D0169375

DIFFERENTIATION RULES

General Formulas

Assume u and v are differentiable functions of x.

Constant: $\frac{d}{dx}(c) = 0$

Sum: $\frac{d}{dx}(u + v) = \frac{du}{dx} + \frac{dv}{dx}$

Difference: $\frac{d}{dx}(u - v) = \frac{du}{dx} - \frac{dv}{dx}$

Constant Multiple: $\frac{d}{dx}(cu) = c\frac{du}{dx}$

Product: $\frac{d}{dx}(uv) = u\frac{dv}{dx} + v\frac{du}{dx}$

Quotient: $\frac{d}{dx}\left(\frac{u}{v}\right) = \frac{v\frac{du}{dx} - u\frac{dv}{dx}}{v^2}$

Power: $\frac{d}{dx}x^n = nx^{n-1}$

Chain Rule: $\frac{d}{dx}(f(g(x)) = f'(g(x)) \cdot g'(x)$

Trigonometric Functions

$\frac{d}{dx}(\sin x) = \cos x$ $\qquad$ $\frac{d}{dx}(\cos x) = -\sin x$

$\frac{d}{dx}(\tan x) = \sec^2 x$ $\qquad$ $\frac{d}{dx}(\sec x) = \sec x \tan x$

$\frac{d}{dx}(\cot x) = -\csc^2 x$ $\qquad$ $\frac{d}{dx}(\csc x) = -\csc x \cot x$

Exponential and Logarithmic Functions

$\frac{d}{dx}e^x = e^x$ $\qquad$ $\frac{d}{dx}\ln x = \frac{1}{x}$

$\frac{d}{dx}a^x = a^x \ln a$ $\qquad$ $\frac{d}{dx}(\log_a x) = \frac{1}{x \ln a}$

Inverse Trigonometric Functions

$\frac{d}{dx}(\sin^{-1} x) = \frac{1}{\sqrt{1 - x^2}}$ $\qquad$ $\frac{d}{dx}(\cos^{-1} x) = -\frac{1}{\sqrt{1 - x^2}}$

$\frac{d}{dx}(\tan^{-1} x) = \frac{1}{1 + x^2}$ $\qquad$ $\frac{d}{dx}(\sec^{-1} x) = \frac{1}{|x|\sqrt{x^2 - 1}}$

$\frac{d}{dx}(\cot^{-1} x) = -\frac{1}{1 + x^2}$ $\qquad$ $\frac{d}{dx}(\csc^{-1} x) = -\frac{1}{|x|\sqrt{x^2 - 1}}$

Hyperbolic Functions

$\frac{d}{dx}(\sinh x) = \cosh x$ $\qquad$ $\frac{d}{dx}(\cosh x) = \sinh x$

$\frac{d}{dx}(\tanh x) = \operatorname{sech}^2 x$ $\qquad$ $\frac{d}{dx}(\operatorname{sech} x) = -\operatorname{sech} x \tanh x$

$\frac{d}{dx}(\coth x) = -\operatorname{csch}^2 x$ $\qquad$ $\frac{d}{dx}(\operatorname{csch} x) = -\operatorname{csch} x \coth x$

Inverse Hyperbolic Functions

$\frac{d}{dx}(\sinh^{-1} x) = \frac{1}{\sqrt{1 + x^2}}$ $\qquad$ $\frac{d}{dx}(\cosh^{-1} x) = \frac{1}{\sqrt{x^2 - 1}}$

$\frac{d}{dx}(\tanh^{-1} x) = \frac{1}{1 - x^2}$ $\qquad$ $\frac{d}{dx}(\operatorname{sech}^{-1} x) = -\frac{1}{x\sqrt{1 - x^2}}$

$\frac{d}{dx}(\coth^{-1} x) = \frac{1}{1 - x^2}$ $\qquad$ $\frac{d}{dx}(\operatorname{csch}^{-1} x) = -\frac{1}{|x|\sqrt{1 + x^2}}$

Parametric Equations

If $x = f(t)$ and $y = g(t)$ are differentiable, then

$$y' = \frac{dy}{dx} = \frac{dy/dt}{dx/dt} \quad \text{and} \quad \frac{d^2y}{dx^2} = \frac{dy'/dt}{dx/dt}$$

THOMAS'
CALCULUS

ELEVENTH EDITION

MEDIA UPGRADE

Based on the original work by
George B. Thomas, Jr.
Massachusetts Institute of Technology

as revised by

Maurice D. Weir
Naval Postgraduate School

Joel Hass
University of California, Davis

Frank R. Giordano
Naval Postgraduate School

PEARSON
Addison Wesley

Boston San Francisco New York
London Toronto Sydney Tokyo Singapore Madrid
Mexico City Munich Paris Cape Town Hong Kong Montreal

Publisher: Greg Tobin
Acquisitions Editor: William Hoffman
Managing Editor: Karen Wernholm
Senior Project Editor: Rachel S. Reeve
Editorial Assistants: Mary Reynolds, Emily Portwood
Production Supervisor: Julie LaChance James
Marketing Manager: Phyllis Hubbard
Marketing Assistant: Heather Peck
Senior Manufacturing Buyer: Evelyn Beaton
Senior Prepress Supervisor: Caroline Fell
Associate Media Producer: Sara Anderson
Software Editors: David Malone, Bob Carroll
Senior Author Support/ Technology Specialist: Joe Vetere
Supplements Production Supervisor: Sheila Spinney
Composition and Production Services: Nesbitt Graphics, Inc.
Illustrations: Techsetters, Inc.
Senior Designer: Barbara T. Atkinson
Interior Design: Geri Davis/The Davis Group, Inc.
Cover Design: Barbara T. Atkinson
Cover Photograph: © Benjamin Mendlowitz

Dedicated to

Ross Lee Finney III

(1933–2000)

Scholar, Educator, Author,

Humanitarian, Friend to all

Library of Congress Cataloging-in-Publication Data

Thomas' calculus.—Media upgrade, 11th ed. / Joel Hass . . . [et al.].
p. cm
Updated ed. of: Thomas' calculus. 11th ed. / . . . as revised by Maurice D. Weir, Joel Hass, Frank R. Giordano. c2005.
Includes index.
ISBN 0-321-48987-X
1. Calculus—Textbooks. I. Hass, Joel. II. Weir, Maurice D. Thomas' calculus. 11th ed.

QA303.2. W45 2008
515–dc22 2006051369

2 3 4 5 6 7 8 9 10-QWD-09 08 07

CONTENTS

10 Conic Sections and Polar Coordinates 685

11 Infinite Sequences and Series 746

12 Vectors and the Geometry of Space 848

13 Vector-Valued Functions and Motion in Space 906

14 Partial Derivatives 965

15 Multiple Integrals 1067

PREFACE

OVERVIEW In preparing the eleventh edition of *Thomas' Calculus*, we have worked to capture the style and strengths of earlier editions. Our goal has been to revisit the best features of the *Thomas' Calculus* classic editions while listening carefully to the suggestions of our many users and reviewers. With these high standards in mind, we have reconstructed the exercises and clarified some difficult topics. In the words of George Thomas, "(We) have tried to write the book as clearly and precisely as is possible." In addition, we have restructured the contents to be more logical and in alignment with the standard syllabus. In looking backward, we have learned much to help us create a useful and appealing calculus text for the next generation of engineers and scientists.

In the eleventh edition the text introduces students not just to the methods and applications of calculus, but also to a mathematical way of thinking. From the exercises to the examples to the narrative that develops the concepts and reveals the theory in readable language, this book is about thinking and communicating mathematical ideas. Calculus contains many of the key paradigms of mathematics, and it marks the real beginnings of how to think about physical and mathematical subjects in a precise and logical way. We try to help students achieve the mathematical maturity required to master the material and apply its power. The insights that come from a deep understanding are well worth the effort.

After completing this book, students should be well versed in the mathematical language needed for applying the concepts of calculus to numerous applications in science and engineering. They should also be well prepared for courses in differential equations, linear algebra, or advanced calculus.

Changes for the Eleventh Edition

EXERCISES Exercises and examples play a crucial role in learning calculus. We have included in this new edition many of the exercises that appeared in previous editions of *Thomas' Calculus*, and which constituted a great strength of those editions. Within each section we have organized and grouped the exercises by topic, progressing from computational problems to applied and theoretical problems. This arrangement gives students the opportunity to develop skills in using the methods of calculus and to deepen their appreciation and understanding of its applications and coherent mathematical structure.

RIGOR The level of rigor, while comparable to earlier editions, is more consistent throughout. We give both formal and informal discussions, making clear the distinction between the two, and we include precise definitions and accessible proofs for the students.

The text is organized so the material can be covered informally, giving the instructor a degree of flexibility. For example, while we do not prove that a continuous function on a closed and bounded interval has a maximum there, we do state this theorem carefully and use it to prove several subsequent results. Moreover, the chapter on limits has been substantially reorganized, with greater attention to both clarity and precision. As in previous editions, the limit concept is still motivated by the important idea of obtaining the slope of the line tangent to a curve at a point on it.

CONTENT During the preparation of this edition, we have paid considerable attention to the suggestions and comments from users of previous *Thomas' Calculus* editions and from our reviewers. This has resulted in extensive revisions and changes to several chapters.

- **Preliminaries** We have rewritten Chapter 1 as a brief review of the elementary functions. While many instructors may choose to skip the chapter, it allows for easy reference and review by the student, standardizes notation, and indicates what is assumed as background material. It also contains some helpful material that many students may not have seen, such as the pitfalls of relying entirely on a calculator or computer to give the graph of a function.
- **Limits** Included in Chapter 2 are epsilon-delta definitions, proofs of many theorems, limits at infinity and infinite limits (and their relationship to asymptotes of a graph).
- **Antiderivatives** We present the derivative and its important applications in Chapters 3 and 4, concluding with the antiderivative concept, which sets the stage for integration.
- **Integration** After discussing several examples of finite sums, we introduce in Chapter 5 the definite integral in its traditional setting of the area under a curve. Following the treatment of the Fundamental Theorem of Calculus, bridging derivatives and antiderivatives, we present the indefinite integral, along with the Substitution Rule for integration. The traditional chapter on applications of definite integrals follows.
- **Techniques of integration** The main techniques of integration, including numerical integration, are presented in Chapter 8. This follows the introduction of the transcendental functions, where we define the natural logarithm as an integral and the exponential function as its inverse.
- **Differential equations** The bulk of the material on solving basic differential equations is now organized into a single Chapter 9. This organization allows for greater instructor flexibility in the coverage of those topics.
- **Conics** At the request of many users, Chapter 10 on the conic sections has been fully restored. This chapter also completes the material on parametric equations by giving parametrizations of parabolas, hyperbolas, and cycloids.
- **Series** In Chapter 11 we have restored the more complete development of the series' convergence tests that appeared in the ninth edition. We also include a brief section introducing Fourier series (which may be omitted) at the end of the chapter.
- **Vectors** To avoid repetition of the central algebraic and geometric ideas, we have combined the treatment of two- and three-dimensional vectors into a single Chapter 12. This presentation is followed by a chapter on vector-valued functions in the plane and in space.
- **The real numbers** We have written a brief new appendix on the theory of real numbers as it applies to calculus.

ART We realize that figures and illustrations are a critical component to learning calculus, so we have taken a fresh look at all of the figures in the book. When revising existing figures and creating new ones, we worked to improve the clarity with which the figures illustrate their associated concepts. This is especially evident with the three-dimensional graphics, where we were able to better indicate depth, layering, and rotation (see figures below). We also attempted to ensure a consistent and pedagogical use of color and assembled a team dedicated to proofreading the completed pieces.

FIGURE 6.11, page 402
Finding the volume of the solid generated by revolving the region (a) about the y-axis.

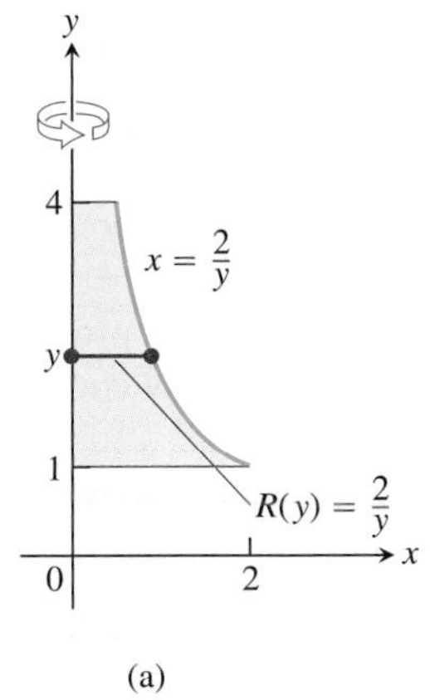

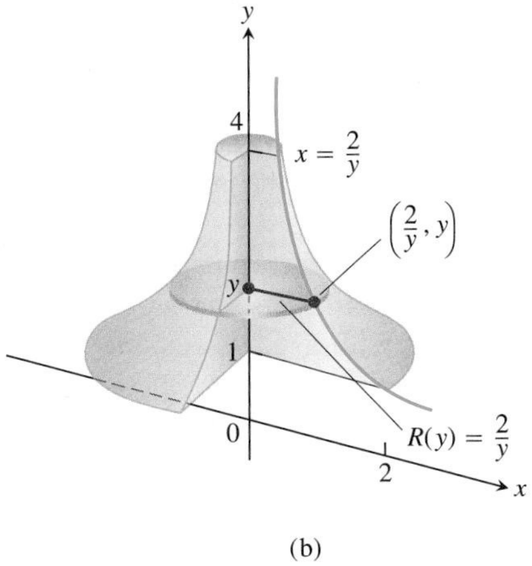

FIGURE 6.13, page 403
The cross-sections of the solid of revolution generated here are washers, not disks.

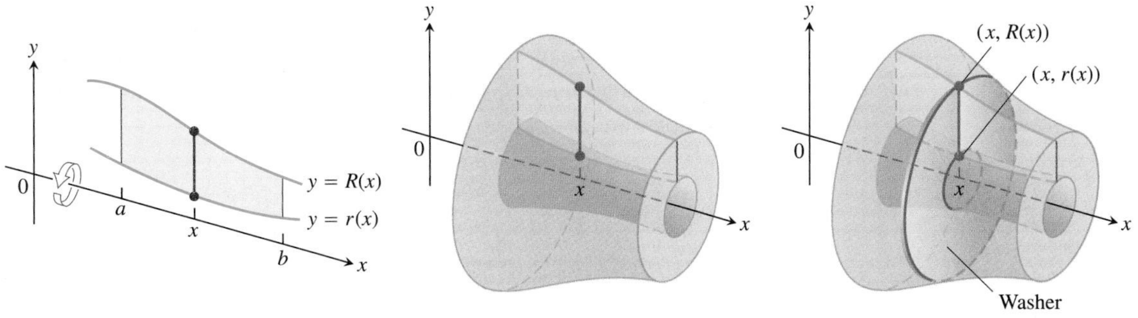

Continuing Features

END-OF-CHAPTER REVIEWS AND PROJECTS In addition to problems appearing after each section, each chapter culminates with review questions, practice exercises covering the entire chapter, and a series of Additional and Advanced Exercises serving to include more challenging or synthesizing problems. Most chapters also include descriptions of several student projects that can be worked on by individual students, or groups of students, over a longer period of time. These projects require the use of a computer and additional material that is available over the Internet at **www.aw-bc.com/thomas**.

WRITING EXERCISES Writing exercises placed throughout the text ask students to explore and explain a variety of calculus concepts and applications. In addition, each chapter end contains a list of questions for students to review and summarize what they have learned. Many of these exercises make good writing assignments.

ANSWERS Answers are provided for all odd-numbered exercises when appropriate, and these have been carefully checked for correctness.

MATHEMATICAL CORRECTNESS As in previous editions, we have been careful to say only what is true and mathematically sound. Every definition, theorem, corollary, and proof has been reviewed for clarity and mathematical correctness.

WRITING AND APPLICATIONS As always, this text continues to be easy to read, conversational, and mathematically rich. Each new topic is motivated by clear, easy-to-understand examples and is then reinforced by its application to real-world problems of immediate interest to students. A hallmark of this book has been the application of calculus to science and engineering. These applied problems have been updated, improved, and extended continually over the last several editions.

TECHNOLOGY In a course using the text, technology can be incorporated according to the taste of the instructor. Each section contains exercises requiring the use of technology; these are marked with a T if suitable for calculator or computer usage or are labeled **Computer Explorations** if a computer algebra system (CAS, such as *Maple* or *Mathematica*) is required. While we continue to provide support for technology, we have toned down its visibility within the chapters from the tenth edition.

EARLY TRANSCENDENTALS For instructors who require an earlier treatment of the calculus of transcendental functions, we have prepared an *Early Transcendentals* version of this text, in which the exponential and logarithmic functions are introduced in the first chapter. Their limits, derivatives, and integrals are given in Chapters 2 through 5, along with those for polynomials and other algebraic functions. Examples and exercises involving the transcendental functions are then interlaced throughout those chapters as the various calculus topics are developed.

Text Versions

STUDENT EDITION OF THOMAS' CALCULUS, Eleventh Edition
Complete (Chapters 1–16), ISBN 0-321-18558-7
Part One, Single Variable Calculus (Chapters 1–11), ISBN 0-321-22642-9
Part Two, Multivariable Calculus (Chapters 11–16), ISBN 0-321-22651-8

STUDENT EDITION OF THOMAS' CALCULUS: EARLY TRANSCENDENTALS, Eleventh Edition
Complete (Chapters 1–16), ISBN 0-321-19800-X
Part One, Single Variable Calculus (Chapters 1–11), ISBN 0-321-22633-X
Part Two, Multivariable Calculus (Chapters 11–16), ISBN 0-321-22651-8
The *Early Transcendentals* version of *Thomas' Calculus* introduces and integrates transcendental functions (such as inverse trigonometric, exponential, and logarithmic functions) into the exposition, examples, and exercises of the early chapters alongside the algebraic functions. Part Two for *Thomas' Calculus: Early Transcendentals* is the same text as Part Two for *Thomas' Calculus*.

Print Supplements

INSTRUCTOR'S SOLUTIONS MANUAL
Part One (Chapters 1–11), ISBN 0-321-22653-4
Part Two (Chapters 11–16), ISBN 0-321-22650-X
The *Instructor's Solutions Manual* by William Ardis, Joseph Borzellino, Linda Buchanan, Alexis T. Mogill, and Patricia Nelson contains complete worked-out solutions to all of the exercises in the text.

ANSWER BOOK
ISBN 0-321-22649-6
The *Answer Book* by William Ardis, Joseph Borzellino, Linda Buchanan, Alexis T. Mogill, and Patricia Nelson contains short answers to most of the exercises in the text.

STUDENT OUTLINES
Part One (Chapters 1–11), ISBN 0-321-22640-2
Part Two (Chapters 11–16), ISBN 0-321-22641-0
Organized to correspond to the text, the *Student Outlines* by Joseph Borzellino and Patricia Nelson reinforces important concepts and provides an outline of the important topics, theorems, and definitions, as well as study tips and additional practice problems.

STUDENT'S SOLUTIONS MANUAL
Part One (Chapters 1–11), ISBN 0-321-22646-1
Part Two (Chapters 11–16), ISBN 0-321-22647-X
The *Student's Solutions Manual* by William Ardis, Joseph Borzellino, Linda Buchanan, Alexis T. Mogill, and Patricia Nelson is designed for the student and contains carefully worked-out solutions to all the odd-numbered exercises in the text.

JUST-IN-TIME ALGEBRA AND TRIGONOMETRY FOR CALCULUS, Third Edition
ISBN 0-321-26943-8
Sharp algebra and trigonometry skills are critical to mastering calculus, and *Just-in-Time Algebra and Trigonometry for Calculus,* Third Edition, by Guntram Mueller and Ronald I. Brent is designed to bolster these skills while students study calculus. As students make their way through calculus, this text is with them every step of the way, showing them the necessary algebra or trigonometry topics and pointing out potential problem spots. The easy-to-use contents has algebra and trigonometry topics arranged in the order in which students will need them as they study calculus.

ADDISON-WESLEY'S CALCULUS REVIEW CARD
The Calculus Review Card is a resource for students containing important formulas, functions, definitions, and theorems that correspond precisely to *Thomas' Calculus.* This card

can work as a reference for completing homework assignments or as an aid in studying and is available bundled with a new text. Contact your Addison-Wesley sales representative for more information.

Media and Online Supplements

TECHNOLOGY RESOURCE MANUALS

Maple Manual by Donald Hartig, California Polytechnic State University
Mathematica Manual by Marie Vanisko, California State University Stanislaus, and Lyle Cochran, Whitworth College
TI-Graphing Calculator Manual by Luz DeAlba, Drake University
These manuals cover *Maple* 9, *Mathematica* 5, and the TI-83 Plus/TI-84 Plus, TI-85/TI-86, and TI-89/TI-92 Plus, respectively. Each manual provides detailed guidance for integrating a specific software package or graphing calculator throughout the course, including syntax and commands. These manuals are available to qualified instructors through **http://suppscentral.aw.com.**

MYMATHLAB™

MyMathLab is a series of text-specific, easily customizable online courses for Addison-Wesley textbooks in mathematics and statistics. MyMathLab is powered by CourseCompass™—Pearson Education's online teaching and learning environment—and by MathXL—Addison-Wesley's online homework, tutorial, and assessment system. MyMathLab gives you the tools you need to deliver all or a portion of your course online, whether your students are in a lab setting or working from home. MyMathLab provides a rich and flexible set of course materials, featuring free-response exercises that are algorithmically generated for unlimited practice and mastery. Students can also use online tools, such as video lectures, animations, multimedia textbook, and Maple/Mathematica projects, to independently improve their understanding and performance. Instructors can use MyMathLab's homework and test managers to select and assign online exercises correlated directly to the textbook, and they can import TestGen tests into MyMathLab for added flexibility. MyMathLab's online gradebook—designed specifically for mathematics and statistics—automatically tracks students' homework and test results and gives the instructor control over how to calculate final grades. MyMathLab is available to qualified adopters. For more information, visit our Web site at **www.mymathlab.com** or contact your Addison-Wesley sales representative for a product demonstration.

MATHXL®

MathXL is a powerful online homework, tutorial, and assessment system that accompanies your Addison-Wesley textbook in mathematics or statistics. With MathXL, instructors can create, edit, and assign online homework and tests using algorithmically generated exercises correlated at the objective level to the textbook. All student work is tracked in MathXL's online gradebook. Students can take chapter tests in MathXL and receive personalized study plans based on their test results. The study plan diagnoses weaknesses and links students directly to tutorial exercises for the objectives they need to study and retest. Students can also access supplemental animations and video clips directly from selected exercises. MathXL is available to qualified adopters. For more information, visit our Web site at **www.mathxl.com** or contact your Addison-Wesley sales representative for a product demonstration.

WebAssign®

WebAssign is an online homework, quizzing, and testing management system. As a hosted application service, WebAssign gives professors the tools they need to assign their students algorithmically generated questions. WebAssign also includes a grade book that is automatically populated with student answers. WebAssign works with personal computers using any recent operating system and browser.

WeBWorK®
WeBWorK is an Internet-based method for delivering homework problems to students online. It automatically grades homework and provides immediate feedback, allowing students to correct mistakes while they are still working on the problem. WeBWorK works with personal computers using any recent operating system and browser.

TESTGEN WITH QUIZMASTER
TestGen enables instructors to build, edit, print, and administer tests using a computerized bank of questions developed to cover all the objectives of the text. TestGen is algorithmically based, allowing instructors to create multiple but equivalent versions of the same question or test with the click of a button. Instructors can also modify test bank questions or add new questions by using the built-in question editor, which allows users to create graphs, import graphics, and insert math notation, variable numbers, or text. Tests can be printed or administered online via the Internet or another network. TestGen comes packaged with QuizMaster, which allows students to take tests on a local area network. The software is available on a dual-platform Windows/Macintosh CD-ROM.

DIGITAL VIDEO TUTOR
The Digital Video Tutor features an engaging team of mathematics instructors who present comprehensive coverage of topics in the text. The lecturers' presentations include examples and exercises from the text and support an approach that emphasizes visualization and problem solving. The video lectures are available on CD-ROM, making it easy and convenient for students to watch the videos from a computer at home or on campus. The complete digitized video set, affordable and portable for students, is ideal for distance learning or supplemental instruction.

WEB SITE **www.aw-bc.com/thomas**
The *Thomas' Calculus* Web site provides the expanded historical biographies and essays referenced in the text. Also available is a collection of *Maple* and *Mathematica* modules that can be used as projects by individual students or groups of students.

ADDISON-WESLEY MATH TUTOR CENTER
The Addison-Wesley Math Tutor Center is staffed by qualified mathematics and statistics instructors who provide students with tutoring on examples and odd-numbered exercises from the textbook. Tutoring is available via toll-free telephone, toll-free fax, e-mail, and the Internet. Interactive, Web-based technology allows tutors and students to view and work through problems together in real time over the Internet. The Addison-Wesley Math Tutor Center is available to qualified adopters. For more information, please visit our Web site at **www.aw-bc.com/tutorcenter** or call us at 1-888-777-0463.

Acknowledgments

We would like to express our thanks to the people who made many valuable contributions to this edition as it developed through its various stages:

Development Editors
Elka Block
David Chelton
Frank Purcell

Accuracy Checkers
William Ardis
Karl Kattchee
Douglas B. Meade
Robert Pierce
Frank Purcell
Marie Vanisko
Thomas Wegleitner

Super Reviewers

Harry Allen, *Ohio State University*
Rebecca Goldin, *George Mason University*
Christopher Heil, *Georgia Institute of Technology*
Dominic Naughton, *Purdue University*
Maria Terrell, *Cornell University*
Clifford Weil, *Michigan State University*

Reviewers

Robert Andersen, *University of Wisconsin–Eau Claire*
Charles Ashley, *Villanova University*
David Bachman, *California Polytechnic State University*
Elizabeth Bator, *University of North Texas*
William Bogley, *Oregon State University*
Kaddour Boukaabar, *California University of Pennsylvania*
Deborah Brandon, *Carnegie Mellon University*
Mark Bridger, *Northeastern University*
Sean Cleary, *City College of New York*
Edward Crotty, *University of Pennsylvania*
Mark Davidson, *Louisiana State University*
Richard Davitt, *University of Louisville*
Elias Deeba, *University of Houston, Downtown Campus*
Anne Dougherty, *University of Colorado*
Rafael Espericueta, *Bakersfield College*
Klaus Fischer, *George Mason University*
William Fitzgibbon, *University of Houston*
Carol Flakus, *Lower Columbia College*
Tim Flood, *Pittsburg State University*
Robert Gardner, *East Tennessee State University*
John Gilbert, *University of Texas at Austin*
Mark Hanisch, *Calvin College*
Zahid Hasan, *California State University, San Bernardino*
Jo W. Heath, *Auburn University*
Ken Holladay, *University of New Orleans*
Hugh Howards, *Wake Forest University*
Dwayne Jennings, *Union University*
Jennifer M. Johnson, *Princeton University*
Matthias Kawski, *Arizona State University*
Bill Kincaid, *Wilmington College*
Mark M. Maxwell, *Robert Morris University*
Jack Mealy, *Austin College*
Richard Mercer, *Wright State University*
Victor Nistor, *Pennsylvania State University*
Michael O'Leary, *Towson University*
Bogdan Oporowski, *Louisiana State University*
Troy Riggs, *Union University*
Ferdinand Rivera, *San Jose State University*
Mohammed Saleem, *San Jose State University*
Tatiana Shubin, *San Jose State University*
Alex Smith, *University of Wisconsin-Eau Claire*
Donald Solomon, *University of Wisconsin–Milwaukee*
Chia Chi Tung, *Minnesota State University*
William L. Van Alstine, *Aiken Technical College*
Bobby Winters, *Pittsburg State University*
Dennis Wortman, *University of Massachusetts at Boston*

Survey Participants

Omar Adawi, *Parkland College*
Siham Alfred, *Raritan Valley Community College*
Donna J. Bailey, *Truman State University*
Rajesh K. Barnwal, *Middle Tennessee State University*
Robert C. Brigham, *University of Central Florida* (retired)
Thomas A. Carnevale, *Valdosta State University*
Leonard Chastkofsky, *University of Georgia*
Richard Dalrymple, *Minnesota West Community & Technical College*
W. Lloyd Davis, *College of San Mateo*
Will-Matthis Dunn III, *Montgomery College*
George F. Feissner, *SUNY College at Cortland*
Bruno Harris, *Brown University*
Celeste Hernandez, *Richland College*
Wei-Min Huang, *Lehigh University*
Herbert E. Kasube, *Bradley University*
Frederick W. Keene, *Pasadena City College*
Michael Kent, *Borough of Manhattan Community College*
Robert Levine, *Community College of Allegheny County, Boyce Campus*
John Martin, *Santa Rosa Junior College*
Michael Scott McClendon, *University of Central Oklahoma*
Ching-Tsuan Pan, *Northern Illinois University*
Emma Previato, *Boston University*
S.S. Ravindran, *University of Alabama*
Dan Rothe, *Alpena Community College*
John T. Saccoman, *Seton Hall University*
Mansour Samimi, *Winston-Salem State University*
Ned W. Schillow, *Lehigh Carbon Community College*
W.R. Schrank, *Angelina College*
Mark R. Woodard, *Furman University*

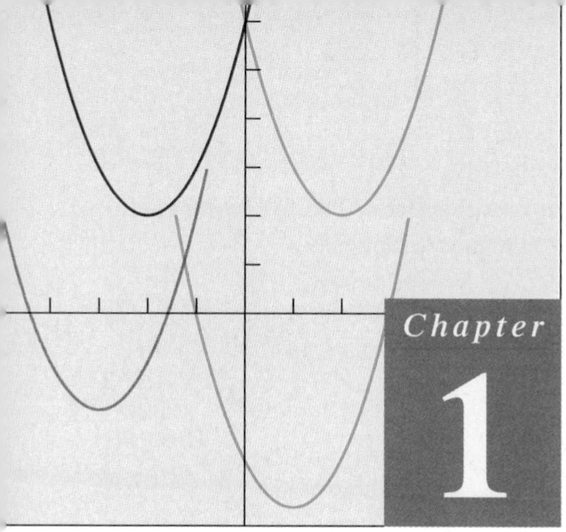

Chapter 1

PRELIMINARIES

OVERVIEW This chapter reviews the basic ideas you need to start calculus. The topics include the real number system, Cartesian coordinates in the plane, straight lines, parabolas, circles, functions, and trigonometry. We also discuss the use of graphing calculators and computer graphing software.

1.1 Real Numbers and the Real Line

This section reviews real numbers, inequalities, intervals, and absolute values.

Real Numbers

Much of calculus is based on properties of the real number system. **Real numbers** are numbers that can be expressed as decimals, such as

$$-\frac{3}{4} = -0.75000\ldots$$

$$\frac{1}{3} = 0.33333\ldots$$

$$\sqrt{2} = 1.4142\ldots$$

The dots ... in each case indicate that the sequence of decimal digits goes on forever. Every conceivable decimal expansion represents a real number, although some numbers have two representations. For instance, the infinite decimals .999 ... and 1.000 ... represent the same real number 1. A similar statement holds for any number with an infinite tail of 9's.

The real numbers can be represented geometrically as points on a number line called the **real line**.

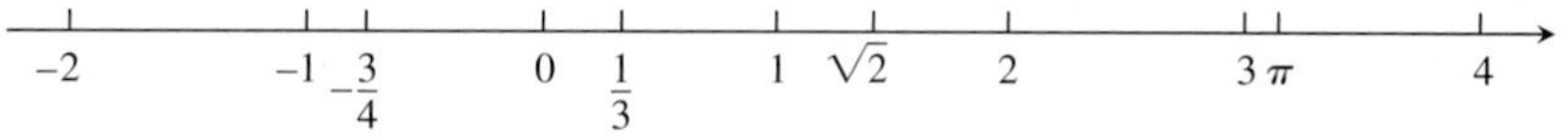

The symbol $\mathbb{R}$ denotes either the real number system or, equivalently, the real line.

The properties of the real number system fall into three categories: algebraic properties, order properties, and completeness. The **algebraic properties** say that the real numbers can be added, subtracted, multiplied, and divided (except by 0) to produce more real numbers under the usual rules of arithmetic. *You can never divide by* 0.

The **order properties** of real numbers are given in Appendix 4. The following useful rules can be derived from them, where the symbol $\Rightarrow$ means "implies."

Rules for Inequalities

If a, b, and c are real numbers, then:

1. $a < b \Rightarrow a + c < b + c$
2. $a < b \Rightarrow a - c < b - c$
3. $a < b$ and $c > 0 \Rightarrow ac < bc$
4. $a < b$ and $c < 0 \Rightarrow bc < ac$
   Special case: $a < b \Rightarrow -b < -a$
5. $a > 0 \Rightarrow \dfrac{1}{a} > 0$
6. If a and b are both positive or both negative, then $a < b \Rightarrow \dfrac{1}{b} < \dfrac{1}{a}$

Notice the rules for multiplying an inequality by a number. Multiplying by a positive number preserves the inequality; multiplying by a negative number reverses the inequality. Also, reciprocation reverses the inequality for numbers of the same sign. For example, $2 < 5$ but $-2 > -5$ and $1/2 > 1/5$.

The **completeness property** of the real number system is deeper and harder to define precisely. However, the property is essential to the idea of a limit (Chapter 2). Roughly speaking, it says that there are enough real numbers to "complete" the real number line, in the sense that there are no "holes" or "gaps" in it. Many theorems of calculus would fail if the real number system were not complete. The topic is best saved for a more advanced course, but Appendix 4 hints about what is involved and how the real numbers are constructed.

We distinguish three special subsets of real numbers.

1. The **natural numbers**, namely $1, 2, 3, 4, \ldots$
2. The **integers**, namely $0, \pm 1, \pm 2, \pm 3, \ldots$
3. The **rational numbers**, namely the numbers that can be expressed in the form of a fraction m/n, where m and n are integers and $n \neq 0$. Examples are

$$\frac{1}{3}, \quad -\frac{4}{9} = \frac{-4}{9} = \frac{4}{-9}, \quad \frac{200}{13}, \quad \text{and} \quad 57 = \frac{57}{1}.$$

The rational numbers are precisely the real numbers with decimal expansions that are either

(a) terminating (ending in an infinite string of zeros), for example,

$$\frac{3}{4} = 0.75000\ldots = 0.75 \qquad \text{or}$$

(b) eventually repeating (ending with a block of digits that repeats over and over), for example

$$\frac{23}{11} = 2.090909\ldots = 2.\overline{09}$$

The bar indicates the block of repeating digits.

A terminating decimal expansion is a special type of repeating decimal since the ending zeros repeat.

The set of rational numbers has all the algebraic and order properties of the real numbers but lacks the completeness property. For example, there is no rational number whose square is 2; there is a "hole" in the rational line where $\sqrt{2}$ should be.

Real numbers that are not rational are called **irrational numbers**. They are characterized by having nonterminating and nonrepeating decimal expansions. Examples are π, $\sqrt{2}$, $\sqrt[3]{5}$, and $\log_{10} 3$. Since every decimal expansion represents a real number, it should be clear that there are infinitely many irrational numbers. Both rational and irrational numbers are found arbitrarily close to any point on the real line.

Set notation is very useful for specifying a particular subset of real numbers. A **set** is a collection of objects, and these objects are the **elements** of the set. If S is a set, the notation $a \in S$ means that a is an element of S, and $a \notin S$ means that a is not an element of S. If S and T are sets, then $S \cup T$ is their **union** and consists of all elements belonging either to S or T (or to both S and T). The **intersection** $S \cap T$ consists of all elements belonging to both S and T. The **empty set** $\varnothing$ is the set that contains no elements. For example, the intersection of the rational numbers and the irrational numbers is the empty set.

Some sets can be described by *listing* their elements in braces. For instance, the set A consisting of the natural numbers (or positive integers) less than 6 can be expressed as

$$A = \{1, 2, 3, 4, 5\}.$$

The entire set of integers is written as

$$\{0, \pm 1, \pm 2, \pm 3, \dots\}.$$

Another way to describe a set is to enclose in braces a rule that generates all the elements of the set. For instance, the set

$$A = \{x \mid x \text{ is an integer and } 0 < x < 6\}$$

is the set of positive integers less than 6.

Intervals

A subset of the real line is called an **interval** if it contains at least two numbers and contains all the real numbers lying between any two of its elements. For example, the set of all real numbers x such that $x > 6$ is an interval, as is the set of all x such that $-2 \le x \le 5$. The set of all nonzero real numbers is not an interval; since 0 is absent, the set fails to contain every real number between -1 and 1 (for example).

Geometrically, intervals correspond to rays and line segments on the real line, along with the real line itself. Intervals of numbers corresponding to line segments are **finite intervals**; intervals corresponding to rays and the real line are **infinite intervals**.

A finite interval is said to be **closed** if it contains both of its endpoints, **half-open** if it contains one endpoint but not the other, and **open** if it contains neither endpoint. The endpoints are also called **boundary points**; they make up the interval's **boundary**. The remaining points of the interval are **interior points** and together comprise the interval's **interior**. Infinite intervals are closed if they contain a finite endpoint, and open otherwise. The entire real line $\mathbb{R}$ is an infinite interval that is both open and closed.

Solving Inequalities

The process of finding the interval or intervals of numbers that satisfy an inequality in x is called **solving** the inequality.

TABLE 1.1 Types of intervals

	Notation	Set description	Type	Picture
Finite:	(a, b)	$\{x \mid a < x < b\}$	Open	
	$[a, b]$	$\{x \mid a \le x \le b\}$	Closed	
	$[a, b)$	$\{x \mid a \le x < b\}$	Half-open	
	$(a, b]$	$\{x \mid a < x \le b\}$	Half-open	
Infinite:	(a, ∞)	$\{x \mid x > a\}$	Open	
	$[a, \infty)$	$\{x \mid x \ge a\}$	Closed	
	$(-\infty, b)$	$\{x \mid x < b\}$	Open	
	$(-\infty, b]$	$\{x \mid x \le b\}$	Closed	
	$(-\infty, \infty)$	$\mathbb{R}$ (set of all real numbers)	Both open and closed	

EXAMPLE 1 Solve the following inequalities and show their solution sets on the real line.

(a) $2x - 1 < x + 3$ **(b)** $-\dfrac{x}{3} < 2x + 1$ **(c)** $\dfrac{6}{x - 1} \ge 5$

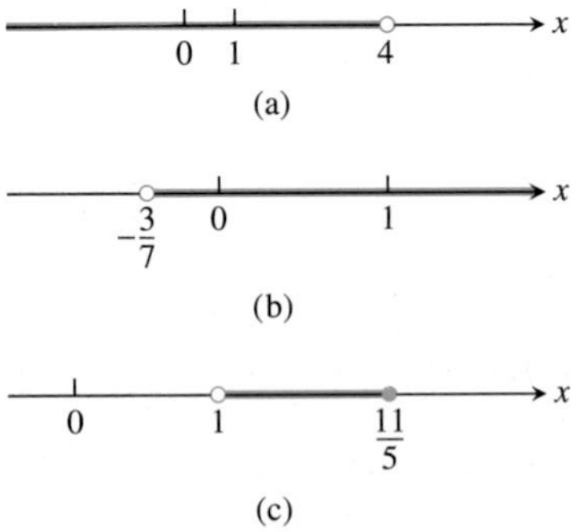

FIGURE 1.1 Solution sets for the inequalities in Example 1.

Solution

(a)

$$2x - 1 < x + 3$$
$$2x < x + 4 \qquad \text{Add 1 to both sides.}$$
$$x < 4 \qquad \text{Subtract } x \text{ from both sides.}$$

The solution set is the open interval $(-\infty, 4)$ (Figure 1.1a).

(b)

$$-\frac{x}{3} < 2x + 1$$
$$-x < 6x + 3 \qquad \text{Multiply both sides by 3.}$$
$$0 < 7x + 3 \qquad \text{Add } x \text{ to both sides.}$$
$$-3 < 7x \qquad \text{Subtract 3 from both sides.}$$
$$-\frac{3}{7} < x \qquad \text{Divide by 7.}$$

The solution set is the open interval $(-3/7, \infty)$ (Figure 1.1b).

(c) The inequality $6/(x-1) \geq 5$ can hold only if $x > 1$, because otherwise $6/(x-1)$ is undefined or negative. Therefore, $(x-1)$ is positive and the inequality will be preserved if we multiply both sides by $(x-1)$, and we have

$$\begin{aligned} \frac{6}{x-1} &\geq 5 \\ 6 &\geq 5x - 5 && \text{Multiply both sides by } (x-1). \\ 11 &\geq 5x && \text{Add 5 to both sides.} \\ \frac{11}{5} &\geq x. && \text{Or } x \leq \frac{11}{5}. \end{aligned}$$

The solution set is the half-open interval $(1, 11/5]$ (Figure 1.1c). ■

Absolute Value

The **absolute value** of a number x, denoted by $|x|$, is defined by the formula

$$|x| = \begin{cases} x, & x \geq 0 \\ -x, & x < 0. \end{cases}$$

EXAMPLE 2 Finding Absolute Values

$$|3| = 3, \quad |0| = 0, \quad |-5| = -(-5) = 5, \quad |-|a|| = |a|$$ ■

Geometrically, the absolute value of x is the distance from x to 0 on the real number line. Since distances are always positive or 0, we see that $|x| \geq 0$ for every real number x, and $|x| = 0$ if and only if $x = 0$. Also,

$$|x - y| = \text{the distance between } x \text{ and } y$$

on the real line (Figure 1.2).

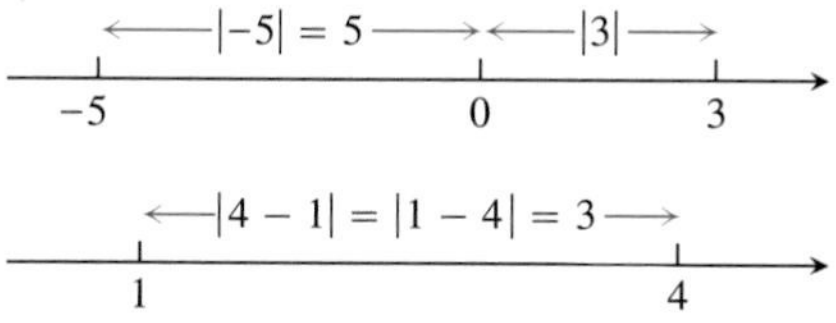

FIGURE 1.2 Absolute values give distances between points on the number line.

Since the symbol $\sqrt{a}$ always denotes the *nonnegative* square root of a, an alternate definition of $|x|$ is

$$|x| = \sqrt{x^2}.$$

It is important to remember that $\sqrt{a^2} = |a|$. Do not write $\sqrt{a^2} = a$ unless you already know that $a \geq 0$.

The absolute value has the following properties. (You are asked to prove these properties in the exercises.)

Absolute Value Properties

1. $\lvert -a \rvert = \lvert a \rvert$	A number and its additive inverse or negative have the same absolute value.
2. $\lvert ab \rvert = \lvert a \rvert \lvert b \rvert$	The absolute value of a product is the product of the absolute values.
3. $\left\lvert \frac{a}{b} \right\rvert = \frac{\lvert a \rvert}{\lvert b \rvert}$	The absolute value of a quotient is the quotient of the absolute values.
4. $\lvert a + b \rvert \leq \lvert a \rvert + \lvert b \rvert$	The **triangle inequality**. The absolute value of the sum of two numbers is less than or equal to the sum of their absolute values.

Note that $|-a| \neq -|a|$. For example, $|-3| = 3$, whereas $-|3| = -3$. If a and b differ in sign, then $|a + b|$ is less than $|a| + |b|$. In all other cases, $|a + b|$ equals $|a| + |b|$. Absolute value bars in expressions like $|-3 + 5|$ work like parentheses: We do the arithmetic inside *before* taking the absolute value.

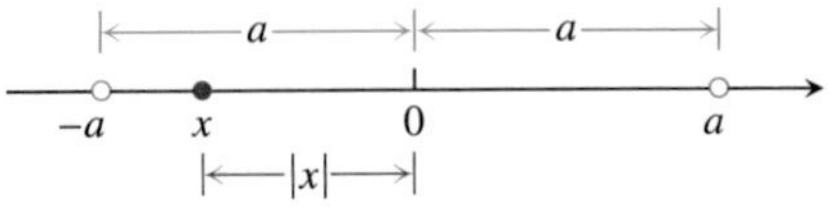

FIGURE 1.3 $|x| < a$ means x lies between $-a$ and a.

EXAMPLE 3 Illustrating the Triangle Inequality

$$|-3 + 5| = |2| = 2 < |-3| + |5| = 8$$
$$|3 + 5| = |8| = |3| + |5|$$
$$|-3 - 5| = |-8| = 8 = |-3| + |-5|$$

■

The inequality $|x| < a$ says that the distance from x to 0 is less than the positive number a. This means that x must lie between $-a$ and a, as we can see from Figure 1.3.

The following statements are all consequences of the definition of absolute value and are often helpful when solving equations or inequalities involving absolute values.

Absolute Values and Intervals

If a is any positive number, then

5. $|x| = a$ if and only if $x = \pm a$
6. $|x| < a$ if and only if $-a < x < a$
7. $|x| > a$ if and only if $x > a$ or $x < -a$
8. $|x| \leq a$ if and only if $-a \leq x \leq a$
9. $|x| \geq a$ if and only if $x \geq a$ or $x \leq -a$

The symbol $\Leftrightarrow$ is often used by mathematicians to denote the "if and only if" logical relationship. It also means "implies and is implied by."

EXAMPLE 4 Solving an Equation with Absolute Values

Solve the equation $|2x - 3| = 7$.

Solution By Property 5, $2x - 3 = \pm 7$, so there are two possibilities:

$2x - 3 = 7$	$2x - 3 = -7$	Equivalent equations without absolute values
$2x = 10$	$2x = -4$	Solve as usual.
$x = 5$	$x = -2$	

The solutions of $|2x - 3| = 7$ are $x = 5$ and $x = -2$. ■

EXAMPLE 5 Solving an Inequality Involving Absolute Values

Solve the inequality $\left|5 - \frac{2}{x}\right| < 1$.

Solution We have

$$\begin{aligned}
\left|5 - \frac{2}{x}\right| < 1 &\Leftrightarrow -1 < 5 - \frac{2}{x} < 1 && \text{Property 6}\\
&\Leftrightarrow -6 < -\frac{2}{x} < -4 && \text{Subtract 5.}\\
&\Leftrightarrow 3 > \frac{1}{x} > 2 && \text{Multiply by } -\tfrac{1}{2}.\\
&\Leftrightarrow \frac{1}{3} < x < \frac{1}{2}. && \text{Take reciprocals.}
\end{aligned}$$

Notice how the various rules for inequalities were used here. Multiplying by a negative number reverses the inequality. So does taking reciprocals in an inequality in which both sides are positive. The original inequality holds if and only if $(1/3) < x < (1/2)$. The solution set is the open interval $(1/3, 1/2)$. ■

EXAMPLE 6 Solve the inequality and show the solution set on the real line:

(a) $|2x - 3| \le 1$ **(b)** $|2x - 3| \ge 1$

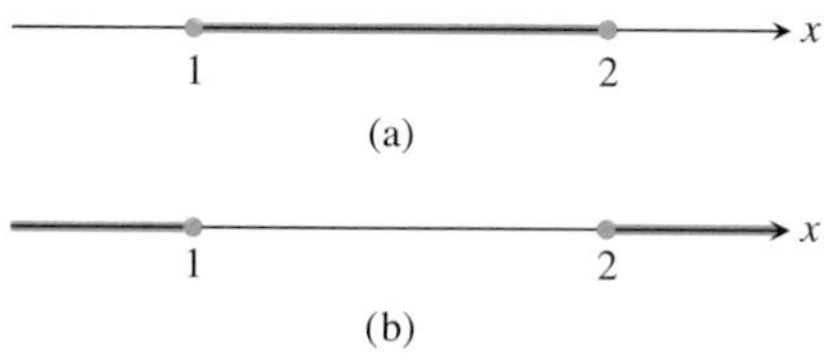

FIGURE 1.4 The solution sets (a) [1, 2] and (b) $(-\infty, 1] \cup [2, \infty)$ in Example 6.

Solution

(a)

$$\begin{aligned}
|2x - 3| &\le 1 &&\\
-1 \le 2x - 3 &\le 1 && \text{Property 8}\\
2 \le 2x &\le 4 && \text{Add 3.}\\
1 \le x &\le 2 && \text{Divide by 2.}
\end{aligned}$$

The solution set is the closed interval [1, 2] (Figure 1.4a).

(b)

$$|2x - 3| \ge 1$$

$$\begin{aligned}
2x - 3 \ge 1 \quad &\text{or} \quad 2x - 3 \le -1 && \text{Property 9}\\
x - \frac{3}{2} \ge \frac{1}{2} \quad &\text{or} \quad x - \frac{3}{2} \le -\frac{1}{2} && \text{Divide by 2.}\\
x \ge 2 \quad &\text{or} \quad x \le 1 && \text{Add } \tfrac{3}{2}.
\end{aligned}$$

The solution set is $(-\infty, 1] \cup [2, \infty)$ (Figure 1.4b). ■

EXERCISES 1.1

Decimal Representations

1. Express 1/9 as a repeating decimal, using a bar to indicate the repeating digits. What are the decimal representations of 2/9? 3/9? 8/9? 9/9?

2. Express 1/11 as a repeating decimal, using a bar to indicate the repeating digits. What are the decimal representations of 2/11? 3/11? 9/11? 11/11?

Inequalities

3. If $2 < x < 6$, which of the following statements about x are necessarily true, and which are not necessarily true?

 a. $0 < x < 4$
 b. $0 < x - 2 < 4$
 c. $1 < \frac{x}{2} < 3$
 d. $\frac{1}{6} < \frac{1}{x} < \frac{1}{2}$
 e. $1 < \frac{6}{x} < 3$
 f. $|x - 4| < 2$
 g. $-6 < -x < 2$
 h. $-6 < -x < -2$

4. If $-1 < y - 5 < 1$, which of the following statements about y are necessarily true, and which are not necessarily true?

a. $4 < y < 6$
b. $-6 < y < -4$
c. $y > 4$
d. $y < 6$
e. $0 < y - 4 < 2$
f. $2 < \frac{y}{2} < 3$
g. $\frac{1}{6} < \frac{1}{y} < \frac{1}{4}$
h. $|y - 5| < 1$

In Exercises 5–12, solve the inequalities and show the solution sets on the real line.

5. $-2x > 4$
6. $8 - 3x \geq 5$
7. $5x - 3 \leq 7 - 3x$
8. $3(2 - x) > 2(3 + x)$
9. $2x - \frac{1}{2} \geq 7x + \frac{7}{6}$
10. $\frac{6 - x}{4} < \frac{3x - 4}{2}$
11. $\frac{4}{5}(x - 2) < \frac{1}{3}(x - 6)$
12. $-\frac{x + 5}{2} \leq \frac{12 + 3x}{4}$

Absolute Value

Solve the equations in Exercises 13–18.

13. $|y| = 3$
14. $|y - 3| = 7$
15. $|2t + 5| = 4$
16. $|1 - t| = 1$
17. $|8 - 3s| = \frac{9}{2}$
18. $\left|\frac{s}{2} - 1\right| = 1$

Solve the inequalities in Exercises 19–34, expressing the solution sets as intervals or unions of intervals. Also, show each solution set on the real line.

19. $|x| < 2$
20. $|x| \leq 2$
21. $|t - 1| \leq 3$
22. $|t + 2| < 1$
23. $|3y - 7| < 4$
24. $|2y + 5| < 1$
25. $\left|\frac{z}{5} - 1\right| \leq 1$
26. $\left|\frac{3}{2}z - 1\right| \leq 2$
27. $\left|3 - \frac{1}{x}\right| < \frac{1}{2}$
28. $\left|\frac{2}{x} - 4\right| < 3$
29. $|2s| \geq 4$
30. $|s + 3| \geq \frac{1}{2}$
31. $|1 - x| > 1$
32. $|2 - 3x| > 5$
33. $\left|\frac{r + 1}{2}\right| \geq 1$
34. $\left|\frac{3r}{5} - 1\right| > \frac{2}{5}$

Quadratic Inequalities

Solve the inequalities in Exercises 35–42. Express the solution sets as intervals or unions of intervals and show them on the real line. Use the result $\sqrt{a^2} = |a|$ as appropriate.

35. $x^2 < 2$
36. $4 \leq x^2$
37. $4 < x^2 < 9$
38. $\frac{1}{9} < x^2 < \frac{1}{4}$
39. $(x - 1)^2 < 4$
40. $(x + 3)^2 < 2$
41. $x^2 - x < 0$
42. $x^2 - x - 2 \geq 0$

Theory and Examples

43. Do not fall into the trap $|-a| = a$. For what real numbers a is this equation true? For what real numbers is it false?

44. Solve the equation $|x - 1| = 1 - x$.

45. **A proof of the triangle inequality** Give the reason justifying each of the numbered steps in the following proof of the triangle inequality.

$$\begin{aligned} |a + b|^2 &= (a + b)^2 && (1) \\ &= a^2 + 2ab + b^2 \\ &\leq a^2 + 2|a||b| + b^2 && (2) \\ &= |a|^2 + 2|a||b| + |b|^2 && (3) \\ &= (|a| + |b|)^2 \\ |a + b| &\leq |a| + |b| && (4) \end{aligned}$$

46. Prove that $|ab| = |a||b|$ for any numbers a and b.

47. If $|x| \leq 3$ and $x > -1/2$, what can you say about x?

48. Graph the inequality $|x| + |y| \leq 1$.

49. Let $f(x) = 2x + 1$ and let $\delta > 0$ be any positive number. Prove that $|x - 1| < \delta$ implies $|f(x) - f(1)| < 2\delta$. Here the notation $f(a)$ means the value of the expression $2x + 1$ when $x = a$. This *function notation* is explained in Section 1.3.

50. Let $f(x) = 2x + 3$ and let $\epsilon > 0$ be any positive number. Prove that $|f(x) - f(0)| < \epsilon$ whenever $|x - 0| < \frac{\epsilon}{2}$. Here the notation $f(a)$ means the value of the expression $2x + 3$ when $x = a$. (See Section 1.3.)

51. For any number a, prove that $|-a| = |a|$.

52. Let a be any positive number. Prove that $|x| > a$ if and only if $x > a$ or $x < -a$.

53. a. If b is any nonzero real number, prove that $|1/b| = 1/|b|$.

b. Prove that $\left|\frac{a}{b}\right| = \frac{|a|}{|b|}$ for any numbers a and $b \neq 0$.

54. Using mathematical induction (see Appendix 1), prove that $|a^n| = |a|^n$ for any number a and positive integer n.

1.2 Lines, Circles, and Parabolas

This section reviews coordinates, lines, distance, circles, and parabolas in the plane. The notion of increment is also discussed.

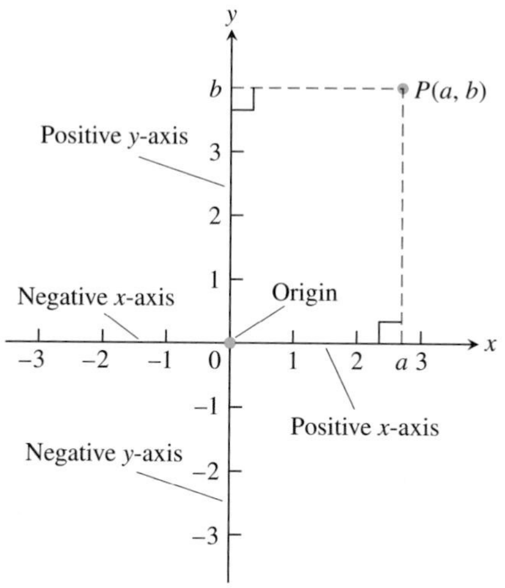

FIGURE 1.5 Cartesian coordinates in the plane are based on two perpendicular axes intersecting at the origin.

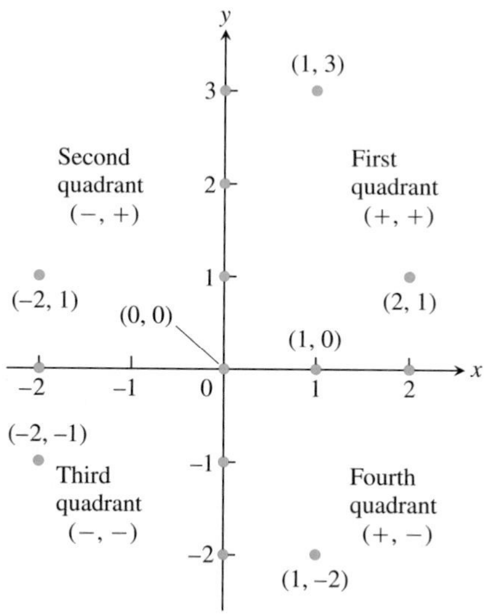

FIGURE 1.6 Points labeled in the xy-coordinate or Cartesian plane. The points on the axes all have coordinate pairs but are usually labeled with single real numbers, (so (1, 0) on the x-axis is labeled as 1). Notice the coordinate sign patterns of the quadrants.

Cartesian Coordinates in the Plane

In the previous section we identified the points on the line with real numbers by assigning them coordinates. Points in the plane can be identified with ordered pairs of real numbers. To begin, we draw two perpendicular coordinate lines that intersect at the 0-point of each line. These lines are called **coordinate axes** in the plane. On the horizontal x-axis, numbers are denoted by x and increase to the right. On the vertical y-axis, numbers are denoted by y and increase upward (Figure 1.5). Thus "upward" and "to the right" are positive directions, whereas "downward" and "to the left" are considered as negative. The **origin** O, also labeled 0, of the coordinate system is the point in the plane where x and y are both zero.

If P is any point in the plane, it can be located by exactly one ordered pair of real numbers in the following way. Draw lines through P perpendicular to the two coordinate axes. These lines intersect the axes at points with coordinates a and b (Figure 1.5). The ordered pair (a, b) is assigned to the point P and is called its **coordinate pair**. The first number a is the **x-coordinate** (or **abscissa**) of P; the second number b is the **y-coordinate** (or **ordinate**) of P. The x-coordinate of every point on the y-axis is 0. The y-coordinate of every point on the x-axis is 0. The origin is the point (0, 0).

Starting with an ordered pair (a, b), we can reverse the process and arrive at a corresponding point P in the plane. Often we identify P with the ordered pair and write $P(a, b)$. We sometimes also refer to "the point (a, b)" and it will be clear from the context when (a, b) refers to a point in the plane and not to an open interval on the real line. Several points labeled by their coordinates are shown in Figure 1.6.

This coordinate system is called the **rectangular coordinate system** or **Cartesian coordinate system** (after the sixteenth century French mathematician René Descartes). The coordinate axes of this coordinate or Cartesian plane divide the plane into four regions called **quadrants**, numbered counterclockwise as shown in Figure 1.6.

The **graph** of an equation or inequality in the variables x and y is the set of all points $P(x, y)$ in the plane whose coordinates satisfy the equation or inequality. When we plot data in the coordinate plane or graph formulas whose variables have different units of measure, we do not need to use the same scale on the two axes. If we plot time vs. thrust for a rocket motor, for example, there is no reason to place the mark that shows 1 sec on the time axis the same distance from the origin as the mark that shows 1 lb on the thrust axis.

Usually when we graph functions whose variables do not represent physical measurements and when we draw figures in the coordinate plane to study their geometry and trigonometry, we try to make the scales on the axes identical. A vertical unit of distance then looks the same as a horizontal unit. As on a surveyor's map or a scale drawing, line segments that are supposed to have the same length will look as if they do and angles that are supposed to be congruent will look congruent.

Computer displays and calculator displays are another matter. The vertical and horizontal scales on machine-generated graphs usually differ, and there are corresponding distortions in distances, slopes, and angles. Circles may look like ellipses, rectangles may look like squares, right angles may appear to be acute or obtuse, and so on. We discuss these displays and distortions in greater detail in Section 1.7.

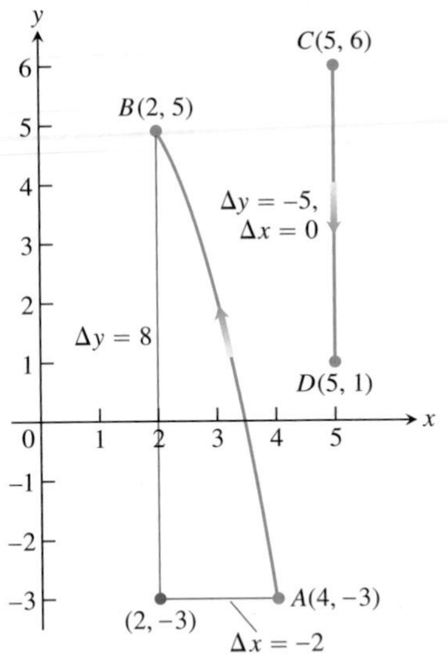

FIGURE 1.7 Coordinate increments may be positive, negative, or zero (Example 1).

HISTORICAL BIOGRAPHY*

René Descartes
(1596–1650)

Increments and Straight Lines

When a particle moves from one point in the plane to another, the net changes in its coordinates are called *increments*. They are calculated by subtracting the coordinates of the starting point from the coordinates of the ending point. If x changes from x_1 to x_2, the **increment** in x is

$$\Delta x = x_2 - x_1.$$

EXAMPLE 1 In going from the point $A(4, -3)$ to the point $B(2, 5)$ the increments in the x- and y-coordinates are

$$\Delta x = 2 - 4 = -2, \qquad \Delta y = 5 - (-3) = 8.$$

From $C(5, 6)$ to $D(5, 1)$ the coordinate increments are

$$\Delta x = 5 - 5 = 0, \qquad \Delta y = 1 - 6 = -5.$$

See Figure 1.7. ■

Given two points $P_1(x_1, y_1)$ and $P_2(x_2, y_2)$ in the plane, we call the increments $\Delta x = x_2 - x_1$ and $\Delta y = y_2 - y_1$ the **run** and the **rise**, respectively, between P_1 and P_2. Two such points always determine a unique straight line (usually called simply a line) passing through them both. We call the line P_1P_2.

Any nonvertical line in the plane has the property that the ratio

$$m = \frac{\text{rise}}{\text{run}} = \frac{\Delta y}{\Delta x} = \frac{y_2 - y_1}{x_2 - x_1}$$

has the same value for every choice of the two points $P_1(x_1, y_1)$ and $P_2(x_2, y_2)$ on the line (Figure 1.8). This is because the ratios of corresponding sides for similar triangles are equal.

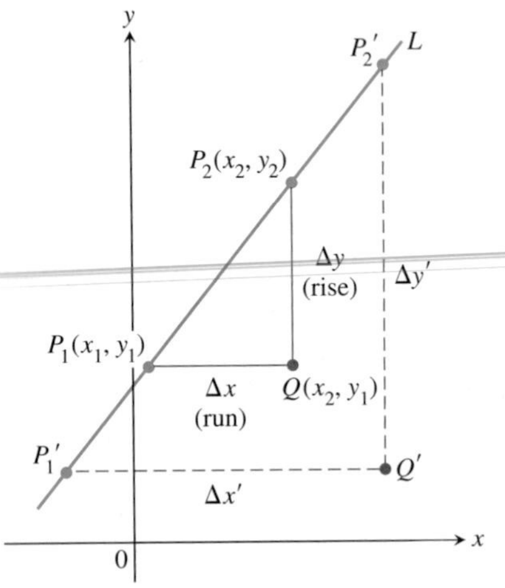

FIGURE 1.8 Triangles P_1QP_2 and $P_1'Q'P_2'$ are similar, so the ratio of their sides has the same value for any two points on the line. This common value is the line's slope.

> **DEFINITION Slope**
> The constant
> $$m = \frac{\text{rise}}{\text{run}} = \frac{\Delta y}{\Delta x} = \frac{y_2 - y_1}{x_2 - x_1}$$
> is the **slope** of the nonvertical line P_1P_2.

The slope tells us the direction (uphill, downhill) and steepness of a line. A line with positive slope rises uphill to the right; one with negative slope falls downhill to the right (Figure 1.9). The greater the absolute value of the slope, the more rapid the rise or fall. The slope of a vertical line is *undefined*. Since the run Δx is zero for a vertical line, we cannot evaluate the slope ratio m.

The direction and steepness of a line can also be measured with an angle. The **angle of inclination** of a line that crosses the x-axis is the smallest counterclockwise angle from the x-axis to the line (Figure 1.10). The inclination of a horizontal line is 0°. The inclination of a vertical line is 90°. If ϕ (the Greek letter phi) is the inclination of a line, then $0 \leq \phi < 180°$.

*To learn more about the historical figures and the development of the major elements and topics of calculus, visit **www.aw-bc.com/thomas**.

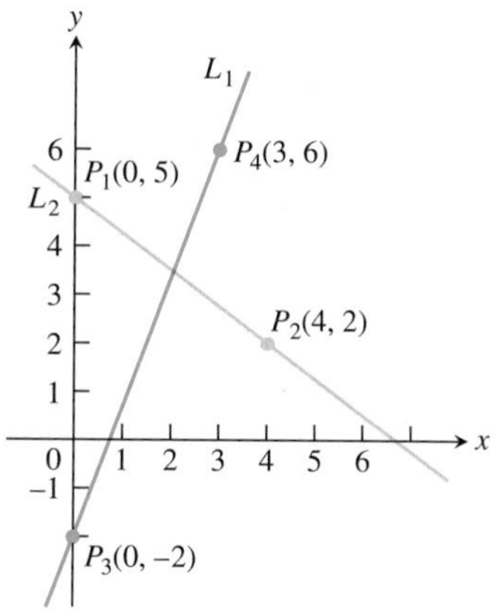

FIGURE 1.9 The slope of L_1 is

$$m = \frac{\Delta y}{\Delta x} = \frac{6 - (-2)}{3 - 0} = \frac{8}{3}.$$

That is, y increases 8 units every time x increases 3 units. The slope of L_2 is

$$m = \frac{\Delta y}{\Delta x} = \frac{2 - 5}{4 - 0} = \frac{-3}{4}.$$

That is, y decreases 3 units every time x increases 4 units.

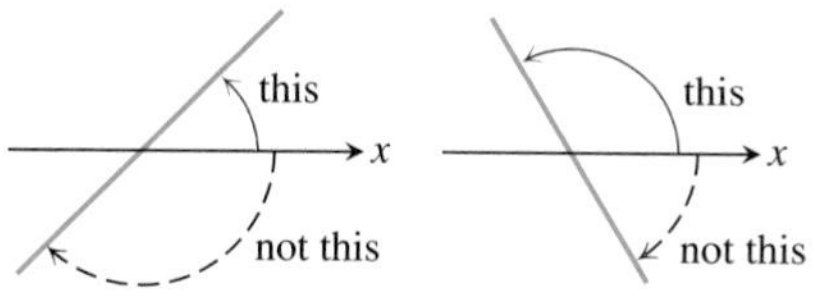

FIGURE 1.10 Angles of inclination are measured counterclockwise from the x-axis.

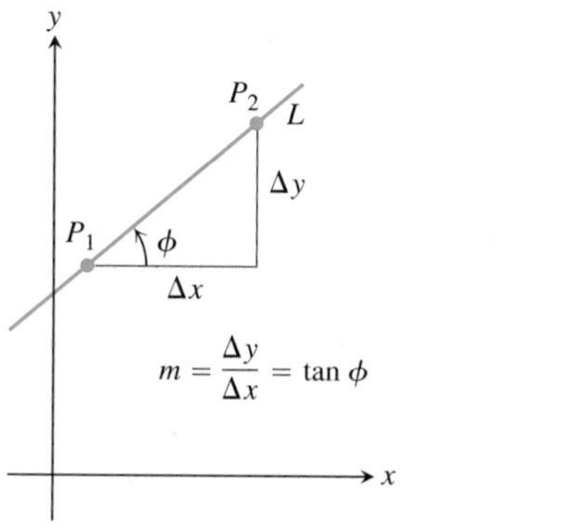

FIGURE 1.11 The slope of a nonvertical line is the tangent of its angle of inclination.

The relationship between the slope m of a nonvertical line and the line's angle of inclination ϕ is shown in Figure 1.11:

$$m = \tan \phi.$$

Straight lines have relatively simple equations. All points on the *vertical line* through the point a on the x-axis have x-coordinates equal to a. Thus, $x = a$ is an equation for the vertical line. Similarly, $y = b$ is an equation for the *horizontal line* meeting the y-axis at b. (See Figure 1.12.)

We can write an equation for a nonvertical straight line L if we know its slope m and the coordinates of one point $P_1(x_1, y_1)$ on it. If $P(x, y)$ is *any* other point on L, then we can use the two points P_1 and P to compute the slope,

$$m = \frac{y - y_1}{x - x_1}$$

so that

$$y - y_1 = m(x - x_1) \qquad \text{or} \qquad y = y_1 + m(x - x_1).$$

> The equation
>
> $$y = y_1 + m(x - x_1)$$
>
> is the **point-slope equation** of the line that passes through the point (x_1, y_1) and has slope m.

EXAMPLE 2 Write an equation for the line through the point (2, 3) with slope $-3/2$.

Solution We substitute $x_1 = 2$, $y_1 = 3$, and $m = -3/2$ into the point-slope equation and obtain

$$y = 3 - \frac{3}{2}(x - 2), \qquad \text{or} \qquad y = -\frac{3}{2}x + 6.$$

When $x = 0$, $y = 6$ so the line intersects the y-axis at $y = 6$. ■

EXAMPLE 3 A Line Through Two Points

Write an equation for the line through $(-2, -1)$ and $(3, 4)$.

Solution The line's slope is

$$m = \frac{-1 - 4}{-2 - 3} = \frac{-5}{-5} = 1.$$

We can use this slope with either of the two given points in the point-slope equation:

With $(x_1, y_1) = (-2, -1)$	**With** $(x_1, y_1) = (3, 4)$
$y = -1 + 1 \cdot (x - (-2))$	$y = 4 + 1 \cdot (x - 3)$
$y = -1 + x + 2$	$y = 4 + x - 3$
$y = x + 1$	$y = x + 1$

Same result

Either way, $y = x + 1$ is an equation for the line (Figure 1.13). ■

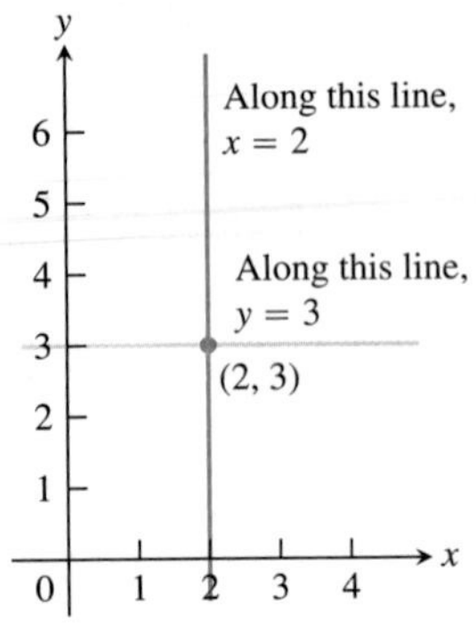

FIGURE 1.12 The standard equations for the vertical and horizontal lines through (2, 3) are $x = 2$ and $y = 3$.

The y-coordinate of the point where a nonvertical line intersects the y-axis is called the **y-intercept** of the line. Similarly, the **x-intercept** of a nonhorizontal line is the x-coordinate of the point where it crosses the x-axis (Figure 1.14). A line with slope m and y-intercept b passes through the point $(0, b)$, so it has equation

$$y = b + m(x - 0), \qquad \text{or, more simply,} \qquad y = mx + b.$$

> The equation
>
> $$y = mx + b$$
>
> is called the **slope-intercept equation** of the line with slope m and y-intercept b.

Lines with equations of the form $y = mx$ have y-intercept 0 and so pass through the origin. Equations of lines are called **linear** equations.

The equation

$$Ax + By = C \qquad (A \text{ and } B \text{ not both } 0)$$

is called the **general linear equation** in x and y because its graph always represents a line and every line has an equation in this form (including lines with undefined slope).

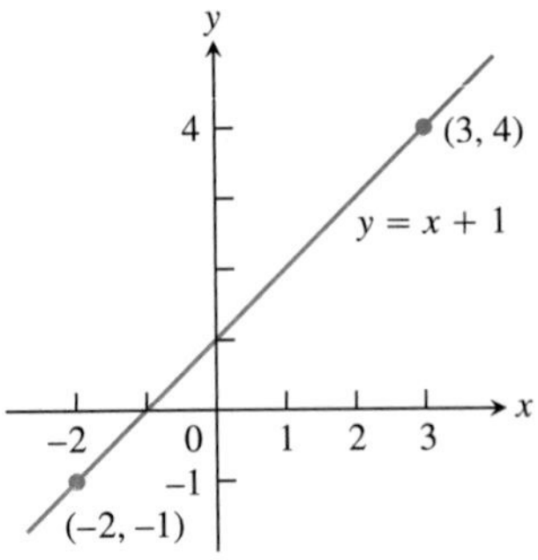

FIGURE 1.13 The line in Example 3.

EXAMPLE 4 Finding the Slope and y-Intercept

Find the slope and y-intercept of the line $8x + 5y = 20$.

Solution Solve the equation for y to put it in slope-intercept form:

$$8x + 5y = 20$$
$$5y = -8x + 20$$
$$y = -\frac{8}{5}x + 4.$$

The slope is $m = -8/5$. The y-intercept is $b = 4$. ■

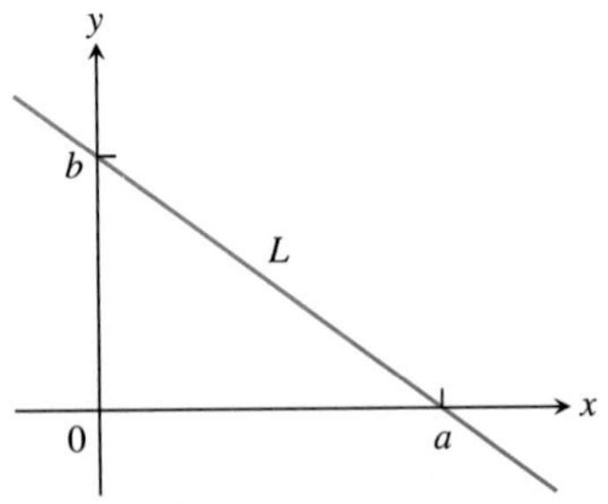

FIGURE 1.14 Line L has x-intercept a and y-intercept b.

Parallel and Perpendicular Lines

Lines that are parallel have equal angles of inclination, so they have the same slope (if they are not vertical). Conversely, lines with equal slopes have equal angles of inclination and so are parallel.

If two nonvertical lines L_1 and L_2 are perpendicular, their slopes m_1 and m_2 satisfy $m_1 m_2 = -1$, so each slope is the *negative reciprocal* of the other:

$$m_1 = -\frac{1}{m_2}, \qquad m_2 = -\frac{1}{m_1}.$$

To see this, notice by inspecting similar triangles in Figure 1.15 that $m_1 = a/h$, and $m_2 = -h/a$. Hence, $m_1 m_2 = (a/h)(-h/a) = -1$.

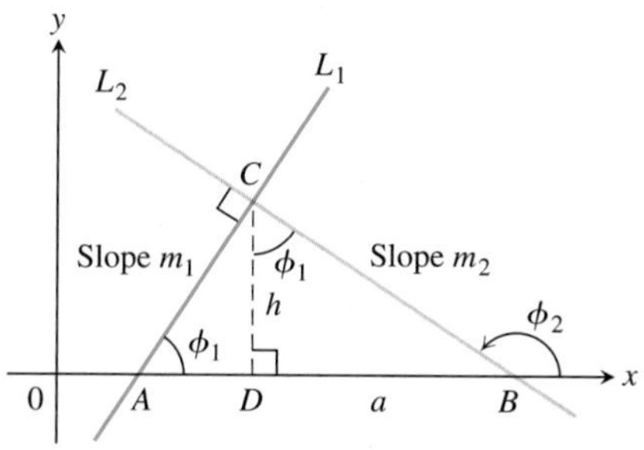

FIGURE 1.15 ΔADC is similar to ΔCDB. Hence ϕ_1 is also the upper angle in ΔCDB. From the sides of ΔCDB, we read $\tan \phi_1 = a/h$.

Distance and Circles in the Plane

The distance between points in the plane is calculated with a formula that comes from the Pythagorean theorem (Figure 1.16).

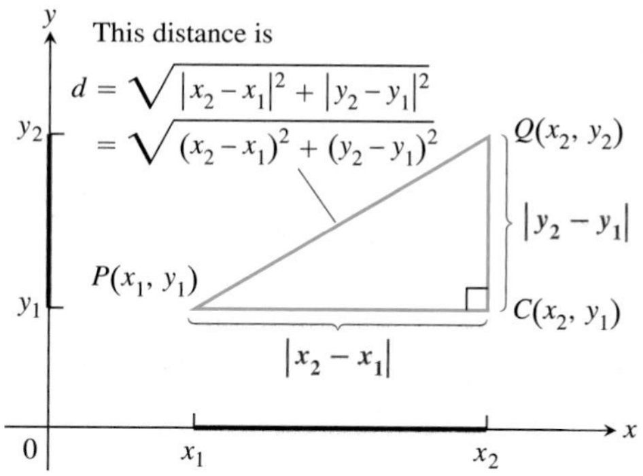

FIGURE 1.16 To calculate the distance between $P(x_1, y_1)$ and $Q(x_2, y_2)$, apply the Pythagorean theorem to triangle PCQ.

Distance Formula for Points in the Plane

The distance between $P(x_1, y_1)$ and $Q(x_2, y_2)$ is

$$d = \sqrt{(\Delta x)^2 + (\Delta y)^2} = \sqrt{(x_2 - x_1)^2 + (y_2 - y_1)^2}.$$

EXAMPLE 5 Calculating Distance

(a) The distance between $P(-1, 2)$ and $Q(3, 4)$ is

$$\sqrt{(3 - (-1))^2 + (4 - 2)^2} = \sqrt{(4)^2 + (2)^2} = \sqrt{20} = \sqrt{4 \cdot 5} = 2\sqrt{5}.$$

(b) The distance from the origin to $P(x, y)$ is

$$\sqrt{(x - 0)^2 + (y - 0)^2} = \sqrt{x^2 + y^2}.$$

■

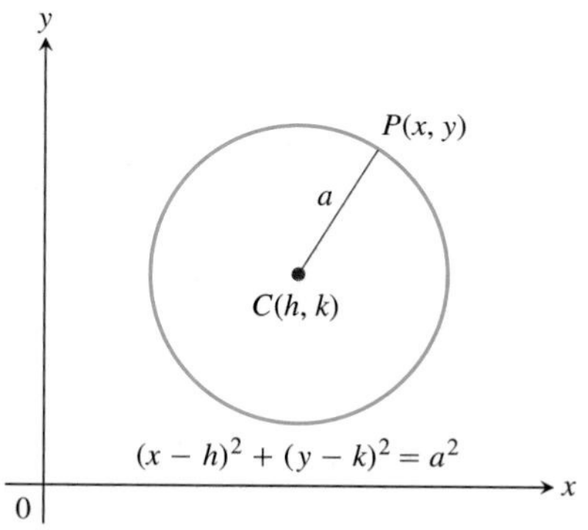

FIGURE 1.17 A circle of radius a in the xy-plane, with center at (h, k).

By definition, a **circle** of radius a is the set of all points $P(x, y)$ whose distance from some center $C(h, k)$ equals a (Figure 1.17). From the distance formula, P lies on the circle if and only if

$$\sqrt{(x - h)^2 + (y - k)^2} = a,$$

so

$$(x - h)^2 + (y - k)^2 = a^2. \tag{1}$$

Equation (1) is the **standard equation** of a circle with center (h, k) and radius a. The circle of radius $a = 1$ and centered at the origin is the **unit circle** with equation

$$x^2 + y^2 = 1.$$

EXAMPLE 6

(a) The standard equation for the circle of radius 2 centered at (3, 4) is

$$(x - 3)^2 + (y - 4)^2 = 2^2 = 4.$$

(b) The circle

$$(x - 1)^2 + (y + 5)^2 = 3$$

has $h = 1$, $k = -5$, and $a = \sqrt{3}$. The center is the point $(h, k) = (1, -5)$ and the radius is $a = \sqrt{3}$. ■

If an equation for a circle is not in standard form, we can find the circle's center and radius by first converting the equation to standard form. The algebraic technique for doing so is *completing the square* (see Appendix 9).

EXAMPLE 7 Finding a Circle's Center and Radius

Find the center and radius of the circle

$$x^2 + y^2 + 4x - 6y - 3 = 0.$$

Solution We convert the equation to standard form by completing the squares in x and y:

$$x^2 + y^2 + 4x - 6y - 3 = 0$$ Start with the given equation.

$$(x^2 + 4x \quad) + (y^2 - 6y \quad) = 3$$ Gather terms. Move the constant to the right-hand side.

$$\left(x^2 + 4x + \left(\frac{4}{2}\right)^2\right) + \left(y^2 - 6y + \left(\frac{-6}{2}\right)^2\right) = 3 + \left(\frac{4}{2}\right)^2 + \left(\frac{-6}{2}\right)^2$$ Add the square of half the coefficient of x to each side of the equation. Do the same for y. The parenthetical expressions on the left-hand side are now perfect squares.

$$(x^2 + 4x + 4) + (y^2 - 6y + 9) = 3 + 4 + 9$$

$$(x + 2)^2 + (y - 3)^2 = 16$$ Write each quadratic as a squared linear expression.

The center is $(-2, 3)$ and the radius is $a = 4$. ■

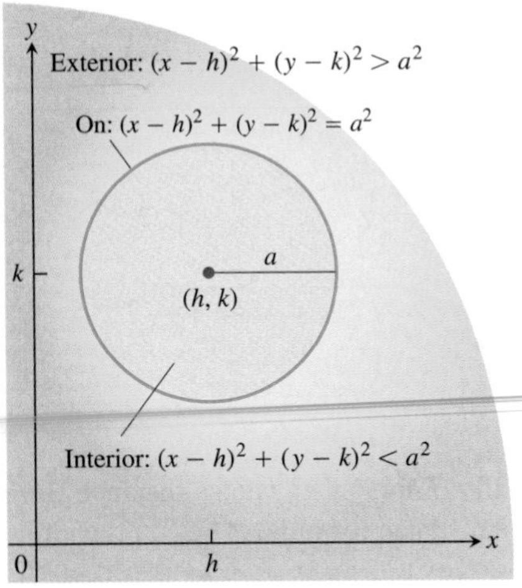

FIGURE 1.18 The interior and exterior of the circle $(x - h)^2 + (y - k)^2 = a^2$.

The points (x, y) satisfying the inequality

$$(x - h)^2 + (y - k)^2 < a^2$$

make up the **interior** region of the circle with center (h, k) and radius a (Figure 1.18). The circle's **exterior** consists of the points (x, y) satisfying

$$(x - h)^2 + (y - k)^2 > a^2.$$

Parabolas

The geometric definition and properties of general parabolas are reviewed in Section 10.1. Here we look at parabolas arising as the graphs of equations of the form $y = ax^2 + bx + c$.

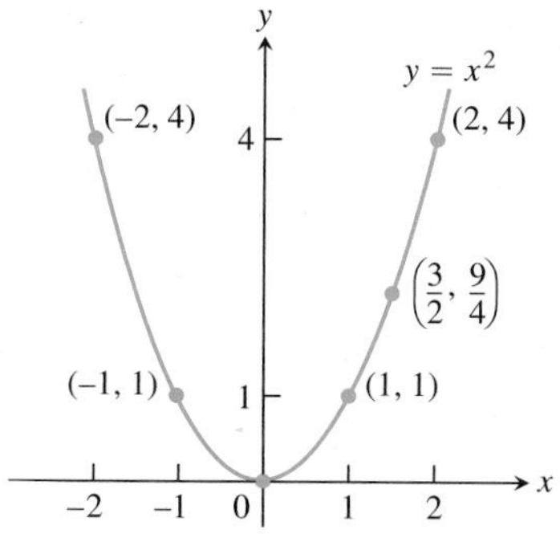

FIGURE 1.19 The parabola $y = x^2$ (Example 8).

EXAMPLE 8 The Parabola $y = x^2$

Consider the equation $y = x^2$. Some points whose coordinates satisfy this equation are $(0, 0)$, $(1, 1)$, $\left(\frac{3}{2}, \frac{9}{4}\right)$, $(-1, 1)$, $(2, 4)$, and $(-2, 4)$. These points (and all others satisfying the equation) make up a smooth curve called a parabola (Figure 1.19). ■

The graph of an equation of the form

$$y = ax^2$$

is a **parabola** whose **axis** (axis of symmetry) is the y-axis. The parabola's **vertex** (point where the parabola and axis cross) lies at the origin. The parabola opens upward if $a > 0$ and downward if $a < 0$. The larger the value of $|a|$, the narrower the parabola (Figure 1.20).

Generally, the graph of $y = ax^2 + bx + c$ is a shifted and scaled version of the parabola $y = x^2$. We discuss shifting and scaling of graphs in more detail in Section 1.5.

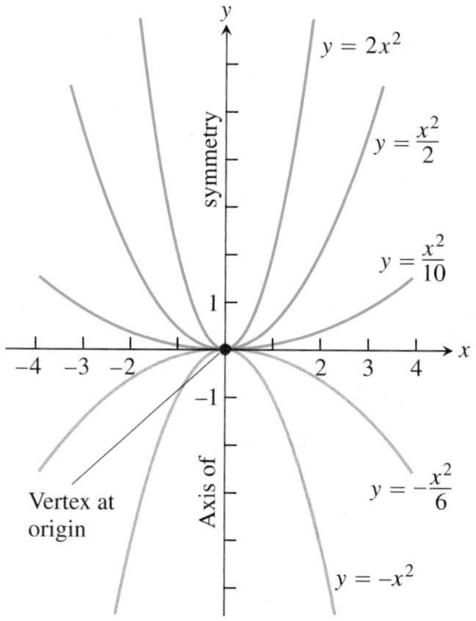

FIGURE 1.20 Besides determining the direction in which the parabola $y = ax^2$ opens, the number a is a scaling factor. The parabola widens as a approaches zero and narrows as $|a|$ becomes large.

The Graph of $y = ax^2 + bx + c, \quad a \neq 0$

The graph of the equation $y = ax^2 + bx + c$, $a \neq 0$, is a parabola. The parabola opens upward if $a > 0$ and downward if $a < 0$. The **axis** is the line

$$x = -\frac{b}{2a}. \tag{2}$$

The **vertex** of the parabola is the point where the axis and parabola intersect. Its x-coordinate is $x = -b/2a$; its y-coordinate is found by substituting $x = -b/2a$ in the parabola's equation.

Notice that if $a = 0$, then we have $y = bx + c$ which is an equation for a line. The axis, given by Equation (2), can be found by completing the square or by using a technique we study in Section 4.1.

EXAMPLE 9 Graphing a Parabola

Graph the equation $y = -\frac{1}{2}x^2 - x + 4$.

Solution Comparing the equation with $y = ax^2 + bx + c$ we see that

$$a = -\frac{1}{2}, \qquad b = -1, \qquad c = 4.$$

Since $a < 0$, the parabola opens downward. From Equation (2) the axis is the vertical line

$$x = -\frac{b}{2a} = -\frac{(-1)}{2(-1/2)} = -1.$$

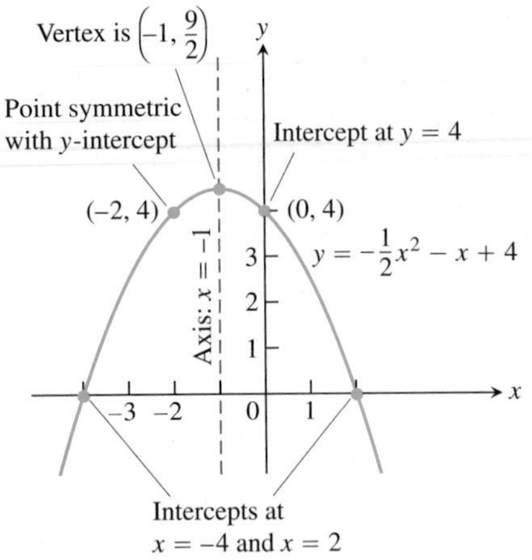

FIGURE 1.21 The parabola in Example 9.

When $x = -1$, we have

$$y = -\frac{1}{2}(-1)^2 - (-1) + 4 = \frac{9}{2}.$$

The vertex is $(-1, 9/2)$.

The x-intercepts are where $y = 0$:

$$-\frac{1}{2}x^2 - x + 4 = 0$$

$$x^2 + 2x - 8 = 0$$

$$(x - 2)(x + 4) = 0$$

$$x = 2, \qquad x = -4$$

We plot some points, sketch the axis, and use the direction of opening to complete the graph in Figure 1.21. ■

EXERCISES 1.2

Increments and Distance

In Exercises 1–4, a particle moves from A to B in the coordinate plane. Find the increments Δx and Δy in the particle's coordinates. Also find the distance from A to B.

1. $A(-3, 2)$, $B(-1, -2)$ **2.** $A(-1, -2)$, $B(-3, 2)$

3. $A(-3.2, -2)$, $B(-8.1, -2)$ **4.** $A(\sqrt{2}, 4)$, $B(0, 1.5)$

Describe the graphs of the equations in Exercises 5–8.

5. $x^2 + y^2 = 1$ **6.** $x^2 + y^2 = 2$

7. $x^2 + y^2 \le 3$ **8.** $x^2 + y^2 = 0$

Slopes, Lines, and Intercepts

Plot the points in Exercises 9–12 and find the slope (if any) of the line they determine. Also find the common slope (if any) of the lines perpendicular to line AB.

9. $A(-1, 2)$, $B(-2, -1)$ **10.** $A(-2, 1)$, $B(2, -2)$

11. $A(2, 3)$, $B(-1, 3)$ **12.** $A(-2, 0)$, $B(-2, -2)$

In Exercises 13–16, find an equation for (a) the vertical line and (b) the horizontal line through the given point.

13. $(-1, 4/3)$ **14.** $\left(\sqrt{2}, -1.3\right)$

15. $\left(0, -\sqrt{2}\right)$ **16.** $(-\pi, 0)$

In Exercises 17–30, write an equation for each line described.

17. Passes through $(-1, 1)$ with slope -1

18. Passes through $(2, -3)$ with slope $1/2$

19. Passes through $(3, 4)$ and $(-2, 5)$

20. Passes through $(-8, 0)$ and $(-1, 3)$

21. Has slope $-5/4$ and y-intercept 6

22. Has slope $1/2$ and y-intercept -3

23. Passes through $(-12, -9)$ and has slope 0

24. Passes through $(1/3, 4)$, and has no slope

25. Has y-intercept 4 and x-intercept -1

26. Has y-intercept -6 and x-intercept 2

27. Passes through $(5, -1)$ and is parallel to the line $2x + 5y = 15$

28. Passes through $\left(-\sqrt{2}, 2\right)$ parallel to the line $\sqrt{2}x + 5y = \sqrt{3}$

29. Passes through $(4, 10)$ and is perpendicular to the line $6x - 3y = 5$

30. Passes through $(0, 1)$ and is perpendicular to the line $8x - 13y = 13$

In Exercises 31–34, find the line's x- and y-intercepts and use this information to graph the line.

31. $3x + 4y = 12$ **32.** $x + 2y = -4$

33. $\sqrt{2}x - \sqrt{3}y = \sqrt{6}$ **34.** $1.5x - y = -3$

35. Is there anything special about the relationship between the lines $Ax + By = C_1$ and $Bx - Ay = C_2$ ($A \ne 0, B \ne 0$)? Give reasons for your answer.

36. Is there anything special about the relationship between the lines $Ax + By = C_1$ and $Ax + By = C_2$ ($A \ne 0, B \ne 0$)? Give reasons for your answer.

Increments and Motion

37. A particle starts at $A(-2, 3)$ and its coordinates change by increments $\Delta x = 5$, $\Delta y = -6$. Find its new position.

38. A particle starts at $A(6, 0)$ and its coordinates change by increments $\Delta x = -6$, $\Delta y = 0$. Find its new position.

39. The coordinates of a particle change by $\Delta x = 5$ and $\Delta y = 6$ as it moves from $A(x, y)$ to $B(3, -3)$. Find x and y.

40. A particle started at $A(1, 0)$, circled the origin once counterclockwise, and returned to $A(1, 0)$. What were the net changes in its coordinates?

Circles

In Exercises 41–46, find an equation for the circle with the given center $C(h, k)$ and radius a. Then sketch the circle in the xy-plane. Include the circle's center in your sketch. Also, label the circle's x- and y-intercepts, if any, with their coordinate pairs.

41. $C(0, 2), \quad a = 2$
42. $C(-3, 0), \quad a = 3$
43. $C(-1, 5), \quad a = \sqrt{10}$
44. $C(1, 1), \quad a = \sqrt{2}$
45. $C\left(-\sqrt{3}, -2\right), \quad a = 2$
46. $C(3, 1/2), \quad a = 5$

Graph the circles whose equations are given in Exercises 47–52. Label each circle's center and intercepts (if any) with their coordinate pairs.

47. $x^2 + y^2 + 4x - 4y + 4 = 0$
48. $x^2 + y^2 - 8x + 4y + 16 = 0$
49. $x^2 + y^2 - 3y - 4 = 0$
50. $x^2 + y^2 - 4x - (9/4) = 0$
51. $x^2 + y^2 - 4x + 4y = 0$
52. $x^2 + y^2 + 2x = 3$

Parabolas

Graph the parabolas in Exercises 53–60. Label the vertex, axis, and intercepts in each case.

53. $y = x^2 - 2x - 3$
54. $y = x^2 + 4x + 3$
55. $y = -x^2 + 4x$
56. $y = -x^2 + 4x - 5$
57. $y = -x^2 - 6x - 5$
58. $y = 2x^2 - x + 3$
59. $y = \frac{1}{2}x^2 + x + 4$
60. $y = -\frac{1}{4}x^2 + 2x + 4$

Inequalities

Describe the regions defined by the inequalities and pairs of inequalities in Exercises 61–68.

61. $x^2 + y^2 > 7$
62. $x^2 + y^2 < 5$
63. $(x - 1)^2 + y^2 \le 4$
64. $x^2 + (y - 2)^2 \ge 4$
65. $x^2 + y^2 > 1, \quad x^2 + y^2 < 4$
66. $x^2 + y^2 \le 4, \quad (x + 2)^2 + y^2 \le 4$
67. $x^2 + y^2 + 6y < 0, \quad y > -3$
68. $x^2 + y^2 - 4x + 2y > 4, \quad x > 2$

69. Write an inequality that describes the points that lie inside the circle with center $(-2, 1)$ and radius $\sqrt{6}$.

70. Write an inequality that describes the points that lie outside the circle with center $(-4, 2)$ and radius 4.

71. Write a pair of inequalities that describe the points that lie inside or on the circle with center $(0, 0)$ and radius $\sqrt{2}$, and on or to the right of the vertical line through $(1, 0)$.

72. Write a pair of inequalities that describe the points that lie outside the circle with center $(0, 0)$ and radius 2, and inside the circle that has center $(1, 3)$ and passes through the origin.

Intersecting Lines, Circles, and Parabolas

In Exercises 73–80, graph the two equations and find the points in which the graphs intersect.

73. $y = 2x, \quad x^2 + y^2 = 1$
74. $x + y = 1, \quad (x - 1)^2 + y^2 = 1$
75. $y - x = 1, \quad y = x^2$
76. $x + y = 0, \quad y = -(x - 1)^2$
77. $y = -x^2, \quad y = 2x^2 - 1$
78. $y = \frac{1}{4}x^2, \quad y = (x - 1)^2$
79. $x^2 + y^2 = 1, \quad (x - 1)^2 + y^2 = 1$
80. $x^2 + y^2 = 1, \quad x^2 + y = 1$

Applications

81. **Insulation** By measuring slopes in the accompanying figure, estimate the temperature change in degrees per inch for (a) the gypsum wallboard; (b) the fiberglass insulation; (c) the wood sheathing.

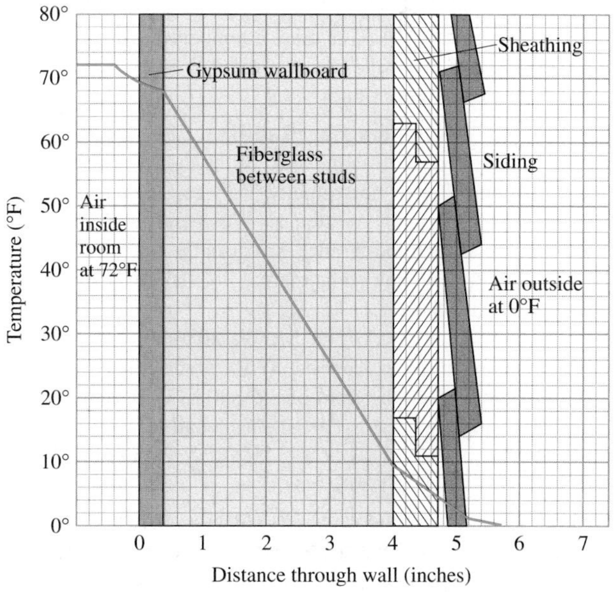

The temperature changes in the wall in Exercises 81 and 82.

82. Insulation According to the figure in Exercise 81, which of the materials is the best insulator? the poorest? Explain.

83. Pressure under water The pressure p experienced by a diver under water is related to the diver's depth d by an equation of the form $p = kd + 1$ (k a constant). At the surface, the pressure is 1 atmosphere. The pressure at 100 meters is about 10.94 atmospheres. Find the pressure at 50 meters.

84. Reflected light A ray of light comes in along the line $x + y = 1$ from the second quadrant and reflects off the x-axis (see the accompanying figure). The angle of incidence is equal to the angle of reflection. Write an equation for the line along which the departing light travels.

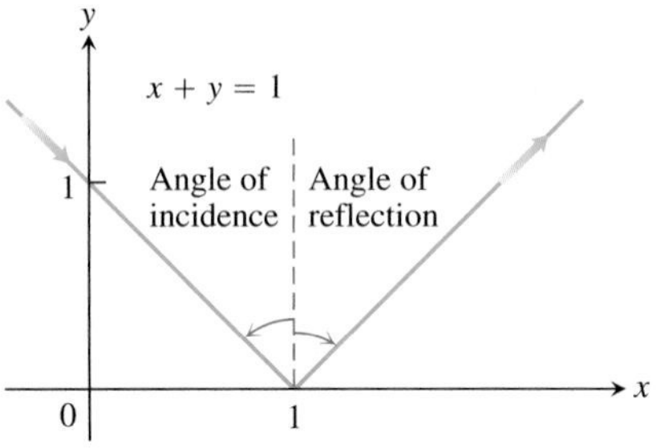

The path of the light ray in Exercise 84. Angles of incidence and reflection are measured from the perpendicular.

85. Fahrenheit vs. Celsius In the FC-plane, sketch the graph of the equation

$$C = \frac{5}{9}(F - 32)$$

linking Fahrenheit and Celsius temperatures. On the same graph sketch the line $C = F$. Is there a temperature at which a Celsius thermometer gives the same numerical reading as a Fahrenheit thermometer? If so, find it.

86. The Mt. Washington Cog Railway Civil engineers calculate the slope of roadbed as the ratio of the distance it rises or falls to the distance it runs horizontally. They call this ratio the **grade** of the roadbed, usually written as a percentage. Along the coast, commercial railroad grades are usually less than 2%. In the mountains, they may go as high as 4%. Highway grades are usually less than 5%.

The steepest part of the Mt. Washington Cog Railway in New Hampshire has an exceptional 37.1% grade. Along this part of the track, the seats in the front of the car are 14 ft above those in the rear. About how far apart are the front and rear rows of seats?

Theory and Examples

87. By calculating the lengths of its sides, show that the triangle with vertices at the points $A(1, 2)$, $B(5, 5)$, and $C(4, -2)$ is isosceles but not equilateral.

88. Show that the triangle with vertices $A(0, 0)$, $B\left(1, \sqrt{3}\right)$, and $C(2, 0)$ is equilateral.

89. Show that the points $A(2, -1)$, $B(1, 3)$, and $C(-3, 2)$ are vertices of a square, and find the fourth vertex.

90. The rectangle shown here has sides parallel to the axes. It is three times as long as it is wide, and its perimeter is 56 units. Find the coordinates of the vertices A, B, and C.

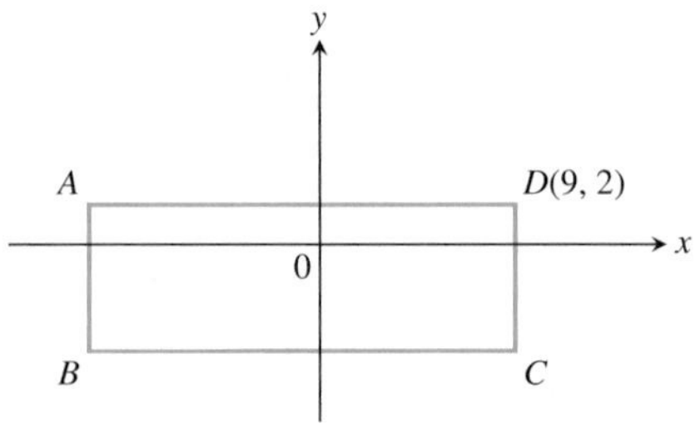

91. Three different parallelograms have vertices at $(-1, 1)$, $(2, 0)$, and $(2, 3)$. Sketch them and find the coordinates of the fourth vertex of each.

92. A 90° rotation counterclockwise about the origin takes $(2, 0)$ to $(0, 2)$, and $(0, 3)$ to $(-3, 0)$, as shown in the accompanying figure. Where does it take each of the following points?

a. $(4, 1)$ **b.** $(-2, -3)$ **c.** $(2, -5)$
d. $(x, 0)$ **e.** $(0, y)$ **f.** (x, y)
g. What point is taken to $(10, 3)$?

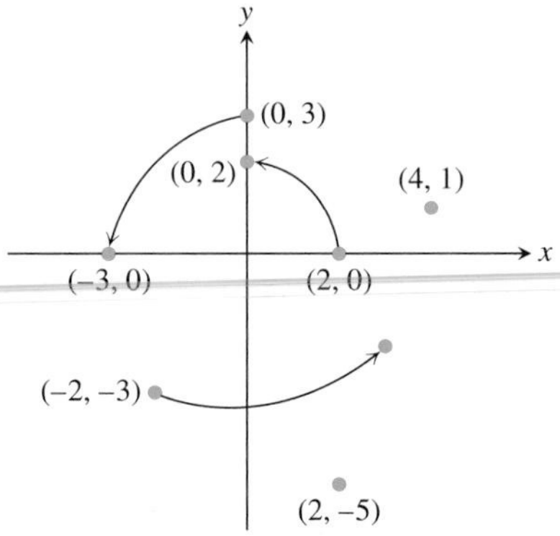

93. For what value of k is the line $2x + ky = 3$ perpendicular to the line $4x + y = 1$? For what value of k are the lines parallel?

94. Find the line that passes through the point $(1, 2)$ and through the point of intersection of the two lines $x + 2y = 3$ and $2x - 3y = -1$.

95. Midpoint of a line segment Show that the point with coordinates

$$\left(\frac{x_1 + x_2}{2}, \frac{y_1 + y_2}{2}\right)$$

is the midpoint of the line segment joining $P(x_1, y_1)$ to $Q(x_2, y_2)$.

96. The distance from a point to a line We can find the distance from a point $P(x_0, y_0)$ to a line $L: Ax + By = C$ by taking the following steps (there is a somewhat faster method in Section 12.5):

1. Find an equation for the line M through P perpendicular to L.
2. Find the coordinates of the point Q in which M and L intersect.
3. Find the distance from P to Q.

Use these steps to find the distance from P to L in each of the following cases.

a. $P(2, 1),\quad L: y = x + 2$

b. $P(4, 6),\quad L: 4x + 3y = 12$

c. $P(a, b),\quad L: x = -1$

d. $P(x_0, y_0),\quad L: Ax + By = C$

1.3 Functions and Their Graphs

Functions are the key to describing the real world in mathematical terms. This section reviews the ideas of functions, their graphs, and ways of representing them.

Functions; Domain and Range

The temperature at which water boils depends on the elevation above sea level (the boiling point drops as you ascend). The interest paid on a cash investment depends on the length of time the investment is held. The area of a circle depends on the radius of the circle. The distance an object travels at constant speed from an initial location along a straight line path depends on the elapsed time.

In each case, the value of one variable quantity, which we might call y, depends on the value of another variable quantity, which we might call x. Since the value of y is completely determined by the value of x, we say that y is a function of x. Often the value of y is given by a *rule* or formula that says how to calculate it from the variable x. For instance, the equation $A = \pi r^2$ is a rule that calculates the area A of a circle from its radius r.

In calculus we may want to refer to an unspecified function without having any particular formula in mind. A symbolic way to say "y is a function of x" is by writing

$$y = f(x) \qquad (\text{“}y \text{ equals } f \text{ of } x\text{”})$$

In this notation, the symbol f represents the function. The letter x, called the **independent variable**, represents the input value of f, and y, the **dependent variable**, represents the corresponding output **value** of f at x.

DEFINITION Function

A **function** from a set D to a set Y is a rule that assigns a *unique* (single) element $f(x) \in Y$ to each element $x \in D$.

The set D of all possible input values is called the **domain** of the function. The set of all values of $f(x)$ as x varies throughout D is called the **range** of the function. The range may not include every element in the set Y.

The domain and range of a function can be any sets of objects, but often in calculus they are sets of real numbers. (In Chapters 13–16 many variables may be involved.)

Think of a function f as a kind of machine that produces an output value $f(x)$ in its range whenever we feed it an input value x from its domain (Figure 1.22). The function

FIGURE 1.22 A diagram showing a function as a kind of machine.

keys on a calculator give an example of a function as a machine. For instance, the $\sqrt{x}$ key on a calculator gives an output value (the square root) whenever you enter a nonnegative number x and press the $\sqrt{x}$ key. The output value appearing in the display is usually a decimal approximation to the square root of x. If you input a number $x < 0$, then the calculator will indicate an error because $x < 0$ is not in the domain of the function and cannot be accepted as an input. The $\sqrt{x}$ key on a calculator is not the same as the exact mathematical function f defined by $f(x) = \sqrt{x}$ because it is limited to decimal outputs and has only finitely many inputs.

A function can also be pictured as an **arrow diagram** (Figure 1.23). Each arrow associates an element of the domain D to a unique or single element in the set Y. In Figure 1.23, the arrows indicate that $f(a)$ is associated with a, $f(x)$ is associated with x, and so on.

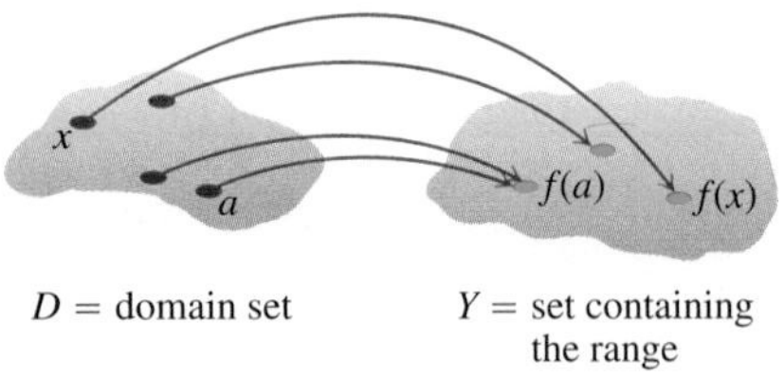

FIGURE 1.23 A function from a set D to a set Y assigns a unique element of Y to each element in D.

The domain of a function may be restricted by context. For example, the domain of the area function given by $A = \pi r^2$ only allows the radius r to be positive. When we define a function $y = f(x)$ with a formula and the domain is not stated explicitly or restricted by context, the domain is assumed to be the largest set of real x-values for which the formula gives real y-values, the so-called **natural domain**. If we want to restrict the domain in some way, we must say so. The domain of $y = x^2$ is the entire set of real numbers. To restrict the function to, say, positive values of x, we would write "$y = x^2, x > 0$."

Changing the domain to which we apply a formula usually changes the range as well. The range of $y = x^2$ is $[0, \infty)$. The range of $y = x^2, x \geq 2$, is the set of all numbers obtained by squaring numbers greater than or equal to 2. In set notation, the range is $\{x^2 | x \geq 2\}$ or $\{y | y \geq 4\}$ or $[4, \infty)$.

When the range of a function is a set of real numbers, the function is said to be **real-valued**. The domains and ranges of many real-valued functions of a real variable are intervals or combinations of intervals. The intervals may be open, closed, or half open, and may be finite or infinite.

EXAMPLE 1 Identifying Domain and Range

Verify the domains and ranges of these functions.

Function	**Domain (x)**	**Range (y)**
$y = x^2$	$(-\infty, \infty)$	$[0, \infty)$
$y = 1/x$	$(-\infty, 0) \cup (0, \infty)$	$(-\infty, 0) \cup (0, \infty)$
$y = \sqrt{x}$	$[0, \infty)$	$[0, \infty)$
$y = \sqrt{4 - x}$	$(-\infty, 4]$	$[0, \infty)$
$y = \sqrt{1 - x^2}$	$[-1, 1]$	$[0, 1]$

Solution The formula $y = x^2$ gives a real y-value for any real number x, so the domain is $(-\infty, \infty)$. The range of $y = x^2$ is $[0, \infty)$ because the square of any real number is nonnegative and every nonnegative number y is the square of its own square root, $y = \left(\sqrt{y}\right)^2$ for $y \geq 0$.

The formula $y = 1/x$ gives a real y-value for every x except $x = 0$. *We cannot divide any number by zero.* The range of $y = 1/x$, the set of reciprocals of all nonzero real numbers, is the set of all nonzero real numbers, since $y = 1/(1/y)$.

The formula $y = \sqrt{x}$ gives a real y-value only if $x \geq 0$. The range of $y = \sqrt{x}$ is $[0, \infty)$ because every nonnegative number is some number's square root (namely, it is the square root of its own square).

In $y = \sqrt{4 - x}$, the quantity $4 - x$ cannot be negative. That is, $4 - x \geq 0$, or $x \leq 4$. The formula gives real y-values for all $x \leq 4$. The range of $\sqrt{4 - x}$ is $[0, \infty)$, the set of all nonnegative numbers.

The formula $y = \sqrt{1 - x^2}$ gives a real y-value for every x in the closed interval from -1 to 1. Outside this domain, $1 - x^2$ is negative and its square root is not a real number. The values of $1 - x^2$ vary from 0 to 1 on the given domain, and the square roots of these values do the same. The range of $\sqrt{1 - x^2}$ is [0, 1]. ■

Graphs of Functions

Another way to visualize a function is its graph. If f is a function with domain D, its **graph** consists of the points in the Cartesian plane whose coordinates are the input-output pairs for f. In set notation, the graph is

$$\{(x, f(x)) \mid x \in D\}.$$

The graph of the function $f(x) = x + 2$ is the set of points with coordinates (x, y) for which $y = x + 2$. Its graph is sketched in Figure 1.24.

The graph of a function f is a useful picture of its behavior. If (x, y) is a point on the graph, then $y = f(x)$ is the height of the graph above the point x. The height may be positive or negative, depending on the sign of $f(x)$ (Figure 1.25).

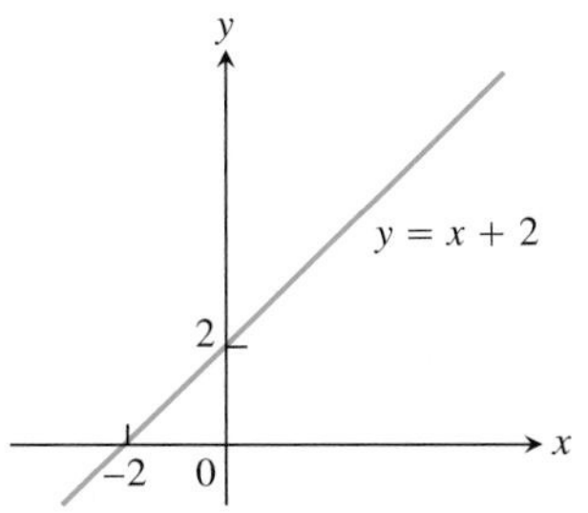

FIGURE 1.24 The graph of $f(x) = x + 2$ is the set of points (x, y) for which y has the value $x + 2$.

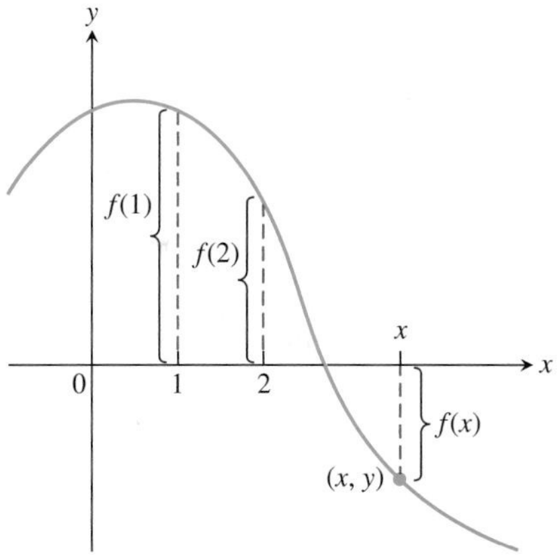

FIGURE 1.25 If (x, y) lies on the graph of f, then the value $y = f(x)$ is the height of the graph above the point x (or below x if $f(x)$ is negative).

x	$y = x^2$
-2	4
-1	1
0	0
1	1
$\frac{3}{2}$	$\frac{9}{4}$
2	4

EXAMPLE 2 Sketching a Graph

Graph the function $y = x^2$ over the interval $[-2, 2]$.

Solution

1. Make a table of xy-pairs that satisfy the function rule, in this case the equation $y = x^2$.

2. Plot the points (x, y) whose coordinates appear in the table. Use fractions when they are convenient computationally.

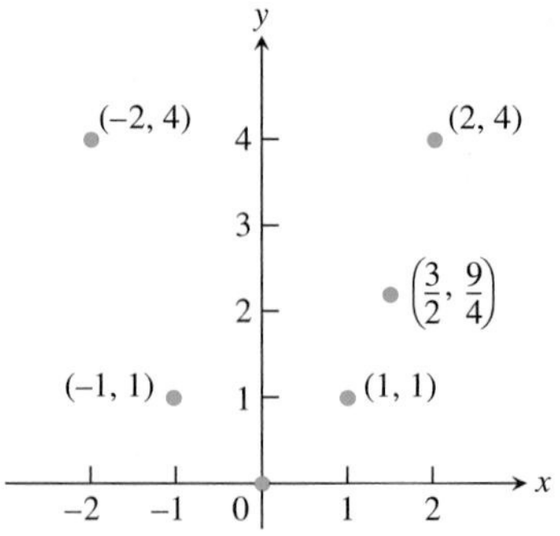

3. Draw a smooth curve through the plotted points. Label the curve with its equation.

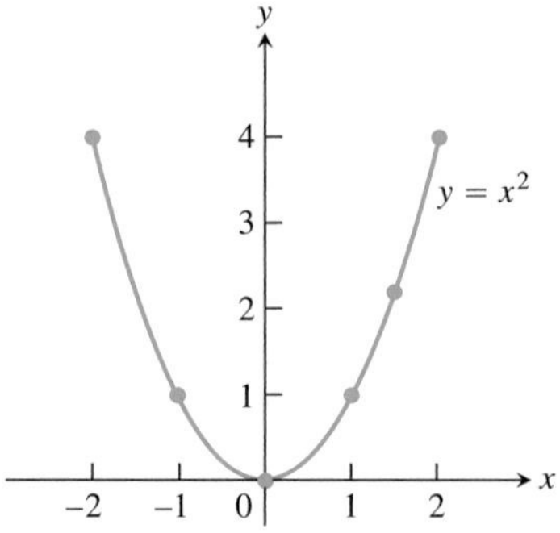

■

How do we know that the graph of $y = x^2$ doesn't look like one of these curves?

Computers and graphing calculators graph functions in much this way—by stringing together plotted points—and the same question arises.

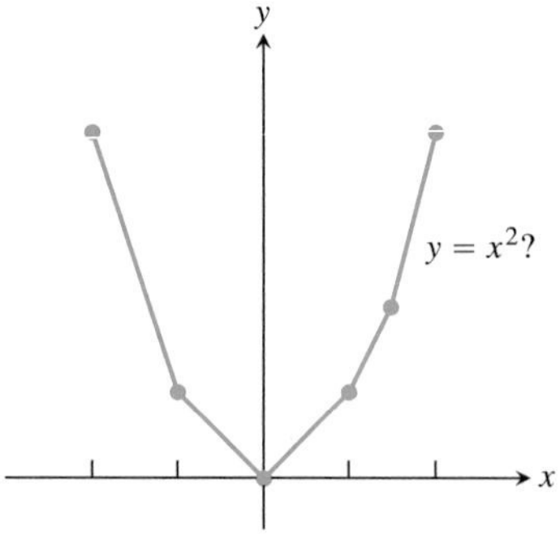

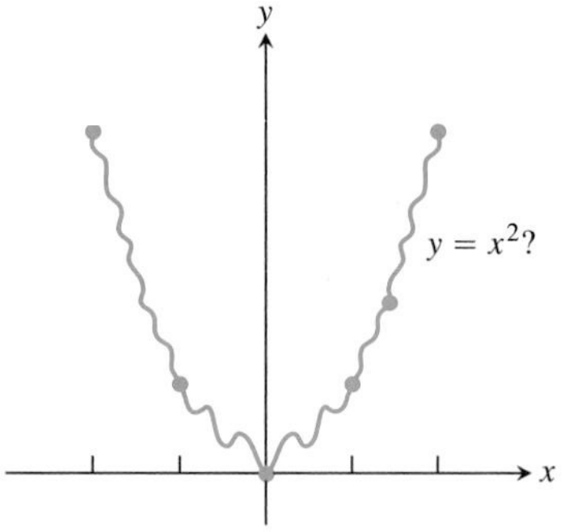

To find out, we could plot more points. But how would we then connect *them*? The basic question still remains: How do we know for sure what the graph looks like between the points we plot? The answer lies in calculus, as we will see in Chapter 4. There we will use the *derivative* to find a curve's shape between plotted points. Meanwhile we will have to settle for plotting points and connecting them as best we can.

EXAMPLE 3 Evaluating a Function from Its Graph

The graph of a fruit fly population p is shown in Figure 1.26.

(a) Find the populations after 20 and 45 days.

(b) What is the (approximate) range of the population function over the time interval $0 \le t \le 50$?

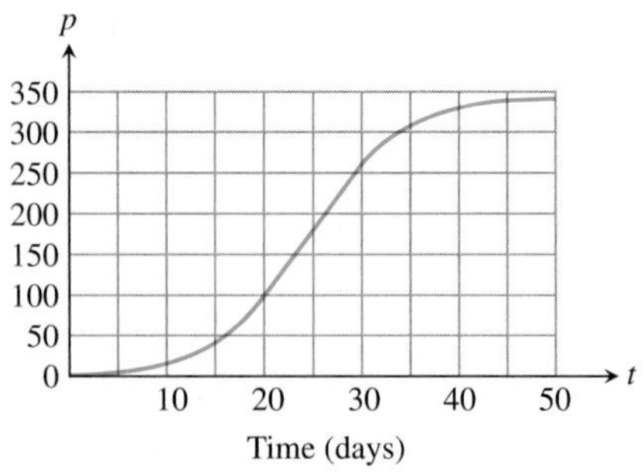

FIGURE 1.26 Graph of a fruit fly population versus time (Example 3).

Solution

(a) We see from Figure 1.26 that the point (20, 100) lies on the graph, so the value of the population p at 20 is $p(20) = 100$. Likewise, $p(45)$ is about 340.

(b) The range of the population function over $0 \le t \le 50$ is approximately [0, 345]. We also observe that the population appears to get closer and closer to the value $p = 350$ as time advances. ■

Representing a Function Numerically

We have seen how a function may be represented algebraically by a formula (the area function) and visually by a graph (Examples 2 and 3). Another way to represent a function is **numerically**, through a table of values. Numerical representations are often used by engineers and applied scientists. From an appropriate table of values, a graph of the function can be obtained using the method illustrated in Example 2, possibly with the aid of a computer. The graph of only the tabled points is called a **scatterplot**.

EXAMPLE 4 A Function Defined by a Table of Values

Musical notes are pressure waves in the air that can be recorded. The data in Table 1.2 give recorded pressure displacement versus time in seconds of a musical note produced by a tuning fork. The table provides a representation of the pressure function over time. If we first make a scatterplot and then connect the data points (t, p) from the table, we obtain the graph shown in Figure 1.27.

TABLE 1.2 Tuning fork data

Time	Pressure	Time	Pressure
0.00091	−0.080	0.00362	0.217
0.00108	0.200	0.00379	0.480
0.00125	0.480	0.00398	0.681
0.00144	0.693	0.00416	0.810
0.00162	0.816	0.00435	0.827
0.00180	0.844	0.00453	0.749
0.00198	0.771	0.00471	0.581
0.00216	0.603	0.00489	0.346
0.00234	0.368	0.00507	0.077
0.00253	0.099	0.00525	−0.164
0.00271	−0.141	0.00543	−0.320
0.00289	−0.309	0.00562	−0.354
0.00307	−0.348	0.00579	−0.248
0.00325	−0.248	0.00598	−0.035
0.00344	−0.041		

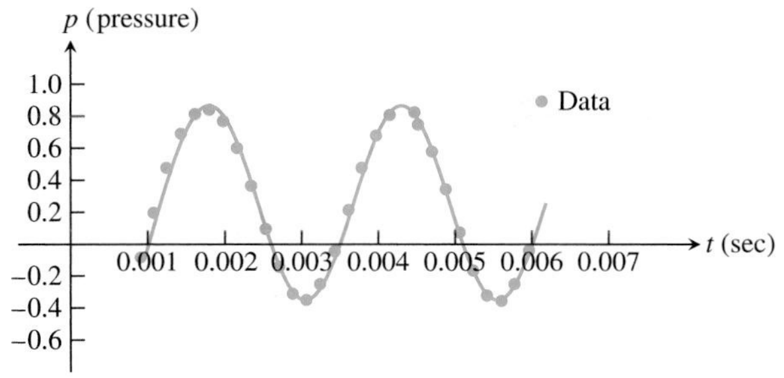

FIGURE 1.27 A smooth curve through the plotted points gives a graph of the pressure function represented by Table 1.2.

■

The Vertical Line Test

Not every curve you draw is the graph of a function. A function f can have only one value $f(x)$ for each x in its domain, so no *vertical line* can intersect the graph of a function more than once. Thus, a circle cannot be the graph of a function since some vertical lines intersect the circle twice (Figure 1.28a). If a is in the domain of a function f, then the vertical line $x = a$ will intersect the graph of f in the single point $(a, f(a))$.

The circle in Figure 1.28a, however, does contain the graphs of *two* functions of x; the upper semicircle defined by the function $f(x) = \sqrt{1 - x^2}$ and the lower semicircle defined by the function $g(x) = -\sqrt{1 - x^2}$ (Figures 1.28b and 1.28c).

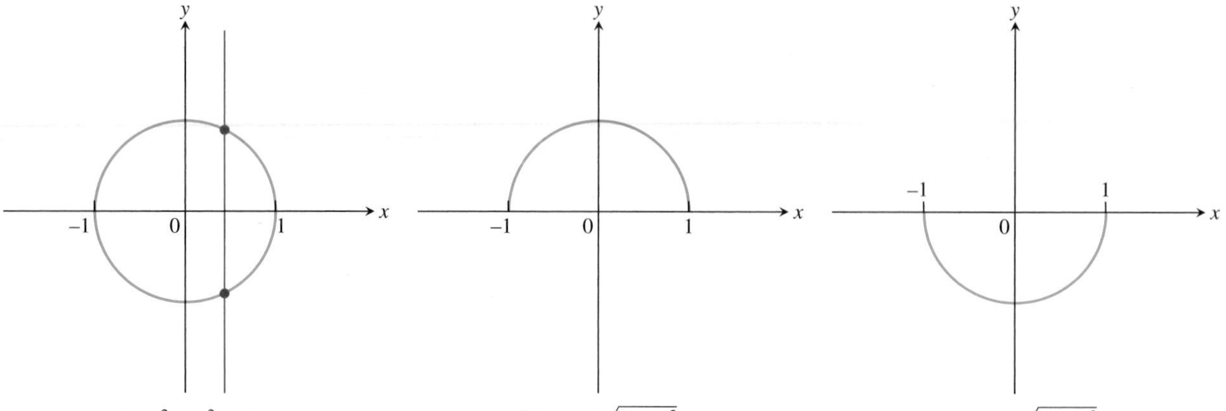

(a) $x^2 + y^2 = 1$ (b) $y = \sqrt{1 - x^2}$ (c) $y = -\sqrt{1 - x^2}$

FIGURE 1.28 (a) The circle is not the graph of a function; it fails the vertical line test. (b) The upper semicircle is the graph of a function $f(x) = \sqrt{1 - x^2}$. (c) The lower semicircle is the graph of a function $g(x) = -\sqrt{1 - x^2}$.

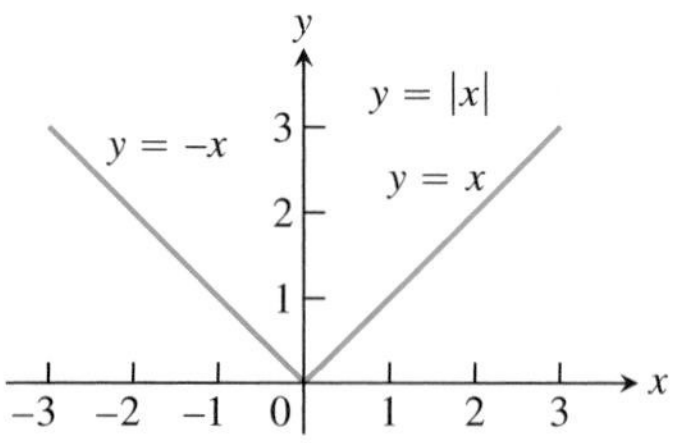

FIGURE 1.29 The absolute value function has domain $(-\infty, \infty)$ and range $[0, \infty)$.

Piecewise-Defined Functions

Sometimes a function is described by using different formulas on different parts of its domain. One example is the **absolute value function**

$$|x| = \begin{cases} x, & x \geq 0 \\ -x, & x < 0, \end{cases}$$

whose graph is given in Figure 1.29. Here are some other examples.

EXAMPLE 5 Graphing Piecewise-Defined Functions

The function

$$f(x) = \begin{cases} -x, & x < 0 \\ x^2, & 0 \leq x \leq 1 \\ 1, & x > 1 \end{cases}$$

is defined on the entire real line but has values given by different formulas depending on the position of x. The values of f are given by: $y = -x$ when $x < 0$, $y = x^2$ when $0 \leq x \leq 1$, and $y = 1$ when $x > 1$. The function, however, is *just one function* whose domain is the entire set of real numbers (Figure 1.30). ■

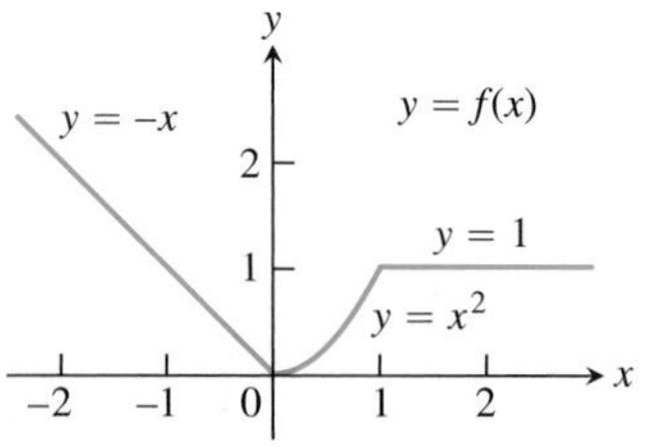

FIGURE 1.30 To graph the function $y = f(x)$ shown here, we apply different formulas to different parts of its domain (Example 5).

EXAMPLE 6 The Greatest Integer Function

The function whose value at any number x is the *greatest integer less than or equal to* x is called the **greatest integer function** or the **integer floor function**. It is denoted $\lfloor x \rfloor$, or, in some books, $[x]$ or $[[x]]$ or int x. Figure 1.31 shows the graph. Observe that

$$\lfloor 2.4 \rfloor = 2, \quad \lfloor 1.9 \rfloor = 1, \quad \lfloor 0 \rfloor = 0, \quad \lfloor -1.2 \rfloor = -2,$$
$$\lfloor 2 \rfloor = 2, \quad \lfloor 0.2 \rfloor = 0, \quad \lfloor -0.3 \rfloor = -1 \quad \lfloor -2 \rfloor = -2.$$

■

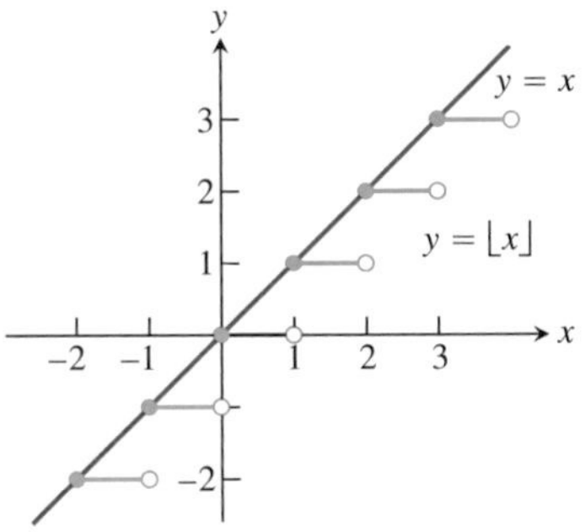

FIGURE 1.31 The graph of the greatest integer function $y = \lfloor x \rfloor$ lies on or below the line $y = x$, so it provides an integer floor for x (Example 6).

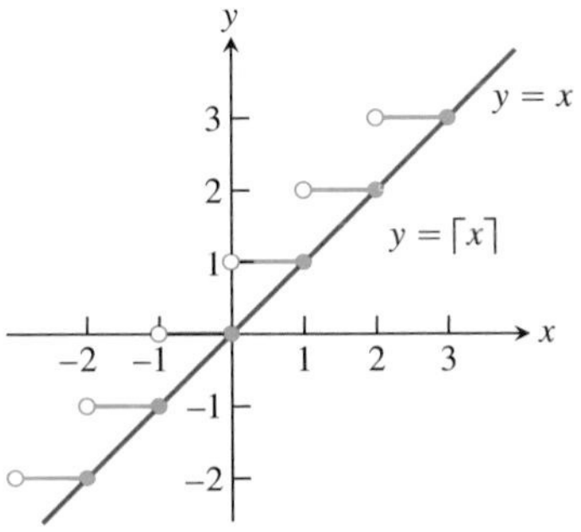

FIGURE 1.32 The graph of the least integer function $y = \lceil x \rceil$ lies on or above the line $y = x$, so it provides an integer ceiling for x (Example 7).

EXAMPLE 7 The Least Integer Function

The function whose value at any number x is the *smallest integer greater than or equal to* x is called the **least integer function** or the **integer ceiling function**. It is denoted $\lceil x \rceil$. Figure 1.32 shows the graph. For positive values of x, this function might represent, for example, the cost of parking x hours in a parking lot which charges \$1 for each hour or part of an hour. ■

EXAMPLE 8 Writing Formulas for Piecewise-Defined Functions

Write a formula for the function $y = f(x)$ whose graph consists of the two line segments in Figure 1.33.

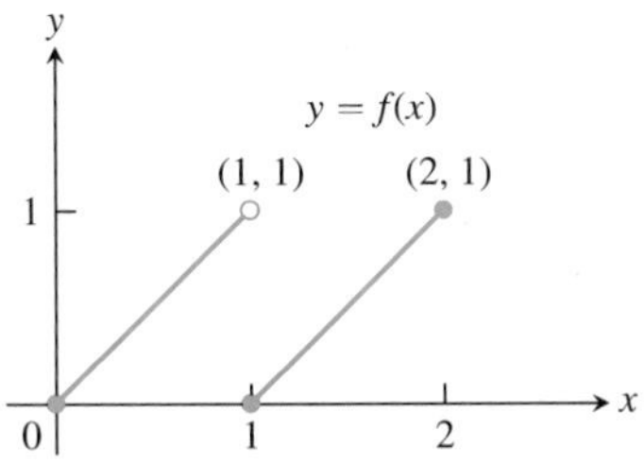

FIGURE 1.33 The segment on the left contains (0, 0) but not (1, 1). The segment on the right contains both of its endpoints (Example 8).

Solution We find formulas for the segments from (0, 0) to (1, 1), and from (1, 0) to (2, 1) and piece them together in the manner of Example 5.

Segment from (0, 0) to (1, 1) The line through (0, 0) and (1, 1) has slope $m = (1 - 0)/(1 - 0) = 1$ and y-intercept $b = 0$. Its slope-intercept equation is $y = x$. The segment from (0, 0) to (1, 1) that includes the point (0, 0) but not the point (1, 1) is the graph of the function $y = x$ restricted to the half-open interval $0 \le x < 1$, namely,

$$y = x, \qquad 0 \le x < 1.$$

Segment from (1, 0) to (2, 1) The line through (1, 0) and (2, 1) has slope $m = (1 - 0)/(2 - 1) = 1$ and passes through the point (1, 0). The corresponding point-slope equation for the line is

$$y = 0 + 1(x - 1), \qquad \text{or} \qquad y = x - 1.$$

The segment from (1, 0) to (2, 1) that includes both endpoints is the graph of $y = x - 1$ restricted to the closed interval $1 \le x \le 2$, namely,

$$y = x - 1, \qquad 1 \le x \le 2.$$

Piecewise formula Combining the formulas for the two pieces of the graph, we obtain

$$f(x) = \begin{cases} x, & 0 \le x < 1 \\ x - 1, & 1 \le x \le 2. \end{cases}$$ ■

EXERCISES 1.3

Functions

In Exercises 1–6, find the domain and range of each function.

1. $f(x) = 1 + x^2$
2. $f(x) = 1 - \sqrt{x}$
3. $F(t) = \dfrac{1}{\sqrt{t}}$
4. $F(t) = \dfrac{1}{1 + \sqrt{t}}$
5. $g(z) = \sqrt{4 - z^2}$
6. $g(z) = \dfrac{1}{\sqrt{4 - z^2}}$

In Exercises 7 and 8, which of the graphs are graphs of functions of x, and which are not? Give reasons for your answers.

7. **a.**

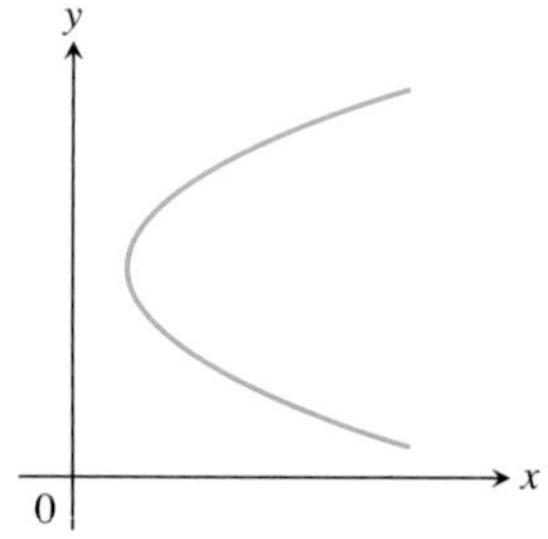

b.

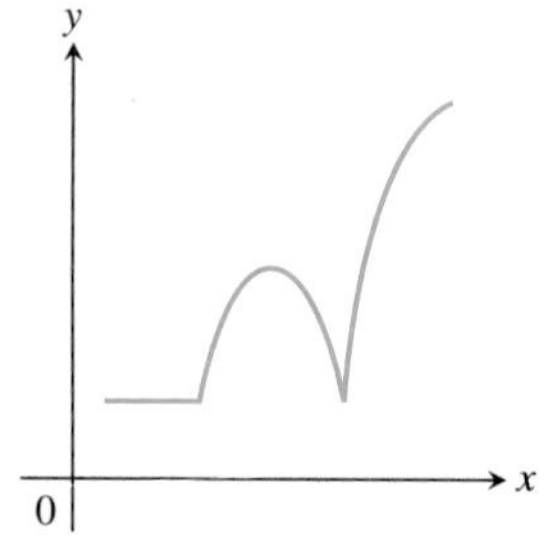

8. **a.**

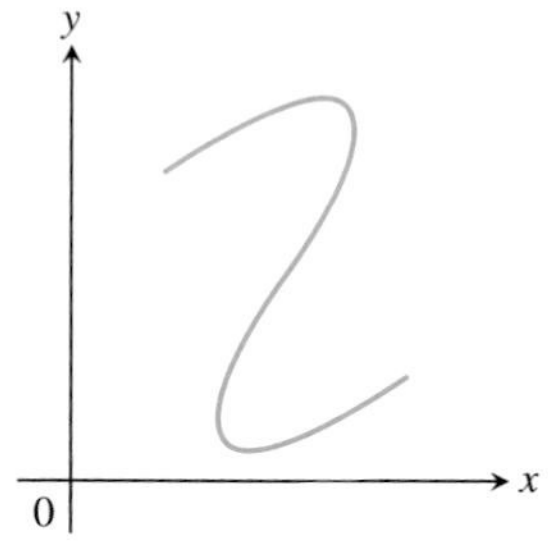

b.

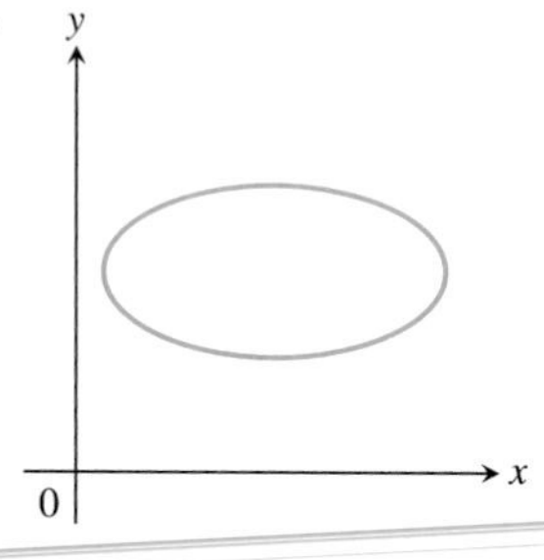

9. Consider the function $y = \sqrt{(1/x) - 1}$.
   - **a.** Can x be negative?
   - **b.** Can $x = 0$?
   - **c.** Can x be greater than 1?
   - **d.** What is the domain of the function?
10. Consider the function $y = \sqrt{2 - \sqrt{x}}$.
    - **a.** Can x be negative?
    - **b.** Can $\sqrt{x}$ be greater than 2?
    - **c.** What is the domain of the function?

Finding Formulas for Functions

11. Express the area and perimeter of an equilateral triangle as a function of the triangle's side length x.
12. Express the side length of a square as a function of the length d of the square's diagonal. Then express the area as a function of the diagonal length.
13. Express the edge length of a cube as a function of the cube's diagonal length d. Then express the surface area and volume of the cube as a function of the diagonal length.
14. A point P in the first quadrant lies on the graph of the function $f(x) = \sqrt{x}$. Express the coordinates of P as functions of the slope of the line joining P to the origin.

Functions and Graphs

Find the domain and graph the functions in Exercises 15–20.

15. $f(x) = 5 - 2x$
16. $f(x) = 1 - 2x - x^2$
17. $g(x) = \sqrt{|x|}$
18. $g(x) = \sqrt{-x}$
19. $F(t) = t/|t|$
20. $G(t) = 1/|t|$
21. Graph the following equations and explain why they are not graphs of functions of x.
    - **a.** $|y| = x$
    - **b.** $y^2 = x^2$
22. Graph the following equations and explain why they are not graphs of functions of x.
    - **a.** $|x| + |y| = 1$
    - **b.** $|x + y| = 1$

Piecewise-Defined Functions

Graph the functions in Exercises 23–26.

23. $f(x) = \begin{cases} x, & 0 \le x \le 1 \\ 2 - x, & 1 < x \le 2 \end{cases}$
24. $g(x) = \begin{cases} 1 - x, & 0 \le x \le 1 \\ 2 - x, & 1 < x \le 2 \end{cases}$
25. $F(x) = \begin{cases} 3 - x, & x \le 1 \\ 2x, & x > 1 \end{cases}$
26. $G(x) = \begin{cases} 1/x, & x < 0 \\ x, & 0 \le x \end{cases}$
27. Find a formula for each function graphed.

a.

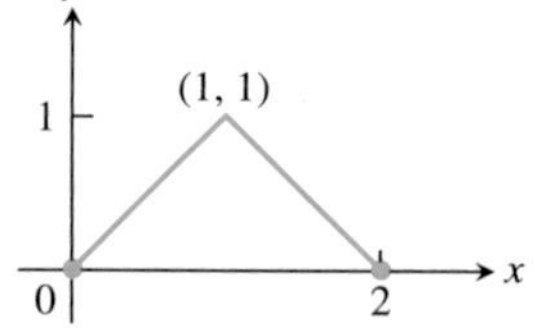

b.

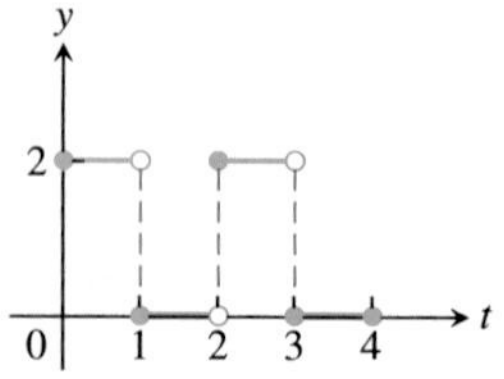

28. a.

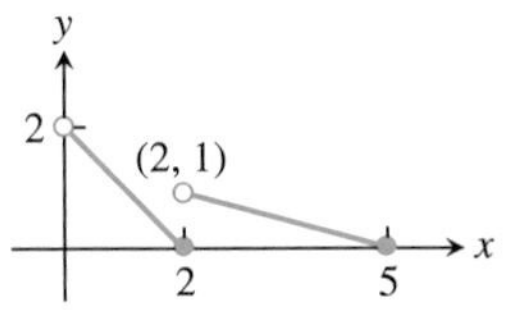

b.

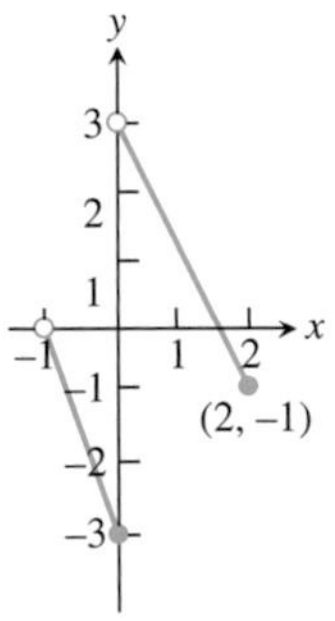

29. a.

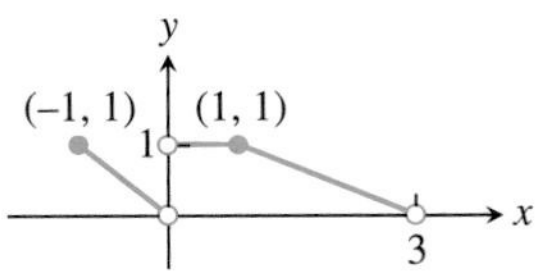

b.

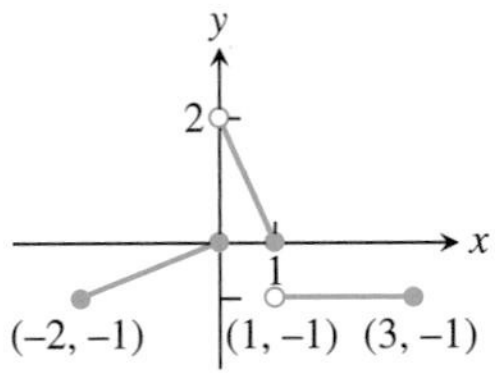

30. a.

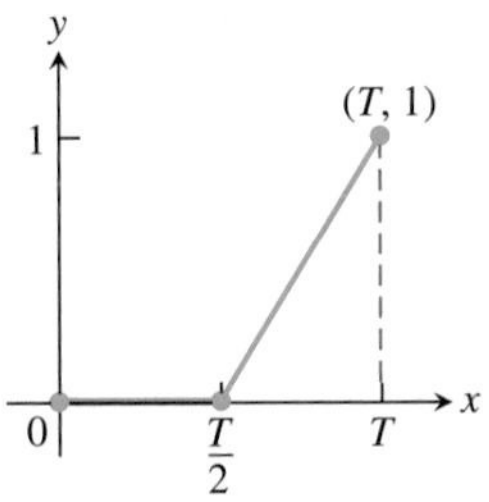

b.

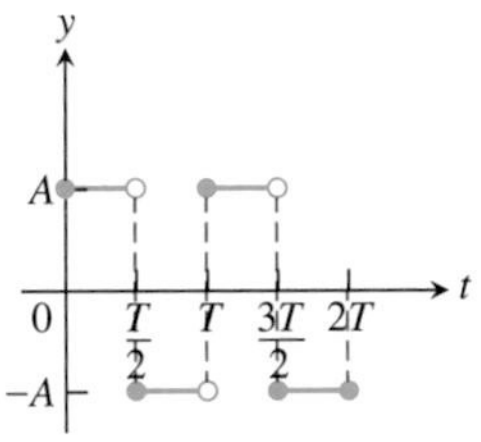

31. a. Graph the functions $f(x) = x/2$ and $g(x) = 1 + (4/x)$ together to identify the values of x for which

$$\frac{x}{2} > 1 + \frac{4}{x}.$$

b. Confirm your findings in part (a) algebraically.

32. a. Graph the functions $f(x) = 3/(x - 1)$ and $g(x) = 2/(x + 1)$ together to identify the values of x for which

$$\frac{3}{x - 1} < \frac{2}{x + 1}.$$

b. Confirm your findings in part (a) algebraically.

The Greatest and Least Integer Functions

33. For what values of x is

a. $\lfloor x \rfloor = 0$? **b.** $\lceil x \rceil = 0$?

34. What real numbers x satisfy the equation $\lfloor x \rfloor = \lceil x \rceil$?

35. Does $\lceil -x \rceil = -\lfloor x \rfloor$ for all real x? Give reasons for your answer.

36. Graph the function

$$f(x) = \begin{cases} \lfloor x \rfloor, & x \ge 0 \\ \lceil x \rceil, & x < 0 \end{cases}$$

Why is $f(x)$ called the *integer part* of x?

Theory and Examples

37. A box with an open top is to be constructed from a rectangular piece of cardboard with dimensions 14 in. by 22 in. by cutting out equal squares of side x at each corner and then folding up the sides as in the figure. Express the volume V of the box as a function of x.

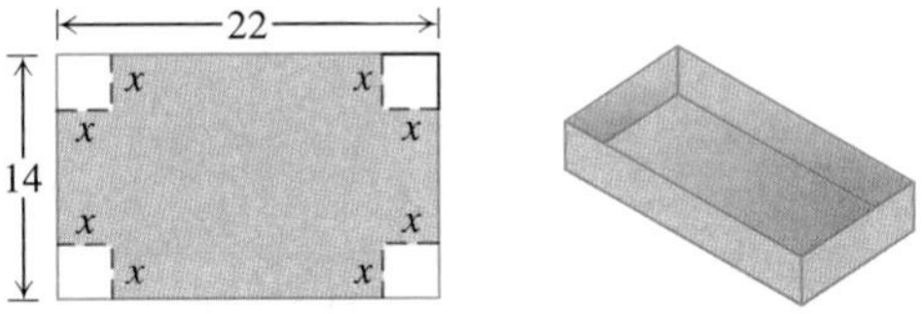

38. The figure shown here shows a rectangle inscribed in an isosceles right triangle whose hypotenuse is 2 units long.

a. Express the y-coordinate of P in terms of x. (You might start by writing an equation for the line AB.)

b. Express the area of the rectangle in terms of x.

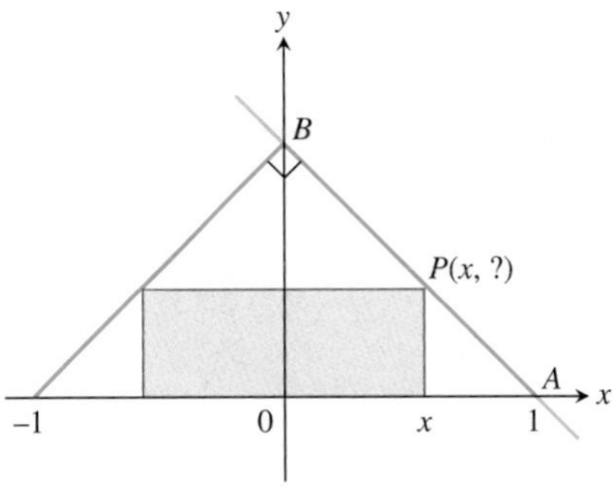

39. A cone problem Begin with a circular piece of paper with a 4 in. radius as shown in part (a). Cut out a sector with an arc length of x. Join the two edges of the remaining portion to form a cone with radius r and height h, as shown in part (b).

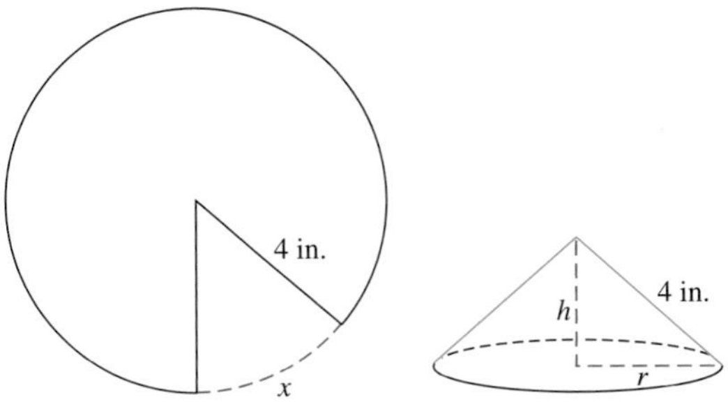

(a) (b)

a. Explain why the circumference of the base of the cone is $8\pi - x$.

b. Express the radius r as a function of x.

c. Express the height h as a function of x.

d. Express the volume V of the cone as a function of x.

40. Industrial costs Dayton Power and Light, Inc., has a power plant on the Miami River where the river is 800 ft wide. To lay a new cable from the plant to a location in the city 2 mi downstream on the opposite side costs \$180 per foot across the river and \$100 per foot along the land.

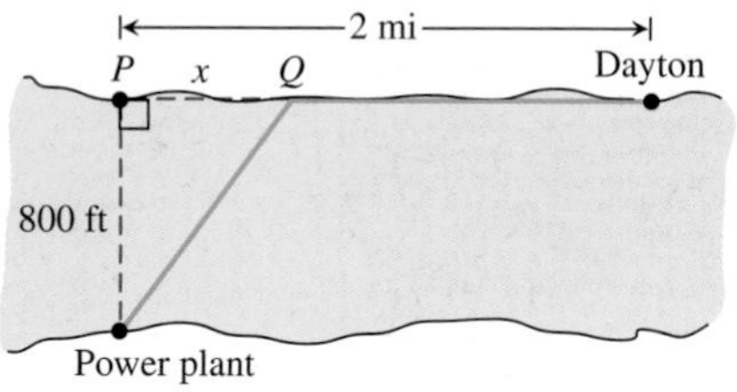

a. Suppose that the cable goes from the plant to a point Q on the opposite side that is x ft from the point P directly opposite the plant. Write a function $C(x)$ that gives the cost of laying the cable in terms of the distance x.

b. Generate a table of values to determine if the least expensive location for point Q is less than 2000 ft or greater than 2000 ft from point P.

41. For a curve to be *symmetric about the x-axis*, the point (x, y) must lie on the curve if and only if the point $(x, -y)$ lies on the curve. Explain why a curve that is symmetric about the x-axis is not the graph of a function, unless the function is $y = 0$.

42. A magic trick You may have heard of a magic trick that goes like this: Take any number. Add 5. Double the result. Subtract 6. Divide by 2. Subtract 2. Now tell me your answer, and I'll tell you what you started with. Pick a number and try it.

You can see what is going on if you let x be your original number and follow the steps to make a formula $f(x)$ for the number you end up with.

1.4 Identifying Functions; Mathematical Models

There are a number of important types of functions frequently encountered in calculus. We identify and briefly summarize them here.

Linear Functions A function of the form $f(x) = mx + b$, for constants m and b, is called a **linear function**. Figure 1.34 shows an array of lines $f(x) = mx$ where $b = 0$, so these lines pass through the origin. Constant functions result when the slope $m = 0$ (Figure 1.35).

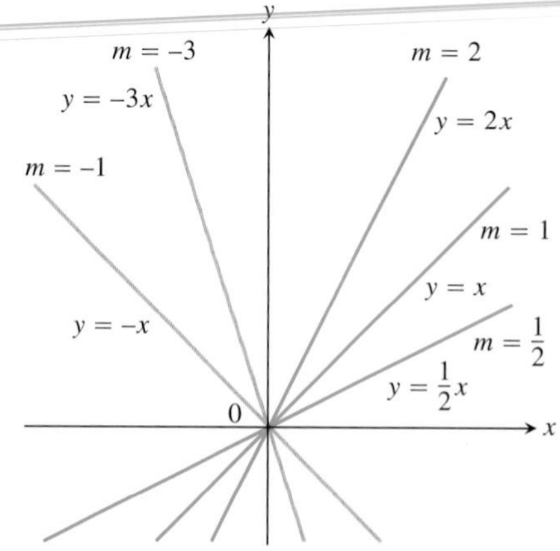

FIGURE 1.34 The collection of lines $y = mx$ has slope m and all lines pass through the origin.

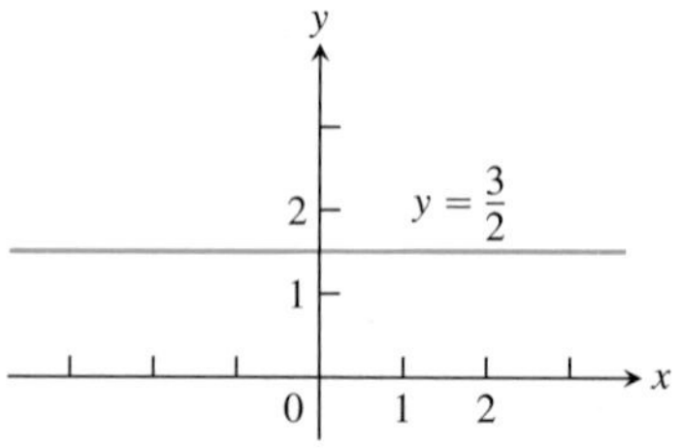

FIGURE 1.35 A constant function has slope $m = 0$.

Power Functions A function $f(x) = x^a$, where a is a constant, is called a **power function**. There are several important cases to consider.

(a) $a = n$, a positive integer.

The graphs of $f(x) = x^n$, for $n = 1, 2, 3, 4, 5$, are displayed in Figure 1.36. These functions are defined for all real values of x. Notice that as the power n gets larger, the curves tend to flatten toward the x-axis on the interval $(-1, 1)$, and also rise more steeply for $|x| > 1$. Each curve passes through the point $(1, 1)$ and through the origin.

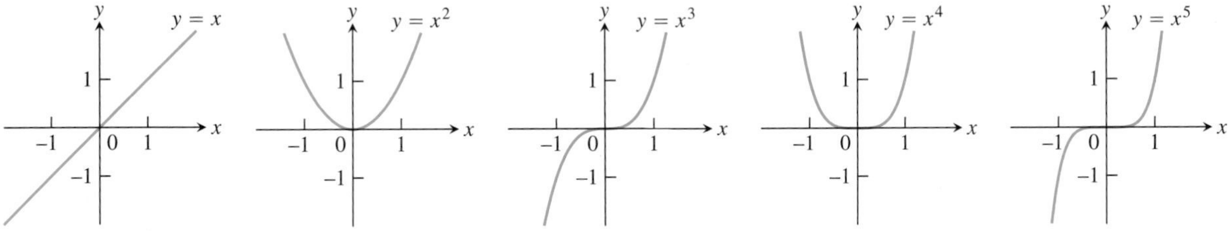

FIGURE 1.36 Graphs of $f(x) = x^n$, $n = 1, 2, 3, 4, 5$ defined for $-\infty < x < \infty$.

(b) $a = -1$ or $a = -2$.

The graphs of the functions $f(x) = x^{-1} = 1/x$ and $g(x) = x^{-2} = 1/x^2$ are shown in Figure 1.37. Both functions are defined for all $x \neq 0$ (you can never divide by zero). The graph of $y = 1/x$ is the hyperbola $xy = 1$ which approaches the coordinate axes far from the origin. The graph of $y = 1/x^2$ also approaches the coordinate axes.

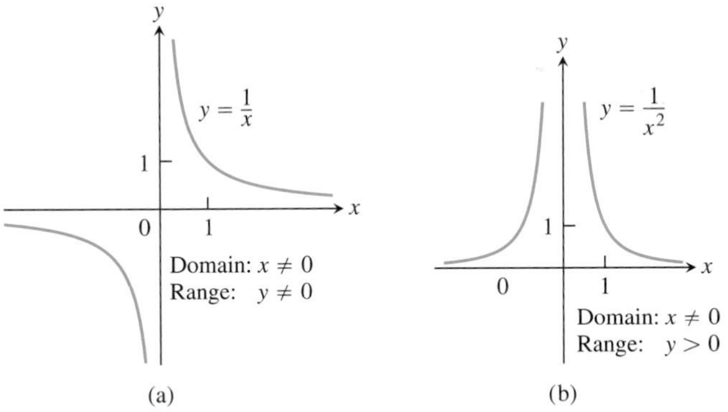

FIGURE 1.37 Graphs of the power functions $f(x) = x^a$ for part (a) $a = -1$ and for part (b) $a = -2$.

(c) $a = \frac{1}{2}, \frac{1}{3}, \frac{3}{2}$, and $\frac{2}{3}$.

The functions $f(x) = x^{1/2} = \sqrt{x}$ and $g(x) = x^{1/3} = \sqrt[3]{x}$ are the **square root** and **cube root** functions, respectively. The domain of the square root function is $[0, \infty)$, but the cube root function is defined for all real x. Their graphs are displayed in Figure 1.38 along with the graphs of $y = x^{3/2}$ and $y = x^{2/3}$. (Recall that $x^{3/2} = (x^{1/2})^3$ and $x^{2/3} = (x^{1/3})^2$.)

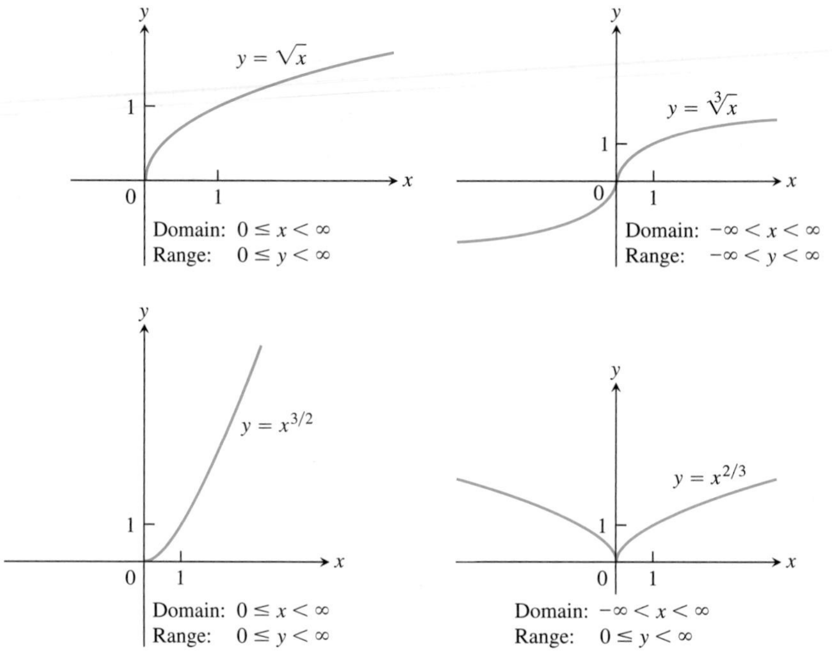

FIGURE 1.38 Graphs of the power functions $f(x) = x^a$ for $a = \frac{1}{2}, \frac{1}{3}, \frac{3}{2}$, and $\frac{2}{3}$.

Polynomials A function p is a **polynomial** if

$$p(x) = a_n x^n + a_{n-1}x^{n-1} + \cdots + a_1 x + a_0$$

where n is a nonnegative integer and the numbers $a_0, a_1, a_2, \ldots, a_n$ are real constants (called the **coefficients** of the polynomial). All polynomials have domain $(-\infty, \infty)$. If the leading coefficient $a_n \neq 0$ and $n > 0$, then n is called the **degree** of the polynomial. Linear functions with $m \neq 0$ are polynomials of degree 1. Polynomials of degree 2, usually written as $p(x) = ax^2 + bx + c$, are called **quadratic functions**. Likewise, **cubic functions** are polynomials $p(x) = ax^3 + bx^2 + cx + d$ of degree 3. Figure 1.39 shows the graphs of three polynomials. You will learn how to graph polynomials in Chapter 4.

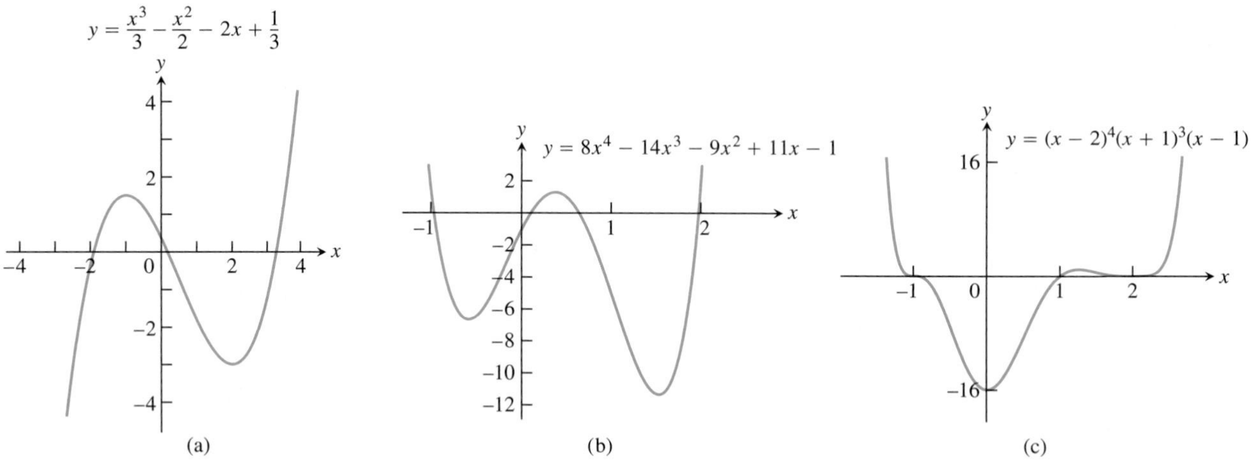

FIGURE 1.39 Graphs of three polynomial functions.

Rational Functions A **rational function** is a quotient or ratio of two polynomials:

$$f(x) = \frac{p(x)}{q(x)}$$

where p and q are polynomials. The domain of a rational function is the set of all real x for which $q(x) \neq 0$. For example, the function

$$f(x) = \frac{2x^2 - 3}{7x + 4}$$

is a rational function with domain $\{x \mid x \neq -4/7\}$. Its graph is shown in Figure 1.40a with the graphs of two other rational functions in Figures 1.40b and 1.40c.

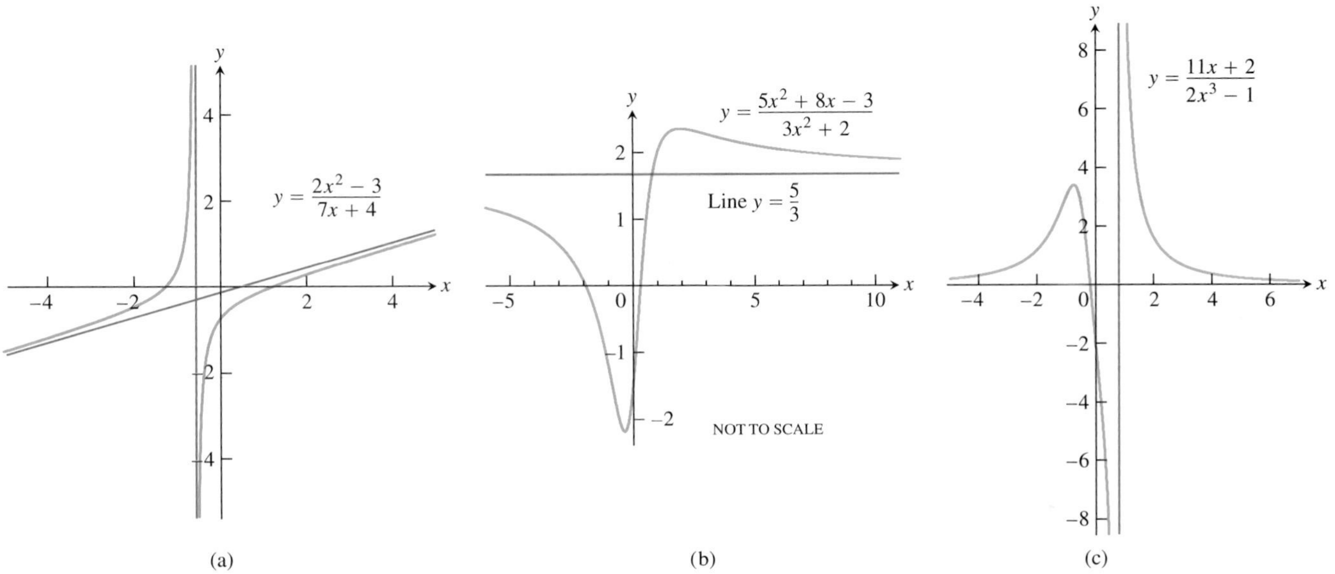

FIGURE 1.40 Graphs of three rational functions.

Algebraic Functions Any function constructed from polynomials using algebraic operations (addition, subtraction, multiplication, division, and taking roots) lies within the class of **algebraic functions**. All rational functions are algebraic, but also included are more complicated functions (such as those satisfying an equation like $y^3 - 9xy + x^3 = 0$, studied in Section 3.6). Figure 1.41 displays the graphs of three algebraic functions.

Trigonometric Functions We review trigonometric functions in Section 1.6. The graphs of the sine and cosine functions are shown in Figure 1.42.

Exponential Functions Functions of the form $f(x) = a^x$, where the base $a > 0$ is a positive constant and $a \neq 1$, are called **exponential functions**. All exponential functions have domain $(-\infty, \infty)$ and range $(0, \infty)$. So an exponential function never assumes the value 0. The graphs of some exponential functions are shown in Figure 1.43. The calculus of exponential functions is studied in Chapter 7.

Logarithmic Functions These are the functions $f(x) = \log_a x$, where the base $a \neq 1$ is a positive constant. They are the *inverse functions* of the exponential functions, and the

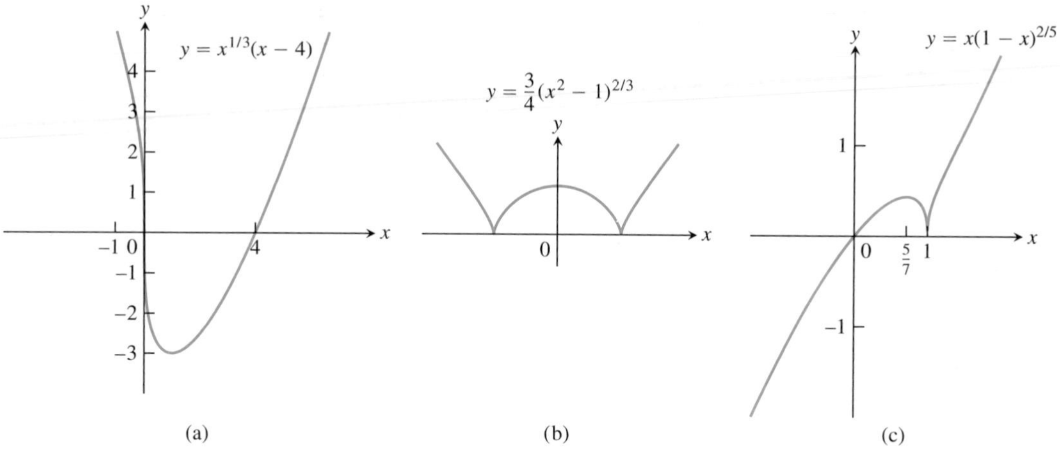

FIGURE 1.41 Graphs of three algebraic functions.

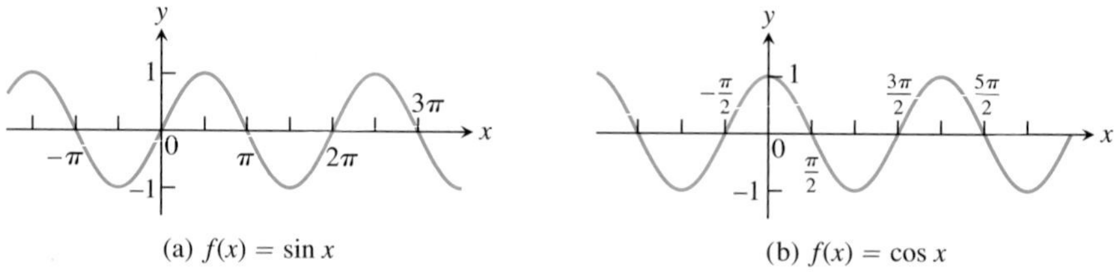

FIGURE 1.42 Graphs of the sine and cosine functions.

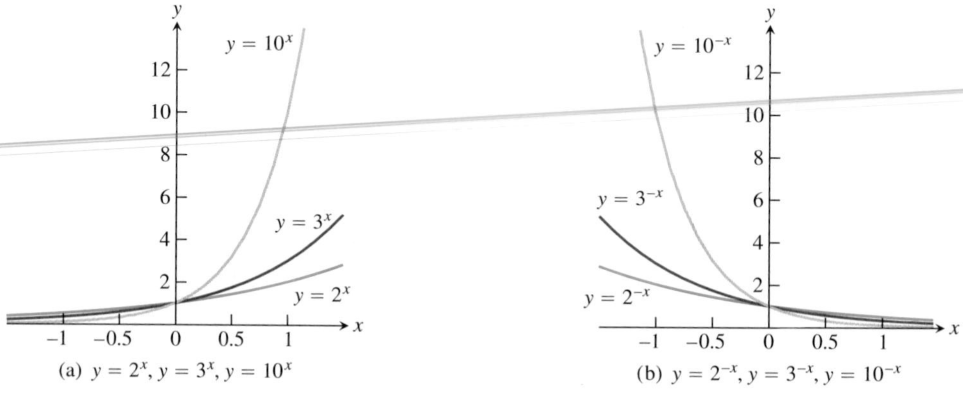

FIGURE 1.43 Graphs of exponential functions.

calculus of these functions is studied in Chapter 7. Figure 1.44 shows the graphs of four logarithmic functions with various bases. In each case the domain is $(0, \infty)$ and the range is $(-\infty, \infty)$.

Transcendental Functions These are functions that are not algebraic. They include the trigonometric, inverse trigonometric, exponential, and logarithmic functions, and many

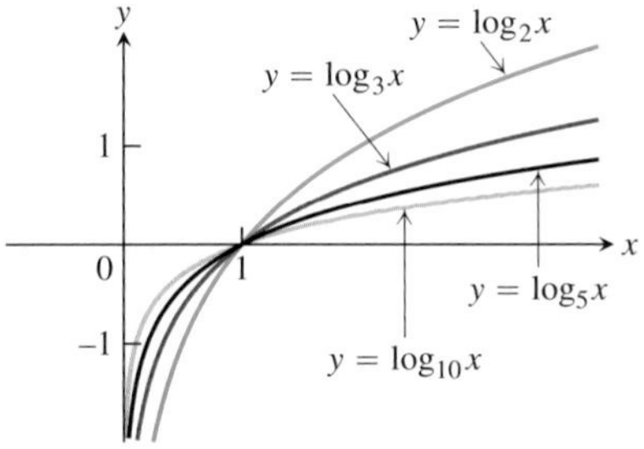

FIGURE 1.44 Graphs of four logarithmic functions.

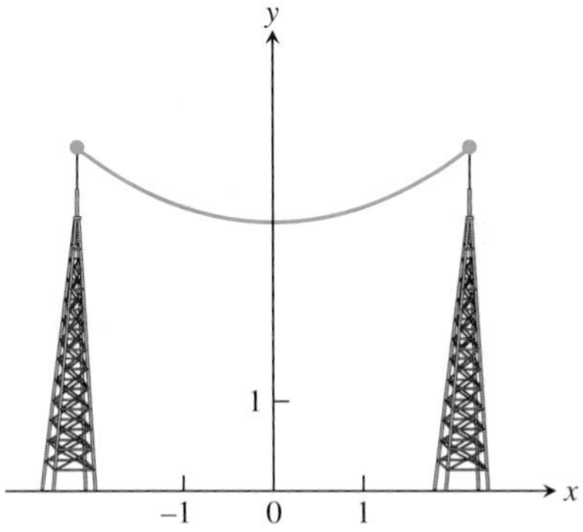

FIGURE 1.45 Graph of a catenary or hanging cable. (The Latin word *catena* means "chain.")

other functions as well (such as the hyperbolic functions studied in Chapter 7). An example of a transcendental function is a **catenary**. Its graph takes the shape of a cable, like a telephone line or TV cable, strung from one support to another and hanging freely under its own weight (Figure 1.45).

EXAMPLE 1 Recognizing Functions

Identify each function given here as one of the types of functions we have discussed. Keep in mind that some functions can fall into more than one category. For example, $f(x) = x^2$ is both a power function and a polynomial of second degree.

(a) $f(x) = 1 + x - \frac{1}{2}x^5$ **(b)** $g(x) = 7^x$ **(c)** $h(z) = z^7$

(d) $y(t) = \sin\left(t - \frac{\pi}{4}\right)$

Solution

(a) $f(x) = 1 + x - \frac{1}{2}x^5$ is a polynomial of degree 5.

(b) $g(x) = 7^x$ is an exponential function with base 7. Notice that the variable x is the exponent.

(c) $h(z) = z^7$ is a power function. (The variable z is the base.)

(d) $y(t) = \sin\left(t - \frac{\pi}{4}\right)$ is a trigonometric function. ■

Increasing Versus Decreasing Functions

If the graph of a function *climbs* or *rises* as you move from left to right, we say that the function is *increasing*. If the graph *descends* or *falls* as you move from left to right, the function is *decreasing*. We give formal definitions of increasing functions and decreasing functions in Section 4.3. In that section, you will learn how to find the intervals over which a function is increasing and the intervals where it is decreasing. Here are examples from Figures 1.36, 1.37, and 1.38.

Function	Where increasing	Where decreasing
$y = x^2$	$0 \le x < \infty$	$-\infty < x \le 0$
$y = x^3$	$-\infty < x < \infty$	Nowhere
$y = 1/x$	Nowhere	$-\infty < x < 0$ and $0 < x < \infty$
$y = 1/x^2$	$-\infty < x < 0$	$0 < x < \infty$
$y = \sqrt{x}$	$0 \le x < \infty$	Nowhere
$y = x^{2/3}$	$0 \le x < \infty$	$-\infty < x \le 0$

Even Functions and Odd Functions: Symmetry

The graphs of *even* and *odd* functions have characteristic symmetry properties.

DEFINITIONS Even Function, Odd Function

A function $y = f(x)$ is an

$$\textbf{even function of } x \quad \text{if } f(-x) = f(x),$$
$$\textbf{odd function of } x \quad \text{if } f(-x) = -f(x),$$

for every x in the function's domain.

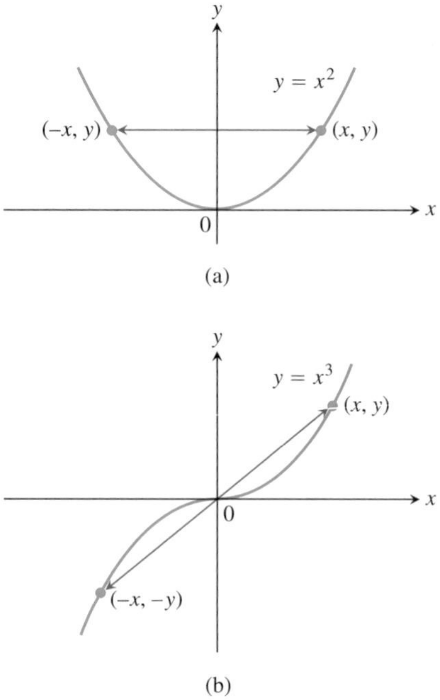

FIGURE 1.46 In part (a) the graph of $y = x^2$ (an even function) is symmetric about the y-axis. The graph of $y = x^3$ (an odd function) in part (b) is symmetric about the origin.

The names even and odd come from powers of x. If y is an even power of x, as in $y = x^2$ or $y = x^4$, it is an even function of x (because $(-x)^2 = x^2$ and $(-x)^4 = x^4$). If y is an odd power of x, as in $y = x$ or $y = x^3$, it is an odd function of x (because $(-x)^1 = -x$ and $(-x)^3 = -x^3$).

The graph of an even function is **symmetric about the y-axis**. Since $f(-x) = f(x)$, a point (x, y) lies on the graph if and only if the point $(-x, y)$ lies on the graph (Figure 1.46a). A reflection across the y-axis leaves the graph unchanged.

The graph of an odd function is **symmetric about the origin**. Since $f(-x) = -f(x)$, a point (x, y) lies on the graph if and only if the point $(-x, -y)$ lies on the graph (Figure 1.46b). Equivalently, a graph is symmetric about the origin if a rotation of 180° about the origin leaves the graph unchanged. Notice that the definitions imply both x and $-x$ must be in the domain of f.

EXAMPLE 2 Recognizing Even and Odd Functions

$f(x) = x^2$ Even function: $(-x)^2 = x^2$ for all x; symmetry about y-axis.

$f(x) = x^2 + 1$ Even function: $(-x)^2 + 1 = x^2 + 1$ for all x; symmetry about y-axis (Figure 1.47a).

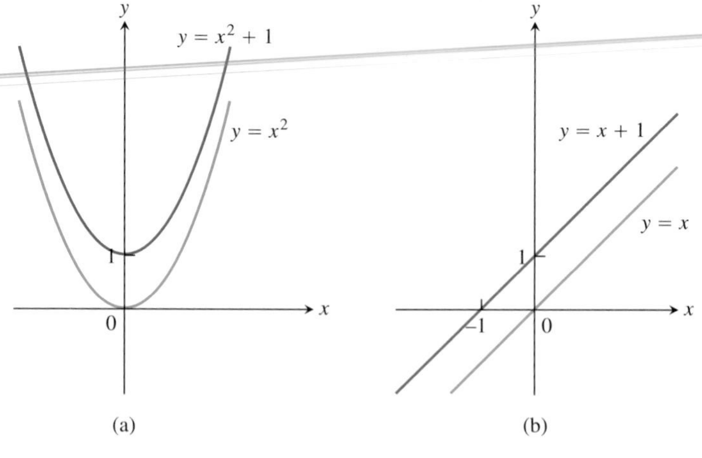

FIGURE 1.47 (a) When we add the constant term 1 to the function $y = x^2$, the resulting function $y = x^2 + 1$ is still even and its graph is still symmetric about the y-axis. (b) When we add the constant term 1 to the function $y = x$, the resulting function $y = x + 1$ is no longer odd. The symmetry about the origin is lost (Example 2).

$f(x) = x$ Odd function: $(-x) = -x$ for all x; symmetry about the origin.

$f(x) = x + 1$ Not odd: $f(-x) = -x + 1$, but $-f(x) = -x - 1$. The two are not equal.
Not even: $(-x) + 1 \neq x + 1$ for all $x \neq 0$ (Figure 1.47b). ∎

Mathematical Models

To help us better understand our world, we often describe a particular phenomenon mathematically (by means of a function or an equation, for instance). Such a **mathematical model** is an idealization of the real-world phenomenon and is seldom a completely accurate representation. Although any model has its limitations, a good one can provide valuable results and conclusions. A model allows us to reach conclusions, as illustrated in Figure 1.48.

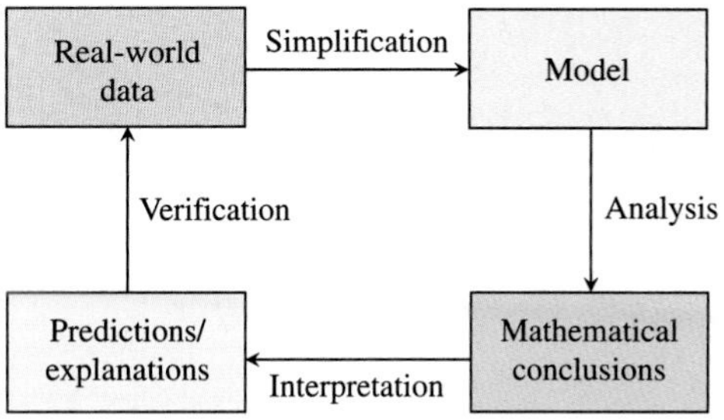

FIGURE 1.48 A flow of the modeling process beginning with an examination of real-world data.

Most models simplify reality and can only *approximate* real-world behavior. One simplifying relationship is *proportionality*.

> **DEFINITION Proportionality**
> Two variables y and x are **proportional** (to one another) if one is always a constant multiple of the other; that is, if
> $$y = kx$$
> for some nonzero constant k.

The definition means that the graph of y versus x lies along a straight line through the origin. This graphical observation is useful in testing whether a given data collection reasonably assumes a proportionality relationship. If a proportionality is reasonable, a plot of one variable against the other should approximate a straight line through the origin.

EXAMPLE 3 Kepler's Third Law

A famous proportionality, postulated by the German astronomer Johannes Kepler in the early seventeenth century, is his third law. If T is the period in days for a planet to complete one full orbit around the sun, and R is the mean distance of the planet to the sun, then Kepler postulated that T is proportional to R raised to the $3/2$ power. That is, for some constant k,

$$T = kR^{3/2}.$$

Let's compare his law to the data in Table 1.3 taken from the *1993 World Almanac*.

TABLE 1.3 Orbital periods and mean distances of planets from the sun

Planet	T Period (days)	R Mean distance (millions of miles)
Mercury	88.0	36
Venus	224.7	67.25
Earth	365.3	93
Mars	687.0	141.75
Jupiter	4,331.8	483.80
Saturn	10,760.0	887.97
Uranus	30,684.0	1,764.50
Neptune	60,188.3	2,791.05
Pluto	90,466.8	3,653.90

The graphing principle in this example may be new to you. To plot T versus $R^{3/2}$ we first calculate the value of $R^{3/2}$ for each value in Table 1.3. For example, $3653.90^{3/2} \approx 220{,}869.1$ and $36^{3/2} = 216$. The horizontal axis represents $R^{3/2}$ (not R values) and we plot the ordered pairs $(R^{3/2}, T)$ in the coordinate system in Figure 1.49. This plot of ordered pairs or scatterplot gives a graph of the period versus the mean distance to the $3/2$ power. We observe that the scatterplot in the figure does lie approximately along a straight line that projects through the origin. By picking two points that lie on that line we can easily estimate the slope, which is the constant of proportionality (in days per miles $\times 10^{-4}$).

$$k = slope = \frac{90,466.8 - 88}{220{,}869.1 - 216} \approx 0.410$$

We estimate the model of Kepler's third law to be $T = 0.410R^{3/2}$ (which depends on our choice of units). We need to be careful to point out that this is *not a proof* of Kepler's third

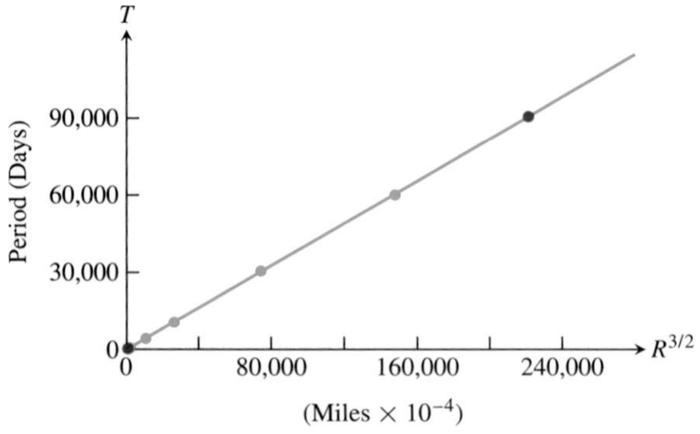

FIGURE 1.49 Graph of Kepler's third law as a proportionality: $T = 0.410R^{3/2}$ (Example 3).

law. We cannot prove or verify a theorem by just looking at some examples. Nevertheless, Figure 1.49 suggests that Kepler's third law is reasonable. ■

The concept of proportionality is one way to test the reasonableness of a conjectured relationship between two variables, as in Example 3. It can also provide the basis for an **empirical model** which comes entirely from a table of collected data.

EXERCISES 1.4

Recognizing Functions

In Exercises 1–4, identify each function as a constant function, linear function, power function, polynomial (state its degree), rational function, algebraic function, trigonometric function, exponential function, or logarithmic function. Remember that some functions can fall into more than one category.

1. a. $f(x) = 7 - 3x$ **b.** $g(x) = \sqrt[5]{x}$

c. $h(x) = \dfrac{x^2 - 1}{x^2 + 1}$ **d.** $r(x) = 8^x$

2. a. $F(t) = t^4 - t$ **b.** $G(t) = 5^t$

c. $H(z) = \sqrt{z^3 + 1}$ **d.** $R(z) = \sqrt[3]{z^7}$

3. a. $y = \dfrac{3 + 2x}{x - 1}$ **b.** $y = x^{5/2} - 2x + 1$

c. $y = \tan \pi x$ **d.** $y = \log_7 x$

4. a. $y = \log_5\left(\dfrac{1}{t}\right)$ **b.** $f(z) = \dfrac{z^5}{\sqrt{z} + 1}$

c. $g(x) = 2^{1/x}$ **d.** $w = 5\cos\left(\dfrac{t}{2} + \dfrac{\pi}{6}\right)$

In Exercises 5 and 6, match each equation with its graph. Do not use a graphing device, and give reasons for your answer.

5. a. $y = x^4$ **b.** $y = x^7$ **c.** $y = x^{10}$

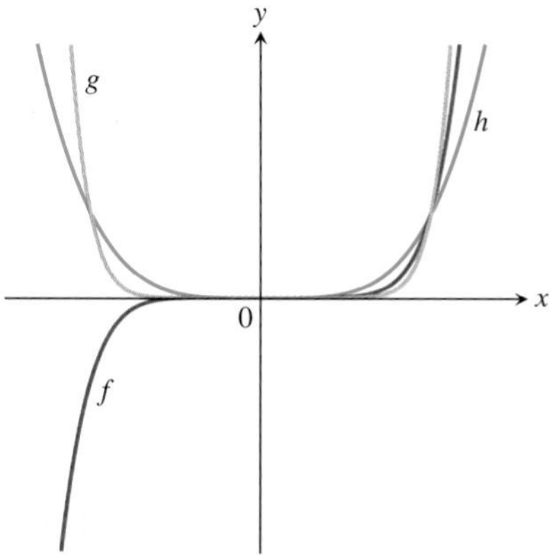

6. a. $y = 5x$ **b.** $y = 5^x$ **c.** $y = x^5$

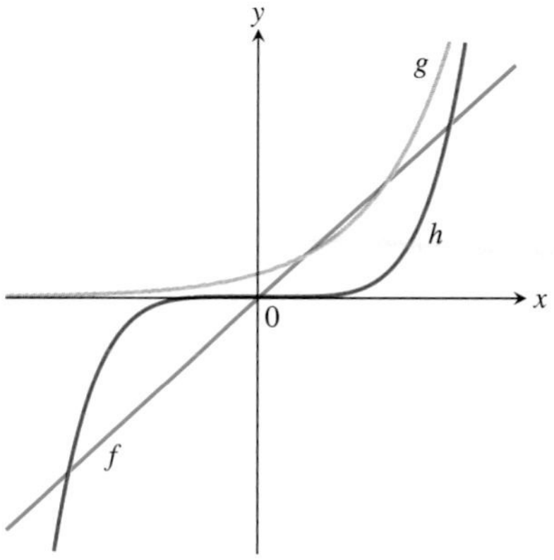

Increasing and Decreasing Functions

Graph the functions in Exercises 7–18. What symmetries, if any, do the graphs have? Specify the intervals over which the function is increasing and the intervals where it is decreasing.

7. $y = -x^3$ **8.** $y = -\dfrac{1}{x^2}$

9. $y = -\dfrac{1}{x}$ **10.** $y = \dfrac{1}{|x|}$

11. $y = \sqrt{|x|}$ **12.** $y = \sqrt{-x}$

13. $y = x^3/8$ **14.** $y = -4\sqrt{x}$

15. $y = -x^{3/2}$ **16.** $y = (-x)^{3/2}$

17. $y = (-x)^{2/3}$ **18.** $y = -x^{2/3}$

Even and Odd Functions

In Exercises 19–30, say whether the function is even, odd, or neither. Give reasons for your answer.

19. $f(x) = 3$ **20.** $f(x) = x^{-5}$

21. $f(x) = x^2 + 1$ **22.** $f(x) = x^2 + x$

23. $g(x) = x^3 + x$ **24.** $g(x) = x^4 + 3x^2 - 1$

25. $g(x) = \dfrac{1}{x^2 - 1}$

26. $g(x) = \dfrac{x}{x^2 - 1}$

27. $h(t) = \dfrac{1}{t - 1}$

28. $h(t) = |t^3|$

29. $h(t) = 2t + 1$

30. $h(t) = 2|t| + 1$

Proportionality

In Exercises 31 and 32, assess whether the given data sets reasonably support the stated proportionality assumption. Graph an appropriate scatterplot for your investigation and, if the proportionality assumption seems reasonable, estimate the constant of proportionality.

31. **a.** y is proportional to x

y	1	2	3	4	5	6	7	8
x	5.9	12.1	17.9	23.9	29.9	36.2	41.8	48.2

b. y is proportional to $x^{1/2}$

y	3.5	5	6	7	8
x	3	6	9	12	15

32. **a.** y is proportional to 3^x

y	5	15	45	135	405	1215	3645	10,935
x	0	1	2	3	4	5	6	7

b. y is proportional to $\ln x$

y	2	4.8	5.3	6.5	8.0	10.5	14.4	15.0
x	2.0	5.0	6.0	9.0	14.0	35.0	120.0	150.0

T 33. The accompanying table shows the distance a car travels during the time the driver is reacting before applying the brakes, and the distance the car travels after the brakes are applied. The distances (in feet) depend on the speed of the car (in miles per hour). Test the reasonableness of the following proportionality assumptions and estimate the constants of proportionality.

a. reaction distance is proportional to speed.

b. braking distance is proportional to the square of the speed.

Speed (mph)	20	25	30	35	40	45	50	55	60	65	70	75	80
Reaction distance (ft)	22	28	33	39	44	50	55	61	66	72	77	83	88
Braking distance (ft)	20	28	41	53	72	93	118	149	182	221	266	318	376

34. In October 2002, astronomers discovered a rocky, icy mini-planet tentatively named "Quaoar" circling the sun far beyond Neptune. The new planet is about 4 billion miles from Earth in an outer fringe of the solar system known as the Kuiper Belt. Using Kepler's third law, estimate the time T it takes Quaoar to complete one full orbit around the sun.

T 35. **Spring elongation** The response of a spring to various loads must be modeled to design a vehicle such as a dump truck, utility vehicle, or a luxury car that responds to road conditions in a desired way. We conducted an experiment to measure the stretch y of a spring in inches as a function of the number x of units of mass placed on the spring.

x (number of units of mass)	0	1	2	3	4	5
y (elongation in inches)	0	0.875	1.721	2.641	3.531	4.391

x (number of units of mass)	6	7	8	9	10
y (elongation in inches)	5.241	6.120	6.992	7.869	8.741

a. Make a scatterplot of the data to test the reasonableness of the hypothesis that stretch y is proportional to the mass x.

b. Estimate the constant of proportionality from your graph obtained in part (a).

c. Predict the elongation of the spring for 13 units of mass.

36. **Ponderosa pines** In the table, x represents the girth (distance around) of a pine tree measured in inches (in.) at shoulder height; y represents the board feet (bf) of lumber finally obtained.

x (in.)	17	19	20	23	25	28	32	38	39	41
y (bf)	19	25	32	57	71	113	123	252	259	294

Formulate and test the following two models: that usable board feet is proportional to **(a)** the square of the girth and **(b)** the cube of the girth. Does one model provide a better "explanation" than the other?

1.5 Combining Functions; Shifting and Scaling Graphs

In this section we look at the main ways functions are combined or transformed to form new functions.

Sums, Differences, Products, and Quotients

Like numbers, functions can be added, subtracted, multiplied, and divided (except where the denominator is zero) to produce new functions. If f and g are functions, then for every x that belongs to the domains of both f and g (that is, for $x \in D(f) \cap D(g)$), we define functions $f + g$, $f - g$, and fg by the formulas

$$(f + g)(x) = f(x) + g(x).$$
$$(f - g)(x) = f(x) - g(x).$$
$$(fg)(x) = f(x)g(x).$$

Notice that the $+$ sign on the left-hand side of the first equation represents the operation of addition of *functions*, whereas the $+$ on the right-hand side of the equation means addition of the real numbers $f(x)$ and $g(x)$.

At any point of $D(f) \cap D(g)$ at which $g(x) \neq 0$, we can also define the function f/g by the formula

$$\left(\frac{f}{g}\right)(x) = \frac{f(x)}{g(x)} \qquad (\text{where } g(x) \neq 0).$$

Functions can also be multiplied by constants: If c is a real number, then the function cf is defined for all x in the domain of f by

$$(cf)(x) = cf(x).$$

EXAMPLE 1 Combining Functions Algebraically

The functions defined by the formulas

$$f(x) = \sqrt{x} \qquad \text{and} \qquad g(x) = \sqrt{1 - x},$$

have domains $D(f) = [0, \infty)$ and $D(g) = (-\infty, 1]$. The points common to these domains are the points

$$[0, \infty) \cap (-\infty, 1] = [0, 1].$$

The following table summarizes the formulas and domains for the various algebraic combinations of the two functions. We also write $f \cdot g$ for the product function fg.

Function	**Formula**	**Domain**
$f + g$	$(f + g)(x) = \sqrt{x} + \sqrt{1 - x}$	$[0, 1] = D(f) \cap D(g)$
$f - g$	$(f - g)(x) = \sqrt{x} - \sqrt{1 - x}$	$[0, 1]$
$g - f$	$(g - f)(x) = \sqrt{1 - x} - \sqrt{x}$	$[0, 1]$
$f \cdot g$	$(f \cdot g)(x) = f(x)g(x) = \sqrt{x(1 - x)}$	$[0, 1]$
f/g	$\frac{f}{g}(x) = \frac{f(x)}{g(x)} = \sqrt{\frac{x}{1 - x}}$	$[0, 1)$ ($x = 1$ excluded)
g/f	$\frac{g}{f}(x) = \frac{g(x)}{f(x)} = \sqrt{\frac{1 - x}{x}}$	$(0, 1]$ ($x = 0$ excluded)

■

The graph of the function $f + g$ is obtained from the graphs of f and g by adding the corresponding y-coordinates $f(x)$ and $g(x)$ at each point $x \in D(f) \cap D(g)$, as in Figure 1.50. The graphs of $f + g$ and $f \cdot g$ from Example 1 are shown in Figure 1.51.

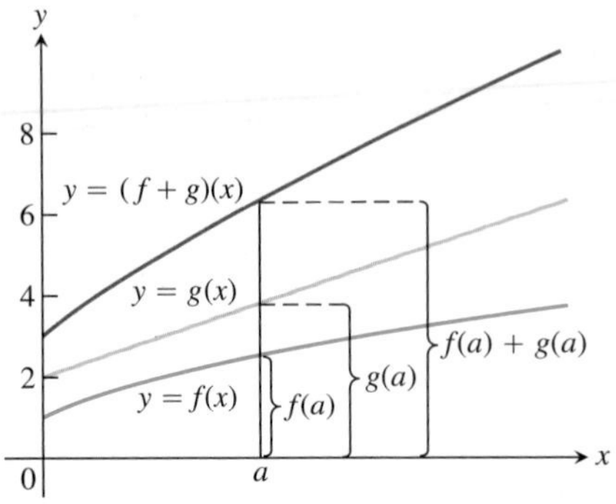

FIGURE 1.50 Graphical addition of two functions.

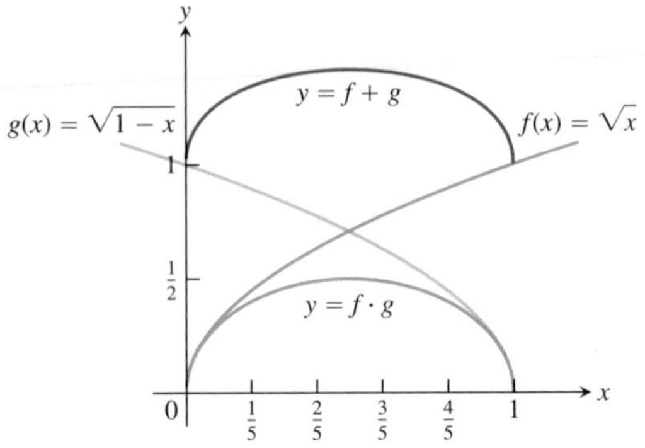

FIGURE 1.51 The domain of the function $f + g$ is the intersection of the domains of f and g, the interval $[0, 1]$ on the x-axis where these domains overlap. This interval is also the domain of the function $f \cdot g$ (Example 1).

Composite Functions

Composition is another method for combining functions.

> **DEFINITION Composition of Functions**
>
> If f and g are functions, the **composite** function $f \circ g$ ("f composed with g") is defined by
>
> $$(f \circ g)(x) = f(g(x)).$$
>
> The domain of $f \circ g$ consists of the numbers x in the domain of g for which $g(x)$ lies in the domain of f.

The definition says that $f \circ g$ can be formed when the range of g lies in the domain of f. To find $(f \circ g)(x)$, *first* find $g(x)$ and *second* find $f(g(x))$. Figure 1.52 pictures $f \circ g$ as a machine diagram and Figure 1.53 shows the composite as an arrow diagram.

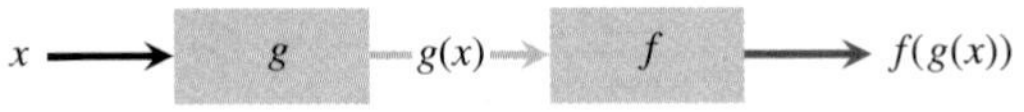

FIGURE 1.52 Two functions can be composed at x whenever the value of one function at x lies in the domain of the other. The composite is denoted by $f \circ g$.

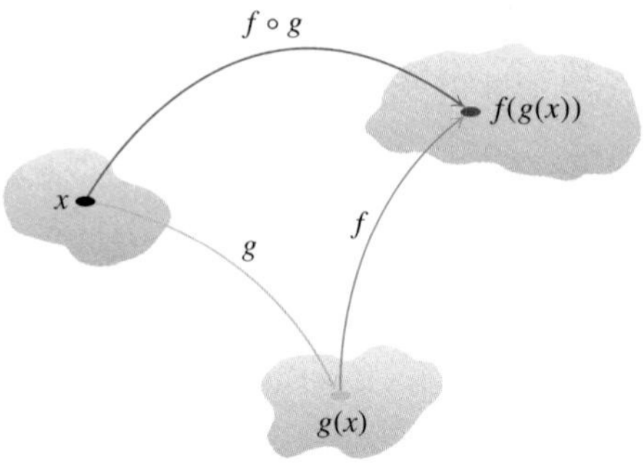

FIGURE 1.53 Arrow diagram for $f \circ g$.

EXAMPLE 2 Viewing a Function as a Composite

The function $y = \sqrt{1 - x^2}$ can be thought of as first calculating $1 - x^2$ and then taking the square root of the result. The function y is the composite of the function $g(x) = 1 - x^2$ and the function $f(x) = \sqrt{x}$. Notice that $1 - x^2$ cannot be negative. The domain of the composite is $[-1, 1]$. ■

To evaluate the composite function $g \circ f$ (when defined), we reverse the order, finding $f(x)$ first and then $g(f(x))$. The domain of $g \circ f$ is the set of numbers x in the domain of f such that $f(x)$ lies in the domain of g.

The functions $f \circ g$ and $g \circ f$ are usually quite different.

EXAMPLE 3 Finding Formulas for Composites

If $f(x) = \sqrt{x}$ and $g(x) = x + 1$, find

(a) $(f \circ g)(x)$ **(b)** $(g \circ f)(x)$ **(c)** $(f \circ f)(x)$ **(d)** $(g \circ g)(x)$.

Solution

Composite	**Domain**
(a) $(f \circ g)(x) = f(g(x)) = \sqrt{g(x)} = \sqrt{x + 1}$	$[-1, \infty)$
(b) $(g \circ f)(x) = g(f(x)) = f(x) + 1 = \sqrt{x} + 1$	$[0, \infty)$
(c) $(f \circ f)(x) = f(f(x)) = \sqrt{f(x)} = \sqrt{\sqrt{x}} = x^{1/4}$	$[0, \infty)$
(d) $(g \circ g)(x) = g(g(x)) = g(x) + 1 = (x + 1) + 1 = x + 2$	$(-\infty, \infty)$

To see why the domain of $f \circ g$ is $[-1, \infty)$, notice that $g(x) = x + 1$ is defined for all real x but belongs to the domain of f only if $x + 1 \geq 0$, that is to say, when $x \geq -1$. ■

Notice that if $f(x) = x^2$ and $g(x) = \sqrt{x}$, then $(f \circ g)(x) = \left(\sqrt{x}\right)^2 = x$. However, the domain of $f \circ g$ is $[0, \infty)$, not $(-\infty, \infty)$.

Shifting a Graph of a Function

To shift the graph of a function $y = f(x)$ straight up, add a positive constant to the right-hand side of the formula $y = f(x)$.

To shift the graph of a function $y = f(x)$ straight down, add a negative constant to the right-hand side of the formula $y = f(x)$.

To shift the graph of $y = f(x)$ to the left, add a positive constant to x. To shift the graph of $y = f(x)$ to the right, add a negative constant to x.

Shift Formulas

Vertical Shifts

$y = f(x) + k$	Shifts the graph of f *up* k units if $k > 0$ Shifts it *down* $\lvert k \rvert$ units if $k < 0$

Horizontal Shifts

$y = f(x + h)$	Shifts the graph of f *left* h units if $h > 0$ Shifts it *right* $\lvert h \rvert$ units if $h < 0$

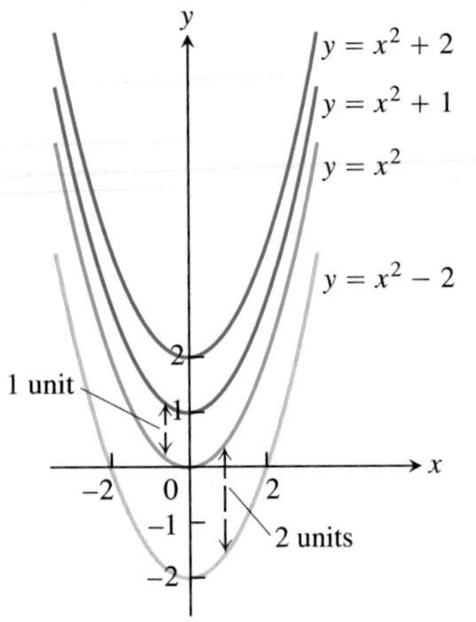

FIGURE 1.54 To shift the graph of $f(x) = x^2$ up (or down), we add positive (or negative) constants to the formula for f (Example 4a and b).

EXAMPLE 4 Shifting a Graph

(a) Adding 1 to the right-hand side of the formula $y = x^2$ to get $y = x^2 + 1$ shifts the graph up 1 unit (Figure 1.54).

(b) Adding -2 to the right-hand side of the formula $y = x^2$ to get $y = x^2 - 2$ shifts the graph down 2 units (Figure 1.54).

(c) Adding 3 to x in $y = x^2$ to get $y = (x + 3)^2$ shifts the graph 3 units to the left (Figure 1.55).

(d) Adding -2 to x in $y = |x|$, and then adding -1 to the result, gives $y = |x - 2| - 1$ and shifts the graph 2 units to the right and 1 unit down (Figure 1.56).

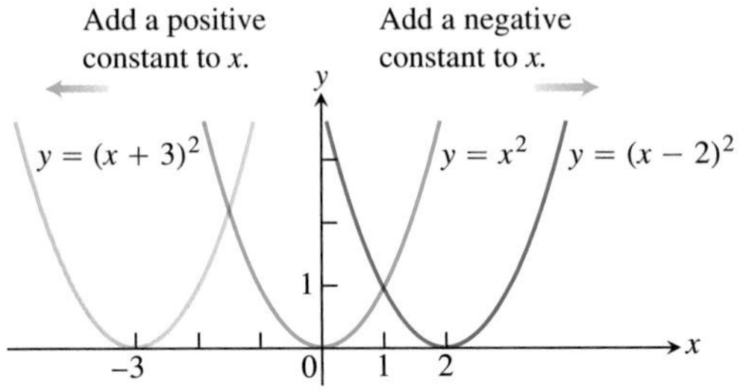

FIGURE 1.55 To shift the graph of $y = x^2$ to the left, we add a positive constant to x. To shift the graph to the right, we add a negative constant to x (Example 4c).

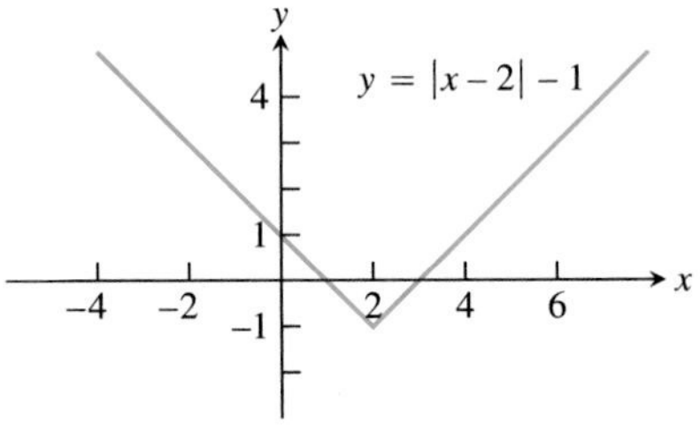

FIGURE 1.56 Shifting the graph of $y = |x|$ 2 units to the right and 1 unit down (Example 4d). ■

Scaling and Reflecting a Graph of a Function

To scale the graph of a function $y = f(x)$ is to stretch or compress it, vertically or horizontally. This is accomplished by multiplying the function f, or the independent variable x, by an appropriate constant c. Reflections across the coordinate axes are special cases where $c = -1$.

Vertical and Horizontal Scaling and Reflecting Formulas

For $c > 1$,

$y = cf(x)$	Stretches the graph of f vertically by a factor of c.
$y = \frac{1}{c} f(x)$	Compresses the graph of f vertically by a factor of c.
$y = f(cx)$	Compresses the graph of f horizontally by a factor of c.
$y = f(x/c)$	Stretches the graph of f horizontally by a factor of c.

For $c = -1$,

$y = -f(x)$	Reflects the graph of f across the x-axis.
$y = f(-x)$	Reflects the graph of f across the y-axis.

EXAMPLE 5 Scaling and Reflecting a Graph

(a) **Vertical:** Multiplying the right-hand side of $y = \sqrt{x}$ by 3 to get $y = 3\sqrt{x}$ stretches the graph vertically by a factor of 3, whereas multiplying by $1/3$ compresses the graph by a factor of 3 (Figure 1.57).

(b) **Horizontal:** The graph of $y = \sqrt{3x}$ is a horizontal compression of the graph of $y = \sqrt{x}$ by a factor of 3, and $y = \sqrt{x/3}$ is a horizontal stretching by a factor of 3 (Figure 1.58). Note that $y = \sqrt{3x} = \sqrt{3}\sqrt{x}$ so a horizontal compression *may* correspond to a vertical stretching by a different scaling factor. Likewise, a horizontal stretching may correspond to a vertical compression by a different scaling factor.

(c) **Reflection:** The graph of $y = -\sqrt{x}$ is a reflection of $y = \sqrt{x}$ across the x-axis, and $y = \sqrt{-x}$ is a reflection across the y-axis (Figure 1.59).

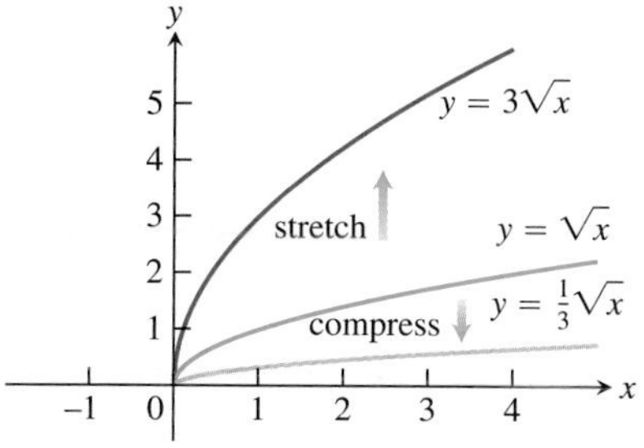

FIGURE 1.57 Vertically stretching and compressing the graph $y = \sqrt{x}$ by a factor of 3 (Example 5a).

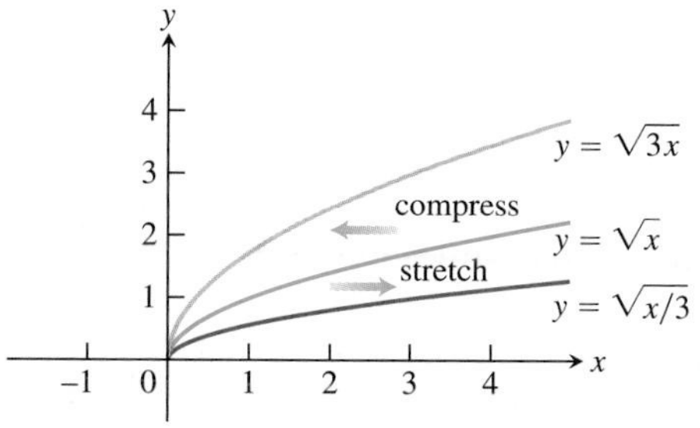

FIGURE 1.58 Horizontally stretching and compressing the graph $y = \sqrt{x}$ by a factor of 3 (Example 5b).

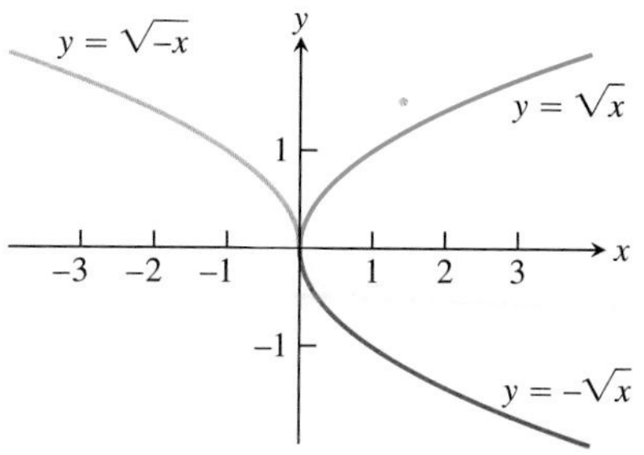

FIGURE 1.59 Reflections of the graph $y = \sqrt{x}$ across the coordinate axes (Example 5c). ■

EXAMPLE 6 Combining Scalings and Reflections

Given the function $f(x) = x^4 - 4x^3 + 10$ (Figure 1.60a), find formulas to

(a) compress the graph horizontally by a factor of 2 followed by a reflection across the y-axis (Figure 1.60b).

(b) compress the graph vertically by a factor of 2 followed by a reflection across the x-axis (Figure 1.60c).

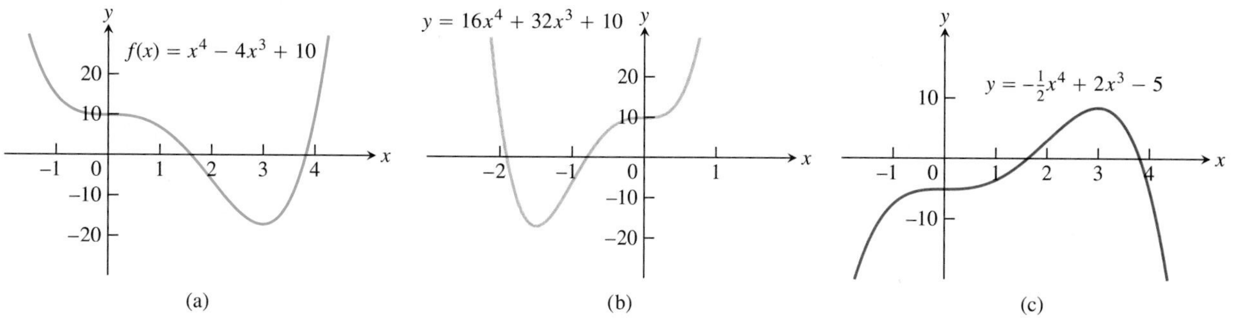

FIGURE 1.60 (a) The original graph of f. (b) The horizontal compression of $y = f(x)$ in part (a) by a factor of 2, followed by a reflection across the y-axis. (c) The vertical compression of $y = f(x)$ in part (a) by a factor of 2, followed by a reflection across the x-axis (Example 6).

Solution

(a) The formula is obtained by substituting $-2x$ for x in the right-hand side of the equation for f

$$\begin{aligned} y = f(-2x) &= (-2x)^4 - 4(-2x)^3 + 10 \\ &= 16x^4 + 32x^3 + 10. \end{aligned}$$

(b) The formula is

$$y = -\frac{1}{2}f(x) = -\frac{1}{2}x^4 + 2x^3 - 5.$$

Ellipses

Substituting cx for x in the standard equation for a circle of radius r centered at the origin gives

$$c^2x^2 + y^2 = r^2. \tag{1}$$

If $0 < c < 1$, the graph of Equation (1) horizontally stretches the circle; if $c > 1$ the circle is compressed horizontally. In either case, the graph of Equation (1) is an ellipse (Figure 1.61). Notice in Figure 1.61 that the y-intercepts of all three graphs are always $-r$ and r. In Figure 1.61b, the line segment joining the points $(\pm r/c, 0)$ is called the **major axis** of the ellipse; the **minor axis** is the line segment joining $(0, \pm r)$. The axes of the ellipse are reversed in Figure 1.61c: the major axis is the line segment joining the points $(0, \pm r)$ and the minor axis is the line segment joining the points $(\pm r/c, 0)$. In both cases, the major axis is the line segment having the longer length.

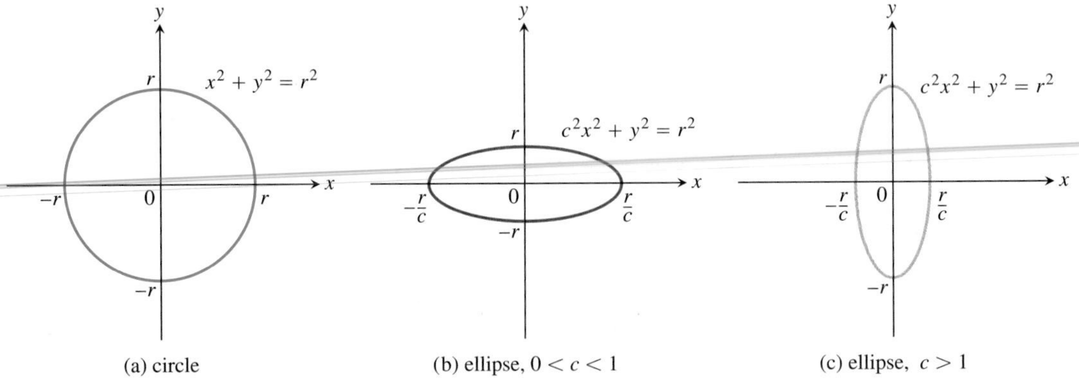

FIGURE 1.61 Horizontal stretchings or compressions of a circle produce graphs of ellipses.

If we divide both sides of Equation (1) by r^2, we obtain

$$\frac{x^2}{a^2} + \frac{y^2}{b^2} = 1. \tag{2}$$

where $a = r/c$ and $b = r$. If $a > b$, the major axis is horizontal; if $a < b$, the major axis is vertical. The **center** of the ellipse given by Equation (2) is the origin (Figure 1.62).

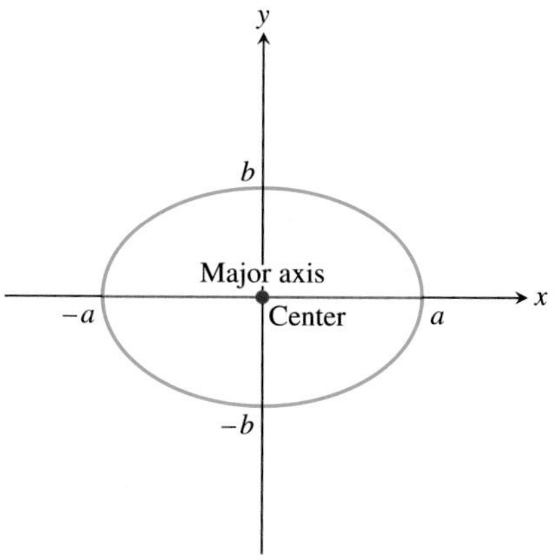

FIGURE 1.62 Graph of the ellipse $\frac{x^2}{a^2} + \frac{y^2}{b^2} = 1$, $a > b$, where the major axis is horizontal.

Substituting $x - h$ for x, and $y - k$ for y, in Equation (2) results in

$$\frac{(x-h)^2}{a^2} + \frac{(y-k)^2}{b^2} = 1. \qquad (3)$$

Equation (3) is the **standard equation of an ellipse** with center at (h, k). The geometric definition and properties of ellipses are reviewed in Section 10.1.

EXERCISES 1.5

Sums, Differences, Products, and Quotients

In Exercises 1 and 2, find the domains and ranges of f, g, $f + g$, and $f \cdot g$.

1. $f(x) = x, \quad g(x) = \sqrt{x - 1}$
2. $f(x) = \sqrt{x + 1}, \quad g(x) = \sqrt{x - 1}$

In Exercises 3 and 4, find the domains and ranges of f, g, f/g, and g/f.

3. $f(x) = 2, \quad g(x) = x^2 + 1$
4. $f(x) = 1, \quad g(x) = 1 + \sqrt{x}$

Composites of Functions

5. If $f(x) = x + 5$ and $g(x) = x^2 - 3$, find the following.
   a. $f(g(0))$ b. $g(f(0))$
   c. $f(g(x))$ d. $g(f(x))$
   e. $f(f(-5))$ f. $g(g(2))$
   g. $f(f(x))$ h. $g(g(x))$
6. If $f(x) = x - 1$ and $g(x) = 1/(x + 1)$, find the following.
   a. $f(g(1/2))$ b. $g(f(1/2))$
   c. $f(g(x))$ d. $g(f(x))$
   e. $f(f(2))$ f. $g(g(2))$
   g. $f(f(x))$ h. $g(g(x))$
7. If $u(x) = 4x - 5$, $v(x) = x^2$, and $f(x) = 1/x$, find formulas for the following.
   a. $u(v(f(x)))$ b. $u(f(v(x)))$
   c. $v(u(f(x)))$ d. $v(f(u(x)))$
   e. $f(u(v(x)))$ f. $f(v(u(x)))$
8. If $f(x) = \sqrt{x}$, $g(x) = x/4$, and $h(x) = 4x - 8$, find formulas for the following.
   a. $h(g(f(x)))$ b. $h(f(g(x)))$
   c. $g(h(f(x)))$ d. $g(f(h(x)))$
   e. $f(g(h(x)))$ f. $f(h(g(x)))$

Let $f(x) = x - 3$, $g(x) = \sqrt{x}$, $h(x) = x^3$, and $j(x) = 2x$. Express each of the functions in Exercises 9 and 10 as a composite involving one or more of f, g, h, and j.

9. a. $y = \sqrt{x} - 3$ b. $y = 2\sqrt{x}$
   c. $y = x^{1/4}$ d. $y = 4x$
   e. $y = \sqrt{(x - 3)^3}$ f. $y = (2x - 6)^3$
10. a. $y = 2x - 3$ b. $y = x^{3/2}$
    c. $y = x^9$ d. $y = x - 6$
    e. $y = 2\sqrt{x - 3}$ f. $y = \sqrt{x^3 - 3}$

11. Copy and complete the following table.

	$g(x)$	$f(x)$	$(f \circ g)(x)$
a.	$x - 7$	$\sqrt{x}$	
b.	$x + 2$	$3x$	
c.		$\sqrt{x - 5}$	$\sqrt{x^2 - 5}$
d.	$\dfrac{x}{x - 1}$	$\dfrac{x}{x - 1}$	
e.		$1 + \dfrac{1}{x}$	x
f.	$\dfrac{1}{x}$		x

12. Copy and complete the following table.

	$g(x)$	$f(x)$	$(f \circ g)(x)$
a.	$\dfrac{1}{x - 1}$	$\lvert x \rvert$	?
b.	?	$\dfrac{x - 1}{x}$	$\dfrac{x}{x + 1}$
c.	?	$\sqrt{x}$	$\lvert x \rvert$
d.	$\sqrt{x}$	?	$\lvert x \rvert$

In Exercises 13 and 14, **(a)** write a formula for $f \circ g$ and $g \circ f$ and find the **(b)** domain and **(c)** range of each.

13. $f(x) = \sqrt{x + 1}, \quad g(x) = \dfrac{1}{x}$

14. $f(x) = x^2, \quad g(x) = 1 - \sqrt{x}$

Shifting Graphs

15. The accompanying figure shows the graph of $y = -x^2$ shifted to two new positions. Write equations for the new graphs.

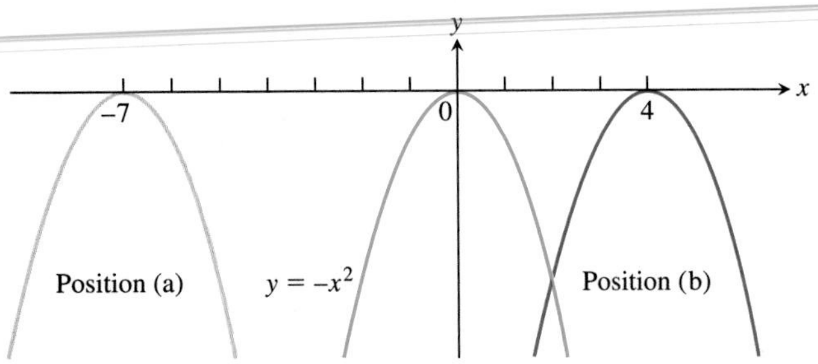

16. The accompanying figure shows the graph of $y = x^2$ shifted to two new positions. Write equations for the new graphs.

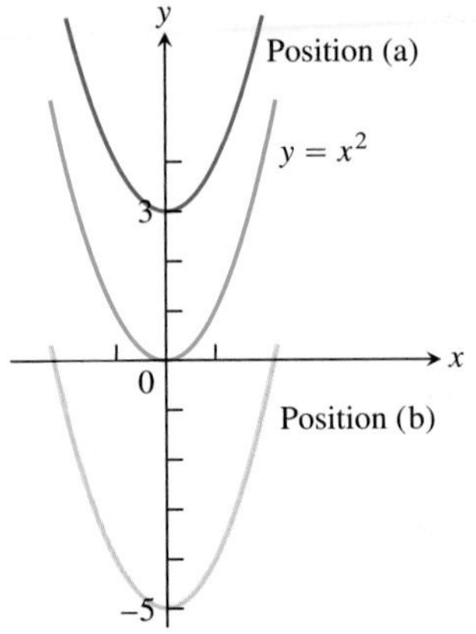

17. Match the equations listed in parts (a)–(d) to the graphs in the accompanying figure.

a. $y = (x - 1)^2 - 4$ **b.** $y = (x - 2)^2 + 2$

c. $y = (x + 2)^2 + 2$ **d.** $y = (x + 3)^2 - 2$

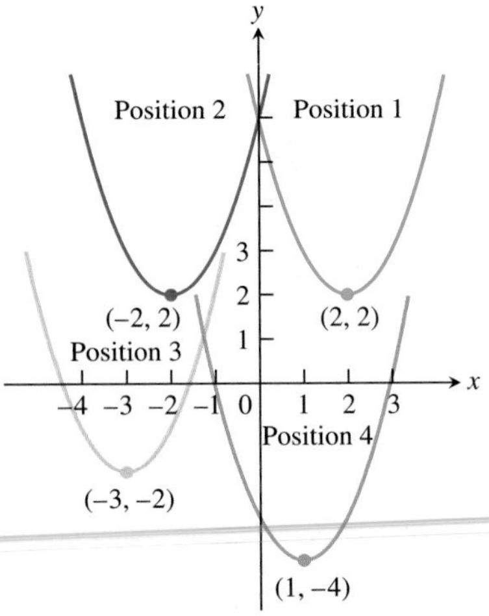

18. The accompanying figure shows the graph of $y = -x^2$ shifted to four new positions. Write an equation for each new graph.

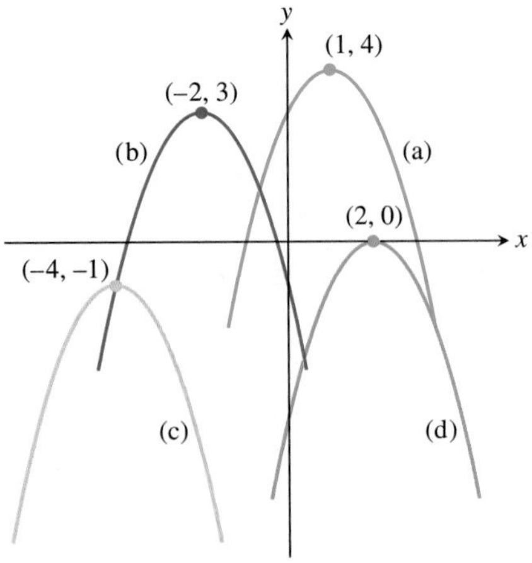

Exercises 19–28 tell how many units and in what directions the graphs of the given equations are to be shifted. Give an equation for the shifted graph. Then sketch the original and shifted graphs together, labeling each graph with its equation.

19. $x^2 + y^2 = 49$ Down 3, left 2

20. $x^2 + y^2 = 25$ Up 3, left 4

21. $y = x^3$ Left 1, down 1

22. $y = x^{2/3}$ Right 1, down 1

23. $y = \sqrt{x}$ Left 0.81

24. $y = -\sqrt{x}$ Right 3

25. $y = 2x - 7$ Up 7

26. $y = \frac{1}{2}(x + 1) + 5$ Down 5, right 1

27. $y = 1/x$ Up 1, right 1

28. $y = 1/x^2$ Left 2, down 1

Graph the functions in Exercises 29–48.

29. $y = \sqrt{x + 4}$

30. $y = \sqrt{9 - x}$

31. $y = |x - 2|$

32. $y = |1 - x| - 1$

33. $y = 1 + \sqrt{x - 1}$

34. $y = 1 - \sqrt{x}$

35. $y = (x + 1)^{2/3}$

36. $y = (x - 8)^{2/3}$

37. $y = 1 - x^{2/3}$

38. $y + 4 = x^{2/3}$

39. $y = \sqrt[3]{x - 1} - 1$

40. $y = (x + 2)^{3/2} + 1$

41. $y = \frac{1}{x - 2}$

42. $y = \frac{1}{x} - 2$

43. $y = \frac{1}{x} + 2$

44. $y = \frac{1}{x + 2}$

45. $y = \frac{1}{(x - 1)^2}$

46. $y = \frac{1}{x^2} - 1$

47. $y = \frac{1}{x^2} + 1$

48. $y = \frac{1}{(x + 1)^2}$

49. The accompanying figure shows the graph of a function $f(x)$ with domain $[0, 2]$ and range $[0, 1]$. Find the domains and ranges of the following functions, and sketch their graphs.

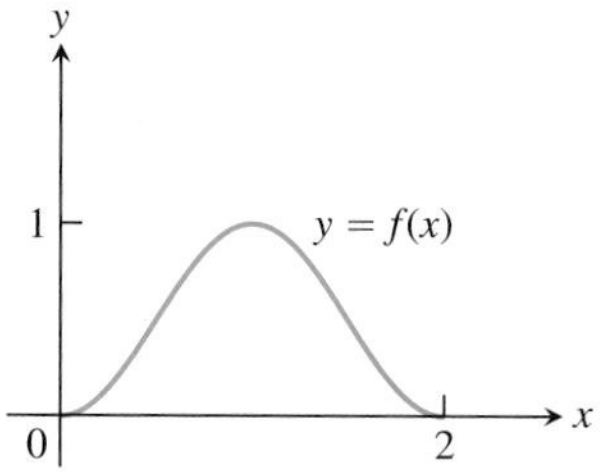

a. $f(x) + 2$

b. $f(x) - 1$

c. $2f(x)$

d. $-f(x)$

e. $f(x + 2)$

f. $f(x - 1)$

g. $f(-x)$

h. $-f(x + 1) + 1$

50. The accompanying figure shows the graph of a function $g(t)$ with domain $[-4, 0]$ and range $[-3, 0]$. Find the domains and ranges of the following functions, and sketch their graphs.

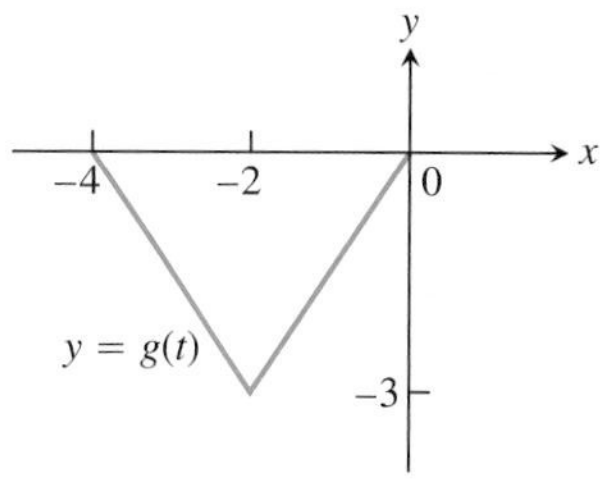

a. $g(-t)$

b. $-g(t)$

c. $g(t) + 3$

d. $1 - g(t)$

e. $g(-t + 2)$

f. $g(t - 2)$

g. $g(1 - t)$

h. $-g(t - 4)$

Vertical and Horizontal Scaling

Exercises 51–60 tell by what factor and direction the graphs of the given functions are to be stretched or compressed. Give an equation for the stretched or compressed graph.

51. $y = x^2 - 1$, stretched vertically by a factor of 3

52. $y = x^2 - 1$, compressed horizontally by a factor of 2

53. $y = 1 + \frac{1}{x^2}$, compressed vertically by a factor of 2

54. $y = 1 + \frac{1}{x^2}$, stretched horizontally by a factor of 3

55. $y = \sqrt{x + 1}$, compressed horizontally by a factor of 4

56. $y = \sqrt{x + 1}$, stretched vertically by a factor of 3

57. $y = \sqrt{4 - x^2}$, stretched horizontally by a factor of 2

58. $y = \sqrt{4 - x^2}$, compressed vertically by a factor of 3

59. $y = 1 - x^3$, compressed horizontally by a factor of 3

60. $y = 1 - x^3$, stretched horizontally by a factor of 2

Graphing

In Exercises 61–68, graph each function, not by plotting points, but by starting with the graph of one of the standard functions presented in Figures 1.36–1.38, and applying an appropriate transformation.

61. $y = -\sqrt{2x + 1}$

62. $y = \sqrt{1 - \frac{x}{2}}$

63. $y = (x - 1)^3 + 2$

64. $y = (1 - x)^3 + 2$

65. $y = \frac{1}{2x} - 1$

66. $y = \frac{2}{x^2} + 1$

67. $y = -\sqrt[3]{x}$

68. $y = (-2x)^{2/3}$

69. Graph the function $y = |x^2 - 1|$.

70. Graph the function $y = \sqrt{|x|}$.

Ellipses

Exercises 71–76 give equations of ellipses. Put each equation in standard form and sketch the ellipse.

71. $9x^2 + 25y^2 = 225$ **72.** $16x^2 + 7y^2 = 112$

73. $3x^2 + (y - 2)^2 = 3$ **74.** $(x + 1)^2 + 2y^2 = 4$

75. $3(x - 1)^2 + 2(y + 2)^2 = 6$

76. $6\left(x + \frac{3}{2}\right)^2 + 9\left(y - \frac{1}{2}\right)^2 = 54$

77. Write an equation for the ellipse $(x^2/16) + (y^2/9) = 1$ shifted 4 units to the left and 3 units up. Sketch the ellipse and identify its center and major axis.

78. Write an equation for the ellipse $(x^2/4) + (y^2/25) = 1$ shifted 3 units to the right and 2 units down. Sketch the ellipse and identify its center and major axis.

Even and Odd Functions

79. Assume that f is an even function, g is an odd function, and both f and g are defined on the entire real line $\mathbb{R}$. Which of the following (where defined) are even? odd?

a. fg **b.** f/g **c.** g/f

d. $f^2 = ff$ **e.** $g^2 = gg$ **f.** $f \circ g$

g. $g \circ f$ **h.** $f \circ f$ **i.** $g \circ g$

80. Can a function be both even and odd? Give reasons for your answer.

T **81.** (*Continuation of Example* 1.) Graph the functions $f(x) = \sqrt{x}$ and $g(x) = \sqrt{1 - x}$ together with their (a) sum, (b) product, (c) two differences, (d) two quotients.

T **82.** Let $f(x) = x - 7$ and $g(x) = x^2$. Graph f and g together with $f \circ g$ and $g \circ f$.

1.6 Trigonometric Functions

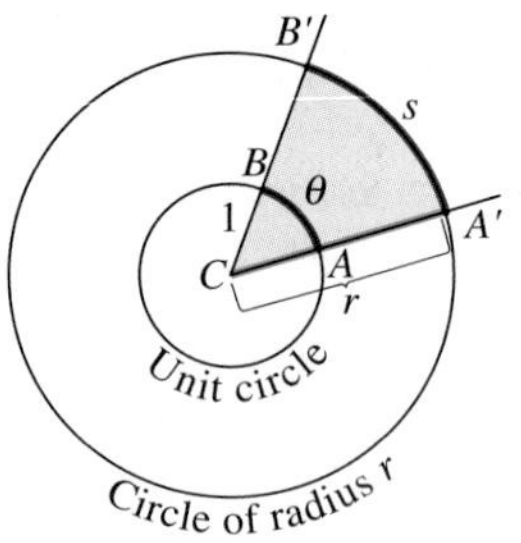

FIGURE 1.63 The radian measure of angle ACB is the length θ of arc AB on the unit circle centered at C. The value of θ can be found from any other circle, however, as the ratio s/r. Thus $s = r\theta$ is the length of arc on a circle of radius r when θ is measured in radians.

This section reviews the basic trigonometric functions. The trigonometric functions are important because they are periodic, or repeating, and therefore model many naturally occurring periodic processes.

Radian Measure

In navigation and astronomy, angles are measured in degrees, but in calculus it is best to use units called *radians* because of the way they simplify later calculations.

The **radian measure** of the angle ACB at the center of the unit circle (Figure 1.63) equals the length of the arc that ACB cuts from the unit circle. Figure 1.63 shows that $s = r\theta$ is the **length of arc** cut from a circle of radius r when the subtending angle θ producing the arc is measured in radians.

Since the circumference of the circle is 2π and one complete revolution of a circle is 360°, the relation between radians and degrees is given by

$$\pi \text{ radians} = 180°.$$

For example, 45° in radian measure is

$$45 \cdot \frac{\pi}{180} = \frac{\pi}{4} \text{ rad},$$

and $\pi/6$ radians is

$$\frac{\pi}{6} \cdot \frac{180}{\pi} = 30°.$$

Conversion Formulas

$$1 \text{ degree} = \frac{\pi}{180} (\approx 0.02) \text{ radians}$$

Degrees to radians: multiply by $\frac{\pi}{180}$

$$1 \text{ radian} = \frac{180}{\pi} (\approx 57) \text{ degrees}$$

Radians to degrees: multiply by $\frac{180}{\pi}$

Figure 1.64 shows the angles of two common triangles in both measures.

An angle in the xy-plane is said to be in **standard position** if its vertex lies at the origin and its initial ray lies along the positive x-axis (Figure 1.65). Angles measured counterclockwise from the positive x-axis are assigned positive measures; angles measured clockwise are assigned negative measures.

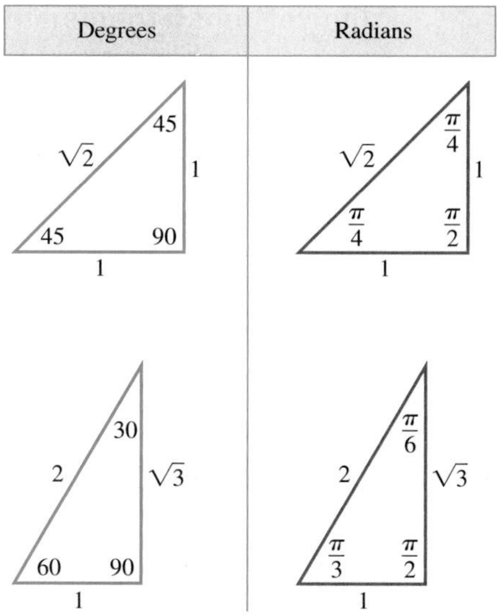

FIGURE 1.64 The angles of two common triangles, in degrees and radians.

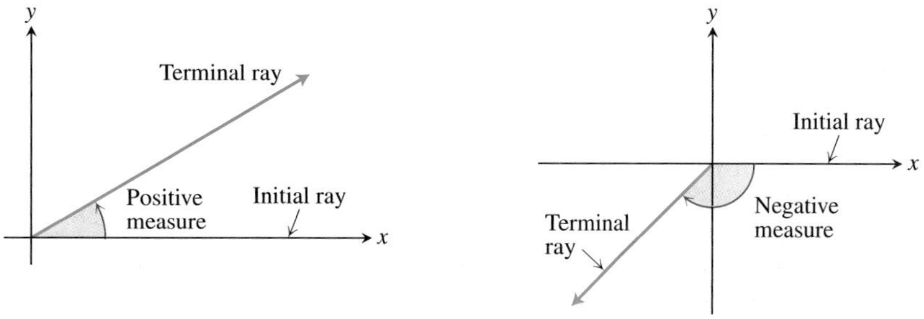

FIGURE 1.65 Angles in standard position in the xy-plane.

When angles are used to describe counterclockwise rotations, our measurements can go arbitrarily far beyond 2π radians or 360°. Similarly, angles describing clockwise rotations can have negative measures of all sizes (Figure 1.66).

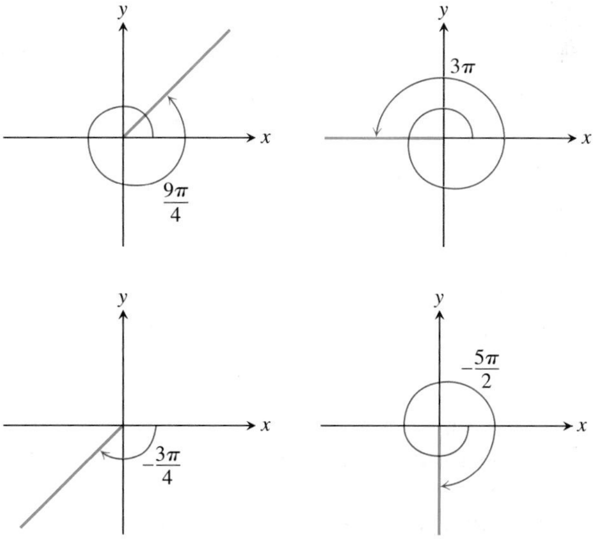

FIGURE 1.66 Nonzero radian measures can be positive or negative and can go beyond 2π.

Angle Convention: Use Radians

From now on in this book it is assumed that all angles are measured in radians unless degrees or some other unit is stated explicitly. When we talk about the angle $\pi/3$, we mean $\pi/3$ radians (which is 60°), not $\pi/3$ degrees. When you do calculus, keep your calculator in radian mode.

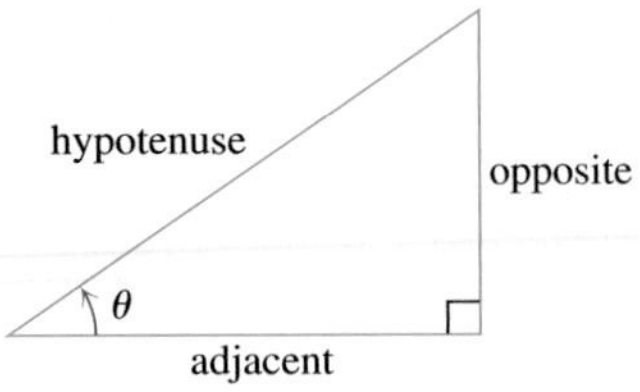

$$\sin\theta = \frac{\text{opp}}{\text{hyp}} \qquad \csc\theta = \frac{\text{hyp}}{\text{opp}}$$

$$\cos\theta = \frac{\text{adj}}{\text{hyp}} \qquad \sec\theta = \frac{\text{hyp}}{\text{adj}}$$

$$\tan\theta = \frac{\text{opp}}{\text{adj}} \qquad \cot\theta = \frac{\text{adj}}{\text{opp}}$$

FIGURE 1.67 Trigonometric ratios of an acute angle.

The Six Basic Trigonometric Functions

You are probably familiar with defining the trigonometric functions of an acute angle in terms of the sides of a right triangle (Figure 1.67). We extend this definition to obtuse and negative angles by first placing the angle in standard position in a circle of radius r. We then define the trigonometric functions in terms of the coordinates of the point $P(x, y)$ where the angle's terminal ray intersects the circle (Figure 1.68).

$$\textbf{sine:}\quad \sin\theta = \frac{y}{r} \qquad \textbf{cosecant:}\quad \csc\theta = \frac{r}{y}$$

$$\textbf{cosine:}\quad \cos\theta = \frac{x}{r} \qquad \textbf{secant:}\quad \sec\theta = \frac{r}{x}$$

$$\textbf{tangent:}\quad \tan\theta = \frac{y}{x} \qquad \textbf{cotangent:}\quad \cot\theta = \frac{x}{y}$$

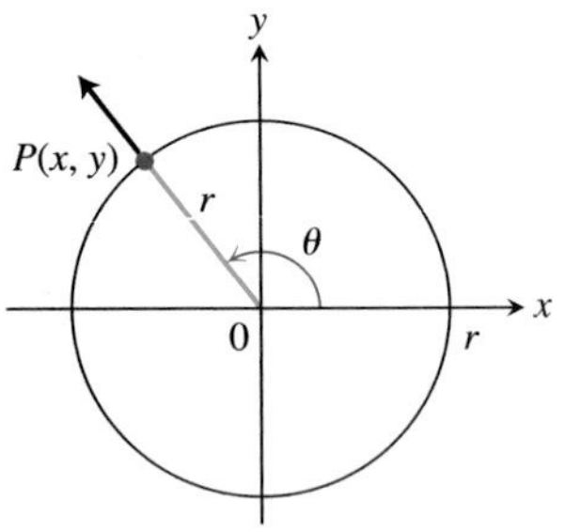

FIGURE 1.68 The trigonometric functions of a general angle θ are defined in terms of x, y, and r.

These extended definitions agree with the right-triangle definitions when the angle is acute (Figure 1.69).

Notice also the following definitions, whenever the quotients are defined.

$$\tan\theta = \frac{\sin\theta}{\cos\theta} \qquad \cot\theta = \frac{1}{\tan\theta}$$

$$\sec\theta = \frac{1}{\cos\theta} \qquad \csc\theta = \frac{1}{\sin\theta}$$

As you can see, $\tan\theta$ and $\sec\theta$ are not defined if $x = 0$. This means they are not defined if θ is $\pm\pi/2, \pm3\pi/2, \ldots$. Similarly, $\cot\theta$ and $\csc\theta$ are not defined for values of θ for which $y = 0$, namely $\theta = 0, \pm\pi, \pm2\pi, \ldots$.

The exact values of these trigonometric ratios for some angles can be read from the triangles in Figure 1.64. For instance,

$$\sin\frac{\pi}{4} = \frac{1}{\sqrt{2}} \qquad \sin\frac{\pi}{6} = \frac{1}{2} \qquad \sin\frac{\pi}{3} = \frac{\sqrt{3}}{2}$$

$$\cos\frac{\pi}{4} = \frac{1}{\sqrt{2}} \qquad \cos\frac{\pi}{6} = \frac{\sqrt{3}}{2} \qquad \cos\frac{\pi}{3} = \frac{1}{2}$$

$$\tan\frac{\pi}{4} = 1 \qquad \tan\frac{\pi}{6} = \frac{1}{\sqrt{3}} \qquad \tan\frac{\pi}{3} = \sqrt{3}$$

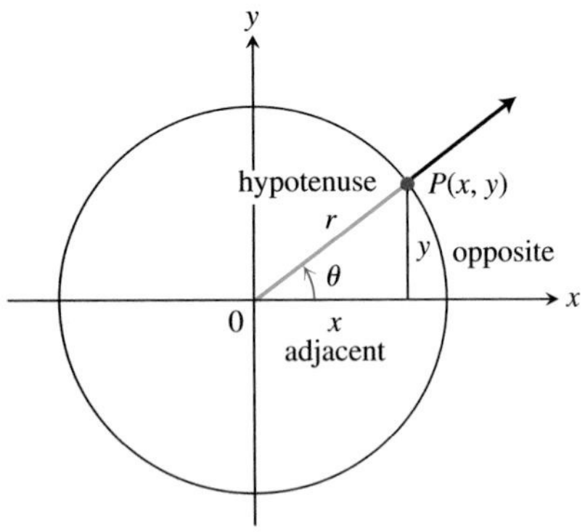

FIGURE 1.69 The new and old definitions agree for acute angles.

The CAST rule (Figure 1.70) is useful for remembering when the basic trigonometric functions are positive or negative. For instance, from the triangle in Figure 1.71, we see that

$$\sin\frac{2\pi}{3} = \frac{\sqrt{3}}{2}, \qquad \cos\frac{2\pi}{3} = -\frac{1}{2}, \qquad \tan\frac{2\pi}{3} = -\sqrt{3}.$$

Using a similar method we determined the values of $\sin\theta$, $\cos\theta$, and $\tan\theta$ shown in Table 1.4.

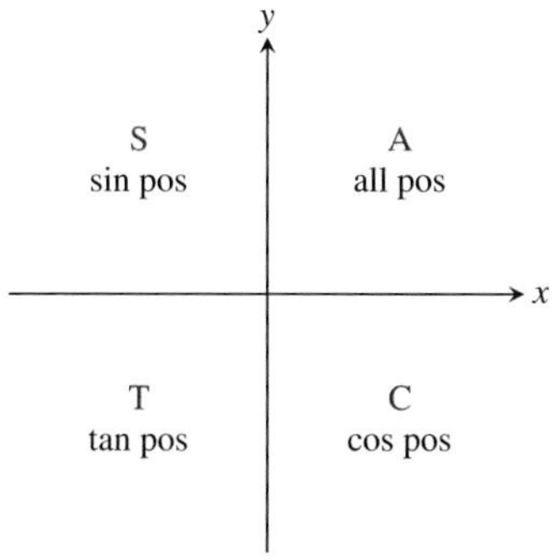

FIGURE 1.70 The CAST rule, remembered by the statement "All Students Take Calculus," tells which trigonometric functions are positive in each quadrant.

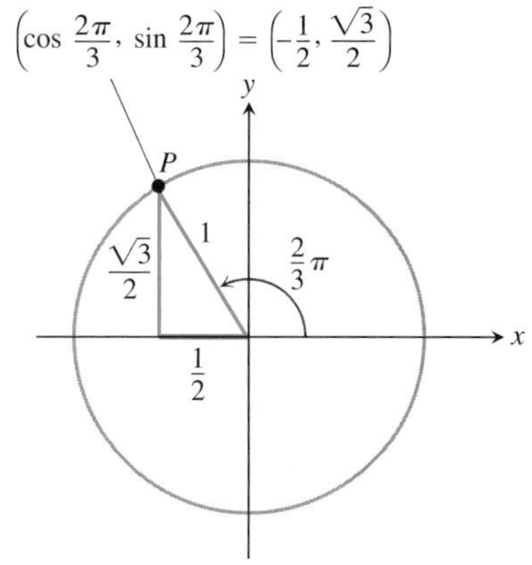

FIGURE 1.71 The triangle for calculating the sine and cosine of $2\pi/3$ radians. The side lengths come from the geometry of right triangles.

Most calculators and computers readily provide values of the trigonometric functions for angles given in either radians or degrees.

TABLE 1.4 Values of $\sin\theta$, $\cos\theta$, and $\tan\theta$ for selected values of θ

Degrees	−180	−135	−90	−45	0	30	45	60	90	120	135	150	180	270	360
θ (radians)	$-\pi$	$\frac{-3\pi}{4}$	$\frac{-\pi}{2}$	$\frac{-\pi}{4}$	0	$\frac{\pi}{6}$	$\frac{\pi}{4}$	$\frac{\pi}{3}$	$\frac{\pi}{2}$	$\frac{2\pi}{3}$	$\frac{3\pi}{4}$	$\frac{5\pi}{6}$	π	$\frac{3\pi}{2}$	2π
$\sin\theta$	0	$\frac{-\sqrt{2}}{2}$	-1	$\frac{-\sqrt{2}}{2}$	0	$\frac{1}{2}$	$\frac{\sqrt{2}}{2}$	$\frac{\sqrt{3}}{2}$	1	$\frac{\sqrt{3}}{2}$	$\frac{\sqrt{2}}{2}$	$\frac{1}{2}$	0	-1	0
$\cos\theta$	-1	$\frac{-\sqrt{2}}{2}$	0	$\frac{\sqrt{2}}{2}$	1	$\frac{\sqrt{3}}{2}$	$\frac{\sqrt{2}}{2}$	$\frac{1}{2}$	0	$-\frac{1}{2}$	$\frac{-\sqrt{2}}{2}$	$\frac{-\sqrt{3}}{2}$	-1	0	1
$\tan\theta$	0	1		-1	0	$\frac{\sqrt{3}}{3}$	1	$\sqrt{3}$		$-\sqrt{3}$	-1	$\frac{-\sqrt{3}}{3}$	0		0

EXAMPLE 1 Finding Trigonometric Function Values

If $\tan\theta = 3/2$ and $0 < \theta < \pi/2$, find the five other trigonometric functions of θ.

Solution From $\tan\theta = 3/2$, we construct the right triangle of height 3 (opposite) and base 2 (adjacent) in Figure 1.72. The Pythagorean theorem gives the length of the hypotenuse, $\sqrt{4 + 9} = \sqrt{13}$. From the triangle we write the values of the other five trigonometric functions:

$$\cos\theta = \frac{2}{\sqrt{13}}, \quad \sin\theta = \frac{3}{\sqrt{13}}, \quad \sec\theta = \frac{\sqrt{13}}{2}, \quad \csc\theta = \frac{\sqrt{13}}{3}, \quad \cot\theta = \frac{2}{3}$$

■

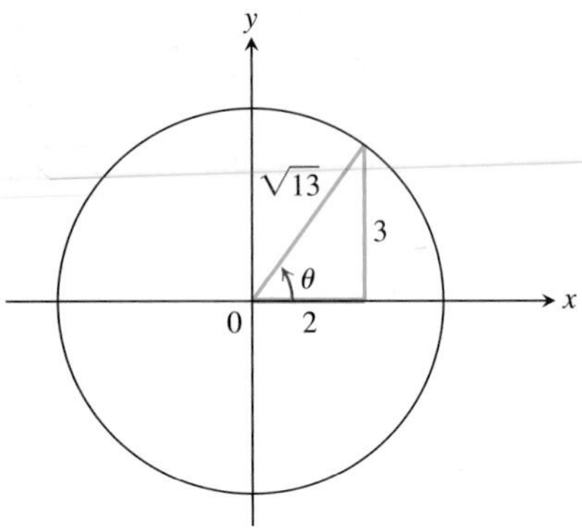

FIGURE 1.72 The triangle for calculating the trigonometric functions in Example 1.

Periodicity and Graphs of the Trigonometric Functions

When an angle of measure θ and an angle of measure $\theta + 2\pi$ are in standard position, their terminal rays coincide. The two angles therefore have the same trigonometric function values:

$$\cos(\theta + 2\pi) = \cos\theta \qquad \sin(\theta + 2\pi) = \sin\theta \qquad \tan(\theta + 2\pi) = \tan\theta$$
$$\sec(\theta + 2\pi) = \sec\theta \qquad \csc(\theta + 2\pi) = \csc\theta \qquad \cot(\theta + 2\pi) = \cot\theta$$

Similarly, $\cos(\theta - 2\pi) = \cos\theta$, $\sin(\theta - 2\pi) = \sin\theta$, and so on. We describe this repeating behavior by saying that the six basic trigonometric functions are *periodic*.

DEFINITION Periodic Function
A function $f(x)$ is **periodic** if there is a positive number p such that $f(x + p) = f(x)$ for every value of x. The smallest such value of p is the **period** of f.

When we graph trigonometric functions in the coordinate plane, we usually denote the independent variable by x instead of θ. See Figure 1.73.

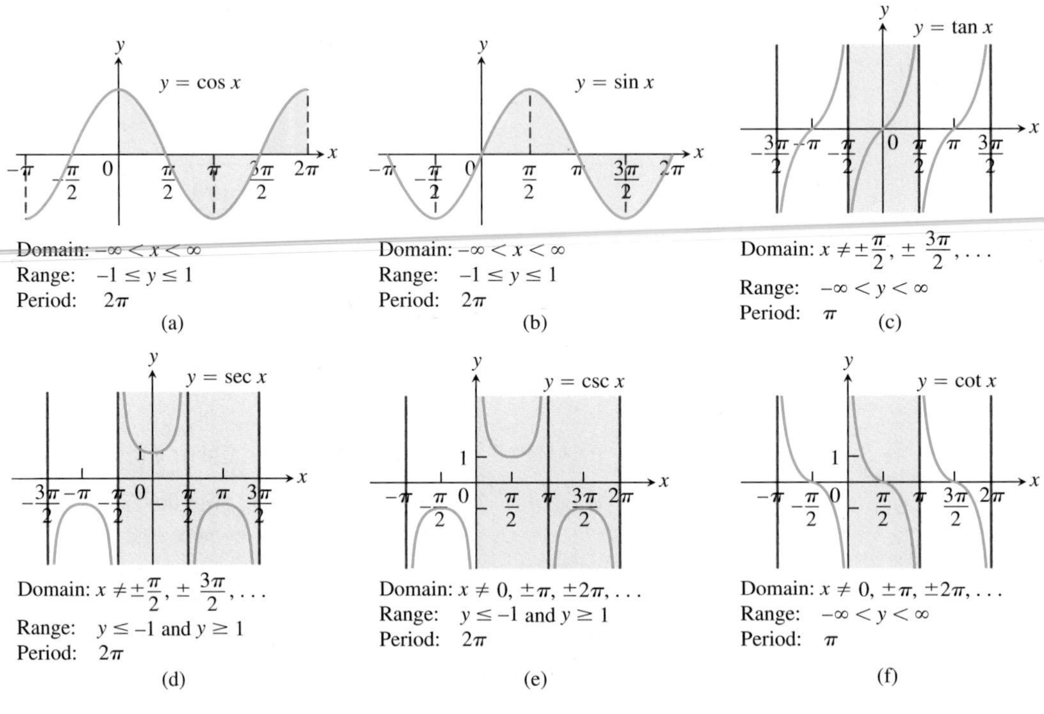

FIGURE 1.73 Graphs of the (a) cosine, (b) sine, (c) tangent, (d) secant, (e) cosecant, and (f) cotangent functions using radian measure. The shading for each trigonometric function indicates its periodicity.

Periods of Trigonometric Functions

Period π: $\tan(x + \pi) = \tan x$
$\cot(x + \pi) = \cot x$

Period 2π: $\sin(x + 2\pi) = \sin x$
$\cos(x + 2\pi) = \cos x$
$\sec(x + 2\pi) = \sec x$
$\csc(x + 2\pi) = \csc x$

As we can see in Figure 1.73, the tangent and cotangent functions have period $p = \pi$. The other four functions have period 2π. Periodic functions are important because many behaviors studied in science are approximately periodic. A theorem from advanced calculus says that every periodic function we want to use in mathematical modeling can be written as an algebraic combination of sines and cosines. We show how to do this in Section 11.11.

The symmetries in the graphs in Figure 1.73 reveal that the cosine and secant functions are even and the other four functions are odd:

Even	**Odd**
$\cos(-x) = \cos x$	$\sin(-x) = -\sin x$
$\sec(-x) = \sec x$	$\tan(-x) = -\tan x$
	$\csc(-x) = -\csc x$
	$\cot(-x) = -\cot x$

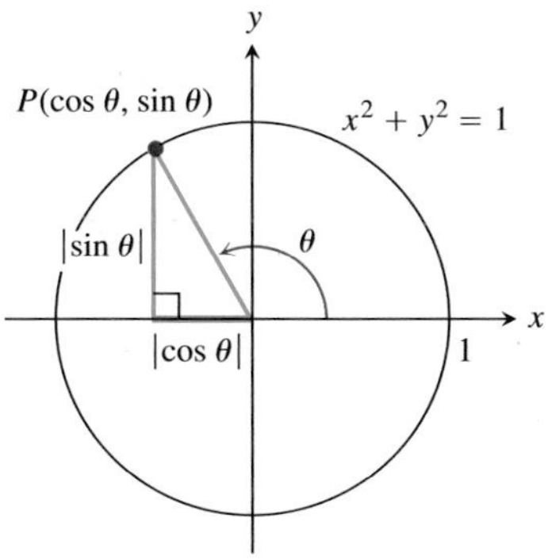

FIGURE 1.74 The reference triangle for a general angle θ.

Identities

The coordinates of any point $P(x, y)$ in the plane can be expressed in terms of the point's distance r from the origin and the angle θ that ray OP makes with the positive x-axis (Figure 1.69). Since $x/r = \cos\theta$ and $y/r = \sin\theta$, we have

$$x = r\cos\theta, \qquad y = r\sin\theta.$$

When $r = 1$ we can apply the Pythagorean theorem to the reference right triangle in Figure 1.74 and obtain the equation

$$\cos^2\theta + \sin^2\theta = 1. \tag{1}$$

This equation, true for all values of θ, is the most frequently used identity in trigonometry. Dividing this identity in turn by $\cos^2\theta$ and $\sin^2\theta$ gives

$$1 + \tan^2\theta = \sec^2\theta.$$
$$1 + \cot^2\theta = \csc^2\theta.$$

The following formulas hold for all angles A and B (Exercises 53 and 54).

Addition Formulas

$$\cos(A + B) = \cos A\cos B - \sin A\sin B$$
$$\sin(A + B) = \sin A\cos B + \cos A\sin B \tag{2}$$

There are similar formulas for $\cos(A - B)$ and $\sin(A - B)$ (Exercises 35 and 36). All the trigonometric identities needed in this book derive from Equations (1) and (2). For example, substituting θ for both A and B in the addition formulas gives

Double-Angle Formulas

$$\begin{aligned} \cos 2\theta &= \cos^2\theta - \sin^2\theta \\ \sin 2\theta &= 2\sin\theta\cos\theta \end{aligned} \tag{3}$$

Additional formulas come from combining the equations

$$\cos^2\theta + \sin^2\theta = 1, \qquad \cos^2\theta - \sin^2\theta = \cos 2\theta.$$

We add the two equations to get $2\cos^2\theta = 1 + \cos 2\theta$ and subtract the second from the first to get $2\sin^2\theta = 1 - \cos 2\theta$. This results in the following identities, which are useful in integral calculus.

Half-Angle Formulas

$$\cos^2\theta = \frac{1 + \cos 2\theta}{2} \tag{4}$$

$$\sin^2\theta = \frac{1 - \cos 2\theta}{2} \tag{5}$$

The Law of Cosines

If a, b, and c are sides of a triangle ABC and if θ is the angle opposite c, then

$$c^2 = a^2 + b^2 - 2ab\cos\theta. \tag{6}$$

This equation is called the **law of cosines**.

We can see why the law holds if we introduce coordinate axes with the origin at C and the positive x-axis along one side of the triangle, as in Figure 1.75. The coordinates of A are $(b, 0)$; the coordinates of B are $(a\cos\theta, a\sin\theta)$. The square of the distance between A and B is therefore

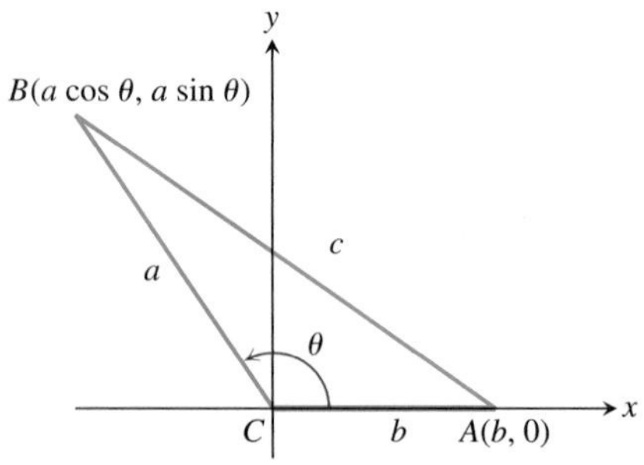

FIGURE 1.75 The square of the distance between A and B gives the law of cosines.

$$\begin{aligned} c^2 &= (a\cos\theta - b)^2 + (a\sin\theta)^2 \\ &= a^2\underbrace{(\cos^2\theta + \sin^2\theta)}_{1} + b^2 - 2ab\cos\theta \\ &= a^2 + b^2 - 2ab\cos\theta. \end{aligned}$$

The law of cosines generalizes the Pythagorean theorem. If $\theta = \pi/2$, then $\cos\theta = 0$ and $c^2 = a^2 + b^2$.

Transformations of Trigonometric Graphs

The rules for shifting, stretching, compressing, and reflecting the graph of a function apply to the trigonometric functions. The following diagram will remind you of the controlling parameters.

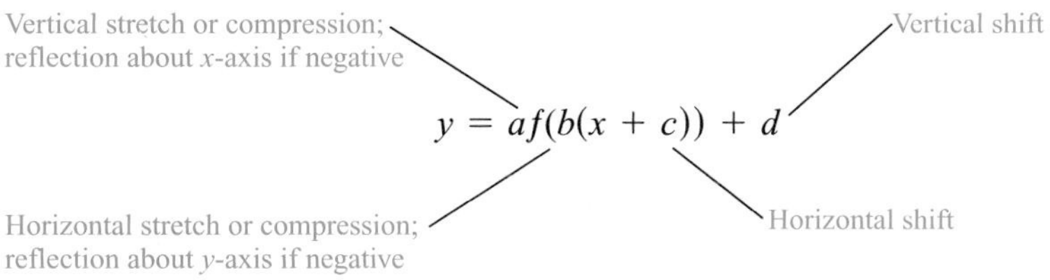

EXAMPLE 2 Modeling Temperature in Alaska

The builders of the Trans-Alaska Pipeline used insulated pads to keep the pipeline heat from melting the permanently frozen soil beneath. To design the pads, it was necessary to take into account the variation in air temperature throughout the year. The variation was represented in the calculations by a **general sine function** or **sinusoid** of the form

$$f(x) = A \sin\left[\frac{2\pi}{B}(x - C)\right] + D,$$

where $|A|$ is the *amplitude*, $|B|$ is the *period*, C is the *horizontal shift*, and D is the *vertical shift* (Figure 1.76).

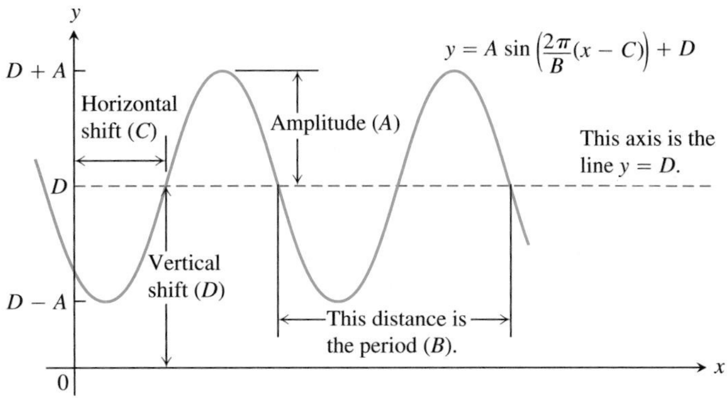

FIGURE 1.76 The general sine curve $y = A \sin[(2\pi/B)(x - C)] + D$, shown for A, B, C, and D positive (Example 2).

Figure 1.77 shows how to use such a function to represent temperature data. The data points in the figure are plots of the mean daily air temperatures for Fairbanks, Alaska, based on records of the National Weather Service from 1941 to 1970. The sine function used to fit the data is

$$f(x) = 37 \sin\left[\frac{2\pi}{365}(x - 101)\right] + 25,$$

where f is temperature in degrees Fahrenheit and x is the number of the day counting from the beginning of the year. The fit, obtained by using the sinusoidal regression feature on a calculator or computer, as we discuss in the next section, is very good at capturing the trend of the data.

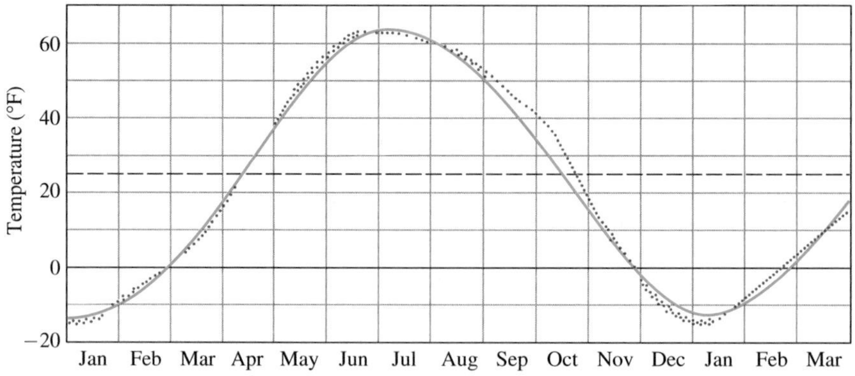

FIGURE 1.77 Normal mean air temperatures for Fairbanks, Alaska, plotted as data points (red). The approximating sine function (blue) is

$$f(x) = 37 \sin [(2\pi/365)(x - 101)] + 25.$$

■

EXERCISES 1.6

Radians, Degrees, and Circular Arcs

1. On a circle of radius 10 m, how long is an arc that subtends a central angle of (a) $4\pi/5$ radians? (b) $110°$?
2. A central angle in a circle of radius 8 is subtended by an arc of length 10π. Find the angle's radian and degree measures.
3. You want to make an $80°$ angle by marking an arc on the perimeter of a 12-in.-diameter disk and drawing lines from the ends of the arc to the disk's center. To the nearest tenth of an inch, how long should the arc be?
4. If you roll a 1-m-diameter wheel forward 30 cm over level ground, through what angle will the wheel turn? Answer in radians (to the nearest tenth) and degrees (to the nearest degree).

Evaluating Trigonometric Functions

5. Copy and complete the following table of function values. If the function is undefined at a given angle, enter "UND." Do not use a calculator or tables.

θ	$-\pi$	$-2\pi/3$	0	$\pi/2$	$3\pi/4$
$\sin\theta$					
$\cos\theta$					
$\tan\theta$					
$\cot\theta$					
$\sec\theta$					
$\csc\theta$					

6. Copy and complete the following table of function values. If the function is undefined at a given angle, enter "UND." Do not use a calculator or tables.

θ	$-3\pi/2$	$-\pi/3$	$-\pi/6$	$\pi/4$	$5\pi/6$
$\sin\theta$					
$\cos\theta$					
$\tan\theta$					
$\cot\theta$					
$\sec\theta$					
$\csc\theta$					

In Exercises 7–12, one of $\sin x$, $\cos x$, and $\tan x$ is given. Find the other two if x lies in the specified interval.

7. $\sin x = \frac{3}{5}, \quad x \in \left[\frac{\pi}{2}, \pi\right]$

8. $\tan x = 2, \quad x \in \left[0, \frac{\pi}{2}\right]$

9. $\cos x = \frac{1}{3}, \quad x \in \left[-\frac{\pi}{2}, 0\right]$

10. $\cos x = -\frac{5}{13}, \quad x \in \left[\frac{\pi}{2}, \pi\right]$

11. $\tan x = \frac{1}{2}, \quad x \in \left[\pi, \frac{3\pi}{2}\right]$

12. $\sin x = -\frac{1}{2}, \quad x \in \left[\pi, \frac{3\pi}{2}\right]$

Graphing Trigonometric Functions

Graph the functions in Exercises 13–22. What is the period of each function?

13. $\sin 2x$

14. $\sin (x/2)$

15. $\cos \pi x$

16. $\cos \frac{\pi x}{2}$

17. $-\sin \frac{\pi x}{3}$

18. $-\cos 2\pi x$

19. $\cos\left(x - \frac{\pi}{2}\right)$

20. $\sin\left(x + \frac{\pi}{2}\right)$

21. $\sin\left(x - \frac{\pi}{4}\right) + 1$

22. $\cos\left(x + \frac{\pi}{4}\right) - 1$

Graph the functions in Exercises 23–26 in the ts-plane (t-axis horizontal, s-axis vertical). What is the period of each function? What symmetries do the graphs have?

23. $s = \cot 2t$

24. $s = -\tan \pi t$

25. $s = \sec\left(\frac{\pi t}{2}\right)$

26. $s = \csc\left(\frac{t}{2}\right)$

T **27.** **a.** Graph $y = \cos x$ and $y = \sec x$ together for $-3\pi/2 \le x \le 3\pi/2$. Comment on the behavior of sec x in relation to the signs and values of cos x.

b. Graph $y = \sin x$ and $y = \csc x$ together for $-\pi \le x \le 2\pi$. Comment on the behavior of csc x in relation to the signs and values of sin x.

T **28.** Graph $y = \tan x$ and $y = \cot x$ together for $-7 \le x \le 7$. Comment on the behavior of cot x in relation to the signs and values of tan x.

29. Graph $y = \sin x$ and $y = \lfloor \sin x \rfloor$ together. What are the domain and range of $\lfloor \sin x \rfloor$?

30. Graph $y = \sin x$ and $y = \lceil \sin x \rceil$ together. What are the domain and range of $\lceil \sin x \rceil$?

Additional Trigonometric Identities

Use the addition formulas to derive the identities in Exercises 31–36.

31. $\cos\left(x - \frac{\pi}{2}\right) = \sin x$

32. $\cos\left(x + \frac{\pi}{2}\right) = -\sin x$

33. $\sin\left(x + \frac{\pi}{2}\right) = \cos x$

34. $\sin\left(x - \frac{\pi}{2}\right) = -\cos x$

35. $\cos(A - B) = \cos A \cos B + \sin A \sin B$ (Exercise 53 provides a different derivation.)

36. $\sin(A - B) = \sin A \cos B - \cos A \sin B$

37. What happens if you take $B = A$ in the identity $\cos(A - B) = \cos A \cos B + \sin A \sin B$? Does the result agree with something you already know?

38. What happens if you take $B = 2\pi$ in the addition formulas? Do the results agree with something you already know?

Using the Addition Formulas

In Exercises 39–42, express the given quantity in terms of sin x and cos x.

39. $\cos(\pi + x)$

40. $\sin(2\pi - x)$

41. $\sin\left(\frac{3\pi}{2} - x\right)$

42. $\cos\left(\frac{3\pi}{2} + x\right)$

43. Evaluate $\sin \frac{7\pi}{12}$ as $\sin\left(\frac{\pi}{4} + \frac{\pi}{3}\right)$.

44. Evaluate $\cos \frac{11\pi}{12}$ as $\cos\left(\frac{\pi}{4} + \frac{2\pi}{3}\right)$.

45. Evaluate $\cos \frac{\pi}{12}$.

46. Evaluate $\sin \frac{5\pi}{12}$.

Using the Double-Angle Formulas

Find the function values in Exercises 47–50.

47. $\cos^2 \frac{\pi}{8}$

48. $\cos^2 \frac{\pi}{12}$

49. $\sin^2 \frac{\pi}{12}$

50. $\sin^2 \frac{\pi}{8}$

Theory and Examples

51. The tangent sum formula The standard formula for the tangent of the sum of two angles is

$$\tan(A + B) = \frac{\tan A + \tan B}{1 - \tan A \tan B}.$$

Derive the formula.

52. *(Continuation of Exercise 51.)* Derive a formula for $\tan(A - B)$.

53. Apply the law of cosines to the triangle in the accompanying figure to derive the formula for $\cos(A - B)$.

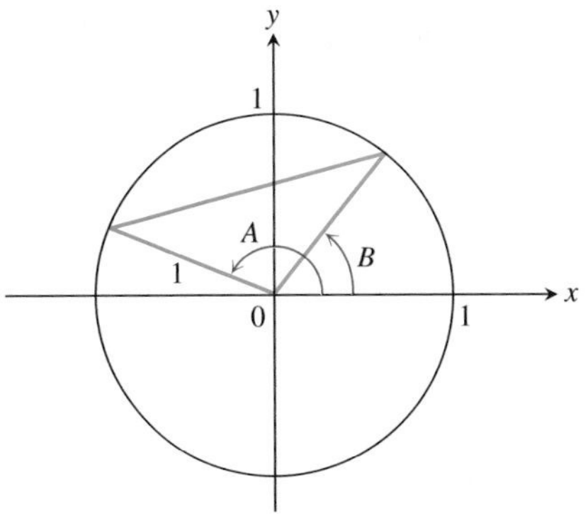

54. **a.** Apply the formula for $\cos(A - B)$ to the identity $\sin\theta = \cos\left(\frac{\pi}{2} - \theta\right)$ to obtain the addition formula for $\sin(A + B)$.

b. Derive the formula for $\cos(A + B)$ by substituting $-B$ for B in the formula for $\cos(A - B)$ from Exercise 35.

55. A triangle has sides $a = 2$ and $b = 3$ and angle $C = 60°$. Find the length of side c.

56. A triangle has sides $a = 2$ and $b = 3$ and angle $C = 40°$. Find the length of side c.

57. The law of sines The *law of sines* says that if a, b, and c are the sides opposite the angles A, B, and C in a triangle, then

$$\frac{\sin A}{a} = \frac{\sin B}{b} = \frac{\sin C}{c}.$$

Use the accompanying figures and the identity $\sin(\pi - \theta) = \sin\theta$, if required, to derive the law.

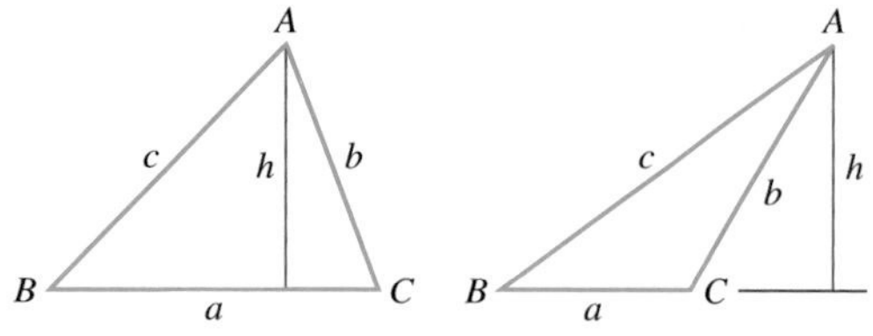

58. A triangle has sides $a = 2$ and $b = 3$ and angle $C = 60°$ (as in Exercise 55). Find the sine of angle B using the law of sines.

59. A triangle has side $c = 2$ and angles $A = \pi/4$ and $B = \pi/3$. Find the length a of the side opposite A.

T **60. The approximation $\sin x \approx x$** It is often useful to know that, when x is measured in radians, $\sin x \approx x$ for numerically small values of x. In Section 3.8, we will see why the approximation holds. The approximation error is less than 1 in 5000 if $|x| < 0.1$.

a. With your grapher in radian mode, graph $y = \sin x$ and $y = x$ together in a viewing window about the origin. What do you see happening as x nears the origin?

b. With your grapher in degree mode, graph $y = \sin x$ and $y = x$ together about the origin again. How is the picture different from the one obtained with radian mode?

c. A quick radian mode check Is your calculator in radian mode? Evaluate $\sin x$ at a value of x near the origin, say $x = 0.1$. If $\sin x \approx x$, the calculator is in radian mode; if not, it isn't. Try it.

General Sine Curves

For

$$f(x) = A\sin\left(\frac{2\pi}{B}(x - C)\right) + D,$$

identify A, B, C, and D for the sine functions in Exercises 61–64 and sketch their graphs (see Figure 1.76).

61. $y = 2\sin(x + \pi) - 1$

62. $y = \frac{1}{2}\sin(\pi x - \pi) + \frac{1}{2}$

63. $y = -\frac{2}{\pi}\sin\left(\frac{\pi}{2}t\right) + \frac{1}{\pi}$

64. $y = \frac{L}{2\pi}\sin\frac{2\pi t}{L},\quad L > 0$

65. Temperature in Fairbanks, Alaska Find the **(a)** amplitude, **(b)** period, **(c)** horizontal shift, and **(d)** vertical shift of the general sine function

$$f(x) = 37\sin\left(\frac{2\pi}{365}(x - 101)\right) + 25.$$

66. Temperature in Fairbanks, Alaska Use the equation in Exercise 65 to approximate the answers to the following questions about the temperature in Fairbanks, Alaska, shown in Figure 1.77. Assume that the year has 365 days.

a. What are the highest and lowest mean daily temperatures shown?

b. What is the average of the highest and lowest mean daily temperatures shown? Why is this average the vertical shift of the function?

COMPUTER EXPLORATIONS

In Exercises 67–70, you will explore graphically the general sine function

$$f(x) = A\sin\left(\frac{2\pi}{B}(x - C)\right) + D$$

as you change the values of the constants A, B, C, and D. Use a CAS or computer grapher to perform the steps in the exercises.

67. The period B Set the constants $A = 3$, $C = D = 0$.

a. Plot $f(x)$ for the values $B = 1, 3, 2\pi, 5\pi$ over the interval $-4\pi \le x \le 4\pi$. Describe what happens to the graph of the general sine function as the period increases.

b. What happens to the graph for negative values of B? Try it with $B = -3$ and $B = -2\pi$.

68. The horizontal shift C Set the constants $A = 3$, $B = 6$, $D = 0$.

a. Plot $f(x)$ for the values $C = 0$, 1, and 2 over the interval $-4\pi \le x \le 4\pi$. Describe what happens to the graph of the general sine function as C increases through positive values.

b. What happens to the graph for negative values of C?

c. What smallest positive value should be assigned to C so the graph exhibits no horizontal shift? Confirm your answer with a plot.

69. The vertical shift D Set the constants $A = 3$, $B = 6$, $C = 0$.

a. Plot $f(x)$ for the values $D = 0$, 1, and 3 over the interval $-4\pi \le x \le 4\pi$. Describe what happens to the graph of the general sine function as D increases through positive values.

b. What happens to the graph for negative values of D?

70. The amplitude A Set the constants $B = 6$, $C = D = 0$.

a. Describe what happens to the graph of the general sine function as A increases through positive values. Confirm your answer by plotting $f(x)$ for the values $A = 1, 5$, and 9.

b. What happens to the graph for negative values of A?

1.7 Graphing with Calculators and Computers

A graphing calculator or a computer with graphing software enables us to graph very complicated functions with high precision. Many of these functions could not otherwise be easily graphed. However, care must be taken when using such devices for graphing purposes and we address those issues in this section. In Chapter 4 we will see how calculus helps us to be certain we are viewing accurately all the important features of a function's graph.

Graphing Windows

When using a graphing calculator or computer as a graphing tool, a portion of the graph is displayed in a rectangular **display** or **viewing window**. Often the default window gives an incomplete or misleading picture of the graph. We use the term *square window* when the units or scales on both axis are the same. This term does not mean that the display window itself is square in shape (usually it is rectangular), but means instead that the x-unit is the same as the y-unit.

When a graph is displayed in the default window, the x-unit may differ from the y-unit of scaling in order to fit the graph in the display. The viewing window in the display is set by specifying the minimum and maximum values of the independent and dependent variables. That is, an interval $a \leq x \leq b$ is specified as well as a range $c \leq y \leq d$. The machine selects a certain number of equally spaced values of x between a and b. Starting with a first value for x, if it lies within the domain of the function f being graphed, and if $f(x)$ lies inside the range $[c, d]$, then the point $(x, f(x))$ is plotted. If x lies outside the domain of f, or $f(x)$ lies outside the specified range $[c, d]$, the machine just moves on to the next x-value since it cannot plot $(x, f(x))$ in that case. The machine plots a large number of points $(x, f(x))$ in this way and approximates the curve representing the graph by drawing a short line segment between each plotted point and its next neighboring point, as we might do by hand. Usually, adjacent points are so close together that the graphical representation has the appearance of a smooth curve. Things can go wrong with this procedure and we illustrate the most common problems through the following examples.

EXAMPLE 1 Choosing a Viewing Window

Graph the function $f(x) = x^3 - 7x^2 + 28$ in each of the following display or viewing windows:

(a) $[-10, 10]$ by $[-10, 10]$ **(b)** $[-4, 4]$ by $[-50, 10]$ **(c)** $[-4, 10]$ by $[-60, 60]$

Solution

(a) We select $a = -10$, $b = 10$, $c = -10$, and $d = 10$ to specify the interval of x-values and the range of y-values for the window. The resulting graph is shown in Figure 1.78a. It appears that the window is cutting off the bottom part of the graph and that the interval of x-values is too large. Let's try the next window.

(b) Now we see more features of the graph (Figure 1.78b), but the top is missing and we need to view more to the right of $x = 4$ as well. The next window should help.

(c) Figure 1.78c shows the graph in this new viewing window. Observe that we get a more complete picture of the graph in this window and it is a reasonable graph of a third-degree polynomial. Choosing a good viewing window is a trial-and-error process which may require some troubleshooting as well.

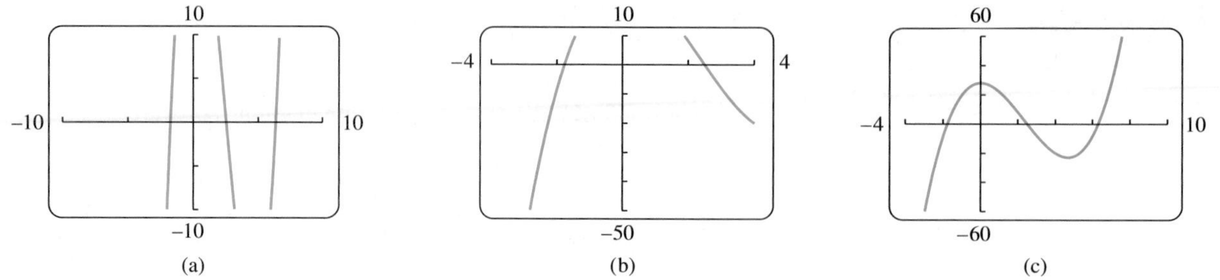

FIGURE 1.78 The graph of $f(x) = x^3 - 7x^2 + 28$ in different viewing windows (Example 1). ■

EXAMPLE 2 Square Windows

When a graph is displayed, the x-unit may differ from the y-unit, as in the graphs shown in Figures 1.78b and 1.78c. The result is distortion in the picture, which may be misleading. The display window can be made square by compressing or stretching the units on one axis to match the scale on the other, giving the true graph. Many systems have built-in functions to make the window "square." If yours does not, you will have to do some calculations and set the window size manually to get a square window, or bring to your viewing some foreknowledge of the true picture.

Figure 1.79a shows the graphs of the perpendicular lines $y = x$ and $y = -x + 3\sqrt{2}$, together with the semicircle $y = \sqrt{9 - x^2}$, in a nonsquare $[-6, 6]$ by $[-6, 8]$ display window. Notice the distortion. The lines do not appear to be perpendicular, and the semicircle appears to be elliptical in shape.

Figure 1.79b shows the graphs of the same functions in a square window in which the x-units are scaled to be the same as the y-units. Notice that the $[-6, 6]$ by $[-4, 4]$ viewing window has the same x-axis in both figures, but the scaling on the x-axis has been compressed in Figure 1.79b to make the window square. Figure 1.79c gives an enlarged view with a square $[-3, 3]$ by $[0, 4]$ window.

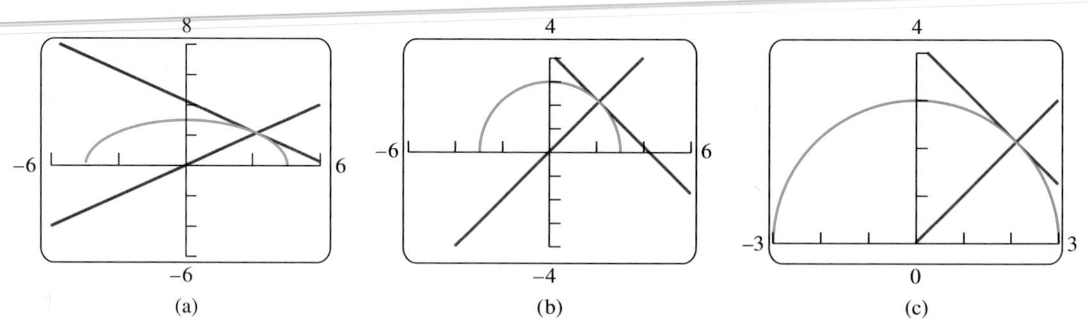

FIGURE 1.79 Graphs of the perpendicular lines $y = x$ and $y = -x + 3\sqrt{2}$, and the semicircle $y = \sqrt{9 - x^2}$, in (a) a nonsquare window, and (b) and (c) square windows (Example 2). ■

If the denominator of a rational function is zero at some x-value within the viewing window, a calculator or graphing computer software may produce a steep near-vertical line segment from the top to the bottom of the window. Here is an example.

EXAMPLE 3 Graph of a Rational Function

Graph the function $y = \dfrac{1}{2 - x}$.

Solution Figure 1.80a shows the graph in the $[-10, 10]$ by $[-10, 10]$ default square window with our computer graphing software. Notice the near-vertical line segment at $x = 2$. It is not truly a part of the graph and $x = 2$ does not belong to the domain of the function. By trial and error we can eliminate the line by changing the viewing window to the smaller $[-6, 6]$ by $[-4, 4]$ view, revealing a better graph (Figure 1.80b).

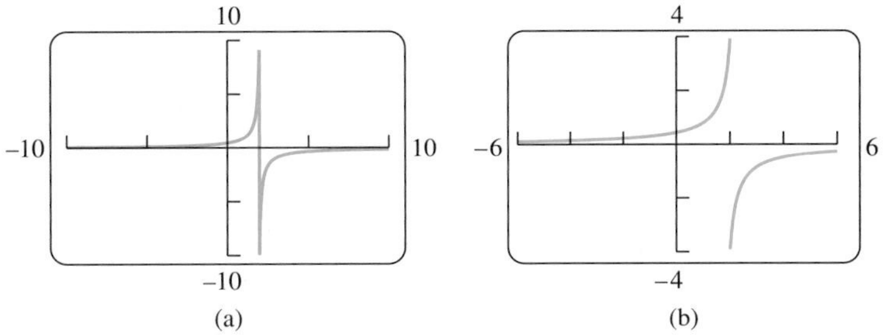

FIGURE 1.80 Graphs of the function $y = \dfrac{1}{2 - x}$ (Example 3). ■

Sometimes the graph of a trigonometric function oscillates very rapidly. When a calculator or computer software plots the points of the graph and connects them, many of the maximum and minimum points are actually missed. The resulting graph is then very misleading.

EXAMPLE 4 Graph of a Rapidly Oscillating Function

Graph the function $f(x) = \sin 100x$.

Solution Figure 1.81a shows the graph of f in the viewing window $[-12, 12]$ by $[-1, 1]$. We see that the graph looks very strange because the sine curve should oscillate periodically between -1 and 1. This behavior is not exhibited in Figure 1.81a. We might experiment with a smaller viewing window, say $[-6, 6]$ by $[-1, 1]$, but the graph is not better (Figure 1.81b). The difficulty is that the period of the trigonometric function $y = \sin 100x$ is very small $(2\pi/100 \approx 0.063)$. If we choose the much smaller viewing window $[-0.1, 0.1]$ by $[-1, 1]$ we get the graph shown in Figure 1.81c. This graph reveals the expected oscillations of a sine curve.

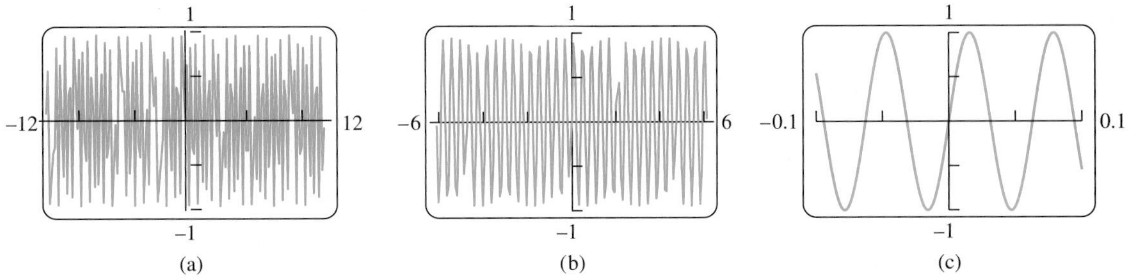

FIGURE 1.81 Graphs of the function $y = \sin 100x$ in three viewing windows. Because the period is $2\pi/100 \approx 0.063$, the smaller window in (c) best displays the true aspects of this rapidly oscillating function (Example 4). ■

EXAMPLE 5 Another Rapidly Oscillating Function

Graph the function $y = \cos x + \frac{1}{50}\sin 50x$.

Solution In the viewing window $[-6, 6]$ by $[-1, 1]$ the graph appears much like the cosine function with some small sharp wiggles on it (Figure 1.82a). We get a better look when we significantly reduce the window to $[-0.6, 0.6]$ by $[0.8, 1.02]$, obtaining the graph in Figure 1.82b. We now see the small but rapid oscillations of the second term, $1/50 \sin 50x$, added to the comparatively larger values of the cosine curve.

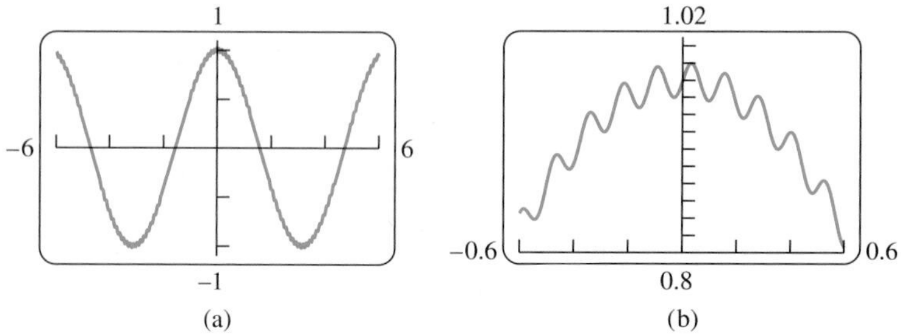

FIGURE 1.82 In (b) we see a close-up view of the function $y = \cos x + \frac{1}{50}\sin 50x$ graphed in (a). The term $\cos x$ clearly dominates the second term, $\frac{1}{50}\sin 50x$, which produces the rapid oscillations along the cosine curve (Example 5).

■

EXAMPLE 6 Graphing an Odd Fractional Power

Graph the function $y = x^{1/3}$.

Solution Many graphing devices display the graph shown in Figure 1.83a. When we compare it with the graph of $y = x^{1/3} = \sqrt[3]{x}$ in Figure 1.38, we see that the left branch for $x < 0$ is missing. The reason the graphs differ is that many calculators and computer soft-

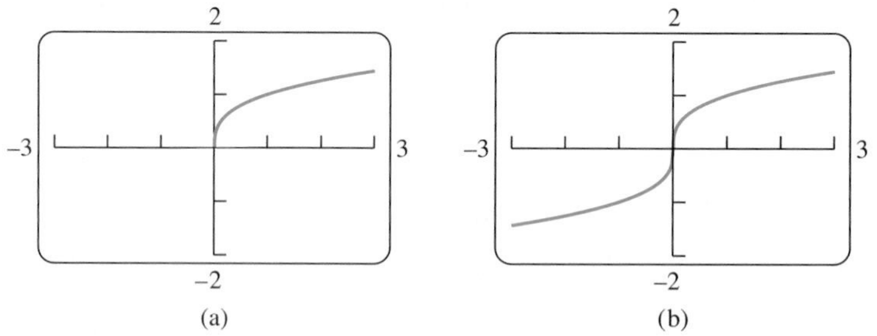

FIGURE 1.83 The graph of $y = x^{1/3}$ is missing the left branch in (a). In (b) we graph the function $f(x) = \frac{x}{|x|} \cdot |x|^{1/3}$ obtaining both branches. (See Example 6.)

ware programs calculate $x^{1/3}$ as $e^{(1/3)\ln x}$. (The exponential and logarithmic functions are studied in Chapter 7.) Since the logarithmic function is not defined for negative values of x, the computing device can only produce the right branch where $x > 0$.

To obtain the full picture showing both branches, we can graph the function

$$f(x) = \frac{x}{|x|} \cdot |x|^{1/3}.$$

This function equals $x^{1/3}$ except at $x = 0$ (where f is undefined, although $0^{1/3} = 0$). The graph of f is shown in Figure 1.83b. ■

Empirical Modeling: Capturing the Trend of Collected Data

In Example 3 of Section 1.4, we verified the reasonableness of Kepler's hypothesis that the period of a planet's orbit is proportional to its mean distance from the sun raised to the 3/2 power. If we cannot hypothesize a relationship between a dependent variable and an independent variable, we might collect data points and try to find a curve that "fits" the data and captures the trend of the scatterplot. The process of finding a curve to fit data is called **regression analysis** and the curve is called a **regression curve**. A computer or graphing calculator finds the regression curve by finding the particular curve which minimizes the sum of the squares of the vertical distances between the data points and the curve. This method of **least squares** is discussed in the Section 14.7 exercises.

There are many useful types of regression curves, such as straight lines, power, polynomial, exponential, logarithmic, and sinusoidal curves. Many computers or graphing calculators have a regression analysis feature to fit a variety of regression curve types. The next example illustrates using a graphing calculator's linear regression feature to fit data from Table 1.5 with a linear equation.

EXAMPLE 7 Fitting a Regression Line

Starting with the data in Table 1.5, build a model for the price of a postage stamp as a function of time. After verifying that the model is "reasonable," use it to predict the price in 2010.

TABLE 1.5 Price of a U.S. postage stamp

Year x	Cost y
1968	0.06
1971	0.08
1974	0.10
1975	0.13
1977	0.15
1981	0.18
1981	0.20
1985	0.22
1987	0.25
1991	0.29
1995	0.32
1998	0.33
2002	0.37

Solution We are building a model for the price of a stamp since 1968. There were two increases in 1981, one of three cents followed by another of two cents. To make 1981 comparable with the other listed years, we lump them together as a single five-cent increase, giving the data in Table 1.6. Figure 1.84a gives the scatterplot for Table 1.6.

TABLE 1.6 Price of a U.S postage stamp since 1968

x	0	3	6	7	9	13	17	19	23	27	30	34
y	6	8	10	13	15	20	22	25	29	32	33	37

Since the scatterplot is fairly linear, we investigate a linear model. Upon entering the data into a graphing calculator (or computer software) and selecting the linear regression option, we find the regression line to be

$$y = 0.94x + 6.10.$$

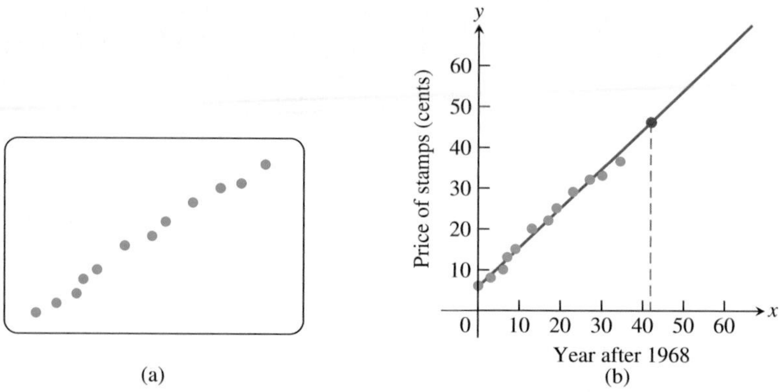

FIGURE 1.84 (a) Scatterplot of (x, y) data in Table 1.6. (b) Using the regression line to estimate the price of a stamp in 2010. (Example 7).

Figure 1.84b shows the line and scatterplot together. The fit is remarkably good, so the model seems reasonable.

Evaluating the regression line, we conclude that in 2010 $(x = 42)$, the price of a stamp will be

$$y = 0.94(42) + 6.10 \approx 46 \text{ cents}.$$

The prediction is shown as the red point on the regression line in Figure 1.84b. ■

EXAMPLE 8 Finding a Curve to Predict Population Levels

We may want to predict the future size of a population, such as the number of trout or catfish living in a fish farm. Figure 1.85 shows a scatterplot of the data collected by R. Pearl for a collection of yeast cells (measured as **biomass**) growing over time (measured in hours) in a nutrient.

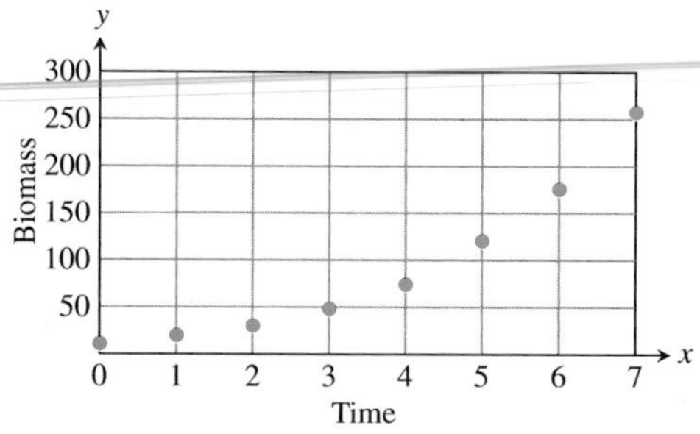

FIGURE 1.85 Biomass of a yeast culture versus elapsed time (Example 8).
(Data from R. Pearl, "The Growth of Population," *Quart. Rev. Biol.*, Vol. 2 (1927), pp. 532–548.)

The plot of points appears to be reasonably smooth with an upward curving trend. We might attempt to capture this trend by fitting a polynomial (for example, a quadratic $y = ax^2 + bx + c$), a power curve $(y = ax^b)$, or an exponential curve $(y = ae^{bx})$. Figure 1.86 shows the result of using a calculator to fit a quadratic model.

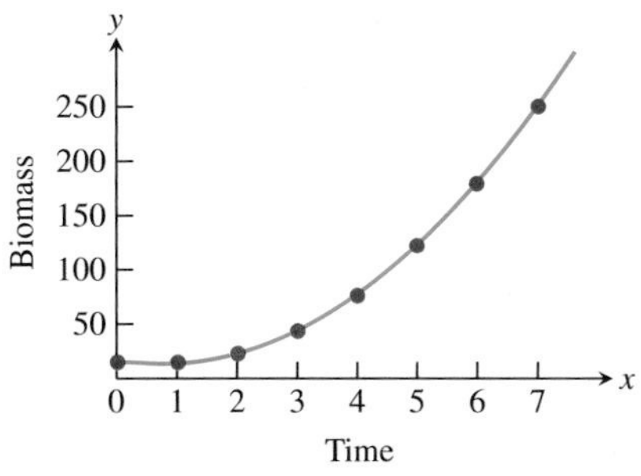

FIGURE 1.86 Fitting a quadratic to Pearl's data gives the equation $y = 6.10x^2 - 9.28x + 16.43$ and the prediction $y(17) = 1622.65$ (Example 8).

The quadratic model $y = 6.10x^2 - 9.28x + 16.43$ appears to fit the collected data reasonably well (Figure 1.86). Using this model, we predict the population after 17 hours as $y(17) = 1622.65$. Let us examine more of Pearl's data to see if our quadratic model continues to be a good one.

In Figure 1.87, we display all of Pearl's data. Now you see that the prediction of $y(17) = 1622.65$ grossly overestimates the observed population of 659.6. Why did the quadratic model fail to predict a more accurate value?

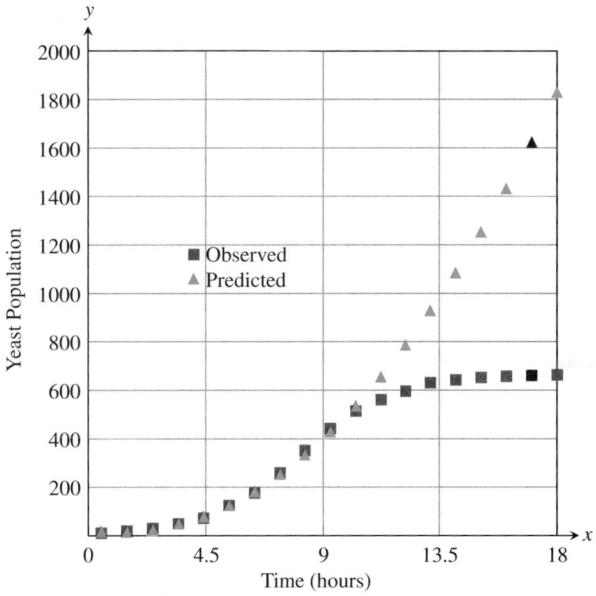

FIGURE 1.87 The rest of Pearl's data (Example 8).

The problem lies in the danger of predicting beyond the range of data used to build the empirical model. (The range of data creating our model was $0 \le x \le 7$.) Such *extrapolation* is especially dangerous when the model selected is not supported by some underlying rationale suggesting the form of the model. In our yeast example, why would we expect a quadratic function as underlying population growth? Why not an exponential function? In the face of this, how then do we predict future values? Often, calculus can help, and in Chapter 9 we use it to model population growth. ■

Regression Analysis

Regression analysis has four steps:

1. Plot the data (scatterplot).
2. Find a regression equation. For a line, it has the form $y = mx + b$, and for a quadratic, the form $y = ax^2 + bx + c$.
3. Superimpose the graph of the regression equation on the scatterplot to see the fit.
4. If the fit is satisfactory, use the regression equation to predict y-values for values of x not in the table.

EXERCISES 1.7

Choosing a Viewing Window

In Exercises 1–4, use a graphing calculator or computer to determine which of the given viewing windows displays the most appropriate graph of the specified function.

1. $f(x) = x^4 - 7x^2 + 6x$
   a. $[-1, 1]$ by $[-1, 1]$ b. $[-2, 2]$ by $[-5, 5]$
   c. $[-10, 10]$ by $[-10, 10]$ d. $[-5, 5]$ by $[-25, 15]$
2. $f(x) = x^3 - 4x^2 - 4x + 16$
   a. $[-1, 1]$ by $[-5, 5]$ b. $[-3, 3]$ by $[-10, 10]$
   c. $[-5, 5]$ by $[-10, 20]$ d. $[-20, 20]$ by $[-100, 100]$
3. $f(x) = 5 + 12x - x^3$
   a. $[-1, 1]$ by $[-1, 1]$ b. $[-5, 5]$ by $[-10, 10]$
   c. $[-4, 4]$ by $[-20, 20]$ d. $[-4, 5]$ by $[-15, 25]$
4. $f(x) = \sqrt{5 + 4x - x^2}$
   a. $[-2, 2]$ by $[-2, 2]$ b. $[-2, 6]$ by $[-1, 4]$
   c. $[-3, 7]$ by $[0, 10]$ d. $[-10, 10]$ by $[-10, 10]$

Determining a Viewing Window

In Exercises 5–30, determine an appropriate viewing window for the given function and use it to display its graph.

5. $f(x) = x^4 - 4x^3 + 15$
6. $f(x) = \frac{x^3}{3} - \frac{x^2}{2} - 2x + 1$
7. $f(x) = x^5 - 5x^4 + 10$
8. $f(x) = 4x^3 - x^4$
9. $f(x) = x\sqrt{9 - x^2}$
10. $f(x) = x^2(6 - x^3)$
11. $y = 2x - 3x^{2/3}$
12. $y = x^{1/3}(x^2 - 8)$
13. $y = 5x^{2/5} - 2x$
14. $y = x^{2/3}(5 - x)$
15. $y = |x^2 - 1|$
16. $y = |x^2 - x|$
17. $y = \frac{x + 3}{x + 2}$
18. $y = 1 - \frac{1}{x + 3}$
19. $f(x) = \frac{x^2 + 2}{x^2 + 1}$
20. $f(x) = \frac{x^2 - 1}{x^2 + 1}$
21. $f(x) = \frac{x - 1}{x^2 - x - 6}$
22. $f(x) = \frac{8}{x^2 - 9}$
23. $f(x) = \frac{6x^2 - 15x + 6}{4x^2 - 10x}$
24. $f(x) = \frac{x^2 - 3}{x - 2}$
25. $y = \sin 250x$
26. $y = 3\cos 60x$
27. $y = \cos\left(\frac{x}{50}\right)$
28. $y = \frac{1}{10}\sin\left(\frac{x}{10}\right)$
29. $y = x + \frac{1}{10}\sin 30x$
30. $y = x^2 + \frac{1}{50}\cos 100x$
31. Graph the lower half of the circle defined by the equation $x^2 + 2x = 4 + 4y - y^2$.
32. Graph the upper branch of the hyperbola $y^2 - 16x^2 = 1$.
33. Graph four periods of the function $f(x) = -\tan 2x$.
34. Graph two periods of the function $f(x) = 3\cot\frac{x}{2} + 1$.
35. Graph the function $f(x) = \sin 2x + \cos 3x$.
36. Graph the function $f(x) = \sin^3 x$.

Graphing in Dot Mode

Another way to avoid incorrect connections when using a graphing device is through the use of a "dot mode," which plots only the points. If your graphing utility allows that mode, use it to plot the functions in Exercises 37–40.

37. $y = \frac{1}{x - 3}$
38. $y = \sin\frac{1}{x}$
39. $y = x\lfloor x \rfloor$
40. $y = \frac{x^3 - 1}{x^2 - 1}$

Regression Analysis

T 41. Table 1.7 shows the mean annual compensation of construction workers.

TABLE 1.7 Construction workers' average annual compensation

Year	Annual compensation (dollars)
1980	22,033
1985	27,581
1988	30,466
1990	32,836
1992	34,815
1995	37,996
1999	42,236
2002	45,413

Source: U.S. Bureau of Economic Analysis.

a. Find a linear regression equation for the data.
b. Find the slope of the regression line. What does the slope represent?

c. Superimpose the graph of the linear regression equation on a scatterplot of the data.

d. Use the regression equation to predict the construction workers' average annual compensation in 2010.

T **42.** The median price of existing single-family homes has increased consistently since 1970. The data in Table 1.8, however, show that there have been differences in various parts of the country.

a. Find a linear regression equation for home cost in the Northeast.

b. What does the slope of the regression line represent?

c. Find a linear regression equation for home cost in the Midwest.

d. Where is the median price increasing more rapidly, in the Northeast or the Midwest?

TABLE 1.8 Median price of single-family homes

Year	Northeast (dollars)	Midwest (dollars)
1970	25,200	20,100
1975	39,300	30,100
1980	60,800	51,900
1985	88,900	58,900
1990	141,200	74,000
1995	197,100	88,300
2000	264,700	97,000

Source: National Association of Realtors®

T **43. Vehicular stopping distance** Table 1.9 shows the total stopping distance of a car as a function of its speed.

a. Find the quadratic regression equation for the data in Table 1.9.

b. Superimpose the graph of the quadratic regression equation on a scatterplot of the data.

c. Use the graph of the quadratic regression equation to predict the average total stopping distance for speeds of 72 and 85 mph. Confirm algebraically.

d. Now use *linear* regression to predict the average total stopping distance for speeds of 72 and 85 mph. Superimpose the regression line on a scatterplot of the data. Which gives the better fit, the line here or the graph in part (b)?

TABLE 1.9 Vehicular stopping distance

Speed (mph)	Average total stopping distance (ft)
20	42
25	56
30	73.5
35	91.5
40	116
45	142.5
50	173
55	209.5
60	248
65	292.5
70	343
75	401
80	464

Source: U.S. Bureau of Public Roads.

T **44. Stern waves** Observations of the stern waves that follow a boat at right angles to its course have disclosed that the distance between the crests of these waves (their *wave length*) increases with the speed of the boat. Table 1.10 shows the relationship between wave length and the speed of the boat.

TABLE 1.10 Wave lengths

Wave length (m)	Speed (km/h)
0.20	1.8
0.65	3.6
1.13	5.4
2.55	7.2
4.00	9.0
5.75	10.8
7.80	12.6
10.20	14.4
12.90	16.2
16.00	18.0
18.40	19.8

a. Find a power regression equation $y = ax^b$ for the data in Table 1.10, where x is the wave length, and y the speed of the boat.

b. Superimpose the graph of the power regression equation on a scatterplot of the data.

c. Use the graph of the power regression equation to predict the speed of the boat when the wave length is 11 m. Confirm algebraically.

d. Now use *linear* regression to predict the speed when the wave length is 11 m. Superimpose the regression line on a scatterplot of the data. Which gives the better fit, the line here or the curve in part (b)?

Chapter 1 Questions to Guide Your Review

1. How are the real numbers represented? What are the main categories characterizing the properties of the real number system? What are the primary subsets of the real numbers?
2. How are the rational numbers described in terms of decimal expansions? What are the irrational numbers? Give examples.
3. What are the order properties of the real numbers? How are they used in solving equations?
4. What is a number's absolute value? Give examples? How are $|-a|$, $|ab|$, $|a/b|$, and $|a + b|$ related to $|a|$ and $|b|$?
5. How are absolute values used to describe intervals or unions of intervals? Give examples.
6. How do we identify points in the plane using the Cartesian coordinate system? What is the graph of an equation in the variables x and y?
7. How can you write an equation for a line if you know the coordinates of two points on the line? The line's slope and the coordinates of one point on the line? The line's slope and y-intercept? Give examples.
8. What are the standard equations for lines perpendicular to the coordinate axes?
9. How are the slopes of mutually perpendicular lines related? What about parallel lines? Give examples.
10. When a line is not vertical, what is the relation between its slope and its angle of inclination?
11. How do you find the distance between two points in the coordinate plane?
12. What is the standard equation of a circle with center (h, k) and radius a? What is the unit circle and what is its equation?
13. Describe the steps you would take to graph the circle $x^2 + y^2 + 4x - 6y + 12 = 0$.
14. What inequality describes the points in the coordinate plane that lie inside the circle of radius a centered at the point (h, k)? That lie inside or on the circle? That lie outside the circle? That lie outside or on the circle?
15. If a, b, and c are constants and $a \neq 0$, what can you say about the graph of the equation $y = ax^2 + bx + c$? In particular, how would you go about sketching the curve $y = 2x^2 + 4x$?
16. What is a function? What is its domain? Its range? What is an arrow diagram for a function? Give examples.
17. What is the graph of a real-valued function of a real variable? What is the vertical line test?
18. What is a piecewise-defined function? Give examples.
19. What are the important types of functions frequently encountered in calculus? Give an example of each type.
20. In terms of its graph, what is meant by an increasing function? A decreasing function? Give an example of each.
21. What is an even function? An odd function? What symmetry properties do the graphs of such functions have? What advantage can we take of this? Given an example of a function that is neither even nor odd.
22. What does it mean to say that y is proportional to x? To $x^{3/2}$? What is the geometric interpretation of proportionality? How can this interpretation be used to test a proposed proportionality?
23. If f and g are real-valued functions, how are the domains of $f + g$, $f - g$, fg, and f/g related to the domains of f and g? Give examples.
24. When is it possible to compose one function with another? Give examples of composites and their values at various points. Does the order in which functions are composed ever matter?
25. How do you change the equation $y = f(x)$ to shift its graph vertically up or down by a factor $k > 0$? Horizontally to the left or right? Give examples.
26. How do you change the equation $y = f(x)$ to compress or stretch the graph by $c > 1$? Reflect the graph across a coordinate axis? Give examples.
27. What is the standard equation of an ellipse with center (h, k)? What is its major axis? Its minor axis? Give examples.
28. What is radian measure? How do you convert from radians to degrees? Degrees to radians?
29. Graph the six basic trigonometric functions. What symmetries do the graphs have?
30. What is a periodic function? Give examples. What are the periods of the six basic trigonometric functions?

31. Starting with the identity $\sin^2\theta + \cos^2\theta = 1$ and the formulas for $\cos(A + B)$ and $\sin(A + B)$, show how a variety of other trigonometric identities may be derived.

32. How does the formula for the general sine function $f(x) = A\sin((2\pi/B)(x - C)) + D$ relate to the shifting, stretching, compressing, and reflection of its graph? Give examples. Graph the general sine curve and identify the constants A, B, C, and D.

33. Name three issues that arise when functions are graphed using a calculator or computer with graphing software. Give examples.

Chapter 1 Practice Exercises

Inequalities

In Exercises 1–4, solve the inequalities and show the solution sets on the real line.

1. $7 + 2x \geq 3$
2. $-3x < 10$
3. $\frac{1}{5}(x - 1) < \frac{1}{4}(x - 2)$
4. $\frac{x - 3}{2} \geq -\frac{4 + x}{3}$

Absolute Value

Solve the equations or inequalities in Exercises 5–8.

5. $|x + 1| = 7$
6. $|y - 3| < 4$
7. $\left|1 - \frac{x}{2}\right| > \frac{3}{2}$
8. $\left|\frac{2x + 7}{3}\right| \leq 5$

Coordinates

9. A particle in the plane moved from $A(-2, 5)$ to the y-axis in such a way that Δy equaled $3\Delta x$. What were the particle's new coordinates?

10. **a.** Plot the points $A(8, 1)$, $B(2, 10)$, $C(-4, 6)$, $D(2, -3)$, and $E(14/3, 6)$.
    b. Find the slopes of the lines AB, BC, CD, DA, CE, and BD.
    c. Do any four of the five points A, B, C, D, and E form a parallelogram?
    d. Are any three of the five points collinear? How do you know?
    e. Which of the lines determined by the five points pass through the origin?

11. Do the points $A(6, 4)$, $B(4, -3)$, and $C(-2, 3)$ form an isosceles triangle? A right triangle? How do you know?

12. Find the coordinates of the point on the line $y = 3x + 1$ that is equidistant from $(0, 0)$ and $(-3, 4)$.

Lines

In Exercises 13–24, write an equation for the specified line.

13. through $(1, -6)$ with slope 3
14. through $(-1, 2)$ with slope $-1/2$
15. the vertical line through $(0, -3)$
16. through $(-3, 6)$ and $(1, -2)$
17. the horizontal line through $(0, 2)$
18. through $(3, 3)$ and $(-2, 5)$
19. with slope -3 and y-intercept 3
20. through $(3, 1)$ and parallel to $2x - y = -2$
21. through $(4, -12)$ and parallel to $4x + 3y = 12$
22. through $(-2, -3)$ and perpendicular to $3x - 5y = 1$
23. through $(-1, 2)$ and perpendicular to $(1/2)x + (1/3)y = 1$
24. with x-intercept 3 and y-intercept -5

Functions and Graphs

25. Express the area and circumference of a circle as functions of the circle's radius. Then express the area as a function of the circumference.

26. Express the radius of a sphere as a function of the sphere's surface area. Then express the surface area as a function of the volume.

27. A point P in the first quadrant lies on the parabola $y = x^2$. Express the coordinates of P as functions of the angle of inclination of the line joining P to the origin.

28. A hot-air balloon rising straight up from a level field is tracked by a range finder located 500 ft from the point of liftoff. Express the balloon's height as a function of the angle the line from the range finder to the balloon makes with the ground.

In Exercises 29–32, determine whether the graph of the function is symmetric about the y-axis, the origin, or neither.

29. $y = x^{1/5}$
30. $y = x^{2/5}$
31. $y = x^2 - 2x - 1$
32. $y = e^{-x^2}$

In Exercises 33–40, determine whether the function is even, odd, or neither.

33. $y = x^2 + 1$
34. $y = x^5 - x^3 - x$
35. $y = 1 - \cos x$
36. $y = \sec x \tan x$
37. $y = \frac{x^4 + 1}{x^3 - 2x}$
38. $y = 1 - \sin x$
39. $y = x + \cos x$
40. $y = \sqrt{x^4 - 1}$

In Exercises 41–50, find the **(a)** domain and **(b)** range.

41. $y = |x| - 2$

42. $y = -2 + \sqrt{1 - x}$

43. $y = \sqrt{16 - x^2}$

44. $y = 3^{2-x} + 1$

45. $y = 2e^{-x} - 3$

46. $y = \tan(2x - \pi)$

47. $y = 2\sin(3x + \pi) - 1$

48. $y = x^{2/5}$

49. $y = \ln(x - 3) + 1$

50. $y = -1 + \sqrt[3]{2 - x}$

Piecewise-Defined Functions

In Exercises 51 and 52, find the **(a)** domain and **(b)** range.

51. $y = \begin{cases} \sqrt{-x}, & -4 \le x \le 0 \\ \sqrt{x}, & 0 < x \le 4 \end{cases}$

52. $y = \begin{cases} -x - 2, & -2 \le x \le -1 \\ x, & -1 < x \le 1 \\ -x + 2, & 1 < x \le 2 \end{cases}$

In Exercises 53 and 54, write a piecewise formula for the function.

53.

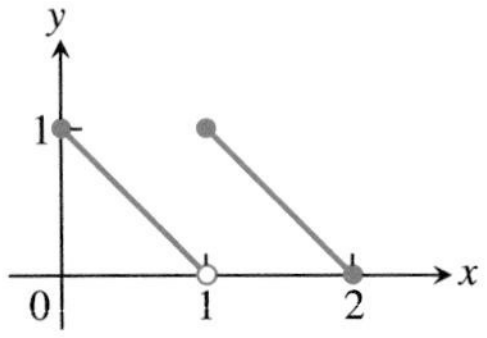

54.

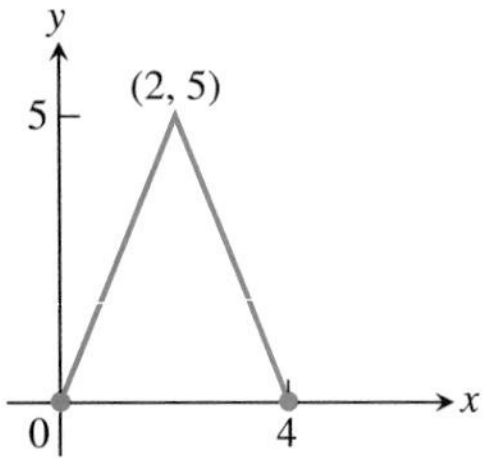

Composition of Functions

In Exercises 55 and 56, find

a. $(f \circ g)(-1)$.

b. $(g \circ f)(2)$.

c. $(f \circ f)(x)$.

d. $(g \circ g)(x)$.

55. $f(x) = \dfrac{1}{x}, \quad g(x) = \dfrac{1}{\sqrt{x + 2}}$

56. $f(x) = 2 - x, \quad g(x) = \sqrt[3]{x + 1}$

In Exercises 57 and 58, **(a)** write a formula for $f \circ g$ and $g \circ f$ and find the **(b)** domain and **(c)** range of each.

57. $f(x) = 2 - x^2, \quad g(x) = \sqrt{x + 2}$

58. $f(x) = \sqrt{x}, \quad g(x) = \sqrt{1 - x}$

Composition with absolute values In Exercises 59–64, graph f_1 and f_2 together. Then describe how applying the absolute value function before applying f_1 affects the graph.

	$f_1(x)$	$f_2(x) = f_1(\lvert x \rvert)$
59.	x	$\lvert x \rvert$
60.	x^3	$\lvert x \rvert^3$
61.	x^2	$\lvert x \rvert^2$
62.	$\dfrac{1}{x}$	$\dfrac{1}{\lvert x \rvert}$
63.	$\sqrt{x}$	$\sqrt{\lvert x \rvert}$
64.	$\sin x$	$\sin \lvert x \rvert$

Composition with absolute values In Exercises 65–68, graph g_1 and g_2 together. Then describe how taking absolute values after applying g_1 affects the graph.

	$g_1(x)$	$g_2(x) = \lvert g_1(x) \rvert$
65.	x^3	$\lvert x^3 \rvert$
66.	$\sqrt{x}$	$\lvert \sqrt{x} \rvert$
67.	$4 - x^2$	$\lvert 4 - x^2 \rvert$
68.	$x^2 + x$	$\lvert x^2 + x \rvert$

Trigonometry

In Exercises 69–72, sketch the graph of the given function. What is the period of the function?

69. $y = \cos 2x$

70. $y = \sin \dfrac{x}{2}$

71. $y = \sin \pi x$

72. $y = \cos \dfrac{\pi x}{2}$

73. Sketch the graph $y = 2\cos\left(x - \dfrac{\pi}{3}\right)$.

74. Sketch the graph $y = 1 + \sin\left(x + \dfrac{\pi}{4}\right)$.

In Exercises 75–78, ABC is a right triangle with the right angle at C. The sides opposite angles A, B, and C are a, b, and c, respectively.

75. a. Find a and b if $c = 2$, $B = \pi/3$.

b. Find a and c if $b = 2$, $B = \pi/3$.

76. a. Express a in terms of A and c.

b. Express a in terms of A and b.

77. a. Express a in terms of B and b.

b. Express c in terms of A and a.

78. a. Express $\sin A$ in terms of a and c.

b. Express $\sin A$ in terms of b and c.

79. Height of a pole Two wires stretch from the top T of a vertical pole to points B and C on the ground, where C is 10 m closer to the base of the pole than is B. If wire BT makes an angle of 35° with the horizontal and wire CT makes an angle of 50° with the horizontal, how high is the pole?

80. Height of a weather balloon Observers at positions A and B 2 km apart simultaneously measure the angle of elevation of a weather balloon to be 40° and 70°, respectively. If the balloon is directly above a point on the line segment between A and B, find the height of the balloon.

T **81. a.** Graph the function $f(x) = \sin x + \cos(x/2)$.

b. What appears to be the period of this function?

c. Confirm your finding in part (b) algebraically.

T **82. a.** Graph $f(x) = \sin(1/x)$.

b. What are the domain and range of f?

c. Is f periodic? Give reasons for your answer.

Chapter 1 Additional and Advanced Exercises

Functions and Graphs

1. The graph of f is shown. Draw the graph of each function.

a. $y = f(-x)$ **b.** $y = -f(x)$

c. $y = -2f(x + 1) + 1$ **d.** $y = 3f(x - 2) - 2$

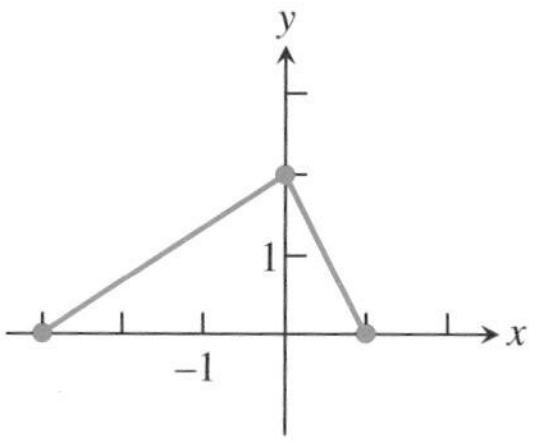

2. A portion of the graph of a function defined on $[-3, 3]$ is shown. Complete the graph assuming that the function is

a. even. **b.** odd.

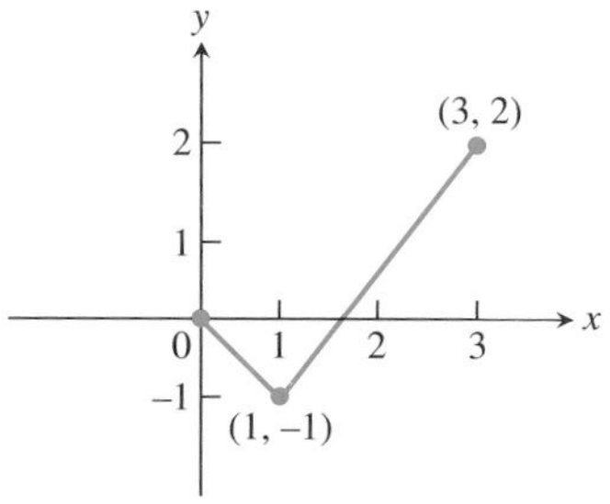

3. Are there two functions f and g such that $f \circ g = g \circ f$? Give reasons for your answer.

4. Are there two functions f and g with the following property? The graphs of f and g are not straight lines but the graph of $f \circ g$ is a straight line. Give reasons for your answer.

5. If $f(x)$ is odd, can anything be said of $g(x) = f(x) - 2$? What if f is even instead? Give reasons for your answer.

6. If $g(x)$ is an odd function defined for all values of x, can anything be said about $g(0)$? Give reasons for your answer.

7. Graph the equation $|x| + |y| = 1 + x$.

8. Graph the equation $y + |y| = x + |x|$.

Trigonometry

In Exercises 9–14, ABC is an arbitrary triangle with sides a, b, and c opposite angles A, B, and C, respectively.

9. Find b if $a = \sqrt{3}, A = \pi/3, B = \pi/4$.

10. Find $\sin B$ if $a = 4, b = 3, A = \pi/4$.

11. Find $\cos A$ if $a = 2, b = 2, c = 3$.

12. Find c if $a = 2, b = 3, C = \pi/4$.

13. Find $\sin B$ if $a = 2, b = 3, c = 4$.

14. Find $\sin C$ if $a = 2, b = 4, c = 5$.

Derivations and Proofs

15. Prove the following identities.

a. $\dfrac{1 - \cos x}{\sin x} = \dfrac{\sin x}{1 + \cos x}$

b. $\dfrac{1 - \cos x}{1 + \cos x} = \tan^2 \dfrac{x}{2}$

16. Explain the following "proof without words" of the law of cosines. (Source: "Proof without Words: The Law of Cosines," Sidney H. Kung, *Mathematics Magazine*, Vol. 63, No. 5, Dec. 1990, p. 342.)

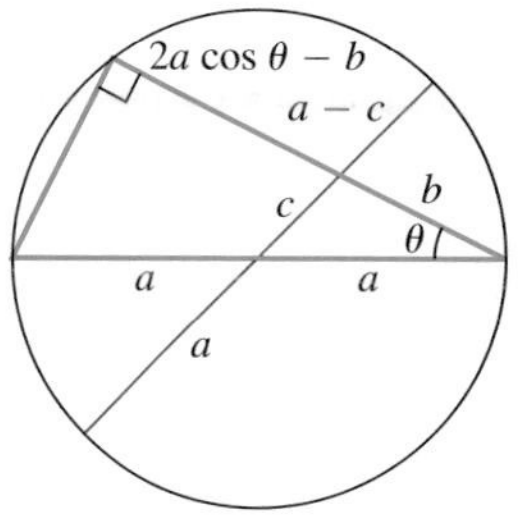

17. Show that the area of triangle ABC is given by $(1/2)ab \sin C = (1/2)bc \sin A = (1/2)ca \sin B$.

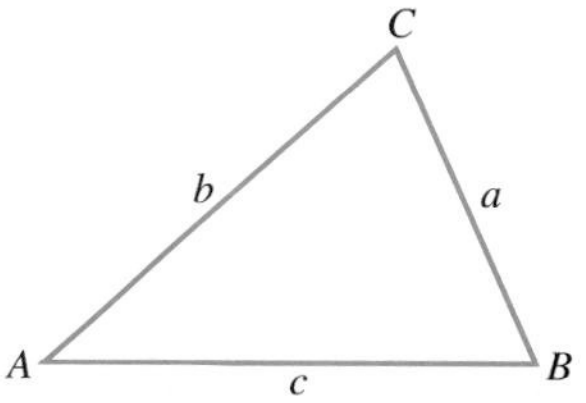

18. Show that the area of triangle ABC is given by $\sqrt{s(s - a)(s - b)(s - c)}$ where $s = (a + b + c)/2$ is the semiperimeter of the triangle.

19. Properties of inequalities If a and b are real numbers, we say that ***a* is less than *b*** and write $a < b$ if (and only if) $b - a$ is positive. Use this definition to prove the following properties of inequalities.

If a, b, and c are real numbers, then:

1. $a < b \Rightarrow a + c < b + c$

2. $a < b \Rightarrow a - c < b - c$

3. $a < b$ and $c > 0 \Rightarrow ac < bc$

4. $a < b$ and $c < 0 \Rightarrow bc < ac$
(Special case: $a < b \Rightarrow -b < -a$)

5. $a > 0 \Rightarrow \frac{1}{a} > 0$

6. $0 < a < b \Rightarrow \frac{1}{b} < \frac{1}{a}$

7. $a < b < 0 \Rightarrow \frac{1}{b} < \frac{1}{a}$

20. Prove that the following inequalities hold for any real numbers a and b.

a. $|a| < |b|$ if and only if $a^2 < b^2$

b. $|a - b| \geq ||a| - |b||$

Generalizing the triangle inequality Prove by mathematical induction that the inequalities in Exercises 21 and 22 hold for any n real numbers $a_1, a_2, \ldots, a_n$. (Mathematical induction is reviewed in Appendix 1.)

21. $|a_1 + a_2 + \cdots + a_n| \leq |a_1| + |a_2| + \cdots + |a_n|$

22. $|a_1 + a_2 + \cdots + a_n| \geq |a_1| - |a_2| - \cdots - |a_n|$

23. Show that if f is both even and odd, then $f(x) = 0$ for every x in the domain of f.

24. a. Even-odd decompositions Let f be a function whose domain is symmetric about the origin, that is, $-x$ belongs to the domain whenever x does. Show that f is the sum of an even function and an odd function:

$$f(x) = E(x) + O(x),$$

where E is an even function and O is an odd function. (*Hint:* Let $E(x) = (f(x) + f(-x))/2$. Show that $E(-x) = E(x)$, so that E is even. Then show that $O(x) = f(x) - E(x)$ is odd.)

b. Uniqueness Show that there is only one way to write f as the sum of an even and an odd function. (*Hint:* One way is given in part (a). If also $f(x) = E_1(x) + O_1(x)$ where E_1 is even and O_1 is odd, show that $E - E_1 = O_1 - O$. Then use Exercise 23 to show that $E = E_1$ and $O = O_1$.)

Grapher Explorations—Effects of Parameters

25. What happens to the graph of $y = ax^2 + bx + c$ as

a. a changes while b and c remain fixed?

b. b changes (a and c fixed, $a \neq 0$)?

c. c changes (a and b fixed, $a \neq 0$)?

26. What happens to the graph of $y = a(x + b)^3 + c$ as

a. a changes while b and c remain fixed?

b. b changes (a and c fixed, $a \neq 0$)?

c. c changes (a and b fixed, $a \neq 0$)?

27. Find all values of the slope of the line $y = mx + 2$ for which the x-intercept exceeds $1/2$.

Geometry

28. An object's center of mass moves at a constant velocity v along a straight line past the origin. The accompanying figure shows the coordinate system and the line of motion. The dots show positions that are 1 sec apart. Why are the areas $A_1, A_2, \ldots, A_5$ in the figure all equal? As in Kepler's equal area law (see Section 13.6), the line that joins the object's center of mass to the origin sweeps out equal areas in equal times.

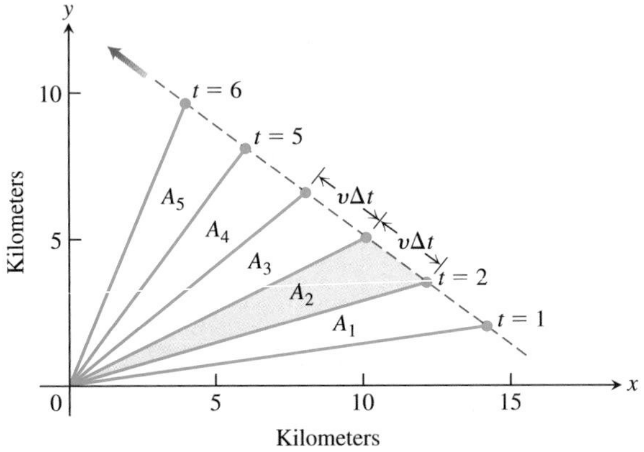

29. a. Find the slope of the line from the origin to the midpoint P, of side AB in the triangle in the accompanying figure ($a, b > 0$).

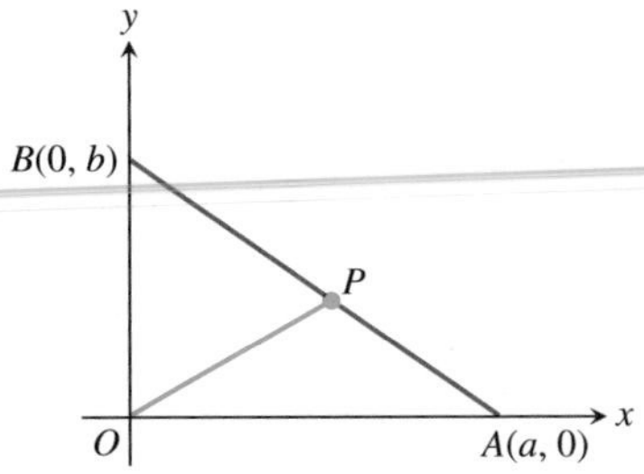

b. When is OP perpendicular to AB?

Chapter 1 Technology Application Projects

An Overview of Mathematica
An overview of *Mathematica* sufficient to complete the *Mathematica* modules appearing on the Web site.

Mathematica/Maple Module
Modeling Change: Springs, Driving Safety, Radioactivity, Trees, Fish, and Mammals.
Construct and interpret mathematical models, analyze and improve them, and make predictions using them.

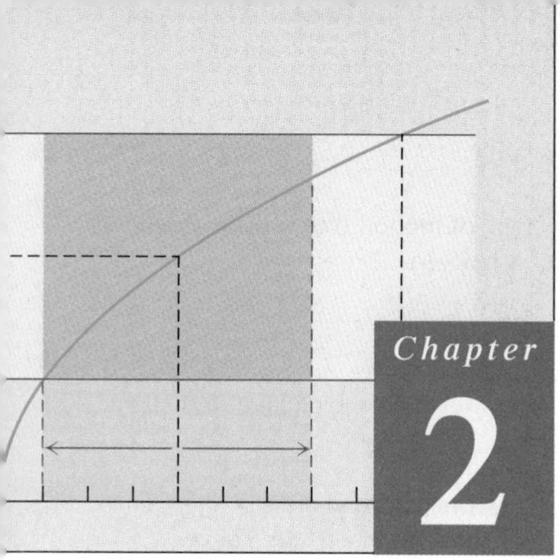

Chapter 2 LIMITS AND CONTINUITY

OVERVIEW The concept of a limit is a central idea that distinguishes calculus from algebra and trigonometry. It is fundamental to finding the tangent to a curve or the velocity of an object.

In this chapter we develop the limit, first intuitively and then formally. We use limits to describe the way a function f varies. Some functions vary continuously; small changes in x produce only small changes in $f(x)$. Other functions can have values that jump or vary erratically. The notion of limit gives a precise way to distinguish between these behaviors. The geometric application of using limits to define the tangent to a curve leads at once to the important concept of the derivative of a function. The derivative, which we investigate thoroughly in Chapter 3, quantifies the way a function's values change.

2.1 Rates of Change and Limits

In this section, we introduce average and instantaneous rates of change. These lead to the main idea of the section, the idea of limit.

Average and Instantaneous Speed

A moving body's **average speed** during an interval of time is found by dividing the distance covered by the time elapsed. The unit of measure is length per unit time: kilometers per hour, feet per second, or whatever is appropriate to the problem at hand.

EXAMPLE 1 Finding an Average Speed

A rock breaks loose from the top of a tall cliff. What is its average speed

(a) during the first 2 sec of fall?

(b) during the 1-sec interval between second 1 and second 2?

Solution In solving this problem we use the fact, discovered by Galileo in the late sixteenth century, that a solid object dropped from rest (not moving) to fall freely near the surface of the earth will fall a distance proportional to the square of the time it has been falling. (This assumes negligible air resistance to slow the object down and that gravity is

HISTORICAL BIOGRAPHY*

Galileo Galilei
(1564–1642)

the only force acting on the falling body. We call this type of motion **free fall**.) If y denotes the distance fallen in feet after t seconds, then Galileo's law is

$$y = 16t^2,$$

where 16 is the constant of proportionality.

The average speed of the rock during a given time interval is the change in distance, Δy, divided by the length of the time interval, Δt.

(a) For the first 2 sec: $\dfrac{\Delta y}{\Delta t} = \dfrac{16(2)^2 - 16(0)^2}{2 - 0} = 32 \dfrac{\text{ft}}{\text{sec}}$

(b) From sec 1 to sec 2: $\dfrac{\Delta y}{\Delta t} = \dfrac{16(2)^2 - 16(1)^2}{2 - 1} = 48 \dfrac{\text{ft}}{\text{sec}}$ ■

The next example examines what happens when we look at the average speed of a falling object over shorter and shorter time intervals.

EXAMPLE 2 Finding an Instantaneous Speed

Find the speed of the falling rock at $t = 1$ and $t = 2$ sec.

Solution We can calculate the average speed of the rock over a time interval $[t_0, t_0 + h]$, having length $\Delta t = h$, as

$$\frac{\Delta y}{\Delta t} = \frac{16(t_0 + h)^2 - 16t_0^2}{h}. \tag{1}$$

We cannot use this formula to calculate the "instantaneous" speed at t_0 by substituting $h = 0$, because we cannot divide by zero. But we *can* use it to calculate average speeds over increasingly short time intervals starting at $t_0 = 1$ and $t_0 = 2$. When we do so, we see a pattern (Table 2.1).

TABLE 2.1 Average speeds over short time intervals

Average speed: $\dfrac{\Delta y}{\Delta t} = \dfrac{16(t_0 + h)^2 - 16t_0^2}{h}$

Length of time interval h	Average speed over interval of length h starting at $t_0 = 1$	Average speed over interval of length h starting at $t_0 = 2$
1	48	80
0.1	33.6	65.6
0.01	32.16	64.16
0.001	32.016	64.016
0.0001	32.0016	64.0016

The average speed on intervals starting at $t_0 = 1$ seems to approach a limiting value of 32 as the length of the interval decreases. This suggests that the rock is falling at a speed of 32 ft/sec at $t_0 = 1$ sec. Let's confirm this algebraically.

---

*To learn more about the historical figures and the development of the major elements and topics of calculus, visit **www.aw-bc.com/thomas**.

If we set $t_0 = 1$ and then expand the numerator in Equation (1) and simplify, we find that

$$\frac{\Delta y}{\Delta t} = \frac{16(1 + h)^2 - 16(1)^2}{h} = \frac{16(1 + 2h + h^2) - 16}{h}$$
$$= \frac{32h + 16h^2}{h} = 32 + 16h.$$

For values of h different from 0, the expressions on the right and left are equivalent and the average speed is $32 + 16h$ ft/sec. We can now see why the average speed has the limiting value $32 + 16(0) = 32$ ft/sec as h approaches 0.

Similarly, setting $t_0 = 2$ in Equation (1), the procedure yields

$$\frac{\Delta y}{\Delta t} = 64 + 16h$$

for values of h different from 0. As h gets closer and closer to 0, the average speed has the limiting value 64 ft/sec. at $t_0 = 2$ sec. ■

Average Rates of Change and Secant Lines

Given an arbitrary function $y = f(x)$, we calculate the average rate of change of y with respect to x over the interval $[x_1, x_2]$ by dividing the change in the value of y, $\Delta y = f(x_2) - f(x_1)$, by the length $\Delta x = x_2 - x_1 = h$ of the interval over which the change occurs.

DEFINITION Average Rate of Change over an Interval

The **average rate of change** of $y = f(x)$ with respect to x over the interval $[x_1, x_2]$ is

$$\frac{\Delta y}{\Delta x} = \frac{f(x_2) - f(x_1)}{x_2 - x_1} = \frac{f(x_1 + h) - f(x_1)}{h}, \qquad h \neq 0.$$

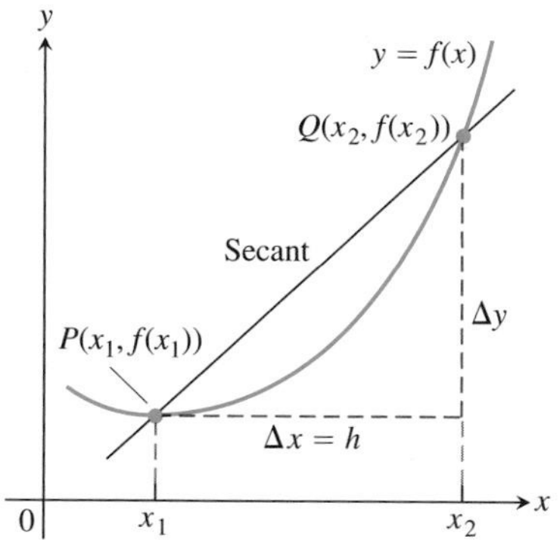

FIGURE 2.1 A secant to the graph $y = f(x)$. Its slope is $\Delta y/\Delta x$, the average rate of change of f over the interval $[x_1, x_2]$.

Geometrically, the rate of change of f over $[x_1, x_2]$ is the slope of the line through the points $P(x_1, f(x_1))$ and $Q(x_2, f(x_2))$ (Figure 2.1). In geometry, a line joining two points of a curve is a **secant** to the curve. Thus, the average rate of change of f from x_1 to x_2 is identical with the slope of secant PQ.

Experimental biologists often want to know the rates at which populations grow under controlled laboratory conditions.

EXAMPLE 3 The Average Growth Rate of a Laboratory Population

Figure 2.2 shows how a population of fruit flies (*Drosophila*) grew in a 50-day experiment. The number of flies was counted at regular intervals, the counted values plotted with respect to time, and the points joined by a smooth curve (colored blue in Figure 2.2). Find the average growth rate from day 23 to day 45.

Solution There were 150 flies on day 23 and 340 flies on day 45. Thus the number of flies increased by $340 - 150 = 190$ in $45 - 23 = 22$ days. The average rate of change of the population from day 23 to day 45 was

$$\text{Average rate of change: } \frac{\Delta p}{\Delta t} = \frac{340 - 150}{45 - 23} = \frac{190}{22} \approx 8.6 \text{ flies/day}.$$

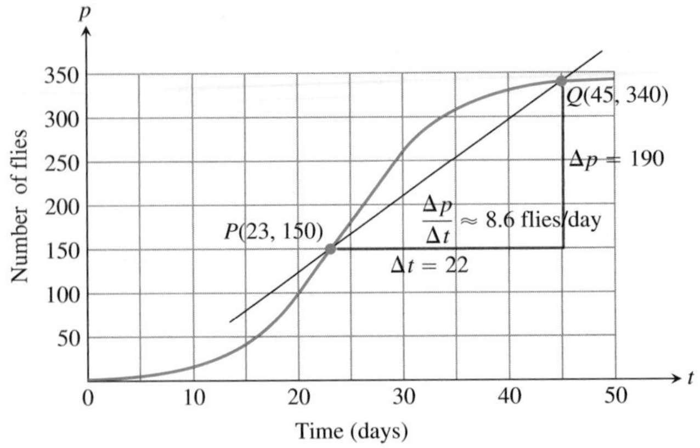

FIGURE 2.2 Growth of a fruit fly population in a controlled experiment. The average rate of change over 22 days is the slope $\Delta p/\Delta t$ of the secant line.

This average is the slope of the secant through the points P and Q on the graph in Figure 2.2. ■

The average rate of change from day 23 to day 45 calculated in Example 3 does not tell us how fast the population was changing on day 23 itself. For that we need to examine time intervals closer to the day in question.

EXAMPLE 4 The Growth Rate on Day 23

How fast was the number of flies in the population of Example 3 growing on day 23?

Solution To answer this question, we examine the average rates of change over increasingly short time intervals starting at day 23. In geometric terms, we find these rates by calculating the slopes of secants from P to Q, for a sequence of points Q approaching P along the curve (Figure 2.3).

Q	Slope of $PQ = \Delta p/\Delta t$ (flies/day)
(45, 340)	$\frac{340 - 150}{45 - 23} \approx 8.6$
(40, 330)	$\frac{330 - 150}{40 - 23} \approx 10.6$
(35, 310)	$\frac{310 - 150}{35 - 23} \approx 13.3$
(30, 265)	$\frac{265 - 150}{30 - 23} \approx 16.4$

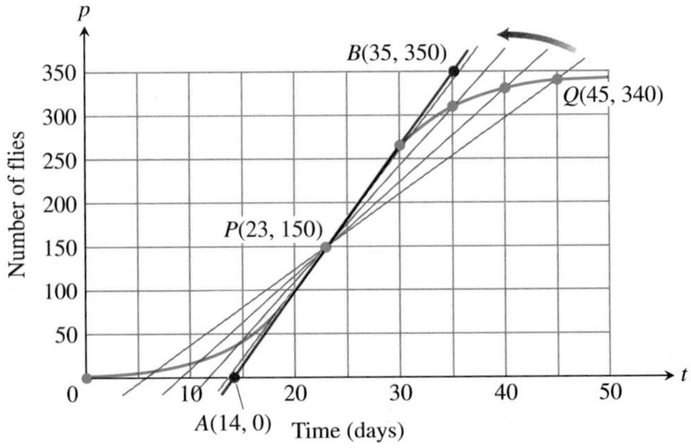

FIGURE 2.3 The positions and slopes of four secants through the point P on the fruit fly graph (Example 4).

The values in the table show that the secant slopes rise from 8.6 to 16.4 as the t-coordinate of Q decreases from 45 to 30, and we would expect the slopes to rise slightly higher as t continued on toward 23. Geometrically, the secants rotate about P and seem to approach the red line in the figure, a line that goes through P in the same direction that the curve goes through P. We will see that this line is called the *tangent* to the curve at P. Since the line appears to pass through the points (14, 0) and (35, 350), it has slope

$$\frac{350 - 0}{35 - 14} = 16.7 \text{ flies/day (approximately).}$$

On day 23 the population was increasing at a rate of about 16.7 flies/day. ■

The rates at which the rock in Example 2 was falling at the instants $t = 1$ and $t = 2$ and the rate at which the population in Example 4 was changing on day $t = 23$ are called *instantaneous rates of change*. As the examples suggest, we find instantaneous rates as limiting values of average rates. In Example 4, we also pictured the tangent line to the population curve on day 23 as a limiting position of secant lines. Instantaneous rates and tangent lines, intimately connected, appear in many other contexts. To talk about the two constructively, and to understand the connection further, we need to investigate the process by which we determine limiting values, or *limits*, as we will soon call them.

Limits of Function Values

Our examples have suggested the limit idea. Let's begin with an informal definition of limit, postponing the precise definition until we've gained more insight.

Let $f(x)$ be defined on an open interval about x_0, *except possibly at x_0 itself.* If $f(x)$ gets arbitrarily close to L (as close to L as we like) for all x sufficiently close to x_0, we say that f approaches the **limit** L as x approaches x_0, and we write

$$\lim_{x \to x_0} f(x) = L,$$

which is read "the limit of $f(x)$ as x approaches x_0 is L". Essentially, the definition says that the values of $f(x)$ are close to the number L whenever x is close to x_0 (on either side of x_0). This definition is "informal" because phrases like *arbitrarily close* and *sufficiently close* are imprecise; their meaning depends on the context. To a machinist manufacturing a piston, *close* may mean *within a few thousandths of an inch*. To an astronomer studying distant galaxies, *close* may mean *within a few thousand light-years*. The definition is clear enough, however, to enable us to recognize and evaluate limits of specific functions. We will need the precise definition of Section 2.3, however, when we set out to prove theorems about limits.

EXAMPLE 5 Behavior of a Function Near a Point

How does the function

$$f(x) = \frac{x^2 - 1}{x - 1}$$

behave near $x = 1$?

Solution The given formula defines f for all real numbers x except $x = 1$ (we cannot divide by zero). For any $x \neq 1$, we can simplify the formula by factoring the numerator and canceling common factors:

$$f(x) = \frac{(x - 1)(x + 1)}{x - 1} = x + 1 \qquad \text{for} \qquad x \neq 1.$$

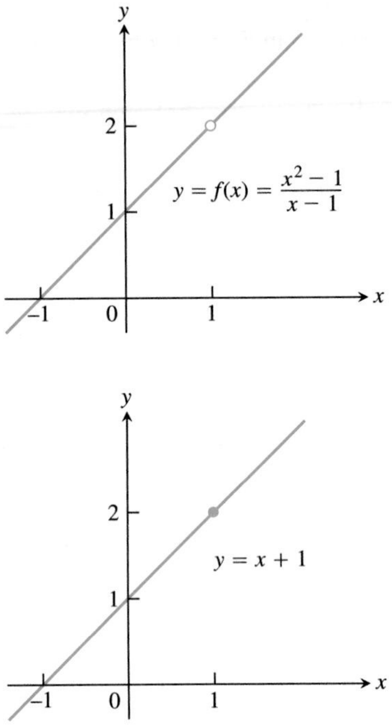

FIGURE 2.4 The graph of f is identical with the line $y = x + 1$ except at $x = 1$, where f is not defined (Example 5).

The graph of f is thus the line $y = x + 1$ with the point (1, 2) *removed*. This removed point is shown as a "hole" in Figure 2.4. Even though $f(1)$ is not defined, it is clear that we can make the value of $f(x)$ *as close as we want* to 2 by choosing x *close enough* to 1 (Table 2.2).

TABLE 2.2 The closer x gets to 1, the closer $f(x) = (x^2 - 1)/(x - 1)$ seems to get to 2

Values of x below and above 1	$f(x) = \frac{x^2 - 1}{x - 1} = x + 1, \quad x \neq 1$
0.9	1.9
1.1	2.1
0.99	1.99
1.01	2.01
0.999	1.999
1.001	2.001
0.999999	1.999999
1.000001	2.000001

We say that $f(x)$ approaches the *limit* 2 as x approaches 1, and write

$$\lim_{x \to 1} f(x) = 2, \qquad \text{or} \qquad \lim_{x \to 1} \frac{x^2 - 1}{x - 1} = 2.$$ ■

EXAMPLE 6 The Limit Value Does Not Depend on How the Function Is Defined at x_0

The function f in Figure 2.5 has limit 2 as $x \to 1$ even though f is not defined at $x = 1$. The function g has limit 2 as $x \to 1$ even though $2 \neq g(1)$. The function h is the only one

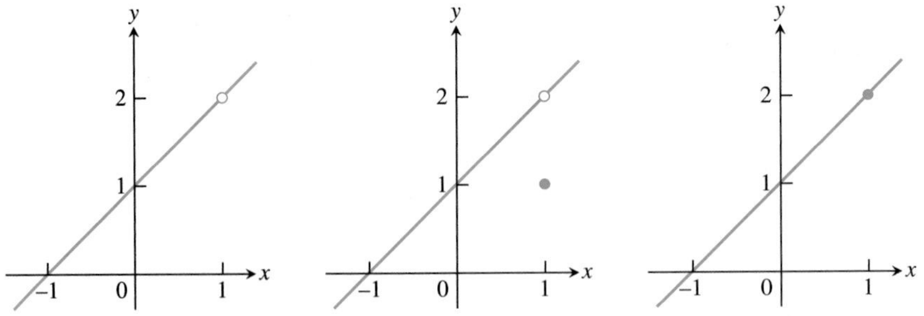

(a) $f(x) = \frac{x^2 - 1}{x - 1}$ (b) $g(x) = \begin{cases} \frac{x^2 - 1}{x - 1}, & x \neq 1 \\ 1, & x = 1 \end{cases}$ (c) $h(x) = x + 1$

FIGURE 2.5 The limits of $f(x)$, $g(x)$, and $h(x)$ all equal 2 as x approaches 1. However, only $h(x)$ has the same function value as its limit at $x = 1$ (Example 6).

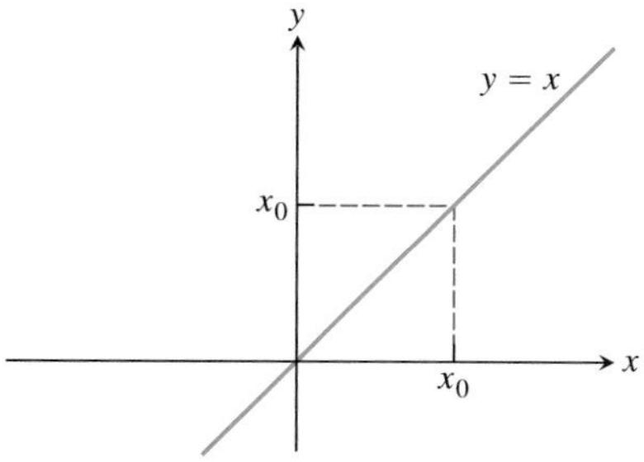

(a) Identity function

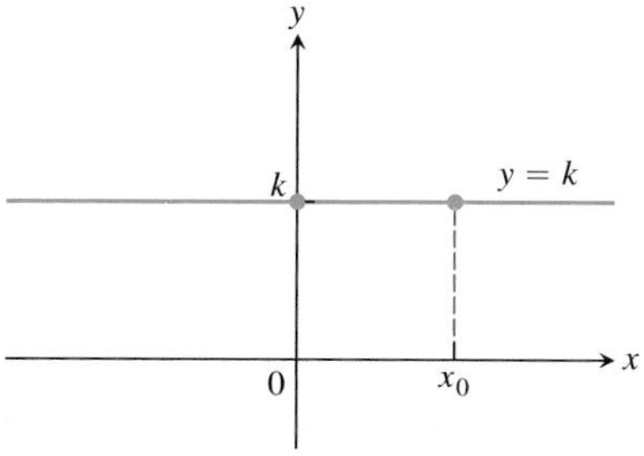

(b) Constant function

FIGURE 2.6 The functions in Example 8.

whose limit as $x \to 1$ equals its value at $x = 1$. For h, we have $\lim_{x\to 1} h(x) = h(1)$. This equality of limit and function value is special, and we return to it in Section 2.6. ■

Sometimes $\lim_{x\to x_0} f(x)$ can be evaluated by calculating $f(x_0)$. This holds, for example, whenever $f(x)$ is an algebraic combination of polynomials and trigonometric functions for which $f(x_0)$ is defined. (We will say more about this in Sections 2.2 and 2.6.)

EXAMPLE 7 Finding Limits by Calculating $f(x_0)$

(a) $\lim_{x\to 2}(4) = 4$

(b) $\lim_{x\to -13}(4) = 4$

(c) $\lim_{x\to 3} x = 3$

(d) $\lim_{x\to 2}(5x - 3) = 10 - 3 = 7$

(e) $\lim_{x\to -2} \dfrac{3x + 4}{x + 5} = \dfrac{-6 + 4}{-2 + 5} = -\dfrac{2}{3}$ ■

EXAMPLE 8 The Identity and Constant Functions Have Limits at Every Point

(a) If f is the **identity function** $f(x) = x$, then for any value of x_0 (Figure 2.6a),

$$\lim_{x\to x_0} f(x) = \lim_{x\to x_0} x = x_0.$$

(b) If f is the **constant function** $f(x) = k$ (function with the constant value k), then for any value of x_0 (Figure 2.6b),

$$\lim_{x\to x_0} f(x) = \lim_{x\to x_0} k = k.$$

For instance,

$$\lim_{x\to 3} x = 3 \quad \text{and} \quad \lim_{x\to -7}(4) = \lim_{x\to 2}(4) = 4.$$

We prove these results in Example 3 in Section 2.3. ■

Some ways that limits can fail to exist are illustrated in Figure 2.7 and described in the next example.

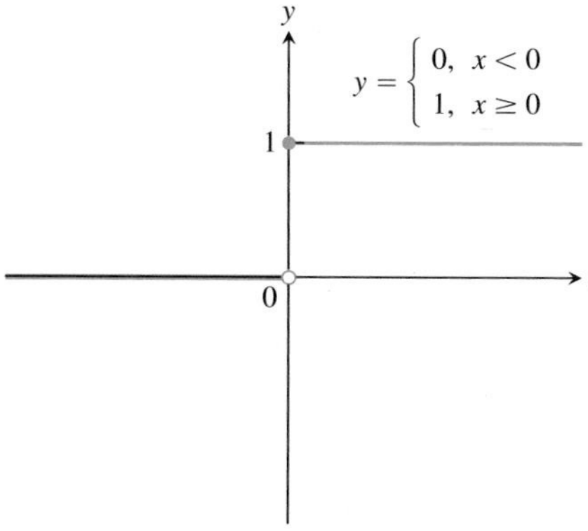

(a) Unit step function $U(x)$

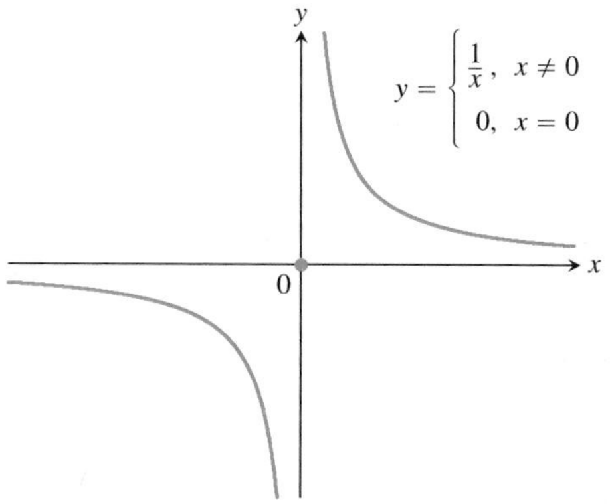

(b) $g(x)$

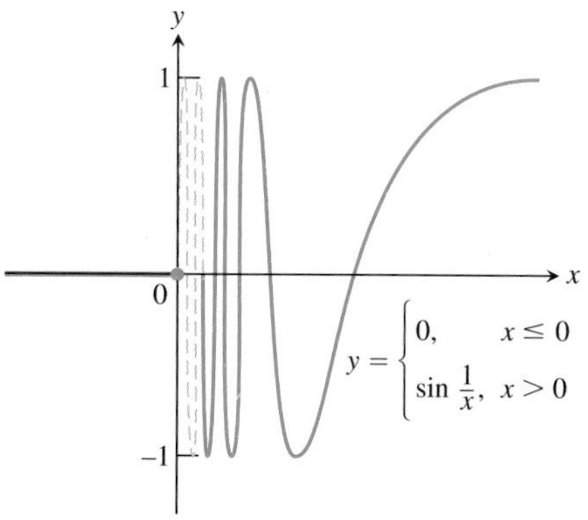

(c) $f(x)$

FIGURE 2.7 None of these functions has a limit as x approaches 0 (Example 9).

EXAMPLE 9 A Function May Fail to Have a Limit at a Point in Its Domain

Discuss the behavior of the following functions as $x \to 0$.

(a) $U(x) = \begin{cases} 0, & x < 0 \\ 1, & x \geq 0 \end{cases}$

(b) $g(x) = \begin{cases} \frac{1}{x}, & x \neq 0 \\ 0, & x = 0 \end{cases}$

(c) $f(x) = \begin{cases} 0, & x \leq 0 \\ \sin \frac{1}{x}, & x > 0 \end{cases}$

Solution

(a) It *jumps:* The **unit step function** $U(x)$ has no limit as $x \to 0$ because its values jump at $x = 0$. For negative values of x arbitrarily close to zero, $U(x) = 0$. For positive values of x arbitrarily close to zero, $U(x) = 1$. There is no *single* value L approached by $U(x)$ as $x \to 0$ (Figure 2.7a).

(b) It *grows too large to have a limit*: $g(x)$ has no limit as $x \to 0$ because the values of g grow arbitrarily large in absolute value as $x \to 0$ and do not stay close to *any* real number (Figure 2.7b).

(c) It *oscillates too much to have a limit*: $f(x)$ has no limit as $x \to 0$ because the function's values oscillate between $+1$ and -1 in every open interval containing 0. The values do not stay close to any one number as $x \to 0$ (Figure 2.7c). ■

Using Calculators and Computers to Estimate Limits

Tables 2.1 and 2.2 illustrate using a calculator or computer to guess a limit numerically as x gets closer and closer to x_0. That procedure would also be successful for the limits of functions like those in Example 7 (these are *continuous* functions and we study them in Section 2.6). However, calculators and computers can give *false values and misleading impressions* for functions that are undefined at a point or fail to have a limit there. The differential calculus will help us know when a calculator or computer is providing strange or ambiguous information about a function's behavior near some point (see Sections 4.4 and 4.6). For now, we simply need to be attentive to the fact that pitfalls may occur when using computing devices to guess the value of a limit. Here's one example.

EXAMPLE 10 Guessing a Limit

Guess the value of $\lim_{x \to 0} \frac{\sqrt{x^2 + 100} - 10}{x^2}$.

Solution Table 2.3 lists values of the function for several values near $x = 0$. As x approaches 0 through the values ± 1, ± 0.5, ± 0.10, and ± 0.01, the function seems to approach the number 0.05.

As we take even smaller values of x, ± 0.0005, ± 0.0001, ± 0.00001, and ± 0.000001, the function appears to approach the value 0.

So what is the answer? Is it 0.05 or 0, or some other value? The calculator/computer values are ambiguous, but the theorems on limits presented in the next section will confirm the correct limit value to be $0.05 (= 1/20)$. Problems such as these demonstrate the

TABLE 2.3 Computer values of $f(x) = \dfrac{\sqrt{x^2 + 100} - 10}{x^2}$ Near $x = 0$

x	$f(x)$	
± 1	0.049876	
± 0.5	0.049969	approaches 0.05?
± 0.1	0.049999	
± 0.01	0.050000	
± 0.0005	0.080000	
± 0.0001	0.000000	approaches 0?
± 0.00001	0.000000	
± 0.000001	0.000000	

power of mathematical reasoning, once it is developed, over the conclusions we might draw from making a few observations. Both approaches have advantages and disadvantages in revealing nature's realities. ■

EXERCISES 2.1

Limits from Graphs

1. For the function $g(x)$ graphed here, find the following limits or explain why they do not exist.

   a. $\lim_{x\to 1} g(x)$ **b.** $\lim_{x\to 2} g(x)$ **c.** $\lim_{x\to 3} g(x)$

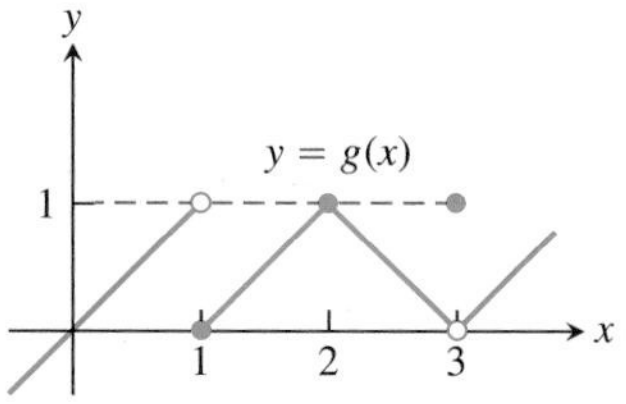

2. For the function $f(t)$ graphed here, find the following limits or explain why they do not exist.

   a. $\lim_{t\to -2} f(t)$ **b.** $\lim_{t\to -1} f(t)$ **c.** $\lim_{t\to 0} f(t)$

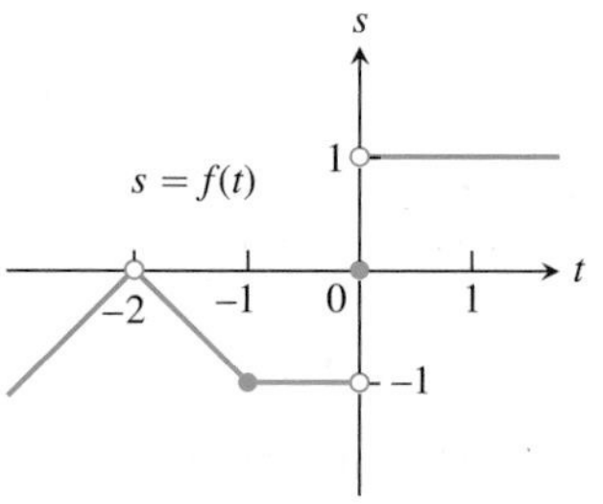

3. Which of the following statements about the function $y = f(x)$ graphed here are true, and which are false?

   a. $\lim_{x\to 0} f(x)$ exists.

   b. $\lim_{x\to 0} f(x) = 0$.

   c. $\lim_{x\to 0} f(x) = 1$.

   d. $\lim_{x\to 1} f(x) = 1$.

   e. $\lim_{x\to 1} f(x) = 0$.

   f. $\lim_{x\to x_0} f(x)$ exists at every point x_0 in $(-1, 1)$.

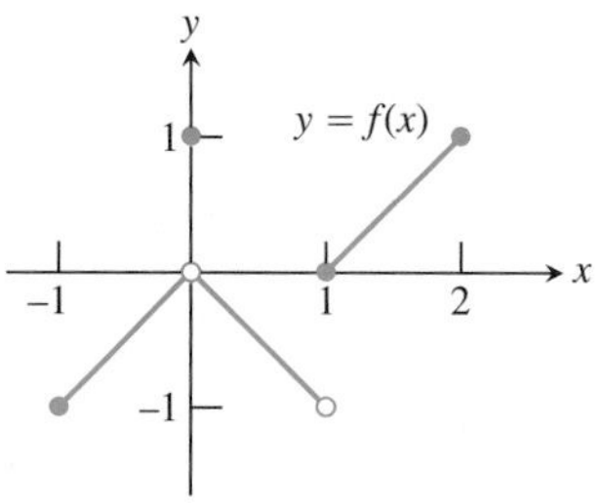

4. Which of the following statements about the function $y = f(x)$ graphed here are true, and which are false?

   a. $\lim_{x\to 2} f(x)$ does not exist.

   b. $\lim_{x\to 2} f(x) = 2$.

c. $\lim_{x\to 1} f(x)$ does not exist.

d. $\lim_{x\to x_0} f(x)$ exists at every point x_0 in $(-1, 1)$.

e. $\lim_{x\to x_0} f(x)$ exists at every point x_0 in $(1, 3)$.

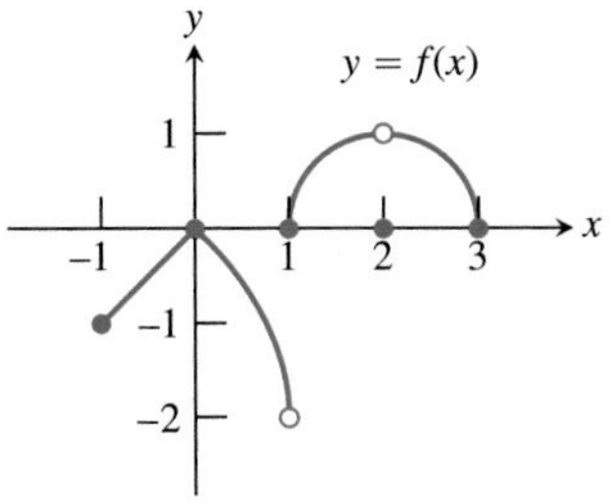

Existence of Limits

In Exercises 5 and 6, explain why the limits do not exist.

5. $\lim_{x\to 0} \dfrac{x}{|x|}$

6. $\lim_{x\to 1} \dfrac{1}{x-1}$

7. Suppose that a function $f(x)$ is defined for all real values of x except $x = x_0$. Can anything be said about the existence of $\lim_{x\to x_0} f(x)$? Give reasons for your answer.

8. Suppose that a function $f(x)$ is defined for all x in $[-1, 1]$. Can anything be said about the existence of $\lim_{x\to 0} f(x)$? Give reasons for your answer.

9. If $\lim_{x\to 1} f(x) = 5$, must f be defined at $x = 1$? If it is, must $f(1) = 5$? Can we conclude *anything* about the values of f at $x = 1$? Explain.

10. If $f(1) = 5$, must $\lim_{x\to 1} f(x)$ exist? If it does, then must $\lim_{x\to 1} f(x) = 5$? Can we conclude *anything* about $\lim_{x\to 1} f(x)$? Explain.

Estimating Limits

T You will find a graphing calculator useful for Exercises 11–20.

11. Let $f(x) = (x^2 - 9)/(x + 3)$.

a. Make a table of the values of f at the points $x = -3.1$, -3.01, -3.001, and so on as far as your calculator can go. Then estimate $\lim_{x\to -3} f(x)$. What estimate do you arrive at if you evaluate f at $x = -2.9, -2.99, -2.999, \ldots$ instead?

b. Support your conclusions in part (a) by graphing f near $x_0 = -3$ and using Zoom and Trace to estimate y-values on the graph as $x \to -3$.

c. Find $\lim_{x\to -3} f(x)$ algebraically, as in Example 5.

12. Let $g(x) = (x^2 - 2)/\left(x - \sqrt{2}\right)$.

a. Make a table of the values of g at the points $x = 1.4, 1.41$, 1.414, and so on through successive decimal approximations of $\sqrt{2}$. Estimate $\lim_{x\to\sqrt{2}} g(x)$.

b. Support your conclusion in part (a) by graphing g near $x_0 = \sqrt{2}$ and using Zoom and Trace to estimate y-values on the graph as $x \to \sqrt{2}$.

c. Find $\lim_{x\to\sqrt{2}} g(x)$ algebraically.

13. Let $G(x) = (x + 6)/(x^2 + 4x - 12)$.

a. Make a table of the values of G at $x = -5.9, -5.99, -5.999$, and so on. Then estimate $\lim_{x\to -6} G(x)$. What estimate do you arrive at if you evaluate G at $x = -6.1, -6.01, -6.001, \ldots$ instead?

b. Support your conclusions in part (a) by graphing G and using Zoom and Trace to estimate y-values on the graph as $x \to -6$.

c. Find $\lim_{x\to -6} G(x)$ algebraically.

14. Let $h(x) = (x^2 - 2x - 3)/(x^2 - 4x + 3)$.

a. Make a table of the values of h at $x = 2.9, 2.99, 2.999$, and so on. Then estimate $\lim_{x\to 3} h(x)$. What estimate do you arrive at if you evaluate h at $x = 3.1, 3.01, 3.001, \ldots$ instead?

b. Support your conclusions in part (a) by graphing h near $x_0 = 3$ and using Zoom and Trace to estimate y-values on the graph as $x \to 3$.

c. Find $\lim_{x\to 3} h(x)$ algebraically.

15. Let $f(x) = (x^2 - 1)/(|x| - 1)$.

a. Make tables of the values of f at values of x that approach $x_0 = -1$ from above and below. Then estimate $\lim_{x\to -1} f(x)$.

b. Support your conclusion in part (a) by graphing f near $x_0 = -1$ and using Zoom and Trace to estimate y-values on the graph as $x \to -1$.

c. Find $\lim_{x\to -1} f(x)$ algebraically.

16. Let $F(x) = (x^2 + 3x + 2)/(2 - |x|)$.

a. Make tables of values of F at values of x that approach $x_0 = -2$ from above and below. Then estimate $\lim_{x\to -2} F(x)$.

b. Support your conclusion in part (a) by graphing F near $x_0 = -2$ and using Zoom and Trace to estimate y-values on the graph as $x \to -2$.

c. Find $\lim_{x\to -2} F(x)$ algebraically.

17. Let $g(\theta) = (\sin\theta)/\theta$.

a. Make a table of the values of g at values of θ that approach $\theta_0 = 0$ from above and below. Then estimate $\lim_{\theta\to 0} g(\theta)$.

b. Support your conclusion in part (a) by graphing g near $\theta_0 = 0$.

18. Let $G(t) = (1 - \cos t)/t^2$.

a. Make tables of values of G at values of t that approach $t_0 = 0$ from above and below. Then estimate $\lim_{t\to 0} G(t)$.

b. Support your conclusion in part (a) by graphing G near $t_0 = 0$.

19. Let $f(x) = x^{1/(1-x)}$.

a. Make tables of values of f at values of x that approach $x_0 = 1$ from above and below. Does f appear to have a limit as $x \to 1$? If so, what is it? If not, why not?

b. Support your conclusions in part (a) by graphing f near $x_0 = 1$.

20. Let $f(x) = (3^x - 1)/x$.

a. Make tables of values of f at values of x that approach $x_0 = 0$ from above and below. Does f appear to have a limit as $x \to 0$? If so, what is it? If not, why not?

b. Support your conclusions in part (a) by graphing f near $x_0 = 0$.

Limits by Substitution

In Exercises 21–28, find the limits by substitution. *Support your answers with a computer or calculator if available.*

21. $\lim_{x \to 2} 2x$

22. $\lim_{x \to 0} 2x$

23. $\lim_{x \to 1/3} (3x - 1)$

24. $\lim_{x \to 1} \dfrac{-1}{(3x - 1)}$

25. $\lim_{x \to -1} 3x(2x - 1)$

26. $\lim_{x \to -1} \dfrac{3x^2}{2x - 1}$

27. $\lim_{x \to \pi/2} x \sin x$

28. $\lim_{x \to \pi} \dfrac{\cos x}{1 - \pi}$

Average Rates of Change

In Exercises 29–34, find the average rate of change of the function over the given interval or intervals.

29. $f(x) = x^3 + 1$;

a. $[2, 3]$ **b.** $[-1, 1]$

30. $g(x) = x^2$;

a. $[-1, 1]$ **b.** $[-2, 0]$

31. $h(t) = \cot t$;

a. $[\pi/4, 3\pi/4]$ **b.** $[\pi/6, \pi/2]$

32. $g(t) = 2 + \cos t$;

a. $[0, \pi]$ **b.** $[-\pi, \pi]$

33. $R(\theta) = \sqrt{4\theta + 1}$; $[0, 2]$

34. $P(\theta) = \theta^3 - 4\theta^2 + 5\theta$; $[1, 2]$

35. A Ford Mustang Cobra's speed The accompanying figure shows the time-to-distance graph for a 1994 Ford Mustang Cobra accelerating from a standstill.

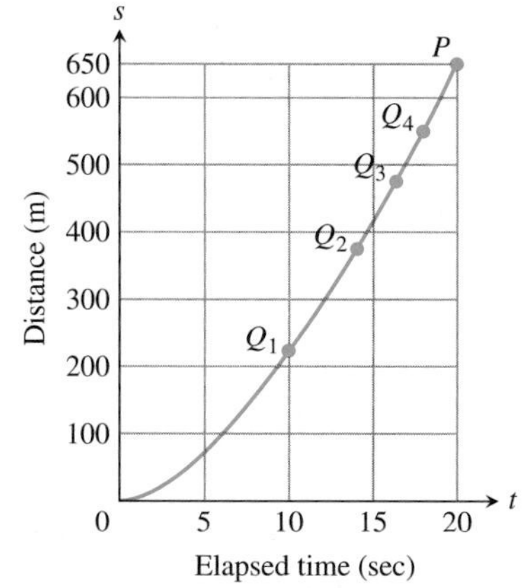

a. Estimate the slopes of secants PQ_1, PQ_2, PQ_3, and PQ_4, arranging them in order in a table like the one in Figure 2.3. What are the appropriate units for these slopes?

b. Then estimate the Cobra's speed at time $t = 20$ sec.

36. The accompanying figure shows the plot of distance fallen versus time for an object that fell from the lunar landing module a distance 80 m to the surface of the moon.

a. Estimate the slopes of the secants PQ_1, PQ_2, PQ_3, and PQ_4, arranging them in a table like the one in Figure 2.3.

b. About how fast was the object going when it hit the surface?

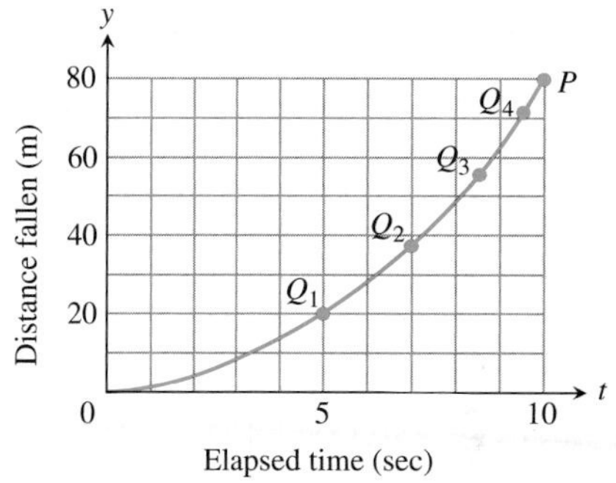

T 37. The profits of a small company for each of the first five years of its operation are given in the following table:

Year	Profit in $1000s
1990	6
1991	27
1992	62
1993	111
1994	174

a. Plot points representing the profit as a function of year, and join them by as smooth a curve as you can.

b. What is the average rate of increase of the profits between 1992 and 1994?

c. Use your graph to estimate the rate at which the profits were changing in 1992.

T 38. Make a table of values for the function $F(x) = (x + 2)/(x - 2)$ at the points $x = 1.2$, $x = 11/10$, $x = 101/100$, $x = 1001/1000$, $x = 10001/10000$, and $x = 1$.

a. Find the average rate of change of $F(x)$ over the intervals $[1, x]$ for each $x \neq 1$ in your table.

b. Extending the table if necessary, try to determine the rate of change of $F(x)$ at $x = 1$.

T 39. Let $g(x) = \sqrt{x}$ for $x \geq 0$.

a. Find the average rate of change of $g(x)$ with respect to x over the intervals $[1, 2]$, $[1, 1.5]$ and $[1, 1 + h]$.

b. Make a table of values of the average rate of change of g with respect to x over the interval $[1, 1 + h]$ for some values of h

approaching zero, say $h = 0.1, 0.01, 0.001, 0.0001, 0.00001$, and 0.000001.

c. What does your table indicate is the rate of change of $g(x)$ with respect to x at $x = 1$?

d. Calculate the limit as h approaches zero of the average rate of change of $g(x)$ with respect to x over the interval $[1, 1 + h]$.

T **40.** Let $f(t) = 1/t$ for $t \neq 0$.

a. Find the average rate of change of f with respect to t over the intervals (i) from $t = 2$ to $t = 3$, and (ii) from $t = 2$ to $t = T$.

b. Make a table of values of the average rate of change of f with respect to t over the interval $[2, T]$, for some values of T approaching 2, say $T = 2.1, 2.01, 2.001, 2.0001, 2.00001$, and 2.000001.

c. What does your table indicate is the rate of change of f with respect to t at $t = 2$?

d. Calculate the limit as T approaches 2 of the average rate of change of f with respect to t over the interval from 2 to T. You will have to do some algebra before you can substitute $T = 2$.

COMPUTER EXPLORATIONS

Graphical Estimates of Limits

In Exercises 41–46, use a CAS to perform the following steps:

a. Plot the function near the point x_0 being approached.

b. From your plot guess the value of the limit.

41. $\lim_{x \to 2} \dfrac{x^4 - 16}{x - 2}$

42. $\lim_{x \to -1} \dfrac{x^3 - x^2 - 5x - 3}{(x + 1)^2}$

43. $\lim_{x \to 0} \dfrac{\sqrt[3]{1 + x} - 1}{x}$

44. $\lim_{x \to 3} \dfrac{x^2 - 9}{\sqrt{x^2 + 7} - 4}$

45. $\lim_{x \to 0} \dfrac{1 - \cos x}{x \sin x}$

46. $\lim_{x \to 0} \dfrac{2x^2}{3 - 3\cos x}$

2.2 Calculating Limits Using the Limit Laws

HISTORICAL ESSAY*

Limits

In Section 2.1 we used graphs and calculators to guess the values of limits. This section presents theorems for calculating limits. The first three let us build on the results of Example 8 in the preceding section to find limits of polynomials, rational functions, and powers. The fourth and fifth prepare for calculations later in the text.

The Limit Laws

The next theorem tells how to calculate limits of functions that are arithmetic combinations of functions whose limits we already know.

THEOREM 1 Limit Laws

If L, M, c and k are real numbers and

$$\lim_{x \to c} f(x) = L \quad \text{and} \quad \lim_{x \to c} g(x) = M, \quad \text{then}$$

1. *Sum Rule:* $\lim_{x \to c} (f(x) + g(x)) = L + M$

The limit of the sum of two functions is the sum of their limits.

2. *Difference Rule:* $\lim_{x \to c} (f(x) - g(x)) = L - M$

The limit of the difference of two functions is the difference of their limits.

3. *Product Rule:* $\lim_{x \to c} (f(x) \cdot g(x)) = L \cdot M$

The limit of a product of two functions is the product of their limits.

---

*To learn more about the historical figures and the development of the major elements and topics of calculus, visit **www.aw-bc.com/thomas**.

4. *Constant Multiple Rule:* $\lim_{x\to c}(k \cdot f(x)) = k \cdot L$

The limit of a constant times a function is the constant times the limit of the function.

5. *Quotient Rule*: $\lim_{x\to c} \frac{f(x)}{g(x)} = \frac{L}{M}, \quad M \neq 0$

The limit of a quotient of two functions is the quotient of their limits, provided the limit of the denominator is not zero.

6. *Power Rule*: If r and s are integers with no common factor and $s \neq 0$, then

$$\lim_{x\to c}(f(x))^{r/s} = L^{r/s}$$

provided that $L^{r/s}$ is a real number. (If s is even, we assume that $L > 0$.)

The limit of a rational power of a function is that power of the limit of the function, provided the latter is a real number.

It is easy to convince ourselves that the properties in Theorem 1 are true (although these intuitive arguments do not constitute proofs). If x is sufficiently close to c, then $f(x)$ is close to L and $g(x)$ is close to M, from our informal definition of a limit. It is then reasonable that $f(x) + g(x)$ is close to $L + M$; $f(x) - g(x)$ is close to $L - M$; $f(x)g(x)$ is close to LM; $kf(x)$ is close to kL; and that $f(x)/g(x)$ is close to L/M if M is not zero. We prove the Sum Rule in Section 2.3, based on a precise definition of limit. Rules 2–5 are proved in Appendix 2. Rule 6 is proved in more advanced texts.

Here are some examples of how Theorem 1 can be used to find limits of polynomial and rational functions.

EXAMPLE 1 Using the Limit Laws

Use the observations $\lim_{x\to c} k = k$ and $\lim_{x\to c} x = c$ (Example 8 in Section 2.1) and the properties of limits to find the following limits.

(a) $\lim_{x\to c}(x^3 + 4x^2 - 3)$ **(b)** $\lim_{x\to c}\frac{x^4 + x^2 - 1}{x^2 + 5}$ **(c)** $\lim_{x\to -2}\sqrt{4x^2 - 3}$

Solution

(a)
$$\begin{aligned}\lim_{x\to c}(x^3 + 4x^2 - 3) &= \lim_{x\to c} x^3 + \lim_{x\to c} 4x^2 - \lim_{x\to c} 3 && \text{Sum and Difference Rules}\\ &= c^3 + 4c^2 - 3 && \text{Product and Multiple Rules}\end{aligned}$$

(b)
$$\begin{aligned}\lim_{x\to c}\frac{x^4 + x^2 - 1}{x^2 + 5} &= \frac{\lim_{x\to c}(x^4 + x^2 - 1)}{\lim_{x\to c}(x^2 + 5)} && \text{Quotient Rule}\\ &= \frac{\lim_{x\to c} x^4 + \lim_{x\to c} x^2 - \lim_{x\to c} 1}{\lim_{x\to c} x^2 + \lim_{x\to c} 5} && \text{Sum and Difference Rules}\\ &= \frac{c^4 + c^2 - 1}{c^2 + 5} && \text{Power or Product Rule}\end{aligned}$$

$$
\begin{aligned}
\textbf{(c)}\quad \lim_{x\to-2}\sqrt{4x^2-3} &= \sqrt{\lim_{x\to-2}(4x^2-3)} && \text{Power Rule with } r/s = 1/2\\
&= \sqrt{\lim_{x\to-2}4x^2 - \lim_{x\to-2}3} && \text{Difference Rule}\\
&= \sqrt{4(-2)^2-3} && \text{Product and Multiple Rules}\\
&= \sqrt{16-3}\\
&= \sqrt{13}
\end{aligned}
$$

■

Two consequences of Theorem 1 further simplify the task of calculating limits of polynomials and rational functions. To evaluate the limit of a polynomial function as x approaches c, merely substitute c for x in the formula for the function. To evaluate the limit of a rational function as x approaches a point c *at which the denominator is not zero*, substitute c for x in the formula for the function. (See Examples 1a and 1b.)

THEOREM 2 Limits of Polynomials Can Be Found by Substitution

If $P(x) = a_nx^n + a_{n-1}x^{n-1} + \cdots + a_0$, then

$$\lim_{x\to c} P(x) = P(c) = a_nc^n + a_{n-1}c^{n-1} + \cdots + a_0.$$

THEOREM 3 Limits of Rational Functions Can Be Found by Substitution If the Limit of the Denominator Is Not Zero

If $P(x)$ and $Q(x)$ are polynomials and $Q(c) \neq 0$, then

$$\lim_{x\to c}\frac{P(x)}{Q(x)} = \frac{P(c)}{Q(c)}.$$

EXAMPLE 2 Limit of a Rational Function

$$\lim_{x\to-1}\frac{x^3+4x^2-3}{x^2+5} = \frac{(-1)^3+4(-1)^2-3}{(-1)^2+5} = \frac{0}{6} = 0$$

This result is similar to the second limit in Example 1 with $c = -1$, now done in one step.

■

Identifying Common Factors
It can be shown that if $Q(x)$ is a polynomial and $Q(c) = 0$, then $(x - c)$ is a factor of $Q(x)$. Thus, if the numerator and denominator of a rational function of x are both zero at $x = c$, they have $(x - c)$ as a common factor.

Eliminating Zero Denominators Algebraically

Theorem 3 applies only if the denominator of the rational function is not zero at the limit point c. If the denominator is zero, canceling common factors in the numerator and denominator may reduce the fraction to one whose denominator is no longer zero at c. If this happens, we can find the limit by substitution in the simplified fraction.

EXAMPLE 3 Canceling a Common Factor

Evaluate

$$\lim_{x\to1}\frac{x^2+x-2}{x^2-x}.$$

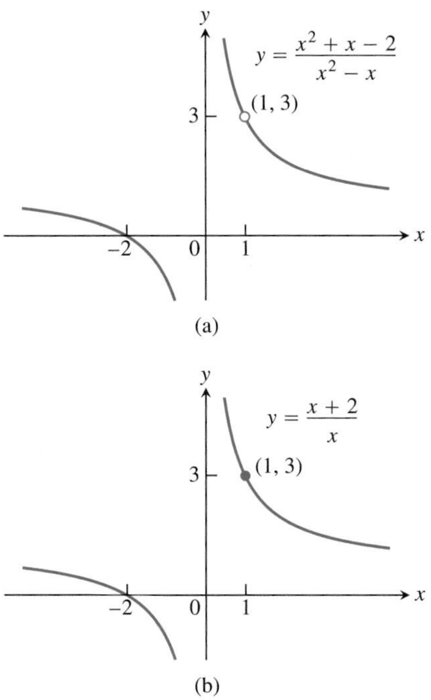

FIGURE 2.8 The graph of $f(x) = (x^2 + x - 2)/(x^2 - x)$ in part (a) is the same as the graph of $g(x) = (x + 2)/x$ in part (b) except at $x = 1$, where f is undefined. The functions have the same limit as $x \to 1$ (Example 3).

Solution We cannot substitute $x = 1$ because it makes the denominator zero. We test the numerator to see if it, too, is zero at $x = 1$. It is, so it has a factor of $(x - 1)$ in common with the denominator. Canceling the $(x - 1)$'s gives a simpler fraction with the same values as the original for $x \neq 1$:

$$\frac{x^2 + x - 2}{x^2 - x} = \frac{(x - 1)(x + 2)}{x(x - 1)} = \frac{x + 2}{x}, \qquad \text{if } x \neq 1.$$

Using the simpler fraction, we find the limit of these values as $x \to 1$ by substitution:

$$\lim_{x \to 1} \frac{x^2 + x - 2}{x^2 - x} = \lim_{x \to 1} \frac{x + 2}{x} = \frac{1 + 2}{1} = 3.$$

See Figure 2.8. ■

EXAMPLE 4 Creating and Canceling a Common Factor

Evaluate

$$\lim_{x \to 0} \frac{\sqrt{x^2 + 100} - 10}{x^2}.$$

Solution This is the limit we considered in Example 10 of the preceding section. We cannot substitute $x = 0$, and the numerator and denominator have no obvious common factors. We can create a common factor by multiplying both numerator and denominator by the expression $\sqrt{x^2 + 100} + 10$ (obtained by changing the sign after the square root). The preliminary algebra rationalizes the numerator:

$$\begin{aligned}
\frac{\sqrt{x^2 + 100} - 10}{x^2} &= \frac{\sqrt{x^2 + 100} - 10}{x^2} \cdot \frac{\sqrt{x^2 + 100} + 10}{\sqrt{x^2 + 100} + 10} \\
&= \frac{x^2 + 100 - 100}{x^2\left(\sqrt{x^2 + 100} + 10\right)} \\
&= \frac{x^2}{x^2\left(\sqrt{x^2 + 100} + 10\right)} && \text{Common factor } x^2 \\
&= \frac{1}{\sqrt{x^2 + 100} + 10}. && \text{Cancel } x^2 \text{ for } x \neq 0
\end{aligned}$$

Therefore,

$$\begin{aligned}
\lim_{x \to 0} \frac{\sqrt{x^2 + 100} - 10}{x^2} &= \lim_{x \to 0} \frac{1}{\sqrt{x^2 + 100} + 10} \\
&= \frac{1}{\sqrt{0^2 + 100} + 10} && \text{Denominator not 0 at } x = 0\text{; substitute} \\
&= \frac{1}{20} = 0.05.
\end{aligned}$$

This calculation provides the correct answer to the ambiguous computer results in Example 10 of the preceding section. ■

The Sandwich Theorem

The following theorem will enable us to calculate a variety of limits in subsequent chapters. It is called the Sandwich Theorem because it refers to a function f whose values are

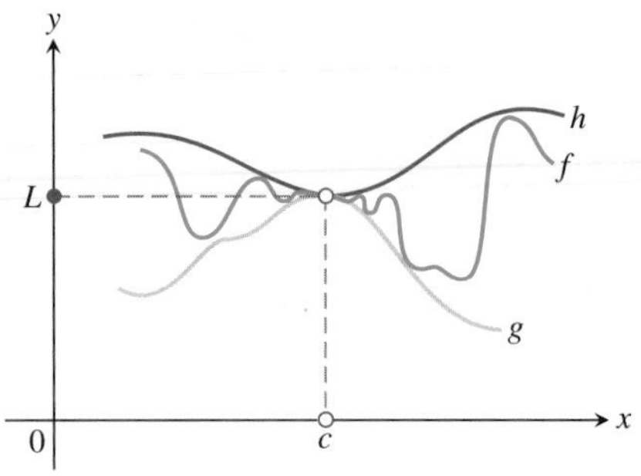

FIGURE 2.9 The graph of f is sandwiched between the graphs of g and h.

sandwiched between the values of two other functions g and h that have the same limit L at a point c. Being trapped between the values of two functions that approach L, the values of f must also approach L (Figure 2.9). You will find a proof in Appendix 2.

THEOREM 4 The Sandwich Theorem
Suppose that $g(x) \leq f(x) \leq h(x)$ for all x in some open interval containing c, except possibly at $x = c$ itself. Suppose also that

$$\lim_{x \to c} g(x) = \lim_{x \to c} h(x) = L.$$

Then $\lim_{x \to c} f(x) = L$.

The Sandwich Theorem is sometimes called the Squeeze Theorem or the Pinching Theorem.

EXAMPLE 5 Applying the Sandwich Theorem

Given that

$$1 - \frac{x^2}{4} \leq u(x) \leq 1 + \frac{x^2}{2} \qquad \text{for all } x \neq 0,$$

find $\lim_{x \to 0} u(x)$, no matter how complicated u is.

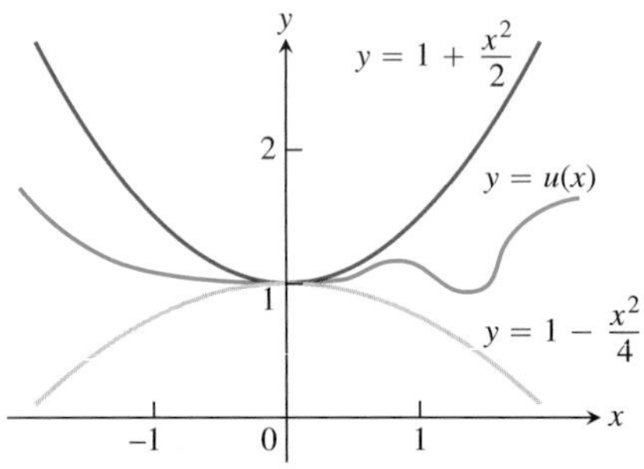

FIGURE 2.10 Any function $u(x)$ whose graph lies in the region between $y = 1 + (x^2/2)$ and $y = 1 - (x^2/4)$ has limit 1 as $x \to 0$ (Example 5).

Solution Since

$$\lim_{x \to 0} (1 - (x^2/4)) = 1 \qquad \text{and} \qquad \lim_{x \to 0} (1 + (x^2/2)) = 1,$$

the Sandwich Theorem implies that $\lim_{x \to 0} u(x) = 1$ (Figure 2.10). ■

EXAMPLE 6 More Applications of the Sandwich Theorem

(a) (Figure 2.11a). It follows from the definition of $\sin\theta$ that $-|\theta| \leq \sin\theta \leq |\theta|$ for all θ, and since $\lim_{\theta \to 0} (-|\theta|) = \lim_{\theta \to 0} |\theta| = 0$, we have

$$\lim_{\theta \to 0} \sin\theta = 0.$$

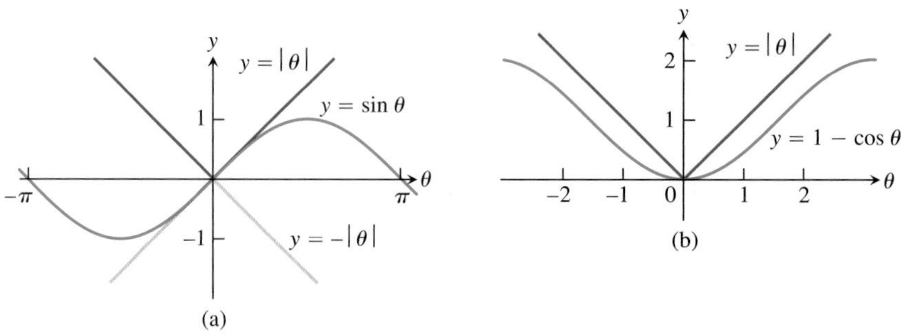

FIGURE 2.11 The Sandwich Theorem confirms that (a) $\lim_{\theta \to 0} \sin\theta = 0$ and (b) $\lim_{\theta \to 0} (1 - \cos\theta) = 0$ (Example 6).

(b) (Figure 2.11b). From the definition of $\cos\theta$, $0 \le 1 - \cos\theta \le |\theta|$ for all θ, and we have $\lim_{\theta\to 0}(1 - \cos\theta) = 0$ or

$$\lim_{\theta\to 0}\cos\theta = 1.$$

(c) For any function $f(x)$, if $\lim_{x\to c}|f(x)| = 0$, then $\lim_{x\to c} f(x) = 0$. The argument: $-|f(x)| \le f(x) \le |f(x)|$ and $-|f(x)|$ and $|f(x)|$ have limit 0 as $x \to c$. ■

Another important property of limits is given by the next theorem. A proof is given in the next section.

> **THEOREM 5** If $f(x) \le g(x)$ for all x in some open interval containing c, except possibly at $x = c$ itself, and the limits of f and g both exist as x approaches c, then
>
> $$\lim_{x\to c} f(x) \le \lim_{x\to c} g(x).$$

The assertion resulting from replacing the less than or equal to $\le$ inequality by the strict $<$ inequality in Theorem 5 is false. Figure 2.11a shows that for $\theta \ne 0$, $-|\theta| < \sin\theta < |\theta|$, but in the limit as $\theta \to 0$, equality holds.

EXERCISES 2.2

Limit Calculations

Find the limits in Exercises 1–18.

1. $\lim_{x\to -7}(2x + 5)$
2. $\lim_{x\to 12}(10 - 3x)$
3. $\lim_{x\to 2}(-x^2 + 5x - 2)$
4. $\lim_{x\to -2}(x^3 - 2x^2 + 4x + 8)$
5. $\lim_{t\to 6} 8(t - 5)(t - 7)$
6. $\lim_{s\to 2/3} 3s(2s - 1)$
7. $\lim_{x\to 2}\dfrac{x + 3}{x + 6}$
8. $\lim_{x\to 5}\dfrac{4}{x - 7}$
9. $\lim_{y\to -5}\dfrac{y^2}{5 - y}$
10. $\lim_{y\to 2}\dfrac{y + 2}{y^2 + 5y + 6}$
11. $\lim_{x\to -1} 3(2x - 1)^2$
12. $\lim_{x\to -4}(x + 3)^{1984}$
13. $\lim_{y\to -3}(5 - y)^{4/3}$
14. $\lim_{z\to 0}(2z - 8)^{1/3}$
15. $\lim_{h\to 0}\dfrac{3}{\sqrt{3h + 1} + 1}$
16. $\lim_{h\to 0}\dfrac{5}{\sqrt{5h + 4} + 2}$
17. $\lim_{h\to 0}\dfrac{\sqrt{3h + 1} - 1}{h}$
18. $\lim_{h\to 0}\dfrac{\sqrt{5h + 4} - 2}{h}$

Find the limits in Exercises 19–36.

19. $\lim_{x\to 5}\dfrac{x - 5}{x^2 - 25}$
20. $\lim_{x\to -3}\dfrac{x + 3}{x^2 + 4x + 3}$
21. $\lim_{x\to -5}\dfrac{x^2 + 3x - 10}{x + 5}$
22. $\lim_{x\to 2}\dfrac{x^2 - 7x + 10}{x - 2}$
23. $\lim_{t\to 1}\dfrac{t^2 + t - 2}{t^2 - 1}$
24. $\lim_{t\to -1}\dfrac{t^2 + 3t + 2}{t^2 - t - 2}$
25. $\lim_{x\to -2}\dfrac{-2x - 4}{x^3 + 2x^2}$
26. $\lim_{y\to 0}\dfrac{5y^3 + 8y^2}{3y^4 - 16y^2}$
27. $\lim_{u\to 1}\dfrac{u^4 - 1}{u^3 - 1}$
28. $\lim_{v\to 2}\dfrac{v^3 - 8}{v^4 - 16}$
29. $\lim_{x\to 9}\dfrac{\sqrt{x} - 3}{x - 9}$
30. $\lim_{x\to 4}\dfrac{4x - x^2}{2 - \sqrt{x}}$
31. $\lim_{x\to 1}\dfrac{x - 1}{\sqrt{x + 3} - 2}$
32. $\lim_{x\to -1}\dfrac{\sqrt{x^2 + 8} - 3}{x + 1}$
33. $\lim_{x\to 2}\dfrac{\sqrt{x^2 + 12} - 4}{x - 2}$
34. $\lim_{x\to -2}\dfrac{x + 2}{\sqrt{x^2 + 5} - 3}$

35. $\lim_{x\to-3} \dfrac{2-\sqrt{x^2-5}}{x+3}$

36. $\lim_{x\to4} \dfrac{4-x}{5-\sqrt{x^2+9}}$

Using Limit Rules

37. Suppose $\lim_{x\to0} f(x) = 1$ and $\lim_{x\to0} g(x) = -5$. Name the rules in Theorem 1 that are used to accomplish steps (a), (b), and (c) of the following calculation.

$$\lim_{x\to0} \frac{2f(x)-g(x)}{(f(x)+7)^{2/3}} = \frac{\lim_{x\to0}(2f(x)-g(x))}{\lim_{x\to0}(f(x)+7)^{2/3}} \qquad \text{(a)}$$

$$= \frac{\lim_{x\to0} 2f(x) - \lim_{x\to0} g(x)}{\left(\lim_{x\to0}\left(f(x)+7\right)\right)^{2/3}} \qquad \text{(b)}$$

$$= \frac{2\lim_{x\to0} f(x) - \lim_{x\to0} g(x)}{\left(\lim_{x\to0} f(x) + \lim_{x\to0} 7\right)^{2/3}} \qquad \text{(c)}$$

$$= \frac{(2)(1)-(-5)}{(1+7)^{2/3}} = \frac{7}{4}$$

38. Let $\lim_{x\to1} h(x) = 5$, $\lim_{x\to1} p(x) = 1$, and $\lim_{x\to1} r(x) = 2$. Name the rules in Theorem 1 that are used to accomplish steps (a), (b), and (c) of the following calculation.

$$\lim_{x\to1} \frac{\sqrt{5h(x)}}{p(x)(4-r(x))} = \frac{\lim_{x\to1}\sqrt{5h(x)}}{\lim_{x\to1}(p(x)(4-r(x)))} \qquad \text{(a)}$$

$$= \frac{\sqrt{\lim_{x\to1} 5h(x)}}{\left(\lim_{x\to1} p(x)\right)\left(\lim_{x\to1}\left(4-r(x)\right)\right)} \qquad \text{(b)}$$

$$= \frac{\sqrt{5\lim_{x\to1} h(x)}}{\left(\lim_{x\to1} p(x)\right)\left(\lim_{x\to1} 4 - \lim_{x\to1} r(x)\right)} \qquad \text{(c)}$$

$$= \frac{\sqrt{(5)(5)}}{(1)(4-2)} = \frac{5}{2}$$

39. Suppose $\lim_{x\to c} f(x) = 5$ and $\lim_{x\to c} g(x) = -2$. Find

a. $\lim_{x\to c} f(x)g(x)$ **b.** $\lim_{x\to c} 2f(x)g(x)$

c. $\lim_{x\to c} (f(x)+3g(x))$ **d.** $\lim_{x\to c} \dfrac{f(x)}{f(x)-g(x)}$

40. Suppose $\lim_{x\to4} f(x) = 0$ and $\lim_{x\to4} g(x) = -3$. Find

a. $\lim_{x\to4} (g(x)+3)$ **b.** $\lim_{x\to4} xf(x)$

c. $\lim_{x\to4} (g(x))^2$ **d.** $\lim_{x\to4} \dfrac{g(x)}{f(x)-1}$

41. Suppose $\lim_{x\to b} f(x) = 7$ and $\lim_{x\to b} g(x) = -3$. Find

a. $\lim_{x\to b} (f(x)+g(x))$ **b.** $\lim_{x\to b} f(x)\cdot g(x)$

c. $\lim_{x\to b} 4g(x)$ **d.** $\lim_{x\to b} f(x)/g(x)$

42. Suppose that $\lim_{x\to-2} p(x) = 4$, $\lim_{x\to-2} r(x) = 0$, and $\lim_{x\to-2} s(x) = -3$. Find

a. $\lim_{x\to-2} (p(x)+r(x)+s(x))$

b. $\lim_{x\to-2} p(x)\cdot r(x)\cdot s(x)$

c. $\lim_{x\to-2} (-4p(x)+5r(x))/s(x)$

Limits of Average Rates of Change

Because of their connection with secant lines, tangents, and instantaneous rates, limits of the form

$$\lim_{h\to0} \frac{f(x+h)-f(x)}{h}$$

occur frequently in calculus. In Exercises 43–48, evaluate this limit for the given value of x and function f.

43. $f(x) = x^2$, $x = 1$

44. $f(x) = x^2$, $x = -2$

45. $f(x) = 3x-4$, $x = 2$

46. $f(x) = 1/x$, $x = -2$

47. $f(x) = \sqrt{x}$, $x = 7$

48. $f(x) = \sqrt{3x+1}$, $x = 0$

Using the Sandwich Theorem

49. If $\sqrt{5-2x^2} \le f(x) \le \sqrt{5-x^2}$ for $-1 \le x \le 1$, find $\lim_{x\to0} f(x)$.

50. If $2 - x^2 \le g(x) \le 2\cos x$ for all x, find $\lim_{x\to0} g(x)$.

51. a. It can be shown that the inequalities

$$1 - \frac{x^2}{6} < \frac{x\sin x}{2-2\cos x} < 1$$

hold for all values of x close to zero. What, if anything, does this tell you about

$$\lim_{x\to0} \frac{x\sin x}{2-2\cos x}?$$

Give reasons for your answer.

T **b.** Graph

$$y = 1-(x^2/6),\ y = (x\sin x)/(2-2\cos x),\ \text{and } y = 1$$

together for $-2 \le x \le 2$. Comment on the behavior of the graphs as $x \to 0$.

52. a. Suppose that the inequalities

$$\frac{1}{2} - \frac{x^2}{24} < \frac{1-\cos x}{x^2} < \frac{1}{2}$$

hold for values of x close to zero. (They do, as you will see in Section 11.9.) What, if anything, does this tell you about

$$\lim_{x\to0} \frac{1-\cos x}{x^2}?$$

Give reasons for your answer.

b. Graph the equations $y = (1/2) - (x^2/24)$, $y = (1-\cos x)/x^2$, and $y = 1/2$ together for $-2 \le x \le 2$. Comment on the behavior of the graphs as $x \to 0$.

Theory and Examples

53. If $x^4 \le f(x) \le x^2$ for x in $[-1, 1]$ and $x^2 \le f(x) \le x^4$ for $x < -1$ and $x > 1$, at what points c do you automatically know $\lim_{x\to c} f(x)$? What can you say about the value of the limit at these points?

54. Suppose that $g(x) \le f(x) \le h(x)$ for all $x \ne 2$ and suppose that

$$\lim_{x\to 2} g(x) = \lim_{x\to 2} h(x) = -5.$$

Can we conclude anything about the values of f, g, and h at $x = 2$? Could $f(2) = 0$? Could $\lim_{x\to 2} f(x) = 0$? Give reasons for your answers.

55. If $\lim_{x\to 4} \dfrac{f(x) - 5}{x - 2} = 1$, find $\lim_{x\to 4} f(x)$.

56. If $\lim_{x\to -2} \dfrac{f(x)}{x^2} = 1$, find

a. $\lim_{x\to -2} f(x)$ **b.** $\lim_{x\to -2} \dfrac{f(x)}{x}$

57. a. If $\lim_{x\to 2} \dfrac{f(x) - 5}{x - 2} = 3$, find $\lim_{x\to 2} f(x)$.

b. If $\lim_{x\to 2} \dfrac{f(x) - 5}{x - 2} = 4$, find $\lim_{x\to 2} f(x)$.

58. If $\lim_{x\to 0} \dfrac{f(x)}{x^2} = 1$, find

a. $\lim_{x\to 0} f(x)$ **b.** $\lim_{x\to 0} \dfrac{f(x)}{x}$

T **59. a.** Graph $g(x) = x\sin(1/x)$ to estimate $\lim_{x\to 0} g(x)$, zooming in on the origin as necessary.

b. Confirm your estimate in part (a) with a proof.

T **60. a.** Graph $h(x) = x^2\cos(1/x^3)$ to estimate $\lim_{x\to 0} h(x)$, zooming in on the origin as necessary.

b. Confirm your estimate in part (a) with a proof.

2.3 The Precise Definition of a Limit

Now that we have gained some insight into the limit concept, working intuitively with the informal definition, we turn our attention to its precise definition. We replace vague phrases like "gets arbitrarily close to" in the informal definition with specific conditions that can be applied to any particular example. With a precise definition we will be able to prove conclusively the limit properties given in the preceding section, and we can establish other particular limits important to the study of calculus.

To show that the limit of $f(x)$ as $x \to x_0$ equals the number L, we need to show that the gap between $f(x)$ and L can be made "as small as we choose" if x is kept "close enough" to x_0. Let us see what this would require if we specified the size of the gap between $f(x)$ and L.

EXAMPLE 1 A Linear Function

Consider the function $y = 2x - 1$ near $x_0 = 4$. Intuitively it is clear that y is close to 7 when x is close to 4, so $\lim_{x\to 4}(2x - 1) = 7$. However, how close to $x_0 = 4$ does x have to be so that $y = 2x - 1$ differs from 7 by, say, less than 2 units?

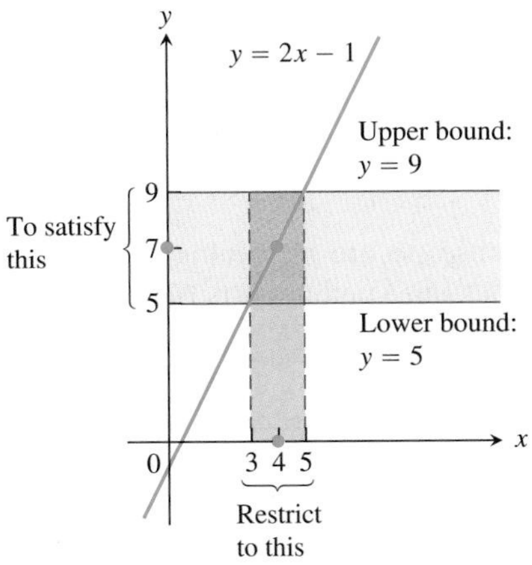

FIGURE 2.12 Keeping x within 1 unit of $x_0 = 4$ will keep y within 2 units of $y_0 = 7$ (Example 1).

Solution We are asked: For what values of x is $|y - 7| < 2$? To find the answer we first express $|y - 7|$ in terms of x:

$$|y - 7| = |(2x - 1) - 7| = |2x - 8|.$$

The question then becomes: what values of x satisfy the inequality $|2x - 8| < 2$? To find out, we solve the inequality:

$$|2x - 8| < 2$$
$$-2 < 2x - 8 < 2$$
$$6 < 2x < 10$$
$$3 < x < 5$$
$$-1 < x - 4 < 1.$$

Keeping x within 1 unit of $x_0 = 4$ will keep y within 2 units of $y_0 = 7$ (Figure 2.12). ■

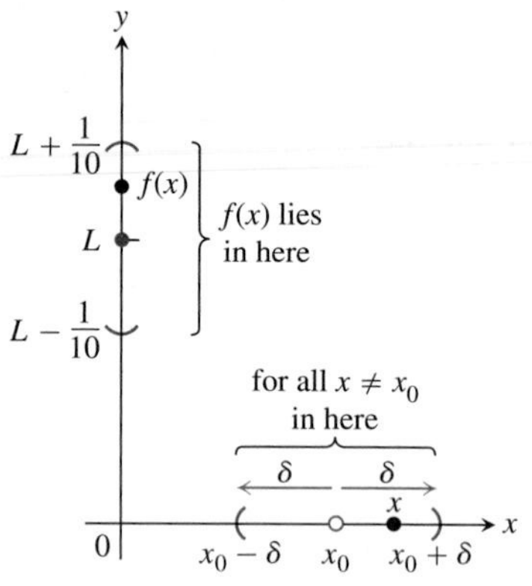

FIGURE 2.13 How should we define $\delta > 0$ so that keeping x within the interval $(x_0 - \delta, x_0 + \delta)$ will keep $f(x)$ within the interval $\left(L - \frac{1}{10}, L + \frac{1}{10}\right)$?

In the previous example we determined how close x must be to a particular value x_0 to ensure that the outputs $f(x)$ of some function lie within a prescribed interval about a limit value L. To show that the limit of $f(x)$ as $x \to x_0$ actually equals L, we must be able to show that the gap between $f(x)$ and L can be made less than *any prescribed error*, no matter how small, by holding x close enough to x_0.

Definition of Limit

Suppose we are watching the values of a function $f(x)$ as x approaches x_0 (without taking on the value of x_0 itself). Certainly we want to be able to say that $f(x)$ stays within one-tenth of a unit of L as soon as x stays within some distance δ of x_0 (Figure 2.13). But that in itself is not enough, because as x continues on its course toward x_0, what is to prevent $f(x)$ from jittering about within the interval from $L - (1/10)$ to $L + (1/10)$ without tending toward L?

We can be told that the error can be no more than $1/100$ or $1/1000$ or $1/100{,}000$. Each time, we find a new δ-interval about x_0 so that keeping x within that interval satisfies the new error tolerance. And each time the possibility exists that $f(x)$ jitters away from L at some stage.

The figures on the next page illustrate the problem. You can think of this as a quarrel between a skeptic and a scholar. The skeptic presents ϵ-challenges to prove that the limit does not exist or, more precisely, that there is room for doubt, and the scholar answers every challenge with a δ-interval around x_0.

How do we stop this seemingly endless series of challenges and responses? By proving that for every error tolerance ϵ that the challenger can produce, we can find, calculate, or conjure a matching distance δ that keeps x "close enough" to x_0 to keep $f(x)$ within that tolerance of L (Figure 2.14). This leads us to the precise definition of a limit.

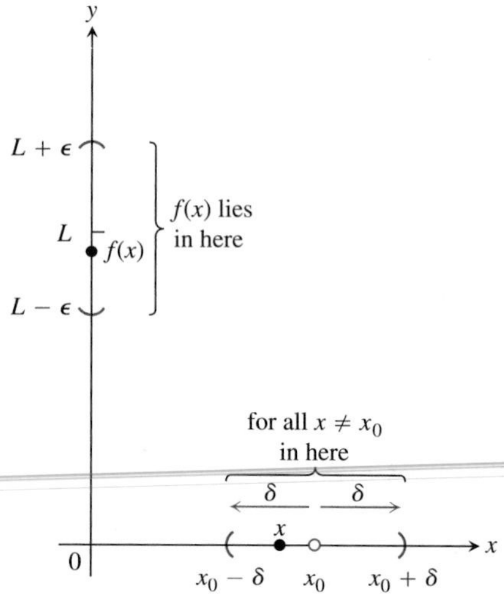

FIGURE 2.14 The relation of δ and ϵ in the definition of limit.

DEFINITION Limit of a Function

Let $f(x)$ be defined on an open interval about x_0, except possibly at x_0 itself. We say that the **limit of $f(x)$ as x approaches x_0 is the number L**, and write

$$\lim_{x \to x_0} f(x) = L,$$

if, for every number $\epsilon > 0$, there exists a corresponding number $\delta > 0$ such that for all x,

$$0 < |x - x_0| < \delta \quad \Rightarrow \quad |f(x) - L| < \epsilon.$$

One way to think about the definition is to suppose we are machining a generator shaft to a close tolerance. We may try for diameter L, but since nothing is perfect, we must be satisfied with a diameter $f(x)$ somewhere between $L - \epsilon$ and $L + \epsilon$. The δ is the measure of how accurate our control setting for x must be to guarantee this degree of accuracy in the diameter of the shaft. Notice that as the tolerance for error becomes stricter, we may have to adjust δ. That is, the value of δ, how tight our control setting must be, depends on the value of ϵ, the error tolerance.

Examples: Testing the Definition

The formal definition of limit does not tell how to find the limit of a function, but it enables us to verify that a suspected limit is correct. The following examples show how the

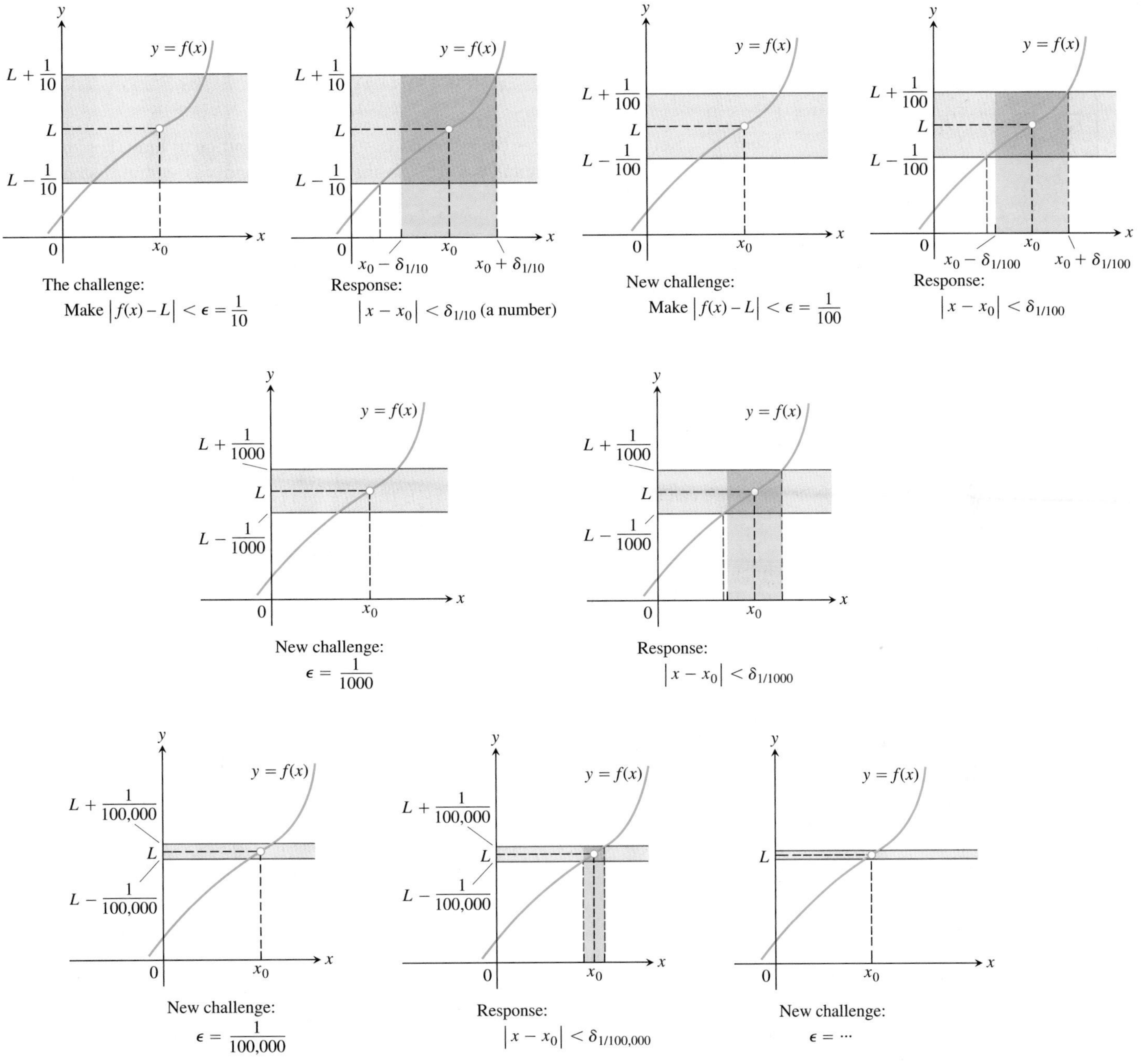

definition can be used to verify limit statements for specific functions. (The first two examples correspond to parts of Examples 7 and 8 in Section 2.1.) However, the real purpose of the definition is not to do calculations like this, but rather to prove general theorems so that the calculation of specific limits can be simplified.

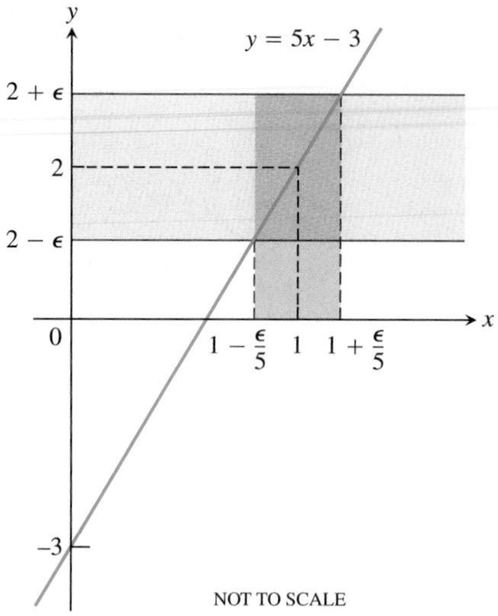

FIGURE 2.15 If $f(x) = 5x - 3$, then $0 < |x - 1| < \epsilon/5$ guarantees that $|f(x) - 2| < \epsilon$ (Example 2).

EXAMPLE 2 Testing the Definition

Show that

$$\lim_{x \to 1} (5x - 3) = 2.$$

Solution Set $x_0 = 1$, $f(x) = 5x - 3$, and $L = 2$ in the definition of limit. For any given $\epsilon > 0$, we have to find a suitable $\delta > 0$ so that if $x \neq 1$ and x is within distance δ of $x_0 = 1$, that is, whenever

$$0 < |x - 1| < \delta,$$

it is true that $f(x)$ is within distance ϵ of $L = 2$, so

$$|f(x) - 2| < \epsilon.$$

We find δ by working backward from the ϵ-inequality:

$$\begin{aligned} |(5x - 3) - 2| = |5x - 5| &< \epsilon \\ 5|x - 1| &< \epsilon \\ |x - 1| &< \epsilon/5. \end{aligned}$$

Thus, we can take $\delta = \epsilon/5$ (Figure 2.15). If $0 < |x - 1| < \delta = \epsilon/5$, then

$$|(5x - 3) - 2| = |5x - 5| = 5|x - 1| < 5(\epsilon/5) = \epsilon,$$

which proves that $\lim_{x \to 1}(5x - 3) = 2$.

The value of $\delta = \epsilon/5$ is not the only value that will make $0 < |x - 1| < \delta$ imply $|5x - 5| < \epsilon$. Any smaller positive δ will do as well. The definition does not ask for a "best" positive δ, just one that will work. ■

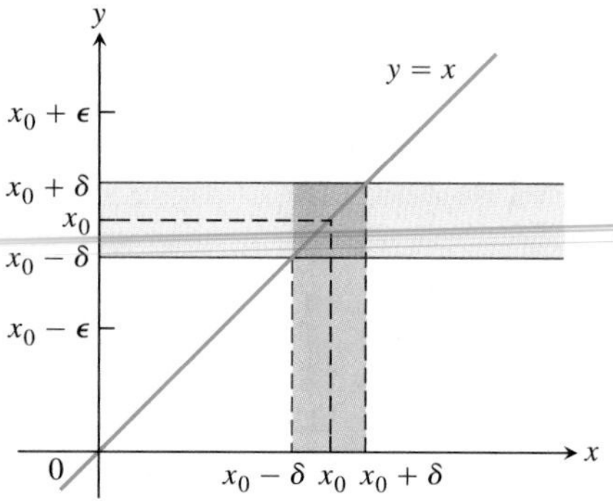

FIGURE 2.16 For the function $f(x) = x$, we find that $0 < |x - x_0| < \delta$ will guarantee $|f(x) - x_0| < \epsilon$ whenever $\delta \leq \epsilon$ (Example 3a).

EXAMPLE 3 Limits of the Identity and Constant Functions

Prove:

(a) $\lim_{x \to x_0} x = x_0$ **(b)** $\lim_{x \to x_0} k = k$ (k constant).

Solution

(a) Let $\epsilon > 0$ be given. We must find $\delta > 0$ such that for all x

$$0 < |x - x_0| < \delta \quad \text{implies} \quad |x - x_0| < \epsilon.$$

The implication will hold if δ equals ϵ or any smaller positive number (Figure 2.16). This proves that $\lim_{x \to x_0} x = x_0$.

(b) Let $\epsilon > 0$ be given. We must find $\delta > 0$ such that for all x

$$0 < |x - x_0| < \delta \quad \text{implies} \quad |k - k| < \epsilon.$$

Since $k - k = 0$, we can use any positive number for δ and the implication will hold (Figure 2.17). This proves that $\lim_{x \to x_0} k = k$. ■

Finding Deltas Algebraically for Given Epsilons

In Examples 2 and 3, the interval of values about x_0 for which $|f(x) - L|$ was less than ϵ was symmetric about x_0 and we could take δ to be half the length of that interval. When

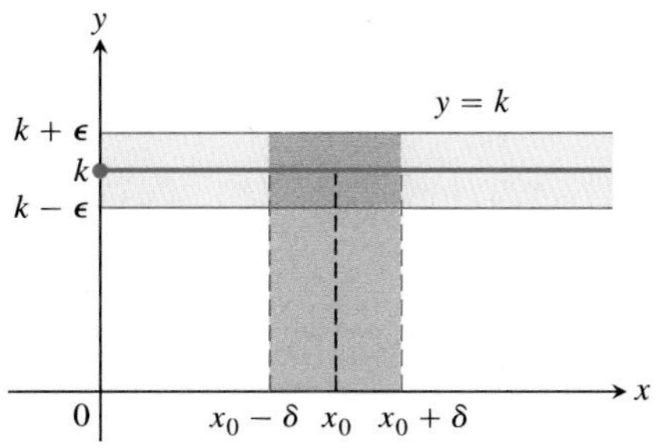

FIGURE 2.17 For the function $f(x) = k$, we find that $|f(x) - k| < \epsilon$ for any positive δ (Example 3b).

such symmetry is absent, as it usually is, we can take δ to be the distance from x_0 to the interval's *nearer* endpoint.

EXAMPLE 4 Finding Delta Algebraically

For the limit $\lim_{x \to 5} \sqrt{x - 1} = 2$, find a $\delta > 0$ that works for $\epsilon = 1$. That is, find a $\delta > 0$ such that for all x

$$0 < |x - 5| < \delta \qquad \Rightarrow \qquad |\sqrt{x - 1} - 2| < 1.$$

Solution We organize the search into two steps, as discussed below.

1. *Solve the inequality* $|\sqrt{x - 1} - 2| < 1$ *to find an interval containing* $x_0 = 5$ *on which the inequality holds for all* $x \neq x_0$.

$$|\sqrt{x - 1} - 2| < 1$$
$$-1 < \sqrt{x - 1} - 2 < 1$$
$$1 < \sqrt{x - 1} < 3$$
$$1 < x - 1 < 9$$
$$2 < x < 10$$

The inequality holds for all x in the open interval (2, 10), so it holds for all $x \neq 5$ in this interval as well (see Figure 2.19).

2. *Find a value of* $\delta > 0$ *to place the centered interval* $5 - \delta < x < 5 + \delta$ (centered at $x_0 = 5$) *inside the interval* (2, 10). The distance from 5 to the nearer endpoint of (2, 10) is 3 (Figure 2.18). If we take $\delta = 3$ or any smaller positive number, then the inequality $0 < |x - 5| < \delta$ will automatically place x between 2 and 10 to make $|\sqrt{x - 1} - 2| < 1$ (Figure 2.19)

$$0 < |x - 5| < 3 \qquad \Rightarrow \qquad |\sqrt{x - 1} - 2| < 1.$$

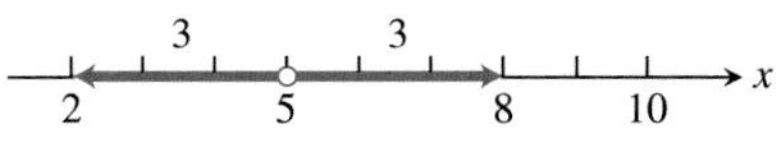

FIGURE 2.18 An open interval of radius 3 about $x_0 = 5$ will lie inside the open interval (2, 10).

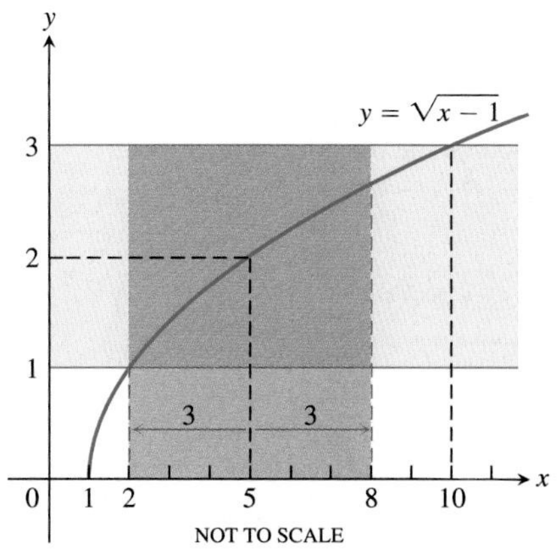

FIGURE 2.19 The function and intervals in Example 4. ■

How to Find Algebraically a δ for a Given f, L, x_0, and $\epsilon > 0$

The process of finding a $\delta > 0$ such that for all x

$$0 < |x - x_0| < \delta \quad \Rightarrow \quad |f(x) - L| < \epsilon$$

can be accomplished in two steps.

1. *Solve the inequality* $|f(x) - L| < \epsilon$ to find an open interval (a, b) containing x_0 on which the inequality holds for all $x \neq x_0$.
2. *Find a value of* $\delta > 0$ that places the open interval $(x_0 - \delta, x_0 + \delta)$ centered at x_0 inside the interval (a, b). The inequality $|f(x) - L| < \epsilon$ will hold for all $x \neq x_0$ in this δ-interval.

EXAMPLE 5 Finding Delta Algebraically

Prove that $\lim_{x \to 2} f(x) = 4$ if

$$f(x) = \begin{cases} x^2, & x \neq 2 \\ 1, & x = 2. \end{cases}$$

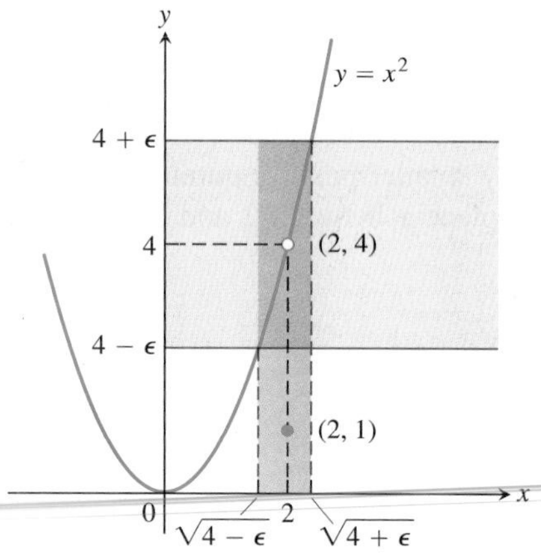

FIGURE 2.20 An interval containing $x = 2$ so that the function in Example 5 satisfies $|f(x) - 4| < \epsilon$.

Solution Our task is to show that given $\epsilon > 0$ there exists a $\delta > 0$ such that for all x

$$0 < |x - 2| < \delta \quad \Rightarrow \quad |f(x) - 4| < \epsilon.$$

1. *Solve the inequality $|f(x) - 4| < \epsilon$ to find an open interval containing $x_0 = 2$ on which the inequality holds for all $x \neq x_0$.*

   For $x \neq x_0 = 2$, we have $f(x) = x^2$, and the inequality to solve is $|x^2 - 4| < \epsilon$:

$$|x^2 - 4| < \epsilon$$
$$-\epsilon < x^2 - 4 < \epsilon$$
$$4 - \epsilon < x^2 < 4 + \epsilon$$
$$\sqrt{4 - \epsilon} < |x| < \sqrt{4 + \epsilon} \qquad \text{Assumes } \epsilon < 4\text{; see below.}$$
$$\sqrt{4 - \epsilon} < x < \sqrt{4 + \epsilon}. \qquad \text{An open interval about } x_0 = 2 \text{ that solves the inequality}$$

   The inequality $|f(x) - 4| < \epsilon$ holds for all $x \neq 2$ in the open interval $\left(\sqrt{4 - \epsilon}, \sqrt{4 + \epsilon}\right)$ (Figure 2.20).

2. *Find a value of $\delta > 0$ that places the centered interval $(2 - \delta, 2 + \delta)$ inside the interval $\left(\sqrt{4 - \epsilon}, \sqrt{4 + \epsilon}\right)$.*

   Take δ to be the distance from $x_0 = 2$ to the nearer endpoint of $\left(\sqrt{4 - \epsilon}, \sqrt{4 + \epsilon}\right)$. In other words, take $\delta = \min\left\{2 - \sqrt{4 - \epsilon}, \sqrt{4 + \epsilon} - 2\right\}$, the *minimum* (the smaller) of the two numbers $2 - \sqrt{4 - \epsilon}$ and $\sqrt{4 + \epsilon} - 2$. If δ has this or any smaller positive value, the inequality $0 < |x - 2| < \delta$ will automatically place x between $\sqrt{4 - \epsilon}$ and $\sqrt{4 + \epsilon}$ to make $|f(x) - 4| < \epsilon$. For all x,

$$0 < |x - 2| < \delta \quad \Rightarrow \quad |f(x) - 4| < \epsilon.$$

This completes the proof for $\epsilon < 4$.

If $\epsilon \geq 4$, then we take δ to be the distance from $x_0 = 2$ to the nearer endpoint of the interval $\left(0, \sqrt{4 + \epsilon}\right)$. In other words, take $\delta = \min\left\{2\sqrt{4 + \epsilon} - 2\right\}$. (See Figure 2.20) ■

Using the Definition to Prove Theorems

We do not usually rely on the formal definition of limit to verify specific limits such as those in the preceding examples. Rather we appeal to general theorems about limits, in particular the theorems of Section 2.2. The definition is used to prove these theorems (Appendix 2). As an example, we prove part 1 of Theorem 1, the Sum Rule.

EXAMPLE 6 Proving the Rule for the Limit of a Sum

Given that $\lim_{x \to c} f(x) = L$ and $\lim_{x \to c} g(x) = M$, prove that

$$\lim_{x \to c} (f(x) + g(x)) = L + M.$$

Solution Let $\epsilon > 0$ be given. We want to find a positive number δ such that for all x

$$0 < |x - c| < \delta \quad \Rightarrow \quad |f(x) + g(x) - (L + M)| < \epsilon.$$

Regrouping terms, we get

$$\begin{aligned} |f(x) + g(x) - (L + M)| &= |(f(x) - L) + (g(x) - M)| \\ &\leq |f(x) - L| + |g(x) - M|. \end{aligned}$$

Triangle Inequality: $|a + b| \leq |a| + |b|$

Since $\lim_{x \to c} f(x) = L$, there exists a number $\delta_1 > 0$ such that for all x

$$0 < |x - c| < \delta_1 \quad \Rightarrow \quad |f(x) - L| < \epsilon/2.$$

Similarly, since $\lim_{x \to c} g(x) = M$, there exists a number $\delta_2 > 0$ such that for all x

$$0 < |x - c| < \delta_2 \quad \Rightarrow \quad |g(x) - M| < \epsilon/2.$$

Let $\delta = \min\{\delta_1, \delta_2\}$, the smaller of δ_1 and δ_2. If $0 < |x - c| < \delta$ then $|x - c| < \delta_1$, so $|f(x) - L| < \epsilon/2$, and $|x - c| < \delta_2$, so $|g(x) - M| < \epsilon/2$. Therefore

$$|f(x) + g(x) - (L + M)| < \frac{\epsilon}{2} + \frac{\epsilon}{2} = \epsilon.$$

This shows that $\lim_{x \to c}(f(x) + g(x)) = L + M$. ■

Let's also prove Theorem 5 of Section 2.2.

EXAMPLE 7 Given that $\lim_{x \to c} f(x) = L$ and $\lim_{x \to c} g(x) = M$, and that $f(x) \leq g(x)$ for all x in an open interval containing c (except possibly c itself), prove that $L \leq M$.

Solution We use the method of proof by contradiction. Suppose, on the contrary, that $L > M$. Then by the limit of a difference property in Theorem 1,

$$\lim_{x \to c} (g(x) - f(x)) = M - L.$$

Therefore, for any $\epsilon > 0$, there exists $\delta > 0$ such that

$$|(g(x) - f(x)) - (M - L)| < \epsilon \qquad \text{whenever} \quad 0 < |x - c| < \delta.$$

Since $L - M > 0$ by hypothesis, we take $\epsilon = L - M$ in particular and we have a number $\delta > 0$ such that

$$|(g(x) - f(x)) - (M - L)| < L - M \qquad \text{whenever} \quad 0 < |x - c| < \delta.$$

Since $a \le |a|$ for any number a, we have

$$(g(x) - f(x)) - (M - L) < L - M \qquad \text{whenever} \quad 0 < |x - c| < \delta$$

which simplifies to

$$g(x) < f(x) \qquad \text{whenever} \quad 0 < |x - c| < \delta.$$

But this contradicts $f(x) \le g(x)$. Thus the inequality $L > M$ must be false. Therefore $L \le M$. ■

EXERCISES 2.3

Centering Intervals About a Point

In Exercises 1–6, sketch the interval (a, b) on the x-axis with the point x_0 inside. Then find a value of $\delta > 0$ such that for all x, $0 < |x - x_0| < \delta \Rightarrow a < x < b$.

1. $a = 1, \quad b = 7, \quad x_0 = 5$
2. $a = 1, \quad b = 7, \quad x_0 = 2$
3. $a = -7/2, \quad b = -1/2, \quad x_0 = -3$
4. $a = -7/2, \quad b = -1/2, \quad x_0 = -3/2$
5. $a = 4/9, \quad b = 4/7, \quad x_0 = 1/2$
6. $a = 2.7591, \quad b = 3.2391, \quad x_0 = 3$

Finding Deltas Graphically

In Exercises 7–14, use the graphs to find a $\delta > 0$ such that for all x

$$0 < |x - x_0| < \delta \quad \Rightarrow \quad |f(x) - L| < \epsilon.$$

7.

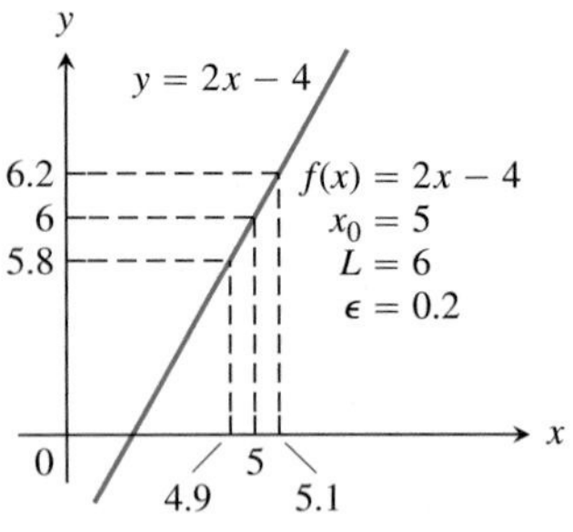

NOT TO SCALE

8.

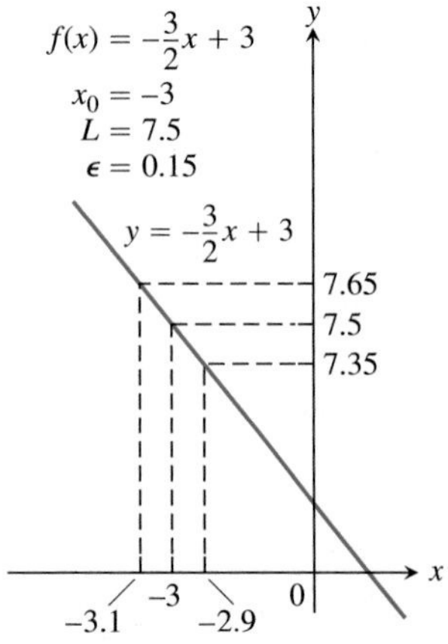

NOT TO SCALE

9.

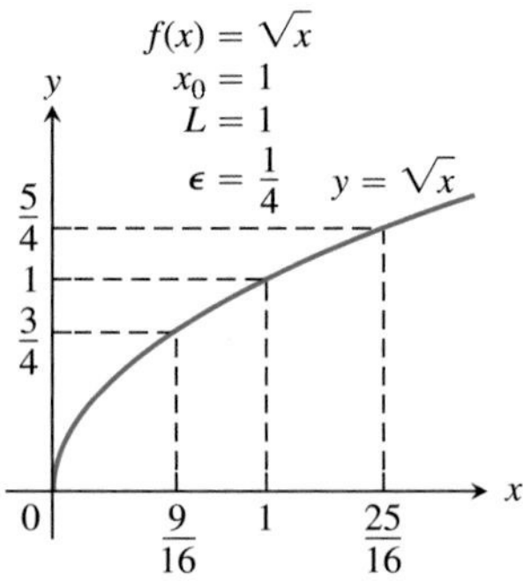

10.

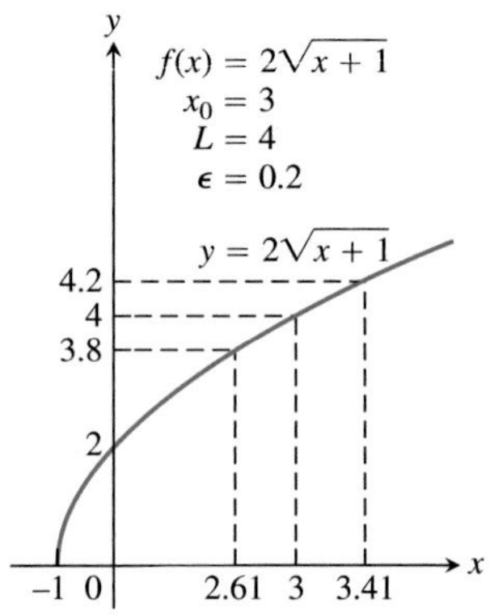

NOT TO SCALE

11.

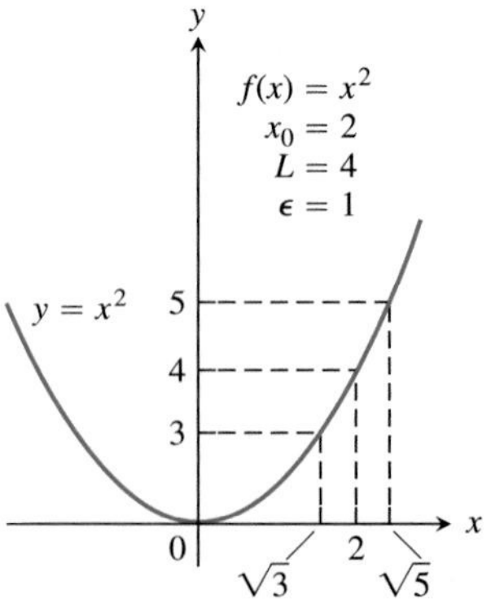

NOT TO SCALE

12.

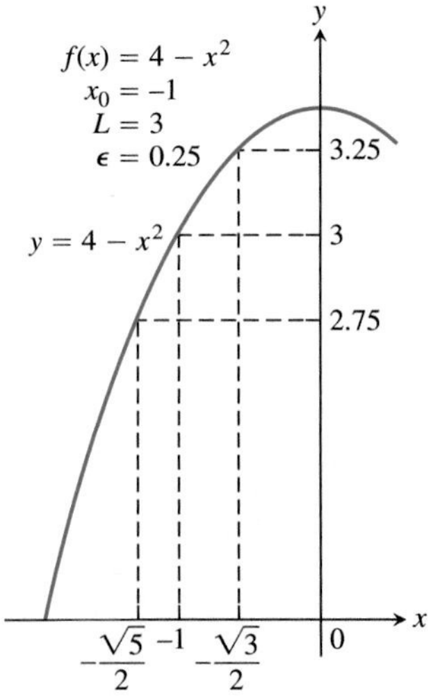

NOT TO SCALE

13.

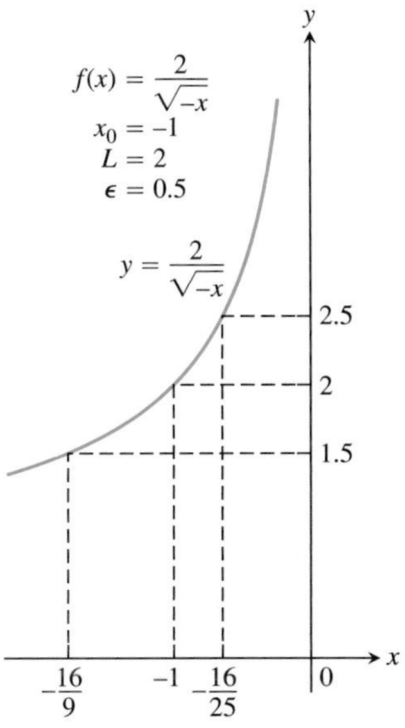

14.

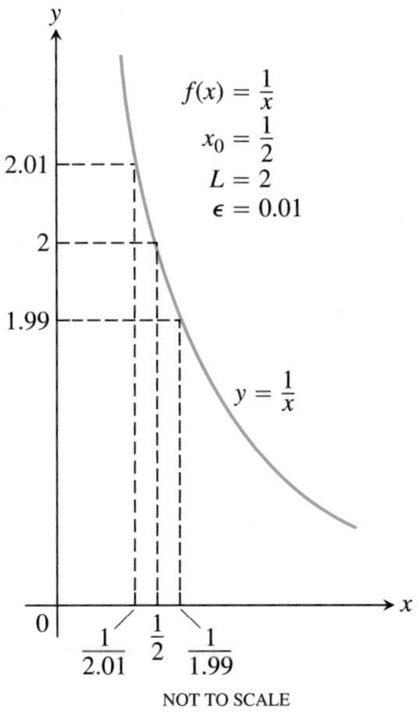

Finding Deltas Algebraically

Each of Exercises 15–30 gives a function $f(x)$ and numbers L, x_0 and $\epsilon > 0$. In each case, find an open interval about x_0 on which the inequality $|f(x) - L| < \epsilon$ holds. Then give a value for $\delta > 0$ such that for all x satisfying $0 < |x - x_0| < \delta$ the inequality $|f(x) - L| < \epsilon$ holds.

15. $f(x) = x + 1, \quad L = 5, \quad x_0 = 4, \quad \epsilon = 0.01$

16. $f(x) = 2x - 2, \quad L = -6, \quad x_0 = -2, \quad \epsilon = 0.02$

17. $f(x) = \sqrt{x + 1}, \quad L = 1, \quad x_0 = 0, \quad \epsilon = 0.1$

18. $f(x) = \sqrt{x}, \quad L = 1/2, \quad x_0 = 1/4, \quad \epsilon = 0.1$

19. $f(x) = \sqrt{19 - x}, \quad L = 3, \quad x_0 = 10, \quad \epsilon = 1$

20. $f(x) = \sqrt{x - 7}, \quad L = 4, \quad x_0 = 23, \quad \epsilon = 1$

21. $f(x) = 1/x, \quad L = 1/4, \quad x_0 = 4, \quad \epsilon = 0.05$

22. $f(x) = x^2, \quad L = 3, \quad x_0 = \sqrt{3}, \quad \epsilon = 0.1$

23. $f(x) = x^2, \quad L = 4, \quad x_0 = -2, \quad \epsilon = 0.5$

24. $f(x) = 1/x, \quad L = -1, \quad x_0 = -1, \quad \epsilon = 0.1$

25. $f(x) = x^2 - 5, \quad L = 11, \quad x_0 = 4, \quad \epsilon = 1$

26. $f(x) = 120/x, \quad L = 5, \quad x_0 = 24, \quad \epsilon = 1$

27. $f(x) = mx, \quad m > 0, \quad L = 2m, \quad x_0 = 2, \quad \epsilon = 0.03$

28. $f(x) = mx, \quad m > 0, \quad L = 3m, \quad x_0 = 3,$
$\epsilon = c > 0$

29. $f(x) = mx + b, \quad m > 0, \quad L = (m/2) + b,$
$x_0 = 1/2, \quad \epsilon = c > 0$

30. $f(x) = mx + b, \quad m > 0, \quad L = m + b, \quad x_0 = 1,$
$\epsilon = 0.05$

More on Formal Limits

Each of Exercises 31–36 gives a function $f(x)$, a point x_0, and a positive number ϵ. Find $L = \lim_{x \to x_0} f(x)$. Then find a number $\delta > 0$ such that for all x

$$0 < |x - x_0| < \delta \quad \Rightarrow \quad |f(x) - L| < \epsilon.$$

31. $f(x) = 3 - 2x, \quad x_0 = 3, \quad \epsilon = 0.02$

32. $f(x) = -3x - 2, \quad x_0 = -1, \quad \epsilon = 0.03$

33. $f(x) = \dfrac{x^2 - 4}{x - 2}, \quad x_0 = 2, \quad \epsilon = 0.05$

34. $f(x) = \dfrac{x^2 + 6x + 5}{x + 5}, \quad x_0 = -5, \quad \epsilon = 0.05$

35. $f(x) = \sqrt{1 - 5x}, \quad x_0 = -3, \quad \epsilon = 0.5$

36. $f(x) = 4/x, \quad x_0 = 2, \quad \epsilon = 0.4$

Prove the limit statements in Exercises 37–50.

37. $\lim_{x \to 4} (9 - x) = 5$

38. $\lim_{x \to 3} (3x - 7) = 2$

39. $\lim_{x \to 9} \sqrt{x - 5} = 2$

40. $\lim_{x \to 0} \sqrt{4 - x} = 2$

41. $\lim_{x \to 1} f(x) = 1 \quad \text{if} \quad f(x) = \begin{cases} x^2, & x \neq 1 \\ 2, & x = 1 \end{cases}$

42. $\lim_{x \to -2} f(x) = 4 \quad \text{if} \quad f(x) = \begin{cases} x^2, & x \neq -2 \\ 1, & x = -2 \end{cases}$

43. $\lim_{x \to 1} \dfrac{1}{x} = 1$

44. $\lim_{x \to \sqrt{3}} \dfrac{1}{x^2} = \dfrac{1}{3}$

45. $\lim_{x \to -3} \dfrac{x^2 - 9}{x + 3} = -6$

46. $\lim_{x \to 1} \dfrac{x^2 - 1}{x - 1} = 2$

47. $\lim_{x \to 1} f(x) = 2 \quad \text{if} \quad f(x) = \begin{cases} 4 - 2x, & x < 1 \\ 6x - 4, & x \geq 1 \end{cases}$

48. $\lim_{x \to 0} f(x) = 0 \quad \text{if} \quad f(x) = \begin{cases} 2x, & x < 0 \\ x/2, & x \geq 0 \end{cases}$

49. $\lim_{x \to 0} x \sin \dfrac{1}{x} = 0$

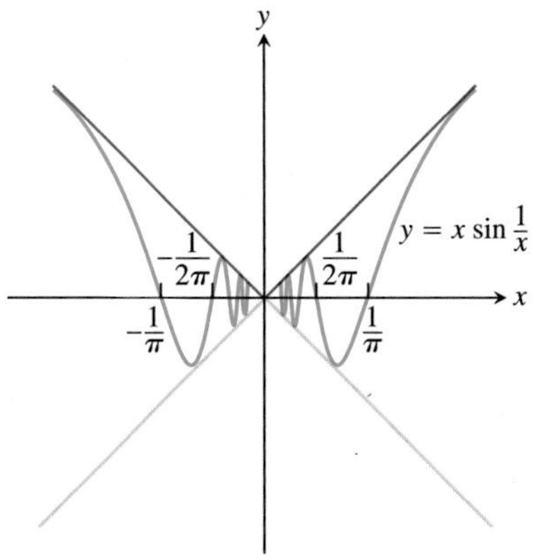

50. $\lim_{x \to 0} x^2 \sin \frac{1}{x} = 0$

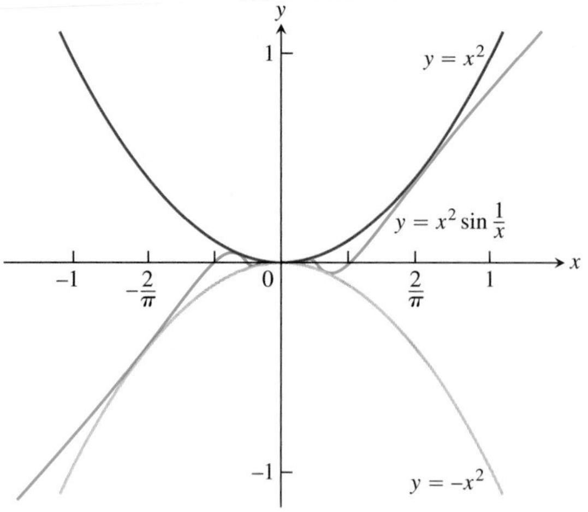

Theory and Examples

51. Define what it means to say that $\lim_{x \to 0} g(x) = k$.

52. Prove that $\lim_{x \to c} f(x) = L$ if and only if $\lim_{h \to 0} f(h + c) = L$.

53. **A wrong statement about limits** Show by example that the following statement is wrong.

    The number L is the limit of $f(x)$ as x approaches x_0 if $f(x)$ gets closer to L as x approaches x_0.

    Explain why the function in your example does not have the given value of L as a limit as $x \to x_0$.

54. **Another wrong statement about limits** Show by example that the following statement is wrong.

    The number L is the limit of $f(x)$ as x approaches x_0 if, given any $\epsilon > 0$, there exists a value of x for which $|f(x) - L| < \epsilon$.

    Explain why the function in your example does not have the given value of L as a limit as $x \to x_0$.

T 55. **Grinding engine cylinders** Before contracting to grind engine cylinders to a cross-sectional area of 9 in^2, you need to know how much deviation from the ideal cylinder diameter of $x_0 = 3.385$ in. you can allow and still have the area come within 0.01 in^2 of the required 9 in^2. To find out, you let $A = \pi(x/2)^2$ and look for the interval in which you must hold x to make $|A - 9| \le 0.01$. What interval do you find?

56. **Manufacturing electrical resistors** Ohm's law for electrical circuits like the one shown in the accompanying figure states that $V = RI$. In this equation, V is a constant voltage, I is the current in amperes, and R is the resistance in ohms. Your firm has been asked to supply the resistors for a circuit in which V will be 120 volts and I is to be 5 ± 0.1 amp. In what interval does R have to lie for I to be within 0.1 amp of the value $I_0 = 5$?

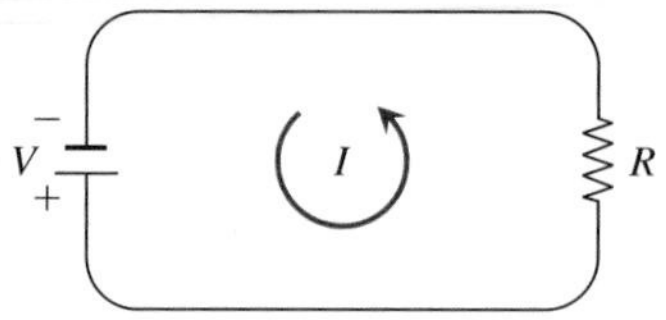

When Is a Number L Not the Limit of $f(x)$ as $x \to x_0$?

We can prove that $\lim_{x \to x_0} f(x) \ne L$ by providing an $\epsilon > 0$ such that no possible $\delta > 0$ satisfies the condition

$$\text{For all } x, \quad 0 < |x - x_0| < \delta \quad \Rightarrow \quad |f(x) - L| < \epsilon.$$

We accomplish this for our candidate ϵ by showing that for each $\delta > 0$ there exists a value of x such that

$$0 < |x - x_0| < \delta \quad \text{and} \quad |f(x) - L| \ge \epsilon.$$

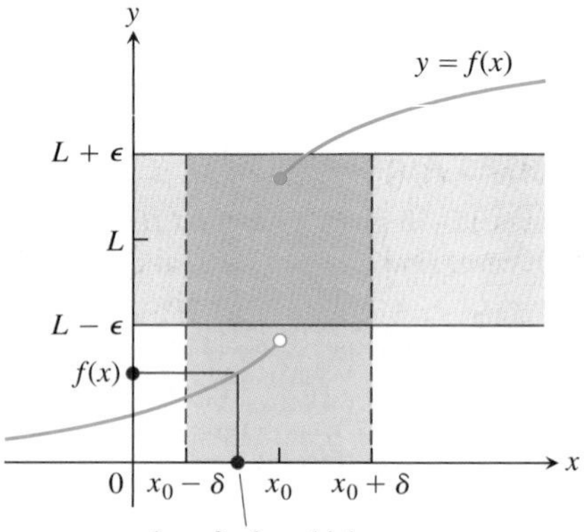

a value of x for which $0 < |x - x_0| < \delta$ and $|f(x) - L| \ge \epsilon$

57. Let $f(x) = \begin{cases} x, & x < 1 \\ x + 1, & x > 1. \end{cases}$

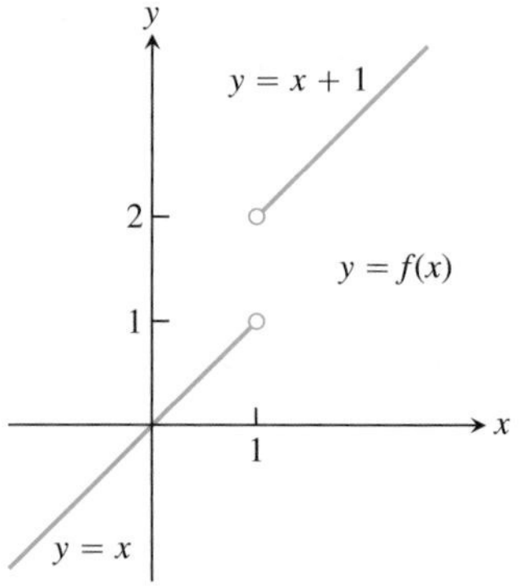

a. Let $\epsilon = 1/2$. Show that no possible $\delta > 0$ satisfies the following condition:

For all x, $0 < |x - 1| < \delta \quad \Rightarrow \quad |f(x) - 2| < 1/2.$

That is, for each $\delta > 0$ show that there is a value of x such that

$$0 < |x - 1| < \delta \quad \text{and} \quad |f(x) - 2| \geq 1/2.$$

This will show that $\lim_{x \to 1} f(x) \neq 2$.

b. Show that $\lim_{x \to 1} f(x) \neq 1$.

c. Show that $\lim_{x \to 1} f(x) \neq 1.5$.

58. Let $h(x) = \begin{cases} x^2, & x < 2 \\ 3, & x = 2 \\ 2, & x > 2. \end{cases}$

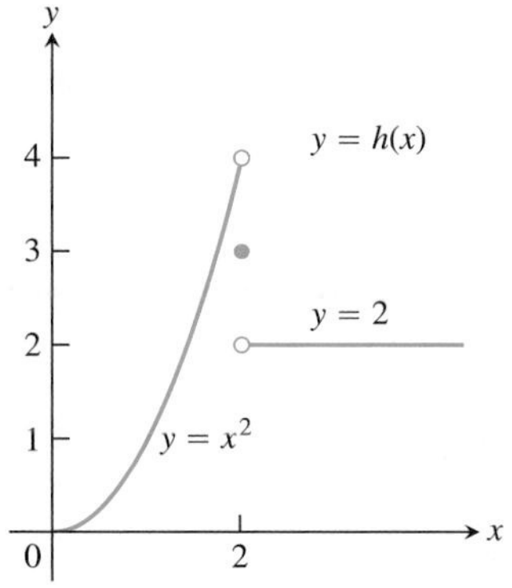

Show that

a. $\lim_{x \to 2} h(x) \neq 4$

b. $\lim_{x \to 2} h(x) \neq 3$

c. $\lim_{x \to 2} h(x) \neq 2$

59. For the function graphed here, explain why

a. $\lim_{x \to 3} f(x) \neq 4$

b. $\lim_{x \to 3} f(x) \neq 4.8$

c. $\lim_{x \to 3} f(x) \neq 3$

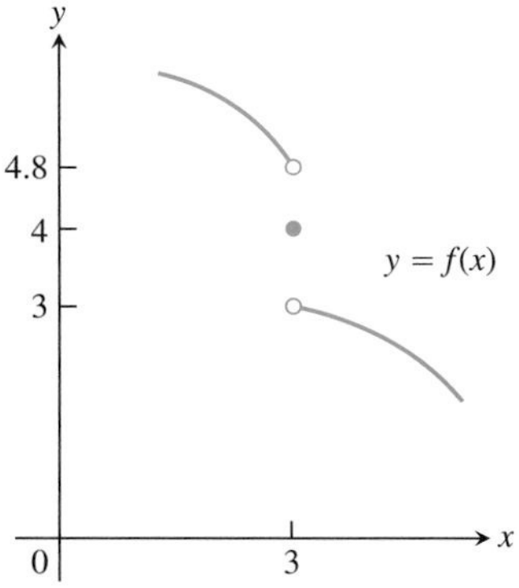

60. a. For the function graphed here, show that $\lim_{x \to -1} g(x) \neq 2$.

b. Does $\lim_{x \to -1} g(x)$ appear to exist? If so, what is the value of the limit? If not, why not?

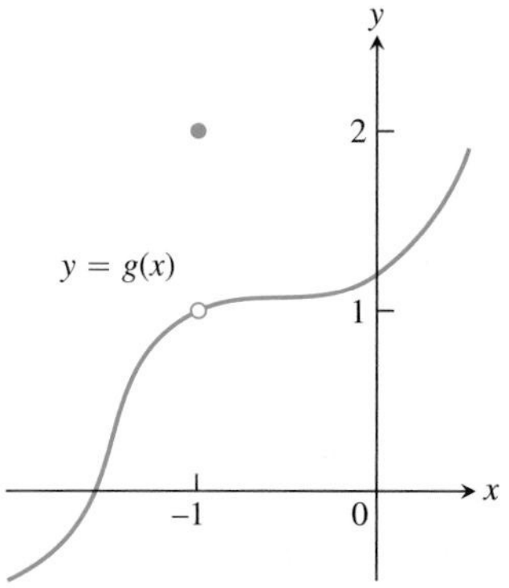

COMPUTER EXPLORATIONS

In Exercises 61–66, you will further explore finding deltas graphically. Use a CAS to perform the following steps:

a. Plot the function $y = f(x)$ near the point x_0 being approached.

b. Guess the value of the limit L and then evaluate the limit symbolically to see if you guessed correctly.

c. Using the value $\epsilon = 0.2$, graph the banding lines $y_1 = L - \epsilon$ and $y_2 = L + \epsilon$ together with the function f near x_0.

d. From your graph in part (c), estimate a $\delta > 0$ such that for all x

$$0 < |x - x_0| < \delta \quad \Rightarrow \quad |f(x) - L| < \epsilon.$$

Test your estimate by plotting f, y_1, and y_2 over the interval $0 < |x - x_0| < \delta$. For your viewing window use $x_0 - 2\delta \leq x \leq x_0 + 2\delta$ and $L - 2\epsilon \leq y \leq L + 2\epsilon$. If any function values lie outside the interval $[L - \epsilon, L + \epsilon]$, your choice of δ was too large. Try again with a smaller estimate.

e. Repeat parts (c) and (d) successively for $\epsilon = 0.1, 0.05$, and 0.001.

61. $f(x) = \dfrac{x^4 - 81}{x - 3}, \quad x_0 = 3$

62. $f(x) = \dfrac{5x^3 + 9x^2}{2x^5 + 3x^2}, \quad x_0 = 0$

63. $f(x) = \dfrac{\sin 2x}{3x}, \quad x_0 = 0$

64. $f(x) = \dfrac{x(1 - \cos x)}{x - \sin x}, \quad x_0 = 0$

65. $f(x) = \dfrac{\sqrt[3]{x} - 1}{x - 1}, \quad x_0 = 1$

66. $f(x) = \dfrac{3x^2 - (7x + 1)\sqrt{x} + 5}{x - 1}, \quad x_0 = 1$

2.4 One-Sided Limits and Limits at Infinity

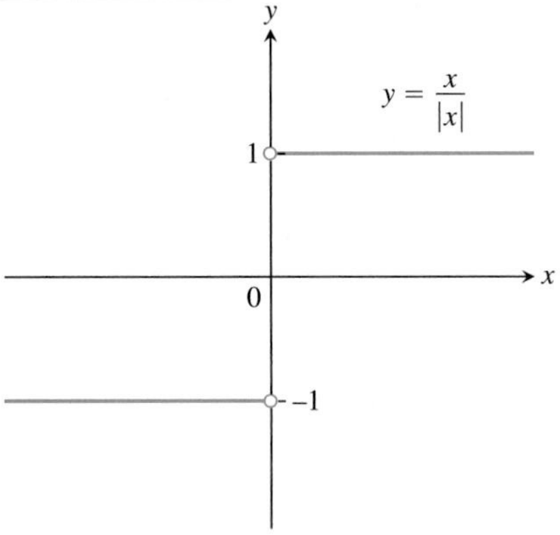

FIGURE 2.21 Different right-hand and left-hand limits at the origin.

In this section we extend the limit concept to *one-sided limits*, which are limits as x approaches the number x_0 from the left-hand side (where $x < x_0$) or the right-hand side ($x > x_0$) only. We also analyze the graphs of certain rational functions as well as other functions with limit behavior as $x \to \pm\infty$.

One-Sided Limits

To have a limit L as x approaches c, a function f must be defined on *both sides* of c and its values $f(x)$ must approach L as x approaches c from either side. Because of this, ordinary limits are called **two-sided**.

If f fails to have a two-sided limit at c, it may still have a one-sided limit, that is, a limit if the approach is only from one side. If the approach is from the right, the limit is a **right-hand limit**. From the left, it is a **left-hand limit**.

The function $f(x) = x/|x|$ (Figure 2.21) has limit 1 as x approaches 0 from the right, and limit -1 as x approaches 0 from the left. Since these one-sided limit values are not the same, there is no single number that $f(x)$ approaches as x approaches 0. So $f(x)$ does not have a (two-sided) limit at 0.

Intuitively, if $f(x)$ is defined on an interval (c, b), where $c < b$, and approaches arbitrarily close to L as x approaches c from within that interval, then f has **right-hand limit** L at c. We write

$$\lim_{x \to c^+} f(x) = L.$$

The symbol "$x \to c^+$" means that we consider only values of x greater than c.

Similarly, if $f(x)$ is defined on an interval (a, c), where $a < c$ and approaches arbitrarily close to M as x approaches c from within that interval, then f has **left-hand limit** M at c. We write

$$\lim_{x \to c^-} f(x) = M.$$

The symbol "$x \to c^-$" means that we consider only x values less than c.

These informal definitions are illustrated in Figure 2.22. For the function $f(x) = x/|x|$ in Figure 2.21 we have

$$\lim_{x \to 0^+} f(x) = 1 \qquad \text{and} \qquad \lim_{x \to 0^-} f(x) = -1.$$

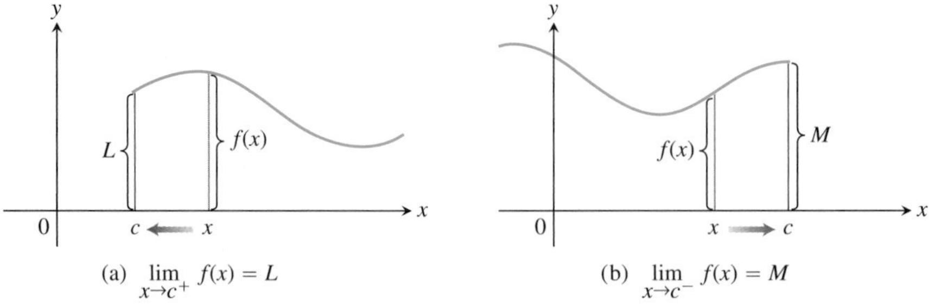

FIGURE 2.22 (a) Right-hand limit as x approaches c. (b) Left-hand limit as x approaches c.

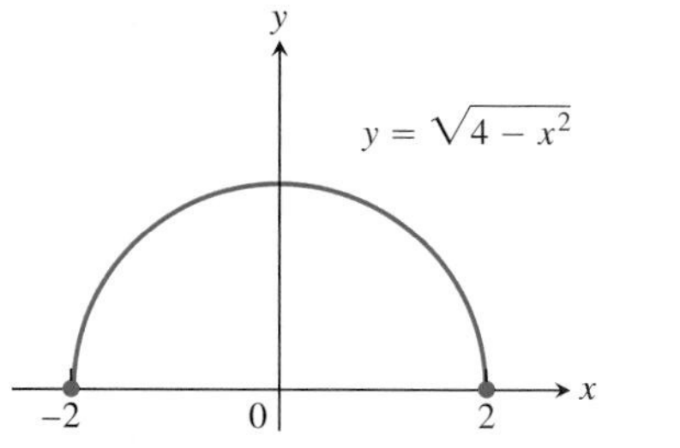

FIGURE 2.23 $\lim_{x\to 2^-} \sqrt{4 - x^2} = 0$ and $\lim_{x\to -2^+} \sqrt{4 - x^2} = 0$ (Example 1).

EXAMPLE 1 One-Sided Limits for a Semicircle

The domain of $f(x) = \sqrt{4 - x^2}$ is $[-2, 2]$; its graph is the semicircle in Figure 2.23. We have

$$\lim_{x\to -2^+} \sqrt{4 - x^2} = 0 \quad \text{and} \quad \lim_{x\to 2^-} \sqrt{4 - x^2} = 0.$$

The function does not have a left-hand limit at $x = -2$ or a right-hand limit at $x = 2$. It does not have ordinary two-sided limits at either -2 or 2. ■

One-sided limits have all the properties listed in Theorem 1 in Section 2.2. The right-hand limit of the sum of two functions is the sum of their right-hand limits, and so on. The theorems for limits of polynomials and rational functions hold with one-sided limits, as does the Sandwich Theorem and Theorem 5. One-sided limits are related to limits in the following way.

THEOREM 6
A function $f(x)$ has a limit as x approaches c if and only if it has left-hand and right-hand limits there and these one-sided limits are equal:

$$\lim_{x\to c} f(x) = L \quad \Leftrightarrow \quad \lim_{x\to c^-} f(x) = L \quad \text{and} \quad \lim_{x\to c^+} f(x) = L.$$

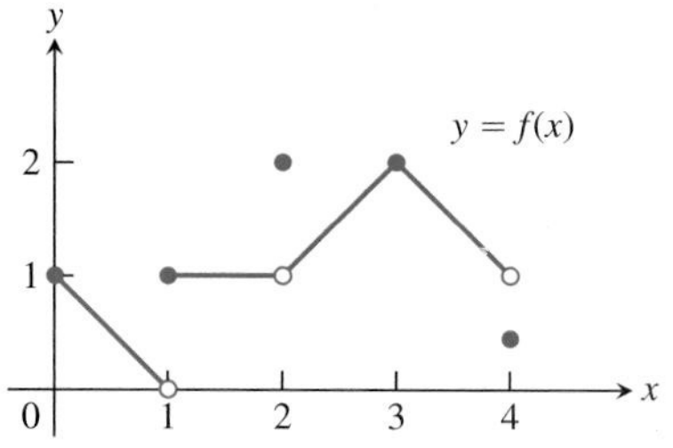

FIGURE 2.24 Graph of the function in Example 2.

EXAMPLE 2 Limits of the Function Graphed in Figure 2.24

At $x = 0$: $\lim_{x\to 0^+} f(x) = 1$,
$\lim_{x\to 0^-} f(x)$ and $\lim_{x\to 0} f(x)$ do not exist. The function is not defined to the left of $x = 0$.

At $x = 1$: $\lim_{x\to 1^-} f(x) = 0$ even though $f(1) = 1$,
$\lim_{x\to 1^+} f(x) = 1$,
$\lim_{x\to 1} f(x)$ does not exist. The right- and left-hand limits are not equal.

At $x = 2$: $\lim_{x\to 2^-} f(x) = 1$,
$\lim_{x\to 2^+} f(x) = 1$,
$\lim_{x\to 2} f(x) = 1$ even though $f(2) = 2$.

At $x = 3$: $\lim_{x\to 3^-} f(x) = \lim_{x\to 3^+} f(x) = \lim_{x\to 3} f(x) = f(3) = 2$.

At $x = 4$: $\lim_{x\to 4^-} f(x) = 1$ even though $f(4) \neq 1$,
$\lim_{x\to 4^+} f(x)$ and $\lim_{x\to 4} f(x)$ do not exist. The function is not defined to the right of $x = 4$.

At every other point c in $[0, 4]$, $f(x)$ has limit $f(c)$. ■

Precise Definitions of One-Sided Limits

The formal definition of the limit in Section 2.3 is readily modified for one-sided limits.

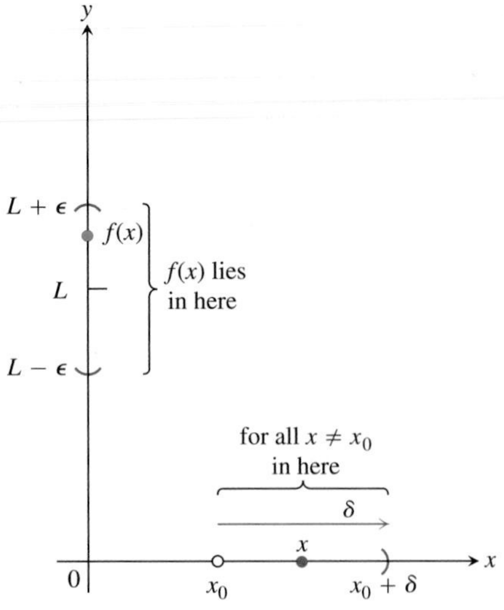

FIGURE 2.25 Intervals associated with the definition of right-hand limit.

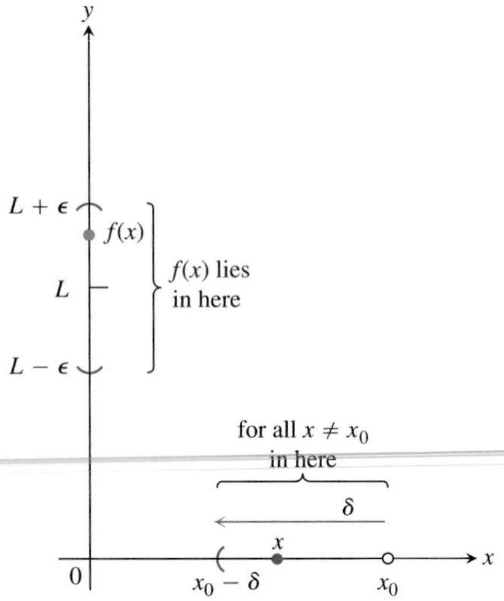

FIGURE 2.26 Intervals associated with the definition of left-hand limit.

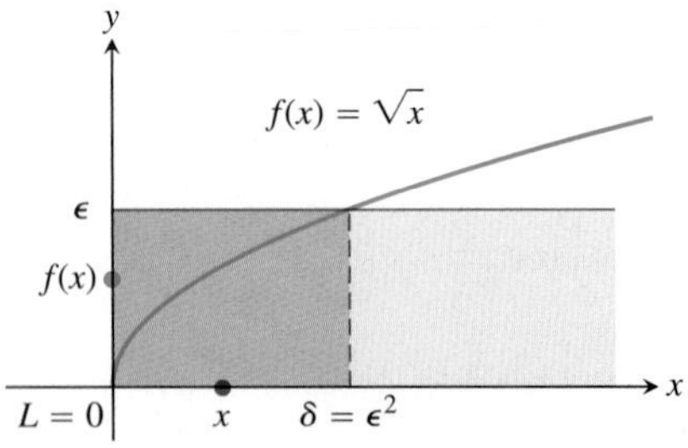

FIGURE 2.27 $\lim_{x \to 0^+} \sqrt{x} = 0$ in Example 3.

DEFINITIONS Right-Hand, Left-Hand Limits

We say that $f(x)$ has **right-hand limit L at x_0**, and write

$$\lim_{x \to x_0^+} f(x) = L \qquad \text{(See Figure 2.25)}$$

if for every number $\epsilon > 0$ there exists a corresponding number $\delta > 0$ such that for all x

$$x_0 < x < x_0 + \delta \quad \Rightarrow \quad |f(x) - L| < \epsilon.$$

We say that f has **left-hand limit L at x_0**, and write

$$\lim_{x \to x_0^-} f(x) = L \qquad \text{(See Figure 2.26)}$$

if for every number $\epsilon > 0$ there exists a corresponding number $\delta > 0$ such that for all x

$$x_0 - \delta < x < x_0 \quad \Rightarrow \quad |f(x) - L| < \epsilon.$$

EXAMPLE 3 Applying the Definition to Find Delta

Prove that

$$\lim_{x \to 0^+} \sqrt{x} = 0.$$

Solution Let $\epsilon > 0$ be given. Here $x_0 = 0$ and $L = 0$, so we want to find a $\delta > 0$ such that for all x

$$0 < x < \delta \quad \Rightarrow \quad |\sqrt{x} - 0| < \epsilon,$$

or

$$0 < x < \delta \quad \Rightarrow \quad \sqrt{x} < \epsilon.$$

Squaring both sides of this last inequality gives

$$x < \epsilon^2 \quad \text{if} \quad 0 < x < \delta.$$

If we choose $\delta = \epsilon^2$ we have

$$0 < x < \delta = \epsilon^2 \quad \Rightarrow \quad \sqrt{x} < \epsilon,$$

or

$$0 < x < \epsilon^2 \quad \Rightarrow \quad |\sqrt{x} - 0| < \epsilon.$$

According to the definition, this shows that $\lim_{x \to 0^+} \sqrt{x} = 0$ (Figure 2.27). ■

The functions examined so far have had some kind of limit at each point of interest. In general, that need not be the case.

EXAMPLE 4 A Function Oscillating Too Much

Show that $y = \sin(1/x)$ has no limit as x approaches zero from either side (Figure 2.28).

Solution As x approaches zero, its reciprocal, $1/x$, grows without bound and the values of $\sin(1/x)$ cycle repeatedly from -1 to 1. There is no single number L that the function's

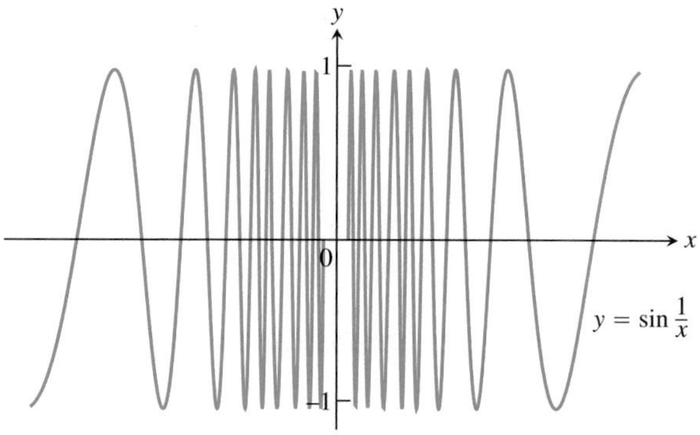

FIGURE 2.28 The function $y = \sin(1/x)$ has neither a right-hand nor a left-hand limit as x approaches zero (Example 4).

values stay increasingly close to as x approaches zero. This is true even if we restrict x to positive values or to negative values. The function has neither a right-hand limit nor a left-hand limit at $x = 0$. ■

Limits Involving $(\sin \theta)/\theta$

A central fact about $(\sin \theta)/\theta$ is that in radian measure its limit as $\theta \to 0$ is 1. We can see this in Figure 2.29 and confirm it algebraically using the Sandwich Theorem. You will see the importance of this limit in Section 3.4, where instantaneous rates of change of the trigonometric functions are studied.

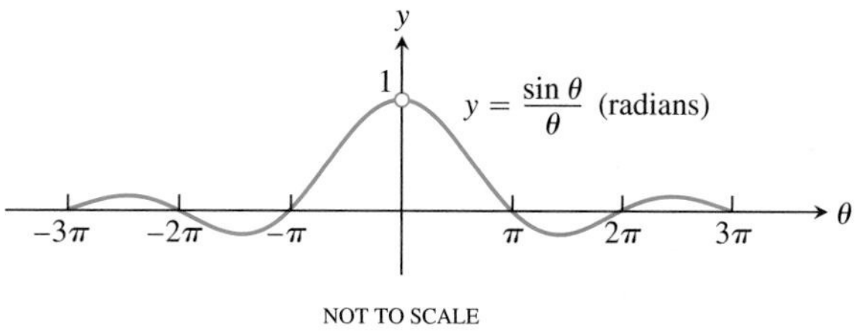

FIGURE 2.29 The graph of $f(\theta) = (\sin \theta)/\theta$.

THEOREM 7

$$\lim_{\theta \to 0} \frac{\sin \theta}{\theta} = 1 \qquad (\theta \text{ in radians}) \tag{1}$$

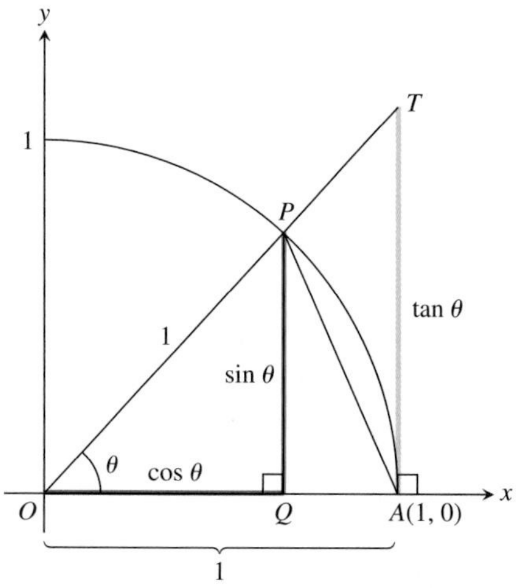

FIGURE 2.30 The figure for the proof of Theorem 7. $TA/OA = \tan \theta$, but $OA = 1$, so $TA = \tan \theta$.

Proof The plan is to show that the right-hand and left-hand limits are both 1. Then we will know that the two-sided limit is 1 as well.

To show that the right-hand limit is 1, we begin with positive values of θ less than $\pi/2$ (Figure 2.30). Notice that

$$\text{Area } \Delta OAP < \text{ area sector } OAP < \text{ area } \Delta OAT.$$

Equation (2) is where radian measure comes in: The area of sector OAP is $\theta/2$ only if θ is measured in radians.

We can express these areas in terms of θ as follows:

$$\text{Area } \Delta OAP = \frac{1}{2}\text{base} \times \text{height} = \frac{1}{2}(1)(\sin\theta) = \frac{1}{2}\sin\theta$$

$$\text{Area sector } OAP = \frac{1}{2}r^2\theta = \frac{1}{2}(1)^2\theta = \frac{\theta}{2} \tag{2}$$

$$\text{Area } \Delta OAT = \frac{1}{2}\text{base} \times \text{height} = \frac{1}{2}(1)(\tan\theta) = \frac{1}{2}\tan\theta.$$

Thus,

$$\frac{1}{2}\sin\theta < \frac{1}{2}\theta < \frac{1}{2}\tan\theta.$$

This last inequality goes the same way if we divide all three terms by the number $(1/2)\sin\theta$, which is positive since $0 < \theta < \pi/2$:

$$1 < \frac{\theta}{\sin\theta} < \frac{1}{\cos\theta}.$$

Taking reciprocals reverses the inequalities:

$$1 > \frac{\sin\theta}{\theta} > \cos\theta.$$

Since $\lim_{\theta\to 0^+}\cos\theta = 1$ (Example 6b, Section 2.2), the Sandwich Theorem gives

$$\lim_{\theta\to 0^+}\frac{\sin\theta}{\theta} = 1.$$

Recall that $\sin\theta$ and θ are both *odd functions* (Section 1.4). Therefore, $f(\theta) = (\sin\theta)/\theta$ is an *even function*, with a graph symmetric about the y-axis (see Figure 2.29). This symmetry implies that the left-hand limit at 0 exists and has the same value as the right-hand limit:

$$\lim_{\theta\to 0^-}\frac{\sin\theta}{\theta} = 1 = \lim_{\theta\to 0^+}\frac{\sin\theta}{\theta},$$

so $\lim_{\theta\to 0}(\sin\theta)/\theta = 1$ by Theorem 6. ■

EXAMPLE 5 Using $\lim_{\theta\to 0}\frac{\sin\theta}{\theta} = 1$

Show that **(a)** $\lim_{h\to 0}\frac{\cos h - 1}{h} = 0$ and **(b)** $\lim_{x\to 0}\frac{\sin 2x}{5x} = \frac{2}{5}$.

Solution

(a) Using the half-angle formula $\cos h = 1 - 2\sin^2(h/2)$, we calculate

$$\begin{aligned}\lim_{h\to 0}\frac{\cos h - 1}{h} &= \lim_{h\to 0} -\frac{2\sin^2(h/2)}{h}\\ &= -\lim_{\theta\to 0}\frac{\sin\theta}{\theta}\sin\theta \qquad \text{Let } \theta = h/2.\\ &= -(1)(0) = 0.\end{aligned}$$

(b) Equation (1) does not apply to the original fraction. We need a $2x$ in the denominator, not a $5x$. We produce it by multiplying numerator and denominator by $2/5$:

$$\begin{aligned}\lim_{x\to 0}\frac{\sin 2x}{5x} &= \lim_{x\to 0}\frac{(2/5)\cdot\sin 2x}{(2/5)\cdot 5x}\\ &= \frac{2}{5}\lim_{x\to 0}\frac{\sin 2x}{2x} \qquad \text{Now, Eq. (1) applies with } \theta = 2x.\\ &= \frac{2}{5}(1) = \frac{2}{5}\end{aligned}$$

■

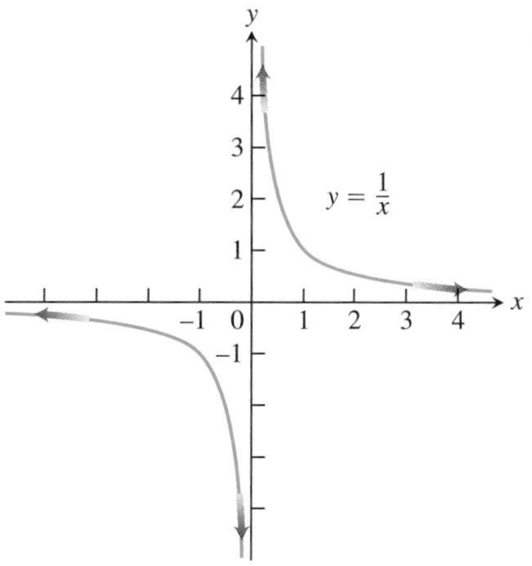

FIGURE 2.31 The graph of $y = 1/x$.

Finite Limits as $x \to \pm\infty$

The symbol for infinity (∞) does not represent a real number. We use ∞ to describe the behavior of a function when the values in its domain or range outgrow all finite bounds. For example, the function $f(x) = 1/x$ is defined for all $x \neq 0$ (Figure 2.31). When x is positive and becomes increasingly large, $1/x$ becomes increasingly small. When x is negative and its magnitude becomes increasingly large, $1/x$ again becomes small. We summarize these observations by saying that $f(x) = 1/x$ has limit 0 as $x \to \pm\infty$ or that 0 is a *limit of* $f(x) = 1/x$ *at infinity and negative infinity*. Here is a precise definition.

DEFINITIONS Limit as x approaches ∞ or $-\infty$

1. We say that $f(x)$ has the **limit L as x approaches infinity** and write

$$\lim_{x\to\infty} f(x) = L$$

if, for every number $\epsilon > 0$, there exists a corresponding number M such that for all x

$$x > M \quad \Rightarrow \quad |f(x) - L| < \epsilon.$$

2. We say that $f(x)$ has the **limit L as x approaches minus infinity** and write

$$\lim_{x\to-\infty} f(x) = L$$

if, for every number $\epsilon > 0$, there exists a corresponding number N such that for all x

$$x < N \quad \Rightarrow \quad |f(x) - L| < \epsilon.$$

Intuitively, $\lim_{x\to\infty} f(x) = L$ if, as x moves increasingly far from the origin in the positive direction, $f(x)$ gets arbitrarily close to L. Similarly, $\lim_{x\to-\infty} f(x) = L$ if, as x moves increasingly far from the origin in the negative direction, $f(x)$ gets arbitrarily close to L.

The strategy for calculating limits of functions as $x \to \pm\infty$ is similar to the one for finite limits in Section 2.2. There we first found the limits of the constant and identity functions $y = k$ and $y = x$. We then extended these results to other functions by applying a theorem about limits of algebraic combinations. Here we do the same thing, except that the starting functions are $y = k$ and $y = 1/x$ instead of $y = k$ and $y = x$.

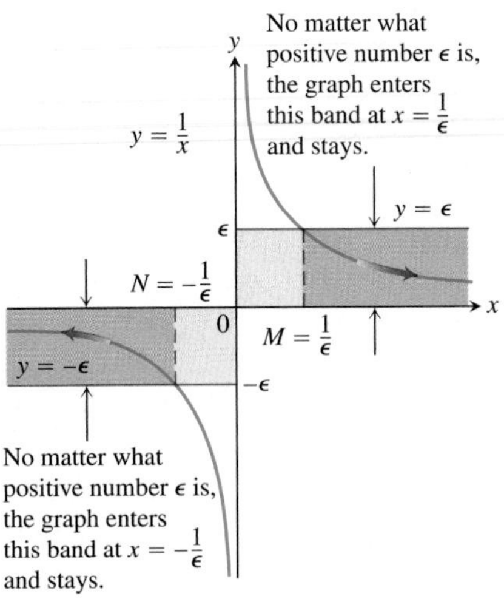

FIGURE 2.32 The geometry behind the argument in Example 6.

The basic facts to be verified by applying the formal definition are

$$\lim_{x\to\pm\infty} k = k \qquad \text{and} \qquad \lim_{x\to\pm\infty} \frac{1}{x} = 0. \tag{3}$$

We prove the latter and leave the former to Exercises 71 and 72.

EXAMPLE 6 Limits at Infinity for $f(x) = \dfrac{1}{x}$

Show that

(a) $\displaystyle\lim_{x\to\infty} \frac{1}{x} = 0$ **(b)** $\displaystyle\lim_{x\to-\infty} \frac{1}{x} = 0.$

Solution

(a) Let $\epsilon > 0$ be given. We must find a number M such that for all x

$$x > M \quad \Rightarrow \quad \left|\frac{1}{x} - 0\right| = \left|\frac{1}{x}\right| < \epsilon.$$

The implication will hold if $M = 1/\epsilon$ or any larger positive number (Figure 2.32). This proves $\lim_{x\to\infty}(1/x) = 0$.

(b) Let $\epsilon > 0$ be given. We must find a number N such that for all x

$$x < N \quad \Rightarrow \quad \left|\frac{1}{x} - 0\right| = \left|\frac{1}{x}\right| < \epsilon.$$

The implication will hold if $N = -1/\epsilon$ or any number less than $-1/\epsilon$ (Figure 2.32). This proves $\lim_{x\to-\infty}(1/x) = 0$. ■

Limits at infinity have properties similar to those of finite limits.

THEOREM 8 Limit Laws as $x \to \pm\infty$

If L, M, and k, are real numbers and

$$\lim_{x\to\pm\infty} f(x) = L \quad \text{and} \quad \lim_{x\to\pm\infty} g(x) = M, \quad \text{then}$$

1. *Sum Rule:* $\displaystyle\lim_{x\to\pm\infty}(f(x) + g(x)) = L + M$
2. *Difference Rule:* $\displaystyle\lim_{x\to\pm\infty}(f(x) - g(x)) = L - M$
3. *Product Rule:* $\displaystyle\lim_{x\to\pm\infty}(f(x)\cdot g(x)) = L\cdot M$
4. *Constant Multiple Rule:* $\displaystyle\lim_{x\to\pm\infty}(k\cdot f(x)) = k\cdot L$
5. *Quotient Rule:* $\displaystyle\lim_{x\to\pm\infty}\frac{f(x)}{g(x)} = \frac{L}{M}, \quad M \neq 0$
6. *Power Rule:* If r and s are integers with no common factors, $s \neq 0$, then

$$\lim_{x\to\pm\infty}(f(x))^{r/s} = L^{r/s}$$

provided that $L^{r/s}$ is a real number. (If s is even, we assume that $L > 0$.)

These properties are just like the properties in Theorem 1, Section 2.2, and we use them the same way.

EXAMPLE 7 Using Theorem 8

(a) $$\lim_{x\to\infty}\left(5+\frac{1}{x}\right) = \lim_{x\to\infty} 5 + \lim_{x\to\infty}\frac{1}{x} \qquad \text{Sum Rule}$$
$$= 5 + 0 = 5 \qquad \text{Known limits}$$

(b) $$\lim_{x\to-\infty}\frac{\pi\sqrt{3}}{x^2} = \lim_{x\to-\infty}\pi\sqrt{3}\cdot\frac{1}{x}\cdot\frac{1}{x}$$
$$= \lim_{x\to-\infty}\pi\sqrt{3}\cdot\lim_{x\to-\infty}\frac{1}{x}\cdot\lim_{x\to-\infty}\frac{1}{x} \qquad \text{Product rule}$$
$$= \pi\sqrt{3}\cdot 0\cdot 0 = 0 \qquad \text{Known limits}$$ ■

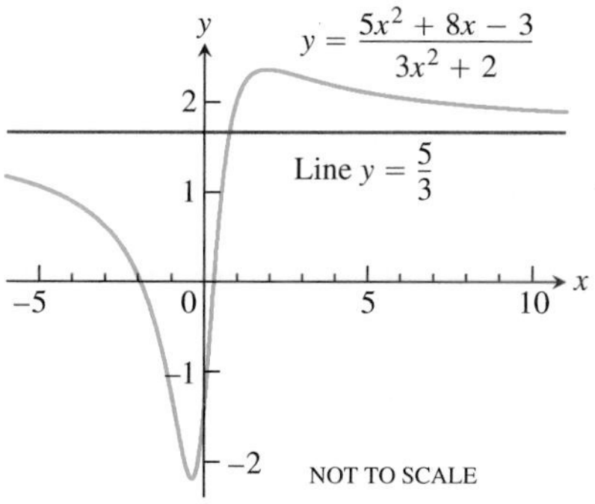

FIGURE 2.33 The graph of the function in Example 8. The graph approaches the line $y = 5/3$ as $|x|$ increases.

Limits at Infinity of Rational Functions

To determine the limit of a rational function as $x \to \pm\infty$, we can divide the numerator and denominator by the highest power of x in the denominator. What happens then depends on the degrees of the polynomials involved.

EXAMPLE 8 Numerator and Denominator of Same Degree

$$\lim_{x\to\infty}\frac{5x^2+8x-3}{3x^2+2} = \lim_{x\to\infty}\frac{5+(8/x)-(3/x^2)}{3+(2/x^2)} \qquad \text{Divide numerator and denominator by } x^2.$$
$$= \frac{5+0-0}{3+0} = \frac{5}{3} \qquad \text{See Fig. 2.33.}$$ ■

EXAMPLE 9 Degree of Numerator Less Than Degree of Denominator

$$\lim_{x\to-\infty}\frac{11x+2}{2x^3-1} = \lim_{x\to-\infty}\frac{(11/x^2)+(2/x^3)}{2-(1/x^3)} \qquad \text{Divide numerator and denominator by } x^3.$$
$$= \frac{0+0}{2-0} = 0 \qquad \text{See Fig. 2.34.}$$ ■

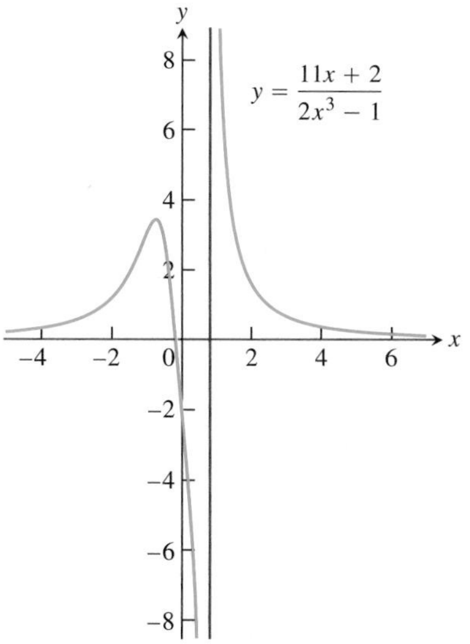

FIGURE 2.34 The graph of the function in Example 9. The graph approaches the x-axis as $|x|$ increases.

We give an example of the case when the degree of the numerator is greater than the degree of the denominator in the next section (Example 8, Section 2.5).

Horizontal Asymptotes

If the distance between the graph of a function and some fixed line approaches zero as a point on the graph moves increasingly far from the origin, we say that the graph approaches the line asymptotically and that the line is an *asymptote* of the graph.

Looking at $f(x) = 1/x$ (See Figure 2.31), we observe that the x-axis is an asymptote of the curve on the right because

$$\lim_{x\to\infty}\frac{1}{x} = 0$$

and on the left because

$$\lim_{x\to-\infty}\frac{1}{x} = 0.$$

We say that the x-axis is a *horizontal asymptote* of the graph of $f(x) = 1/x$.

DEFINITION Horizontal Asymptote

A line $y = b$ is a **horizontal asymptote** of the graph of a function $y = f(x)$ if either

$$\lim_{x\to\infty} f(x) = b \quad \text{or} \quad \lim_{x\to-\infty} f(x) = b.$$

The curve

$$f(x) = \frac{5x^2 + 8x - 3}{3x^2 + 2}$$

sketched in Figure 2.33 (Example 8) has the line $y = 5/3$ as a horizontal asymptote on both the right and the left because

$$\lim_{x\to\infty} f(x) = \frac{5}{3} \quad \text{and} \quad \lim_{x\to-\infty} f(x) = \frac{5}{3}.$$

EXAMPLE 10 Substituting a New Variable

Find $\lim_{x\to\infty} \sin(1/x)$.

Solution We introduce the new variable $t = 1/x$. From Example 6, we know that $t \to 0^+$ as $x \to \infty$ (see Figure 2.31). Therefore,

$$\lim_{x\to\infty} \sin\frac{1}{x} = \lim_{t\to 0^+} \sin t = 0.$$

■

The Sandwich Theorem Revisited

The Sandwich Theorem also holds for limits as $x \to \pm\infty$.

EXAMPLE 11 A Curve May Cross Its Horizontal Asymptote

Using the Sandwich Theorem, find the horizontal asymptote of the curve

$$y = 2 + \frac{\sin x}{x}.$$

Solution We are interested in the behavior as $x \to \pm\infty$. Since

$$0 \le \left|\frac{\sin x}{x}\right| \le \left|\frac{1}{x}\right|$$

and $\lim_{x\to\pm\infty} |1/x| = 0$, we have $\lim_{x\to\pm\infty} (\sin x)/x = 0$ by the Sandwich Theorem. Hence,

$$\lim_{x\to\pm\infty}\left(2 + \frac{\sin x}{x}\right) = 2 + 0 = 2,$$

and the line $y = 2$ is a horizontal asymptote of the curve on both left and right (Figure 2.35).

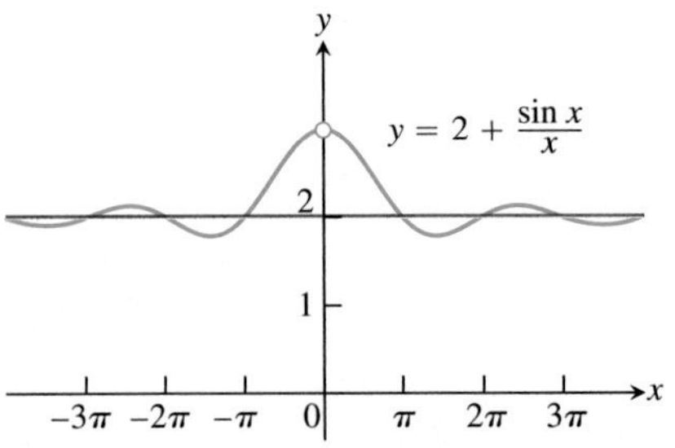

FIGURE 2.35 A curve may cross one of its asymptotes infinitely often (Example 11).

This example illustrates that a curve may cross one of its horizontal asymptotes, perhaps many times. ■

Oblique Asymptotes

If the degree of the numerator of a rational function is one greater than the degree of the denominator, the graph has an **oblique (slanted) asymptote**. We find an equation for the asymptote by dividing numerator by denominator to express f as a linear function plus a remainder that goes to zero as $x \to \pm\infty$. Here's an example.

EXAMPLE 12 Finding an Oblique Asymptote

Find the oblique asymptote for the graph of

$$f(x) = \frac{2x^2 - 3}{7x + 4}$$

in Figure 2.36.

Solution By long division, we find

$$f(x) = \frac{2x^2 - 3}{7x + 4}$$

$$= \underbrace{\left(\frac{2}{7}x - \frac{8}{49}\right)}_{\text{linear function } g(x)} + \underbrace{\frac{-115}{49(7x + 4)}}_{\text{remainder}}$$

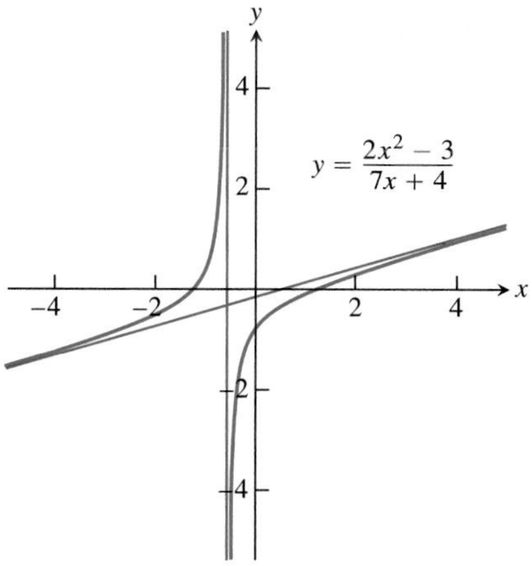

FIGURE 2.36 The function in Example 12 has an oblique asymptote.

As $x \to \pm\infty$, the remainder, whose magnitude gives the vertical distance between the graphs of f and g, goes to zero, making the (slanted) line

$$g(x) = \frac{2}{7}x - \frac{8}{49}$$

an asymptote of the graph of f (Figure 2.36). The line $y = g(x)$ is an asymptote both to the right and to the left. In the next section you will see that the function $f(x)$ grows arbitrarily large in absolute value as x approaches $-4/7$, where the denominator becomes zero (Figure 2.36). ■

EXERCISES 2.4

Finding Limits Graphically

1. Which of the following statements about the function $y = f(x)$ graphed here are true, and which are false?

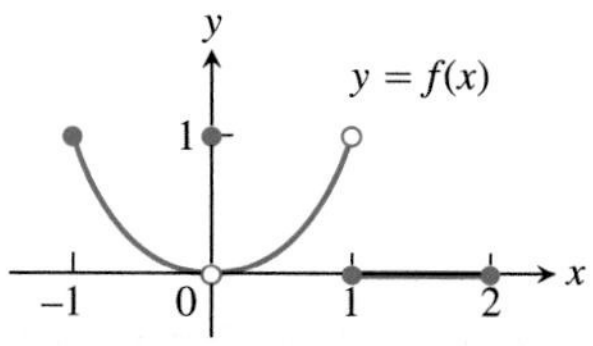

a. $\lim_{x \to -1^+} f(x) = 1$

b. $\lim_{x \to 0^-} f(x) = 0$

c. $\lim_{x \to 0^-} f(x) = 1$

d. $\lim_{x \to 0^-} f(x) = \lim_{x \to 0^+} f(x)$

e. $\lim_{x \to 0} f(x)$ exists

f. $\lim_{x \to 0} f(x) = 0$

g. $\lim_{x \to 0} f(x) = 1$

h. $\lim_{x \to 1} f(x) = 1$

i. $\lim_{x \to 1} f(x) = 0$

j. $\lim_{x \to 2^-} f(x) = 2$

k. $\lim_{x \to -1^-} f(x)$ does not exist.

l. $\lim_{x \to 2^+} f(x) = 0$

2. Which of the following statements about the function $y = f(x)$ graphed here are true, and which are false?

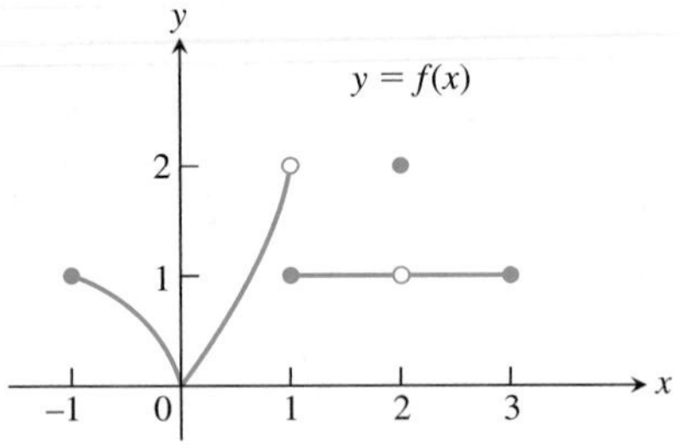

a. $\lim_{x \to -1^+} f(x) = 1$

b. $\lim_{x \to 2} f(x)$ does not exist.

c. $\lim_{x \to 2} f(x) = 2$

d. $\lim_{x \to 1^-} f(x) = 2$

e. $\lim_{x \to 1^+} f(x) = 1$

f. $\lim_{x \to 1} f(x)$ does not exist.

g. $\lim_{x \to 0^+} f(x) = \lim_{x \to 0^-} f(x)$

h. $\lim_{x \to c} f(x)$ exists at every c in the open interval $(-1, 1)$.

i. $\lim_{x \to c} f(x)$ exists at every c in the open interval $(1, 3)$.

j. $\lim_{x \to -1^-} f(x) = 0$

k. $\lim_{x \to 3^+} f(x)$ does not exist.

3. Let $f(x) = \begin{cases} 3 - x, & x < 2 \\ \dfrac{x}{2} + 1, & x > 2. \end{cases}$

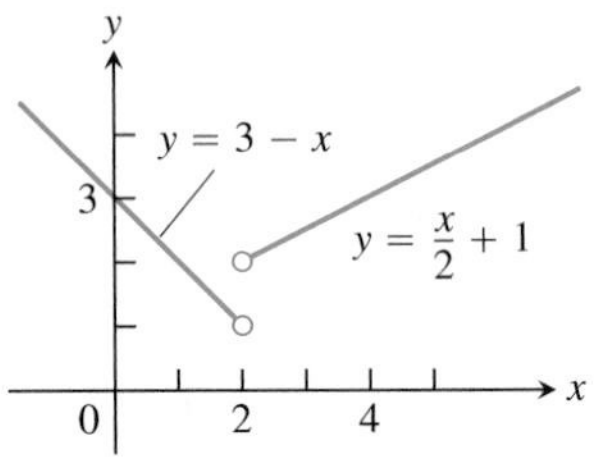

a. Find $\lim_{x \to 2^+} f(x)$ and $\lim_{x \to 2^-} f(x)$.

b. Does $\lim_{x \to 2} f(x)$ exist? If so, what is it? If not, why not?

c. Find $\lim_{x \to 4^-} f(x)$ and $\lim_{x \to 4^+} f(x)$.

d. Does $\lim_{x \to 4} f(x)$ exist? If so, what is it? If not, why not?

4. Let $f(x) = \begin{cases} 3 - x, & x < 2 \\ 2, & x = 2 \\ \dfrac{x}{2}, & x > 2. \end{cases}$

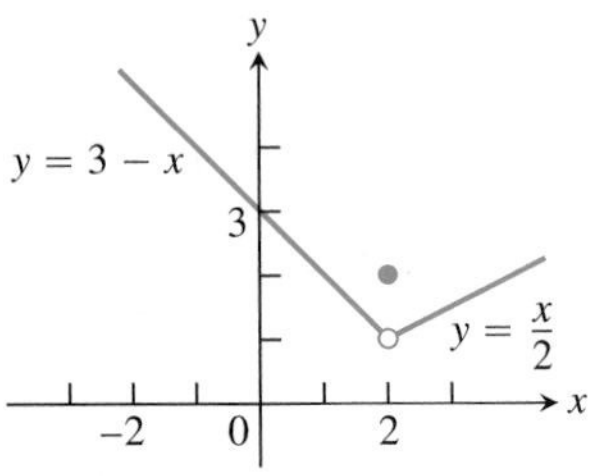

a. Find $\lim_{x \to 2^+} f(x)$, $\lim_{x \to 2^-} f(x)$, and $f(2)$.

b. Does $\lim_{x \to 2} f(x)$ exist? If so, what is it? If not, why not?

c. Find $\lim_{x \to -1^-} f(x)$ and $\lim_{x \to -1^+} f(x)$.

d. Does $\lim_{x \to -1} f(x)$ exist? If so, what is it? If not, why not?

5. Let $f(x) = \begin{cases} 0, & x \le 0 \\ \sin \dfrac{1}{x}, & x > 0. \end{cases}$

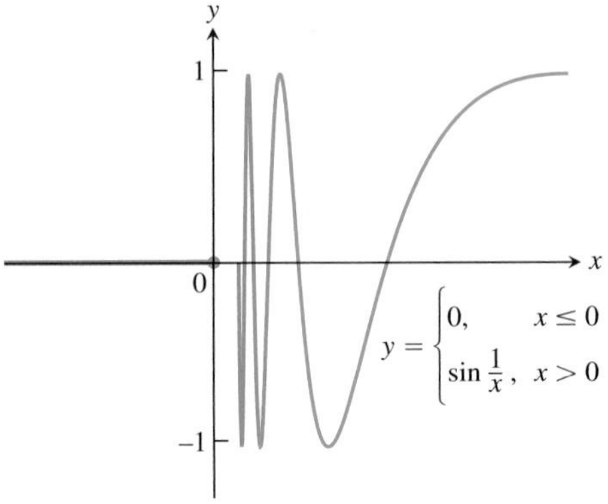

a. Does $\lim_{x \to 0^+} f(x)$ exist? If so, what is it? If not, why not?

b. Does $\lim_{x \to 0^-} f(x)$ exist? If so, what is it? If not, why not?

c. Does $\lim_{x \to 0} f(x)$ exist? If so, what is it? If not, why not?

6. Let $g(x) = \sqrt{x} \sin(1/x)$.

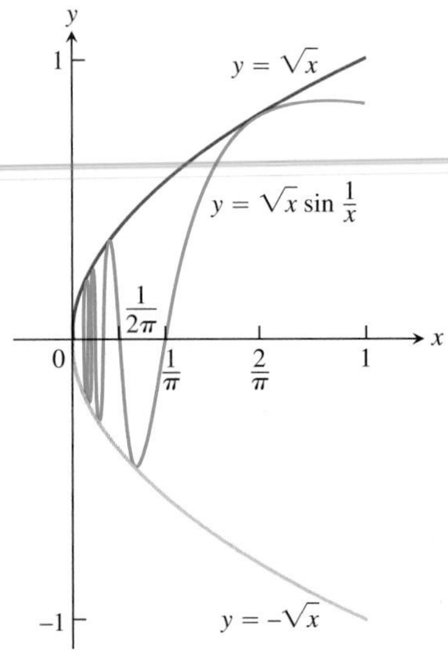

a. Does $\lim_{x \to 0^+} g(x)$ exist? If so, what is it? If not, why not?

b. Does $\lim_{x \to 0^-} g(x)$ exist? If so, what is it? If not, why not?

c. Does $\lim_{x \to 0} g(x)$ exist? If so, what is it? If not, why not?

7. a. Graph $f(x) = \begin{cases} x^3, & x \neq 1 \\ 0, & x = 1. \end{cases}$

b. Find $\lim_{x\to 1^-} f(x)$ and $\lim_{x\to 1^+} f(x)$.

c. Does $\lim_{x\to 1} f(x)$ exist? If so, what is it? If not, why not?

8. a. Graph $f(x) = \begin{cases} 1 - x^2, & x \neq 1 \\ 2, & x = 1. \end{cases}$

b. Find $\lim_{x\to 1^+} f(x)$ and $\lim_{x\to 1^-} f(x)$.

c. Does $\lim_{x\to 1} f(x)$ exist? If so, what is it? If not, why not?

Graph the functions in Exercises 9 and 10. Then answer these questions.

a. What are the domain and range of f?

b. At what points c, if any, does $\lim_{x\to c} f(x)$ exist?

c. At what points does only the left-hand limit exist?

d. At what points does only the right-hand limit exist?

9. $f(x) = \begin{cases} \sqrt{1 - x^2}, & 0 \le x < 1 \\ 1, & 1 \le x < 2 \\ 2, & x = 2 \end{cases}$

10. $f(x) = \begin{cases} x, & -1 \le x < 0, \quad \text{or} \quad 0 < x \le 1 \\ 1, & x = 0 \\ 0, & x < -1, \quad \text{or} \quad x > 1 \end{cases}$

Finding One-Sided Limits Algebraically

Find the limits in Exercises 11–18.

11. $\displaystyle\lim_{x\to -0.5^-} \sqrt{\frac{x+2}{x+1}}$

12. $\displaystyle\lim_{x\to 1^+} \sqrt{\frac{x-1}{x+2}}$

13. $\displaystyle\lim_{x\to -2^+} \left(\frac{x}{x+1}\right)\left(\frac{2x+5}{x^2+x}\right)$

14. $\displaystyle\lim_{x\to 1^-} \left(\frac{1}{x+1}\right)\left(\frac{x+6}{x}\right)\left(\frac{3-x}{7}\right)$

15. $\displaystyle\lim_{h\to 0^+} \frac{\sqrt{h^2+4h+5} - \sqrt{5}}{h}$

16. $\displaystyle\lim_{h\to 0^-} \frac{\sqrt{6} - \sqrt{5h^2+11h+6}}{h}$

17. a. $\displaystyle\lim_{x\to -2^+} (x+3)\frac{|x+2|}{x+2}$ **b.** $\displaystyle\lim_{x\to -2^-} (x+3)\frac{|x+2|}{x+2}$

18. a. $\displaystyle\lim_{x\to 1^+} \frac{\sqrt{2x}\,(x-1)}{|x-1|}$ **b.** $\displaystyle\lim_{x\to 1^-} \frac{\sqrt{2x}\,(x-1)}{|x-1|}$

Use the graph of the greatest integer function $y = \lfloor x \rfloor$ (sometimes written $y = \text{int}\, x$), Figure 1.31 in Section 1.3, to help you find the limits in Exercises 19 and 20.

19. a. $\displaystyle\lim_{\theta\to 3^+} \frac{\lfloor \theta \rfloor}{\theta}$ **b.** $\displaystyle\lim_{\theta\to 3^-} \frac{\lfloor \theta \rfloor}{\theta}$

20. a. $\displaystyle\lim_{t\to 4^+} (t - \lfloor t \rfloor)$ **b.** $\displaystyle\lim_{t\to 4^-} (t - \lfloor t \rfloor)$

Using $\displaystyle\lim_{\theta\to 0} \frac{\sin\theta}{\theta} = 1$

Find the limits in Exercises 21–36.

21. $\displaystyle\lim_{\theta\to 0} \frac{\sin\sqrt{2}\theta}{\sqrt{2}\theta}$

22. $\displaystyle\lim_{t\to 0} \frac{\sin kt}{t}$ (k constant)

23. $\displaystyle\lim_{y\to 0} \frac{\sin 3y}{4y}$

24. $\displaystyle\lim_{h\to 0^-} \frac{h}{\sin 3h}$

25. $\displaystyle\lim_{x\to 0} \frac{\tan 2x}{x}$

26. $\displaystyle\lim_{t\to 0} \frac{2t}{\tan t}$

27. $\displaystyle\lim_{x\to 0} \frac{x \csc 2x}{\cos 5x}$

28. $\displaystyle\lim_{x\to 0} 6x^2(\cot x)(\csc 2x)$

29. $\displaystyle\lim_{x\to 0} \frac{x + x\cos x}{\sin x \cos x}$

30. $\displaystyle\lim_{x\to 0} \frac{x^2 - x + \sin x}{2x}$

31. $\displaystyle\lim_{t\to 0} \frac{\sin(1-\cos t)}{1-\cos t}$

32. $\displaystyle\lim_{h\to 0} \frac{\sin(\sin h)}{\sin h}$

33. $\displaystyle\lim_{\theta\to 0} \frac{\sin\theta}{\sin 2\theta}$

34. $\displaystyle\lim_{x\to 0} \frac{\sin 5x}{\sin 4x}$

35. $\displaystyle\lim_{x\to 0} \frac{\tan 3x}{\sin 8x}$

36. $\displaystyle\lim_{y\to 0} \frac{\sin 3y \cot 5y}{y \cot 4y}$

Calculating Limits as $x \to \pm\infty$

In Exercises 37–42, find the limit of each function **(a)** as $x \to \infty$ and **(b)** as $x \to -\infty$. (You may wish to visualize your answer with a graphing calculator or computer.)

37. $f(x) = \dfrac{2}{x} - 3$

38. $f(x) = \pi - \dfrac{2}{x^2}$

39. $g(x) = \dfrac{1}{2 + (1/x)}$

40. $g(x) = \dfrac{1}{8 - (5/x^2)}$

41. $h(x) = \dfrac{-5 + (7/x)}{3 - (1/x^2)}$

42. $h(x) = \dfrac{3 - (2/x)}{4 + \left(\sqrt{2}/x^2\right)}$

Find the limits in Exercises 43–46.

43. $\displaystyle\lim_{x\to\infty} \frac{\sin 2x}{x}$

44. $\displaystyle\lim_{\theta\to -\infty} \frac{\cos\theta}{3\theta}$

45. $\displaystyle\lim_{t\to -\infty} \frac{2 - t + \sin t}{t + \cos t}$

46. $\displaystyle\lim_{r\to\infty} \frac{r + \sin r}{2r + 7 - 5\sin r}$

Limits of Rational Functions

In Exercises 47–56, find the limit of each rational function **(a)** as $x \to \infty$ and **(b)** as $x \to -\infty$.

47. $f(x) = \dfrac{2x+3}{5x+7}$

48. $f(x) = \dfrac{2x^3+7}{x^3 - x^2 + x + 7}$

49. $f(x) = \dfrac{x+1}{x^2+3}$

50. $f(x) = \dfrac{3x+7}{x^2-2}$

51. $h(x) = \dfrac{7x^3}{x^3 - 3x^2 + 6x}$

52. $g(x) = \dfrac{1}{x^3 - 4x + 1}$

53. $g(x) = \dfrac{10x^5 + x^4 + 31}{x^6}$

54. $h(x) = \dfrac{9x^4 + x}{2x^4 + 5x^2 - x + 6}$

55. $h(x) = \dfrac{-2x^3 - 2x + 3}{3x^3 + 3x^2 - 5x}$

56. $h(x) = \dfrac{-x^4}{x^4 - 7x^3 + 7x^2 + 9}$

Limits with Noninteger or Negative Powers

The process by which we determine limits of rational functions applies equally well to ratios containing noninteger or negative powers of x: divide numerator and denominator by the highest power of x in the denominator and proceed from there. Find the limits in Exercises 57–62.

57. $\lim\limits_{x\to\infty} \dfrac{2\sqrt{x} + x^{-1}}{3x - 7}$

58. $\lim\limits_{x\to\infty} \dfrac{2 + \sqrt{x}}{2 - \sqrt{x}}$

59. $\lim\limits_{x\to-\infty} \dfrac{\sqrt[3]{x} - \sqrt[5]{x}}{\sqrt[3]{x} + \sqrt[5]{x}}$

60. $\lim\limits_{x\to\infty} \dfrac{x^{-1} + x^{-4}}{x^{-2} - x^{-3}}$

61. $\lim\limits_{x\to\infty} \dfrac{2x^{5/3} - x^{1/3} + 7}{x^{8/5} + 3x + \sqrt{x}}$

62. $\lim\limits_{x\to-\infty} \dfrac{\sqrt[3]{x} - 5x + 3}{2x + x^{2/3} - 4}$

Theory and Examples

63. Once you know $\lim_{x\to a^+} f(x)$ and $\lim_{x\to a^-} f(x)$ at an interior point of the domain of f, do you then know $\lim_{x\to a} f(x)$? Give reasons for your answer.

64. If you know that $\lim_{x\to c} f(x)$ exists, can you find its value by calculating $\lim_{x\to c^+} f(x)$? Give reasons for your answer.

65. Suppose that f is an odd function of x. Does knowing that $\lim_{x\to 0^+} f(x) = 3$ tell you anything about $\lim_{x\to 0^-} f(x)$? Give reasons for your answer.

66. Suppose that f is an even function of x. Does knowing that $\lim_{x\to 2^-} f(x) = 7$ tell you anything about either $\lim_{x\to -2^-} f(x)$ or $\lim_{x\to -2^+} f(x)$? Give reasons for your answer.

67. Suppose that $f(x)$ and $g(x)$ are polynomials in x and that $\lim_{x\to\infty} (f(x)/g(x)) = 2$. Can you conclude anything about $\lim_{x\to-\infty} (f(x)/g(x))$? Give reasons for your answer.

68. Suppose that $f(x)$ and $g(x)$ are polynomials in x. Can the graph of $f(x)/g(x)$ have an asymptote if $g(x)$ is never zero? Give reasons for your answer.

69. How many horizontal asymptotes can the graph of a given rational function have? Give reasons for your answer.

70. Find $\lim\limits_{x\to\infty} \left(\sqrt{x^2 + x} - \sqrt{x^2 - x}\right)$.

Use the formal definitions of limits as $x \to \pm\infty$ to establish the limits in Exercises 71 and 72.

71. If f has the constant value $f(x) = k$, then $\lim\limits_{x\to\infty} f(x) = k$.

72. If f has the constant value $f(x) = k$, then $\lim\limits_{x\to-\infty} f(x) = k$.

Formal Definitions of One-Sided Limits

73. Given $\epsilon > 0$, find an interval $I = (5, 5 + \delta)$, $\delta > 0$, such that if x lies in I, then $\sqrt{x - 5} < \epsilon$. What limit is being verified and what is its value?

74. Given $\epsilon > 0$, find an interval $I = (4 - \delta, 4)$, $\delta > 0$, such that if x lies in I, then $\sqrt{4 - x} < \epsilon$. What limit is being verified and what is its value?

Use the definitions of right-hand and left-hand limits to prove the limit statements in Exercises 75 and 76.

75. $\lim\limits_{x\to 0^-} \dfrac{x}{|x|} = -1$

76. $\lim\limits_{x\to 2^+} \dfrac{x - 2}{|x - 2|} = 1$

77. **Greatest integer function** Find **(a)** $\lim_{x\to 400^+} \lfloor x \rfloor$ and **(b)** $\lim_{x\to 400^-} \lfloor x \rfloor$; then use limit definitions to verify your findings. **(c)** Based on your conclusions in parts (a) and (b), can anything be said about $\lim_{x\to 400} \lfloor x \rfloor$? Give reasons for your answers.

78. **One-sided limits** Let $f(x) = \begin{cases} x^2 \sin(1/x), & x < 0 \\ \sqrt{x}, & x > 0. \end{cases}$

Find **(a)** $\lim_{x\to 0^+} f(x)$ and **(b)** $\lim_{x\to 0^-} f(x)$; then use limit definitions to verify your findings. **(c)** Based on your conclusions in parts (a) and (b), can anything be said about $\lim_{x\to 0} f(x)$? Give reasons for your answer.

Grapher Explorations—"Seeing" Limits at Infinity

Sometimes a change of variable can change an unfamiliar expression into one whose limit we know how to find. For example,

$$\lim_{x\to\infty} \sin\frac{1}{x} = \lim_{\theta\to 0^+} \sin\theta \qquad \text{Substitute } \theta = 1/x$$
$$= 0.$$

This suggests a creative way to "see" limits at infinity. Describe the procedure and use it to picture and determine limits in Exercises 79–84.

79. $\lim\limits_{x\to\pm\infty} x\sin\dfrac{1}{x}$

80. $\lim\limits_{x\to-\infty} \dfrac{\cos(1/x)}{1 + (1/x)}$

81. $\lim\limits_{x\to\pm\infty} \dfrac{3x + 4}{2x - 5}$

82. $\lim\limits_{x\to\infty} \left(\dfrac{1}{x}\right)^{1/x}$

83. $\lim\limits_{x\to\pm\infty} \left(3 + \dfrac{2}{x}\right)\left(\cos\dfrac{1}{x}\right)$

84. $\lim\limits_{x\to\infty} \left(\dfrac{3}{x^2} - \cos\dfrac{1}{x}\right)\left(1 + \sin\dfrac{1}{x}\right)$

2.5 Infinite Limits and Vertical Asymptotes

In this section we extend the concept of limit to *infinite limits*, which are not limits as before, but rather an entirely new use of the term limit. Infinite limits provide useful symbols and language for describing the behavior of functions whose values become arbitrarily large, positive or negative. We continue our analysis of graphs of rational functions from the last section, using vertical asymptotes and dominant terms for numerically large values of x.

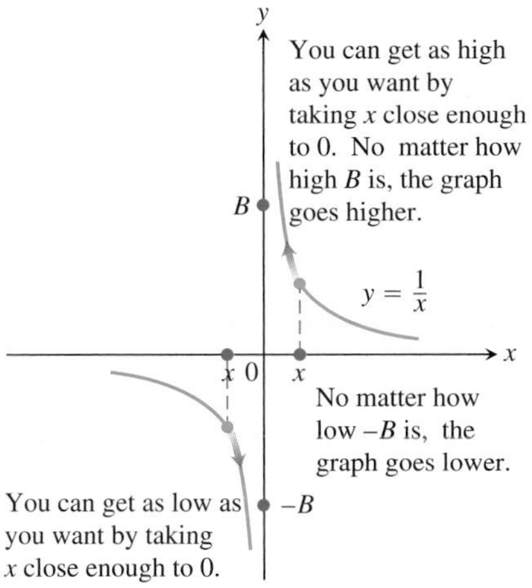

FIGURE 2.37 One-sided infinite limits:

$$\lim_{x\to 0^+}\frac{1}{x}=\infty \quad\text{and}\quad \lim_{x\to 0^-}\frac{1}{x}=-\infty$$

Infinite Limits

Let us look again at the function $f(x) = 1/x$. As $x \to 0^+$, the values of f grow without bound, eventually reaching and surpassing every positive real number. That is, given any positive real number B, however large, the values of f become larger still (Figure 2.37). Thus, f has no limit as $x \to 0^+$. It is nevertheless convenient to describe the behavior of f by saying that $f(x)$ approaches ∞ as $x \to 0^+$. We write

$$\lim_{x\to 0^+} f(x) = \lim_{x\to 0^+}\frac{1}{x} = \infty .$$

In writing this, we are *not* saying that the limit exists. Nor are we saying that there is a real number ∞, for there is no such number. Rather, we are saying that $\lim_{x\to 0^+}(1/x)$ *does not exist because* $1/x$ *becomes arbitrarily large and positive as* $x \to 0^+$.

As $x \to 0^-$, the values of $f(x) = 1/x$ become arbitrarily large and negative. Given any negative real number $-B$, the values of f eventually lie below $-B$. (See Figure 2.37.) We write

$$\lim_{x\to 0^-} f(x) = \lim_{x\to 0^-}\frac{1}{x} = -\infty .$$

Again, we are not saying that the limit exists and equals the number $-\infty$. There *is* no real number $-\infty$. We are describing the behavior of a function whose limit as $x \to 0^-$ *does not exist because its values become arbitrarily large and negative.*

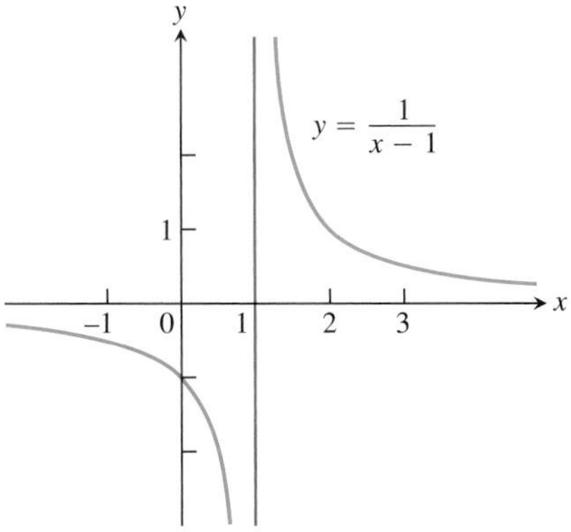

FIGURE 2.38 Near $x = 1$, the function $y = 1/(x-1)$ behaves the way the function $y = 1/x$ behaves near $x = 0$. Its graph is the graph of $y = 1/x$ shifted 1 unit to the right (Example 1).

EXAMPLE 1 One-Sided Infinite Limits

Find $\displaystyle\lim_{x\to 1^+}\frac{1}{x-1}$ and $\displaystyle\lim_{x\to 1^-}\frac{1}{x-1}$.

Geometric Solution The graph of $y = 1/(x-1)$ is the graph of $y = 1/x$ shifted 1 unit to the right (Figure 2.38). Therefore, $y = 1/(x-1)$ behaves near 1 exactly the way $y = 1/x$ behaves near 0:

$$\lim_{x\to 1^+}\frac{1}{x-1} = \infty \quad\text{and}\quad \lim_{x\to 1^-}\frac{1}{x-1} = -\infty .$$

Analytic Solution Think about the number $x-1$ and its reciprocal. As $x \to 1^+$, we have $(x-1) \to 0^+$ and $1/(x-1) \to \infty$. As $x \to 1^-$, we have $(x-1) \to 0^-$ and $1/(x-1) \to -\infty$. ■

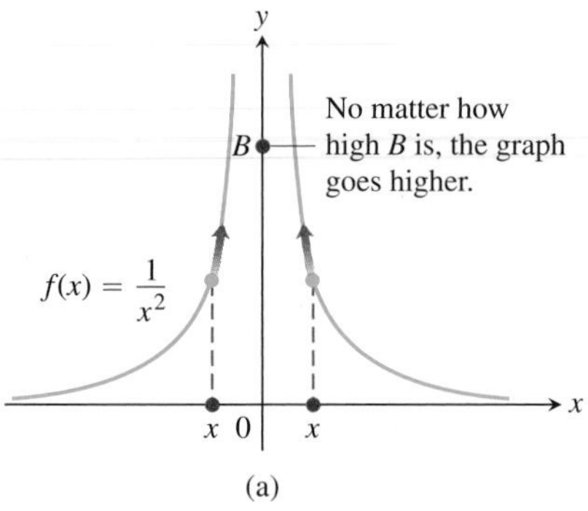

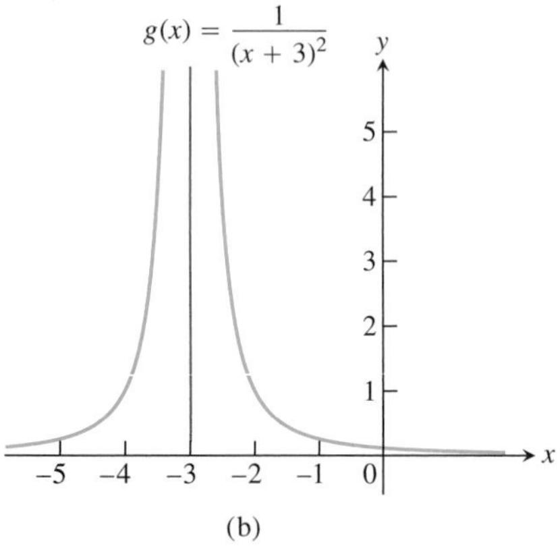

FIGURE 2.39 The graphs of the functions in Example 2. (a) $f(x)$ approaches infinity as $x \to 0$. (b) $g(x)$ approaches infinity as $x \to -3$.

EXAMPLE 2 Two-Sided Infinite Limits

Discuss the behavior of

(a) $f(x) = \dfrac{1}{x^2}$ near $x = 0$,

(b) $g(x) = \dfrac{1}{(x+3)^2}$ near $x = -3$.

Solution

(a) As x approaches zero from either side, the values of $1/x^2$ are positive and become arbitrarily large (Figure 2.39a):

$$\lim_{x \to 0} f(x) = \lim_{x \to 0} \frac{1}{x^2} = \infty.$$

(b) The graph of $g(x) = 1/(x+3)^2$ is the graph of $f(x) = 1/x^2$ shifted 3 units to the left (Figure 2.39b). Therefore, g behaves near -3 exactly the way f behaves near 0.

$$\lim_{x \to -3} g(x) = \lim_{x \to -3} \frac{1}{(x+3)^2} = \infty.$$ ■

The function $y = 1/x$ shows no consistent behavior as $x \to 0$. We have $1/x \to \infty$ if $x \to 0^+$, but $1/x \to -\infty$ if $x \to 0^-$. All we can say about $\lim_{x \to 0}(1/x)$ is that it does not exist. The function $y = 1/x^2$ is different. Its values approach infinity as x approaches zero from either side, so we can say that $\lim_{x \to 0}(1/x^2) = \infty$.

EXAMPLE 3 Rational Functions Can Behave in Various Ways Near Zeros of Their Denominators

(a) $$\lim_{x \to 2} \frac{(x-2)^2}{x^2-4} = \lim_{x \to 2} \frac{(x-2)^2}{(x-2)(x+2)} = \lim_{x \to 2} \frac{x-2}{x+2} = 0$$

(b) $$\lim_{x \to 2} \frac{x-2}{x^2-4} = \lim_{x \to 2} \frac{x-2}{(x-2)(x+2)} = \lim_{x \to 2} \frac{1}{x+2} = \frac{1}{4}$$

(c) $$\lim_{x \to 2^+} \frac{x-3}{x^2-4} = \lim_{x \to 2^+} \frac{x-3}{(x-2)(x+2)} = -\infty$$ The values are negative for $x > 2$, x near 2.

(d) $$\lim_{x \to 2^-} \frac{x-3}{x^2-4} = \lim_{x \to 2^-} \frac{x-3}{(x-2)(x+2)} = \infty$$ The values are positive for $x < 2$, x near 2.

(e) $$\lim_{x \to 2} \frac{x-3}{x^2-4} = \lim_{x \to 2} \frac{x-3}{(x-2)(x+2)} \text{ does not exist.}$$ See parts (c) and (d).

(f) $$\lim_{x \to 2} \frac{2-x}{(x-2)^3} = \lim_{x \to 2} \frac{-(x-2)}{(x-2)^3} = \lim_{x \to 2} \frac{-1}{(x-2)^2} = -\infty$$

In parts (a) and (b) the effect of the zero in the denominator at $x = 2$ is canceled because the numerator is zero there also. Thus a finite limit exists. This is not true in part (f), where cancellation still leaves a zero in the denominator. ■

Precise Definitions of Infinite Limits

Instead of requiring $f(x)$ to lie arbitrarily close to a finite number L for all x sufficiently close to x_0, the definitions of infinite limits require $f(x)$ to lie arbitrarily far from the origin. Except for this change, the language is identical with what we have seen before. Figures 2.40 and 2.41 accompany these definitions.

DEFINITIONS Infinity, Negative Infinity as Limits

1. We say that **$f(x)$ approaches infinity as x approaches x_0**, and write
$$\lim_{x \to x_0} f(x) = \infty,$$
if for every positive real number B there exists a corresponding $\delta > 0$ such that for all x
$$0 < |x - x_0| < \delta \quad \Rightarrow \quad f(x) > B.$$

2. We say that **$f(x)$ approaches negative infinity as x approaches x_0**, and write
$$\lim_{x \to x_0} f(x) = -\infty,$$
if for every negative real number $-B$ there exists a corresponding $\delta > 0$ such that for all x
$$0 < |x - x_0| < \delta \quad \Rightarrow \quad f(x) < -B.$$

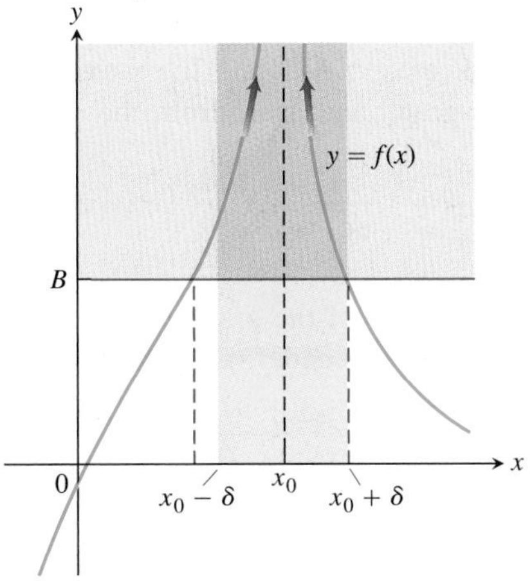

FIGURE 2.40 For $x_0 - \delta < x < x_0 + \delta$, the graph of $f(x)$ lies above the line $y = B$.

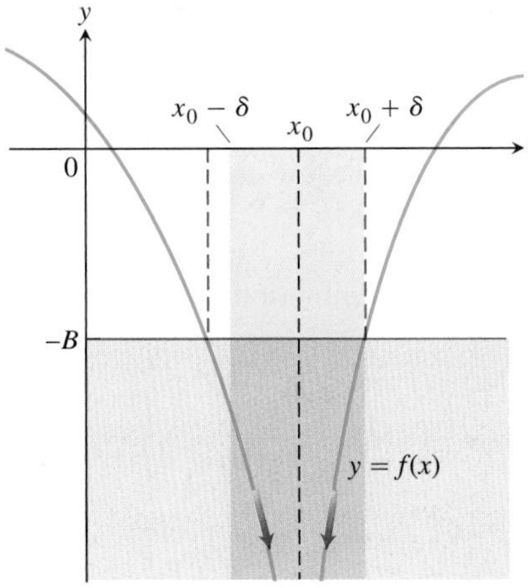

FIGURE 2.41 For $x_0 - \delta < x < x_0 + \delta$, the graph of $f(x)$ lies below the line $y = -B$.

The precise definitions of one-sided infinite limits at x_0 are similar and are stated in the exercises.

EXAMPLE 4 Using the Definition of Infinite Limits

Prove that $\lim_{x \to 0} \frac{1}{x^2} = \infty$.

Solution Given $B > 0$, we want to find $\delta > 0$ such that
$$0 < |x - 0| < \delta \quad \text{implies} \quad \frac{1}{x^2} > B.$$

Now,
$$\frac{1}{x^2} > B \qquad \text{if and only if } x^2 < \frac{1}{B}$$

or, equivalently,
$$|x| < \frac{1}{\sqrt{B}}.$$

Thus, choosing $\delta = 1/\sqrt{B}$ (or any smaller positive number), we see that
$$|x| < \delta \quad \text{implies} \quad \frac{1}{x^2} > \frac{1}{\delta^2} \geq B.$$

Therefore, by definition,
$$\lim_{x \to 0} \frac{1}{x^2} = \infty.$$
■

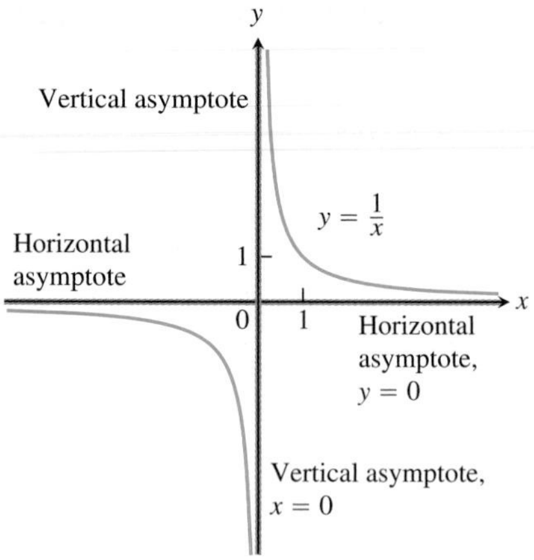

FIGURE 2.42 The coordinate axes are asymptotes of both branches of the hyperbola $y = 1/x$.

Vertical Asymptotes

Notice that the distance between a point on the graph of $y = 1/x$ and the y-axis approaches zero as the point moves vertically along the graph and away from the origin (Figure 2.42). This behavior occurs because

$$\lim_{x \to 0^+} \frac{1}{x} = \infty \quad \text{and} \quad \lim_{x \to 0^-} \frac{1}{x} = -\infty.$$

We say that the line $x = 0$ (the y-axis) is a *vertical asymptote* of the graph of $y = 1/x$. Observe that the denominator is zero at $x = 0$ and the function is undefined there.

DEFINITION Vertical Asymptote

A line $x = a$ is a **vertical asymptote** of the graph of a function $y = f(x)$ if either

$$\lim_{x \to a^+} f(x) = \pm\infty \quad \text{or} \quad \lim_{x \to a^-} f(x) = \pm\infty.$$

EXAMPLE 5 Looking for Asymptotes

Find the horizontal and vertical asymptotes of the curve

$$y = \frac{x + 3}{x + 2}.$$

Solution We are interested in the behavior as $x \to \pm\infty$ and as $x \to -2$, where the denominator is zero.

The asymptotes are quickly revealed if we recast the rational function as a polynomial with a remainder, by dividing $(x + 2)$ into $(x + 3)$.

$$\begin{array}{r} 1 \\ x + 2\overline{)x + 3} \\ \underline{x + 2} \\ 1 \end{array}$$

This result enables us to rewrite y:

$$y = 1 + \frac{1}{x + 2}.$$

We now see that the curve in question is the graph of $y = 1/x$ shifted 1 unit up and 2 units left (Figure 2.43). The asymptotes, instead of being the coordinate axes, are now the lines $y = 1$ and $x = -2$. ■

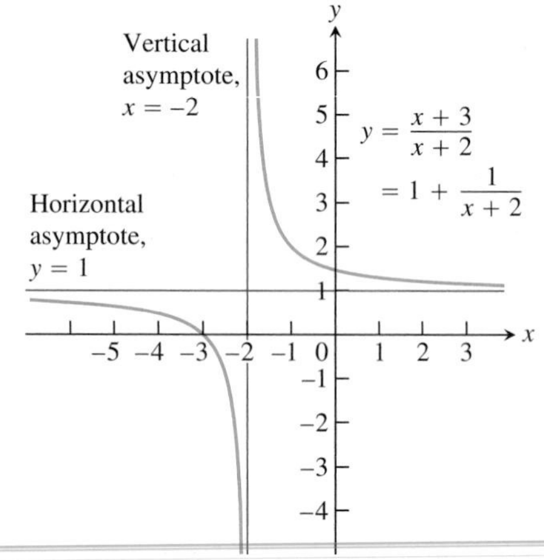

FIGURE 2.43 The lines $y = 1$ and $x = -2$ are asymptotes of the curve $y = (x + 3)/(x + 2)$ (Example 5).

EXAMPLE 6 Asymptotes Need Not Be Two-Sided

Find the horizontal and vertical asymptotes of the graph of

$$f(x) = -\frac{8}{x^2 - 4}.$$

Solution We are interested in the behavior as $x \to \pm\infty$ and as $x \to \pm 2$, where the denominator is zero. Notice that f is an even function of x, so its graph is symmetric with respect to the y-axis.

(a) *The behavior as* $x \to \pm\infty$. Since $\lim_{x \to \infty} f(x) = 0$, the line $y = 0$ is a horizontal asymptote of the graph to the right. By symmetry it is an asymptote to the left as well

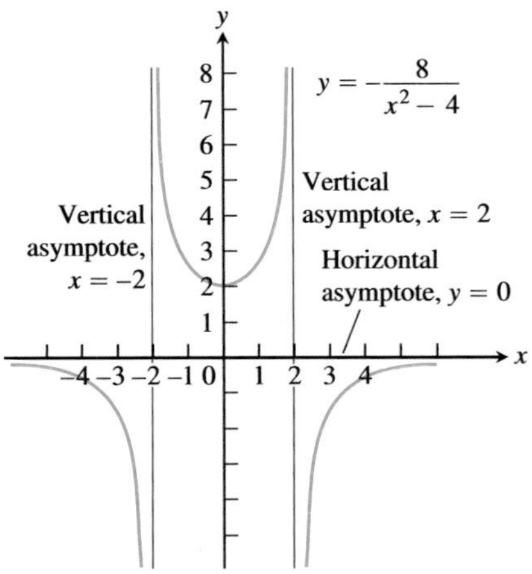

FIGURE 2.44 Graph of $y = -8/(x^2 - 4)$. Notice that the curve approaches the x-axis from only one side. Asymptotes do not have to be two-sided (Example 6).

(Figure 2.44). Notice that the curve approaches the x-axis from only the negative side (or from below).

(b) *The behavior as* $x \rightarrow \pm 2$. Since

$$\lim_{x\to 2^+} f(x) = -\infty \qquad \text{and} \qquad \lim_{x\to 2^-} f(x) = \infty,$$

the line $x = 2$ is a vertical asymptote both from the right and from the left. By symmetry, the same holds for the line $x = -2$.

There are no other asymptotes because f has a finite limit at every other point. ■

EXAMPLE 7 Curves with Infinitely Many Asymptotes

The curves

$$y = \sec x = \frac{1}{\cos x} \qquad \text{and} \qquad y = \tan x = \frac{\sin x}{\cos x}$$

both have vertical asymptotes at odd-integer multiples of $\pi/2$, where $\cos x = 0$ (Figure 2.45).

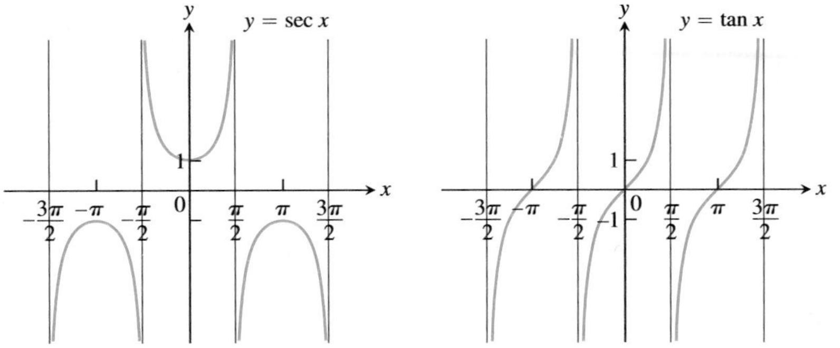

FIGURE 2.45 The graphs of $\sec x$ and $\tan x$ have infinitely many vertical asymptotes (Example 7).

The graphs of

$$y = \csc x = \frac{1}{\sin x} \qquad \text{and} \qquad y = \cot x = \frac{\cos x}{\sin x}$$

have vertical asymptotes at integer multiples of π, where $\sin x = 0$ (Figure 2.46).

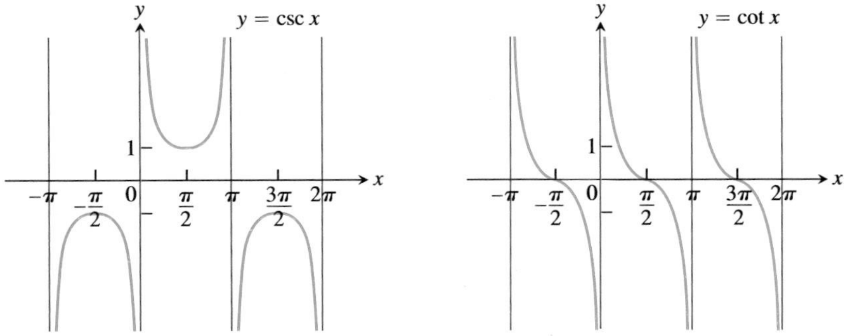

FIGURE 2.46 The graphs of $\csc x$ and $\cot x$ (Example 7). ■

EXAMPLE 8 A Rational Function with Degree of Numerator Greater than Degree of Denominator

Find the asymptotes of the graph of

$$f(x) = \frac{x^2 - 3}{2x - 4}.$$

Solution We are interested in the behavior as $x \to \pm\infty$ and also as $x \to 2$, where the denominator is zero. We divide $(2x - 4)$ into $(x^2 - 3)$:

$$\begin{array}{r} \frac{x}{2} + 1 \\ 2x - 4\overline{)x^2 - 3} \\ \underline{x^2 - 2x} \\ 2x - 3 \\ \underline{2x - 4} \\ 1 \end{array}$$

This tells us that

$$f(x) = \frac{x^2 - 3}{2x - 4} = \underbrace{\frac{x}{2} + 1}_{\text{linear}} + \underbrace{\frac{1}{2x - 4}}_{\text{remainder}}.$$

Since $\lim_{x\to 2^+} f(x) = \infty$ and $\lim_{x\to 2^-} f(x) = -\infty$, the line $x = 2$ is a two-sided vertical asymptote. As $x \to \pm\infty$, the remainder approaches 0 and $f(x) \to (x/2) + 1$. The line $y = (x/2) + 1$ is an oblique asymptote both to the right and to the left (Figure 2.47). ■

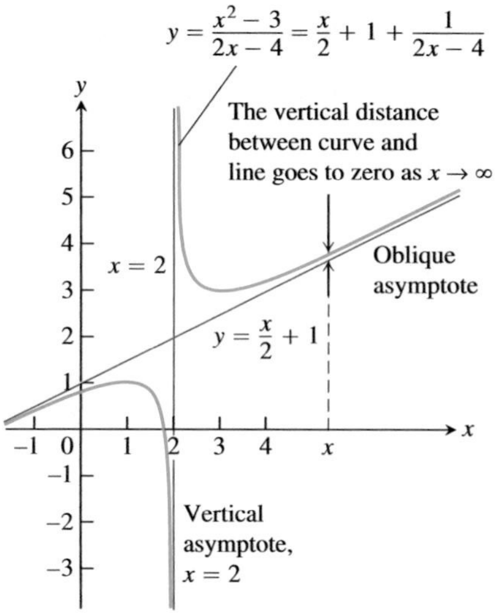

FIGURE 2.47 The graph of $f(x) = (x^2 - 3)/(2x - 4)$ has a vertical asymptote and an oblique asymptote (Example 8).

Notice in Example 8, that if the degree of the numerator in a rational function is greater than the degree of the denominator, then the limit is $+\infty$ or $-\infty$, depending on the signs assumed by the numerator and denominator as $|x|$ becomes large.

Dominant Terms

Of all the observations we can make quickly about the function

$$f(x) = \frac{x^2 - 3}{2x - 4}$$

in Example 8, probably the most useful is that

$$f(x) = \frac{x}{2} + 1 + \frac{1}{2x - 4}.$$

This tells us immediately that

$$f(x) \approx \frac{x}{2} + 1 \qquad \text{For } x \text{ numerically large}$$

$$f(x) \approx \frac{1}{2x - 4} \qquad \text{For } x \text{ near 2}$$

If we want to know how f behaves, this is the way to find out. It behaves like $y = (x/2) + 1$ when x is numerically large and the contribution of $1/(2x - 4)$ to the total value of f is insignificant. It behaves like $1/(2x - 4)$ when x is so close to 2 that $1/(2x - 4)$ makes the dominant contribution.

We say that $(x/2) + 1$ **dominates** when x is numerically large, and we say that $1/(2x - 4)$ dominates when x is near 2. **Dominant terms** like these are the key to predicting a function's behavior. Here's another example.

EXAMPLE 9 Two Graphs Appearing Identical on a Large Scale

Let $f(x) = 3x^4 - 2x^3 + 3x^2 - 5x + 6$ and $g(x) = 3x^4$. Show that although f and g are quite different for numerically small values of x, they are virtually identical for $|x|$ very large.

Solution The graphs of f and g behave quite differently near the origin (Figure 2.48a), but appear as virtually identical on a larger scale (Figure 2.48b).

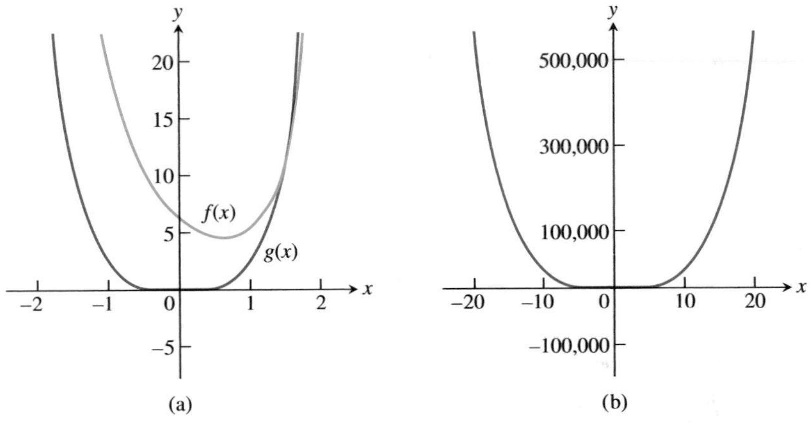

FIGURE 2.48 The graphs of f and g, (a) are distinct for $|x|$ small, and (b) nearly identical for $|x|$ large (Example 9).

We can test that the term $3x^4$ in f, represented graphically by g, dominates the polynomial f for numerically large values of x by examining the ratio of the two functions as $x \to \pm\infty$. We find that

$$\begin{aligned}\lim_{x\to\pm\infty} \frac{f(x)}{g(x)} &= \lim_{x\to\pm\infty} \frac{3x^4 - 2x^3 + 3x^2 - 5x + 6}{3x^4} \\ &= \lim_{x\to\pm\infty} \left(1 - \frac{2}{3x} + \frac{1}{x^2} - \frac{5}{3x^3} + \frac{2}{x^4}\right) \\ &= 1,\end{aligned}$$

so that f and g are nearly identical for $|x|$ large. ■

EXERCISES 2.5

Infinite Limits

Find the limits in Exercises 1–12.

1. $\lim_{x\to 0^+} \frac{1}{3x}$

2. $\lim_{x\to 0^-} \frac{5}{2x}$

3. $\lim_{x\to 2^-} \frac{3}{x-2}$

4. $\lim_{x\to 3^+} \frac{1}{x-3}$

5. $\lim_{x\to -8^+} \frac{2x}{x+8}$

6. $\lim_{x\to -5^-} \frac{3x}{2x+10}$

7. $\lim_{x\to 7} \frac{4}{(x-7)^2}$

8. $\lim_{x\to 0} \frac{-1}{x^2(x+1)}$

9. **a.** $\lim_{x\to 0^+} \frac{2}{3x^{1/3}}$ **b.** $\lim_{x\to 0^-} \frac{2}{3x^{1/3}}$

10. **a.** $\lim_{x\to 0^+} \frac{2}{x^{1/5}}$ **b.** $\lim_{x\to 0^-} \frac{2}{x^{1/5}}$

11. $\lim_{x\to 0} \frac{4}{x^{2/5}}$

12. $\lim_{x\to 0} \frac{1}{x^{2/3}}$

Find the limits in Exercises 13–16.

13. $\lim_{x\to(\pi/2)^-} \tan x$

14. $\lim_{x\to(-\pi/2)^+} \sec x$

15. $\lim_{\theta\to 0^-} (1 + \csc\theta)$

16. $\lim_{\theta\to 0} (2 - \cot\theta)$

Additional Calculations

Find the limits in Exercises 17–22.

17. $\lim \frac{1}{x^2-4}$ as

a. $x\to 2^+$ **b.** $x\to 2^-$

c. $x\to -2^+$ **d.** $x\to -2^-$

18. $\lim \frac{x}{x^2-1}$ as

a. $x\to 1^+$ **b.** $x\to 1^-$

c. $x\to -1^+$ **d.** $x\to -1^-$

19. $\lim \left(\frac{x^2}{2} - \frac{1}{x}\right)$ as

a. $x\to 0^+$ **b.** $x\to 0^-$

c. $x\to \sqrt[3]{2}$ **d.** $x\to -1$

20. $\lim \frac{x^2-1}{2x+4}$ as

a. $x\to -2^+$ **b.** $x\to -2^-$

c. $x\to 1^+$ **d.** $x\to 0^-$

21. $\lim \frac{x^2-3x+2}{x^3-2x^2}$ as

a. $x\to 0^+$ **b.** $x\to 2^+$

c. $x\to 2^-$ **d.** $x\to 2$

e. What, if anything, can be said about the limit as $x\to 0$?

22. $\lim \frac{x^2-3x+2}{x^3-4x}$ as

a. $x\to 2^+$ **b.** $x\to -2^+$

c. $x\to 0^-$ **d.** $x\to 1^+$

e. What, if anything, can be said about the limit as $x\to 0$?

Find the limits in Exercises 23–26.

23. $\lim \left(2 - \frac{3}{t^{1/3}}\right)$ as

a. $t\to 0^+$ **b.** $t\to 0^-$

24. $\lim \left(\frac{1}{t^{3/5}} + 7\right)$ as

a. $t\to 0^+$ **b.** $t\to 0^-$

25. $\lim \left(\frac{1}{x^{2/3}} + \frac{2}{(x-1)^{2/3}}\right)$ as

a. $x\to 0^+$ **b.** $x\to 0^-$

c. $x\to 1^+$ **d.** $x\to 1^-$

26. $\lim \left(\frac{1}{x^{1/3}} - \frac{1}{(x-1)^{4/3}}\right)$ as

a. $x\to 0^+$ **b.** $x\to 0^-$

c. $x\to 1^+$ **d.** $x\to 1^-$

Graphing Rational Functions

Graph the rational functions in Exercises 27–38. Include the graphs and equations of the asymptotes and dominant terms.

27. $y = \frac{1}{x-1}$

28. $y = \frac{1}{x+1}$

29. $y = \frac{1}{2x+4}$

30. $y = \frac{-3}{x-3}$

31. $y = \frac{x+3}{x+2}$

32. $y = \frac{2x}{x+1}$

33. $y = \frac{x^2}{x-1}$

34. $y = \frac{x^2+1}{x-1}$

35. $y = \frac{x^2-4}{x-1}$

36. $y = \frac{x^2-1}{2x+4}$

37. $y = \frac{x^2-1}{x}$

38. $y = \frac{x^3+1}{x^2}$

Inventing Graphs from Values and Limits

In Exercises 39–42, sketch the graph of a function $y = f(x)$ that satisfies the given conditions. No formulas are required—just label the coordinate axes and sketch an appropriate graph. (The answers are not unique, so your graphs may not be exactly like those in the answer section.)

39. $f(0) = 0, f(1) = 2, f(-1) = -2, \lim_{x\to-\infty} f(x) = -1$, and $\lim_{x\to\infty} f(x) = 1$

40. $f(0) = 0, \lim_{x\to\pm\infty} f(x) = 0, \lim_{x\to 0^+} f(x) = 2$, and $\lim_{x\to 0^-} f(x) = -2$

41. $f(0) = 0, \lim_{x\to\pm\infty} f(x) = 0, \lim_{x\to 1^-} f(x) = \lim_{x\to -1^+} f(x) = \infty$, $\lim_{x\to 1^+} f(x) = -\infty$, and $\lim_{x\to -1^-} f(x) = -\infty$

42. $f(2) = 1, f(-1) = 0, \lim_{x\to\infty} f(x) = 0, \lim_{x\to 0^+} f(x) = \infty$, $\lim_{x\to 0^-} f(x) = -\infty$, and $\lim_{x\to-\infty} f(x) = 1$

Inventing Functions

In Exercises 43–46, find a function that satisfies the given conditions and sketch its graph. (The answers here are not unique. Any function that satisfies the conditions is acceptable. Feel free to use formulas defined in pieces if that will help.)

43. $\lim_{x\to\pm\infty} f(x) = 0, \lim_{x\to 2^-} f(x) = \infty$, and $\lim_{x\to 2^+} f(x) = \infty$

44. $\lim_{x\to\pm\infty} g(x) = 0, \lim_{x\to 3^-} g(x) = -\infty$, and $\lim_{x\to 3^+} g(x) = \infty$

45. $\lim_{x\to-\infty} h(x) = -1, \lim_{x\to\infty} h(x) = 1, \lim_{x\to 0^-} h(x) = -1$, and $\lim_{x\to 0^+} h(x) = 1$

46. $\lim_{x\to\pm\infty} k(x) = 1, \lim_{x\to 1^-} k(x) = \infty$, and $\lim_{x\to 1^+} k(x) = -\infty$

The Formal Definition of Infinite Limit

Use formal definitions to prove the limit statements in Exercises 47–50.

47. $\lim_{x\to 0} \dfrac{-1}{x^2} = -\infty$

48. $\lim_{x\to 0} \dfrac{1}{|x|} = \infty$

49. $\lim_{x\to 3} \dfrac{-2}{(x-3)^2} = -\infty$

50. $\lim_{x\to -5} \dfrac{1}{(x+5)^2} = \infty$

Formal Definitions of Infinite One-Sided Limits

51. Here is the definition of **infinite right-hand limit**.

> We say that $f(x)$ approaches infinity as x approaches x_0 from the right, and write
>
> $$\lim_{x\to x_0^+} f(x) = \infty,$$
>
> if, for every positive real number B, there exists a corresponding number $\delta > 0$ such that for all x
>
> $$x_0 < x < x_0 + \delta \quad \Rightarrow \quad f(x) > B.$$

Modify the definition to cover the following cases.

a. $\lim_{x\to x_0^-} f(x) = \infty$

b. $\lim_{x\to x_0^+} f(x) = -\infty$

c. $\lim_{x\to x_0^-} f(x) = -\infty$

Use the formal definitions from Exercise 51 to prove the limit statements in Exercises 52–56.

52. $\lim_{x\to 0^+} \dfrac{1}{x} = \infty$

53. $\lim_{x\to 0^-} \dfrac{1}{x} = -\infty$

54. $\lim_{x\to 2^-} \dfrac{1}{x-2} = -\infty$

55. $\lim_{x\to 2^+} \dfrac{1}{x-2} = \infty$

56. $\lim_{x\to 1^-} \dfrac{1}{1-x^2} = \infty$

Graphing Terms

Each of the functions in Exercises 57–60 is given as the sum or difference of two terms. First graph the terms (with the same set of axes). Then, using these graphs as guides, sketch in the graph of the function.

57. $y = \sec x + \dfrac{1}{x}, \quad -\dfrac{\pi}{2} < x < \dfrac{\pi}{2}$

58. $y = \sec x - \dfrac{1}{x^2}, \quad -\dfrac{\pi}{2} < x < \dfrac{\pi}{2}$

59. $y = \tan x + \dfrac{1}{x^2}, \quad -\dfrac{\pi}{2} < x < \dfrac{\pi}{2}$

60. $y = \dfrac{1}{x} - \tan x, \quad -\dfrac{\pi}{2} < x < \dfrac{\pi}{2}$

Grapher Explorations—Comparing Graphs with Formulas

Graph the curves in Exercises 61–64. Explain the relation between the curve's formula and what you see.

61. $y = \dfrac{x}{\sqrt{4-x^2}}$

62. $y = \dfrac{-1}{\sqrt{4-x^2}}$

63. $y = x^{2/3} + \dfrac{1}{x^{1/3}}$

64. $y = \sin\left(\dfrac{\pi}{x^2+1}\right)$

2.6 Continuity

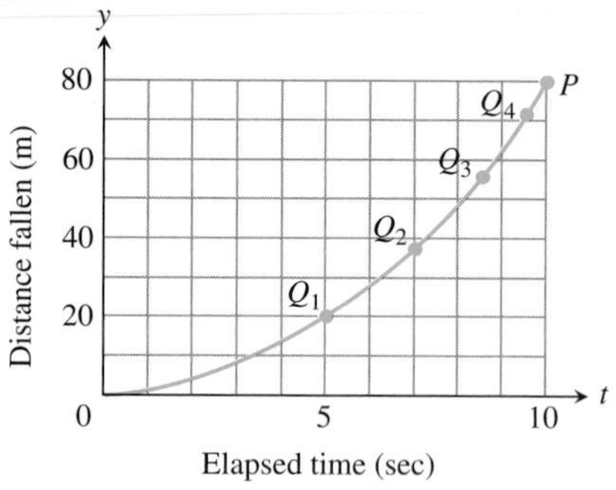

FIGURE 2.49 Connecting plotted points by an unbroken curve from experimental data $Q_1, Q_2, Q_3, \ldots$ for a falling object.

When we plot function values generated in a laboratory or collected in the field, we often connect the plotted points with an unbroken curve to show what the function's values are likely to have been at the times we did not measure (Figure 2.49). In doing so, we are assuming that we are working with a *continuous function*, so its outputs vary continuously with the inputs and do not jump from one value to another without taking on the values in between. The limit of a continuous function as x approaches c can be found simply by calculating the value of the function at c. (We found this to be true for polynomials in Section 2.2.)

Any function $y = f(x)$ whose graph can be sketched over its domain in one continuous motion without lifting the pencil is an example of a continuous function. In this section we investigate more precisely what it means for a function to be continuous. We also study the properties of continuous functions, and see that many of the function types presented in Section 1.4 are continuous.

Continuity at a Point

To understand continuity, we need to consider a function like the one in Figure 2.50 whose limits we investigated in Example 2, Section 2.4.

EXAMPLE 1 Investigating Continuity

Find the points at which the function f in Figure 2.50 is continuous and the points at which f is discontinuous.

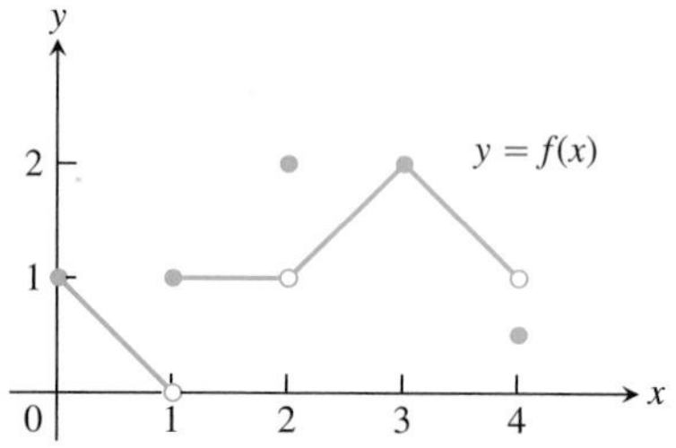

FIGURE 2.50 The function is continuous on $[0, 4]$ except at $x = 1, x = 2$, and $x = 4$ (Example 1).

Solution The function f is continuous at every point in its domain $[0, 4]$ except at $x = 1, x = 2$, and $x = 4$. At these points, there are breaks in the graph. Note the relationship between the limit of f and the value of f at each point of the function's domain.

Points at which *f* is continuous:

At $x = 0$,	$\lim_{x \to 0^+} f(x) = f(0)$.
At $x = 3$,	$\lim_{x \to 3} f(x) = f(3)$.
At $0 < c < 4, c \neq 1, 2$,	$\lim_{x \to c} f(x) = f(c)$.

Points at which *f* is discontinuous:

At $x = 1$,	$\lim_{x \to 1} f(x)$ does not exist.
At $x = 2$,	$\lim_{x \to 2} f(x) = 1$, but $1 \neq f(2)$.
At $x = 4$,	$\lim_{x \to 4^-} f(x) = 1$, but $1 \neq f(4)$.
At $c < 0, c > 4$,	these points are not in the domain of f. ■

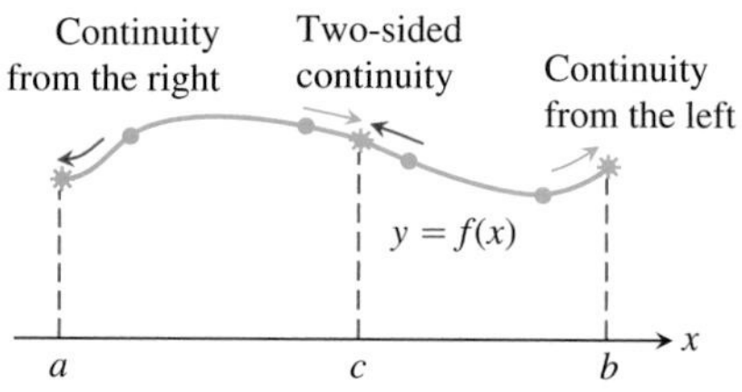

FIGURE 2.51 Continuity at points a, b, and c.

To define continuity at a point in a function's domain, we need to define continuity at an interior point (which involves a two-sided limit) and continuity at an endpoint (which involves a one-sided limit) (Figure 2.51).

DEFINITION Continuous at a Point

Interior point: A function $y = f(x)$ is **continuous at an interior point c** of its domain if

$$\lim_{x \to c} f(x) = f(c).$$

Endpoint: A function $y = f(x)$ is **continuous at a left endpoint a** or is **continuous at a right endpoint b** of its domain if

$$\lim_{x \to a^+} f(x) = f(a) \quad \text{or} \quad \lim_{x \to b^-} f(x) = f(b), \quad \text{respectively.}$$

If a function f is not continuous at a point c, we say that f is **discontinuous** at c and c is a **point of discontinuity** of f. Note that c need not be in the domain of f.

A function f is **right-continuous (continuous from the right)** at a point $x = c$ in its domain if $\lim_{x \to c^+} f(x) = f(c)$. It is **left-continuous (continuous from the left)** at c if $\lim_{x \to c^-} f(x) = f(c)$. Thus, a function is continuous at a left endpoint a of its domain if it is right-continuous at a and continuous at a right endpoint b of its domain if it is left-continuous at b. A function is continuous at an interior point c of its domain if and only if it is both right-continuous and left-continuous at c (Figure 2.51).

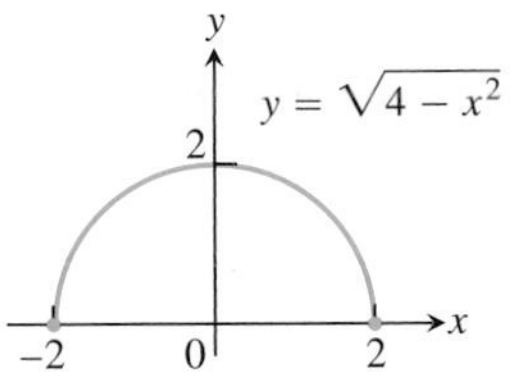

FIGURE 2.52 A function that is continuous at every domain point (Example 2).

EXAMPLE 2 A Function Continuous Throughout Its Domain

The function $f(x) = \sqrt{4 - x^2}$ is continuous at every point of its domain, $[-2, 2]$ (Figure 2.52), including $x = -2$, where f is right-continuous, and $x = 2$, where f is left-continuous. ■

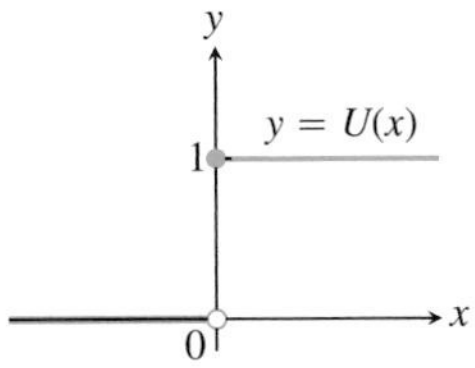

FIGURE 2.53 A function that is right-continuous, but not left-continuous, at the origin. It has a jump discontinuity there (Example 3).

EXAMPLE 3 The Unit Step Function Has a Jump Discontinuity

The unit step function $U(x)$, graphed in Figure 2.53, is right-continuous at $x = 0$, but is neither left-continuous nor continuous there. It has a jump discontinuity at $x = 0$. ■

We summarize continuity at a point in the form of a test.

Continuity Test

A function $f(x)$ is continuous at an interior point of its domain $x = c$ if and only if it meets the following three conditions.

1. $f(c)$ exists (c lies in the domain of f)
2. $\lim_{x \to c} f(x)$ exists (f has a limit as $x \to c$)
3. $\lim_{x \to c} f(x) = f(c)$ (the limit equals the function value)

For one-sided continuity and continuity at an endpoint, the limits in parts 2 and 3 of the test should be replaced by the appropriate one-sided limits.

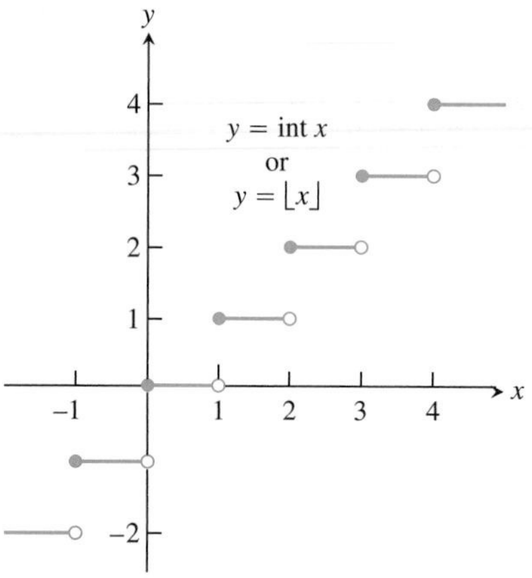

FIGURE 2.54 The greatest integer function is continuous at every noninteger point. It is right-continuous, but not left-continuous, at every integer point (Example 4).

EXAMPLE 4 The Greatest Integer Function

The function $y = \lfloor x \rfloor$ or $y = \operatorname{int} x$, introduced in Chapter 1, is graphed in Figure 2.54. It is discontinuous at every integer because the limit does not exist at any integer n:

$$\lim_{x \to n^-} \operatorname{int} x = n - 1 \quad \text{and} \quad \lim_{x \to n^+} \operatorname{int} x = n$$

so the left-hand and right-hand limits are not equal as $x \to n$. Since $\operatorname{int} n = n$, the greatest integer function is right-continuous at every integer n (but not left-continuous).

The greatest integer function is continuous at every real number other than the integers. For example,

$$\lim_{x \to 1.5} \operatorname{int} x = 1 = \operatorname{int} 1.5.$$

In general, if $n - 1 < c < n$, n an integer, then

$$\lim_{x \to c} \operatorname{int} x = n - 1 = \operatorname{int} c.$$ ■

Figure 2.55 is a catalog of discontinuity types. The function in Figure 2.55a is continuous at $x = 0$. The function in Figure 2.55b would be continuous if it had $f(0) = 1$. The function in Figure 2.55c would be continuous if $f(0)$ were 1 instead of 2. The discontinuities in Figure 2.55b and c are **removable**. Each function has a limit as $x \to 0$, and we can remove the discontinuity by setting $f(0)$ equal to this limit.

The discontinuities in Figure 2.55d through f are more serious: $\lim_{x \to 0} f(x)$ does not exist, and there is no way to improve the situation by changing f at 0. The step function in Figure 2.55d has a **jump discontinuity**: The one-sided limits exist but have different values. The function $f(x) = 1/x^2$ in Figure 2.55e has an **infinite discontinuity**. The function in Figure 2.55f has an **oscillating discontinuity**: It oscillates too much to have a limit as $x \to 0$.

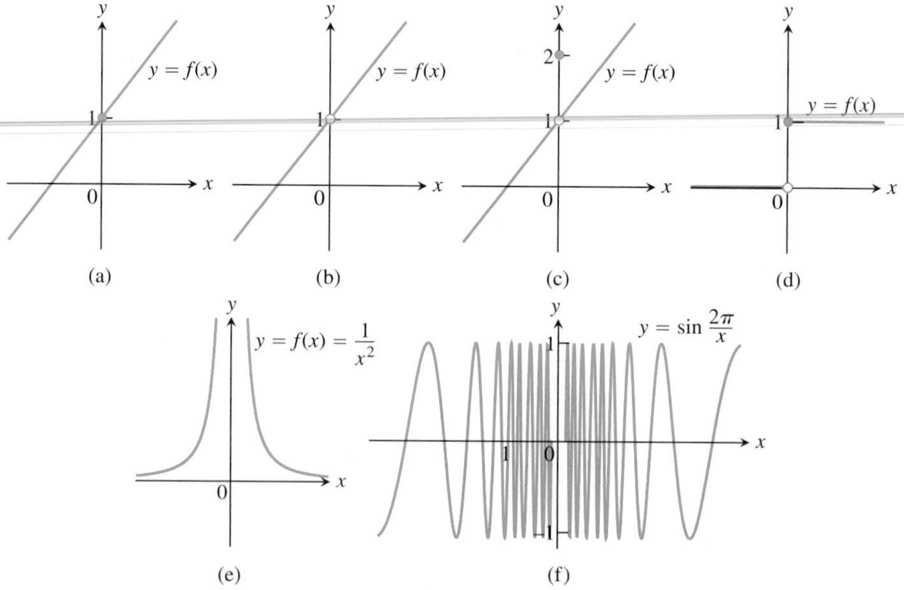

FIGURE 2.55 The function in (a) is continuous at $x = 0$; the functions in (b) through (f) are not.

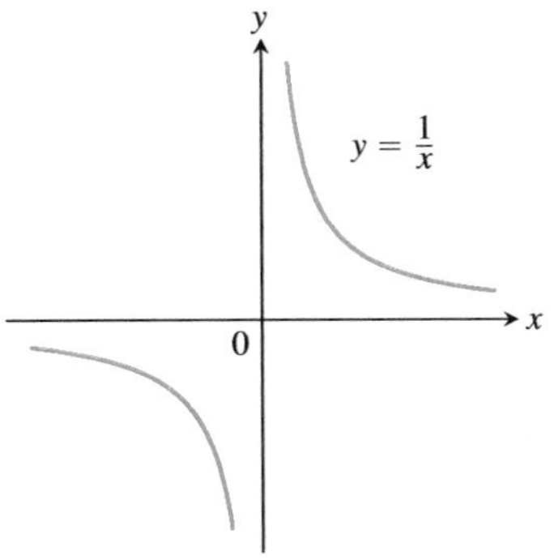

FIGURE 2.56 The function $y = 1/x$ is continuous at every value of x except $x = 0$. It has a point of discontinuity at $x = 0$ (Example 5).

Continuous Functions

A function is **continuous on an interval** if and only if it is continuous at every point of the interval. For example, the semicircle function graphed in Figure 2.52 is continuous on the interval $[-2, 2]$, which is its domain. A **continuous function** is one that is continuous at every point of its domain. A continuous function need not be continuous on every interval. For example, $y = 1/x$ is not continuous on $[-1, 1]$ (Figure 2.56), but it is continuous over its domain $(-\infty, 0) \cup (0, \infty)$.

EXAMPLE 5 Identifying Continuous Functions

(a) The function $y = 1/x$ (Figure 2.56) is a continuous function because it is continuous at every point of its domain. It has a point of discontinuity at $x = 0$, however, because it is not defined there.

(b) The identity function $f(x) = x$ and constant functions are continuous everywhere by Example 3, Section 2.3. ■

Algebraic combinations of continuous functions are continuous wherever they are defined.

THEOREM 9 Properties of Continuous Functions

If the functions f and g are continuous at $x = c$, then the following combinations are continuous at $x = c$.

1. *Sums*: $f + g$
2. *Differences*: $f - g$
3. *Products*: $f \cdot g$
4. *Constant multiples*: $k \cdot f$, for any number k
5. *Quotients*: f/g provided $g(c) \neq 0$
6. *Powers*: $f^{r/s}$, provided it is defined on an open interval containing c, where r and s are integers

Most of the results in Theorem 9 are easily proved from the limit rules in Theorem 1, Section 2.2. For instance, to prove the sum property we have

$$\begin{aligned}\lim_{x \to c}(f + g)(x) &= \lim_{x \to c}(f(x) + g(x)) \\ &= \lim_{x \to c} f(x) + \lim_{x \to c} g(x), && \text{Sum Rule, Theorem 1} \\ &= f(c) + g(c) && \text{Continuity of } f, g \text{ at } c \\ &= (f + g)(c).\end{aligned}$$

This shows that $f + g$ is continuous.

EXAMPLE 6 Polynomial and Rational Functions Are Continuous

(a) Every polynomial $P(x) = a_n x^n + a_{n-1}x^{n-1} + \cdots + a_0$ is continuous because $\lim_{x \to c} P(x) = P(c)$ by Theorem 2, Section 2.2.

(b) If $P(x)$ and $Q(x)$ are polynomials, then the rational function $P(x)/Q(x)$ is continuous wherever it is defined ($Q(c) \neq 0$) by the Quotient Rule in Theorem 9.

EXAMPLE 7 Continuity of the Absolute Value Function

The function $f(x) = |x|$ is continuous at every value of x. If $x > 0$, we have $f(x) = x$, a polynomial. If $x < 0$, we have $f(x) = -x$, another polynomial. Finally, at the origin, $\lim_{x \to 0} |x| = 0 = |0|$. ■

The functions $y = \sin x$ and $y = \cos x$ are continuous at $x = 0$ by Example 6 of Section 2.2. Both functions are, in fact, continuous everywhere (see Exercise 62). It follows from Theorem 9 that all six trigonometric functions are then continuous wherever they are defined. For example, $y = \tan x$ is continuous on $\cdots \cup (-\pi/2, \pi/2) \cup (\pi/2, 3\pi/2) \cup \cdots$.

Composites

All composites of continuous functions are continuous. The idea is that if $f(x)$ is continuous at $x = c$ and $g(x)$ is continuous at $x = f(c)$, then $g \circ f$ is continuous at $x = c$ (Figure 2.57). In this case, the limit as $x \to c$ is $g(f(c))$.

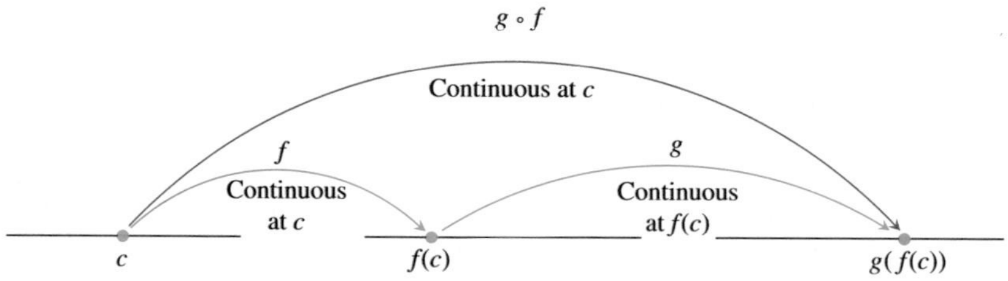

FIGURE 2.57 Composites of continuous functions are continuous.

> **THEOREM 10 Composite of Continuous Functions**
> If f is continuous at c and g is continuous at $f(c)$, then the composite $g \circ f$ is continuous at c.

Intuitively, Theorem 10 is reasonable because if x is close to c, then $f(x)$ is close to $f(c)$, and since g is continuous at $f(c)$, it follows that $g(f(x))$ is close to $g(f(c))$.

The continuity of composites holds for any finite number of functions. The only requirement is that each function be continuous where it is applied. For an outline of the proof of Theorem 10, see Exercise 6 in Appendix 2.

EXAMPLE 8 Applying Theorems 9 and 10

Show that the following functions are continuous everywhere on their respective domains.

(a) $y = \sqrt{x^2 - 2x - 5}$

(b) $y = \dfrac{x^{2/3}}{1 + x^4}$

(c) $y = \left|\dfrac{x - 2}{x^2 - 2}\right|$

(d) $y = \left|\dfrac{x \sin x}{x^2 + 2}\right|$

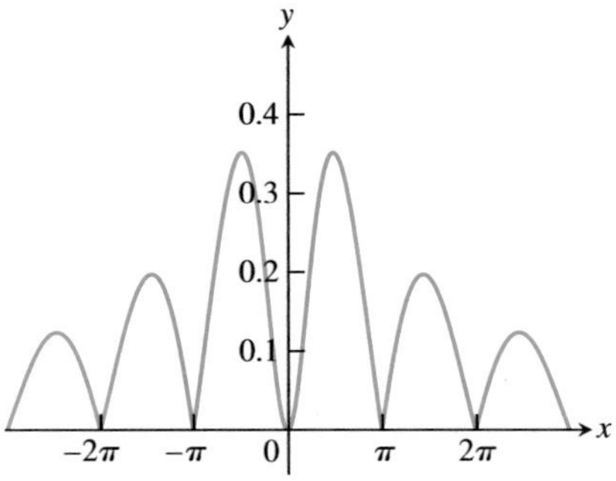

FIGURE 2.58 The graph suggests that $y = |(x \sin x)/(x^2 + 2)|$ is continuous (Example 8d).

Solution

(a) The square root function is continuous on $[0, \infty)$ because it is a rational power of the continuous identity function $f(x) = x$ (Part 6, Theorem 9). The given function is then the composite of the polynomial $f(x) = x^2 - 2x - 5$ with the square root function $g(t) = \sqrt{t}$.

(b) The numerator is a rational power of the identity function; the denominator is an everywhere-positive polynomial. Therefore, the quotient is continuous.

(c) The quotient $(x - 2)/(x^2 - 2)$ is continuous for all $x \neq \pm\sqrt{2}$, and the function is the composition of this quotient with the continuous absolute value function (Example 7).

(d) Because the sine function is everywhere-continuous (Exercise 62), the numerator term $x \sin x$ is the product of continuous functions, and the denominator term $x^2 + 2$ is an everywhere-positive polynomial. The given function is the composite of a quotient of continuous functions with the continuous absolute value function (Figure 2.58). ■

Continuous Extension to a Point

The function $y = (\sin x)/x$ is continuous at every point except $x = 0$. In this it is like the function $y = 1/x$. But $y = (\sin x)/x$ is different from $y = 1/x$ in that it has a finite limit as $x \to 0$ (Theorem 7). It is therefore possible to extend the function's domain to include the point $x = 0$ in such a way that the extended function is continuous at $x = 0$. We define

$$F(x) = \begin{cases} \dfrac{\sin x}{x}, & x \neq 0 \\ 1, & x = 0. \end{cases}$$

The function $F(x)$ is continuous at $x = 0$ because

$$\lim_{x \to 0} \frac{\sin x}{x} = F(0)$$

(Figure 2.59).

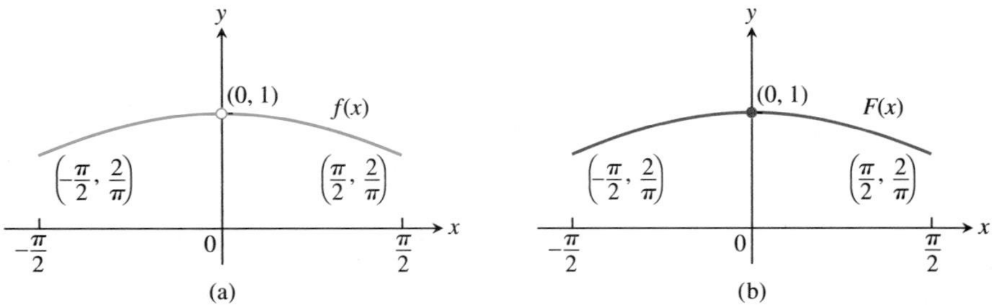

FIGURE 2.59 The graph (a) of $f(x) = (\sin x)/x$ for $-\pi/2 \leq x \leq \pi/2$ does not include the point (0, 1) because the function is not defined at $x = 0$. (b) We can remove the discontinuity from the graph by defining the new function $F(x)$ with $F(0) = 1$ and $F(x) = f(x)$ everywhere else. Note that $F(0) = \lim_{x \to 0} f(x)$.

More generally, a function (such as a rational function) may have a limit even at a point where it is not defined. If $f(c)$ is not defined, but $\lim_{x \to c} f(x) = L$ exists, we can define a new function $F(x)$ by the rule

$$F(x) = \begin{cases} f(x), & \text{if } x \text{ is in the domain of } f \\ L, & \text{if } x = c. \end{cases}$$

The function F is continuous at $x = c$. It is called the **continuous extension** of f to $x = c$. For rational functions f, continuous extensions are usually found by canceling common factors.

EXAMPLE 9 A Continuous Extension

Show that

$$f(x) = \frac{x^2 + x - 6}{x^2 - 4}$$

has a continuous extension to $x = 2$, and find that extension.

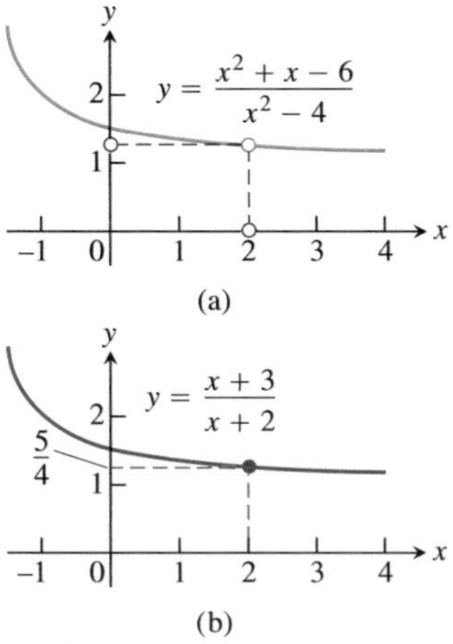

FIGURE 2.60 (a) The graph of $f(x)$ and (b) the graph of its continuous extension $F(x)$ (Example 9).

Solution Although $f(2)$ is not defined, if $x \neq 2$ we have

$$f(x) = \frac{x^2 + x - 6}{x^2 - 4} = \frac{(x - 2)(x + 3)}{(x - 2)(x + 2)} = \frac{x + 3}{x + 2}.$$

The new function

$$F(x) = \frac{x + 3}{x + 2}$$

is equal to $f(x)$ for $x \neq 2$, but is continuous at $x = 2$, having there the value of $5/4$. Thus F is the continuous extension of f to $x = 2$, and

$$\lim_{x \to 2} \frac{x^2 + x - 6}{x^2 - 4} = \lim_{x \to 2} f(x) = \frac{5}{4}.$$

The graph of f is shown in Figure 2.60. The continuous extension F has the same graph except with no hole at $(2, 5/4)$. Effectively, F is the function f with its point of discontinuity at $x = 2$ removed. ■

Intermediate Value Theorem for Continuous Functions

Functions that are continuous on intervals have properties that make them particularly useful in mathematics and its applications. One of these is the *Intermediate Value Property*. A function is said to have the **Intermediate Value Property** if whenever it takes on two values, it also takes on all the values in between.

THEOREM 11 The Intermediate Value Theorem for Continuous Functions
A function $y = f(x)$ that is continuous on a closed interval $[a, b]$ takes on every value between $f(a)$ and $f(b)$. In other words, if y_0 is any value between $f(a)$ and $f(b)$, then $y_0 = f(c)$ for some c in $[a, b]$.

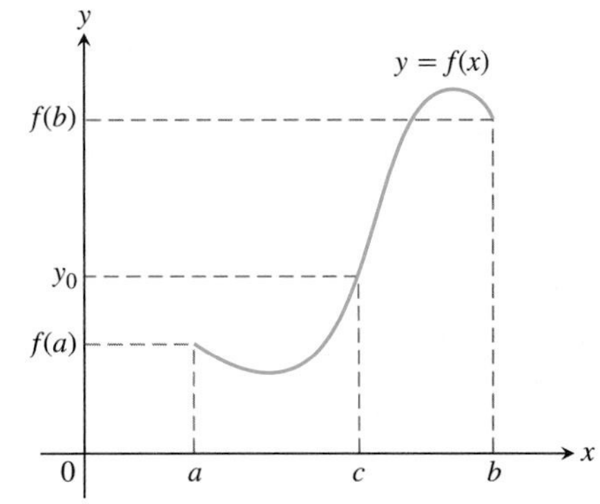

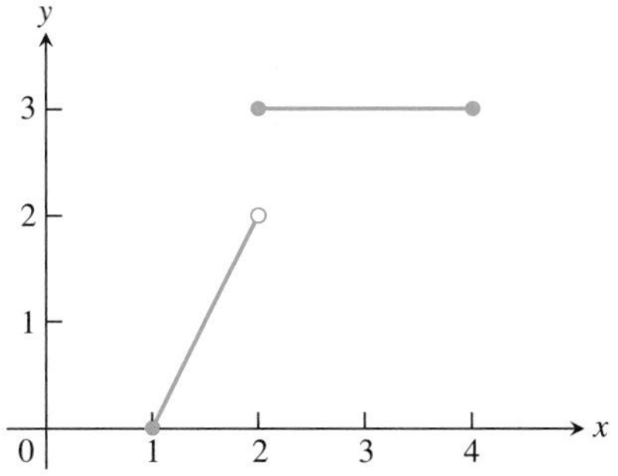

FIGURE 2.61 The function
$$f(x) = \begin{cases} 2x - 2, & 1 \le x < 2 \\ 3, & 2 \le x \le 4 \end{cases}$$
does not take on all values between $f(1) = 0$ and $f(4) = 3$; it misses all the values between 2 and 3.

Geometrically, the Intermediate Value Theorem says that any horizontal line $y = y_0$ crossing the y-axis between the numbers $f(a)$ and $f(b)$ will cross the curve $y = f(x)$ at least once over the interval $[a, b]$.

The proof of the Intermediate Value Theorem depends on the completeness property of the real number system and can be found in more advanced texts.

The continuity of f on the interval is essential to Theorem 11. If f is discontinuous at even one point of the interval, the theorem's conclusion may fail, as it does for the function graphed in Figure 2.61.

A Consequence for Graphing: Connectivity Theorem 11 is the reason the graph of a function continuous on an interval cannot have any breaks over the interval. It will be **connected**, a single, unbroken curve, like the graph of $\sin x$. It will not have jumps like the graph of the greatest integer function (Figure 2.54) or separate branches like the graph of $1/x$ (Figure 2.56).

A Consequence for Root Finding We call a solution of the equation $f(x) = 0$ a **root** of the equation or **zero** of the function f. The Intermediate Value Theorem tells us that if f is continuous, then any interval on which f changes sign contains a zero of the function.

In practical terms, when we see the graph of a continuous function cross the horizontal axis on a computer screen, we know it is not stepping across. There really is a point where the function's value is zero. This consequence leads to a procedure for estimating the zeros of any continuous function we can graph:

1. Graph the function over a large interval to see roughly where the zeros are.
2. Zoom in on each zero to estimate its x-coordinate value.

You can practice this procedure on your graphing calculator or computer in some of the exercises. Figure 2.62 shows a typical sequence of steps in a graphical solution of the equation $x^3 - x - 1 = 0$.

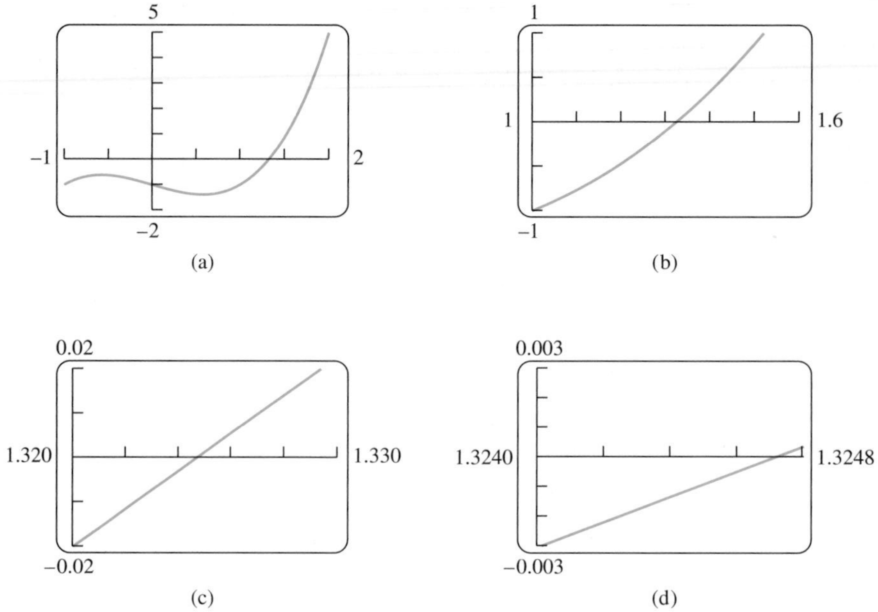

FIGURE 2.62 Zooming in on a zero of the function $f(x) = x^3 - x - 1$. The zero is near $x = 1.3247$.

EXERCISES 2.6

Continuity from Graphs

In Exercises 1–4, say whether the function graphed is continuous on $[-1, 3]$. If not, where does it fail to be continuous and why?

1.

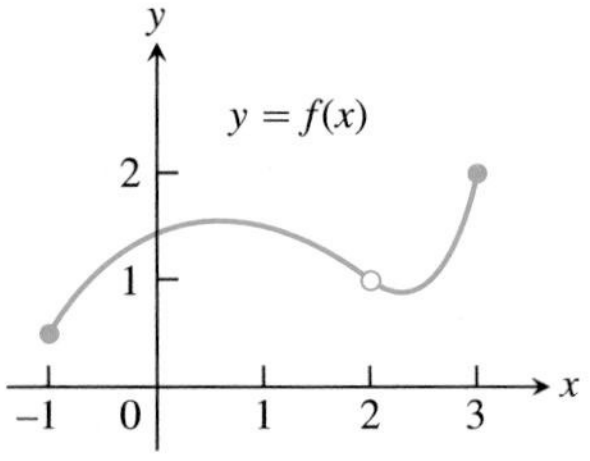

2.

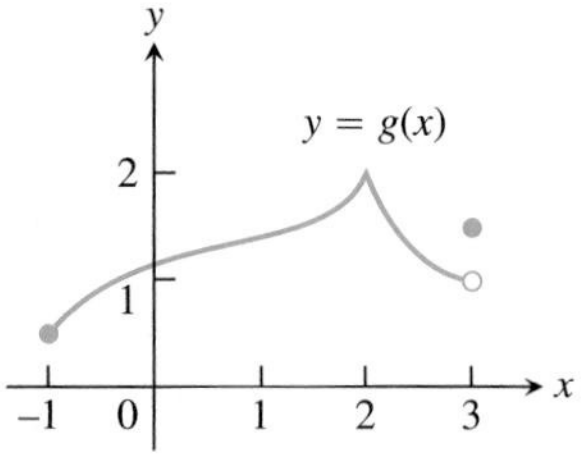

3.

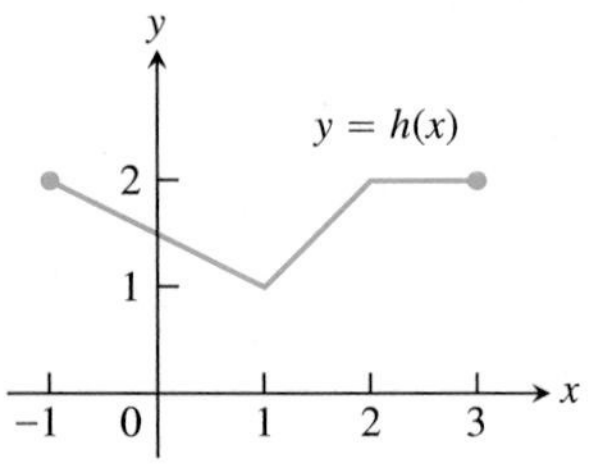

4.

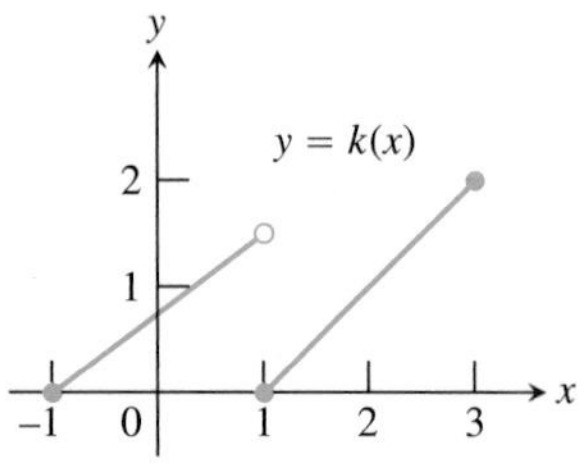

Exercises 5–10 are about the function

$$f(x) = \begin{cases} x^2 - 1, & -1 \le x < 0 \\ 2x, & 0 < x < 1 \\ 1, & x = 1 \\ -2x + 4, & 1 < x < 2 \\ 0, & 2 < x < 3 \end{cases}$$

graphed in the accompanying figure.

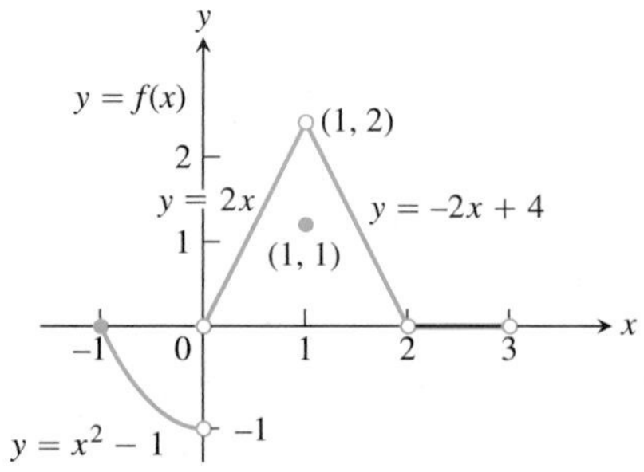

The graph for Exercises 5–10.

5. **a.** Does $f(-1)$ exist?
   b. Does $\lim_{x\to -1^+} f(x)$ exist?
   c. Does $\lim_{x\to -1^+} f(x) = f(-1)$?
   d. Is f continuous at $x = -1$?

6. **a.** Does $f(1)$ exist?
   b. Does $\lim_{x\to 1} f(x)$ exist?
   c. Does $\lim_{x\to 1} f(x) = f(1)$?
   d. Is f continuous at $x = 1$?

7. **a.** Is f defined at $x = 2$? (Look at the definition of f.)
   b. Is f continuous at $x = 2$?

8. At what values of x is f continuous?

9. What value should be assigned to $f(2)$ to make the extended function continuous at $x = 2$?

10. To what new value should $f(1)$ be changed to remove the discontinuity?

Applying the Continuity Test

At which points do the functions in Exercises 11 and 12 fail to be continuous? At which points, if any, are the discontinuities removable? Not removable? Give reasons for your answers.

11. Exercise 1, Section 2.4

12. Exercise 2, Section 2.4

At what points are the functions in Exercises 13–28 continuous?

13. $y = \dfrac{1}{x-2} - 3x$

14. $y = \dfrac{1}{(x+2)^2} + 4$

15. $y = \dfrac{x+1}{x^2 - 4x + 3}$

16. $y = \dfrac{x+3}{x^2 - 3x - 10}$

17. $y = |x - 1| + \sin x$

18. $y = \dfrac{1}{|x| + 1} - \dfrac{x^2}{2}$

19. $y = \dfrac{\cos x}{x}$

20. $y = \dfrac{x+2}{\cos x}$

21. $y = \csc 2x$

22. $y = \tan \dfrac{\pi x}{2}$

23. $y = \dfrac{x \tan x}{x^2 + 1}$

24. $y = \dfrac{\sqrt{x^4 + 1}}{1 + \sin^2 x}$

25. $y = \sqrt{2x + 3}$

26. $y = \sqrt[4]{3x - 1}$

27. $y = (2x - 1)^{1/3}$

28. $y = (2 - x)^{1/5}$

Composite Functions

Find the limits in Exercises 29–34. Are the functions continuous at the point being approached?

29. $\lim_{x\to\pi} \sin(x - \sin x)$

30. $\lim_{t\to 0} \sin\left(\dfrac{\pi}{2}\cos(\tan t)\right)$

31. $\lim_{y\to 1} \sec(y \sec^2 y - \tan^2 y - 1)$

32. $\lim_{x\to 0} \tan\left(\dfrac{\pi}{4}\cos(\sin x^{1/3})\right)$

33. $\lim_{t\to 0} \cos\left(\dfrac{\pi}{\sqrt{19 - 3\sec 2t}}\right)$

34. $\lim_{x\to\pi/6} \sqrt{\csc^2 x + 5\sqrt{3}\tan x}$

Continuous Extensions

35. Define $g(3)$ in a way that extends $g(x) = (x^2 - 9)/(x - 3)$ to be continuous at $x = 3$.

36. Define $h(2)$ in a way that extends $h(t) = (t^2 + 3t - 10)/(t - 2)$ to be continuous at $t = 2$.

37. Define $f(1)$ in a way that extends $f(s) = (s^3 - 1)/(s^2 - 1)$ to be continuous at $s = 1$.

38. Define $g(4)$ in a way that extends $g(x) = (x^2 - 16)/(x^2 - 3x - 4)$ to be continuous at $x = 4$.

39. For what value of a is
$$f(x) = \begin{cases} x^2 - 1, & x < 3 \\ 2ax, & x \ge 3 \end{cases}$$
continuous at every x?

40. For what value of b is
$$g(x) = \begin{cases} x, & x < -2 \\ bx^2, & x \ge -2 \end{cases}$$
continuous at every x?

T In Exercises 41–44, graph the function f to see whether it appears to have a continuous extension to the origin. If it does, use Trace and Zoom to find a good candidate for the extended function's value at $x = 0$. If the function does not appear to have a continuous extension, can it be extended to be continuous at the origin from the right or from the left? If so, what do you think the extended function's value(s) should be?

41. $f(x) = \dfrac{10^x - 1}{x}$

42. $f(x) = \dfrac{10^{|x|} - 1}{x}$

43. $f(x) = \dfrac{\sin x}{|x|}$

44. $f(x) = (1 + 2x)^{1/x}$

Theory and Examples

45. A continuous function $y = f(x)$ is known to be negative at $x = 0$ and positive at $x = 1$. Why does the equation $f(x) = 0$ have at least one solution between $x = 0$ and $x = 1$? Illustrate with a sketch.

46. Explain why the equation $\cos x = x$ has at least one solution.

47. **Roots of a cubic** Show that the equation $x^3 - 15x + 1 = 0$ has three solutions in the interval $[-4, 4]$.

48. **A function value** Show that the function $F(x) = (x - a)^2 \cdot (x - b)^2 + x$ takes on the value $(a + b)/2$ for some value of x.

49. **Solving an equation** If $f(x) = x^3 - 8x + 10$, show that there are values c for which $f(c)$ equals **(a)** π; **(b)** $-\sqrt{3}$; **(c)** 5,000,000.

50. Explain why the following five statements ask for the same information.

a. Find the roots of $f(x) = x^3 - 3x - 1$.

b. Find the x-coordinates of the points where the curve $y = x^3$ crosses the line $y = 3x + 1$.

c. Find all the values of x for which $x^3 - 3x = 1$.

d. Find the x-coordinates of the points where the cubic curve $y = x^3 - 3x$ crosses the line $y = 1$.

e. Solve the equation $x^3 - 3x - 1 = 0$.

51. **Removable discontinuity** Give an example of a function $f(x)$ that is continuous for all values of x except $x = 2$, where it has a removable discontinuity. Explain how you know that f is discontinuous at $x = 2$, and how you know the discontinuity is removable.

52. **Nonremovable discontinuity** Give an example of a function $g(x)$ that is continuous for all values of x except $x = -1$, where it has a nonremovable discontinuity. Explain how you know that g is discontinuous there and why the discontinuity is not removable.

53. **A function discontinuous at every point**

a. Use the fact that every nonempty interval of real numbers contains both rational and irrational numbers to show that the function

$$f(x) = \begin{cases} 1, & \text{if } x \text{ is rational} \\ 0, & \text{if } x \text{ is irrational} \end{cases}$$

is discontinuous at every point.

b. Is f right-continuous or left-continuous at any point?

54. If functions $f(x)$ and $g(x)$ are continuous for $0 \le x \le 1$, could $f(x)/g(x)$ possibly be discontinuous at a point of $[0, 1]$? Give reasons for your answer.

55. If the product function $h(x) = f(x) \cdot g(x)$ is continuous at $x = 0$, must $f(x)$ and $g(x)$ be continuous at $x = 0$? Give reasons for your answer.

56. **Discontinuous composite of continuous functions** Give an example of functions f and g, both continuous at $x = 0$, for which the composite $f \circ g$ is discontinuous at $x = 0$. Does this contradict Theorem 10? Give reasons for your answer.

57. **Never-zero continuous functions** Is it true that a continuous function that is never zero on an interval never changes sign on that interval? Give reasons for your answer.

58. **Stretching a rubber band** Is it true that if you stretch a rubber band by moving one end to the right and the other to the left, some point of the band will end up in its original position? Give reasons for your answer.

59. **A fixed point theorem** Suppose that a function f is continuous on the closed interval $[0, 1]$ and that $0 \le f(x) \le 1$ for every x in $[0, 1]$. Show that there must exist a number c in $[0, 1]$ such that $f(c) = c$ (c is called a **fixed point** of f).

60. **The sign-preserving property of continuous functions** Let f be defined on an interval (a, b) and suppose that $f(c) \ne 0$ at some c where f is continuous. Show that there is an interval $(c - \delta, c + \delta)$ about c where f has the same sign as $f(c)$. Notice how remarkable this conclusion is. Although f is defined throughout (a, b), it is not required to be continuous at any point except c. That and the condition $f(c) \ne 0$ are enough to make f different from zero (positive or negative) throughout an entire interval.

61. Prove that f is continuous at c if and only if

$$\lim_{h \to 0} f(c + h) = f(c).$$

62. Use Exercise 61 together with the identities

$$\sin(h + c) = \sin h \cos c + \cos h \sin c,$$
$$\cos(h + c) = \cos h \cos c - \sin h \sin c$$

to prove that $f(x) = \sin x$ and $g(x) = \cos x$ are continuous at every point $x = c$.

Solving Equations Graphically

T Use a graphing calculator or computer grapher to solve the equations in Exercises 63–70.

63. $x^3 - 3x - 1 = 0$

64. $2x^3 - 2x^2 - 2x + 1 = 0$

65. $x(x - 1)^2 = 1$ (one root)

66. $x^x = 2$

67. $\sqrt{x} + \sqrt{1 + x} = 4$

68. $x^3 - 15x + 1 = 0$ (three roots)

69. $\cos x = x$ (one root). Make sure you are using radian mode.

70. $2 \sin x = x$ (three roots). Make sure you are using radian mode.

2.7 Tangents and Derivatives

This section continues the discussion of secants and tangents begun in Section 2.1. We calculate limits of secant slopes to find tangents to curves.

What *Is* a Tangent to a Curve?

For circles, tangency is straightforward. A line L is tangent to a circle at a point P if L passes through P perpendicular to the radius at P (Figure 2.63). Such a line just *touches*

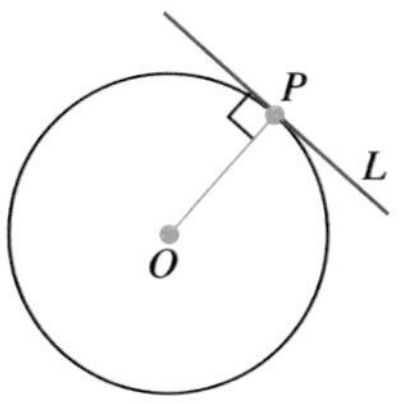

FIGURE 2.63 L is tangent to the circle at P if it passes through P perpendicular to radius OP.

the circle. But what does it mean to say that a line L is tangent to some other curve C at a point P? Generalizing from the geometry of the circle, we might say that it means one of the following:

1. L passes through P perpendicular to the line from P to the center of C.
2. L passes through only one point of C, namely P.
3. L passes through P and lies on one side of C only.

Although these statements are valid if C is a circle, none of them works consistently for more general curves. Most curves do not have centers, and a line we may want to call tangent may intersect C at other points or cross C at the point of tangency (Figure 2.64).

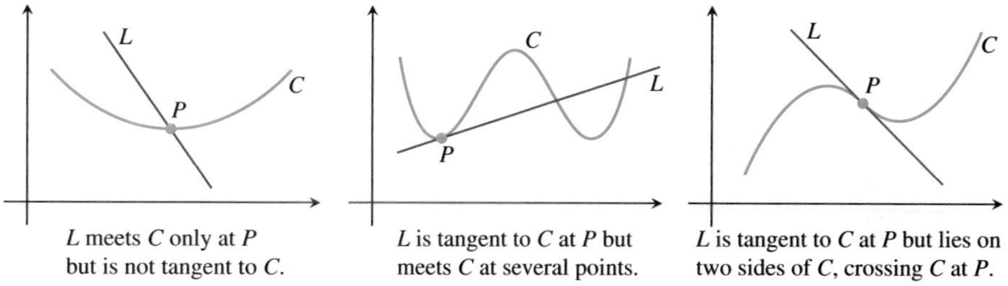

FIGURE 2.64 Exploding myths about tangent lines.

HISTORICAL BIOGRAPHY

Pierre de Fermat
(1601–1665)

To define tangency for general curves, we need a *dynamic* approach that takes into account the behavior of the secants through P and nearby points Q as Q moves toward P along the curve (Figure 2.65). It goes like this:

1. We start with what we *can* calculate, namely the slope of the secant PQ.
2. Investigate the limit of the secant slope as Q approaches P along the curve.
3. If the limit exists, take it to be the slope of the curve at P and define the tangent to the curve at P to be the line through P with this slope.

This approach is what we were doing in the falling-rock and fruit fly examples in Section 2.1.

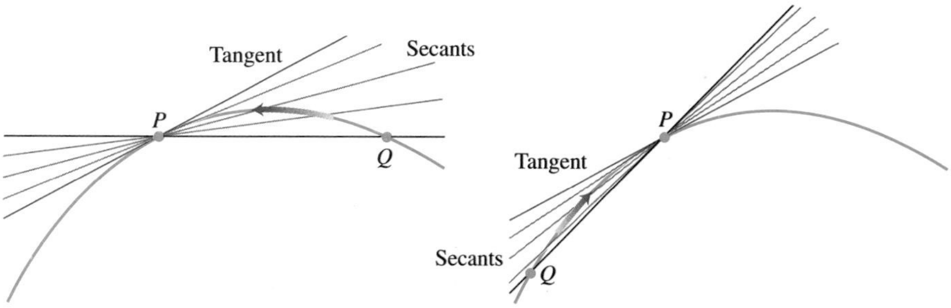

FIGURE 2.65 The dynamic approach to tangency. The tangent to the curve at P is the line through P whose slope is the limit of the secant slopes as $Q \to P$ from either side.

EXAMPLE 1 Tangent Line to a Parabola

Find the slope of the parabola $y = x^2$ at the point $P(2, 4)$. Write an equation for the tangent to the parabola at this point.

Solution We begin with a secant line through $P(2, 4)$ and $Q(2 + h, (2 + h)^2)$ nearby. We then write an expression for the slope of the secant PQ and investigate what happens to the slope as Q approaches P along the curve:

$$\text{Secant slope} = \frac{\Delta y}{\Delta x} = \frac{(2 + h)^2 - 2^2}{h} = \frac{h^2 + 4h + 4 - 4}{h}$$

$$= \frac{h^2 + 4h}{h} = h + 4.$$

If $h > 0$, then Q lies above and to the right of P, as in Figure 2.66. If $h < 0$, then Q lies to the left of P (not shown). In either case, as Q approaches P along the curve, h approaches zero and the secant slope approaches 4:

$$\lim_{h \to 0} (h + 4) = 4.$$

We take 4 to be the parabola's slope at P.

The tangent to the parabola at P is the line through P with slope 4:

$$y = 4 + 4(x - 2) \quad \text{Point-slope equation}$$

$$y = 4x - 4.$$

■

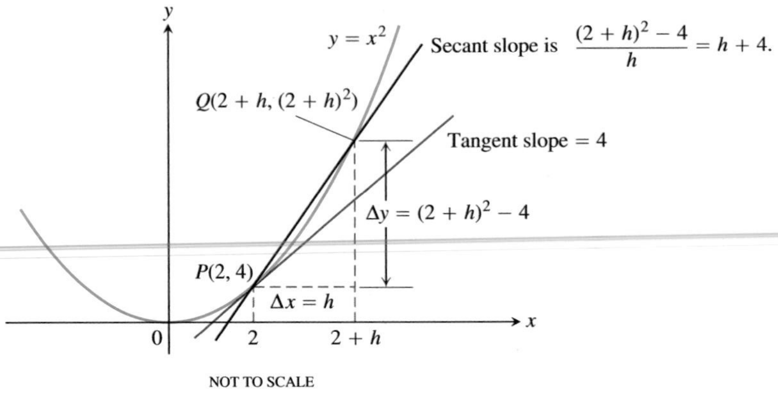

FIGURE 2.66 Finding the slope of the parabola $y = x^2$ at the point $P(2, 4)$ (Example 1).

Finding a Tangent to the Graph of a Function

The problem of finding a tangent to a curve was the dominant mathematical problem of the early seventeenth century. In optics, the tangent determined the angle at which a ray of light entered a curved lens. In mechanics, the tangent determined the direction of a body's motion at every point along its path. In geometry, the tangents to two curves at a point of intersection determined the angles at which they intersected. To find a tangent to an arbitrary curve $y = f(x)$ at a point $P(x_0, f(x_0))$, we use the same dynamic procedure. We calculate the slope of the secant through P and a point $Q(x_0 + h, f(x_0 + h))$. We then investigate the limit of the slope as $h \to 0$ (Figure 2.67). If the limit exists, we call it the slope of the curve at P and define the tangent at P to be the line through P having this slope.

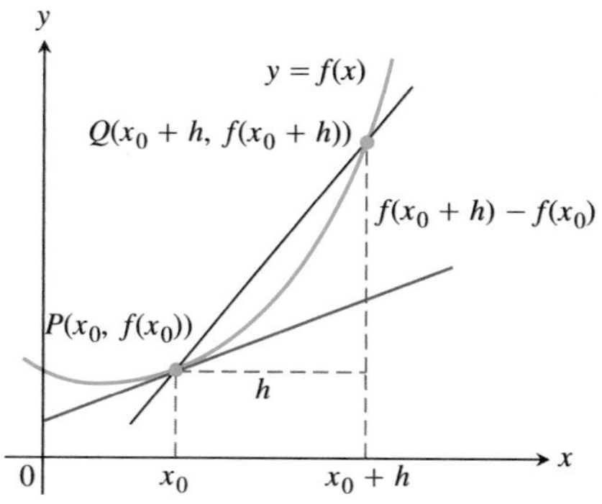

FIGURE 2.67 The slope of the tangent line at P is $\lim_{h\to 0} \dfrac{f(x_0 + h) - f(x_0)}{h}$.

DEFINITIONS Slope, Tangent Line

The **slope of the curve** $y = f(x)$ at the point $P(x_0, f(x_0))$ is the number

$$m = \lim_{h\to 0} \frac{f(x_0 + h) - f(x_0)}{h} \qquad \text{(provided the limit exists).}$$

The **tangent line** to the curve at P is the line through P with this slope.

Whenever we make a new definition, we try it on familiar objects to be sure it is consistent with results we expect in familiar cases. Example 2 shows that the new definition of slope agrees with the old definition from Section 1.2 when we apply it to nonvertical lines.

EXAMPLE 2 Testing the Definition

Show that the line $y = mx + b$ is its own tangent at any point $(x_0, mx_0 + b)$.

Solution We let $f(x) = mx + b$ and organize the work into three steps.

1. *Find $f(x_0)$ and $f(x_0 + h)$.*

$$\begin{aligned} f(x_0) &= mx_0 + b \\ f(x_0 + h) &= m(x_0 + h) + b = mx_0 + mh + b \end{aligned}$$

2. *Find the slope $\lim_{h\to 0}(f(x_0 + h) - f(x_0))/h$.*

$$\begin{aligned} \lim_{h\to 0} \frac{f(x_0 + h) - f(x_0)}{h} &= \lim_{h\to 0} \frac{(mx_0 + mh + b) - (mx_0 + b)}{h} \\ &= \lim_{h\to 0} \frac{mh}{h} = m \end{aligned}$$

3. *Find the tangent line using the point-slope equation.* The tangent line at the point $(x_0, mx_0 + b)$ is

$$\begin{aligned} y &= (mx_0 + b) + m(x - x_0) \\ y &= mx_0 + b + mx - mx_0 \\ y &= mx + b. \end{aligned}$$

■

Let's summarize the steps in Example 2.

Finding the Tangent to the Curve $y = f(x)$ at (x_0, y_0)

1. Calculate $f(x_0)$ and $f(x_0 + h)$.
2. Calculate the slope
$$m = \lim_{h\to 0} \frac{f(x_0 + h) - f(x_0)}{h}.$$
3. If the limit exists, find the tangent line as
$$y = y_0 + m(x - x_0).$$

EXAMPLE 3 Slope and Tangent to $y = 1/x,\ x \neq 0$

(a) Find the slope of the curve $y = 1/x$ at $x = a \neq 0$.

(b) Where does the slope equal $-1/4$?

(c) What happens to the tangent to the curve at the point $(a, 1/a)$ as a changes?

Solution

(a) Here $f(x) = 1/x$. The slope at $(a, 1/a)$ is

$$\begin{aligned}\lim_{h\to 0}\frac{f(a+h)-f(a)}{h} &= \lim_{h\to 0}\frac{\dfrac{1}{a+h}-\dfrac{1}{a}}{h}\\ &= \lim_{h\to 0}\frac{1}{h}\,\frac{a-(a+h)}{a(a+h)}\\ &= \lim_{h\to 0}\frac{-h}{ha(a+h)}\\ &= \lim_{h\to 0}\frac{-1}{a(a+h)} = -\frac{1}{a^2}.\end{aligned}$$

Notice how we had to keep writing "$\lim_{h\to 0}$" before each fraction until the stage where we could evaluate the limit by substituting $h = 0$. The number a may be positive or negative, but not 0.

(b) The slope of $y = 1/x$ at the point where $x = a$ is $-1/a^2$. It will be $-1/4$ provided that

$$-\frac{1}{a^2} = -\frac{1}{4}.$$

This equation is equivalent to $a^2 = 4$, so $a = 2$ or $a = -2$. The curve has slope $-1/4$ at the two points $(2, 1/2)$ and $(-2, -1/2)$ (Figure 2.68).

(c) Notice that the slope $-1/a^2$ is always negative if $a \neq 0$. As $a \to 0^+$, the slope approaches $-\infty$ and the tangent becomes increasingly steep (Figure 2.69). We see this situation again as $a \to 0^-$. As a moves away from the origin in either direction, the slope approaches 0^- and the tangent levels off to become horizontal. ■

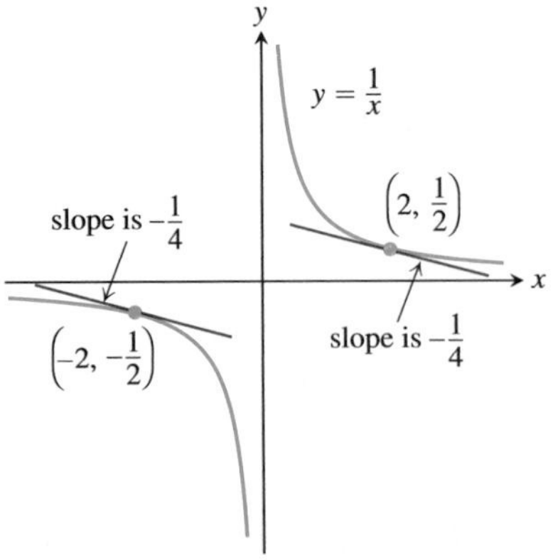

FIGURE 2.68 The two tangent lines to $y = 1/x$ having slope $-1/4$ (Example 3).

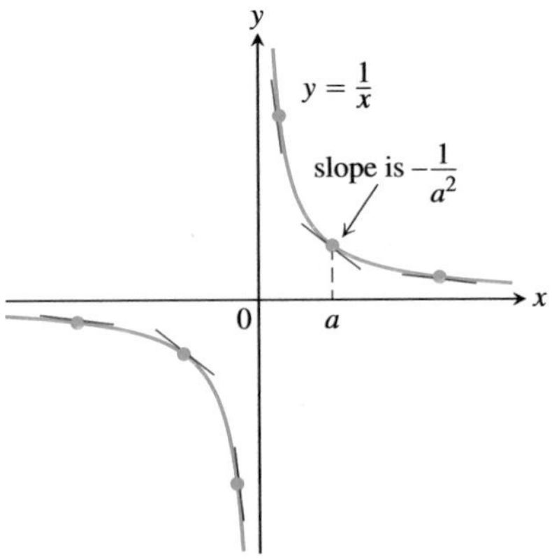

FIGURE 2.69 The tangent slopes, steep near the origin, become more gradual as the point of tangency moves away.

Rates of Change: Derivative at a Point

The expression

$$\frac{f(x_0 + h) - f(x_0)}{h}$$

is called the **difference quotient of f at x_0 with increment h**. If the difference quotient has a limit as h approaches zero, that limit is called the **derivative of f at x_0**. If we interpret the difference quotient as a secant slope, the derivative gives the slope of the curve and tangent at the point where $x = x_0$. If we interpret the difference quotient as an average rate of change, as we did in Section 2.1, the derivative gives the function's rate of change with respect to x at the point $x = x_0$. The derivative is one of the two most important mathematical objects considered in calculus. We begin a thorough study of it in Chapter 3. The other important object is the integral, and we initiate its study in Chapter 5.

EXAMPLE 4 Instantaneous Speed (Continuation of Section 2.1, Examples 1 and 2)

In Examples 1 and 2 in Section 2.1, we studied the speed of a rock falling freely from rest near the surface of the earth. We knew that the rock fell $y = 16t^2$ feet during the first t sec, and we used a sequence of average rates over increasingly short intervals to estimate the rock's speed at the instant $t = 1$. Exactly what *was* the rock's speed at this time?

Solution We let $f(t) = 16t^2$. The average speed of the rock over the interval between $t = 1$ and $t = 1 + h$ seconds was

$$\frac{f(1 + h) - f(1)}{h} = \frac{16(1 + h)^2 - 16(1)^2}{h} = \frac{16(h^2 + 2h)}{h} = 16(h + 2).$$

The rock's speed at the instant $t = 1$ was

$$\lim_{h \to 0} 16(h + 2) = 16(0 + 2) = 32 \text{ ft/sec}.$$

Our original estimate of 32 ft/ sec was right. ■

Summary

We have been discussing slopes of curves, lines tangent to a curve, the rate of change of a function, the limit of the difference quotient, and the derivative of a function at a point. All of these ideas refer to the same thing, summarized here:

1. The slope of $y = f(x)$ at $x = x_0$
2. The slope of the tangent to the curve $y = f(x)$ at $x = x_0$
3. The rate of change of $f(x)$ with respect to x at $x = x_0$
4. The derivative of f at $x = x_0$
5. The limit of the difference quotient, $\lim_{h \to 0} \dfrac{f(x_0 + h) - f(x_0)}{h}$

EXERCISES 2.7

Slopes and Tangent Lines

In Exercises 1–4, use the grid and a straight edge to make a rough estimate of the slope of the curve (in y-units per x-unit) at the points P_1 and P_2. Graphs can shift during a press run, so your estimates may be somewhat different from those in the back of the book.

1.

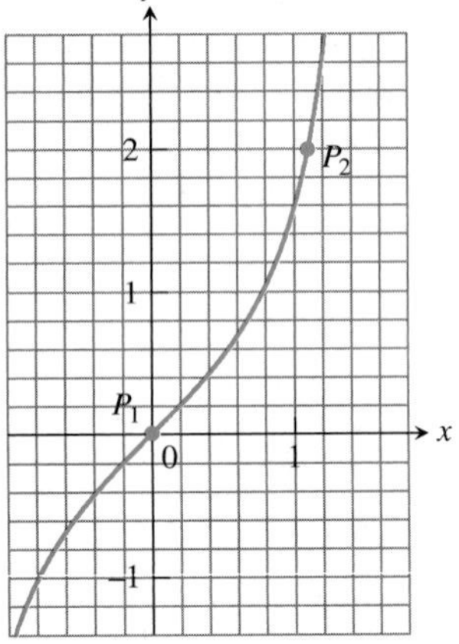

2.

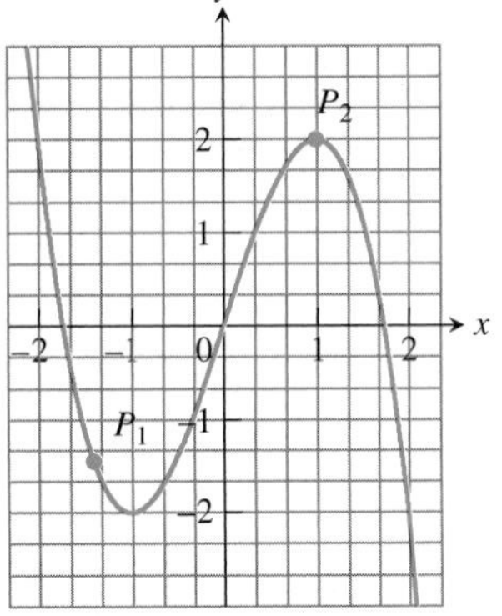

3.

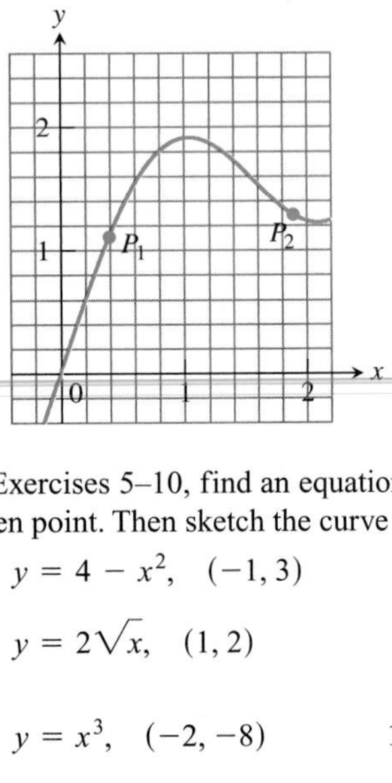

4.

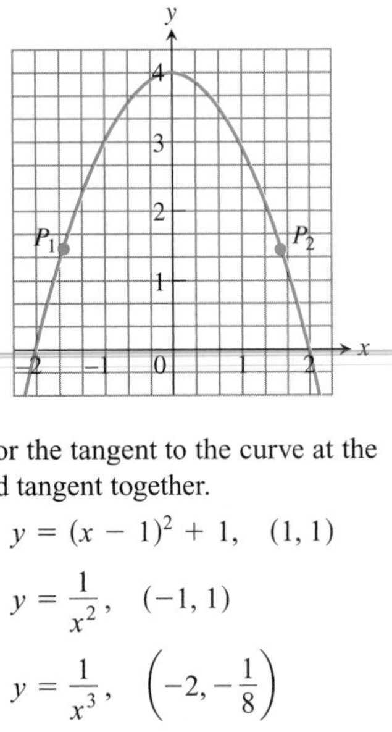

In Exercises 5–10, find an equation for the tangent to the curve at the given point. Then sketch the curve and tangent together.

5. $y = 4 - x^2, \quad (-1, 3)$

6. $y = (x - 1)^2 + 1, \quad (1, 1)$

7. $y = 2\sqrt{x}, \quad (1, 2)$

8. $y = \dfrac{1}{x^2}, \quad (-1, 1)$

9. $y = x^3, \quad (-2, -8)$

10. $y = \dfrac{1}{x^3}, \quad \left(-2, -\dfrac{1}{8}\right)$

In Exercises 11–18, find the slope of the function's graph at the given point. Then find an equation for the line tangent to the graph there.

11. $f(x) = x^2 + 1, \quad (2, 5)$

12. $f(x) = x - 2x^2, \quad (1, -1)$

13. $g(x) = \dfrac{x}{x - 2}, \quad (3, 3)$

14. $g(x) = \dfrac{8}{x^2}, \quad (2, 2)$

15. $h(t) = t^3, \quad (2, 8)$

16. $h(t) = t^3 + 3t, \quad (1, 4)$

17. $f(x) = \sqrt{x}, \quad (4, 2)$

18. $f(x) = \sqrt{x + 1}, \quad (8, 3)$

In Exercises 19–22, find the slope of the curve at the point indicated.

19. $y = 5x^2, \quad x = -1$

20. $y = 1 - x^2, \quad x = 2$

21. $y = \dfrac{1}{x - 1}, \quad x = 3$

22. $y = \dfrac{x - 1}{x + 1}, \quad x = 0$

Tangent Lines with Specified Slopes

At what points do the graphs of the functions in Exercises 23 and 24 have horizontal tangents?

23. $f(x) = x^2 + 4x - 1$

24. $g(x) = x^3 - 3x$

25. Find equations of all lines having slope -1 that are tangent to the curve $y = 1/(x - 1)$.

26. Find an equation of the straight line having slope $1/4$ that is tangent to the curve $y = \sqrt{x}$.

Rates of Change

27. Object dropped from a tower An object is dropped from the top of a 100-m-high tower. Its height above ground after t sec is $100 - 4.9t^2$ m. How fast is it falling 2 sec after it is dropped?

28. Speed of a rocket At t sec after liftoff, the height of a rocket is $3t^2$ ft. How fast is the rocket climbing 10 sec after liftoff?

29. Circle's changing area What is the rate of change of the area of a circle ($A = \pi r^2$) with respect to the radius when the radius is $r = 3$?

30. Ball's changing volume What is the rate of change of the volume of a ball ($V = (4/3)\pi r^3$) with respect to the radius when the radius is $r = 2$?

Testing for Tangents

31. Does the graph of

$$f(x) = \begin{cases} x^2 \sin(1/x), & x \neq 0 \\ 0, & x = 0 \end{cases}$$

have a tangent at the origin? Give reasons for your answer.

32. Does the graph of

$$g(x) = \begin{cases} x \sin(1/x), & x \neq 0 \\ 0, & x = 0 \end{cases}$$

have a tangent at the origin? Give reasons for your answer.

Vertical Tangents

We say that the curve $y = f(x)$ has a **vertical tangent** at the point where $x = x_0$ if $\lim_{h \to 0} (f(x_0 + h) - f(x_0))/h = \infty$ or $-\infty$.

Vertical tangent at $x = 0$ (see accompanying figure):

$$\lim_{h \to 0} \frac{f(0 + h) - f(0)}{h} = \lim_{h \to 0} \frac{h^{1/3} - 0}{h}$$

$$= \lim_{h \to 0} \frac{1}{h^{2/3}} = \infty$$

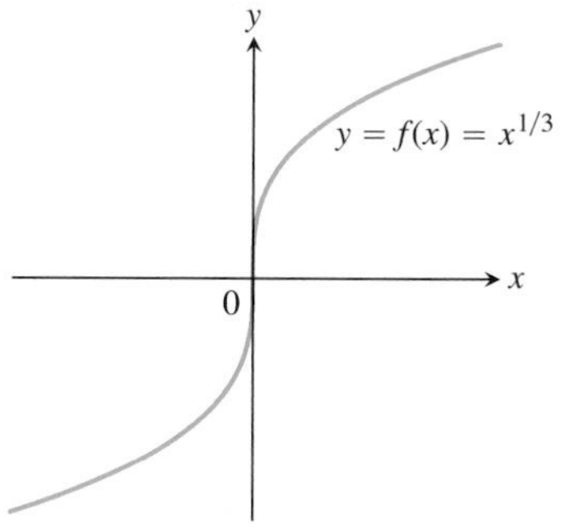

VERTICAL TANGENT AT ORIGIN

No vertical tangent at $x = 0$ (see next figure):

$$\lim_{h\to 0} \frac{g(0 + h) - g(0)}{h} = \lim_{h\to 0} \frac{h^{2/3} - 0}{h} = \lim_{h\to 0} \frac{1}{h^{1/3}}$$

does not exist, because the limit is ∞ from the right and $-\infty$ from the left.

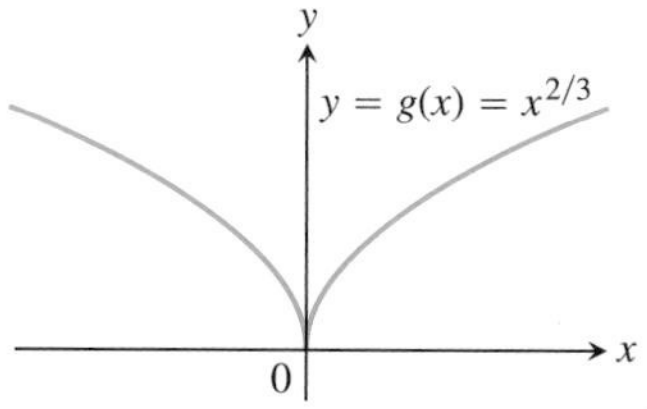

NO VERTICAL TANGENT AT ORIGIN

33. Does the graph of

$$f(x) = \begin{cases} -1, & x < 0 \\ 0, & x = 0 \\ 1, & x > 0 \end{cases}$$

have a vertical tangent at the origin? Give reasons for your answer.

34. Does the graph of

$$U(x) = \begin{cases} 0, & x < 0 \\ 1, & x \ge 0 \end{cases}$$

have a vertical tangent at the point (0, 1)? Give reasons for your answer.

T **a.** Graph the curves in Exercises 35–44. Where do the graphs appear to have vertical tangents?

b. Confirm your findings in part (a) with limit calculations. But before you do, read the introduction to Exercises 33 and 34.

35. $y = x^{2/5}$

36. $y = x^{4/5}$

37. $y = x^{1/5}$

38. $y = x^{3/5}$

39. $y = 4x^{2/5} - 2x$

40. $y = x^{5/3} - 5x^{2/3}$

41. $y = x^{2/3} - (x - 1)^{1/3}$

42. $y = x^{1/3} + (x - 1)^{1/3}$

43. $y = \begin{cases} -\sqrt{|x|}, & x \le 0 \\ \sqrt{x}, & x > 0 \end{cases}$

44. $y = \sqrt{|4 - x|}$

COMPUTER EXPLORATIONS

Graphing Secant and Tangent Lines

Use a CAS to perform the following steps for the functions in Exercises 45–48.

a. Plot $y = f(x)$ over the interval $(x_0 - 1/2) \le x \le (x_0 + 3)$.

b. Holding x_0 fixed, the difference quotient

$$q(h) = \frac{f(x_0 + h) - f(x_0)}{h}$$

at x_0 becomes a function of the step size h. Enter this function into your CAS workspace.

c. Find the limit of q as $h \to 0$.

d. Define the secant lines $y = f(x_0) + q \cdot (x - x_0)$ for $h = 3, 2$, and 1. Graph them together with f and the tangent line over the interval in part (a).

45. $f(x) = x^3 + 2x, \quad x_0 = 0$

46. $f(x) = x + \dfrac{5}{x}, \quad x_0 = 1$

47. $f(x) = x + \sin(2x), \quad x_0 = \pi/2$

48. $f(x) = \cos x + 4\sin(2x), \quad x_0 = \pi$

Chapter 2 Questions to Guide Your Review

1. What is the average rate of change of the function $g(t)$ over the interval from $t = a$ to $t = b$? How is it related to a secant line?
2. What limit must be calculated to find the rate of change of a function $g(t)$ at $t = t_0$?
3. What is an informal or intuitive definition of the limit

$$\lim_{x\to x_0} f(x) = L?$$

Why is the definition "informal"? Give examples.

4. Does the existence and value of the limit of a function $f(x)$ as x approaches x_0 ever depend on what happens at $x = x_0$? Explain and give examples.
5. What function behaviors might occur for which the limit may fail to exist? Give examples.
6. What theorems are available for calculating limits? Give examples of how the theorems are used.

7. How are one-sided limits related to limits? How can this relationship sometimes be used to calculate a limit or prove it does not exist? Give examples.

8. What is the value of $\lim_{\theta \to 0}((\sin\theta)/\theta)$? Does it matter whether θ is measured in degrees or radians? Explain.

9. What exactly does $\lim_{x \to x_0} f(x) = L$ mean? Give an example in which you find a $\delta > 0$ for a given f, L, x_0, and $\epsilon > 0$ in the precise definition of limit.

10. Give precise definitions of the following statements.

 a. $\lim_{x \to 2^-} f(x) = 5$ **b.** $\lim_{x \to 2^+} f(x) = 5$

 c. $\lim_{x \to 2} f(x) = \infty$ **d.** $\lim_{x \to 2} f(x) = -\infty$

11. What exactly do $\lim_{x \to \infty} f(x) = L$ and $\lim_{x \to -\infty} f(x) = L$ mean? Give examples.

12. What are $\lim_{x \to \pm\infty} k$ (k a constant) and $\lim_{x \to \pm\infty}(1/x)$? How do you extend these results to other functions? Give examples.

13. How do you find the limit of a rational function as $x \to \pm\infty$? Give examples.

14. What are horizontal, vertical, and oblique asymptotes? Give examples.

15. What conditions must be satisfied by a function if it is to be continuous at an interior point of its domain? At an endpoint?

16. How can looking at the graph of a function help you tell where the function is continuous?

17. What does it mean for a function to be right-continuous at a point? Left-continuous? How are continuity and one-sided continuity related?

18. What can be said about the continuity of polynomials? Of rational functions? Of trigonometric functions? Of rational powers and algebraic combinations of functions? Of composites of functions? Of absolute values of functions?

19. Under what circumstances can you extend a function $f(x)$ to be continuous at a point $x = c$? Give an example.

20. What does it mean for a function to be continuous on an interval?

21. What does it mean for a function to be continuous? Give examples to illustrate the fact that a function that is not continuous on its entire domain may still be continuous on selected intervals within the domain.

22. What are the basic types of discontinuity? Give an example of each. What is a removable discontinuity? Give an example.

23. What does it mean for a function to have the Intermediate Value Property? What conditions guarantee that a function has this property over an interval? What are the consequences for graphing and solving the equation $f(x) = 0$?

24. It is often said that a function is continuous if you can draw its graph without having to lift your pen from the paper. Why is that?

25. What does it mean for a line to be tangent to a curve C at a point P?

26. What is the significance of the formula

$$\lim_{h \to 0} \frac{f(x + h) - f(x)}{h}?$$

Interpret the formula geometrically and physically.

27. How do you find the tangent to the curve $y = f(x)$ at a point (x_0, y_0) on the curve?

28. How does the slope of the curve $y = f(x)$ at $x = x_0$ relate to the function's rate of change with respect to x at $x = x_0$? To the derivative of f at x_0?

Chapter 2 Practice Exercises

Limits and Continuity

1. Graph the function

$$f(x) = \begin{cases} 1, & x \le -1 \\ -x, & -1 < x < 0 \\ 1, & x = 0 \\ -x, & 0 < x < 1 \\ 1, & x \ge 1. \end{cases}$$

Then discuss, in detail, limits, one-sided limits, continuity, and one-sided continuity of f at $x = -1, 0$, and 1. Are any of the discontinuities removable? Explain.

2. Repeat the instructions of Exercise 1 for

$$f(x) = \begin{cases} 0, & x \le -1 \\ 1/x, & 0 < |x| < 1 \\ 0, & x = 1 \\ 1, & x > 1. \end{cases}$$

3. Suppose that $f(t)$ and $g(t)$ are defined for all t and that $\lim_{t \to t_0} f(t) = -7$ and $\lim_{t \to t_0} g(t) = 0$. Find the limit as $t \to t_0$ of the following functions.

 a. $3f(t)$ **b.** $(f(t))^2$

 c. $f(t) \cdot g(t)$ **d.** $\dfrac{f(t)}{g(t) - 7}$

 e. $\cos(g(t))$ **f.** $|f(t)|$

 g. $f(t) + g(t)$ **h.** $1/f(t)$

4. Suppose that $f(x)$ and $g(x)$ are defined for all x and that $\lim_{x\to 0} f(x) = 1/2$ and $\lim_{x\to 0} g(x) = \sqrt{2}$. Find the limits as $x \to 0$ of the following functions.

   a. $-g(x)$ b. $g(x) \cdot f(x)$
   c. $f(x) + g(x)$ d. $1/f(x)$
   e. $x + f(x)$ f. $\dfrac{f(x) \cdot \cos x}{x - 1}$

In Exercises 5 and 6, find the value that $\lim_{x\to 0} g(x)$ must have if the given limit statements hold.

5. $\lim_{x\to 0}\left(\dfrac{4 - g(x)}{x}\right) = 1$ 6. $\lim_{x\to -4}\left(x \lim_{x\to 0} g(x)\right) = 2$

7. On what intervals are the following functions continuous?

   a. $f(x) = x^{1/3}$ b. $g(x) = x^{3/4}$
   c. $h(x) = x^{-2/3}$ d. $k(x) = x^{-1/6}$

8. On what intervals are the following functions continuous?

   a. $f(x) = \tan x$ b. $g(x) = \csc x$
   c. $h(x) = \dfrac{\cos x}{x - \pi}$ d. $k(x) = \dfrac{\sin x}{x}$

Finding Limits

In Exercises 9–16, find the limit or explain why it does not exist.

9. $\lim \dfrac{x^2 - 4x + 4}{x^3 + 5x^2 - 14x}$

   a. as $x \to 0$ b. as $x \to 2$

10. $\lim \dfrac{x^2 + x}{x^5 + 2x^4 + x^3}$

   a. as $x \to 0$ b. as $x \to -1$

11. $\lim_{x\to 1} \dfrac{1 - \sqrt{x}}{1 - x}$ 12. $\lim_{x\to a} \dfrac{x^2 - a^2}{x^4 - a^4}$

13. $\lim_{h\to 0} \dfrac{(x + h)^2 - x^2}{h}$ 14. $\lim_{x\to 0} \dfrac{(x + h)^2 - x^2}{h}$

15. $\lim_{x\to 0} \dfrac{\frac{1}{2 + x} - \frac{1}{2}}{x}$ 16. $\lim_{x\to 0} \dfrac{(2 + x)^3 - 8}{x}$

In Exercises 17–20, find the limit of $g(x)$ as x approaches the indicated value.

17. $\lim_{x\to 0^+} (4g(x))^{1/3} = 2$ 18. $\lim_{x\to \sqrt{5}} \dfrac{1}{x + g(x)} = 2$

19. $\lim_{x\to 1} \dfrac{3x^2 + 1}{g(x)} = \infty$ 20. $\lim_{x\to -2} \dfrac{5 - x^2}{\sqrt{g(x)}} = 0$

Limits at Infinity

Find the limits in Exercises 21–30.

21. $\lim_{x\to \infty} \dfrac{2x + 3}{5x + 7}$ 22. $\lim_{x\to -\infty} \dfrac{2x^2 + 3}{5x^2 + 7}$

23. $\lim_{x\to -\infty} \dfrac{x^2 - 4x + 8}{3x^3}$ 24. $\lim_{x\to \infty} \dfrac{1}{x^2 - 7x + 1}$

25. $\lim_{x\to -\infty} \dfrac{x^2 - 7x}{x + 1}$ 26. $\lim_{x\to \infty} \dfrac{x^4 + x^3}{12x^3 + 128}$

27. $\lim_{x\to \infty} \dfrac{\sin x}{\lfloor x \rfloor}$ (If you have a grapher, try graphing the function for $-5 \le x \le 5$.)

28. $\lim_{\theta\to \infty} \dfrac{\cos \theta - 1}{\theta}$ (If you have a grapher, try graphing $f(x) = x(\cos(1/x) - 1)$ near the origin to "see" the limit at infinity.)

29. $\lim_{x\to \infty} \dfrac{x + \sin x + 2\sqrt{x}}{x + \sin x}$ 30. $\lim_{x\to \infty} \dfrac{x^{2/3} + x^{-1}}{x^{2/3} + \cos^2 x}$

Continuous Extension

31. Can $f(x) = x(x^2 - 1)/|x^2 - 1|$ be extended to be continuous at $x = 1$ or -1? Give reasons for your answers. (Graph the function—you will find the graph interesting.)

32. Explain why the function $f(x) = \sin(1/x)$ has no continuous extension to $x = 0$.

T In Exercises 33–36, graph the function to see whether it appears to have a continuous extension to the given point a. If it does, use Trace and Zoom to find a good candidate for the extended function's value at a. If the function does not appear to have a continuous extension, can it be extended to be continuous from the right or left? If so, what do you think the extended function's value should be?

33. $f(x) = \dfrac{x - 1}{x - \sqrt[4]{x}}$, $a = 1$ 34. $g(\theta) = \dfrac{5\cos\theta}{4\theta - 2\pi}$, $a = \pi/2$

35. $h(t) = (1 + |t|)^{1/t}$, $a = 0$ 36. $k(x) = \dfrac{x}{1 - 2^{|x|}}$, $a = 0$

Roots

T 37. Let $f(x) = x^3 - x - 1$.

   a. Show that f has a zero between -1 and 2.
   b. Solve the equation $f(x) = 0$ graphically with an error of magnitude at most 10^{-8}.
   c. It can be shown that the exact value of the solution in part (b) is
   $$\left(\frac{1}{2} + \frac{\sqrt{69}}{18}\right)^{1/3} + \left(\frac{1}{2} - \frac{\sqrt{69}}{18}\right)^{1/3}$$
   Evaluate this exact answer and compare it with the value you found in part (b).

T 38. Let $f(\theta) = \theta^3 - 2\theta + 2$.

   a. Show that f has a zero between -2 and 0.
   b. Solve the equation $f(\theta) = 0$ graphically with an error of magnitude at most 10^{-4}.
   c. It can be shown that the exact value of the solution in part (b) is
   $$\left(\sqrt{\frac{19}{27}} - 1\right)^{1/3} - \left(\sqrt{\frac{19}{27}} + 1\right)^{1/3}$$
   Evaluate this exact answer and compare it with the value you found in part (b).

Chapter 2 Additional and Advanced Exercises

T **1. Assigning a value to 0^0** The rules of exponents (see Appendix 9) tell us that $a^0 = 1$ if a is any number different from zero. They also tell us that $0^n = 0$ if n is any positive number.

If we tried to extend these rules to include the case 0^0, we would get conflicting results. The first rule would say $0^0 = 1$, whereas the second would say $0^0 = 0$.

We are not dealing with a question of right or wrong here. Neither rule applies as it stands, so there is no contradiction. We could, in fact, define 0^0 to have any value we wanted as long as we could persuade others to agree.

What value would you like 0^0 to have? Here is an example that might help you to decide. (See Exercise 2 below for another example.)

a. Calculate x^x for $x = 0.1, 0.01, 0.001$, and so on as far as your calculator can go. Record the values you get. What pattern do you see?

b. Graph the function $y = x^x$ for $0 < x \le 1$. Even though the function is not defined for $x \le 0$, the graph will approach the y-axis from the right. Toward what y-value does it seem to be headed? Zoom in to further support your idea.

T **2. A reason you might want 0^0 to be something other than 0 or 1** As the number x increases through positive values, the numbers $1/x$ and $1/(\ln x)$ both approach zero. What happens to the number

$$f(x) = \left(\frac{1}{x}\right)^{1/(\ln x)}$$

as x increases? Here are two ways to find out.

a. Evaluate f for $x = 10, 100, 1000$, and so on as far as your calculator can reasonably go. What pattern do you see?

b. Graph f in a variety of graphing windows, including windows that contain the origin. What do you see? Trace the y-values along the graph. What do you find?

3. Lorentz contraction In relativity theory, the length of an object, say a rocket, appears to an observer to depend on the speed at which the object is traveling with respect to the observer. If the observer measures the rocket's length as L_0 at rest, then at speed v the length will appear to be

$$L = L_0\sqrt{1 - \frac{v^2}{c^2}}.$$

This equation is the Lorentz contraction formula. Here, c is the speed of light in a vacuum, about 3×10^8 m/sec. What happens to L as v increases? Find $\lim_{v \to c^-} L$. Why was the left-hand limit needed?

4. Controlling the flow from a draining tank Torricelli's law says that if you drain a tank like the one in the figure shown, the rate y at which water runs out is a constant times the square root of the water's depth x. The constant depends on the size and shape of the exit valve.

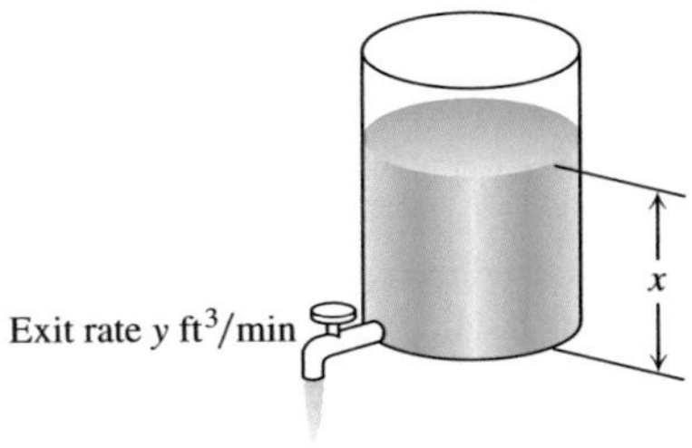

Suppose that $y = \sqrt{x}/2$ for a certain tank. You are trying to maintain a fairly constant exit rate by adding water to the tank with a hose from time to time. How deep must you keep the water if you want to maintain the exit rate

a. within 0.2 ft^3/min of the rate $y_0 = 1$ ft^3/min?

b. within 0.1 ft^3/min of the rate $y_0 = 1$ ft^3/min?

5. Thermal expansion in precise equipment As you may know, most metals expand when heated and contract when cooled. The dimensions of a piece of laboratory equipment are sometimes so critical that the shop where the equipment is made must be held at the same temperature as the laboratory where the equipment is to be used. A typical aluminum bar that is 10 cm wide at 70°F will be

$$y = 10 + (t - 70) \times 10^{-4}$$

centimeters wide at a nearby temperature t. Suppose that you are using a bar like this in a gravity wave detector, where its width must stay within 0.0005 cm of the ideal 10 cm. How close to $t_0 = 70°\text{F}$ must you maintain the temperature to ensure that this tolerance is not exceeded?

6. Stripes on a measuring cup The interior of a typical 1-L measuring cup is a right circular cylinder of radius 6 cm (see accompanying figure). The volume of water we put in the cup is therefore a function of the level h to which the cup is filled, the formula being

$$V = \pi 6^2 h = 36\pi h.$$

How closely must we measure h to measure out 1 L of water (1000 cm^3) with an error of no more than 1% (10 cm^3)?

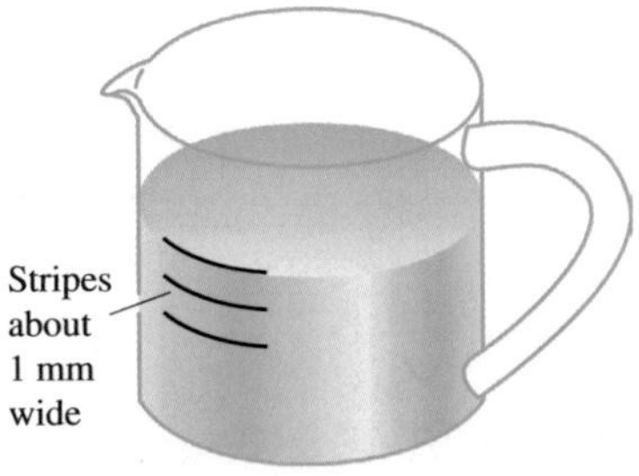

(a)

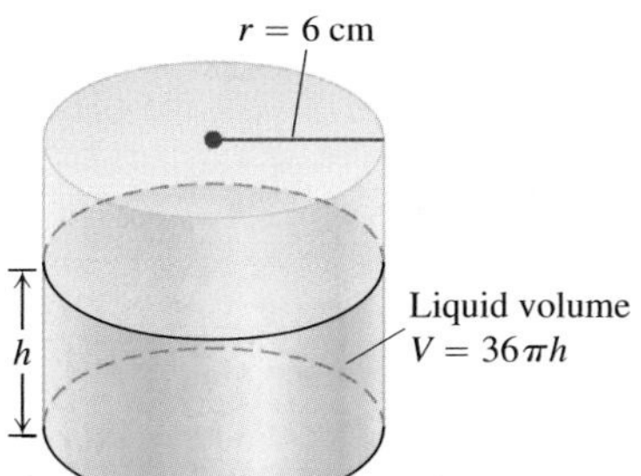

(b)

A 1-L measuring cup (a), modeled as a right circular cylinder (b) of radius $r = 6$ cm

Precise Definition of Limit

In Exercises 7–10, use the formal definition of limit to prove that the function is continuous at x_0.

7. $f(x) = x^2 - 7, \quad x_0 = 1$ **8.** $g(x) = 1/(2x), \quad x_0 = 1/4$

9. $h(x) = \sqrt{2x - 3}, \quad x_0 = 2$ **10.** $F(x) = \sqrt{9 - x}, \quad x_0 = 5$

11. Uniqueness of limits Show that a function cannot have two different limits at the same point. That is, if $\lim_{x\to x_0} f(x) = L_1$ and $\lim_{x\to x_0} f(x) = L_2$, then $L_1 = L_2$.

12. Prove the limit Constant Multiple Rule:
$\lim_{x\to c} kf(x) = k \lim_{x\to c} f(x)$ for any constant k.

13. One-sided limits If $\lim_{x\to 0^+} f(x) = A$ and $\lim_{x\to 0^-} f(x) = B$, find

a. $\lim_{x\to 0^+} f(x^3 - x)$ **b.** $\lim_{x\to 0^-} f(x^3 - x)$

c. $\lim_{x\to 0^+} f(x^2 - x^4)$ **d.** $\lim_{x\to 0^-} f(x^2 - x^4)$

14. Limits and continuity Which of the following statements are true, and which are false? If true, say why; if false, give a counterexample (that is, an example confirming the falsehood).

a. If $\lim_{x\to a} f(x)$ exists but $\lim_{x\to a} g(x)$ does not exist, then $\lim_{x\to a}(f(x) + g(x))$ does not exist.

b. If neither $\lim_{x\to a} f(x)$ nor $\lim_{x\to a} g(x)$ exists, then $\lim_{x\to a} (f(x) + g(x))$ does not exist.

c. If f is continuous at x, then so is $|f|$.

d. If $|f|$ is continuous at a, then so is f.

In Exercises 15 and 16, use the formal definition of limit to prove that the function has a continuous extension to the given value of x.

15. $f(x) = \dfrac{x^2 - 1}{x + 1}, \quad x = -1$ **16.** $g(x) = \dfrac{x^2 - 2x - 3}{2x - 6}, \quad x = 3$

17. A function continuous at only one point Let

$$f(x) = \begin{cases} x, & \text{if } x \text{ is rational} \\ 0, & \text{if } x \text{ is irrational.} \end{cases}$$

a. Show that f is continuous at $x = 0$.

b. Use the fact that every nonempty open interval of real numbers contains both rational and irrational numbers to show that f is not continuous at any nonzero value of x.

18. The Dirichlet ruler function If x is a rational number, then x can be written in a unique way as a quotient of integers m/n where $n > 0$ and m and n have no common factors greater than 1. (We say that such a fraction is in *lowest terms*. For example, 6/4 written in lowest terms is 3/2.) Let $f(x)$ be defined for all x in the interval [0, 1] by

$$f(x) = \begin{cases} 1/n, & \text{if } x = m/n \text{ is a rational number in lowest terms} \\ 0, & \text{if } x \text{ is irrational.} \end{cases}$$

For instance, $f(0) = f(1) = 1$, $f(1/2) = 1/2$, $f(1/3) = f(2/3) = 1/3$, $f(1/4) = f(3/4) = 1/4$, and so on.

a. Show that f is discontinuous at every rational number in [0, 1].

b. Show that f is continuous at every irrational number in [0, 1]. (*Hint*: If ϵ is a given positive number, show that there are only finitely many rational numbers r in [0, 1] such that $f(r) \geq \epsilon$.)

c. Sketch the graph of f. Why do you think f is called the "ruler function"?

19. Antipodal points Is there any reason to believe that there is always a pair of antipodal (diametrically opposite) points on Earth's equator where the temperatures are the same? Explain.

20. If $\lim_{x\to c} (f(x) + g(x)) = 3$ and $\lim_{x\to c} (f(x) - g(x)) = -1$, find $\lim_{x\to c} f(x)g(x)$.

21. Roots of a quadratic equation that is almost linear The equation $ax^2 + 2x - 1 = 0$, where a is a constant, has two roots if $a > -1$ and $a \neq 0$, one positive and one negative:

$$r_+(a) = \frac{-1 + \sqrt{1 + a}}{a}, \qquad r_-(a) = \frac{-1 - \sqrt{1 + a}}{a}.$$

a. What happens to $r_+(a)$ as $a \to 0$? As $a \to -1^+$?

b. What happens to $r_-(a)$ as $a \to 0$? As $a \to -1^+$?

c. Support your conclusions by graphing $r_+(a)$ and $r_-(a)$ as functions of a. Describe what you see.

d. For added support, graph $f(x) = ax^2 + 2x - 1$ simultaneously for $a = 1, 0.5, 0.2, 0.1$, and 0.05.

22. Root of an equation Show that the equation $x + 2\cos x = 0$ has at least one solution.

23. Bounded functions A real-valued function f is **bounded from above** on a set D if there exists a number N such that $f(x) \leq N$ for all x in D. We call N, when it exists, an **upper bound** for f on D and say that f is bounded from above by N. In a similar manner, we say that f is **bounded from below** on D if there exists a number M such that $f(x) \geq M$ for all x in D. We call M, when it exists, a **lower bound** for f on D and say that f is bounded from below by M. We say that f is **bounded** on D if it is bounded from both above and below.

a. Show that f is bounded on D if and only if there exists a number B such that $|f(x)| \leq B$ for all x in D.

b. Suppose that f is bounded from above by N. Show that if $\lim_{x\to x_0} f(x) = L$, then $L \leq N$.

c. Suppose that f is bounded from below by M. Show that if $\lim_{x\to x_0} f(x) = L$, then $L \geq M$.

24. ***Max $\{a, b\}$ and min $\{a, b\}$***

a. Show that the expression

$$\max\{a, b\} = \frac{a+b}{2} + \frac{|a-b|}{2}$$

equals a if $a \geq b$ and equals b if $b \geq a$. In other words, $\max\{a, b\}$ gives the larger of the two numbers a and b.

b. Find a similar expression for $\min\{a, b\}$, the smaller of a and b.

Generalized Limits Involving $\frac{\sin\theta}{\theta}$

The formula $\lim_{\theta\to 0}(\sin\theta)/\theta = 1$ can be generalized. If $\lim_{x\to c} f(x) = 0$ and $f(x)$ is never zero in an open interval containing the point $x = c$, except possibly c itself, then

$$\lim_{x\to c} \frac{\sin f(x)}{f(x)} = 1.$$

Here are several examples.

a. $\displaystyle \lim_{x\to 0} \frac{\sin x^2}{x^2} = 1.$

b. $\displaystyle \lim_{x\to 0} \frac{\sin x^2}{x} = \lim_{x\to 0} \frac{\sin x^2}{x^2} \lim_{x\to 0} \frac{x^2}{x} = 1 \cdot 0 = 0.$

c. $\displaystyle \lim_{x\to -1} \frac{\sin(x^2 - x - 2)}{x+1} = \lim_{x\to -1} \frac{\sin(x^2 - x - 2)}{(x^2 - x - 2)} \cdot$

$\displaystyle \lim_{x\to -1} \frac{(x^2 - x - 2)}{x+1} = 1 \cdot \lim_{x\to -1} \frac{(x+1)(x-2)}{x+1} = -3.$

d. $\displaystyle \lim_{x\to 1} \frac{\sin\left(1-\sqrt{x}\right)}{x-1} = \lim_{x\to 1} \frac{\sin\left(1-\sqrt{x}\right)}{1-\sqrt{x}} \frac{1-\sqrt{x}}{x-1} =$

$\displaystyle 1 \cdot \lim_{x\to 1} \frac{\left(1-\sqrt{x}\right)\left(1+\sqrt{x}\right)}{(x-1)\left(1+\sqrt{x}\right)} = \lim_{x\to 1} \frac{1-x}{(x-1)\left(1+\sqrt{x}\right)} = -\frac{1}{2}.$

Find the limits in Exercises 25–30.

25. $\displaystyle \lim_{x\to 0} \frac{\sin(1-\cos x)}{x}$

26. $\displaystyle \lim_{x\to 0^+} \frac{\sin x}{\sin\sqrt{x}}$

27. $\displaystyle \lim_{x\to 0} \frac{\sin(\sin x)}{x}$

28. $\displaystyle \lim_{x\to 0} \frac{\sin(x^2 + x)}{x}$

29. $\displaystyle \lim_{x\to 2} \frac{\sin(x^2 - 4)}{x-2}$

30. $\displaystyle \lim_{x\to 9} \frac{\sin\left(\sqrt{x} - 3\right)}{x-9}$

Chapter 2 Technology Application Projects

Mathematica-Maple Module

Take It to the Limit

Part I

Part II (Zero Raised to the Power Zero: What Does it Mean?)

Part III (One-Sided Limits)

Visualize and interpret the limit concept through graphical and numerical explorations.

Part IV (What a Difference a Power Makes)

See how sensitive limits can be with various powers of x.

Mathematica-Maple Module

Going to Infinity

Part I (Exploring Function Behavior as $x \to \infty$ or $x \to -\infty$)

This module provides four examples to explore the behavior of a function as $x \to \infty$ or $x \to -\infty$.

Part II (Rates of Growth)

Observe graphs that *appear* to be continuous, yet the function is not continuous. Several issues of continuity are explored to obtain results that you may find surprising.

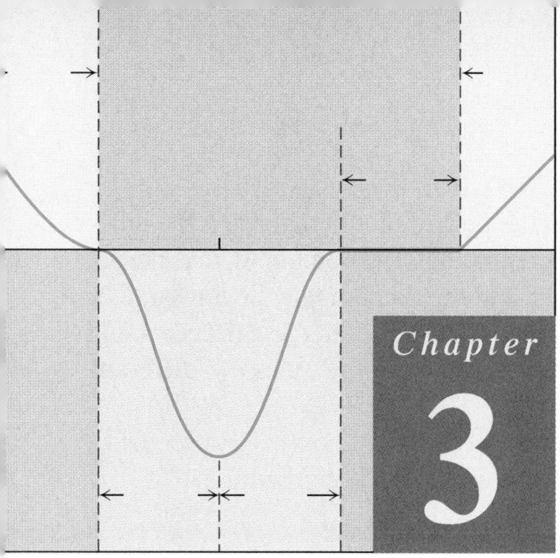

Chapter

3 DIFFERENTIATION

OVERVIEW In Chapter 2, we defined the slope of a curve at a point as the limit of secant slopes. This limit, called a derivative, measures the rate at which a function changes, and it is one of the most important ideas in calculus. Derivatives are used to calculate velocity and acceleration, to estimate the rate of spread of a disease, to set levels of production so as to maximize efficiency, to find the best dimensions of a cylindrical can, to find the age of a prehistoric artifact, and for many other applications. In this chapter, we develop techniques to calculate derivatives easily and learn how to use derivatives to approximate complicated functions.

3.1 The Derivative as a Function

HISTORICAL ESSAY

The Derivative

At the end of Chapter 2, we defined the slope of a curve $y = f(x)$ at the point where $x = x_0$ to be

$$\lim_{h \to 0} \frac{f(x_0 + h) - f(x_0)}{h}.$$

We called this limit, when it existed, the derivative of f at x_0. We now investigate the derivative as a *function* derived from f by considering the limit at each point of the domain of f.

DEFINITION Derivative Function

The **derivative** of the function $f(x)$ with respect to the variable x is the function f' whose value at x is

$$f'(x) = \lim_{h \to 0} \frac{f(x + h) - f(x)}{h},$$

provided the limit exists.

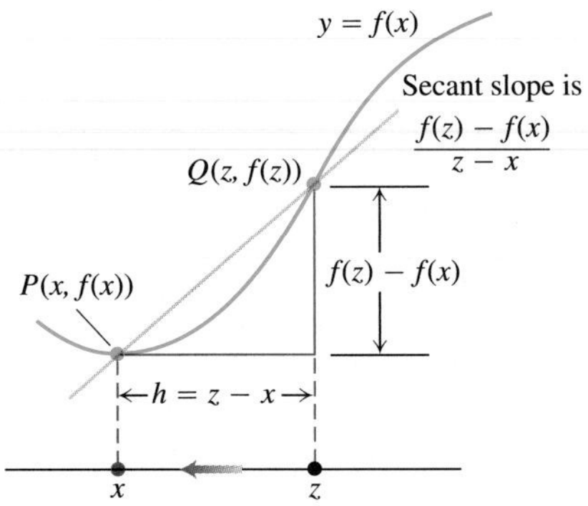

Derivative of f at x is

$$f'(x) = \lim_{h \to 0} \frac{f(x+h) - f(x)}{h}$$

$$= \lim_{z \to x} \frac{f(z) - f(x)}{z - x}$$

FIGURE 3.1 The way we write the difference quotient for the derivative of a function f depends on how we label the points involved.

We use the notation $f(x)$ rather than simply f in the definition to emphasize the independent variable x, with respect to which we are differentiating. The domain of f' is the set of points in the domain of f for which the limit exists, and the domain may be the same or smaller than the domain of f. If f' exists at a particular x, we say that f is **differentiable (has a derivative)** at x. If f' exists at every point in the domain of f, we call f **differentiable**.

If we write $z = x + h$, then $h = z - x$ and h approaches 0 if and only if z approaches x. Therefore, an equivalent definition of the derivative is as follows (see Figure 3.1).

Alternative Formula for the Derivative

$$f'(x) = \lim_{z \to x} \frac{f(z) - f(x)}{z - x}.$$

Calculating Derivatives from the Definition

The process of calculating a derivative is called **differentiation**. To emphasize the idea that differentiation is an operation performed on a function $y = f(x)$, we use the notation

$$\frac{d}{dx} f(x)$$

as another way to denote the derivative $f'(x)$. Examples 2 and 3 of Section 2.7 illustrate the differentiation process for the functions $y = mx + b$ and $y = 1/x$. Example 2 shows that

$$\frac{d}{dx}(mx + b) = m.$$

For instance,

$$\frac{d}{dx}\left(\frac{3}{2}x - 4\right) = \frac{3}{2}.$$

In Example 3, we see that

$$\frac{d}{dx}\left(\frac{1}{x}\right) = -\frac{1}{x^2}.$$

Here are two more examples.

EXAMPLE 1 Applying the Definition

Differentiate $f(x) = \dfrac{x}{x - 1}$.

Solution Here we have $f(x) = \dfrac{x}{x - 1}$

and

$$f(x+h) = \frac{(x+h)}{(x+h)-1}, \text{ so}$$

$$\begin{aligned}
f'(x) &= \lim_{h\to 0} \frac{f(x+h)-f(x)}{h} \\
&= \lim_{h\to 0} \frac{\dfrac{x+h}{x+h-1} - \dfrac{x}{x-1}}{h} \\
&= \lim_{h\to 0} \frac{1}{h} \cdot \frac{(x+h)(x-1) - x(x+h-1)}{(x+h-1)(x-1)} \qquad \frac{a}{b} - \frac{c}{d} = \frac{ad-cb}{bd} \\
&= \lim_{h\to 0} \frac{1}{h} \cdot \frac{-h}{(x+h-1)(x-1)} \\
&= \lim_{h\to 0} \frac{-1}{(x+h-1)(x-1)} = \frac{-1}{(x-1)^2}.
\end{aligned}$$

■

EXAMPLE 2 Derivative of the Square Root Function

(a) Find the derivative of $y = \sqrt{x}$ for $x > 0$.

(b) Find the tangent line to the curve $y = \sqrt{x}$ at $x = 4$.

You will often need to know the derivative of $\sqrt{x}$ for $x > 0$:

$$\frac{d}{dx}\sqrt{x} = \frac{1}{2\sqrt{x}}.$$

Solution

(a) We use the equivalent form to calculate f':

$$\begin{aligned}
f'(x) &= \lim_{z\to x} \frac{f(z)-f(x)}{z-x} \\
&= \lim_{z\to x} \frac{\sqrt{z}-\sqrt{x}}{z-x} \\
&= \lim_{z\to x} \frac{\sqrt{z}-\sqrt{x}}{\left(\sqrt{z}-\sqrt{x}\right)\left(\sqrt{z}+\sqrt{x}\right)} \\
&= \lim_{z\to x} \frac{1}{\sqrt{z}+\sqrt{x}} = \frac{1}{2\sqrt{x}}.
\end{aligned}$$

(b) The slope of the curve at $x = 4$ is

$$f'(4) = \frac{1}{2\sqrt{4}} = \frac{1}{4}.$$

The tangent is the line through the point (4, 2) with slope $1/4$ (Figure 3.2):

$$y = 2 + \frac{1}{4}(x-4)$$

$$y = \frac{1}{4}x + 1.$$

■

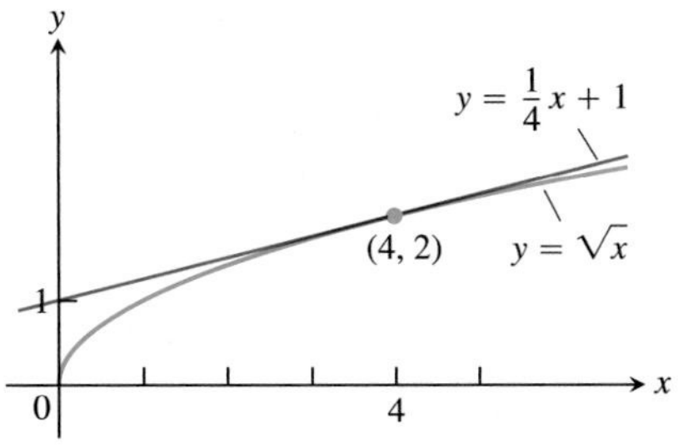

FIGURE 3.2 The curve $y = \sqrt{x}$ and its tangent at (4, 2). The tangent's slope is found by evaluating the derivative at $x = 4$ (Example 2).

We consider the derivative of $y = \sqrt{x}$ when $x = 0$ in Example 6.

Notations

There are many ways to denote the derivative of a function $y = f(x)$, where the independent variable is x and the dependent variable is y. Some common alternative notations for the derivative are

$$f'(x) = y' = \frac{dy}{dx} = \frac{df}{dx} = \frac{d}{dx}f(x) = D(f)(x) = D_x f(x).$$

The symbols d/dx and D indicate the operation of differentiation and are called **differentiation operators**. We read dy/dx as "the derivative of y with respect to x," and df/dx and $(d/dx)f(x)$ as "the derivative of f with respect to x." The "prime" notations y' and f' come from notations that Newton used for derivatives. The d/dx notations are similar to those used by Leibniz. The symbol dy/dx should not be regarded as a ratio (until we introduce the idea of "differentials" in Section 3.8).

Be careful not to confuse the notation $D(f)$ as meaning the domain of the function f instead of the derivative function f'. The distinction should be clear from the context.

To indicate the value of a derivative at a specified number $x = a$, we use the notation

$$f'(a) = \left.\frac{dy}{dx}\right|_{x=a} = \left.\frac{df}{dx}\right|_{x=a} = \left.\frac{d}{dx}f(x)\right|_{x=a}.$$

For instance, in Example 2b we could write

$$f'(4) = \left.\frac{d}{dx}\sqrt{x}\right|_{x=4} = \left.\frac{1}{2\sqrt{x}}\right|_{x=4} = \frac{1}{2\sqrt{4}} = \frac{1}{4}.$$

To evaluate an expression, we sometimes use the right bracket] in place of the vertical bar $|$.

Graphing the Derivative

We can often make a reasonable plot of the derivative of $y = f(x)$ by estimating the slopes on the graph of f. That is, we plot the points $(x, f'(x))$ in the xy-plane and connect them with a smooth curve, which represents $y = f'(x)$.

EXAMPLE 3 Graphing a Derivative

Graph the derivative of the function $y = f(x)$ in Figure 3.3a.

Solution We sketch the tangents to the graph of f at frequent intervals and use their slopes to estimate the values of $f'(x)$ at these points. We plot the corresponding $(x, f'(x))$ pairs and connect them with a smooth curve as sketched in Figure 3.3b. ■

What can we learn from the graph of $y = f'(x)$? At a glance we can see

1. where the rate of change of f is positive, negative, or zero;
2. the rough size of the growth rate at any x and its size in relation to the size of $f(x)$;
3. where the rate of change itself is increasing or decreasing.

Here's another example.

EXAMPLE 4 Concentration of Blood Sugar

On April 23, 1988, the human-powered airplane *Daedalus* flew a record-breaking 119 km from Crete to the island of Santorini in the Aegean Sea, southeast of mainland Greece. Dur-

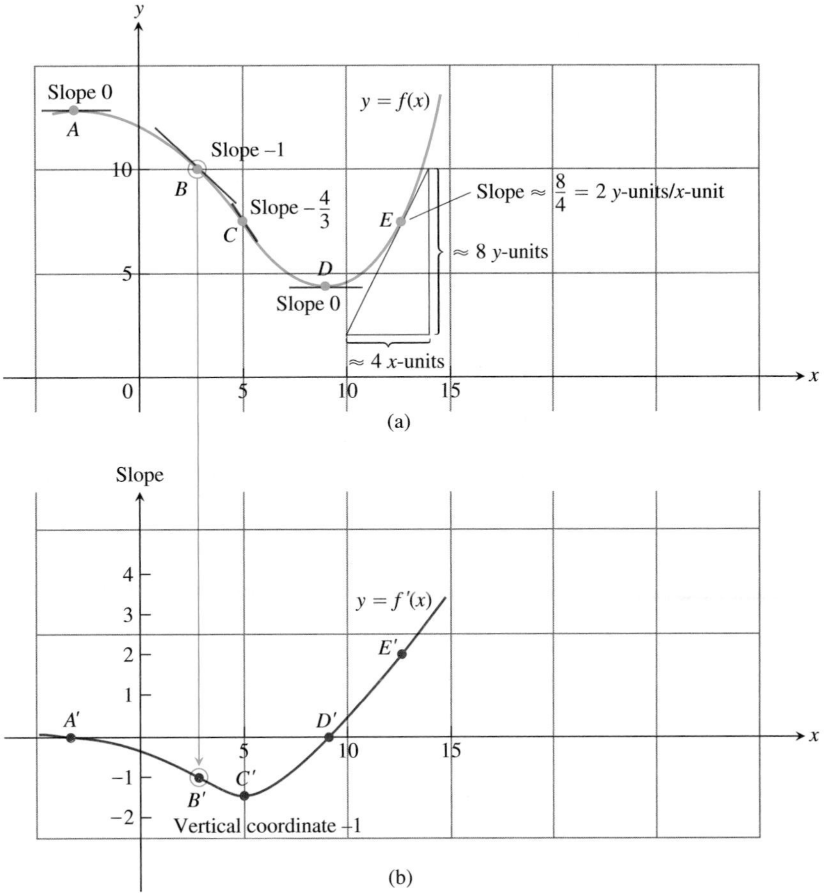

FIGURE 3.3 We made the graph of $y = f'(x)$ in (b) by plotting slopes from the graph of $y = f(x)$ in (a). The vertical coordinate of B' is the slope at B and so on. The graph of f' is a visual record of how the slope of f changes with x.

ing the 6-hour endurance tests before the flight, researchers monitored the prospective pilots' blood-sugar concentrations. The concentration graph for one of the athlete-pilots is shown in Figure 3.4a, where the concentration in milligrams/deciliter is plotted against time in hours.

The graph consists of line segments connecting data points. The constant slope of each segment gives an estimate of the derivative of the concentration between measurements. We calculated the slope of each segment from the coordinate grid and plotted the derivative as a step function in Figure 3.4b. To make the plot for the first hour, for instance, we observed that the concentration increased from about 79 mg/dL to 93 mg/dL. The net increase was $\Delta y = 93 - 79 = 14$ mg/dL. Dividing this by $\Delta t = 1$ hour gave the rate of change as

$$\frac{\Delta y}{\Delta t} = \frac{14}{1} = 14 \text{ mg/dL per hour}.$$

Notice that we can make no estimate of the concentration's rate of change at times $t = 1, 2, \ldots, 5$, where the graph we have drawn for the concentration has a corner and no slope. The derivative step function is not defined at these times. ■

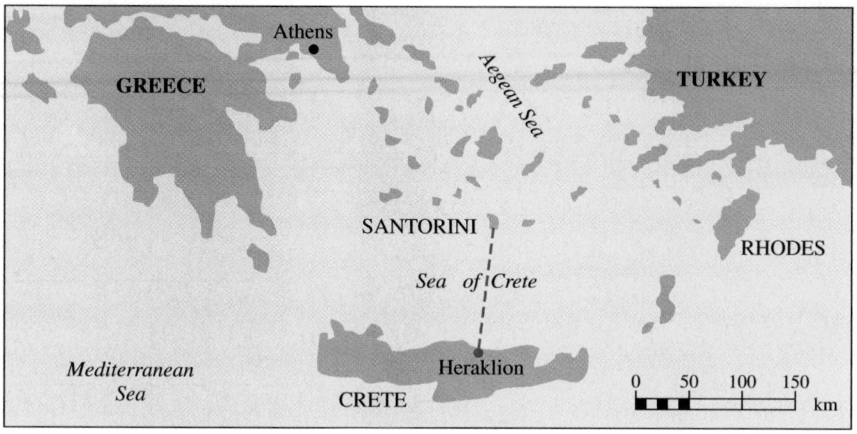

Daedalus's flight path on April 23, 1988

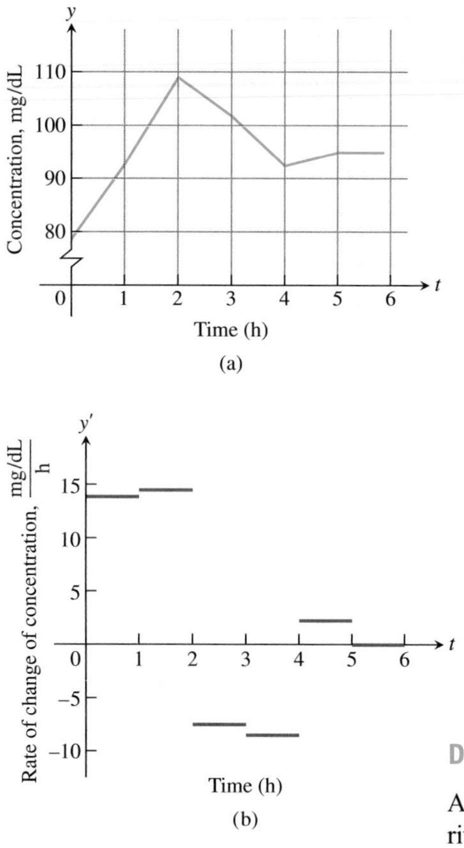

FIGURE 3.4 (a) Graph of the sugar concentration in the blood of a *Daedalus* pilot during a 6-hour preflight endurance test. (b) The derivative of the pilot's blood-sugar concentration shows how rapidly the concentration rose and fell during various portions of the test.

Differentiable on an Interval; One-Sided Derivatives

A function $y = f(x)$ is **differentiable** on an open interval (finite or infinite) if it has a derivative at each point of the interval. It is differentiable on a closed interval $[a, b]$ if it is differentiable on the interior (a, b) and if the limits

$$\lim_{h \to 0^+} \frac{f(a + h) - f(a)}{h} \qquad \textbf{Right-hand derivative at } \boldsymbol{a}$$

$$\lim_{h \to 0^-} \frac{f(b + h) - f(b)}{h} \qquad \textbf{Left-hand derivative at } \boldsymbol{b}$$

exist at the endpoints (Figure 3.5).

Right-hand and left-hand derivatives may be defined at any point of a function's domain. The usual relation between one-sided and two-sided limits holds for these derivatives. Because of Theorem 6, Section 2.4, a function has a derivative at a point if and only if it has left-hand and right-hand derivatives there, and these one-sided derivatives are equal.

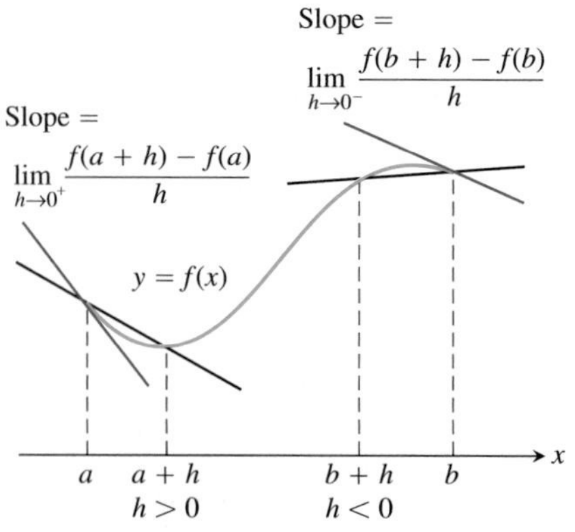

FIGURE 3.5 Derivatives at endpoints are one-sided limits.

EXAMPLE 5 $y = |x|$ Is Not Differentiable at the Origin

Show that the function $y = |x|$ is differentiable on $(-\infty, 0)$ and $(0, \infty)$ but has no derivative at $x = 0$.

Solution To the right of the origin,

$$\frac{d}{dx}(|x|) = \frac{d}{dx}(x) = \frac{d}{dx}(1 \cdot x) = 1. \qquad \frac{d}{dx}(mx + b) = m, |x| = x$$

To the left,

$$\frac{d}{dx}(|x|) = \frac{d}{dx}(-x) = \frac{d}{dx}(-1 \cdot x) = -1 \qquad |x| = -x$$

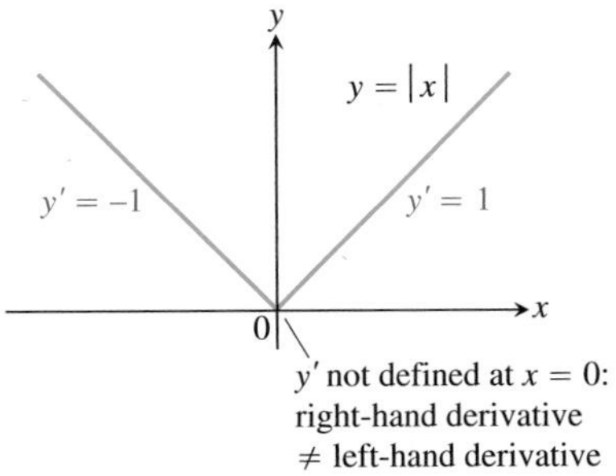

FIGURE 3.6 The function $y = |x|$ is not differentiable at the origin where the graph has a "corner."

(Figure 3.6). There can be no derivative at the origin because the one-sided derivatives differ there:

$$\begin{aligned}
\text{Right-hand derivative of } |x| \text{ at zero} &= \lim_{h\to 0^+} \frac{|0+h| - |0|}{h} = \lim_{h\to 0^+} \frac{|h|}{h} \\
&= \lim_{h\to 0^+} \frac{h}{h} \qquad |h| = h \text{ when } h > 0. \\
&= \lim_{h\to 0^+} 1 = 1 \\
\text{Left-hand derivative of } |x| \text{ at zero} &= \lim_{h\to 0^-} \frac{|0+h| - |0|}{h} = \lim_{h\to 0^-} \frac{|h|}{h} \\
&= \lim_{h\to 0^-} \frac{-h}{h} \qquad |h| = -h \text{ when } h < 0. \\
&= \lim_{h\to 0^-} -1 = -1.
\end{aligned}$$

■

EXAMPLE 6 $y = \sqrt{x}$ **Is Not Differentiable at** $x = 0$

In Example 2 we found that for $x > 0$,

$$\frac{d}{dx}\sqrt{x} = \frac{1}{2\sqrt{x}}.$$

We apply the definition to examine if the derivative exists at $x = 0$:

$$\lim_{h\to 0^+} \frac{\sqrt{0+h} - \sqrt{0}}{h} = \lim_{h\to 0^+} \frac{1}{\sqrt{h}} = \infty.$$

Since the (right-hand) limit is not finite, there is no derivative at $x = 0$. Since the slopes of the secant lines joining the origin to the points $(h, \sqrt{h})$ on a graph of $y = \sqrt{x}$ approach ∞, the graph has a *vertical tangent* at the origin. ■

When Does a Function *Not* Have a Derivative at a Point?

A function has a derivative at a point x_0 if the slopes of the secant lines through $P(x_0, f(x_0))$ and a nearby point Q on the graph approach a limit as Q approaches P. Whenever the secants fail to take up a limiting position or become vertical as Q approaches P, the derivative does not exist. Thus differentiability is a "smoothness" condition on the graph of f. A function whose graph is otherwise smooth will fail to have a derivative at a point for several reasons, such as at points where the graph has

1. a *corner*, where the one-sided derivatives differ.

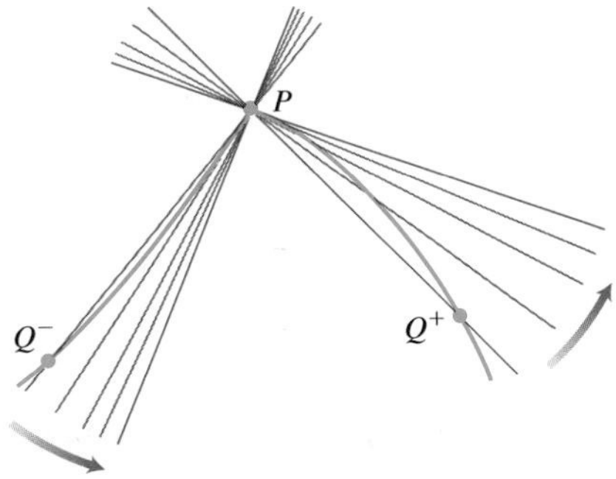

2. a *cusp*, where the slope of PQ approaches ∞ from one side and $-\infty$ from the other.

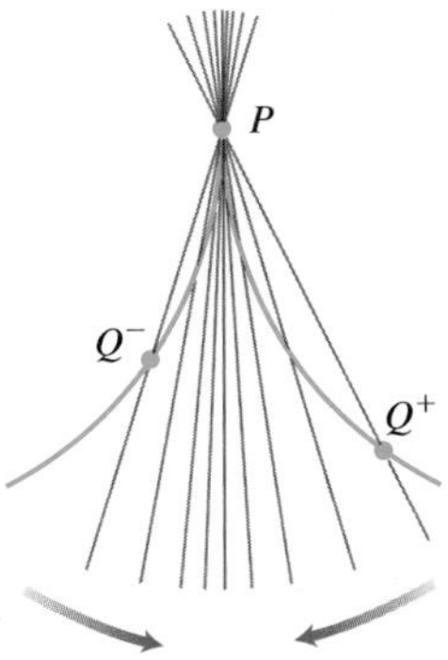

3. a *vertical tangent*, where the slope of PQ approaches ∞ from both sides or approaches $-\infty$ from both sides (here, $-\infty$).

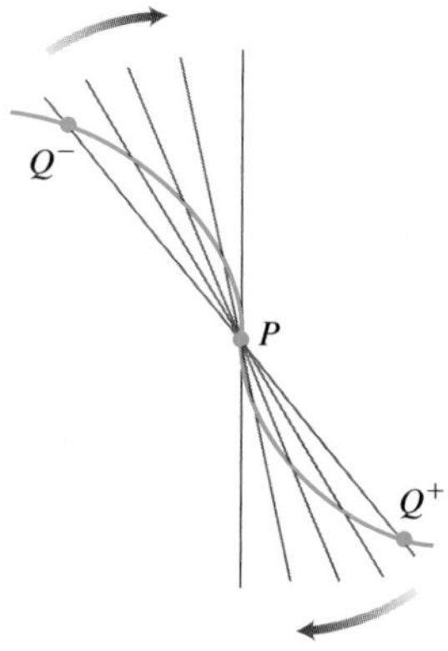

4. a *discontinuity*.

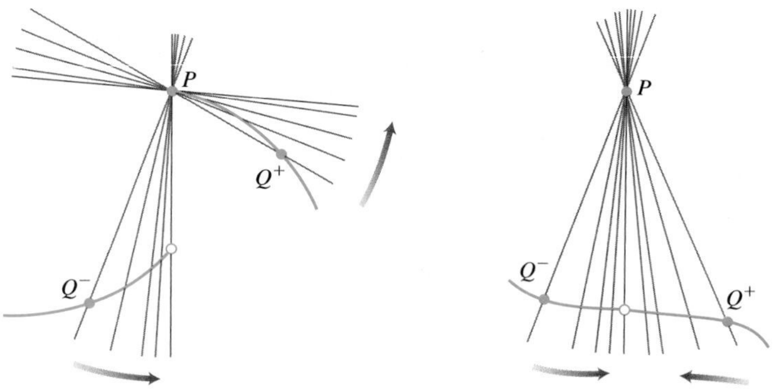

Differentiable Functions Are Continuous

A function is continuous at every point where it has a derivative.

> **THEOREM 1 Differentiability Implies Continuity**
> If f has a derivative at $x = c$, then f is continuous at $x = c$.

Proof Given that $f'(c)$ exists, we must show that $\lim_{x\to c} f(x) = f(c)$, or equivalently, that $\lim_{h\to 0} f(c + h) = f(c)$. If $h \neq 0$, then

$$\begin{aligned} f(c + h) &= f(c) + (f(c + h) - f(c)) \\ &= f(c) + \frac{f(c + h) - f(c)}{h} \cdot h. \end{aligned}$$

Now take limits as $h \to 0$. By Theorem 1 of Section 2.2,

$$\begin{aligned}\lim_{h\to 0} f(c+h) &= \lim_{h\to 0} f(c) + \lim_{h\to 0} \frac{f(c+h)-f(c)}{h} \cdot \lim_{h\to 0} h \\ &= f(c) + f'(c)\cdot 0 \\ &= f(c) + 0 \\ &= f(c).\end{aligned}$$

■

Similar arguments with one-sided limits show that if f has a derivative from one side (right or left) at $x = c$ then f is continuous from that side at $x = c$.

Theorem 1 on page 154 says that if a function has a discontinuity at a point (for instance, a jump discontinuity), then it cannot be differentiable there. The greatest integer function $y = \lfloor x \rfloor = \operatorname{int} x$ fails to be differentiable at every integer $x = n$ (Example 4, Section 2.6).

CAUTION The converse of Theorem 1 is false. A function need not have a derivative at a point where it is continuous, as we saw in Example 5.

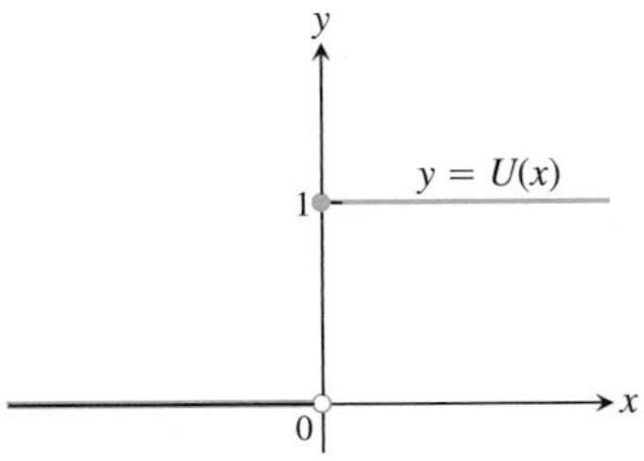

FIGURE 3.7 The unit step function does not have the Intermediate Value Property and cannot be the derivative of a function on the real line.

The Intermediate Value Property of Derivatives

Not every function can be some function's derivative, as we see from the following theorem, first proved in 1875 by the French mathematician Jean Gaston Darboux (1842–1917).

> **THEOREM 2 Darboux's Theorem**
> If a and b are any two points in an interval on which f is differentiable, then f' takes on every value between $f'(a)$ and $f'(b)$.

Theorem 2 (which we will not prove) says that a function cannot *be* a derivative on an interval unless it has the Intermediate Value Property there. For example, the unit step function in Figure 3.7 cannot be the derivative of any real-valued function on the real line. In Chapter 5 we will see that every continuous function is a derivative of some function.

In Section 4.4, we invoke Theorem 2 to analyze what happens at a point on the graph of a twice-differentiable function where it changes its "bending" behavior.

EXERCISES 3.1

Finding Derivative Functions and Values

Using the definition, calculate the derivatives of the functions in Exercises 1–6. Then find the values of the derivatives as specified.

1. $f(x) = 4 - x^2;\quad f'(-3), f'(0), f'(1)$
2. $F(x) = (x-1)^2 + 1;\quad F'(-1), F'(0), F'(2)$
3. $g(t) = \dfrac{1}{t^2};\quad g'(-1), g'(2), g'(\sqrt{3})$
4. $k(z) = \dfrac{1-z}{2z};\quad k'(-1), k'(1), k'(\sqrt{2})$
5. $p(\theta) = \sqrt{3\theta};\quad p'(1), p'(3), p'(2/3)$
6. $r(s) = \sqrt{2s+1};\quad r'(0), r'(1), r'(1/2)$

In Exercises 7–12, find the indicated derivatives.

7. $\dfrac{dy}{dx}$ if $y = 2x^3$
8. $\dfrac{dr}{ds}$ if $r = \dfrac{s^3}{2} + 1$

9. $\dfrac{ds}{dt}$ if $s = \dfrac{t}{2t + 1}$

10. $\dfrac{dv}{dt}$ if $v = t - \dfrac{1}{t}$

11. $\dfrac{dp}{dq}$ if $p = \dfrac{1}{\sqrt{q + 1}}$

12. $\dfrac{dz}{dw}$ if $z = \dfrac{1}{\sqrt{3w - 2}}$

Slopes and Tangent Lines

In Exercises 13–16, differentiate the functions and find the slope of the tangent line at the given value of the independent variable.

13. $f(x) = x + \dfrac{9}{x}, \quad x = -3$

14. $k(x) = \dfrac{1}{2 + x}, \quad x = 2$

15. $s = t^3 - t^2, \quad t = -1$

16. $y = (x + 1)^3, \quad x = -2$

In Exercises 17–18, differentiate the functions. Then find an equation of the tangent line at the indicated point on the graph of the function.

17. $y = f(x) = \dfrac{8}{\sqrt{x - 2}}, \quad (x, y) = (6, 4)$

18. $w = g(z) = 1 + \sqrt{4 - z}, \quad (z, w) = (3, 2)$

In Exercises 19–22, find the values of the derivatives.

19. $\left.\dfrac{ds}{dt}\right|_{t=-1}$ if $s = 1 - 3t^2$

20. $\left.\dfrac{dy}{dx}\right|_{x=\sqrt{3}}$ if $y = 1 - \dfrac{1}{x}$

21. $\left.\dfrac{dr}{d\theta}\right|_{\theta=0}$ if $r = \dfrac{2}{\sqrt{4 - \theta}}$

22. $\left.\dfrac{dw}{dz}\right|_{z=4}$ if $w = z + \sqrt{z}$

Using the Alternative Formula for Derivatives

Use the formula

$$f'(x) = \lim_{z \to x} \frac{f(z) - f(x)}{z - x}$$

to find the derivative of the functions in Exercises 23–26.

23. $f(x) = \dfrac{1}{x + 2}$

24. $f(x) = \dfrac{1}{(x - 1)^2}$

25. $g(x) = \dfrac{x}{x - 1}$

26. $g(x) = 1 + \sqrt{x}$

Graphs

Match the functions graphed in Exercises 27–30 with the derivatives graphed in the accompanying figures (a)–(d).

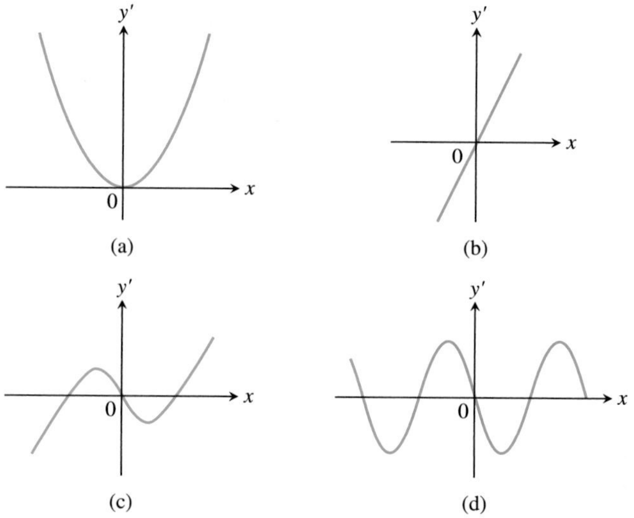

27.

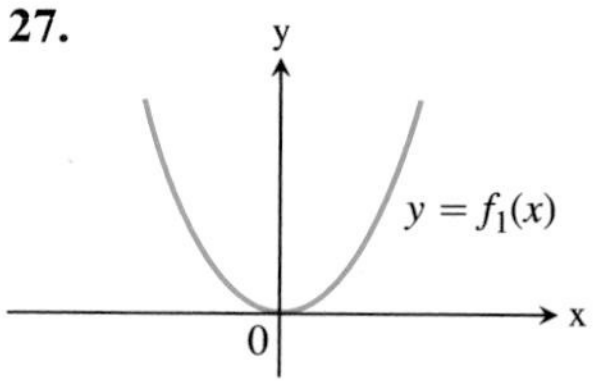

28.

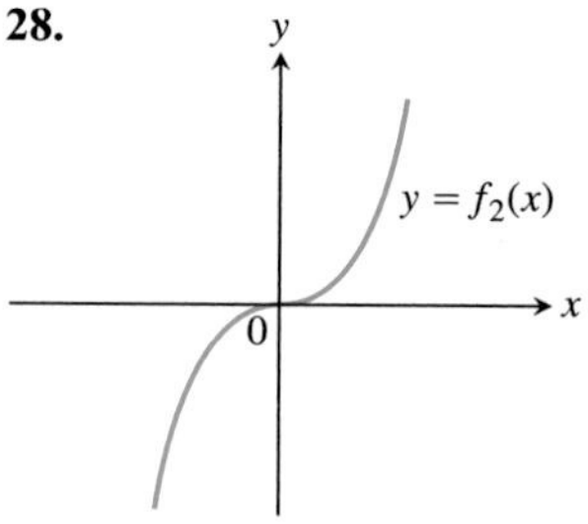

29.

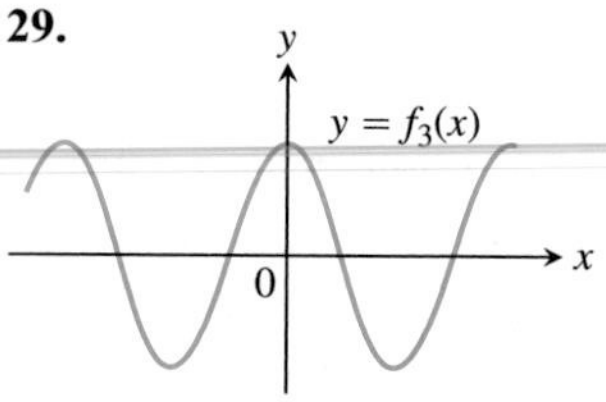

30.

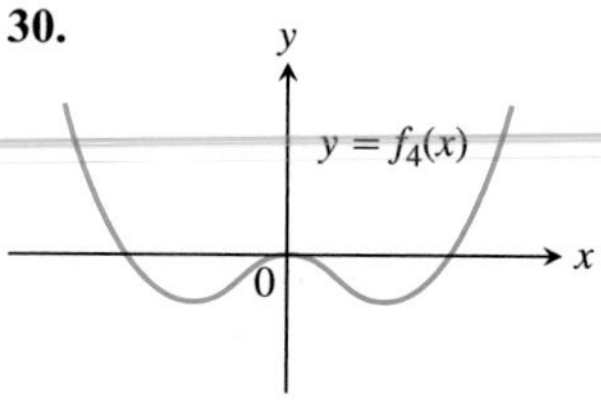

31. **a.** The graph in the accompanying figure is made of line segments joined end to end. At which points of the interval $[-4, 6]$ is f' not defined? Give reasons for your answer.

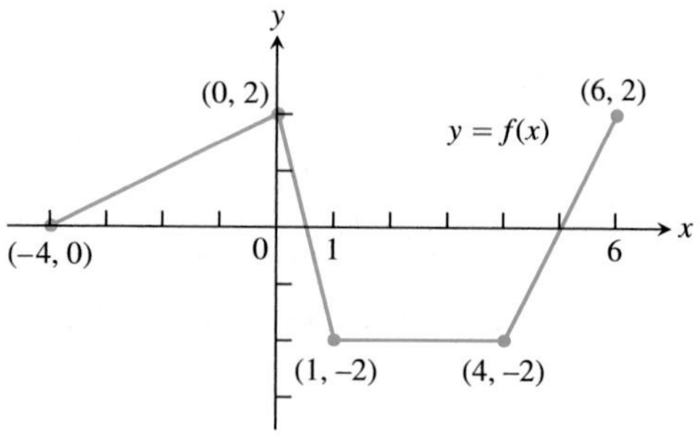

b. Graph the derivative of f.
The graph should show a step function.

32. Recovering a function from its derivative

a. Use the following information to graph the function f over the closed interval $[-2, 5]$.

i) The graph of f is made of closed line segments joined end to end.

ii) The graph starts at the point $(-2, 3)$.

iii) The derivative of f is the step function in the figure shown here.

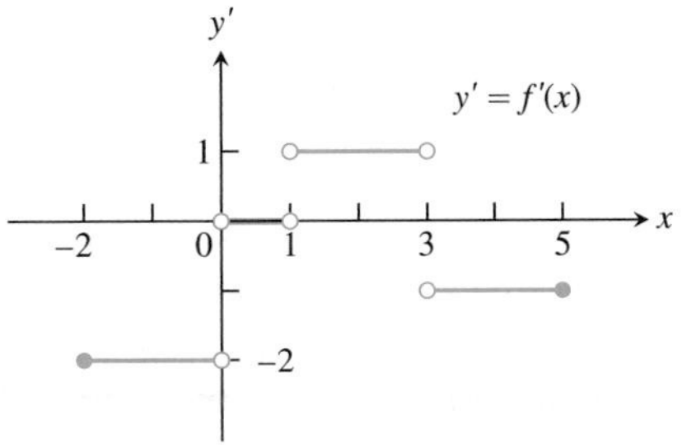

b. Repeat part (a) assuming that the graph starts at $(-2, 0)$ instead of $(-2, 3)$.

33. Growth in the economy The graph in the accompanying figure shows the average annual percentage change $y = f(t)$ in the U.S. gross national product (GNP) for the years 1983–1988. Graph dy/dt (where defined). (*Source*: *Statistical Abstracts of the United States*, 110th Edition, U.S. Department of Commerce, p. 427.)

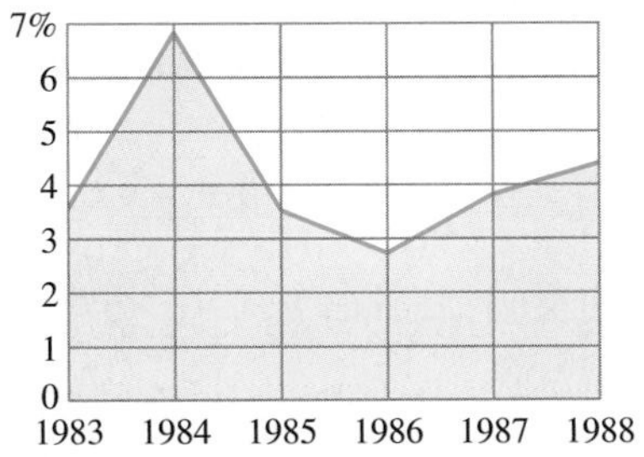

34. Fruit flies (*Continuation of Example 3, Section 2.1.*) Populations starting out in closed environments grow slowly at first, when there are relatively few members, then more rapidly as the number of reproducing individuals increases and resources are still abundant, then slowly again as the population reaches the carrying capacity of the environment.

a. Use the graphical technique of Example 3 to graph the derivative of the fruit fly population introduced in Section 2.1. The graph of the population is reproduced here.

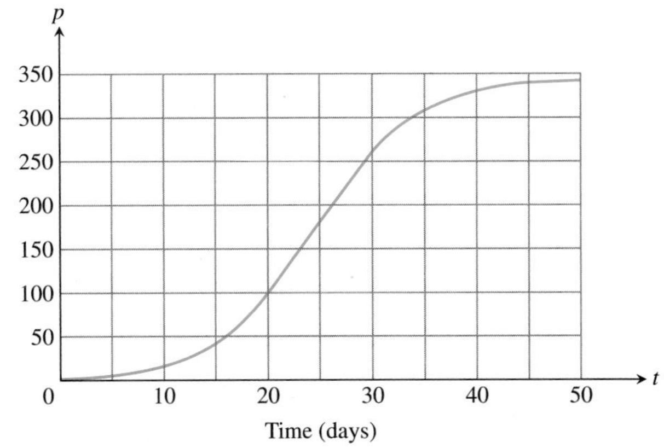

b. During what days does the population seem to be increasing fastest? Slowest?

One-Sided Derivatives

Compare the right-hand and left-hand derivatives to show that the functions in Exercises 35–38 are not differentiable at the point P.

35.

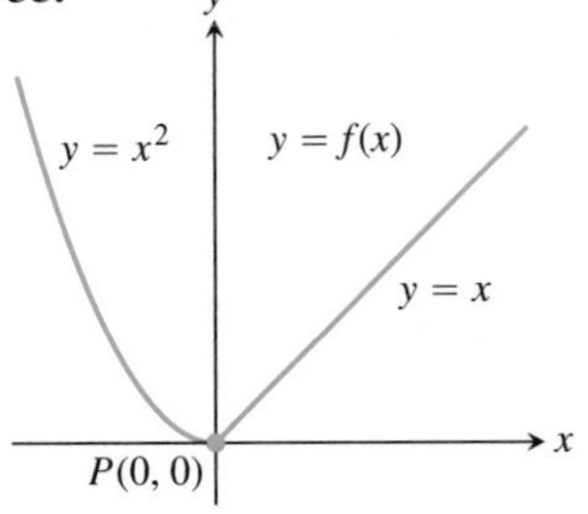

36.

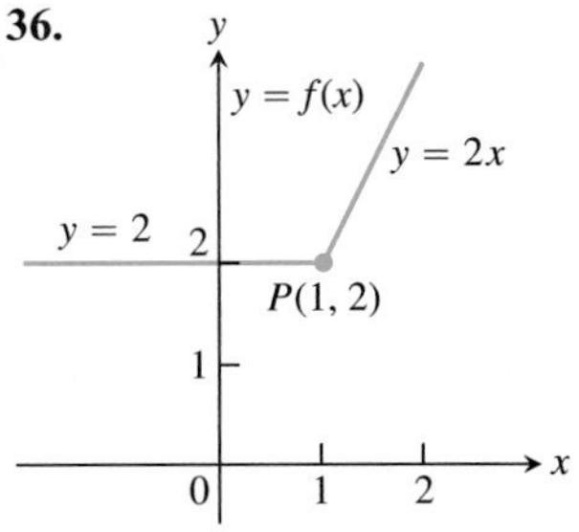

37.

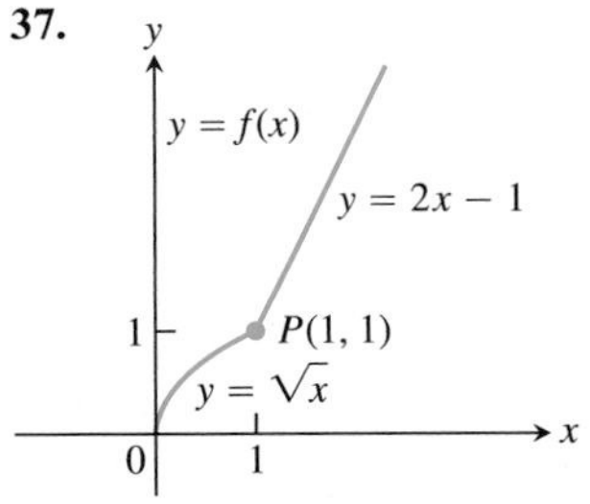

38.

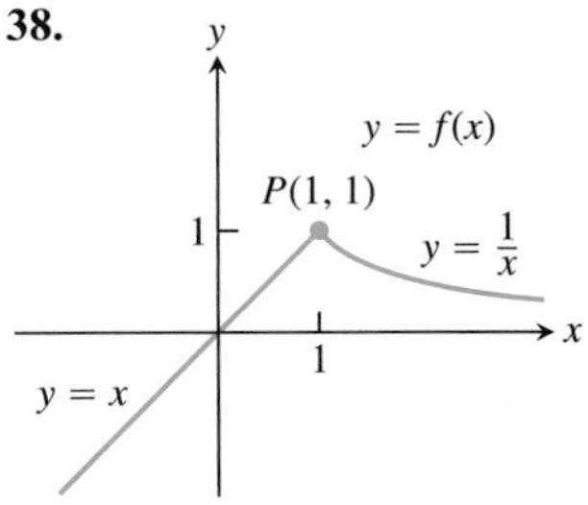

Differentiability and Continuity on an Interval

Each figure in Exercises 39–44 shows the graph of a function over a closed interval D. At what domain points does the function appear to be

a. differentiable?

b. continuous but not differentiable?

c. neither continuous nor differentiable?

Give reasons for your answers.

39.

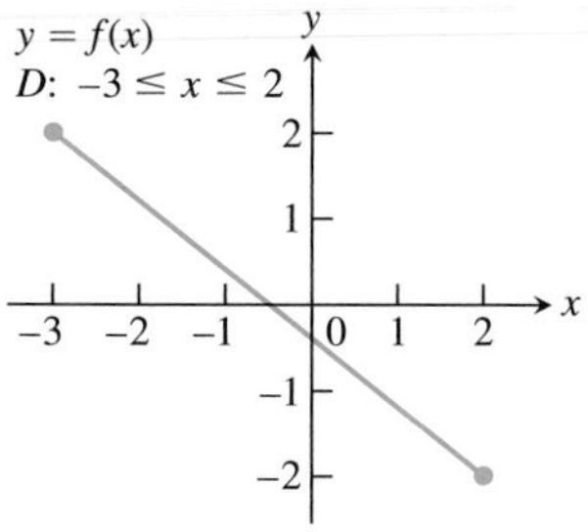

40.

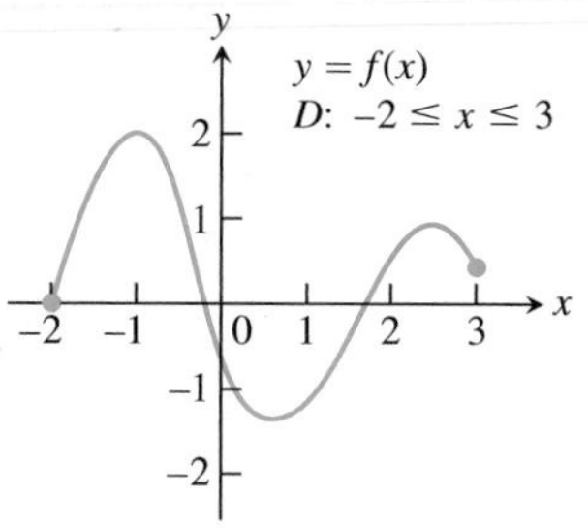

41.

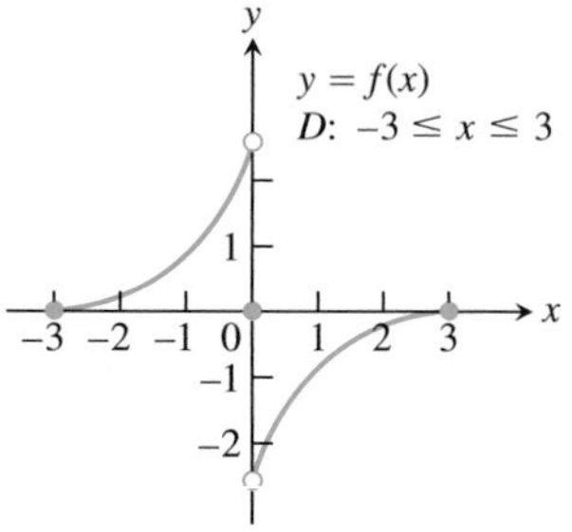

42.

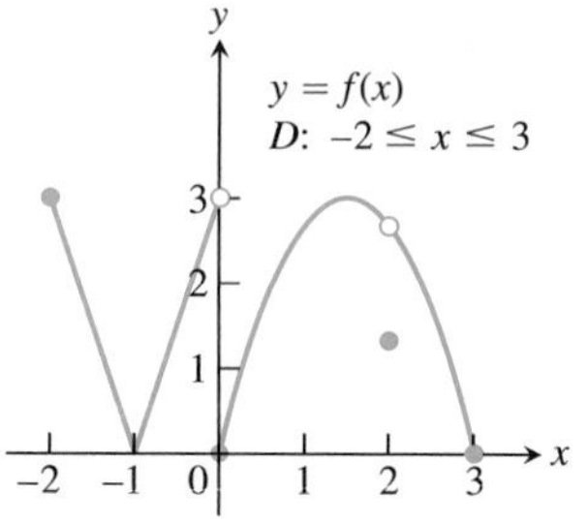

43.

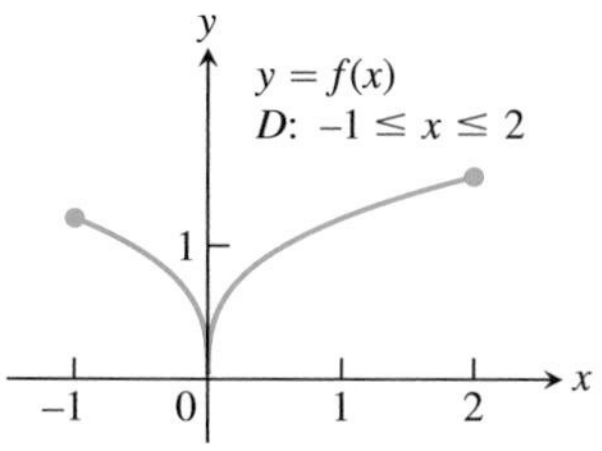

44.

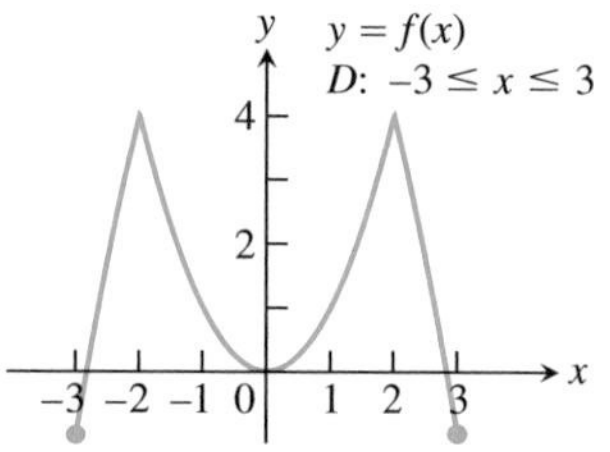

Theory and Examples

In Exercises 45–48,

a. Find the derivative $f'(x)$ of the given function $y = f(x)$.

b. Graph $y = f(x)$ and $y = f'(x)$ side by side using separate sets of coordinate axes, and answer the following questions.

c. For what values of x, if any, is f' positive? Zero? Negative?

d. Over what intervals of x-values, if any, does the function $y = f(x)$ increase as x increases? Decrease as x increases? How is this related to what you found in part (c)? (We will say more about this relationship in Chapter 4.)

45. $y = -x^2$

46. $y = -1/x$

47. $y = x^3/3$

48. $y = x^4/4$

49. Does the curve $y = x^3$ ever have a negative slope? If so, where? Give reasons for your answer.

50. Does the curve $y = 2\sqrt{x}$ have any horizontal tangents? If so, where? Give reasons for your answer.

51. Tangent to a parabola Does the parabola $y = 2x^2 - 13x + 5$ have a tangent whose slope is -1? If so, find an equation for the line and the point of tangency. If not, why not?

52. Tangent to $y = \sqrt{x}$ Does any tangent to the curve $y = \sqrt{x}$ cross the x-axis at $x = -1$? If so, find an equation for the line and the point of tangency. If not, why not?

53. Greatest integer in x Does any function differentiable on $(-\infty, \infty)$ have $y = \text{int}\, x$, the greatest integer in x (see Figure 2.55), as its derivative? Give reasons for your answer.

54. Derivative of $y = |x|$ Graph the derivative of $f(x) = |x|$. Then graph $y = (|x| - 0)/(x - 0) = |x|/x$. What can you conclude?

55. Derivative of $-f$ Does knowing that a function $f(x)$ is differentiable at $x = x_0$ tell you anything about the differentiability of the function $-f$ at $x = x_0$? Give reasons for your answer.

56. Derivative of multiples Does knowing that a function $g(t)$ is differentiable at $t = 7$ tell you anything about the differentiability of the function $3g$ at $t = 7$? Give reasons for your answer.

57. Limit of a quotient Suppose that functions $g(t)$ and $h(t)$ are defined for all values of t and $g(0) = h(0) = 0$. Can $\lim_{t\to 0} (g(t))/(h(t))$ exist? If it does exist, must it equal zero? Give reasons for your answers.

58. a. Let $f(x)$ be a function satisfying $|f(x)| \le x^2$ for $-1 \le x \le 1$. Show that f is differentiable at $x = 0$ and find $f'(0)$.

b. Show that

$$f(x) = \begin{cases} x^2 \sin \dfrac{1}{x}, & x \neq 0 \\ 0, & x = 0 \end{cases}$$

is differentiable at $x = 0$ and find $f'(0)$.

T **59.** Graph $y = 1/(2\sqrt{x})$ in a window that has $0 \le x \le 2$. Then, on the same screen, graph

$$y = \frac{\sqrt{x + h} - \sqrt{x}}{h}$$

for $h = 1, 0.5, 0.1$. Then try $h = -1, -0.5, -0.1$. Explain what is going on.

T **60.** Graph $y = 3x^2$ in a window that has $-2 \le x \le 2$, $0 \le y \le 3$. Then, on the same screen, graph

$$y = \frac{(x + h)^3 - x^3}{h}$$

for $h = 2, 1, 0.2$. Then try $h = -2, -1, -0.2$. Explain what is going on.

T **61. Weierstrass's nowhere differentiable continuous function** The sum of the first eight terms of the Weierstrass function $f(x) = \sum_{n=0}^{\infty} (2/3)^n \cos(9^n \pi x)$ is

$$g(x) = \cos(\pi x) + (2/3)^1 \cos(9\pi x) + (2/3)^2 \cos(9^2 \pi x) + (2/3)^3 \cos(9^3 \pi x) + \cdots + (2/3)^7 \cos(9^7 \pi x).$$

Graph this sum. Zoom in several times. How wiggly and bumpy is this graph? Specify a viewing window in which the displayed portion of the graph is smooth.

COMPUTER EXPLORATIONS

Use a CAS to perform the following steps for the functions in Exercises 62–67.

a. Plot $y = f(x)$ to see that function's global behavior.

b. Define the difference quotient q at a general point x, with general step size h.

c. Take the limit as $h \to 0$. What formula does this give?

d. Substitute the value $x = x_0$ and plot the function $y = f(x)$ together with its tangent line at that point.

e. Substitute various values for x larger and smaller than x_0 into the formula obtained in part (c). Do the numbers make sense with your picture?

f. Graph the formula obtained in part (c). What does it mean when its values are negative? Zero? Positive? Does this make sense with your plot from part (a)? Give reasons for your answer.

62. $f(x) = x^3 + x^2 - x, \quad x_0 = 1$

63. $f(x) = x^{1/3} + x^{2/3}, \quad x_0 = 1$

64. $f(x) = \dfrac{4x}{x^2 + 1}, \quad x_0 = 2$ **65.** $f(x) = \dfrac{x - 1}{3x^2 + 1}, \quad x_0 = -1$

66. $f(x) = \sin 2x, \quad x_0 = \pi/2$ **67.** $f(x) = x^2 \cos x, \quad x_0 = \pi/4$

3.2 Differentiation Rules

This section introduces a few rules that allow us to differentiate a great variety of functions. By proving these rules here, we can differentiate functions without having to apply the definition of the derivative each time.

Powers, Multiples, Sums, and Differences

The first rule of differentiation is that the derivative of every constant function is zero.

> **RULE 1 Derivative of a Constant Function**
> If f has the constant value $f(x) = c$, then
> $$\frac{df}{dx} = \frac{d}{dx}(c) = 0.$$

EXAMPLE 1

If f has the constant value $f(x) = 8$, then

$$\frac{df}{dx} = \frac{d}{dx}(8) = 0.$$

Similarly,

$$\frac{d}{dx}\left(-\frac{\pi}{2}\right) = 0 \quad \text{and} \quad \frac{d}{dx}\left(\sqrt{3}\right) = 0. \quad \blacksquare$$

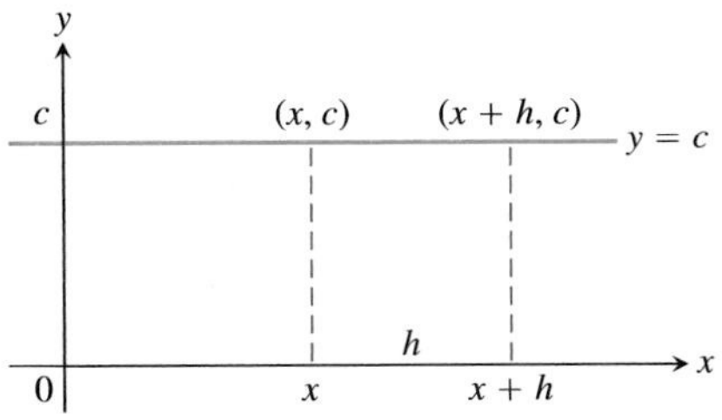

FIGURE 3.8 The rule $(d/dx)(c) = 0$ is another way to say that the values of constant functions never change and that the slope of a horizontal line is zero at every point.

Proof of Rule 1 We apply the definition of derivative to $f(x) = c$, the function whose outputs have the constant value c (Figure 3.8). At every value of x, we find that

$$f'(x) = \lim_{h \to 0} \frac{f(x + h) - f(x)}{h} = \lim_{h \to 0} \frac{c - c}{h} = \lim_{h \to 0} 0 = 0. \quad \blacksquare$$

The second rule tells how to differentiate x^n if n is a positive integer.

RULE 2 Power Rule for Positive Integers
If n is a positive integer, then

$$\frac{d}{dx}x^n = nx^{n-1}.$$

To apply the Power Rule, we subtract 1 from the original exponent (n) and multiply the result by n.

EXAMPLE 2 Interpreting Rule 2

f	x	x^2	x^3	x^4	...
f'	1	$2x$	$3x^2$	$4x^3$	...

■

HISTORICAL BIOGRAPHY

Richard Courant
(1888–1972)

First Proof of Rule 2 The formula

$$z^n - x^n = (z - x)(z^{n-1} + z^{n-2}x + \cdots + zx^{n-2} + x^{n-1})$$

can be verified by multiplying out the right-hand side. Then from the alternative form for the definition of the derivative,

$$\begin{aligned} f'(x) &= \lim_{z \to x} \frac{f(z) - f(x)}{z - x} = \lim_{z \to x} \frac{z^n - x^n}{z - x} \\ &= \lim_{z \to x}(z^{n-1} + z^{n-2}x + \cdots + zx^{n-2} + x^{n-1}) \\ &= nx^{n-1} \end{aligned}$$

Second Proof of Rule 2 If $f(x) = x^n$, then $f(x + h) = (x + h)^n$. Since n is a positive integer, we can expand $(x + h)^n$ by the Binomial Theorem to get

$$\begin{aligned} f'(x) &= \lim_{h \to 0} \frac{f(x + h) - f(x)}{h} = \lim_{h \to 0} \frac{(x + h)^n - x^n}{h} \\ &= \lim_{h \to 0} \frac{\left[x^n + nx^{n-1}h + \frac{n(n-1)}{2}x^{n-2}h^2 + \cdots + nxh^{n-1} + h^n\right] - x^n}{h} \\ &= \lim_{h \to 0} \frac{nx^{n-1}h + \frac{n(n-1)}{2}x^{n-2}h^2 + \cdots + nxh^{n-1} + h^n}{h} \\ &= \lim_{h \to 0} \left[nx^{n-1} + \frac{n(n-1)}{2}x^{n-2}h + \cdots + nxh^{n-2} + h^{n-1}\right] \\ &= nx^{n-1} \end{aligned}$$

■

The third rule says that when a differentiable function is multiplied by a constant, its derivative is multiplied by the same constant.

RULE 3 Constant Multiple Rule
If u is a differentiable function of x, and c is a constant, then

$$\frac{d}{dx}(cu) = c\frac{du}{dx}.$$

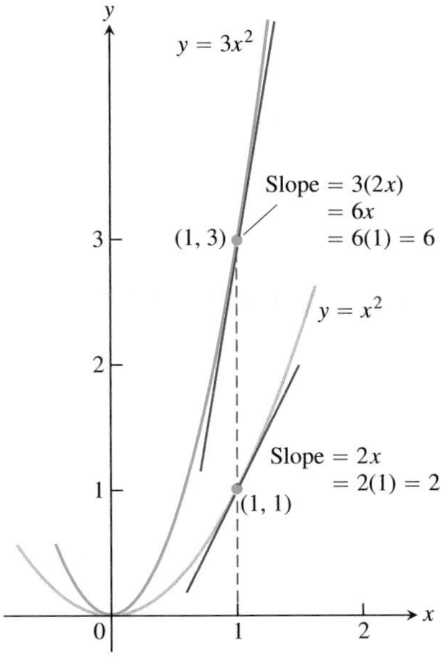

FIGURE 3.9 The graphs of $y = x^2$ and $y = 3x^2$. Tripling the y-coordinates triples the slope (Example 3).

In particular, if n is a positive integer, then

$$\frac{d}{dx}(cx^n) = cnx^{n-1}.$$

EXAMPLE 3

(a) The derivative formula

$$\frac{d}{dx}(3x^2) = 3 \cdot 2x = 6x$$

says that if we rescale the graph of $y = x^2$ by multiplying each y-coordinate by 3, then we multiply the slope at each point by 3 (Figure 3.9).

(b) A useful special case

The derivative of the negative of a differentiable function u is the negative of the function's derivative. Rule 3 with $c = -1$ gives

$$\frac{d}{dx}(-u) = \frac{d}{dx}(-1 \cdot u) = -1 \cdot \frac{d}{dx}(u) = -\frac{du}{dx}. \quad \blacksquare$$

Proof of Rule 3

$$\begin{aligned}
\frac{d}{dx}cu &= \lim_{h\to 0}\frac{cu(x+h) - cu(x)}{h} && \text{Derivative definition with } f(x) = cu(x) \\
&= c\lim_{h\to 0}\frac{u(x+h) - u(x)}{h} && \text{Limit property} \\
&= c\frac{du}{dx} && u \text{ is differentiable.} \quad \blacksquare
\end{aligned}$$

The next rule says that the derivative of the sum of two differentiable functions is the sum of their derivatives.

Denoting Functions by u and v
The functions we are working with when we need a differentiation formula are likely to be denoted by letters like f and g. When we apply the formula, we do not want to find it using these same letters in some other way. To guard against this problem, we denote the functions in differentiation rules by letters like u and v that are not likely to be already in use.

RULE 4 Derivative Sum Rule
If u and v are differentiable functions of x, then their sum $u + v$ is differentiable at every point where u and v are both differentiable. At such points,

$$\frac{d}{dx}(u + v) = \frac{du}{dx} + \frac{dv}{dx}.$$

EXAMPLE 4 Derivative of a Sum

$$\begin{aligned} y &= x^4 + 12x \\ \frac{dy}{dx} &= \frac{d}{dx}(x^4) + \frac{d}{dx}(12x) \\ &= 4x^3 + 12 \end{aligned}$$

■

Proof of Rule 4 We apply the definition of derivative to $f(x) = u(x) + v(x)$:

$$\begin{aligned} \frac{d}{dx}[u(x) + v(x)] &= \lim_{h \to 0} \frac{[u(x+h) + v(x+h)] - [u(x) + v(x)]}{h} \\ &= \lim_{h \to 0} \left[\frac{u(x+h) - u(x)}{h} + \frac{v(x+h) - v(x)}{h} \right] \\ &= \lim_{h \to 0} \frac{u(x+h) - u(x)}{h} + \lim_{h \to 0} \frac{v(x+h) - v(x)}{h} = \frac{du}{dx} + \frac{dv}{dx}. \end{aligned}$$

■

Combining the Sum Rule with the Constant Multiple Rule gives the **Difference Rule,** which says that the derivative of a *difference* of differentiable functions is the difference of their derivatives.

$$\frac{d}{dx}(u - v) = \frac{d}{dx}[u + (-1)v] = \frac{du}{dx} + (-1)\frac{dv}{dx} = \frac{du}{dx} - \frac{dv}{dx}$$

The Sum Rule also extends to sums of more than two functions, as long as there are only finitely many functions in the sum. If $u_1, u_2, \ldots, u_n$ are differentiable at x, then so is $u_1 + u_2 + \cdots + u_n$, and

$$\frac{d}{dx}(u_1 + u_2 + \cdots + u_n) = \frac{du_1}{dx} + \frac{du_2}{dx} + \cdots + \frac{du_n}{dx}.$$

EXAMPLE 5 Derivative of a Polynomial

$$\begin{aligned} y &= x^3 + \frac{4}{3}x^2 - 5x + 1 \\ \frac{dy}{dx} &= \frac{d}{dx}x^3 + \frac{d}{dx}\left(\frac{4}{3}x^2\right) - \frac{d}{dx}(5x) + \frac{d}{dx}(1) \\ &= 3x^2 + \frac{4}{3} \cdot 2x - 5 + 0 \\ &= 3x^2 + \frac{8}{3}x - 5 \end{aligned}$$

■

Notice that we can differentiate any polynomial term by term, the way we differentiated the polynomial in Example 5. All polynomials are differentiable everywhere.

Proof of the Sum Rule for Sums of More Than Two Functions We prove the statement

$$\frac{d}{dx}(u_1 + u_2 + \cdots + u_n) = \frac{du_1}{dx} + \frac{du_2}{dx} + \cdots + \frac{du_n}{dx}$$

by mathematical induction (see Appendix 1). The statement is true for $n = 2$, as was just proved. This is Step 1 of the induction proof.

Step 2 is to show that if the statement is true for any positive integer $n = k$, where $k \geq n_0 = 2$, then it is also true for $n = k + 1$. So suppose that

$$\frac{d}{dx}(u_1 + u_2 + \cdots + u_k) = \frac{du_1}{dx} + \frac{du_2}{dx} + \cdots + \frac{du_k}{dx}. \tag{1}$$

Then

$$\frac{d}{dx}(\underbrace{u_1 + u_2 + \cdots + u_k}_{\text{Call the function defined by this sum } u.} + \underbrace{u_{k+1}}_{\text{Call this function } v.})$$

$$= \frac{d}{dx}(u_1 + u_2 + \cdots + u_k) + \frac{du_{k+1}}{dx} \qquad \text{Rule 4 for } \frac{d}{dx}(u + v)$$

$$= \frac{du_1}{dx} + \frac{du_2}{dx} + \cdots + \frac{du_k}{dx} + \frac{du_{k+1}}{dx}. \qquad \text{Eq. (1)}$$

With these steps verified, the mathematical induction principle now guarantees the Sum Rule for every integer $n \geq 2$. ■

EXAMPLE 6 Finding Horizontal Tangents

Does the curve $y = x^4 - 2x^2 + 2$ have any horizontal tangents? If so, where?

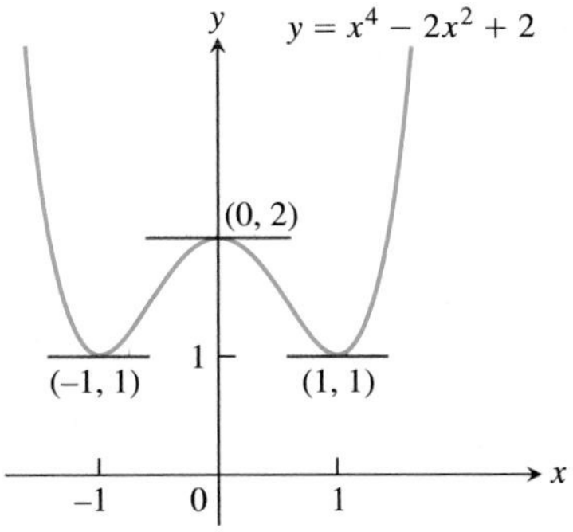

FIGURE 3.10 The curve $y = x^4 - 2x^2 + 2$ and its horizontal tangents (Example 6).

Solution The horizontal tangents, if any, occur where the slope dy/dx is zero. We have,

$$\frac{dy}{dx} = \frac{d}{dx}(x^4 - 2x^2 + 2) = 4x^3 - 4x.$$

Now solve the equation $\dfrac{dy}{dx} = 0$ for x:

$$4x^3 - 4x = 0$$
$$4x(x^2 - 1) = 0$$
$$x = 0, 1, -1.$$

The curve $y = x^4 - 2x^2 + 2$ has horizontal tangents at $x = 0, 1$, and -1. The corresponding points on the curve are (0, 2), (1, 1) and $(-1, 1)$. See Figure 3.10. ■

Products and Quotients

While the derivative of the sum of two functions is the sum of their derivatives, the derivative of the product of two functions is *not* the product of their derivatives. For instance,

$$\frac{d}{dx}(x \cdot x) = \frac{d}{dx}(x^2) = 2x, \qquad \text{while} \qquad \frac{d}{dx}(x) \cdot \frac{d}{dx}(x) = 1 \cdot 1 = 1.$$

The derivative of a product of two functions is the sum of *two* products, as we now explain.

RULE 5 Derivative Product Rule

If u and v are differentiable at x, then so is their product uv, and

$$\frac{d}{dx}(uv) = u\frac{dv}{dx} + v\frac{du}{dx}.$$

Picturing the Product Rule

Suppose $u(x)$ and $v(x)$ are positive and increase when x increases, and $h > 0$.

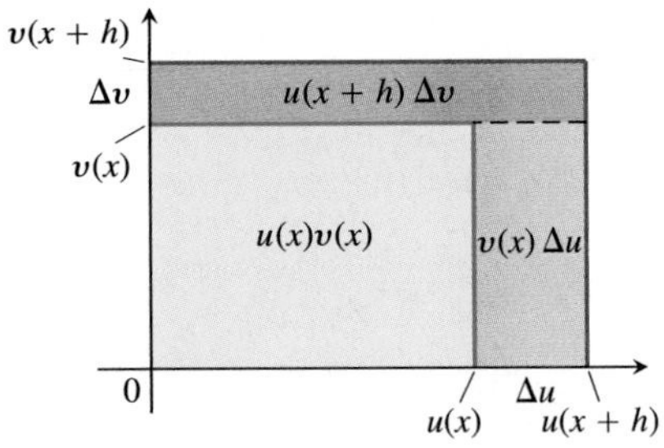

Then the change in the product uv is the difference in areas of the larger and smaller "squares," which is the sum of the upper and right-hand reddish-shaded rectangles. That is,

$$\begin{aligned}\Delta(uv) &= u(x+h)v(x+h) - u(x)v(x)\\ &= u(x+h)\Delta v + v(x)\Delta u.\end{aligned}$$

Division by h gives

$$\frac{\Delta(uv)}{h} = u(x+h)\frac{\Delta v}{h} + v(x)\frac{\Delta u}{h}.$$

The limit as $h \to 0^+$ gives the Product Rule.

The derivative of the product uv is u times the derivative of v plus v times the derivative of u. In *prime notation*, $(uv)' = uv' + vu'$. In function notation,

$$\frac{d}{dx}[f(x)g(x)] = f(x)g'(x) + g(x)f'(x).$$

EXAMPLE 7 Using the Product Rule

Find the derivative of

$$y = \frac{1}{x}\left(x^2 + \frac{1}{x}\right).$$

Solution We apply the Product Rule with $u = 1/x$ and $v = x^2 + (1/x)$:

$$\begin{aligned}\frac{d}{dx}\left[\frac{1}{x}\left(x^2 + \frac{1}{x}\right)\right] &= \frac{1}{x}\left(2x - \frac{1}{x^2}\right) + \left(x^2 + \frac{1}{x}\right)\left(-\frac{1}{x^2}\right)\\ &= 2 - \frac{1}{x^3} - 1 - \frac{1}{x^3}\\ &= 1 - \frac{2}{x^3}.\end{aligned}$$

$\frac{d}{dx}(uv) = u\frac{dv}{dx} + v\frac{du}{dx}$, and $\frac{d}{dx}\left(\frac{1}{x}\right) = -\frac{1}{x^2}$ by Example 3, Section 2.7.

Proof of Rule 5

$$\frac{d}{dx}(uv) = \lim_{h\to 0}\frac{u(x+h)v(x+h) - u(x)v(x)}{h}$$

To change this fraction into an equivalent one that contains difference quotients for the derivatives of u and v, we subtract and add $u(x+h)v(x)$ in the numerator:

$$\begin{aligned}\frac{d}{dx}(uv) &= \lim_{h\to 0}\frac{u(x+h)v(x+h) - u(x+h)v(x) + u(x+h)v(x) - u(x)v(x)}{h}\\ &= \lim_{h\to 0}\left[u(x+h)\frac{v(x+h) - v(x)}{h} + v(x)\frac{u(x+h) - u(x)}{h}\right]\\ &= \lim_{h\to 0}u(x+h)\cdot\lim_{h\to 0}\frac{v(x+h) - v(x)}{h} + v(x)\cdot\lim_{h\to 0}\frac{u(x+h) - u(x)}{h}.\end{aligned}$$

As h approaches zero, $u(x+h)$ approaches $u(x)$ because u, being differentiable at x, is continuous at x. The two fractions approach the values of dv/dx at x and du/dx at x. In short,

$$\frac{d}{dx}(uv) = u\frac{dv}{dx} + v\frac{du}{dx}.$$

In the following example, we have only numerical values with which to work.

EXAMPLE 8 Derivative from Numerical Values

Let $y = uv$ be the product of the functions u and v. Find $y'(2)$ if

$$u(2) = 3, \quad u'(2) = -4, \quad v(2) = 1, \quad \text{and} \quad v'(2) = 2.$$

Solution From the Product Rule, in the form

$$y' = (uv)' = uv' + vu',$$

we have

$$\begin{aligned} y'(2) &= u(2)v'(2) + v(2)u'(2) \\ &= (3)(2) + (1)(-4) = 6 - 4 = 2. \end{aligned}$$

■

EXAMPLE 9 Differentiating a Product in Two Ways

Find the derivative of $y = (x^2 + 1)(x^3 + 3)$.

Solution

(a) From the Product Rule with $u = x^2 + 1$ and $v = x^3 + 3$, we find

$$\begin{aligned} \frac{d}{dx}\left[\left(x^2 + 1\right)\left(x^3 + 3\right)\right] &= (x^2 + 1)(3x^2) + (x^3 + 3)(2x) \\ &= 3x^4 + 3x^2 + 2x^4 + 6x \\ &= 5x^4 + 3x^2 + 6x. \end{aligned}$$

(b) This particular product can be differentiated as well (perhaps better) by multiplying out the original expression for y and differentiating the resulting polynomial:

$$\begin{aligned} y &= (x^2 + 1)(x^3 + 3) = x^5 + x^3 + 3x^2 + 3 \\ \frac{dy}{dx} &= 5x^4 + 3x^2 + 6x. \end{aligned}$$

This is in agreement with our first calculation. ■

Just as the derivative of the product of two differentiable functions is not the product of their derivatives, the derivative of the quotient of two functions is not the quotient of their derivatives. What happens instead is the Quotient Rule.

RULE 6 **Derivative Quotient Rule**

If u and v are differentiable at x and if $v(x) \neq 0$, then the quotient u/v is differentiable at x, and

$$\frac{d}{dx}\left(\frac{u}{v}\right) = \frac{v\dfrac{du}{dx} - u\dfrac{dv}{dx}}{v^2}.$$

In function notation,

$$\frac{d}{dx}\left[\frac{f(x)}{g(x)}\right] = \frac{g(x)f'(x) - f(x)g'(x)}{g^2(x)}.$$

EXAMPLE 10 Using the Quotient Rule

Find the derivative of

$$y = \frac{t^2 - 1}{t^2 + 1}.$$

Solution

We apply the Quotient Rule with $u = t^2 - 1$ and $v = t^2 + 1$:

$$\frac{dy}{dt} = \frac{(t^2+1)\cdot 2t - (t^2-1)\cdot 2t}{(t^2+1)^2} \qquad \frac{d}{dt}\left(\frac{u}{v}\right) = \frac{v(du/dt) - u(dv/dt)}{v^2}$$

$$= \frac{2t^3 + 2t - 2t^3 + 2t}{(t^2+1)^2}$$

$$= \frac{4t}{(t^2+1)^2}.$$

■

Proof of Rule 6

$$\frac{d}{dx}\left(\frac{u}{v}\right) = \lim_{h\to 0} \frac{\dfrac{u(x+h)}{v(x+h)} - \dfrac{u(x)}{v(x)}}{h}$$

$$= \lim_{h\to 0} \frac{v(x)u(x+h) - u(x)v(x+h)}{hv(x+h)v(x)}$$

To change the last fraction into an equivalent one that contains the difference quotients for the derivatives of u and v, we subtract and add $v(x)u(x)$ in the numerator. We then get

$$\frac{d}{dx}\left(\frac{u}{v}\right) = \lim_{h\to 0} \frac{v(x)u(x+h) - v(x)u(x) + v(x)u(x) - u(x)v(x+h)}{hv(x+h)v(x)}$$

$$= \lim_{h\to 0} \frac{v(x)\dfrac{u(x+h) - u(x)}{h} - u(x)\dfrac{v(x+h) - v(x)}{h}}{v(x+h)v(x)}.$$

Taking the limit in the numerator and denominator now gives the Quotient Rule. ■

Negative Integer Powers of x

The Power Rule for negative integers is the same as the rule for positive integers.

RULE 7 Power Rule for Negative Integers

If n is a negative integer and $x \neq 0$, then

$$\frac{d}{dx}(x^n) = nx^{n-1}.$$

EXAMPLE 11

(a) $\dfrac{d}{dx}\left(\dfrac{1}{x}\right) = \dfrac{d}{dx}(x^{-1}) = (-1)x^{-2} = -\dfrac{1}{x^2}$ Agrees with Example 3, Section 2.7

(b) $\dfrac{d}{dx}\left(\dfrac{4}{x^3}\right) = 4\dfrac{d}{dx}(x^{-3}) = 4(-3)x^{-4} = -\dfrac{12}{x^4}$ ■

Proof of Rule 7 The proof uses the Quotient Rule. If n is a negative integer, then $n = -m$, where m is a positive integer. Hence, $x^n = x^{-m} = 1/x^m$, and

$$\begin{aligned}
\frac{d}{dx}(x^n) &= \frac{d}{dx}\left(\frac{1}{x^m}\right) \\
&= \frac{x^m \cdot \frac{d}{dx}(1) - 1 \cdot \frac{d}{dx}(x^m)}{(x^m)^2} && \text{Quotient Rule with } u = 1 \text{ and } v = x^m \\
&= \frac{0 - mx^{m-1}}{x^{2m}} && \text{Since } m > 0, \frac{d}{dx}(x^m) = mx^{m-1} \\
&= -mx^{-m-1} \\
&= nx^{n-1}. && \text{Since } -m = n
\end{aligned}$$

■

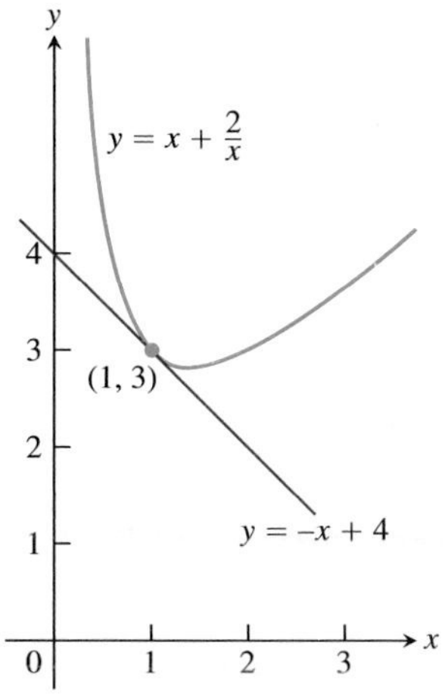

FIGURE 3.11 The tangent to the curve $y = x + (2/x)$ at (1, 3) in Example 12. The curve has a third-quadrant portion not shown here. We see how to graph functions like this one in Chapter 4.

EXAMPLE 12 Tangent to a Curve

Find an equation for the tangent to the curve

$$y = x + \frac{2}{x}$$

at the point (1, 3) (Figure 3.11).

Solution The slope of the curve is

$$\frac{dy}{dx} = \frac{d}{dx}(x) + 2\frac{d}{dx}\left(\frac{1}{x}\right) = 1 + 2\left(-\frac{1}{x^2}\right) = 1 - \frac{2}{x^2}.$$

The slope at $x = 1$ is

$$\left.\frac{dy}{dx}\right|_{x=1} = \left[1 - \frac{2}{x^2}\right]_{x=1} = 1 - 2 = -1.$$

The line through (1, 3) with slope $m = -1$ is

$$\begin{aligned}
y - 3 &= (-1)(x - 1) && \text{Point-slope equation} \\
y &= -x + 1 + 3 \\
y &= -x + 4.
\end{aligned}$$

■

The choice of which rules to use in solving a differentiation problem can make a difference in how much work you have to do. Here is an example.

EXAMPLE 13 Choosing Which Rule to Use

Rather than using the Quotient Rule to find the derivative of

$$y = \frac{(x - 1)(x^2 - 2x)}{x^4},$$

expand the numerator and divide by x^4:

$$y = \frac{(x - 1)(x^2 - 2x)}{x^4} = \frac{x^3 - 3x^2 + 2x}{x^4} = x^{-1} - 3x^{-2} + 2x^{-3}.$$

Then use the Sum and Power Rules:

$$\begin{aligned}\frac{dy}{dx} &= -x^{-2} - 3(-2)x^{-3} + 2(-3)x^{-4}\\ &= -\frac{1}{x^2} + \frac{6}{x^3} - \frac{6}{x^4}.\end{aligned}$$

■

Second- and Higher-Order Derivatives

If $y = f(x)$ is a differentiable function, then its derivative $f'(x)$ is also a function. If f' is also differentiable, then we can differentiate f' to get a new function of x denoted by f''. So $f'' = (f')'$. The function f'' is called the **second derivative** of f because it is the derivative of the first derivative. Notationally,

$$f''(x) = \frac{d^2y}{dx^2} = \frac{d}{dx}\left(\frac{dy}{dx}\right) = \frac{dy'}{dx} = y'' = D^2(f)(x) = D_x^{\,2} f(x).$$

The symbol D^2 means the operation of differentiation is performed twice.

If $y = x^6$, then $y' = 6x^5$ and we have

$$y'' = \frac{dy'}{dx} = \frac{d}{dx}\left(6x^5\right) = 30x^4.$$

Thus $D^2\left(x^6\right) = 30x^4$.

If y'' is differentiable, its derivative, $y''' = dy''/dx = d^3y/dx^3$ is the **third derivative** of y with respect to x. The names continue as you imagine, with

$$y^{(n)} = \frac{d}{dx}y^{(n-1)} = \frac{d^ny}{dx^n} = D^ny$$

denoting the ***n*th derivative** of y with respect to x for any positive integer n.

We can interpret the second derivative as the rate of change of the slope of the tangent to the graph of $y = f(x)$ at each point. You will see in the next chapter that the second derivative reveals whether the graph bends upward or downward from the tangent line as we move off the point of tangency. In the next section, we interpret both the second and third derivatives in terms of motion along a straight line.

How to Read the Symbols for Derivatives

y'	"y prime"
y''	"y double prime"
$\frac{d^2y}{dx^2}$	"d squared y dx squared"
y'''	"y triple prime"
$y^{(n)}$	"y super n"
$\frac{d^ny}{dx^n}$	"d to the n of y by dx to the n"
D^n	"D to the n"

EXAMPLE 14 Finding Higher Derivatives

The first four derivatives of $y = x^3 - 3x^2 + 2$ are

First derivative:	$y' = 3x^2 - 6x$
Second derivative:	$y'' = 6x - 6$
Third derivative:	$y''' = 6$
Fourth derivative:	$y^{(4)} = 0.$

The function has derivatives of all orders, the fifth and later derivatives all being zero.

■

EXERCISES 3.2

Derivative Calculations

In Exercises 1–12, find the first and second derivatives.

1. $y = -x^2 + 3$
2. $y = x^2 + x + 8$
3. $s = 5t^3 - 3t^5$
4. $w = 3z^7 - 7z^3 + 21z^2$
5. $y = \dfrac{4x^3}{3} - x$
6. $y = \dfrac{x^3}{3} + \dfrac{x^2}{2} + \dfrac{x}{4}$
7. $w = 3z^{-2} - \dfrac{1}{z}$
8. $s = -2t^{-1} + \dfrac{4}{t^2}$
9. $y = 6x^2 - 10x - 5x^{-2}$
10. $y = 4 - 2x - x^{-3}$
11. $r = \dfrac{1}{3s^2} - \dfrac{5}{2s}$
12. $r = \dfrac{12}{\theta} - \dfrac{4}{\theta^3} + \dfrac{1}{\theta^4}$

In Exercises 13–16, find y' **(a)** by applying the Product Rule and **(b)** by multiplying the factors to produce a sum of simpler terms to differentiate.

13. $y = (3 - x^2)(x^3 - x + 1)$
14. $y = (x - 1)(x^2 + x + 1)$
15. $y = (x^2 + 1)\left(x + 5 + \dfrac{1}{x}\right)$
16. $y = \left(x + \dfrac{1}{x}\right)\left(x - \dfrac{1}{x} + 1\right)$

Find the derivatives of the functions in Exercises 17–28.

17. $y = \dfrac{2x + 5}{3x - 2}$
18. $z = \dfrac{2x + 1}{x^2 - 1}$
19. $g(x) = \dfrac{x^2 - 4}{x + 0.5}$
20. $f(t) = \dfrac{t^2 - 1}{t^2 + t - 2}$
21. $v = (1 - t)(1 + t^2)^{-1}$
22. $w = (2x - 7)^{-1}(x + 5)$
23. $f(s) = \dfrac{\sqrt{s} - 1}{\sqrt{s} + 1}$
24. $u = \dfrac{5x + 1}{2\sqrt{x}}$
25. $v = \dfrac{1 + x - 4\sqrt{x}}{x}$
26. $r = 2\left(\dfrac{1}{\sqrt{\theta}} + \sqrt{\theta}\right)$
27. $y = \dfrac{1}{(x^2 - 1)(x^2 + x + 1)}$
28. $y = \dfrac{(x + 1)(x + 2)}{(x - 1)(x - 2)}$

Find the derivatives of all orders of the functions in Exercises 29 and 30.

29. $y = \dfrac{x^4}{2} - \dfrac{3}{2}x^2 - x$
30. $y = \dfrac{x^5}{120}$

Find the first and second derivatives of the functions in Exercises 31–38.

31. $y = \dfrac{x^3 + 7}{x}$
32. $s = \dfrac{t^2 + 5t - 1}{t^2}$
33. $r = \dfrac{(\theta - 1)(\theta^2 + \theta + 1)}{\theta^3}$
34. $u = \dfrac{(x^2 + x)(x^2 - x + 1)}{x^4}$
35. $w = \left(\dfrac{1 + 3z}{3z}\right)(3 - z)$
36. $w = (z + 1)(z - 1)(z^2 + 1)$
37. $p = \left(\dfrac{q^2 + 3}{12q}\right)\left(\dfrac{q^4 - 1}{q^3}\right)$
38. $p = \dfrac{q^2 + 3}{(q - 1)^3 + (q + 1)^3}$

Using Numerical Values

39. Suppose u and v are functions of x that are differentiable at $x = 0$ and that

$$u(0) = 5, \quad u'(0) = -3, \quad v(0) = -1, \quad v'(0) = 2.$$

Find the values of the following derivatives at $x = 0$.

a. $\dfrac{d}{dx}(uv)$ **b.** $\dfrac{d}{dx}\left(\dfrac{u}{v}\right)$ **c.** $\dfrac{d}{dx}\left(\dfrac{v}{u}\right)$ **d.** $\dfrac{d}{dx}(7v - 2u)$

40. Suppose u and v are differentiable functions of x and that

$$u(1) = 2, \quad u'(1) = 0, \quad v(1) = 5, \quad v'(1) = -1.$$

Find the values of the following derivatives at $x = 1$.

a. $\dfrac{d}{dx}(uv)$ **b.** $\dfrac{d}{dx}\left(\dfrac{u}{v}\right)$ **c.** $\dfrac{d}{dx}\left(\dfrac{v}{u}\right)$ **d.** $\dfrac{d}{dx}(7v - 2u)$

Slopes and Tangents

41. **a. Normal to a curve** Find an equation for the line perpendicular to the tangent to the curve $y = x^3 - 4x + 1$ at the point $(2, 1)$.

b. Smallest slope What is the smallest slope on the curve? At what point on the curve does the curve have this slope?

c. Tangents having specified slope Find equations for the tangents to the curve at the points where the slope of the curve is 8.

42. **a. Horizontal tangents** Find equations for the horizontal tangents to the curve $y = x^3 - 3x - 2$. Also find equations for the lines that are perpendicular to these tangents at the points of tangency.

b. Smallest slope What is the smallest slope on the curve? At what point on the curve does the curve have this slope? Find an equation for the line that is perpendicular to the curve's tangent at this point.

43. Find the tangents to *Newton's serpentine* (graphed here) at the origin and the point $(1, 2)$.

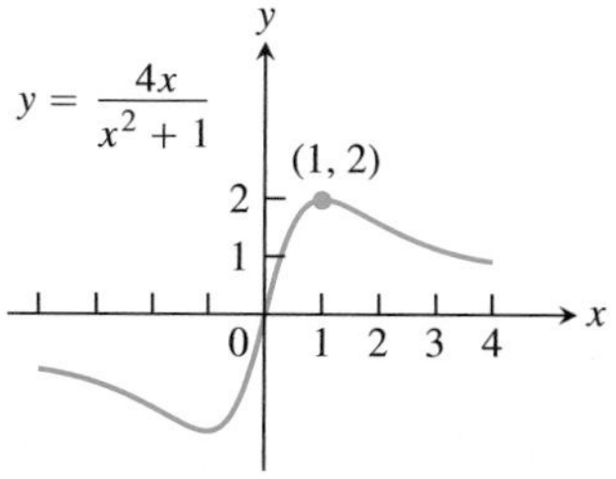

44. Find the tangent to the *Witch of Agnesi* (graphed here) at the point (2, 1).

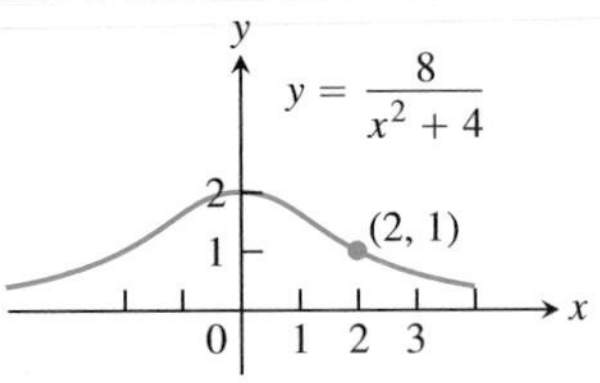

45. Quadratic tangent to identity function The curve $y = ax^2 + bx + c$ passes through the point (1, 2) and is tangent to the line $y = x$ at the origin. Find a, b, and c.

46. Quadratics having a common tangent The curves $y = x^2 + ax + b$ and $y = cx - x^2$ have a common tangent line at the point (1, 0). Find a, b, and c.

47. a. Find an equation for the line that is tangent to the curve $y = x^3 - x$ at the point $(-1, 0)$.

T **b.** Graph the curve and tangent line together. The tangent intersects the curve at another point. Use Zoom and Trace to estimate the point's coordinates.

T **c.** Confirm your estimates of the coordinates of the second intersection point by solving the equations for the curve and tangent simultaneously (Solver key).

48. a. Find an equation for the line that is tangent to the curve $y = x^3 - 6x^2 + 5x$ at the origin.

T **b.** Graph the curve and tangent together. The tangent intersects the curve at another point. Use Zoom and Trace to estimate the point's coordinates.

T **c.** Confirm your estimates of the coordinates of the second intersection point by solving the equations for the curve and tangent simultaneously (Solver key).

Theory and Examples

49. The general polynomial of degree n has the form

$$P(x) = a_n x^n + a_{n-1}x^{n-1} + \cdots + a_2 x^2 + a_1 x + a_0$$

where $a_n \neq 0$. Find $P'(x)$.

50. The body's reaction to medicine The reaction of the body to a dose of medicine can sometimes be represented by an equation of the form

$$R = M^2\left(\frac{C}{2} - \frac{M}{3}\right),$$

where C is a positive constant and M is the amount of medicine absorbed in the blood. If the reaction is a change in blood pressure, R is measured in millimeters of mercury. If the reaction is a change in temperature, R is measured in degrees, and so on.

Find dR/dM. This derivative, as a function of M, is called the sensitivity of the body to the medicine. In Section 4.5, we will see how to find the amount of medicine to which the body is most sensitive.

51. Suppose that the function v in the Product Rule has a constant value c. What does the Product Rule then say? What does this say about the Constant Multiple Rule?

52. The Reciprocal Rule

a. The *Reciprocal Rule* says that at any point where the function $v(x)$ is differentiable and different from zero,

$$\frac{d}{dx}\left(\frac{1}{v}\right) = -\frac{1}{v^2}\frac{dv}{dx}.$$

Show that the Reciprocal Rule is a special case of the Quotient Rule.

b. Show that the Reciprocal Rule and the Product Rule together imply the Quotient Rule.

53. Generalizing the Product Rule The Product Rule gives the formula

$$\frac{d}{dx}(uv) = u\frac{dv}{dx} + v\frac{du}{dx}$$

for the derivative of the product uv of two differentiable functions of x.

a. What is the analogous formula for the derivative of the product uvw of *three* differentiable functions of x?

b. What is the formula for the derivative of the product $u_1 u_2 u_3 u_4$ of *four* differentiable functions of x?

c. What is the formula for the derivative of a product $u_1 u_2 u_3 \ldots u_n$ of a finite number n of differentiable functions of x?

54. Rational Powers

a. Find $\frac{d}{dx}\left(x^{3/2}\right)$ by writing $x^{3/2}$ as $x \cdot x^{1/2}$ and using the Product Rule. Express your answer as a rational number times a rational power of x. Work parts (b) and (c) by a similar method.

b. Find $\frac{d}{dx}(x^{5/2})$.

c. Find $\frac{d}{dx}(x^{7/2})$.

d. What patterns do you see in your answers to parts (a), (b), and (c)? Rational powers are one of the topics in Section 3.6.

55. Cylinder pressure If gas in a cylinder is maintained at a constant temperature T, the pressure P is related to the volume V by a formula of the form

$$P = \frac{nRT}{V - nb} - \frac{an^2}{V^2},$$

in which a, b, n, and R are constants. Find dP/dV. (See accompanying figure.)

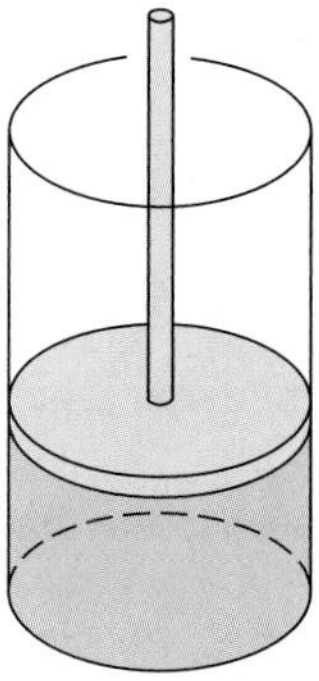

56. The best quantity to order One of the formulas for inventory management says that the average weekly cost of ordering, paying for, and holding merchandise is

$$A(q) = \frac{km}{q} + cm + \frac{hq}{2},$$

where q is the quantity you order when things run low (shoes, radios, brooms, or whatever the item might be); k is the cost of placing an order (the same, no matter how often you order); c is the cost of one item (a constant); m is the number of items sold each week (a constant); and h is the weekly holding cost per item (a constant that takes into account things such as space, utilities, insurance, and security). Find dA/dq and d^2A/dq^2.

3.3 The Derivative as a Rate of Change

In Section 2.1, we initiated the study of average and instantaneous rates of change. In this section, we continue our investigations of applications in which derivatives are used to model the rates at which things change in the world around us. We revisit the study of motion along a line and examine other applications.

It is natural to think of change as change with respect to time, but other variables can be treated in the same way. For example, a physician may want to know how change in dosage affects the body's response to a drug. An economist may want to study how the cost of producing steel varies with the number of tons produced.

Instantaneous Rates of Change

If we interpret the difference quotient $(f(x + h) - f(x))/h$ as the average rate of change in f over the interval from x to $x + h$, we can interpret its limit as $h \to 0$ as the rate at which f is changing at the point x.

DEFINITION Instantaneous Rate of Change

The **instantaneous rate of change** of f with respect to x at x_0 is the derivative

$$f'(x_0) = \lim_{h \to 0} \frac{f(x_0 + h) - f(x_0)}{h},$$

provided the limit exists.

Thus, instantaneous rates are limits of average rates.

It is conventional to use the word *instantaneous* even when x does not represent time. The word is, however, frequently omitted. When we say *rate of change*, we mean *instantaneous rate of change*.

EXAMPLE 1 How a Circle's Area Changes with Its Diameter

The area A of a circle is related to its diameter by the equation

$$A = \frac{\pi}{4} D^2.$$

How fast does the area change with respect to the diameter when the diameter is 10 m?

Solution The rate of change of the area with respect to the diameter is

$$\frac{dA}{dD} = \frac{\pi}{4} \cdot 2D = \frac{\pi D}{2}.$$

When $D = 10$ m, the area is changing at rate $(\pi/2)10 = 5\pi \text{ m}^2/\text{m}$. ■

Motion Along a Line: Displacement, Velocity, Speed, Acceleration, and Jerk

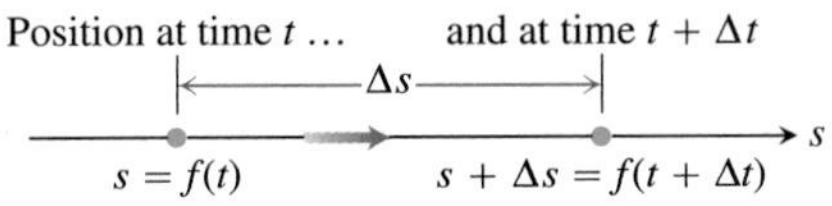

FIGURE 3.12 The positions of a body moving along a coordinate line at time t and shortly later at time $t + \Delta t$.

Suppose that an object is moving along a coordinate line (say an s-axis) so that we know its position s on that line as a function of time t:

$$s = f(t).$$

The **displacement** of the object over the time interval from t to $t + \Delta t$ (Figure 3.12) is

$$\Delta s = f(t + \Delta t) - f(t),$$

and the **average velocity** of the object over that time interval is

$$v_{av} = \frac{\text{displacement}}{\text{travel time}} = \frac{\Delta s}{\Delta t} = \frac{f(t + \Delta t) - f(t)}{\Delta t}.$$

To find the body's velocity at the exact instant t, we take the limit of the average velocity over the interval from t to $t + \Delta t$ as Δt shrinks to zero. This limit is the derivative of f with respect to t.

DEFINITION Velocity

Velocity (instantaneous velocity) is the derivative of position with respect to time. If a body's position at time t is $s = f(t)$, then the body's velocity at time t is

$$v(t) = \frac{ds}{dt} = \lim_{\Delta t \to 0} \frac{f(t + \Delta t) - f(t)}{\Delta t}.$$

EXAMPLE 2 Finding the Velocity of a Race Car

Figure 3.13 shows the time-to-distance graph of a 1996 Riley & Scott Mk III-Olds WSC race car. The slope of the secant PQ is the average velocity for the 3-sec interval from $t = 2$ to $t = 5$ sec; in this case, it is about 100 ft/sec or 68 mph.

The slope of the tangent at P is the speedometer reading at $t = 2$ sec, about 57 ft/sec or 39 mph. The acceleration for the period shown is a nearly constant 28.5 ft/sec^2 during

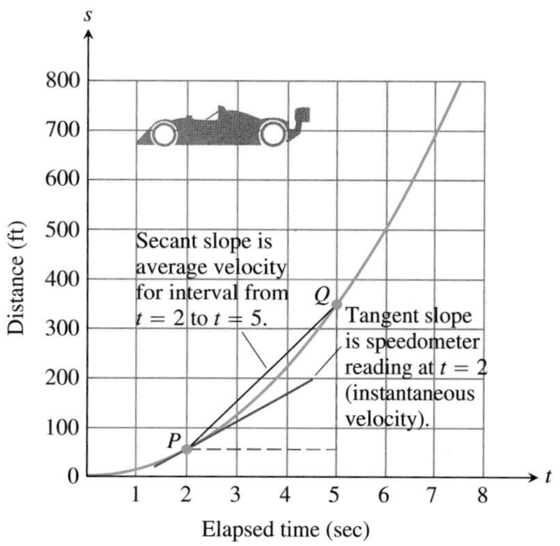

FIGURE 3.13 The time-to-distance graph for Example 2. The slope of the tangent line at P is the instantaneous velocity at $t = 2$ sec.

each second, which is about $0.89g$, where g is the acceleration due to gravity. The race car's top speed is an estimated 190 mph. (*Source: Road and Track*, March 1997.) ■

Besides telling how fast an object is moving, its velocity tells the direction of motion. When the object is moving forward (s increasing), the velocity is positive; when the body is moving backward (s decreasing), the velocity is negative (Figure 3.14).

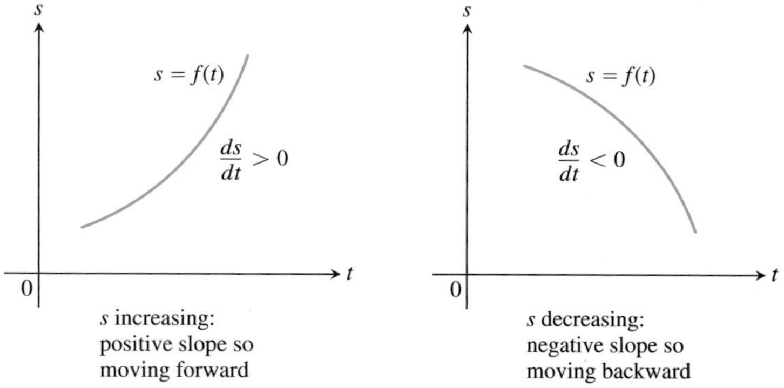

FIGURE 3.14 For motion $s = f(t)$ along a straight line, $v = ds/dt$ is positive when s increases and negative when s decreases.

If we drive to a friend's house and back at 30 mph, say, the speedometer will show 30 on the way over but it will not show -30 on the way back, even though our distance from home is decreasing. The speedometer always shows *speed*, which is the absolute value of velocity. Speed measures the rate of progress regardless of direction.

DEFINITION Speed

Speed is the absolute value of velocity.

$$\text{Speed} = |v(t)| = \left|\frac{ds}{dt}\right|$$

EXAMPLE 3 Horizontal Motion

Figure 3.15 shows the velocity $v = f'(t)$ of a particle moving on a coordinate line. The particle moves forward for the first 3 sec, moves backward for the next 2 sec, stands still for a second, and moves forward again. The particle achieves its greatest speed at time $t = 4$, while moving backward. ■

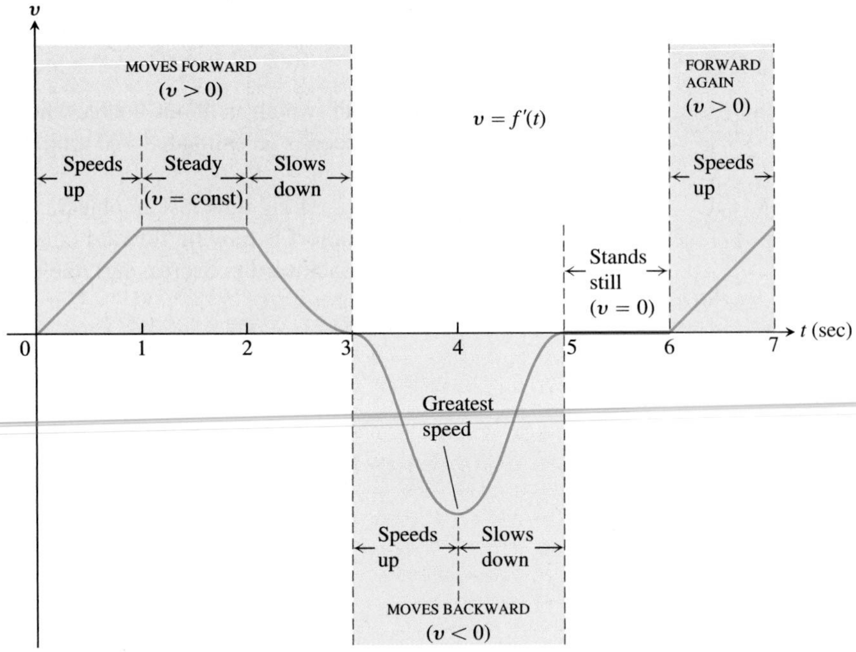

FIGURE 3.15 The velocity graph for Example 3.

HISTORICAL BIOGRAPHY

Bernard Bolzano
(1781–1848)

The rate at which a body's velocity changes is the body's *acceleration*. The acceleration measures how quickly the body picks up or loses speed.

A sudden change in acceleration is called a *jerk*. When a ride in a car or a bus is jerky, it is not that the accelerations involved are necessarily large but that the changes in acceleration are abrupt.

DEFINITIONS Acceleration, Jerk

Acceleration is the derivative of velocity with respect to time. If a body's position at time t is $s = f(t)$, then the body's acceleration at time t is

$$a(t) = \frac{dv}{dt} = \frac{d^2s}{dt^2}.$$

Jerk is the derivative of acceleration with respect to time:

$$j(t) = \frac{da}{dt} = \frac{d^3s}{dt^3}.$$

Near the surface of the Earth all bodies fall with the same constant acceleration. Galileo's experiments with free fall (Example 1, Section 2.1) lead to the equation

$$s = \frac{1}{2}gt^2,$$

where s is distance and g is the acceleration due to Earth's gravity. This equation holds in a vacuum, where there is no air resistance, and closely models the fall of dense, heavy objects, such as rocks or steel tools, for the first few seconds of their fall, before air resistance starts to slow them down.

The value of g in the equation $s = (1/2)gt^2$ depends on the units used to measure t and s. With t in seconds (the usual unit), the value of g determined by measurement at sea level is approximately 32 ft/sec^2 (feet per second squared) in English units, and $g = 9.8$ m/sec^2 (meters per second squared) in metric units. (These gravitational constants depend on the distance from Earth's center of mass, and are slightly lower on top of Mt. Everest, for example.)

The jerk of the constant acceleration of gravity ($g = 32$ ft/sec^2) is zero:

$$j = \frac{d}{dt}(g) = 0.$$

An object does not exhibit jerkiness during free fall.

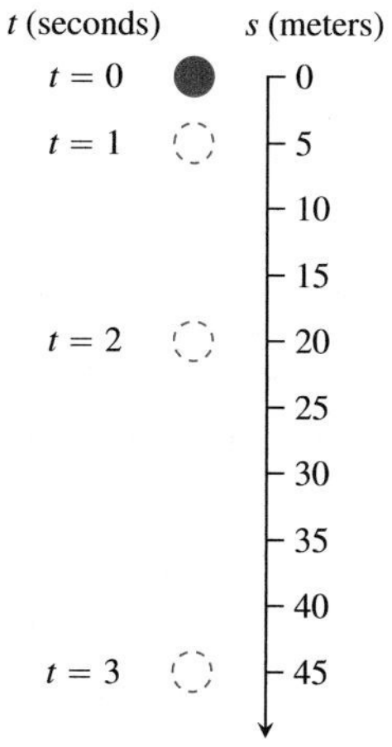

FIGURE 3.16 A ball bearing falling from rest (Example 4).

EXAMPLE 4 Modeling Free Fall

Figure 3.16 shows the free fall of a heavy ball bearing released from rest at time $t = 0$ sec.

(a) How many meters does the ball fall in the first 2 sec?

(b) What is its velocity, speed, and acceleration then?

Solution

(a) The metric free-fall equation is $s = 4.9t^2$. During the first 2 sec, the ball falls

$$s(2) = 4.9(2)^2 = 19.6 \text{ m}.$$

(b) At any time t, *velocity* is the derivative of position:

$$v(t) = s'(t) = \frac{d}{dt}(4.9t^2) = 9.8t.$$

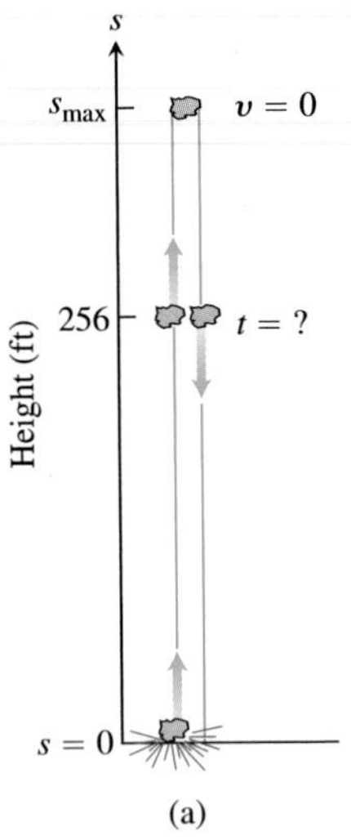

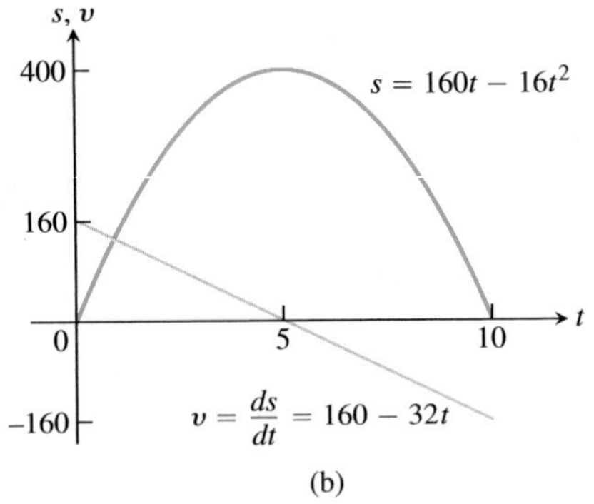

FIGURE 3.17 (a) The rock in Example 5. (b) The graphs of s and v as functions of time; s is largest when $v = ds/dt = 0$. The graph of s is *not* the path of the rock: It is a plot of height versus time. The slope of the plot is the rock's velocity, graphed here as a straight line.

At $t = 2$, the velocity is

$$v(2) = 19.6 \text{ m/sec}$$

in the downward (increasing s) direction. The *speed* at $t = 2$ is

$$\text{Speed} = |v(2)| = 19.6 \text{ m/sec}.$$

The *acceleration* at any time t is

$$a(t) = v'(t) = s''(t) = 9.8 \text{ m/sec}^2.$$

At $t = 2$, the acceleration is 9.8 m/sec^2. ■

EXAMPLE 5 Modeling Vertical Motion

A dynamite blast blows a heavy rock straight up with a launch velocity of 160 ft/sec (about 109 mph) (Figure 3.17a). It reaches a height of $s = 160t - 16t^2$ ft after t sec.

(a) How high does the rock go?

(b) What are the velocity and speed of the rock when it is 256 ft above the ground on the way up? On the way down?

(c) What is the acceleration of the rock at any time t during its flight (after the blast)?

(d) When does the rock hit the ground again?

Solution

(a) In the coordinate system we have chosen, s measures height from the ground up, so the velocity is positive on the way up and negative on the way down. The instant the rock is at its highest point is the one instant during the flight when the velocity is 0. To find the maximum height, all we need to do is to find when $v = 0$ and evaluate s at this time.

At any time t, the velocity is

$$v = \frac{ds}{dt} = \frac{d}{dt}(160t - 16t^2) = 160 - 32t \text{ ft/sec}.$$

The velocity is zero when

$$160 - 32t = 0 \quad \text{or} \quad t = 5 \text{ sec}.$$

The rock's height at $t = 5$ sec is

$$s_{max} = s(5) = 160(5) - 16(5)^2 = 800 - 400 = 400 \text{ ft}.$$

See Figure 3.17b.

(b) To find the rock's velocity at 256 ft on the way up and again on the way down, we first find the two values of t for which

$$s(t) = 160t - 16t^2 = 256.$$

To solve this equation, we write

$$\begin{aligned} 16t^2 - 160t + 256 &= 0 \\ 16(t^2 - 10t + 16) &= 0 \\ (t - 2)(t - 8) &= 0 \\ t = 2 \text{ sec}, t &= 8 \text{ sec}. \end{aligned}$$

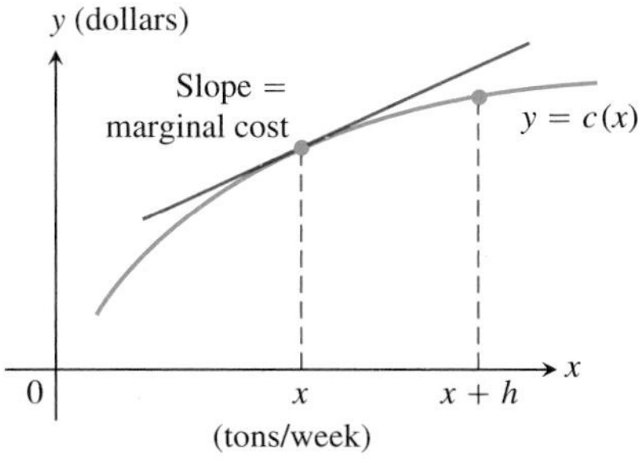

FIGURE 3.18 Weekly steel production: $c(x)$ is the cost of producing x tons per week. The cost of producing an additional h tons is $c(x + h) - c(x)$.

The rock is 256 ft above the ground 2 sec after the explosion and again 8 sec after the explosion. The rock's velocities at these times are

$$v(2) = 160 - 32(2) = 160 - 64 = 96 \text{ ft/sec}.$$
$$v(8) = 160 - 32(8) = 160 - 256 = -96 \text{ ft/sec}.$$

At both instants, the rock's speed is 96 ft/sec. Since $v(2) > 0$, the rock is moving upward (s is increasing) at $t = 2$ sec; it is moving downward (s is decreasing) at $t = 8$ because $v(8) < 0$.

(c) At any time during its flight following the explosion, the rock's acceleration is a constant

$$a = \frac{dv}{dt} = \frac{d}{dt}(160 - 32t) = -32 \text{ ft/sec}^2.$$

The acceleration is always downward. As the rock rises, it slows down; as it falls, it speeds up.

(d) The rock hits the ground at the positive time t for which $s = 0$. The equation $160t - 16t^2 = 0$ factors to give $16t(10 - t) = 0$, so it has solutions $t = 0$ and $t = 10$. At $t = 0$, the blast occurred and the rock was thrown upward. It returned to the ground 10 sec later. ■

Derivatives in Economics

Engineers use the terms *velocity* and *acceleration* to refer to the derivatives of functions describing motion. Economists, too, have a specialized vocabulary for rates of change and derivatives. They call them *marginals*.

In a manufacturing operation, the *cost of production* $c(x)$ is a function of x, the number of units produced. The **marginal cost of production** is the rate of change of cost with respect to level of production, so it is dc/dx.

Suppose that $c(x)$ represents the dollars needed to produce x tons of steel in one week. It costs more to produce $x + h$ units per week, and the cost difference, divided by h, is the average cost of producing each additional ton:

$$\frac{c(x + h) - c(x)}{h} = \begin{array}{l}\text{average cost of each of the additional} \\ h \text{ tons of steel produced.}\end{array}$$

The limit of this ratio as $h \to 0$ is the *marginal cost* of producing more steel per week when the current weekly production is x tons (Figure 3.18).

$$\frac{dc}{dx} = \lim_{h \to 0} \frac{c(x + h) - c(x)}{h} = \text{marginal cost of production}.$$

Sometimes the marginal cost of production is loosely defined to be the extra cost of producing one unit:

$$\frac{\Delta c}{\Delta x} = \frac{c(x + 1) - c(x)}{1},$$

which is approximated by the value of dc/dx at x. This approximation is acceptable if the slope of the graph of c does not change quickly near x. Then the difference quotient will be close to its limit dc/dx, which is the rise in the tangent line if $\Delta x = 1$ (Figure 3.19). The approximation works best for large values of x.

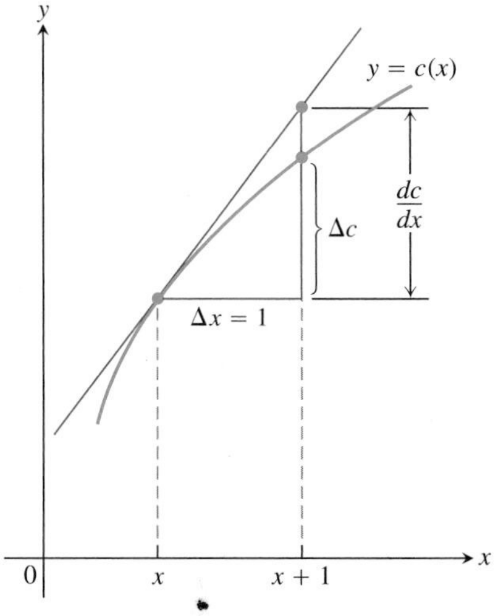

FIGURE 3.19 The marginal cost dc/dx is approximately the extra cost Δc of producing $\Delta x = 1$ more unit.

Economists often represent a total cost function by a cubic polynomial

$$c(x) = \alpha x^3 + \beta x^2 + \gamma x + \delta$$

where δ represents *fixed costs* such as rent, heat, equipment capitalization, and management costs. The other terms represent *variable costs* such as the costs of raw materials, taxes, and labor. Fixed costs are independent of the number of units produced, whereas variable costs depend on the quantity produced. A cubic polynomial is usually complicated enough to capture the cost behavior on a relevant quantity interval.

EXAMPLE 6 Marginal Cost and Marginal Revenue

Suppose that it costs

$$c(x) = x^3 - 6x^2 + 15x$$

dollars to produce x radiators when 8 to 30 radiators are produced and that

$$r(x) = x^3 - 3x^2 + 12x$$

gives the dollar revenue from selling x radiators. Your shop currently produces 10 radiators a day. About how much extra will it cost to produce one more radiator a day, and what is your estimated increase in revenue for selling 11 radiators a day?

Solution The cost of producing one more radiator a day when 10 are produced is about $c'(10)$:

$$c'(x) = \frac{d}{dx}\left(x^3 - 6x^2 + 15x\right) = 3x^2 - 12x + 15$$

$$c'(10) = 3(100) - 12(10) + 15 = 195.$$

The additional cost will be about \$195. The marginal revenue is

$$r'(x) = \frac{d}{dx}\left(x^3 - 3x^2 + 12x\right) = 3x^2 - 6x + 12.$$

The marginal revenue function estimates the increase in revenue that will result from selling one additional unit. If you currently sell 10 radiators a day, you can expect your revenue to increase by about

$$r'(10) = 3(100) - 6(10) + 12 = \$252$$

if you increase sales to 11 radiators a day. ■

EXAMPLE 7 Marginal Tax Rate

To get some feel for the language of marginal rates, consider marginal tax rates. If your marginal income tax rate is 28% and your income increases by \$1000, you can expect to pay an extra \$280 in taxes. This does not mean that you pay 28% of your entire income in taxes. It just means that at your current income level I, the rate of increase of taxes T with respect to income is $dT/dI = 0.28$. You will pay \$0.28 out of every extra dollar you earn in taxes. Of course, if you earn a lot more, you may land in a higher tax bracket and your marginal rate will increase. ■

Sensitivity to Change

When a small change in x produces a large change in the value of a function $f(x)$, we say that the function is relatively **sensitive** to changes in x. The derivative $f'(x)$ is a measure of this sensitivity.

EXAMPLE 8 Genetic Data and Sensitivity to Change

The Austrian monk Gregor Johann Mendel (1822–1884), working with garden peas and other plants, provided the first scientific explanation of hybridization.

His careful records showed that if p (a number between 0 and 1) is the frequency of the gene for smooth skin in peas (dominant) and $(1 - p)$ is the frequency of the gene for wrinkled skin in peas, then the proportion of smooth-skinned peas in the next generation will be

$$y = 2p(1 - p) + p^2 = 2p - p^2.$$

The graph of y versus p in Figure 3.20a suggests that the value of y is more sensitive to a change in p when p is small than when p is large. Indeed, this fact is borne out by the derivative graph in Figure 3.20b, which shows that dy/dp is close to 2 when p is near 0 and close to 0 when p is near 1.

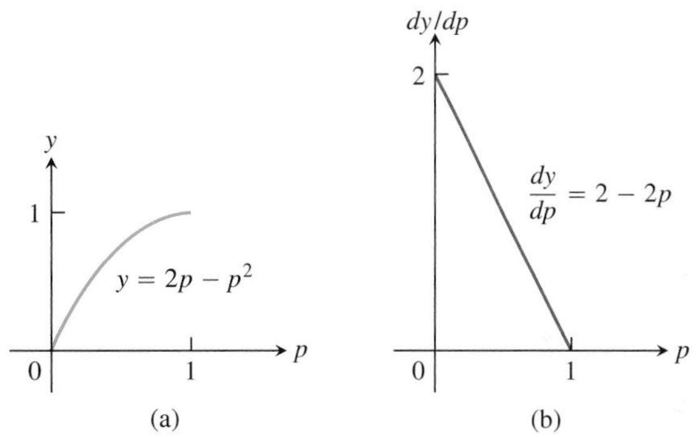

FIGURE 3.20 (a) The graph of $y = 2p - p^2$, describing the proportion of smooth-skinned peas. (b) The graph of dy/dp (Example 8).

The implication for genetics is that introducing a few more dominant genes into a highly recessive population (where the frequency of wrinkled skin peas is small) will have a more dramatic effect on later generations than will a similar increase in a highly dominant population. ■

EXERCISES 3.3

Motion Along a Coordinate Line

Exercises 1–6 give the positions $s = f(t)$ of a body moving on a coordinate line, with s in meters and t in seconds.

a. Find the body's displacement and average velocity for the given time interval.

b. Find the body's speed and acceleration at the endpoints of the interval.

c. When, if ever, during the interval does the body change direction?

1. $s = t^2 - 3t + 2, \quad 0 \le t \le 2$

2. $s = 6t - t^2, \quad 0 \le t \le 6$

3. $s = -t^3 + 3t^2 - 3t, \quad 0 \le t \le 3$

4. $s = (t^4/4) - t^3 + t^2, \quad 0 \le t \le 3$

5. $s = \dfrac{25}{t^2} - \dfrac{5}{t}, \quad 1 \le t \le 5$

6. $s = \dfrac{25}{t+5}, \quad -4 \le t \le 0$

7. Particle motion At time t, the position of a body moving along the s-axis is $s = t^3 - 6t^2 + 9t$ m.

a. Find the body's acceleration each time the velocity is zero.

b. Find the body's speed each time the acceleration is zero.

c. Find the total distance traveled by the body from $t = 0$ to $t = 2$.

8. Particle motion At time $t \ge 0$, the velocity of a body moving along the s-axis is $v = t^2 - 4t + 3$.

a. Find the body's acceleration each time the velocity is zero.

b. When is the body moving forward? Backward?

c. When is the body's velocity increasing? Decreasing?

Free-Fall Applications

9. Free fall on Mars and Jupiter The equations for free fall at the surfaces of Mars and Jupiter (s in meters, t in seconds) are $s = 1.86t^2$ on Mars and $s = 11.44t^2$ on Jupiter. How long does it take a rock falling from rest to reach a velocity of 27.8 m/sec (about 100 km/h) on each planet?

10. Lunar projectile motion A rock thrown vertically upward from the surface of the moon at a velocity of 24 m/sec (about 86 km/h) reaches a height of $s = 24t - 0.8t^2$ meters in t sec.

a. Find the rock's velocity and acceleration at time t. (The acceleration in this case is the acceleration of gravity on the moon.)

b. How long does it take the rock to reach its highest point?

c. How high does the rock go?

d. How long does it take the rock to reach half its maximum height?

e. How long is the rock aloft?

11. Finding g on a small airless planet Explorers on a small airless planet used a spring gun to launch a ball bearing vertically upward from the surface at a launch velocity of 15 m/sec. Because the acceleration of gravity at the planet's surface was g_s m/sec^2, the explorers expected the ball bearing to reach a height of $s = 15t - (1/2)g_s t^2$ meters t sec later. The ball bearing reached its maximum height 20 sec after being launched. What was the value of g_s?

12. Speeding bullet A 45-caliber bullet fired straight up from the surface of the moon would reach a height of $s = 832t - 2.6t^2$ feet after t sec. On Earth, in the absence of air, its height would be $s = 832t - 16t^2$ ft after t sec. How long will the bullet be aloft in each case? How high will the bullet go?

13. Free fall from the Tower of Pisa Had Galileo dropped a cannonball from the Tower of Pisa, 179 ft above the ground, the ball's height above ground t sec into the fall would have been $s = 179 - 16t^2$.

a. What would have been the ball's velocity, speed, and acceleration at time t?

b. About how long would it have taken the ball to hit the ground?

c. What would have been the ball's velocity at the moment of impact?

14. Galileo's free-fall formula Galileo developed a formula for a body's velocity during free fall by rolling balls from rest down increasingly steep inclined planks and looking for a limiting formula that would predict a ball's behavior when the plank was vertical and the ball fell freely; see part (a) of the accompanying figure. He found that, for any given angle of the plank, the ball's velocity t sec into motion was a constant multiple of t. That is, the velocity was given by a formula of the form $v = kt$. The value of the constant k depended on the inclination of the plank.

In modern notation—part (b) of the figure—with distance in meters and time in seconds, what Galileo determined by experiment was that, for any given angle θ, the ball's velocity t sec into the roll was

$$v = 9.8(\sin\theta)t \text{ m/sec}.$$

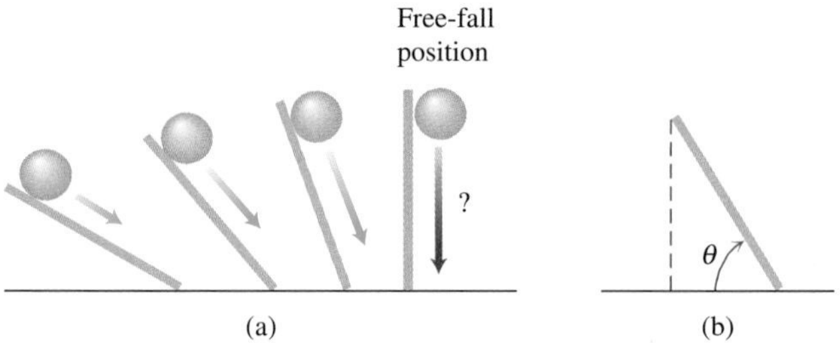

a. What is the equation for the ball's velocity during free fall?

b. Building on your work in part (a), what constant acceleration does a freely falling body experience near the surface of Earth?

Conclusions About Motion from Graphs

15. The accompanying figure shows the velocity $v = ds/dt = f(t)$ (m/sec) of a body moving along a coordinate line.

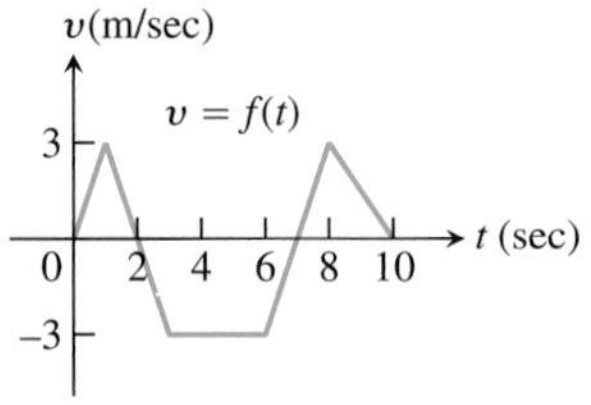

a. When does the body reverse direction?

b. When (approximately) is the body moving at a constant speed?

c. Graph the body's speed for $0 \le t \le 10$.

d. Graph the acceleration, where defined.

16. A particle P moves on the number line shown in part (a) of the accompanying figure. Part (b) shows the position of P as a function of time t.

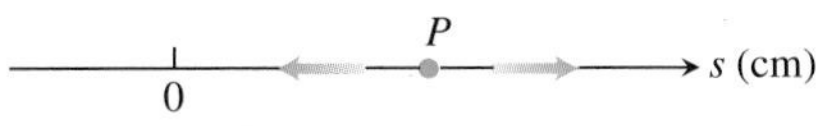

(a)

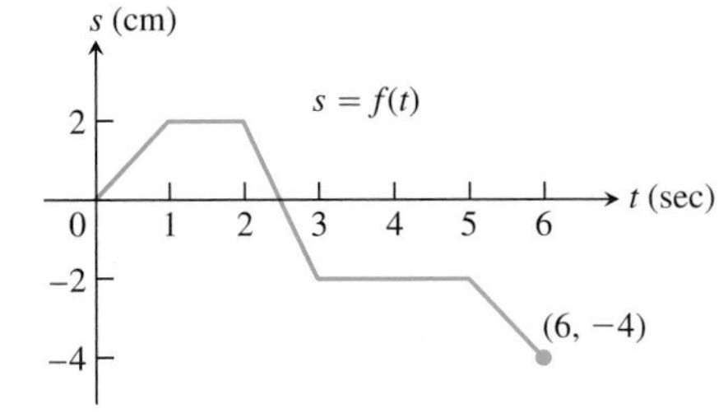

(b)

a. When is P moving to the left? Moving to the right? Standing still?

b. Graph the particle's velocity and speed (where defined).

17. **Launching a rocket** When a model rocket is launched, the propellant burns for a few seconds, accelerating the rocket upward. After burnout, the rocket coasts upward for a while and then begins to fall. A small explosive charge pops out a parachute shortly after the rocket starts down. The parachute slows the rocket to keep it from breaking when it lands.

The figure here shows velocity data from the flight of the model rocket. Use the data to answer the following.

a. How fast was the rocket climbing when the engine stopped?

b. For how many seconds did the engine burn?

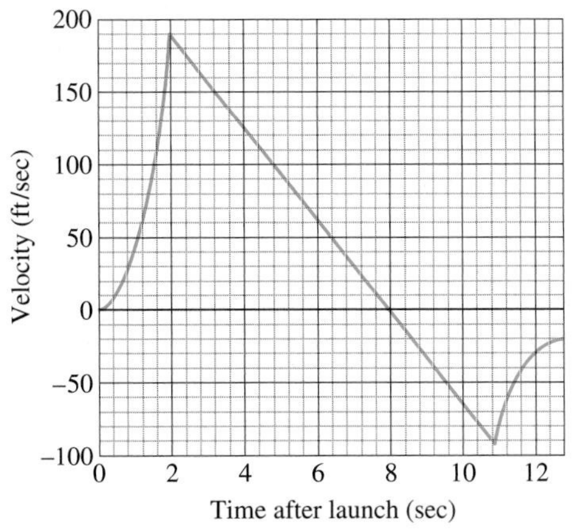

c. When did the rocket reach its highest point? What was its velocity then?

d. When did the parachute pop out? How fast was the rocket falling then?

e. How long did the rocket fall before the parachute opened?

f. When was the rocket's acceleration greatest?

g. When was the acceleration constant? What was its value then (to the nearest integer)?

18. The accompanying figure shows the velocity $v = f(t)$ of a particle moving on a coordinate line.

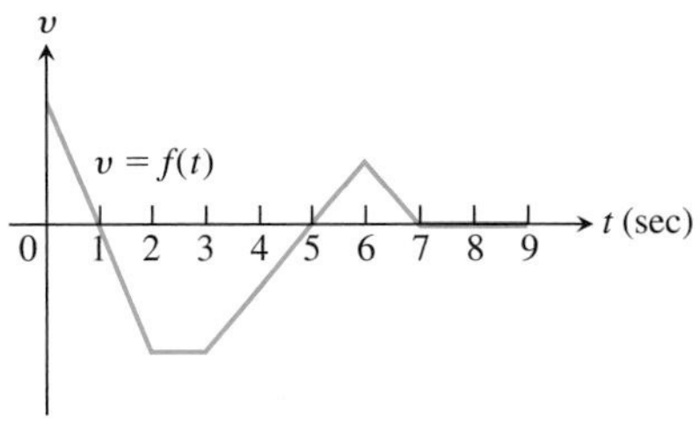

a. When does the particle move forward? Move backward? Speed up? Slow down?

b. When is the particle's acceleration positive? Negative? Zero?

c. When does the particle move at its greatest speed?

d. When does the particle stand still for more than an instant?

19. **Two falling balls** The multiflash photograph in the accompanying figure shows two balls falling from rest. The vertical rulers are marked in centimeters. Use the equation $s = 490t^2$ (the free-fall equation for s in centimeters and t in seconds) to answer the following questions.

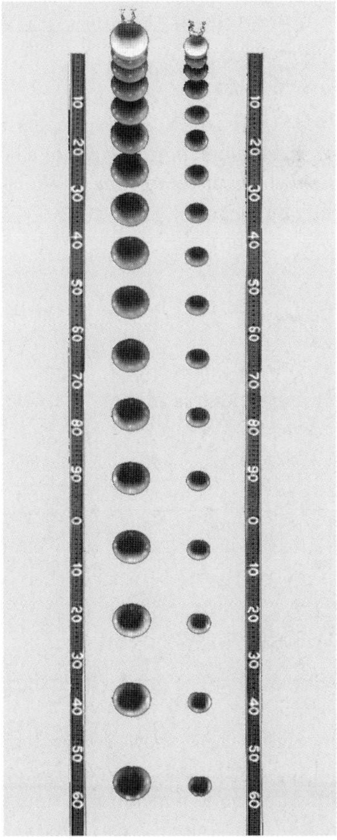

a. How long did it take the balls to fall the first 160 cm? What was their average velocity for the period?

b. How fast were the balls falling when they reached the 160-cm mark? What was their acceleration then?

c. About how fast was the light flashing (flashes per second)?

20. A traveling truck The accompanying graph shows the position s of a truck traveling on a highway. The truck starts at $t = 0$ and returns 15 h later at $t = 15$.

a. Use the technique described in Section 3.1, Example 3, to graph the truck's velocity $v = ds/dt$ for $0 \le t \le 15$. Then repeat the process, with the velocity curve, to graph the truck's acceleration dv/dt.

b. Suppose that $s = 15t^2 - t^3$. Graph ds/dt and d^2s/dt^2 and compare your graphs with those in part (a).

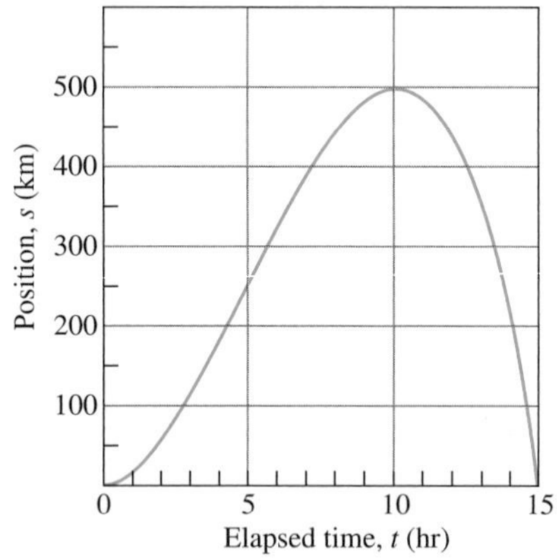

21. The graphs in Figure 3.21 show the position s, velocity $v = ds/dt$, and acceleration $a = d^2s/dt^2$ of a body moving along a coordinate line as functions of time t. Which graph is which? Give reasons for your answers.

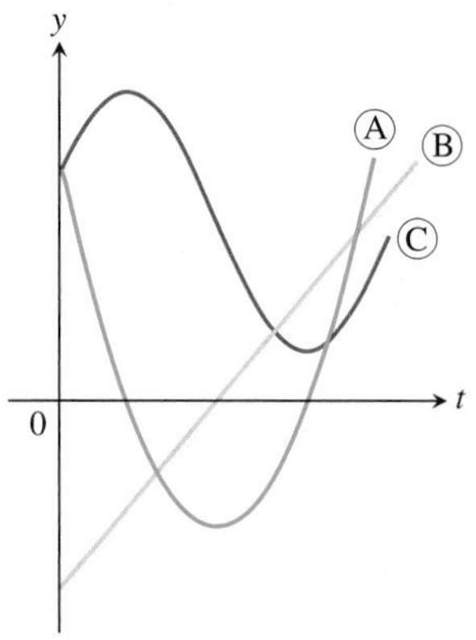

FIGURE 3.21 The graphs for Exercise 21.

22. The graphs in Figure 3.22 show the position s, the velocity $v = ds/dt$, and the acceleration $a = d^2s/dt^2$ of a body moving along the coordinate line as functions of time t. Which graph is which? Give reasons for your answers.

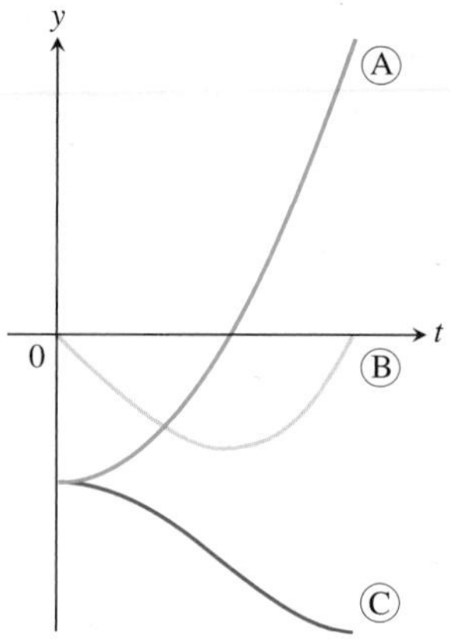

FIGURE 3.22 The graphs for Exercise 22.

Economics

23. Marginal cost Suppose that the dollar cost of producing x washing machines is $c(x) = 2000 + 100x - 0.1x^2$.

a. Find the average cost per machine of producing the first 100 washing machines.

b. Find the marginal cost when 100 washing machines are produced.

c. Show that the marginal cost when 100 washing machines are produced is approximately the cost of producing one more washing machine after the first 100 have been made, by calculating the latter cost directly.

24. Marginal revenue Suppose that the revenue from selling x washing machines is

$$r(x) = 20{,}000\left(1 - \frac{1}{x}\right)$$

dollars.

a. Find the marginal revenue when 100 machines are produced.

b. Use the function $r'(x)$ to estimate the increase in revenue that will result from increasing production from 100 machines a week to 101 machines a week.

c. Find the limit of $r'(x)$ as $x \to \infty$. How would you interpret this number?

Additional Applications

25. Bacterium population When a bactericide was added to a nutrient broth in which bacteria were growing, the bacterium population continued to grow for a while, but then stopped growing and began to decline. The size of the population at time t (hours) was $b = 10^6 + 10^4t - 10^3t^2$. Find the growth rates at

a. $t = 0$ hours.

b. $t = 5$ hours.

c. $t = 10$ hours.

26. Draining a tank The number of gallons of water in a tank t minutes after the tank has started to drain is $Q(t) = 200(30 - t)^2$. How fast is the water running out at the end of 10 min? What is the average rate at which the water flows out during the first 10 min?

T **27. Draining a tank** It takes 12 hours to drain a storage tank by opening the valve at the bottom. The depth y of fluid in the tank t hours after the valve is opened is given by the formula

$$y = 6\left(1 - \frac{t}{12}\right)^2 \text{ m}.$$

a. Find the rate dy/dt (m/h) at which the tank is draining at time t.

b. When is the fluid level in the tank falling fastest? Slowest? What are the values of dy/dt at these times?

c. Graph y and dy/dt together and discuss the behavior of y in relation to the signs and values of dy/dt.

28. Inflating a balloon The volume $V = (4/3)\pi r^3$ of a spherical balloon changes with the radius.

a. At what rate (ft^3/ft) does the volume change with respect to the radius when $r = 2$ ft?

b. By approximately how much does the volume increase when the radius changes from 2 to 2.2 ft?

29. Airplane takeoff Suppose that the distance an aircraft travels along a runway before takeoff is given by $D = (10/9)t^2$, where D is measured in meters from the starting point and t is measured in seconds from the time the brakes are released. The aircraft will become airborne when its speed reaches 200 km/h. How long will it take to become airborne, and what distance will it travel in that time?

30. Volcanic lava fountains Although the November 1959 Kilauea Iki eruption on the island of Hawaii began with a line of fountains along the wall of the crater, activity was later confined to a single vent in the crater's floor, which at one point shot lava 1900 ft straight into the air (a world record). What was the lava's exit velocity in feet per second? In miles per hour? (*Hint*: If v_0 is the exit velocity of a particle of lava, its height t sec later will be $s = v_0 t - 16t^2$ ft. Begin by finding the time at which $ds/dt = 0$. Neglect air resistance.)

T Exercises 31–34 give the position function $s = f(t)$ of a body moving along the s-axis as a function of time t. Graph f together with the velocity function $v(t) = ds/dt = f'(t)$ and the acceleration function $a(t) = d^2s/dt^2 = f''(t)$. Comment on the body's behavior in relation to the signs and values of v and a. Include in your commentary such topics as the following:

a. When is the body momentarily at rest?

b. When does it move to the left (down) or to the right (up)?

c. When does it change direction?

d. When does it speed up and slow down?

e. When is it moving fastest (highest speed)? Slowest?

f. When is it farthest from the axis origin?

31. $s = 200t - 16t^2, \quad 0 \le t \le 12.5$ (a heavy object fired straight up from Earth's surface at 200 ft/sec)

32. $s = t^2 - 3t + 2, \quad 0 \le t \le 5$

33. $s = t^3 - 6t^2 + 7t, \quad 0 \le t \le 4$

34. $s = 4 - 7t + 6t^2 - t^3, \quad 0 \le t \le 4$

35. Thoroughbred racing A racehorse is running a 10-furlong race. (A furlong is 220 yards, although we will use furlongs and seconds as our units in this exercise.) As the horse passes each furlong marker (F), a steward records the time elapsed (t) since the beginning of the race, as shown in the table:

F	0	1	2	3	4	5	6	7	8	9	10
t	0	20	33	46	59	73	86	100	112	124	135

a. How long does it take the horse to finish the race?

b. What is the average speed of the horse over the first 5 furlongs?

c. What is the approximate speed of the horse as it passes the 3-furlong marker?

d. During which portion of the race is the horse running the fastest?

e. During which portion of the race is the horse accelerating the fastest?

3.4 Derivatives of Trigonometric Functions

Many of the phenomena we want information about are approximately periodic (electromagnetic fields, heart rhythms, tides, weather). The derivatives of sines and cosines play a key role in describing periodic changes. This section shows how to differentiate the six basic trigonometric functions.

Derivative of the Sine Function

To calculate the derivative of $f(x) = \sin x$, for x measured in radians, we combine the limits in Example 5a and Theorem 7 in Section 2.4 with the angle sum identity for the sine:

$$\sin(x + h) = \sin x \cos h + \cos x \sin h.$$

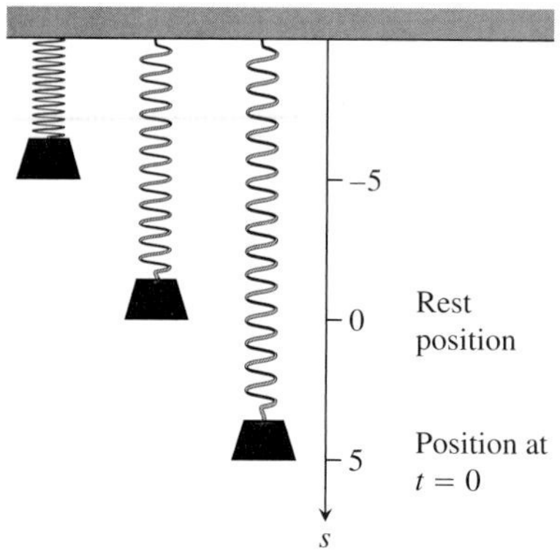

FIGURE 3.24 A body hanging from a vertical spring and then displaced oscillates above and below its rest position. Its motion is described by trigonometric functions (Example 3).

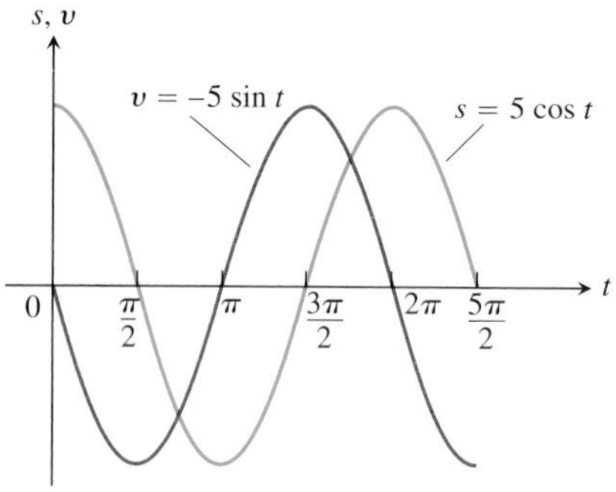

FIGURE 3.25 The graphs of the position and velocity of the body in Example 3.

Simple Harmonic Motion

The motion of a body bobbing freely up and down on the end of a spring or bungee cord is an example of *simple harmonic motion*. The next example describes a case in which there are no opposing forces such as friction or buoyancy to slow the motion down.

EXAMPLE 3 Motion on a Spring

A body hanging from a spring (Figure 3.24) is stretched 5 units beyond its rest position and released at time $t = 0$ to bob up and down. Its position at any later time t is

$$s = 5 \cos t.$$

What are its velocity and acceleration at time t?

Solution We have

Position: $s = 5 \cos t$

Velocity: $v = \frac{ds}{dt} = \frac{d}{dt}(5 \cos t) = -5 \sin t$

Acceleration: $a = \frac{dv}{dt} = \frac{d}{dt}(-5 \sin t) = -5 \cos t.$

Notice how much we can learn from these equations:

1. As time passes, the weight moves down and up between $s = -5$ and $s = 5$ on the s-axis. The amplitude of the motion is 5. The period of the motion is 2π.
2. The velocity $v = -5 \sin t$ attains its greatest magnitude, 5, when $\cos t = 0$, as the graphs show in Figure 3.25. Hence, the speed of the weight, $|v| = 5|\sin t|$, is greatest when $\cos t = 0$, that is, when $s = 0$ (the rest position). The speed of the weight is zero when $\sin t = 0$. This occurs when $s = 5 \cos t = \pm 5$, at the endpoints of the interval of motion.
3. The acceleration value is always the exact opposite of the position value. When the weight is above the rest position, gravity is pulling it back down; when the weight is below the rest position, the spring is pulling it back up.
4. The acceleration, $a = -5 \cos t$, is zero only at the rest position, where $\cos t = 0$ and the force of gravity and the force from the spring offset each other. When the weight is anywhere else, the two forces are unequal and acceleration is nonzero. The acceleration is greatest in magnitude at the points farthest from the rest position, where $\cos t = \pm 1$. ■

EXAMPLE 4 Jerk

The jerk of the simple harmonic motion in Example 3 is

$$j = \frac{da}{dt} = \frac{d}{dt}(-5 \cos t) = 5 \sin t.$$

It has its greatest magnitude when $\sin t = \pm 1$, not at the extremes of the displacement but at the rest position, where the acceleration changes direction and sign. ■

Derivatives of the Other Basic Trigonometric Functions

Because $\sin x$ and $\cos x$ are differentiable functions of x, the related functions

$$\tan x = \frac{\sin x}{\cos x}, \quad \cot x = \frac{\cos x}{\sin x}, \quad \sec x = \frac{1}{\cos x}, \quad \text{and} \quad \csc x = \frac{1}{\sin x}$$

are differentiable at every value of x at which they are defined. Their derivatives, calculated from the Quotient Rule, are given by the following formulas. Notice the negative signs in the derivative formulas for the cofunctions.

Derivatives of the Other Trigonometric Functions

$$\frac{d}{dx}(\tan x) = \sec^2 x$$

$$\frac{d}{dx}(\sec x) = \sec x \tan x$$

$$\frac{d}{dx}(\cot x) = -\csc^2 x$$

$$\frac{d}{dx}(\csc x) = -\csc x \cot x$$

To show a typical calculation, we derive the derivative of the tangent function. The other derivations are left to Exercise 50.

EXAMPLE 5

Find $d(\tan x)/dx$.

Solution

$$\begin{aligned}
\frac{d}{dx}(\tan x) = \frac{d}{dx}\left(\frac{\sin x}{\cos x}\right) &= \frac{\cos x \frac{d}{dx}(\sin x) - \sin x \frac{d}{dx}(\cos x)}{\cos^2 x} && \text{Quotient Rule}\\
&= \frac{\cos x \cos x - \sin x(-\sin x)}{\cos^2 x}\\
&= \frac{\cos^2 x + \sin^2 x}{\cos^2 x}\\
&= \frac{1}{\cos^2 x} = \sec^2 x
\end{aligned}$$

■

EXAMPLE 6

Find y'' if $y = \sec x$.

Solution

$$\begin{aligned}
y &= \sec x\\
y' &= \sec x \tan x\\
y'' &= \frac{d}{dx}(\sec x \tan x)\\
&= \sec x \frac{d}{dx}(\tan x) + \tan x \frac{d}{dx}(\sec x) && \text{Product Rule}\\
&= \sec x(\sec^2 x) + \tan x(\sec x \tan x)\\
&= \sec^3 x + \sec x \tan^2 x
\end{aligned}$$

■

The differentiability of the trigonometric functions throughout their domains gives another proof of their continuity at every point in their domains (Theorem 1, Section 3.1). So we can calculate limits of algebraic combinations and composites of trigonometric functions by direct substitution.

EXAMPLE 7 Finding a Trigonometric Limit

$$\lim_{x \to 0} \frac{\sqrt{2 + \sec x}}{\cos(\pi - \tan x)} = \frac{\sqrt{2 + \sec 0}}{\cos(\pi - \tan 0)} = \frac{\sqrt{2 + 1}}{\cos(\pi - 0)} = \frac{\sqrt{3}}{-1} = -\sqrt{3}$$ ■

EXERCISES 3.4

Derivatives

In Exercises 1–12, find dy/dx.

1. $y = -10x + 3\cos x$

2. $y = \frac{3}{x} + 5\sin x$

3. $y = \csc x - 4\sqrt{x} + 7$

4. $y = x^2 \cot x - \frac{1}{x^2}$

5. $y = (\sec x + \tan x)(\sec x - \tan x)$

6. $y = (\sin x + \cos x)\sec x$

7. $y = \frac{\cot x}{1 + \cot x}$

8. $y = \frac{\cos x}{1 + \sin x}$

9. $y = \frac{4}{\cos x} + \frac{1}{\tan x}$

10. $y = \frac{\cos x}{x} + \frac{x}{\cos x}$

11. $y = x^2 \sin x + 2x\cos x - 2\sin x$

12. $y = x^2 \cos x - 2x\sin x - 2\cos x$

In Exercises 13–16, find ds/dt.

13. $s = \tan t - t$

14. $s = t^2 - \sec t + 1$

15. $s = \frac{1 + \csc t}{1 - \csc t}$

16. $s = \frac{\sin t}{1 - \cos t}$

In Exercises 17–20, find $dr/d\theta$.

17. $r = 4 - \theta^2 \sin\theta$

18. $r = \theta\sin\theta + \cos\theta$

19. $r = \sec\theta\csc\theta$

20. $r = (1 + \sec\theta)\sin\theta$

In Exercises 21–24, find dp/dq.

21. $p = 5 + \frac{1}{\cot q}$

22. $p = (1 + \csc q)\cos q$

23. $p = \frac{\sin q + \cos q}{\cos q}$

24. $p = \frac{\tan q}{1 + \tan q}$

25. Find y'' if

a. $y = \csc x$.

b. $y = \sec x$.

26. Find $y^{(4)} = d^4y/dx^4$ if

a. $y = -2\sin x$.

b. $y = 9\cos x$.

Tangent Lines

In Exercises 27–30, graph the curves over the given intervals, together with their tangents at the given values of x. Label each curve and tangent with its equation.

27. $y = \sin x, \quad -3\pi/2 \le x \le 2\pi$
$x = -\pi, 0, 3\pi/2$

28. $y = \tan x, \quad -\pi/2 < x < \pi/2$
$x = -\pi/3, 0, \pi/3$

29. $y = \sec x, \quad -\pi/2 < x < \pi/2$
$x = -\pi/3, \pi/4$

30. $y = 1 + \cos x, \quad -3\pi/2 \le x \le 2\pi$
$x = -\pi/3, 3\pi/2$

T Do the graphs of the functions in Exercises 31–34 have any horizontal tangents in the interval $0 \le x \le 2\pi$? If so, where? If not, why not? Visualize your findings by graphing the functions with a grapher.

31. $y = x + \sin x$

32. $y = 2x + \sin x$

33. $y = x - \cot x$

34. $y = x + 2\cos x$

35. Find all points on the curve $y = \tan x$, $-\pi/2 < x < \pi/2$, where the tangent line is parallel to the line $y = 2x$. Sketch the curve and tangent(s) together, labeling each with its equation.

36. Find all points on the curve $y = \cot x$, $0 < x < \pi$, where the tangent line is parallel to the line $y = -x$. Sketch the curve and tangent(s) together, labeling each with its equation.

In Exercises 37 and 38, find an equation for **(a)** the tangent to the curve at P and **(b)** the horizontal tangent to the curve at Q.

37.

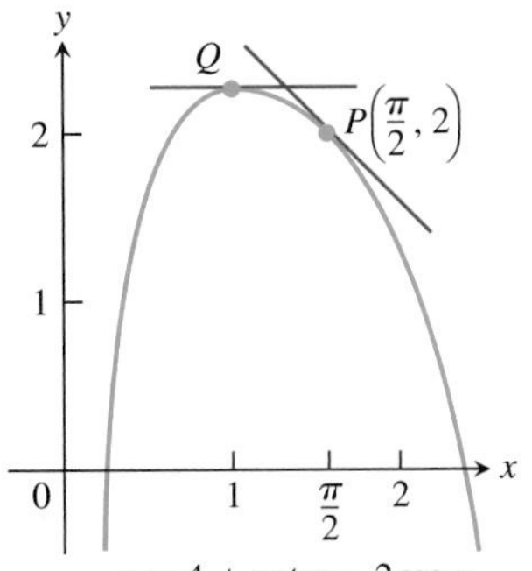

$y = 4 + \cot x - 2\csc x$

38.

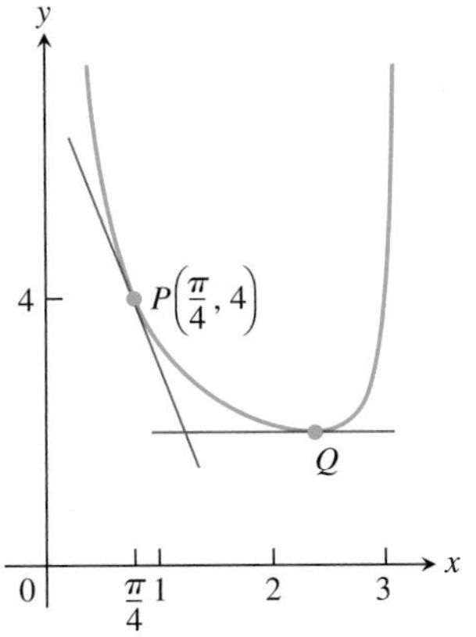

$y = 1 + \sqrt{2}\csc x + \cot x$

Trigonometric Limits

Find the limits in Exercises 39–44.

39. $\lim_{x\to 2} \sin\left(\frac{1}{x} - \frac{1}{2}\right)$

40. $\lim_{x\to -\pi/6} \sqrt{1 + \cos(\pi \csc x)}$

41. $\lim_{x\to 0} \sec\left[\cos x + \pi \tan\left(\frac{\pi}{4 \sec x}\right) - 1\right]$

42. $\lim_{x\to 0} \sin\left(\frac{\pi + \tan x}{\tan x - 2\sec x}\right)$

43. $\lim_{t\to 0} \tan\left(1 - \frac{\sin t}{t}\right)$

44. $\lim_{\theta\to 0} \cos\left(\frac{\pi\theta}{\sin\theta}\right)$

Simple Harmonic Motion

The equations in Exercises 45 and 46 give the position $s = f(t)$ of a body moving on a coordinate line (s in meters, t in seconds). Find the body's velocity, speed, acceleration, and jerk at time $t = \pi/4$ sec.

45. $s = 2 - 2\sin t$

46. $s = \sin t + \cos t$

Theory and Examples

47. Is there a value of c that will make

$$f(x) = \begin{cases} \dfrac{\sin^2 3x}{x^2}, & x \neq 0 \\ c, & x = 0 \end{cases}$$

continuous at $x = 0$? Give reasons for your answer.

48. Is there a value of b that will make

$$g(x) = \begin{cases} x + b, & x < 0 \\ \cos x, & x \geq 0 \end{cases}$$

continuous at $x = 0$? Differentiable at $x = 0$? Give reasons for your answers.

49. Find $d^{999}/dx^{999}(\cos x)$.

50. Derive the formula for the derivative with respect to x of

a. $\sec x$. **b.** $\csc x$. **c.** $\cot x$.

T **51.** Graph $y = \cos x$ for $-\pi \leq x \leq 2\pi$. On the same screen, graph

$$y = \frac{\sin(x + h) - \sin x}{h}$$

for $h = 1, 0.5, 0.3$, and 0.1. Then, in a new window, try $h = -1, -0.5$, and -0.3. What happens as $h \to 0^+$? As $h \to 0^-$? What phenomenon is being illustrated here?

T **52.** Graph $y = -\sin x$ for $-\pi \leq x \leq 2\pi$. On the same screen, graph

$$y = \frac{\cos(x + h) - \cos x}{h}$$

for $h = 1, 0.5, 0.3$, and 0.1. Then, in a new window, try $h = -1, -0.5$, and -0.3. What happens as $h \to 0^+$? As $h \to 0^-$? What phenomenon is being illustrated here?

T **53. Centered difference quotients** The *centered difference quotient*

$$\frac{f(x + h) - f(x - h)}{2h}$$

is used to approximate $f'(x)$ in numerical work because (1) its limit as $h \to 0$ equals $f'(x)$ when $f'(x)$ exists, and (2) it usually gives a better approximation of $f'(x)$ for a given value of h than Fermat's difference quotient

$$\frac{f(x + h) - f(x)}{h}.$$

See the accompanying figure.

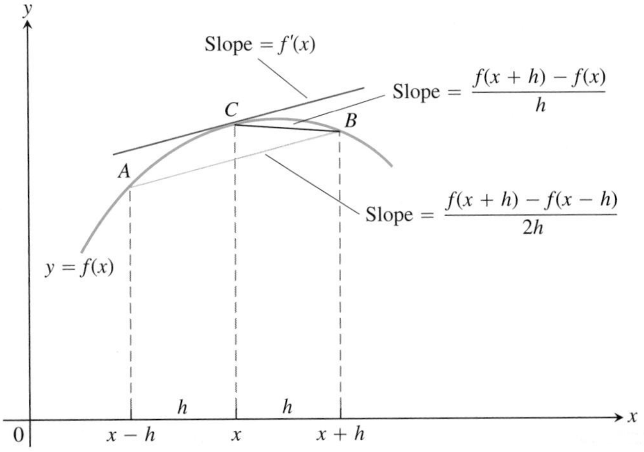

a. To see how rapidly the centered difference quotient for $f(x) = \sin x$ converges to $f'(x) = \cos x$, graph $y = \cos x$ together with

$$y = \frac{\sin(x + h) - \sin(x - h)}{2h}$$

over the interval $[-\pi, 2\pi]$ for $h = 1, 0.5$, and 0.3. Compare the results with those obtained in Exercise 51 for the same values of h.

b. To see how rapidly the centered difference quotient for $f(x) = \cos x$ converges to $f'(x) = -\sin x$, graph $y = -\sin x$ together with

$$y = \frac{\cos(x + h) - \cos(x - h)}{2h}$$

over the interval $[-\pi, 2\pi]$ for $h = 1, 0.5$, and 0.3. Compare the results with those obtained in Exercise 52 for the same values of h.

54. A caution about centered difference quotients (*Continuation of Exercise 53.*) The quotient

$$\frac{f(x + h) - f(x - h)}{2h}$$

may have a limit as $h \to 0$ when f has no derivative at x. As a case in point, take $f(x) = |x|$ and calculate

$$\lim_{h \to 0} \frac{|0 + h| - |0 - h|}{2h}.$$

As you will see, the limit exists even though $f(x) = |x|$ has no derivative at $x = 0$. *Moral:* Before using a centered difference quotient, be sure the derivative exists.

T **55. Slopes on the graph of the tangent function** Graph $y = \tan x$ and its derivative together on $(-\pi/2, \pi/2)$. Does the graph of the tangent function appear to have a smallest slope? a largest slope? Is the slope ever negative? Give reasons for your answers.

T **56. Slopes on the graph of the cotangent function** Graph $y = \cot x$ and its derivative together for $0 < x < \pi$. Does the graph of the cotangent function appear to have a smallest slope? A largest slope? Is the slope ever positive? Give reasons for your answers.

T **57. Exploring $(\sin kx)/x$** Graph $y = (\sin x)/x$, $y = (\sin 2x)/x$, and $y = (\sin 4x)/x$ together over the interval $-2 \le x \le 2$. Where does each graph appear to cross the y-axis? Do the graphs really intersect the axis? What would you expect the graphs of $y = (\sin 5x)/x$ and $y = (\sin(-3x))/x$ to do as $x \to 0$? Why? What about the graph of $y = (\sin kx)/x$ for other values of k? Give reasons for your answers.

T **58. Radians versus degrees: degree mode derivatives** What happens to the derivatives of $\sin x$ and $\cos x$ if x is measured in degrees instead of radians? To find out, take the following steps.

a. With your graphing calculator or computer grapher in *degree mode*, graph

$$f(h) = \frac{\sin h}{h}$$

and estimate $\lim_{h \to 0} f(h)$. Compare your estimate with $\pi/180$. Is there any reason to believe the limit *should* be $\pi/180$?

b. With your grapher still in degree mode, estimate

$$\lim_{h \to 0} \frac{\cos h - 1}{h}.$$

c. Now go back to the derivation of the formula for the derivative of $\sin x$ in the text and carry out the steps of the derivation using degree-mode limits. What formula do you obtain for the derivative?

d. Work through the derivation of the formula for the derivative of $\cos x$ using degree-mode limits. What formula do you obtain for the derivative?

e. The disadvantages of the degree-mode formulas become apparent as you start taking derivatives of higher order. Try it. What are the second and third degree-mode derivatives of $\sin x$ and $\cos x$?

3.5 The Chain Rule and Parametric Equations

We know how to differentiate $y = f(u) = \sin u$ and $u = g(x) = x^2 - 4$, but how do we differentiate a composite like $F(x) = f(g(x)) = \sin(x^2 - 4)$? The differentiation formulas we have studied so far do not tell us how to calculate $F'(x)$. So how do we find the derivative of $F = f \circ g$? The answer is, with the Chain Rule, which says that the derivative of the composite of two differentiable functions is the product of their derivatives evaluated at appropriate points. The Chain Rule is one of the most important and widely used rules of differentiation. This section describes the rule and how to use it. We then apply the rule to describe curves in the plane and their tangent lines in another way.

Derivative of a Composite Function

We begin with examples.

EXAMPLE 1 Relating Derivatives

The function $y = \frac{3}{2}x = \frac{1}{2}(3x)$ is the composite of the functions $y = \frac{1}{2}u$ and $u = 3x$. How are the derivatives of these functions related?

Solution We have

$$\frac{dy}{dx} = \frac{3}{2}, \quad \frac{dy}{du} = \frac{1}{2}, \quad \text{and} \quad \frac{du}{dx} = 3.$$

Since $\frac{3}{2} = \frac{1}{2} \cdot 3$, we see that

$$\frac{dy}{dx} = \frac{dy}{du} \cdot \frac{du}{dx}.$$

Is it an accident that

$$\frac{dy}{dx} = \frac{dy}{du} \cdot \frac{du}{dx}?$$

If we think of the derivative as a rate of change, our intuition allows us to see that this relationship is reasonable. If $y = f(u)$ changes half as fast as u and $u = g(x)$ changes three times as fast as x, then we expect y to change $3/2$ times as fast as x. This effect is much like that of a multiple gear train (Figure 3.26). ■

FIGURE 3.26 When gear A makes x turns, gear B makes u turns and gear C makes y turns. By comparing circumferences or counting teeth, we see that $y = u/2$ (C turns one-half turn for each B turn) and $u = 3x$ (B turns three times for A's one), so $y = 3x/2$. Thus, $dy/dx = 3/2 = (1/2)(3) = (dy/du)(du/dx)$.

EXAMPLE 2

The function

$$y = 9x^4 + 6x^2 + 1 = (3x^2 + 1)^2$$

is the composite of $y = u^2$ and $u = 3x^2 + 1$. Calculating derivatives, we see that

$$\begin{aligned} \frac{dy}{du} \cdot \frac{du}{dx} &= 2u \cdot 6x \\ &= 2(3x^2 + 1) \cdot 6x \\ &= 36x^3 + 12x. \end{aligned}$$

Calculating the derivative from the expanded formula, we get

$$\begin{aligned} \frac{dy}{dx} &= \frac{d}{dx}\left(9x^4 + 6x^2 + 1\right) \\ &= 36x^3 + 12x. \end{aligned}$$

Once again,

$$\frac{dy}{du} \cdot \frac{du}{dx} = \frac{dy}{dx}.$$

■

The derivative of the composite function $f(g(x))$ at x is the derivative of f at $g(x)$ times the derivative of g at x. This is known as the Chain Rule (Figure 3.27).

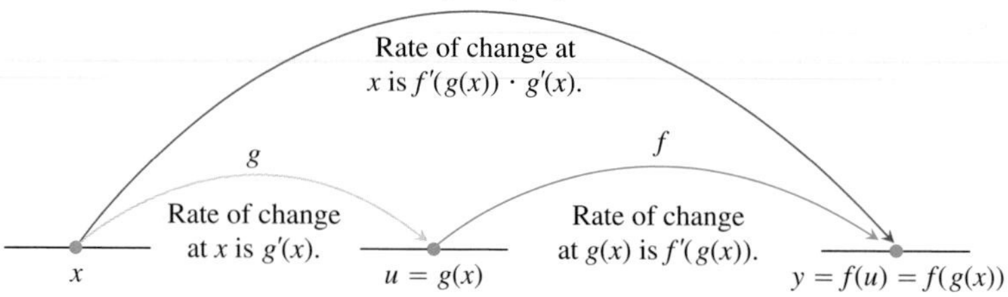

FIGURE 3.27 Rates of change multiply: The derivative of $f \circ g$ at x is the derivative of f at $g(x)$ times the derivative of g at x.

THEOREM 3 The Chain Rule

If $f(u)$ is differentiable at the point $u = g(x)$ and $g(x)$ is differentiable at x, then the composite function $(f \circ g)(x) = f(g(x))$ is differentiable at x, and

$$(f \circ g)'(x) = f'(g(x)) \cdot g'(x).$$

In Leibniz's notation, if $y = f(u)$ and $u = g(x)$, then

$$\frac{dy}{dx} = \frac{dy}{du} \cdot \frac{du}{dx},$$

where dy/du is evaluated at $u = g(x)$.

Intuitive "Proof" of the Chain Rule:

Let Δu be the change in u corresponding to a change of Δx in x, that is

$$\Delta u = g(x + \Delta x) - g(x)$$

Then the corresponding change in y is

$$\Delta y = f(u + \Delta u) - f(u).$$

It would be tempting to write

$$\frac{\Delta y}{\Delta x} = \frac{\Delta y}{\Delta u} \cdot \frac{\Delta u}{\Delta x} \tag{1}$$

and take the limit as $\Delta x \to 0$:

$$\begin{aligned} \frac{dy}{dx} &= \lim_{\Delta x \to 0} \frac{\Delta y}{\Delta x} \\ &= \lim_{\Delta x \to 0} \frac{\Delta y}{\Delta u} \cdot \frac{\Delta u}{\Delta x} \\ &= \lim_{\Delta x \to 0} \frac{\Delta y}{\Delta u} \cdot \lim_{\Delta x \to 0} \frac{\Delta u}{\Delta x} \\ &= \lim_{\Delta u \to 0} \frac{\Delta y}{\Delta u} \cdot \lim_{\Delta x \to 0} \frac{\Delta u}{\Delta x} \\ &= \frac{dy}{du} \frac{du}{dx}. \end{aligned}$$

(Note that $\Delta u \to 0$ as $\Delta x \to 0$ since g is continuous.)

The only flaw in this reasoning is that in Equation (1) it might happen that $\Delta u = 0$ (even when $\Delta x \neq 0$) and, of course, we can't divide by 0. The proof requires a different approach to overcome this flaw, and we give a precise proof in Section 3.8. ■

EXAMPLE 3 Applying the Chain Rule

An object moves along the x-axis so that its position at any time $t \geq 0$ is given by $x(t) = \cos(t^2 + 1)$. Find the velocity of the object as a function of t.

Solution We know that the velocity is dx/dt. In this instance, x is a composite function: $x = \cos(u)$ and $u = t^2 + 1$. We have

$$\frac{dx}{du} = -\sin(u) \qquad x = \cos(u)$$

$$\frac{du}{dt} = 2t. \qquad u = t^2 + 1$$

By the Chain Rule,

$$\begin{aligned} \frac{dx}{dt} &= \frac{dx}{du} \cdot \frac{du}{dt} \\ &= -\sin(u) \cdot 2t \qquad \frac{dx}{du} \text{ evaluated at } u \\ &= -\sin(t^2 + 1) \cdot 2t \\ &= -2t \sin(t^2 + 1). \end{aligned}$$

■

As we see from Example 3, a difficulty with the Leibniz notation is that it doesn't state specifically where the derivatives are supposed to be evaluated.

"Outside-Inside" Rule

It sometimes helps to think about the Chain Rule this way: If $y = f(g(x))$, then

$$\frac{dy}{dx} = f'(g(x)) \cdot g'(x).$$

In words, differentiate the "outside" function f and evaluate it at the "inside" function $g(x)$ left alone; then multiply by the derivative of the "inside function."

EXAMPLE 4 Differentiating from the Outside In

Differentiate $\sin(x^2 + x)$ with respect to x.

Solution

$$\frac{d}{dx} \sin \underbrace{(x^2 + x)}_{\text{inside}} = \cos \underbrace{(x^2 + x)}_{\substack{\text{inside} \\ \text{left alone}}} \cdot \underbrace{(2x + 1)}_{\substack{\text{derivative of} \\ \text{the inside}}}$$

■

Repeated Use of the Chain Rule

We sometimes have to use the Chain Rule two or more times to find a derivative. Here is an example.

HISTORICAL BIOGRAPHY

Johann Bernoulli
(1667–1748)

EXAMPLE 5 A Three-Link "Chain"

Find the derivative of $g(t) = \tan(5 - \sin 2t)$.

Solution Notice here that the tangent is a function of $5 - \sin 2t$, whereas the sine is a function of $2t$, which is itself a function of t. Therefore, by the Chain Rule,

$$\begin{aligned} g'(t) &= \frac{d}{dt}\left(\tan\left(5 - \sin 2t\right)\right) \\ &= \sec^2(5 - \sin 2t)\cdot\frac{d}{dt}\left(5 - \sin 2t\right) && \text{Derivative of } \tan u \text{ with } u = 5 - \sin 2t \\ &= \sec^2(5 - \sin 2t)\cdot\left(0 - \cos 2t\cdot\frac{d}{dt}\left(2t\right)\right) && \text{Derivative of } 5 - \sin u \text{ with } u = 2t \\ &= \sec^2(5 - \sin 2t)\cdot(-\cos 2t)\cdot 2 \\ &= -2(\cos 2t)\sec^2(5 - \sin 2t). \end{aligned}$$

■

The Chain Rule with Powers of a Function

If f is a differentiable function of u and if u is a differentiable function of x, then substituting $y = f(u)$ into the Chain Rule formula

$$\frac{dy}{dx} = \frac{dy}{du}\cdot\frac{du}{dx}$$

leads to the formula

$$\frac{d}{dx}f(u) = f'(u)\frac{du}{dx}.$$

Here's an example of how it works: If n is a positive or negative integer and $f(u) = u^n$, the Power Rules (Rules 2 and 7) tell us that $f'(u) = nu^{n-1}$. If u is a differentiable function of x, then we can use the Chain Rule to extend this to the **Power Chain Rule**:

$$\frac{d}{dx}u^n = nu^{n-1}\frac{du}{dx}. \qquad \frac{d}{du}\left(u^n\right) = nu^{n-1}$$

EXAMPLE 6 Applying the Power Chain Rule

(a)
$$\begin{aligned} \frac{d}{dx}(5x^3 - x^4)^7 &= 7(5x^3 - x^4)^6\frac{d}{dx}\left(5x^3 - x^4\right) && \text{Power Chain Rule with } u = 5x^3 - x^4,\ n = 7 \\ &= 7(5x^3 - x^4)^6(5\cdot 3x^2 - 4x^3) \\ &= 7(5x^3 - x^4)^6(15x^2 - 4x^3) \end{aligned}$$

(b)
$$\begin{aligned} \frac{d}{dx}\left(\frac{1}{3x - 2}\right) &= \frac{d}{dx}(3x - 2)^{-1} \\ &= -1(3x - 2)^{-2}\frac{d}{dx}(3x - 2) && \text{Power Chain Rule with } u = 3x - 2,\ n = -1 \\ &= -1(3x - 2)^{-2}(3) \\ &= -\frac{3}{(3x - 2)^2} \end{aligned}$$

In part (b) we could also have found the derivative with the Quotient Rule. ■

$\sin^n x$ means $(\sin x)^n$, $n \neq -1$.

EXAMPLE 7 Finding Tangent Slopes

(a) Find the slope of the line tangent to the curve $y = \sin^5 x$ at the point where $x = \pi/3$.

(b) Show that the slope of every line tangent to the curve $y = 1/(1 - 2x)^3$ is positive.

Solution

(a)
$$\frac{dy}{dx} = 5\sin^4 x \cdot \frac{d}{dx}\sin x \qquad \text{Power Chain Rule with } u = \sin x,\ n = 5$$
$$= 5\sin^4 x \cos x$$

The tangent line has slope

$$\left.\frac{dy}{dx}\right|_{x=\pi/3} = 5\left(\frac{\sqrt{3}}{2}\right)^4\left(\frac{1}{2}\right) = \frac{45}{32}.$$

(b)
$$\frac{dy}{dx} = \frac{d}{dx}(1 - 2x)^{-3}$$
$$= -3(1 - 2x)^{-4} \cdot \frac{d}{dx}(1 - 2x) \qquad \text{Power Chain Rule with } u = (1 - 2x),\ n = -3$$
$$= -3(1 - 2x)^{-4} \cdot (-2)$$
$$= \frac{6}{(1 - 2x)^4}$$

At any point (x, y) on the curve, $x \neq 1/2$ and the slope of the tangent line is

$$\frac{dy}{dx} = \frac{6}{(1 - 2x)^4},$$

the quotient of two positive numbers. ■

EXAMPLE 8 Radians Versus Degrees

It is important to remember that the formulas for the derivatives of both $\sin x$ and $\cos x$ were obtained under the assumption that x is measured in radians, *not* degrees. The Chain Rule gives us new insight into the difference between the two. Since $180° = \pi$ radians, $x° = \pi x/180$ radians where $x°$ means the angle x measured in degrees.

By the Chain Rule,

$$\frac{d}{dx}\sin(x°) = \frac{d}{dx}\sin\left(\frac{\pi x}{180}\right) = \frac{\pi}{180}\cos\left(\frac{\pi x}{180}\right) = \frac{\pi}{180}\cos(x°).$$

See Figure 3.28. Similarly, the derivative of $\cos(x°)$ is $-(\pi/180)\sin(x°)$.

The factor $\pi/180$, annoying in the first derivative, would compound with repeated differentiation. We see at a glance the compelling reason for the use of radian measure. ■

Parametric Equations

Instead of describing a curve by expressing the y-coordinate of a point $P(x, y)$ on the curve as a function of x, it is sometimes more convenient to describe the curve by expressing *both* coordinates as functions of a third variable t. Figure 3.29 shows the path of a moving particle described by a pair of equations, $x = f(t)$ and $y = g(t)$. For studying motion,

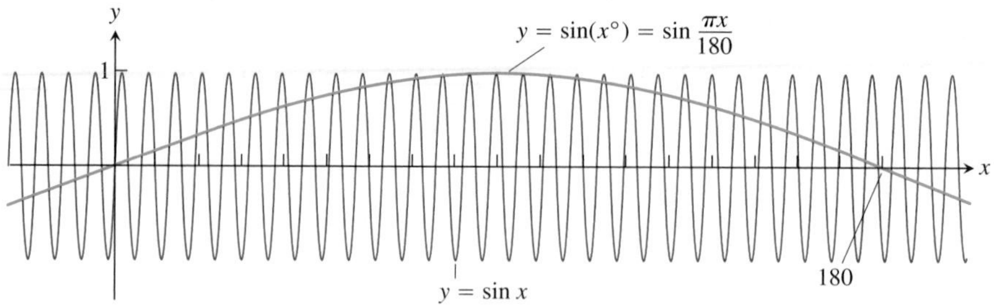

FIGURE 3.28 Sin ($x°$) oscillates only $\pi/180$ times as often as $\sin x$ oscillates. Its maximum slope is $\pi/180$ at $x = 0$ (Example 8).

t usually denotes time. Equations like these are better than a Cartesian formula because they tell us the particle's position $(x, y) = (f(t), g(t))$ at any time t.

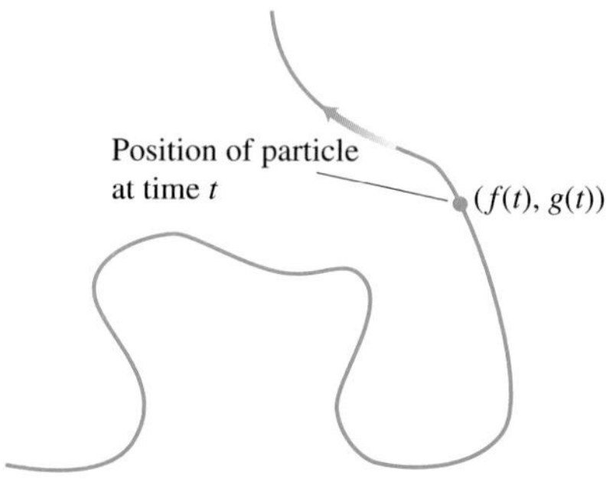

FIGURE 3.29 The path traced by a particle moving in the xy-plane is not always the graph of a function of x or a function of y.

DEFINITION Parametric Curve

If x and y are given as functions

$$x = f(t), \qquad y = g(t)$$

over an interval of t-values, then the set of points $(x, y) = (f(t), g(t))$ defined by these equations is a **parametric curve**. The equations are **parametric equations** for the curve.

The variable t is a **parameter** for the curve, and its domain I is the **parameter interval**. If I is a closed interval, $a \le t \le b$, the point $(f(a), g(a))$ is the **initial point** of the curve. The point $(f(b), g(b))$ is the **terminal point**. When we give parametric equations and a parameter interval for a curve, we say that we have **parametrized** the curve. The equations and interval together constitute a **parametrization** of the curve.

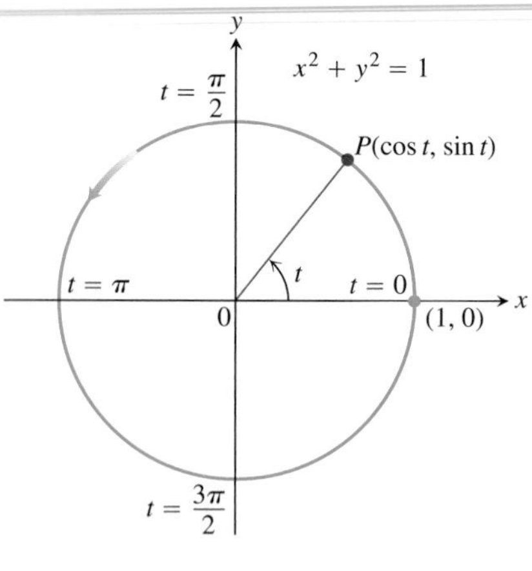

FIGURE 3.30 The equations $x = \cos t$ and $y = \sin t$ describe motion on the circle $x^2 + y^2 = 1$. The arrow shows the direction of increasing t (Example 9).

EXAMPLE 9 Moving Counterclockwise on a Circle

Graph the parametric curves

(a) $x = \cos t, \qquad y = \sin t, \qquad 0 \le t \le 2\pi.$

(b) $x = a\cos t, \qquad y = a\sin t, \qquad 0 \le t \le 2\pi.$

Solution

(a) Since $x^2 + y^2 = \cos^2 t + \sin^2 t = 1$, the parametric curve lies along the unit circle $x^2 + y^2 = 1$. As t increases from 0 to 2π, the point $(x, y) = (\cos t, \sin t)$ starts at $(1, 0)$ and traces the entire circle once counterclockwise (Figure 3.30).

(b) For $x = a\cos t$, $y = a\sin t$, $0 \le t \le 2\pi$, we have $x^2 + y^2 = a^2\cos^2 t + a^2\sin^2 t = a^2$. The parametrization describes a motion that begins at the point $(a, 0)$ and traverses the circle $x^2 + y^2 = a^2$ once counterclockwise, returning to $(a, 0)$ at $t = 2\pi$. ■

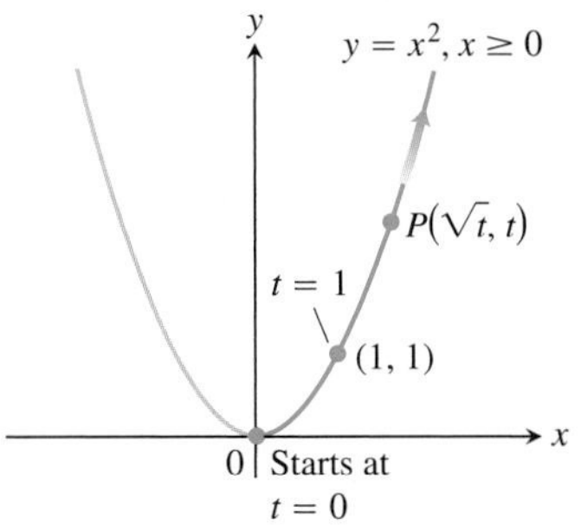

FIGURE 3.31 The equations $x = \sqrt{t}$ and $y = t$ and the interval $t \geq 0$ describe the motion of a particle that traces the right-hand half of the parabola $y = x^2$ (Example 10).

EXAMPLE 10 Moving Along a Parabola

The position $P(x, y)$ of a particle moving in the xy-plane is given by the equations and parameter interval

$$x = \sqrt{t}, \qquad y = t, \qquad t \geq 0.$$

Identify the path traced by the particle and describe the motion.

Solution We try to identify the path by eliminating t between the equations $x = \sqrt{t}$ and $y = t$. With any luck, this will produce a recognizable algebraic relation between x and y. We find that

$$y = t = \left(\sqrt{t}\right)^2 = x^2.$$

Thus, the particle's position coordinates satisfy the equation $y = x^2$, so the particle moves along the parabola $y = x^2$.

It would be a mistake, however, to conclude that the particle's path is the entire parabola $y = x^2$; it is only half the parabola. The particle's x-coordinate is never negative. The particle starts at (0, 0) when $t = 0$ and rises into the first quadrant as t increases (Figure 3.31). The parameter interval is $[0, \infty)$ and there is no terminal point. ■

EXAMPLE 11 Parametrizing a Line Segment

Find a parametrization for the line segment with endpoints $(-2, 1)$ and (3, 5).

Solution Using $(-2, 1)$ we create the parametric equations

$$x = -2 + at, \qquad y = 1 + bt.$$

These represent a line, as we can see by solving each equation for t and equating to obtain

$$\frac{x + 2}{a} = \frac{y - 1}{b}.$$

This line goes through the point $(-2, 1)$ when $t = 0$. We determine a and b so that the line goes through (3, 5) when $t = 1$.

$$3 = -2 + a \quad \Rightarrow \quad a = 5 \qquad x = 3 \text{ when } t = 1.$$
$$5 = 1 + b \quad \Rightarrow \quad b = 4 \qquad y = 5 \text{ when } t = 1.$$

Therefore,

$$x = -2 + 5t, \qquad y = 1 + 4t, \qquad 0 \leq t \leq 1$$

is a parametrization of the line segment with initial point $(-2, 1)$ and terminal point (3, 5). ■

Slopes of Parametrized Curves

A parametrized curve $x = f(t)$ and $y = g(t)$ is **differentiable** at t if f and g are differentiable at t. At a point on a differentiable parametrized curve where y is also a differentiable function of x, the derivatives dy/dt, dx/dt, and dy/dx are related by the Chain Rule:

$$\frac{dy}{dt} = \frac{dy}{dx} \cdot \frac{dx}{dt}.$$

If $dx/dt \neq 0$, we may divide both sides of this equation by dx/dt to solve for dy/dx.

Parametric Formula for dy/dx
If all three derivatives exist and $dx/dt \neq 0$,

$$\frac{dy}{dx} = \frac{dy/dt}{dx/dt}. \tag{2}$$

EXAMPLE 12 Differentiating with a Parameter

If $x = 2t + 3$ and $y = t^2 - 1$, find the value of dy/dx at $t = 6$.

Solution Equation (2) gives dy/dx as a function of t:

$$\frac{dy}{dx} = \frac{dy/dt}{dx/dt} = \frac{2t}{2} = t = \frac{x-3}{2}.$$

When $t = 6$, $dy/dx = 6$. Notice that we are also able to find the derivative dy/dx as a function of x. ■

EXAMPLE 13 Moving Along the Ellipse $x^2/a^2 + y^2/b^2 = 1$

Describe the motion of a particle whose position $P(x, y)$ at time t is given by

$$x = a\cos t, \qquad y = b\sin t, \qquad 0 \leq t \leq 2\pi.$$

Find the line tangent to the curve at the point $\left(a/\sqrt{2}, b/\sqrt{2}\right)$, where $t = \pi/4$. (The constants a and b are both positive.)

Solution We find a Cartesian equation for the particle's coordinates by eliminating t between the equations

$$\cos t = \frac{x}{a}, \qquad \sin t = \frac{y}{b}.$$

The identity $\cos^2 t + \sin^2 t = 1$, yields

$$\left(\frac{x}{a}\right)^2 + \left(\frac{y}{b}\right)^2 = 1, \qquad \text{or} \qquad \frac{x^2}{a^2} + \frac{y^2}{b^2} = 1.$$

The particle's coordinates (x, y) satisfy the equation $(x^2/a^2) + (y^2/b^2) = 1$, so the particle moves along this ellipse. When $t = 0$, the particle's coordinates are

$$x = a\cos(0) = a, \qquad y = b\sin(0) = 0,$$

so the motion starts at $(a, 0)$. As t increases, the particle rises and moves toward the left, moving counterclockwise. It traverses the ellipse once, returning to its starting position $(a, 0)$ at $t = 2\pi$.

The slope of the tangent line to the ellipse when $t = \pi/4$ is

$$\begin{aligned}
\left.\frac{dy}{dx}\right|_{t=\pi/4} &= \left.\frac{dy/dt}{dx/dt}\right|_{t=\pi/4} && \text{Eq. (2)} \\
&= \left.\frac{b\cos t}{-a\sin t}\right|_{t=\pi/4} \\
&= \frac{b/\sqrt{2}}{-a/\sqrt{2}} = -\frac{b}{a}.
\end{aligned}$$

The tangent line is

$$y - \frac{b}{\sqrt{2}} = -\frac{b}{a}\left(x - \frac{a}{\sqrt{2}}\right)$$
$$y = \frac{b}{\sqrt{2}} - \frac{b}{a}\left(x - \frac{a}{\sqrt{2}}\right)$$

or

$$y = -\frac{b}{a}x + \sqrt{2}b.$$ ■

If parametric equations define y as a twice-differentiable function of x, we can apply Equation (2) to the function $dy/dx = y'$ to calculate d^2y/dx^2 as a function of t:

$$\frac{d^2y}{dx^2} = \frac{d}{dx}(y') = \frac{dy'/dt}{dx/dt}. \qquad \text{Eq. (2) with } y' \text{ in place of } y$$

Parametric Formula for d^2y/dx^2

If the equations $x = f(t)$, $y = g(t)$ define y as a twice-differentiable function of x, then at any point where $dx/dt \neq 0$,

$$\frac{d^2y}{dx^2} = \frac{dy'/dt}{dx/dt}. \qquad (3)$$

EXAMPLE 14 Finding d^2y/dx^2 for a Parametrized Curve

Find d^2y/dx^2 as a function of t if $x = t - t^2$, $y = t - t^3$.

Finding d^2y/dx^2 in Terms of t

1. Express $y' = dy/dx$ in terms of t.
2. Find dy'/dt.
3. Divide dy'/dt by dx/dt.

Solution

1. Express $y' = dy/dx$ in terms of t.

$$y' = \frac{dy}{dx} = \frac{dy/dt}{dx/dt} = \frac{1 - 3t^2}{1 - 2t}$$

2. Differentiate y' with respect to t.

$$\frac{dy'}{dt} = \frac{d}{dt}\left(\frac{1 - 3t^2}{1 - 2t}\right) = \frac{2 - 6t + 6t^2}{(1 - 2t)^2} \qquad \text{Quotient Rule}$$

3. Divide dy'/dt by dx/dt.

$$\frac{d^2y}{dx^2} = \frac{dy'/dt}{dx/dt} = \frac{(2 - 6t + 6t^2)/(1 - 2t)^2}{1 - 2t} = \frac{2 - 6t + 6t^2}{(1 - 2t)^3} \qquad \text{Eq. (3)}$$ ■

EXAMPLE 15 Dropping Emergency Supplies

A Red Cross aircraft is dropping emergency food and medical supplies into a disaster area. If the aircraft releases the supplies immediately above the edge of an open field 700 ft long and if the cargo moves along the path

$$x = 120t \quad \text{and} \quad y = -16t^2 + 500, \quad t \geq 0$$

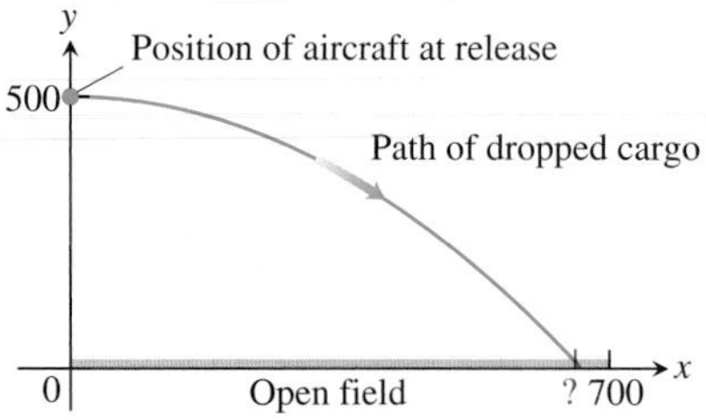

FIGURE 3.32 The path of the dropped cargo of supplies in Example 15.

does the cargo land in the field? The coordinates x and y are measured in feet, and the parameter t (time since release) in seconds. Find a Cartesian equation for the path of the falling cargo (Figure 3.32) and the cargo's rate of descent relative to its forward motion when it hits the ground.

Solution The cargo hits the ground when $y = 0$, which occurs at time t when

$$-16t^2 + 500 = 0 \qquad \text{Set } y = 0.$$

$$t = \sqrt{\frac{500}{16}} = \frac{5\sqrt{5}}{2} \text{ sec}. \qquad t \geq 0$$

The x-coordinate at the time of the release is $x = 0$. At the time the cargo hits the ground, the x-coordinate is

$$x = 120t = 120\left(\frac{5\sqrt{5}}{2}\right) = 300\sqrt{5} \text{ ft}.$$

Since $300\sqrt{5} \approx 670.8 < 700$, the cargo does land in the field.

We find a Cartesian equation for the cargo's coordinates by eliminating t between the parametric equations:

$$y = -16t^2 + 500 \qquad \text{Parametric equation for } y$$

$$= -16\left(\frac{x}{120}\right)^2 + 500 \qquad \text{Substitute for } t \text{ from the equation } x = 120t.$$

$$= -\frac{1}{900}x^2 + 500. \qquad \text{A parabola}$$

The rate of descent relative to its forward motion when the cargo hits the ground is

$$\left.\frac{dy}{dx}\right|_{t=5\sqrt{5}/2} = \left.\frac{dy/dt}{dx/dt}\right|_{t=5\sqrt{5}/2} \qquad \text{Eq. (2)}$$

$$= \left.\frac{-32t}{120}\right|_{t=5\sqrt{5}/2}$$

$$= -\frac{2\sqrt{5}}{3} \approx -1.49.$$

Thus, it is falling about 1.5 feet for every foot of forward motion when it hits the ground. ■

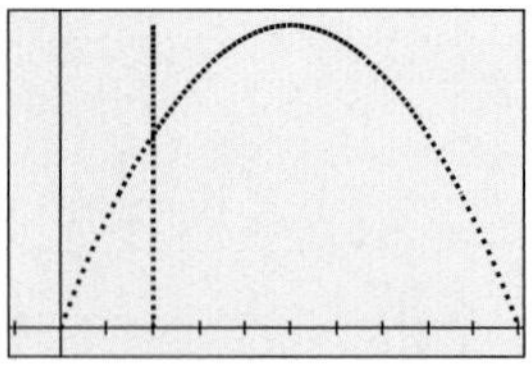

$\begin{cases} x(t) = 2 \\ y(t) = 160t - 16t^2 \end{cases}$

and

$\begin{cases} x(t) = t \\ y(t) = 160t - 16t^2 \end{cases}$

in dot mode

USING TECHNOLOGY Simulation of Motion on a Vertical Line

The parametric equations

$$x(t) = c, \qquad y(t) = f(t)$$

will illuminate pixels along the vertical line $x = c$. If $f(t)$ denotes the height of a moving body at time t, graphing $(x(t), y(t)) = (c, f(t))$ will simulate the actual motion. Try it for the rock in Example 5, Section 3.3 with $x(t) = 2$, say, and $y(t) = 160t - 16t^2$, in dot mode with tStep $= 0.1$. Why does the spacing of the dots vary? Why does the grapher seem to stop after it reaches the top? (Try the plots for $0 \leq t \leq 5$ and $5 \leq t \leq 10$ separately.)

For a second experiment, plot the parametric equations

$$x(t) = t, \qquad y(t) = 160t - 16t^2$$

together with the vertical line simulation of the motion, again in dot mode. Use what you know about the behavior of the rock from the calculations of Example 5 to select a window size that will display all the interesting behavior.

Standard Parametrizations and Derivative Rules

CIRCLE $x^2 + y^2 = a^2$:

$$x = a\cos t$$
$$y = a\sin t$$
$$0 \le t \le 2\pi$$

ELLIPSE $\frac{x^2}{a^2} + \frac{y^2}{b^2} = 1$:

$$x = a\cos t$$
$$y = b\sin t$$
$$0 \le t \le 2\pi$$

FUNCTION $y = f(x)$:

$$x = t$$
$$y = f(t)$$

DERIVATIVES

$$y' = \frac{dy}{dx} = \frac{dy/dt}{dx/dt}, \qquad \frac{d^2y}{dx^2} = \frac{dy'/dt}{dx/dt}$$

EXERCISES 3.5

Derivative Calculations

In Exercises 1–8, given $y = f(u)$ and $u = g(x)$, find $dy/dx = f'(g(x))g'(x)$.

1. $y = 6u - 9, \quad u = (1/2)x^4$ **2.** $y = 2u^3, \quad u = 8x - 1$

3. $y = \sin u, \quad u = 3x + 1$ **4.** $y = \cos u, \quad u = -x/3$

5. $y = \cos u, \quad u = \sin x$ **6.** $y = \sin u, \quad u = x - \cos x$

7. $y = \tan u, \quad u = 10x - 5$ **8.** $y = -\sec u, \quad u = x^2 + 7x$

In Exercises 9–18, write the function in the form $y = f(u)$ and $u = g(x)$. Then find dy/dx as a function of x.

9. $y = (2x + 1)^5$ **10.** $y = (4 - 3x)^9$

11. $y = \left(1 - \frac{x}{7}\right)^{-7}$ **12.** $y = \left(\frac{x}{2} - 1\right)^{-10}$

13. $y = \left(\frac{x^2}{8} + x - \frac{1}{x}\right)^4$ **14.** $y = \left(\frac{x}{5} + \frac{1}{5x}\right)^5$

15. $y = \sec(\tan x)$ **16.** $y = \cot\left(\pi - \frac{1}{x}\right)$

17. $y = \sin^3 x$ **18.** $y = 5\cos^{-4} x$

Find the derivatives of the functions in Exercises 19–38.

19. $p = \sqrt{3 - t}$ **20.** $q = \sqrt{2r - r^2}$

21. $s = \frac{4}{3\pi}\sin 3t + \frac{4}{5\pi}\cos 5t$

22. $s = \sin\left(\frac{3\pi t}{2}\right) + \cos\left(\frac{3\pi t}{2}\right)$

23. $r = (\csc\theta + \cot\theta)^{-1}$ **24.** $r = -(\sec\theta + \tan\theta)^{-1}$

25. $y = x^2\sin^4 x + x\cos^{-2} x$ **26.** $y = \frac{1}{x}\sin^{-5} x - \frac{x}{3}\cos^3 x$

27. $y = \frac{1}{21}(3x - 2)^7 + \left(4 - \frac{1}{2x^2}\right)^{-1}$

28. $y = (5 - 2x)^{-3} + \frac{1}{8}\left(\frac{2}{x} + 1\right)^4$

29. $y = (4x + 3)^4(x + 1)^{-3}$ **30.** $y = (2x - 5)^{-1}(x^2 - 5x)^6$

31. $h(x) = x\tan\left(2\sqrt{x}\right) + 7$ **32.** $k(x) = x^2\sec\left(\frac{1}{x}\right)$

33. $f(\theta) = \left(\frac{\sin\theta}{1 + \cos\theta}\right)^2$ **34.** $g(t) = \left(\frac{1 + \cos t}{\sin t}\right)^{-1}$

35. $r = \sin(\theta^2)\cos(2\theta)$ **36.** $r = \sec\sqrt{\theta}\tan\left(\frac{1}{\theta}\right)$

37. $q = \sin\left(\frac{t}{\sqrt{t + 1}}\right)$ **38.** $q = \cot\left(\frac{\sin t}{t}\right)$

In Exercises 39–48, find dy/dt.

39. $y = \sin^2(\pi t - 2)$ **40.** $y = \sec^2 \pi t$

41. $y = (1 + \cos 2t)^{-4}$ **42.** $y = (1 + \cot(t/2))^{-2}$

43. $y = \sin(\cos(2t - 5))$

44. $y = \cos\left(5 \sin\left(\frac{t}{3}\right)\right)$

45. $y = \left(1 + \tan^4\left(\frac{t}{12}\right)\right)^3$

46. $y = \frac{1}{6}\left(1 + \cos^2(7t)\right)^3$

47. $y = \sqrt{1 + \cos(t^2)}$

48. $y = 4 \sin\left(\sqrt{1 + \sqrt{t}}\right)$

Second Derivatives

Find y'' in Exercises 49–52.

49. $y = \left(1 + \frac{1}{x}\right)^3$

50. $y = \left(1 - \sqrt{x}\right)^{-1}$

51. $y = \frac{1}{9}\cot(3x - 1)$

52. $y = 9 \tan\left(\frac{x}{3}\right)$

Finding Numerical Values of Derivatives

In Exercises 53–58, find the value of $(f \circ g)'$ at the given value of x.

53. $f(u) = u^5 + 1, \quad u = g(x) = \sqrt{x}, \quad x = 1$

54. $f(u) = 1 - \frac{1}{u}, \quad u = g(x) = \frac{1}{1 - x}, \quad x = -1$

55. $f(u) = \cot\frac{\pi u}{10}, \quad u = g(x) = 5\sqrt{x}, \quad x = 1$

56. $f(u) = u + \frac{1}{\cos^2 u}, \quad u = g(x) = \pi x, \quad x = 1/4$

57. $f(u) = \frac{2u}{u^2 + 1}, \quad u = g(x) = 10x^2 + x + 1, \quad x = 0$

58. $f(u) = \left(\frac{u - 1}{u + 1}\right)^2, \quad u = g(x) = \frac{1}{x^2} - 1, \quad x = -1$

59. Suppose that functions f and g and their derivatives with respect to x have the following values at $x = 2$ and $x = 3$.

x	$f(x)$	$g(x)$	$f'(x)$	$g'(x)$
2	8	2	1/3	−3
3	3	−4	2π	5

Find the derivatives with respect to x of the following combinations at the given value of x.

a. $2f(x), \quad x = 2$

b. $f(x) + g(x), \quad x = 3$

c. $f(x) \cdot g(x), \quad x = 3$

d. $f(x)/g(x), \quad x = 2$

e. $f(g(x)), \quad x = 2$

f. $\sqrt{f(x)}, \quad x = 2$

g. $1/g^2(x), \quad x = 3$

h. $\sqrt{f^2(x) + g^2(x)}, \quad x = 2$

60. Suppose that the functions f and g and their derivatives with respect to x have the following values at $x = 0$ and $x = 1$.

x	$f(x)$	$g(x)$	$f'(x)$	$g'(x)$
0	1	1	5	1/3
1	3	−4	−1/3	−8/3

Find the derivatives with respect to x of the following combinations at the given value of x,

a. $5f(x) - g(x), \quad x = 1$

b. $f(x)g^3(x), \quad x = 0$

c. $\frac{f(x)}{g(x) + 1}, \quad x = 1$

d. $f(g(x)), \quad x = 0$

e. $g(f(x)), \quad x = 0$

f. $(x^{11} + f(x))^{-2}, \quad x = 1$

g. $f(x + g(x)), \quad x = 0$

61. Find ds/dt when $\theta = 3\pi/2$ if $s = \cos\theta$ and $d\theta/dt = 5$.

62. Find dy/dt when $x = 1$ if $y = x^2 + 7x - 5$ and $dx/dt = 1/3$.

Choices in Composition

What happens if you can write a function as a composite in different ways? Do you get the same derivative each time? The Chain Rule says you should. Try it with the functions in Exercises 63 and 64.

63. Find dy/dx if $y = x$ by using the Chain Rule with y as a composite of

a. $y = (u/5) + 7$ and $u = 5x - 35$

b. $y = 1 + (1/u)$ and $u = 1/(x - 1)$.

64. Find dy/dx if $y = x^{3/2}$ by using the Chain Rule with y as a composite of

a. $y = u^3$ and $u = \sqrt{x}$

b. $y = \sqrt{u}$ and $u = x^3$.

Tangents and Slopes

65. a. Find the tangent to the curve $y = 2\tan(\pi x/4)$ at $x = 1$.

b. **Slopes on a tangent curve** What is the smallest value the slope of the curve can ever have on the interval $-2 < x < 2$? Give reasons for your answer.

66. **Slopes on sine curves**

a. Find equations for the tangents to the curves $y = \sin 2x$ and $y = -\sin(x/2)$ at the origin. Is there anything special about how the tangents are related? Give reasons for your answer.

b. Can anything be said about the tangents to the curves $y = \sin mx$ and $y = -\sin(x/m)$ at the origin (m a constant $\neq 0$)? Give reasons for your answer.

c. For a given m, what are the largest values the slopes of the curves $y = \sin mx$ and $y = -\sin(x/m)$ can ever have? Give reasons for your answer.

d. The function $y = \sin x$ completes one period on the interval $[0, 2\pi]$, the function $y = \sin 2x$ completes two periods, the function $y = \sin(x/2)$ completes half a period, and so on. Is there any relation between the number of periods $y = \sin mx$ completes on $[0, 2\pi]$ and the slope of the curve $y = \sin mx$ at the origin? Give reasons for your answer.

Finding Cartesian Equations from Parametric Equations

Exercises 67–78 give parametric equations and parameter intervals for the motion of a particle in the xy-plane. Identify the particle's path by finding a Cartesian equation for it. Graph the Cartesian equation. (The graphs will vary with the equation used.) Indicate the portion of the graph traced by the particle and the direction of motion.

67. $x = \cos 2t, \quad y = \sin 2t, \quad 0 \le t \le \pi$

68. $x = \cos(\pi - t), \quad y = \sin(\pi - t), \quad 0 \le t \le \pi$

69. $x = 4\cos t, \quad y = 2\sin t, \quad 0 \le t \le 2\pi$

70. $x = 4\sin t, \quad y = 5\cos t, \quad 0 \le t \le 2\pi$

71. $x = 3t, \quad y = 9t^2, \quad -\infty < t < \infty$

72. $x = -\sqrt{t}, \quad y = t, \quad t \ge 0$

73. $x = 2t - 5, \quad y = 4t - 7, \quad -\infty < t < \infty$

74. $x = 3 - 3t, \quad y = 2t, \quad 0 \le t \le 1$

75. $x = t, \quad y = \sqrt{1 - t^2}, \quad -1 \le t \le 0$

76. $x = \sqrt{t + 1}, \quad y = \sqrt{t}, \quad t \ge 0$

77. $x = \sec^2 t - 1, \quad y = \tan t, \quad -\pi/2 < t < \pi/2$

78. $x = -\sec t, \quad y = \tan t, \quad -\pi/2 < t < \pi/2$

Determining Parametric Equations

79. Find parametric equations and a parameter interval for the motion of a particle that starts at $(a, 0)$ and traces the circle $x^2 + y^2 = a^2$

a. once clockwise. **b.** once counterclockwise.

c. twice clockwise. **d.** twice counterclockwise.

(There are many ways to do these, so your answers may not be the same as the ones in the back of the book.)

80. Find parametric equations and a parameter interval for the motion of a particle that starts at $(a, 0)$ and traces the ellipse $(x^2/a^2) + (y^2/b^2) = 1$

a. once clockwise. **b.** once counterclockwise.

c. twice clockwise. **d.** twice counterclockwise.

(As in Exercise 79, there are many correct answers.)

In Exercises 81–86, find a parametrization for the curve.

81. the line segment with endpoints $(-1, -3)$ and $(4, 1)$

82. the line segment with endpoints $(-1, 3)$ and $(3, -2)$

83. the lower half of the parabola $x - 1 = y^2$

84. the left half of the parabola $y = x^2 + 2x$

85. the ray (half line) with initial point $(2, 3)$ that passes through the point $(-1, -1)$

86. the ray (half line) with initial point $(-1, 2)$ that passes through the point $(0, 0)$

Tangents to Parametrized Curves

In Exercises 87–94, find an equation for the line tangent to the curve at the point defined by the given value of t. Also, find the value of d^2y/dx^2 at this point.

87. $x = 2\cos t, \quad y = 2\sin t, \quad t = \pi/4$

88. $x = \cos t, \quad y = \sqrt{3}\cos t, \quad t = 2\pi/3$

89. $x = t, \quad y = \sqrt{t}, \quad t = 1/4$

90. $x = -\sqrt{t + 1}, \quad y = \sqrt{3t}, \quad t = 3$

91. $x = 2t^2 + 3, \quad y = t^4, \quad t = -1$

92. $x = t - \sin t, \quad y = 1 - \cos t, \quad t = \pi/3$

93. $x = \cos t, \quad y = 1 + \sin t, \quad t = \pi/2$

94. $x = \sec^2 t - 1, \quad y = \tan t, \quad t = -\pi/4$

Theory, Examples, and Applications

95. Running machinery too fast Suppose that a piston is moving straight up and down and that its position at time t sec is

$$s = A\cos(2\pi bt),$$

with A and b positive. The value of A is the amplitude of the motion, and b is the frequency (number of times the piston moves up and down each second). What effect does doubling the frequency have on the piston's velocity, acceleration, and jerk? (Once you find out, you will know why machinery breaks when you run it too fast.)

96. Temperatures in Fairbanks, Alaska The graph in Figure 3.33 shows the average Fahrenheit temperature in Fairbanks, Alaska, during a typical 365-day year. The equation that approximates the temperature on day x is

$$y = 37\sin\left[\frac{2\pi}{365}(x - 101)\right] + 25.$$

a. On what day is the temperature increasing the fastest?

b. About how many degrees per day is the temperature increasing when it is increasing at its fastest?

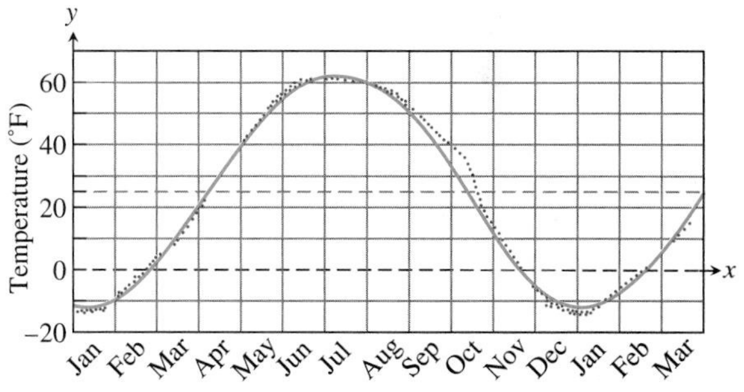

FIGURE 3.33 Normal mean air temperatures at Fairbanks, Alaska, plotted as data points, and the approximating sine function (Exercise 96).

97. Particle motion The position of a particle moving along a coordinate line is $s = \sqrt{1 + 4t}$, with s in meters and t in seconds. Find the particle's velocity and acceleration at $t = 6$ sec.

98. Constant acceleration Suppose that the velocity of a falling body is $v = k\sqrt{s}$ m/sec (k a constant) at the instant the body has fallen s m from its starting point. Show that the body's acceleration is constant.

99. Falling meteorite The velocity of a heavy meteorite entering Earth's atmosphere is inversely proportional to $\sqrt{s}$ when it is s km from Earth's center. Show that the meteorite's acceleration is inversely proportional to s^2.

100. Particle acceleration A particle moves along the x-axis with velocity $dx/dt = f(x)$. Show that the particle's acceleration is $f(x)f'(x)$.

101. Temperature and the period of a pendulum For oscillations of small amplitude (short swings), we may safely model the relationship between the period T and the length L of a simple pendulum with the equation

$$T = 2\pi\sqrt{\frac{L}{g}},$$

where g is the constant acceleration of gravity at the pendulum's location. If we measure g in centimeters per second squared, we measure L in centimeters and T in seconds. If the pendulum is made of metal, its length will vary with temperature, either increasing or decreasing at a rate that is roughly proportional to L. In symbols, with u being temperature and k the proportionality constant,

$$\frac{dL}{du} = kL.$$

Assuming this to be the case, show that the rate at which the period changes with respect to temperature is $kT/2$.

102. Chain Rule Suppose that $f(x) = x^2$ and $g(x) = |x|$. Then the composites

$$(f \circ g)(x) = |x|^2 = x^2 \quad \text{and} \quad (g \circ f)(x) = |x^2| = x^2$$

are both differentiable at $x = 0$ even though g itself is not differentiable at $x = 0$. Does this contradict the Chain Rule? Explain.

103. Tangents Suppose that $u = g(x)$ is differentiable at $x = 1$ and that $y = f(u)$ is differentiable at $u = g(1)$. If the graph of $y = f(g(x))$ has a horizontal tangent at $x = 1$, can we conclude anything about the tangent to the graph of g at $x = 1$ or the tangent to the graph of f at $u = g(1)$? Give reasons for your answer.

104. Suppose that $u = g(x)$ is differentiable at $x = -5$, $y = f(u)$ is differentiable at $u = g(-5)$, and $(f \circ g)'(-5)$ is negative. What, if anything, can be said about the values of $g'(-5)$ and $f'(g(-5))$?

T **105. The derivative of sin 2x** Graph the function $y = 2\cos 2x$ for $-2 \le x \le 3.5$. Then, on the same screen, graph

$$y = \frac{\sin 2(x + h) - \sin 2x}{h}$$

for $h = 1.0, 0.5$, and 0.2. Experiment with other values of h, including negative values. What do you see happening as $h \to 0$? Explain this behavior.

T **106. The derivative of cos (x^2)** Graph $y = -2x\sin(x^2)$ for $-2 \le x \le 3$. Then, on the same screen, graph

$$y = \frac{\cos((x + h)^2) - \cos(x^2)}{h}$$

for $h = 1.0, 0.7$, and 0.3. Experiment with other values of h. What do you see happening as $h \to 0$? Explain this behavior.

T The curves in Exercises 107 and 108 are called *Bowditch curves* or *Lissajous figures*. In each case, find the point in the interior of the first quadrant where the tangent to the curve is horizontal, and find the equations of the two tangents at the origin.

107.

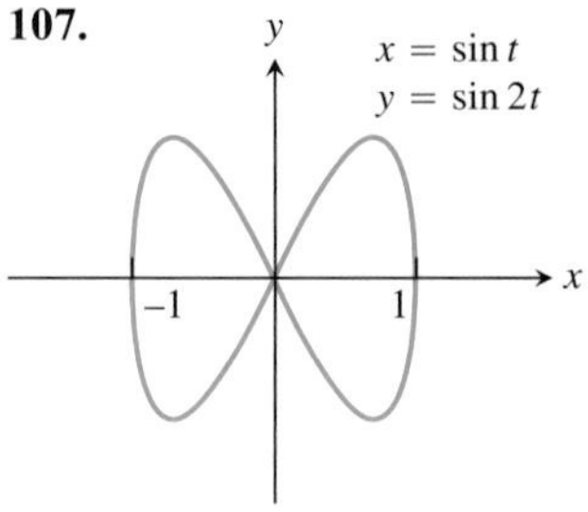

108.

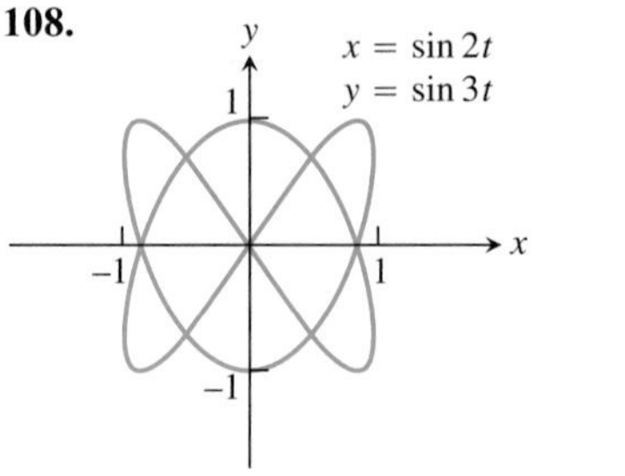

Using the Chain Rule, show that the power rule $(d/dx)x^n = nx^{n-1}$ holds for the functions x^n in Exercises 109 and 110.

109. $x^{1/4} = \sqrt{\sqrt{x}}$

110. $x^{3/4} = \sqrt{x\sqrt{x}}$

COMPUTER EXPLORATIONS

Trigonometric Polynomials

111. As Figure 3.34 shows, the trigonometric "polynomial"

$$s = f(t) = 0.78540 - 0.63662\cos 2t - 0.07074\cos 6t$$
$$-0.02546\cos 10t - 0.01299\cos 14t$$

gives a good approximation of the sawtooth function $s = g(t)$ on the interval $[-\pi, \pi]$. How well does the derivative of f approximate the derivative of g at the points where dg/dt is defined? To find out, carry out the following steps.

a. Graph dg/dt (where defined) over $[-\pi, \pi]$.

b. Find df/dt.

c. Graph df/dt. Where does the approximation of dg/dt by df/dt seem to be best? Least good? Approximations by trigonometric polynomials are important in the theories of heat and oscillation, but we must not expect too much of them, as we see in the next exercise.

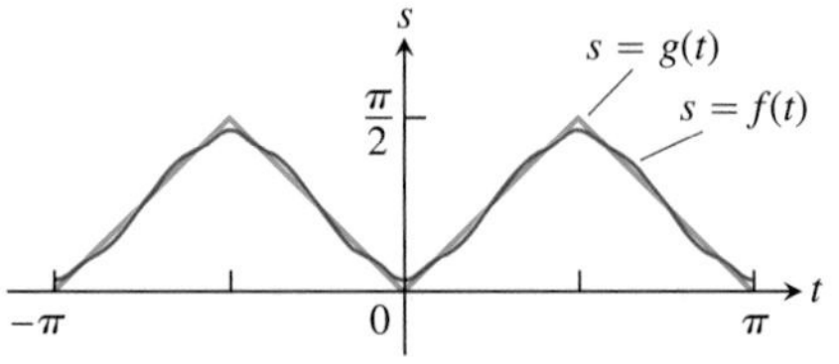

FIGURE 3.34 The approximation of a sawtooth function by a trigonometric "polynomial" (Exercise 111).

112. (*Continuation of Exercise 111.*) In Exercise 111, the trigonometric polynomial $f(t)$ that approximated the sawtooth function $g(t)$ on $[-\pi, \pi]$ had a derivative that approximated the derivative of the sawtooth function. It is possible, however, for a trigonometric polynomial to approximate a function in a reasonable way without its derivative approximating the function's derivative at all well. As a case in point, the "polynomial"

$$s = h(t) = 1.2732 \sin 2t + 0.4244 \sin 6t + 0.25465 \sin 10t + 0.18189 \sin 14t + 0.14147 \sin 18t$$

graphed in Figure 3.35 approximates the step function $s = k(t)$ shown there. Yet the derivative of h is nothing like the derivative of k.

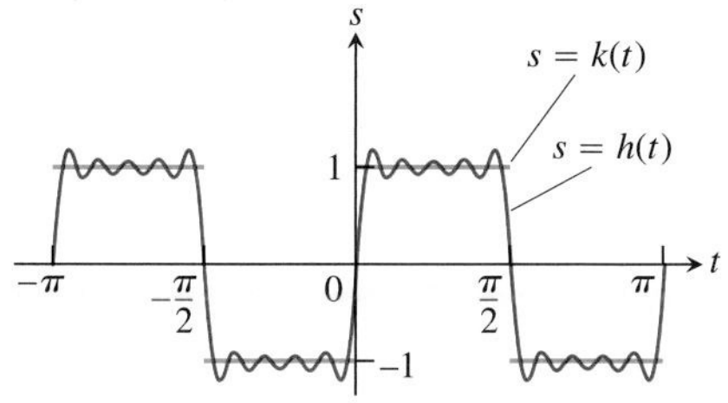

FIGURE 3.35 The approximation of a step function by a trigonometric "polynomial" (Exercise 112).

a. Graph dk/dt (where defined) over $[-\pi, \pi]$.

b. Find dh/dt.

c. Graph dh/dt to see how badly the graph fits the graph of dk/dt. Comment on what you see.

Parametrized Curves

Use a CAS to perform the following steps on the parametrized curves in Exercises 113–116.

a. Plot the curve for the given interval of t values.

b. Find dy/dx and d^2y/dx^2 at the point t_0.

c. Find an equation for the tangent line to the curve at the point defined by the given value t_0. Plot the curve together with the tangent line on a single graph.

113. $x = \frac{1}{3}t^3, \quad y = \frac{1}{2}t^2, \quad 0 \le t \le 1, \quad t_0 = 1/2$

114. $x = 2t^3 - 16t^2 + 25t + 5, \quad y = t^2 + t - 3, \quad 0 \le t \le 6, \quad t_0 = 3/2$

115. $x = t - \cos t, \quad y = 1 + \sin t, \quad -\pi \le t \le \pi, \quad t_0 = \pi/4$

116. $x = e^t \cos t, \quad y = e^t \sin t, \quad 0 \le t \le \pi, \quad t_0 = \pi/2$

3.6 Implicit Differentiation

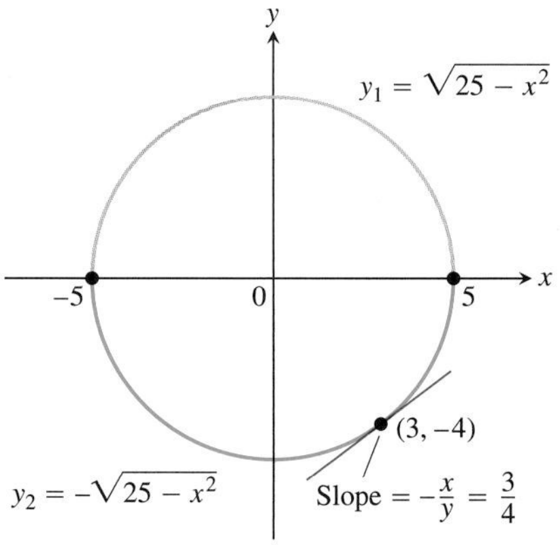

FIGURE 3.36 The circle combines the graphs of two functions. The graph of y_2 is the lower semicircle and passes through $(3, -4)$.

Most of the functions we have dealt with so far have been described by an equation of the form $y = f(x)$ that expresses y explicitly in terms of the variable x. We have learned rules for differentiating functions defined in this way. In Section 3.5 we also learned how to find the derivative dy/dx when a curve is defined parametrically by equations $x = x(t)$ and $y = y(t)$. A third situation occurs when we encounter equations like

$$x^2 + y^2 - 25 = 0, \qquad y^2 - x = 0, \qquad \text{or} \quad x^3 + y^3 - 9xy = 0.$$

(See Figures 3.36, 3.37, and 3.38.) These equations define an *implicit* relation between the variables x and y. In some cases we may be able to solve such an equation for y as an explicit function (or even several functions) of x. When we cannot put an equation $F(x, y) = 0$ in the form $y = f(x)$ to differentiate it in the usual way, we may still be able to find dy/dx by *implicit differentiation*. This consists of differentiating both sides of the equation with respect to x and then solving the resulting equation for y'. This section describes the technique and uses it to extend the Power Rule for differentiation to include rational exponents. In the examples and exercises of this section it is always assumed that the given equation determines y implicitly as a differentiable function of x.

Implicitly Defined Functions

We begin with an example.

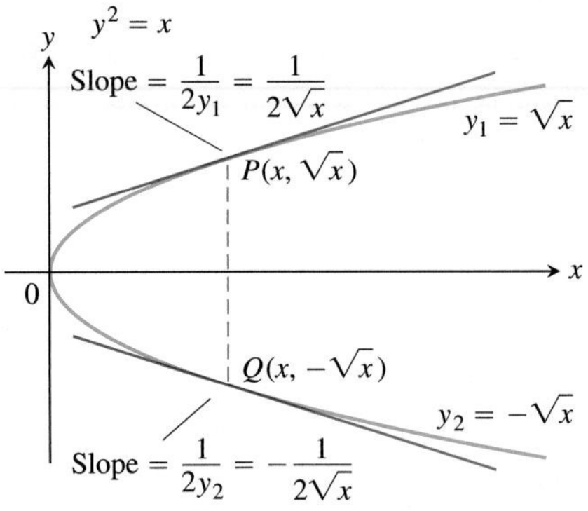

FIGURE 3.37 The equation $y^2 - x = 0$, or $y^2 = x$ as it is usually written, defines two differentiable functions of x on the interval $x \geq 0$. Example 1 shows how to find the derivatives of these functions without solving the equation $y^2 = x$ for y.

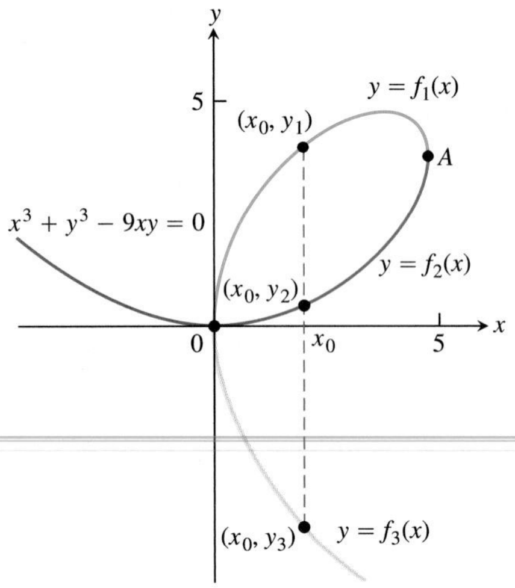

FIGURE 3.38 The curve $x^3 + y^3 - 9xy = 0$ is not the graph of any one function of x. The curve can, however, be divided into separate arcs that *are* the graphs of functions of x. This particular curve, called a *folium*, dates to Descartes in 1638.

EXAMPLE 1 Differentiating Implicitly

Find dy/dx if $y^2 = x$.

Solution The equation $y^2 = x$ defines two differentiable functions of x that we can actually find, namely $y_1 = \sqrt{x}$ and $y_2 = -\sqrt{x}$ (Figure 3.37). We know how to calculate the derivative of each of these for $x > 0$:

$$\frac{dy_1}{dx} = \frac{1}{2\sqrt{x}} \quad \text{and} \quad \frac{dy_2}{dx} = -\frac{1}{2\sqrt{x}}.$$

But suppose that we knew only that the equation $y^2 = x$ defined y as one or more differentiable functions of x for $x > 0$ without knowing exactly what these functions were. Could we still find dy/dx?

The answer is yes. To find dy/dx, we simply differentiate both sides of the equation $y^2 = x$ with respect to x, treating $y = f(x)$ as a differentiable function of x:

$$y^2 = x$$

$$2y\frac{dy}{dx} = 1 \qquad \text{The Chain Rule gives } \frac{d}{dx}(y^2) = \frac{d}{dx}[f(x)]^2 = 2f(x)f'(x) = 2y\frac{dy}{dx}.$$

$$\frac{dy}{dx} = \frac{1}{2y}.$$

This one formula gives the derivatives we calculated for *both* explicit solutions $y_1 = \sqrt{x}$ and $y_2 = -\sqrt{x}$:

$$\frac{dy_1}{dx} = \frac{1}{2y_1} = \frac{1}{2\sqrt{x}} \quad \text{and} \quad \frac{dy_2}{dx} = \frac{1}{2y_2} = \frac{1}{2(-\sqrt{x})} = -\frac{1}{2\sqrt{x}}.$$

EXAMPLE 2 Slope of a Circle at a Point

Find the slope of circle $x^2 + y^2 = 25$ at the point $(3, -4)$.

Solution The circle is not the graph of a single function of x. Rather it is the combined graphs of two differentiable functions, $y_1 = \sqrt{25 - x^2}$ and $y_2 = -\sqrt{25 - x^2}$ (Figure 3.36). The point $(3, -4)$ lies on the graph of y_2, so we can find the slope by calculating explicitly:

$$\left.\frac{dy_2}{dx}\right|_{x=3} = -\left.\frac{-2x}{2\sqrt{25 - x^2}}\right|_{x=3} = -\frac{-6}{2\sqrt{25 - 9}} = \frac{3}{4}.$$

But we can also solve the problem more easily by differentiating the given equation of the circle implicitly with respect to x:

$$\frac{d}{dx}(x^2) + \frac{d}{dx}(y^2) = \frac{d}{dx}(25)$$

$$2x + 2y\frac{dy}{dx} = 0$$

$$\frac{dy}{dx} = -\frac{x}{y}.$$

The slope at $(3, -4)$ is $-\left.\frac{x}{y}\right|_{(3,-4)} = -\frac{3}{-4} = \frac{3}{4}.$

Notice that unlike the slope formula for dy_2/dx, which applies only to points below the x-axis, the formula $dy/dx = -x/y$ applies everywhere the circle has a slope. Notice also that the derivative involves *both* variables x and y, not just the independent variable x. ■

To calculate the derivatives of other implicitly defined functions, we proceed as in Examples 1 and 2: We treat y as a differentiable implicit function of x and apply the usual rules to differentiate both sides of the defining equation.

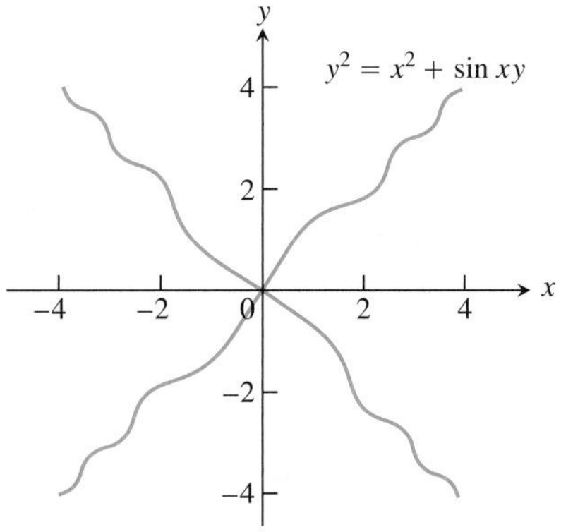

FIGURE 3.39 The graph of $y^2 = x^2 + \sin xy$ in Example 3. The example shows how to find slopes on this implicitly defined curve.

EXAMPLE 3 Differentiating Implicitly

Find dy/dx if $y^2 = x^2 + \sin xy$ (Figure 3.39).

Solution

$$y^2 = x^2 + \sin xy$$

$$\frac{d}{dx}\left(y^2\right) = \frac{d}{dx}\left(x^2\right) + \frac{d}{dx}\left(\sin xy\right) \qquad \text{Differentiate both sides with respect to } x \ldots$$

$$2y\frac{dy}{dx} = 2x + (\cos xy)\frac{d}{dx}(xy) \qquad \ldots \text{treating } y \text{ as a function of } x \text{ and using the Chain Rule.}$$

$$2y\frac{dy}{dx} = 2x + (\cos xy)\left(y + x\frac{dy}{dx}\right) \qquad \text{Treat } xy \text{ as a product.}$$

$$2y\frac{dy}{dx} - (\cos xy)\left(x\frac{dy}{dx}\right) = 2x + (\cos xy)y \qquad \text{Collect terms with } dy/dx \ldots$$

$$(2y - x\cos xy)\frac{dy}{dx} = 2x + y\cos xy \qquad \ldots \text{and factor out } dy/dx.$$

$$\frac{dy}{dx} = \frac{2x + y\cos xy}{2y - x\cos xy} \qquad \text{Solve for } dy/dx \text{ by dividing.}$$

Notice that the formula for dy/dx applies everywhere that the implicitly defined curve has a slope. Notice again that the derivative involves *both* variables x and y, not just the independent variable x. ■

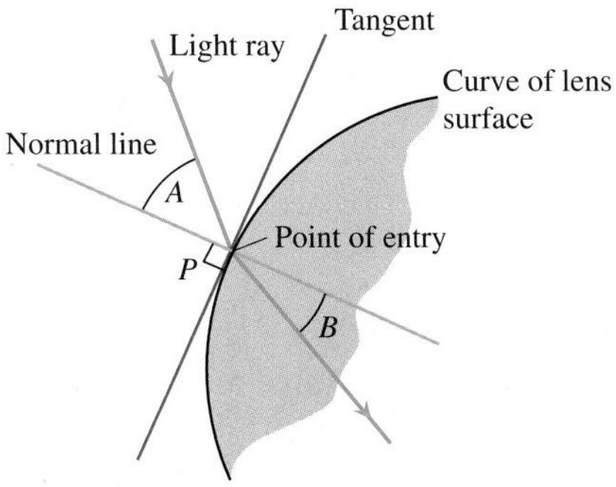

FIGURE 3.40 The profile of a lens, showing the bending (refraction) of a ray of light as it passes through the lens surface.

Implicit Differentiation

1. Differentiate both sides of the equation with respect to x, treating y as a differentiable function of x.
2. Collect the terms with dy/dx on one side of the equation.
3. Solve for dy/dx.

Lenses, Tangents, and Normal Lines

In the law that describes how light changes direction as it enters a lens, the important angles are the angles the light makes with the line perpendicular to the surface of the lens at the point of entry (angles A and B in Figure 3.40). This line is called the *normal* to the surface at the point of entry. In a profile view of a lens like the one in Figure 3.40, the **normal** is the line perpendicular to the tangent to the profile curve at the point of entry.

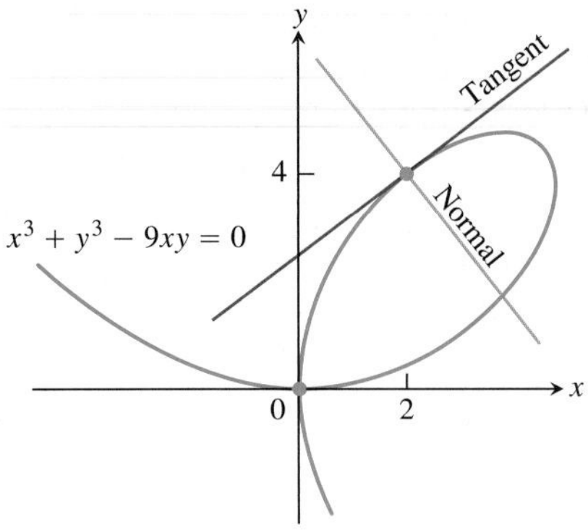

FIGURE 3.41 Example 4 shows how to find equations for the tangent and normal to the folium of Descartes at (2, 4).

EXAMPLE 4 Tangent and Normal to the Folium of Descartes

Show that the point (2, 4) lies on the curve $x^3 + y^3 - 9xy = 0$. Then find the tangent and normal to the curve there (Figure 3.41).

Solution The point (2, 4) lies on the curve because its coordinates satisfy the equation given for the curve: $2^3 + 4^3 - 9(2)(4) = 8 + 64 - 72 = 0$.

To find the slope of the curve at (2, 4), we first use implicit differentiation to find a formula for dy/dx:

$$x^3 + y^3 - 9xy = 0$$

$$\frac{d}{dx}\left(x^3\right) + \frac{d}{dx}\left(y^3\right) - \frac{d}{dx}\left(9xy\right) = \frac{d}{dx}\left(0\right) \qquad \text{Differentiate both sides with respect to } x.$$

$$3x^2 + 3y^2\frac{dy}{dx} - 9\left(x\frac{dy}{dx} + y\frac{dx}{dx}\right) = 0 \qquad \text{Treat } xy \text{ as a product and } y \text{ as a function of } x.$$

$$(3y^2 - 9x)\frac{dy}{dx} + 3x^2 - 9y = 0$$

$$3(y^2 - 3x)\frac{dy}{dx} = 9y - 3x^2$$

$$\frac{dy}{dx} = \frac{3y - x^2}{y^2 - 3x}. \qquad \text{Solve for } dy/dx.$$

We then evaluate the derivative at $(x, y) = (2, 4)$:

$$\left.\frac{dy}{dx}\right|_{(2,4)} = \left.\frac{3y - x^2}{y^2 - 3x}\right|_{(2,4)} = \frac{3(4) - 2^2}{4^2 - 3(2)} = \frac{8}{10} = \frac{4}{5}.$$

The tangent at (2, 4) is the line through (2, 4) with slope 4/5:

$$y = 4 + \frac{4}{5}\left(x - 2\right)$$

$$y = \frac{4}{5}x + \frac{12}{5}.$$

The normal to the curve at (2, 4) is the line perpendicular to the tangent there, the line through (2, 4) with slope $-5/4$:

$$y = 4 - \frac{5}{4}(x - 2)$$

$$y = -\frac{5}{4}x + \frac{13}{2}.$$

■

The quadratic formula enables us to solve a second-degree equation like $y^2 - 2xy + 3x^2 = 0$ for y in terms of x. There is a formula for the three roots of a cubic equation that is like the quadratic formula but much more complicated. If this formula is used to solve the equation $x^3 + y^3 = 9xy$ for y in terms of x, then three functions determined by the equation are

$$y = f(x) = \sqrt[3]{-\frac{x^3}{2} + \sqrt{\frac{x^6}{4} - 27x^3}} + \sqrt[3]{-\frac{x^3}{2} - \sqrt{\frac{x^6}{4} - 27x^3}}$$

and

$$y = \frac{1}{2}\left[-f(x) \pm \sqrt{-3}\left(\sqrt[3]{-\frac{x^3}{2} + \sqrt{\frac{x^6}{4} - 27x^3}} - \sqrt[3]{-\frac{x^3}{2} - \sqrt{\frac{x^6}{4} - 27x^3}}\right)\right].$$

Using implicit differentiation in Example 4 was much simpler than calculating dy/dx directly from any of the above formulas. Finding slopes on curves defined by higher-degree equations usually requires implicit differentiation.

Derivatives of Higher Order

Implicit differentiation can also be used to find higher derivatives. Here is an example.

EXAMPLE 5 Finding a Second Derivative Implicitly

Find d^2y/dx^2 if $2x^3 - 3y^2 = 8$.

Solution To start, we differentiate both sides of the equation with respect to x in order to find $y' = dy/dx$.

$$\frac{d}{dx}\left(2x^3 - 3y^2\right) = \frac{d}{dx}(8)$$

$$6x^2 - 6yy' = 0 \qquad \text{Treat } y \text{ as a function of } x.$$

$$x^2 - yy' = 0$$

$$y' = \frac{x^2}{y}, \quad \text{when } y \neq 0 \qquad \text{Solve for } y'.$$

We now apply the Quotient Rule to find y''.

$$y'' = \frac{d}{dx}\left(\frac{x^2}{y}\right) = \frac{2xy - x^2y'}{y^2} = \frac{2x}{y} - \frac{x^2}{y^2}\cdot y'$$

Finally, we substitute $y' = x^2/y$ to express y'' in terms of x and y.

$$y'' = \frac{2x}{y} - \frac{x^2}{y^2}\left(\frac{x^2}{y}\right) = \frac{2x}{y} - \frac{x^4}{y^3}, \quad \text{when } y \neq 0$$

■

Rational Powers of Differentiable Functions

We know that the rule

$$\frac{d}{dx}x^n = nx^{n-1}$$

holds when n is an integer. Using implicit differentiation we can show that it holds when n is any rational number.

THEOREM 4 Power Rule for Rational Powers

If p/q is a rational number, then $x^{p/q}$ is differentiable at every interior point of the domain of $x^{(p/q)-1}$, and

$$\frac{d}{dx}x^{p/q} = \frac{p}{q}x^{(p/q)-1}.$$

EXAMPLE 6 Using the Rational Power Rule

(a) $\frac{d}{dx}\left(x^{1/2}\right) = \frac{1}{2}x^{-1/2} = \frac{1}{2\sqrt{x}}$ for $x > 0$

(b) $\frac{d}{dx}\left(x^{2/3}\right) = \frac{2}{3}x^{-1/3}$ for $x \neq 0$

(c) $\frac{d}{dx}\left(x^{-4/3}\right) = -\frac{4}{3}x^{-7/3}$ for $x \neq 0$ ■

Proof of Theorem 4 Let p and q be integers with $q > 0$ and suppose that $y = \sqrt[q]{x^p} = x^{p/q}$. Then

$$y^q = x^p.$$

Since p and q are integers (for which we already have the Power Rule), and assuming that y is a differentiable function of x, we can differentiate both sides of the equation with respect to x and get

$$qy^{q-1}\frac{dy}{dx} = px^{p-1}.$$

If $y \neq 0$, we can divide both sides of the equation by qy^{q-1} to solve for dy/dx, obtaining

$$\begin{aligned}
\frac{dy}{dx} &= \frac{px^{p-1}}{qy^{q-1}} \\
&= \frac{p}{q}\cdot\frac{x^{p-1}}{(x^{p/q})^{q-1}} && y = x^{p/q} \\
&= \frac{p}{q}\cdot\frac{x^{p-1}}{x^{p-p/q}} && \frac{p}{q}(q-1) = p - \frac{p}{q} \\
&= \frac{p}{q}\cdot x^{(p-1)-(p-p/q)} && \text{A law of exponents} \\
&= \frac{p}{q}\cdot x^{(p/q)-1},
\end{aligned}$$

which proves the rule. ■

We will drop the assumption of differentiability used in the proof of Theorem 4 in Chapter 7, where we prove the Power Rule for any nonzero real exponent. (See Section 7.3.)

By combining the result of Theorem 4 with the Chain Rule, we get an extension of the Power Chain Rule to rational powers of u: If p/q is a rational number and u is a differentiable function of x, then $u^{p/q}$ is a differentiable function of x and

$$\frac{d}{dx}u^{p/q} = \frac{p}{q}u^{(p/q)-1}\frac{du}{dx},$$

provided that $u \neq 0$ if $(p/q) < 1$. This restriction is necessary because 0 might be in the domain of $u^{p/q}$ but not in the domain of $u^{(p/q)-1}$, as we see in the next example.

EXAMPLE 7 Using the Rational Power and Chain Rules

function defined on $[-1, 1]$

(a) $\dfrac{d}{dx}\left(1 - x^2\right)^{1/4} = \dfrac{1}{4}\left(1 - x^2\right)^{-3/4}(-2x)$ Power Chain Rule with $u = 1 - x^2$

$$= \frac{-x}{2\left(1 - x^2\right)^{3/4}}$$

derivative defined only on $(-1, 1)$

(b) $\dfrac{d}{dx}(\cos x)^{-1/5} = -\dfrac{1}{5}(\cos x)^{-6/5}\dfrac{d}{dx}(\cos x)$

$$= -\frac{1}{5}(\cos x)^{-6/5}(-\sin x)$$

$$= \frac{1}{5}(\sin x)(\cos x)^{-6/5}$$

■

EXERCISES 3.6

Derivatives of Rational Powers

Find dy/dx in Exercises 1-10.

1. $y = x^{9/4}$
2. $y = x^{-3/5}$
3. $y = \sqrt[3]{2x}$
4. $y = \sqrt[4]{5x}$
5. $y = 7\sqrt{x + 6}$
6. $y = -2\sqrt{x - 1}$
7. $y = (2x + 5)^{-1/2}$
8. $y = (1 - 6x)^{2/3}$
9. $y = x(x^2 + 1)^{1/2}$
10. $y = x(x^2 + 1)^{-1/2}$

Find the first derivatives of the functions in Exercises 11-18.

11. $s = \sqrt[7]{t^2}$
12. $r = \sqrt[4]{\theta^{-3}}$
13. $y = \sin\left[(2t + 5)^{-2/3}\right]$
14. $z = \cos\left[(1 - 6t)^{2/3}\right]$
15. $f(x) = \sqrt{1 - \sqrt{x}}$
16. $g(x) = 2(2x^{-1/2} + 1)^{-1/3}$
17. $h(\theta) = \sqrt[3]{1 + \cos(2\theta)}$
18. $k(\theta) = (\sin(\theta + 5))^{5/4}$

Differentiating Implicitly

Use implicit differentiation to find dy/dx in Exercises 19-32.

19. $x^2y + xy^2 = 6$
20. $x^3 + y^3 = 18xy$
21. $2xy + y^2 = x + y$
22. $x^3 - xy + y^3 = 1$
23. $x^2(x - y)^2 = x^2 - y^2$
24. $(3xy + 7)^2 = 6y$
25. $y^2 = \dfrac{x - 1}{x + 1}$
26. $x^2 = \dfrac{x - y}{x + y}$
27. $x = \tan y$
28. $xy = \cot(xy)$
29. $x + \tan(xy) = 0$
30. $x + \sin y = xy$
31. $y\sin\left(\dfrac{1}{y}\right) = 1 - xy$
32. $y^2\cos\left(\dfrac{1}{y}\right) = 2x + 2y$

Find $dr/d\theta$ in Exercises 33-36.

33. $\theta^{1/2} + r^{1/2} = 1$
34. $r - 2\sqrt{\theta} = \dfrac{3}{2}\theta^{2/3} + \dfrac{4}{3}\theta^{3/4}$
35. $\sin(r\theta) = \dfrac{1}{2}$
36. $\cos r + \cot\theta = r\theta$

Second Derivatives

In Exercises 37-42, use implicit differentiation to find dy/dx and then d^2y/dx^2.

37. $x^2 + y^2 = 1$
38. $x^{2/3} + y^{2/3} = 1$
39. $y^2 = x^2 + 2x$
40. $y^2 - 2x = 1 - 2y$
41. $2\sqrt{y} = x - y$
42. $xy + y^2 = 1$
43. If $x^3 + y^3 = 16$, find the value of d^2y/dx^2 at the point $(2, 2)$.
44. If $xy + y^2 = 1$, find the value of d^2y/dx^2 at the point $(0, -1)$.

Slopes, Tangents, and Normals

In Exercises 45 and 46, find the slope of the curve at the given points.

45. $y^2 + x^2 = y^4 - 2x$ at $(-2, 1)$ and $(-2, -1)$
46. $(x^2 + y^2)^2 = (x - y)^2$ at $(1, 0)$ and $(1, -1)$

In Exercises 47-56, verify that the given point is on the curve and find the lines that are **(a)** tangent and **(b)** normal to the curve at the given point.

47. $x^2 + xy - y^2 = 1,\quad (2, 3)$
48. $x^2 + y^2 = 25,\quad (3, -4)$
49. $x^2y^2 = 9,\quad (-1, 3)$

50. $y^2 - 2x - 4y - 1 = 0, \quad (-2, 1)$

51. $6x^2 + 3xy + 2y^2 + 17y - 6 = 0, \quad (-1, 0)$

52. $x^2 - \sqrt{3}xy + 2y^2 = 5, \quad (\sqrt{3}, 2)$

53. $2xy + \pi \sin y = 2\pi, \quad (1, \pi/2)$

54. $x \sin 2y = y \cos 2x, \quad (\pi/4, \pi/2)$

55. $y = 2 \sin(\pi x - y), \quad (1, 0)$

56. $x^2 \cos^2 y - \sin y = 0, \quad (0, \pi)$

57. Parallel tangents Find the two points where the curve $x^2 + xy + y^2 = 7$ crosses the x-axis, and show that the tangents to the curve at these points are parallel. What is the common slope of these tangents?

58. Tangents parallel to the coordinate axes Find points on the curve $x^2 + xy + y^2 = 7$ **(a)** where the tangent is parallel to the x-axis and **(b)** where the tangent is parallel to the y-axis. In the latter case, dy/dx is not defined, but dx/dy is. What value does dx/dy have at these points?

59. The eight curve Find the slopes of the curve $y^4 = y^2 - x^2$ at the two points shown here.

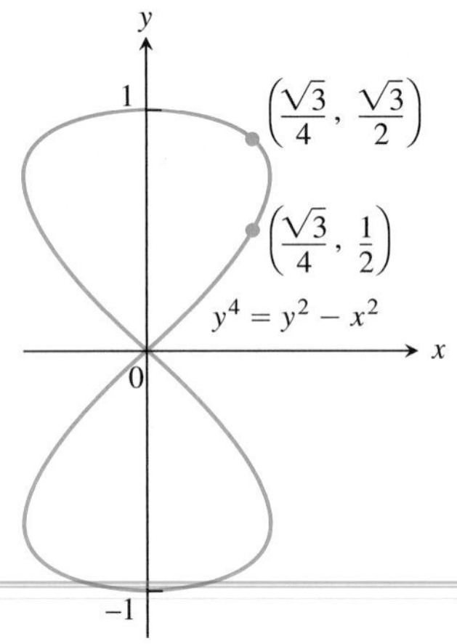

60. The cissoid of Diocles (from about 200 B.C.) Find equations for the tangent and normal to the cissoid of Diocles $y^2(2 - x) = x^3$ at $(1, 1)$.

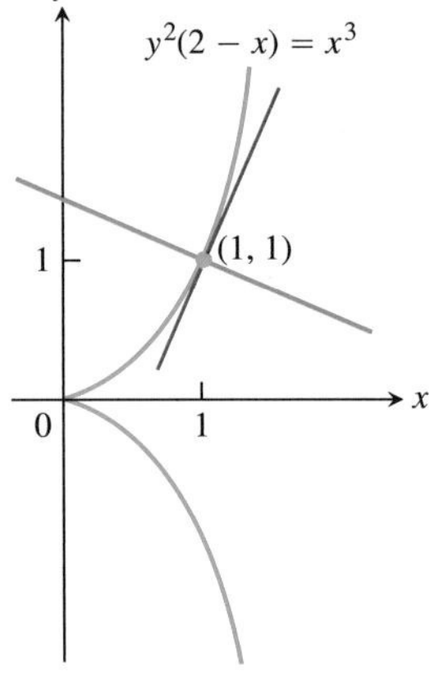

61. The devil's curve (Gabriel Cramer [the Cramer of Cramer's rule], 1750) Find the slopes of the devil's curve $y^4 - 4y^2 = x^4 - 9x^2$ at the four indicated points.

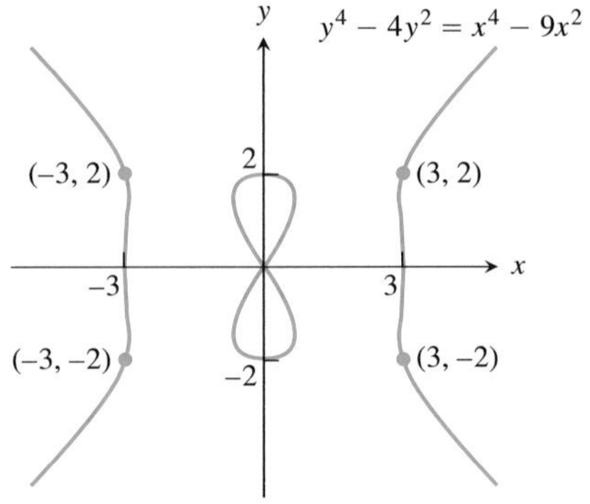

62. The folium of Descartes (See Figure 3.38)

a. Find the slope of the folium of Descartes, $x^3 + y^3 - 9xy = 0$ at the points (4, 2) and (2, 4).

b. At what point other than the origin does the folium have a horizontal tangent?

c. Find the coordinates of the point A in Figure 3.38, where the folium has a vertical tangent.

Implicitly Defined Parametrizations

Assuming that the equations in Exercises 63–66 define x and y implicitly as differentiable functions $x = f(t), y = g(t)$, find the slope of the curve $x = f(t), y = g(t)$ at the given value of t.

63. $x^2 - 2tx + 2t^2 = 4, \quad 2y^3 - 3t^2 = 4, \quad t = 2$

64. $x = \sqrt{5 - \sqrt{t}}, \quad y(t - 1) = \sqrt{t}, \quad t = 4$

65. $x + 2x^{3/2} = t^2 + t, \quad y\sqrt{t + 1} + 2t\sqrt{y} = 4, \quad t = 0$

66. $x \sin t + 2x = t, \quad t \sin t - 2t = y, \quad t = \pi$

Theory and Examples

67. Which of the following could be true if $f''(x) = x^{-1/3}$?

a. $f(x) = \frac{3}{2}x^{2/3} - 3$ **b.** $f(x) = \frac{9}{10}x^{5/3} - 7$

c. $f'''(x) = -\frac{1}{3}x^{-4/3}$ **d.** $f'(x) = \frac{3}{2}x^{2/3} + 6$

68. Is there anything special about the tangents to the curves $y^2 = x^3$ and $2x^2 + 3y^2 = 5$ at the points $(1, \pm 1)$? Give reasons for your answer.

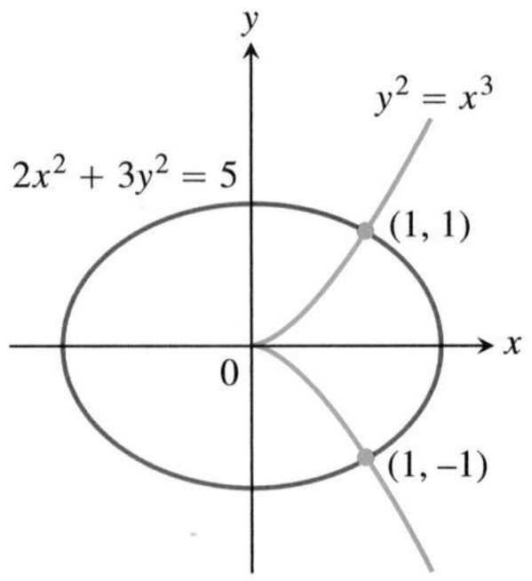

69. Intersecting normal The line that is normal to the curve $x^2 + 2xy - 3y^2 = 0$ at $(1, 1)$ intersects the curve at what other point?

70. Normals parallel to a line Find the normals to the curve $xy + 2x - y = 0$ that are parallel to the line $2x + y = 0$.

71. Normals to a parabola Show that if it is possible to draw three normals from the point $(a, 0)$ to the parabola $x = y^2$ shown here, then a must be greater than $1/2$. One of the normals is the x-axis. For what value of a are the other two normals perpendicular?

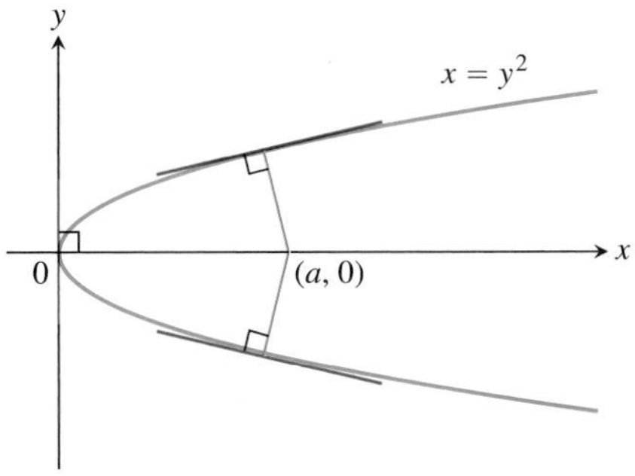

72. What is the geometry behind the restrictions on the domains of the derivatives in Example 6(b) and Example 7(a)?

T In Exercises 73 and 74, find both dy/dx (treating y as a differentiable function of x) and dx/dy (treating x as a differentiable function of y). How do dy/dx and dx/dy seem to be related? Explain the relationship geometrically in terms of the graphs.

73. $xy^3 + x^2y = 6$ **74.** $x^3 + y^2 = \sin^2 y$

COMPUTER EXPLORATIONS

75. a. Given that $x^4 + 4y^2 = 1$, find dy/dx two ways: (1) by solving for y and differentiating the resulting functions in the usual way and (2) by implicit differentiation. Do you get the same result each way?

b. Solve the equation $x^4 + 4y^2 = 1$ for y and graph the resulting functions together to produce a complete graph of the equation $x^4 + 4y^2 = 1$. Then add the graphs of the first derivatives of these functions to your display. Could you have predicted the general behavior of the derivative graphs from looking at the graph of $x^4 + 4y^2 = 1$? Could you have predicted the general behavior of the graph of $x^4 + 4y^2 = 1$ by looking at the derivative graphs? Give reasons for your answers.

76. a. Given that $(x - 2)^2 + y^2 = 4$ find dy/dx two ways: (1) by solving for y and differentiating the resulting functions with respect to x and (2) by implicit differentiation. Do you get the same result each way?

b. Solve the equation $(x - 2)^2 + y^2 = 4$ for y and graph the resulting functions together to produce a complete graph of the equation $(x - 2)^2 + y^2 = 4$. Then add the graphs of the functions' first derivatives to your picture. Could you have predicted the general behavior of the derivative graphs from looking at the graph of $(x - 2)^2 + y^2 = 4$? Could you have predicted the general behavior of the graph of $(x - 2)^2 + y^2 = 4$ by looking at the derivative graphs? Give reasons for your answers.

Use a CAS to perform the following steps in Exercises 77–84.

a. Plot the equation with the implicit plotter of a CAS. Check to see that the given point P satisfies the equation.

b. Using implicit differentiation, find a formula for the derivative dy/dx and evaluate it at the given point P.

c. Use the slope found in part (b) to find an equation for the tangent line to the curve at P. Then plot the implicit curve and tangent line together on a single graph.

77. $x^3 - xy + y^3 = 7, \quad P(2, 1)$

78. $x^5 + y^3x + yx^2 + y^4 = 4, \quad P(1, 1)$

79. $y^2 + y = \dfrac{2 + x}{1 - x}, \quad P(0, 1)$

80. $y^3 + \cos xy = x^2, \quad P(1, 0)$

81. $x + \tan\left(\dfrac{y}{x}\right) = 2, \quad P\left(1, \dfrac{\pi}{4}\right)$

82. $xy^3 + \tan(x + y) = 1, \quad P\left(\dfrac{\pi}{4}, 0\right)$

83. $2y^2 + (xy)^{1/3} = x^2 + 2, \quad P(1, 1)$

84. $x\sqrt{1 + 2y} + y = x^2, \quad P(1, 0)$

3.7 Related Rates

In this section we look at problems that ask for the rate at which some variable changes. In each case the rate is a derivative that has to be computed from the rate at which some other variable (or perhaps several variables) is known to change. To find it, we write an equation that relates the variables involved and differentiate it to get an equation that relates the rate we seek to the rates we know. The problem of finding a rate you cannot measure easily from some other rates that you can is called a *related rates problem*.

Related Rates Equations

Suppose we are pumping air into a spherical balloon. Both the volume and radius of the balloon are increasing over time. If V is the volume and r is the radius of the balloon at an instant of time, then

$$V = \frac{4}{3}\pi r^3.$$

Using the Chain Rule, we differentiate to find the related rates equation

$$\frac{dV}{dt} = \frac{dV}{dr}\frac{dr}{dt} = 4\pi r^2 \frac{dr}{dt}.$$

So if we know the radius r of the balloon and the rate dV/dt at which the volume is increasing at a given instant of time, then we can solve this last equation for dr/dt to find how fast the radius is increasing at that instant. Note that it is easier to measure directly the rate of increase of the volume than it is to measure the increase in the radius. The related rates equation allows us to calculate dr/dt from dV/dt.

Very often the key to relating the variables in a related rates problem is drawing a picture that shows the geometric relations between them, as illustrated in the following example.

EXAMPLE 1 Pumping Out a Tank

How rapidly will the fluid level inside a vertical cylindrical tank drop if we pump the fluid out at the rate of 3000 L/min?

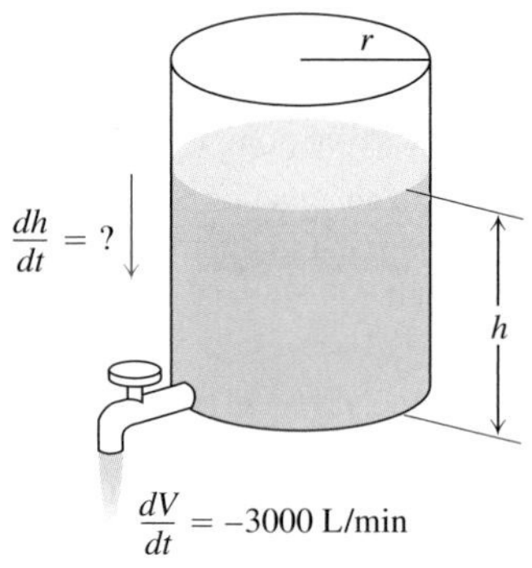

FIGURE 3.42 The rate of change of fluid volume in a cylindrical tank is related to the rate of change of fluid level in the tank (Example 1).

Solution We draw a picture of a partially filled vertical cylindrical tank, calling its radius r and the height of the fluid h (Figure 3.42). Call the volume of the fluid V.

As time passes, the radius remains constant, but V and h change. We think of V and h as differentiable functions of time and use t to represent time. We are told that

$$\frac{dV}{dt} = -3000. \qquad \text{We pump out at the rate of 3000 L/min. The rate is negative because the volume is decreasing.}$$

We are asked to find

$$\frac{dh}{dt}. \qquad \text{How fast will the fluid level drop?}$$

To find dh/dt, we first write an equation that relates h to V. The equation depends on the units chosen for V, r, and h. With V in liters and r and h in meters, the appropriate equation for the cylinder's volume is

$$V = 1000\pi r^2 h$$

because a cubic meter contains 1000 L.

Since V and h are differentiable functions of t, we can differentiate both sides of the equation $V = 1000\pi r^2 h$ with respect to t to get an equation that relates dh/dt to dV/dt:

$$\frac{dV}{dt} = 1000\pi r^2 \frac{dh}{dt}. \qquad r \text{ is a constant.}$$

We substitute the known value $dV/dt = -3000$ and solve for dh/dt:

$$\frac{dh}{dt} = \frac{-3000}{1000\pi r^2} = -\frac{3}{\pi r^2}.$$

The fluid level will drop at the rate of $3/(\pi r^2)$ m/min.

The equation $dh/dt = -3/\pi r^2$ shows how the rate at which the fluid level drops depends on the tank's radius. If r is small, dh/dt will be large; if r is large, dh/dt will be small.

$$\text{If } r = 1 \text{ m:} \qquad \frac{dh}{dt} = -\frac{3}{\pi} \approx -0.95 \text{ m/min} = -95 \text{ cm/min}.$$

$$\text{If } r = 10 \text{ m:} \qquad \frac{dh}{dt} = -\frac{3}{100\pi} \approx -0.0095 \text{ m/min} = -0.95 \text{ cm/min}.$$

■

Related Rates Problem Strategy

1. *Draw a picture and name the variables and constants.* Use t for time. Assume that all variables are differentiable functions of t.
2. *Write down the numerical information* (in terms of the symbols you have chosen).
3. *Write down what you are asked to find* (usually a rate, expressed as a derivative).
4. *Write an equation that relates the variables.* You may have to combine two or more equations to get a single equation that relates the variable whose rate you want to the variables whose rates you know.
5. *Differentiate with respect to t.* Then express the rate you want in terms of the rate and variables whose values you know.
6. *Evaluate.* Use known values to find the unknown rate.

EXAMPLE 2 A Rising Balloon

A hot air balloon rising straight up from a level field is tracked by a range finder 500 ft from the liftoff point. At the moment the range finder's elevation angle is $\pi/4$, the angle is increasing at the rate of 0.14 rad/min. How fast is the balloon rising at that moment?

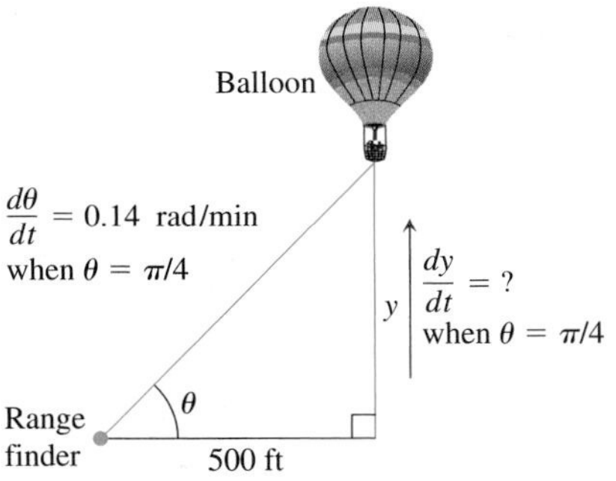

FIGURE 3.43 The rate of change of the balloon's height is related to the rate of change of the angle the range finder makes with the ground (Example 2).

Solution We answer the question in six steps.

1. *Draw a picture and name the variables and constants* (Figure 3.43). The variables in the picture are

   $\theta =$ the angle in radians the range finder makes with the ground.

   $y =$ the height in feet of the balloon.

We let t represent time in minutes and assume that θ and y are differentiable functions of t.

The one constant in the picture is the distance from the range finder to the liftoff point (500 ft). There is no need to give it a special symbol.

2. *Write down the additional numerical information.*

$$\frac{d\theta}{dt} = 0.14 \text{ rad/min} \qquad \text{when} \qquad \theta = \frac{\pi}{4}$$

3. *Write down what we are to find.* We want dy/dt when $\theta = \pi/4$.

4. *Write an equation that relates the variables y and θ.*

$$\frac{y}{500} = \tan\theta \qquad \text{or} \qquad y = 500\tan\theta$$

5. *Differentiate with respect to t using the Chain Rule.* The result tells how dy/dt (which we want) is related to $d\theta/dt$ (which we know).

$$\frac{dy}{dt} = 500\,(\sec^2\theta)\frac{d\theta}{dt}$$

6. *Evaluate with $\theta = \pi/4$ and $d\theta/dt = 0.14$ to find dy/dt.*

$$\frac{dy}{dt} = 500(\sqrt{2})^2(0.14) = 140 \qquad \sec\frac{\pi}{4} = \sqrt{2}$$

At the moment in question, the balloon is rising at the rate of 140 ft/min. ■

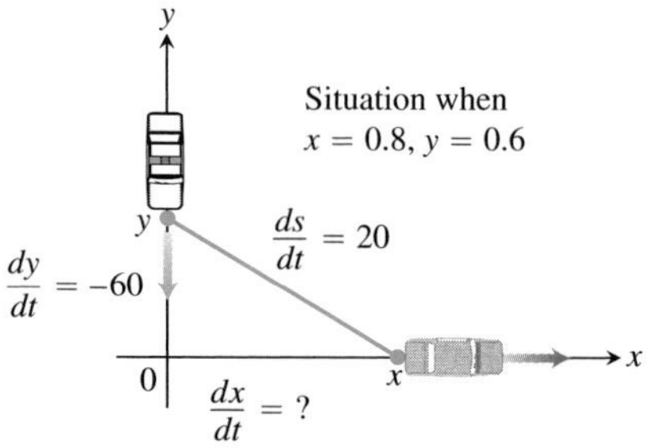

FIGURE 3.44 The speed of the car is related to the speed of the police cruiser and the rate of change of the distance between them (Example 3).

EXAMPLE 3 A Highway Chase

A police cruiser, approaching a right-angled intersection from the north, is chasing a speeding car that has turned the corner and is now moving straight east. When the cruiser is 0.6 mi north of the intersection and the car is 0.8 mi to the east, the police determine with radar that the distance between them and the car is increasing at 20 mph. If the cruiser is moving at 60 mph at the instant of measurement, what is the speed of the car?

Solution We picture the car and cruiser in the coordinate plane, using the positive x-axis as the eastbound highway and the positive y-axis as the southbound highway (Figure 3.44). We let t represent time and set

x = position of car at time t
y = position of cruiser at time t
s = distance between car and cruiser at time t.

We assume that x, y, and s are differentiable functions of t.

We want to find dx/dt when

$$x = 0.8 \text{ mi}, \qquad y = 0.6 \text{ mi}, \qquad \frac{dy}{dt} = -60 \text{ mph}, \qquad \frac{ds}{dt} = 20 \text{ mph}.$$

Note that dy/dt is negative because y is decreasing.

We differentiate the distance equation

$$s^2 = x^2 + y^2$$

(we could also use $s = \sqrt{x^2 + y^2}$), and obtain

$$2s\frac{ds}{dt} = 2x\frac{dx}{dt} + 2y\frac{dy}{dt}$$

$$\frac{ds}{dt} = \frac{1}{s}\left(x\frac{dx}{dt} + y\frac{dy}{dt}\right)$$

$$= \frac{1}{\sqrt{x^2 + y^2}}\left(x\frac{dx}{dt} + y\frac{dy}{dt}\right).$$

Finally, use $x = 0.8$, $y = 0.6$, $dy/dt = -60$, $ds/dt = 20$, and solve for dx/dt.

$$20 = \frac{1}{\sqrt{(0.8)^2 + (0.6)^2}}\left(0.8\frac{dx}{dt} + (0.6)(-60)\right)$$

$$\frac{dx}{dt} = \frac{20\sqrt{(0.8)^2 + (0.6)^2} + (0.6)(60)}{0.8} = 70$$

At the moment in question, the car's speed is 70 mph. ■

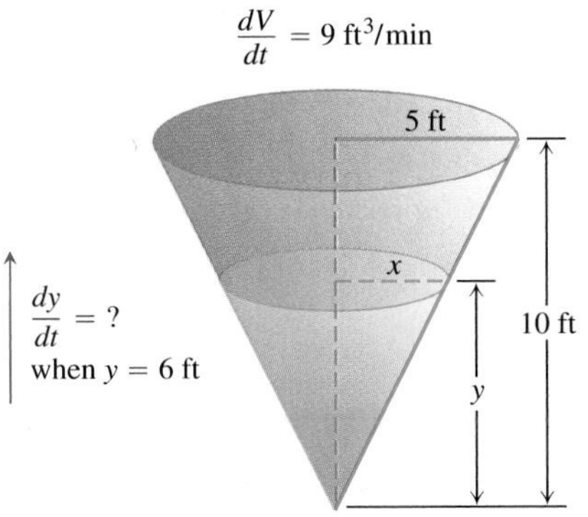

FIGURE 3.45 The geometry of the conical tank and the rate at which water fills the tank determine how fast the water level rises (Example 4).

EXAMPLE 4 Filling a Conical Tank

Water runs into a conical tank at the rate of $9\ \text{ft}^3/\text{min}$. The tank stands point down and has a height of 10 ft and a base radius of 5 ft. How fast is the water level rising when the water is 6 ft deep?

Solution Figure 3.45 shows a partially filled conical tank. The variables in the problem are

V = volume (ft^3) of the water in the tank at time t (min)
x = radius (ft) of the surface of the water at time t
y = depth (ft) of water in tank at time t.

We assume that V, x, and y are differentiable functions of t. The constants are the dimensions of the tank. We are asked for dy/dt when

$$y = 6\ \text{ft} \quad \text{and} \quad \frac{dV}{dt} = 9\ \text{ft}^3/\text{min}.$$

The water forms a cone with volume

$$V = \frac{1}{3}\pi x^2 y.$$

This equation involves x as well as V and y. Because no information is given about x and dx/dt at the time in question, we need to eliminate x. The similar triangles in Figure 3.45 give us a way to express x in terms of y:

$$\frac{x}{y} = \frac{5}{10} \quad \text{or} \quad x = \frac{y}{2}.$$

Therefore,

$$V = \frac{1}{3}\pi\left(\frac{y}{2}\right)^2 y = \frac{\pi}{12}y^3$$

to give the derivative

$$\frac{dV}{dt} = \frac{\pi}{12}\cdot 3y^2\frac{dy}{dt} = \frac{\pi}{4}y^2\frac{dy}{dt}.$$

Finally, use $y = 6$ and $dV/dt = 9$ to solve for dy/dt.

$$9 = \frac{\pi}{4}(6)^2\frac{dy}{dt}$$

$$\frac{dy}{dt} = \frac{1}{\pi} \approx 0.32$$

At the moment in question, the water level is rising at about 0.32 ft/min. ■

EXERCISES 3.7

1. **Area** Suppose that the radius r and area $A = \pi r^2$ of a circle are differentiable functions of t. Write an equation that relates dA/dt to dr/dt.

2. **Surface area** Suppose that the radius r and surface area $S = 4\pi r^2$ of a sphere are differentiable functions of t. Write an equation that relates dS/dt to dr/dt.

3. **Volume** The radius r and height h of a right circular cylinder are related to the cylinder's volume V by the formula $V = \pi r^2 h$.
   a. How is dV/dt related to dh/dt if r is constant?
   b. How is dV/dt related to dr/dt if h is constant?
   c. How is dV/dt related to dr/dt and dh/dt if neither r nor h is constant?

4. **Volume** The radius r and height h of a right circular cone are related to the cone's volume V by the equation $V = (1/3)\pi r^2 h$.
   a. How is dV/dt related to dh/dt if r is constant?
   b. How is dV/dt related to dr/dt if h is constant?
   c. How is dV/dt related to dr/dt and dh/dt if neither r nor h is constant?

5. **Changing voltage** The voltage V (volts), current I (amperes), and resistance R (ohms) of an electric circuit like the one shown here are related by the equation $V = IR$. Suppose that V is increasing at the rate of 1 volt/sec while I is decreasing at the rate of 1/3 amp/sec. Let t denote time in seconds.

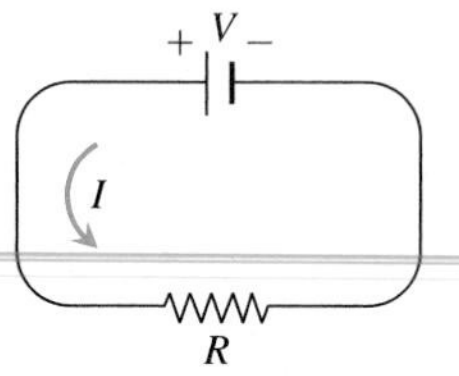

   a. What is the value of dV/dt?
   b. What is the value of dI/dt?
   c. What equation relates dR/dt to dV/dt and dI/dt?
   d. Find the rate at which R is changing when $V = 12$ volts and $I = 2$ amp. Is R increasing, or decreasing?

6. **Electrical power** The power P (watts) of an electric circuit is related to the circuit's resistance R (ohms) and current I (amperes) by the equation $P = RI^2$.
   a. How are dP/dt, dR/dt, and dI/dt related if none of P, R, and I are constant?
   b. How is dR/dt related to dI/dt if P is constant?

7. **Distance** Let x and y be differentiable functions of t and let $s = \sqrt{x^2 + y^2}$ be the distance between the points $(x, 0)$ and $(0, y)$ in the xy-plane.
   a. How is ds/dt related to dx/dt if y is constant?
   b. How is ds/dt related to dx/dt and dy/dt if neither x nor y is constant?
   c. How is dx/dt related to dy/dt if s is constant?

8. **Diagonals** If x, y, and z are lengths of the edges of a rectangular box, the common length of the box's diagonals is $s = \sqrt{x^2 + y^2 + z^2}$.
   a. Assuming that x, y, and z are differentiable functions of t, how is ds/dt related to dx/dt, dy/dt, and dz/dt?
   b. How is ds/dt related to dy/dt and dz/dt if x is constant?
   c. How are dx/dt, dy/dt, and dz/dt related if s is constant?

9. **Area** The area A of a triangle with sides of lengths a and b enclosing an angle of measure θ is

$$A = \frac{1}{2} ab \sin \theta.$$

   a. How is dA/dt related to $d\theta/dt$ if a and b are constant?
   b. How is dA/dt related to $d\theta/dt$ and da/dt if only b is constant?
   c. How is dA/dt related to $d\theta/dt$, da/dt, and db/dt if none of a, b, and θ are constant?

10. **Heating a plate** When a circular plate of metal is heated in an oven, its radius increases at the rate of 0.01 cm/min. At what rate is the plate's area increasing when the radius is 50 cm?

11. **Changing dimensions in a rectangle** The length l of a rectangle is decreasing at the rate of 2 cm/sec while the width w is increasing at the rate of 2 cm/sec. When $l = 12$ cm and $w = 5$ cm, find the rates of change of **(a)** the area, **(b)** the perimeter, and **(c)** the lengths of the diagonals of the rectangle. Which of these quantities are decreasing, and which are increasing?

12. **Changing dimensions in a rectangular box** Suppose that the edge lengths x, y, and z of a closed rectangular box are changing at the following rates:

$$\frac{dx}{dt} = 1 \text{ m/sec}, \quad \frac{dy}{dt} = -2 \text{ m/sec}, \quad \frac{dz}{dt} = 1 \text{ m/sec}.$$

Find the rates at which the box's **(a)** volume, **(b)** surface area, and **(c)** diagonal length $s = \sqrt{x^2 + y^2 + z^2}$ are changing at the instant when $x = 4$, $y = 3$, and $z = 2$.

13. **A sliding ladder** A 13-ft ladder is leaning against a house when its base starts to slide away. By the time the base is 12 ft from the house, the base is moving at the rate of 5 ft/sec.
   a. How fast is the top of the ladder sliding down the wall then?
   b. At what rate is the area of the triangle formed by the ladder, wall, and ground changing then?
   c. At what rate is the angle θ between the ladder and the ground changing then?

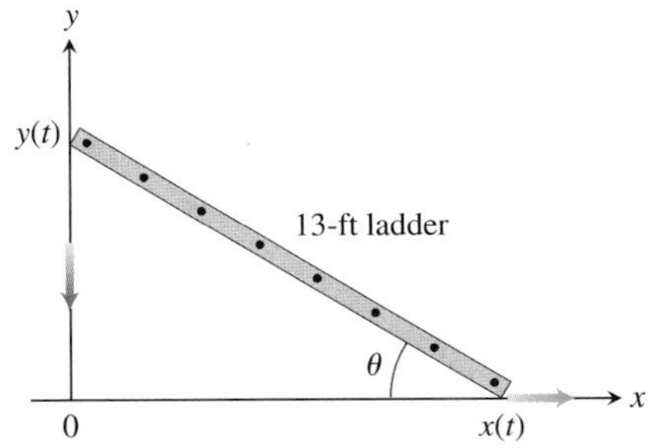

14. Commercial air traffic Two commercial airplanes are flying at 40,000 ft along straight-line courses that intersect at right angles. Plane A is approaching the intersection point at a speed of 442 knots (nautical miles per hour; a nautical mile is 2000 yd). Plane B is approaching the intersection at 481 knots. At what rate is the distance between the planes changing when A is 5 nautical miles from the intersection point and B is 12 nautical miles from the intersection point?

15. Flying a kite A girl flies a kite at a height of 300 ft, the wind carrying the kite horizontally away from her at a rate of 25 ft/sec. How fast must she let out the string when the kite is 500 ft away from her?

16. Boring a cylinder The mechanics at Lincoln Automotive are reboring a 6-in.-deep cylinder to fit a new piston. The machine they are using increases the cylinder's radius one-thousandth of an inch every 3 min. How rapidly is the cylinder volume increasing when the bore (diameter) is 3.800 in.?

17. A growing sand pile Sand falls from a conveyor belt at the rate of $10\ \text{m}^3/\text{min}$ onto the top of a conical pile. The height of the pile is always three-eighths of the base diameter. How fast are the **(a)** height and **(b)** radius changing when the pile is 4 m high? Answer in centimeters per minute.

18. A draining conical reservoir Water is flowing at the rate of $50\ \text{m}^3/\text{min}$ from a shallow concrete conical reservoir (vertex down) of base radius 45 m and height 6 m.

a. How fast (centimeters per minute) is the water level falling when the water is 5 m deep?

b. How fast is the radius of the water's surface changing then? Answer in centimeters per minute.

19. A draining hemispherical reservoir Water is flowing at the rate of $6\ \text{m}^3/\text{min}$ from a reservoir shaped like a hemispherical bowl of radius 13 m, shown here in profile. Answer the following questions, given that the volume of water in a hemispherical bowl of radius R is $V = (\pi/3)y^2(3R - y)$ when the water is y meters deep.

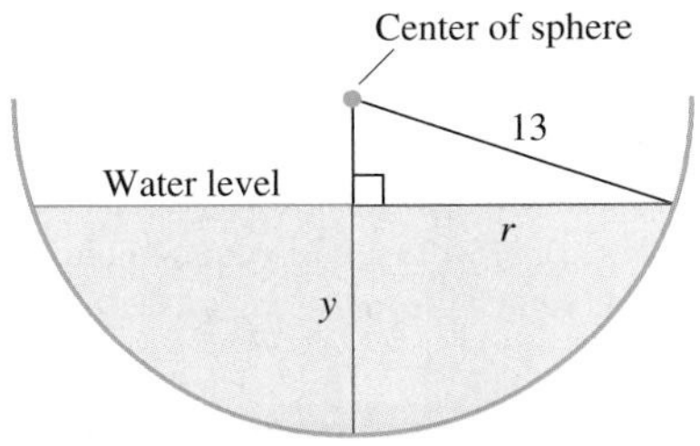

a. At what rate is the water level changing when the water is 8 m deep?

b. What is the radius r of the water's surface when the water is y m deep?

c. At what rate is the radius r changing when the water is 8 m deep?

20. A growing raindrop Suppose that a drop of mist is a perfect sphere and that, through condensation, the drop picks up moisture at a rate proportional to its surface area. Show that under these circumstances the drop's radius increases at a constant rate.

21. The radius of an inflating balloon A spherical balloon is inflated with helium at the rate of $100\pi\ \text{ft}^3/\text{min}$. How fast is the balloon's radius increasing at the instant the radius is 5 ft? How fast is the surface area increasing?

22. Hauling in a dinghy A dinghy is pulled toward a dock by a rope from the bow through a ring on the dock 6 ft above the bow. The rope is hauled in at the rate of 2 ft/sec.

a. How fast is the boat approaching the dock when 10 ft of rope are out?

b. At what rate is the angle θ changing then (see the figure)?

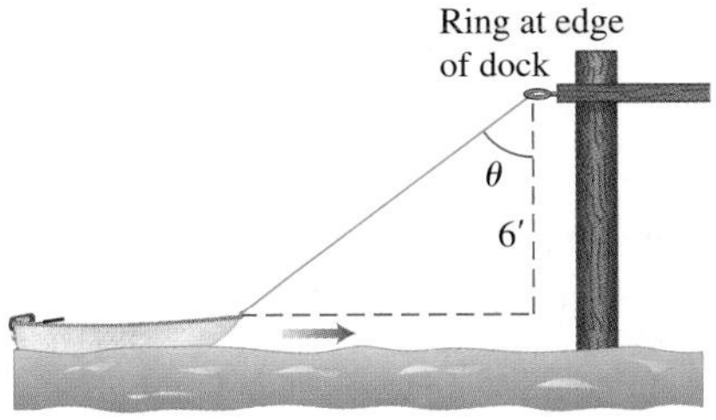

23. A balloon and a bicycle A balloon is rising vertically above a level, straight road at a constant rate of 1 ft/sec. Just when the balloon is 65 ft above the ground, a bicycle moving at a constant rate of 17 ft/sec passes under it. How fast is the distance $s(t)$ between the bicycle and balloon increasing 3 sec later?

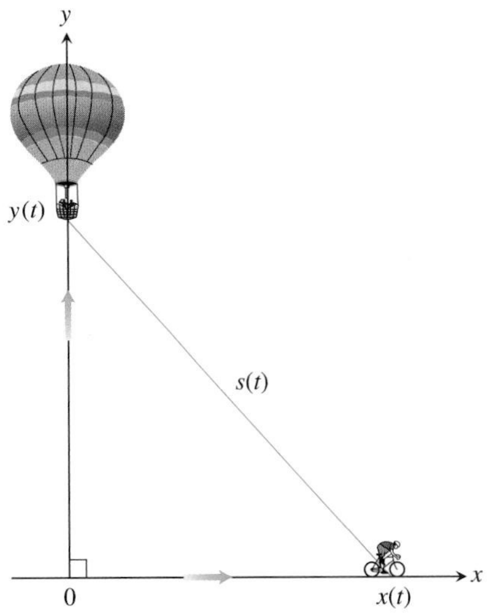

24. Making coffee Coffee is draining from a conical filter into a cylindrical coffeepot at the rate of $10 \text{ in}^3/\text{min}$.

a. How fast is the level in the pot rising when the coffee in the cone is 5 in. deep?

b. How fast is the level in the cone falling then?

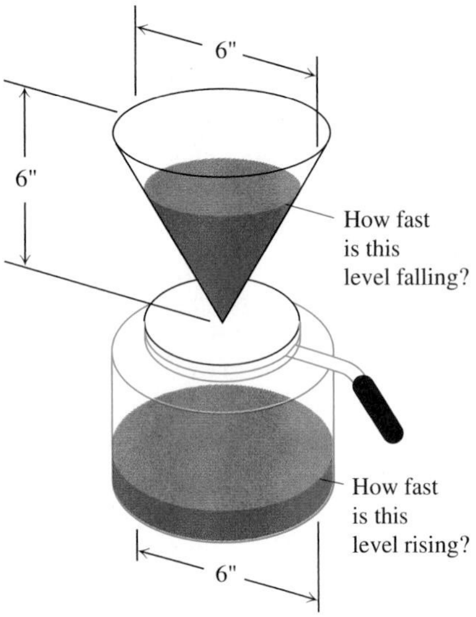

25. Cardiac output In the late 1860s, Adolf Fick, a professor of physiology in the Faculty of Medicine in Würzberg, Germany, developed one of the methods we use today for measuring how much blood your heart pumps in a minute. Your cardiac output as you read this sentence is probably about 7 L/min. At rest it is likely to be a bit under 6 L/min. If you are a trained marathon runner running a marathon, your cardiac output can be as high as 30 L/min.

Your cardiac output can be calculated with the formula

$$y = \frac{Q}{D},$$

where Q is the number of milliliters of CO_2 you exhale in a minute and D is the difference between the CO_2 concentration (ml/L) in the blood pumped to the lungs and the CO_2 concentration in the blood returning from the lungs. With $Q = 233$ ml/min and $D = 97 - 56 = 41$ ml/L,

$$y = \frac{233 \text{ ml/min}}{41 \text{ ml/L}} \approx 5.68 \text{ L/min},$$

fairly close to the 6 L/min that most people have at basal (resting) conditions. (Data courtesy of J. Kenneth Herd, M.D., Quillan College of Medicine, East Tennessee State University.)

Suppose that when $Q = 233$ and $D = 41$, we also know that D is decreasing at the rate of 2 units a minute but that Q remains unchanged. What is happening to the cardiac output?

26. Cost, revenue, and profit A company can manufacture x items at a cost of $c(x)$ thousand dollars, a sales revenue of $r(x)$ thousand dollars, and a profit of $p(x) = r(x) - c(x)$ thousand dollars. Find dc/dt, dr/dt, and dp/dt for the following values of x and dx/dt.

a. $r(x) = 9x$, $c(x) = x^3 - 6x^2 + 15x$, and $dx/dt = 0.1$ when $x = 2$

b. $r(x) = 70x$, $c(x) = x^3 - 6x^2 + 45/x$, and $dx/dt = 0.05$ when $x = 1.5$

27. Moving along a parabola A particle moves along the parabola $y = x^2$ in the first quadrant in such a way that its x-coordinate (measured in meters) increases at a steady 10 m/sec. How fast is the angle of inclination θ of the line joining the particle to the origin changing when $x = 3$ m?

28. Moving along another parabola A particle moves from right to left along the parabolic curve $y = \sqrt{-x}$ in such a way that its x-coordinate (measured in meters) decreases at the rate of 8 m/sec. How fast is the angle of inclination θ of the line joining the particle to the origin changing when $x = -4$?

29. Motion in the plane The coordinates of a particle in the metric xy-plane are differentiable functions of time t with $dx/dt = -1$ m/sec and $dy/dt = -5$ m/sec. How fast is the particle's distance from the origin changing as it passes through the point (5, 12)?

30. A moving shadow A man 6 ft tall walks at the rate of 5 ft/sec toward a streetlight that is 16 ft above the ground. At what rate is the tip of his shadow moving? At what rate is the length of his shadow changing when he is 10 ft from the base of the light?

31. Another moving shadow A light shines from the top of a pole 50 ft high. A ball is dropped from the same height from a point 30 ft away from the light. (See accompanying figure.) How fast is the shadow of the ball moving along the ground 1/2 sec later? (Assume the ball falls a distance $s = 16t^2$ ft in t sec.)

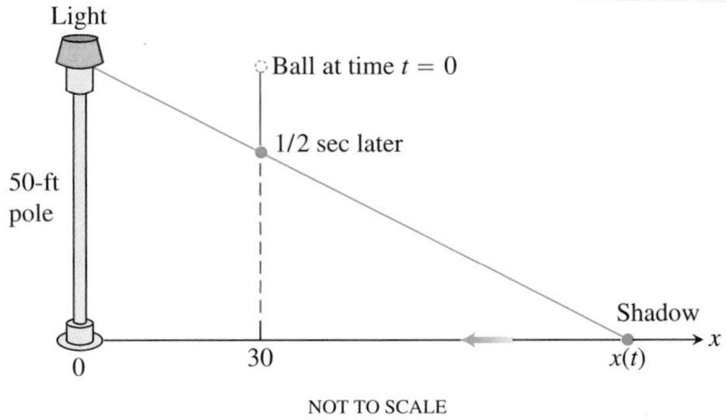

32. Videotaping a moving car You are videotaping a race from a stand 132 ft from the track, following a car that is moving at 180 mi/h (264 ft/sec). How fast will your camera angle θ be changing when the car is right in front of you? A half second later?

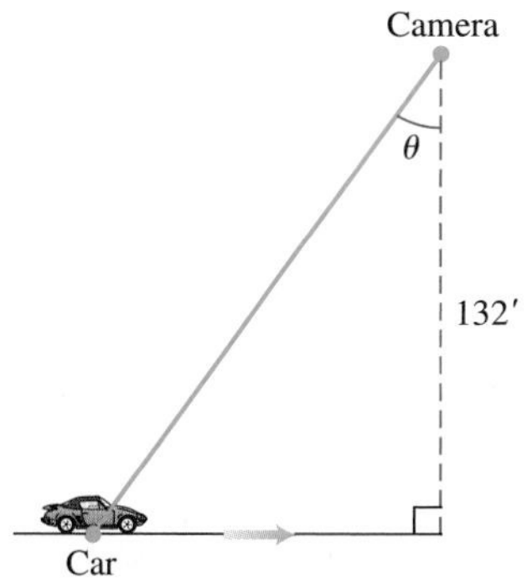

33. A melting ice layer A spherical iron ball 8 in. in diameter is coated with a layer of ice of uniform thickness. If the ice melts at the rate of $10 \text{ in}^3/\text{min}$, how fast is the thickness of the ice decreasing when it is 2 in. thick? How fast is the outer surface area of ice decreasing?

34. Highway patrol A highway patrol plane flies 3 mi above a level, straight road at a steady 120 mi/h. The pilot sees an oncoming car and with radar determines that at the instant the line-of-sight distance from plane to car is 5 mi, the line-of-sight distance is decreasing at the rate of 160 mi/h. Find the car's speed along the highway.

35. A building's shadow On a morning of a day when the sun will pass directly overhead, the shadow of an 80-ft building on level ground is 60 ft long. At the moment in question, the angle θ the sun makes with the ground is increasing at the rate of $0.27°/\text{min}$. At what rate is the shadow decreasing? (Remember to use radians. Express your answer in inches per minute, to the nearest tenth.)

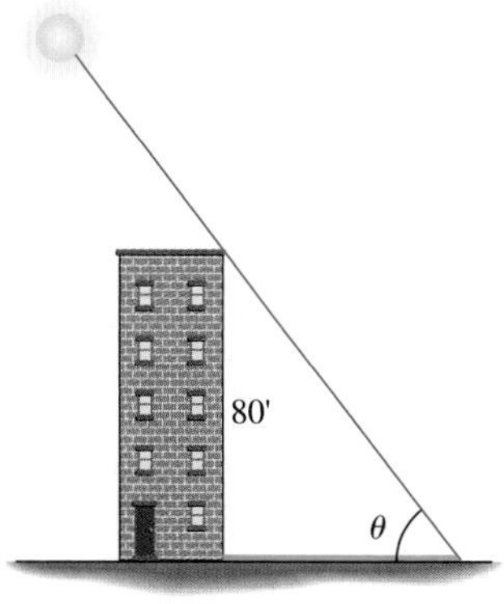

36. Walkers A and B are walking on straight streets that meet at right angles. A approaches the intersection at 2 m/sec; B moves away from the intersection 1 m/sec. At what rate is the angle θ changing when A is 10 m from the intersection and B is 20 m from the intersection? Express your answer in degrees per second to the nearest degree.

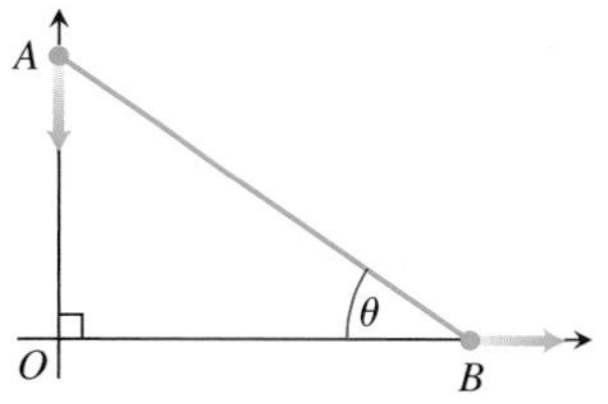

37. Baseball players A baseball diamond is a square 90 ft on a side. A player runs from first base to second at a rate of 16 ft/sec.

a. At what rate is the player's distance from third base changing when the player is 30 ft from first base?

b. At what rates are angles θ_1 and θ_2 (see the figure) changing at that time?

c. The player slides into second base at the rate of 15 ft/sec. At what rates are angles θ_1 and θ_2 changing as the player touches base?

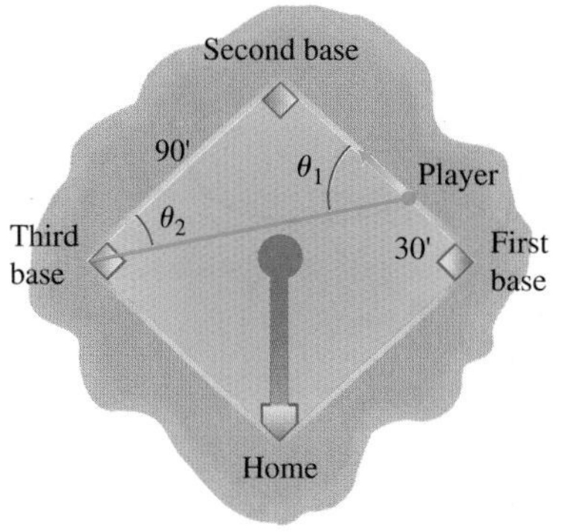

38. Ships Two ships are steaming straight away from a point O along routes that make a 120° angle. Ship A moves at 14 knots (nautical miles per hour; a nautical mile is 2000 yd). Ship B moves at 21 knots. How fast are the ships moving apart when $OA = 5$ and $OB = 3$ nautical miles?

3.8 Linearization and Differentials

Sometimes we can approximate complicated functions with simpler ones that give the accuracy we want for specific applications and are easier to work with. The approximating functions discussed in this section are called *linearizations*, and they are based on tangent lines. Other approximating functions, such as polynomials, are discussed in Chapter 11.

We introduce new variables dx and dy, called *differentials*, and define them in a way that makes Leibniz's notation for the derivative dy/dx a true ratio. We use dy to estimate error in measurement and sensitivity of a function to change. Application of these ideas then provides for a precise proof of the Chain Rule (Section 3.5).

Linearization

As you can see in Figure 3.46, the tangent to the curve $y = x^2$ lies close to the curve near the point of tangency. For a brief interval to either side, the y-values along the tangent line give good approximations to the y-values on the curve. We observe this phenomenon by zooming in on the two graphs at the point of tangency or by looking at tables of values for the difference between $f(x)$ and its tangent line near the x-coordinate of the point of tangency. Locally, every differentiable curve behaves like a straight line.

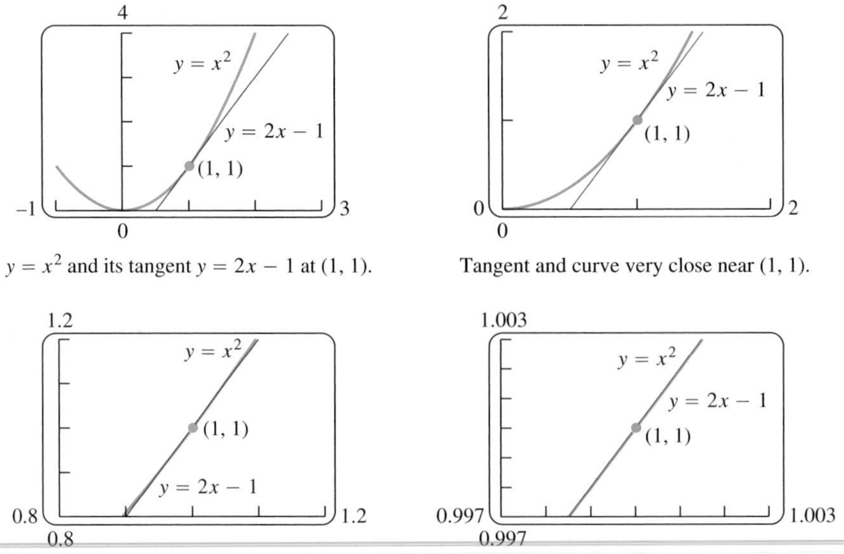

$y = x^2$ and its tangent $y = 2x - 1$ at (1, 1).

Tangent and curve very close near (1, 1).

Tangent and curve very close throughout entire x-interval shown.

Tangent and curve closer still. Computer screen cannot distinguish tangent from curve on this x-interval.

FIGURE 3.46 The more we magnify the graph of a function near a point where the function is differentiable, the flatter the graph becomes and the more it resembles its tangent.

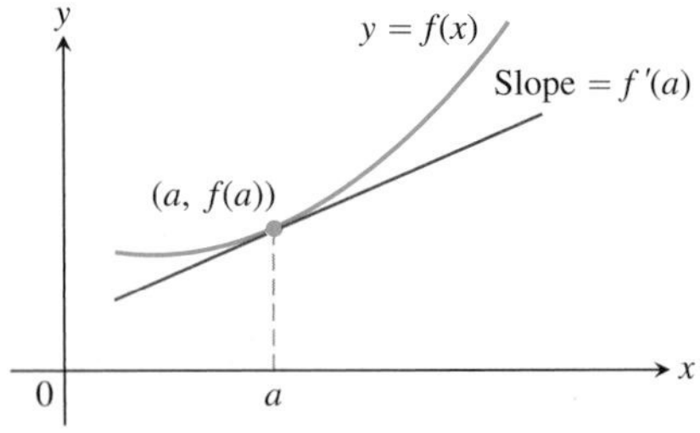

FIGURE 3.47 The tangent to the curve $y = f(x)$ at $x = a$ is the line $L(x) = f(a) + f'(a)(x - a)$.

In general, the tangent to $y = f(x)$ at a point $x = a$, where f is differentiable (Figure 3.47), passes through the point $(a, f(a))$, so its point-slope equation is

$$y = f(a) + f'(a)(x - a).$$

Thus, this tangent line is the graph of the linear function

$$L(x) = f(a) + f'(a)(x - a).$$

For as long as this line remains close to the graph of f, $L(x)$ gives a good approximation to $f(x)$.

DEFINITIONS Linearization, Standard Linear Approximation

If f is differentiable at $x = a$, then the approximating function

$$L(x) = f(a) + f'(a)(x - a)$$

is the **linearization** of f at a. The approximation

$$f(x) \approx L(x)$$

of f by L is the **standard linear approximation** of f at a. The point $x = a$ is the **center** of the approximation.

EXAMPLE 1 Finding a Linearization

Find the linearization of $f(x) = \sqrt{1 + x}$ at $x = 0$ (Figure 3.48).

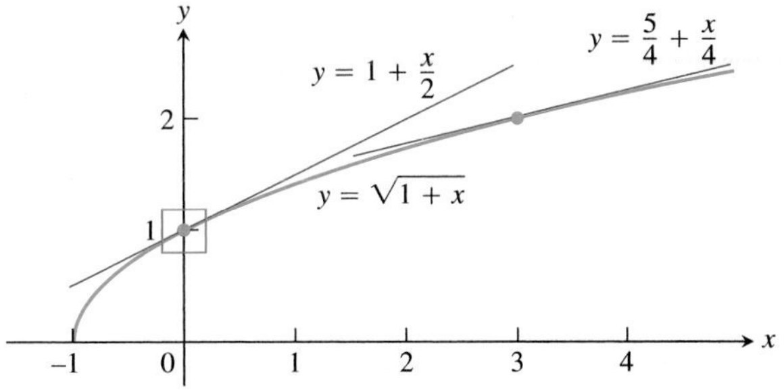

FIGURE 3.48 The graph of $y = \sqrt{1 + x}$ and its linearizations at $x = 0$ and $x = 3$. Figure 3.49 shows a magnified view of the small window about 1 on the y-axis.

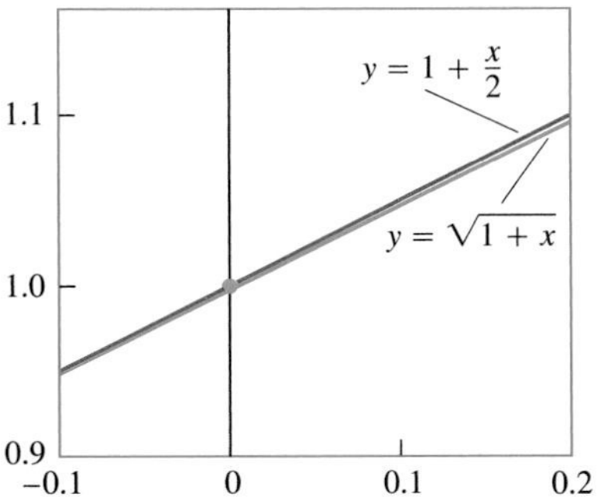

FIGURE 3.49 Magnified view of the window in Figure 3.48.

Solution Since

$$f'(x) = \frac{1}{2}\left(1 + x\right)^{-1/2},$$

we have $f(0) = 1$ and $f'(0) = 1/2$, giving the linearization

$$L(x) = f(a) + f'(a)(x - a) = 1 + \frac{1}{2}\left(x - 0\right) = 1 + \frac{x}{2}.$$

See Figure 3.49. ■

Look at how accurate the approximation $\sqrt{1 + x} \approx 1 + (x/2)$ from Example 1 is for values of x near 0.

As we move away from zero, we lose accuracy. For example, for $x = 2$, the linearization gives 2 as the approximation for $\sqrt{3}$, which is not even accurate to one decimal place.

Do not be misled by the preceding calculations into thinking that whatever we do with a linearization is better done with a calculator. In practice, we would never use a linearization to find a particular square root. The utility of a linearization is its ability to replace a complicated formula by a simpler one over an entire interval of values. If we have to work with $\sqrt{1 + x}$ for x close to 0 and can tolerate the small amount of error involved, we can

Approximation	True value	$\lvert$True value − approximation$\rvert$
$\sqrt{1.2} \approx 1 + \frac{0.2}{2} = 1.10$	1.095445	$<10^{-2}$
$\sqrt{1.05} \approx 1 + \frac{0.05}{2} = 1.025$	1.024695	$<10^{-3}$
$\sqrt{1.005} \approx 1 + \frac{0.005}{2} = 1.00250$	1.002497	$<10^{-5}$

work with $1 + (x/2)$ instead. Of course, we then need to know how much error there is. We have more to say on the estimation of error in Chapter 11.

A linear approximation normally loses accuracy away from its center. As Figure 3.48 suggests, the approximation $\sqrt{1 + x} \approx 1 + (x/2)$ will probably be too crude to be useful near $x = 3$. There, we need the linearization at $x = 3$.

EXAMPLE 2 Finding a Linearization at Another Point

Find the linearization of $f(x) = \sqrt{1 + x}$ at $x = 3$.

Solution We evaluate the equation defining $L(x)$ at $a = 3$. With

$$f(3) = 2, \qquad f'(3) = \frac{1}{2}(1 + x)^{-1/2}\bigg|_{x=3} = \frac{1}{4},$$

we have

$$L(x) = 2 + \frac{1}{4}(x - 3) = \frac{5}{4} + \frac{x}{4}. \qquad \blacksquare$$

At $x = 3.2$, the linearization in Example 2 gives

$$\sqrt{1 + x} = \sqrt{1 + 3.2} \approx \frac{5}{4} + \frac{3.2}{4} = 1.250 + 0.800 = 2.050,$$

which differs from the true value $\sqrt{4.2} \approx 2.04939$ by less than one one-thousandth. The linearization in Example 1 gives

$$\sqrt{1 + x} = \sqrt{1 + 3.2} \approx 1 + \frac{3.2}{2} = 1 + 1.6 = 2.6,$$

a result that is off by more than 25%.

EXAMPLE 3 Finding a Linearization for the Cosine Function

Find the linearization of $f(x) = \cos x$ at $x = \pi/2$ (Figure 3.50).

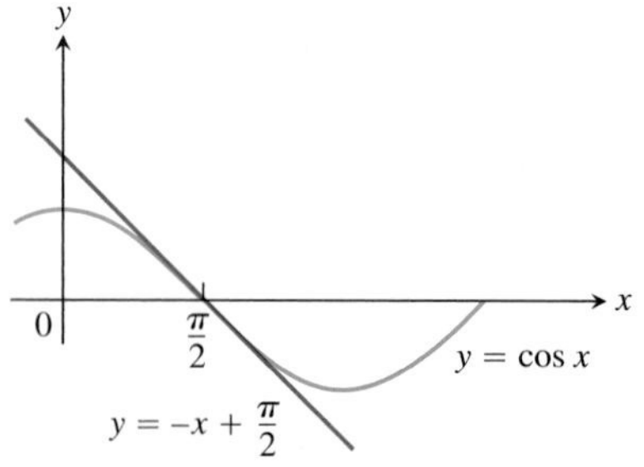

FIGURE 3.50 The graph of $f(x) = \cos x$ and its linearization at $x = \pi/2$. Near $x = \pi/2$, $\cos x \approx -x + (\pi/2)$ (Example 3).

Solution Since $f(\pi/2) = \cos(\pi/2) = 0$, $f'(x) = -\sin x$, and $f'(\pi/2) = -\sin(\pi/2) = -1$, we have

$$\begin{aligned} L(x) &= f(a) + f'(a)(x - a) \\ &= 0 + (-1)\left(x - \frac{\pi}{2}\right) \\ &= -x + \frac{\pi}{2}. \end{aligned} \qquad \blacksquare$$

An important linear approximation for roots and powers is

$$(1 + x)^k \approx 1 + kx \qquad (x \text{ near } 0; \text{ any number } k)$$

(Exercise 15). This approximation, good for values of x sufficiently close to zero, has broad application. For example, when x is small,

$$\sqrt{1 + x} \approx 1 + \frac{1}{2}x \qquad k = 1/2$$

$$\frac{1}{1 - x} = (1 - x)^{-1} \approx 1 + (-1)(-x) = 1 + x \qquad k = -1; \text{ replace } x \text{ by } -x.$$

$$\sqrt[3]{1 + 5x^4} = (1 + 5x^4)^{1/3} \approx 1 + \frac{1}{3}\left(5x^4\right) = 1 + \frac{5}{3}x^4 \qquad k = 1/3; \text{ replace } x \text{ by } 5x^4.$$

$$\frac{1}{\sqrt{1 - x^2}} = (1 - x^2)^{-1/2} \approx 1 + \left(-\frac{1}{2}\right)(-x^2) = 1 + \frac{1}{2}x^2 \qquad k = -1/2; \text{ replace } x \text{ by } -x^2.$$

Differentials

We sometimes use the Leibniz notation dy/dx to represent the derivative of y with respect to x. Contrary to its appearance, it is not a ratio. We now introduce two new variables dx and dy with the property that if their ratio exists, it will be equal to the derivative.

DEFINITION Differential

Let $y = f(x)$ be a differentiable function. The **differential *dx*** is an independent variable. The **differential *dy*** is

$$dy = f'(x)\,dx.$$

Unlike the independent variable dx, the variable dy is always a dependent variable. It depends on both x and dx. If dx is given a specific value and x is a particular number in the domain of the function f, then the numerical value of dy is determined.

EXAMPLE 4 Finding the Differential *dy*

(a) Find dy if $y = x^5 + 37x$.

(b) Find the value of dy when $x = 1$ and $dx = 0.2$.

Solution

(a) $dy = (5x^4 + 37)\,dx$

(b) Substituting $x = 1$ and $dx = 0.2$ in the expression for dy, we have

$$dy = (5 \cdot 1^4 + 37)0.2 = 8.4.$$

■

The geometric meaning of differentials is shown in Figure 3.51. Let $x = a$ and set $dx = \Delta x$. The corresponding change in $y = f(x)$ is

$$\Delta y = f(a + dx) - f(a).$$

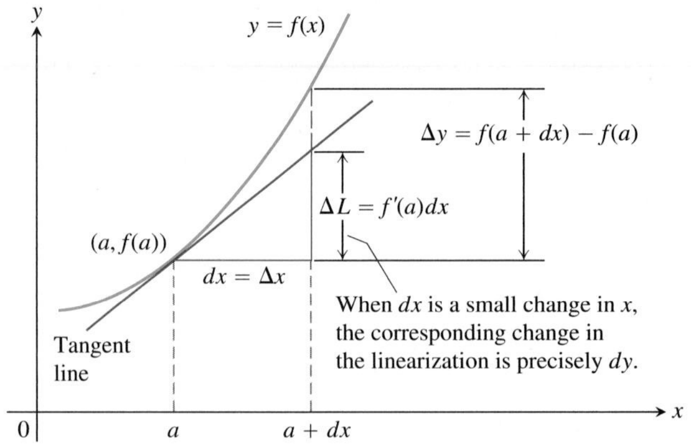

FIGURE 3.51 Geometrically, the differential dy is the change ΔL in the linearization of f when $x = a$ changes by an amount $dx = \Delta x$.

The corresponding change in the tangent line L is

$$\begin{aligned}\Delta L &= L(a + dx) - L(a)\\ &= \underbrace{f(a) + f'(a)[(a + dx) - a]}_{L(a+dx)} - \underbrace{f(a)}_{L(a)}\\ &= f'(a)\,dx.\end{aligned}$$

That is, the change in the linearization of f is precisely the value of the differential dy when $x = a$ and $dx = \Delta x$. Therefore, dy represents the amount the tangent line rises or falls when x changes by an amount $dx = \Delta x$.

If $dx \neq 0$, then the quotient of the differential dy by the differential dx is equal to the derivative $f'(x)$ because

$$dy \div dx = \frac{f'(x)\,dx}{dx} = f'(x) = \frac{dy}{dx}.$$

We sometimes write

$$df = f'(x)\,dx$$

in place of $dy = f'(x)\,dx$, calling df the **differential of f**. For instance, if $f(x) = 3x^2 - 6$, then

$$df = d(3x^2 - 6) = 6x\,dx.$$

Every differentiation formula like

$$\frac{d(u + v)}{dx} = \frac{du}{dx} + \frac{dv}{dx} \quad \text{or} \quad \frac{d(\sin u)}{dx} = \cos u\,\frac{du}{dx}$$

has a corresponding differential form like

$$d(u + v) = du + dv \quad \text{or} \quad d(\sin u) = \cos u\,du.$$

EXAMPLE 5 Finding Differentials of Functions

(a) $d(\tan 2x) = \sec^2(2x)\, d(2x) = 2\sec^2 2x\, dx$

(b) $d\left(\dfrac{x}{x+1}\right) = \dfrac{(x+1)\,dx - x\,d(x+1)}{(x+1)^2} = \dfrac{x\,dx + dx - x\,dx}{(x+1)^2} = \dfrac{dx}{(x+1)^2}$ ■

Estimating with Differentials

Suppose we know the value of a differentiable function $f(x)$ at a point a and want to predict how much this value will change if we move to a nearby point $a + dx$. If dx is small, then we can see from Figure 3.51 that Δy is approximately equal to the differential dy. Since

$$f(a + dx) = f(a) + \Delta y,$$

the differential approximation gives

$$f(a + dx) \approx f(a) + dy$$

where $dx = \Delta x$. Thus the approximation $\Delta y \approx dy$ can be used to calculate $f(a + dx)$ when $f(a)$ is known and dx is small.

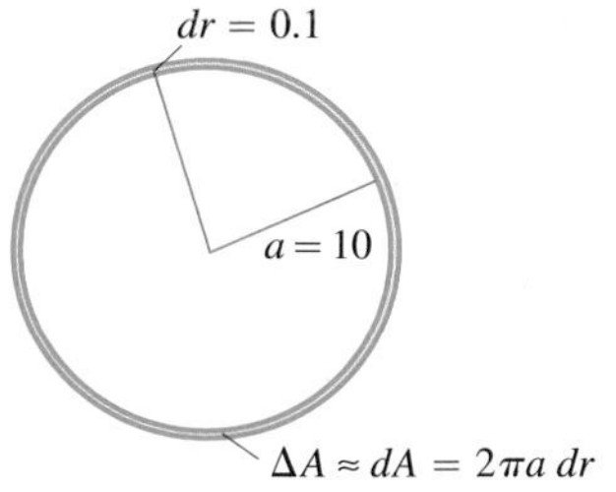

FIGURE 3.52 When dr is small compared with a, as it is when $dr = 0.1$ and $a = 10$, the differential $dA = 2\pi a\, dr$ gives a way to estimate the area of the circle with radius $r = a + dr$ (Example 6).

EXAMPLE 6 Estimating with Differentials

The radius r of a circle increases from $a = 10$ m to 10.1 m (Figure 3.52). Use dA to estimate the increase in the circle's area A. Estimate the area of the enlarged circle and compare your estimate to the true area.

Solution Since $A = \pi r^2$, the estimated increase is

$$dA = A'(a)\, dr = 2\pi a\, dr = 2\pi(10)(0.1) = 2\pi \text{ m}^2.$$

Thus,

$$\begin{aligned} A(10 + 0.1) &\approx A(10) + 2\pi \\ &= \pi(10)^2 + 2\pi = 102\pi. \end{aligned}$$

The area of a circle of radius 10.1 m is approximately 102π m^2.

The true area is

$$\begin{aligned} A(10.1) &= \pi(10.1)^2 \\ &= 102.01\pi \text{ m}^2. \end{aligned}$$

The error in our estimate is 0.01π m^2, which is the difference $\Delta A - dA$. ■

Error in Differential Approximation

Let $f(x)$ be differentiable at $x = a$ and suppose that $dx = \Delta x$ is an increment of x. We have two ways to describe the change in f as x changes from a to $a + \Delta x$:

The true change:	$\Delta f = f(a + \Delta x) - f(a)$
The differential estimate:	$df = f'(a)\,\Delta x$.

How well does df approximate Δf?

We measure the approximation error by subtracting df from Δf:

$$\begin{aligned}\text{Approximation error} &= \Delta f - df \\ &= \Delta f - f'(a)\Delta x \\ &= \underbrace{f(a + \Delta x) - f(a)}_{\Delta f} - f'(a)\Delta x \\ &= \underbrace{\left(\frac{f(a + \Delta x) - f(a)}{\Delta x} - f'(a)\right)}_{\text{Call this part } \epsilon} \cdot \Delta x \\ &= \epsilon \cdot \Delta x.\end{aligned}$$

As $\Delta x \to 0$, the difference quotient

$$\frac{f(a + \Delta x) - f(a)}{\Delta x}$$

approaches $f'(a)$ (remember the definition of $f'(a)$), so the quantity in parentheses becomes a very small number (which is why we called it ϵ). In fact, $\epsilon \to 0$ as $\Delta x \to 0$. When Δx is small, the approximation error $\epsilon\, \Delta x$ is smaller still.

$$\underbrace{\Delta f}_{\substack{\text{true}\\ \text{change}}} = \underbrace{f'(a)\Delta x}_{\substack{\text{estimated}\\ \text{change}}} + \underbrace{\epsilon\, \Delta x}_{\text{error}}$$

Although we do not know exactly how small the error is and will not be able to make much progress on this front until Chapter 11, there is something worth noting here, namely the *form* taken by the equation.

Change in $y = f(x)$ near $x = a$

If $y = f(x)$ is differentiable at $x = a$ and x changes from a to $a + \Delta x$, the change Δy in f is given by an equation of the form

$$\Delta y = f'(a)\, \Delta x + \epsilon\, \Delta x \tag{1}$$

in which $\epsilon \to 0$ as $\Delta x \to 0$.

In Example 6 we found that

$$\Delta A = \pi(10.1)^2 - \pi(10)^2 = (102.01 - 100)\pi = (\underbrace{2\pi}_{dA} + \underbrace{0.01\pi}_{\text{error}})\ \text{m}^2$$

so the approximation error is $\Delta A - dA = \epsilon\, \Delta r = 0.01\pi$ and $\epsilon = 0.01\pi/\Delta r = 0.01\pi/0.1 = 0.1\pi$ m.

Equation (1) enables us to bring the proof of the Chain Rule to a successful conclusion.

Proof of the Chain Rule

Our goal is to show that if $f(u)$ is a differentiable function of u and $u = g(x)$ is a differentiable function of x, then the composite $y = f(g(x))$ is a differentiable function of x.

More precisely, if g is differentiable at x_0 and f is differentiable at $g(x_0)$, then the composite is differentiable at x_0 and

$$\left.\frac{dy}{dx}\right|_{x=x_0} = f'(g(x_0)) \cdot g'(x_0).$$

Let Δx be an increment in x and let Δu and Δy be the corresponding increments in u and y. Applying Equation (1) we have,

$$\Delta u = g'(x_0)\Delta x + \epsilon_1 \Delta x = (g'(x_0) + \epsilon_1)\Delta x,$$

where $\epsilon_1 \to 0$ as $\Delta x \to 0$. Similarly,

$$\Delta y = f'(u_0)\Delta u + \epsilon_2 \Delta u = (f'(u_0) + \epsilon_2)\Delta u,$$

where $\epsilon_2 \to 0$ as $\Delta u \to 0$. Notice also that $\Delta u \to 0$ as $\Delta x \to 0$. Combining the equations for Δu and Δy gives

$$\Delta y = (f'(u_0) + \epsilon_2)(g'(x_0) + \epsilon_1)\Delta x,$$

so

$$\frac{\Delta y}{\Delta x} = f'(u_0)g'(x_0) + \epsilon_2 g'(x_0) + f'(u_0)\epsilon_1 + \epsilon_2\epsilon_1.$$

Since ϵ_1 and ϵ_2 go to zero as Δx goes to zero, three of the four terms on the right vanish in the limit, leaving

$$\left.\frac{dy}{dx}\right|_{x=x_0} = \lim_{\Delta x \to 0} \frac{\Delta y}{\Delta x} = f'(u_0)g'(x_0) = f'(g(x_0)) \cdot g'(x_0).$$

This concludes the proof. ■

Sensitivity to Change

The equation $df = f'(x)\,dx$ tells how *sensitive* the output of f is to a change in input at different values of x. The larger the value of f' at x, the greater the effect of a given change dx. As we move from a to a nearby point $a + dx$, we can describe the change in f in three ways:

	True	**Estimated**
Absolute change	$\Delta f = f(a + dx) - f(a)$	$df = f'(a)\,dx$
Relative change	$\dfrac{\Delta f}{f(a)}$	$\dfrac{df}{f(a)}$
Percentage change	$\dfrac{\Delta f}{f(a)} \times 100$	$\dfrac{df}{f(a)} \times 100$

EXAMPLE 7 Finding the Depth of a Well

You want to calculate the depth of a well from the equation $s = 16t^2$ by timing how long it takes a heavy stone you drop to splash into the water below. How sensitive will your calculations be to a 0.1-sec error in measuring the time?

Solution The size of ds in the equation

$$ds = 32t\,dt$$

depends on how big t is. If $t = 2$ sec, the change caused by $dt = 0.1$ is about

$$ds = 32(2)(0.1) = 6.4 \text{ ft}.$$

Three seconds later at $t = 5$ sec, the change caused by the same dt is

$$ds = 32(5)(0.1) = 16 \text{ ft}.$$

The estimated depth of the well differs from its true depth by a greater distance the longer the time it takes the stone to splash into the water below, for a given error in measuring the time. ■

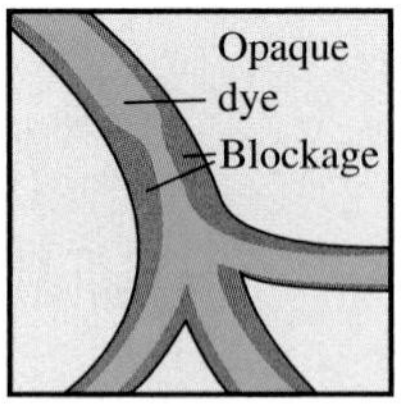

Angiography

An opaque dye is injected into a partially blocked artery to make the inside visible under X-rays. This reveals the location and severity of the blockage.

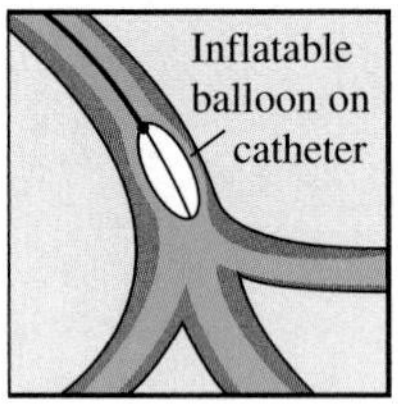

Angioplasty

A balloon-tipped catheter is inflated inside the artery to widen it at the blockage site.

EXAMPLE 8 Unclogging Arteries

In the late 1830s, French physiologist Jean Poiseuille ("pwa-ZOY") discovered the formula we use today to predict how much the radius of a partially clogged artery has to be expanded to restore normal flow. His formula,

$$V = kr^4,$$

says that the volume V of fluid flowing through a small pipe or tube in a unit of time at a fixed pressure is a constant times the fourth power of the tube's radius r. How will a 10% increase in r affect V?

Solution The differentials of r and V are related by the equation

$$dV = \frac{dV}{dr}\,dr = 4kr^3\,dr.$$

The relative change in V is

$$\frac{dV}{V} = \frac{4kr^3\,dr}{kr^4} = 4\frac{dr}{r}.$$

The relative change in V is 4 times the relative change in r, so a 10% increase in r will produce a 40% increase in the flow. ■

EXAMPLE 9 Converting Mass to Energy

Newton's second law,

$$F = \frac{d}{dt}(mv) = m\frac{dv}{dt} = ma,$$

is stated with the assumption that mass is constant, but we know this is not strictly true because the mass of a body increases with velocity. In Einstein's corrected formula, mass has the value

$$m = \frac{m_0}{\sqrt{1 - v^2/c^2}},$$

where the "rest mass" m_0 represents the mass of a body that is not moving and c is the speed of light, which is about 300,000 km/sec. Use the approximation

$$\frac{1}{\sqrt{1 - x^2}} \approx 1 + \frac{1}{2}x^2 \tag{2}$$

to estimate the increase Δm in mass resulting from the added velocity v.

Solution When v is very small compared with c, v^2/c^2 is close to zero and it is safe to use the approximation

$$\frac{1}{\sqrt{1 - v^2/c^2}} \approx 1 + \frac{1}{2}\left(\frac{v^2}{c^2}\right) \qquad \text{Eq. (2) with } x = \frac{v}{c}$$

to obtain

$$m = \frac{m_0}{\sqrt{1 - v^2/c^2}} \approx m_0\left[1 + \frac{1}{2}\left(\frac{v^2}{c^2}\right)\right] = m_0 + \frac{1}{2}m_0 v^2\left(\frac{1}{c^2}\right),$$

or

$$m \approx m_0 + \frac{1}{2}m_0 v^2\left(\frac{1}{c^2}\right). \tag{3}$$

Equation (3) expresses the increase in mass that results from the added velocity v.

Energy Interpretation

In Newtonian physics, $(1/2)m_0 v^2$ is the kinetic energy (KE) of the body, and if we rewrite Equation (3) in the form

$$(m - m_0)c^2 \approx \frac{1}{2}m_0 v^2,$$

we see that

$$(m - m_0)c^2 \approx \frac{1}{2}m_0 v^2 = \frac{1}{2}m_0 v^2 - \frac{1}{2}m_0(0)^2 = \Delta(\text{KE}),$$

or

$$(\Delta m)c^2 \approx \Delta(\text{KE}).$$

So the change in kinetic energy $\Delta(\text{KE})$ in going from velocity 0 to velocity v is approximately equal to $(\Delta m)c^2$, the change in mass times the square of the speed of light. Using $c \approx 3 \times 10^8$ m/sec, we see that a small change in mass can create a large change in energy. ■

EXERCISES 3.8

Finding Linearizations

In Exercises 1–4, find the linearization $L(x)$ of $f(x)$ at $x = a$.

1. $f(x) = x^3 - 2x + 3, \quad a = 2$
2. $f(x) = \sqrt{x^2 + 9}, \quad a = -4$
3. $f(x) = x + \frac{1}{x}, \quad a = 1$
4. $f(x) = \sqrt[3]{x}, \quad a = -8$

Linearization for Approximation

You want linearizations that will replace the functions in Exercises 5–10 over intervals that include the given points x_0. To make your subsequent work as simple as possible, you want to center each linearization not at x_0 but at a nearby integer $x = a$ at which the given function and its derivative are easy to evaluate. What linearization do you use in each case?

5. $f(x) = x^2 + 2x, \quad x_0 = 0.1$
6. $f(x) = x^{-1}, \quad x_0 = 0.9$
7. $f(x) = 2x^2 + 4x - 3, \quad x_0 = -0.9$
8. $f(x) = 1 + x, \quad x_0 = 8.1$
9. $f(x) = \sqrt[3]{x}, \quad x_0 = 8.5$
10. $f(x) = \frac{x}{x + 1}, \quad x_0 = 1.3$

Linearizing Trigonometric Functions

In Exercises 11–14, find the linearization of f at $x = a$. Then graph the linearization and f together.

11. $f(x) = \sin x$ at (a) $x = 0$, (b) $x = \pi$

12. $f(x) = \cos x$ at (a) $x = 0$, (b) $x = -\pi/2$

13. $f(x) = \sec x$ at (a) $x = 0$, (b) $x = -\pi/3$

14. $f(x) = \tan x$ at (a) $x = 0$, (b) $x = \pi/4$

The Approximation $(1 + x)^k \approx 1 + kx$

15. Show that the linearization of $f(x) = (1 + x)^k$ at $x = 0$ is $L(x) = 1 + kx$.

16. Use the linear approximation $(1 + x)^k \approx 1 + kx$ to find an approximation for the function $f(x)$ for values of x near zero.

a. $f(x) = (1 - x)^6$ **b.** $f(x) = \dfrac{2}{1 - x}$

c. $f(x) = \dfrac{1}{\sqrt{1 + x}}$ **d.** $f(x) = \sqrt{2 + x^2}$

e. $f(x) = (4 + 3x)^{1/3}$ **f.** $f(x) = \sqrt[3]{\left(1 - \dfrac{1}{2 + x}\right)^2}$

17. Faster than a calculator Use the approximation $(1 + x)^k \approx 1 + kx$ to estimate the following.

a. $(1.0002)^{50}$ **b.** $\sqrt[3]{1.009}$

18. Find the linearization of $f(x) = \sqrt{x + 1} + \sin x$ at $x = 0$. How is it related to the individual linearizations of $\sqrt{x + 1}$ and $\sin x$ at $x = 0$?

Derivatives in Differential Form

In Exercises 19–30, find dy.

19. $y = x^3 - 3\sqrt{x}$ **20.** $y = x\sqrt{1 - x^2}$

21. $y = \dfrac{2x}{1 + x^2}$ **22.** $y = \dfrac{2\sqrt{x}}{3(1 + \sqrt{x})}$

23. $2y^{3/2} + xy - x = 0$ **24.** $xy^2 - 4x^{3/2} - y = 0$

25. $y = \sin(5\sqrt{x})$ **26.** $y = \cos(x^2)$

27. $y = 4\tan(x^3/3)$ **28.** $y = \sec(x^2 - 1)$

29. $y = 3\csc(1 - 2\sqrt{x})$ **30.** $y = 2\cot\left(\dfrac{1}{\sqrt{x}}\right)$

Approximation Error

In Exercises 31–36, each function $f(x)$ changes value when x changes from x_0 to $x_0 + dx$. Find

a. the change $\Delta f = f(x_0 + dx) - f(x_0)$;

b. the value of the estimate $df = f'(x_0)\,dx$; and

c. the approximation error $|\Delta f - df|$.

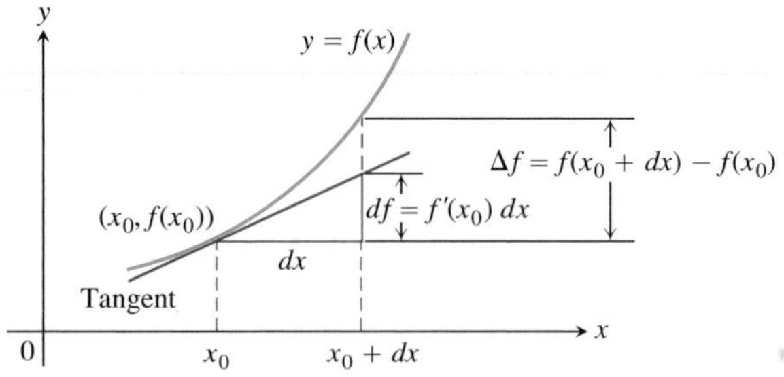

31. $f(x) = x^2 + 2x$, $x_0 = 1$, $dx = 0.1$

32. $f(x) = 2x^2 + 4x - 3$, $x_0 = -1$, $dx = 0.1$

33. $f(x) = x^3 - x$, $x_0 = 1$, $dx = 0.1$

34. $f(x) = x^4$, $x_0 = 1$, $dx = 0.1$

35. $f(x) = x^{-1}$, $x_0 = 0.5$, $dx = 0.1$

36. $f(x) = x^3 - 2x + 3$, $x_0 = 2$, $dx = 0.1$

Differential Estimates of Change

In Exercises 37–42, write a differential formula that estimates the given change in volume or surface area.

37. The change in the volume $V = (4/3)\pi r^3$ of a sphere when the radius changes from r_0 to $r_0 + dr$

38. The change in the volume $V = x^3$ of a cube when the edge lengths change from x_0 to $x_0 + dx$

39. The change in the surface area $S = 6x^2$ of a cube when the edge lengths change from x_0 to $x_0 + dx$

40. The change in the lateral surface area $S = \pi r\sqrt{r^2 + h^2}$ of a right circular cone when the radius changes from r_0 to $r_0 + dr$ and the height does not change

41. The change in the volume $V = \pi r^2 h$ of a right circular cylinder when the radius changes from r_0 to $r_0 + dr$ and the height does not change

42. The change in the lateral surface area $S = 2\pi rh$ of a right circular cylinder when the height changes from h_0 to $h_0 + dh$ and the radius does not change

Applications

43. The radius of a circle is increased from 2.00 to 2.02 m.

a. Estimate the resulting change in area.

b. Express the estimate as a percentage of the circle's original area.

44. The diameter of a tree was 10 in. During the following year, the circumference increased 2 in. About how much did the tree's diameter increase? The tree's cross-section area?

45. Estimating volume Estimate the volume of material in a cylindrical shell with height 30 in., radius 6 in., and shell thickness 0.5 in.

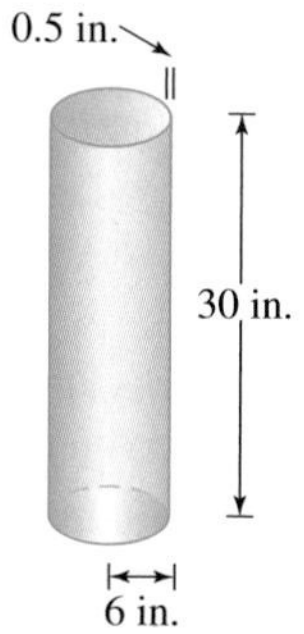

46. **Estimating height of a building** A surveyor, standing 30 ft from the base of a building, measures the angle of elevation to the top of the building to be 75°. How accurately must the angle be measured for the percentage error in estimating the height of the building to be less than 4%?

47. **Tolerance** The height and radius of a right circular cylinder are equal, so the cylinder's volume is $V = \pi h^3$. The volume is to be calculated with an error of no more than 1% of the true value. Find approximately the greatest error that can be tolerated in the measurement of h, expressed as a percentage of h.

48. **Tolerance**

    a. About how accurately must the interior diameter of a 10-m-high cylindrical storage tank be measured to calculate the tank's volume to within 1% of its true value?

    b. About how accurately must the tank's exterior diameter be measured to calculate the amount of paint it will take to paint the side of the tank to within 5% of the true amount?

49. **Minting coins** A manufacturer contracts to mint coins for the federal government. How much variation dr in the radius of the coins can be tolerated if the coins are to weigh within $1/1000$ of their ideal weight? Assume that the thickness does not vary.

50. **Sketching the change in a cube's volume** The volume $V = x^3$ of a cube with edges of length x increases by an amount ΔV when x increases by an amount Δx. Show with a sketch how to represent ΔV geometrically as the sum of the volumes of

    a. three slabs of dimensions x by x by Δx

    b. three bars of dimensions x by Δx by Δx

    c. one cube of dimensions Δx by Δx by Δx.

    The differential formula $dV = 3x^2\,dx$ estimates the change in V with the three slabs.

51. **The effect of flight maneuvers on the heart** The amount of work done by the heart's main pumping chamber, the left ventricle, is given by the equation

$$W = PV + \frac{V\delta v^2}{2g},$$

where W is the work per unit time, P is the average blood pressure, V is the volume of blood pumped out during the unit of time, δ ("delta") is the weight density of the blood, v is the average velocity of the exiting blood, and g is the acceleration of gravity.

When P, V, δ, and v remain constant, W becomes a function of g, and the equation takes the simplified form

$$W = a + \frac{b}{g} \quad (a, b \text{ constant}).$$

As a member of NASA's medical team, you want to know how sensitive W is to apparent changes in g caused by flight maneuvers, and this depends on the initial value of g. As part of your investigation, you decide to compare the effect on W of a given change dg on the moon, where $g = 5.2 \text{ ft/sec}^2$, with the effect the same change dg would have on Earth, where $g = 32 \text{ ft/sec}^2$. Use the simplified equation above to find the ratio of dW_{moon} to dW_{Earth}.

52. **Measuring acceleration of gravity** When the length L of a clock pendulum is held constant by controlling its temperature, the pendulum's period T depends on the acceleration of gravity g. The period will therefore vary slightly as the clock is moved from place to place on the earth's surface, depending on the change in g. By keeping track of ΔT, we can estimate the variation in g from the equation $T = 2\pi(L/g)^{1/2}$ that relates T, g, and L.

    a. With L held constant and g as the independent variable, calculate dT and use it to answer parts (b) and (c).

    b. If g increases, will T increase or decrease? Will a pendulum clock speed up or slow down? Explain.

    c. A clock with a 100-cm pendulum is moved from a location where $g = 980 \text{ cm/sec}^2$ to a new location. This increases the period by $dT = 0.001$ sec. Find dg and estimate the value of g at the new location.

53. The edge of a cube is measured as 10 cm with an error of 1%. The cube's volume is to be calculated from this measurement. Estimate the percentage error in the volume calculation.

54. About how accurately should you measure the side of a square to be sure of calculating the area within 2% of its true value?

55. The diameter of a sphere is measured as 100 ± 1 cm and the volume is calculated from this measurement. Estimate the percentage error in the volume calculation.

56. Estimate the allowable percentage error in measuring the diameter D of a sphere if the volume is to be calculated correctly to within 3%.

57. (*Continuation of Example 7.*) Show that a 5% error in measuring t will cause about a 10% error in calculating s from the equation $s = 16t^2$.

58. (*Continuation of Example 8.*) By what percentage should r be increased to increase V by 50%?

Theory and Examples

59. Show that the approximation of $\sqrt{1+x}$ by its linearization at the origin must improve as $x \to 0$ by showing that

$$\lim_{x \to 0} \frac{\sqrt{1+x}}{1 + (x/2)} = 1.$$

60. Show that the approximation of tan x by its linearization at the origin must improve as $x \to 0$ by showing that

$$\lim_{x\to 0} \frac{\tan x}{x} = 1.$$

61. The linearization is the best linear approximation (This is why we use the linearization.) Suppose that $y = f(x)$ is differentiable at $x = a$ and that $g(x) = m(x - a) + c$ is a linear function in which m and c are constants. If the error $E(x) = f(x) - g(x)$ were small enough near $x = a$, we might think of using g as a linear approximation of f instead of the linearization $L(x) = f(a) + f'(a)(x - a)$. Show that if we impose on g the conditions

1. $E(a) = 0$ — The approximation error is zero at $x = a$.
2. $\lim_{x\to a} \frac{E(x)}{x - a} = 0$ — The error is negligible when compared with $x - a$.

then $g(x) = f(a) + f'(a)(x - a)$. Thus, the linearization $L(x)$ gives the only linear approximation whose error is both zero at $x = a$ and negligible in comparison with $x - a$.

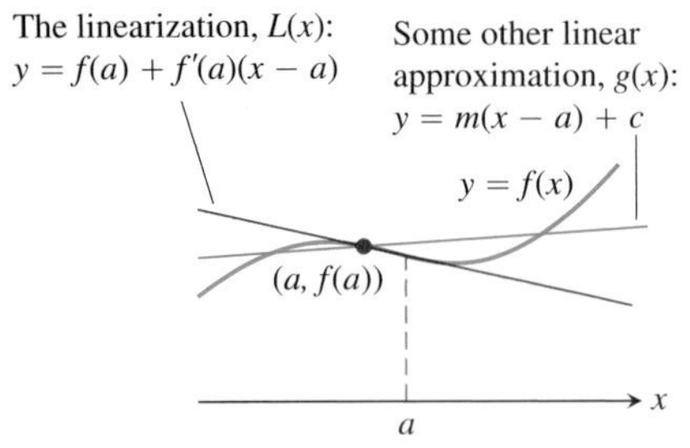

62. Quadratic approximations

a. Let $Q(x) = b_0 + b_1(x - a) + b_2(x - a)^2$ be a quadratic approximation to $f(x)$ at $x = a$ with the properties:

i. $Q(a) = f(a)$

ii. $Q'(a) = f'(a)$

iii. $Q''(a) = f''(a)$

Determine the coefficients b_0, b_1, and b_2.

b. Find the quadratic approximation to $f(x) = 1/(1 - x)$ at $x = 0$.

c. Graph $f(x) = 1/(1 - x)$ and its quadratic approximation at $x = 0$. Then zoom in on the two graphs at the point (0, 1). Comment on what you see.

d. Find the quadratic approximation to $g(x) = 1/x$ at $x = 1$. Graph g and its quadratic approximation together. Comment on what you see.

e. Find the quadratic approximation to $h(x) = \sqrt{1 + x}$ at $x = 0$. Graph h and its quadratic approximation together. Comment on what you see.

f. What are the linearizations of f, g, and h at the respective points in parts (b), (d), and (e)?

63. Reading derivatives from graphs The idea that differentiable curves flatten out when magnified can be used to estimate the values of the derivatives of functions at particular points. We magnify the curve until the portion we see looks like a straight line through the point in question, and then we use the screen's coordinate grid to read the slope of the curve as the slope of the line it resembles.

a. To see how the process works, try it first with the function $y = x^2$ at $x = 1$. The slope you read should be 2.

b. Then try it with the curve $y = e^x$ at $x = 1$, $x = 0$, and $x = -1$. In each case, compare your estimate of the derivative with the value of e^x at the point. What pattern do you see? Test it with other values of x. Chapter 7 will explain what is going on.

64. Suppose that the graph of a differentiable function $f(x)$ has a horizontal tangent at $x = a$. Can anything be said about the linearization of f at $x = a$? Give reasons for your answer.

65. To what relative speed should a body at rest be accelerated to increase its mass by 1%?

66. Repeated root-taking

a. Enter 2 in your calculator and take successive square roots by pressing the square root key repeatedly (or raising the displayed number repeatedly to the 0.5 power). What pattern do you see emerging? Explain what is going on. What happens if you take successive tenth roots instead?

b. Repeat the procedure with 0.5 in place of 2 as the original entry. What happens now? Can you use any positive number x in place of 2? Explain what is going on.

COMPUTER EXPLORATIONS

Comparing Functions with Their Linearizations

In Exercises 67–70, use a CAS to estimate the magnitude of the error in using the linearization in place of the function over a specified interval I. Perform the following steps:

a. Plot the function f over I.

b. Find the linearization L of the function at the point a.

c. Plot f and L together on a single graph.

d. Plot the absolute error $|f(x) - L(x)|$ over I and find its maximum value.

e. From your graph in part (d), estimate as large a $\delta > 0$ as you can, satisfying

$$|x - a| < \delta \quad \Rightarrow \quad |f(x) - L(x)| < \epsilon$$

for $\epsilon = 0.5, 0.1$, and 0.01. Then check graphically to see if your δ-estimate holds true.

67. $f(x) = x^3 + x^2 - 2x, \quad [-1, 2], \quad a = 1$

68. $f(x) = \frac{x - 1}{4x^2 + 1}, \quad \left[-\frac{3}{4}, 1\right], \quad a = \frac{1}{2}$

69. $f(x) = x^{2/3}(x - 2), \quad [-2, 3], \quad a = 2$

70. $f(x) = \sqrt{x} - \sin x, \quad [0, 2\pi], \quad a = 2$

Chapter 3 Questions to Guide Your Review

1. What is the derivative of a function f? How is its domain related to the domain of f? Give examples.
2. What role does the derivative play in defining slopes, tangents, and rates of change?
3. How can you sometimes graph the derivative of a function when all you have is a table of the function's values?
4. What does it mean for a function to be differentiable on an open interval? On a closed interval?
5. How are derivatives and one-sided derivatives related?
6. Describe geometrically when a function typically does *not* have a derivative at a point.
7. How is a function's differentiability at a point related to its continuity there, if at all?
8. Could the unit step function

$$U(x) = \begin{cases} 0, & x < 0 \\ 1, & x \geq 0 \end{cases}$$

possibly be the derivative of some other function on $[-1, 1]$? Explain.

9. What rules do you know for calculating derivatives? Give some examples.
10. Explain how the three formulas

a. $\frac{d}{dx}(x^n) = nx^{n-1}$

b. $\frac{d}{dx}(cu) = c\frac{du}{dx}$

c. $\frac{d}{dx}(u_1 + u_2 + \cdots + u_n) = \frac{du_1}{dx} + \frac{du_2}{dx} + \cdots + \frac{du_n}{dx}$

enable us to differentiate any polynomial.

11. What formula do we need, in addition to the three listed in Question 10, to differentiate rational functions?
12. What is a second derivative? A third derivative? How many derivatives do the functions you know have? Give examples.
13. What is the relationship between a function's average and instantaneous rates of change? Give an example.
14. How do derivatives arise in the study of motion? What can you learn about a body's motion along a line by examining the derivatives of the body's position function? Give examples.
15. How can derivatives arise in economics?
16. Give examples of still other applications of derivatives.
17. What do the limits $\lim_{h\to 0}((\sin h)/h)$ and $\lim_{h\to 0}((\cos h - 1)/h)$ have to do with the derivatives of the sine and cosine functions? What *are* the derivatives of these functions?
18. Once you know the derivatives of $\sin x$ and $\cos x$, how can you find the derivatives of $\tan x$, $\cot x$, $\sec x$, and $\csc x$? What *are* the derivatives of these functions?
19. At what points are the six basic trigonometric functions continuous? How do you know?
20. What is the rule for calculating the derivative of a composite of two differentiable functions? How is such a derivative evaluated? Give examples.
21. What is the formula for the slope dy/dx of a parametrized curve $x = f(t), y = g(t)$? When does the formula apply? When can you expect to be able to find d^2y/dx^2 as well? Give examples.
22. If u is a differentiable function of x, how do you find $(d/dx)(u^n)$ if n is an integer? If n is a rational number? Give examples.
23. What is implicit differentiation? When do you need it? Give examples.
24. How do related rates problems arise? Give examples.
25. Outline a strategy for solving related rates problems. Illustrate with an example.
26. What is the linearization $L(x)$ of a function $f(x)$ at a point $x = a$? What is required of f at a for the linearization to exist? How are linearizations used? Give examples.
27. If x moves from a to a nearby value $a + dx$, how do you estimate the corresponding change in the value of a differentiable function $f(x)$? How do you estimate the relative change? The percentage change? Give an example.

Chapter 3 Practice Exercises

Derivatives of Functions

Find the derivatives of the functions in Exercises 1-40.

1. $y = x^5 - 0.125x^2 + 0.25x$
2. $y = 3 - 0.7x^3 + 0.3x^7$
3. $y = x^3 - 3(x^2 + \pi^2)$
4. $y = x^7 + \sqrt{7x} - \frac{1}{\pi + 1}$
5. $y = (x + 1)^2(x^2 + 2x)$
6. $y = (2x - 5)(4 - x)^{-1}$
7. $y = (\theta^2 + \sec\theta + 1)^3$
8. $y = \left(-1 - \frac{\csc\theta}{2} - \frac{\theta^2}{4}\right)^2$
9. $s = \frac{\sqrt{t}}{1 + \sqrt{t}}$
10. $s = \frac{1}{\sqrt{t} - 1}$

11. $y = 2\tan^2 x - \sec^2 x$

12. $y = \dfrac{1}{\sin^2 x} - \dfrac{2}{\sin x}$

13. $s = \cos^4(1 - 2t)$

14. $s = \cot^3\left(\dfrac{2}{t}\right)$

15. $s = (\sec t + \tan t)^5$

16. $s = \csc^5(1 - t + 3t^2)$

17. $r = \sqrt{2\theta \sin\theta}$

18. $r = 2\theta\sqrt{\cos\theta}$

19. $r = \sin\sqrt{2\theta}$

20. $r = \sin\left(\theta + \sqrt{\theta + 1}\right)$

21. $y = \dfrac{1}{2}x^2 \csc\dfrac{2}{x}$

22. $y = 2\sqrt{x}\sin\sqrt{x}$

23. $y = x^{-1/2}\sec(2x)^2$

24. $y = \sqrt{x}\csc(x + 1)^3$

25. $y = 5\cot x^2$

26. $y = x^2\cot 5x$

27. $y = x^2\sin^2(2x^2)$

28. $y = x^{-2}\sin^2(x^3)$

29. $s = \left(\dfrac{4t}{t + 1}\right)^{-2}$

30. $s = \dfrac{-1}{15(15t - 1)^3}$

31. $y = \left(\dfrac{\sqrt{x}}{1 + x}\right)^2$

32. $y = \left(\dfrac{2\sqrt{x}}{2\sqrt{x} + 1}\right)^2$

33. $y = \sqrt{\dfrac{x^2 + x}{x^2}}$

34. $y = 4x\sqrt{x + \sqrt{x}}$

35. $r = \left(\dfrac{\sin\theta}{\cos\theta - 1}\right)^2$

36. $r = \left(\dfrac{1 + \sin\theta}{1 - \cos\theta}\right)^2$

37. $y = (2x + 1)\sqrt{2x + 1}$

38. $y = 20(3x - 4)^{1/4}(3x - 4)^{-1/5}$

39. $y = \dfrac{3}{(5x^2 + \sin 2x)^{3/2}}$

40. $y = (3 + \cos^3 3x)^{-1/3}$

Implicit Differentiation

In Exercises 41–48, find dy/dx.

41. $xy + 2x + 3y = 1$

42. $x^2 + xy + y^2 - 5x = 2$

43. $x^3 + 4xy - 3y^{4/3} = 2x$

44. $5x^{4/5} + 10y^{6/5} = 15$

45. $\sqrt{xy} = 1$

46. $x^2y^2 = 1$

47. $y^2 = \dfrac{x}{x + 1}$

48. $y^2 = \sqrt{\dfrac{1 + x}{1 - x}}$

In Exercises 49 and 50, find dp/dq.

49. $p^3 + 4pq - 3q^2 = 2$

50. $q = (5p^2 + 2p)^{-3/2}$

In Exercises 51 and 52, find dr/ds.

51. $r\cos 2s + \sin^2 s = \pi$

52. $2rs - r - s + s^2 = -3$

53. Find d^2y/dx^2 by implicit differentiation:
   a. $x^3 + y^3 = 1$
   b. $y^2 = 1 - \dfrac{2}{x}$

54. a. By differentiating $x^2 - y^2 = 1$ implicitly, show that $dy/dx = x/y$.
   b. Then show that $d^2y/dx^2 = -1/y^3$.

Numerical Values of Derivatives

55. Suppose that functions $f(x)$ and $g(x)$ and their first derivatives have the following values at $x = 0$ and $x = 1$.

x	$f(x)$	$g(x)$	$f'(x)$	$g'(x)$
0	1	1	−3	1/2
1	3	5	1/2	−4

Find the first derivatives of the following combinations at the given value of x.

a. $6f(x) - g(x), \quad x = 1$
b. $f(x)g^2(x), \quad x = 0$
c. $\dfrac{f(x)}{g(x) + 1}, \quad x = 1$
d. $f(g(x)), \quad x = 0$
e. $g(f(x)), \quad x = 0$
f. $(x + f(x))^{3/2}, \quad x = 1$
g. $f(x + g(x)), \quad x = 0$

56. Suppose that the function $f(x)$ and its first derivative have the following values at $x = 0$ and $x = 1$.

x	$f(x)$	$f'(x)$
0	9	−2
1	−3	1/5

Find the first derivatives of the following combinations at the given value of x.

a. $\sqrt{x}\,f(x), \quad x = 1$
b. $\sqrt{f(x)}, \quad x = 0$
c. $f(\sqrt{x}), \quad x = 1$
d. $f(1 - 5\tan x), \quad x = 0$
e. $\dfrac{f(x)}{2 + \cos x}, \quad x = 0$
f. $10\sin\left(\dfrac{\pi x}{2}\right)f^2(x), \quad x = 1$

57. Find the value of dy/dt at $t = 0$ if $y = 3\sin 2x$ and $x = t^2 + \pi$.

58. Find the value of ds/du at $u = 2$ if $s = t^2 + 5t$ and $t = (u^2 + 2u)^{1/3}$.

59. Find the value of dw/ds at $s = 0$ if $w = \sin\left(\sqrt{r} - 2\right)$ and $r = 8\sin(s + \pi/6)$.

60. Find the value of dr/dt at $t = 0$ if $r = (\theta^2 + 7)^{1/3}$ and $\theta^2 t + \theta = 1$.

61. If $y^3 + y = 2\cos x$, find the value of d^2y/dx^2 at the point (0, 1).

62. If $x^{1/3} + y^{1/3} = 4$, find d^2y/dx^2 at the point (8, 8).

Derivative Definition

In Exercises 63 and 64, find the derivative using the definition.

63. $f(t) = \dfrac{1}{2t + 1}$

64. $g(x) = 2x^2 + 1$

65. a. Graph the function

$$f(x) = \begin{cases} x^2, & -1 \le x < 0 \\ -x^2, & 0 \le x \le 1. \end{cases}$$

   b. Is f continuous at $x = 0$?
   c. Is f differentiable at $x = 0$?

Give reasons for your answers.

66. a. Graph the function

$$f(x) = \begin{cases} x, & -1 \le x < 0 \\ \tan x, & 0 \le x \le \pi/4. \end{cases}$$

b. Is f continuous at $x = 0$?

c. Is f differentiable at $x = 0$?

Give reasons for your answers.

67. a. Graph the function

$$f(x) = \begin{cases} x, & 0 \le x \le 1 \\ 2 - x, & 1 < x \le 2. \end{cases}$$

b. Is f continuous at $x = 1$?

c. Is f differentiable at $x = 1$?

Give reasons for your answers.

68. For what value or values of the constant m, if any, is

$$f(x) = \begin{cases} \sin 2x, & x \le 0 \\ mx, & x > 0 \end{cases}$$

a. continuous at $x = 0$?

b. differentiable at $x = 0$?

Give reasons for your answers.

Slopes, Tangents, and Normals

69. Tangents with specified slope Are there any points on the curve $y = (x/2) + 1/(2x - 4)$ where the slope is $-3/2$? If so, find them.

70. Tangents with specified slope Are there any points on the curve $y = x - 1/(2x)$ where the slope is 3? If so, find them.

71. Horizontal tangents Find the points on the curve $y = 2x^3 - 3x^2 - 12x + 20$ where the tangent is parallel to the x-axis.

72. Tangent intercepts Find the x- and y-intercepts of the line that is tangent to the curve $y = x^3$ at the point $(-2, -8)$.

73. Tangents perpendicular or parallel to lines Find the points on the curve $y = 2x^3 - 3x^2 - 12x + 20$ where the tangent is

a. perpendicular to the line $y = 1 - (x/24)$.

b. parallel to the line $y = \sqrt{2} - 12x$.

74. Intersecting tangents Show that the tangents to the curve $y = (\pi \sin x)/x$ at $x = \pi$ and $x = -\pi$ intersect at right angles.

75. Normals parallel to a line Find the points on the curve $y = \tan x$, $-\pi/2 < x < \pi/2$, where the normal is parallel to the line $y = -x/2$. Sketch the curve and normals together, labeling each with its equation.

76. Tangent and normal lines Find equations for the tangent and normal to the curve $y = 1 + \cos x$ at the point $(\pi/2, 1)$. Sketch the curve, tangent, and normal together, labeling each with its equation.

77. Tangent parabola The parabola $y = x^2 + C$ is to be tangent to the line $y = x$. Find C.

78. Slope of tangent Show that the tangent to the curve $y = x^3$ at any point (a, a^3) meets the curve again at a point where the slope is four times the slope at (a, a^3).

79. Tangent curve For what value of c is the curve $y = c/(x + 1)$ tangent to the line through the points $(0, 3)$ and $(5, -2)$?

80. Normal to a circle Show that the normal line at any point of the circle $x^2 + y^2 = a^2$ passes through the origin.

Tangents and Normals to Implicitly Defined Curves

In Exercises 81–86, find equations for the lines that are tangent and normal to the curve at the given point.

81. $x^2 + 2y^2 = 9, \quad (1, 2)$

82. $x^3 + y^2 = 2, \quad (1, 1)$

83. $xy + 2x - 5y = 2, \quad (3, 2)$

84. $(y - x)^2 = 2x + 4, \quad (6, 2)$

85. $x + \sqrt{xy} = 6, \quad (4, 1)$

86. $x^{3/2} + 2y^{3/2} = 17, \quad (1, 4)$

87. Find the slope of the curve $x^3y^3 + y^2 = x + y$ at the points $(1, 1)$ and $(1, -1)$.

88. The graph shown suggests that the curve $y = \sin(x - \sin x)$ might have horizontal tangents at the x-axis. Does it? Give reasons for your answer.

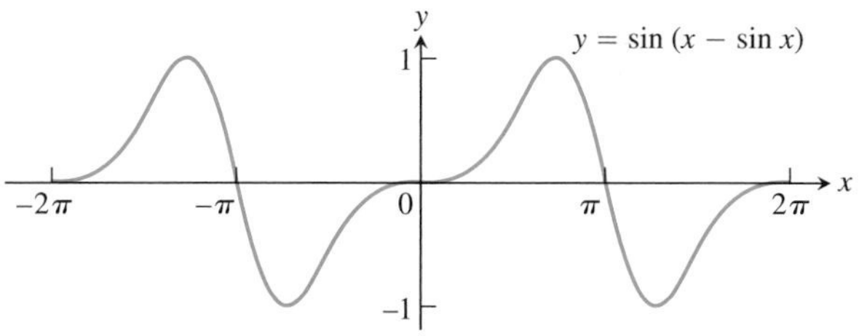

Tangents to Parametrized Curves

In Exercises 89 and 90, find an equation for the line in the xy-plane that is tangent to the curve at the point corresponding to the given value of t. Also, find the value of d^2y/dx^2 at this point.

89. $x = (1/2)\tan t, \quad y = (1/2)\sec t, \quad t = \pi/3$

90. $x = 1 + 1/t^2, \quad y = 1 - 3/t, \quad t = 2$

Analyzing Graphs

Each of the figures in Exercises 91 and 92 shows two graphs, the graph of a function $y = f(x)$ together with the graph of its derivative $f'(x)$. Which graph is which? How do you know?

91.

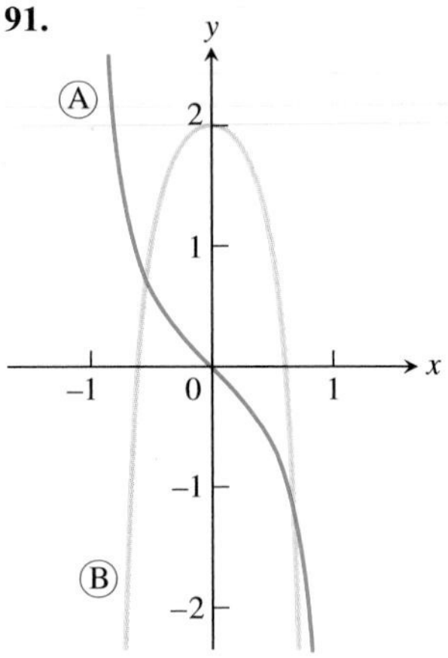

92.

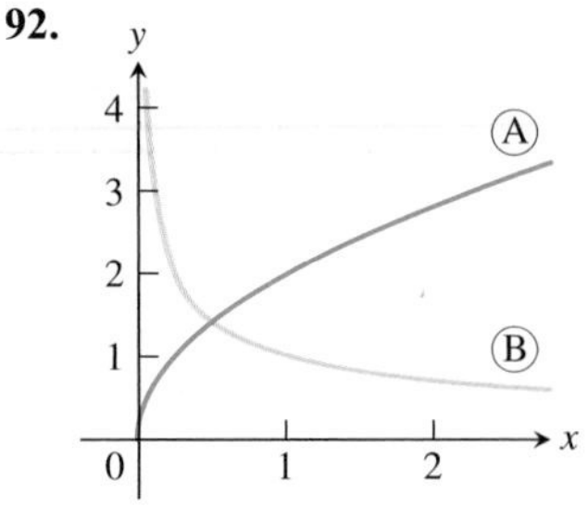

93. Use the following information to graph the function $y = f(x)$ for $-1 \le x \le 6$.

i. The graph of f is made of line segments joined end to end.

ii. The graph starts at the point $(-1, 2)$.

iii. The derivative of f, where defined, agrees with the step function shown here.

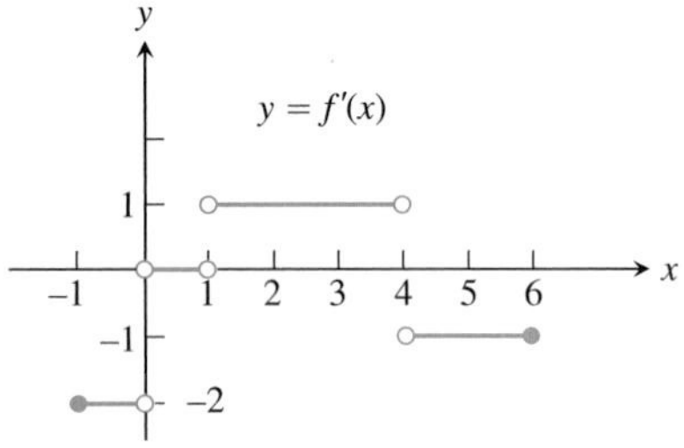

94. Repeat Exercise 93, supposing that the graph starts at $(-1, 0)$ instead of $(-1, 2)$.

Exercises 95 and 96 are about the graphs in Figure 3.53 (right-hand column). The graphs in part (a) show the numbers of rabbits and foxes in a small arctic population. They are plotted as functions of time for 200 days. The number of rabbits increases at first, as the rabbits reproduce. But the foxes prey on rabbits and, as the number of foxes increases, the rabbit population levels off and then drops. Figure 3.53b shows the graph of the derivative of the rabbit population. We made it by plotting slopes.

95. a. What is the value of the derivative of the rabbit population in Figure 3.53 when the number of rabbits is largest? Smallest?

b. What is the size of the rabbit population in Figure 3.53 when its derivative is largest? Smallest (negative value)?

96. In what units should the slopes of the rabbit and fox population curves be measured?

Trigonometric Limits

97. $\lim_{x\to 0} \dfrac{\sin x}{2x^2 - x}$

98. $\lim_{x\to 0} \dfrac{3x - \tan 7x}{2x}$

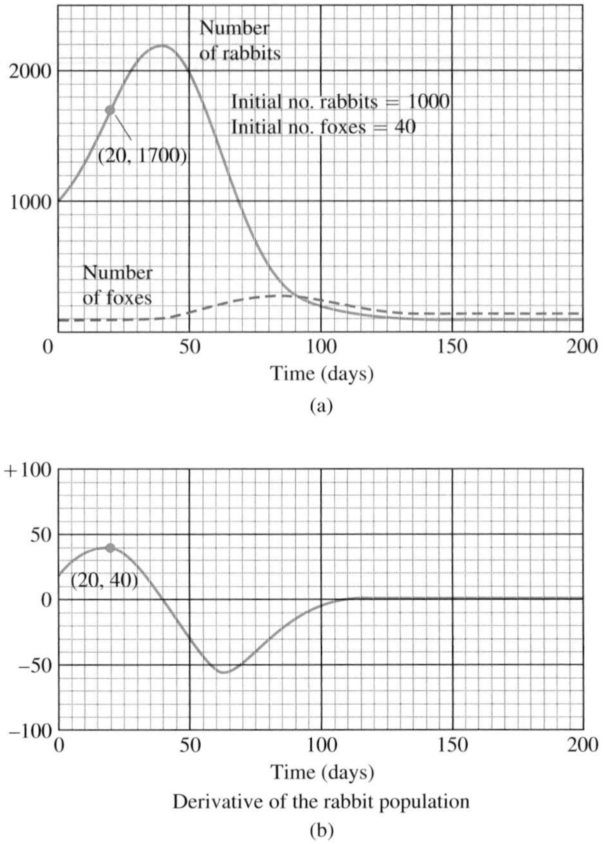

FIGURE 3.53 Rabbits and foxes in an arctic predator-prey food chain.

99. $\lim_{r\to 0} \dfrac{\sin r}{\tan 2r}$

100. $\lim_{\theta\to 0} \dfrac{\sin(\sin\theta)}{\theta}$

101. $\lim_{\theta\to(\pi/2)^-} \dfrac{4\tan^2\theta + \tan\theta + 1}{\tan^2\theta + 5}$

102. $\lim_{\theta\to 0^+} \dfrac{1 - 2\cot^2\theta}{5\cot^2\theta - 7\cot\theta - 8}$

103. $\lim_{x\to 0} \dfrac{x\sin x}{2 - 2\cos x}$

104. $\lim_{\theta\to 0} \dfrac{1 - \cos\theta}{\theta^2}$

Show how to extend the functions in Exercises 105 and 106 to be continuous at the origin.

105. $g(x) = \dfrac{\tan(\tan x)}{\tan x}$

106. $f(x) = \dfrac{\tan(\tan x)}{\sin(\sin x)}$

Related Rates

107. Right circular cylinder The total surface area S of a right circular cylinder is related to the base radius r and height h by the equation $S = 2\pi r^2 + 2\pi rh$.

a. How is dS/dt related to dr/dt if h is constant?

b. How is dS/dt related to dh/dt if r is constant?

c. How is dS/dt related to dr/dt and dh/dt if neither r nor h is constant?

d. How is dr/dt related to dh/dt if S is constant?

108. Right circular cone The lateral surface area S of a right circular cone is related to the base radius r and height h by the equation $S = \pi r\sqrt{r^2 + h^2}$.

a. How is dS/dt related to dr/dt if h is constant?

b. How is dS/dt related to dh/dt if r is constant?

c. How is dS/dt related to dr/dt and dh/dt if neither r nor h is constant?

109. Circle's changing area The radius of a circle is changing at the rate of $-2/\pi$ m/sec. At what rate is the circle's area changing when $r = 10$ m?

110. Cube's changing edges The volume of a cube is increasing at the rate of 1200 cm^3/min at the instant its edges are 20 cm long. At what rate are the lengths of the edges changing at that instant?

111. Resistors connected in parallel If two resistors of R_1 and R_2 ohms are connected in parallel in an electric circuit to make an R-ohm resistor, the value of R can be found from the equation

$$\frac{1}{R} = \frac{1}{R_1} + \frac{1}{R_2}.$$

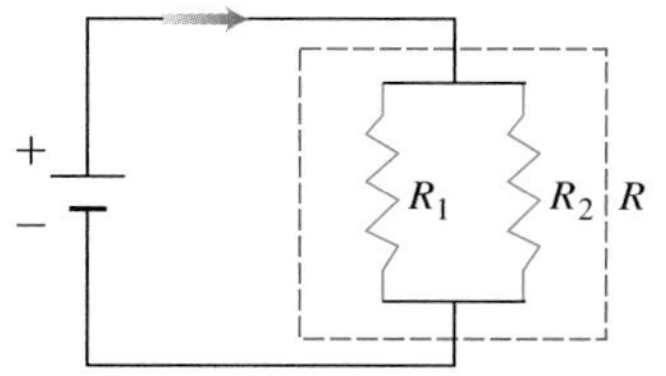

If R_1 is decreasing at the rate of 1 ohm/sec and R_2 is increasing at the rate of 0.5 ohm/sec, at what rate is R changing when $R_1 = 75$ ohms and $R_2 = 50$ ohms?

112. Impedance in a series circuit The impedance Z (ohms) in a series circuit is related to the resistance R (ohms) and reactance X (ohms) by the equation $Z = \sqrt{R^2 + X^2}$. If R is increasing at 3 ohms/sec and X is decreasing at 2 ohms/sec, at what rate is Z changing when $R = 10$ ohms and $X = 20$ ohms?

113. Speed of moving particle The coordinates of a particle moving in the metric xy-plane are differentiable functions of time t with $dx/dt = 10$ m/sec and $dy/dt = 5$ m/sec. How fast is the particle moving away from the origin as it passes through the point $(3, -4)$?

114. Motion of a particle A particle moves along the curve $y = x^{3/2}$ in the first quadrant in such a way that its distance from the origin increases at the rate of 11 units per second. Find dx/dt when $x = 3$.

115. Draining a tank Water drains from the conical tank shown in the accompanying figure at the rate of 5 ft^3/min.

a. What is the relation between the variables h and r in the figure?

b. How fast is the water level dropping when $h = 6$ ft?

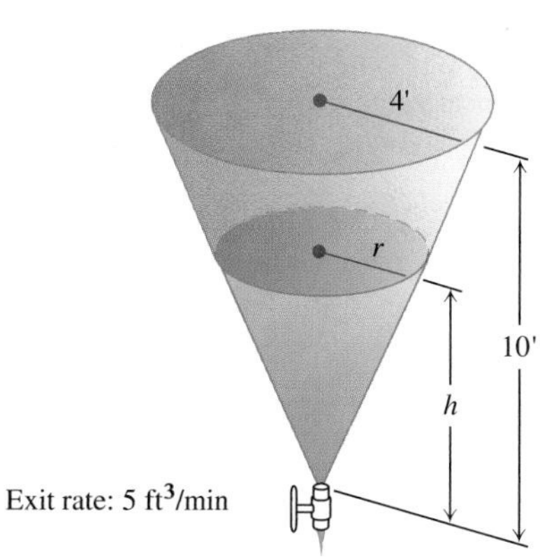

116. Rotating spool As television cable is pulled from a large spool to be strung from the telephone poles along a street, it unwinds from the spool in layers of constant radius (see accompanying figure). If the truck pulling the cable moves at a steady 6 ft/sec (a touch over 4 mph), use the equation $s = r\theta$ to find how fast (radians per second) the spool is turning when the layer of radius 1.2 ft is being unwound.

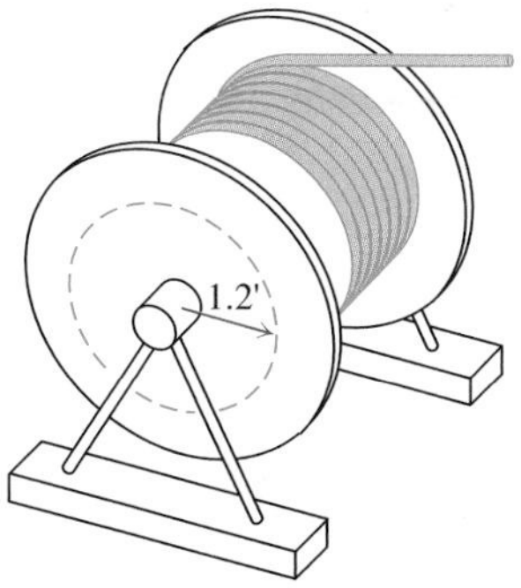

117. Moving searchlight beam The figure shows a boat 1 km offshore, sweeping the shore with a searchlight. The light turns at a constant rate, $d\theta/dt = -0.6$ rad/sec.

a. How fast is the light moving along the shore when it reaches point A?

b. How many revolutions per minute is 0.6 rad/sec?

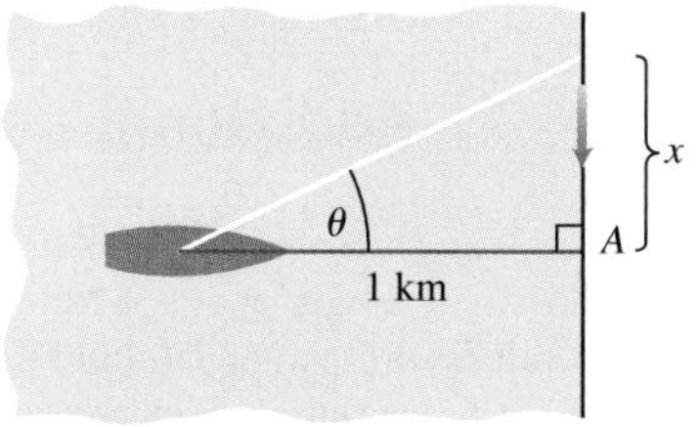

118. Points moving on coordinate axes Points A and B move along the x- and y-axes, respectively, in such a way that the distance r (meters) along the perpendicular from the origin to the line AB remains constant. How fast is OA changing, and is it increasing, or decreasing, when $OB = 2r$ and B is moving toward O at the rate of $0.3r$ m/sec?

Linearization

119. Find the linearizations of

a. $\tan x$ at $x = -\pi/4$ **b.** $\sec x$ at $x = -\pi/4$.

Graph the curves and linearizations together.

120. We can obtain a useful linear approximation of the function $f(x) = 1/(1 + \tan x)$ at $x = 0$ by combining the approximations

$$\frac{1}{1+x} \approx 1 - x \quad \text{and} \quad \tan x \approx x$$

to get

$$\frac{1}{1+\tan x} \approx 1 - x.$$

Show that this result is the standard linear approximation of $1/(1 + \tan x)$ at $x = 0$.

121. Find the linearization of $f(x) = \sqrt{1+x} + \sin x - 0.5$ at $x = 0$.

122. Find the linearization of $f(x) = 2/(1 - x) + \sqrt{1+x} - 3.1$ at $x = 0$.

Differential Estimates of Change

123. Surface area of a cone Write a formula that estimates the change that occurs in the lateral surface area of a right circular cone when the height changes from h_0 to $h_0 + dh$ and the radius does not change.

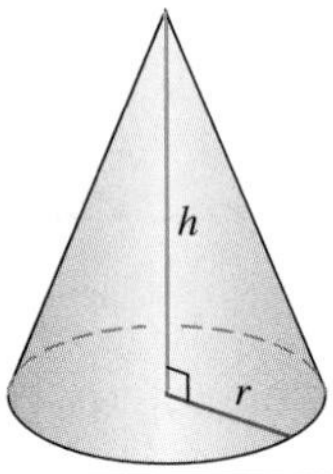

$V = \frac{1}{3}\pi r^2 h$

$S = \pi r\sqrt{r^2 + h^2}$

(Lateral surface area)

124. Controlling error

a. How accurately should you measure the edge of a cube to be reasonably sure of calculating the cube's surface area with an error of no more than 2%?

b. Suppose that the edge is measured with the accuracy required in part (a). About how accurately can the cube's volume be calculated from the edge measurement? To find out, estimate the percentage error in the volume calculation that might result from using the edge measurement.

125. Compounding error The circumference of the equator of a sphere is measured as 10 cm with a possible error of 0.4 cm. This measurement is then used to calculate the radius. The radius is then used to calculate the surface area and volume of the sphere. Estimate the percentage errors in the calculated values of

a. the radius.

b. the surface area.

c. the volume.

126. Finding height To find the height of a lamppost (see accompanying figure), you stand a 6 ft pole 20 ft from the lamp and measure the length a of its shadow, finding it to be 15 ft, give or take an inch. Calculate the height of the lamppost using the value $a = 15$ and estimate the possible error in the result.

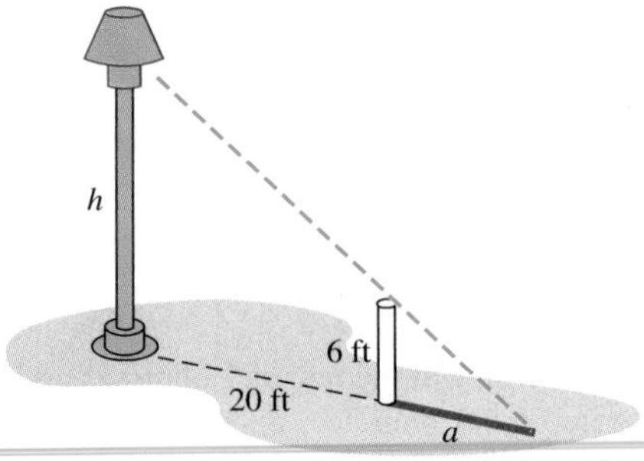

Chapter 3 Additional and Advanced Exercises

1. An equation like $\sin^2\theta + \cos^2\theta = 1$ is called an **identity** because it holds for all values of θ. An equation like $\sin\theta = 0.5$ is not an identity because it holds only for selected values of θ, not all. If you differentiate both sides of a trigonometric identity in θ with respect to θ, the resulting new equation will also be an identity.

Differentiate the following to show that the resulting equations hold for all θ.

a. $\sin 2\theta = 2\sin\theta\cos\theta$

b. $\cos 2\theta = \cos^2\theta - \sin^2\theta$

2. If the identity $\sin(x + a) = \sin x\cos a + \cos x\sin a$ is differentiated with respect to x, is the resulting equation also an identity? Does this principle apply to the equation $x^2 - 2x - 8 = 0$? Explain.

3. a. Find values for the constants a, b, and c that will make

$$f(x) = \cos x \quad \text{and} \quad g(x) = a + bx + cx^2$$

satisfy the conditions

$$f(0) = g(0), \quad f'(0) = g'(0), \quad \text{and} \quad f''(0) = g''(0).$$

b. Find values for b and c that will make

$$f(x) = \sin(x + a) \quad \text{and} \quad g(x) = b\sin x + c\cos x$$

satisfy the conditions

$$f(0) = g(0) \quad \text{and} \quad f'(0) = g'(0).$$

c. For the determined values of a, b, and c, what happens for the third and fourth derivatives of f and g in each of parts (a) and (b)?

4. Solutions to differential equations

a. Show that $y = \sin x$, $y = \cos x$, and $y = a\cos x + b\sin x$ (a and b constants) all satisfy the equation

$$y'' + y = 0.$$

b. How would you modify the functions in part (a) to satisfy the equation

$$y'' + 4y = 0?$$

Generalize this result.

5. An osculating circle Find the values of h, k, and a that make the circle $(x - h)^2 + (y - k)^2 = a^2$ tangent to the parabola $y = x^2 + 1$ at the point (1, 2) and that also make the second derivatives d^2y/dx^2 have the same value on both curves there. Circles like this one that are tangent to a curve and have the same second derivative as the curve at the point of tangency are called *osculating circles* (from the Latin *osculari*, meaning "to kiss"). We encounter them again in Chapter 13.

6. Marginal revenue A bus will hold 60 people. The number x of people per trip who use the bus is related to the fare charged (p dollars) by the law $p = [3 - (x/40)]^2$. Write an expression for the total revenue $r(x)$ per trip received by the bus company. What number of people per trip will make the marginal revenue dr/dx equal to zero? What is the corresponding fare? (This fare is the one that maximizes the revenue, so the bus company should probably rethink its fare policy.)

7. Industrial production

a. Economists often use the expression "rate of growth" in relative rather than absolute terms. For example, let $u = f(t)$ be the number of people in the labor force at time t in a given industry. (We treat this function as though it were differentiable even though it is an integer-valued step function.)

Let $v = g(t)$ be the average production per person in the labor force at time t. The total production is then $y = uv$. If the labor force is growing at the rate of 4% per year ($du/dt = 0.04u$) and the production per worker is growing at the rate of 5% per year ($dv/dt = 0.05v$), find the rate of growth of the total production, y.

b. Suppose that the labor force in part (a) is decreasing at the rate of 2% per year while the production per person is increasing at the rate of 3% per year. Is the total production increasing, or is it decreasing, and at what rate?

8. Designing a gondola The designer of a 30-ft-diameter spherical hot air balloon wants to suspend the gondola 8 ft below the bottom of the balloon with cables tangent to the surface of the balloon, as shown. Two of the cables are shown running from the top edges of the gondola to their points of tangency, $(-12, -9)$ and $(12, -9)$. How wide should the gondola be?

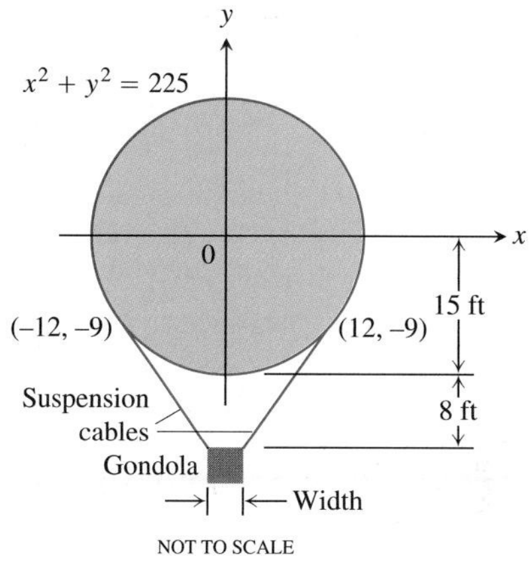

9. Pisa by parachute The photograph shows Mike McCarthy parachuting from the top of the Tower of Pisa on August 5, 1988. Make a rough sketch to show the shape of the graph of his speed during the jump.

Mike McCarthy of London jumped from the Tower of Pisa and then opened his parachute in what he said was a world record low-level parachute jump of 179 ft. (*Source: Boston Globe*, Aug. 6, 1988.)

10. Motion of a particle The position at time $t \geq 0$ of a particle moving along a coordinate line is

$$s = 10\cos(t + \pi/4).$$

a. What is the particle's starting position ($t = 0$)?

b. What are the points farthest to the left and right of the origin reached by the particle?

c. Find the particle's velocity and acceleration at the points in part (b).

d. When does the particle first reach the origin? What are its velocity, speed, and acceleration then?

11. Shooting a paper clip On Earth, you can easily shoot a paper clip 64 ft straight up into the air with a rubber band. In t sec after firing, the paper clip is $s = 64t - 16t^2$ ft above your hand.

a. How long does it take the paper clip to reach its maximum height? With what velocity does it leave your hand?

b. On the moon, the same acceleration will send the paper clip to a height of $s = 64t - 2.6t^2$ ft in t sec. About how long will it take the paper clip to reach its maximum height, and how high will it go?

12. Velocities of two particles At time t sec, the positions of two particles on a coordinate line are $s_1 = 3t^3 - 12t^2 + 18t + 5$ m and $s_2 = -t^3 + 9t^2 - 12t$ m. When do the particles have the same velocities?

13. Velocity of a particle A particle of constant mass m moves along the x-axis. Its velocity v and position x satisfy the equation

$$\frac{1}{2}m(v^2 - v_0{}^2) = \frac{1}{2}k(x_0{}^2 - x^2),$$

where k, v_0, and x_0 are constants. Show that whenever $v \neq 0$,

$$m\frac{dv}{dt} = -kx.$$

14. Average and instantaneous velocity

a. Show that if the position x of a moving point is given by a quadratic function of t, $x = At^2 + Bt + C$, then the average velocity over any time interval $[t_1, t_2]$ is equal to the instantaneous velocity at the midpoint of the time interval.

b. What is the geometric significance of the result in part (a)?

15. Find all values of the constants m and b for which the function

$$y = \begin{cases} \sin x, & x < \pi \\ mx + b, & x \geq \pi \end{cases}$$

is

a. continuous at $x = \pi$.

b. differentiable at $x = \pi$.

16. Does the function

$$f(x) = \begin{cases} \dfrac{1 - \cos x}{x}, & x \neq 0 \\ 0, & x = 0 \end{cases}$$

have a derivative at $x = 0$? Explain.

17. a. For what values of a and b will

$$f(x) = \begin{cases} ax, & x < 2 \\ ax^2 - bx + 3, & x \geq 2 \end{cases}$$

be differentiable for all values of x?

b. Discuss the geometry of the resulting graph of f.

18. a. For what values of a and b will

$$g(x) = \begin{cases} ax + b, & x \leq -1 \\ ax^3 + x + 2b, & x > -1 \end{cases}$$

be differentiable for all values of x?

b. Discuss the geometry of the resulting graph of g.

19. Odd differentiable functions Is there anything special about the derivative of an odd differentiable function of x? Give reasons for your answer.

20. Even differentiable functions Is there anything special about the derivative of an even differentiable function of x? Give reasons for your answer.

21. Suppose that the functions f and g are defined throughout an open interval containing the point x_0, that f is differentiable at x_0, that $f(x_0) = 0$, and that g is continuous at x_0. Show that the product fg is differentiable at x_0. This process shows, for example, that although $|x|$ is not differentiable at $x = 0$, the product $x|x|$ *is* differentiable at $x = 0$.

22. *(Continuation of Exercise 21.)* Use the result of Exercise 21 to show that the following functions are differentiable at $x = 0$.

a. $|x|\sin x$ **b.** $x^{2/3}\sin x$ **c.** $\sqrt[3]{x}(1 - \cos x)$

d. $h(x) = \begin{cases} x^2\sin(1/x), & x \neq 0 \\ 0, & x = 0 \end{cases}$

23. Is the derivative of

$$h(x) = \begin{cases} x^2\sin(1/x), & x \neq 0 \\ 0, & x = 0 \end{cases}$$

continuous at $x = 0$? How about the derivative of $k(x) = xh(x)$? Give reasons for your answers.

24. Suppose that a function f satisfies the following conditions for all real values of x and y:

i. $f(x + y) = f(x) \cdot f(y)$.

ii. $f(x) = 1 + xg(x)$, where $\lim_{x \to 0} g(x) = 1$.

Show that the derivative $f'(x)$ exists at every value of x and that $f'(x) = f(x)$.

25. The generalized product rule Use mathematical induction to prove that if $y = u_1u_2 \cdots u_n$ is a finite product of differentiable functions, then y is differentiable on their common domain and

$$\frac{dy}{dx} = \frac{du_1}{dx}u_2 \cdots u_n + u_1\frac{du_2}{dx} \cdots u_n + \cdots + u_1u_2 \cdots u_{n-1}\frac{du_n}{dx}.$$

26. Leibniz's rule for higher-order derivatives of products Leibniz's rule for higher-order derivatives of products of differentiable functions says that

a. $$\frac{d^2(uv)}{dx^2} = \frac{d^2u}{dx^2}v + 2\frac{du}{dx}\frac{dv}{dx} + u\frac{d^2v}{dx^2}$$

b. $$\frac{d^3(uv)}{dx^3} = \frac{d^3u}{dx^3}v + 3\frac{d^2u}{dx^2}\frac{dv}{dx} + 3\frac{du}{dx}\frac{d^2v}{dx^2} + u\frac{d^3v}{dx^3}$$

c. $$\frac{d^n(uv)}{dx^n} = \frac{d^nu}{dx^n}v + n\frac{d^{n-1}u}{dx^{n-1}}\frac{dv}{dx} + \cdots + \frac{n(n-1)\cdots(n-k+1)}{k!}\frac{d^{n-k}u}{dx^{n-k}}\frac{d^kv}{dx^k} + \cdots + u\frac{d^nv}{dx^n}.$$

The equations in parts (a) and (b) are special cases of the equation in part (c). Derive the equation in part (c) by mathematical induction, using

$$\binom{m}{k} + \binom{m}{k+1} = \frac{m!}{k!(m-k)!} + \frac{m!}{(k+1)!(m-k-1)!}.$$

27. The period of a clock pendulum The period T of a clock pendulum (time for one full swing and back) is given by the formula $T^2 = 4\pi^2 L/g$, where T is measured in seconds, $g = 32.2$ ft/sec^2, and L, the length of the pendulum, is measured in feet. Find approximately

a. the length of a clock pendulum whose period is $T = 1$ sec.

b. the change dT in T if the pendulum in part (a) is lengthened 0.01 ft.

c. the amount the clock gains or loses in a day as a result of the period's changing by the amount dT found in part (b).

28. The melting ice cube Assume an ice cube retains its cubical shape as it melts. If we call its edge length s, its volume is $V = s^3$ and its surface area is $6s^2$. We assume that V and s are differentiable functions of time t. We assume also that the cube's volume decreases at a rate that is proportional to its surface area. (This latter assumption seems reasonable enough when we think that the melting takes place at the surface: Changing the amount of surface changes the amount of ice exposed to melt.) In mathematical terms,

$$\frac{dV}{dt} = -k(6s^2), \qquad k > 0.$$

The minus sign indicates that the volume is decreasing. We assume that the proportionality factor k is constant. (It probably depends on many things, such as the relative humidity of the surrounding air, the air temperature, and the incidence or absence of sunlight, to name only a few.) Assume a particular set of conditions in which the cube lost $1/4$ of its volume during the first hour, and that the volume is V_0 when $t = 0$. How long will it take the ice cube to melt?

Chapter 3 Technology Application Projects

Mathematica/Maple Module
Convergence of Secant Slopes to the Derivative Function

You will visualize the secant line between successive points on a curve and observe what happens as the distance between them becomes small. The function, sample points, and secant lines are plotted on a single graph, while a second graph compares the slopes of the secant lines with the derivative function.

Mathematica/Maple Module
Derivatives, Slopes, Tangent Lines, and Making Movies

Parts I–III. You will visualize the derivative at a point, the linearization of a function, and the derivative of a function. You learn how to plot the function and selected tangents on the same graph.

Part IV (Plotting Many Tangents)

Part V (Making Movies). Parts IV and V of the module can be used to animate tangent lines as one moves along the graph of a function.

Mathematica/Maple Module
Convergence of Secant Slopes to the Derivative Function

You will visualize right-hand and left-hand derivatives.

Mathematica/Maple Module
Motion Along a Straight Line: Position → Velocity → Acceleration

Observe dramatic animated visualizations of the derivative relations among the position, velocity, and acceleration functions. Figures in the text can be animated.

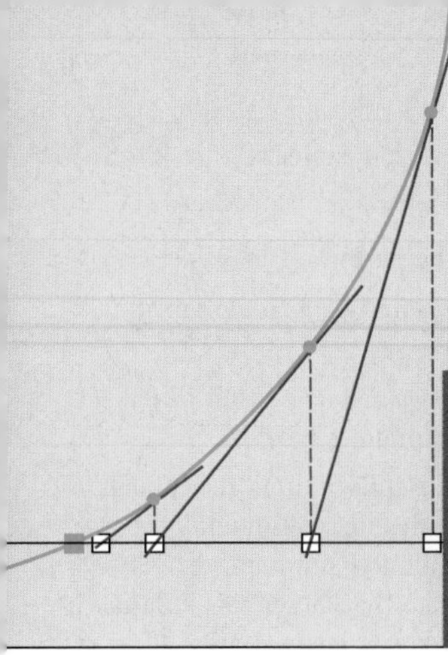

Chapter 4

APPLICATIONS OF DERIVATIVES

OVERVIEW This chapter studies some of the important applications of derivatives. We learn how derivatives are used to find extreme values of functions, to determine and analyze the shapes of graphs, to calculate limits of fractions whose numerators and denominators both approach zero or infinity, and to find numerically where a function equals zero. We also consider the process of recovering a function from its derivative. The key to many of these accomplishments is the Mean Value Theorem, a theorem whose corollaries provide the gateway to integral calculus in Chapter 5.

4.1 Extreme Values of Functions

This section shows how to locate and identify extreme values of a continuous function from its derivative. Once we can do this, we can solve a variety of *optimization problems* in which we find the optimal (best) way to do something in a given situation.

DEFINITIONS **Absolute Maximum, Absolute Minimum**

Let f be a function with domain D. Then f has an **absolute maximum** value on D at a point c if

$$f(x) \leq f(c) \qquad \text{for all } x \text{ in } D$$

and an **absolute minimum** value on D at c if

$$f(x) \geq f(c) \qquad \text{for all } x \text{ in } D.$$

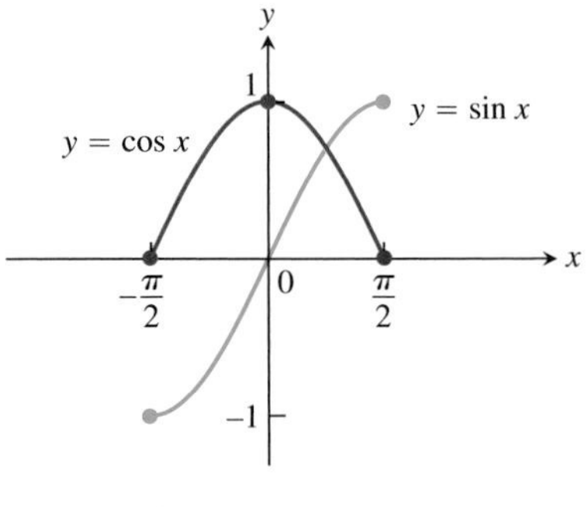

FIGURE 4.1 Absolute extrema for the sine and cosine functions on $[-\pi/2, \pi/2]$. These values can depend on the domain of a function.

Absolute maximum and minimum values are called absolute **extrema** (plural of the Latin *extremum*). Absolute extrema are also called **global** extrema, to distinguish them from *local extrema* defined below.

For example, on the closed interval $[-\pi/2, \pi/2]$ the function $f(x) = \cos x$ takes on an absolute maximum value of 1 (once) and an absolute minimum value of 0 (twice). On the same interval, the function $g(x) = \sin x$ takes on a maximum value of 1 and a minimum value of -1 (Figure 4.1).

Functions with the same defining rule can have different extrema, depending on the domain.

EXAMPLE 1 Exploring Absolute Extrema

The absolute extrema of the following functions on their domains can be seen in Figure 4.2.

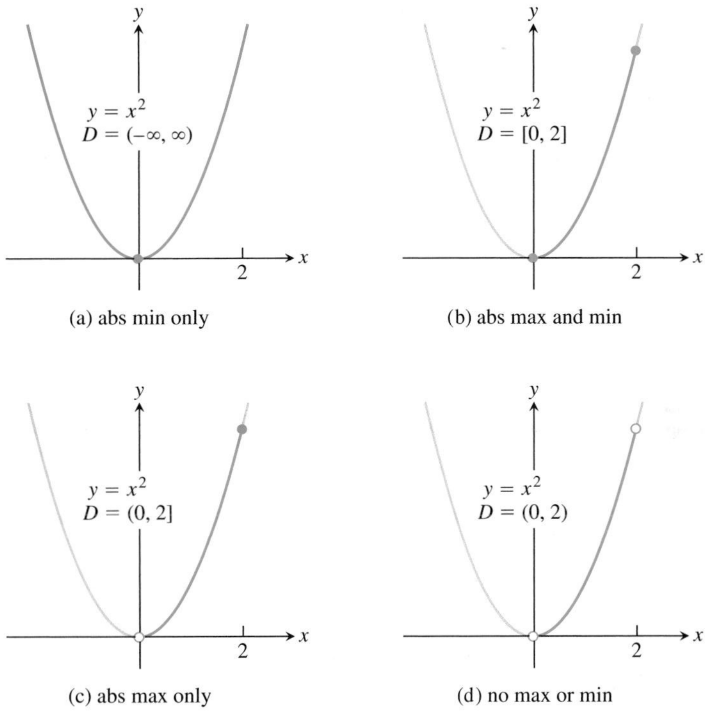

FIGURE 4.2 Graphs for Example 1.

Function rule	**Domain *D***	**Absolute extrema on *D***
(a) $y = x^2$	$(-\infty, \infty)$	No absolute maximum. Absolute minimum of 0 at $x = 0$.
(b) $y = x^2$	$[0, 2]$	Absolute maximum of 4 at $x = 2$. Absolute minimum of 0 at $x = 0$.
(c) $y = x^2$	$(0, 2]$	Absolute maximum of 4 at $x = 2$. No absolute minimum.
(d) $y = x^2$	$(0, 2)$	No absolute extrema.

■

HISTORICAL BIOGRAPHY

Daniel Bernoulli
(1700–1782)

The following theorem asserts that a function which is continuous at every point of a closed interval $[a, b]$ has an absolute maximum and an absolute minimum value on the interval. We always look for these values when we graph a function.

THEOREM 1 The Extreme Value Theorem
If f is continuous on a closed interval $[a, b]$, then f attains both an absolute maximum value M and an absolute minimum value m in $[a, b]$. That is, there are numbers x_1 and x_2 in $[a, b]$ with $f(x_1) = m$, $f(x_2) = M$, and $m \le f(x) \le M$ for every other x in $[a, b]$ (Figure 4.3).

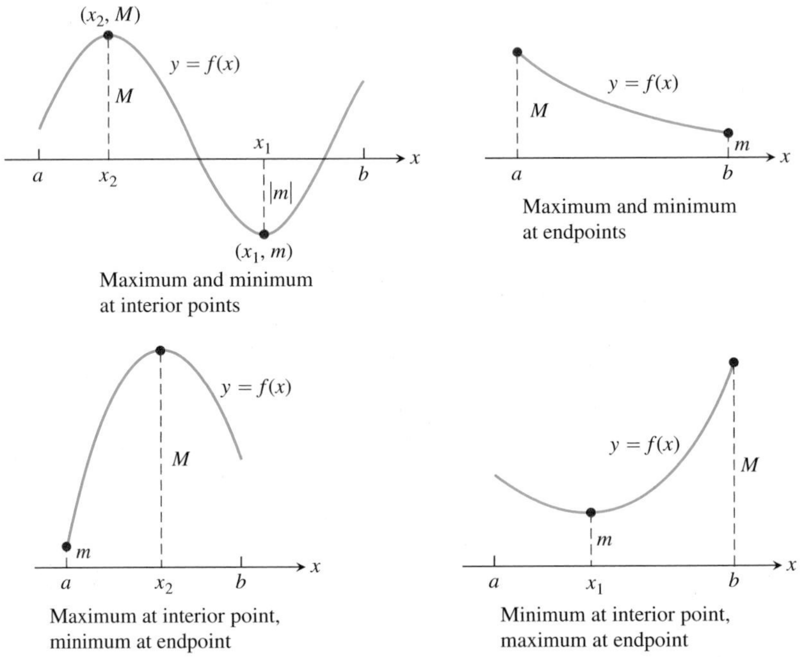

FIGURE 4.3 Some possibilities for a continuous function's maximum and minimum on a closed interval $[a, b]$.

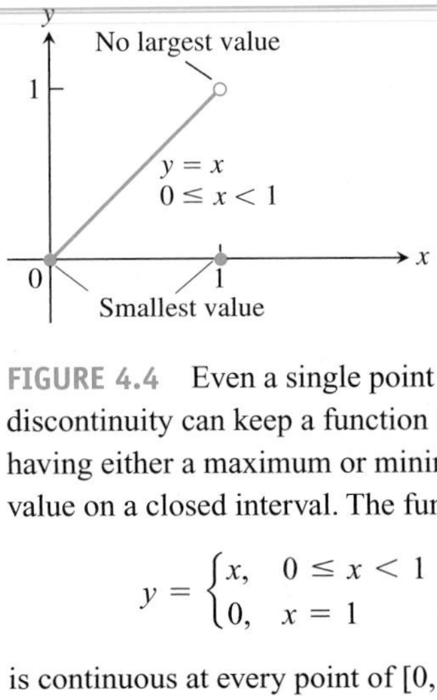

FIGURE 4.4 Even a single point of discontinuity can keep a function from having either a maximum or minimum value on a closed interval. The function

$$y = \begin{cases} x, & 0 \le x < 1 \\ 0, & x = 1 \end{cases}$$

is continuous at every point of $[0, 1]$ except $x = 1$, yet its graph over $[0, 1]$ does not have a highest point.

The proof of The Extreme Value Theorem requires a detailed knowledge of the real number system (see Appendix 4) and we will not give it here. Figure 4.3 illustrates possible locations for the absolute extrema of a continuous function on a closed interval $[a, b]$. As we observed for the function $y = \cos x$, it is possible that an absolute minimum (or absolute maximum) may occur at two or more different points of the interval.

The requirements in Theorem 1 that the interval be closed and finite, and that the function be continuous, are key ingredients. Without them, the conclusion of the theorem need not hold. Example 1 shows that an absolute extreme value may not exist if the interval fails to be both closed and finite. Figure 4.4 shows that the continuity requirement cannot be omitted.

Local (Relative) Extreme Values

Figure 4.5 shows a graph with five points where a function has extreme values on its domain $[a, b]$. The function's absolute minimum occurs at a even though at e the function's value is

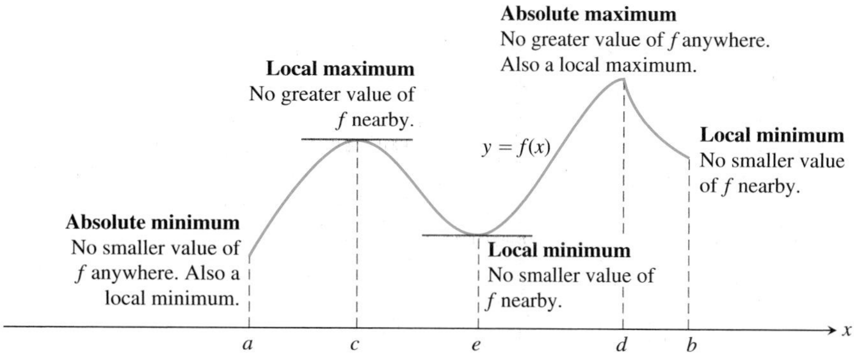

FIGURE 4.5 How to classify maxima and minima.

smaller than at any other point *nearby*. The curve rises to the left and falls to the right around c, making $f(c)$ a maximum locally. The function attains its absolute maximum at d.

DEFINITIONS Local Maximum, Local Minimum

A function f has a **local maximum** value at an interior point c of its domain if

$$f(x) \leq f(c) \qquad \text{for all } x \text{ in some open interval containing } c.$$

A function f has a **local minimum** value at an interior point c of its domain if

$$f(x) \geq f(c) \qquad \text{for all } x \text{ in some open interval containing } c.$$

We can extend the definitions of local extrema to the endpoints of intervals by defining f to have a **local maximum** or **local minimum** value *at an endpoint* c if the appropriate inequality holds for all x in some half-open interval in its domain containing c. In Figure 4.5, the function f has local maxima at c and d and local minima at a, e, and b. Local extrema are also called **relative extrema**.

An absolute maximum is also a local maximum. Being the largest value overall, it is also the largest value in its immediate neighborhood. Hence, *a list of all local maxima will automatically include the absolute maximum if there is one*. Similarly, *a list of all local minima will include the absolute minimum if there is one.*

Finding Extrema

The next theorem explains why we usually need to investigate only a few values to find a function's extrema.

THEOREM 2 The First Derivative Theorem for Local Extreme Values

If f has a local maximum or minimum value at an interior point c of its domain, and if f' is defined at c, then

$$f'(c) = 0.$$

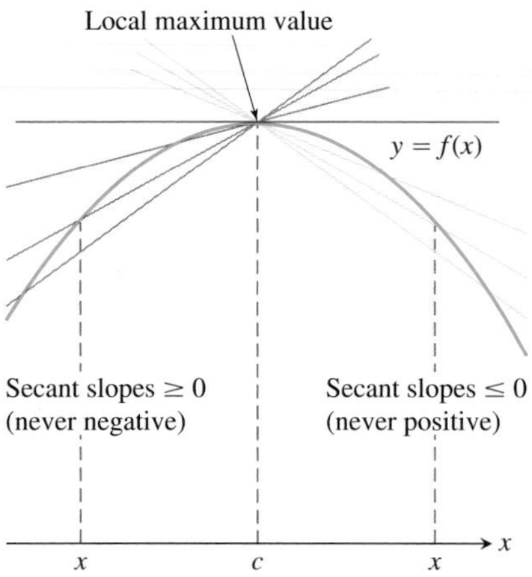

FIGURE 4.6 A curve with a local maximum value. The slope at c, simultaneously the limit of nonpositive numbers and nonnegative numbers, is zero.

Proof To prove that $f'(c)$ is zero at a local extremum, we show first that $f'(c)$ cannot be positive and second that $f'(c)$ cannot be negative. The only number that is neither positive nor negative is zero, so that is what $f'(c)$ must be.

To begin, suppose that f has a local maximum value at $x = c$ (Figure 4.6) so that $f(x) - f(c) \leq 0$ for all values of x near enough to c. Since c is an interior point of f's domain, $f'(c)$ is defined by the two-sided limit

$$\lim_{x \to c} \frac{f(x) - f(c)}{x - c}.$$

This means that the right-hand and left-hand limits both exist at $x = c$ and equal $f'(c)$. When we examine these limits separately, we find that

$$f'(c) = \lim_{x \to c^+} \frac{f(x) - f(c)}{x - c} \leq 0. \qquad \text{Because } (x - c) > 0 \text{ and } f(x) \leq f(c) \qquad (1)$$

Similarly,

$$f'(c) = \lim_{x \to c^-} \frac{f(x) - f(c)}{x - c} \geq 0. \qquad \text{Because } (x - c) < 0 \text{ and } f(x) \leq f(c) \qquad (2)$$

Together, Equations (1) and (2) imply $f'(c) = 0$.

This proves the theorem for local maximum values. To prove it for local minimum values, we simply use $f(x) \geq f(c)$, which reverses the inequalities in Equations (1) and (2). ■

Theorem 2 says that a function's first derivative is always zero at an interior point where the function has a local extreme value and the derivative is defined. Hence the only places where a function f can possibly have an extreme value (local or global) are

1. interior points where $f' = 0$,
2. interior points where f' is undefined,
3. endpoints of the domain of f.

The following definition helps us to summarize.

DEFINITION **Critical Point**
An interior point of the domain of a function f where f' is zero or undefined is a **critical point** of f.

Thus the only domain points where a function can assume extreme values are critical points and endpoints.

Be careful not to misinterpret Theorem 2 because its converse is false. A differentiable function may have a critical point at $x = c$ without having a local extreme value there. For instance, the function $f(x) = x^3$ has a critical point at the origin and zero value there, but is positive to the right of the origin and negative to the left. So it cannot have a local extreme value at the origin. Instead, it has a *point of inflection* there. This idea is defined and discussed further in Section 4.4.

Most quests for extreme values call for finding the absolute extrema of a continuous function on a closed and finite interval. Theorem 1 assures us that such values exist; Theorem 2 tells us that they are taken on only at critical points and endpoints. Often we can simply list these points and calculate the corresponding function values to find what the largest and smallest values are, and where they are located.

How to Find the Absolute Extrema of a Continuous Function f on a Finite Closed Interval

1. Evaluate f at all critical points and endpoints.
2. Take the largest and smallest of these values.

EXAMPLE 2 Finding Absolute Extrema

Find the absolute maximum and minimum values of $f(x) = x^2$ on $[-2, 1]$.

Solution The function is differentiable over its entire domain, so the only critical point is where $f'(x) = 2x = 0$, namely $x = 0$. We need to check the function's values at $x = 0$ and at the endpoints $x = -2$ and $x = 1$:

Critical point value: $f(0) = 0$

Endpoint values: $f(-2) = 4$

$f(1) = 1$

The function has an absolute maximum value of 4 at $x = -2$ and an absolute minimum value of 0 at $x = 0$. ■

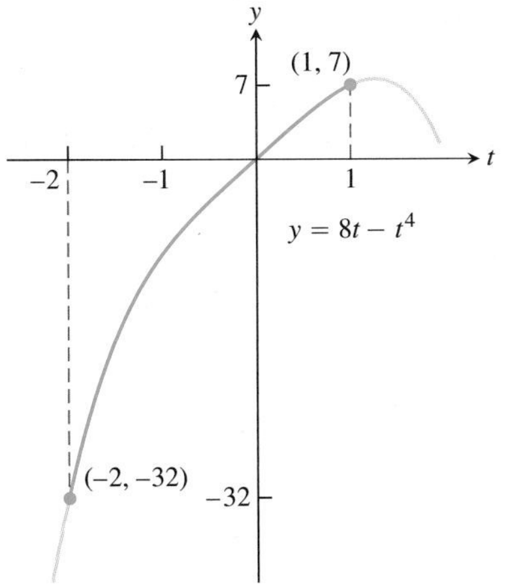

FIGURE 4.7 The extreme values of $g(t) = 8t - t^4$ on $[-2, 1]$ (Example 3).

EXAMPLE 3 Absolute Extrema at Endpoints

Find the absolute extrema values of $g(t) = 8t - t^4$ on $[-2, 1]$.

Solution The function is differentiable on its entire domain, so the only critical points occur where $g'(t) = 0$. Solving this equation gives

$$8 - 4t^3 = 0 \quad \text{or} \quad t = \sqrt[3]{2} > 1,$$

a point not in the given domain. The function's absolute extrema therefore occur at the endpoints, $g(-2) = -32$ (absolute minimum), and $g(1) = 7$ (absolute maximum). See Figure 4.7. ■

EXAMPLE 4 Finding Absolute Extrema on a Closed Interval

Find the absolute maximum and minimum values of $f(x) = x^{2/3}$ on the interval $[-2, 3]$.

Solution We evaluate the function at the critical points and endpoints and take the largest and smallest of the resulting values.

The first derivative

$$f'(x) = \frac{2}{3}x^{-1/3} = \frac{2}{3\sqrt[3]{x}}$$

has no zeros but is undefined at the interior point $x = 0$. The values of f at this one critical point and at the endpoints are

Critical point value: $f(0) = 0$

Endpoint values: $f(-2) = (-2)^{2/3} = \sqrt[3]{4}$

$f(3) = (3)^{2/3} = \sqrt[3]{9}.$

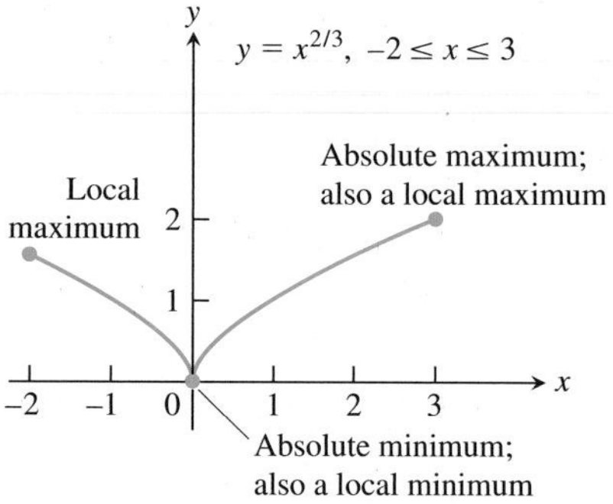

FIGURE 4.8 The extreme values of $f(x) = x^{2/3}$ on $[-2, 3]$ occur at $x = 0$ and $x = 3$ (Example 4).

We can see from this list that the function's absolute maximum value is $\sqrt[3]{9} \approx 2.08$, and it occurs at the right endpoint $x = 3$. The absolute minimum value is 0, and it occurs at the interior point $x = 0$. (Figure 4.8). ■

While a function's extrema can occur only at critical points and endpoints, not every critical point or endpoint signals the presence of an extreme value. Figure 4.9 illustrates this for interior points.

We complete this section with an example illustrating how the concepts we studied are used to solve a real-world optimization problem.

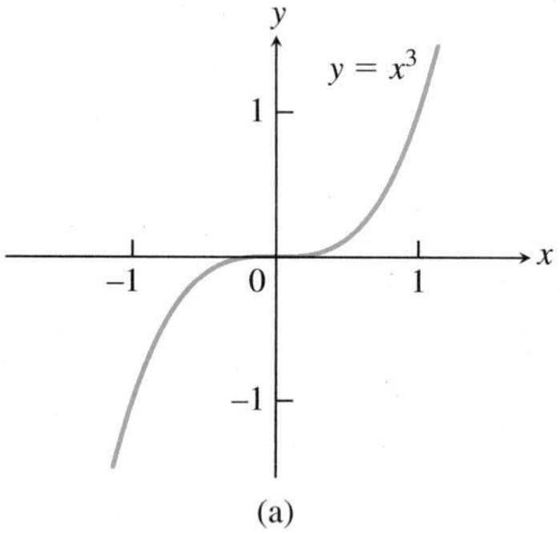

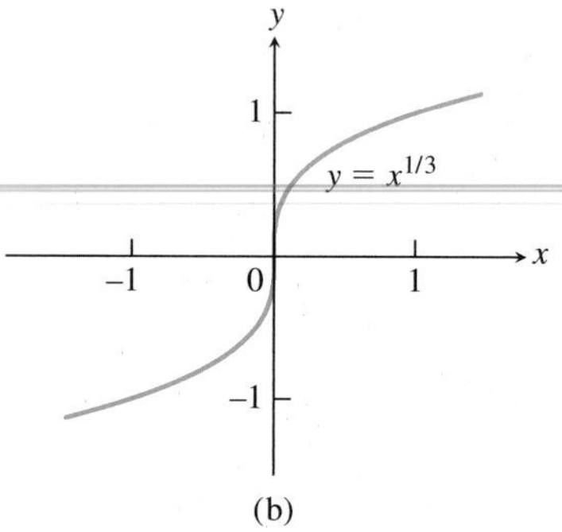

FIGURE 4.9 Critical points without extreme values. (a) $y' = 3x^2$ is 0 at $x = 0$, but $y = x^3$ has no extremum there. (b) $y' = (1/3)x^{-2/3}$ is undefined at $x = 0$, but $y = x^{1/3}$ has no extremum there.

EXAMPLE 5 Piping Oil from a Drilling Rig to a Refinery

A drilling rig 12 mi offshore is to be connected by pipe to a refinery onshore, 20 mi straight down the coast from the rig. If underwater pipe costs \$500,000 per mile and land-based pipe costs \$300,000 per mile, what combination of the two will give the least expensive connection?

Solution We try a few possibilities to get a feel for the problem:

(a) *Smallest amount of underwater pipe*

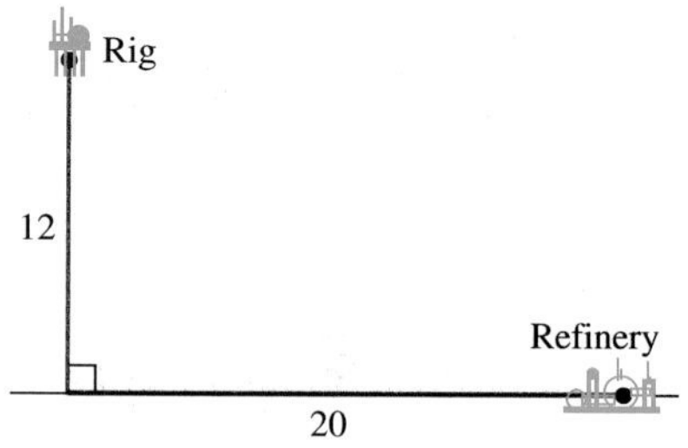

Underwater pipe is more expensive, so we use as little as we can. We run straight to shore (12 mi) and use land pipe for 20 mi to the refinery.

$$\text{Dollar cost} = 12(500{,}000) + 20(300{,}000) = 12{,}000{,}000$$

(b) *All pipe underwater (most direct route)*

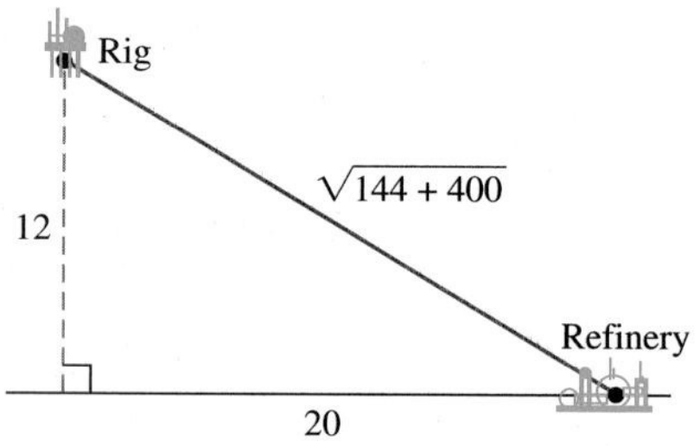

We go straight to the refinery underwater.

$$\text{Dollar cost} = \sqrt{544}\,(500{,}000) \approx 11{,}661{,}900$$

This is less expensive than plan (a).

(c) *Something in between*

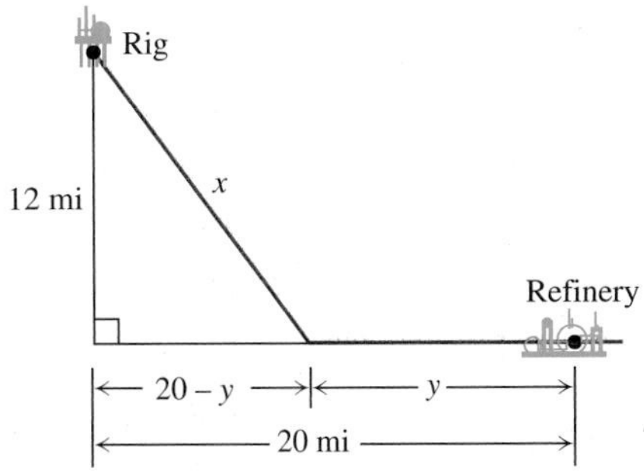

Now we introduce the length x of underwater pipe and the length y of land-based pipe as variables. The right angle opposite the rig is the key to expressing the relationship between x and y, for the Pythagorean theorem gives

$$x^2 = 12^2 + (20 - y)^2$$
$$x = \sqrt{144 + (20 - y)^2}. \qquad (3)$$

Only the positive root has meaning in this model.

The dollar cost of the pipeline is

$$c = 500{,}000x + 300{,}000y.$$

To express c as a function of a single variable, we can substitute for x, using Equation (3):

$$c(y) = 500{,}000\sqrt{144 + (20 - y)^2} + 300{,}000y.$$

Our goal now is to find the minimum value of $c(y)$ on the interval $0 \le y \le 20$. The first derivative of $c(y)$ with respect to y according to the Chain Rule is

$$c'(y) = 500{,}000 \cdot \frac{1}{2} \cdot \frac{2(20 - y)(-1)}{\sqrt{144 + (20 - y)^2}} + 300{,}000$$
$$= -500{,}000 \frac{20 - y}{\sqrt{144 + (20 - y)^2}} + 300{,}000.$$

Setting c' equal to zero gives

$$500{,}000\,(20 - y) = 300{,}000\sqrt{144 + (20 - y)^2}$$
$$\frac{5}{3}\left(20 - y\right) = \sqrt{144 + (20 - y)^2}$$
$$\frac{25}{9}\left(20 - y\right)^2 = 144 + (20 - y)^2$$
$$\frac{16}{9}\left(20 - y\right)^2 = 144$$
$$(20 - y) = \pm\frac{3}{4} \cdot 12 = \pm 9$$
$$y = 20 \pm 9$$
$$y = 11 \quad \text{or} \quad y = 29.$$

Only $y = 11$ lies in the interval of interest. The values of c at this one critical point and at the endpoints are

$$c(11) = 10{,}800{,}000$$
$$c(0) = 11{,}661{,}900$$
$$c(20) = 12{,}000{,}000$$

The least expensive connection costs \$10,800,000, and we achieve it by running the line underwater to the point on shore 11 mi from the refinery. ■

EXERCISES 4.1

Finding Extrema from Graphs

In Exercises 1–6, determine from the graph whether the function has any absolute extreme values on $[a, b]$. Then explain how your answer is consistent with Theorem 1.

1.

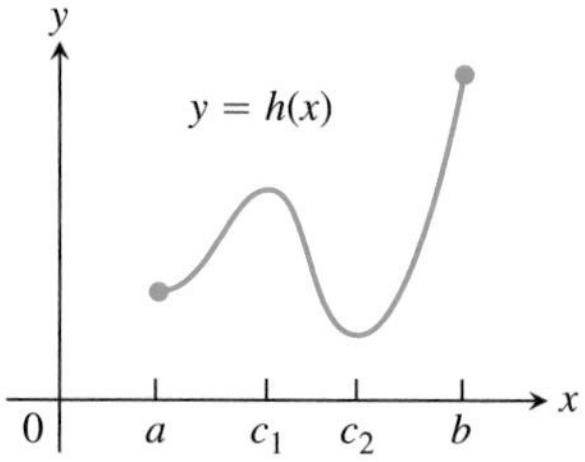

2.

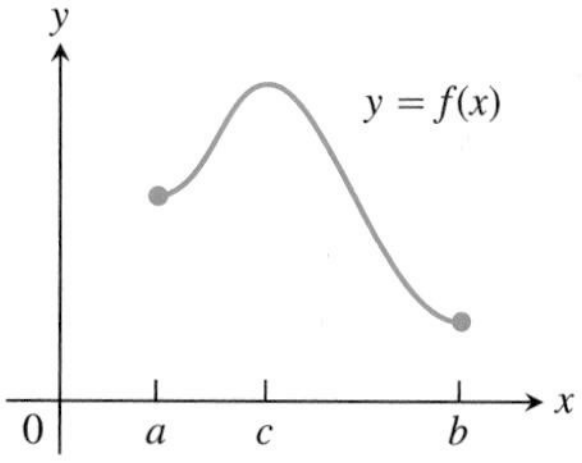

3.

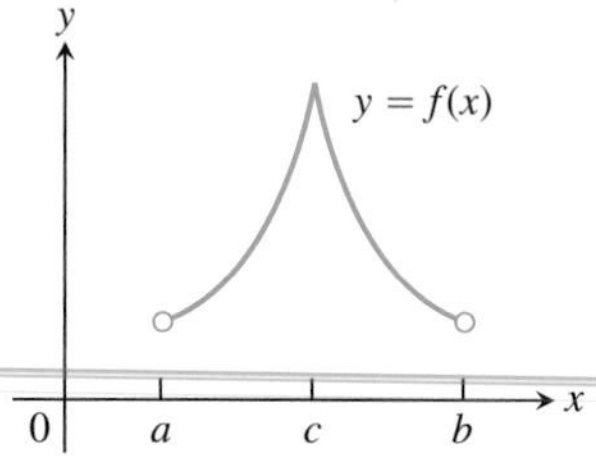

4.

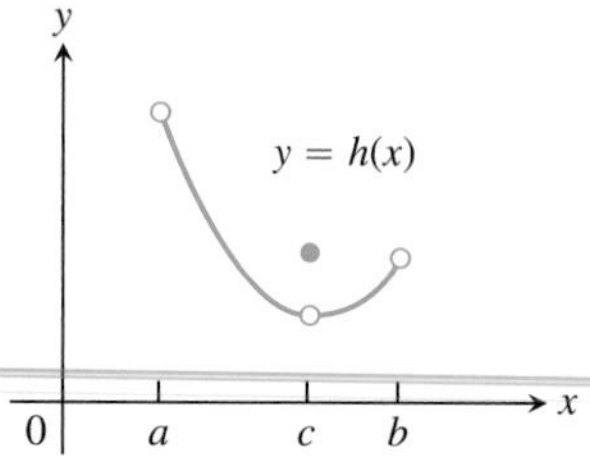

5.

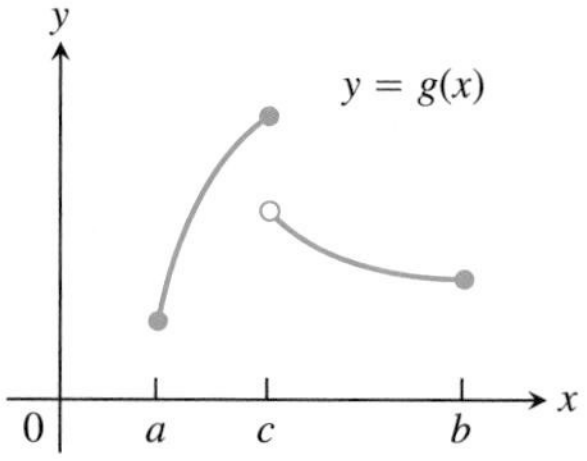

6.

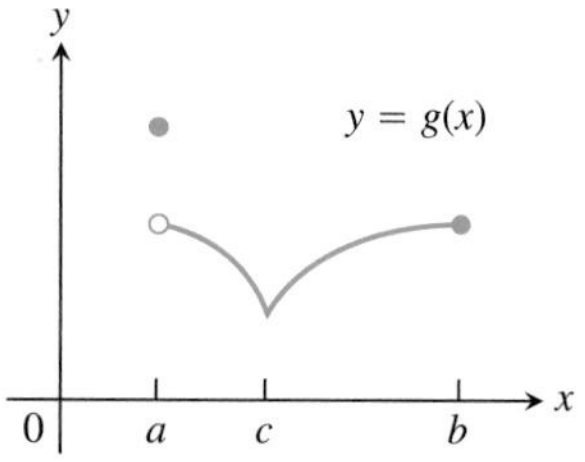

In Exercises 7–10, find the extreme values and where they occur.

7.

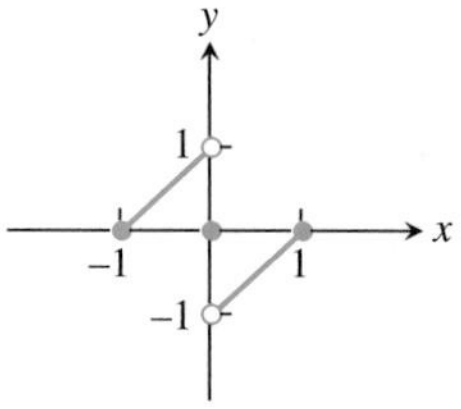

8.

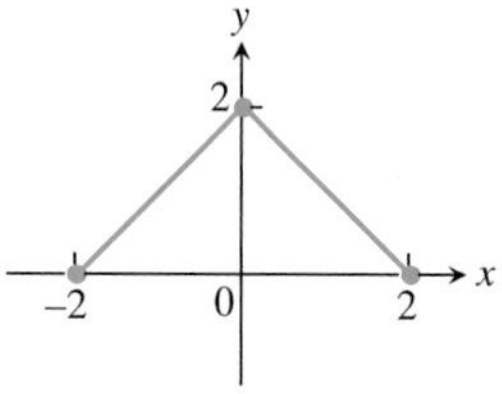

9.

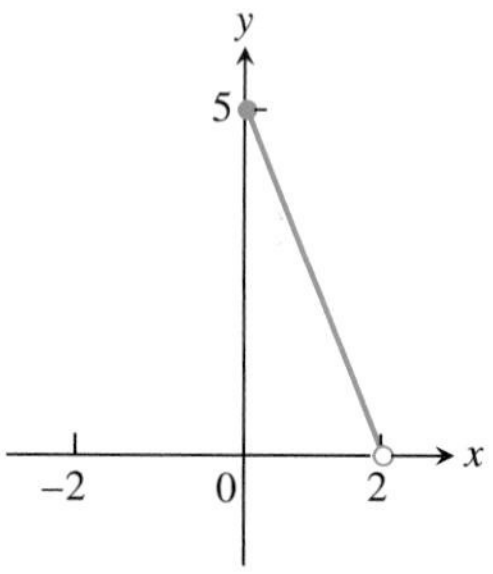

10.

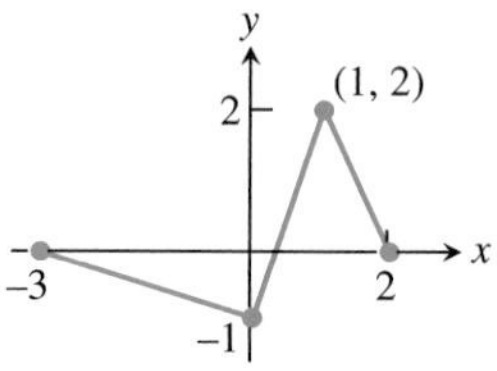

In Exercises 11-14, match the table with a graph.

11.

x	$f'(x)$
a	0
b	0
c	5

12.

x	$f'(x)$
a	0
b	0
c	-5

13.

x	$f'(x)$
a	does not exist
b	0
c	-2

14.

x	$f'(x)$
a	does not exist
b	does not exist
c	-1.7

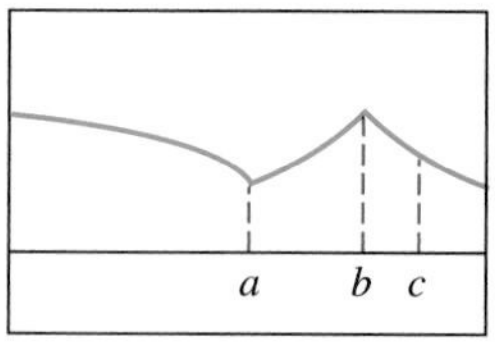

(a)

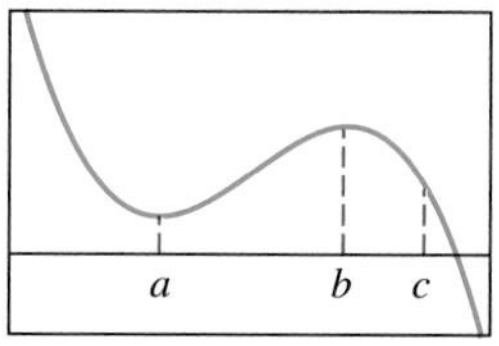

(b)

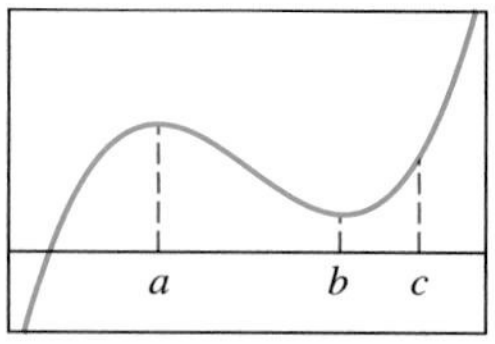

(c)

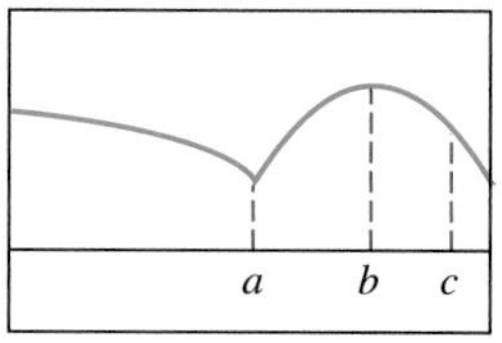

(d)

Absolute Extrema on Finite Closed Intervals

In Exercises 15–30, find the absolute maximum and minimum values of each function on the given interval. Then graph the function. Identify the points on the graph where the absolute extrema occur, and include their coordinates.

15. $f(x) = \frac{2}{3}x - 5, \quad -2 \le x \le 3$
16. $f(x) = -x - 4, \quad -4 \le x \le 1$
17. $f(x) = x^2 - 1, \quad -1 \le x \le 2$
18. $f(x) = 4 - x^2, \quad -3 \le x \le 1$
19. $F(x) = -\frac{1}{x^2}, \quad 0.5 \le x \le 2$
20. $F(x) = -\frac{1}{x}, \quad -2 \le x \le -1$
21. $h(x) = \sqrt[3]{x}, \quad -1 \le x \le 8$
22. $h(x) = -3x^{2/3}, \quad -1 \le x \le 1$
23. $g(x) = \sqrt{4 - x^2}, \quad -2 \le x \le 1$
24. $g(x) = -\sqrt{5 - x^2}, \quad -\sqrt{5} \le x \le 0$
25. $f(\theta) = \sin\theta, \quad -\frac{\pi}{2} \le \theta \le \frac{5\pi}{6}$
26. $f(\theta) = \tan\theta, \quad -\frac{\pi}{3} \le \theta \le \frac{\pi}{4}$
27. $g(x) = \csc x, \quad \frac{\pi}{3} \le x \le \frac{2\pi}{3}$
28. $g(x) = \sec x, \quad -\frac{\pi}{3} \le x \le \frac{\pi}{6}$
29. $f(t) = 2 - |t|, \quad -1 \le t \le 3$
30. $f(t) = |t - 5|, \quad 4 \le t \le 7$

In Exercises 31–34, find the function's absolute maximum and minimum values and say where they are assumed.

31. $f(x) = x^{4/3}, \quad -1 \le x \le 8$
32. $f(x) = x^{5/3}, \quad -1 \le x \le 8$
33. $g(\theta) = \theta^{3/5}, \quad -32 \le \theta \le 1$
34. $h(\theta) = 3\theta^{2/3}, \quad -27 \le \theta \le 8$

Finding Extreme Values

In Exercises 35–44, find the extreme values of the function and where they occur.

35. $y = 2x^2 - 8x + 9$
36. $y = x^3 - 2x + 4$
37. $y = x^3 + x^2 - 8x + 5$
38. $y = x^3 - 3x^2 + 3x - 2$
39. $y = \sqrt{x^2 - 1}$
40. $y = \frac{1}{\sqrt{1 - x^2}}$
41. $y = \frac{1}{\sqrt[3]{1 - x^2}}$
42. $y = \sqrt{3 + 2x - x^2}$
43. $y = \frac{x}{x^2 + 1}$
44. $y = \frac{x + 1}{x^2 + 2x + 2}$

Local Extrema and Critical Points

In Exercises 45–52, find the derivative at each critical point and determine the local extreme values.

45. $y = x^{2/3}(x + 2)$
46. $y = x^{2/3}(x^2 - 4)$
47. $y = x\sqrt{4 - x^2}$
48. $y = x^2\sqrt{3 - x}$
49. $y = \begin{cases} 4 - 2x, & x \le 1 \\ x + 1, & x > 1 \end{cases}$
50. $y = \begin{cases} 3 - x, & x < 0 \\ 3 + 2x - x^2, & x \ge 0 \end{cases}$
51. $y = \begin{cases} -x^2 - 2x + 4, & x \le 1 \\ -x^2 + 6x - 4, & x > 1 \end{cases}$
52. $y = \begin{cases} -\frac{1}{4}x^2 - \frac{1}{2}x + \frac{15}{4}, & x \le 1 \\ x^3 - 6x^2 + 8x, & x > 1 \end{cases}$

In Exercises 53 and 54, give reasons for your answers.

53. Let $f(x) = (x - 2)^{2/3}$.
    - **a.** Does $f'(2)$ exist?
    - **b.** Show that the only local extreme value of f occurs at $x = 2$.
    - **c.** Does the result in part (b) contradict the Extreme Value Theorem?
    - **d.** Repeat parts (a) and (b) for $f(x) = (x - a)^{2/3}$, replacing 2 by a.
54. Let $f(x) = |x^3 - 9x|$.
    - **a.** Does $f'(0)$ exist?
    - **b.** Does $f'(3)$ exist?
    - **c.** Does $f'(-3)$ exist?
    - **d.** Determine all extrema of f.

Optimization Applications

Whenever you are maximizing or minimizing a function of a single variable, we urge you to graph the function over the domain that is appropriate to the problem you are solving. The graph will provide insight before you begin to calculate and will furnish a visual context for understanding your answer.

55. **Constructing a pipeline** Supertankers off-load oil at a docking facility 4 mi offshore. The nearest refinery is 9 mi east of the shore point nearest the docking facility. A pipeline must be constructed connecting the docking facility with the refinery. The pipeline costs $300,000 per mile if constructed underwater and $200,000 per mile if overland.

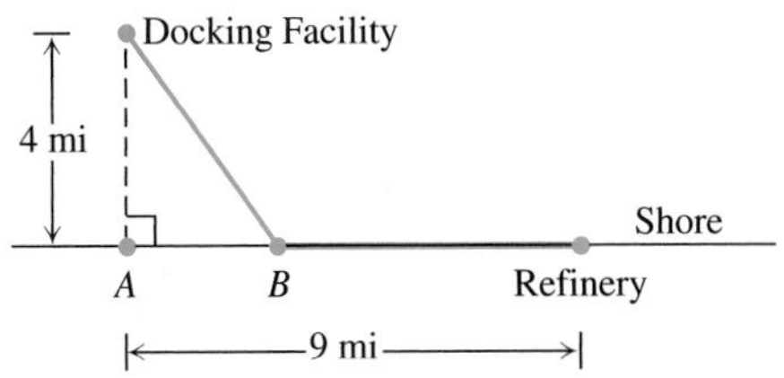

a. Locate Point B to minimize the cost of the construction.

b. The cost of underwater construction is expected to increase, whereas the cost of overland construction is expected to stay constant. At what cost does it become optimal to construct the pipeline directly to Point A?

56. Upgrading a highway A highway must be constructed to connect Village A with Village B. There is a rudimentary roadway that can be upgraded 50 mi south of the line connecting the two villages. The cost of upgrading the existing roadway is \$300,000 per mile, whereas the cost of constructing a new highway is \$500,000 per mile. Find the combination of upgrading and new construction that minimizes the cost of connecting the two villages. Clearly define the location of the proposed highway.

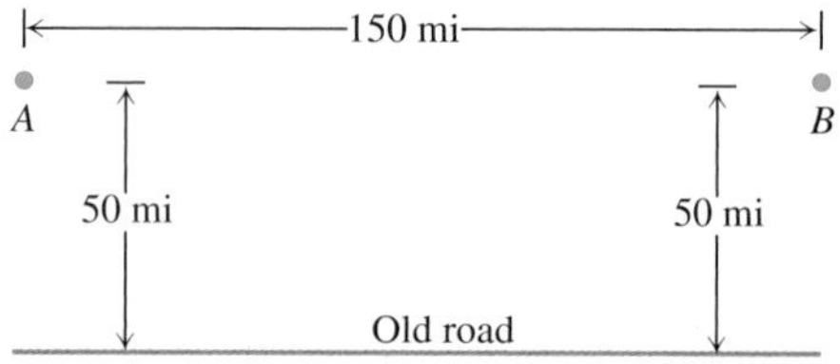

57. Locating a pumping station Two towns lie on the south side of a river. A pumping station is to be located to serve the two towns. A pipeline will be constructed from the pumping station to each of the towns along the line connecting the town and the pumping station. Locate the pumping station to minimize the amount of pipeline that must be constructed.

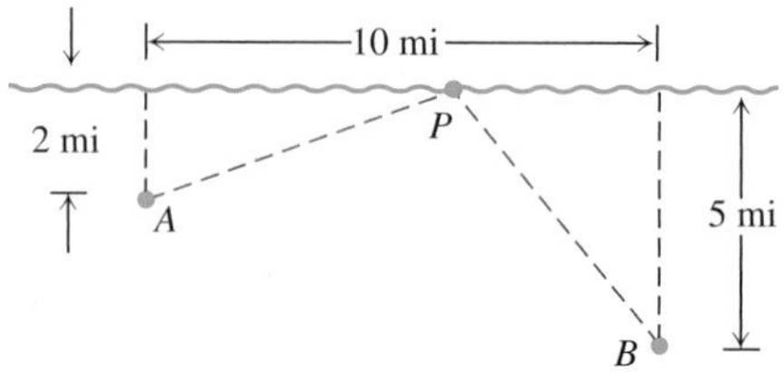

58. Length of a guy wire One tower is 50 ft high and another tower is 30 ft high. The towers are 150 ft apart. A guy wire is to run from Point A to the top of each tower.

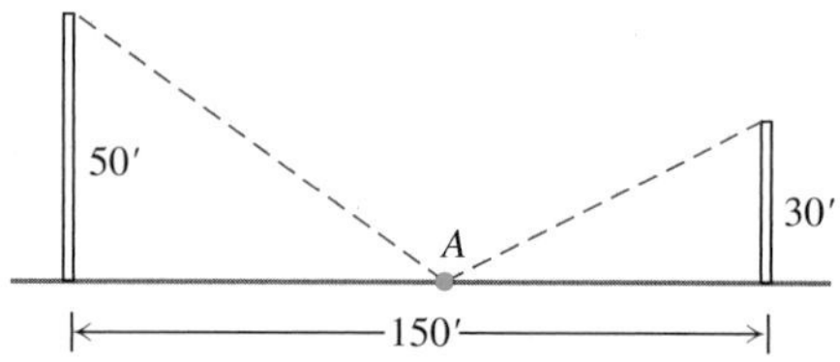

a. Locate Point A so that the total length of guy wire is minimal.

b. Show in general that regardless of the height of the towers, the length of guy wire is minimized if the angles at A are equal.

59. The function

$$V(x) = x(10 - 2x)(16 - 2x), \qquad 0 < x < 5,$$

models the volume of a box.

a. Find the extreme values of V.

b. Interpret any values found in part (a) in terms of volume of the box.

60. The function

$$P(x) = 2x + \frac{200}{x}, \qquad 0 < x < \infty,$$

models the perimeter of a rectangle of dimensions x by $100/x$.

a. Find any extreme values of P.

b. Give an interpretation in terms of perimeter of the rectangle for any values found in part (a).

61. Area of a right triangle What is the largest possible area for a right triangle whose hypotenuse is 5 cm long?

62. Area of an athletic field An athletic field is to be built in the shape of a rectangle x units long capped by semicircular regions of radius r at the two ends. The field is to be bounded by a 400-m racetrack.

a. Express the area of the rectangular portion of the field as a function of x alone or r alone (your choice).

b. What values of x and r give the rectangular portion the largest possible area?

63. Maximum height of a vertically moving body The height of a body moving vertically is given by

$$s = -\frac{1}{2}gt^2 + v_0t + s_0, \qquad g > 0,$$

with s in meters and t in seconds. Find the body's maximum height.

64. Peak alternating current Suppose that at any given time t (in seconds) the current i (in amperes) in an alternating current circuit is $i = 2\cos t + 2\sin t$. What is the peak current for this circuit (largest magnitude)?

Theory and Examples

65. A minimum with no derivative The function $f(x) = |x|$ has an absolute minimum value at $x = 0$ even though f is not differentiable at $x = 0$. Is this consistent with Theorem 2? Give reasons for your answer.

66. Even functions If an even function $f(x)$ has a local maximum value at $x = c$, can anything be said about the value of f at $x = -c$? Give reasons for your answer.

67. Odd functions If an odd function $g(x)$ has a local minimum value at $x = c$, can anything be said about the value of g at $x = -c$? Give reasons for your answer.

68. We know how to find the extreme values of a continuous function $f(x)$ by investigating its values at critical points and endpoints. But what if there *are* no critical points or endpoints? What happens then? Do such functions really exist? Give reasons for your answers.

69. Cubic functions Consider the cubic function

$$f(x) = ax^3 + bx^2 + cx + d.$$

a. Show that f can have 0, 1, or 2 critical points. Give examples and graphs to support your argument.

b. How many local extreme values can f have?

T **70. Functions with no extreme values at endpoints**

a. Graph the function

$$f(x) = \begin{cases} \sin \frac{1}{x}, & x > 0 \\ 0, & x = 0. \end{cases}$$

Explain why $f(0) = 0$ is not a local extreme value of f.

b. Construct a function of your own that fails to have an extreme value at a domain endpoint.

T Graph the functions in Exercises 71–74. Then find the extreme values of the function on the interval and say where they occur.

71. $f(x) = |x - 2| + |x + 3|, \quad -5 \le x \le 5$

72. $g(x) = |x - 1| - |x - 5|, \quad -2 \le x \le 7$

73. $h(x) = |x + 2| - |x - 3|, \quad -\infty < x < \infty$

74. $k(x) = |x + 1| + |x - 3|, \quad -\infty < x < \infty$

COMPUTER EXPLORATIONS

In Exercises 75–80, you will use a CAS to help find the absolute extrema of the given function over the specified closed interval. Perform the following steps.

a. Plot the function over the interval to see its general behavior there.

b. Find the interior points where $f' = 0$. (In some exercises, you may have to use the numerical equation solver to approximate a solution.) You may want to plot f' as well.

c. Find the interior points where f' does not exist.

d. Evaluate the function at all points found in parts (b) and (c) and at the endpoints of the interval.

e. Find the function's absolute extreme values on the interval and identify where they occur.

75. $f(x) = x^4 - 8x^2 + 4x + 2, \quad [-20/25, 64/25]$

76. $f(x) = -x^4 + 4x^3 - 4x + 1, \quad [-3/4, 3]$

77. $f(x) = x^{2/3}(3 - x), \quad [-2, 2]$

78. $f(x) = 2 + 2x - 3x^{2/3}, \quad [-1, 10/3]$

79. $f(x) = \sqrt{x} + \cos x, \quad [0, 2\pi]$

80. $f(x) = x^{3/4} - \sin x + \frac{1}{2}, \quad [0, 2\pi]$

4.2 The Mean Value Theorem

We know that constant functions have zero derivatives, but could there be a complicated function, with many terms, the derivatives of which all cancel to give zero? What is the relationship between two functions that have identical derivatives over an interval? What we are really asking here is what functions can have a particular *kind* of derivative. These and many other questions we study in this chapter are answered by applying the Mean Value Theorem. To arrive at this theorem we first need Rolle's Theorem.

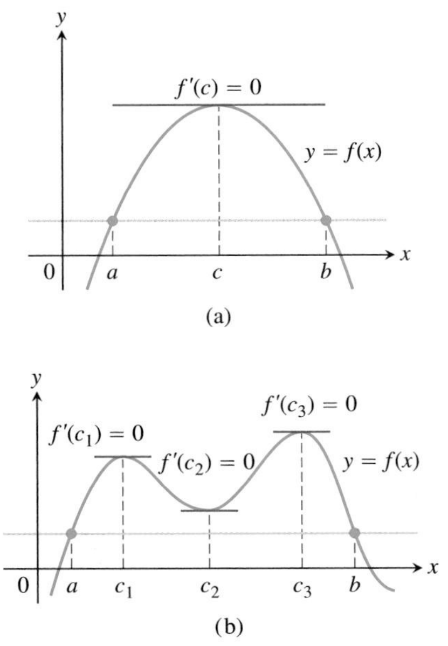

FIGURE 4.10 Rolle's Theorem says that a differentiable curve has at least one horizontal tangent between any two points where it crosses a horizontal line. It may have just one (a), or it may have more (b).

Rolle's Theorem

Drawing the graph of a function gives strong geometric evidence that between any two points where a differentiable function crosses a horizontal line there is at least one point on the curve where the tangent is horizontal (Figure 4.10). More precisely, we have the following theorem.

THEOREM 3 Rolle's Theorem

Suppose that $y = f(x)$ is continuous at every point of the closed interval $[a, b]$ and differentiable at every point of its interior (a, b). If

$$f(a) = f(b),$$

then there is at least one number c in (a, b) at which

$$f'(c) = 0.$$

Proof Being continuous, f assumes absolute maximum and minimum values on $[a, b]$. These can occur only

HISTORICAL BIOGRAPHY

Michel Rolle
(1652–1719)

1. at interior points where f' is zero,
2. at interior points where f' does not exist,
3. at the endpoints of the function's domain, in this case a and b.

By hypothesis, f has a derivative at every interior point. That rules out possibility (2), leaving us with interior points where $f' = 0$ and with the two endpoints a and b.

If either the maximum or the minimum occurs at a point c between a and b, then $f'(c) = 0$ by Theorem 2 in Section 4.1, and we have found a point for Rolle's theorem.

If both the absolute maximum and the absolute minimum occur at the endpoints, then because $f(a) = f(b)$ it must be the case that f is a constant function with $f(x) = f(a) = f(b)$ for every $x \in [a, b]$. Therefore $f'(x) = 0$ and the point c can be taken anywhere in the interior (a, b). ■

The hypotheses of Theorem 3 are essential. If they fail at even one point, the graph may not have a horizontal tangent (Figure 4.11).

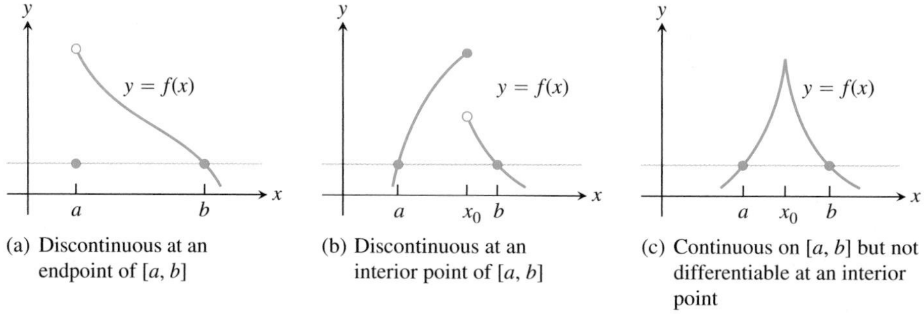

FIGURE 4.11 There may be no horizontal tangent if the hypotheses of Rolle's Theorem do not hold.

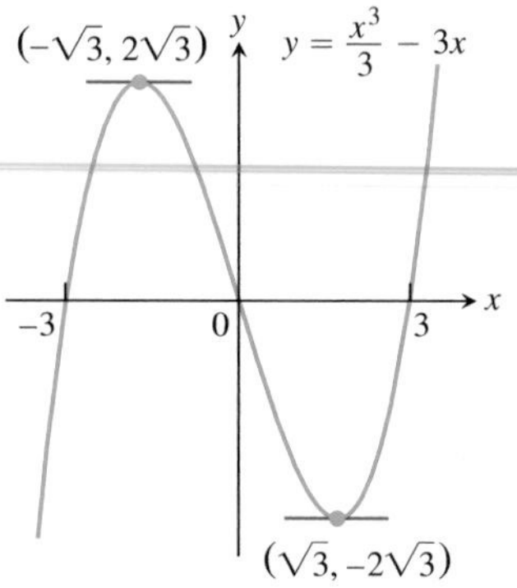

FIGURE 4.12 As predicted by Rolle's Theorem, this curve has horizontal tangents between the points where it crosses the x-axis (Example 1).

EXAMPLE 1 Horizontal Tangents of a Cubic Polynomial

The polynomial function

$$f(x) = \frac{x^3}{3} - 3x$$

graphed in Figure 4.12 is continuous at every point of $[-3, 3]$ and is differentiable at every point of $(-3, 3)$. Since $f(-3) = f(3) = 0$, Rolle's Theorem says that f' must be zero at least once in the open interval between $a = -3$ and $b = 3$. In fact, $f'(x) = x^2 - 3$ is zero twice in this interval, once at $x = -\sqrt{3}$ and again at $x = \sqrt{3}$. ■

EXAMPLE 2 Solution of an Equation $f(x) = 0$

Show that the equation

$$x^3 + 3x + 1 = 0$$

has exactly one real solution.

Solution Let

$$y = f(x) = x^3 + 3x + 1.$$

Then the derivative

$$f'(x) = 3x^2 + 3$$

is never zero (because it is always positive). Now, if there were even two points $x = a$ and $x = b$ where $f(x)$ was zero, Rolle's Theorem would guarantee the existence of a point $x = c$ in between them where f' was zero. Therefore, f has no more than one zero. It does in fact have one zero, because the Intermediate Value Theorem tells us that the graph of $y = f(x)$ crosses the x-axis somewhere between $x = -1$ (where $y = -3$) and $x = 0$ (where $y = 1$). (See Figure 4.13.) ■

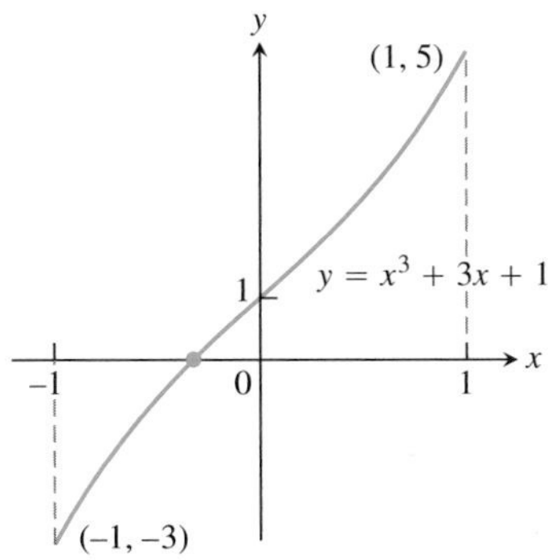

FIGURE 4.13 The only real zero of the polynomial $y = x^3 + 3x + 1$ is the one shown here where the curve crosses the x-axis between -1 and 0 (Example 2).

Our main use of Rolle's Theorem is in proving the Mean Value Theorem.

The Mean Value Theorem

The Mean Value Theorem, which was first stated by Joseph-Louis Lagrange, is a slanted version of Rolle's Theorem (Figure 4.14). There is a point where the tangent is parallel to chord AB.

THEOREM 4 The Mean Value Theorem
Suppose $y = f(x)$ is continuous on a closed interval $[a, b]$ and differentiable on the interval's interior (a, b). Then there is at least one point c in (a, b) at which

$$\frac{f(b) - f(a)}{b - a} = f'(c). \tag{1}$$

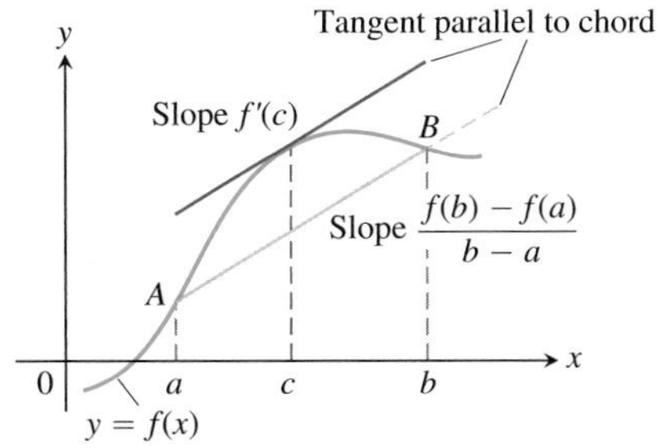

FIGURE 4.14 Geometrically, the Mean Value Theorem says that somewhere between A and B the curve has at least one tangent parallel to chord AB.

Proof We picture the graph of f as a curve in the plane and draw a line through the points $A(a, f(a))$ and $B(b, f(b))$ (see Figure 4.15). The line is the graph of the function

$$g(x) = f(a) + \frac{f(b) - f(a)}{b - a}(x - a) \tag{2}$$

(point-slope equation). The vertical difference between the graphs of f and g at x is

$$\begin{aligned} h(x) &= f(x) - g(x) \\ &= f(x) - f(a) - \frac{f(b) - f(a)}{b - a}(x - a). \end{aligned} \tag{3}$$

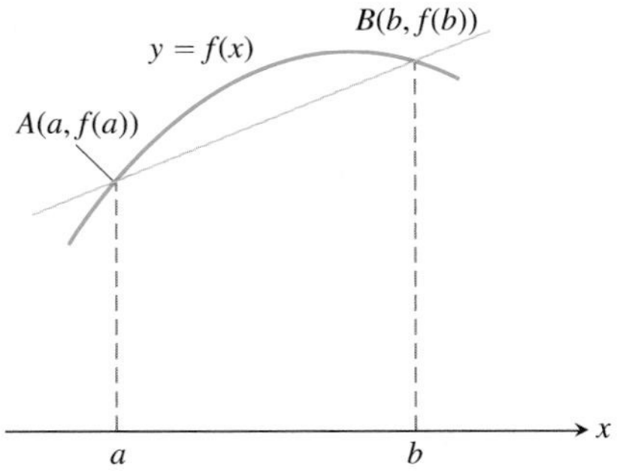

FIGURE 4.15 The graph of f and the chord AB over the interval $[a, b]$.

Figure 4.16 shows the graphs of f, g, and h together.

The function h satisfies the hypotheses of Rolle's Theorem on $[a, b]$. It is continuous on $[a, b]$ and differentiable on (a, b) because both f and g are. Also, $h(a) = h(b) = 0$ because the graphs of f and g both pass through A and B. Therefore $h'(c) = 0$ at some point $c \in (a, b)$. This is the point we want for Equation (1).

To verify Equation (1), we differentiate both sides of Equation (3) with respect to x and then set $x = c$:

$$h'(x) = f'(x) - \frac{f(b) - f(a)}{b - a} \qquad \text{Derivative of Eq. (3)} \ldots$$

$$h'(c) = f'(c) - \frac{f(b) - f(a)}{b - a} \qquad \ldots \text{with } x = c$$

$$0 = f'(c) - \frac{f(b) - f(a)}{b - a} \qquad h'(c) = 0$$

$$f'(c) = \frac{f(b) - f(a)}{b - a}, \qquad \text{Rearranged}$$

which is what we set out to prove. ■

HISTORICAL BIOGRAPHY

Joseph-Louis Lagrange (1736–1813)

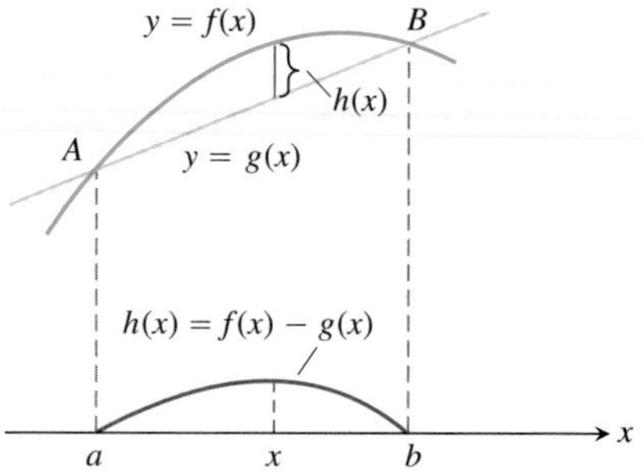

FIGURE 4.16 The chord AB is the graph of the function $g(x)$. The function $h(x) = f(x) - g(x)$ gives the vertical distance between the graphs of f and g at x.

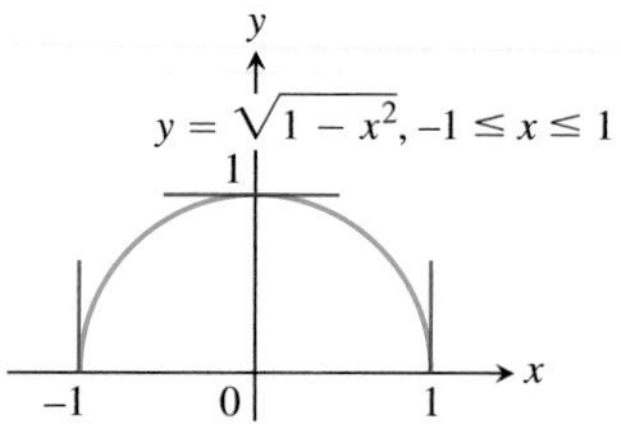

FIGURE 4.17 The function $f(x) = \sqrt{1 - x^2}$ satisfies the hypotheses (and conclusion) of the Mean Value Theorem on $[-1, 1]$ even though f is not differentiable at -1 and 1.

The hypotheses of the Mean Value Theorem do not require f to be differentiable at either a or b. Continuity at a and b is enough (Figure 4.17).

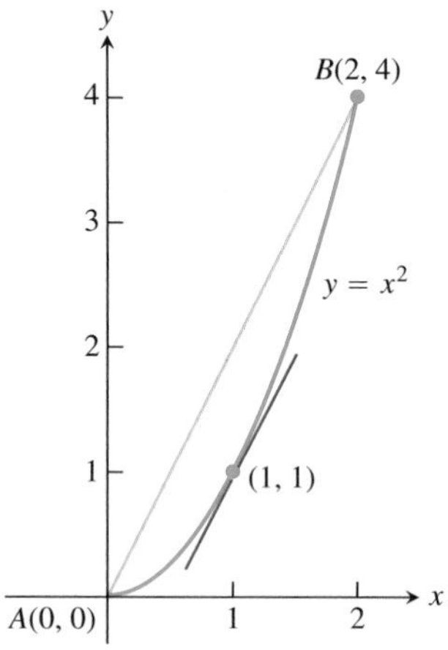

FIGURE 4.18 As we find in Example 3, $c = 1$ is where the tangent is parallel to the chord.

EXAMPLE 3 The function $f(x) = x^2$ (Figure 4.18) is continuous for $0 \le x \le 2$ and differentiable for $0 < x < 2$. Since $f(0) = 0$ and $f(2) = 4$, the Mean Value Theorem says that at some point c in the interval, the derivative $f'(x) = 2x$ must have the value $(4 - 0)/(2 - 0) = 2$. In this (exceptional) case we can identify c by solving the equation $2c = 2$ to get $c = 1$. ■

A Physical Interpretation

If we think of the number $(f(b) - f(a))/(b - a)$ as the average change in f over $[a, b]$ and $f'(c)$ as an instantaneous change, then the Mean Value Theorem says that at some interior point the instantaneous change must equal the average change over the entire interval.

EXAMPLE 4 If a car accelerating from zero takes 8 sec to go 352 ft, its average velocity for the 8-sec interval is $352/8 = 44$ ft/sec. At some point during the acceleration, the Mean Value Theorem says, the speedometer must read exactly 30 mph (44 ft/sec) (Figure 4.19). ■

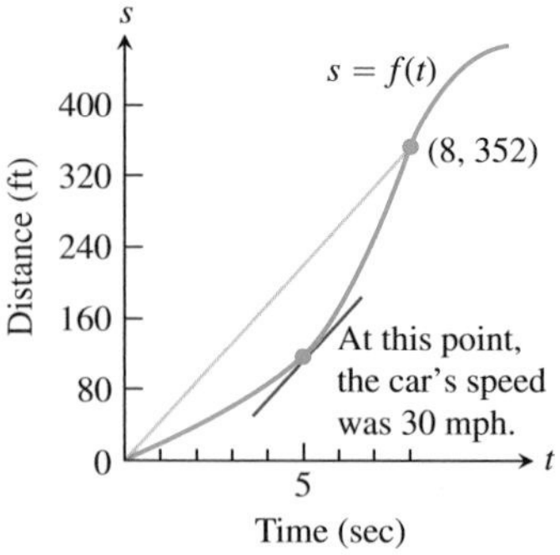

FIGURE 4.19 Distance versus elapsed time for the car in Example 4.

Mathematical Consequences

At the beginning of the section, we asked what kind of function has a zero derivative over an interval. The first corollary of the Mean Value Theorem provides the answer.

COROLLARY 1 Functions with Zero Derivatives Are Constant
If $f'(x) = 0$ at each point x of an open interval (a, b), then $f(x) = C$ for all $x \in (a, b)$, where C is a constant.

Proof We want to show that f has a constant value on the interval (a, b). We do so by showing that if x_1 and x_2 are any two points in (a, b), then $f(x_1) = f(x_2)$. Numbering x_1 and x_2 from left to right, we have $x_1 < x_2$. Then f satisfies the hypotheses of the Mean Value Theorem on $[x_1, x_2]$: It is differentiable at every point of $[x_1, x_2]$ and hence continuous at every point as well. Therefore,

$$\frac{f(x_2) - f(x_1)}{x_2 - x_1} = f'(c)$$

at some point c between x_1 and x_2. Since $f' = 0$ throughout (a, b), this equation translates successively into

$$\frac{f(x_2) - f(x_1)}{x_2 - x_1} = 0, \qquad f(x_2) - f(x_1) = 0, \qquad \text{and} \qquad f(x_1) = f(x_2). \qquad \blacksquare$$

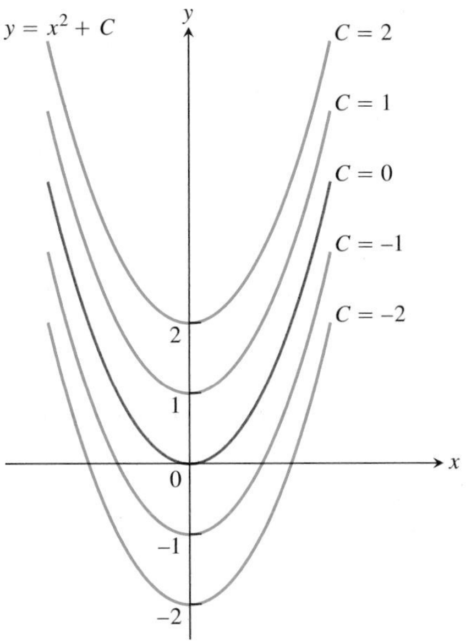

FIGURE 4.20 From a geometric point of view, Corollary 2 of the Mean Value Theorem says that the graphs of functions with identical derivatives on an interval can differ only by a vertical shift there. The graphs of the functions with derivative $2x$ are the parabolas $y = x^2 + C$, shown here for selected values of C.

At the beginning of this section, we also asked about the relationship between two functions that have identical derivatives over an interval. The next corollary tells us that their values on the interval have a constant difference.

> **COROLLARY 2 Functions with the Same Derivative Differ by a Constant**
> If $f'(x) = g'(x)$ at each point x in an open interval (a, b), then there exists a constant C such that $f(x) = g(x) + C$ for all $x \in (a, b)$. That is, $f - g$ is a constant on (a, b).

Proof At each point $x \in (a, b)$ the derivative of the difference function $h = f - g$ is

$$h'(x) = f'(x) - g'(x) = 0.$$

Thus, $h(x) = C$ on (a, b) by Corollary 1. That is, $f(x) - g(x) = C$ on (a, b), so $f(x) = g(x) + C$. $\blacksquare$

Corollaries 1 and 2 are also true if the open interval (a, b) fails to be finite. That is, they remain true if the interval is (a, ∞), $(-\infty, b)$, or $(-\infty, \infty)$.

Corollary 2 plays an important role when we discuss antiderivatives in Section 4.8. It tells us, for instance, that since the derivative of $f(x) = x^2$ on $(-\infty, \infty)$ is $2x$, any other function with derivative $2x$ on $(-\infty, \infty)$ must have the formula $x^2 + C$ for some value of C (Figure 4.20).

EXAMPLE 5 Find the function $f(x)$ whose derivative is $\sin x$ and whose graph passes through the point $(0, 2)$.

Solution Since $f(x)$ has the same derivative as $g(x) = -\cos x$, we know that $f(x) = -\cos x + C$ for some constant C. The value of C can be determined from the condition that $f(0) = 2$ (the graph of f passes through $(0, 2)$):

$$f(0) = -\cos(0) + C = 2, \qquad \text{so} \qquad C = 3.$$

The function is $f(x) = -\cos x + 3$. $\blacksquare$

Finding Velocity and Position from Acceleration

Here is how to find the velocity and displacement functions of a body falling freely from rest with acceleration 9.8 m/sec^2.

We know that $v(t)$ is some function whose derivative is 9.8. We also know that the derivative of $g(t) = 9.8t$ is 9.8. By Corollary 2,

$$v(t) = 9.8t + C$$

for some constant C. Since the body falls from rest, $v(0) = 0$. Thus

$$9.8(0) + C = 0, \qquad \text{and} \qquad C = 0.$$

The velocity function must be $v(t) = 9.8t$. How about the position function $s(t)$?

We know that $s(t)$ is some function whose derivative is $9.8t$. We also know that the derivative of $f(t) = 4.9t^2$ is $9.8t$. By Corollary 2,

$$s(t) = 4.9t^2 + C$$

for some constant C. If the initial height is $s(0) = h$, measured positive downward from the rest position, then

$$4.9(0)^2 + C = h, \qquad \text{and} \qquad C = h.$$

The position function must be $s(t) = 4.9t^2 + h$.

The ability to find functions from their rates of change is one of the very powerful tools of calculus. As we will see, it lies at the heart of the mathematical developments in Chapter 5.

EXERCISES 4.2

Finding c in the Mean Value Theorem

Find the value or values of c that satisfy the equation

$$\frac{f(b) - f(a)}{b - a} = f'(c)$$

in the conclusion of the Mean Value Theorem for the functions and intervals in Exercises 1–4.

1. $f(x) = x^2 + 2x - 1, \quad [0, 1]$

2. $f(x) = x^{2/3}, \quad [0, 1]$

3. $f(x) = x + \frac{1}{x}, \quad \left[\frac{1}{2}, 2\right]$

4. $f(x) = \sqrt{x - 1}, \quad [1, 3]$

Checking and Using Hypotheses

Which of the functions in Exercises 5–8 satisfy the hypotheses of the Mean Value Theorem on the given interval, and which do not? Give reasons for your answers.

5. $f(x) = x^{2/3}, \quad [-1, 8]$

6. $f(x) = x^{4/5}, \quad [0, 1]$

7. $f(x) = \sqrt{x(1 - x)}, \quad [0, 1]$

8. $f(x) = \begin{cases} \dfrac{\sin x}{x}, & -\pi \le x < 0 \\ 0, & x = 0 \end{cases}$

9. The function

$$f(x) = \begin{cases} x, & 0 \le x < 1 \\ 0, & x = 1 \end{cases}$$

is zero at $x = 0$ and $x = 1$ and differentiable on $(0, 1)$, but its derivative on $(0, 1)$ is never zero. How can this be? Doesn't Rolle's Theorem say the derivative has to be zero somewhere in $(0, 1)$? Give reasons for your answer.

10. For what values of a, m and b does the function

$$f(x) = \begin{cases} 3, & x = 0 \\ -x^2 + 3x + a, & 0 < x < 1 \\ mx + b, & 1 \le x \le 2 \end{cases}$$

satisfy the hypotheses of the Mean Value Theorem on the interval $[0, 2]$?

Roots (Zeros)

11. a. Plot the zeros of each polynomial on a line together with the zeros of its first derivative.

i) $y = x^2 - 4$

ii) $y = x^2 + 8x + 15$

iii) $y = x^3 - 3x^2 + 4 = (x + 1)(x - 2)^2$

iv) $y = x^3 - 33x^2 + 216x = x(x - 9)(x - 24)$

b. Use Rolle's Theorem to prove that between every two zeros of $x^n + a_{n-1}x^{n-1} + \cdots + a_1x + a_0$ there lies a zero of

$$nx^{n-1} + (n-1)a_{n-1}x^{n-2} + \cdots + a_1.$$

12. Suppose that f'' is continuous on $[a, b]$ and that f has three zeros in the interval. Show that f'' has at least one zero in (a, b). Generalize this result.

13. Show that if $f'' > 0$ throughout an interval $[a, b]$, then f' has at most one zero in $[a, b]$. What if $f'' < 0$ throughout $[a, b]$ instead?

14. Show that a cubic polynomial can have at most three real zeros.

Show that the functions in Exercises 15–22 have exactly one zero in the given interval.

15. $f(x) = x^4 + 3x + 1, \quad [-2, -1]$

16. $f(x) = x^3 + \dfrac{4}{x^2} + 7, \quad (-\infty, 0)$

17. $g(t) = \sqrt{t} + \sqrt{1+t} - 4, \quad (0, \infty)$

18. $g(t) = \dfrac{1}{1-t} + \sqrt{1+t} - 3.1, \quad (-1, 1)$

19. $r(\theta) = \theta + \sin^2\left(\dfrac{\theta}{3}\right) - 8, \quad (-\infty, \infty)$

20. $r(\theta) = 2\theta - \cos^2\theta + \sqrt{2}, \quad (-\infty, \infty)$

21. $r(\theta) = \sec\theta - \dfrac{1}{\theta^3} + 5, \quad (0, \pi/2)$

22. $r(\theta) = \tan\theta - \cot\theta - \theta, \quad (0, \pi/2)$

Finding Functions from Derivatives

23. Suppose that $f(-1) = 3$ and that $f'(x) = 0$ for all x. Must $f(x) = 3$ for all x? Give reasons for your answer.

24. Suppose that $f(0) = 5$ and that $f'(x) = 2$ for all x. Must $f(x) = 2x + 5$ for all x. Give reasons for your answer.

25. Suppose that $f'(x) = 2x$ for all x. Find $f(2)$ if

a. $f(0) = 0$ **b.** $f(1) = 0$ **c.** $f(-2) = 3$.

26. What can be said about functions whose derivatives are constant? Give reasons for your answer.

In Exercises 27–32, find all possible functions with the given derivative.

27. a. $y' = x$ **b.** $y' = x^2$ **c.** $y' = x^3$

28. a. $y' = 2x$ **b.** $y' = 2x - 1$ **c.** $y' = 3x^2 + 2x - 1$

29. a. $y' = -\dfrac{1}{x^2}$ **b.** $y' = 1 - \dfrac{1}{x^2}$ **c.** $y' = 5 + \dfrac{1}{x^2}$

30. a. $y' = \dfrac{1}{2\sqrt{x}}$ **b.** $y' = \dfrac{1}{\sqrt{x}}$ **c.** $y' = 4x - \dfrac{1}{\sqrt{x}}$

31. a. $y' = \sin 2t$ **b.** $y' = \cos\dfrac{t}{2}$ **c.** $y' = \sin 2t + \cos\dfrac{t}{2}$

32. a. $y' = \sec^2\theta$ **b.** $y' = \sqrt{\theta}$ **c.** $y' = \sqrt{\theta} - \sec^2\theta$

In Exercises 33–36, find the function with the given derivative whose graph passes through the point P.

33. $f'(x) = 2x - 1, \quad P(0, 0)$

34. $g'(x) = \dfrac{1}{x^2} + 2x, \quad P(-1, 1)$

35. $r'(\theta) = 8 - \csc^2\theta, \quad P\left(\dfrac{\pi}{4}, 0\right)$

36. $r'(t) = \sec t \tan t - 1, \quad P(0, 0)$

Finding Position from Velocity

Exercises 37–40 give the velocity $v = ds/dt$ and initial position of a body moving along a coordinate line. Find the body's position at time t.

37. $v = 9.8t + 5, \quad s(0) = 10$

38. $v = 32t - 2, \quad s(0.5) = 4$

39. $v = \sin \pi t, \quad s(0) = 0$

40. $v = \dfrac{2}{\pi}\cos\dfrac{2t}{\pi}, \quad s(\pi^2) = 1$

Finding Position from Acceleration

Exercises 41–44 give the acceleration $a = d^2s/dt^2$, initial velocity, and initial position of a body moving on a coordinate line. Find the body's position at time t.

41. $a = 32, \quad v(0) = 20, \quad s(0) = 5$

42. $a = 9.8, \quad v(0) = -3, \quad s(0) = 0$

43. $a = -4\sin 2t, \quad v(0) = 2, \quad s(0) = -3$

44. $a = \dfrac{9}{\pi^2}\cos\dfrac{3t}{\pi}, \quad v(0) = 0, \quad s(0) = -1$

Applications

45. Temperature change It took 14 sec for a mercury thermometer to rise from $-19°$C to $100°$C when it was taken from a freezer and placed in boiling water. Show that somewhere along the way the mercury was rising at the rate of $8.5°$C/sec.

46. A trucker handed in a ticket at a toll booth showing that in 2 hours she had covered 159 mi on a toll road with speed limit 65 mph. The trucker was cited for speeding. Why?

47. Classical accounts tell us that a 170-oar trireme (ancient Greek or Roman warship) once covered 184 sea miles in 24 hours. Explain why at some point during this feat the trireme's speed exceeded 7.5 knots (sea miles per hour).

48. A marathoner ran the 26.2-mi New York City Marathon in 2.2 hours. Show that at least twice the marathoner was running at exactly 11 mph.

49. Show that at some instant during a 2-hour automobile trip the car's speedometer reading will equal the average speed for the trip.

50. Free fall on the moon On our moon, the acceleration of gravity is 1.6 m/sec^2. If a rock is dropped into a crevasse, how fast will it be going just before it hits bottom 30 sec later?

Theory and Examples

51. The geometric mean of *a* and *b* The *geometric mean* of two positive numbers a and b is the number $\sqrt{ab}$. Show that the value of c in the conclusion of the Mean Value Theorem for $f(x) = 1/x$ on an interval of positive numbers $[a, b]$ is $c = \sqrt{ab}$.

52. The arithmetic mean of *a* and *b* The *arithmetic mean* of two numbers a and b is the number $(a + b)/2$. Show that the value of c in the conclusion of the Mean Value Theorem for $f(x) = x^2$ on any interval $[a, b]$ is $c = (a + b)/2$.

T **53.** Graph the function

$$f(x) = \sin x \sin(x + 2) - \sin^2(x + 1).$$

What does the graph do? Why does the function behave this way? Give reasons for your answers.

54. Rolle's Theorem

a. Construct a polynomial $f(x)$ that has zeros at $x = -2, -1, 0, 1$, and 2.

b. Graph f and its derivative f' together. How is what you see related to Rolle's Theorem?

c. Do $g(x) = \sin x$ and its derivative g' illustrate the same phenomenon?

55. Unique solution Assume that f is continuous on $[a, b]$ and differentiable on (a, b). Also assume that $f(a)$ and $f(b)$ have opposite signs and that $f' \neq 0$ between a and b. Show that $f(x) = 0$ exactly once between a and b.

56. Parallel tangents Assume that f and g are differentiable on $[a, b]$ and that $f(a) = g(a)$ and $f(b) = g(b)$. Show that there is at least one point between a and b where the tangents to the graphs of f and g are parallel or the same line. Illustrate with a sketch.

57. If the graphs of two differentiable functions $f(x)$ and $g(x)$ start at the same point in the plane and the functions have the same rate of change at every point, do the graphs have to be identical? Give reasons for your answer.

58. Show that for any numbers a and b, the inequality $|\sin b - \sin a| \leq |b - a|$ is true.

59. Assume that f is differentiable on $a \leq x \leq b$ and that $f(b) < f(a)$. Show that f' is negative at some point between a and b.

60. Let f be a function defined on an interval $[a, b]$. What conditions could you place on f to guarantee that

$$\min f' \leq \frac{f(b) - f(a)}{b - a} \leq \max f',$$

where $\min f'$ and $\max f'$ refer to the minimum and maximum values of f' on $[a, b]$? Give reasons for your answers.

T **61.** Use the inequalities in Exercise 60 to estimate $f(0.1)$ if $f'(x) = 1/(1 + x^4 \cos x)$ for $0 \leq x \leq 0.1$ and $f(0) = 1$.

T **62.** Use the inequalities in Exercise 60 to estimate $f(0.1)$ if $f'(x) = 1/(1 - x^4)$ for $0 \leq x \leq 0.1$ and $f(0) = 2$.

63. Let f be differentiable at every value of x and suppose that $f(1) = 1$, that $f' < 0$ on $(-\infty, 1)$, and that $f' > 0$ on $(1, \infty)$.

a. Show that $f(x) \geq 1$ for all x.

b. Must $f'(1) = 0$? Explain.

64. Let $f(x) = px^2 + qx + r$ be a quadratic function defined on a closed interval $[a, b]$. Show that there is exactly one point c in (a, b) at which f satisfies the conclusion of the Mean Value Theorem.

4.3 Monotonic Functions and The First Derivative Test

In sketching the graph of a differentiable function it is useful to know where it increases (rises from left to right) and where it decreases (falls from left to right) over an interval. This section defines precisely what it means for a function to be increasing or decreasing over an interval, and gives a test to determine where it increases and where it decreases. We also show how to test the critical points of a function for the presence of local extreme values.

Increasing Functions and Decreasing Functions

What kinds of functions have positive derivatives or negative derivatives? The answer, provided by the Mean Value Theorem's third corollary, is this: The only functions with positive derivatives are increasing functions; the only functions with negative derivatives are decreasing functions.

DEFINITIONS Increasing, Decreasing Function

Let f be a function defined on an interval I and let x_1 and x_2 be any two points in I.

1. If $f(x_1) < f(x_2)$ whenever $x_1 < x_2$, then f is said to be **increasing** on I.
2. If $f(x_2) < f(x_1)$ whenever $x_1 < x_2$, then f is said to be **decreasing** on I.

A function that is increasing or decreasing on I is called **monotonic** on I.

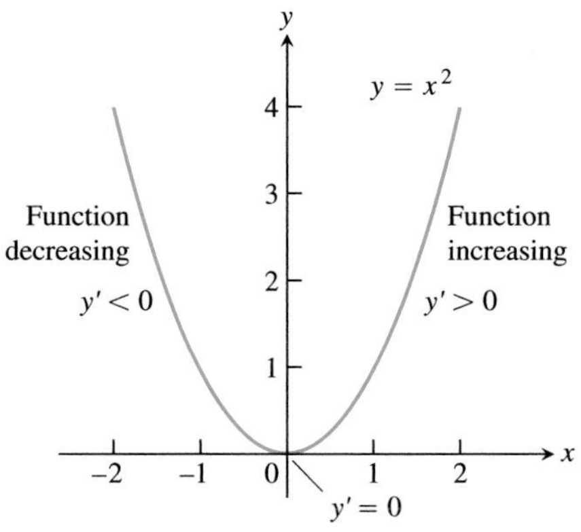

FIGURE 4.21 The function $f(x) = x^2$ is monotonic on the intervals $(-\infty, 0]$ and $[0, \infty)$, but it is not monotonic on $(-\infty, \infty)$.

It is important to realize that the definitions of increasing and decreasing functions must be satisfied for *every* pair of points x_1 and x_2 in I with $x_1 < x_2$. Because of the inequality $<$ comparing the function values, and not $\leq$, some books say that f is *strictly* increasing or decreasing on I. The interval I may be finite or infinite.

The function $f(x) = x^2$ decreases on $(-\infty, 0]$ and increases on $[0, \infty)$ as can be seen from its graph (Figure 4.21). The function f is monotonic on $(-\infty, 0]$ and $[0, \infty)$, but it is not monotonic on $(-\infty, \infty)$. Notice that on the interval $(-\infty, 0)$ the tangents have negative slopes, so the first derivative is always negative there; for $(0, \infty)$ the tangents have positive slopes and the first derivative is positive. The following result confirms these observations.

COROLLARY 3 First Derivative Test for Monotonic Functions

Suppose that f is continuous on $[a, b]$ and differentiable on (a, b).

If $f'(x) > 0$ at each point $x \in (a, b)$, then f is increasing on $[a, b]$.
If $f'(x) < 0$ at each point $x \in (a, b)$, then f is decreasing on $[a, b]$.

Proof Let x_1 and x_2 be any two points in $[a, b]$ with $x_1 < x_2$. The Mean Value Theorem applied to f on $[x_1, x_2]$ says that

$$f(x_2) - f(x_1) = f'(c)(x_2 - x_1)$$

for some c between x_1 and x_2. The sign of the right-hand side of this equation is the same as the sign of $f'(c)$ because $x_2 - x_1$ is positive. Therefore, $f(x_2) > f(x_1)$ if f' is positive on (a, b) and $f(x_2) < f(x_1)$ if f' is negative on (a, b). ■

Here is how to apply the First Derivative Test to find where a function is increasing and decreasing. If $a < b$ are two critical points for a function f, and if f' exists but is not zero on the interval (a, b), then f' must be positive on (a, b) or negative there (Theorem 2, Section 3.1). One way we can determine the sign of f' on the interval is simply by evaluating f' for some point x in (a, b). Then we apply Corollary 3.

EXAMPLE 1 Using the First Derivative Test for Monotonic Functions

Find the critical points of $f(x) = x^3 - 12x - 5$ and identify the intervals on which f is increasing and decreasing.

Solution The function f is everywhere continuous and differentiable. The first derivative

$$\begin{aligned} f'(x) &= 3x^2 - 12 = 3(x^2 - 4) \\ &= 3(x + 2)(x - 2) \end{aligned}$$

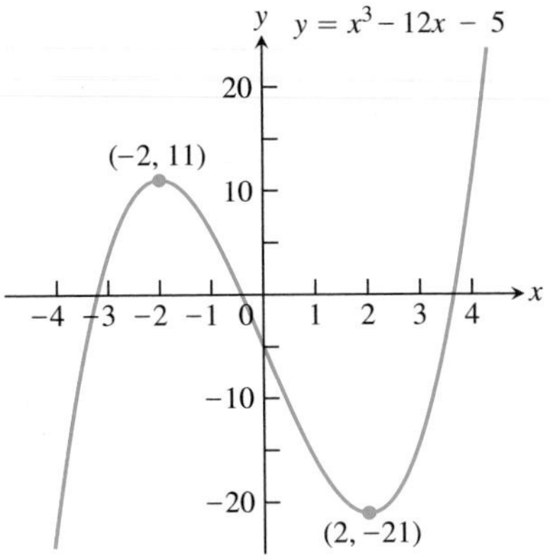

FIGURE 4.22 The function $f(x) = x^3 - 12x - 5$ is monotonic on three separate intervals (Example 1).

is zero at $x = -2$ and $x = 2$. These critical points subdivide the domain of f into intervals $(-\infty, -2)$, $(-2, 2)$, and $(2, \infty)$ on which f' is either positive or negative. We determine the sign of f' by evaluating f at a convenient point in each subinterval. The behavior of f is determined by then applying Corollary 3 to each subinterval. The results are summarized in the following table, and the graph of f is given in Figure 4.22.

Intervals	$-\infty < x < -2$	$-2 < x < 2$	$2 < x < \infty$
f' Evaluated	$f'(-3) = 15$	$f'(0) = -12$	$f'(3) = 15$
Sign of f'	+	−	+
Behavior of f	increasing	decreasing	increasing

■

Corollary 3 is valid for infinite as well as finite intervals, and we used that fact in our analysis in Example 1.

Knowing where a function increases and decreases also tells us how to test for the nature of local extreme values.

HISTORICAL BIOGRAPHY

Edmund Halley
(1656–1742)

First Derivative Test for Local Extrema

In Figure 4.23, at the points where f has a minimum value, $f' < 0$ immediately to the left and $f' > 0$ immediately to the right. (If the point is an endpoint, there is only one side to consider.) Thus, the function is decreasing on the left of the minimum value and it is increasing on its right. Similarly, at the points where f has a maximum value, $f' > 0$ immediately to the left and $f' < 0$ immediately to the right. Thus, the function is increasing on the left of the maximum value and decreasing on its right. In summary, at a local extreme point, the sign of $f'(x)$ changes.

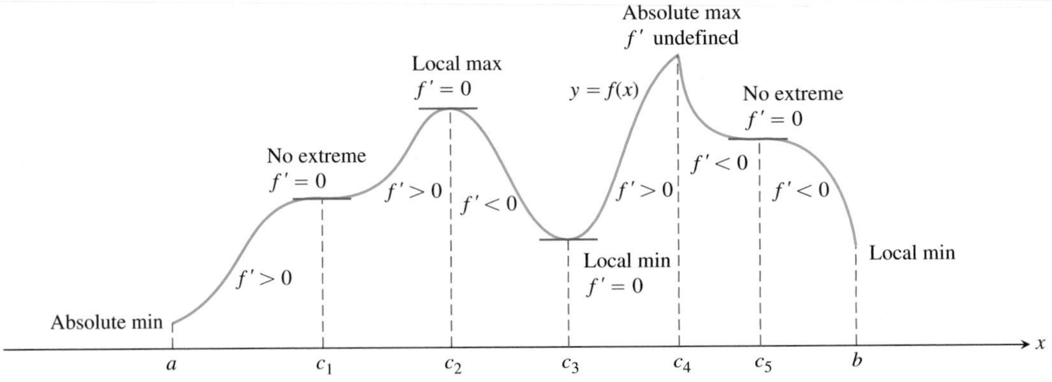

FIGURE 4.23 A function's first derivative tells how the graph rises and falls.

These observations lead to a test for the presence and nature of local extreme values of differentiable functions.

First Derivative Test for Local Extrema

Suppose that c is a critical point of a continuous function f, and that f is differentiable at every point in some interval containing c except possibly at c itself. Moving across c from left to right,

1. if f' changes from negative to positive at c, then f has a local minimum at c;
2. if f' changes from positive to negative at c, then f has a local maximum at c;
3. if f' does not change sign at c (that is, f' is positive on both sides of c or negative on both sides), then f has no local extremum at c.

The test for local extrema at endpoints is similar, but there is only one side to consider.

Proof Part (1). Since the sign of f' changes from negative to positive at c, these are numbers a and b such that $f' < 0$ on (a, c) and $f' > 0$ on (c, b). If $x \in (a, c)$, then $f(c) < f(x)$ because $f' < 0$ implies that f is decreasing on $[a, c]$. If $x \in (c, b)$, then $f(c) < f(x)$ because $f' > 0$ implies that f is increasing on $[c, b]$. Therefore, $f(x) \geq f(c)$ for every $x \in (a, b)$. By definition, f has a local minimum at c.

Parts (2) and (3) are proved similarly. ■

EXAMPLE 2 Using the First Derivative Test for Local Extrema

Find the critical points of

$$f(x) = x^{1/3}(x - 4) = x^{4/3} - 4x^{1/3}.$$

Identify the intervals on which f is increasing and decreasing. Find the function's local and absolute extreme values.

Solution The function f is continuous at all x since it is the product of two continuous functions, $x^{1/3}$ and $(x - 4)$. The first derivative

$$\begin{aligned} f'(x) &= \frac{d}{dx}\left(x^{4/3} - 4x^{1/3}\right) = \frac{4}{3}x^{1/3} - \frac{4}{3}x^{-2/3} \\ &= \frac{4}{3}x^{-2/3}\left(x - 1\right) = \frac{4(x - 1)}{3x^{2/3}} \end{aligned}$$

is zero at $x = 1$ and undefined at $x = 0$. There are no endpoints in the domain, so the critical points $x = 0$ and $x = 1$ are the only places where f might have an extreme value.

The critical points partition the x-axis into intervals on which f' is either positive or negative. The sign pattern of f' reveals the behavior of f between and at the critical points. We can display the information in a table like the following:

Intervals	$x < 0$	$0 < x < 1$	$x > 1$
Sign of f'	$-$	$-$	$+$
Behavior of f	decreasing	decreasing	increasing

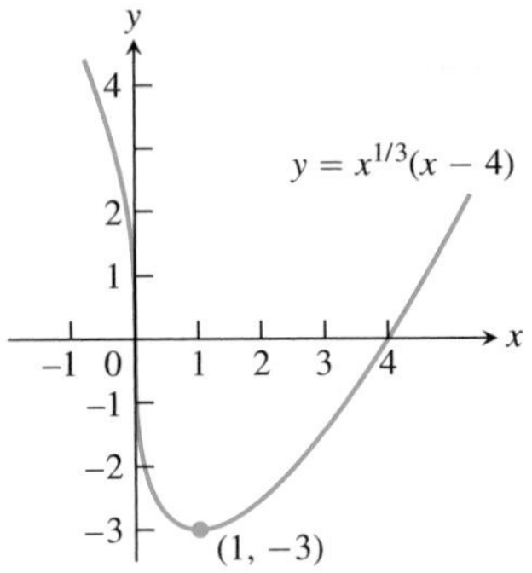

FIGURE 4.24 The function $f(x) = x^{1/3}(x - 4)$ decreases when $x < 1$ and increases when $x > 1$ (Example 2).

Corollary 3 to the Mean Value Theorem tells us that f decreases on $(-\infty, 0)$, decreases on $(0, 1)$, and increases on $(1, \infty)$. The First Derivative Test for Local Extrema tells us that f does not have an extreme value at $x = 0$ (f' does not change sign) and that f has a local minimum at $x = 1$ (f' changes from negative to positive).

The value of the local minimum is $f(1) = 1^{1/3}(1 - 4) = -3$. This is also an absolute minimum because the function's values fall toward it from the left and rise away from it on the right. Figure 4.24 shows this value in relation to the function's graph.

Note that $\lim_{x\to 0} f'(x) = -\infty$, so the graph of f has a vertical tangent at the origin. ■

EXERCISES 4.3

Analyzing f Given f'

Answer the following questions about the functions whose derivatives are given in Exercises 1–8:

a. What are the critical points of f?

b. On what intervals is f increasing or decreasing?

c. At what points, if any, does f assume local maximum and minimum values?

1. $f'(x) = x(x - 1)$ **2.** $f'(x) = (x - 1)(x + 2)$

3. $f'(x) = (x - 1)^2(x + 2)$ **4.** $f'(x) = (x - 1)^2(x + 2)^2$

5. $f'(x) = (x - 1)(x + 2)(x - 3)$

6. $f'(x) = (x - 7)(x + 1)(x + 5)$

7. $f'(x) = x^{-1/3}(x + 2)$ **8.** $f'(x) = x^{-1/2}(x - 3)$

Extremes of Given Functions

In Exercises 9–28:

a. Find the intervals on which the function is increasing and decreasing.

b. Then identify the function's local extreme values, if any, saying where they are taken on.

c. Which, if any, of the extreme values are absolute?

T **d.** Support your findings with a graphing calculator or computer grapher.

9. $g(t) = -t^2 - 3t + 3$ **10.** $g(t) = -3t^2 + 9t + 5$

11. $h(x) = -x^3 + 2x^2$ **12.** $h(x) = 2x^3 - 18x$

13. $f(\theta) = 3\theta^2 - 4\theta^3$ **14.** $f(\theta) = 6\theta - \theta^3$

15. $f(r) = 3r^3 + 16r$ **16.** $h(r) = (r + 7)^3$

17. $f(x) = x^4 - 8x^2 + 16$ **18.** $g(x) = x^4 - 4x^3 + 4x^2$

19. $H(t) = \frac{3}{2}t^4 - t^6$ **20.** $K(t) = 15t^3 - t^5$

21. $g(x) = x\sqrt{8 - x^2}$ **22.** $g(x) = x^2\sqrt{5 - x}$

23. $f(x) = \frac{x^2 - 3}{x - 2}, \quad x \neq 2$ **24.** $f(x) = \frac{x^3}{3x^2 + 1}$

25. $f(x) = x^{1/3}(x + 8)$ **26.** $g(x) = x^{2/3}(x + 5)$

27. $h(x) = x^{1/3}(x^2 - 4)$ **28.** $k(x) = x^{2/3}(x^2 - 4)$

Extreme Values on Half-Open Intervals

In Exercises 29–36:

a. Identify the function's local extreme values in the given domain, and say where they are assumed.

b. Which of the extreme values, if any, are absolute?

T **c.** Support your findings with a graphing calculator or computer grapher.

29. $f(x) = 2x - x^2, \quad -\infty < x \le 2$

30. $f(x) = (x + 1)^2, \quad -\infty < x \le 0$

31. $g(x) = x^2 - 4x + 4, \quad 1 \le x < \infty$

32. $g(x) = -x^2 - 6x - 9, \quad -4 \le x < \infty$

33. $f(t) = 12t - t^3, \quad -3 \le t < \infty$

34. $f(t) = t^3 - 3t^2, \quad -\infty < t \le 3$

35. $h(x) = \frac{x^3}{3} - 2x^2 + 4x, \quad 0 \le x < \infty$

36. $k(x) = x^3 + 3x^2 + 3x + 1, \quad -\infty < x \le 0$

Graphing Calculator or Computer Grapher

In Exercises 37–40:

a. Find the local extrema of each function on the given interval, and say where they are assumed.

T **b.** Graph the function and its derivative together. Comment on the behavior of f in relation to the signs and values of f'.

37. $f(x) = \frac{x}{2} - 2\sin\frac{x}{2}, \quad 0 \le x \le 2\pi$

38. $f(x) = -2\cos x - \cos^2 x, \quad -\pi \le x \le \pi$

39. $f(x) = \csc^2 x - 2\cot x, \quad 0 < x < \pi$

40. $f(x) = \sec^2 x - 2\tan x, \quad \frac{-\pi}{2} < x < \frac{\pi}{2}$

Theory and Examples

Show that the functions in Exercises 41 and 42 have local extreme values at the given values of θ, and say which kind of local extreme the function has.

41. $h(\theta) = 3\cos\frac{\theta}{2}, \quad 0 \le \theta \le 2\pi, \quad \text{at } \theta = 0 \text{ and } \theta = 2\pi$

42. $h(\theta) = 5\sin\frac{\theta}{2}, \quad 0 \le \theta \le \pi, \quad \text{at } \theta = 0 \text{ and } \theta = \pi$

43. Sketch the graph of a differentiable function $y = f(x)$ through the point $(1, 1)$ if $f'(1) = 0$ and

a. $f'(x) > 0$ for $x < 1$ and $f'(x) < 0$ for $x > 1$;

b. $f'(x) < 0$ for $x < 1$ and $f'(x) > 0$ for $x > 1$;

c. $f'(x) > 0$ for $x \ne 1$;

d. $f'(x) < 0$ for $x \ne 1$.

44. Sketch the graph of a differentiable function $y = f(x)$ that has

a. a local minimum at $(1, 1)$ and a local maximum at $(3, 3)$;

b. a local maximum at $(1, 1)$ and a local minimum at $(3, 3)$;

c. local maxima at $(1, 1)$ and $(3, 3)$;

d. local minima at $(1, 1)$ and $(3, 3)$.

45. Sketch the graph of a continuous function $y = g(x)$ such that

a. $g(2) = 2$, $0 < g' < 1$ for $x < 2$, $g'(x) \to 1^-$ as $x \to 2^-$, $-1 < g' < 0$ for $x > 2$, and $g'(x) \to -1^+$ as $x \to 2^+$;

b. $g(2) = 2$, $g' < 0$ for $x < 2$, $g'(x) \to -\infty$ as $x \to 2^-$, $g' > 0$ for $x > 2$, and $g'(x) \to \infty$ as $x \to 2^+$.

46. Sketch the graph of a continuous function $y = h(x)$ such that

a. $h(0) = 0$, $-2 \le h(x) \le 2$ for all x, $h'(x) \to \infty$ as $x \to 0^-$, and $h'(x) \to \infty$ as $x \to 0^+$;

b. $h(0) = 0$, $-2 \le h(x) \le 0$ for all x, $h'(x) \to \infty$ as $x \to 0^-$, and $h'(x) \to -\infty$ as $x \to 0^+$.

47. As x moves from left to right through the point $c = 2$, is the graph of $f(x) = x^3 - 3x + 2$ rising, or is it falling? Give reasons for your answer.

48. Find the intervals on which the function $f(x) = ax^2 + bx + c$, $a \ne 0$, is increasing and decreasing. Describe the reasoning behind your answer.

4.4 Concavity and Curve Sketching

In Section 4.3 we saw how the first derivative tells us where a function is increasing and where it is decreasing. At a critical point of a differentiable function, the First Derivative Test tells us whether there is a local maximum or a local minimum, or whether the graph just continues to rise or fall there.

In this section we see how the second derivative gives information about the way the graph of a differentiable function bends or turns. This additional information enables us to capture key aspects of the behavior of a function and its graph, and then present these features in a sketch of the graph.

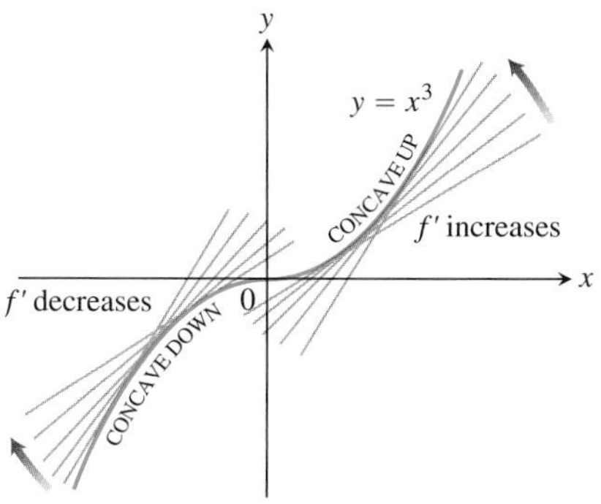

FIGURE 4.25 The graph of $f(x) = x^3$ is concave down on $(-\infty, 0)$ and concave up on $(0, \infty)$ (Example 1a).

Concavity

As you can see in Figure 4.25, the curve $y = x^3$ rises as x increases, but the portions defined on the intervals $(-\infty, 0)$ and $(0, \infty)$ turn in different ways. As we approach the origin from the left along the curve, the curve turns to our right and falls below its tangents. The slopes of the tangents are decreasing on the interval $(-\infty, 0)$. As we move away from the origin along the curve to the right, the curve turns to our left and rises above its tangents. The slopes of the tangents are increasing on the interval $(0, \infty)$. This turning or bending behavior defines the *concavity* of the curve.

DEFINITION Concave Up, Concave Down

The graph of a differentiable function $y = f(x)$ is

(a) concave up on an open interval I if f' is increasing on I

(b) concave down on an open interval I if f' is decreasing on I.

If $y = f(x)$ has a second derivative, we can apply Corollary 3 of the Mean Value Theorem to conclude that f' increases if $f'' > 0$ on I, and decreases if $f'' < 0$.

The Second Derivative Test for Concavity

Let $y = f(x)$ be twice-differentiable on an interval I.

1. If $f'' > 0$ on I, the graph of f over I is concave up.
2. If $f'' < 0$ on I, the graph of f over I is concave down.

If $y = f(x)$ is twice-differentiable, we will use the notations f'' and y'' interchangeably when denoting the second derivative.

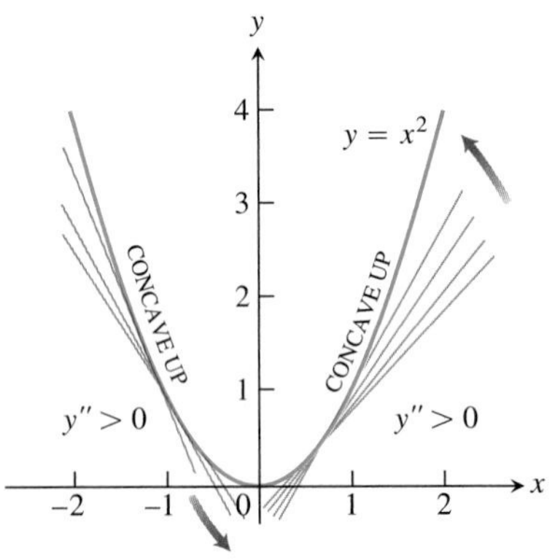

FIGURE 4.26 The graph of $f(x) = x^2$ is concave up on every interval (Example 1b).

EXAMPLE 1 Applying the Concavity Test

(a) The curve $y = x^3$ (Figure 4.25) is concave down on $(-\infty, 0)$ where $y'' = 6x < 0$ and concave up on $(0, \infty)$ where $y'' = 6x > 0$.

(b) The curve $y = x^2$ (Figure 4.26) is concave up on $(-\infty, \infty)$ because its second derivative $y'' = 2$ is always positive. ■

EXAMPLE 2 Determining Concavity

Determine the concavity of $y = 3 + \sin x$ on $[0, 2\pi]$.

Solution The graph of $y = 3 + \sin x$ is concave down on $(0, \pi)$, where $y'' = -\sin x$ is negative. It is concave up on $(\pi, 2\pi)$, where $y'' = -\sin x$ is positive (Figure 4.27). ■

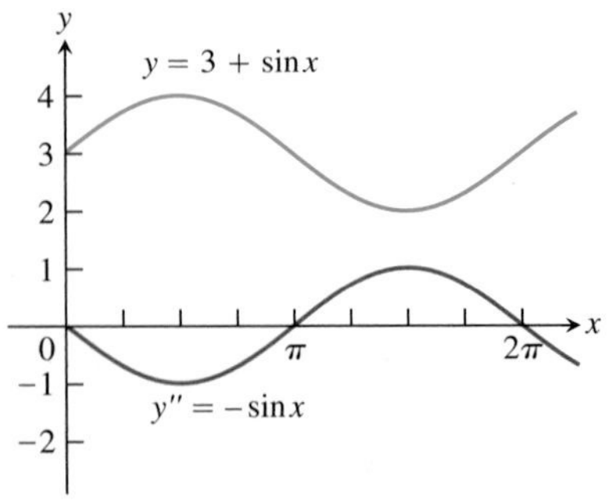

FIGURE 4.27 Using the graph of y'' to determine the concavity of y (Example 2).

Points of Inflection

The curve $y = 3 + \sin x$ in Example 2 changes concavity at the point $(\pi, 3)$. We call $(\pi, 3)$ a *point of inflection* of the curve.

DEFINITION Point of Inflection

A point where the graph of a function has a tangent line and where the concavity changes is a **point of inflection**.

A point on a curve where y'' is positive on one side and negative on the other is a point of inflection. At such a point, y'' is either zero (because derivatives have the Intermediate Value Property) or undefined. If y is a twice-differentiable function, $y'' = 0$ at a point of inflection and y' has a local maximum or minimum.

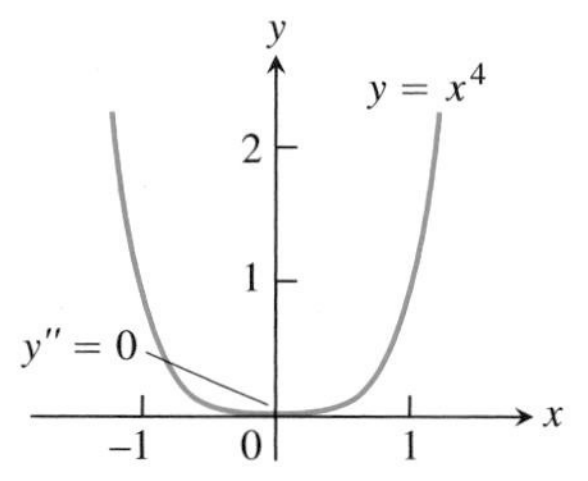

FIGURE 4.28 The graph of $y = x^4$ has no inflection point at the origin, even though $y'' = 0$ there (Example 3).

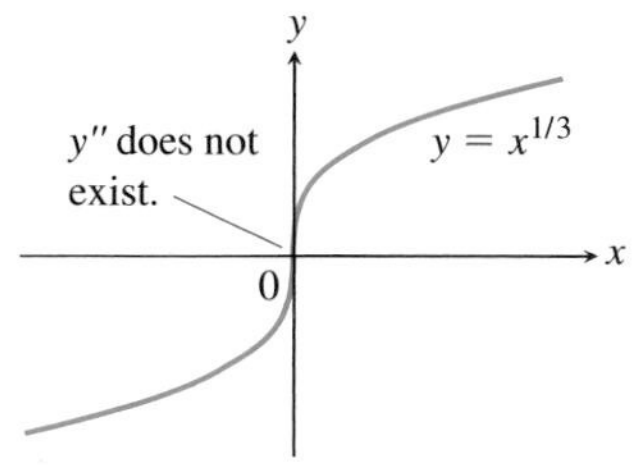

FIGURE 4.29 A point where y'' fails to exist can be a point of inflection (Example 4).

EXAMPLE 3 An Inflection Point May Not Exist Where $y'' = 0$

The curve $y = x^4$ has no inflection point at $x = 0$ (Figure 4.28). Even though $y'' = 12x^2$ is zero there, it does not change sign. ■

EXAMPLE 4 An Inflection Point May Occur Where y'' Does Not Exist

The curve $y = x^{1/3}$ has a point of inflection at $x = 0$ (Figure 4.29), but y'' does not exist there.

$$y'' = \frac{d^2}{dx^2}\left(x^{1/3}\right) = \frac{d}{dx}\left(\frac{1}{3}x^{-2/3}\right) = -\frac{2}{9}x^{-5/3}.$$ ■

We see from Example 3 that a zero second derivative does not always produce a point of inflection. From Example 4, we see that inflection points can also occur where there *is* no second derivative.

To study the motion of a body moving along a line as a function of time, we often are interested in knowing when the body's acceleration, given by the second derivative, is positive or negative. The points of inflection on the graph of the body's position function reveal where the acceleration changes sign.

EXAMPLE 5 Studying Motion Along a Line

A particle is moving along a horizontal line with position function

$$s(t) = 2t^3 - 14t^2 + 22t - 5, \qquad t \geq 0.$$

Find the velocity and acceleration, and describe the motion of the particle.

Solution The velocity is

$$v(t) = s'(t) = 6t^2 - 28t + 22 = 2(t - 1)(3t - 11),$$

and the acceleration is

$$a(t) = v'(t) = s''(t) = 12t - 28 = 4(3t - 7).$$

When the function $s(t)$ is increasing, the particle is moving to the right; when $s(t)$ is decreasing, the particle is moving to the left.

Notice that the first derivative ($v = s'$) is zero when $t = 1$ and $t = 11/3$.

Intervals	$0 < t < 1$	$1 < t < 11/3$	$11/3 < t$
Sign of $v = s'$	+	−	+
Behavior of s	increasing	decreasing	increasing
Particle motion	right	left	right

The particle is moving to the right in the time intervals $[0, 1)$ and $(11/3, \infty)$, and moving to the left in $(1, 11/3)$. It is momentarily stationary (at rest), at $t = 1$ and $t = 11/3$.

The acceleration $a(t) = s''(t) = 4(3t - 7)$ is zero when $t = 7/3$.

Intervals	$0 < t < 7/3$	$7/3 < t$
Sign of $a = s''$	−	+
Graph of s	concave down	concave up

The accelerating force is directed toward the left during the time interval [0, 7/3], is momentarily zero at $t = 7/3$, and is directed toward the right thereafter. ■

Second Derivative Test for Local Extrema

Instead of looking for sign changes in f' at critical points, we can sometimes use the following test to determine the presence and character of local extrema.

THEOREM 5 Second Derivative Test for Local Extrema

Suppose f'' is continuous on an open interval that contains $x = c$.

1. If $f'(c) = 0$ and $f''(c) < 0$, then f has a local maximum at $x = c$.
2. If $f'(c) = 0$ and $f''(c) > 0$, then f has a local minimum at $x = c$.
3. If $f'(c) = 0$ and $f''(c) = 0$, then the test fails. The function f may have a local maximum, a local minimum, or neither.

$f' = 0, f'' < 0$ $\Rightarrow$ local max $f' = 0, f'' > 0$ $\Rightarrow$ local min

Proof Part (1). If $f''(c) < 0$, then $f''(x) < 0$ on some open interval I containing the point c, since f'' is continuous. Therefore, f' is decreasing on I. Since $f'(c) = 0$, the sign of f' changes from positive to negative at c so f has a local maximum at c by the First Derivative Test.

The proof of Part (2) is similar.

For Part (3), consider the three functions $y = x^4$, $y = -x^4$, and $y = x^3$. For each function, the first and second derivatives are zero at $x = 0$. Yet the function $y = x^4$ has a local minimum there, $y = -x^4$ has a local maximum, and $y = x^3$ is increasing in any open interval containing $x = 0$ (having neither a maximum nor a minimum there). Thus the test fails. ■

This test requires us to know f'' *only at c itself* and not in an interval about c. This makes the test easy to apply. That's the good news. The bad news is that the test is inconclusive if $f'' = 0$ or if f'' does not exist at $x = c$. When this happens, use the First Derivative Test for local extreme values.

Together f' and f'' tell us the shape of the function's graph, that is, where the critical points are located and what happens at a critical point, where the function is increasing and where it is decreasing, and how the curve is turning or bending as defined by its concavity. We use this information to sketch a graph of the function that captures its key features.

EXAMPLE 6 Using f' and f'' to Graph f

Sketch a graph of the function

$$f(x) = x^4 - 4x^3 + 10$$

using the following steps.

(a) Identify where the extrema of f occur.

(b) Find the intervals on which f is increasing and the intervals on which f is decreasing.

(c) Find where the graph of f is concave up and where it is concave down.

(d) Sketch the general shape of the graph for f.

(e) Plot some specific points, such as local maximum and minimum points, points of inflection, and intercepts. Then sketch the curve.

Solution f is continuous since $f'(x) = 4x^3 - 12x^2$ exists. The domain of f is $(-\infty, \infty)$, and the domain of f' is also $(-\infty, \infty)$. Thus, the critical points of f occur only at the zeros of f'. Since

$$f'(x) = 4x^3 - 12x^2 = 4x^2(x - 3)$$

the first derivative is zero at $x = 0$ and $x = 3$.

Intervals	$x < 0$	$0 < x < 3$	$3 < x$
Sign of f'	$-$	$-$	$+$
Behavior of f	decreasing	decreasing	increasing

(a) Using the First Derivative Test for local extrema and the table above, we see that there is no extremum at $x = 0$ and a local minimum at $x = 3$.

(b) Using the table above, we see that f is decreasing on $(-\infty, 0]$ and $[0, 3]$, and increasing on $[3, \infty)$.

(c) $f''(x) = 12x^2 - 24x = 12x(x - 2)$ is zero at $x = 0$ and $x = 2$.

Intervals	$x < 0$	$0 < x < 2$	$2 < x$
Sign of f''	$+$	$-$	$+$
Behavior of f	concave up	concave down	concave up

We see that f is concave up on the intervals $(-\infty, 0)$ and $(2, \infty)$, and concave down on $(0, 2)$.

(d) Summarizing the information in the two tables above, we obtain

$x < 0$	$0 < x < 2$	$2 < x < 3$	$3 < x$
decreasing concave up	decreasing concave down	decreasing concave up	increasing concave up

The general shape of the curve is

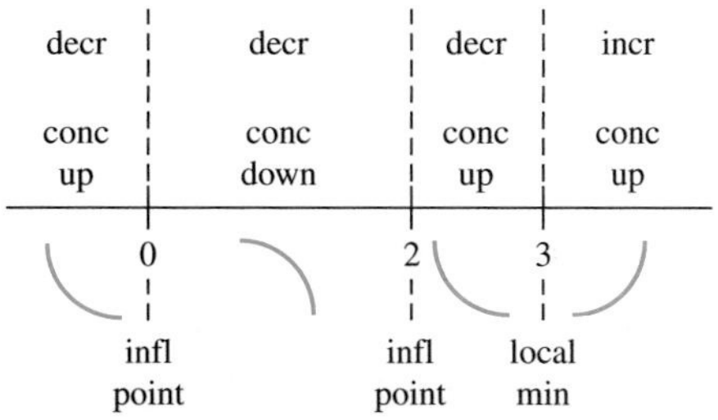

General shape.

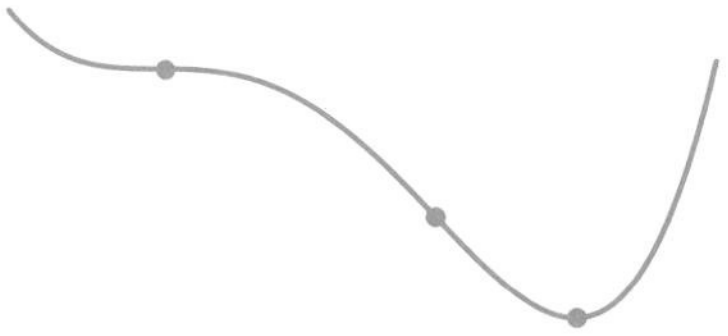

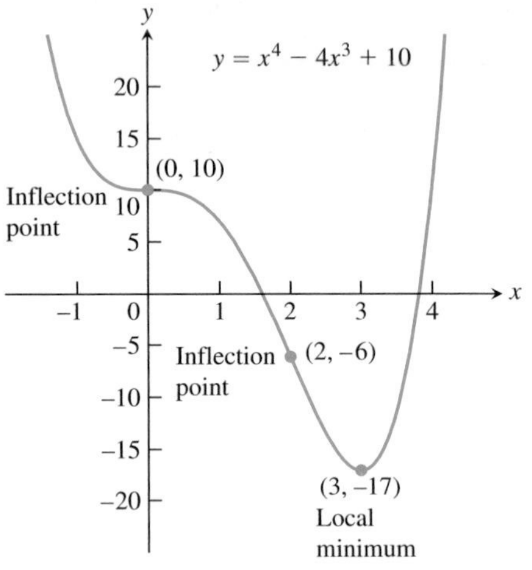

FIGURE 4.30 The graph of $f(x) = x^4 - 4x^3 + 10$ (Example 6).

(e) Plot the curve's intercepts (if possible) and the points where y' and y'' are zero. Indicate any local extreme values and inflection points. Use the general shape as a guide to sketch the curve. (Plot additional points as needed.) Figure 4.30 shows the graph of f. ■

The steps in Example 6 help in giving a procedure for graphing to capture the key features of a function and its graph.

Strategy for Graphing $y = f(x)$

1. Identify the domain of f and any symmetries the curve may have.
2. Find y' and y''.
3. Find the critical points of f, and identify the function's behavior at each one.
4. Find where the curve is increasing and where it is decreasing.
5. Find the points of inflection, if any occur, and determine the concavity of the curve.
6. Identify any asymptotes.
7. Plot key points, such as the intercepts and the points found in Steps 3–5, and sketch the curve.

EXAMPLE 7 Using the Graphing Strategy

Sketch the graph of $f(x) = \dfrac{(x+1)^2}{1+x^2}$.

Solution

1. The domain of f is $(-\infty, \infty)$ and there are no symmetries about either axis or the origin (Section 1.4).
2. *Find f' and f''.*

$$f(x) = \frac{(x+1)^2}{1+x^2}$$ x-intercept at $x = -1$, y-intercept ($y = 1$) at $x = 0$

$$f'(x) = \frac{(1+x^2)\cdot 2(x+1) - (x+1)^2 \cdot 2x}{(1+x^2)^2}$$

$$= \frac{2(1-x^2)}{(1+x^2)^2}$$ Critical points: $x = -1, x = 1$

$$f''(x) = \frac{(1+x^2)^2 \cdot 2(-2x) - 2(1-x^2)[2(1+x^2)\cdot 2x]}{(1+x^2)^4}$$

$$= \frac{4x(x^2-3)}{(1+x^2)^3}$$ After some algebra

3. *Behavior at critical points.* The critical points occur only at $x = \pm 1$ where $f'(x) = 0$ (Step 2) since f' exists everywhere over the domain of f. At $x = -1$, $f''(-1) = 1 > 0$ yielding a relative minimum by the Second Derivative Test. At $x = 1$, $f''(1) = -1 < 0$ yielding a relative maximum by the Second Derivative Test. We will see in Step 6 that both are absolute extrema as well.

4. *Increasing and decreasing*. We see that on the interval $(-\infty, -1)$ the derivative $f'(x) < 0$, and the curve is decreasing. On the interval $(-1, 1)$, $f'(x) > 0$ and the curve is increasing; it is decreasing on $(1, \infty)$ where $f'(x) < 0$ again.
5. *Inflection points*. Notice that the denominator of the second derivative (Step 2) is always positive. The second derivative f'' is zero when $x = -\sqrt{3}, 0$, and $\sqrt{3}$. The second derivative changes sign at each of these points: negative on $\left(-\infty, -\sqrt{3}\right)$, positive on $\left(-\sqrt{3}, 0\right)$, negative on $\left(0, \sqrt{3}\right)$, and positive again on $\left(\sqrt{3}, \infty\right)$. Thus each point is a point of inflection. The curve is concave down on the interval $\left(-\infty, -\sqrt{3}\right)$, concave up on $\left(-\sqrt{3}, 0\right)$, concave down on $\left(0, \sqrt{3}\right)$, and concave up again on $\left(\sqrt{3}, \infty\right)$.
6. *Asymptotes*. Expanding the numerator of $f(x)$ and then dividing both numerator and denominator by x^2 gives

$$f(x) = \frac{(x+1)^2}{1+x^2} = \frac{x^2+2x+1}{1+x^2} \qquad \text{Expanding numerator}$$

$$= \frac{1 + (2/x) + (1/x^2)}{(1/x^2) + 1}. \qquad \text{Dividing by } x^2$$

We see that $f(x) \to 1^+$ as $x \to \infty$ and that $f(x) \to 1^-$ as $x \to -\infty$. Thus, the line $y = 1$ is a horizontal asymptote.

Since f decreases on $(-\infty, -1)$ and then increases on $(-1, 1)$, we know that $f(-1) = 0$ is a local minimum. Although f decreases on $(1, \infty)$, it never crosses the horizontal asymptote $y = 1$ on that interval (it approaches the asymptote from above). So the graph never becomes negative, and $f(-1) = 0$ is an absolute minimum as well. Likewise, $f(1) = 2$ is an absolute maximum because the graph never crosses the asymptote $y = 1$ on the interval $(-\infty, -1)$, approaching it from below. Therefore, there are no vertical asymptotes (the range of f is $0 \le y \le 2$).

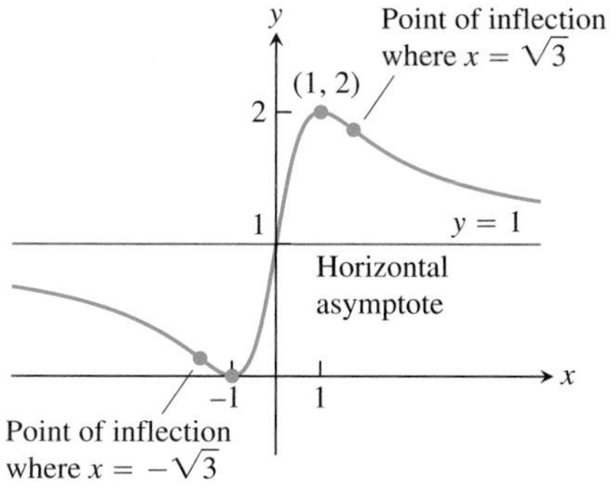

FIGURE 4.31 The graph of $y = \dfrac{(x+1)^2}{1+x^2}$ (Example 7).

7. The graph of f is sketched in Figure 4.31. Notice how the graph is concave down as it approaches the horizontal asymptote $y = 1$ as $x \to -\infty$, and concave up in its approach to $y = 1$ as $x \to \infty$. ■

Learning About Functions from Derivatives

As we saw in Examples 6 and 7, we can learn almost everything we need to know about a twice-differentiable function $y = f(x)$ by examining its first derivative. We can find where the function's graph rises and falls and where any local extrema are assumed. We can differentiate y' to learn how the graph bends as it passes over the intervals of rise and fall. We can determine the shape of the function's graph. Information we cannot get from the derivative is how to place the graph in the xy-plane. But, as we discovered in Section 4.2, the only additional information we need to position the graph is the value of f at one point. The derivative does not give us information about the asymptotes, which are found using limits (Sections 2.4 and 2.5).

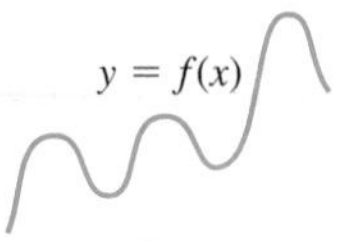 Differentiable ⇒ smooth, connected; graph may rise and fall	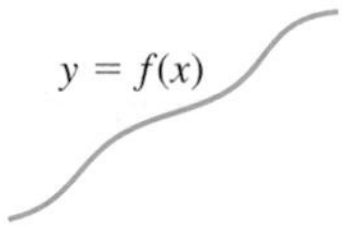 $y' > 0 \Rightarrow$ rises from left to right; may be wavy	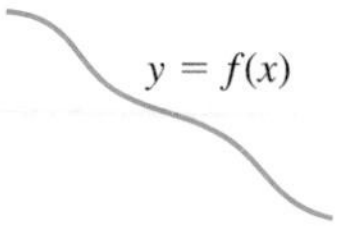 $y' < 0 \Rightarrow$ falls from left to right; may be wavy
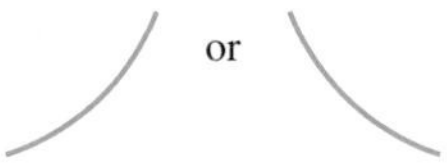 $y'' > 0 \Rightarrow$ concave up throughout; no waves; graph may rise or fall	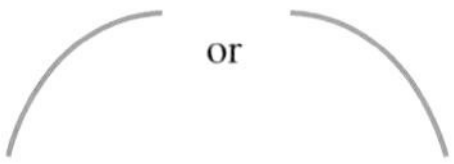 $y'' < 0 \Rightarrow$ concave down throughout; no waves; graph may rise or fall	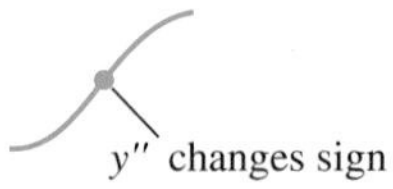 Inflection point
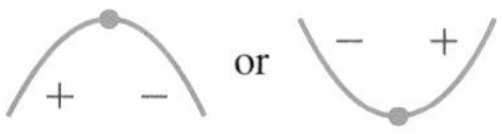 y' changes sign ⇒ graph has local maximum or local minimum	 $y' = 0$ and $y'' < 0$ at a point; graph has local maximum	 $y' = 0$ and $y'' > 0$ at a point; graph has local minimum

EXERCISES 4.4

Analyzing Graphed Functions

Identify the inflection points and local maxima and minima of the functions graphed in Exercises 1–8. Identify the intervals on which the functions are concave up and concave down.

1. $y = \frac{x^3}{3} - \frac{x^2}{2} - 2x + \frac{1}{3}$

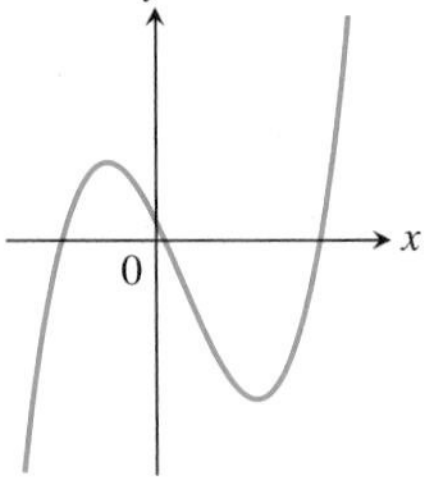

2. $y = \frac{x^4}{4} - 2x^2 + 4$

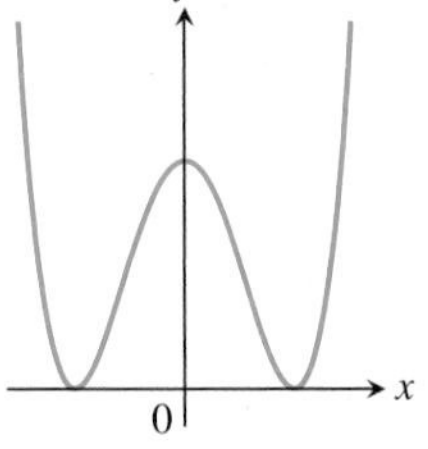

3. $y = \frac{3}{4}(x^2 - 1)^{2/3}$

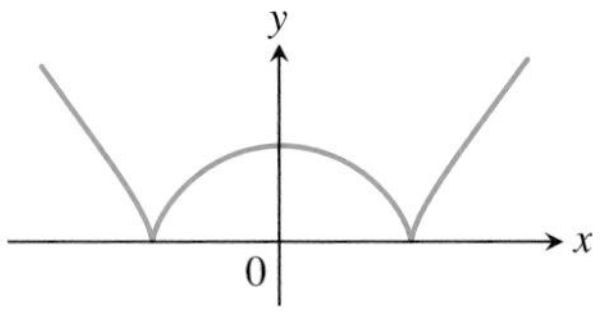

4. $y = \frac{9}{14}x^{1/3}(x^2 - 7)$

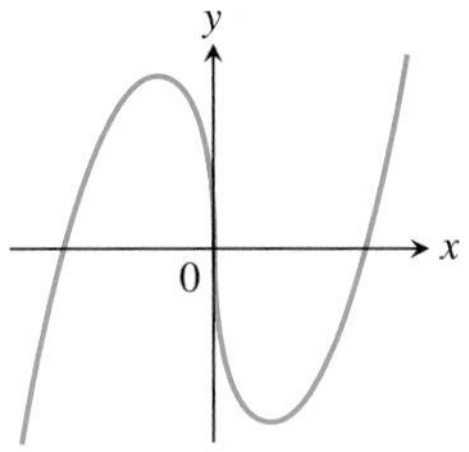

5. $y = x + \sin 2x,\ -\frac{2\pi}{3} \le x \le \frac{2\pi}{3}$

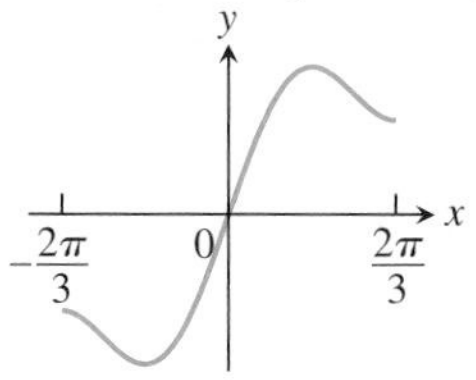

6. $y = \tan x - 4x,\ -\frac{\pi}{2} < x < \frac{\pi}{2}$

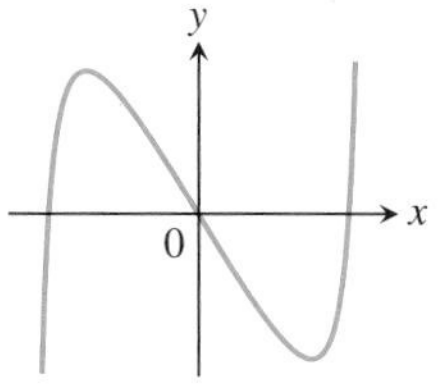

7. $y = \sin |x|,\ -2\pi \le x \le 2\pi$

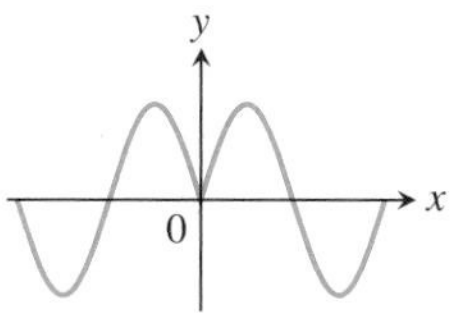

NOT TO SCALE

8. $y = 2\cos x - \sqrt{2}x,\ -\pi \le x \le \frac{3\pi}{2}$

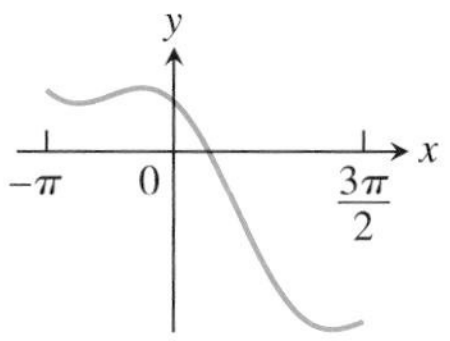

Graphing Equations

Use the steps of the graphing procedure on page 272 to graph the equations in Exercises 9–40. Include the coordinates of any local extreme points and inflection points.

9. $y = x^2 - 4x + 3$

10. $y = 6 - 2x - x^2$

11. $y = x^3 - 3x + 3$

12. $y = x(6 - 2x)^2$

13. $y = -2x^3 + 6x^2 - 3$

14. $y = 1 - 9x - 6x^2 - x^3$

15. $y = (x - 2)^3 + 1$

16. $y = 1 - (x + 1)^3$

17. $y = x^4 - 2x^2 = x^2(x^2 - 2)$

18. $y = -x^4 + 6x^2 - 4 = x^2(6 - x^2) - 4$

19. $y = 4x^3 - x^4 = x^3(4 - x)$

20. $y = x^4 + 2x^3 = x^3(x + 2)$

21. $y = x^5 - 5x^4 = x^4(x - 5)$

22. $y = x\left(\frac{x}{2} - 5\right)^4$

23. $y = x + \sin x,\quad 0 \le x \le 2\pi$

24. $y = x - \sin x,\quad 0 \le x \le 2\pi$

25. $y = x^{1/5}$

26. $y = x^{3/5}$

27. $y = x^{2/5}$

28. $y = x^{4/5}$

29. $y = 2x - 3x^{2/3}$

30. $y = 5x^{2/5} - 2x$

31. $y = x^{2/3}\left(\frac{5}{2} - x\right)$

32. $y = x^{2/3}(x - 5)$

33. $y = x\sqrt{8 - x^2}$

34. $y = (2 - x^2)^{3/2}$

35. $y = \dfrac{x^2 - 3}{x - 2},\quad x \ne 2$

36. $y = \dfrac{x^3}{3x^2 + 1}$

37. $y = |x^2 - 1|$

38. $y = |x^2 - 2x|$

39. $y = \sqrt{|x|} = \begin{cases} \sqrt{-x}, & x < 0 \\ \sqrt{x}, & x \ge 0 \end{cases}$

40. $y = \sqrt{|x - 4|}$

Sketching the General Shape Knowing *y′*

Each of Exercises 41–62 gives the first derivative of a continuous function $y = f(x)$. Find y'' and then use steps 2–4 of the graphing procedure on page 272 to sketch the general shape of the graph of f.

41. $y' = 2 + x - x^2$

42. $y' = x^2 - x - 6$

43. $y' = x(x - 3)^2$

44. $y' = x^2(2 - x)$

45. $y' = x(x^2 - 12)$

46. $y' = (x - 1)^2(2x + 3)$

47. $y' = (8x - 5x^2)(4 - x)^2$

48. $y' = (x^2 - 2x)(x - 5)^2$

49. $y' = \sec^2 x,\quad -\frac{\pi}{2} < x < \frac{\pi}{2}$

50. $y' = \tan x,\quad -\frac{\pi}{2} < x < \frac{\pi}{2}$

51. $y' = \cot \frac{\theta}{2},\quad 0 < \theta < 2\pi$

52. $y' = \csc^2 \frac{\theta}{2},\quad 0 < \theta < 2\pi$

53. $y' = \tan^2 \theta - 1,\quad -\frac{\pi}{2} < \theta < \frac{\pi}{2}$

54. $y' = 1 - \cot^2 \theta,\quad 0 < \theta < \pi$

55. $y' = \cos t,\quad 0 \le t \le 2\pi$

56. $y' = \sin t,\quad 0 \le t \le 2\pi$

57. $y' = (x + 1)^{-2/3}$

58. $y' = (x - 2)^{-1/3}$

59. $y' = x^{-2/3}(x - 1)$

60. $y' = x^{-4/5}(x + 1)$

61. $y' = 2|x| = \begin{cases} -2x, & x < 0 \\ 2x, & x \ge 0 \end{cases}$

62. $y' = \begin{cases} -x^2, & x \le 0 \\ x^2, & x > 0 \end{cases}$

Sketching *y* from Graphs of *y′* and *y″*

Each of Exercises 63–66 shows the graphs of the first and second derivatives of a function $y = f(x)$. Copy the picture and add to it a sketch of the approximate graph of f, given that the graph passes through the point P.

63.

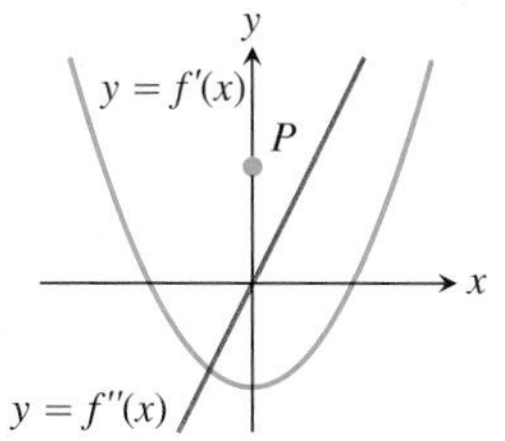

64.

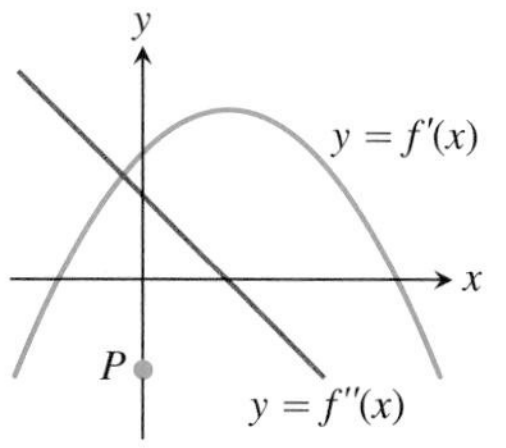

65.

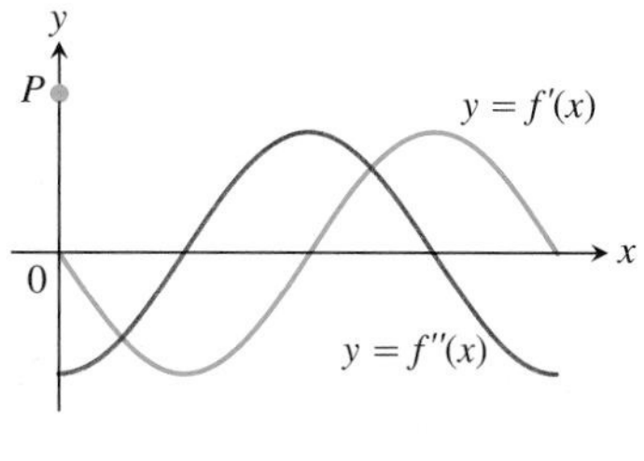

66.

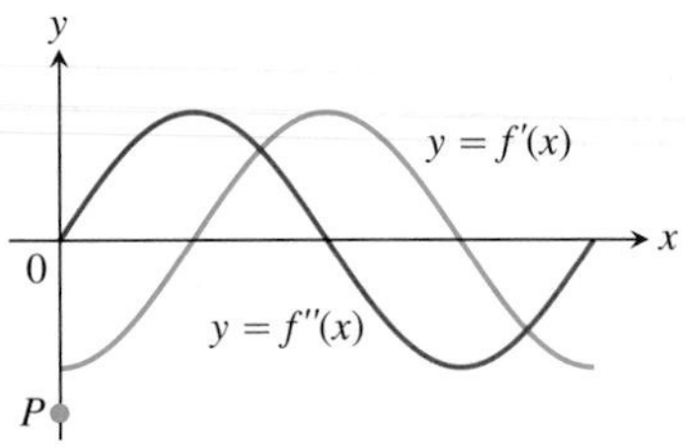

Theory and Examples

67. The accompanying figure shows a portion of the graph of a twice-differentiable function $y = f(x)$. At each of the five labeled points, classify y' and y'' as positive, negative, or zero.

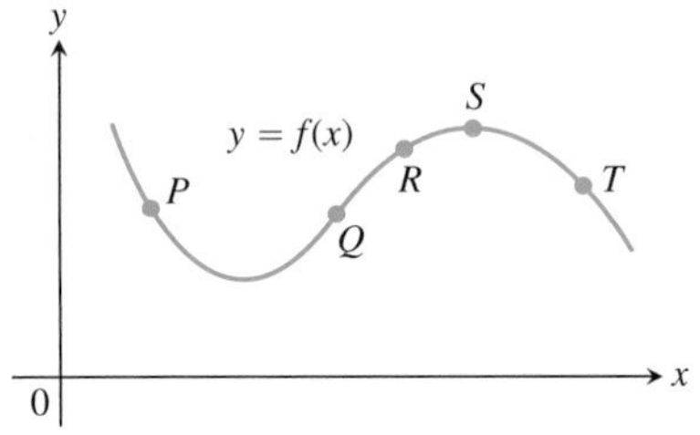

68. Sketch a smooth connected curve $y = f(x)$ with

$f(-2) = 8,$ $\quad f'(2) = f'(-2) = 0,$

$f(0) = 4,$ $\quad f'(x) < 0$ for $|x| < 2,$

$f(2) = 0,$ $\quad f''(x) < 0$ for $x < 0,$

$f'(x) > 0$ for $|x| > 2,$ $\quad f''(x) > 0$ for $x > 0.$

69. Sketch the graph of a twice-differentiable function $y = f(x)$ with the following properties. Label coordinates where possible.

x	y	Derivatives
$x < 2$		$y' < 0,\ y'' > 0$
2	1	$y' = 0,\ y'' > 0$
$2 < x < 4$		$y' > 0,\ y'' > 0$
4	4	$y' > 0,\ y'' = 0$
$4 < x < 6$		$y' > 0,\ y'' < 0$
6	7	$y' = 0,\ y'' < 0$
$x > 6$		$y' < 0,\ y'' < 0$

70. Sketch the graph of a twice-differentiable function $y = f(x)$ that passes through the points $(-2, 2), (-1, 1), (0, 0), (1, 1)$ and $(2, 2)$ and whose first two derivatives have the following sign patterns:

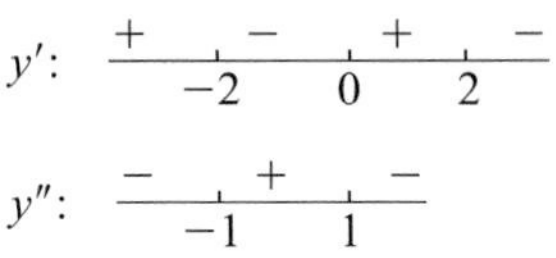

Motion Along a Line The graphs in Exercises 71 and 72 show the position $s = f(t)$ of a body moving back and forth on a coordinate line. (**a**) When is the body moving away from the origin? Toward the origin? At approximately what times is the (**b**) velocity equal to zero? (**c**) Acceleration equal to zero? (**d**) When is the acceleration positive? Negative?

71.

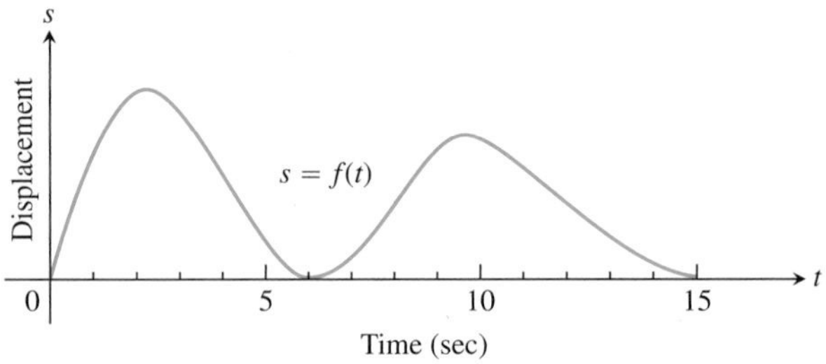

72.

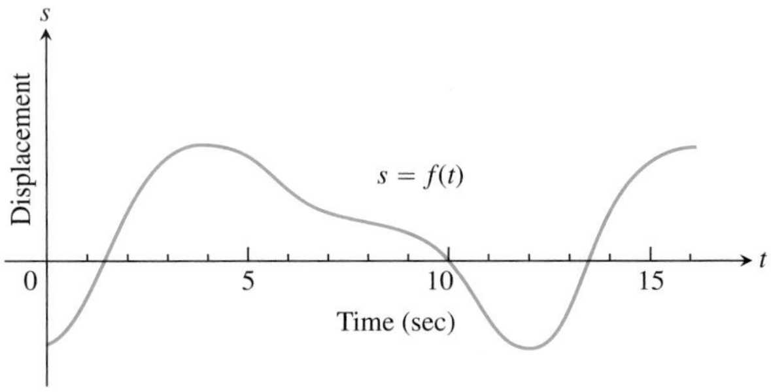

73. Marginal cost The accompanying graph shows the hypothetical cost $c = f(x)$ of manufacturing x items. At approximately what production level does the marginal cost change from decreasing to increasing?

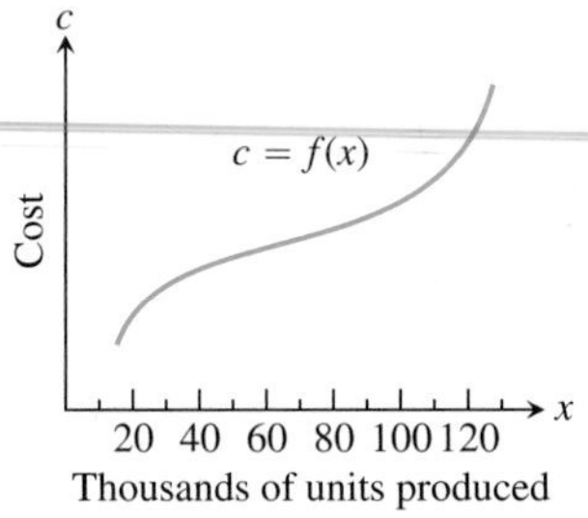

74. The accompanying graph shows the monthly revenue of the Widget Corporation for the last 12 years. During approximately what time intervals was the marginal revenue increasing? decreasing?

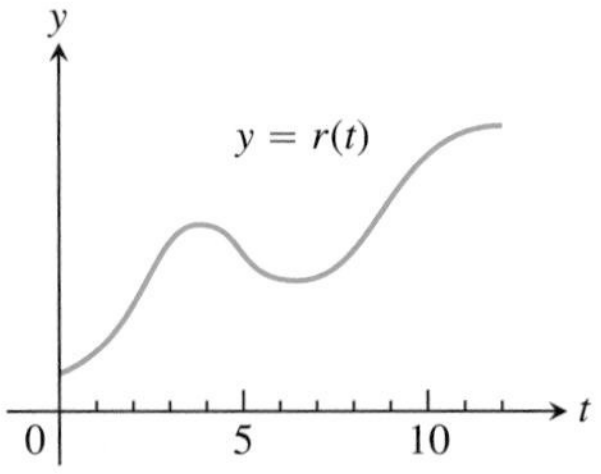

75. Suppose the derivative of the function $y = f(x)$ is

$$y' = (x - 1)^2(x - 2).$$

At what points, if any, does the graph of f have a local minimum, local maximum, or point of inflection? (*Hint:* Draw the sign pattern for y'.)

76. Suppose the derivative of the function $y = f(x)$ is

$$y' = (x - 1)^2(x - 2)(x - 4).$$

At what points, if any, does the graph of f have a local minimum, local maximum, or point of inflection?

77. For $x > 0$, sketch a curve $y = f(x)$ that has $f(1) = 0$ and $f'(x) = 1/x$. Can anything be said about the concavity of such a curve? Give reasons for your answer.

78. Can anything be said about the graph of a function $y = f(x)$ that has a continuous second derivative that is never zero? Give reasons for your answer.

79. If b, c, and d are constants, for what value of b will the curve $y = x^3 + bx^2 + cx + d$ have a point of inflection at $x = 1$? Give reasons for your answer.

80. Horizontal tangents True, or false? Explain.

a. The graph of every polynomial of even degree (largest exponent even) has at least one horizontal tangent.

b. The graph of every polynomial of odd degree (largest exponent odd) has at least one horizontal tangent.

81. Parabolas

a. Find the coordinates of the vertex of the parabola $y = ax^2 + bx + c, a \neq 0$.

b. When is the parabola concave up? Concave down? Give reasons for your answers.

82. Is it true that the concavity of the graph of a twice-differentiable function $y = f(x)$ changes every time $f''(x) = 0$? Give reasons for your answer.

83. Quadratic curves What can you say about the inflection points of a quadratic curve $y = ax^2 + bx + c, a \neq 0$? Give reasons for your answer.

84. Cubic curves What can you say about the inflection points of a cubic curve $y = ax^3 + bx^2 + cx + d, a \neq 0$? Give reasons for your answer.

COMPUTER EXPLORATIONS

In Exercises 85–88, find the inflection points (if any) on the graph of the function and the coordinates of the points on the graph where the function has a local maximum or local minimum value. Then graph the function in a region large enough to show all these points simultaneously. Add to your picture the graphs of the function's first and second derivatives. How are the values at which these graphs intersect the x-axis related to the graph of the function? In what other ways are the graphs of the derivatives related to the graph of the function?

85. $y = x^5 - 5x^4 - 240$

86. $y = x^3 - 12x^2$

87. $y = \frac{4}{5}x^5 + 16x^2 - 25$

88. $y = \frac{x^4}{4} - \frac{x^3}{3} - 4x^2 + 12x + 20$

89. Graph $f(x) = 2x^4 - 4x^2 + 1$ and its first two derivatives together. Comment on the behavior of f in relation to the signs and values of f' and f''.

90. Graph $f(x) = x \cos x$ and its second derivative together for $0 \leq x \leq 2\pi$. Comment on the behavior of the graph of f in relation to the signs and values of f''.

91. a. On a common screen, graph $f(x) = x^3 + kx$ for $k = 0$ and nearby positive and negative values of k. How does the value of k seem to affect the shape of the graph?

b. Find $f'(x)$. As you will see, $f'(x)$ is a quadratic function of x. Find the discriminant of the quadratic (the discriminant of $ax^2 + bx + c$ is $b^2 - 4ac$). For what values of k is the discriminant positive? Zero? Negative? For what values of k does f' have two zeros? One or no zeros? Now explain what the value of k has to do with the shape of the graph of f.

c. Experiment with other values of k. What appears to happen as $k \to -\infty$? as $k \to \infty$?

92. a. On a common screen, graph $f(x) = x^4 + kx^3 + 6x^2$, $-2 \leq x \leq 2$ for $k = -4$, and some nearby integer values of k. How does the value of k seem to affect the shape of the graph?

b. Find $f''(x)$. As you will see, $f''(x)$ is a quadratic function of x. What is the discriminant of this quadratic (see Exercise 91(b))? For what values of k is the discriminant positive? Zero? Negative? For what values of k does $f''(x)$ have two zeros? One or no zeros? Now explain what the value of k has to do with the shape of the graph of f.

93. a. Graph $y = x^{2/3}(x^2 - 2)$ for $-3 \leq x \leq 3$. Then use calculus to confirm what the screen shows about concavity, rise, and fall. (Depending on your grapher, you may have to enter $x^{2/3}$ as $(x^2)^{1/3}$ to obtain a plot for negative values of x.)

b. Does the curve have a cusp at $x = 0$, or does it just have a corner with different right-hand and left-hand derivatives?

94. a. Graph $y = 9x^{2/3}(x - 1)$ for $-0.5 \leq x \leq 1.5$. Then use calculus to confirm what the screen shows about concavity, rise, and fall. What concavity does the curve have to the left of the origin? (Depending on your grapher, you may have to enter $x^{2/3}$ as $(x^2)^{1/3}$ to obtain a plot for negative values of x.)

b. Does the curve have a cusp at $x = 0$, or does it just have a corner with different right-hand and left-hand derivatives?

95. Does the curve $y = x^2 + 3 \sin 2x$ have a horizontal tangent near $x = -3$? Give reasons for your answer.

4.5 Applied Optimization Problems

To optimize something means to maximize or minimize some aspect of it. What are the dimensions of a rectangle with fixed perimeter having maximum area? What is the least expensive shape for a cylindrical can? What is the size of the most profitable production run? The differential calculus is a powerful tool for solving problems that call for maximizing or minimizing a function. In this section we solve a variety of optimization problems from business, mathematics, physics, and economics.

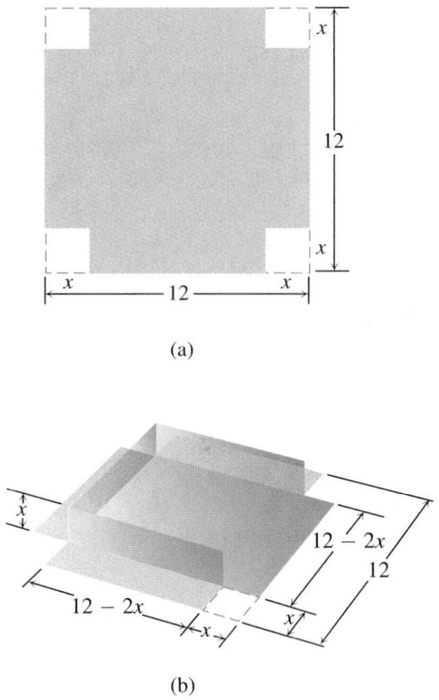

FIGURE 4.32 An open box made by cutting the corners from a square sheet of tin. What size corners maximize the box's volume (Example 1)?

Examples from Business and Industry

EXAMPLE 1 Fabricating a Box

An open-top box is to be made by cutting small congruent squares from the corners of a 12-in.-by-12-in. sheet of tin and bending up the sides. How large should the squares cut from the corners be to make the box hold as much as possible?

Solution We start with a picture (Figure 4.32). In the figure, the corner squares are x in. on a side. The volume of the box is a function of this variable:

$$V(x) = x(12 - 2x)^2 = 144x - 48x^2 + 4x^3. \qquad V = hlw$$

Since the sides of the sheet of tin are only 12 in. long, $x \le 6$ and the domain of V is the interval $0 \le x \le 6$.

A graph of V (Figure 4.33) suggests a minimum value of 0 at $x = 0$ and $x = 6$ and a maximum near $x = 2$. To learn more, we examine the first derivative of V with respect to x:

$$\frac{dV}{dx} = 144 - 96x + 12x^2 = 12(12 - 8x + x^2) = 12(2 - x)(6 - x).$$

Of the two zeros, $x = 2$ and $x = 6$, only $x = 2$ lies in the interior of the function's domain and makes the critical-point list. The values of V at this one critical point and two endpoints are

Critical-point value: $V(2) = 128$

Endpoint values: $V(0) = 0, \qquad V(6) = 0.$

The maximum volume is 128 in.^{3}. The cutout squares should be 2 in. on a side. ■

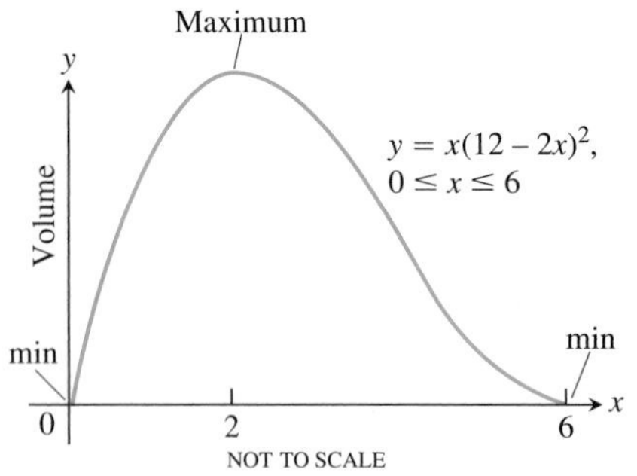

FIGURE 4.33 The volume of the box in Figure 4.32 graphed as a function of x.

EXAMPLE 2 Designing an Efficient Cylindrical Can

You have been asked to design a 1-liter can shaped like a right circular cylinder (Figure 4.34). What dimensions will use the least material?

Solution *Volume of can:* If r and h are measured in centimeters, then the volume of the can in cubic centimeters is

$$\pi r^2 h = 1000. \qquad 1 \text{ liter} = 1000 \text{ cm}^3$$

Surface area of can: $A = \underbrace{2\pi r^2}_{\text{circular ends}} + \underbrace{2\pi rh}_{\text{circular wall}}$

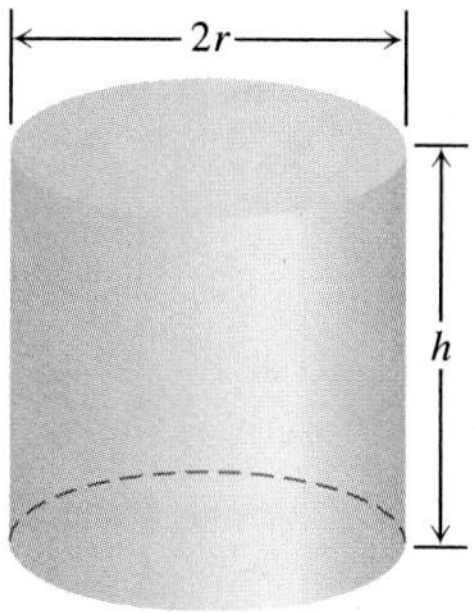

FIGURE 4.34 This 1-L can uses the least material when $h = 2r$ (Example 2).

How can we interpret the phrase "least material"? First, it is customary to ignore the thickness of the material and the waste in manufacturing. Then we ask for dimensions r and h that make the total surface area as small as possible while satisfying the constraint $\pi r^2 h = 1000$.

To express the surface area as a function of one variable, we solve for one of the variables in $\pi r^2 h = 1000$ and substitute that expression into the surface area formula. Solving for h is easier:

$$h = \frac{1000}{\pi r^2}.$$

Thus,

$$\begin{aligned} A &= 2\pi r^2 + 2\pi r h \\ &= 2\pi r^2 + 2\pi r\left(\frac{1000}{\pi r^2}\right) \\ &= 2\pi r^2 + \frac{2000}{r}. \end{aligned}$$

Our goal is to find a value of $r > 0$ that minimizes the value of A. Figure 4.35 suggests that such a value exists.

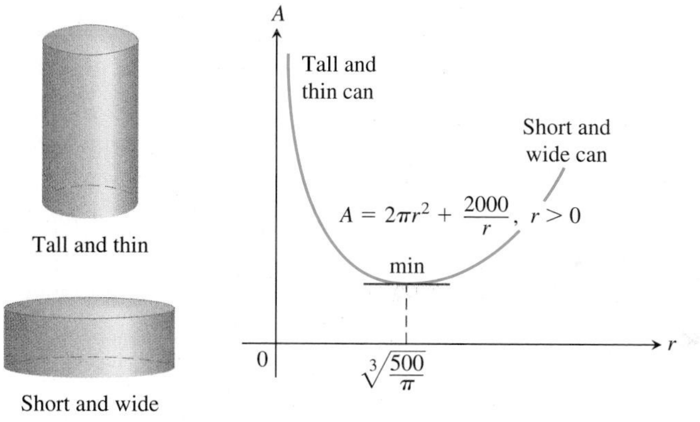

FIGURE 4.35 The graph of $A = 2\pi r^2 + 2000/r$ is concave up.

Notice from the graph that for small r (a tall thin container, like a piece of pipe), the term $2000/r$ dominates and A is large. For large r (a short wide container, like a pizza pan), the term $2\pi r^2$ dominates and A again is large.

Since A is differentiable on $r > 0$, an interval with no endpoints, it can have a minimum value only where its first derivative is zero.

$$\begin{aligned} \frac{dA}{dr} &= 4\pi r - \frac{2000}{r^2} & \\ 0 &= 4\pi r - \frac{2000}{r^2} & \text{Set } dA/dr = 0. \\ 4\pi r^3 &= 2000 & \text{Multiply by } r^2. \\ r &= \sqrt[3]{\frac{500}{\pi}} \approx 5.42 & \text{Solve for } r. \end{aligned}$$

What happens at $r = \sqrt[3]{500/\pi}$?

The second derivative

$$\frac{d^2A}{dr^2} = 4\pi + \frac{4000}{r^3}$$

is positive throughout the domain of A. The graph is therefore everywhere concave up and the value of A at $r = \sqrt[3]{500/\pi}$ an absolute minimum.

The corresponding value of h (after a little algebra) is

$$h = \frac{1000}{\pi r^2} = 2\sqrt[3]{\frac{500}{\pi}} = 2r.$$

The 1-L can that uses the least material has height equal to the diameter, here with $r \approx 5.42$ cm and $h \approx 10.84$ cm. ■

Solving Applied Optimization Problems

1. *Read the problem.* Read the problem until you understand it. What is given? What is the unknown quantity to be optimized?
2. *Draw a picture.* Label any part that may be important to the problem.
3. *Introduce variables.* List every relation in the picture and in the problem as an equation or algebraic expression, and identify the unknown variable.
4. *Write an equation for the unknown quantity.* If you can, express the unknown as a function of a single variable or in two equations in two unknowns. This may require considerable manipulation.
5. *Test the critical points and endpoints in the domain of the unknown.* Use what you know about the shape of the function's graph. Use the first and second derivatives to identify and classify the function's critical points.

Examples from Mathematics and Physics

EXAMPLE 3 Inscribing Rectangles

A rectangle is to be inscribed in a semicircle of radius 2. What is the largest area the rectangle can have, and what are its dimensions?

Solution Let $(x, \sqrt{4 - x^2})$ be the coordinates of the corner of the rectangle obtained by placing the circle and rectangle in the coordinate plane (Figure 4.36). The length, height, and area of the rectangle can then be expressed in terms of the position x of the lower right-hand corner:

$$\text{Length: } 2x, \qquad \text{Height: } \sqrt{4 - x^2}, \qquad \text{Area: } 2x \cdot \sqrt{4 - x^2}.$$

Notice that the values of x are to be found in the interval $0 \le x \le 2$, where the selected corner of the rectangle lies.

Our goal is to find the absolute maximum value of the function

$$A(x) = 2x\sqrt{4 - x^2}$$

on the domain [0, 2].

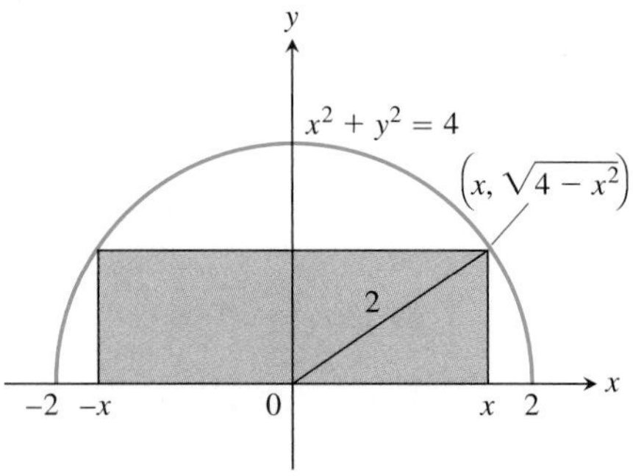

FIGURE 4.36 The rectangle inscribed in the semicircle in Example 3.

The derivative

$$\frac{dA}{dx} = \frac{-2x^2}{\sqrt{4 - x^2}} + 2\sqrt{4 - x^2}$$

is not defined when $x = 2$ and is equal to zero when

$$\begin{aligned} \frac{-2x^2}{\sqrt{4 - x^2}} + 2\sqrt{4 - x^2} &= 0 \\ -2x^2 + 2(4 - x^2) &= 0 \\ 8 - 4x^2 &= 0 \\ x^2 = 2 \text{ or } x &= \pm\sqrt{2}. \end{aligned}$$

Of the two zeros, $x = \sqrt{2}$ and $x = -\sqrt{2}$, only $x = \sqrt{2}$ lies in the interior of A's domain and makes the critical-point list. The values of A at the endpoints and at this one critical point are

Critical-point value: $A\left(\sqrt{2}\right) = 2\sqrt{2}\sqrt{4 - 2} = 4$

Endpoint values: $A(0) = 0, \quad A(2) = 0.$

The area has a maximum value of 4 when the rectangle is $\sqrt{4 - x^2} = \sqrt{2}$ units high and $2x = 2\sqrt{2}$ unit long. ■

HISTORICAL BIOGRAPHY

Willebrord Snell van Royen (1580–1626)

EXAMPLE 4 Fermat's Principle and Snell's Law

The speed of light depends on the medium through which it travels, and is generally slower in denser media.

Fermat's principle in optics states that light travels from one point to another along a path for which the time of travel is a minimum. Find the path that a ray of light will follow in going from a point A in a medium where the speed of light is c_1 to a point B in a second medium where its speed is c_2.

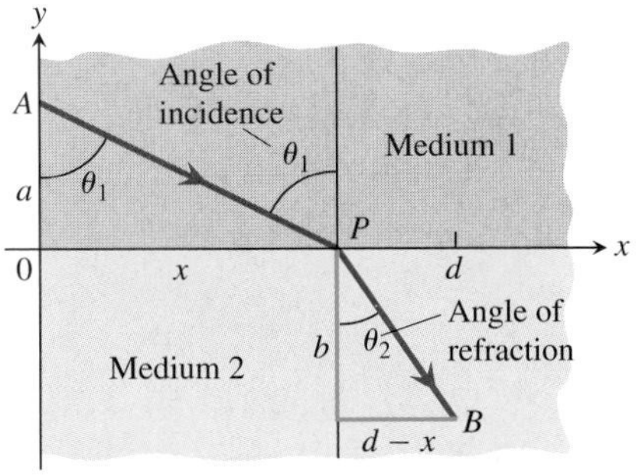

FIGURE 4.37 A light ray refracted (deflected from its path) as it passes from one medium to a denser medium (Example 4).

Solution Since light traveling from A to B follows the quickest route, we look for a path that will minimize the travel time. We assume that A and B lie in the xy-plane and that the line separating the two media is the x-axis (Figure 4.37).

In a uniform medium, where the speed of light remains constant, "shortest time" means "shortest path," and the ray of light will follow a straight line. Thus the path from A to B will consist of a line segment from A to a boundary point P, followed by another line segment from P to B. Distance equals rate times time, so

$$\text{Time} = \frac{\text{distance}}{\text{rate}}.$$

The time required for light to travel from A to P is

$$t_1 = \frac{AP}{c_1} = \frac{\sqrt{a^2 + x^2}}{c_1}.$$

From P to B, the time is

$$t_2 = \frac{PB}{c_2} = \frac{\sqrt{b^2 + (d - x)^2}}{c_2}.$$

The time from A to B is the sum of these:

$$t = t_1 + t_2 = \frac{\sqrt{a^2 + x^2}}{c_1} + \frac{\sqrt{b^2 + (d - x)^2}}{c_2}.$$

This equation expresses t as a differentiable function of x whose domain is $[0, d]$. We want to find the absolute minimum value of t on this closed interval. We find the derivative

$$\frac{dt}{dx} = \frac{x}{c_1\sqrt{a^2 + x^2}} - \frac{d - x}{c_2\sqrt{b^2 + (d - x)^2}}.$$

In terms of the angles θ_1 and θ_2 in Figure 4.37,

$$\frac{dt}{dx} = \frac{\sin\theta_1}{c_1} - \frac{\sin\theta_2}{c_2}.$$

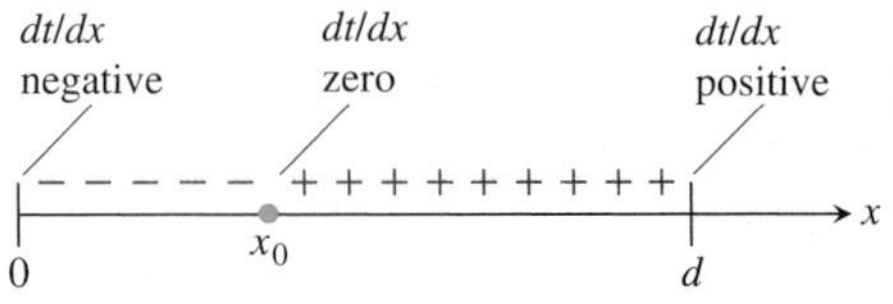

FIGURE 4.38 The sign pattern of dt/dx in Example 4.

If we restrict x to the interval $0 \le x \le d$, then t has a negative derivative at $x = 0$ and a positive derivative at $x = d$. By the Intermediate Value Theorem for Derivatives (Section 3.1), there is a point $x_0 \in [0, d]$ where $dt/dx = 0$ (Figure 4.38). There is only one such point because dt/dx is an increasing function of x (Exercise 54). At this point

$$\frac{\sin\theta_1}{c_1} = \frac{\sin\theta_2}{c_2}.$$

This equation is **Snell's Law** or the **Law of Refraction**, and is an important principle in the theory of optics. It describes the path the ray of light follows. ■

Examples from Economics

In these examples we point out two ways that calculus makes a contribution to economics. The first has to do with maximizing profit. The second has to do with minimizing average cost.

Suppose that

$$r(x) = \text{the revenue from selling } x \text{ items}$$
$$c(x) = \text{the cost of producing the } x \text{ items}$$
$$p(x) = r(x) - c(x) = \text{the profit from producing and selling } x \text{ items}.$$

The **marginal revenue, marginal cost,** and **marginal profit** when producing and selling x items are

$$\frac{dr}{dx} = \text{marginal revenue},$$

$$\frac{dc}{dx} = \text{marginal cost},$$

$$\frac{dp}{dx} = \text{marginal profit}.$$

The first observation is about the relationship of p to these derivatives.

If $r(x)$ and $c(x)$ are differentiable for all $x > 0$, and if $p(x) = r(x) - c(x)$ has a maximum value, it occurs at a production level at which $p'(x) = 0$. Since $p'(x) = r'(x) - c'(x)$, $p'(x) = 0$ implies that

$$r'(x) - c'(x) = 0 \quad \text{or} \quad r'(x) = c'(x).$$

Therefore

> At a production level yielding maximum profit, marginal revenue equals marginal cost (Figure 4.39).

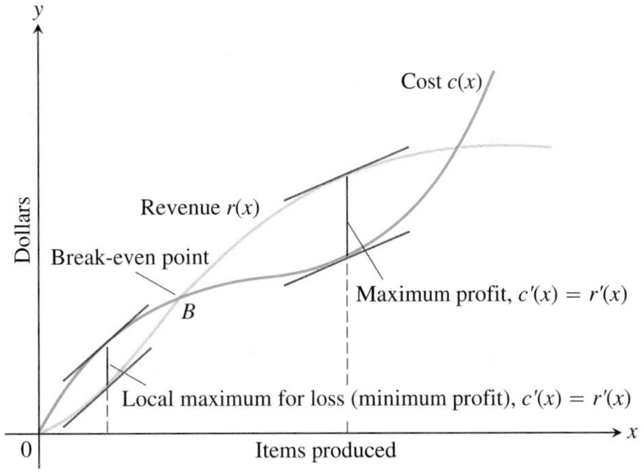

FIGURE 4.39 The graph of a typical cost function starts concave down and later turns concave up. It crosses the revenue curve at the break-even point B. To the left of B, the company operates at a loss. To the right, the company operates at a profit, with the maximum profit occurring where $c'(x) = r'(x)$. Farther to the right, cost exceeds revenue (perhaps because of a combination of rising labor and material costs and market saturation) and production levels become unprofitable again.

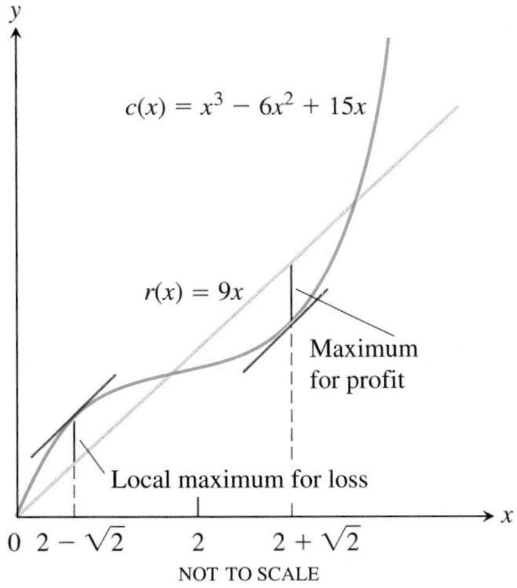

FIGURE 4.40 The cost and revenue curves for Example 5.

EXAMPLE 5 Maximizing Profit

Suppose that $r(x) = 9x$ and $c(x) = x^3 - 6x^2 + 15x$, where x represents thousands of units. Is there a production level that maximizes profit? If so, what is it?

Solution Notice that $r'(x) = 9$ and $c'(x) = 3x^2 - 12x + 15$.

$$3x^2 - 12x + 15 = 9 \qquad \text{Set } c'(x) = r'(x).$$

$$3x^2 - 12x + 6 = 0$$

The two solutions of the quadratic equation are

$$x_1 = \frac{12 - \sqrt{72}}{6} = 2 - \sqrt{2} \approx 0.586 \qquad \text{and}$$

$$x_2 = \frac{12 + \sqrt{72}}{6} = 2 + \sqrt{2} \approx 3.414.$$

The possible production levels for maximum profit are $x \approx 0.586$ thousand units or $x \approx 3.414$ thousand units. The second derivative of $p(x) = r(x) - c(x)$ is $p''(x) = -c''(x)$ since $r''(x)$ is everywhere zero. Thus, $p''(x) = 6(2 - x)$ which is negative at $x = 2 + \sqrt{2}$ and positive at $x = 2 - \sqrt{2}$. By the Second Derivative Test, a maximum profit occurs at about $x = 3.414$ (where revenue exceeds costs) and maximum loss occurs at about $x = 0.586$. The graph of $r(x)$ is shown in Figure 4.40. ■

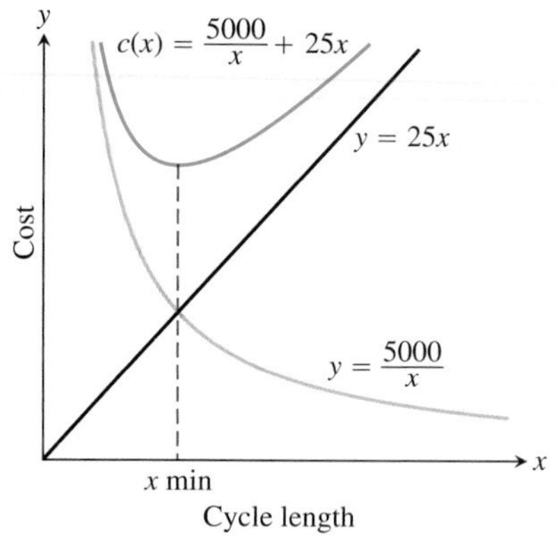

FIGURE 4.41 The average daily cost $c(x)$ is the sum of a hyperbola and a linear function (Example 6).

EXAMPLE 6 Minimizing Costs

A cabinetmaker uses plantation-farmed mahogany to produce 5 furnishings each day. Each delivery of one container of wood is \$5000, whereas the storage of that material is \$10 per day per unit stored, where a unit is the amount of material needed by her to produce 1 furnishing. How much material should be ordered each time and how often should the material be delivered to minimize her average daily cost in the production cycle between deliveries?

Solution If she asks for a delivery every x days, then she must order $5x$ units to have enough material for that delivery cycle. The *average* amount in storage is approximately one-half of the delivery amount, or $5x/2$. Thus, the cost of delivery and storage for each cycle is approximately

$$\text{Cost per cycle} = \text{delivery costs} + \text{storage costs}$$

$$\text{Cost per cycle} = \underbrace{5000}_{\substack{\text{delivery}\\ \text{cost}}} + \underbrace{\left(\frac{5x}{2}\right)}_{\substack{\text{average}\\ \text{amount stored}}} \cdot \underbrace{x}_{\substack{\text{number of}\\ \text{days stored}}} \cdot \underbrace{10}_{\substack{\text{storage cost}\\ \text{per day}}}$$

We compute the *average daily cost* $c(x)$ by dividing the cost per cycle by the number of days x in the cycle (see Figure 4.41).

$$c(x) = \frac{5000}{x} + 25x, \qquad x > 0.$$

As $x \to 0$ and as $x \to \infty$, the average daily cost becomes large. So we expect a minimum to exist, but where? Our goal is to determine the number of days x between deliveries that provides the absolute minimum cost.

We find the critical points by determining where the derivative is equal to zero:

$$c'(x) = -\frac{5000}{x^2} + 25 = 0$$
$$x = \pm\sqrt{200} \approx \pm 14.14.$$

Of the two critical points, only $\sqrt{200}$ lies in the domain of $c(x)$. The critical-point value of the average daily cost is

$$c\left(\sqrt{200}\right) = \frac{5000}{\sqrt{200}} + 25\sqrt{200} = 500\sqrt{2} \approx \$707.11.$$

We note that $c(x)$ is defined over the open interval $(0, \infty)$ with $c''(x) = 10000/x^3 > 0$. Thus, an absolute minimum exists at $x = \sqrt{200} \approx 14.14$ days.

The cabinetmaker should schedule a delivery of $5(14) = 70$ units of the exotic wood every 14 days. ■

In Examples 5 and 6 we allowed the number of items x to be any positive real number. In reality it usually only makes sense for x to be a positive integer (or zero). If we must round our answers, should we round up or down?

EXAMPLE 7 Sensitivity of the Minimum Cost

Should we round the number of days between deliveries up or down for the best solution in Example 6?

Solution The average daily cost will increase by about \$0.03 if we round down from 14.14 to 14 days:

$$c(14) = \frac{5000}{14} + 25(14) = \$707.14$$

and

$$c(14) - c(14.14) = \$707.14 - \$707.11 = \$0.03.$$

On the other hand, $c(15) = \$708.33$, and our cost would increase by $\$708.33 - \$707.11 = \$1.22$ if we round up. Thus, it is better that we round x down to 14 days. ■

EXERCISES 4.5

Whenever you are maximizing or minimizing a function of a single variable, we urge you to graph it over the domain that is appropriate to the problem you are solving. The graph will provide insight before you calculate and will furnish a visual context for understanding your answer.

Applications in Geometry

1. **Minimizing perimeter** What is the smallest perimeter possible for a rectangle whose area is 16 in.^{2}, and what are its dimensions?

2. Show that among all rectangles with an 8-m perimeter, the one with largest area is a square.

3. The figure shows a rectangle inscribed in an isosceles right triangle whose hypotenuse is 2 units long.
   a. Express the y-coordinate of P in terms of x. (*Hint*: Write an equation for the line AB.)
   b. Express the area of the rectangle in terms of x.
   c. What is the largest area the rectangle can have, and what are its dimensions?

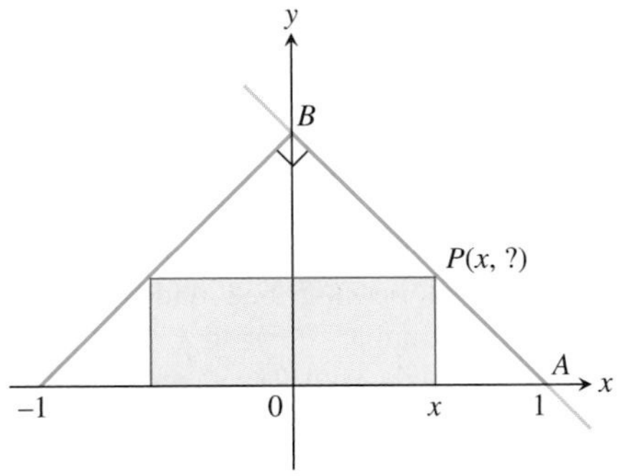

4. A rectangle has its base on the x-axis and its upper two vertices on the parabola $y = 12 - x^2$. What is the largest area the rectangle can have, and what are its dimensions?

5. You are planning to make an open rectangular box from an 8-in.-by-15-in. piece of cardboard by cutting congruent squares from the corners and folding up the sides. What are the dimensions of the box of largest volume you can make this way, and what is its volume?

6. You are planning to close off a corner of the first quadrant with a line segment 20 units long running from $(a, 0)$ to $(0, b)$. Show that the area of the triangle enclosed by the segment is largest when $a = b$.

7. **The best fencing plan** A rectangular plot of farmland will be bounded on one side by a river and on the other three sides by a single-strand electric fence. With 800m of wire at your disposal, what is the largest area you can enclose, and what are its dimensions?

8. **The shortest fence** A 216 m^2 rectangular pea patch is to be enclosed by a fence and divided into two equal parts by another fence parallel to one of the sides. What dimensions for the outer rectangle will require the smallest total length of fence? How much fence will be needed?

9. **Designing a tank** Your iron works has contracted to design and build a 500 ft^3, square-based, open-top, rectangular steel holding tank for a paper company. The tank is to be made by welding thin stainless steel plates together along their edges. As the production engineer, your job is to find dimensions for the base and height that will make the tank weigh as little as possible.
   a. What dimensions do you tell the shop to use?
   b. Briefly describe how you took weight into account.

10. **Catching rainwater** A 1125 ft^3 open-top rectangular tank with a square base x ft on a side and y ft deep is to be built with its top flush with the ground to catch runoff water. The costs associated with the tank involve not only the material from which the tank is made but also an excavation charge proportional to the product xy.
    a. If the total cost is
    $$c = 5(x^2 + 4xy) + 10xy,$$
    what values of x and y will minimize it?
    b. Give a possible scenario for the cost function in part (a).

11. Designing a poster You are designing a rectangular poster to contain 50 in.^{2} of printing with a 4-in. margin at the top and bottom and a 2-in. margin at each side. What overall dimensions will minimize the amount of paper used?

12. Find the volume of the largest right circular cone that can be inscribed in a sphere of radius 3.

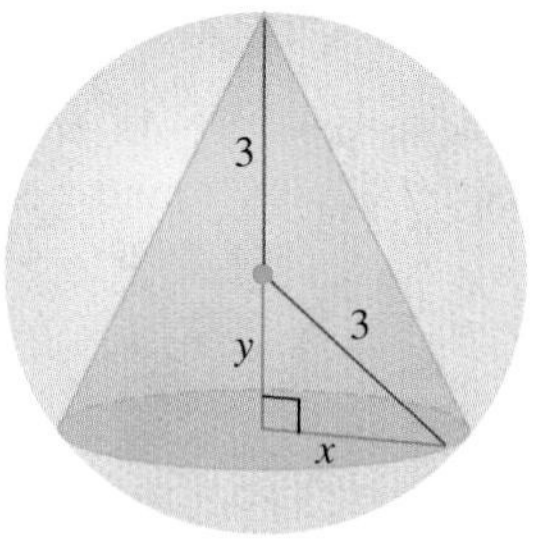

13. Two sides of a triangle have lengths a and b, and the angle between them is θ. What value of θ will maximize the triangle's area? (*Hint:* $A = (1/2)ab \sin \theta$.)

14. Designing a can What are the dimensions of the lightest open-top right circular cylindrical can that will hold a volume of 1000 cm^3? Compare the result here with the result in Example 2.

15. Designing a can You are designing a 1000 cm^3 right circular cylindrical can whose manufacture will take waste into account. There is no waste in cutting the aluminum for the side, but the top and bottom of radius r will be cut from squares that measure $2r$ units on a side. The total amount of aluminum used up by the can will therefore be

$$A = 8r^2 + 2\pi rh$$

rather than the $A = 2\pi r^2 + 2\pi rh$ in Example 2. In Example 2, the ratio of h to r for the most economical can was 2 to 1. What is the ratio now?

T **16. Designing a box with a lid** A piece of cardboard measures 10 in. by 15 in. Two equal squares are removed from the corners of a 10-in. side as shown in the figure. Two equal rectangles are removed from the other corners so that the tabs can be folded to form a rectangular box with lid.

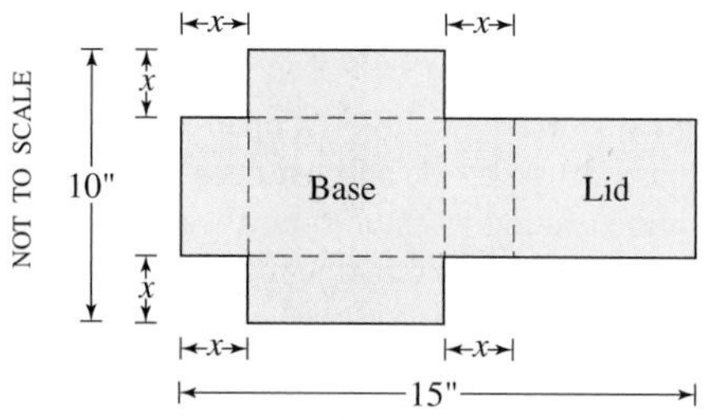

a. Write a formula $V(x)$ for the volume of the box.

b. Find the domain of V for the problem situation and graph V over this domain.

c. Use a graphical method to find the maximum volume and the value of x that gives it.

d. Confirm your result in part (c) analytically.

T **17. Designing a suitcase** A 24-in.-by-36-in. sheet of cardboard is folded in half to form a 24-in.-by-18-in. rectangle as shown in the accompanying figure. Then four congruent squares of side length x are cut from the corners of the folded rectangle. The sheet is unfolded, and the six tabs are folded up to form a box with sides and a lid.

a. Write a formula $V(x)$ for the volume of the box.

b. Find the domain of V for the problem situation and graph V over this domain.

c. Use a graphical method to find the maximum volume and the value of x that gives it.

d. Confirm your result in part (c) analytically.

e. Find a value of x that yields a volume of 1120 in.^{3}.

f. Write a paragraph describing the issues that arise in part (b).

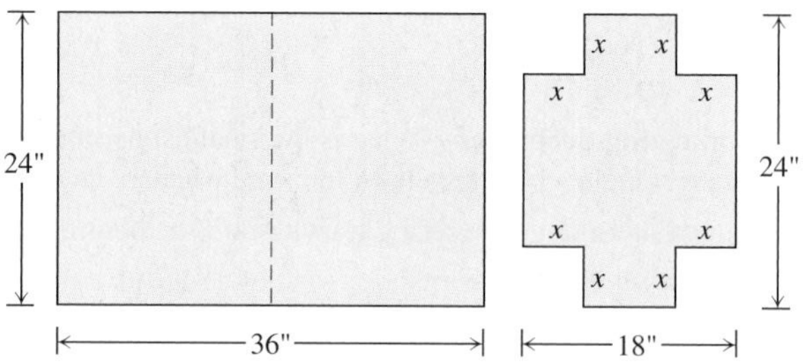

The sheet is then unfolded.

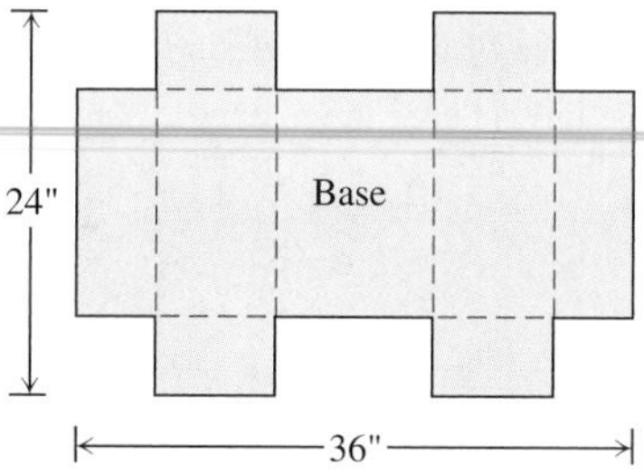

18. A rectangle is to be inscribed under the arch of the curve $y = 4\cos(0.5x)$ from $x = -\pi$ to $x = \pi$. What are the dimensions of the rectangle with largest area, and what is the largest area?

19. Find the dimensions of a right circular cylinder of maximum volume that can be inscribed in a sphere of radius 10 cm. What is the maximum volume?

20. a. The U.S. Postal Service will accept a box for domestic shipment only if the sum of its length and girth (distance around) does not exceed 108 in. What dimensions will give a box with a square end the largest possible volume?

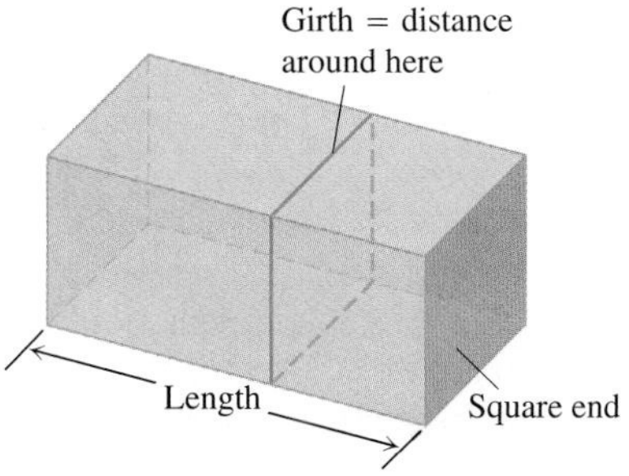

b. Graph the volume of a 108-in. box (length plus girth equals 108 in.) as a function of its length and compare what you see with your answer in part (a).

21. (*Continuation of Exercise 20.*)

a. Suppose that instead of having a box with square ends you have a box with square sides so that its dimensions are h by h by w and the girth is $2h + 2w$. What dimensions will give the box its largest volume now?

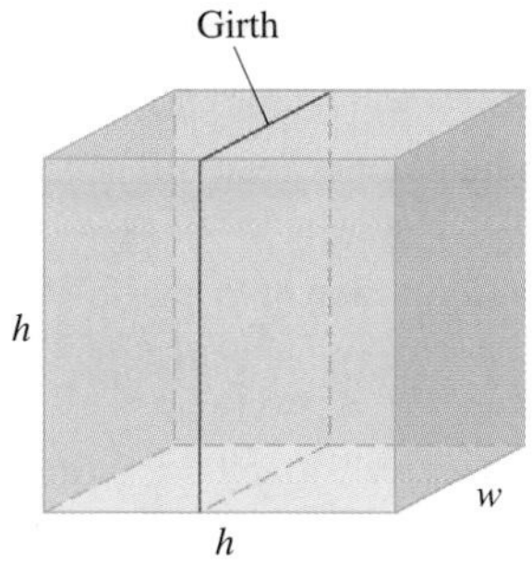

b. Graph the volume as a function of h and compare what you see with your answer in part (a).

22. A window is in the form of a rectangle surmounted by a semicircle. The rectangle is of clear glass, whereas the semicircle is of tinted glass that transmits only half as much light per unit area as clear glass does. The total perimeter is fixed. Find the proportions of the window that will admit the most light. Neglect the thickness of the frame.

23. A silo (base not included) is to be constructed in the form of a cylinder surmounted by a hemisphere. The cost of construction per square unit of surface area is twice as great for the hemisphere as it is for the cylindrical sidewall. Determine the dimensions to be used if the volume is fixed and the cost of construction is to be kept to a minimum. Neglect the thickness of the silo and waste in construction.

24. The trough in the figure is to be made to the dimensions shown. Only the angle θ can be varied. What value of θ will maximize the trough's volume?

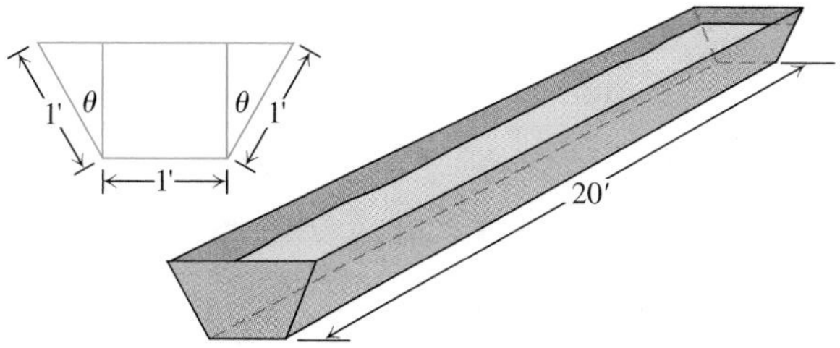

25. Paper folding A rectangular sheet of 8.5-in.-by-11-in. paper is placed on a flat surface. One of the corners is placed on the opposite longer edge, as shown in the figure, and held there as the paper is smoothed flat. The problem is to make the length of the crease as small as possible. Call the length L. Try it with paper.

a. Show that $L^2 = 2x^3/(2x - 8.5)$.

b. What value of x minimizes L^2?

c. What is the minimum value of L?

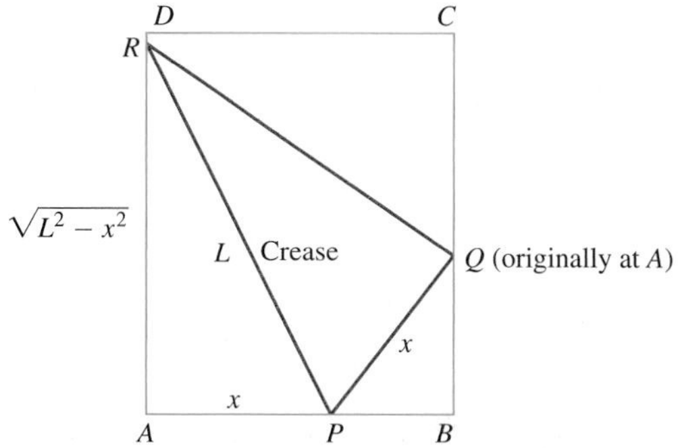

26. Constructing cylinders Compare the answers to the following two construction problems.

a. A rectangular sheet of perimeter 36 cm and dimensions x cm by y cm to be rolled into a cylinder as shown in part (a) of the figure. What values of x and y give the largest volume?

b. The same sheet is to be revolved about one of the sides of length y to sweep out the cylinder as shown in part (b) of the figure. What values of x and y give the largest volume?

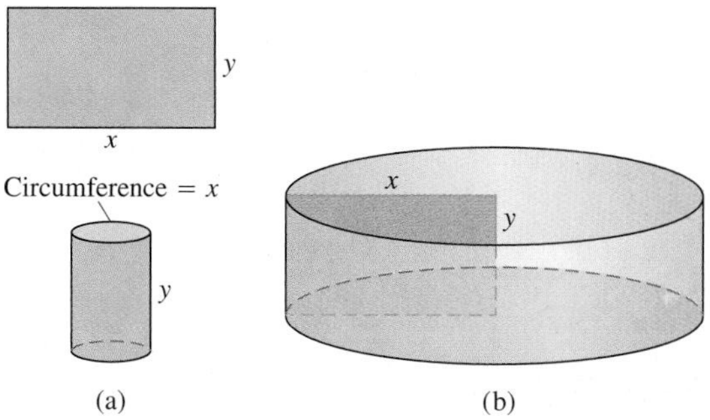

27. Constructing cones A right triangle whose hypotenuse is $\sqrt{3}$ m long is revolved about one of its legs to generate a right circular cone. Find the radius, height, and volume of the cone of greatest volume that can be made this way.

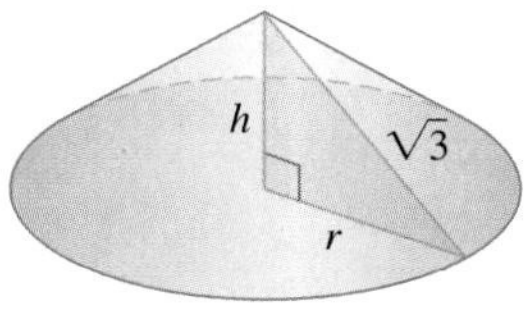

28. What value of a makes $f(x) = x^2 + (a/x)$ have

a. a local minimum at $x = 2$?

b. a point of inflection at $x = 1$?

29. Show that $f(x) = x^2 + (a/x)$ cannot have a local maximum for any value of a.

30. What values of a and b make $f(x) = x^3 + ax^2 + bx$ have

a. a local maximum at $x = -1$ and a local minimum at $x = 3$?

b. a local minimum at $x = 4$ and a point of inflection at $x = 1$?

Physical Applications

31. Vertical motion The height of an object moving vertically is given by

$$s = -16t^2 + 96t + 112,$$

with s in feet and t in seconds. Find

a. the object's velocity when $t = 0$

b. its maximum height and when it occurs

c. its velocity when $s = 0$.

32. Quickest route Jane is 2 mi offshore in a boat and wishes to reach a coastal village 6 mi down a straight shoreline from the point nearest the boat. She can row 2 mph and can walk 5 mph. Where should she land her boat to reach the village in the least amount of time?

33. Shortest beam The 8-ft wall shown here stands 27 ft from the building. Find the length of the shortest straight beam that will reach to the side of the building from the ground outside the wall.

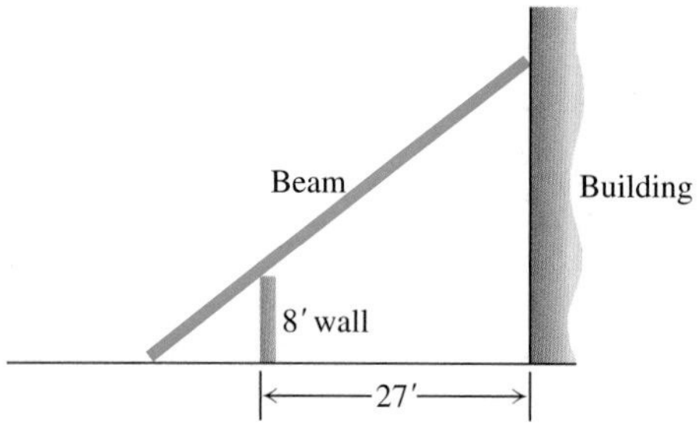

T **34. Strength of a beam** The strength S of a rectangular wooden beam is proportional to its width times the square of its depth. (See accompanying figure.)

a. Find the dimensions of the strongest beam that can be cut from a 12-in.-diameter cylindrical log.

b. Graph S as a function of the beam's width w, assuming the proportionality constant to be $k = 1$. Reconcile what you see with your answer in part (a).

c. On the same screen, graph S as a function of the beam's depth d, again taking $k = 1$. Compare the graphs with one another and with your answer in part (a). What would be the effect of changing to some other value of k? Try it.

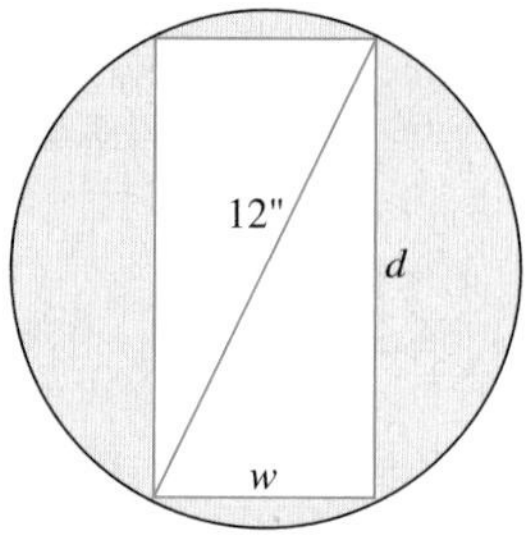

T **35. Stiffness of a beam** The stiffness S of a rectangular beam is proportional to its width times the cube of its depth.

a. Find the dimensions of the stiffest beam that can be cut from a 12-in.-diameter cylindrical log.

b. Graph S as a function of the beam's width w, assuming the proportionality constant to be $k = 1$. Reconcile what you see with your answer in part (a).

c. On the same screen, graph S as a function of the beam's depth d, again taking $k = 1$. Compare the graphs with one another and with your answer in part (a). What would be the effect of changing to some other value of k? Try it.

36. Motion on a line The positions of two particles on the s-axis are $s_1 = \sin t$ and $s_2 = \sin(t + \pi/3)$, with s_1 and s_2 in meters and t in seconds.

a. At what time(s) in the interval $0 \le t \le 2\pi$ do the particles meet?

b. What is the farthest apart that the particles ever get?

c. When in the interval $0 \le t \le 2\pi$ is the distance between the particles changing the fastest?

37. Frictionless cart A small frictionless cart, attached to the wall by a spring, is pulled 10 cm from its rest position and released at time $t = 0$ to roll back and forth for 4 sec. Its position at time t is $s = 10 \cos \pi t$.

a. What is the cart's maximum speed? When is the cart moving that fast? Where is it then? What is the magnitude of the acceleration then?

b. Where is the cart when the magnitude of the acceleration is greatest? What is the cart's speed then?

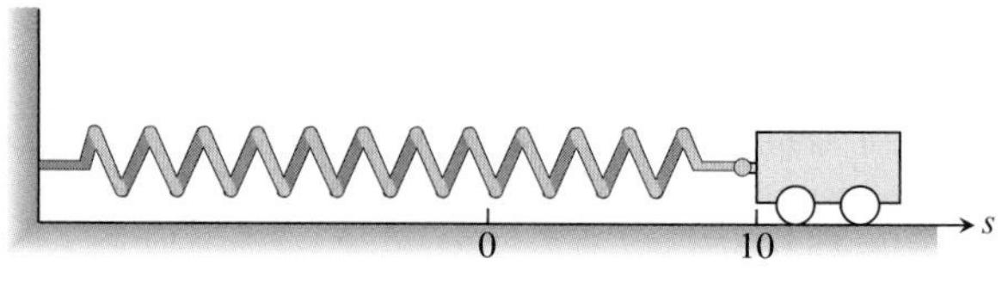

38. Two masses hanging side by side from springs have positions $s_1 = 2 \sin t$ and $s_2 = \sin 2t$, respectively.

a. At what times in the interval $0 < t$ do the masses pass each other? (*Hint:* $\sin 2t = 2 \sin t \cos t$.)

b. When in the interval $0 \le t \le 2\pi$ is the vertical distance between the masses the greatest? What is this distance? (*Hint:* $\cos 2t = 2\cos^2 t - 1$.)

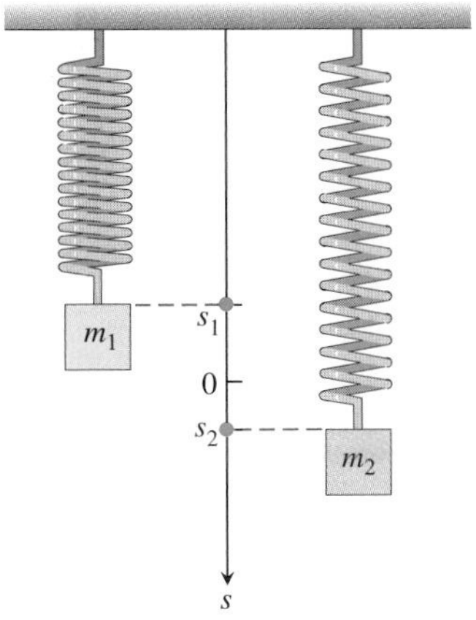

39. Distance between two ships At noon, ship A was 12 nautical miles due north of ship B. Ship A was sailing south at 12 knots (nautical miles per hour; a nautical mile is 2000 yd) and continued to do so all day. Ship B was sailing east at 8 knots and continued to do so all day.

a. Start counting time with $t = 0$ at noon and express the distance s between the ships as a function of t.

b. How rapidly was the distance between the ships changing at noon? One hour later?

c. The visibility that day was 5 nautical miles. Did the ships ever sight each other?

T **d.** Graph s and ds/dt together as functions of t for $-1 \le t \le 3$, using different colors if possible. Compare the graphs and reconcile what you see with your answers in parts (b) and (c).

e. The graph of ds/dt looks as if it might have a horizontal asymptote in the first quadrant. This in turn suggests that ds/dt approaches a limiting value as $t \to \infty$. What is this value? What is its relation to the ships' individual speeds?

40. Fermat's principle in optics Fermat's principle in optics states that light always travels from one point to another along a path that minimizes the travel time. Light from a source A is reflected by a plane mirror to a receiver at point B, as shown in the figure. Show that for the light to obey Fermat's principle, the angle of incidence must equal the angle of reflection, both measured from the line normal to the reflecting surface. (This result can also be derived without calculus. There is a purely geometric argument, which you may prefer.)

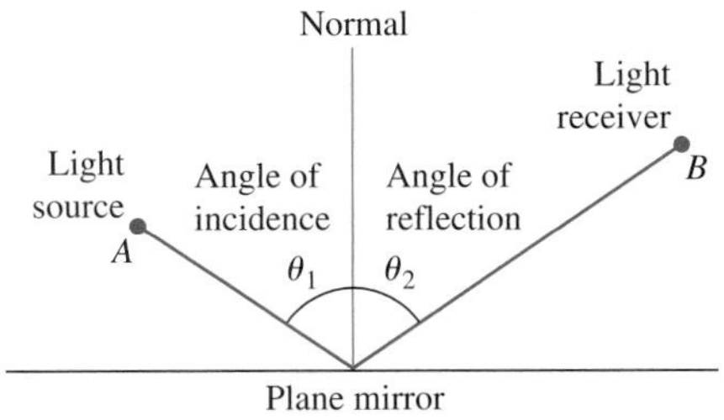

41. Tin pest When metallic tin is kept below 13.2°C, it slowly becomes brittle and crumbles to a gray powder. Tin objects eventually crumble to this gray powder spontaneously if kept in a cold climate for years. The Europeans who saw tin organ pipes in their churches crumble away years ago called the change *tin pest* because it seemed to be contagious, and indeed it was, for the gray powder is a catalyst for its own formation.

A *catalyst* for a chemical reaction is a substance that controls the rate of reaction without undergoing any permanent change in itself. An *autocatalytic reaction* is one whose product is a catalyst for its own formation. Such a reaction may proceed slowly at first if the amount of catalyst present is small and slowly again at the end, when most of the original substance is used up. But in between, when both the substance and its catalyst product are abundant, the reaction proceeds at a faster pace.

In some cases, it is reasonable to assume that the rate $v = dx/dt$ of the reaction is proportional both to the amount of the original substance present and to the amount of product. That is, v may be considered to be a function of x alone, and

$$v = kx(a - x) = kax - kx^2,$$

where

$x =$ the amount of product

$a =$ the amount of substance at the beginning

$k =$ a positive constant.

At what value of x does the rate v have a maximum? What is the maximum value of v?

42. Airplane landing path An airplane is flying at altitude H when it begins its descent to an airport runway that is at horizontal ground distance L from the airplane, as shown in the figure. Assume that the landing path of the airplane is the graph of a cubic polynomial function $y = ax^3 + bx^2 + cx + d$, where $y(-L) = H$ and $y(0) = 0$.

a. What is dy/dx at $x = 0$?

b. What is dy/dx at $x = -L$?

c. Use the values for dy/dx at $x = 0$ and $x = -L$ together with $y(0) = 0$ and $y(-L) = H$ to show that

$$y(x) = H\left[2\left(\frac{x}{L}\right)^3 + 3\left(\frac{x}{L}\right)^2\right].$$

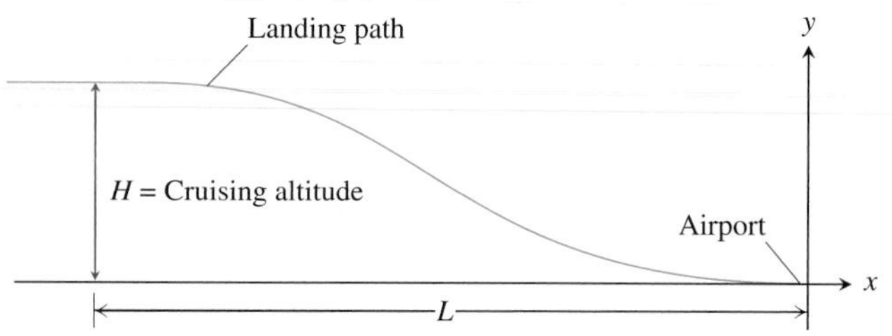

Business and Economics

43. It costs you c dollars each to manufacture and distribute backpacks. If the backpacks sell at x dollars each, the number sold is given by

$$n = \frac{a}{x - c} + b(100 - x),$$

where a and b are positive constants. What selling price will bring a maximum profit?

44. You operate a tour service that offers the following rates:

$200 per person if 50 people (the minimum number to book the tour) go on the tour.

For each additional person, up to a maximum of 80 people total, the rate per person is reduced by $2.

It costs $6000 (a fixed cost) plus $32 per person to conduct the tour. How many people does it take to maximize your profit?

45. Wilson lot size formula One of the formulas for inventory management says that the average weekly cost of ordering, paying for, and holding merchandise is

$$A(q) = \frac{km}{q} + cm + \frac{hq}{2},$$

where q is the quantity you order when things run low (shoes, radios, brooms, or whatever the item might be), k is the cost of placing an order (the same, no matter how often you order), c is the cost of one item (a constant), m is the number of items sold each week (a constant), and h is the weekly holding cost per item (a constant that takes into account things such as space, utilities, insurance, and security).

a. Your job, as the inventory manager for your store, is to find the quantity that will minimize $A(q)$. What is it? (The formula you get for the answer is called the *Wilson lot size formula*.)

b. Shipping costs sometimes depend on order size. When they do, it is more realistic to replace k by $k + bq$, the sum of k and a constant multiple of q. What is the most economical quantity to order now?

46. Production level Prove that the production level (if any) at which average cost is smallest is a level at which the average cost equals marginal cost.

47. Show that if $r(x) = 6x$ and $c(x) = x^3 - 6x^2 + 15x$ are your revenue and cost functions, then the best you can do is break even (have revenue equal cost).

48. Production level Suppose that $c(x) = x^3 - 20x^2 + 20{,}000x$ is the cost of manufacturing x items. Find a production level that will minimize the average cost of making x items.

49. Average daily cost In Example 6, assume for any material that a cost of d is incurred per delivery, the storage cost is s dollars per unit stored per day, and the production rate is p units per day.

a. How much should be delivered every x days?

b. Show that

$$\text{cost per cycle} = d + \frac{px}{2}sx.$$

c. Find the time between deliveries x^* and the amount to deliver that minimizes the *average daily cost* of delivery and storage.

d. Show that x^* occurs at the intersection of the hyperbola $y = d/x$ and the line $y = psx/2$.

50. Minimizing average cost Suppose that $c(x) = 2000 + 96x + 4x^{3/2}$, where x represents thousands of units. Is there a production level that minimizes average cost? If so, what is it?

Medicine

51. Sensitivity to medicine (*Continuation of Exercise 50, Section 3.2.*) Find the amount of medicine to which the body is most sensitive by finding the value of M that maximizes the derivative dR/dM, where

$$R = M^2\left(\frac{C}{2} - \frac{M}{3}\right)$$

and C is a constant.

52. How we cough

a. When we cough, the trachea (windpipe) contracts to increase the velocity of the air going out. This raises the questions of how much it should contract to maximize the velocity and whether it really contracts that much when we cough.

Under reasonable assumptions about the elasticity of the tracheal wall and about how the air near the wall is slowed by friction, the average flow velocity v can be modeled by the equation

$$v = c(r_0 - r)r^2 \text{ cm/sec}, \qquad \frac{r_0}{2} \le r \le r_0,$$

where r_0 is the rest radius of the trachea in centimeters and c is a positive constant whose value depends in part on the length of the trachea.

Show that v is greatest when $r = (2/3)r_0$, that is, when the trachea is about 33% contracted. The remarkable fact is that X-ray photographs confirm that the trachea contracts about this much during a cough.

T **b.** Take r_0 to be 0.5 and c to be 1 and graph v over the interval $0 \le r \le 0.5$. Compare what you see with the claim that v is at a maximum when $r = (2/3)r_0$.

Theory and Examples

53. An inequality for positive integers Show that if a, b, c, and d are positive integers, then

$$\frac{(a^2+1)(b^2+1)(c^2+1)(d^2+1)}{abcd} \geq 16.$$

54. The derivative dt/dx in Example 4

a. Show that

$$f(x) = \frac{x}{\sqrt{a^2+x^2}}$$

is an increasing function of x.

b. Show that

$$g(x) = \frac{d-x}{\sqrt{b^2+(d-x)^2}}$$

is a decreasing function of x.

c. Show that

$$\frac{dt}{dx} = \frac{x}{c_1\sqrt{a^2+x^2}} - \frac{d-x}{c_2\sqrt{b^2+(d-x)^2}}$$

is an increasing function of x.

55. Let $f(x)$ and $g(x)$ be the differentiable functions graphed here. Point c is the point where the vertical distance between the curves is the greatest. Is there anything special about the tangents to the two curves at c? Give reasons for your answer.

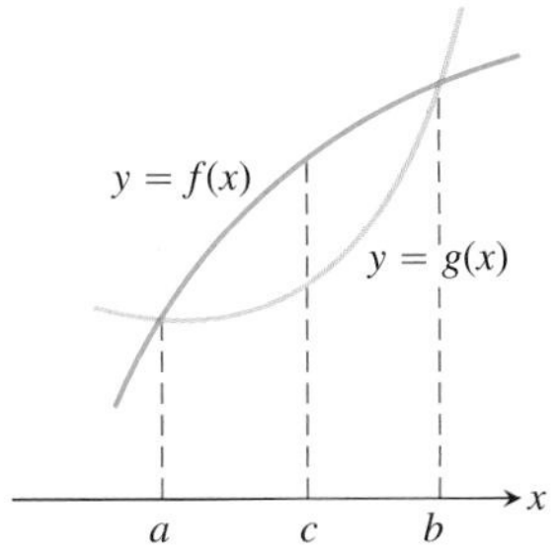

56. You have been asked to determine whether the function $f(x) = 3 + 4\cos x + \cos 2x$ is ever negative.

a. Explain why you need to consider values of x only in the interval $[0, 2\pi]$.

b. Is f ever negative? Explain.

57. a. The function $y = \cot x - \sqrt{2}\csc x$ has an absolute maximum value on the interval $0 < x < \pi$. Find it.

T **b.** Graph the function and compare what you see with your answer in part (a).

58. a. The function $y = \tan x + 3\cot x$ has an absolute minimum value on the interval $0 < x < \pi/2$. Find it.

T **b.** Graph the function and compare what you see with your answer in part (a).

59. a. How close does the curve $y = \sqrt{x}$ come to the point $(3/2, 0)$? (*Hint*: If you minimize the *square* of the distance, you can avoid square roots.)

T **b.** Graph the distance function and $y = \sqrt{x}$ together and reconcile what you see with your answer in part (a).

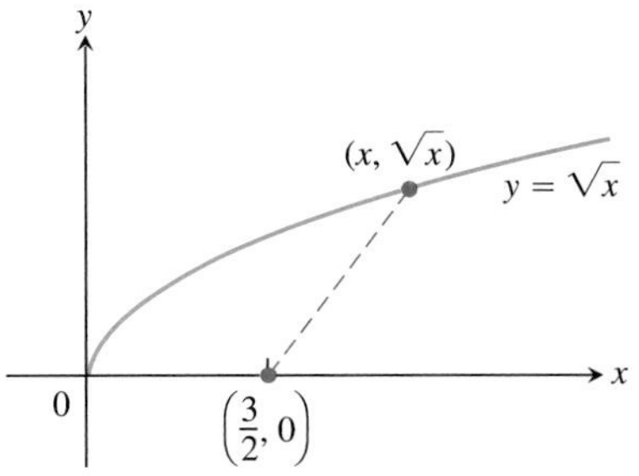

60. a. How close does the semicircle $y = \sqrt{16 - x^2}$ come to the point $(1, \sqrt{3})$?

T **b.** Graph the distance function and $y = \sqrt{16 - x^2}$ together and reconcile what you see with your answer in part (a).

COMPUTER EXPLORATIONS

In Exercises 61 and 62, you may find it helpful to use a CAS.

61. Generalized cone problem A cone of height h and radius r is constructed from a flat, circular disk of radius a in. by removing a sector AOC of arc length x in. and then connecting the edges OA and OC.

a. Find a formula for the volume V of the cone in terms of x and a.

b. Find r and h in the cone of maximum volume for $a = 4, 5, 6, 8$.

c. Find a simple relationship between r and h that is independent of a for the cone of maximum volume. Explain how you arrived at your relationship.

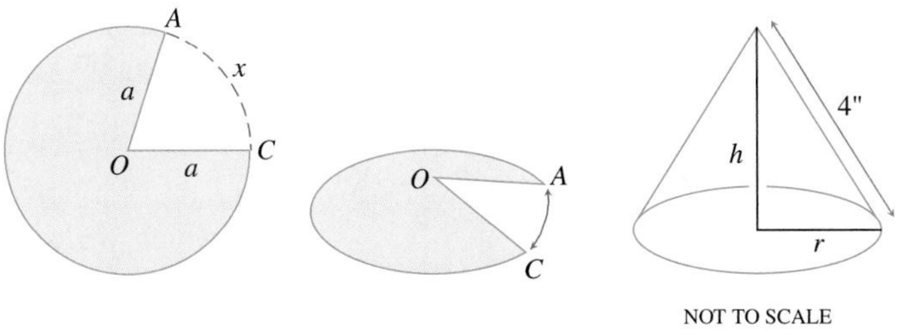

62. Circumscribing an ellipse Let $P(x, a)$ and $Q(-x, a)$ be two points on the upper half of the ellipse

$$\frac{x^2}{100} + \frac{(y-5)^2}{25} = 1$$

centered at $(0, 5)$. A triangle RST is formed by using the tangent lines to the ellipse at Q and P as shown in the figure.

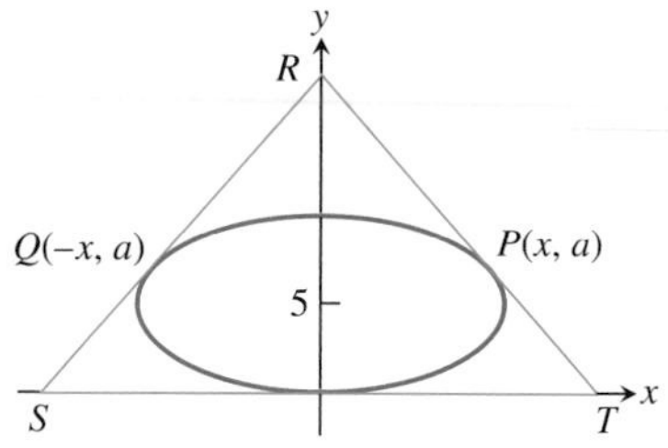

a. Show that the area of the triangle is

$$A(x) = -f'(x)\left[x - \frac{f(x)}{f'(x)}\right]^2,$$

where $y = f(x)$ is the function representing the upper half of the ellipse.

b. What is the domain of A? Draw the graph of A. How are the asymptotes of the graph related to the problem situation?

c. Determine the height of the triangle with minimum area. How is it related to the y coordinate of the center of the ellipse?

d. Repeat parts (a) through (c) for the ellipse

$$\frac{x^2}{C^2} + \frac{(y - B)^2}{B^2} = 1$$

centered at $(0, B)$. Show that the triangle has minimum area when its height is $3B$.

4.6 Indeterminate Forms and L'Hôpital's Rule

HISTORICAL BIOGRAPHY

Guillaume François Antoine de l'Hôpital (1661–1704)

John Bernoulli discovered a rule for calculating limits of fractions whose numerators and denominators both approach zero or $+\infty$. The rule is known today as **l'Hôpital's Rule**, after Guillaume de l'Hôpital. He was a French nobleman who wrote the first introductory differential calculus text, where the rule first appeared in print.

Indeterminate Form 0/0

If the continuous functions $f(x)$ and $g(x)$ are both zero at $x = a$, then

$$\lim_{x \to a} \frac{f(x)}{g(x)}$$

cannot be found by substituting $x = a$. The substitution produces $0/0$, a meaningless expression, which we cannot evaluate. We use $0/0$ as a notation for an expression known as an **indeterminate form**. Sometimes, but not always, limits that lead to indeterminate forms may be found by cancellation, rearrangement of terms, or other algebraic manipulations. This was our experience in Chapter 2. It took considerable analysis in Section 2.4 to find $\lim_{x \to 0} (\sin x)/x$. But we have had success with the limit

$$f'(a) = \lim_{x \to a} \frac{f(x) - f(a)}{x - a},$$

from which we calculate derivatives and which always produces the equivalent of $0/0$ when we substitute $x = a$. L'Hôpital's Rule enables us to draw on our success with derivatives to evaluate limits that otherwise lead to indeterminate forms.

THEOREM 6 L'Hôpital's Rule (First Form)
Suppose that $f(a) = g(a) = 0$, that $f'(a)$ and $g'(a)$ exist, and that $g'(a) \neq 0$. Then

$$\lim_{x \to a} \frac{f(x)}{g(x)} = \frac{f'(a)}{g'(a)}.$$

Caution
To apply l'Hôpital's Rule to f/g, divide the derivative of f by the derivative of g. Do not fall into the trap of taking the derivative of f/g. The quotient to use is f'/g', not $(f/g)'$.

Proof Working backward from $f'(a)$ and $g'(a)$, which are themselves limits, we have

$$\frac{f'(a)}{g'(a)} = \frac{\lim\limits_{x\to a} \dfrac{f(x) - f(a)}{x - a}}{\lim\limits_{x\to a} \dfrac{g(x) - g(a)}{x - a}} = \lim_{x\to a} \frac{\dfrac{f(x) - f(a)}{x - a}}{\dfrac{g(x) - g(a)}{x - a}}$$

$$= \lim_{x\to a} \frac{f(x) - f(a)}{g(x) - g(a)} = \lim_{x\to a} \frac{f(x) - 0}{g(x) - 0} = \lim_{x\to a} \frac{f(x)}{g(x)}.$$ ■

EXAMPLE 1 Using L'Hôpital's Rule

(a) $$\lim_{x\to 0} \frac{3x - \sin x}{x} = \left.\frac{3 - \cos x}{1}\right|_{x=0} = 2$$

(b) $$\lim_{x\to 0} \frac{\sqrt{1 + x} - 1}{x} = \left.\frac{\dfrac{1}{2\sqrt{1 + x}}}{1}\right|_{x=0} = \frac{1}{2}$$ ■

Sometimes after differentiation, the new numerator and denominator both equal zero at $x = a$, as we see in Example 2. In these cases, we apply a stronger form of l'Hôpital's Rule.

> **THEOREM 7** L'Hôpital's Rule (Stronger Form)
>
> Suppose that $f(a) = g(a) = 0$, that f and g are differentiable on an open interval I containing a, and that $g'(x) \neq 0$ on I if $x \neq a$. Then
>
> $$\lim_{x\to a} \frac{f(x)}{g(x)} = \lim_{x\to a} \frac{f'(x)}{g'(x)},$$
>
> assuming that the limit on the right side exists.

Before we give a proof of Theorem 7, let's consider an example.

EXAMPLE 2 Applying the Stronger Form of L'Hôpital's Rule

(a) $$\lim_{x\to 0} \frac{\sqrt{1 + x} - 1 - x/2}{x^2} \qquad \frac{0}{0}$$

$$= \lim_{x\to 0} \frac{(1/2)(1 + x)^{-1/2} - 1/2}{2x} \qquad \text{Still } \frac{0}{0}\text{; differentiate again.}$$

$$= \lim_{x\to 0} \frac{-(1/4)(1 + x)^{-3/2}}{2} = -\frac{1}{8} \qquad \text{Not } \frac{0}{0}\text{; limit is found.}$$

(b) $$\lim_{x\to 0} \frac{x - \sin x}{x^3} \qquad \frac{0}{0}$$

$$= \lim_{x\to 0} \frac{1 - \cos x}{3x^2} \qquad \text{Still } \frac{0}{0}$$

$$= \lim_{x\to 0} \frac{\sin x}{6x} \qquad \text{Still } \frac{0}{0}$$

$$= \lim_{x\to 0} \frac{\cos x}{6} = \frac{1}{6} \qquad \text{Not } \frac{0}{0}\text{; limit is found.}$$ ■

The proof of the stronger form of l'Hôpital's Rule is based on Cauchy's Mean Value Theorem, a Mean Value Theorem that involves two functions instead of one. We prove Cauchy's Theorem first and then show how it leads to l'Hôpital's Rule.

HISTORICAL BIOGRAPHY

Augustin-Louis Cauchy (1789–1857)

THEOREM 8 Cauchy's Mean Value Theorem

Suppose functions f and g are continuous on $[a, b]$ and differentiable throughout (a, b) and also suppose $g'(x) \neq 0$ throughout (a, b). Then there exists a number c in (a, b) at which

$$\frac{f'(c)}{g'(c)} = \frac{f(b) - f(a)}{g(b) - g(a)}.$$

Proof We apply the Mean Value Theorem of Section 4.2 twice. First we use it to show that $g(a) \neq g(b)$. For if $g(b)$ did equal $g(a)$, then the Mean Value Theorem would give

$$g'(c) = \frac{g(b) - g(a)}{b - a} = 0$$

for some c between a and b, which cannot happen because $g'(x) \neq 0$ in (a, b).

We next apply the Mean Value Theorem to the function

$$F(x) = f(x) - f(a) - \frac{f(b) - f(a)}{g(b) - g(a)}[g(x) - g(a)].$$

This function is continuous and differentiable where f and g are, and $F(b) = F(a) = 0$. Therefore, there is a number c between a and b for which $F'(c) = 0$. When expressed in terms of f and g, this equation becomes

$$F'(c) = f'(c) - \frac{f(b) - f(a)}{g(b) - g(a)}[g'(c)] = 0$$

or

$$\frac{f'(c)}{g'(c)} = \frac{f(b) - f(a)}{g(b) - g(a)}.$$ ■

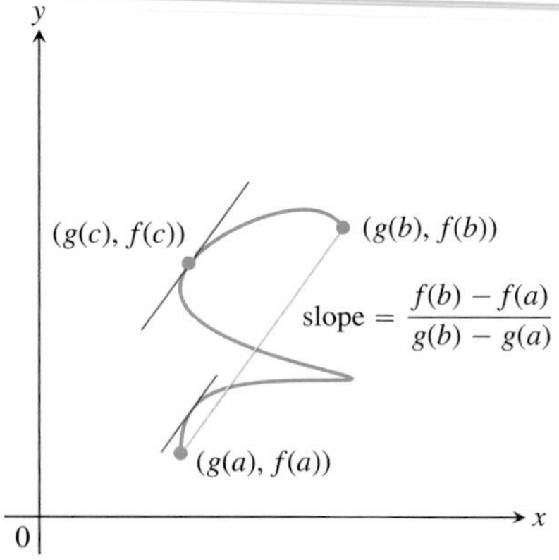

FIGURE 4.42 There is at least one value of the parameter $t = c$, $a < c < b$, for which the slope of the tangent to the curve at $(g(c), f(c))$ is the same as the slope of the secant line joining the points $(g(a), f(a))$ and $(g(b), f(b))$.

Notice that the Mean Value Theorem in Section 4.2 is Theorem 8 with $g(x) = x$.

Cauchy's Mean Value Theorem has a geometric interpretation for a curve C defined by the parametric equations $x = g(t)$ and $y = f(t)$. From Equation (2) in Section 3.5, the slope of the parametric curve at t is given by

$$\frac{dy/dt}{dx/dt} = \frac{f'(t)}{g'(t)},$$

so $f'(c)/g'(c)$ is the slope of the tangent to the curve when $t = c$. The secant line joining the two points $(g(a), f(a))$ and $(g(b), f(b))$ on C has slope

$$\frac{f(b) - f(a)}{g(b) - g(a)}.$$

Theorem 8 says that there is a parameter value c in the interval (a, b) for which the slope of the tangent to the curve at the point $(g(c), f(c))$ is the same as the slope of the secant line joining the points $(g(a), f(a))$ and $(g(b), f(b))$. This geometric result is shown in Figure 4.42. Note that more than one such value c of the parameter may exist.

We now prove Theorem 7.

Proof of the Stronger Form of l'Hôpital's Rule We first establish the limit equation for the case $x \to a^+$. The method needs almost no change to apply to $x \to a^-$, and the combination of these two cases establishes the result.

Suppose that x lies to the right of a. Then $g'(x) \neq 0$, and we can apply Cauchy's Mean Value Theorem to the closed interval from a to x. This step produces a number c between a and x such that

$$\frac{f'(c)}{g'(c)} = \frac{f(x) - f(a)}{g(x) - g(a)}.$$

But $f(a) = g(a) = 0$, so

$$\frac{f'(c)}{g'(c)} = \frac{f(x)}{g(x)}.$$

As x approaches a, c approaches a because it always lies between a and x. Therefore,

$$\lim_{x \to a^+} \frac{f(x)}{g(x)} = \lim_{c \to a^+} \frac{f'(c)}{g'(c)} = \lim_{x \to a^+} \frac{f'(x)}{g'(x)},$$

which establishes l'Hôpital's Rule for the case where x approaches a from above. The case where x approaches a from below is proved by applying Cauchy's Mean Value Theorem to the closed interval $[x, a]$, $x < a$. ■

Most functions encountered in the real world and most functions in this book satisfy the conditions of l'Hôpital's Rule.

> **Using L'Hôpital's Rule**
>
> To find
>
> $$\lim_{x \to a} \frac{f(x)}{g(x)}$$
>
> by l'Hôpital's Rule, continue to differentiate f and g, so long as we still get the form $0/0$ at $x = a$. But as soon as one or the other of these derivatives is different from zero at $x = a$ we stop differentiating. L'Hôpital's Rule does not apply when either the numerator or denominator has a finite nonzero limit.

EXAMPLE 3 Incorrectly Applying the Stronger Form of L'Hôpital's Rule

$$\lim_{x \to 0} \frac{1 - \cos x}{x + x^2} \qquad \frac{0}{0}$$

$$= \lim_{x \to 0} \frac{\sin x}{1 + 2x} = \frac{0}{1} = 0 \qquad \text{Not } \frac{0}{0}\text{; limit is found.}$$

Up to now the calculation is correct, but if we continue to differentiate in an attempt to apply l'Hôpital's Rule once more, we get

$$\lim_{x \to 0} \frac{1 - \cos x}{x + x^2} = \lim_{x \to 0} \frac{\sin x}{1 + 2x} = \lim_{x \to 0} \frac{\cos x}{2} = \frac{1}{2},$$

which is wrong. L'Hôpital's Rule can only be applied to limits which give indeterminate forms, and $0/1$ is not an indeterminate form. ■

L'Hôpital's Rule applies to one-sided limits as well, which is apparent from the proof of Theorem 7.

EXAMPLE 4 Using L'Hôpital's Rule with One-Sided Limits

Recall that ∞ and $+\infty$ mean the same thing.

(a) $\lim_{x \to 0^+} \frac{\sin x}{x^2}$ $\quad \frac{0}{0}$

$= \lim_{x \to 0^+} \frac{\cos x}{2x} = \infty$ $\quad$ Positive for $x > 0$.

(b) $\lim_{x \to 0^-} \frac{\sin x}{x^2}$ $\quad \frac{0}{0}$

$= \lim_{x \to 0^-} \frac{\cos x}{2x} = -\infty$ $\quad$ Negative for $x < 0$. ■

Indeterminate Forms ∞/∞, $\infty \cdot 0$, $\infty - \infty$

Sometimes when we try to evaluate a limit as $x \to a$ by substituting $x = a$ we get an ambiguous expression like ∞/∞, $\infty \cdot 0$, or $\infty - \infty$, instead of $0/0$. We first consider the form ∞/∞.

In more advanced books it is proved that l'Hôpital's Rule applies to the indeterminate form ∞/∞ as well as to $0/0$. If $f(x) \to \pm\infty$ and $g(x) \to \pm\infty$ as $x \to a$, then

$$\lim_{x \to a} \frac{f(x)}{g(x)} = \lim_{x \to a} \frac{f'(x)}{g'(x)}$$

provided the limit on the right exists. In the notation $x \to a$, a may be either finite or infinite. Moreover $x \to a$ may be replaced by the one-sided limits $x \to a^+$ or $x \to a^-$.

EXAMPLE 5 Working with the Indeterminate Form ∞/∞

Find

(a) $\lim_{x \to \pi/2} \frac{\sec x}{1 + \tan x}$

(b) $\lim_{x \to \infty} \frac{x - 2x^2}{3x^2 + 5x}$

Solution

(a) The numerator and denominator are discontinuous at $x = \pi/2$, so we investigate the one-sided limits there. To apply l'Hôpital's Rule, we can choose I to be any open interval with $x = \pi/2$ as an endpoint.

$\lim_{x \to (\pi/2)^-} \frac{\sec x}{1 + \tan x}$ $\quad \frac{\infty}{\infty}$ from the left

$$= \lim_{x \to (\pi/2)^-} \frac{\sec x \tan x}{\sec^2 x} = \lim_{x \to (\pi/2)^-} \sin x = 1$$

The right-hand limit is 1 also, with $(-\infty)/(-\infty)$ as the indeterminate form. Therefore, the two-sided limit is equal to 1.

(b) $\lim_{x \to \infty} \frac{x - 2x^2}{3x^2 + 5x} = \lim_{x \to \infty} \frac{1 - 4x}{6x + 5} = \lim_{x \to \infty} \frac{-4}{6} = -\frac{2}{3}$. ■

Next we turn our attention to the indeterminate forms $\infty \cdot 0$ and $\infty - \infty$. Sometimes these forms can be handled by using algebra to convert them to a $0/0$ or ∞/∞ form. Here again we do not mean to suggest that $\infty \cdot 0$ or $\infty - \infty$ is a number. They are only notations for functional behaviors when considering limits. Here are examples of how we might work with these indeterminate forms.

EXAMPLE 6 Working with the Indeterminate Form $\infty \cdot 0$

Find

$$\lim_{x\to\infty}\left(x \sin\frac{1}{x}\right)$$

Solution

$$\begin{aligned}
&\lim_{x\to\infty}\left(x \sin\frac{1}{x}\right) && \infty \cdot 0\\
&\qquad = \lim_{h\to 0^+}\left(\frac{1}{h}\sin h\right) && \text{Let } h = 1/x.\\
&\qquad = 1
\end{aligned}$$

■

EXAMPLE 7 Working with the Indeterminate Form $\infty - \infty$

Find

$$\lim_{x\to 0}\left(\frac{1}{\sin x} - \frac{1}{x}\right).$$

Solution If $x \to 0^+$, then $\sin x \to 0^+$ and

$$\frac{1}{\sin x} - \frac{1}{x} \to \infty - \infty.$$

Similarly, if $x \to 0^-$, then $\sin x \to 0^-$ and

$$\frac{1}{\sin x} - \frac{1}{x} \to -\infty - (-\infty) = -\infty + \infty.$$

Neither form reveals what happens in the limit. To find out, we first combine the fractions:

$$\frac{1}{\sin x} - \frac{1}{x} = \frac{x - \sin x}{x \sin x} \qquad \text{Common denominator is } x \sin x$$

Then apply l'Hôpital's Rule to the result:

$$\begin{aligned}
\lim_{x\to 0}\left(\frac{1}{\sin x} - \frac{1}{x}\right) &= \lim_{x\to 0}\frac{x - \sin x}{x \sin x} && \frac{0}{0}\\
&= \lim_{x\to 0}\frac{1 - \cos x}{\sin x + x\cos x} && \text{Still } \frac{0}{0}\\
&= \lim_{x\to 0}\frac{\sin x}{2\cos x - x\sin x} = \frac{0}{2} = 0.
\end{aligned}$$

■

EXERCISES 4.6

Finding Limits

In Exercises 1–6, use l'Hôpital's Rule to evaluate the limit. Then evaluate the limit using a method studied in Chapter 2.

1. $\lim_{x\to 2} \dfrac{x-2}{x^2-4}$

2. $\lim_{x\to 0} \dfrac{\sin 5x}{x}$

3. $\lim_{x\to\infty} \dfrac{5x^2-3x}{7x^2+1}$

4. $\lim_{x\to 1} \dfrac{x^3-1}{4x^3-x-3}$

5. $\lim_{x\to 0} \dfrac{1-\cos x}{x^2}$

6. $\lim_{x\to\infty} \dfrac{2x^2+3x}{x^3+x+1}$

Applying l'Hôpital's Rule

Use l'Hôpital's Rule to find the limits in Exercises 7–26.

7. $\lim_{t\to 0} \dfrac{\sin t^2}{t}$

8. $\lim_{x\to\pi/2} \dfrac{2x-\pi}{\cos x}$

9. $\lim_{\theta\to\pi} \dfrac{\sin\theta}{\pi-\theta}$

10. $\lim_{x\to\pi/2} \dfrac{1-\sin x}{1+\cos 2x}$

11. $\lim_{x\to\pi/4} \dfrac{\sin x-\cos x}{x-\pi/4}$

12. $\lim_{x\to\pi/3} \dfrac{\cos x-0.5}{x-\pi/3}$

13. $\lim_{x\to(\pi/2)} -\left(x-\dfrac{\pi}{2}\right)\tan x$

14. $\lim_{x\to 0} \dfrac{2x}{x+7\sqrt{x}}$

15. $\lim_{x\to 1} \dfrac{2x^2-(3x+1)\sqrt{x}+2}{x-1}$

16. $\lim_{x\to 2} \dfrac{\sqrt{x^2+5}-3}{x^2-4}$

17. $\lim_{x\to 0} \dfrac{\sqrt{a(a+x)}-a}{x}, \quad a>0$

18. $\lim_{t\to 0} \dfrac{10(\sin t-t)}{t^3}$

19. $\lim_{x\to 0} \dfrac{x(\cos x-1)}{\sin x-x}$

20. $\lim_{h\to 0} \dfrac{\sin(a+h)-\sin a}{h}$

21. $\lim_{r\to 1} \dfrac{a(r^n-1)}{r-1}, \quad n$ a positive integer

22. $\lim_{x\to 0^+} \left(\dfrac{1}{x}-\dfrac{1}{\sqrt{x}}\right)$

23. $\lim_{x\to\infty} (x-\sqrt{x^2+x})$

24. $\lim_{x\to\infty} x\tan\dfrac{1}{x}$

25. $\lim_{x\to\pm\infty} \dfrac{3x-5}{2x^2-x+2}$

26. $\lim_{x\to 0} \dfrac{\sin 7x}{\tan 11x}$

Theory and Applications

L'Hôpital's Rule does not help with the limits in Exercises 27–30. Try it; you just keep on cycling. Find the limits some other way.

27. $\lim_{x\to\infty} \dfrac{\sqrt{9x+1}}{\sqrt{x+1}}$

28. $\lim_{x\to 0^+} \dfrac{\sqrt{x}}{\sqrt{\sin x}}$

29. $\lim_{x\to(\pi/2)^-} \dfrac{\sec x}{\tan x}$

30. $\lim_{x\to 0^+} \dfrac{\cot x}{\csc x}$

31. Which one is correct, and which one is wrong? Give reasons for your answers.

 a. $\lim_{x\to 3} \dfrac{x-3}{x^2-3} = \lim_{x\to 3} \dfrac{1}{2x} = \dfrac{1}{6}$

 b. $\lim_{x\to 3} \dfrac{x-3}{x^2-3} = \dfrac{0}{6} = 0$

32. **∞/∞ Form** Give an example of two differentiable functions f and g with $\lim_{x\to\infty} f(x) = \lim_{x\to\infty} g(x) = \infty$ that satisfy the following.

 a. $\lim_{x\to\infty} \dfrac{f(x)}{g(x)} = 3$ b. $\lim_{x\to\infty} \dfrac{f(x)}{g(x)} = 0$

 c. $\lim_{x\to\infty} \dfrac{f(x)}{g(x)} = \infty$

33. **Continuous extension** Find a value of c that makes the function

$$f(x) = \begin{cases} \dfrac{9x-3\sin 3x}{5x^3}, & x \neq 0 \\ c, & x = 0 \end{cases}$$

continuous at $x = 0$. Explain why your value of c works.

34. Let

$$f(x) = \begin{cases} x+2, & x\neq 0 \\ 0, & x = 0 \end{cases} \quad \text{and} \quad g(x) = \begin{cases} x+1, & x\neq 0 \\ 0, & x = 0. \end{cases}$$

 a. Show that

$$\lim_{x\to 0} \frac{f'(x)}{g'(x)} = 1 \quad \text{but} \quad \lim_{x\to 0} \frac{f(x)}{g(x)} = 2.$$

 b. Explain why this does not contradict l'Hôpital's Rule.

T 35. **0/0 Form** Estimate the value of

$$\lim_{x\to 1} \frac{2x^2-(3x+1)\sqrt{x}+2}{x-1}$$

by graphing. Then confirm your estimate with l'Hôpital's Rule.

T 36. **∞ − ∞ Form**

 a. Estimate the value of

$$\lim_{x\to\infty} \left(x-\sqrt{x^2+x}\right)$$

 by graphing $f(x) = x-\sqrt{x^2+x}$ over a suitably large interval of x-values.

 b. Now confirm your estimate by finding the limit with l'Hôpital's Rule. As the first step, multiply $f(x)$ by the fraction $\left(x+\sqrt{x^2+x}\right)/\left(x+\sqrt{x^2+x}\right)$ and simplify the new numerator.

T **37.** Let

$$f(x) = \frac{1 - \cos x^6}{x^{12}}.$$

Explain why some graphs of f may give false information about $\lim_{x \to 0} f(x)$. (*Hint:* Try the window $[-1, 1]$ by $[-0.5, 1]$.)

38. Find all values of c, that satisfy the conclusion of Cauchy's Mean Value Theorem for the given functions and interval.

a. $f(x) = x,\quad g(x) = x^2,\quad (a, b) = (-2, 0)$

b. $f(x) = x,\quad g(x) = x^2,\quad (a, b)$ arbitrary

c. $f(x) = x^3/3 - 4x,\quad g(x) = x^2,\quad (a, b) = (0, 3)$

39. In the accompanying figure, the circle has radius OA equal to 1, and AB is tangent to the circle at A. The arc AC has radian measure θ and the segment AB also has length θ. The line through B and C crosses the x-axis at $P(x, 0)$.

a. Show that the length of PA is

$$1 - x = \frac{\theta(1 - \cos\theta)}{\theta - \sin\theta}.$$

b. Find $\lim_{\theta \to 0} (1 - x)$.

c. Show that $\lim_{\theta \to \infty} [(1 - x) - (1 - \cos\theta)] = 0.$

Interpret this geometrically.

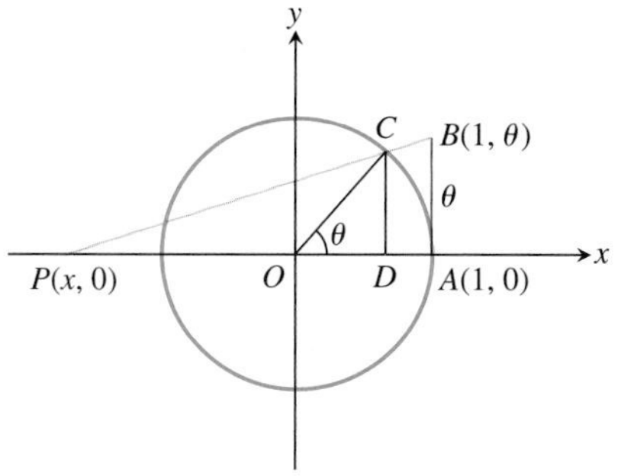

40. A right triangle has one leg of length 1, another of length y, and a hypotenuse of length r. The angle opposite y has radian measure θ. Find the limits as $\theta \to \pi/2$ of

a. $r - y$.

b. $r^2 - y^2$.

c. $r^3 - y^3$.

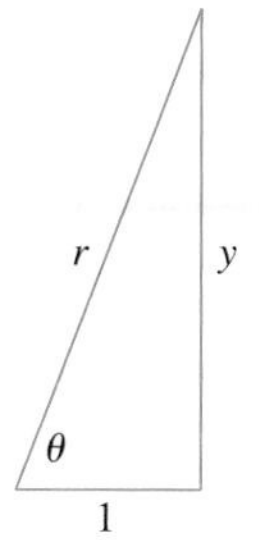

4.7 Newton's Method

HISTORICAL BIOGRAPHY

Niels Henrik Abel (1802–1829)

One of the basic problems of mathematics is solving equations. Using the quadratic root formula, we know how to find a point (solution) where $x^2 - 3x + 2 = 0$. There are more complicated formulas to solve cubic or quartic equations (polynomials of degree 3 or 4), but the Norwegian mathematician Niels Abel showed that no simple formulas exist to solve polynomials of degree equal to five. There is also no simple formula for solving equations like $\sin x = x^2$, which involve transcendental functions as well as polynomials or other algebraic functions.

In this section we study a numerical method, called *Newton's method* or the *Newton–Raphson method*, which is a technique to approximate the solution to an equation $f(x) = 0$. Essentially it uses tangent lines in place of the graph of $y = f(x)$ near the points where f is zero. (A value of x where f is zero is a *root* of the function f and a *solution* of the equation $f(x) = 0$.)

Procedure for Newton's Method

The goal of Newton's method for estimating a solution of an equation $f(x) = 0$ is to produce a sequence of approximations that approach the solution. We pick the first number x_0 of the sequence. Then, under favorable circumstances, the method does the rest by moving step by step toward a point where the graph of f crosses the x-axis (Figure 4.43). At each

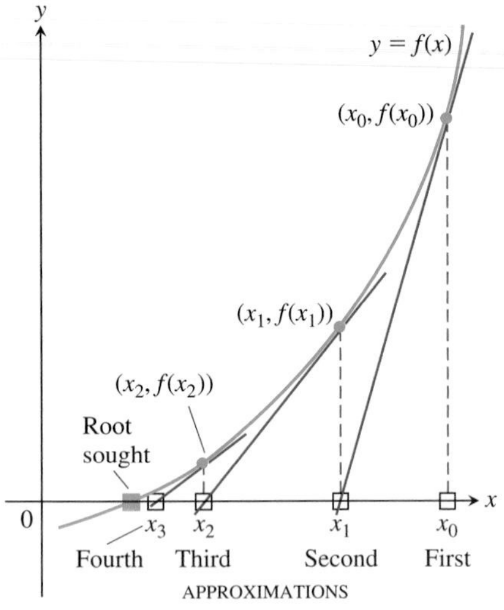

FIGURE 4.43 Newton's method starts with an initial guess x_0 and (under favorable circumstances) improves the guess one step at a time.

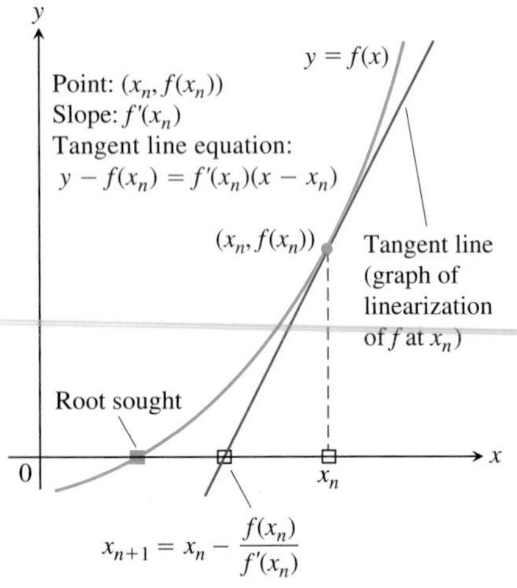

FIGURE 4.44 The geometry of the successive steps of Newton's method. From x_n we go up to the curve and follow the tangent line down to find x_{n+1}.

step the method approximates a zero of f with a zero of one of its linearizations. Here is how it works.

The initial estimate, x_0, may be found by graphing or just plain guessing. The method then uses the tangent to the curve $y = f(x)$ at $(x_0, f(x_0))$ to approximate the curve, calling the point x_1 where the tangent meets the x-axis (Figure 4.43). The number x_1 is usually a better approximation to the solution than is x_0. The point x_2 where the tangent to the curve at $(x_1, f(x_1))$ crosses the x-axis is the next approximation in the sequence. We continue on, using each approximation to generate the next, until we are close enough to the root to stop.

We can derive a formula for generating the successive approximations in the following way. Given the approximation x_n, the point-slope equation for the tangent to the curve at $(x_n, f(x_n))$ is

$$y = f(x_n) + f'(x_n)(x - x_n).$$

We can find where it crosses the x-axis by setting $y = 0$ (Figure 4.44).

$$0 = f(x_n) + f'(x_n)(x - x_n)$$

$$-\frac{f(x_n)}{f'(x_n)} = x - x_n$$

$$x = x_n - \frac{f(x_n)}{f'(x_n)} \qquad \text{If } f'(x_n) \neq 0$$

This value of x is the next approximation x_{n+1}. Here is a summary of Newton's method.

Procedure for Newton's Method

1. Guess a first approximation to a solution of the equation $f(x) = 0$. A graph of $y = f(x)$ may help.
2. Use the first approximation to get a second, the second to get a third, and so on, using the formula

$$x_{n+1} = x_n - \frac{f(x_n)}{f'(x_n)}, \qquad \text{if } f'(x_n) \neq 0 \tag{1}$$

Applying Newton's Method

Applications of Newton's method generally involve many numerical computations, making them well suited for computers or calculators. Nevertheless, even when the calculations are done by hand (which may be very tedious), they give a powerful way to find solutions of equations.

In our first example, we find decimal approximations to $\sqrt{2}$ by estimating the positive root of the equation $f(x) = x^2 - 2 = 0$.

EXAMPLE 1 Finding the Square Root of 2

Find the positive root of the equation

$$f(x) = x^2 - 2 = 0.$$

Solution With $f(x) = x^2 - 2$ and $f'(x) = 2x$, Equation (1) becomes

$$x_{n+1} = x_n - \frac{x_n{}^2 - 2}{2x_n}$$
$$= x_n - \frac{x_n}{2} + \frac{1}{x_n}$$
$$= \frac{x_n}{2} + \frac{1}{x_n}.$$

The equation

$$x_{n+1} = \frac{x_n}{2} + \frac{1}{x_n}$$

enables us to go from each approximation to the next with just a few keystrokes. With the starting value $x_0 = 1$, we get the results in the first column of the following table. (To five decimal places, $\sqrt{2} = 1.41421$.)

	Error	Number of correct digits
$x_0 = 1$	−0.41421	1
$x_1 = 1.5$	0.08579	1
$x_2 = 1.41667$	0.00246	3
$x_3 = 1.41422$	0.00001	5

■

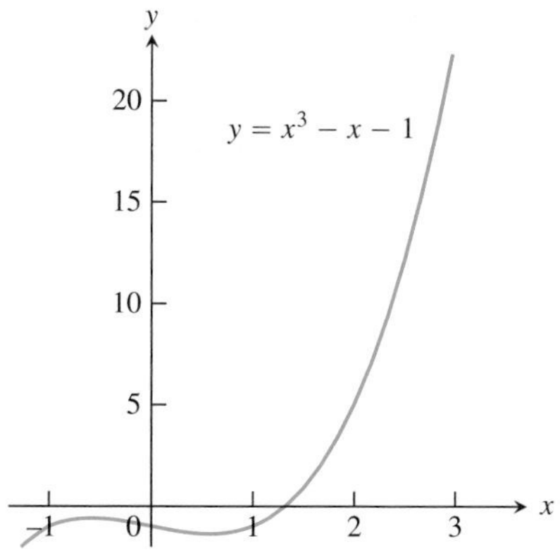

FIGURE 4.45 The graph of $f(x) = x^3 - x - 1$ crosses the x-axis once; this is the root we want to find (Example 2).

Newton's method is the method used by most calculators to calculate roots because it converges so fast (more about this later). If the arithmetic in the table in Example 1 had been carried to 13 decimal places instead of 5, then going one step further would have given $\sqrt{2}$ correctly to more than 10 decimal places.

EXAMPLE 2 Using Newton's Method

Find the x-coordinate of the point where the curve $y = x^3 - x$ crosses the horizontal line $y = 1$.

Solution The curve crosses the line when $x^3 - x = 1$ or $x^3 - x - 1 = 0$. When does $f(x) = x^3 - x - 1$ equal zero? Since $f(1) = -1$ and $f(2) = 5$, we know by the Intermediate Value Theorem there is a root in the interval (1, 2) (Figure 4.45).

We apply Newton's method to f with the starting value $x_0 = 1$. The results are displayed in Table 4.1 and Figure 4.46.

At $n = 5$, we come to the result $x_6 = x_5 = 1.3247\ 17957$. When $x_{n+1} = x_n$, Equation (1) shows that $f(x_n) = 0$. We have found a solution of $f(x) = 0$ to nine decimals. ■

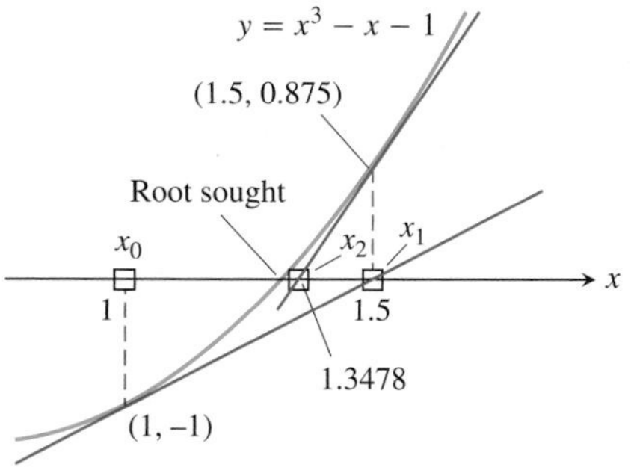

FIGURE 4.46 The first three x-values in Table 4.1 (four decimal places).

In Figure 4.47 we have indicated that the process in Example 2 might have started at the point $B_0(3, 23)$ on the curve, with $x_0 = 3$. Point B_0 is quite far from the x-axis, but the tangent at B_0 crosses the x-axis at about (2.12, 0), so x_1 is still an improvement over x_0. If we use Equation (1) repeatedly as before, with $f(x) = x^3 - x - 1$ and $f'(x) = 3x^2 - 1$, we confirm the nine-place solution $x_7 = x_6 = 1.3247\ 17957$ in seven steps.

TABLE 4.1 The result of applying Newton's method to $f(x) = x^3 - x - 1$ with $x_0 = 1$

n	x_n	$f(x_n)$	$f'(x_n)$	$x_{n+1} = x_n - \dfrac{f(x_n)}{f'(x_n)}$
0	1	−1	2	1.5
1	1.5	0.875	5.75	1.3478 26087
2	1.3478 26087	0.1006 82173	4.4499 05482	1.3252 00399
3	1.3252 00399	0.0020 58362	4.2684 68292	1.3247 18174
4	1.3247 18174	0.0000 00924	4.2646 34722	1.3247 17957
5	1.3247 17957	−1.8672E-13	4.2646 32999	1.3247 17957

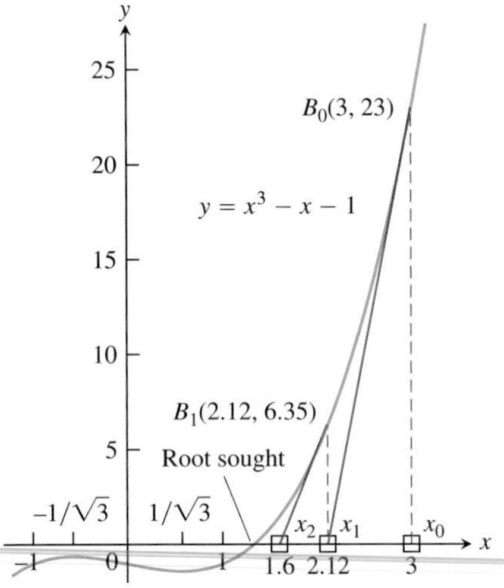

FIGURE 4.47 Any starting value x_0 to the right of $x = 1/\sqrt{3}$ will lead to the root.

The curve in Figure 4.47 has a local maximum at $x = -1/\sqrt{3}$ and a local minimum at $x = 1/\sqrt{3}$. We would not expect good results from Newton's method if we were to start with x_0 between these points, but we can start any place to the right of $x = 1/\sqrt{3}$ and get the answer. It would not be very clever to do so, but we could even begin far to the right of B_0, for example with $x_0 = 10$. It takes a bit longer, but the process still converges to the same answer as before.

Convergence of Newton's Method

In practice, Newton's method usually converges with impressive speed, but this is not guaranteed. One way to test convergence is to begin by graphing the function to estimate a good starting value for x_0. You can test that you are getting closer to a zero of the function by evaluating $|f(x_n)|$ and check that the method is converging by evaluating $|x_n - x_{n+1}|$.

Theory does provide some help. A theorem from advanced calculus says that if

$$\left|\frac{f(x)f''(x)}{(f'(x))^2}\right| < 1 \tag{2}$$

for all x in an interval about a root r, then the method will converge to r for any starting value x_0 in that interval. Note that this condition is satisfied if the graph of f is not too horizontal near where it crosses the x-axis.

Newton's method always converges if, between r and x_0, the graph of f is concave up when $f(x_0) > 0$ and concave down when $f(x_0) < 0$. (See Figure 4.48.) In most cases, the speed of the convergence to the root r is expressed by the advanced calculus formula

$$\underbrace{|x_{n+1} - r|}_{\text{error } e_{n+1}} \le \frac{\max|f''|}{2\min|f'|}|x_n - r|^2 = \text{constant}\cdot\underbrace{|x_n - r|^2}_{\text{error } e_n}, \tag{3}$$

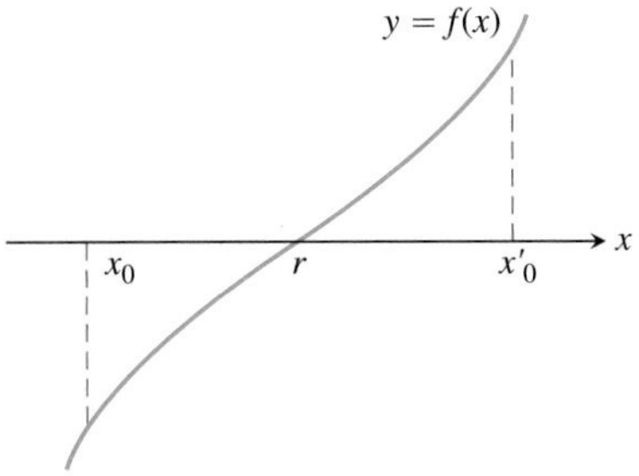

FIGURE 4.48 Newton's method will converge to r from either starting point.

where max and min refer to the maximum and minimum values in an interval surrounding r. The formula says that the error in step $n + 1$ is no greater than a constant times the square of the error in step n. This may not seem like much, but think of what it says. If the constant is less than or equal to 1 and $|x_n - r| < 10^{-3}$, then $|x_{n+1} - r| < 10^{-6}$. *In a single step*, the method moves from three decimal places of accuracy to six, and the number of decimals of accuracy continues to double with each successive step.

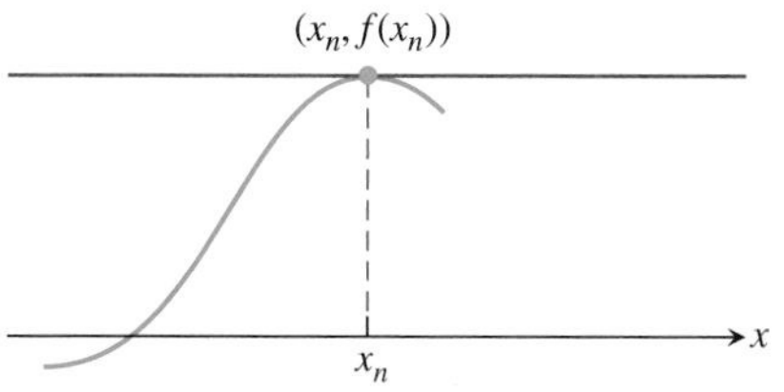

FIGURE 4.49 If $f'(x_n) = 0$, there is no intersection point to define x_{n+1}.

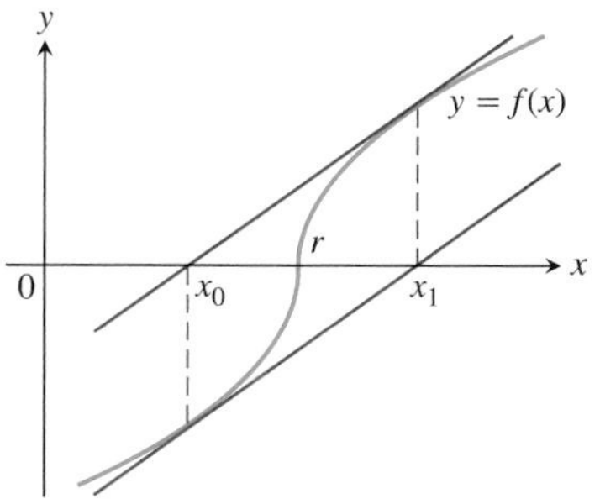

FIGURE 4.50 Newton's method fails to converge. You go from x_0 to x_1 and back to x_0, never getting any closer to r.

But Things Can Go Wrong

Newton's method stops if $f'(x_n) = 0$ (Figure 4.49). In that case, try a new starting point. Of course, f and f' may have the same root. To detect whether this is so, you could first find the solutions of $f'(x) = 0$ and check f at those values, or you could graph f and f' together.

Newton's method does not always converge. For instance, if

$$f(x) = \begin{cases} -\sqrt{r - x}, & x < r \\ \sqrt{x - r}, & x \geq r, \end{cases}$$

the graph will be like the one in Figure 4.50. If we begin with $x_0 = r - h$, we get $x_1 = r + h$, and successive approximations go back and forth between these two values. No amount of iteration brings us closer to the root than our first guess.

If Newton's method does converge, it converges to a root. Be careful, however. There are situations in which the method appears to converge but there is no root there. Fortunately, such situations are rare.

When Newton's method converges to a root, it may not be the root you have in mind. Figure 4.51 shows two ways this can happen.

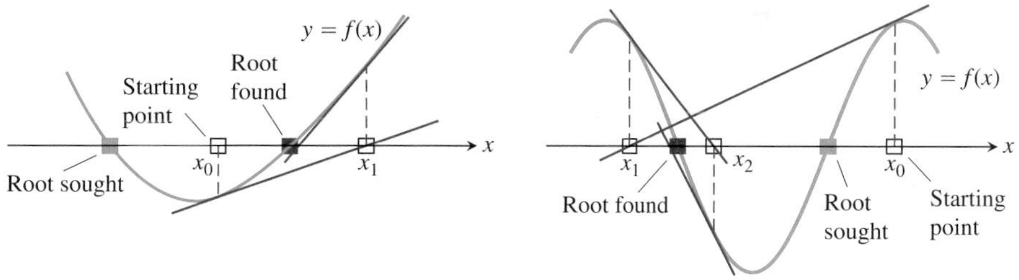

FIGURE 4.51 If you start too far away, Newton's method may miss the root you want.

Fractal Basins and Newton's Method

The process of finding roots by Newton's method can be uncertain in the sense that for some equations, the final outcome can be extremely sensitive to the starting value's location.

The equation $4x^4 - 4x^2 = 0$ is a case in point (Figure 4.52a). Starting values in the blue zone on the x-axis lead to root A. Starting values in the black lead to root B, and starting values in the red zone lead to root C. The points $\pm\sqrt{2}/2$ give horizontal tangents. The points $\pm\sqrt{21}/7$ "cycle," each leading to the other, and back (Figure 4.52b).

The interval between $\sqrt{21}/7$ and $\sqrt{2}/2$ contains infinitely many open intervals of points leading to root A, alternating with intervals of points leading to root C (Figure 4.52c). The boundary points separating consecutive intervals (there are infinitely many) do not lead to roots, but cycle back and forth from one to another. Moreover, as we select points that approach $\sqrt{21}/7$ from the right, it becomes increasingly difficult to distinguish which lead to root A and which to root C. On the same side of $\sqrt{21}/7$, we find arbitrarily close together points whose ultimate destinations are far apart.

If we think of the roots as "attractors" of other points, the coloring in Figure 4.52 shows the intervals of the points they attract (the "intervals of attraction"). You might think that points between roots A and B would be attracted to either A or B, but, as we see, that is not the case. Between A and B there are infinitely many intervals of points attracted to C. Similarly between B and C lie infinitely many intervals of points attracted to A.

We encounter an even more dramatic example of such behavior when we apply Newton's method to solve the complex-number equation $z^6 - 1 = 0$. It has six solutions: 1, -1, and the four numbers $\pm(1/2) \pm \left(\sqrt{3}/2\right)i$. As Figure 4.53 suggests, each of the

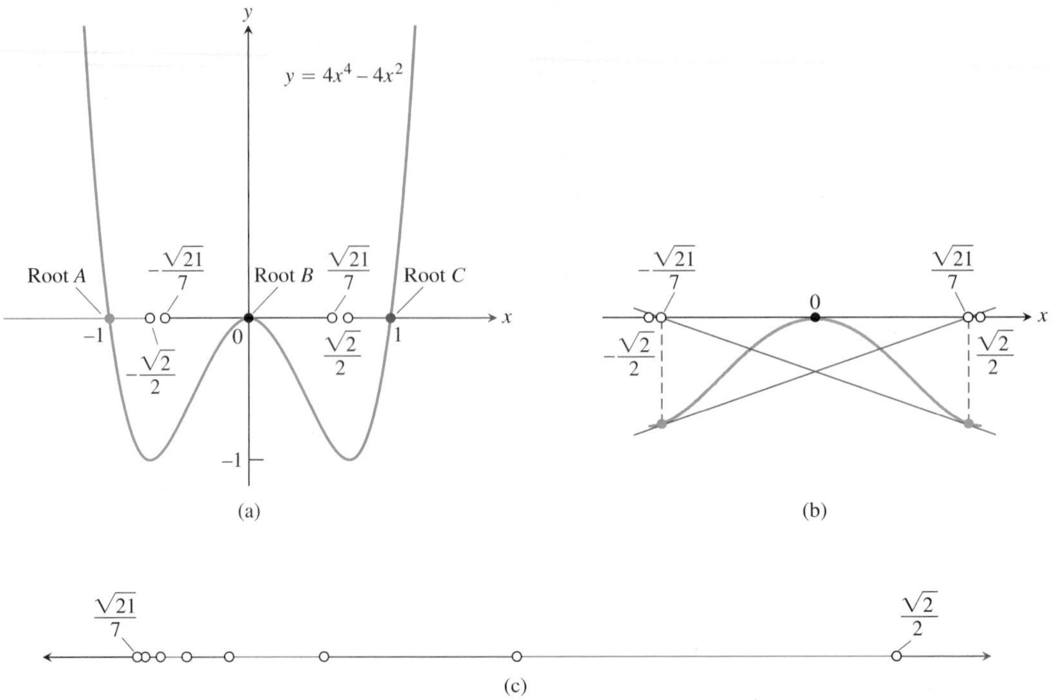

FIGURE 4.52 (a) Starting values in $\left(-\infty, -\sqrt{2}/2\right)$, $\left(-\sqrt{21}/7, \sqrt{21}/7\right)$, and $\left(\sqrt{2}/2, \infty\right)$ lead respectively to roots A, B, and C. (b) The values $x = \pm\sqrt{21}/7$ lead only to each other. (c) Between $\sqrt{21}/7$ and $\sqrt{2}/2$, there are infinitely many open intervals of points attracted to A alternating with open intervals of points attracted to C. This behavior is mirrored in the interval $\left(-\sqrt{2}/2, -\sqrt{21}/7\right)$.

FIGURE 4.53 This computer-generated initial value portrait uses color to show where different points in the complex plane end up when they are used as starting values in applying Newton's method to solve the equation $z^6 - 1 = 0$. Red points go to 1, green points to $(1/2) + \left(\sqrt{3}/2\right)i$, dark blue points to $(-1/2) + \left(\sqrt{3}/2\right)i$, and so on. Starting values that generate sequences that do not arrive within 0.1 unit of a root after 32 steps are colored black.

six roots has infinitely many "basins" of attraction in the complex plane (Appendix 5). Starting points in red basins are attracted to the root 1, those in the green basin to the root $(1/2) + \left(\sqrt{3}/2\right)i$, and so on. Each basin has a boundary whose complicated pattern repeats without end under successive magnifications. These basins are called **fractal basins.**

EXERCISES 4.7

Root-Finding

1. Use Newton's method to estimate the solutions of the equation $x^2 + x - 1 = 0$. Start with $x_0 = -1$ for the left-hand solution and with $x_0 = 1$ for the solution on the right. Then, in each case, find x_2.
2. Use Newton's method to estimate the one real solution of $x^3 + 3x + 1 = 0$. Start with $x_0 = 0$ and then find x_2.
3. Use Newton's method to estimate the two zeros of the function $f(x) = x^4 + x - 3$. Start with $x_0 = -1$ for the left-hand zero and with $x_0 = 1$ for the zero on the right. Then, in each case, find x_2.
4. Use Newton's method to estimate the two zeros of the function $f(x) = 2x - x^2 + 1$. Start with $x_0 = 0$ for the left-hand zero and with $x_0 = 2$ for the zero on the right. Then, in each case, find x_2.
5. Use Newton's method to find the positive fourth root of 2 by solving the equation $x^4 - 2 = 0$. Start with $x_0 = 1$ and find x_2.
6. Use Newton's method to find the negative fourth root of 2 by solving the equation $x^4 - 2 = 0$. Start with $x_0 = -1$ and find x_2.

Theory, Examples, and Applications

7. **Guessing a root** Suppose that your first guess is lucky, in the sense that x_0 is a root of $f(x) = 0$. Assuming that $f'(x_0)$ is defined and not 0, what happens to x_1 and later approximations?
8. **Estimating pi** You plan to estimate $\pi/2$ to five decimal places by using Newton's method to solve the equation $\cos x = 0$. Does it matter what your starting value is? Give reasons for your answer.
9. **Oscillation** Show that if $h > 0$, applying Newton's method to

$$f(x) = \begin{cases} \sqrt{x}, & x \geq 0 \\ \sqrt{-x}, & x < 0 \end{cases}$$

leads to $x_1 = -h$ if $x_0 = h$ and to $x_1 = h$ if $x_0 = -h$. Draw a picture that shows what is going on.

10. **Approximations that get worse and worse** Apply Newton's method to $f(x) = x^{1/3}$ with $x_0 = 1$ and calculate x_1, x_2, x_3, and x_4. Find a formula for $|x_n|$. What happens to $|x_n|$ as $n \to \infty$? Draw a picture that shows what is going on.
11. Explain why the following four statements ask for the same information:

 i) Find the roots of $f(x) = x^3 - 3x - 1$.

 ii) Find the x-coordinates of the intersections of the curve $y = x^3$ with the line $y = 3x + 1$.

 iii) Find the x-coordinates of the points where the curve $y = x^3 - 3x$ crosses the horizontal line $y = 1$.

 iv) Find the values of x where the derivative of $g(x) = (1/4)x^4 - (3/2)x^2 - x + 5$ equals zero.

12. **Locating a planet** To calculate a planet's space coordinates, we have to solve equations like $x = 1 + 0.5 \sin x$. Graphing the function $f(x) = x - 1 - 0.5 \sin x$ suggests that the function has a root near $x = 1.5$. Use one application of Newton's method to improve this estimate. That is, start with $x_0 = 1.5$ and find x_1. (The value of the root is 1.49870 to five decimal places.) Remember to use radians.

T 13. **A program for using Newton's method on a grapher** Let $f(x) = x^3 + 3x + 1$. Here is a home screen program to perform the computations in Newton's method.

 a. Let $y_0 = f(x)$ and $y_1 = \text{NDER } f(x)$.

 b. Store $x_0 = -0.3$ into x.

 c. Then store $x - (y_0/y_1)$ into x and press the Enter key over and over. Watch as the numbers converge to the zero of f.

 d. Use different values for x_0 and repeat steps (b) and (c).

 e. Write your own equation and use this approach to solve it using Newton's method. Compare your answer with the answer given by the built-in feature of your calculator that gives zeros of functions.

T 14. (*Continuation of Exercise 11.*)

 a. Use Newton's method to find the two negative zeros of $f(x) = x^3 - 3x - 1$ to five decimal places.

 b. Graph $f(x) = x^3 - 3x - 1$ for $-2 \leq x \leq 2.5$. Use the Zoom and Trace features to estimate the zeros of f to five decimal places.

 c. Graph $g(x) = 0.25x^4 - 1.5x^2 - x + 5$. Use the Zoom and Trace features with appropriate rescaling to find, to five decimal places, the values of x where the graph has horizontal tangents.

T 15. **Intersecting curves** The curve $y = \tan x$ crosses the line $y = 2x$ between $x = 0$ and $x = \pi/2$. Use Newton's method to find where.

T 16. **Real solutions of a quartic** Use Newton's method to find the two real solutions of the equation $x^4 - 2x^3 - x^2 - 2x + 2 = 0$.

T 17. **a.** How many solutions does the equation $\sin 3x = 0.99 - x^2$ have?

 b. Use Newton's method to find them.

T **18. Intersection of curves**

a. Does cos 3x ever equal x? Give reasons for your answer.

b. Use Newton's method to find where.

T **19.** Find the four real zeros of the function $f(x) = 2x^4 - 4x^2 + 1$.

T **20. Estimating pi** Estimate π to as many decimal places as your calculator will display by using Newton's method to solve the equation $\tan x = 0$ with $x_0 = 3$.

21. At what values(s) of x does $\cos x = 2x$?

22. At what value(s) of x does $\cos x = -x$?

23. Use the Intermediate Value Theorem from Section 2.6 to show that $f(x) = x^3 + 2x - 4$ has a root between $x = 1$ and $x = 2$. Then find the root to five decimal places.

24. Factoring a quartic Find the approximate values of r_1 through r_4 in the factorization

$$8x^4 - 14x^3 - 9x^2 + 11x - 1 = 8(x - r_1)(x - r_2)(x - r_3)(x - r_4).$$

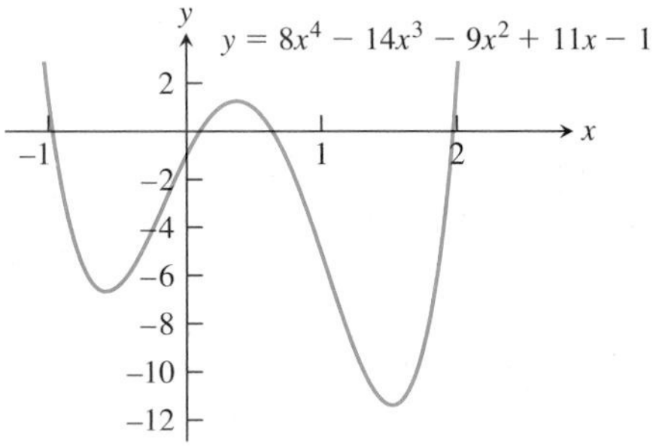

T **25. Converging to different zeros** Use Newton's method to find the zeros of $f(x) = 4x^4 - 4x^2$ using the given starting values (Figure 4.52).

a. $x_0 = -2$ and $x_0 = -0.8$, lying in $\left(-\infty, -\sqrt{2}/2\right)$

b. $x_0 = -0.5$ and $x_0 = 0.25$, lying in $\left(-\sqrt{21}/7, \sqrt{21}/7\right)$

c. $x_0 = 0.8$ and $x_0 = 2$, lying in $\left(\sqrt{2}/2, \infty\right)$

d. $x_0 = -\sqrt{21}/7$ and $x_0 = \sqrt{21}/7$

26. The sonobuoy problem In submarine location problems, it is often necessary to find a submarine's closest point of approach (CPA) to a sonobuoy (sound detector) in the water. Suppose that the submarine travels on the parabolic path $y = x^2$ and that the buoy is located at the point $(2, -1/2)$.

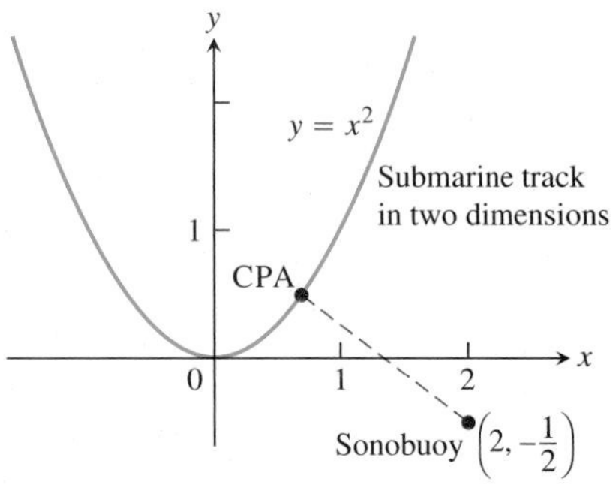

a. Show that the value of x that minimizes the distance between the submarine and the buoy is a solution of the equation $x = 1/(x^2 + 1)$.

b. Solve the equation $x = 1/(x^2 + 1)$ with Newton's method.

27. Curves that are nearly flat at the root Some curves are so flat that, in practice, Newton's method stops too far from the root to give a useful estimate. Try Newton's method on $f(x) = (x - 1)^{40}$ with a starting value of $x_0 = 2$ to see how close your machine comes to the root $x = 1$.

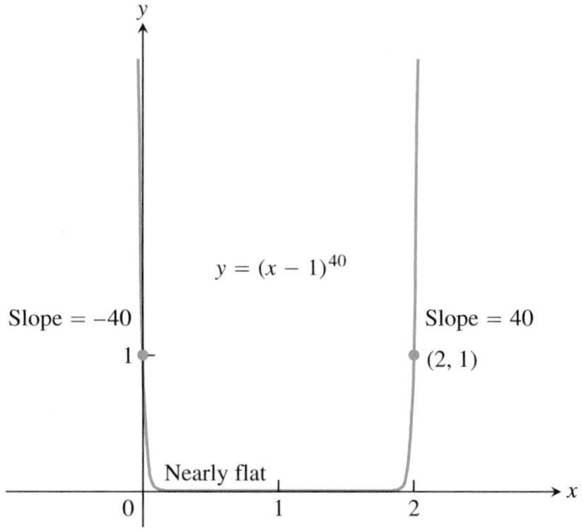

28. Finding a root different from the one sought All three roots of $f(x) = 4x^4 - 4x^2$ can be found by starting Newton's method near $x = \sqrt{21}/7$. Try it. (See Figure 4.52.)

29. Finding an ion concentration While trying to find the acidity of a saturated solution of magnesium hydroxide in hydrochloric acid, you derive the equation

$$\frac{3.64 \times 10^{-11}}{[H_3O^+]^2} = [H_3O^+] + 3.6 \times 10^{-4}$$

for the hydronium ion concentration $[H_3O^+]$. To find the value of $[H_3O^+]$, you set $x = 10^4[H_3O^+]$ and convert the equation to

$$x^3 + 3.6x^2 - 36.4 = 0.$$

You then solve this by Newton's method. What do you get for x? (Make it good to two decimal places.) For $[H_3O^+]$?

T **30. Complex roots** If you have a computer or a calculator that can be programmed to do complex-number arithmetic, experiment with Newton's method to solve the equation $z^6 - 1 = 0$. The recursion relation to use is

$$z_{n+1} = z_n - \frac{z_n^{\,6} - 1}{6z_n^{\,5}} \quad \text{or} \quad z_{n+1} = \frac{5}{6}z_n + \frac{1}{6z_n^{\,5}}.$$

Try these starting values (among others): 2, i, $\sqrt{3} + i$.

4.8 Antiderivatives

We have studied how to find the derivative of a function. However, many problems require that we recover a function from its known derivative (from its known rate of change). For instance, we may know the velocity function of an object falling from an initial height and need to know its height at any time over some period. More generally, we want to find a function F from its derivative f. If such a function F exists, it is called an *antiderivative* of f.

Finding Antiderivatives

> **DEFINITION Antiderivative**
> A function F is an **antiderivative** of f on an interval I if $F'(x) = f(x)$ for all x in I.

The process of recovering a function $F(x)$ from its derivative $f(x)$ is called *antidifferentiation*. We use capital letters such as F to represent an antiderivative of a function f, G to represent an antiderivative of g, and so forth.

EXAMPLE 1 Finding Antiderivatives

Find an antiderivative for each of the following functions.

(a) $f(x) = 2x$

(b) $g(x) = \cos x$

(c) $h(x) = 2x + \cos x$

Solution

(a) $F(x) = x^2$

(b) $G(x) = \sin x$

(c) $H(x) = x^2 + \sin x$

Each answer can be checked by differentiating. The derivative of $F(x) = x^2$ is $2x$. The derivative of $G(x) = \sin x$ is $\cos x$ and the derivative of $H(x) = x^2 + \sin x$ is $2x + \cos x$. ■

The function $F(x) = x^2$ is not the only function whose derivative is $2x$. The function $x^2 + 1$ has the same derivative. So does $x^2 + C$ for any constant C. Are there others?

Corollary 2 of the Mean Value Theorem in Section 4.2 gives the answer: Any two antiderivatives of a function differ by a constant. So the functions $x^2 + C$, where C is an **arbitrary constant**, form *all* the antiderivatives of $f(x) = 2x$. More generally, we have the following result.

> If F is an antiderivative of f on an interval I, then the most general antiderivative of f on I is
> $$F(x) + C$$
> where C is an arbitrary constant.

Thus the most general antiderivative of f on I is a *family* of functions $F(x) + C$ whose graphs are vertical translates of one another. We can select a particular antiderivative from this family by assigning a specific value to C. Here is an example showing how such an assignment might be made.

EXAMPLE 2 Finding a Particular Antiderivative

Find an antiderivative of $f(x) = \sin x$ that satisfies $F(0) = 3$.

Solution Since the derivative of $-\cos x$ is $\sin x$, the general antiderivative

$$F(x) = -\cos x + C$$

gives all the antiderivatives of $f(x)$. The condition $F(0) = 3$ determines a specific value for C. Substituting $x = 0$ into $F(x) = -\cos x + C$ gives

$$F(0) = -\cos 0 + C = -1 + C.$$

Since $F(0) = 3$, solving for C gives $C = 4$. So

$$F(x) = -\cos x + 4$$

is the antiderivative satisfying $F(0) = 3$. ■

By working backward from assorted differentiation rules, we can derive formulas and rules for antiderivatives. In each case there is an arbitrary constant C in the general expression representing all antiderivatives of a given function. Table 4.2 gives antiderivative formulas for a number of important functions.

TABLE 4.2 Antiderivative formulas

	Function	General antiderivative
1.	x^n	$\dfrac{x^{n+1}}{n+1} + C, \quad n \neq -1$, n rational
2.	$\sin kx$	$-\dfrac{\cos kx}{k} + C, \quad k$ a constant, $k \neq 0$
3.	$\cos kx$	$\dfrac{\sin kx}{k} + C, \quad k$ a constant, $k \neq 0$
4.	$\sec^2 x$	$\tan x + C$
5.	$\csc^2 x$	$-\cot x + C$
6.	$\sec x \tan x$	$\sec x + C$
7.	$\csc x \cot x$	$-\csc x + C$

The rules in Table 4.2 are easily verified by differentiating the general antiderivative formula to obtain the function to its left. For example, the derivative of $\tan x + C$ is $\sec^2 x$, whatever the value of the constant C, and this establishes the formula for the most general antiderivative of $\sec^2 x$.

EXAMPLE 3 Finding Antiderivatives Using Table 4.2

Find the general antiderivative of each of the following functions.

(a) $f(x) = x^5$

(b) $g(x) = \dfrac{1}{\sqrt{x}}$

(c) $h(x) = \sin 2x$

(d) $i(x) = \cos \dfrac{x}{2}$

Solution

(a) $F(x) = \dfrac{x^6}{6} + C$ Formula 1 with $n = 5$

(b) $g(x) = x^{-1/2}$, so

$$G(x) = \frac{x^{1/2}}{1/2} + C = 2\sqrt{x} + C$$ Formula 1 with $n = -1/2$

(c) $H(x) = \dfrac{-\cos 2x}{2} + C$ Formula 2 with $k = 2$

(d) $I(x) = \dfrac{\sin (x/2)}{1/2} + C = 2 \sin \dfrac{x}{2} + C$ Formula 3 with $k = 1/2$ ■

Other derivative rules also lead to corresponding antiderivative rules. We can add and subtract antiderivatives, and multiply them by constants.

The formulas in Table 4.3 are easily proved by differentiating the antiderivatives and verifying that the result agrees with the original function. Formula 2 is the special case $k = -1$ in Formula 1.

TABLE 4.3 Antiderivative linearity rules

		Function	General antiderivative
1.	*Constant Multiple Rule:*	$kf(x)$	$kF(x) + C$, k a constant
2.	*Negative Rule:*	$-f(x)$	$-F(x) + C$,
3.	*Sum or Difference Rule:*	$f(x) \pm g(x)$	$F(x) \pm G(x) + C$

EXAMPLE 4 Using the Linearity Rules for Antiderivatives

Find the general antiderivative of

$$f(x) = \frac{3}{\sqrt{x}} + \sin 2x.$$

Solution We have that $f(x) = 3g(x) + h(x)$ for the functions g and h in Example 3. Since $G(x) = 2\sqrt{x}$ is an antiderivative of $g(x)$ from Example 3b, it follows from the Constant Multiple Rule for antiderivatives that $3G(x) = 3 \cdot 2\sqrt{x} = 6\sqrt{x}$ is an antiderivative of $3g(x) = 3/\sqrt{x}$. Likewise, from Example 3c we know that $H(x) = (-1/2)\cos 2x$ is an antiderivative of $h(x) = \sin 2x$. From the Sum Rule for antiderivatives, we then get that

$$\begin{aligned} F(x) &= 3G(x) + H(x) + C \\ &= 6\sqrt{x} - \frac{1}{2}\cos 2x + C \end{aligned}$$

is the general antiderivative formula for $f(x)$, where C is an arbitrary constant. ■

Antiderivatives play several important roles, and methods and techniques for finding them are a major part of calculus. (This is the subject of Chapter 8.)

Initial Value Problems and Differential Equations

Finding an antiderivative for a function $f(x)$ is the same problem as finding a function $y(x)$ that satisfies the equation

$$\frac{dy}{dx} = f(x).$$

This is called a **differential equation**, since it is an equation involving an unknown function y that is being differentiated. To solve it, we need a function $y(x)$ that satisfies the equation. This function is found by taking the antiderivative of $f(x)$. We fix the arbitrary constant arising in the antidifferentiation process by specifying an initial condition

$$y(x_0) = y_0.$$

This condition means the function $y(x)$ has the value y_0 when $x = x_0$. The combination of a differential equation and an initial condition is called an **initial value problem**. Such problems play important roles in all branches of science. Here's an example of solving an initial value problem.

EXAMPLE 5 Finding a Curve from Its Slope Function and a Point

Find the curve whose slope at the point (x, y) is $3x^2$ if the curve is required to pass through the point $(1, -1)$.

Solution In mathematical language, we are asked to solve the initial value problem that consists of the following.

The differential equation: $\dfrac{dy}{dx} = 3x^2$ The curve's slope is $3x^2$.

The initial condition: $y(1) = -1$

1. *Solve the differential equation*: The function y is an antiderivative of $f(x) = 3x^2$, so

$$y = x^3 + C.$$

This result tells us that y equals $x^3 + C$ for some value of C. We find that value from the initial condition $y(1) = -1$.

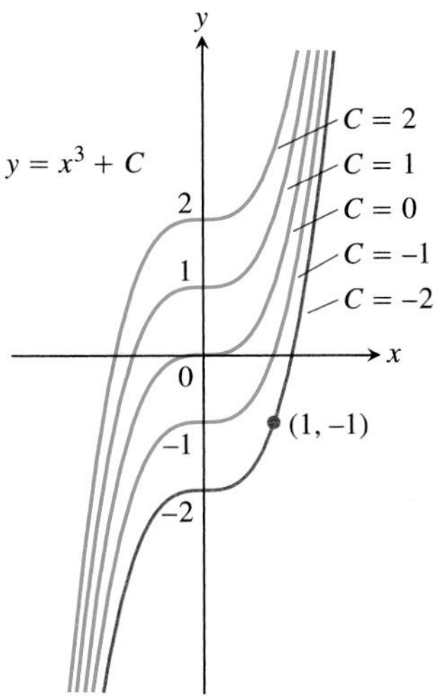

FIGURE 4.54 The curves $y = x^3 + C$ fill the coordinate plane without overlapping. In Example 5, we identify the curve $y = x^3 - 2$ as the one that passes through the given point $(1, -1)$.

2. *Evaluate C:*

$$\begin{aligned} y &= x^3 + C \\ -1 &= (1)^3 + C \qquad \text{Initial condition } y(1) = -1 \\ C &= -2. \end{aligned}$$

The curve we want is $y = x^3 - 2$ (Figure 4.54). ■

The most general antiderivative $F(x) + C$ (which is $x^3 + C$ in Example 5) of the function $f(x)$ gives the **general solution** $y = F(x) + C$ of the differential equation $dy/dx = f(x)$. The general solution gives *all* the solutions of the equation (there are infinitely many, one for each value of C). We **solve** the differential equation by finding its general solution. We then solve the initial value problem by finding the **particular solution** that satisfies the initial condition $y(x_0) = y_0$.

Antiderivatives and Motion

We have seen that the derivative of the position of an object gives its velocity, and the derivative of its velocity gives its acceleration. If we know an object's acceleration, then by finding an antiderivative we can recover the velocity, and from an antiderivative of the velocity we can recover its position function. This procedure was used as an application of Corollary 2 in Section 4.2. Now that we have a terminology and conceptual framework in terms of antiderivatives, we revisit the problem from the point of view of differential equations.

EXAMPLE 6 Dropping a Package from an Ascending Balloon

A balloon ascending at the rate of 12 ft/sec is at a height 80 ft above the ground when a package is dropped. How long does it take the package to reach the ground?

Solution Let $v(t)$ denote the velocity of the package at time t, and let $s(t)$ denote its height above the ground. The acceleration of gravity near the surface of the earth is 32 ft/sec^2. Assuming no other forces act on the dropped package, we have

$$\frac{dv}{dt} = -32. \qquad \text{Negative because gravity acts in the direction of decreasing } s.$$

This leads to the initial value problem.

$$\begin{aligned} &\textit{Differential equation:} && \frac{dv}{dt} = -32 \\ &\textit{Initial condition:} && v(0) = 12, \end{aligned}$$

which is our mathematical model for the package's motion. We solve the initial value problem to obtain the velocity of the package.

1. *Solve the differential equation*: The general formula for an antiderivative of -32 is

$$v = -32t + C.$$

Having found the general solution of the differential equation, we use the initial condition to find the particular solution that solves our problem.

2. *Evaluate C:*

$$\begin{aligned} 12 &= -32(0) + C \qquad \text{Initial condition } v(0) = 12 \\ C &= 12. \end{aligned}$$

The solution of the initial value problem is

$$v = -32t + 12.$$

Since velocity is the derivative of height and the height of the package is 80 ft at the time $t = 0$ when it is dropped, we now have a second initial value problem.

$$\textit{Differential equation:} \quad \frac{ds}{dt} = -32t + 12 \qquad \text{Set } v = ds/dt \text{ in the last equation.}$$

$$\textit{Initial condition:} \quad s(0) = 80$$

We solve this initial value problem to find the height as a function of t.

1. *Solve the differential equation*: Finding the general antiderivative of $-32t + 12$ gives

$$s = -16t^2 + 12t + C.$$

2. *Evaluate C*:

$$80 = -16(0)^2 + 12(0) + C \qquad \text{Initial condition } s(0) = 80$$

$$C = 80.$$

The package's height above ground at time t is

$$s = -16t^2 + 12t + 80.$$

Use the solution: To find how long it takes the package to reach the ground, we set s equal to 0 and solve for t:

$$-16t^2 + 12t + 80 = 0$$

$$-4t^2 + 3t + 20 = 0$$

$$t = \frac{-3 \pm \sqrt{329}}{-8} \qquad \text{Quadratic formula}$$

$$t \approx -1.89, \qquad t \approx 2.64.$$

The package hits the ground about 2.64 sec after it is dropped from the balloon. (The negative root has no physical meaning.) ■

Indefinite Integrals

A special symbol is used to denote the collection of all antiderivatives of a function f.

DEFINITION Indefinite Integral, Integrand

The set of all antiderivatives of f is the **indefinite integral** of f with respect to x, denoted by

$$\int f(x)\,dx.$$

The symbol $\int$ is an **integral sign**. The function f is the **integrand** of the integral, and x is the **variable of integration**.

Using this notation, we restate the solutions of Example 1, as follows:

$$\int 2x\,dx = x^2 + C,$$

$$\int \cos x\,dx = \sin x + C,$$

$$\int (2x + \cos x)\,dx = x^2 + \sin x + C.$$

This notation is related to the main application of antiderivatives, which will be explored in Chapter 5. Antiderivatives play a key role in computing limits of infinite sums, an unexpected and wonderfully useful role that is described in a central result of Chapter 5, called the Fundamental Theorem of Calculus.

EXAMPLE 7 Indefinite Integration Done Term-by-Term and Rewriting the Constant of Integration

Evaluate

$$\int (x^2 - 2x + 5)\,dx.$$

Solution If we recognize that $(x^3/3) - x^2 + 5x$ is an antiderivative of $x^2 - 2x + 5$, we can evaluate the integral as

$$\int (x^2 - 2x + 5)\,dx = \overbrace{\frac{x^3}{3} - x^2 + 5x}^{\text{antiderivative}} + \underbrace{C}_{\text{arbitrary constant}}.$$

If we do not recognize the antiderivative right away, we can generate it term-by-term with the Sum, Difference, and Constant Multiple Rules:

$$\begin{aligned}
\int (x^2 - 2x + 5)\,dx &= \int x^2\,dx - \int 2x\,dx + \int 5\,dx \\
&= \int x^2\,dx - 2\int x\,dx + 5\int 1\,dx \\
&= \left(\frac{x^3}{3} + C_1\right) - 2\left(\frac{x^2}{2} + C_2\right) + 5(x + C_3) \\
&= \frac{x^3}{3} + C_1 - x^2 - 2C_2 + 5x + 5C_3.
\end{aligned}$$

This formula is more complicated than it needs to be. If we combine C_1, $-2C_2$, and $5C_3$ into a single arbitrary constant $C = C_1 - 2C_2 + 5C_3$, the formula simplifies to

$$\frac{x^3}{3} - x^2 + 5x + C$$

and *still* gives all the antiderivatives there are. For this reason, we recommend that you go right to the final form even if you elect to integrate term-by-term. Write

$$\int (x^2 - 2x + 5)\,dx = \int x^2\,dx - \int 2x\,dx + \int 5\,dx$$
$$= \frac{x^3}{3} - x^2 + 5x + C.$$

Find the simplest antiderivative you can for each part and add the arbitrary constant of integration at the end. ■

EXERCISES 4.8

Finding Antiderivatives

In Exercises 1–16, find an antiderivative for each function. Do as many as you can mentally. Check your answers by differentiation.

1. a. $2x$ b. x^2 c. $x^2 - 2x + 1$
2. a. $6x$ b. x^7 c. $x^7 - 6x + 8$
3. a. $-3x^{-4}$ b. x^{-4} c. $x^{-4} + 2x + 3$
4. a. $2x^{-3}$ b. $\frac{x^{-3}}{2} + x^2$ c. $-x^{-3} + x - 1$
5. a. $\frac{1}{x^2}$ b. $\frac{5}{x^2}$ c. $2 - \frac{5}{x^2}$
6. a. $-\frac{2}{x^3}$ b. $\frac{1}{2x^3}$ c. $x^3 - \frac{1}{x^3}$
7. a. $\frac{3}{2}\sqrt{x}$ b. $\frac{1}{2\sqrt{x}}$ c. $\sqrt{x} + \frac{1}{\sqrt{x}}$
8. a. $\frac{4}{3}\sqrt[3]{x}$ b. $\frac{1}{3\sqrt[3]{x}}$ c. $\sqrt[3]{x} + \frac{1}{\sqrt[3]{x}}$
9. a. $\frac{2}{3}x^{-1/3}$ b. $\frac{1}{3}x^{-2/3}$ c. $-\frac{1}{3}x^{-4/3}$
10. a. $\frac{1}{2}x^{-1/2}$ b. $-\frac{1}{2}x^{-3/2}$ c. $-\frac{3}{2}x^{-5/2}$
11. a. $-\pi \sin \pi x$ b. $3 \sin x$ c. $\sin \pi x - 3 \sin 3x$
12. a. $\pi \cos \pi x$ b. $\frac{\pi}{2} \cos \frac{\pi x}{2}$ c. $\cos \frac{\pi x}{2} + \pi \cos x$
13. a. $\sec^2 x$ b. $\frac{2}{3} \sec^2 \frac{x}{3}$ c. $-\sec^2 \frac{3x}{2}$
14. a. $\csc^2 x$ b. $-\frac{3}{2} \csc^2 \frac{3x}{2}$ c. $1 - 8 \csc^2 2x$
15. a. $\csc x \cot x$ b. $-\csc 5x \cot 5x$ c. $-\pi \csc \frac{\pi x}{2} \cot \frac{\pi x}{2}$
16. a. $\sec x \tan x$ b. $4 \sec 3x \tan 3x$ c. $\sec \frac{\pi x}{2} \tan \frac{\pi x}{2}$

Finding Indefinite Integrals

In Exercises 17–54, find the most general antiderivative or indefinite integral. Check your answers by differentiation.

17. $\int (x + 1)\,dx$
18. $\int (5 - 6x)\,dx$
19. $\int \left(3t^2 + \frac{t}{2}\right) dt$
20. $\int \left(\frac{t^2}{2} + 4t^3\right) dt$
21. $\int (2x^3 - 5x + 7)\,dx$
22. $\int (1 - x^2 - 3x^5)\,dx$
23. $\int \left(\frac{1}{x^2} - x^2 - \frac{1}{3}\right) dx$
24. $\int \left(\frac{1}{5} - \frac{2}{x^3} + 2x\right) dx$
25. $\int x^{-1/3}\,dx$
26. $\int x^{-5/4}\,dx$
27. $\int \left(\sqrt{x} + \sqrt[3]{x}\right) dx$
28. $\int \left(\frac{\sqrt{x}}{2} + \frac{2}{\sqrt{x}}\right) dx$
29. $\int \left(8y - \frac{2}{y^{1/4}}\right) dy$
30. $\int \left(\frac{1}{7} - \frac{1}{y^{5/4}}\right) dy$
31. $\int 2x(1 - x^{-3})\,dx$
32. $\int x^{-3}(x + 1)\,dx$
33. $\int \frac{t\sqrt{t} + \sqrt{t}}{t^2}\,dt$
34. $\int \frac{4 + \sqrt{t}}{t^3}\,dt$
35. $\int (-2 \cos t)\,dt$
36. $\int (-5 \sin t)\,dt$
37. $\int 7 \sin \frac{\theta}{3}\,d\theta$
38. $\int 3 \cos 5\theta\,d\theta$
39. $\int (-3 \csc^2 x)\,dx$
40. $\int \left(-\frac{\sec^2 x}{3}\right) dx$
41. $\int \frac{\csc \theta \cot \theta}{2}\,d\theta$
42. $\int \frac{2}{5} \sec \theta \tan \theta\,d\theta$

43. $\int (4 \sec x \tan x - 2 \sec^2 x)\, dx$

44. $\int \frac{1}{2}(\csc^2 x - \csc x \cot x)\, dx$

45. $\int (\sin 2x - \csc^2 x)\, dx$

46. $\int (2 \cos 2x - 3 \sin 3x)\, dx$

47. $\int \frac{1 + \cos 4t}{2}\, dt$

48. $\int \frac{1 - \cos 6t}{2}\, dt$

49. $\int (1 + \tan^2 \theta)\, d\theta$

(*Hint:* $1 + \tan^2 \theta = \sec^2 \theta$)

50. $\int (2 + \tan^2 \theta)\, d\theta$

51. $\int \cot^2 x\, dx$

(*Hint:* $1 + \cot^2 x = \csc^2 x$)

52. $\int (1 - \cot^2 x)\, dx$

53. $\int \cos \theta (\tan \theta + \sec \theta)\, d\theta$

54. $\int \frac{\csc \theta}{\csc \theta - \sin \theta}\, d\theta$

Checking Antiderivative Formulas

Verify the formulas in Exercises 55–60 by differentiation.

55. $\int (7x - 2)^3\, dx = \frac{(7x - 2)^4}{28} + C$

56. $\int (3x + 5)^{-2}\, dx = -\frac{(3x + 5)^{-1}}{3} + C$

57. $\int \sec^2 (5x - 1)\, dx = \frac{1}{5} \tan (5x - 1) + C$

58. $\int \csc^2 \left(\frac{x - 1}{3}\right) dx = -3 \cot \left(\frac{x - 1}{3}\right) + C$

59. $\int \frac{1}{(x + 1)^2}\, dx = -\frac{1}{x + 1} + C$

60. $\int \frac{1}{(x + 1)^2}\, dx = \frac{x}{x + 1} + C$

61. Right, or wrong? Say which for each formula and give a brief reason for each answer.

a. $\int x \sin x\, dx = \frac{x^2}{2} \sin x + C$

b. $\int x \sin x\, dx = -x \cos x + C$

c. $\int x \sin x\, dx = -x \cos x + \sin x + C$

62. Right, or wrong? Say which for each formula and give a brief reason for each answer.

a. $\int \tan \theta \sec^2 \theta\, d\theta = \frac{\sec^3 \theta}{3} + C$

b. $\int \tan \theta \sec^2 \theta\, d\theta = \frac{1}{2} \tan^2 \theta + C$

c. $\int \tan \theta \sec^2 \theta\, d\theta = \frac{1}{2} \sec^2 \theta + C$

63. Right, or wrong? Say which for each formula and give a brief reason for each answer.

a. $\int (2x + 1)^2\, dx = \frac{(2x + 1)^3}{3} + C$

b. $\int 3(2x + 1)^2\, dx = (2x + 1)^3 + C$

c. $\int 6(2x + 1)^2\, dx = (2x + 1)^3 + C$

64. Right, or wrong? Say which for each formula and give a brief reason for each answer.

a. $\int \sqrt{2x + 1}\, dx = \sqrt{x^2 + x + C}$

b. $\int \sqrt{2x + 1}\, dx = \sqrt{x^2 + x} + C$

c. $\int \sqrt{2x + 1}\, dx = \frac{1}{3}\left(\sqrt{2x + 1}\right)^3 + C$

Initial Value Problems

65. Which of the following graphs shows the solution of the initial value problem

$$\frac{dy}{dx} = 2x, \quad y = 4 \text{ when } x = 1?$$

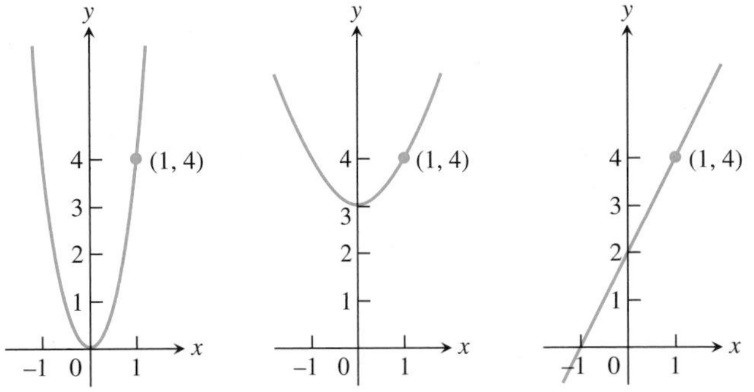

(a) (b) (c)

Give reasons for your answer.

66. Which of the following graphs shows the solution of the initial value problem

$$\frac{dy}{dx} = -x, \quad y = 1 \text{ when } x = -1?$$

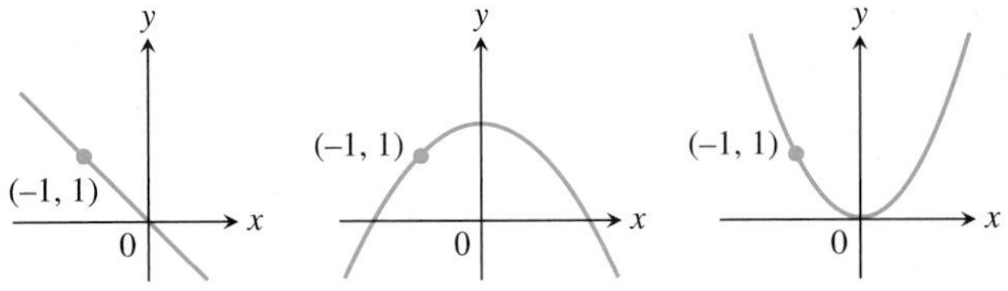

(a) (b) (c)

Give reasons for your answer.

Solve the initial value problems in Exercises 67–86.

67. $\dfrac{dy}{dx} = 2x - 7, \quad y(2) = 0$

68. $\dfrac{dy}{dx} = 10 - x, \quad y(0) = -1$

69. $\dfrac{dy}{dx} = \dfrac{1}{x^2} + x, \quad x > 0; \quad y(2) = 1$

70. $\dfrac{dy}{dx} = 9x^2 - 4x + 5, \quad y(-1) = 0$

71. $\dfrac{dy}{dx} = 3x^{-2/3}, \quad y(-1) = -5$

72. $\dfrac{dy}{dx} = \dfrac{1}{2\sqrt{x}}, \quad y(4) = 0$

73. $\dfrac{ds}{dt} = 1 + \cos t, \quad s(0) = 4$

74. $\dfrac{ds}{dt} = \cos t + \sin t, \quad s(\pi) = 1$

75. $\dfrac{dr}{d\theta} = -\pi \sin \pi\theta, \quad r(0) = 0$

76. $\dfrac{dr}{d\theta} = \cos \pi\theta, \quad r(0) = 1$

77. $\dfrac{dv}{dt} = \dfrac{1}{2} \sec t \tan t, \quad v(0) = 1$

78. $\dfrac{dv}{dt} = 8t + \csc^2 t, \quad v\left(\dfrac{\pi}{2}\right) = -7$

79. $\dfrac{d^2y}{dx^2} = 2 - 6x; \quad y'(0) = 4, \quad y(0) = 1$

80. $\dfrac{d^2y}{dx^2} = 0; \quad y'(0) = 2, \quad y(0) = 0$

81. $\dfrac{d^2r}{dt^2} = \dfrac{2}{t^3}; \quad \left.\dfrac{dr}{dt}\right|_{t=1} = 1, \quad r(1) = 1$

82. $\dfrac{d^2s}{dt^2} = \dfrac{3t}{8}; \quad \left.\dfrac{ds}{dt}\right|_{t=4} = 3, \quad s(4) = 4$

83. $\dfrac{d^3y}{dx^3} = 6; \quad y''(0) = -8, \quad y'(0) = 0, \quad y(0) = 5$

84. $\dfrac{d^3\theta}{dt^3} = 0; \quad \theta''(0) = -2, \quad \theta'(0) = -\dfrac{1}{2}, \quad \theta(0) = \sqrt{2}$

85. $y^{(4)} = -\sin t + \cos t;$
$y'''(0) = 7, \quad y''(0) = y'(0) = -1, \quad y(0) = 0$

86. $y^{(4)} = -\cos x + 8 \sin 2x;$
$y'''(0) = 0, \quad y''(0) = y'(0) = 1, \quad y(0) = 3$

Finding Curves

87. Find the curve $y = f(x)$ in the xy-plane that passes through the point (9, 4) and whose slope at each point is $3\sqrt{x}$.

88. a. Find a curve $y = f(x)$ with the following properties:

i) $\dfrac{d^2y}{dx^2} = 6x$

ii) Its graph passes through the point (0, 1), and has a horizontal tangent there.

b. How many curves like this are there? How do you know?

Solution (Integral) Curves

Exercises 89–92 show solution curves of differential equations. In each exercise, find an equation for the curve through the labeled point.

89.

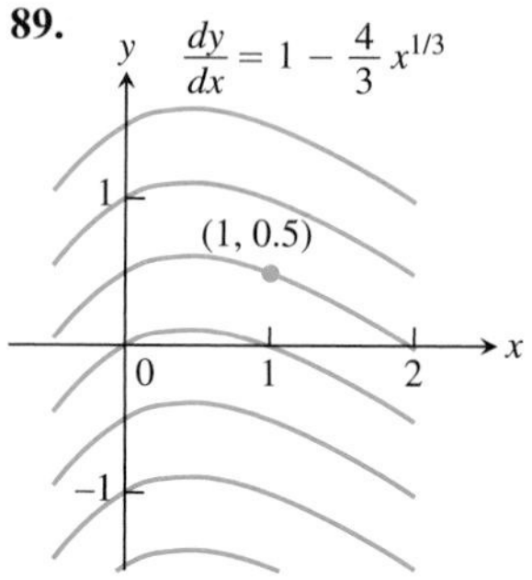

90.

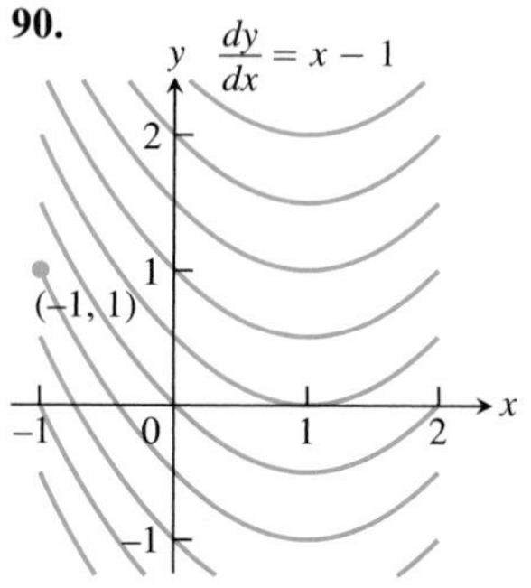

91.

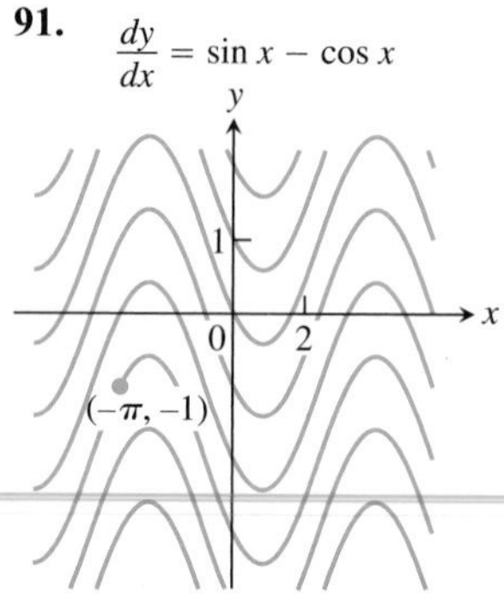

92.

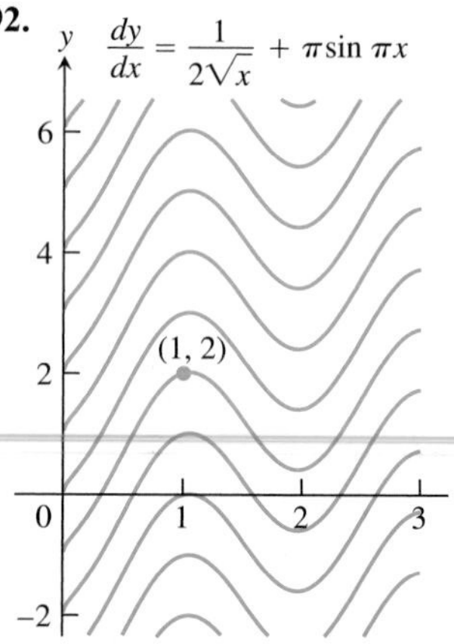

Applications

93. Finding displacement from an antiderivative of velocity

a. Suppose that the velocity of a body moving along the s-axis is

$$\frac{ds}{dt} = v = 9.8t - 3.$$

i) Find the body's displacement over the time interval from $t = 1$ to $t = 3$ given that $s = 5$ when $t = 0$.

ii) Find the body's displacement from $t = 1$ to $t = 3$ given that $s = -2$ when $t = 0$.

iii) Now find the body's displacement from $t = 1$ to $t = 3$ given that $s = s_0$ when $t = 0$.

b. Suppose that the position s of a body moving along a coordinate line is a differentiable function of time t. Is it true that once you know an antiderivative of the velocity function ds/dt you can find the body's displacement from $t = a$ to $t = b$ even if you do not know the body's exact position at either of those times? Give reasons for your answer.

94. Liftoff from Earth A rocket lifts off the surface of Earth with a constant acceleration of $20\ \text{m/sec}^2$. How fast will the rocket be going 1 min later?

95. Stopping a car in time You are driving along a highway at a steady 60 mph (88 ft/sec) when you see an accident ahead and slam on the brakes. What constant deceleration is required to stop your car in 242 ft? To find out, carry out the following steps.

1. Solve the initial value problem

Differential equation: $\dfrac{d^2s}{dt^2} = -k \quad (k \text{ constant})$

Initial conditions: $\dfrac{ds}{dt} = 88$ and $s = 0$ when $t = 0$.

Measuring time and distance from when the brakes are applied.

2. Find the value of t that makes $ds/dt = 0$. (The answer will involve k.)

3. Find the value of k that makes $s = 242$ for the value of t you found in Step 2.

96. Stopping a motorcycle The State of Illinois Cycle Rider Safety Program requires riders to be able to brake from 30 mph (44 ft/sec) to 0 in 45 ft. What constant deceleration does it take to do that?

97. Motion along a coordinate line A particle moves on a coordinate line with acceleration $a = d^2s/dt^2 = 15\sqrt{t} - \left(3/\sqrt{t}\right)$, subject to the conditions that $ds/dt = 4$ and $s = 0$ when $t = 1$. Find

a. the velocity $v = ds/dt$ in terms of t

b. the position s in terms of t.

T **98. The hammer and the feather** When *Apollo 15* astronaut David Scott dropped a hammer and a feather on the moon to demonstrate that in a vacuum all bodies fall with the same (constant) acceleration, he dropped them from about 4 ft above the ground. The television footage of the event shows the hammer and the feather falling more slowly than on Earth, where, in a vacuum, they would have taken only half a second to fall the 4 ft. How long did it take the hammer and feather to fall 4 ft on the moon? To find out, solve the following initial value problem for s as a function of t. Then find the value of t that makes s equal to 0.

Differential equation: $\dfrac{d^2s}{dt^2} = -5.2\ \text{ft/sec}^2$

Initial conditions: $\dfrac{ds}{dt} = 0$ and $s = 4$ when $t = 0$

99. Motion with constant acceleration The standard equation for the position s of a body moving with a constant acceleration a along a coordinate line is

$$s = \frac{a}{2}t^2 + v_0 t + s_0, \qquad (1)$$

where v_0 and s_0 are the body's velocity and position at time $t = 0$. Derive this equation by solving the initial value problem

Differential equation: $\dfrac{d^2s}{dt^2} = a$

Initial conditions: $\dfrac{ds}{dt} = v_0$ and $s = s_0$ when $t = 0$.

100. Free fall near the surface of a planet For free fall near the surface of a planet where the acceleration due to gravity has a constant magnitude of g length-units/sec^2, Equation (1) in Exercise 99 takes the form

$$s = -\frac{1}{2}gt^2 + v_0 t + s_0, \qquad (2)$$

where s is the body's height above the surface. The equation has a minus sign because the acceleration acts downward, in the direction of decreasing s. The velocity v_0 is positive if the object is rising at time $t = 0$ and negative if the object is falling.

Instead of using the result of Exercise 99, you can derive Equation (2) directly by solving an appropriate initial value problem. What initial value problem? Solve it to be sure you have the right one, explaining the solution steps as you go along.

Theory and Examples

101. Suppose that

$$f(x) = \frac{d}{dx}\left(1 - \sqrt{x}\right) \quad \text{and} \quad g(x) = \frac{d}{dx}(x + 2).$$

Find:

a. $\displaystyle\int f(x)\,dx$ **b.** $\displaystyle\int g(x)\,dx$

c. $\displaystyle\int [-f(x)]\,dx$ **d.** $\displaystyle\int [-g(x)]\,dx$

e. $\displaystyle\int [f(x) + g(x)]\,dx$ **f.** $\displaystyle\int [f(x) - g(x)]\,dx$

102. Uniqueness of solutions If differentiable functions $y = F(x)$ and $y = G(x)$ both solve the initial value problem

$$\frac{dy}{dx} = f(x), \qquad y(x_0) = y_0,$$

on an interval I, must $F(x) = G(x)$ for every x in I? Give reasons for your answer.

Chapter 4 Additional and Advanced Exercises

1. What can you say about a function whose maximum and minimum values on an interval are equal? Give reasons for your answer.

2. Is it true that a discontinuous function cannot have both an absolute maximum and an absolute minimum value on a closed interval? Give reasons for your answer.

3. Can you conclude anything about the extreme values of a continuous function on an open interval? On a half-open interval? Give reasons for your answer.

4. **Local extrema** Use the sign pattern for the derivative

$$\frac{df}{dx} = 6(x-1)(x-2)^2(x-3)^3(x-4)^4$$

to identify the points where f has local maximum and minimum values.

5. **Local extrema**

   a. Suppose that the first derivative of $y = f(x)$ is

$$y' = 6(x+1)(x-2)^2.$$

   At what points, if any, does the graph of f have a local maximum, local minimum, or point of inflection?

   b. Suppose that the first derivative of $y = f(x)$ is

$$y' = 6x(x+1)(x-2).$$

   At what points, if any, does the graph of f have a local maximum, local minimum, or point of inflection?

6. If $f'(x) \le 2$ for all x, what is the most the values of f can increase on $[0, 6]$? Give reasons for your answer.

7. **Bounding a function** Suppose that f is continuous on $[a, b]$ and that c is an interior point of the interval. Show that if $f'(x) \le 0$ on $[a, c)$ and $f'(x) \ge 0$ on $(c, b]$, then $f(x)$ is never less than $f(c)$ on $[a, b]$.

8. **An inequality**

   a. Show that $-1/2 \le x/(1+x^2) \le 1/2$ for every value of x.

   b. Suppose that f is a function whose derivative is $f'(x) = x/(1+x^2)$. Use the result in part (a) to show that

$$\left|f(b) - f(a)\right| \le \frac{1}{2}\left|b - a\right|$$

   for any a and b.

9. The derivative of $f(x) = x^2$ is zero at $x = 0$, but f is not a constant function. Doesn't this contradict the corollary of the Mean Value Theorem that says that functions with zero derivatives are constant? Give reasons for your answer.

10. **Extrema and inflection points** Let $h = fg$ be the product of two differentiable functions of x.

    a. If f and g are positive, with local maxima at $x = a$, and if f' and g' change sign at a, does h have a local maximum at a?

    b. If the graphs of f and g have inflection points at $x = a$, does the graph of h have an inflection point at a?

    In either case, if the answer is yes, give a proof. If the answer is no, give a counterexample.

11. **Finding a function** Use the following information to find the values of a, b, and c in the formula $f(x) = (x+a)/(bx^2 + cx + 2)$.

    i) The values of a, b, and c are either 0 or 1.

    ii) The graph of f passes through the point $(-1, 0)$.

    iii) The line $y = 1$ is an asymptote of the graph of f.

12. **Horizontal tangent** For what value or values of the constant k will the curve $y = x^3 + kx^2 + 3x - 4$ have exactly one horizontal tangent?

13. **Largest inscribed triangle** Points A and B lie at the ends of a diameter of a unit circle and point C lies on the circumference. Is it true that the area of triangle ABC is largest when the triangle is isosceles? How do you know?

14. **Proving the second derivative test** The Second Derivative Test for Local Maxima and Minima (Section 4.4) says:

    a. f has a local maximum value at $x = c$ if $f'(c) = 0$ and $f''(c) < 0$

    b. f has a local minimum value at $x = c$ if $f'(c) = 0$ and $f''(c) > 0$.

    To prove statement (a), let $\epsilon = (1/2)|f''(c)|$. Then use the fact that

$$f''(c) = \lim_{h\to 0} \frac{f'(c+h) - f'(c)}{h} = \lim_{h\to 0} \frac{f'(c+h)}{h}$$

    to conclude that for some $\delta > 0$,

$$0 < |h| < \delta \quad \Rightarrow \quad \frac{f'(c+h)}{h} < f''(c) + \epsilon < 0.$$

    Thus, $f'(c+h)$ is positive for $-\delta < h < 0$ and negative for $0 < h < \delta$. Prove statement (b) in a similar way.

15. **Hole in a water tank** You want to bore a hole in the side of the tank shown here at a height that will make the stream of water coming out hit the ground as far from the tank as possible. If you drill the hole near the top, where the pressure is low, the water will exit slowly but spend a relatively long time in the air. If you drill the hole near the bottom, the water will exit at a higher velocity but have only a short time to fall. Where is the best place, if any, for the hole? (*Hint:* How long will it take an exiting particle of water to fall from height y to the ground?)

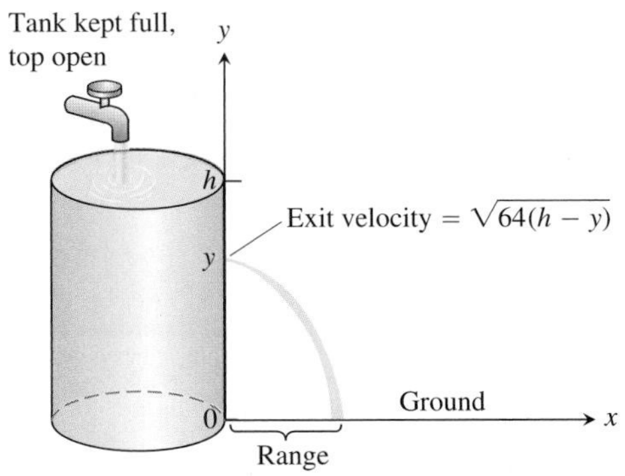

16. Kicking a field goal An American football player wants to kick a field goal with the ball being on a right hash mark. Assume that the goal posts are b feet apart and that the hash mark line is a distance $a > 0$ feet from the right goal post. (See the accompanying figure.) Find the distance h from the goal post line that gives the kicker his largest angle β. Assume that the football field is flat.

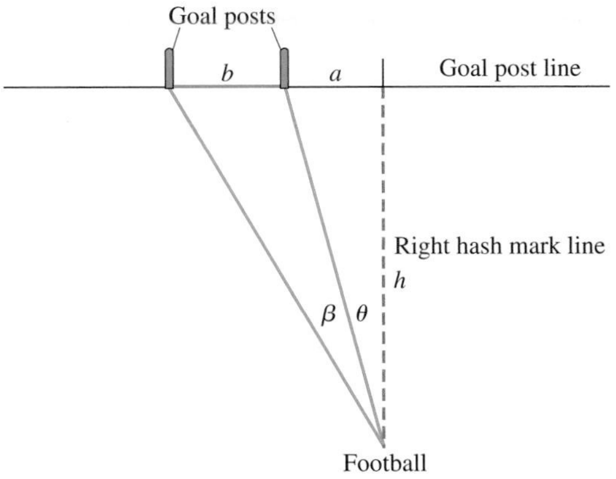

17. A max-min problem with a variable answer Sometimes the solution of a max-min problem depends on the proportions of the shapes involved. As a case in point, suppose that a right circular cylinder of radius r and height h is inscribed in a right circular cone of radius R and height H, as shown here. Find the value of r (in terms of R and H) that maximizes the total surface area of the cylinder (including top and bottom). As you will see, the solution depends on whether $H \le 2R$ or $H > 2R$.

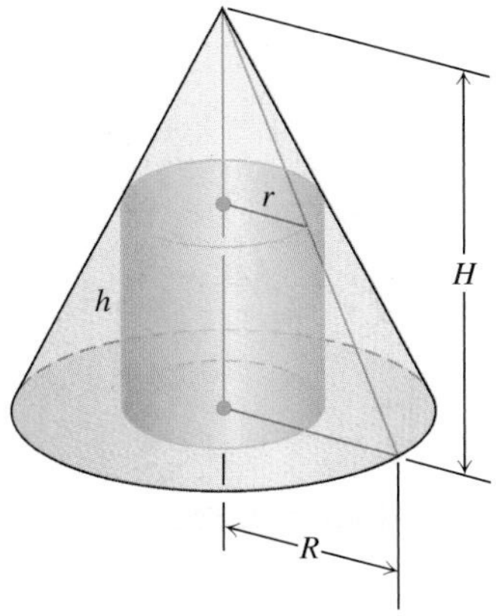

18. Minimizing a parameter Find the smallest value of the positive constant m that will make $mx - 1 + (1/x)$ greater than or equal to zero for all positive values of x.

19. Evaluate the following limits.

a. $\lim_{x\to 0} \dfrac{2\sin 5x}{3x}$ **b.** $\lim_{x\to 0} \sin 5x \cot 3x$

c. $\lim_{x\to 0} x \csc^2 \sqrt{2x}$ **d.** $\lim_{x\to \pi/2} (\sec x - \tan x)$

e. $\lim_{x\to 0} \dfrac{x - \sin x}{x - \tan x}$ **f.** $\lim_{x\to 0} \dfrac{\sin x^2}{x \sin x}$

g. $\lim_{x\to 0} \dfrac{\sec x - 1}{x^2}$ **h.** $\lim_{x\to 2} \dfrac{x^3 - 8}{x^2 - 4}$

20. L'Hôpital's Rule does not help with the following limits. Find them some other way.

a. $\lim_{x\to\infty} \dfrac{\sqrt{x + 5}}{\sqrt{x} + 5}$ **b.** $\lim_{x\to\infty} \dfrac{2x}{x + 7\sqrt{x}}$

21. Suppose that it costs a company $y = a + bx$ dollars to produce x units per week. It can sell x units per week at a price of $P = c - ex$ dollars per unit. Each of a, b, c, and e represents a positive constant. **(a)** What production level maximizes the profit? **(b)** What is the corresponding price? **(c)** What is the weekly profit at this level of production? **(d)** At what price should each item be sold to maximize profits if the government imposes a tax of t dollars per item sold? Comment on the difference between this price and the price before the tax.

22. Estimating reciprocals without division You can estimate the value of the reciprocal of a number a without ever dividing by a if you apply Newton's method to the function $f(x) = (1/x) - a$. For example, if $a = 3$, the function involved is $f(x) = (1/x) - 3$.

a. Graph $y = (1/x) - 3$. Where does the graph cross the x-axis?

b. Show that the recursion formula in this case is

$$x_{n+1} = x_n(2 - 3x_n),$$

so there is no need for division.

23. To find $x = \sqrt[q]{a}$, we apply Newton's method to $f(x) = x^q - a$. Here we assume that a is a positive real number and q is a positive integer. Show that x_1 is a "weighted average" of x_0 and a/x_0^{q-1}, and find the coefficients m_0, m_1 such that

$$x_1 = m_0 x_0 + m_1\left(\frac{a}{x_0^{q-1}}\right), \qquad \begin{aligned} m_0 > 0,\ m_1 > 0, \\ m_0 + m_1 = 1. \end{aligned}$$

What conclusion would you reach if x_0 and a/x_0^{q-1} were equal? What would be the value of x_1 in that case?

24. The family of straight lines $y = ax + b$ (a, b arbitrary constants) can be characterized by the relation $y'' = 0$. Find a similar relation satisfied by the family of all circles

$$(x - h)^2 + (y - h)^2 = r^2,$$

where h and r are arbitrary constants. (*Hint:* Eliminate h and r from the set of three equations including the given one and two obtained by successive differentiation.)

25. Assume that the brakes of an automobile produce a constant deceleration of k ft/sec^2. **(a)** Determine what k must be to bring an automobile traveling 60 mi/hr (88 ft/sec) to rest in a distance of 100 ft from the point where the brakes are applied. **(b)** With the same k, how far would a car traveling 30 mi/hr travel before being brought to a stop?

26. Let $f(x)$, $g(x)$ be two continuously differentiable functions satisfying the relationships $f'(x) = g(x)$ and $f''(x) = -f(x)$. Let $h(x) = f^2(x) + g^2(x)$. If $h(0) = 5$, find $h(10)$.

27. Can there be a curve satisfying the following conditions? d^2y/dx^2 is everywhere equal to zero and, when $x = 0$, $y = 0$ and $dy/dx = 1$. Give a reason for your answer.

28. Find the equation for the curve in the xy-plane that passes through the point $(1, -1)$ if its slope at x is always $3x^2 + 2$.

29. A particle moves along the x-axis. Its acceleration is $a = -t^2$. At $t = 0$, the particle is at the origin. In the course of its motion, it reaches the point $x = b$, where $b > 0$, but no point beyond b. Determine its velocity at $t = 0$.

30. A particle moves with acceleration $a = \sqrt{t} - \left(1/\sqrt{t}\right)$. Assuming that the velocity $v = 4/3$ and the position $s = -4/15$ when $t = 0$, find

a. the velocity v in terms of t.

b. the position s in terms of t.

31. Given $f(x) = ax^2 + 2bx + c$ with $a > 0$. By considering the minimum, prove that $f(x) \geq 0$ for all real x if, and only if, $b^2 - ac \leq 0$.

32. Schwarz's inequality

a. In Exercise 31, let

$$f(x) = (a_1x + b_1)^2 + (a_2x + b_2)^2 + \cdots + (a_nx + b_n)^2,$$

and deduce Schwarz's inequality:

$$(a_1b_1 + a_2b_2 + \cdots + a_nb_n)^2 \leq (a_1^2 + a_2^2 + \cdots + a_n^2)(b_1^2 + b_2^2 + \cdots + b_n^2).$$

b. Show that equality holds in Schwarz's inequality only if there exists a real number x that makes a_ix equal $-b_i$ for every value of i from 1 to n.

Chapter 4 Technology Application Projects

Mathematica/Maple Module

Motion Along a Straight Line: Position → Velocity → Acceleration

You will observe the shape of a graph through dramatic animated visualizations of the derivative relations among the position, velocity, and acceleration. Figures in the text can be animated.

Mathematica/Maple Module

Newton's Method: Estimate π to How Many Places?

Plot a function, observe a root, pick a starting point near the root, and use Newton's Iteration Procedure to approximate the root to a desired accuracy. The numbers π, e, and $\sqrt{2}$ are approximated.

Chapter 5 INTEGRATION

OVERVIEW One of the great achievements of classical geometry was to obtain formulas for the areas and volumes of triangles, spheres, and cones. In this chapter we study a method to calculate the areas and volumes of these and other more general shapes. The method we develop, called *integration*, is a tool for calculating much more than areas and volumes. The *integral* has many applications in statistics, economics, the sciences, and engineering. It allows us to calculate quantities ranging from probabilities and averages to energy consumption and the forces against a dam's floodgates.

The idea behind integration is that we can effectively compute many quantities by breaking them into small pieces, and then summing the contributions from each small part. We develop the theory of the integral in the setting of area, where it most clearly reveals its nature. We begin with examples involving finite sums. These lead naturally to the question of what happens when more and more terms are summed. Passing to the limit, as the number of terms goes to infinity, then gives an integral. While integration and differentiation are closely connected, we will not see the roles of the derivative and antiderivative emerge until Section 5.4. The nature of their connection, contained in the Fundamental Theorem of Calculus, is one of the most important ideas in calculus.

5.1 Estimating with Finite Sums

This section shows how area, average values, and the distance traveled by an object over time can all be approximated by finite sums. Finite sums are the basis for defining the integral in Section 5.3.

Area

The area of a region with a curved boundary can be approximated by summing the areas of a collection of rectangles. Using more rectangles can increase the accuracy of the approximation.

EXAMPLE 1 Approximating Area

What is the area of the shaded region R that lies above the x-axis, below the graph of $y = 1 - x^2$, and between the vertical lines $x = 0$ and $x = 1$? (See Figure 5.1.) An architect might want to know this area to calculate the weight of a custom window with a shape described by R. Unfortunately, there is no simple geometric formula for calculating the areas of shapes having curved boundaries like the region R.

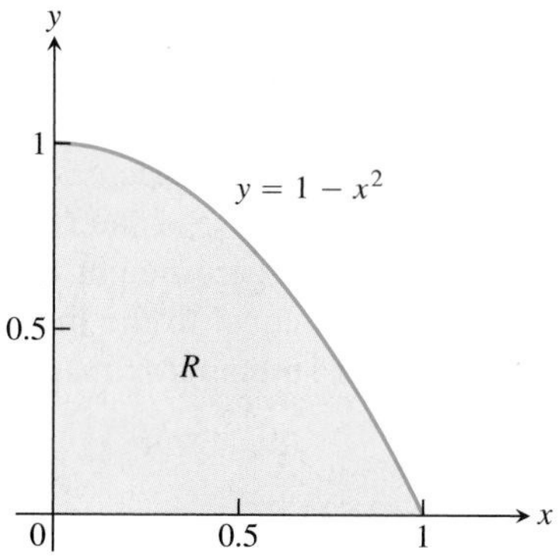

FIGURE 5.1 The area of the region R cannot be found by a simple geometry formula (Example 1).

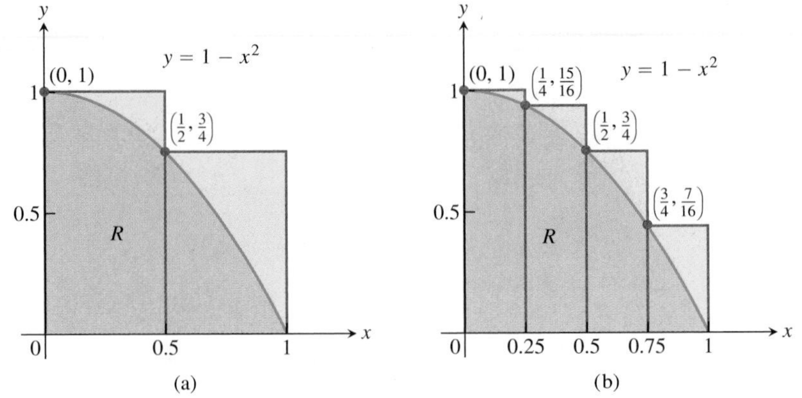

FIGURE 5.2 (a) We get an upper estimate of the area of R by using two rectangles containing R. (b) Four rectangles give a better upper estimate. Both estimates overshoot the true value for the area.

While we do not yet have a method for determining the exact area of R, we can approximate it in a simple way. Figure 5.2a shows two rectangles that together contain the region R. Each rectangle has width $1/2$ and they have heights 1 and $3/4$, moving from left to right. The height of each rectangle is the maximum value of the function f, obtained by evaluating f at the left endpoint of the subinterval of $[0, 1]$ forming the base of the rectangle. The total area of the two rectangles approximates the area A of the region R,

$$A \approx 1 \cdot \frac{1}{2} + \frac{3}{4} \cdot \frac{1}{2} = \frac{7}{8} = 0.875.$$

This estimate is larger than the true area A, since the two rectangles contain R. We say that 0.875 is an **upper sum** because it is obtained by taking the height of each rectangle as the maximum (uppermost) value of $f(x)$ for x a point in the base interval of the rectangle. In Figure 5.2b, we improve our estimate by using four thinner rectangles, each of width $1/4$, which taken together contain the region R. These four rectangles give the approximation

$$A \approx 1 \cdot \frac{1}{4} + \frac{15}{16} \cdot \frac{1}{4} + \frac{3}{4} \cdot \frac{1}{4} + \frac{7}{16} \cdot \frac{1}{4} = \frac{25}{32} = 0.78125,$$

which is still greater than A since the four rectangles contain R.

Suppose instead we use four rectangles contained *inside* the region R to estimate the area, as in Figure 5.3a. Each rectangle has width $1/4$ as before, but the rectangles are shorter and lie entirely beneath the graph of f. The function $f(x) = 1 - x^2$ is decreasing on $[0, 1]$, so the height of each of these rectangles is given by the value of f at the right endpoint of the subinterval forming its base. The fourth rectangle has zero height and therefore contributes no area. Summing these rectangles with heights equal to the minimum value of $f(x)$ for x a point in each base subinterval, gives a **lower sum** approximation to the area,

$$A \approx \frac{15}{16} \cdot \frac{1}{4} + \frac{3}{4} \cdot \frac{1}{4} + \frac{7}{16} \cdot \frac{1}{4} + 0 \cdot \frac{1}{4} = \frac{17}{32} = 0.53125.$$

This estimate is smaller than the area A since the rectangles all lie inside of the region R. The true value of A lies somewhere between these lower and upper sums:

$$0.53125 < A < 0.78125.$$

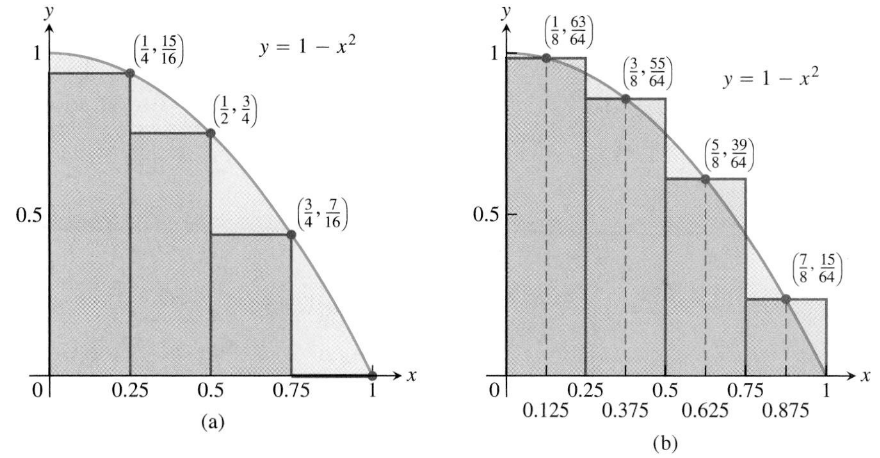

FIGURE 5.3 (a) Rectangles contained in R give an estimate for the area that undershoots the true value. (b) The midpoint rule uses rectangles whose height is the value of $y = f(x)$ at the midpoints of their bases.

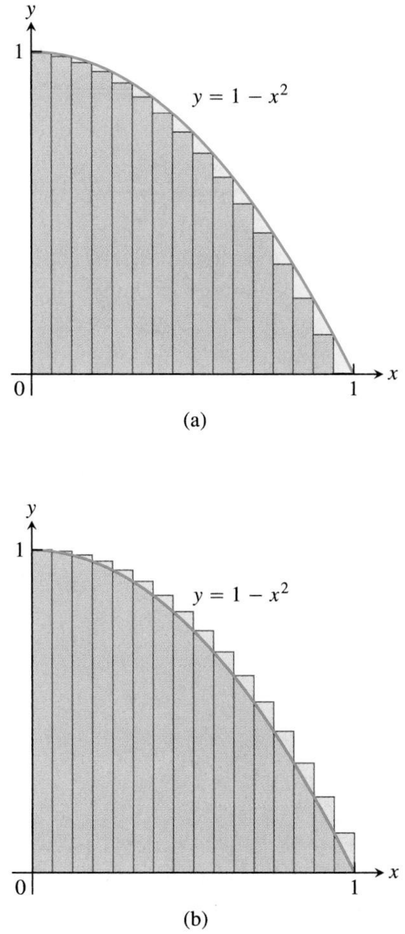

FIGURE 5.4 (a) A lower sum using 16 rectangles of equal width $\Delta x = 1/16$. (b) An upper sum using 16 rectangles.

By considering both lower and upper sum approximations we get not only estimates for the area, but also a bound on the size of the possible error in these estimates since the true value of the area lies somewhere between them. Here the error cannot be greater than the difference $0.78125 - 0.53125 = 0.25$.

Yet another estimate can be obtained by using rectangles whose heights are the values of f at the midpoints of their bases (Figure 5.3b). This method of estimation is called the **midpoint rule** for approximating the area. The midpoint rule gives an estimate that is between a lower sum and an upper sum, but it is not clear whether it overestimates or underestimates the true area. With four rectangles of width $1/4$ as before, the midpoint rule estimates the area of R to be

$$A \approx \frac{63}{64} \cdot \frac{1}{4} + \frac{55}{64} \cdot \frac{1}{4} + \frac{39}{64} \cdot \frac{1}{4} + \frac{15}{64} \cdot \frac{1}{4} = \frac{172}{64} \cdot \frac{1}{4} = 0.671875.$$

In each of our computed sums, the interval $[a, b]$ over which the function f is defined was subdivided into n subintervals of equal width (also called length) $\Delta x = (b - a)/n$, and f was evaluated at a point in each subinterval: c_1 in the first subinterval, c_2 in the second subinterval, and so on. The finite sums then all take the form

$$f(c_1)\,\Delta x + f(c_2)\,\Delta x + f(c_3)\,\Delta x + \cdots + f(c_n)\,\Delta x.$$

By taking more and more rectangles, with each rectangle thinner than before, it appears that these finite sums give better and better approximations to the true area of the region R.

Figure 5.4a shows a lower sum approximation for the area of R using 16 rectangles of equal width. The sum of their areas is 0.634765625, which appears close to the true area, but is still smaller since the rectangles lie inside R.

Figure 5.4b shows an upper sum approximation using 16 rectangles of equal width. The sum of their areas is 0.697265625, which is somewhat larger than the true area because the rectangles taken together contain R. The midpoint rule for 16 rectangles gives a total area approximation of 0.6669921875, but it is not immediately clear whether this estimate is larger or smaller than the true area.

TABLE 5.1 Finite approximations for the area of R

Number of subintervals	Lower sum	Midpoint rule	Upper sum
2	.375	.6875	.875
4	.53125	.671875	.78125
16	.634765625	.6669921875	.697265625
50	.6566	.6667	.6766
100	.66165	.666675	.67165
1000	.6661665	.66666675	.6671665

Table 5.1 shows the values of upper and lower sum approximations to the area of R using up to 1000 rectangles. In Section 5.2 we will see how to get an exact value of the areas of regions such as R by taking a limit as the base width of each rectangle goes to zero and the number of rectangles goes to infinity. With the techniques developed there, we will be able to show that the area of R is exactly $2/3$. ■

Distance Traveled

Suppose we know the velocity function $v(t)$ of a car moving down a highway, without changing direction, and want to know how far it traveled between times $t = a$ and $t = b$. If we already know an antiderivative $F(t)$ of $v(t)$ we can find the car's position function $s(t)$ by setting $s(t) = F(t) + C$. The distance traveled can then be found by calculating the change in position, $s(b) - s(a)$ (see Exercise 93, Section 4.8). If the velocity function is determined by recording a speedometer reading at various times on the car, then we have no formula from which to obtain an antiderivative function for velocity. So what do we do in this situation?

When we don't know an antiderivative for the velocity function $v(t)$, we can approximate the distance traveled in the following way. Subdivide the interval $[a, b]$ into short time intervals on each of which the velocity is considered to be fairly constant. Then approximate the distance traveled on each time subinterval with the usual distance formula

$$\text{distance} = \text{velocity} \times \text{time}$$

and add the results across $[a, b]$.

Suppose the subdivided interval looks like

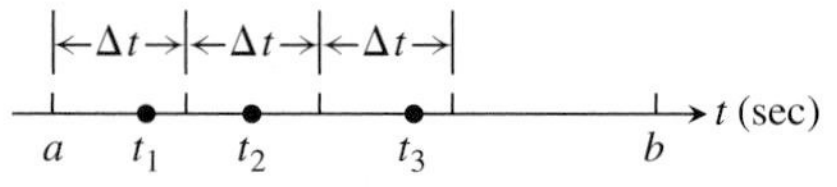

with the subintervals all of equal length Δt. Pick a number t_1 in the first interval. If Δt is so small that the velocity barely changes over a short time interval of duration Δt, then the distance traveled in the first time interval is about $v(t_1)\,\Delta t$. If t_2 is a number in the second interval, the distance traveled in the second time interval is about $v(t_2)\,\Delta t$. The sum of the distances traveled over all the time intervals is

$$D \approx v(t_1)\,\Delta t + v(t_2)\,\Delta t + \cdots + v(t_n)\,\Delta t,$$

where n is the total number of subintervals.

EXAMPLE 2 Estimating the Height of a Projectile

The velocity function of a projectile fired straight into the air is $f(t) = 160 - 9.8t$ m/sec. Use the summation technique just described to estimate how far the projectile rises during the first 3 sec. How close do the sums come to the exact figure of 435.9 m?

Solution We explore the results for different numbers of intervals and different choices of evaluation points. Notice that $f(t)$ is decreasing, so choosing left endpoints gives an upper sum estimate; choosing right endpoints gives a lower sum estimate.

(a) *Three subintervals of length* 1, *with* f *evaluated at left endpoints giving an upper sum*:

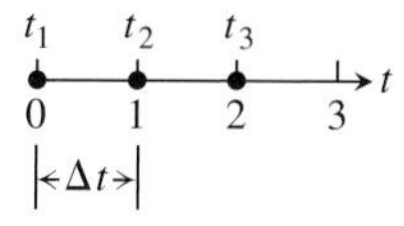

With f evaluated at $t = 0, 1$, and 2, we have

$$\begin{aligned} D &\approx f(t_1)\,\Delta t + f(t_2)\,\Delta t + f(t_3)\,\Delta t \\ &= [160 - 9.8(0)](1) + [160 - 9.8(1)](1) + [160 - 9.8(2)](1) \\ &= 450.6. \end{aligned}$$

(b) *Three subintervals of length* 1, *with* f *evaluated at right endpoints giving a lower sum*:

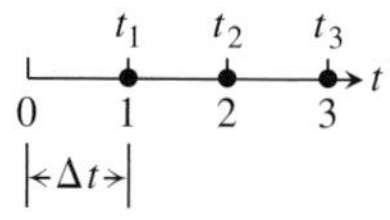

With f evaluated at $t = 1, 2$, and 3, we have

$$\begin{aligned} D &\approx f(t_1)\,\Delta t + f(t_2)\,\Delta t + f(t_3)\,\Delta t \\ &= [160 - 9.8(1)](1) + [160 - 9.8(2)](1) + [160 - 9.8(3)](1) \\ &= 421.2. \end{aligned}$$

(c) *With six subintervals of length* $1/2$, *we get*

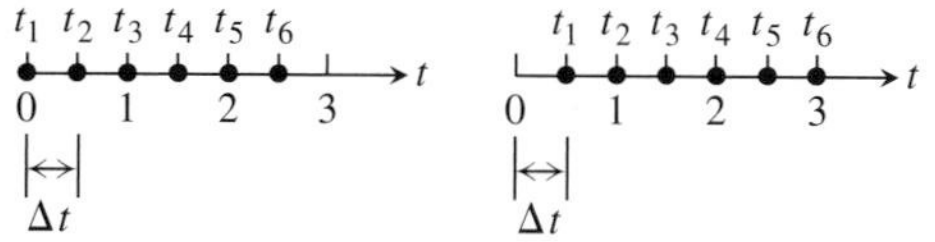

An upper sum using left endpoints: $D \approx 443.25$; a lower sum using right endpoints: $D \approx 428.55$.

These six-interval estimates are somewhat closer than the three-interval estimates. The results improve as the subintervals get shorter.

As we can see in Table 5.2, the left-endpoint upper sums approach the true value 435.9 from above, whereas the right-endpoint lower sums approach it from below. The true

TABLE 5.2 Travel-distance estimates

Number of subintervals	Length of each subinterval	Upper sum	Lower sum
3	1	450.6	421.2
6	1/2	443.25	428.55
12	1/4	439.57	432.22
24	1/8	437.74	434.06
48	1/16	436.82	434.98
96	1/32	436.36	435.44
192	1/64	436.13	435.67

value lies between these upper and lower sums. The magnitude of the error in the closest entries is 0.23, a small percentage of the true value.

$$\begin{aligned}\text{Error magnitude} &= |\text{true value} - \text{calculated value}| \\ &= |435.9 - 435.67| = 0.23.\end{aligned}$$

$$\text{Error percentage} = \frac{0.23}{435.9} \approx 0.05\%.$$

It would be reasonable to conclude from the table's last entries that the projectile rose about 436 m during its first 3 sec of flight. ■

Displacement Versus Distance Traveled

If a body with position function $s(t)$ moves along a coordinate line without changing direction, we can calculate the total distance it travels from $t = a$ to $t = b$ by summing the distance traveled over small intervals, as in Example 2. If the body changes direction one or more times during the trip, then we need to use the body's *speed* $|v(t)|$, which is the absolute value of its velocity function, $v(t)$, to find the total distance traveled. Using the velocity itself, as in Example 2, only gives an estimate to the body's **displacement**, $s(b) - s(a)$, the difference between its initial and final positions.

To see why, partition the time interval $[a, b]$ into small enough equal subintervals Δt so that the body's velocity does not change very much from time t_{k-1} to t_k. Then $v(t_k)$ gives a good approximation of the velocity throughout the interval. Accordingly, the change in the body's position coordinate during the time interval is about

$$v(t_k)\,\Delta t.$$

The change is positive if $v(t_k)$ is positive and negative if $v(t_k)$ is negative.

In either case, the distance traveled during the subinterval is about

$$|v(t_k)|\,\Delta t.$$

The **total distance traveled** is approximately the sum

$$|v(t_1)|\,\Delta t + |v(t_2)|\,\Delta t + \cdots + |v(t_n)|\,\Delta t.$$

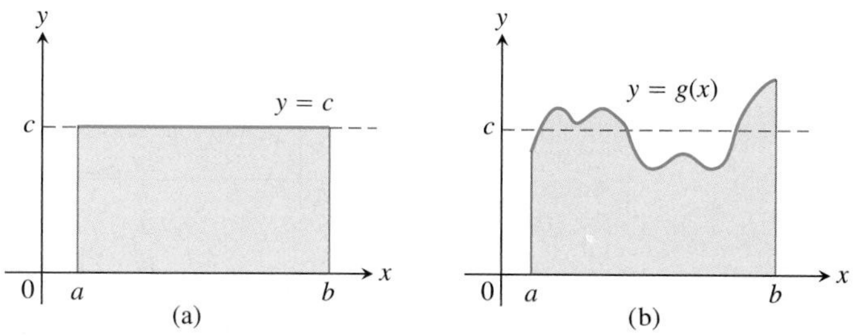

FIGURE 5.5 (a) The average value of $f(x) = c$ on $[a, b]$ is the area of the rectangle divided by $b - a$. (b) The average value of $g(x)$ on $[a, b]$ is the area beneath its graph divided by $b - a$.

Average Value of a Nonnegative Function

The average value of a collection of n numbers $x_1, x_2, \ldots, x_n$ is obtained by adding them together and dividing by n. But what is the average value of a continuous function f on an interval $[a, b]$? Such a function can assume infinitely many values. For example, the temperature at a certain location in a town is a continuous function that goes up and down each day. What does it mean to say that the average temperature in the town over the course of a day is 73 degrees?

When a function is constant, this question is easy to answer. A function with constant value c on an interval $[a, b]$ has average value c. When c is positive, its graph over $[a, b]$ gives a rectangle of height c. The average value of the function can then be interpreted geometrically as the area of this rectangle divided by its width $b - a$ (Figure 5.5a).

What if we want to find the average value of a nonconstant function, such as the function g in Figure 5.5b? We can think of this graph as a snapshot of the height of some water that is sloshing around in a tank, between enclosing walls at $x = a$ and $x = b$. As the water moves, its height over each point changes, but its average height remains the same. To get the average height of the water, we let it settle down until it is level and its height is constant. The resulting height c equals the area under the graph of g divided by $b - a$. We are led to *define* the average value of a nonnegative function on an interval $[a, b]$ to be the area under its graph divided by $b - a$. For this definition to be valid, we need a precise understanding of what is meant by the area under a graph. This will be obtained in Section 5.3, but for now we look at two simple examples.

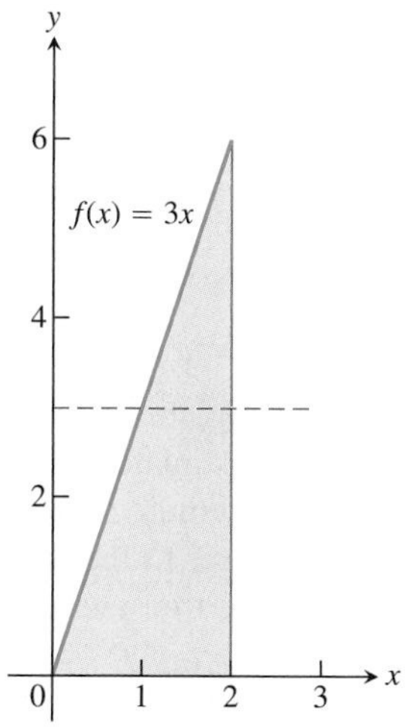

FIGURE 5.6 The average value of $f(x) = 3x$ over $[0, 2]$ is 3 (Example 3).

EXAMPLE 3 The Average Value of a Linear Function

What is the average value of the function $f(x) = 3x$ on the interval $[0, 2]$?

Solution The average equals the area under the graph divided by the width of the interval. In this case we do not need finite approximation to estimate the area of the region under the graph: a triangle of height 6 and base 2 has area 6 (Figure 5.6). The width of the interval is $b - a = 2 - 0 = 2$. The average value of the function is $6/2 = 3$. ■

EXAMPLE 4 The Average Value of sin x

Estimate the average value of the function $f(x) = \sin x$ on the interval $[0, \pi]$.

Solution Looking at the graph of sin x between 0 and π in Figure 5.7, we can see that its average height is somewhere between 0 and 1. To find the average we need to

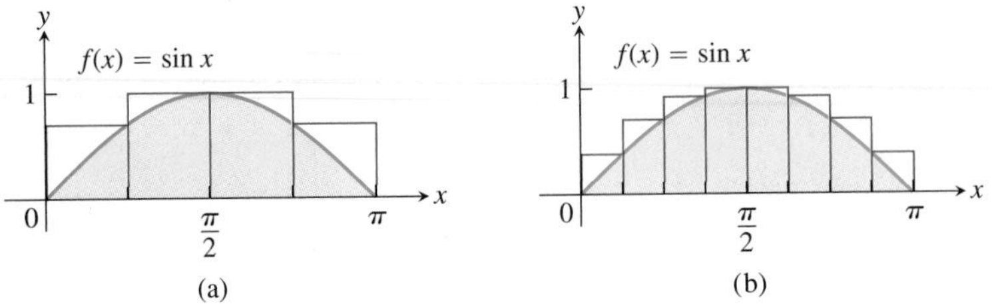

FIGURE 5.7 Approximating the area under $f(x) = \sin x$ between 0 and π to compute the average value of $\sin x$ over $[0, \pi]$, using (a) four rectangles; (b) eight rectangles (Example 4).

calculate the area A under the graph and then divide this area by the length of the interval, $\pi - 0 = \pi$.

We do not have a simple way to determine the area, so we approximate it with finite sums. To get an upper sum estimate, we add the areas of four rectangles of equal width $\pi/4$ that together contain the region beneath the graph of $y = \sin x$ and above the x-axis on $[0, \pi]$. We choose the heights of the rectangles to be the largest value of $\sin x$ on each subinterval. Over a particular subinterval, this largest value may occur at the left endpoint, the right endpoint, or somewhere between them. We evaluate $\sin x$ at this point to get the height of the rectangle for an upper sum. The sum of the rectangle areas then estimates the total area (Figure 5.7a):

$$A \approx \left(\sin\frac{\pi}{4}\right)\cdot\frac{\pi}{4} + \left(\sin\frac{\pi}{2}\right)\cdot\frac{\pi}{4} + \left(\sin\frac{\pi}{2}\right)\cdot\frac{\pi}{4} + \left(\sin\frac{3\pi}{4}\right)\cdot\frac{\pi}{4}$$

$$= \left(\frac{1}{\sqrt{2}} + 1 + 1 + \frac{1}{\sqrt{2}}\right)\cdot\frac{\pi}{4} \approx (3.42)\cdot\frac{\pi}{4} \approx 2.69.$$

To estimate the average value of $\sin x$ we divide the estimated area by π and obtain the approximation $2.69/\pi \approx 0.86$.

If we use eight rectangles of equal width $\pi/8$ all lying above the graph of $y = \sin x$ (Figure 5.7b), we get the area estimate

$$A \approx \left(\sin\frac{\pi}{8} + \sin\frac{\pi}{4} + \sin\frac{3\pi}{8} + \sin\frac{\pi}{2} + \sin\frac{\pi}{2} + \sin\frac{5\pi}{8} + \sin\frac{3\pi}{4} + \sin\frac{7\pi}{8}\right)\cdot\frac{\pi}{8}$$

$$\approx (.38 + .71 + .92 + 1 + 1 + .92 + .71 + .38)\cdot\frac{\pi}{8} = (6.02)\cdot\frac{\pi}{8} \approx 2.365.$$

Dividing this result by the length π of the interval gives a more accurate estimate of 0.753 for the average. Since we used an upper sum to approximate the area, this estimate is still greater than the actual average value of $\sin x$ over $[0, \pi]$. If we use more and more rectangles, with each rectangle getting thinner and thinner, we get closer and closer to the true average value. Using the techniques of Section 5.3, we will show that the true average value is $2/\pi \approx 0.64$.

As before, we could just as well have used rectangles lying under the graph of $y = \sin x$ and calculated a lower sum approximation, or we could have used the midpoint rule. In Section 5.3, we will see that it doesn't matter whether our approximating rectangles are chosen to give upper sums, lower sums, or a sum in between. In each case, the approximations are close to the true area if all the rectangles are sufficiently thin. ■

Summary

The area under the graph of a positive function, the distance traveled by a moving object that doesn't change direction, and the average value of a nonnegative function over an interval can all be approximated by finite sums. First we subdivide the interval into subintervals, treating the appropriate function f as if it were constant over each particular subinterval. Then we multiply the width of each subinterval by the value of f at some point within it, and add these products together. If the interval $[a, b]$ is subdivided into n subintervals of equal widths $\Delta x = (b - a)/n$, and if $f(c_k)$ is the value of f at the chosen point c_k in the kth subinterval, this process gives a finite sum of the form

$$f(c_1)\,\Delta x + f(c_2)\,\Delta x + f(c_3)\,\Delta x + \cdots + f(c_n)\,\Delta x.$$

The choices for the c_k could maximize or minimize the value of f in the kth subinterval, or give some value in between. The true value lies somewhere between the approximations given by upper sums and lower sums. The finite sum approximations we looked at improved as we took more subintervals of thinner width.

EXERCISES 5.1

Area

In Exercises 1–4 use finite approximations to estimate the area under the graph of the function using

a. a lower sum with two rectangles of equal width.

b. a lower sum with four rectangles of equal width.

c. an upper sum with two rectangles of equal width.

d. an upper sum with four rectangles of equal width.

1. $f(x) = x^2$ between $x = 0$ and $x = 1$.

2. $f(x) = x^3$ between $x = 0$ and $x = 1$.

3. $f(x) = 1/x$ between $x = 1$ and $x = 5$.

4. $f(x) = 4 - x^2$ between $x = -2$ and $x = 2$.

Using rectangles whose height is given by the value of the function at the midpoint of the rectangle's base (*the midpoint rule*) estimate the area under the graphs of the following functions, using first two and then four rectangles.

5. $f(x) = x^2$ between $x = 0$ and $x = 1$.

6. $f(x) = x^3$ between $x = 0$ and $x = 1$.

7. $f(x) = 1/x$ between $x = 1$ and $x = 5$.

8. $f(x) = 4 - x^2$ between $x = -2$ and $x = 2$.

Distance

9. Distance traveled The accompanying table shows the velocity of a model train engine moving along a track for 10 sec. Estimate the distance traveled by the engine using 10 subintervals of length 1 with

a. left-endpoint values.

b. right-endpoint values.

Time (sec)	Velocity (in./sec)	Time (sec)	Velocity (in./sec)
0	0	6	11
1	12	7	6
2	22	8	2
3	10	9	6
4	5	10	0
5	13		

10. Distance traveled upstream You are sitting on the bank of a tidal river watching the incoming tide carry a bottle upstream. You record the velocity of the flow every 5 minutes for an hour, with the results shown in the accompanying table. About how far upstream did the bottle travel during that hour? Find an estimate using 12 subintervals of length 5 with

a. left-endpoint values.

b. right-endpoint values.

Time (min)	Velocity (m/sec)	Time (min)	Velocity (m/sec)
0	1	35	1.2
5	1.2	40	1.0
10	1.7	45	1.8
15	2.0	50	1.5
20	1.8	55	1.2
25	1.6	60	0
30	1.4		

11. Length of a road You and a companion are about to drive a twisty stretch of dirt road in a car whose speedometer works but whose odometer (mileage counter) is broken. To find out how long this particular stretch of road is, you record the car's velocity at 10-sec intervals, with the results shown in the accompanying table. Estimate the length of the road using

a. left-endpoint values.

b. right-endpoint values.

Time (sec)	Velocity (converted to ft/sec) (30 mi/h = 44 ft/sec)	Time (sec)	Velocity (converted to ft/sec) (30 mi/h = 44 ft/sec)
0	0	70	15
10	44	80	22
20	15	90	35
30	35	100	44
40	30	110	30
50	44	120	35
60	35		

12. Distance from velocity data The accompanying table gives data for the velocity of a vintage sports car accelerating from 0 to 142 mi/h in 36 sec (10 thousandths of an hour).

Time (h)	Velocity (mi/h)	Time (h)	Velocity (mi/h)
0.0	0	0.006	116
0.001	40	0.007	125
0.002	62	0.008	132
0.003	82	0.009	137
0.004	96	0.010	142
0.005	108		

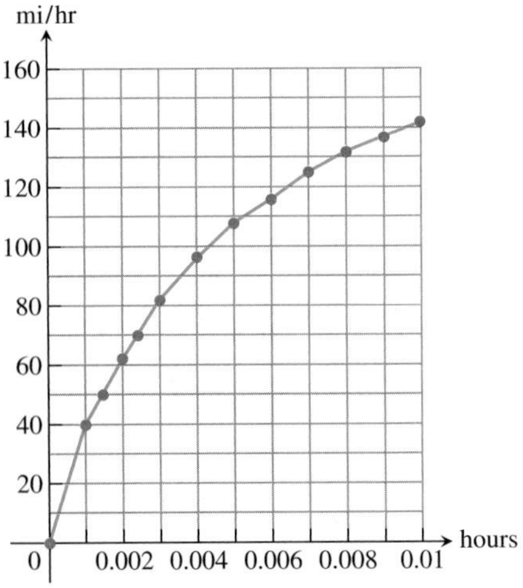

a. Use rectangles to estimate how far the car traveled during the 36 sec it took to reach 142 mi/h.

b. Roughly how many seconds did it take the car to reach the halfway point? About how fast was the car going then?

Velocity and Distance

13. Free fall with air resistance An object is dropped straight down from a helicopter. The object falls faster and faster but its acceleration (rate of change of its velocity) decreases over time because of air resistance. The acceleration is measured in ft/sec^2 and recorded every second after the drop for 5 sec, as shown:

t	0	1	2	3	4	5
a	32.00	19.41	11.77	7.14	4.33	2.63

a. Find an upper estimate for the speed when $t = 5$.

b. Find a lower estimate for the speed when $t = 5$.

c. Find an upper estimate for the distance fallen when $t = 3$.

14. Distance traveled by a projectile An object is shot straight upward from sea level with an initial velocity of 400 ft/sec.

a. Assuming that gravity is the only force acting on the object, give an upper estimate for its velocity after 5 sec have elapsed. Use $g = 32$ ft/sec^2 for the gravitational acceleration.

b. Find a lower estimate for the height attained after 5 sec.

Average Value of a Function

In Exercises 15–18, use a finite sum to estimate the average value of f on the given interval by partitioning the interval into four subintervals of equal length and evaluating f at the subinterval midpoints.

15. $f(x) = x^3$ on $[0, 2]$ **16.** $f(x) = 1/x$ on $[1, 9]$

17. $f(t) = (1/2) + \sin^2 \pi t$ on $[0, 2]$

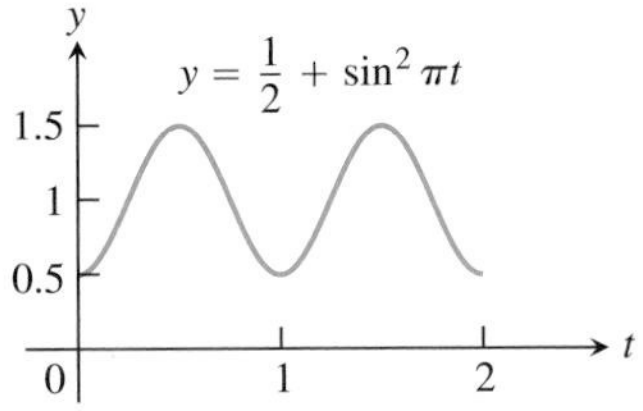

18. $f(t) = 1 - \left(\cos \frac{\pi t}{4}\right)^4$ on $[0, 4]$

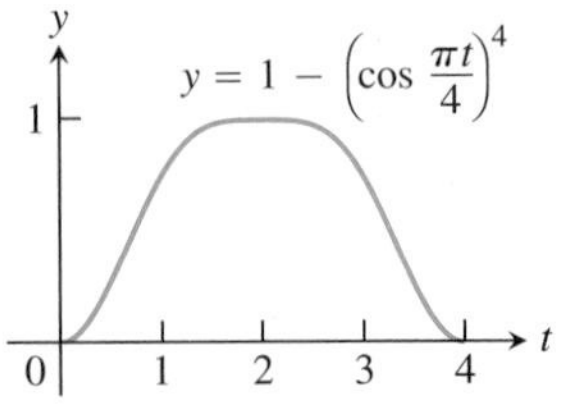

Pollution Control

19. Water pollution Oil is leaking out of a tanker damaged at sea. The damage to the tanker is worsening as evidenced by the increased leakage each hour, recorded in the following table.

Time (h)	0	1	2	3	4
Leakage (gal/h)	50	70	97	136	190

Time (h)	5	6	7	8
Leakage (gal/h)	265	369	516	720

a. Give an upper and a lower estimate of the total quantity of oil that has escaped after 5 hours.

b. Repeat part (a) for the quantity of oil that has escaped after 8 hours.

c. The tanker continues to leak 720 gal/h after the first 8 hours. If the tanker originally contained 25,000 gal of oil, approximately how many more hours will elapse in the worst case before all the oil has spilled? In the best case?

20. Air pollution A power plant generates electricity by burning oil. Pollutants produced as a result of the burning process are removed by scrubbers in the smokestacks. Over time, the scrubbers become less efficient and eventually they must be replaced when the amount of pollution released exceeds government standards. Measurements are taken at the end of each month determining the rate at which pollutants are released into the atmosphere, recorded as follows.

Month	Jan	Feb	Mar	Apr	May	Jun
Pollutant Release rate (tons/day)	0.20	0.25	0.27	0.34	0.45	0.52

Month	Jul	Aug	Sep	Oct	Nov	Dec
Pollutant Release rate (tons/day)	0.63	0.70	0.81	0.85	0.89	0.95

a. Assuming a 30-day month and that new scrubbers allow only 0.05 ton/day released, give an upper estimate of the total tonnage of pollutants released by the end of June. What is a lower estimate?

b. In the best case, approximately when will a total of 125 tons of pollutants have been released into the atmosphere?

Area of a Circle

21. Inscribe a regular n-sided polygon inside a circle of radius 1 and compute the area of the polygon for the following values of n:

a. 4 (square) **b.** 8 (octagon) **c.** 16

d. Compare the areas in parts (a), (b), and (c) with the area of the circle.

22. (*Continuation of Exercise 21*)

a. Inscribe a regular n-sided polygon inside a circle of radius 1 and compute the area of one of the n congruent triangles formed by drawing radii to the vertices of the polygon.

b. Compute the limit of the area of the inscribed polygon as $n \to \infty$.

c. Repeat the computations in parts (a) and (b) for a circle of radius r.

COMPUTER EXPLORATIONS

In Exercises 23–26, use a CAS to perform the following steps.

a. Plot the functions over the given interval.

b. Subdivide the interval into $n = 100, 200$, and 1000 subintervals of equal length and evaluate the function at the midpoint of each subinterval.

c. Compute the average value of the function values generated in part (b).

d. Solve the equation $f(x) = (\text{average value})$ for x using the average value calculated in part (c) for the $n = 1000$ partitioning.

23. $f(x) = \sin x \quad \text{on} \quad [0, \pi]$ **24.** $f(x) = \sin^2 x \quad \text{on} \quad [0, \pi]$

25. $f(x) = x \sin \frac{1}{x} \quad \text{on} \quad \left[\frac{\pi}{4}, \pi\right]$

26. $f(x) = x \sin^2 \frac{1}{x} \quad \text{on} \quad \left[\frac{\pi}{4}, \pi\right]$

5.2 Sigma Notation and Limits of Finite Sums

In estimating with finite sums in Section 5.1, we often encountered sums with many terms (up to 1000 in Table 5.1, for instance). In this section we introduce a notation to write sums with a large number of terms. After describing the notation and stating several of its properties, we look at what happens to a finite sum approximation as the number of terms approaches infinity.

Finite Sums and Sigma Notation

Sigma notation enables us to write a sum with many terms in the compact form

$$\sum_{k=1}^{n} a_k = a_1 + a_2 + a_3 + \cdots + a_{n-1} + a_n.$$

The Greek letter Σ (capital sigma, corresponding to our letter S), stands for "sum." The **index of summation** k tells us where the sum begins (at the number below the Σ symbol) and where it ends (at the number above Σ). Any letter can be used to denote the index, but the letters i, j, and k are customary.

$$\sum_{k=1}^{n} a_k$$

The index k ends at $k = n$.

The summation symbol (Greek letter sigma)

a_k is a formula for the kth term.

The index k starts at $k = 1$.

Thus we can write

$$1^2 + 2^2 + 3^2 + 4^2 + 5^2 + 6^2 + 7^2 + 8^2 + 9^2 + 10^2 + 11^2 = \sum_{k=1}^{11} k^2,$$

and

$$f(1) + f(2) + f(3) + \cdots + f(100) = \sum_{i=1}^{100} f(i).$$

The sigma notation used on the right side of these equations is much more compact than the summation expressions on the left side.

EXAMPLE 1 Using Sigma Notation

The sum in sigma notation	**The sum written out, one term for each value of k**	**The value of the sum**
$\sum_{k=1}^{5} k$	$1 + 2 + 3 + 4 + 5$	15
$\sum_{k=1}^{3} (-1)^k k$	$(-1)^1(1) + (-1)^2(2) + (-1)^3(3)$	$-1 + 2 - 3 = -2$
$\sum_{k=1}^{2} \frac{k}{k+1}$	$\frac{1}{1+1} + \frac{2}{2+1}$	$\frac{1}{2} + \frac{2}{3} = \frac{7}{6}$
$\sum_{k=4}^{5} \frac{k^2}{k-1}$	$\frac{4^2}{4-1} + \frac{5^2}{5-1}$	$\frac{16}{3} + \frac{25}{4} = \frac{139}{12}$

■

The lower limit of summation does not have to be 1; it can be any integer.

EXAMPLE 2 Using Different Index Starting Values

Express the sum $1 + 3 + 5 + 7 + 9$ in sigma notation.

Solution The formula generating the terms changes with the lower limit of summation, but the terms generated remain the same. It is often simplest to start with $k = 0$ or $k = 1$.

Starting with $k = 0$: $$1 + 3 + 5 + 7 + 9 = \sum_{k=0}^{4} (2k + 1)$$

Starting with $k = 1$: $$1 + 3 + 5 + 7 + 9 = \sum_{k=1}^{5} (2k - 1)$$

Starting with $k = 2$: $$1 + 3 + 5 + 7 + 9 = \sum_{k=2}^{6} (2k - 3)$$

Starting with $k = -3$: $$1 + 3 + 5 + 7 + 9 = \sum_{k=-3}^{1} (2k + 7)$$ ■

When we have a sum such as

$$\sum_{k=1}^{3} (k + k^2)$$

we can rearrange its terms,

$$\begin{aligned} \sum_{k=1}^{3} (k + k^2) &= (1 + 1^2) + (2 + 2^2) + (3 + 3^2) \\ &= (1 + 2 + 3) + (1^2 + 2^2 + 3^2) \qquad \text{Regroup terms.} \\ &= \sum_{k=1}^{3} k + \sum_{k=1}^{3} k^2 \end{aligned}$$

This illustrates a general rule for finite sums:

$$\sum_{k=1}^{n} (a_k + b_k) = \sum_{k=1}^{n} a_k + \sum_{k=1}^{n} b_k$$

Four such rules are given below. A proof that they are valid can be obtained using mathematical induction (see Appendix 1).

Algebra Rules for Finite Sums

1. *Sum Rule*: $$\sum_{k=1}^{n} (a_k + b_k) = \sum_{k=1}^{n} a_k + \sum_{k=1}^{n} b_k$$

2. *Difference Rule*: $$\sum_{k=1}^{n} (a_k - b_k) = \sum_{k=1}^{n} a_k - \sum_{k=1}^{n} b_k$$

3. *Constant Multiple Rule*: $$\sum_{k=1}^{n} ca_k = c \cdot \sum_{k=1}^{n} a_k \qquad \text{(Any number } c\text{)}$$

4. *Constant Value Rule*: $$\sum_{k=1}^{n} c = n \cdot c \qquad (c \text{ is any constant value.})$$

EXAMPLE 3 Using the Finite Sum Algebra Rules

(a) $\sum_{k=1}^{n}(3k - k^2) = 3\sum_{k=1}^{n} k - \sum_{k=1}^{n} k^2$ — Difference Rule and Constant Multiple Rule

(b) $\sum_{k=1}^{n}(-a_k) = \sum_{k=1}^{n}(-1)\cdot a_k = -1\cdot\sum_{k=1}^{n} a_k = -\sum_{k=1}^{n} a_k$ — Constant Multiple Rule

(c) $\sum_{k=1}^{3}(k + 4) = \sum_{k=1}^{3} k + \sum_{k=1}^{3} 4$ — Sum Rule

$= (1 + 2 + 3) + (3\cdot 4)$ — Constant Value Rule

$= 6 + 12 = 18$

(d) $\sum_{k=1}^{n}\frac{1}{n} = n\cdot\frac{1}{n} = 1$ — Constant Value Rule ($1/n$ is constant) ■

HISTORICAL BIOGRAPHY

Carl Friedrich Gauss
(1777–1855)

Over the years people have discovered a variety of formulas for the values of finite sums. The most famous of these are the formula for the sum of the first n integers (Gauss may have discovered it at age 8) and the formulas for the sums of the squares and cubes of the first n integers.

EXAMPLE 4 The Sum of the First n Integers

Show that the sum of the first n integers is

$$\sum_{k=1}^{n} k = \frac{n(n+1)}{2}.$$

Solution: The formula tells us that the sum of the first 4 integers is

$$\frac{(4)(5)}{2} = 10.$$

Addition verifies this prediction:

$$1 + 2 + 3 + 4 = 10.$$

To prove the formula in general, we write out the terms in the sum twice, once forward and once backward.

$$\begin{array}{ccccccccc} 1 & + & 2 & + & 3 & + & \cdots & + & n \\ n & + & (n-1) & + & (n-2) & + & \cdots & + & 1 \end{array}$$

If we add the two terms in the first column we get $1 + n = n + 1$. Similarly, if we add the two terms in the second column we get $2 + (n - 1) = n + 1$. The two terms in any column sum to $n + 1$. When we add the n columns together we get n terms, each equal to $n + 1$, for a total of $n(n + 1)$. Since this is twice the desired quantity, the sum of the first n integers is $(n)(n + 1)/2$. ■

Formulas for the sums of the squares and cubes of the first n integers are proved using mathematical induction (see Appendix 1). We state them here.

The first n squares: $\sum_{k=1}^{n} k^2 = \frac{n(n+1)(2n+1)}{6}$

The first n cubes: $\sum_{k=1}^{n} k^3 = \left(\frac{n(n+1)}{2}\right)^2$

Limits of Finite Sums

The finite sum approximations we considered in Section 5.1 got more accurate as the number of terms increased and the subinterval widths (lengths) became thinner. The next example shows how to calculate a limiting value as the widths of the subintervals go to zero and their number grows to infinity.

EXAMPLE 5 The Limit of Finite Approximations to an Area

Find the limiting value of lower sum approximations to the area of the region R below the graph of $y = 1 - x^2$ and above the interval $[0, 1]$ on the x-axis using equal width rectangles whose widths approach zero and whose number approaches infinity. (See Figure 5.4a.)

Solution We compute a lower sum approximation using n rectangles of equal width $\Delta x = (1 - 0)/n$, and then we see what happens as $n \to \infty$. We start by subdividing $[0, 1]$ into n equal width subintervals

$$\left[0, \frac{1}{n}\right], \left[\frac{1}{n}, \frac{2}{n}\right], \ldots, \left[\frac{n-1}{n}, \frac{n}{n}\right].$$

Each subinterval has width $1/n$. The function $1 - x^2$ is decreasing on $[0, 1]$, and its smallest value in a subinterval occurs at the subinterval's right endpoint. So a lower sum is constructed with rectangles whose height over the subinterval $[(k - 1)/n, k/n]$ is $f(k/n) = 1 - (k/n)^2$, giving the sum

$$f\left(\frac{1}{n}\right)\left(\frac{1}{n}\right) + f\left(\frac{2}{n}\right)\left(\frac{1}{n}\right) + \cdots + f\left(\frac{k}{n}\right)\left(\frac{1}{n}\right) + \cdots + f\left(\frac{n}{n}\right)\left(\frac{1}{n}\right).$$

We write this in sigma notation and simplify,

$$\begin{aligned}
\sum_{k=1}^{n} f\left(\frac{k}{n}\right)\left(\frac{1}{n}\right) &= \sum_{k=1}^{n}\left(1 - \left(\frac{k}{n}\right)^2\right)\left(\frac{1}{n}\right) \\
&= \sum_{k=1}^{n}\left(\frac{1}{n} - \frac{k^2}{n^3}\right) \\
&= \sum_{k=1}^{n}\frac{1}{n} - \sum_{k=1}^{n}\frac{k^2}{n^3} && \text{Difference Rule} \\
&= n \cdot \frac{1}{n} - \frac{1}{n^3}\sum_{k=1}^{n} k^2 && \text{Constant Value and Constant Multiple Rules} \\
&= 1 - \left(\frac{1}{n^3}\right)\frac{(n)(n+1)(2n+1)}{6} && \text{Sum of the First } n \text{ Squares} \\
&= 1 - \frac{2n^3 + 3n^2 + n}{6n^3}. && \text{Numerator expanded}
\end{aligned}$$

We have obtained an expression for the lower sum that holds for any n. Taking the limit of this expression as $n \to \infty$, we see that the lower sums converge as the number of subintervals increases and the subinterval widths approach zero:

$$\lim_{n\to\infty}\left(1 - \frac{2n^3 + 3n^2 + n}{6n^3}\right) = 1 - \frac{2}{6} = \frac{2}{3}.$$

The lower sum approximations converge to $2/3$. A similar calculation shows that the upper sum approximations also converge to $2/3$ (Exercise 35). Any finite sum approximation, in the sense of our summary at the end of Section 5.1, also converges to the same value

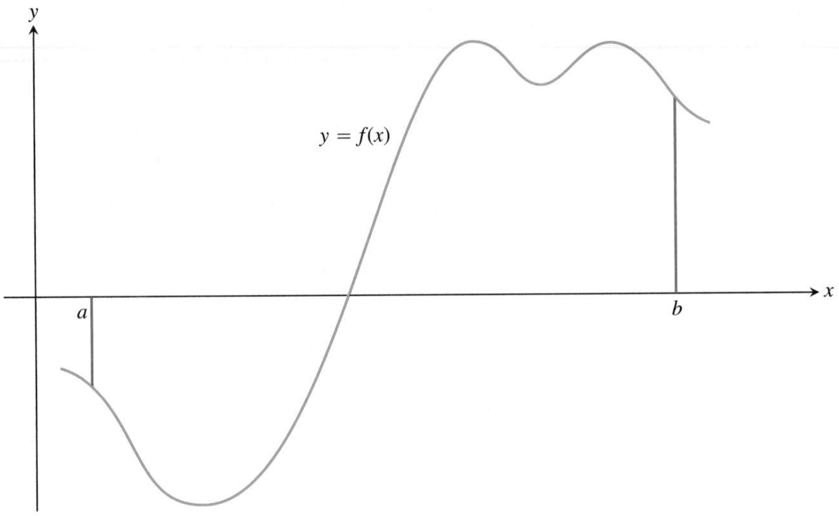

FIGURE 5.8 A typical continuous function $y = f(x)$ over a closed interval $[a, b]$.

2/3. This is because it is possible to show that any finite sum approximation is trapped between the lower and upper sum approximations. For this reason we are led to *define* the area of the region R as this limiting value. In Section 5.3 we study the limits of such finite approximations in their more general setting. ■

Riemann Sums

HISTORICAL BIOGRAPHY

Georg Friedrich Bernhard Riemann (1826–1866)

The theory of limits of finite approximations was made precise by the German mathematician Bernhard Riemann. We now introduce the notion of a *Riemann sum*, which underlies the theory of the definite integral studied in the next section.

We begin with an arbitrary function f defined on a closed interval $[a, b]$. Like the function pictured in Figure 5.8, f may have negative as well as positive values. We subdivide the interval $[a, b]$ into subintervals, not necessarily of equal widths (or lengths), and form sums in the same way as for the finite approximations in Section 5.1. To do so, we choose $n - 1$ points $\{x_1, x_2, x_3, \ldots, x_{n-1}\}$ between a and b and satisfying

$$a < x_1 < x_2 < \cdots < x_{n-1} < b.$$

To make the notation consistent, we denote a by x_0 and b by x_n, so that

$$a = x_0 < x_1 < x_2 < \cdots < x_{n-1} < x_n = b.$$

The set

$$P = \{x_0, x_1, x_2, \ldots, x_{n-1}, x_n\}$$

is called a **partition** of $[a, b]$.

The partition P divides $[a, b]$ into n closed subintervals

$$[x_0, x_1], [x_1, x_2], \ldots, [x_{n-1}, x_n].$$

The first of these subintervals is $[x_0, x_1]$, the second is $[x_1, x_2]$, and the **kth subinterval of** P is $[x_{k-1}, x_k]$, for k an integer between 1 and n.

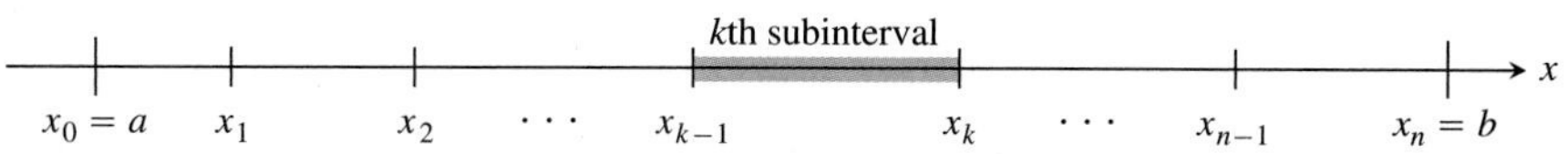

The width of the first subinterval $[x_0, x_1]$ is denoted Δx_1, the width of the second $[x_1, x_2]$ is denoted Δx_2, and the width of the kth subinterval is $\Delta x_k = x_k - x_{k-1}$. If all n subintervals have equal width, then the common width Δx is equal to $(b - a)/n$.

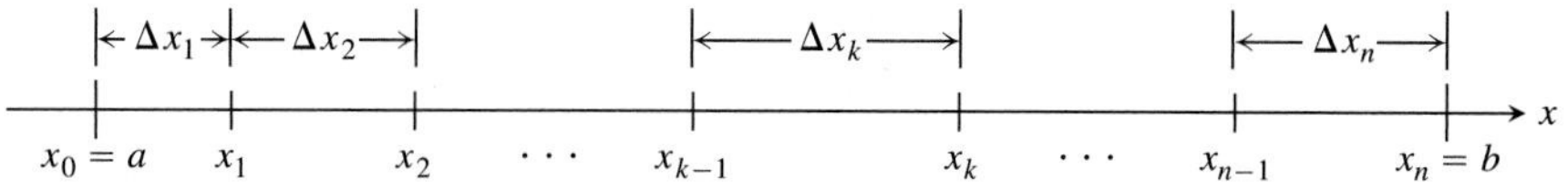

In each subinterval we select some point. The point chosen in the kth subinterval $[x_{k-1}, x_k]$ is called c_k. Then on each subinterval we stand a vertical rectangle that stretches from the x-axis to touch the curve at $(c_k, f(c_k))$. These rectangles can be above or below the x-axis, depending on whether $f(c_k)$ is positive or negative, or on it if $f(c_k) = 0$ (Figure 5.9).

On each subinterval we form the product $f(c_k) \cdot \Delta x_k$. This product is positive, negative or zero, depending on the sign of $f(c_k)$. When $f(c_k) > 0$, the product $f(c_k) \cdot \Delta x_k$ is the area of a rectangle with height $f(c_k)$ and width Δx_k. When $f(c_k) < 0$, the product $f(c_k) \cdot \Delta x_k$ is a negative number, the negative of the area of a rectangle of width Δx_k that drops from the x-axis to the negative number $f(c_k)$.

Finally we sum all these products to get

$$S_P = \sum_{k=1}^{n} f(c_k)\, \Delta x_k .$$

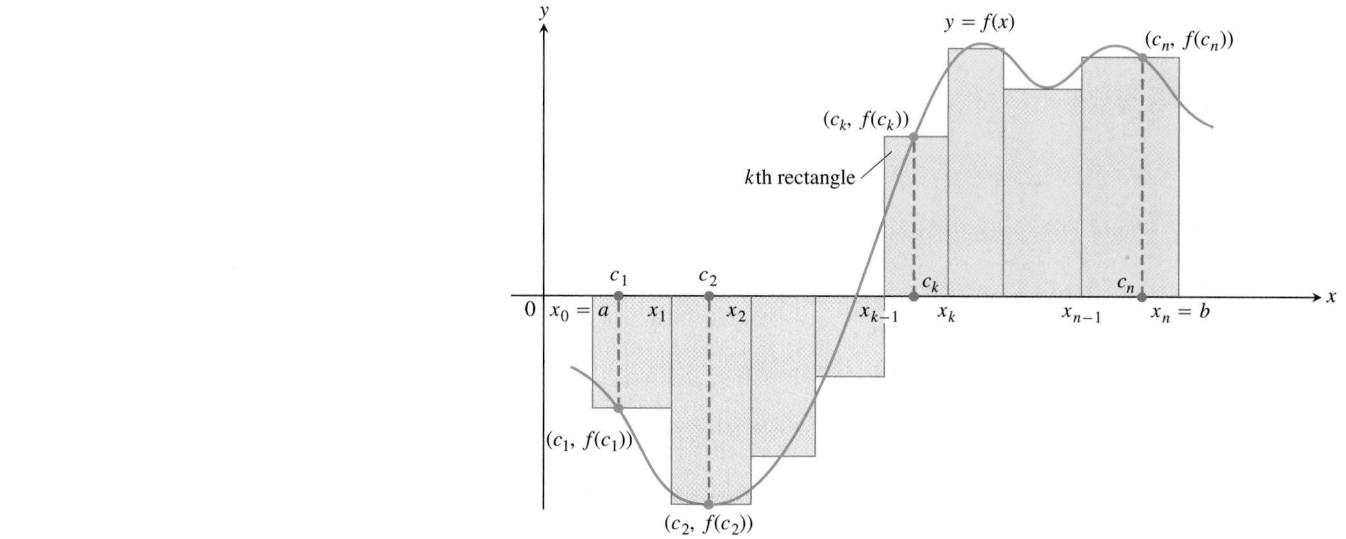

FIGURE 5.9 The rectangles approximate the region between the graph of the function $y = f(x)$ and the x-axis.

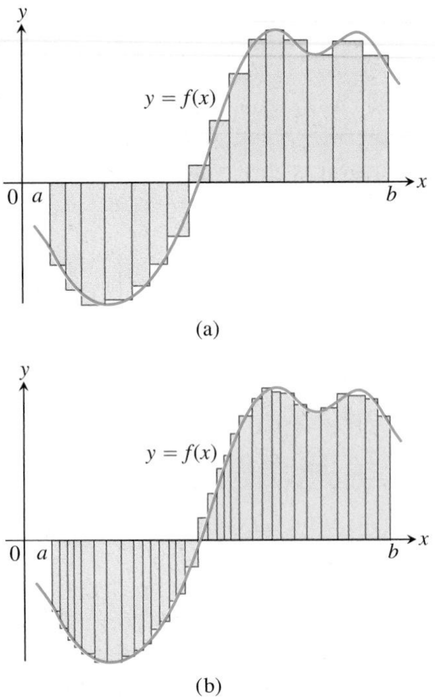

FIGURE 5.10 The curve of Figure 5.9 with rectangles from finer partitions of $[a, b]$. Finer partitions create collections of rectangles with thinner bases that approximate the region between the graph of f and the x-axis with increasing accuracy.

The sum S_P is called a **Riemann sum for f on the interval $[a, b]$**. There are many such sums, depending on the partition P we choose, and the choices of the points c_k in the subintervals.

In Example 5, where the subintervals all had equal widths $\Delta x = 1/n$, we could make them thinner by simply increasing their number n. When a partition has subintervals of varying widths, we can ensure they are all thin by controlling the width of a widest (longest) subinterval. We define the **norm** of a partition P, written $\|P\|$, to be the largest of all the subinterval widths. If $\|P\|$ is a small number, then all of the subintervals in the partition P have a small width. Let's look at an example of these ideas.

EXAMPLE 6 Partitioning a Closed Interval

The set $P = \{0, 0.2, 0.6, 1, 1.5, 2\}$ is a partition of $[0, 2]$. There are five subintervals of P: $[0, 0.2]$, $[0.2, 0.6]$, $[0.6, 1]$, $[1, 1.5]$, and $[1.5, 2]$:

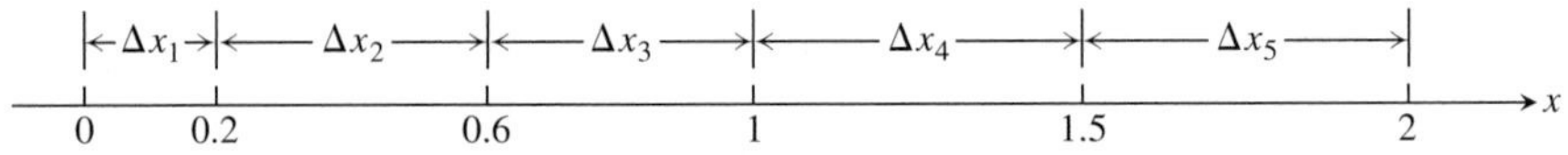

The lengths of the subintervals are $\Delta x_1 = 0.2$, $\Delta x_2 = 0.4$, $\Delta x_3 = 0.4$, $\Delta x_4 = 0.5$, and $\Delta x_5 = 0.5$. The longest subinterval length is 0.5, so the norm of the partition is $\|P\| = 0.5$. In this example, there are two subintervals of this length. ■

Any Riemann sum associated with a partition of a closed interval $[a, b]$ defines rectangles that approximate the region between the graph of a continuous function f and the x-axis. Partitions with norm approaching zero lead to collections of rectangles that approximate this region with increasing accuracy, as suggested by Figure 5.10. We will see in the next section that if the function f is continuous over the closed interval $[a, b]$, then no matter how we choose the partition P and the points c_k in its subintervals to construct a Riemann sum, a single limiting value is approached as the subinterval widths, controlled by the norm of the partition, approach zero.

EXERCISES 5.2

Sigma Notation

Write the sums in Exercises 1–6 without sigma notation. Then evaluate them.

1. $\displaystyle\sum_{k=1}^{2} \frac{6k}{k+1}$

2. $\displaystyle\sum_{k=1}^{3} \frac{k-1}{k}$

3. $\displaystyle\sum_{k=1}^{4} \cos k\pi$

4. $\displaystyle\sum_{k=1}^{5} \sin k\pi$

5. $\displaystyle\sum_{k=1}^{3} (-1)^{k+1} \sin \frac{\pi}{k}$

6. $\displaystyle\sum_{k=1}^{4} (-1)^{k} \cos k\pi$

7. Which of the following express $1 + 2 + 4 + 8 + 16 + 32$ in sigma notation?

   a. $\displaystyle\sum_{k=1}^{6} 2^{k-1}$ **b.** $\displaystyle\sum_{k=0}^{5} 2^{k}$ **c.** $\displaystyle\sum_{k=-1}^{4} 2^{k+1}$

8. Which of the following express $1 - 2 + 4 - 8 + 16 - 32$ in sigma notation?

   a. $\displaystyle\sum_{k=1}^{6} (-2)^{k-1}$ **b.** $\displaystyle\sum_{k=0}^{5} (-1)^{k} 2^{k}$ **c.** $\displaystyle\sum_{k=-2}^{3} (-1)^{k+1} 2^{k+2}$

9. Which formula is not equivalent to the other two?

   a. $\displaystyle\sum_{k=2}^{4} \frac{(-1)^{k-1}}{k-1}$ **b.** $\displaystyle\sum_{k=0}^{2} \frac{(-1)^{k}}{k+1}$ **c.** $\displaystyle\sum_{k=-1}^{1} \frac{(-1)^{k}}{k+2}$

10. Which formula is not equivalent to the other two?

    a. $\displaystyle\sum_{k=1}^{4} (k-1)^2$ **b.** $\displaystyle\sum_{k=-1}^{3} (k+1)^2$ **c.** $\displaystyle\sum_{k=-3}^{-1} k^2$

Express the sums in Exercises 11–16 in sigma notation. The form of your answer will depend on your choice of the lower limit of summation.

11. $1 + 2 + 3 + 4 + 5 + 6$ **12.** $1 + 4 + 9 + 16$

13. $\frac{1}{2} + \frac{1}{4} + \frac{1}{8} + \frac{1}{16}$ **14.** $2 + 4 + 6 + 8 + 10$

15. $1 - \frac{1}{2} + \frac{1}{3} - \frac{1}{4} + \frac{1}{5}$ **16.** $-\frac{1}{5} + \frac{2}{5} - \frac{3}{5} + \frac{4}{5} - \frac{5}{5}$

Values of Finite Sums

17. Suppose that $\sum_{k=1}^{n} a_k = -5$ and $\sum_{k=1}^{n} b_k = 6$. Find the values of

a. $\sum_{k=1}^{n} 3a_k$ **b.** $\sum_{k=1}^{n} \frac{b_k}{6}$ **c.** $\sum_{k=1}^{n} (a_k + b_k)$

d. $\sum_{k=1}^{n} (a_k - b_k)$ **e.** $\sum_{k=1}^{n} (b_k - 2a_k)$

18. Suppose that $\sum_{k=1}^{n} a_k = 0$ and $\sum_{k=1}^{n} b_k = 1$. Find the values of

a. $\sum_{k=1}^{n} 8a_k$ **b.** $\sum_{k=1}^{n} 250b_k$

c. $\sum_{k=1}^{n} (a_k + 1)$ **d.** $\sum_{k=1}^{n} (b_k - 1)$

Evaluate the sums in Exercises 19–28.

19. a. $\sum_{k=1}^{10} k$ **b.** $\sum_{k=1}^{10} k^2$ **c.** $\sum_{k=1}^{10} k^3$

20. a. $\sum_{k=1}^{13} k$ **b.** $\sum_{k=1}^{13} k^2$ **c.** $\sum_{k=1}^{13} k^3$

21. $\sum_{k=1}^{7} (-2k)$ **22.** $\sum_{k=1}^{5} \frac{\pi k}{15}$

23. $\sum_{k=1}^{6} (3 - k^2)$ **24.** $\sum_{k=1}^{6} (k^2 - 5)$

25. $\sum_{k=1}^{5} k(3k + 5)$ **26.** $\sum_{k=1}^{7} k(2k + 1)$

27. $\sum_{k=1}^{5} \frac{k^3}{225} + \left(\sum_{k=1}^{5} k\right)^3$ **28.** $\left(\sum_{k=1}^{7} k\right)^2 - \sum_{k=1}^{7} \frac{k^3}{4}$

Rectangles for Riemann Sums

In Exercises 29–32, graph each function $f(x)$ over the given interval. Partition the interval into four subintervals of equal length. Then add to your sketch the rectangles associated with the Riemann sum $\sum_{k=1}^{4} f(c_k)\,\Delta x_k$, given that c_k is the (a) left-hand endpoint, (b) right-hand endpoint, (c) midpoint of the kth subinterval. (Make a separate sketch for each set of rectangles.)

29. $f(x) = x^2 - 1, \quad [0, 2]$

30. $f(x) = -x^2, \quad [0, 1]$

31. $f(x) = \sin x, \quad [-\pi, \pi]$

32. $f(x) = \sin x + 1, \quad [-\pi, \pi]$

33. Find the norm of the partition $P = \{0, 1.2, 1.5, 2.3, 2.6, 3\}$.

34. Find the norm of the partition $P = \{-2, -1.6, -0.5, 0, 0.8, 1\}$.

Limits of Upper Sums

For the functions in Exercises 35–40 find a formula for the upper sum obtained by dividing the interval $[a, b]$ into n equal subintervals. Then take a limit of these sums as $n \to \infty$ to calculate the area under the curve over $[a, b]$.

35. $f(x) = 1 - x^2$ over the interval $[0, 1]$.

36. $f(x) = 2x$ over the interval $[0, 3]$.

37. $f(x) = x^2 + 1$ over the interval $[0, 3]$.

38. $f(x) = 3x^2$ over the interval $[0, 1]$.

39. $f(x) = x + x^2$ over the interval $[0, 1]$.

40. $f(x) = 3x + 2x^2$ over the interval $[0, 1]$.

5.3 The Definite Integral

In Section 5.2 we investigated the limit of a finite sum for a function defined over a closed interval $[a, b]$ using n subintervals of equal width (or length), $(b - a)/n$. In this section we consider the limit of more general Riemann sums as the norm of the partitions of $[a, b]$ approaches zero. For general Riemann sums the subintervals of the partitions need not have equal widths. The limiting process then leads to the definition of the *definite integral* of a function over a closed interval $[a, b]$.

Limits of Riemann Sums

The definition of the definite integral is based on the idea that for certain functions, as the norm of the partitions of $[a, b]$ approaches zero, the values of the corresponding Riemann sums approach a limiting value I. What we mean by this converging idea is that a Riemann sum will be close to the number I provided that the norm of its partition is sufficiently small (so that all of its subintervals have thin enough widths). We introduce the symbol ϵ as a small positive number that specifies how close to I the Riemann sum must be, and the symbol δ as a second small positive number that specifies how small the norm of a partition must be in order for that to happen. Here is a precise formulation.

DEFINITION The Definite Integral as a Limit of Riemann Sums

Let $f(x)$ be a function defined on a closed interval $[a, b]$. We say that a number I is the **definite integral of f over $[a, b]$** and that I is the limit of the Riemann sums $\sum_{k=1}^{n} f(c_k)\,\Delta x_k$ if the following condition is satisfied:

Given any number $\epsilon > 0$ there is a corresponding number $\delta > 0$ such that for every partition $P = \{x_0, x_1, \dots, x_n\}$ of $[a, b]$ with $\|P\| < \delta$ and any choice of c_k in $[x_{k-1}, x_k]$, we have

$$\left| \sum_{k=1}^{n} f(c_k)\,\Delta x_k - I \right| < \epsilon.$$

Leibniz introduced a notation for the definite integral that captures its construction as a limit of Riemann sums. He envisioned the finite sums $\sum_{k=1}^{n} f(c_k)\,\Delta x_k$ becoming an infinite sum of function values $f(x)$ multiplied by "infinitesimal" subinterval widths dx. The sum symbol $\sum$ is replaced in the limit by the integral symbol $\int$, whose origin is in the letter "S." The function values $f(c_k)$ are replaced by a continuous selection of function values $f(x)$. The subinterval widths Δx_k become the differential dx. It is as if we are summing all products of the form $f(x) \cdot dx$ as x goes from a to b. While this notation captures the process of constructing an integral, it is Riemann's definition that gives a precise meaning to the definite integral.

Notation and Existence of the Definite Integral

The symbol for the number I in the definition of the definite integral is

$$\int_a^b f(x)\,dx$$

which is read as "the integral from a to b of f of x dee x" or sometimes as "the integral from a to b of f of x with respect to x." The component parts in the integral symbol also have names:

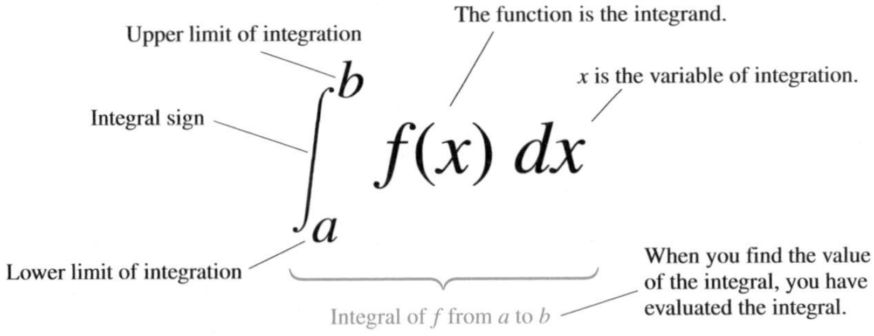

When the definition is satisfied, we say the Riemann sums of f on $[a, b]$ **converge** to the definite integral $I = \int_a^b f(x)\,dx$ and that f is **integrable** over $[a, b]$. We have many choices for a partition P with norm going to zero, and many choices of points c_k for each partition. The definite integral exists when we always get the same limit I, no matter what choices are made. When the limit exists we write it as the definite integral

$$\lim_{\|P\| \to 0} \sum_{k=1}^{n} f(c_k)\,\Delta x_k = I = \int_a^b f(x)\,dx.$$

When each partition has n equal subintervals, each of width $\Delta x = (b - a)/n$, we will also write

$$\lim_{n \to \infty} \sum_{k=1}^{n} f(c_k)\,\Delta x = I = \int_a^b f(x)\,dx.$$

The limit is always taken as the norm of the partitions approaches zero and the number of subintervals goes to infinity.

The value of the definite integral of a function over any particular interval depends on the function, not on the letter we choose to represent its independent variable. If we decide to use t or u instead of x, we simply write the integral as

$$\int_a^b f(t)\,dt \qquad \text{or} \qquad \int_a^b f(u)\,du \qquad \text{instead of} \qquad \int_a^b f(x)\,dx.$$

No matter how we write the integral, it is still the same number, defined as a limit of Riemann sums. Since it does not matter what letter we use, the variable of integration is called a **dummy variable**.

Since there are so many choices to be made in taking a limit of Riemann sums, it might seem difficult to show that such a limit exists. It turns out, however, that no matter what choices are made, the Riemann sums associated with a *continuous* function converge to the same limit.

THEOREM 1 The Existence of Definite Integrals

A continuous function is integrable. That is, if a function f is continuous on an interval $[a, b]$, then its definite integral over $[a, b]$ exists.

By the Extreme Value Theorem (Theorem 1, Section 4.1), when f is continuous we can choose c_k so that $f(c_k)$ gives the maximum value of f on $[x_{k-1}, x_k]$, giving an **upper sum**. We can choose c_k to give the minimum value of f on $[x_{k-1}, x_k]$, giving a **lower sum**. We can pick c_k to be the midpoint of $[x_{k-1}, x_k]$, the rightmost point x_k, or a random point. We can take the partitions of equal or varying widths. In each case we get the same limit for $\sum_{k=1}^{n} f(c_k)\,\Delta x_k$ as $\|P\| \to 0$. The idea behind Theorem 1 is that a Riemann sum associated with a partition is no more than the upper sum of that partition and no less than the lower sum. The upper and lower sums converge to the same value when $\|P\| \to 0$. All other Riemann sums lie between the upper and lower sums and have the same limit. A proof of Theorem 1 involves a careful analysis of functions, partitions, and limits along this line of thinking and is left to a more advanced text. An indication of this proof is given in Exercises 80 and 81.

Theorem 1 says nothing about how to *calculate* definite integrals. A method of calculation will be developed in Section 5.4, through a connection to the process of taking antiderivatives.

Integrable and Nonintegrable Functions

Theorem 1 tells us that functions continuous over the interval $[a, b]$ are integrable there. Functions that are not continuous may or may not be integrable. Discontinuous functions that are integrable include those that are increasing on $[a, b]$ (Exercise 77), and the *piecewise-continuous functions* defined in the Additional Exercises at the end of this chapter. (The latter are continuous except at a finite number of points in $[a, b]$.) For integrability to fail, a function needs to be sufficiently discontinuous so that the region between its graph and the x-axis cannot be approximated well by increasingly thin rectangles. Here is an example of a function that is not integrable.

EXAMPLE 1 A Nonintegrable Function on [0, 1]

The function

$$f(x) = \begin{cases} 1, & \text{if } x \text{ is rational} \\ 0, & \text{if } x \text{ is irrational} \end{cases}$$

has no Riemann integral over $[0, 1]$. Underlying this is the fact that between any two numbers there is both a rational number and an irrational number. Thus the function jumps up and down too erratically over $[0, 1]$ to allow the region beneath its graph and above the x-axis to be approximated by rectangles, no matter how thin they are. We show, in fact, that upper sum approximations and lower sum approximations converge to different limiting values.

If we pick a partition P of $[0, 1]$ and choose c_k to be the maximum value for f on $[x_{k-1}, x_k]$ then the corresponding Riemann sum is

$$U = \sum_{k=1}^{n} f(c_k)\,\Delta x_k = \sum_{k=1}^{n} (1)\,\Delta x_k = 1,$$

since each subinterval $[x_{k-1}, x_k]$ contains a rational number where $f(c_k) = 1$. Note that the lengths of the intervals in the partition sum to 1, $\sum_{k=1}^{n} \Delta x_k = 1$. So each such Riemann sum equals 1, and a limit of Riemann sums using these choices equals 1.

On the other hand, if we pick c_k to be the minimum value for f on $[x_{k-1}, x_k]$, then the Riemann sum is

$$L = \sum_{k=1}^{n} f(c_k)\,\Delta x_k = \sum_{k=1}^{n} (0)\,\Delta x_k = 0,$$

since each subinterval $[x_{k-1}, x_k]$ contains an irrational number c_k where $f(c_k) = 0$. The limit of Riemann sums using these choices equals zero. Since the limit depends on the choices of c_k, the function f is not integrable. ■

Properties of Definite Integrals

In defining $\int_a^b f(x)\,dx$ as a limit of sums $\sum_{k=1}^{n} f(c_k)\,\Delta x_k$, we moved from left to right across the interval $[a, b]$. What would happen if we instead move right to left, starting with $x_0 = b$ and ending at $x_n = a$. Each Δx_k in the Riemann sum would change its sign, with $x_k - x_{k-1}$ now negative instead of positive. With the same choices of c_k in each subinterval, the sign of any Riemann sum would change, as would the sign of the limit, the integral

$\int_b^a f(x)\,dx$. Since we have not previously given a meaning to integrating backward, we are led to define

$$\int_b^a f(x)\,dx = -\int_a^b f(x)\,dx.$$

Another extension of the integral is to an interval of zero width, when $a = b$. Since $f(c_k)\,\Delta x_k$ is zero when the interval width $\Delta x_k = 0$, whenever $f(a)$ exists we define

$$\int_a^a f(x)\,dx = 0.$$

Theorem 2 states seven properties of integrals, given as rules that they satisfy, including the two above. These rules become very useful in the process of computing integrals. We will refer to them repeatedly to simplify our calculations.

Rules 2 through 7 have geometric interpretations, shown in Figure 5.11. The graphs in these figures are of positive functions, but the rules apply to general integrable functions.

THEOREM 2
When f and g are integrable on the interval $[a, b]$, the definite integral satisfies Rules 1 to 7 in Table 5.3.

TABLE 5.3 Rules satisfied by definite integrals

1. *Order of Integration:* $\int_b^a f(x)\,dx = -\int_a^b f(x)\,dx$ — A Definition

2. *Zero Width Interval:* $\int_a^a f(x)\,dx = 0$ — A Definition when $f(a)$ exists

3. *Constant Multiple:* $\int_a^b kf(x)\,dx = k\int_a^b f(x)\,dx$ — Any Number k

   $\int_a^b -f(x)\,dx = -\int_a^b f(x)\,dx$ — $k = -1$

4. *Sum and Difference:* $\int_a^b (f(x) \pm g(x))\,dx = \int_a^b f(x)\,dx \pm \int_a^b g(x)\,dx$

5. *Additivity:* $\int_a^b f(x)\,dx + \int_b^c f(x)\,dx = \int_a^c f(x)\,dx$

6. *Max-Min Inequality:* If f has maximum value $\max f$ and minimum value $\min f$ on $[a, b]$, then

$$\min f \cdot (b - a) \le \int_a^b f(x)\,dx \le \max f \cdot (b - a).$$

7. *Domination:* $f(x) \ge g(x)$ on $[a, b] \Rightarrow \int_a^b f(x)\,dx \ge \int_a^b g(x)\,dx$

   $f(x) \ge 0$ on $[a, b] \Rightarrow \int_a^b f(x)\,dx \ge 0$ (Special Case)

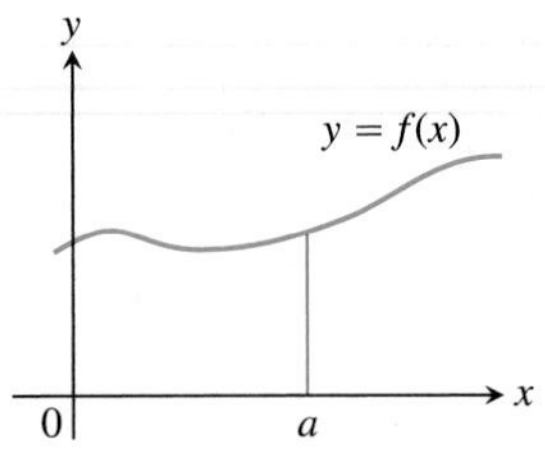

(a) *Zero Width Interval:*

$$\int_a^a f(x)\,dx = 0.$$

(The area over a point is 0.)

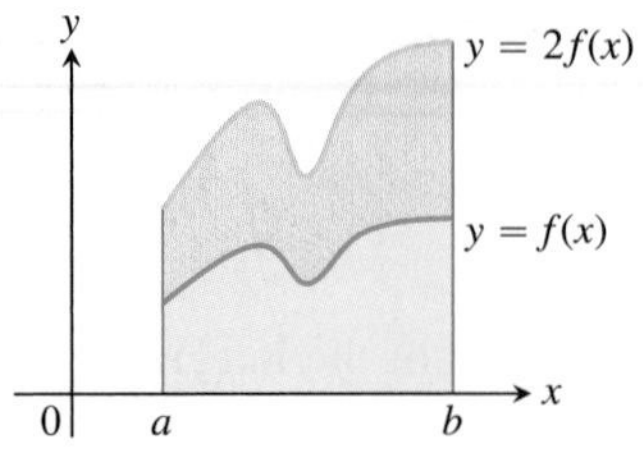

(b) *Constant Multiple:*

$$\int_a^b kf(x)\,dx = k\int_a^b f(x)\,dx.$$

(Shown for $k = 2$.)

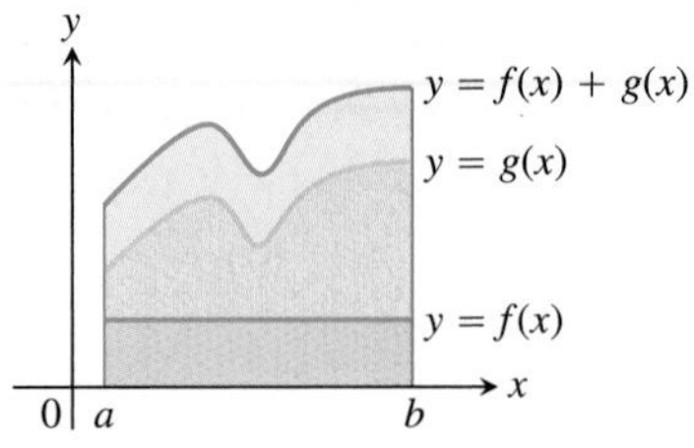

(c) *Sum:*

$$\int_a^b (f(x) + g(x))\,dx = \int_a^b f(x)\,dx + \int_a^b g(x)\,dx$$

(Areas add)

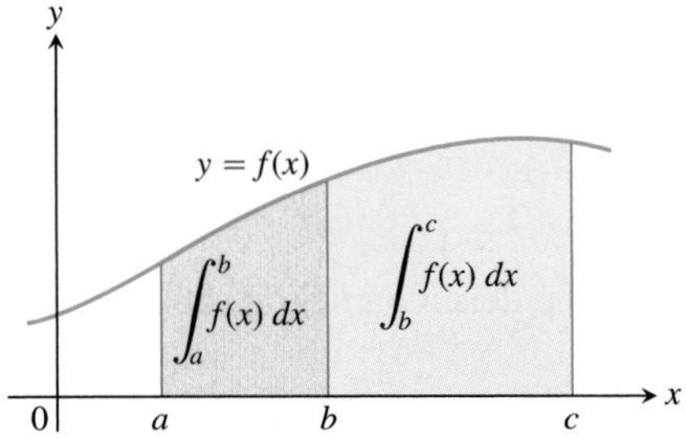

(d) *Additivity for definite integrals:*

$$\int_a^b f(x)\,dx + \int_b^c f(x)\,dx = \int_a^c f(x)\,dx$$

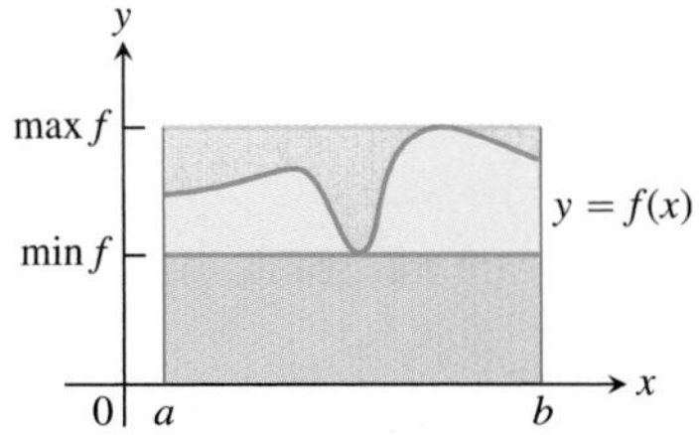

(e) *Max-Min Inequality:*

$$\min f \cdot (b - a) \le \int_a^b f(x)\,dx \le \max f \cdot (b - a)$$

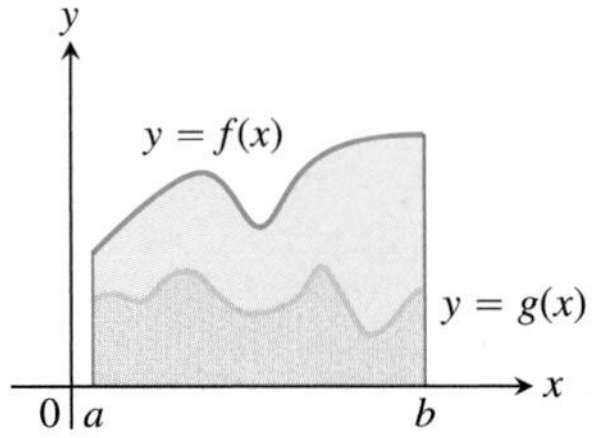

(f) *Domination:*

$$f(x) \ge g(x) \text{ on } [a, b] \Rightarrow \int_a^b f(x)\,dx \ge \int_a^b g(x)\,dx$$

FIGURE 5.11

While Rules 1 and 2 are definitions, Rules 3 to 7 of Table 5.3 must be proved. The proofs are based on the definition of the definite integral as a limit of Riemann sums. The following is a proof of one of these rules. Similar proofs can be given to verify the other properties in Table 5.3.

Proof of Rule 6 Rule 6 says that the integral of f over $[a, b]$ is never smaller than the minimum value of f times the length of the interval and never larger than the maximum value of f times the length of the interval. The reason is that for every partition of $[a, b]$ and for every choice of the points c_k,

$$\begin{aligned}
\min f \cdot (b - a) &= \min f \cdot \sum_{k=1}^{n} \Delta x_k && \sum_{k=1}^{n} \Delta x_k = b - a \\
&= \sum_{k=1}^{n} \min f \cdot \Delta x_k && \text{Constant Multiple Rule} \\
&\le \sum_{k=1}^{n} f(c_k)\,\Delta x_k && \min f \le f(c_k) \\
&\le \sum_{k=1}^{n} \max f \cdot \Delta x_k && f(c_k) \le \max f \\
&= \max f \cdot \sum_{k=1}^{n} \Delta x_k && \text{Constant Multiple Rule} \\
&= \max f \cdot (b - a).
\end{aligned}$$

In short, all Riemann sums for f on $[a, b]$ satisfy the inequality

$$\min f \cdot (b - a) \leq \sum_{k=1}^{n} f(c_k)\,\Delta x_k \leq \max f \cdot (b - a).$$

Hence their limit, the integral, does too. ■

EXAMPLE 2 Using the Rules for Definite Integrals

Suppose that

$$\int_{-1}^{1} f(x)\,dx = 5, \qquad \int_{1}^{4} f(x)\,dx = -2, \qquad \int_{-1}^{1} h(x)\,dx = 7.$$

Then

1. $\displaystyle\int_{4}^{1} f(x)\,dx = -\int_{1}^{4} f(x)\,dx = -(-2) = 2$ Rule 1

2. $\displaystyle\int_{-1}^{1} [2f(x) + 3h(x)]\,dx = 2\int_{-1}^{1} f(x)\,dx + 3\int_{-1}^{1} h(x)\,dx$ Rules 3 and 4

$$= 2(5) + 3(7) = 31$$

3. $\displaystyle\int_{-1}^{4} f(x)\,dx = \int_{-1}^{1} f(x)\,dx + \int_{1}^{4} f(x)\,dx = 5 + (-2) = 3$ Rule 5 ■

EXAMPLE 3 Finding Bounds for an Integral

Show that the value of $\int_0^1 \sqrt{1 + \cos x}\,dx$ is less than $3/2$.

Solution The Max-Min Inequality for definite integrals (Rule 6) says that $\min f \cdot (b - a)$ is a *lower bound* for the value of $\int_a^b f(x)\,dx$ and that $\max f \cdot (b - a)$ is an *upper bound*. The maximum value of $\sqrt{1 + \cos x}$ on $[0, 1]$ is $\sqrt{1 + 1} = \sqrt{2}$, so

$$\int_0^1 \sqrt{1 + \cos x}\,dx \leq \sqrt{2} \cdot (1 - 0) = \sqrt{2}.$$

Since $\int_0^1 \sqrt{1 + \cos x}\,dx$ is bounded from above by $\sqrt{2}$ (which is 1.414 ...), the integral is less than $3/2$. ■

Area Under the Graph of a Nonnegative Function

We now make precise the notion of the area of a region with curved boundary, capturing the idea of approximating a region by increasingly many rectangles. The area under the graph of a nonnegative continuous function is defined to be a definite integral.

DEFINITION Area Under a Curve as a Definite Integral

If $y = f(x)$ is nonnegative and integrable over a closed interval $[a, b]$, then the **area under the curve $y = f(x)$ over $[a, b]$** is the integral of f from a to b,

$$A = \int_a^b f(x)\,dx.$$

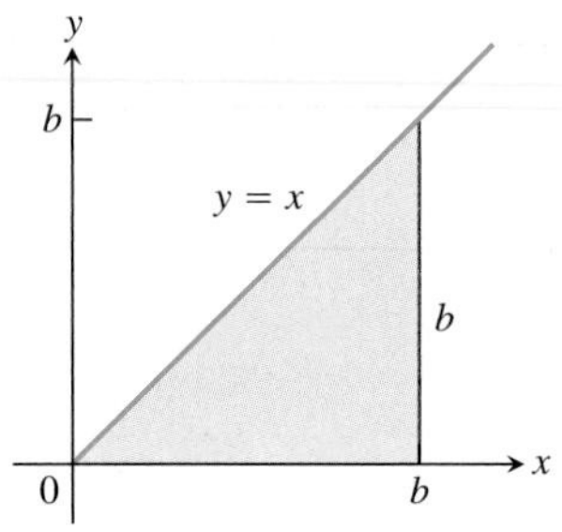

FIGURE 5.12 The region in Example 4 is a triangle.

For the first time we have a rigorous definition for the area of a region whose boundary is the graph of any continuous function. We now apply this to a simple example, the area under a straight line, where we can verify that our new definition agrees with our previous notion of area.

EXAMPLE 4 Area Under the Line $y = x$

Compute $\int_0^b x\,dx$ and find the area A under $y = x$ over the interval $[0, b]$, $b > 0$.

Solution The region of interest is a triangle (Figure 5.12). We compute the area in two ways.

(a) To compute the definite integral as the limit of Riemann sums, we calculate $\lim_{\|P\|\to 0} \sum_{k=1}^{n} f(c_k)\,\Delta x_k$ for partitions whose norms go to zero. Theorem 1 tells us that it does not matter how we choose the partitions or the points c_k as long as the norms approach zero. All choices give the exact same limit. So we consider the partition P that subdivides the interval $[0, b]$ into n subintervals of equal width $\Delta x = (b - 0)/n = b/n$, and we choose c_k to be the right endpoint in each subinterval. The partition is $P = \left\{0, \frac{b}{n}, \frac{2b}{n}, \frac{3b}{n}, \cdots, \frac{nb}{n}\right\}$ and $c_k = \frac{kb}{n}$. So

$$\begin{aligned}
\sum_{k=1}^{n} f(c_k)\,\Delta x &= \sum_{k=1}^{n} \frac{kb}{n} \cdot \frac{b}{n} && f(c_k) = c_k \\
&= \sum_{k=1}^{n} \frac{kb^2}{n^2} \\
&= \frac{b^2}{n^2} \sum_{k=1}^{n} k && \text{Constant Multiple Rule} \\
&= \frac{b^2}{n^2} \cdot \frac{n(n+1)}{2} && \text{Sum of First } n \text{ Integers} \\
&= \frac{b^2}{2}\left(1 + \frac{1}{n}\right)
\end{aligned}$$

As $n \to \infty$ and $\|P\| \to 0$, this last expression on the right has the limit $b^2/2$. Therefore,

$$\int_0^b x\,dx = \frac{b^2}{2}.$$

(b) Since the area equals the definite integral for a nonnegative function, we can quickly derive the definite integral by using the formula for the area of a triangle having base length b and height $y = b$. The area is $A = (1/2)\,b \cdot b = b^2/2$. Again we have that $\int_0^b x\,dx = b^2/2$. ■

Example 4 can be generalized to integrate $f(x) = x$ over any closed interval $[a, b]$, $0 < a < b$.

$$\begin{aligned}
\int_a^b x\,dx &= \int_a^0 x\,dx + \int_0^b x\,dx && \text{Rule 5} \\
&= -\int_0^a x\,dx + \int_0^b x\,dx && \text{Rule 1} \\
&= -\frac{a^2}{2} + \frac{b^2}{2}. && \text{Example 4}
\end{aligned}$$

In conclusion, we have the following rule for integrating $f(x) = x$:

$$\int_a^b x\,dx = \frac{b^2}{2} - \frac{a^2}{2}, \qquad a < b \tag{1}$$

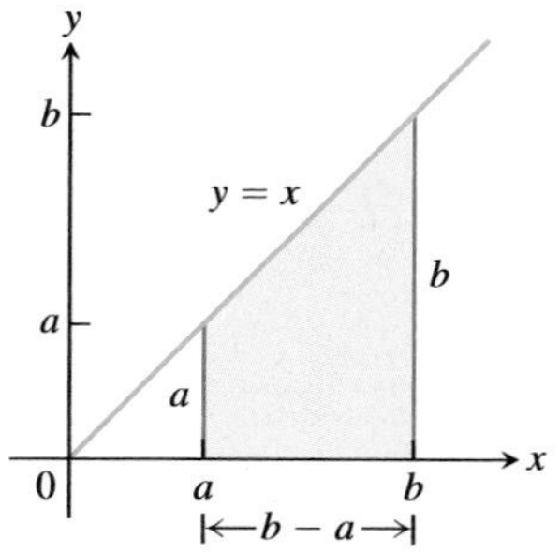

FIGURE 5.13 The area of this trapezoidal region is $A = (b^2 - a^2)/2$.

This computation gives the area of a trapezoid (Figure 5.13). Equation (1) remains valid when a and b are negative. When $a < b < 0$, the definite integral value $(b^2 - a^2)/2$ is a negative number, the negative of the area of a trapezoid dropping down to the line $y = x$ below the x-axis. When $a < 0$ and $b > 0$, Equation (1) is still valid and the definite integral gives the difference between two areas, the area under the graph and above $[0, b]$ minus the area below $[a, 0]$ and over the graph.

The following results can also be established using a Riemann sum calculation similar to that in Example 4 (Exercises 75 and 76).

$$\int_a^b c\,dx = c(b - a), \qquad c \text{ any constant} \tag{2}$$

$$\int_a^b x^2\,dx = \frac{b^3}{3} - \frac{a^3}{3}, \qquad a < b \tag{3}$$

Average Value of a Continuous Function Revisited

In Section 5.1 we introduced informally the average value of a nonnegative continuous function f over an interval $[a, b]$, leading us to define this average as the area under the graph of $y = f(x)$ divided by $b - a$. In integral notation we write this as

$$\text{Average} = \frac{1}{b - a}\int_a^b f(x)\,dx.$$

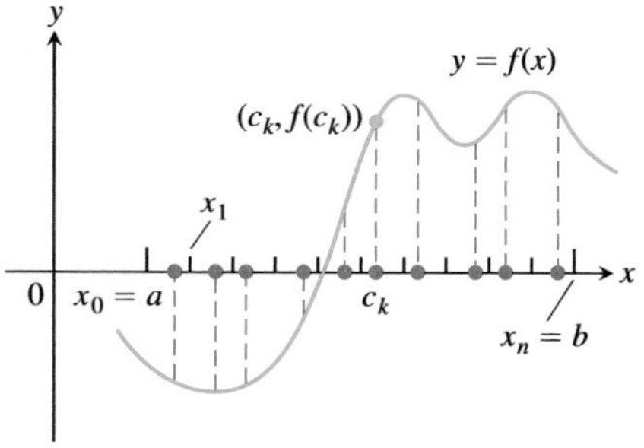

FIGURE 5.14 A sample of values of a function on an interval $[a, b]$.

We can use this formula to give a precise definition of the average value of any continuous (or integrable) function, whether positive, negative or both.

Alternately, we can use the following reasoning. We start with the idea from arithmetic that the average of n numbers is their sum divided by n. A continuous function f on $[a, b]$ may have infinitely many values, but we can still sample them in an orderly way. We divide $[a, b]$ into n subintervals of equal width $\Delta x = (b - a)/n$ and evaluate f at a point c_k in each (Figure 5.14). The average of the n sampled values is

$$\begin{aligned}
\frac{f(c_1) + f(c_2) + \cdots + f(c_n)}{n} &= \frac{1}{n}\sum_{k=1}^{n} f(c_k) \\
&= \frac{\Delta x}{b - a}\sum_{k=1}^{n} f(c_k) \qquad \Delta x = \frac{b-a}{n}, \text{ so } \frac{1}{n} = \frac{\Delta x}{b-a} \\
&= \frac{1}{b - a}\sum_{k=1}^{n} f(c_k)\,\Delta x
\end{aligned}$$

The average is obtained by dividing a Riemann sum for f on $[a, b]$ by $(b - a)$. As we increase the size of the sample and let the norm of the partition approach zero, the average approaches $(1/(b - a))\int_a^b f(x)\,dx$. Both points of view lead us to the following definition.

DEFINITION The Average or Mean Value of a Function

If f is integrable on $[a, b]$, then its **average value on** $[a, b]$, also called its **mean value**, is

$$\text{av}(f) = \frac{1}{b-a}\int_a^b f(x)\,dx.$$

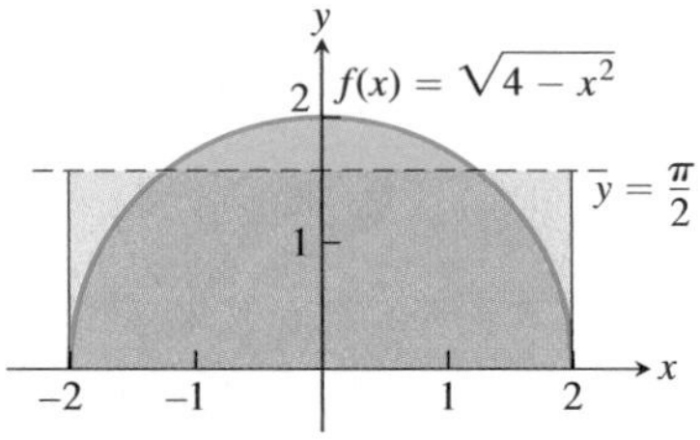

FIGURE 5.15 The average value of $f(x) = \sqrt{4 - x^2}$ on $[-2, 2]$ is $\pi/2$ (Example 5).

EXAMPLE 5 Finding an Average Value

Find the average value of $f(x) = \sqrt{4 - x^2}$ on $[-2, 2]$.

Solution We recognize $f(x) = \sqrt{4 - x^2}$ as a function whose graph is the upper semicircle of radius 2 centered at the origin (Figure 5.15).

The area between the semicircle and the x-axis from -2 to 2 can be computed using the geometry formula

$$\text{Area} = \frac{1}{2}\cdot \pi r^2 = \frac{1}{2}\cdot \pi(2)^2 = 2\pi.$$

Because f is nonnegative, the area is also the value of the integral of f from -2 to 2,

$$\int_{-2}^{2}\sqrt{4 - x^2}\,dx = 2\pi.$$

Therefore, the average value of f is

$$\text{av}(f) = \frac{1}{2-(-2)}\int_{-2}^{2}\sqrt{4 - x^2}\,dx = \frac{1}{4}(2\pi) = \frac{\pi}{2}.$$ ■

EXERCISES 5.3

Expressing Limits as Integrals

Express the limits in Exercises 1–8 as definite integrals.

1. $\lim_{\|P\|\to 0}\sum_{k=1}^{n} c_k^2\,\Delta x_k$, where P is a partition of $[0, 2]$
2. $\lim_{\|P\|\to 0}\sum_{k=1}^{n} 2c_k^3\,\Delta x_k$, where P is a partition of $[-1, 0]$
3. $\lim_{\|P\|\to 0}\sum_{k=1}^{n} (c_k^2 - 3c_k)\,\Delta x_k$, where P is a partition of $[-7, 5]$
4. $\lim_{\|P\|\to 0}\sum_{k=1}^{n} \left(\frac{1}{c_k}\right)\Delta x_k$, where P is a partition of $[1, 4]$
5. $\lim_{\|P\|\to 0}\sum_{k=1}^{n} \frac{1}{1 - c_k}\,\Delta x_k$, where P is a partition of $[2, 3]$
6. $\lim_{\|P\|\to 0}\sum_{k=1}^{n} \sqrt{4 - c_k^2}\,\Delta x_k$, where P is a partition of $[0, 1]$
7. $\lim_{\|P\|\to 0}\sum_{k=1}^{n} (\sec c_k)\,\Delta x_k$, where P is a partition of $[-\pi/4, 0]$
8. $\lim_{\|P\|\to 0}\sum_{k=1}^{n} (\tan c_k)\,\Delta x_k$, where P is a partition of $[0, \pi/4]$

Using Properties and Known Values to Find Other Integrals

9. Suppose that f and g are integrable and that

$$\int_1^2 f(x)\,dx = -4, \quad \int_1^5 f(x)\,dx = 6, \quad \int_1^5 g(x)\,dx = 8.$$

Use the rules in Table 5.3 to find

a. $\int_2^2 g(x)\,dx$ **b.** $\int_5^1 g(x)\,dx$

c. $\int_1^2 3f(x)\,dx$ **d.** $\int_2^5 f(x)\,dx$

e. $\int_1^5 [f(x) - g(x)]\,dx$ **f.** $\int_1^5 [4f(x) - g(x)]\,dx$

10. Suppose that f and h are integrable and that

$$\int_1^9 f(x)\,dx = -1, \quad \int_7^9 f(x)\,dx = 5, \quad \int_7^9 h(x)\,dx = 4.$$

Use the rules in Table 5.3 to find

a. $\int_1^9 -2f(x)\,dx$ **b.** $\int_7^9 [f(x) + h(x)]\,dx$

c. $\int_7^9 [2f(x) - 3h(x)]\,dx$ **d.** $\int_9^1 f(x)\,dx$

e. $\int_1^7 f(x)\,dx$ **f.** $\int_9^7 [h(x) - f(x)]\,dx$

11. Suppose that $\int_1^2 f(x)\,dx = 5$. Find

a. $\int_1^2 f(u)\,du$ **b.** $\int_1^2 \sqrt{3}f(z)\,dz$

c. $\int_2^1 f(t)\,dt$ **d.** $\int_1^2 [-f(x)]\,dx$

12. Suppose that $\int_{-3}^0 g(t)\,dt = \sqrt{2}$. Find

a. $\int_0^{-3} g(t)\,dt$ **b.** $\int_{-3}^0 g(u)\,du$

c. $\int_{-3}^0 [-g(x)]\,dx$ **d.** $\int_{-3}^0 \frac{g(r)}{\sqrt{2}}\,dr$

13. Suppose that f is integrable and that $\int_0^3 f(z)\,dz = 3$ and $\int_0^4 f(z)\,dz = 7$. Find

a. $\int_3^4 f(z)\,dz$ **b.** $\int_4^3 f(t)\,dt$

14. Suppose that h is integrable and that $\int_{-1}^1 h(r)\,dr = 0$ and $\int_{-1}^3 h(r)\,dr = 6$. Find

a. $\int_1^3 h(r)\,dr$ **b.** $-\int_3^1 h(u)\,du$

Using Area to Evaluate Definite Integrals

In Exercises 15–22, graph the integrands and use areas to evaluate the integrals.

15. $\int_{-2}^4 \left(\frac{x}{2} + 3\right) dx$ **16.** $\int_{1/2}^{3/2} (-2x + 4)\,dx$

17. $\int_{-3}^3 \sqrt{9 - x^2}\,dx$ **18.** $\int_{-4}^0 \sqrt{16 - x^2}\,dx$

19. $\int_{-2}^1 |x|\,dx$ **20.** $\int_{-1}^1 (1 - |x|)\,dx$

21. $\int_{-1}^1 (2 - |x|)\,dx$ **22.** $\int_{-1}^1 \left(1 + \sqrt{1 - x^2}\right) dx$

Use areas to evaluate the integrals in Exercises 23–26.

23. $\int_0^b \frac{x}{2}\,dx, \quad b > 0$ **24.** $\int_0^b 4x\,dx, \quad b > 0$

25. $\int_a^b 2s\,ds, \quad 0 < a < b$ **26.** $\int_a^b 3t\,dt, \quad 0 < a < b$

Evaluations

Use the results of Equations (1) and (3) to evaluate the integrals in Exercises 27–38.

27. $\int_1^{\sqrt{2}} x\,dx$ **28.** $\int_{0.5}^{2.5} x\,dx$ **29.** $\int_\pi^{2\pi} \theta\,d\theta$

30. $\int_{\sqrt{2}}^{5\sqrt{2}} r\,dr$ **31.** $\int_0^{\sqrt[3]{7}} x^2\,dx$ **32.** $\int_0^{0.3} s^2\,ds$

33. $\int_0^{1/2} t^2\,dt$ **34.** $\int_0^{\pi/2} \theta^2\,d\theta$ **35.** $\int_a^{2a} x\,dx$

36. $\int_a^{\sqrt{3}a} x\,dx$ **37.** $\int_0^{\sqrt[3]{b}} x^2\,dx$ **38.** $\int_0^{3b} x^2\,dx$

Use the rules in Table 5.3 and Equations (1)–(3) to evaluate the integrals in Exercises 39–50.

39. $\int_3^1 7\,dx$ **40.** $\int_0^{-2} \sqrt{2}\,dx$

41. $\int_0^2 5x\,dx$ **42.** $\int_3^5 \frac{x}{8}\,dx$

43. $\int_0^2 (2t - 3)\,dt$ **44.** $\int_0^{\sqrt{2}} \left(t - \sqrt{2}\right) dt$

45. $\int_2^1 \left(1 + \frac{z}{2}\right) dz$ **46.** $\int_3^0 (2z - 3)\,dz$

47. $\int_1^2 3u^2\,du$ **48.** $\int_{1/2}^1 24u^2\,du$

49. $\int_0^2 (3x^2 + x - 5)\,dx$ **50.** $\int_1^0 (3x^2 + x - 5)\,dx$

Finding Area

In Exercises 51–54 use a definite integral to find the area of the region between the given curve and the x-axis on the interval $[0, b]$.

51. $y = 3x^2$ **52.** $y = \pi x^2$

53. $y = 2x$ **54.** $y = \frac{x}{2} + 1$

Average Value

In Exercises 55–62, graph the function and find its average value over the given interval.

55. $f(x) = x^2 - 1$ on $[0, \sqrt{3}]$

56. $f(x) = -\frac{x^2}{2}$ on $[0, 3]$ **57.** $f(x) = -3x^2 - 1$ on $[0, 1]$

58. $f(x) = 3x^2 - 3$ on $[0, 1]$

59. $f(t) = (t - 1)^2$ on $[0, 3]$

60. $f(t) = t^2 - t$ on $[-2, 1]$

61. $g(x) = |x| - 1$ on **a.** $[-1, 1]$, **b.** $[1, 3]$, and **c.** $[-1, 3]$

62. $h(x) = -|x|$ on **a.** $[-1, 0]$, **b.** $[0, 1]$, and **c.** $[-1, 1]$

Theory and Examples

63. What values of a and b maximize the value of

$$\int_a^b (x - x^2)\, dx?$$

(*Hint:* Where is the integrand positive?)

64. What values of a and b minimize the value of

$$\int_a^b (x^4 - 2x^2)\, dx?$$

65. Use the Max-Min Inequality to find upper and lower bounds for the value of

$$\int_0^1 \frac{1}{1 + x^2}\, dx.$$

66. (*Continuation of Exercise 65*) Use the Max-Min Inequality to find upper and lower bounds for

$$\int_0^{0.5} \frac{1}{1 + x^2}\, dx \quad \text{and} \quad \int_{0.5}^1 \frac{1}{1 + x^2}\, dx.$$

Add these to arrive at an improved estimate of

$$\int_0^1 \frac{1}{1 + x^2}\, dx.$$

67. Show that the value of $\int_0^1 \sin(x^2)\, dx$ cannot possibly be 2.

68. Show that the value of $\int_0^1 \sqrt{x + 8}\, dx$ lies between $2\sqrt{2} \approx 2.8$ and 3.

69. Integrals of nonnegative functions Use the Max-Min Inequality to show that if f is integrable then

$$f(x) \geq 0 \quad \text{on} \quad [a, b] \quad \Rightarrow \quad \int_a^b f(x)\, dx \geq 0.$$

70. Integrals of nonpositive functions Show that if f is integrable then

$$f(x) \leq 0 \quad \text{on} \quad [a, b] \quad \Rightarrow \quad \int_a^b f(x)\, dx \leq 0.$$

71. Use the inequality $\sin x \leq x$, which holds for $x \geq 0$, to find an upper bound for the value of $\int_0^1 \sin x\, dx$.

72. The inequality $\sec x \geq 1 + (x^2/2)$ holds on $(-\pi/2, \pi/2)$. Use it to find a lower bound for the value of $\int_0^1 \sec x\, dx$.

73. If av(f) really is a typical value of the integrable function $f(x)$ on $[a, b]$, then the number av(f) should have the same integral over $[a, b]$ that f does. Does it? That is, does

$$\int_a^b \operatorname{av}(f)\, dx = \int_a^b f(x)\, dx?$$

Give reasons for your answer.

74. It would be nice if average values of integrable functions obeyed the following rules on an interval $[a, b]$.

a. $\operatorname{av}(f + g) = \operatorname{av}(f) + \operatorname{av}(g)$

b. $\operatorname{av}(kf) = k \operatorname{av}(f)$ (any number k)

c. $\operatorname{av}(f) \leq \operatorname{av}(g)$ if $f(x) \leq g(x)$ on $[a, b]$.

Do these rules ever hold? Give reasons for your answers.

75. Use limits of Riemann sums as in Example 4a to establish Equation (2).

76. Use limits of Riemann sums as in Example 4a to establish Equation (3).

77. Upper and lower sums for increasing functions

a. Suppose the graph of a continuous function $f(x)$ rises steadily as x moves from left to right across an interval $[a, b]$. Let P be a partition of $[a, b]$ into n subintervals of length $\Delta x = (b - a)/n$. Show by referring to the accompanying figure that the difference between the upper and lower sums for f on this partition can be represented graphically as the area of a rectangle R whose dimensions are $[f(b) - f(a)]$ by Δx. (*Hint*: The difference $U - L$ is the sum of areas of rectangles whose diagonals $Q_0Q_1, Q_1Q_2, \ldots, Q_{n-1}Q_n$ lie along the curve. There is no overlapping when these rectangles are shifted horizontally onto R.)

b. Suppose that instead of being equal, the lengths Δx_k of the subintervals of the partition of $[a, b]$ vary in size. Show that

$$U - L \leq |f(b) - f(a)|\, \Delta x_{\max},$$

where $\Delta x_{\max}$ is the norm of P, and hence that $\lim_{\|P\| \to 0} (U - L) = 0$.

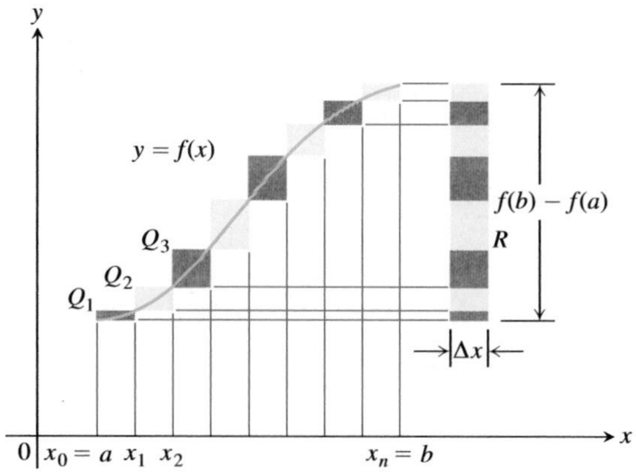

78. Upper and lower sums for decreasing functions (*Continuation of Exercise 77*)

a. Draw a figure like the one in Exercise 77 for a continuous function $f(x)$ whose values decrease steadily as x moves from left to right across the interval $[a, b]$. Let P be a partition of $[a, b]$ into subintervals of equal length. Find an expression for $U - L$ that is analogous to the one you found for $U - L$ in Exercise 77a.

b. Suppose that instead of being equal, the lengths Δx_k of the subintervals of P vary in size. Show that the inequality

$$U - L \leq |f(b) - f(a)|\ \Delta x_{\max}$$

of Exercise 77b still holds and hence that $\lim_{\|P\|\to 0} (U - L) = 0$.

79. Use the formula

$$\sin h + \sin 2h + \sin 3h + \cdots + \sin mh = \frac{\cos(h/2) - \cos((m + (1/2))h)}{2\sin(h/2)}$$

to find the area under the curve $y = \sin x$ from $x = 0$ to $x = \pi/2$ in two steps:

a. Partition the interval $[0, \pi/2]$ into n subintervals of equal length and calculate the corresponding upper sum U; then

b. Find the limit of U as $n \to \infty$ and $\Delta x = (b - a)/n \to 0$.

80. Suppose that f is continuous and nonnegative over $[a, b]$, as in the figure at the right. By inserting points

$$x_1, x_2, \ldots, x_{k-1}, x_k, \ldots, x_{n-1}$$

as shown, divide $[a, b]$ into n subintervals of lengths $\Delta x_1 = x_1 - a$, $\Delta x_2 = x_2 - x_1, \ldots, \Delta x_n = b - x_{n-1}$, which need not be equal.

a. If $m_k = \min\{f(x)$ for x in the kth subinterval$\}$, explain the connection between the *lower sum*

$$L = m_1\,\Delta x_1 + m_2\,\Delta x_2 + \cdots + m_n\,\Delta x_n$$

and the shaded region in the first part of the figure.

b. If $M_k = \max\{f(x)$ for x in the kth subinterval$\}$, explain the connection between the *upper sum*

$$U = M_1\,\Delta x_1 + M_2\,\Delta x_2 + \cdots + M_n\,\Delta x_n$$

and the shaded region in the second part of the figure.

c. Explain the connection between $U - L$ and the shaded regions along the curve in the third part of the figure.

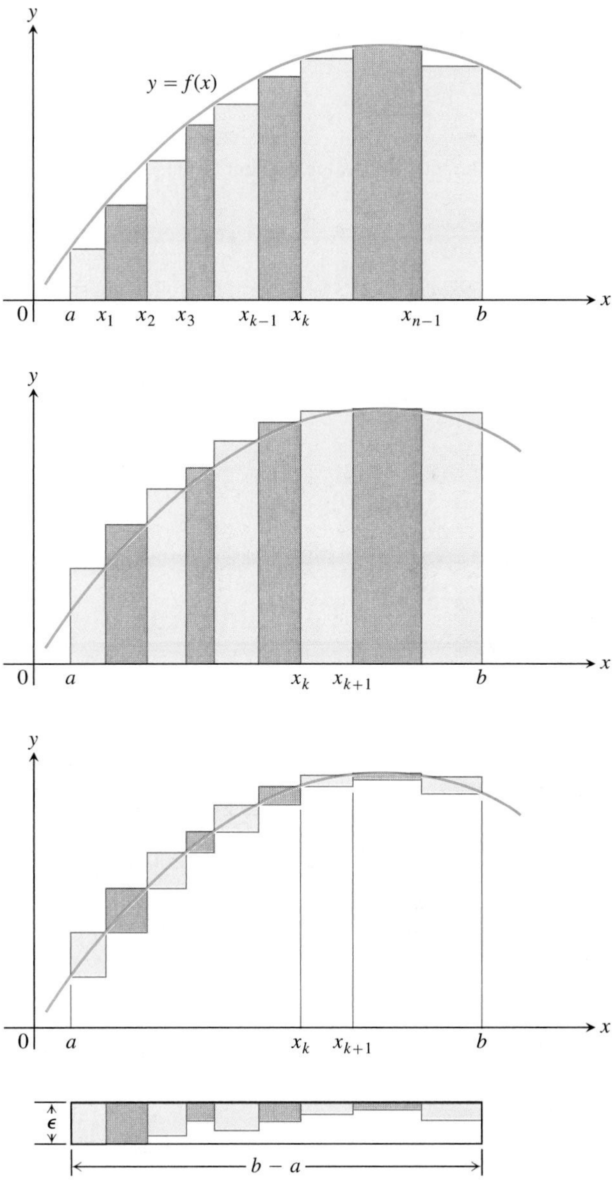

81. We say f is **uniformly continuous** on $[a, b]$ if given any $\epsilon > 0$ there is a $\delta > 0$ such that if x_1, x_2 are in $[a, b]$ and $|x_1 - x_2| < \delta$ then $|f(x_1) - f(x_2)| < \epsilon$. It can be shown that a continuous function on $[a, b]$ is uniformly continuous. Use this and the figure at the right to show that if f is continuous and $\epsilon > 0$ is given, it is possible to make $U - L \leq \epsilon \cdot (b - a)$ by making the largest of the Δx_k's sufficiently small.

82. If you average 30 mi/h on a 150-mi trip and then return over the same 150 mi at the rate of 50 mi/h, what is your average speed for the trip? Give reasons for your answer. (Source: David H. Pleacher, *The Mathematics Teacher*, Vol. 85, No. 6, pp. 445–446, September 1992.)

COMPUTER EXPLORATIONS

Finding Riemann Sums

If your CAS can draw rectangles associated with Riemann sums, use it to draw rectangles associated with Riemann sums that converge to the integrals in Exercises 83–88. Use $n = 4$, 10, 20, and 50 subintervals of equal length in each case.

83. $\displaystyle\int_0^1 (1 - x)\,dx = \frac{1}{2}$

84. $\displaystyle\int_0^1 (x^2 + 1)\,dx = \frac{4}{3}$

85. $\int_{-\pi}^{\pi} \cos x \, dx = 0$

86. $\int_{0}^{\pi/4} \sec^2 x \, dx = 1$

87. $\int_{-1}^{1} |x| \, dx = 1$

88. $\int_{1}^{2} \frac{1}{x} \, dx$ (The integral's value is about 0.693.)

Average Value

In Exercises 89–92, use a CAS to perform the following steps:

a. Plot the functions over the given interval.

b. Partition the interval into $n = 100, 200$, and 1000 subintervals of equal length, and evaluate the function at the midpoint of each subinterval.

c. Compute the average value of the function values generated in part (b).

d. Solve the equation $f(x) =$ (average value) for x using the average value calculated in part (c) for the $n = 1000$ partitioning.

89. $f(x) = \sin x \quad \text{on} \quad [0, \pi]$

90. $f(x) = \sin^2 x \quad \text{on} \quad [0, \pi]$

91. $f(x) = x \sin \frac{1}{x} \quad \text{on} \quad \left[\frac{\pi}{4}, \pi\right]$

92. $f(x) = x \sin^2 \frac{1}{x} \quad \text{on} \quad \left[\frac{\pi}{4}, \pi\right]$

5.4 The Fundamental Theorem of Calculus

In this section we present the Fundamental Theorem of Calculus, which is the central theorem of integral calculus. It connects integration and differentiation, enabling us to compute integrals using an antiderivative of the integrand function rather than by taking limits of Riemann sums as we did in Section 5.3. Leibniz and Newton exploited this relationship and started mathematical developments that fueled the scientific revolution for the next 200 years.

Along the way, we present the integral version of the Mean Value Theorem, which is another important theorem of integral calculus and used to prove the Fundamental Theorem.

HISTORICAL BIOGRAPHY

Sir Isaac Newton (1642–1727)

Mean Value Theorem for Definite Integrals

In the previous section, we defined the average value of a continuous function over a closed interval $[a, b]$ as the definite integral $\int_a^b f(x)\,dx$ divided by the length or width $b - a$ of the interval. The Mean Value Theorem for Definite Integrals asserts that this average value is *always* taken on at least once by the function f in the interval.

The graph in Figure 5.16 shows a *positive* continuous function $y = f(x)$ defined over the interval $[a, b]$. Geometrically, the Mean Value Theorem says that there is a number c in $[a, b]$ such that the rectangle with height equal to the average value $f(c)$ of the function and base width $b - a$ has exactly the same area as the region beneath the graph of f from a to b.

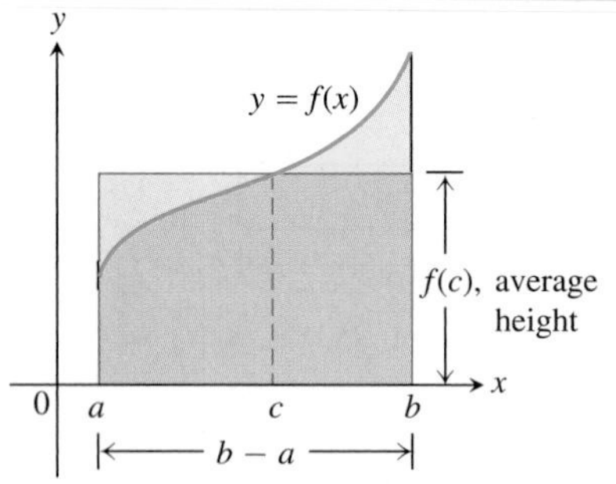

FIGURE 5.16 The value $f(c)$ in the Mean Value Theorem is, in a sense, the average (or *mean*) height of f on $[a, b]$. When $f \geq 0$, the area of the rectangle is the area under the graph of f from a to b,

$$f(c)(b - a) = \int_a^b f(x)\,dx.$$

THEOREM 3 The Mean Value Theorem for Definite Integrals
If f is continuous on $[a, b]$, then at some point c in $[a, b]$,

$$f(c) = \frac{1}{b - a} \int_a^b f(x)\,dx.$$

Proof If we divide both sides of the Max-Min Inequality (Table 5.3, Rule 6) by $(b - a)$, we obtain

$$\min f \le \frac{1}{b-a}\int_a^b f(x)\,dx \le \max f.$$

Since f is continuous, the Intermediate Value Theorem for Continuous Functions (Section 2.6) says that f must assume every value between $\min f$ and $\max f$. It must therefore assume the value $(1/(b-a))\int_a^b f(x)\,dx$ at some point c in $[a, b]$. ■

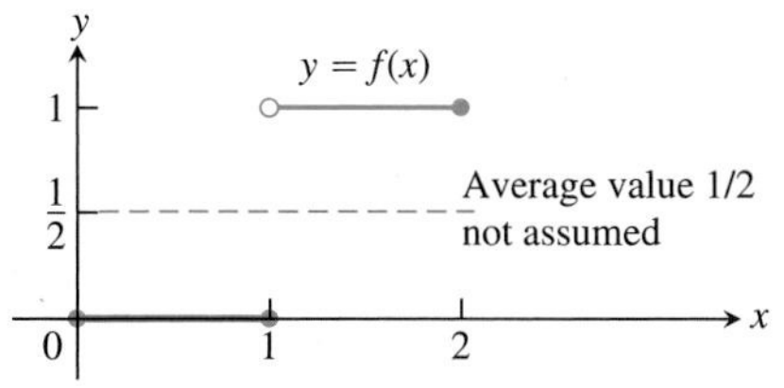

FIGURE 5.17 A discontinuous function need not assume its average value.

The continuity of f is important here. It is possible that a discontinuous function never equals its average value (Figure 5.17).

EXAMPLE 1 Applying the Mean Value Theorem for Integrals

Find the average value of $f(x) = 4 - x$ on $[0, 3]$ and where f actually takes on this value at some point in the given domain.

Solution

$$\begin{aligned}
\text{av}(f) &= \frac{1}{b-a}\int_a^b f(x)\,dx \\
&= \frac{1}{3-0}\int_0^3 (4-x)\,dx = \frac{1}{3}\left(\int_0^3 4\,dx - \int_0^3 x\,dx\right) \\
&= \frac{1}{3}\left(4(3-0) - \left(\frac{3^2}{2} - \frac{0^2}{2}\right)\right) \qquad \text{Section 5.3, Eqs. (1) and (2)} \\
&= 4 - \frac{3}{2} = \frac{5}{2}.
\end{aligned}$$

The average value of $f(x) = 4 - x$ over $[0, 3]$ is $5/2$. The function assumes this value when $4 - x = 5/2$ or $x = 3/2$, which is the point c in Theorem 3. (Figure 5.18) ■

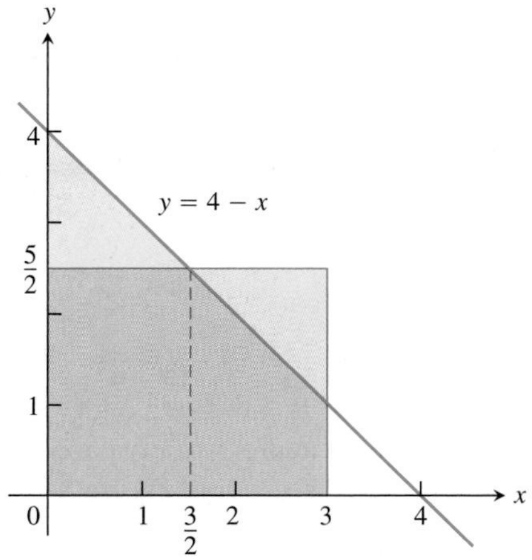

FIGURE 5.18 The area of the rectangle with base $[0, 3]$ and height $5/2$ (the average value of the function $f(x) = 4 - x$) is equal to the area between the graph of f and the x-axis from 0 to 3 (Example 1).

In Example 1, we actually found a point c where f assumed its average value by setting $f(x)$ equal to the calculated average value and solving for x. It's not always possible to solve easily for the value c. What else can we learn from the Mean Value Theorem for integrals? Here's an example.

EXAMPLE 2 Show that if f is continuous on $[a, b]$, $a \ne b$, and if

$$\int_a^b f(x)\,dx = 0,$$

then $f(x) = 0$ at least once in $[a, b]$.

Solution The average value of f on $[a, b]$ is

$$\text{av}(f) = \frac{1}{b-a}\int_a^b f(x)\,dx = \frac{1}{b-a}\cdot 0 = 0.$$

By the Mean Value Theorem, f assumes this value at some point $c \in [a, b]$. ■

Fundamental Theorem, Part 1

If $f(t)$ is an integrable function over a finite interval I, then the integral from any fixed number $a \in I$ to another number $x \in I$ defines a new function F whose value at x is

$$F(x) = \int_a^x f(t)\, dt. \tag{1}$$

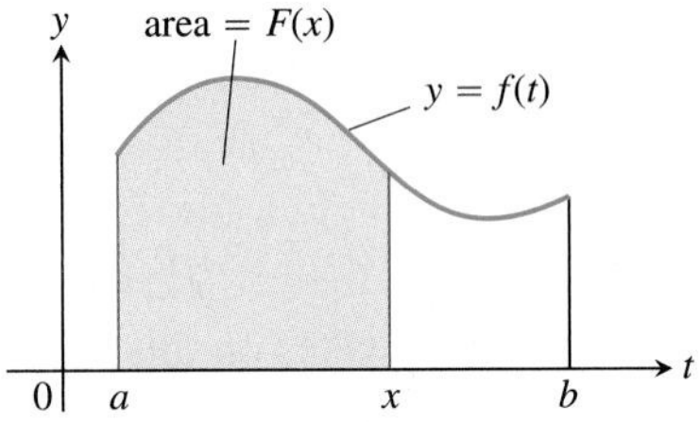

FIGURE 5.19 The function $F(x)$ defined by Equation (1) gives the area under the graph of f from a to x when f is nonnegative and $x > a$.

For example, if f is nonnegative and x lies to the right of a, then $F(x)$ is the area under the graph from a to x (Figure 5.19). The variable x is the upper limit of integration of an integral, but F is just like any other real-valued function of a real variable. For each value of the input x, there is a well-defined numerical output, in this case the definite integral of f from a to x.

Equation (1) gives a way to define new functions, but its importance now is the connection it makes between integrals and derivatives. If f is any continuous function, then the Fundamental Theorem asserts that F is a differentiable function of x whose derivative is f itself. At every value of x,

$$\frac{d}{dx}F(x) = \frac{d}{dx}\int_a^x f(t)\, dt = f(x).$$

To gain some insight into why this result holds, we look at the geometry behind it.

If $f \geq 0$ on $[a, b]$, then the computation of $F'(x)$ from the definition of the derivative means taking the limit as $h \to 0$ of the difference quotient

$$\frac{F(x + h) - F(x)}{h}.$$

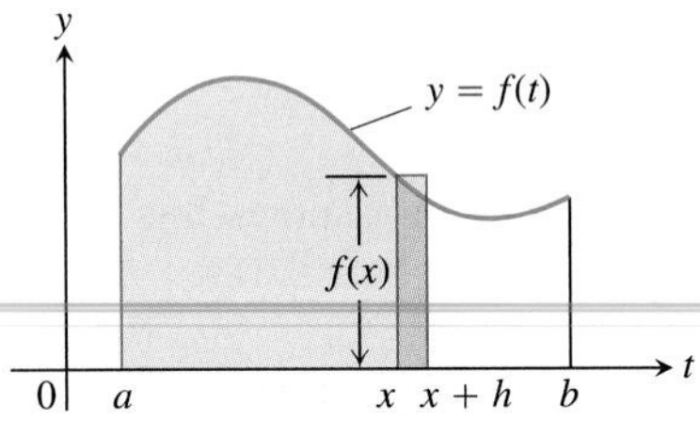

FIGURE 5.20 In Equation (1), $F(x)$ is the area to the left of x. Also, $F(x + h)$ is the area to the left of $x + h$. The difference quotient $[F(x + h) - F(x)]/h$ is then approximately equal to $f(x)$, the height of the rectangle shown here.

For $h > 0$, the numerator is obtained by subtracting two areas, so it is the area under the graph of f from x to $x + h$ (Figure 5.20). If h is small, this area is approximately equal to the area of the rectangle of height $f(x)$ and width h, which can be seen from Figure 5.20. That is,

$$F(x + h) - F(x) \approx hf(x).$$

Dividing both sides of this approximation by h and letting $h \to 0$, it is reasonable to expect that

$$F'(x) = \lim_{h\to 0} \frac{F(x + h) - F(x)}{h} = f(x).$$

This result is true even if the function f is not positive, and it forms the first part of the Fundamental Theorem of Calculus.

THEOREM 4 The Fundamental Theorem of Calculus, Part 1
If f is continuous on $[a, b]$ then $F(x) = \int_a^x f(t)\, dt$ is continuous on $[a, b]$ and differentiable on (a, b) and its derivative is $f(x)$;

$$F'(x) = \frac{d}{dx}\int_a^x f(t)\, dt = f(x). \tag{2}$$

Before proving Theorem 4, we look at several examples to gain a better understanding of what it says.

EXAMPLE 3 Applying the Fundamental Theorem

Use the Fundamental Theorem to find

(a) $\dfrac{d}{dx}\displaystyle\int_a^x \cos t\, dt$

(b) $\dfrac{d}{dx}\displaystyle\int_0^x \frac{1}{1+t^2}\, dt$

(c) $\dfrac{dy}{dx}$ if $y = \displaystyle\int_x^5 3t \sin t\, dt$

(d) $\dfrac{dy}{dx}$ if $y = \displaystyle\int_1^{x^2} \cos t\, dt$

Solution

(a) $\dfrac{d}{dx}\displaystyle\int_a^x \cos t\, dt = \cos x$ Eq. 2 with $f(t) = \cos t$

(b) $\dfrac{d}{dx}\displaystyle\int_0^x \frac{1}{1+t^2}\, dt = \frac{1}{1+x^2}$ Eq. 2 with $f(t) = \dfrac{1}{1+t^2}$

(c) Rule 1 for integrals in Table 5.3 of Section 5.3 sets this up for the Fundamental Theorem.

$$\frac{dy}{dx} = \frac{d}{dx}\int_x^5 3t\sin t\, dt = \frac{d}{dx}\left(-\int_5^x 3t\sin t\, dt\right) \qquad \text{Rule 1}$$

$$= -\frac{d}{dx}\int_5^x 3t\sin t\, dt$$

$$= -3x\sin x$$

(d) The upper limit of integration is not x but x^2. This makes y a composite of the two functions,

$$y = \int_1^u \cos t\, dt \qquad \text{and} \qquad u = x^2.$$

We must therefore apply the Chain Rule when finding dy/dx.

$$\frac{dy}{dx} = \frac{dy}{du}\cdot\frac{du}{dx}$$

$$= \left(\frac{d}{du}\int_1^u \cos t\, dt\right)\cdot\frac{du}{dx}$$

$$= \cos u \cdot \frac{du}{dx}$$

$$= \cos(x^2)\cdot 2x$$

$$= 2x\cos x^2$$

■

EXAMPLE 4 Constructing a Function with a Given Derivative and Value

Find a function $y = f(x)$ on the domain $(-\pi/2, \pi/2)$ with derivative

$$\frac{dy}{dx} = \tan x$$

that satisfies the condition $f(3) = 5$.

Solution The Fundamental Theorem makes it easy to construct a function with derivative $\tan x$ that equals 0 at $x = 3$:

$$y = \int_3^x \tan t\, dt.$$

Since $y(3) = \int_3^3 \tan t\, dt = 0$, we have only to add 5 to this function to construct one with derivative $\tan x$ whose value at $x = 3$ is 5:

$$f(x) = \int_3^x \tan t\, dt + 5.$$

■

Although the solution to the problem in Example 4 satisfies the two required conditions, you might ask whether it is in a useful form. The answer is yes, since today we have computers and calculators that are capable of approximating integrals. In Chapter 7 we will learn to write the solution in Example 4 exactly as

$$y = \ln\left|\frac{\cos 3}{\cos x}\right| + 5.$$

We now give a proof of the Fundamental Theorem for an arbitrary continuous function.

Proof of Theorem 4 We prove the Fundamental Theorem by applying the definition of the derivative directly to the function $F(x)$, when x and $x + h$ are in (a, b). This means writing out the difference quotient

$$\frac{F(x + h) - F(x)}{h} \tag{3}$$

and showing that its limit as $h \to 0$ is the number $f(x)$ for each x in (a, b). Thus,

$$\begin{aligned} F'(x) &= \lim_{h\to 0} \frac{F(x + h) - F(x)}{h} \\ &= \lim_{h\to 0} \frac{1}{h}\left[\int_a^{x+h} f(t)\, dt - \int_a^x f(t)\, dt\right] \\ &= \lim_{h\to 0} \frac{1}{h}\int_x^{x+h} f(t)\, dt \qquad \text{Table 5.3, Rule 5} \end{aligned}$$

According to the Mean Value Theorem for Definite Integrals, the value before taking the limit in the last expression is one of the values taken on by f in the interval between x and $x + h$. That is, for some number c in this interval,

$$\frac{1}{h}\int_x^{x+h} f(t)\, dt = f(c). \tag{4}$$

As $h \to 0$, $x + h$ approaches x, forcing c to approach x also (because c is trapped between x and $x + h$). Since f is continuous at x, $f(c)$ approaches $f(x)$:

$$\lim_{h\to 0} f(c) = f(x). \tag{5}$$

In conclusion, we have

$$\begin{aligned} F'(x) &= \lim_{h\to 0} \frac{1}{h}\int_x^{x+h} f(t)\, dt \\ &= \lim_{h\to 0} f(c) && \text{Eq. (4)} \\ &= f(x). && \text{Eq. (5)} \end{aligned}$$

If $x = a$ or b, then the limit of Equation (3) is interpreted as a one-sided limit with $h \to 0^+$ or $h \to 0^-$, respectively. Then Theorem 1 in Section 3.1 shows that F is continuous for every point of $[a, b]$. This concludes the proof. ■

Fundamental Theorem, Part 2 (The Evaluation Theorem)

We now come to the second part of the Fundamental Theorem of Calculus. This part describes how to evaluate definite integrals without having to calculate limits of Riemann sums. Instead we find and evaluate an antiderivative at the upper and lower limits of integration.

> **THEOREM 4 (Continued) The Fundamental Theorem of Calculus, Part 2**
> If f is continuous at every point of $[a, b]$ and F is any antiderivative of f on $[a, b]$, then
>
> $$\int_a^b f(x)\, dx = F(b) - F(a).$$

Proof Part 1 of the Fundamental Theorem tells us that an antiderivative of f exists, namely

$$G(x) = \int_a^x f(t)\, dt.$$

Thus, if F is *any* antiderivative of f, then $F(x) = G(x) + C$ for some constant C for $a < x < b$ (by Corollary 2 of the Mean Value Theorem for Derivatives, Section 4.2). Since both F and G are continuous on $[a, b]$, we see that $F(x) = G(x) + C$ also holds when $x = a$ and $x = b$ by taking one-sided limits (as $x \to a^+$ and $x \to b^-$).

Evaluating $F(b) - F(a)$, we have

$$\begin{aligned} F(b) - F(a) &= [G(b) + C] - [G(a) + C] \\ &= G(b) - G(a) \\ &= \int_a^b f(t)\,dt - \int_a^a f(t)\,dt \\ &= \int_a^b f(t)\,dt - 0 \\ &= \int_a^b f(t)\,dt. \end{aligned}$$

■

The theorem says that to calculate the definite integral of f over $[a, b]$ all we need to do is:

1. Find an antiderivative F of f, and
2. Calculate the number $\int_a^b f(x)\,dx = F(b) - F(a)$.

The usual notation for $F(b) - F(a)$ is

$$F(x)\Big]_a^b \quad \text{or} \quad \Big[F(x)\Big]_a^b,$$

depending on whether F has one or more terms.

EXAMPLE 5 Evaluating Integrals

(a) $\displaystyle\int_0^{\pi} \cos x\,dx = \sin x\Big]_0^{\pi} = \sin \pi - \sin 0 = 0 - 0 = 0$

(b) $\displaystyle\int_{-\pi/4}^{0} \sec x \tan x\,dx = \sec x\Big]_{-\pi/4}^{0} = \sec 0 - \sec\left(-\frac{\pi}{4}\right) = 1 - \sqrt{2}$

(c) $$\begin{aligned} \int_1^4 \left(\frac{3}{2}\sqrt{x} - \frac{4}{x^2}\right) dx &= \left[x^{3/2} + \frac{4}{x}\right]_1^4 \\ &= \left[(4)^{3/2} + \frac{4}{4}\right] - \left[(1)^{3/2} + \frac{4}{1}\right] \\ &= [8 + 1] - [5] = 4. \end{aligned}$$

■

The process used in Example 5 was much easier than a Riemann sum computation.

The conclusions of the Fundamental Theorem tell us several things. Equation (2) can be rewritten as

$$\frac{d}{dx}\int_a^x f(t)\,dt = \frac{dF}{dx} = f(x),$$

which says that if you first integrate the function f and then differentiate the result, you get the function f back again. Likewise, the equation

$$\int_a^x \frac{dF}{dt}\,dt = \int_a^x f(t)\,dt = F(x) - F(a)$$

says that if you first differentiate the function F and then integrate the result, you get the function F back (adjusted by an integration constant). In a sense, the processes of integration and differentiation are "inverses" of each other. The Fundamental Theorem also says that every continuous function f has an antiderivative F. And it says that the differential equation $dy/dx = f(x)$ has a solution (namely, the function $y = F(x)$) for every continuous function f.

Total Area

The Riemann sum contains terms such as $f(c_k)\,\Delta_k$ which give the area of a rectangle when $f(c_k)$ is positive. When $f(c_k)$ is negative, then the product $f(c_k)\,\Delta_k$ is the negative of the rectangle's area. When we add up such terms for a negative function we get the negative of the area between the curve and the x-axis. If we then take the absolute value, we obtain the correct positive area.

EXAMPLE 6 Finding Area Using Antiderivatives

Calculate the area bounded by the x-axis and the parabola $y = 6 - x - x^2$.

Solution We find where the curve crosses the x-axis by setting

$$y = 0 = 6 - x - x^2 = (3 + x)(2 - x),$$

which gives

$$x = -3 \quad \text{or} \quad x = 2.$$

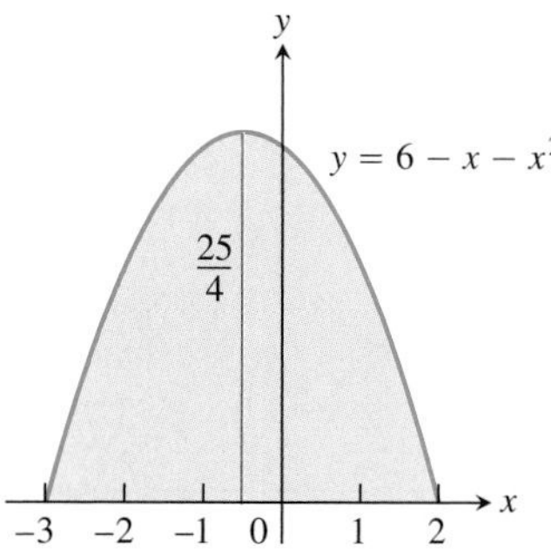

FIGURE 5.21 The area of this parabolic arch is calculated with a definite integral (Example 6).

The curve is sketched in Figure 5.21, and is nonnegative on $[-3, 2]$.

The area is

$$\int_{-3}^{2}(6 - x - x^2)\,dx = \left[6x - \frac{x^2}{2} - \frac{x^3}{3}\right]_{-3}^{2}$$

$$= \left(12 - 2 - \frac{8}{3}\right) - \left(-18 - \frac{9}{2} + \frac{27}{3}\right) = 20\tfrac{5}{6}.$$

The curve in Figure 5.21 is an arch of a parabola, and it is interesting to note that the area under such an arch is exactly equal to two-thirds the base times the altitude:

$$\frac{2}{3}(5)\left(\frac{25}{4}\right) = \frac{125}{6} = 20\tfrac{5}{6}.$$

■

To compute the area of the region bounded by the graph of a function $y = f(x)$ and the x-axis requires more care when the function takes on both positive and negative values. We must be careful to break up the interval $[a, b]$ into subintervals on which the function doesn't change sign. Otherwise we might get cancellation between positive and negative signed areas, leading to an incorrect total. The correct total area is obtained by adding the absolute value of the definite integral over each subinterval where $f(x)$ does not change sign. The term "area" will be taken to mean *total area*.

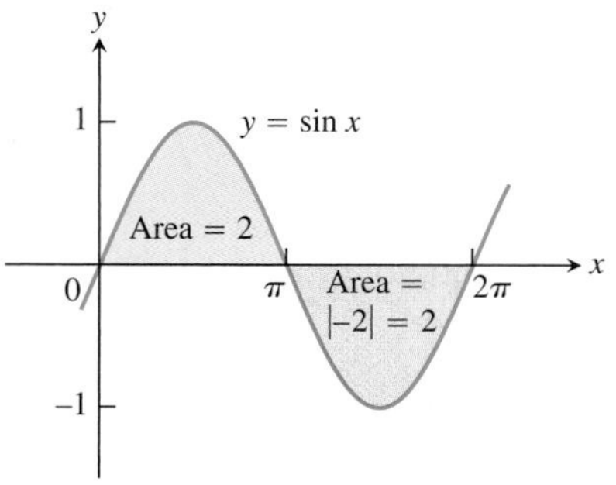

FIGURE 5.22 The total area between $y = \sin x$ and the x-axis for $0 \le x \le 2\pi$ is the sum of the absolute values of two integrals (Example 7).

EXAMPLE 7 Canceling Areas

Figure 5.22 shows the graph of the function $f(x) = \sin x$ between $x = 0$ and $x = 2\pi$. Compute

(a) the definite integral of $f(x)$ over $[0, 2\pi]$.

(b) the area between the graph of $f(x)$ and the x-axis over $[0, 2\pi]$.

Solution The definite integral for $f(x) = \sin x$ is given by

$$\int_0^{2\pi} \sin x\,dx = -\cos x\Big]_0^{2\pi} = -[\cos 2\pi - \cos 0] = -[1 - 1] = 0.$$

The definite integral is zero because the portions of the graph above and below the x-axis make canceling contributions.

The area between the graph of $f(x)$ and the x-axis over $[0, 2\pi]$ is calculated by breaking up the domain of $\sin x$ into two pieces: the interval $[0, \pi]$ over which it is nonnegative and the interval $[\pi, 2\pi]$ over which it is nonpositive.

$$\int_0^{\pi} \sin x\,dx = -\cos x\Big]_0^{\pi} = -[\cos \pi - \cos 0] = -[-1 - 1] = 2.$$

$$\int_{\pi}^{2\pi} \sin x\,dx = -\cos x\Big]_{\pi}^{2\pi} = -[\cos 2\pi - \cos \pi] = -[1 - (-1)] = -2.$$

The second integral gives a negative value. The area between the graph and the axis is obtained by adding the absolute values

$$\text{Area} = |2| + |-2| = 4.$$

■

> **Summary:**
>
> To find the area between the graph of $y = f(x)$ and the x-axis over the interval $[a, b]$, do the following:
>
> 1. Subdivide $[a, b]$ at the zeros of f.
> 2. Integrate f over each subinterval.
> 3. Add the absolute values of the integrals.

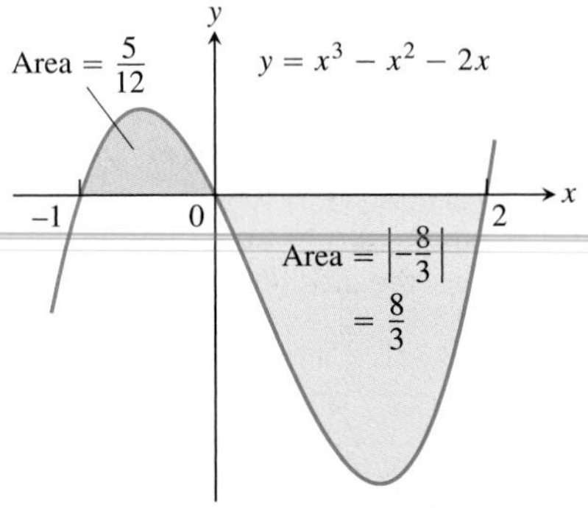

FIGURE 5.23 The region between the curve $y = x^3 - x^2 - 2x$ and the x-axis (Example 8).

EXAMPLE 8 Finding Area Using Antiderivatives

Find the area of the region between the x-axis and the graph of $f(x) = x^3 - x^2 - 2x$, $-1 \le x \le 2$.

Solution First find the zeros of f. Since

$$f(x) = x^3 - x^2 - 2x = x(x^2 - x - 2) = x(x + 1)(x - 2),$$

the zeros are $x = 0, -1$, and 2 (Figure 5.23). The zeros subdivide $[-1, 2]$ into two subintervals: $[-1, 0]$, on which $f \ge 0$, and $[0, 2]$, on which $f \le 0$. We integrate f over each subinterval and add the absolute values of the calculated integrals.

$$\int_{-1}^{0} (x^3 - x^2 - 2x)\,dx = \left[\frac{x^4}{4} - \frac{x^3}{3} - x^2\right]_{-1}^{0} = 0 - \left[\frac{1}{4} + \frac{1}{3} - 1\right] = \frac{5}{12}$$

$$\int_{0}^{2} (x^3 - x^2 - 2x)\,dx = \left[\frac{x^4}{4} - \frac{x^3}{3} - x^2\right]_{0}^{2} = \left[4 - \frac{8}{3} - 4\right] - 0 = -\frac{8}{3}$$

The total enclosed area is obtained by adding the absolute values of the calculated integrals,

$$\text{Total enclosed area} = \frac{5}{12} + \left|-\frac{8}{3}\right| = \frac{37}{12}.$$

■

EXERCISES 5.4

Evaluating Integrals

Evaluate the integrals in Exercises 1–26.

1. $\int_{-2}^{0}(2x+5)\,dx$

2. $\int_{-3}^{4}\left(5-\frac{x}{2}\right)dx$

3. $\int_{0}^{4}\left(3x-\frac{x^3}{4}\right)dx$

4. $\int_{-2}^{2}(x^3-2x+3)\,dx$

5. $\int_{0}^{1}\left(x^2+\sqrt{x}\right)dx$

6. $\int_{0}^{5}x^{3/2}\,dx$

7. $\int_{1}^{32}x^{-6/5}\,dx$

8. $\int_{-2}^{-1}\frac{2}{x^2}\,dx$

9. $\int_{0}^{\pi}\sin x\,dx$

10. $\int_{0}^{\pi}(1+\cos x)\,dx$

11. $\int_{0}^{\pi/3}2\sec^2 x\,dx$

12. $\int_{\pi/6}^{5\pi/6}\csc^2 x\,dx$

13. $\int_{\pi/4}^{3\pi/4}\csc\theta\cot\theta\,d\theta$

14. $\int_{0}^{\pi/3}4\sec u\tan u\,du$

15. $\int_{\pi/2}^{0}\frac{1+\cos 2t}{2}\,dt$

16. $\int_{-\pi/3}^{\pi/3}\frac{1-\cos 2t}{2}\,dt$

17. $\int_{-\pi/2}^{\pi/2}(8y^2+\sin y)\,dy$

18. $\int_{-\pi/3}^{-\pi/4}\left(4\sec^2 t+\frac{\pi}{t^2}\right)dt$

19. $\int_{1}^{-1}(r+1)^2\,dr$

20. $\int_{-\sqrt{3}}^{\sqrt{3}}(t+1)(t^2+4)\,dt$

21. $\int_{\sqrt{2}}^{1}\left(\frac{u^7}{2}-\frac{1}{u^5}\right)du$

22. $\int_{1/2}^{1}\left(\frac{1}{v^3}-\frac{1}{v^4}\right)dv$

23. $\int_{1}^{\sqrt{2}}\frac{s^2+\sqrt{s}}{s^2}\,ds$

24. $\int_{9}^{4}\frac{1-\sqrt{u}}{\sqrt{u}}\,du$

25. $\int_{-4}^{4}|x|\,dx$

26. $\int_{0}^{\pi}\frac{1}{2}(\cos x+|\cos x|)\,dx$

Derivatives of Integrals

Find the derivatives in Exercises 27–30

a. by evaluating the integral and differentiating the result.

b. by differentiating the integral directly.

27. $\frac{d}{dx}\int_{0}^{\sqrt{x}}\cos t\,dt$

28. $\frac{d}{dx}\int_{1}^{\sin x}3t^2\,dt$

29. $\frac{d}{dt}\int_{0}^{t^4}\sqrt{u}\,du$

30. $\frac{d}{d\theta}\int_{0}^{\tan\theta}\sec^2 y\,dy$

Find dy/dx in Exercises 31–36.

31. $y=\int_{0}^{x}\sqrt{1+t^2}\,dt$

32. $y=\int_{1}^{x}\frac{1}{t}\,dt,\quad x>0$

33. $y=\int_{\sqrt{x}}^{0}\sin(t^2)\,dt$

34. $y=\int_{0}^{x^2}\cos\sqrt{t}\,dt$

35. $y=\int_{0}^{\sin x}\frac{dt}{\sqrt{1-t^2}},\quad |x|<\frac{\pi}{2}$

36. $y=\int_{\tan x}^{0}\frac{dt}{1+t^2}$

Area

In Exercises 37–42, find the total area between the region and the x-axis.

37. $y=-x^2-2x,\quad -3\le x\le 2$

38. $y=3x^2-3,\quad -2\le x\le 2$

39. $y=x^3-3x^2+2x,\quad 0\le x\le 2$

40. $y=x^3-4x,\quad -2\le x\le 2$

41. $y=x^{1/3},\quad -1\le x\le 8$

42. $y=x^{1/3}-x,\quad -1\le x\le 8$

Find the areas of the shaded regions in Exercises 43–46.

43.

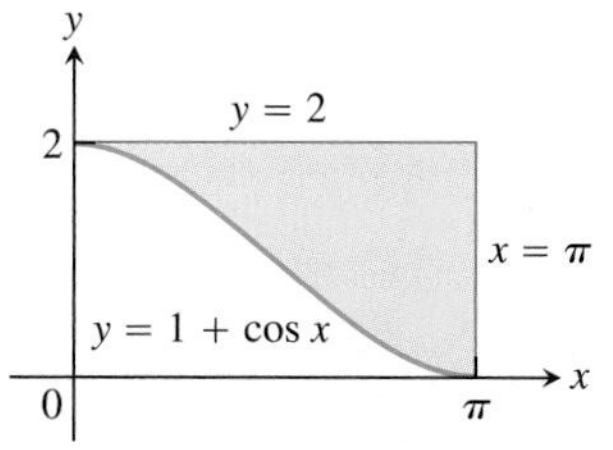

44.

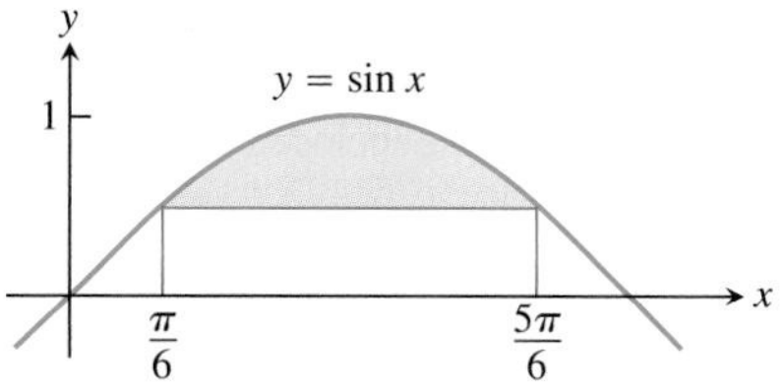

45.

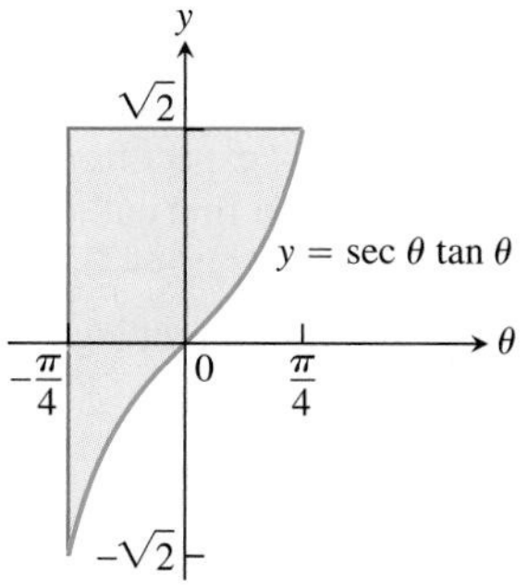

46.

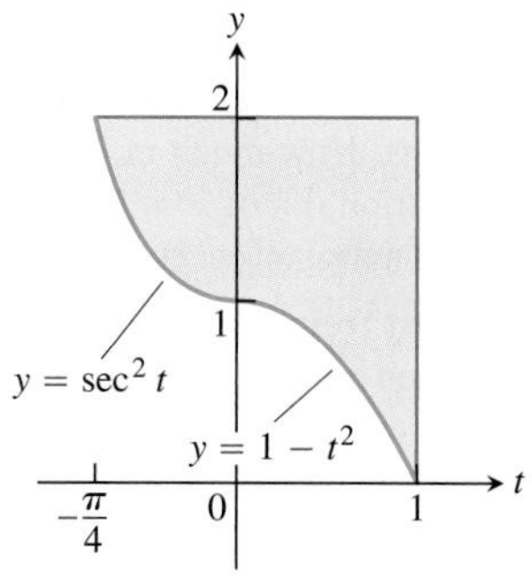

Initial Value Problems

Each of the following functions solves one of the initial value problems in Exercises 47–50. Which function solves which problem? Give brief reasons for your answers.

a. $y = \int_1^x \frac{1}{t}\,dt - 3$ **b.** $y = \int_0^x \sec t\,dt + 4$

c. $y = \int_{-1}^x \sec t\,dt + 4$ **d.** $y = \int_\pi^x \frac{1}{t}\,dt - 3$

47. $\frac{dy}{dx} = \frac{1}{x}, \quad y(\pi) = -3$ **48.** $y' = \sec x, \quad y(-1) = 4$

49. $y' = \sec x, \quad y(0) = 4$ **50.** $y' = \frac{1}{x}, \quad y(1) = -3$

Express the solutions of the initial value problems in Exercises 51–54 in terms of integrals.

51. $\frac{dy}{dx} = \sec x, \quad y(2) = 3$

52. $\frac{dy}{dx} = \sqrt{1 + x^2}, \quad y(1) = -2$

53. $\frac{ds}{dt} = f(t), \quad s(t_0) = s_0$

54. $\frac{dv}{dt} = g(t), \quad v(t_0) = v_0$

Applications

55. Archimedes' area formula for parabolas Archimedes (287–212 B.C.), inventor, military engineer, physicist, and the greatest mathematician of classical times in the Western world, discovered that the area under a parabolic arch is two-thirds the base times the height. Sketch the parabolic arch $y = h - (4h/b^2)x^2$, $-b/2 \le x \le b/2$, assuming that h and b are positive. Then use calculus to find the area of the region enclosed between the arch and the x-axis.

56. Revenue from marginal revenue Suppose that a company's marginal revenue from the manufacture and sale of egg beaters is

$$\frac{dr}{dx} = 2 - 2/(x + 1)^2,$$

where r is measured in thousands of dollars and x in thousands of units. How much money should the company expect from a production run of $x = 3$ thousand egg beaters? To find out, integrate the marginal revenue from $x = 0$ to $x = 3$.

57. Cost from marginal cost The marginal cost of printing a poster when x posters have been printed is

$$\frac{dc}{dx} = \frac{1}{2\sqrt{x}}$$

dollars. Find $c(100) - c(1)$, the cost of printing posters 2–100.

58. (*Continuation of Exercise 57.*) Find $c(400) - c(100)$, the cost of printing posters 101–400.

Drawing Conclusions About Motion from Graphs

59. Suppose that f is the differentiable function shown in the accompanying graph and that the position at time t (sec) of a particle moving along a coordinate axis is

$$s = \int_0^t f(x)\,dx$$

meters. Use the graph to answer the following questions. Give reasons for your answers.

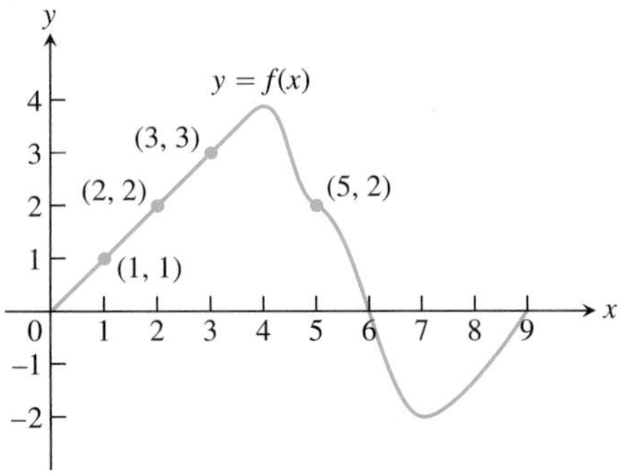

a. What is the particle's velocity at time $t = 5$?

b. Is the acceleration of the particle at time $t = 5$ positive, or negative?

c. What is the particle's position at time $t = 3$?

d. At what time during the first 9 sec does s have its largest value?

e. Approximately when is the acceleration zero?

f. When is the particle moving toward the origin? away from the origin?

g. On which side of the origin does the particle lie at time $t = 9$?

60. Suppose that g is the differentiable function graphed here and that the position at time t (sec) of a particle moving along a coordinate axis is

$$s = \int_0^t g(x)\,dx$$

meters. Use the graph to answer the following questions. Give reasons for your answers.

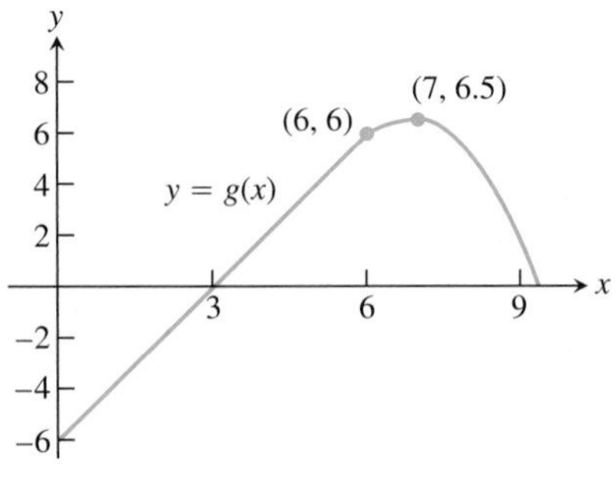

a. What is the particle's velocity at $t = 3$?

b. Is the acceleration at time $t = 3$ positive, or negative?

c. What is the particle's position at time $t = 3$?

d. When does the particle pass through the origin?

e. When is the acceleration zero?

f. When is the particle moving away from the origin? toward the origin?

g. On which side of the origin does the particle lie at $t = 9$?

Theory and Examples

61. Show that if k is a positive constant, then the area between the x-axis and one arch of the curve $y = \sin kx$ is $2/k$.

62. Find

$$\lim_{x \to 0} \frac{1}{x^3} \int_0^x \frac{t^2}{t^4 + 1}\, dt.$$

63. Suppose $\int_1^x f(t)\, dt = x^2 - 2x + 1$. Find $f(x)$.

64. Find $f(4)$ if $\int_0^x f(t)\, dt = x \cos \pi x$.

65. Find the linearization of

$$f(x) = 2 - \int_2^{x+1} \frac{9}{1 + t}\, dt$$

at $x = 1$.

66. Find the linearization of

$$g(x) = 3 + \int_1^{x^2} \sec(t - 1)\, dt$$

at $x = -1$.

67. Suppose that f has a positive derivative for all values of x and that $f(1) = 0$. Which of the following statements must be true of the function

$$g(x) = \int_0^x f(t)\, dt?$$

Give reasons for your answers.

a. g is a differentiable function of x.

b. g is a continuous function of x.

c. The graph of g has a horizontal tangent at $x = 1$.

d. g has a local maximum at $x = 1$.

e. g has a local minimum at $x = 1$.

f. The graph of g has an inflection point at $x = 1$.

g. The graph of dg/dx crosses the x-axis at $x = 1$.

68. Another proof of The Evaluation Theorem

a. Let $a = x_0 < x_1 < x_2 < \cdots < x_n = b$ be any partition of $[a, b]$, and let F be any antiderivative of f. Show that

$$F(b) - F(a) = \sum_{i=1}^{n} [F(x_i) - F(x_{i-1})].$$

b. Apply the Mean Value Theorem to each term to show $F(x_i) - F(x_{i-1}) = f(c_i)(x_i - x_{i-1})$ for some c_i in the interval (x_{i-1}, x_i). Then show that $F(b) - F(a)$ is a Riemann sum for f on $[a, b]$.

c. From part b and the definition of the definite integral, show that

$$F(b) - F(a) = \int_a^b f(x)\, dx.$$

T **69. The Fundamental Theorem** If f is continuous, we expect

$$\lim_{h \to 0} \frac{1}{h} \int_x^{x+h} f(t)\, dt$$

to equal $f(x)$, as in the proof of Part 1 of the Fundamental Theorem. For instance, if $f(t) = \cos t$, then

$$\frac{1}{h} \int_x^{x+h} \cos t\, dt = \frac{\sin(x + h) - \sin x}{h}. \tag{6}$$

The right-hand side of Equation (7) is the difference quotient for the derivative of the sine, and we expect its limit as $h \to 0$ to be $\cos x$.

Graph $\cos x$ for $-\pi \le x \le 2\pi$. Then, in a different color if possible, graph the right-hand side of Equation (6) as a function of x for $h = 2, 1, 0.5$, and 0.1. Watch how the latter curves converge to the graph of the cosine as $h \to 0$.

T **70.** Repeat Exercise 69 for $f(t) = 3t^2$. What is

$$\lim_{h \to 0} \frac{1}{h} \int_x^{x+h} 3t^2\, dt = \lim_{h \to 0} \frac{(x + h)^3 - x^3}{h}?$$

Graph $f(x) = 3x^2$ for $-1 \le x \le 1$. Then graph the quotient $((x + h)^3 - x^3)/h$ as a function of x for $h = 1, 0.5, 0.2$, and 0.1. Watch how the latter curves converge to the graph of $3x^2$ as $h \to 0$.

COMPUTER EXPLORATIONS

In Exercises 71–74, let $F(x) = \int_a^x f(t)\, dt$ for the specified function f and interval $[a, b]$. Use a CAS to perform the following steps and answer the questions posed.

a. Plot the functions f and F together over $[a, b]$.

b. Solve the equation $F'(x) = 0$. What can you see to be true about the graphs of f and F at points where $F'(x) = 0$? Is your observation borne out by Part 1 of the Fundamental Theorem coupled with information provided by the first derivative? Explain your answer.

c. Over what intervals (approximately) is the function F increasing and decreasing? What is true about f over those intervals?

d. Calculate the derivative f' and plot it together with F. What can you see to be true about the graph of F at points where $f'(x) = 0$? Is your observation borne out by Part 1 of the Fundamental Theorem? Explain your answer.

71. $f(x) = x^3 - 4x^2 + 3x, \quad [0, 4]$

72. $f(x) = 2x^4 - 17x^3 + 46x^2 - 43x + 12, \quad \left[0, \frac{9}{2}\right]$

73. $f(x) = \sin 2x \cos \frac{x}{3}, \quad [0, 2\pi]$

74. $f(x) = x \cos \pi x, \quad [0, 2\pi]$

In Exercises 75–78, let $F(x) = \int_a^{u(x)} f(t)\,dt$ for the specified a, u, and f. Use a CAS to perform the following steps and answer the questions posed.

a. Find the domain of F.

b. Calculate $F'(x)$ and determine its zeros. For what points in its domain is F increasing? decreasing?

c. Calculate $F''(x)$ and determine its zero. Identify the local extrema and the points of inflection of F.

d. Using the information from parts (a)–(c), draw a rough hand-sketch of $y = F(x)$ over its domain. Then graph $F(x)$ on your CAS to support your sketch.

75. $a = 1, \quad u(x) = x^2, \quad f(x) = \sqrt{1 - x^2}$

76. $a = 0, \quad u(x) = x^2, \quad f(x) = \sqrt{1 - x^2}$

77. $a = 0, \quad u(x) = 1 - x, \quad f(x) = x^2 - 2x - 3$

78. $a = 0, \quad u(x) = 1 - x^2, \quad f(x) = x^2 - 2x - 3$

In Exercises 79 and 80, assume that f is continuous and $u(x)$ is twice-differentiable.

79. Calculate $\frac{d}{dx}\int_a^{u(x)} f(t)\,dt$ and check your answer using a CAS.

80. Calculate $\frac{d^2}{dx^2}\int_a^{u(x)} f(t)\,dt$ and check your answer using a CAS.

5.5 Indefinite Integrals and the Substitution Rule

A definite integral is a number defined by taking the limit of Riemann sums associated with partitions of a finite closed interval whose norms go to zero. The Fundamental Theorem of Calculus says that a definite integral of a continuous function can be computed easily if we can find an antiderivative of the function. Antiderivatives generally turn out to be more difficult to find than derivatives. However, it is well worth the effort to learn techniques for computing them.

Recall from Section 4.8 that the set of *all* antiderivatives of the function f is called the **indefinite integral** of f with respect to x, and is symbolized by

$$\int f(x)\,dx.$$

The connection between antiderivatives and the definite integral stated in the Fundamental Theorem now explains this notation. When finding the indefinite integral of a function f, remember that it always includes an arbitrary constant C.

We must distinguish carefully between definite and indefinite integrals. A definite integral $\int_a^b f(x)\,dx$ is a *number*. An indefinite integral $\int f(x)\,dx$ is a *function* plus an arbitrary constant C.

So far, we have only been able to find antiderivatives of functions that are clearly recognizable as derivatives. In this section we begin to develop more general techniques for finding antiderivatives. The first integration techniques we develop are obtained by inverting rules for finding derivatives, such as the Power Rule and the Chain Rule.

The Power Rule in Integral Form

If u is a differentiable function of x and n is a rational number different from -1, the Chain Rule tells us that

$$\frac{d}{dx}\left(\frac{u^{n+1}}{n+1}\right) = u^n \frac{du}{dx}.$$

From another point of view, this same equation says that $u^{n+1}/(n+1)$ is one of the antiderivatives of the function $u^n(du/dx)$. Therefore,

$$\int \left(u^n \frac{du}{dx}\right) dx = \frac{u^{n+1}}{n+1} + C.$$

The integral on the left-hand side of this equation is usually written in the simpler "differential" form,

$$\int u^n \, du,$$

obtained by treating the dx's as differentials that cancel. We are thus led to the following rule.

> If u is any differentiable function, then
>
> $$\int u^n \, du = \frac{u^{n+1}}{n+1} + C \qquad (n \neq -1,\ n \text{ rational}). \tag{1}$$

Equation (1) actually holds for any real exponent $n \neq -1$, as we see in Chapter 7.

In deriving Equation (1), we assumed u to be a differentiable function of the variable x, but the name of the variable does not matter and does not appear in the final formula. We could have represented the variable with θ, t, y, or any other letter. Equation (1) says that whenever we can cast an integral in the form

$$\int u^n \, du, \qquad (n \neq -1),$$

with u a differentiable function and du its differential, we can evaluate the integral as $[u^{n+1}/(n+1)] + C$.

EXAMPLE 1 Using the Power Rule

$$\int \sqrt{1+y^2} \cdot 2y \, dy = \int \sqrt{u} \cdot \left(\frac{du}{dy}\right) dy \qquad \text{Let } u = 1 + y^2,\ du/dy = 2y$$

$$= \int u^{1/2} \, du$$

$$= \frac{u^{(1/2)+1}}{(1/2)+1} + C \qquad \text{Integrate, using Eq. (1) with } n = 1/2.$$

$$= \frac{2}{3} u^{3/2} + C \qquad \text{Simpler form}$$

$$= \frac{2}{3}(1+y^2)^{3/2} + C \qquad \text{Replace } u \text{ by } 1 + y^2. \quad \blacksquare$$

EXAMPLE 2 Adjusting the Integrand by a Constant

$$\begin{aligned}
\int \sqrt{4t-1}\,dt &= \int \frac{1}{4}\cdot\sqrt{4t-1}\cdot 4\,dt \\
&= \frac{1}{4}\int \sqrt{u}\cdot\left(\frac{du}{dt}\right)dt && \text{Let } u = 4t-1,\ du/dt = 4. \\
&= \frac{1}{4}\int u^{1/2}\,du && \text{With the 1/4 out front, the integral is now in standard form.} \\
&= \frac{1}{4}\cdot\frac{u^{3/2}}{3/2} + C && \text{Integrate, using Eq. (1) with } n = 1/2. \\
&= \frac{1}{6}u^{3/2} + C && \text{Simpler form} \\
&= \frac{1}{6}(4t-1)^{3/2} + C && \text{Replace } u \text{ by } 4t-1.
\end{aligned}$$

■

Substitution: Running the Chain Rule Backwards

The substitutions in Examples 1 and 2 are instances of the following general rule.

THEOREM 5 The Substitution Rule

If $u = g(x)$ is a differentiable function whose range is an interval I and f is continuous on I, then

$$\int f(g(x))g'(x)\,dx = \int f(u)\,du.$$

Proof The rule is true because, by the Chain Rule, $F(g(x))$ is an antiderivative of $f(g(x))\cdot g'(x)$ whenever F is an antiderivative of f:

$$\begin{aligned}
\frac{d}{dx}F(g(x)) &= F'(g(x))\cdot g'(x) && \text{Chain Rule} \\
&= f(g(x))\cdot g'(x). && \text{Because } F' = f
\end{aligned}$$

If we make the substitution $u = g(x)$ then

$$\begin{aligned}
\int f(g(x))g'(x)\,dx &= \int \frac{d}{dx}F(g(x))\,dx \\
&= F(g(x)) + C && \text{Fundamental Theorem} \\
&= F(u) + C && u = g(x) \\
&= \int F'(u)\,du && \text{Fundamental Theorem} \\
&= \int f(u)\,du && F' = f
\end{aligned}$$

■

The Substitution Rule provides the following method to evaluate the integral

$$\int f(g(x))g'(x)\,dx,$$

when f and g' are continuous functions:

1. Substitute $u = g(x)$ and $du = g'(x)\,dx$ to obtain the integral

$$\int f(u)\,du.$$

2. Integrate with respect to u.
3. Replace u by $g(x)$ in the result.

EXAMPLE 3 Using Substitution

$$\int \cos(7\theta + 5)\,d\theta = \int \cos u \cdot \frac{1}{7}\,du$$

Let $u = 7\theta + 5$, $du = 7\,d\theta$, $(1/7)\,du = d\theta$.

$$= \frac{1}{7}\int \cos u\,du$$

With the (1/7) out front, the integral is now in standard form.

$$= \frac{1}{7}\sin u + C$$

Integrate with respect to u, Table 4.2.

$$= \frac{1}{7}\sin(7\theta + 5) + C$$

Replace u by $7\theta + 5$.

We can verify this solution by differentiating and checking that we obtain the original function $\cos(7\theta + 5)$. ■

EXAMPLE 4 Using Substitution

$$\int x^2 \sin(x^3)\,dx = \int \sin(x^3) \cdot x^2\,dx$$

$$= \int \sin u \cdot \frac{1}{3}\,du$$

Let $u = x^3$, $du = 3x^2\,dx$, $(1/3)\,du = x^2\,dx$.

$$= \frac{1}{3}\int \sin u\,du$$

$$= \frac{1}{3}(-\cos u) + C$$

Integrate with respect to u.

$$= -\frac{1}{3}\cos(x^3) + C$$

Replace u by x^3. ■

EXAMPLE 5 Using Identities and Substitution

$$\begin{aligned}
\int \frac{1}{\cos^2 2x}\,dx &= \int \sec^2 2x\,dx && \frac{1}{\cos 2x} = \sec 2x \\
&= \int \sec^2 u \cdot \frac{1}{2}\,du && u = 2x,\ du = 2\,dx,\ dx = (1/2)\,du \\
&= \frac{1}{2}\int \sec^2 u\,du \\
&= \frac{1}{2}\tan u + C && \frac{d}{du}\tan u = \sec^2 u \\
&= \frac{1}{2}\tan 2x + C && u = 2x
\end{aligned}$$

■

The success of the substitution method depends on finding a substitution that changes an integral we cannot evaluate directly into one that we can. If the first substitution fails, try to simplify the integrand further with an additional substitution or two (see Exercises 49 and 50). Alternatively, we can start fresh. There can be more than one good way to start, as in the next example.

EXAMPLE 6 Using Different Substitutions

Evaluate

$$\int \frac{2z\,dz}{\sqrt[3]{z^2+1}}.$$

Solution We can use the substitution method of integration as an exploratory tool: Substitute for the most troublesome part of the integrand and see how things work out. For the integral here, we might try $u = z^2 + 1$ or we might even press our luck and take u to be the entire cube root. Here is what happens in each case.

Solution 1: Substitute $u = z^2 + 1$.

$$\begin{aligned}
\int \frac{2z\,dz}{\sqrt[3]{z^2+1}} &= \int \frac{du}{u^{1/3}} && \text{Let } u = z^2+1,\ du = 2z\,dz. \\
&= \int u^{-1/3}\,du && \text{In the form } \textstyle\int u^n\,du \\
&= \frac{u^{2/3}}{2/3} + C && \text{Integrate with respect to } u. \\
&= \frac{3}{2}u^{2/3} + C \\
&= \frac{3}{2}(z^2+1)^{2/3} + C && \text{Replace } u \text{ by } z^2+1.
\end{aligned}$$

Solution 2: Substitute $u = \sqrt[3]{z^2 + 1}$ instead.

$$\int \frac{2z\,dz}{\sqrt[3]{z^2+1}} = \int \frac{3u^2\,du}{u} \qquad \begin{array}{l}\text{Let } u = \sqrt[3]{z^2+1},\\ u^3 = z^2 + 1,\\ 3u^2\,du = 2z\,dz.\end{array}$$

$$= 3\int u\,du$$

$$= 3 \cdot \frac{u^2}{2} + C \qquad \text{Integrate with respect to } u.$$

$$= \frac{3}{2}(z^2+1)^{2/3} + C \qquad \text{Replace } u \text{ by } (z^2+1)^{1/3}.$$ ■

The Integrals of $\sin^2 x$ and $\cos^2 x$

Sometimes we can use trigonometric identities to transform integrals we do not know how to evaluate into ones we can using the substitution rule. Here is an example giving the integral formulas for $\sin^2 x$ and $\cos^2 x$ which arise frequently in applications.

EXAMPLE 7

(a) $$\int \sin^2 x\,dx = \int \frac{1-\cos 2x}{2}\,dx \qquad \sin^2 x = \frac{1-\cos 2x}{2}$$

$$= \frac{1}{2}\int (1-\cos 2x)\,dx = \frac{1}{2}\int dx - \frac{1}{2}\int \cos 2x\,dx$$

$$= \frac{1}{2}x - \frac{1}{2}\frac{\sin 2x}{2} + C = \frac{x}{2} - \frac{\sin 2x}{4} + C$$

(b) $$\int \cos^2 x\,dx = \int \frac{1+\cos 2x}{2}\,dx \qquad \cos^2 x = \frac{1+\cos 2x}{2}$$

$$= \frac{x}{2} + \frac{\sin 2x}{4} + C \qquad \text{As in part (a), but with a sign change}$$ ■

EXAMPLE 8 Area Beneath the Curve $y = \sin^2 x$

Figure 5.24 shows the graph of $g(x) = \sin^2 x$ over the interval $[0, 2\pi]$. Find

(a) the definite integral of $g(x)$ over $[0, 2\pi]$.

(b) the area between the graph of the function and the x-axis over $[0, 2\pi]$.

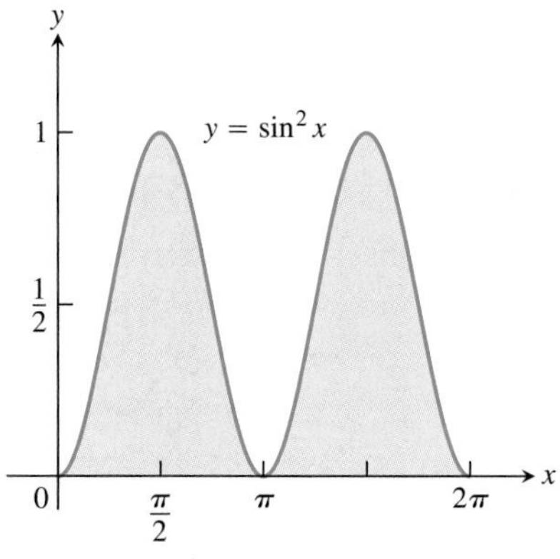

FIGURE 5.24 The area beneath the curve $y = \sin^2 x$ over $[0, 2\pi]$ equals π square units (Example 8).

Solution

(a) From Example 7(a), the definite integral is

$$\int_0^{2\pi} \sin^2 x\,dx = \left[\frac{x}{2} - \frac{\sin 2x}{4}\right]_0^{2\pi} = \left[\frac{2\pi}{2} - \frac{\sin 4\pi}{4}\right] - \left[\frac{0}{2} - \frac{\sin 0}{4}\right]$$

$$= [\pi - 0] - [0 - 0] = \pi.$$

(b) The function $\sin^2 x$ is nonnegative, so the area is equal to the definite integral, or π. ■

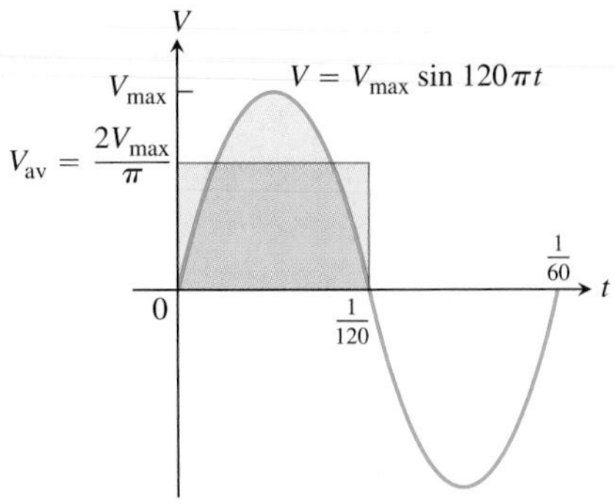

FIGURE 5.25 The graph of the voltage $V = V_{max} \sin 120\pi t$ over a full cycle. Its average value over a half-cycle is $2V_{max}/\pi$. Its average value over a full cycle is zero (Example 9).

EXAMPLE 9 Household Electricity

We can model the voltage in our home wiring with the sine function

$$V = V_{max} \sin 120\pi t,$$

which expresses the voltage V in volts as a function of time t in seconds. The function runs through 60 cycles each second (its frequency is 60 hertz, or 60 Hz). The positive constant V_{max} ("vee max") is the **peak voltage**.

The average value of V over the half-cycle from 0 to $1/120$ sec (see Figure 5.25) is

$$\begin{aligned} V_{av} &= \frac{1}{(1/120) - 0} \int_0^{1/120} V_{max} \sin 120\pi t \, dt \\ &= 120 V_{max} \left[-\frac{1}{120\pi} \cos 120\pi t \right]_0^{1/120} \\ &= \frac{V_{max}}{\pi} [-\cos \pi + \cos 0] \\ &= \frac{2V_{max}}{\pi}. \end{aligned}$$

The average value of the voltage over a full cycle is zero, as we can see from Figure 5.25. (Also see Exercise 63.) If we measured the voltage with a standard moving-coil galvanometer, the meter would read zero.

To measure the voltage effectively, we use an instrument that measures the square root of the average value of the square of the voltage, namely

$$V_{rms} = \sqrt{(V^2)_{av}}.$$

The subscript "rms" (read the letters separately) stands for "root mean square." Since the average value of $V^2 = (V_{max})^2 \sin^2 120\pi t$ over a cycle is

$$(V^2)_{av} = \frac{1}{(1/60) - 0} \int_0^{1/60} (V_{max})^2 \sin^2 120\pi t \, dt = \frac{(V_{max})^2}{2},$$

(Exercise 63, part c), the rms voltage is

$$V_{rms} = \sqrt{\frac{(V_{max})^2}{2}} = \frac{V_{max}}{\sqrt{2}}.$$

The values given for household currents and voltages are always rms values. Thus, "115 volts ac" means that the rms voltage is 115. The peak voltage, obtained from the last equation, is

$$V_{max} = \sqrt{2}\, V_{rms} = \sqrt{2} \cdot 115 \approx 163 \text{ volts},$$

which is considerably higher. ■

EXERCISES 5.5

Evaluating Integrals

Evaluate the indefinite integrals in Exercises 1–12 by using the given substitutions to reduce the integrals to standard form.

1. $\int \sin 3x \, dx, \quad u = 3x$

2. $\int x \sin (2x^2) \, dx, \quad u = 2x^2$

3. $\int \sec 2t \tan 2t \, dt, \quad u = 2t$

4. $\int \left(1 - \cos \frac{t}{2}\right)^2 \sin \frac{t}{2} \, dt, \quad u = 1 - \cos \frac{t}{2}$

5. $\int 28(7x-2)^{-5}\,dx, \quad u = 7x - 2$

6. $\int x^3(x^4-1)^2\,dx, \quad u = x^4 - 1$

7. $\int \frac{9r^2\,dr}{\sqrt{1-r^3}}, \quad u = 1 - r^3$

8. $\int 12(y^4+4y^2+1)^2(y^3+2y)\,dy, \quad u = y^4 + 4y^2 + 1$

9. $\int \sqrt{x}\sin^2(x^{3/2}-1)\,dx, \quad u = x^{3/2} - 1$

10. $\int \frac{1}{x^2}\cos^2\left(\frac{1}{x}\right)dx, \quad u = -\frac{1}{x}$

11. $\int \csc^2 2\theta \cot 2\theta\,d\theta$

a. Using $u = \cot 2\theta$ b. Using $u = \csc 2\theta$

12. $\int \frac{dx}{\sqrt{5x+8}}$

a. Using $u = 5x + 8$ b. Using $u = \sqrt{5x+8}$

Evaluate the integrals in Exercises 13–48.

13. $\int \sqrt{3-2s}\,ds$

14. $\int (2x+1)^3\,dx$

15. $\int \frac{1}{\sqrt{5s+4}}\,ds$

16. $\int \frac{3\,dx}{(2-x)^2}$

17. $\int \theta\sqrt[4]{1-\theta^2}\,d\theta$

18. $\int 8\theta\sqrt[3]{\theta^2-1}\,d\theta$

19. $\int 3y\sqrt{7-3y^2}\,dy$

20. $\int \frac{4y\,dy}{\sqrt{2y^2+1}}$

21. $\int \frac{1}{\sqrt{x}(1+\sqrt{x})^2}\,dx$

22. $\int \frac{(1+\sqrt{x})^3}{\sqrt{x}}\,dx$

23. $\int \cos(3z+4)\,dz$

24. $\int \sin(8z-5)\,dz$

25. $\int \sec^2(3x+2)\,dx$

26. $\int \tan^2 x \sec^2 x\,dx$

27. $\int \sin^5\frac{x}{3}\cos\frac{x}{3}\,dx$

28. $\int \tan^7\frac{x}{2}\sec^2\frac{x}{2}\,dx$

29. $\int r^2\left(\frac{r^3}{18}-1\right)^5 dr$

30. $\int r^4\left(7-\frac{r^5}{10}\right)^3 dr$

31. $\int x^{1/2}\sin(x^{3/2}+1)\,dx$

32. $\int x^{1/3}\sin(x^{4/3}-8)\,dx$

33. $\int \sec\left(v+\frac{\pi}{2}\right)\tan\left(v+\frac{\pi}{2}\right)dv$

34. $\int \csc\left(\frac{v-\pi}{2}\right)\cot\left(\frac{v-\pi}{2}\right)dv$

35. $\int \frac{\sin(2t+1)}{\cos^2(2t+1)}\,dt$

36. $\int \frac{6\cos t}{(2+\sin t)^3}\,dt$

37. $\int \sqrt{\cot y}\csc^2 y\,dy$

38. $\int \frac{\sec z\tan z}{\sqrt{\sec z}}\,dz$

39. $\int \frac{1}{t^2}\cos\left(\frac{1}{t}-1\right)dt$

40. $\int \frac{1}{\sqrt{t}}\cos(\sqrt{t}+3)\,dt$

41. $\int \frac{1}{\theta^2}\sin\frac{1}{\theta}\cos\frac{1}{\theta}\,d\theta$

42. $\int \frac{\cos\sqrt{\theta}}{\sqrt{\theta}\sin^2\sqrt{\theta}}\,d\theta$

43. $\int (s^3+2s^2-5s+5)(3s^2+4s-5)\,ds$

44. $\int (\theta^4-2\theta^2+8\theta-2)(\theta^3-\theta+2)\,d\theta$

45. $\int t^3(1+t^4)^3\,dt$

46. $\int \sqrt{\frac{x-1}{x^5}}\,dx$

47. $\int x^3\sqrt{x^2+1}\,dx$

48. $\int 3x^5\sqrt{x^3+1}\,dx$

Simplifying Integrals Step by Step

If you do not know what substitution to make, try reducing the integral step by step, using a trial substitution to simplify the integral a bit and then another to simplify it some more. You will see what we mean if you try the sequences of substitutions in Exercises 49 and 50.

49. $\int \frac{18\tan^2 x\sec^2 x}{(2+\tan^3 x)^2}\,dx$

a. $u = \tan x$, followed by $v = u^3$, then by $w = 2 + v$

b. $u = \tan^3 x$, followed by $v = 2 + u$

c. $u = 2 + \tan^3 x$

50. $\int \sqrt{1+\sin^2(x-1)}\sin(x-1)\cos(x-1)\,dx$

a. $u = x - 1$, followed by $v = \sin u$, then by $w = 1 + v^2$

b. $u = \sin(x-1)$, followed by $v = 1 + u^2$

c. $u = 1 + \sin^2(x-1)$

Evaluate the integrals in Exercises 51 and 52.

51. $\int \frac{(2r-1)\cos\sqrt{3(2r-1)^2+6}}{\sqrt{3(2r-1)^2+6}}\,dr$

52. $\int \frac{\sin\sqrt{\theta}}{\sqrt{\theta}\cos^3\sqrt{\theta}}\,d\theta$

Initial Value Problems

Solve the initial value problems in Exercises 53–58.

53. $\frac{ds}{dt} = 12t(3t^2-1)^3, \quad s(1) = 3$

54. $\frac{dy}{dx} = 4x(x^2+8)^{-1/3}, \quad y(0) = 0$

55. $\frac{ds}{dt} = 8\sin^2\left(t+\frac{\pi}{12}\right), \quad s(0) = 8$

56. $\frac{dr}{d\theta} = 3\cos^2\left(\frac{\pi}{4}-\theta\right), \quad r(0) = \frac{\pi}{8}$

57. $\frac{d^2s}{dt^2} = -4\sin\left(2t - \frac{\pi}{2}\right), \quad s'(0) = 100, \quad s(0) = 0$

58. $\frac{d^2y}{dx^2} = 4\sec^2 2x \tan 2x, \quad y'(0) = 4, \quad y(0) = -1$

59. The velocity of a particle moving back and forth on a line is $v = ds/dt = 6\sin 2t$ m/sec for all t. If $s = 0$ when $t = 0$, find the value of s when $t = \pi/2$ sec.

60. The acceleration of a particle moving back and forth on a line is $a = d^2s/dt^2 = \pi^2 \cos \pi t$ m/sec^2 for all t. If $s = 0$ and $v = 8$ m/sec when $t = 0$, find s when $t = 1$ sec.

Theory and Examples

61. It looks as if we can integrate $2\sin x \cos x$ with respect to x in three different ways:

a. $\int 2\sin x\cos x\,dx = \int 2u\,du \qquad u = \sin x,$

$= u^2 + C_1 = \sin^2 x + C_1$

b. $\int 2\sin x\cos x\,dx = \int -2u\,du \qquad u = \cos x,$

$= -u^2 + C_2 = -\cos^2 x + C_2$

c. $\int 2\sin x\cos x\,dx = \int \sin 2x\,dx \qquad 2\sin x\cos x = \sin 2x$

$= -\frac{\cos 2x}{2} + C_3.$

Can all three integrations be correct? Give reasons for your answer.

62. The substitution $u = \tan x$ gives

$$\int \sec^2 x \tan x\,dx = \int u\,du = \frac{u^2}{2} + C = \frac{\tan^2 x}{2} + C.$$

The substitution $u = \sec x$ gives

$$\int \sec^2 x \tan x\,dx = \int u\,du = \frac{u^2}{2} + C = \frac{\sec^2 x}{2} + C.$$

Can both integrations be correct? Give reasons for your answer.

63. (*Continuation of Example 9.*)

a. Show by evaluating the integral in the expression

$$\frac{1}{(1/60) - 0}\int_0^{1/60} V_{max}\sin 120\,\pi t\,dt$$

that the average value of $V = V_{max}\sin 120\,\pi t$ over a full cycle is zero.

b. The circuit that runs your electric stove is rated 240 volts rms. What is the peak value of the allowable voltage?

c. Show that

$$\int_0^{1/60} (V_{max})^2 \sin^2 120\,\pi t\,dt = \frac{(V_{max})^2}{120}.$$

5.6 Substitution and Area Between Curves

There are two methods for evaluating a definite integral by substitution. The first method is to find an antiderivative using substitution, and then to evaluate the definite integral by applying the Fundamental Theorem. We used this method in Examples 8 and 9 of the preceding section. The second method extends the process of substitution directly to *definite* integrals. We apply the new formula introduced here to the problem of computing the area between two curves.

Substitution Formula

In the following formula, the limits of integration change when the variable of integration is changed by substitution.

THEOREM 6 Substitution in Definite Integrals
If g' is continuous on the interval $[a, b]$ and f is continuous on the range of g, then

$$\int_a^b f(g(x))\cdot g'(x)\,dx = \int_{g(a)}^{g(b)} f(u)\,du$$

Proof Let F denote any antiderivative of f. Then,

$$\begin{aligned}
\int_a^b f(g(x)) \cdot g'(x)\,dx &= F(g(x))\Big]_{x=a}^{x=b} && \frac{d}{dx}F(g(x)) = F'(g(x))g'(x) = f(g(x))g'(x) \\
&= F(g(b)) - F(g(a)) \\
&= F(u)\Big]_{u=g(a)}^{u=g(b)} \\
&= \int_{g(a)}^{g(b)} f(u)\,du. && \text{Fundamental Theorem, Part 2}
\end{aligned}$$

■

To use the formula, make the same u-substitution $u = g(x)$ and $du = g'(x)\,dx$ you would use to evaluate the corresponding indefinite integral. Then integrate the transformed integral with respect to u from the value $g(a)$ (the value of u at $x = a$) to the value $g(b)$ (the value of u at $x = b$).

EXAMPLE 1 Substitution by Two Methods

Evaluate $\int_{-1}^{1} 3x^2\sqrt{x^3+1}\,dx$.

Solution We have two choices.

Method 1: Transform the integral and evaluate the transformed integral with the transformed limits given in Theorem 6.

$$\begin{aligned}
&\int_{-1}^{1} 3x^2\sqrt{x^3+1}\,dx \\
&= \int_0^2 \sqrt{u}\,du && \text{Let } u = x^3+1,\ du = 3x^2\,dx. \text{ When } x=-1,\ u=(-1)^3+1=0. \text{ When } x=1,\ u=(1)^3+1=2. \\
&= \frac{2}{3}u^{3/2}\Big]_0^2 && \text{Evaluate the new definite integral.} \\
&= \frac{2}{3}\left[2^{3/2} - 0^{3/2}\right] = \frac{2}{3}\left[2\sqrt{2}\right] = \frac{4\sqrt{2}}{3}
\end{aligned}$$

Method 2: Transform the integral as an indefinite integral, integrate, change back to x, and use the original x-limits.

$$\begin{aligned}
\int 3x^2\sqrt{x^3+1}\,dx &= \int \sqrt{u}\,du && \text{Let } u = x^3+1,\ du = 3x^2\,dx. \\
&= \frac{2}{3}u^{3/2} + C && \text{Integrate with respect to } u. \\
&= \frac{2}{3}(x^3+1)^{3/2} + C && \text{Replace } u \text{ by } x^3+1. \\
\int_{-1}^{1} 3x^2\sqrt{x^3+1}\,dx &= \frac{2}{3}(x^3+1)^{3/2}\Big]_{-1}^{1} && \text{Use the integral just found, with limits of integration for } x. \\
&= \frac{2}{3}\left[((1)^3+1)^{3/2} - ((-1)^3+1)^{3/2}\right] \\
&= \frac{2}{3}\left[2^{3/2} - 0^{3/2}\right] = \frac{2}{3}\left[2\sqrt{2}\right] = \frac{4\sqrt{2}}{3}
\end{aligned}$$

■

Which method is better—evaluating the transformed definite integral with transformed limits using Theorem 6, or transforming the integral, integrating, and transforming back to use the original limits of integration? In Example 1, the first method seems easier, but that is not always the case. Generally, it is best to know both methods and to use whichever one seems better at the time.

EXAMPLE 2 Using the Substitution Formula

$$\begin{aligned}\int_{\pi/4}^{\pi/2} \cot\theta \csc^2\theta \, d\theta &= \int_1^0 u \cdot (-du) \\ &= -\int_1^0 u\, du \\ &= -\left[\frac{u^2}{2}\right]_1^0 \\ &= -\left[\frac{(0)^2}{2} - \frac{(1)^2}{2}\right] = \frac{1}{2}\end{aligned}$$

Let $u = \cot\theta$, $du = -\csc^2\theta\, d\theta$,
$-du = \csc^2\theta\, d\theta$.
When $\theta = \pi/4$, $u = \cot(\pi/4) = 1$.
When $\theta = \pi/2$, $u = \cot(\pi/2) = 0$.

Definite Integrals of Symmetric Functions

The Substitution Formula in Theorem 6 simplifies the calculation of definite integrals of even and odd functions (Section 1.4) over a symmetric interval $[-a, a]$ (Figure 5.26).

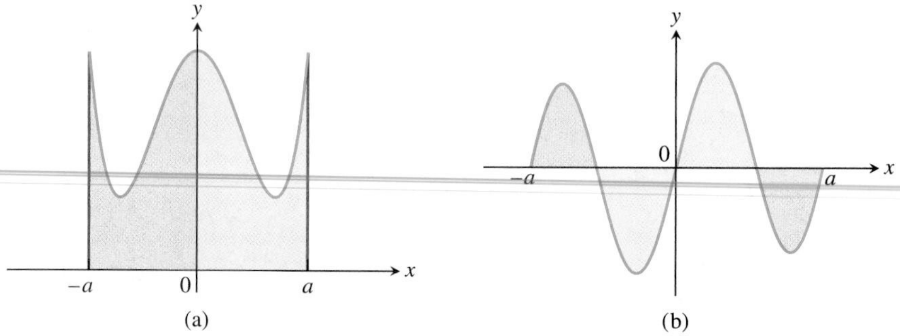

FIGURE 5.26 (a) f even, $\int_{-a}^{a} f(x)\,dx = 2\int_0^a f(x)\,dx$ (b) f odd, $\int_{-a}^{a} f(x)\,dx = 0$

Theorem 7

Let f be continuous on the symmetric interval $[-a, a]$.

(a) If f is even, then $\displaystyle\int_{-a}^{a} f(x)\,dx = 2\int_0^a f(x)\,dx.$

(b) If f is odd, then $\displaystyle\int_{-a}^{a} f(x)\,dx = 0.$

Proof of Part (a)

$$\begin{aligned}\int_{-a}^{a} f(x)\,dx &= \int_{-a}^{0} f(x)\,dx + \int_{0}^{a} f(x)\,dx && \text{Additivity Rule for Definite Integrals}\\ &= -\int_{0}^{-a} f(x)\,dx + \int_{0}^{a} f(x)\,dx && \text{Order of Integration Rule}\\ &= -\int_{0}^{a} f(-u)(-du) + \int_{0}^{a} f(x)\,dx && \text{Let } u=-x,\ du=-dx.\ \text{When } x=0,\ u=0.\ \text{When } x=-a,\ u=a.\\ &= \int_{0}^{a} f(-u)\,du + \int_{0}^{a} f(x)\,dx\\ &= \int_{0}^{a} f(u)\,du + \int_{0}^{a} f(x)\,dx && f \text{ is even, so } f(-u)=f(u).\\ &= 2\int_{0}^{a} f(x)\,dx\end{aligned}$$

The proof of part (b) is entirely similar and you are asked to give it in Exercise 86. ■

The assertions of Theorem 7 remain true when f is an integrable function (rather than having the stronger property of being continuous), but the proof is somewhat more difficult and best left to a more advanced course.

EXAMPLE 3 Integral of an Even Function

Evaluate $\displaystyle\int_{-2}^{2}(x^4 - 4x^2 + 6)\,dx.$

Solution Since $f(x) = x^4 - 4x^2 + 6$ satisfies $f(-x) = f(x)$, it is even on the symmetric interval $[-2, 2]$, so

$$\begin{aligned}\int_{-2}^{2}(x^4 - 4x^2 + 6)\,dx &= 2\int_{0}^{2}(x^4 - 4x^2 + 6)\,dx\\ &= 2\left[\frac{x^5}{5} - \frac{4}{3}x^3 + 6x\right]_0^2\\ &= 2\left(\frac{32}{5} - \frac{32}{3} + 12\right) = \frac{232}{15}.\end{aligned}$$ ■

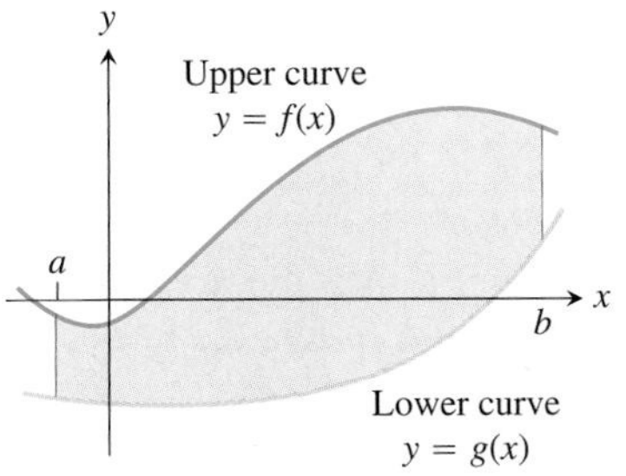

FIGURE 5.27 The region between the curves $y = f(x)$ and $y = g(x)$ and the lines $x = a$ and $x = b$.

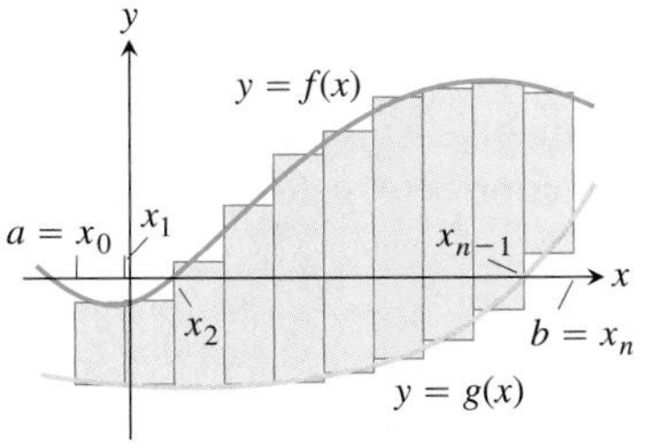

FIGURE 5.28 We approximate the region with rectangles perpendicular to the x-axis.

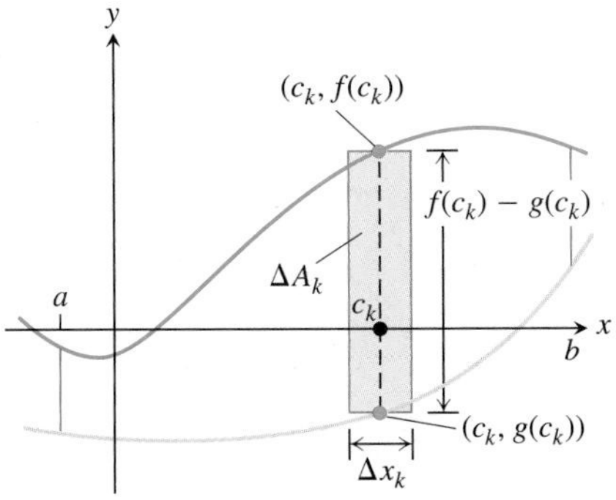

FIGURE 5.29 The area ΔA_k of the kth rectangle is the product of its height, $f(c_k) - g(c_k)$, and its width, Δx_k.

Areas Between Curves

Suppose we want to find the area of a region that is bounded above by the curve $y = f(x)$, below by the curve $y = g(x)$, and on the left and right by the lines $x = a$ and $x = b$ (Figure 5.27). The region might accidentally have a shape whose area we could find with geometry, but if f and g are arbitrary continuous functions, we usually have to find the area with an integral.

To see what the integral should be, we first approximate the region with n vertical rectangles based on a partition $P = \{x_0, x_1, \ldots, x_n\}$ of $[a, b]$ (Figure 5.28). The area of the kth rectangle (Figure 5.29) is

$$\Delta A_k = \text{height} \times \text{width} = [f(c_k) - g(c_k)]\,\Delta x_k.$$

We then approximate the area of the region by adding the areas of the n rectangles:

$$A \approx \sum_{k=1}^{n} \Delta A_k = \sum_{k=1}^{n} [f(c_k) - g(c_k)]\, \Delta x_k. \qquad \text{Riemann Sum}$$

As $\|P\| \to 0$, the sums on the right approach the limit $\int_a^b [f(x) - g(x)]\, dx$ because f and g are continuous. We take the area of the region to be the value of this integral. That is,

$$A = \lim_{\|P\| \to 0} \sum_{k=1}^{n} [f(c_k) - g(c_k)]\, \Delta x_k = \int_a^b [f(x) - g(x)]\, dx.$$

DEFINITION Area Between Curves

If f and g are continuous with $f(x) \ge g(x)$ throughout $[a, b]$, then the **area of the region between the curves $y = f(x)$ and $y = g(x)$ from a to b** is the integral of $(f - g)$ from a to b:

$$A = \int_a^b [f(x) - g(x)]\, dx.$$

When applying this definition it is helpful to graph the curves. The graph reveals which curve is the upper curve f and which is the lower curve g. It also helps you find the limits of integration if they are not already known. You may need to find where the curves intersect to determine the limits of integration, and this may involve solving the equation $f(x) = g(x)$ for values of x. Then you can integrate the function $f - g$ for the area between the intersections.

EXAMPLE 4 Area Between Intersecting Curves

Find the area of the region enclosed by the parabola $y = 2 - x^2$ and the line $y = -x$.

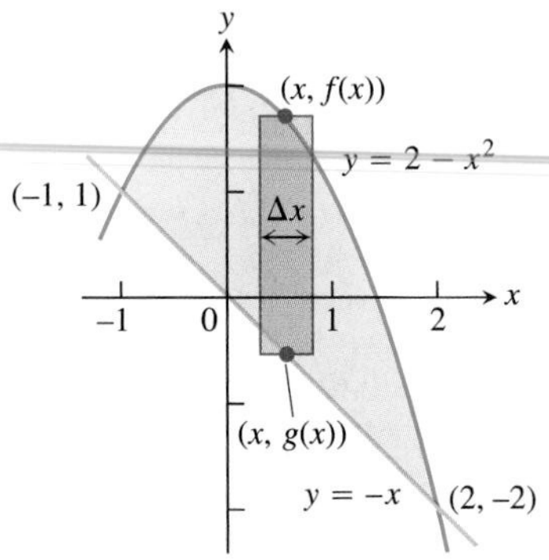

FIGURE 5.30 The region in Example 4 with a typical approximating rectangle.

Solution First we sketch the two curves (Figure 5.30). The limits of integration are found by solving $y = 2 - x^2$ and $y = -x$ simultaneously for x.

$$\begin{aligned} 2 - x^2 &= -x && \text{Equate } f(x) \text{ and } g(x). \\ x^2 - x - 2 &= 0 && \text{Rewrite.} \\ (x + 1)(x - 2) &= 0 && \text{Factor.} \\ x = -1, \quad x &= 2. && \text{Solve.} \end{aligned}$$

The region runs from $x = -1$ to $x = 2$. The limits of integration are $a = -1$, $b = 2$.

The area between the curves is

$$\begin{aligned} A &= \int_a^b [f(x) - g(x)]\, dx = \int_{-1}^{2} [(2 - x^2) - (-x)]\, dx \\ &= \int_{-1}^{2} (2 + x - x^2)\, dx = \left[2x + \frac{x^2}{2} - \frac{x^3}{3}\right]_{-1}^{2} \\ &= \left(4 + \frac{4}{2} - \frac{8}{3}\right) - \left(-2 + \frac{1}{2} + \frac{1}{3}\right) = \frac{9}{2} \end{aligned}$$

■

HISTORICAL BIOGRAPHY

Richard Dedekind
(1831–1916)

If the formula for a bounding curve changes at one or more points, we subdivide the region into subregions that correspond to the formula changes and apply the formula for the area between curves to each subregion.

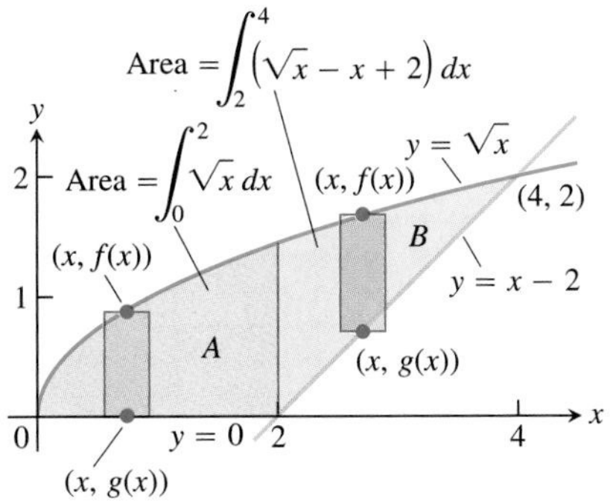

FIGURE 5.31 When the formula for a bounding curve changes, the area integral changes to become the sum of integrals to match, one integral for each of the shaded regions shown here for Example 5.

EXAMPLE 5 Changing the Integral to Match a Boundary Change

Find the area of the region in the first quadrant that is bounded above by $y = \sqrt{x}$ and below by the x-axis and the line $y = x - 2$.

Solution The sketch (Figure 5.31) shows that the region's upper boundary is the graph of $f(x) = \sqrt{x}$. The lower boundary changes from $g(x) = 0$ for $0 \le x \le 2$ to $g(x) = x - 2$ for $2 \le x \le 4$ (there is agreement at $x = 2$). We subdivide the region at $x = 2$ into subregions A and B, shown in Figure 5.31.

The limits of integration for region A are $a = 0$ and $b = 2$. The left-hand limit for region B is $a = 2$. To find the right-hand limit, we solve the equations $y = \sqrt{x}$ and $y = x - 2$ simultaneously for x:

$$\begin{aligned} \sqrt{x} &= x - 2 && \text{Equate } f(x) \text{ and } g(x). \\ x &= (x-2)^2 = x^2 - 4x + 4 && \text{Square both sides.} \\ x^2 - 5x + 4 &= 0 && \text{Rewrite.} \\ (x-1)(x-4) &= 0 && \text{Factor.} \\ x = 1, \quad x &= 4. && \text{Solve.} \end{aligned}$$

Only the value $x = 4$ satisfies the equation $\sqrt{x} = x - 2$. The value $x = 1$ is an extraneous root introduced by squaring. The right-hand limit is $b = 4$.

$$\begin{aligned} &\text{For } 0 \le x \le 2: && f(x) - g(x) = \sqrt{x} - 0 = \sqrt{x} \\ &\text{For } 2 \le x \le 4: && f(x) - g(x) = \sqrt{x} - (x-2) = \sqrt{x} - x + 2 \end{aligned}$$

We add the area of subregions A and B to find the total area:

$$\begin{aligned} \text{Total area} &= \underbrace{\int_0^2 \sqrt{x}\,dx}_{\text{area of } A} + \underbrace{\int_2^4 (\sqrt{x} - x + 2)\,dx}_{\text{area of } B} \\ &= \left[\frac{2}{3}x^{3/2}\right]_0^2 + \left[\frac{2}{3}x^{3/2} - \frac{x^2}{2} + 2x\right]_2^4 \\ &= \frac{2}{3}(2)^{3/2} - 0 + \left(\frac{2}{3}(4)^{3/2} - 8 + 8\right) - \left(\frac{2}{3}(2)^{3/2} - 2 + 4\right) \\ &= \frac{2}{3}(8) - 2 = \frac{10}{3}. \end{aligned}$$

■

Integration with Respect to y

If a region's bounding curves are described by functions of y, the approximating rectangles are horizontal instead of vertical and the basic formula has y in place of x.

For regions like these

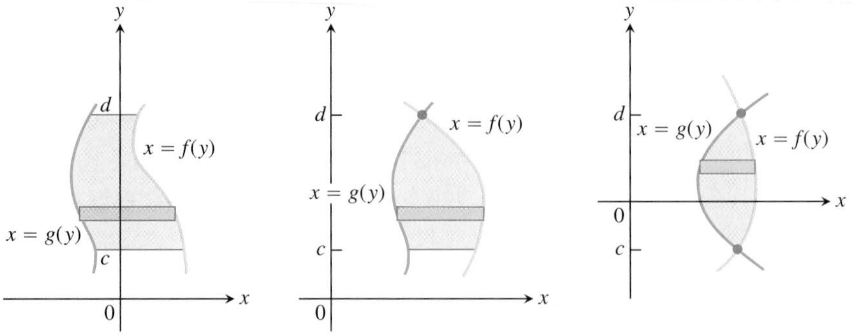

use the formula

$$A = \int_c^d [f(y) - g(y)]\,dy.$$

In this equation f always denotes the right-hand curve and g the left-hand curve, so $f(y) - g(y)$ is nonnegative.

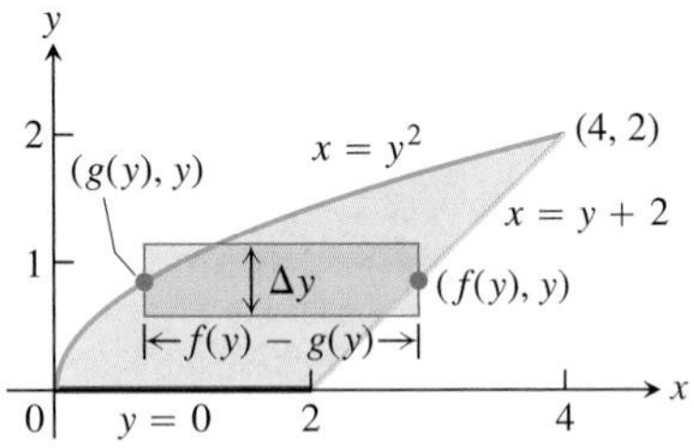

FIGURE 5.32 It takes two integrations to find the area of this region if we integrate with respect to x. It takes only one if we integrate with respect to y (Example 6).

EXAMPLE 6 Find the area of the region in Example 5 by integrating with respect to y.

Solution We first sketch the region and a typical *horizontal* rectangle based on a partition of an interval of y-values (Figure 5.32). The region's right-hand boundary is the line $x = y + 2$, so $f(y) = y + 2$. The left-hand boundary is the curve $x = y^2$, so $g(y) = y^2$. The lower limit of integration is $y = 0$. We find the upper limit by solving $x = y + 2$ and $x = y^2$ simultaneously for y:

$$\begin{aligned} y + 2 &= y^2 && \text{Equate } f(y) = y + 2 \text{ and } g(y) = y^2. \\ y^2 - y - 2 &= 0 && \text{Rewrite.} \\ (y + 1)(y - 2) &= 0 && \text{Factor.} \\ y = -1, \quad y &= 2 && \text{Solve.} \end{aligned}$$

The upper limit of integration is $b = 2$. (The value $y = -1$ gives a point of intersection *below* the x-axis.)

The area of the region is

$$\begin{aligned} A = \int_a^b [f(y) - g(y)]\,dy &= \int_0^2 [y + 2 - y^2]\,dy \\ &= \int_0^2 [2 + y - y^2]\,dy \\ &= \left[2y + \frac{y^2}{2} - \frac{y^3}{3}\right]_0^2 \\ &= 4 + \frac{4}{2} - \frac{8}{3} = \frac{10}{3}. \end{aligned}$$

This is the result of Example 5, found with less work. ■

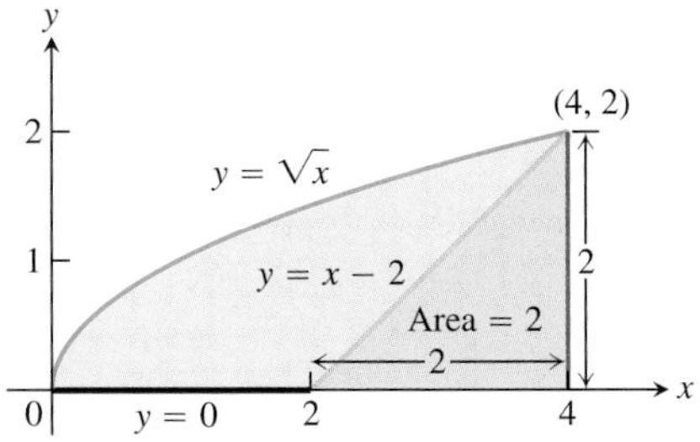

FIGURE 5.33 The area of the blue region is the area under the parabola $y = \sqrt{x}$ minus the area of the triangle (Example 7).

Combining Integrals with Formulas from Geometry

The fastest way to find an area may be to combine calculus and geometry.

EXAMPLE 7 The Area of the Region in Example 5 Found the Fastest Way

Find the area of the region in Example 5.

Solution The area we want is the area between the curve $y = \sqrt{x}$, $0 \le x \le 4$, and the x-axis, *minus* the area of a triangle with base 2 and height 2 (Figure 5.33):

$$\begin{aligned}\text{Area} &= \int_0^4 \sqrt{x}\,dx - \frac{1}{2}(2)(2)\\ &= \frac{2}{3}x^{3/2}\Big]_0^4 - 2\\ &= \frac{2}{3}(8) - 0 - 2 = \frac{10}{3}.\end{aligned}$$

Conclusion from Examples 5–7 It is sometimes easier to find the area between two curves by integrating with respect to y instead of x. Also, it may help to combine geometry and calculus. After sketching the region, take a moment to think about the best way to proceed.

EXERCISES 5.6

Evaluating Definite Integrals

Use the Substitution Formula in Theorem 6 to evaluate the integrals in Exercises 1–24.

1. a. $\int_0^3 \sqrt{y+1}\,dy$ b. $\int_{-1}^0 \sqrt{y+1}\,dy$
2. a. $\int_0^1 r\sqrt{1-r^2}\,dr$ b. $\int_{-1}^1 r\sqrt{1-r^2}\,dr$
3. a. $\int_0^{\pi/4} \tan x \sec^2 x\,dx$ b. $\int_{-\pi/4}^0 \tan x \sec^2 x\,dx$
4. a. $\int_0^{\pi} 3\cos^2 x \sin x\,dx$ b. $\int_{2\pi}^{3\pi} 3\cos^2 x \sin x\,dx$
5. a. $\int_0^1 t^3(1+t^4)^3\,dt$ b. $\int_{-1}^1 t^3(1+t^4)^3\,dt$
6. a. $\int_0^{\sqrt{7}} t(t^2+1)^{1/3}\,dt$ b. $\int_{-\sqrt{7}}^0 t(t^2+1)^{1/3}\,dt$
7. a. $\int_{-1}^1 \frac{5r}{(4+r^2)^2}\,dr$ b. $\int_0^1 \frac{5r}{(4+r^2)^2}\,dr$
8. a. $\int_0^1 \frac{10\sqrt{v}}{(1+v^{3/2})^2}\,dv$ b. $\int_1^4 \frac{10\sqrt{v}}{(1+v^{3/2})^2}\,dv$
9. a. $\int_0^{\sqrt{3}} \frac{4x}{\sqrt{x^2+1}}\,dx$ b. $\int_{-\sqrt{3}}^{\sqrt{3}} \frac{4x}{\sqrt{x^2+1}}\,dx$
10. a. $\int_0^1 \frac{x^3}{\sqrt{x^4+9}}\,dx$ b. $\int_{-1}^0 \frac{x^3}{\sqrt{x^4+9}}\,dx$
11. a. $\int_0^{\pi/6} (1-\cos 3t)\sin 3t\,dt$ b. $\int_{\pi/6}^{\pi/3} (1-\cos 3t)\sin 3t\,dt$
12. a. $\int_{-\pi/2}^0 \left(2+\tan\frac{t}{2}\right)\sec^2\frac{t}{2}\,dt$ b. $\int_{-\pi/2}^{\pi/2} \left(2+\tan\frac{t}{2}\right)\sec^2\frac{t}{2}\,dt$
13. a. $\int_0^{2\pi} \frac{\cos z}{\sqrt{4+3\sin z}}\,dz$ b. $\int_{-\pi}^{\pi} \frac{\cos z}{\sqrt{4+3\sin z}}\,dz$
14. a. $\int_{-\pi/2}^0 \frac{\sin w}{(3+2\cos w)^2}\,dw$ b. $\int_0^{\pi/2} \frac{\sin w}{(3+2\cos w)^2}\,dw$
15. $\int_0^1 \sqrt{t^5+2t}\,(5t^4+2)\,dt$
16. $\int_1^4 \frac{dy}{2\sqrt{y}(1+\sqrt{y})^2}$
17. $\int_0^{\pi/6} \cos^{-3} 2\theta \sin 2\theta\,d\theta$
18. $\int_{\pi}^{3\pi/2} \cot^5\left(\frac{\theta}{6}\right)\sec^2\left(\frac{\theta}{6}\right)d\theta$
19. $\int_0^{\pi} 5(5-4\cos t)^{1/4}\sin t\,dt$
20. $\int_0^{\pi/4} (1-\sin 2t)^{3/2}\cos 2t\,dt$
21. $\int_0^1 (4y-y^2+4y^3+1)^{-2/3}(12y^2-2y+4)\,dy$
22. $\int_0^1 (y^3+6y^2-12y+9)^{-1/2}(y^2+4y-4)\,dy$

23. $\displaystyle\int_0^{\sqrt[3]{\pi^2}} \sqrt{\theta}\cos^2(\theta^{3/2})\,d\theta$

24. $\displaystyle\int_{-1}^{-1/2} t^{-2}\sin^2\left(1+\frac{1}{t}\right)dt$

Area

Find the total areas of the shaded regions in Exercises 25–40.

25.

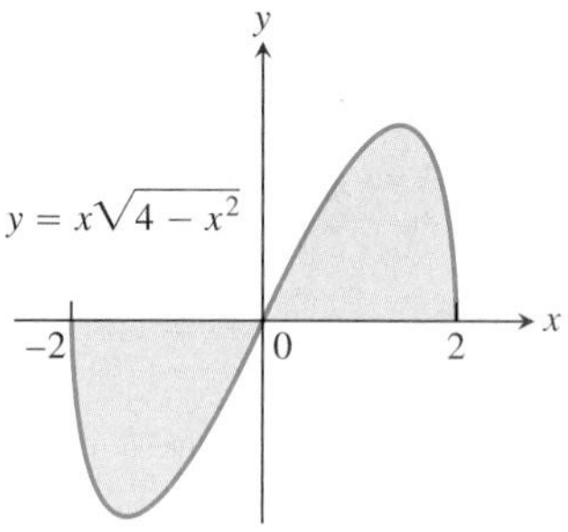

26.

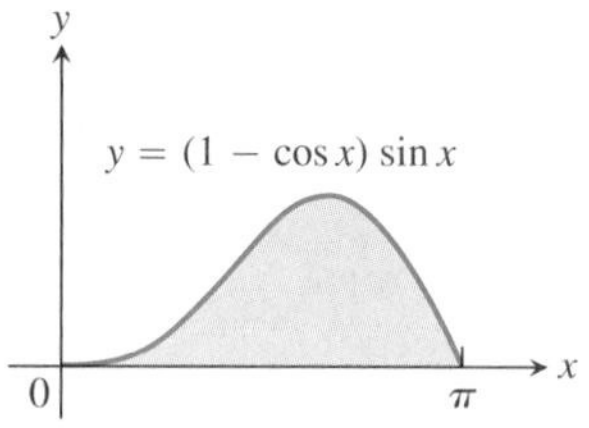

27.

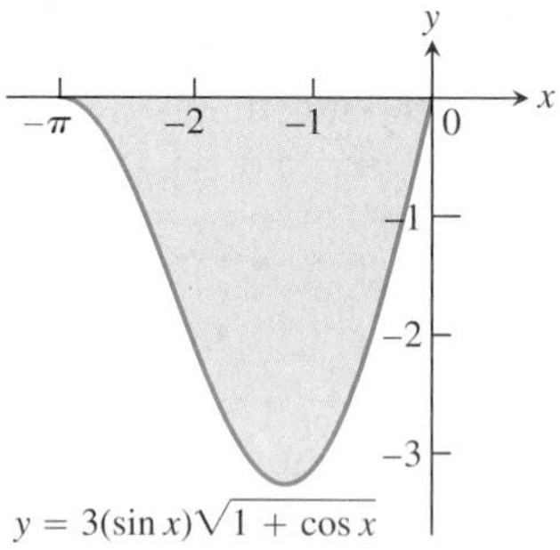

28.

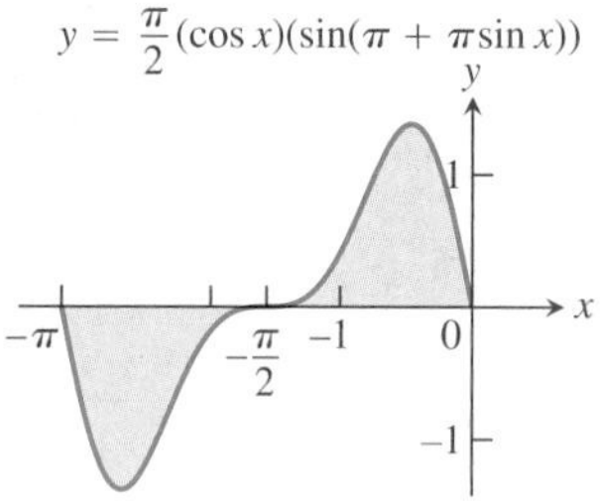

29.

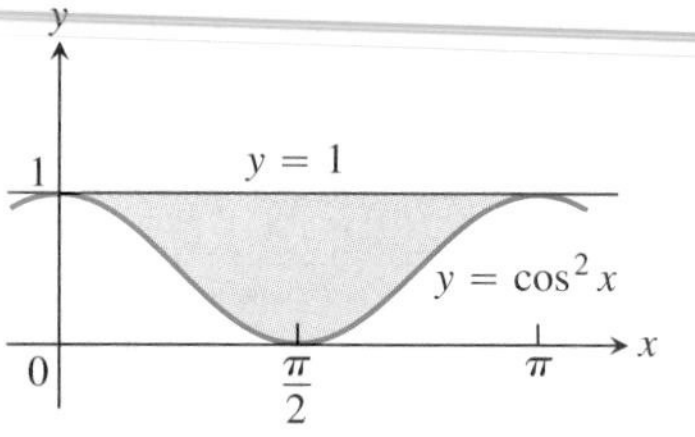

30.

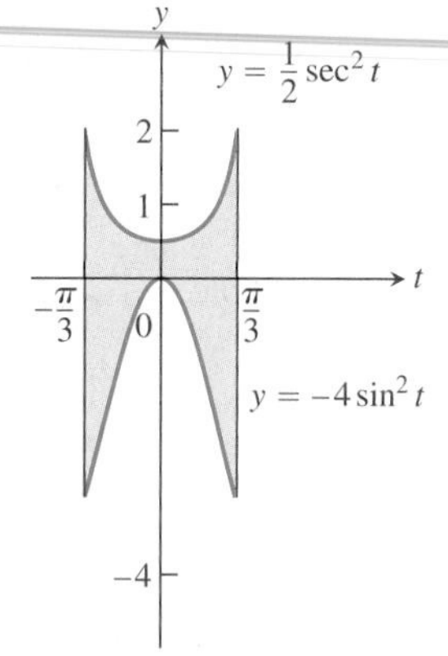

31.

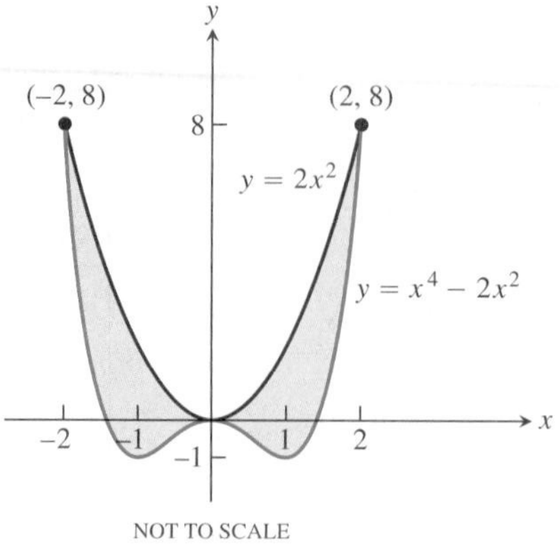

32.

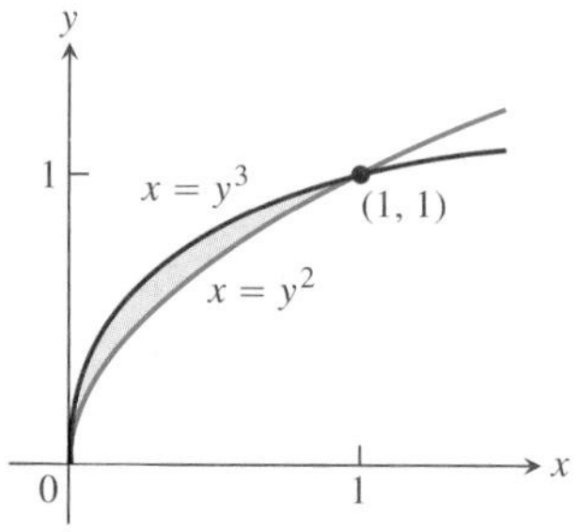

33.

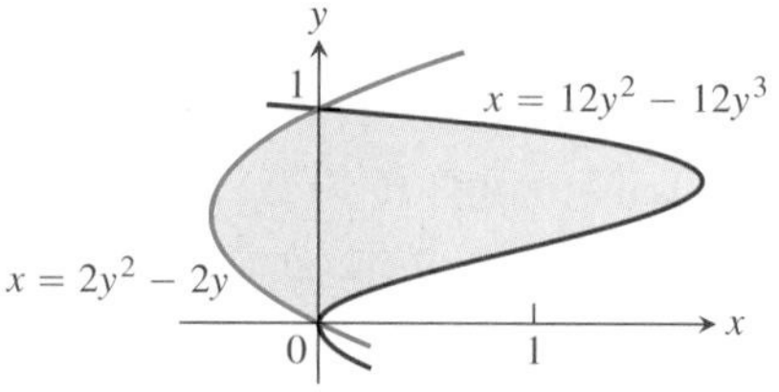

34.

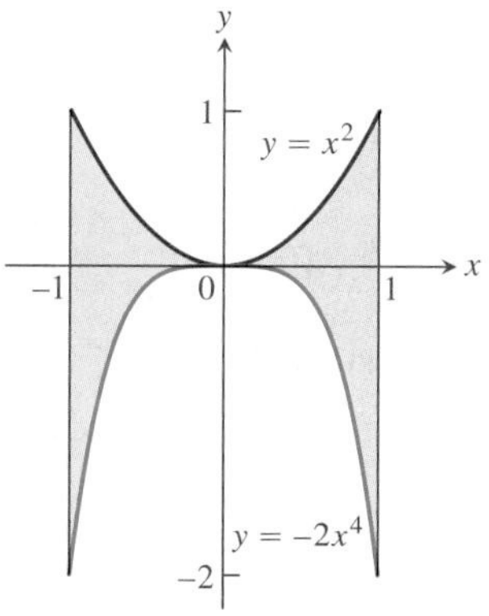

35.

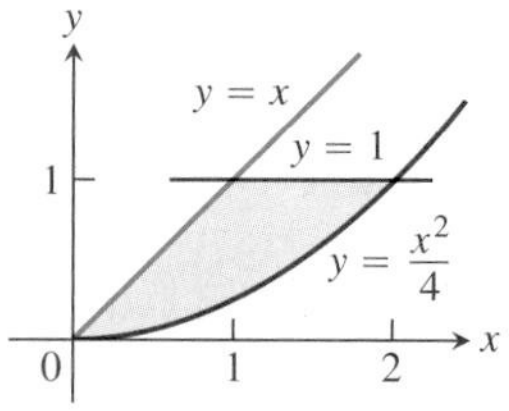

36.
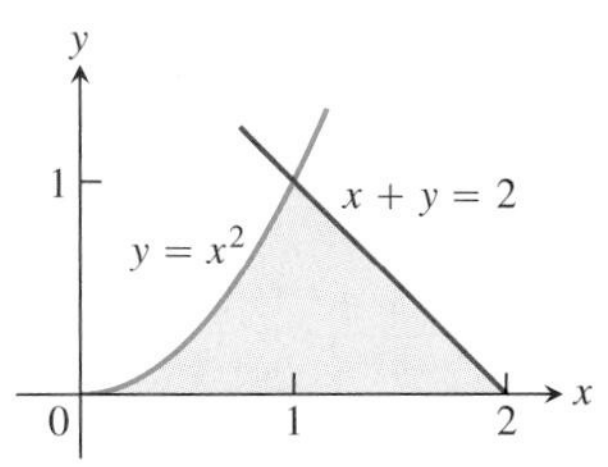

37.
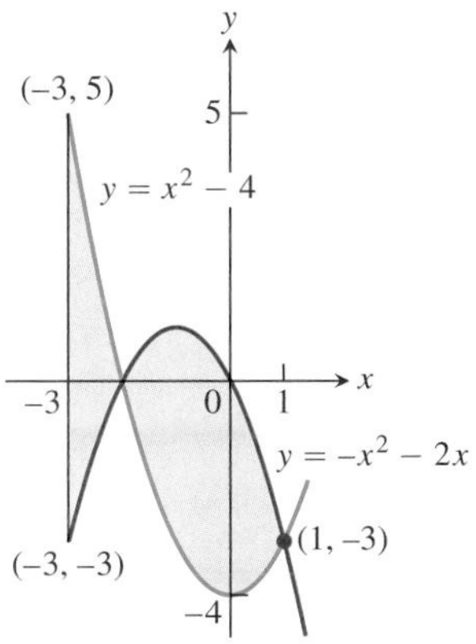

38.
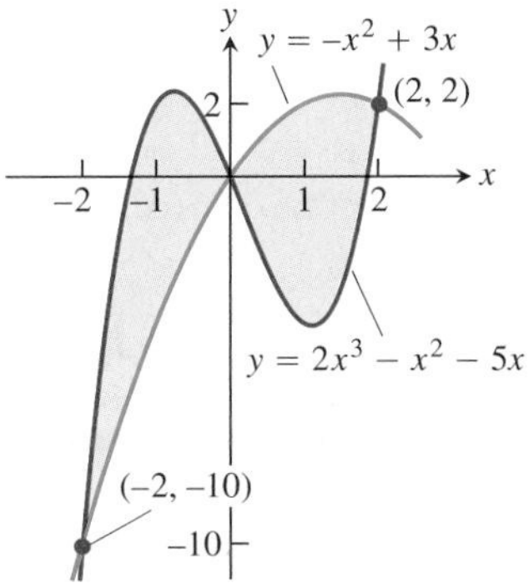

39.
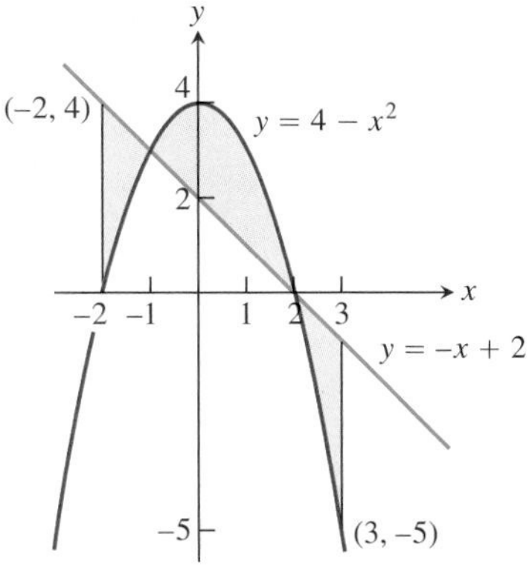

40.
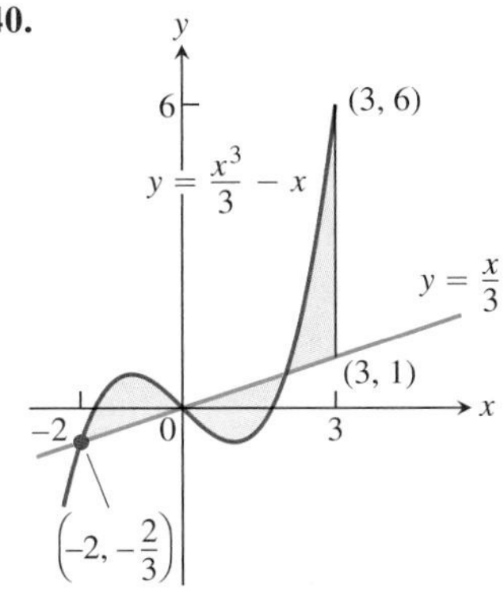

Find the areas of the regions enclosed by the lines and curves in Exercises 41–50.

41. $y = x^2 - 2$ and $y = 2$

42. $y = 2x - x^2$ and $y = -3$

43. $y = x^4$ and $y = 8x$

44. $y = x^2 - 2x$ and $y = x$

45. $y = x^2$ and $y = -x^2 + 4x$

46. $y = 7 - 2x^2$ and $y = x^2 + 4$

47. $y = x^4 - 4x^2 + 4$ and $y = x^2$

48. $y = x\sqrt{a^2 - x^2}$, $a > 0$, and $y = 0$

49. $y = \sqrt{|x|}$ and $5y = x + 6$ (How many intersection points are there?)

50. $y = |x^2 - 4|$ and $y = (x^2/2) + 4$

Find the areas of the regions enclosed by the lines and curves in Exercises 51–58.

51. $x = 2y^2$, $x = 0$, and $y = 3$

52. $x = y^2$ and $x = y + 2$

53. $y^2 - 4x = 4$ and $4x - y = 16$

54. $x - y^2 = 0$ and $x + 2y^2 = 3$

55. $x + y^2 = 0$ and $x + 3y^2 = 2$

56. $x - y^{2/3} = 0$ and $x + y^4 = 2$

57. $x = y^2 - 1$ and $x = |y|\sqrt{1 - y^2}$

58. $x = y^3 - y^2$ and $x = 2y$

Find the areas of the regions enclosed by the curves in Exercises 59–62.

59. $4x^2 + y = 4$ and $x^4 - y = 1$

60. $x^3 - y = 0$ and $3x^2 - y = 4$

61. $x + 4y^2 = 4$ and $x + y^4 = 1$, for $x \geq 0$

62. $x + y^2 = 3$ and $4x + y^2 = 0$

Find the areas of the regions enclosed by the lines and curves in Exercises 63–70.

63. $y = 2\sin x$ and $y = \sin 2x$, $0 \leq x \leq \pi$

64. $y = 8\cos x$ and $y = \sec^2 x$, $-\pi/3 \leq x \leq \pi/3$

65. $y = \cos(\pi x/2)$ and $y = 1 - x^2$

66. $y = \sin(\pi x/2)$ and $y = x$

67. $y = \sec^2 x$, $y = \tan^2 x$, $x = -\pi/4$, and $x = \pi/4$

68. $x = \tan^2 y$ and $x = -\tan^2 y$, $-\pi/4 \leq y \leq \pi/4$

69. $x = 3\sin y\sqrt{\cos y}$ and $x = 0$, $0 \leq y \leq \pi/2$

70. $y = \sec^2(\pi x/3)$ and $y = x^{1/3}$, $-1 \leq x \leq 1$

71. Find the area of the propeller-shaped region enclosed by the curve $x - y^3 = 0$ and the line $x - y = 0$.

72. Find the area of the propeller-shaped region enclosed by the curves $x - y^{1/3} = 0$ and $x - y^{1/5} = 0$.

73. Find the area of the region in the first quadrant bounded by the line $y = x$, the line $x = 2$, the curve $y = 1/x^2$, and the x-axis.

74. Find the area of the "triangular" region in the first quadrant bounded on the left by the y-axis and on the right by the curves $y = \sin x$ and $y = \cos x$.

75. The region bounded below by the parabola $y = x^2$ and above by the line $y = 4$ is to be partitioned into two subsections of equal area by cutting across it with the horizontal line $y = c$.

   a. Sketch the region and draw a line $y = c$ across it that looks about right. In terms of c, what are the coordinates of the points where the line and parabola intersect? Add them to your figure.

b. Find c by integrating with respect to y. (This puts c in the limits of integration.)

c. Find c by integrating with respect to x. (This puts c into the integrand as well.)

76. Find the area of the region between the curve $y = 3 - x^2$ and the line $y = -1$ by integrating with respect to **a.** x, **b.** y.

77. Find the area of the region in the first quadrant bounded on the left by the y-axis, below by the line $y = x/4$, above left by the curve $y = 1 + \sqrt{x}$, and above right by the curve $y = 2/\sqrt{x}$.

78. Find the area of the region in the first quadrant bounded on the left by the y-axis, below by the curve $x = 2\sqrt{y}$, above left by the curve $x = (y - 1)^2$, and above right by the line $x = 3 - y$.

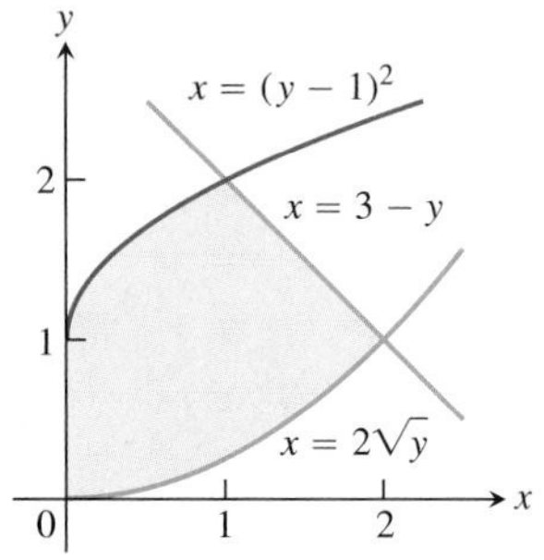

79. The figure here shows triangle AOC inscribed in the region cut from the parabola $y = x^2$ by the line $y = a^2$. Find the limit of the ratio of the area of the triangle to the area of the parabolic region as a approaches zero.

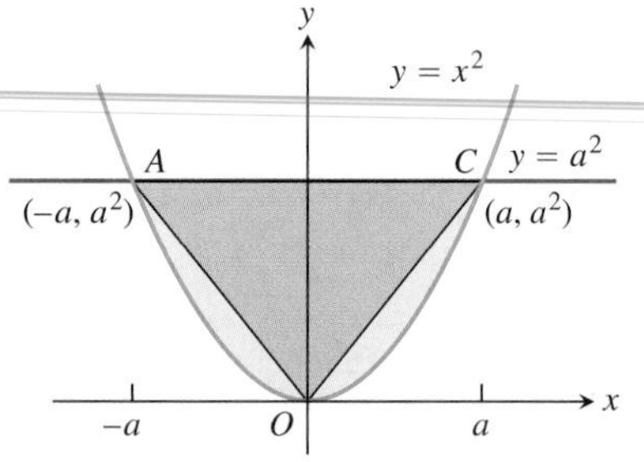

80. Suppose the area of the region between the graph of a positive continuous function f and the x-axis from $x = a$ to $x = b$ is 4 square units. Find the area between the curves $y = f(x)$ and $y = 2f(x)$ from $x = a$ to $x = b$.

81. Which of the following integrals, if either, calculates the area of the shaded region shown here? Give reasons for your answer.

a. $\int_{-1}^{1} (x - (-x))\, dx = \int_{-1}^{1} 2x\, dx$

b. $\int_{-1}^{1} (-x - (x))\, dx = \int_{-1}^{1} -2x\, dx$

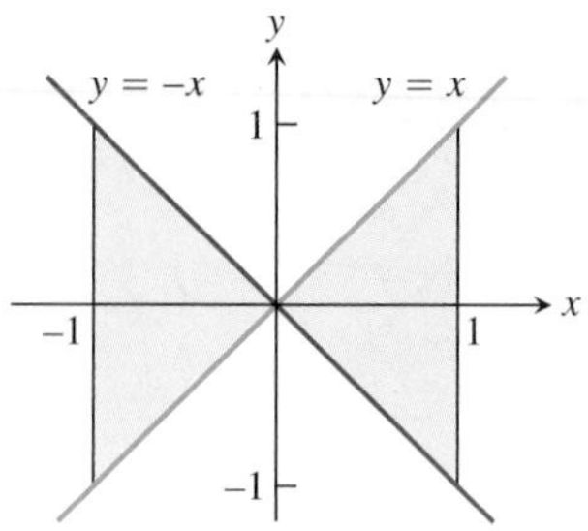

82. True, sometimes true, or never true? The area of the region between the graphs of the continuous functions $y = f(x)$ and $y = g(x)$ and the vertical lines $x = a$ and $x = b$ $(a < b)$ is

$$\int_a^b [f(x) - g(x)]\, dx.$$

Give reasons for your answer.

Theory and Examples

83. Suppose that $F(x)$ is an antiderivative of $f(x) = (\sin x)/x,\ x > 0$. Express

$$\int_1^3 \frac{\sin 2x}{x}\, dx$$

in terms of F.

84. Show that if f is continuous, then

$$\int_0^1 f(x)\, dx = \int_0^1 f(1 - x)\, dx.$$

85. Suppose that

$$\int_0^1 f(x)\, dx = 3.$$

Find

$$\int_{-1}^0 f(x)\, dx$$

if **a.** f is odd, **b.** f is even.

86. a. Show that if f is odd on $[-a, a]$, then

$$\int_{-a}^a f(x)\, dx = 0.$$

b. Test the result in part (a) with $f(x) = \sin x$ and $a = \pi/2$.

87. If f is a continuous function, find the value of the integral

$$I = \int_0^a \frac{f(x)\, dx}{f(x) + f(a - x)}$$

by making the substitution $u = a - x$ and adding the resulting integral to I.

88. By using a substitution, prove that for all positive numbers x and y,

$$\int_x^{xy} \frac{1}{t}\,dt = \int_1^y \frac{1}{t}\,dt.$$

The Shift Property for Definite Integrals

A basic property of definite integrals is their invariance under translation, as expressed by the equation.

$$\int_a^b f(x)\,dx = \int_{a-c}^{b-c} f(x+c)\,dx. \qquad (1)$$

The equation holds whenever f is integrable and defined for the necessary values of x. For example in the accompanying figure, show that

$$\int_{-2}^{-1} (x+2)^3\,dx = \int_0^1 x^3\,dx$$

because the areas of the shaded regions are congruent.

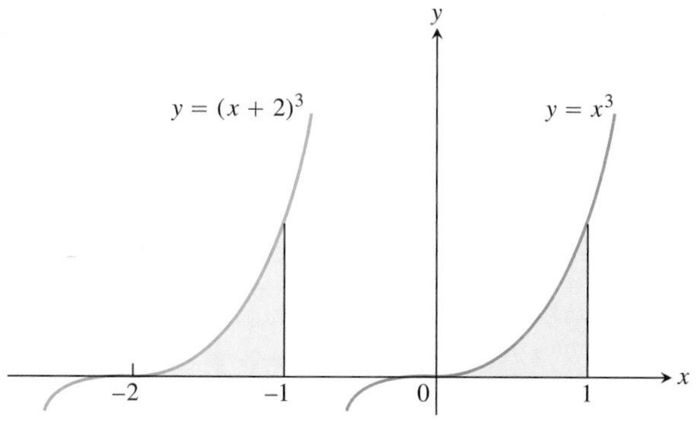

89. Use a substitution to verify Equation (1).

90. For each of the following functions, graph $f(x)$ over $[a, b]$ and $f(x+c)$ over $[a-c, b-c]$ to convince yourself that Equation (1) is reasonable.

a. $f(x) = x^2, \quad a = 0, \quad b = 1, \quad c = 1$

b. $f(x) = \sin x, \quad a = 0, \quad b = \pi, \quad c = \pi/2$

c. $f(x) = \sqrt{x-4}, \quad a = 4, \quad b = 8, \quad c = 5$

COMPUTER EXPLORATIONS

In Exercises 91–94, you will find the area between curves in the plane when you cannot find their points of intersection using simple algebra. Use a CAS to perform the following steps:

a. Plot the curves together to see what they look like and how many points of intersection they have.

b. Use the numerical equation solver in your CAS to find all the points of intersection.

c. Integrate $|f(x) - g(x)|$ over consecutive pairs of intersection values.

d. Sum together the integrals found in part (c).

91. $f(x) = \dfrac{x^3}{3} - \dfrac{x^2}{2} - 2x + \dfrac{1}{3}, \quad g(x) = x - 1$

92. $f(x) = \dfrac{x^4}{2} - 3x^3 + 10, \quad g(x) = 8 - 12x$

93. $f(x) = x + \sin(2x), \quad g(x) = x^3$

94. $f(x) = x^2 \cos x, \quad g(x) = x^3 - x$

Chapter 5 Questions to Guide Your Review

1. How can you sometimes estimate quantities like distance traveled, area, and average value with finite sums? Why might you want to do so?
2. What is sigma notation? What advantage does it offer? Give examples.
3. What is a Riemann sum? Why might you want to consider such a sum?
4. What is the norm of a partition of a closed interval?
5. What is the definite integral of a function f over a closed interval $[a, b]$? When can you be sure it exists?
6. What is the relation between definite integrals and area? Describe some other interpretations of definite integrals.
7. What is the average value of an integrable function over a closed interval? Must the function assume its average value? Explain.
8. Describe the rules for working with definite integrals (Table 5.3). Give examples.
9. What is the Fundamental Theorem of Calculus? Why is it so important? Illustrate each part of the theorem with an example.
10. How does the Fundamental Theorem provide a solution to the initial value problem $dy/dx = f(x), y(x_0) = y_0$, when f is continuous?
11. How is integration by substitution related to the Chain Rule?
12. How can you sometimes evaluate indefinite integrals by substitution? Give examples.
13. How does the method of substitution work for definite integrals? Give examples.
14. How do you define and calculate the area of the region between the graphs of two continuous functions? Give an example.

Chapter 5 Practice Exercises

Finite Sums and Estimates

1. The accompanying figure shows the graph of the velocity (ft/sec) of a model rocket for the first 8 sec after launch. The rocket accelerated straight up for the first 2 sec and then coasted to reach its maximum height at $t = 8$ sec.

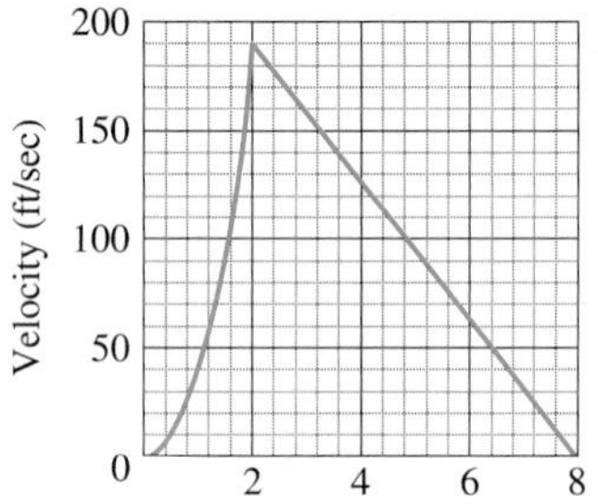

a. Assuming that the rocket was launched from ground level, about how high did it go? (This is the rocket in Section 3.3, Exercise 17, but you do not need to do Exercise 17 to do the exercise here.)

b. Sketch a graph of the rocket's height aboveground as a function of time for $0 \le t \le 8$.

2. a. The accompanying figure shows the velocity (m/sec) of a body moving along the s-axis during the time interval from $t = 0$ to $t = 10$ sec. About how far did the body travel during those 10 sec?

b. Sketch a graph of s as a function of t for $0 \le t \le 10$ assuming $s(0) = 0$.

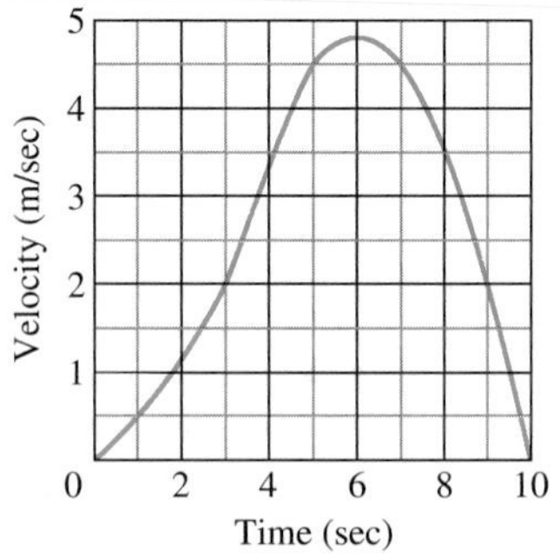

3. Suppose that $\sum_{k=1}^{10} a_k = -2$ and $\sum_{k=1}^{10} b_k = 25$. Find the value of

a. $\sum_{k=1}^{10} \frac{a_k}{4}$ b. $\sum_{k=1}^{10} (b_k - 3a_k)$

c. $\sum_{k=1}^{10} (a_k + b_k - 1)$ d. $\sum_{k=1}^{10} \left(\frac{5}{2} - b_k\right)$

4. Suppose that $\sum_{k=1}^{20} a_k = 0$ and $\sum_{k=1}^{20} b_k = 7$. Find the values of

a. $\sum_{k=1}^{20} 3a_k$ b. $\sum_{k=1}^{20} (a_k + b_k)$

c. $\sum_{k=1}^{20} \left(\frac{1}{2} - \frac{2b_k}{7}\right)$ d. $\sum_{k=1}^{20} (a_k - 2)$

Definite Integrals

In Exercises 5–8, express each limit as a definite integral. Then evaluate the integral to find the value of the limit. In each case, P is a partition of the given interval and the numbers c_k are chosen from the subintervals of P.

5. $\lim_{\|P\|\to 0} \sum_{k=1}^{n} (2c_k - 1)^{-1/2}\,\Delta x_k$, where P is a partition of $[1, 5]$

6. $\lim_{\|P\|\to 0} \sum_{k=1}^{n} c_k(c_k^2 - 1)^{1/3}\,\Delta x_k$, where P is a partition of $[1, 3]$

7. $\lim_{\|P\|\to 0} \sum_{k=1}^{n} \left(\cos\left(\frac{c_k}{2}\right)\right)\Delta x_k$, where P is a partition of $[-\pi, 0]$

8. $\lim_{\|P\|\to 0} \sum_{k=1}^{n} (\sin c_k)(\cos c_k)\,\Delta x_k$, where P is a partition of $[0, \pi/2]$

9. If $\int_{-2}^{2} 3f(x)\,dx = 12$, $\int_{-2}^{5} f(x)\,dx = 6$, and $\int_{-2}^{5} g(x)\,dx = 2$, find the values of the following.

a. $\int_{-2}^{2} f(x)\,dx$ b. $\int_{2}^{5} f(x)\,dx$

c. $\int_{5}^{-2} g(x)\,dx$ d. $\int_{-2}^{5} (-\pi g(x))\,dx$

e. $\int_{-2}^{5} \left(\frac{f(x) + g(x)}{5}\right)dx$

10. If $\int_{0}^{2} f(x)\,dx = \pi$, $\int_{0}^{2} 7g(x)\,dx = 7$, and $\int_{0}^{1} g(x)\,dx = 2$, find the values of the following.

a. $\int_{0}^{2} g(x)\,dx$ b. $\int_{1}^{2} g(x)\,dx$

c. $\int_{2}^{0} f(x)\,dx$ d. $\int_{0}^{2} \sqrt{2}\,f(x)\,dx$

e. $\int_{0}^{2} (g(x) - 3f(x))\,dx$

Area

In Exercise 11–14, find the total area of the region between the graph of f and the x-axis.

11. $f(x) = x^2 - 4x + 3, \quad 0 \le x \le 3$

12. $f(x) = 1 - (x^2/4), \quad -2 \le x \le 3$

13. $f(x) = 5 - 5x^{2/3}, \quad -1 \le x \le 8$

14. $f(x) = 1 - \sqrt{x}, \quad 0 \le x \le 4$

Find the areas of the regions enclosed by the curves and lines in Exercises 15–26.

15. $y = x, \quad y = 1/x^2, \quad x = 2$

16. $y = x, \quad y = 1/\sqrt{x}, \quad x = 2$

17. $\sqrt{x} + \sqrt{y} = 1, \quad x = 0, \quad y = 0$

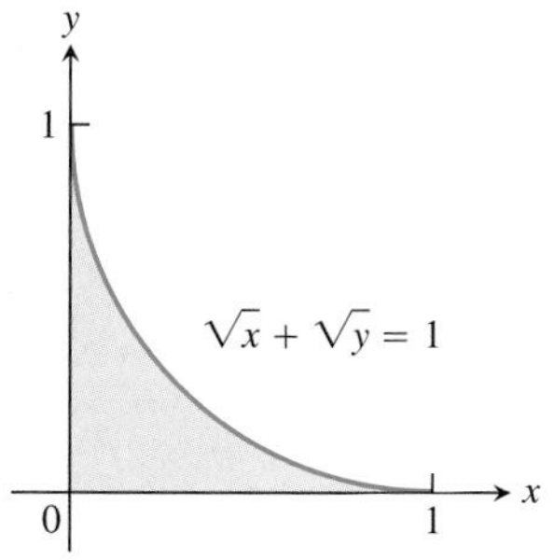

18. $x^3 + \sqrt{y} = 1, \quad x = 0, \quad y = 0, \quad \text{for} \quad 0 \le x \le 1$

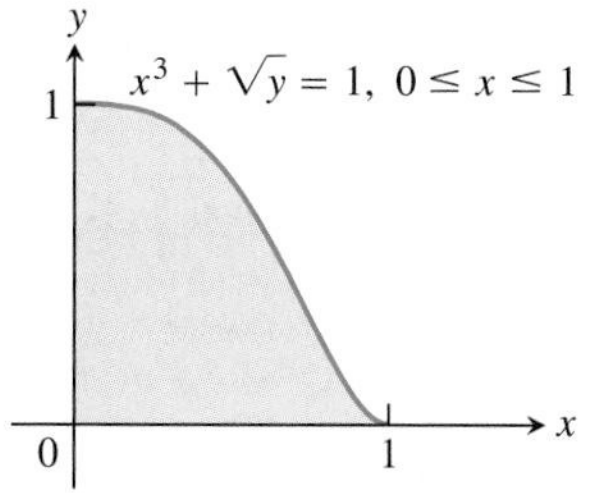

19. $x = 2y^2, \quad x = 0, \quad y = 3$

20. $x = 4 - y^2, \quad x = 0$

21. $y^2 = 4x, \quad y = 4x - 2$

22. $y^2 = 4x + 4, \quad y = 4x - 16$

23. $y = \sin x, \quad y = x, \quad 0 \le x \le \pi/4$

24. $y = |\sin x|, \quad y = 1, \quad -\pi/2 \le x \le \pi/2$

25. $y = 2\sin x, \quad y = \sin 2x, \quad 0 \le x \le \pi$

26. $y = 8\cos x, \quad y = \sec^2 x, \quad -\pi/3 \le x \le \pi/3$

27. Find the area of the "triangular" region bounded on the left by $x + y = 2$, on the right by $y = x^2$, and above by $y = 2$.

28. Find the area of the "triangular" region bounded on the left by $y = \sqrt{x}$, on the right by $y = 6 - x$, and below by $y = 1$.

29. Find the extreme values of $f(x) = x^3 - 3x^2$ and find the area of the region enclosed by the graph of f and the x-axis.

30. Find the area of the region cut from the first quadrant by the curve $x^{1/2} + y^{1/2} = a^{1/2}$.

31. Find the total area of the region enclosed by the curve $x = y^{2/3}$ and the lines $x = y$ and $y = -1$.

32. Find the total area of the region between the curves $y = \sin x$ and $y = \cos x$ for $0 \le x \le 3\pi/2$.

Initial Value Problems

33. Show that $y = x^2 + \int_1^x \frac{1}{t}\,dt$ solves the initial value problem

$$\frac{d^2y}{dx^2} = 2 - \frac{1}{x^2}; \quad y'(1) = 3, \quad y(1) = 1.$$

34. Show that $y = \int_0^x \left(1 + 2\sqrt{\sec t}\right) dt$ solves the initial value problem

$$\frac{d^2y}{dx^2} = \sqrt{\sec x}\tan x; \quad y'(0) = 3, \quad y(0) = 0.$$

Express the solutions of the initial value problems in Exercises 35 and 36 in terms of integrals.

35. $\dfrac{dy}{dx} = \dfrac{\sin x}{x}, \quad y(5) = -3$

36. $\dfrac{dy}{dx} = \sqrt{2 - \sin^2 x}, \quad y(-1) = 2$

Evaluating Indefinite Integrals

Evaluate the integrals in Exercises 37–44.

37. $\int 2(\cos x)^{-1/2} \sin x\,dx$

38. $\int (\tan x)^{-3/2} \sec^2 x\,dx$

39. $\int (2\theta + 1 + 2\cos(2\theta + 1))\,d\theta$

40. $\int \left(\dfrac{1}{\sqrt{2\theta - \pi}} + 2\sec^2(2\theta - \pi)\right) d\theta$

41. $\int \left(t - \dfrac{2}{t}\right)\left(t + \dfrac{2}{t}\right) dt$

42. $\int \dfrac{(t+1)^2 - 1}{t^4}\,dt$

43. $\int \sqrt{t}\sin(2t^{3/2})\,dt$

44. $\int \sec\theta\tan\theta\sqrt{1 + \sec\theta}\,d\theta$

Evaluating Definite Integrals

Evaluate the integrals in Exercises 45–70.

45. $\int_{-1}^{1} (3x^2 - 4x + 7)\,dx$

46. $\int_0^1 (8s^3 - 12s^2 + 5)\,ds$

47. $\int_1^2 \dfrac{4}{v^2}\,dv$

48. $\int_1^{27} x^{-4/3}\,dx$

49. $\int_1^4 \dfrac{dt}{t\sqrt{t}}$

50. $\int_1^4 \dfrac{\left(1 + \sqrt{u}\right)^{1/2}}{\sqrt{u}}\,du$

51. $\int_0^1 \dfrac{36\,dx}{(2x + 1)^3}$

52. $\int_0^1 \dfrac{dr}{\sqrt[3]{(7 - 5r)^2}}$

53. $\int_{1/8}^{1} x^{-1/3}(1 - x^{2/3})^{3/2}\,dx$

54. $\int_0^{1/2} x^3(1 + 9x^4)^{-3/2}\,dx$

55. $\int_0^{\pi} \sin^2 5r\,dr$

56. $\int_0^{\pi/4} \cos^2\left(4t - \dfrac{\pi}{4}\right) dt$

57. $\int_{0}^{\pi/3} \sec^2 \theta \, d\theta$

58. $\int_{\pi/4}^{3\pi/4} \csc^2 x \, dx$

59. $\int_{\pi}^{3\pi} \cot^2 \frac{x}{6} \, dx$

60. $\int_{0}^{\pi} \tan^2 \frac{\theta}{3} \, d\theta$

61. $\int_{-\pi/3}^{0} \sec x \tan x \, dx$

62. $\int_{\pi/4}^{3\pi/4} \csc z \cot z \, dz$

63. $\int_{0}^{\pi/2} 5(\sin x)^{3/2} \cos x \, dx$

64. $\int_{-1}^{1} 2x \sin (1 - x^2) \, dx$

65. $\int_{-\pi/2}^{\pi/2} 15 \sin^4 3x \cos 3x \, dx$

66. $\int_{0}^{2\pi/3} \cos^{-4}\left(\frac{x}{2}\right) \sin\left(\frac{x}{2}\right) dx$

67. $\int_{0}^{\pi/2} \frac{3 \sin x \cos x}{\sqrt{1 + 3 \sin^2 x}} \, dx$

68. $\int_{0}^{\pi/4} \frac{\sec^2 x}{(1 + 7 \tan x)^{2/3}} \, dx$

69. $\int_{0}^{\pi/3} \frac{\tan \theta}{\sqrt{2 \sec \theta}} \, d\theta$

70. $\int_{\pi^2/36}^{\pi^2/4} \frac{\cos \sqrt{t}}{\sqrt{t \sin \sqrt{t}}} \, dt$

Average Values

71. Find the average value of $f(x) = mx + b$

a. over $[-1, 1]$

b. over $[-k, k]$

72. Find the average value of

a. $y = \sqrt{3x}$ over $[0, 3]$

b. $y = \sqrt{ax}$ over $[0, a]$

73. Let f be a function that is differentiable on $[a, b]$. In Chapter 2 we defined the average rate of change of f over $[a, b]$ to be

$$\frac{f(b) - f(a)}{b - a}$$

and the instantaneous rate of change of f at x to be $f'(x)$. In this chapter we defined the average value of a function. For the new definition of average to be consistent with the old one, we should have

$$\frac{f(b) - f(a)}{b - a} = \text{average value of } f' \text{ on } [a, b].$$

Is this the case? Give reasons for your answer.

74. Is it true that the average value of an integrable function over an interval of length 2 is half the function's integral over the interval? Give reasons for your answer.

T 75. Compute the average value of the temperature function

$$f(x) = 37 \sin\left(\frac{2\pi}{365}(x - 101)\right) + 25$$

for a 365-day year. This is one way to estimate the annual mean air temperature in Fairbanks, Alaska. The National Weather Service's official figure, a numerical average of the daily normal mean air temperatures for the year, is 25.7°F, which is slightly higher than the average value of $f(x)$. Figure 3.33 shows why.

T 76. **Specific heat of a gas** Specific heat C_v is the amount of heat required to raise the temperature of a given mass of gas with constant volume by 1°C, measured in units of cal/deg-mole (calories per degree gram molecule). The specific heat of oxygen depends on its temperature T and satisfies the formula

$$C_v = 8.27 + 10^{-5}(26T - 1.87T^2).$$

Find the average value of C_v for $20° \le T \le 675°\text{C}$ and the temperature at which it is attained.

Differentiating Integrals

In Exercises 77–80, find dy/dx.

77. $y = \int_{2}^{x} \sqrt{2 + \cos^3 t} \, dt$

78. $y = \int_{2}^{7x^2} \sqrt{2 + \cos^3 t} \, dt$

79. $y = \int_{x}^{1} \frac{6}{3 + t^4} \, dt$

80. $y = \int_{\sec x}^{2} \frac{1}{t^2 + 1} \, dt$

Theory and Examples

81. Is it true that every function $y = f(x)$ that is differentiable on $[a, b]$ is itself the derivative of some function on $[a, b]$? Give reasons for your answer.

82. Suppose that $F(x)$ is an antiderivative of $f(x) = \sqrt{1 + x^4}$. Express $\int_0^1 \sqrt{1 + x^4}\, dx$ in terms of F and give a reason for your answer.

83. Find dy/dx if $y = \int_x^1 \sqrt{1 + t^2}\, dt$. Explain the main steps in your calculation.

84. Find dy/dx if $y = \int_{\cos x}^0 (1/(1 - t^2))\, dt$. Explain the main steps in your calculation.

85. **A new parking lot** To meet the demand for parking, your town has allocated the area shown here. As the town engineer, you have been asked by the town council to find out if the lot can be built for \$10,000. The cost to clear the land will be \$0.10 a square foot, and the lot will cost \$2.00 a square foot to pave. Can the job be done for \$10,000? Use a lower sum estimate to see. (Answers may vary slightly, depending on the estimate used.)

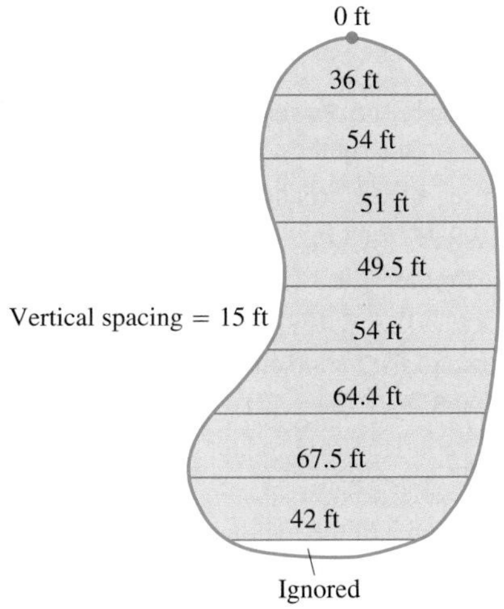

86. Skydivers A and B are in a helicopter hovering at 6400 ft. Skydiver A jumps and descends for 4 sec before opening her parachute. The helicopter then climbs to 7000 ft and hovers there. Forty-five seconds after A leaves the aircraft, B jumps and descends for 13 sec before opening his parachute. Both skydivers descend at 16 ft/sec with parachutes open. Assume that the skydivers fall freely (no effective air resistance) before their parachutes open.

a. At what altitude does A's parachute open?

b. At what altitude does B's parachute open?

c. Which skydiver lands first?

Average Daily Inventory

Average value is used in economics to study such things as average daily inventory. If $I(t)$ is the number of radios, tires, shoes, or whatever product a firm has on hand on day t (we call I an **inventory function**), the average value of I over a time period $[0, T]$ is called the firm's average daily inventory for the period.

$$\textbf{Average daily inventory} = \text{av}(I) = \frac{1}{T}\int_0^T I(t)\,dt.$$

If h is the dollar cost of holding one item per day, the product $\text{av}(I)\cdot h$ is the **average daily holding cost** for the period.

87. As a wholesaler, Tracey Burr Distributors receives a shipment of 1200 cases of chocolate bars every 30 days. TBD sells the chocolate to retailers at a steady rate, and t days after a shipment arrives, its inventory of cases on hand is $I(t) = 1200 - 40t$, $0 \le t \le 30$. What is TBD's average daily inventory for the 30-day period? What is its average daily holding cost if the cost of holding one case is 3¢ a day?

88. Rich Wholesale Foods, a manufacturer of cookies, stores its cases of cookies in an air-conditioned warehouse for shipment every 14 days. Rich tries to keep 600 cases on reserve to meet occasional peaks in demand, so a typical 14-day inventory function is $I(t) = 600 + 600t$, $0 \le t \le 14$. The daily holding cost for each case is 4¢ per day. Find Rich's average daily inventory and average daily holding cost.

89. Solon Container receives 450 drums of plastic pellets every 30 days. The inventory function (drums on hand as a function of days) is $I(t) = 450 - t^2/2$. Find the average daily inventory. If the holding cost for one drum is 2¢ per day, find the average daily holding cost.

90. Mitchell Mailorder receives a shipment of 600 cases of athletic socks every 60 days. The number of cases on hand t days after the shipment arrives is $I(t) = 600 - 20\sqrt{15t}$. Find the average daily inventory. If the holding cost for one case is 1/2¢ per day, find the average daily holding cost.

Chapter 5 Additional and Advanced Exercises

Theory and Examples

1. a. If $\int_0^1 7f(x)\,dx = 7$, does $\int_0^1 f(x)\,dx = 1$?

b. If $\int_0^1 f(x)\,dx = 4$ and $f(x) \ge 0$, does

$$\int_0^1 \sqrt{f(x)}\,dx = \sqrt{4} = 2?$$

Give reasons for your answers.

2. Suppose $\int_{-2}^2 f(x)\,dx = 4$, $\int_2^5 f(x)\,dx = 3$, $\int_{-2}^5 g(x)\,dx = 2$.

Which, if any, of the following statements are true?

a. $\int_5^2 f(x)\,dx = -3$ **b.** $\int_{-2}^5 (f(x) + g(x)) = 9$

c. $f(x) \le g(x)$ on the interval $-2 \le x \le 5$

3. Initial value problem Show that

$$y = \frac{1}{a}\int_0^x f(t)\sin a(x - t)\,dt$$

solves the initial value problem

$$\frac{d^2y}{dx^2} + a^2y = f(x), \qquad \frac{dy}{dx} = 0 \text{ and } y = 0 \text{ when } x = 0.$$

(*Hint:* $\sin(ax - at) = \sin ax \cos at - \cos ax \sin at$.)

4. Proportionality Suppose that x and y are related by the equation

$$x = \int_0^y \frac{1}{\sqrt{1 + 4t^2}}\,dt.$$

Show that d^2y/dx^2 is proportional to y and find the constant of proportionality.

5. Find $f(4)$ if

a. $\int_0^{x^2} f(t)\,dt = x\cos\pi x$ **b.** $\int_0^{f(x)} t^2\,dt = x\cos\pi x$.

6. Find $f(\pi/2)$ from the following information.

i. f is positive and continuous.

ii. The area under the curve $y = f(x)$ from $x = 0$ to $x = a$ is

$$\frac{a^2}{2} + \frac{a}{2}\sin a + \frac{\pi}{2}\cos a.$$

7. The area of the region in the xy-plane enclosed by the x-axis, the curve $y = f(x)$, $f(x) \geq 0$, and the lines $x = 1$ and $x = b$ is equal to $\sqrt{b^2 + 1} - \sqrt{2}$ for all $b > 1$. Find $f(x)$.

8. Prove that

$$\int_0^x \left(\int_0^u f(t)\, dt \right) du = \int_0^x f(u)(x - u)\, du.$$

(*Hint:* Express the integral on the right-hand side as the difference of two integrals. Then show that both sides of the equation have the same derivative with respect to x.)

9. Finding a curve Find the equation for the curve in the xy-plane that passes through the point $(1, -1)$ if its slope at x is always $3x^2 + 2$.

10. Shoveling dirt You sling a shovelful of dirt up from the bottom of a hole with an initial velocity of 32 ft/sec. The dirt must rise 17 ft above the release point to clear the edge of the hole. Is that enough speed to get the dirt out, or had you better duck?

Piecewise Continuous Functions

Although we are mainly interested in continuous functions, many functions in applications are piecewise continuous. A function $f(x)$ is **piecewise continuous on a closed interval I** if f has only finitely many discontinuities in I, the limits

$$\lim_{x \to c^-} f(x) \text{ and } \lim_{x \to c^+} f(x)$$

exist and are finite at every interior point of I, and the appropriate one-sided limits exist and are finite at the endpoints of I. All piecewise continuous functions are integrable. The points of discontinuity subdivide I into open and half-open subintervals on which f is continuous, and the limit criteria above guarantee that f has a continuous extension to the closure of each subinterval. To integrate a piecewise continuous function, we integrate the individual extensions and add the results. The integral of

$$f(x) = \begin{cases} 1 - x, & -1 \leq x < 0 \\ x^2, & 0 \leq x < 2 \\ -1, & 2 \leq x \leq 3 \end{cases}$$

(Figure 5.34) over $[-1, 3]$ is

$$\begin{aligned} \int_{-1}^{3} f(x)\, dx &= \int_{-1}^{0} (1 - x)\, dx + \int_0^2 x^2\, dx + \int_2^3 (-1)\, dx \\ &= \left[x - \frac{x^2}{2} \right]_{-1}^{0} + \left[\frac{x^3}{3} \right]_0^2 + \left[-x \right]_2^3 \\ &= \frac{3}{2} + \frac{8}{3} - 1 = \frac{19}{6}. \end{aligned}$$

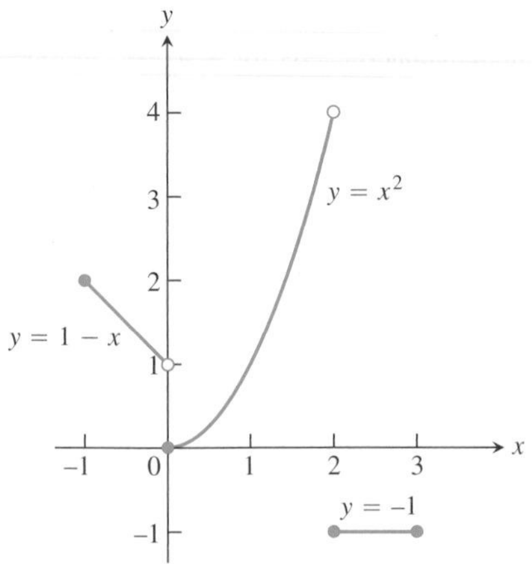

FIGURE 5.34 Piecewise continuous functions like this are integrated piece by piece.

The Fundamental Theorem applies to piecewise continuous functions with the restriction that $(d/dx)\int_a^x f(t)\, dt$ is expected to equal $f(x)$ only at values of x at which f is continuous. There is a similar restriction on Leibniz's Rule below.

Graph the functions in Exercises 11–16 and integrate them over their domains.

11. $f(x) = \begin{cases} x^{2/3}, & -8 \leq x < 0 \\ -4, & 0 \leq x \leq 3 \end{cases}$

12. $f(x) = \begin{cases} \sqrt{-x}, & -4 \leq x < 0 \\ x^2 - 4, & 0 \leq x \leq 3 \end{cases}$

13. $g(t) = \begin{cases} t, & 0 \leq t < 1 \\ \sin \pi t, & 1 \leq t \leq 2 \end{cases}$

14. $h(z) = \begin{cases} \sqrt{1 - z}, & 0 \leq z < 1 \\ (7z - 6)^{-1/3}, & 1 \leq z \leq 2 \end{cases}$

15. $f(x) = \begin{cases} 1, & -2 \leq x < -1 \\ 1 - x^2, & -1 \leq x < 1 \\ 2, & 1 \leq x \leq 2 \end{cases}$

16. $h(r) = \begin{cases} r, & -1 \leq r < 0 \\ 1 - r^2, & 0 \leq r < 1 \\ 1, & 1 \leq r \leq 2 \end{cases}$

17. Find the average value of the function graphed in the accompanying figure.

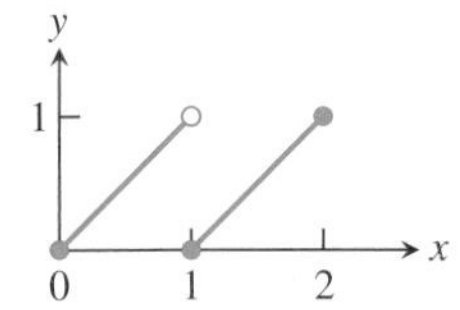

18. Find the average value of the function graphed in the accompanying figure.

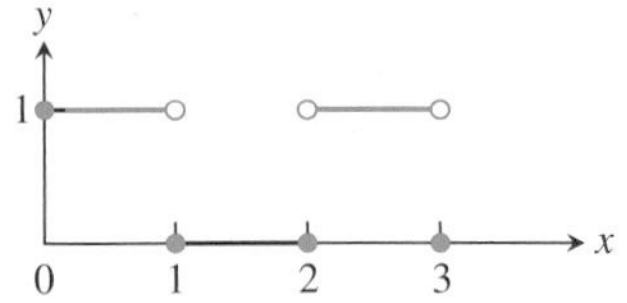

Leibniz's Rule

In applications, we sometimes encounter functions like

$$f(x) = \int_{\sin x}^{x^2} (1 + t)\, dt \qquad \text{and} \qquad g(x) = \int_{\sqrt{x}}^{2\sqrt{x}} \sin t^2\, dt,$$

defined by integrals that have variable upper limits of integration and variable lower limits of integration at the same time. The first integral can be evaluated directly, but the second cannot. We may find the derivative of either integral, however, by a formula called **Leibniz's Rule**.

> **Leibniz's Rule**
>
> If f is continuous on $[a, b]$ and if $u(x)$ and $v(x)$ are differentiable functions of x whose values lie in $[a, b]$, then
>
> $$\frac{d}{dx}\int_{u(x)}^{v(x)} f(t)\, dt = f(v(x))\frac{dv}{dx} - f(u(x))\frac{du}{dx}.$$

Figure 5.35 gives a geometric interpretation of Leibniz's Rule. It shows a carpet of variable width $f(t)$ that is being rolled up at the left at the same time x as it is being unrolled at the right. (In this interpretation, time is x, not t.) At time x, the floor is covered from $u(x)$ to $v(x)$. The rate du/dx at which the carpet is being rolled up need not be the same as the rate dv/dx at which the carpet is being laid down. At any given time x, the area covered by carpet is

$$A(x) = \int_{u(x)}^{v(x)} f(t)\, dt.$$

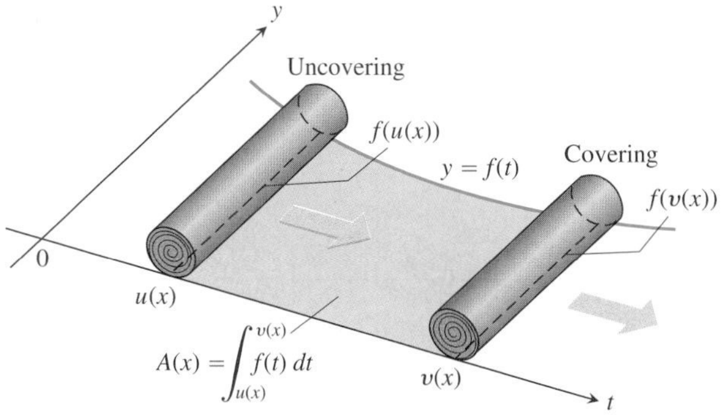

FIGURE 5.35 Rolling and unrolling a carpet: a geometric interpretation of Leibniz's Rule:

$$\frac{dA}{dx} = f(v(x))\frac{dv}{dx} - f(u(x))\frac{du}{dx}.$$

At what rate is the covered area changing? At the instant x, $A(x)$ is increasing by the width $f(v(x))$ of the unrolling carpet times the rate dv/dx at which the carpet is being unrolled. That is, $A(x)$ is being increased at the rate

$$f(v(x))\frac{dv}{dx}.$$

At the same time, A is being decreased at the rate

$$f(u(x))\frac{du}{dx},$$

the width at the end that is being rolled up times the rate du/dx. The net rate of change in A is

$$\frac{dA}{dx} = f(v(x))\frac{dv}{dx} - f(u(x))\frac{du}{dx},$$

which is precisely Leibniz's Rule.

To prove the rule, let F be an antiderivative of f on $[a, b]$. Then

$$\int_{u(x)}^{v(x)} f(t)\, dt = F(v(x)) - F(u(x)).$$

Differentiating both sides of this equation with respect to x gives the equation we want:

$$\begin{aligned}\frac{d}{dx}\int_{u(x)}^{v(x)} f(t)\, dt &= \frac{d}{dx}\Big[F(v(x)) - F(u(x))\Big]\\ &= F'(v(x))\frac{dv}{dx} - F'(u(x))\frac{du}{dx} \qquad \text{Chain Rule}\\ &= f(v(x))\frac{dv}{dx} - f(u(x))\frac{du}{dx}.\end{aligned}$$

Use Leibniz's Rule to find the derivatives of the functions in Exercises 19–21.

19. $f(x) = \int_{1/x}^{x} \frac{1}{t}\, dt$

20. $f(x) = \int_{\cos x}^{\sin x} \frac{1}{1 - t^2}\, dt$

21. $g(y) = \int_{\sqrt{y}}^{2\sqrt{y}} \sin t^2\, dt$

22. Use Leibniz's Rule to find the value of x that maximizes the value of the integral

$$\int_{x}^{x+3} t(5 - t)\, dt.$$

Problems like this arise in the mathematical theory of political elections. See "The Entry Problem in a Political Race," by Steven J. Brams and Philip D. Straffin, Jr., in *Political Equilibrium*, Peter Ordeshook and Kenneth Shepfle, Editors, Kluwer-Nijhoff, Boston, 1982, pp. 181–195.

Approximating Finite Sums with Integrals

In many applications of calculus, integrals are used to approximate finite sums—the reverse of the usual procedure of using finite sums to approximate integrals.

For example, let's estimate the sum of the square roots of the first n positive integers, $\sqrt{1} + \sqrt{2} + \cdots + \sqrt{n}$. The integral

$$\int_0^1 \sqrt{x}\, dx = \frac{2}{3}x^{3/2}\Big]_0^1 = \frac{2}{3}$$

is the limit of the upper sums

$$\begin{aligned} S_n &= \sqrt{\frac{1}{n}} \cdot \frac{1}{n} + \sqrt{\frac{2}{n}} \cdot \frac{1}{n} + \cdots + \sqrt{\frac{n}{n}} \cdot \frac{1}{n} \\ &= \frac{\sqrt{1} + \sqrt{2} + \cdots + \sqrt{n}}{n^{3/2}}. \end{aligned}$$

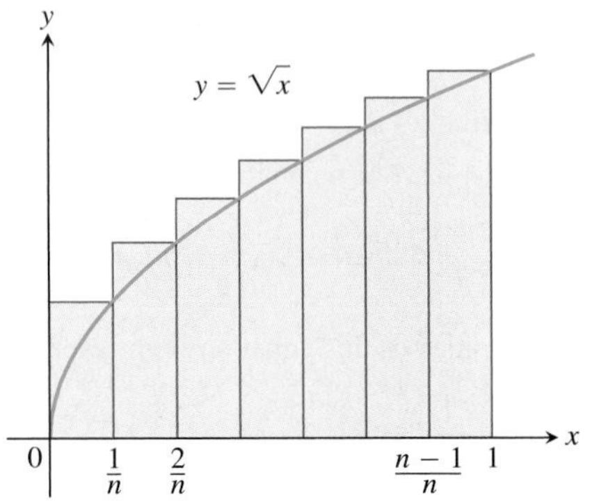

Therefore, when n is large, S_n will be close to $2/3$ and we will have

$$\text{Root sum} = \sqrt{1} + \sqrt{2} + \cdots + \sqrt{n} = S_n \cdot n^{3/2} \approx \frac{2}{3}n^{3/2}.$$

The following table shows how good the approximation can be.

n	**Root sum**	$(2/3)n^{3/2}$	**Relative error**
10	22.468	21.082	$1.386/22.468 \approx 6\%$
50	239.04	235.70	1.4%
100	671.46	666.67	0.7%
1000	21,097	21,082	0.07%

23. Evaluate

$$\lim_{n\to\infty} \frac{1^5 + 2^5 + 3^5 + \cdots + n^5}{n^6}$$

by showing that the limit is

$$\int_0^1 x^5\, dx$$

and evaluating the integral.

24. See Exercise 23. Evaluate

$$\lim_{n\to\infty} \frac{1}{n^4}(1^3 + 2^3 + 3^3 + \cdots + n^3).$$

25. Let $f(x)$ be a continuous function. Express

$$\lim_{n\to\infty} \frac{1}{n}\left[f\left(\frac{1}{n}\right) + f\left(\frac{2}{n}\right) + \cdots + f\left(\frac{n}{n}\right)\right]$$

as a definite integral.

26. Use the result of Exercise 25 to evaluate

a. $\lim_{n\to\infty} \frac{1}{n^2}(2 + 4 + 6 + \cdots + 2n)$,

b. $\lim_{n\to\infty} \frac{1}{n^{16}}(1^{15} + 2^{15} + 3^{15} + \cdots + n^{15})$,

c. $\lim_{n\to\infty} \frac{1}{n}\left(\sin\frac{\pi}{n} + \sin\frac{2\pi}{n} + \sin\frac{3\pi}{n} + \cdots + \sin\frac{n\pi}{n}\right)$.

What can be said about the following limits?

d. $\lim_{n\to\infty} \frac{1}{n^{17}}(1^{15} + 2^{15} + 3^{15} + \cdots + n^{15})$

e. $\lim_{n\to\infty} \frac{1}{n^{15}}(1^{15} + 2^{15} + 3^{15} + \cdots + n^{15})$

27. a. Show that the area A_n of an n-sided regular polygon in a circle of radius r is

$$A_n = \frac{nr^2}{2}\sin\frac{2\pi}{n}.$$

b. Find the limit of A_n as $n \to \infty$. Is this answer consistent with what you know about the area of a circle?

28. A differential equation Show that $y = \sin x + \int_x^\pi \cos 2t\, dt + 1$ satisfies both of the following conditions:

i. $y'' = -\sin x + 2\sin 2x$

ii. $y = 1$ and $y' = -2$ when $x = \pi$.

29. A function defined by an integral The graph of a function f consists of a semicircle and two line segments as shown. Let $g(x) = \int_1^x f(t)\, dt$.

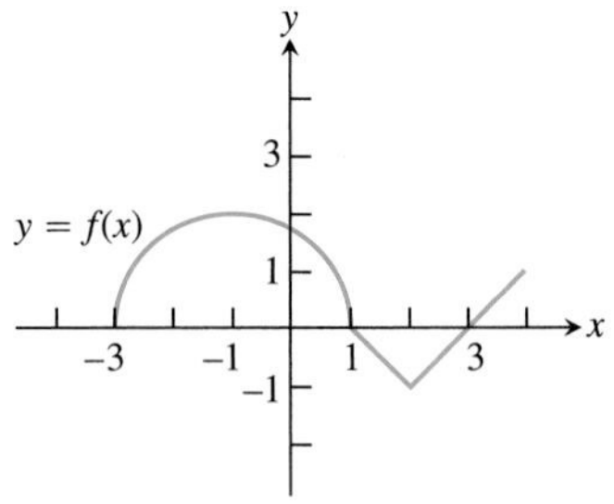

a. Find $g(1)$. **b.** Find $g(3)$. **c.** Find $g(-1)$.

d. Find all values of x on the open interval $(-3, 4)$ at which g has a relative maximum.

e. Write an equation for the line tangent to the graph of g at $x = -1$.

f. Find the x-coordinate of each point of inflection of the graph of g on the open interval $(-3, 4)$.

g. Find the range of g.

Chapter 5 Technology Application Projects

Mathematica/Maple Module

Using Riemann Sums to Estimate Areas, Volumes, and Lengths of Curves

Visualize and approximate areas and volumes in Part I.

Mathematica/Maple Module

Riemann Sums, Definite Integrals, and the Fundamental Theorem of Calculus

Parts I, II, and III develop Riemann sums and definite integrals. Part IV continues the development of the Riemann sum and definite integral using the Fundamental Theorem to solve problems previously investigated.

Mathematica/Maple Module

Rain Catchers, Elevators, and Rockets

Part I illustrates that the area under a curve is the same as the area of an appropriate rectangle for examples taken from the chapter. You will compute the amount of water accumulating in basins of different shapes as the basin is filled and drained.

Mathematica/Maple Module

Motion Along a Straight Line, Part II

You will observe the shape of a graph through dramatic animated visualizations of the derivative relations among the position, velocity, and acceleration. Figures in the text can be animated using this software.

Mathematica/Maple Module

Bending of Beams

Study bent shapes of beams, determine their maximum deflections, concavity and inflection points, and interpret the results in terms of a beam's compression and tension.

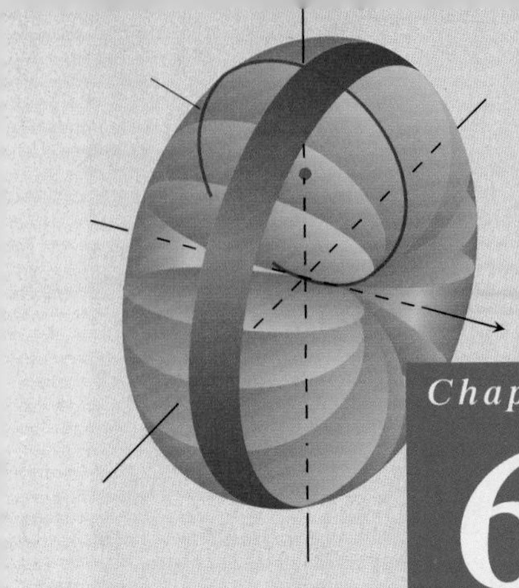

Chapter

6

APPLICATIONS OF DEFINITE INTEGRALS

OVERVIEW In Chapter 5 we discovered the connection between Riemann sums

$$S_P = \sum_{k=1}^{n} f(c_k)\,\Delta x_k$$

associated with a partition P of the finite closed interval $[a, b]$ and the process of integration. We found that for a continuous function f on $[a, b]$, the limit of S_P as the norm of the partition $\|P\|$ approaches zero is the number

$$\int_a^b f(x)\,dx = F(b) - F(a)$$

where F is any antiderivative of f. We applied this to the problems of computing the area between the x-axis and the graph of $y = f(x)$ for $a \le x \le b$, and to finding the area between two curves.

In this chapter we extend the applications to finding volumes, lengths of plane curves, centers of mass, areas of surfaces of revolution, work, and fluid forces against planar walls. We define all these as limits of Riemann sums of continuous functions on closed intervals—that is, as definite integrals which can be evaluated using the Fundamental Theorem of Calculus.

6.1 Volumes by Slicing and Rotation About an Axis

In this section we define volumes of solids whose cross-sections are plane regions. A **cross-section** of a solid S is the plane region formed by intersecting S with a plane (Figure 6.1).

Suppose we want to find the volume of a solid S like the one in Figure 6.1. We begin by extending the definition of a cylinder from classical geometry to cylindrical solids with arbitrary bases (Figure 6.2). If the cylindrical solid has a known base area A and height h, then the volume of the cylindrical solid is

$$\text{Volume} = \text{area} \times \text{height} = A \cdot h.$$

This equation forms the basis for defining the volumes of many solids that are not cylindrical, like the one in Figure 6.1, by the *method of slicing*.

If the cross-section of the solid S at each point x in the interval $[a, b]$ is a region $R(x)$ of area $A(x)$, and A is a continuous function of x, we can define and calculate the volume of the solid S as a definite integral in the following way.

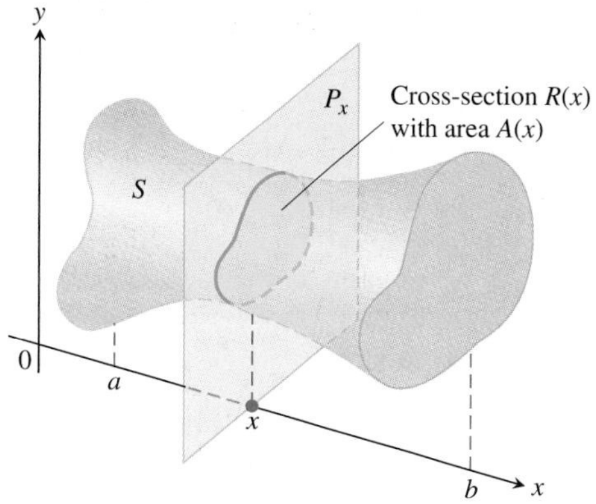

FIGURE 6.1 A cross-section of the solid S formed by intersecting S with a plane P_x perpendicular to the x-axis through the point x in the interval $[a, b]$.

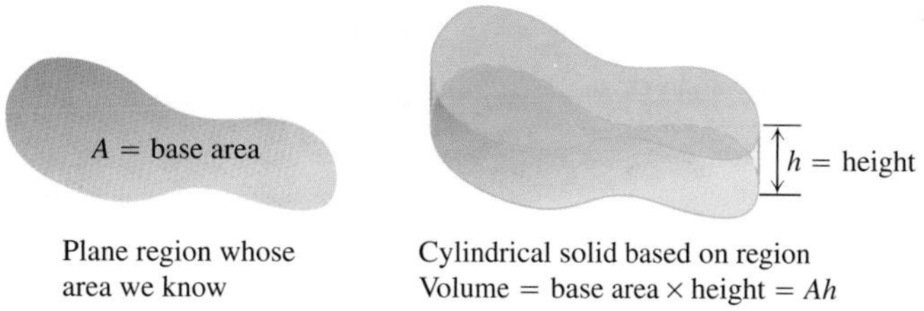

FIGURE 6.2 The volume of a cylindrical solid is always defined to be its base area times its height.

We partition $[a, b]$ into subintervals of width (length) Δx_k and slice the solid, as we would a loaf of bread, by planes perpendicular to the x-axis at the partition points $a = x_0 < x_1 < \cdots < x_n = b$. The planes P_{x_k}, perpendicular to the x-axis at the partition points, slice S into thin "slabs" (like thin slices of a loaf of bread). A typical slab is shown in Figure 6.3. We approximate the slab between the plane at x_{k-1} and the plane at x_k by a cylindrical solid with base area $A(x_k)$ and height $\Delta x_k = x_k - x_{k-1}$ (Figure 6.4). The volume V_k of this cylindrical solid is $A(x_k) \cdot \Delta x_k$, which is approximately the same volume as that of the slab:

$$\text{Volume of the } k\text{th slab} \approx V_k = A(x_k)\, \Delta x_k.$$

The volume V of the entire solid S is therefore approximated by the sum of these cylindrical volumes,

$$V \approx \sum_{k=1}^{n} V_k = \sum_{k=1}^{n} A(x_k)\, \Delta x_k.$$

This is a Riemann sum for the function $A(x)$ on $[a, b]$. We expect the approximations from these sums to improve as the norm of the partition of $[a, b]$ goes to zero. Taking a partition of $[a, b]$ into n subintervals with $\|P\| \to 0$ gives

$$V = \lim_{n \to \infty} \sum_{k=1}^{n} A(x_k)\, \Delta x_k = \int_a^b A(x)\, dx.$$

So we define their limiting definite integral to be the volume of the solid S.

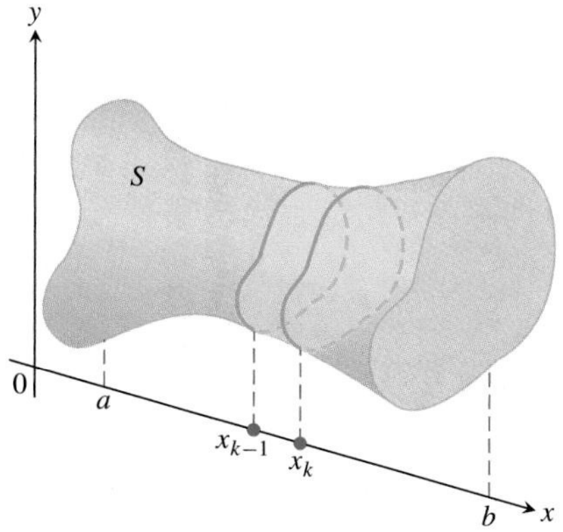

FIGURE 6.3 A typical thin slab in the solid S.

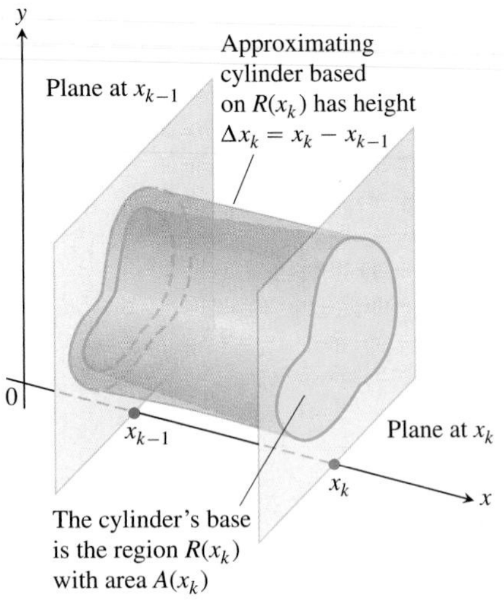

FIGURE 6.4 The solid thin slab in Figure 6.3 is approximated by the cylindrical solid with base $R(x_k)$ having area $A(x_k)$ and height $\Delta x_k = x_k - x_{k-1}$.

DEFINITION Volume

The **volume** of a solid of known integrable cross-sectional area $A(x)$ from $x = a$ to $x = b$ is the integral of A from a to b,

$$V = \int_a^b A(x)\,dx.$$

This definition applies whenever $A(x)$ is continuous, or more generally, when it is integrable. To apply the formula in the definition to calculate the volume of a solid, take the following steps:

Calculating the Volume of a Solid

1. *Sketch the solid and a typical cross-section.*
2. *Find a formula for* $A(x)$, the area of a typical cross-section.
3. *Find the limits of integration.*
4. *Integrate* $A(x)$ using the Fundamental Theorem.

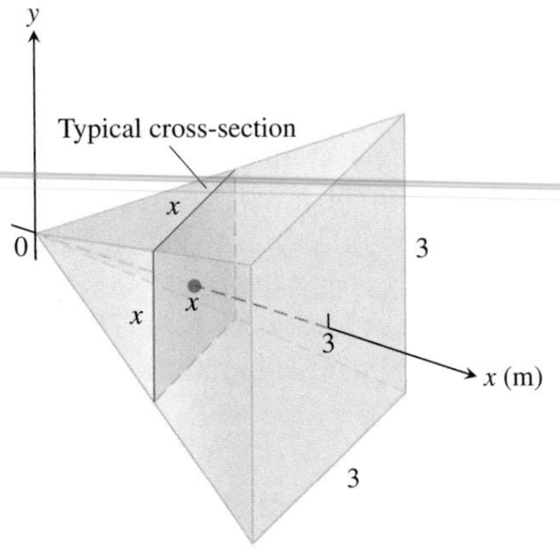

FIGURE 6.5 The cross-sections of the pyramid in Example 1 are squares.

EXAMPLE 1 Volume of a Pyramid

A pyramid 3 m high has a square base that is 3 m on a side. The cross-section of the pyramid perpendicular to the altitude x m down from the vertex is a square x m on a side. Find the volume of the pyramid.

Solution

1. *A sketch.* We draw the pyramid with its altitude along the x-axis and its vertex at the origin and include a typical cross-section (Figure 6.5).
2. *A formula for* $A(x)$. The cross-section at x is a square x meters on a side, so its area is
$$A(x) = x^2.$$
3. *The limits of integration.* The squares lie on the planes from $x = 0$ to $x = 3$.
4. *Integrate to find the volume.*
$$V = \int_0^3 A(x)\,dx = \int_0^3 x^2\,dx = \frac{x^3}{3}\bigg]_0^3 = 9\text{ m}^3$$ ■

HISTORICAL BIOGRAPHY

Bonaventura Cavalieri
(1598–1647)

EXAMPLE 2 Cavalieri's Principle

Cavalieri's principle says that solids with equal altitudes and identical cross-sectional areas at each height have the same volume (Figure 6.6). This follows immediately from the definition of volume, because the cross-sectional area function $A(x)$ and the interval $[a, b]$ are the same for both solids. ■

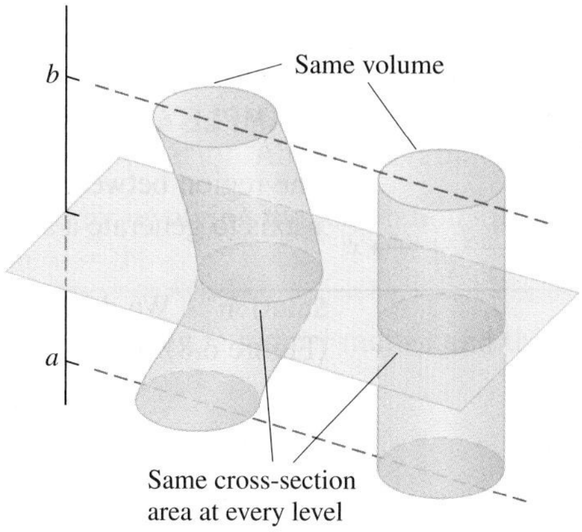

FIGURE 6.6 *Cavalieri's Principle:* These solids have the same volume, which can be illustrated with stacks of coins (Example 2).

EXAMPLE 3 Volume of a Wedge

A curved wedge is cut from a cylinder of radius 3 by two planes. One plane is perpendicular to the axis of the cylinder. The second plane crosses the first plane at a 45° angle at the center of the cylinder. Find the volume of the wedge.

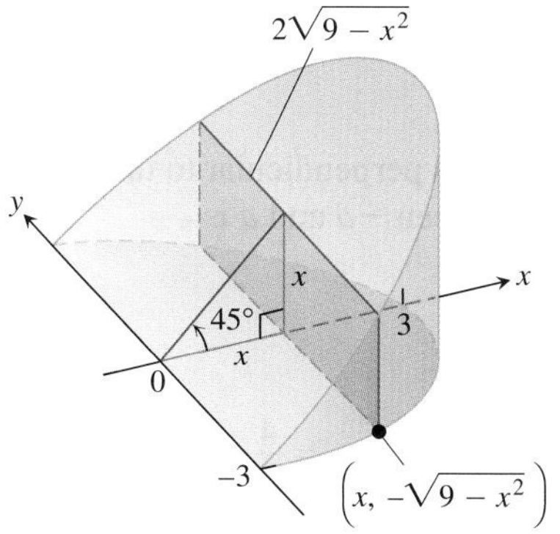

FIGURE 6.7 The wedge of Example 3, sliced perpendicular to the x-axis. The cross-sections are rectangles.

Solution We draw the wedge and sketch a typical cross-section perpendicular to the x-axis (Figure 6.7). The cross-section at x is a rectangle of area

$$\begin{aligned} A(x) &= (\text{height})(\text{width}) = (x)\left(2\sqrt{9-x^2}\right) \\ &= 2x\sqrt{9-x^2}. \end{aligned}$$

The rectangles run from $x = 0$ to $x = 3$, so we have

$$\begin{aligned} V &= \int_a^b A(x)\,dx = \int_0^3 2x\sqrt{9-x^2}\,dx \\ &= -\frac{2}{3}(9-x^2)^{3/2}\Big]_0^3 \\ &= 0 + \frac{2}{3}(9)^{3/2} \\ &= 18. \end{aligned}$$

Let $u = 9 - x^2$, $du = -2x\,dx$, integrate, and substitute back.

■

Solids of Revolution: The Disk Method

The solid generated by rotating a plane region about an axis in its plane is called a **solid of revolution**. To find the volume of a solid like the one shown in Figure 6.8, we need only observe that the cross-sectional area $A(x)$ is the area of a disk of radius $R(x)$, the distance of the planar region's boundary from the axis of revolution. The area is then

$$A(x) = \pi(\text{radius})^2 = \pi[R(x)]^2.$$

So the definition of volume gives

$$V = \int_a^b A(x)\,dx = \int_a^b \pi[R(x)]^2\,dx.$$

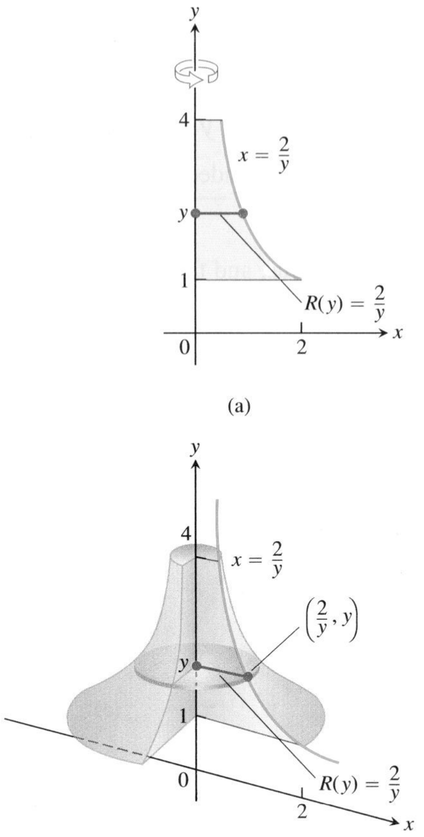

FIGURE 6.11 The region (a) and part of the solid of revolution (b) in Example 7.

Solution We draw figures showing the region, a typical radius, and the generated solid (Figure 6.11). The volume is

$$
\begin{aligned}
V &= \int_1^4 \pi[R(y)]^2\,dy \\
&= \int_1^4 \pi\left(\frac{2}{y}\right)^2 dy \\
&= \pi\int_1^4 \frac{4}{y^2}\,dy = 4\pi\left[-\frac{1}{y}\right]_1^4 = 4\pi\left[\frac{3}{4}\right] \\
&= 3\pi.
\end{aligned}
$$

■

EXAMPLE 8 Rotation About a Vertical Axis

Find the volume of the solid generated by revolving the region between the parabola $x = y^2 + 1$ and the line $x = 3$ about the line $x = 3$.

Solution We draw figures showing the region, a typical radius, and the generated solid (Figure 6.12). Note that the cross-sections are perpendicular to the line $x = 3$. The volume is

$$
\begin{aligned}
V &= \int_{-\sqrt{2}}^{\sqrt{2}} \pi[R(y)]^2\,dy \\
&= \int_{-\sqrt{2}}^{\sqrt{2}} \pi[2 - y^2]^2\,dy \qquad R(y) = 3 - (y^2 + 1) = 2 - y^2 \\
&= \pi\int_{-\sqrt{2}}^{\sqrt{2}} [4 - 4y^2 + y^4]\,dy \\
&= \pi\left[4y - \frac{4}{3}y^3 + \frac{y^5}{5}\right]_{-\sqrt{2}}^{\sqrt{2}} \\
&= \frac{64\pi\sqrt{2}}{15}.
\end{aligned}
$$

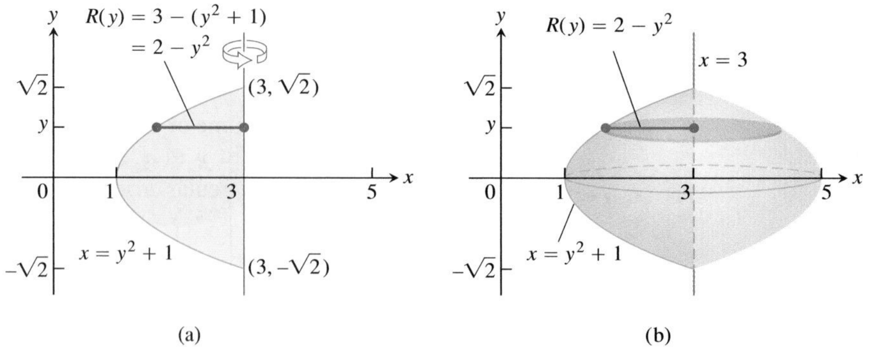

FIGURE 6.12 The region (a) and solid of revolution (b) in Example 8.

■

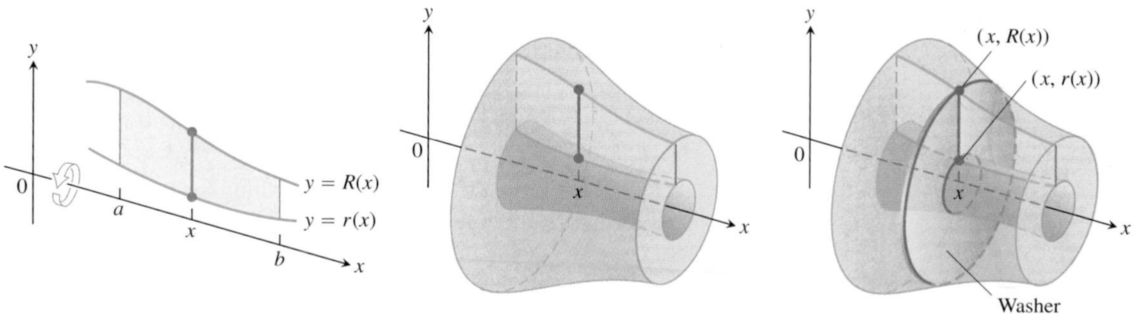

FIGURE 6.13 The cross-sections of the solid of revolution generated here are washers, not disks, so the integral $\int_a^b A(x)\,dx$ leads to a slightly different formula.

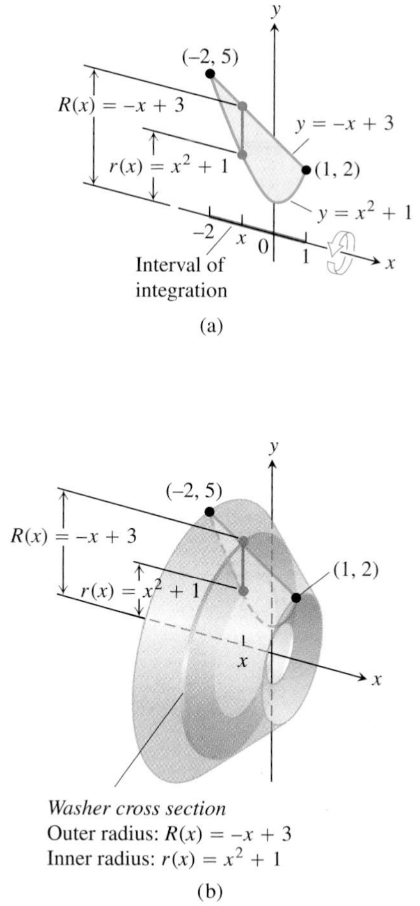

FIGURE 6.14 (a) The region in Example 9 spanned by a line segment perpendicular to the axis of revolution. (b) When the region is revolved about the x-axis, the line segment generates a washer.

Solids of Revolution: The Washer Method

If the region we revolve to generate a solid does not border on or cross the axis of revolution, the solid has a hole in it (Figure 6.13). The cross-sections perpendicular to the axis of revolution are washers (the purplish circular surface in Figure 6.13) instead of disks. The dimensions of a typical washer are

Outer radius:	$R(x)$
Inner radius:	$r(x)$

The washer's area is

$$A(x) = \pi[R(x)]^2 - \pi[r(x)]^2 = \pi([R(x)]^2 - [r(x)]^2).$$

Consequently, the definition of volume gives

$$V = \int_a^b A(x)\,dx = \int_a^b \pi([R(x)]^2 - [r(x)]^2)\,dx.$$

This method for calculating the volume of a solid of revolution is called the **washer method** because a slab is a circular washer of outer radius $R(x)$ and inner radius $r(x)$.

EXAMPLE 9 A Washer Cross-Section (Rotation About the x-Axis)

The region bounded by the curve $y = x^2 + 1$ and the line $y = -x + 3$ is revolved about the x-axis to generate a solid. Find the volume of the solid.

Solution

1. Draw the region and sketch a line segment across it perpendicular to the axis of revolution (the red segment in Figure 6.14).
2. Find the outer and inner radii of the washer that would be swept out by the line segment if it were revolved about the x-axis along with the region.

These radii are the distances of the ends of the line segment from the axis of revolution (Figure 6.14).

$$\text{Outer radius:} \quad R(x) = -x + 3$$
$$\text{Inner radius:} \quad r(x) = x^2 + 1$$

3. Find the limits of integration by finding the x-coordinates of the intersection points of the curve and line in Figure 6.14a.

$$\begin{aligned} x^2 + 1 &= -x + 3 \\ x^2 + x - 2 &= 0 \\ (x + 2)(x - 1) &= 0 \\ x = -2, \quad x &= 1 \end{aligned}$$

4. Evaluate the volume integral.

$$\begin{aligned} V &= \int_a^b \pi([R(x)]^2 - [r(x)]^2)\,dx \\ &= \int_{-2}^{1} \pi((-x + 3)^2 - (x^2 + 1)^2)\,dx \qquad \text{Values from Steps 2 and 3} \\ &= \int_{-2}^{1} \pi(8 - 6x - x^2 - x^4)\,dx \\ &= \pi\left[8x - 3x^2 - \frac{x^3}{3} - \frac{x^5}{5}\right]_{-2}^{1} = \frac{117\pi}{5} \end{aligned}$$

■

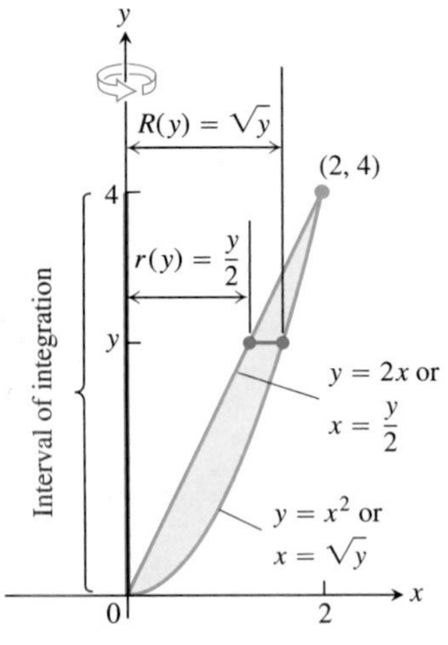

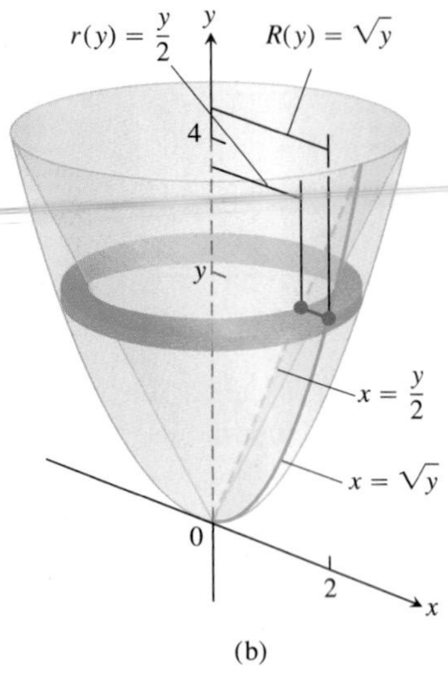

FIGURE 6.15 (a) The region being rotated about the y-axis, the washer radii, and limits of integration in Example 10. (b) The washer swept out by the line segment in part (a).

To find the volume of a solid formed by revolving a region about the y-axis, we use the same procedure as in Example 9, but integrate with respect to y instead of x. In this situation the line segment sweeping out a typical washer is perpendicular to the y-axis (the axis of revolution), and the outer and inner radii of the washer are functions of y.

EXAMPLE 10 A Washer Cross-Section (Rotation About the y-Axis)

The region bounded by the parabola $y = x^2$ and the line $y = 2x$ in the first quadrant is revolved about the y-axis to generate a solid. Find the volume of the solid.

Solution First we sketch the region and draw a line segment across it perpendicular to the axis of revolution (the y-axis). See Figure 6.15a.

The radii of the washer swept out by the line segment are $R(y) = \sqrt{y}$, $r(y) = y/2$ (Figure 6.15).

The line and parabola intersect at $y = 0$ and $y = 4$, so the limits of integration are $c = 0$ and $d = 4$. We integrate to find the volume:

$$\begin{aligned} V &= \int_c^d \pi([R(y)]^2 - [r(y)]^2)\,dy \\ &= \int_0^4 \pi\left(\left[\sqrt{y}\right]^2 - \left[\frac{y}{2}\right]^2\right)dy \\ &= \pi\int_0^4 \left(y - \frac{y^2}{4}\right)dy = \pi\left[\frac{y^2}{2} - \frac{y^3}{12}\right]_0^4 = \frac{8}{3}\pi. \end{aligned}$$

■

Summary

In all of our volume examples, no matter how the cross-sectional area $A(x)$ of a typical slab is determined, the definition of volume as the definite integral $V = \int_a^b A(x)\,dx$ is the heart of the calculations we made.

EXERCISES 6.1

Cross-Sectional Areas

In Exercises 1 and 2, find a formula for the area $A(x)$ of the cross-sections of the solid perpendicular to the x-axis.

1. The solid lies between planes perpendicular to the x-axis at $x = -1$ and $x = 1$. In each case, the cross-sections perpendicular to the x-axis between these planes run from the semicircle $y = -\sqrt{1 - x^2}$ to the semicircle $y = \sqrt{1 - x^2}$.

a. The cross-sections are circular disks with diameters in the xy-plane.

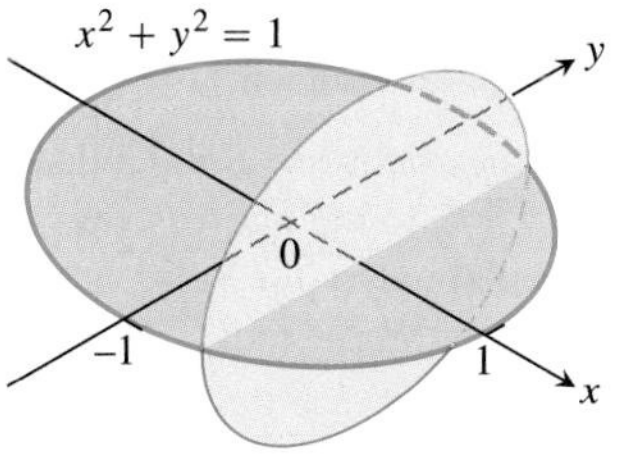

b. The cross-sections are squares with bases in the xy-plane.

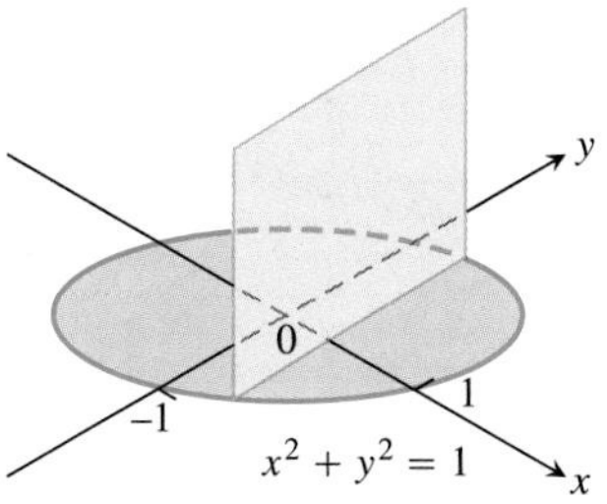

c. The cross-sections are squares with diagonals in the xy-plane. (The length of a square's diagonal is $\sqrt{2}$ times the length of its sides.)

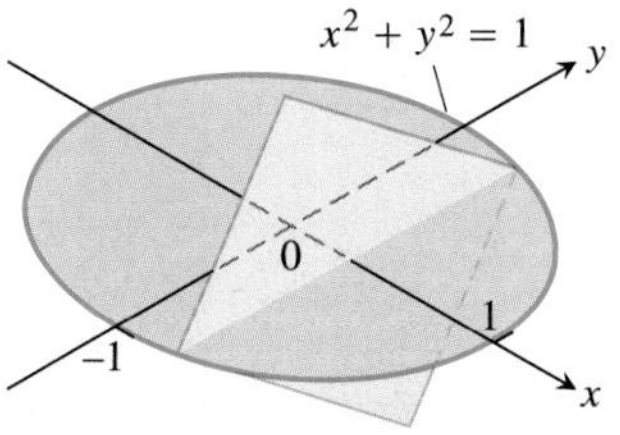

d. The cross-sections are equilateral triangles with bases in the xy-plane.

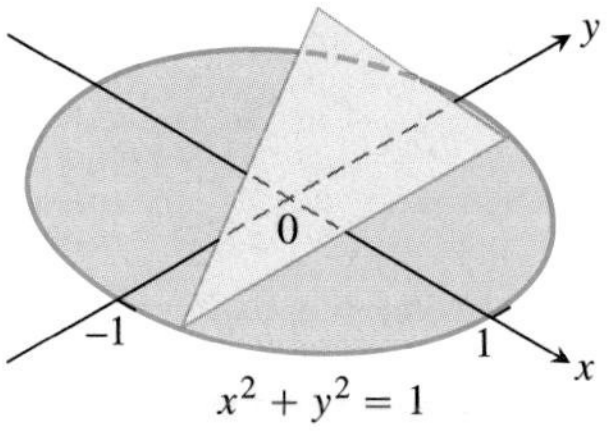

2. The solid lies between planes perpendicular to the x-axis at $x = 0$ and $x = 4$. The cross-sections perpendicular to the x-axis between these planes run from the parabola $y = -\sqrt{x}$ to the parabola $y = \sqrt{x}$.

a. The cross-sections are circular disks with diameters in the xy-plane.

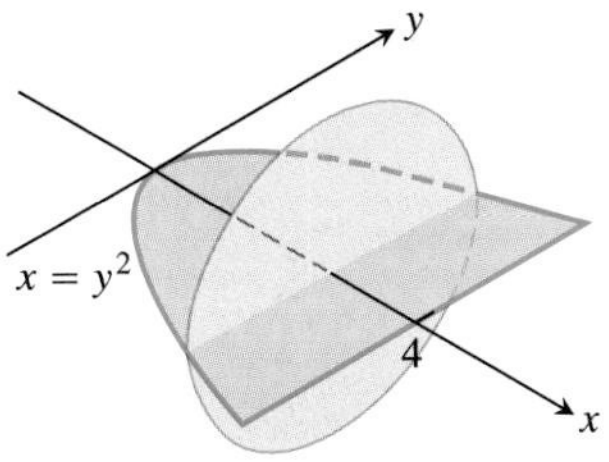

b. The cross-sections are squares with bases in the xy-plane.

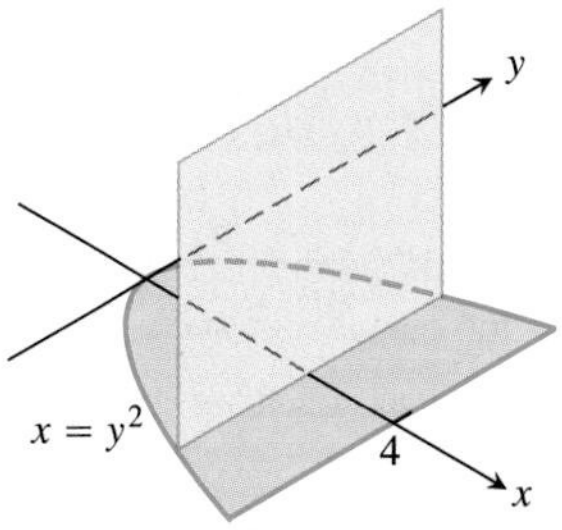

c. The cross-sections are squares with diagonals in the xy-plane.

d. The cross-sections are equilateral triangles with bases in the xy-plane.

Volumes by Slicing

Find the volumes of the solids in Exercises 3–10.

3. The solid lies between planes perpendicular to the x-axis at $x = 0$ and $x = 4$. The cross-sections perpendicular to the axis on the interval $0 \le x \le 4$ are squares whose diagonals run from the parabola $y = -\sqrt{x}$ to the parabola $y = \sqrt{x}$.

4. The solid lies between planes perpendicular to the x-axis at $x = -1$ and $x = 1$. The cross-sections perpendicular to the x-axis are circular disks whose diameters run from the parabola $y = x^2$ to the parabola $y = 2 - x^2$.

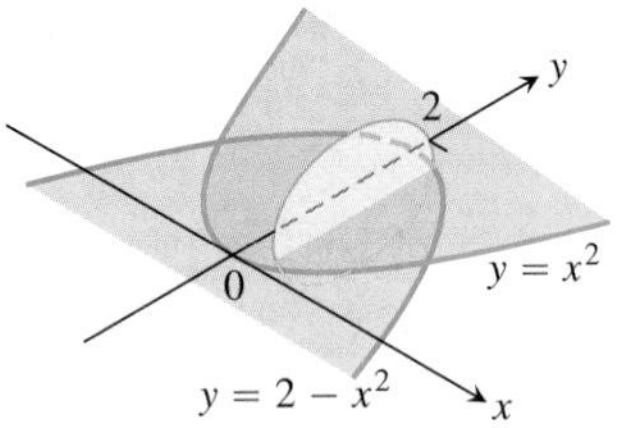

5. The solid lies between planes perpendicular to the x-axis at $x = -1$ and $x = 1$. The cross-sections perpendicular to the x-axis between these planes are squares whose bases run from the semicircle $y = -\sqrt{1 - x^2}$ to the semicircle $y = \sqrt{1 - x^2}$.

6. The solid lies between planes perpendicular to the x-axis at $x = -1$ and $x = 1$. The cross-sections perpendicular to the x-axis between these planes are squares whose diagonals run from the semicircle $y = -\sqrt{1 - x^2}$ to the semicircle $y = \sqrt{1 - x^2}$.

7. The base of a solid is the region between the curve $y = 2\sqrt{\sin x}$ and the interval $[0, \pi]$ on the x-axis. The cross-sections perpendicular to the x-axis are
   a. equilateral triangles with bases running from the x-axis to the curve as shown in the figure.

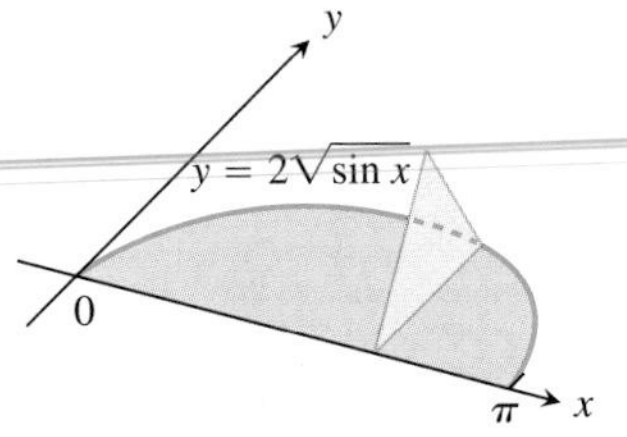

   b. squares with bases running from the x-axis to the curve.

8. The solid lies between planes perpendicular to the x-axis at $x = -\pi/3$ and $x = \pi/3$. The cross-sections perpendicular to the x-axis are
   a. circular disks with diameters running from the curve $y = \tan x$ to the curve $y = \sec x$.
   b. squares whose bases run from the curve $y = \tan x$ to the curve $y = \sec x$.

9. The solid lies between planes perpendicular to the y-axis at $y = 0$ and $y = 2$. The cross-sections perpendicular to the y-axis are circular disks with diameters running from the y-axis to the parabola $x = \sqrt{5}y^2$.

10. The base of the solid is the disk $x^2 + y^2 \le 1$. The cross-sections by planes perpendicular to the y-axis between $y = -1$ and $y = 1$ are isosceles right triangles with one leg in the disk.

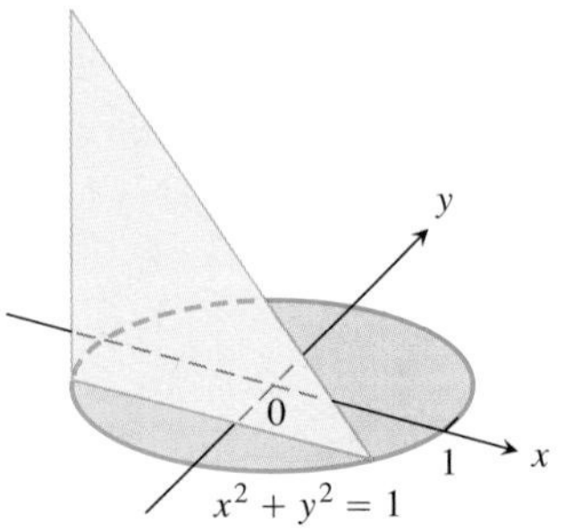

11. **A twisted solid** A square of side length s lies in a plane perpendicular to a line L. One vertex of the square lies on L. As this square moves a distance h along L, the square turns one revolution about L to generate a corkscrew-like column with square cross-sections.
    a. Find the volume of the column.
    b. What will the volume be if the square turns twice instead of once? Give reasons for your answer.

12. **Cavalieri's Principle** A solid lies between planes perpendicular to the x-axis at $x = 0$ and $x = 12$. The cross-sections by planes perpendicular to the x-axis are circular disks whose diameters run from the line $y = x/2$ to the line $y = x$ as shown in the accompanying figure. Explain why the solid has the same volume as a right circular cone with base radius 3 and height 12.

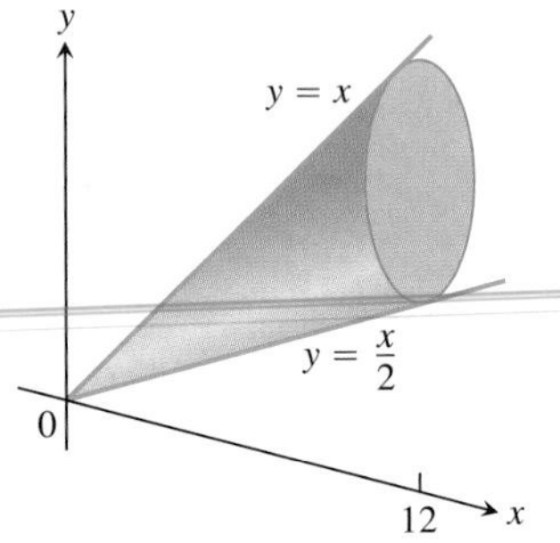

Volumes by the Disk Method

In Exercises 13–16, find the volume of the solid generated by revolving the shaded region about the given axis.

13. About the x-axis

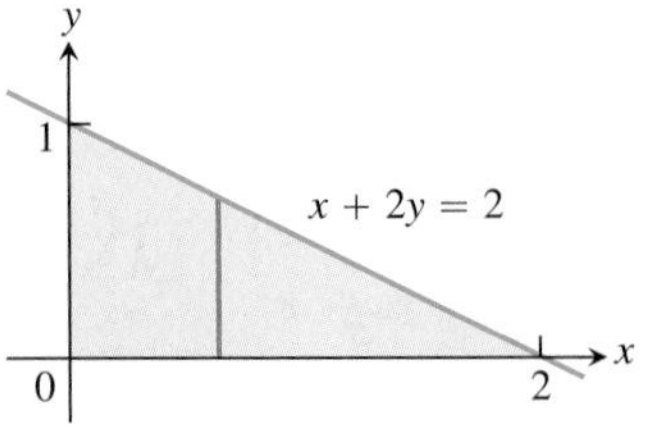

14. About the y-axis

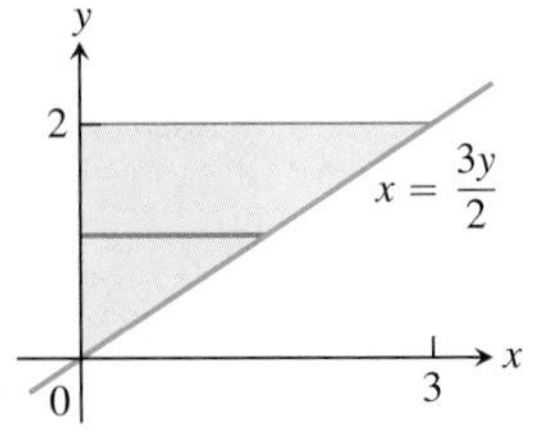

15. About the y-axis

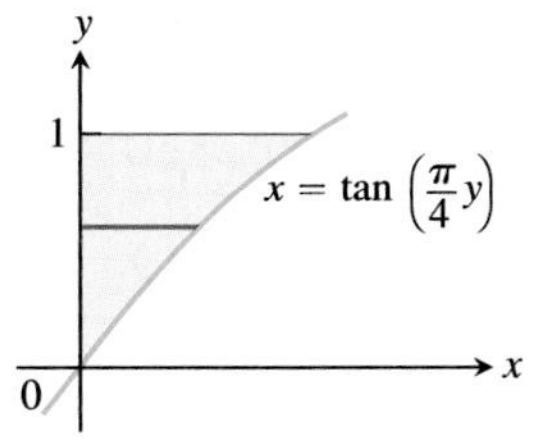

16. About the x-axis

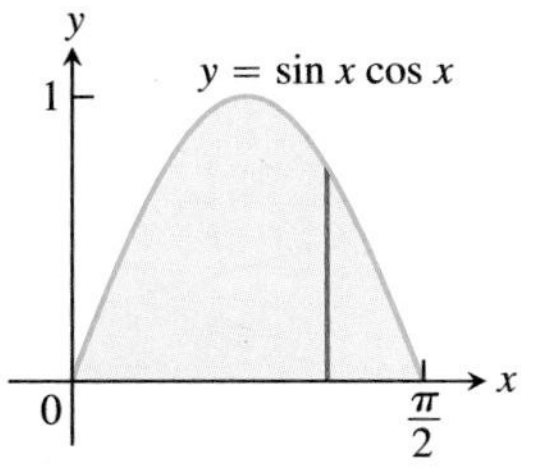

Find the volumes of the solids generated by revolving the regions bounded by the lines and curves in Exercises 17–22 about the x-axis.

17. $y = x^2, \quad y = 0, \quad x = 2$

18. $y = x^3, \quad y = 0, \quad x = 2$

19. $y = \sqrt{9 - x^2}, \quad y = 0$

20. $y = x - x^2, \quad y = 0$

21. $y = \sqrt{\cos x}, \quad 0 \le x \le \pi/2, \quad y = 0, \quad x = 0$

22. $y = \sec x, \quad y = 0, \quad x = -\pi/4, \quad x = \pi/4$

In Exercises 23 and 24, find the volume of the solid generated by revolving the region about the given line.

23. The region in the first quadrant bounded above by the line $y = \sqrt{2}$, below by the curve $y = \sec x \tan x$, and on the left by the y-axis, about the line $y = \sqrt{2}$

24. The region in the first quadrant bounded above by the line $y = 2$, below by the curve $y = 2 \sin x$, $0 \le x \le \pi/2$, and on the left by the y-axis, about the line $y = 2$

Find the volumes of the solids generated by revolving the regions bounded by the lines and curves in Exercises 25–30 about the y-axis.

25. The region enclosed by $x = \sqrt{5}y^2, \quad x = 0, \quad y = -1, \quad y = 1$

26. The region enclosed by $x = y^{3/2}, \quad x = 0, \quad y = 2$

27. The region enclosed by $x = \sqrt{2 \sin 2y}, \quad 0 \le y \le \pi/2, \quad x = 0$

28. The region enclosed by $x = \sqrt{\cos(\pi y/4)}, \quad -2 \le y \le 0, \quad x = 0$

29. $x = 2/(y + 1), \quad x = 0, \quad y = 0, \quad y = 3$

30. $x = \sqrt{2y}/(y^2 + 1), \quad x = 0, \quad y = 1$

Volumes by the Washer Method

Find the volumes of the solids generated by revolving the shaded regions in Exercises 31 and 32 about the indicated axes.

31. The x-axis

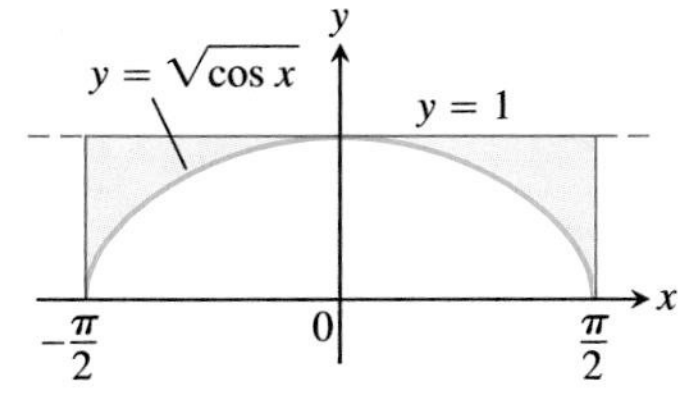

32. The y-axis

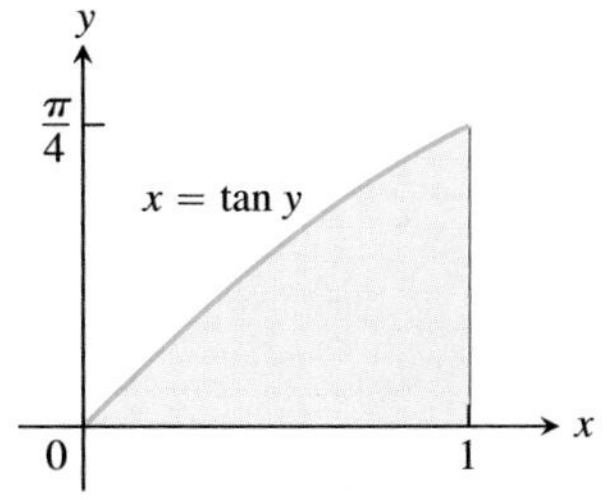

Find the volumes of the solids generated by revolving the regions bounded by the lines and curves in Exercises 33–38 about the x-axis.

33. $y = x, \quad y = 1, \quad x = 0$

34. $y = 2\sqrt{x}, \quad y = 2, \quad x = 0$

35. $y = x^2 + 1, \quad y = x + 3$

36. $y = 4 - x^2, \quad y = 2 - x$

37. $y = \sec x, \quad y = \sqrt{2}, \quad -\pi/4 \le x \le \pi/4$

38. $y = \sec x, \quad y = \tan x, \quad x = 0, \quad x = 1$

In Exercises 39–42, find the volume of the solid generated by revolving each region about the y-axis.

39. The region enclosed by the triangle with vertices (1, 0), (2, 1), and (1, 1)

40. The region enclosed by the triangle with vertices (0, 1), (1, 0), and (1, 1)

41. The region in the first quadrant bounded above by the parabola $y = x^2$, below by the x-axis, and on the right by the line $x = 2$

42. The region in the first quadrant bounded on the left by the circle $x^2 + y^2 = 3$, on the right by the line $x = \sqrt{3}$, and above by the line $y = \sqrt{3}$

In Exercises 43 and 44, find the volume of the solid generated by revolving each region about the given axis.

43. The region in the first quadrant bounded above by the curve $y = x^2$, below by the x-axis, and on the right by the line $x = 1$, about the line $x = -1$

44. The region in the second quadrant bounded above by the curve $y = -x^3$, below by the x-axis, and on the left by the line $x = -1$, about the line $x = -2$

Volumes of Solids of Revolution

45. Find the volume of the solid generated by revolving the region bounded by $y = \sqrt{x}$ and the lines $y = 2$ and $x = 0$ about

a. the x-axis. **b.** the y-axis.

c. the line $y = 2$. **d.** the line $x = 4$.

46. Find the volume of the solid generated by revolving the triangular region bounded by the lines $y = 2x$, $y = 0$, and $x = 1$ about

a. the line $x = 1$. **b.** the line $x = 2$.

47. Find the volume of the solid generated by revolving the region bounded by the parabola $y = x^2$ and the line $y = 1$ about

a. the line $y = 1$. **b.** the line $y = 2$.

c. the line $y = -1$.

48. By integration, find the volume of the solid generated by revolving the triangular region with vertices $(0, 0)$, $(b, 0)$, $(0, h)$ about

a. the x-axis. **b.** the y-axis.

Theory and Applications

49. The volume of a torus The disk $x^2 + y^2 \le a^2$ is revolved about the line $x = b$ $(b > a)$ to generate a solid shaped like a doughnut and called a *torus*. Find its volume. (*Hint:* $\int_{-a}^{a}\sqrt{a^2 - y^2}\,dy = \pi a^2/2$, since it is the area of a semicircle of radius a.)

50. Volume of a bowl A bowl has a shape that can be generated by revolving the graph of $y = x^2/2$ between $y = 0$ and $y = 5$ about the y-axis.

a. Find the volume of the bowl.

b. Related rates If we fill the bowl with water at a constant rate of 3 cubic units per second, how fast will the water level in the bowl be rising when the water is 4 units deep?

51. Volume of a bowl

a. A hemispherical bowl of radius a contains water to a depth h. Find the volume of water in the bowl.

b. Related rates Water runs into a sunken concrete hemispherical bowl of radius 5 m at the rate of $0.2\ \text{m}^3/\text{sec}$. How fast is the water level in the bowl rising when the water is 4 m deep?

52. Explain how you could estimate the volume of a solid of revolution by measuring the shadow cast on a table parallel to its axis of revolution by a light shining directly above it.

53. Volume of a hemisphere Derive the formula $V = (2/3)\pi R^3$ for the volume of a hemisphere of radius R by comparing its cross-sections with the cross-sections of a solid right circular cylinder of radius R and height R from which a solid right circular cone of base radius R and height R has been removed as suggested by the accompanying figure.

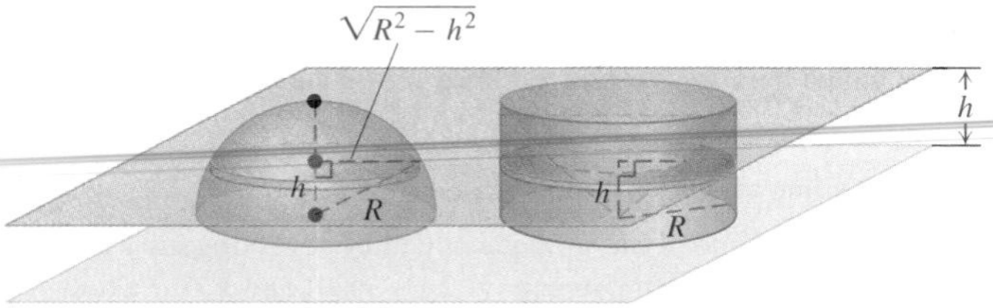

54. Volume of a cone Use calculus to find the volume of a right circular cone of height h and base radius r.

55. Designing a wok You are designing a wok frying pan that will be shaped like a spherical bowl with handles. A bit of experimentation at home persuades you that you can get one that holds about 3 L if you make it 9 cm deep and give the sphere a radius of 16 cm. To be sure, you picture the wok as a solid of revolution, as shown here, and calculate its volume with an integral. To the nearest cubic centimeter, what volume do you really get? ($1\ \text{L} = 1000\ \text{cm}^3$.)

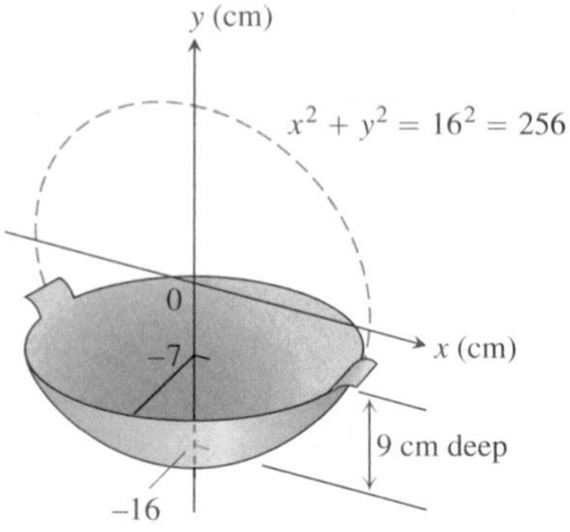

56. Designing a plumb bob Having been asked to design a brass plumb bob that will weigh in the neighborhood of 190 g, you decide to shape it like the solid of revolution shown here. Find the plumb bob's volume. If you specify a brass that weighs $8.5\ \text{g/cm}^3$, how much will the plumb bob weigh (to the nearest gram)?

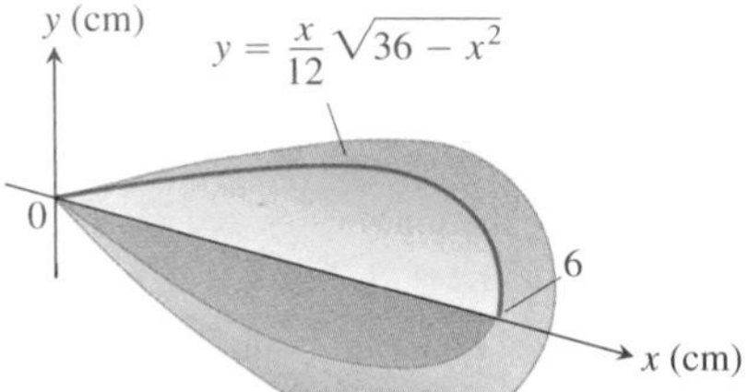

57. Max-min The arch $y = \sin x$, $0 \le x \le \pi$, is revolved about the line $y = c$, $0 \le c \le 1$, to generate the solid in Figure 6.16.

a. Find the value of c that minimizes the volume of the solid. What is the minimum volume?

b. What value of c in $[0, 1]$ maximizes the volume of the solid?

T **c.** Graph the solid's volume as a function of c, first for $0 \le c \le 1$ and then on a larger domain. What happens to the volume of the solid as c moves away from $[0, 1]$? Does this make sense physically? Give reasons for your answers.

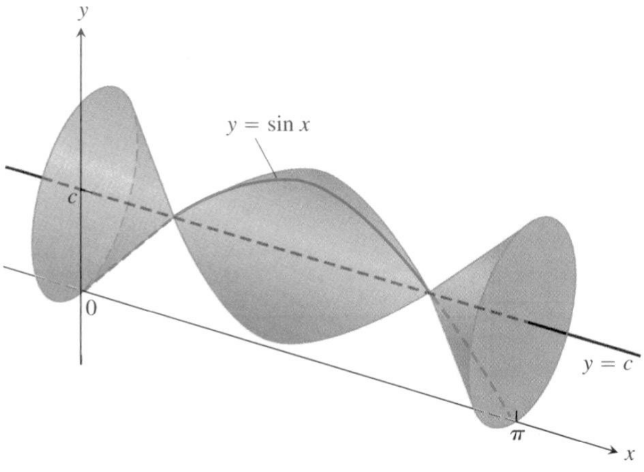

FIGURE 6.16

58. An auxiliary fuel tank You are designing an auxiliary fuel tank that will fit under a helicopter's fuselage to extend its range. After some experimentation at your drawing board, you decide to shape the tank like the surface generated by revolving the curve $y = 1 - (x^2/16)$, $-4 \le x \le 4$, about the x-axis (dimensions in feet).

a. How many cubic feet of fuel will the tank hold (to the nearest cubic foot)?

b. A cubic foot holds 7.481 gal. If the helicopter gets 2 mi to the gallon, how many additional miles will the helicopter be able to fly once the tank is installed (to the nearest mile)?

6.2 Volumes by Cylindrical Shells

In Section 6.1 we defined the volume of a solid S as the definite integral

$$V = \int_a^b A(x)\, dx,$$

where $A(x)$ is an integrable cross-sectional area of S from $x = a$ to $x = b$. The area $A(x)$ was obtained by slicing through the solid with a plane perpendicular to the x-axis. In this section we use the same integral definition for volume, but obtain the area by slicing through the solid in a different way. Now we slice through the solid using circular cylinders of increasing radii, like cookie cutters. We slice straight down through the solid perpendicular to the x-axis, with the axis of the cylinder parallel to the y-axis. The vertical axis of each cylinder is the same line, but the radii of the cylinders increase with each slice. In this way the solid S is sliced up into thin cylindrical shells of constant thickness that grow outward from their common axis, like circular tree rings. Unrolling a cylindrical shell shows that its volume is approximately that of a rectangular slab with area $A(x)$ and thickness Δx. This allows us to apply the same integral definition for volume as before. Before describing the method in general, let's look at an example to gain some insight.

EXAMPLE 1 Finding a Volume Using Shells

The region enclosed by the x-axis and the parabola $y = f(x) = 3x - x^2$ is revolved about the vertical line $x = -1$ to generate the shape of a solid (Figure 6.17). Find the volume of the solid.

Solution Using the washer method from Section 6.1 would be awkward here because we would need to express the x-values of the left and right branches of the parabola in terms

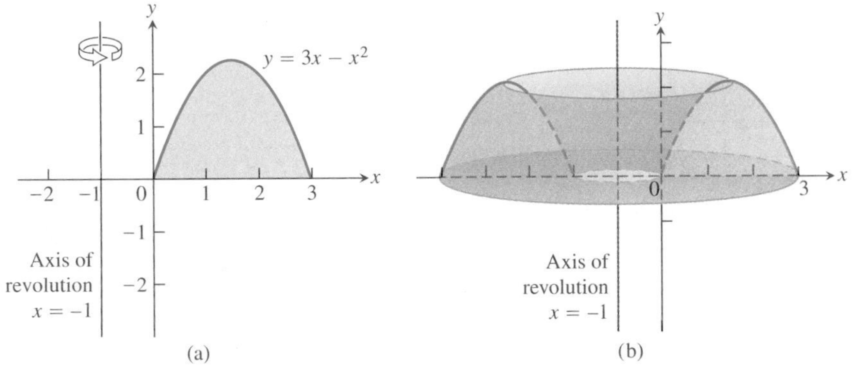

FIGURE 6.17 (a) The graph of the region in Example 1, before revolution. (b) The solid formed when the region in part (a) is revolved about the axis of revolution $x = -1$.

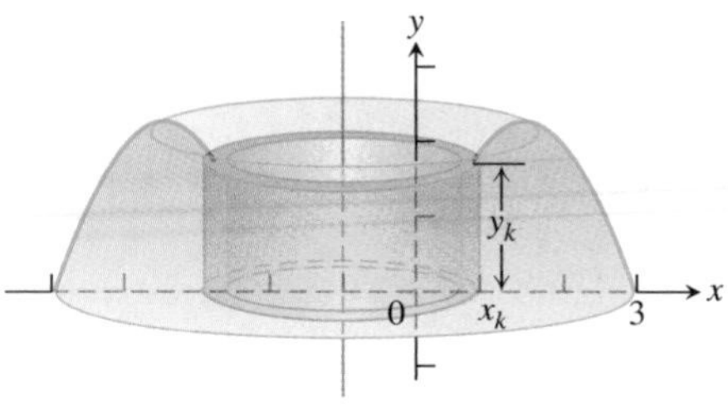

FIGURE 6.18 A cylindrical shell of height y_k obtained by rotating a vertical strip of thickness Δx about the line $x = -1$. The outer radius of the cylinder occurs at x_k, where the height of the parabola is $y_k = 3x_k - x_k^2$ (Example 1).

of y. (These x-values are the inner and outer radii for a typical washer, leading to complicated formulas.) Instead of rotating a horizontal strip of thickness Δy, we rotate a *vertical strip* of thickness Δx. This rotation produces a *cylindrical shell* of height y_k above a point x_k within the base of the vertical strip, and of thickness Δx. An example of a cylindrical shell is shown as the orange-shaded region in Figure 6.18. We can think of the cylindrical shell shown in the figure as approximating a slice of the solid obtained by cutting straight down through it, parallel to the axis of revolution, all the way around close to the inside hole. We then cut another cylindrical slice around the enlarged hole, then another, and so on, obtaining n cylinders. The radii of the cylinders gradually increase, and the heights of the cylinders follow the contour of the parabola: shorter to taller, then back to shorter (Figure 6.17a).

Each slice is sitting over a subinterval of the x-axis of length (width) Δx. Its radius is approximately $(1 + x_k)$, and its height is approximately $3x_k - x_k^2$. If we unroll the cylinder at x_k and flatten it out, it becomes (approximately) a rectangular slab with thickness Δx (Figure 6.19). The outer circumference of the kth cylinder is $2\pi \cdot \text{radius} = 2\pi(1 + x_k)$, and this is the length of the rolled-out rectangular slab. Its volume is approximated by that of a rectangular solid,

$$\begin{aligned}\Delta V_k &= \text{circumference} \times \text{height} \times \text{thickness} \\ &= 2\pi(1 + x_k) \cdot \left(3x_k - x_k^2\right) \cdot \Delta x.\end{aligned}$$

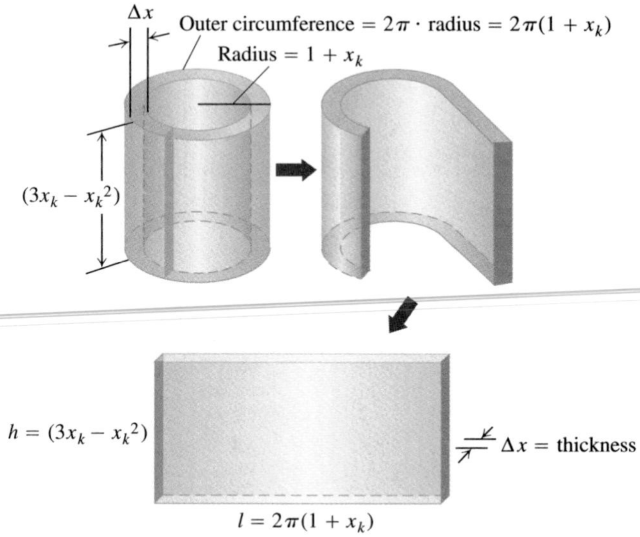

FIGURE 6.19 Imagine cutting and unrolling a cylindrical shell to get a flat (nearly) rectangular solid (Example 1).

Summing together the volumes ΔV_k of the individual cylindrical shells over the interval $[0, 3]$ gives the Riemann sum

$$\sum_{k=1}^{n} \Delta V_k = \sum_{k=1}^{n} 2\pi(x_k + 1)\left(3x_k - x_k^2\right) \Delta x.$$

Taking the limit as the thickness $\Delta x \to 0$ gives the volume integral

$$\begin{aligned} V &= \lim_{n\to\infty} \sum_{k=1}^{n} 2\pi(x_k + 1)(3x_k - x_k^2)\,\Delta x \\ &= \int_0^3 2\pi(x+1)(3x - x^2)\,dx \\ &= \int_0^3 2\pi(3x^2 + 3x - x^3 - x^2)\,dx \\ &= 2\pi \int_0^3 (2x^2 + 3x - x^3)\,dx \\ &= 2\pi\left[\frac{2}{3}x^3 + \frac{3}{2}x^2 - \frac{1}{4}x^4\right]_0^3 = \frac{45\pi}{2}. \end{aligned}$$

We now generalize the procedure used in Example 1.

The Shell Method

Suppose the region bounded by the graph of a nonnegative continuous function $y = f(x)$ and the x-axis over the finite closed interval $[a, b]$ lies to the right of the vertical line $x = L$ (Figure 6.20a). We assume $a \geq L$, so the vertical line may touch the region, but not pass through it. We generate a solid S by rotating this region about the vertical line L.

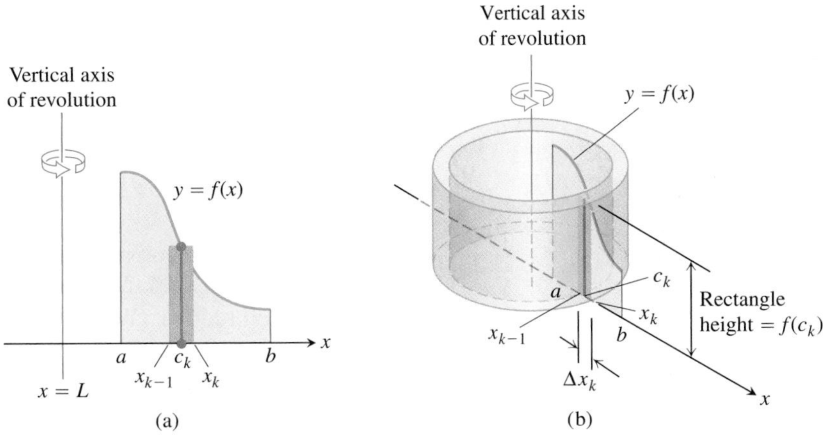

FIGURE 6.20 When the region shown in (a) is revolved about the vertical line $x = L$, a solid is produced which can be sliced into cylindrical shells. A typical shell is shown in (b).

Let P be a partition of the interval $[a, b]$ by the points $a = x_0 < x_1 < \cdots < x_n = b$, and let c_k be the midpoint of the kth subinterval $[x_{k-1}, x_k]$. We approximate the region in Figure 6.20a with rectangles based on this partition of $[a, b]$. A typical approximating rectangle has height $f(c_k)$ and width $\Delta x_k = x_k - x_{k-1}$. If this rectangle is rotated about the vertical line $x = L$, then a shell is swept out, as in Figure 6.20b. A formula from geometry tells us that the volume of the shell swept out by the rectangle is

$$\begin{aligned} \Delta V_k &= 2\pi \times \text{average shell radius} \times \text{shell height} \times \text{thickness} \\ &= 2\pi \cdot (c_k - L) \cdot f(c_k) \cdot \Delta x_k. \end{aligned}$$

We approximate the volume of the solid S by summing the volumes of the shells swept out by the n rectangles based on P:

$$V \approx \sum_{k=1}^{n} \Delta V_k.$$

The limit of this Riemann sum as $\|P\| \to 0$ gives the volume of the solid as a definite integral:

$$V = \lim_{\|P\| \to 0} \sum_{k=1}^{n} \Delta V_k = \int_a^b 2\pi(\text{shell radius})(\text{shell height})\, dx.$$

$$= \int_a^b 2\pi(x - L)f(x)\, dx.$$

We refer to the variable of integration, here x, as the **thickness variable**. We use the first integral, rather than the second containing a formula for the integrand, to emphasize the *process* of the shell method. This will allow for rotations about a horizontal line L as well.

Shell Formula for Revolution About a Vertical Line

The volume of the solid generated by revolving the region between the x-axis and the graph of a continuous function $y = f(x) \geq 0, L \leq a \leq x \leq b$, about a vertical line $x = L$ is

$$V = \int_a^b 2\pi \begin{pmatrix}\text{shell}\\ \text{radius}\end{pmatrix}\begin{pmatrix}\text{shell}\\ \text{height}\end{pmatrix} dx.$$

EXAMPLE 2 Cylindrical Shells Revolving About the y-Axis

The region bounded by the curve $y = \sqrt{x}$, the x-axis, and the line $x = 4$ is revolved about the y-axis to generate a solid. Find the volume of the solid.

Solution Sketch the region and draw a line segment across it *parallel* to the axis of revolution (Figure 6.21a). Label the segment's height (shell height) and distance from the axis of revolution (shell radius). (We drew the shell in Figure 6.21b, but you need not do that.)

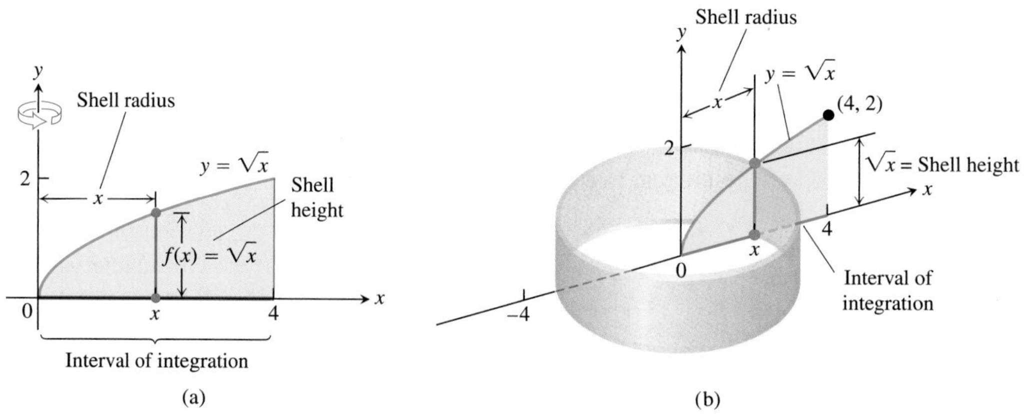

FIGURE 6.21 (a) The region, shell dimensions, and interval of integration in Example 2. (b) The shell swept out by the vertical segment in part (a) with a width Δx.

The shell thickness variable is x, so the limits of integration for the shell formula are $a = 0$ and $b = 4$ (Figure 6.21). The volume is then

$$\begin{aligned} V &= \int_a^b 2\pi \begin{pmatrix}\text{shell}\\ \text{radius}\end{pmatrix}\begin{pmatrix}\text{shell}\\ \text{height}\end{pmatrix} dx \\ &= \int_0^4 2\pi(x)\left(\sqrt{x}\right) dx \\ &= 2\pi \int_0^4 x^{3/2}\, dx = 2\pi \left[\frac{2}{5}x^{5/2}\right]_0^4 = \frac{128\pi}{5}. \end{aligned}$$

■

So far, we have used vertical axes of revolution. For horizontal axes, we replace the x's with y's.

EXAMPLE 3 Cylindrical Shells Revolving About the x-Axis

The region bounded by the curve $y = \sqrt{x}$, the x-axis, and the line $x = 4$ is revolved about the x-axis to generate a solid. Find the volume of the solid.

Solution Sketch the region and draw a line segment across it *parallel* to the axis of revolution (Figure 6.22a). Label the segment's length (shell height) and distance from the axis of revolution (shell radius). (We drew the shell in Figure 6.22b, but you need not do that.)

In this case, the shell thickness variable is y, so the limits of integration for the shell formula method are $a = 0$ and $b = 2$ (along the y-axis in Figure 6.22). The volume of the solid is

$$\begin{aligned} V &= \int_a^b 2\pi \begin{pmatrix}\text{shell}\\ \text{radius}\end{pmatrix}\begin{pmatrix}\text{shell}\\ \text{height}\end{pmatrix} dy \\ &= \int_0^2 2\pi(y)(4 - y^2)\, dy \\ &= \int_0^2 2\pi(4y - y^3)\, dy \\ &= 2\pi\left[2y^2 - \frac{y^4}{4}\right]_0^2 = 8\pi. \end{aligned}$$

■

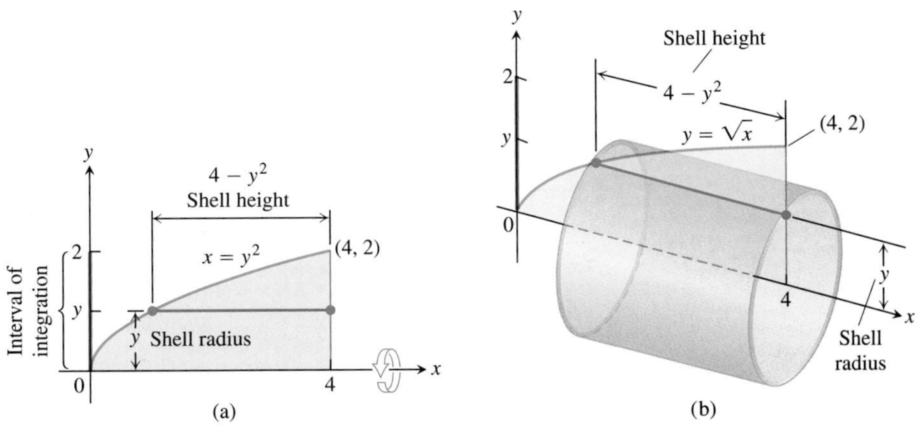

FIGURE 6.22 (a) The region, shell dimensions, and interval of integration in Example 3. (b) The shell swept out by the horizontal segment in part (a) with a width Δy.

> **Summary of the Shell Method**
>
> Regardless of the position of the axis of revolution (horizontal or vertical), the steps for implementing the shell method are these.
>
> 1. *Draw the region and sketch a line segment* across it *parallel* to the axis of revolution. *Label* the segment's height or length (shell height) and distance from the axis of revolution (shell radius).
> 2. *Find* the limits of integration for the thickness variable.
> 3. *Integrate* the product 2π (shell radius) (shell height) with respect to the thickness variable (x or y) to find the volume.

The shell method gives the same answer as the washer method when both are used to calculate the volume of a region. We do not prove that result here, but it is illustrated in Exercises 33 and 34. Both volume formulas are actually special cases of a general volume formula we look at in studying double and triple integrals in Chapter 15. That general formula also allows for computing volumes of solids other than those swept out by regions of revolution.

EXERCISES 6.2

In Exercises 1–6, use the shell method to find the volumes of the solids generated by revolving the shaded region about the indicated axis.

1.

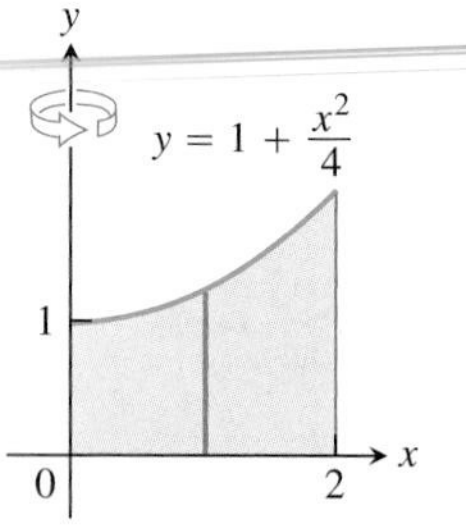

2.

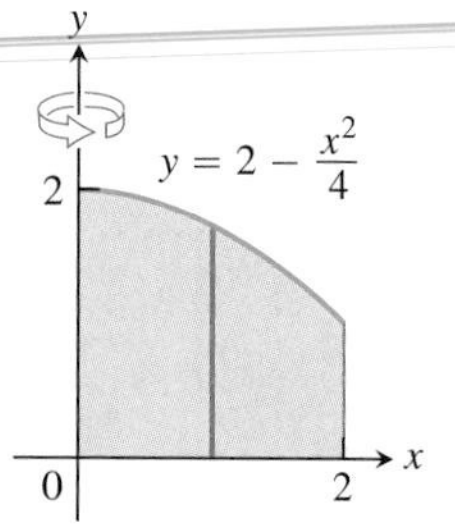

3.

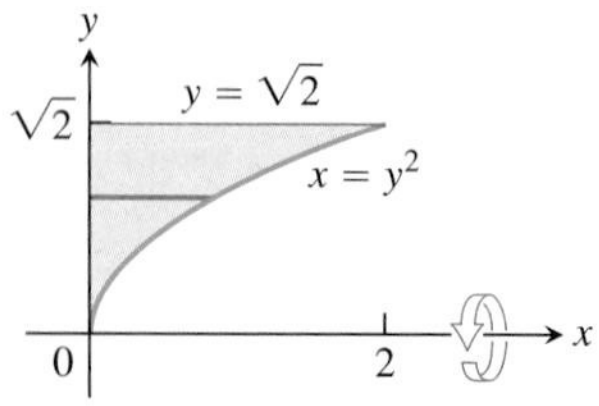

4.

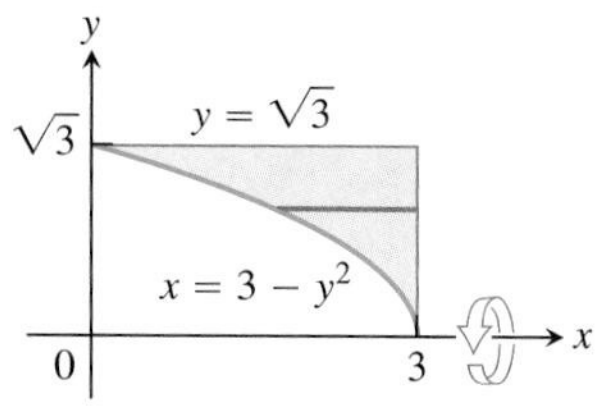

5. The y-axis

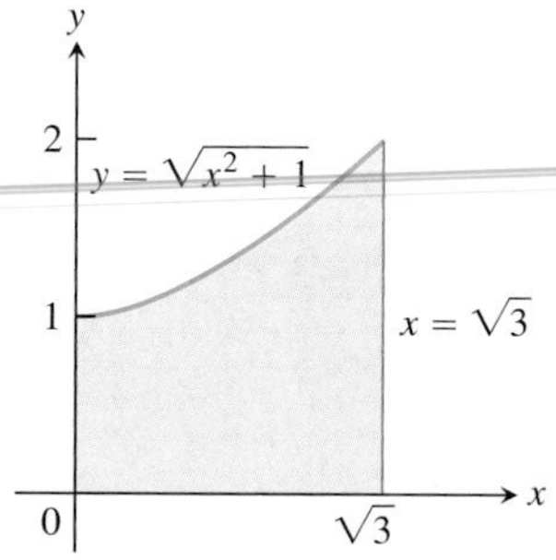

6. The y-axis

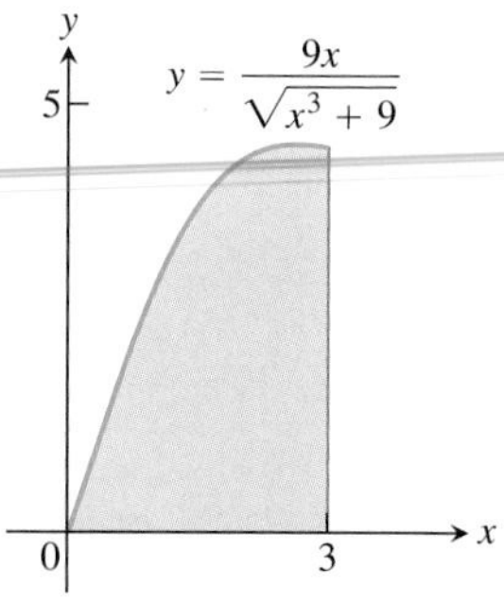

Revolution About the y-Axis

Use the shell method to find the volumes of the solids generated by revolving the regions bounded by the curves and lines in Exercises 7–14 about the y-axis.

7. $y = x, \quad y = -x/2, \quad x = 2$

8. $y = 2x, \quad y = x/2, \quad x = 1$

9. $y = x^2, \quad y = 2 - x, \quad x = 0$, for $x \geq 0$

10. $y = 2 - x^2, \quad y = x^2, \quad x = 0$

11. $y = 2x - 1, \quad y = \sqrt{x}, \quad x = 0$

12. $y = 3/(2\sqrt{x}), \quad y = 0, \quad x = 1, \quad x = 4$

13. Let $f(x) = \begin{cases} (\sin x)/x, & 0 < x \le \pi \\ 1, & x = 0 \end{cases}$

a. Show that $x f(x) = \sin x, 0 \le x \le \pi$.

b. Find the volume of the solid generated by revolving the shaded region about the y-axis.

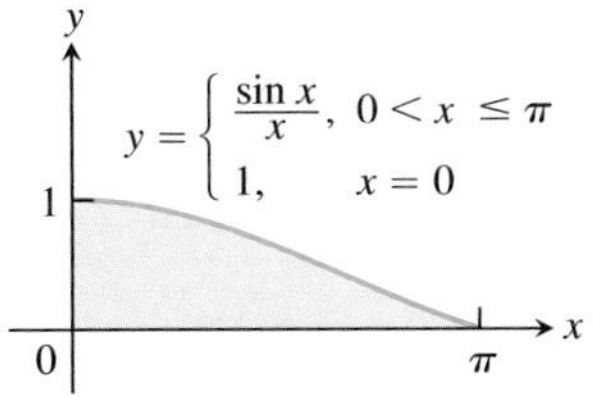

14. Let $g(x) = \begin{cases} (\tan x)^2/x, & 0 < x \le \pi/4 \\ 0, & x = 0 \end{cases}$

a. Show that $x g(x) = (\tan x)^2, 0 \le x \le \pi/4$.

b. Find the volume of the solid generated by revolving the shaded region about the y-axis.

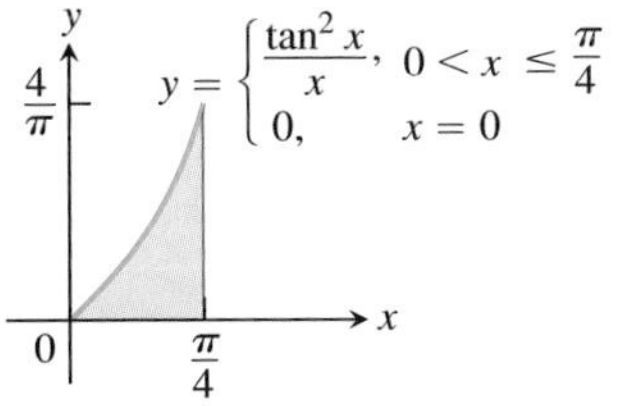

Revolution About the x-Axis

Use the shell method to find the volumes of the solids generated by revolving the regions bounded by the curves and lines in Exercises 15–22 about the x-axis.

15. $x = \sqrt{y}, \quad x = -y, \quad y = 2$

16. $x = y^2, \quad x = -y, \quad y = 2, \quad y \ge 0$

17. $x = 2y - y^2, \quad x = 0$

18. $x = 2y - y^2, \quad x = y$

19. $y = |x|, \quad y = 1$

20. $y = x, \quad y = 2x, \quad y = 2$

21. $y = \sqrt{x}, \quad y = 0, \quad y = x - 2$

22. $y = \sqrt{x}, \quad y = 0, \quad y = 2 - x$

Revolution About Horizontal Lines

In Exercises 23 and 24, use the shell method to find the volumes of the solids generated by revolving the shaded regions about the indicated axes.

23. **a.** The x-axis **b.** The line $y = 1$

c. The line $y = 8/5$ **d.** The line $y = -2/5$

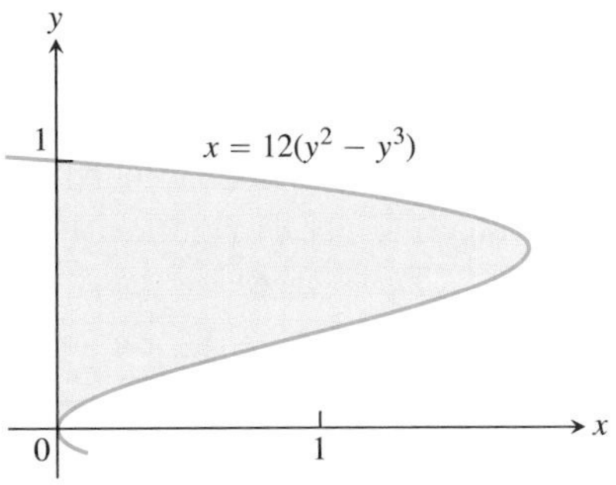

24. **a.** The x-axis **b.** The line $y = 2$

c. The line $y = 5$ **d.** The line $y = -5/8$

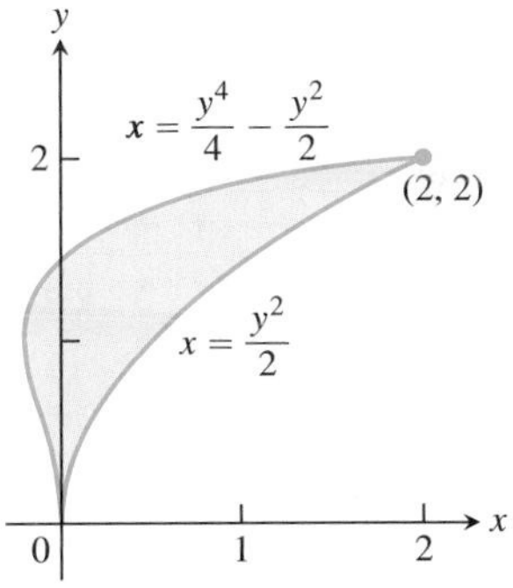

Comparing the Washer and Shell Models

For some regions, both the washer and shell methods work well for the solid generated by revolving the region about the coordinate axes, but this is not always the case. When a region is revolved about the y-axis, for example, and washers are used, we must integrate with respect to y. It may not be possible, however, to express the integrand in terms of y. In such a case, the shell method allows us to integrate with respect to x instead. Exercises 25 and 26 provide some insight.

25. Compute the volume of the solid generated by revolving the region bounded by $y = x$ and $y = x^2$ about each coordinate axis using

a. the shell method. **b.** the washer method.

26. Compute the volume of the solid generated by revolving the triangular region bounded by the lines $2y = x + 4$, $y = x$, and $x = 0$ about

a. the x-axis using the washer method.

b. the y-axis using the shell method.

c. the line $x = 4$ using the shell method.

d. the line $y = 8$ using the washer method.

Choosing Shells or Washers

In Exercises 27–32, find the volumes of the solids generated by revolving the regions about the given axes. If you think it would be better to use washers in any given instance, feel free to do so.

27. The triangle with vertices (1, 1), (1, 2), and (2, 2) about

a. the x-axis **b.** the y-axis

c. the line $x = 10/3$ **d.** the line $y = 1$

28. The region bounded by $y = \sqrt{x}$, $y = 2$, $x = 0$ about

a. the x-axis **b.** the y-axis

c. the line $x = 4$ **d.** the line $y = 2$

29. The region in the first quadrant bounded by the curve $x = y - y^3$ and the y-axis about

a. the x-axis **b.** the line $y = 1$

30. The region in the first quadrant bounded by $x = y - y^3$, $x = 1$, and $y = 1$ about

a. the x-axis **b.** the y-axis

c. the line $x = 1$ **d.** the line $y = 1$

31. The region bounded by $y = \sqrt{x}$ and $y = x^2/8$ about

a. the x-axis **b.** the y-axis

32. The region bounded by $y = 2x - x^2$ and $y = x$ about

a. the y-axis **b.** the line $x = 1$

33. The region in the first quadrant that is bounded above by the curve $y = 1/x^{1/4}$, on the left by the line $x = 1/16$, and below by the line $y = 1$ is revolved about the x-axis to generate a solid. Find the volume of the solid by

a. the washer method. **b.** the shell method.

34. The region in the first quadrant that is bounded above by the curve $y = 1/\sqrt{x}$, on the left by the line $x = 1/4$, and below by the line $y = 1$ is revolved about the y-axis to generate a solid. Find the volume of the solid by

a. the washer method. **b.** the shell method.

Choosing Disks, Washers, or Shells

35. The region shown here is to be revolved about the x-axis to generate a solid. Which of the methods (disk, washer, shell) could you use to find the volume of the solid? How many integrals would be required in each case? Explain.

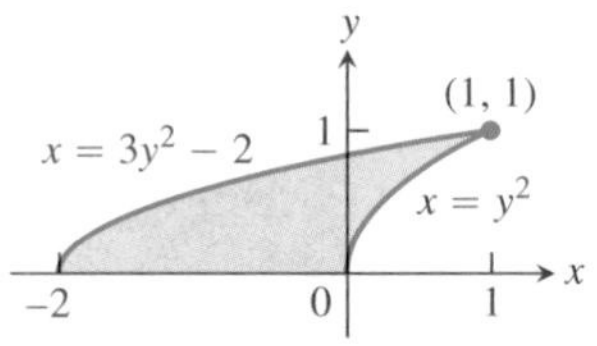

36. The region shown here is to be revolved about the y-axis to generate a solid. Which of the methods (disk, washer, shell) could you use to find the volume of the solid? How many integrals would be required in each case? Give reasons for your answers.

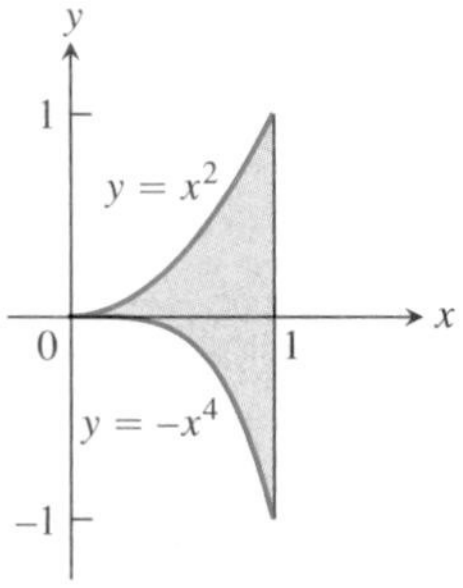

6.3 Lengths of Plane Curves

HISTORICAL BIOGRAPHY

Archimedes
(287–212 B.C.)

We know what is meant by the length of a straight line segment, but without calculus, we have no precise notion of the length of a general winding curve. The idea of approximating the length of a curve running from point A to point B by subdividing the curve into many pieces and joining successive points of division by straight line segments dates back to the ancient Greeks. Archimedes used this method to approximate the circumference of a circle by inscribing a polygon of n sides and then using geometry to compute its perimeter

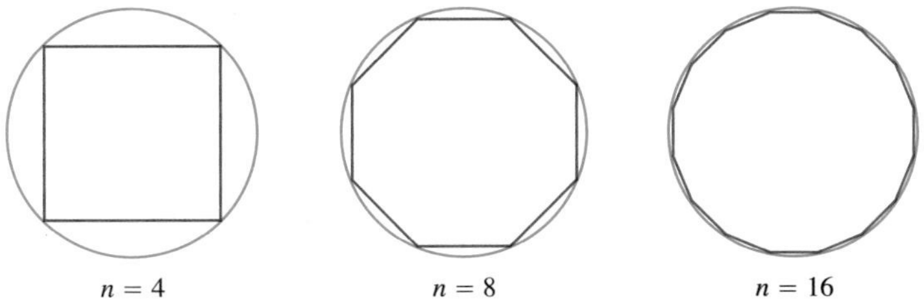

FIGURE 6.23 Archimedes used the perimeters of inscribed polygons to approximate the circumference of a circle. For $n = 96$ the approximation method gives $\pi \approx 3.14103$ as the circumference of the unit circle.

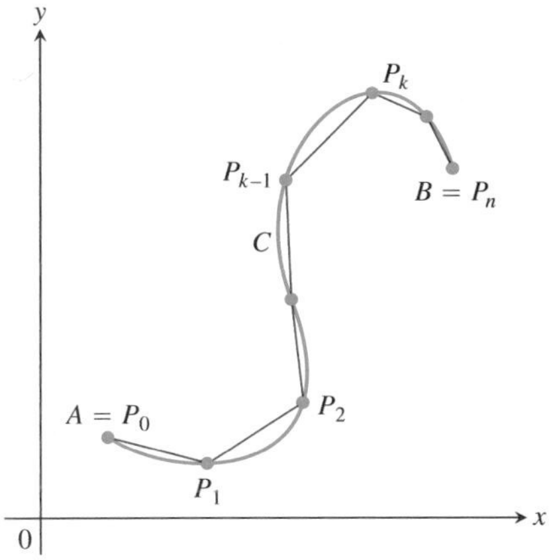

FIGURE 6.24 The curve C defined parametrically by the equations $x = f(t)$ and $y = g(t)$, $a \le t \le b$. The length of the curve from A to B is approximated by the sum of the lengths of the polygonal path (straight line segments) starting at $A = P_0$, then to P_1, and so on, ending at $B = P_n$.

(Figure 6.23). The extension of this idea to a more general curve is displayed in Figure 6.24, and we now describe how that method works.

Length of a Parametrically Defined Curve

Let C be a curve given parametrically by the equations

$$x = f(t) \quad \text{and} \quad y = g(t), \qquad a \le t \le b.$$

We assume the functions f and g have continuous first derivatives on the interval $[a, b]$ that are not simultaneously zero. Such functions are said to be **continuously differentiable**, and the curve C defined by them is called a **smooth curve**. It may be helpful to imagine the curve as the path of a particle moving from point $A = (f(a), g(a))$ at time $t = a$ to point $B = (f(b), g(b))$ in Figure 6.24. We subdivide the path (or arc) AB into n pieces at points $A = P_0, P_1, P_2, \dots, P_n = B$. These points correspond to a partition of the interval $[a, b]$ by $a = t_0 < t_1 < t_2 < \cdots < t_n = b$, where $P_k = (f(t_k), g(t_k))$. Join successive points of this subdivision by straight line segments (Figure 6.24). A representative line segment has length

$$\begin{aligned} L_k &= \sqrt{(\Delta x_k)^2 + (\Delta y_k)^2} \\ &= \sqrt{[f(t_k) - f(t_{k-1})]^2 + [g(t_k) - g(t_{k-1})]^2} \end{aligned}$$

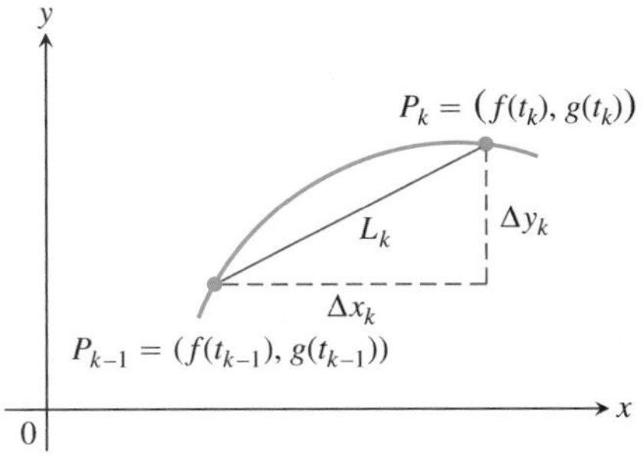

FIGURE 6.25 The arc $P_{k-1}P_k$ is approximated by the straight line segment shown here, which has length $L_k = \sqrt{(\Delta x_k)^2 + (\Delta y_k)^2}$.

(see Figure 6.25). If Δt_k is small, the length L_k is approximately the length of arc $P_{k-1}P_k$. By the Mean Value Theorem there are numbers t_k^* and t_k^{**} in $[t_{k-1}, t_k]$ such that

$$\begin{aligned} \Delta x_k &= f(t_k) - f(t_{k-1}) = f'(t_k^*)\,\Delta t_k, \\ \Delta y_k &= g(t_k) - g(t_{k-1}) = g'(t_k^{**})\,\Delta t_k. \end{aligned}$$

Assuming the path from A to B is traversed exactly once as t increases from $t = a$ to $t = b$, with no doubling back or retracing, an intuitive approximation to the "length" of the curve AB is the sum of all the lengths L_k:

$$\begin{aligned} \sum_{k=1}^{n} L_k &= \sum_{k=1}^{n} \sqrt{(\Delta x_k)^2 + (\Delta y_k)^2} \\ &= \sum_{k=1}^{n} \sqrt{[f'(t_k^*)]^2 + [g'(t_k^{**})]^2}\,\Delta t_k. \end{aligned}$$

Although this last sum on the right is not exactly a Riemann sum (because f' and g' are evaluated at different points), a theorem in advanced calculus guarantees its limit, as the norm of the partition tends to zero, to be the definite integral

$$\lim_{n\to\infty} \sum_{k=1}^{n} \sqrt{[f'(t_k^*)]^2 + [g'(t_k^{**})]^2}\, \Delta t_k = \int_a^b \sqrt{[f'(t)]^2 + [g'(t)]^2}\, dt.$$

Therefore, it is reasonable to define the length of the curve from A to B as this integral.

DEFINITION Length of a Parametric Curve

If a curve C is defined parametrically by $x = f(t)$ and $y = g(t)$, $a \le t \le b$, where f' and g' are continuous and not simultaneously zero on $[a, b]$, and C is traversed exactly once as t increases from $t = a$ to $t = b$, then **the length of C** is the definite integral

$$L = \int_a^b \sqrt{[f'(t)]^2 + [g'(t)]^2}\, dt.$$

A smooth curve C does not double back or reverse the direction of motion over the time interval $[a, b]$ since $(f')^2 + (g')^2 > 0$ throughout the interval.

If $x = f(t)$ and $y = g(t)$, then using the Leibniz notation we have the following result for arc length:

$$L = \int_a^b \sqrt{\left(\frac{dx}{dt}\right)^2 + \left(\frac{dy}{dt}\right)^2}\, dt. \qquad (1)$$

What if there are two different parametrizations for a curve C whose length we want to find; does it matter which one we use? The answer, from advanced calculus, is no, as long as the parametrization we choose meets the conditions stated in the definition of the length of C (see Exercise 29).

EXAMPLE 1 The Circumference of a Circle

Find the length of the circle of radius r defined parametrically by

$$x = r\cos t \quad \text{and} \quad y = r\sin t, \quad 0 \le t \le 2\pi.$$

Solution As t varies from 0 to 2π, the circle is traversed exactly once, so the circumference is

$$L = \int_0^{2\pi} \sqrt{\left(\frac{dx}{dt}\right)^2 + \left(\frac{dy}{dt}\right)^2}\, dt.$$

We find

$$\frac{dx}{dt} = -r\sin t, \quad \frac{dy}{dt} = r\cos t$$

and

$$\left(\frac{dx}{dt}\right)^2 + \left(\frac{dy}{dt}\right)^2 = r^2(\sin^2 t + \cos^2 t) = r^2.$$

So

$$L = \int_0^{2\pi} \sqrt{r^2}\, dt = r\Big[t\Big]_0^{2\pi} = 2\pi r. \qquad \blacksquare$$

EXAMPLE 2 Applying the Parametric Formula for Length of a Curve

Find the length of the astroid (Figure 6.26)

$$x = \cos^3 t, \qquad y = \sin^3 t, \qquad 0 \le t \le 2\pi.$$

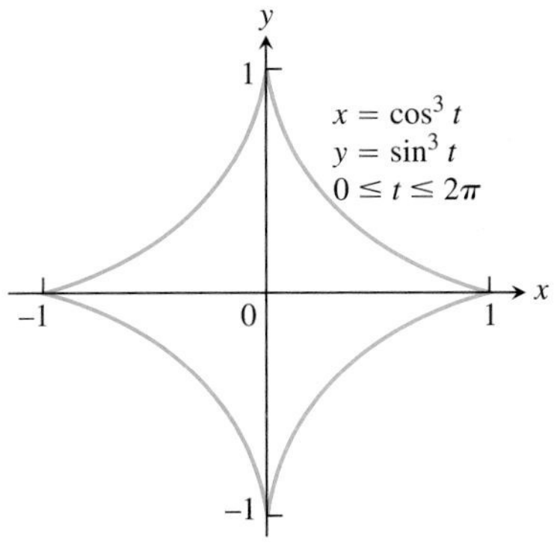

FIGURE 6.26 The astroid in Example 2.

Solution Because of the curve's symmetry with respect to the coordinate axes, its length is four times the length of the first-quadrant portion. We have

$$x = \cos^3 t, \qquad y = \sin^3 t$$

$$\left(\frac{dx}{dt}\right)^2 = [3\cos^2 t(-\sin t)]^2 = 9\cos^4 t \sin^2 t$$

$$\left(\frac{dy}{dt}\right)^2 = [3\sin^2 t(\cos t)]^2 = 9\sin^4 t \cos^2 t$$

$$\sqrt{\left(\frac{dx}{dt}\right)^2 + \left(\frac{dy}{dt}\right)^2} = \sqrt{9\cos^2 t \sin^2 t\underbrace{(\cos^2 t + \sin^2 t)}_{1}}$$

$$= \sqrt{9\cos^2 t \sin^2 t}$$

$$= 3|\cos t \sin t|$$

$$= 3\cos t \sin t. \qquad \begin{array}{l}\cos t \sin t \ge 0 \text{ for}\\ 0 \le t \le \pi/2\end{array}$$

Therefore,

$$\text{Length of first-quadrant portion} = \int_0^{\pi/2} 3\cos t \sin t\, dt$$

$$= \frac{3}{2}\int_0^{\pi/2} \sin 2t\, dt \qquad \begin{array}{l}\cos t \sin t =\\ (1/2)\sin 2t\end{array}$$

$$= -\frac{3}{4}\cos 2t\Big]_0^{\pi/2} = \frac{3}{2}.$$

The length of the astroid is four times this: $4(3/2) = 6$. ■

HISTORICAL BIOGRAPHY

Gregory St. Vincent
(1584–1667)

Length of a Curve $y = f(x)$

Given a continuously differentiable function $y = f(x)$, $a \le x \le b$, we can assign $x = t$ as a parameter. The graph of the function f is then the curve C defined parametrically by

$$x = t \quad \text{and} \quad y = f(t), \qquad a \le t \le b,$$

a special case of what we considered before. Then,

$$\frac{dx}{dt} = 1 \quad \text{and} \quad \frac{dy}{dt} = f'(t).$$

From our calculations in Section 3.5, we have

$$\frac{dy}{dx} = \frac{dy/dt}{dx/dt} = f'(t)$$

giving

$$\begin{aligned}\left(\frac{dx}{dt}\right)^2 + \left(\frac{dy}{dt}\right)^2 &= 1 + [f'(t)]^2 \\ &= 1 + \left(\frac{dy}{dx}\right)^2 \\ &= 1 + [f'(x)]^2.\end{aligned}$$

Substitution into Equation (1) gives the arc length formula for the graph of $y = f(x)$.

Formula for the Length of $y = f(x)$, $a \le x \le b$
If f is continuously differentiable on the closed interval $[a, b]$, the length of the curve (graph) $y = f(x)$ from $x = a$ to $x = b$ is

$$L = \int_a^b \sqrt{1 + \left(\frac{dy}{dx}\right)^2}\, dx = \int_a^b \sqrt{1 + [f'(x)]^2}\, dx. \qquad (2)$$

EXAMPLE 3 Applying the Arc Length Formula for a Graph

Find the length of the curve

$$y = \frac{4\sqrt{2}}{3}x^{3/2} - 1, \qquad 0 \le x \le 1.$$

Solution We use Equation (2) with $a = 0$, $b = 1$, and

$$\begin{aligned} y &= \frac{4\sqrt{2}}{3}x^{3/2} - 1 \\ \frac{dy}{dx} &= \frac{4\sqrt{2}}{3} \cdot \frac{3}{2}x^{1/2} = 2\sqrt{2}x^{1/2} \\ \left(\frac{dy}{dx}\right)^2 &= \left(2\sqrt{2}x^{1/2}\right)^2 = 8x.\end{aligned}$$

The length of the curve from $x = 0$ to $x = 1$ is

$$\begin{aligned} L &= \int_0^1 \sqrt{1 + \left(\frac{dy}{dx}\right)^2}\, dx = \int_0^1 \sqrt{1 + 8x}\, dx && \text{Eq. (2) with } a = 0, b = 1 \\ &= \frac{2}{3} \cdot \frac{1}{8}(1 + 8x)^{3/2}\Big]_0^1 = \frac{13}{6}. && \text{Let } u = 1 + 8x \text{, integrate, and replace } u \text{ by } 1 + 8x.\end{aligned}$$

■

Dealing with Discontinuities in dy/dx

At a point on a curve where dy/dx fails to exist, dx/dy may exist and we may be able to find the curve's length by expressing x as a function of y and applying the following analogue of Equation (2):

> **Formula for the Length of $x = g(y)$, $\quad c \le y \le d$**
> If g is continuously differentiable on $[c, d]$, the length of the curve $x = g(y)$ from $y = c$ to $y = d$ is
>
> $$L = \int_c^d \sqrt{1 + \left(\frac{dx}{dy}\right)^2}\, dy = \int_c^d \sqrt{1 + [g'(y)]^2}\, dy. \qquad (3)$$

EXAMPLE 4 Length of a Graph Which Has a Discontinuity in dy/dx

Find the length of the curve $y = (x/2)^{2/3}$ from $x = 0$ to $x = 2$.

Solution The derivative

$$\frac{dy}{dx} = \frac{2}{3}\left(\frac{x}{2}\right)^{-1/3}\left(\frac{1}{2}\right) = \frac{1}{3}\left(\frac{2}{x}\right)^{1/3}$$

is not defined at $x = 0$, so we cannot find the curve's length with Equation (2).

We therefore rewrite the equation to express x in terms of y:

$$y = \left(\frac{x}{2}\right)^{2/3}$$

$$y^{3/2} = \frac{x}{2} \qquad \text{Raise both sides to the power 3/2.}$$

$$x = 2y^{3/2}. \qquad \text{Solve for } x.$$

From this we see that the curve whose length we want is also the graph of $x = 2y^{3/2}$ from $y = 0$ to $y = 1$ (Figure 6.27).

The derivative

$$\frac{dx}{dy} = 2\left(\frac{3}{2}\right)y^{1/2} = 3y^{1/2}$$

is continuous on $[0, 1]$. We may therefore use Equation (3) to find the curve's length:

$$L = \int_c^d \sqrt{1 + \left(\frac{dx}{dy}\right)^2}\, dy = \int_0^1 \sqrt{1 + 9y}\, dy \qquad \text{Eq. (3) with } c = 0, d = 1.$$

$$= \frac{1}{9}\cdot\frac{2}{3}(1 + 9y)^{3/2}\Big]_0^1 \qquad \text{Let } u = 1 + 9y,\ du/9 = dy, \text{ integrate, and substitute back.}$$

$$= \frac{2}{27}\left(10\sqrt{10} - 1\right) \approx 2.27.$$

■

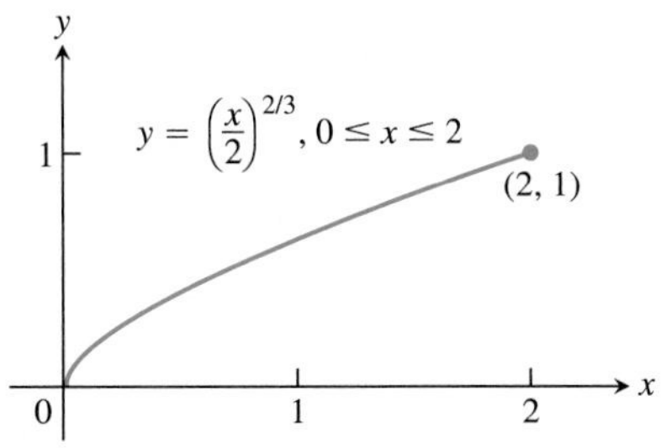

FIGURE 6.27 The graph of $y = (x/2)^{2/3}$ from $x = 0$ to $x = 2$ is also the graph of $x = 2y^{3/2}$ from $y = 0$ to $y = 1$ (Example 4).

HISTORICAL BIOGRAPHY

James Gregory
(1638–1675)

The Short Differential Formula

Equation (1) is frequently written in terms of differentials in place of derivatives. This is done formally by writing $(dt)^2$ under the radical in place of the dt outside the radical, and then writing

$$\left(\frac{dx}{dt}\right)^2(dt)^2 = \left(\frac{dx}{dt}\,dt\right)^2 = (dx)^2$$

and

$$\left(\frac{dy}{dt}\right)^2(dt)^2 = \left(\frac{dy}{dt}\,dt\right)^2 = (dy)^2.$$

It is also customary to eliminate the parentheses in $(dx)^2$ and write dx^2 instead, so that Equation (1) is written

$$L = \int \sqrt{dx^2 + dy^2}. \tag{4}$$

We can think of these differentials as a way to summarize and simplify the properties of integrals. Differentials are given a precise mathematical definition in a more advanced text.

To do an integral computation, dx and dy must both be expressed in terms of one and the same variable, and appropriate limits must be supplied in Equation (4).

A useful way to remember Equation (4) is to write

$$ds = \sqrt{dx^2 + dy^2} \tag{5}$$

and treat ds as the differential of arc length, which can be integrated between appropriate limits to give the total length of a curve. Figure 6.28a gives the exact interpretation of ds corresponding to Equation (5). Figure 6.28b is not strictly accurate but is to be thought of as a simplified approximation of Figure 6.28a.

With Equation (5) in mind, the quickest way to recall the formulas for arc length is to remember the equation

$$\text{Arc length} = \int ds.$$

If we write $L = \int ds$ and have the graph of $y = f(x)$, we can rewrite Equation (5) to get

$$ds = \sqrt{dx^2 + dy^2} = \sqrt{dx^2 + \frac{dy^2}{dx^2}\,dx^2} = \sqrt{1 + \frac{dy^2}{dx^2}}\,dx = \sqrt{1 + \left(\frac{dy}{dx}\right)^2}\,dx,$$

resulting in Equation (2). If we have instead $x = g(y)$, we rewrite Equation (5)

$$ds = \sqrt{dx^2 + dy^2} = \sqrt{dy^2 + \frac{dx^2}{dy^2}\,dy^2} = \sqrt{1 + \frac{dx^2}{dy^2}}\,dy = \sqrt{1 + \left(\frac{dx}{dy}\right)^2}\,dy,$$

and obtain Equation (3).

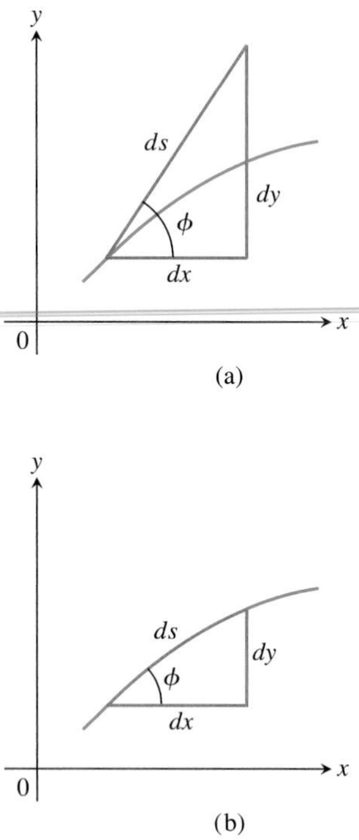

FIGURE 6.28 Diagrams for remembering the equation $ds = \sqrt{dx^2 + dy^2}$.

EXERCISES 6.3

Lengths of Parametrized Curves

Find the lengths of the curves in Exercises 1–6.

1. $x = 1 - t, \quad y = 2 + 3t, \quad -2/3 \le t \le 1$

2. $x = \cos t, \quad y = t + \sin t, \quad 0 \le t \le \pi$

3. $x = t^3, \quad y = 3t^2/2, \quad 0 \le t \le \sqrt{3}$

4. $x = t^2/2, \quad y = (2t + 1)^{3/2}/3, \quad 0 \le t \le 4$

5. $x = (2t + 3)^{3/2}/3, \quad y = t + t^2/2, \quad 0 \le t \le 3$

6. $x = 8\cos t + 8t\sin t, \quad y = 8\sin t - 8t\cos t, \quad 0 \le t \le \pi/2$

Finding Lengths of Curves

Find the lengths of the curves in Exercises 7–16. If you have a grapher, you may want to graph these curves to see what they look like.

7. $y = (1/3)(x^2 + 2)^{3/2}$ from $x = 0$ to $x = 3$

8. $y = x^{3/2}$ from $x = 0$ to $x = 4$

9. $x = (y^3/3) + 1/(4y)$ from $y = 1$ to $y = 3$
(*Hint:* $1 + (dx/dy)^2$ is a perfect square.)

10. $x = (y^{3/2}/3) - y^{1/2}$ from $y = 1$ to $y = 9$
(*Hint:* $1 + (dx/dy)^2$ is a perfect square.)

11. $x = (y^4/4) + 1/(8y^2)$ from $y = 1$ to $y = 2$
(*Hint:* $1 + (dx/dy)^2$ is a perfect square.)

12. $x = (y^3/6) + 1/(2y)$ from $y = 2$ to $y = 3$
(*Hint:* $1 + (dx/dy)^2$ is a perfect square.)

13. $y = (3/4)x^{4/3} - (3/8)x^{2/3} + 5, \quad 1 \le x \le 8$

14. $y = (x^3/3) + x^2 + x + 1/(4x + 4), \quad 0 \le x \le 2$

15. $x = \int_0^y \sqrt{\sec^4 t - 1}\, dt, \quad -\pi/4 \le y \le \pi/4$

16. $y = \int_{-2}^x \sqrt{3t^4 - 1}\, dt, \quad -2 \le x \le -1$

T Finding Integrals for Lengths of Curves

In Exercises 17–24, do the following.

a. Set up an integral for the length of the curve.

b. Graph the curve to see what it looks like.

c. Use your grapher's or computer's integral evaluator to find the curve's length numerically.

17. $y = x^2, \quad -1 \le x \le 2$

18. $y = \tan x, \quad -\pi/3 \le x \le 0$

19. $x = \sin y, \quad 0 \le y \le \pi$

20. $x = \sqrt{1 - y^2}, \quad -1/2 \le y \le 1/2$

21. $y^2 + 2y = 2x + 1$ from $(-1, -1)$ to $(7, 3)$

22. $y = \sin x - x\cos x, \quad 0 \le x \le \pi$

23. $y = \int_0^x \tan t\, dt, \quad 0 \le x \le \pi/6$

24. $x = \int_0^y \sqrt{\sec^2 t - 1}\, dt, \quad -\pi/3 \le y \le \pi/4$

Theory and Applications

25. Is there a smooth (continuously differentiable) curve $y = f(x)$ whose length over the interval $0 \le x \le a$ is always $\sqrt{2}a$? Give reasons for your answer.

26. Using tangent fins to derive the length formula for curves Assume that f is smooth on $[a, b]$ and partition the interval $[a, b]$ in the usual way. In each subinterval $[x_{k-1}, x_k]$, construct the *tangent fin* at the point $(x_{k-1}, f(x_{k-1}))$, as shown in the accompanying figure.

a. Show that the length of the kth tangent fin over the interval $[x_{k-1}, x_k]$ equals $\sqrt{(\Delta x_k)^2 + (f'(x_{k-1})\,\Delta x_k)^2}$.

b. Show that

$$\lim_{n\to\infty} \sum_{k=1}^{n} (\text{length of } k\text{th tangent fin}) = \int_a^b \sqrt{1 + (f'(x))^2}\, dx,$$

which is the length L of the curve $y = f(x)$ from a to b.

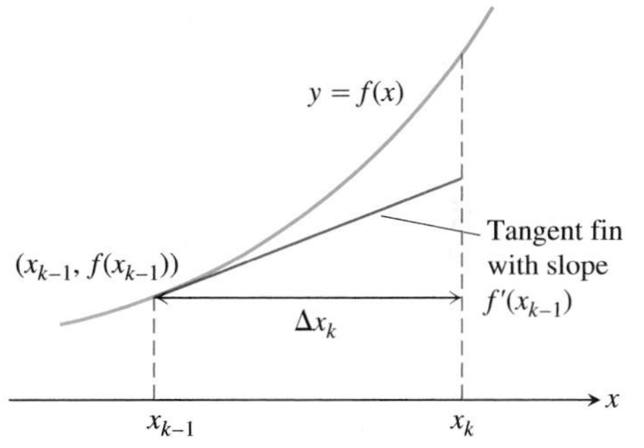

27. a. Find a curve through the point $(1, 1)$ whose length integral is

$$L = \int_1^4 \sqrt{1 + \frac{1}{4x}}\, dx.$$

b. How many such curves are there? Give reasons for your answer.

28. a. Find a curve through the point $(0, 1)$ whose length integral is

$$L = \int_1^2 \sqrt{1 + \frac{1}{y^4}}\, dy.$$

b. How many such curves are there? Give reasons for your answer.

29. Length is independent of parametrization To illustrate the fact that the numbers we get for length do not depend on the way we parametrize our curves (except for the mild restrictions preventing doubling back mentioned earlier), calculate the length of the semicircle $y = \sqrt{1 - x^2}$ with these two different parametrizations:

a. $x = \cos 2t, \quad y = \sin 2t, \quad 0 \le t \le \pi/2$

b. $x = \sin \pi t, \quad y = \cos \pi t, \quad -1/2 \le t \le 1/2$

30. Find the length of one arch of the cycloid $x = a(\theta - \sin \theta)$, $y = a(1 - \cos \theta)$, $0 \le \theta \le 2\pi$, shown in the accompanying figure. A **cycloid** is the curve traced out by a point P on the circumference of a circle rolling along a straight line, such as the x-axis.

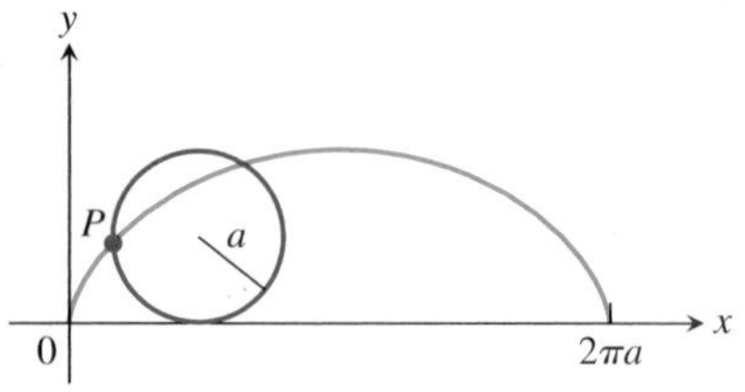

COMPUTER EXPLORATIONS

In Exercises 31–36, use a CAS to perform the following steps for the given curve over the closed interval.

a. Plot the curve together with the polygonal path approximations for $n = 2, 4, 8$ partition points over the interval. (See Figure 6.24.)

b. Find the corresponding approximation to the length of the curve by summing the lengths of the line segments.

c. Evaluate the length of the curve using an integral. Compare your approximations for $n = 2, 4, 8$ with the actual length given by the integral. How does the actual length compare with the approximations as n increases? Explain your answer.

31. $f(x) = \sqrt{1 - x^2}, \quad -1 \le x \le 1$

32. $f(x) = x^{1/3} + x^{2/3}, \quad 0 \le x \le 2$

33. $f(x) = \sin(\pi x^2), \quad 0 \le x \le \sqrt{2}$

34. $f(x) = x^2 \cos x, \quad 0 \le x \le \pi$

35. $f(x) = \dfrac{x - 1}{4x^2 + 1}, \quad -\dfrac{1}{2} \le x \le 1$

36. $f(x) = x^3 - x^2, \quad -1 \le x \le 1$

37. $x = \dfrac{1}{3}t^3, \quad y = \dfrac{1}{2}t^2, \quad 0 \le t \le 1$

38. $x = 2t^3 - 16t^2 + 25t + 5, \quad y = t^2 + t - 3,$ $0 \le t \le 6$

39. $x = t - \cos t, \quad y = 1 + \sin t, \quad -\pi \le t \le \pi$

40. $x = e^t \cos t, \quad y = e^t \sin t, \quad 0 \le t \le \pi$

6.4 Moments and Centers of Mass

Many structures and mechanical systems behave as if their masses were concentrated at a single point, called the *center of mass* (Figure 6.29). It is important to know how to locate this point, and doing so is basically a mathematical enterprise. For the moment, we deal with one- and two-dimensional objects. Three-dimensional objects are best done with the multiple integrals of Chapter 15.

Masses Along a Line

We develop our mathematical model in stages. The first stage is to imagine masses m_1, m_2, and m_3 on a rigid x-axis supported by a fulcrum at the origin.

The resulting system might balance, or it might not. It depends on how large the masses are and how they are arranged.

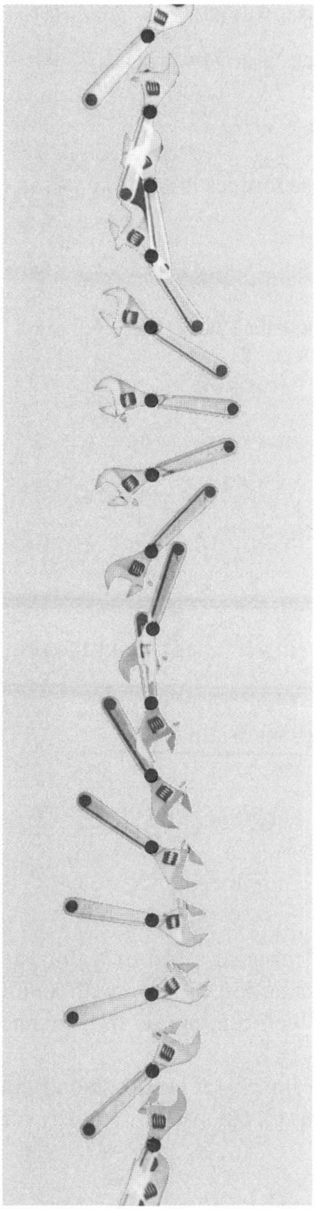
(a)

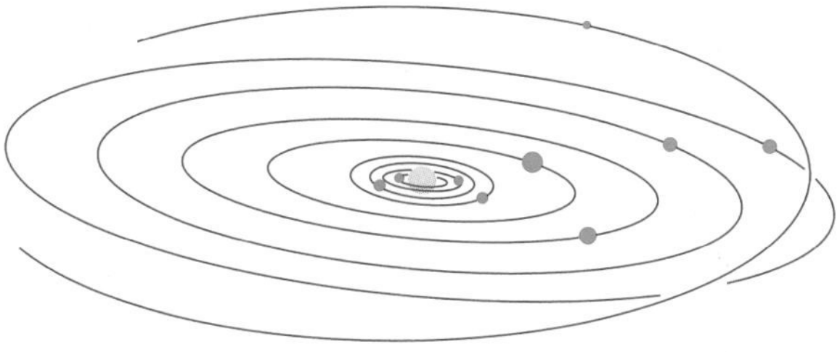
(b)

FIGURE 6.29 (a) The motion of this wrench gliding on ice seems haphazard until we notice that the wrench is simply turning about its center of mass as the center glides in a straight line. (b) The planets, asteroids, and comets of our solar system revolve about their collective center of mass. (It lies inside the sun.)

Each mass m_k exerts a downward force $m_k g$ (the weight of m_k) equal to the magnitude of the mass times the acceleration of gravity. Each of these forces has a tendency to turn the axis about the origin, the way you turn a seesaw. This turning effect, called a **torque**, is measured by multiplying the force $m_k g$ by the signed distance x_k from the point of application to the origin. Masses to the left of the origin exert negative (counterclockwise) torque. Masses to the right of the origin exert positive (clockwise) torque.

The sum of the torques measures the tendency of a system to rotate about the origin. This sum is called the **system torque**.

$$\text{System torque} = m_1 g x_1 + m_2 g x_2 + m_3 g x_3 \tag{1}$$

The system will balance if and only if its torque is zero.

If we factor out the g in Equation (1), we see that the system torque is

$$\underbrace{g}_{\text{a feature of the environment}} \cdot \underbrace{(m_1 x_1 + m_2 x_2 + m_3 x_3)}_{\text{a feature of the system}}$$

Thus, the torque is the product of the gravitational acceleration g, which is a feature of the environment in which the system happens to reside, and the number $(m_1 x_1 + m_2 x_2 + m_3 x_3)$, which is a feature of the system itself, a constant that stays the same no matter where the system is placed.

The number $(m_1 x_1 + m_2 x_2 + m_3 x_3)$ is called the **moment of the system about the origin**. It is the sum of the **moments** $m_1 x_1$, $m_2 x_2$, $m_3 x_3$ of the individual masses.

$$M_0 = \text{Moment of system about origin} = \sum m_k x_k.$$

(We shift to sigma notation here to allow for sums with more terms.)

We usually want to know where to place the fulcrum to make the system balance, that is, at what point $\bar{x}$ to place it to make the torques add to zero.

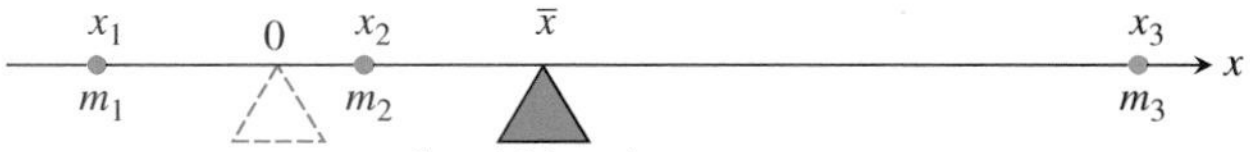

The torque of each mass about the fulcrum in this special location is

$$\begin{aligned}\text{Torque of } m_k \text{ about } \bar{x} &= \begin{pmatrix}\text{signed distance}\\ \text{of } m_k \text{ from } \bar{x}\end{pmatrix}\begin{pmatrix}\text{downward}\\ \text{force}\end{pmatrix}\\ &= (x_k - \bar{x})m_k g.\end{aligned}$$

When we write the equation that says that the sum of these torques is zero, we get an equation we can solve for $\bar{x}$:

$$\begin{aligned}\sum (x_k - \bar{x})m_k g &= 0 && \text{Sum of the torques equals zero}\\ g\sum (x_k - \bar{x})m_k &= 0 && \text{Constant Multiple Rule for Sums}\\ \sum (m_k x_k - \bar{x}m_k) &= 0 && g \text{ divided out, } m_k \text{ distributed}\\ \sum m_k x_k - \sum \bar{x}m_k &= 0 && \text{Difference Rule for Sums}\\ \sum m_k x_k &= \bar{x}\sum m_k && \text{Rearranged, Constant Multiple Rule again}\\ \bar{x} &= \frac{\sum m_k x_k}{\sum m_k}. && \text{Solved for } \bar{x}\end{aligned}$$

This last equation tells us to find $\bar{x}$ by dividing the system's moment about the origin by the system's total mass:

$$\bar{x} = \frac{\sum m_k x_k}{\sum m_k} = \frac{\text{system moment about origin}}{\text{system mass}}.$$

The point $\bar{x}$ is called the system's **center of mass**.

Wires and Thin Rods

In many applications, we want to know the center of mass of a rod or a thin strip of metal. In cases like these where we can model the distribution of mass with a continuous function, the summation signs in our formulas become integrals in a manner we now describe.

Imagine a long, thin strip lying along the x-axis from $x = a$ to $x = b$ and cut into small pieces of mass Δm_k by a partition of the interval $[a, b]$. Choose x_k to be any point in the kth subinterval of the partition.

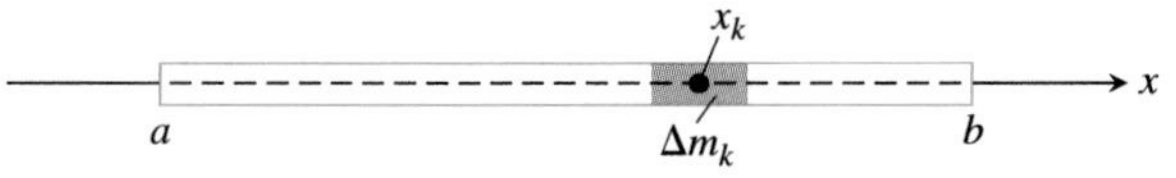

The kth piece is Δx_k units long and lies approximately x_k units from the origin. Now observe three things.

First, the strip's center of mass $\bar{x}$ is nearly the same as that of the system of point masses we would get by attaching each mass Δm_k to the point x_k:

$$\bar{x} \approx \frac{\text{system moment}}{\text{system mass}}.$$

Density
A material's density is its mass per unit volume. In practice, however, we tend to use units we can conveniently measure. For wires, rods, and narrow strips, we use mass per unit length. For flat sheets and plates, we use mass per unit area.

Second, the moment of each piece of the strip about the origin is approximately $x_k \,\Delta m_k$, so the system moment is approximately the sum of the $x_k \,\Delta m_k$:

$$\text{System moment} \approx \sum x_k \,\Delta m_k.$$

Third, if the density of the strip at x_k is $\delta(x_k)$, expressed in terms of mass per unit length and if δ is continuous, then Δm_k is approximately equal to $\delta(x_k)\,\Delta x_k$ (mass per unit length times length):

$$\Delta m_k \approx \delta(x_k)\,\Delta x_k.$$

Combining these three observations gives

$$\bar{x} \approx \frac{\text{system moment}}{\text{system mass}} \approx \frac{\sum x_k \,\Delta m_k}{\sum \Delta m_k} \approx \frac{\sum x_k\,\delta(x_k)\,\Delta x_k}{\sum \delta(x_k)\,\Delta x_k}. \tag{2}$$

The sum in the last numerator in Equation (2) is a Riemann sum for the continuous function $x\delta(x)$ over the closed interval $[a, b]$. The sum in the denominator is a Riemann sum for the function $\delta(x)$ over this interval. We expect the approximations in Equation (2) to improve as the strip is partitioned more finely, and we are led to the equation

$$\bar{x} = \frac{\int_a^b x\delta(x)\,dx}{\int_a^b \delta(x)\,dx}.$$

This is the formula we use to find $\bar{x}$.

Moment, Mass, and Center of Mass of a Thin Rod or Strip Along the x-Axis with Density Function $\delta(x)$

Moment about the origin: $$M_0 = \int_a^b x\delta(x)\,dx \tag{3a}$$

Mass: $$M = \int_a^b \delta(x)\,dx \tag{3b}$$

Center of mass: $$\bar{x} = \frac{M_0}{M} \tag{3c}$$

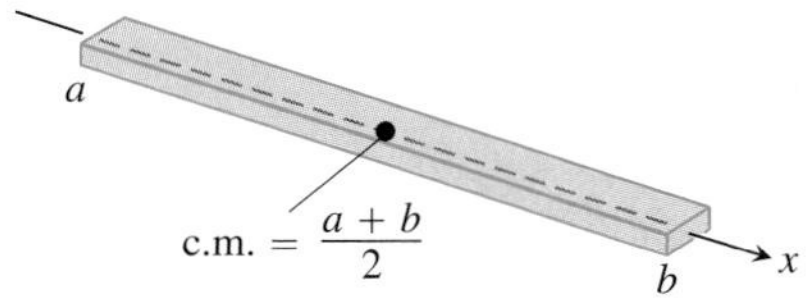

FIGURE 6.30 The center of mass of a straight, thin rod or strip of constant density lies halfway between its ends (Example 1).

EXAMPLE 1 Strips and Rods of Constant Density

Show that the center of mass of a straight, thin strip or rod of constant density lies halfway between its two ends.

Solution We model the strip as a portion of the x-axis from $x = a$ to $x = b$ (Figure 6.30). Our goal is to show that $\bar{x} = (a + b)/2$, the point halfway between a and b.

The key is the density's having a constant value. This enables us to regard the function $\delta(x)$ in the integrals in Equation (3) as a constant (call it δ), with the result that

$$M_0 = \int_a^b \delta x\, dx = \delta \int_a^b x\, dx = \delta\left[\frac{1}{2}x^2\right]_a^b = \frac{\delta}{2}(b^2 - a^2)$$

$$M = \int_a^b \delta\, dx = \delta \int_a^b dx = \delta\big[x\big]_a^b = \delta(b - a)$$

$$\bar{x} = \frac{M_0}{M} = \frac{\frac{\delta}{2}(b^2 - a^2)}{\delta(b - a)}$$

$$= \frac{a + b}{2}.$$

The δ's cancel in the formula for $\bar{x}$.

■

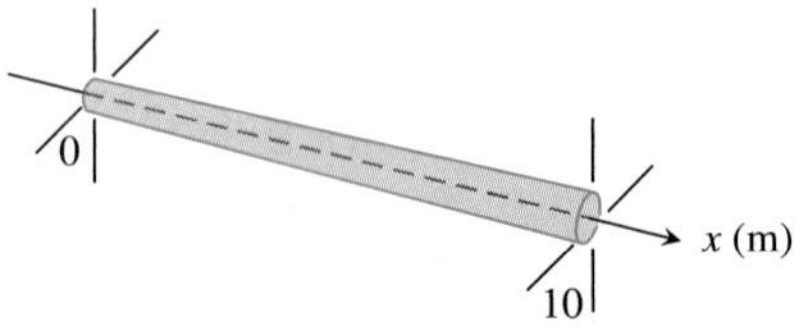

FIGURE 6.31 We can treat a rod of variable thickness as a rod of variable density (Example 2).

EXAMPLE 2 Variable-Density Rod

The 10-m-long rod in Figure 6.31 thickens from left to right so that its density, instead of being constant, is $\delta(x) = 1 + (x/10)$ kg/m. Find the rod's center of mass.

Solution The rod's moment about the origin (Equation 3a) is

$$M_0 = \int_0^{10} x\delta(x)\, dx = \int_0^{10} x\left(1 + \frac{x}{10}\right) dx = \int_0^{10} \left(x + \frac{x^2}{10}\right) dx$$

$$= \left[\frac{x^2}{2} + \frac{x^3}{30}\right]_0^{10} = 50 + \frac{100}{3} = \frac{250}{3}\ \text{kg} \cdot \text{m}.$$

The units of a moment are mass × length.

The rod's mass (Equation 3b) is

$$M = \int_0^{10} \delta(x)\, dx = \int_0^{10} \left(1 + \frac{x}{10}\right) dx = \left[x + \frac{x^2}{20}\right]_0^{10} = 10 + 5 = 15\ \text{kg}.$$

The center of mass (Equation 3c) is located at the point

$$\bar{x} = \frac{M_0}{M} = \frac{250}{3} \cdot \frac{1}{15} = \frac{50}{9} \approx 5.56\ \text{m}.$$

■

Masses Distributed over a Plane Region

Suppose that we have a finite collection of masses located in the plane, with mass m_k at the point (x_k, y_k) (see Figure 6.32). The mass of the system is

$$\text{System mass: } M = \sum m_k.$$

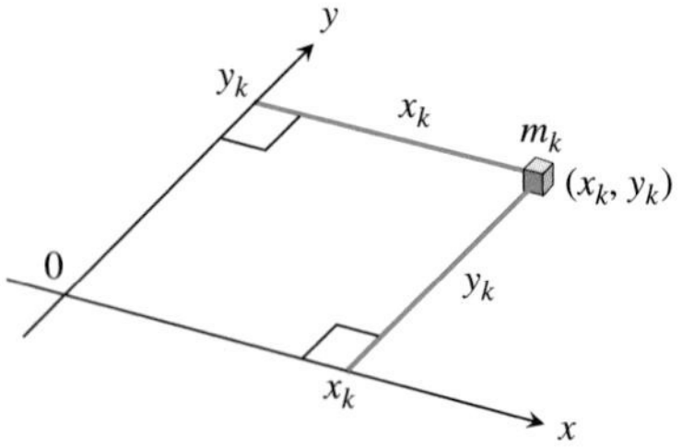

FIGURE 6.32 Each mass m_k has a moment about each axis.

Each mass m_k has a moment about each axis. Its moment about the x-axis is $m_k y_k$, and its moment about the y-axis is $m_k x_k$. The moments of the entire system about the two axes are

$$\text{Moment about } x\text{-axis:} \quad M_x = \sum m_k y_k,$$

$$\text{Moment about } y\text{-axis:} \quad M_y = \sum m_k x_k.$$

The x-coordinate of the system's center of mass is defined to be

$$\bar{x} = \frac{M_y}{M} = \frac{\sum m_k x_k}{\sum m_k}. \tag{4}$$

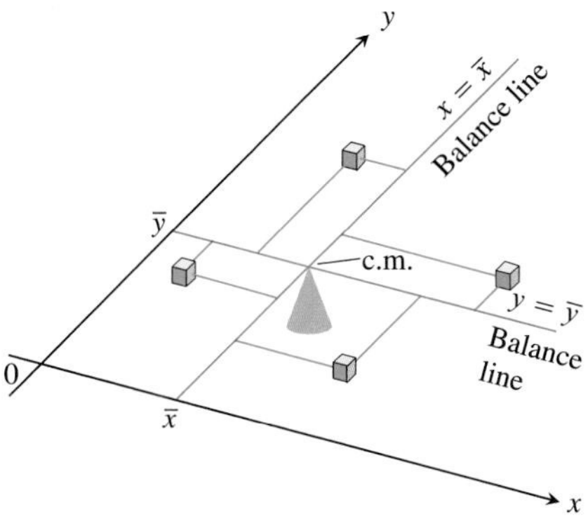

FIGURE 6.33 A two-dimensional array of masses balances on its center of mass.

With this choice of $\bar{x}$, as in the one-dimensional case, the system balances about the line $x = \bar{x}$ (Figure 6.33).

The y-coordinate of the system's center of mass is defined to be

$$\bar{y} = \frac{M_x}{M} = \frac{\sum m_k y_k}{\sum m_k}. \tag{5}$$

With this choice of $\bar{y}$, the system balances about the line $y = \bar{y}$ as well. The torques exerted by the masses about the line $y = \bar{y}$ cancel out. Thus, as far as balance is concerned, the system behaves as if all its mass were at the single point $(\bar{x}, \bar{y})$. We call this point the system's **center of mass**.

Thin, Flat Plates

In many applications, we need to find the center of mass of a thin, flat plate: a disk of aluminum, say, or a triangular sheet of steel. In such cases, we assume the distribution of mass to be continuous, and the formulas we use to calculate $\bar{x}$ and $\bar{y}$ contain integrals instead of finite sums. The integrals arise in the following way.

Imagine the plate occupying a region in the xy-plane, cut into thin strips parallel to one of the axes (in Figure 6.34, the y-axis). The center of mass of a typical strip is $(\tilde{x}, \tilde{y})$. We treat the strip's mass Δm as if it were concentrated at $(\tilde{x}, \tilde{y})$. The moment of the strip about the y-axis is then $\tilde{x}\,\Delta m$. The moment of the strip about the x-axis is $\tilde{y}\,\Delta m$. Equations (4) and (5) then become

$$\bar{x} = \frac{M_y}{M} = \frac{\sum \tilde{x}\,\Delta m}{\sum \Delta m}, \qquad \bar{y} = \frac{M_x}{M} = \frac{\sum \tilde{y}\,\Delta m}{\sum \Delta m}.$$

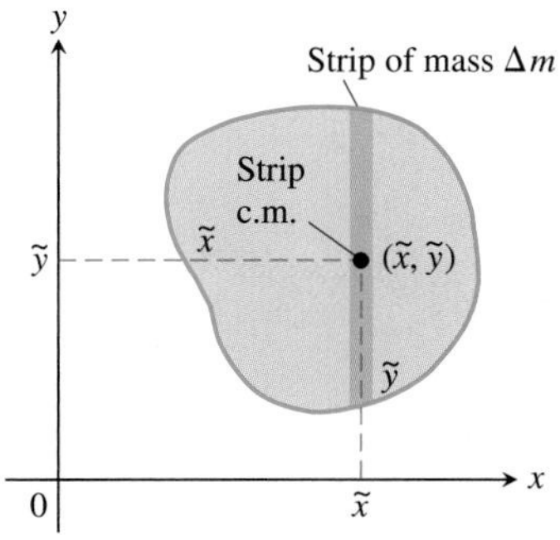

FIGURE 6.34 A plate cut into thin strips parallel to the y-axis. The moment exerted by a typical strip about each axis is the moment its mass Δm would exert if concentrated at the strip's center of mass $(\tilde{x}, \tilde{y})$.

As in the one-dimensional case, the sums are Riemann sums for integrals and approach these integrals as limiting values as the strips into which the plate is cut become narrower and narrower. We write these integrals symbolically as

$$\bar{x} = \frac{\int \tilde{x}\,dm}{\int dm} \quad \text{and} \quad \bar{y} = \frac{\int \tilde{y}\,dm}{\int dm}.$$

Moments, Mass, and Center of Mass of a Thin Plate Covering a Region in the xy-Plane

$$\begin{aligned} &\text{Moment about the } x\text{-axis:} && M_x = \int \tilde{y}\,dm \\ &\text{Moment about the } y\text{-axis:} && M_y = \int \tilde{x}\,dm \\ &\text{Mass:} && M = \int dm \\ &\text{Center of mass:} && \bar{x} = \frac{M_y}{M}, \quad \bar{y} = \frac{M_x}{M} \end{aligned} \tag{6}$$

To evaluate these integrals, we picture the plate in the coordinate plane and sketch a strip of mass parallel to one of the coordinates axes. We then express the strip's mass dm and the coordinates $(\tilde{x}, \tilde{y})$ of the strip's center of mass in terms of x or y. Finally, we integrate $\tilde{y}\,dm$, $\tilde{x}\,dm$, and dm between limits of integration determined by the plate's location in the plane.

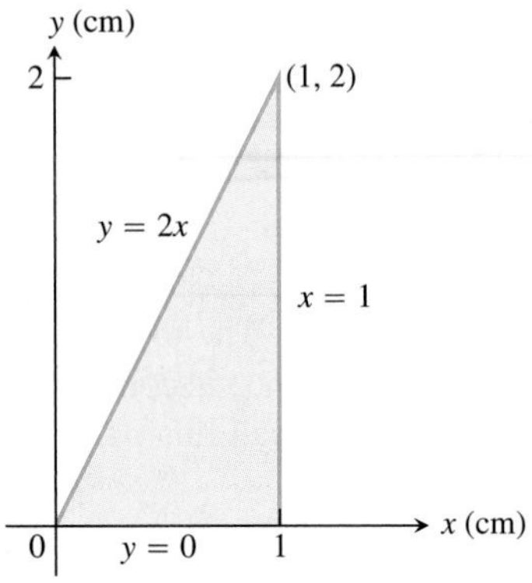

FIGURE 6.35 The plate in Example 3.

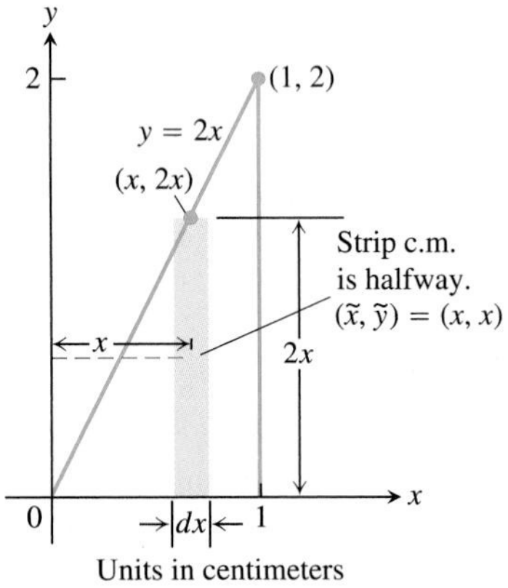

FIGURE 6.36 Modeling the plate in Example 3 with vertical strips.

EXAMPLE 3 Constant-Density Plate

The triangular plate shown in Figure 6.35 has a constant density of $\delta = 3\ \text{g/cm}^2$. Find

(a) the plate's moment M_y about the y-axis.

(b) the plate's mass M.

(c) the x-coordinate of the plate's center of mass (c.m.).

Solution

Method 1: Vertical Strips (Figure 6.36)

(a) The moment M_y: The typical vertical strip has

center of mass (c.m.): $(\tilde{x}, \tilde{y}) = (x, x)$

length: $2x$

width: dx

area: $dA = 2x\,dx$

mass: $dm = \delta\,dA = 3 \cdot 2x\,dx = 6x\,dx$

distance of c.m. from y-axis: $\tilde{x} = x$.

The moment of the strip about the y-axis is

$$\tilde{x}\,dm = x \cdot 6x\,dx = 6x^2\,dx.$$

The moment of the plate about the y-axis is therefore

$$M_y = \int \tilde{x}\,dm = \int_0^1 6x^2\,dx = 2x^3\Big]_0^1 = 2\ \text{g} \cdot \text{cm}.$$

(b) The plate's mass:

$$M = \int dm = \int_0^1 6x\,dx = 3x^2\Big]_0^1 = 3\ \text{g}.$$

(c) The x-coordinate of the plate's center of mass:

$$\bar{x} = \frac{M_y}{M} = \frac{2\ \text{g} \cdot \text{cm}}{3\ \text{g}} = \frac{2}{3}\ \text{cm}.$$

By a similar computation, we could find M_x and $\bar{y} = M_x/M$.

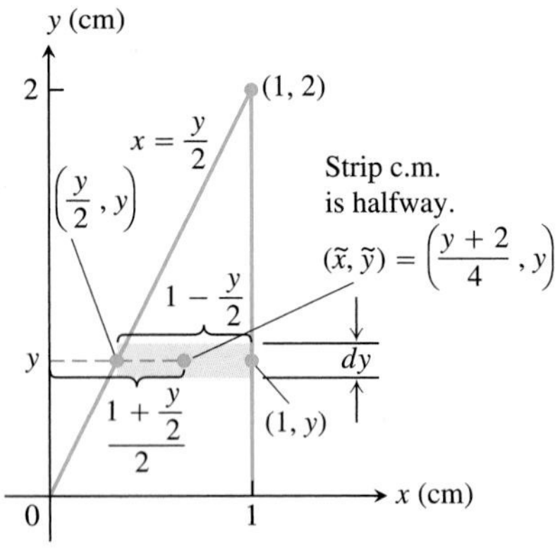

FIGURE 6.37 Modeling the plate in Example 3 with horizontal strips.

Method 2: Horizontal Strips (Figure 6.37)

(a) The moment M_y: The y-coordinate of the center of mass of a typical horizontal strip is y (see the figure), so

$$\tilde{y} = y.$$

The x-coordinate is the x-coordinate of the point halfway across the triangle. This makes it the average of $y/2$ (the strip's left-hand x-value) and 1 (the strip's right-hand x-value):

$$\tilde{x} = \frac{(y/2) + 1}{2} = \frac{y}{4} + \frac{1}{2} = \frac{y + 2}{4}.$$

We also have

$$\text{length:} \quad 1 - \frac{y}{2} = \frac{2-y}{2}$$

$$\text{width:} \quad dy$$

$$\text{area:} \quad dA = \frac{2-y}{2}\,dy$$

$$\text{mass:} \quad dm = \delta\,dA = 3 \cdot \frac{2-y}{2}\,dy$$

$$\text{distance of c.m. to } y\text{-axis:} \quad \widetilde{x} = \frac{y+2}{4}.$$

The moment of the strip about the y-axis is

$$\widetilde{x}\,dm = \frac{y+2}{4} \cdot 3 \cdot \frac{2-y}{2}\,dy = \frac{3}{8}(4 - y^2)\,dy.$$

The moment of the plate about the y-axis is

$$M_y = \int \widetilde{x}\,dm = \int_0^2 \frac{3}{8}(4 - y^2)\,dy = \frac{3}{8}\left[4y - \frac{y^3}{3}\right]_0^2 = \frac{3}{8}\left(\frac{16}{3}\right) = 2\text{ g}\cdot\text{cm}.$$

(b) The plate's mass:

$$M = \int dm = \int_0^2 \frac{3}{2}(2 - y)\,dy = \frac{3}{2}\left[2y - \frac{y^2}{2}\right]_0^2 = \frac{3}{2}(4 - 2) = 3\text{ g}.$$

(c) The x-coordinate of the plate's center of mass:

$$\bar{x} = \frac{M_y}{M} = \frac{2\text{ g}\cdot\text{cm}}{3\text{ g}} = \frac{2}{3}\text{ cm}.$$

By a similar computation, we could find M_x and $\bar{y}$. ■

> **How to Find a Plate's Center of Mass**
>
> 1. Picture the plate in the xy-plane.
> 2. Sketch a strip of mass parallel to one of the coordinate axes and find its dimensions.
> 3. Find the strip's mass dm and center of mass $(\widetilde{x}, \widetilde{y})$.
> 4. Integrate $\widetilde{y}\,dm$, $\widetilde{x}\,dm$, and dm to find M_x, M_y, and M.
> 5. Divide the moments by the mass to calculate $\bar{x}$ and $\bar{y}$.

If the distribution of mass in a thin, flat plate has an axis of symmetry, the center of mass will lie on this axis. If there are two axes of symmetry, the center of mass will lie at their intersection. These facts often help to simplify our work.

EXAMPLE 4 Constant-Density Plate

Find the center of mass of a thin plate of constant density δ covering the region bounded above by the parabola $y = 4 - x^2$ and below by the x-axis (Figure 6.38).

Solution Since the plate is symmetric about the y-axis and its density is constant, the distribution of mass is symmetric about the y-axis and the center of mass lies on the y-axis. Thus, $\bar{x} = 0$. It remains to find $\bar{y} = M_x/M$.

A trial calculation with horizontal strips (Figure 6.38a) leads to an inconvenient integration

$$M_x = \int_0^4 2\delta y\sqrt{4 - y}\,dy.$$

We therefore model the distribution of mass with vertical strips instead (Figure 6.38b).

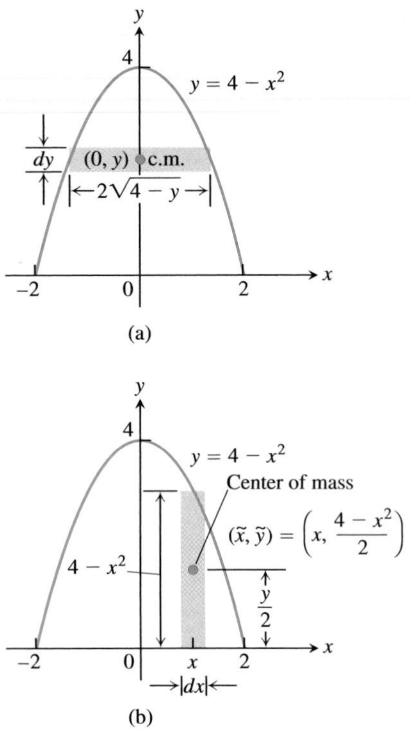

FIGURE 6.38 Modeling the plate in Example 4 with (a) horizontal strips leads to an inconvenient integration, so we model with (b) vertical strips instead.

The typical vertical strip has

center of mass (c.m.): $(\tilde{x}, \tilde{y}) = \left(x, \dfrac{4 - x^2}{2}\right)$

length: $4 - x^2$

width: dx

area: $dA = (4 - x^2)\, dx$

mass: $dm = \delta\, dA = \delta(4 - x^2)\, dx$

distance from c.m. to x-axis: $\tilde{y} = \dfrac{4 - x^2}{2}.$

The moment of the strip about the x-axis is

$$\tilde{y}\, dm = \frac{4 - x^2}{2} \cdot \delta(4 - x^2)\, dx = \frac{\delta}{2}(4 - x^2)^2\, dx.$$

The moment of the plate about the x-axis is

$$M_x = \int \tilde{y}\, dm = \int_{-2}^{2} \frac{\delta}{2}(4 - x^2)^2\, dx$$
$$= \frac{\delta}{2}\int_{-2}^{2} (16 - 8x^2 + x^4)\, dx = \frac{256}{15}\delta. \tag{7}$$

The mass of the plate is

$$M = \int dm = \int_{-2}^{2} \delta(4 - x^2)\, dx = \frac{32}{3}\delta. \tag{8}$$

Therefore,

$$\bar{y} = \frac{M_x}{M} = \frac{(256/15)\,\delta}{(32/3)\,\delta} = \frac{8}{5}.$$

The plate's center of mass is the point

$$(\bar{x}, \bar{y}) = \left(0, \frac{8}{5}\right).$$

■

EXAMPLE 5 Variable-Density Plate

Find the center of mass of the plate in Example 4 if the density at the point (x, y) is $\delta = 2x^2$, twice the square of the distance from the point to the y-axis.

Solution The mass distribution is still symmetric about the y-axis, so $\bar{x} = 0$. With $\delta = 2x^2$, Equations (7) and (8) become

$$M_x = \int \widetilde{y}\, dm = \int_{-2}^{2} \frac{\delta}{2}(4 - x^2)^2\, dx = \int_{-2}^{2} x^2(4 - x^2)^2\, dx$$

$$= \int_{-2}^{2} (16x^2 - 8x^4 + x^6)\, dx = \frac{2048}{105} \tag{7'}$$

$$M = \int dm = \int_{-2}^{2} \delta(4 - x^2)\, dx = \int_{-2}^{2} 2x^2(4 - x^2)\, dx$$

$$= \int_{-2}^{2} (8x^2 - 2x^4)\, dx = \frac{256}{15}. \tag{8'}$$

Therefore,

$$\bar{y} = \frac{M_x}{M} = \frac{2048}{105} \cdot \frac{15}{256} = \frac{8}{7}.$$

The plate's new center of mass is

$$(\bar{x}, \bar{y}) = \left(0, \frac{8}{7}\right).$$

■

EXAMPLE 6 Constant-Density Wire

Find the center of mass of a wire of constant density δ shaped like a semicircle of radius a.

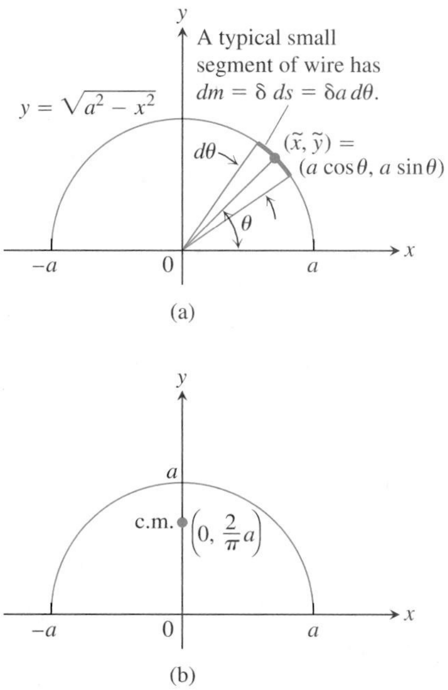

FIGURE 6.39 The semicircular wire in Example 6. (a) The dimensions and variables used in finding the center of mass. (b) The center of mass does not lie on the wire.

Solution We model the wire with the semicircle $y = \sqrt{a^2 - x^2}$ (Figure 6.39). The distribution of mass is symmetric about the y-axis, so $\bar{x} = 0$. To find $\bar{y}$, we imagine the wire divided into short segments. The typical segment (Figure 6.39a) has

length: $ds = a\, d\theta$

mass: $dm = \delta\, ds = \delta a\, d\theta$ — Mass per unit length times length

distance of c.m. to x-axis: $\widetilde{y} = a \sin\theta$.

Hence,

$$\bar{y} = \frac{\int \widetilde{y}\, dm}{\int dm} = \frac{\int_0^{\pi} a \sin\theta \cdot \delta a\, d\theta}{\int_0^{\pi} \delta a\, d\theta} = \frac{\delta a^2\big[-\cos\theta\big]_0^{\pi}}{\delta a \pi} = \frac{2}{\pi} a.$$

The center of mass lies on the axis of symmetry at the point $(0, 2a/\pi)$, about two-thirds of the way up from the origin (Figure 6.39b).

Centroids

When the density function is constant, it cancels out of the numerator and denominator of the formulas for $\bar{x}$ and $\bar{y}$. This happened in nearly every example in this section. As far as $\bar{x}$ and $\bar{y}$ were concerned, δ might as well have been 1. Thus, when the density is constant, the location of the center of mass is a feature of the geometry of the object and not of the material from which it is made. In such cases, engineers may call the center of mass the **centroid** of the shape, as in "Find the centroid of a triangle or a solid cone." To do so, just set δ equal to 1 and proceed to find $\bar{x}$ and $\bar{y}$ as before, by dividing moments by masses.

EXERCISES 6.4

Thin Rods

1. An 80-lb child and a 100-lb child are balancing on a seesaw. The 80-lb child is 5 ft from the fulcrum. How far from the fulcrum is the 100-lb child?

2. The ends of a log are placed on two scales. One scale reads 100 kg and the other 200 kg. Where is the log's center of mass?

3. The ends of two thin steel rods of equal length are welded together to make a right-angled frame. Locate the frame's center of mass. (*Hint:* Where is the center of mass of each rod?)

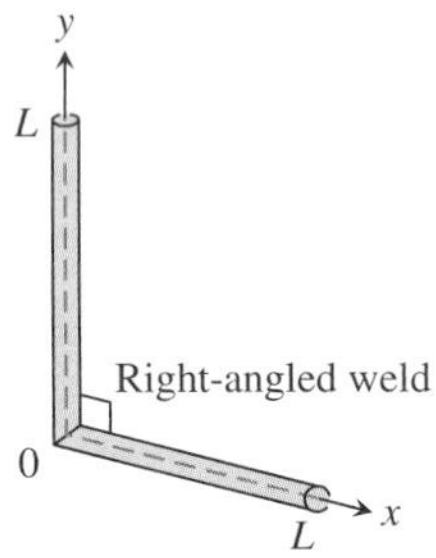

4. You weld the ends of two steel rods into a right-angled frame. One rod is twice the length of the other. Where is the frame's center of mass? (*Hint:* Where is the center of mass of each rod?)

Exercises 5–12 give density functions of thin rods lying along various intervals of the x-axis. Use Equations (3a) through (3c) to find each rod's moment about the origin, mass, and center of mass.

5. $\delta(x) = 4, \quad 0 \le x \le 2$

6. $\delta(x) = 4, \quad 1 \le x \le 3$

7. $\delta(x) = 1 + (x/3), \quad 0 \le x \le 3$

8. $\delta(x) = 2 - (x/4), \quad 0 \le x \le 4$

9. $\delta(x) = 1 + \left(1/\sqrt{x}\right), \quad 1 \le x \le 4$

10. $\delta(x) = 3(x^{-3/2} + x^{-5/2}), \quad 0.25 \le x \le 1$

11. $\delta(x) = \begin{cases} 2 - x, & 0 \le x < 1 \\ x, & 1 \le x \le 2 \end{cases}$

12. $\delta(x) = \begin{cases} x + 1, & 0 \le x < 1 \\ 2, & 1 \le x \le 2 \end{cases}$

Thin Plates with Constant Density

In Exercises 13–24, find the center of mass of a thin plate of constant density δ covering the given region.

13. The region bounded by the parabola $y = x^2$ and the line $y = 4$

14. The region bounded by the parabola $y = 25 - x^2$ and the x-axis

15. The region bounded by the parabola $y = x - x^2$ and the line $y = -x$

16. The region enclosed by the parabolas $y = x^2 - 3$ and $y = -2x^2$

17. The region bounded by the y-axis and the curve $x = y - y^3$, $0 \le y \le 1$

18. The region bounded by the parabola $x = y^2 - y$ and the line $y = x$

19. The region bounded by the x-axis and the curve $y = \cos x$, $-\pi/2 \le x \le \pi/2$

20. The region between the x-axis and the curve $y = \sec^2 x$, $-\pi/4 \le x \le \pi/4$

21. The region bounded by the parabolas $y = 2x^2 - 4x$ and $y = 2x - x^2$

22. **a.** The region cut from the first quadrant by the circle $x^2 + y^2 = 9$

    b. The region bounded by the x-axis and the semicircle $y = \sqrt{9 - x^2}$

    Compare your answer in part (b) with the answer in part (a).

23. The "triangular" region in the first quadrant between the circle $x^2 + y^2 = 9$ and the lines $x = 3$ and $y = 3$. (*Hint:* Use geometry to find the area.)

24. The region bounded above by the curve $y = 1/x^3$, below by the curve $y = -1/x^3$, and on the left and right by the lines $x = 1$ and $x = a > 1$. Also, find $\lim_{a\to\infty} \bar{x}$.

Thin Plates with Varying Density

25. Find the center of mass of a thin plate covering the region between the x-axis and the curve $y = 2/x^2$, $1 \le x \le 2$, if the plate's density at the point (x, y) is $\delta(x) = x^2$.

26. Find the center of mass of a thin plate covering the region bounded below by the parabola $y = x^2$ and above by the line $y = x$ if the plate's density at the point (x, y) is $\delta(x) = 12x$.

27. The region bounded by the curves $y = \pm 4/\sqrt{x}$ and the lines $x = 1$ and $x = 4$ is revolved about the y-axis to generate a solid.

    a. Find the volume of the solid.

    b. Find the center of mass of a thin plate covering the region if the plate's density at the point (x, y) is $\delta(x) = 1/x$.

    c. Sketch the plate and show the center of mass in your sketch.

28. The region between the curve $y = 2/x$ and the x-axis from $x = 1$ to $x = 4$ is revolved about the x-axis to generate a solid.

    a. Find the volume of the solid.

    b. Find the center of mass of a thin plate covering the region if the plate's density at the point (x, y) is $\delta(x) = \sqrt{x}$.

    c. Sketch the plate and show the center of mass in your sketch.

Centroids of Triangles

29. **The centroid of a triangle lies at the intersection of the triangle's medians** (*Figure 6.40a*) You may recall that the point

inside a triangle that lies one-third of the way from each side toward the opposite vertex is the point where the triangle's three medians intersect. Show that the centroid lies at the intersection of the medians by showing that it too lies one-third of the way from each side toward the opposite vertex. To do so, take the following steps.

i. Stand one side of the triangle on the x-axis as in Figure 6.40b. Express dm in terms of L and dy.

ii. Use similar triangles to show that $L = (b/h)(h - y)$. Substitute this expression for L in your formula for dm.

iii. Show that $\bar{y} = h/3$.

iv. Extend the argument to the other sides.

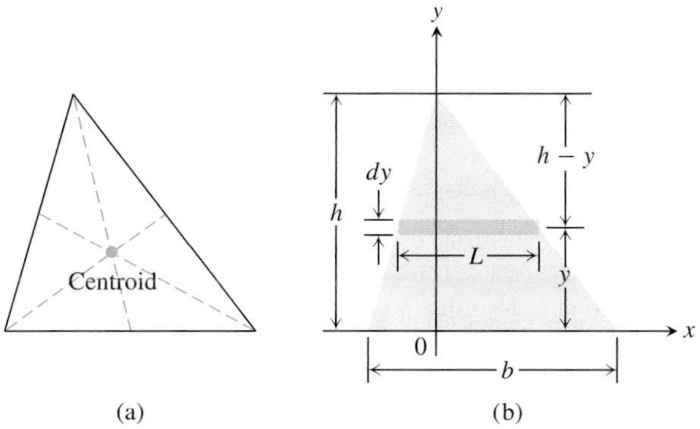

FIGURE 6.40 The triangle in Exercise 29. (a) The centroid. (b) The dimensions and variables to use in locating the center of mass.

Use the result in Exercise 29 to find the centroids of the triangles whose vertices appear in Exercises 30–34. Assume $a, b > 0$.

30. $(-1, 0), (1, 0), (0, 3)$

31. $(0, 0), (1, 0), (0, 1)$

32. $(0, 0), (a, 0), (0, a)$

33. $(0, 0), (a, 0), (0, b)$

34. $(0, 0), (a, 0), (a/2, b)$

Thin Wires

35. Constant density Find the moment about the x-axis of a wire of constant density that lies along the curve $y = \sqrt{x}$ from $x = 0$ to $x = 2$.

36. Constant density Find the moment about the x-axis of a wire of constant density that lies along the curve $y = x^3$ from $x = 0$ to $x = 1$.

37. Variable density Suppose that the density of the wire in Example 6 is $\delta = k \sin \theta$ (k constant). Find the center of mass.

38. Variable density Suppose that the density of the wire in Example 6 is $\delta = 1 + k|\cos \theta|$ (k constant). Find the center of mass.

Engineering Formulas

Verify the statements and formulas in Exercises 39–42.

39. The coordinates of the centroid of a differentiable plane curve are

$$\bar{x} = \frac{\int x\, ds}{\text{length}}, \qquad \bar{y} = \frac{\int y\, ds}{\text{length}}.$$

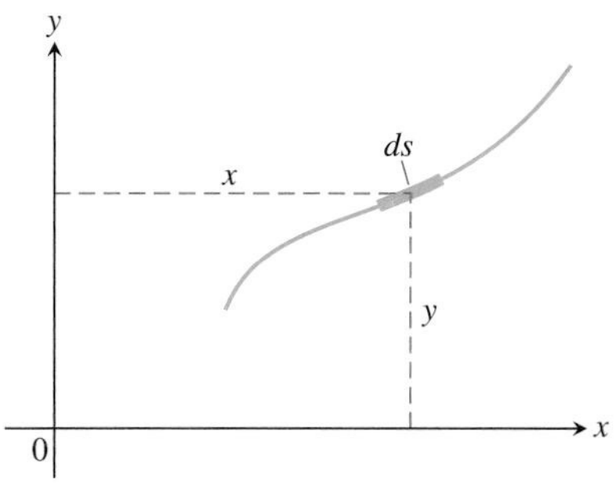

40. Whatever the value of $p > 0$ in the equation $y = x^2/(4p)$, the y-coordinate of the centroid of the parabolic segment shown here is $\bar{y} = (3/5)a$.

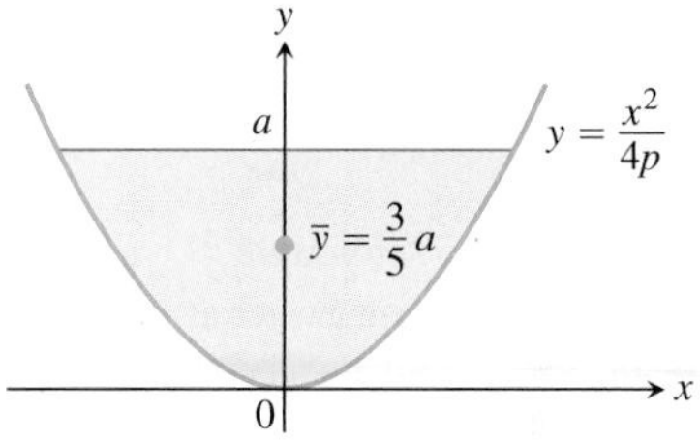

41. For wires and thin rods of constant density shaped like circular arcs centered at the origin and symmetric about the y-axis, the y-coordinate of the center of mass is

$$\bar{y} = \frac{a \sin \alpha}{\alpha} = \frac{ac}{s}.$$

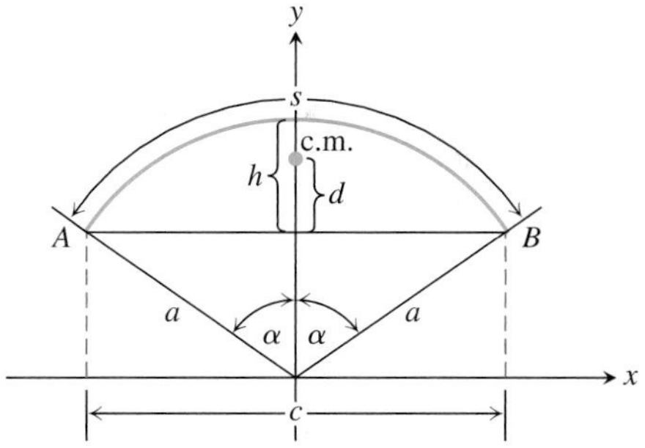

42. (*Continuation of Exercise 41.*)

a. Show that when α is small, the distance d from the centroid to chord AB is about $2h/3$ (in the notation of the figure here) by taking the following steps.

i. Show that

$$\frac{d}{h} = \frac{\sin \alpha - \alpha \cos \alpha}{\alpha - \alpha \cos \alpha}. \tag{9}$$

T **ii.** Graph

$$f(\alpha) = \frac{\sin \alpha - \alpha \cos \alpha}{\alpha - \alpha \cos \alpha}$$

and use the trace feature to show that $\lim_{\alpha \to 0^+} f(\alpha) \approx 2/3$.

b. The error (difference between d and $2h/3$) is small even for angles greater that 45°. See for yourself by evaluating the right-hand side of Equation (9) for $\alpha = 0.2, 0.4, 0.6, 0.8$, and 1.0 rad.

6.5 Areas of Surfaces of Revolution and the Theorems of Pappus

When you jump rope, the rope sweeps out a surface in the space around you called a *surface of revolution*. The "area" of this surface depends on the length of the rope and the distance of each of its segments from the axis of revolution. In this section we define areas of surfaces of revolution. More complicated surfaces are treated in Chapter 16.

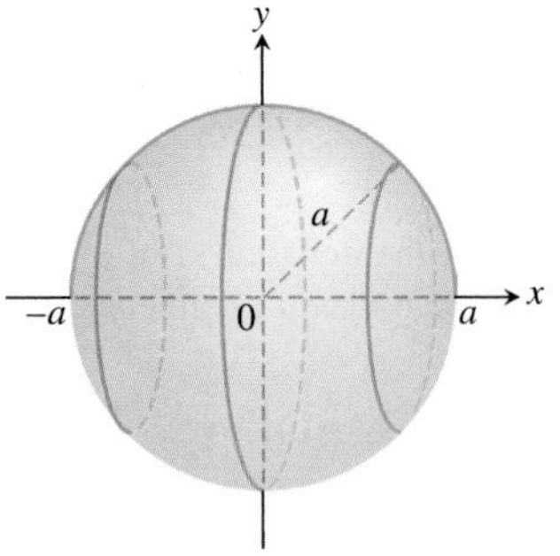

FIGURE 6.41 Rotating the semicircle $y = \sqrt{a^2 - x^2}$ of radius a with center at the origin generates a spherical surface with area $4\pi a^2$.

Defining Surface Area

We want our definition of the area of a surface of revolution to be consistent with known results from classical geometry for the surface areas of spheres, circular cylinders, and cones. So if the jump rope discussed in the introduction takes the shape of a semicircle with radius a rotated about the x-axis (Figure 6.41), it generates a sphere with surface area $4\pi a^2$.

Before considering general curves, we begin by rotating horizontal and slanted line segments about the x-axis. If we rotate the horizontal line segment AB having length Δx about the x-axis (Figure 6.42a), we generate a cylinder with surface area $2\pi y \Delta x$. This area is the same as that of a rectangle with side lengths Δx and $2\pi y$ (Figure 6.42b). The length $2\pi y$ is the circumference of the circle of radius y generated by rotating the point (x, y) on the line AB about the x-axis.

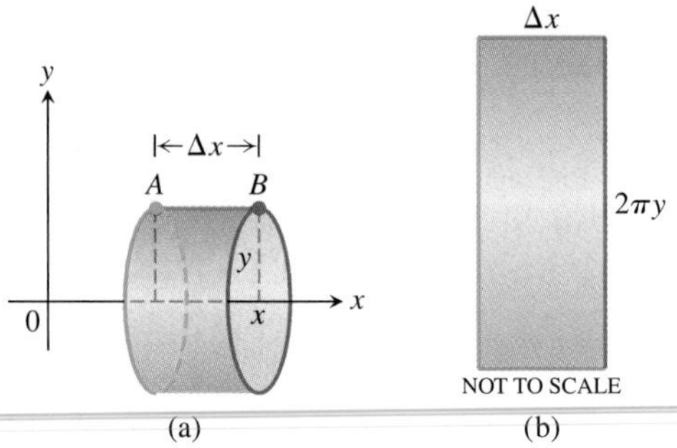

FIGURE 6.42 (a) A cylindrical surface generated by rotating the horizontal line segment AB of length Δx about the x-axis has area $2\pi y \Delta x$. (b) The cut and rolled out cylindrical surface as a rectangle.

Suppose the line segment AB has length Δs and is slanted rather than horizontal. Now when AB is rotated about the x-axis, it generates a frustum of a cone (Figure 6.43a). From classical geometry, the surface area of this frustum is $2\pi y^* \Delta s$, where $y^* = (y_1 + y_2)/2$ is the average height of the slanted segment AB above the x-axis. This surface area is the same as that of a rectangle with side lengths Δs and $2\pi y^*$ (Figure 6.43b).

Let's build on these geometric principles to define the area of a surface swept out by revolving more general curves about the x-axis. Suppose we want to find the area of the surface swept out by revolving the graph of a nonnegative continuous function $y = f(x)$, $a \le x \le b$, about the x-axis. We partition the closed interval $[a, b]$ in the usual way and use the points in the partition to subdivide the graph into short arcs. Figure 6.44 shows a typical arc PQ and the band it sweeps out as part of the graph of f.

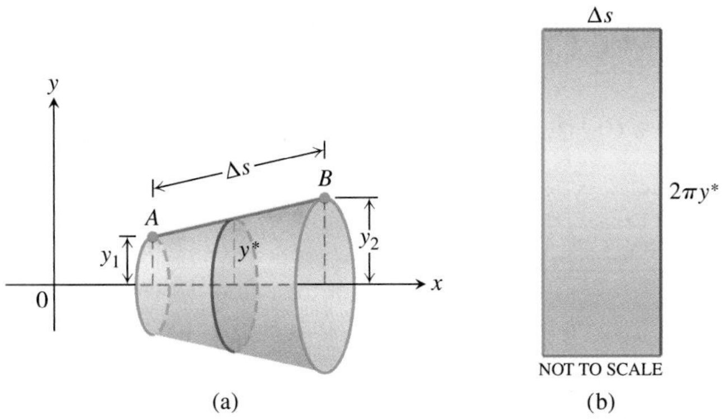

FIGURE 6.43 (a) The frustum of a cone generated by rotating the slanted line segment AB of length Δs about the x-axis has area $2\pi y^* \Delta s$. (b) The area of the rectangle for $y^* = \dfrac{y_1 + y_2}{2}$, the average height of AB above the x-axis.

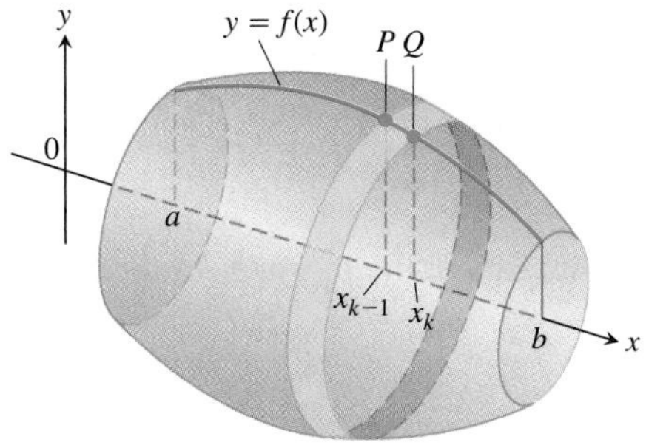

FIGURE 6.44 The surface generated by revolving the graph of a nonnegative function $y = f(x)$, $a \le x \le b$, about the x-axis. The surface is a union of bands like the one swept out by the arc PQ.

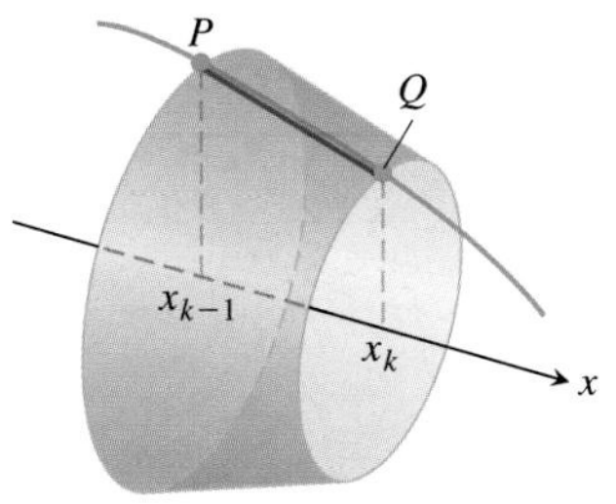

FIGURE 6.45 The line segment joining P and Q sweeps out a frustum of a cone.

As the arc PQ revolves about the x-axis, the line segment joining P and Q sweeps out a frustum of a cone whose axis lies along the x-axis (Figure 6.45). The surface area of this frustum approximates the surface area of the band swept out by the arc PQ. The surface area of the frustum of the cone shown in Figure 6.45 is $2\pi y^* L$, where y^* is the average height of the line segment joining P and Q, and L is its length (just as before). Since $f \ge 0$, from Figure 6.46 we see that the average height of the line segment is $y^* = (f(x_{k-1}) + f(x_k))/2$, and the slant length is $L = \sqrt{(\Delta x_k)^2 + (\Delta y_k)^2}$. Therefore,

$$\text{Frustum surface area} = 2\pi \cdot \frac{f(x_{k-1}) + f(x_k)}{2} \cdot \sqrt{(\Delta x_k)^2 + (\Delta y_k)^2}$$

$$= \pi(f(x_{k-1}) + f(x_k))\sqrt{(\Delta x_k)^2 + (\Delta y_k)^2}.$$

The area of the original surface, being the sum of the areas of the bands swept out by arcs like arc PQ, is approximated by the frustum area sum

$$\sum_{k=1}^{n} \pi(f(x_{k-1}) + f(x_k))\sqrt{(\Delta x_k)^2 + (\Delta y_k)^2}. \tag{1}$$

We expect the approximation to improve as the partition of $[a, b]$ becomes finer. Moreover, if the function f is differentiable, then by the Mean Value Theorem, there is a point $(c_k, f(c_k))$ on the curve between P and Q where the tangent is parallel to the segment PQ (Figure 6.47). At this point,

$$f'(c_k) = \frac{\Delta y_k}{\Delta x_k},$$

$$\Delta y_k = f'(c_k)\,\Delta x_k.$$

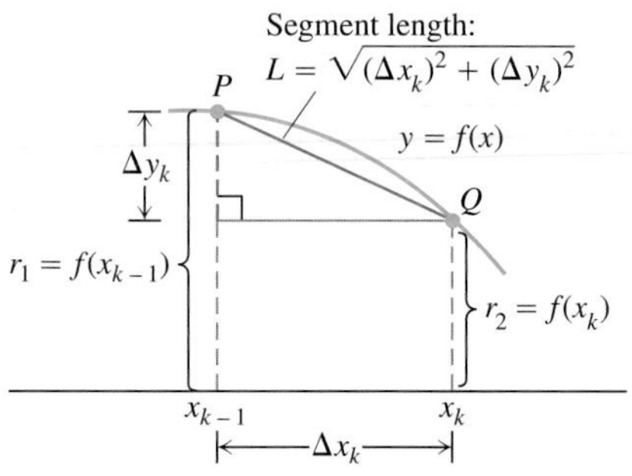

FIGURE 6.46 Dimensions associated with the arc and line segment PQ.

With this substitution for Δy_k, the sums in Equation (1) take the form

$$\sum_{k=1}^{n} \pi(f(x_{k-1}) + f(x_k))\sqrt{(\Delta x_k)^2 + (f'(c_k)\,\Delta x_k)^2}$$

$$= \sum_{k=1}^{n} \pi(f(x_{k-1}) + f(x_k))\sqrt{1 + (f'(c_k))^2}\,\Delta x_k. \tag{2}$$

These sums are not the Riemann sums of any function because the points x_{k-1}, x_k, and c_k are not the same. However, a theorem from advanced calculus assures us that as the norm of the partition of $[a, b]$ goes to zero, the sums in Equation (2) converge to the integral

$$\int_a^b 2\pi f(x)\sqrt{1 + (f'(x))^2}\,dx.$$

We therefore define this integral to be the area of the surface swept out by the graph of f from a to b.

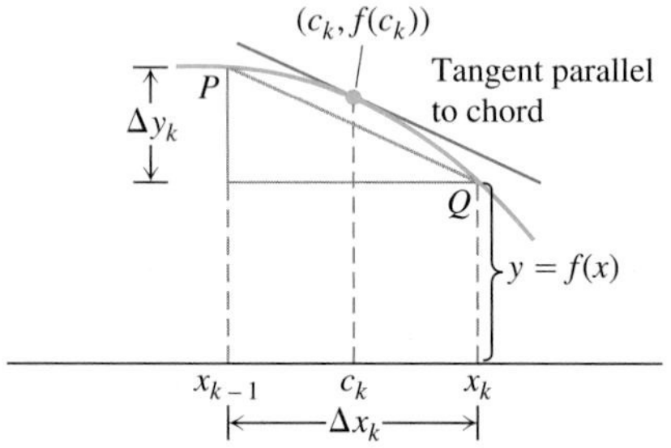

FIGURE 6.47 If f is smooth, the Mean Value Theorem guarantees the existence of a point c_k where the tangent is parallel to segment PQ.

DEFINITION Surface Area for Revolution About the x-Axis

If the function $f(x) \geq 0$ is continuously differentiable on $[a, b]$, the **area** of the surface generated by revolving the curve $y = f(x)$ about the x-axis is

$$S = \int_a^b 2\pi y\sqrt{1 + \left(\frac{dy}{dx}\right)^2}\,dx = \int_a^b 2\pi f(x)\sqrt{1 + (f'(x))^2}\,dx. \tag{3}$$

The square root in Equation (3) is the same one that appears in the formula for the length of the generating curve in Equation (2) of Section 6.3.

EXAMPLE 1 Applying the Surface Area Formula

Find the area of the surface generated by revolving the curve $y = 2\sqrt{x}$, $1 \leq x \leq 2$, about the x-axis (Figure 6.48).

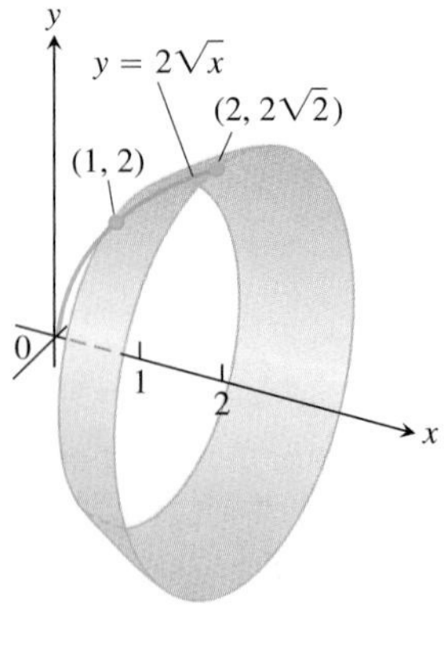

FIGURE 6.48 In Example 1 we calculate the area of this surface.

Solution We evaluate the formula

$$S = \int_a^b 2\pi y\sqrt{1 + \left(\frac{dy}{dx}\right)^2}\,dx \qquad \text{Eq. (3)}$$

with

$$a = 1, \qquad b = 2, \qquad y = 2\sqrt{x}, \qquad \frac{dy}{dx} = \frac{1}{\sqrt{x}},$$

$$\sqrt{1 + \left(\frac{dy}{dx}\right)^2} = \sqrt{1 + \left(\frac{1}{\sqrt{x}}\right)^2}$$

$$= \sqrt{1 + \frac{1}{x}} = \sqrt{\frac{x+1}{x}} = \frac{\sqrt{x+1}}{\sqrt{x}}.$$

With these substitutions,

$$S = \int_1^2 2\pi \cdot 2\sqrt{x}\frac{\sqrt{x+1}}{\sqrt{x}}\,dx = 4\pi\int_1^2 \sqrt{x+1}\,dx$$

$$= 4\pi \cdot \frac{2}{3}(x+1)^{3/2}\Big]_1^2 = \frac{8\pi}{3}\left(3\sqrt{3} - 2\sqrt{2}\right).$$

Revolution About the y-Axis

For revolution about the y-axis, we interchange x and y in Equation (3).

Surface Area for Revolution About the y-Axis

If $x = g(y) \geq 0$ is continuously differentiable on $[c, d]$, the area of the surface generated by revolving the curve $x = g(y)$ about the y-axis is

$$S = \int_c^d 2\pi x\sqrt{1 + \left(\frac{dx}{dy}\right)^2}\,dy = \int_c^d 2\pi g(y)\sqrt{1 + (g'(y))^2}\,dy. \qquad (4)$$

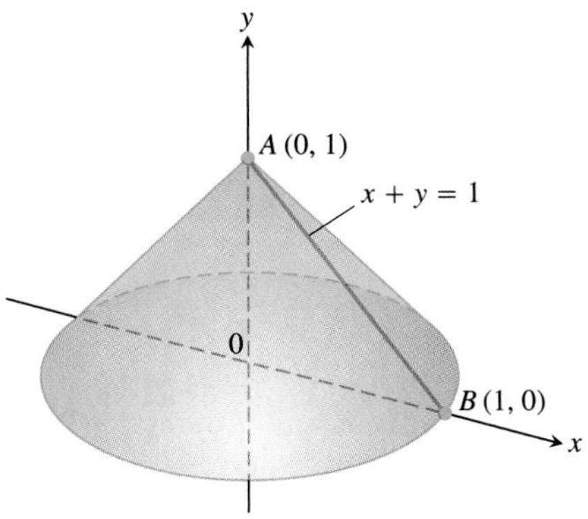

FIGURE 6.49 Revolving line segment AB about the y-axis generates a cone whose lateral surface area we can now calculate in two different ways (Example 2).

EXAMPLE 2 Finding Area for Revolution about the y-Axis

The line segment $x = 1 - y, 0 \leq y \leq 1$, is revolved about the y-axis to generate the cone in Figure 6.49. Find its lateral surface area (which excludes the base area).

Solution Here we have a calculation we can check with a formula from geometry:

$$\text{Lateral surface area} = \frac{\text{base circumference}}{2} \times \text{slant height} = \pi\sqrt{2}.$$

To see how Equation (4) gives the same result, we take

$$c = 0, \qquad d = 1, \qquad x = 1 - y, \qquad \frac{dx}{dy} = -1,$$

$$\sqrt{1 + \left(\frac{dx}{dy}\right)^2} = \sqrt{1 + (-1)^2} = \sqrt{2}$$

and calculate

$$S = \int_c^d 2\pi x\sqrt{1 + \left(\frac{dx}{dy}\right)^2}\,dy = \int_0^1 2\pi(1 - y)\sqrt{2}\,dy$$

$$= 2\pi\sqrt{2}\left[y - \frac{y^2}{2}\right]_0^1 = 2\pi\sqrt{2}\left(1 - \frac{1}{2}\right)$$

$$= \pi\sqrt{2}.$$

The results agree, as they should.

Parametrized Curves

Regardless of the coordinate axis of revolution, the square roots appearing in Equations (3) and (4) are the same ones that appear in the formulas for arc length in Section 6.3. If the curve is parametrized by the equations $x = f(t)$ and $y = g(t)$, $a \le t \le b$, where f and g are continuously differentiable on $[a, b]$, then the corresponding square root appearing in the arc length formula is

$$\sqrt{[f'(t)]^2 + [g'(t)]^2} = \sqrt{\left(\frac{dx}{dt}\right)^2 + \left(\frac{dy}{dt}\right)^2}.$$

This observation leads to the following formulas for area of surfaces of revolution for smooth parametrized curves.

Surface Area of Revolution for Parametrized Curves
If a smooth curve $x = f(t)$, $y = g(t)$, $a \le t \le b$, is traversed exactly once as t increases from a to b, then the areas of the surfaces generated by revolving the curve about the coordinate axes are as follows.

1. Revolution about the x-axis ($y \ge 0$):

$$S = \int_a^b 2\pi y \sqrt{\left(\frac{dx}{dt}\right)^2 + \left(\frac{dy}{dt}\right)^2}\, dt \qquad (5)$$

2. Revolution about the y-axis ($x \ge 0$):

$$S = \int_a^b 2\pi x \sqrt{\left(\frac{dx}{dt}\right)^2 + \left(\frac{dy}{dt}\right)^2}\, dt \qquad (6)$$

As with length, we can calculate surface area from any convenient parametrization that meets the stated criteria.

EXAMPLE 3 Applying Surface Area Formula

The standard parametrization of the circle of radius 1 centered at the point (0, 1) in the xy-plane is

$$x = \cos t, \qquad y = 1 + \sin t, \qquad 0 \le t \le 2\pi.$$

Use this parametrization to find the area of the surface swept out by revolving the circle about the x-axis (Figure 6.50).

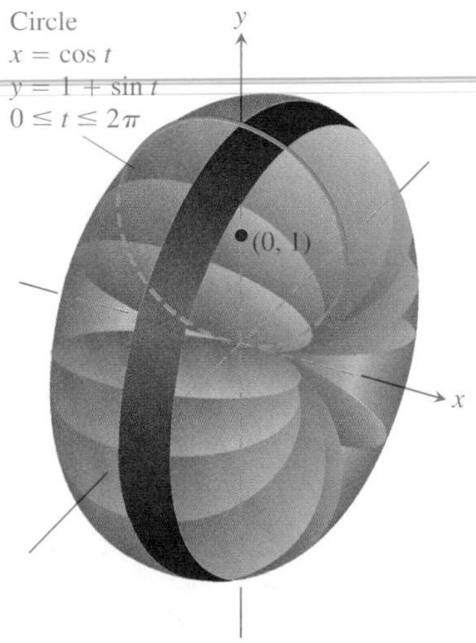

FIGURE 6.50 In Example 3 we calculate the area of the surface of revolution swept out by this parametrized curve.

Solution We evaluate the formula

$$S = \int_a^b 2\pi y \sqrt{\left(\frac{dx}{dt}\right)^2 + \left(\frac{dy}{dt}\right)^2}\, dt$$

Eq. (5) for revolution about the x-axis; $y = 1 + \sin t > 0$

$$= \int_0^{2\pi} 2\pi(1 + \sin t)\underbrace{\sqrt{(-\sin t)^2 + (\cos t)^2}}_{1}\, dt$$

$$= 2\pi \int_0^{2\pi} (1 + \sin t)\, dt$$

$$= 2\pi \Big[t - \cos t\Big]_0^{2\pi} = 4\pi^2.$$

■

The Differential Form

The equations

$$S = \int_a^b 2\pi y \sqrt{1 + \left(\frac{dy}{dx}\right)^2}\, dx \quad \text{and} \quad S = \int_c^d 2\pi x \sqrt{\left(\frac{dx}{dy}\right)^2}\, dy$$

are often written in terms of the arc length differential $ds = \sqrt{dx^2 + dy^2}$ as

$$S = \int_a^b 2\pi y\, ds \quad \text{and} \quad S = \int_c^d 2\pi x\, ds.$$

In the first of these, y is the distance from the x-axis to an element of arc length ds. In the second, x is the distance from the y-axis to an element of arc length ds. Both integrals have the form

$$S = \int 2\pi(\text{radius})(\text{band width}) = \int 2\pi\rho\, ds \qquad (7)$$

where ρ is the radius from the axis of revolution to an element of arc length ds (Figure 6.51).

In any particular problem, you would then express the radius function ρ and the arc length differential ds in terms of a common variable and supply limits of integration for that variable.

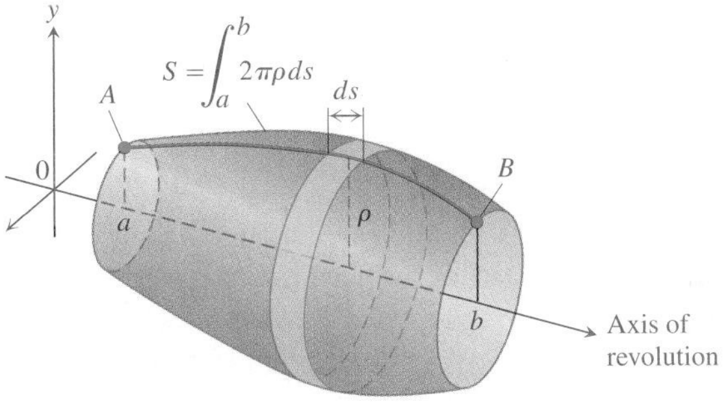

FIGURE 6.51 The area of the surface swept out by revolving arc AB about the axis shown here is $\int_a^b 2\pi\rho\, ds$. The exact expression depends on the formulas for ρ and ds.

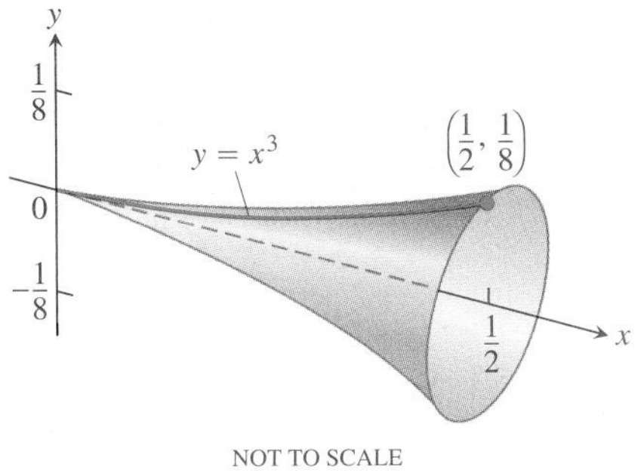

FIGURE 6.52 The surface generated by revolving the curve $y = x^3$, $0 \le x \le 1/2$, about the x-axis could be the design for a champagne glass (Example 4).

EXAMPLE 4 Using the Differential Form for Surface Areas

Find the area of the surface generated by revolving the curve $y = x^3$, $0 \le x \le 1/2$, about the x-axis (Figure 6.52).

Solution We start with the short differential form:

$$S = \int 2\pi\rho\, ds$$

$$= \int 2\pi y\, ds \qquad \text{For revolution about the } x\text{-axis, the radius function is } \rho = y > 0 \text{ on } 0 \le x \le 1/2.$$

$$= \int 2\pi y \sqrt{dx^2 + dy^2}. \qquad ds = \sqrt{dx^2 + dy^2}$$

We then decide whether to express dy in terms of dx or dx in terms of dy. The original form of the equation, $y = x^3$, makes it easier to express dy in terms of dx, so we continue the calculation with

$$y = x^3, \qquad dy = 3x^2\,dx, \qquad \text{and} \qquad \sqrt{dx^2 + dy^2} = \sqrt{dx^2 + (3x^2\,dx)^2} = \sqrt{1 + 9x^4}\,dx.$$

With these substitutions, x becomes the variable of integration and

$$\begin{aligned} S &= \int_{x=0}^{x=1/2} 2\pi y \sqrt{dx^2 + dy^2} \\ &= \int_0^{1/2} 2\pi x^3 \sqrt{1 + 9x^4}\,dx \\ &= 2\pi \left(\frac{1}{36}\right)\left(\frac{2}{3}\right)(1 + 9x^4)^{3/2}\bigg]_0^{1/2} \\ &= \frac{\pi}{27}\left[\left(1 + \frac{9}{16}\right)^{3/2} - 1\right] \\ &= \frac{\pi}{27}\left[\left(\frac{25}{16}\right)^{3/2} - 1\right] = \frac{\pi}{27}\left(\frac{125}{64} - 1\right) \\ &= \frac{61\pi}{1728}. \end{aligned}$$

Substitute $u = 1 + 9x^4$, $du/36 = x^3\,dx$; integrate, and substitute back.

■

Cylindrical Versus Conical Bands

Why not find the surface area by approximating with cylindrical bands instead of conical bands, as suggested in Figure 6.53? The Riemann sums we get this way converge just as nicely as the ones based on conical bands, and the resulting integral is simpler. For revolution about the x-axis in this case, the radius in Equation (7) is $\rho = y$ and the band width is $ds = dx$. This leads to the integral formula

$$S = \int_a^b 2\pi f(x)\,dx \tag{8}$$

rather than the defining Equation (3). The problem with this new formula is that it fails to give results consistent with the surface area formulas from classical geometry, and that was one of our stated goals at the outset. Just because we end up with a nice-looking integral from a Riemann sum derivation does not mean it will calculate what we intend. (See Exercise 40.)

CAUTION Do not use Equation (8) to calculate surface area. It does *not* give the correct result.

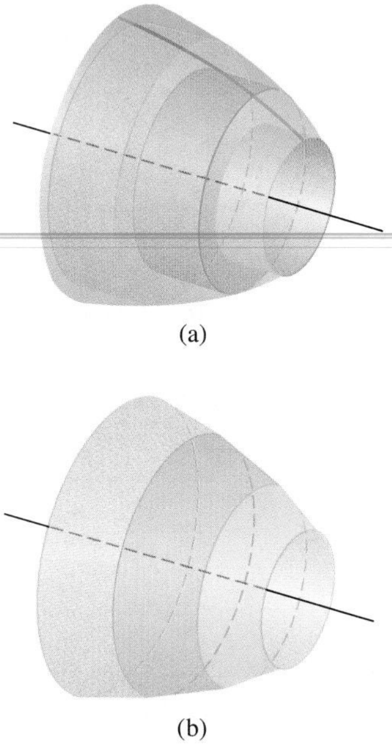

FIGURE 6.53 Why not use (a) cylindrical bands instead of (b) conical bands to approximate surface area?

The Theorems of Pappus

In the third century, an Alexandrian Greek named Pappus discovered two formulas that relate centroids to surfaces and solids of revolution. The formulas provide shortcuts to a number of otherwise lengthy calculations.

THEOREM 1 Pappus's Theorem for Volumes
If a plane region is revolved once about a line in the plane that does not cut through the region's interior, then the volume of the solid it generates is equal to the region's area times the distance traveled by the region's centroid during the revolution. If ρ is the distance from the axis of revolution to the centroid, then

$$V = 2\pi\rho A. \tag{9}$$

Proof We draw the axis of revolution as the x-axis with the region R in the first quadrant (Figure 6.54). We let $L(y)$ denote the length of the cross-section of R perpendicular to the y-axis at y. We assume $L(y)$ to be continuous.

By the method of cylindrical shells, the volume of the solid generated by revolving the region about the x-axis is

$$V = \int_c^d 2\pi(\text{shell radius})(\text{shell height})\,dy = 2\pi\int_c^d y\,L(y)\,dy. \tag{10}$$

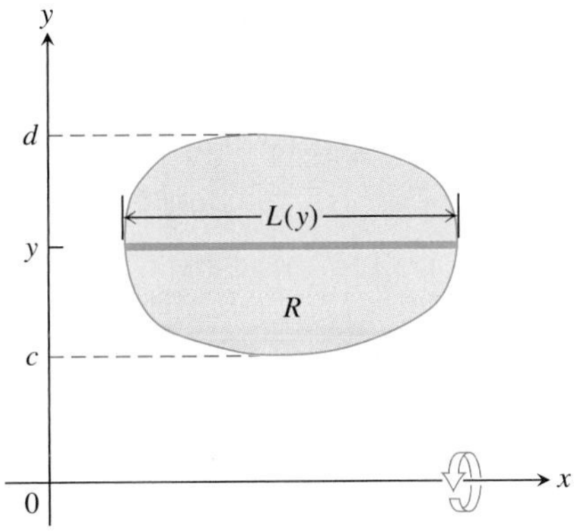

FIGURE 6.54 The region R is to be revolved (once) about the x-axis to generate a solid. A 1700-year-old theorem says that the solid's volume can be calculated by multiplying the region's area by the distance traveled by its centroid during the revolution.

The y-coordinate of R's centroid is

$$\bar{y} = \frac{\int_c^d \tilde{y}\,dA}{A} = \frac{\int_c^d y\,L(y)\,dy}{A}, \qquad \tilde{y} = y,\ dA = L(y)dy$$

so that

$$\int_c^d y\,L(y)\,dy = A\bar{y}.$$

Substituting $A\bar{y}$ for the last integral in Equation (10) gives $V = 2\pi\bar{y}A$. With ρ equal to $\bar{y}$, we have $V = 2\pi\rho A$. ■

EXAMPLE 5 Volume of a Torus

The volume of the torus (doughnut) generated by revolving a circular disk of radius a about an axis in its plane at a distance $b \geq a$ from its center (Figure 6.55) is

$$V = 2\pi(b)(\pi a^2) = 2\pi^2 ba^2.$$ ■

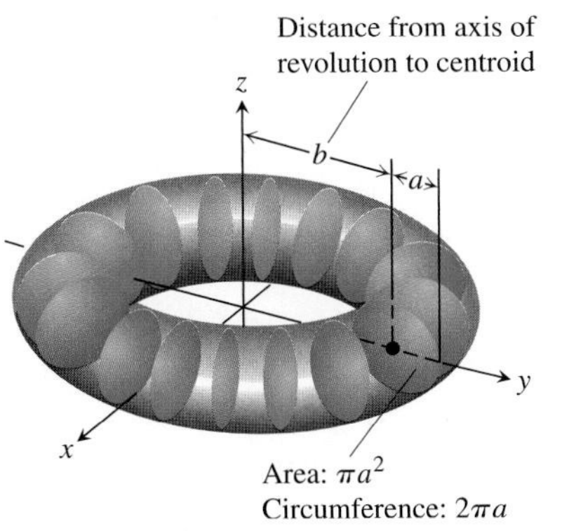

FIGURE 6.55 With Pappus's first theorem, we can find the volume of a torus without having to integrate (Example 5).

EXAMPLE 6 Locate the Centroid of a Semicircular Region

Solution We model the region as the region between the semicircle $y = \sqrt{a^2 - x^2}$ (Figure 6.56) and the x-axis and imagine revolving the region about the x-axis to generate a solid sphere. By symmetry, the x-coordinate of the centroid is $\bar{x} = 0$. With $\bar{y} = \rho$ in Equation (9), we have

$$\bar{y} = \frac{V}{2\pi A} = \frac{(4/3)\pi a^3}{2\pi(1/2)\pi a^2} = \frac{4}{3\pi}a.$$ ■

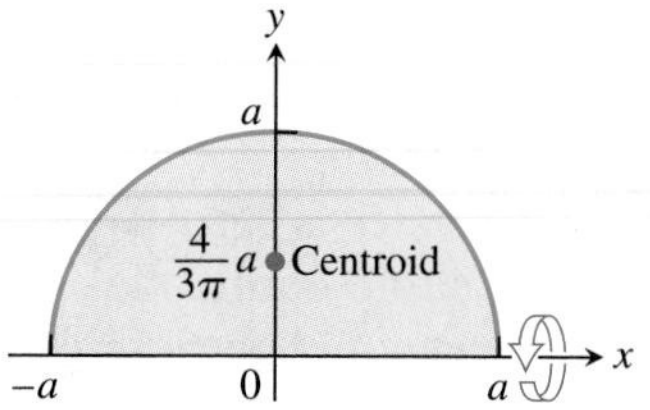

FIGURE 6.56 With Pappus's first theorem, we can locate the centroid of a semicircular region without having to integrate (Example 6).

THEOREM 2 Pappus's Theorem for Surface Areas
If an arc of a smooth plane curve is revolved once about a line in the plane that does not cut through the arc's interior, then the area of the surface generated by the arc equals the length of the arc times the distance traveled by the arc's centroid during the revolution. If ρ is the distance from the axis of revolution to the centroid, then

$$S = 2\pi\rho L. \tag{11}$$

The proof we give assumes that we can model the axis of revolution as the x-axis and the arc as the graph of a continuously differentiable function of x.

Proof We draw the axis of revolution as the x-axis with the arc extending from $x = a$ to $x = b$ in the first quadrant (Figure 6.57). The area of the surface generated by the arc is

$$S = \int_{x=a}^{x=b} 2\pi y\, ds = 2\pi \int_{x=a}^{x=b} y\, ds. \tag{12}$$

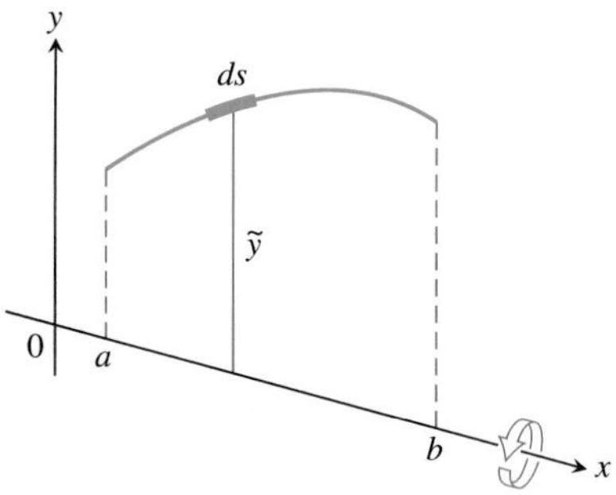

FIGURE 6.57 Figure for proving Pappus's area theorem.

The y-coordinate of the arc's centroid is

$$\bar{y} = \frac{\int_{x=a}^{x=b} \tilde{y}\, ds}{\int_{x=a}^{x=b} ds} = \frac{\int_{x=a}^{x=b} y\, ds}{L}.$$

$L = \int ds$ is the arc's length and $\tilde{y} = y$.

Hence

$$\int_{x=a}^{x=b} y\, ds = \bar{y}L.$$

Substituting $\bar{y}L$ for the last integral in Equation (12) gives $S = 2\pi\bar{y}L$. With ρ equal to $\bar{y}$, we have $S = 2\pi\rho L$. ■

EXAMPLE 7 Surface Area of a Torus

The surface area of the torus in Example 5 is

$$S = 2\pi(b)(2\pi a) = 4\pi^2 ba.$$

■

EXERCISES 6.5

Finding Integrals for Surface Area

In Exercises 1–8:

a. Set up an integral for the area of the surface generated by revolving the given curve about the indicated axis.

T **b.** Graph the curve to see what it looks like. If you can, graph the surface, too.

T **c.** Use your grapher's or computer's integral evaluator to find the surface's area numerically.

1. $y = \tan x, \quad 0 \le x \le \pi/4; \quad x$-axis

2. $y = x^2, \quad 0 \le x \le 2; \quad x$-axis

3. $xy = 1, \quad 1 \le y \le 2; \quad y$-axis

4. $x = \sin y, \quad 0 \le y \le \pi; \quad y$-axis

5. $x^{1/2} + y^{1/2} = 3$ from $(4, 1)$ to $(1, 4)$; x-axis

6. $y + 2\sqrt{y} = x, \quad 1 \le y \le 2; \quad y$-axis

7. $x = \int_0^y \tan t\, dt, \quad 0 \le y \le \pi/3; \quad y$-axis

8. $y = \int_1^x \sqrt{t^2 - 1}\, dt, \quad 1 \le x \le \sqrt{5}; \quad x$-axis

Finding Surface Areas

9. Find the lateral (side) surface area of the cone generated by revolving the line segment $y = x/2, 0 \le x \le 4$, about the x-axis. Check your answer with the geometry formula

$$\text{Lateral surface area} = \frac{1}{2} \times \text{base circumference} \times \text{slant height}.$$

10. Find the lateral surface area of the cone generated by revolving the line segment $y = x/2, 0 \le x \le 4$ about the y-axis. Check your answer with the geometry formula

$$\text{Lateral surface area} = \frac{1}{2} \times \text{base circumference} \times \text{slant height}.$$

11. Find the surface area of the cone frustum generated by revolving the line segment $y = (x/2) + (1/2), 1 \le x \le 3$, about the x-axis. Check your result with the geometry formula

$$\text{Frustum surface area} = \pi(r_1 + r_2) \times \text{slant height}.$$

12. Find the surface area of the cone frustum generated by revolving the line segment $y = (x/2) + (1/2), 1 \le x \le 3$, about the y-axis. Check your result with the geometry formula

$$\text{Frustum surface area} = \pi(r_1 + r_2) \times \text{slant height}.$$

Find the areas of the surfaces generated by revolving the curves in Exercises 13–22 about the indicated axes. If you have a grapher, you may want to graph these curves to see what they look like.

13. $y = x^3/9, \quad 0 \le x \le 2; \quad x$-axis

14. $y = \sqrt{x}, \quad 3/4 \le x \le 15/4; \quad x$-axis

15. $y = \sqrt{2x - x^2}, \quad 0.5 \le x \le 1.5; \quad x$-axis

16. $y = \sqrt{x + 1}, \quad 1 \le x \le 5; \quad x$-axis

17. $x = y^3/3, \quad 0 \le y \le 1; \quad y$-axis

18. $x = (1/3)y^{3/2} - y^{1/2}, \quad 1 \le y \le 3; \quad y$-axis

19. $x = 2\sqrt{4 - y}, \quad 0 \le y \le 15/4; \quad y$-axis

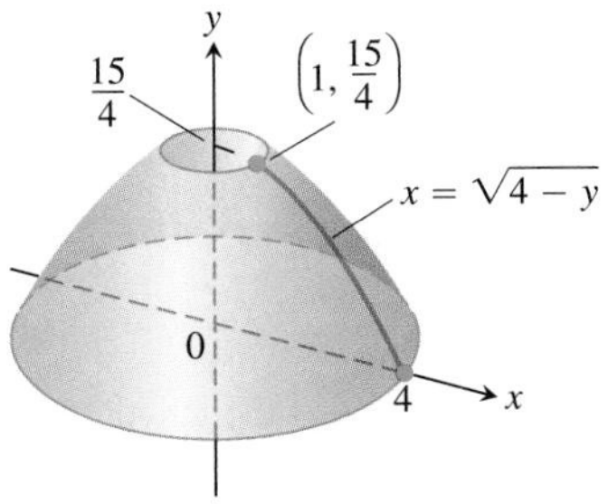

20. $x = \sqrt{2y - 1}, \quad 5/8 \le y \le 1; \quad y$-axis

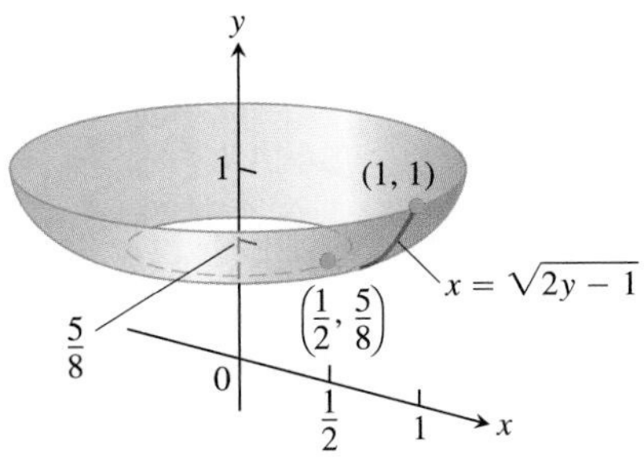

21. $x = (y^4/4) + 1/(8y^2), \quad 1 \le y \le 2; \quad x$-axis (*Hint:* Express $ds = \sqrt{dx^2 + dy^2}$ in terms of dy, and evaluate the integral $S = \int 2\pi y \, ds$ with appropriate limits.)

22. $y = (1/3)(x^2 + 2)^{3/2}, \quad 0 \le x \le \sqrt{2}; \quad y$-axis (*Hint:* Express $ds = \sqrt{dx^2 + dy^2}$ in terms of dx, and evaluate the integral $S = \int 2\pi x \, ds$ with appropriate limits.)

23. Testing the new definition Show that the surface area of a sphere of radius a is still $4\pi a^2$ by using Equation (3) to find the area of the surface generated by revolving the curve $y = \sqrt{a^2 - x^2}, \ -a \le x \le a$, about the x-axis.

24. Testing the new definition The lateral (side) surface area of a cone of height h and base radius r should be $\pi r\sqrt{r^2 + h^2}$, the semiperimeter of the base times the slant height. Show that this is still the case by finding the area of the surface generated by revolving the line segment $y = (r/h)x, 0 \le x \le h$, about the x-axis.

25. Write an integral for the area of the surface generated by revolving the curve $y = \cos x, -\pi/2 \le x \le \pi/2$, about the x-axis. In Section 8.5 we will see how to evaluate such integrals.

26. The surface of an astroid Find the area of the surface generated by revolving about the x-axis the portion of the astroid $x^{2/3} + y^{2/3} = 1$ shown here. (*Hint:* Revolve the first-quadrant portion $y = (1 - x^{2/3})^{3/2}, 0 \le x \le 1$, about the x-axis and double your result.)

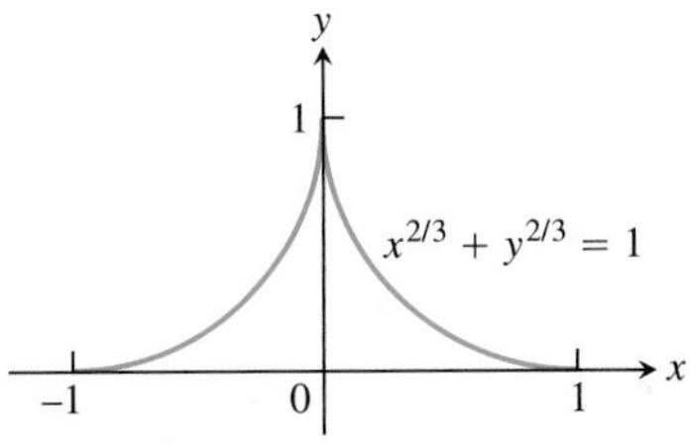

T **27. Enameling woks** Your company decided to put out a deluxe version of the successful wok you designed in Section 6.1, Exercise 55. The plan is to coat it inside with white enamel and outside with blue enamel. Each enamel will be sprayed on 0.5 mm thick before baking. (See diagram here.) Your manufacturing department wants to know how much enamel to have on hand for a production run of 5000 woks. What do you tell them? (Neglect waste and unused material and give your answer in liters. Remember that $1 \text{ cm}^3 = 1 \text{ mL}$, so $1 \text{ L} = 1000 \text{ cm}^3$.)

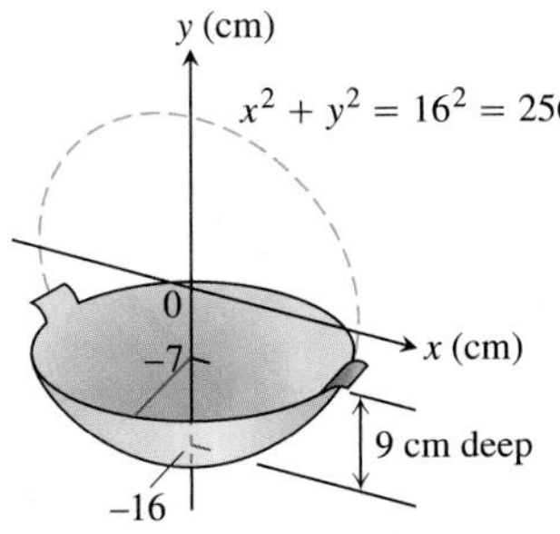

28. Slicing bread Did you know that if you cut a spherical loaf of bread into slices of equal width, each slice will have the same amount of crust? To see why, suppose the semicircle $y = \sqrt{r^2 - x^2}$ shown here is revolved about the x-axis to generate a sphere. Let AB be an arc of the semicircle that lies above an interval of length h on the x-axis. Show that the area swept out by AB does not depend on the location of the interval. (It does depend on the length of the interval.)

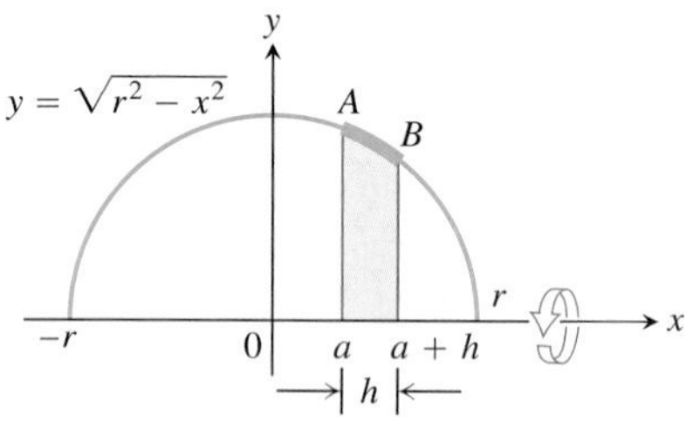

29. The shaded band shown here is cut from a sphere of radius R by parallel planes h units apart. Show that the surface area of the band is $2\pi Rh$.

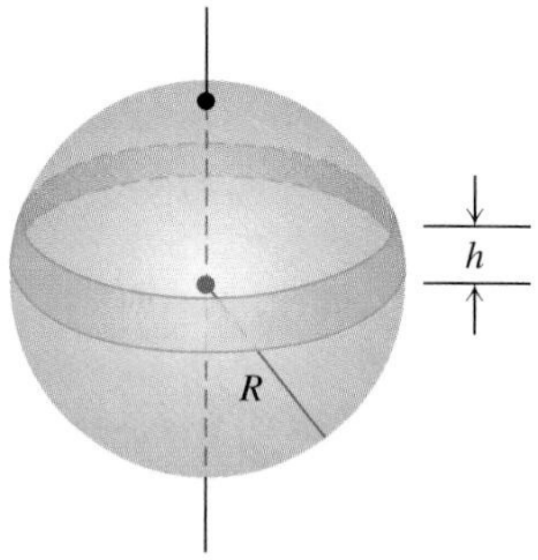

30. Here is a schematic drawing of the 90-ft dome used by the U.S. National Weather Service to house radar in Bozeman, Montana.

a. How much outside surface is there to paint (not counting the bottom)?

T **b.** Express the answer to the nearest square foot.

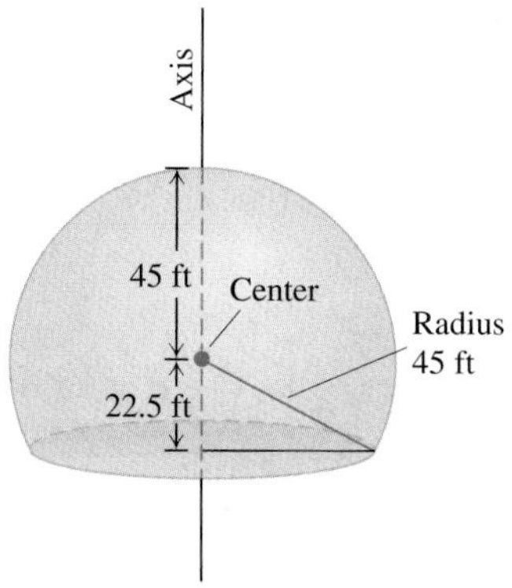

31. Surfaces generated by curves that cross the axis of revolution The surface area formula in Equation (3) was developed under the assumption that the function f whose graph generated the surface was nonnegative over the interval $[a, b]$. For curves that cross the axis of revolution, we replace Equation (3) with the absolute value formula

$$S = \int 2\pi\rho\, ds = \int 2\pi |f(x)|\, ds. \qquad (13)$$

Use Equation (13) to find the surface area of the double cone generated by revolving the line segment $y = x, -1 \le x \le 2$, about the x-axis.

32. (*Continuation of Exercise 31.*) Find the area of the surface generated by revolving the curve $y = x^3/9, -\sqrt{3} \le x \le \sqrt{3}$, about the x-axis. What do you think will happen if you drop the absolute value bars from Equation (13) and attempt to find the surface area with the formula $S = \int 2\pi f(x)\, ds$ instead? Try it.

Parametrizations

Find the areas of the surfaces generated by revolving the curves in Exercises 33–35 about the indicated axes.

33. $x = \cos t, \quad y = 2 + \sin t, \quad 0 \le t \le 2\pi; \quad x$-axis

34. $x = (2/3)t^{3/2}, \quad y = 2\sqrt{t}, \quad 0 \le t \le \sqrt{3}; \quad y$-axis

35. $x = t + \sqrt{2}, \quad y = (t^2/2) + \sqrt{2}t, \quad -\sqrt{2} \le t \le \sqrt{2}; \quad y$-axis

36. Set up, but do not evaluate, an integral that represents the area of the surface obtained by rotating the curve $x = a\,(t - \sin t)$, $y = a\,(1 - \cos t), 0 \le t \le 2\pi$, about the x-axis.

37. A cone frustum The line segment joining the points (0, 1) and (2, 2) is revolved about the x-axis to generate a frustum of a cone. Find the surface area of the frustum using the parametrization $x = 2t, y = t + 1, 0 \le t \le 1$. Check your result with the geometry formula: Area $= \pi(r_1 + r_2)$(slant height).

38. A cone The line segment joining the origin to the point (h, r) is revolved about the x-axis to generate a cone of height h and base radius r. Find the cone's surface area with the parametric equations $x = ht, y = rt, 0 \le t \le 1$. Check your result with the geometry formula: Area $= \pi r$(slant height).

39. An alternative derivation of the surface area formula Assume f is smooth on $[a, b]$ and partition $[a, b]$ in the usual way. In the kth subinterval $[x_{k-1}, x_k]$ construct the tangent line to the curve at the midpoint $m_k = (x_{k-1} + x_k)/2$, as in the figure here.

a. Show that $r_1 = f(m_k) - f'(m_k)\dfrac{\Delta x_k}{2}$ and $r_2 = f(m_k) + f'(m_k)\dfrac{\Delta x_k}{2}$.

b. Show that the length L_k of the tangent line segment in the kth subinterval is $L_k = \sqrt{(\Delta x_k)^2 + (f'(m_k)\,\Delta x_k)^2}$.

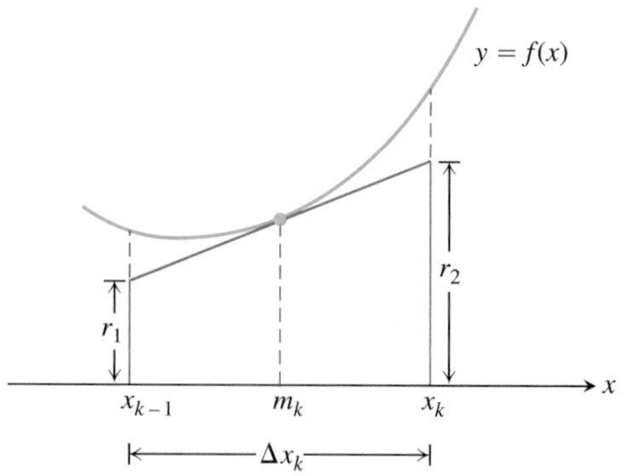

c. Show that the lateral surface area of the frustum of the cone swept out by the tangent line segment as it revolves about the x-axis is $2\pi f(m_k)\sqrt{1 + (f'(m_k))^2}\,\Delta x_k$.

d. Show that the area of the surface generated by revolving $y = f(x)$ about the x-axis over $[a, b]$ is

$$\lim_{n\to\infty} \sum_{k=1}^{n} \begin{pmatrix}\text{lateral surface area} \\ \text{of } k\text{th frustum}\end{pmatrix} = \int_a^b 2\pi f(x)\sqrt{1 + (f'(x))^2}\,dx.$$

40. Modeling surface area The lateral surface area of the cone swept out by revolving the line segment $y = x/\sqrt{3},\ 0 \le x \le \sqrt{3}$, about the x-axis should be $(1/2)$(base circumference)(slant height) $= (1/2)(2\pi)(2) = 2\pi$. What do you get if you use Equation (8) with $f(x) = x/\sqrt{3}$?

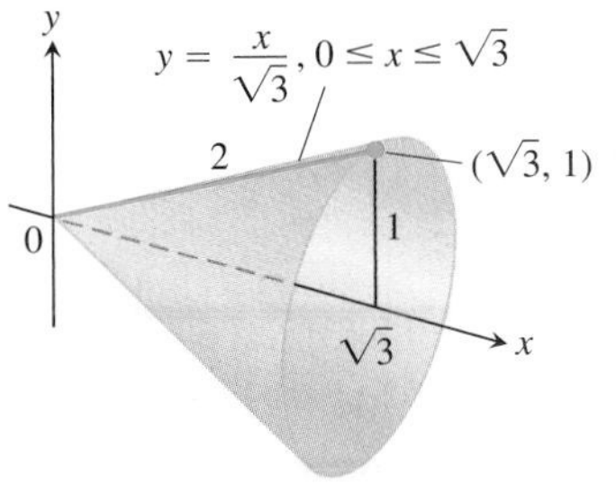

The Theorems of Pappus

41. The square region with vertices (0, 2), (2, 0), (4, 2), and (2, 4) is revolved about the x-axis to generate a solid. Find the volume and surface area of the solid.

42. Use a theorem of Pappus to find the volume generated by revolving about the line $x = 5$ the triangular region bounded by the coordinate axes and the line $2x + y = 6$. (As you saw in Exercise 29 of Section 6.4, the centroid of a triangle lies at the intersection of the medians, one-third of the way from the midpoint of each side toward the opposite vertex.)

43. Find the volume of the torus generated by revolving the circle $(x - 2)^2 + y^2 = 1$ about the y-axis.

44. Use the theorems of Pappus to find the lateral surface area and the volume of a right circular cone.

45. Use the Second Theorem of Pappus and the fact that the surface area of a sphere of radius a is $4\pi a^2$ to find the centroid of the semicircle $y = \sqrt{a^2 - x^2}$.

46. As found in Exercise 45, the centroid of the semicircle $y = \sqrt{a^2 - x^2}$ lies at the point $(0, 2a/\pi)$. Find the area of the surface swept out by revolving the semicircle about the line $y = a$.

47. The area of the region R enclosed by the semiellipse $y = (b/a)\sqrt{a^2 - x^2}$ and the x-axis is $(1/2)\pi ab$ and the volume of the ellipsoid generated by revolving R about the x-axis is $(4/3)\pi ab^2$. Find the centroid of R. Notice that the location is independent of a.

48. As found in Example 6, the centroid of the region enclosed by the x-axis and the semicircle $y = \sqrt{a^2 - x^2}$ lies at the point $(0, 4a/3\pi)$. Find the volume of the solid generated by revolving this region about the line $y = -a$.

49. The region of Exercise 48 is revolved about the line $y = x - a$ to generate a solid. Find the volume of the solid.

50. As found in Exercise 45, the centroid of the semicircle $y = \sqrt{a^2 - x^2}$ lies at the point $(0, 2a/\pi)$. Find the area of the surface generated by revolving the semicircle about the line $y = x - a$.

51. Find the moment about the x-axis of the semicircular region in Example 6. If you use results already known, you will not need to integrate.

6.6 Work

In everyday life, *work* means an activity that requires muscular or mental effort. In science, the term refers specifically to a force acting on a body and the body's subsequent displacement. This section shows how to calculate work. The applications run from compressing railroad car springs and emptying subterranean tanks to forcing electrons together and lifting satellites into orbit.

Work Done by a Constant Force

When a body moves a distance d along a straight line as a result of being acted on by a force of constant magnitude F in the direction of motion, we define the **work** W done by the force on the body with the formula

$$W = Fd \qquad \text{(Constant-force formula for work).} \tag{1}$$

Joules
The joule, abbreviated J and pronounced "jewel," is named after the English physicist James Prescott Joule (1818–1889). The defining equation is

$$1 \text{ joule} = (1 \text{ newton})(1 \text{ meter}).$$

In symbols, $1 \text{ J} = 1 \text{ N} \cdot \text{m}$.

From Equation (1) we see that the unit of work in any system is the unit of force multiplied by the unit of distance. In SI units (SI stands for *Système International*, or International System), the unit of force is a newton, the unit of distance is a meter, and the unit of work is a newton-meter $(\text{N} \cdot \text{m})$. This combination appears so often it has a special name, the **joule**. In the British system, the unit of work is the foot-pound, a unit frequently used by engineers.

EXAMPLE 1 Jacking Up a Car

If you jack up the side of a 2000-lb car 1.25 ft to change a tire (you have to apply a constant vertical force of about 1000 lb) you will perform $1000 \times 1.25 = 1250$ ft-lb of work on the car. In SI units, you have applied a force of 4448 N through a distance of 0.381 m to do $4448 \times 0.381 \approx 1695$ J of work. ■

Work Done by a Variable Force Along a Line

If the force you apply varies along the way, as it will if you are compressing a spring, the formula $W = Fd$ has to be replaced by an integral formula that takes the variation in F into account.

Suppose that the force performing the work acts along a line that we take to be the x-axis and that its magnitude F is a continuous function of the position. We want to find the work done over the interval from $x = a$ to $x = b$. We partition $[a, b]$ in the usual way and choose an arbitrary point c_k in each subinterval $[x_{k-1}, x_k]$. If the subinterval is short enough, F, being continuous, will not vary much from x_{k-1} to x_k. The amount of work done across the interval will be about $F(c_k)$ times the distance Δx_k, the same as it would be if F were constant and we could apply Equation (1). The total work done from a to b is therefore approximated by the Riemann sum

$$\text{Work} \approx \sum_{k=1}^{n} F(c_k)\,\Delta x_k .$$

We expect the approximation to improve as the norm of the partition goes to zero, so we define the work done by the force from a to b to be the integral of F from a to b:

$$\lim_{n\to\infty} \sum_{k=1}^{n} F(c_k)\,\Delta x_k = \int_a^b F(x)\,dx.$$

DEFINITION Work
The **work** done by a variable force $F(x)$ directed along the x-axis from $x = a$ to $x = b$ is

$$W = \int_a^b F(x)\,dx. \tag{2}$$

The units of the integral are joules if F is in newtons and x is in meters, and foot-pounds if F is in pounds and x in feet. So, the work done by a force of $F(x) = 1/x^2$ newtons along the x-axis from $x = 1$ m to $x = 10$ m is

$$W = \int_1^{10} \frac{1}{x^2}\,dx = -\frac{1}{x}\bigg]_1^{10} = -\frac{1}{10} + 1 = 0.9 \text{ J}.$$

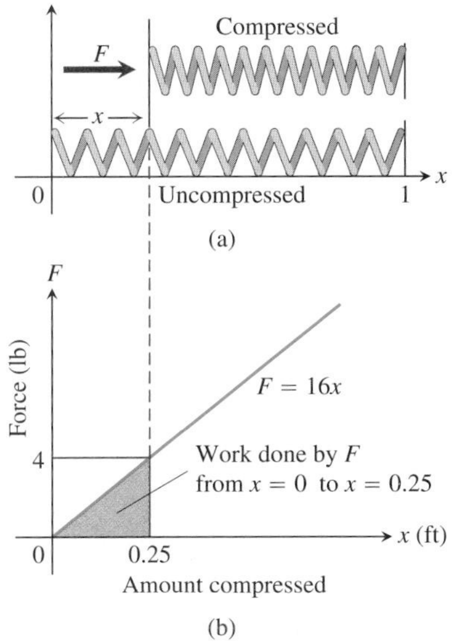

FIGURE 6.58 The force F needed to hold a spring under compression increases linearly as the spring is compressed (Example 2).

Hooke's Law for Springs: $F = kx$

Hooke's Law says that the force it takes to stretch or compress a spring x length units from its natural (unstressed) length is proportional to x. In symbols,

$$F = kx. \tag{3}$$

The constant k, measured in force units per unit length, is a characteristic of the spring, called the **force constant** (or **spring constant**) of the spring. Hooke's Law, Equation (3), gives good results as long as the force doesn't distort the metal in the spring. We assume that the forces in this section are too small to do that.

EXAMPLE 2 Compressing a Spring

Find the work required to compress a spring from its natural length of 1 ft to a length of 0.75 ft if the force constant is $k = 16$ lb/ft.

Solution We picture the uncompressed spring laid out along the x-axis with its movable end at the origin and its fixed end at $x = 1$ ft (Figure 6.58). This enables us to describe the force required to compress the spring from 0 to x with the formula $F = 16x$. To compress the spring from 0 to 0.25 ft, the force must increase from

$$F(0) = 16 \cdot 0 = 0 \text{ lb} \qquad \text{to} \qquad F(0.25) = 16 \cdot 0.25 = 4 \text{ lb}.$$

The work done by F over this interval is

$$W = \int_0^{0.25} 16x\,dx = 8x^2\Big]_0^{0.25} = 0.5 \text{ ft-lb}. \qquad \text{Eq. (2) with } a = 0,\ b = 0.25,\ F(x) = 16x$$

■

EXAMPLE 3 Stretching a Spring

A spring has a natural length of 1 m. A force of 24 N stretches the spring to a length of 1.8 m.

(a) Find the force constant k.

(b) How much work will it take to stretch the spring 2 m beyond its natural length?

(c) How far will a 45-N force stretch the spring?

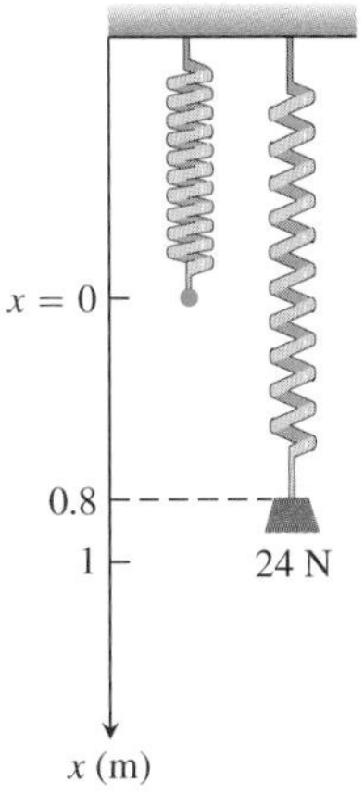

FIGURE 6.59 A 24-N weight stretches this spring 0.8 m beyond its unstressed length (Example 3).

Solution

(a) *The force constant.* We find the force constant from Equation (3). A force of 24 N stretches the spring 0.8 m, so

$$24 = k(0.8) \qquad \text{Eq. (3) with } F = 24,\ x = 0.8$$
$$k = 24/0.8 = 30 \text{ N/m}.$$

(b) *The work to stretch the spring* 2 m. We imagine the unstressed spring hanging along the x-axis with its free end at $x = 0$ (Figure 6.59). The force required to stretch the spring x m beyond its natural length is the force required to pull the free end of the spring x units from the origin. Hooke's Law with $k = 30$ says that this force is

$$F(x) = 30x.$$

The work done by F on the spring from $x = 0$ m to $x = 2$ m is

$$W = \int_0^2 30x\,dx = 15x^2\Big]_0^2 = 60\text{ J}.$$

(c) *How far will a* 45-N *force stretch the spring?* We substitute $F = 45$ in the equation $F = 30x$ to find

$$45 = 30x, \qquad \text{or} \qquad x = 1.5\text{ m}.$$

A 45-N force will stretch the spring 1.5 m. No calculus is required to find this. ■

The work integral is useful to calculate the work done in lifting objects whose weights vary with their elevation.

EXAMPLE 4 Lifting a Rope and Bucket

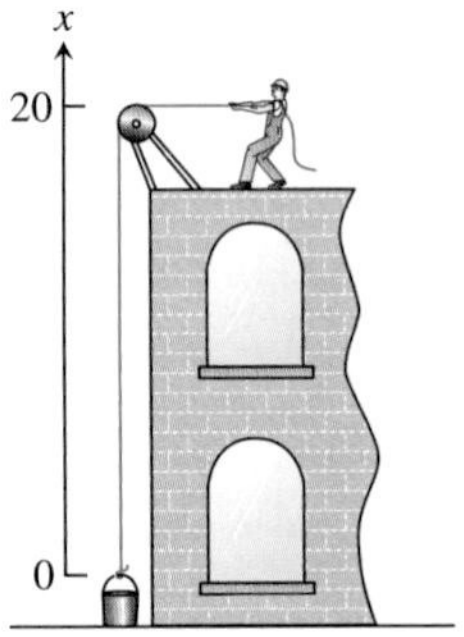

FIGURE 6.60 Lifting the bucket in Example 4.

A 5-lb bucket is lifted from the ground into the air by pulling in 20 ft of rope at a constant speed (Figure 6.60). The rope weighs 0.08 lb/ft. How much work was spent lifting the bucket and rope?

Solution The bucket has constant weight so the work done lifting it alone is weight × distance = $5 \cdot 20 = 100$ ft-lb.

The weight of the rope varies with the bucket's elevation, because less of it is freely hanging. When the bucket is x ft off the ground, the remaining proportion of the rope still being lifted weighs $(0.08) \cdot (20 - x)$ lb. So the work in lifting the rope is

$$\begin{aligned}\text{Work on rope} &= \int_0^{20} (0.08)(20 - x)\,dx = \int_0^{20} (1.6 - 0.08x)\,dx \\ &= \Big[1.6x - 0.04x^2\Big]_0^{20} = 32 - 16 = 16\text{ ft-lb}.\end{aligned}$$

The total work for the bucket and rope combined is

$$100 + 16 = 116\text{ ft-lb}.$$ ■

Pumping Liquids from Containers

How much work does it take to pump all or part of the liquid from a container? To find out, we imagine lifting the liquid out one thin horizontal slab at a time and applying the equation $W = Fd$ to each slab. We then evaluate the integral this leads to as the slabs become thinner and more numerous. The integral we get each time depends on the weight of the liquid and the dimensions of the container, but the way we find the integral is always the same. The next examples show what to do.

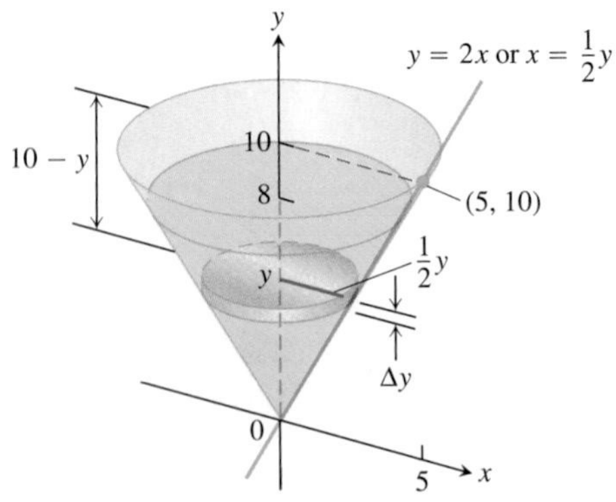

FIGURE 6.61 The olive oil and tank in Example 5.

EXAMPLE 5 Pumping Oil from a Conical Tank

The conical tank in Figure 6.61 is filled to within 2 ft of the top with olive oil weighing 57 lb/ft^3. How much work does it take to pump the oil to the rim of the tank?

Solution We imagine the oil divided into thin slabs by planes perpendicular to the y-axis at the points of a partition of the interval [0, 8].

The typical slab between the planes at y and $y + \Delta y$ has a volume of about

$$\Delta V = \pi(\text{radius})^2(\text{thickness}) = \pi\left(\frac{1}{2}y\right)^2 \Delta y = \frac{\pi}{4}y^2\,\Delta y\text{ ft}^3.$$

The force $F(y)$ required to lift this slab is equal to its weight,

$$F(y) = 57\,\Delta V = \frac{57\pi}{4}y^2\,\Delta y \text{ lb}.$$

Weight = weight per unit volume × volume

The distance through which $F(y)$ must act to lift this slab to the level of the rim of the cone is about $(10 - y)$ ft, so the work done lifting the slab is about

$$\Delta W = \frac{57\pi}{4}(10 - y)y^2\,\Delta y \text{ ft-lb}.$$

Assuming there are n slabs associated with the partition of $[0, 8]$, and that $y = y_k$ denotes the plane associated with the kth slab of thickness Δy_k, we can approximate the work done lifting all of the slabs with the Riemann sum

$$W \approx \sum_{k=1}^{n} \frac{57\pi}{4}(10 - y_k)y_k{}^2\,\Delta y_k \text{ ft-lb}.$$

The work of pumping the oil to the rim is the limit of these sums as the norm of the partition goes to zero.

$$\begin{aligned} W = \lim_{n\to\infty}\sum_{k=1}^{n}\frac{57\pi}{4}(10 - y_k)y_k^2\,\Delta y_k &= \int_0^8 \frac{57\pi}{4}(10 - y)y^2\,dy \\ &= \frac{57\pi}{4}\int_0^8 (10y^2 - y^3)\,dy \\ &= \frac{57\pi}{4}\left[\frac{10y^3}{3} - \frac{y^4}{4}\right]_0^8 \approx 30{,}561 \text{ ft-lb}. \end{aligned}$$

■

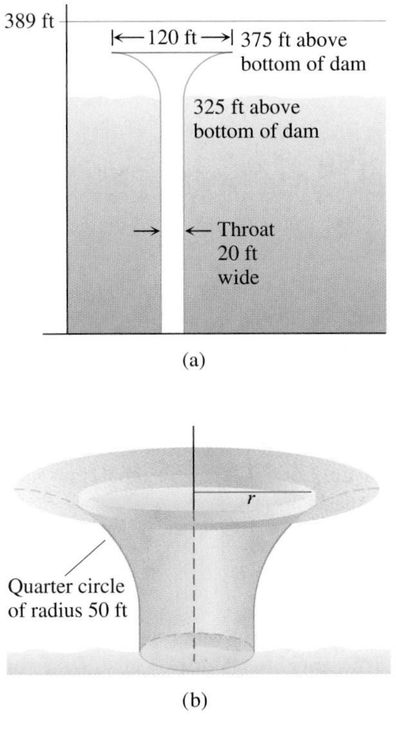

FIGURE 6.62 (a) Cross-section of the glory hole for a dam and (b) the top of the glory hole (Example 6).

EXAMPLE 6 Pumping Water from a Glory Hole

A glory hole is a vertical drain pipe that keeps the water behind a dam from getting too high. The top of the glory hole for a dam is 14 ft below the top of the dam and 375 ft above the bottom (Figure 6.62). The hole needs to be pumped out from time to time to permit the removal of seasonal debris.

From the cross-section in Figure 6.62a, we see that the glory hole is a funnel-shaped drain. The throat of the funnel is 20 ft wide and the head is 120 ft across. The outside boundary of the head cross-section are quarter circles formed with 50-ft radii, shown in Figure 6.62b. The glory hole is formed by rotating a cross-section around its center. Consequently, all horizontal cross-sections are circular disks throughout the entire glory hole. We calculate the work required to pump water from

(a) the throat of the hole.

(b) the funnel portion.

Solution

(a) *Pumping from the throat.* A typical slab in the throat between the planes at y and $y + \Delta y$ has a volume of about

$$\Delta V = \pi(\text{radius})^2(\text{thickness}) = \pi(10)^2\,\Delta y \text{ ft}^3.$$

The force $F(y)$ required to lift this slab is equal to its weight (about 62.4 lb/ft^3 for water),

$$F(y) = 62.4\,\Delta V = 6240\pi\,\Delta y \text{ lb}.$$

The distance through which $F(y)$ must act to lift this slab to the top of the hole is $(375 - y)$ ft, so the work done lifting the slab is

$$\Delta W = 6240\pi(375 - y)\,\Delta y \text{ ft-lb}.$$

We can approximate the work done in pumping the water from the throat by summing the work done lifting all the slabs individually, and then taking the limit of this Riemann sum as the norm of the partition goes to zero. This gives the integral

$$\begin{aligned} W &= \int_0^{325} 6240\pi(375 - y)\,dy \\ &= 6240\pi\left[375y - \frac{y^2}{2}\right]_0^{325} \\ &\approx 1{,}353{,}869{,}354 \text{ ft-lb}. \end{aligned}$$

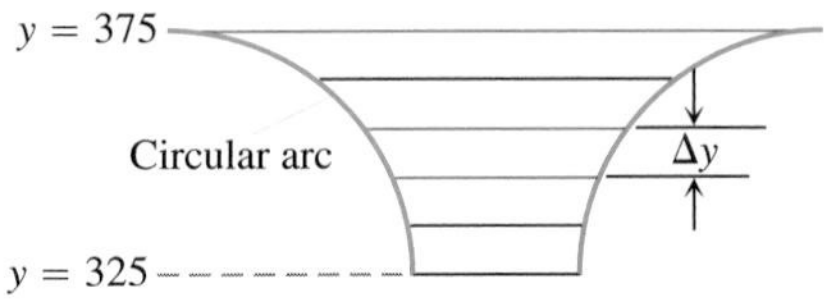

FIGURE 6.63 The glory hole funnel portion.

(b) *Pumping from the funnel.* To compute the work necessary to pump water from the funnel portion of the glory hole, from $y = 325$ to $y = 375$, we need to compute ΔV for approximating elements in the funnel as shown in Figure 6.63. As can be seen from the figure, the radii of the slabs vary with height y.

In Exercises 33 and 34, you are asked to complete the analysis to determine the total work required to pump the water and to find the power of the pumps necessary to pump out the glory hole. ■

EXERCISES 6.6

Springs

1. **Spring constant** It took 1800 J of work to stretch a spring from its natural length of 2 m to a length of 5 m. Find the spring's force constant.

2. **Stretching a spring** A spring has a natural length of 10 in. An 800-lb force stretches the spring to 14 in.
   a. Find the force constant.
   b. How much work is done in stretching the spring from 10 in. to 12 in.?
   c. How far beyond its natural length will a 1600-lb force stretch the spring?

3. **Stretching a rubber band** A force of 2 N will stretch a rubber band 2 cm (0.02 m). Assuming that Hooke's Law applies, how far will a 4-N force stretch the rubber band? How much work does it take to stretch the rubber band this far?

4. **Stretching a spring** If a force of 90 N stretches a spring 1 m beyond its natural length, how much work does it take to stretch the spring 5 m beyond its natural length?

5. **Subway car springs** It takes a force of 21,714 lb to compress a coil spring assembly on a New York City Transit Authority subway car from its free height of 8 in. to its fully compressed height of 5 in.
   a. What is the assembly's force constant?
   b. How much work does it take to compress the assembly the first half inch? the second half inch? Answer to the nearest in.-lb.

   (Data courtesy of Bombardier, Inc., Mass Transit Division, for spring assemblies in subway cars delivered to the New York City Transit Authority from 1985 to 1987.)

6. **Bathroom scale** A bathroom scale is compressed $1/16$ in. when a 150-lb person stands on it. Assuming that the scale behaves like a spring that obeys Hooke's Law, how much does someone who compresses the scale $1/8$ in. weigh? How much work is done compressing the scale $1/8$ in.?

Work Done By a Variable Force

7. **Lifting a rope** A mountain climber is about to haul up a 50 m length of hanging rope. How much work will it take if the rope weighs 0.624 N/m?

8. **Leaky sandbag** A bag of sand originally weighing 144 lb was lifted at a constant rate. As it rose, sand also leaked out at a constant rate. The sand was half gone by the time the bag had been

lifted to 18 ft. How much work was done lifting the sand this far? (Neglect the weight of the bag and lifting equipment.)

9. **Lifting an elevator cable** An electric elevator with a motor at the top has a multistrand cable weighing 4.5 lb/ft. When the car is at the first floor, 180 ft of cable are paid out, and effectively 0 ft are out when the car is at the top floor. How much work does the motor do just lifting the cable when it takes the car from the first floor to the top?

10. **Force of attraction** When a particle of mass m is at $(x, 0)$, it is attracted toward the origin with a force whose magnitude is k/x^2. If the particle starts from rest at $x = b$ and is acted on by no other forces, find the work done on it by the time it reaches $x = a$, $0 < a < b$.

11. **Compressing gas** Suppose that the gas in a circular cylinder of cross-sectional area A is being compressed by a piston. If p is the pressure of the gas in pounds per square inch and V is the volume in cubic inches, show that the work done in compressing the gas from state (p_1, V_1) to state (p_2, V_2) is given by the equation

$$\text{Work} = \int_{(p_1, V_1)}^{(p_2, V_2)} p\, dV.$$

(*Hint:* In the coordinates suggested in the figure here, $dV = A\, dx$. The force against the piston is pA.)

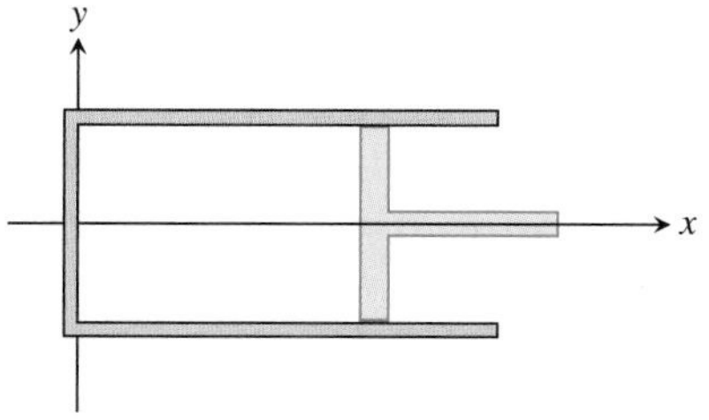

12. (*Continuation of Exercise 11.*) Use the integral in Exercise 11 to find the work done in compressing the gas from $V_1 = 243$ in.^{3} to $V_2 = 32$ in.^{3} if $p_1 = 50$ lb/in.^{3} and p and V obey the gas law $pV^{1.4} =$ constant (for adiabatic processes).

13. **Leaky bucket** Assume the bucket in Example 4 is leaking. It starts with 2 gal of water (16 lb) and leaks at a constant rate. It finishes draining just as it reaches the top. How much work was spent lifting the water alone? (*Hint:* Do not include the rope and bucket, and find the proportion of water left at elevation x ft.)

14. (*Continuation of Exercise 13.*) The workers in Example 4 and Exercise 13 changed to a larger bucket that held 5 gal (40 lb) of water, but the new bucket had an even larger leak so that it, too, was empty by the time it reached the top. Assuming that the water leaked out at a steady rate, how much work was done lifting the water alone? (Do not include the rope and bucket.)

Pumping Liquids from Containers

> **The Weight of Water**
> Because of Earth's rotation and variations in its gravitational field, the weight of a cubic foot of water at sea level can vary from about 62.26 lb at the equator to as much as 62.59 lb near the poles, a variation of about 0.5%. A cubic foot that weighs about 62.4 lb in Melbourne and New York City will weigh 62.5 lb in Juneau and Stockholm. Although 62.4 is a typical figure and common textbook value, there is considerable variation.

15. **Pumping water** The rectangular tank shown here, with its top at ground level, is used to catch runoff water. Assume that the water weighs 62.4 lb/ft^3.
    a. How much work does it take to empty the tank by pumping the water back to ground level once the tank is full?
    b. If the water is pumped to ground level with a $(5/11)$-horsepower (hp) motor (work output 250 ft-lb/sec), how long will it take to empty the full tank (to the nearest minute)?
    c. Show that the pump in part (b) will lower the water level 10 ft (halfway) during the first 25 min of pumping.
    d. **The weight of water** What are the answers to parts (a) and (b) in a location where water weighs 62.26 lb/ft^3? 62.59 lb/ft^3?

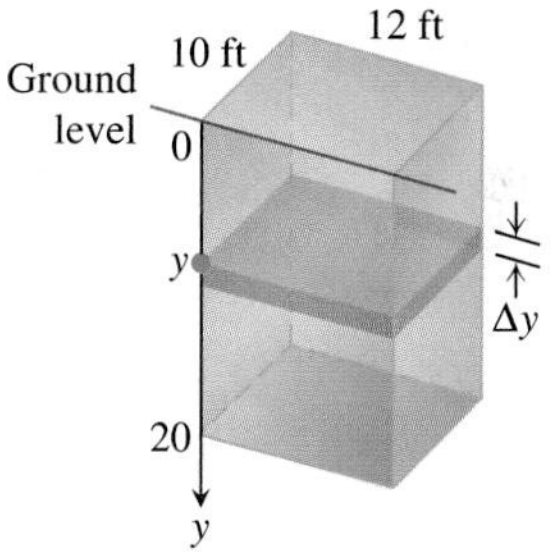

16. **Emptying a cistern** The rectangular cistern (storage tank for rainwater) shown below has its top 10 ft below ground level. The cistern, currently full, is to be emptied for inspection by pumping its contents to ground level.
    a. How much work will it take to empty the cistern?
    b. How long will it take a 1/2 hp pump, rated at 275 ft-lb/sec, to pump the tank dry?
    c. How long will it take the pump in part (b) to empty the tank halfway? (It will be less than half the time required to empty the tank completely.)
    d. **The weight of water** What are the answers to parts (a) through (c) in a location where water weighs 62.26 lb/ft^3? 62.59 lb/ft^3?

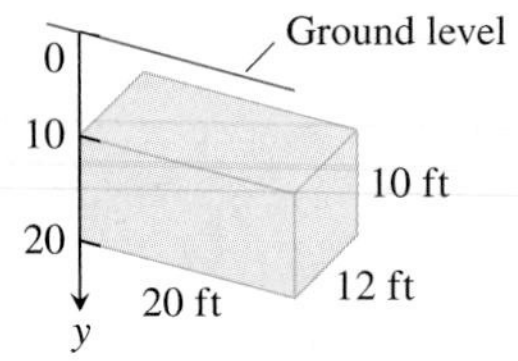

17. Pumping oil How much work would it take to pump oil from the tank in Example 5 to the level of the top of the tank if the tank were completely full?

18. Pumping a half-full tank Suppose that, instead of being full, the tank in Example 5 is only half full. How much work does it take to pump the remaining oil to a level 4 ft above the top of the tank?

19. Emptying a tank A vertical right circular cylindrical tank measures 30 ft high and 20 ft in diameter. It is full of kerosene weighing 51.2 lb/ft^3. How much work does it take to pump the kerosene to the level of the top of the tank?

20. The cylindrical tank shown here can be filled by pumping water from a lake 15 ft below the bottom of the tank. There are two ways to go about it. One is to pump the water through a hose attached to a valve in the bottom of the tank. The other is to attach the hose to the rim of the tank and let the water pour in. Which way will be faster? Give reasons for your answer.

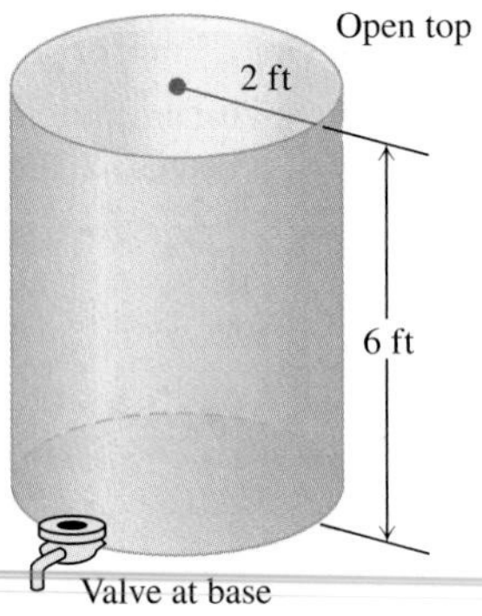

21. a. Pumping milk Suppose that the conical container in Example 5 contains milk (weighing 64.5 lb/ft^3) instead of olive oil. How much work will it take to pump the contents to the rim?

b. Pumping oil How much work will it take to pump the oil in Example 5 to a level 3 ft above the cone's rim?

22. Pumping seawater To design the interior surface of a huge stainless-steel tank, you revolve the curve $y = x^2, 0 \le x \le 4$, about the y-axis. The container, with dimensions in meters, is to be filled with seawater, which weighs 10,000 N/m^3. How much work will it take to empty the tank by pumping the water to the tank's top?

23. Emptying a water reservoir We model pumping from spherical containers the way we do from other containers, with the axis of integration along the vertical axis of the sphere. Use the figure here to find how much work it takes to empty a full hemispherical water reservoir of radius 5 m by pumping the water to a height of 4 m above the top of the reservoir. Water weighs 9800 N/m^3.

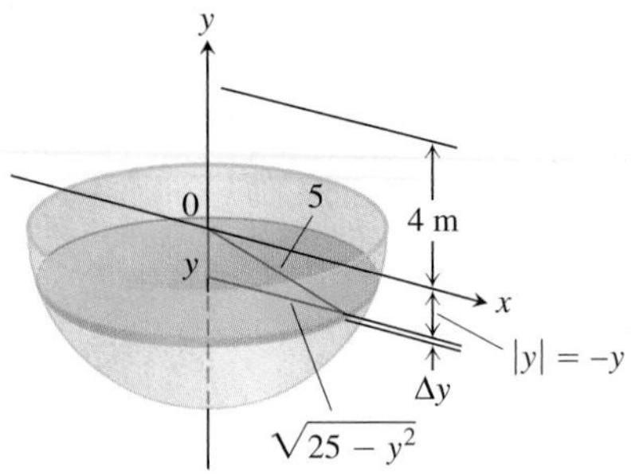

24. You are in charge of the evacuation and repair of the storage tank shown here. The tank is a hemisphere of radius 10 ft and is full of benzene weighing 56 lb/ft^3. A firm you contacted says it can empty the tank for 1/2¢ per foot-pound of work. Find the work required to empty the tank by pumping the benzene to an outlet 2 ft above the top of the tank. If you have \$5000 budgeted for the job, can you afford to hire the firm?

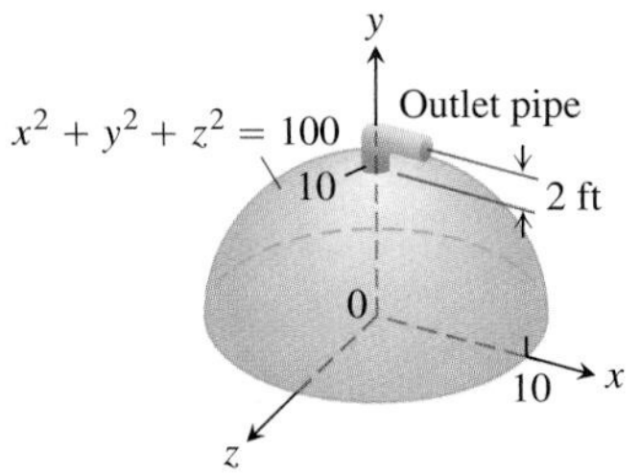

Work and Kinetic Energy

25. Kinetic energy If a variable force of magnitude $F(x)$ moves a body of mass m along the x-axis from x_1 to x_2, the body's velocity v can be written as dx/dt (where t represents time). Use Newton's second law of motion $F = m(dv/dt)$ and the Chain Rule

$$\frac{dv}{dt} = \frac{dv}{dx}\frac{dx}{dt} = v\frac{dv}{dx}$$

to show that the net work done by the force in moving the body from x_1 to x_2 is

$$W = \int_{x_1}^{x_2} F(x)\,dx = \frac{1}{2}mv_2{}^2 - \frac{1}{2}mv_1{}^2,$$

where v_1 and v_2 are the body's velocities at x_1 and x_2. In physics, the expression $(1/2)mv^2$ is called the *kinetic energy* of a body of mass m moving with velocity v. Therefore, *the work done by the force equals the change in the body's kinetic energy*, and we can find the work by calculating this change.

In Exercises 26–32, use the result of Exercise 25.

26. Tennis A 2-oz tennis ball was served at 160 ft/sec (about 109 mph). How much work was done on the ball to make it go this fast? (To find the ball's mass from its weight, express the weight in pounds and divide by 32 ft/sec^2, the acceleration of gravity.)

27. Baseball How many foot-pounds of work does it take to throw a baseball 90 mph? A baseball weighs 5 oz, or 0.3125 lb.

28. **Golf** A 1.6-oz golf ball is driven off the tee at a speed of 280 ft/sec (about 191 mph). How many foot-pounds of work are done on the ball getting it into the air?

29. **Tennis** During the match in which Pete Sampras won the 1990 U.S. Open men's tennis championship, Sampras hit a serve that was clocked at a phenomenal 124 mph. How much work did Sampras have to do on the 2-oz ball to get it to that speed?

30. **Football** A quarterback threw a 14.5-oz football 88 ft/sec (60 mph). How many foot-pounds of work were done on the ball to get it to this speed?

31. **Softball** How much work has to be performed on a 6.5-oz softball to pitch it 132 ft/sec (90 mph)?

32. **A ball bearing** A 2-oz steel ball bearing is placed on a vertical spring whose force constant is $k = 18$ lb/ft. The spring is compressed 2 in. and released. About how high does the ball bearing go?

33. **Pumping the funnel of the glory hole** (*Continuation of Example 6.*)
   a. Find the radius of the cross-section (funnel portion) of the glory hole in Example 6 as a function of the height y above the floor of the dam (from $y = 325$ to $y = 375$).
   b. Find ΔV for the funnel section of the glory hole (from $y = 325$ to $y = 375$).
   c. Find the work necessary to pump out the funnel section by formulating and evaluating the appropriate definite integral.

34. **Pumping water from a glory hole** (*Continuation of Exercise 33.*)
   a. Find the total work necessary to pump out the glory hole, by adding the work necessary to pump both the throat and funnel sections.
   b. Your answer to part (a) is in foot-pounds. A more useful form is horsepower-hours, since motors are rated in horsepower. To convert from foot-pounds to horsepower-hours, divide by 1.98×10^6. How many hours would it take a 1000-horsepower motor to pump out the glory hole, assuming that the motor was fully efficient?

35. **Drinking a milkshake** The truncated conical container shown here is full of strawberry milkshake that weighs 4/9 oz/in.^{3} As you can see, the container is 7 in. deep, 2.5 in. across at the base, and 3.5 in. across at the top (a standard size at Brigham's in Boston). The straw sticks up an inch above the top. About how much work does it take to suck up the milkshake through the straw (neglecting friction)? Answer in inch-ounces.

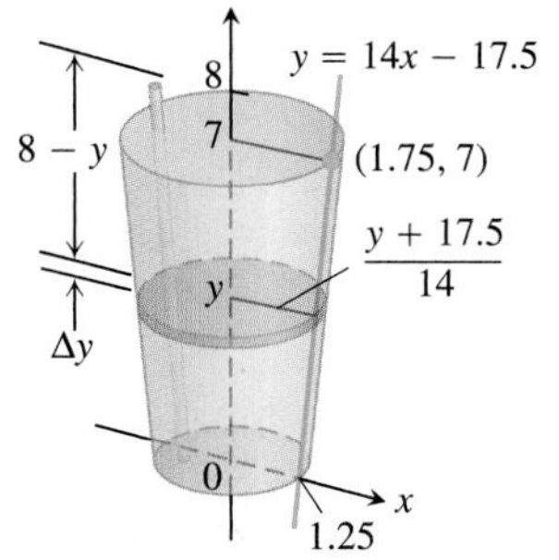

Dimensions in inches

36. **Water tower** Your town has decided to drill a well to increase its water supply. As the town engineer, you have determined that a water tower will be necessary to provide the pressure needed for distribution, and you have designed the system shown here. The water is to be pumped from a 300 ft well through a vertical 4 in. pipe into the base of a cylindrical tank 20 ft in diameter and 25 ft high. The base of the tank will be 60 ft aboveground. The pump is a 3 hp pump, rated at 1650 ft · lb/sec. To the nearest hour, how long will it take to fill the tank the first time? (Include the time it takes to fill the pipe.) Assume that water weighs 62.4 lb/ft^3.

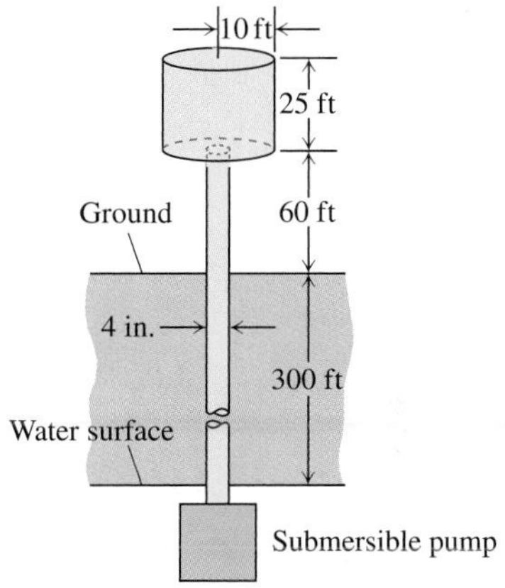

NOT TO SCALE

37. **Putting a satellite in orbit** The strength of Earth's gravitational field varies with the distance r from Earth's center, and the magnitude of the gravitational force experienced by a satellite of mass m during and after launch is

$$F(r) = \frac{mMG}{r^2}.$$

Here, $M = 5.975 \times 10^{24}$ kg is Earth's mass, $G = 6.6720 \times 10^{-11}\ \mathrm{N \cdot m^2\, kg^{-2}}$ is the universal gravitational constant, and r is measured in meters. The work it takes to lift a 1000-kg satellite from Earth's surface to a circular orbit 35,780 km above Earth's center is therefore given by the integral

$$\text{Work} = \int_{6{,}370{,}000}^{35{,}780{,}000} \frac{1000MG}{r^2}\, dr \text{ joules}.$$

Evaluate the integral. The lower limit of integration is Earth's radius in meters at the launch site. (This calculation does not take into account energy spent lifting the launch vehicle or energy spent bringing the satellite to orbit velocity.)

38. **Forcing electrons together** Two electrons r meters apart repel each other with a force of

$$F = \frac{23 \times 10^{-29}}{r^2} \text{ newtons}.$$

   a. Suppose one electron is held fixed at the point $(1, 0)$ on the x-axis (units in meters). How much work does it take to move a second electron along the x-axis from the point $(-1, 0)$ to the origin?
   b. Suppose an electron is held fixed at each of the points $(-1, 0)$ and $(1, 0)$. How much work does it take to move a third electron along the x-axis from $(5, 0)$ to $(3, 0)$?

6.7 Fluid Pressures and Forces

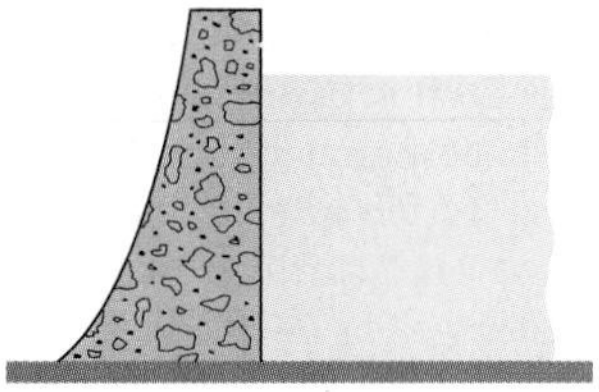

FIGURE 6.64 To withstand the increasing pressure, dams are built thicker as they go down.

We make dams thicker at the bottom than at the top (Figure 6.64) because the pressure against them increases with depth. The pressure at any point on a dam depends only on how far below the surface the point is and not on how much the surface of the dam happens to be tilted at that point. The pressure, in pounds per square foot at a point h feet below the surface, is always $62.4h$. The number 62.4 is the weight-density of water in pounds per cubic foot. The pressure h feet below the surface of any fluid is the fluid's *weight-density* times h.

The Pressure-Depth Equation

In a fluid that is standing still, the pressure p at depth h is the fluid's weight-density w times h:

$$p = wh. \tag{1}$$

In this section we use the equation $p = wh$ to derive a formula for the total force exerted by a fluid against all or part of a vertical or horizontal containing wall.

Weight-density

A fluid's weight-density is its weight per unit volume. Typical values (lb/ft^3) are

Gasoline	42
Mercury	849
Milk	64.5
Molasses	100
Olive oil	57
Seawater	64
Water	62.4

The Constant-Depth Formula for Fluid Force

In a container of fluid with a flat horizontal base, the total force exerted by the fluid against the base can be calculated by multiplying the area of the base by the pressure at the base. We can do this because total force equals force per unit area (pressure) times area. (See Figure 6.65.) If F, p, and A are the total force, pressure, and area, then

$$\begin{aligned} F &= \text{total force} = \text{force per unit area} \times \text{area} \\ &= \text{pressure} \times \text{area} = pA \\ &= whA. \qquad p = wh \text{ from Eq. (1)} \end{aligned}$$

Fluid Force on a Constant-Depth Surface

$$F = pA = whA \tag{2}$$

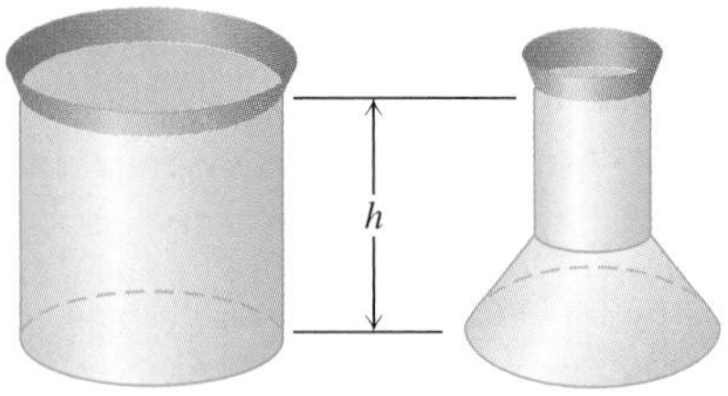

FIGURE 6.65 These containers are filled with water to the same depth and have the same base area. The total force is therefore the same on the bottom of each container. The containers' shapes do not matter here.

For example, the weight-density of water is $62.4\ \text{lb/ft}^3$, so the fluid force at the bottom of a 10 ft $\times$ 20 ft rectangular swimming pool 3 ft deep is

$$\begin{aligned} F = whA &= (62.4\ \text{lb/ft}^3)(3\ \text{ft})(10 \cdot 20\ \text{ft}^2) \\ &= 37{,}440\ \text{lb}. \end{aligned}$$

For a flat plate submerged *horizontally*, like the bottom of the swimming pool just discussed, the downward force acting on its upper face due to liquid pressure is given by Equation (2). If the plate is submerged *vertically*, however, then the pressure against it will be different at different depths and Equation (2) no longer is usable in that form (because h varies). By dividing the plate into many narrow horizontal bands or strips, we can create a Riemann sum whose limit is the fluid force against the side of the submerged vertical plate. Here is the procedure.

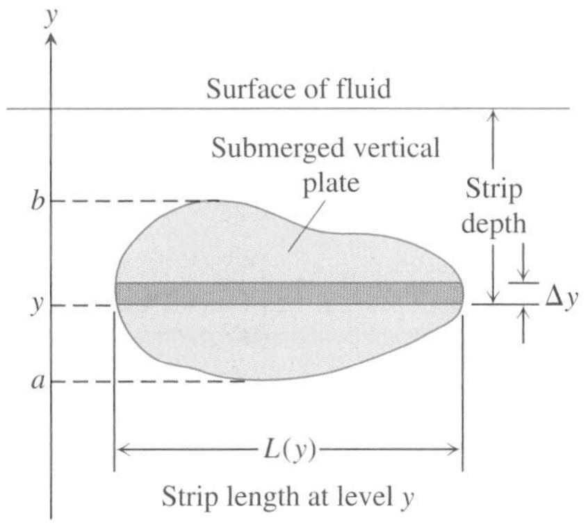

FIGURE 6.66 The force exerted by a fluid against one side of a thin, flat horizontal strip is about $\Delta F = \text{pressure} \times \text{area} = w \times (\text{strip depth}) \times L(y)\,\Delta y$.

The Variable-Depth Formula

Suppose we want to know the force exerted by a fluid against one side of a vertical plate submerged in a fluid of weight-density w. To find it, we model the plate as a region extending from $y = a$ to $y = b$ in the xy-plane (Figure 6.66). We partition $[a, b]$ in the usual way and imagine the region to be cut into thin horizontal strips by planes perpendicular to the y-axis at the partition points. The typical strip from y to $y + \Delta y$ is Δy units wide by $L(y)$ units long. We assume $L(y)$ to be a continuous function of y.

The pressure varies across the strip from top to bottom. If the strip is narrow enough, however, the pressure will remain close to its bottom-edge value of $w \times (\text{strip depth})$. The force exerted by the fluid against one side of the strip will be about

$$\begin{aligned}\Delta F &= (\text{pressure along bottom edge}) \times (\text{area})\\ &= w \cdot (\text{strip depth}) \cdot L(y)\,\Delta y.\end{aligned}$$

Assume there are n strips associated with the partition of $a \le y \le b$ and that y_k is the bottom edge of the kth strip having length $L(y_k)$ and width Δy_k. The force against the entire plate is approximated by summing the forces against each strip, giving the Riemann sum

$$F \approx \sum_{k=1}^{n} (w \cdot (\text{strip depth})_k \cdot L(y_k))\,\Delta y_k. \tag{3}$$

The sum in Equation (3) is a Riemann sum for a continuous function on $[a, b]$, and we expect the approximations to improve as the norm of the partition goes to zero. The force against the plate is the limit of these sums:

$$\lim_{n\to\infty} \sum_{k=1}^{n} (w \cdot (\text{strip depth})_k \cdot L(y_k))\,\Delta y_k = \int_a^b w \cdot (\text{strip depth}) \cdot L(y)\,dy.$$

The Integral for Fluid Force Against a Vertical Flat Plate

Suppose that a plate submerged vertically in fluid of weight-density w runs from $y = a$ to $y = b$ on the y-axis. Let $L(y)$ be the length of the horizontal strip measured from left to right along the surface of the plate at level y. Then the force exerted by the fluid against one side of the plate is

$$F = \int_a^b w \cdot (\text{strip depth}) \cdot L(y)\,dy. \tag{4}$$

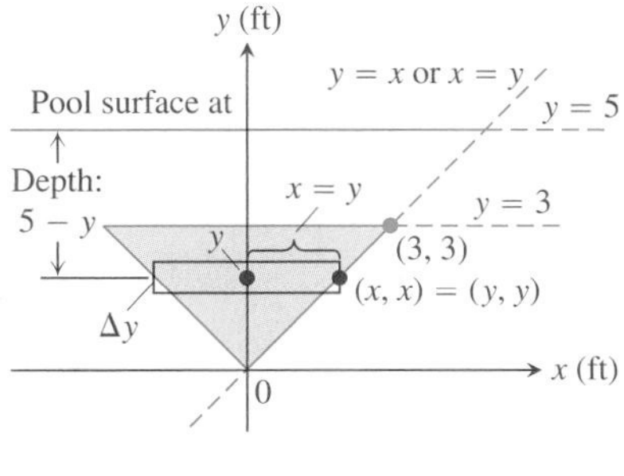

FIGURE 6.67 To find the force on one side of the submerged plate in Example 1, we can use a coordinate system like the one here.

EXAMPLE 1 Applying the Integral for Fluid Force

A flat isosceles right triangular plate with base 6 ft and height 3 ft is submerged vertically, base up, 2 ft below the surface of a swimming pool. Find the force exerted by the water against one side of the plate.

Solution We establish a coordinate system to work in by placing the origin at the plate's bottom vertex and running the y-axis upward along the plate's axis of symmetry (Figure 6.67). The surface of the pool lies along the line $y = 5$ and the plate's top edge along the line $y = 3$. The plate's right-hand edge lies along the line $y = x$, with the upper right vertex at $(3, 3)$. The length of a thin strip at level y is

$$L(y) = 2x = 2y.$$

The depth of the strip beneath the surface is $(5 - y)$. The force exerted by the water against one side of the plate is therefore

$$\begin{aligned} F &= \int_a^b w \cdot \begin{pmatrix}\text{strip} \\ \text{depth}\end{pmatrix} \cdot L(y)\, dy \qquad \text{Eq. (4)} \\ &= \int_0^3 62.4(5 - y)2y\, dy \\ &= 124.8 \int_0^3 (5y - y^2)\, dy \\ &= 124.8 \left[\frac{5}{2}y^2 - \frac{y^3}{3}\right]_0^3 = 1684.8 \text{ lb}. \end{aligned}$$

■

Fluid Forces and Centroids

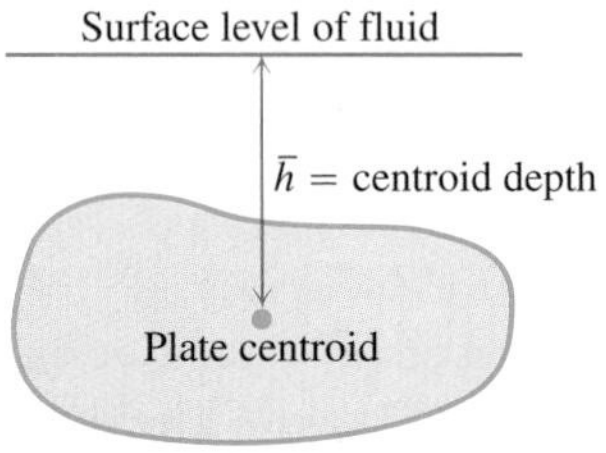

FIGURE 6.68 The force against one side of the plate is $w \cdot \bar{h} \cdot$ plate area.

If we know the location of the centroid of a submerged flat vertical plate (Figure 6.68), we can take a shortcut to find the force against one side of the plate. From Equation (4),

$$\begin{aligned} F &= \int_a^b w \times (\text{strip depth}) \times L(y)\, dy \\ &= w \int_a^b (\text{strip depth}) \times L(y)\, dy \\ &= w \times (\text{moment about surface level line of region occupied by plate}) \\ &= w \times (\text{depth of plate's centroid}) \times (\text{area of plate}). \end{aligned}$$

Fluid Forces and Centroids

The force of a fluid of weight-density w against one side of a submerged flat vertical plate is the product of w, the distance $\bar{h}$ from the plate's centroid to the fluid surface, and the plate's area:

$$F = w\bar{h}A. \qquad (5)$$

EXAMPLE 2 Finding Fluid Force Using Equation (5)

Use Equation (5) to find the force in Example 1.

Solution The centroid of the triangle (Figure 6.67) lies on the y-axis, one-third of the way from the base to the vertex, so $\bar{h} = 3$. The triangle's area is

$$\begin{aligned} A &= \frac{1}{2}(\text{base})(\text{height}) \\ &= \frac{1}{2}(6)(3) = 9. \end{aligned}$$

Hence,

$$\begin{aligned} F &= w\bar{h}A = (62.4)(3)(9) \\ &= 1684.8 \text{ lb}. \end{aligned}$$

■

EXERCISES 6.7

The weight-densities of the fluids in the following exercises can be found in the table on page 456.

1. **Triangular plate** Calculate the fluid force on one side of the plate in Example 1 using the coordinate system shown here.

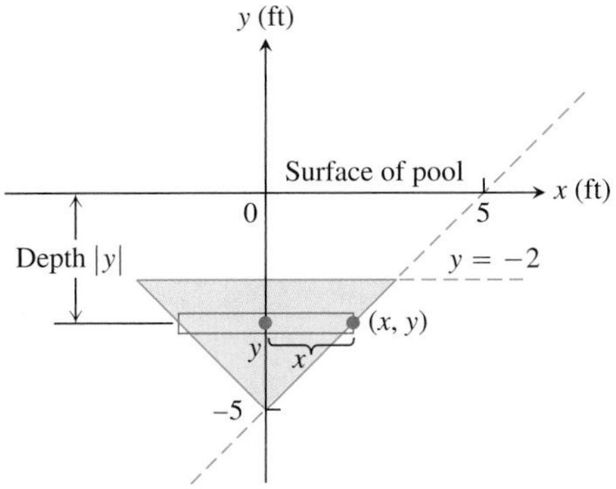

2. **Triangular plate** Calculate the fluid force on one side of the plate in Example 1 using the coordinate system shown here.

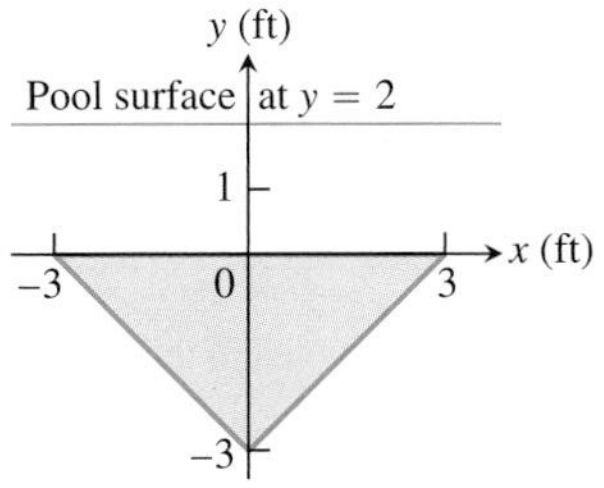

3. **Lowered triangular plate** The plate in Example 1 is lowered another 2 ft into the water. What is the fluid force on one side of the plate now?

4. **Raised triangular plate** The plate in Example 1 is raised to put its top edge at the surface of the pool. What is the fluid force on one side of the plate now?

5. **Triangular plate** The isosceles triangular plate shown here is submerged vertically 1 ft below the surface of a freshwater lake.

   a. Find the fluid force against one face of the plate.

   b. What would be the fluid force on one side of the plate if the water were seawater instead of freshwater?

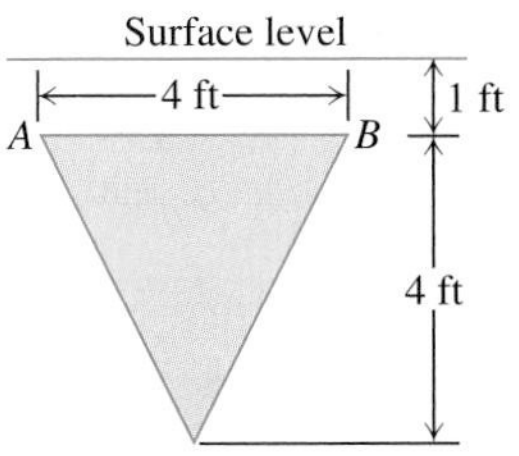

6. **Rotated triangular plate** The plate in Exercise 5 is revolved $180°$ about line AB so that part of the plate sticks out of the lake, as shown here. What force does the water exert on one face of the plate now?

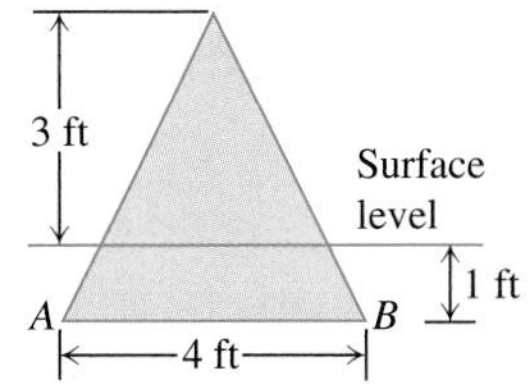

7. **New England Aquarium** The viewing portion of the rectangular glass window in a typical fish tank at the New England Aquarium in Boston is 63 in. wide and runs from 0.5 in. below the water's surface to 33.5 in. below the surface. Find the fluid force against this portion of the window. The weight-density of seawater is $64\ \text{lb/ft}^3$. (In case you were wondering, the glass is $3/4$ in. thick and the tank walls extend 4 in. above the water to keep the fish from jumping out.)

8. **Fish tank** A horizontal rectangular freshwater fish tank with base $2\ \text{ft} \times 4\ \text{ft}$ and height 2 ft (interior dimensions) is filled to within 2 in. of the top.

   a. Find the fluid force against each side and end of the tank.

   b. If the tank is sealed and stood on end (without spilling), so that one of the square ends is the base, what does that do to the fluid forces on the rectangular sides?

9. **Semicircular plate** A semicircular plate 2 ft in diameter sticks straight down into freshwater with the diameter along the surface. Find the force exerted by the water on one side of the plate.

10. **Milk truck** A tank truck hauls milk in a 6-ft-diameter horizontal right circular cylindrical tank. How much force does the milk exert on each end of the tank when the tank is half full?

11. The cubical metal tank shown here has a parabolic gate, held in place by bolts and designed to withstand a fluid force of 160 lb without rupturing. The liquid you plan to store has a weight-density of $50\ \text{lb/ft}^3$.

    a. What is the fluid force on the gate when the liquid is 2 ft deep?

    b. What is the maximum height to which the container can be filled without exceeding its design limitation?

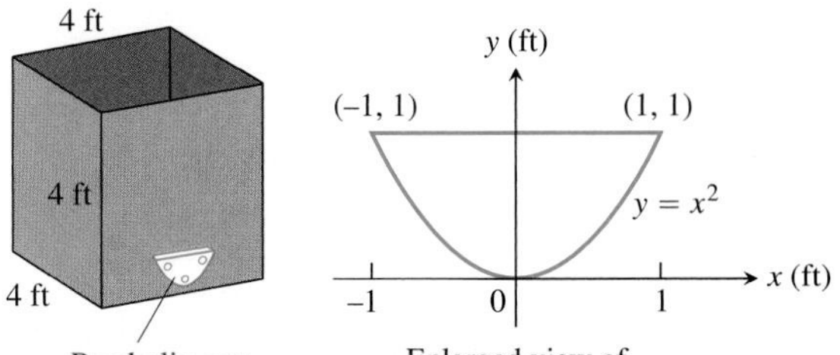

Parabolic gate

Enlarged view of parabolic gate

12. The rectangular tank shown here has a 1 ft × 1 ft square window 1 ft above the base. The window is designed to withstand a fluid force of 312 lb without cracking.

a. What fluid force will the window have to withstand if the tank is filled with water to a depth of 3 ft?

b. To what level can the tank be filled with water without exceeding the window's design limitation?

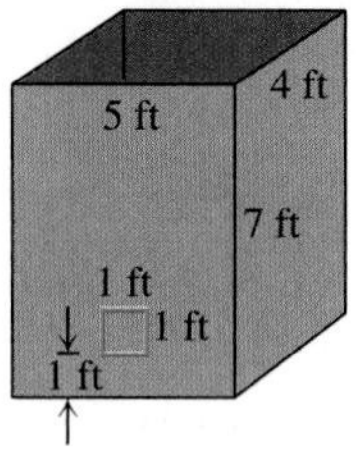

13. The end plates of the trough shown here were designed to withstand a fluid force of 6667 lb. How many cubic feet of water can the tank hold without exceeding this limitation? Round down to the nearest cubic foot.

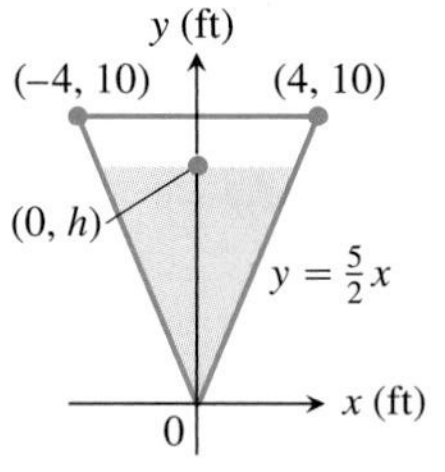

End view of trough

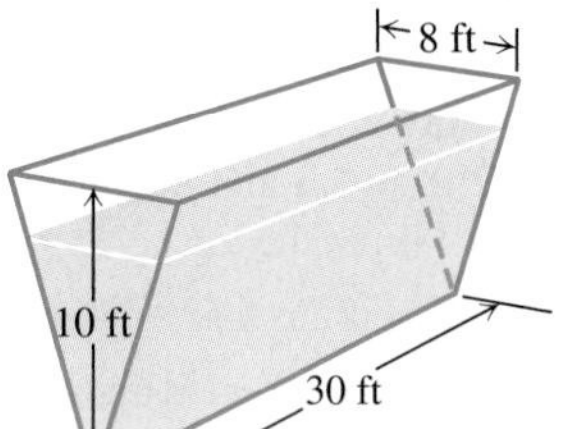

Dimensional view of trough

14. Water is running into the rectangular swimming pool shown here at the rate of 1000 ft^3/h.

a. Find the fluid force against the triangular drain plate after 9 h of filling.

b. The drain plate is designed to withstand a fluid force of 520 lb. How high can you fill the pool without exceeding this limitation?

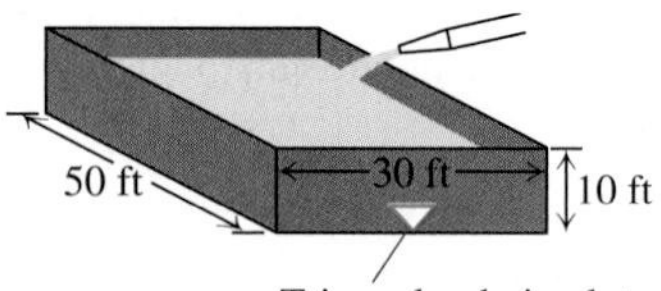

Triangular drain plate

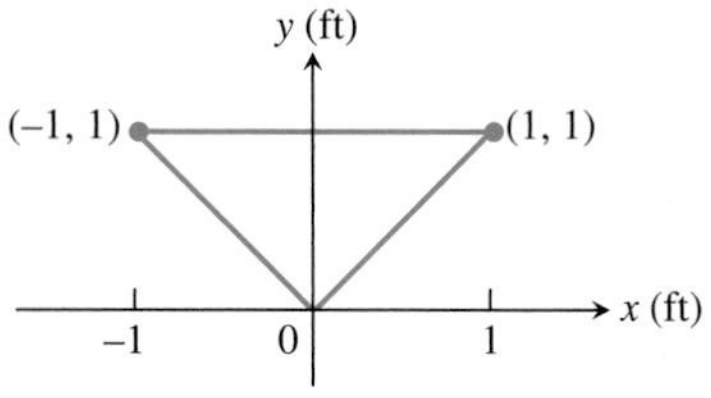

Enlarged view of drain plate

15. A vertical rectangular plate a units long by b units wide is submerged in a fluid of weight-density w with its long edges parallel to the fluid's surface. Find the average value of the pressure along the vertical dimension of the plate. Explain your answer.

16. (*Continuation of Exercise 15.*) Show that the force exerted by the fluid on one side of the plate is the average value of the pressure (found in Exercise 15) times the area of the plate.

17. Water pours into the tank here at the rate of 4 ft^3/min. The tank's cross-sections are 4-ft-diameter semicircles. One end of the tank is movable, but moving it to increase the volume compresses a spring. The spring constant is $k = 100$ lb/ft. If the end of the tank moves 5 ft against the spring, the water will drain out of a safety hole in the bottom at the rate of 5 ft^3/min. Will the movable end reach the hole before the tank overflows?

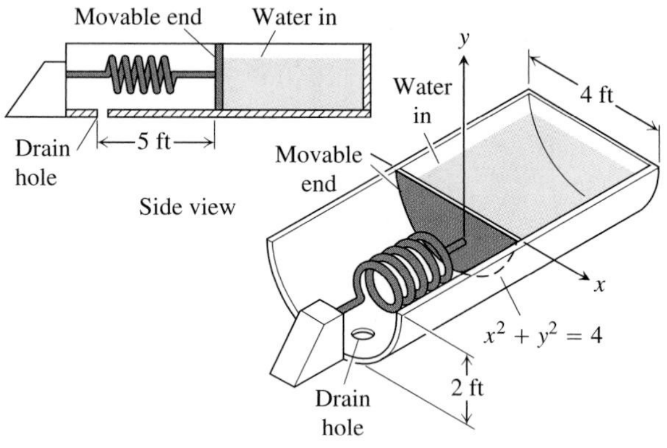

18. **Watering trough** The vertical ends of a watering trough are squares 3 ft on a side.

a. Find the fluid force against the ends when the trough is full.

b. How many inches do you have to lower the water level in the trough to reduce the fluid force by 25%?

19. **Milk carton** A rectangular milk carton measures 3.75 in. × 3.75 in. at the base and is 7.75 in. tall. Find the force of the milk on one side when the carton is full.

20. **Olive oil can** A standard olive oil can measures 5.75 in. × 3.5 in. at the base and is 10 in. tall. Find the fluid force against the base and each side when the can is full.

21. **Watering trough** The vertical ends of a watering trough are isosceles triangles like the one shown here (dimensions in feet).

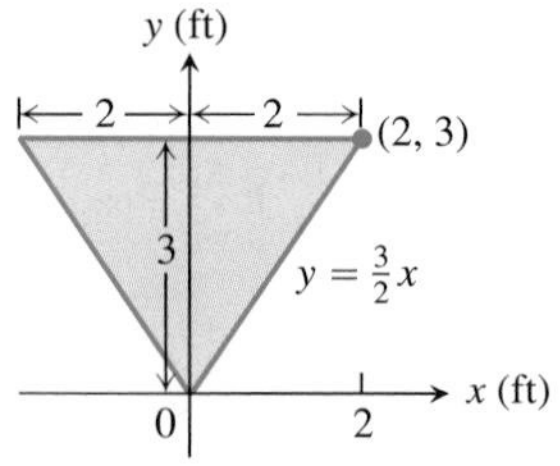

a. Find the fluid force against the ends when the trough is full.

b. How many inches do you have to lower the water level in the trough to cut the fluid force on the ends in half? (Answer to the nearest half-inch.)

c. Does it matter how long the trough is? Give reasons for your answer.

22. The face of a dam is a rectangle, *ABCD*, of dimensions $AB = CD = 100$ ft, $AD = BC = 26$ ft. Instead of being vertical, the plane *ABCD* is inclined as indicated in the accompanying figure, so that the top of the dam is 24 ft higher than the bottom. Find the force due to water pressure on the dam when the surface of the water is level with the top of the dam.

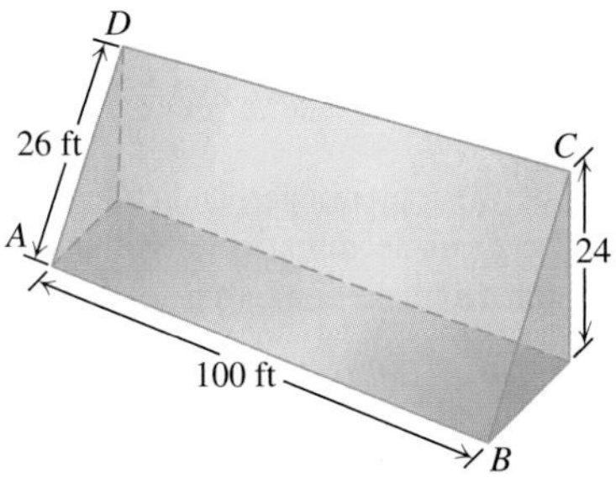

Chapter 6 Questions to Guide Your Review

1. How do you define and calculate the volumes of solids by the method of slicing? Give an example.

2. How are the disk and washer methods for calculating volumes derived from the method of slicing? Give examples of volume calculations by these methods.

3. Describe the method of cylindrical shells. Give an example.

4. How do you define the length of a smooth parametrized curve $x = f(t), y = g(t), a \le t \le b$? What does smoothness have to do with length? What else do you need to know about the parametrization to find the curve's length? Give examples.

5. How do you find the length of the graph of a smooth function over a closed interval? Give an example. What about functions that do not have continuous first derivatives?

6. What is a center of mass?

7. How do you locate the center of mass of a straight, narrow rod or strip of material? Give an example. If the density of the material is constant, you can tell right away where the center of mass is. Where is it?

8. How do you locate the center of mass of a thin flat plate of material? Give an example.

9. How do you define and calculate the area of the surface swept out by revolving the graph of a smooth function $y = f(x)$, $a \le x \le b$, about the *x*-axis? Give an example.

10. Under what conditions can you find the area of the surface generated by revolving a curve $x = f(t), y = g(t), a \le t \le b$, about the *x*-axis? The *y*-axis? Give examples.

11. What do Pappus's two theorems say? Give examples of how they are used to calculate surface areas and volumes and to locate centroids.

12. How do you define and calculate the work done by a variable force directed along a portion of the *x*-axis? How do you calculate the work it takes to pump a liquid from a tank? Give examples.

13. How do you calculate the force exerted by a liquid against a portion of a vertical wall? Give an example.

Chapter 6 Practice Exercises

Volumes

Find the volumes of the solids in Exercises 1–16.

1. The solid lies between planes perpendicular to the *x*-axis at $x = 0$ and $x = 1$. The cross-sections perpendicular to the *x*-axis between these planes are circular disks whose diameters run from the parabola $y = x^2$ to the parabola $y = \sqrt{x}$.

2. The base of the solid is the region in the first quadrant between the line $y = x$ and the parabola $y = 2\sqrt{x}$. The cross-sections of the solid perpendicular to the x-axis are equilateral triangles whose bases stretch from the line to the curve.

3. The solid lies between planes perpendicular to the x-axis at $x = \pi/4$ and $x = 5\pi/4$. The cross-sections between these planes are circular disks whose diameters run from the curve $y = 2\cos x$ to the curve $y = 2\sin x$.

4. The solid lies between planes perpendicular to the x-axis at $x = 0$ and $x = 6$. The cross-sections between these planes are squares whose bases run from the x-axis up to the curve $x^{1/2} + y^{1/2} = \sqrt{6}$.

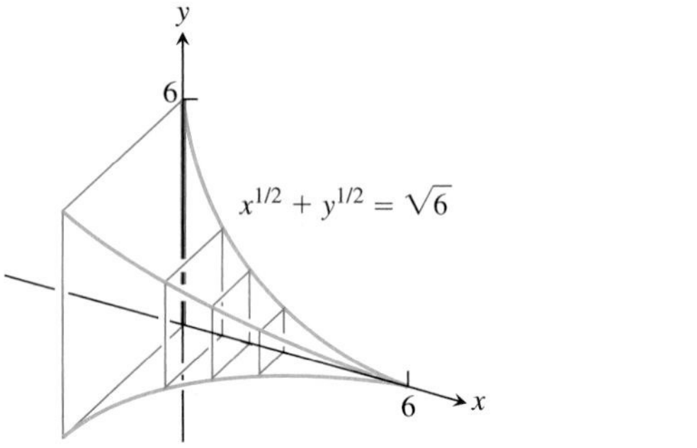

5. The solid lies between planes perpendicular to the x-axis at $x = 0$ and $x = 4$. The cross-sections of the solid perpendicular to the x-axis between these planes are circular disks whose diameters run from the curve $x^2 = 4y$ to the curve $y^2 = 4x$.

6. The base of the solid is the region bounded by the parabola $y^2 = 4x$ and the line $x = 1$ in the xy-plane. Each cross-section perpendicular to the x-axis is an equilateral triangle with one edge in the plane. (The triangles all lie on the same side of the plane.)

7. Find the volume of the solid generated by revolving the region bounded by the x-axis, the curve $y = 3x^4$, and the lines $x = 1$ and $x = -1$ about **(a)** the x-axis; **(b)** the y-axis; **(c)** the line $x = 1$; **(d)** the line $y = 3$.

8. Find the volume of the solid generated by revolving the "triangular" region bounded by the curve $y = 4/x^3$ and the lines $x = 1$ and $y = 1/2$ about **(a)** the x-axis; **(b)** the y-axis; **(c)** the line $x = 2$; **(d)** the line $y = 4$.

9. Find the volume of the solid generated by revolving the region bounded on the left by the parabola $x = y^2 + 1$ and on the right by the line $x = 5$ about **(a)** the x-axis; **(b)** the y-axis; **(c)** the line $x = 5$.

10. Find the volume of the solid generated by revolving the region bounded by the parabola $y^2 = 4x$ and the line $y = x$ about **(a)** the x-axis; **(b)** the y-axis; **(c)** the line $x = 4$; **(d)** the line $y = 4$.

11. Find the volume of the solid generated by revolving the "triangular" region bounded by the x-axis, the line $x = \pi/3$, and the curve $y = \tan x$ in the first quadrant about the x-axis.

12. Find the volume of the solid generated by revolving the region bounded by the curve $y = \sin x$ and the lines $x = 0$, $x = \pi$, and $y = 2$ about the line $y = 2$.

13. Find the volume of the solid generated by revolving the region between the x-axis and the curve $y = x^2 - 2x$ about **(a)** the x-axis; **(b)** the line $y = -1$; **(c)** the line $x = 2$; **(d)** the line $y = 2$.

14. Find the volume of the solid generated by revolving about the x-axis the region bounded by $y = 2\tan x$, $y = 0$, $x = -\pi/4$, and $x = \pi/4$. (The region lies in the first and third quadrants and resembles a skewed bowtie.)

15. **Volume of a solid sphere hole** A round hole of radius $\sqrt{3}$ ft is bored through the center of a solid sphere of a radius 2 ft. Find the volume of material removed from the sphere.

16. **Volume of a football** The profile of a football resembles the ellipse shown here. Find the football's volume to the nearest cubic inch.

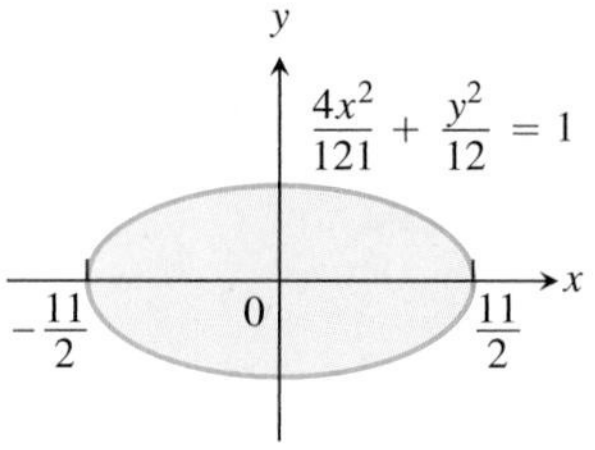

Lengths of Curves

Find the lengths of the curves in Exercises 17–23.

17. $y = x^{1/2} - (1/3)x^{3/2}, \quad 1 \le x \le 4$

18. $x = y^{2/3}, \quad 1 \le y \le 8$

19. $y = (5/12)x^{6/5} - (5/8)x^{4/5}, \quad 1 \le x \le 32$

20. $x = (y^3/12) + (1/y), \quad 1 \le y \le 2$

21. $x = 5\cos t - \cos 5t, \quad y = 5\sin t - \sin 5t, \quad 0 \le t \le \pi/2$

22. $x = t^3 - 6t^2, \quad y = t^3 + 6t^2, \quad 0 \le t \le 1$

23. $x = 3\cos\theta, \quad y = 3\sin\theta, \quad 0 \le \theta \le \dfrac{3\pi}{2}$

24. Find the length of the enclosed loop $x = t^2$, $y = (t^3/3) - t$ shown here. The loop starts at $t = -\sqrt{3}$ and ends at $t = \sqrt{3}$.

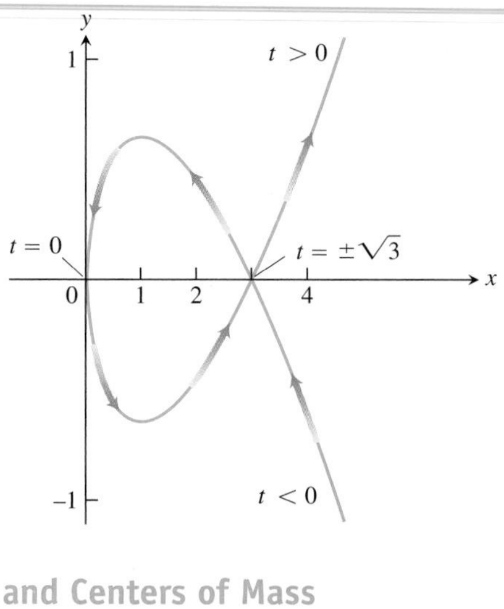

Centroids and Centers of Mass

25. Find the centroid of a thin, flat plate covering the region enclosed by the parabolas $y = 2x^2$ and $y = 3 - x^2$.

26. Find the centroid of a thin, flat plate covering the region enclosed by the x-axis, the lines $x = 2$ and $x = -2$, and the parabola $y = x^2$.

27. Find the centroid of a thin, flat plate covering the "triangular" region in the first quadrant bounded by the y-axis, the parabola $y = x^2/4$, and the line $y = 4$.

28. Find the centroid of a thin, flat plate covering the region enclosed by the parabola $y^2 = x$ and the line $x = 2y$.

29. Find the center of mass of a thin, flat plate covering the region enclosed by the parabola $y^2 = x$ and the line $x = 2y$ if the density function is $\delta(y) = 1 + y$. (Use horizontal strips.)

30. **a.** Find the center of mass of a thin plate of constant density covering the region between the curve $y = 3/x^{3/2}$ and the x-axis from $x = 1$ to $x = 9$.

b. Find the plate's center of mass if, instead of being constant, the density is $\delta(x) = x$. (Use vertical strips.)

Areas of Surfaces of Revolution

In Exercises 31–36, find the areas of the surfaces generated by revolving the curves about the given axes.

31. $y = \sqrt{2x + 1}, \quad 0 \le x \le 3; \quad x\text{-axis}$

32. $y = x^3/3, \quad 0 \le x \le 1; \quad x\text{-axis}$

33. $x = \sqrt{4y - y^2}, \quad 1 \le y \le 2; \quad y\text{-axis}$

34. $x = \sqrt{y}, \quad 2 \le y \le 6; \quad y\text{-axis}$

35. $x = t^2/2, \quad y = 2t, \quad 0 \le t \le \sqrt{5}; \quad x\text{-axis}$

36. $x = t^2 + 1/(2t), \quad y = 4\sqrt{t}, \quad 1/\sqrt{2} \le t \le 1; \quad y\text{-axis}$

Work

37. Lifting equipment A rock climber is about to haul up 100 N (about 22.5 lb) of equipment that has been hanging beneath her on 40 m of rope that weighs 0.8 newton per meter. How much work will it take? (*Hint:* Solve for the rope and equipment separately, then add.)

38. Leaky tank truck You drove an 800-gal tank truck of water from the base of Mt. Washington to the summit and discovered on arrival that the tank was only half full. You started with a full tank, climbed at a steady rate, and accomplished the 4750-ft elevation change in 50 min. Assuming that the water leaked out at a steady rate, how much work was spent in carrying water to the top? Do not count the work done in getting yourself and the truck there. Water weights 8 lb/U.S. gal.

39. Stretching a spring If a force of 20 lb is required to hold a spring 1 ft beyond its unstressed length, how much work does it take to stretch the spring this far? An additional foot?

40. Garage door spring A force of 200 N will stretch a garage door spring 0.8 m beyond its unstressed length. How far will a 300-N force stretch the spring? How much work does it take to stretch the spring this far from its unstressed length?

41. Pumping a reservoir A reservoir shaped like a right circular cone, point down, 20 ft across the top and 8 ft deep, is full of water. How much work does it take to pump the water to a level 6 ft above the top?

42. Pumping a reservoir (*Continuation of Exercise 41.*) The reservoir is filled to a depth of 5 ft, and the water is to be pumped to the same level as the top. How much work does it take?

43. Pumping a conical tank A right circular conical tank, point down, with top radius 5 ft and height 10 ft is filled with a liquid whose weight-density is 60 lb/ft^3. How much work does it take to pump the liquid to a point 2 ft above the tank? If the pump is driven by a motor rated at 275 ft-lb/sec (1/2 hp), how long will it take to empty the tank?

44. Pumping a cylindrical tank A storage tank is a right circular cylinder 20 ft long and 8 ft in diameter with its axis horizontal. If the tank is half full of olive oil weighing 57 lb/ft^3, find the work done in emptying it through a pipe that runs from the bottom of the tank to an outlet that is 6 ft above the top of the tank.

Fluid Force

45. Trough of water The vertical triangular plate shown here is the end plate of a trough full of water ($w = 62.4$). What is the fluid force against the plate?

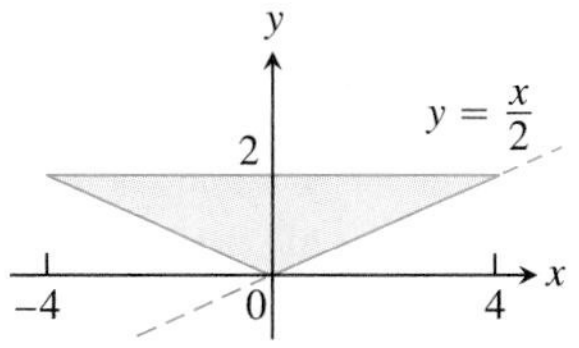

46. Trough of maple syrup The vertical trapezoid plate shown here is the end plate of a trough full of maple syrup weighing 75 lb/ft^3. What is the force exerted by the syrup against the end plate of the trough when the syrup is 10 in. deep?

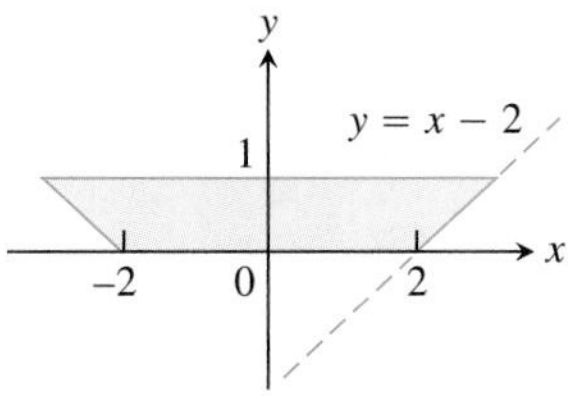

47. Force on a parabolic gate A flat vertical gate in the face of a dam is shaped like the parabolic region between the curve $y = 4x^2$ and the line $y = 4$, with measurements in feet. The top of the gate lies 5 ft below the surface of the water. Find the force exerted by the water against the gate ($w = 62.4$).

T **48.** You plan to store mercury ($w = 849$ lb/ft^3) in a vertical rectangular tank with a 1 ft square base side whose interior side wall can withstand a total fluid force of 40,000 lb. About how many cubic feet of mercury can you store in the tank at any one time?

49. The container profiled in the accompanying figure is filled with two nonmixing liquids of weight-density w_1 and w_2. Find the fluid force on one side of the vertical square plate $ABCD$. The points B and D lie in the boundary layer and the square is $6\sqrt{2}$ ft on a side.

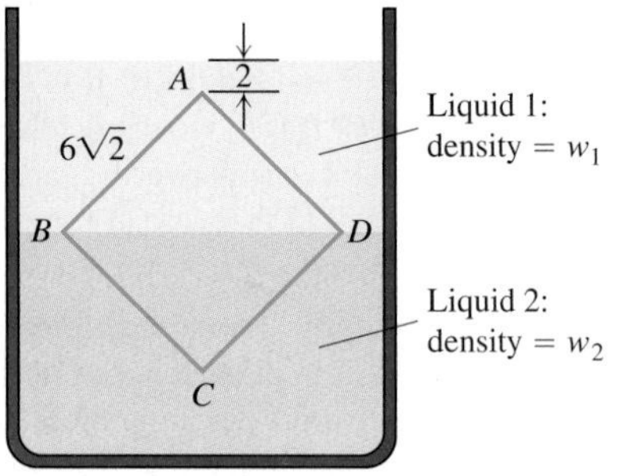

50. The isosceles trapezoidal plate shown here is submerged vertically in water ($w = 62.4$) with its upper edge 4 ft below the surface. Find the fluid force on one side of the plate in two different ways:

a. By evaluating an integral.

b. By dividing the plate into a parallelogram and an isosceles triangle, locating their centroids, and using the equation $F = w\bar{h}A$ from Section 6.7.

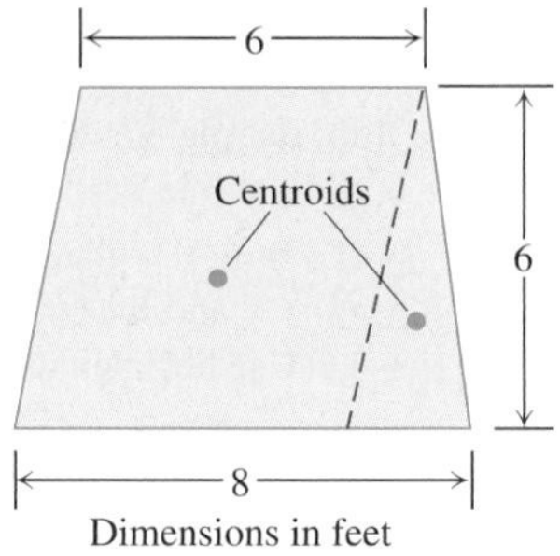

Chapter 6 Additional and Advanced Exercises

Volume and Length

1. A solid is generated by revolving about the x-axis the region bounded by the graph of the positive continuous function $y = f(x)$, the x-axis, and the fixed line $x = a$ and the variable line $x = b$, $b > a$. Its volume, for all b, is $b^2 - ab$. Find $f(x)$.

2. A solid is generated by revolving about the x-axis the region bounded by the graph of the positive continuous function $y = f(x)$, the x-axis, and the lines $x = 0$ and $x = a$. Its volume, for all $a > 0$, is $a^2 + a$. Find $f(x)$.

3. Suppose that the increasing function $f(x)$ is smooth for $x \geq 0$ and that $f(0) = a$. Let $s(x)$ denote the length of the graph of f from $(0, a)$ to $(x, f(x))$, $x > 0$. Find $f(x)$ if $s(x) = Cx$ for some constant C. What are the allowable values for C?

4. a. Show that for $0 < \alpha \leq \pi/2$,

$$\int_0^{\alpha} \sqrt{1 + \cos^2 \theta}\, d\theta > \sqrt{\alpha^2 + \sin^2 \alpha}.$$

b. Generalize the result in part (a).

Moments and Centers of Mass

5. Find the centroid of the region bounded below by the x-axis and above by the curve $y = 1 - x^n$, n an even positive integer. What is the limiting position of the centroid as $n \to \infty$?

6. If you haul a telephone pole on a two-wheeled carriage behind a truck, you want the wheels to be 3 ft or so behind the pole's center of mass to provide an adequate "tongue" weight. NYNEX's class 1.40-ft wooden poles have a 27-in. circumference at the top and a 43.5-in. circumference at the base. About how far from the top is the center of mass?

7. Suppose that a thin metal plate of area A and constant density δ occupies a region R in the xy-plane, and let M_y be the plate's moment about the y-axis. Show that the plate's moment about the line $x = b$ is

a. $M_y - b\delta A$ if the plate lies to the right of the line, and

b. $b\delta A - M_y$ if the plate lies to the left of the line.

8. Find the center of mass of a thin plate covering the region bounded by the curve $y^2 = 4ax$ and the line $x = a$, $a =$ positive constant, if the density at (x, y) is directly proportional to **(a)** x, **(b)** $|y|$.

9. a. Find the centroid of the region in the first quadrant bounded by two concentric circles and the coordinate axes, if the circles have radii a and b, $0 < a < b$, and their centers are at the origin.

b. Find the limits of the coordinates of the centroid as a approaches b and discuss the meaning of the result.

10. A triangular corner is cut from a square 1 ft on a side. The area of the triangle removed is 36 in.^{2}. If the centroid of the remaining region is 7 in. from one side of the original square, how far is it from the remaining sides?

Surface Area

11. At points on the curve $y = 2\sqrt{x}$, line segments of length $h = y$ are drawn perpendicular to the xy-plane. (See accompanying figure.) Find the area of the surface formed by these perpendiculars from $(0, 0)$ to $(3, 2\sqrt{3})$.

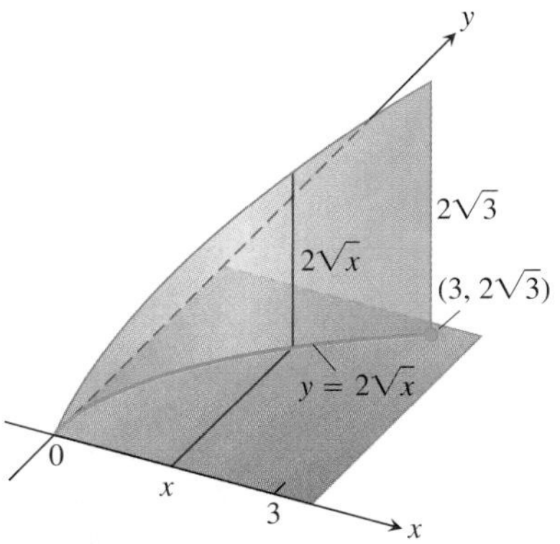

12. At points on a circle of radius a, line segments are drawn perpendicular to the plane of the circle, the perpendicular at each point P being of length ks, where s is the length of the arc of the circle measured counterclockwise from $(a, 0)$ to P and k is a positive constant, as shown here. Find the area of the surface formed by the perpendiculars along the arc beginning at $(a, 0)$ and extending once around the circle.

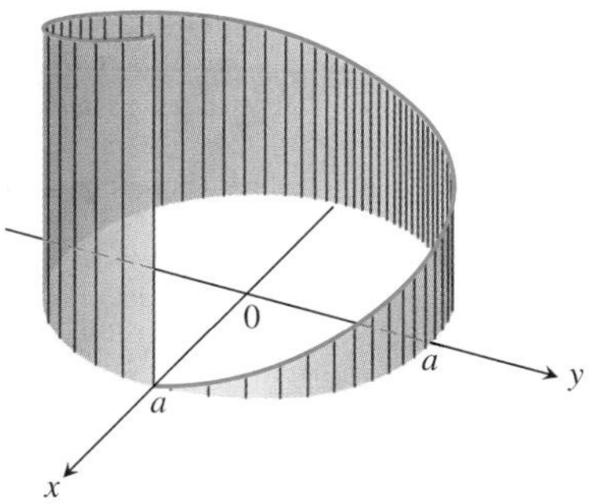

Work

13. A particle of mass m starts from rest at time $t = 0$ and is moved along the x-axis with constant acceleration a from $x = 0$ to $x = h$ against a variable force of magnitude $F(t) = t^2$. Find the work done.

14. Work and kinetic energy Suppose a 1.6-oz golf ball is placed on a vertical spring with force constant $k = 2$ lb/in. The spring is compressed 6 in. and released. About how high does the ball go (measured from the spring's rest position)?

Fluid Force

15. A triangular plate ABC is submerged in water with its plane vertical. The side AB, 4 ft long, is 6 ft below the surface of the water, while the vertex C is 2 ft below the surface. Find the force exerted by the water on one side of the plate.

16. A vertical rectangular plate is submerged in a fluid with its top edge parallel to the fluid's surface. Show that the force exerted by the fluid on one side of the plate equals the average value of the pressure up and down the plate times the area of the plate.

17. The *center of pressure* on one side of a plane region submerged in a fluid is defined to be the point at which the total force exerted by the fluid can be applied without changing its total moment about any axis in the plane. Find the depth to the center of pressure **(a)** on a vertical rectangle of height h and width b if its upper edge is in the surface of the fluid; **(b)** on a vertical triangle of height h and base b if the vertex opposite b is a ft and the base b is $(a + h)$ ft below the surface of the fluid.

Chapter 6 Technology Application Projects

Mathematica/Maple Module

Using Riemann Sums to Estimate Areas, Volumes, and Lengths of Curves

Visualize and approximate areas and volumes in **Part I** and **Part II**: Volumes of Revolution; and **Part III**: Lengths of Curves.

Mathematica/Maple Module

Modeling a Bungee Cord Jump

Collect data (or use data previously collected) to build and refine a model for the force exerted by a jumper's bungee cord. Use the work-energy theorem to compute the distance fallen for a given jumper and a given length of bungee cord.

Chapter 7

TRANSCENDENTAL FUNCTIONS

OVERVIEW Functions can be classified into two broad groups (see Section 1.4). Polynomial functions are called *algebraic*, as are functions obtained from them by addition, multiplication, division, or taking powers and roots. Functions that are not algebraic are called *transcendental.* The trigonometric, exponential, logarithmic, and hyperbolic functions are transcendental, as are their inverses.

Transcendental functions occur frequently in many calculus settings and applications, including growths of populations, vibrations and waves, efficiencies of computer algorithms, and the stability of engineered structures. In this chapter we introduce several important transcendental functions and investigate their graphs, properties, derivatives, and integrals.

7.1 Inverse Functions and Their Derivatives

A function that undoes, or inverts, the effect of a function f is called the *inverse* of f. Many common functions, though not all, are paired with an inverse. Important inverse functions often show up in formulas for antiderivatives and solutions of differential equations. Inverse functions also play a key role in the development and properties of the logarithmic and exponential functions, as we will see in Section 7.3.

One-to-One Functions

A function is a rule that assigns a value from its range to each element in its domain. Some functions assign the same range value to more than one element in the domain. The function $f(x) = x^2$ assigns the same value, 1, to both of the numbers -1 and $+1$; the sines of $\pi/3$ and $2\pi/3$ are both $\sqrt{3}/2$. Other functions assume each value in their range no more than once. The square roots and cubes of different numbers are always different. A function that has distinct values at distinct elements in its domain is called one-to-one. These functions take on any one value in their range exactly once.

> **DEFINITION One-to-One Function**
> A function $f(x)$ is **one-to-one** on a domain D if $f(x_1) \neq f(x_2)$ whenever $x_1 \neq x_2$ in D.

EXAMPLE 1 Domains of One-to-One Functions

(a) $f(x) = \sqrt{x}$ is one-to-one on any domain of nonnegative numbers because $\sqrt{x_1} \neq \sqrt{x_2}$ whenever $x_1 \neq x_2$.

(b) $g(x) = \sin x$ is *not* one-to-one on the interval $[0, \pi]$ because $\sin(\pi/6) = \sin(5\pi/6)$. The sine *is* one-to-one on $[0, \pi/2]$, however, because it is a strictly increasing function on $[0, \pi/2]$. ■

The graph of a one-to-one function $y = f(x)$ can intersect a given horizontal line at most once. If it intersects the line more than once, it assumes the same y-value more than once, and is therefore not one-to-one (Figure 7.1).

> **The Horizontal Line Test for One-to-One Functions**
> A function $y = f(x)$ is one-to-one if and only if its graph intersects each horizontal line at most once.

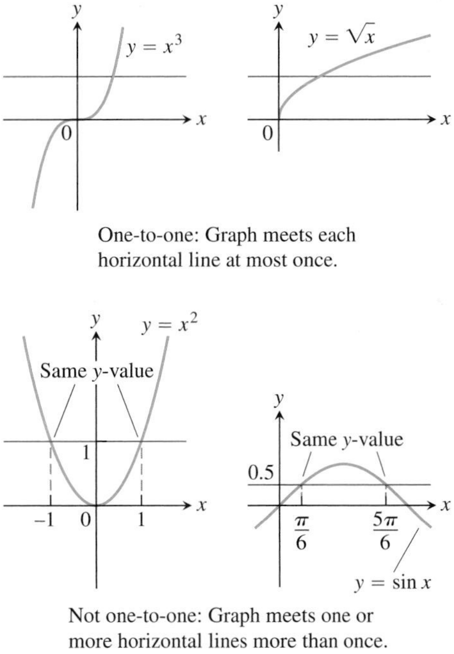

FIGURE 7.1 Using the horizontal line test, we see that $y = x^3$ and $y = \sqrt{x}$ are one-to-one on their domains $(-\infty, \infty)$ and $[0, \infty)$, but $y = x^2$ and $y = \sin x$ are not one-to-one on their domains $(-\infty, \infty)$.

Inverse Functions

Since each output of a one-to-one function comes from just one input, the effect of the function can be inverted to send an output back to the input from which it came.

DEFINITION Inverse Function

Suppose that f is a one-to-one function on a domain D with range R. The **inverse function** f^{-1} is defined by

$$f^{-1}(a) = b \text{ if } f(b) = a.$$

The domain of f^{-1} is R and the range of f^{-1} is D.

The domains and ranges of f and f^{-1} are interchanged. The symbol f^{-1} for the inverse of f is read "f inverse." The "-1" in f^{-1} is *not* an exponent: $f^{-1}(x)$ does not mean $1/f(x)$.

If we apply f to send an input x to the output $f(x)$ and follow by applying f^{-1} to $f(x)$ we get right back to x, just where we started. Similarly, if we take some number y in the range of f, apply f^{-1} to it, and then apply f to the resulting value $f^{-1}(y)$, we get back the value y with which we began. Composing a function and its inverse has the same effect as doing nothing.

$$(f^{-1} \circ f)(x) = x, \qquad \text{for all } x \text{ in the domain of } f$$

$$(f \circ f^{-1})(y) = y, \qquad \text{for all } y \text{ in the domain of } f^{-1} \text{ (or range of } f)$$

Only a one-to-one function can have an inverse. The reason is that if $f(x_1) = y$ and $f(x_2) = y$ for two distinct inputs x_1 and x_2, then there is no way to assign a value to $f^{-1}(y)$ that satisfies both $f^{-1}(f(x_1)) = x_1$ and $f^{-1}(f(x_2)) = x_2$.

A function that is increasing on an interval, satisfying $f(x_2) > f(x_1)$ when $x_2 > x_1$, is one-to-one and has an inverse. Decreasing functions also have an inverse (Exercise 39). Functions that have positive derivatives at all x are increasing (Corollary 3 of the Mean Value Theorem, Section 4.2), and so they have inverses. Similarly, functions with negative derivatives at all x are decreasing and have inverses. Functions that are neither increasing nor decreasing may still be one-to-one and have an inverse, as with the function $\sec^{-1} x$ in Section 7.7.

Finding Inverses

The graphs of a function and its inverse are closely related. To read the value of a function from its graph, we start at a point x on the x-axis, go vertically to the graph, and then move horizontally to the y-axis to read the value of y. The inverse function can be read from the graph by reversing this process. Start with a point y on the y-axis, go horizontally to the graph, and then move vertically to the x-axis to read the value of $x = f^{-1}(y)$ (Figure 7.2).

We want to set up the graph of f^{-1} so that its input values lie along the x-axis, as is usually done for functions, rather then on the y-axis. To achieve this we interchange the x and y axes by reflecting across the 45° line $y = x$. After this reflection we have a new graph that represents f^{-1}. The value of $f^{-1}(x)$ can now be read from the graph in the usual way, by starting with a point x on the x-axis, going vertically to the graph and then horizontally to the y-axis to get the value of $f^{-1}(x)$. Figure 7.2 indicates the relation between the graphs of f and f^{-1}. The graphs are interchanged by reflection through the line $y = x$.

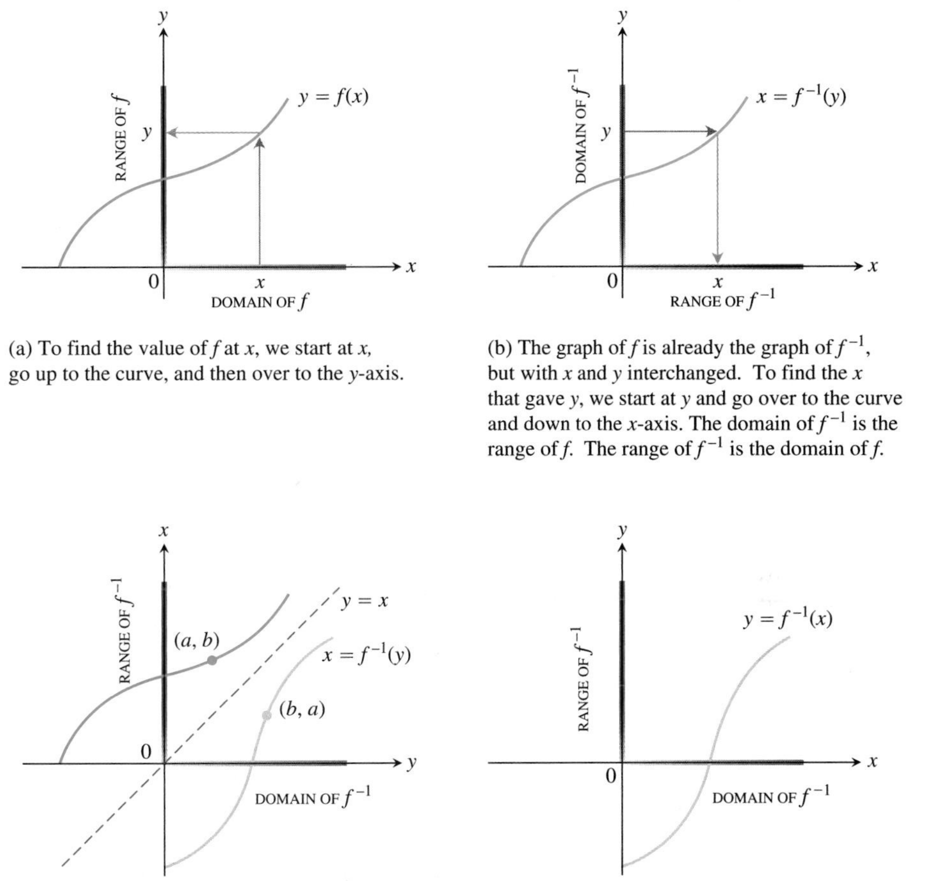

(a) To find the value of f at x, we start at x, go up to the curve, and then over to the y-axis.

(b) The graph of f is already the graph of f^{-1}, but with x and y interchanged. To find the x that gave y, we start at y and go over to the curve and down to the x-axis. The domain of f^{-1} is the range of f. The range of f^{-1} is the domain of f.

(c) To draw the graph of f^{-1} in the more usual way, we reflect the system in the line $y = x$.

(d) Then we interchange the letters x and y. We now have a normal-looking graph of f^{-1} as a function of x.

FIGURE 7.2 Determining the graph of $y = f^{-1}(x)$ from the graph of $y = f(x)$.

The process of passing from f to f^{-1} can be summarized as a two-step process.

1. Solve the equation $y = f(x)$ for x. This gives a formula $x = f^{-1}(y)$ where x is expressed as a function of y.
2. Interchange x and y, obtaining a formula $y = f^{-1}(x)$ where f^{-1} is expressed in the conventional format with x as the independent variable and y as the dependent variable.

EXAMPLE 2 Finding an Inverse Function

Find the inverse of $y = \frac{1}{2}x + 1$, expressed as a function of x.

Solution

1. *Solve for x in terms of y:*

$$\begin{aligned} y &= \frac{1}{2}x + 1 \\ 2y &= x + 2 \\ x &= 2y - 2. \end{aligned}$$

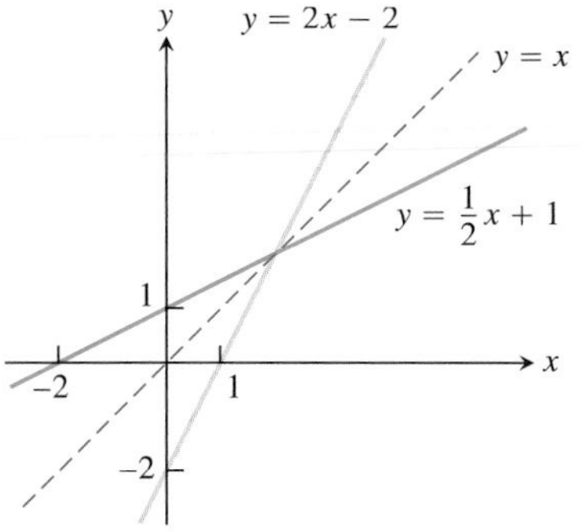

FIGURE 7.3 Graphing $f(x) = (1/2)x + 1$ and $f^{-1}(x) = 2x - 2$ together shows the graphs' symmetry with respect to the line $y = x$. The slopes are reciprocals of each other (Example 2).

2. *Interchange x and y:* $y = 2x - 2$.

The inverse of the function $f(x) = (1/2)x + 1$ is the function $f^{-1}(x) = 2x - 2$. To check, we verify that both composites give the identity function:

$$f^{-1}(f(x)) = 2\left(\frac{1}{2}x + 1\right) - 2 = x + 2 - 2 = x$$

$$f(f^{-1}(x)) = \frac{1}{2}(2x - 2) + 1 = x - 1 + 1 = x.$$

See Figure 7.3. ■

EXAMPLE 3 Finding an Inverse Function

Find the inverse of the function $y = x^2, x \geq 0$, expressed as a function of x.

Solution We first solve for x in terms of y:

$$y = x^2$$

$$\sqrt{y} = \sqrt{x^2} = |x| = x \qquad |x| = x \text{ because } x \geq 0$$

We then interchange x and y, obtaining

$$y = \sqrt{x}.$$

The inverse of the function $y = x^2, x \geq 0$, is the function $y = \sqrt{x}$ (Figure 7.4).

Notice that, unlike the restricted function $y = x^2, x \geq 0$, the unrestricted function $y = x^2$ is not one-to-one and therefore has no inverse. ■

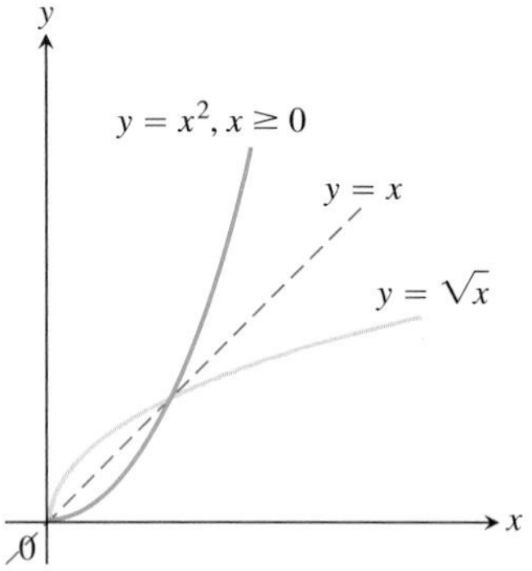

FIGURE 7.4 The functions $y = \sqrt{x}$ and $y = x^2, x \geq 0$, are inverses of one another (Example 3).

Derivatives of Inverses of Differentiable Functions

If we calculate the derivatives of $f(x) = (1/2)x + 1$ and its inverse $f^{-1}(x) = 2x - 2$ from Example 2, we see that

$$\frac{d}{dx}f(x) = \frac{d}{dx}\left(\frac{1}{2}x + 1\right) = \frac{1}{2}$$

$$\frac{d}{dx}f^{-1}(x) = \frac{d}{dx}(2x - 2) = 2.$$

The derivatives are reciprocals of one another. The graph of f is the line $y = (1/2)x + 1$, and the graph of f^{-1} is the line $y = 2x - 2$ (Figure 7.3). Their slopes are reciprocals of one another.

This is not a special case. Reflecting any nonhorizontal or nonvertical line across the line $y = x$ always inverts the line's slope. If the original line has slope $m \neq 0$ (Figure 7.5), the reflected line has slope $1/m$ (Exercise 36).

The reciprocal relationship between the slopes of f and f^{-1} holds for other functions as well, but we must be careful to compare slopes at corresponding points. If the slope of $y = f(x)$ at the point $(a, f(a))$ is $f'(a)$ and $f'(a) \neq 0$, then the slope of $y = f^{-1}(x)$ at the point $(f(a), a)$ is the reciprocal $1/f'(a)$ (Figure 7.6). If we set $b = f(a)$, then

$$(f^{-1})'(b) = \frac{1}{f'(a)} = \frac{1}{f'(f^{-1}(b))}.$$

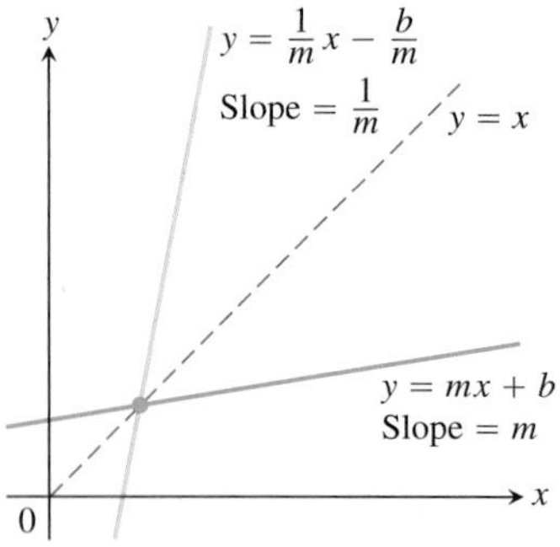

FIGURE 7.5 The slopes of nonvertical lines reflected through the line $y = x$ are reciprocals of each other.

If $y = f(x)$ has a horizontal tangent line at $(a, f(a))$ then the inverse function f^{-1} has a vertical tangent line at $(f(a), a)$, and this infinite slope implies that f^{-1} is not differentiable

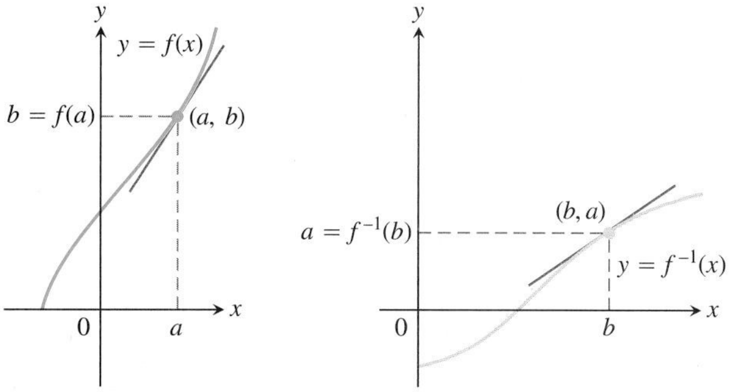

The slopes are reciprocal: $(f^{-1})'(b) = \dfrac{1}{f'(a)}$ or $(f^{-1})'(b) = \dfrac{1}{f'(f^{-1}(b))}$

FIGURE 7.6 The graphs of inverse functions have reciprocal slopes at corresponding points.

at $f(a)$. Theorem 1 gives the conditions under which f^{-1} is differentiable in its domain, which is the same as the range of f.

THEOREM 1 The Derivative Rule for Inverses

If f has an interval I as domain and $f'(x)$ exists and is never zero on I, then f^{-1} is differentiable at every point in its domain. The value of $(f^{-1})'$ at a point b in the domain of f^{-1} is the reciprocal of the value of f' at the point $a = f^{-1}(b)$:

$$(f^{-1})'(b) = \frac{1}{f'(f^{-1}(b))}$$

or

$$\left.\frac{df^{-1}}{dx}\right|_{x=b} = \frac{1}{\left.\dfrac{df}{dx}\right|_{x=f^{-1}(b)}} \tag{1}$$

The proof of Theorem 1 is omitted, but here is another way to view it. When $y = f(x)$ is differentiable at $x = a$ and we change x by a small amount dx, the corresponding change in y is approximately

$$dy = f'(a)\,dx.$$

This means that y changes about $f'(a)$ times as fast as x when $x = a$ and that x changes about $1/f'(a)$ times as fast as y when $y = b$. It is reasonable that the derivative of f^{-1} at b is the reciprocal of the derivative of f at a.

EXAMPLE 4 Applying Theorem 1

The function $f(x) = x^2, x \geq 0$ and its inverse $f^{-1}(x) = \sqrt{x}$ have derivatives $f'(x) = 2x$ and $(f^{-1})'(x) = 1/(2\sqrt{x})$.

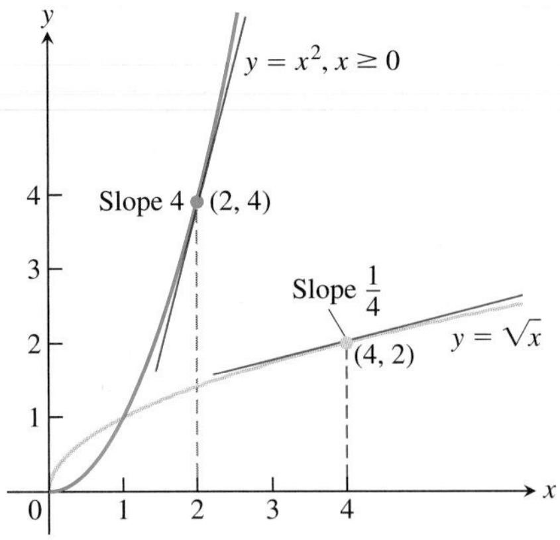

FIGURE 7.7 The derivative of $f^{-1}(x) = \sqrt{x}$ at the point (4, 2) is the reciprocal of the derivative of $f(x) = x^2$ at (2, 4) (Example 4).

Theorem 1 predicts that the derivative of $f^{-1}(x)$ is

$$\begin{aligned}(f^{-1})'(x) &= \frac{1}{f'(f^{-1}(x))}\\ &= \frac{1}{2(f^{-1}(x))}\\ &= \frac{1}{2(\sqrt{x})}.\end{aligned}$$

Theorem 1 gives a derivative that agrees with our calculation using the Power Rule for the derivative of the square root function.

Let's examine Theorem 1 at a specific point. We pick $x = 2$ (the number a) and $f(2) = 4$ (the value b). Theorem 1 says that the derivative of f at 2, $f'(2) = 4$, and the derivative of f^{-1} at $f(2)$, $(f^{-1})'(4)$, are reciprocals. It states that

$$(f^{-1})'(4) = \frac{1}{f'(f^{-1}(4))} = \frac{1}{f'(2)} = \frac{1}{2x}\bigg|_{x=2} = \frac{1}{4}.$$

See Figure 7.7. ■

Equation (1) sometimes enables us to find specific values of df^{-1}/dx without knowing a formula for f^{-1}.

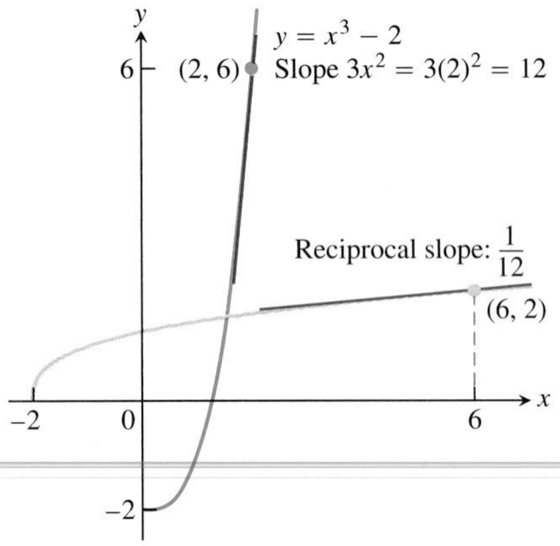

FIGURE 7.8 The derivative of $f(x) = x^3 - 2$ at $x = 2$ tells us the derivative of f^{-1} at $x = 6$ (Example 5).

EXAMPLE 5 Finding a Value of the Inverse Derivative

Let $f(x) = x^3 - 2$. Find the value of df^{-1}/dx at $x = 6 = f(2)$ without finding a formula for $f^{-1}(x)$.

Solution

$$\frac{df}{dx}\bigg|_{x=2} = 3x^2\bigg|_{x=2} = 12$$

$$\frac{df^{-1}}{dx}\bigg|_{x=f(2)} = \frac{1}{\left.\dfrac{df}{dx}\right|_{x=2}} = \frac{1}{12} \qquad \text{Eq. (1)}$$

See Figure 7.8. ■

Parametrizing Inverse Functions

We can graph or represent any function $y = f(x)$ parametrically as

$$x = t \quad \text{and} \quad y = f(t).$$

Interchanging t and $f(t)$ produces parametric equations for the inverse:

$$x = f(t) \quad \text{and} \quad y = t$$

(see Section 3.5).

For example, to graph the one-to-one function $f(x) = x^2, x \geq 0$, on a grapher together with its inverse and the line $y = x, x \geq 0$, use the parametric graphing option with

$$\begin{aligned}&\text{Graph of } f: && x_1 = t, && y_1 = t^2, && t \geq 0\\ &\text{Graph of } f^{-1}: && x_2 = t^2, && y_2 = t\\ &\text{Graph of } y = x: && x_3 = t, && y_3 = t\end{aligned}$$

EXERCISES 7.1

Identifying One-to-One Functions Graphically

Which of the functions graphed in Exercises 1–6 are one-to-one, and which are not?

1.

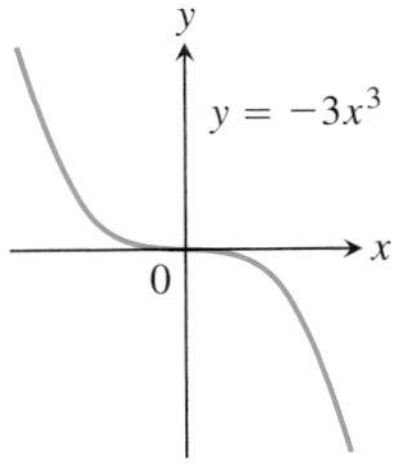

2.

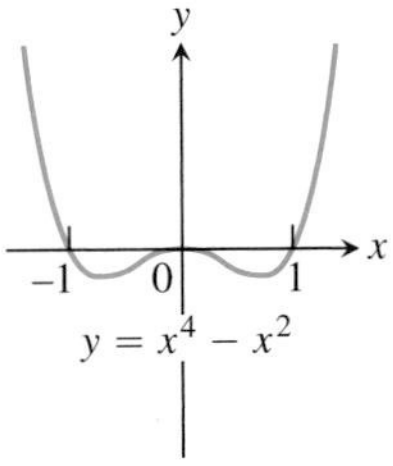

3.

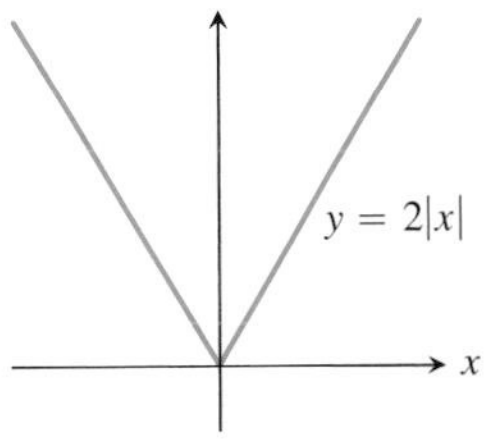

4.

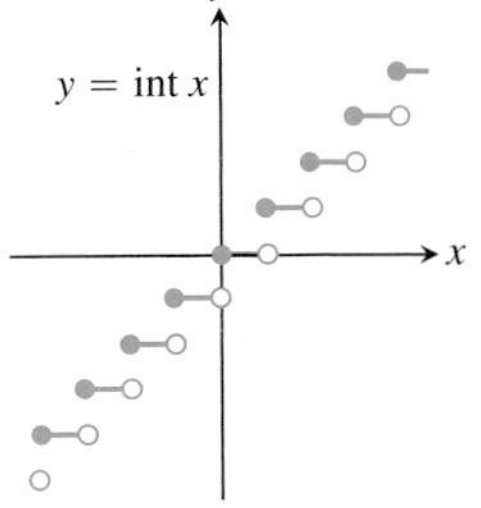

5.

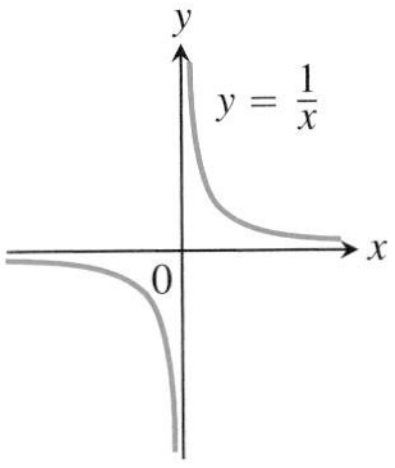

6.

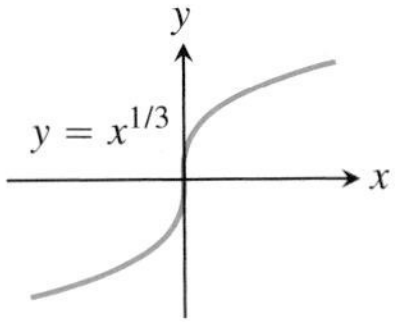

Graphing Inverse Functions

Each of Exercises 7–10 shows the graph of a function $y = f(x)$. Copy the graph and draw in the line $y = x$. Then use symmetry with respect to the line $y = x$ to add the graph of f^{-1} to your sketch. (It is not necessary to find a formula for f^{-1}.) Identify the domain and range of f^{-1}.

7.

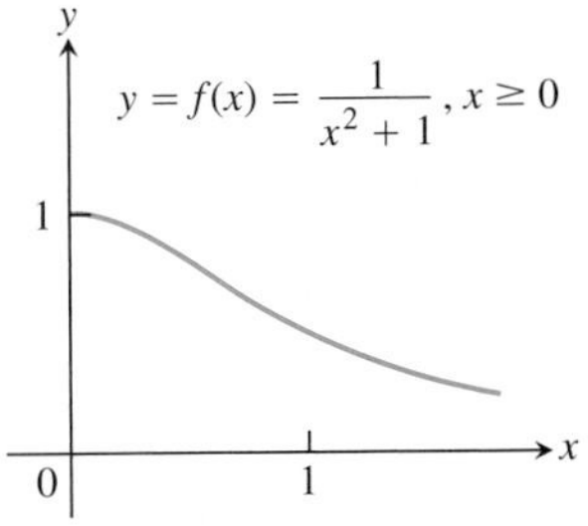

8.

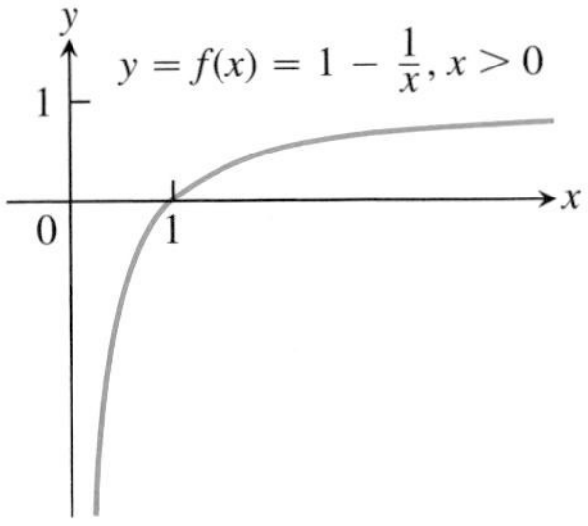

9.

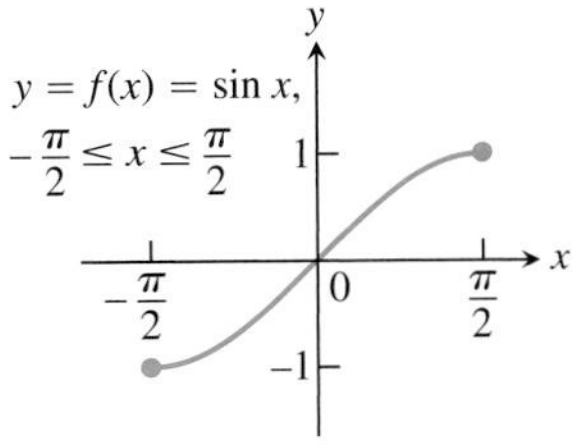

10.

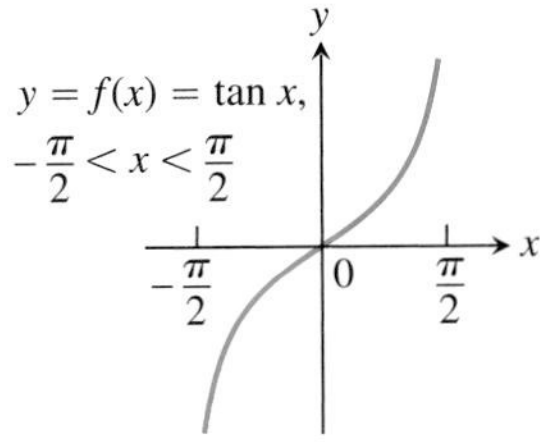

11. a. Graph the function $f(x) = \sqrt{1 - x^2}, 0 \leq x \leq 1$. What symmetry does the graph have?

b. Show that f is its own inverse. (Remember that $\sqrt{x^2} = x$ if $x \geq 0$.)

12. a. Graph the function $f(x) = 1/x$. What symmetry does the graph have?

b. Show that f is its own inverse.

Formulas for Inverse Functions

Each of Exercises 13–18 gives a formula for a function $y = f(x)$ and shows the graphs of f and f^{-1}. Find a formula for f^{-1} in each case.

13. $f(x) = x^2 + 1, \quad x \geq 0$

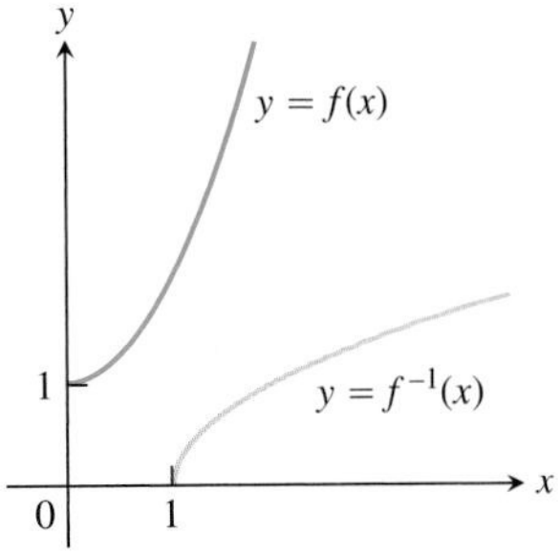

14. $f(x) = x^2, \quad x \leq 0$

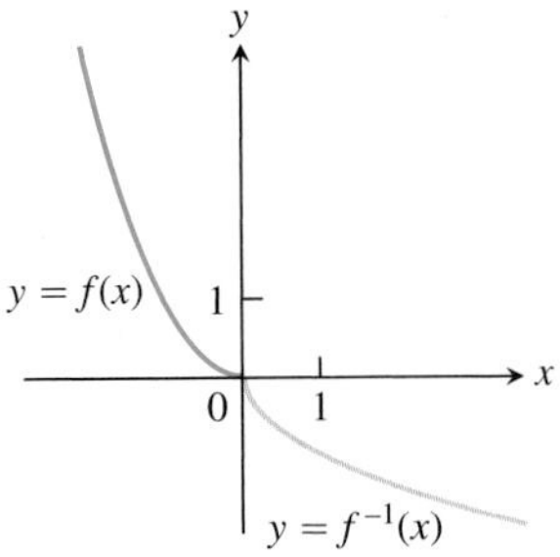

15. $f(x) = x^3 - 1$

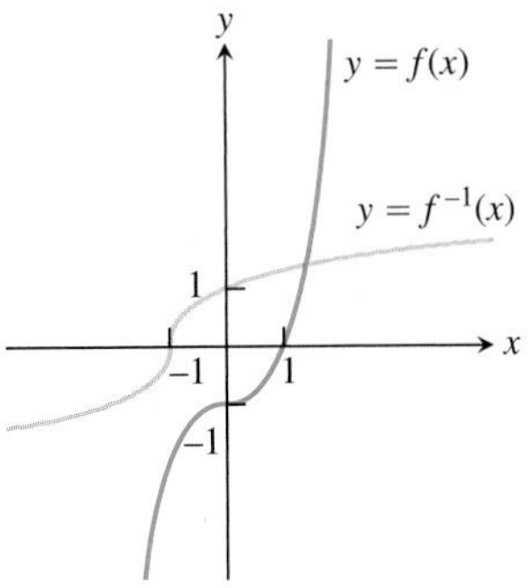

16. $f(x) = x^2 - 2x + 1, \quad x \geq 1$

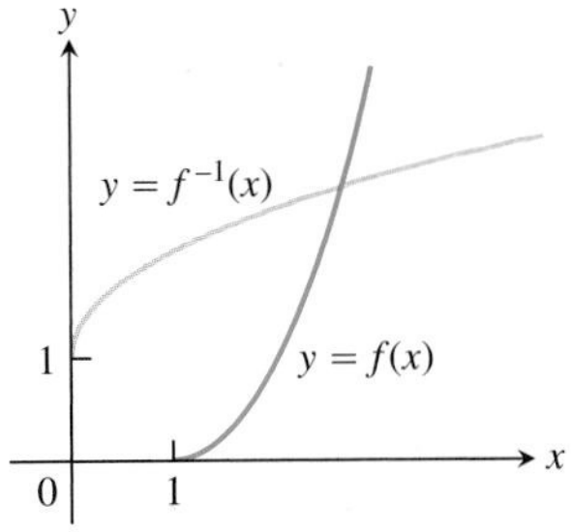

17. $f(x) = (x + 1)^2, \quad x \geq -1$

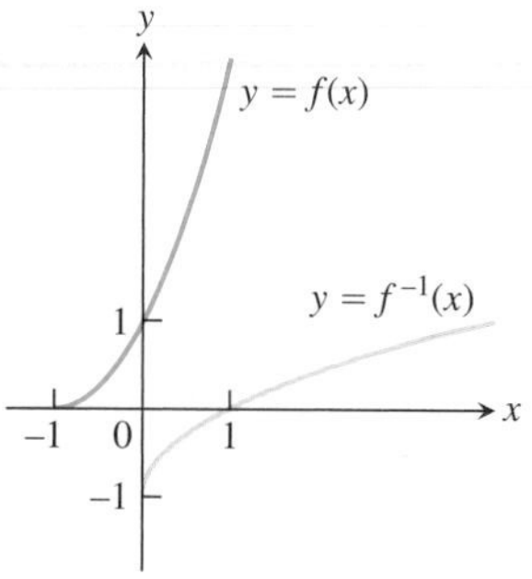

18. $f(x) = x^{2/3}, \quad x \geq 0$

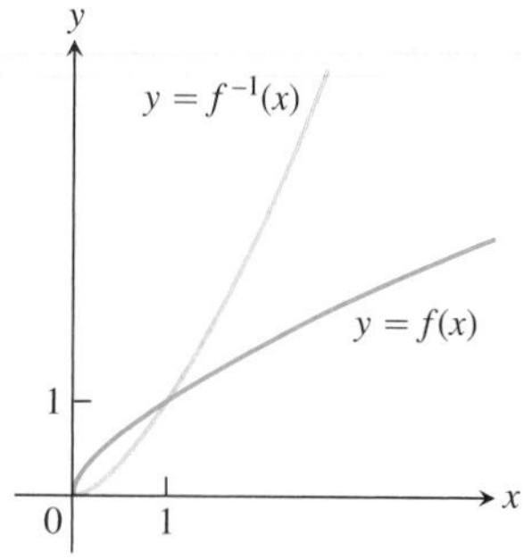

Each of Exercises 19–24 gives a formula for a function $y = f(x)$. In each case, find $f^{-1}(x)$ and identify the domain and range of f^{-1}. As a check, show that $f(f^{-1}(x)) = f^{-1}(f(x)) = x$.

19. $f(x) = x^5$

20. $f(x) = x^4, \quad x \geq 0$

21. $f(x) = x^3 + 1$

22. $f(x) = (1/2)x - 7/2$

23. $f(x) = 1/x^2, \quad x > 0$

24. $f(x) = 1/x^3, \quad x \neq 0$

Derivatives of Inverse Functions

In Exercises 25–28:

a. Find $f^{-1}(x)$.

b. Graph f and f^{-1} together.

c. Evaluate df/dx at $x = a$ and df^{-1}/dx at $x = f(a)$ to show that at these points $df^{-1}/dx = 1/(df/dx)$.

25. $f(x) = 2x + 3, \quad a = -1$

26. $f(x) = (1/5)x + 7, \quad a = -1$

27. $f(x) = 5 - 4x, \quad a = 1/2$

28. $f(x) = 2x^2, \quad x \geq 0, \quad a = 5$

29. a. Show that $f(x) = x^3$ and $g(x) = \sqrt[3]{x}$ are inverses of one another.

b. Graph f and g over an x-interval large enough to show the graphs intersecting at $(1, 1)$ and $(-1, -1)$. Be sure the picture shows the required symmetry about the line $y = x$.

c. Find the slopes of the tangents to the graphs of f and g at $(1, 1)$ and $(-1, -1)$ (four tangents in all).

d. What lines are tangent to the curves at the origin?

30. a. Show that $h(x) = x^3/4$ and $k(x) = (4x)^{1/3}$ are inverses of one another.

b. Graph h and k over an x-interval large enough to show the graphs intersecting at $(2, 2)$ and $(-2, -2)$. Be sure the picture shows the required symmetry about the line $y = x$.

c. Find the slopes of the tangents to the graphs at h and k at $(2, 2)$ and $(-2, -2)$.

d. What lines are tangent to the curves at the origin?

31. Let $f(x) = x^3 - 3x^2 - 1, x \geq 2$. Find the value of df^{-1}/dx at the point $x = -1 = f(3)$.

32. Let $f(x) = x^2 - 4x - 5, x > 2$. Find the value of df^{-1}/dx at the point $x = 0 = f(5)$.

33. Suppose that the differentiable function $y = f(x)$ has an inverse and that the graph of f passes through the point $(2, 4)$ and has a slope of $1/3$ there. Find the value of df^{-1}/dx at $x = 4$.

34. Suppose that the differentiable function $y = g(x)$ has an inverse and that the graph of g passes through the origin with slope 2. Find the slope of the graph of g^{-1} at the origin.

Inverses of Lines

35. a. Find the inverse of the function $f(x) = mx$, where m is a constant different from zero.

b. What can you conclude about the inverse of a function $y = f(x)$ whose graph is a line through the origin with a nonzero slope m?

36. Show that the graph of the inverse of $f(x) = mx + b$, where m and b are constants and $m \neq 0$, is a line with slope $1/m$ and y-intercept $-b/m$.

37. a. Find the inverse of $f(x) = x + 1$. Graph f and its inverse together. Add the line $y = x$ to your sketch, drawing it with dashes or dots for contrast.

b. Find the inverse of $f(x) = x + b$ (b constant). How is the graph of f^{-1} related to the graph of f?

c. What can you conclude about the inverses of functions whose graphs are lines parallel to the line $y = x$?

38. a. Find the inverse of $f(x) = -x + 1$. Graph the line $y = -x + 1$ together with the line $y = x$. At what angle do the lines intersect?

b. Find the inverse of $f(x) = -x + b$ (b constant). What angle does the line $y = -x + b$ make with the line $y = x$?

c. What can you conclude about the inverses of functions whose graphs are lines perpendicular to the line $y = x$?

Increasing and Decreasing Functions

39. As in Section 4.3, a function $f(x)$ increases on an interval I if for any two points x_1 and x_2 in I,

$$x_2 > x_1 \quad \Rightarrow \quad f(x_2) > f(x_1).$$

Similarly, a function decreases on I if for any two points x_1 and x_2 in I,

$$x_2 > x_1 \quad \Rightarrow \quad f(x_2) < f(x_1).$$

Show that increasing functions and decreasing functions are one-to-one. That is, show that for any x_1 and x_2 in I, $x_2 \neq x_1$ implies $f(x_2) \neq f(x_1)$.

Use the results of Exercise 39 to show that the functions in Exercises 40–44 have inverses over their domains. Find a formula for df^{-1}/dx using Theorem 1.

40. $f(x) = (1/3)x + (5/6)$

41. $f(x) = 27x^3$

42. $f(x) = 1 - 8x^3$

43. $f(x) = (1 - x)^3$

44. $f(x) = x^{5/3}$

Theory and Applications

45. If $f(x)$ is one-to-one, can anything be said about $g(x) = -f(x)$? Is it also one-to-one? Give reasons for your answer.

46. If $f(x)$ is one-to-one and $f(x)$ is never zero, can anything be said about $h(x) = 1/f(x)$? Is it also one-to-one? Give reasons for your answer.

47. Suppose that the range of g lies in the domain of f so that the composite $f \circ g$ is defined. If f and g are one-to-one, can anything be said about $f \circ g$? Give reasons for your answer.

48. If a composite $f \circ g$ is one-to-one, must g be one-to-one? Give reasons for your answer.

49. Suppose $f(x)$ is positive, continuous, and increasing over the interval $[a, b]$. By interpreting the graph of f show that

$$\int_a^b f(x)\,dx + \int_{f(a)}^{f(b)} f^{-1}(y)\,dy = bf(b) - af(a).$$

50. Determine conditions on the constants a, b, c, and d so that the rational function

$$f(x) = \frac{ax + b}{cx + d}$$

has an inverse.

51. If we write $g(x)$ for $f^{-1}(x)$, Equation (1) can be written as

$$g'(f(a)) = \frac{1}{f'(a)}, \quad \text{or} \quad g'(f(a)) \cdot f'(a) = 1.$$

If we then write x for a, we get

$$g'(f(x)) \cdot f'(x) = 1.$$

The latter equation may remind you of the Chain Rule, and indeed there is a connection.

Assume that f and g are differentiable functions that are inverses of one another, so that $(g \circ f)(x) = x$. Differentiate both sides of this equation with respect to x, using the Chain Rule to express $(g \circ f)'(x)$ as a product of derivatives of g and f. What do you find? (This is not a proof of Theorem 1 because we assume here the theorem's conclusion that $g = f^{-1}$ is differentiable.)

52. Equivalence of the washer and shell methods for finding volume Let f be differentiable and increasing on the interval $a \le x \le b$, with $a > 0$, and suppose that f has a differentiable inverse, f^{-1}. Revolve about the y-axis the region bounded by the graph of f and the lines $x = a$ and $y = f(b)$ to generate a solid. Then the values of the integrals given by the washer and shell methods for the volume have identical values:

$$\int_{f(a)}^{f(b)} \pi((f^{-1}(y))^2 - a^2)\,dy = \int_a^b 2\pi x(f(b) - f(x))\,dx.$$

To prove this equality, define

$$W(t) = \int_{f(a)}^{f(t)} \pi((f^{-1}(y))^2 - a^2)\,dy$$

$$S(t) = \int_a^t 2\pi x(f(t) - f(x))\,dx.$$

Then show that the functions W and S agree at a point of $[a, b]$ and have identical derivatives on $[a, b]$. As you saw in Section 4.8, Exercise 102, this will guarantee $W(t) = S(t)$ for all t in $[a, b]$. In particular, $W(b) = S(b)$. (*Source:* "Disks and Shells Revisited," by Walter Carlip, *American Mathematical Monthly*, Vol. 98, No. 2, Feb. 1991, pp. 154–156.)

COMPUTER EXPLORATIONS

In Exercises 53–60, you will explore some functions and their inverses together with their derivatives and linear approximating functions at specified points. Perform the following steps using your CAS:

a. Plot the function $y = f(x)$ together with its derivative over the given interval. Explain why you know that f is one-to-one over the interval.

b. Solve the equation $y = f(x)$ for x as a function of y, and name the resulting inverse function g.

c. Find the equation for the tangent line to f at the specified point $(x_0, f(x_0))$.

d. Find the equation for the tangent line to g at the point $(f(x_0), x_0)$ located symmetrically across the 45° line $y = x$ (which is the graph of the identity function). Use Theorem 1 to find the slope of this tangent line.

e. Plot the functions f and g, the identity, the two tangent lines, and the line segment joining the points $(x_0, f(x_0))$ and $(f(x_0), x_0)$. Discuss the symmetries you see across the main diagonal.

53. $y = \sqrt{3x - 2}, \quad \frac{2}{3} \le x \le 4, \quad x_0 = 3$

54. $y = \dfrac{3x + 2}{2x - 11}, \quad -2 \le x \le 2, \quad x_0 = 1/2$

55. $y = \dfrac{4x}{x^2 + 1}, \quad -1 \le x \le 1, \quad x_0 = 1/2$

56. $y = \dfrac{x^3}{x^2 + 1}, \quad -1 \le x \le 1, \quad x_0 = 1/2$

57. $y = x^3 - 3x^2 - 1, \quad 2 \le x \le 5, \quad x_0 = \dfrac{27}{10}$

58. $y = 2 - x - x^3, \quad -2 \le x \le 2, \quad x_0 = \dfrac{3}{2}$

59. $y = e^x, \quad -3 \le x \le 5, \quad x_0 = 1$

60. $y = \sin x, \quad -\dfrac{\pi}{2} \le x \le \dfrac{\pi}{2}, \quad x_0 = 1$

In Exercises 61 and 62, repeat the steps above to solve for the functions $y = f(x)$ and $x = f^{-1}(y)$ defined implicitly by the given equations over the interval.

61. $y^{1/3} - 1 = (x + 2)^3, \quad -5 \le x \le 5, \quad x_0 = -3/2$

62. $\cos y = x^{1/5}, \quad 0 \le x \le 1, \quad x_0 = 1/2$

7.2 Natural Logarithms

For any positive number a, the function value $f(x) = a^x$ is easy to define when x is an integer or rational number. When x is irrational, the meaning of a^x is not so clear. Similarly, the definition of the logarithm $\log_a x$, the inverse function of $f(x) = a^x$, is not completely obvious. In this section we use integral calculus to define the *natural logarithm* function, for which the number a is a particularly important value. This function allows us to define and analyze general exponential and logarithmic functions, $y = a^x$ and $y = \log_a x$.

Logarithms originally played important roles in arithmetic computations. Historically, considerable labor went into producing long tables of logarithms, correct to five, eight, or even more, decimal places of accuracy. Prior to the modern age of electronic calculators and computers, every engineer owned slide rules marked with logarithmic scales. Calculations with logarithms made possible the great seventeenth-century advances in offshore navigation and celestial mechanics. Today we know such calculations are done using calculators or computers, but the properties and numerous applications of logarithms are as important as ever.

Definition of the Natural Logarithm Function

One solid approach to defining and understanding logarithms begins with a study of the natural logarithm function defined as an integral through the Fundamental Theorem of Calculus. While this approach may seem indirect, it enables us to derive quickly the familiar properties of logarithmic and exponential functions. The functions we have studied so far were analyzed using the techniques of calculus, but here we do something more fundamental. We use calculus for the very definition of the logarithmic and exponential functions.

The natural logarithm of a positive number x, written as $\ln x$, is the value of an integral.

DEFINITION The Natural Logarithm Function

$$\ln x = \int_1^x \frac{1}{t}\,dt, \qquad x > 0$$

If $x > 1$, then $\ln x$ is the area under the curve $y = 1/t$ from $t = 1$ to $t = x$ (Figure 7.9). For $0 < x < 1$, $\ln x$ gives the negative of the area under the curve from x to 1. The function is not defined for $x \leq 0$. From the Zero Width Interval Rule for definite integrals, we also have

$$\ln 1 = \int_1^1 \frac{1}{t}\,dt = 0.$$

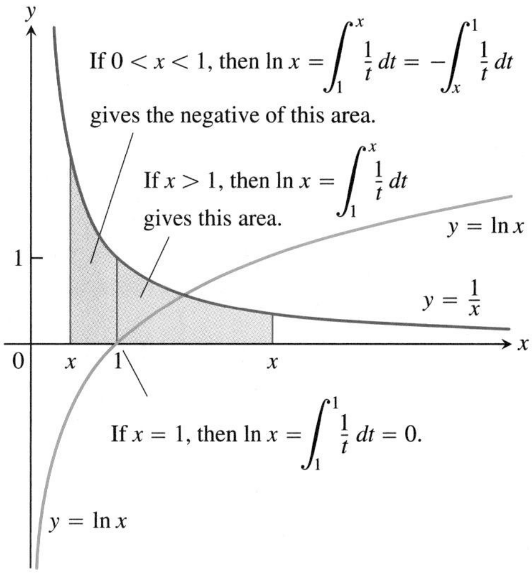

FIGURE 7.9 The graph of $y = \ln x$ and its relation to the function $y = 1/x, x > 0$. The graph of the logarithm rises above the x-axis as x moves from 1 to the right, and it falls below the axis as x moves from 1 to the left.

TABLE 7.1 Typical 2-place values of $\ln x$

x	$\ln x$
0	undefined
0.05	−3.00
0.5	−0.69
1	0
2	0.69
3	1.10
4	1.39
10	2.30

Notice that we show the graph of $y = 1/x$ in Figure 7.9 but use $y = 1/t$ in the integral. Using x for everything would have us writing

$$\ln x = \int_1^x \frac{1}{x}\,dx,$$

with x meaning two different things. So we change the variable of integration to t.

By using rectangles to obtain finite approximations of the area under the graph of $y = 1/t$ and over the interval between $t = 1$ and $t = x$, as in Section 5.1, we can approximate the values of the function $\ln x$. Several values are given in Table 7.1. There is an important number whose natural logarithm equals 1.

DEFINITION The Number e

The number e is that number in the domain of the natural logarithm satisfying

$$\ln(e) = 1$$

Geometrically, the number e corresponds to the point on the x-axis for which the area under the graph of $y = 1/t$ and above the interval $[1, e]$ is the exact area of the unit square. The area of the region shaded blue in Figure 7.9 is 1 sq unit when $x = e$.

The Derivative of $y = \ln x$

By the first part of the Fundamental Theorem of Calculus (Section 5.4),

$$\frac{d}{dx}\ln x = \frac{d}{dx}\int_1^x \frac{1}{t}\,dt = \frac{1}{x}.$$

For every positive value of x, we have

$$\frac{d}{dx}\ln x = \frac{1}{x}.$$

Therefore, the function $y = \ln x$ is a solution to the initial value problem $dy/dx = 1/x$, $x > 0$, with $y(1) = 0$. Notice that the derivative is always positive so the natural logarithm is an increasing function, hence it is one-to-one and invertible. Its inverse is studied in Section 7.3.

If u is a differentiable function of x whose values are positive, so that $\ln u$ is defined, then applying the Chain Rule

$$\frac{dy}{dx} = \frac{dy}{du}\frac{du}{dx}$$

to the function $y = \ln u$ gives

$$\frac{d}{dx}\ln u = \frac{d}{du}\ln u \cdot \frac{du}{dx} = \frac{1}{u}\frac{du}{dx}.$$

$$\frac{d}{dx}\ln u = \frac{1}{u}\frac{du}{dx}, \qquad u > 0 \tag{1}$$

EXAMPLE 1 Derivatives of Natural Logarithms

(a) $\frac{d}{dx}\ln 2x = \frac{1}{2x}\frac{d}{dx}(2x) = \frac{1}{2x}(2) = \frac{1}{x}$

(b) Equation (1) with $u = x^2 + 3$ gives

$$\frac{d}{dx}\ln(x^2+3) = \frac{1}{x^2+3}\cdot\frac{d}{dx}(x^2+3) = \frac{1}{x^2+3}\cdot 2x = \frac{2x}{x^2+3}.$$

■

Notice the remarkable occurrence in Example 1a. The function $y = \ln 2x$ has the same derivative as the function $y = \ln x$. This is true of $y = \ln ax$ for any positive number a:

$$\frac{d}{dx}\ln ax = \frac{1}{ax}\cdot\frac{d}{dx}(ax) = \frac{1}{ax}(a) = \frac{1}{x}. \tag{2}$$

Since they have the same derivative, the functions $y = \ln ax$ and $y = \ln x$ differ by a constant.

HISTORICAL BIOGRAPHY

John Napier
(1550–1617)

Properties of Logarithms

Logarithms were invented by John Napier and were the single most important improvement in arithmetic calculation before the modern electronic computer. What made them so useful is that the properties of logarithms enable multiplication of positive numbers by addition of their logarithms, division of positive numbers by subtraction of their logarithms, and exponentiation of a number by multiplying its logarithm by the exponent. We summarize these properties as a series of rules in Theorem 2. For the moment, we restrict the exponent r in Rule 4 to be a rational number; you will see why when we prove the rule.

THEOREM 2 Properties of Logarithms

For any numbers $a > 0$ and $x > 0$, the natural logarithm satisfies the following rules:

1. *Product Rule*: $\ln ax = \ln a + \ln x$
2. *Quotient Rule*: $\ln \frac{a}{x} = \ln a - \ln x$
3. *Reciprocal Rule*: $\ln \frac{1}{x} = -\ln x$ — Rule 2 with $a = 1$
4. *Power Rule*: $\ln x^r = r \ln x$ — r rational

We illustrate how these rules apply.

EXAMPLE 2 Interpreting the Properties of Logarithms

(a) $\ln 6 = \ln (2 \cdot 3) = \ln 2 + \ln 3$ — Product

(b) $\ln 4 - \ln 5 = \ln \frac{4}{5} = \ln 0.8$ — Quotient

(c) $\ln \frac{1}{8} = -\ln 8$ — Reciprocal

$= -\ln 2^3 = -3 \ln 2$ — Power ■

EXAMPLE 3 Applying the Properties to Function Formulas

(a) $\ln 4 + \ln \sin x = \ln (4 \sin x)$ — Product

(b) $\ln \frac{x+1}{2x-3} = \ln (x+1) - \ln (2x-3)$ — Quotient

(c) $\ln \sec x = \ln \dfrac{1}{\cos x} = -\ln \cos x$ Reciprocal

(d) $\ln \sqrt[3]{x+1} = \ln (x+1)^{1/3} = \dfrac{1}{3} \ln (x+1)$ Power ■

We now give the proof of Theorem 2. The steps in the proof are similar to those used in solving problems involving logarithms.

Proof that $\ln ax = \ln a + \ln x$ The argument is unusual—and elegant. It starts by observing that $\ln ax$ and $\ln x$ have the same derivative (Equation 2). According to Corollary 2 of the Mean Value Theorem, then, the functions must differ by a constant, which means that

$$\ln ax = \ln x + C$$

for some C.

Since this last equation holds for all positive values of x, it must hold for $x = 1$. Hence,

$$\begin{aligned} \ln (a \cdot 1) &= \ln 1 + C \\ \ln a &= 0 + C && \ln 1 = 0 \\ C &= \ln a. \end{aligned}$$

By substituting we conclude,

$$\ln ax = \ln a + \ln x.$$

Proof that $\ln x^r = r \ln x$ (assuming r rational) We use the same-derivative argument again. For all positive values of x,

$$\begin{aligned} \frac{d}{dx} \ln x^r &= \frac{1}{x^r} \frac{d}{dx}(x^r) && \text{Eq. (1) with } u = x^r \\ &= \frac{1}{x^r} r x^{r-1} && \text{Here is where we need } r \text{ to be rational, at least for now. We have proved the Power Rule only for rational exponents.} \\ &= r \cdot \frac{1}{x} = \frac{d}{dx}(r \ln x). \end{aligned}$$

Since $\ln x^r$ and $r \ln x$ have the same derivative,

$$\ln x^r = r \ln x + C$$

for some constant C. Taking x to be 1 identifies C as zero, and we're done.

You are asked to prove Rule 2 in Exercise 84. Rule 3 is a special case of Rule 2, obtained by setting $a = 1$ and noting that $\ln 1 = 0$. So we have established all cases of Theorem 2. ■

We have not yet proved Rule 4 for r irrational; we will return to this case in Section 7.3. The rule does hold for all r, rational or irrational.

The Graph and Range of $\ln x$

The derivative $d(\ln x)/dx = 1/x$ is positive for $x > 0$, so $\ln x$ is an increasing function of x. The second derivative, $-1/x^2$, is negative, so the graph of $\ln x$ is concave down.

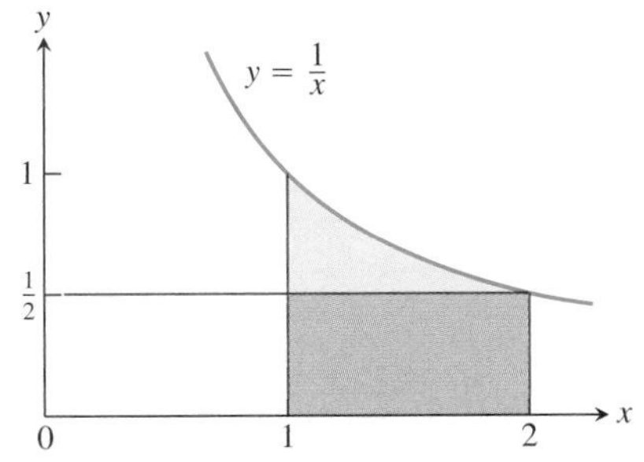

FIGURE 7.10 The rectangle of height $y = 1/2$ fits beneath the graph of $y = 1/x$ for the interval $1 \le x \le 2$.

We can estimate the value of ln 2 by considering the area under the graph of $y = 1/x$ and above the interval $[1, 2]$. In Figure 7.10 a rectangle of height $1/2$ over the interval $[1, 2]$ fits under the graph. Therefore the area under the graph, which is ln 2, is greater than the area, $1/2$, of the rectangle. So $\ln 2 > 1/2$. Knowing this we have,

$$\ln 2^n = n \ln 2 > n\left(\frac{1}{2}\right) = \frac{n}{2}$$

and

$$\ln 2^{-n} = -n \ln 2 < -n\left(\frac{1}{2}\right) = -\frac{n}{2}.$$

It follows that

$$\lim_{x \to \infty} \ln x = \infty \qquad \text{and} \qquad \lim_{x \to 0^+} \ln x = -\infty.$$

We defined $\ln x$ for $x > 0$, so the domain of $\ln x$ is the set of positive real numbers. The above discussion and the Intermediate Value Theorem show that its range is the entire real line giving the graph of $y = \ln x$ shown in Figure 7.9.

The Integral $\int (1/u)\, du$

Equation (1) leads to the integral formula

$$\int \frac{1}{u}\, du = \ln u + C \tag{3}$$

when u is a positive differentiable function, but what if u is negative? If u is negative, then $-u$ is positive and

$$\begin{aligned} \int \frac{1}{u}\, du &= \int \frac{1}{(-u)}\, d(-u) && \text{Eq. (3) with } u \text{ replaced by } -u \\ &= \ln(-u) + C. && (4) \end{aligned}$$

We can combine Equations (3) and (4) into a single formula by noticing that in each case the expression on the right is $\ln|u| + C$. In Equation (3), $\ln u = \ln|u|$ because $u > 0$; in Equation (4), $\ln(-u) = \ln|u|$ because $u < 0$. Whether u is positive or negative, the integral of $(1/u)\, du$ is $\ln|u| + C$.

> If u is a differentiable function that is never zero,
>
> $$\int \frac{1}{u}\, du = \ln|u| + C. \tag{5}$$

Equation (5) applies anywhere on the domain of $1/u$, the points where $u \neq 0$.

We know that

$$\int u^n\, du = \frac{u^{n+1}}{n+1} + C, \qquad n \neq -1 \text{ and rational}$$

Equation (5) explains what to do when n equals -1. Equation (5) says integrals of a certain *form* lead to logarithms. If $u = f(x)$, then $du = f'(x)\,dx$ and

$$\int \frac{1}{u}\,du = \int \frac{f'(x)}{f(x)}\,dx.$$

So Equation (5) gives

$$\int \frac{f'(x)}{f(x)}\,dx = \ln|f(x)| + C$$

whenever $f(x)$ is a differentiable function that maintains a constant sign on the domain given for it.

EXAMPLE 4 Applying Equation (5)

(a) $$\int_0^2 \frac{2x}{x^2-5}\,dx = \int_{-5}^{-1} \frac{du}{u} = \ln|u|\Big]_{-5}^{-1}$$ $u = x^2 - 5,\quad du = 2x\,dx,$ $u(0) = -5,\quad u(2) = -1$

$$= \ln|-1| - \ln|-5| = \ln 1 - \ln 5 = -\ln 5$$

(b) $$\int_{-\pi/2}^{\pi/2} \frac{4\cos\theta}{3 + 2\sin\theta}\,d\theta = \int_1^5 \frac{2}{u}\,du$$ $u = 3 + 2\sin\theta,\quad du = 2\cos\theta\,d\theta,$ $u(-\pi/2) = 1,\quad u(\pi/2) = 5$

$$= 2\ln|u|\Big]_1^5$$

$$= 2\ln|5| - 2\ln|1| = 2\ln 5$$

Note that $u = 3 + 2\sin\theta$ is always positive on $[-\pi/2, \pi/2]$, so Equation (5) applies. ■

The Integrals of tan x and cot x

Equation (5) tells us at last how to integrate the tangent and cotangent functions. For the tangent function,

$$\int \tan x\,dx = \int \frac{\sin x}{\cos x}\,dx = \int \frac{-du}{u}$$ $u = \cos x > 0$ on $(-\pi/2, \pi/2)$, $du = -\sin x\,dx$

$$= -\int \frac{du}{u} = -\ln|u| + C$$

$$= -\ln|\cos x| + C = \ln\frac{1}{|\cos x|} + C$$ Reciprocal Rule

$$= \ln|\sec x| + C.$$

For the cotangent,

$$\int \cot x\,dx = \int \frac{\cos x\,dx}{\sin x} = \int \frac{du}{u}$$ $u = \sin x,$ $du = \cos x\,dx$

$$= \ln|u| + C = \ln|\sin x| + C = -\ln|\csc x| + C.$$

$$\int \tan u \, du = -\ln|\cos u| + C = \ln|\sec u| + C$$

$$\int \cot u \, du = \ln|\sin u| + C = -\ln|\csc x| + C$$

EXAMPLE 5

$$\int_0^{\pi/6} \tan 2x \, dx = \int_0^{\pi/3} \tan u \cdot \frac{du}{2} = \frac{1}{2}\int_0^{\pi/3} \tan u \, du$$

$$= \frac{1}{2}\ln|\sec u|\Big]_0^{\pi/3} = \frac{1}{2}(\ln 2 - \ln 1) = \frac{1}{2}\ln 2$$

Substitute $u = 2x$, $dx = du/2$, $u(0) = 0$, $u(\pi/6) = \pi/3$ ■

Logarithmic Differentiation

The derivatives of positive functions given by formulas that involve products, quotients, and powers can often be found more quickly if we take the natural logarithm of both sides before differentiating. This enables us to use the laws of logarithms to simplify the formulas before differentiating. The process, called **logarithmic differentiation**, is illustrated in the next example.

EXAMPLE 6 Using Logarithmic Differentiation

Find dy/dx if

$$y = \frac{(x^2 + 1)(x + 3)^{1/2}}{x - 1}, \qquad x > 1.$$

Solution We take the natural logarithm of both sides and simplify the result with the properties of logarithms:

$$\ln y = \ln \frac{(x^2 + 1)(x + 3)^{1/2}}{x - 1}$$

$$= \ln((x^2 + 1)(x + 3)^{1/2}) - \ln(x - 1) \qquad \text{Rule 2}$$

$$= \ln(x^2 + 1) + \ln(x + 3)^{1/2} - \ln(x - 1) \qquad \text{Rule 1}$$

$$= \ln(x^2 + 1) + \frac{1}{2}\ln(x + 3) - \ln(x - 1). \qquad \text{Rule 3}$$

We then take derivatives of both sides with respect to x, using Equation (1) on the left:

$$\frac{1}{y}\frac{dy}{dx} = \frac{1}{x^2 + 1} \cdot 2x + \frac{1}{2} \cdot \frac{1}{x + 3} - \frac{1}{x - 1}.$$

Next we solve for dy/dx:

$$\frac{dy}{dx} = y\left(\frac{2x}{x^2 + 1} + \frac{1}{2x + 6} - \frac{1}{x - 1}\right).$$

Finally, we substitute for y:

$$\frac{dy}{dx} = \frac{(x^2+1)(x+3)^{1/2}}{x-1}\left(\frac{2x}{x^2+1} + \frac{1}{2x+6} - \frac{1}{x-1}\right).$$ ■

A direct computation in Example 6, using the Quotient and Product Rules, would be much longer.

EXERCISES 7.2

Using the Properties of Logarithms

1. Express the following logarithms in terms of ln 2 and ln 3.

a. $\ln 0.75$ **b.** $\ln(4/9)$ **c.** $\ln(1/2)$

d. $\ln \sqrt[3]{9}$ **e.** $\ln 3\sqrt{2}$ **f.** $\ln \sqrt{13.5}$

2. Express the following logarithms in terms of ln 5 and ln 7.

a. $\ln(1/125)$ **b.** $\ln 9.8$ **c.** $\ln 7\sqrt{7}$

d. $\ln 1225$ **e.** $\ln 0.056$

f. $(\ln 35 + \ln(1/7))/(\ln 25)$

Use the properties of logarithms to simplify the expressions in Exercises 3 and 4.

3. a. $\ln \sin\theta - \ln\left(\frac{\sin\theta}{5}\right)$ **b.** $\ln(3x^2 - 9x) + \ln\left(\frac{1}{3x}\right)$

c. $\frac{1}{2}\ln(4t^4) - \ln 2$

4. a. $\ln \sec\theta + \ln\cos\theta$ **b.** $\ln(8x+4) - 2\ln 2$

c. $3\ln\sqrt[3]{t^2 - 1} - \ln(t+1)$

Derivatives of Logarithms

In Exercises 5–36, find the derivative of y with respect to x, t, or θ, as appropriate.

5. $y = \ln 3x$ **6.** $y = \ln kx$, k constant

7. $y = \ln(t^2)$ **8.** $y = \ln(t^{3/2})$

9. $y = \ln\frac{3}{x}$ **10.** $y = \ln\frac{10}{x}$

11. $y = \ln(\theta + 1)$ **12.** $y = \ln(2\theta + 2)$

13. $y = \ln x^3$ **14.** $y = (\ln x)^3$

15. $y = t(\ln t)^2$ **16.** $y = t\sqrt{\ln t}$

17. $y = \frac{x^4}{4}\ln x - \frac{x^4}{16}$ **18.** $y = \frac{x^3}{3}\ln x - \frac{x^3}{9}$

19. $y = \frac{\ln t}{t}$ **20.** $y = \frac{1 + \ln t}{t}$

21. $y = \frac{\ln x}{1 + \ln x}$ **22.** $y = \frac{x \ln x}{1 + \ln x}$

23. $y = \ln(\ln x)$ **24.** $y = \ln(\ln(\ln x))$

25. $y = \theta(\sin(\ln\theta) + \cos(\ln\theta))$

26. $y = \ln(\sec\theta + \tan\theta)$

27. $y = \ln\frac{1}{x\sqrt{x+1}}$ **28.** $y = \frac{1}{2}\ln\frac{1+x}{1-x}$

29. $y = \frac{1 + \ln t}{1 - \ln t}$ **30.** $y = \sqrt{\ln\sqrt{t}}$

31. $y = \ln(\sec(\ln\theta))$ **32.** $y = \ln\left(\frac{\sqrt{\sin\theta\cos\theta}}{1 + 2\ln\theta}\right)$

33. $y = \ln\left(\frac{(x^2+1)^5}{\sqrt{1-x}}\right)$ **34.** $y = \ln\sqrt{\frac{(x+1)^5}{(x+2)^{20}}}$

35. $y = \int_{x^2/2}^{x^2} \ln\sqrt{t}\,dt$ **36.** $y = \int_{\sqrt{x}}^{\sqrt[3]{x}} \ln t\,dt$

Integration

Evaluate the integrals in Exercises 37–54.

37. $\int_{-3}^{-2}\frac{dx}{x}$ **38.** $\int_{-1}^{0}\frac{3\,dx}{3x-2}$

39. $\int\frac{2y\,dy}{y^2 - 25}$ **40.** $\int\frac{8r\,dr}{4r^2 - 5}$

41. $\int_0^{\pi}\frac{\sin t}{2 - \cos t}\,dt$ **42.** $\int_0^{\pi/3}\frac{4\sin\theta}{1 - 4\cos\theta}\,d\theta$

43. $\int_1^2\frac{2\ln x}{x}\,dx$ **44.** $\int_2^4\frac{dx}{x\ln x}$

45. $\int_2^4\frac{dx}{x(\ln x)^2}$ **46.** $\int_2^{16}\frac{dx}{2x\sqrt{\ln x}}$

47. $\int\frac{3\sec^2 t}{6 + 3\tan t}\,dt$ **48.** $\int\frac{\sec y\tan y}{2 + \sec y}\,dy$

49. $\int_0^{\pi/2}\tan\frac{x}{2}\,dx$ **50.** $\int_{\pi/4}^{\pi/2}\cot t\,dt$

51. $\int_{\pi/2}^{\pi}2\cot\frac{\theta}{3}\,d\theta$ **52.** $\int_0^{\pi/12}6\tan 3x\,dx$

53. $\int\frac{dx}{2\sqrt{x} + 2x}$ **54.** $\int\frac{\sec x\,dx}{\sqrt{\ln(\sec x + \tan x)}}$

Logarithmic Differentiation

In Exercises 55–68, use logarithmic differentiation to find the derivative of y with respect to the given independent variable.

55. $y = \sqrt{x(x+1)}$

56. $y = \sqrt{(x^2+1)(x-1)^2}$

57. $y = \sqrt{\frac{t}{t+1}}$

58. $y = \sqrt{\frac{1}{t(t+1)}}$

59. $y = \sqrt{\theta + 3}\sin\theta$

60. $y = (\tan\theta)\sqrt{2\theta + 1}$

61. $y = t(t+1)(t+2)$

62. $y = \frac{1}{t(t+1)(t+2)}$

63. $y = \frac{\theta + 5}{\theta\cos\theta}$

64. $y = \frac{\theta\sin\theta}{\sqrt{\sec\theta}}$

65. $y = \frac{x\sqrt{x^2+1}}{(x+1)^{2/3}}$

66. $y = \sqrt{\frac{(x+1)^{10}}{(2x+1)^5}}$

67. $y = \sqrt[3]{\frac{x(x-2)}{x^2+1}}$

68. $y = \sqrt[3]{\frac{x(x+1)(x-2)}{(x^2+1)(2x+3)}}$

Theory and Applications

69. Locate and identify the absolute extreme values of

a. $\ln(\cos x)$ on $[-\pi/4, \pi/3]$,

b. $\cos(\ln x)$ on $[1/2, 2]$.

70. a. Prove that $f(x) = x - \ln x$ is increasing for $x > 1$.

b. Using part (a), show that $\ln x < x$ if $x > 1$.

71. Find the area between the curves $y = \ln x$ and $y = \ln 2x$ from $x = 1$ to $x = 5$.

72. Find the area between the curve $y = \tan x$ and the x-axis from $x = -\pi/4$ to $x = \pi/3$.

73. The region in the first quadrant bounded by the coordinate axes, the line $y = 3$, and the curve $x = 2/\sqrt{y+1}$ is revolved about the y-axis to generate a solid. Find the volume of the solid.

74. The region between the curve $y = \sqrt{\cot x}$ and the x-axis from $x = \pi/6$ to $x = \pi/2$ is revolved about the x-axis to generate a solid. Find the volume of the solid.

75. The region between the curve $y = 1/x^2$ and the x-axis from $x = 1/2$ to $x = 2$ is revolved about the y-axis to generate a solid. Find the volume of the solid.

76. In Section 6.2, Exercise 6, we revolved about the y-axis the region between the curve $y = 9x/\sqrt{x^3+9}$ and the x-axis from $x = 0$ to $x = 3$ to generate a solid of volume 36π. What volume do you get if you revolve the region about the x-axis instead? (See Section 6.2, Exercise 6, for a graph.)

77. Find the lengths of the following curves.

a. $y = (x^2/8) - \ln x, \quad 4 \le x \le 8$

b. $x = (y/4)^2 - 2\ln(y/4), \quad 4 \le y \le 12$

78. Find a curve through the point $(1, 0)$ whose length from $x = 1$ to $x = 2$ is

$$L = \int_1^2 \sqrt{1 + \frac{1}{x^2}}\,dx.$$

T **79. a.** Find the centroid of the region between the curve $y = 1/x$ and the x-axis from $x = 1$ to $x = 2$. Give the coordinates to two decimal places.

b. Sketch the region and show the centroid in your sketch.

80. a. Find the center of mass of a thin plate of constant density covering the region between the curve $y = 1/\sqrt{x}$ and the x-axis from $x = 1$ to $x = 16$.

b. Find the center of mass if, instead of being constant, the density function is $\delta(x) = 4/\sqrt{x}$.

Solve the initial value problems in Exercises 81 and 82.

81. $\frac{dy}{dx} = 1 + \frac{1}{x}, \quad y(1) = 3$

82. $\frac{d^2y}{dx^2} = \sec^2 x, \quad y(0) = 0 \quad \text{and} \quad y'(0) = 1$

T **83. The linearization of $\ln(1+x)$ at $x = 0$** Instead of approximating $\ln x$ near $x = 1$, we approximate $\ln(1+x)$ near $x = 0$. We get a simpler formula this way.

a. Derive the linearization $\ln(1+x) \approx x$ at $x = 0$.

b. Estimate to five decimal places the error involved in replacing $\ln(1+x)$ by x on the interval $[0, 0.1]$.

c. Graph $\ln(1+x)$ and x together for $0 \le x \le 0.5$. Use different colors, if available. At what points does the approximation of $\ln(1+x)$ seem best? Least good? By reading coordinates from the graphs, find as good an upper bound for the error as your grapher will allow.

84. Use the same-derivative argument, as was done to prove Rules 1 and 4 of Theorem 2, to prove the Quotient Rule property of logarithms.

Grapher Explorations

85. Graph $\ln x$, $\ln 2x$, $\ln 4x$, $\ln 8x$, and $\ln 16x$ (as many as you can) together for $0 < x \le 10$. What is going on? Explain.

86. Graph $y = \ln|\sin x|$ in the window $0 \le x \le 22$, $-2 \le y \le 0$. Explain what you see. How could you change the formula to turn the arches upside down?

87. a. Graph $y = \sin x$ and the curves $y = \ln(a + \sin x)$ for $a = 2$, 4, 8, 20, and 50 together for $0 \le x \le 23$.

b. Why do the curves flatten as a increases? (*Hint:* Find an a-dependent upper bound for $|y'|$.)

88. Does the graph of $y = \sqrt{x} - \ln x$, $x > 0$, have an inflection point? Try to answer the question **(a)** by graphing, **(b)** by using calculus.

7.3 The Exponential Function

Having developed the theory of the function $\ln x$, we introduce the exponential function $\exp x = e^x$ as the inverse of $\ln x$. We study its properties and compute its derivative and integral. Knowing its derivative, we prove the power rule to differentiate x^n when n is *any* real number, rational or irrational.

The Inverse of ln *x* and the Number *e*

The function $\ln x$, being an increasing function of x with domain $(0, \infty)$ and range $(-\infty, \infty)$, has an inverse $\ln^{-1} x$ with domain $(-\infty, \infty)$ and range $(0, \infty)$. The graph of $\ln^{-1} x$ is the graph of $\ln x$ reflected across the line $y = x$. As you can see in Figure 7.11,

$$\lim_{x\to\infty} \ln^{-1} x = \infty \qquad \text{and} \qquad \lim_{x\to-\infty} \ln^{-1} x = 0.$$

The function $\ln^{-1} x$ is also denoted by $\exp x$.

In Section 7.2 we defined the number e by the equation $\ln(e) = 1$, so $e = \ln^{-1}(1) = \exp(1)$. Although e is not a rational number, later in this section we see one way to express it as a limit. In Chapter 11, we will calculate its value with a computer to as many places of accuracy as we want with a different formula (Section 11.9, Example 6). To 15 places,

$$e = 2.718281828459045.$$

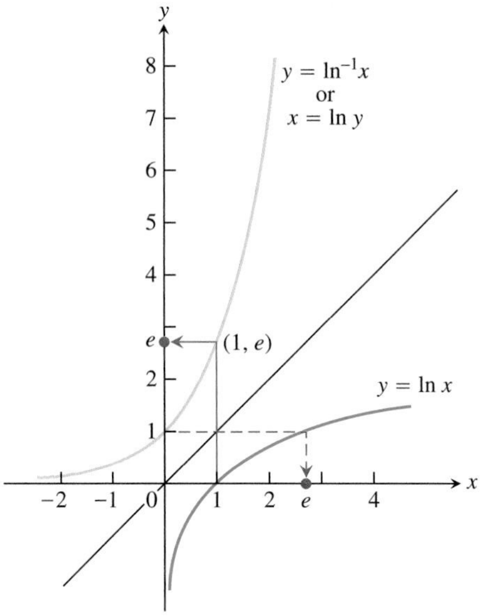

FIGURE 7.11 The graphs of $y = \ln x$ and $y = \ln^{-1} x = \exp x$. The number e is $\ln^{-1} 1 = \exp(1)$.

The Function $y = e^x$

We can raise the number e to a rational power r in the usual way:

$$e^2 = e \cdot e, \qquad e^{-2} = \frac{1}{e^2}, \qquad e^{1/2} = \sqrt{e},$$

and so on. Since e is positive, e^r is positive too. Thus, e^r has a logarithm. When we take the logarithm, we find that

$$\ln e^r = r \ln e = r \cdot 1 = r.$$

Since $\ln x$ is one-to-one and $\ln(\ln^{-1} r) = r$, this equation tells us that

$$e^r = \ln^{-1} r = \exp r \qquad \text{for } r \text{ rational}. \tag{1}$$

We have not yet found a way to give an obvious meaning to e^x for x irrational. But $\ln^{-1} x$ has meaning for any x, rational or irrational. So Equation (1) provides a way to extend the definition of e^x to irrational values of x. The function $\ln^{-1} x$ is defined for all x, so we use it to assign a value to e^x at every point where e^x had no previous definition.

Typical values of e^x

x	e^x (rounded)
-1	0.37
0	1
1	2.72
2	7.39
10	22026
100	2.6881×10^{43}

DEFINITION The Natural Exponential Function
For every real number x, $e^x = \ln^{-1} x = \exp x$.

For the first time we have a precise meaning for an irrational exponent. Usually the exponential function is denoted by e^x rather than $\exp x$. Since $\ln x$ and e^x are inverses of one another, we have

Inverse Equations for e^x and $\ln x$

$$e^{\ln x} = x \qquad (\text{all } x > 0) \tag{2}$$

$$\ln(e^x) = x \qquad (\text{all } x) \tag{3}$$

Transcendental Numbers and Transcendental Functions

Numbers that are solutions of polynomial equations with rational coefficients are called **algebraic**: -2 is algebraic because it satisfies the equation $x + 2 = 0$, and $\sqrt{3}$ is algebraic because it satisfies the equation $x^2 - 3 = 0$. Numbers that are not algebraic are called **transcendental**, like e and π. In 1873, Charles Hermite proved the transcendence of e in the sense that we describe. In 1882, C.L.F. Lindemann proved the transcendence of π.

Today, we call a function $y = f(x)$ algebraic if it satisfies an equation of the form

$$P_n y^n + \cdots + P_1 y + P_0 = 0$$

in which the P's are polynomials in x with rational coefficients. The function $y = 1/\sqrt{x+1}$ is algebraic because it satisfies the equation $(x + 1)y^2 - 1 = 0$. Here the polynomials are $P_2 = x + 1$, $P_1 = 0$, and $P_0 = -1$. Functions that are not algebraic are called transcendental.

The domain of $\ln x$ is $(0, \infty)$ and its range is $(-\infty, \infty)$. So the domain of e^x is $(-\infty, \infty)$ and its range is $(0, \infty)$.

EXAMPLE 1 Using the Inverse Equations

(a) $\ln e^2 = 2$

(b) $\ln e^{-1} = -1$

(c) $\ln \sqrt{e} = \frac{1}{2}$

(d) $\ln e^{\sin x} = \sin x$

(e) $e^{\ln 2} = 2$

(f) $e^{\ln(x^2+1)} = x^2 + 1$

(g) $e^{3\ln 2} = e^{\ln 2^3} = e^{\ln 8} = 8$ One way

(h) $e^{3\ln 2} = (e^{\ln 2})^3 = 2^3 = 8$ Another way ■

EXAMPLE 2 Solving for an Exponent

Find k if $e^{2k} = 10$.

Solution Take the natural logarithm of both sides:

$$\begin{aligned} e^{2k} &= 10 \\ \ln e^{2k} &= \ln 10 \\ 2k &= \ln 10 && \text{Eq. (3)} \\ k &= \frac{1}{2}\ln 10. \end{aligned}$$

■

The General Exponential Function a^x

Since $a = e^{\ln a}$ for any positive number a, we can think of a^x as $(e^{\ln a})^x = e^{x \ln a}$. We therefore make the following definition.

DEFINITION General Exponential Functions

For any numbers $a > 0$ and x, the exponential function with base a is

$$a^x = e^{x \ln a}.$$

When $a = e$, the definition gives $a^x = e^{x \ln a} = e^{x \ln e} = e^{x \cdot 1} = e^x$.

HISTORICAL BIOGRAPHY

Siméon Denis Poisson
(1781–1840)

EXAMPLE 3 Evaluating Exponential Functions

(a) $2^{\sqrt{3}} = e^{\sqrt{3}\ln 2} \approx e^{1.20} \approx 3.32$

(b) $2^{\pi} = e^{\pi \ln 2} \approx e^{2.18} \approx 8.8$ ■

We study the calculus of general exponential functions and their inverses in the next section. Here we need the definition in order to discuss the laws of exponents for e^x.

Laws of Exponents

Even though e^x is defined in a seemingly roundabout way as $\ln^{-1} x$, it obeys the familiar laws of exponents from algebra. Theorem 3 shows us that these laws are consequences of the definitions of $\ln x$ and e^x.

THEOREM 3 Laws of Exponents for e^x

For all numbers x, x_1, and x_2, the natural exponential e^x obeys the following laws:

1. $e^{x_1} \cdot e^{x_2} = e^{x_1 + x_2}$

2. $e^{-x} = \dfrac{1}{e^x}$

3. $\dfrac{e^{x_1}}{e^{x_2}} = e^{x_1 - x_2}$

4. $(e^{x_1})^{x_2} = e^{x_1 x_2} = (e^{x_2})^{x_1}$

Proof of Law 1 Let

$$y_1 = e^{x_1} \quad \text{and} \quad y_2 = e^{x_2}. \tag{4}$$

Then

$$\begin{aligned}
x_1 &= \ln y_1 \quad \text{and} \quad x_2 = \ln y_2 && \text{Take logs of both sides of Eqs. (4).}\\
x_1 + x_2 &= \ln y_1 + \ln y_2 \\
&= \ln y_1 y_2 && \text{Product Rule for logarithms}\\
e^{x_1 + x_2} &= e^{\ln y_1 y_2} && \text{Exponentiate.}\\
&= y_1 y_2 && e^{\ln u} = u\\
&= e^{x_1} e^{x_2}.
\end{aligned}$$ ■

The proof of Law 4 is similar. Laws 2 and 3 follow from Law 1 (Exercise 78).

EXAMPLE 4 Applying the Exponent Laws

(a) $e^{x + \ln 2} = e^x \cdot e^{\ln 2} = 2e^x$ Law 1

(b) $e^{-\ln x} = \dfrac{1}{e^{\ln x}} = \dfrac{1}{x}$ Law 2

(c) $\dfrac{e^{2x}}{e} = e^{2x-1}$ Law 3

(d) $(e^3)^x = e^{3x} = (e^x)^3$ Law 4 ■

Theorem 3 is also valid for a^x, the exponential function with base a. For example,

$$\begin{aligned} a^{x_1} \cdot a^{x_2} &= e^{x_1 \ln a} \cdot e^{x_2 \ln a} && \text{Definition of } a^x \\ &= e^{x_1 \ln a + x_2 \ln a} && \text{Law 1} \\ &= e^{(x_1 + x_2)\ln a} && \text{Factor } \ln a \\ &= a^{x_1 + x_2}. && \text{Definition of } a^x \end{aligned}$$

The Derivative and Integral of e^x

The exponential function is differentiable because it is the inverse of a differentiable function whose derivative is never zero (Theorem 1). We calculate its derivative using Theorem 1 and our knowledge of the derivative of $\ln x$. Let

$$f(x) = \ln x \qquad \text{and} \qquad y = e^x = \ln^{-1} x = f^{-1}(x).$$

Then,

$$\begin{aligned} \frac{dy}{dx} = \frac{d}{dx}(e^x) &= \frac{d}{dx}\ln^{-1} x \\ &= \frac{d}{dx} f^{-1}(x) \\ &= \frac{1}{f'(f^{-1}(x))} && \text{Theorem 1} \\ &= \frac{1}{f'(e^x)} && f^{-1}(x) = e^x \\ &= \frac{1}{\left(\dfrac{1}{e^x}\right)} && f'(z) = \frac{1}{z} \text{ with } z = e^x \\ &= e^x. \end{aligned}$$

That is, for $y = e^x$, we find that $dy/dx = e^x$ so the natural exponential function e^x is its own derivative. We will see in Section 7.5 that the only functions that behave this way are constant multiples of e^x. In summary,

$$\frac{d}{dx} e^x = e^x \tag{5}$$

EXAMPLE 5 Differentiating an Exponential

$$\begin{aligned} \frac{d}{dx}(5e^x) &= 5\frac{d}{dx} e^x \\ &= 5e^x \end{aligned}$$

■

The Chain Rule extends Equation (5) in the usual way to a more general form.

If u is any differentiable function of x, then

$$\frac{d}{dx} e^u = e^u \frac{du}{dx}. \tag{6}$$

EXAMPLE 6 Applying the Chain Rule with Exponentials

(a) $\frac{d}{dx} e^{-x} = e^{-x} \frac{d}{dx}(-x) = e^{-x}(-1) = -e^{-x}$ Eq. (6) with $u = -x$

(b) $\frac{d}{dx} e^{\sin x} = e^{\sin x} \frac{d}{dx}(\sin x) = e^{\sin x} \cdot \cos x$ Eq. (6) with $u = \sin x$

The integral equivalent of Equation (6) is

$$\int e^u \, du = e^u + C.$$

EXAMPLE 7 Integrating Exponentials

(a)
$$\begin{aligned} \int_0^{\ln 2} e^{3x} \, dx &= \int_0^{\ln 8} e^u \cdot \frac{1}{3} \, du \\ &= \frac{1}{3} \int_0^{\ln 8} e^u \, du \\ &= \frac{1}{3} e^u \Big]_0^{\ln 8} \\ &= \frac{1}{3}(8 - 1) = \frac{7}{3} \end{aligned}$$

$u = 3x, \quad \frac{1}{3} du = dx, \quad u(0) = 0,$
$u(\ln 2) = 3 \ln 2 = \ln 2^3 = \ln 8$

(b)
$$\begin{aligned} \int_0^{\pi/2} e^{\sin x} \cos x \, dx &= e^{\sin x} \Big]_0^{\pi/2} \\ &= e^1 - e^0 = e - 1 \end{aligned}$$

Antiderivative from Example 6

EXAMPLE 8 Solving an Initial Value Problem

Solve the initial value problem

$$e^y \frac{dy}{dx} = 2x, \qquad x > \sqrt{3}; \qquad y(2) = 0.$$

Solution We integrate both sides of the differential equation with respect to x to obtain

$$e^y = x^2 + C.$$

We use the initial condition $y(2) = 0$ to determine C:

$$C = e^0 - (2)^2$$
$$= 1 - 4 = -3.$$

This completes the formula for e^y:

$$e^y = x^2 - 3.$$

To find y, we take logarithms of both sides:

$$\ln e^y = \ln (x^2 - 3)$$
$$y = \ln (x^2 - 3).$$

Notice that the solution is valid for $x > \sqrt{3}$.

Let's check the solution in the original equation.

$$e^y \frac{dy}{dx} = e^y \frac{d}{dx} \ln (x^2 - 3)$$
$$= e^y \frac{2x}{x^2 - 3} \qquad \text{Derivative of } \ln (x^2 - 3)$$
$$= e^{\ln (x^2 - 3)} \frac{2x}{x^2 - 3} \qquad y = \ln (x^2 - 3)$$
$$= (x^2 - 3) \frac{2x}{x^2 - 3} \qquad e^{\ln y} = y$$
$$= 2x.$$

The solution checks. ■

The Number e Expressed as a Limit

We have defined the number e as the number for which $\ln e = 1$, or the value $\exp (1)$. We see that e is an important constant for the logarithmic and exponential functions, but what is its numerical value? The next theorem shows one way to calculate e as a limit.

THEOREM 4 The Number e as a Limit
The number e can be calculated as the limit

$$e = \lim_{x \to 0} (1 + x)^{1/x}.$$

Proof If $f(x) = \ln x$, then $f'(x) = 1/x$, so $f'(1) = 1$. But, by the definition of derivative,

$$f'(1) = \lim_{h \to 0} \frac{f(1 + h) - f(1)}{h} = \lim_{x \to 0} \frac{f(1 + x) - f(1)}{x}$$
$$= \lim_{x \to 0} \frac{\ln (1 + x) - \ln 1}{x} = \lim_{x \to 0} \frac{1}{x} \ln (1 + x) \qquad \ln 1 = 0$$
$$= \lim_{x \to 0} \ln (1 + x)^{1/x} = \ln \left[\lim_{x \to 0} (1 + x)^{1/x} \right] \qquad \ln \text{ is continuous.}$$

Because $f'(1) = 1$, we have

$$\ln\left[\lim_{x\to 0}(1 + x)^{1/x}\right] = 1$$

Therefore,

$$\lim_{x\to 0}(1 + x)^{1/x} = e \qquad \ln e = 1 \text{ and } \ln \text{ is one-to-one}$$

■

By substituting $y = 1/x$, we can also express the limit in Theorem 4 as

$$e = \lim_{y\to\infty}\left(1 + \frac{1}{y}\right)^{y}. \tag{7}$$

At the beginning of the section we noted that $e = 2.718281828459045$ to 15 decimal places.

The Power Rule (General Form)

We can now define x^n for any $x > 0$ and any real number n as $x^n = e^{n \ln x}$. Therefore, the n in the equation $\ln x^n = n \ln x$ no longer needs to be rational—it can be any number as long as $x > 0$:

$$\ln x^n = \ln(e^{n \ln x}) = n \ln x \qquad \ln e^u = u, \text{ any } u$$

Together, the law $a^x/a^y = a^{x-y}$ and the definition $x^n = e^{n \ln x}$ enable us to establish the Power Rule for differentiation in its final form. Differentiating x^n with respect to x gives

$$\begin{aligned}
\frac{d}{dx}x^n &= \frac{d}{dx}e^{n \ln x} && \text{Definition of } x^n,\ x > 0 \\
&= e^{n \ln x} \cdot \frac{d}{dx}(n \ln x) && \text{Chain Rule for } e^u \\
&= x^n \cdot \frac{n}{x} && \text{The definition again} \\
&= nx^{n-1}.
\end{aligned}$$

In short, as long as $x > 0$,

$$\frac{d}{dx}x^n = nx^{n-1}.$$

The Chain Rule extends this equation to the Power Rule's general form.

Power Rule (General Form)
If u is a positive differentiable function of x and n is any real number, then u^n is a differentiable function of x and

$$\frac{d}{dx}u^n = nu^{n-1}\frac{du}{dx}.$$

EXAMPLE 9 Using the Power Rule with Irrational Powers

(a) $\frac{d}{dx}x^{\sqrt{2}} = \sqrt{2}x^{\sqrt{2}-1} \qquad (x > 0)$

(b) $$\frac{d}{dx}(2 + \sin 3x)^{\pi} = \pi(2 + \sin 3x)^{\pi-1}(\cos 3x)\cdot 3$$
$$= 3\pi(2 + \sin 3x)^{\pi-1}(\cos 3x).$$

■

EXERCISES 7.3

Algebraic Calculations with the Exponential and Logarithm

Find simpler expressions for the quantities in Exercises 1–4.

1. a. $e^{\ln 7.2}$ **b.** $e^{-\ln x^2}$ **c.** $e^{\ln x - \ln y}$

2. a. $e^{\ln(x^2+y^2)}$ **b.** $e^{-\ln 0.3}$ **c.** $e^{\ln \pi x - \ln 2}$

3. a. $2\ln\sqrt{e}$ **b.** $\ln(\ln e^e)$ **c.** $\ln(e^{-x^2-y^2})$

4. a. $\ln(e^{\sec\theta})$ **b.** $\ln(e^{(e^x)})$ **c.** $\ln(e^{2\ln x})$

Solving Equations with Logarithmic or Exponential Terms

In Exercises 5–10, solve for y in terms of t or x, as appropriate.

5. $\ln y = 2t + 4$

6. $\ln y = -t + 5$

7. $\ln(y - 40) = 5t$

8. $\ln(1 - 2y) = t$

9. $\ln(y - 1) - \ln 2 = x + \ln x$

10. $\ln(y^2 - 1) - \ln(y + 1) = \ln(\sin x)$

In Exercises 11 and 12, solve for k.

11. a. $e^{2k} = 4$ **b.** $100e^{10k} = 200$ **c.** $e^{k/1000} = a$

12. a. $e^{5k} = \frac{1}{4}$ **b.** $80e^k = 1$ **c.** $e^{(\ln 0.8)k} = 0.8$

In Exercises 13–16, solve for t.

13. a. $e^{-0.3t} = 27$ **b.** $e^{kt} = \frac{1}{2}$ **c.** $e^{(\ln 0.2)t} = 0.4$

14. a. $e^{-0.01t} = 1000$ **b.** $e^{kt} = \frac{1}{10}$ **c.** $e^{(\ln 2)t} = \frac{1}{2}$

15. $e^{\sqrt{t}} = x^2$

16. $e^{(x^2)}e^{(2x+1)} = e^t$

Derivatives

In Exercises 17–36, find the derivative of y with respect to x, t, or θ, as appropriate.

17. $y = e^{-5x}$

18. $y = e^{2x/3}$

19. $y = e^{5-7x}$

20. $y = e^{(4\sqrt{x}+x^2)}$

21. $y = xe^x - e^x$

22. $y = (1 + 2x)e^{-2x}$

23. $y = (x^2 - 2x + 2)e^x$

24. $y = (9x^2 - 6x + 2)e^{3x}$

25. $y = e^{\theta}(\sin\theta + \cos\theta)$

26. $y = \ln(3\theta e^{-\theta})$

27. $y = \cos(e^{-\theta^2})$

28. $y = \theta^3 e^{-2\theta}\cos 5\theta$

29. $y = \ln(3te^{-t})$

30. $y = \ln(2e^{-t}\sin t)$

31. $y = \ln\left(\frac{e^{\theta}}{1 + e^{\theta}}\right)$

32. $y = \ln\left(\frac{\sqrt{\theta}}{1 + \sqrt{\theta}}\right)$

33. $y = e^{(\cos t + \ln t)}$

34. $y = e^{\sin t}(\ln t^2 + 1)$

35. $y = \int_0^{\ln x} \sin e^t\, dt$

36. $y = \int_{e^{4\sqrt{x}}}^{e^{2x}} \ln t\, dt$

In Exercises 37–40, find dy/dx.

37. $\ln y = e^y \sin x$

38. $\ln xy = e^{x+y}$

39. $e^{2x} = \sin(x + 3y)$

40. $\tan y = e^x + \ln x$

Integrals

Evaluate the integrals in Exercises 41–62.

41. $\int (e^{3x} + 5e^{-x})\, dx$

42. $\int (2e^x - 3e^{-2x})\, dx$

43. $\int_{\ln 2}^{\ln 3} e^x\, dx$

44. $\int_{-\ln 2}^{0} e^{-x}\, dx$

45. $\int 8e^{(x+1)}\, dx$

46. $\int 2e^{(2x-1)}\, dx$

47. $\int_{\ln 4}^{\ln 9} e^{x/2}\, dx$

48. $\int_0^{\ln 16} e^{x/4}\, dx$

49. $\int \frac{e^{\sqrt{r}}}{\sqrt{r}}\, dr$

50. $\int \frac{e^{-\sqrt{r}}}{\sqrt{r}}\, dr$

51. $\int 2t\, e^{-t^2}\, dt$

52. $\int t^3 e^{(t^4)}\, dt$

53. $\int \frac{e^{1/x}}{x^2}\, dx$

54. $\int \frac{e^{-1/x^2}}{x^3}\, dx$

55. $\int_0^{\pi/4} (1 + e^{\tan\theta})\sec^2\theta\, d\theta$

56. $\int_{\pi/4}^{\pi/2} (1 + e^{\cot\theta})\csc^2\theta\, d\theta$

57. $\int e^{\sec \pi t}\sec \pi t \tan \pi t\, dt$

58. $\int e^{\csc(\pi+t)}\csc(\pi + t)\cot(\pi + t)\, dt$

59. $\int_{\ln(\pi/6)}^{\ln(\pi/2)} 2e^v \cos e^v \, dv$

60. $\int_0^{\sqrt{\ln \pi}} 2x\, e^{x^2} \cos(e^{x^2})\, dx$

61. $\int \frac{e^r}{1 + e^r}\, dr$

62. $\int \frac{dx}{1 + e^x}$

Initial Value Problems

Solve the initial value problems in Exercises 63–66.

63. $\frac{dy}{dt} = e^t \sin(e^t - 2), \quad y(\ln 2) = 0$

64. $\frac{dy}{dt} = e^{-t} \sec^2(\pi e^{-t}), \quad y(\ln 4) = 2/\pi$

65. $\frac{d^2y}{dx^2} = 2e^{-x}, \quad y(0) = 1 \quad \text{and} \quad y'(0) = 0$

66. $\frac{d^2y}{dt^2} = 1 - e^{2t}, \quad y(1) = -1 \quad \text{and} \quad y'(1) = 0$

Theory and Applications

67. Find the absolute maximum and minimum values of $f(x) = e^x - 2x$ on $[0, 1]$.

68. Where does the periodic function $f(x) = 2e^{\sin(x/2)}$ take on its extreme values and what are these values?

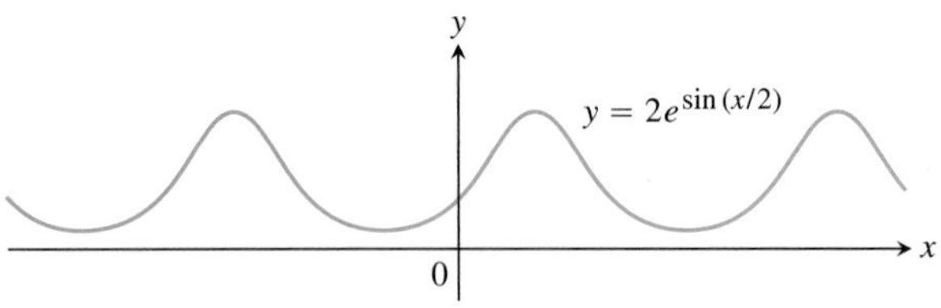

69. Find the absolute maximum value of $f(x) = x^2 \ln(1/x)$ and say where it is assumed.

T 70. Graph $f(x) = (x - 3)^2 e^x$ and its first derivative together. Comment on the behavior of f in relation to the signs and values of f'. Identify significant points on the graphs with calculus, as necessary.

71. Find the area of the "triangular" region in the first quadrant that is bounded above by the curve $y = e^{2x}$, below by the curve $y = e^x$, and on the right by the line $x = \ln 3$.

72. Find the area of the "triangular" region in the first quadrant that is bounded above by the curve $y = e^{x/2}$, below by the curve $y = e^{-x/2}$, and on the right by the line $x = 2\ln 2$.

73. Find a curve through the origin in the xy-plane whose length from $x = 0$ to $x = 1$ is

$$L = \int_0^1 \sqrt{1 + \frac{1}{4}e^x}\, dx.$$

74. Find the area of the surface generated by revolving the curve $x = (e^y + e^{-y})/2$, $0 \le y \le \ln 2$, about the y-axis.

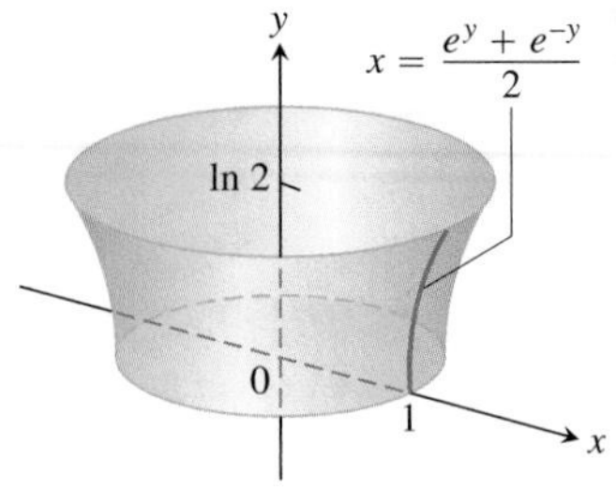

75. **a.** Show that $\int \ln x\, dx = x \ln x - x + C$.

b. Find the average value of $\ln x$ over $[1, e]$.

76. Find the average value of $f(x) = 1/x$ on $[1, 2]$.

77. **The linearization of e^x at $x = 0$**

a. Derive the linear approximation $e^x \approx 1 + x$ at $x = 0$.

T **b.** Estimate to five decimal places the magnitude of the error involved in replacing e^x by $1 + x$ on the interval $[0, 0.2]$.

T **c.** Graph e^x and $1 + x$ together for $-2 \le x \le 2$. Use different colors, if available. On what intervals does the approximation appear to overestimate e^x? Underestimate e^x?

78. **Laws of Exponents**

a. Starting with the equation $e^{x_1}e^{x_2} = e^{x_1 + x_2}$, derived in the text, show that $e^{-x} = 1/e^x$ for any real number x. Then show that $e^{x_1}/e^{x_2} = e^{x_1 - x_2}$ for any numbers x_1 and x_2.

b. Show that $(e^{x_1})^{x_2} = e^{x_1 x_2} = (e^{x_2})^{x_1}$ for any numbers x_1 and x_2.

T 79. **A decimal representation of e** Find e to as many decimal places as your calculator allows by solving the equation $\ln x = 1$.

T 80. **The inverse relation between e^x and $\ln x$** Find out how good your calculator is at evaluating the composites

$$e^{\ln x} \quad \text{and} \quad \ln(e^x).$$

81. Show that for any number $a > 1$

$$\int_1^a \ln x\, dx + \int_0^{\ln a} e^y\, dy = a \ln a.$$

(See accompanying figure.)

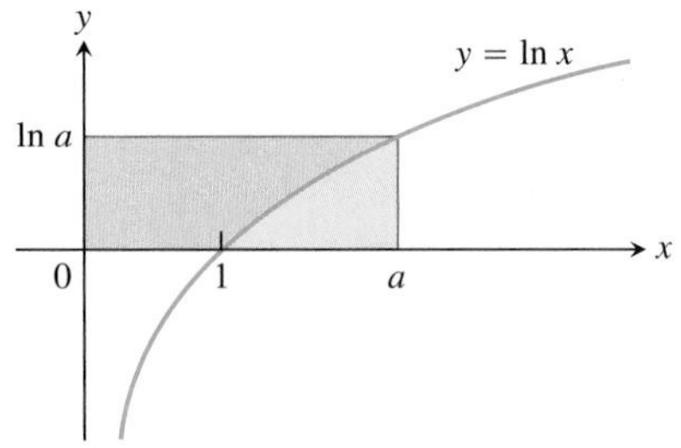

82. **The geometric, logarithmic, and arithmetic mean inequality**

a. Show that the graph of e^x is concave up over every interval of x-values.

b. Show, by reference to the accompanying figure, that if $0 < a < b$ then

$$e^{(\ln a+\ln b)/2}\cdot(\ln b - \ln a) < \int_{\ln a}^{\ln b} e^x\,dx < \frac{e^{\ln a} + e^{\ln b}}{2}\cdot(\ln b - \ln a).$$

c. Use the inequality in part (b) to conclude that

$$\sqrt{ab} < \frac{b - a}{\ln b - \ln a} < \frac{a + b}{2}.$$

This inequality says that the geometric mean of two positive numbers is less than their logarithmic mean, which in turn is less than their arithmetic mean.

(For more about this inequality, see "The Geometric, Logarithmic, and Arithmetic Mean Inequality" by Frank Burk, *American Mathematical Monthly*, Vol. 94, No. 6, June–July 1987, pp. 527–528.)

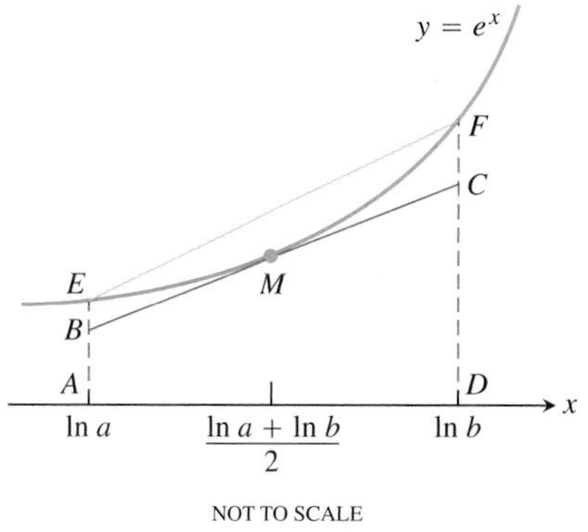

7.4 a^x and $\log_a x$

We have defined general exponential functions such as 2^x, 10^x, and π^x. In this section we compute their derivatives and integrals. We also define the general logarithmic functions such as $\log_2 x$, $\log_{10} x$, and $\log_\pi x$, and find their derivatives and integrals as well.

The Derivative of a^u

We start with the definition $a^x = e^{x \ln a}$:

$$\frac{d}{dx}a^x = \frac{d}{dx}e^{x\ln a} = e^{x\ln a}\cdot\frac{d}{dx}(x\ln a) \qquad \frac{d}{dx}e^u = e^u\frac{du}{dx}$$

$$= a^x \ln a.$$

If $a > 0$, then

$$\frac{d}{dx}a^x = a^x \ln a.$$

With the Chain Rule, we get a more general form.

> If $a > 0$ and u is a differentiable function of x, then a^u is a differentiable function of x and
>
> $$\frac{d}{dx}a^u = a^u \ln a \,\frac{du}{dx}. \tag{1}$$

These equations show why e^x is the exponential function preferred in calculus. If $a = e$, then $\ln a = 1$ and the derivative of a^x simplifies to

$$\frac{d}{dx}e^x = e^x \ln e = e^x.$$

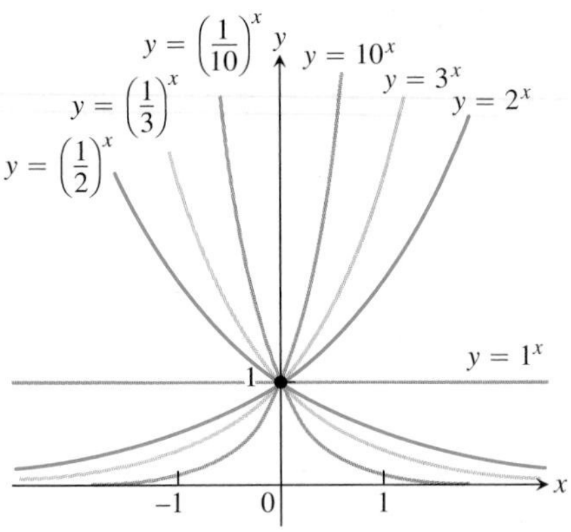

FIGURE 7.12 Exponential functions decrease if $0 < a < 1$ and increase if $a > 1$. As $x \to \infty$, we have $a^x \to 0$ if $0 < a < 1$ and $a^x \to \infty$ if $a > 1$. As $x \to -\infty$, we have $a^x \to \infty$ if $0 < a < 1$ and $a^x \to 0$ if $a > 1$.

EXAMPLE 1 Differentiating General Exponential Functions

(a) $\dfrac{d}{dx} 3^x = 3^x \ln 3$

(b) $\dfrac{d}{dx} 3^{-x} = 3^{-x}(\ln 3)\dfrac{d}{dx}(-x) = -3^{-x}\ln 3$

(c) $\dfrac{d}{dx} 3^{\sin x} = 3^{\sin x}(\ln 3)\dfrac{d}{dx}(\sin x) = 3^{\sin x}(\ln 3)\cos x$ ■

From Equation (1), we see that the derivative of a^x is positive if $\ln a > 0$, or $a > 1$, and negative if $\ln a < 0$, or $0 < a < 1$. Thus, a^x is an increasing function of x if $a > 1$ and a decreasing function of x if $0 < a < 1$. In each case, a^x is one-to-one. The second derivative

$$\frac{d^2}{dx^2}(a^x) = \frac{d}{dx}(a^x \ln a) = (\ln a)^2 a^x$$

is positive for all x, so the graph of a^x is concave up on every interval of the real line (Figure 7.12).

Other Power Functions

The ability to raise positive numbers to arbitrary real powers makes it possible to define functions like x^x and $x^{\ln x}$ for $x > 0$. We find the derivatives of such functions by rewriting the functions as powers of e.

EXAMPLE 2 Differentiating a General Power Function

Find dy/dx if $y = x^x$, $x > 0$.

Solution Write x^x as a power of e:

$$y = x^x = e^{x \ln x}. \qquad a^x \text{ with } a = x.$$

Then differentiate as usual:

$$\begin{aligned}\frac{dy}{dx} &= \frac{d}{dx}e^{x\ln x}\\ &= e^{x\ln x}\frac{d}{dx}(x\ln x)\\ &= x^x\left(x\cdot\frac{1}{x} + \ln x\right)\\ &= x^x(1 + \ln x).\end{aligned}$$ ■

The Integral of a^u

If $a \neq 1$, so that $\ln a \neq 0$, we can divide both sides of Equation (1) by $\ln a$ to obtain

$$a^u \frac{du}{dx} = \frac{1}{\ln a}\frac{d}{dx}(a^u).$$

Integrating with respect to x then gives

$$\int a^u \frac{du}{dx}\,dx = \int \frac{1}{\ln a}\frac{d}{dx}(a^u)\,dx = \frac{1}{\ln a}\int \frac{d}{dx}(a^u)\,dx = \frac{1}{\ln a}a^u + C.$$

Writing the first integral in differential form gives

$$\int a^u\,du = \frac{a^u}{\ln a} + C. \tag{2}$$

EXAMPLE 3 Integrating General Exponential Functions

(a) $\displaystyle\int 2^x\,dx = \frac{2^x}{\ln 2} + C$ Eq. (2) with $a = 2$, $u = x$

(b) $\displaystyle\int 2^{\sin x}\cos x\,dx$

$\displaystyle= \int 2^u\,du = \frac{2^u}{\ln 2} + C$ $u = \sin x$, $du = \cos x\,dx$, and Eq. (2)

$\displaystyle= \frac{2^{\sin x}}{\ln 2} + C$ u replaced by $\sin x$ ■

Logarithms with Base *a*

As we saw earlier, if a is any positive number other than 1, the function a^x is one-to-one and has a nonzero derivative at every point. It therefore has a differentiable inverse. We call the inverse the **logarithm of x with base a** and denote it by $\log_a x$.

DEFINITION $\log_a x$

For any positive number $a \neq 1$,

$\log_a x$ is the inverse function of a^x.

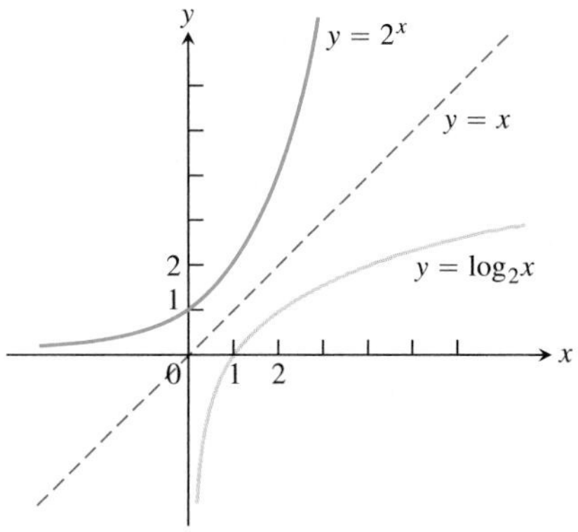

FIGURE 7.13 The graph of 2^x and its inverse, $\log_2 x$.

The graph of $y = \log_a x$ can be obtained by reflecting the graph of $y = a^x$ across the 45° line $y = x$ (Figure 7.13). When $a = e$, we have $\log_e x =$ inverse of $e^x = \ln x$. Since $\log_a x$ and a^x are inverses of one another, composing them in either order gives the identity function.

Inverse Equations for a^x and $\log_a x$

$$a^{\log_a x} = x \qquad (x > 0) \tag{3}$$

$$\log_a(a^x) = x \qquad (\text{all } x) \tag{4}$$

EXAMPLE 4 Applying the Inverse Equations

(a) $\log_2(2^5) = 5$ **(b)** $\log_{10}(10^{-7}) = -7$

(c) $2^{\log_2(3)} = 3$ **(d)** $10^{\log_{10}(4)} = 4$ ■

Evaluation of $\log_a x$

The evaluation of $\log_a x$ is simplified by the observation that $\log_a x$ is a numerical multiple of $\ln x$.

$$\log_a x = \frac{1}{\ln a} \cdot \ln x = \frac{\ln x}{\ln a} \tag{5}$$

We can derive this equation from Equation (3):

$$\begin{aligned} a^{\log_a(x)} &= x && \text{Eq. (3)} \\ \ln a^{\log_a(x)} &= \ln x && \text{Take the natural logarithm of both sides.} \\ \log_a(x) \cdot \ln a &= \ln x && \text{The Power Rule in Theorem 2} \\ \log_a x &= \frac{\ln x}{\ln a} && \text{Solve for } \log_a x. \end{aligned}$$

For example,

$$\log_{10} 2 = \frac{\ln 2}{\ln 10} \approx \frac{0.69315}{2.30259} \approx 0.30103$$

The arithmetic rules satisfied by $\log_a x$ are the same as the ones for $\ln x$ (Theorem 2). These rules, given in Table 7.2, can be proved by dividing the corresponding rules for the natural logarithm function by $\ln a$. For example,

$$\begin{aligned} \ln xy &= \ln x + \ln y && \text{Rule 1 for natural logarithms } \ldots \\ \frac{\ln xy}{\ln a} &= \frac{\ln x}{\ln a} + \frac{\ln y}{\ln a} && \ldots \text{ divided by } \ln a \ldots \\ \log_a xy &= \log_a x + \log_a y. && \ldots \text{ gives Rule 1 for base } a \text{ logarithms.} \end{aligned}$$

TABLE 7.2 Rules for base a logarithms

For any numbers $x > 0$ and $y > 0$,

1. *Product Rule:*
   $\log_a xy = \log_a x + \log_a y$
2. *Quotient Rule:*
   $\log_a \frac{x}{y} = \log_a x - \log_a y$
3. *Reciprocal Rule:*
   $\log_a \frac{1}{y} = -\log_a y$
4. *Power Rule:*
   $\log_a x^y = y \log_a x$

Derivatives and Integrals Involving $\log_a x$

To find derivatives or integrals involving base a logarithms, we convert them to natural logarithms.

If u is a positive differentiable function of x, then

$$\frac{d}{dx}(\log_a u) = \frac{d}{dx}\left(\frac{\ln u}{\ln a}\right) = \frac{1}{\ln a}\frac{d}{dx}(\ln u) = \frac{1}{\ln a} \cdot \frac{1}{u}\frac{du}{dx}.$$

$$\frac{d}{dx}(\log_a u) = \frac{1}{\ln a} \cdot \frac{1}{u}\frac{du}{dx}$$

EXAMPLE 5

(a) $$\frac{d}{dx}\log_{10}(3x+1) = \frac{1}{\ln 10}\cdot\frac{1}{3x+1}\frac{d}{dx}(3x+1) = \frac{3}{(\ln 10)(3x+1)}$$

(b) $$\int \frac{\log_2 x}{x}\,dx = \frac{1}{\ln 2}\int\frac{\ln x}{x}\,dx \qquad \log_2 x = \frac{\ln x}{\ln 2}$$

$$= \frac{1}{\ln 2}\int u\,du \qquad u = \ln x,\quad du = \frac{1}{x}dx$$

$$= \frac{1}{\ln 2}\frac{u^2}{2} + C = \frac{1}{\ln 2}\frac{(\ln x)^2}{2} + C = \frac{(\ln x)^2}{2\ln 2} + C$$ ■

Indeterminate Powers

Limits that lead to the indeterminate forms 1^∞, 0^0, and ∞^0 can sometimes be handled by first taking the logarithm of the function. We use l'Hôpital's Rule to find the limit of the logarithm expression and then exponentiate the result to find the original function limit. This procedure is formulated as follows.

> If $\lim_{x\to a}\ln f(x) = L$, then
>
> $$\lim_{x\to a} f(x) = \lim_{x\to a} e^{\ln f(x)} = e^L.$$
>
> Here a may be either finite or infinite.

EXAMPLE 6 Working with the Indeterminate Form 1^∞

Apply l'Hôpital's Rule to show that $\lim_{x\to 0^+}(1+x)^{1/x} = e$.

Solution The limit leads to the indeterminate form 1^∞. We let $f(x) = (1+x)^{1/x}$ and find $\lim_{x\to 0^+}\ln f(x)$. Since

$$\ln f(x) = \ln(1+x)^{1/x} = \frac{1}{x}\ln(1+x),$$

l' Hôpital's Rule now applies to give

$$\lim_{x\to 0^+}\ln f(x) = \lim_{x\to 0^+}\frac{\ln(1+x)}{x} \qquad \frac{0}{0}$$

$$= \lim_{x\to 0^+}\frac{\frac{1}{1+x}}{1}$$

$$= \frac{1}{1} = 1.$$

Therefore,

$$\lim_{x\to 0^+}(1+x)^{1/x} = \lim_{x\to 0^+} f(x) = \lim_{x\to 0^+} e^{\ln f(x)} = e^1 = e.$$ ■

EXAMPLE 7 An Indeterminant Form ∞^0

Find $\lim_{x\to\infty} x^{1/x}$.

Solution The limit leads to the indeterminate form ∞^0. We let $f(x) = x^{1/x}$ and find $\lim_{x\to\infty} \ln f(x)$. Since

$$\ln f(x) = \ln x^{1/x} = \frac{\ln x}{x},$$

l'Hôpital's Rule gives

$$\begin{aligned}\lim_{x\to\infty} \ln f(x) &= \lim_{x\to\infty} \frac{\ln x}{x} \qquad \frac{\infty}{\infty}\\ &= \lim_{x\to\infty} \frac{1/x}{1}\\ &= \frac{0}{1} = 0.\end{aligned}$$

Therefore,

$$\lim_{x\to\infty} x^{1/x} = \lim_{x\to\infty} f(x) = \lim_{x\to\infty} e^{\ln f(x)} = e^0 = 1. \qquad \blacksquare$$

EXERCISES 7.4

Algebraic Calculations With a^x and $\log_a x$

Simplify the expressions in Exercises 1–4.

1. a. $5^{\log_5 7}$ **b.** $8^{\log_8 \sqrt{2}}$ **c.** $1.3^{\log_{1.3} 75}$

d. $\log_4 16$ **e.** $\log_3 \sqrt{3}$ **f.** $\log_4 \left(\frac{1}{4}\right)$

2. a. $2^{\log_2 3}$ **b.** $10^{\log_{10}(1/2)}$ **c.** $\pi^{\log_\pi 7}$

d. $\log_{11} 121$ **e.** $\log_{121} 11$ **f.** $\log_3 \left(\frac{1}{9}\right)$

3. a. $2^{\log_4 x}$ **b.** $9^{\log_3 x}$ **c.** $\log_2 \left(e^{(\ln 2)(\sin x)}\right)$

4. a. $25^{\log_5 (3x^2)}$ **b.** $\log_e (e^x)$ **c.** $\log_4 \left(2^{e^x \sin x}\right)$

Express the ratios in Exercises 5 and 6 as ratios of natural logarithms and simplify.

5. a. $\dfrac{\log_2 x}{\log_3 x}$ **b.** $\dfrac{\log_2 x}{\log_8 x}$ **c.** $\dfrac{\log_x a}{\log_{x^2} a}$

6. a. $\dfrac{\log_9 x}{\log_3 x}$ **b.** $\dfrac{\log_{\sqrt{10}} x}{\log_{\sqrt{2}} x}$ **c.** $\dfrac{\log_a b}{\log_b a}$

Solve the equations in Exercises 7–10 for x.

7. $3^{\log_3 (7)} + 2^{\log_2 (5)} = 5^{\log_5 (x)}$

8. $8^{\log_8 (3)} - e^{\ln 5} = x^2 - 7^{\log_7 (3x)}$

9. $3^{\log_3 (x^2)} = 5e^{\ln x} - 3 \cdot 10^{\log_{10} (2)}$

10. $\ln e + 4^{-2 \log_4 (x)} = \dfrac{1}{x} \log_{10} (100)$

Derivatives

In Exercises 11–38, find the derivative of y with respect to the given independent variable.

11. $y = 2^x$ **12.** $y = 3^{-x}$

13. $y = 5^{\sqrt{s}}$ **14.** $y = 2^{(s^2)}$

15. $y = x^\pi$ **16.** $y = t^{1-e}$

17. $y = (\cos \theta)^{\sqrt{2}}$ **18.** $y = (\ln \theta)^\pi$

19. $y = 7^{\sec \theta} \ln 7$ **20.** $y = 3^{\tan \theta} \ln 3$

21. $y = 2^{\sin 3t}$ **22.** $y = 5^{-\cos 2t}$

23. $y = \log_2 5\theta$ **24.** $y = \log_3(1 + \theta \ln 3)$

25. $y = \log_4 x + \log_4 x^2$ **26.** $y = \log_{25} e^x - \log_5 \sqrt{x}$

27. $y = \log_2 r \cdot \log_4 r$ **28.** $y = \log_3 r \cdot \log_9 r$

29. $y = \log_3 \left(\left(\dfrac{x+1}{x-1}\right)^{\ln 3}\right)$ **30.** $y = \log_5 \sqrt{\left(\dfrac{7x}{3x+2}\right)^{\ln 5}}$

31. $y = \theta \sin (\log_7 \theta)$ **32.** $y = \log_7 \left(\dfrac{\sin \theta \cos \theta}{e^\theta 2^\theta}\right)$

33. $y = \log_5 e^x$ **34.** $y = \log_2 \left(\dfrac{x^2 e^2}{2\sqrt{x+1}}\right)$

35. $y = 3^{\log_2 t}$ **36.** $y = 3 \log_8 (\log_2 t)$

37. $y = \log_2 (8t^{\ln 2})$ **38.** $y = t \log_3 \left(e^{(\sin t)(\ln 3)}\right)$

Logarithmic Differentiation

In Exercises 39–46, use logarithmic differentiation to find the derivative of y with respect to the given independent variable.

39. $y = (x+1)^x$ **40.** $y = x^{(x+1)}$

41. $y = (\sqrt{t})^t$ **42.** $y = t^{\sqrt{t}}$

43. $y = (\sin x)^x$ **44.** $y = x^{\sin x}$

45. $y = x^{\ln x}$ **46.** $y = (\ln x)^{\ln x}$

Integration

Evaluate the integrals in Exercises 47–56.

47. $\displaystyle\int 5^x \, dx$ **48.** $\displaystyle\int (1.3)^x \, dx$

49. $\int_0^1 2^{-\theta}\,d\theta$

50. $\int_{-2}^0 5^{-\theta}\,d\theta$

51. $\int_1^{\sqrt{2}} x2^{(x^2)}\,dx$

52. $\int_1^4 \frac{2^{\sqrt{x}}}{\sqrt{x}}\,dx$

53. $\int_0^{\pi/2} 7^{\cos t}\sin t\,dt$

54. $\int_0^{\pi/4}\left(\frac{1}{3}\right)^{\tan t}\sec^2 t\,dt$

55. $\int_2^4 x^{2x}(1+\ln x)\,dx$

56. $\int_1^2 \frac{2^{\ln x}}{x}\,dx$

Evaluate the integrals in Exercises 57–60.

57. $\int 3x^{\sqrt{3}}\,dx$

58. $\int x^{\sqrt{2}-1}\,dx$

59. $\int_0^3 (\sqrt{2}+1)x^{\sqrt{2}}\,dx$

60. $\int_1^e x^{(\ln 2)-1}\,dx$

Evaluate the integrals in Exercises 61–70.

61. $\int \frac{\log_{10} x}{x}\,dx$

62. $\int_1^4 \frac{\log_2 x}{x}\,dx$

63. $\int_1^4 \frac{\ln 2 \log_2 x}{x}\,dx$

64. $\int_1^e \frac{2\ln 10 \log_{10} x}{x}\,dx$

65. $\int_0^2 \frac{\log_2(x+2)}{x+2}\,dx$

66. $\int_{1/10}^{10} \frac{\log_{10}(10x)}{x}\,dx$

67. $\int_0^9 \frac{2\log_{10}(x+1)}{x+1}\,dx$

68. $\int_2^3 \frac{2\log_2(x-1)}{x-1}\,dx$

69. $\int \frac{dx}{x\log_{10} x}$

70. $\int \frac{dx}{x(\log_8 x)^2}$

Evaluate the integrals in Exercises 71–74.

71. $\int_1^{\ln x}\frac{1}{t}\,dt,\quad x>1$

72. $\int_1^{e^x}\frac{1}{t}\,dt$

73. $\int_1^{1/x}\frac{1}{t}\,dt,\quad x>0$

74. $\frac{1}{\ln a}\int_1^x \frac{1}{t}\,dt,\quad x>0$

Theory and Applications

75. Find the area of the region between the curve $y = 2x/(1+x^2)$ and the interval $-2 \le x \le 2$ of the x-axis.

76. Find the area of the region between the curve $y = 2^{1-x}$ and the interval $-1 \le x \le 1$ of the x-axis.

77. Blood pH The pH of human blood normally falls between 7.37 and 7.44. Find the corresponding bounds for $[H_3O^+]$.

78. Brain fluid pH The cerebrospinal fluid in the brain has a hydronium ion concentration of about $[H_3O^+] = 4.8 \times 10^{-8}$ moles per liter. What is the pH?

79. Audio amplifiers By what factor k do you have to multiply the intensity of I of the sound from your audio amplifier to add 10 dB to the sound level?

80. Audio amplifiers You multiplied the intensity of the sound of your audio system by a factor of 10. By how many decibels did this increase the sound level?

81. In any solution, the product of the hydronium ion concentration $[H_3O^+]$ (moles/L) and the hydroxyl ion concentration $[OH^-]$ (moles/L) is about 10^{-14}.

a. What value of $[H_3O^+]$ minimizes the sum of the concentrations, $S = [H_3O^+] + [OH^-]$? (*Hint:* Change notation. Let $x = [H_3O^+]$.)

b. What is the pH of a solution in which S has this minimum value?

c. What ratio of $[H_3O^+]$ to $[OH^-]$ minimizes S?

82. Could $\log_a b$ possibly equal $1/\log_b a$? Give reasons for your answer.

T **83.** The equation $x^2 = 2^x$ has three solutions: $x = 2$, $x = 4$, and one other. Estimate the third solution as accurately as you can by graphing.

T **84.** Could $x^{\ln 2}$ possibly be the same as $2^{\ln x}$ for $x > 0$? Graph the two functions and explain what you see.

85. The linearization of 2^x

a. Find the linearization of $f(x) = 2^x$ at $x = 0$. Then round its coefficients to two decimal places.

T b. Graph the linearization and function together for $-3 \le x \le 3$ and $-1 \le x \le 1$.

86. The linearization of $\log_3 x$

a. Find the linearization of $f(x) = \log_3 x$ at $x = 3$. Then round its coefficients to two decimal places.

T b. Graph the linearization and function together in the window $0 \le x \le 8$ and $2 \le x \le 4$.

Calculations with Other Bases

T **87.** Most scientific calculators have keys for $\log_{10} x$ and $\ln x$. To find logarithms to other bases, we use the Equation (5), $\log_a x = (\ln x)/(\ln a)$.

Find the following logarithms to five decimal places.

a. $\log_3 8$

b. $\log_7 0.5$

c. $\log_{20} 17$

d. $\log_{0.5} 7$

e. $\ln x$, given that $\log_{10} x = 2.3$

f. $\ln x$, given that $\log_2 x = 1.4$

g. $\ln x$, given that $\log_2 x = -1.5$

h. $\ln x$, given that $\log_{10} x = -0.7$

88. Conversion factors

a. Show that the equation for converting base 10 logarithms to base 2 logarithms is

$$\log_2 x = \frac{\ln 10}{\ln 2}\log_{10} x.$$

b. Show that the equation for converting base a logarithms to base b logarithms is

$$\log_b x = \frac{\ln a}{\ln b}\log_a x.$$

89. Orthogonal families of curves Prove that all curves in the family

$$y = -\frac{1}{2}x^2 + k$$

(k any constant) are perpendicular to all curves in the family $y = \ln x + c$ (c any constant) at their points of intersection. (See the accompanying figure.)

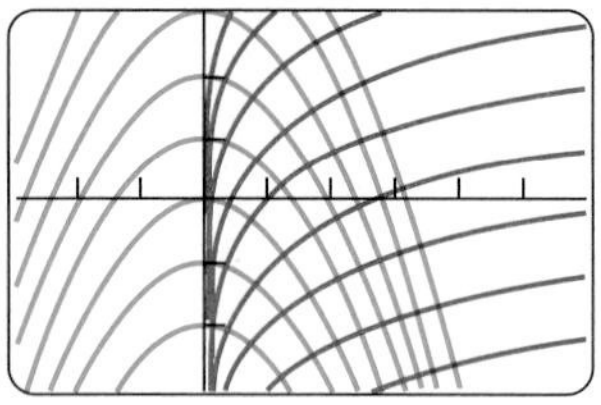

T **90. Which is bigger, π^e or e^π?** Calculators have taken some of the mystery out of this once-challenging question. (Go ahead and check; you will see that it is a surprisingly close call.) You can answer the question without a calculator, though.

a. Find an equation for the line through the origin tangent to the graph of $y = \ln x$.

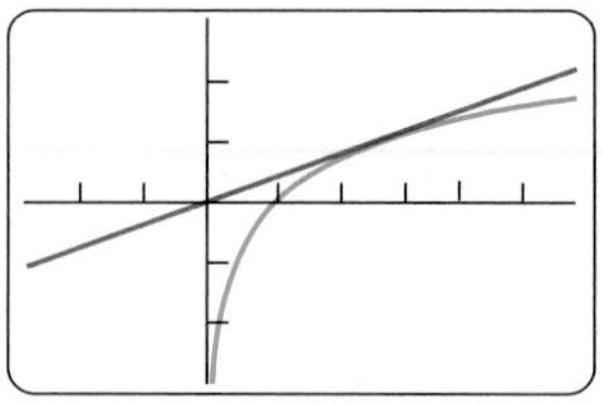

[–3, 6] by [–3, 3]

b. Give an argument based on the graphs of $y = \ln x$ and the tangent line to explain why $\ln x < x/e$ for all positive $x \neq e$.

c. Show that $\ln(x^e) < x$ for all positive $x \neq e$.

d. Conclude that $x^e < e^x$ for all positive $x \neq e$.

e. So which bigger, π^e or e^π?

Find the limits in Exercise 91–100.

91. $\lim_{x \to 1^+} x^{1/(1-x)}$

92. $\lim_{x \to 1^+} x^{1/(x-1)}$

93. $\lim_{x \to \infty} (\ln x)^{1/x}$

94. $\lim_{x \to e^+} (\ln x)^{1/(x-e)}$

95. $\lim_{x \to 0^+} x^{-1/\ln x}$

96. $\lim_{x \to \infty} x^{1/\ln x}$

97. $\lim_{x \to \infty} (1 + 2x)^{1/(2 \ln x)}$

98. $\lim_{x \to 0} (e^x + x)^{1/x}$

99. $\lim_{x \to 0^+} x^x$

100. $\lim_{x \to 0^+} \left(1 + \frac{1}{x}\right)^x$

7.5 Exponential Growth and Decay

Exponential functions increase or decrease very rapidly with changes in the independent variable. They describe growth or decay in a wide variety of natural and industrial situations. The variety of models based on these functions partly accounts for their importance.

The Law of Exponential Change

In modeling many real-world situations, a quantity y increases or decreases at a rate proportional to its size at a given time t. Examples of such quantities include the amount of a decaying radioactive material, funds earning interest in a bank account, the size of a population, and the temperature difference between a hot cup of coffee and the room in which it sits. Such quantities change according to the *law of exponential change*, which we derive in this section.

If the amount present at time $t = 0$ is called y_0, then we can find y as a function of t by solving the following initial value problem:

$$\text{Differential equation:} \quad \frac{dy}{dt} = ky$$
$$\text{Initial condition:} \quad y = y_0 \quad \text{when} \quad t = 0. \tag{1}$$

If y is positive and increasing, then k is positive, and we use Equation (1) to say that the rate of growth is proportional to what has already been accumulated. If y is positive and decreasing, then k is negative, and we use Equation (1) to say that the rate of decay is proportional to the amount still left.

We see right away that the constant function $y = 0$ is a solution of Equation (1) if $y_0 = 0$. To find the nonzero solutions, we divide Equation (1) by y:

$$\frac{1}{y} \cdot \frac{dy}{dt} = k$$

$$\int \frac{1}{y}\frac{dy}{dt}\,dt = \int k\,dt \qquad \text{Integrate with respect to } t;$$

$$\ln|y| = kt + C \qquad \int (1/u)\,du = \ln|u| + C.$$

$$|y| = e^{kt+C} \qquad \text{Exponentiate.}$$

$$|y| = e^{C} \cdot e^{kt} \qquad e^{a+b} = e^{a} \cdot e^{b}$$

$$y = \pm e^{C} e^{kt} \qquad \text{If } |y| = r, \text{ then } y = \pm r.$$

$$y = Ae^{kt}. \qquad A \text{ is a shorter name for } \pm e^{C}.$$

By allowing A to take on the value 0 in addition to all possible values $\pm e^{C}$, we can include the solution $y = 0$ in the formula.

We find the value of A for the initial value problem by solving for A when $y = y_0$ and $t = 0$:

$$y_0 = Ae^{k \cdot 0} = A.$$

The solution of the initial value problem is therefore $y = y_0 e^{kt}$.

Quantities changing in this way are said to undergo **exponential growth** if $k > 0$, and **exponential decay** if $k < 0$.

The Law of Exponential Change

$$y = y_0 e^{kt} \tag{2}$$

Growth: $k > 0$ Decay: $k < 0$

The number k is the **rate constant** of the equation.

The derivation of Equation (2) shows that the only functions that are their own derivatives are constant multiples of the exponential function.

Unlimited Population Growth

Strictly speaking, the number of individuals in a population (of people, plants, foxes, or bacteria, for example) is a discontinuous function of time because it takes on discrete values. However, when the number of individuals becomes large enough, the population can be approximated by a continuous function. Differentiability of the approximating function is another reasonable hypothesis in many settings, allowing for the use of calculus to model and predict population sizes.

If we assume that the proportion of reproducing individuals remains constant and assume a constant fertility, then at any instant t the birth rate is proportional to the number $y(t)$ of individuals present. Let's assume, too, that the death rate of the population is stable and proportional to $y(t)$. If, further, we neglect departures and arrivals, the growth rate

dy/dt is the birth rate minus the death rate, which is the difference of the two proportionalities under our assumptions. In other words, $dy/dt = ky$, so that $y = y_0 e^{kt}$, where y_0 is the size of the population at time $t = 0$. As with all kinds of growth, there may be limitations imposed by the surrounding environment, but we will not go into these here. (This situation is analyzed in Section 9.5.)

In the following example we assume this population model to look at how the number of individuals infected by a disease within a given population decreases as the disease is appropriately treated.

EXAMPLE 1 Reducing the Cases of an Infectious Disease

One model for the way diseases die out when properly treated assumes that the rate dy/dt at which the number of infected people changes is proportional to the number y. The number of people cured is proportional to the number that have the disease. Suppose that in the course of any given year the number of cases of a disease is reduced by 20%. If there are 10,000 cases today, how many years will it take to reduce the number to 1000?

Solution We use the equation $y = y_0 e^{kt}$. There are three things to find: the value of y_0, the value of k, and the time t when $y = 1000$.

The value of y_0. We are free to count time beginning anywhere we want. If we count from today, then $y = 10{,}000$ when $t = 0$, so $y_0 = 10{,}000$. Our equation is now

$$y = 10{,}000 e^{kt}. \tag{3}$$

The value of k. When $t = 1$ year, the number of cases will be 80% of its present value, or 8000. Hence,

$$\begin{aligned} 8000 &= 10{,}000 e^{k(1)} && \text{Eq. (3) with } t = 1 \text{ and } y = 8000 \\ e^k &= 0.8 \\ \ln(e^k) &= \ln 0.8 && \text{Logs of both sides} \\ k &= \ln 0.8 < 0. \end{aligned}$$

At any given time t,

$$y = 10{,}000 e^{(\ln 0.8)t}. \tag{4}$$

The value of t that makes $y = 1000$. We set y equal to 1000 in Equation (4) and solve for t:

$$\begin{aligned} 1000 &= 10{,}000 e^{(\ln 0.8)t} \\ e^{(\ln 0.8)t} &= 0.1 \\ (\ln 0.8)t &= \ln 0.1 && \text{Logs of both sides} \\ t &= \frac{\ln 0.1}{\ln 0.8} \approx 10.32 \text{ years}. \end{aligned}$$

It will take a little more than 10 years to reduce the number of cases to 1000. ■

Continuously Compounded Interest

If you invest an amount A_0 of money at a fixed annual interest rate r (expressed as a decimal) and if interest is added to your account k times a year, the formula for the amount of money you will have at the end of t years is

$$A_t = A_0 \left(1 + \frac{r}{k}\right)^{kt}. \tag{5}$$

The interest might be added ("compounded," bankers say) monthly ($k = 12$), weekly ($k = 52$), daily ($k = 365$), or even more frequently, say by the hour or by the minute. By taking the limit as interest is compounded more and more often, we arrive at the following formula for the amount after t years,

$$\begin{aligned}
\lim_{k\to\infty} A_t &= \lim_{k\to\infty} A_0\left(1+\frac{r}{k}\right)^{kt} \\
&= A_0 \lim_{k\to\infty}\left(1+\frac{r}{k}\right)^{\frac{k}{r}\cdot rt} \\
&= A_0\left[\lim_{\frac{r}{k}\to 0}\left(1+\frac{r}{k}\right)^{\frac{k}{r}}\right]^{rt} && \text{As } k\to\infty,\ \frac{r}{k}\to 0 \\
&= A_0\left[\lim_{x\to 0}(1+x)^{1/x}\right]^{rt} && \text{Substitute } x=\frac{r}{k} \\
&= A_0 e^{rt} && \text{Theorem 4}
\end{aligned}$$

The resulting formula for the amount of money in your account after t years is

$$A(t) = A_0 e^{rt}. \tag{6}$$

Interest paid according to this formula is said to be **compounded continuously**. The number r is called the **continuous interest rate.** The amount of money after t years is calculated with the law of exponential change given in Equation (6).

EXAMPLE 2 A Savings Account

Suppose you deposit \$621 in a bank account that pays 6% compounded continuously. How much money will you have 8 years later?

Solution We use Equation (6) with $A_0 = 621$, $r = 0.06$, and $t = 8$:

$$A(8) = 621e^{(0.06)(8)} = 621e^{0.48} = 1003.58 \qquad \text{Nearest cent}$$

Had the bank paid interest quarterly ($k = 4$ in Equation 5), the amount in your account would have been \$1000.01. Thus the effect of continuous compounding, as compared with quarterly compounding, has been an addition of \$3.57. A bank might decide it would be worth this additional amount to be able to advertise, "We compound interest every second, night and day—better yet, we compound the interest continuously." ■

Radioactivity

Some atoms are unstable and can spontaneously emit mass or radiation. This process is called **radioactive decay**, and an element whose atoms go spontaneously through this process is called **radioactive**. Sometimes when an atom emits some of its mass through this process of radioactivity, the remainder of the atom re-forms to make an atom of some new element. For example, radioactive carbon-14 decays into nitrogen; radium, through a number of intermediate radioactive steps, decays into lead.

Experiments have shown that at any given time the rate at which a radioactive element decays (as measured by the number of nuclei that change per unit time) is approximately proportional to the number of radioactive nuclei present. Thus, the decay of a radioactive element is described by the equation $dy/dt = -ky$, $k > 0$. It is conventional to use

For radon-222 gas, t is measured in days and $k = 0.18$. For radium-226, which used to be painted on watch dials to make them glow at night (a dangerous practice), t is measured in years and $k = 4.3 \times 10^{-4}$.

$-k(k > 0)$ here instead of $k(k < 0)$ to emphasize that y is decreasing. If y_0 is the number of radioactive nuclei present at time zero, the number still present at any later time t will be

$$y = y_0 e^{-kt}, \qquad k > 0.$$

EXAMPLE 3 Half-Life of a Radioactive Element

The **half-life** of a radioactive element is the time required for half of the radioactive nuclei present in a sample to decay. It is an interesting fact that the half-life is a constant that does not depend on the number of radioactive nuclei initially present in the sample, but only on the radioactive substance.

To see why, let y_0 be the number of radioactive nuclei initially present in the sample. Then the number y present at any later time t will be $y = y_0 e^{-kt}$. We seek the value of t at which the number of radioactive nuclei present equals half the original number:

$$\begin{aligned} y_0 e^{-kt} &= \frac{1}{2} y_0 \\ e^{-kt} &= \frac{1}{2} \\ -kt &= \ln \frac{1}{2} = -\ln 2 && \text{Reciprocal Rule for logarithms} \\ t &= \frac{\ln 2}{k} \end{aligned}$$

This value of t is the half-life of the element. It depends only on the value of k; the number y_0 does not enter in.

$$\text{Half-life} = \frac{\ln 2}{k} \tag{7}$$

■

EXAMPLE 4 Half-Life of Polonium-210

The effective radioactive lifetime of polonium-210 is so short we measure it in days rather than years. The number of radioactive atoms remaining after t days in a sample that starts with y_0 radioactive atoms is

$$y = y_0 e^{-5 \times 10^{-3} t}.$$

Find the element's half-life.

Solution

$$\begin{aligned} \text{Half-life} &= \frac{\ln 2}{k} && \text{Eq. (7)} \\ &= \frac{\ln 2}{5 \times 10^{-3}} && \text{The } k \text{ from polonium's decay equation} \\ &\approx 139 \text{ days} \end{aligned}$$

■

EXAMPLE 5 Carbon-14 Dating

The decay of radioactive elements can sometimes be used to date events from the Earth's past. In a living organism, the ratio of radioactive carbon, carbon-14, to ordinary carbon stays fairly constant during the lifetime of the organism, being approximately equal to the ratio in the organism's surroundings at the time. After the organism's death, however, no new carbon is ingested, and the proportion of carbon-14 in the organism's remains decreases as the carbon-14 decays.

Scientists who do carbon-14 dating use a figure of 5700 years for its half-life (more about carbon-14 dating in the exercises). Find the age of a sample in which 10% of the radioactive nuclei originally present have decayed.

Solution We use the decay equation $y = y_0 e^{-kt}$. There are two things to find: the value of k and the value of t when y is $0.9y_0$ (90% of the radioactive nuclei are still present). That is, find t when $y_0 e^{-kt} = 0.9y_0$, or $e^{-kt} = 0.9$.

The value of k. We use the half-life Equation (7):

$$k = \frac{\ln 2}{\text{half-life}} = \frac{\ln 2}{5700} \qquad (\text{about } 1.2 \times 10^{-4})$$

The value of t that makes $e^{-kt} = 0.9$:

$$e^{-kt} = 0.9$$

$$e^{-(\ln 2/5700)t} = 0.9$$

$$-\frac{\ln 2}{5700} t = \ln 0.9 \qquad \text{Logs of both sides}$$

$$t = -\frac{5700 \ln 0.9}{\ln 2} \approx 866 \text{ years}.$$

The sample is about 866 years old. ■

Heat Transfer: Newton's Law of Cooling

Hot soup left in a tin cup cools to the temperature of the surrounding air. A hot silver ingot immersed in a large tub of water cools to the temperature of the surrounding water. In situations like these, the rate at which an object's temperature is changing at any given time is roughly proportional to the difference between its temperature and the temperature of the surrounding medium. This observation is called *Newton's law of cooling*, although it applies to warming as well, and there is an equation for it.

If H is the temperature of the object at time t and H_S is the constant surrounding temperature, then the differential equation is

$$\frac{dH}{dt} = -k(H - H_S). \tag{8}$$

If we substitute y for $(H - H_S)$, then

$$\begin{aligned} \frac{dy}{dt} &= \frac{d}{dt}(H - H_S) = \frac{dH}{dt} - \frac{d}{dt}(H_S) \\ &= \frac{dH}{dt} - 0 && H_S \text{ is a constant.} \\ &= \frac{dH}{dt} \\ &= -k(H - H_S) && \text{Eq. (8)} \\ &= -ky. && H - H_S = y. \end{aligned}$$

Now we know that the solution of $dy/dt = -ky$ is $y = y_0 e^{-kt}$, where $y(0) = y_0$. Substituting $(H - H_S)$ for y, this says that

$$H - H_S = (H_0 - H_S)e^{-kt}, \tag{9}$$

where H_0 is the temperature at $t = 0$. This is the equation for Newton's Law of Cooling.

EXAMPLE 6 Cooling a Hard-Boiled Egg

A hard-boiled egg at 98°C is put in a sink of 18°C water. After 5 min, the egg's temperature is 38°C. Assuming that the water has not warmed appreciably, how much longer will it take the egg to reach 20°C?

Solution We find how long it would take the egg to cool from 98°C to 20°C and subtract the 5 min that have already elapsed. Using Equation (9) with $H_S = 18$ and $H_0 = 98$, the egg's temperature t min after it is put in the sink is

$$H = 18 + (98 - 18)e^{-kt} = 18 + 80e^{-kt}.$$

To find k, we use the information that $H = 38$ when $t = 5$:

$$\begin{aligned}
38 &= 18 + 80e^{-5k} \\
e^{-5k} &= \frac{1}{4} \\
-5k &= \ln\frac{1}{4} = -\ln 4 \\
k &= \frac{1}{5}\ln 4 = 0.2\ln 4 \qquad \text{(about 0.28)}.
\end{aligned}$$

The egg's temperature at time t is $H = 18 + 80e^{-(0.2\ln 4)t}$. Now find the time t when $H = 20$:

$$\begin{aligned}
20 &= 18 + 80e^{-(0.2\ln 4)t} \\
80e^{-(0.2\ln 4)t} &= 2 \\
e^{-(0.2\ln 4)t} &= \frac{1}{40} \\
-(0.2\ln 4)t &= \ln\frac{1}{40} = -\ln 40 \\
t &= \frac{\ln 40}{0.2\ln 4} \approx 13 \text{ min}.
\end{aligned}$$

The egg's temperature will reach 20°C about 13 min after it is put in the water to cool. Since it took 5 min to reach 38°C, it will take about 8 min more to reach 20°C. ■

EXERCISES 7.5

The answers to most of the following exercises are in terms of logarithms and exponentials. A calculator can be helpful, enabling you to express the answers in decimal form.

1. **Human evolution continues** The analysis of tooth shrinkage by C. Loring Brace and colleagues at the University of Michigan's Museum of Anthropology indicates that human tooth size is continuing to decrease and that the evolutionary process did not come to a halt some 30,000 years ago as many scientists contend. In northern Europeans, for example, tooth size reduction now has a rate of 1% per 1000 years.

a. If t represents time in years and y represents tooth size, use the condition that $y = 0.99y_0$ when $t = 1000$ to find the value of k in the equation $y = y_0 e^{kt}$. Then use this value of k to answer the following questions.

b. In about how many years will human teeth be 90% of their present size?

c. What will be our descendants' tooth size 20,000 years from now (as a percentage of our present tooth size)?

(*Source: LSA Magazine*, Spring 1989, Vol. 12, No. 2, p. 19, Ann Arbor, MI.)

2. **Atmospheric pressure** The earth's atmospheric pressure p is often modeled by assuming that the rate dp/dh at which p changes with the altitude h above sea level is proportional to p. Suppose that the pressure at sea level is 1013 millibars (about 14.7 pounds per square inch) and that the pressure at an altitude of 20 km is 90 millibars.

a. Solve the initial value problem

Differential equation: $dp/dh = kp$ (k a constant)
Initial condition: $p = p_0$ when $h = 0$

to express p in terms of h. Determine the values of p_0 and k from the given altitude-pressure data.

b. What is the atmospheric pressure at $h = 50$ km?

c. At what altitude does the pressure equal 900 millibars?

3. **First-order chemical reactions** In some chemical reactions, the rate at which the amount of a substance changes with time is proportional to the amount present. For the change of δ-glucono lactone into gluconic acid, for example,

$$\frac{dy}{dt} = -0.6y$$

when t is measured in hours. If there are 100 grams of δ-glucono lactone present when $t = 0$, how many grams will be left after the first hour?

4. **The inversion of sugar** The processing of raw sugar has a step called "inversion" that changes the sugar's molecular structure. Once the process has begun, the rate of change of the amount of raw sugar is proportional to the amount of raw sugar remaining. If 1000 kg of raw sugar reduces to 800 kg of raw sugar during the first 10 hours, how much raw sugar will remain after another 14 hours?

5. **Working underwater** The intensity $L(x)$ of light x feet beneath the surface of the ocean satisfies the differential equation

$$\frac{dL}{dx} = -kL.$$

As a diver, you know from experience that diving to 18 ft in the Caribbean Sea cuts the intensity in half. You cannot work without artificial light when the intensity falls below one-tenth of the surface value. About how deep can you expect to work without artificial light?

6. **Voltage in a discharging capacitor** Suppose that electricity is draining from a capacitor at a rate that is proportional to the voltage V across its terminals and that, if t is measured in seconds,

$$\frac{dV}{dt} = -\frac{1}{40}V.$$

Solve this equation for V, using V_0 to denote the value of V when $t = 0$. How long will it take the voltage to drop to 10% of its original value?

7. **Cholera bacteria** Suppose that the bacteria in a colony can grow unchecked, by the law of exponential change. The colony starts with 1 bacterium and doubles every half-hour. How many bacteria will the colony contain at the end of 24 hours? (Under favorable laboratory conditions, the number of cholera bacteria can double every 30 min. In an infected person, many bacteria are destroyed, but this example helps explain why a person who feels well in the morning may be dangerously ill by evening.)

8. **Growth of bacteria** A colony of bacteria is grown under ideal conditions in a laboratory so that the population increases exponentially with time. At the end of 3 hours there are 10,000 bacteria. At the end of 5 hours there are 40,000. How many bacteria were present initially?

9. **The incidence of a disease** (*Continuation of Example 1.*) Suppose that in any given year the number of cases can be reduced by 25% instead of 20%.

a. How long will it take to reduce the number of cases to 1000?

b. How long will it take to eradicate the disease, that is, reduce the number of cases to less than 1?

10. **The U.S. population** The Museum of Science in Boston displays a running total of the U.S. population. On May 11, 1993, the total was increasing at the rate of 1 person every 14 sec. The displayed population figure for 3:45 P.M. that day was 257,313,431.

a. Assuming exponential growth at a constant rate, find the rate constant for the population's growth (people per 365-day year).

b. At this rate, what will the U.S. population be at 3:45 P.M. Boston time on May 11, 2008?

11. **Oil depletion** Suppose the amount of oil pumped from one of the canyon wells in Whittier, California, decreases at the continuous rate of 10% per year. When will the well's output fall to one-fifth of its present value?

12. **Continuous price discounting** To encourage buyers to place 100-unit orders, your firm's sales department applies a continuous discount that makes the unit price a function $p(x)$ of the number of units x ordered. The discount decreases the price at the rate of \$0.01 per unit ordered. The price per unit for a 100-unit order is $p(100) = \$20.09$.

a. Find $p(x)$ by solving the following initial value problem:

Differential equation: $\dfrac{dp}{dx} = -\dfrac{1}{100}p$

Initial condition: $p(100) = 20.09.$

b. Find the unit price $p(10)$ for a 10-unit order and the unit price $p(90)$ for a 90-unit order.

c. The sales department has asked you to find out if it is discounting so much that the firm's revenue, $r(x) = x \cdot p(x)$, will actually be less for a 100-unit order than, say, for a 90-unit order. Reassure them by showing that r has its maximum value at $x = 100$.

T **d.** Graph the revenue function $r(x) = xp(x)$ for $0 \le x \le 200$.

13. Continuously compounded interest You have just placed A_0 dollars in a bank account that pays 4% interest, compounded continuously.

a. How much money will you have in the account in 5 years?

b. How long will it take your money to double? To triple?

14. John Napier's question John Napier (1550–1617), the Scottish laird who invented logarithms, was the first person to answer the question, What happens if you invest an amount of money at 100% interest, compounded continuously?

a. What does happen?

b. How long does it take to triple your money?

c. How much can you earn in a year?

Give reasons for your answers.

15. Benjamin Franklin's will The Franklin Technical Institute of Boston owes its existence to a provision in a codicil to Benjamin Franklin's will. In part the codicil reads:

> I wish to be useful even after my Death, if possible, in forming and advancing other young men that may be serviceable to their Country in both Boston and Philadelphia. To this end I devote Two thousand Pounds Sterling, which I give, one thousand thereof to the Inhabitants of the Town of Boston in Massachusetts, and the other thousand to the inhabitants of the City of Philadelphia, in Trust and for the Uses, Interests and Purposes hereinafter mentioned and declared.

Franklin's plan was to lend money to young apprentices at 5% interest with the provision that each borrower should pay each year along

> . . . with the yearly Interest, one tenth part of the Principal, which sums of Principal and Interest shall be again let to fresh Borrowers. . . . If this plan is executed and succeeds as projected without interruption for one hundred Years, the Sum will then be one hundred and thirty-one thousand Pounds of which I would have the Managers of the Donation to the Inhabitants of the Town of Boston, then lay out at their discretion one hundred thousand Pounds in Public Works. . . . The remaining thirty-one thousand Pounds, I would have continued to be let out on Interest in the manner above directed for another hundred Years. . . . At the end of this second term if no unfortunate accident has prevented the operation the sum will be Four Millions and Sixty-one Thousand Pounds.

It was not always possible to find as many borrowers as Franklin had planned, but the managers of the trust did the best they could. At the end of 100 years from the reception of the Franklin gift, in January 1894, the fund had grown from 1000 pounds to almost exactly 90,000 pounds. In 100 years the original capital had multiplied about 90 times instead of the 131 times Franklin had imagined.

What rate of interest, compounded continuously for 100 years, would have multiplied Benjamin Franklin's original capital by 90?

16. (*Continuation of Exercise 15.*) In Benjamin Franklin's estimate that the original 1000 pounds would grow to 131,000 in 100 years, he was using an annual rate of 5% and compounding once each year. What rate of interest per year when compounded continuously for 100 years would multiply the original amount by 131?

17. Radon-222 The decay equation for radon-222 gas is known to be $y = y_0 e^{-0.18t}$, with t in days. About how long will it take the radon in a sealed sample of air to fall to 90% of its original value?

18. Polonium-210 The half-life of polonium is 139 days, but your sample will not be useful to you after 95% of the radioactive nuclei present on the day the sample arrives has disintegrated. For about how many days after the sample arrives will you be able to use the polonium?

19. The mean life of a radioactive nucleus Physicists using the radioactivity equation $y = y_0 e^{-kt}$ call the number $1/k$ the *mean life* of a radioactive nucleus. The mean life of a radon nucleus is about $1/0.18 = 5.6$ days. The mean life of a carbon-14 nucleus is more than 8000 years. Show that 95% of the radioactive nuclei originally present in a sample will disintegrate within three mean lifetimes, i.e., by time $t = 3/k$. Thus, the mean life of a nucleus gives a quick way to estimate how long the radioactivity of a sample will last.

20. Californium-252 What costs \$27 million per gram and can be used to treat brain cancer, analyze coal for its sulfur content, and detect explosives in luggage? The answer is californium-252, a radioactive isotope so rare that only 8 g of it have been made in the western world since its discovery by Glenn Seaborg in 1950. The half-life of the isotope is 2.645 years—long enough for a useful service life and short enough to have a high radioactivity per unit mass. One microgram of the isotope releases 170 million neutrons per second.

a. What is the value of k in the decay equation for this isotope?

b. What is the isotope's mean life? (See Exercise 19.)

c. How long will it take 95% of a sample's radioactive nuclei to disintegrate?

21. Cooling soup Suppose that a cup of soup cooled from 90°C to 60°C after 10 min in a room whose temperature was 20°C. Use Newton's law of cooling to answer the following questions.

a. How much longer would it take the soup to cool to 35°C?

b. Instead of being left to stand in the room, the cup of 90°C soup is put in a freezer whose temperature is −15°C. How long will it take the soup to cool from 90°C to 35°C?

22. **A beam of unknown temperature** An aluminum beam was brought from the outside cold into a machine shop where the temperature was held at 65°F. After 10 min, the beam warmed to 35°F and after another 10 min it was 50°F. Use Newton's law of cooling to estimate the beam's initial temperature.

23. **Surrounding medium of unknown temperature** A pan of warm water (46°C) was put in a refrigerator. Ten minutes later, the water's temperature was 39°C; 10 min after that, it was 33°C. Use Newton's law of cooling to estimate how cold the refrigerator was.

24. **Silver cooling in air** The temperature of an ingot of silver is 60°C above room temperature right now. Twenty minutes ago, it was 70°C above room temperature. How far above room temperature will the silver be
    a. 15 min from now?
    b. 2 hours from now?
    c. When will the silver be 10°C above room temperature?

25. **The age of Crater Lake** The charcoal from a tree killed in the volcanic eruption that formed Crater Lake in Oregon contained 44.5% of the carbon-14 found in living matter. About how old is Crater Lake?

26. **The sensitivity of carbon-14 dating to measurement** To see the effect of a relatively small error in the estimate of the amount of carbon-14 in a sample being dated, consider this hypothetical situation:
    a. A fossilized bone found in central Illinois in the year A.D. 2000 contains 17% of its original carbon-14 content. Estimate the year the animal died.
    b. Repeat part (a) assuming 18% instead of 17%.
    c. Repeat part (a) assuming 16% instead of 17%.

27. **Art forgery** A painting attributed to Vermeer (1632–1675), which should contain no more than 96.2% of its original carbon-14, contains 99.5% instead. About how old is the forgery?

7.6 Relative Rates of Growth

It is often important in mathematics, computer science, and engineering to compare the rates at which functions of x grow as x becomes large. Exponential functions are important in these comparisons because of their very fast growth, and logarithmic functions because of their very slow growth. In this section we introduce the *little-oh* and *big-oh* notation used to describe the results of these comparisons. We restrict our attention to functions whose values eventually become and remain positive as $x \to \infty$.

Growth Rates of Functions

You may have noticed that exponential functions like 2^x and e^x seem to grow more rapidly as x gets large than do polynomials and rational functions. These exponentials certainly grow more rapidly than x itself, and you can see 2^x outgrowing x^2 as x increases in Figure 7.14. In fact, as $x \to \infty$, the functions 2^x and e^x grow faster than any power of x, even $x^{1{,}000{,}000}$ (Exercise 19).

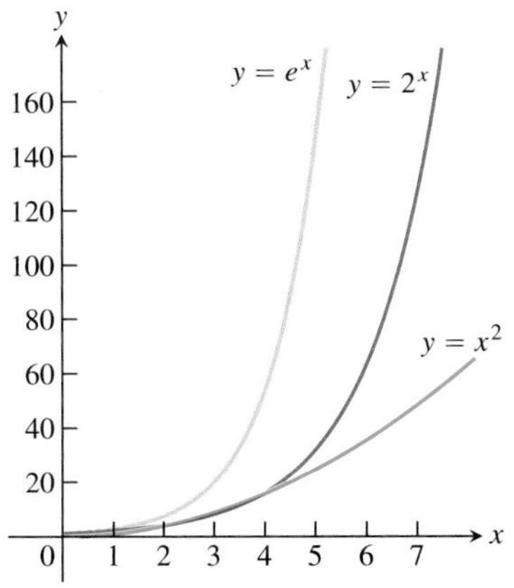

FIGURE 7.14 The graphs of e^x, 2^x, and x^2.

To get a feeling for how rapidly the values of $y = e^x$ grow with increasing x, think of graphing the function on a large blackboard, with the axes scaled in centimeters. At $x = 1$ cm, the graph is $e^1 \approx 3$ cm above the x-axis. At $x = 6$ cm, the graph is $e^6 \approx 403$ cm ≈ 4 m high (it is about to go through the ceiling if it hasn't done so already). At $x = 10$ cm, the graph is $e^{10} \approx 22{,}026$ cm ≈ 220 m high, higher than most buildings. At $x = 24$ cm, the graph is more than halfway to the moon, and at $x = 43$ cm from the origin, the graph is high enough to reach past the sun's closest stellar neighbor, the red dwarf star Proxima Centauri:

$$\begin{aligned} e^{43} &\approx 4.73 \times 10^{18} \text{ cm} \\ &= 4.73 \times 10^{13} \text{ km} \\ &\approx 1.58 \times 10^{8} \text{ light-seconds} \\ &\approx 5.0 \text{ light-years} \end{aligned}$$

In a vacuum, light travels at 300,000 km/sec.

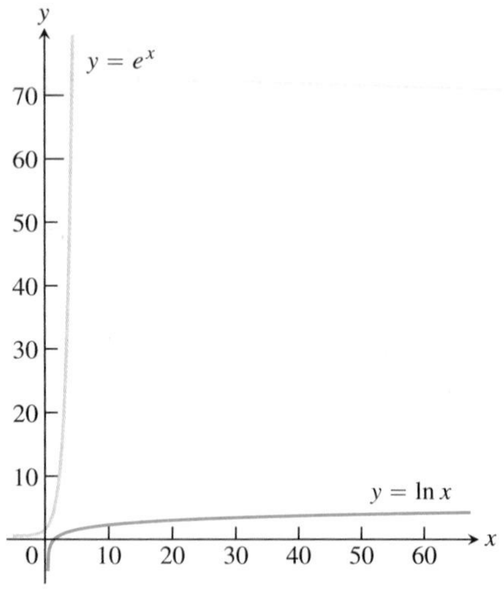

FIGURE 7.15 Scale drawings of the graphs of e^x and $\ln x$.

The distance to Proxima Centauri is about 4.22 light-years. Yet with $x = 43$ cm from the origin, the graph is still less than 2 feet to the right of the y-axis.

In contrast, logarithmic functions like $y = \log_2 x$ and $y = \ln x$ grow more slowly as $x \to \infty$ than any positive power of x (Exercise 21). With axes scaled in centimeters, you have to go nearly 5 light-years out on the x-axis to find a point where the graph of $y = \ln x$ is even $y = 43$ cm high. See Figure 7.15.

These important comparisons of exponential, polynomial, and logarithmic functions can be made precise by defining what it means for a function $f(x)$ to grow faster than another function $g(x)$ as $x \to \infty$.

DEFINITION **Rates of Growth as $x \to \infty$**

Let $f(x)$ and $g(x)$ be positive for x sufficiently large.

1. f **grows faster than** g as $x \to \infty$ if

$$\lim_{x\to\infty} \frac{f(x)}{g(x)} = \infty$$

or, equivalently, if

$$\lim_{x\to\infty} \frac{g(x)}{f(x)} = 0.$$

We also say that g **grows slower than** f as $x \to \infty$.

2. f and g **grow at the same rate** as $x \to \infty$ if

$$\lim_{x\to\infty} \frac{f(x)}{g(x)} = L$$

where L is finite and positive.

According to these definitions, $y = 2x$ does not grow faster than $y = x$. The two functions grow at the same rate because

$$\lim_{x\to\infty} \frac{2x}{x} = \lim_{x\to\infty} 2 = 2,$$

which is a finite, nonzero limit. The reason for this apparent disregard of common sense is that we want "f grows faster than g" to mean that for large x-values g is negligible when compared with f.

EXAMPLE 1 Several Useful Comparisons of Growth Rates

(a) e^x grows faster than x^2 as $x \to \infty$ because

$$\underbrace{\lim_{x\to\infty} \frac{e^x}{x^2}}_{\infty/\infty} = \underbrace{\lim_{x\to\infty} \frac{e^x}{2x}}_{\infty/\infty} = \lim_{x\to\infty} \frac{e^x}{2} = \infty. \qquad \text{Using l'Hôpital's Rule twice}$$

(b) 3^x grows faster than 2^x as $x \to \infty$ because

$$\lim_{x\to\infty} \frac{3^x}{2^x} = \lim_{x\to\infty} \left(\frac{3}{2}\right)^x = \infty.$$

(c) x^2 grows faster than $\ln x$ as $x \to \infty$, because

$$\lim_{x\to\infty} \frac{x^2}{\ln x} = \lim_{x\to\infty} \frac{2x}{1/x} = \lim_{x\to\infty} 2x^2 = \infty. \qquad \text{l'Hôpital's Rule}$$

(d) $\ln x$ grows slower than x as $x \to \infty$ because

$$\lim_{x\to\infty} \frac{\ln x}{x} = \lim_{x\to\infty} \frac{1/x}{1} \qquad \text{l'Hôpital's Rule}$$

$$= \lim_{x\to\infty} \frac{1}{x} = 0.$$

■

EXAMPLE 2 Exponential and Logarithmic Functions with Different Bases

(a) As Example 1b suggests, exponential functions with different bases never grow at the same rate as $x \to \infty$. If $a > b > 0$, then a^x grows faster than b^x. Since $(a/b) > 1$,

$$\lim_{x\to\infty} \frac{a^x}{b^x} = \lim_{x\to\infty} \left(\frac{a}{b}\right)^x = \infty.$$

(b) In contrast to exponential functions, logarithmic functions with different bases a and b always grow at the same rate as $x \to \infty$:

$$\lim_{x\to\infty} \frac{\log_a x}{\log_b x} = \lim_{x\to\infty} \frac{\ln x/\ln a}{\ln x/\ln b} = \frac{\ln b}{\ln a}.$$

The limiting ratio is always finite and never zero. ■

If f grows at the same rate as g as $x \to \infty$, and g grows at the same rate as h as $x \to \infty$, then f grows at the same rate as h as $x \to \infty$. The reason is that

$$\lim_{x\to\infty} \frac{f}{g} = L_1 \qquad \text{and} \qquad \lim_{x\to\infty} \frac{g}{h} = L_2$$

together imply

$$\lim_{x\to\infty} \frac{f}{h} = \lim_{x\to\infty} \frac{f}{g} \cdot \frac{g}{h} = L_1 L_2.$$

If L_1 and L_2 are finite and nonzero, then so is $L_1 L_2$.

EXAMPLE 3 Functions Growing at the Same Rate

Show that $\sqrt{x^2 + 5}$ and $(2\sqrt{x} - 1)^2$ grow at the same rate as $x \to \infty$.

Solution We show that the functions grow at the same rate by showing that they both grow at the same rate as the function $g(x) = x$:

$$\lim_{x\to\infty} \frac{\sqrt{x^2+5}}{x} = \lim_{x\to\infty} \sqrt{1 + \frac{5}{x^2}} = 1,$$

$$\lim_{x\to\infty} \frac{(2\sqrt{x} - 1)^2}{x} = \lim_{x\to\infty} \left(\frac{2\sqrt{x} - 1}{\sqrt{x}}\right)^2 = \lim_{x\to\infty} \left(2 - \frac{1}{\sqrt{x}}\right)^2 = 4.$$

■

Order and Oh-Notation

Here we introduce the "little-oh" and "big-oh" notation invented by number theorists a hundred years ago and now commonplace in mathematical analysis and computer science.

DEFINITION Little-oh

A function f is **of smaller order than** g as $x \to \infty$ if $\lim_{x \to \infty} \frac{f(x)}{g(x)} = 0$. We indicate this by writing $\boldsymbol{f = o(g)}$ ("f is little-oh of g").

Notice that saying $f = o(g)$ as $x \to \infty$ is another way to say that f grows slower than g as $x \to \infty$.

EXAMPLE 4 Using Little-oh Notation

(a) $\ln x = o(x)$ as $x \to \infty$ because $\lim_{x \to \infty} \frac{\ln x}{x} = 0$

(b) $x^2 = o(x^3 + 1)$ as $x \to \infty$ because $\lim_{x \to \infty} \frac{x^2}{x^3 + 1} = 0$ ■

DEFINITION Big-oh

Let $f(x)$ and $g(x)$ be positive for x sufficiently large. Then f is **of at most the order of** g as $x \to \infty$ if there is a positive integer M for which

$$\frac{f(x)}{g(x)} \leq M,$$

for x sufficiently large. We indicate this by writing $\boldsymbol{f = O(g)}$ ("f is big-oh of g").

EXAMPLE 5 Using Big-oh Notation

(a) $x + \sin x = O(x)$ as $x \to \infty$ because $\frac{x + \sin x}{x} \leq 2$ for x sufficiently large.

(b) $e^x + x^2 = O(e^x)$ as $x \to \infty$ because $\frac{e^x + x^2}{e^x} \to 1$ as $x \to \infty$.

(c) $x = O(e^x)$ as $x \to \infty$ because $\frac{x}{e^x} \to 0$ as $x \to \infty$. ■

If you look at the definitions again, you will see that $f = o(g)$ implies $f = O(g)$ for functions that are positive for x sufficiently large. Also, if f and g grow at the same rate, then $f = O(g)$ and $g = O(f)$ (Exercise 11).

Sequential vs. Binary Search

Computer scientists often measure the efficiency of an algorithm by counting the number of steps a computer must take to execute the algorithm. There can be significant differences in how efficiently algorithms perform, even if they are designed to accomplish the same task. These differences are often described in big-oh notation. Here is an example.

Webster's Third New International Dictionary lists about 26,000 words that begin with the letter a. One way to look up a word, or to learn if it is not there, is to read through the list one word at a time until you either find the word or determine that it is not there. This method, called sequential search, makes no particular use of the words' alphabetical arrangement. You are sure to get an answer, but it might take 26,000 steps.

Another way to find the word or to learn it is not there is to go straight to the middle of the list (give or take a few words). If you do not find the word, then go to the middle of the half that contains it and forget about the half that does not. (You know which half contains it because you know the list is ordered alphabetically.) This method eliminates roughly 13,000 words in a single step. If you do not find the word on the second try, then jump to the middle of the half that contains it. Continue this way until you have either found the word or divided the list in half so many times there are no words left. How many times do you have to divide the list to find the word or learn that it is not there? At most 15, because

$$(26{,}000/2^{15}) < 1.$$

That certainly beats a possible 26,000 steps.

For a list of length n, a sequential search algorithm takes on the order of n steps to find a word or determine that it is not in the list. A binary search, as the second algorithm is called, takes on the order of $\log_2 n$ steps. The reason is that if $2^{m-1} < n \le 2^m$, then $m - 1 < \log_2 n \le m$, and the number of bisections required to narrow the list to one word will be at most $m = \lceil \log_2 n \rceil$, the integer ceiling for $\log_2 n$.

Big-oh notation provides a compact way to say all this. The number of steps in a sequential search of an ordered list is $O(n)$; the number of steps in a binary search is $O(\log_2 n)$. In our example, there is a big difference between the two (26,000 vs. 15), and the difference can only increase with n because n grows faster than $\log_2 n$ as $n \to \infty$ (as in Example 1d).

EXERCISES 7.6

Comparisons with the Exponential e^x

1. Which of the following functions grow faster than e^x as $x \to \infty$? Which grow at the same rate as e^x? Which grow slower?

a. $x + 3$ **b.** $x^3 + \sin^2 x$

c. $\sqrt{x}$ **d.** 4^x

e. $(3/2)^x$ **f.** $e^{x/2}$

g. $e^x/2$ **h.** $\log_{10} x$

2. Which of the following functions grow faster than e^x as $x \to \infty$? Which grow at the same rate as e^x? Which grow slower?

a. $10x^4 + 30x + 1$ **b.** $x \ln x - x$

c. $\sqrt{1 + x^4}$ **d.** $(5/2)^x$

e. e^{-x} **f.** xe^x

g. $e^{\cos x}$ **h.** e^{x-1}

Comparisons with the Power x^2

3. Which of the following functions grow faster than x^2 as $x \to \infty$? Which grow at the same rate as x^2? Which grow slower?

a. $x^2 + 4x$ **b.** $x^5 - x^2$

c. $\sqrt{x^4 + x^3}$ **d.** $(x + 3)^2$

e. $x \ln x$ **f.** 2^x

g. $x^3 e^{-x}$ **h.** $8x^2$

4. Which of the following functions grow faster than x^2 as $x \to \infty$? Which grow at the same rate as x^2? Which grow slower?

a. $x^2 + \sqrt{x}$ **b.** $10x^2$

c. $x^2 e^{-x}$ **d.** $\log_{10}(x^2)$

e. $x^3 - x^2$ **f.** $(1/10)^x$

g. $(1.1)^x$ **h.** $x^2 + 100x$

Comparisons with the Logarithm ln x

5. Which of the following functions grow faster than $\ln x$ as $x \to \infty$? Which grow at the same rate as $\ln x$? Which grow slower?

a. $\log_3 x$ b. $\ln 2x$
c. $\ln \sqrt{x}$ d. $\sqrt{x}$
e. x f. $5 \ln x$
g. $1/x$ h. e^x

6. Which of the following functions grow faster than $\ln x$ as $x \to \infty$? Which grow at the same rate as $\ln x$? Which grow slower?

a. $\log_2 (x^2)$ b. $\log_{10} 10x$
c. $1/\sqrt{x}$ d. $1/x^2$
e. $x - 2 \ln x$ f. e^{-x}
g. $\ln (\ln x)$ h. $\ln (2x + 5)$

Ordering Functions by Growth Rates

7. Order the following functions from slowest growing to fastest growing as $x \to \infty$.

a. e^x b. x^x
c. $(\ln x)^x$ d. $e^{x/2}$

8. Order the following functions from slowest growing to fastest growing as $x \to \infty$.

a. 2^x b. x^2
c. $(\ln 2)^x$ d. e^x

Big-oh and Little-oh; Order

9. True, or false? As $x \to \infty$,

a. $x = o(x)$ b. $x = o(x + 5)$
c. $x = O(x + 5)$ d. $x = O(2x)$
e. $e^x = o(e^{2x})$ f. $x + \ln x = O(x)$
g. $\ln x = o(\ln 2x)$ h. $\sqrt{x^2 + 5} = O(x)$

10. True, or false? As $x \to \infty$,

a. $\dfrac{1}{x + 3} = O\left(\dfrac{1}{x}\right)$ b. $\dfrac{1}{x} + \dfrac{1}{x^2} = O\left(\dfrac{1}{x}\right)$
c. $\dfrac{1}{x} - \dfrac{1}{x^2} = o\left(\dfrac{1}{x}\right)$ d. $2 + \cos x = O(2)$
e. $e^x + x = O(e^x)$ f. $x \ln x = o(x^2)$
g. $\ln (\ln x) = O(\ln x)$ h. $\ln (x) = o(\ln (x^2 + 1))$

11. Show that if positive functions $f(x)$ and $g(x)$ grow at the same rate as $x \to \infty$, then $f = O(g)$ and $g = O(f)$.

12. When is a polynomial $f(x)$ of smaller order than a polynomial $g(x)$ as $x \to \infty$? Give reasons for your answer.

13. When is a polynomial $f(x)$ of at most the order of a polynomial $g(x)$ as $x \to \infty$? Give reasons for your answer.

14. What do the conclusions we drew in Section 2.4 about the limits of rational functions tell us about the relative growth of polynomials as $x \to \infty$?

Other Comparisons

T 15. Investigate

$$\lim_{x\to\infty} \frac{\ln (x + 1)}{\ln x} \quad \text{and} \quad \lim_{x\to\infty} \frac{\ln (x + 999)}{\ln x}.$$

Then use l'Hôpital's Rule to explain what you find.

16. (*Continuation of Exercise 15.*) Show that the value of

$$\lim_{x\to\infty} \frac{\ln (x + a)}{\ln x}$$

is the same no matter what value you assign to the constant a. What does this say about the relative rates at which the functions $f(x) = \ln (x + a)$ and $g(x) = \ln x$ grow?

17. Show that $\sqrt{10x + 1}$ and $\sqrt{x + 1}$ grow at the same rate as $x \to \infty$ by showing that they both grow at the same rate as $\sqrt{x}$ as $x \to \infty$.

18. Show that $\sqrt{x^4 + x}$ and $\sqrt{x^4 - x^3}$ grow at the same rate as $x \to \infty$ by showing that they both grow at the same rate as x^2 as $x \to \infty$.

19. Show that e^x grows faster as $x \to \infty$ than x^n for any positive integer n, even $x^{1{,}000{,}000}$. (*Hint:* What is the nth derivative of x^n?)

20. **The function e^x outgrows any polynomial** Show that e^x grows faster as $x \to \infty$ than any polynomial

$$a_n x^n + a_{n-1} x^{n-1} + \cdots + a_1 x + a_0.$$

21. a. Show that $\ln x$ grows slower as $x \to \infty$ than $x^{1/n}$ for any positive integer n, even $x^{1/1{,}000{,}000}$.

T b. Although the values of $x^{1/1{,}000{,}000}$ eventually overtake the values of $\ln x$, you have to go way out on the x-axis before this happens. Find a value of x greater than 1 for which $x^{1/1{,}000{,}000} > \ln x$. You might start by observing that when $x > 1$ the equation $\ln x = x^{1/1{,}000{,}000}$ is equivalent to the equation $\ln (\ln x) = (\ln x)/1{,}000{,}000$.

T c. Even $x^{1/10}$ takes a long time to overtake $\ln x$. Experiment with a calculator to find the value of x at which the graphs of $x^{1/10}$ and $\ln x$ cross, or, equivalently, at which $\ln x = 10 \ln (\ln x)$. Bracket the crossing point between powers of 10 and then close in by successive halving.

T d. (*Continuation of part (c).*) The value of x at which $\ln x = 10 \ln (\ln x)$ is too far out for some graphers and root finders to identify. Try it on the equipment available to you and see what happens.

22. **The function ln x grows slower than any polynomial** Show that $\ln x$ grows slower as $x \to \infty$ than any nonconstant polynomial.

Algorithms and Searches

23. a. Suppose you have three different algorithms for solving the same problem and each algorithm takes a number of steps that is of the order of one of the functions listed here:

$$n \log_2 n, \quad n^{3/2}, \quad n(\log_2 n)^2.$$

Which of the algorithms is the most efficient in the long run? Give reasons for your answer.

T **b.** Graph the functions in part (a) together to get a sense of how rapidly each one grows.

24. Repeat Exercise 23 for the functions

$$n, \quad \sqrt{n}\log_2 n, \quad (\log_2 n)^2.$$

T **25.** Suppose you are looking for an item in an ordered list one million items long. How many steps might it take to find that item with a sequential search? A binary search?

T **26.** You are looking for an item in an ordered list 450,000 items long (the length of *Webster's Third New International Dictionary*). How many steps might it take to find the item with a sequential search? A binary search?

7.7 Inverse Trigonometric Functions

Inverse trigonometric functions arise when we want to calculate angles from side measurements in triangles. They also provide useful antiderivatives and appear frequently in the solutions of differential equations. This section shows how these functions are defined, graphed, and evaluated, how their derivatives are computed, and why they appear as important antiderivatives.

Defining the Inverses

The six basic trigonometric functions are not one-to-one (their values repeat periodically). However we can restrict their domains to intervals on which they are one-to-one. The sine function increases from -1 at $x = -\pi/2$ to $+1$ at $x = \pi/2$. By restricting its domain to the interval $[-\pi/2, \pi/2]$ we make it one-to-one, so that it has an inverse $\sin^{-1} x$ (Figure 7.16). Similar domain restrictions can be applied to all six trigonometric functions.

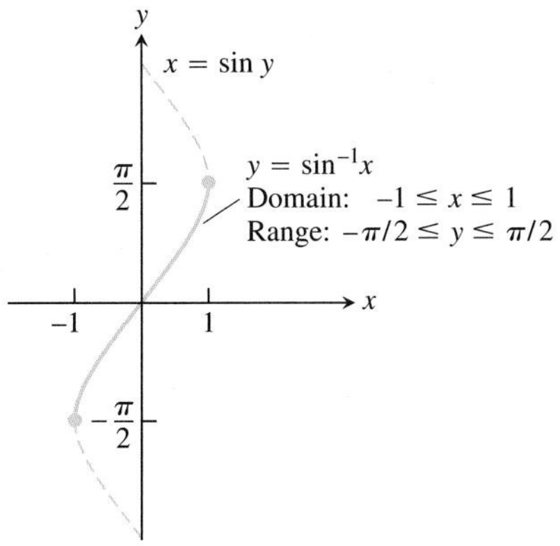

FIGURE 7.16 The graph of $y = \sin^{-1} x$.

Domain restrictions that make the trigonometric functions one-to-one

Function	**Domain**	**Range**	
$\sin x$	$[-\pi/2, \pi/2]$	$[-1, 1]$	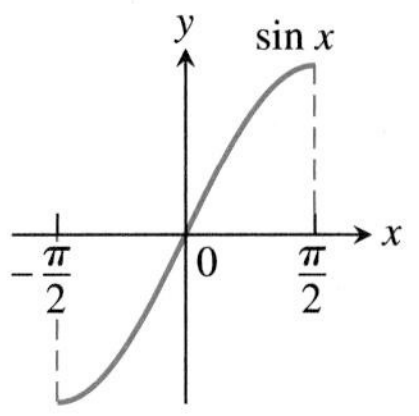
$\cos x$	$[0, \pi]$	$[-1, 1]$	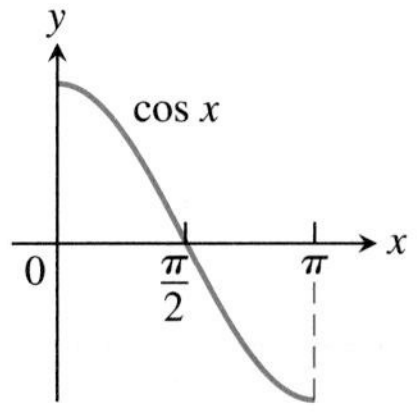

$\tan x$	$(-\pi/2, \pi/2)$	$(-\infty, \infty)$	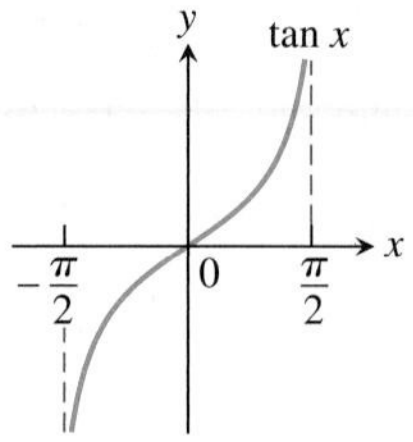
$\cot x$	$(0, \pi)$	$(-\infty, \infty)$	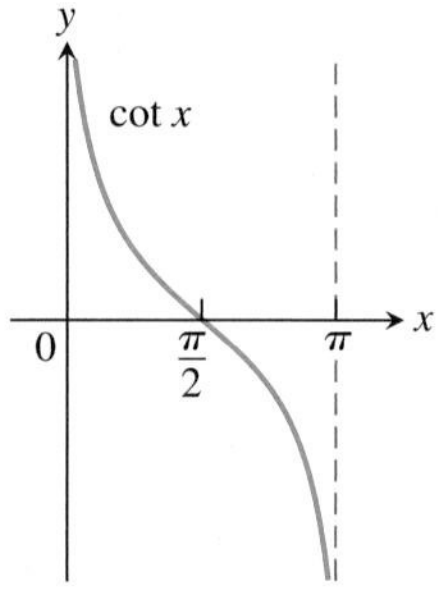
$\sec x$	$[0, \pi/2) \cup (\pi/2, \pi]$	$(-\infty, -1] \cup [1, \infty)$	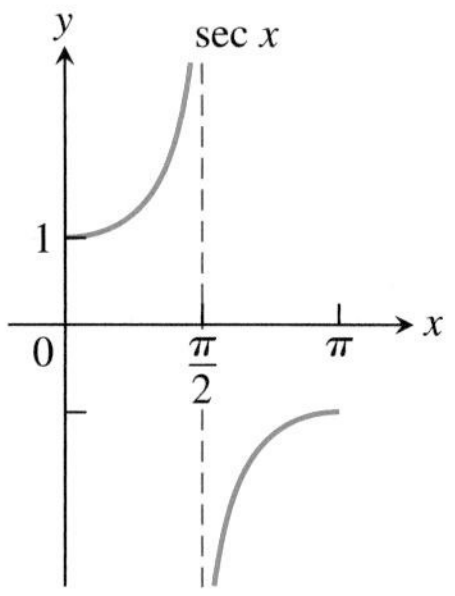
$\csc x$	$[-\pi/2, 0) \cup (0, \pi/2]$	$(-\infty, -1] \cup [1, \infty)$	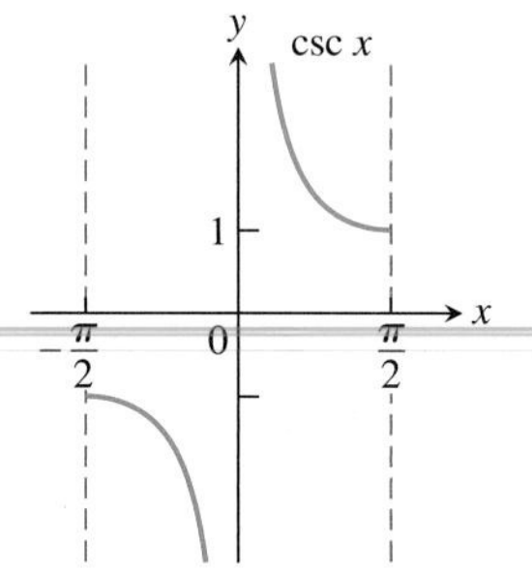

Since these restricted functions are now one-to-one, they have inverses, which we denote by

$$y = \sin^{-1} x \quad \text{or} \quad y = \arcsin x$$
$$y = \cos^{-1} x \quad \text{or} \quad y = \arccos x$$
$$y = \tan^{-1} x \quad \text{or} \quad y = \arctan x$$
$$y = \cot^{-1} x \quad \text{or} \quad y = \operatorname{arccot} x$$
$$y = \sec^{-1} x \quad \text{or} \quad y = \operatorname{arcsec} x$$
$$y = \csc^{-1} x \quad \text{or} \quad y = \operatorname{arccsc} x$$

These equations are read "y equals the arcsine of x" or "y equals arcsin x" and so on.

CAUTION The -1 in the expressions for the inverse means "inverse." It does *not* mean reciprocal. For example, the *reciprocal* of $\sin x$ is $(\sin x)^{-1} = 1/\sin x = \csc x$.

The graphs of the six inverse trigonometric functions are shown in Figure 7.17. We can obtain these graphs by reflecting the graphs of the restricted trigonometric functions through the line $y = x$, as in Section 7.1. We now take a closer look at these functions and their derivatives.

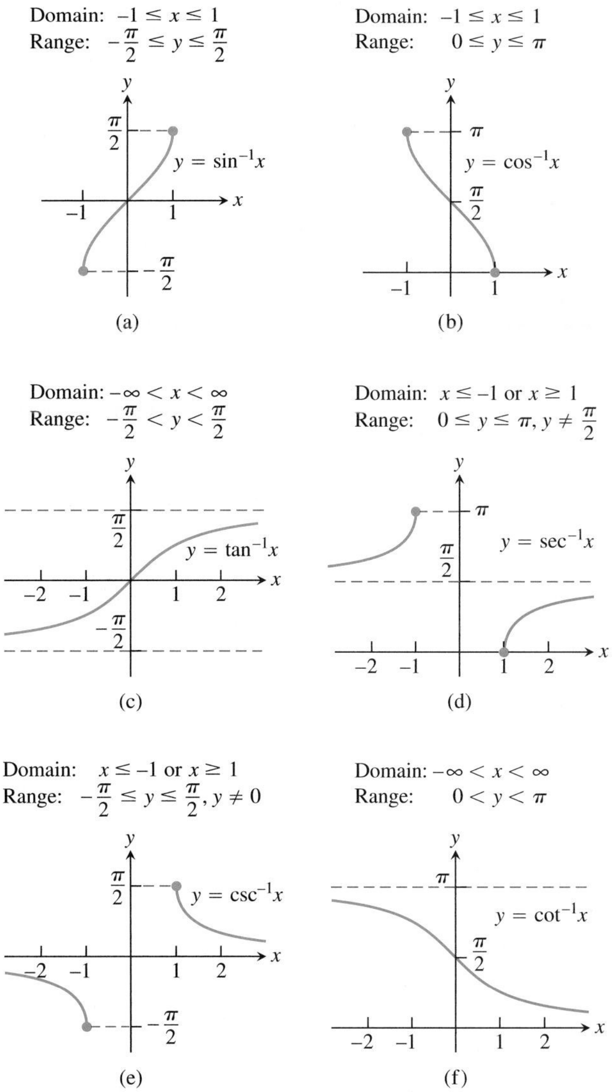

FIGURE 7.17 Graphs of the six basic inverse trigonometric functions.

The Arcsine and Arccosine Functions

The arcsine of x is the angle in $[-\pi/2, \pi/2]$ whose sine is x. The arccosine is an angle in $[0, \pi]$ whose cosine is x.

DEFINITION Arcsine and Arccosine Functions

$y = \sin^{-1} x$ is the number in $[-\pi/2, \pi/2]$ for which $\sin y = x$.

$y = \cos^{-1} x$ is the number in $[0, \pi]$ for which $\cos y = x$.

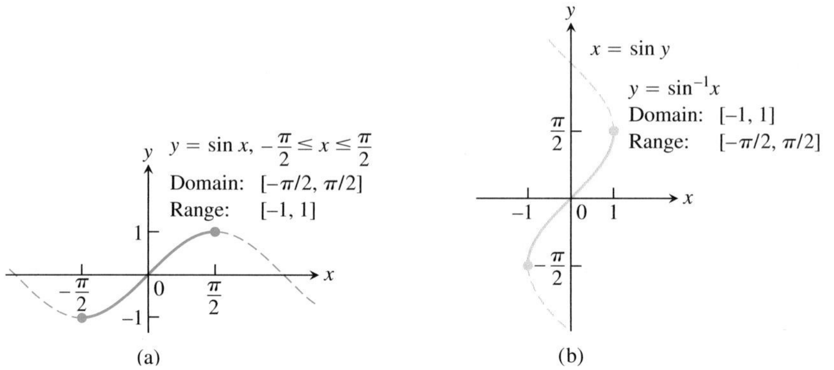

FIGURE 7.18 The graphs of (a) $y = \sin x$, $-\pi/2 \le x \le \pi/2$, and (b) its inverse, $y = \sin^{-1} x$. The graph of $\sin^{-1} x$, obtained by reflection across the line $y = x$, is a portion of the curve $x = \sin y$.

The "Arc" in Arc Sine and Arc Cosine

The accompanying figure gives a geometric interpretation of $y = \sin^{-1} x$ and $y = \cos^{-1} x$ for radian angles in the first quadrant. For a unit circle, the equation $s = r\theta$ becomes $s = \theta$, so central angles and the arcs they subtend have the same measure. If $x = \sin y$, then, in addition to being the angle whose sine is x, y is also the length of arc on the unit circle that subtends an angle whose sine is x. So we call y "the arc whose sine is x."

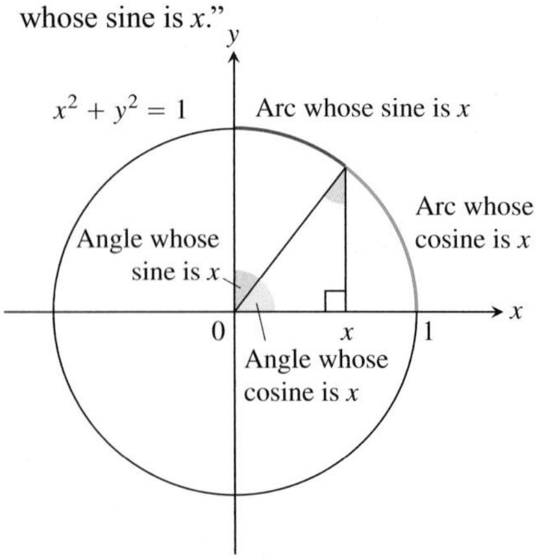

The graph of $y = \sin^{-1} x$ (Figure 7.18) is symmetric about the origin (it lies along the graph of $x = \sin y$). The arcsine is therefore an odd function:

$$\sin^{-1}(-x) = -\sin^{-1} x. \tag{1}$$

The graph of $y = \cos^{-1} x$ (Figure 7.19) has no such symmetry.

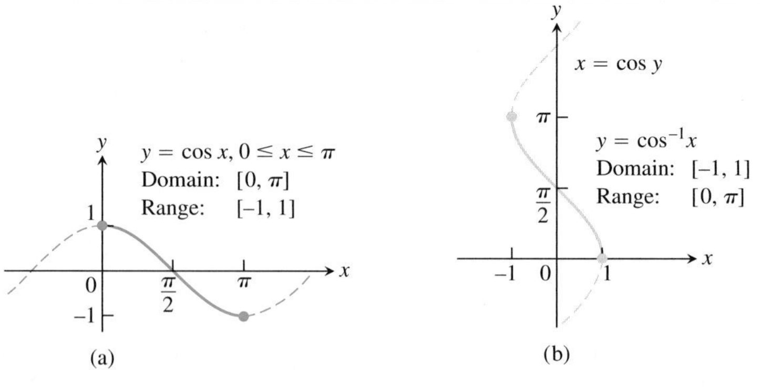

FIGURE 7.19 The graphs of (a) $y = \cos x$, $0 \le x \le \pi$, and (b) its inverse, $y = \cos^{-1} x$. The graph of $\cos^{-1} x$, obtained by reflection across the line $y = x$, is a portion of the curve $x = \cos y$.

Known values of $\sin x$ and $\cos x$ can be inverted to find values of $\sin^{-1} x$ and $\cos^{-1} x$.

x	$\sin^{-1} x$
$\sqrt{3}/2$	$\pi/3$
$\sqrt{2}/2$	$\pi/4$
$1/2$	$\pi/6$
$-1/2$	$-\pi/6$
$-\sqrt{2}/2$	$-\pi/4$
$-\sqrt{3}/2$	$-\pi/3$

EXAMPLE 1 Common Values of $\sin^{-1} x$

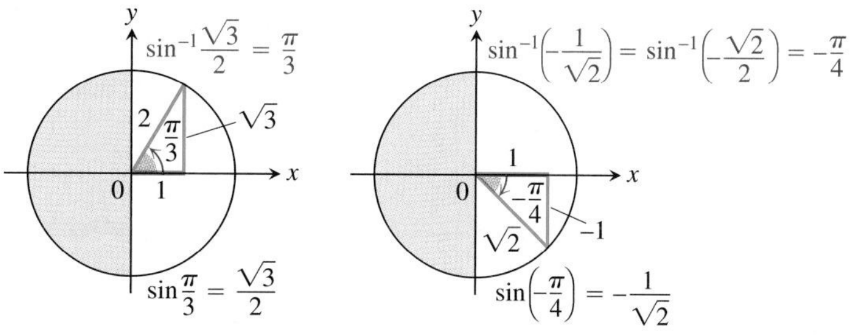

The angles come from the first and fourth quadrants because the range of $\sin^{-1} x$ is $[-\pi/2, \pi/2]$. ■

x	$\cos^{-1} x$
$\sqrt{3}/2$	$\pi/6$
$\sqrt{2}/2$	$\pi/4$
$1/2$	$\pi/3$
$-1/2$	$2\pi/3$
$-\sqrt{2}/2$	$3\pi/4$
$-\sqrt{3}/2$	$5\pi/6$

EXAMPLE 2 Common Values of $\cos^{-1} x$

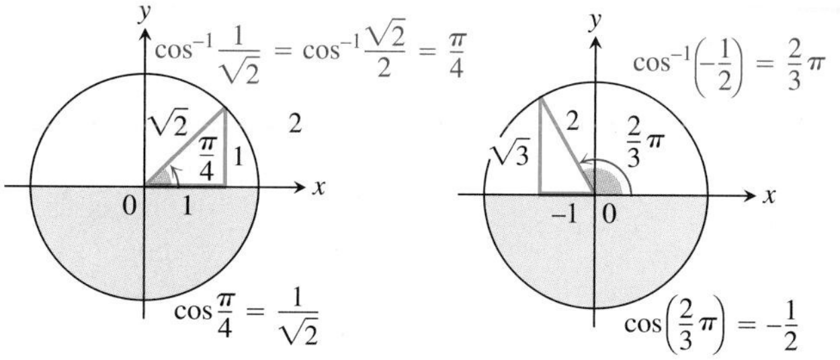

The angles come from the first and second quadrants because the range of $\cos^{-1} x$ is $[0, \pi]$. ■

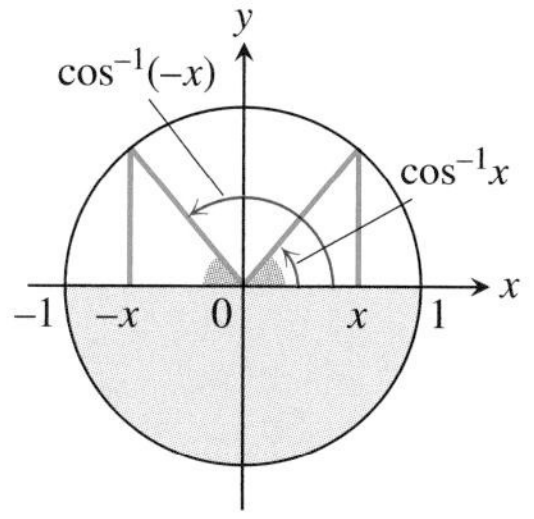

FIGURE 7.20 $\cos^{-1} x$ and $\cos^{-1}(-x)$ are supplementary angles (so their sum is π).

Identities Involving Arcsine and Arccosine

As we can see from Figure 7.20, the arccosine of x satisfies the identity

$$\cos^{-1} x + \cos^{-1}(-x) = \pi, \tag{2}$$

or

$$\cos^{-1}(-x) = \pi - \cos^{-1} x. \tag{3}$$

Also, we can see from the triangle in Figure 7.21 that for $x > 0$,

$$\sin^{-1} x + \cos^{-1} x = \pi/2. \tag{4}$$

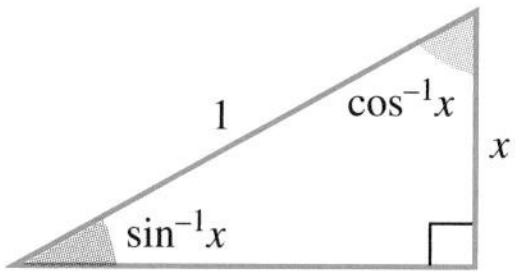

FIGURE 7.21 $\sin^{-1} x$ and $\cos^{-1} x$ are complementary angles (so their sum is $\pi/2$).

Equation (4) holds for the other values of x in $[-1, 1]$ as well, but we cannot conclude this from the triangle in Figure 7.21. It is, however, a consequence of Equations (1) and (3) (Exercise 131).

Inverses of tan x, cot x, sec x, and csc x

The arctangent of x is an angle whose tangent is x. The arccotangent of x is an angle whose cotangent is x.

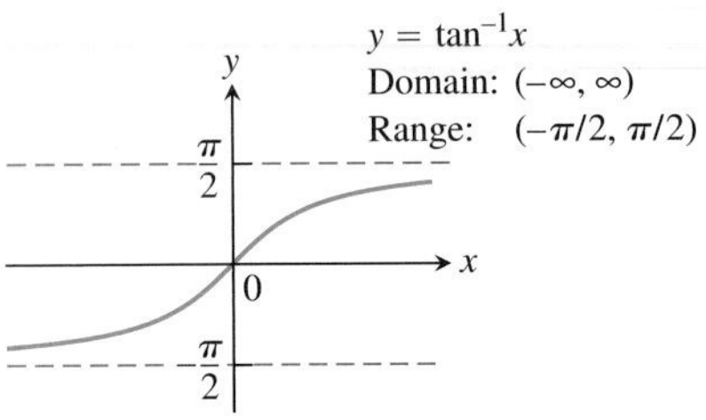

FIGURE 7.22 The graph of $y = \tan^{-1} x$.

DEFINITION **Arctangent and Arccotangent Functions**

$y = \tan^{-1} x$ is the number in $(-\pi/2, \pi/2)$ for which $\tan y = x$.

$y = \cot^{-1} x$ is the number in $(0, \pi)$ for which $\cot y = x$.

We use open intervals to avoid values where the tangent and cotangent are undefined.

The graph of $y = \tan^{-1} x$ is symmetric about the origin because it is a branch of the graph $x = \tan y$ that is symmetric about the origin (Figure 7.22). Algebraically this means that

$$\tan^{-1}(-x) = -\tan^{-1} x;$$

the arctangent is an odd function. The graph of $y = \cot^{-1} x$ has no such symmetry (Figure 7.23).

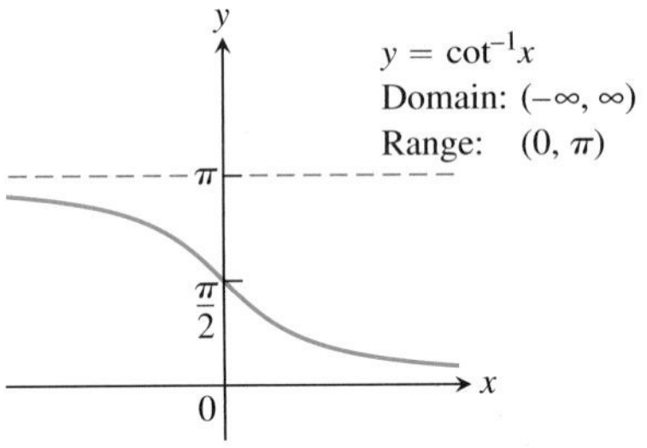

FIGURE 7.23 The graph of $y = \cot^{-1} x$.

The inverses of the restricted forms of $\sec x$ and $\csc x$ are chosen to be the functions graphed in Figures 7.24 and 7.25.

CAUTION There is no general agreement about how to define $\sec^{-1} x$ for negative values of x. We chose angles in the second quadrant between $\pi/2$ and π. This choice makes $\sec^{-1} x = \cos^{-1}(1/x)$. It also makes $\sec^{-1} x$ an increasing function on each interval of its domain. Some tables choose $\sec^{-1} x$ to lie in $[-\pi, -\pi/2)$ for $x < 0$ and some texts choose it to lie in $[\pi, 3\pi/2)$ (Figure 7.26). These choices simplify the formula for the derivative (our formula needs absolute value signs) but fail to satisfy the computational equation $\sec^{-1} x = \cos^{-1}(1/x)$. From this, we can derive the identity

$$\sec^{-1} x = \cos^{-1}\left(\frac{1}{x}\right) = \frac{\pi}{2} - \sin^{-1}\left(\frac{1}{x}\right) \tag{5}$$

by applying Equation (4).

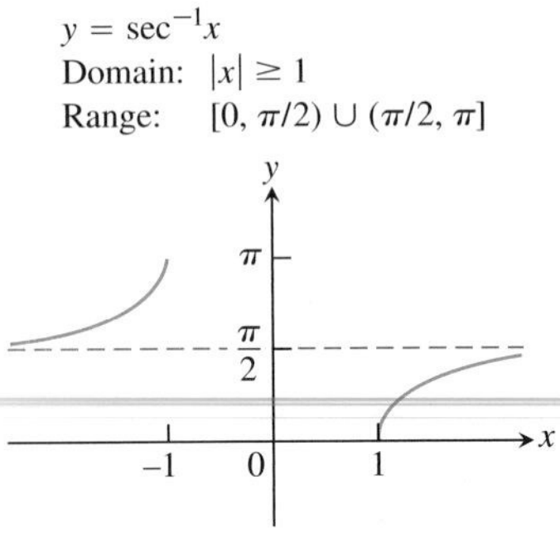

FIGURE 7.24 The graph of $y = \sec^{-1} x$.

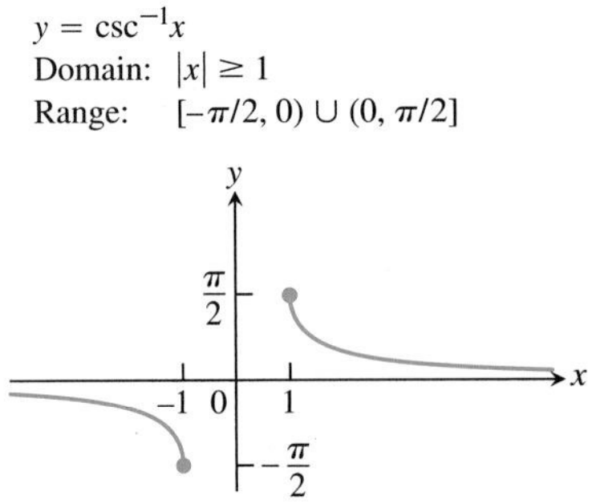

FIGURE 7.25 The graph of $y = \csc^{-1} x$.

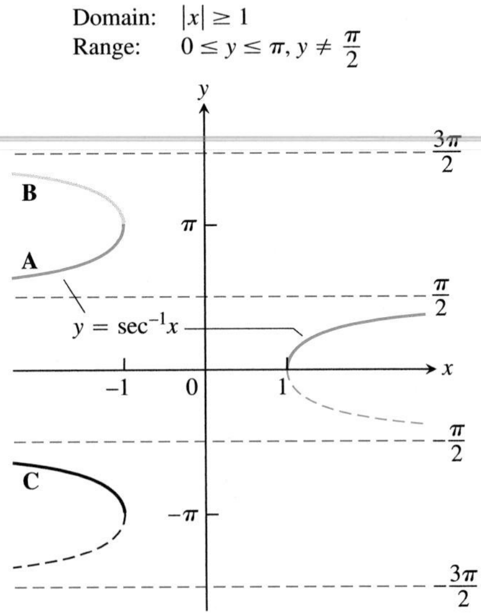

FIGURE 7.26 There are several logical choices for the left-hand branch of $y = \sec^{-1} x$. With choice **A**, $\sec^{-1} x = \cos^{-1}(1/x)$, a useful identity employed by many calculators.

x	$\tan^{-1} x$
$\sqrt{3}$	$\pi/3$
1	$\pi/4$
$\sqrt{3}/3$	$\pi/6$
$-\sqrt{3}/3$	$-\pi/6$
-1	$-\pi/4$
$-\sqrt{3}$	$-\pi/3$

EXAMPLE 3 Common Values of $\tan^{-1} x$

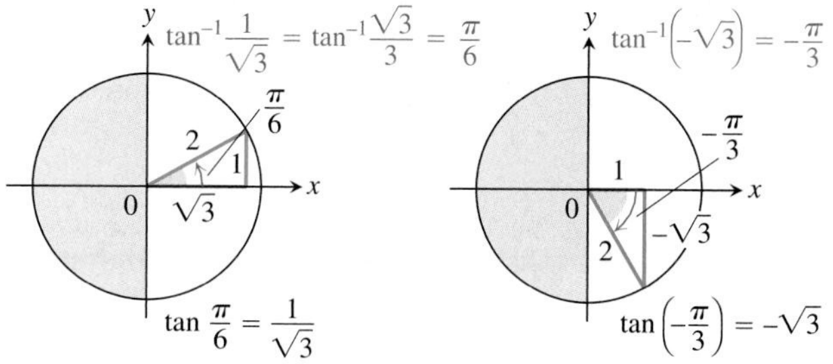

The angles come from the first and fourth quadrants because the range of $\tan^{-1} x$ is $(-\pi/2, \pi/2)$. ■

EXAMPLE 4 Find $\cos\alpha$, $\tan\alpha$, $\sec\alpha$, $\csc\alpha$, and $\cot\alpha$ if

$$\alpha = \sin^{-1}\frac{2}{3}.$$

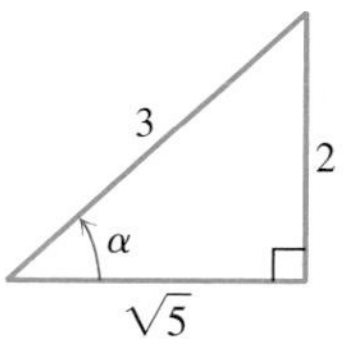

FIGURE 7.27 If $\alpha = \sin^{-1}(2/3)$, then the values of the other basic trigonometric functions of α can be read from this triangle (Example 4).

Solution This equation says that $\sin\alpha = 2/3$. We picture α as an angle in a right triangle with opposite side 2 and hypotenuse 3 (Figure 7.27). The length of the remaining side is

$$\sqrt{(3)^2 - (2)^2} = \sqrt{9 - 4} = \sqrt{5}. \qquad \text{Pythagorean theorem}$$

We add this information to the figure and then read the values we want from the completed triangle:

$$\cos\alpha = \frac{\sqrt{5}}{3}, \quad \tan\alpha = \frac{2}{\sqrt{5}}, \quad \sec\alpha = \frac{3}{\sqrt{5}}, \quad \csc\alpha = \frac{3}{2}, \quad \cot\alpha = \frac{\sqrt{5}}{2}.$$ ■

EXAMPLE 5 Find $\sec\left(\tan^{-1}\frac{x}{3}\right)$.

Solution We let $\theta = \tan^{-1}(x/3)$ (to give the angle a name) and picture θ in a right triangle with

$$\tan\theta = \text{opposite}/\text{adjacent} = x/3.$$

The length of the triangle's hypotenuse is

$$\sqrt{x^2 + 3^2} = \sqrt{x^2 + 9}.$$

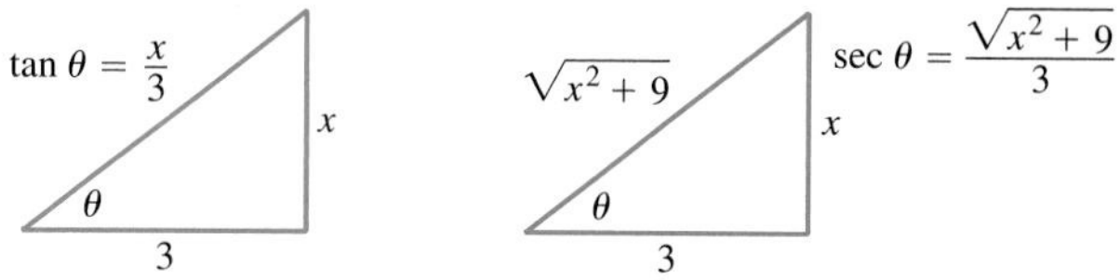

Thus,

$$\sec\left(\tan^{-1}\frac{x}{3}\right) = \sec\theta$$

$$= \frac{\sqrt{x^2 + 9}}{3}. \qquad \sec\theta = \frac{\text{hypotenuse}}{\text{adjacent}}$$ ■

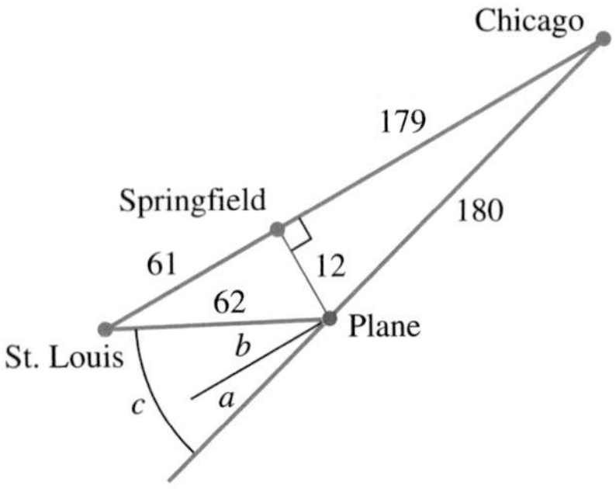

FIGURE 7.28 Diagram for drift correction (Example 6), with distances rounded to the nearest mile (drawing not to scale).

EXAMPLE 6 Drift Correction

During an airplane flight from Chicago to St. Louis the navigator determines that the plane is 12 mi off course, as shown in Figure 7.28. Find the angle a for a course parallel to the original, correct course, the angle b, and the correction angle $c = a + b$.

Solution

$$a = \sin^{-1}\frac{12}{180} \approx 0.067 \text{ radian} \approx 3.8°$$

$$b = \sin^{-1}\frac{12}{62} \approx 0.195 \text{ radian} \approx 11.2°$$

$$c = a + b \approx 15°.$$

■

The Derivative of $y = \sin^{-1} u$

We know that the function $x = \sin y$ is differentiable in the interval $-\pi/2 < y < \pi/2$ and that its derivative, the cosine, is positive there. Theorem 1 in Section 7.1 therefore assures us that the inverse function $y = \sin^{-1} x$ is differentiable throughout the interval $-1 < x < 1$. We cannot expect it to be differentiable at $x = 1$ or $x = -1$ because the tangents to the graph are vertical at these points (see Figure 7.29).

We find the derivative of $y = \sin^{-1} x$ by applying Theorem 1 with $f(x) = \sin x$ and $f^{-1}(x) = \sin^{-1} x$.

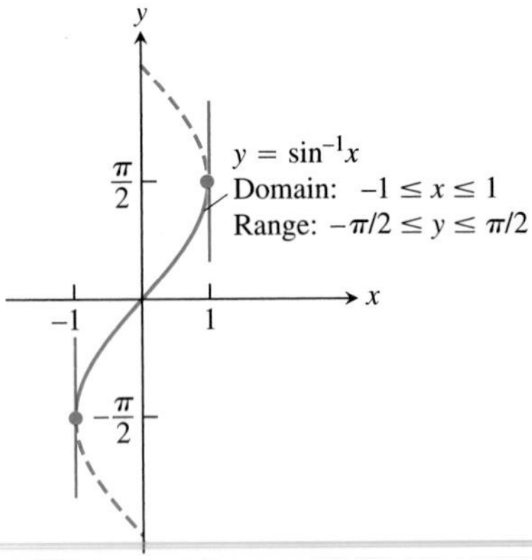

FIGURE 7.29 The graph of $y = \sin^{-1} x$.

$$\begin{aligned}
(f^{-1})'(x) &= \frac{1}{f'(f^{-1}(x))} && \text{Theorem 1}\\
&= \frac{1}{\cos(\sin^{-1} x)} && f'(u) = \cos u\\
&= \frac{1}{\sqrt{1 - \sin^2(\sin^{-1} x)}} && \cos u = \sqrt{1 - \sin^2 u}\\
&= \frac{1}{\sqrt{1 - x^2}} && \sin(\sin^{-1} x) = x
\end{aligned}$$

Alternate Derivation: Instead of applying Theorem 1 directly, we can find the derivative of $y = \sin^{-1} x$ using implicit differentiation as follows:

$$\begin{aligned}
\sin y &= x && y = \sin^{-1} x \Leftrightarrow \sin y = x\\
\frac{d}{dx}(\sin y) &= 1 && \text{Derivative of both sides with respect to } x\\
\cos y \frac{dy}{dx} &= 1 && \text{Chain Rule}\\
\frac{dy}{dx} &= \frac{1}{\cos y} && \text{We can divide because } \cos y > 0 \text{ for } -\pi/2 < y < \pi/2.\\
&= \frac{1}{\sqrt{1 - x^2}} && \cos y = \sqrt{1 - \sin^2 y}
\end{aligned}$$

No matter which derivation we use, we have that the derivative of $y = \sin^{-1} x$ with respect to x is

$$\frac{d}{dx}(\sin^{-1} x) = \frac{1}{\sqrt{1 - x^2}}.$$

If u is a differentiable function of x with $|u| < 1$, we apply the Chain Rule to get

$$\frac{d}{dx}(\sin^{-1} u) = \frac{1}{\sqrt{1 - u^2}} \frac{du}{dx}, \qquad |u| < 1.$$

EXAMPLE 7 Applying the Derivative Formula

$$\frac{d}{dx}(\sin^{-1} x^2) = \frac{1}{\sqrt{1 - (x^2)^2}} \cdot \frac{d}{dx}(x^2) = \frac{2x}{\sqrt{1 - x^4}}$$ ■

The Derivative of $y = \tan^{-1} u$

We find the derivative of $y = \tan^{-1} x$ by applying Theorem 1 with $f(x) = \tan x$ and $f^{-1}(x) = \tan^{-1} x$. Theorem 1 can be applied because the derivative of $\tan x$ is positive for $-\pi/2 < x < \pi/2$.

$$\begin{aligned}
(f^{-1})'(x) &= \frac{1}{f'(f^{-1}(x))} && \text{Theorem 1} \\
&= \frac{1}{\sec^2(\tan^{-1} x)} && f'(u) = \sec^2 u \\
&= \frac{1}{1 + \tan^2(\tan^{-1} x)} && \sec^2 u = 1 + \tan^2 u \\
&= \frac{1}{1 + x^2} && \tan(\tan^{-1} x) = x
\end{aligned}$$

The derivative is defined for all real numbers. If u is a differentiable function of x, we get the Chain Rule form:

$$\frac{d}{dx}(\tan^{-1} u) = \frac{1}{1 + u^2} \frac{du}{dx}.$$

EXAMPLE 8 A Moving Particle

A particle moves along the x-axis so that its position at any time $t \geq 0$ is $x(t) = \tan^{-1}\sqrt{t}$. What is the velocity of the particle when $t = 16$?

Solution

$$v(t) = \frac{d}{dt}\tan^{-1}\sqrt{t} = \frac{1}{1 + (\sqrt{t})^2} \cdot \frac{d}{dt}\sqrt{t} = \frac{1}{1 + t} \cdot \frac{1}{2\sqrt{t}}$$

When $t = 16$, the velocity is

$$v(16) = \frac{1}{1+16} \cdot \frac{1}{2\sqrt{16}} = \frac{1}{136}.$$

The Derivative of $y = \sec^{-1} u$

Since the derivative of sec x is positive for $0 < x < \pi/2$ and $\pi/2 < x < \pi$, Theorem 1 says that the inverse function $y = \sec^{-1} x$ is differentiable. Instead of applying the formula in Theorem 1 directly, we find the derivative of $y = \sec^{-1} x$, $|x| > 1$, using implicit differentiation and the Chain Rule as follows:

$$y = \sec^{-1} x$$

$$\sec y = x \qquad \text{Inverse function relationship}$$

$$\frac{d}{dx}(\sec y) = \frac{d}{dx}x \qquad \text{Differentiate both sides.}$$

$$\sec y \tan y \frac{dy}{dx} = 1 \qquad \text{Chain Rule}$$

$$\frac{dy}{dx} = \frac{1}{\sec y \tan y} \qquad \text{Since } |x| > 1,\ y \text{ lies in } (0, \pi/2) \cup (\pi/2, \pi) \text{ and } \sec y \tan y \neq 0.$$

To express the result in terms of x, we use the relationships

$$\sec y = x \qquad \text{and} \qquad \tan y = \pm\sqrt{\sec^2 y - 1} = \pm\sqrt{x^2 - 1}$$

to get

$$\frac{dy}{dx} = \pm\frac{1}{x\sqrt{x^2-1}}.$$

Can we do anything about the $\pm$ sign? A glance at Figure 7.30 shows that the slope of the graph $y = \sec^{-1} x$ is always positive. Thus,

$$\frac{d}{dx}\sec^{-1} x = \begin{cases} +\dfrac{1}{x\sqrt{x^2-1}} & \text{if } x > 1 \\ -\dfrac{1}{x\sqrt{x^2-1}} & \text{if } x < -1. \end{cases}$$

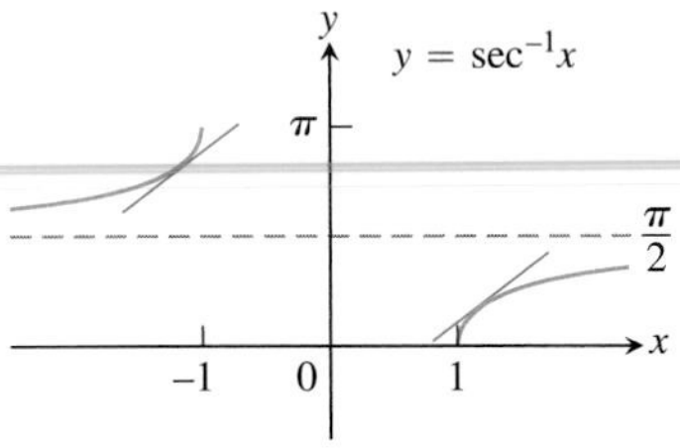

FIGURE 7.30 The slope of the curve $y = \sec^{-1} x$ is positive for both $x < -1$ and $x > 1$.

With the absolute value symbol, we can write a single expression that eliminates the "±" ambiguity:

$$\frac{d}{dx}\sec^{-1} x = \frac{1}{|x|\sqrt{x^2-1}}.$$

If u is a differentiable function of x with $|u| > 1$, we have the formula

$$\frac{d}{dx}(\sec^{-1} u) = \frac{1}{|u|\sqrt{u^2-1}}\frac{du}{dx}, \qquad |u| > 1.$$

EXAMPLE 9 Using the Formula

$$\frac{d}{dx}\sec^{-1}(5x^4) = \frac{1}{|5x^4|\sqrt{(5x^4)^2 - 1}}\frac{d}{dx}(5x^4)$$
$$= \frac{1}{5x^4\sqrt{25x^8 - 1}}(20x^3) \qquad 5x^4 > 0$$
$$= \frac{4}{x\sqrt{25x^8 - 1}}$$

Derivatives of the Other Three

We could use the same techniques to find the derivatives of the other three inverse trigonometric functions—arccosine, arccotangent, and arccosecant—but there is a much easier way, thanks to the following identities.

Inverse Function–Inverse Cofunction Identities

$$\cos^{-1}x = \pi/2 - \sin^{-1}x$$
$$\cot^{-1}x = \pi/2 - \tan^{-1}x$$
$$\csc^{-1}x = \pi/2 - \sec^{-1}x$$

We saw the first of these identities in Equation (4). The others are derived in a similar way. It follows easily that the derivatives of the inverse cofunctions are the negatives of the derivatives of the corresponding inverse functions. For example, the derivative of $\cos^{-1}x$ is calculated as follows:

$$\frac{d}{dx}(\cos^{-1}x) = \frac{d}{dx}\left(\frac{\pi}{2} - \sin^{-1}x\right) \qquad \text{Identity}$$
$$= -\frac{d}{dx}(\sin^{-1}x)$$
$$= -\frac{1}{\sqrt{1 - x^2}} \qquad \text{Derivative of arcsine}$$

EXAMPLE 10 A Tangent Line to the Arccotangent Curve

Find an equation for the line tangent to the graph of $y = \cot^{-1}x$ at $x = -1$.

Solution First we note that

$$\cot^{-1}(-1) = \pi/2 - \tan^{-1}(-1) = \pi/2 - (-\pi/4) = 3\pi/4.$$

The slope of the tangent line is

$$\left.\frac{dy}{dx}\right|_{x=-1} = -\left.\frac{1}{1 + x^2}\right|_{x=-1} = -\frac{1}{1 + (-1)^2} = -\frac{1}{2},$$

so the tangent line has equation $y - 3\pi/4 = (-1/2)(x + 1)$.

The derivatives of the inverse trigonometric functions are summarized in Table 7.3.

TABLE 7.3 Derivatives of the inverse trigonometric functions

1. $\dfrac{d(\sin^{-1} u)}{dx} = \dfrac{du/dx}{\sqrt{1 - u^2}}, \quad |u| < 1$

2. $\dfrac{d(\cos^{-1} u)}{dx} = -\dfrac{du/dx}{\sqrt{1 - u^2}}, \quad |u| < 1$

3. $\dfrac{d(\tan^{-1} u)}{dx} = \dfrac{du/dx}{1 + u^2}$

4. $\dfrac{d(\cot^{-1} u)}{dx} = -\dfrac{du/dx}{1 + u^2}$

5. $\dfrac{d(\sec^{-1} u)}{dx} = \dfrac{du/dx}{|u|\sqrt{u^2 - 1}}, \quad |u| > 1$

6. $\dfrac{d(\csc^{-1} u)}{dx} = \dfrac{-du/dx}{|u|\sqrt{u^2 - 1}}, \quad |u| > 1$

Integration Formulas

The derivative formulas in Table 7.3 yield three useful integration formulas in Table 7.4. The formulas are readily verified by differentiating the functions on the right-hand sides.

TABLE 7.4 Integrals evaluated with inverse trigonometric functions

The following formulas hold for any constant $a \neq 0$.

1. $\displaystyle\int \frac{du}{\sqrt{a^2 - u^2}} = \sin^{-1}\left(\frac{u}{a}\right) + C \qquad$ (Valid for $u^2 < a^2$)

2. $\displaystyle\int \frac{du}{a^2 + u^2} = \frac{1}{a}\tan^{-1}\left(\frac{u}{a}\right) + C \qquad$ (Valid for all u)

3. $\displaystyle\int \frac{du}{u\sqrt{u^2 - a^2}} = \frac{1}{a}\sec^{-1}\left|\frac{u}{a}\right| + C \qquad$ (Valid for $|u| > a > 0$)

The derivative formulas in Table 7.3 have $a = 1$, but in most integrations $a \neq 1$, and the formulas in Table 7.4 are more useful.

EXAMPLE 11 Using the Integral Formulas

(a) $$\int_{\sqrt{2}/2}^{\sqrt{3}/2} \frac{dx}{\sqrt{1 - x^2}} = \sin^{-1} x \Big]_{\sqrt{2}/2}^{\sqrt{3}/2}$$

$$= \sin^{-1}\left(\frac{\sqrt{3}}{2}\right) - \sin^{-1}\left(\frac{\sqrt{2}}{2}\right) = \frac{\pi}{3} - \frac{\pi}{4} = \frac{\pi}{12}$$

(b) $\displaystyle\int_0^1 \frac{dx}{1+x^2} = \tan^{-1} x\Big]_0^1 = \tan^{-1}(1) - \tan^{-1}(0) = \frac{\pi}{4} - 0 = \frac{\pi}{4}$

(c) $\displaystyle\int_{2/\sqrt{3}}^{\sqrt{2}} \frac{dx}{x\sqrt{x^2-1}} = \sec^{-1} x\Big]_{2/\sqrt{3}}^{\sqrt{2}} = \frac{\pi}{4} - \frac{\pi}{6} = \frac{\pi}{12}$ ■

EXAMPLE 12 Using Substitution and Table 7.4

(a) $\displaystyle\int \frac{dx}{\sqrt{9-x^2}} = \int \frac{dx}{\sqrt{(3)^2 - x^2}} = \sin^{-1}\left(\frac{x}{3}\right) + C$ Table 7.4 Formula 1, with $a = 3$, $u = x$

(b) $\displaystyle\int \frac{dx}{\sqrt{3-4x^2}} = \frac{1}{2}\int \frac{du}{\sqrt{a^2-u^2}}$ $a = \sqrt{3}$, $u = 2x$, and $du/2 = dx$

$\displaystyle = \frac{1}{2}\sin^{-1}\left(\frac{u}{a}\right) + C$ Formula 1

$\displaystyle = \frac{1}{2}\sin^{-1}\left(\frac{2x}{\sqrt{3}}\right) + C$ ■

EXAMPLE 13 Completing the Square

Evaluate

$$\int \frac{dx}{\sqrt{4x - x^2}}.$$

Solution The expression $\sqrt{4x - x^2}$ does not match any of the formulas in Table 7.4, so we first rewrite $4x - x^2$ by completing the square:

$$4x - x^2 = -(x^2 - 4x) = -(x^2 - 4x + 4) + 4 = 4 - (x-2)^2.$$

Then we substitute $a = 2$, $u = x - 2$, and $du = dx$ to get

$$\int \frac{dx}{\sqrt{4x-x^2}} = \int \frac{dx}{\sqrt{4-(x-2)^2}}$$

$\displaystyle = \int \frac{du}{\sqrt{a^2-u^2}}$ $a = 2$, $u = x - 2$, and $du = dx$

$\displaystyle = \sin^{-1}\left(\frac{u}{a}\right) + C$ Table 7.4, Formula 1

$\displaystyle = \sin^{-1}\left(\frac{x-2}{2}\right) + C$ ■

EXAMPLE 14 Completing the Square

Evaluate

$$\int \frac{dx}{4x^2 + 4x + 2}.$$

Solution We complete the square on the binomial $4x^2 + 4x$:

$$4x^2 + 4x + 2 = 4(x^2 + x) + 2 = 4\left(x^2 + x + \frac{1}{4}\right) + 2 - \frac{4}{4}$$

$$= 4\left(x + \frac{1}{2}\right)^2 + 1 = (2x + 1)^2 + 1.$$

Then,

$$\int \frac{dx}{4x^2 + 4x + 2} = \int \frac{dx}{(2x + 1)^2 + 1} = \frac{1}{2}\int \frac{du}{u^2 + a^2} \qquad \begin{array}{l} a = 1, u = 2x + 1, \\ \text{and } du/2 = dx \end{array}$$

$$= \frac{1}{2} \cdot \frac{1}{a} \tan^{-1}\left(\frac{u}{a}\right) + C \qquad \text{Table 7.4, Formula 2}$$

$$= \frac{1}{2} \tan^{-1}(2x + 1) + C \qquad a = 1, u = 2x + 1$$

■

EXAMPLE 15 Using Substitution

Evaluate

$$\int \frac{dx}{\sqrt{e^{2x} - 6}}.$$

Solution

$$\int \frac{dx}{\sqrt{e^{2x} - 6}} = \int \frac{du/u}{\sqrt{u^2 - a^2}} \qquad \begin{array}{l} u = e^x, du = e^x\,dx, \\ dx = du/e^x = du/u, \\ a = \sqrt{6} \end{array}$$

$$= \int \frac{du}{u\sqrt{u^2 - a^2}}$$

$$= \frac{1}{a} \sec^{-1}\left|\frac{u}{a}\right| + C \qquad \text{Table 7.4, Formula 3}$$

$$= \frac{1}{\sqrt{6}} \sec^{-1}\left(\frac{e^x}{\sqrt{6}}\right) + C$$

■

EXERCISES 7.7

Common Values of Inverse Trignonometric Functions

Use reference triangles like those in Examples 1–3 to find the angles in Exercises 1–12.

1. **a.** $\tan^{-1} 1$ **b.** $\tan^{-1}(-\sqrt{3})$ **c.** $\tan^{-1}\left(\frac{1}{\sqrt{3}}\right)$

2. **a.** $\tan^{-1}(-1)$ **b.** $\tan^{-1}\sqrt{3}$ **c.** $\tan^{-1}\left(\frac{-1}{\sqrt{3}}\right)$

3. **a.** $\sin^{-1}\left(\frac{-1}{2}\right)$ **b.** $\sin^{-1}\left(\frac{1}{\sqrt{2}}\right)$ **c.** $\sin^{-1}\left(\frac{-\sqrt{3}}{2}\right)$

4. **a.** $\sin^{-1}\left(\frac{1}{2}\right)$ **b.** $\sin^{-1}\left(\frac{-1}{\sqrt{2}}\right)$ **c.** $\sin^{-1}\left(\frac{\sqrt{3}}{2}\right)$

5. **a.** $\cos^{-1}\left(\frac{1}{2}\right)$ **b.** $\cos^{-1}\left(\frac{-1}{\sqrt{2}}\right)$ **c.** $\cos^{-1}\left(\frac{\sqrt{3}}{2}\right)$

6. **a.** $\cos^{-1}\left(\frac{-1}{2}\right)$ **b.** $\cos^{-1}\left(\frac{1}{\sqrt{2}}\right)$ **c.** $\cos^{-1}\left(\frac{-\sqrt{3}}{2}\right)$

7. **a.** $\sec^{-1}(-\sqrt{2})$ **b.** $\sec^{-1}\left(\frac{2}{\sqrt{3}}\right)$ **c.** $\sec^{-1}(-2)$

8. **a.** $\sec^{-1}\sqrt{2}$ **b.** $\sec^{-1}\left(\frac{-2}{\sqrt{3}}\right)$ **c.** $\sec^{-1} 2$

9. **a.** $\csc^{-1}\sqrt{2}$ **b.** $\csc^{-1}\left(\frac{-2}{\sqrt{3}}\right)$ **c.** $\csc^{-1} 2$

10. **a.** $\csc^{-1}(-\sqrt{2})$ **b.** $\csc^{-1}\left(\frac{2}{\sqrt{3}}\right)$ **c.** $\csc^{-1}(-2)$

11. **a.** $\cot^{-1}(-1)$ **b.** $\cot^{-1}(\sqrt{3})$ **c.** $\cot^{-1}\left(\frac{-1}{\sqrt{3}}\right)$

12. **a.** $\cot^{-1}(1)$ **b.** $\cot^{-1}(-\sqrt{3})$ **c.** $\cot^{-1}\left(\frac{1}{\sqrt{3}}\right)$

Trigonometric Function Values

13. Given that $\alpha = \sin^{-1}(5/13)$, find $\cos\alpha$, $\tan\alpha$, $\sec\alpha$, $\csc\alpha$, and $\cot\alpha$.

14. Given that $\alpha = \tan^{-1}(4/3)$, find $\sin\alpha$, $\cos\alpha$, $\sec\alpha$, $\csc\alpha$, and $\cot\alpha$.

15. Given that $\alpha = \sec^{-1}(-\sqrt{5})$, find $\sin\alpha$, $\cos\alpha$, $\tan\alpha$, $\csc\alpha$, and $\cot\alpha$.

16. Given that $\alpha = \sec^{-1}(-\sqrt{13}/2)$, find $\sin\alpha$, $\cos\alpha$, $\tan\alpha$, $\csc\alpha$, and $\cot\alpha$.

Evaluating Trigonometric and Inverse Trigonometric Terms

Find the values in Exercises 17–28.

17. $\sin\left(\cos^{-1}\left(\frac{\sqrt{2}}{2}\right)\right)$ **18.** $\sec\left(\cos^{-1}\frac{1}{2}\right)$

19. $\tan\left(\sin^{-1}\left(-\frac{1}{2}\right)\right)$ **20.** $\cot\left(\sin^{-1}\left(-\frac{\sqrt{3}}{2}\right)\right)$

21. $\csc(\sec^{-1} 2) + \cos(\tan^{-1}(-\sqrt{3}))$

22. $\tan(\sec^{-1} 1) + \sin(\csc^{-1}(-2))$

23. $\sin\left(\sin^{-1}\left(-\frac{1}{2}\right) + \cos^{-1}\left(-\frac{1}{2}\right)\right)$

24. $\cot\left(\sin^{-1}\left(-\frac{1}{2}\right) - \sec^{-1} 2\right)$

25. $\sec(\tan^{-1} 1 + \csc^{-1} 1)$ **26.** $\sec(\cot^{-1}\sqrt{3} + \csc^{-1}(-1))$

27. $\sec^{-1}\left(\sec\left(-\frac{\pi}{6}\right)\right)$ (The answer is *not* $-\pi/6$.)

28. $\cot^{-1}\left(\cot\left(-\frac{\pi}{4}\right)\right)$ (The answer is *not* $-\pi/4$.)

Finding Trigonometric Expressions

Evaluate the expressions in Exercises 29–40.

29. $\sec\left(\tan^{-1}\frac{x}{2}\right)$ **30.** $\sec(\tan^{-1} 2x)$

31. $\tan(\sec^{-1} 3y)$ **32.** $\tan\left(\sec^{-1}\frac{y}{5}\right)$

33. $\cos(\sin^{-1} x)$ **34.** $\tan(\cos^{-1} x)$

35. $\sin(\tan^{-1}\sqrt{x^2 - 2x}), \quad x \geq 2$

36. $\sin\left(\tan^{-1}\frac{x}{\sqrt{x^2+1}}\right)$ **37.** $\cos\left(\sin^{-1}\frac{2y}{3}\right)$

38. $\cos\left(\sin^{-1}\frac{y}{5}\right)$ **39.** $\sin\left(\sec^{-1}\frac{x}{4}\right)$

40. $\sin\sec^{-1}\left(\frac{\sqrt{x^2+4}}{x}\right)$

Limits

Find the limits in Exercises 41–48. (If in doubt, look at the function's graph.)

41. $\lim_{x\to 1^-}\sin^{-1} x$ **42.** $\lim_{x\to -1^+}\cos^{-1} x$

43. $\lim_{x\to\infty}\tan^{-1} x$ **44.** $\lim_{x\to -\infty}\tan^{-1} x$

45. $\lim_{x\to\infty}\sec^{-1} x$ **46.** $\lim_{x\to -\infty}\sec^{-1} x$

47. $\lim_{x\to\infty}\csc^{-1} x$ **48.** $\lim_{x\to -\infty}\csc^{-1} x$

Finding Derivatives

In Exercises 49–70, find the derivative of y with respect to the appropriate variable.

49. $y = \cos^{-1}(x^2)$ **50.** $y = \cos^{-1}(1/x)$

51. $y = \sin^{-1}\sqrt{2}\,t$ **52.** $y = \sin^{-1}(1 - t)$

53. $y = \sec^{-1}(2s + 1)$ **54.** $y = \sec^{-1} 5s$

55. $y = \csc^{-1}(x^2 + 1), \quad x > 0$

56. $y = \csc^{-1}\frac{x}{2}$

57. $y = \sec^{-1}\frac{1}{t}, \quad 0 < t < 1$ **58.** $y = \sin^{-1}\frac{3}{t^2}$

59. $y = \cot^{-1}\sqrt{t}$ **60.** $y = \cot^{-1}\sqrt{t - 1}$

61. $y = \ln(\tan^{-1} x)$ **62.** $y = \tan^{-1}(\ln x)$

63. $y = \csc^{-1}(e^t)$ **64.** $y = \cos^{-1}(e^{-t})$

65. $y = s\sqrt{1 - s^2} + \cos^{-1} s$ **66.** $y = \sqrt{s^2 - 1} - \sec^{-1} s$

67. $y = \tan^{-1}\sqrt{x^2 - 1} + \csc^{-1} x, \quad x > 1$

68. $y = \cot^{-1}\frac{1}{x} - \tan^{-1} x$ **69.** $y = x\sin^{-1} x + \sqrt{1 - x^2}$

70. $y = \ln(x^2 + 4) - x\tan^{-1}\left(\frac{x}{2}\right)$

Evaluating Integrals

Evaluate the integrals in Exercises 71–94.

71. $\int \frac{dx}{\sqrt{9 - x^2}}$ **72.** $\int \frac{dx}{\sqrt{1 - 4x^2}}$

73. $\displaystyle\int \frac{dx}{17 + x^2}$

74. $\displaystyle\int \frac{dx}{9 + 3x^2}$

75. $\displaystyle\int \frac{dx}{x\sqrt{25x^2 - 2}}$

76. $\displaystyle\int \frac{dx}{x\sqrt{5x^2 - 4}}$

77. $\displaystyle\int_0^1 \frac{4\, ds}{\sqrt{4 - s^2}}$

78. $\displaystyle\int_0^{3\sqrt{2}/4} \frac{ds}{\sqrt{9 - 4s^2}}$

79. $\displaystyle\int_0^2 \frac{dt}{8 + 2t^2}$

80. $\displaystyle\int_{-2}^2 \frac{dt}{4 + 3t^2}$

81. $\displaystyle\int_{-1}^{-\sqrt{2}/2} \frac{dy}{y\sqrt{4y^2 - 1}}$

82. $\displaystyle\int_{-2/3}^{-\sqrt{2}/3} \frac{dy}{y\sqrt{9y^2 - 1}}$

83. $\displaystyle\int \frac{3\, dr}{\sqrt{1 - 4(r - 1)^2}}$

84. $\displaystyle\int \frac{6\, dr}{\sqrt{4 - (r + 1)^2}}$

85. $\displaystyle\int \frac{dx}{2 + (x - 1)^2}$

86. $\displaystyle\int \frac{dx}{1 + (3x + 1)^2}$

87. $\displaystyle\int \frac{dx}{(2x - 1)\sqrt{(2x - 1)^2 - 4}}$

88. $\displaystyle\int \frac{dx}{(x + 3)\sqrt{(x + 3)^2 - 25}}$

89. $\displaystyle\int_{-\pi/2}^{\pi/2} \frac{2\cos\theta\, d\theta}{1 + (\sin\theta)^2}$

90. $\displaystyle\int_{\pi/6}^{\pi/4} \frac{\csc^2 x\, dx}{1 + (\cot x)^2}$

91. $\displaystyle\int_0^{\ln\sqrt{3}} \frac{e^x\, dx}{1 + e^{2x}}$

92. $\displaystyle\int_1^{e^{\pi/4}} \frac{4\, dt}{t(1 + \ln^2 t)}$

93. $\displaystyle\int \frac{y\, dy}{\sqrt{1 - y^4}}$

94. $\displaystyle\int \frac{\sec^2 y\, dy}{\sqrt{1 - \tan^2 y}}$

Evaluate the integrals in Exercises 95–104.

95. $\displaystyle\int \frac{dx}{\sqrt{-x^2 + 4x - 3}}$

96. $\displaystyle\int \frac{dx}{\sqrt{2x - x^2}}$

97. $\displaystyle\int_{-1}^0 \frac{6\, dt}{\sqrt{3 - 2t - t^2}}$

98. $\displaystyle\int_{1/2}^1 \frac{6\, dt}{\sqrt{3 + 4t - 4t^2}}$

99. $\displaystyle\int \frac{dy}{y^2 - 2y + 5}$

100. $\displaystyle\int \frac{dy}{y^2 + 6y + 10}$

101. $\displaystyle\int_1^2 \frac{8\, dx}{x^2 - 2x + 2}$

102. $\displaystyle\int_2^4 \frac{2\, dx}{x^2 - 6x + 10}$

103. $\displaystyle\int \frac{dx}{(x + 1)\sqrt{x^2 + 2x}}$

104. $\displaystyle\int \frac{dx}{(x - 2)\sqrt{x^2 - 4x + 3}}$

Evaluate the integrals in Exercises 105–112.

105. $\displaystyle\int \frac{e^{\sin^{-1} x}\, dx}{\sqrt{1 - x^2}}$

106. $\displaystyle\int \frac{e^{\cos^{-1} x}\, dx}{\sqrt{1 - x^2}}$

107. $\displaystyle\int \frac{(\sin^{-1} x)^2\, dx}{\sqrt{1 - x^2}}$

108. $\displaystyle\int \frac{\sqrt{\tan^{-1} x}\, dx}{1 + x^2}$

109. $\displaystyle\int \frac{dy}{(\tan^{-1} y)(1 + y^2)}$

110. $\displaystyle\int \frac{dy}{(\sin^{-1} y)\sqrt{1 - y^2}}$

111. $\displaystyle\int_{\sqrt{2}}^2 \frac{\sec^2(\sec^{-1} x)\, dx}{x\sqrt{x^2 - 1}}$

112. $\displaystyle\int_{2/\sqrt{3}}^2 \frac{\cos(\sec^{-1} x)\, dx}{x\sqrt{x^2 - 1}}$

Limits

Find the limits in Exercises 113–116.

113. $\displaystyle\lim_{x\to 0} \frac{\sin^{-1} 5x}{x}$

114. $\displaystyle\lim_{x\to 1^+} \frac{\sqrt{x^2 - 1}}{\sec^{-1} x}$

115. $\displaystyle\lim_{x\to\infty} x\tan^{-1}\frac{2}{x}$

116. $\displaystyle\lim_{x\to 0} \frac{2\tan^{-1} 3x^2}{7x^2}$

Integration Formulas

Verify the integration formulas in Exercises 117–120.

117. $\displaystyle\int \frac{\tan^{-1} x}{x^2}\, dx = \ln x - \frac{1}{2}\ln(1 + x^2) - \frac{\tan^{-1} x}{x} + C$

118. $\displaystyle\int x^3\cos^{-1} 5x\, dx = \frac{x^4}{4}\cos^{-1} 5x + \frac{5}{4}\int \frac{x^4\, dx}{\sqrt{1 - 25x^2}}$

119. $\displaystyle\int (\sin^{-1} x)^2\, dx = x(\sin^{-1} x)^2 - 2x + 2\sqrt{1 - x^2}\,\sin^{-1} x + C$

120. $\displaystyle\int \ln(a^2 + x^2)\, dx = x\ln(a^2 + x^2) - 2x + 2a\tan^{-1}\frac{x}{a} + C$

Initial Value Problems

Solve the initial value problems in Exercises 121–124.

121. $\displaystyle\frac{dy}{dx} = \frac{1}{\sqrt{1 - x^2}}, \quad y(0) = 0$

122. $\displaystyle\frac{dy}{dx} = \frac{1}{x^2 + 1} - 1, \quad y(0) = 1$

123. $\displaystyle\frac{dy}{dx} = \frac{1}{x\sqrt{x^2 - 1}}, \quad x > 1; \quad y(2) = \pi$

124. $\displaystyle\frac{dy}{dx} = \frac{1}{1 + x^2} - \frac{2}{\sqrt{1 - x^2}}, \quad y(0) = 2$

Applications and Theory

125. You are sitting in a classroom next to the wall looking at the blackboard at the front of the room. The blackboard is 12 ft long and starts 3 ft from the wall you are sitting next to. Show that your viewing angle is

$$\alpha = \cot^{-1}\frac{x}{15} - \cot^{-1}\frac{x}{3}$$

if you are x ft from the front wall.

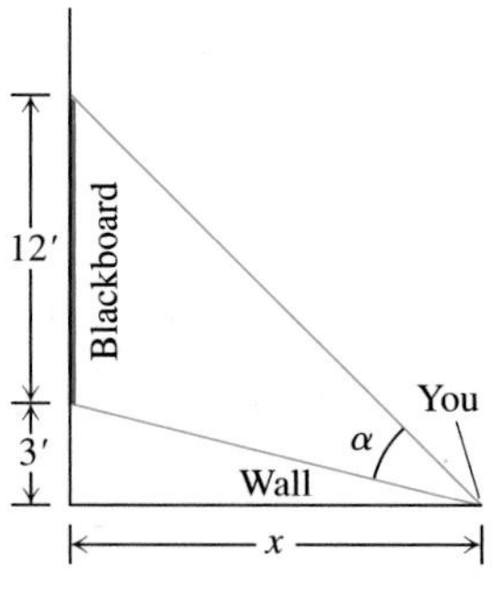

126. The region between the curve $y = \sec^{-1} x$ and the x-axis from $x = 1$ to $x = 2$ (shown here) is revolved about the y-axis to generate a solid. Find the volume of the solid.

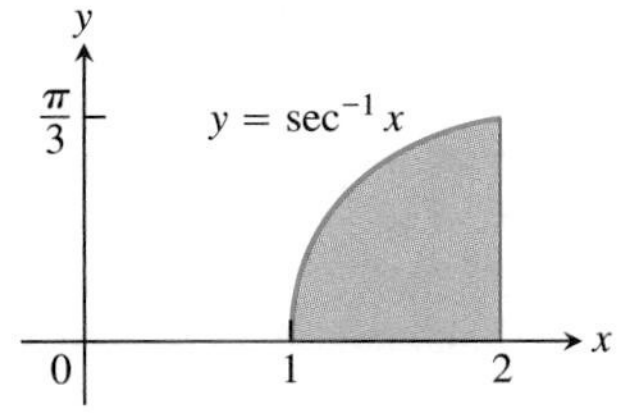

127. The slant height of the cone shown here is 3 m. How large should the indicated angle be to maximize the cone's volume?

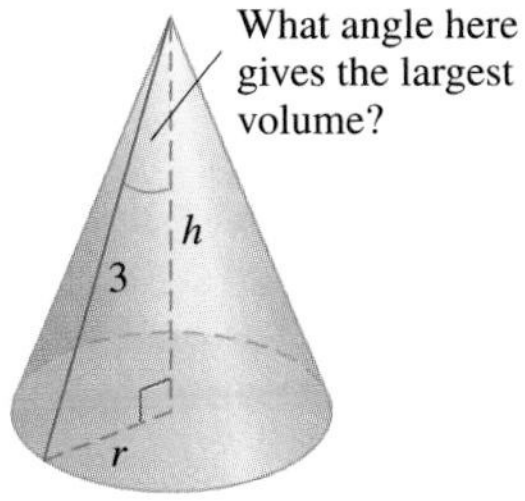

128. Find the angle α.

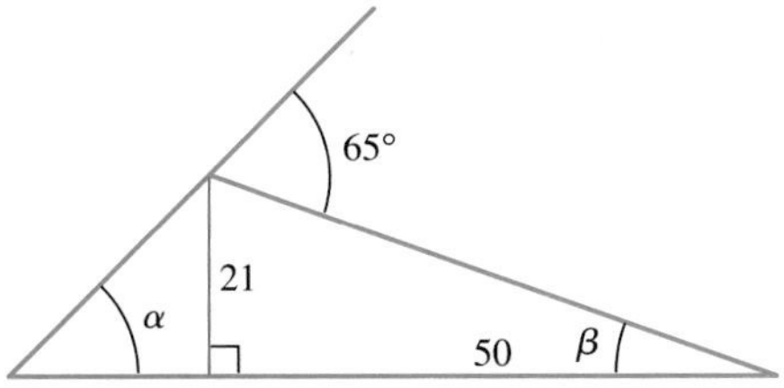

129. Here is an informal proof that $\tan^{-1} 1 + \tan^{-1} 2 + \tan^{-1} 3 = \pi$. Explain what is going on.

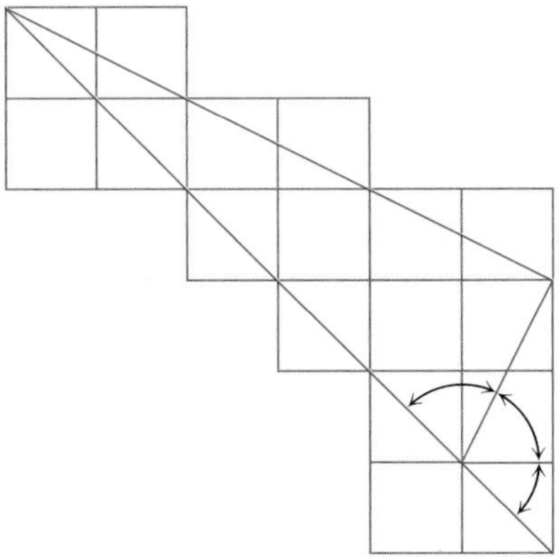

130. Two derivations of the identity $\sec^{-1}(-x) = \pi - \sec^{-1} x$

a. (*Geometric*) Here is a pictorial proof that $\sec^{-1}(-x) = \pi - \sec^{-1} x$. See if you can tell what is going on.

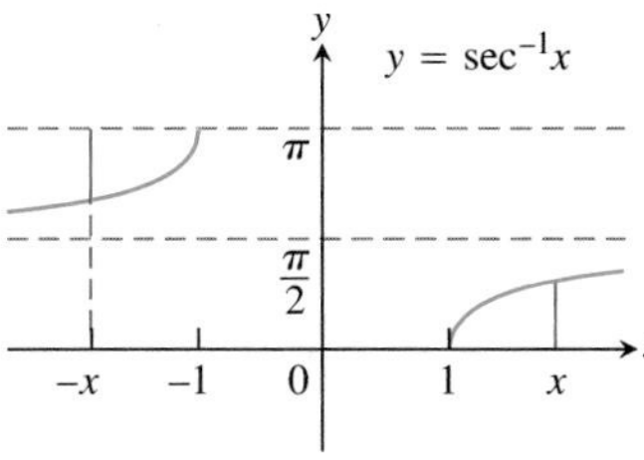

b. (*Algebraic*) Derive the identity $\sec^{-1}(-x) = \pi - \sec^{-1} x$ by combining the following two equations from the text:

$$\cos^{-1}(-x) = \pi - \cos^{-1} x \qquad \text{Eq. (3)}$$

$$\sec^{-1} x = \cos^{-1}(1/x) \qquad \text{Eq. (5)}$$

131. The identity $\sin^{-1} x + \cos^{-1} x = \pi/2$ Figure 7.21 establishes the identity for $0 < x < 1$. To establish it for the rest of $[-1, 1]$, verify by direct calculation that it holds for $x = 1, 0$, and -1. Then, for values of x in $(-1, 0)$, let $x = -a, a > 0$, and apply Eqs. (1) and (3) to the sum $\sin^{-1}(-a) + \cos^{-1}(-a)$.

132. Show that the sum $\tan^{-1} x + \tan^{-1}(1/x)$ is constant.

Which of the expressions in Exercises 133–136 are defined, and which are not? Give reasons for your answers.

133. a. $\tan^{-1} 2$ **b.** $\cos^{-1} 2$

134. a. $\csc^{-1}(1/2)$ **b.** $\csc^{-1} 2$

135. a. $\sec^{-1} 0$ **b.** $\sin^{-1}\sqrt{2}$

136. a. $\cot^{-1}(-1/2)$ **b.** $\cos^{-1}(-5)$

137. (*Continuation of Exercise 125.*) You want to position your chair along the wall to maximize your viewing angle α. How far from the front of the room should you sit?

138. What value of x maximizes the angle θ shown here? How large is θ at that point? Begin by showing that $\theta = \pi - \cot^{-1} x - \cot^{-1}(2 - x)$.

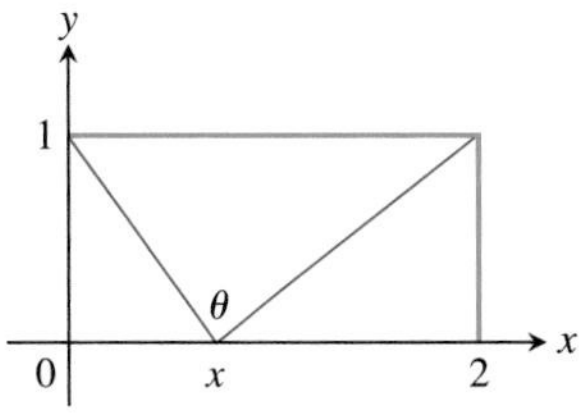

139. Can the integrations in (a) and (b) both be correct? Explain.

a. $$\int \frac{dx}{\sqrt{1 - x^2}} = \sin^{-1} x + C$$

b. $$\int \frac{dx}{\sqrt{1 - x^2}} = -\int -\frac{dx}{\sqrt{1 - x^2}} = -\cos^{-1} x + C$$

140. Can the integrations in (a) and (b) both be correct? Explain.

a. $$\int \frac{dx}{\sqrt{1 - x^2}} = -\int -\frac{dx}{\sqrt{1 - x^2}} = -\cos^{-1} x + C$$

b. $\displaystyle\int \frac{dx}{\sqrt{1-x^2}} = \int \frac{-du}{\sqrt{1-(-u)^2}}$ $x = -u,$ $dx = -du$

$\displaystyle = \int \frac{-du}{\sqrt{1-u^2}}$

$= \cos^{-1} u + C$

$= \cos^{-1}(-x) + C$ $u = -x$

141. Use the identity

$$\csc^{-1} u = \frac{\pi}{2} - \sec^{-1} u$$

to derive the formula for the derivative of $\csc^{-1} u$ in Table 7.3 from the formula for the derivative of $\sec^{-1} u$.

142. Derive the formula

$$\frac{dy}{dx} = \frac{1}{1+x^2}$$

for the derivative of $y = \tan^{-1} x$ by differentiating both sides of the equivalent equation $\tan y = x$.

143. Use the Derivative Rule in Section 7.1, Theorem 1, to derive

$$\frac{d}{dx}\sec^{-1} x = \frac{1}{|x|\sqrt{x^2-1}}, \quad |x| > 1.$$

144. Use the identity

$$\cot^{-1} u = \frac{\pi}{2} - \tan^{-1} u$$

to derive the formula for the derivative of $\cot^{-1} u$ in Table 7.3 from the formula for the derivative of $\tan^{-1} u$.

145. What is special about the functions

$$f(x) = \sin^{-1}\frac{x-1}{x+1}, \quad x \ge 0, \quad \text{and} \quad g(x) = 2\tan^{-1}\sqrt{x}?$$

Explain.

146. What is special about the functions

$$f(x) = \sin^{-1}\frac{1}{\sqrt{x^2+1}} \quad \text{and} \quad g(x) = \tan^{-1}\frac{1}{x}?$$

Explain.

147. Find the volume of the solid of revolution shown here.

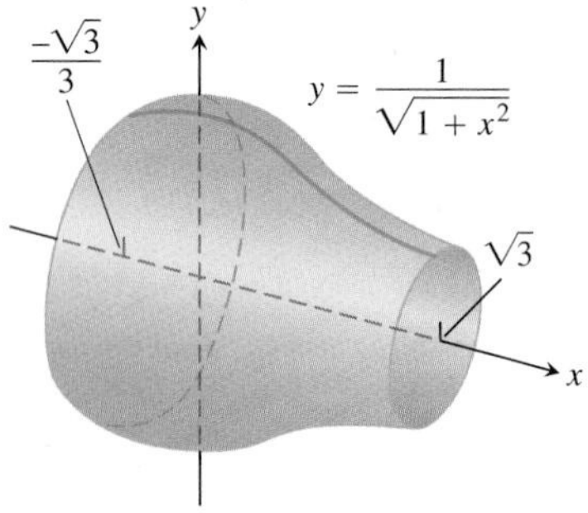

148. Arc length Find the length of the curve $y = \sqrt{1-x^2}$, $-1/2 \le x \le 1/2$.

Volumes by Slicing

Find the volumes of the solids in Exercises 149 and 150.

149. The solid lies between planes perpendicular to the x-axis at $x = -1$ and $x = 1$. The cross-sections perpendicular to the x-axis are

a. circles whose diameters stretch from the curve $y = -1/\sqrt{1+x^2}$ to the curve $y = 1/\sqrt{1+x^2}$.

b. vertical squares whose base edges run from the curve $y = -1/\sqrt{1+x^2}$ to the curve $y = 1/\sqrt{1+x^2}$.

150. The solid lies between planes perpendicular to the x-axis at $x = -\sqrt{2}/2$ and $x = \sqrt{2}/2$. The cross-sections perpendicular to the x-axis are

a. circles whose diameters stretch from the x-axis to the curve $y = 2/\sqrt[4]{1-x^2}$.

b. squares whose diagonals stretch from the x-axis to the curve $y = 2/\sqrt[4]{1-x^2}$.

T Calculator and Grapher Explorations

151. Find the values of

a. $\sec^{-1} 1.5$ **b.** $\csc^{-1}(-1.5)$ **c.** $\cot^{-1} 2$

152. Find the values of

a. $\sec^{-1}(-3)$ **b.** $\csc^{-1} 1.7$ **c.** $\cot^{-1}(-2)$

In Exercises 153–155, find the domain and range of each composite function. Then graph the composites on separate screens. Do the graphs make sense in each case? Give reasons for your answers. Comment on any differences you see.

153. a. $y = \tan^{-1}(\tan x)$ **b.** $y = \tan(\tan^{-1} x)$

154. a. $y = \sin^{-1}(\sin x)$ **b.** $y = \sin(\sin^{-1} x)$

155. a. $y = \cos^{-1}(\cos x)$ **b.** $y = \cos(\cos^{-1} x)$

156. Graph $y = \sec(\sec^{-1} x) = \sec(\cos^{-1}(1/x))$. Explain what you see.

157. Newton's serpentine Graph Newton's serpentine, $y = 4x/(x^2+1)$. Then graph $y = 2\sin(2\tan^{-1} x)$ in the same graphing window. What do you see? Explain.

158. Graph the rational function $y = (2-x^2)/x^2$. Then graph $y = \cos(2\sec^{-1} x)$ in the same graphing window. What do you see? Explain.

159. Graph $f(x) = \sin^{-1} x$ together with its first two derivatives. Comment on the behavior of f and the shape of its graph in relation to the signs and values of f' and f''.

160. Graph $f(x) = \tan^{-1} x$ together with its first two derivatives. Comment on the behavior of f and the shape of its graph in relation to the signs and values of f' and f''.

7.8 Hyperbolic Functions

The hyperbolic functions are formed by taking combinations of the two exponential functions e^x and e^{-x}. The hyperbolic functions simplify many mathematical expressions and they are important in applications. For instance, they are used in problems such as computing the tension in a cable suspended by its two ends, as in an electric transmission line. They also play an important role in finding solutions to differential equations. In this section, we give a brief introduction to hyperbolic functions, their graphs, how their derivatives are calculated, and why they appear as important antiderivatives.

Even and Odd Parts of the Exponential Function

Recall the definitions of even and odd functions from Section 1.4, and the symmetries of their graphs. An even function f satisfies $f(-x) = f(x)$, while an odd function satisfies $f(-x) = -f(x)$. Every function f that is defined on an interval centered at the origin can be written in a unique way as the sum of one even function and one odd function. The decomposition is

$$f(x) = \underbrace{\frac{f(x) + f(-x)}{2}}_{\text{even part}} + \underbrace{\frac{f(x) - f(-x)}{2}}_{\text{odd part}}.$$

If we write e^x this way, we get

$$e^x = \underbrace{\frac{e^x + e^{-x}}{2}}_{\text{even part}} + \underbrace{\frac{e^x - e^{-x}}{2}}_{\text{odd part}}.$$

The even and odd parts of e^x, called the hyperbolic cosine and hyperbolic sine of x, respectively, are useful in their own right. They describe the motions of waves in elastic solids and the temperature distributions in metal cooling fins. The centerline of the Gateway Arch to the West in St. Louis is a weighted hyperbolic cosine curve.

Definitions and Identities

The hyperbolic cosine and hyperbolic sine functions are defined by the first two equations in Table 7.5. The table also lists the definitions of the hyperbolic tangent, cotangent, secant, and cosecant. As we will see, the hyperbolic functions bear a number of similarities to the trigonometric functions after which they are named. (See Exercise 84 as well.)

The notation $\cosh x$ is often read "kosh x," rhyming with "gosh x," and $\sinh x$ is pronounced as if spelled "cinch x," rhyming with "pinch x."

Hyperbolic functions satisfy the identities in Table 7.6. Except for differences in sign, these resemble identities we already know for trigonometric functions.

The second equation is obtained as follows:

$$\begin{aligned} 2\sinh x \cosh x &= 2\left(\frac{e^x - e^{-x}}{2}\right)\left(\frac{e^x + e^{-x}}{2}\right) \\ &= \frac{e^{2x} - e^{-2x}}{2} \\ &= \sinh 2x. \end{aligned}$$

TABLE 7.5 The six basic hyperbolic functions

Hyperbolic sine of x:	$\sinh x = \dfrac{e^x - e^{-x}}{2}$
Hyperbolic cosine of x:	$\cosh x = \dfrac{e^x + e^{-x}}{2}$
Hyperbolic tangent:	$\tanh x = \dfrac{\sinh x}{\cosh x} = \dfrac{e^x - e^{-x}}{e^x + e^{-x}}$
Hyperbolic cotangent:	$\coth x = \dfrac{\cosh x}{\sinh x} = \dfrac{e^x + e^{-x}}{e^x - e^{-x}}$
Hyperbolic secant:	$\operatorname{sech} x = \dfrac{1}{\cosh x} = \dfrac{2}{e^x + e^{-x}}$
Hyperbolic cosecant:	$\operatorname{csch} x = \dfrac{1}{\sinh x} = \dfrac{2}{e^x - e^{-x}}$

FIGURE 7.31

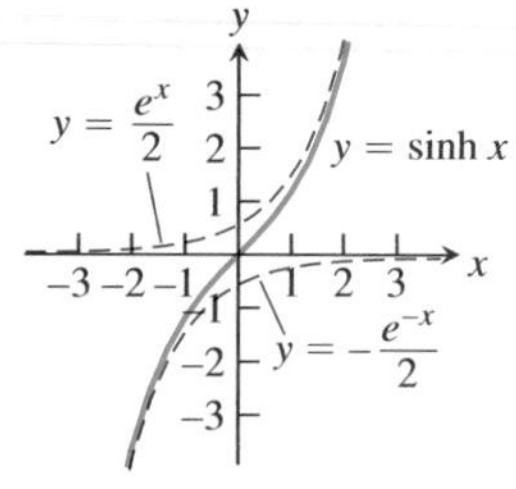

(a)

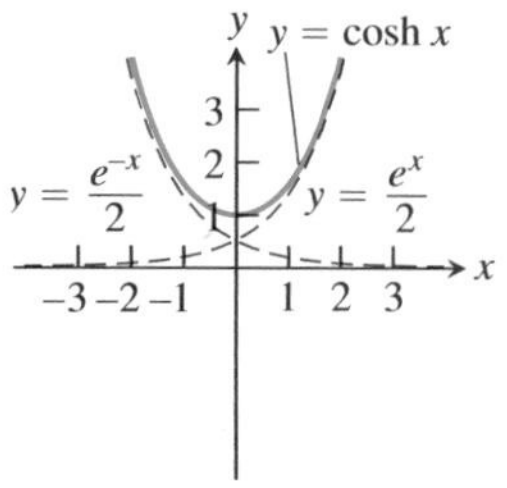

(b)

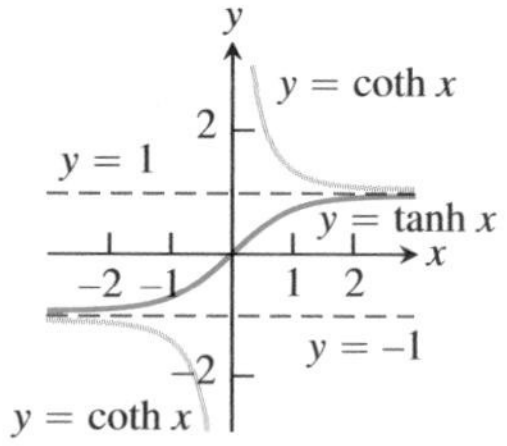

(c)

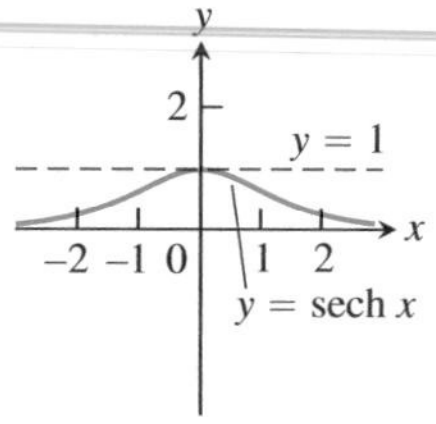

(d)

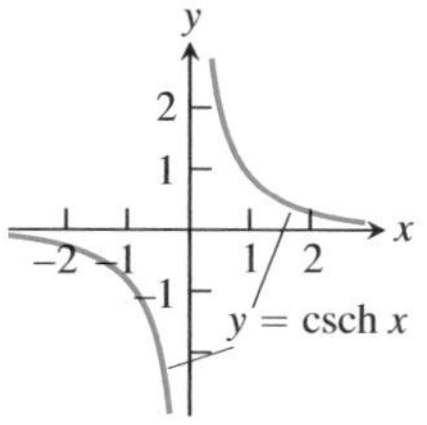

(e)

TABLE 7.6 Identities for hyperbolic functions

$$\cosh^2 x - \sinh^2 x = 1$$
$$\sinh 2x = 2 \sinh x \cosh x$$
$$\cosh 2x = \cosh^2 x + \sinh^2 x$$
$$\cosh^2 x = \frac{\cosh 2x + 1}{2}$$
$$\sinh^2 x = \frac{\cosh 2x - 1}{2}$$
$$\tanh^2 x = 1 - \operatorname{sech}^2 x$$
$$\coth^2 x = 1 + \operatorname{csch}^2 x$$

The other identities are obtained similarly, by substituting in the definitions of the hyperbolic functions and using algebra. Like many standard functions, hyperbolic functions and their inverses are easily evaluated with calculators, which have special keys or keystroke sequences for that purpose.

Derivatives and Integrals

The six hyperbolic functions, being rational combinations of the differentiable functions e^x and e^{-x}, have derivatives at every point at which they are defined (Table 7.7). Again, there are similarities with trigonometric functions. The derivative formulas in Table 7.7 lead to the integral formulas in Table 7.8.

TABLE 7.7 Derivatives of hyperbolic functions

$$\frac{d}{dx}(\sinh u) = \cosh u \frac{du}{dx}$$

$$\frac{d}{dx}(\cosh u) = \sinh u \frac{du}{dx}$$

$$\frac{d}{dx}(\tanh u) = \operatorname{sech}^2 u \frac{du}{dx}$$

$$\frac{d}{dx}(\coth u) = -\operatorname{csch}^2 u \frac{du}{dx}$$

$$\frac{d}{dx}(\operatorname{sech} u) = -\operatorname{sech} u \tanh u \frac{du}{dx}$$

$$\frac{d}{dx}(\operatorname{csch} u) = -\operatorname{csch} u \coth u \frac{du}{dx}$$

TABLE 7.8 Integral formulas for hyperbolic functions

$$\int \sinh u \, du = \cosh u + C$$

$$\int \cosh u \, du = \sinh u + C$$

$$\int \operatorname{sech}^2 u \, du = \tanh u + C$$

$$\int \operatorname{csch}^2 u \, du = -\coth u + C$$

$$\int \operatorname{sech} u \tanh u \, du = -\operatorname{sech} u + C$$

$$\int \operatorname{csch} u \coth u \, du = -\operatorname{csch} u + C$$

The derivative formulas are derived from the derivative of e^u:

$$\begin{aligned} \frac{d}{dx}(\sinh u) &= \frac{d}{dx}\left(\frac{e^u - e^{-u}}{2}\right) && \text{Definition of } \sinh u \\ &= \frac{e^u \, du/dx + e^{-u} \, du/dx}{2} && \text{Derivative of } e^u \\ &= \cosh u \frac{du}{dx} && \text{Definition of } \cosh u \end{aligned}$$

This gives the first derivative formula. The calculation

$$\begin{aligned} \frac{d}{dx}(\operatorname{csch} u) &= \frac{d}{dx}\left(\frac{1}{\sinh u}\right) && \text{Definition of } \operatorname{csch} u \\ &= -\frac{\cosh u}{\sinh^2 u}\frac{du}{dx} && \text{Quotient Rule} \\ &= -\frac{1}{\sinh u}\frac{\cosh u}{\sinh u}\frac{du}{dx} && \text{Rearrange terms.} \\ &= -\operatorname{csch} u \coth u \frac{du}{dx} && \text{Definitions of } \operatorname{csch} u \text{ and } \coth u \end{aligned}$$

gives the last formula. The others are obtained similarly.

EXAMPLE 1 Finding Derivatives and Integrals

(a) $$\frac{d}{dt}\left(\tanh \sqrt{1 + t^2}\right) = \operatorname{sech}^2 \sqrt{1 = t^2} \cdot \frac{d}{dt}\left(\sqrt{1 + t^2}\right)$$
$$= \frac{t}{\sqrt{1 + t^2}} \operatorname{sech}^2 \sqrt{1 + t^2}$$

(b) $$\int \coth 5x \, dx = \int \frac{\cosh 5x}{\sinh 5x} dx = \frac{1}{5}\int \frac{du}{u}$$
$u = \sinh 5x$, $du = 5 \cosh 5x \, dx$
$$= \frac{1}{5} \ln |u| + C = \frac{1}{5} \ln |\sinh 5x| + C$$

(c) $$\int_0^1 \sinh^2 x \, dx = \int_0^1 \frac{\cosh 2x - 1}{2} dx$$
Table 7.6
$$= \frac{1}{2}\int_0^1 (\cosh 2x - 1)\, dx = \frac{1}{2}\left[\frac{\sinh 2x}{2} - x\right]_0^1$$
$$= \frac{\sinh 2}{4} - \frac{1}{2} \approx 0.40672$$
Evaluate with a calculator

(d) $$\int_0^{\ln 2} 4e^x \sinh x \, dx = \int_0^{\ln 2} 4e^x \frac{e^x - e^{-x}}{2} dx = \int_0^{\ln 2} (2e^{2x} - 2)\, dx$$
$$= \left[e^{2x} - 2x\right]_0^{\ln 2} = (e^{2\ln 2} - 2\ln 2) - (1 - 0)$$
$$= 4 - 2\ln 2 - 1$$
$$\approx 1.6137$$

Inverse Hyperbolic Functions

The inverses of the six basic hyperbolic functions are very useful in integration. Since $d(\sinh x)/dx = \cosh x > 0$, the hyperbolic sine is an increasing function of x. We denote its inverse by

$$y = \sinh^{-1} x.$$

For every value of x in the interval $-\infty < x < \infty$, the value of $y = \sinh^{-1} x$ is the number whose hyperbolic sine is x. The graphs of $y = \sinh x$ and $y = \sinh^{-1} x$ are shown in Figure 7.32a.

The function $y = \cosh x$ is not one-to-one, as we can see from the graph in Figure 7.31b. The restricted function $y = \cosh x, x \geq 0$, however, is one-to-one and therefore has an inverse, denoted by

$$y = \cosh^{-1} x.$$

For every value of $x \geq 1, y = \cosh^{-1} x$ is the number in the interval $0 \leq y < \infty$ whose hyperbolic cosine is x. The graphs of $y = \cosh x, x \geq 0$, and $y = \cosh^{-1} x$ are shown in Figure 7.32b.

Like $y = \cosh x$, the function $y = \operatorname{sech} x = 1/\cosh x$ fails to be one-to-one, but its restriction to nonnegative values of x does have an inverse, denoted by

$$y = \operatorname{sech}^{-1} x.$$

For every value of x in the interval $(0, 1]$, $y = \operatorname{sech}^{-1} x$ is the nonnegative number whose hyperbolic secant is x. The graphs of $y = \operatorname{sech} x, x \geq 0$, and $y = \operatorname{sech}^{-1} x$ are shown in Figure 7.32c.

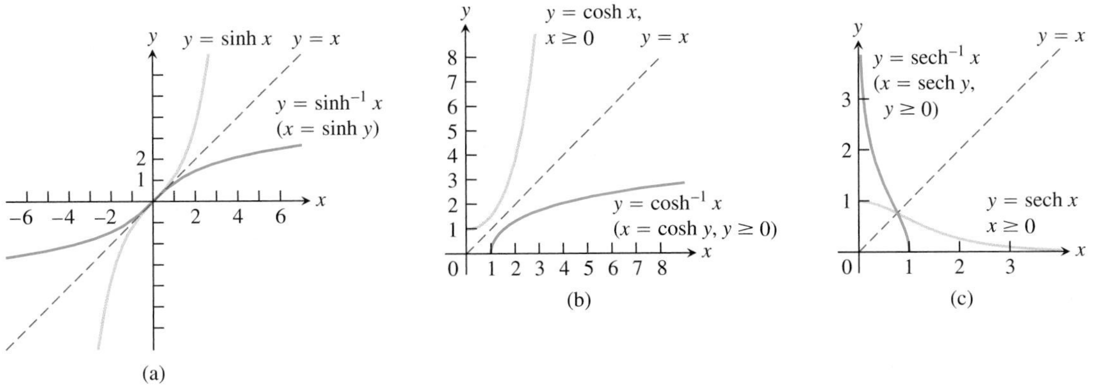

FIGURE 7.32 The graphs of the inverse hyperbolic sine, cosine, and secant of x. Notice the symmetries about the line $y = x$.

The hyperbolic tangent, cotangent, and cosecant are one-to-one on their domains and therefore have inverses, denoted by

$$y = \tanh^{-1} x, \qquad y = \coth^{-1} x, \qquad y = \operatorname{csch}^{-1} x.$$

These functions are graphed in Figure 7.33.

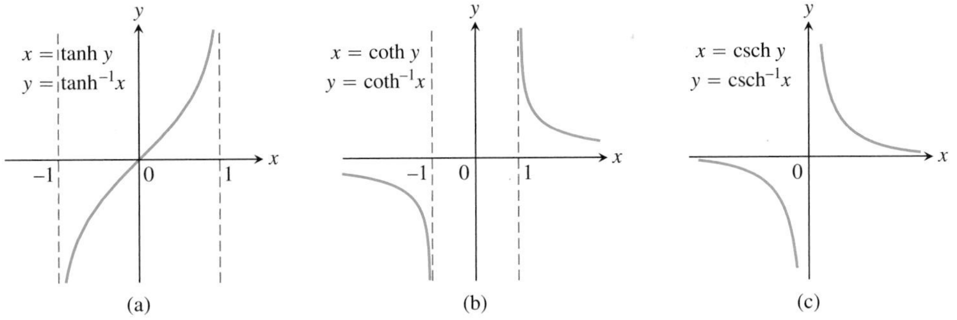

FIGURE 7.33 The graphs of the inverse hyperbolic tangent, cotangent, and cosecant of x.

TABLE 7.9 Identities for inverse hyperbolic functions

$$\operatorname{sech}^{-1} x = \cosh^{-1} \frac{1}{x}$$

$$\operatorname{csch}^{-1} x = \sinh^{-1} \frac{1}{x}$$

$$\coth^{-1} x = \tanh^{-1} \frac{1}{x}$$

Useful Identities

We use the identities in Table 7.9 to calculate the values of $\operatorname{sech}^{-1} x$, $\operatorname{csch}^{-1} x$, and $\coth^{-1} x$ on calculators that give only $\cosh^{-1} x$, $\sinh^{-1} x$, and $\tanh^{-1} x$. These identities are direct consequences of the definitions. For example, if $0 < x \leq 1$, then

$$\operatorname{sech}\left(\cosh^{-1}\left(\frac{1}{x}\right)\right) = \frac{1}{\cosh\left(\cosh^{-1}\left(\frac{1}{x}\right)\right)} = \frac{1}{\left(\frac{1}{x}\right)} = x$$

so

$$\cosh^{-1}\left(\frac{1}{x}\right) = \operatorname{sech}^{-1} x$$

since the hyperbolic secant is one-to-one on $(0, 1]$.

Derivatives and Integrals

The chief use of inverse hyperbolic functions lies in integrations that reverse the derivative formulas in Table 7.10.

TABLE 7.10 Derivatives of inverse hyperbolic functions

$$\frac{d(\sinh^{-1} u)}{dx} = \frac{1}{\sqrt{1 + u^2}}\frac{du}{dx}$$

$$\frac{d(\cosh^{-1} u)}{dx} = \frac{1}{\sqrt{u^2 - 1}}\frac{du}{dx}, \qquad u > 1$$

$$\frac{d(\tanh^{-1} u)}{dx} = \frac{1}{1 - u^2}\frac{du}{dx}, \qquad |u| < 1$$

$$\frac{d(\coth^{-1} u)}{dx} = \frac{1}{1 - u^2}\frac{du}{dx}, \qquad |u| > 1$$

$$\frac{d(\operatorname{sech}^{-1} u)}{dx} = \frac{-du/dx}{u\sqrt{1 - u^2}}, \qquad 0 < u < 1$$

$$\frac{d(\operatorname{csch}^{-1} u)}{dx} = \frac{-du/dx}{|u|\sqrt{1 + u^2}}, \qquad u \neq 0$$

The restrictions $|u| < 1$ and $|u| > 1$ on the derivative formulas for $\tanh^{-1} u$ and $\coth^{-1} u$ come from the natural restrictions on the values of these functions. (See Figure 7.33a and b.) The distinction between $|u| < 1$ and $|u| > 1$ becomes important when we convert the derivative formulas into integral formulas. If $|u| < 1$, the integral of $1/(1 - u^2)$ is $\tanh^{-1} u + C$. If $|u| > 1$, the integral is $\coth^{-1} u + C$.

We illustrate how the derivatives of the inverse hyperbolic functions are found in Example 2, where we calculate $d(\cosh^{-1} u)/dx$. The other derivatives are obtained by similar calculations.

HISTORICAL BIOGRAPHY

Sonya Kovalevsky
(1850–1891)

EXAMPLE 2 Derivative of the Inverse Hyperbolic Cosine

Show that if u is a differentiable function of x whose values are greater than 1, then

$$\frac{d}{dx}(\cosh^{-1} u) = \frac{1}{\sqrt{u^2 - 1}}\frac{du}{dx}.$$

Solution First we find the derivative of $y = \cosh^{-1} x$ for $x > 1$ by applying Theorem 1 with $f(x) = \cosh x$ and $f^{-1}(x) = \cosh^{-1} x$. Theorem 1 can be applied because the derivative of $\cosh x$ is positive for $0 < x$.

$$\begin{aligned}
(f^{-1})'(x) &= \frac{1}{f'(f^{-1}(x))} && \text{Theorem 1} \\
&= \frac{1}{\sinh(\cosh^{-1} x)} && f'(u) = \sinh u \\
&= \frac{1}{\sqrt{\cosh^2(\cosh^{-1} x) - 1}} && \cosh^2 u - \sinh^2 u = 1,\ \sinh u = \sqrt{\cosh^2 u - 1} \\
&= \frac{1}{\sqrt{x^2 - 1}} && \cosh(\cosh^{-1} x) = x
\end{aligned}$$

In short,

$$\frac{d}{dx}(\cosh^{-1} x) = \frac{1}{\sqrt{x^2 - 1}}.$$

The Chain Rule gives the final result:

$$\frac{d}{dx}(\cosh^{-1} u) = \frac{1}{\sqrt{u^2 - 1}}\frac{du}{dx}.$$

■

Instead of applying Theorem 1 directly, as in Example 2, we could also find the derivative of $y = \cosh^{-1} x,\ x > 1$, using implicit differentiation and the Chain Rule:

$$\begin{aligned}
y &= \cosh^{-1} x \\
x &= \cosh y && \text{Equivalent equation} \\
1 &= \sinh y \frac{dy}{dx} && \text{Implicit differentiation with respect to } x \text{, and the Chain Rule} \\
\frac{dy}{dx} &= \frac{1}{\sinh y} = \frac{1}{\sqrt{\cosh^2 y - 1}} && \text{Since } x > 1,\ y > 0 \text{ and } \sinh y > 0 \\
&= \frac{1}{\sqrt{x^2 - 1}}. && \cosh y = x
\end{aligned}$$

With appropriate substitutions, the derivative formulas in Table 7.10 lead to the integration formulas in Table 7.11. Each of the formulas in Table 7.11 can be verified by differentiating the expression on the right-hand side.

EXAMPLE 3 Using Table 7.11

Evaluate

$$\int_0^1 \frac{2\,dx}{\sqrt{3 + 4x^2}}.$$

TABLE 7.11 Integrals leading to inverse hyperbolic functions

1.	$\displaystyle\int \frac{du}{\sqrt{a^2+u^2}} = \sinh^{-1}\left(\frac{u}{a}\right) + C,$	$a > 0$
2.	$\displaystyle\int \frac{du}{\sqrt{u^2-a^2}} = \cosh^{-1}\left(\frac{u}{a}\right) + C,$	$u > a > 0$
3.	$\displaystyle\int \frac{du}{a^2-u^2} = \frac{1}{a}\tanh^{-1}\left(\frac{u}{a}\right) + C$	if $u^2 < a^2$
	$\displaystyle\phantom{\int \frac{du}{a^2-u^2}} = \frac{1}{a}\coth^{-1}\left(\frac{u}{a}\right) + C,$	if $u^2 > a^2$
4.	$\displaystyle\int \frac{du}{u\sqrt{a^2-u^2}} = -\frac{1}{a}\operatorname{sech}^{-1}\left(\frac{u}{a}\right) + C,$	$0 < u < a$
5.	$\displaystyle\int \frac{du}{u\sqrt{a^2+u^2}} = -\frac{1}{a}\operatorname{csch}^{-1}\left\lvert\frac{u}{a}\right\rvert + C,$	$u \neq 0$ and $a > 0$

Solution The indefinite integral is

$$\int \frac{2\,dx}{\sqrt{3+4x^2}} = \int \frac{du}{\sqrt{a^2+u^2}} \qquad u = 2x,\quad du = 2\,dx,\quad a = \sqrt{3}$$

$$= \sinh^{-1}\left(\frac{u}{a}\right) + C \qquad \text{Formula from Table 7.11}$$

$$= \sinh^{-1}\left(\frac{2x}{\sqrt{3}}\right) + C.$$

Therefore,

$$\int_0^1 \frac{2\,dx}{\sqrt{3+4x^2}} = \sinh^{-1}\left(\frac{2x}{\sqrt{3}}\right)\Bigg]_0^1 = \sinh^{-1}\left(\frac{2}{\sqrt{3}}\right) - \sinh^{-1}(0)$$

$$= \sinh^{-1}\left(\frac{2}{\sqrt{3}}\right) - 0 \approx 0.98665.$$

■

EXERCISES 7.8

Hyperbolic Function Values and Identities

Each of Exercises 1–4 gives a value of $\sinh x$ or $\cosh x$. Use the definitions and the identity $\cosh^2 x - \sinh^2 x = 1$ to find the values of the remaining five hyperbolic functions.

1. $\sinh x = -\dfrac{3}{4}$

2. $\sinh x = \dfrac{4}{3}$

3. $\cosh x = \dfrac{17}{15}, \quad x > 0$

4. $\cosh x = \dfrac{13}{5}, \quad x > 0$

Rewrite the expressions in Exercises 5–10 in terms of exponentials and simplify the results as much as you can.

5. $2\cosh(\ln x)$

6. $\sinh(2\ln x)$

7. $\cosh 5x + \sinh 5x$

8. $\cosh 3x - \sinh 3x$

9. $(\sinh x + \cosh x)^4$

10. $\ln(\cosh x + \sinh x) + \ln(\cosh x - \sinh x)$

11. Use the identities

$$\sinh(x + y) = \sinh x \cosh y + \cosh x \sinh y$$
$$\cosh(x + y) = \cosh x \cosh y + \sinh x \sinh y$$

to show that

a. $\sinh 2x = 2 \sinh x \cosh x$

b. $\cosh 2x = \cosh^2 x + \sinh^2 x$.

12. Use the definitions of $\cosh x$ and $\sinh x$ to show that

$$\cosh^2 x - \sinh^2 x = 1.$$

Derivatives

In Exercises 13–24, find the derivative of y with respect to the appropriate variable.

13. $y = 6 \sinh \frac{x}{3}$

14. $y = \frac{1}{2} \sinh(2x + 1)$

15. $y = 2\sqrt{t} \tanh \sqrt{t}$

16. $y = t^2 \tanh \frac{1}{t}$

17. $y = \ln(\sinh z)$

18. $y = \ln(\cosh z)$

19. $y = \operatorname{sech} \theta(1 - \ln \operatorname{sech} \theta)$

20. $y = \operatorname{csch} \theta(1 - \ln \operatorname{csch} \theta)$

21. $y = \ln \cosh v - \frac{1}{2} \tanh^2 v$

22. $y = \ln \sinh v - \frac{1}{2} \coth^2 v$

23. $y = (x^2 + 1) \operatorname{sech}(\ln x)$

(*Hint:* Before differentiating, express in terms of exponentials and simplify.)

24. $y = (4x^2 - 1) \operatorname{csch}(\ln 2x)$

In Exercises 25–36, find the derivative of y with respect to the appropriate variable.

25. $y = \sinh^{-1} \sqrt{x}$

26. $y = \cosh^{-1} 2\sqrt{x + 1}$

27. $y = (1 - \theta) \tanh^{-1} \theta$

28. $y = (\theta^2 + 2\theta) \tanh^{-1}(\theta + 1)$

29. $y = (1 - t) \coth^{-1} \sqrt{t}$

30. $y = (1 - t^2) \coth^{-1} t$

31. $y = \cos^{-1} x - x \operatorname{sech}^{-1} x$

32. $y = \ln x + \sqrt{1 - x^2} \operatorname{sech}^{-1} x$

33. $y = \operatorname{csch}^{-1} \left(\frac{1}{2}\right)^\theta$

34. $y = \operatorname{csch}^{-1} 2^\theta$

35. $y = \sinh^{-1}(\tan x)$

36. $y = \cosh^{-1}(\sec x), \quad 0 < x < \pi/2$

Integration Formulas

Verify the integration formulas in Exercises 37–40.

37. a. $\int \operatorname{sech} x \, dx = \tan^{-1}(\sinh x) + C$

b. $\int \operatorname{sech} x \, dx = \sin^{-1}(\tanh x) + C$

38. $\int x \operatorname{sech}^{-1} x \, dx = \frac{x^2}{2} \operatorname{sech}^{-1} x - \frac{1}{2}\sqrt{1 - x^2} + C$

39. $\int x \coth^{-1} x \, dx = \frac{x^2 - 1}{2} \coth^{-1} x + \frac{x}{2} + C$

40. $\int \tanh^{-1} x \, dx = x \tanh^{-1} x + \frac{1}{2} \ln(1 - x^2) + C$

Indefinite Integrals

Evaluate the integrals in Exercises 41–50.

41. $\int \sinh 2x \, dx$

42. $\int \sinh \frac{x}{5} \, dx$

43. $\int 6 \cosh \left(\frac{x}{2} - \ln 3\right) dx$

44. $\int 4 \cosh(3x - \ln 2) \, dx$

45. $\int \tanh \frac{x}{7} \, dx$

46. $\int \coth \frac{\theta}{\sqrt{3}} \, d\theta$

47. $\int \operatorname{sech}^2 \left(x - \frac{1}{2}\right) dx$

48. $\int \operatorname{csch}^2(5 - x) \, dx$

49. $\int \frac{\operatorname{sech} \sqrt{t} \tanh \sqrt{t} \, dt}{\sqrt{t}}$

50. $\int \frac{\operatorname{csch}(\ln t) \coth(\ln t) \, dt}{t}$

Definite Integrals

Evaluate the integrals in Exercises 51–60.

51. $\int_{\ln 2}^{\ln 4} \coth x \, dx$

52. $\int_0^{\ln 2} \tanh 2x \, dx$

53. $\int_{-\ln 4}^{-\ln 2} 2e^\theta \cosh \theta \, d\theta$

54. $\int_0^{\ln 2} 4e^{-\theta} \sinh \theta \, d\theta$

55. $\int_{-\pi/4}^{\pi/4} \cosh(\tan \theta) \sec^2 \theta \, d\theta$

56. $\int_0^{\pi/2} 2 \sinh(\sin \theta) \cos \theta \, d\theta$

57. $\int_1^2 \frac{\cosh(\ln t)}{t} \, dt$

58. $\int_1^4 \frac{8 \cosh \sqrt{x}}{\sqrt{x}} \, dx$

59. $\int_{-\ln 2}^0 \cosh^2 \left(\frac{x}{2}\right) dx$

60. $\int_0^{\ln 10} 4 \sinh^2 \left(\frac{x}{2}\right) dx$

Evaluating Inverse Hyperbolic Functions and Related Integrals

When hyperbolic function keys are not available on a calculator, it is still possible to evaluate the inverse hyperbolic functions by expressing them as logarithms, as shown here.

$\sinh^{-1} x = \ln\left(x + \sqrt{x^2 + 1}\right),$	$-\infty < x < \infty$
$\cosh^{-1} x = \ln\left(x + \sqrt{x^2 + 1}\right),$	$x \geq 1$
$\tanh^{-1} x = \frac{1}{2} \ln \frac{1 + x}{1 - x},$	$\lvert x \rvert < 1$
$\operatorname{sech}^{-1} x = \ln\left(\frac{1 + \sqrt{1 - x^2}}{x}\right),$	$0 < x \leq 1$
$\operatorname{csch}^{-1} x = \ln\left(\frac{1}{x} + \frac{\sqrt{1 + x^2}}{\lvert x \rvert}\right),$	$x \neq 0$
$\coth^{-1} x = \frac{1}{2} \ln \frac{x + 1}{x - 1},$	$\lvert x \rvert > 1$

Use the formulas in the box here to express the numbers in Exercises 61–66 in terms of natural logarithms.

61. $\sinh^{-1}(-5/12)$

62. $\cosh^{-1}(5/3)$

63. $\tanh^{-1}(-1/2)$

64. $\coth^{-1}(5/4)$

65. $\operatorname{sech}^{-1}(3/5)$

66. $\operatorname{csch}^{-1}(-1/\sqrt{3})$

Evaluate the integrals in Exercises 67–74 in terms of

a. inverse hyperbolic functions.

b. natural logarithms.

67. $\displaystyle\int_0^{2\sqrt{3}} \frac{dx}{\sqrt{4+x^2}}$

68. $\displaystyle\int_0^{1/3} \frac{6\,dx}{\sqrt{1+9x^2}}$

69. $\displaystyle\int_{5/4}^{2} \frac{dx}{1-x^2}$

70. $\displaystyle\int_0^{1/2} \frac{dx}{1-x^2}$

71. $\displaystyle\int_{1/5}^{3/13} \frac{dx}{x\sqrt{1-16x^2}}$

72. $\displaystyle\int_1^{2} \frac{dx}{x\sqrt{4+x^2}}$

73. $\displaystyle\int_0^{\pi} \frac{\cos x\,dx}{\sqrt{1+\sin^2 x}}$

74. $\displaystyle\int_1^{e} \frac{dx}{x\sqrt{1+(\ln x)^2}}$

Applications and Theory

75. a. Show that if a function f is defined on an interval symmetric about the origin (so that f is defined at $-x$ whenever it is defined at x), then

$$f(x) = \frac{f(x)+f(-x)}{2} + \frac{f(x)-f(-x)}{2}. \qquad (1)$$

Then show that $(f(x)+f(-x))/2$ is even and that $(f(x)-f(-x))/2$ is odd.

b. Equation (1) simplifies considerably if f itself is (i) even or (ii) odd. What are the new equations? Give reasons for your answers.

76. Derive the formula $\sinh^{-1} x = \ln\left(x + \sqrt{x^2+1}\right)$, $-\infty < x < \infty$. Explain in your derivation why the plus sign is used with the square root instead of the minus sign.

77. Skydiving If a body of mass m falling from rest under the action of gravity encounters an air resistance proportional to the square of the velocity, then the body's velocity t sec into the fall satisfies the differential equation

$$m\frac{dv}{dt} = mg - kv^2,$$

where k is a constant that depends on the body's aerodynamic properties and the density of the air. (We assume that the fall is short enough so that the variation in the air's density will not affect the outcome significantly.)

a. Show that

$$v = \sqrt{\frac{mg}{k}}\tanh\left(\sqrt{\frac{gk}{m}}\,t\right)$$

satisfies the differential equation and the initial condition that $v = 0$ when $t = 0$.

b. Find the body's *limiting velocity*, $\lim_{t\to\infty} v$.

c. For a 160-lb skydiver ($mg = 160$), with time in seconds and distance in feet, a typical value for k is 0.005. What is the diver's limiting velocity?

78. Accelerations whose magnitudes are proportional to displacement Suppose that the position of a body moving along a coordinate line at time t is

a. $s = a\cos kt + b\sin kt$

b. $s = a\cosh kt + b\sinh kt$.

Show in both cases that the acceleration d^2s/dt^2 is proportional to s but that in the first case it is directed toward the origin, whereas in the second case it is directed away from the origin.

79. Tractor trailers and the tractrix When a tractor trailer turns into a cross street or driveway, its rear wheels follow a curve like the one shown here. (This is why the rear wheels sometimes ride up over the curb.) We can find an equation for the curve if we picture the rear wheels as a mass M at the point $(1, 0)$ on the x-axis attached by a rod of unit length to a point P representing the cab at the origin. As the point P moves up the y-axis, it drags M along behind it. The curve traced by M—called a *tractrix* from the Latin word *tractum*, for "drag"—can be shown to be the graph of the function $y = f(x)$ that solves the initial value problem

Differential equation: $\displaystyle\frac{dy}{dx} = -\frac{1}{x\sqrt{1-x^2}} + \frac{x}{\sqrt{1-x^2}}$

Initial condition: $y = 0$ when $x = 1$.

Solve the initial value problem to find an equation for the curve. (You need an inverse hyperbolic function.)

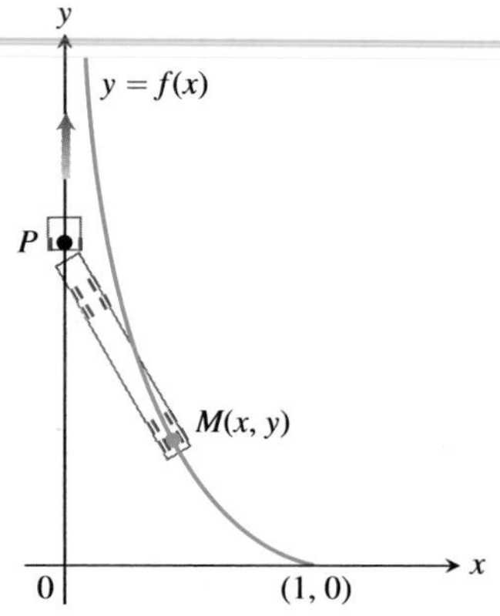

80. Area Show that the area of the region in the first quadrant enclosed by the curve $y = (1/a)\cosh ax$, the coordinate axes, and the line $x = b$ is the same as the area of a rectangle of height $1/a$ and length s, where s is the length of the curve from $x = 0$ to $x = b$. (See accompanying figure.)

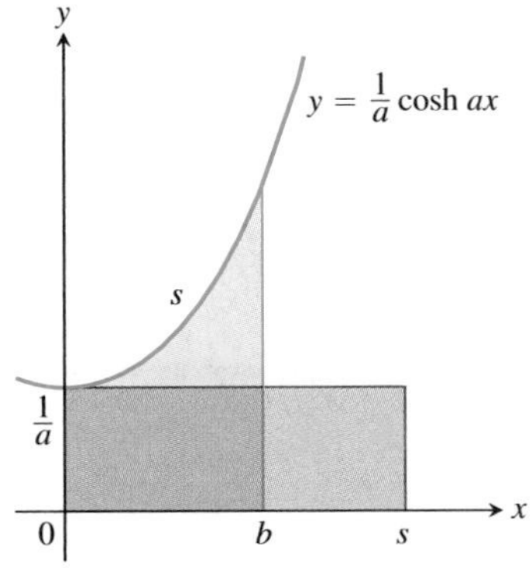

81. Volume A region in the first quadrant is bounded above by the curve $y = \cosh x$, below by the curve $y = \sinh x$, and on the left and right by the y-axis and the line $x = 2$, respectively. Find the volume of the solid generated by revolving the region about the x-axis.

82. Volume The region enclosed by the curve $y = \operatorname{sech} x$, the x-axis, and the lines $x = \pm \ln \sqrt{3}$ is revolved about the x-axis to generate a solid. Find the volume of the solid.

83. Arc length Find the length of the segment of the curve $y = (1/2)\cosh 2x$ from $x = 0$ to $x = \ln \sqrt{5}$.

84. The hyperbolic in hyperbolic functions In case you are wondering where the name *hyperbolic* comes from, here is the answer: Just as $x = \cos u$ and $y = \sin u$ are identified with points (x, y) on the unit circle, the functions $x = \cosh u$ and $y = \sinh u$ are identified with points (x, y) on the right-hand branch of the unit hyperbola, $x^2 - y^2 = 1$.

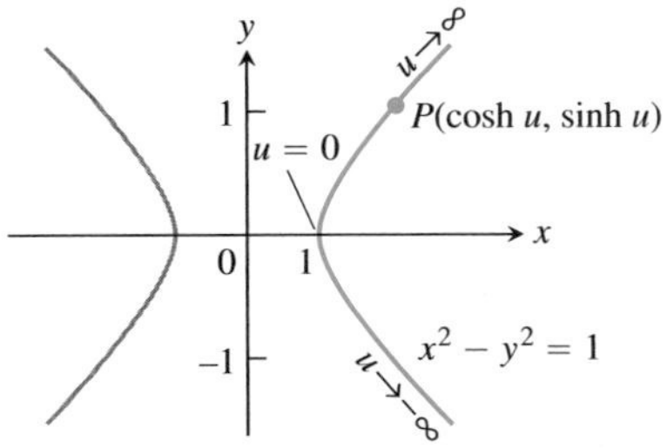

Since $\cosh^2 u - \sinh^2 u = 1$, the point $(\cosh u, \sinh u)$ lies on the right-hand branch of the hyperbola $x^2 - y^2 = 1$ for every value of u (Exercise 84).

Another analogy between hyperbolic and circular functions is that the variable u in the coordinates $(\cosh u, \sinh u)$ for the points of the right-hand branch of the hyperbola $x^2 - y^2 = 1$ is twice the area of the sector AOP pictured in the accompanying figure. To see why this is so, carry out the following steps.

a. Show that the area $A(u)$ of sector AOP is

$$A(u) = \frac{1}{2}\cosh u \sinh u - \int_1^{\cosh u} \sqrt{x^2 - 1}\, dx.$$

b. Differentiate both sides of the equation in part (a) with respect to u to show that

$$A'(u) = \frac{1}{2}.$$

c. Solve this last equation for $A(u)$. What is the value of $A(0)$? What is the value of the constant of integration C in your solution? With C determined, what does your solution say about the relationship of u to $A(u)$?

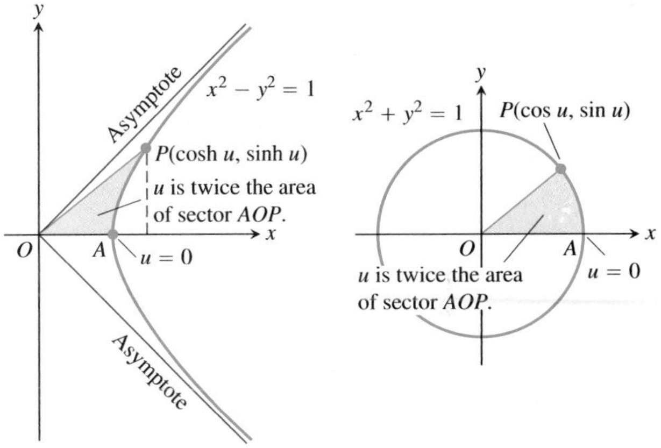

One of the analogies between hyperbolic and circular functions is revealed by these two diagrams (Exercise 84).

85. A minimal surface Find the area of the surface swept out by revolving about the x-axis the curve $y = 4\cosh(x/4)$, $-\ln 16 \le x \le \ln 81$.

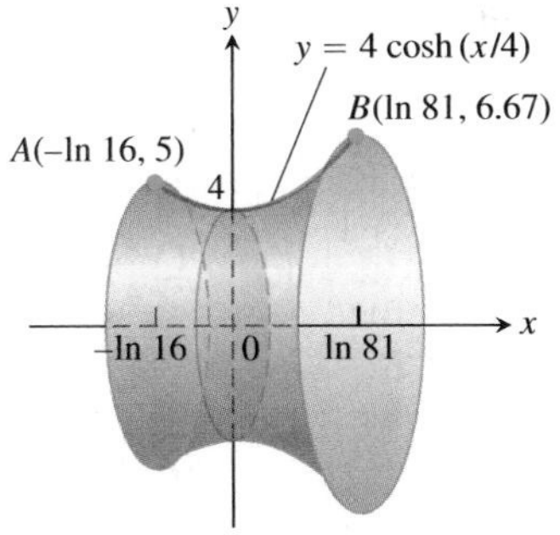

It can be shown that, of all continuously differentiable curves joining points A and B in the figure, the curve $y = 4\cosh(x/4)$ generates the surface of least area. If you made a rigid wire frame of the end-circles through A and B and dipped them in a soap-film solution, the surface spanning the circles would be the one generated by the curve.

T **86. a.** Find the centroid of the curve $y = \cosh x$, $-\ln 2 \le x \le \ln 2$.

b. Evaluate the coordinates to two decimal places. Then sketch the curve and plot the centroid to show its relation to the curve.

Hanging Cables

87. Imagine a cable, like a telephone line or TV cable, strung from one support to another and hanging freely. The cable's weight per unit length is w and the horizontal tension at its lowest point is a vector of length H. If we choose a coordinate system for the plane of the cable in which the x-axis is horizontal, the force of gravity is straight down, the positive y-axis points straight up, and the lowest point of the cable lies at the point $y = H/w$ on the y-axis (see accompanying figure), then it can be shown that the cable lies along the graph of the hyperbolic cosine

$$y = \frac{H}{w}\cosh\frac{w}{H}x.$$

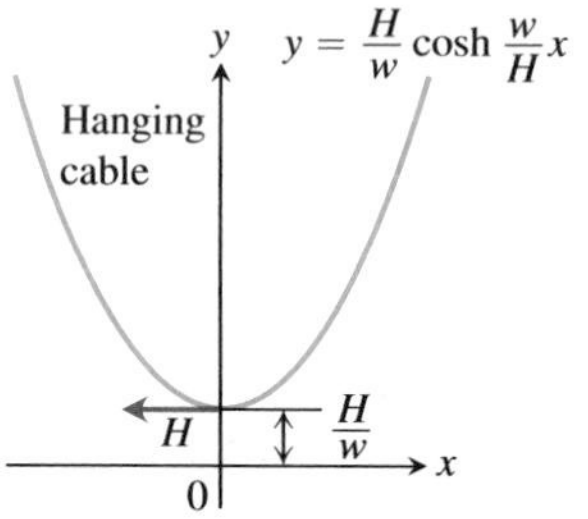

Such a curve is sometimes called a **chain curve** or a **catenary**, the latter deriving from the Latin *catena*, meaning "chain."

a. Let $P(x, y)$ denote an arbitrary point on the cable. The next accompanying figure displays the tension at P as a vector of length (magnitude) T, as well as the tension H at the lowest point A. Show that the cable's slope at P is

$$\tan\phi = \frac{dy}{dx} = \sinh\frac{w}{H}x.$$

b. Using the result from part (a) and the fact that the horizontal tension at P must equal H (the cable is not moving), show that $T = wy$. Hence, the magnitude of the tension at $P(x, y)$ is exactly equal to the weight of y units of cable.

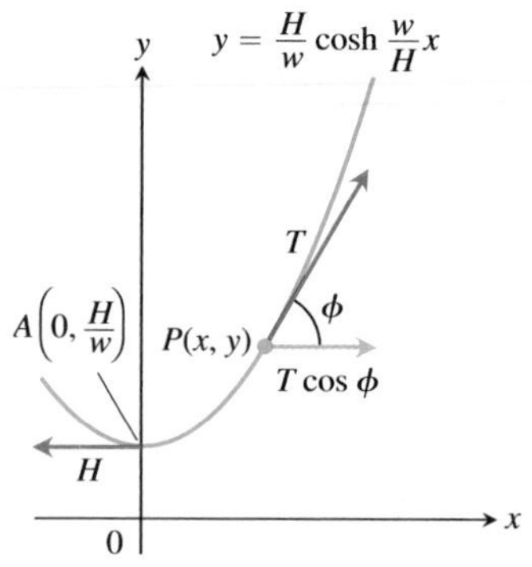

88. (*Continuation of Exercise 87.*) The length of arc AP in the Exercise 87 figure is $s = (1/a)\sinh ax$, where $a = w/H$. Show that the coordinates of P may be expressed in terms of s as

$$x = \frac{1}{a}\sinh^{-1} as, \qquad y = \sqrt{s^2 + \frac{1}{a^2}}.$$

89. The sag and horizontal tension in a cable The ends of a cable 32 ft long and weighing 2 lb/ft are fastened at the same level to posts 30 ft apart.

a. Model the cable with the equation

$$y = \frac{1}{a}\cosh ax, \quad -15 \le x \le 15.$$

Use information from Exercise 88 to show that a satisfies the equation

$$16a = \sinh 15a. \tag{2}$$

T **b.** Solve Equation (2) graphically by estimating the coordinates of the points where the graphs of the equations $y = 16a$ and $y = \sinh 15a$ intersect in the ay-plane.

T **c.** Solve Equation (2) for a numerically. Compare your solution with the value you found in part (b).

d. Estimate the horizontal tension in the cable at the cable's lowest point.

T **e.** Using the value found for a in part (c), graph the catenary

$$y = \frac{1}{a}\cosh ax$$

over the interval $-15 \le x \le 15$. Estimate the sag in the cable at its center.

Chapter 7 Questions to Guide Your Review

1. What functions have inverses? How do you know if two functions f and g are inverses of one another? Give examples of functions that are (are not) inverses of one another.
2. How are the domains, ranges, and graphs of functions and their inverses related? Give an example.
3. How can you sometimes express the inverse of a function of x as a function of x?
4. Under what circumstances can you be sure that the inverse of a function f is differentiable? How are the derivatives of f and f^{-1} related?

5. What is the natural logarithm function? What are its domain, range, and derivative? What arithmetic properties does it have? Comment on its graph.
6. What is logarithmic differentiation? Give an example.
7. What integrals lead to logarithms? Give examples. What are the integrals of $\tan x$ and $\cot x$?
8. How is the exponential function e^x defined? What are its domain, range, and derivative? What laws of exponents does it obey? Comment on its graph.
9. How are the functions a^x and $\log_a x$ defined? Are there any restrictions on a? How is the graph of $\log_a x$ related to the graph of $\ln x$? What truth is there in the statement that there is really only one exponential function and one logarithmic function?
10. Describe some of the applications of base 10 logarithms.
11. What is the law of exponential change? How can it be derived from an initial value problem? What are some of the applications of the law?
12. How do you compare the growth rates of positive functions as $x \to \infty$?
13. What roles do the functions e^x and $\ln x$ play in growth comparisons?
14. Describe big-oh and little-oh notation. Give examples.
15. Which is more efficient—a sequential search or a binary search? Explain.
16. How are the inverse trigonometric functions defined? How can you sometimes use right triangles to find values of these functions? Give examples.
17. What are the derivatives of the inverse trigonometric functions? How do the domains of the derivatives compare with the domains of the functions?
18. What integrals lead to inverse trigonometric functions? How do substitution and completing the square broaden the application of these integrals?
19. What are the six basic hyperbolic functions? Comment on their domains, ranges, and graphs. What are some of the identities relating them?
20. What are the derivatives of the six basic hyperbolic functions? What are the corresponding integral formulas? What similarities do you see here with the six basic trigonometric functions?
21. How are the inverse hyperbolic functions defined? Comment on their domains, ranges, and graphs. How can you find values of $\operatorname{sech}^{-1} x$, $\operatorname{csch}^{-1} x$, and $\coth^{-1} x$ using a calculator's keys for $\cosh^{-1} x$, $\sinh^{-1} x$, and $\tanh^{-1} x$?
22. What integrals lead naturally to inverse hyperbolic functions?

Chapter 7 Practice Exercises

Differentiation

In Exercises 1–24, find the derivative of y with respect to the appropriate variable.

1. $y = 10e^{-x/5}$
2. $y = \sqrt{2}e^{\sqrt{2}x}$
3. $y = \frac{1}{4}xe^{4x} - \frac{1}{16}e^{4x}$
4. $y = x^2e^{-2/x}$
5. $y = \ln(\sin^2\theta)$
6. $y = \ln(\sec^2\theta)$
7. $y = \log_2(x^2/2)$
8. $y = \log_5(3x - 7)$
9. $y = 8^{-t}$
10. $y = 9^{2t}$
11. $y = 5x^{3.6}$
12. $y = \sqrt{2}x^{-\sqrt{2}}$
13. $y = (x + 2)^{x+2}$
14. $y = 2(\ln x)^{x/2}$
15. $y = \sin^{-1}\sqrt{1 - u^2}, \quad 0 < u < 1$
16. $y = \sin^{-1}\left(\frac{1}{\sqrt{v}}\right), \quad v > 1$
17. $y = \ln \cos^{-1} x$
18. $y = z\cos^{-1} z - \sqrt{1 - z^2}$
19. $y = t\tan^{-1} t - \frac{1}{2}\ln t$
20. $y = (1 + t^2)\cot^{-1} 2t$
21. $y = z\sec^{-1} z - \sqrt{z^2 - 1}, \quad z > 1$
22. $y = 2\sqrt{x - 1}\sec^{-1}\sqrt{x}$
23. $y = \csc^{-1}(\sec\theta), \quad 0 < \theta < \pi/2$
24. $y = (1 + x^2)e^{\tan^{-1} x}$

Logarithmic Differentiation

In Exercises 25–30, use logarithmic differentiation to find the derivative of y with respect to the appropriate variable.

25. $y = \dfrac{2(x^2 + 1)}{\sqrt{\cos 2x}}$
26. $y = \sqrt[10]{\dfrac{3x + 4}{2x - 4}}$
27. $y = \left(\dfrac{(t + 1)(t - 1)}{(t - 2)(t + 3)}\right)^5, \quad t > 2$
28. $y = \dfrac{2u2^u}{\sqrt{u^2 + 1}}$
29. $y = (\sin\theta)^{\sqrt{\theta}}$
30. $y = (\ln x)^{1/(\ln x)}$

Integration

Evaluate the integrals in Exercises 31–78.

31. $\int e^x \sin(e^x)\,dx$
32. $\int e^t \cos(3e^t - 2)\,dt$

solving the initial value problem implicit in his model. Solve the problem and compare your solution graphically with the equation $s = 16t^2$. You will see that it describes a motion that starts too slowly at first and then becomes too fast too soon to be realistic.

25. The best branching angles for blood vessels and pipes When a smaller pipe branches off from a larger one in a flow system, we may want it to run off at an angle that is best from some energy-saving point of view. We might require, for instance, that energy loss due to friction be minimized along the section AOB shown in the accompanying figure. In this diagram, B is a given point to be reached by the smaller pipe, A is a point in the larger pipe upstream from B, and O is the point where the branching occurs. A law due to Poiseuille states that the loss of energy due to friction in nonturbulent flow is proportional to the length of the path and inversely proportional to the fourth power of the radius. Thus, the loss along AO is $(kd_1)/R^4$ and along OB is $(kd_2)/r^4$, where k is a constant, d_1 is the length of AO, d_2 is the length of OB, R is the radius of the larger pipe, and r is the radius of the smaller pipe. The angle θ is to be chosen to minimize the sum of these two losses:

$$L = k\frac{d_1}{R^4} + k\frac{d_2}{r^4}.$$

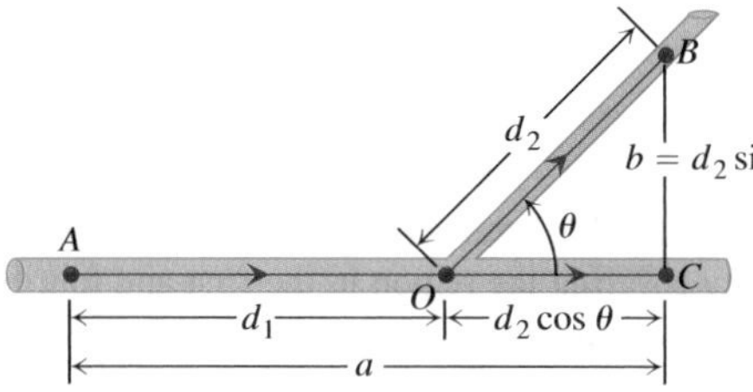

In our model, we assume that $AC = a$ and $BC = b$ are fixed. Thus we have the relations

$$d_1 + d_2\cos\theta = a \qquad d_2\sin\theta = b,$$

so that

$$d_2 = b\csc\theta,$$
$$d_1 = a - d_2\cos\theta = a - b\cot\theta.$$

We can express the total loss L as a function of θ:

$$L = k\left(\frac{a - b\cot\theta}{R^4} + \frac{b\csc\theta}{r^4}\right).$$

a. Show that the critical value of θ for which $dL/d\theta$ equals zero is

$$\theta_c = \cos^{-1}\frac{r^4}{R^4}.$$

b. If the ratio of the pipe radii is $r/R = 5/6$, estimate to the nearest degree the optimal branching angle given in part (a).

The mathematical analysis described here is also used to explain the angles at which arteries branch in an animal's body. (See *Introduction to Mathematics for Life Scientists*, Second Edition, by E. Batschelet [New York: Springer-Verlag, 1976].)

T **26. Group blood testing** During World War II it was necessary to administer blood tests to large numbers of recruits. There are two standard ways to administer a blood test to N people. In method 1, each person is tested separately. In method 2, the blood samples of x people are pooled and tested as one large sample. If the test is negative, this one test is enough for all x people. If the test is positive, then each of the x people is tested separately, requiring a total of $x + 1$ tests. Using the second method and some probability theory it can be shown that, on the average, the total number of tests y will be

$$y = N\left(1 - q^x + \frac{1}{x}\right).$$

With $q = 0.99$ and $N = 1000$, find the integer value of x that minimizes y. Also find the integer value of x that maximizes y. (This second result is not important to the real-life situation.) The group testing method was used in World War II with a savings of 80% over the individual testing method, but not with the given value of q.

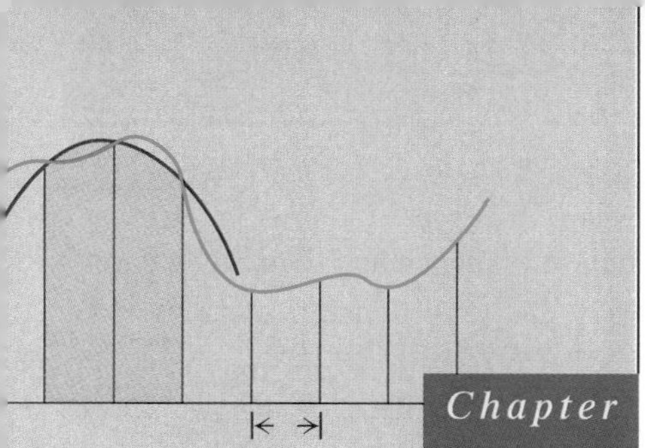

Chapter

8 TECHNIQUES OF INTEGRATION

OVERVIEW The Fundamental Theorem connects antiderivatives and the definite integral. Evaluating the indefinite integral

$$\int f(x)\,dx$$

is equivalent to finding a function F such that $F'(x) = f(x)$, and then adding an arbitrary constant C:

$$\int f(x)\,dx = F(x) + C.$$

In this chapter we study a number of important techniques for finding indefinite integrals of more complicated functions than those seen before. The goal of this chapter is to show how to change unfamiliar integrals into integrals we can recognize, find in a table, or evaluate with a computer. We also extend the idea of the definite integral to *improper integrals* for which the integrand may be unbounded over the interval of integration, or the interval itself may no longer be finite.

8.1 Basic Integration Formulas

To help us in the search for finding indefinite integrals, it is useful to build up a table of integral formulas by inverting formulas for derivatives, as we have done in previous chapters. Then we try to match any integral that confronts us against one of the standard types. This usually involves a certain amount of algebraic manipulation as well as use of the Substitution Rule.

Recall the Substitution Rule from Section 5.5:

$$\int f(g(x))g'(x)\,dx = \int f(u)\,du$$

where $u = g(x)$ is a differentiable function whose range is an interval I and f is continuous on I. Success in integration often hinges on the ability to spot what part of the integrand should be called u in order that one will also have du, so that a known formula can be applied. This means that the first requirement for skill in integration is a thorough mastery of the formulas for differentiation.

Table 8.1 shows the basic forms of integrals we have evaluated so far. In this section we present several algebraic or substitution methods to help us use this table. There is a more extensive table at the back of the book; we discuss its use in Section 8.6.

TABLE 8.1 Basic integration formulas

1. $\int du = u + C$
2. $\int k\,du = ku + C \quad$ (any number k)
3. $\int (du + dv) = \int du + \int dv$
4. $\int u^n\,du = \dfrac{u^{n+1}}{n+1} + C \quad (n \neq -1)$
5. $\int \dfrac{du}{u} = \ln|u| + C$
6. $\int \sin u\,du = -\cos u + C$
7. $\int \cos u\,du = \sin u + C$
8. $\int \sec^2 u\,du = \tan u + C$
9. $\int \csc^2 u\,du = -\cot u + C$
10. $\int \sec u \tan u\,du = \sec u + C$
11. $\int \csc u \cot u\,du = -\csc u + C$
12. $\int \tan u\,du = -\ln|\cos u| + C$
    $= \ln|\sec u| + C$
13. $\int \cot u\,du = \ln|\sin u| + C$
    $= -\ln|\csc u| + C$
14. $\int e^u\,du = e^u + C$
15. $\int a^u\,du = \dfrac{a^u}{\ln a} + C \quad (a > 0, a \neq 1)$
16. $\int \sinh u\,du = \cosh u + C$
17. $\int \cosh u\,du = \sinh u + C$
18. $\int \dfrac{du}{\sqrt{a^2 - u^2}} = \sin^{-1}\left(\dfrac{u}{a}\right) + C$
19. $\int \dfrac{du}{a^2 + u^2} = \dfrac{1}{a}\tan^{-1}\left(\dfrac{u}{a}\right) + C$
20. $\int \dfrac{du}{u\sqrt{u^2 - a^2}} = \dfrac{1}{a}\sec^{-1}\left|\dfrac{u}{a}\right| + C$
21. $\int \dfrac{du}{\sqrt{a^2 + u^2}} = \sinh^{-1}\left(\dfrac{u}{a}\right) + C \quad (a > 0)$
22. $\int \dfrac{du}{\sqrt{u^2 - a^2}} = \cosh^{-1}\left(\dfrac{u}{a}\right) + C \quad (u > a > 0)$

We often have to rewrite an integral to match it to a standard formula.

EXAMPLE 1 Making a Simplifying Substitution

Evaluate

$$\int \frac{2x - 9}{\sqrt{x^2 - 9x + 1}}\,dx.$$

Solution

$$\begin{aligned}\int \frac{2x-9}{\sqrt{x^2-9x+1}}\,dx &= \int \frac{du}{\sqrt{u}} && u = x^2-9x+1,\ du = (2x-9)\,dx. \\ &= \int u^{-1/2}\,du \\ &= \frac{u^{(-1/2)+1}}{(-1/2)+1} + C && \text{Table 8.1 Formula 4, with } n = -1/2 \\ &= 2u^{1/2} + C \\ &= 2\sqrt{x^2-9x+1} + C\end{aligned}$$

■

EXAMPLE 2 Completing the Square

Evaluate

$$\int \frac{dx}{\sqrt{8x-x^2}}.$$

Solution We complete the square to simplify the denominator:

$$\begin{aligned}8x-x^2 &= -(x^2-8x) = -(x^2-8x+16-16) \\ &= -(x^2-8x+16)+16 = 16-(x-4)^2.\end{aligned}$$

Then

$$\begin{aligned}\int \frac{dx}{\sqrt{8x-x^2}} &= \int \frac{dx}{\sqrt{16-(x-4)^2}} \\ &= \int \frac{du}{\sqrt{a^2-u^2}} && a = 4,\ u = (x-4),\ du = dx \\ &= \sin^{-1}\left(\frac{u}{a}\right) + C && \text{Table 8.1, Formula 18} \\ &= \sin^{-1}\left(\frac{x-4}{4}\right) + C.\end{aligned}$$

■

EXAMPLE 3 Expanding a Power and Using a Trigonometric Identity

Evaluate

$$\int (\sec x + \tan x)^2\,dx.$$

Solution We expand the integrand and get

$$(\sec x + \tan x)^2 = \sec^2 x + 2\sec x\tan x + \tan^2 x.$$

The first two terms on the right-hand side of this equation are familiar; we can integrate them at once. How about $\tan^2 x$? There is an identity that connects it with $\sec^2 x$:

$$\tan^2 x + 1 = \sec^2 x, \qquad \tan^2 x = \sec^2 x - 1.$$

We replace $\tan^2 x$ by $\sec^2 x - 1$ and get

$$\begin{aligned}\int (\sec x + \tan x)^2\, dx &= \int (\sec^2 x + 2\sec x \tan x + \sec^2 x - 1)\, dx \\ &= 2\int \sec^2 x\, dx + 2\int \sec x \tan x\, dx - \int 1\, dx \\ &= 2\tan x + 2\sec x - x + C.\end{aligned}$$

■

EXAMPLE 4 Eliminating a Square Root

Evaluate

$$\int_0^{\pi/4} \sqrt{1 + \cos 4x}\, dx.$$

Solution We use the identity

$$\cos^2 \theta = \frac{1 + \cos 2\theta}{2}, \qquad \text{or} \qquad 1 + \cos 2\theta = 2\cos^2 \theta.$$

With $\theta = 2x$, this identity becomes

$$1 + \cos 4x = 2\cos^2 2x.$$

Hence,

$$\begin{aligned}\int_0^{\pi/4} \sqrt{1 + \cos 4x}\, dx &= \int_0^{\pi/4} \sqrt{2}\sqrt{\cos^2 2x}\, dx \\ &= \sqrt{2}\int_0^{\pi/4} |\cos 2x|\, dx && \sqrt{u^2} = |u| \\ &= \sqrt{2}\int_0^{\pi/4} \cos 2x\, dx && \text{On } [0, \pi/4],\ \cos 2x \geq 0, \text{ so } |\cos 2x| = \cos 2x. \\ &= \sqrt{2}\left[\frac{\sin 2x}{2}\right]_0^{\pi/4} && \text{Table 8.1, Formula 7, with } u = 2x \text{ and } du = 2\,dx \\ &= \sqrt{2}\left[\frac{1}{2} - 0\right] = \frac{\sqrt{2}}{2}.\end{aligned}$$

■

EXAMPLE 5 Reducing an Improper Fraction

Evaluate

$$\int \frac{3x^2 - 7x}{3x + 2}\, dx.$$

Solution The integrand is an improper fraction (degree of numerator greater than or equal to degree of denominator). To integrate it, we divide first, getting a quotient plus a remainder that is a proper fraction:

$$\frac{3x^2 - 7x}{3x + 2} = x - 3 + \frac{6}{3x + 2}.$$

$$\begin{array}{r} x - 3 \\ 3x + 2\,\overline{)\,3x^2 - 7x} \\ \underline{3x^2 + 2x} \\ -9x \\ \underline{-9x - 6} \\ +\,6 \end{array}$$

Therefore,

$$\int \frac{3x^2 - 7x}{3x + 2}\, dx = \int \left(x - 3 + \frac{6}{3x + 2}\right) dx = \frac{x^2}{2} - 3x + 2\ln|3x + 2| + C.$$

■

Reducing an improper fraction by long division (Example 5) does not always lead to an expression we can integrate directly. We see what to do about that in Section 8.5.

EXAMPLE 6 Separating a Fraction

Evaluate

$$\int \frac{3x + 2}{\sqrt{1 - x^2}}\, dx.$$

Solution We first separate the integrand to get

$$\int \frac{3x + 2}{\sqrt{1 - x^2}}\, dx = 3\int \frac{x\, dx}{\sqrt{1 - x^2}} + 2\int \frac{dx}{\sqrt{1 - x^2}}.$$

In the first of these new integrals, we substitute

$$u = 1 - x^2, \qquad du = -2x\, dx, \qquad \text{and} \qquad x\, dx = -\frac{1}{2}\, du.$$

$$\begin{aligned} 3\int \frac{x\, dx}{\sqrt{1 - x^2}} &= 3\int \frac{(-1/2)\, du}{\sqrt{u}} = -\frac{3}{2}\int u^{-1/2}\, du \\ &= -\frac{3}{2}\cdot\frac{u^{1/2}}{1/2} + C_1 = -3\sqrt{1 - x^2} + C_1 \end{aligned}$$

The second of the new integrals is a standard form,

$$2\int \frac{dx}{\sqrt{1 - x^2}} = 2\sin^{-1}x + C_2.$$

Combining these results and renaming $C_1 + C_2$ as C gives

$$\int \frac{3x + 2}{\sqrt{1 - x^2}}\, dx = -3\sqrt{1 - x^2} + 2\sin^{-1}x + C.$$ ■

The final example of this section calculates an important integral by the algebraic technique of multiplying the integrand by a form of 1 to change the integrand into one we can integrate.

HISTORICAL BIOGRAPHY

George David Birkhoff (1884–1944)

EXAMPLE 7 Integral of $y = \sec x$—Multiplying by a Form of 1

Evaluate

$$\int \sec x\, dx.$$

Solution

$$\begin{aligned} \int \sec x\, dx &= \int (\sec x)(1)\, dx = \int \sec x \cdot \frac{\sec x + \tan x}{\sec x + \tan x}\, dx \\ &= \int \frac{\sec^2 x + \sec x \tan x}{\sec x + \tan x}\, dx \\ &= \int \frac{du}{u} \\ &= \ln|u| + C = \ln|\sec x + \tan x| + C. \end{aligned}$$ ■

$u = \tan x + \sec x,$
$du = (\sec^2 x + \sec x \tan x)\, dx$

With cosecants and cotangents in place of secants and tangents, the method of Example 7 leads to a companion formula for the integral of the cosecant (see Exercise 95).

TABLE 8.2 The secant and cosecant integrals

1. $\int \sec u \, du = \ln |\sec u + \tan u| + C$

2. $\int \csc u \, du = -\ln |\csc u + \cot u| + C$

Procedures for Matching Integrals to Basic Formulas

Procedure	**Example**
Making a simplifying substitution	$\dfrac{2x - 9}{\sqrt{x^2 - 9x + 1}} dx = \dfrac{du}{\sqrt{u}}$
Completing the square	$\sqrt{8x - x^2} = \sqrt{16 - (x - 4)^2}$
Using a trigonometric identity	$(\sec x + \tan x)^2 = \sec^2 x + 2 \sec x \tan x + \tan^2 x$ $= \sec^2 x + 2 \sec x \tan x + (\sec^2 x - 1)$ $= 2 \sec^2 x + 2 \sec x \tan x - 1$
Eliminating a square root	$\sqrt{1 + \cos 4x} = \sqrt{2 \cos^2 2x} = \sqrt{2}\, |\cos 2x|$
Reducing an improper fraction	$\dfrac{3x^2 - 7x}{3x + 2} = x - 3 + \dfrac{6}{3x + 2}$
Separating a fraction	$\dfrac{3x + 2}{\sqrt{1 - x^2}} = \dfrac{3x}{\sqrt{1 - x^2}} + \dfrac{2}{\sqrt{1 - x^2}}$
Multiplying by a form of 1	$\sec x = \sec x \cdot \dfrac{\sec x + \tan x}{\sec x + \tan x}$ $= \dfrac{\sec^2 x + \sec x \tan x}{\sec x + \tan x}$

EXERCISES 8.1

Basic Substitutions

Evaluate each integral in Exercises 1–36 by using a substitution to reduce it to standard form.

1. $\displaystyle\int \frac{16x \, dx}{\sqrt{8x^2 + 1}}$

2. $\displaystyle\int \frac{3 \cos x \, dx}{\sqrt{1 + 3 \sin x}}$

3. $\displaystyle\int 3\sqrt{\sin v} \cos v \, dv$

4. $\displaystyle\int \cot^3 y \csc^2 y \, dy$

5. $\displaystyle\int_0^1 \frac{16x \, dx}{8x^2 + 2}$

6. $\displaystyle\int_{\pi/4}^{\pi/3} \frac{\sec^2 z}{\tan z} dz$

7. $\int \frac{dx}{\sqrt{x}(\sqrt{x}+1)}$

8. $\int \frac{dx}{x-\sqrt{x}}$

9. $\int \cot(3-7x)\,dx$

10. $\int \csc(\pi x-1)\,dx$

11. $\int e^{\theta}\csc(e^{\theta}+1)\,d\theta$

12. $\int \frac{\cot(3+\ln x)}{x}\,dx$

13. $\int \sec\frac{t}{3}\,dt$

14. $\int x\sec(x^2-5)\,dx$

15. $\int \csc(s-\pi)\,ds$

16. $\int \frac{1}{\theta^2}\csc\frac{1}{\theta}\,d\theta$

17. $\int_0^{\sqrt{\ln 2}} 2x\,e^{x^2}\,dx$

18. $\int_{\pi/2}^{\pi} (\sin y)e^{\cos y}\,dy$

19. $\int e^{\tan v}\sec^2 v\,dv$

20. $\int \frac{e^{\sqrt{t}}\,dt}{\sqrt{t}}$

21. $\int 3^{x+1}\,dx$

22. $\int \frac{2^{\ln x}}{x}\,dx$

23. $\int \frac{2^{\sqrt{w}}\,dw}{2\sqrt{w}}$

24. $\int 10^{2\theta}\,d\theta$

25. $\int \frac{9\,du}{1+9u^2}$

26. $\int \frac{4\,dx}{1+(2x+1)^2}$

27. $\int_0^{1/6} \frac{dx}{\sqrt{1-9x^2}}$

28. $\int_0^1 \frac{dt}{\sqrt{4-t^2}}$

29. $\int \frac{2s\,ds}{\sqrt{1-s^4}}$

30. $\int \frac{2\,dx}{x\sqrt{1-4\ln^2 x}}$

31. $\int \frac{6\,dx}{x\sqrt{25x^2-1}}$

32. $\int \frac{dr}{r\sqrt{r^2-9}}$

33. $\int \frac{dx}{e^x+e^{-x}}$

34. $\int \frac{dy}{\sqrt{e^{2y}-1}}$

35. $\int_1^{e^{\pi/3}} \frac{dx}{x\cos(\ln x)}$

36. $\int \frac{\ln x\,dx}{x+4x\ln^2 x}$

Completing the Square

Evaluate each integral in Exercises 37–42 by completing the square and using a substitution to reduce it to standard form.

37. $\int_1^2 \frac{8\,dx}{x^2-2x+2}$

38. $\int_2^4 \frac{2\,dx}{x^2-6x+10}$

39. $\int \frac{dt}{\sqrt{-t^2+4t-3}}$

40. $\int \frac{d\theta}{\sqrt{2\theta-\theta^2}}$

41. $\int \frac{dx}{(x+1)\sqrt{x^2+2x}}$

42. $\int \frac{dx}{(x-2)\sqrt{x^2-4x+3}}$

Trigonometric Identities

Evaluate each integral in Exercises 43–46 by using trigonometric identities and substitutions to reduce it to standard form.

43. $\int (\sec x+\cot x)^2\,dx$

44. $\int (\csc x-\tan x)^2\,dx$

45. $\int \csc x\sin 3x\,dx$

46. $\int (\sin 3x\cos 2x-\cos 3x\sin 2x)\,dx$

Improper Fractions

Evaluate each integral in Exercises 47–52 by reducing the improper fraction and using a substitution (if necessary) to reduce it to standard form.

47. $\int \frac{x}{x+1}\,dx$

48. $\int \frac{x^2}{x^2+1}\,dx$

49. $\int_{\sqrt{2}}^3 \frac{2x^3}{x^2-1}\,dx$

50. $\int_{-1}^3 \frac{4x^2-7}{2x+3}\,dx$

51. $\int \frac{4t^3-t^2+16t}{t^2+4}\,dt$

52. $\int \frac{2\theta^3-7\theta^2+7\theta}{2\theta-5}\,d\theta$

Separating Fractions

Evaluate each integral in Exercises 53–56 by separating the fraction and using a substitution (if necessary) to reduce it to standard form.

53. $\int \frac{1-x}{\sqrt{1-x^2}}\,dx$

54. $\int \frac{x+2\sqrt{x-1}}{2x\sqrt{x-1}}\,dx$

55. $\int_0^{\pi/4} \frac{1+\sin x}{\cos^2 x}\,dx$

56. $\int_0^{1/2} \frac{2-8x}{1+4x^2}\,dx$

Multiplying by a Form of 1

Evaluate each integral in Exercises 57–62 by multiplying by a form of 1 and using a substitution (if necessary) to reduce it to standard form.

57. $\int \frac{1}{1+\sin x}\,dx$

58. $\int \frac{1}{1+\cos x}\,dx$

59. $\int \frac{1}{\sec\theta+\tan\theta}\,d\theta$

60. $\int \frac{1}{\csc\theta+\cot\theta}\,d\theta$

61. $\int \frac{1}{1-\sec x}\,dx$

62. $\int \frac{1}{1-\csc x}\,dx$

Eliminating Square Roots

Evaluate each integral in Exercises 63–70 by eliminating the square root.

63. $\int_0^{2\pi} \sqrt{\frac{1-\cos x}{2}}\,dx$

64. $\int_0^{\pi} \sqrt{1-\cos 2x}\,dx$

65. $\int_{\pi/2}^{\pi} \sqrt{1 + \cos 2t}\, dt$

66. $\int_{-\pi}^{0} \sqrt{1 + \cos t}\, dt$

67. $\int_{-\pi}^{0} \sqrt{1 - \cos^2 \theta}\, d\theta$

68. $\int_{\pi/2}^{\pi} \sqrt{1 - \sin^2 \theta}\, d\theta$

69. $\int_{-\pi/4}^{\pi/4} \sqrt{1 + \tan^2 y}\, dy$

70. $\int_{-\pi/4}^{0} \sqrt{\sec^2 y - 1}\, dy$

Assorted Integrations

Evaluate each integral in Exercises 71–82 by using any technique you think is appropriate.

71. $\int_{\pi/4}^{3\pi/4} (\csc x - \cot x)^2\, dx$

72. $\int_{0}^{\pi/4} (\sec x + 4 \cos x)^2\, dx$

73. $\int \cos \theta \csc (\sin \theta)\, d\theta$

74. $\int \left(1 + \frac{1}{x}\right) \cot (x + \ln x)\, dx$

75. $\int (\csc x - \sec x)(\sin x + \cos x)\, dx$

76. $\int 3 \sinh \left(\frac{x}{2} + \ln 5\right) dx$

77. $\int \frac{6\, dy}{\sqrt{y}(1 + y)}$

78. $\int \frac{dx}{x\sqrt{4x^2 - 1}}$

79. $\int \frac{7\, dx}{(x - 1)\sqrt{x^2 - 2x - 48}}$

80. $\int \frac{dx}{(2x + 1)\sqrt{4x^2 + 4x}}$

81. $\int \sec^2 t \tan (\tan t)\, dt$

82. $\int \frac{dx}{x\sqrt{3 + x^2}}$

Trigonometric Powers

83. a. Evaluate $\int \cos^3 \theta\, d\theta$. (*Hint:* $\cos^2 \theta = 1 - \sin^2 \theta$.)

b. Evaluate $\int \cos^5 \theta\, d\theta$.

c. Without actually evaluating the integral, explain how you would evaluate $\int \cos^9 \theta\, d\theta$.

84. a. Evaluate $\int \sin^3 \theta\, d\theta$. (*Hint:* $\sin^2 \theta = 1 - \cos^2 \theta$.)

b. Evaluate $\int \sin^5 \theta\, d\theta$.

c. Evaluate $\int \sin^7 \theta\, d\theta$.

d. Without actually evaluating the integral, explain how you would evaluate $\int \sin^{13} \theta\, d\theta$.

85. a. Express $\int \tan^3 \theta\, d\theta$ in terms of $\int \tan \theta\, d\theta$. Then evaluate $\int \tan^3 \theta\, d\theta$. (*Hint:* $\tan^2 \theta = \sec^2 \theta - 1$.)

b. Express $\int \tan^5 \theta\, d\theta$ in terms of $\int \tan^3 \theta\, d\theta$.

c. Express $\int \tan^7 \theta\, d\theta$ in terms of $\int \tan^5 \theta\, d\theta$.

d. Express $\int \tan^{2k+1} \theta\, d\theta$, where k is a positive integer, in terms of $\int \tan^{2k-1} \theta\, d\theta$.

86. a. Express $\int \cot^3 \theta\, d\theta$ in terms of $\int \cot \theta\, d\theta$. Then evaluate $\int \cot^3 \theta\, d\theta$. (*Hint:* $\cot^2 \theta = \csc^2 \theta - 1$.)

b. Express $\int \cot^5 \theta\, d\theta$ in terms of $\int \cot^3 \theta\, d\theta$.

c. Express $\int \cot^7 \theta\, d\theta$ in terms of $\int \cot^5 \theta\, d\theta$.

d. Express $\int \cot^{2k+1} \theta\, d\theta$, where k is a positive integer, in terms of $\int \cot^{2k-1} \theta\, d\theta$.

Theory and Examples

87. **Area** Find the area of the region bounded above by $y = 2 \cos x$ and below by $y = \sec x$, $-\pi/4 \le x \le \pi/4$.

88. **Area** Find the area of the "triangular" region that is bounded from above and below by the curves $y = \csc x$ and $y = \sin x$, $\pi/6 \le x \le \pi/2$, and on the left by the line $x = \pi/6$.

89. **Volume** Find the volume of the solid generated by revolving the region in Exercise 87 about the x-axis.

90. **Volume** Find the volume of the solid generated by revolving the region in Exercise 88 about the x-axis.

91. **Arc length** Find the length of the curve $y = \ln (\cos x)$, $0 \le x \le \pi/3$.

92. **Arc length** Find the length of the curve $y = \ln (\sec x)$, $0 \le x \le \pi/4$.

93. **Centroid** Find the centroid of the region bounded by the x-axis, the curve $y = \sec x$, and the lines $x = -\pi/4$, $x = \pi/4$.

94. **Centroid** Find the centroid of the region that is bounded by the x-axis, the curve $y = \csc x$, and the lines $x = \pi/6$, $x = 5\pi/6$.

95. **The integral of csc x** Repeat the derivation in Example 7, using cofunctions, to show that

$$\int \csc x\, dx = -\ln |\csc x + \cot x| + C.$$

96. **Using different substitutions** Show that the integral

$$\int ((x^2 - 1)(x + 1))^{-2/3}\, dx$$

can be evaluated with any of the following substitutions.

a. $u = 1/(x + 1)$

b. $u = ((x - 1)/(x + 1))^k$ for $k = 1, 1/2, 1/3, -1/3, -2/3$, and -1

c. $u = \tan^{-1} x$

d. $u = \tan^{-1} \sqrt{x}$

e. $u = \tan^{-1} ((x - 1)/2)$

f. $u = \cos^{-1} x$

g. $u = \cosh^{-1} x$

What is the value of the integral? (*Source:* "Problems and Solutions," *College Mathematics Journal*, Vol. 21, No. 5 (Nov. 1990), pp. 425–426.)

8.2 Integration by Parts

Since

$$\int x\,dx = \frac{1}{2}x^2 + C$$

and

$$\int x^2\,dx = \frac{1}{3}x^3 + C,$$

it is apparent that

$$\int x \cdot x\,dx \neq \int x\,dx \cdot \int x\,dx.$$

In other words, the integral of a product is generally *not* the product of the individual-integrals:

$$\int f(x)g(x)\,dx \text{ is not equal to } \int f(x)\,dx \cdot \int g(x)\,dx.$$

Integration by parts is a technique for simplifying integrals of the form

$$\int f(x)g(x)\,dx.$$

It is useful when f can be differentiated repeatedly and g can be integrated repeatedly without difficulty. The integral

$$\int xe^x\,dx$$

is such an integral because $f(x) = x$ can be differentiated twice to become zero and $g(x) = e^x$ can be integrated repeatedly without difficulty. Integration by parts also applies to integrals like

$$\int e^x \sin x\,dx$$

in which each part of the integrand appears again after repeated differentiation or integration.

In this section, we describe integration by parts and show how to apply it.

Product Rule in Integral Form

If f and g are differentiable functions of x, the Product Rule says

$$\frac{d}{dx}[f(x)g(x)] = f'(x)g(x) + f(x)g'(x).$$

In terms of indefinite integrals, this equation becomes

$$\int \frac{d}{dx}[f(x)g(x)]\,dx = \int [f'(x)g(x) + f(x)g'(x)]\,dx$$

or

$$\int \frac{d}{dx}[f(x)g(x)]\,dx = \int f'(x)g(x)\,dx + \int f(x)g'(x)\,dx.$$

Rearranging the terms of this last equation, we get

$$\int f(x)g'(x)\,dx = \int \frac{d}{dx}[f(x)g(x)]\,dx - \int f'(x)g(x)\,dx$$

leading to the **integration by parts** formula

$$\int f(x)g'(x)\,dx = f(x)g(x) - \int f'(x)g(x)\,dx \qquad (1)$$

Sometimes it is easier to remember the formula if we write it in differential form. Let $u = f(x)$ and $v = g(x)$. Then $du = f'(x)\,dx$ and $dv = g'(x)\,dx$. Using the Substitution Rule, the integration by parts formula becomes

Integration by Parts Formula

$$\int u\,dv = uv - \int v\,du \qquad (2)$$

This formula expresses one integral, $\int u\,dv$, in terms of a second integral, $\int v\,du$. With a proper choice of u and v, the second integral may be easier to evaluate than the first. In using the formula, various choices may be available for u and dv. The next examples illustrate the technique.

EXAMPLE 1 Using Integration by Parts

Find

$$\int x \cos x\,dx.$$

Solution We use the formula $\int u\,dv = uv - \int v\,du$ with

$$u = x, \qquad dv = \cos x\,dx,$$
$$du = dx, \qquad v = \sin x. \qquad \text{Simplest antiderivative of } \cos x$$

Then

$$\int x\cos x\,dx = x\sin x - \int \sin x\,dx = x\sin x + \cos x + C.$$ ■

Let us examine the choices available for u and dv in Example 1.

EXAMPLE 2 Example 1 Revisited

To apply integration by parts to

$$\int x\cos x\,dx = \int u\,dv$$

we have four possible choices:

1. Let $u = 1$ and $dv = x \cos x \, dx$.
2. Let $u = x$ and $dv = \cos x \, dx$.
3. Let $u = x \cos x$ and $dv = dx$.
4. Let $u = \cos x$ and $dv = x \, dx$.

Let's examine these one at a time.

Choice 1 won't do because we don't know how to integrate $dv = x \cos x \, dx$ to get v.

Choice 2 works well, as we saw in Example 1.

Choice 3 leads to

$$u = x \cos x, \qquad dv = dx,$$
$$du = (\cos x - x \sin x)\, dx, \qquad v = x,$$

and the new integral

$$\int v \, du = \int (x \cos x - x^2 \sin x)\, dx.$$

This is worse than the integral we started with.

Choice 4 leads to

$$u = \cos x, \qquad dv = x \, dx,$$
$$du = -\sin x \, dx, \qquad v = x^2/2,$$

so the new integral is

$$\int v \, du = -\int \frac{x^2}{2} \sin x \, dx.$$

This, too, is worse. ■

The goal of integration by parts is to go from an integral $\int u \, dv$ that we don't see how to evaluate to an integral $\int v \, du$ that we can evaluate. Generally, you choose dv first to be as much of the integrand, including dx, as you can readily integrate; u is the leftover part. Keep in mind that integration by parts does not always work.

EXAMPLE 3 Integral of the Natural Logarithm

Find

$$\int \ln x \, dx.$$

Solution Since $\int \ln x \, dx$ can be written as $\int \ln x \cdot 1 \, dx$, we use the formula $\int u \, dv = uv - \int v \, du$ with

$u = \ln x$ Simplifies when differentiated $\qquad dv = dx$ Easy to integrate

$du = \frac{1}{x} dx,$ $\qquad v = x.$ Simplest antiderivative

Then

$$\int \ln x \, dx = x \ln x - \int x \cdot \frac{1}{x} dx = x \ln x - \int dx = x \ln x - x + C.$$ ■

Sometimes we have to use integration by parts more than once.

EXAMPLE 4 Repeated Use of Integration by Parts

Evaluate

$$\int x^2 e^x\,dx.$$

Solution With $u = x^2$, $dv = e^x\,dx$, $du = 2x\,dx$, and $v = e^x$, we have

$$\int x^2 e^x\,dx = x^2 e^x - 2\int x e^x\,dx.$$

The new integral is less complicated than the original because the exponent on x is reduced by one. To evaluate the integral on the right, we integrate by parts again with $u = x$, $dv = e^x\,dx$. Then $du = dx$, $v = e^x$, and

$$\int x e^x\,dx = x e^x - \int e^x\,dx = x e^x - e^x + C.$$

Hence,

$$\begin{aligned}\int x^2 e^x\,dx &= x^2 e^x - 2\int x e^x\,dx \\ &= x^2 e^x - 2x e^x + 2e^x + C.\end{aligned}$$

■

The technique of Example 4 works for any integral $\int x^n e^x\,dx$ in which n is a positive integer, because differentiating x^n will eventually lead to zero and integrating e^x is easy. We say more about this later in this section when we discuss *tabular integration.*

Integrals like the one in the next example occur in electrical engineering. Their evaluation requires two integrations by parts, followed by solving for the unknown integral.

EXAMPLE 5 Solving for the Unknown Integral

Evaluate

$$\int e^x \cos x\,dx.$$

Solution Let $u = e^x$ and $dv = \cos x\,dx$. Then $du = e^x\,dx$, $v = \sin x$, and

$$\int e^x \cos x\,dx = e^x \sin x - \int e^x \sin x\,dx.$$

The second integral is like the first except that it has $\sin x$ in place of $\cos x$. To evaluate it, we use integration by parts with

$$u = e^x, \qquad dv = \sin x\,dx, \qquad v = -\cos x, \qquad du = e^x\,dx.$$

Then

$$\begin{aligned}\int e^x \cos x\,dx &= e^x \sin x - \left(-e^x \cos x - \int (-\cos x)(e^x\,dx)\right) \\ &= e^x \sin x + e^x \cos x - \int e^x \cos x\,dx.\end{aligned}$$

The unknown integral now appears on both sides of the equation. Adding the integral to both sides and adding the constant of integration gives

$$2\int e^x \cos x\, dx = e^x \sin x + e^x \cos x + C_1.$$

Dividing by 2 and renaming the constant of integration gives

$$\int e^x \cos x\, dx = \frac{e^x \sin x + e^x \cos x}{2} + C.$$ ■

Evaluating Definite Integrals by Parts

The integration by parts formula in Equation (1) can be combined with Part 2 of the Fundamental Theorem in order to evaluate definite integrals by parts. Assuming that both f' and g' are continuous over the interval $[a, b]$, Part 2 of the Fundamental Theorem gives

Integration by Parts Formula for Definite Integrals

$$\int_a^b f(x)g'(x)\, dx = f(x)g(x)\Big]_a^b - \int_a^b f'(x)g(x)\, dx \qquad (3)$$

In applying Equation (3), we normally use the u and v notation from Equation (2) because it is easier to remember. Here is an example.

EXAMPLE 6 Finding Area

Find the area of the region bounded by the curve $y = xe^{-x}$ and the x-axis from $x = 0$ to $x = 4$.

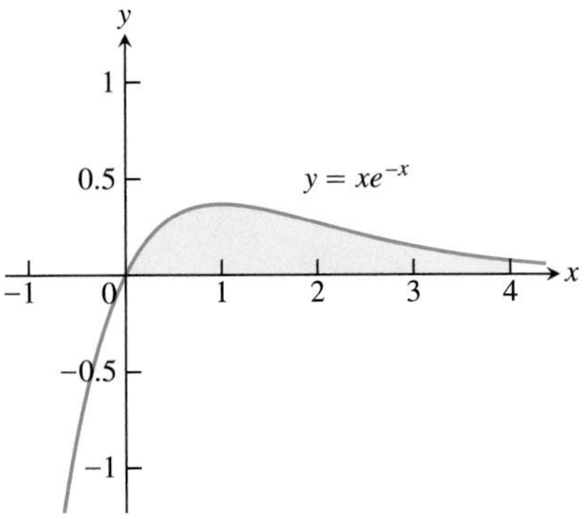

FIGURE 8.1 The region in Example 6.

Solution The region is shaded in Figure 8.1. Its area is

$$\int_0^4 xe^{-x}\, dx.$$

Let $u = x$, $dv = e^{-x}\, dx$, $v = -e^{-x}$, and $du = dx$. Then,

$$\begin{aligned}
\int_0^4 xe^{-x}\, dx &= -xe^{-x}\Big]_0^4 - \int_0^4 (-e^{-x})\, dx \\
&= [-4e^{-4} - (0)] + \int_0^4 e^{-x}\, dx \\
&= -4e^{-4} - e^{-x}\Big]_0^4 \\
&= -4e^{-4} - e^{-4} - (-e^0) = 1 - 5e^{-4} \approx 0.91.
\end{aligned}$$ ■

Tabular Integration

We have seen that integrals of the form $\int f(x)g(x)\, dx$, in which f can be differentiated repeatedly to become zero and g can be integrated repeatedly without difficulty, are natural candidates for integration by parts. However, if many repetitions are required, the calculations can be cumbersome. In situations like this, there is a way to organize the calculations that saves a great deal of work. It is called **tabular integration** and is illustrated in the following examples.

EXAMPLE 7 Using Tabular Integration

Evaluate

$$\int x^2 e^x \, dx.$$

Solution With $f(x) = x^2$ and $g(x) = e^x$, we list:

$f(x)$ and its derivatives		$g(x)$ and its integrals
x^2	$(+)$	e^x
$2x$	$(-)$	e^x
2	$(+)$	e^x
0		e^x

We combine the products of the functions connected by the arrows according to the operation signs above the arrows to obtain

$$\int x^2 e^x \, dx = x^2 e^x - 2xe^x + 2e^x + C.$$

Compare this with the result in Example 4. ■

EXAMPLE 8 Using Tabular Integration

Evaluate

$$\int x^3 \sin x \, dx.$$

Solution With $f(x) = x^3$ and $g(x) = \sin x$, we list:

$f(x)$ and its derivatives		$g(x)$ and its integrals
x^3	$(+)$	$\sin x$
$3x^2$	$(-)$	$-\cos x$
$6x$	$(+)$	$-\sin x$
6	$(-)$	$\cos x$
0		$\sin x$

Again we combine the products of the functions connected by the arrows according to the operation signs above the arrows to obtain

$$\int x^3 \sin x \, dx = -x^3 \cos x + 3x^2 \sin x + 6x \cos x - 6 \sin x + C.$$ ■

The Additional Exercises at the end of this chapter show how tabular integration can be used when neither function f nor g can be differentiated repeatedly to become zero.

Summary

When substitution doesn't work, try integration by parts. Start with an integral in which the integrand is the product of two functions,

$$\int f(x)g(x)\,dx.$$

(Remember that g may be the constant function 1, as in Example 3.) Match the integral with the form

$$\int u\,dv$$

by choosing dv to be part of the integrand including dx and either $f(x)$ or $g(x)$. Remember that we must be able to readily integrate dv to get v in order to obtain the right side of the formula

$$\int u\,dv = uv - \int v\,du.$$

If the new integral on the right side is more complex than the original one, try a different choice for u and dv.

EXAMPLE 9 A Reduction Formula

Obtain a "reduction" formula that expresses the integral

$$\int \cos^n x\,dx$$

in terms of an integral of a lower power of $\cos x$.

Solution We may think of $\cos^n x$ as $\cos^{n-1} x \cdot \cos x$. Then we let

$$u = \cos^{n-1} x \qquad \text{and} \qquad dv = \cos x\,dx,$$

so that

$$du = (n-1)\cos^{n-2} x\,(-\sin x\,dx) \qquad \text{and} \qquad v = \sin x.$$

Hence

$$\begin{aligned}
\int \cos^n x\,dx &= \cos^{n-1} x \sin x + (n-1)\int \sin^2 x \cos^{n-2} x\,dx \\
&= \cos^{n-1} x \sin x + (n-1)\int (1-\cos^2 x)\cos^{n-2} x\,dx, \\
&= \cos^{n-1} x \sin x + (n-1)\int \cos^{n-2} x\,dx - (n-1)\int \cos^n x\,dx.
\end{aligned}$$

If we add

$$(n-1)\int \cos^n x\,dx$$

to both sides of this equation, we obtain

$$n\int \cos^n x\,dx = \cos^{n-1} x \sin x + (n-1)\int \cos^{n-2} x\,dx.$$

We then divide through by n, and the final result is

$$\int \cos^n x\,dx = \frac{\cos^{n-1} x \sin x}{n} + \frac{n-1}{n}\int \cos^{n-2} x\,dx.$$

This allows us to reduce the exponent on $\cos x$ by 2 and is a very useful formula. When n is a positive integer, we may apply the formula repeatedly until the remaining integral is either

$$\int \cos x\,dx = \sin x + C \qquad \text{or} \qquad \int \cos^0 x\,dx = \int dx = x + C.$$

■

EXAMPLE 10 Using a Reduction Formula

Evaluate

$$\int \cos^3 x\,dx.$$

Solution From the result in Example 9,

$$\int \cos^3 x\,dx = \frac{\cos^2 x \sin x}{3} + \frac{2}{3}\int \cos x\,dx$$

$$= \frac{1}{3}\cos^2 x \sin x + \frac{2}{3}\sin x + C.$$

■

EXERCISES 8.2

Integration by Parts

Evaluate the integrals in Exercises 1–24.

1. $\int x \sin \frac{x}{2}\,dx$
2. $\int \theta \cos \pi\theta\,d\theta$
3. $\int t^2 \cos t\,dt$
4. $\int x^2 \sin x\,dx$
5. $\int_1^2 x \ln x\,dx$
6. $\int_1^e x^3 \ln x\,dx$
7. $\int \tan^{-1} y\,dy$
8. $\int \sin^{-1} y\,dy$
9. $\int x \sec^2 x\,dx$
10. $\int 4x \sec^2 2x\,dx$
11. $\int x^3 e^x\,dx$
12. $\int p^4 e^{-p}\,dp$
13. $\int (x^2 - 5x)e^x\,dx$
14. $\int (r^2 + r + 1)e^r\,dr$
15. $\int x^5 e^x\,dx$
16. $\int t^2 e^{4t}\,dt$
17. $\int_0^{\pi/2} \theta^2 \sin 2\theta\,d\theta$
18. $\int_0^{\pi/2} x^3 \cos 2x\,dx$
19. $\int_{2/\sqrt{3}}^{2} t \sec^{-1} t\,dt$
20. $\int_0^{1/\sqrt{2}} 2x \sin^{-1}(x^2)\,dx$
21. $\int e^\theta \sin\theta\,d\theta$
22. $\int e^{-y} \cos y\,dy$
23. $\int e^{2x} \cos 3x\,dx$
24. $\int e^{-2x} \sin 2x\,dx$

Substitution and Integration by Parts

Evaluate the integrals in Exercises 25–30 by using a substitution prior to integration by parts.

25. $\int e^{\sqrt{3s+9}}\, ds$

26. $\int_0^1 x\sqrt{1-x}\, dx$

27. $\int_0^{\pi/3} x \tan^2 x\, dx$

28. $\int \ln (x + x^2)\, dx$

29. $\int \sin (\ln x)\, dx$

30. $\int z(\ln z)^2\, dz$

Theory and Examples

31. Finding area Find the area of the region enclosed by the curve $y = x \sin x$ and the x-axis (see the accompanying figure) for

a. $0 \le x \le \pi$ **b.** $\pi \le x \le 2\pi$ **c.** $2\pi \le x \le 3\pi$.

d. What pattern do you see here? What is the area between the curve and the x-axis for $n\pi \le x \le (n + 1)\pi$, n an arbitrary nonnegative integer? Give reasons for your answer.

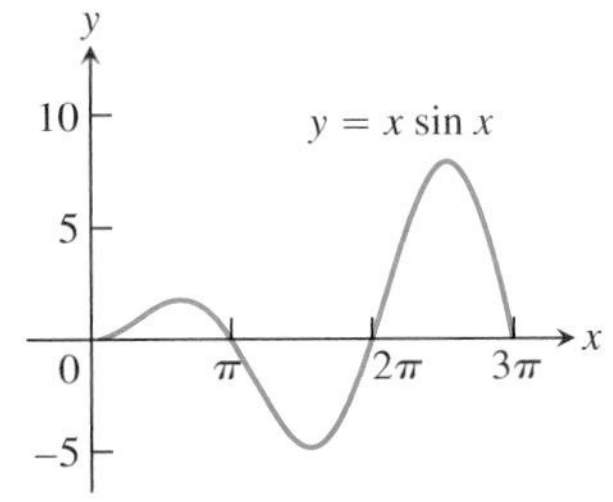

32. Finding area Find the area of the region enclosed by the curve $y = x \cos x$ and the x-axis (see the accompanying figure) for

a. $\pi/2 \le x \le 3\pi/2$ **b.** $3\pi/2 \le x \le 5\pi/2$

c. $5\pi/2 \le x \le 7\pi/2$.

d. What pattern do you see? What is the area between the curve and the x-axis for

$$\left(\frac{2n-1}{2}\right)\pi \le x \le \left(\frac{2n+1}{2}\right)\pi,$$

n an arbitrary positive integer? Give reasons for your answer.

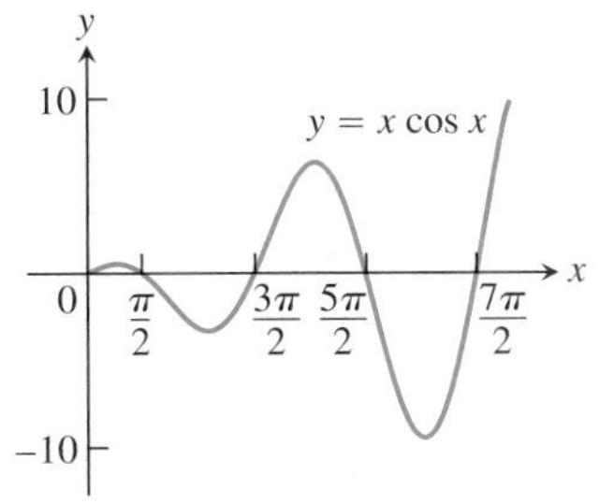

33. Finding volume Find the volume of the solid generated by revolving the region in the first quadrant bounded by the coordinate axes, the curve $y = e^x$, and the line $x = \ln 2$ about the line $x = \ln 2$.

34. Finding volume Find the volume of the solid generated by revolving the region in the first quadrant bounded by the coordinate axes, the curve $y = e^{-x}$, and the line $x = 1$

a. about the y-axis. **b.** about the line $x = 1$.

35. Finding volume Find the volume of the solid generated by revolving the region in the first quadrant bounded by the coordinate axes and the curve $y = \cos x$, $0 \le x \le \pi/2$, about

a. the y-axis. **b.** the line $x = \pi/2$.

36. Finding volume Find the volume of the solid generated by revolving the region bounded by the x-axis and the curve $y = x \sin x$, $0 \le x \le \pi$, about

a. the y-axis. **b.** the line $x = \pi$.

(See Exercise 31 for a graph.)

37. Average value A retarding force, symbolized by the dashpot in the figure, slows the motion of the weighted spring so that the mass's position at time t is

$$y = 2e^{-t} \cos t, \qquad t \ge 0.$$

Find the average value of y over the interval $0 \le t \le 2\pi$.

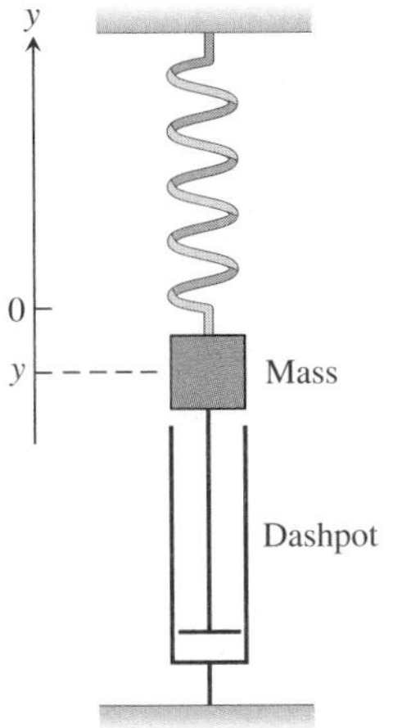

38. Average value In a mass-spring-dashpot system like the one in Exercise 37, the mass's position at time t is

$$y = 4e^{-t} (\sin t - \cos t), \qquad t \ge 0.$$

Find the average value of y over the interval $0 \le t \le 2\pi$.

Reduction Formulas

In Exercises 39–42, use integration by parts to establish the *reduction formula*.

39. $\int x^n \cos x\,dx = x^n \sin x - n\int x^{n-1}\sin x\,dx$

40. $\int x^n \sin x\,dx = -x^n \cos x + n\int x^{n-1}\cos x\,dx$

41. $\int x^n e^{ax}\,dx = \frac{x^n e^{ax}}{a} - \frac{n}{a}\int x^{n-1}e^{ax}\,dx, \quad a \neq 0$

42. $\int (\ln x)^n\,dx = x(\ln x)^n - n\int (\ln x)^{n-1}\,dx$

Integrating Inverses of Functions

Integration by parts leads to a rule for integrating inverses that usually gives good results:

$$\int f^{-1}(x)\,dx = \int yf'(y)\,dy \qquad \begin{array}{l} y = f^{-1}(x), \quad x = f(y) \\ dx = f'(y)\,dy \end{array}$$

$$= yf(y) - \int f(y)\,dy \qquad \begin{array}{l} \text{Integration by parts with} \\ u = y,\ dv = f'(y)\,dy \end{array}$$

$$= xf^{-1}(x) - \int f(y)\,dy$$

The idea is to take the most complicated part of the integral, in this case $f^{-1}(x)$, and simplify it first. For the integral of $\ln x$, we get

$$\int \ln x\,dx = \int ye^y\,dy \qquad \begin{array}{l} y = \ln x, \quad x = e^y \\ dx = e^y\,dy \end{array}$$

$$= ye^y - e^y + C$$

$$= x\ln x - x + C.$$

For the integral of $\cos^{-1} x$ we get

$$\int \cos^{-1} x\,dx = x\cos^{-1} x - \int \cos y\,dy \qquad y = \cos^{-1} x$$

$$= x\cos^{-1} x - \sin y + C$$

$$= x\cos^{-1} x - \sin(\cos^{-1} x) + C.$$

Use the formula

$$\int f^{-1}(x)\,dx = xf^{-1}(x) - \int f(y)\,dy \qquad y = f^{-1}(x) \qquad (4)$$

to evaluate the integrals in Exercises 43–46. Express your answers in terms of x.

43. $\int \sin^{-1} x\,dx$

44. $\int \tan^{-1} x\,dx$

45. $\int \sec^{-1} x\,dx$

46. $\int \log_2 x\,dx$

Another way to integrate $f^{-1}(x)$ (when f^{-1} is integrable, of course) is to use integration by parts with $u = f^{-1}(x)$ and $dv = dx$ to rewrite the integral of f^{-1} as

$$\int f^{-1}(x)\,dx = xf^{-1}(x) - \int x\left(\frac{d}{dx}f^{-1}(x)\right)dx. \qquad (5)$$

Exercises 47 and 48 compare the results of using Equations (4) and (5).

47. Equations (4) and (5) give different formulas for the integral of $\cos^{-1} x$:

a. $\int \cos^{-1} x\,dx = x\cos^{-1} x - \sin(\cos^{-1} x) + C$ Eq. (4)

b. $\int \cos^{-1} x\,dx = x\cos^{-1} x - \sqrt{1 - x^2} + C$ Eq. (5)

Can both integrations be correct? Explain.

48. Equations (4) and (5) lead to different formulas for the integral of $\tan^{-1} x$:

a. $\int \tan^{-1} x\,dx = x\tan^{-1} x - \ln\sec(\tan^{-1} x) + C$ Eq. (4)

b. $\int \tan^{-1} x\,dx = x\tan^{-1} x - \ln\sqrt{1 + x^2} + C$ Eq. (5)

Can both integrations be correct? Explain.

Evaluate the integrals in Exercises 49 and 50 with **(a)** Eq. (4) and **(b)** Eq. (5). In each case, check your work by differentiating your answer with respect to x.

49. $\int \sinh^{-1} x\,dx$

50. $\int \tanh^{-1} x\,dx$

8.3 Integration of Rational Functions by Partial Fractions

This section shows how to express a rational function (a quotient of polynomials) as a sum of simpler fractions, called *partial fractions*, which are easily integrated. For instance, the rational function $(5x - 3)/(x^2 - 2x - 3)$ can be rewritten as

$$\frac{5x - 3}{x^2 - 2x - 3} = \frac{2}{x + 1} + \frac{3}{x - 3},$$

which can be verified algebraically by placing the fractions on the right side over a common denominator $(x+1)(x-3)$. The skill acquired in writing rational functions as such a sum is useful in other settings as well (for instance, when using certain transform methods to solve differential equations). To integrate the rational function $(5x-3)/(x+1)(x-3)$ on the left side of our previous expression, we simply sum the integrals of the fractions on the right side:

$$\int \frac{5x-3}{(x+1)(x-3)}\,dx = \int \frac{2}{x+1}\,dx + \int \frac{3}{x-3}\,dx$$

$$= 2\ln|x+1| + 3\ln|x-3| + C.$$

The method for rewriting rational functions as a sum of simpler fractions is called **the method of partial fractions**. In the case of the above example, it consists of finding constants A and B such that

$$\frac{5x-3}{x^2-2x-3} = \frac{A}{x+1} + \frac{B}{x-3}. \tag{1}$$

(Pretend for a moment that we do not know that $A = 2$ and $B = 3$ will work.) We call the fractions $A/(x+1)$ and $B/(x-3)$ **partial fractions** because their denominators are only part of the original denominator $x^2 - 2x - 3$. We call A and B **undetermined coefficients** until proper values for them have been found.

To find A and B, we first clear Equation (1) of fractions, obtaining

$$5x - 3 = A(x-3) + B(x+1) = (A+B)x - 3A + B.$$

This will be an identity in x if and only if the coefficients of like powers of x on the two sides are equal:

$$A + B = 5, \qquad -3A + B = -3.$$

Solving these equations simultaneously gives $A = 2$ and $B = 3$.

General Description of the Method

Success in writing a rational function $f(x)/g(x)$ as a sum of partial fractions depends on two things:

- *The degree of $f(x)$ must be less than the degree of $g(x)$.* That is, the fraction must be proper. If it isn't, divide $f(x)$ by $g(x)$ and work with the remainder term. See Example 3 of this section.
- *We must know the factors of $g(x)$.* In theory, any polynomial with real coefficients can be written as a product of real linear factors and real quadratic factors. In practice, the factors may be hard to find.

Here is how we find the partial fractions of a proper fraction $f(x)/g(x)$ when the factors of g are known.

Method of Partial Fractions ($f(x)/g(x)$ Proper)

1. Let $x - r$ be a linear factor of $g(x)$. Suppose that $(x - r)^m$ is the highest power of $x - r$ that divides $g(x)$. Then, to this factor, assign the sum of the m partial fractions:
$$\frac{A_1}{x - r} + \frac{A_2}{(x - r)^2} + \cdots + \frac{A_m}{(x - r)^m}.$$
Do this for each distinct linear factor of $g(x)$.
2. Let $x^2 + px + q$ be a quadratic factor of $g(x)$. Suppose that $(x^2 + px + q)^n$ is the highest power of this factor that divides $g(x)$. Then, to this factor, assign the sum of the n partial fractions:
$$\frac{B_1x + C_1}{x^2 + px + q} + \frac{B_2x + C_2}{(x^2 + px + q)^2} + \cdots + \frac{B_nx + C_n}{(x^2 + px + q)^n}.$$
Do this for each distinct quadratic factor of $g(x)$ that cannot be factored into linear factors with real coefficients.
3. Set the original fraction $f(x)/g(x)$ equal to the sum of all these partial fractions. Clear the resulting equation of fractions and arrange the terms in decreasing powers of x.
4. Equate the coefficients of corresponding powers of x and solve the resulting equations for the undetermined coefficients.

EXAMPLE 1 Distinct Linear Factors

Evaluate
$$\int \frac{x^2 + 4x + 1}{(x - 1)(x + 1)(x + 3)}\,dx$$
using partial fractions.

Solution The partial fraction decomposition has the form
$$\frac{x^2 + 4x + 1}{(x - 1)(x + 1)(x + 3)} = \frac{A}{x - 1} + \frac{B}{x + 1} + \frac{C}{x + 3}.$$

To find the values of the undetermined coefficients A, B, and C we clear fractions and get
$$\begin{aligned} x^2 + 4x + 1 &= A(x + 1)(x + 3) + B(x - 1)(x + 3) + C(x - 1)(x + 1) \\ &= (A + B + C)x^2 + (4A + 2B)x + (3A - 3B - C). \end{aligned}$$

The polynomials on both sides of the above equation are identical, so we equate coefficients of like powers of x obtaining
$$\begin{aligned} &\text{Coefficient of } x^2: && A + B + C = 1 \\ &\text{Coefficient of } x^1: && 4A + 2B \phantom{{}- C} = 4 \\ &\text{Coefficient of } x^0: && 3A - 3B - C = 1 \end{aligned}$$

There are several ways for solving such a system of linear equations for the unknowns A, B, and C, including elimination of variables, or the use of a calculator or computer. Whatever method is used, the solution is $A = 3/4$, $B = 1/2$, and $C = -1/4$. Hence we have

$$\int \frac{x^2 + 4x + 1}{(x-1)(x+1)(x+3)}\,dx = \int \left[\frac{3}{4}\frac{1}{x-1} + \frac{1}{2}\frac{1}{x+1} - \frac{1}{4}\frac{1}{x+3}\right] dx$$

$$= \frac{3}{4}\ln|x-1| + \frac{1}{2}\ln|x+1| - \frac{1}{4}\ln|x+3| + K,$$

where K is the arbitrary constant of integration (to avoid confusion with the undetermined coefficient we labeled as C). ■

EXAMPLE 2 A Repeated Linear Factor

Evaluate

$$\int \frac{6x+7}{(x+2)^2}\,dx.$$

Solution First we express the integrand as a sum of partial fractions with undetermined coefficients.

$$\frac{6x+7}{(x+2)^2} = \frac{A}{x+2} + \frac{B}{(x+2)^2}$$

$$6x + 7 = A(x+2) + B \qquad \text{Multiply both sides by } (x+2)^2.$$

$$= Ax + (2A + B)$$

Equating coefficients of corresponding powers of x gives

$$A = 6 \quad \text{and} \quad 2A + B = 12 + B = 7, \quad \text{or} \quad A = 6 \quad \text{and} \quad B = -5.$$

Therefore,

$$\int \frac{6x+7}{(x+2)^2}\,dx = \int \left(\frac{6}{x+2} - \frac{5}{(x+2)^2}\right) dx$$

$$= 6\int \frac{dx}{x+2} - 5\int (x+2)^{-2}\,dx$$

$$= 6\ln|x+2| + 5(x+2)^{-1} + C$$ ■

EXAMPLE 3 Integrating an Improper Fraction

Evaluate

$$\int \frac{2x^3 - 4x^2 - x - 3}{x^2 - 2x - 3}\,dx.$$

Solution First we divide the denominator into the numerator to get a polynomial plus a proper fraction.

$$\begin{array}{r} 2x \\ x^2 - 2x - 3\overline{)\,2x^3 - 4x^2 - x - 3} \\ \underline{2x^3 - 4x^2 - 6x} \\ 5x - 3 \end{array}$$

Then we write the improper fraction as a polynomial plus a proper fraction.

$$\frac{2x^3 - 4x^2 - x - 3}{x^2 - 2x - 3} = 2x + \frac{5x - 3}{x^2 - 2x - 3}$$

We found the partial fraction decomposition of the fraction on the right in the opening example, so

$$\begin{aligned}\int \frac{2x^3 - 4x^2 - x - 3}{x^2 - 2x - 3}\,dx &= \int 2x\,dx + \int \frac{5x - 3}{x^2 - 2x - 3}\,dx \\ &= \int 2x\,dx + \int \frac{2}{x + 1}\,dx + \int \frac{3}{x - 3}\,dx \\ &= x^2 + 2\ln|x + 1| + 3\ln|x - 3| + C.\end{aligned}$$

■

A quadratic polynomial is **irreducible** if it cannot be written as the product of two linear factors with real coefficients.

EXAMPLE 4 Integrating with an Irreducible Quadratic Factor in the Denominator

Evaluate

$$\int \frac{-2x + 4}{(x^2 + 1)(x - 1)^2}\,dx$$

using partial fractions.

Solution The denominator has an irreducible quadratic factor as well as a repeated linear factor, so we write

$$\frac{-2x + 4}{(x^2 + 1)(x - 1)^2} = \frac{Ax + B}{x^2 + 1} + \frac{C}{x - 1} + \frac{D}{(x - 1)^2}. \qquad (2)$$

Clearing the equation of fractions gives

$$\begin{aligned}-2x + 4 &= (Ax + B)(x - 1)^2 + C(x - 1)(x^2 + 1) + D(x^2 + 1) \\ &= (A + C)x^3 + (-2A + B - C + D)x^2 \\ &\quad + (A - 2B + C)x + (B - C + D).\end{aligned}$$

Equating coefficients of like terms gives

Coefficients of x^3:	$0 = A + C$
Coefficients of x^2:	$0 = -2A + B - C + D$
Coefficients of x^1:	$-2 = A - 2B + C$
Coefficients of x^0:	$4 = B - C + D$

We solve these equations simultaneously to find the values of A, B, C, and D:

$-4 = -2A, \quad A = 2$	Subtract fourth equation from second.
$C = -A = -2$	From the first equation
$B = 1$	$A = 2$ and $C = -2$ in third equation.
$D = 4 - B + C = 1.$	From the fourth equation

We substitute these values into Equation (2), obtaining

$$\frac{-2x+4}{(x^2+1)(x-1)^2} = \frac{2x+1}{x^2+1} - \frac{2}{x-1} + \frac{1}{(x-1)^2}.$$

Finally, using the expansion above we can integrate:

$$\begin{aligned}
\int \frac{-2x+4}{(x^2+1)(x-1)^2}\,dx &= \int \left(\frac{2x+1}{x^2+1} - \frac{2}{x-1} + \frac{1}{(x-1)^2}\right) dx \\
&= \int \left(\frac{2x}{x^2+1} + \frac{1}{x^2+1} - \frac{2}{x-1} + \frac{1}{(x-1)^2}\right) dx \\
&= \ln(x^2+1) + \tan^{-1} x - 2\ln|x-1| - \frac{1}{x-1} + C.
\end{aligned}$$

EXAMPLE 5 A Repeated Irreducible Quadratic Factor

Evaluate

$$\int \frac{dx}{x(x^2+1)^2}.$$

Solution The form of the partial fraction decomposition is

$$\frac{1}{x(x^2+1)^2} = \frac{A}{x} + \frac{Bx+C}{x^2+1} + \frac{Dx+E}{(x^2+1)^2}$$

Multiplying by $x(x^2+1)^2$, we have

$$\begin{aligned}
1 &= A(x^2+1)^2 + (Bx+C)x(x^2+1) + (Dx+E)x \\
&= A(x^4+2x^2+1) + B(x^4+x^2) + C(x^3+x) + Dx^2 + Ex \\
&= (A+B)x^4 + Cx^3 + (2A+B+D)x^2 + (C+E)x + A
\end{aligned}$$

If we equate coefficients, we get the system

$$A+B=0, \quad C=0, \quad 2A+B+D=0, \quad C+E=0, \quad A=1.$$

Solving this system gives $A=1$, $B=-1$, $C=0$, $D=-1$, and $E=0$. Thus,

$$\begin{aligned}
\int \frac{dx}{x(x^2+1)^2} &= \int \left[\frac{1}{x} + \frac{-x}{x^2+1} + \frac{-x}{(x^2+1)^2}\right] dx \\
&= \int \frac{dx}{x} - \int \frac{x\,dx}{x^2+1} - \int \frac{x\,dx}{(x^2+1)^2} \\
&= \int \frac{dx}{x} - \frac{1}{2}\int \frac{du}{u} - \frac{1}{2}\int \frac{du}{u^2} \qquad u = x^2+1,\ du = 2x\,dx \\
&= \ln|x| - \frac{1}{2}\ln|u| + \frac{1}{2u} + K \\
&= \ln|x| - \frac{1}{2}\ln(x^2+1) + \frac{1}{2(x^2+1)} + K \\
&= \ln\frac{|x|}{\sqrt{x^2+1}} + \frac{1}{2(x^2+1)} + K.
\end{aligned}$$

Historical Biography

Oliver Heaviside
(1850–1925)

The Heaviside "Cover-up" Method for Linear Factors

When the degree of the polynomial $f(x)$ is less than the degree of $g(x)$ and

$$g(x) = (x - r_1)(x - r_2)\cdots(x - r_n)$$

is a product of n distinct linear factors, each raised to the first power, there is a quick way to expand $f(x)/g(x)$ by partial fractions.

EXAMPLE 6 Using the Heaviside Method

Find A, B, and C in the partial-fraction expansion

$$\frac{x^2 + 1}{(x - 1)(x - 2)(x - 3)} = \frac{A}{x - 1} + \frac{B}{x - 2} + \frac{C}{x - 3}. \tag{3}$$

Solution If we multiply both sides of Equation (3) by $(x - 1)$ to get

$$\frac{x^2 + 1}{(x - 2)(x - 3)} = A + \frac{B(x - 1)}{x - 2} + \frac{C(x - 1)}{x - 3}$$

and set $x = 1$, the resulting equation gives the value of A:

$$\frac{(1)^2 + 1}{(1 - 2)(1 - 3)} = A + 0 + 0,$$

$$A = 1.$$

Thus, the value of A is the number we would have obtained if we had covered the factor $(x - 1)$ in the denominator of the original fraction

$$\frac{x^2 + 1}{(x - 1)(x - 2)(x - 3)} \tag{4}$$

and evaluated the rest at $x = 1$:

$$A = \frac{(1)^2 + 1}{\boxed{(x - 1)}\,(1 - 2)(1 - 3)} = \frac{2}{(-1)(-2)} = 1.$$

↑ Cover

Similarly, we find the value of B in Equation (3) by covering the factor $(x - 2)$ in Equation (4) and evaluating the rest at $x = 2$:

$$B = \frac{(2)^2 + 1}{(2 - 1)\boxed{(x - 2)}(2 - 3)} = \frac{5}{(1)(-1)} = -5.$$

↑ Cover

Finally, C is found by covering the $(x - 3)$ in Equation (4) and evaluating the rest at $x = 3$:

$$C = \frac{(3)^2 + 1}{(3 - 1)(3 - 2)\boxed{(x - 3)}} = \frac{10}{(2)(1)} = 5.$$

↑ Cover

■

Heaviside Method

1. *Write the quotient with $g(x)$ factored:*

$$\frac{f(x)}{g(x)} = \frac{f(x)}{(x - r_1)(x - r_2)\cdots(x - r_n)}.$$

2. *Cover the factors $(x - r_i)$ of $g(x)$ one at a time*, each time replacing all the uncovered x's by the number r_i. This gives a number A_i for each root r_i:

$$A_1 = \frac{f(r_1)}{(r_1 - r_2)\cdots(r_1 - r_n)}$$

$$A_2 = \frac{f(r_2)}{(r_2 - r_1)(r_2 - r_3)\cdots(r_2 - r_n)}$$

$$\vdots$$

$$A_n = \frac{f(r_n)}{(r_n - r_1)(r_n - r_2)\cdots(r_n - r_{n-1})}.$$

3. *Write the partial-fraction expansion of $f(x)/g(x)$* as

$$\frac{f(x)}{g(x)} = \frac{A_1}{(x - r_1)} + \frac{A_2}{(x - r_2)} + \cdots + \frac{A_n}{(x - r_n)}.$$

EXAMPLE 7 Integrating with the Heaviside Method

Evaluate

$$\int \frac{x + 4}{x^3 + 3x^2 - 10x}\, dx.$$

Solution The degree of $f(x) = x + 4$ is less than the degree of $g(x) = x^3 + 3x^2 - 10x$, and, with $g(x)$ factored,

$$\frac{x + 4}{x^3 + 3x^2 - 10x} = \frac{x + 4}{x(x - 2)(x + 5)}.$$

The roots of $g(x)$ are $r_1 = 0$, $r_2 = 2$, and $r_3 = -5$. We find

$$A_1 = \frac{0 + 4}{\boxed{x}\,(0 - 2)(0 + 5)} = \frac{4}{(-2)(5)} = -\frac{2}{5}$$

$\Uparrow$ Cover (under x)

$$A_2 = \frac{2 + 4}{2\,\boxed{(x - 2)}\,(2 + 5)} = \frac{6}{(2)(7)} = \frac{3}{7}$$

$\Uparrow$ Cover (under $(x-2)$)

$$A_3 = \frac{-5 + 4}{(-5)(-5 - 2)\,\boxed{(x + 5)}} = \frac{-1}{(-5)(-7)} = -\frac{1}{35}.$$

$\Uparrow$ Cover (under $(x+5)$)

Therefore,

$$\frac{x+4}{x(x-2)(x+5)} = -\frac{2}{5x} + \frac{3}{7(x-2)} - \frac{1}{35(x+5)},$$

and

$$\int \frac{x+4}{x(x-2)(x+5)}\,dx = -\frac{2}{5}\ln|x| + \frac{3}{7}\ln|x-2| - \frac{1}{35}\ln|x+5| + C.$$ ■

Other Ways to Determine the Coefficients

Another way to determine the constants that appear in partial fractions is to differentiate, as in the next example. Still another is to assign selected numerical values to x.

EXAMPLE 8 Using Differentiation

Find A, B, and C in the equation

$$\frac{x-1}{(x+1)^3} = \frac{A}{x+1} + \frac{B}{(x+1)^2} + \frac{C}{(x+1)^3}.$$

Solution We first clear fractions:

$$x - 1 = A(x+1)^2 + B(x+1) + C.$$

Substituting $x = -1$ shows $C = -2$. We then differentiate both sides with respect to x, obtaining

$$1 = 2A(x+1) + B.$$

Substituting $x = -1$ shows $B = 1$. We differentiate again to get $0 = 2A$, which shows $A = 0$. Hence,

$$\frac{x-1}{(x+1)^3} = \frac{1}{(x+1)^2} - \frac{2}{(x+1)^3}.$$ ■

In some problems, assigning small values to x such as $x = 0, \pm 1, \pm 2$, to get equations in A, B, and C provides a fast alternative to other methods.

EXAMPLE 9 Assigning Numerical Values to x

Find A, B, and C in

$$\frac{x^2+1}{(x-1)(x-2)(x-3)} = \frac{A}{x-1} + \frac{B}{x-2} + \frac{C}{x-3}.$$

Solution Clear fractions to get

$$x^2 + 1 = A(x-2)(x-3) + B(x-1)(x-3) + C(x-1)(x-2).$$

Then let $x = 1, 2, 3$ successively to find A, B, and C:

$$\begin{aligned} x = 1: \quad (1)^2 + 1 &= A(-1)(-2) + B(0) + C(0) \\ 2 &= 2A \\ A &= 1 \\ x = 2: \quad (2)^2 + 1 &= A(0) + B(1)(-1) + C(0) \\ 5 &= -B \\ B &= -5 \\ x = 3: \quad (3)^2 + 1 &= A(0) + B(0) + C(2)(1) \\ 10 &= 2C \\ C &= 5. \end{aligned}$$

Conclusion:

$$\frac{x^2 + 1}{(x - 1)(x - 2)(x - 3)} = \frac{1}{x - 1} - \frac{5}{x - 2} + \frac{5}{x - 3}.$$

■

EXERCISES 8.3

Expanding Quotients into Partial Fractions

Expand the quotients in Exercises 1–8 by partial fractions.

1. $\dfrac{5x - 13}{(x - 3)(x - 2)}$
2. $\dfrac{5x - 7}{x^2 - 3x + 2}$
3. $\dfrac{x + 4}{(x + 1)^2}$
4. $\dfrac{2x + 2}{x^2 - 2x + 1}$
5. $\dfrac{z + 1}{z^2(z - 1)}$
6. $\dfrac{z}{z^3 - z^2 - 6z}$
7. $\dfrac{t^2 + 8}{t^2 - 5t + 6}$
8. $\dfrac{t^4 + 9}{t^4 + 9t^2}$

Nonrepeated Linear Factors

In Exercises 9–16, express the integrands as a sum of partial fractions and evaluate the integrals.

9. $\displaystyle\int \frac{dx}{1 - x^2}$
10. $\displaystyle\int \frac{dx}{x^2 + 2x}$
11. $\displaystyle\int \frac{x + 4}{x^2 + 5x - 6}\,dx$
12. $\displaystyle\int \frac{2x + 1}{x^2 - 7x + 12}\,dx$
13. $\displaystyle\int_4^8 \frac{y\,dy}{y^2 - 2y - 3}$
14. $\displaystyle\int_{1/2}^1 \frac{y + 4}{y^2 + y}\,dy$
15. $\displaystyle\int \frac{dt}{t^3 + t^2 - 2t}$
16. $\displaystyle\int \frac{x + 3}{2x^3 - 8x}\,dx$

Repeated Linear Factors

In Exercises 17–20, express the integrands as a sum of partial fractions and evaluate the integrals.

17. $\displaystyle\int_0^1 \frac{x^3\,dx}{x^2 + 2x + 1}$
18. $\displaystyle\int_{-1}^0 \frac{x^3\,dx}{x^2 - 2x + 1}$
19. $\displaystyle\int \frac{dx}{(x^2 - 1)^2}$
20. $\displaystyle\int \frac{x^2\,dx}{(x - 1)(x^2 + 2x + 1)}$

Irreducible Quadratic Factors

In Exercises 21–28, express the integrands as a sum of partial fractions and evaluate the integrals.

21. $\displaystyle\int_0^1 \frac{dx}{(x + 1)(x^2 + 1)}$
22. $\displaystyle\int_1^{\sqrt{3}} \frac{3t^2 + t + 4}{t^3 + t}\,dt$
23. $\displaystyle\int \frac{y^2 + 2y + 1}{(y^2 + 1)^2}\,dy$
24. $\displaystyle\int \frac{8x^2 + 8x + 2}{(4x^2 + 1)^2}\,dx$
25. $\displaystyle\int \frac{2s + 2}{(s^2 + 1)(s - 1)^3}\,ds$
26. $\displaystyle\int \frac{s^4 + 81}{s(s^2 + 9)^2}\,ds$
27. $\displaystyle\int \frac{2\theta^3 + 5\theta^2 + 8\theta + 4}{(\theta^2 + 2\theta + 2)^2}\,d\theta$
28. $\displaystyle\int \frac{\theta^4 - 4\theta^3 + 2\theta^2 - 3\theta + 1}{(\theta^2 + 1)^3}\,d\theta$

Improper Fractions

In Exercises 29–34, perform long division on the integrand, write the proper fraction as a sum of partial fractions, and then evaluate the integral.

29. $\displaystyle\int \frac{2x^3 - 2x^2 + 1}{x^2 - x}\,dx$
30. $\displaystyle\int \frac{x^4}{x^2 - 1}\,dx$
31. $\displaystyle\int \frac{9x^3 - 3x + 1}{x^3 - x^2}\,dx$
32. $\displaystyle\int \frac{16x^3}{4x^2 - 4x + 1}\,dx$
33. $\displaystyle\int \frac{y^4 + y^2 - 1}{y^3 + y}\,dy$
34. $\displaystyle\int \frac{2y^4}{y^3 - y^2 + y - 1}\,dy$

Evaluating Integrals

Evaluate the integrals in Exercises 35–40.

35. $\displaystyle\int \frac{e^t\,dt}{e^{2t} + 3e^t + 2}$

36. $\displaystyle\int \frac{e^{4t} + 2e^{2t} - e^t}{e^{2t} + 1}\,dt$

37. $\displaystyle\int \frac{\cos y\,dy}{\sin^2 y + \sin y - 6}$

38. $\displaystyle\int \frac{\sin\theta\,d\theta}{\cos^2\theta + \cos\theta - 2}$

39. $\displaystyle\int \frac{(x-2)^2\tan^{-1}(2x) - 12x^3 - 3x}{(4x^2+1)(x-2)^2}\,dx$

40. $\displaystyle\int \frac{(x+1)^2\tan^{-1}(3x) + 9x^3 + x}{(9x^2+1)(x+1)^2}\,dx$

Initial Value Problems

Solve the initial value problems in Exercises 41–44 for x as a function of t.

41. $(t^2 - 3t + 2)\dfrac{dx}{dt} = 1 \quad (t > 2), \quad x(3) = 0$

42. $(3t^4 + 4t^2 + 1)\dfrac{dx}{dt} = 2\sqrt{3}, \quad x(1) = -\pi\sqrt{3}/4$

43. $(t^2 + 2t)\dfrac{dx}{dt} = 2x + 2 \quad (t, x > 0), \quad x(1) = 1$

44. $(t + 1)\dfrac{dx}{dt} = x^2 + 1 \quad (t > -1), \quad x(0) = \pi/4$

Applications and Examples

In Exercises 45 and 46, find the volume of the solid generated by revolving the shaded region about the indicated axis.

45. The x-axis

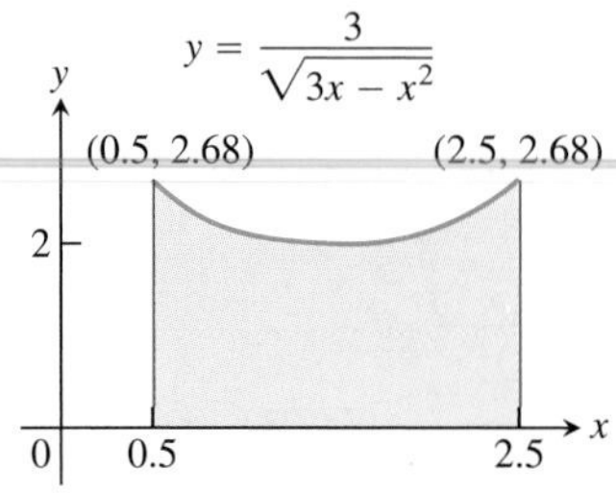

46. The y-axis

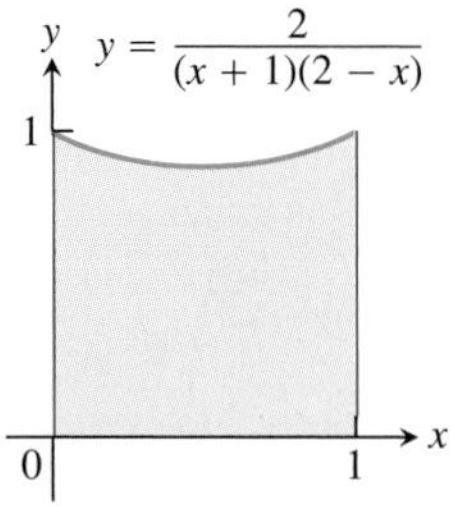

T 47. Find, to two decimal places, the x-coordinate of the centroid of the region in the first quadrant bounded by the x-axis, the curve $y = \tan^{-1} x$, and the line $x = \sqrt{3}$.

T 48. Find the x-coordinate of the centroid of this region to two decimal places.

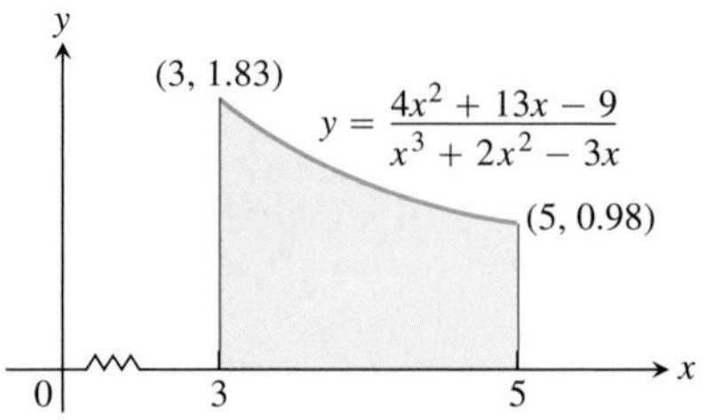

T 49. Social diffusion Sociologists sometimes use the phrase "social diffusion" to describe the way information spreads through a population. The information might be a rumor, a cultural fad, or news about a technical innovation. In a sufficiently large population, the number of people x who have the information is treated as a differentiable function of time t, and the rate of diffusion, dx/dt, is assumed to be proportional to the number of people who have the information times the number of people who do not. This leads to the equation

$$\frac{dx}{dt} = kx(N - x),$$

where N is the number of people in the population.

Suppose t is in days, $k = 1/250$, and two people start a rumor at time $t = 0$ in a population of $N = 1000$ people.

a. Find x as a function of t.

b. When will half the population have heard the rumor? (This is when the rumor will be spreading the fastest.)

T 50. Second-order chemical reactions Many chemical reactions are the result of the interaction of two molecules that undergo a change to produce a new product. The rate of the reaction typically depends on the concentrations of the two kinds of molecules. If a is the amount of substance A and b is the amount of substance B at time $t = 0$, and if x is the amount of product at time t, then the rate of formation of x may be given by the differential equation

$$\frac{dx}{dt} = k(a - x)(b - x),$$

or

$$\frac{1}{(a - x)(b - x)}\frac{dx}{dt} = k,$$

where k is a constant for the reaction. Integrate both sides of this equation to obtain a relation between x and t **(a)** if $a = b$, and **(b)** if $a \neq b$. Assume in each case that $x = 0$ when $t = 0$.

51. An integral connecting π to the approximation 22/7

a. Evaluate $\displaystyle\int_0^1 \frac{x^4(x-1)^4}{x^2+1}\,dx$.

b. How good is the approximation $\pi \approx 22/7$? Find out by expressing $\left(\dfrac{22}{7} - \pi\right)$ as a percentage of π.

c. Graph the function $y = \dfrac{x^4(x-1)^4}{x^2+1}$ for $0 \le x \le 1$. Experiment with the range on the y-axis set between 0 and 1, then between 0 and 0.5, and then decreasing the range until the graph can be seen. What do you conclude about the area under the curve?

52. Find the second-degree polynomial $P(x)$ such that $P(0) = 1$, $P'(0) = 0$, and

$$\int \frac{P(x)}{x^3(x-1)^2}\,dx$$

is a rational function.

8.4 Trigonometric Integrals

Trigonometric integrals involve algebraic combinations of the six basic trigonometric functions. In principle, we can always express such integrals in terms of sines and cosines, but it is often simpler to work with other functions, as in the integral

$$\int \sec^2 x\,dx = \tan x + C.$$

The general idea is to use identities to transform the integrals we have to find into integrals that are easier to work with.

Products of Powers of Sines and Cosines

We begin with integrals of the form:

$$\int \sin^m x \cos^n x\,dx,$$

where m and n are nonnegative integers (positive or zero). We can divide the work into three cases.

Case 1 If m is odd, we write m as $2k+1$ and use the identity $\sin^2 x = 1 - \cos^2 x$ to obtain

$$\sin^m x = \sin^{2k+1} x = (\sin^2 x)^k \sin x = (1 - \cos^2 x)^k \sin x. \tag{1}$$

Then we combine the single $\sin x$ with dx in the integral and set $\sin x\,dx$ equal to $-d(\cos x)$.

Case 2 If m is even and n is odd in $\int \sin^m x \cos^n x\,dx$, we write n as $2k+1$ and use the identity $\cos^2 x = 1 - \sin^2 x$ to obtain

$$\cos^n x = \cos^{2k+1} x = (\cos^2 x)^k \cos x = (1 - \sin^2 x)^k \cos x.$$

We then combine the single $\cos x$ with dx and set $\cos x\,dx$ equal to $d(\sin x)$.

Case 3 If both m and n are even in $\int \sin^m x \cos^n x\,dx$, we substitute

$$\sin^2 x = \frac{1 - \cos 2x}{2}, \qquad \cos^2 x = \frac{1 + \cos 2x}{2} \tag{2}$$

to reduce the integrand to one in lower powers of $\cos 2x$.

Here are some examples illustrating each case.

EXAMPLE 1 *m* is Odd

Evaluate

$$\int \sin^3 x \cos^2 x\,dx.$$

Solution

$$\begin{aligned}
\int \sin^3 x \cos^2 x \, dx &= \int \sin^2 x \cos^2 x \sin x \, dx \\
&= \int (1 - \cos^2 x) \cos^2 x \, (-d(\cos x)) \\
&= \int (1 - u^2)(u^2)(-du) && u = \cos x \\
&= \int (u^4 - u^2) \, du \\
&= \frac{u^5}{5} - \frac{u^3}{3} + C \\
&= \frac{\cos^5 x}{5} - \frac{\cos^3 x}{3} + C.
\end{aligned}$$

■

EXAMPLE 2 *m* is Even and *n* is Odd

Evaluate

$$\int \cos^5 x \, dx.$$

Solution

$$\begin{aligned}
\int \cos^5 x \, dx &= \int \cos^4 x \cos x \, dx = \int (1 - \sin^2 x)^2 \, d(\sin x) && m = 0 \\
&= \int (1 - u^2)^2 \, du && u = \sin x \\
&= \int (1 - 2u^2 + u^4) \, du \\
&= u - \frac{2}{3}u^3 + \frac{1}{5}u^5 + C = \sin x - \frac{2}{3}\sin^3 x + \frac{1}{5}\sin^5 x + C.
\end{aligned}$$

■

EXAMPLE 3 *m* and *n* are Both Even

Evaluate

$$\int \sin^2 x \cos^4 x \, dx.$$

Solution

$$\begin{aligned}
\int \sin^2 x \cos^4 x \, dx &= \int \left(\frac{1 - \cos 2x}{2}\right)\left(\frac{1 + \cos 2x}{2}\right)^2 dx \\
&= \frac{1}{8}\int (1 - \cos 2x)(1 + 2\cos 2x + \cos^2 2x) \, dx \\
&= \frac{1}{8}\int (1 + \cos 2x - \cos^2 2x - \cos^3 2x) \, dx \\
&= \frac{1}{8}\left[x + \frac{1}{2}\sin 2x - \int (\cos^2 2x + \cos^3 2x) \, dx\right].
\end{aligned}$$

For the term involving $\cos^2 2x$ we use

$$\int \cos^2 2x\,dx = \frac{1}{2}\int (1 + \cos 4x)\,dx$$

$$= \frac{1}{2}\left(x + \frac{1}{4}\sin 4x\right).$$

Omitting the constant of integration until the final result

For the $\cos^3 2x$ term we have

$$\int \cos^3 2x\,dx = \int (1 - \sin^2 2x)\cos 2x\,dx$$

$u = \sin 2x$, $du = 2\cos 2x\,dx$

$$= \frac{1}{2}\int (1 - u^2)\,du = \frac{1}{2}\left(\sin 2x - \frac{1}{3}\sin^3 2x\right).$$

Again omitting C

Combining everything and simplifying we get

$$\int \sin^2 x \cos^4 x\,dx = \frac{1}{16}\left(x - \frac{1}{4}\sin 4x + \frac{1}{3}\sin^3 2x\right) + C. \qquad \blacksquare$$

Eliminating Square Roots

In the next example, we use the identity $\cos^2\theta = (1 + \cos 2\theta)/2$ to eliminate a square root.

EXAMPLE 4 Evaluate

$$\int_0^{\pi/4} \sqrt{1 + \cos 4x}\,dx.$$

Solution To eliminate the square root we use the identity

$$\cos^2\theta = \frac{1 + \cos 2\theta}{2}, \qquad \text{or} \qquad 1 + \cos 2\theta = 2\cos^2\theta.$$

With $\theta = 2x$, this becomes

$$1 + \cos 4x = 2\cos^2 2x.$$

Therefore,

$$\int_0^{\pi/4} \sqrt{1 + \cos 4x}\,dx = \int_0^{\pi/4} \sqrt{2\cos^2 2x}\,dx = \int_0^{\pi/4} \sqrt{2}\sqrt{\cos^2 2x}\,dx$$

$$= \sqrt{2}\int_0^{\pi/4} |\cos 2x|\,dx = \sqrt{2}\int_0^{\pi/4} \cos 2x\,dx$$

$\cos 2x \geq 0$ on $[0, \pi/4]$

$$= \sqrt{2}\left[\frac{\sin 2x}{2}\right]_0^{\pi/4} = \frac{\sqrt{2}}{2}[1 - 0] = \frac{\sqrt{2}}{2}. \qquad \blacksquare$$

Integrals of Powers of tan *x* and sec *x*

We know how to integrate the tangent and secant and their squares. To integrate higher powers we use the identities $\tan^2 x = \sec^2 x - 1$ and $\sec^2 x = \tan^2 x + 1$, and integrate by parts when necessary to reduce the higher powers to lower powers.

EXAMPLE 5 Evaluate

$$\int \tan^4 x\, dx.$$

Solution

$$\begin{aligned}
\int \tan^4 x\, dx &= \int \tan^2 x \cdot \tan^2 x\, dx = \int \tan^2 x \cdot (\sec^2 x - 1)\, dx \\
&= \int \tan^2 x \sec^2 x\, dx - \int \tan^2 x\, dx \\
&= \int \tan^2 x \sec^2 x\, dx - \int (\sec^2 x - 1)\, dx \\
&= \int \tan^2 x \sec^2 x\, dx - \int \sec^2 x\, dx + \int dx.
\end{aligned}$$

In the first integral, we let

$$u = \tan x, \qquad du = \sec^2 x\, dx$$

and have

$$\int u^2\, du = \frac{1}{3}u^3 + C_1.$$

The remaining integrals are standard forms, so

$$\int \tan^4 x\, dx = \frac{1}{3}\tan^3 x - \tan x + x + C.$$

■

EXAMPLE 6 Evaluate

$$\int \sec^3 x\, dx.$$

Solution We integrate by parts, using

$$u = \sec x, \qquad dv = \sec^2 x\, dx, \qquad v = \tan x, \qquad du = \sec x \tan x\, dx.$$

Then

$$\begin{aligned}
\int \sec^3 x\, dx &= \sec x \tan x - \int (\tan x)(\sec x \tan x\, dx) \\
&= \sec x \tan x - \int (\sec^2 x - 1) \sec x\, dx \qquad \tan^2 x = \sec^2 x - 1 \\
&= \sec x \tan x + \int \sec x\, dx - \int \sec^3 x\, dx.
\end{aligned}$$

Combining the two secant-cubed integrals gives

$$2\int \sec^3 x\, dx = \sec x \tan x + \int \sec x\, dx$$

and

$$\int \sec^3 x\,dx = \frac{1}{2}\sec x \tan x + \frac{1}{2}\ln|\sec x + \tan x| + C.$$ ■

Products of Sines and Cosines

The integrals

$$\int \sin mx \sin nx\,dx, \qquad \int \sin mx \cos nx\,dx, \qquad \text{and} \qquad \int \cos mx \cos nx\,dx$$

arise in many places where trigonometric functions are applied to problems in mathematics and science. We can evaluate these integrals through integration by parts, but two such integrations are required in each case. It is simpler to use the identities

$$\sin mx \sin nx = \frac{1}{2}[\cos(m-n)x - \cos(m+n)x], \tag{3}$$

$$\sin mx \cos nx = \frac{1}{2}[\sin(m-n)x + \sin(m+n)x], \tag{4}$$

$$\cos mx \cos nx = \frac{1}{2}[\cos(m-n)x + \cos(m+n)x]. \tag{5}$$

These come from the angle sum formulas for the sine and cosine functions (Section 1.6). They give functions whose antiderivatives are easily found.

EXAMPLE 7 Evaluate

$$\int \sin 3x \cos 5x\,dx.$$

Solution From Equation (4) with $m = 3$ and $n = 5$ we get

$$\begin{aligned}\int \sin 3x \cos 5x\,dx &= \frac{1}{2}\int [\sin(-2x) + \sin 8x]\,dx \\ &= \frac{1}{2}\int (\sin 8x - \sin 2x)\,dx \\ &= -\frac{\cos 8x}{16} + \frac{\cos 2x}{4} + C.\end{aligned}$$ ■

EXERCISES 8.4

Products of Powers of Sines and Cosines

Evaluate the integrals in Exercises 1–14.

1. $\int_0^{\pi/2} \sin^5 x\,dx$

2. $\int_0^{\pi} \sin^5 \frac{x}{2}\,dx$

3. $\int_{-\pi/2}^{\pi/2} \cos^3 x\,dx$

4. $\int_0^{\pi/6} 3\cos^5 3x\,dx$

5. $\int_0^{\pi/2} \sin^7 y\,dy$

6. $\int_0^{\pi/2} 7\cos^7 t\,dt$

7. $\int_0^{\pi} 8 \sin^4 x \, dx$

8. $\int_0^{1} 8 \cos^4 2\pi x \, dx$

9. $\int_{-\pi/4}^{\pi/4} 16 \sin^2 x \cos^2 x \, dx$

10. $\int_0^{\pi} 8 \sin^4 y \cos^2 y \, dy$

11. $\int_0^{\pi/2} 35 \sin^4 x \cos^3 x \, dx$

12. $\int_0^{\pi} \sin 2x \cos^2 2x \, dx$

13. $\int_0^{\pi/4} 8 \cos^3 2\theta \sin 2\theta \, d\theta$

14. $\int_0^{\pi/2} \sin^2 2\theta \cos^3 2\theta \, d\theta$

Integrals with Square Roots

Evaluate the integrals in Exercises 15–22.

15. $\int_0^{2\pi} \sqrt{\frac{1 - \cos x}{2}} \, dx$

16. $\int_0^{\pi} \sqrt{1 - \cos 2x} \, dx$

17. $\int_0^{\pi} \sqrt{1 - \sin^2 t} \, dt$

18. $\int_0^{\pi} \sqrt{1 - \cos^2 \theta} \, d\theta$

19. $\int_{-\pi/4}^{\pi/4} \sqrt{1 + \tan^2 x} \, dx$

20. $\int_{-\pi/4}^{\pi/4} \sqrt{\sec^2 x - 1} \, dx$

21. $\int_0^{\pi/2} \theta \sqrt{1 - \cos 2\theta} \, d\theta$

22. $\int_{-\pi}^{\pi} (1 - \cos^2 t)^{3/2} \, dt$

Powers of Tan *x* and Sec *x*

Evaluate the integrals in Exercises 23–32.

23. $\int_{-\pi/3}^{0} 2 \sec^3 x \, dx$

24. $\int e^x \sec^3 e^x \, dx$

25. $\int_0^{\pi/4} \sec^4 \theta \, d\theta$

26. $\int_0^{\pi/12} 3 \sec^4 3x \, dx$

27. $\int_{\pi/4}^{\pi/2} \csc^4 \theta \, d\theta$

28. $\int_{\pi/2}^{\pi} 3 \csc^4 \frac{\theta}{2} \, d\theta$

29. $\int_0^{\pi/4} 4 \tan^3 x \, dx$

30. $\int_{-\pi/4}^{\pi/4} 6 \tan^4 x \, dx$

31. $\int_{\pi/6}^{\pi/3} \cot^3 x \, dx$

32. $\int_{\pi/4}^{\pi/2} 8 \cot^4 t \, dt$

Products of Sines and Cosines

Evaluate the integrals in Exercises 33–38.

33. $\int_{-\pi}^{0} \sin 3x \cos 2x \, dx$

34. $\int_0^{\pi/2} \sin 2x \cos 3x \, dx$

35. $\int_{-\pi}^{\pi} \sin 3x \sin 3x \, dx$

36. $\int_0^{\pi/2} \sin x \cos x \, dx$

37. $\int_0^{\pi} \cos 3x \cos 4x \, dx$

38. $\int_{-\pi/2}^{\pi/2} \cos x \cos 7x \, dx$

Theory and Examples

39. **Surface area** Find the area of the surface generated by revolving the arc

$$x = t^{2/3}, \quad y = t^2/2, \quad 0 \le t \le 2,$$

about the x-axis.

40. **Arc length** Find the length of the curve

$$y = \ln(\cos x), \quad 0 \le x \le \pi/3.$$

41. **Arc length** Find the length of the curve

$$y = \ln(\sec x), \quad 0 \le x \le \pi/4.$$

42. **Center of gravity** Find the center of gravity of the region bounded by the x-axis, the curve $y = \sec x$, and the lines $x = -\pi/4, x = \pi/4$.

43. **Volume** Find the volume generated by revolving one arch of the curve $y = \sin x$ about the x-axis.

44. **Area** Find the area between the x-axis and the curve $y = \sqrt{1 + \cos 4x}, 0 \le x \le \pi$.

45. **Orthogonal functions** Two functions f and g are said to be **orthogonal** on an interval $a \le x \le b$ if $\int_a^b f(x)g(x) \, dx = 0$.

   a. Prove that $\sin mx$ and $\sin nx$ are orthogonal on any interval of length 2π provided m and n are integers such that $m^2 \ne n^2$.

   b. Prove the same for $\cos mx$ and $\cos nx$.

   c. Prove the same for $\sin mx$ and $\cos nx$ even if $m = n$.

46. **Fourier series** A finite Fourier series is given by the sum

$$f(x) = \sum_{n=1}^{N} a_n \sin nx = a_1 \sin x + a_2 \sin 2x + \cdots + a_N \sin Nx$$

Show that the mth coefficient a_m is given by the formula

$$a_m = \frac{1}{\pi} \int_{-\pi}^{\pi} f(x) \sin mx \, dx.$$

8.5 Trigonometric Substitutions

Trigonometric substitutions can be effective in transforming integrals involving $\sqrt{a^2 - x^2}$, $\sqrt{a^2 + x^2}$, and $\sqrt{x^2 - a^2}$ into integrals we can evaluate directly.

Three Basic Substitutions

The most common substitutions are $x = a \tan\theta$, $x = a\sin\theta$, and $x = a\sec\theta$. They come from the reference right triangles in Figure 8.2.

With $x = a\tan\theta$,

$$a^2 + x^2 = a^2 + a^2\tan^2\theta = a^2(1 + \tan^2\theta) = a^2\sec^2\theta.$$

With $x = a\sin\theta$,

$$a^2 - x^2 = a^2 - a^2\sin^2\theta = a^2(1 - \sin^2\theta) = a^2\cos^2\theta.$$

With $x = a\sec\theta$,

$$x^2 - a^2 = a^2\sec^2\theta - a^2 = a^2(\sec^2\theta - 1) = a^2\tan^2\theta.$$

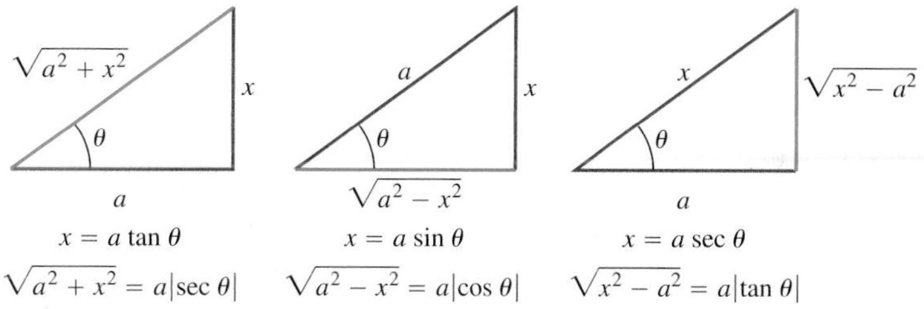

FIGURE 8.2 Reference triangles for the three basic substitutions identifying the sides labeled x and a for each substitution.

We want any substitution we use in an integration to be reversible so that we can change back to the original variable afterward. For example, if $x = a\tan\theta$, we want to be able to set $\theta = \tan^{-1}(x/a)$ after the integration takes place. If $x = a\sin\theta$, we want to be able to set $\theta = \sin^{-1}(x/a)$ when we're done, and similarly for $x = a\sec\theta$.

As we know from Section 7.7, the functions in these substitutions have inverses only for selected values of θ (Figure 8.3). For reversibility,

$$x = a\tan\theta \quad \text{requires} \quad \theta = \tan^{-1}\left(\frac{x}{a}\right) \quad \text{with} \quad -\frac{\pi}{2} < \theta < \frac{\pi}{2},$$

$$x = a\sin\theta \quad \text{requires} \quad \theta = \sin^{-1}\left(\frac{x}{a}\right) \quad \text{with} \quad -\frac{\pi}{2} \le \theta \le \frac{\pi}{2},$$

$$x = a\sec\theta \quad \text{requires} \quad \theta = \sec^{-1}\left(\frac{x}{a}\right) \quad \text{with} \quad \begin{cases} 0 \le \theta < \dfrac{\pi}{2} & \text{if } \dfrac{x}{a} \ge 1, \\[2ex] \dfrac{\pi}{2} < \theta \le \pi & \text{if } \dfrac{x}{a} \le -1. \end{cases}$$

To simplify calculations with the substitution $x = a\sec\theta$, we will restrict its use to integrals in which $x/a \ge 1$. This will place θ in $[0, \pi/2)$ and make $\tan\theta \ge 0$. We will then have $\sqrt{x^2 - a^2} = \sqrt{a^2\tan^2\theta} = |a\tan\theta| = a\tan\theta$, free of absolute values, provided $a > 0$.

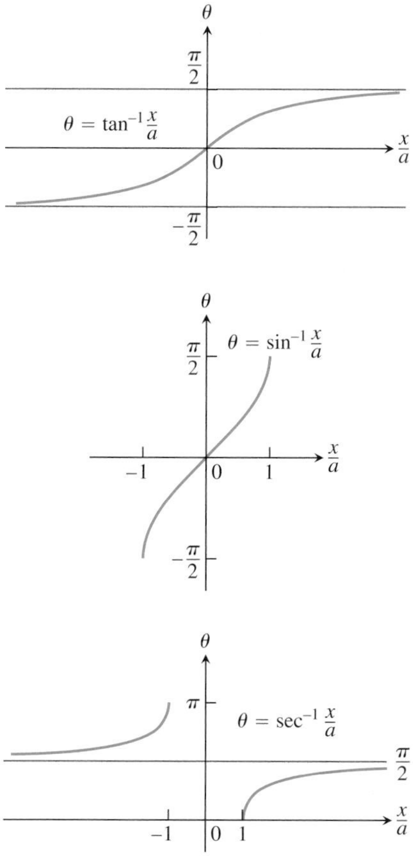

FIGURE 8.3 The arctangent, arcsine, and arcsecant of x/a, graphed as functions of x/a.

EXAMPLE 1 Using the Substitution $x = a\tan\theta$

Evaluate

$$\int \frac{dx}{\sqrt{4 + x^2}}.$$

Solution We set

$$x = 2\tan\theta, \qquad dx = 2\sec^2\theta\,d\theta, \qquad -\frac{\pi}{2} < \theta < \frac{\pi}{2},$$
$$4 + x^2 = 4 + 4\tan^2\theta = 4(1 + \tan^2\theta) = 4\sec^2\theta.$$

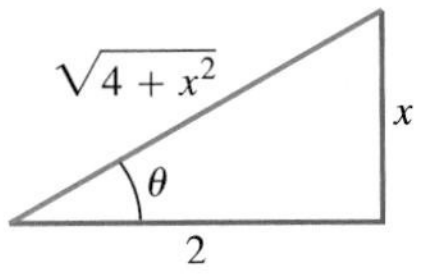

FIGURE 8.4 Reference triangle for $x = 2\tan\theta$ (Example 1):

$$\tan\theta = \frac{x}{2}$$

and

$$\sec\theta = \frac{\sqrt{4+x^2}}{2}.$$

Then

$$\begin{aligned}
\int \frac{dx}{\sqrt{4+x^2}} &= \int \frac{2\sec^2\theta\,d\theta}{\sqrt{4\sec^2\theta}} = \int \frac{\sec^2\theta\,d\theta}{|\sec\theta|} && \sqrt{\sec^2\theta} = |\sec\theta| \\
&= \int \sec\theta\,d\theta && \sec\theta > 0 \text{ for } -\frac{\pi}{2} < \theta < \frac{\pi}{2} \\
&= \ln|\sec\theta + \tan\theta| + C \\
&= \ln\left|\frac{\sqrt{4+x^2}}{2} + \frac{x}{2}\right| + C && \text{From Fig. 8.4} \\
&= \ln\left|\sqrt{4+x^2} + x\right| + C'. && \text{Taking } C' = C - \ln 2
\end{aligned}$$

Notice how we expressed $\ln|\sec\theta + \tan\theta|$ in terms of x: We drew a reference triangle for the original substitution $x = 2\tan\theta$ (Figure 8.4) and read the ratios from the triangle. ■

EXAMPLE 2 Using the Substitution $x = a\sin\theta$

Evaluate

$$\int \frac{x^2\,dx}{\sqrt{9-x^2}}.$$

Solution We set

$$x = 3\sin\theta, \qquad dx = 3\cos\theta\,d\theta, \qquad -\frac{\pi}{2} < \theta < \frac{\pi}{2}$$
$$9 - x^2 = 9 - 9\sin^2\theta = 9(1 - \sin^2\theta) = 9\cos^2\theta.$$

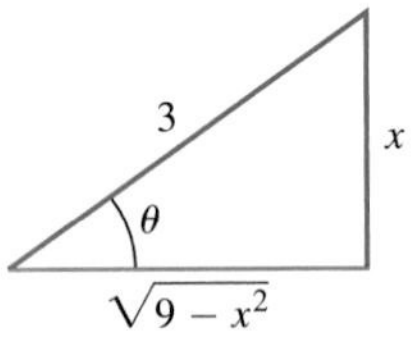

FIGURE 8.5 Reference triangle for $x = 3\sin\theta$ (Example 2):

$$\sin\theta = \frac{x}{3}$$

and

$$\cos\theta = \frac{\sqrt{9-x^2}}{3}.$$

Then

$$\begin{aligned}
\int \frac{x^2\,dx}{\sqrt{9-x^2}} &= \int \frac{9\sin^2\theta \cdot 3\cos\theta\,d\theta}{|3\cos\theta|} \\
&= 9\int \sin^2\theta\,d\theta && \cos\theta > 0 \text{ for } -\frac{\pi}{2} < \theta < \frac{\pi}{2} \\
&= 9\int \frac{1 - \cos 2\theta}{2}\,d\theta \\
&= \frac{9}{2}\left(\theta - \frac{\sin 2\theta}{2}\right) + C \\
&= \frac{9}{2}(\theta - \sin\theta\cos\theta) + C && \sin 2\theta = 2\sin\theta\cos\theta \\
&= \frac{9}{2}\left(\sin^{-1}\frac{x}{3} - \frac{x}{3}\cdot\frac{\sqrt{9-x^2}}{3}\right) + C && \text{Fig. 8.5} \\
&= \frac{9}{2}\sin^{-1}\frac{x}{3} - \frac{x}{2}\sqrt{9-x^2} + C.
\end{aligned}$$

■

EXAMPLE 3 Using the Substitution $x = a\sec\theta$

Evaluate

$$\int \frac{dx}{\sqrt{25x^2 - 4}}, \qquad x > \frac{2}{5}.$$

Solution We first rewrite the radical as

$$\sqrt{25x^2 - 4} = \sqrt{25\left(x^2 - \frac{4}{25}\right)} = 5\sqrt{x^2 - \left(\frac{2}{5}\right)^2}$$

to put the radicand in the form $x^2 - a^2$. We then substitute

$$x = \frac{2}{5}\sec\theta, \qquad dx = \frac{2}{5}\sec\theta\tan\theta\, d\theta, \qquad 0 < \theta < \frac{\pi}{2}$$

$$x^2 - \left(\frac{2}{5}\right)^2 = \frac{4}{25}\sec^2\theta - \frac{4}{25} = \frac{4}{25}(\sec^2\theta - 1) = \frac{4}{25}\tan^2\theta$$

$$\sqrt{x^2 - \left(\frac{2}{5}\right)^2} = \frac{2}{5}|\tan\theta| = \frac{2}{5}\tan\theta.$$

$\tan\theta > 0$ for $0 < \theta < \pi/2$

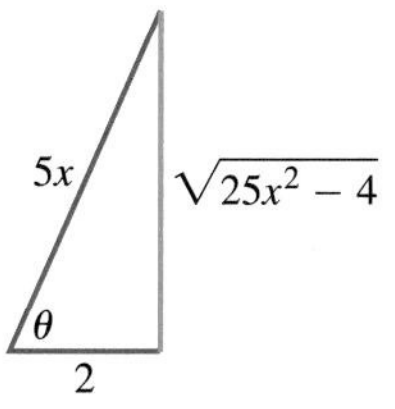

FIGURE 8.6 If $x = (2/5)\sec\theta$, $0 < \theta < \pi/2$, then $\theta = \sec^{-1}(5x/2)$, and we can read the values of the other trigonometric functions of θ from this right triangle (Example 3).

With these substitutions, we have

$$\int \frac{dx}{\sqrt{25x^2 - 4}} = \int \frac{dx}{5\sqrt{x^2 - (4/25)}} = \int \frac{(2/5)\sec\theta\tan\theta\, d\theta}{5\cdot(2/5)\tan\theta}$$

$$= \frac{1}{5}\int \sec\theta\, d\theta = \frac{1}{5}\ln|\sec\theta + \tan\theta| + C$$

$$= \frac{1}{5}\ln\left|\frac{5x}{2} + \frac{\sqrt{25x^2 - 4}}{2}\right| + C.$$

Fig. 8.6 ■

A trigonometric substitution can sometimes help us to evaluate an integral containing an integer power of a quadratic binomial, as in the next example.

EXAMPLE 4 Finding the Volume of a Solid of Revolution

Find the volume of the solid generated by revolving about the x-axis the region bounded by the curve $y = 4/(x^2 + 4)$, the x-axis, and the lines $x = 0$ and $x = 2$.

Solution We sketch the region (Figure 8.7) and use the disk method:

$$V = \int_0^2 \pi[R(x)]^2\, dx = 16\pi\int_0^2 \frac{dx}{(x^2 + 4)^2}.$$

$R(x) = \dfrac{4}{x^2 + 4}$

To evaluate the integral, we set

$$x = 2\tan\theta, \qquad dx = 2\sec^2\theta\, d\theta, \qquad \theta = \tan^{-1}\frac{x}{2},$$

$$x^2 + 4 = 4\tan^2\theta + 4 = 4(\tan^2\theta + 1) = 4\sec^2\theta$$

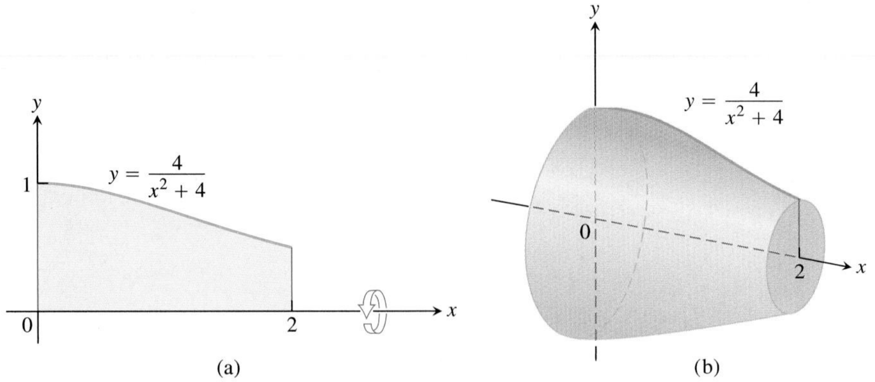

FIGURE 8.7 The region (a) and solid (b) in Example 4.

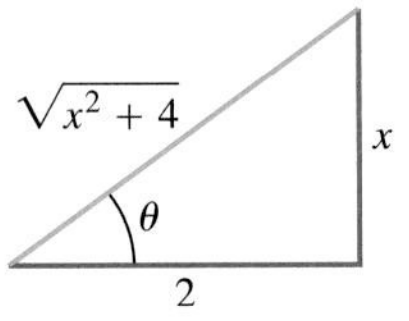

FIGURE 8.8 Reference triangle for $x = 2 \tan \theta$ (Example 4).

(Figure 8.8). With these substitutions,

$$V = 16\pi \int_0^2 \frac{dx}{(x^2 + 4)^2}$$

$$= 16\pi \int_0^{\pi/4} \frac{2 \sec^2 \theta \, d\theta}{(4 \sec^2 \theta)^2} \qquad \theta = 0 \text{ when } x = 0; \; \theta = \pi/4 \text{ when } x = 2$$

$$= 16\pi \int_0^{\pi/4} \frac{2 \sec^2 \theta \, d\theta}{16 \sec^4 \theta} = \pi \int_0^{\pi/4} 2 \cos^2 \theta \, d\theta$$

$$= \pi \int_0^{\pi/4} (1 + \cos 2\theta) \, d\theta = \pi \left[\theta + \frac{\sin 2\theta}{2} \right]_0^{\pi/4} \qquad 2\cos^2\theta = 1 + \cos 2\theta$$

$$= \pi \left[\frac{\pi}{4} + \frac{1}{2} \right] \approx 4.04.$$

■

EXAMPLE 5 Finding the Area of an Ellipse

Find the area enclosed by the ellipse

$$\frac{x^2}{a^2} + \frac{y^2}{b^2} = 1$$

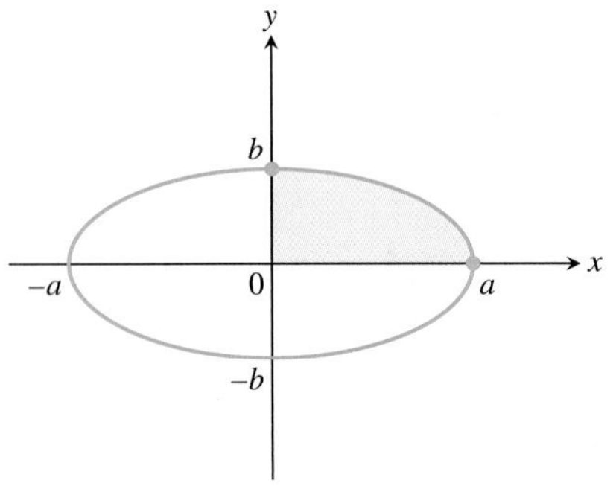

FIGURE 8.9 The ellipse $\frac{x^2}{a^2} + \frac{y^2}{b^2} = 1$ in Example 5.

Solution Because the ellipse is symmetric with respect to both axes, the total area A is four times the area in the first quadrant (Figure 8.9). Solving the equation of the ellipse for $y \geq 0$, we get

$$\frac{y^2}{b^2} = 1 - \frac{x^2}{a^2} = \frac{a^2 - x^2}{a^2},$$

or

$$y = \frac{b}{a}\sqrt{a^2 - x^2} \qquad 0 \leq x \leq a$$

The area of the ellipse is

$$\begin{aligned} A &= 4\int_0^a \frac{b}{a}\sqrt{a^2 - x^2}\,dx \\ &= 4\frac{b}{a}\int_0^{\pi/2} a\cos\theta \cdot a\cos\theta\,d\theta \qquad \begin{array}{l} x = a\sin\theta,\ dx = a\cos\theta\,d\theta, \\ \theta = 0 \text{ when } x = 0; \\ \theta = \pi/2 \text{ when } x = a \end{array} \\ &= 4ab\int_0^{\pi/2} \cos^2\theta\,d\theta \\ &= 4ab\int_0^{\pi/2} \frac{1 + \cos 2\theta}{2}\,d\theta \\ &= 2ab\left[\theta + \frac{\sin 2\theta}{2}\right]_0^{\pi/2} \\ &= 2ab\left[\frac{\pi}{2} + 0 - 0\right] = \pi ab. \end{aligned}$$

If $a = b = r$ we get that the area of a circle with radius r is πr^2. ■

EXERCISES 8.5

Basic Trigonometric Substitutions

Evaluate the integrals in Exercises 1–28.

1. $\displaystyle\int \frac{dy}{\sqrt{9 + y^2}}$

2. $\displaystyle\int \frac{3\,dy}{\sqrt{1 + 9y^2}}$

3. $\displaystyle\int_{-2}^{2} \frac{dx}{4 + x^2}$

4. $\displaystyle\int_0^2 \frac{dx}{8 + 2x^2}$

5. $\displaystyle\int_0^{3/2} \frac{dx}{\sqrt{9 - x^2}}$

6. $\displaystyle\int_0^{1/2\sqrt{2}} \frac{2\,dx}{\sqrt{1 - 4x^2}}$

7. $\displaystyle\int \sqrt{25 - t^2}\,dt$

8. $\displaystyle\int \sqrt{1 - 9t^2}\,dt$

9. $\displaystyle\int \frac{dx}{\sqrt{4x^2 - 49}},\quad x > \frac{7}{2}$

10. $\displaystyle\int \frac{5\,dx}{\sqrt{25x^2 - 9}},\quad x > \frac{3}{5}$

11. $\displaystyle\int \frac{\sqrt{y^2 - 49}}{y}\,dy,\quad y > 7$

12. $\displaystyle\int \frac{\sqrt{y^2 - 25}}{y^3}\,dy,\quad y > 5$

13. $\displaystyle\int \frac{dx}{x^2\sqrt{x^2 - 1}},\quad x > 1$

14. $\displaystyle\int \frac{2\,dx}{x^3\sqrt{x^2 - 1}},\quad x > 1$

15. $\displaystyle\int \frac{x^3\,dx}{\sqrt{x^2 + 4}}$

16. $\displaystyle\int \frac{dx}{x^2\sqrt{x^2 + 1}}$

17. $\displaystyle\int \frac{8\,dw}{w^2\sqrt{4 - w^2}}$

18. $\displaystyle\int \frac{\sqrt{9 - w^2}}{w^2}\,dw$

19. $\displaystyle\int_0^{\sqrt{3}/2} \frac{4x^2\,dx}{(1 - x^2)^{3/2}}$

20. $\displaystyle\int_0^1 \frac{dx}{(4 - x^2)^{3/2}}$

21. $\displaystyle\int \frac{dx}{(x^2 - 1)^{3/2}},\quad x > 1$

22. $\displaystyle\int \frac{x^2\,dx}{(x^2 - 1)^{5/2}},\quad x > 1$

23. $\displaystyle\int \frac{(1 - x^2)^{3/2}}{x^6}\,dx$

24. $\displaystyle\int \frac{(1 - x^2)^{1/2}}{x^4}\,dx$

25. $\displaystyle\int \frac{8\,dx}{(4x^2 + 1)^2}$

26. $\displaystyle\int \frac{6\,dt}{(9t^2 + 1)^2}$

27. $\displaystyle\int \frac{v^2\,dv}{(1 - v^2)^{5/2}}$

28. $\displaystyle\int \frac{(1 - r^2)^{5/2}}{r^8}\,dr$

In Exercises 29–36, use an appropriate substitution and then a trigonometric substitution to evaluate the integrals.

29. $\displaystyle\int_0^{\ln 4} \frac{e^t\,dt}{\sqrt{e^{2t} + 9}}$

30. $\displaystyle\int_{\ln(3/4)}^{\ln(4/3)} \frac{e^t\,dt}{(1 + e^{2t})^{3/2}}$

31. $\displaystyle\int_{1/12}^{1/4} \frac{2\,dt}{\sqrt{t} + 4t\sqrt{t}}$

32. $\displaystyle\int_1^e \frac{dy}{y\sqrt{1 + (\ln y)^2}}$

33. $\displaystyle\int \frac{dx}{x\sqrt{x^2 - 1}}$

34. $\displaystyle\int \frac{dx}{1 + x^2}$

35. $\displaystyle\int \frac{x\,dx}{\sqrt{x^2 - 1}}$

36. $\displaystyle\int \frac{dx}{\sqrt{1 - x^2}}$

Initial Value Problems

Solve the initial value problems in Exercises 37–40 for y as a function of x.

37. $x\dfrac{dy}{dx} = \sqrt{x^2 - 4},\quad x \ge 2,\quad y(2) = 0$

38. $\sqrt{x^2 - 9}\dfrac{dy}{dx} = 1,\quad x > 3,\quad y(5) = \ln 3$

39. $(x^2 + 4)\dfrac{dy}{dx} = 3,\quad y(2) = 0$

40. $(x^2 + 1)^2\dfrac{dy}{dx} = \sqrt{x^2 + 1},\quad y(0) = 1$

Applications

41. Find the area of the region in the first quadrant that is enclosed by the coordinate axes and the curve $y = \sqrt{9 - x^2}/3$.

42. Find the volume of the solid generated by revolving about the x-axis the region in the first quadrant enclosed by the coordinate axes, the curve $y = 2/(1 + x^2)$, and the line $x = 1$.

The Substitution $z = \tan(x/2)$

The substitution

$$z = \tan\frac{x}{2} \qquad (1)$$

reduces the problem of integrating a rational expression in $\sin x$ and $\cos x$ to a problem of integrating a rational function of z. This in turn can be integrated by partial fractions.

From the accompanying figure

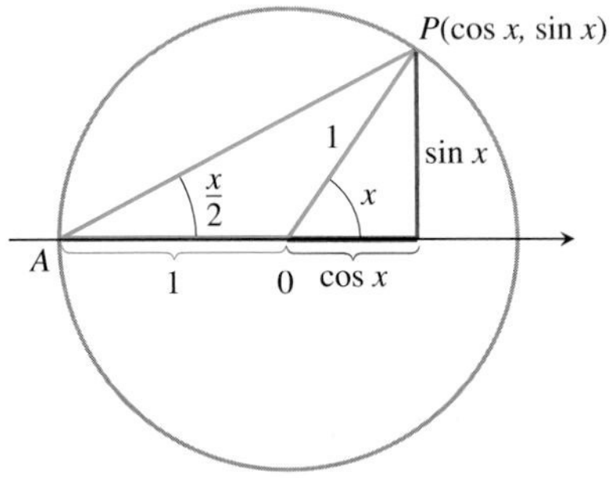

we can read the relation

$$\tan\frac{x}{2} = \frac{\sin x}{1 + \cos x}.$$

To see the effect of the substitution, we calculate

$$\cos x = 2\cos^2\left(\frac{x}{2}\right) - 1 = \frac{2}{\sec^2(x/2)} - 1$$

$$= \frac{2}{1 + \tan^2(x/2)} - 1 = \frac{2}{1 + z^2} - 1$$

$$\cos x = \frac{1 - z^2}{1 + z^2}, \qquad (2)$$

and

$$\sin x = 2\sin\frac{x}{2}\cos\frac{x}{2} = 2\frac{\sin(x/2)}{\cos(x/2)} \cdot \cos^2\left(\frac{x}{2}\right)$$

$$= 2\tan\frac{x}{2} \cdot \frac{1}{\sec^2(x/2)} = \frac{2\tan(x/2)}{1 + \tan^2(x/2)}$$

$$\sin x = \frac{2z}{1 + z^2}. \qquad (3)$$

Finally, $x = 2\tan^{-1} z$, so

$$dx = \frac{2\,dz}{1 + z^2}. \qquad (4)$$

Examples

a. $$\int \frac{1}{1 + \cos x}\,dx = \int \frac{1 + z^2}{2}\,\frac{2\,dz}{1 + z^2}$$

$$= \int dz = z + C$$

$$= \tan\left(\frac{x}{2}\right) + C$$

b. $$\int \frac{1}{2 + \sin x}\,dx = \int \frac{1 + z^2}{2 + 2z + 2z^2}\,\frac{2\,dz}{1 + z^2}$$

$$= \int \frac{dz}{z^2 + z + 1} = \int \frac{dz}{(z + (1/2))^2 + 3/4}$$

$$= \int \frac{du}{u^2 + a^2}$$

$$= \frac{1}{a}\tan^{-1}\left(\frac{u}{a}\right) + C$$

$$= \frac{2}{\sqrt{3}}\tan^{-1}\frac{2z + 1}{\sqrt{3}} + C$$

$$= \frac{2}{\sqrt{3}}\tan^{-1}\frac{1 + 2\tan(x/2)}{\sqrt{3}} + C$$

Use the substitutions in Equations (1)–(4) to evaluate the integrals in Exercises 43–50. Integrals like these arise in calculating the average angular velocity of the output shaft of a universal joint when the input and output shafts are not aligned.

43. $\displaystyle\int \frac{dx}{1 - \sin x}$

44. $\displaystyle\int \frac{dx}{1 + \sin x + \cos x}$

45. $\displaystyle\int_0^{\pi/2} \frac{dx}{1 + \sin x}$

46. $\displaystyle\int_{\pi/3}^{\pi/2} \frac{dx}{1 - \cos x}$

47. $\displaystyle\int_0^{\pi/2} \frac{d\theta}{2 + \cos\theta}$

48. $\displaystyle\int_{\pi/2}^{2\pi/3} \frac{\cos\theta\,d\theta}{\sin\theta\cos\theta + \sin\theta}$

49. $\displaystyle\int \frac{dt}{\sin t - \cos t}$

50. $\displaystyle\int \frac{\cos t\,dt}{1 - \cos t}$

Use the substitution $z = \tan(\theta/2)$ to evaluate the integrals in Exercises 51 and 52.

51. $\displaystyle\int \sec\theta\,d\theta$

52. $\displaystyle\int \csc\theta\,d\theta$

8.6 Integral Tables and Computer Algebra Systems

As we have studied, the basic techniques of integration are substitution and integration by parts. We apply these techniques to transform unfamiliar integrals into integrals whose forms we recognize or can find in a table. But where do the integrals in the tables come from? They come from applying substitutions and integration by parts, or by differentiating important functions that arise in practice or applications and tabling the results (as we did in creating Table 8.1). When an integral matches an integral in the table or can be changed into one of the tabulated integrals with some appropriate combination of algebra, trigonometry, substitution, and calculus, the matched result can be used to solve the integration problem at hand.

Computer Algebra Systems (CAS) can also be used to evaluate an integral, if such a system is available. However, beware that there are many relatively simple functions, like $\sin(x^2)$ or $1/\ln x$ for which even the most powerful computer algebra systems cannot find explicit antiderivative formulas because no such formulas exist.

In this section we discuss how to use tables and computer algebra systems to evaluate integrals.

Integral Tables

A Brief Table of Integrals is provided at the back of the book, after the index. (More extensive tables appear in compilations such as *CRC Mathematical Tables*, which contain thousands of integrals.) The integration formulas are stated in terms of constants a, b, c, m, n, and so on. These constants can usually assume any real value and need not be integers. Occasional limitations on their values are stated with the formulas. Formula 5 requires $n \neq -1$, for example, and Formula 11 requires $n \neq 2$.

The formulas also assume that the constants do not take on values that require dividing by zero or taking even roots of negative numbers. For example, Formula 8 assumes that $a \neq 0$, and Formulas 13(a) and (b) cannot be used unless b is positive.

The integrals in Examples 1–5 of this section can be evaluated using algebraic manipulation, substitution, or integration by parts. Here we illustrate how the integrals are found using the Brief Table of Integrals.

EXAMPLE 1 Find

$$\int x(2x+5)^{-1}\,dx.$$

Solution We use Formula 8 (not 7, which requires $n \neq -1$):

$$\int x(ax+b)^{-1}\,dx = \frac{x}{a} - \frac{b}{a^2}\ln|ax+b| + C.$$

With $a = 2$ and $b = 5$, we have

$$\int x(2x+5)^{-1}\,dx = \frac{x}{2} - \frac{5}{4}\ln|2x+5| + C.$$

■

EXAMPLE 2 Find

$$\int \frac{dx}{x\sqrt{2x+4}}.$$

Solution We use Formula 13(b):

$$\int \frac{dx}{x\sqrt{ax+b}} = \frac{1}{\sqrt{b}} \ln \left| \frac{\sqrt{ax+b} - \sqrt{b}}{\sqrt{ax+b} + \sqrt{b}} \right| + C, \qquad \text{if } b > 0.$$

With $a = 2$ and $b = 4$, we have

$$\int \frac{dx}{x\sqrt{2x+4}} = \frac{1}{\sqrt{4}} \ln \left| \frac{\sqrt{2x+4} - \sqrt{4}}{\sqrt{2x+4} + \sqrt{4}} \right| + C$$

$$= \frac{1}{2} \ln \left| \frac{\sqrt{2x+4} - 2}{\sqrt{2x+4} + 2} \right| + C.$$

Formula 13(a), which would require $b < 0$ here, is not appropriate in Example 2. It *is* appropriate, however, in the next example.

EXAMPLE 3 Find

$$\int \frac{dx}{x\sqrt{2x-4}}.$$

Solution We use Formula 13(a):

$$\int \frac{dx}{x\sqrt{ax-b}} = \frac{2}{\sqrt{b}} \tan^{-1} \sqrt{\frac{ax-b}{b}} + C.$$

With $a = 2$ and $b = 4$, we have

$$\int \frac{dx}{x\sqrt{2x-4}} = \frac{2}{\sqrt{4}} \tan^{-1} \sqrt{\frac{2x-4}{4}} + C = \tan^{-1} \sqrt{\frac{x-2}{2}} + C.$$

EXAMPLE 4 Find

$$\int \frac{dx}{x^2\sqrt{2x-4}}.$$

Solution We begin with Formula 15:

$$\int \frac{dx}{x^2\sqrt{ax+b}} = -\frac{\sqrt{ax+b}}{bx} - \frac{a}{2b} \int \frac{dx}{x\sqrt{ax+b}} + C.$$

With $a = 2$ and $b = -4$, we have

$$\int \frac{dx}{x^2\sqrt{2x-4}} = -\frac{\sqrt{2x-4}}{-4x} + \frac{2}{2 \cdot 4} \int \frac{dx}{x\sqrt{2x-4}} + C.$$

We then use Formula 13(a) to evaluate the integral on the right (Example 3) to obtain

$$\int \frac{dx}{x^2\sqrt{2x-4}} = \frac{\sqrt{2x-4}}{4x} + \frac{1}{4} \tan^{-1} \sqrt{\frac{x-2}{2}} + C.$$

EXAMPLE 5 Find

$$\int x \sin^{-1} x \, dx.$$

Solution We use Formula 99:

$$\int x^n \sin^{-1} ax\, dx = \frac{x^{n+1}}{n+1}\sin^{-1} ax - \frac{a}{n+1}\int \frac{x^{n+1}\, dx}{\sqrt{1-a^2x^2}}, \qquad n \neq -1.$$

With $n = 1$ and $a = 1$, we have

$$\int x \sin^{-1} x\, dx = \frac{x^2}{2}\sin^{-1} x - \frac{1}{2}\int \frac{x^2\, dx}{\sqrt{1-x^2}}.$$

The integral on the right is found in the table as Formula 33:

$$\int \frac{x^2}{\sqrt{a^2-x^2}}\, dx = \frac{a^2}{2}\sin^{-1}\left(\frac{x}{a}\right) - \frac{1}{2}x\sqrt{a^2-x^2} + C.$$

With $a = 1$,

$$\int \frac{x^2\, dx}{\sqrt{1-x^2}} = \frac{1}{2}\sin^{-1} x - \frac{1}{2}x\sqrt{1-x^2} + C.$$

The combined result is

$$\begin{aligned}\int x \sin^{-1} x\, dx &= \frac{x^2}{2}\sin^{-1} x - \frac{1}{2}\left(\frac{1}{2}\sin^{-1} x - \frac{1}{2}x\sqrt{1-x^2} + C\right)\\ &= \left(\frac{x^2}{2} - \frac{1}{4}\right)\sin^{-1} x + \frac{1}{4}x\sqrt{1-x^2} + C'.\end{aligned}$$

■

Reduction Formulas

The time required for repeated integrations by parts can sometimes be shortened by applying formulas like

$$\int \tan^n x\, dx = \frac{1}{n-1}\tan^{n-1} x - \int \tan^{n-2} x\, dx \tag{1}$$

$$\int (\ln x)^n\, dx = x(\ln x)^n - n\int (\ln x)^{n-1}\, dx \tag{2}$$

$$\int \sin^n x \cos^m x\, dx = -\frac{\sin^{n-1} x \cos^{m+1} x}{m+n} + \frac{n-1}{m+n}\int \sin^{n-2} x \cos^m x\, dx \qquad (n \neq -m). \tag{3}$$

Formulas like these are called **reduction formulas** because they replace an integral containing some power of a function with an integral of the same form with the power reduced. By applying such a formula repeatedly, we can eventually express the original integral in terms of a power low enough to be evaluated directly.

EXAMPLE 6 Using a Reduction Formula

Find

$$\int \tan^5 x\, dx.$$

Solution We apply Equation (1) with $n = 5$ to get

$$\int \tan^5 x\,dx = \frac{1}{4}\tan^4 x - \int \tan^3 x\,dx.$$

We then apply Equation (1) again, with $n = 3$, to evaluate the remaining integral:

$$\int \tan^3 x\,dx = \frac{1}{2}\tan^2 x - \int \tan x\,dx = \frac{1}{2}\tan^2 x + \ln|\cos x| + C.$$

The combined result is

$$\int \tan^5 x\,dx = \frac{1}{4}\tan^4 x - \frac{1}{2}\tan^2 x - \ln|\cos x| + C'.$$ ■

As their form suggests, reduction formulas are derived by integration by parts.

EXAMPLE 7 Deriving a Reduction Formula

Show that for any positive integer n,

$$\int (\ln x)^n\,dx = x(\ln x)^n - n\int (\ln x)^{n-1}\,dx.$$

Solution We use the integration by parts formula

$$\int u\,dv = uv - \int v\,du$$

with

$$u = (\ln x)^n, \qquad du = n(\ln x)^{n-1}\frac{dx}{x}, \qquad dv = dx, \qquad v = x,$$

to obtain

$$\int (\ln x)^n\,dx = x(\ln x)^n - n\int (\ln x)^{n-1}\,dx.$$ ■

Sometimes two reduction formulas come into play.

EXAMPLE 8 Find

$$\int \sin^2 x\cos^3 x\,dx.$$

Solution 1 We apply Equation (3) with $n = 2$ and $m = 3$ to get

$$\begin{aligned}\int \sin^2 x\cos^3 x\,dx &= -\frac{\sin x\cos^4 x}{2+3} + \frac{1}{2+3}\int \sin^0 x\cos^3 x\,dx\\ &= -\frac{\sin x\cos^4 x}{5} + \frac{1}{5}\int \cos^3 x\,dx.\end{aligned}$$

We can evaluate the remaining integral with Formula 61 (another reduction formula):

$$\int \cos^n ax\,dx = \frac{\cos^{n-1}ax\sin ax}{na} + \frac{n-1}{n}\int \cos^{n-2}ax\,dx.$$

With $n = 3$ and $a = 1$, we have

$$\begin{aligned}\int \cos^3 x\,dx &= \frac{\cos^2 x \sin x}{3} + \frac{2}{3}\int \cos x\,dx \\ &= \frac{\cos^2 x \sin x}{3} + \frac{2}{3}\sin x + C.\end{aligned}$$

The combined result is

$$\begin{aligned}\int \sin^2 x \cos^3 x\,dx &= -\frac{\sin x \cos^4 x}{5} + \frac{1}{5}\left(\frac{\cos^2 x \sin x}{3} + \frac{2}{3}\sin x + C\right) \\ &= -\frac{\sin x \cos^4 x}{5} + \frac{\cos^2 x \sin x}{15} + \frac{2}{15}\sin x + C'.\end{aligned}$$

Solution 2 Equation (3) corresponds to Formula 68 in the table, but there is another formula we might use, namely Formula 69. With $a = 1$, Formula 69 gives

$$\int \sin^n x \cos^m x\,dx = \frac{\sin^{n+1} x \cos^{m-1} x}{m+n} + \frac{m-1}{m+n}\int \sin^n x \cos^{m-2} x\,dx.$$

In our case, $n = 2$ and $m = 3$, so that

$$\begin{aligned}\int \sin^2 x \cos^3 x\,dx &= \frac{\sin^3 x \cos^2 x}{5} + \frac{2}{5}\int \sin^2 x \cos x\,dx \\ &= \frac{\sin^3 x \cos^2 x}{5} + \frac{2}{5}\left(\frac{\sin^3 x}{3}\right) + C \\ &= \frac{\sin^3 x \cos^2 x}{5} + \frac{2}{15}\sin^3 x + C.\end{aligned}$$

As you can see, it is faster to use Formula 69, but we often cannot tell beforehand how things will work out. Do not spend a lot of time looking for the "best" formula. Just find one that will work and forge ahead.

Notice also that Formulas 68 (Solution 1) and 69 (Solution 2) lead to different-looking answers. That is often the case with trigonometric integrals and is no cause for concern. The results are equivalent, and we may use whichever one we please. ■

Nonelementary Integrals

The development of computers and calculators that find antiderivatives by symbolic manipulation has led to a renewed interest in determining which antiderivatives can be expressed as finite combinations of elementary functions (the functions we have been studying) and which cannot. Integrals of functions that do not have elementary antiderivatives are called **nonelementary** integrals. They require infinite series (Chapter 11) or numerical methods for their evaluation. Examples of the latter include the error function (which measures the probability of random errors)

$$\operatorname{erf}(x) = \frac{2}{\sqrt{\pi}}\int_0^x e^{-t^2}\,dt$$

and integrals such as

$$\int \sin x^2\,dx \qquad \text{and} \qquad \int \sqrt{1 + x^4}\,dx$$

that arise in engineering and physics. These and a number of others, such as

$$\int \frac{e^x}{x}\,dx, \qquad \int e^{(e^x)}\,dx, \qquad \int \frac{1}{\ln x}\,dx, \qquad \int \ln(\ln x)\,dx, \qquad \int \frac{\sin x}{x}\,dx,$$

$$\int \sqrt{1 - k^2 \sin^2 x}\,dx, \qquad 0 < k < 1,$$

look so easy they tempt us to try them just to see how they turn out. It can be proved, however, that there is no way to express these integrals as finite combinations of elementary functions. The same applies to integrals that can be changed into these by substitution. The integrands all have antiderivatives, as a consequence of the Fundamental Theorem of the Calculus, Part 1, because they are continuous. However, none of the antiderivatives is elementary.

None of the integrals you are asked to evaluate in the present chapter falls into this category, but you may encounter nonelementary integrals in your other work.

Integration with a CAS

A powerful capability of computer algebra systems is their ability to integrate symbolically. This is performed with the **integrate command** specified by the particular system (for example, **int** in Maple, **Integrate** in Mathematica).

EXAMPLE 9 Using a CAS with a Named Function

Suppose that you want to evaluate the indefinite integral of the function

$$f(x) = x^2\sqrt{a^2 + x^2}.$$

Using Maple, you first define or name the function:

```
> f:= x^2 * sqrt (a^2 + x^2);
```

Then you use the integrate command on f, identifying the variable of integration:

```
> int(f, x);
```

Maple returns the answer

$$\frac{1}{4}x(a^2 + x^2)^{3/2} - \frac{1}{8}a^2x\sqrt{a^2 + x^2} - \frac{1}{8}a^4 \ln\left(x + \sqrt{a^2 + x^2}\right).$$

If you want to see if the answer can be simplified, enter

```
> simplify(%);
```

Maple returns

$$\frac{1}{8}a^2x\sqrt{a^2 + x^2} + \frac{1}{4}x^3\sqrt{a^2 + x^2} - \frac{1}{8}a^4 \ln\left(x + \sqrt{a^2 + x^2}\right).$$

If you want the definite integral for $0 \le x \le \pi/2$, you can use the format

```
> int(f, x = 0..Pi/2);
```

Maple (Version 5.1) will return the expression

$$\frac{1}{64}\pi(4a^2 + \pi^2)^{(3/2)} - \frac{1}{32}a^2\pi\sqrt{4a^2 + \pi^2} + \frac{1}{8}a^4\ln(2)$$
$$- \frac{1}{8}a^4\ln\left(\pi + \sqrt{4a^2 + \pi^2}\right) + \frac{1}{16}a^4\ln(a^2).$$

You can also find the definite integral for a particular value of the constant a:

$$> a := 1;$$

$$> \text{int}(f, x = 0..1);$$

Maple returns the numerical answer

$$\frac{3}{8}\sqrt{2} + \frac{1}{8}\ln\left(\sqrt{2} - 1\right).$$

EXAMPLE 10 Using a CAS Without Naming the Function

Use a CAS to find

$$\int \sin^2 x \cos^3 x \, dx.$$

Solution With Maple, we have the entry

$$> \text{int}\,((\sin\char`^2)(x) * (\cos\char`^3)(x), x);$$

with the immediate return

$$-\frac{1}{5}\sin(x)\cos(x)^4 + \frac{1}{15}\cos(x)^2\sin(x) + \frac{2}{15}\sin(x).$$

EXAMPLE 11 A CAS May Not Return a Closed Form Solution

Use a CAS to find

$$\int (\cos^{-1} ax)^2 \, dx.$$

Solution Using Maple, we enter

$$> \text{int}((\arccos(a * x))\char`^2, x);$$

and Maple returns the expression

$$\int \arccos(ax)^2 \, dx,$$

indicating that it does not have a closed form solution known by Maple. In Chapter 11, you will see how series expansion may help to evaluate such an integral.

Computer algebra systems vary in how they process integrations. We used Maple 5.1 in Examples 9–11. Mathematica 4 would have returned somewhat different results:

1. In Example 9, given

   $$\textit{In [1]:=} \text{ Integrate } [x\char`^2 * \text{Sqrt } [a\char`^2 + x\char`^2], x]$$

   Mathematica returns

   $$\textit{Out [1]=} \sqrt{a^2 + x^2}\left(\frac{a^2 x}{8} + \frac{x^3}{4}\right) - \frac{1}{8}a^4 \,\text{Log}\left[x + \sqrt{a^2 + x^2}\right]$$

   without having to simplify an intermediate result. The answer is close to Formula 22 in the integral tables.

2. The Mathematica answer to the integral

In [2]:= Integrate [Sin [x]^2 * Cos [x]^3, x]

in Example 10 is

$$\textit{Out [2]}= \frac{\text{Sin}[x]}{8} - \frac{1}{48}\text{Sin}[3\,x] - \frac{1}{80}\text{Sin}[5\,x]$$

differing from both the Maple answer and the answers in Example 8.

3. Mathematica does give a result for the integration

In [3]:= Integrate [ArcCos [a * x]^2, x]

in Example 11, provided $a \neq 0$:

$$\textit{Out [3]}= -2x - \frac{2\sqrt{1 - a^2 x^2}\,\text{ArcCos}[a\,x]}{a} + x\,\text{ArcCos}[a\,x]^2$$

Although a CAS is very powerful and can aid us in solving difficult problems, each CAS has its own limitations. There are even situations where a CAS may further complicate a problem (in the sense of producing an answer that is extremely difficult to use or interpret). Note, too, that neither Maple nor Mathematica return an arbitrary constant $+C$. On the other hand, a little mathematical thinking on your part may reduce the problem to one that is quite easy to handle. We provide an example in Exercise 111.

EXERCISES 8.6

Using Integral Tables

Use the table of integrals at the back of the book to evaluate the integrals in Exercises 1–38.

1. $\int \frac{dx}{x\sqrt{x-3}}$
2. $\int \frac{dx}{x\sqrt{x+4}}$
3. $\int \frac{x\,dx}{\sqrt{x-2}}$
4. $\int \frac{x\,dx}{(2x+3)^{3/2}}$
5. $\int x\sqrt{2x-3}\,dx$
6. $\int x(7x+5)^{3/2}\,dx$
7. $\int \frac{\sqrt{9-4x}}{x^2}\,dx$
8. $\int \frac{dx}{x^2\sqrt{4x-9}}$
9. $\int x\sqrt{4x-x^2}\,dx$
10. $\int \frac{\sqrt{x-x^2}}{x}\,dx$
11. $\int \frac{dx}{x\sqrt{7+x^2}}$
12. $\int \frac{dx}{x\sqrt{7-x^2}}$
13. $\int \frac{\sqrt{4-x^2}}{x}\,dx$
14. $\int \frac{\sqrt{x^2-4}}{x}\,dx$
15. $\int \sqrt{25-p^2}\,dp$
16. $\int q^2\sqrt{25-q^2}\,dq$
17. $\int \frac{r^2}{\sqrt{4-r^2}}\,dr$
18. $\int \frac{ds}{\sqrt{s^2-2}}$
19. $\int \frac{d\theta}{5+4\sin 2\theta}$
20. $\int \frac{d\theta}{4+5\sin 2\theta}$
21. $\int e^{2t}\cos 3t\,dt$
22. $\int e^{-3t}\sin 4t\,dt$
23. $\int x\cos^{-1}x\,dx$
24. $\int x\tan^{-1}x\,dx$
25. $\int \frac{ds}{(9-s^2)^2}$
26. $\int \frac{d\theta}{(2-\theta^2)^2}$
27. $\int \frac{\sqrt{4x+9}}{x^2}\,dx$
28. $\int \frac{\sqrt{9x-4}}{x^2}\,dx$
29. $\int \frac{\sqrt{3t-4}}{t}\,dt$
30. $\int \frac{\sqrt{3t+9}}{t}\,dt$
31. $\int x^2\tan^{-1}x\,dx$
32. $\int \frac{\tan^{-1}x}{x^2}\,dx$
33. $\int \sin 3x\cos 2x\,dx$
34. $\int \sin 2x\cos 3x\,dx$

35. $\int 8 \sin 4t \sin \frac{t}{2}\, dt$

36. $\int \sin \frac{t}{3} \sin \frac{t}{6}\, dt$

37. $\int \cos \frac{\theta}{3} \cos \frac{\theta}{4}\, d\theta$

38. $\int \cos \frac{\theta}{2} \cos 7\theta\, d\theta$

Substitution and Integral Tables

In Exercises 39–52, use a substitution to change the integral into one you can find in the table. Then evaluate the integral.

39. $\int \frac{x^3 + x + 1}{(x^2 + 1)^2}\, dx$

40. $\int \frac{x^2 + 6x}{(x^2 + 3)^2}\, dx$

41. $\int \sin^{-1} \sqrt{x}\, dx$

42. $\int \frac{\cos^{-1} \sqrt{x}}{\sqrt{x}}\, dx$

43. $\int \frac{\sqrt{x}}{\sqrt{1 - x}}\, dx$

44. $\int \frac{\sqrt{2 - x}}{\sqrt{x}}\, dx$

45. $\int \cot t \sqrt{1 - \sin^2 t}\, dt, \quad 0 < t < \pi/2$

46. $\int \frac{dt}{\tan t \sqrt{4 - \sin^2 t}}$

47. $\int \frac{dy}{y \sqrt{3 + (\ln y)^2}}$

48. $\int \frac{\cos \theta\, d\theta}{\sqrt{5 + \sin^2 \theta}}$

49. $\int \frac{3\, dr}{\sqrt{9r^2 - 1}}$

50. $\int \frac{3\, dy}{\sqrt{1 + 9y^2}}$

51. $\int \cos^{-1} \sqrt{x}\, dx$

52. $\int \tan^{-1} \sqrt{y}\, dy$

Using Reduction Formulas

Use reduction formulas to evaluate the integrals in Exercises 53–72.

53. $\int \sin^5 2x\, dx$

54. $\int \sin^5 \frac{\theta}{2}\, d\theta$

55. $\int 8 \cos^4 2\pi t\, dt$

56. $\int 3 \cos^5 3y\, dy$

57. $\int \sin^2 2\theta \cos^3 2\theta\, d\theta$

58. $\int 9 \sin^3 \theta \cos^{3/2} \theta\, d\theta$

59. $\int 2 \sin^2 t \sec^4 t\, dt$

60. $\int \csc^2 y \cos^5 y\, dy$

61. $\int 4 \tan^3 2x\, dx$

62. $\int \tan^4 \left(\frac{x}{2}\right) dx$

63. $\int 8 \cot^4 t\, dt$

64. $\int 4 \cot^3 2t\, dt$

65. $\int 2 \sec^3 \pi x\, dx$

66. $\int \frac{1}{2} \csc^3 \frac{x}{2}\, dx$

67. $\int 3 \sec^4 3x\, dx$

68. $\int \csc^4 \frac{\theta}{3}\, d\theta$

69. $\int \csc^5 x\, dx$

70. $\int \sec^5 x\, dx$

71. $\int 16x^3 (\ln x)^2\, dx$

72. $\int (\ln x)^3\, dx$

Powers of x Times Exponentials

Evaluate the integrals in Exercises 73–80 using table Formulas 103–106. These integrals can also be evaluated using tabular integration (Section 8.2).

73. $\int x e^{3x}\, dx$

74. $\int x e^{-2x}\, dx$

75. $\int x^3 e^{x/2}\, dx$

76. $\int x^2 e^{\pi x}\, dx$

77. $\int x^2\, 2^x\, dx$

78. $\int x^2\, 2^{-x}\, dx$

79. $\int x \pi^x\, dx$

80. $\int x 2^{\sqrt{2}x}\, dx$

Substitutions with Reduction Formulas

Evaluate the integrals in Exercises 81–86 by making a substitution (possibly trigonometric) and then applying a reduction formula.

81. $\int e^t \sec^3 (e^t - 1)\, dt$

82. $\int \frac{\csc^3 \sqrt{\theta}}{\sqrt{\theta}}\, d\theta$

83. $\int_0^1 2\sqrt{x^2 + 1}\, dx$

84. $\int_0^{\sqrt{3}/2} \frac{dy}{(1 - y^2)^{5/2}}$

85. $\int_1^2 \frac{(r^2 - 1)^{3/2}}{r}\, dr$

86. $\int_0^{1/\sqrt{3}} \frac{dt}{(t^2 + 1)^{7/2}}$

Hyperbolic Functions

Use the integral tables to evaluate the integrals in Exercises 87–92.

87. $\int \frac{1}{8} \sinh^5 3x\, dx$

88. $\int \frac{\cosh^4 \sqrt{x}}{\sqrt{x}}\, dx$

89. $\int x^2 \cosh 3x\, dx$

90. $\int x \sinh 5x\, dx$

91. $\int \operatorname{sech}^7 x \tanh x\, dx$

92. $\int \operatorname{csch}^3 2x \coth 2x\, dx$

Theory and Examples

Exercises 93–100 refer to formulas in the table of integrals at the back of the book.

93. Derive Formula 9 by using the substitution $u = ax + b$ to evaluate

$$\int \frac{x}{(ax + b)^2}\, dx.$$

94. Derive Formula 17 by using a trigonometric substitution to evaluate

$$\int \frac{dx}{(a^2 + x^2)^2}.$$

95. Derive Formula 29 by using a trigonometric substitution to evaluate

$$\int \sqrt{a^2 - x^2}\, dx.$$

96. Derive Formula 46 by using a trigonometric substitution to evaluate

$$\int \frac{dx}{x^2 \sqrt{x^2 - a^2}}.$$

97. Derive Formula 80 by evaluating

$$\int x^n \sin ax\, dx$$

by integration by parts.

98. Derive Formula 110 by evaluating

$$\int x^n (\ln ax)^m\, dx$$

by integration by parts.

99. Derive Formula 99 by evaluating

$$\int x^n \sin^{-1} ax\, dx$$

by integration by parts.

100. Derive Formula 101 by evaluating

$$\int x^n \tan^{-1} ax\, dx$$

by integration by parts.

101. Surface area Find the area of the surface generated by revolving the curve $y = \sqrt{x^2 + 2}$, $0 \le x \le \sqrt{2}$, about the x-axis.

102. Arc length Find the length of the curve $y = x^2$, $0 \le x \le \sqrt{3}/2$.

103. Centroid Find the centroid of the region cut from the first quadrant by the curve $y = 1/\sqrt{x + 1}$ and the line $x = 3$.

104. Moment about y-axis A thin plate of constant density $\delta = 1$ occupies the region enclosed by the curve $y = 36/(2x + 3)$ and the line $x = 3$ in the first quadrant. Find the moment of the plate about the y-axis.

T **105.** Use the integral table and a calculator to find to two decimal places the area of the surface generated by revolving the curve $y = x^2$, $-1 \le x \le 1$, about the x-axis.

106. Volume The head of your firm's accounting department has asked you to find a formula she can use in a computer program to calculate the year-end inventory of gasoline in the company's tanks. A typical tank is shaped like a right circular cylinder of radius r and length L, mounted horizontally, as shown here. The data come to the accounting office as depth measurements taken with a vertical measuring stick marked in centimeters.

a. Show, in the notation of the figure here, that the volume of gasoline that fills the tank to a depth d is

$$V = 2L \int_{-r}^{-r+d} \sqrt{r^2 - y^2}\, dy.$$

b. Evaluate the integral.

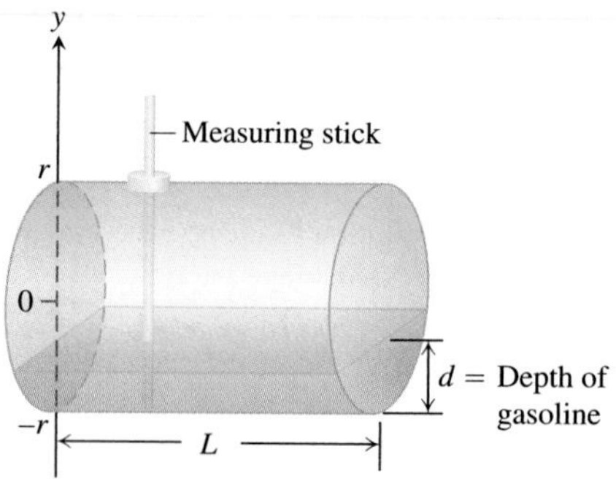

107. What is the largest value

$$\int_a^b \sqrt{x - x^2}\, dx$$

can have for any a and b? Give reasons for your answer.

108. What is the largest value

$$\int_a^b x\sqrt{2x - x^2}\, dx$$

can have for any a and b? Give reasons for your answer.

COMPUTER EXPLORATIONS

In Exercises 109 and 110, use a CAS to perform the integrations.

109. Evaluate the integrals

a. $\int x \ln x\, dx$ **b.** $\int x^2 \ln x\, dx$ **c.** $\int x^3 \ln x\, dx.$

d. What pattern do you see? Predict the formula for $\int x^4 \ln x\, dx$ and then see if you are correct by evaluating it with a CAS.

e. What is the formula for $\int x^n \ln x\, dx$, $n \ge 1$? Check your answer using a CAS.

110. Evaluate the integrals

a. $\int \frac{\ln x}{x^2}\, dx$ **b.** $\int \frac{\ln x}{x^3}\, dx$ **c.** $\int \frac{\ln x}{x^4}\, dx.$

d. What pattern do you see? Predict the formula for

$$\int \frac{\ln x}{x^5}\, dx$$

and then see if you are correct by evaluating it with a CAS.

e. What is the formula for

$$\int \frac{\ln x}{x^n}\, dx, \quad n \ge 2?$$

Check your answer using a CAS.

111. a. Use a CAS to evaluate

$$\int_0^{\pi/2} \frac{\sin^n x}{\sin^n x + \cos^n x}\,dx$$

where n is an arbitrary positive integer. Does your CAS find the result?

b. In succession, find the integral when $n = 1, 2, 3, 5, 7$. Comment on the complexity of the results.

c. Now substitute $x = (\pi/2) - u$ and add the new and old integrals. What is the value of

$$\int_0^{\pi/2} \frac{\sin^n x}{\sin^n x + \cos^n x}\,dx?$$

This exercise illustrates how a little mathematical ingenuity solves a problem not immediately amenable to solution by a CAS.

8.7 Numerical Integration

As we have seen, the ideal way to evaluate a definite integral $\int_a^b f(x)\,dx$ is to find a formula $F(x)$ for one of the antiderivatives of $f(x)$ and calculate the number $F(b) - F(a)$. But some antiderivatives require considerable work to find, and still others, like the antiderivatives of $\sin(x^2)$, $1/\ln x$, and $\sqrt{1 + x^4}$, have no elementary formulas.

Another situation arises when a function is defined by a table whose entries were obtained experimentally through instrument readings. In this case a formula for the function may not even exist.

Whatever the reason, when we cannot evaluate a definite integral with an antiderivative, we turn to numerical methods such as the *Trapezoidal Rule* and *Simpson's Rule* developed in this section. These rules usually require far fewer subdivisions of the integration interval to get accurate results compared to the various rectangle rules presented in Sections 5.1 and 5.2. We also estimate the error obtained when using these approximation methods.

Trapezoidal Approximations

When we cannot find a workable antiderivative for a function f that we have to integrate, we partition the interval of integration, replace f by a closely fitting polynomial on each subinterval, integrate the polynomials, and add the results to approximate the integral of f. In our presentation we assume that f is positive, but the only requirement is for f to be continuous over the interval of integration $[a, b]$.

The Trapezoidal Rule for the value of a definite integral is based on approximating the region between a curve and the x-axis with trapezoids instead of rectangles, as in Figure 8.10. It is not necessary for the subdivision points $x_0, x_1, x_2, \ldots, x_n$ in the figure to be evenly spaced, but the resulting formula is simpler if they are. We therefore assume that the length of each subinterval is

$$\Delta x = \frac{b - a}{n}.$$

The length $\Delta x = (b - a)/n$ is called the **step size** or **mesh size**. The area of the trapezoid that lies above the ith subinterval is

$$\Delta x\left(\frac{y_{i-1} + y_i}{2}\right) = \frac{\Delta x}{2}(y_{i-1} + y_i),$$

Using T to approximate $\int_a^b f(x)\,dx$, we have

$$\text{av}(f) \approx \frac{1}{b-a} \cdot T = \frac{1}{12} \cdot 782 \approx 65.17.$$

Rounding to the accuracy of the given data, we estimate the average temperature as 65 degrees. ■

Error Estimates for the Trapezoidal Rule

As n increases and the step size $\Delta x = (b-a)/n$ approaches zero, T approaches the exact value of $\int_a^b f(x)\,dx$. To see why, write

$$\begin{aligned} T &= \Delta x \left(\frac{1}{2} y_0 + y_1 + y_2 + \cdots + y_{n-1} + \frac{1}{2} y_n \right) \\ &= (y_1 + y_2 + \cdots + y_n)\Delta x + \frac{1}{2}(y_0 - y_n)\,\Delta x \\ &= \sum_{k=1}^{n} f(x_k)\Delta x + \frac{1}{2}[f(a) - f(b)]\,\Delta x. \end{aligned}$$

As $n \to \infty$ and $\Delta x \to 0$,

$$\sum_{k=1}^{n} f(x_k)\Delta x \to \int_a^b f(x)\,dx \quad \text{and} \quad \frac{1}{2}[f(a) - f(b)]\Delta x \to 0.$$

Therefore,

$$\lim_{n\to\infty} T = \int_a^b f(x)\,dx + 0 = \int_a^b f(x)\,dx.$$

This means that in theory we can make the difference between T and the integral as small as we want by taking n large enough, assuming only that f is integrable. In practice, though, how do we tell how large n should be for a given tolerance?

We answer this question with a result from advanced calculus, which says that if f'' is continuous on $[a, b]$, then

$$\int_a^b f(x)\,dx = T - \frac{b-a}{12} \cdot f''(c)(\Delta x)^2$$

for some number c between a and b. Thus, as Δx approaches zero, the error defined by

$$E_T = -\frac{b-a}{12} \cdot f''(c)(\Delta x)^2$$

approaches zero as the *square* of Δx.

The inequality

$$|E_T| \le \frac{b-a}{12} \max|f''(x)|(\Delta x)^2,$$

where max refers to the interval $[a, b]$, gives an upper bound for the magnitude of the error. In practice, we usually cannot find the exact value of $\max|f''(x)|$ and have to estimate an upper bound or "worst case" value for it instead. If M is any upper bound for the values of $|f''(x)|$ on $[a, b]$, so that $|f''(x)| \le M$ on $[a, b]$, then

$$|E_T| \le \frac{b-a}{12} M(\Delta x)^2.$$

If we substitute $(b - a)/n$ for Δx, we get

$$|E_T| \le \frac{M(b-a)^3}{12n^2}.$$

This is the inequality we normally use in estimating $|E_T|$. We find the best M we can and go on to estimate $|E_T|$ from there. This may sound careless, but it works. To make $|E_T|$ small for a given M, we just make n large.

The Error Estimate for the Trapezoidal Rule

If f'' is continuous and M is any upper bound for the values of $|f''|$ on $[a, b]$, then the error E_T in the trapezoidal approximation of the integral of f from a to b for n steps satisfies the inequality

$$|E_T| \le \frac{M(b-a)^3}{12n^2}.$$

EXAMPLE 3 Bounding the Trapezoidal Rule Error

Find an upper bound for the error incurred in estimating

$$\int_0^{\pi} x \sin x \, dx$$

with the Trapezoidal Rule with $n = 10$ steps (Figure 8.12).

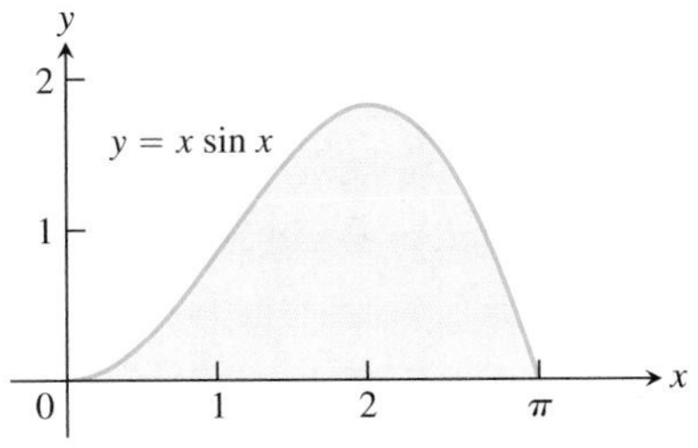

FIGURE 8.12 Graph of the integrand in Example 3.

Solution With $a = 0$, $b = \pi$, and $n = 10$, the error estimate gives

$$|E_T| \le \frac{M(b-a)^3}{12n^2} = \frac{\pi^3}{1200}M.$$

The number M can be any upper bound for the magnitude of the second derivative of $f(x) = x \sin x$ on $[0, \pi]$. A routine calculation gives

$$f''(x) = 2\cos x - x \sin x,$$

so

$$\begin{aligned} |f''(x)| &= |2\cos x - x\sin x| \\ &\le 2|\cos x| + |x||\sin x| \\ &\le 2 \cdot 1 + \pi \cdot 1 = 2 + \pi. \end{aligned}$$

$|\cos x|$ and $|\sin x|$ never exceed 1, and $0 \le x \le \pi$.

We can safely take $M = 2 + \pi$. Therefore,

$$|E_T| \le \frac{\pi^3}{1200}M = \frac{\pi^3(2+\pi)}{1200} < 0.133.$$

Rounded up to be safe

The absolute error is no greater than 0.133.

For greater accuracy, we would not try to improve M but would take more steps. With $n = 100$ steps, for example, we get

$$|E_T| \le \frac{(2+\pi)\pi^3}{120{,}000} < 0.00133 = 1.33 \times 10^{-3}.$$

■

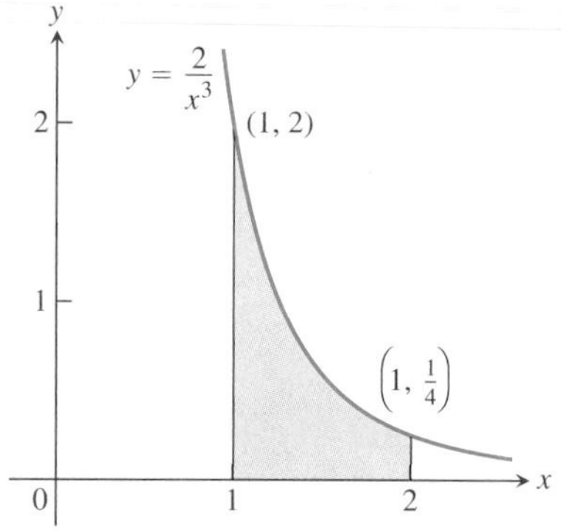

FIGURE 8.13 The continuous function $y = 2/x^3$ has its maximum value on [1, 2] at $x = 1$.

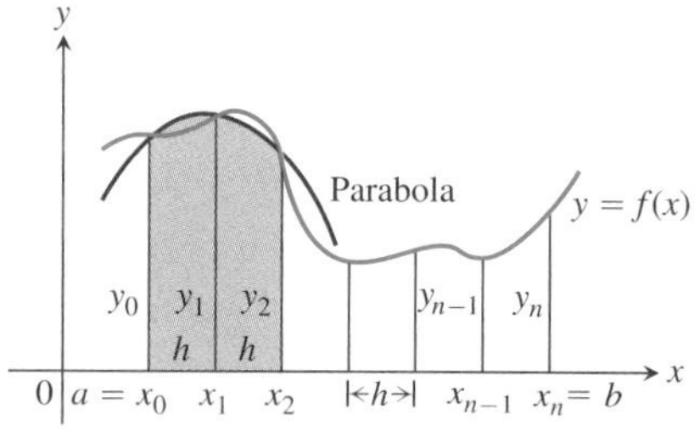

FIGURE 8.14 Simpson's Rule approximates short stretches of the curve with parabolas.

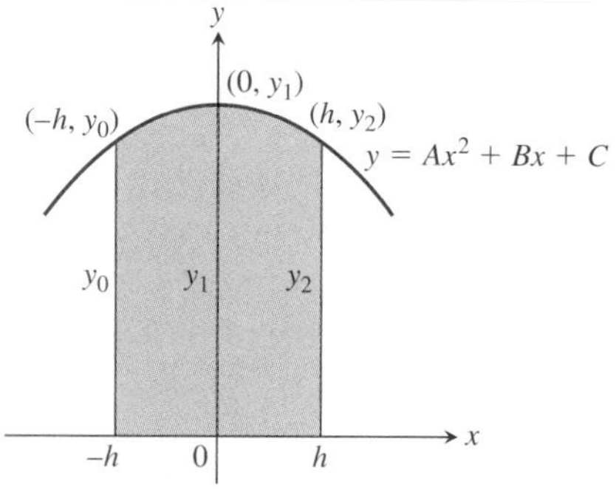

FIGURE 8.15 By integrating from $-h$ to h, we find the shaded area to be

$$\frac{h}{3}(y_0 + 4y_1 + y_2).$$

EXAMPLE 4 Finding How Many Steps Are Needed for a Specific Accuracy

How many subdivisions should be used in the Trapezoidal Rule to approximate

$$\ln 2 = \int_1^2 \frac{1}{x}\,dx$$

with an error whose absolute value is less than 10^{-4}?

Solution With $a = 1$ and $b = 2$, the error estimate is

$$|E_T| \le \frac{M(2-1)^3}{12n^2} = \frac{M}{12n^2}.$$

This is one of the rare cases in which we actually can find $\max|f''|$ rather than having to settle for an upper bound M. With $f(x) = 1/x$, we find

$$f''(x) = \frac{d^2}{dx^2}(x^{-1}) = 2x^{-3} = \frac{2}{x^3}.$$

On [1, 2], $y = 2/x^3$ decreases steadily from a maximum of $y = 2$ to a minimum of $y = 1/4$ (Figure 8.13). Therefore, $M = 2$ and

$$|E_T| \le \frac{2}{12n^2} = \frac{1}{6n^2}.$$

The error's absolute value will therefore be less than 10^{-4} if

$$\frac{1}{6n^2} < 10^{-4}, \qquad \frac{10^4}{6} < n^2, \qquad \frac{100}{\sqrt{6}} < n, \qquad \text{or} \qquad 40.83 < n.$$

The first integer beyond 40.83 is $n = 41$. With $n = 41$ subdivisions we can guarantee calculating ln 2 with an error of magnitude less than 10^{-4}. Any larger n will work, too. ■

Simpson's Rule: Approximations Using Parabolas

Riemann sums and the Trapezoidal Rule both give reasonable approximations to the integral of a continuous function over a closed interval. The Trapezoidal Rule is more efficient, giving a better approximation for small values of n, which makes it a faster algorithm for numerical integration.

Another rule for approximating the definite integral of a continuous function results from using parabolas instead of the straight line segments which produced trapezoids. As before, we partition the interval $[a, b]$ into n subintervals of equal length $h = \Delta x = (b - a)/n$, but this time we require that n be an *even* number. On each consecutive pair of intervals we approximate the curve $y = f(x) \ge 0$ by a parabola, as shown in Figure 8.14. A typical parabola passes through three consecutive points (x_{i-1}, y_{i-1}), (x_i, y_i), and (x_{i+1}, y_{i+1}) on the curve.

Let's calculate the shaded area beneath a parabola passing through three consecutive points. To simplify our calculations, we first take the case where $x_0 = -h$, $x_1 = 0$, and $x_2 = h$ (Figure 8.15), where $h = \Delta x = (b - a)/n$. The area under the parabola will be the same if we shift the y-axis to the left or right. The parabola has an equation of the form

$$y = Ax^2 + Bx + C,$$

so the area under it from $x = -h$ to $x = h$ is

$$\begin{aligned} A_p &= \int_{-h}^{h} (Ax^2 + Bx + C)\,dx \\ &= \frac{Ax^3}{3} + \frac{Bx^2}{2} + Cx\bigg]_{-h}^{h} \\ &= \frac{2Ah^3}{3} + 2Ch = \frac{h}{3}(2Ah^2 + 6C). \end{aligned}$$

Since the curve passes through the three points $(-h, y_0)$, $(0, y_1)$, and (h, y_2), we also have

$$y_0 = Ah^2 - Bh + C, \qquad y_1 = C, \qquad y_2 = Ah^2 + Bh + C,$$

from which we obtain

$$\begin{aligned} C &= y_1, \\ Ah^2 - Bh &= y_0 - y_1, \\ Ah^2 + Bh &= y_2 - y_1, \\ 2Ah^2 &= y_0 + y_2 - 2y_1. \end{aligned}$$

Hence, expressing the area A_p in terms of the ordinates y_0, y_1, and y_2, we have

$$A_p = \frac{h}{3}(2Ah^2 + 6C) = \frac{h}{3}((y_0 + y_2 - 2y_1) + 6y_1) = \frac{h}{3}(y_0 + 4y_1 + y_2).$$

Now shifting the parabola horizontally to its shaded position in Figure 8.14 does not change the area under it. Thus the area under the parabola through (x_0, y_0), (x_1, y_1), and (x_2, y_2) in Figure 8.14 is still

$$\frac{h}{3}(y_0 + 4y_1 + y_2).$$

Similarly, the area under the parabola through the points (x_2, y_2), (x_3, y_3), and (x_4, y_4) is

$$\frac{h}{3}(y_2 + 4y_3 + y_4).$$

Computing the areas under all the parabolas and adding the results gives the approximation

$$\begin{aligned} \int_a^b f(x)\,dx &\approx \frac{h}{3}(y_0 + 4y_1 + y_2) + \frac{h}{3}(y_2 + 4y_3 + y_4) + \cdots \\ &\quad + \frac{h}{3}(y_{n-2} + 4y_{n-1} + y_n) \\ &= \frac{h}{3}(y_0 + 4y_1 + 2y_2 + 4y_3 + 2y_4 + \cdots + 2y_{n-2} + 4y_{n-1} + y_n). \end{aligned}$$

HISTORICAL BIOGRAPHY

Thomas Simpson
(1720–1761)

The result is known as Simpson's Rule, and it is again valid for any continuous function $y = f(x)$ (Exercise 38). The function need not be positive, as in our derivation. The number n of subintervals must be even to apply the rule because each parabolic arc uses two subintervals.

Simpson's Rule

To approximate $\int_a^b f(x)\,dx$, use

$$S = \frac{\Delta x}{3}\left(y_0 + 4y_1 + 2y_2 + 4y_3 + \cdots + 2y_{n-2} + 4y_{n-1} + y_n\right).$$

The y's are the values of f at the partition points

$$x_0 = a, x_1 = a + \Delta x, x_2 = a + 2\Delta x, \ldots, x_{n-1} = a + (n-1)\Delta x, x_n = b.$$

The number n is even, and $\Delta x = (b - a)/n$.

Note the pattern of the coefficients in the above rule: $1, 4, 2, 4, 2, 4, 2, \ldots, 4, 2, 1$.

TABLE 8.4

x	$y = 5x^4$
0	0
$\frac{1}{2}$	$\frac{5}{16}$
1	5
$\frac{3}{2}$	$\frac{405}{16}$
2	80

EXAMPLE 5 Applying Simpson's Rule

Use Simpson's Rule with $n = 4$ to approximate $\int_0^2 5x^4\,dx$.

Solution Partition [0, 2] into four subintervals and evaluate $y = 5x^4$ at the partition points (Table 8.4). Then apply Simpson's Rule with $n = 4$ and $\Delta x = 1/2$:

$$\begin{aligned} S &= \frac{\Delta x}{3}\left(y_0 + 4y_1 + 2y_2 + 4y_3 + y_4\right) \\ &= \frac{1}{6}\left(0 + 4\left(\frac{5}{16}\right) + 2(5) + 4\left(\frac{405}{16}\right) + 80\right) \\ &= 32\frac{1}{12}. \end{aligned}$$

This estimate differs from the exact value (32) by only $1/12$, a percentage error of less than three-tenths of one percent, and this was with just four subintervals. ■

Error Estimates for Simpson's Rule

To estimate the error in Simpson's rule, we start with a result from advanced calculus that says that if the fourth derivative $f^{(4)}$ is continuous, then

$$\int_a^b f(x)\,dx = S - \frac{b-a}{180} \cdot f^{(4)}(c)(\Delta x)^4$$

for some point c between a and b. Thus, as Δx approaches zero, the error,

$$E_S = -\frac{b-a}{180} \cdot f^{(4)}(c)(\Delta x)^4,$$

approaches zero as the *fourth power* of Δx (This helps to explain why Simpson's Rule is likely to give better results than the Trapezoidal Rule.)

The inequality

$$|E_S| \le \frac{b-a}{180} \max|f^{(4)}(x)|\,(\Delta x)^4$$

where max refers to the interval $[a, b]$, gives an upper bound for the magnitude of the error. As with $\max|f''|$ in the error formula for the Trapezoidal Rule, we usually cannot find the exact value of $\max|f^{(4)}(x)|$ and have to replace it with an upper bound. If M is any upper bound for the values of $|f^{(4)}|$ on $[a, b]$, then

$$|E_S| \leq \frac{b-a}{180} M(\Delta x)^4.$$

Substituting $(b - a)/n$ for Δx in this last expression gives

$$|E_S| \leq \frac{M(b-a)^5}{180n^4}.$$

This is the formula we usually use in estimating the error in Simpson's Rule. We find a reasonable value for M and go on to estimate $|E_S|$ from there.

The Error Estimate for Simpson's Rule

If $f^{(4)}$ is continuous and M is any upper bound for the values of $|f^{(4)}|$ on $[a, b]$, then the error E_S in the Simpson's Rule approximation of the integral of f from a to b for n steps satisfies the inequality

$$|E_S| \leq \frac{M(b-a)^5}{180n^4}.$$

As with the Trapezoidal Rule, we often cannot find the smallest possible value of M. We just find the best value we can and go on from there.

EXAMPLE 6 Bounding the Error in Simpson's Rule

Find an upper bound for the error in estimating $\int_0^2 5x^4\,dx$ using Simpson's Rule with $n = 4$ (Example 5).

Solution To estimate the error, we first find an upper bound M for the magnitude of the fourth derivative of $f(x) = 5x^4$ on the interval $0 \leq x \leq 2$. Since the fourth derivative has the constant value $f^{(4)}(x) = 120$, we take $M = 120$. With $b - a = 2$ and $n = 4$, the error estimate for Simpson's Rule gives

$$|E_S| \leq \frac{M(b-a)^5}{180n^4} = \frac{120(2)^5}{180 \cdot 4^4} = \frac{1}{12}.$$

■

EXAMPLE 7 Comparing the Trapezoidal Rule and Simpson's Rule Approximations

As we saw in Chapter 7, the value of ln 2 can be calculated from the integral

$$\ln 2 = \int_1^2 \frac{1}{x}\,dx.$$

Table 8.5 shows T and S values for approximations of $\int_1^2 (1/x)\,dx$ using various values of n. Notice how Simpson's Rule dramatically improves over the Trapezoidal Rule. In particular, notice that when we double the value of n (thereby halving the value of $h = \Delta x$), the T error is divided by 2 *squared*, whereas the S error is divided by 2 *to the fourth*.

TABLE 8.5 Trapezoidal Rule approximations (T_n) and Simpson's Rule approximations (S_n) of $\ln 2 = \int_1^2 (1/x)\,dx$

n	T_n	|Error| less than . . .	S_n	|Error| less than . . .
10	0.6937714032	0.0006242227	0.6931502307	0.0000030502
20	0.6933033818	0.0001562013	0.6931473747	0.0000001942
30	0.6932166154	0.0000694349	0.6931472190	0.0000000385
40	0.6931862400	0.0000390595	0.6931471927	0.0000000122
50	0.6931721793	0.0000249988	0.6931471856	0.0000000050
100	0.6931534305	0.0000062500	0.6931471809	0.0000000004

This has a dramatic effect as $\Delta x = (2 - 1)/n$ gets very small. The Simpson approximation for $n = 50$ rounds accurately to seven places and for $n = 100$ agrees to nine decimal places (billionths)! ■

If $f(x)$ is a polynomial of degree less than four, then its fourth derivative is zero, and

$$E_S = -\frac{b-a}{180} f^{(4)}(c)(\Delta x)^4 = -\frac{b-a}{180}(0)(\Delta x)^4 = 0.$$

Thus, there will be no error in the Simpson approximation of any integral of f. In other words, if f is a constant, a linear function, or a quadratic or cubic polynomial, Simpson's Rule will give the value of any integral of f exactly, whatever the number of subdivisions. Similarly, if f is a constant or a linear function, then its second derivative is zero and

$$E_T = -\frac{b-a}{12} f''(c)(\Delta x)^2 = -\frac{b-a}{12}(0)(\Delta x)^2 = 0.$$

The Trapezoidal Rule will therefore give the exact value of any integral of f. This is no surprise, for the trapezoids fit the graph perfectly. Although decreasing the step size Δx reduces the error in the Simpson and Trapezoidal approximations in theory, it may fail to do so in practice.

When Δx is very small, say $\Delta x = 10^{-5}$, computer or calculator round-off errors in the arithmetic required to evaluate S and T may accumulate to such an extent that the error formulas no longer describe what is going on. Shrinking Δx below a certain size can actually make things worse. Although this is not an issue in this book, you should consult a text on numerical analysis for alternative methods if you are having problems with round-off.

EXAMPLE 8 Estimate

$$\int_0^2 x^3\,dx$$

with Simpson's Rule.

Solution The fourth derivative of $f(x) = x^3$ is zero, so we expect Simpson's Rule to give the integral's exact value with any (even) number of steps. Indeed, with $n = 2$ and $\Delta x = (2 - 0)/2 = 1$,

$$S = \frac{\Delta x}{3}(y_0 + 4y_1 + y_2)$$
$$= \frac{1}{3}((0)^3 + 4(1)^3 + (2)^3) = \frac{12}{3} = 4,$$

while

$$\int_0^2 x^3\,dx = \frac{x^4}{4}\bigg]_0^2 = \frac{16}{4} - 0 = 4.$$

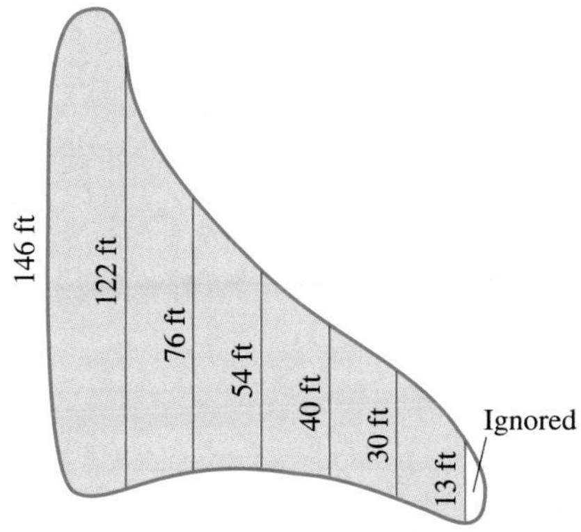

FIGURE 8.16 The dimensions of the swamp in Example 9.

EXAMPLE 9 Draining a Swamp

A town wants to drain and fill a small polluted swamp (Figure 8.16). The swamp averages 5 ft deep. About how many cubic yards of dirt will it take to fill the area after the swamp is drained?

Solution To calculate the volume of the swamp, we estimate the surface area and multiply by 5. To estimate the area, we use Simpson's Rule with $\Delta x = 20$ ft and the y's equal to the distances measured across the swamp, as shown in Figure 8.16.

$$S = \frac{\Delta x}{3}(y_0 + 4y_1 + 2y_2 + 4y_3 + 2y_4 + 4y_5 + y_6)$$
$$= \frac{20}{3}(146 + 488 + 152 + 216 + 80 + 120 + 13) = 8100$$

The volume is about $(8100)(5) = 40{,}500\ \text{ft}^3$ or $1500\ \text{yd}^3$.

EXERCISES 8.7

Estimating Integrals

The instructions for the integrals in Exercises 1–10 have two parts, one for the Trapezoidal Rule and one for Simpson's Rule.

I. Using the Trapezoidal Rule

a. Estimate the integral with $n = 4$ steps and find an upper bound for $|E_T|$.

b. Evaluate the integral directly and find $|E_T|$.

c. Use the formula $(|E_T|/(\text{true value})) \times 100$ to express $|E_T|$ as a percentage of the integral's true value.

II. Using Simpson's Rule

a. Estimate the integral with $n = 4$ steps and find an upper bound for $|E_S|$.

b. Evaluate the integral directly and find $|E_S|$.

c. Use the formula $(|E_S|/(\text{true value})) \times 100$ to express $|E_S|$ as a percentage of the integral's true value.

1. $\int_1^2 x\,dx$

2. $\int_1^3 (2x - 1)\,dx$

3. $\int_{-1}^1 (x^2 + 1)\,dx$

4. $\int_{-2}^0 (x^2 - 1)\,dx$

5. $\int_0^2 (t^3 + t)\,dt$

6. $\int_{-1}^1 (t^3 + 1)\,dt$

7. $\int_1^2 \frac{1}{s^2}\,ds$

8. $\int_2^4 \frac{1}{(s - 1)^2}\,ds$

9. $\int_0^\pi \sin t\,dt$

10. $\int_0^1 \sin \pi t\,dt$

In Exercises 11–14, use the tabulated values of the integrand to estimate the integral with **(a)** the Trapezoidal Rule and **(b)** Simpson's Rule with $n = 8$ steps. Round your answers to five decimal places. Then **(c)** find the integral's exact value and the approximation error E_T or E_S, as appropriate.

11. $\displaystyle\int_0^1 x\sqrt{1 - x^2}\, dx$

x	$x\sqrt{1 - x^2}$
0	0.0
0.125	0.12402
0.25	0.24206
0.375	0.34763
0.5	0.43301
0.625	0.48789
0.75	0.49608
0.875	0.42361
1.0	0

12. $\displaystyle\int_0^3 \frac{\theta}{\sqrt{16 + \theta^2}}\, d\theta$

θ	$\theta/\sqrt{16 + \theta^2}$
0	0.0
0.375	0.09334
0.75	0.18429
1.125	0.27075
1.5	0.35112
1.875	0.42443
2.25	0.49026
2.625	0.58466
3.0	0.6

13. $\displaystyle\int_{-\pi/2}^{\pi/2} \frac{3\cos t}{(2 + \sin t)^2}\, dt$

t	$(3\cos t)/(2 + \sin t)^2$
−1.57080	0.0
−1.17810	0.99138
−0.78540	1.26906
−0.39270	1.05961
0	0.75
0.39270	0.48821
0.78540	0.28946
1.17810	0.13429
1.57080	0

14. $\displaystyle\int_{\pi/4}^{\pi/2} (\csc^2 y)\sqrt{\cot y}\, dy$

y	$(\csc^2 y)\sqrt{\cot y}$
0.78540	2.0
0.88357	1.51606
0.98175	1.18237
1.07992	0.93998
1.17810	0.75402
1.27627	0.60145
1.37445	0.46364
1.47262	0.31688
1.57080	0

The Minimum Number of Subintervals

In Exercises 15–26, estimate the minimum number of subintervals needed to approximate the integrals with an error of magnitude less than 10^{-4} by **(a)** the Trapezoidal Rule and **(b)** Simpson's Rule. (The integrals in Exercises 15–22 are the integrals from Exercises 1–8.)

15. $\displaystyle\int_1^2 x\, dx$

16. $\displaystyle\int_1^3 (2x - 1)\, dx$

17. $\displaystyle\int_{-1}^1 (x^2 + 1)\, dx$

18. $\displaystyle\int_{-2}^0 (x^2 - 1)\, dx$

19. $\displaystyle\int_0^2 (t^3 + t)\, dt$

20. $\displaystyle\int_{-1}^1 (t^3 + 1)\, dt$

21. $\displaystyle\int_1^2 \frac{1}{s^2}\, ds$

22. $\displaystyle\int_2^4 \frac{1}{(s - 1)^2}\, ds$

23. $\displaystyle\int_0^3 \sqrt{x + 1}\, dx$

24. $\displaystyle\int_0^3 \frac{1}{\sqrt{x + 1}}\, dx$

25. $\displaystyle\int_0^2 \sin(x + 1)\, dx$

26. $\displaystyle\int_{-1}^1 \cos(x + \pi)\, dx$

Applications

27. Volume of water in a swimming pool A rectangular swimming pool is 30 ft wide and 50 ft long. The table shows the depth $h(x)$ of the water at 5-ft intervals from one end of the pool to the other. Estimate the volume of water in the pool using the Trapezoidal Rule with $n = 10$, applied to the integral

$$V = \int_0^{50} 30 \cdot h(x)\, dx.$$

Position (ft) x	Depth (ft) $h(x)$	Position (ft) x	Depth (ft) $h(x)$
0	6.0	30	11.5
5	8.2	35	11.9
10	9.1	40	12.3
15	9.9	45	12.7
20	10.5	50	13.0
25	11.0		

28. Stocking a fish pond As the fish and game warden of your township, you are responsible for stocking the town pond with fish before the fishing season. The average depth of the pond is 20 ft. Using a scaled map, you measure distances across the pond at 200-ft intervals, as shown in the accompanying diagram.

a. Use the Trapezoidal Rule to estimate the volume of the pond.

b. You plan to start the season with one fish per 1000 cubic feet. You intend to have at least 25% of the opening day's fish population left at the end of the season. What is the maximum number of licenses the town can sell if the average seasonal catch is 20 fish per license?

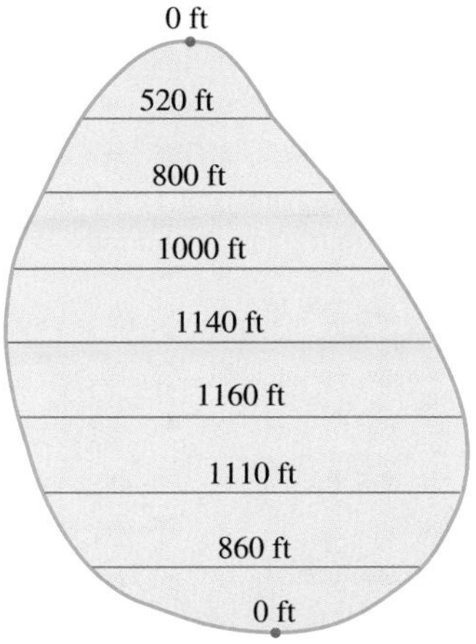

Vertical spacing = 200 ft

29. Ford® Mustang Cobra™ The accompanying table shows time-to-speed data for a 1994 Ford Mustang Cobra accelerating from rest to 130 mph. How far had the Mustang traveled by the time it reached this speed? (Use trapezoids to estimate the area under the velocity curve, but be careful: The time intervals vary in length.)

Speed change	**Time (sec)**
Zero to 30 mph	2.2
40 mph	3.2
50 mph	4.5
60 mph	5.9
70 mph	7.8
80 mph	10.2
90 mph	12.7
100 mph	16.0
110 mph	20.6
120 mph	26.2
130 mph	37.1

Source: Car and Driver, April 1994.

30. Aerodynamic drag A vehicle's aerodynamic drag is determined in part by its cross-sectional area, so, all other things being equal, engineers try to make this area as small as possible. Use Simpson's Rule to estimate the cross-sectional area of the body of James Worden's solar-powered Solectria® automobile at MIT from the diagram.

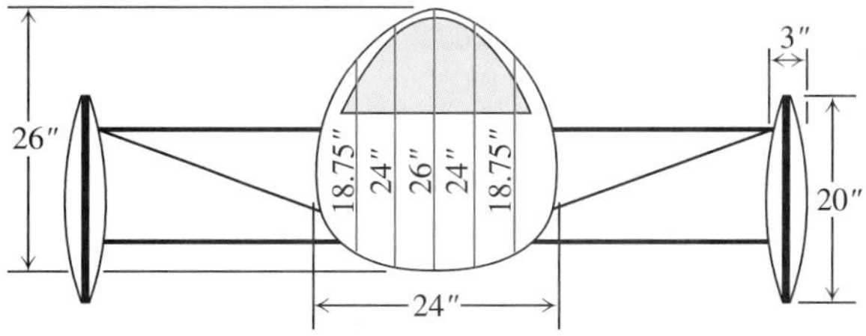

31. Wing design The design of a new airplane requires a gasoline tank of constant cross-sectional area in each wing. A scale drawing of a cross-section is shown here. The tank must hold 5000 lb of gasoline, which has a density of 42 lb/ft^3. Estimate the length of the tank.

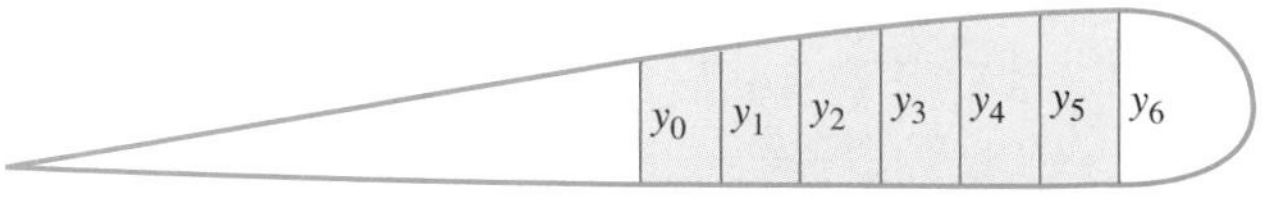

$y_0 = 1.5$ ft, $y_1 = 1.6$ ft, $y_2 = 1.8$ ft, $y_3 = 1.9$ ft,
$y_4 = 2.0$ ft, $y_5 = y_6 = 2.1$ ft Horizontal spacing = 1 ft

32. Oil consumption on Pathfinder Island A diesel generator runs continuously, consuming oil at a gradually increasing rate until it must be temporarily shut down to have the filters replaced.

Use the Trapezoidal Rule to estimate the amount of oil consumed by the generator during that week.

Day	Oil consumption rate (liters/h)
Sun	0.019
Mon	0.020
Tue	0.021
Wed	0.023
Thu	0.025
Fri	0.028
Sat	0.031
Sun	0.035

Theory and Examples

33. Usable values of the sine-integral function *The sine-integral function,*

$$\mathrm{Si}(x) = \int_0^x \frac{\sin t}{t}\,dt, \qquad \text{“Sine integral of } x\text{”}$$

is one of the many functions in engineering whose formulas cannot be simplified. There is no elementary formula for the antiderivative of $(\sin t)/t$. The values of $\mathrm{Si}(x)$, however, are readily estimated by numerical integration.

Although the notation does not show it explicitly, the function being integrated is

$$f(t) = \begin{cases} \dfrac{\sin t}{t}, & t \neq 0 \\ 1, & t = 0, \end{cases}$$

the continuous extension of $(\sin t)/t$ to the interval $[0, x]$. The function has derivatives of all orders at every point of its domain. Its graph is smooth, and you can expect good results from Simpson's Rule.

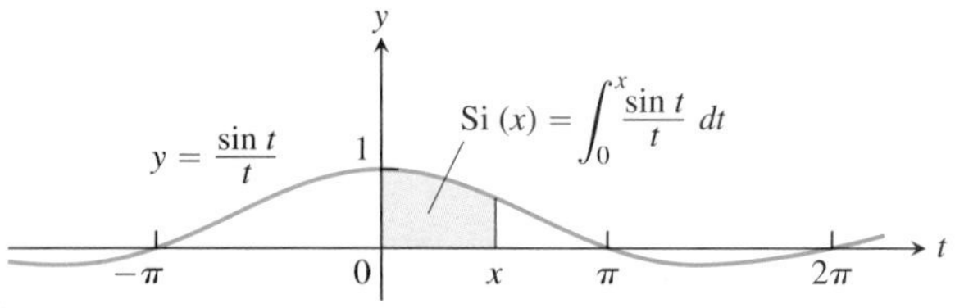

a. Use the fact that $|f^{(4)}| \leq 1$ on $[0, \pi/2]$ to give an upper bound for the error that will occur if

$$\mathrm{Si}\left(\frac{\pi}{2}\right) = \int_0^{\pi/2} \frac{\sin t}{t}\,dt$$

is estimated by Simpson's Rule with $n = 4$.

b. Estimate $\mathrm{Si}(\pi/2)$ by Simpson's Rule with $n = 4$.

c. Express the error bound you found in part (a) as a percentage of the value you found in part (b).

34. The error function *The error function,*

$$\mathrm{erf}(x) = \frac{2}{\sqrt{\pi}} \int_0^x e^{-t^2}\,dt,$$

important in probability and in the theories of heat flow and signal transmission, must be evaluated numerically because there is no elementary expression for the antiderivative of e^{-t^2}.

a. Use Simpson's Rule with $n = 10$ to estimate $\mathrm{erf}(1)$.

b. In $[0, 1]$,

$$\left| \frac{d^4}{dt^4}\left(e^{-t^2}\right) \right| \leq 12.$$

Give an upper bound for the magnitude of the error of the estimate in part (a).

35. (*Continuation of Example 3.*) The error bounds for E_T and E_S are "worst case" estimates, and the Trapezoidal and Simpson Rules are often more accurate than the bounds suggest. The Trapezoidal Rule estimate of

$$\int_0^{\pi} x \sin x\,dx$$

in Example 3 is a case in point.

a. Use the Trapezoidal Rule with $n = 10$ to approximate the value of the integral. The table to the right gives the necessary y-values.

x	$x \sin x$
0	0
$(0.1)\pi$	0.09708
$(0.2)\pi$	0.36932
$(0.3)\pi$	0.76248
$(0.4)\pi$	1.19513
$(0.5)\pi$	1.57080
$(0.6)\pi$	1.79270
$(0.7)\pi$	1.77912
$(0.8)\pi$	1.47727
$(0.9)\pi$	0.87372
π	0

b. Find the magnitude of the difference between π, the integral's value, and your approximation in part (a). You will find the difference to be considerably less than the upper bound of 0.133 calculated with $n = 10$ in Example 3.

T **c.** The upper bound of 0.133 for $|E_T|$ in Example 3 could have been improved somewhat by having a better bound for

$$|f''(x)| = |2 \cos x - x \sin x|$$

on $[0, \pi]$. The upper bound we used was $2 + \pi$. Graph f'' over $[0, \pi]$ and use Trace or Zoom to improve this upper bound.

Use the improved upper bound as M to make an improved estimate of $|E_T|$. Notice that the Trapezoidal Rule approximation in part (a) is also better than this improved estimate would suggest.

T **36.** (*Continuation of Exercise 35.*)

a. Show that the fourth derivative of $f(x) = x \sin x$ is

$$f^{(4)}(x) = -4 \cos x + x \sin x.$$

Use Trace or Zoom to find an upper bound M for the values of $|f^{(4)}|$ on $[0, \pi]$.

b. Use the value of M from part (a) to obtain an upper bound for the magnitude of the error in estimating the value of

$$\int_0^{\pi} x \sin x \, dx$$

with Simpson's Rule with $n = 10$ steps.

c. Use the data in the table in Exercise 35 to estimate $\int_0^{\pi} x \sin x \, dx$ with Simpson's Rule with $n = 10$ steps.

d. To six decimal places, find the magnitude of the difference between your estimate in part (c) and the integral's true value, π. You will find the error estimate obtained in part (b) to be quite good.

37. Prove that the sum T in the Trapezoidal Rule for $\int_a^b f(x)\,dx$ is a Riemann sum for f continuous on $[a, b]$. (*Hint:* Use the Intermediate Value Theorem to show the existence of c_k in the subinterval $[x_{k-1}, x_k]$ satisfying $f(c_k) = (f(x_{k-1}) + f(x_k))/2$.)

38. Prove that the sum S in Simpson's Rule for $\int_a^b f(x)\,dx$ is a Riemann sum for f continuous on $[a, b]$. (See Exercise 37.)

Numerical Integration

As we mentioned at the beginning of the section, the definite integrals of many continuous functions cannot be evaluated with the Fundamental Theorem of Calculus because their antiderivatives lack elementary formulas. Numerical integration offers a practical way to estimate the values of these so-called *nonelementary integrals*. If your calculator or computer has a numerical integration routine, try it on the integrals in Exercises 39–42.

39. $\displaystyle\int_0^1 \sqrt{1 + x^4}\, dx$ — A nonelementary integral that came up in Newton's research

40. $\displaystyle\int_0^{\pi/2} \frac{\sin x}{x}\, dx$ — The integral from Exercise 33. To avoid division by zero, you may have to start the integration at a small positive number like 10^{-6} instead of 0.

41. $\displaystyle\int_0^{\pi/2} \sin(x^2)\, dx$ — An integral associated with the diffraction of light

42. $\displaystyle\int_0^{\pi/2} 40\sqrt{1 - 0.64\cos^2 t}\, dt$ — The length of the ellipse $(x^2/25) + (y^2/9) = 1$

43. Consider the integral $\int_0^{\pi} \sin x\, dx$.

a. Find the Trapezoidal Rule approximations for $n = 10$, 100, and 1000.

b. Record the errors with as many decimal places of accuracy as you can.

c. What pattern do you see?

d. Explain how the error bound for E_T accounts for the pattern.

44. (*Continuation of Exercise 43.*) Repeat Exercise 43 with Simpson's Rule and E_S.

45. Consider the integral $\int_{-1}^{1} \sin(x^2)\, dx$.

a. Find f'' for $f(x) = \sin(x^2)$.

b. Graph $y = f''(x)$ in the viewing window $[-1, 1]$ by $[-3, 3]$.

c. Explain why the graph in part (b) suggests that $|f''(x)| \le 3$ for $-1 \le x \le 1$.

d. Show that the error estimate for the Trapezoidal Rule in this case becomes

$$|E_T| \le \frac{(\Delta x)^2}{2}.$$

e. Show that the Trapezoidal Rule error will be less than or equal to 0.01 in magnitude if $\Delta x \le 0.1$.

f. How large must n be for $\Delta x \le 0.1$?

46. Consider the integral $\int_{-1}^{1} \sin(x^2)\, dx$.

a. Find $f^{(4)}$ for $f(x) = \sin(x^2)$. (You may want to check your work with a CAS if you have one available.)

b. Graph $y = f^{(4)}(x)$ in the viewing window $[-1, 1]$ by $[-30, 10]$.

c. Explain why the graph in part (b) suggests that $|f^{(4)}(x)| \le 30$ for $-1 \le x \le 1$.

d. Show that the error estimate for Simpson's Rule in this case becomes

$$|E_S| \le \frac{(\Delta x)^4}{3}.$$

e. Show that the Simpson's Rule error will be less than or equal to 0.01 in magnitude if $\Delta x \le 0.4$.

f. How large must n be for $\Delta x \le 0.4$?

47. A vase We wish to estimate the volume of a flower vase using only a calculator, a string, and a ruler. We measure the height of the vase to be 6 in. We then use the string and the ruler to find circumferences of the vase (in inches) at half-inch intervals. (We list them from the top down to correspond with the picture of the vase.)

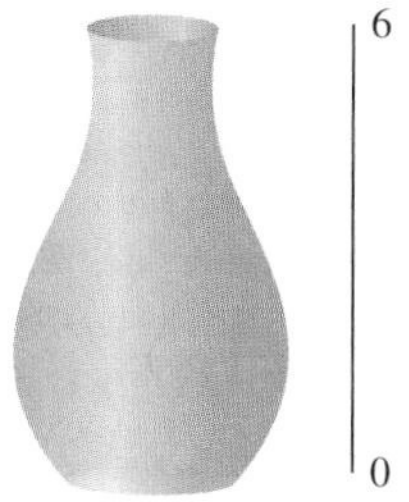

Circumferences	
5.4	10.8
4.5	11.6
4.4	11.6
5.1	10.8
6.3	9.0
7.8	6.3
9.4	

a. Find the areas of the cross-sections that correspond to the given circumferences.

b. Express the volume of the vase as an integral with respect to y over the interval $[0, 6]$.

c. Approximate the integral using the Trapezoidal Rule with $n = 12$.

d. Approximate the integral using Simpson's Rule with $n = 12$. Which result do you think is more accurate? Give reasons for your answer.

T **48. A sailboat's displacement** To find the volume of water displaced by a sailboat, the common practice is to partition the waterline into 10 subintervals of equal length, measure the cross-sectional area $A(x)$ of the submerged portion of the hull at each partition point, and then use Simpson's Rule to estimate the integral of $A(x)$ from one end of the waterline to the other. The table here lists the area measurements at "Stations" 0 through 10, as the partition points are called, for the cruising sloop *Pipedream*, shown here. The common subinterval length (distance between consecutive stations) is $\Delta x = 2.54$ ft (about 2 ft 6-1/2 in., chosen for the convenience of the builder).

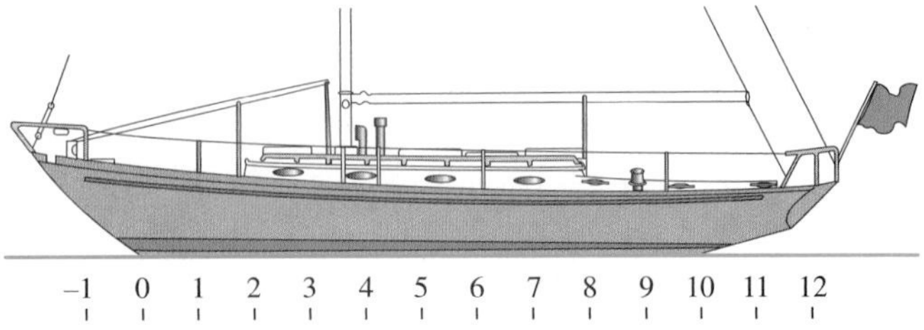

a. Estimate *Pipedream*'s displacement volume to the nearest cubic foot.

Station	Submerged area (ft^2)
0	0
1	1.07
2	3.84
3	7.82
4	12.20
5	15.18
6	16.14
7	14.00
8	9.21
9	3.24
10	0

b. The figures in the table are for seawater, which weighs $64\ \text{lb/ft}^3$. How many pounds of water does *Pipedream* displace? (Displacement is given in pounds for small craft and in long tons (1 long ton $=$ 2240 lb) for larger vessels.) (Data from *Skene's Elements of Yacht Design* by Francis S. Kinney (Dodd, Mead, 1962.)

c. Prismatic coefficients A boat's prismatic coefficient is the ratio of the displacement volume to the volume of a prism whose height equals the boat's waterline length and whose base equals the area of the boat's largest submerged cross-section. The best sailboats have prismatic coefficients between 0.51 and 0.54. Find *Pipedream*'s prismatic coefficient, given a waterline length of 25.4 ft and a largest submerged cross-sectional area of 16.14 ft^2 (at Station 6).

T **49. Elliptic integrals** The length of the ellipse

$$x = a\cos t, \quad y = b\sin t, \quad 0 \le t \le 2\pi$$

turns out to be

$$\text{Length} = 4a\int_0^{\pi/2}\sqrt{1 - e^2\cos^2 t}\,dt,$$

where e is the ellipse's eccentricity. The integral in this formula, called an *elliptic integral*, is nonelementary except when $e = 0$ or 1.

a. Use the Trapezoidal Rule with $n = 10$ to estimate the length of the ellipse when $a = 1$ and $e = 1/2$.

b. Use the fact that the absolute value of the second derivative of $f(t) = \sqrt{1 - e^2\cos^2 t}$ is less than 1 to find an upper bound for the error in the estimate you obtained in part (a).

T **50.** The length of one arch of the curve $y = \sin x$ is given by

$$L = \int_0^{\pi}\sqrt{1 + \cos^2 x}\,dx.$$

Estimate L by Simpson's Rule with $n = 8$.

T **51.** Your metal fabrication company is bidding for a contract to make sheets of corrugated iron roofing like the one shown here. The cross-sections of the corrugated sheets are to conform to the curve

$$y = \sin\frac{3\pi}{20}x, \quad 0 \le x \le 20\text{ in}.$$

If the roofing is to be stamped from flat sheets by a process that does not stretch the material, how wide should the original material be? To find out, use numerical integration to approximate the length of the sine curve to two decimal places.

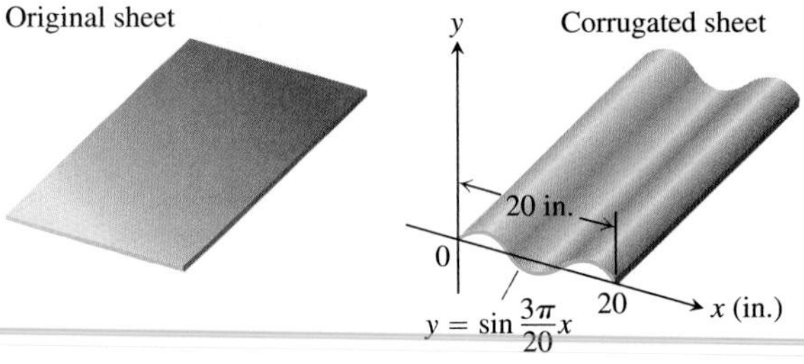

T **52.** Your engineering firm is bidding for the contract to construct the tunnel shown here. The tunnel is 300 ft long and 50 ft wide at the base. The cross-section is shaped like one arch of the curve $y = 25\cos(\pi x/50)$. Upon completion, the tunnel's inside surface (excluding the roadway) will be treated with a waterproof sealer that costs \$1.75 per square foot to apply. How much will it cost to apply the sealer? (*Hint:* Use numerical integration to find the length of the cosine curve.)

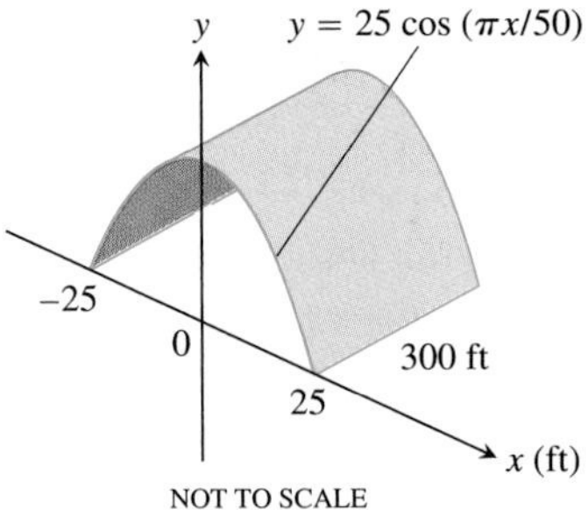

Surface Area

Find, to two decimal places, the areas of the surfaces generated by revolving the curves in Exercises 53–56 about the x-axis.

53. $y = \sin x, \quad 0 \le x \le \pi$

54. $y = x^2/4, \quad 0 \le x \le 2$

55. $y = x + \sin 2x, \quad -2\pi/3 \le x \le 2\pi/3$ (the curve in Section 4.4, Exercise 5)

56. $y = \frac{x}{12}\sqrt{36 - x^2}, 0 \le x \le 6$ (the surface of the plumb bob in Section 6.1, Exercise 56)

Estimating Function Values

57. Use numerical integration to estimate the value of

$$\sin^{-1} 0.6 = \int_0^{0.6} \frac{dx}{\sqrt{1 - x^2}}.$$

For reference, $\sin^{-1} 0.6 = 0.64350$ to five decimal places.

58. Use numerical integration to estimate the value of

$$\pi = 4\int_0^1 \frac{1}{1 + x^2}\,dx.$$

8.8 Improper Integrals

Up to now, definite integrals have been required to have two properties. First, that the domain of integration $[a, b]$ be finite. Second, that the range of the integrand be finite on this domain. In practice, we may encounter problems that fail to meet one or both of these conditions. The integral for the area under the curve $y = (\ln x)/x^2$ from $x = 1$ to $x = \infty$ is an example for which the domain is infinite (Figure 8.17a). The integral for the area under the curve of $y = 1/\sqrt{x}$ between $x = 0$ and $x = 1$ is an example for which the range of the integrand is infinite (Figure 8.17b). In either case, the integrals are said to be *improper* and are calculated as limits. We will see that improper integrals play an important role when investigating the convergence of certain infinite series in Chapter 11.

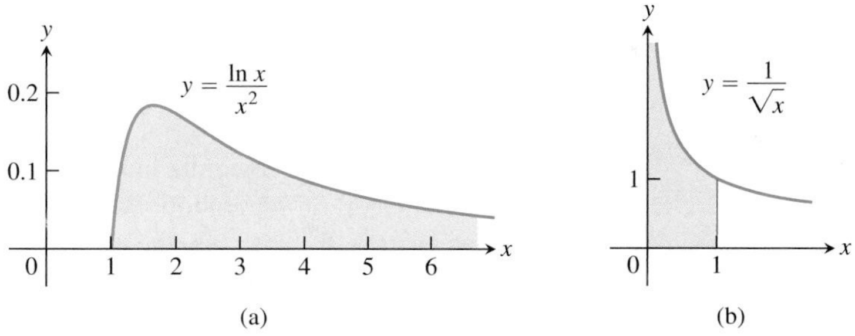

FIGURE 8.17 Are the areas under these infinite curves finite?

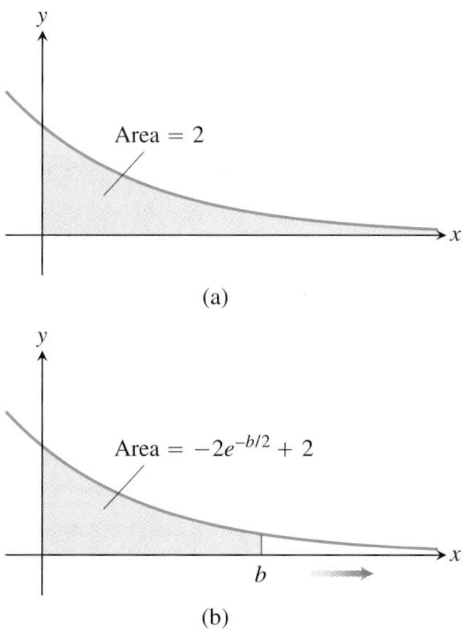

FIGURE 8.18 (a) The area in the first quadrant under the curve $y = e^{-x/2}$ is (b) an improper integral of the first type.

Infinite Limits of Integration

Consider the infinite region that lies under the curve $y = e^{-x/2}$ in the first quadrant (Figure 8.18a). You might think this region has infinite area, but we will see that the natural value to assign is finite. Here is how to assign a value to the area. First find the area $A(b)$ of the portion of the region that is bounded on the right by $x = b$ (Figure 8.18b).

$$A(b) = \int_0^b e^{-x/2}\,dx = -2e^{-x/2}\Big]_0^b = -2e^{-b/2} + 2$$

Then find the limit of $A(b)$ as $b \to \infty$

$$\lim_{b\to\infty} A(b) = \lim_{b\to\infty} (-2e^{-b/2} + 2) = 2.$$

The value we assign to the area under the curve from 0 to ∞ is

$$\int_0^\infty e^{-x/2}\,dx = \lim_{b\to\infty}\int_0^b e^{-x/2}\,dx = 2.$$

DEFINITION Type I Improper Integrals

Integrals with infinite limits of integration are **improper integrals of Type I**.

1. If $f(x)$ is continuous on $[a, \infty)$, then

$$\int_a^\infty f(x)\,dx = \lim_{b\to\infty}\int_a^b f(x)\,dx.$$

2. If $f(x)$ is continuous on $(-\infty, b]$, then

$$\int_{-\infty}^b f(x)\,dx = \lim_{a\to-\infty}\int_a^b f(x)\,dx.$$

3. If $f(x)$ is continuous on $(-\infty, \infty)$, then

$$\int_{-\infty}^\infty f(x)\,dx = \int_{-\infty}^c f(x)\,dx + \int_c^\infty f(x)\,dx,$$

where c is any real number.

In each case, if the limit is finite we say that the improper integral **converges** and that the limit is the **value** of the improper integral. If the limit fails to exist, the improper integral **diverges**.

It can be shown that the choice of c in Part 3 of the definition is unimportant. We can evaluate or determine the convergence or divergence of $\int_{-\infty}^{\infty} f(x)\,dx$ with any convenient choice.

Any of the integrals in the above definition can be interpreted as an area if $f \ge 0$ on the interval of integration. For instance, we interpreted the improper integral in Figure 8.18 as an area. In that case, the area has the finite value 2. If $f \ge 0$ and the improper integral diverges, we say the area under the curve is **infinite**.

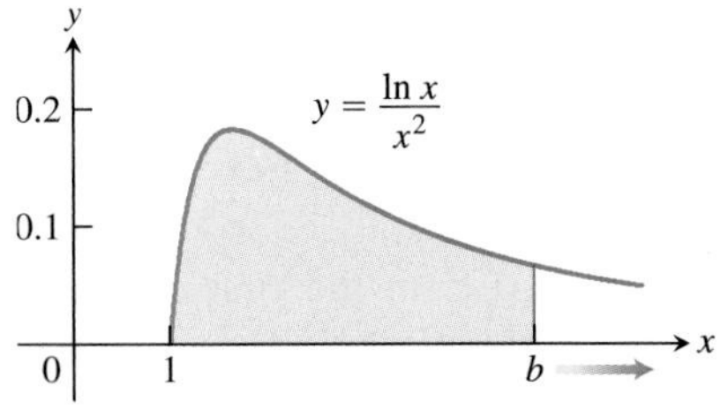

FIGURE 8.19 The area under this curve is an improper integral (Example 1).

EXAMPLE 1 Evaluating an Improper Integral on $[1, \infty)$

Is the area under the curve $y = (\ln x)/x^2$ from $x = 1$ to $x = \infty$ finite? If so, what is it?

Solution We find the area under the curve from $x = 1$ to $x = b$ and examine the limit as $b \to \infty$. If the limit is finite, we take it to be the area under the curve (Figure 8.19). The area from 1 to b is

$$\int_1^b \frac{\ln x}{x^2}\,dx = \left[(\ln x)\left(-\frac{1}{x}\right)\right]_1^b - \int_1^b \left(-\frac{1}{x}\right)\left(\frac{1}{x}\right)dx$$

Integration by parts with $u = \ln x$, $dv = dx/x^2$, $du = dx/x$, $v = -1/x$.

$$= -\frac{\ln b}{b} - \left[\frac{1}{x}\right]_1^b$$

$$= -\frac{\ln b}{b} - \frac{1}{b} + 1.$$

The limit of the area as $b \to \infty$ is

$$\begin{aligned}\int_1^\infty \frac{\ln x}{x^2}\,dx &= \lim_{b\to\infty}\int_1^b \frac{\ln x}{x^2}\,dx \\ &= \lim_{b\to\infty}\left[-\frac{\ln b}{b} - \frac{1}{b} + 1\right] \\ &= -\left[\lim_{b\to\infty}\frac{\ln b}{b}\right] - 0 + 1 \\ &= -\left[\lim_{b\to\infty}\frac{1/b}{1}\right] + 1 = 0 + 1 = 1. \qquad \text{l'Hôpital's Rule}\end{aligned}$$

Thus, the improper integral converges and the area has finite value 1. ■

EXAMPLE 2 Evaluating an Integral on $(-\infty, \infty)$

Evaluate

$$\int_{-\infty}^{\infty}\frac{dx}{1+x^2}.$$

Solution According to the definition (Part 3), we can write

$$\int_{-\infty}^{\infty}\frac{dx}{1+x^2} = \int_{-\infty}^{0}\frac{dx}{1+x^2} + \int_{0}^{\infty}\frac{dx}{1+x^2}.$$

HISTORICAL BIOGRAPHY

Lejeune Dirichlet (1805–1859)

Next we evaluate each improper integral on the right side of the equation above.

$$\begin{aligned}\int_{-\infty}^{0}\frac{dx}{1+x^2} &= \lim_{a\to-\infty}\int_a^0 \frac{dx}{1+x^2} \\ &= \lim_{a\to-\infty}\tan^{-1}x\Big]_a^0 \\ &= \lim_{a\to-\infty}(\tan^{-1}0 - \tan^{-1}a) = 0 - \left(-\frac{\pi}{2}\right) = \frac{\pi}{2}\end{aligned}$$

$$\begin{aligned}\int_{0}^{\infty}\frac{dx}{1+x^2} &= \lim_{b\to\infty}\int_0^b \frac{dx}{1+x^2} \\ &= \lim_{b\to\infty}\tan^{-1}x\Big]_0^b \\ &= \lim_{b\to\infty}(\tan^{-1}b - \tan^{-1}0) = \frac{\pi}{2} - 0 = \frac{\pi}{2}\end{aligned}$$

Thus,

$$\int_{-\infty}^{\infty}\frac{dx}{1+x^2} = \frac{\pi}{2} + \frac{\pi}{2} = \pi.$$

Since $1/(1+x^2) > 0$, the improper integral can be interpreted as the (finite) area beneath the curve and above the x-axis (Figure 8.20). ■

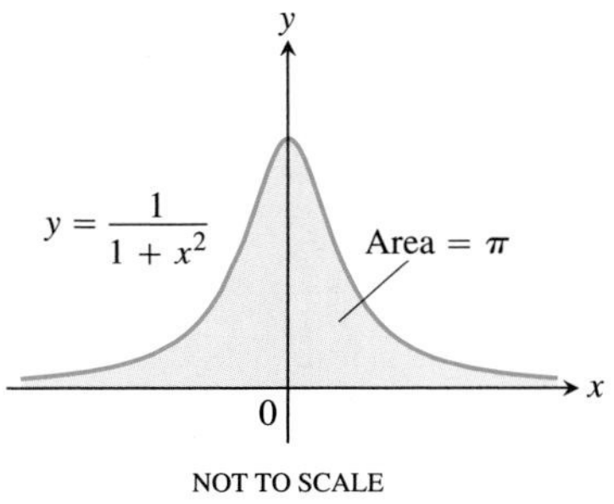

FIGURE 8.20 The area under this curve is finite (Example 2).

The Integral $\int_1^\infty \frac{dx}{x^p}$

The function $y = 1/x$ is the boundary between the convergent and divergent improper integrals with integrands of the form $y = 1/x^p$. As the next example shows, the improper integral converges if $p > 1$ and diverges if $p \leq 1$.

EXAMPLE 3 Determining Convergence

For what values of p does the integral $\int_1^\infty dx/x^p$ converge? When the integral does converge, what is its value?

Solution If $p \neq 1$,

$$\int_1^b \frac{dx}{x^p} = \frac{x^{-p+1}}{-p+1}\bigg]_1^b = \frac{1}{1-p}\left(b^{-p+1} - 1\right) = \frac{1}{1-p}\left(\frac{1}{b^{p-1}} - 1\right).$$

Thus,

$$\int_1^\infty \frac{dx}{x^p} = \lim_{b\to\infty} \int_1^b \frac{dx}{x^p}$$

$$= \lim_{b\to\infty}\left[\frac{1}{1-p}\left(\frac{1}{b^{p-1}} - 1\right)\right] = \begin{cases} \dfrac{1}{p-1}, & p > 1 \\ \infty, & p < 1 \end{cases}$$

because

$$\lim_{b\to\infty} \frac{1}{b^{p-1}} = \begin{cases} 0, & p > 1 \\ \infty, & p < 1. \end{cases}$$

Therefore, the integral converges to the value $1/(p-1)$ if $p > 1$ and it diverges if $p < 1$.

If $p = 1$, the integral also diverges:

$$\int_1^\infty \frac{dx}{x^p} = \int_1^\infty \frac{dx}{x}$$

$$= \lim_{b\to\infty} \int_1^b \frac{dx}{x}$$

$$= \lim_{b\to\infty} \ln x\bigg]_1^b$$

$$= \lim_{b\to\infty} (\ln b - \ln 1) = \infty.$$

■

Integrands with Vertical Asymptotes

Another type of improper integral arises when the integrand has a vertical asymptote—an infinite discontinuity—at a limit of integration or at some point between the limits of integration. If the integrand f is positive over the interval of integration, we can again interpret the improper integral as the area under the graph of f and above the x-axis between the limits of integration.

Consider the region in the first quadrant that lies under the curve $y = 1/\sqrt{x}$ from $x = 0$ to $x = 1$ (Figure 8.17b). First we find the area of the portion from a to 1 (Figure 8.21).

$$\int_a^1 \frac{dx}{\sqrt{x}} = 2\sqrt{x}\Big]_a^1 = 2 - 2\sqrt{a}$$

Then we find the limit of this area as $a \to 0^+$:

$$\lim_{a\to 0^+} \int_a^1 \frac{dx}{\sqrt{x}} = \lim_{a\to 0^+} \left(2 - 2\sqrt{a}\right) = 2.$$

The area under the curve from 0 to 1 is finite and equals

$$\int_0^1 \frac{dx}{\sqrt{x}} = \lim_{a\to 0^+} \int_a^1 \frac{dx}{\sqrt{x}} = 2.$$

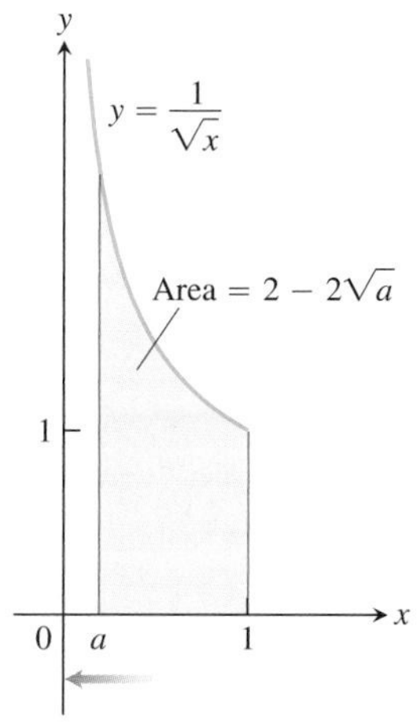

FIGURE 8.21 The area under this curve is

$$\lim_{a\to 0^+} \int_a^1 \left(\frac{1}{\sqrt{x}}\right) dx = 2,$$

an improper integral of the second kind.

DEFINITION Type II Improper Integrals

Integrals of functions that become infinite at a point within the interval of integration are **improper integrals of Type II**.

1. If $f(x)$ is continuous on $(a, b]$ and is discontinuous at a then
$$\int_a^b f(x)\,dx = \lim_{c\to a^+} \int_c^b f(x)\,dx.$$

2. If $f(x)$ is continuous on $[a, b)$ and is discontinuous at b, then
$$\int_a^b f(x)\,dx = \lim_{c\to b^-} \int_a^c f(x)\,dx.$$

3. If $f(x)$ is discontinuous at c, where $a < c < b$, and continuous on $[a, c) \cup (c, b]$, then
$$\int_a^b f(x)\,dx = \int_a^c f(x)\,dx + \int_c^b f(x)\,dx.$$

In each case, if the limit is finite we say the improper integral **converges** and that the limit is the **value** of the improper integral. If the limit does not exist, the integral **diverges**.

In Part 3 of the definition, the integral on the left side of the equation converges if *both* integrals on the right side converge; otherwise it diverges.

EXAMPLE 4 A Divergent Improper Integral

Investigate the convergence of

$$\int_0^1 \frac{1}{1 - x}\,dx.$$

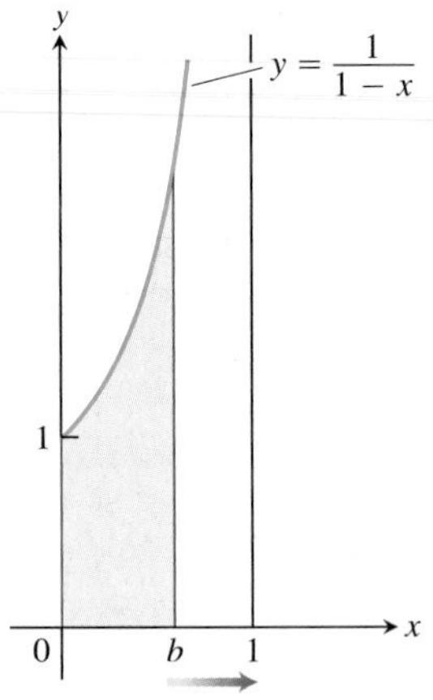

FIGURE 8.22 The limit does not exist:

$$\int_0^1 \left(\frac{1}{1-x}\right) dx = \lim_{b\to 1^-} \int_0^b \frac{1}{1-x}\, dx = \infty.$$

The area beneath the curve and above the x-axis for $[0, 1)$ is not a real number (Example 4).

Solution The integrand $f(x) = 1/(1 - x)$ is continuous on $[0, 1)$ but is discontinuous at $x = 1$ and becomes infinite as $x \to 1^-$ (Figure 8.22). We evaluate the integral as

$$\lim_{b\to 1^-} \int_0^b \frac{1}{1-x}\, dx = \lim_{b\to 1^-} \Big[-\ln|1 - x|\Big]_0^b$$

$$= \lim_{b\to 1^-} [-\ln(1 - b) + 0] = \infty.$$

The limit is infinite, so the integral diverges. ■

EXAMPLE 5 Vertical Asympote at an Interior Point

Evaluate

$$\int_0^3 \frac{dx}{(x-1)^{2/3}}.$$

Solution The integrand has a vertical asymptote at $x = 1$ and is continuous on $[0, 1)$ and $(1, 3]$ (Figure 8.23). Thus, by Part 3 of the definition above,

$$\int_0^3 \frac{dx}{(x-1)^{2/3}} = \int_0^1 \frac{dx}{(x-1)^{2/3}} + \int_1^3 \frac{dx}{(x-1)^{2/3}}.$$

Next, we evaluate each improper integral on the right-hand side of this equation.

$$\int_0^1 \frac{dx}{(x-1)^{2/3}} = \lim_{b\to 1^-} \int_0^b \frac{dx}{(x-1)^{2/3}}$$

$$= \lim_{b\to 1^-} 3(x-1)^{1/3}\Big]_0^b$$

$$= \lim_{b\to 1^-} [3(b-1)^{1/3} + 3] = 3$$

$$\int_1^3 \frac{dx}{(x-1)^{2/3}} = \lim_{c\to 1^+} \int_c^3 \frac{dx}{(x-1)^{2/3}}$$

$$= \lim_{c\to 1^+} 3(x-1)^{1/3}\Big]_c^3$$

$$= \lim_{c\to 1^+} \left[3(3-1)^{1/3} - 3(c-1)^{1/3}\right] = 3\sqrt[3]{2}$$

We conclude that

$$\int_0^3 \frac{dx}{(x-1)^{2/3}} = 3 + 3\sqrt[3]{2}.$$ ■

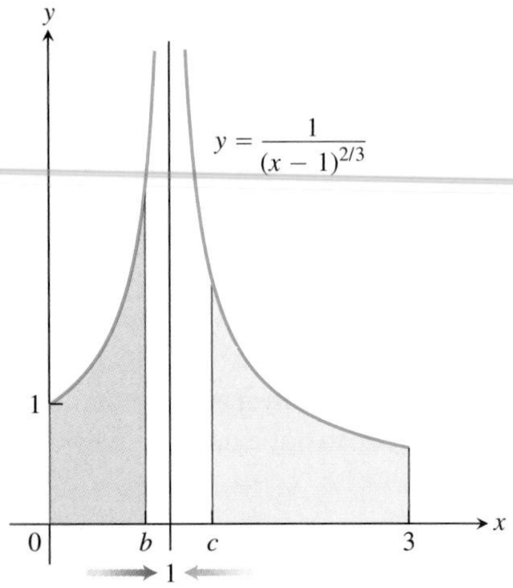

FIGURE 8.23 Example 5 shows the convergence of

$$\int_0^3 \frac{1}{(x-1)^{2/3}}\, dx = 3 + 3\sqrt[3]{2},$$

so the area under the curve exists (so it is a real number).

EXAMPLE 6 A Convergent Improper Integral

Evaluate

$$\int_2^\infty \frac{x+3}{(x-1)(x^2+1)}\, dx.$$

Solution

$$
\begin{aligned}
\int_2^{\infty} \frac{x+3}{(x-1)(x^2+1)}\,dx &= \lim_{b\to\infty}\int_2^{b} \frac{x+3}{(x-1)(x^2+1)}\,dx \\
&= \lim_{b\to\infty}\int_2^{b} \left(\frac{2}{x-1} - \frac{2x+1}{x^2+1}\right) dx \qquad \text{Partial fractions} \\
&= \lim_{b\to\infty}\left[2\ln(x-1) - \ln(x^2+1) - \tan^{-1}x\right]_2^b \\
&= \lim_{b\to\infty}\left[\ln\frac{(x-1)^2}{x^2+1} - \tan^{-1}x\right]_2^b \qquad \text{Combine the logarithms.} \\
&= \lim_{b\to\infty}\left[\ln\left(\frac{(b-1)^2}{b^2+1}\right) - \tan^{-1}b\right] - \ln\left(\frac{1}{5}\right) + \tan^{-1}2 \\
&= 0 - \frac{\pi}{2} + \ln 5 + \tan^{-1}2 \approx 1.1458
\end{aligned}
$$

Notice that we combined the logarithms in the antiderivative *before* we calculated the limit as $b \to \infty$. Had we not done so, we would have encountered the indeterminate form

$$\lim_{b\to\infty}(2\ln(b-1) - \ln(b^2+1)) = \infty - \infty.$$

The way to evaluate the indeterminate form, of course, is to combine the logarithms, so we would have arrived at the same answer in the end. ■

Computer algebra systems can evaluate many convergent improper integrals. To evaluate the integral in Example 6 using Maple, enter

> *f*:= (*x* + 3)/((*x* − 1) * (*x*^2 + 1));

Then use the integration command

> int(*f*, *x* = 2..infinity);

Maple returns the answer

$$-\frac{1}{2}\pi + \ln(5) + \arctan(2).$$

To obtain a numerical result, use the evaluation command **evalf** and specify the number of digits, as follows:

> evalf(%, 6);

The symbol % instructs the computer to evaluate the last expression on the screen, in this case $(-1/2)\pi + \ln(5) + \arctan(2)$. Maple returns 1.14579.

Using Mathematica, entering

In [1]:= Integrate [(*x* + 3)/((*x* − 1)(*x*^2 + 1)), {*x*, 2, Infinity}]

returns

Out [1]= $\frac{-\text{Pi}}{2}$ + ArcTan [2] + Log [5].

To obtain a numerical result with six digits, use the command "N[%, 6]"; it also yields 1.14579.

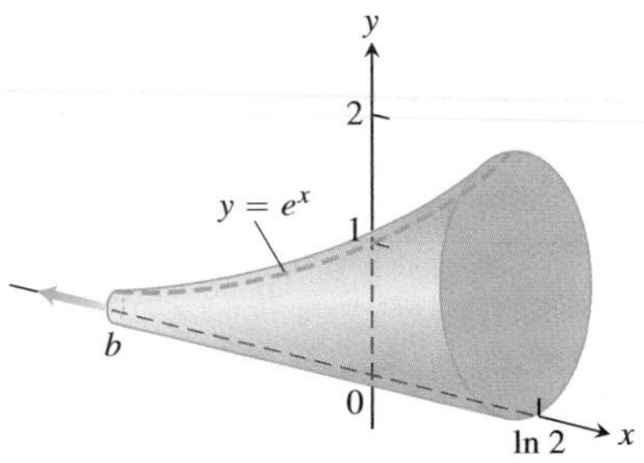

FIGURE 8.24 The calculation in Example 7 shows that this infinite horn has a finite volume.

EXAMPLE 7 Finding the Volume of an Infinite Solid

The cross-sections of the solid horn in Figure 8.24 perpendicular to the x-axis are circular disks with diameters reaching from the x-axis to the curve $y = e^x$, $-\infty < x \leq \ln 2$. Find the volume of the horn.

Solution The area of a typical cross-section is

$$A(x) = \pi(\text{radius})^2 = \pi\left(\frac{1}{2}y\right)^2 = \frac{\pi}{4}e^{2x}.$$

We define the volume of the horn to be the limit as $b \to -\infty$ of the volume of the portion from b to $\ln 2$. As in Section 6.1 (the method of slicing), the volume of this portion is

$$V = \int_b^{\ln 2} A(x)\,dx = \int_b^{\ln 2} \frac{\pi}{4}e^{2x}\,dx = \frac{\pi}{8}e^{2x}\Big]_b^{\ln 2}$$

$$= \frac{\pi}{8}\left(e^{\ln 4} - e^{2b}\right) = \frac{\pi}{8}\left(4 - e^{2b}\right).$$

As $b \to -\infty$, $e^{2b} \to 0$ and $V \to (\pi/8)(4 - 0) = \pi/2$. The volume of the horn is $\pi/2$. ∎

EXAMPLE 8 An Incorrect Calculation

Evaluate

$$\int_0^3 \frac{dx}{x-1}.$$

Solution Suppose we fail to notice the discontinuity of the integrand at $x = 1$, interior to the interval of integration. If we evaluate the integral as an ordinary integral we get

$$\int_0^3 \frac{dx}{x-1} = \ln|x-1|\Big]_0^3 = \ln 2 - \ln 1 = \ln 2.$$

This result is *wrong* because the integral is improper. The correct evaluation uses limits:

$$\int_0^3 \frac{dx}{x-1} = \int_0^1 \frac{dx}{x-1} + \int_1^3 \frac{dx}{x-1}$$

where

$$\int_0^1 \frac{dx}{x-1} = \lim_{b\to 1^-}\int_0^b \frac{dx}{x-1} = \lim_{b\to 1^-} \ln|x-1|\Big]_0^b$$

$$= \lim_{b\to 1^-}\left(\ln|b-1| - \ln|-1|\right)$$

$$= \lim_{b\to 1^-} \ln(1-b) = -\infty. \qquad 1 - b \to 0^+ \text{ as } b \to 1^-$$

Since $\int_0^1 dx/(x-1)$ is divergent, the original integral $\int_0^3 dx/(x-1)$ is divergent. ∎

Example 8 illustrates what can go wrong if you mistake an improper integral for an ordinary integral. Whenever you encounter an integral $\int_a^b f(x)\,dx$ you must examine the function f on $[a, b]$ and then decide if the integral is improper. If f is continuous on $[a, b]$, it will be proper, an ordinary integral.

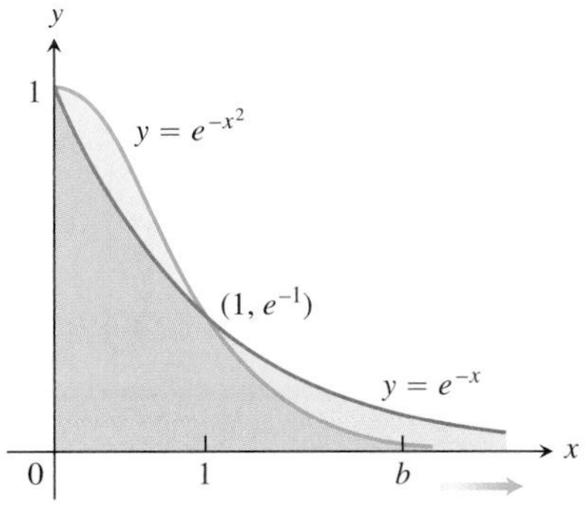

FIGURE 8.25 The graph of e^{-x^2} lies below the graph of e^{-x} for $x > 1$ (Example 9).

Tests for Convergence and Divergence

When we cannot evaluate an improper integral directly, we try to determine whether it converges or diverges. If the integral diverges, that's the end of the story. If it converges, we can use numerical methods to approximate its value. The principal tests for convergence or divergence are the Direct Comparison Test and the Limit Comparison Test.

EXAMPLE 9 Investigating Convergence

Does the integral $\int_1^{\infty} e^{-x^2}\,dx$ converge?

Solution By definition,

$$\int_1^{\infty} e^{-x^2}\,dx = \lim_{b\to\infty}\int_1^b e^{-x^2}\,dx.$$

We cannot evaluate the latter integral directly because it is nonelementary. But we *can* show that its limit as $b \to \infty$ is finite. We know that $\int_1^b e^{-x^2}\,dx$ is an increasing function of b. Therefore either it becomes infinite as $b \to \infty$ or it has a finite limit as $b \to \infty$. It does not become infinite: For every value of $x \geq 1$ we have $e^{-x^2} \leq e^{-x}$ (Figure 8.25), so that

$$\int_1^b e^{-x^2}\,dx \leq \int_1^b e^{-x}\,dx = -e^{-b} + e^{-1} < e^{-1} \approx 0.36788.$$

Hence

$$\int_1^{\infty} e^{-x^2}\,dx = \lim_{b\to\infty}\int_1^b e^{-x^2}\,dx$$

converges to some definite finite value. We do not know exactly what the value is except that it is something positive and less than 0.37. Here we are relying on the completeness property of the real numbers, discussed in Appendix 4. ■

The comparison of e^{-x^2} and e^{-x} in Example 9 is a special case of the following test.

HISTORICAL BIOGRAPHY

Karl Weierstrass (1815–1897)

THEOREM 1 Direct Comparison Test
Let f and g be continuous on $[a, \infty)$ with $0 \leq f(x) \leq g(x)$ for all $x \geq a$. Then

1. $\int_a^{\infty} f(x)\,dx$ converges if $\int_a^{\infty} g(x)\,dx$ converges

2. $\int_a^{\infty} g(x)\,dx$ diverges if $\int_a^{\infty} f(x)\,dx$ diverges.

The reasoning behind the argument establishing Theorem 1 is similar to that in Example 9.

If $0 \leq f(x) \leq g(x)$ for $x \geq a$, then

$$\int_a^b f(x)\,dx \leq \int_a^b g(x)\,dx, \qquad b > a.$$

From this it can be argued, as in Example 9, that

$$\int_a^\infty f(x)\,dx \text{ converges if } \int_a^\infty g(x)\,dx \text{ converges.}$$

Turning this around says that

$$\int_a^\infty g(x)\,dx \text{ diverges if } \int_a^\infty f(x)\,dx \text{ diverges.}$$

EXAMPLE 10 Using the Direct Comparison Test

(a) $\displaystyle\int_1^\infty \frac{\sin^2 x}{x^2}\,dx$ converges because

$$0 \le \frac{\sin^2 x}{x^2} \le \frac{1}{x^2} \quad \text{on} \quad [1, \infty) \quad \text{and} \quad \int_1^\infty \frac{1}{x^2}\,dx \text{ converges.}$$ Example 3

(b) $\displaystyle\int_1^\infty \frac{1}{\sqrt{x^2 - 0.1}}\,dx$ diverges because

$$\frac{1}{\sqrt{x^2 - 0.1}} \ge \frac{1}{x} \quad \text{on} \quad [1, \infty) \quad \text{and} \quad \int_1^\infty \frac{1}{x}\,dx \text{ diverges.}$$ Example 3

■

THEOREM 2 Limit Comparison Test
If the positive functions f and g are continuous on $[a, \infty)$ and if

$$\lim_{x\to\infty} \frac{f(x)}{g(x)} = L, \qquad 0 < L < \infty,$$

then

$$\int_a^\infty f(x)\,dx \quad \text{and} \quad \int_a^\infty g(x)\,dx$$

both converge or both diverge.

A proof of Theorem 2 is given in advanced calculus.

Although the improper integrals of two functions from a to ∞ may both converge, this does not mean that their integrals necessarily have the same value, as the next example shows.

EXAMPLE 11 Using the Limit Comparison Test

Show that

$$\int_1^\infty \frac{dx}{1+x^2}$$

converges by comparison with $\int_1^\infty (1/x^2)\,dx$. Find and compare the two integral values.

Solution The functions $f(x) = 1/x^2$ and $g(x) = 1/(1+x^2)$ are positive and continuous on $[1, \infty)$. Also,

$$\lim_{x\to\infty} \frac{f(x)}{g(x)} = \lim_{x\to\infty} \frac{1/x^2}{1/(1+x^2)} = \lim_{x\to\infty} \frac{1+x^2}{x^2}$$

$$= \lim_{x\to\infty} \left(\frac{1}{x^2} + 1\right) = 0 + 1 = 1,$$

a positive finite limit (Figure 8.26). Therefore, $\int_1^\infty \frac{dx}{1+x^2}$ converges because $\int_1^\infty \frac{dx}{x^2}$ converges.

The integrals converge to different values, however.

$$\int_1^\infty \frac{dx}{x^2} = \frac{1}{2-1} = 1 \qquad \text{Example 3}$$

and

$$\int_1^\infty \frac{dx}{1+x^2} = \lim_{b\to\infty} \int_1^b \frac{dx}{1+x^2}$$

$$= \lim_{b\to\infty} [\tan^{-1} b - \tan^{-1} 1] = \frac{\pi}{2} - \frac{\pi}{4} = \frac{\pi}{4}$$ ■

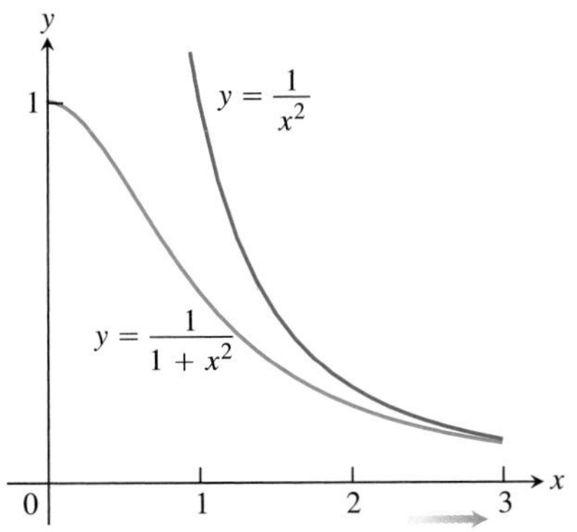

FIGURE 8.26 The functions in Example 11.

EXAMPLE 12 Using the Limit Comparison Test

Show that

$$\int_1^\infty \frac{3}{e^x + 5}\,dx$$

converges.

Solution From Example 9, it is easy to see that $\int_1^\infty e^{-x}\,dx = \int_1^\infty (1/e^x)\,dx$ converges. Moreover, we have

$$\lim_{x\to\infty} \frac{1/e^x}{3/(e^x+5)} = \lim_{x\to\infty} \frac{e^x+5}{3e^x} = \lim_{x\to\infty} \left(\frac{1}{3} + \frac{5}{3e^x}\right) = \frac{1}{3},$$

a positive finite limit. As far as the convergence of the improper integral is concerned, $3/(e^x+5)$ behaves like $1/e^x$. ■

Types of Improper Integrals Discussed in This Section

Infinite Limits of Integration: Type I

1. Upper limit

$$\int_1^{\infty} \frac{\ln x}{x^2}\, dx = \lim_{b \to \infty} \int_1^b \frac{\ln x}{x^2}\, dx$$

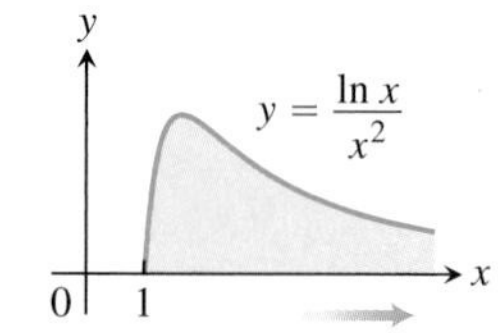

2. Lower limit

$$\int_{-\infty}^{0} \frac{dx}{1 + x^2} = \lim_{a \to -\infty} \int_a^0 \frac{dx}{1 + x^2}$$

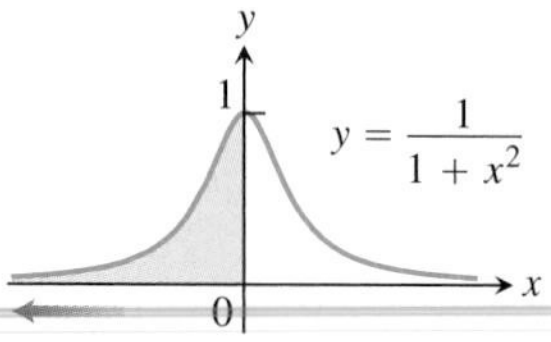

3. Both limits

$$\int_{-\infty}^{\infty} \frac{dx}{1 + x^2} = \lim_{b \to -\infty} \int_b^0 \frac{dx}{1 + x^2} + \lim_{c \to \infty} \int_0^c \frac{dx}{1 + x^2}$$

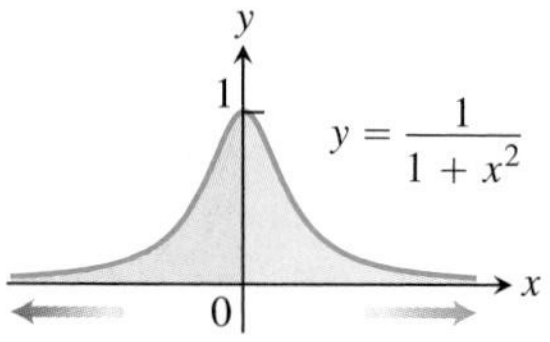

Integrand Becomes Infinite: Type II

4. Upper endpoint

$$\int_0^1 \frac{dx}{(x - 1)^{2/3}} = \lim_{b \to 1^-} \int_0^b \frac{dx}{(x - 1)^{2/3}}$$

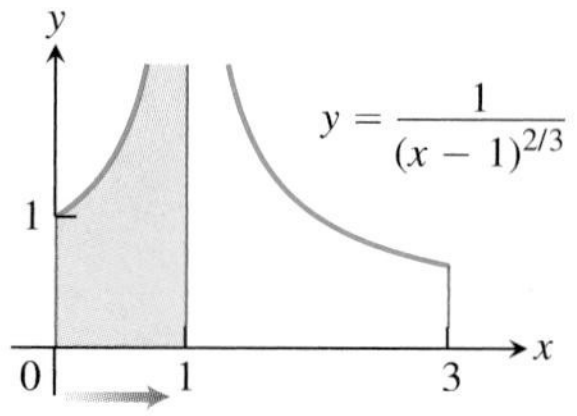

5. Lower endpoint

$$\int_1^3 \frac{dx}{(x - 1)^{2/3}} = \lim_{d \to 1^+} \int_d^3 \frac{dx}{(x - 1)^{2/3}}$$

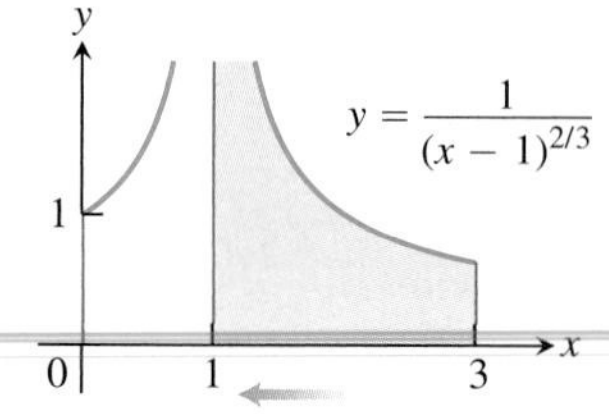

6. Interior point

$$\int_0^3 \frac{dx}{(x - 1)^{2/3}} = \int_0^1 \frac{dx}{(x - 1)^{2/3}} + \int_1^3 \frac{dx}{(x - 1)^{2/3}}$$

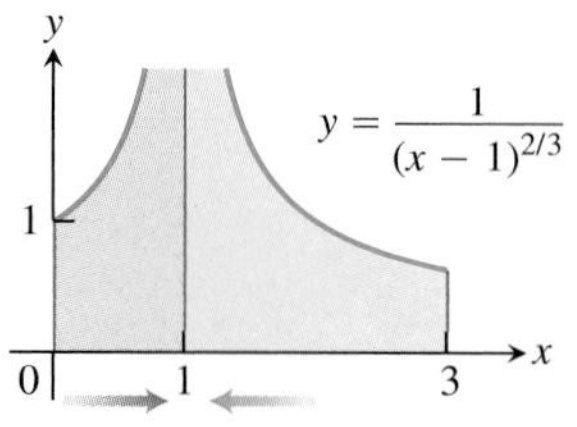

EXERCISES 8.8

Evaluating Improper Integrals

Evaluate the integrals in Exercises 1–34 without using tables.

1. $\displaystyle\int_0^\infty \frac{dx}{x^2+1}$

2. $\displaystyle\int_1^\infty \frac{dx}{x^{1.001}}$

3. $\displaystyle\int_0^1 \frac{dx}{\sqrt{x}}$

4. $\displaystyle\int_0^4 \frac{dx}{\sqrt{4-x}}$

5. $\displaystyle\int_{-1}^1 \frac{dx}{x^{2/3}}$

6. $\displaystyle\int_{-8}^1 \frac{dx}{x^{1/3}}$

7. $\displaystyle\int_0^1 \frac{dx}{\sqrt{1-x^2}}$

8. $\displaystyle\int_0^1 \frac{dr}{r^{0.999}}$

9. $\displaystyle\int_{-\infty}^{-2} \frac{2\,dx}{x^2-1}$

10. $\displaystyle\int_{-\infty}^{2} \frac{2\,dx}{x^2+4}$

11. $\displaystyle\int_2^\infty \frac{2}{v^2-v}\,dv$

12. $\displaystyle\int_2^\infty \frac{2\,dt}{t^2-1}$

13. $\displaystyle\int_{-\infty}^\infty \frac{2x\,dx}{(x^2+1)^2}$

14. $\displaystyle\int_{-\infty}^\infty \frac{x\,dx}{(x^2+4)^{3/2}}$

15. $\displaystyle\int_0^1 \frac{\theta+1}{\sqrt{\theta^2+2\theta}}\,d\theta$

16. $\displaystyle\int_0^2 \frac{s+1}{\sqrt{4-s^2}}\,ds$

17. $\displaystyle\int_0^\infty \frac{dx}{(1+x)\sqrt{x}}$

18. $\displaystyle\int_1^\infty \frac{1}{x\sqrt{x^2-1}}\,dx$

19. $\displaystyle\int_0^\infty \frac{dv}{(1+v^2)(1+\tan^{-1}v)}$

20. $\displaystyle\int_0^\infty \frac{16\tan^{-1}x}{1+x^2}\,dx$

21. $\displaystyle\int_{-\infty}^0 \theta e^\theta\,d\theta$

22. $\displaystyle\int_0^\infty 2e^{-\theta}\sin\theta\,d\theta$

23. $\displaystyle\int_{-\infty}^0 e^{-|x|}\,dx$

24. $\displaystyle\int_{-\infty}^\infty 2xe^{-x^2}\,dx$

25. $\displaystyle\int_0^1 x\ln x\,dx$

26. $\displaystyle\int_0^1 (-\ln x)\,dx$

27. $\displaystyle\int_0^2 \frac{ds}{\sqrt{4-s^2}}$

28. $\displaystyle\int_0^1 \frac{4r\,dr}{\sqrt{1-r^4}}$

29. $\displaystyle\int_1^2 \frac{ds}{s\sqrt{s^2-1}}$

30. $\displaystyle\int_2^4 \frac{dt}{t\sqrt{t^2-4}}$

31. $\displaystyle\int_{-1}^4 \frac{dx}{\sqrt{|x|}}$

32. $\displaystyle\int_0^2 \frac{dx}{\sqrt{|x-1|}}$

33. $\displaystyle\int_{-1}^\infty \frac{d\theta}{\theta^2+5\theta+6}$

34. $\displaystyle\int_0^\infty \frac{dx}{(x+1)(x^2+1)}$

Testing for Convergence

In Exercises 35–64, use integration, the Direct Comparison Test, or the Limit Comparison Test to test the integrals for convergence. If more than one method applies, use whatever method you prefer.

35. $\displaystyle\int_0^{\pi/2} \tan\theta\,d\theta$

36. $\displaystyle\int_0^{\pi/2} \cot\theta\,d\theta$

37. $\displaystyle\int_0^\pi \frac{\sin\theta\,d\theta}{\sqrt{\pi-\theta}}$

38. $\displaystyle\int_{-\pi/2}^{\pi/2} \frac{\cos\theta\,d\theta}{(\pi-2\theta)^{1/3}}$

39. $\displaystyle\int_0^{\ln 2} x^{-2}e^{-1/x}\,dx$

40. $\displaystyle\int_0^1 \frac{e^{-\sqrt{x}}}{\sqrt{x}}\,dx$

41. $\displaystyle\int_0^\pi \frac{dt}{\sqrt{t}+\sin t}$

42. $\displaystyle\int_0^1 \frac{dt}{t-\sin t}$ (*Hint:* $t \ge \sin t$ for $t \ge 0$)

43. $\displaystyle\int_0^2 \frac{dx}{1-x^2}$

44. $\displaystyle\int_0^2 \frac{dx}{1-x}$

45. $\displaystyle\int_{-1}^1 \ln|x|\,dx$

46. $\displaystyle\int_{-1}^1 -x\ln|x|\,dx$

47. $\displaystyle\int_1^\infty \frac{dx}{x^3+1}$

48. $\displaystyle\int_4^\infty \frac{dx}{\sqrt{x}-1}$

49. $\displaystyle\int_2^\infty \frac{dv}{\sqrt{v-1}}$

50. $\displaystyle\int_0^\infty \frac{d\theta}{1+e^\theta}$

51. $\displaystyle\int_0^\infty \frac{dx}{\sqrt{x^6+1}}$

52. $\displaystyle\int_2^\infty \frac{dx}{\sqrt{x^2-1}}$

53. $\displaystyle\int_1^\infty \frac{\sqrt{x+1}}{x^2}\,dx$

54. $\displaystyle\int_2^\infty \frac{x\,dx}{\sqrt{x^4-1}}$

55. $\displaystyle\int_\pi^\infty \frac{2+\cos x}{x}\,dx$

56. $\displaystyle\int_\pi^\infty \frac{1+\sin x}{x^2}\,dx$

57. $\displaystyle\int_4^\infty \frac{2\,dt}{t^{3/2}-1}$

58. $\displaystyle\int_2^\infty \frac{1}{\ln x}\,dx$

59. $\displaystyle\int_1^\infty \frac{e^x}{x}\,dx$

60. $\displaystyle\int_{e^e}^\infty \ln(\ln x)\,dx$

61. $\displaystyle\int_1^\infty \frac{1}{\sqrt{e^x-x}}\,dx$

62. $\displaystyle\int_1^\infty \frac{1}{e^x-2^x}\,dx$

63. $\displaystyle\int_{-\infty}^\infty \frac{dx}{\sqrt{x^4+1}}$

64. $\displaystyle\int_{-\infty}^\infty \frac{dx}{e^x+e^{-x}}$

Theory and Examples

65. Find the values of p for which each integral converges.

    a. $\displaystyle\int_1^2 \frac{dx}{x(\ln x)^p}$

    b. $\displaystyle\int_2^\infty \frac{dx}{x(\ln x)^p}$

66. **$\int_{-\infty}^\infty f(x)\,dx$ may not equal $\lim_{b\to\infty}\int_{-b}^b f(x)\,dx$** Show that

$$\int_0^\infty \frac{2x\,dx}{x^2+1}$$

diverges and hence that

$$\int_{-\infty}^\infty \frac{2x\,dx}{x^2+1}$$

diverges. Then show that

$$\lim_{b\to\infty} \int_{-b}^{b} \frac{2x\,dx}{x^2+1} = 0.$$

Exercises 67–70 are about the infinite region in the first quadrant between the curve $y = e^{-x}$ and the x-axis.

67. Find the area of the region.

68. Find the centroid of the region.

69. Find the volume of the solid generated by revolving the region about the y-axis.

70. Find the volume of the solid generated by revolving the region about the x-axis.

71. Find the area of the region that lies between the curves $y = \sec x$ and $y = \tan x$ from $x = 0$ to $x = \pi/2$.

72. The region in Exercise 71 is revolved about the x-axis to generate a solid.

a. Find the volume of the solid.

b. Show that the inner and outer surfaces of the solid have infinite area.

73. Estimating the value of a convergent improper integral whose domain is infinite

a. Show that

$$\int_3^\infty e^{-3x}\,dx = \frac{1}{3}e^{-9} < 0.000042,$$

and hence that $\int_3^\infty e^{-x^2}\,dx < 0.000042$. Explain why this means that $\int_0^\infty e^{-x^2}\,dx$ can be replaced by $\int_0^3 e^{-x^2}\,dx$ without introducing an error of magnitude greater than 0.000042.

T **b.** Evaluate $\int_0^3 e^{-x^2}\,dx$ numerically.

74. The infinite paint can or Gabriel's horn As Example 3 shows, the integral $\int_1^\infty (dx/x)$ diverges. This means that the integral

$$\int_1^\infty 2\pi \frac{1}{x}\sqrt{1+\frac{1}{x^4}}\,dx,$$

which measures the *surface area* of the solid of revolution traced out by revolving the curve $y = 1/x$, $1 \le x$, about the x-axis, diverges also. By comparing the two integrals, we see that, for every finite value $b > 1$,

$$\int_1^b 2\pi \frac{1}{x}\sqrt{1+\frac{1}{x^4}}\,dx > 2\pi \int_1^b \frac{1}{x}\,dx.$$

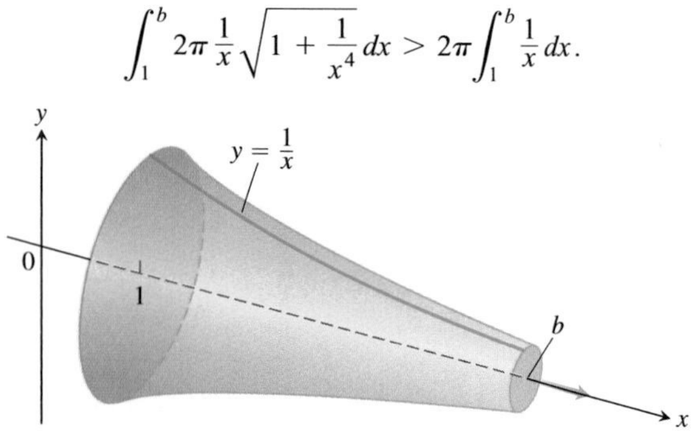

However, the integral

$$\int_1^\infty \pi\left(\frac{1}{x}\right)^2 dx$$

for the *volume* of the solid converges. **(a)** Calculate it. **(b)** This solid of revolution is sometimes described as a can that does not hold enough paint to cover its own interior. Think about that for a moment. It is common sense that a finite amount of paint cannot cover an infinite surface. But if we fill the horn with paint (a finite amount), then we *will* have covered an infinite surface. Explain the apparent contradiction.

75. Sine-integral function The integral

$$\text{Si}\,(x) = \int_0^x \frac{\sin t}{t}\,dt,$$

called the *sine-integral function,* has important applications in optics.

T **a.** Plot the integrand $(\sin t)/t$ for $t > 0$. Is the Si function everywhere increasing or decreasing? Do you think $\text{Si}\,(x) = 0$ for $x > 0$? Check your answers by graphing the function $\text{Si}\,(x)$ for $0 \le x \le 25$.

b. Explore the convergence of

$$\int_0^\infty \frac{\sin t}{t}\,dt.$$

If it converges, what is its value?

76. Error function The function

$$\text{erf}\,(x) = \int_0^x \frac{2e^{-t^2}}{\sqrt{\pi}}\,dt,$$

called the *error function*, has important applications in probability and statistics.

T **a.** Plot the error function for $0 \le x \le 25$.

b. Explore the convergence of

$$\int_0^\infty \frac{2e^{-t^2}}{\sqrt{\pi}}\,dt.$$

If it converges, what appears to be its value? You will see how to confirm your estimate in Section 15.3, Exercise 37.

77. Normal probability distribution function The function

$$f(x) = \frac{1}{\sigma\sqrt{2\pi}}e^{-\frac{1}{2}\left(\frac{x-\mu}{\sigma}\right)^2}$$

is called the *normal probability density function* with mean μ and standard deviation σ. The number μ tells where the distribution is centered, and σ measures the "scatter" around the mean.

From the theory of probability, it is known that

$$\int_{-\infty}^{\infty} f(x)\,dx = 1.$$

In what follows, let $\mu = 0$ and $\sigma = 1$.

T **a.** Draw the graph of f. Find the intervals on which f is increasing, the intervals on which f is decreasing, and any local extreme values and where they occur.

b. Evaluate

$$\int_{-n}^{n} f(x)\,dx$$

for $n = 1, 2, 3$.

c. Give a convincing argument that

$$\int_{-\infty}^{\infty} f(x)\,dx = 1.$$

(*Hint:* Show that $0 < f(x) < e^{-x/2}$ for $x > 1$, and for $b > 1$,

$$\int_{b}^{\infty} e^{-x/2}\,dx \to 0 \quad \text{as} \quad b \to \infty.)$$

78. Here is an argument that $\ln 3$ equals $\infty - \infty$. Where does the argument go wrong? Give reasons for your answer.

$$\begin{aligned}
\ln 3 &= \ln 1 + \ln 3 = \ln 1 - \ln\frac{1}{3} \\
&= \lim_{b\to\infty} \ln\left(\frac{b-2}{b}\right) - \ln\frac{1}{3} \\
&= \lim_{b\to\infty}\left[\ln\frac{x-2}{x}\right]_3^b \\
&= \lim_{b\to\infty}\left[\ln(x-2) - \ln x\right]_3^b \\
&= \lim_{b\to\infty}\int_3^b \left(\frac{1}{x-2} - \frac{1}{x}\right)dx \\
&= \int_3^\infty \left(\frac{1}{x-2} - \frac{1}{x}\right)dx \\
&= \int_3^\infty \frac{1}{x-2}\,dx - \int_3^\infty \frac{1}{x}\,dx \\
&= \lim_{b\to\infty}\left[\ln(x-2)\right]_3^b - \lim_{b\to\infty}\left[\ln x\right]_3^b \\
&= \infty - \infty.
\end{aligned}$$

79. Show that if $f(x)$ is integrable on every interval of real numbers and a and b are real numbers with $a < b$, then

a. $\int_{-\infty}^{a} f(x)\,dx$ and $\int_{a}^{\infty} f(x)\,dx$ both converge if and only if $\int_{-\infty}^{b} f(x)\,dx$ and $\int_{b}^{\infty} f(x)\,dx$ both converge.

b. $\int_{-\infty}^{a} f(x)\,dx + \int_{a}^{\infty} f(x)\,dx = \int_{-\infty}^{b} f(x)\,dx + \int_{b}^{\infty} f(x)\,dx$ when the integrals involved converge.

80. a. Show that if f is even and the necessary integrals exist, then

$$\int_{-\infty}^{\infty} f(x)\,dx = 2\int_{0}^{\infty} f(x)\,dx.$$

b. Show that if f is odd and the necessary integrals exist, then

$$\int_{-\infty}^{\infty} f(x)\,dx = 0.$$

Use direct evaluation, the comparison tests, and the results in Exercise 80, as appropriate, to determine the convergence or divergence of the integrals in Exercises 81–88. If more than one method applies, use whatever method you prefer.

81. $\displaystyle\int_{-\infty}^{\infty} \frac{dx}{\sqrt{x^2+1}}$

82. $\displaystyle\int_{-\infty}^{\infty} \frac{dx}{\sqrt{x^6+1}}$

83. $\displaystyle\int_{-\infty}^{\infty} \frac{dx}{e^x + e^{-x}}$

84. $\displaystyle\int_{-\infty}^{\infty} \frac{e^{-x}\,dx}{x^2+1}$

85. $\displaystyle\int_{-\infty}^{\infty} e^{-|x|}\,dx$

86. $\displaystyle\int_{-\infty}^{\infty} \frac{dx}{(x+1)^2}$

87. $\displaystyle\int_{-\infty}^{\infty} \frac{|\sin x| + |\cos x|}{|x|+1}\,dx$

(*Hint:* $|\sin\theta| + |\cos\theta| \ge \sin^2\theta + \cos^2\theta$.)

88. $\displaystyle\int_{-\infty}^{\infty} \frac{x\,dx}{(x^2+1)(x^2+2)}$

COMPUTER EXPLORATIONS

Exploring Integrals of $x^p \ln x$

In Exercises 89–92, use a CAS to explore the integrals for various values of p (include noninteger values). For what values of p does the integral converge? What is the value of the integral when it does converge? Plot the integrand for various values of p.

89. $\displaystyle\int_{0}^{e} x^p \ln x\,dx$

90. $\displaystyle\int_{e}^{\infty} x^p \ln x\,dx$

91. $\displaystyle\int_{0}^{\infty} x^p \ln x\,dx$

92. $\displaystyle\int_{-\infty}^{\infty} x^p \ln|x|\,dx$

Chapter 8 Questions to Guide Your Review

1. What basic integration formulas do you know?
2. What procedures do you know for matching integrals to basic formulas?
3. What is the formula for integration by parts? Where does it come from? Why might you want to use it?
4. When applying the formula for integration by parts, how do you choose the u and dv? How can you apply integration by parts to an integral of the form $\int f(x)\,dx$?
5. What is tabular integration? Give an example.
6. What is the goal of the method of partial fractions?

7. When the degree of a polynomial $f(x)$ is less than the degree of a polynomial $g(x)$, how do you write $f(x)/g(x)$ as a sum of partial fractions if $g(x)$
   a. is a product of distinct linear factors?
   b. consists of a repeated linear factor?
   c. contains an irreducible quadratic factor?

   What do you do if the degree of f is *not* less than the degree of g?
8. If an integrand is a product of the form $\sin^n x \cos^m x$, where m and n are nonnegative integers, how do you evaluate the integral? Give a specific example of each case.
9. What substitutions are made to evaluate integrals of $\sin mx \sin nx$, $\sin mx \cos nx$, and $\cos mx \cos nx$? Give an example of each case.
10. What substitutions are sometimes used to transform integrals involving $\sqrt{a^2 - x^2}$, $\sqrt{a^2 + x^2}$, and $\sqrt{x^2 - a^2}$ into integrals that can be evaluated directly? Give an example of each case.
11. What restrictions can you place on the variables involved in the three basic trigonometric substitutions to make sure the substitutions are reversible (have inverses)?
12. How are integral tables typically used? What do you do if a particular integral you want to evaluate is not listed in the table?
13. What is a reduction formula? How are reduction formulas typically derived? How are reduction formulas used? Give an example.
14. You are collaborating to produce a short "how-to" manual for numerical integration, and you are writing about the Trapezoidal Rule. **(a)** What would you say about the rule itself and how to use it? How to achieve accuracy? **(b)** What would you say if you were writing about Simpson's Rule instead?
15. How would you compare the relative merits of Simpson's Rule and the Trapezoidal Rule?
16. What is an improper integral of Type I? Type II? How are the values of various types of improper integrals defined? Give examples.
17. What tests are available for determining the convergence and divergence of improper integrals that cannot be evaluated directly? Give examples of their use.

Chapter 8 Practice Exercises

Integration Using Substitutions

Evaluate the integrals in Exercises 1–82. To transform each integral into a recognizable basic form, it may be necessary to use one or more of the techniques of algebraic substitution, completing the square, separating fractions, long division, or trigonometric substitution.

1. $\int x\sqrt{4x^2 - 9}\, dx$
2. $\int 6x\sqrt{3x^2 + 5}\, dx$
3. $\int x(2x + 1)^{1/2}\, dx$
4. $\int x(1 - x)^{-1/2}\, dx$
5. $\int \frac{x\, dx}{\sqrt{8x^2 + 1}}$
6. $\int \frac{x\, dx}{\sqrt{9 - 4x^2}}$
7. $\int \frac{y\, dy}{25 + y^2}$
8. $\int \frac{y^3\, dy}{4 + y^4}$
9. $\int \frac{t^3\, dt}{\sqrt{9 - 4t^4}}$
10. $\int \frac{2t\, dt}{t^4 + 1}$
11. $\int z^{2/3}(z^{5/3} + 1)^{2/3}\, dz$
12. $\int z^{-1/5}(1 + z^{4/5})^{-1/2}\, dz$
13. $\int \frac{\sin 2\theta\, d\theta}{(1 - \cos 2\theta)^2}$
14. $\int \frac{\cos \theta\, d\theta}{(1 + \sin \theta)^{1/2}}$
15. $\int \frac{\sin t}{3 + 4\cos t}\, dt$
16. $\int \frac{\cos 2t}{1 + \sin 2t}\, dt$
17. $\int \sin 2x\, e^{\cos 2x}\, dx$
18. $\int \sec x \tan x\, e^{\sec x}\, dx$
19. $\int e^\theta \sin(e^\theta) \cos^2(e^\theta)\, d\theta$
20. $\int e^\theta \sec^2(e^\theta)\, d\theta$
21. $\int 2^{x-1}\, dx$
22. $\int 5^{x\sqrt{2}}\, dx$
23. $\int \frac{dv}{v \ln v}$
24. $\int \frac{dv}{v(2 + \ln v)}$
25. $\int \frac{dx}{(x^2 + 1)(2 + \tan^{-1} x)}$
26. $\int \frac{\sin^{-1} x}{\sqrt{1 - x^2}}\, dx$
27. $\int \frac{2\, dx}{\sqrt{1 - 4x^2}}$
28. $\int \frac{dx}{\sqrt{49 - x^2}}$
29. $\int \frac{dt}{\sqrt{16 - 9t^2}}$
30. $\int \frac{dt}{\sqrt{9 - 4t^2}}$
31. $\int \frac{dt}{9 + t^2}$
32. $\int \frac{dt}{1 + 25t^2}$
33. $\int \frac{4\, dx}{5x\sqrt{25x^2 - 16}}$
34. $\int \frac{6\, dx}{x\sqrt{4x^2 - 9}}$
35. $\int \frac{dx}{\sqrt{4x - x^2}}$
36. $\int \frac{dx}{\sqrt{4x - x^2 - 3}}$
37. $\int \frac{dy}{y^2 - 4y + 8}$
38. $\int \frac{dt}{t^2 + 4t + 5}$
39. $\int \frac{dx}{(x - 1)\sqrt{x^2 - 2x}}$
40. $\int \frac{dv}{(v + 1)\sqrt{v^2 + 2v}}$
41. $\int \sin^2 x\, dx$
42. $\int \cos^2 3x\, dx$

43. $\int \sin^3 \frac{\theta}{2}\, d\theta$

44. $\int \sin^3 \theta \cos^2 \theta\, d\theta$

45. $\int \tan^3 2t\, dt$

46. $\int 6 \sec^4 t\, dt$

47. $\int \frac{dx}{2 \sin x \cos x}$

48. $\int \frac{2\, dx}{\cos^2 x - \sin^2 x}$

49. $\int_{\pi/4}^{\pi/2} \sqrt{\csc^2 y - 1}\, dy$

50. $\int_{\pi/4}^{3\pi/4} \sqrt{\cot^2 t + 1}\, dt$

51. $\int_0^{\pi} \sqrt{1 - \cos^2 2x}\, dx$

52. $\int_0^{2\pi} \sqrt{1 - \sin^2 \frac{x}{2}}\, dx$

53. $\int_{-\pi/2}^{\pi/2} \sqrt{1 - \cos 2t}\, dt$

54. $\int_{\pi}^{2\pi} \sqrt{1 + \cos 2t}\, dt$

55. $\int \frac{x^2}{x^2 + 4}\, dx$

56. $\int \frac{x^3}{9 + x^2}\, dx$

57. $\int \frac{4x^2 + 3}{2x - 1}\, dx$

58. $\int \frac{2x}{x - 4}\, dx$

59. $\int \frac{2y - 1}{y^2 + 4}\, dy$

60. $\int \frac{y + 4}{y^2 + 1}\, dy$

61. $\int \frac{t + 2}{\sqrt{4 - t^2}}\, dt$

62. $\int \frac{2t^2 + \sqrt{1 - t^2}}{t\sqrt{1 - t^2}}\, dt$

63. $\int \frac{\tan x\, dx}{\tan x + \sec x}$

64. $\int \frac{\cot x}{\cot x + \csc x}\, dx$

65. $\int \sec (5 - 3x)\, dx$

66. $\int x \csc (x^2 + 3)\, dx$

67. $\int \cot \left(\frac{x}{4}\right) dx$

68. $\int \tan (2x - 7)\, dx$

69. $\int x\sqrt{1 - x}\, dx$

70. $\int 3x\sqrt{2x + 1}\, dx$

71. $\int \sqrt{z^2 + 1}\, dz$

72. $\int (16 + z^2)^{-3/2}\, dz$

73. $\int \frac{dy}{\sqrt{25 + y^2}}$

74. $\int \frac{dy}{\sqrt{25 + 9y^2}}$

75. $\int \frac{dx}{x^2\sqrt{1 - x^2}}$

76. $\int \frac{x^3\, dx}{\sqrt{1 - x^2}}$

77. $\int \frac{x^2\, dx}{\sqrt{1 - x^2}}$

78. $\int \sqrt{4 - x^2}\, dx$

79. $\int \frac{dx}{\sqrt{x^2 - 9}}$

80. $\int \frac{12\, dx}{(x^2 - 1)^{3/2}}$

81. $\int \frac{\sqrt{w^2 - 1}}{w}\, dw$

82. $\int \frac{\sqrt{z^2 - 16}}{z}\, dz$

Integration by Parts

Evaluate the integrals in Exercises 83–90 using integration by parts.

83. $\int \ln (x + 1)\, dx$

84. $\int x^2 \ln x\, dx$

85. $\int \tan^{-1} 3x\, dx$

86. $\int \cos^{-1} \left(\frac{x}{2}\right) dx$

87. $\int (x + 1)^2 e^x\, dx$

88. $\int x^2 \sin (1 - x)\, dx$

89. $\int e^x \cos 2x\, dx$

90. $\int e^{-2x} \sin 3x\, dx$

Partial Fractions

Evaluate the integrals in Exercises 91–110. It may be necessary to use a substitution first.

91. $\int \frac{x\, dx}{x^2 - 3x + 2}$

92. $\int \frac{x\, dx}{x^2 + 4x + 3}$

93. $\int \frac{dx}{x(x + 1)^2}$

94. $\int \frac{x + 1}{x^2(x - 1)}\, dx$

95. $\int \frac{\sin \theta\, d\theta}{\cos^2 \theta + \cos \theta - 2}$

96. $\int \frac{\cos \theta\, d\theta}{\sin^2 \theta + \sin \theta - 6}$

97. $\int \frac{3x^2 + 4x + 4}{x^3 + x}\, dx$

98. $\int \frac{4x\, dx}{x^3 + 4x}$

99. $\int \frac{v + 3}{2v^3 - 8v}\, dv$

100. $\int \frac{(3v - 7)\, dv}{(v - 1)(v - 2)(v - 3)}$

101. $\int \frac{dt}{t^4 + 4t^2 + 3}$

102. $\int \frac{t\, dt}{t^4 - t^2 - 2}$

103. $\int \frac{x^3 + x^2}{x^2 + x - 2}\, dx$

104. $\int \frac{x^3 + 1}{x^3 - x}\, dx$

105. $\int \frac{x^3 + 4x^2}{x^2 + 4x + 3}\, dx$

106. $\int \frac{2x^3 + x^2 - 21x + 24}{x^2 + 2x - 8}\, dx$

107. $\int \frac{dx}{x\left(3\sqrt{x + 1}\right)}$

108. $\int \frac{dx}{x\left(1 + \sqrt[3]{x}\right)}$

109. $\int \frac{ds}{e^s - 1}$

110. $\int \frac{ds}{\sqrt{e^s + 1}}$

Trigonometric Substitutions

Evaluate the integrals in Exercises 111–114 **(a)** without using a trigonometric substitution, **(b)** using a trigonometric substitution.

111. $\int \frac{y\, dy}{\sqrt{16 - y^2}}$

112. $\int \frac{x\, dx}{\sqrt{4 + x^2}}$

113. $\int \frac{x\, dx}{4 - x^2}$

114. $\int \frac{t\, dt}{\sqrt{4t^2 - 1}}$

Quadratic Terms

Evaluate the integrals in Exercises 115–118.

115. $\int \frac{x\, dx}{9 - x^2}$

116. $\int \frac{dx}{x(9 - x^2)}$

117. $\int \frac{dx}{9 - x^2}$

118. $\int \frac{dx}{\sqrt{9 - x^2}}$

Trigonometric Integrals

Evaluate the integrals in Exercises 119–126.

119. $\int \sin^3 x \cos^4 x\, dx$

120. $\int \cos^5 x \sin^5 x\, dx$

121. $\int \tan^4 x \sec^2 x\, dx$

122. $\int \tan^3 x \sec^3 x\, dx$

123. $\int \sin 5\theta \cos 6\theta\, d\theta$

124. $\int \cos 3\theta \cos 3\theta\, d\theta$

125. $\int \sqrt{1 + \cos(t/2)}\, dt$

126. $\int e^t \sqrt{\tan^2 e^t + 1}\, dt$

Numerical Integration

127. According to the error-bound formula for Simpson's Rule, how many subintervals should you use to be sure of estimating the value of

$$\ln 3 = \int_1^3 \frac{1}{x}\, dx$$

by Simpson's Rule with an error of no more than 10^{-4} in absolute value? (Remember that for Simpson's Rule, the number of subintervals has to be even.)

128. A brief calculation shows that if $0 \le x \le 1$, then the second derivative of $f(x) = \sqrt{1 + x^4}$ lies between 0 and 8. Based on this, about how many subdivisions would you need to estimate the integral of f from 0 to 1 with an error no greater than 10^{-3} in absolute value using the Trapezoidal Rule?

129. A direct calculation shows that

$$\int_0^{\pi} 2 \sin^2 x\, dx = \pi.$$

How close do you come to this value by using the Trapezoidal Rule with $n = 6$? Simpson's Rule with $n = 6$? Try them and find out.

130. You are planning to use Simpson's Rule to estimate the value of the integral

$$\int_1^2 f(x)\, dx$$

with an error magnitude less than 10^{-5}. You have determined that $|f^{(4)}(x)| \le 3$ throughout the interval of integration. How many subintervals should you use to assure the required accuracy? (Remember that for Simpson's Rule the number has to be even.)

131. Mean temperature Compute the average value of the temperature function

$$f(x) = 37 \sin\left(\frac{2\pi}{365}(x - 101)\right) + 25$$

for a 365-day year. This is one way to estimate the annual mean air temperature in Fairbanks, Alaska. The National Weather Service's official figure, a numerical average of the daily normal mean air temperatures for the year, is 25.7°F, which is slightly higher than the average value of $f(x)$.

132. Heat capacity of a gas Heat capacity C_v is the amount of heat required to raise the temperature of a given mass of gas with constant volume by 1°C, measured in units of cal/deg-mol (calories per degree gram molecular weight). The heat capacity of oxygen depends on its temperature T and satisfies the formula

$$C_v = 8.27 + 10^{-5}(26T - 1.87T^2).$$

Find the average value of C_v for $20° \le T \le 675°$C and the temperature at which it is attained.

133. Fuel efficiency An automobile computer gives a digital readout of fuel consumption in gallons per hour. During a trip, a passenger recorded the fuel consumption every 5 min for a full hour of travel.

Time	Gal/h	Time	Gal/h
0	2.5	35	2.5
5	2.4	40	2.4
10	2.3	45	2.3
15	2.4	50	2.4
20	2.4	55	2.4
25	2.5	60	2.3
30	2.6		

a. Use the Trapezoidal Rule to approximate the total fuel consumption during the hour.

b. If the automobile covered 60 mi in the hour, what was its fuel efficiency (in miles per gallon) for that portion of the trip?

134. A new parking lot To meet the demand for parking, your town has allocated the area shown here. As the town engineer, you have been asked by the town council to find out if the lot can be built for \$11,000. The cost to clear the land will be \$0.10 a square foot, and the lot will cost \$2.00 a square foot to pave. Use Simpson's Rule to find out if the job can be done for \$11,000.

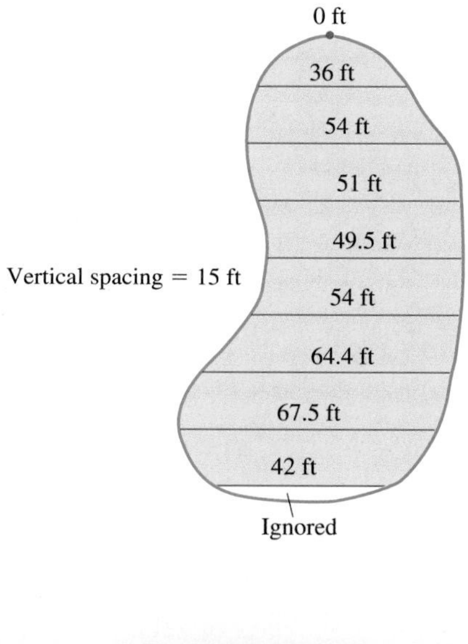

Improper Integrals

Evaluate the improper integrals in Exercises 135–144.

135. $\int_0^3 \frac{dx}{\sqrt{9-x^2}}$

136. $\int_0^1 \ln x\, dx$

137. $\int_{-1}^1 \frac{dy}{y^{2/3}}$

138. $\int_{-2}^0 \frac{d\theta}{(\theta+1)^{3/5}}$

139. $\int_3^\infty \frac{2\,du}{u^2-2u}$

140. $\int_1^\infty \frac{3v-1}{4v^3-v^2}\,dv$

141. $\int_0^\infty x^2 e^{-x}\,dx$

142. $\int_{-\infty}^0 xe^{3x}\,dx$

143. $\int_{-\infty}^\infty \frac{dx}{4x^2+9}$

144. $\int_{-\infty}^\infty \frac{4dx}{x^2+16}$

Convergence or Divergence

Which of the improper integrals in Exercises 145–150 converge and which diverge?

145. $\int_6^\infty \frac{d\theta}{\sqrt{\theta^2+1}}$

146. $\int_0^\infty e^{-u}\cos u\, du$

147. $\int_1^\infty \frac{\ln z}{z}\,dz$

148. $\int_1^\infty \frac{e^{-t}}{\sqrt{t}}\,dt$

149. $\int_{-\infty}^\infty \frac{2\,dx}{e^x+e^{-x}}$

150. $\int_{-\infty}^\infty \frac{dx}{x^2(1+e^x)}$

Assorted Integrations

Evaluate the integrals in Exercises 151–218. The integrals are listed in random order.

151. $\int \frac{x\,dx}{1+\sqrt{x}}$

152. $\int \frac{x^3+2}{4-x^2}\,dx$

153. $\int \frac{dx}{x(x^2+1)^2}$

154. $\int \frac{\cos\sqrt{x}}{\sqrt{x}}\,dx$

155. $\int \frac{dx}{\sqrt{-2x-x^2}}$

156. $\int \frac{(t-1)\,dt}{\sqrt{t^2-2t}}$

157. $\int \frac{du}{\sqrt{1+u^2}}$

158. $\int e^t \cos e^t\, dt$

159. $\int \frac{2-\cos x+\sin x}{\sin^2 x}\,dx$

160. $\int \frac{\sin^2\theta}{\cos^2\theta}\,d\theta$

161. $\int \frac{9\,dv}{81-v^4}$

162. $\int \frac{\cos x\,dx}{1+\sin^2 x}$

163. $\int \theta\cos(2\theta+1)\,d\theta$

164. $\int_2^\infty \frac{dx}{(x-1)^2}$

165. $\int \frac{x^3\,dx}{x^2-2x+1}$

166. $\int \frac{d\theta}{\sqrt{1+\sqrt{\theta}}}$

167. $\int \frac{2\sin\sqrt{x}\,dx}{\sqrt{x}\sec\sqrt{x}}$

168. $\int \frac{x^5\,dx}{x^4-16}$

169. $\int \frac{dy}{\sin y\cos y}$

170. $\int \frac{d\theta}{\theta^2-2\theta+4}$

171. $\int \frac{\tan x}{\cos^2 x}\,dx$

172. $\int \frac{dr}{(r+1)\sqrt{r^2+2r}}$

173. $\int \frac{(r+2)\,dr}{\sqrt{-r^2-4r}}$

174. $\int \frac{y\,dy}{4+y^4}$

175. $\int \frac{\sin 2\theta\,d\theta}{(1+\cos 2\theta)^2}$

176. $\int \frac{dx}{(x^2-1)^2}$

177. $\int_{\pi/4}^{\pi/2} \sqrt{1+\cos 4x}\,dx$

178. $\int (15)^{2x+1}\,dx$

179. $\int \frac{x\,dx}{\sqrt{2-x}}$

180. $\int \frac{\sqrt{1-v^2}}{v^2}\,dv$

181. $\int \frac{dy}{y^2-2y+2}$

182. $\int \ln\sqrt{x-1}\,dx$

183. $\int \theta^2\tan(\theta^3)\,d\theta$

184. $\int \frac{x\,dx}{\sqrt{8-2x^2-x^4}}$

185. $\int \frac{z+1}{z^2(z^2+4)}\,dz$

186. $\int x^3 e^{(x^2)}\,dx$

187. $\int \frac{t\,dt}{\sqrt{9-4t^2}}$

188. $\int_0^{\pi/10} \sqrt{1+\cos 5\theta}\,d\theta$

189. $\int \frac{\cot\theta\,d\theta}{1+\sin^2\theta}$

190. $\int \frac{\tan^{-1}x}{x^2}\,dx$

191. $\int \frac{\tan\sqrt{y}}{2\sqrt{y}}\,dy$

192. $\int \frac{e^t\,dt}{e^{2t}+3e^t+2}$

193. $\int \frac{\theta^2 d\theta}{4-\theta^2}$

194. $\int \frac{1-\cos 2x}{1+\cos 2x}\,dx$

195. $\int \frac{\cos(\sin^{-1}x)}{\sqrt{1-x^2}}\,dx$

196. $\int \frac{\cos x\,dx}{\sin^3 x-\sin x}$

197. $\int \sin\frac{x}{2}\cos\frac{x}{2}\,dx$

198. $\int \frac{x^2-x+2}{(x^2+2)^2}\,dx$

199. $\int \frac{e^t\,dt}{1+e^t}$

200. $\int \tan^3 t\,dt$

201. $\int_1^\infty \frac{\ln y}{y^3}\,dy$

202. $\int \frac{3+\sec^2 x+\sin x}{\tan x}\,dx$

203. $\int \frac{\cot v\,dv}{\ln\sin v}$

204. $\int \frac{dx}{(2x-1)\sqrt{x^2-x}}$

205. $\int e^{\ln\sqrt{x}}\,dx$

206. $\int e^\theta\sqrt{3+4e^\theta}\,d\theta$

207. $\int \frac{\sin 5t\,dt}{1+(\cos 5t)^2}$

208. $\int \frac{dv}{\sqrt{e^{2v}-1}}$

209. $\int (27)^{3\theta+1}\,d\theta$

210. $\int x^5\sin x\,dx$

211. $\int \frac{dr}{1+\sqrt{r}}$

212. $\int \frac{4x^3-20x}{x^4-10x^2+9}\,dx$

213. $\int \frac{8\,dy}{y^3(y+2)}$

214. $\int \frac{(t+1)\,dt}{(t^2+2t)^{2/3}}$

215. $\int \frac{8\,dm}{m\sqrt{49m^2-4}}$

216. $\displaystyle\int \frac{dt}{t(1+\ln t)\sqrt{(\ln t)(2+\ln t)}}$

217. $\displaystyle\int_0^1 3(x-1)^2\left(\int_0^x \sqrt{1+(t-1)^4}\,dt\right)dx$

218. $\displaystyle\int_2^\infty \frac{4v^3+v-1}{v^2(v-1)(v^2+1)}\,dv$

219. Suppose for a certain function f it is known that

$$f'(x)=\frac{\cos x}{x},\quad f(\pi/2)=a,\quad \text{and}\quad f(3\pi/2)=b.$$

Use integration by parts to evaluate

$$\int_{\pi/2}^{3\pi/2} f(x)\,dx.$$

220. Find a positive number a satisfying

$$\int_0^a \frac{dx}{1+x^2}=\int_a^\infty \frac{dx}{1+x^2}.$$

Chapter 8 Additional and Advanced Exercises

Challenging Integrals

Evaluate the integrals in Exercises 1–10.

1. $\displaystyle\int (\sin^{-1} x)^2\,dx$

2. $\displaystyle\int \frac{dx}{x(x+1)(x+2)\cdots(x+m)}$

3. $\displaystyle\int x\sin^{-1} x\,dx$

4. $\displaystyle\int \sin^{-1}\sqrt{y}\,dy$

5. $\displaystyle\int \frac{d\theta}{1-\tan^2\theta}$

6. $\displaystyle\int \ln\left(\sqrt{x}+\sqrt{1+x}\right)dx$

7. $\displaystyle\int \frac{dt}{t-\sqrt{1-t^2}}$

8. $\displaystyle\int \frac{(2e^{2x}-e^x)\,dx}{\sqrt{3e^{2x}-6e^x-1}}$

9. $\displaystyle\int \frac{dx}{x^4+4}$

10. $\displaystyle\int \frac{dx}{x^6-1}$

Limits

Evaluate the limits in Exercises 11 and 12.

11. $\displaystyle\lim_{x\to\infty}\int_{-x}^{x} \sin t\,dt$

12. $\displaystyle\lim_{x\to 0^+} x\int_x^1 \frac{\cos t}{t^2}\,dt$

Evaluate the limits in Exercises 13 and 14 by identifying them with definite integrals and evaluating the integrals.

13. $\displaystyle\lim_{n\to\infty}\sum_{k=1}^{n} \ln\sqrt[n]{1+\frac{k}{n}}$

14. $\displaystyle\lim_{n\to\infty}\sum_{k=0}^{n-1} \frac{1}{\sqrt{n^2-k^2}}$

Theory and Applications

15. Finding arc length Find the length of the curve

$$y=\int_0^x \sqrt{\cos 2t}\,dt,\quad 0\le x\le \pi/4.$$

16. Finding arc length Find the length of the curve $y=\ln(1-x^2)$, $0\le x\le 1/2$.

17. Finding volume The region in the first quadrant that is enclosed by the x-axis and the curve $y=3x\sqrt{1-x}$ is revolved about the y-axis to generate a solid. Find the volume of the solid.

18. Finding volume The region in the first quadrant that is enclosed by the x-axis, the curve $y=5/\left(x\sqrt{5-x}\right)$, and the lines $x=1$ and $x=4$ is revolved about the x-axis to generate a solid. Find the volume of the solid.

19. Finding volume The region in the first quadrant enclosed by the coordinate axes, the curve $y=e^x$, and the line $x=1$ is revolved about the y-axis to generate a solid. Find the volume of the solid.

20. Finding volume The region in the first quadrant that is bounded above by the curve $y=e^x-1$, below by the x-axis, and on the right by the line $x=\ln 2$ is revolved about the line $x=\ln 2$ to generate a solid. Find the volume of the solid.

21. Finding volume Let R be the "triangular" region in the first quadrant that is bounded above by the line $y=1$, below by the curve $y=\ln x$, and on the left by the line $x=1$. Find the volume of the solid generated by revolving R about

a. the x-axis. **b.** the line $y=1$.

22. Finding volume (*Continuation of Exercise 21.*) Find the volume of the solid generated by revolving the region R about

a. the y-axis. **b.** the line $x=1$.

23. Finding volume The region between the x-axis and the curve

$$y=f(x)=\begin{cases}0, & x=0\\ x\ln x, & 0<x\le 2\end{cases}$$

is revolved about the x-axis to generate the solid shown here.

a. Show that f is continuous at $x=0$.

b. Find the volume of the solid.

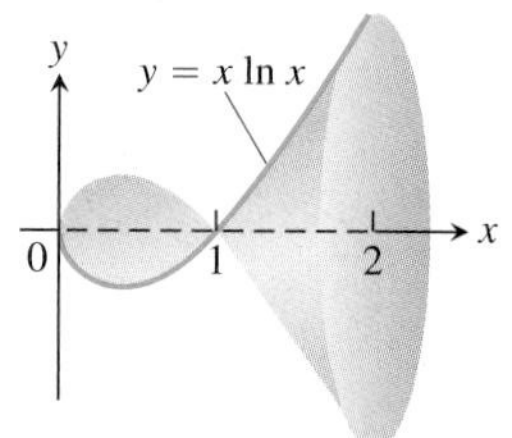

24. **Finding volume** The infinite region bounded by the coordinate axes and the curve $y = -\ln x$ in the first quadrant is revolved about the x-axis to generate a solid. Find the volume of the solid.

25. **Centroid of a region** Find the centroid of the region in the first quadrant that is bounded below by the x-axis, above by the curve $y = \ln x$, and on the right by the line $x = e$.

26. **Centroid of a region** Find the centroid of the region in the plane enclosed by the curves $y = \pm(1 - x^2)^{-1/2}$ and the lines $x = 0$ and $x = 1$.

27. **Length of a curve** Find the length of the curve $y = \ln x$ from $x = 1$ to $x = e$.

28. **Finding surface area** Find the area of the surface generated by revolving the curve in Exercise 27 about the y-axis.

29. **The length of an astroid** The graph of the equation $x^{2/3} + y^{2/3} = 1$ is one of a family of curves called *astroids* (not "asteroids") because of their starlike appearance (see accompanying figure). Find the length of this particular astroid.

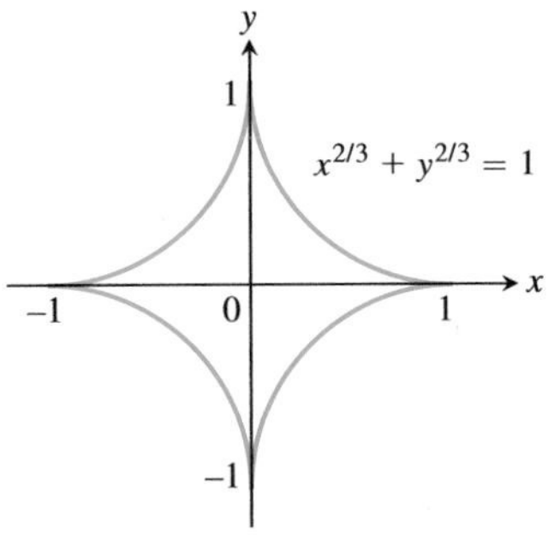

30. **The surface generated by an astroid** Find the area of the surface generated by revolving the curve in Exercise 29 about the x-axis.

31. Find a curve through the origin whose length is

$$\int_0^4 \sqrt{1 + \frac{1}{4x}}\, dx.$$

32. Without evaluating either integral, explain why

$$2\int_{-1}^{1} \sqrt{1 - x^2}\, dx = \int_{-1}^{1} \frac{dx}{\sqrt{1 - x^2}}.$$

T 33. **a.** Graph the function $f(x) = e^{(x - e^x)}$, $-5 \le x \le 3$.

b. Show that $\int_{-\infty}^{\infty} f(x)\, dx$ converges and find its value.

34. Find $\lim_{n\to\infty} \int_0^1 \frac{ny^{n-1}}{1 + y}\, dy$.

35. Derive the integral formula

$$\int x\left(\sqrt{x^2 - a^2}\right)^n dx = \frac{\left(\sqrt{x^2 - a^2}\right)^{n+2}}{n + 2} + C, \quad n \ne -2.$$

36. Prove that

$$\frac{\pi}{6} < \int_0^1 \frac{dx}{\sqrt{4 - x^2 - x^3}} < \frac{\pi\sqrt{2}}{8}.$$

(*Hint:* Observe that for $0 < x < 1$, we have $4 - x^2 > 4 - x^2 - x^3 > 4 - 2x^2$, with the left-hand side becoming an equality for $x = 0$ and the right-hand side becoming an equality for $x = 1$.)

37. For what value or values of a does

$$\int_1^{\infty} \left(\frac{ax}{x^2 + 1} - \frac{1}{2x}\right) dx$$

converge? Evaluate the corresponding integral(s).

38. For each $x > 0$, let $G(x) = \int_0^{\infty} e^{-xt}\, dt$. Prove that $xG(x) = 1$ for each $x > 0$.

39. **Infinite area and finite volume** What values of p have the following property: The area of the region between the curve $y = x^{-p}$, $1 \le x < \infty$, and the x-axis is infinite but the volume of the solid generated by revolving the region about the x-axis is finite.

40. **Infinite area and finite volume** What values of p have the following property: The area of the region in the first quadrant enclosed by the curve $y = x^{-p}$, the y-axis, the line $x = 1$, and the interval $[0, 1]$ on the x-axis is infinite but the volume of the solid generated by revolving the region about one of the coordinate axes is finite.

Tabular Integration

The technique of tabular integration also applies to integrals of the form $\int f(x)g(x)\, dx$ when neither function can be differentiated repeatedly to become zero. For example, to evaluate

$$\int e^{2x} \cos x\, dx$$

we begin as before with a table listing successive derivatives of e^{2x} and integrals of $\cos x$:

e^{2x} and its derivatives		$\cos x$ and its integrals	
e^{2x}	(+)	$\cos x$	
$2e^{2x}$	(−)	$\sin x$	
$4e^{2x}$	(+)	$-\cos x$	← *Stop here:* Row is same as first row except for multiplicative constants (4 on the left, −1 on the right)

We stop differentiating and integrating as soon as we reach a row that is the same as the first row except for multiplicative constants. We interpret the table as saying

$$\int e^{2x} \cos x\, dx = +(e^{2x} \sin x) - (2e^{2x}(-\cos x)) + \int (4e^{2x})(-\cos x)\, dx.$$

We take signed products from the diagonal arrows and a signed integral for the last horizontal arrow. Transposing the integral on the right-hand side over to the left-hand side now gives

$$5\int e^{2x}\cos x\,dx = e^{2x}\sin x + 2e^{2x}\cos x$$

or

$$\int e^{2x}\cos x\,dx = \frac{e^{2x}\sin x + 2e^{2x}\cos x}{5} + C,$$

after dividing by 5 and adding the constant of integration.

Use tabular integration to evaluate the integrals in Exercises 41–48.

41. $\int e^{2x}\cos 3x\,dx$ **42.** $\int e^{3x}\sin 4x\,dx$

43. $\int \sin 3x \sin x\,dx$ **44.** $\int \cos 5x \sin 4x\,dx$

45. $\int e^{ax}\sin bx\,dx$ **46.** $\int e^{ax}\cos bx\,dx$

47. $\int \ln(ax)\,dx$ **48.** $\int x^2\ln(ax)\,dx$

The Gamma Function and Stirling's Formula

Euler's gamma function $\Gamma(x)$ ("gamma of x"; Γ is a Greek capital g) uses an integral to extend the factorial function from the nonnegative integers to other real values. The formula is

$$\Gamma(x) = \int_0^\infty t^{x-1}e^{-t}\,dt, \quad x > 0.$$

For each positive x, the number $\Gamma(x)$ is the integral of $t^{x-1}e^{-t}$ with respect to t from 0 to ∞. Figure 8.27 shows the graph of Γ near the origin. You will see how to calculate $\Gamma(1/2)$ if you do Additional Exercise 31 in Chapter 15.

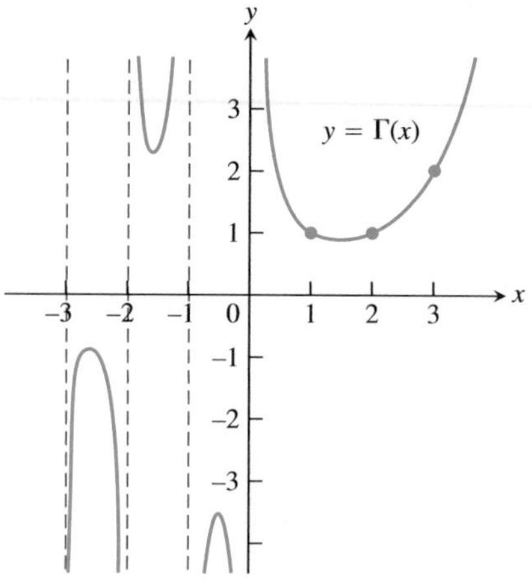

FIGURE 8.27 Euler's gamma function $\Gamma(x)$ is a continuous function of x whose value at each positive integer $n + 1$ is $n!$. The defining integral formula for Γ is valid only for $x > 0$, but we can extend Γ to negative noninteger values of x with the formula $\Gamma(x) = (\Gamma(x + 1))/x$, which is the subject of Exercise 49.

49. If n is a nonnegative integer, $\Gamma(n + 1) = n!$

a. Show that $\Gamma(1) = 1$.

b. Then apply integration by parts to the integral for $\Gamma(x + 1)$ to show that $\Gamma(x + 1) = x\Gamma(x)$. This gives

$$\begin{aligned}\Gamma(2) &= 1\Gamma(1) = 1\\ \Gamma(3) &= 2\Gamma(2) = 2\\ \Gamma(4) &= 3\Gamma(3) = 6\\ &\vdots\\ \Gamma(n + 1) &= n\,\Gamma(n) = n! \qquad (1)\end{aligned}$$

c. Use mathematical induction to verify Equation (1) for every nonnegative integer n.

50. Stirling's formula Scottish mathematician James Stirling (1692–1770) showed that

$$\lim_{x\to\infty}\left(\frac{e}{x}\right)^x\sqrt{\frac{x}{2\pi}}\,\Gamma(x) = 1,$$

so for large x,

$$\Gamma(x) = \left(\frac{x}{e}\right)^x\sqrt{\frac{2\pi}{x}}(1 + \epsilon(x)), \qquad \epsilon(x)\to 0 \text{ as } x\to\infty. \qquad (2)$$

Dropping $\epsilon(x)$ leads to the approximation

$$\Gamma(x) \approx \left(\frac{x}{e}\right)^x\sqrt{\frac{2\pi}{x}} \qquad \textbf{(Stirling's formula)}. \qquad (3)$$

a. Stirling's approximation for $n!$ Use Equation (3) and the fact that $n! = n\Gamma(n)$ to show that

$$n! \approx \left(\frac{n}{e}\right)^n\sqrt{2n\pi} \qquad \textbf{(Stirling's approximation)}. \qquad (4)$$

As you will see if you do Exercise 64 in Section 11.1, Equation (4) leads to the approximation

$$\sqrt[n]{n!} \approx \frac{n}{e}. \qquad (5)$$

T **b.** Compare your calculator's value for $n!$ with the value given by Stirling's approximation for $n = 10, 20, 30, \ldots$, as far as your calculator can go.

T **c.** A refinement of Equation (2) gives

$$\Gamma(x) = \left(\frac{x}{e}\right)^x\sqrt{\frac{2\pi}{x}}\,e^{1/(12x)}(1 + \epsilon(x)),$$

or

$$\Gamma(x) \approx \left(\frac{x}{e}\right)^x \sqrt{\frac{2\pi}{x}}\, e^{1/(12x)}$$

which tells us that

$$n! \approx \left(\frac{n}{e}\right)^n \sqrt{2n\pi}\, e^{1/(12n)}. \qquad (6)$$

Compare the values given for 10! by your calculator, Stirling's approximation, and Equation (6).

Chapter 8 Technology Application Projects

Mathematica/Maple Module
Riemann, Trapezoidal, and Simpson Approximations

Part I: Visualize the error involved in using Riemann sums to approximate the area under a curve.

Part II: Build a table of values and compute the relative magnitude of the error as a function of the step size Δx.

Part III: Investigate the effect of the derivative function on the error.

Parts IV and **V**: Trapezoidal Rule approximations.

Part VI: Simpson's Rule approximations.

Mathematica/Maple Module
Games of Chance: Exploring the Monte Carlo Probabilistic Technique for Numerical Integration

Graphically explore the Monte Carlo method for approximating definite integrals.

Mathematica/Maple Module
Computing Probabilities with Improper Integrals

Graphically explore the Monte Carlo method for approximating definite integrals.

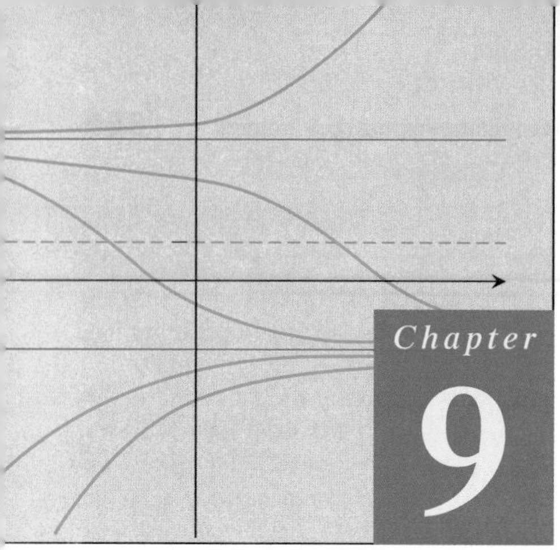

Chapter 9

FURTHER APPLICATIONS OF INTEGRATION

HISTORICAL BIOGRAPHY

Carl Friedrich Gauss
(1777–1855)

OVERVIEW In Section 4.8 we introduced differential equations of the form $dy/dx = f(x)$, where y is an unknown function being differentiated. For a continuous function f, we found the general solution $y(x)$ by integration: $y(x) = \int f(x)\, dx$. (Remember that the indefinite integral represents *all* the antiderivatives of f, so it contains an arbitrary constant $+C$ which must be shown once an antiderivative is found.) Many applications in the sciences, engineering, and economics involve a model formulated by even more general differential equations. In Section 7.5, for example, we found that exponential growth and decay is modeled by a differential equation of the form $dy/dx = ky$, for some constant $k \neq 0$. We have not yet considered differential equations such as $dy/dx = y - x$, yet such equations arise frequently in applications. In this chapter, we study several differential equations having the form $dy/dx = f(x, y)$, where f is a function of *both* the independent and dependent variables. We use the theory of indefinite integration to solve these differential equations, and investigate analytic, graphical, and numerical solution methods.

9.1 Slope Fields and Separable Differential Equations

HISTORICAL BIOGRAPHY

Jules Henri Poincaré
(1854–1912)

In calculating derivatives by implicit differentiation (Section 3.6), we found that the expression for the derivative dy/dx often contained both variables x and y, not just the independent variable x. We begin this section by considering the general differential equation $dy/dx = f(x, y)$ and what is meant by a solution to it. Then we investigate equations having a special form for which the function f can be expressed as a product of a function of x and a function of y.

General First-Order Differential Equations and Solutions

A **first-order differential equation** is an equation

$$\frac{dy}{dx} = f(x, y) \tag{1}$$

in which $f(x, y)$ is a function of two variables defined on a region in the xy-plane. The equation is of *first-order* because it involves only the first derivative dy/dx (and not higher-order derivatives). We point out that the equations

$$y' = f(x, y) \quad \text{and} \quad \frac{d}{dx} y = f(x, y),$$

are equivalent to Equation (1) and all three forms will be used interchangeably in the text.

A **solution** of Equation (1) is a differentiable function $y = y(x)$ defined on an interval I of x-values (perhaps infinite) such that

$$\frac{d}{dx}y(x) = f(x, y(x))$$

on that interval. That is, when $y(x)$ and its derivative $y'(x)$ are substituted into Equation (1), the resulting equation is true for all x over the interval I. The **general solution** to a first-order differential equation is a solution that contains all possible solutions. The general solution always contains an arbitrary constant, but having this property doesn't mean a solution is the general solution. That is, a solution may contain an arbitrary constant without being the general solution. Establishing that a solution *is* the general solution may require deeper results from the theory of differential equations and is best studied in a more advanced course.

EXAMPLE 1 Verifying Solution Functions

Show that every member of the family of functions

$$y = \frac{C}{x} + 2$$

is a solution of the first-order differential equation

$$\frac{dy}{dx} = \frac{1}{x}(2 - y)$$

on the interval $(0, \infty)$, where C is any constant.

Solution Differentiating $y = C/x + 2$ gives

$$\frac{dy}{dx} = C\frac{d}{dx}\left(\frac{1}{x}\right) + 0 = -\frac{C}{x^2}.$$

Thus we need only verify that for all $x \in (0, \infty)$,

$$-\frac{C}{x^2} = \frac{1}{x}\left[2 - \left(\frac{C}{x} + 2\right)\right].$$

This last equation follows immediately by expanding the expression on the right side:

$$\frac{1}{x}\left[2 - \left(\frac{C}{x} + 2\right)\right] = \frac{1}{x}\left(-\frac{C}{x}\right) = -\frac{C}{x^2}.$$

Therefore, for every value of C, the function $y = C/x + 2$ is a solution of the differential equation. ■

As was the case in finding antiderivatives, we often need a *particular* rather than the general solution to a first-order differential equation $y' = f(x, y)$. The **particular solution** satisfying the initial condition $y(x_0) = y_0$ is the solution $y = y(x)$ whose value is y_0 when $x = x_0$. Thus the graph of the particular solution passes through the point (x_0, y_0) in the xy-plane. A **first-order initial value problem** is a differential equation $y' = f(x, y)$ whose solution must satisfy an initial condition $y(x_0) = y_0$.

EXAMPLE 2 Verifying That a Function Is a Particular Solution

Show that the function

$$y = (x + 1) - \frac{1}{3}e^x$$

Substitution into the equation for y gives the particular solution

$$y = 2x^{1/3} - \ln x - 4. \qquad \blacksquare$$

In solving the linear equation in Example 2, we integrated both sides of the equation after multiplying each side by the integrating factor. However, we can shorten the amount of work, as in Example 4, by remembering that the left side *always* integrates into the product $v(x) \cdot y$ of the integrating factor times the solution function. From Equation (3) this means that

$$v(x)y = \int v(x)Q(x)\,dx.$$

We need only integrate the product of the integrating factor $v(x)$ with the right side $Q(x)$ of Equation (1) and then equate the result with $v(x)y$ to obtain the general solution. Nevertheless, to emphasize the role of $v(x)$ in the solution process, we sometimes follow the complete procedure as illustrated in Example 2.

Observe that if the function $Q(x)$ is identically zero in the standard form given by Equation (1), the linear equation is separable:

$$\frac{dy}{dx} + P(x)y = Q(x)$$

$$\frac{dy}{dx} + P(x)y = 0 \qquad Q(x) \equiv 0$$

$$dy = -P(x)\,dx \qquad \text{Separating the variables}$$

We now present two applied problems modeled by a first-order linear differential equation.

RL Circuits

The diagram in Figure 9.5 represents an electrical circuit whose total resistance is a constant R ohms and whose self-inductance, shown as a coil, is L henries, also a constant. There is a switch whose terminals at a and b can be closed to connect a constant electrical source of V volts.

Ohm's Law, $V = RI$, has to be modified for such a circuit. The modified form is

$$L\frac{di}{dt} + Ri = V, \tag{5}$$

where i is the intensity of the current in amperes and t is the time in seconds. By solving this equation, we can predict how the current will flow after the switch is closed.

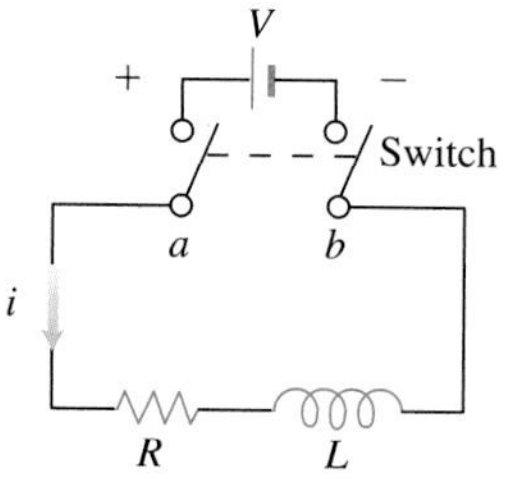

FIGURE 9.5 The *RL* circuit in Example 5.

EXAMPLE 5 Electric Current Flow

The switch in the *RL* circuit in Figure 9.5 is closed at time $t = 0$. How will the current flow as a function of time?

Solution Equation (5) is a first-order linear differential equation for i as a function of t. Its standard form is

$$\frac{di}{dt} + \frac{R}{L}i = \frac{V}{L}, \tag{6}$$

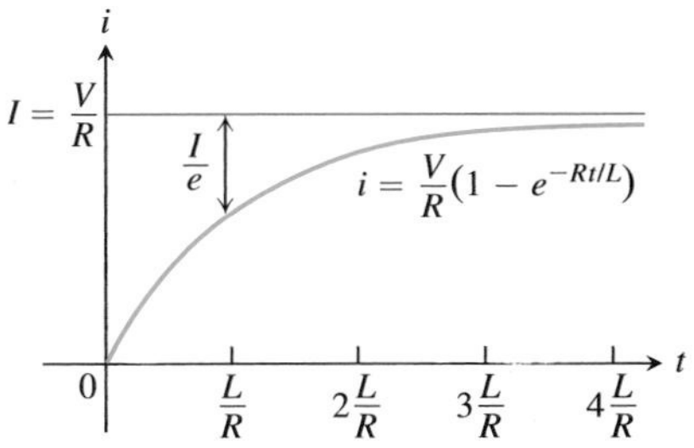

FIGURE 9.6 The growth of the current in the *RL* circuit in Example 5. *I* is the current's steady-state value. The number $t = L/R$ is the time constant of the circuit. The current gets to within 5% of its steady-state value in 3 time constants (Exercise 31).

and the corresponding solution, given that $i = 0$ when $t = 0$, is

$$i = \frac{V}{R} - \frac{V}{R}e^{-(R/L)t} \tag{7}$$

(Exercise 32). Since R and L are positive, $-(R/L)$ is negative and $e^{-(R/L)t} \to 0$ as $t \to \infty$. Thus,

$$\lim_{t\to\infty} i = \lim_{t\to\infty}\left(\frac{V}{R} - \frac{V}{R}e^{-(R/L)t}\right) = \frac{V}{R} - \frac{V}{R}\cdot 0 = \frac{V}{R}.$$

At any given time, the current is theoretically less than V/R, but as time passes, the current approaches the **steady-state value** V/R. According to the equation

$$L\frac{di}{dt} + Ri = V,$$

$I = V/R$ is the current that will flow in the circuit if either $L = 0$ (no inductance) or $di/dt = 0$ (steady current, $i =$ constant) (Figure 9.6).

Equation (7) expresses the solution of Equation (6) as the sum of two terms: a **steady-state solution** V/R and a **transient solution** $-(V/R)e^{-(R/L)t}$ that tends to zero as $t \to \infty$. ■

Mixture Problems

A chemical in a liquid solution (or dispersed in a gas) runs into a container holding the liquid (or the gas) with, possibly, a specified amount of the chemical dissolved as well. The mixture is kept uniform by stirring and flows out of the container at a known rate. In this process, it is often important to know the concentration of the chemical in the container at any given time. The differential equation describing the process is based on the formula

$$\begin{matrix}\text{Rate of change} \\ \text{of amount} \\ \text{in container}\end{matrix} = \begin{pmatrix}\text{rate at which} \\ \text{chemical} \\ \text{arrives}\end{pmatrix} - \begin{pmatrix}\text{rate at which} \\ \text{chemical} \\ \text{departs.}\end{pmatrix} \tag{8}$$

If $y(t)$ is the amount of chemical in the container at time t and $V(t)$ is the total volume of liquid in the container at time t, then the departure rate of the chemical at time t is

$$\begin{aligned}\text{Departure rate} &= \frac{y(t)}{V(t)}\cdot(\text{outflow rate}) \\ &= \begin{pmatrix}\text{concentration in} \\ \text{container at time } t\end{pmatrix}\cdot(\text{outflow rate}).\end{aligned} \tag{9}$$

Accordingly, Equation (8) becomes

$$\frac{dy}{dt} = (\text{chemical's arrival rate}) - \frac{y(t)}{V(t)}\cdot(\text{outflow rate}). \tag{10}$$

If, say, y is measured in pounds, V in gallons, and t in minutes, the units in Equation (10) are

$$\frac{\text{pounds}}{\text{minutes}} = \frac{\text{pounds}}{\text{minutes}} - \frac{\text{pounds}}{\text{gallons}}\cdot\frac{\text{gallons}}{\text{minutes}}.$$

EXAMPLE 6 Oil Refinery Storage Tank

In an oil refinery, a storage tank contains 2000 gal of gasoline that initially has 100 lb of an additive dissolved in it. In preparation for winter weather, gasoline containing 2 lb of

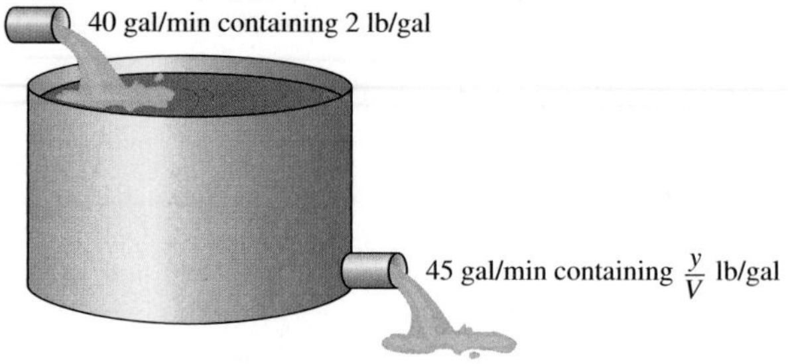

FIGURE 9.7 The storage tank in Example 6 mixes input liquid with stored liquid to produce an output liquid.

additive per gallon is pumped into the tank at a rate of 40 gal/min. The well-mixed solution is pumped out at a rate of 45 gal/min. How much of the additive is in the tank 20 min after the pumping process begins (Figure 9.7)?

Solution Let y be the amount (in pounds) of additive in the tank at time t. We know that $y = 100$ when $t = 0$. The number of gallons of gasoline and additive in solution in the tank at any time t is

$$\begin{aligned} V(t) &= 2000 \text{ gal} + \left(40 \frac{\text{gal}}{\text{min}} - 45 \frac{\text{gal}}{\text{min}}\right)(t \text{ min}) \\ &= (2000 - 5t) \text{ gal}. \end{aligned}$$

Therefore,

$$\begin{aligned} \text{Rate out} &= \frac{y(t)}{V(t)} \cdot \text{outflow rate} && \text{Eq. (9)} \\ &= \left(\frac{y}{2000 - 5t}\right) 45 && \text{Outflow rate is 45 gal/min. and } v = 2000 - 5t. \\ &= \frac{45y}{2000 - 5t} \frac{\text{lb}}{\text{min}}. \end{aligned}$$

Also,

$$\begin{aligned} \text{Rate in} &= \left(2 \frac{\text{lb}}{\text{gal}}\right)\left(40 \frac{\text{gal}}{\text{min}}\right) \\ &= 80 \frac{\text{lb}}{\text{min}}. && \text{Eq. (10)} \end{aligned}$$

The differential equation modeling the mixture process is

$$\frac{dy}{dt} = 80 - \frac{45y}{2000 - 5t}$$

in pounds per minute.

To solve this differential equation, we first write it in standard form:

$$\frac{dy}{dt} + \frac{45}{2000 - 5t} y = 80.$$

Thus, $P(t) = 45/(2000 - 5t)$ and $Q(t) = 80$.

The integrating factor is

$$v(t) = e^{\int P\,dt} = e^{\int \frac{45}{2000-5t}\,dt}$$
$$= e^{-9\ln(2000-5t)} \qquad 2000 - 5t > 0$$
$$= (2000 - 5t)^{-9}.$$

Multiplying both sides of the standard equation by $v(t)$ and integrating both sides gives,

$$(2000 - 5t)^{-9}\cdot\left(\frac{dy}{dt} + \frac{45}{2000 - 5t}\,y\right) = 80(2000 - 5t)^{-9}$$
$$(2000 - 5t)^{-9}\frac{dy}{dt} + 45(2000 - 5t)^{-10}\,y = 80(2000 - 5t)^{-9}$$
$$\frac{d}{dt}\left[(2000 - 5t)^{-9}y\right] = 80(2000 - 5t)^{-9}$$
$$(2000 - 5t)^{-9}y = \int 80(2000 - 5t)^{-9}\,dt$$
$$(2000 - 5t)^{-9}y = 80\cdot\frac{(2000 - 5t)^{-8}}{(-8)(-5)} + C.$$

The general solution is

$$y = 2(2000 - 5t) + C(2000 - 5t)^9.$$

Because $y = 100$ when $t = 0$, we can determine the value of C:

$$100 = 2(2000 - 0) + C(2000 - 0)^9$$
$$C = -\frac{3900}{(2000)^9}.$$

The particular solution of the initial value problem is

$$y = 2(2000 - 5t) - \frac{3900}{(2000)^9}(2000 - 5t)^9.$$

The amount of additive 20 min after the pumping begins is

$$y(20) = 2[2000 - 5(20)] - \frac{3900}{(2000)^9}[2000 - 5(20)]^9 \approx 1342 \text{ lb}.$$ ■

EXERCISES 9.2

First-Order Linear Equations

Solve the differential equations in Exercises 1–14.

1. $x\dfrac{dy}{dx} + y = e^x, \quad x > 0$

2. $e^x\dfrac{dy}{dx} + 2e^x y = 1$

3. $xy' + 3y = \dfrac{\sin x}{x^2}, \quad x > 0$

4. $y' + (\tan x)y = \cos^2 x, \quad -\pi/2 < x < \pi/2$

5. $x\dfrac{dy}{dx} + 2y = 1 - \dfrac{1}{x}, \quad x > 0$

6. $(1 + x)y' + y = \sqrt{x}$

7. $2y' = e^{x/2} + y$

8. $e^{2x}y' + 2e^{2x}y = 2x$

9. $xy' - y = 2x\ln x$

10. $x\dfrac{dy}{dx} = \dfrac{\cos x}{x} - 2y, \quad x > 0$

11. $(t - 1)^3\dfrac{ds}{dt} + 4(t - 1)^2 s = t + 1, \quad t > 1$

12. $(t+1)\dfrac{ds}{dt} + 2s = 3(t+1) + \dfrac{1}{(t+1)^2}, \quad t > -1$

13. $\sin\theta \dfrac{dr}{d\theta} + (\cos\theta)r = \tan\theta, \quad 0 < \theta < \pi/2$

14. $\tan\theta \dfrac{dr}{d\theta} + r = \sin^2\theta, \quad 0 < \theta < \pi/2$

Solving Initial Value Problems

Solve the initial value problems in Exercises 15–20.

15. $\dfrac{dy}{dt} + 2y = 3, \quad y(0) = 1$

16. $t\dfrac{dy}{dt} + 2y = t^3, \quad t > 0, \quad y(2) = 1$

17. $\theta\dfrac{dy}{d\theta} + y = \sin\theta, \quad \theta > 0, \quad y(\pi/2) = 1$

18. $\theta\dfrac{dy}{d\theta} - 2y = \theta^3 \sec\theta \tan\theta, \quad \theta > 0, \quad y(\pi/3) = 2$

19. $(x+1)\dfrac{dy}{dx} - 2(x^2 + x)y = \dfrac{e^{x^2}}{x+1}, \quad x > -1, \quad y(0) = 5$

20. $\dfrac{dy}{dx} + xy = x, \quad y(0) = -6$

21. Solve the exponential growth/decay initial value problem for y as a function of t thinking of the differential equation as a first-order linear equation with $P(x) = -k$ and $Q(x) = 0$:

$$\frac{dy}{dt} = ky \quad (k \text{ constant}), \quad y(0) = y_0$$

22. Solve the following initial value problem for u as a function of t:

$$\frac{du}{dt} + \frac{k}{m}u = 0 \quad (k \text{ and } m \text{ positive constants}), \quad u(0) = u_0$$

a. as a first-order linear equation.

b. as a separable equation.

Theory and Examples

23. Is either of the following equations correct? Give reasons for your answers.

a. $x\displaystyle\int \frac{1}{x}\,dx = x\ln|x| + C$ b. $x\displaystyle\int \frac{1}{x}\,dx = x\ln|x| + Cx$

24. Is either of the following equations correct? Give reasons for your answers.

a. $\dfrac{1}{\cos x}\displaystyle\int \cos x\,dx = \tan x + C$

b. $\dfrac{1}{\cos x}\displaystyle\int \cos x\,dx = \tan x + \dfrac{C}{\cos x}$

25. **Salt mixture** A tank initially contains 100 gal of brine in which 50 lb of salt are dissolved. A brine containing 2 lb/gal of salt runs into the tank at the rate of 5 gal/min. The mixture is kept uniform by stirring and flows out of the tank at the rate of 4 gal/min.

a. At what rate (pounds per minute) does salt enter the tank at time t?

b. What is the volume of brine in the tank at time t?

c. At what rate (pounds per minute) does salt leave the tank at time t?

d. Write down and solve the initial value problem describing the mixing process.

e. Find the concentration of salt in the tank 25 min after the process starts.

26. **Mixture problem** A 200-gal tank is half full of distilled water. At time $t = 0$, a solution containing 0.5 lb/gal of concentrate enters the tank at the rate of 5 gal/min, and the well-stirred mixture is withdrawn at the rate of 3 gal/min.

a. At what time will the tank be full?

b. At the time the tank is full, how many pounds of concentrate will it contain?

27. **Fertilizer mixture** A tank contains 100 gal of fresh water. A solution containing 1 lb/gal of soluble lawn fertilizer runs into the tank at the rate of 1 gal/min, and the mixture is pumped out of the tank at the rate of 3 gal/min. Find the maximum amount of fertilizer in the tank and the time required to reach the maximum.

28. **Carbon monoxide pollution** An executive conference room of a corporation contains 4500 ft^3 of air initially free of carbon monoxide. Starting at time $t = 0$, cigarette smoke containing 4% carbon monoxide is blown into the room at the rate of 0.3 ft^3/min. A ceiling fan keeps the air in the room well circulated and the air leaves the room at the same rate of 0.3 ft^3/min. Find the time when the concentration of carbon monoxide in the room reaches 0.01%.

29. **Current in a closed *RL* circuit** How many seconds after the switch in an RL circuit is closed will it take the current i to reach half of its steady state value? Notice that the time depends on R and L and not on how much voltage is applied.

30. **Current in an open *RL* circuit** If the switch is thrown open after the current in an RL circuit has built up to its steady-state value $I = V/R$, the decaying current (graphed here) obeys the equation

$$L\frac{di}{dt} + Ri = 0,$$

which is Equation (5) with $V = 0$.

a. Solve the equation to express i as a function of t.

b. How long after the switch is thrown will it take the current to fall to half its original value?

c. Show that the value of the current when $t = L/R$ is I/e. (The significance of this time is explained in the next exercise.)

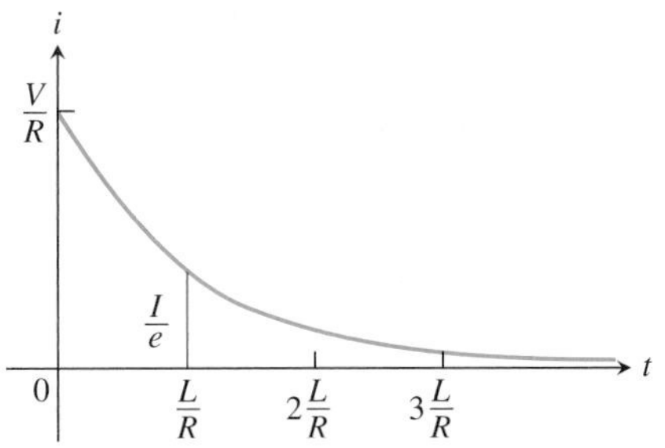

31. Time constants Engineers call the number L/R the *time constant* of the *RL* circuit in Figure 9.6. The significance of the time constant is that the current will reach 95% of its final value within 3 time constants of the time the switch is closed (Figure 9.6). Thus, the time constant gives a built-in measure of how rapidly an individual circuit will reach equilibrium.

a. Find the value of i in Equation (7) that corresponds to $t = 3L/R$ and show that it is about 95% of the steady-state value $I = V/R$.

b. Approximately what percentage of the steady-state current will be flowing in the circuit 2 time constants after the switch is closed (i.e., when $t = 2L/R$)?

32. Derivation of Equation (7) in Example 5

a. Show that the solution of the equation

$$\frac{di}{dt} + \frac{R}{L}i = \frac{V}{L}$$

is

$$i = \frac{V}{R} + Ce^{-(R/L)t}.$$

b. Then use the initial condition $i(0) = 0$ to determine the value of C. This will complete the derivation of Equation (7).

c. Show that $i = V/R$ is a solution of Equation (6) and that $i = Ce^{-(R/L)t}$ satisfies the equation

$$\frac{di}{dt} + \frac{R}{L}i = 0.$$

HISTORICAL BIOGRAPHY

James Bernoulli
(1654–1705)

A **Bernoulli differential equation** is of the form

$$\frac{dy}{dx} + P(x)y = Q(x)y^n.$$

Observe that, if $n = 0$ or 1, the Bernoulli equation is linear. For other values of n, the substitution $u = y^{1-n}$ transforms the Bernoulli equation into the linear equation

$$\frac{du}{dx} + (1 - n)P(x)u = (1 - n)Q(x).$$

For example, in the equation

$$\frac{dy}{dx} - y = e^{-x}y^2$$

we have $n = 2$, so that $u = y^{1-2} = y^{-1}$ and $du/dx = -y^{-2}\,dy/dx$. Then $dy/dx = -y^2\,du/dx = -u^{-2}\,du/dx$. Substitution into the original equation gives

$$-u^{-2}\frac{du}{dx} - u^{-1} = e^{-x}u^{-2}$$

or, equivalently,

$$\frac{du}{dx} + u = -e^{-x}.$$

This last equation is linear in the (unknown) dependent variable u.

Solve the differential equations in Exercises 33–36.

33. $y' - y = -y^2$

34. $y' - y = xy^2$

35. $xy' + y = y^{-2}$

36. $x^2y' + 2xy = y^3$

9.3 Euler's Method

HISTORICAL BIOGRAPHY

Leonhard Euler
(1703–1783)

If we do not require or cannot immediately find an *exact* solution for an initial value problem $y' = f(x, y)$, $y(x_0) = y_0$ we can often use a computer to generate a table of approximate numerical values of y for values of x in an appropriate interval. Such a table is called a **numerical solution** of the problem, and the method by which we generate the table is called a **numerical method**. Numerical methods are generally fast and accurate, and they are often the methods of choice when exact formulas are unnecessary, unavailable, or overly complicated. In this section, we study one such method, called Euler's method, upon which many other numerical methods are based.

Euler's Method

Given a differential equation $dy/dx = f(x, y)$ and an initial condition $y(x_0) = y_0$, we can approximate the solution $y = y(x)$ by its linearization

$$L(x) = y(x_0) + y'(x_0)(x - x_0) \qquad \text{or} \qquad L(x) = y_0 + f(x_0, y_0)(x - x_0).$$

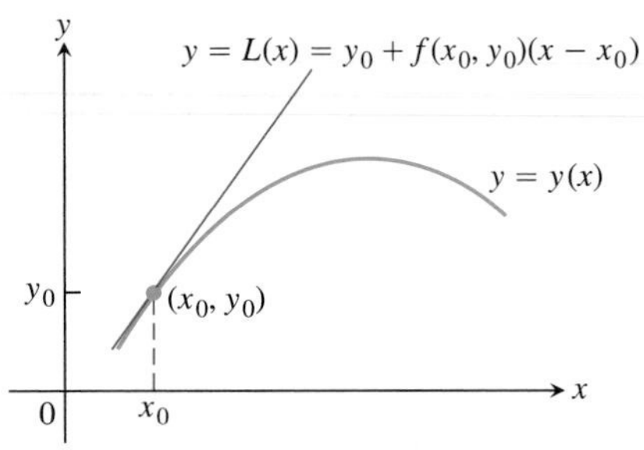

FIGURE 9.8 The linearization $L(x)$ of $y = y(x)$ at $x = x_0$.

The function $L(x)$ gives a good approximation to the solution $y(x)$ in a short interval about x_0 (Figure 9.8). The basis of Euler's method is to patch together a string of linearizations to approximate the curve over a longer stretch. Here is how the method works.

We know the point (x_0, y_0) lies on the solution curve. Suppose that we specify a new value for the independent variable to be $x_1 = x_0 + dx$. (Recall that $dx = \Delta x$ in the definition of differentials.) If the increment dx is small, then

$$y_1 = L(x_1) = y_0 + f(x_0, y_0)\,dx$$

is a good approximation to the exact solution value $y = y(x_1)$. So from the point (x_0, y_0), which lies *exactly* on the solution curve, we have obtained the point (x_1, y_1), which lies very close to the point $(x_1, y(x_1))$ on the solution curve (Figure 9.9).

Using the point (x_1, y_1) and the slope $f(x_1, y_1)$ of the solution curve through (x_1, y_1), we take a second step. Setting $x_2 = x_1 + dx$, we use the linearization of the solution curve through (x_1, y_1) to calculate

$$y_2 = y_1 + f(x_1, y_1)\,dx.$$

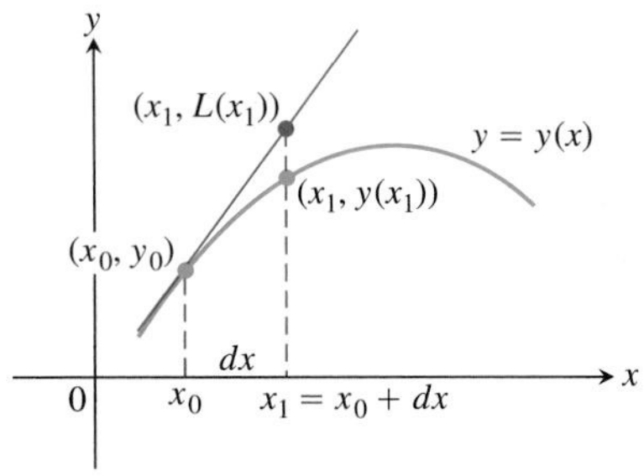

FIGURE 9.9 The first Euler step approximates $y(x_1)$ with $y_1 = L(x_1)$.

This gives the next approximation (x_2, y_2) to values along the solution curve $y = y(x)$ (Figure 9.10). Continuing in this fashion, we take a third step from the point (x_2, y_2) with slope $f(x_2, y_2)$ to obtain the third approximation

$$y_3 = y_2 + f(x_2, y_2)\,dx,$$

and so on. We are literally building an approximation to one of the solutions by following the direction of the slope field of the differential equation.

The steps in Figure 9.10 are drawn large to illustrate the construction process, so the approximation looks crude. In practice, dx would be small enough to make the red curve hug the blue one and give a good approximation throughout.

EXAMPLE 1 Using Euler's Method

Find the first three approximations y_1, y_2, y_3 using Euler's method for the initial value problem

$$y' = 1 + y, \qquad y(0) = 1,$$

starting at $x_0 = 0$ with $dx = 0.1$.

Solution We have $x_0 = 0, y_0 = 1, x_1 = x_0 + dx = 0.1, x_2 = x_0 + 2dx = 0.2$, and $x_3 = x_0 + 3dx = 0.3$.

$$\begin{aligned}
\textit{First:}\quad y_1 &= y_0 + f(x_0, y_0)\,dx \\
&= y_0 + (1 + y_0)\,dx \\
&= 1 + (1 + 1)(0.1) = 1.2 \\
\textit{Second:}\quad y_2 &= y_1 + f(x_1, y_1)\,dx \\
&= y_1 + (1 + y_1)\,dx \\
&= 1.2 + (1 + 1.2)(0.1) = 1.42 \\
\textit{Third:}\quad y_3 &= y_2 + f(x_2, y_2)\,dx \\
&= y_2 + (1 + y_2)\,dx \\
&= 1.42 + (1 + 1.42)(0.1) = 1.662
\end{aligned}$$

■

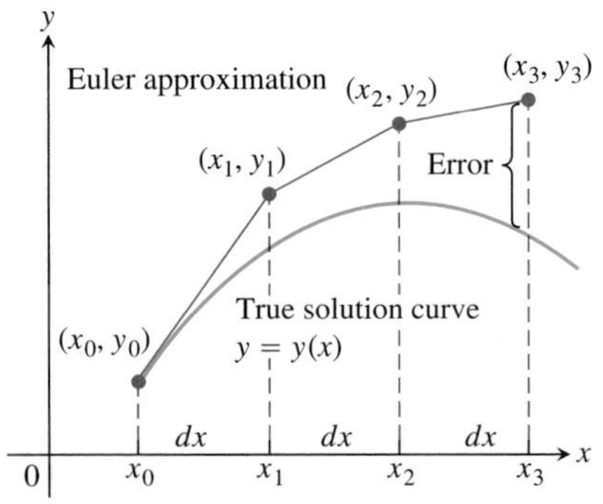

FIGURE 9.10 Three steps in the Euler approximation to the solution of the initial value problem $y' = f(x, y), y(x_0) = y_0$. As we take more steps, the errors involved usually accumulate, but not in the exaggerated way shown here.

The step-by-step process used in Example 1 can be continued easily. Using equally spaced values for the independent variable in the table and generating n of them, set

$$\begin{aligned}
x_1 &= x_0 + dx \\
x_2 &= x_1 + dx \\
&\vdots \\
x_n &= x_{n-1} + dx.
\end{aligned}$$

Then calculate the approximations to the solution,

$$y_1 = y_0 + f(x_0, y_0)\,dx$$
$$y_2 = y_1 + f(x_1, y_1)\,dx$$
$$\vdots$$
$$y_n = y_{n-1} + f(x_{n-1}, y_{n-1})\,dx.$$

The number of steps n can be as large as we like, but errors can accumulate if n is too large.

Euler's method is easy to implement on a computer or calculator. A computer program generates a table of numerical solutions to an initial value problem, allowing us to input x_0 and y_0, the number of steps n, and the step size dx. It then calculates the approximate solution values $y_1, y_2, \ldots, y_n$ in iterative fashion, as just described.

Solving the separable equation in Example 1, we find that the exact solution to the initial value problem is $y = 2e^x - 1$. We use this information in Example 2.

EXAMPLE 2 Investigating the Accuracy of Euler's Method

Use Euler's method to solve

$$y' = 1 + y, \qquad y(0) = 1,$$

on the interval $0 \le x \le 1$, starting at $x_0 = 0$ and taking

(a) $dx = 0.1$

(b) $dx = 0.05$.

Compare the approximations with the values of the exact solution $y = 2e^x - 1$.

Solution

(a) We used a computer to generate the approximate values in Table 9.1. The "error" column is obtained by subtracting the unrounded Euler values from the unrounded values found using the exact solution. All entries are then rounded to four decimal places.

TABLE 9.1 Euler solution of $y' = 1 + y$, $y(0) = 1$, step size $dx = 0.1$

x	y (Euler)	y (exact)	Error
0	1	1	0
0.1	1.2	1.2103	0.0103
0.2	1.42	1.4428	0.0228
0.3	1.662	1.6997	0.0377
0.4	1.9282	1.9836	0.0554
0.5	2.2210	2.2974	0.0764
0.6	2.5431	2.6442	0.1011
0.7	2.8974	3.0275	0.1301
0.8	3.2872	3.4511	0.1639
0.9	3.7159	3.9192	0.2033
1.0	4.1875	4.4366	0.2491

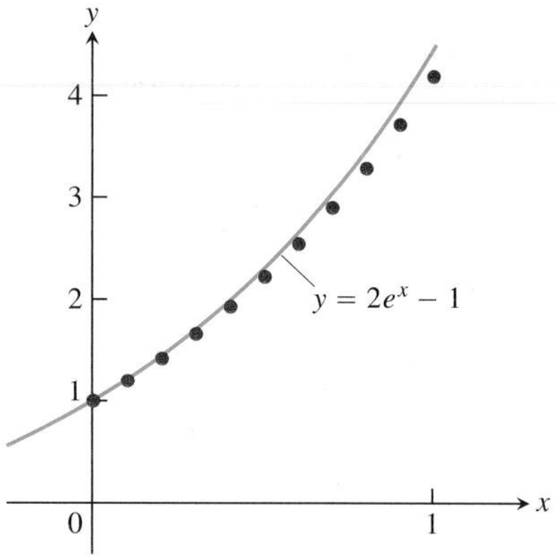

FIGURE 9.11 The graph of $y = 2e^x - 1$ superimposed on a scatterplot of the Euler approximations shown in Table 9.1 (Example 2).

By the time we reach $x = 1$ (after 10 steps), the error is about 5.6% of the exact solution. A plot of the exact solution curve with the scatterplot of Euler solution points from Table 9.1 is shown in Figure 9.11.

(b) One way to try to reduce the error is to decrease the step size. Table 9.2 shows the results and their comparisons with the exact solutions when we decrease the step size to 0.05, doubling the number of steps to 20. As in Table 9.1, all computations are performed before rounding. This time when we reach $x = 1$, the relative error is only about 2.9%.

TABLE 9.2 Euler solution of $y' = 1 + y$, $y(0) = 1$, step size $dx = 0.05$

x	y (Euler)	y (exact)	Error
0	1	1	0
0.05	1.1	1.1025	0.0025
0.10	1.205	1.2103	0.0053
0.15	1.3153	1.3237	0.0084
0.20	1.4310	1.4428	0.0118
0.25	1.5526	1.5681	0.0155
0.30	1.6802	1.6997	0.0195
0.35	1.8142	1.8381	0.0239
0.40	1.9549	1.9836	0.0287
0.45	2.1027	2.1366	0.0340
0.50	2.2578	2.2974	0.0397
0.55	2.4207	2.4665	0.0458
0.60	2.5917	2.6442	0.0525
0.65	2.7713	2.8311	0.0598
0.70	2.9599	3.0275	0.0676
0.75	3.1579	3.2340	0.0761
0.80	3.3657	3.4511	0.0853
0.85	3.5840	3.6793	0.0953
0.90	3.8132	3.9192	0.1060
0.95	4.0539	4.1714	0.1175
1.00	4.3066	4.4366	0.1300

■

It might be tempting to reduce the step size even further in Example 2 to obtain greater accuracy. Each additional calculation, however, not only requires additional computer time but more importantly adds to the buildup of round-off errors due to the approximate representations of numbers inside the computer.

The analysis of error and the investigation of methods to reduce it when making numerical calculations are important but are appropriate for a more advanced course. There are numerical methods more accurate than Euler's method, as you can see in a further study of differential equations. We study one improvement here.

HISTORICAL BIOGRAPHY

Carl Runge
(1856–1927)

Improved Euler's Method

We can improve on Euler's method by taking an average of two slopes. We first estimate y_n as in the original Euler method, but denote it by z_n. We then take the average of $f(x_{n-1}, y_{n-1})$ and $f(x_n, z_n)$ in place of $f(x_{n-1}, y_{n-1})$ in the next step. Thus, we calculate the next approximation y_n using

$$z_n = y_{n-1} + f(x_{n-1}, y_{n-1})\,dx$$

$$y_n = y_{n-1} + \left[\frac{f(x_{n-1}, y_{n-1}) + f(x_n, z_n)}{2}\right] dx.$$

EXAMPLE 3 Investigating the Accuracy of the Improved Euler's Method

Use the improved Euler's method to solve

$$y' = 1 + y, \qquad y(0) = 1,$$

on the interval $0 \leq x \leq 1$, starting at $x_0 = 0$ and taking $dx = 0.1$. Compare the approximations with the values of the exact solution $y = 2e^x - 1$.

Solution We used a computer to generate the approximate values in Table 9.3. The "error" column is obtained by subtracting the unrounded improved Euler values from the unrounded values found using the exact solution. All entries are then rounded to four decimal places.

TABLE 9.3 Improved Euler solution of $y' = 1 + y$, $y(0) = 1$, step size $dx = 0.1$

x	y (improved Euler)	y (exact)	Error
0	1	1	0
0.1	1.21	1.2103	0.0003
0.2	1.4421	1.4428	0.0008
0.3	1.6985	1.6997	0.0013
0.4	1.9818	1.9836	0.0018
0.5	2.2949	2.2974	0.0025
0.6	2.6409	2.6442	0.0034
0.7	3.0231	3.0275	0.0044
0.8	3.4456	3.4511	0.0055
0.9	3.9124	3.9192	0.0068
1.0	4.4282	4.4366	0.0084

By the time we reach $x = 1$ (after 10 steps), the relative error is about 0.19%. ■

By comparing Tables 9.1 and 9.3, we see that the improved Euler's method is considerably more accurate than the regular Euler's method, at least for the initial value problem $y' = 1 + y,\ \ y(0) = 1$.

EXAMPLE 4 Oil Refinery Storage Tank Revisited

In Example 6, Section 9.2, we looked at a problem involving an additive mixture entering a 2000-gallon gasoline tank that was simultaneously being pumped. The analysis gave the initial value problem

$$\frac{dy}{dt} = 80 - \frac{45y}{2000 - 5t}, \qquad y(0) = 100$$

where $y(t)$ is the amount of additive in the tank at time t. The question was to find $y(20)$. Using Euler's method with an increment of $dt = 0.2$ (or 100 steps) gives the approximations

$$y(0.2) \approx 115.55, \qquad y(0.4) \approx 131.0298, \ldots$$

ending with $y(20) \approx 1344.3616$. The relative error from the exact solution $y(20) = 1342$ is about 0.18%. ■

EXERCISES 9.3

Calculating Euler Approximations

In Exercises 1–6, use Euler's method to calculate the first three approximations to the given initial value problem for the specified increment size. Calculate the exact solution and investigate the accuracy of your approximations. Round your results to four decimal places.

1. $y' = 1 - \frac{y}{x}, \quad y(2) = -1, \quad dx = 0.5$
2. $y' = x(1 - y), \quad y(1) = 0, \quad dx = 0.2$
3. $y' = 2xy + 2y, \quad y(0) = 3, \quad dx = 0.2$
4. $y' = y^2(1 + 2x), \quad y(-1) = 1, \quad dx = 0.5$
5. T $y' = 2xe^{x^2}, \quad y(0) = 2, \quad dx = 0.1$
6. T $y' = y + e^x - 2, \quad y(0) = 2, \quad dx = 0.5$
7. Use the Euler method with $dx = 0.2$ to estimate $y(1)$ if $y' = y$ and $y(0) = 1$. What is the exact value of $y(1)$?
8. Use the Euler method with $dx = 0.2$ to estimate $y(2)$ if $y' = y/x$ and $y(1) = 2$. What is the exact value of $y(2)$?
9. T Use the Euler method with $dx = 0.5$ to estimate $y(5)$ if $y' = y^2/\sqrt{x}$ and $y(1) = -1$. What is the exact value of $y(5)$?
10. T Use the Euler method with $dx = 1/3$ to estimate $y(2)$ if $y' = y - e^{2x}$ and $y(0) = 1$. What is the exact value of $y(2)$?

Improved Euler's Method

In Exercises 11 and 12, use the improved Euler's method to calculate the first three approximations to the given initial value problem. Compare the approximations with the values of the exact solution.

11. $y' = 2y(x + 1), \quad y(0) = 3, \quad dx = 0.2$

    (See Exercise 3 for the exact solution.)
12. $y' = x(1 - y), \quad y(1) = 0, \quad dx = 0.2$

    (See Exercise 2 for the exact solution.)

COMPUTER EXPLORATIONS

Euler's Method

In Exercises 13–16, use Euler's method with the specified step size to estimate the value of the solution at the given point x^*. Find the value of the exact solution at x^*.

13. $y' = 2xe^{x^2}, \quad y(0) = 2, \quad dx = 0.1, \quad x^* = 1$
14. $y' = y + e^x - 2, \quad y(0) = 2, \quad dx = 0.5, \quad x^* = 2$
15. $y' = \sqrt{x}/y, \quad y > 0, \quad y(0) = 1, \quad dx = 0.1, \quad x^* = 1$
16. $y' = 1 + y^2, \quad y(0) = 0, \quad dx = 0.1, \quad x^* = 1$

In Exercises 17 and 18, **(a)** find the exact solution of the initial value problem. Then compare the accuracy of the approximation with $y(x^*)$ using Euler's method starting at x_0 with step size **(b)** 0.2, **(c)** 0.1, and **(d)** 0.05.

17. $y' = 2y^2(x - 1), \quad y(2) = -1/2, \quad x_0 = 2, \quad x^* = 3$
18. $y' = y - 1, \quad y(0) = 3, \quad x_0 = 0, \quad x^* = 1$

Improved Euler's Method

In Exercises 19 and 20, compare the accuracy of the approximation with $y(x^*)$ using the improved Euler's method starting at x_0 with step size

a. 0.2 **b.** 0.1 **c.** 0.05

d. Describe what happens to the error as the step size decreases.

19. $y' = 2y^2(x - 1), \quad y(2) = -1/2, \quad x_0 = 2, \quad x^* = 3$

    (See Exercise 17 for the exact solution.)
20. $y' = y - 1, \quad y(0) = 3, \quad x_0 = 0, \quad x^* = 1$

    (See Exercise 18 for the exact solution.)

Exploring Differential Equations Graphically

Use a CAS to explore graphically each of the differential equations in Exercises 21–24. Perform the following steps to help with your explorations.

a. Plot a slope field for the differential equation in the given xy-window.

b. Find the general solution of the differential equation using your CAS DE solver.

c. Graph the solutions for the values of the arbitrary constant $C = -2, -1, 0, 1, 2$ superimposed on your slope field plot.

d. Find and graph the solution that satisfies the specified initial condition over the interval $[0, b]$.

e. Find the Euler numerical approximation to the solution of the initial value problem with 4 subintervals of the x-interval and plot the Euler approximation superimposed on the graph produced in part (d).

f. Repeat part (e) for 8, 16, and 32 subintervals. Plot these three Euler approximations superimposed on the graph from part (e).

g. Find the error $(y(\text{exact}) - y(\text{Euler}))$ at the specified point $x = b$ for each of your four Euler approximations. Discuss the improvement in the percentage error.

21. $y' = x + y,\ \ y(0) = -7/10;\ \ -4 \le x \le 4,\ \ -4 \le y \le 4;\ \ b = 1$

22. $y' = -x/y,\ \ y(0) = 2;\ \ -3 \le x \le 3,\ \ -3 \le y \le 3;\ \ b = 2$

23. A logistic equation $y' = y(2 - y),\ \ y(0) = 1/2;\ \ 0 \le x \le 4,\ \ 0 \le y \le 3;\ \ b = 3$

24. $y' = (\sin x)(\sin y),\ \ y(0) = 2;\ \ -6 \le x \le 6,\ \ -6 \le y \le 6;\ \ b = 3\pi/2$

9.4 Graphical Solutions of Autonomous Differential Equations

In Chapter 4 we learned that the sign of the first derivative tells where the graph of a function is increasing and where it is decreasing. The sign of the second derivative tells the concavity of the graph. We can build on our knowledge of how derivatives determine the shape of a graph to solve differential equations graphically. The starting ideas for doing so are the notions of *phase line* and *equilibrium value*. We arrive at these notions by investigating what happens when the derivative of a differentiable function is zero from a point of view different from that studied in Chapter 4.

Equilibrium Values and Phase Lines

When we differentiate implicitly the equation

$$\frac{1}{5}\ln(5y - 15) = x + 1$$

we obtain

$$\frac{1}{5}\left(\frac{5}{5y - 15}\right)\frac{dy}{dx} = 1.$$

Solving for $y' = dy/dx$ we find $y' = 5y - 15 = 5(y - 3)$. In this case the derivative y' is a function of y only (the dependent variable) and is zero when $y = 3$.

A differential equation for which dy/dx is a function of y only is called an **autonomous** differential equation. Let's investigate what happens when the derivative in an autonomous equation equals zero.

DEFINITION Equilibrium Values

If $dy/dx = g(y)$ is an autonomous differential equation, then the values of y for which $dy/dx = 0$ are called **equilibrium values** or **rest points**.

Thus, equilibrium values are those at which no change occurs in the dependent variable, so y is at *rest*. The emphasis is on the value of y where $dy/dx = 0$, not the value of x, as we studied in Chapter 4.

EXAMPLE 1 Finding Equilibrium Values

The equilibrium values for the autonomous differential equation

$$\frac{dy}{dx} = (y + 1)(y - 2)$$

are $y = -1$ and $y = 2$. ■

To construct a graphical solution to an autonomous differential equation like the one in Example 1, we first make a **phase line** for the equation, a plot on the y-axis that shows the equation's equilibrium values along with the intervals where dy/dx and d^2y/dx^2 are positive and negative. Then we know where the solutions are increasing and decreasing, and the concavity of the solution curves. These are the essential features we found in Section 4.4, so we can determine the shapes of the solution curves without having to find formulas for them.

EXAMPLE 2 Drawing a Phase Line and Sketching Solution Curves

Draw a phase line for the equation

$$\frac{dy}{dx} = (y + 1)(y - 2)$$

and use it to sketch solutions to the equation.

Solution

1. *Draw a number line for y and mark the equilibrium values $y = -1$ and $y = 2$, where $dy/dx = 0$.*

2. *Identify and label the intervals where $y' > 0$ and $y' < 0$.* This step resembles what we did in Section 4.3, only now we are marking the y-axis instead of the x-axis.

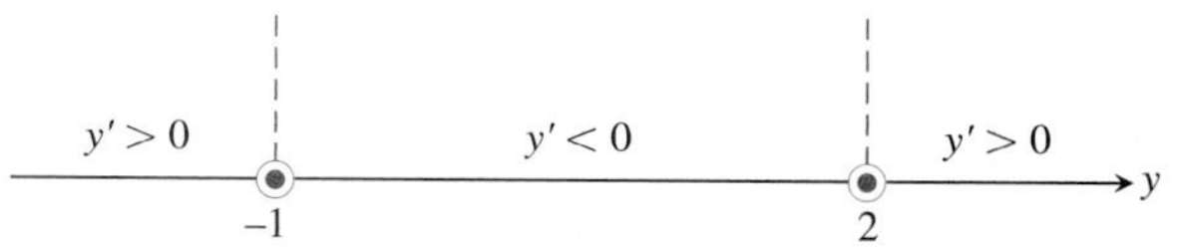

We can encapsulate the information about the sign of y' on the phase line itself. Since $y' > 0$ on the interval to the left of $y = -1$, a solution of the differential equation with a y-value less than -1 will increase from there toward $y = -1$. We display this information by drawing an arrow on the interval pointing to -1.

Similarly, $y' < 0$ between $y = -1$ and $y = 2$, so any solution with a value in this interval will decrease toward $y = -1$.

For $y > 2$, we have $y' > 0$, so a solution with a y-value greater than 2 will increase from there without bound.

In short, solution curves below the horizontal line $y = -1$ in the xy-plane rise toward $y = -1$. Solution curves between the lines $y = -1$ and $y = 2$ fall away from $y = 2$ toward $y = -1$. Solution curves above $y = 2$ rise away from $y = 2$ and keep going up.

3. *Calculate y'' and mark the intervals where $y'' > 0$ and $y'' < 0$.* To find y'', we differentiate y' *with respect to* x, using implicit differentiation.

$$\begin{aligned} y' &= (y+1)(y-2) = y^2 - y - 2 && \text{Formula for } y' \ldots \\ y'' &= \frac{d}{dx}(y') = \frac{d}{dx}(y^2 - y - 2) \\ &= 2yy' - y' && \text{differentiated implicitly with respect to } x. \\ &= (2y-1)y' \\ &= (2y-1)(y+1)(y-2). \end{aligned}$$

From this formula, we see that y'' changes sign at $y = -1, y = 1/2$, and $y = 2$. We add the sign information to the phase line.

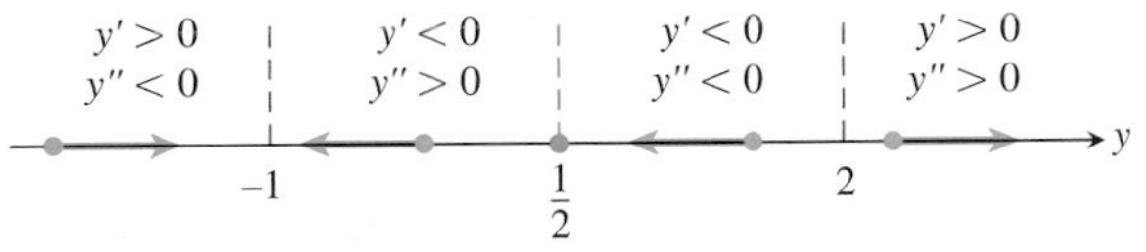

4. *Sketch an assortment of solution curves in the xy-plane.* The horizontal lines $y = -1, y = 1/2$, and $y = 2$ partition the plane into horizontal bands in which we know the signs of y' and y''. In each band, this information tells us whether the solution curves rise or fall and how they bend as x increases (Figure 9.12).

The "equilibrium lines" $y = -1$ and $y = 2$ are also solution curves. (The constant functions $y = -1$ and $y = 2$ satisfy the differential equation.) Solution curves that cross the line $y = 1/2$ have an inflection point there. The concavity changes from concave down (above the line) to concave up (below the line).

As predicted in Step 2, solutions in the middle and lower bands approach the equilibrium value $y = -1$ as x increases. Solutions in the upper band rise steadily away from the value $y = 2$. ■

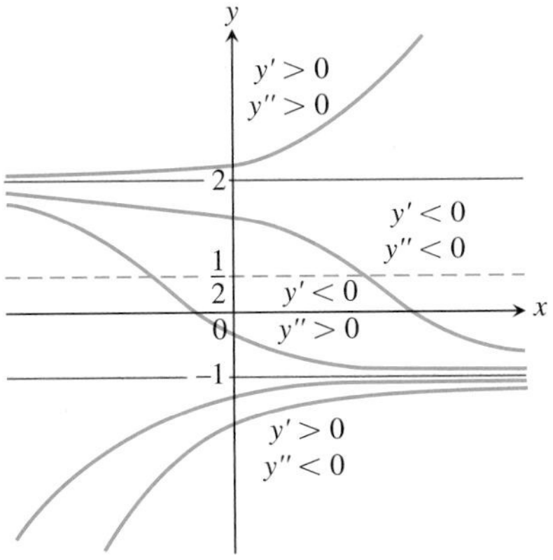

FIGURE 9.12 Graphical solutions from Example 2 include the horizontal lines $y = -1$ and $y = 2$ through the equilibrium values.

Stable and Unstable Equilibria

Look at Figure 9.12 once more, in particular at the behavior of the solution curves near the equilibrium values. Once a solution curve has a value near $y = -1$, it tends steadily toward that value; $y = -1$ is a **stable equilibrium**. The behavior near $y = 2$ is just the opposite: all solutions except the equilibrium solution $y = 2$ itself move *away* from it as x increases. We call $y = 2$ an **unstable equilibrium**. If the solution is *at* that value, it stays, but if it is off by any amount, no matter how small, it moves away. (Sometimes an equilibrium value is unstable because a solution moves away from it only on one side of the point.)

Now that we know what to look for, we can already see this behavior on the initial phase line. The arrows lead away from $y = 2$ and, once to the left of $y = 2$, toward $y = -1$.

We now present several applied examples for which we can sketch a family of solution curves to the differential equation models using the method in Example 2.

In Section 7.5 we solved analytically the differential equation

$$\frac{dH}{dt} = -k(H - H_S), \qquad k > 0$$

modeling Newton's law of cooling. Here H is the temperature (amount of heat) of an object at time t and H_S is the constant temperature of the surrounding medium. Our first example uses a phase line analysis to understand the graphical behavior of this temperature model over time.

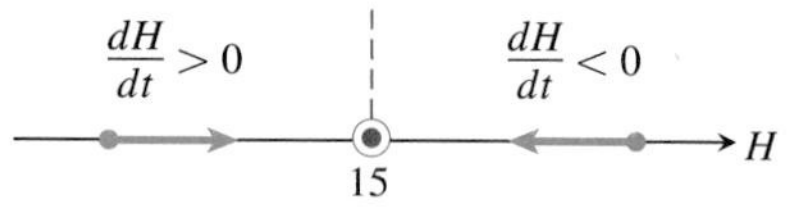

FIGURE 9.13 First step in constructing the phase line for Newton's law of cooling in Example 3. The temperature tends towards the equilibrium (surrounding-medium) value in the long run.

EXAMPLE 3 Cooling Soup

What happens to the temperature of the soup when a cup of hot soup is placed on a table in a room? We know the soup cools down, but what does a typical temperature curve look like as a function of time?

Solution Suppose that the surrounding medium has a constant Celsius temperature of 15°C. We can then express the difference in temperature as $H(t) - 15$. Assuming H is a differentiable function of time t, by Newton's law of cooling, there is a constant of proportionality $k > 0$ such that

$$\frac{dH}{dt} = -k(H - 15) \tag{1}$$

(*minus* k to give a negative derivative when $H > 15$).

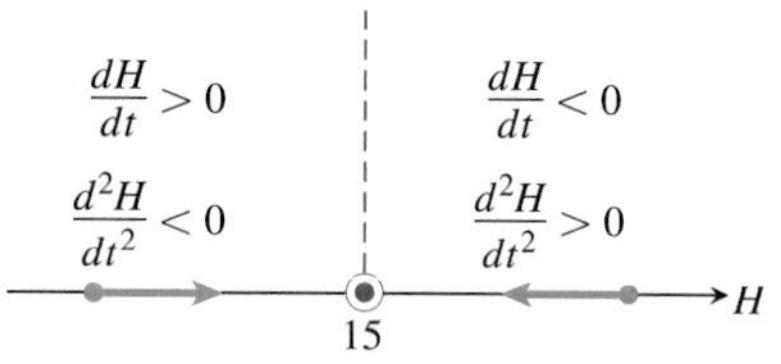

FIGURE 9.14 The complete phase line for Newton's law of cooling (Example 3).

Since $dH/dt = 0$ at $H = 15$, the temperature 15°C is an equilibrium value. If $H > 15$, Equation (1) tells us that $(H - 15) > 0$ and $dH/dt < 0$. If the object is hotter than the room, it will get cooler. Similarly, if $H < 15$, then $(H - 15) < 0$ and $dH/dt > 0$. An object cooler than the room will warm up. Thus, the behavior described by Equation (1) agrees with our intuition of how temperature should behave. These observations are captured in the initial phase line diagram in Figure 9.13. The value $H = 15$ is a stable equilibrium.

We determine the concavity of the solution curves by differentiating both sides of Equation (1) with respect to t:

$$\frac{d}{dt}\left(\frac{dH}{dt}\right) = \frac{d}{dt}(-k(H - 15))$$

$$\frac{d^2H}{dt^2} = -k\frac{dH}{dt}.$$

Since $-k$ is negative, we see that d^2H/dt^2 is positive when $dH/dt < 0$ and negative when $dH/dt > 0$. Figure 9.14 adds this information to the phase line.

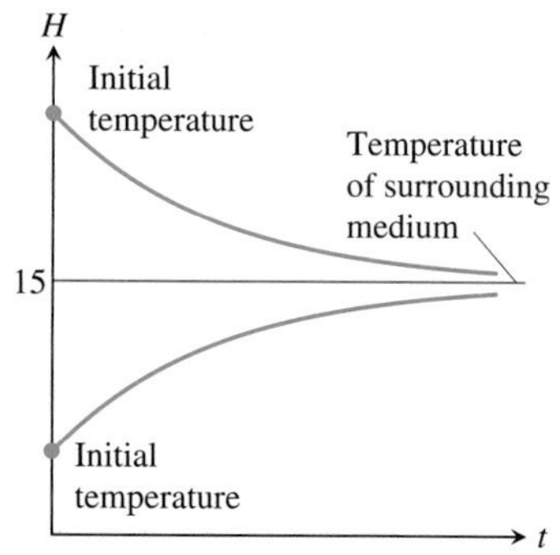

FIGURE 9.15 Temperature versus time. Regardless of initial temperature, the object's temperature $H(t)$ tends toward 15°C, the temperature of the surrounding medium.

The completed phase line shows that if the temperature of the object is above the equilibrium value of 15°C, the graph of $H(t)$ will be decreasing and concave upward. If the temperature is below 15°C (the temperature of the surrounding medium), the graph of $H(t)$ will be increasing and concave downward. We use this information to sketch typical solution curves (Figure 9.15).

From the upper solution curve in Figure 9.15, we see that as the object cools down, the rate at which it cools slows down because dH/dt approaches zero. This observation is implicit in Newton's law of cooling and contained in the differential equation, but the flattening of the graph as time advances gives an immediate visual representation of the phenomenon. The ability to discern physical behavior from graphs is a powerful tool in understanding real-world systems. ■

EXAMPLE 4 Analyzing the Fall of a Body Encountering a Resistive Force

Galileo and Newton both observed that the rate of change in momentum encountered by a moving object is equal to the net force applied to it. In mathematical terms,

$$F = \frac{d}{dt}(mv) \tag{2}$$

where F is the force and m and v the object's mass and velocity. If m varies with time, as it will if the object is a rocket burning fuel, the right-hand side of Equation (2) expands to

$$m\frac{dv}{dt} + v\frac{dm}{dt}$$

using the Product Rule. In many situations, however, m is constant, $dm/dt = 0$, and Equation (2) takes the simpler form

$$F = m\frac{dv}{dt} \quad \text{or} \quad F = ma, \tag{3}$$

known as *Newton's second law of motion*.

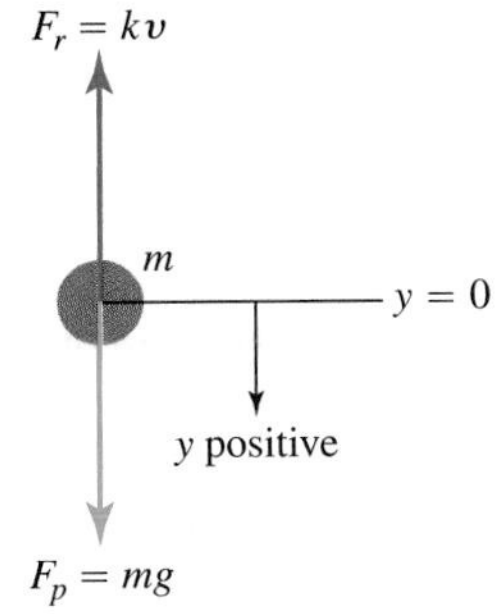

FIGURE 9.16 An object falling under the influence of gravity with a resistive force assumed to be proportional to the velocity.

In free fall, the constant acceleration due to gravity is denoted by g and the one force acting downward on the falling body is

$$F_p = mg,$$

the propulsion due to gravity. If, however, we think of a real body falling through the air—say, a penny from a great height or a parachutist from an even greater height—we know that at some point air resistance is a factor in the speed of the fall. A more realistic model of free fall would include air resistance, shown as a force F_r in the schematic diagram in Figure 9.16.

For low speeds well below the speed of sound, physical experiments have shown that F_r is approximately proportional to the body's velocity. The net force on the falling body is therefore

$$F = F_p - F_r,$$

giving

$$m\frac{dv}{dt} = mg - kv$$

$$\frac{dv}{dt} = g - \frac{k}{m}v. \tag{4}$$

We can use a phase line to analyze the velocity functions that solve this differential equation.

The equilibrium point, obtained by setting the right-hand side of Equation (4) equal to zero, is

$$v = \frac{mg}{k}.$$

If the body is initially moving faster than this, dv/dt is negative and the body slows down. If the body is moving at a velocity below mg/k, then $dv/dt > 0$ and the body speeds up. These observations are captured in the initial phase line diagram in Figure 9.17.

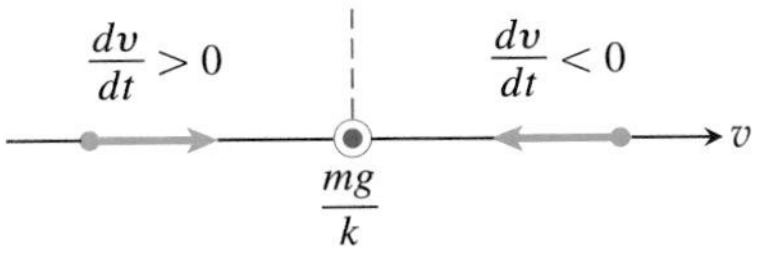

FIGURE 9.17 Initial phase line for Example 4.

We determine the concavity of the solution curves by differentiating both sides of Equation (4) with respect to t:

$$\frac{d^2v}{dt^2} = \frac{d}{dt}\left(g - \frac{k}{m}v\right) = -\frac{k}{m}\frac{dv}{dt}.$$

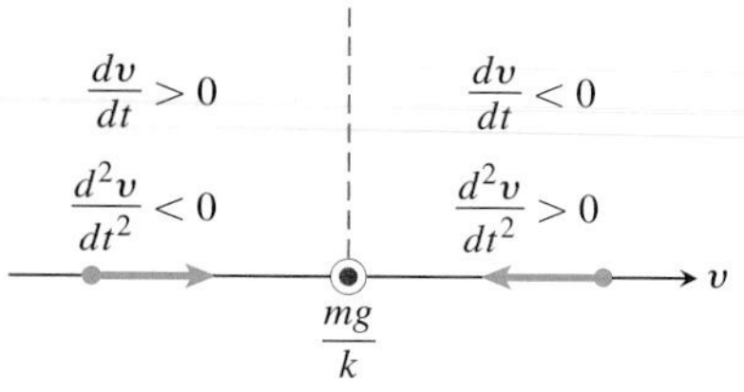

FIGURE 9.18 The completed phase line for Example 4.

We see that $d^2v/dt^2 < 0$ when $v < mg/k$ and $d^2v/dt^2 > 0$ when $v > mg/k$. Figure 9.18 adds this information to the phase line. Notice the similarity to the phase line for Newton's law of cooling (Figure 9.14). The solution curves are similar as well (Figure 9.19).

Figure 9.19 shows two typical solution curves. Regardless of the initial velocity, we see the body's velocity tending toward the limiting value $v = mg/k$. This value, a stable equilibrium point, is called the body's **terminal velocity**. Skydivers can vary their terminal velocity from 95 mph to 180 mph by changing the amount of body area opposing the fall. ■

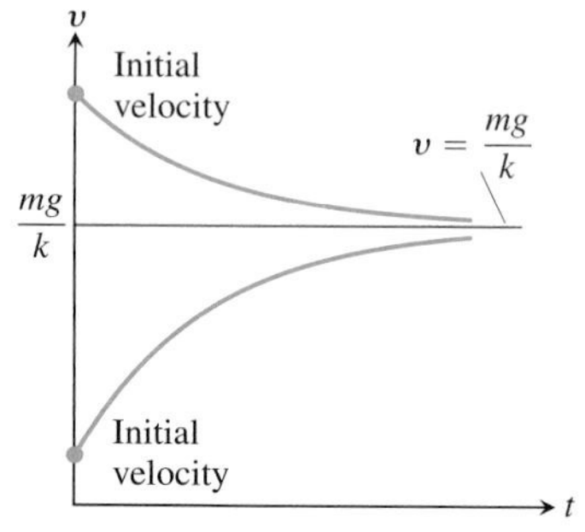

FIGURE 9.19 Typical velocity curves in Example 4. The value $v = mg/k$ is the terminal velocity.

EXAMPLE 5 Analyzing Population Growth in a Limiting Environment

In Section 7.5 we examined population growth using the model of exponential change. That is, if P represents the number of individuals and we neglect departures and arrivals, then

$$\frac{dP}{dt} = kP, \tag{5}$$

where $k > 0$ is the birthrate minus the death rate per individual per unit time.

Because the natural environment has only a limited number of resources to sustain life, it is reasonable to assume that only a maximum population M can be accommodated. As the population approaches this **limiting population** or **carrying capacity**, resources become less abundant and the growth rate k decreases. A simple relationship exhibiting this behavior is

$$k = r(M - P),$$

where $r > 0$ is a constant. Notice that k decreases as P increases toward M and that k is negative if P is greater than M. Substituting $r(M - P)$ for k in Equation (5) gives the differential equation

$$\frac{dP}{dt} = r(M - P)P = rMP - rP^2. \tag{6}$$

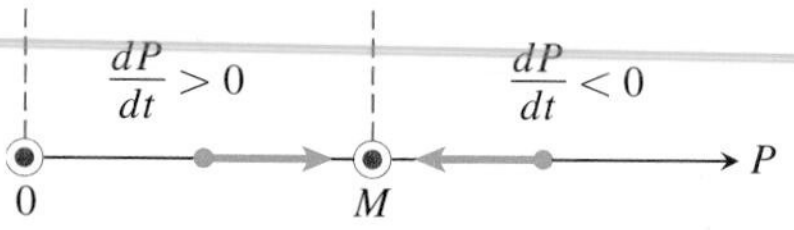

FIGURE 9.20 The initial phase line for Equation 6.

The model given by Equation (6) is referred to as **logistic growth**.

We can forecast the behavior of the population over time by analyzing the phase line for Equation (6). The equilibrium values are $P = M$ and $P = 0$, and we can see that $dP/dt > 0$ if $0 < P < M$ and $dP/dt < 0$ if $P > M$. These observations are recorded on the phase line in Figure 9.20.

We determine the concavity of the population curves by differentiating both sides of Equation (6) with respect to t:

$$\begin{aligned}\frac{d^2P}{dt^2} &= \frac{d}{dt}(rMP - rP^2)\\ &= rM\frac{dP}{dt} - 2rP\frac{dP}{dt}\\ &= r(M - 2P)\frac{dP}{dt}.\end{aligned} \tag{7}$$

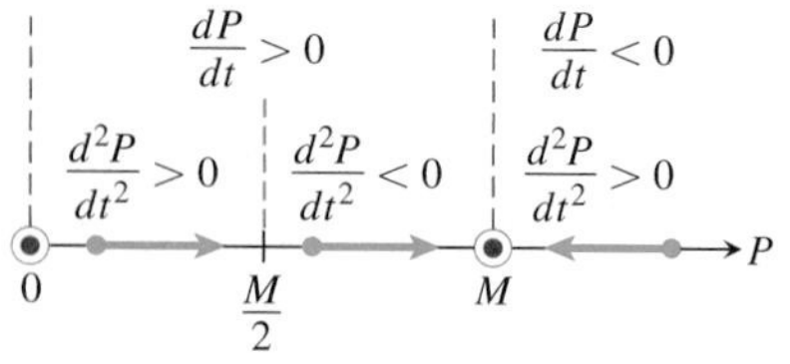

FIGURE 9.21 The completed phase line for logistic growth (Equation 6).

If $P = M/2$, then $d^2P/dt^2 = 0$. If $P < M/2$, then $(M - 2P)$ and dP/dt are positive and $d^2P/dt^2 > 0$. If $M/2 < P < M$, then $(M - 2P) < 0$, $dP/dt > 0$, and $d^2P/dt^2 < 0$. If $P > M$, then $(M - 2P)$ and dP/dt are both negative and $d^2P/dt^2 > 0$. We add this information to the phase line (Figure 9.21).

The lines $P = M/2$ and $P = M$ divide the first quadrant of the tP-plane into horizontal bands in which we know the signs of both dP/dt and d^2P/dt^2. In each band, we know how the solution curves rise and fall, and how they bend as time passes. The equilibrium lines $P = 0$ and $P = M$ are both population curves. Population curves crossing the line $P = M/2$ have an inflection point there, giving them a **sigmoid** shape (curved in two directions like a letter S). Figure 9.22 displays typical population curves. ■

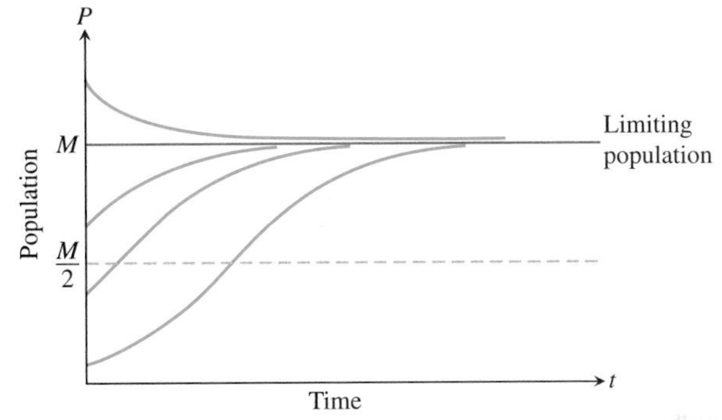

FIGURE 9.22 Population curves in Example 5.

EXERCISES 9.4

Phase Lines and Solution Curves

In Exercises 1–8,

a. Identify the equilibrium values. Which are stable and which are unstable?

b. Construct a phase line. Identify the signs of y' and y''.

c. Sketch several solution curves.

1. $\dfrac{dy}{dx} = (y + 2)(y - 3)$

2. $\dfrac{dy}{dx} = y^2 - 4$

3. $\dfrac{dy}{dx} = y^3 - y$

4. $\dfrac{dy}{dx} = y^2 - 2y$

5. $y' = \sqrt{y}, \quad y > 0$

6. $y' = y - \sqrt{y}, \quad y > 0$

7. $y' = (y - 1)(y - 2)(y - 3)$

8. $y' = y^3 - y^2$

Models of Population Growth

The autonomous differential equations in Exercises 9–12 represent models for population growth. For each exercise, use a phase line analysis to sketch solution curves for $P(t)$, selecting different starting values $P(0)$ (as in Example 5). Which equilibria are stable, and which are unstable?

9. $\dfrac{dP}{dt} = 1 - 2P$

10. $\dfrac{dP}{dt} = P(1 - 2P)$

11. $\dfrac{dP}{dt} = 2P(P - 3)$

12. $\dfrac{dP}{dt} = 3P(1 - P)\left(P - \dfrac{1}{2}\right)$

13. Catastrophic continuation of Example 5 Suppose that a healthy population of some species is growing in a limited environment and that the current population P_0 is fairly close to the carrying capacity M_0. You might imagine a population of fish living in a freshwater lake in a wilderness area. Suddenly a catastrophe such as the Mount St. Helens volcanic eruption contaminates the lake and destroys a significant part of the food and oxygen on which the fish depend. The result is a new environment with a carrying capacity M_1 considerably less than M_0 and, in fact, less than the current population P_0. Starting at some time before the catastrophe, sketch a "before-and-after" curve that shows how the fish population responds to the change in environment.

14. Controlling a population The fish and game department in a certain state is planning to issue hunting permits to control the deer population (one deer per permit). It is known that if the deer population falls below a certain level m, the deer will become extinct. It is also known that if the deer population rises above the carrying capacity M, the population will decrease back to M through disease and malnutrition.

a. Discuss the reasonableness of the following model for the growth rate of the deer population as a function of time:

$$\frac{dP}{dt} = rP(M - P)(P - m),$$

where P is the population of the deer and r is a positive constant of proportionality. Include a phase line.

b. Explain how this model differs from the logistic model $dP/dt = rP(M - P)$. Is it better or worse than the logistic model?

c. Show that if $P > M$ for all t, then $\lim_{t\to\infty} P(t) = M$.

d. What happens if $P < m$ for all t?

e. Discuss the solutions to the differential equation. What are the equilibrium points of the model? Explain the dependence of the steady-state value of P on the initial values of P. About how many permits should be issued?

Applications and Examples

15. Skydiving If a body of mass m falling from rest under the action of gravity encounters an air resistance proportional to the square of velocity, then the body's velocity t seconds into the fall satisfies the equation.

$$m\frac{dv}{dt} = mg - kv^2, \qquad k > 0$$

where k is a constant that depends on the body's aerodynamic properties and the density of the air. (We assume that the fall is too short to be affected by changes in the air's density.)

a. Draw a phase line for the equation.

b. Sketch a typical velocity curve.

c. For a 160-lb skydiver ($mg = 160$) and with time in seconds and distance in feet, a typical value of k is 0.005. What is the diver's terminal velocity?

16. Resistance proportional to $\sqrt{v}$ A body of mass m is projected vertically downward with initial velocity v_0. Assume that the resisting force is proportional to the square root of the velocity and find the terminal velocity from a graphical analysis.

17. Sailing A sailboat is running along a straight course with the wind providing a constant forward force of 50 lb. The only other force acting on the boat is resistance as the boat moves through the water. The resisting force is numerically equal to five times the boat's speed, and the initial velocity is 1 ft/sec. What is the maximum velocity in feet per second of the boat under this wind?

18. The spread of information Sociologists recognize a phenomenon called *social diffusion,* which is the spreading of a piece of information, technological innovation, or cultural fad among a population. The members of the population can be divided into two classes: those who have the information and those who do not. In a fixed population whose size is known, it is reasonable to assume that the rate of diffusion is proportional to the number who have the information times the number yet to receive it. If X denotes the number of individuals who have the information in a population of N people, then a mathematical model for social diffusion is given by

$$\frac{dX}{dt} = kX(N - X),$$

where t represents time in days and k is a positive constant.

a. Discuss the reasonableness of the model.

b. Construct a phase line identifying the signs of X' and X''.

c. Sketch representative solution curves.

d. Predict the value of X for which the information is spreading most rapidly. How many people eventually receive the information?

19. Current in an *RL*-circuit The accompanying diagram represents an electrical circuit whose total resistance is a constant R ohms and whose self-inductance, shown as a coil, is L henries, also a constant. There is a switch whose terminals at a and b can be closed to connect a constant electrical source of V volts.

Ohm's Law, $V = Ri$, has to be modified for such a circuit. The modified form is

$$L\frac{di}{dt} + Ri = V,$$

where i is the intensity of the current in amperes and t is the time in seconds. By solving this equation, we can predict how the current will flow after the switch is closed.

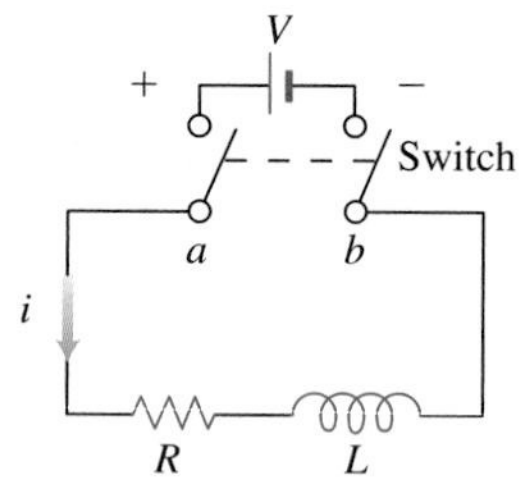

Use a phase line analysis to sketch the solution curve assuming that the switch in the *RL*-circuit is closed at time $t = 0$. What happens to the current as $t \to \infty$? This value is called the *steady-state solution.*

20. A pearl in shampoo Suppose that a pearl is sinking in a thick fluid, like shampoo, subject to a frictional force opposing its fall and proportional to its velocity. Suppose that there is also a resistive buoyant force exerted by the shampoo. According to *Archimedes' principle,* the buoyant force equals the weight of the fluid displaced by the pearl. Using m for the mass of the pearl and P for the mass of the shampoo displaced by the pearl as it descends, complete the following steps.

a. Draw a schematic diagram showing the forces acting on the pearl as it sinks, as in Figure 9.16.

b. Using $v(t)$ for the pearl's velocity as a function of time t, write a differential equation modeling the velocity of the pearl as a falling body.

c. Construct a phase line displaying the signs of v' and v''.

d. Sketch typical solution curves.

e. What is the terminal velocity of the pearl?

9.5 Applications of First-Order Differential Equations

We now look at three applications of the differential equations we have been studying. The first application analyzes an object moving along a straight line while subject to a force opposing its motion. The second is a model of population growth which takes into account factors in the environment placing limits on growth, such as the availability of food or other vital resources. The last application considers a curve or curves intersecting each curve in a second family of curves *orthogonally* (that is, at right angles).

Resistance Proportional to Velocity

In some cases it is reasonable to assume that the resistance encountered by a moving object, such as a car coasting to a stop, is proportional to the object's velocity. The faster the object moves, the more its forward progress is resisted by the air through which it passes. To describe this in mathematical terms, we picture the object as a mass m moving along a coordinate line with position function s and velocity v at time t. From Newton's second law of motion, the resisting force opposing the motion is

$$\text{Force} = \text{mass} \times \text{acceleration} = m\frac{dv}{dt}.$$

We can express the assumption that the resisting force is proportional to velocity by writing

$$m\frac{dv}{dt} = -kv \quad \text{or} \quad \frac{dv}{dt} = -\frac{k}{m}v \qquad (k > 0).$$

This is a separable differential equation representing exponential change. The solution to the equation with initial condition $v = v_0$ at $t = 0$ is (Section 7.5)

$$v = v_0 e^{-(k/m)t}. \tag{1}$$

What can we learn from Equation (1)? For one thing, we can see that if m is something large, like the mass of a 20,000-ton ore boat in Lake Erie, it will take a long time for the velocity to approach zero (because t must be large in the exponent of the equation in order to make kt/m large enough for v to be small). We can learn even more if we integrate Equation (1) to find the position s as a function of time t.

Suppose that a body is coasting to a stop and the only force acting on it is a resistance proportional to its speed. How far will it coast? To find out, we start with Equation (1) and solve the initial value problem

$$\frac{ds}{dt} = v_0 e^{-(k/m)t}, \qquad s(0) = 0.$$

Integrating with respect to t gives

$$s = -\frac{v_0 m}{k} e^{-(k/m)t} + C.$$

Substituting $s = 0$ when $t = 0$ gives

$$0 = -\frac{v_0 m}{k} + C \quad \text{and} \quad C = \frac{v_0 m}{k}.$$

The body's position at time t is therefore

$$s(t) = -\frac{v_0 m}{k} e^{-(k/m)t} + \frac{v_0 m}{k} = \frac{v_0 m}{k}\left(1 - e^{-(k/m)t}\right). \tag{2}$$

To find how far the body will coast, we find the limit of $s(t)$ as $t \to \infty$. Since $-(k/m) < 0$, we know that $e^{-(k/m)t} \to 0$ as $t \to \infty$, so that

$$\begin{aligned}\lim_{t\to\infty} s(t) &= \lim_{t\to\infty} \frac{v_0 m}{k}\left(1 - e^{-(k/m)t}\right) \\ &= \frac{v_0 m}{k}(1 - 0) = \frac{v_0 m}{k}.\end{aligned}$$

Thus,

$$\text{Distance coasted} = \frac{v_0 m}{k}. \tag{3}$$

This is an ideal figure, of course. Only in mathematics can time stretch to infinity. The number $v_0 m/k$ is only an upper bound (albeit a useful one). It is true to life in one respect, at least: if m is large, it will take a lot of energy to stop the body. That is why ocean liners have to be docked by tugboats. Any liner of conventional design entering a slip with enough speed to steer would smash into the pier before it could stop.

In the English system, where weight is measured in pounds, mass is measured in **slugs**. Thus,

$$\text{Pounds} = \text{slugs} \times 32,$$

assuming the gravitational constant is 32 ft/sec^2.

EXAMPLE 1 A Coasting Ice Skater

For a 192-lb ice skater, the k in Equation (1) is about $1/3$ slug/sec and $m = 192/32 = 6$ slugs. How long will it take the skater to coast from 11 ft/sec (7.5 mph) to 1 ft/sec? How far will the skater coast before coming to a complete stop?

Solution We answer the first question by solving Equation (1) for t:

$$\begin{aligned} 11e^{-t/18} &= 1 && \text{Eq. (1) with } k = 1/3,\ m = 6,\ v_0 = 11,\ v = 1 \\ e^{-t/18} &= 1/11 \\ -t/18 &= \ln(1/11) = -\ln 11 \\ t &= 18 \ln 11 \approx 43 \text{ sec}. \end{aligned}$$

We answer the second question with Equation (3):

$$\begin{aligned}\text{Distance coasted} &= \frac{v_0 m}{k} = \frac{11 \cdot 6}{1/3} \\ &= 198 \text{ ft}.\end{aligned}$$

■

Modeling Population Growth

In Section 7.5 we modeled population growth with the Law of Exponential Change:

$$\frac{dP}{dt} = kP, \qquad P(0) = P_0$$

where P is the population at time t, $k > 0$ is a constant growth rate, and P_0 is the size of the population at time $t = 0$. In Section 7.5 we found the solution $P = P_0 e^{kt}$ to this model. However, an issue to be addressed is "how good is the model?"

To begin an assessment of the model, notice that the exponential growth differential equation says that

$$\frac{dP/dt}{P} = k \tag{4}$$

is constant. This rate is called the **relative growth rate**. Now, Table 9.4 gives the world population at midyear for the years 1980 to 1989. Taking $dt = 1$ and $dP \approx \Delta P$, we see from the table that the relative growth rate in Equation (4) is approximately the constant 0.017. Thus, based on the tabled data with $t = 0$ representing 1980, $t = 1$ representing 1981, and so forth, the world population could be modeled by

$$\text{Differential equation:} \quad \frac{dP}{dt} = 0.017P$$

$$\text{Initial condition:} \quad P(0) = 4454.$$

TABLE 9.4 World population (midyear)

Year	Population (millions)	$\Delta P/P$
1980	4454	$76/4454 \approx 0.0171$
1981	4530	$80/4530 \approx 0.0177$
1982	4610	$80/4610 \approx 0.0174$
1983	4690	$80/4690 \approx 0.0171$
1984	4770	$81/4770 \approx 0.0170$
1985	4851	$82/4851 \approx 0.0169$
1986	4933	$85/4933 \approx 0.0172$
1987	5018	$87/5018 \approx 0.0173$
1988	5105	$85/5105 \approx 0.0167$
1989	5190	

Source: U.S. Bureau of the Census (Sept., 1999): www.census.gov/ipc/www/worldpop.html.

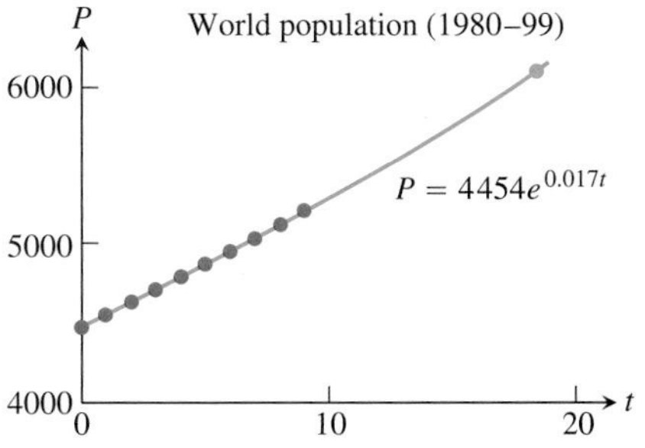

FIGURE 9.23 Notice that the value of the solution $P = 4454e^{0.017t}$ is 6152.16 when $t = 19$, which is slightly higher than the actual population in 1999.

The solution to this initial value problem gives the population function $P = 4454e^{0.017t}$. In year 1999 (so $t = 19$), the solution predicts the world population in midyear to be about 6152 million, or 6.15 billion (Figure 9.23), which is more than the actual population of 6001 million given by the U.S. Bureau of the Census (Table 9.5). Let's examine more recent data to see if there is a change in the growth rate.

Table 9.5 shows the world population for the years 1990 to 2002. From the table we see that the relative growth rate is positive but decreases as the population increases due to environmental, economic, and other factors. On average, the growth rate decreases by about 0.0003 per year over the years 1990 to 2002. That is, the graph of k in Equation (4) is closer to being a line with a negative slope $-r = -0.0003$. In Example 5 of Section 9.4 we proposed the more realistic **logistic growth model**

$$\frac{dP}{dt} = r(M - P)P, \tag{5}$$

where M is the maximum population, or **carrying capacity**, that the environment is capable of sustaining in the long run. Comparing Equation (5) with the exponential model, we see that $k = r(M - P)$ is a linearly decreasing function of the population rather than a constant. The graphical solution curves to the logistic model of Equation (5) were obtained in Section 9.4 and are displayed (again) in Figure 9.24. Notice from the graphs that if $P < M$, the population grows toward M; if $P > M$, the growth rate will be negative (as $r > 0, M > 0$) and the population decreasing.

TABLE 9.5 Recent world population

Year	Population (millions)	$\Delta P/P$
1990	5275	$84/5275 \approx 0.0159$
1991	5359	$84/5359 \approx 0.0157$
1992	5443	$81/5443 \approx 0.0149$
1993	5524	$81/5524 \approx 0.0147$
1994	5605	$80/5605 \approx 0.0143$
1995	5685	$79/5685 \approx 0.0139$
1996	5764	$80/5764 \approx 0.0139$
1997	5844	$79/5844 \approx 0.0135$
1998	5923	$78/5923 \approx 0.0132$
1999	6001	$78/6001 \approx 0.0130$
2000	6079	$73/6079 \approx 0.0120$
2001	6152	$76/6152 \approx 0.0124$
2002	6228	?
2003	?	

Source: U.S. Bureau of the Census (Sept., 2003): www.census.gov/ipc/www/worldpop.html.

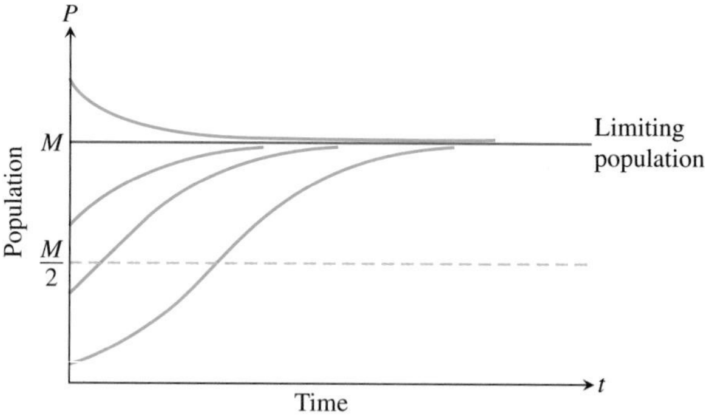

FIGURE 9.24 Solution curves to the logistic population model $dP/dt = r(M - P)P$.

EXAMPLE 2 Modeling a Bear Population

A national park is known to be capable of supporting 100 grizzly bears, but no more. Ten bears are in the park at present. We model the population with a logistic differential equation with $r = 0.001$ (although the model may not give reliable results for very small population levels).

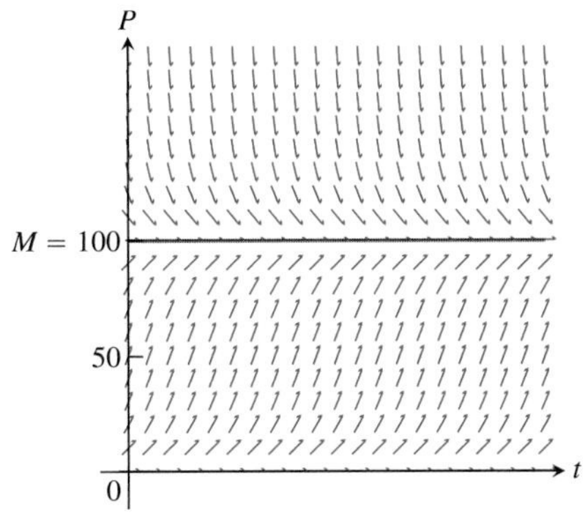

FIGURE 9.25 A slope field for the logistic differential equation $dP/dt = 0.001(100 - P)P$ (Example 2).

(a) Draw and describe a slope field for the differential equation.

(b) Use Euler's method with step size $dt = 1$ to estimate the population size in 20 years.

(c) Find a logistic growth analytic solution $P(t)$ for the population and draw its graph.

(d) When will the bear population reach 50?

Solution

(a) *Slope field.* The carrying capacity is 100, so $M = 100$. The solution we seek is a solution to the following differential equation.

$$\frac{dP}{dt} = 0.001(100 - P)P$$

Figure 9.25 shows a slope field for this differential equation. There appears to be a horizontal asymptote at $P = 100$. The solution curves fall toward this level from above and rise toward it from below.

(b) *Euler's method.* With step size $dt = 1$, $t_0 = 0$, $P(0) = 10$, and

$$\frac{dP}{dt} = f(t, P) = 0.001(100 - P)P,$$

we obtain the approximations in Table 9.6, using the iteration formula

$$P_n = P_{n-1} + 0.001(100 - P_{n-1})P_{n-1}.$$

TABLE 9.6 Euler solution of $dP/dt = 0.001(100 - P)P$, $P(0) = 10$, step size $dt = 1$

t	P (Euler)	t	P (Euler)
0	10		
1	10.9	11	24.3629
2	11.8712	12	26.2056
3	12.9174	13	28.1395
4	14.0423	14	30.1616
5	15.2493	15	32.2680
6	16.5417	16	34.4536
7	17.9222	17	36.7119
8	19.3933	18	39.0353
9	20.9565	19	41.4151
10	22.6130	20	43.8414

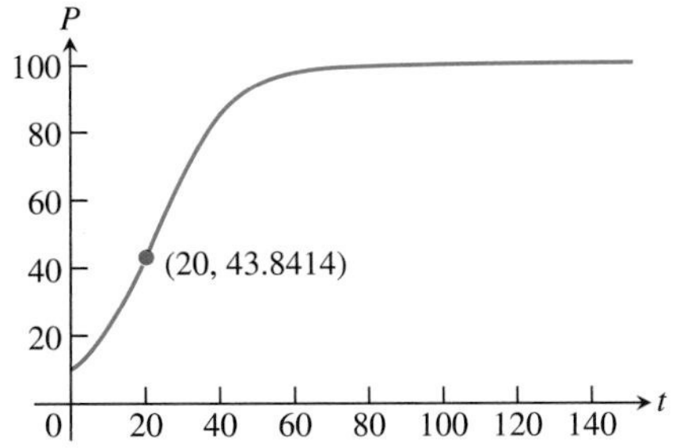

FIGURE 9.26 Euler approximations of the solution to $dP/dt = 0.001(100 - P)P$, $P(0) = 10$, step size $dt = 1$.

There are approximately 44 grizzly bears after 20 years. Figure 9.26 shows a graph of the Euler approximation over the interval $0 \le t \le 150$ with step size $dt = 1$. It looks like the lower curves we sketched in Figure 9.24.

(c) *Analytic solution.* We can assume that $t = 0$ when the bear population is 10, so $P(0) = 10$. The logistic growth model we seek is the solution to the following initial value problem.

$$\text{Differential equation:} \quad \frac{dP}{dt} = 0.001(100 - P)P$$

$$\text{Initial condition:} \quad P(0) = 10$$

To prepare for integration, we rewrite the differential equation in the form

$$\frac{1}{P(100 - P)} \frac{dP}{dt} = 0.001.$$

Using partial fraction decomposition on the left-hand side and multiplying both sides by 100, we get

$$\left(\frac{1}{P} + \frac{1}{100 - P}\right)\frac{dP}{dt} = 0.1$$

$$\ln|P| - \ln|100 - P| = 0.1t + C \qquad \text{Integrate with respect to } t.$$

$$\ln\left|\frac{P}{100 - P}\right| = 0.1t + C$$

$$\ln\left|\frac{100 - P}{P}\right| = -0.1t - C \qquad \ln\frac{a}{b} = -\ln\frac{b}{a}$$

$$\left|\frac{100 - P}{P}\right| = e^{-0.1t - C} \qquad \text{Exponentiate.}$$

$$\frac{100 - P}{P} = (\pm e^{-C})e^{-0.1t}$$

$$\frac{100}{P} - 1 = Ae^{-0.1t} \qquad \text{Let } A = \pm e^{-C}.$$

$$P = \frac{100}{1 + Ae^{-0.1t}}. \qquad \text{Solve for } P.$$

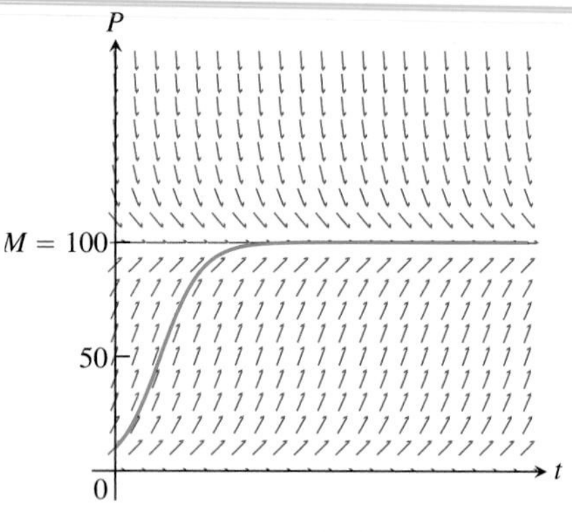

FIGURE 9.27 The graph of

$$P = \frac{100}{1 + 9e^{-0.1t}}$$

superimposed on the slope field in Figure 9.25 (Example 2).

This is the general solution to the differential equation. When $t = 0$, $P = 10$, and we obtain

$$10 = \frac{100}{1 + Ae^0}$$

$$1 + A = 10$$

$$A = 9.$$

Thus, the logistic growth model is

$$P = \frac{100}{1 + 9e^{-0.1t}}.$$

Its graph (Figure 9.27) is superimposed on the slope field from Figure 9.25.

(d) When will the bear population reach 50? For this model,

$$\begin{aligned} 50 &= \frac{100}{1 + 9e^{-0.1t}} \\ 1 + 9e^{-0.1t} &= 2 \\ e^{-0.1t} &= \frac{1}{9} \\ e^{0.1t} &= 9 \\ t &= \frac{\ln 9}{0.1} \approx 22 \text{ years}. \end{aligned}$$

The solution of the general logistic differential equation

$$\frac{dP}{dt} = r(M - P)P$$

can be obtained as in Example 2. In Exercise 10, we ask you to show that the solution is

$$P = \frac{M}{1 + Ae^{-rMt}}.$$

The value of A is determined by an appropriate initial condition.

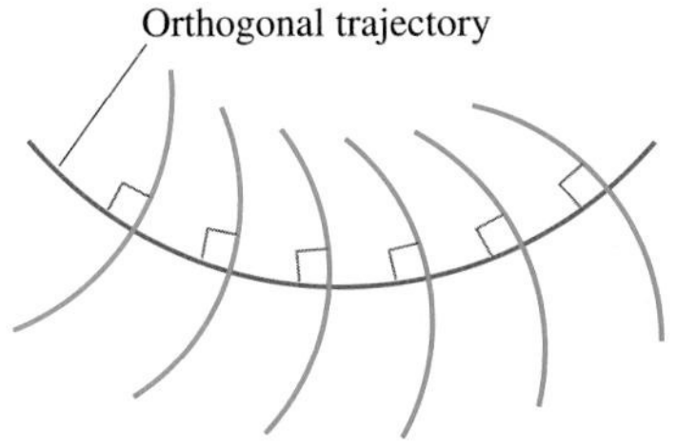

FIGURE 9.28 An orthogonal trajectory intersects the family of curves at right angles, or orthogonally.

Orthogonal Trajectories

An **orthogonal trajectory** of a family of curves is a curve that intersects each curve of the family at right angles, or *orthogonally* (Figure 9.28). For instance, each straight line through the origin is an orthogonal trajectory of the family of circles $x^2 + y^2 = a^2$, centered at the origin (Figure 9.29). Such mutually orthogonal systems of curves are of particular importance in physical problems related to electrical potential, where the curves in one family correspond to flow of electric current and those in the other family correspond to curves of constant potential. They also occur in hydrodynamics and heat-flow problems.

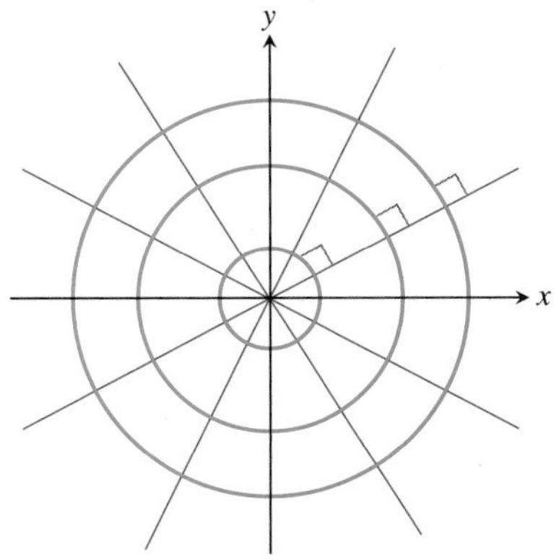

FIGURE 9.29 Every straight line through the origin is orthogonal to the family of circles centered at the origin.

EXAMPLE 3 Finding Orthogonal Trajectories

Find the orthogonal trajectories of the family of curves $xy = a$, where $a \neq 0$ is an arbitrary constant.

Solution The curves $xy = a$ form a family of hyperbolas with asymptotes $y = \pm x$. First we find the slopes of each curve in this family, or their dy/dx values. Differentiating $xy = a$ implicitly gives

$$x\frac{dy}{dx} + y = 0 \qquad \text{or} \qquad \frac{dy}{dx} = -\frac{y}{x}.$$

Thus the slope of the tangent line at any point (x, y) on one of the hyperbolas $xy = a$ is $y' = -y/x$. On an orthogonal trajectory the slope of the tangent line at this same point must be the negative reciprocal, or x/y. Therefore, the orthogonal trajectories must satisfy the differential equation

$$\frac{dy}{dx} = \frac{x}{y}.$$

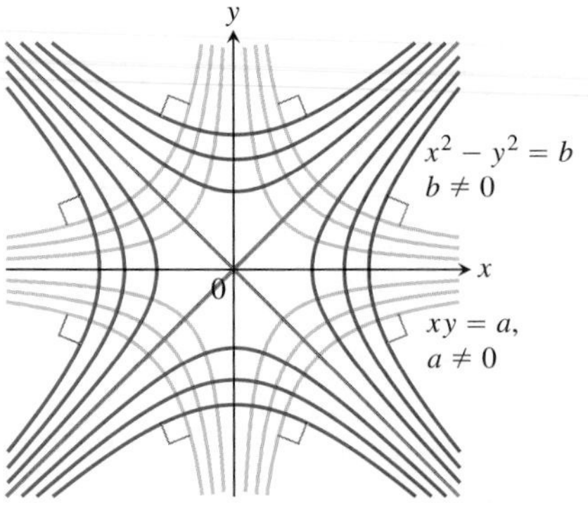

FIGURE 9.30 Each curve is orthogonal to every curve it meets in the other family (Example 3).

This differential equation is separable and we solve it as in Section 9.1:

$$y\,dy = x\,dx \qquad \text{Separate variables.}$$

$$\int y\,dy = \int x\,dx \qquad \text{Integrate both sides.}$$

$$\frac{1}{2}y^2 = \frac{1}{2}x^2 + C$$

$$y^2 - x^2 = b, \tag{6}$$

where $b = 2C$ is an arbitrary constant. The orthogonal trajectories are the family of hyperbolas given by Equation (6) and sketched in Figure 9.30. ■

EXERCISES 9.5

1. **Coasting bicycle** A 66-kg cyclist on a 7-kg bicycle starts coasting on level ground at 9 m/sec. The k in Equation (1) is about 3.9 kg/sec.
   a. About how far will the cyclist coast before reaching a complete stop?
   b. How long will it take the cyclist's speed to drop to 1 m/sec?
2. **Coasting battleship** Suppose that an Iowa class battleship has mass around 51,000 metric tons (51,000,000 kg) and a k value in Equation (1) of about 59,000 kg/sec. Assume that the ship loses power when it is moving at a speed of 9 m/sec.
   a. About how far will the ship coast before it is dead in the water?
   b. About how long will it take the ship's speed to drop to 1 m/sec?
3. The data in Table 9.7 were collected with a motion detector and a CBL™ by Valerie Sharritts, a mathematics teacher at St. Francis DeSales High School in Columbus, Ohio. The table shows the distance s (meters) coasted on in-line skates in t sec by her daughter Ashley when she was 10 years old. Find a model for Ashley's position given by the data in Table 9.7 in the form of Equation (2). Her initial velocity was $v_0 = 2.75$ m/sec, her mass $m = 39.92$ kg (she weighed 88 lb), and her total coasting distance was 4.91 m.
4. **Coasting to a stop** Table 9.8 shows the distance s (meters) coasted on in-line skates in terms of time t (seconds) by Kelly Schmitzer. Find a model for her position in the form of Equation (2). Her initial velocity was $v_0 = 0.80$ m/sec, her mass $m = 49.90$ kg (110 lb), and her total coasting distance was 1.32 m.
5. **Guppy population** A 2000-gal tank can support no more than 150 guppies. Six guppies are introduced into the tank. Assume that the rate of growth of the population is

$$\frac{dP}{dt} = 0.0015(150 - P)P,$$

where time t is in weeks.

TABLE 9.7 Ashley Sharritts skating data

t (sec)	s (m)	t (sec)	s (m)	t (sec)	s (m)
0	0	2.24	3.05	4.48	4.77
0.16	0.31	2.40	3.22	4.64	4.82
0.32	0.57	2.56	3.38	4.80	4.84
0.48	0.80	2.72	3.52	4.96	4.86
0.64	1.05	2.88	3.67	5.12	4.88
0.80	1.28	3.04	3.82	5.28	4.89
0.96	1.50	3.20	3.96	5.44	4.90
1.12	1.72	3.36	4.08	5.60	4.90
1.28	1.93	3.52	4.18	5.76	4.91
1.44	2.09	3.68	4.31	5.92	4.90
1.60	2.30	3.84	4.41	6.08	4.91
1.76	2.53	4.00	4.52	6.24	4.90
1.92	2.73	4.16	4.63	6.40	4.91
2.08	2.89	4.32	4.69	6.56	4.91

   a. Find a formula for the guppy population in terms of t.
   b. How long will it take for the guppy population to be 100? 125?
6. **Gorilla population** A certain wild animal preserve can support no more than 250 lowland gorillas. Twenty-eight gorillas were known to be in the preserve in 1970. Assume that the rate of growth of the population is

$$\frac{dP}{dt} = 0.0004(250 - P)P,$$

where time t is in years.

TABLE 9.8 Kelly Schmitzer skating data

t (sec)	s (m)	t (sec)	s (m)	t (sec)	s (m)
0	0	1.5	0.89	3.1	1.30
0.1	0.07	1.7	0.97	3.3	1.31
0.3	0.22	1.9	1.05	3.5	1.32
0.5	0.36	2.1	1.11	3.7	1.32
0.7	0.49	2.3	1.17	3.9	1.32
0.9	0.60	2.5	1.22	4.1	1.32
1.1	0.71	2.7	1.25	4.3	1.32
1.3	0.81	2.9	1.28	4.5	1.32

a. Find a formula for the gorilla population in terms of t.

b. How long will it take for the gorilla population to reach the carrying capacity of the preserve?

7. **Pacific halibut fishery** The Pacific halibut fishery has been modeled by the logistic equation

$$\frac{dy}{dt} = r(M - y)y$$

where $y(t)$ is the total weight of the halibut population in kilograms at time t (measured in years), the carrying capacity is estimated to be $M = 8 \times 10^7$ kg, and $r = 0.08875 \times 10^{-7}$ per year.

a. If $y(0) = 1.6 \times 10^7$ kg, what is the total weight of the halibut population after 1 year?

b. When will the total weight in the halibut fishery reach 4×10^7 kg?

8. **Modified logistic model** Suppose that the logistic differential equation in Example 2 is modified to

$$\frac{dP}{dt} = 0.001(100 - P)P - c$$

for some constant c.

a. Explain the meaning of the constant c. What values for c might be realistic for the grizzly bear population?

T b. Draw a direction field for the differential equation when $c = 1$. What are the equilibrium solutions (Section 9.4)?

c. Sketch several solution curves in your direction field from part (a). Describe what happens to the grizzly bear population for various initial populations.

9. **Exact solutions** Find the exact solutions to the following initial value problems.

a. $y' = 1 + y, \quad y(0) = 1$

b. $y' = 0.5(400 - y)y, \quad y(0) = 2$

10. **Logistic differential equation** Show that the solution of the differential equation

$$\frac{dP}{dt} = r(M - P)P$$

is

$$P = \frac{M}{1 + Ae^{-rMt}},$$

where A is an arbitrary constant.

11. **Catastrophic solution** Let k and P_0 be positive constants.

a. Solve the initial value problem?

$$\frac{dP}{dt} = kP^2, \quad P(0) = P_0$$

T b. Show that the graph of the solution in part (a) has a vertical asymptote at a positive value of t. What is that value of t?

12. **Extinct populations** Consider the population model

$$\frac{dP}{dt} = r(M - P)(P - m),$$

where $r > 0$, M is the maximum sustainable population, and m is the minimum population below which the species becomes extinct.

a. Let $m = 100$, and $M = 1200$, and assume that $m < P < M$. Show that the differential equation can be rewritten in the form

$$\left[\frac{1}{1200 - P} + \frac{1}{P - 100}\right]\frac{dP}{dt} = 1100r$$

and solve this separable equation.

b. Find the solution to part (a) that satisfies $P(0) = 300$.

c. Solve the differential equation with the restriction $m < P < M$.

Orthogonal Trajectories

In Exercises 13–18, find the orthogonal trajectories of the family of curves. Sketch several members of each family.

13. $y = mx$

14. $y = cx^2$

15. $kx^2 + y^2 = 1$

16. $2x^2 + y^2 = c^2$

17. $y = ce^{-x}$

18. $y = e^{kx}$

19. Show that the curves $2x^2 + 3y^2 = 5$ and $y^2 = x^3$ are orthogonal.

20. Find the family of solutions of the given differential equation and the family of orthogonal trajectories. Sketch both families.

a. $x\,dx + y\,dy = 0$ b. $x\,dy - 2y\,dx = 0$

21. Suppose a and b are positive numbers. Sketch the parabolas

$$y^2 = 4a^2 - 4ax \quad \text{and} \quad y^2 = 4b^2 + 4bx$$

in the same diagram. Show that they intersect at $\left(a - b, \pm 2\sqrt{ab}\right)$, and that each "$a$-parabola" is orthogonal to every "b-parabola."

Chapter 9 Questions to Guide Your Review

1. What is a first-order differential equation? When is a function a solution of such an equation?
2. How do you solve separable first-order differential equations?
3. What is the law of exponential change? How can it be derived from an initial value problem? What are some of the applications of the law?
4. What is the slope field of a differential equation $y' = f(x, y)$? What can we learn from such fields?
5. How do you solve linear first-order differential equations?
6. Describe Euler's method for solving the initial value problem $y' = f(x, y), y(x_0) = y_0$ numerically. Give an example. Comment on the method's accuracy. Why might you want to solve an initial value problem numerically?
7. Describe the improved Euler's method for solving the initial value problem $y' = f(x, y), y(x_0) = y_0$ numerically. How does it compare with Euler's method?
8. What is an autonomous differential equation? What are its equilibrium values? How do they differ from critical points? What is a stable equilibrium value? Unstable?
9. How do you construct the phase line for an autonomous differential equation? How does the phase line help you produce a graph which qualitatively depicts a solution to the differential equation?
10. Why is the exponential model unrealistic for predicting long-term population growth? How does the logistic model correct for the deficiency in the exponential model for population growth? What is the logistic differential equation? What is the form of its solution? Describe the graph of the logistic solution.

Chapter 9 Practice Exercises

In Exercises 1–20 solve the differential equation.

1. $\dfrac{dy}{dx} = \sqrt{y}\cos^2\sqrt{y}$
2. $y' = \dfrac{3y(x+1)^2}{y-1}$
3. $yy' = \sec y^2 \sec^2 x$
4. $y\cos^2 x\,dy + \sin x\,dx = 0$
5. $y' = xe^y\sqrt{x-2}$
6. $y' = xye^{x^2}$
7. $\sec x\,dy + x\cos^2 y\,dx = 0$
8. $2x^2\,dx - 3\sqrt{y}\csc x\,dy = 0$
9. $y' = \dfrac{e^y}{xy}$
10. $y' = xe^{x-y}\csc y$
11. $x(x-1)\,dy - y\,dx = 0$
12. $y' = (y^2-1)x^{-1}$
13. $2y' - y = xe^{x/2}$
14. $\dfrac{y'}{2} + y = e^{-x}\sin x$
15. $xy' + 2y = 1 - x^{-1}$
16. $xy' - y = 2x\ln x$
17. $(1+e^x)\,dy + (ye^x + e^{-x})\,dx = 0$
18. $e^{-x}\,dy + (e^{-x}y - 4x)\,dx = 0$
19. $(x+3y^2)\,dy + y\,dx = 0$ (*Hint:* $d(xy) = y\,dx + x\,dy$)
20. $x\,dy + (3y - x^{-2}\cos x)\,dx = 0, \quad x > 0$

Initial Value Problems

In Exercises 21–30 solve the initial value problem.

21. $\dfrac{dy}{dx} = e^{-x-y-2}, \quad y(0) = -2$
22. $\dfrac{dy}{dx} = \dfrac{y\ln y}{1+x^2}, \quad y(0) = e^2$
23. $(x+1)\dfrac{dy}{dx} + 2y = x, \quad x > -1, \quad y(0) = 1$
24. $x\dfrac{dy}{dx} + 2y = x^2 + 1, \quad x > 0, \quad y(1) = 1$
25. $\dfrac{dy}{dx} + 3x^2y = x^2, \quad y(0) = -1$
26. $x\,dy + (y - \cos x)\,dx = 0, \quad y\left(\dfrac{\pi}{2}\right) = 0$
27. $x\,dy - \left(y + \sqrt{y}\right)dx = 0, \quad y(1) = 1$
28. $y^{-2}\dfrac{dx}{dy} = \dfrac{e^x}{e^{2x}+1}, \quad y(0) = 1$
29. $xy' + (x-2)y = 3x^3e^{-x}, \quad y(1) = 0$
30. $y\,dx + (3x - xy + 2)\,dy = 0, \quad y(2) = -1, \quad y < 0$

Euler's Method

In Exercises 31 and 32, use the stated method to solve the initial value problem on the given interval starting at x_0 with $dx = 0.1$.

T 31. **Euler:** $y' = y + \cos x, \quad y(0) = 0; \quad 0 \le x \le 2; \quad x_0 = 0$

T 32. **Improved Euler:** $y' = (2-y)(2x+3), \quad y(-3) = 1;$ $-3 \le x \le -1; \quad x_0 = -3$

In Exercises 33 and 34, use the stated method with $dx = 0.05$ to estimate $y(c)$ where y is the solution to the given initial value problem.

T 33. **Improved Euler:**

$$c = 3; \quad \frac{dy}{dx} = \frac{x-2y}{x+1}, \quad y(0) = 1$$

34. Euler:

$$c = 4; \quad \frac{dy}{dx} = \frac{x^2 - 2y + 1}{x}, \quad y(1) = 1$$

In Exercises 35 and 36, use the stated method to solve the initial value problem graphically, starting at $x_0 = 0$ with

a. $dx = 0.1$. **b.** $dx = -0.1$.

35. Euler:

$$\frac{dy}{dx} = \frac{1}{e^{x+y+2}}, \quad y(0) = -2$$

36. Improved Euler:

$$\frac{dy}{dx} = -\frac{x^2 + y}{e^y + x}, \quad y(0) = 0$$

Slope Fields

In Exercises 37–40, sketch part of the equation's slope field. Then add to your sketch the solution curve that passes through the point $P(1, -1)$. Use Euler's method with $x_0 = 1$ and $dx = 0.2$ to estimate $y(2)$. Round your answers to four decimal places. Find the exact value of $y(2)$ for comparison.

37. $y' = x$ **38.** $y' = 1/x$

39. $y' = xy$ **40.** $y' = 1/y$

Autonomous Differential Equations and Phase Lines

In Exercises 41 and 42.

a. Identify the equilibrium values. Which are stable and which are unstable?

b. Construct a phase line. Identify the signs of y' and y''.

c. Sketch a representative selection of solution curves.

41. $\frac{dy}{dx} = y^2 - 1$ **42.** $\frac{dy}{dx} = y - y^2$

Applications

43. Escape velocity The gravitational attraction F exerted by an airless moon on a body of mass m at a distance s from the moon's center is given by the equation $F = -mg\,R^2 s^{-2}$, where g is the acceleration of gravity at the moon's surface and R is the moon's radius (see accompanying figure). The force F is negative because it acts in the direction of decreasing s.

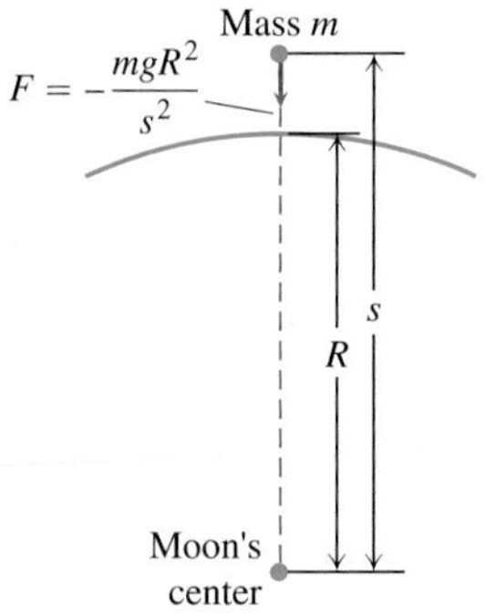

a. If the body is projected vertically upward from the moon's surface with an initial velocity v_0 at time $t = 0$, use Newton's second law, $F = ma$, to show that the body's velocity at position s is given by the equation

$$v^2 = \frac{2gR^2}{s} + v_0{}^2 - 2gR.$$

Thus, the velocity remains positive as long as $v_0 \geq \sqrt{2gR}$. The velocity $v_0 = \sqrt{2gR}$ is the moon's **escape velocity**. A body projected upward with this velocity or a greater one will escape from the moon's gravitational pull.

b. Show that if $v_0 = \sqrt{2gR}$, then

$$s = R\left(1 + \frac{3v_0}{2R}t\right)^{2/3}.$$

44. Coasting to a stop Table 9.9 shows the distance s (meters) coasted on in-line skates in t sec by Johnathon Krueger. Find a model for his position in the form of Equation (2) of Section 9.5. His initial velocity was $v_0 = 0.86$ m/sec, his mass $m = 30.84$ kg (he weighed 68 lb), and his total coasting distance 0.97 m.

TABLE 9.9 Johnathon Krueger skating data

t (sec)	*s* (m)	*t* (sec)	*s* (m)	*t* (sec)	*s* (m)
0	0	0.93	0.61	1.86	0.93
0.13	0.08	1.06	0.68	2.00	0.94
0.27	0.19	1.20	0.74	2.13	0.95
0.40	0.28	1.33	0.79	2.26	0.96
0.53	0.36	1.46	0.83	2.39	0.96
0.67	0.45	1.60	0.87	2.53	0.97
0.80	0.53	1.73	0.90	2.66	0.97

Chapter 9 Additional and Advanced Exercises

Theory and Applications

1. Transport through a cell membrane Under some conditions, the result of the movement of a dissolved substance across a cell's membrane is described by the equation

$$\frac{dy}{dt} = k\frac{A}{V}(c - y).$$

In this equation, y is the concentration of the substance inside the cell and dy/dt is the rate at which y changes over time. The letters

k, A, V, and c stand for constants, k being the *permeability coefficient* (a property of the membrane), A the surface area of the membrane, V the cell's volume, and c the concentration of the substance outside the cell. The equation says that the rate at which the concentration changes within the cell is proportional to the difference between it and the outside concentration.

a. Solve the equation for $y(t)$, using y_0 to denote $y(0)$.

b. Find the steady-state concentration, $\lim_{t\to\infty} y(t)$. (Based on *Some Mathematical Models in Biology*, edited by R. M. Thrall, J. A. Mortimer, K. R. Rebman, and R. F. Baum, rev. ed., Dec. 1967, PB-202 364, pp. 101–103; distributed by N.T.I.S., U.S. Department of Commerce.)

2. Oxygen flow mixture Oxygen flows through one tube into a liter flask filled with air, and the mixture of oxygen and air (considered well stirred) escapes through another tube. Assuming that air contains 21% oxygen, what percentage of oxygen will the flask contain after 5 L have passed through the intake tube?

3. Carbon dioxide in a classroom If the average person breathes 20 times per minute, exhaling each time 100 in^3 of air containing 4% carbon dioxide, find the percentage of carbon dioxide in the air of a 10,000 ft^3 closed room 1 hour after a class of 30 students enters. Assume that the air is fresh at the start, that the ventilators admit 1000 ft^3 of fresh air per minute, and that the fresh air contains 0.04% carbon dioxide.

4. Height of a rocket If an external force F acts upon a system whose mass varies with time, Newton's law of motion is

$$\frac{d(mv)}{dt} = F + (v + u)\frac{dm}{dt}.$$

In this equation, m is the mass of the system at time t, v its velocity, and $v + u$ is the velocity of the mass that is entering (or leaving) the system at the rate dm/dt. Suppose that a rocket of initial mass m_0 starts from rest, but is driven upward by firing some of its mass directly backward at the constant rate of $dm/dt = -b$ units per second and at constant speed relative to the rocket $u = -c$. The only external force acting on the rocket is $F = -mg$ due to gravity. Under these assumptions, show that the height of the rocket above the ground at the end of t seconds (t small compared to m_0/b) is

$$y = c\left[t + \frac{m_0 - bt}{b}\ln\frac{m_0 - bt}{m_0}\right] - \frac{1}{2}gt^2.$$

5. a. Assume that $P(x)$ and $Q(x)$ are continuous over the interval $[a, b]$. Use the Fundamental Theorem of Calculus, Part 1 to show that any function y satisfying the equation

$$v(x)y = \int v(x)Q(x)\,dx + C$$

for $v(x) = e^{\int P(x)\,dx}$ is a solution to the first-order linear equation

$$\frac{dy}{dx} + P(x)y = Q(x).$$

b. If $C = y_0 v(x_0) - \int_{x_0}^{x} v(t)Q(t)\,dt$, then show that any solution y in part (a) satisfies the initial condition $y(x_0) = y_0$.

6. (*Continuation of Exercise 5.*) Assume the hypotheses of Exercise 5, and assume that $y_1(x)$ and $y_2(x)$ are both solutions to the first-order linear equation satisfying the initial condition $y(x_0) = y_0$.

a. Verify that $y(x) = y_1(x) - y_2(x)$ satisfies the initial value problem

$$y' + P(x)y = 0, \quad y(x_0) = 0.$$

b. For the integrating factor $v(x) = e^{\int P(x)\,dx}$, show that

$$\frac{d}{dx}\left(v(x)[y_1(x) - y_2(x)]\right) = 0.$$

Conclude that $v(x)[y_1(x) - y_2(x)] \equiv \text{constant}$.

c. From part (a), we have $y_1(x_0) - y_2(x_0) = 0$. Since $v(x) > 0$ for $a < x < b$, use part (b) to establish that $y_1(x) - y_2(x) \equiv 0$ on the interval (a, b). Thus $y_1(x) = y_2(x)$ for all $a < x < b$.

Chapter 9 Technology Application Projects

Mathematica/Maple Module
Drug Dosages: Are They Effective? Are They Safe?

Formulate and solve an initial value model for the absorption of blood in the bloodstream.

Mathematica/Maple Module
First-Order Differential Equations and Slope Fields

Plot slope fields and solution curves for various initial conditions to selected first-order differential equations.

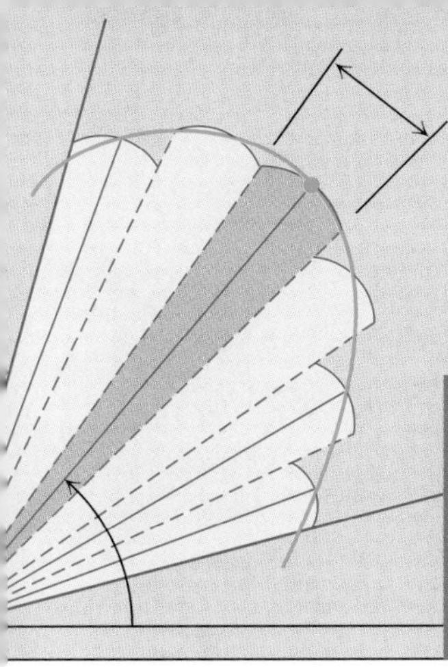

Chapter

10 Conic Sections and Polar Coordinates

OVERVIEW In this chapter we give geometric definitions of parabolas, ellipses, and hyperbolas and derive their standard equations. These curves are called *conic sections*, or *conics*, and model the paths traveled by planets, satellites, and other bodies whose motions are driven by inverse square forces. In Chapter 13 we will see that once the path of a moving body is known to be a conic, we immediately have information about the body's velocity and the force that drives it. Planetary motion is best described with the help of polar coordinates, so we also investigate curves, derivatives, and integrals in this new coordinate system.

10.1 Conic Sections and Quadratic Equations

In Chapter 1 we defined a **circle** as the set of points in a plane whose distance from some fixed center point is a constant radius value. If the center is (h, k) and the radius is a, the standard equation for the circle is $(x - h)^2 + (y - k)^2 = a^2$. It is an example of a conic section, which are the curves formed by cutting a double cone with a plane (Figure 10.1); hence the name *conic section.*

We now describe parabolas, ellipses, and hyperbolas as the graphs of quadratic equations in the coordinate plane.

Parabolas

DEFINITIONS Parabola, Focus, Directrix

A set that consists of all the points in a plane equidistant from a given fixed point and a given fixed line in the plane is a **parabola**. The fixed point is the **focus** of the parabola. The fixed line is the **directrix**.

If the focus F lies on the directrix L, the parabola is the line through F perpendicular to L. We consider this to be a degenerate case and assume henceforth that F does not lie on L.

A parabola has its simplest equation when its focus and directrix straddle one of the coordinate axes. For example, suppose that the focus lies at the point $F(0, p)$ on the positive y-axis and that the directrix is the line $y = -p$ (Figure 10.2). In the notation of the figure,

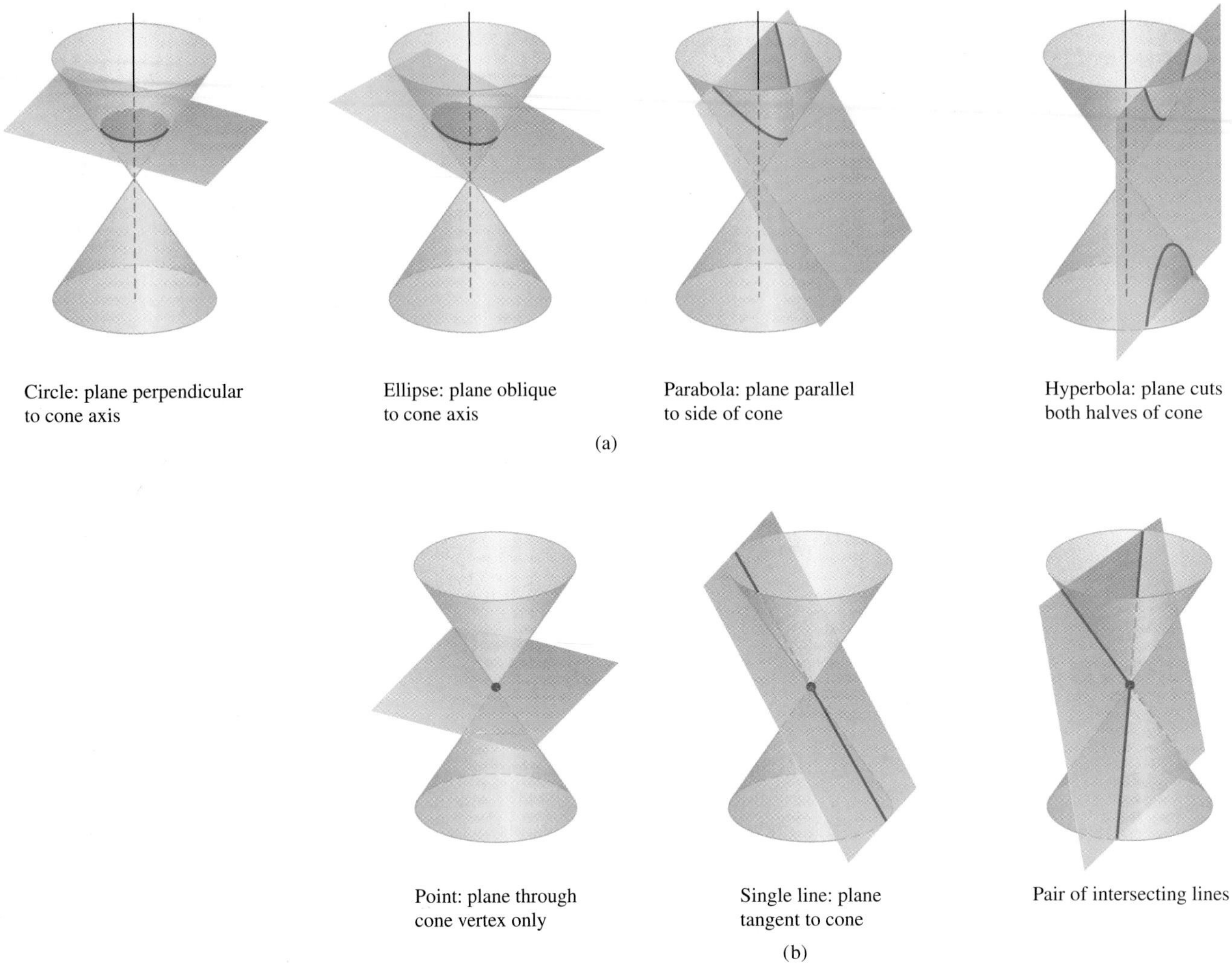

FIGURE 10.1 The standard conic sections (a) are the curves in which a plane cuts a double cone. Hyperbolas come in two parts, called *branches*. The point and lines obtained by passing the plane through the cone's vertex (b) are *degenerate* conic sections.

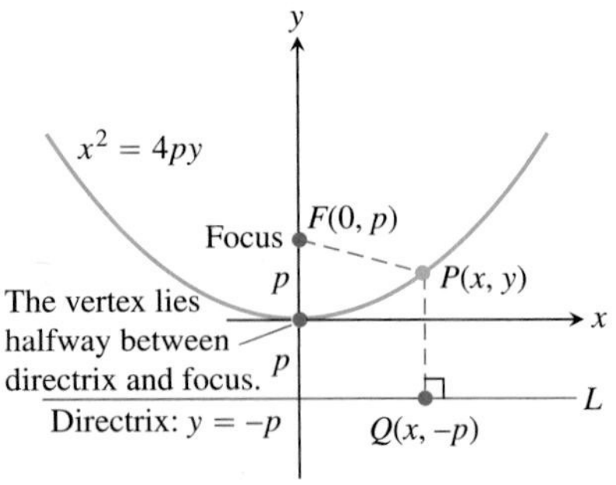

FIGURE 10.2 The standard form of the parabola $x^2 = 4py, p > 0$.

a point $P(x, y)$ lies on the parabola if and only if $PF = PQ$. From the distance formula,

$$PF = \sqrt{(x-0)^2 + (y-p)^2} = \sqrt{x^2 + (y-p)^2}$$

$$PQ = \sqrt{(x-x)^2 + (y-(-p))^2} = \sqrt{(y+p)^2}.$$

When we equate these expressions, square, and simplify, we get

$$y = \frac{x^2}{4p} \quad \text{or} \quad x^2 = 4py. \qquad \text{Standard form} \qquad (1)$$

These equations reveal the parabola's symmetry about the y-axis. We call the y-axis the **axis** of the parabola (short for "axis of symmetry").

The point where a parabola crosses its axis is the **vertex**. The vertex of the parabola $x^2 = 4py$ lies at the origin (Figure 10.2). The positive number p is the parabola's **focal length**.

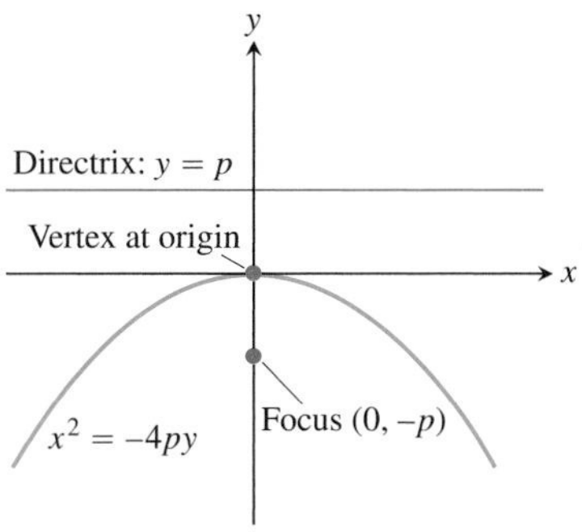

FIGURE 10.3 The parabola $x^2 = -4py, p > 0$.

If the parabola opens downward, with its focus at $(0, -p)$ and its directrix the line $y = p$, then Equations (1) become

$$y = -\frac{x^2}{4p} \quad \text{and} \quad x^2 = -4py$$

(Figure 10.3). We obtain similar equations for parabolas opening to the right or to the left (Figure 10.4 and Table 10.1).

TABLE 10.1 Standard-form equations for parabolas with vertices at the origin $(p > 0)$

Equation	Focus	Directrix	Axis	Opens
$x^2 = 4py$	$(0, p)$	$y = -p$	y-axis	Up
$x^2 = -4py$	$(0, -p)$	$y = p$	y-axis	Down
$y^2 = 4px$	$(p, 0)$	$x = -p$	x-axis	To the right
$y^2 = -4px$	$(-p, 0)$	$x = p$	x-axis	To the left

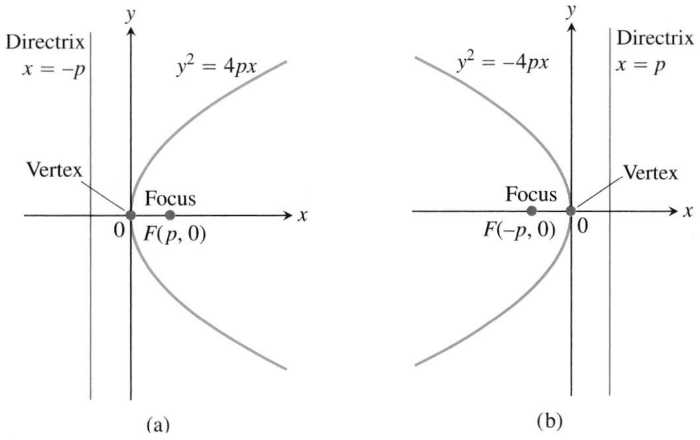

FIGURE 10.4 (a) The parabola $y^2 = 4px$. (b) The parabola $y^2 = -4px$.

EXAMPLE 1 Find the focus and directrix of the parabola $y^2 = 10x$.

Solution We find the value of p in the standard equation $y^2 = 4px$:

$$4p = 10, \qquad \text{so} \qquad p = \frac{10}{4} = \frac{5}{2}.$$

Then we find the focus and directrix for this value of p:

$$\text{Focus:} \qquad (p, 0) = \left(\frac{5}{2}, 0\right)$$

$$\text{Directrix:} \qquad x = -p \qquad \text{or} \qquad x = -\frac{5}{2}.$$

■

The horizontal and vertical shift formulas in Section 1.5, can be applied to the equations in Table 10.1 to give equations for a variety of parabolas in other locations (see Exercises 39, 40, and 45–48).

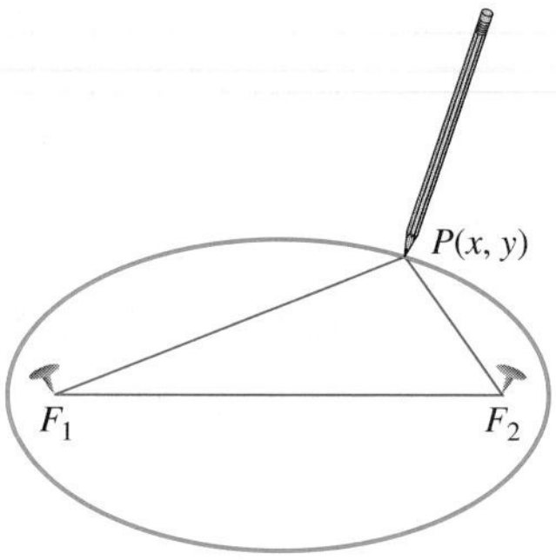

FIGURE 10.5 One way to draw an ellipse uses two tacks and a loop of string to guide the pencil.

Ellipses

> **DEFINITIONS Ellipse, Foci**
>
> An **ellipse** is the set of points in a plane whose distances from two fixed points in the plane have a constant sum. The two fixed points are the **foci** of the ellipse.

The quickest way to construct an ellipse uses the definition. Put a loop of string around two tacks F_1 and F_2, pull the string taut with a pencil point P, and move the pencil around to trace a closed curve (Figure 10.5). The curve is an ellipse because the sum $PF_1 + PF_2$, being the length of the loop minus the distance between the tacks, remains constant. The ellipse's foci lie at F_1 and F_2.

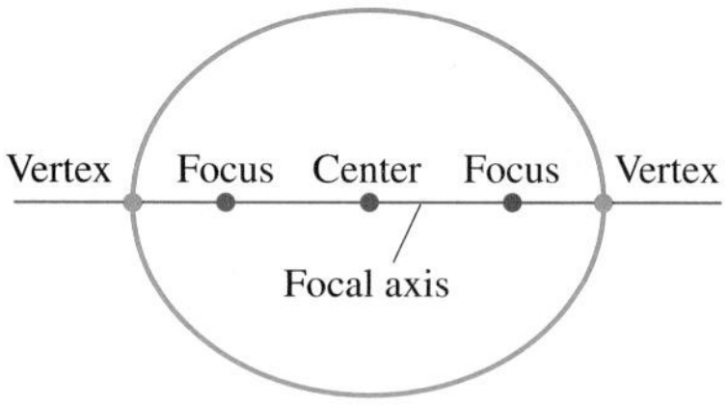

FIGURE 10.6 Points on the focal axis of an ellipse.

> **DEFINITIONS Focal Axis, Center, Vertices**
>
> The line through the foci of an ellipse is the ellipse's **focal axis**. The point on the axis halfway between the foci is the **center**. The points where the focal axis and ellipse cross are the ellipse's **vertices** (Figure 10.6).

If the foci are $F_1(-c, 0)$ and $F_2(c, 0)$ (Figure 10.7), and $PF_1 + PF_2$ is denoted by $2a$, then the coordinates of a point P on the ellipse satisfy the equation

$$\sqrt{(x + c)^2 + y^2} + \sqrt{(x - c)^2 + y^2} = 2a.$$

To simplify this equation, we move the second radical to the right-hand side, square, isolate the remaining radical, and square again, obtaining

$$\frac{x^2}{a^2} + \frac{y^2}{a^2 - c^2} = 1. \tag{2}$$

Since $PF_1 + PF_2$ is greater than the length F_1F_2 (triangle inequality for triangle PF_1F_2), the number $2a$ is greater than $2c$. Accordingly, $a > c$ and the number $a^2 - c^2$ in Equation (2) is positive.

The algebraic steps leading to Equation (2) can be reversed to show that every point P whose coordinates satisfy an equation of this form with $0 < c < a$ also satisfies the equation $PF_1 + PF_2 = 2a$. A point therefore lies on the ellipse if and only if its coordinates satisfy Equation (2).

If

$$b = \sqrt{a^2 - c^2}, \tag{3}$$

then $a^2 - c^2 = b^2$ and Equation (2) takes the form

$$\frac{x^2}{a^2} + \frac{y^2}{b^2} = 1. \tag{4}$$

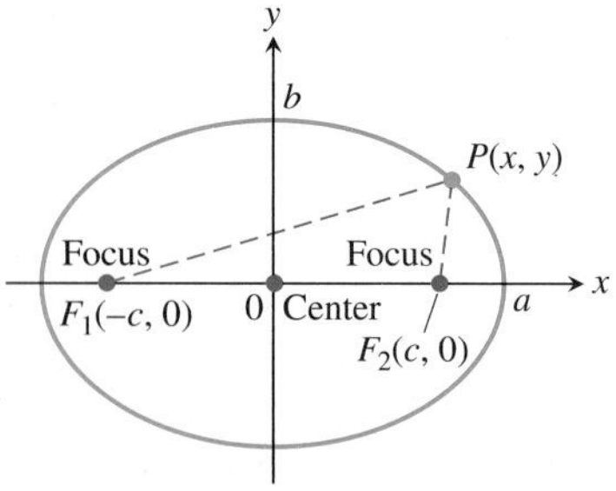

FIGURE 10.7 The ellipse defined by the equation $PF_1 + PF_2 = 2a$ is the graph of the equation $(x^2/a^2) + (y^2/b^2) = 1$, where $b^2 = a^2 - c^2$.

Equation (4) reveals that this ellipse is symmetric with respect to the origin and both coordinate axes. It lies inside the rectangle bounded by the lines $x = \pm a$ and $y = \pm b$. It crosses the axes at the points $(\pm a, 0)$ and $(0, \pm b)$. The tangents at these points are perpendicular to the axes because

$$\frac{dy}{dx} = -\frac{b^2 x}{a^2 y} \qquad \text{Obtained from Equation (4) by implicit differentiation}$$

is zero if $x = 0$ and infinite if $y = 0$.

The **major axis** of the ellipse in Equation (4) is the line segment of length $2a$ joining the points $(\pm a, 0)$. The **minor axis** is the line segment of length $2b$ joining the points $(0, \pm b)$. The number a itself is the **semimajor axis**, the number b the **semiminor axis**. The number c, found from Equation (3) as

$$c = \sqrt{a^2 - b^2},$$

is the **center-to-focus distance** of the ellipse.

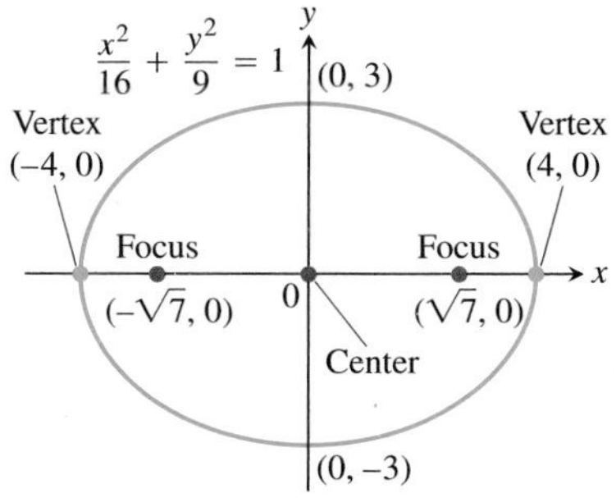

FIGURE 10.8 An ellipse with its major axis horizontal (Example 2).

EXAMPLE 2 Major Axis Horizontal

The ellipse

$$\frac{x^2}{16} + \frac{y^2}{9} = 1 \tag{5}$$

(Figure 10.8) has

Semimajor axis: $a = \sqrt{16} = 4$, Semiminor axis: $b = \sqrt{9} = 3$

Center-to-focus distance: $c = \sqrt{16 - 9} = \sqrt{7}$

Foci: $(\pm c, 0) = \left(\pm\sqrt{7}, 0\right)$

Vertices: $(\pm a, 0) = (\pm 4, 0)$

Center: $(0, 0)$. ■

EXAMPLE 3 Major Axis Vertical

The ellipse

$$\frac{x^2}{9} + \frac{y^2}{16} = 1, \tag{6}$$

obtained by interchanging x and y in Equation (5), has its major axis vertical instead of horizontal (Figure 10.9). With a^2 still equal to 16 and b^2 equal to 9, we have

Semimajor axis: $a = \sqrt{16} = 4$, Semiminor axis: $b = \sqrt{9} = 3$

Center-to-focus distance: $c = \sqrt{16 - 9} = \sqrt{7}$

Foci: $(0, \pm c) = \left(0, \pm\sqrt{7}\right)$

Vertices: $(0, \pm a) = (0, \pm 4)$

Center: $(0, 0)$. ■

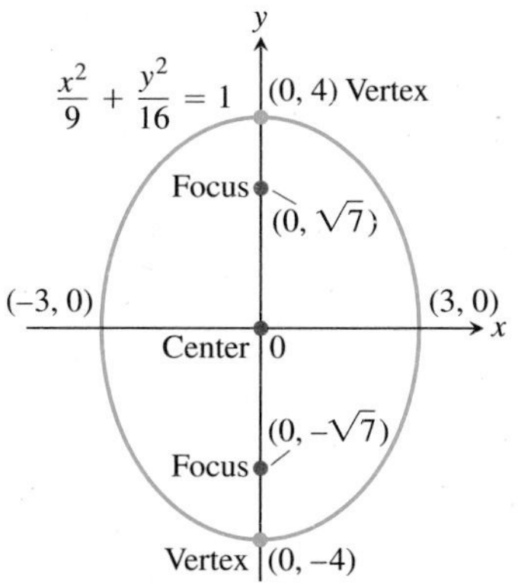

FIGURE 10.9 An ellipse with its major axis vertical (Example 3).

There is never any cause for confusion in analyzing Equations (5) and (6). We simply find the intercepts on the coordinate axes; then we know which way the major axis runs because it is the longer of the two axes. The center always lies at the origin and the foci and vertices lie on the major axis.

Standard-Form Equations for Ellipses Centered at the Origin

Foci on the x-axis: $\frac{x^2}{a^2} + \frac{y^2}{b^2} = 1 \quad (a > b)$

Center-to-focus distance: $c = \sqrt{a^2 - b^2}$

Foci: $(\pm c, 0)$

Vertices: $(\pm a, 0)$

Foci on the y-axis: $\frac{x^2}{b^2} + \frac{y^2}{a^2} = 1 \quad (a > b)$

Center-to-focus distance: $c = \sqrt{a^2 - b^2}$

Foci: $(0, \pm c)$

Vertices: $(0, \pm a)$

In each case, a is the semimajor axis and b is the semiminor axis.

Hyperbolas

DEFINITIONS Hyperbola, Foci

A **hyperbola** is the set of points in a plane whose distances from two fixed points in the plane have a constant difference. The two fixed points are the **foci** of the hyperbola.

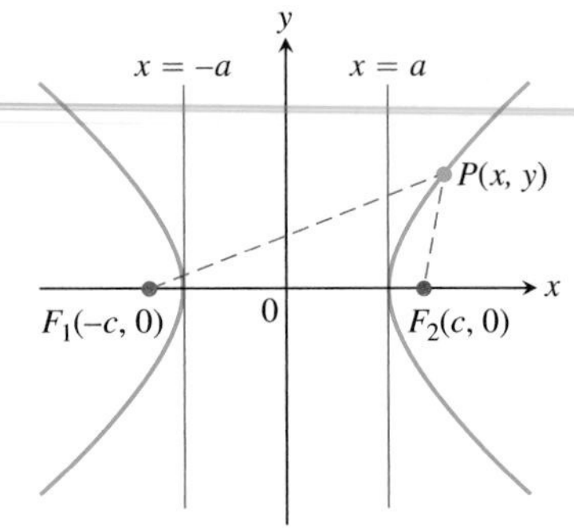

FIGURE 10.10 Hyperbolas have two branches. For points on the right-hand branch of the hyperbola shown here, $PF_1 - PF_2 = 2a$. For points on the left-hand branch, $PF_2 - PF_1 = 2a$. We then let $b = \sqrt{c^2 - a^2}$.

If the foci are $F_1(-c, 0)$ and $F_2(c, 0)$ (Figure 10.10) and the constant difference is $2a$, then a point (x, y) lies on the hyperbola if and only if

$$\sqrt{(x + c)^2 + y^2} - \sqrt{(x - c)^2 + y^2} = \pm 2a. \tag{7}$$

To simplify this equation, we move the second radical to the right-hand side, square, isolate the remaining radical, and square again, obtaining

$$\frac{x^2}{a^2} + \frac{y^2}{a^2 - c^2} = 1. \tag{8}$$

So far, this looks just like the equation for an ellipse. But now $a^2 - c^2$ is negative because $2a$, being the difference of two sides of triangle PF_1F_2, is less than $2c$, the third side.

The algebraic steps leading to Equation (8) can be reversed to show that every point P whose coordinates satisfy an equation of this form with $0 < a < c$ also satisfies Equation (7). A point therefore lies on the hyperbola if and only if its coordinates satisfy Equation (8).

If we let b denote the positive square root of $c^2 - a^2$,

$$b = \sqrt{c^2 - a^2}, \tag{9}$$

then $a^2 - c^2 = -b^2$ and Equation (8) takes the more compact form

$$\frac{x^2}{a^2} - \frac{y^2}{b^2} = 1. \tag{10}$$

The differences between Equation (10) and the equation for an ellipse (Equation 4) are the minus sign and the new relation

$$c^2 = a^2 + b^2. \qquad \text{From Equation (9)}$$

Like the ellipse, the hyperbola is symmetric with respect to the origin and coordinate axes. It crosses the x-axis at the points $(\pm a, 0)$. The tangents at these points are vertical because

$$\frac{dy}{dx} = \frac{b^2 x}{a^2 y} \qquad \text{Obtained from Equation (10) by implicit differentiation}$$

is infinite when $y = 0$. The hyperbola has no y-intercepts; in fact, no part of the curve lies between the lines $x = -a$ and $x = a$.

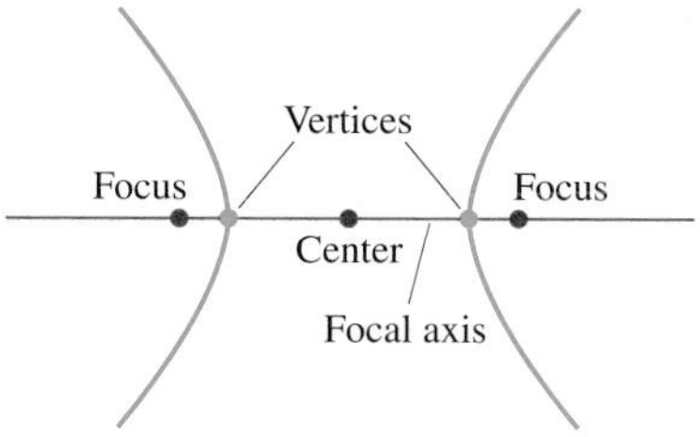

FIGURE 10.11 Points on the focal axis of a hyperbola.

DEFINITIONS **Focal Axis, Center, Vertices**

The line through the foci of a hyperbola is the **focal axis**. The point on the axis halfway between the foci is the hyperbola's **center**. The points where the focal axis and hyperbola cross are the **vertices** (Figure 10.11).

Asymptotes of Hyperbolas and Graphing

If we solve Equation (10) for y we obtain

$$\begin{aligned} y^2 &= b^2\left(\frac{x^2}{a^2} - 1\right) \\ &= \frac{b^2}{a^2}x^2\left(1 - \frac{a^2}{x^2}\right) \end{aligned}$$

or, taking square roots,

$$y = \pm\frac{b}{a}x\sqrt{1 - \frac{a^2}{x^2}}.$$

As $x \to \pm\infty$, the factor $\sqrt{1 - a^2/x^2}$ approaches 1, and the factor $\pm(b/a)x$ is dominant. Thus the lines

$$y = \pm\frac{b}{a}x$$

are the two **asymptotes** of the hyperbola defined by Equation (10). The asymptotes give the guidance we need to graph hyperbolas quickly. The fastest way to find the equations of the asymptotes is to replace the 1 in Equation (10) by 0 and solve the new equation for y:

$$\underbrace{\frac{x^2}{a^2} - \frac{y^2}{b^2} = 1}_{\text{hyperbola}} \rightarrow \underbrace{\frac{x^2}{a^2} - \frac{y^2}{b^2} = 0}_{\text{0 for 1}} \rightarrow \underbrace{y = \pm\frac{b}{a}x.}_{\text{asymptotes}}$$

Standard-Form Equations for Hyperbolas Centered at the Origin

Foci on the x-axis: $\dfrac{x^2}{a^2} - \dfrac{y^2}{b^2} = 1$

Center-to-focus distance: $c = \sqrt{a^2 + b^2}$

Foci: $(\pm c, 0)$

Vertices: $(\pm a, 0)$

Asymptotes: $\dfrac{x^2}{a^2} - \dfrac{y^2}{b^2} = 0$ or $y = \pm\dfrac{b}{a}x$

Foci on the y-axis: $\dfrac{y^2}{a^2} - \dfrac{x^2}{b^2} = 1$

Center-to-focus distance: $c = \sqrt{a^2 + b^2}$

Foci: $(0, \pm c)$

Vertices: $(0, \pm a)$

Asymptotes: $\dfrac{y^2}{a^2} - \dfrac{x^2}{b^2} = 0$ or $y = \pm\dfrac{a}{b}x$

Notice the difference in the asymptote equations (b/a in the first, a/b in the second).

EXAMPLE 4 Foci on the x-axis

The equation

$$\frac{x^2}{4} - \frac{y^2}{5} = 1 \tag{11}$$

is Equation (10) with $a^2 = 4$ and $b^2 = 5$ (Figure 10.12). We have

Center-to-focus distance: $c = \sqrt{a^2 + b^2} = \sqrt{4 + 5} = 3$

Foci: $(\pm c, 0) = (\pm 3, 0)$, Vertices: $(\pm a, 0) = (\pm 2, 0)$

Center: $(0, 0)$

Asymptotes: $\dfrac{x^2}{4} - \dfrac{y^2}{5} = 0$ or $y = \pm\dfrac{\sqrt{5}}{2}x$. ■

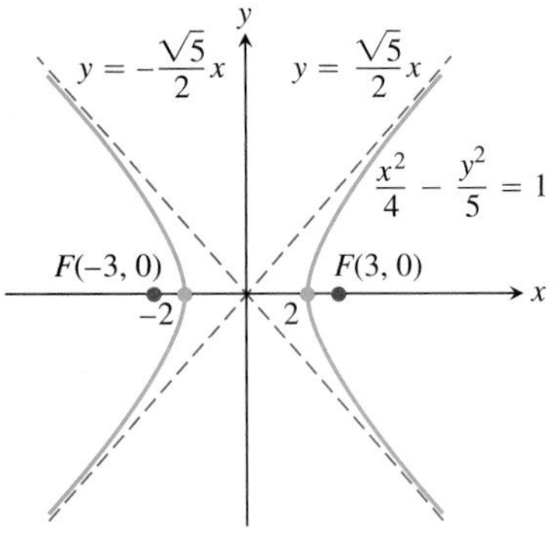

FIGURE 10.12 The hyperbola and its asymptotes in Example 4.

EXAMPLE 5 Foci on the y-axis

The hyperbola

$$\frac{y^2}{4} - \frac{x^2}{5} = 1,$$

obtained by interchanging x and y in Equation (11), has its vertices on the y-axis instead of the x-axis (Figure 10.13). With a^2 still equal to 4 and b^2 equal to 5, we have

Center-to-focus distance: $c = \sqrt{a^2 + b^2} = \sqrt{4 + 5} = 3$

Foci: $(0, \pm c) = (0, \pm 3)$, Vertices: $(0, \pm a) = (0, \pm 2)$

Center: $(0, 0)$

Asymptotes: $\dfrac{y^2}{4} - \dfrac{x^2}{5} = 0$ or $y = \pm\dfrac{2}{\sqrt{5}}x$. ■

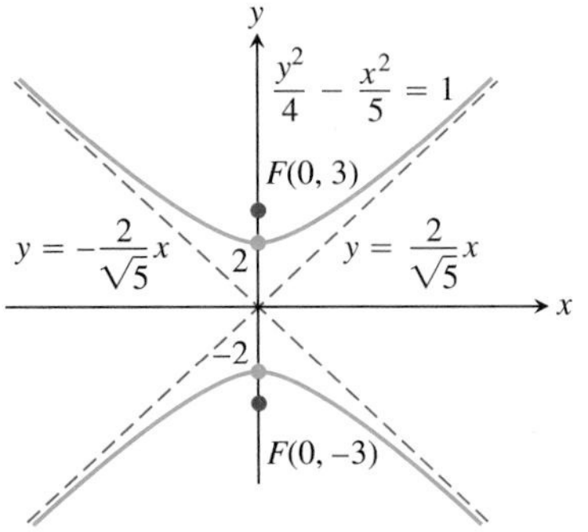

FIGURE 10.13 The hyperbola and its asymptotes in Example 5.

Reflective Properties

The chief applications of parabolas involve their use as reflectors of light and radio waves. Rays originating at a parabola's focus are reflected out of the parabola parallel to the parabola's axis (Figure 10.14 and Exercise 90). Moreover, the time any ray takes from the focus to a line parallel to the parabola's directrix (thus perpendicular to its axis) is the same for each of the rays. These properties are used by flashlight, headlight, and spotlight reflectors and by microwave broadcast antennas.

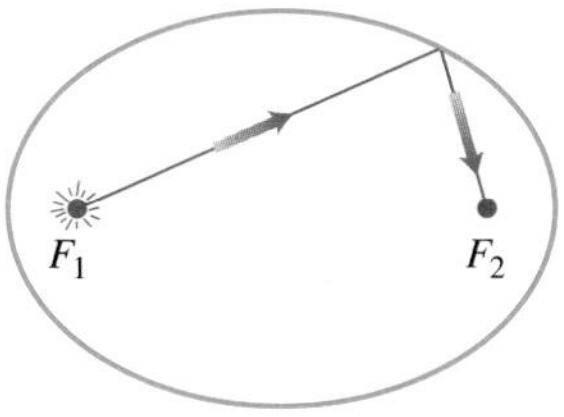

FIGURE 10.15 An elliptical mirror (shown here in profile) reflects light from one focus to the other.

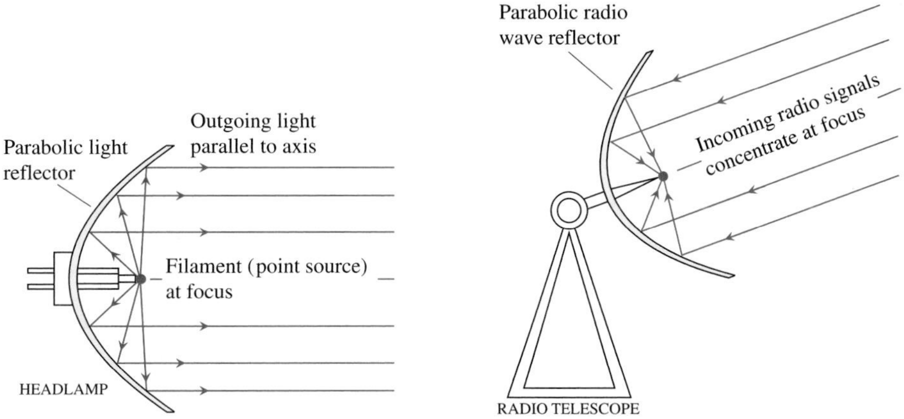

FIGURE 10.14 Parabolic reflectors can generate a beam of light parallel to the parabola's axis from a source at the focus; or they can receive rays parallel to the axis and concentrate them at the focus.

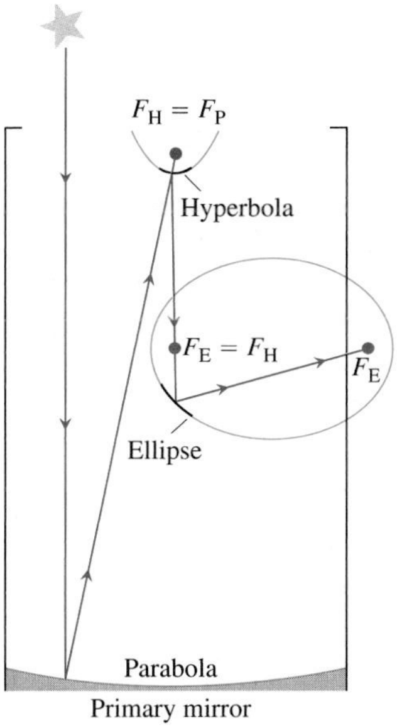

FIGURE 10.16 Schematic drawing of a reflecting telescope.

If an ellipse is revolved about its major axis to generate a surface (the surface is called an *ellipsoid*) and the interior is silvered to produce a mirror, light from one focus will be reflected to the other focus (Figure 10.15). Ellipsoids reflect sound the same way, and this property is used to construct *whispering galleries*, rooms in which a person standing at one focus can hear a whisper from the other focus. (Statuary Hall in the U.S. Capitol building is a whispering gallery.)

Light directed toward one focus of a hyperbolic mirror is reflected toward the other focus. This property of hyperbolas is combined with the reflective properties of parabolas and ellipses in designing some modern telescopes. In Figure 10.16 starlight reflects off a primary parabolic mirror toward the mirror's focus F_P. It is then reflected by a small hyperbolic mirror, whose focus is $F_H = F_P$, toward the second focus of the hyperbola, $F_E = F_H$. Since this focus is shared by an ellipse, the light is reflected by the elliptical mirror to the ellipse's second focus to be seen by an observer.

EXERCISES 10.1

Identifying Graphs

Match the parabolas in Exercises 1–4 with the following equations:

$$x^2 = 2y, \quad x^2 = -6y, \quad y^2 = 8x, \quad y^2 = -4x.$$

Then find the parabola's focus and directrix.

1.

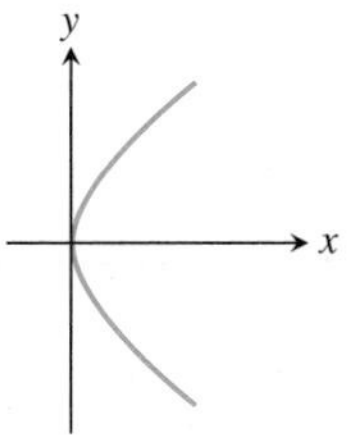

2.

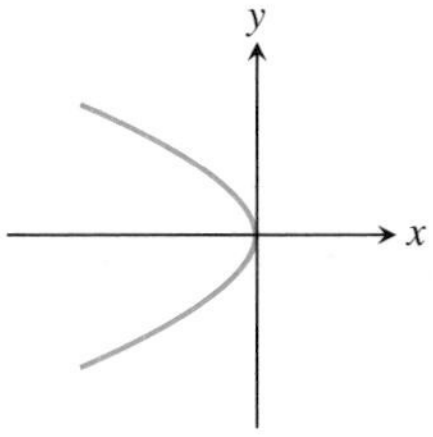

3.

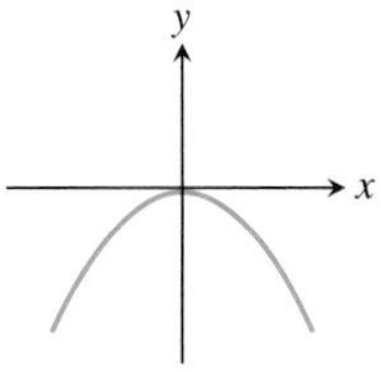

4.

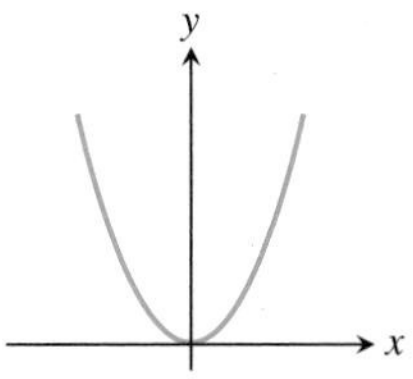

Match each conic section in Exercises 5–8 with one of these equations:

$$\frac{x^2}{4} + \frac{y^2}{9} = 1, \quad \frac{x^2}{2} + y^2 = 1,$$

$$\frac{y^2}{4} - x^2 = 1, \quad \frac{x^2}{4} - \frac{y^2}{9} = 1.$$

Then find the conic section's foci and vertices. If the conic section is a hyperbola, find its asymptotes as well.

5.

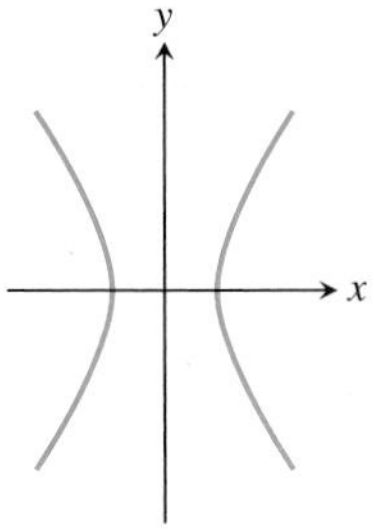

6.

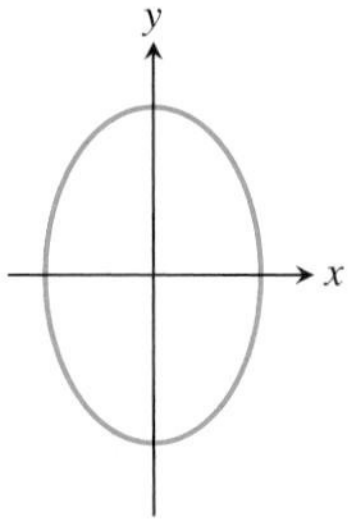

7.

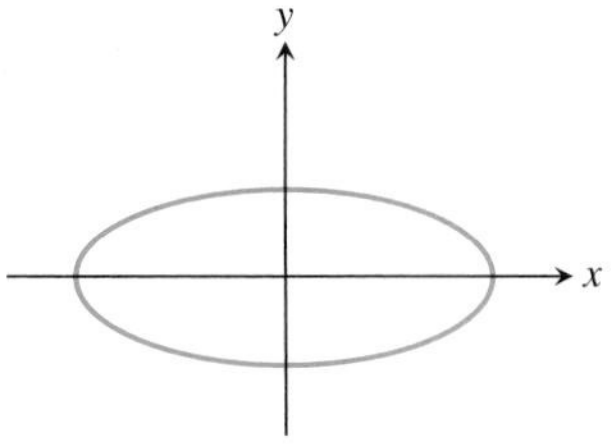

8.

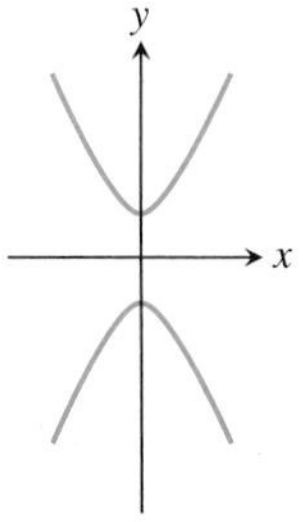

Parabolas

Exercises 9–16 give equations of parabolas. Find each parabola's focus and directrix. Then sketch the parabola. Include the focus and directrix in your sketch.

9. $y^2 = 12x$ **10.** $x^2 = 6y$ **11.** $x^2 = -8y$

12. $y^2 = -2x$ **13.** $y = 4x^2$ **14.** $y = -8x^2$

15. $x = -3y^2$ **16.** $x = 2y^2$

Ellipses

Exercises 17–24 give equations for ellipses. Put each equation in standard form. Then sketch the ellipse. Include the foci in your sketch.

17. $16x^2 + 25y^2 = 400$ **18.** $7x^2 + 16y^2 = 112$

19. $2x^2 + y^2 = 2$ **20.** $2x^2 + y^2 = 4$

21. $3x^2 + 2y^2 = 6$ **22.** $9x^2 + 10y^2 = 90$

23. $6x^2 + 9y^2 = 54$ **24.** $169x^2 + 25y^2 = 4225$

Exercises 25 and 26 give information about the foci and vertices of ellipses centered at the origin of the xy-plane. In each case, find the ellipse's standard-form equation from the given information.

25. Foci: $\left(\pm\sqrt{2}, 0\right)$
Vertices: $(\pm 2, 0)$

26. Foci: $(0, \pm 4)$
Vertices: $(0, \pm 5)$

Hyperbolas

Exercises 27–34 give equations for hyperbolas. Put each equation in standard form and find the hyperbola's asymptotes. Then sketch the hyperbola. Include the asymptotes and foci in your sketch.

27. $x^2 - y^2 = 1$ **28.** $9x^2 - 16y^2 = 144$

29. $y^2 - x^2 = 8$ **30.** $y^2 - x^2 = 4$

31. $8x^2 - 2y^2 = 16$ **32.** $y^2 - 3x^2 = 3$

33. $8y^2 - 2x^2 = 16$ **34.** $64x^2 - 36y^2 = 2304$

Exercises 35–38 give information about the foci, vertices, and asymptotes of hyperbolas centered at the origin of the xy-plane. In each case, find the hyperbola's standard-form equation from the information given.

35. Foci: $\left(0, \pm\sqrt{2}\right)$
Asymptotes: $y = \pm x$

36. Foci: $(\pm 2, 0)$
Asymptotes: $y = \pm \dfrac{1}{\sqrt{3}}x$

37. Vertices: $(\pm 3, 0)$
Asymptotes: $y = \pm \dfrac{4}{3}x$

38. Vertices: $(0, \pm 2)$
Asymptotes: $y = \pm \dfrac{1}{2}x$

Shifting Conic Sections

39. The parabola $y^2 = 8x$ is shifted down 2 units and right 1 unit to generate the parabola $(y + 2)^2 = 8(x - 1)$.

a. Find the new parabola's vertex, focus, and directrix.

b. Plot the new vertex, focus, and directrix, and sketch in the parabola.

40. The parabola $x^2 = -4y$ is shifted left 1 unit and up 3 units to generate the parabola $(x + 1)^2 = -4(y - 3)$.

a. Find the new parabola's vertex, focus, and directrix.

b. Plot the new vertex, focus, and directrix, and sketch in the parabola.

41. The ellipse $(x^2/16) + (y^2/9) = 1$ is shifted 4 units to the right and 3 units up to generate the ellipse

$$\frac{(x - 4)^2}{16} + \frac{(y - 3)^2}{9} = 1.$$

a. Find the foci, vertices, and center of the new ellipse.

b. Plot the new foci, vertices, and center, and sketch in the new ellipse.

42. The ellipse $(x^2/9) + (y^2/25) = 1$ is shifted 3 units to the left and 2 units down to generate the ellipse

$$\frac{(x + 3)^2}{9} + \frac{(y + 2)^2}{25} = 1.$$

a. Find the foci, vertices, and center of the new ellipse.

b. Plot the new foci, vertices, and center, and sketch in the new ellipse.

43. The hyperbola $(x^2/16) - (y^2/9) = 1$ is shifted 2 units to the right to generate the hyperbola

$$\frac{(x - 2)^2}{16} - \frac{y^2}{9} = 1.$$

a. Find the center, foci, vertices, and asymptotes of the new hyperbola.

b. Plot the new center, foci, vertices, and asymptotes, and sketch in the hyperbola.

44. The hyperbola $(y^2/4) - (x^2/5) = 1$ is shifted 2 units down to generate the hyperbola

$$\frac{(y+2)^2}{4} - \frac{x^2}{5} = 1.$$

a. Find the center, foci, vertices, and asymptotes of the new hyperbola.

b. Plot the new center, foci, vertices, and asymptotes, and sketch in the hyperbola.

Exercises 45–48 give equations for parabolas and tell how many units up or down and to the right or left each parabola is to be shifted. Find an equation for the new parabola, and find the new vertex, focus, and directrix.

45. $y^2 = 4x$, left 2, down 3 **46.** $y^2 = -12x$, right 4, up 3

47. $x^2 = 8y$, right 1, down 7 **48.** $x^2 = 6y$, left 3, down 2

Exercises 49–52 give equations for ellipses and tell how many units up or down and to the right or left each ellipse is to be shifted. Find an equation for the new ellipse, and find the new foci, vertices, and center.

49. $\frac{x^2}{6} + \frac{y^2}{9} = 1$, left 2, down 1

50. $\frac{x^2}{2} + y^2 = 1$, right 3, up 4

51. $\frac{x^2}{3} + \frac{y^2}{2} = 1$, right 2, up 3

52. $\frac{x^2}{16} + \frac{y^2}{25} = 1$, left 4, down 5

Exercises 53–56 give equations for hyperbolas and tell how many units up or down and to the right or left each hyperbola is to be shifted. Find an equation for the new hyperbola, and find the new center, foci, vertices, and asymptotes.

53. $\frac{x^2}{4} - \frac{y^2}{5} = 1$, right 2, up 2

54. $\frac{x^2}{16} - \frac{y^2}{9} = 1$, left 2, down 1

55. $y^2 - x^2 = 1$, left 1, down 1

56. $\frac{y^2}{3} - x^2 = 1$, right 1, up 3

Find the center, foci, vertices, asymptotes, and radius, as appropriate, of the conic sections in Exercises 57–68.

57. $x^2 + 4x + y^2 = 12$

58. $2x^2 + 2y^2 - 28x + 12y + 114 = 0$

59. $x^2 + 2x + 4y - 3 = 0$ **60.** $y^2 - 4y - 8x - 12 = 0$

61. $x^2 + 5y^2 + 4x = 1$ **62.** $9x^2 + 6y^2 + 36y = 0$

63. $x^2 + 2y^2 - 2x - 4y = -1$

64. $4x^2 + y^2 + 8x - 2y = -1$

65. $x^2 - y^2 - 2x + 4y = 4$ **66.** $x^2 - y^2 + 4x - 6y = 6$

67. $2x^2 - y^2 + 6y = 3$ **68.** $y^2 - 4x^2 + 16x = 24$

Inequalities

Sketch the regions in the xy-plane whose coordinates satisfy the inequalities or pairs of inequalities in Exercises 69–74.

69. $9x^2 + 16y^2 \le 144$

70. $x^2 + y^2 \ge 1$ and $4x^2 + y^2 \le 4$

71. $x^2 + 4y^2 \ge 4$ and $4x^2 + 9y^2 \le 36$

72. $(x^2 + y^2 - 4)(x^2 + 9y^2 - 9) \le 0$

73. $4y^2 - x^2 \ge 4$ **74.** $|x^2 - y^2| \le 1$

Theory and Examples

75. Archimedes' formula for the volume of a parabolic solid The region enclosed by the parabola $y = (4h/b^2)x^2$ and the line $y = h$ is revolved about the y-axis to generate the solid shown here. Show that the volume of the solid is $3/2$ the volume of the corresponding cone.

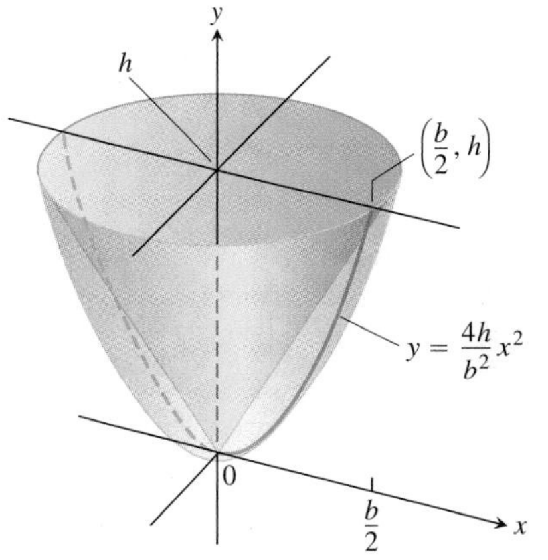

76. Suspension bridge cables hang in parabolas The suspension bridge cable shown here supports a uniform load of w pounds per horizontal foot. It can be shown that if H is the horizontal tension of the cable at the origin, then the curve of the cable satisfies the equation

$$\frac{dy}{dx} = \frac{w}{H}x.$$

Show that the cable hangs in a parabola by solving this differential equation subject to the initial condition that $y = 0$ when $x = 0$.

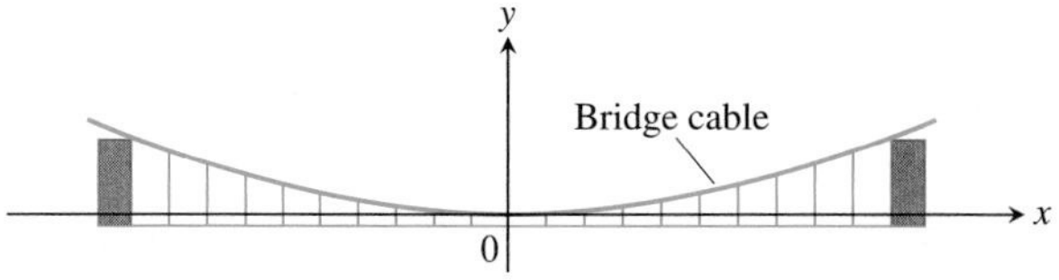

77. Find an equation for the circle through the points (1, 0), (0, 1), and (2, 2).

78. Find an equation for the circle through the points (2, 3), (3, 2), and $(-4, 3)$.

79. Find an equation for the circle centered at $(-2, 1)$ that passes through the point (1, 3). Is the point (1.1, 2.8) inside, outside, or on the circle?

80. Find equations for the tangents to the circle $(x - 2)^2 + (y - 1)^2 = 5$ at the points where the circle crosses the coordinate axes. (*Hint:* Use implicit differentiation.)

81. If lines are drawn parallel to the coordinate axes through a point P on the parabola $y^2 = kx$, $k > 0$, the parabola partitions the rectangular region bounded by these lines and the coordinate axes into two smaller regions, A and B.

a. If the two smaller regions are revolved about the y-axis, show that they generate solids whose volumes have the ratio 4:1.

b. What is the ratio of the volumes generated by revolving the regions about the x-axis?

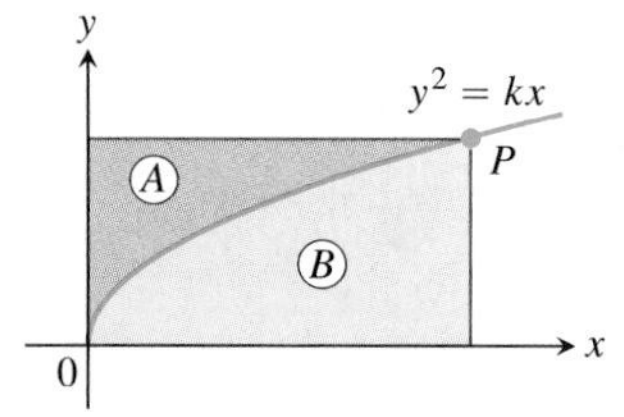

82. Show that the tangents to the curve $y^2 = 4px$ from any point on the line $x = -p$ are perpendicular.

83. Find the dimensions of the rectangle of largest area that can be inscribed in the ellipse $x^2 + 4y^2 = 4$ with its sides parallel to the coordinate axes. What is the area of the rectangle?

84. Find the volume of the solid generated by revolving the region enclosed by the ellipse $9x^2 + 4y^2 = 36$ about the **(a)** x-axis, **(b)** y-axis.

85. The "triangular" region in the first quadrant bounded by the x-axis, the line $x = 4$, and the hyperbola $9x^2 - 4y^2 = 36$ is revolved about the x-axis to generate a solid. Find the volume of the solid.

86. The region bounded on the left by the y-axis, on the right by the hyperbola $x^2 - y^2 = 1$, and above and below by the lines $y = \pm 3$ is revolved about the y-axis to generate a solid. Find the volume of the solid.

87. Find the centroid of the region that is bounded below by the x-axis and above by the ellipse $(x^2/9) + (y^2/16) = 1$.

88. The curve $y = \sqrt{x^2 + 1}$, $0 \le x \le \sqrt{2}$, which is part of the upper branch of the hyperbola $y^2 - x^2 = 1$, is revolved about the x-axis to generate a surface. Find the area of the surface.

89. The circular waves in the photograph here were made by touching the surface of a ripple tank, first at A and then at B. As the waves expanded, their point of intersection appeared to trace a hyperbola. Did it really do that? To find out, we can model the waves with circles centered at A and B.

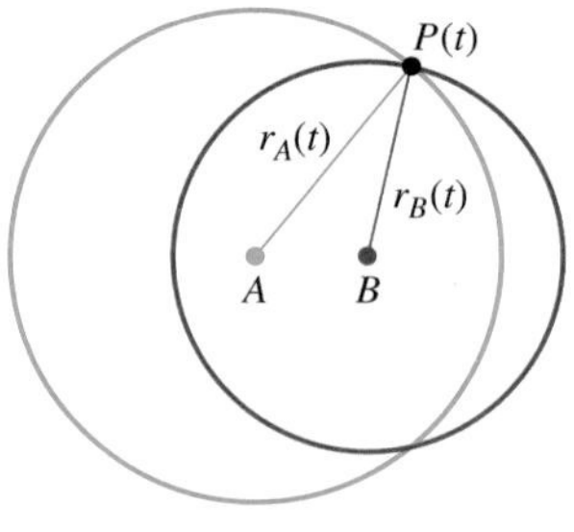

At time t, the point P is $r_A(t)$ units from A and $r_B(t)$ units from B. Since the radii of the circles increase at a constant rate, the rate at which the waves are traveling is

$$\frac{dr_A}{dt} = \frac{dr_B}{dt}.$$

Conclude from this equation that $r_A - r_B$ has a constant value, so that P must lie on a hyperbola with foci at A and B.

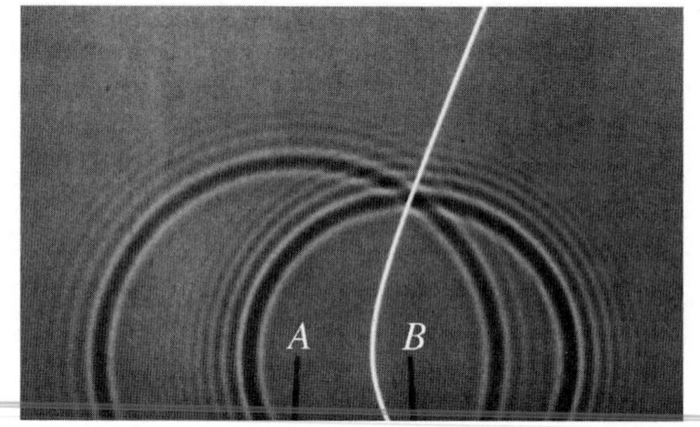

90. The reflective property of parabolas The figure here shows a typical point $P(x_0, y_0)$ on the parabola $y^2 = 4px$. The line L is tangent to the parabola at P. The parabola's focus lies at $F(p, 0)$. The ray L' extending from P to the right is parallel to the x-axis. We show that light from F to P will be reflected out along L' by showing that β equals α. Establish this equality by taking the following steps.

a. Show that $\tan \beta = 2p/y_0$.

b. Show that $\tan \phi = y_0/(x_0 - p)$.

c. Use the identity

$$\tan \alpha = \frac{\tan \phi - \tan \beta}{1 + \tan \phi \tan \beta}$$

to show that $\tan \alpha = 2p/y_0$.

Since α and β are both acute, $\tan \beta = \tan \alpha$ implies $\beta = \alpha$.

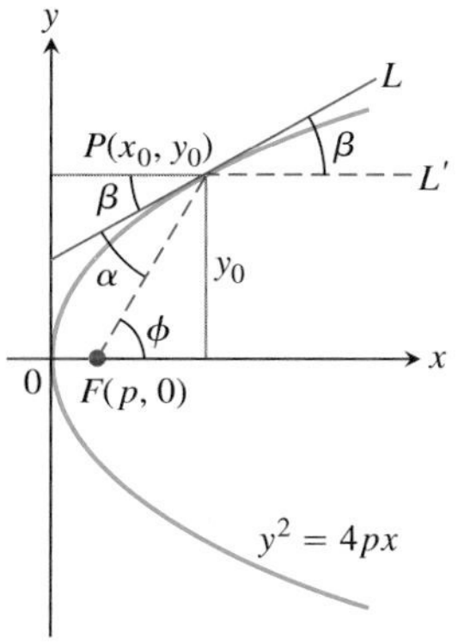

91. How the astronomer Kepler used string to draw parabolas Kepler's method for drawing a parabola (with more modern tools) requires a string the length of a T square and a table whose edge can serve as the parabola's directrix. Pin one end of the string to the point where you want the focus to be and the other end to the upper end of the T square. Then, holding the string taut against the T square with a pencil, slide the T square along the table's edge. As the T square moves, the pencil will trace a parabola. Why?

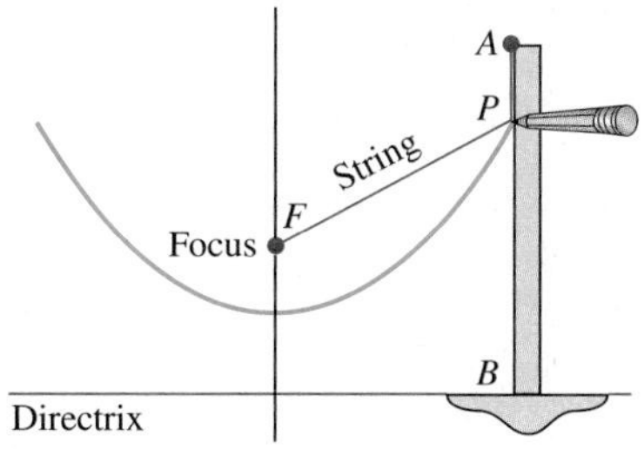

92. Construction of a hyperbola The following diagrams appeared (unlabeled) in Ernest J. Eckert, "Constructions Without Words," *Mathematics Magazine*, Vol. 66, No. 2, Apr. 1993, p. 113. Explain the constructions by finding the coordinates of the point P.

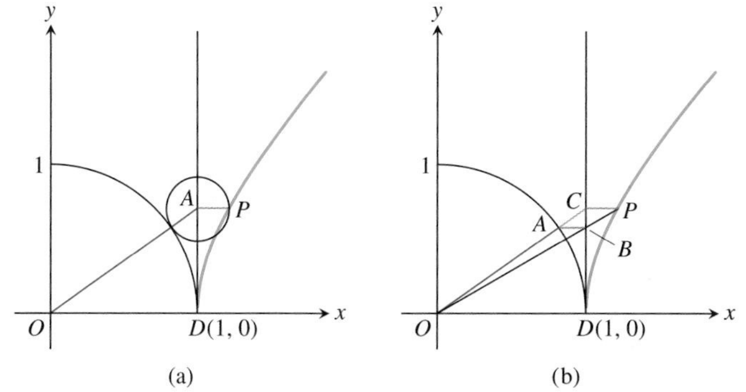

(a) (b)

93. The width of a parabola at the focus Show that the number $4p$ is the *width* of the parabola $x^2 = 4py$ $(p > 0)$ at the focus by showing that the line $y = p$ cuts the parabola at points that are $4p$ units apart.

94. The asymptotes of $(x^2/a^2) - (y^2/b^2) = 1$ Show that the vertical distance between the line $y = (b/a)x$ and the upper half of the right-hand branch $y = (b/a)\sqrt{x^2 - a^2}$ of the hyperbola $(x^2/a^2) - (y^2/b^2) = 1$ approaches 0 by showing that

$$\lim_{x \to \infty} \left(\frac{b}{a}x - \frac{b}{a}\sqrt{x^2 - a^2} \right) = \frac{b}{a} \lim_{x \to \infty} \left(x - \sqrt{x^2 - a^2} \right) = 0.$$

Similar results hold for the remaining portions of the hyperbola and the lines $y = \pm(b/a)x$.

10.2 Classifying Conic Sections by Eccentricity

We now show how to associate with each conic section a number called the conic section's *eccentricity*. The eccentricity reveals the conic section's type (circle, ellipse, parabola, or hyperbola) and, in the case of ellipses and hyperbolas, describes the conic section's general proportions.

Eccentricity

Although the center-to-focus distance c does not appear in the equation

$$\frac{x^2}{a^2} + \frac{y^2}{b^2} = 1, \qquad (a > b)$$

for an ellipse, we can still determine c from the equation $c = \sqrt{a^2 - b^2}$. If we fix a and vary c over the interval $0 \le c \le a$, the resulting ellipses will vary in shape (Figure 10.17). They are circles if $c = 0$ (so that $a = b$) and flatten as c increases. If $c = a$, the foci and vertices overlap and the ellipse degenerates into a line segment.

We use the ratio of c to a to describe the various shapes the ellipse can take. We call this ratio the ellipse's eccentricity.

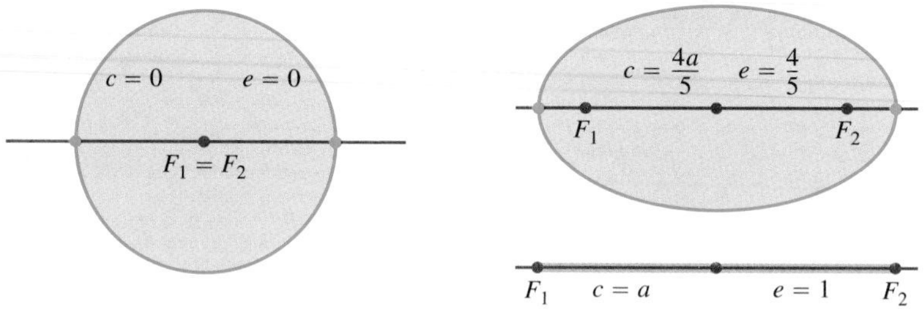

FIGURE 10.17 The ellipse changes from a circle to a line segment as c increases from 0 to a.

DEFINITION Eccentricity of an Ellipse

The **eccentricity** of the ellipse $(x^2/a^2) + (y^2/b^2) = 1$ $(a > b)$ is

$$e = \frac{c}{a} = \frac{\sqrt{a^2 - b^2}}{a}.$$

TABLE 10.2 Eccentricities of planetary orbits

Mercury	0.21	Saturn	0.06
Venus	0.01	Uranus	0.05
Earth	0.02	Neptune	0.01
Mars	0.09	Pluto	0.25
Jupiter	0.05		

The planets in the solar system revolve around the sun in (approximate) elliptical orbits with the sun at one focus. Most of the orbits are nearly circular, as can be seen from the eccentricities in Table 10.2. Pluto has a fairly eccentric orbit, with $e = 0.25$, as does Mercury, with $e = 0.21$. Other members of the solar system have orbits that are even more eccentric. Icarus, an asteroid about 1 mile wide that revolves around the sun every 409 Earth days, has an orbital eccentricity of 0.83 (Figure 10.18).

EXAMPLE 1 Halley's Comet

The orbit of Halley's comet is an ellipse 36.18 astronomical units long by 9.12 astronomical units wide. (One *astronomical unit* [AU] is 149,597,870 km, the semimajor axis of Earth's orbit.) Its eccentricity is

$$e = \frac{\sqrt{a^2 - b^2}}{a} = \frac{\sqrt{(36.18/2)^2 - (9.12/2)^2}}{(1/2)(36.18)} = \frac{\sqrt{(18.09)^2 - (4.56)^2}}{18.09} \approx 0.97. \quad \blacksquare$$

HISTORICAL BIOGRAPHY

Edmund Halley (1656–1742)

Whereas a parabola has one focus and one directrix, each **ellipse** has two foci and two **directrices**. These are the lines perpendicular to the major axis at distances $\pm a/e$ from the center. The parabola has the property that

$$PF = 1 \cdot PD \tag{1}$$

for any point P on it, where F is the focus and D is the point nearest P on the directrix. For an ellipse, it can be shown that the equations that replace Equation (1) are

$$PF_1 = e \cdot PD_1, \qquad PF_2 = e \cdot PD_2. \tag{2}$$

Here, e is the eccentricity, P is any point on the ellipse, F_1 and F_2 are the foci, and D_1 and D_2 are the points on the directrices nearest P (Figure 10.19).

In both Equations (2) the directrix and focus must correspond; that is, if we use the distance from P to F_1, we must also use the distance from P to the directrix at the same end of the ellipse. The directrix $x = -a/e$ corresponds to $F_1(-c, 0)$, and the directrix $x = a/e$ corresponds to $F_2(c, 0)$.

The eccentricity of a hyperbola is also $e = c/a$, only in this case c equals $\sqrt{a^2 + b^2}$ instead of $\sqrt{a^2 - b^2}$. In contrast to the eccentricity of an ellipse, the eccentricity of a hyperbola is always greater than 1.

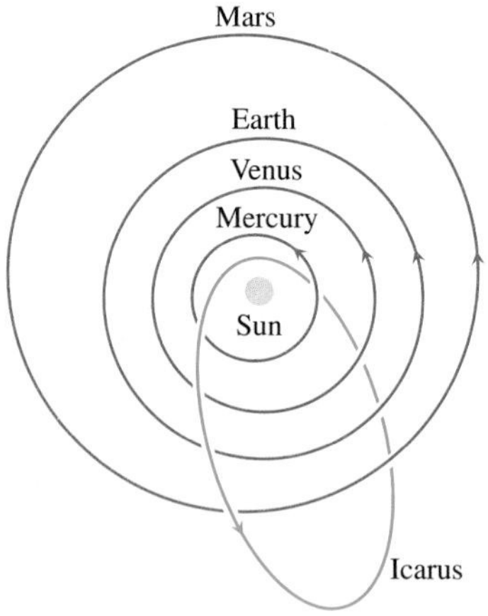

FIGURE 10.18 The orbit of the asteroid Icarus is highly eccentric. Earth's orbit is so nearly circular that its foci lie inside the sun.

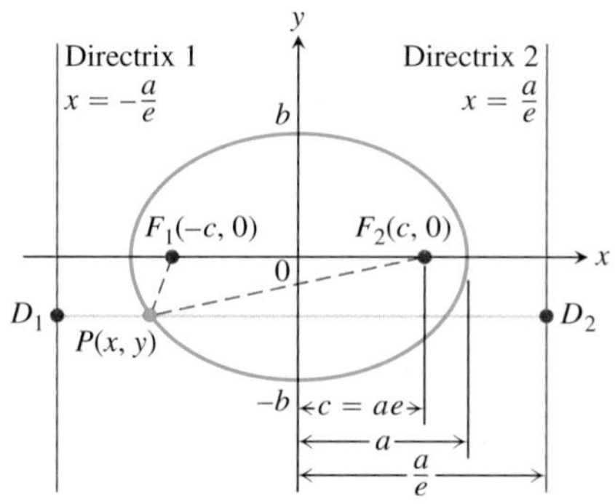

FIGURE 10.19 The foci and directrices of the ellipse $(x^2/a^2) + (y^2/b^2) = 1$. Directrix 1 corresponds to focus F_1, and directrix 2 to focus F_2.

DEFINITION Eccentricity of a Hyperbola

The **eccentricity** of the hyperbola $(x^2/a^2) - (y^2/b^2) = 1$ is

$$e = \frac{c}{a} = \frac{\sqrt{a^2 + b^2}}{a}.$$

In both ellipse and hyperbola, the eccentricity is the ratio of the distance between the foci to the distance between the vertices (because $c/a = 2c/2a$).

$$\text{Eccentricity} = \frac{\text{distance between foci}}{\text{distance between vertices}}$$

In an ellipse, the foci are closer together than the vertices and the ratio is less than 1. In a hyperbola, the foci are farther apart than the vertices and the ratio is greater than 1.

EXAMPLE 2 Finding the Vertices of an Ellipse

Locate the vertices of an ellipse of eccentricity 0.8 whose foci lie at the points $(0, \pm 7)$.

Solution Since $e = c/a$, the vertices are the points $(0, \pm a)$ where

$$a = \frac{c}{e} = \frac{7}{0.8} = 8.75,$$

or $(0, \pm 8.75)$. ■

EXAMPLE 3 Eccentricity of a Hyperbola

Find the eccentricity of the hyperbola $9x^2 - 16y^2 = 144$.

Solution We divide both sides of the hyperbola's equation by 144 to put it in standard form, obtaining

$$\frac{9x^2}{144} - \frac{16y^2}{144} = 1 \quad \text{and} \quad \frac{x^2}{16} - \frac{y^2}{9} = 1.$$

With $a^2 = 16$ and $b^2 = 9$, we find that $c = \sqrt{a^2 + b^2} = \sqrt{16 + 9} = 5$, so

$$e = \frac{c}{a} = \frac{5}{4}.$$ ■

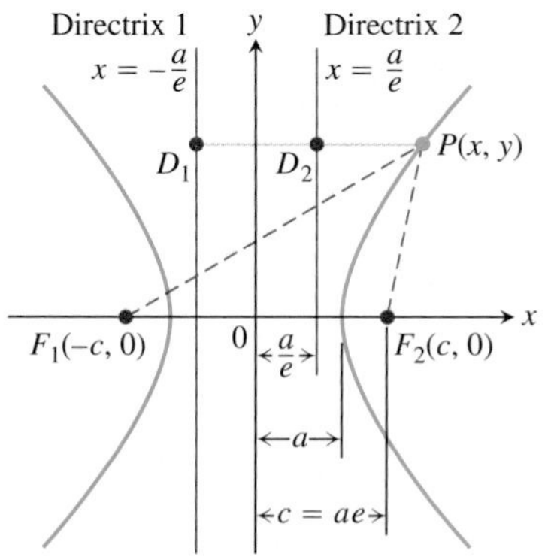

FIGURE 10.20 The foci and directrices of the hyperbola $(x^2/a^2) - (y^2/b^2) = 1$. No matter where P lies on the hyperbola, $PF_1 = e \cdot PD_1$ and $PF_2 = e \cdot PD_2$.

As with the ellipse, it can be shown that the lines $x = \pm a/e$ act as **directrices** for the **hyperbola** and that

$$PF_1 = e \cdot PD_1 \quad \text{and} \quad PF_2 = e \cdot PD_2. \tag{3}$$

Here P is any point on the hyperbola, F_1 and F_2 are the foci, and D_1 and D_2 are the points nearest P on the directrices (Figure 10.20).

To complete the picture, we define the eccentricity of a parabola to be $e = 1$. Equations (1) to (3) then have the common form $PF = e \cdot PD$.

DEFINITION Eccentricity of a Parabola

The **eccentricity** of a parabola is $e = 1$.

The "focus–directrix" equation $PF = e \cdot PD$ unites the parabola, ellipse, and hyperbola in the following way. Suppose that the distance PF of a point P from a fixed point F (the focus) is a constant multiple of its distance from a fixed line (the directrix). That is, suppose

$$PF = e \cdot PD, \tag{4}$$

where e is the constant of proportionality. Then the path traced by P is

(a) a *parabola* if $e = 1$,

(b) an *ellipse* of eccentricity e if $e < 1$, and

(c) a *hyperbola* of eccentricity e if $e > 1$.

There are no coordinates in Equation (4) and when we try to translate it into coordinate form it translates in different ways, depending on the size of e. At least, that is what happens in Cartesian coordinates. However, in polar coordinates, as we will see in Section 10.8, the equation $PF = e \cdot PD$ translates into a single equation regardless of the value of e, an equation so simple that it has been the equation of choice of astronomers and space scientists for nearly 300 years.

Given the focus and corresponding directrix of a hyperbola centered at the origin and with foci on the x-axis, we can use the dimensions shown in Figure 10.20 to find e. Knowing e, we can derive a Cartesian equation for the hyperbola from the equation $PF = e \cdot PD$, as in the next example. We can find equations for ellipses centered at the origin and with foci on the x-axis in a similar way, using the dimensions shown in Figure 10.19.

EXAMPLE 4 Cartesian Equation for a Hyperbola

Find a Cartesian equation for the hyperbola centered at the origin that has a focus at $(3, 0)$ and the line $x = 1$ as the corresponding directrix.

Solution We first use the dimensions shown in Figure 10.20 to find the hyperbola's eccentricity. The focus is

$$(c, 0) = (3, 0) \quad \text{so} \quad c = 3.$$

The directrix is the line

$$x = \frac{a}{e} = 1, \quad \text{so} \quad a = e.$$

When combined with the equation $e = c/a$ that defines eccentricity, these results give

$$e = \frac{c}{a} = \frac{3}{e}, \quad \text{so} \quad e^2 = 3 \quad \text{and} \quad e = \sqrt{3}.$$

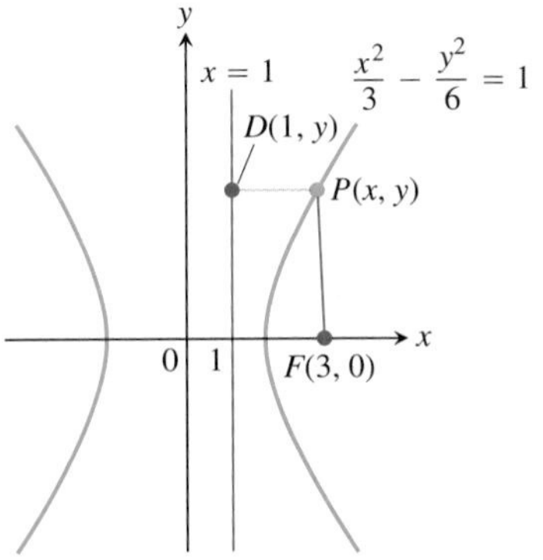

FIGURE 10.21 The hyperbola and directrix in Example 4.

Knowing e, we can now derive the equation we want from the equation $PF = e \cdot PD$. In the notation of Figure 10.21, we have

$$PF = e \cdot PD \qquad \text{Equation (4)}$$
$$\sqrt{(x-3)^2 + (y-0)^2} = \sqrt{3}\,|x-1| \qquad e = \sqrt{3}$$
$$x^2 - 6x + 9 + y^2 = 3(x^2 - 2x + 1)$$
$$2x^2 - y^2 = 6$$
$$\frac{x^2}{3} - \frac{y^2}{6} = 1.$$

■

EXERCISES 10.2

Ellipses

In Exercises 1–8, find the eccentricity of the ellipse. Then find and graph the ellipse's foci and directrices.

1. $16x^2 + 25y^2 = 400$

2. $7x^2 + 16y^2 = 112$

3. $2x^2 + y^2 = 2$

4. $2x^2 + y^2 = 4$

5. $3x^2 + 2y^2 = 6$

6. $9x^2 + 10y^2 = 90$

7. $6x^2 + 9y^2 = 54$

8. $169x^2 + 25y^2 = 4225$

Exercises 9–12 give the foci or vertices and the eccentricities of ellipses centered at the origin of the xy-plane. In each case, find the ellipse's standard-form equation.

9. Foci: $(0, \pm 3)$
Eccentricity: 0.5

10. Foci: $(\pm 8, 0)$
Eccentricity: 0.2

11. Vertices: $(0, \pm 70)$
Eccentricity: 0.1

12. Vertices: $(\pm 10, 0)$
Eccentricity: 0.24

Exercises 13–16 give foci and corresponding directrices of ellipses centered at the origin of the xy-plane. In each case, use the dimensions in Figure 10.19 to find the eccentricity of the ellipse. Then find the ellipse's standard-form equation.

13. Focus: $(\sqrt{5}, 0)$
Directrix: $x = \frac{9}{\sqrt{5}}$

14. Focus: $(4, 0)$
Directrix: $x = \frac{16}{3}$

15. Focus: $(-4, 0)$
Directrix: $x = -16$

16. Focus: $(-\sqrt{2}, 0)$
Directrix: $x = -2\sqrt{2}$

17. Draw an ellipse of eccentricity $4/5$. Explain your procedure.

18. Draw the orbit of Pluto (eccentricity 0.25) to scale. Explain your procedure.

19. The endpoints of the major and minor axes of an ellipse are $(1, 1)$, $(3, 4)$, $(1, 7)$, and $(-1, 4)$. Sketch the ellipse, give its equation in standard form, and find its foci, eccentricity, and directrices.

20. Find an equation for the ellipse of eccentricity $2/3$ that has the line $x = 9$ as a directrix and the point $(4, 0)$ as the corresponding focus.

21. What values of the constants a, b, and c make the ellipse

$$4x^2 + y^2 + ax + by + c = 0$$

lie tangent to the x-axis at the origin and pass through the point $(-1, 2)$? What is the eccentricity of the ellipse?

22. The reflective property of ellipses An ellipse is revolved about its major axis to generate an ellipsoid. The inner surface of the ellipsoid is silvered to make a mirror. Show that a ray of light emanating from one focus will be reflected to the other focus. Sound waves also follow such paths, and this property is used in constructing "whispering galleries." (*Hint:* Place the ellipse in standard position in the xy-plane and show that the lines from a point P on the ellipse to the two foci make congruent angles with the tangent to the ellipse at P.)

Hyperbolas

In Exercises 23–30, find the eccentricity of the hyperbola. Then find and graph the hyperbola's foci and directrices.

23. $x^2 - y^2 = 1$

24. $9x^2 - 16y^2 = 144$

25. $y^2 - x^2 = 8$

26. $y^2 - x^2 = 4$

27. $8x^2 - 2y^2 = 16$

28. $y^2 - 3x^2 = 3$

29. $8y^2 - 2x^2 = 16$

30. $64x^2 - 36y^2 = 2304$

Exercises 31–34 give the eccentricities and the vertices or foci of hyperbolas centered at the origin of the xy-plane. In each case, find the hyperbola's standard-form equation.

31. Eccentricity: 3
Vertices: $(0, \pm 1)$

32. Eccentricity: 2
Vertices: $(\pm 2, 0)$

33. Eccentricity: 3
Foci: $(\pm 3, 0)$

34. Eccentricity: 1.25
Foci: $(0, \pm 5)$

Exercises 35–38 give foci and corresponding directrices of hyperbolas centered at the origin of the xy-plane. In each case, find the hyperbola's eccentricity. Then find the hyperbola's standard-form equation.

35. Focus: $(4, 0)$
Directrix: $x = 2$

36. Focus: $(\sqrt{10}, 0)$
Directrix: $x = \sqrt{2}$

37. Focus: $(-2, 0)$
Directrix: $x = -\frac{1}{2}$

38. Focus: $(-6, 0)$
Directrix: $x = -2$

39. A hyperbola of eccentricity $3/2$ has one focus at $(1, -3)$. The corresponding directrix is the line $y = 2$. Find an equation for the hyperbola.

T **40. The effect of eccentricity on a hyperbola's shape** What happens to the graph of a hyperbola as its eccentricity increases? To find out, rewrite the equation $(x^2/a^2) - (y^2/b^2) = 1$ in terms of a and e instead of a and b. Graph the hyperbola for various values of e and describe what you find.

41. The reflective property of hyperbolas Show that a ray of light directed toward one focus of a hyperbolic mirror, as in the accompanying figure, is reflected toward the other focus. (*Hint:* Show that the tangent to the hyperbola at P bisects the angle made by segments PF_1 and PF_2.)

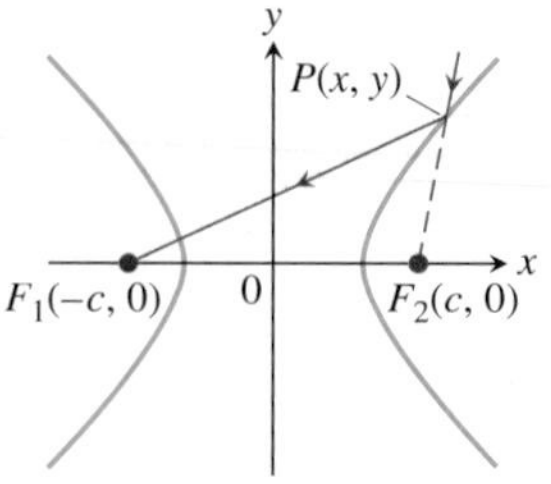

42. A confocal ellipse and hyperbola Show that an ellipse and a hyperbola that have the same foci A and B, as in the accompanying figure, cross at right angles at their point of intersection. (*Hint:* A ray of light from focus A that met the hyperbola at P would be reflected from the hyperbola as if it came directly from B (Exercise 41). The same ray would be reflected off the ellipse to pass through B (Exercise 22).)

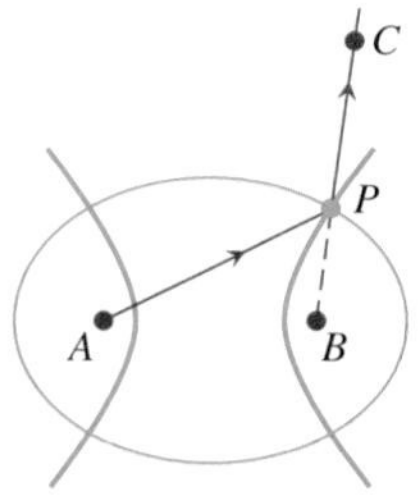

10.3 Quadratic Equations and Rotations

In this section, we examine the Cartesian graph of any equation

$$Ax^2 + Bxy + Cy^2 + Dx + Ey + F = 0, \tag{1}$$

in which A, B, and C are not all zero, and show that it is nearly always a conic section. The exceptions are the cases in which there is no graph at all or the graph consists of two parallel lines. It is conventional to call all graphs of Equation (1), curved or not, **quadratic curves**.

The Cross Product Term

You may have noticed that the term Bxy did not appear in the equations for the conic sections in Section 10.1. This happened because the axes of the conic sections ran parallel to (in fact, coincided with) the coordinate axes.

To see what happens when the parallelism is absent, let us write an equation for a hyperbola with $a = 3$ and foci at $F_1(-3, -3)$ and $F_2(3, 3)$ (Figure 10.22). The equation $|PF_1 - PF_2| = 2a$ becomes $|PF_1 - PF_2| = 2(3) = 6$ and

$$\sqrt{(x+3)^2 + (y+3)^2} - \sqrt{(x-3)^2 + (y-3)^2} = \pm 6.$$

When we transpose one radical, square, solve for the radical that still appears, and square again, the equation reduces to

$$2xy = 9, \tag{2}$$

a case of Equation (1) in which the cross product term is present. The asymptotes of the hyperbola in Equation (2) are the x- and y-axes, and the focal axis makes an angle of $\pi/4$

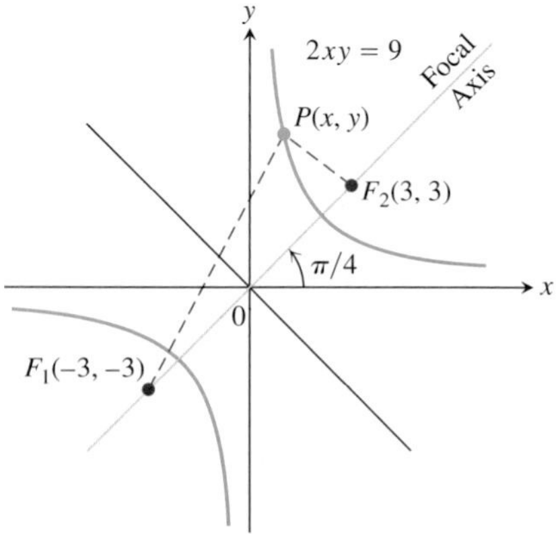

FIGURE 10.22 The focal axis of the hyperbola $2xy = 9$ makes an angle of $\pi/4$ radians with the positive x-axis.

radians with the positive x-axis. As in this example, the cross product term is present in Equation (1) only when the axes of the conic are tilted.

To eliminate the xy-term from the equation of a conic, we rotate the coordinate axes to eliminate the "tilt" in the axes of the conic. The equations for the rotations we use are derived in the following way. In the notation of Figure 10.23, which shows a counterclockwise rotation about the origin through an angle α,

$$\begin{aligned} x &= OM = OP\cos(\theta+\alpha) = OP\cos\theta\cos\alpha - OP\sin\theta\sin\alpha \\ y &= MP = OP\sin(\theta+\alpha) = OP\cos\theta\sin\alpha + OP\sin\theta\cos\alpha. \end{aligned} \tag{3}$$

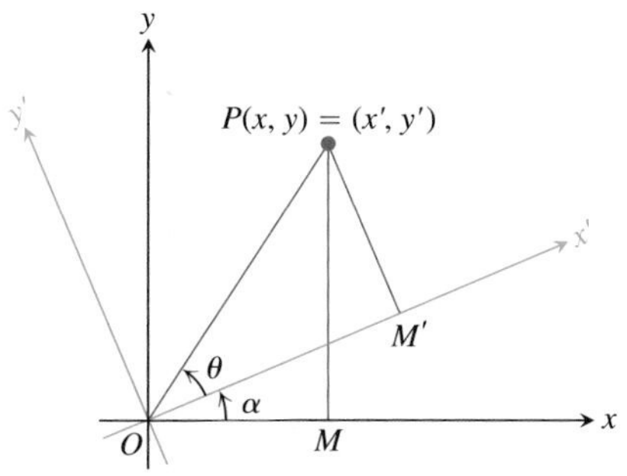

FIGURE 10.23 A counterclockwise rotation through angle α about the origin.

Since

$$OP\cos\theta = OM' = x'$$

and

$$OP\sin\theta = M'P = y',$$

Equations (3) reduce to the following.

Equations for Rotating Coordinate Axes

$$\begin{aligned} x &= x'\cos\alpha - y'\sin\alpha \\ y &= x'\sin\alpha + y'\cos\alpha \end{aligned} \tag{4}$$

EXAMPLE 1 Finding an Equation for a Hyperbola

The x- and y-axes are rotated through an angle of $\pi/4$ radians about the origin. Find an equation for the hyperbola $2xy = 9$ in the new coordinates.

Solution Since $\cos\pi/4 = \sin\pi/4 = 1/\sqrt{2}$, we substitute

$$x = \frac{x' - y'}{\sqrt{2}}, \qquad y = \frac{x' + y'}{\sqrt{2}}$$

from Equations (4) into the equation $2xy = 9$ and obtain

$$\begin{aligned} 2\left(\frac{x'-y'}{\sqrt{2}}\right)\left(\frac{x'+y'}{\sqrt{2}}\right) &= 9 \\ x'^2 - y'^2 &= 9 \\ \frac{x'^2}{9} - \frac{y'^2}{9} &= 1. \end{aligned}$$

See Figure 10.24. ■

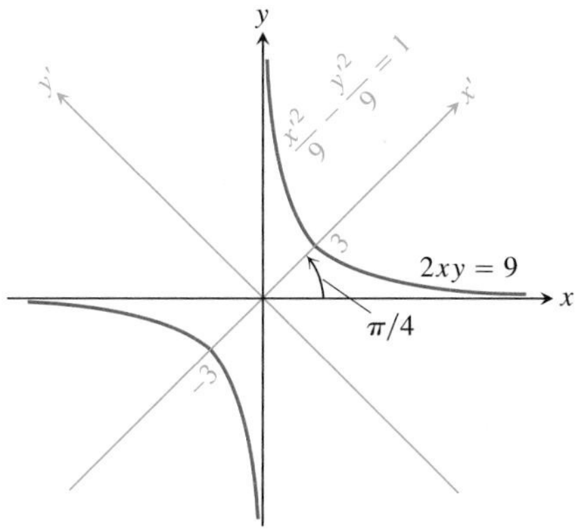

FIGURE 10.24 The hyperbola in Example 1 (x' and y' are the coordinates).

If we apply Equations (4) to the quadratic equation (1), we obtain a new quadratic equation

$$A'x'^2 + B'x'y' + C'y'^2 + D'x' + E'y' + F' = 0. \tag{5}$$

The new and old coefficients are related by the equations

$$\begin{aligned} A' &= A\cos^2\alpha + B\cos\alpha\sin\alpha + C\sin^2\alpha \\ B' &= B\cos 2\alpha + (C - A)\sin 2\alpha \\ C' &= A\sin^2\alpha - B\sin\alpha\cos\alpha + C\cos^2\alpha \\ D' &= D\cos\alpha + E\sin\alpha \\ E' &= -D\sin\alpha + E\cos\alpha \\ F' &= F. \end{aligned} \tag{6}$$

These equations show, among other things, that if we start with an equation for a curve in which the cross product term is present ($B \neq 0$), we can find a rotation angle α that produces an equation in which no cross product term appears ($B' = 0$). To find α, we set $B' = 0$ in the second equation in (6) and solve the resulting equation,

$$B\cos 2\alpha + (C - A)\sin 2\alpha = 0,$$

for α. In practice, this means determining α from one of the two equations

Angle of Rotation

$$\cot 2\alpha = \frac{A - C}{B} \qquad \text{or} \qquad \tan 2\alpha = \frac{B}{A - C}. \tag{7}$$

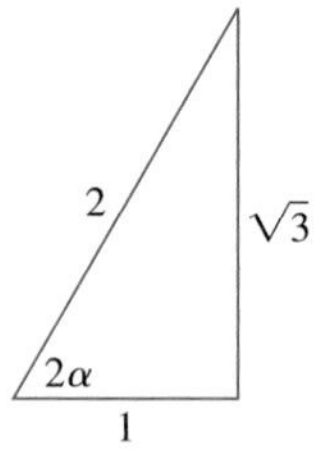

FIGURE 10.25 This triangle identifies $2\alpha = \cot^{-1}(1/\sqrt{3})$ as $\pi/3$ (Example 2).

EXAMPLE 2 Finding the Angle of Rotation

The coordinate axes are to be rotated through an angle α to produce an equation for the curve

$$2x^2 + \sqrt{3}\,xy + y^2 - 10 = 0$$

that has no cross product term. Find α and the new equation. Identify the curve.

Solution The equation $2x^2 + \sqrt{3}\,xy + y^2 - 10 = 0$ has $A = 2, B = \sqrt{3}$, and $C = 1$. We substitute these values into Equation (7) to find α:

$$\cot 2\alpha = \frac{A - C}{B} = \frac{2 - 1}{\sqrt{3}} = \frac{1}{\sqrt{3}}.$$

From the right triangle in Figure 10.25, we see that one appropriate choice of angle is $2\alpha = \pi/3$, so we take $\alpha = \pi/6$. Substituting $\alpha = \pi/6$, $A = 2$, $B = \sqrt{3}$, $C = 1$, $D = E = 0$, and $F = -10$ into Equations (6) gives

$$A' = \frac{5}{2}, \qquad B' = 0, \qquad C' = \frac{1}{2}, \qquad D' = E' = 0, \qquad F' = -10.$$

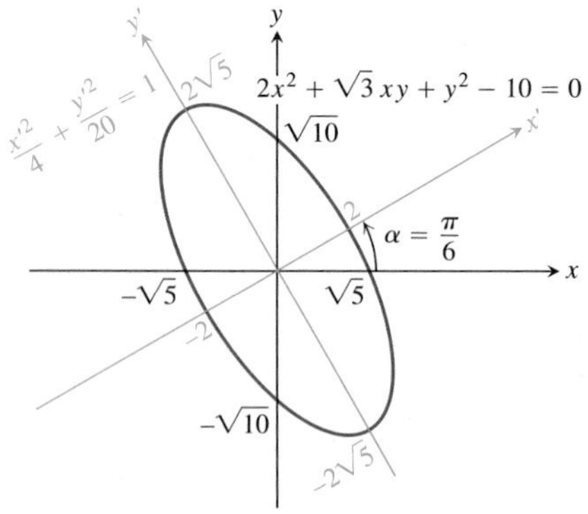

FIGURE 10.26 The conic section in Example 2.

Equation (5) then gives

$$\frac{5}{2}x'^2 + \frac{1}{2}y'^2 - 10 = 0, \qquad \text{or} \qquad \frac{x'^2}{4} + \frac{y'^2}{20} = 1.$$

The curve is an ellipse with foci on the new y'-axis (Figure 10.26). ■

Possible Graphs of Quadratic Equations

We now return to the graph of the general quadratic equation.

Since axes can always be rotated to eliminate the cross product term, there is no loss of generality in assuming that this has been done and that our equation has the form

$$Ax^2 + Cy^2 + Dx + Ey + F = 0. \tag{8}$$

Equation (8) represents

(a) a *circle* if $A = C \neq 0$ (special cases: the graph is a point or there is no graph at all);

(b) a *parabola* if Equation (8) is quadratic in one variable and linear in the other;

(c) an *ellipse* if A and C are both positive or both negative (special cases: circles, a single point, or no graph at all);

(d) a *hyperbola* if A and C have opposite signs (special case: a pair of intersecting lines);

(e) a *straight line* if A and C are zero and at least one of D and E is different from zero;

(f) *one or two straight lines* if the left-hand side of Equation (8) can be factored into the product of two linear factors.

See Table 10.3 for examples.

TABLE 10.3 Examples of quadratic curves $Ax^2 + Bxy + Cy^2 + Dx + Ey + F = 0$

	A	B	C	D	E	F	Equation	Remarks
Circle	1		1			-4	$x^2 + y^2 = 4$	$A = C; F < 0$
Parabola			1	-9			$y^2 = 9x$	Quadratic in y, linear in x
Ellipse	4		9			-36	$4x^2 + 9y^2 = 36$	A, C have same sign, $A \neq C; F < 0$
Hyperbola	1		-1			-1	$x^2 - y^2 = 1$	A, C have opposite signs
One line (still a conic section)	1						$x^2 = 0$	y-axis
Intersecting lines (still a conic section)		1		1	-1	-1	$xy + x - y - 1 = 0$	Factors to $(x - 1)(y + 1) = 0$, so $x = 1, y = -1$
Parallel lines (not a conic section)	1			-3		2	$x^2 - 3x + 2 = 0$	Factors to $(x - 1)(x - 2) = 0$, so $x = 1, x = 2$
Point	1		1				$x^2 + y^2 = 0$	The origin
No graph	1					1	$x^2 = -1$	No graph

The Discriminant Test

We do not need to eliminate the xy-term from the equation

$$Ax^2 + Bxy + Cy^2 + Dx + Ey + F = 0 \tag{9}$$

to tell what kind of conic section the equation represents. If this is the only information we want, we can apply the following test instead.

As we have seen, if $B \neq 0$, then rotating the coordinate axes through an angle α that satisfies the equation

$$\cot 2\alpha = \frac{A - C}{B} \tag{10}$$

will change Equation (9) into an equivalent form

$$A'x'^2 + C'y'^2 + D'x' + E'y' + F' = 0 \tag{11}$$

without a cross product term.

Now, the graph of Equation (11) is a (real or degenerate)

(a) *parabola* if A' or $C' = 0$; that is, if $A'C' = 0$;

(b) *ellipse* if A' and C' have the same sign; that is, if $A'C' > 0$;

(c) *hyperbola* if A' and C' have opposite signs; that is, if $A'C' < 0$.

It can also be verified from Equations (6) that for any rotation of axes,

$$B^2 - 4AC = B'^2 - 4A'C'. \tag{12}$$

This means that the quantity $B^2 - 4AC$ is not changed by a rotation. But when we rotate through the angle α given by Equation (10), B' becomes zero, so

$$B^2 - 4AC = -4A'C'.$$

Since the curve is a parabola if $A'C' = 0$, an ellipse if $A'C' > 0$, and a hyperbola if $A'C' < 0$, the curve must be a parabola if $B^2 - 4AC = 0$, an ellipse if $B^2 - 4AC < 0$, and a hyperbola if $B^2 - 4AC > 0$. The number $B^2 - 4AC$ is called the **discriminant** of Equation (9).

The Discriminant Test

With the understanding that occasional degenerate cases may arise, the quadratic curve $Ax^2 + Bxy + Cy^2 + Dx + Ey + F = 0$ is

(a) a **parabola** if $B^2 - 4AC = 0$,
(b) an **ellipse** if $B^2 - 4AC < 0$,
(c) a **hyperbola** if $B^2 - 4AC > 0$.

EXAMPLE 3 Applying the Discriminant Test

(a) $3x^2 - 6xy + 3y^2 + 2x - 7 = 0$ represents a parabola because

$$B^2 - 4AC = (-6)^2 - 4 \cdot 3 \cdot 3 = 36 - 36 = 0.$$

(b) $x^2 + xy + y^2 - 1 = 0$ represents an ellipse because

$$B^2 - 4AC = (1)^2 - 4 \cdot 1 \cdot 1 = -3 < 0.$$

(c) $xy - y^2 - 5y + 1 = 0$ represents a hyperbola because

$$B^2 - 4AC = (1)^2 - 4(0)(-1) = 1 > 0.$$

■

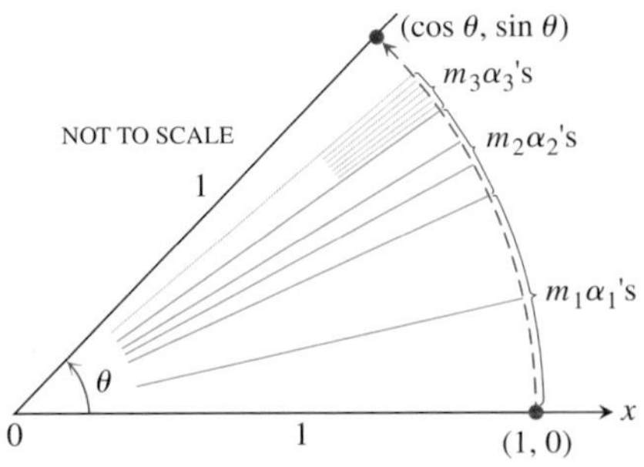

FIGURE 10.27 To calculate the sine and cosine of an angle θ between 0 and 2π, the calculator rotates the point $(1, 0)$ to an appropriate location on the unit circle and displays the resulting coordinates.

USING TECHNOLOGY How Calculators Use Rotations to Evaluate Sines and Cosines

Some calculators use rotations to calculate sines and cosines of arbitrary angles. The procedure goes something like this: The calculator has, stored,

1. ten angles or so, say

$$\alpha_1 = \sin^{-1}(10^{-1}), \qquad \alpha_2 = \sin^{-1}(10^{-2}), \qquad \ldots, \qquad \alpha_{10} = \sin^{-1}(10^{-10}),$$

and

2. twenty numbers, the sines and cosines of the angles $\alpha_1, \alpha_2, \ldots, \alpha_{10}$.

To calculate the sine and cosine of an arbitrary angle θ, we enter θ (in radians) into the calculator. The calculator subtracts or adds multiples of 2π to θ to replace θ by the angle between 0 and 2π that has the same sine and cosine as θ (we continue to call the angle θ). The calculator then "writes" θ as a sum of multiples of α_1 (as many as possible without overshooting) plus multiples of α_2 (again, as many as possible), and so on, working its way to α_{10}. This gives

$$\theta \approx m_1\alpha_1 + m_2\alpha_2 + \cdots + m_{10}\alpha_{10}.$$

The calculator then rotates the point $(1, 0)$ through m_1 copies of α_1 (through α_1, m_1 times in succession), plus m_2 copies of α_2, and so on, finishing off with m_{10} copies of α_{10} (Figure 10.27). The coordinates of the final position of $(1, 0)$ on the unit circle are the values the calculator gives for $(\cos\theta, \sin\theta)$.

EXERCISES 10.3

Using the Discriminant

Use the discriminant $B^2 - 4AC$ to decide whether the equations in Exercises 1–16 represent parabolas, ellipses, or hyperbolas.

1. $x^2 - 3xy + y^2 - x = 0$
2. $3x^2 - 18xy + 27y^2 - 5x + 7y = -4$
3. $3x^2 - 7xy + \sqrt{17}\,y^2 = 1$
4. $2x^2 - \sqrt{15}\,xy + 2y^2 + x + y = 0$
5. $x^2 + 2xy + y^2 + 2x - y + 2 = 0$
6. $2x^2 - y^2 + 4xy - 2x + 3y = 6$
7. $x^2 + 4xy + 4y^2 - 3x = 6$
8. $x^2 + y^2 + 3x - 2y = 10$
9. $xy + y^2 - 3x = 5$
10. $3x^2 + 6xy + 3y^2 - 4x + 5y = 12$
11. $3x^2 - 5xy + 2y^2 - 7x - 14y = -1$
12. $2x^2 - 4.9xy + 3y^2 - 4x = 7$
13. $x^2 - 3xy + 3y^2 + 6y = 7$
14. $25x^2 + 21xy + 4y^2 - 350x = 0$
15. $6x^2 + 3xy + 2y^2 + 17y + 2 = 0$
16. $3x^2 + 12xy + 12y^2 + 435x - 9y + 72 = 0$

Rotating Coordinate Axes

In Exercises 17–26, rotate the coordinate axes to change the given equation into an equation that has no cross product (xy) term. Then identify the graph of the equation. (The new equations will vary with the size and direction of the rotation you use.)

17. $xy = 2$
18. $x^2 + xy + y^2 = 1$
19. $3x^2 + 2\sqrt{3}\,xy + y^2 - 8x + 8\sqrt{3}\,y = 0$
20. $x^2 - \sqrt{3}\,xy + 2y^2 = 1$
21. $x^2 - 2xy + y^2 = 2$
22. $3x^2 - 2\sqrt{3}\,xy + y^2 = 1$
23. $\sqrt{2}\,x^2 + 2\sqrt{2}\,xy + \sqrt{2}\,y^2 - 8x + 8y = 0$
24. $xy - y - x + 1 = 0$
25. $3x^2 + 2xy + 3y^2 = 19$
26. $3x^2 + 4\sqrt{3}\,xy - y^2 = 7$
27. Find the sine and cosine of an angle in Quadrant I through which the coordinate axes can be rotated to eliminate the cross product term from the equation

$$14x^2 + 16xy + 2y^2 - 10x + 26{,}370y - 17 = 0.$$

Do not carry out the rotation.

28. Find the sine and cosine of an angle in Quadrant II through which the coordinate axes can be rotated to eliminate the cross product term from the equation

$$4x^2 - 4xy + y^2 - 8\sqrt{5}\,x - 16\sqrt{5}\,y = 0.$$

Do not carry out the rotation.

T The conic sections in Exercises 17–26 were chosen to have rotation angles that were "nice" in the sense that once we knew $\cot 2\alpha$ or $\tan 2\alpha$ we could identify 2α and find $\sin\alpha$ and $\cos\alpha$ from familiar triangles.

In Exercises 29–34, use a calculator to find an angle α through which the coordinate axes can be rotated to change the given equation into a quadratic equation that has no cross product term. Then find $\sin\alpha$ and $\cos\alpha$ to two decimal places and use Equations (6) to find the coefficients of the new equation to the nearest decimal place. In each case, say whether the conic section is an ellipse, a hyperbola, or a parabola.

29. $x^2 - xy + 3y^2 + x - y - 3 = 0$

30. $2x^2 + xy - 3y^2 + 3x - 7 = 0$

31. $x^2 - 4xy + 4y^2 - 5 = 0$

32. $2x^2 - 12xy + 18y^2 - 49 = 0$

33. $3x^2 + 5xy + 2y^2 - 8y - 1 = 0$

34. $2x^2 + 7xy + 9y^2 + 20x - 86 = 0$

Theory and Examples

35. What effect does a 90° rotation about the origin have on the equations of the following conic sections? Give the new equation in each case.

a. The ellipse $(x^2/a^2) + (y^2/b^2) = 1 \quad (a > b)$

b. The hyperbola $(x^2/a^2) = (y^2/b^2) = 1$

c. The circle $x^2 + y^2 = a^2$

d. The line $y = mx$ **e.** The line $y = mx + b$

36. What effect does a 180° rotation about the origin have on the equations of the following conic sections? Give the new equation in each case.

a. The ellipse $(x^2/a^2) + (y^2/b^2) = 1 \quad (a > b)$

b. The hyperbola $(x^2/a^2) = (y^2/b^2) = 1$

c. The circle $x^2 + y^2 = a^2$

d. The line $y = mx$ **e.** The line $y = mx + b$

37. The Hyperbola $xy = a$ The hyperbola $xy = 1$ is one of many hyperbolas of the form $xy = a$ that appear in science and mathematics.

a. Rotate the coordinate axes through an angle of 45° to change the equation $xy = 1$ into an equation with no xy-term. What is the new equation?

b. Do the same for the equation $xy = a$.

38. Find the eccentricity of the hyperbola $xy = 2$.

39. Can anything be said about the graph of the equation $Ax^2 + Bxy + Cy^2 + Dx + Ey + F = 0$ if $AC < 0$? Give reasons for your answer.

40. Degenerate conics Does any nondegenerate conic section $Ax^2 + Bxy + Cy^2 + Dx + Ey + F = 0$ have all of the following properties?

a. It is symmetric with respect to the origin.

b. It passes through the point $(1, 0)$.

c. It is tangent to the line $y = 1$ at the point $(-2, 1)$.

Give reasons for your answer.

41. Show that the equation $x^2 + y^2 = a^2$ becomes $x'^2 + y'^2 = a^2$ for every choice of the angle α in the rotation equations (4).

42. Show that rotating the axes through an angle of $\pi/4$ radians will eliminate the xy-term from Equation (1) whenever $A = C$.

43. a. Decide whether the equation

$$x^2 + 4xy + 4y^2 + 6x + 12y + 9 = 0$$

represents an ellipse, a parabola, or a hyperbola.

b. Show that the graph of the equation in part (a) is the line $2y = -x - 3$.

44. a. Decide whether the conic section with equation

$$9x^2 + 6xy + y^2 - 12x - 4y + 4 = 0$$

represents a parabola, an ellipse, or a hyperbola.

b. Show that the graph of the equation in part (a) is the line $y = -3x + 2$.

45. a. What kind of conic section is the curve $xy + 2x - y = 0$?

b. Solve the equation $xy + 2x - y = 0$ for y and sketch the curve as the graph of a rational function of x.

c. Find equations for the lines parallel to the line $y = -2x$ that are normal to the curve. Add the lines to your sketch.

46. Prove or find counterexamples to the following statements about the graph of $Ax^2 + Bxy + Cy^2 + Dx + Ey + F = 0$.

a. If $AC > 0$, the graph is an ellipse.

b. If $AC > 0$, the graph is a hyperbola.

c. If $AC < 0$, the graph is a hyperbola.

47. A nice area formula for ellipses When $B^2 - 4AC$ is negative, the equation

$$Ax^2 + Bxy + Cy^2 = 1$$

represents an ellipse. If the ellipse's semi-axes are a and b, its area is πab (a standard formula). Show that the area is also given by the formula $2\pi/\sqrt{4AC - B^2}$. (*Hint:* Rotate the coordinate axes to eliminate the xy-term and apply Equation (12) to the new equation.)

48. Other invariants We describe the fact that $B'^2 - 4A'C'$ equals $B^2 - 4AC$ after a rotation about the origin by saying that the discriminant of a quadratic equation is an *invariant* of the equation.

Use Equations (6) to show that the numbers **(a)** $A + C$ and **(b)** $D^2 + E^2$ are also invariants, in the sense that

$$A' + C' = A + C \quad \text{and} \quad D'^2 + E'^2 = D^2 + E^2.$$

We can use these equalities to check against numerical errors when we rotate axes.

49. A proof that $B'^2 - 4A'C' = B^2 - 4AC$ Use Equations (6) to show that $B'^2 - 4A'C' = B^2 - 4AC$ for any rotation of axes about the origin.

10.4 Conics and Parametric Equations; The Cycloid

Curves in the Cartesian plane defined by parametric equations, and the calculation of their derivatives, were introduced in Section 3.5. There we studied parametrizations of lines, circles, and ellipses. In this section we discuss parametrization of parabolas, hyperbolas, cycloids, brachistocrones, and tautocrones.

Parabolas and Hyperbolas

In Section 3.5 we used the parametrization

$$x = \sqrt{t}, \qquad y = t, \qquad t > 0$$

to describe the motion of a particle moving along the right branch of the parabola $y = x^2$. In the following example we obtain a parametrization of the entire parabola, not just its right branch.

EXAMPLE 1 An Entire Parabola

The position $P(x, y)$ of a particle moving in the xy-plane is given by the equations and parameter interval

$$x = t, \qquad y = t^2, \qquad -\infty < t < \infty.$$

Identify the particle's path and describe the motion.

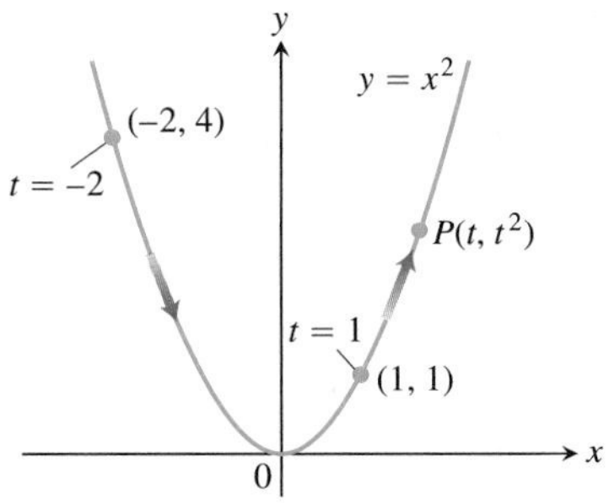

FIGURE 10.28 The path defined by $x = t, y = t^2, -\infty < t < \infty$ is the entire parabola $y = x^2$ (Example 1).

Solution We identify the path by eliminating t between the equations $x = t$ and $y = t^2$, obtaining

$$y = (t)^2 = x^2.$$

The particle's position coordinates satisfy the equation $y = x^2$, so the particle moves along this curve.

In contrast to Example 10 in Section 3.5, the particle now traverses the entire parabola. As t increases from $-\infty$ to ∞, the particle comes down the left-hand side, passes through the origin, and moves up the right-hand side (Figure 10.28). ■

As Example 1 illustrates, any curve $y = f(x)$ has the parametrization $x = t$, $y = f(t)$. This is so simple we usually do not use it, but the point of view is occasionally helpful.

EXAMPLE 2 A Parametrization of the Right-hand Branch of the Hyperbola $x^2 - y^2 = 1$

Describe the motion of the particle whose position $P(x, y)$ at time t is given by

$$x = \sec t, \qquad y = \tan t, \qquad -\frac{\pi}{2} < t < \frac{\pi}{2}.$$

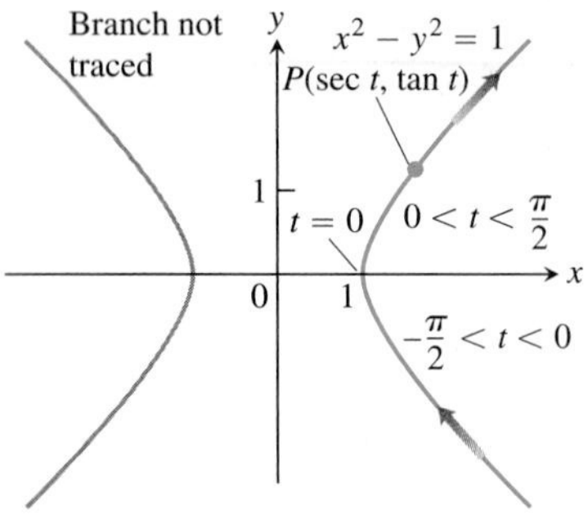

FIGURE 10.29 The equations $x = \sec t, y = \tan t$ and interval $-\pi/2 < t < \pi/2$ describe the right-hand branch of the hyperbola $x^2 - y^2 = 1$ (Example 2).

Solution We find a Cartesian equation for the coordinates of P by eliminating t between the equations

$$\sec t = x, \qquad \tan t = y.$$

We accomplish this with the identity $\sec^2 t - \tan^2 t = 1$, which yields

$$x^2 - y^2 = 1.$$

Since the particle's coordinates (x, y) satisfy the equation $x^2 - y^2 = 1$, the motion takes place somewhere on this hyperbola. As t runs between $-\pi/2$ and $\pi/2$, $x = \sec t$ remains positive and $y = \tan t$ runs between $-\infty$ and ∞, so P traverses the hyperbola's right-hand branch. It comes in along the branch's lower half as $t \to 0^-$, reaches $(1, 0)$ at $t = 0$, and moves out into the first quadrant as t increases toward $\pi/2$ (Figure 10.29). ■

Cycloids

The problem with a pendulum clock whose bob swings in a circular arc is that the frequency of the swing depends on the amplitude of the swing. The wider the swing, the longer it takes the bob to return to center (its lowest position).

This does not happen if the bob can be made to swing in a *cycloid*. In 1673, Christiaan Huygens designed a pendulum clock whose bob would swing in a cycloid, a curve we define in Example 3. He hung the bob from a fine wire constrained by guards that caused it to draw up as it swung away from center (Figure 10.30).

HISTORICAL BIOGRAPHY

Christiaan Huygens (1629–1695)

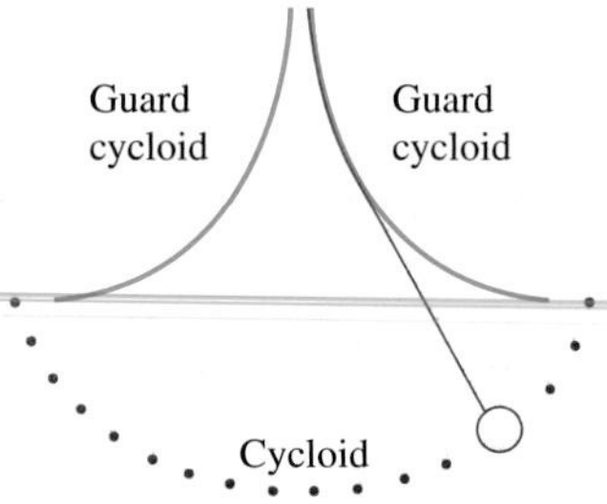

FIGURE 10.30 In Huygens' pendulum clock, the bob swings in a cycloid, so the frequency is independent of the amplitude.

EXAMPLE 3 Parametrizing a Cycloid

A wheel of radius a rolls along a horizontal straight line. Find parametric equations for the path traced by a point P on the wheel's circumference. The path is called a **cycloid**.

Solution We take the line to be the x-axis, mark a point P on the wheel, start the wheel with P at the origin, and roll the wheel to the right. As parameter, we use the angle t through which the wheel turns, measured in radians. Figure 10.31 shows the wheel a short while later, when its base lies at units from the origin. The wheel's center C lies at (at, a) and the coordinates of P are

$$x = at + a\cos\theta, \qquad y = a + a\sin\theta.$$

To express θ in terms of t, we observe that $t + \theta = 3\pi/2$ in the figure, so that

$$\theta = \frac{3\pi}{2} - t.$$

This makes

$$\cos\theta = \cos\left(\frac{3\pi}{2} - t\right) = -\sin t, \qquad \sin\theta = \sin\left(\frac{3\pi}{2} - t\right) = -\cos t.$$

The equations we seek are

$$x = at - a\sin t, \qquad y = a - a\cos t.$$

These are usually written with the a factored out:

$$x = a(t - \sin t), \qquad y = a(1 - \cos t). \tag{1}$$

Figure 10.32 shows the first arch of the cycloid and part of the next. ■

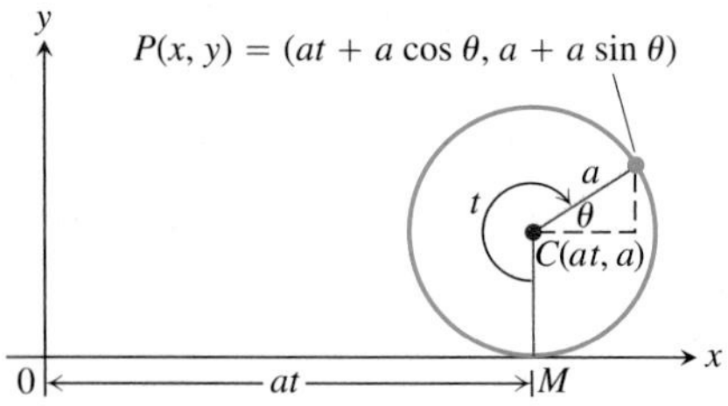

FIGURE 10.31 The position of $P(x, y)$ on the rolling wheel at angle t (Example 3).

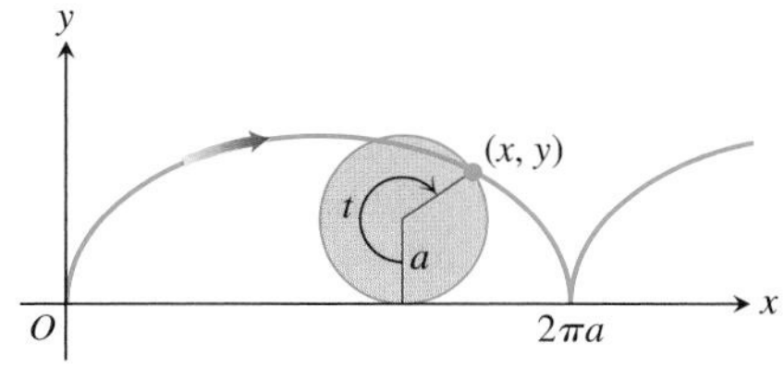

FIGURE 10.32 The cycloid $x = a(t - \sin t), y = a(1 - \cos t)$, for $t \geq 0$.

Brachistochrones and Tautochrones

If we turn Figure 10.32 upside down, Equations (1) still apply and the resulting curve (Figure 10.33) has two interesting physical properties. The first relates to the origin O and the point B at the bottom of the first arch. Among all smooth curves joining these points, the cycloid is the curve along which a frictionless bead, subject only to the force of gravity, will slide from O to B the fastest. This makes the cycloid a **brachistochrone** ("brah-*kiss*-toe-krone"), or shortest time curve for these points. The second property is that even if you start the bead partway down the curve toward B, it will still take the bead the same amount of time to reach B. This makes the cycloid a **tautochrone** ("*taw*-toe-krone"), or same-time curve for O and B.

Are there any other brachistochrones joining O and B, or is the cycloid the only one? We can formulate this as a mathematical question in the following way. At the start, the kinetic energy of the bead is zero, since its velocity is zero. The work done by gravity in moving the bead from (0, 0) to any other point (x, y) in the plane is mgy, and this must equal the change in kinetic energy. That is,

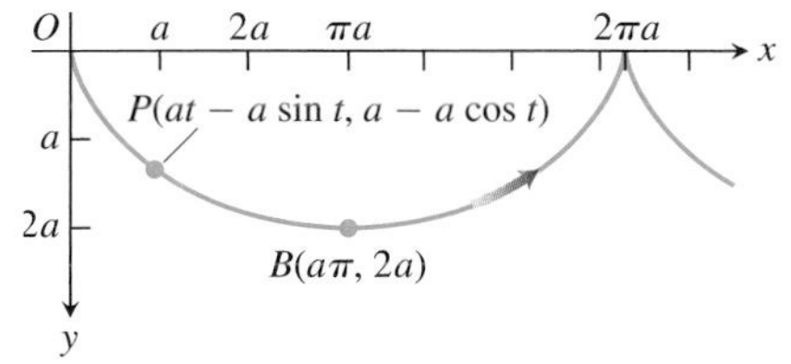

FIGURE 10.33 To study motion along an upside-down cycloid under the influence of gravity, we turn Figure 10.32 upside down. This points the y-axis in the direction of the gravitational force and makes the downward y-coordinates positive. The equations and parameter interval for the cycloid are still

$$x = a(t - \sin t),$$
$$y = a(1 - \cos t), \quad t \geq 0.$$

The arrow shows the direction of increasing t.

$$mgy = \frac{1}{2}mv^2 - \frac{1}{2}m(0)^2.$$

Thus, the velocity of the bead when it reaches (x, y) has to be

$$v = \sqrt{2gy}.$$

That is,

$$\frac{ds}{dt} = \sqrt{2gy}$$

ds is the arc length differential along the bead's path.

or

$$dt = \frac{ds}{\sqrt{2gy}} = \frac{\sqrt{1 + (dy/dx)^2}\,dx}{\sqrt{2gy}}.$$

The time T_f it takes the bead to slide along a particular path $y = f(x)$ from O to $B(a\pi, 2a)$ is

$$T_f = \int_{x=0}^{x=a\pi} \sqrt{\frac{1 + (dy/dx)^2}{2gy}}\,dx. \tag{2}$$

What curves $y = f(x)$, if any, minimize the value of this integral?

At first sight, we might guess that the straight line joining O and B would give the shortest time, but perhaps not. There might be some advantage in having the bead fall vertically at first to build up its velocity faster. With a higher velocity, the bead could travel a longer path and still reach B first. Indeed, this is the right idea. The solution, from a branch of mathematics known as the *calculus of variations,* is that the original cycloid from O to B is the one and only brachistochrone for O and B.

While the solution of the brachistrochrone problem is beyond our present reach, we can still show why the cycloid is a tautochrone. For the cycloid, Equation (2) takes the form

$$\begin{aligned} T_{\text{cycloid}} &= \int_{x=0}^{x=a\pi} \sqrt{\frac{dx^2 + dy^2}{2gy}} \\ &= \int_{t=0}^{t=\pi} \sqrt{\frac{a^2(2 - 2\cos t)}{2ga(1 - \cos t)}}\,dt \\ &= \int_0^{\pi} \sqrt{\frac{a}{g}}\,dt = \pi\sqrt{\frac{a}{g}}. \end{aligned}$$

From Equations (1), $dx = a(1 - \cos t)\,dt$, $dy = a \sin t\,dt$, and $y = a(1 - \cos t)$

Thus, the amount of time it takes the frictionless bead to slide down the cycloid to B after it is released from rest at O is $\pi\sqrt{a/g}$.

Suppose that instead of starting the bead at O we start it at some lower point on the cycloid, a point (x_0, y_0) corresponding to the parameter value $t_0 > 0$. The bead's velocity at any later point (x, y) on the cycloid is

$$v = \sqrt{2g(y - y_0)} = \sqrt{2ga(\cos t_0 - \cos t)}. \qquad y = a(1 - \cos t)$$

Accordingly, the time required for the bead to slide from (x_0, y_0) down to B is

$$T = \int_{t_0}^{\pi} \sqrt{\frac{a^2(2 - 2\cos t)}{2ga(\cos t_0 - \cos t)}}\,dt = \sqrt{\frac{a}{g}} \int_{t_0}^{\pi} \sqrt{\frac{1 - \cos t}{\cos t_0 - \cos t}}\,dt$$

$$= \sqrt{\frac{a}{g}} \int_{t_0}^{\pi} \sqrt{\frac{2\sin^2(t/2)}{(2\cos^2(t_0/2) - 1) - (2\cos^2(t/2) - 1)}}\,dt$$

$$= \sqrt{\frac{a}{g}} \int_{t_0}^{\pi} \frac{\sin(t/2)\,dt}{\sqrt{\cos^2(t_0/2) - \cos^2(t/2)}}$$

$$= \sqrt{\frac{a}{g}} \int_{t=t_0}^{t=\pi} \frac{-2\,du}{\sqrt{a^2 - u^2}} \qquad \begin{aligned} u &= \cos(t/2) \\ -2\,du &= \sin(t/2)\,dt \\ c &= \cos(t_0/2) \end{aligned}$$

$$= 2\sqrt{\frac{a}{g}} \left[-\sin^{-1}\frac{u}{c}\right]_{t=t_0}^{t=\pi}$$

$$= 2\sqrt{\frac{a}{g}} \left[-\sin^{-1}\frac{\cos(t/2)}{\cos(t_0/2)}\right]_{t_0}^{\pi}$$

$$= 2\sqrt{\frac{a}{g}}(-\sin^{-1}0 + \sin^{-1}1) = \pi\sqrt{\frac{a}{g}}.$$

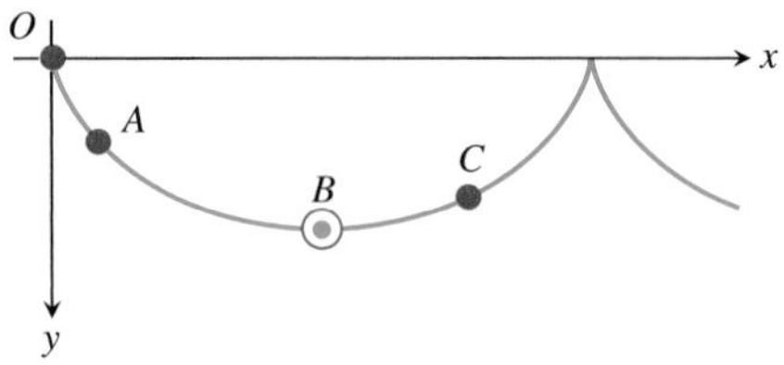

FIGURE 10.34 Beads released simultaneously on the cycloid at O, A, and C will reach B at the same time.

This is precisely the time it takes the bead to slide to B from O. It takes the bead the same amount of time to reach B no matter where it starts. Beads starting simultaneously from O, A, and C in Figure 10.34, for instance, will all reach B at the same time. This is the reason that Huygens' pendulum clock is independent of the amplitude of the swing.

EXERCISES 10.4

Parametric Equations for Conics

Exercises 1–12 give parametric equations and parameter intervals for the motion of a particle in the xy-plane. Identify the particle's path by finding a Cartesian equation for it. Graph the Cartesian equation. (The graphs will vary with the equation used.) Indicate the portion of the graph traced by the particle and the direction of motion.

1. $x = \cos t, \quad y = \sin t, \quad 0 \le t \le \pi$
2. $x = \sin(2\pi(1 - t)), \quad y = \cos(2\pi(1 - t)); \quad 0 \le t \le 1$
3. $x = 4\cos t, \quad y = 5\sin t; \quad 0 \le t \le \pi$
4. $x = 4\sin t, \quad y = 5\cos t; \quad 0 \le t \le 2\pi$
5. $x = t, \quad y = \sqrt{t}; \quad t \ge 0$
6. $x = \sec^2 t - 1, \quad y = \tan t; \quad -\pi/2 < t < \pi/2$
7. $x = -\sec t, \quad y = \tan t; \quad -\pi/2 < t < \pi/2$
8. $x = \csc t, \quad y = \cot t; \quad 0 < t < \pi$
9. $x = t, \quad y = \sqrt{4 - t^2}; \quad 0 \le t \le 2$
10. $x = t^2, \quad y = \sqrt{t^4 + 1}; \quad t \ge 0$
11. $x = -\cosh t, \quad y = \sinh t; \quad -\infty < t < \infty$
12. $x = 2\sinh t, \quad y = 2\cosh t; \quad -\infty < t < \infty$
13. **Hypocycloids** When a circle rolls on the inside of a fixed circle, any point P on the circumference of the rolling circle describes a *hypocycloid*. Let the fixed circle be $x^2 + y^2 = a^2$, let the radius of the rolling circle be b, and let the initial position of the tracing point P be $A(a, 0)$. Find parametric equations for the hypocycloid, using as the parameter the angle θ from the positive x-axis to the line joining the circles' centers. In particular, if

$b = a/4$, as in the accompanying figure, show that the hypocycloid is the astroid

$$x = a\cos^3\theta, \quad y = a\sin^3\theta.$$

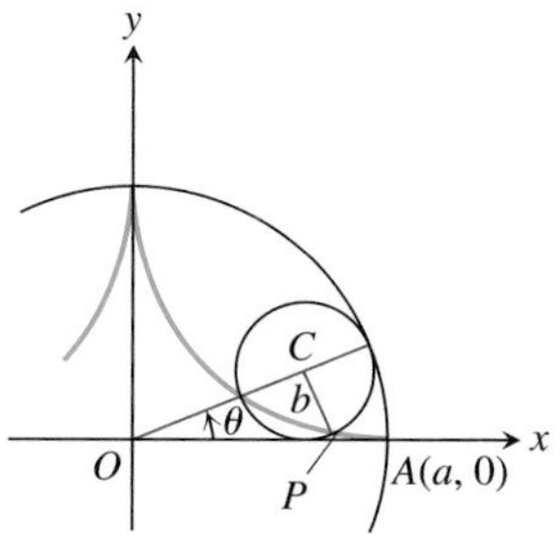

14. More about hypocycloids The accompanying figure shows a circle of radius a tangent to the inside of a circle of radius $2a$. The point P, shown as the point of tangency in the figure, is attached to the smaller circle. What path does P trace as the smaller circle rolls around the inside of the larger circle?

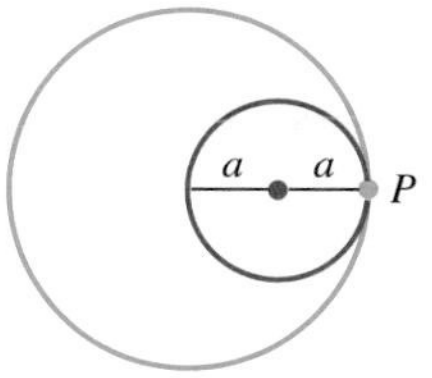

15. As the point N moves along the line $y = a$ in the accompanying figure, P moves in such a way that $OP = MN$. Find parametric equations for the coordinates of P as functions of the angle t that the line ON makes with the positive y-axis.

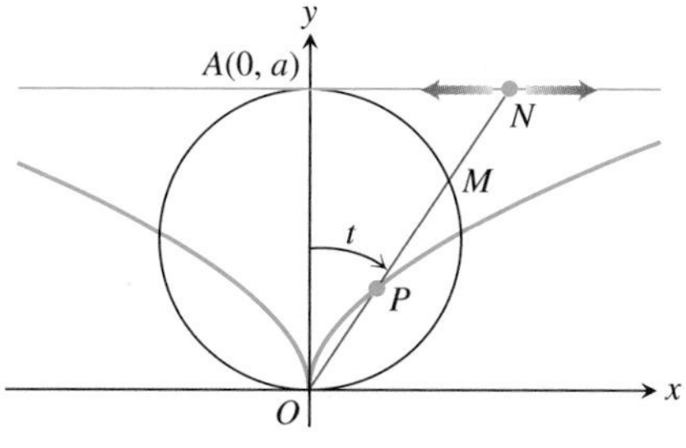

16. Trochoids A wheel of radius a rolls along a horizontal straight line without slipping. Find parametric equations for the curve traced out by a point P on a spoke of the wheel b units from its center. As parameter, use the angle θ through which the wheel turns. The curve is called a *trochoid*, which is a cycloid when $b = a$.

Distance Using Parametric Equations

17. Find the point on the parabola $x = t, y = t^2, -\infty < t < \infty$, closest to the point $(2, 1/2)$. (*Hint:* Minimize the square of the distance as a function of t.)

18. Find the point on the ellipse $x = 2\cos t, y = \sin t, 0 \le t \le 2\pi$ closest to the point $(3/4, 0)$. (*Hint:* Minimize the square of the distance as a function of t.)

T GRAPHER EXPLORATIONS

If you have a parametric equation grapher, graph the following equations over the given intervals.

19. Ellipse $x = 4\cos t, \quad y = 2\sin t, \quad$ over

a. $0 \le t \le 2\pi$ **b.** $0 \le t \le \pi$

c. $-\pi/2 \le t \le \pi/2$.

20. Hyperbola branch $x = \sec t$ (enter as $1/\cos(t)$), $y = \tan t$ (enter as $\sin(t)/\cos(t)$), over

a. $-1.5 \le t \le 1.5$ **b.** $-0.5 \le t \le 0.5$

c. $-0.1 \le t \le 0.1$.

21. Parabola $x = 2t + 3, \quad y = t^2 - 1, \quad -2 \le t \le 2$

22. Cycloid $x = t - \sin t, \quad y = 1 - \cos t, \quad$ over

a. $0 \le t \le 2\pi$ **b.** $0 \le t \le 4\pi$

c. $\pi \le t \le 3\pi$.

23. A nice curve (a deltoid)

$$x = 2\cos t + \cos 2t, \quad y = 2\sin t - \sin 2t; \quad 0 \le t \le 2\pi$$

What happens if you replace 2 with -2 in the equations for x and y? Graph the new equations and find out.

24. An even nicer curve

$$x = 3\cos t + \cos 3t, \quad y = 3\sin t - \sin 3t; \quad 0 \le t \le 2\pi$$

What happens if you replace 3 with -3 in the equations for x and y? Graph the new equations and find out.

25. Three beautiful curves

a. *Epicycloid:*

$$x = 9\cos t - \cos 9t, \quad y = 9\sin t - \sin 9t; \quad 0 \le t \le 2\pi$$

b. *Hypocycloid:*

$$x = 8\cos t + 2\cos 4t, \quad y = 8\sin t - 2\sin 4t; \quad 0 \le t \le 2\pi$$

c. *Hypotrochoid:*

$$x = \cos t + 5\cos 3t, \quad y = 6\cos t - 5\sin 3t; \quad 0 \le t \le 2\pi$$

26. More beautiful curves

a. $x = 6\cos t + 5\cos 3t, \quad y = 6\sin t - 5\sin 3t;$
$0 \le t \le 2\pi$

b. $x = 6\cos 2t + 5\cos 6t, \quad y = 6\sin 2t - 5\sin 6t;$
$0 \le t \le \pi$

c. $x = 6\cos t + 5\cos 3t, \quad y = 6\sin 2t - 5\sin 3t;$
$0 \le t \le 2\pi$

d. $x = 6\cos 2t + 5\cos 6t, \quad y = 6\sin 4t - 5\sin 6t;$
$0 \le t \le \pi$

10.5 Polar Coordinates

In this section, we study polar coordinates and their relation to Cartesian coordinates. While a point in the plane has just one pair of Cartesian coordinates, it has infinitely many pairs of polar coordinates. This has interesting consequences for graphing, as we will see in the next section.

Definition of Polar Coordinates

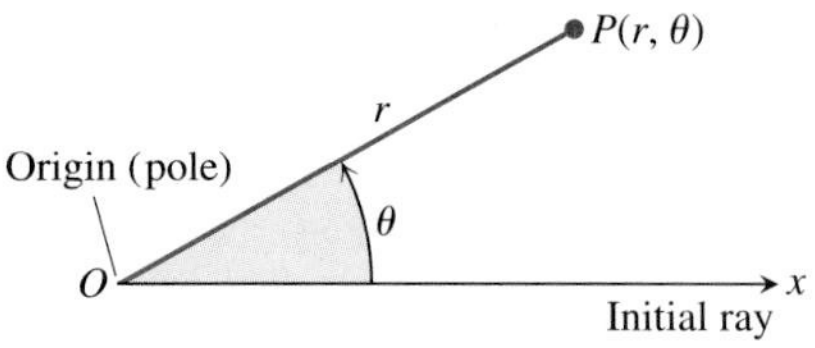

FIGURE 10.35 To define polar coordinates for the plane, we start with an origin, called the pole, and an initial ray.

To define polar coordinates, we first fix an **origin** O (called the **pole**) and an **initial ray** from O (Figure 10.35). Then each point P can be located by assigning to it a **polar coordinate pair** (r, θ) in which r gives the directed distance from O to P and θ gives the directed angle from the initial ray to ray OP.

> **Polar Coordinates**
>
> $$P(r, \theta)$$
>
> Directed distance from O to P — Directed angle from initial ray to OP

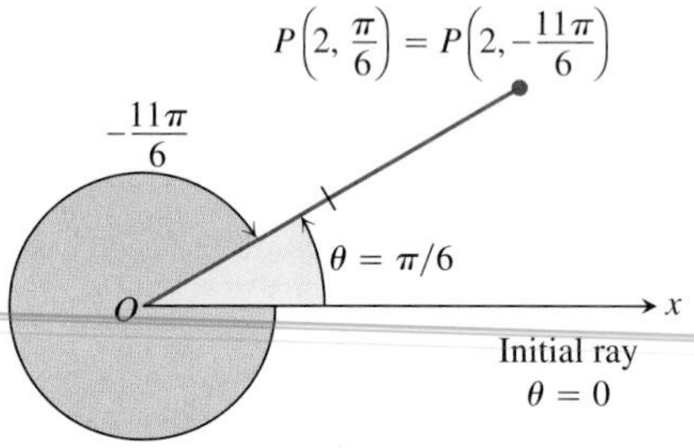

FIGURE 10.36 Polar coordinates are not unique.

As in trigonometry, θ is positive when measured counterclockwise and negative when measured clockwise. The angle associated with a given point is not unique. For instance, the point 2 units from the origin along the ray $\theta = \pi/6$ has polar coordinates $r = 2$, $\theta = \pi/6$. It also has coordinates $r = 2, \theta = -11\pi/6$ (Figure 10.36). There are occasions when we wish to allow r to be negative. That is why we use directed distance in defining $P(r, \theta)$. The point $P(2, 7\pi/6)$ can be reached by turning $7\pi/6$ radians counterclockwise from the initial ray and going forward 2 units (Figure 10.37). It can also be reached by turning $\pi/6$ radians counterclockwise from the initial ray and going *backward* 2 units. So the point also has polar coordinates $r = -2, \theta = \pi/6$.

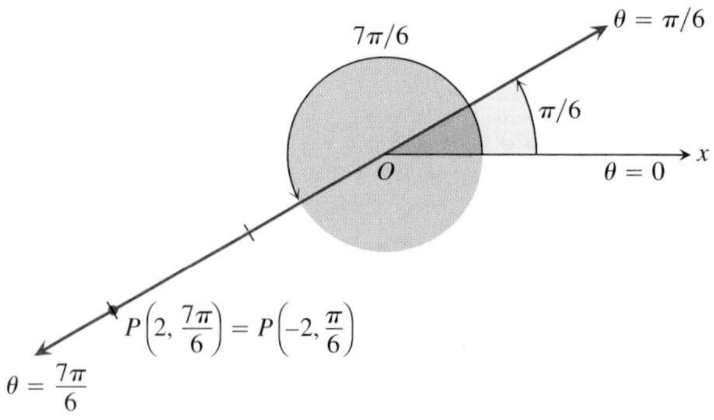

FIGURE 10.37 Polar coordinates can have negative r-values.

EXAMPLE 1 Finding Polar Coordinates

Find all the polar coordinates of the point $P(2, \pi/6)$.

Solution We sketch the initial ray of the coordinate system, draw the ray from the origin that makes an angle of $\pi/6$ radians with the initial ray, and mark the point $(2, \pi/6)$ (Figure 10.38). We then find the angles for the other coordinate pairs of P in which $r = 2$ and $r = -2$.

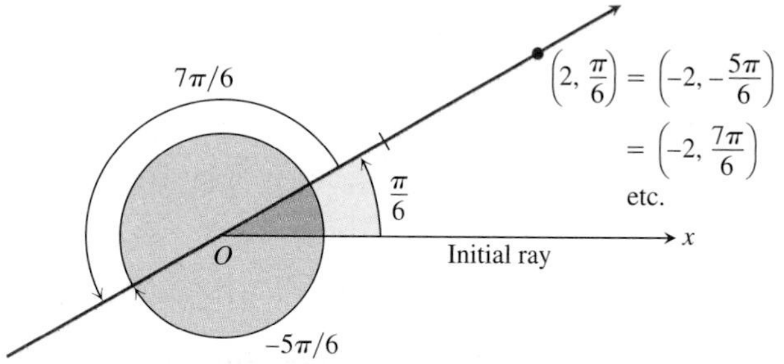

FIGURE 10.38 The point $P(2, \pi/6)$ has infinitely many polar coordinate pairs (Example 1).

For $r = 2$, the complete list of angles is

$$\frac{\pi}{6}, \quad \frac{\pi}{6} \pm 2\pi, \quad \frac{\pi}{6} \pm 4\pi, \quad \frac{\pi}{6} \pm 6\pi, \quad \ldots.$$

For $r = -2$, the angles are

$$-\frac{5\pi}{6}, \quad -\frac{5\pi}{6} \pm 2\pi, \quad -\frac{5\pi}{6} \pm 4\pi, \quad -\frac{5\pi}{6} \pm 6\pi, \quad \ldots.$$

The corresponding coordinate pairs of P are

$$\left(2, \frac{\pi}{6} + 2n\pi\right), \qquad n = 0, \pm 1, \pm 2, \ldots$$

and

$$\left(-2, -\frac{5\pi}{6} + 2n\pi\right), \qquad n = 0, \pm 1, \pm 2, \ldots.$$

When $n = 0$, the formulas give $(2, \pi/6)$ and $(-2, -5\pi/6)$. When $n = 1$, they give $(2, 13\pi/6)$ and $(-2, 7\pi/6)$, and so on. ■

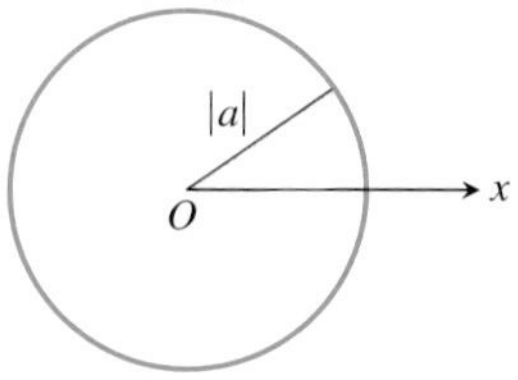

FIGURE 10.39 The polar equation for a circle is $r = a$.

Polar Equations and Graphs

If we hold r fixed at a constant value $r = a \neq 0$, the point $P(r, \theta)$ will lie $|a|$ units from the origin O. As θ varies over any interval of length 2π, P then traces a circle of radius $|a|$ centered at O (Figure 10.39).

If we hold θ fixed at a constant value $\theta = \theta_0$ and let r vary between $-\infty$ and ∞, the point $P(r, \theta)$ traces the line through O that makes an angle of measure θ_0 with the initial ray.

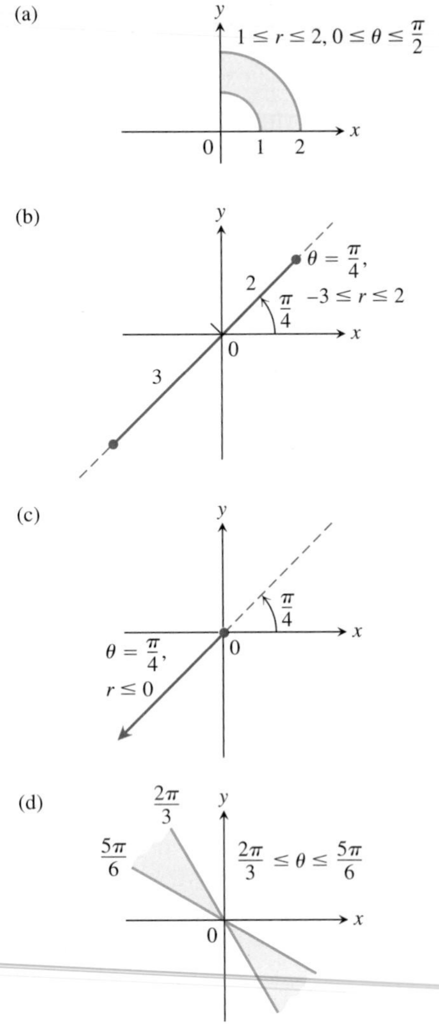

FIGURE 10.40 The graphs of typical inequalities in r and θ (Example 3).

Equation	Graph
$r = a$	Circle radius $\lvert a \rvert$ centered at O
$\theta = \theta_0$	Line through O making an angle θ_0 with the initial ray

EXAMPLE 2 Finding Polar Equations for Graphs

(a) $r = 1$ and $r = -1$ are equations for the circle of radius 1 centered at O.

(b) $\theta = \pi/6, \theta = 7\pi/6$, and $\theta = -5\pi/6$ are equations for the line in Figure 10.38. ■

Equations of the form $r = a$ and $\theta = \theta_0$ can be combined to define regions, segments, and rays.

EXAMPLE 3 Identifying Graphs

Graph the sets of points whose polar coordinates satisfy the following conditions.

(a) $1 \le r \le 2$ and $0 \le \theta \le \dfrac{\pi}{2}$

(b) $-3 \le r \le 2$ and $\theta = \dfrac{\pi}{4}$

(c) $r \le 0$ and $\theta = \dfrac{\pi}{4}$

(d) $\dfrac{2\pi}{3} \le \theta \le \dfrac{5\pi}{6}$ (no restriction on r)

Solution The graphs are shown in Figure 10.40. ■

Relating Polar and Cartesian Coordinates

When we use both polar and Cartesian coordinates in a plane, we place the two origins together and take the initial polar ray as the positive x-axis. The ray $\theta = \pi/2, r > 0$, becomes the positive y-axis (Figure 10.41). The two coordinate systems are then related by the following equations.

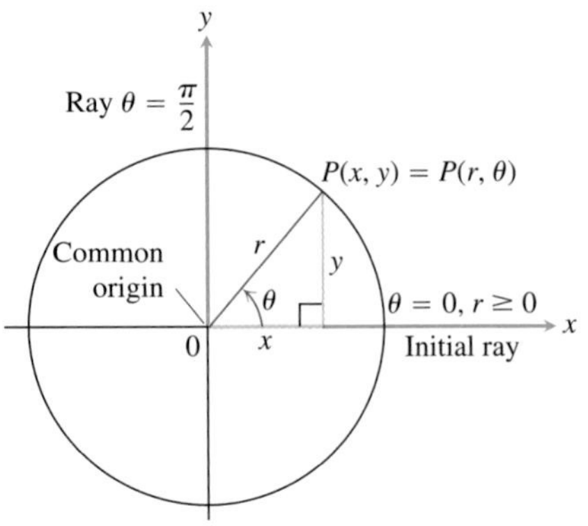

FIGURE 10.41 The usual way to relate polar and Cartesian coordinates.

Equations Relating Polar and Cartesian Coordinates

$$x = r\cos\theta, \qquad y = r\sin\theta, \qquad x^2 + y^2 = r^2$$

The first two of these equations uniquely determine the Cartesian coordinates x and y given the polar coordinates r and θ. On the other hand, if x and y are given, the third equation gives two possible choices for r (a positive and a negative value). For each selection, there is a unique $\theta \in [0, 2\pi)$ satisfying the first two equations, each then giving a polar coordinate representation of the Cartesian point (x, y). The other polar coordinate representations for the point can be determined from these two, as in Example 1.

EXAMPLE 4 Equivalent Equations

Polar equation	Cartesian equivalent
$r\cos\theta = 2$	$x = 2$
$r^2\cos\theta\sin\theta = 4$	$xy = 4$
$r^2\cos^2\theta - r^2\sin^2\theta = 1$	$x^2 - y^2 = 1$
$r = 1 + 2r\cos\theta$	$y^2 - 3x^2 - 4x - 1 = 0$
$r = 1 - \cos\theta$	$x^4 + y^4 + 2x^2y^2 + 2x^3 + 2xy^2 - y^2 = 0$

With some curves, we are better off with polar coordinates; with others, we aren't. ■

EXAMPLE 5 Converting Cartesian to Polar

Find a polar equation for the circle $x^2 + (y - 3)^2 = 9$ (Figure 10.42).

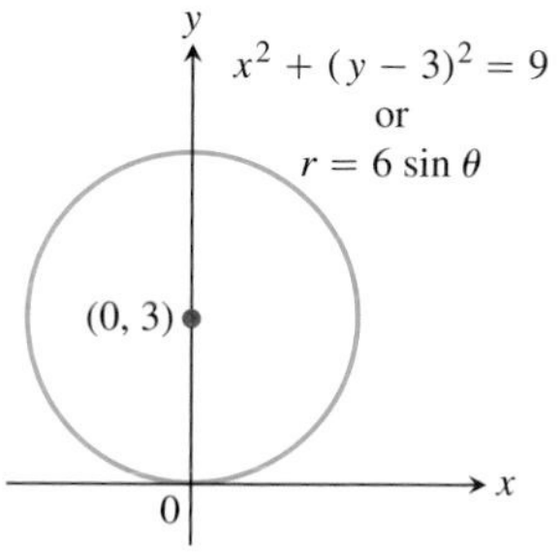

FIGURE 10.42 The circle in Example 5.

Solution

$$
\begin{aligned}
x^2 + y^2 - 6y + 9 &= 9 && \text{Expand } (y-3)^2. \\
x^2 + y^2 - 6y &= 0 && \text{The 9's cancel.} \\
r^2 - 6r\sin\theta &= 0 && x^2 + y^2 = r^2 \\
r = 0 \quad \text{or} \quad r - 6\sin\theta &= 0 \\
r &= 6\sin\theta && \text{Includes both possibilities}
\end{aligned}
$$

We will say more about polar equations of conic sections in Section 10.8. ■

EXAMPLE 6 Converting Polar to Cartesian

Replace the following polar equations by equivalent Cartesian equations, and identify their graphs.

(a) $r\cos\theta = -4$

(b) $r^2 = 4r\cos\theta$

(c) $r = \dfrac{4}{2\cos\theta - \sin\theta}$

Solution We use the substitutions $r\cos\theta = x$, $r\sin\theta = y$, $r^2 = x^2 + y^2$.

(a) $r\cos\theta = -4$

The Cartesian equation:
$$
\begin{aligned}
r\cos\theta &= -4 \\
x &= -4
\end{aligned}
$$

The graph: Vertical line through $x = -4$ on the x-axis

(b) $r^2 = 4r\cos\theta$

The Cartesian equation:
$$
\begin{aligned}
r^2 &= 4r\cos\theta \\
x^2 + y^2 &= 4x \\
x^2 - 4x + y^2 &= 0 \\
x^2 - 4x + 4 + y^2 &= 4 && \text{Completing the square} \\
(x - 2)^2 + y^2 &= 4
\end{aligned}
$$

The graph: Circle, radius 2, center $(h, k) = (2, 0)$

(c) $r = \dfrac{4}{2\cos\theta - \sin\theta}$

The Cartesian equation:

$$r(2\cos\theta - \sin\theta) = 4$$
$$2r\cos\theta - r\sin\theta = 4$$
$$2x - y = 4$$
$$y = 2x - 4$$

The graph: Line, slope $m = 2$, y-intercept $b = -4$ ■

EXERCISES 10.5

Polar Coordinate Pairs

1. Which polar coordinate pairs label the same point?

a. $(3, 0)$ **b.** $(-3, 0)$ **c.** $(2, 2\pi/3)$
d. $(2, 7\pi/3)$ **e.** $(-3, \pi)$ **f.** $(2, \pi/3)$
g. $(-3, 2\pi)$ **h.** $(-2, -\pi/3)$

2. Which polar coordinate pairs label the same point?

a. $(-2, \pi/3)$ **b.** $(2, -\pi/3)$ **c.** (r, θ)
d. $(r, \theta + \pi)$ **e.** $(-r, \theta)$ **f.** $(2, -2\pi/3)$
g. $(-r, \theta + \pi)$ **h.** $(-2, 2\pi/3)$

3. Plot the following points (given in polar coordinates). Then find all the polar coordinates of each point.

a. $(2, \pi/2)$ **b.** $(2, 0)$
c. $(-2, \pi/2)$ **d.** $(-2, 0)$

4. Plot the following points (given in polar coordinates). Then find all the polar coordinates of each point.

a. $(3, \pi/4)$ **b.** $(-3, \pi/4)$
c. $(3, -\pi/4)$ **d.** $(-3, -\pi/4)$

Polar to Cartesian Coordinates

5. Find the Cartesian coordinates of the points in Exercise 1.

6. Find the Cartesian coordinates of the following points (given in polar coordinates).

a. $(\sqrt{2}, \pi/4)$ **b.** $(1, 0)$
c. $(0, \pi/2)$ **d.** $(-\sqrt{2}, \pi/4)$
e. $(-3, 5\pi/6)$ **f.** $(5, \tan^{-1}(4/3))$
g. $(-1, 7\pi)$ **h.** $(2\sqrt{3}, 2\pi/3)$

Graphing Polar Equations and Inequalities

Graph the sets of points whose polar coordinates satisfy the equations and inequalities in Exercises 7–22.

7. $r = 2$ **8.** $0 \le r \le 2$
9. $r \ge 1$ **10.** $1 \le r \le 2$
11. $0 \le \theta \le \pi/6, \quad r \ge 0$ **12.** $\theta = 2\pi/3, \quad r \le -2$
13. $\theta = \pi/3, \quad -1 \le r \le 3$ **14.** $\theta = 11\pi/4, \quad r \ge -1$
15. $\theta = \pi/2, \quad r \ge 0$ **16.** $\theta = \pi/2, \quad r \le 0$
17. $0 \le \theta \le \pi, \quad r = 1$ **18.** $0 \le \theta \le \pi, \quad r = -1$
19. $\pi/4 \le \theta \le 3\pi/4, \quad 0 \le r \le 1$
20. $-\pi/4 \le \theta \le \pi/4, \quad -1 \le r \le 1$
21. $-\pi/2 \le \theta \le \pi/2, \quad 1 \le r \le 2$
22. $0 \le \theta \le \pi/2, \quad 1 \le |r| \le 2$

Polar to Cartesian Equations

Replace the polar equations in Exercises 23–48 by equivalent Cartesian equations. Then describe or identify the graph.

23. $r\cos\theta = 2$ **24.** $r\sin\theta = -1$
25. $r\sin\theta = 0$ **26.** $r\cos\theta = 0$
27. $r = 4\csc\theta$ **28.** $r = -3\sec\theta$
29. $r\cos\theta + r\sin\theta = 1$ **30.** $r\sin\theta = r\cos\theta$
31. $r^2 = 1$ **32.** $r^2 = 4r\sin\theta$
33. $r = \dfrac{5}{\sin\theta - 2\cos\theta}$ **34.** $r^2\sin 2\theta = 2$
35. $r = \cot\theta\csc\theta$ **36.** $r = 4\tan\theta\sec\theta$
37. $r = \csc\theta\, e^{r\cos\theta}$ **38.** $r\sin\theta = \ln r + \ln\cos\theta$
39. $r^2 + 2r^2\cos\theta\sin\theta = 1$ **40.** $\cos^2\theta = \sin^2\theta$
41. $r^2 = -4r\cos\theta$ **42.** $r^2 = -6r\sin\theta$
43. $r = 8\sin\theta$ **44.** $r = 3\cos\theta$
45. $r = 2\cos\theta + 2\sin\theta$ **46.** $r = 2\cos\theta - \sin\theta$
47. $r\sin\left(\theta + \dfrac{\pi}{6}\right) = 2$ **48.** $r\sin\left(\dfrac{2\pi}{3} - \theta\right) = 5$

Cartesian to Polar Equations

Replace the Cartesian equations in Exercises 49–62 by equivalent polar equations.

49. $x = 7$ **50.** $y = 1$ **51.** $x = y$
52. $x - y = 3$ **53.** $x^2 + y^2 = 4$ **54.** $x^2 - y^2 = 1$
55. $\dfrac{x^2}{9} + \dfrac{y^2}{4} = 1$ **56.** $xy = 2$

57. $y^2 = 4x$

58. $x^2 + xy + y^2 = 1$

59. $x^2 + (y - 2)^2 = 4$

60. $(x - 5)^2 + y^2 = 25$

61. $(x - 3)^2 + (y + 1)^2 = 4$

62. $(x + 2)^2 + (y - 5)^2 = 16$

Theory and Examples

63. Find all polar coordinates of the origin.

64. Vertical and horizontal lines

a. Show that every vertical line in the xy-plane has a polar equation of the form $r = a \sec \theta$.

b. Find the analogous polar equation for horizontal lines in the xy-plane.

10.6 Graphing in Polar Coordinates

This section describes techniques for graphing equations in polar coordinates.

Symmetry

Figure 10.43 illustrates the standard polar coordinate tests for symmetry.

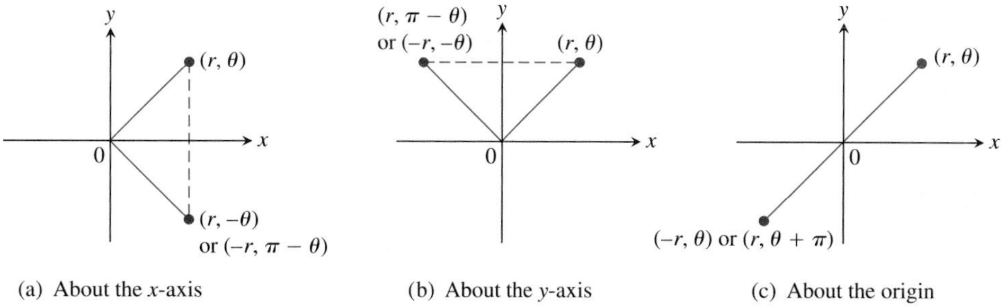

FIGURE 10.43 Three tests for symmetry in polar coordinates.

Symmetry Tests for Polar Graphs

1. *Symmetry about the x-axis:* If the point (r, θ) lies on the graph, the point $(r, -\theta)$ or $(-r, \pi - \theta)$ lies on the graph (Figure 10.43a).
2. *Symmetry about the y-axis:* If the point (r, θ) lies on the graph, the point $(r, \pi - \theta)$ or $(-r, -\theta)$ lies on the graph (Figure 10.43b).
3. *Symmetry about the origin:* If the point (r, θ) lies on the graph, the point $(-r, \theta)$ or $(r, \theta + \pi)$ lies on the graph (Figure 10.43c).

Slope

The slope of a polar curve $r = f(\theta)$ is given by dy/dx, not by $r' = df/d\theta$. To see why, think of the graph of f as the graph of the parametric equations

$$x = r\cos\theta = f(\theta)\cos\theta, \qquad y = r\sin\theta = f(\theta)\sin\theta.$$

If f is a differentiable function of θ, then so are x and y and, when $dx/d\theta \neq 0$, we can calculate dy/dx from the parametric formula

$$\frac{dy}{dx} = \frac{dy/d\theta}{dx/d\theta} \qquad \text{Section 3.5, Equation (2) with } t = \theta$$

$$= \frac{\dfrac{d}{d\theta}(f(\theta)\cdot \sin\theta)}{\dfrac{d}{d\theta}(f(\theta)\cdot \cos\theta)}$$

$$= \frac{\dfrac{df}{d\theta}\sin\theta + f(\theta)\cos\theta}{\dfrac{df}{d\theta}\cos\theta - f(\theta)\sin\theta} \qquad \text{Product Rule for derivatives}$$

Slope of the Curve $r = f(\theta)$

$$\left.\frac{dy}{dx}\right|_{(r,\,\theta)} = \frac{f'(\theta)\sin\theta + f(\theta)\cos\theta}{f'(\theta)\cos\theta - f(\theta)\sin\theta},$$

provided $dx/d\theta \neq 0$ at (r, θ).

If the curve $r = f(\theta)$ passes through the origin at $\theta = \theta_0$, then $f(\theta_0) = 0$, and the slope equation gives

$$\left.\frac{dy}{dx}\right|_{(0,\,\theta_0)} = \frac{f'(\theta_0)\sin\theta_0}{f'(\theta_0)\cos\theta_0} = \tan\theta_0.$$

If the graph of $r = f(\theta)$ passes through the origin at the value $\theta = \theta_0$, the slope of the curve there is $\tan\theta_0$. The reason we say "slope at $(0, \theta_0)$" and not just "slope at the origin" is that a polar curve may pass through the origin (or any point) more than once, with different slopes at different θ-values. This is not the case in our first example, however.

θ	$r = 1 - \cos\theta$
0	0
$\frac{\pi}{3}$	$\frac{1}{2}$
$\frac{\pi}{2}$	1
$\frac{2\pi}{3}$	$\frac{3}{2}$
π	2

(a)

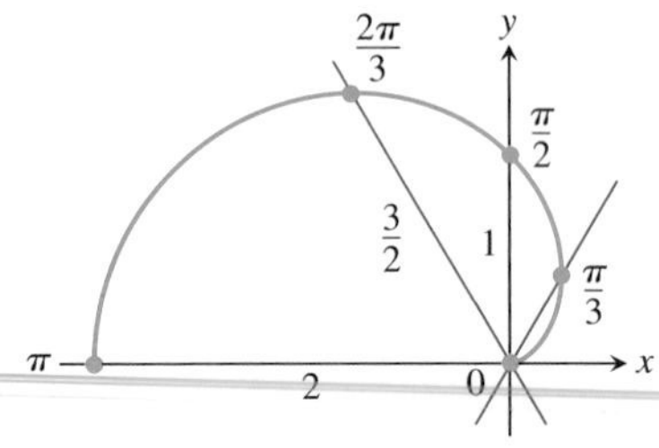

(b)

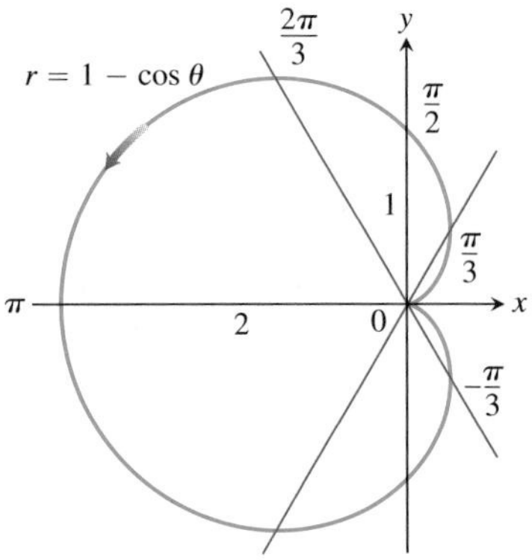

(c)

FIGURE 10.44 The steps in graphing the cardioid $r = 1 - \cos\theta$ (Example 1). The arrow shows the direction of increasing θ.

EXAMPLE 1 A Cardioid

Graph the curve $r = 1 - \cos\theta$.

Solution The curve is symmetric about the x-axis because

$$(r, \theta) \text{ on the graph} \Rightarrow r = 1 - \cos\theta$$
$$\Rightarrow r = 1 - \cos(-\theta) \qquad \cos\theta = \cos(-\theta)$$
$$\Rightarrow (r, -\theta) \text{ on the graph.}$$

As θ increases from 0 to π, $\cos\theta$ decreases from 1 to -1, and $r = 1 - \cos\theta$ increases from a minimum value of 0 to a maximum value of 2. As θ continues on from π to 2π, $\cos\theta$ increases from -1 back to 1 and r decreases from 2 back to 0. The curve starts to repeat when $\theta = 2\pi$ because the cosine has period 2π.

The curve leaves the origin with slope $\tan(0) = 0$ and returns to the origin with slope $\tan(2\pi) = 0$.

We make a table of values from $\theta = 0$ to $\theta = \pi$, plot the points, draw a smooth curve through them with a horizontal tangent at the origin, and reflect the curve across the x-axis to complete the graph (Figure 10.44). The curve is called a *cardioid* because of its heart shape. Cardioid shapes appear in the cams that direct the even layering of thread on bobbins and reels, and in the signal-strength pattern of certain radio antennas. ■

EXAMPLE 2 Graph the Curve $r^2 = 4\cos\theta$.

Solution The equation $r^2 = 4\cos\theta$ requires $\cos\theta \geq 0$, so we get the entire graph by running θ from $-\pi/2$ to $\pi/2$. The curve is symmetric about the x-axis because

$$\begin{aligned}(r, \theta) \text{ on the graph} &\Rightarrow r^2 = 4\cos\theta \\ &\Rightarrow r^2 = 4\cos(-\theta) \qquad \cos\theta = \cos(-\theta) \\ &\Rightarrow (r, -\theta) \text{ on the graph.}\end{aligned}$$

The curve is also symmetric about the origin because

$$\begin{aligned}(r, \theta) \text{ on the graph} &\Rightarrow r^2 = 4\cos\theta \\ &\Rightarrow (-r)^2 = 4\cos\theta \\ &\Rightarrow (-r, \theta) \text{ on the graph.}\end{aligned}$$

Together, these two symmetries imply symmetry about the y-axis.

The curve passes through the origin when $\theta = -\pi/2$ and $\theta = \pi/2$. It has a vertical tangent both times because $\tan\theta$ is infinite.

For each value of θ in the interval between $-\pi/2$ and $\pi/2$, the formula $r^2 = 4\cos\theta$ gives two values of r:

$$r = \pm 2\sqrt{\cos\theta}.$$

We make a short table of values, plot the corresponding points, and use information about symmetry and tangents to guide us in connecting the points with a smooth curve (Figure 10.45). ■

θ	$\cos\theta$	$r = \pm 2\sqrt{\cos\theta}$
0	1	± 2
$\pm\frac{\pi}{6}$	$\frac{\sqrt{3}}{2}$	$\approx \pm 1.9$
$\pm\frac{\pi}{4}$	$\frac{1}{\sqrt{2}}$	$\approx \pm 1.7$
$\pm\frac{\pi}{3}$	$\frac{1}{2}$	$\approx \pm 1.4$
$\pm\frac{\pi}{2}$	0	0

(a)

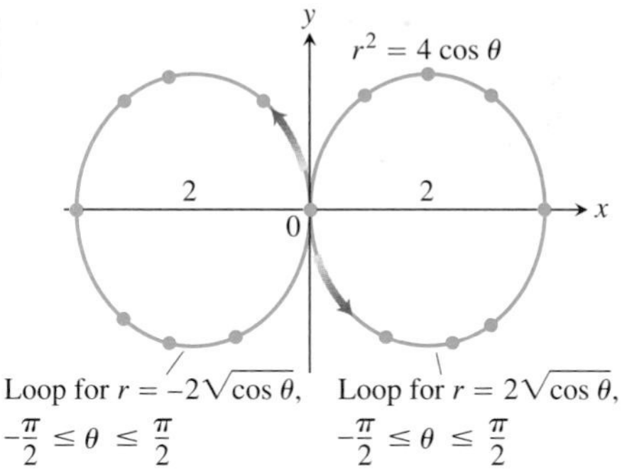

(b)

FIGURE 10.45 The graph of $r^2 = 4\cos\theta$. The arrows show the direction of increasing θ. The values of r in the table are rounded (Example 2).

A Technique for Graphing

One way to graph a polar equation $r = f(\theta)$ is to make a table of (r, θ)-values, plot the corresponding points, and connect them in order of increasing θ. This can work well if enough points have been plotted to reveal all the loops and dimples in the graph. Another method of graphing that is usually quicker and more reliable is to

1. first graph $r = f(\theta)$ in the *Cartesian* $r\theta$-plane,
2. then use the Cartesian graph as a "table" and guide to sketch the *polar* coordinate graph.

This method is better than simple point plotting because the first Cartesian graph, even when hastily drawn, shows at a glance where r is positive, negative, and nonexistent, as well as where r is increasing and decreasing. Here's an example.

EXAMPLE 3 A Lemniscate

Graph the curve

$$r^2 = \sin 2\theta.$$

Solution Here we begin by plotting r^2 (not r) as a function of θ in the Cartesian $r^2\theta$-plane. See Figure 10.46a. We pass from there to the graph of $r = \pm\sqrt{\sin 2\theta}$ in the $r\theta$-plane (Figure 10.46b), and then draw the polar graph (Figure 10.46c). The graph in Figure 10.46b "covers" the final polar graph in Figure 10.46c twice. We could have managed with either loop alone, with the two upper halves, or with the two lower halves. The double covering does no harm, however, and we actually learn a little more about the behavior of the function this way. ■

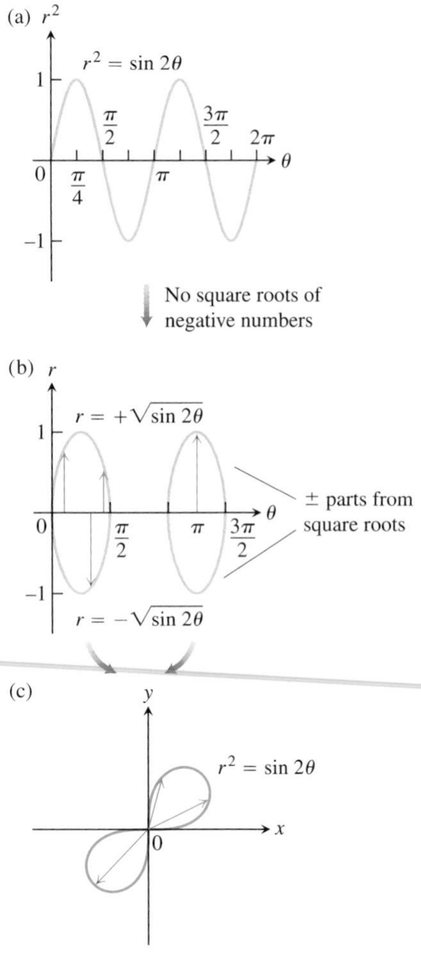

FIGURE 10.46 To plot $r = f(\theta)$ in the Cartesian $r\theta$-plane in (b), we first plot $r^2 = \sin 2\theta$ in the $r^2\theta$-plane in (a) and then ignore the values of θ for which $\sin 2\theta$ is negative. The radii from the sketch in (b) cover the polar graph of the lemniscate in (c) twice (Example 3).

Finding Points Where Polar Graphs Intersect

The fact that we can represent a point in different ways in polar coordinates makes extra care necessary in deciding when a point lies on the graph of a polar equation and in determining the points in which polar graphs intersect. The problem is that a point of intersection may satisfy the equation of one curve with polar coordinates that are different from the ones with which it satisfies the equation of another curve. Thus, solving the equations of two curves simultaneously may not identify all their points of intersection. One sure way to identify all the points of intersection is to graph the equations.

EXAMPLE 4 Deceptive Polar Coordinates

Show that the point $(2, \pi/2)$ lies on the curve $r = 2\cos 2\theta$.

Solution It may seem at first that the point $(2, \pi/2)$ does not lie on the curve because substituting the given coordinates into the equation gives

$$2 = 2\cos 2\left(\frac{\pi}{2}\right) = 2\cos \pi = -2,$$

which is not a true equality. The magnitude is right, but the sign is wrong. This suggests looking for a pair of coordinates for the same given point in which r is negative, for example, $(-2, -(\pi/2))$. If we try these in the equation $r = 2\cos 2\theta$, we find

$$-2 = 2\cos 2\left(-\frac{\pi}{2}\right) = 2(-1) = -2,$$

and the equation is satisfied. The point $(2, \pi/2)$ does lie on the curve. ■

EXAMPLE 5 Elusive Intersection Points

Find the points of intersection of the curves

$$r^2 = 4\cos\theta \qquad \text{and} \qquad r = 1 - \cos\theta.$$

HISTORICAL BIOGRAPHY

Johannes Kepler
(1571–1630)

Solution In Cartesian coordinates, we can always find the points where two curves cross by solving their equations simultaneously. In polar coordinates, the story is different. Simultaneous solution may reveal some intersection points without revealing others. In this example, simultaneous solution reveals only two of the four intersection points. The others are found by graphing. (Also, see Exercise 49.)

If we substitute $\cos\theta = r^2/4$ in the equation $r = 1 - \cos\theta$, we get

$$\begin{aligned} r = 1 - \cos\theta &= 1 - \frac{r^2}{4} \\ 4r &= 4 - r^2 \\ r^2 + 4r - 4 &= 0 \\ r &= -2 \pm 2\sqrt{2}. && \text{Quadratic formula} \end{aligned}$$

The value $r = -2 - 2\sqrt{2}$ has too large an absolute value to belong to either curve. The values of θ corresponding to $r = -2 + 2\sqrt{2}$ are

$$\begin{aligned} \theta &= \cos^{-1}(1 - r) && \text{From } r = 1 - \cos\theta \\ &= \cos^{-1}\left(1 - \left(2\sqrt{2} - 2\right)\right) && \text{Set } r = 2\sqrt{2} - 2. \\ &= \cos^{-1}\left(3 - 2\sqrt{2}\right) \\ &= \pm 80^\circ. && \text{Rounded to the nearest degree} \end{aligned}$$

We have thus identified two intersection points: $(r, \theta) = (2\sqrt{2} - 2, \pm 80^\circ)$.

If we graph the equations $r^2 = 4\cos\theta$ and $r = 1 - \cos\theta$ together (Figure 10.47), as we can now do by combining the graphs in Figures 10.44 and 10.45, we see that the curves also intersect at the point $(2, \pi)$ and the origin. Why weren't the r-values of these points revealed by the simultaneous solution? The answer is that the points $(0, 0)$ and $(2, \pi)$ are not on the curves "simultaneously." They are not reached at the same value of θ. On the curve $r = 1 - \cos\theta$, the point $(2, \pi)$ is reached when $\theta = \pi$. On the curve $r^2 = 4\cos\theta$, it is reached when $\theta = 0$, where it is identified not by the coordinates $(2, \pi)$, which do not satisfy the equation, but by the coordinates $(-2, 0)$, which do. Similarly, the cardioid reaches the origin when $\theta = 0$, but the curve $r^2 = 4\cos\theta$ reaches the origin when $\theta = \pi/2$. ■

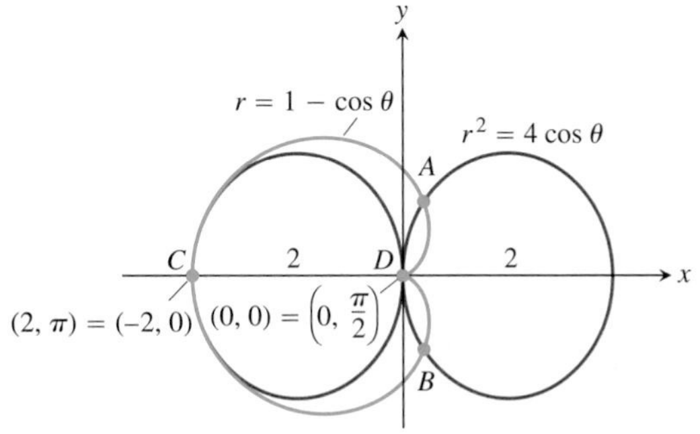

FIGURE 10.47 The four points of intersection of the curves $r = 1 - \cos\theta$ and $r^2 = 4\cos\theta$ (Example 5). Only A and B were found by simultaneous solution. The other two were disclosed by graphing.

USING TECHNOLOGY Graphing Polar Curves Parametrically

For complicated polar curves we may need to use a graphing calculator or computer to graph the curve. If the device does not plot polar graphs directly, we can convert $r = f(\theta)$ into parametric form using the equations

$$x = r\cos\theta = f(\theta)\cos\theta, \qquad y = r\sin\theta = f(\theta)\sin\theta.$$

Then we use the device to draw a parametrized curve in the Cartesian xy-plane. It may be required to use the parameter t rather than θ for the graphing device.

EXERCISES 10.6

Symmetries and Polar Graphs

Identify the symmetries of the curves in Exercises 1–12. Then sketch the curves.

1. $r = 1 + \cos\theta$
2. $r = 2 - 2\cos\theta$
3. $r = 1 - \sin\theta$
4. $r = 1 + \sin\theta$
5. $r = 2 + \sin\theta$
6. $r = 1 + 2\sin\theta$
7. $r = \sin(\theta/2)$
8. $r = \cos(\theta/2)$
9. $r^2 = \cos\theta$
10. $r^2 = \sin\theta$
11. $r^2 = -\sin\theta$
12. $r^2 = -\cos\theta$

Graph the lemniscates in Exercises 13–16. What symmetries do these curves have?

13. $r^2 = 4\cos 2\theta$
14. $r^2 = 4\sin 2\theta$
15. $r^2 = -\sin 2\theta$
16. $r^2 = -\cos 2\theta$

Slopes of Polar Curves

Find the slopes of the curves in Exercises 17–20 at the given points. Sketch the curves along with their tangents at these points.

17. **Cardioid** $r = -1 + \cos\theta;\quad \theta = \pm\pi/2$
18. **Cardioid** $r = -1 + \sin\theta;\quad \theta = 0, \pi$
19. **Four-leaved rose** $r = \sin 2\theta;\quad \theta = \pm\pi/4, \pm 3\pi/4$
20. **Four-leaved rose** $r = \cos 2\theta;\quad \theta = 0, \pm\pi/2, \pi$

Limaçons

Graph the limaçons in Exercises 21–24. Limaçon ("*lee*-ma-sahn") is Old French for "snail." You will understand the name when you graph the limaçons in Exercise 21. Equations for limaçons have the form $r = a \pm b\cos\theta$ or $r = a \pm b\sin\theta$. There are four basic shapes.

21. **Limaçons with an inner loop**

 a. $r = \dfrac{1}{2} + \cos\theta$ **b.** $r = \dfrac{1}{2} + \sin\theta$

22. **Cardioids**

 a. $r = 1 - \cos\theta$ **b.** $r = -1 + \sin\theta$

23. **Dimpled limaçons**

 a. $r = \dfrac{3}{2} + \cos\theta$ **b.** $r = \dfrac{3}{2} - \sin\theta$

24. **Oval limaçons**

 a. $r = 2 + \cos\theta$ **b.** $r = -2 + \sin\theta$

Graphing Polar Inequalities

25. Sketch the region defined by the inequalities $-1 \le r \le 2$ and $-\pi/2 \le \theta \le \pi/2$.
26. Sketch the region defined by the inequalities $0 \le r \le 2\sec\theta$ and $-\pi/4 \le \theta \le \pi/4$.

In Exercises 27 and 28, sketch the region defined by the inequality.

27. $0 \le r \le 2 - 2\cos\theta$
28. $0 \le r^2 \le \cos\theta$

Intersections

29. Show that the point $(2, 3\pi/4)$ lies on the curve $r = 2\sin 2\theta$.
30. Show that $(1/2, 3\pi/2)$ lies on the curve $r = -\sin(\theta/3)$.

Find the points of intersection of the pairs of curves in Exercises 31–38.

31. $r = 1 + \cos\theta,\quad r = 1 - \cos\theta$
32. $r = 1 + \sin\theta,\quad r = 1 - \sin\theta$
33. $r = 2\sin\theta,\quad r = 2\sin 2\theta$
34. $r = \cos\theta,\quad r = 1 - \cos\theta$
35. $r = \sqrt{2},\quad r^2 = 4\sin\theta$
36. $r^2 = \sqrt{2}\sin\theta,\quad r^2 = \sqrt{2}\cos\theta$
37. $r = 1,\quad r^2 = 2\sin 2\theta$
38. $r^2 = \sqrt{2}\cos 2\theta,\quad r^2 = \sqrt{2}\sin 2\theta$

T Find the points of intersection of the pairs of curves in Exercises 39–42.

39. $r^2 = \sin 2\theta,\quad r^2 = \cos 2\theta$
40. $r = 1 + \cos\dfrac{\theta}{2},\quad r = 1 - \sin\dfrac{\theta}{2}$
41. $r = 1,\quad r = 2\sin 2\theta$
42. $r = 1,\quad r^2 = 2\sin 2\theta$

T Grapher Explorations

43. Which of the following has the same graph as $r = 1 - \cos\theta$?

a. $r = -1 - \cos\theta$ **b.** $r = 1 + \cos\theta$

Confirm your answer with algebra.

44. Which of the following has the same graph as $r = \cos 2\theta$?

a. $r = -\sin(2\theta + \pi/2)$ **b.** $r = -\cos(\theta/2)$

Confirm your answer with algebra.

45. A rose within a rose Graph the equation $r = 1 - 2\sin 3\theta$.

46. The nephroid of Freeth Graph the nephroid of Freeth:

$$r = 1 + 2\sin\frac{\theta}{2}.$$

47. Roses Graph the roses $r = \cos m\theta$ for $m = 1/3, 2, 3$, and 7.

48. Spirals Polar coordinates are just the thing for defining spirals. Graph the following spirals.

a. $r = \theta$ **b.** $r = -\theta$

c. *A logarithmic spiral:* $r = e^{\theta/10}$

d. *A hyperbolic spiral:* $r = 8/\theta$

e. *An equilateral hyperbola:* $r = \pm 10/\sqrt{\theta}$

(Use different colors for the two branches.)

Theory and Examples

49. (*Continuation of Example 5.*) The simultaneous solution of the equations

$$r^2 = 4\cos\theta \qquad (1)$$
$$r = 1 - \cos\theta \qquad (2)$$

in the text did not reveal the points $(0, 0)$ and $(2, \pi)$ in which their graphs intersected.

a. We could have found the point $(2, \pi)$, however, by replacing the (r, θ) in Equation (1) by the equivalent $(-r, \theta + \pi)$ to obtain

$$\begin{aligned} r^2 &= 4\cos\theta \\ (-r)^2 &= 4\cos(\theta + \pi) \qquad (3) \\ r^2 &= -4\cos\theta. \end{aligned}$$

Solve Equations (2) and (3) simultaneously to show that $(2, \pi)$ is a common solution. (This will still not reveal that the graphs intersect at $(0, 0)$.)

b. The origin is still a special case. (It often is.) Here is one way to handle it: Set $r = 0$ in Equations (1) and (2) and solve each equation for a corresponding value of θ. Since $(0, \theta)$ is the origin for *any* θ, this will show that both curves pass through the origin even if they do so for different θ-values.

50. If a curve has any two of the symmetries listed at the beginning of the section, can anything be said about its having or not having the third symmetry? Give reasons for your answer.

***51.** Find the maximum width of the petal of the four-leaved rose $r = \cos 2\theta$, which lies along the x-axis.

***52.** Find the maximum height above the x-axis of the cardioid $r = 2(1 + \cos\theta)$.

10.7 Areas and Lengths in Polar Coordinates

This section shows how to calculate areas of plane regions, lengths of curves, and areas of surfaces of revolution in polar coordinates.

Area in the Plane

The region *OTS* in Figure 10.48 is bounded by the rays $\theta = \alpha$ and $\theta = \beta$ and the curve $r = f(\theta)$. We approximate the region with n nonoverlapping fan-shaped circular sectors based on a partition P of angle *TOS*. The typical sector has radius $r_k = f(\theta_k)$ and central angle of radian measure $\Delta\theta_k$. Its area is $\Delta\theta_k/2\pi$ times the area of a circle of radius r_k, or

$$A_k = \frac{1}{2}r_k^2\,\Delta\theta_k = \frac{1}{2}\left(f(\theta_k)\right)^2\Delta\theta_k.$$

The area of region *OTS* is approximately

$$\sum_{k=1}^{n} A_k = \sum_{k=1}^{n}\frac{1}{2}\left(f(\theta_k)\right)^2\Delta\theta_k.$$

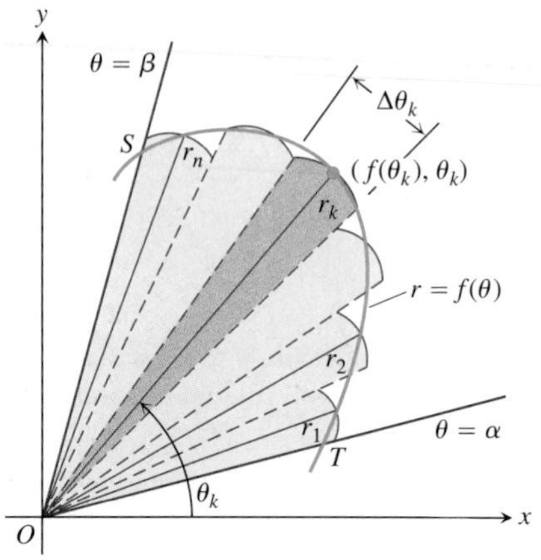

FIGURE 10.48 To derive a formula for the area of region *OTS*, we approximate the region with fan-shaped circular sectors.

If f is continuous, we expect the approximations to improve as the norm of the partition $\|P\| \to 0$, and we are led to the following formula for the region's area:

$$\begin{aligned} A &= \lim_{\|P\|\to 0} \sum_{k=1}^{n} \frac{1}{2}\left(f(\theta_k)\right)^2 \Delta\theta_k \\ &= \int_{\alpha}^{\beta} \frac{1}{2}\left(f(\theta)\right)^2 d\theta. \end{aligned}$$

Area of the Fan-Shaped Region Between the Origin and the Curve $r = f(\theta), \alpha \le \theta \le \beta$

$$A = \int_{\alpha}^{\beta} \frac{1}{2} r^2 \, d\theta.$$

This is the integral of the **area differential** (Figure 10.49)

$$dA = \frac{1}{2} r^2 \, d\theta = \frac{1}{2}(f(\theta))^2 \, d\theta.$$

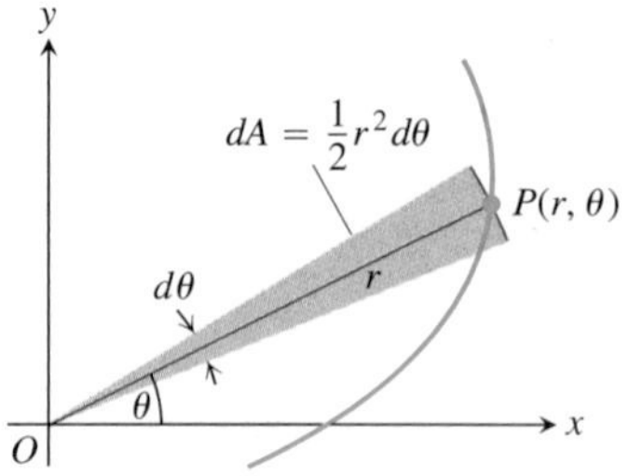

FIGURE 10.49 The area differential dA for the curve $n = f(\theta)$.

EXAMPLE 1 Finding Area

Find the area of the region in the plane enclosed by the cardioid $r = 2(1 + \cos\theta)$.

Solution We graph the cardioid (Figure 10.50) and determine that the radius OP sweeps out the region exactly once as θ runs from 0 to 2π. The area is therefore

$$\begin{aligned} \int_{\theta=0}^{\theta=2\pi} \frac{1}{2} r^2 \, d\theta &= \int_0^{2\pi} \frac{1}{2} \cdot 4(1 + \cos\theta)^2 \, d\theta \\ &= \int_0^{2\pi} 2(1 + 2\cos\theta + \cos^2\theta) \, d\theta \\ &= \int_0^{2\pi} \left(2 + 4\cos\theta + 2\,\frac{1 + \cos 2\theta}{2}\right) d\theta \\ &= \int_0^{2\pi} (3 + 4\cos\theta + \cos 2\theta) \, d\theta \\ &= \left[3\theta + 4\sin\theta + \frac{\sin 2\theta}{2}\right]_0^{2\pi} = 6\pi - 0 = 6\pi. \end{aligned}$$

■

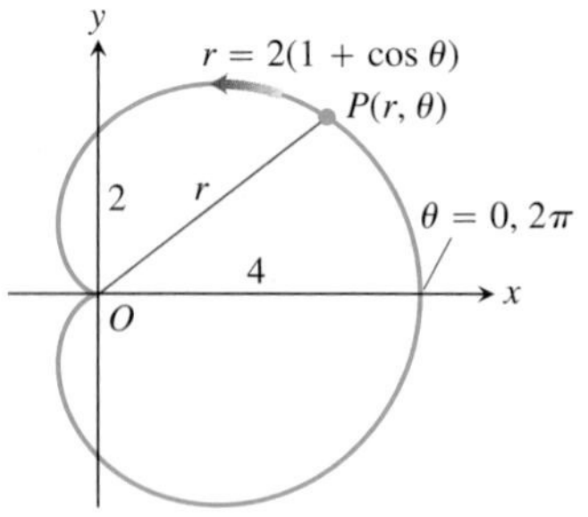

FIGURE 10.50 The cardioid in Example 1.

EXAMPLE 2 Finding Area

Find the area inside the smaller loop of the limaçon

$$r = 2\cos\theta + 1.$$

Solution After sketching the curve (Figure 10.51), we see that the smaller loop is traced out by the point (r, θ) as θ increases from $\theta = 2\pi/3$ to $\theta = 4\pi/3$. Since the curve is symmetric about the x-axis (the equation is unaltered when we replace θ by $-\theta$), we may calculate the area of the shaded half of the inner loop by integrating from $\theta = 2\pi/3$ to $\theta = \pi$. The area we seek will be twice the resulting integral:

$$A = 2\int_{2\pi/3}^{\pi} \frac{1}{2} r^2 \, d\theta = \int_{2\pi/3}^{\pi} r^2 \, d\theta.$$

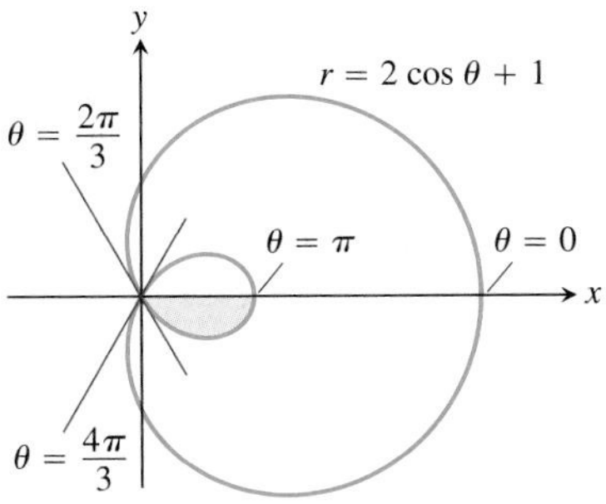

FIGURE 10.51 The limaçon in Example 2. Limaçon (pronounced LEE-ma-sahn) is an old French word for *snail*.

Since

$$r^2 = (2\cos\theta + 1)^2 = 4\cos^2\theta + 4\cos\theta + 1$$
$$= 4\cdot\frac{1 + \cos 2\theta}{2} + 4\cos\theta + 1$$
$$= 2 + 2\cos 2\theta + 4\cos\theta + 1$$
$$= 3 + 2\cos 2\theta + 4\cos\theta,$$

we have

$$A = \int_{2\pi/3}^{\pi} (3 + 2\cos 2\theta + 4\cos\theta)\, d\theta$$
$$= \Big[3\theta + \sin 2\theta + 4\sin\theta\Big]_{2\pi/3}^{\pi}$$
$$= (3\pi) - \left(2\pi - \frac{\sqrt{3}}{2} + 4\cdot\frac{\sqrt{3}}{2}\right)$$
$$= \pi - \frac{3\sqrt{3}}{2}.$$

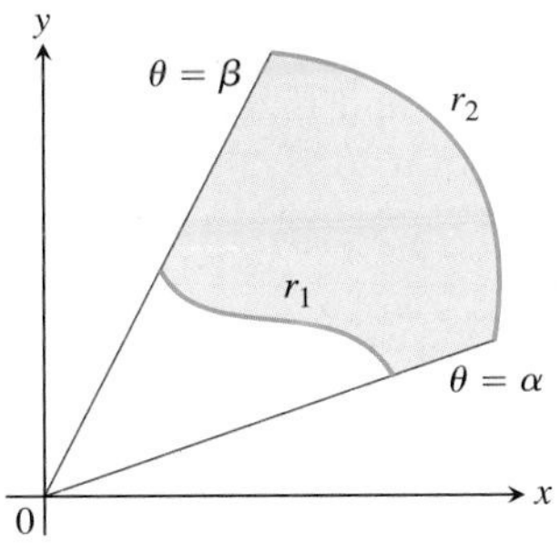

FIGURE 10.52 The area of the shaded region is calculated by subtracting the area of the region between r_1 and the origin from the area of the region between r_2 and the origin.

To find the area of a region like the one in Figure 10.52, which lies between two polar curves $r_1 = r_1(\theta)$ and $r_2 = r_2(\theta)$ from $\theta = \alpha$ to $\theta = \beta$, we subtract the integral of $(1/2)r_1^2\, d\theta$ from the integral of $(1/2)r_2^2\, d\theta$. This leads to the following formula.

Area of the Region $0 \le r_1(\theta) \le r \le r_2(\theta), \quad \alpha \le \theta \le \beta$

$$A = \int_\alpha^\beta \frac{1}{2} r_2^2\, d\theta - \int_\alpha^\beta \frac{1}{2} r_1^2\, d\theta = \int_\alpha^\beta \frac{1}{2}\left(r_2^2 - r_1^2\right) d\theta \qquad (1)$$

EXAMPLE 3 Finding Area Between Polar Curves

Find the area of the region that lies inside the circle $r = 1$ and outside the cardioid $r = 1 - \cos\theta$.

Solution We sketch the region to determine its boundaries and find the limits of integration (Figure 10.53). The outer curve is $r_2 = 1$, the inner curve is $r_1 = 1 - \cos\theta$, and θ runs from $-\pi/2$ to $\pi/2$. The area, from Equation (1), is

$$A = \int_{-\pi/2}^{\pi/2} \frac{1}{2}\left(r_2^2 - r_1^2\right) d\theta$$
$$= 2\int_0^{\pi/2} \frac{1}{2}\left(r_2^2 - r_1^2\right) d\theta \qquad \text{Symmetry}$$
$$= \int_0^{\pi/2} (1 - (1 - 2\cos\theta + \cos^2\theta))\, d\theta$$
$$= \int_0^{\pi/2} (2\cos\theta - \cos^2\theta)\, d\theta = \int_0^{\pi/2}\left(2\cos\theta - \frac{1 + \cos 2\theta}{2}\right) d\theta$$
$$= \left[2\sin\theta - \frac{\theta}{2} - \frac{\sin 2\theta}{4}\right]_0^{\pi/2} = 2 - \frac{\pi}{4}.$$

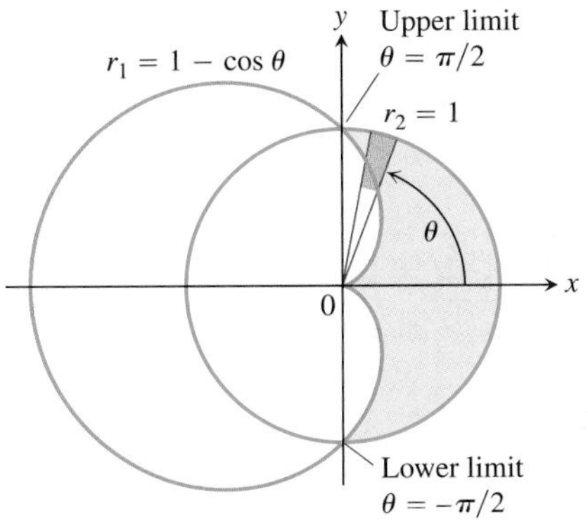

FIGURE 10.53 The region and limits of integration in Example 3.

Length of a Polar Curve

We can obtain a polar coordinate formula for the length of a curve $r = f(\theta)$, $\alpha \le \theta \le \beta$, by parametrizing the curve as

$$x = r\cos\theta = f(\theta)\cos\theta, \qquad y = r\sin\theta = f(\theta)\sin\theta, \qquad \alpha \le \theta \le \beta. \tag{2}$$

The parametric length formula, Equation (1) from Section 6.3, then gives the length as

$$L = \int_\alpha^\beta \sqrt{\left(\frac{dx}{d\theta}\right)^2 + \left(\frac{dy}{d\theta}\right)^2}\, d\theta.$$

This equation becomes

$$L = \int_\alpha^\beta \sqrt{r^2 + \left(\frac{dr}{d\theta}\right)^2}\, d\theta$$

when Equations (2) are substituted for x and y (Exercise 33).

> **Length of a Polar Curve**
> If $r = f(\theta)$ has a continuous first derivative for $\alpha \le \theta \le \beta$ and if the point $P(r, \theta)$ traces the curve $r = f(\theta)$ exactly once as θ runs from α to β, then the length of the curve is
>
> $$L = \int_\alpha^\beta \sqrt{r^2 + \left(\frac{dr}{d\theta}\right)^2}\, d\theta. \tag{3}$$

EXAMPLE 4 Finding the Length of a Cardioid

Find the length of the cardioid $r = 1 - \cos\theta$.

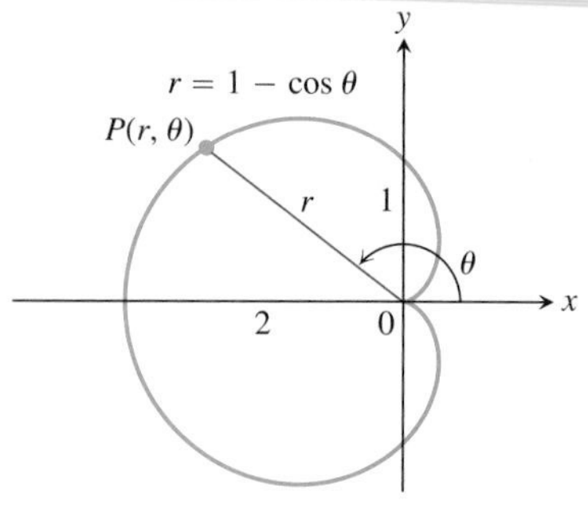

FIGURE 10.54 Calculating the length of a cardioid (Example 4).

Solution We sketch the cardioid to determine the limits of integration (Figure 10.54). The point $P(r, \theta)$ traces the curve once, counterclockwise as θ runs from 0 to 2π, so these are the values we take for α and β.

With

$$r = 1 - \cos\theta, \qquad \frac{dr}{d\theta} = \sin\theta,$$

we have

$$\begin{aligned} r^2 + \left(\frac{dr}{d\theta}\right)^2 &= (1 - \cos\theta)^2 + (\sin\theta)^2 \\ &= 1 - 2\cos\theta + \underbrace{\cos^2\theta + \sin^2\theta}_{1} = 2 - 2\cos\theta \end{aligned}$$

and

$$\begin{aligned} L &= \int_\alpha^\beta \sqrt{r^2 + \left(\frac{dr}{d\theta}\right)^2}\, d\theta = \int_0^{2\pi} \sqrt{2 - 2\cos\theta}\, d\theta \\ &= \int_0^{2\pi} \sqrt{4\sin^2\frac{\theta}{2}}\, d\theta \qquad 1 - \cos\theta = 2\sin^2\frac{\theta}{2} \end{aligned}$$

$$= \int_0^{2\pi} 2\left|\sin\frac{\theta}{2}\right| d\theta$$

$$= \int_0^{2\pi} 2\sin\frac{\theta}{2}\, d\theta \qquad \sin\frac{\theta}{2} \geq 0 \quad \text{for} \quad 0 \leq \theta \leq 2\pi$$

$$= \left[-4\cos\frac{\theta}{2}\right]_0^{2\pi} = 4 + 4 = 8.$$

■

Area of a Surface of Revolution

To derive polar coordinate formulas for the area of a surface of revolution, we parametrize the curve $r = f(\theta)$, $\alpha \leq \theta \leq \beta$, with Equations (2) and apply the surface area equations for parametrized curves in Section 6.5.

> **Area of a Surface of Revolution of a Polar Curve**
>
> If $r = f(\theta)$ has a continuous first derivative for $\alpha \leq \theta \leq \beta$ and if the point $P(r, \theta)$ traces the curve $r = f(\theta)$ exactly once as θ runs from α to β, then the areas of the surfaces generated by revolving the curve about the x- and y-axes are given by the following formulas:
>
> **1.** Revolution about the x-axis $(y \geq 0)$:
>
> $$S = \int_\alpha^\beta 2\pi r \sin\theta \sqrt{r^2 + \left(\frac{dr}{d\theta}\right)^2}\, d\theta \tag{4}$$
>
> **2.** Revolution about the y-axis $(x \geq 0)$:
>
> $$S = \int_\alpha^\beta 2\pi r \cos\theta \sqrt{r^2 + \left(\frac{dr}{d\theta}\right)^2}\, d\theta \tag{5}$$

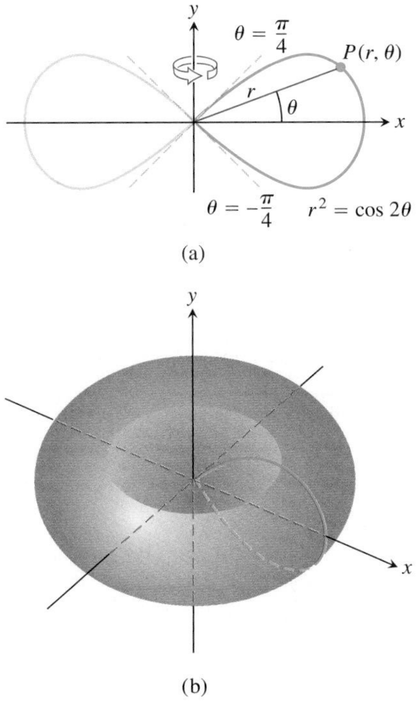

FIGURE 10.55 The right-hand half of a lemniscate (a) is revolved about the y-axis to generate a surface (b), whose area is calculated in Example 5.

EXAMPLE 5 Finding Surface Area

Find the area of the surface generated by revolving the right-hand loop of the lemniscate $r^2 = \cos 2\theta$ about the y-axis.

Solution We sketch the loop to determine the limits of integration (Figure 10.55a). The point $P(r, \theta)$ traces the curve once, counterclockwise as θ runs from $-\pi/4$ to $\pi/4$, so these are the values we take for α and β.

We evaluate the area integrand in Equation (5) in stages. First,

$$2\pi r \cos\theta \sqrt{r^2 + \left(\frac{dr}{d\theta}\right)^2} = 2\pi\cos\theta\sqrt{r^4 + \left(r\frac{dr}{d\theta}\right)^2}. \tag{6}$$

Next, $r^2 = \cos 2\theta$, so

$$2r\frac{dr}{d\theta} = -2\sin 2\theta$$

$$r\frac{dr}{d\theta} = -\sin 2\theta$$

$$\left(r\frac{dr}{d\theta}\right)^2 = \sin^2 2\theta.$$

Finally, $r^4 = (r^2)^2 = \cos^2 2\theta$, so the square root on the right-hand side of Equation (6) simplifies to

$$\sqrt{r^4 + \left(r\frac{dr}{d\theta}\right)^2} = \sqrt{\cos^2 2\theta + \sin^2 2\theta} = 1.$$

All together, we have

$$\begin{aligned} S &= \int_\alpha^\beta 2\pi r\cos\theta\sqrt{r^2 + \left(\frac{dr}{d\theta}\right)^2}\,d\theta \qquad \text{Equation (5)} \\ &= \int_{-\pi/4}^{\pi/4} 2\pi\cos\theta\cdot(1)\,d\theta \\ &= 2\pi\Big[\sin\theta\Big]_{-\pi/4}^{\pi/4} \\ &= 2\pi\left[\frac{\sqrt{2}}{2} + \frac{\sqrt{2}}{2}\right] = 2\pi\sqrt{2}. \end{aligned}$$

■

EXERCISES 10.7

Areas Inside Polar Curves

Find the areas of the regions in Exercises 1–6.

1. Inside the oval limaçon $r = 4 + 2\cos\theta$
2. Inside the cardioid $r = a(1 + \cos\theta), \quad a > 0$
3. Inside one leaf of the four-leaved rose $r = \cos 2\theta$
4. Inside the lemniscate $r^2 = 2a^2\cos 2\theta, \quad a > 0$
5. Inside one loop of the lemniscate $r^2 = 4\sin 2\theta$
6. Inside the six-leaved rose $r^2 = 2\sin 3\theta$

Areas Shared by Polar Regions

Find the areas of the regions in Exercises 7–16.

7. Shared by the circles $r = 2\cos\theta$ and $r = 2\sin\theta$
8. Shared by the circles $r = 1$ and $r = 2\sin\theta$
9. Shared by the circle $r = 2$ and the cardioid $r = 2(1 - \cos\theta)$
10. Shared by the cardioids $r = 2(1 + \cos\theta)$ and $r = 2(1 - \cos\theta)$
11. Inside the lemniscate $r^2 = 6\cos 2\theta$ and outside the circle $r = \sqrt{3}$
12. Inside the circle $r = 3a\cos\theta$ and outside the cardioid $r = a(1 + \cos\theta), a > 0$
13. Inside the circle $r = -2\cos\theta$ and outside the circle $r = 1$
14. **a.** Inside the outer loop of the limaçon $r = 2\cos\theta + 1$ (See Figure 10.51.)

    b. Inside the outer loop and outside the inner loop of the limaçon $r = 2\cos\theta + 1$
15. Inside the circle $r = 6$ above the line $r = 3\csc\theta$
16. Inside the lemniscate $r^2 = 6\cos 2\theta$ to the right of the line $r = (3/2)\sec\theta$
17. **a.** Find the area of the shaded region in the accompanying figure.

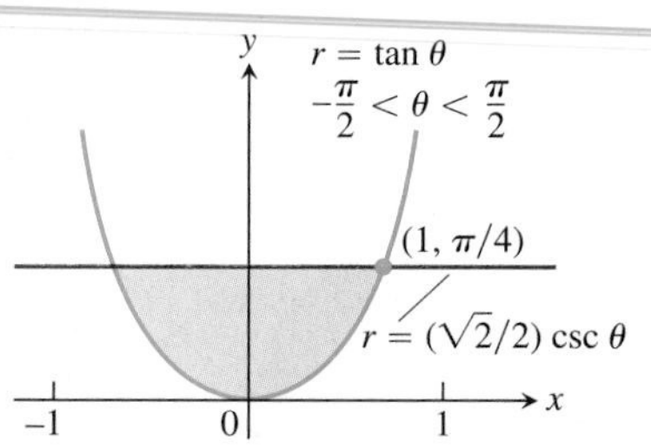

    b. It looks as if the graph of $r = \tan\theta, -\pi/2 < \theta < \pi/2$, could be asymptotic to the lines $x = 1$ and $x = -1$. Is it? Give reasons for your answer.
18. The area of the region that lies inside the cardioid curve $r = \cos\theta + 1$ and outside the circle $r = \cos\theta$ is not

$$\frac{1}{2}\int_0^{2\pi}[(\cos\theta + 1)^2 - \cos^2\theta]\,d\theta = \pi.$$

Why not? What *is* the area? Give reasons for your answers.

Lengths of Polar Curves

Find the lengths of the curves in Exercises 19–27.

19. The spiral $r = \theta^2$, $0 \le \theta \le \sqrt{5}$

20. The spiral $r = e^{\theta}/\sqrt{2}$, $0 \le \theta \le \pi$

21. The cardioid $r = 1 + \cos\theta$

22. The curve $r = a\sin^2(\theta/2)$, $0 \le \theta \le \pi$, $a > 0$

23. The parabolic segment $r = 6/(1 + \cos\theta)$, $0 \le \theta \le \pi/2$

24. The parabolic segment $r = 2/(1 - \cos\theta)$, $\pi/2 \le \theta \le \pi$

25. The curve $r = \cos^3(\theta/3)$, $0 \le \theta \le \pi/4$

26. The curve $r = \sqrt{1 + \sin 2\theta}$, $0 \le \theta \le \pi\sqrt{2}$

27. The curve $r = \sqrt{1 + \cos 2\theta}$, $0 \le \theta \le \pi\sqrt{2}$

28. Circumferences of circles As usual, when faced with a new formula, it is a good idea to try it on familiar objects to be sure it gives results consistent with past experience. Use the length formula in Equation (3) to calculate the circumferences of the following circles $(a > 0)$:

a. $r = a$ **b.** $r = a\cos\theta$ **c.** $r = a\sin\theta$

Surface Area

Find the areas of the surfaces generated by revolving the curves in Exercises 29–32 about the indicated axes.

29. $r = \sqrt{\cos 2\theta}$, $0 \le \theta \le \pi/4$, y-axis

30. $r = \sqrt{2}e^{\theta/2}$, $0 \le \theta \le \pi/2$, x-axis

31. $r^2 = \cos 2\theta$, x-axis

32. $r = 2a\cos\theta$, $a > 0$, y-axis

Theory and Examples

33. The length of the curve $r = f(\theta), \alpha \le \theta \le \beta$ Assuming that the necessary derivatives are continuous, show how the substitutions

$$x = f(\theta)\cos\theta, \quad y = f(\theta)\sin\theta$$

(Equations 2 in the text) transform

$$L = \int_{\alpha}^{\beta} \sqrt{\left(\frac{dx}{d\theta}\right)^2 + \left(\frac{dy}{d\theta}\right)^2}\, d\theta$$

into

$$L = \int_{\alpha}^{\beta} \sqrt{r^2 + \left(\frac{dr}{d\theta}\right)^2}\, d\theta.$$

34. Average value If f is continuous, the average value of the polar coordinate r over the curve $r = f(\theta)$, $\alpha \le \theta \le \beta$, with respect to θ is given by the formula

$$r_{\text{av}} = \frac{1}{\beta - \alpha}\int_{\alpha}^{\beta} f(\theta)\, d\theta.$$

Use this formula to find the average value of r with respect to θ over the following curves $(a > 0)$.

a. The cardioid $r = a(1 - \cos\theta)$

b. The circle $r = a$

c. The circle $r = a\cos\theta$, $-\pi/2 \le \theta \le \pi/2$

35. $r = f(\theta)$ vs. $r = 2f(\theta)$ Can anything be said about the relative lengths of the curves $r = f(\theta)$, $\alpha \le \theta \le \beta$, and $r = 2f(\theta)$, $\alpha \le \theta \le \beta$? Give reasons for your answer.

36. $r = f(\theta)$ vs. $r = 2f(\theta)$ The curves $r = f(\theta)$, $\alpha \le \theta \le \beta$, and $r = 2f(\theta)$, $\alpha \le \theta \le \beta$, are revolved about the x-axis to generate surfaces. Can anything be said about the relative areas of these surfaces? Give reasons for your answer.

Centroids of Fan-Shaped Regions

Since the centroid of a triangle is located on each median, two-thirds of the way from the vertex to the opposite base, the lever arm for the moment about the x-axis of the thin triangular region in the accompanying figure is about $(2/3)r\sin\theta$. Similarly, the lever arm for the moment of the triangular region about the y-axis is about $(2/3)r\cos\theta$. These approximations improve as $\Delta\theta \to 0$ and lead to the following formulas for the coordinates of the centroid of region AOB:

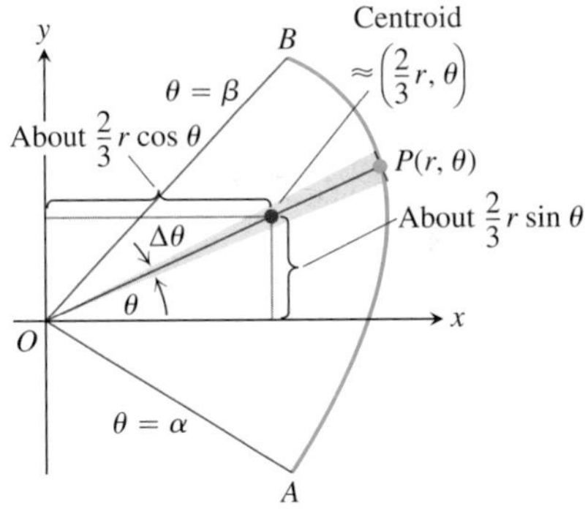

$$\bar{x} = \frac{\displaystyle\int \frac{2}{3}r\cos\theta \cdot \frac{1}{2}r^2\, d\theta}{\displaystyle\int \frac{1}{2}r^2\, d\theta} = \frac{\displaystyle\frac{2}{3}\int r^3\cos\theta\, d\theta}{\displaystyle\int r^2\, d\theta},$$

$$\bar{y} = \frac{\displaystyle\int \frac{2}{3}r\sin\theta \cdot \frac{1}{2}r^2\, d\theta}{\displaystyle\int \frac{1}{2}r^2\, d\theta} = \frac{\displaystyle\frac{2}{3}\int r^3\sin\theta\, d\theta}{\displaystyle\int r^2\, d\theta},$$

with limits $\theta = \alpha$ to $\theta = \beta$ on all integrals.

37. Find the centroid of the region enclosed by the cardioid $r = a(1 + \cos\theta)$.

38. Find the centroid of the semicircular region $0 \le r \le a$, $0 \le \theta \le \pi$.

10.8 Conic Sections in Polar Coordinates

Polar coordinates are important in astronomy and astronautical engineering because the ellipses, parabolas, and hyperbolas along which satellites, moons, planets, and comets approximately move can all be described with a single relatively simple coordinate equation. We develop that equation here.

Lines

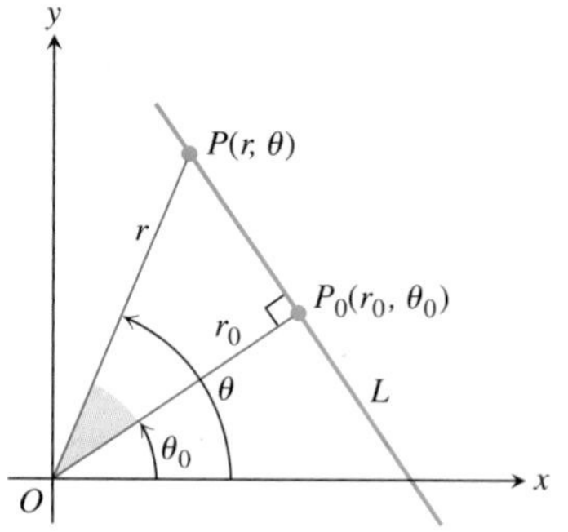

FIGURE 10.56 We can obtain a polar equation for line L by reading the relation $r_0 = r\cos(\theta - \theta_0)$ from the right triangle OP_0P.

Suppose the perpendicular from the origin to line L meets L at the point $P_0(r_0, \theta_0)$, with $r_0 \geq 0$ (Figure 10.56). Then, if $P(r, \theta)$ is any other point on L, the points P, P_0, and O are the vertices of a right triangle, from which we can read the relation

$$r_0 = r\cos(\theta - \theta_0).$$

The Standard Polar Equation for Lines

If the point $P_0(r_0, \theta_0)$ is the foot of the perpendicular from the origin to the line L, and $r_0 \geq 0$, then an equation for L is

$$r\cos(\theta - \theta_0) = r_0. \tag{1}$$

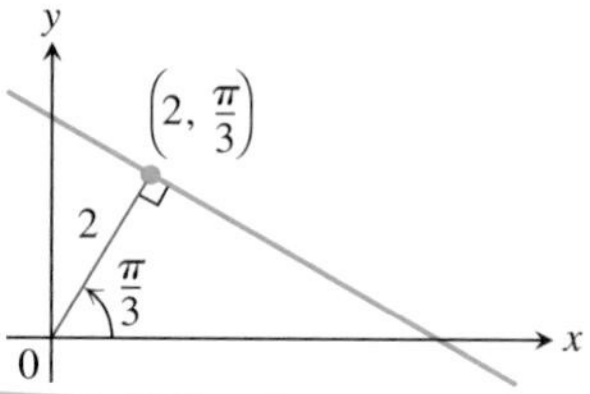

FIGURE 10.57 The standard polar equation of this line converts to the Cartesian equation $x + \sqrt{3}y = 4$ (Example 1).

EXAMPLE 1 Converting a Line's Polar Equation to Cartesian Form

Use the identity $\cos(A - B) = \cos A\cos B + \sin A\sin B$ to find a Cartesian equation for the line in Figure 10.57.

Solution

$$r\cos\left(\theta - \frac{\pi}{3}\right) = 2$$

$$r\left(\cos\theta\cos\frac{\pi}{3} + \sin\theta\sin\frac{\pi}{3}\right) = 2$$

$$\frac{1}{2}r\cos\theta + \frac{\sqrt{3}}{2}r\sin\theta = 2$$

$$\frac{1}{2}x + \frac{\sqrt{3}}{2}y = 2$$

$$x + \sqrt{3}y = 4$$

■

Circles

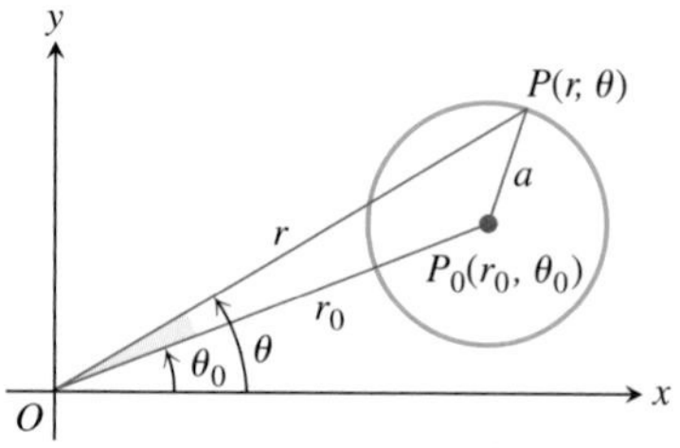

FIGURE 10.58 We can get a polar equation for this circle by applying the Law of Cosines to triangle OP_0P.

To find a polar equation for the circle of radius a centered at $P_0(r_0, \theta_0)$, we let $P(r, \theta)$ be a point on the circle and apply the Law of Cosines to triangle OP_0P (Figure 10.58). This gives

$$a^2 = r_0^2 + r^2 - 2r_0 r\cos(\theta - \theta_0).$$

If the circle passes through the origin, then $r_0 = a$ and this equation simplifies to

$$a^2 = a^2 + r^2 - 2ar\cos(\theta - \theta_0)$$

$$r^2 = 2ar\cos(\theta - \theta_0)$$

$$r = 2a\cos(\theta - \theta_0).$$

If the circle's center lies on the positive x-axis, $\theta_0 = 0$ and we get the further simplification

$$r = 2a\cos\theta$$

(see Figure 10.59a).

If the center lies on the positive y-axis, $\theta = \pi/2$, $\cos(\theta - \pi/2) = \sin\theta$, and the equation $r = 2a\cos(\theta - \theta_0)$ becomes

$$r = 2a\sin\theta$$

(see Figure 10.59b).

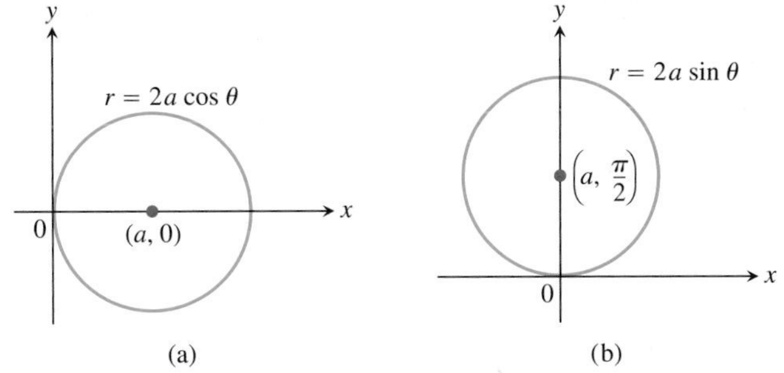

FIGURE 10.59 Polar equation of a circle of radius a through the origin with center on (a) the positive x-axis, and (b) the positive y-axis.

Equations for circles through the origin centered on the negative x- and y-axes can be obtained by replacing r with $-r$ in the above equations (Figure 10.60).

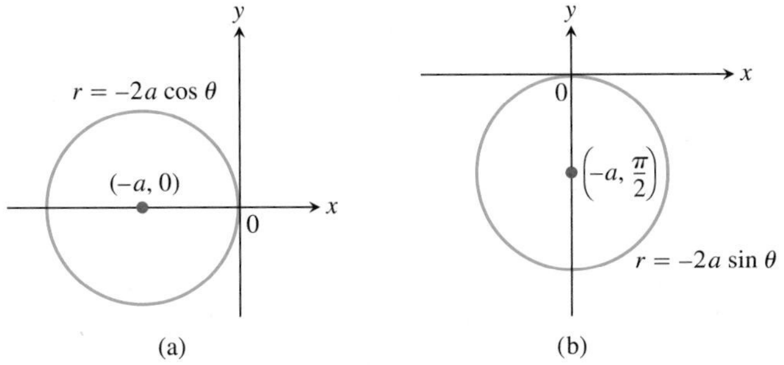

FIGURE 10.60 Polar equation of a circle of radius a through the origin with center on (a) the negative x-axis, and (b) the negative y-axis.

EXAMPLE 2 Circles Through the Origin

Radius	Center (polar coordinates)	Polar equation
3	$(3, 0)$	$r = 6 \cos \theta$
2	$(2, \pi/2)$	$r = 4 \sin \theta$
1/2	$(-1/2, 0)$	$r = -\cos \theta$
1	$(-1, \pi/2)$	$r = -2 \sin \theta$

■

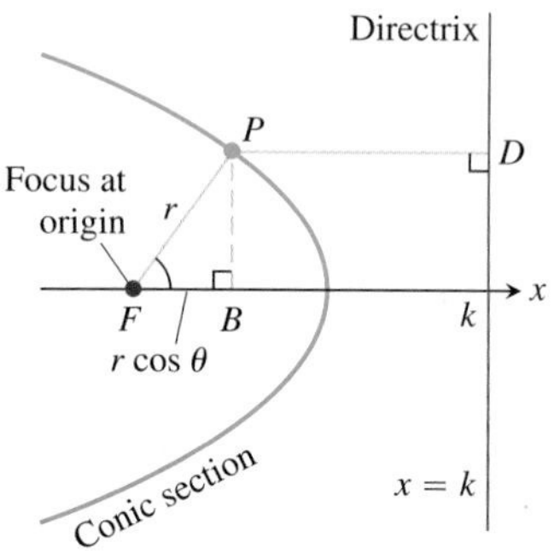

FIGURE 10.61 If a conic section is put in the position with its focus placed at the origin and a directrix perpendicular to the initial ray and right of the origin, we can find its polar equation from the conic's focus–directrix equation.

Ellipses, Parabolas, and Hyperbolas

To find polar equations for ellipses, parabolas, and hyperbolas, we place one focus at the origin and the corresponding directrix to the right of the origin along the vertical line $x = k$ (Figure 10.61). This makes

$$PF = r$$

and

$$PD = k - FB = k - r \cos \theta.$$

The conic's focus–directrix equation $PF = e \cdot PD$ then becomes

$$r = e(k - r \cos \theta),$$

which can be solved for r to obtain

Polar Equation for a Conic with Eccentricity e

$$r = \frac{ke}{1 + e \cos \theta}, \qquad (2)$$

where $x = k > 0$ is the vertical directrix.

This equation represents an ellipse if $0 < e < 1$, a parabola if $e = 1$, and a hyperbola if $e > 1$. That is, ellipses, parabolas, and hyperbolas all have the same basic equation expressed in terms of eccentricity and location of the directrix.

EXAMPLE 3 Polar Equations of Some Conics

$e = \frac{1}{2}$: ellipse $\quad r = \dfrac{k}{2 + \cos \theta}$

$e = 1$: parabola $\quad r = \dfrac{k}{1 + \cos \theta}$

$e = 2$: hyperbola $\quad r = \dfrac{2k}{1 + 2 \cos \theta}$

■

You may see variations of Equation (2) from time to time, depending on the location of the directrix. If the directrix is the line $x = -k$ to the left of the origin (the origin is still a focus), we replace Equation (2) by

$$r = \frac{ke}{1 - e\cos\theta}.$$

The denominator now has a $(-)$ instead of a $(+)$. If the directrix is either of the lines $y = k$ or $y = -k$, the equations have sines in them instead of cosines, as shown in Figure 10.62.

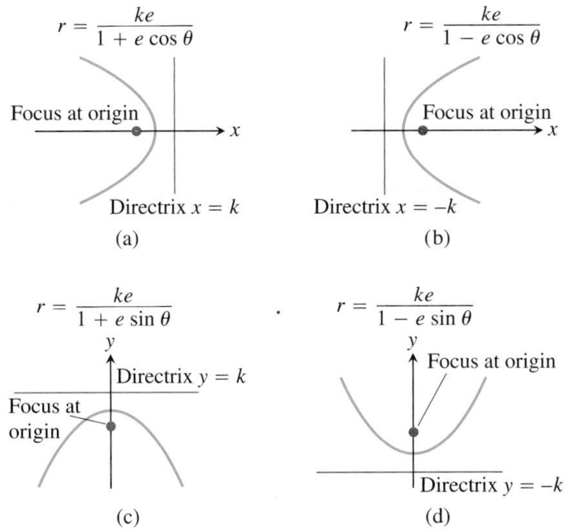

FIGURE 10.62 Equations for conic sections with eccentricity $e > 0$, but different locations of the directrix. The graphs here show a parabola, so $e = 1$.

EXAMPLE 4 Polar Equation of a Hyperbola

Find an equation for the hyperbola with eccentricity $3/2$ and directrix $x = 2$.

Solution We use Equation (2) with $k = 2$ and $e = 3/2$:

$$r = \frac{2(3/2)}{1 + (3/2)\cos\theta} \quad \text{or} \quad r = \frac{6}{2 + 3\cos\theta}.$$ ∎

EXAMPLE 5 Finding a Directrix

Find the directrix of the parabola

$$r = \frac{25}{10 + 10\cos\theta}.$$

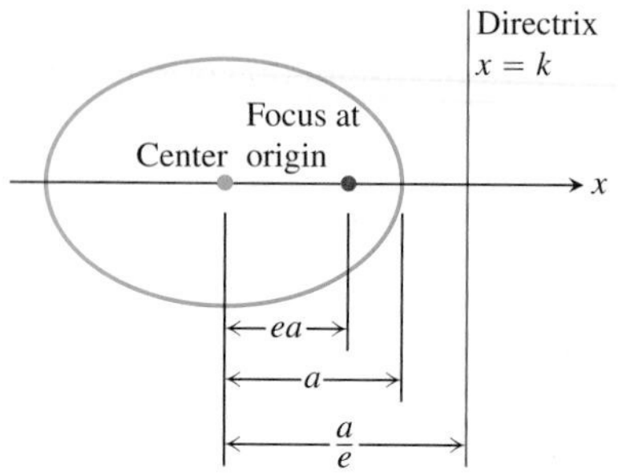

FIGURE 10.63 In an ellipse with semimajor axis a, the focus–directrix distance is $k = (a/e) - ea$, so $ke = a(1 - e^2)$.

Solution We divide the numerator and denominator by 10 to put the equation in standard form:

$$r = \frac{5/2}{1 + \cos\theta}.$$

This is the equation

$$r = \frac{ke}{1 + e\cos\theta}$$

with $k = 5/2$ and $e = 1$. The equation of the directrix is $x = 5/2$. ■

From the ellipse diagram in Figure 10.63, we see that k is related to the eccentricity e and the semimajor axis a by the equation

$$k = \frac{a}{e} - ea.$$

From this, we find that $ke = a(1 - e^2)$. Replacing ke in Equation (2) by $a(1 - e^2)$ gives the standard polar equation for an ellipse.

Polar Equation for the Ellipse with Eccentricity e and Semimajor Axis a

$$r = \frac{a(1 - e^2)}{1 + e\cos\theta} \qquad (3)$$

Notice that when $e = 0$, Equation (3) becomes $r = a$, which represents a circle.

Equation (3) is the starting point for calculating planetary orbits.

EXAMPLE 6 The Planet Pluto's Orbit

Find a polar equation for an ellipse with semimajor axis 39.44 AU (astronomical units) and eccentricity 0.25. This is the approximate size of Pluto's orbit around the sun.

Solution We use Equation (3) with $a = 39.44$ and $e = 0.25$ to find

$$r = \frac{39.44(1 - (0.25)^2)}{1 + 0.25\cos\theta} = \frac{147.9}{4 + \cos\theta}.$$

At its point of closest approach (perihelion) where $\theta = 0$, Pluto is

$$r = \frac{147.9}{4 + 1} = 29.58 \text{ AU}$$

from the sun. At its most distant point (aphelion) where $\theta = \pi$, Pluto is

$$r = \frac{147.9}{4 - 1} = 49.3 \text{ AU}$$

from the sun (Figure 10.64). ■

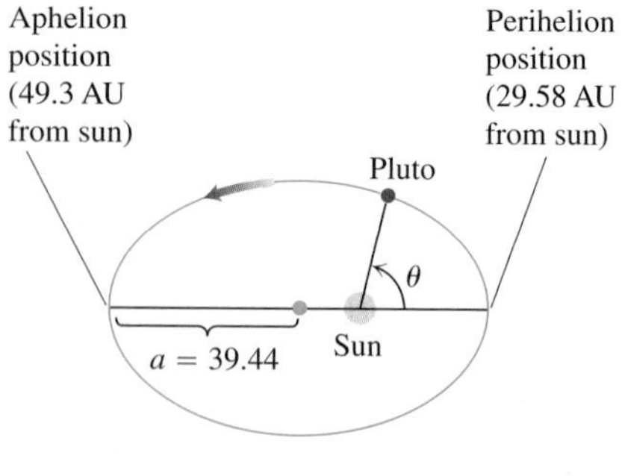

FIGURE 10.64 The orbit of Pluto (Example 6).

EXERCISES 10.8

Lines

Find polar and Cartesian equations for the lines in Exercises 1–4.

1.

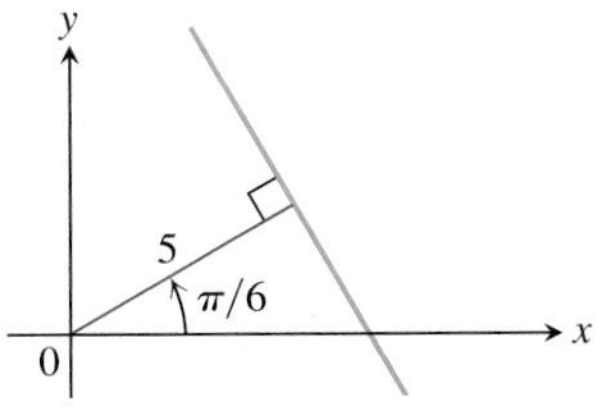

2.

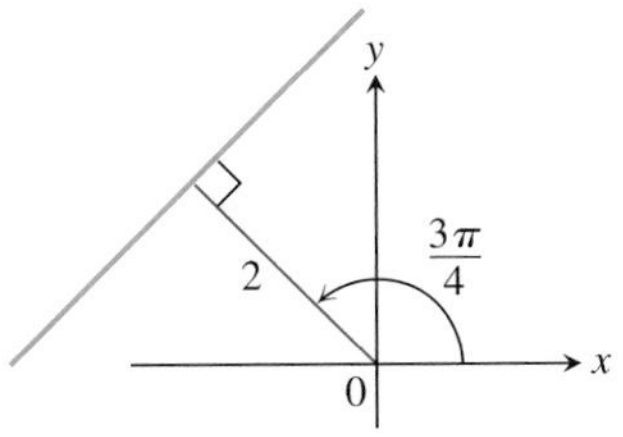

3.

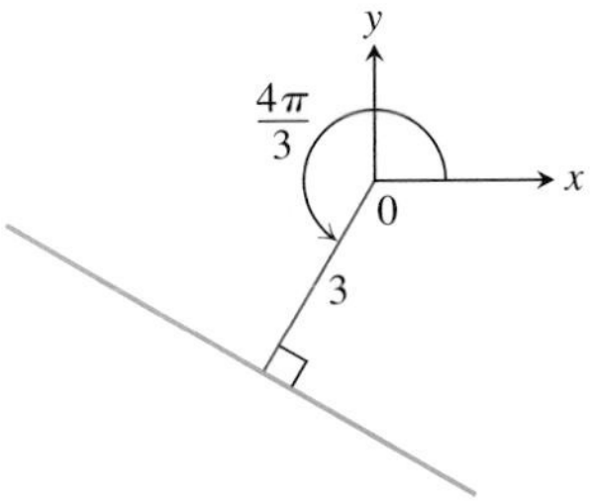

4.

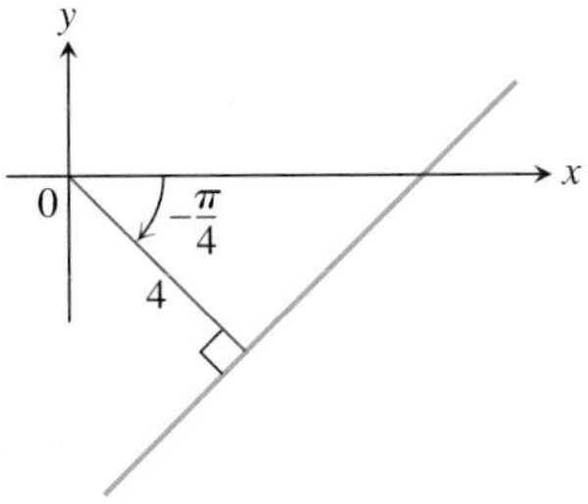

Sketch the lines in Exercises 5–8 and find Cartesian equations for them.

5. $r\cos\left(\theta - \frac{\pi}{4}\right) = \sqrt{2}$ **6.** $r\cos\left(\theta + \frac{3\pi}{4}\right) = 1$

7. $r\cos\left(\theta - \frac{2\pi}{3}\right) = 3$ **8.** $r\cos\left(\theta + \frac{\pi}{3}\right) = 2$

Find a polar equation in the form $r\cos(\theta - \theta_0) = r_0$ for each of the lines in Exercises 9–12.

9. $\sqrt{2}x + \sqrt{2}y = 6$ **10.** $\sqrt{3}x - y = 1$

11. $y = -5$ **12.** $x = -4$

Circles

Find polar equations for the circles in Exercises 13–16.

13.

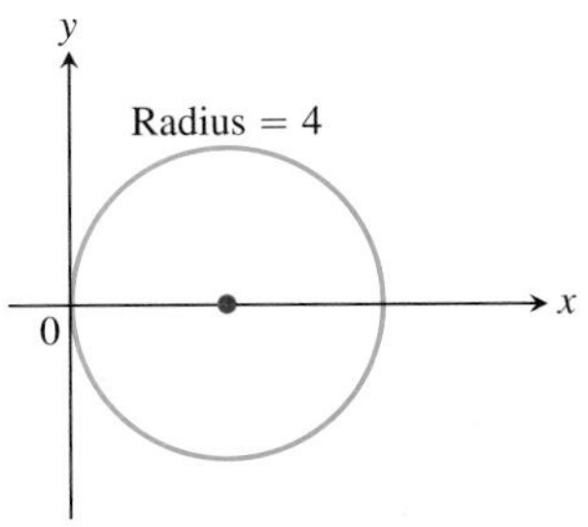

14.

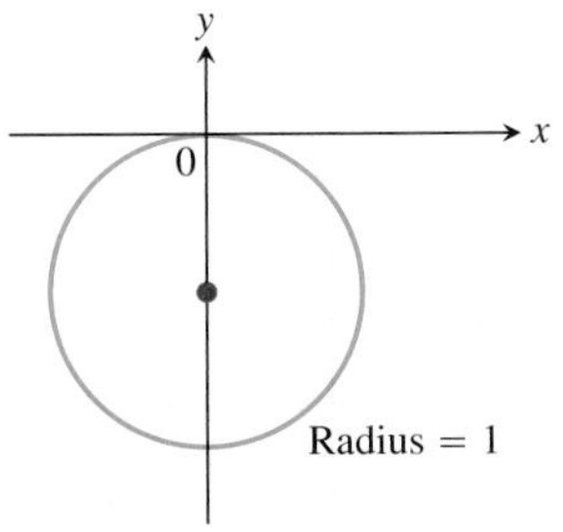

15.

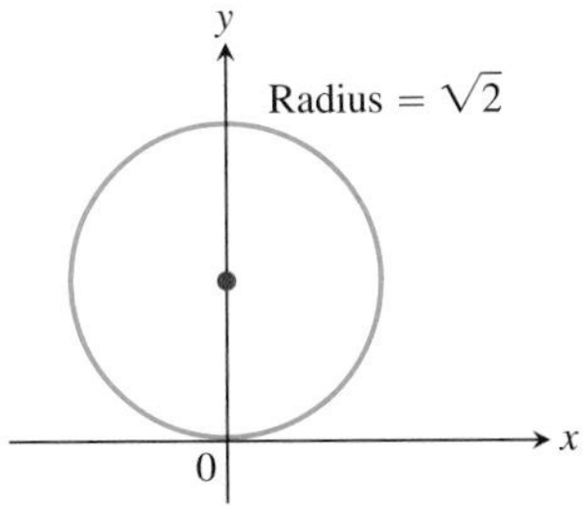

16.

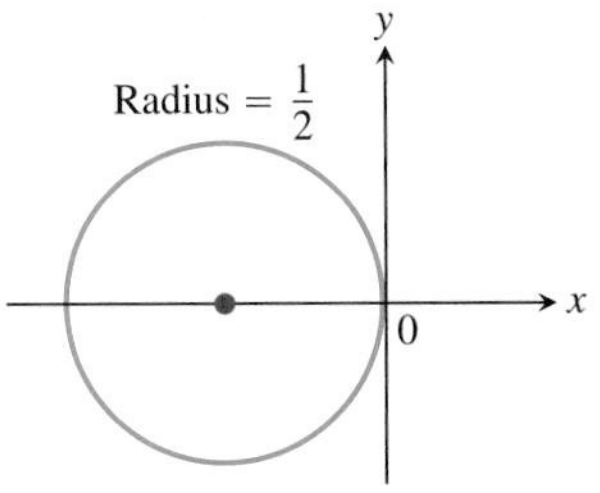

Sketch the circles in Exercises 17–20. Give polar coordinates for their centers and identify their radii.

17. $r = 4\cos\theta$ **18.** $r = 6\sin\theta$

19. $r = -2\cos\theta$ **20.** $r = -8\sin\theta$

Find polar equations for the circles in Exercises 21–28. Sketch each circle in the coordinate plane and label it with both its Cartesian and polar equations.

21. $(x - 6)^2 + y^2 = 36$ **22.** $(x + 2)^2 + y^2 = 4$

23. $x^2 + (y - 5)^2 = 25$ **24.** $x^2 + (y + 7)^2 = 49$

25. $x^2 + 2x + y^2 = 0$ **26.** $x^2 - 16x + y^2 = 0$

27. $x^2 + y^2 + y = 0$ **28.** $x^2 + y^2 - \frac{4}{3}y = 0$

Conic Sections from Eccentricities and Directrices

Exercises 29–36 give the eccentricities of conic sections with one focus at the origin, along with the directrix corresponding to that focus. Find a polar equation for each conic section.

29. $e = 1, \quad x = 2$ **30.** $e = 1, \quad y = 2$

31. $e = 5, \quad y = -6$ **32.** $e = 2, \quad x = 4$

33. $e = 1/2, \quad x = 1$ **34.** $e = 1/4, \quad x = -2$

35. $e = 1/5, \quad y = -10$ **36.** $e = 1/3, \quad y = 6$

Parabolas and Ellipses

Sketch the parabolas and ellipses in Exercises 37–44. Include the directrix that corresponds to the focus at the origin. Label the vertices with appropriate polar coordinates. Label the centers of the ellipses as well.

37. $r = \dfrac{1}{1 + \cos\theta}$ **38.** $r = \dfrac{6}{2 + \cos\theta}$

39. $r = \dfrac{25}{10 - 5\cos\theta}$ **40.** $r = \dfrac{4}{2 - 2\cos\theta}$

41. $r = \dfrac{400}{16 + 8\sin\theta}$ **42.** $r = \dfrac{12}{3 + 3\sin\theta}$

43. $r = \dfrac{8}{2 - 2\sin\theta}$ **44.** $r = \dfrac{4}{2 - \sin\theta}$

Graphing Inequalities

Sketch the regions defined by the inequalities in Exercises 45 and 46.

45. $0 \le r \le 2\cos\theta$

46. $-3\cos\theta \le r \le 0$

T Grapher Explorations

Graph the lines and conic sections in Exercises 47–56.

47. $r = 3\sec(\theta - \pi/3)$

48. $r = 4\sec(\theta + \pi/6)$

49. $r = 4\sin\theta$

50. $r = -2\cos\theta$

51. $r = 8/(4 + \cos\theta)$

52. $r = 8/(4 + \sin\theta)$

53. $r = 1/(1 - \sin\theta)$

54. $r = 1/(1 + \cos\theta)$

55. $r = 1/(1 + 2\sin\theta)$

56. $r = 1/(1 + 2\cos\theta)$

Theory and Examples

57. Perihelion and aphelion A planet travels about its sun in an ellipse whose semimajor axis has length a. (See accompanying figure.)

a. Show that $r = a(1 - e)$ when the planet is closest to the sun and that $r = a(1 + e)$ when the planet is farthest from the sun.

b. Use the data in the table in Exercise 58 to find how close each planet in our solar system comes to the sun and how far away each planet gets from the sun.

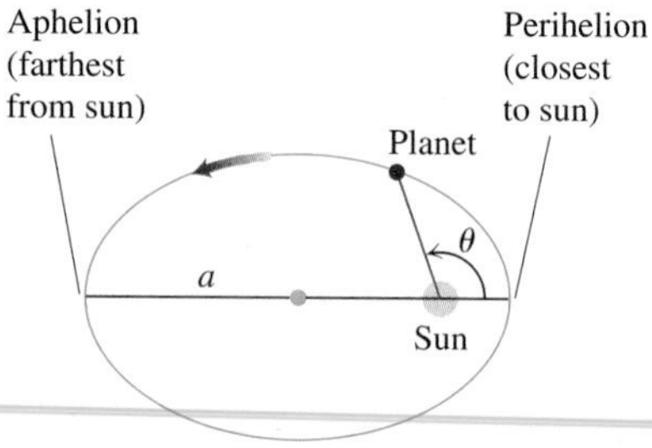

58. Planetary orbits In Example 6, we found a polar equation for the orbit of Pluto. Use the data in the table below to find polar equations for the orbits of the other planets.

Planet	Semimajor axis (astronomical units)	Eccentricity
Mercury	0.3871	0.2056
Venus	0.7233	0.0068
Earth	1.000	0.0167
Mars	1.524	0.0934
Jupiter	5.203	0.0484
Saturn	9.539	0.0543
Uranus	19.18	0.0460
Neptune	30.06	0.0082
Pluto	39.44	0.2481

59. a. Find Cartesian equations for the curves $r = 4\sin\theta$ and $r = \sqrt{3}\sec\theta$.

b. Sketch the curves together and label their points of intersection in both Cartesian and polar coordinates.

60. Repeat Exercise 59 for $r = 8\cos\theta$ and $r = 2\sec\theta$.

61. Find a polar equation for the parabola with focus (0, 0) and directrix $r\cos\theta = 4$.

62. Find a polar equation for the parabola with focus (0, 0) and directrix $r\cos(\theta - \pi/2) = 2$.

63. a. The space engineer's formula for eccentricity The space engineer's formula for the eccentricity of an elliptical orbit is

$$e = \frac{r_{max} - r_{min}}{r_{max} + r_{min}},$$

where r is the distance from the space vehicle to the attracting focus of the ellipse along which it travels. Why does the formula work?

b. Drawing ellipses with string You have a string with a knot in each end that can be pinned to a drawing board. The string is 10 in. long from the center of one knot to the center of the other. How far apart should the pins be to use the method illustrated in Figure 10.5 (Section 10.1) to draw an ellipse of eccentricity 0.2? The resulting ellipse would resemble the orbit of Mercury.

64. Halley's comet (See Section 10.2, Example 1.)

a. Write an equation for the orbit of Halley's comet in a coordinate system in which the sun lies at the origin and the other focus lies on the negative x-axis, scaled in astronomical units.

b. How close does the comet come to the sun in astronomical units? In kilometers?

c. What is the farthest the comet gets from the sun in astronomical units? In kilometers?

In Exercises 65–68, find a polar equation for the given curve. In each case, sketch a typical curve.

65. $x^2 + y^2 - 2ay = 0$

66. $y^2 = 4ax + 4a^2$

67. $x\cos\alpha + y\sin\alpha = p$ (α, p constant)

68. $(x^2 + y^2)^2 + 2ax(x^2 + y^2) - a^2y^2 = 0$

COMPUTER EXPLORATIONS

69. Use a CAS to plot the polar equation

$$r = \frac{ke}{1 + e\cos\theta}$$

for various values of k and e, $-\pi \le \theta \le \pi$. Answer the following questions.

a. Take $k = -2$. Describe what happens to the plots as you take e to be 3/4, 1, and 5/4. Repeat for $k = 2$.

b. Take $k = -1$. Describe what happens to the plots as you take e to be $7/6, 5/4, 4/3, 3/2, 2, 3, 5, 10$, and 20. Repeat for $e = 1/2, 1/3, 1/4, 1/10$, and $1/20$.

c. Now keep $e > 0$ fixed and describe what happens as you take k to be $-1, -2, -3, -4$, and -5. Be sure to look at graphs for parabolas, ellipses, and hyperbolas.

70. Use a CAS to plot the polar ellipse

$$r = \frac{a(1 - e^2)}{1 + e\cos\theta}$$

for various values of $a > 0$ and $0 < e < 1$, $-\pi \le \theta \le \pi$.

a. Take $e = 9/10$. Describe what happens to the plots as you let a equal $1, 3/2, 2, 3, 5$, and 10. Repeat with $e = 1/4$.

b. Take $a = 2$. Describe what happens as you take e to be $9/10, 8/10, 7/10, \ldots, 1/10, 1/20$, and $1/50$.

Chapter 10 Questions to Guide Your Review

1. What is a parabola? What are the Cartesian equations for parabolas whose vertices lie at the origin and whose foci lie on the coordinate axes? How can you find the focus and directrix of such a parabola from its equation?
2. What is an ellipse? What are the Cartesian equations for ellipses centered at the origin with foci on one of the coordinate axes? How can you find the foci, vertices, and directrices of such an ellipse from its equation?
3. What is a hyperbola? What are the Cartesian equations for hyperbolas centered at the origin with foci on one of the coordinate axes? How can you find the foci, vertices, and directrices of such an ellipse from its equation?
4. What is the eccentricity of a conic section? How can you classify conic sections by eccentricity? How are an ellipse's shape and eccentricity related?
5. Explain the equation $PF = e \cdot PD$.
6. What is a quadratic curve in the xy-plane? Give examples of degenerate and nondegenerate quadratic curves.
7. How can you find a Cartesian coordinate system in which the new equation for a conic section in the plane has no xy-term? Give an example.
8. How can you tell what kind of graph to expect from a quadratic equation in x and y? Give examples.
9. What are some typical parametrizations for conic sections?
10. What is a cycloid? What are typical parametric equations for cycloids? What physical properties account for the importance of cycloids?
11. What are polar coordinates? What equations relate polar coordinates to Cartesian coordinates? Why might you want to change from one coordinate system to the other?
12. What consequence does the lack of uniqueness of polar coordinates have for graphing? Give an example.
13. How do you graph equations in polar coordinates? Include in your discussion symmetry, slope, behavior at the origin, and the use of Cartesian graphs. Give examples.
14. How do you find the area of a region $0 \le r_1(\theta) \le r \le r_2(\theta)$, $\alpha \le \theta \le \beta$, in the polar coordinate plane? Give examples.
15. Under what conditions can you find the length of a curve $r = f(\theta)$, $\alpha \le \theta \le \beta$, in the polar coordinate plane? Give an example of a typical calculation.
16. Under what conditions can you find the area of the surface generated by revolving a curve $r = f(\theta)$, $\alpha \le \theta \le \beta$, about the x-axis? The y-axis? Give examples of typical calculations.
17. What are the standard equations for lines and conic sections in polar coordinates? Give examples.

Chapter 10 Practice Exercises

Graphing Conic Sections

Sketch the parabolas in Exercises 1–4. Include the focus and directrix in each sketch.

1. $x^2 = -4y$
2. $x^2 = 2y$
3. $y^2 = 3x$
4. $y^2 = -(8/3)x$

Find the eccentricities of the ellipses and hyperbolas in Exercises 5–8. Sketch each conic section. Include the foci, vertices, and asymptotes (as appropriate) in your sketch.

5. $16x^2 + 7y^2 = 112$
6. $x^2 + 2y^2 = 4$
7. $3x^2 - y^2 = 3$
8. $5y^2 - 4x^2 = 20$

EXAMPLE 2 A Divergent Sequence

Show that the sequence $\{1, -1, 1, -1, 1, -1, \ldots, (-1)^{n+1}, \ldots\}$ diverges.

Solution Suppose the sequence converges to some number L. By choosing $\epsilon = 1/2$ in the definition of the limit, all terms a_n of the sequence with index n larger than some N must lie within $\epsilon = 1/2$ of L. Since the number 1 appears repeatedly as every other term of the sequence, we must have that the number 1 lies within the distance $\epsilon = 1/2$ of L. It follows that $|L - 1| < 1/2$, or equivalently, $1/2 < L < 3/2$. Likewise, the number -1 appears repeatedly in the sequence with arbitrarily high index. So we must also have that $|L - (-1)| < 1/2$, or equivalently, $-3/2 < L < -1/2$. But the number L cannot lie in both of the intervals $(1/2, 3/2)$ and $(-3/2, -1/2)$ because they have no overlap. Therefore, no such limit L exists and so the sequence diverges.

Note that the same argument works for any positive number ϵ smaller than 1, not just $1/2$. ■

The sequence $\{\sqrt{n}\}$ also diverges, but for a different reason. As n increases, its terms become larger than any fixed number. We describe the behavior of this sequence by writing

$$\lim_{n\to\infty} \sqrt{n} = \infty.$$

In writing infinity as the limit of a sequence, we are not saying that the differences between the terms a_n and ∞ become small as n increases. Nor are we asserting that there is some number infinity that the sequence approaches. We are merely using a notation that captures the idea that a_n eventually gets and stays larger than any fixed number as n gets large.

DEFINITION Diverges to Infinity

The sequence $\{a_n\}$ **diverges to infinity** if for every number M there is an integer N such that for all n larger than N, $a_n > M$. If this condition holds we write

$$\lim_{n\to\infty} a_n = \infty \qquad \text{or} \qquad a_n \to \infty.$$

Similarly if for every number m there is an integer N such that for all $n > N$ we have $a_n < m$, then we say $\{a_n\}$ **diverges to negative infinity** and write

$$\lim_{n\to\infty} a_n = -\infty \qquad \text{or} \qquad a_n \to -\infty.$$

A sequence may diverge without diverging to infinity or negative infinity. We saw this in Example 2, and the sequences $\{1, -2, 3, -4, 5, -6, 7, -8, \ldots\}$ and $\{1, 0, 2, 0, 3, 0, \ldots\}$ are also examples of such divergence.

Calculating Limits of Sequences

If we always had to use the formal definition of the limit of a sequence, calculating with ϵ's and N's, then computing limits of sequences would be a formidable task. Fortunately we can derive a few basic examples, and then use these to quickly analyze the limits of many more sequences. We will need to understand how to combine and compare sequences. Since sequences are functions with domain restricted to the positive integers, it is not too surprising that the theorems on limits of functions given in Chapter 2 have versions for sequences.

THEOREM 1

Let $\{a_n\}$ and $\{b_n\}$ be sequences of real numbers and let A and B be real numbers. The following rules hold if $\lim_{n\to\infty} a_n = A$ and $\lim_{n\to\infty} b_n = B$.

1. *Sum Rule:* $\lim_{n\to\infty}(a_n + b_n) = A + B$
2. *Difference Rule:* $\lim_{n\to\infty}(a_n - b_n) = A - B$
3. *Product Rule:* $\lim_{n\to\infty}(a_n \cdot b_n) = A \cdot B$
4. *Constant Multiple Rule:* $\lim_{n\to\infty}(k \cdot b_n) = k \cdot B$ (Any number k)
5. *Quotient Rule:* $\lim_{n\to\infty} \frac{a_n}{b_n} = \frac{A}{B}$ if $B \neq 0$

The proof is similar to that of Theorem 1 of Section 2.2, and is omitted.

EXAMPLE 3 Applying Theorem 1

By combining Theorem 1 with the limits of Example 1, we have:

(a) $\lim_{n\to\infty}\left(-\frac{1}{n}\right) = -1 \cdot \lim_{n\to\infty}\frac{1}{n} = -1 \cdot 0 = 0$ Constant Multiple Rule and Example 1a

(b) $\lim_{n\to\infty}\left(\frac{n-1}{n}\right) = \lim_{n\to\infty}\left(1 - \frac{1}{n}\right) = \lim_{n\to\infty} 1 - \lim_{n\to\infty}\frac{1}{n} = 1 - 0 = 1$ Difference Rule and Example 1a

(c) $\lim_{n\to\infty}\frac{5}{n^2} = 5 \cdot \lim_{n\to\infty}\frac{1}{n} \cdot \lim_{n\to\infty}\frac{1}{n} = 5 \cdot 0 \cdot 0 = 0$ Product Rule

(d) $\lim_{n\to\infty}\frac{4 - 7n^6}{n^6 + 3} = \lim_{n\to\infty}\frac{(4/n^6) - 7}{1 + (3/n^6)} = \frac{0 - 7}{1 + 0} = -7.$ Sum and Quotient Rules ■

Be cautious in applying Theorem 1. It does not say, for example, that each of the sequences $\{a_n\}$ and $\{b_n\}$ have limits if their sum $\{a_n + b_n\}$ has a limit. For instance, $\{a_n\} = \{1, 2, 3, \dots\}$ and $\{b_n\} = \{-1, -2, -3, \dots\}$ both diverge, but their sum $\{a_n + b_n\} = \{0, 0, 0, \dots\}$ clearly converges to 0.

One consequence of Theorem 1 is that every nonzero multiple of a divergent sequence $\{a_n\}$ diverges. For suppose, to the contrary, that $\{ca_n\}$ converges for some number $c \neq 0$. Then, by taking $k = 1/c$ in the Constant Multiple Rule in Theorem 1, we see that the sequence

$$\left\{\frac{1}{c} \cdot ca_n\right\} = \{a_n\}$$

converges. Thus, $\{ca_n\}$ cannot converge unless $\{a_n\}$ also converges. If $\{a_n\}$ does not converge, then $\{ca_n\}$ does not converge.

The next theorem is the sequence version of the Sandwich Theorem in Section 2.2. You are asked to prove the theorem in Exercise 95.

THEOREM 2 The Sandwich Theorem for Sequences

Let $\{a_n\}$, $\{b_n\}$, and $\{c_n\}$ be sequences of real numbers. If $a_n \le b_n \le c_n$ holds for all n beyond some index N, and if $\lim_{n\to\infty} a_n = \lim_{n\to\infty} c_n = L$, then $\lim_{n\to\infty} b_n = L$ also.

An immediate consequence of Theorem 2 is that, if $|b_n| \le c_n$ and $c_n \to 0$, then $b_n \to 0$ because $-c_n \le b_n \le c_n$. We use this fact in the next example.

EXAMPLE 4 Applying the Sandwich Theorem

Since $1/n \to 0$, we know that

(a) $\dfrac{\cos n}{n} \to 0$ because $-\dfrac{1}{n} \le \dfrac{\cos n}{n} \le \dfrac{1}{n}$;

(b) $\dfrac{1}{2^n} \to 0$ because $0 \le \dfrac{1}{2^n} \le \dfrac{1}{n}$;

(c) $(-1)^n \dfrac{1}{n} \to 0$ because $-\dfrac{1}{n} \le (-1)^n \dfrac{1}{n} \le \dfrac{1}{n}$. ■

The application of Theorems 1 and 2 is broadened by a theorem stating that applying a continuous function to a convergent sequence produces a convergent sequence. We state the theorem without proof (Exercise 96).

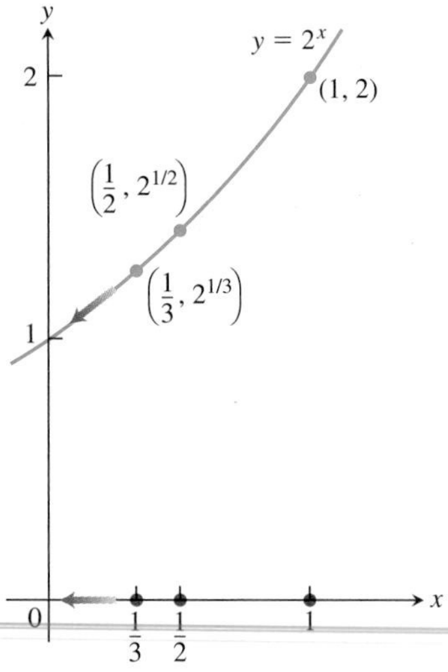

FIGURE 11.3 As $n \to \infty$, $1/n \to 0$ and $2^{1/n} \to 2^0$ (Example 6).

THEOREM 3 The Continuous Function Theorem for Sequences

Let $\{a_n\}$ be a sequence of real numbers. If $a_n \to L$ and if f is a function that is continuous at L and defined at all a_n, then $f(a_n) \to f(L)$.

EXAMPLE 5 Applying Theorem 3

Show that $\sqrt{(n+1)/n} \to 1$.

Solution We know that $(n+1)/n \to 1$. Taking $f(x) = \sqrt{x}$ and $L = 1$ in Theorem 3 gives $\sqrt{(n+1)/n} \to \sqrt{1} = 1$. ■

EXAMPLE 6 The Sequence $\{2^{1/n}\}$

The sequence $\{1/n\}$ converges to 0. By taking $a_n = 1/n$, $f(x) = 2^x$, and $L = 0$ in Theorem 3, we see that $2^{1/n} = f(1/n) \to f(L) = 2^0 = 1$. The sequence $\{2^{1/n}\}$ converges to 1 (Figure 11.3). ■

Using l'Hôpital's Rule

The next theorem enables us to use l'Hôpital's Rule to find the limits of some sequences. It formalizes the connection between $\lim_{n\to\infty} a_n$ and $\lim_{x\to\infty} f(x)$.

THEOREM 4

Suppose that $f(x)$ is a function defined for all $x \ge n_0$ and that $\{a_n\}$ is a sequence of real numbers such that $a_n = f(n)$ for $n \ge n_0$. Then

$$\lim_{x\to\infty} f(x) = L \quad \Rightarrow \quad \lim_{n\to\infty} a_n = L.$$

Proof Suppose that $\lim_{x\to\infty} f(x) = L$. Then for each positive number ϵ there is a number M such that for all x,

$$x > M \quad \Rightarrow \quad |f(x) - L| < \epsilon.$$

Let N be an integer greater than M and greater than or equal to n_0. Then

$$n > N \quad \Rightarrow \quad a_n = f(n) \quad \text{and} \quad |a_n - L| = |f(n) - L| < \epsilon. \qquad \blacksquare$$

EXAMPLE 7 Applying L'Hôpital's Rule

Show that

$$\lim_{n\to\infty} \frac{\ln n}{n} = 0.$$

Solution The function $(\ln x)/x$ is defined for all $x \geq 1$ and agrees with the given sequence at positive integers. Therefore, by Theorem 4, $\lim_{n\to\infty}(\ln n)/n$ will equal $\lim_{x\to\infty}(\ln x)/x$ if the latter exists. A single application of l'Hôpital's Rule shows that

$$\lim_{x\to\infty} \frac{\ln x}{x} = \lim_{x\to\infty} \frac{1/x}{1} = \frac{0}{1} = 0.$$

We conclude that $\lim_{n\to\infty}(\ln n)/n = 0$. ■

When we use l'Hôpital's Rule to find the limit of a sequence, we often treat n as a continuous real variable and differentiate directly with respect to n. This saves us from having to rewrite the formula for a_n as we did in Example 7.

EXAMPLE 8 Applying L'Hôpital's Rule

Find

$$\lim_{n\to\infty} \frac{2^n}{5n}.$$

Solution By l'Hôpital's Rule (differentiating with respect to n),

$$\begin{aligned}\lim_{n\to\infty} \frac{2^n}{5n} &= \lim_{n\to\infty} \frac{2^n \cdot \ln 2}{5} \\ &= \infty. \end{aligned}$$ ■

EXAMPLE 9 Applying L'Hôpital's Rule to Determine Convergence

Does the sequence whose nth term is

$$a_n = \left(\frac{n+1}{n-1}\right)^n$$

converge? If so, find $\lim_{n\to\infty} a_n$.

Solution The limit leads to the indeterminate form 1^∞. We can apply l'Hôpital's Rule if we first change the form to $\infty \cdot 0$ by taking the natural logarithm of a_n:

$$\begin{aligned}\ln a_n &= \ln\left(\frac{n+1}{n-1}\right)^n \\ &= n \ln\left(\frac{n+1}{n-1}\right).\end{aligned}$$

Then,

$$\begin{aligned}
\lim_{n\to\infty} \ln a_n &= \lim_{n\to\infty} n \ln\left(\frac{n+1}{n-1}\right) && \infty \cdot 0 \\
&= \lim_{n\to\infty} \frac{\ln\left(\frac{n+1}{n-1}\right)}{1/n} && \frac{0}{0} \\
&= \lim_{n\to\infty} \frac{-2/(n^2-1)}{-1/n^2} && \text{l'Hôpital's Rule} \\
&= \lim_{n\to\infty} \frac{2n^2}{n^2-1} = 2.
\end{aligned}$$

Since $\ln a_n \to 2$ and $f(x) = e^x$ is continuous, Theorem 4 tells us that

$$a_n = e^{\ln a_n} \to e^2.$$

The sequence $\{a_n\}$ converges to e^2. ■

Commonly Occurring Limits

The next theorem gives some limits that arise frequently.

THEOREM 5
The following six sequences converge to the limits listed below:

1. $\displaystyle\lim_{n\to\infty} \frac{\ln n}{n} = 0$
2. $\displaystyle\lim_{n\to\infty} \sqrt[n]{n} = 1$
3. $\displaystyle\lim_{n\to\infty} x^{1/n} = 1 \qquad (x > 0)$
4. $\displaystyle\lim_{n\to\infty} x^n = 0 \qquad (|x| < 1)$
5. $\displaystyle\lim_{n\to\infty} \left(1 + \frac{x}{n}\right)^n = e^x \qquad (\text{any } x)$
6. $\displaystyle\lim_{n\to\infty} \frac{x^n}{n!} = 0 \qquad (\text{any } x)$

In Formulas (3) through (6), x remains fixed as $n \to \infty$.

Factorial Notation
The notation $n!$ ("n factorial") means the product $1 \cdot 2 \cdot 3 \cdots n$ of the integers from 1 to n. Notice that $(n+1)! = (n+1) \cdot n!$. Thus, $4! = 1 \cdot 2 \cdot 3 \cdot 4 = 24$ and $5! = 1 \cdot 2 \cdot 3 \cdot 4 \cdot 5 = 5 \cdot 4! = 120$. We define 0! to be 1. Factorials grow even faster than exponentials, as the table suggests.

n	e^n (rounded)	$n!$
1	3	1
5	148	120
10	22,026	3,628,800
20	4.9×10^8	2.4×10^{18}

Proof The first limit was computed in Example 7. The next two can be proved by taking logarithms and applying Theorem 4 (Exercises 93 and 94). The remaining proofs are given in Appendix 3. ■

EXAMPLE 10 Applying Theorem 5

(a) $\dfrac{\ln(n^2)}{n} = \dfrac{2\ln n}{n} \to 2 \cdot 0 = 0$ Formula 1

(b) $\sqrt[n]{n^2} = n^{2/n} = (n^{1/n})^2 \to (1)^2 = 1$ Formula 2

(c) $\sqrt[n]{3n} = 3^{1/n}(n^{1/n}) \to 1 \cdot 1 = 1$ Formula 3 with $x = 3$ and Formula 2

(d) $\left(-\frac{1}{2}\right)^n \to 0$ Formula 4 with $x = -\frac{1}{2}$

(e) $\left(\frac{n-2}{n}\right)^n = \left(1 + \frac{-2}{n}\right)^n \to e^{-2}$ Formula 5 with $x = -2$

(f) $\frac{100^n}{n!} \to 0$ Formula 6 with $x = 100$ ■

Recursive Definitions

So far, we have calculated each a_n directly from the value of n. But sequences are often defined **recursively** by giving

1. The value(s) of the initial term or terms, and
2. A rule, called a **recursion formula**, for calculating any later term from terms that precede it.

EXAMPLE 11 Sequences Constructed Recursively

(a) The statements $a_1 = 1$ and $a_n = a_{n-1} + 1$ define the sequence $1, 2, 3, \ldots, n, \ldots$ of positive integers. With $a_1 = 1$, we have $a_2 = a_1 + 1 = 2$, $a_3 = a_2 + 1 = 3$, and so on.

(b) The statements $a_1 = 1$ and $a_n = n \cdot a_{n-1}$ define the sequence $1, 2, 6, 24, \ldots, n!, \ldots$ of factorials. With $a_1 = 1$, we have $a_2 = 2 \cdot a_1 = 2$, $a_3 = 3 \cdot a_2 = 6$, $a_4 = 4 \cdot a_3 = 24$, and so on.

(c) The statements $a_1 = 1, a_2 = 1$, and $a_{n+1} = a_n + a_{n-1}$ define the sequence $1, 1, 2, 3, 5, \ldots$ of **Fibonacci numbers**. With $a_1 = 1$ and $a_2 = 1$, we have $a_3 = 1 + 1 = 2$, $a_4 = 2 + 1 = 3$, $a_5 = 3 + 2 = 5$, and so on.

(d) As we can see by applying Newton's method, the statements $x_0 = 1$ and $x_{n+1} = x_n - [(\sin x_n - x_n^2)/(\cos x_n - 2x_n)]$ define a sequence that converges to a solution of the equation $\sin x - x^2 = 0$. ■

Bounded Nondecreasing Sequences

The terms of a general sequence can bounce around, sometimes getting larger, sometimes smaller. An important special kind of sequence is one for which each term is at least as large as its predecessor.

> **DEFINITION Nondecreasing Sequence**
> A sequence $\{a_n\}$ with the property that $a_n \leq a_{n+1}$ for all n is called a **nondecreasing sequence**.

EXAMPLE 12 Nondecreasing Sequences

(a) The sequence $1, 2, 3, \ldots, n, \ldots$ of natural numbers

(b) The sequence $\frac{1}{2}, \frac{2}{3}, \frac{3}{4}, \ldots, \frac{n}{n+1}, \ldots$

(c) The constant sequence $\{3\}$ ■

There are two kinds of nondecreasing sequences—those whose terms increase beyond any finite bound and those whose terms do not.

DEFINITIONS Bounded, Upper Bound, Least Upper Bound

A sequence $\{a_n\}$ is **bounded from above** if there exists a number M such that $a_n \leq M$ for all n. The number M is an **upper bound** for $\{a_n\}$. If M is an upper bound for $\{a_n\}$ but no number less than M is an upper bound for $\{a_n\}$, then M is the **least upper bound** for $\{a_n\}$.

EXAMPLE 13 Applying the Definition for Boundedness

(a) The sequence $1, 2, 3, \ldots, n, \ldots$ has no upper bound.

(b) The sequence $\frac{1}{2}, \frac{2}{3}, \frac{3}{4}, \ldots, \frac{n}{n+1}, \ldots$ is bounded above by $M = 1$.

No number less than 1 is an upper bound for the sequence, so 1 is the least upper bound (Exercise 113). ■

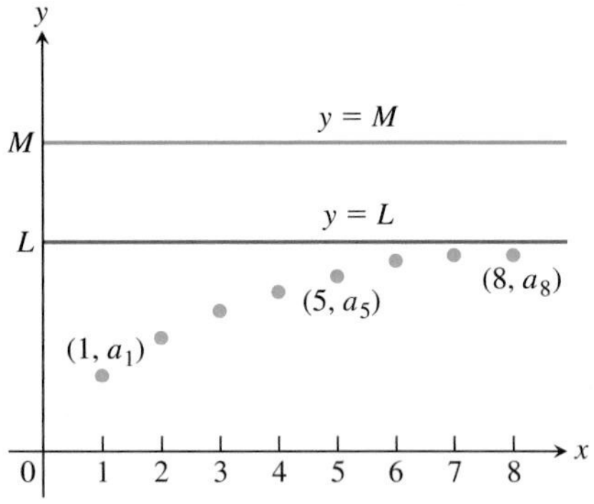

FIGURE 11.4 If the terms of a nondecreasing sequence have an upper bound M, they have a limit $L \leq M$.

A nondecreasing sequence that is bounded from above always has a least upper bound. This is the completeness property of the real numbers, discussed in Appendix 4. We will prove that if L is the least upper bound then the sequence converges to L.

Suppose we plot the points $(1, a_1), (2, a_2), \ldots, (n, a_n), \ldots$ in the xy-plane. If M is an upper bound of the sequence, all these points will lie on or below the line $y = M$ (Figure 11.4). The line $y = L$ is the lowest such line. None of the points (n, a_n) lies above $y = L$, but some do lie above any lower line $y = L - \epsilon$, if ϵ is a positive number. The sequence converges to L because

(a) $a_n \leq L$ for *all* values of n and

(b) given any $\epsilon > 0$, there exists at least one integer N for which $a_N > L - \epsilon$.

The fact that $\{a_n\}$ is nondecreasing tells us further that

$$a_n \geq a_N > L - \epsilon \qquad \text{for all } n \geq N.$$

Thus, *all* the numbers a_n beyond the Nth number lie within ϵ of L. This is precisely the condition for L to be the limit of the sequence $\{a_n\}$.

The facts for nondecreasing sequences are summarized in the following theorem. A similar result holds for nonincreasing sequences (Exercise 107).

THEOREM 6 The Nondecreasing Sequence Theorem

A nondecreasing sequence of real numbers converges if and only if it is bounded from above. If a nondecreasing sequence converges, it converges to its least upper bound.

Theorem 6 implies that a nondecreasing sequence converges when it is bounded from above. It diverges to infinity if it is not bounded from above.

EXERCISES 11.1

Finding Terms of a Sequence

Each of Exercises 1–6 gives a formula for the nth term a_n of a sequence $\{a_n\}$. Find the values of a_1, a_2, a_3, and a_4.

1. $a_n = \dfrac{1-n}{n^2}$
2. $a_n = \dfrac{1}{n!}$
3. $a_n = \dfrac{(-1)^{n+1}}{2n-1}$
4. $a_n = 2 + (-1)^n$
5. $a_n = \dfrac{2^n}{2^{n+1}}$
6. $a_n = \dfrac{2^n - 1}{2^n}$

Each of Exercises 7–12 gives the first term or two of a sequence along with a recursion formula for the remaining terms. Write out the first ten terms of the sequence.

7. $a_1 = 1, \quad a_{n+1} = a_n + (1/2^n)$
8. $a_1 = 1, \quad a_{n+1} = a_n/(n+1)$
9. $a_1 = 2, \quad a_{n+1} = (-1)^{n+1}a_n/2$
10. $a_1 = -2, \quad a_{n+1} = na_n/(n+1)$
11. $a_1 = a_2 = 1, \quad a_{n+2} = a_{n+1} + a_n$
12. $a_1 = 2, \quad a_2 = -1, \quad a_{n+2} = a_{n+1}/a_n$

Finding a Sequence's Formula

In Exercises 13–22, find a formula for the nth term of the sequence.

13. The sequence $1, -1, 1, -1, 1, \ldots$ 1's with alternating signs
14. The sequence $-1, 1, -1, 1, -1, \ldots$ 1's with alternating signs
15. The sequence $1, -4, 9, -16, 25, \ldots$ Squares of the positive integers; with alternating signs
16. The sequence $1, -\frac{1}{4}, \frac{1}{9}, -\frac{1}{16}, \frac{1}{25}, \ldots$ Reciprocals of squares of the positive integers, with alternating signs
17. The sequence $0, 3, 8, 15, 24, \ldots$ Squares of the positive integers diminished by 1
18. The sequence $-3, -2, -1, 0, 1, \ldots$ Integers beginning with -3
19. The sequence $1, 5, 9, 13, 17, \ldots$ Every other odd positive integer
20. The sequence $2, 6, 10, 14, 18, \ldots$ Every other even positive integer
21. The sequence $1, 0, 1, 0, 1, \ldots$ Alternating 1's and 0's
22. The sequence $0, 1, 1, 2, 2, 3, 3, 4, \ldots$ Each positive integer repeated

Finding Limits

Which of the sequences $\{a_n\}$ in Exercises 23–84 converge, and which diverge? Find the limit of each convergent sequence.

23. $a_n = 2 + (0.1)^n$
24. $a_n = \dfrac{n + (-1)^n}{n}$
25. $a_n = \dfrac{1-2n}{1+2n}$
26. $a_n = \dfrac{2n+1}{1-3\sqrt{n}}$
27. $a_n = \dfrac{1-5n^4}{n^4+8n^3}$
28. $a_n = \dfrac{n+3}{n^2+5n+6}$
29. $a_n = \dfrac{n^2-2n+1}{n-1}$
30. $a_n = \dfrac{1-n^3}{70-4n^2}$
31. $a_n = 1 + (-1)^n$
32. $a_n = (-1)^n\left(1 - \frac{1}{n}\right)$
33. $a_n = \left(\dfrac{n+1}{2n}\right)\left(1 - \dfrac{1}{n}\right)$
34. $a_n = \left(2 - \frac{1}{2^n}\right)\left(3 + \frac{1}{2^n}\right)$
35. $a_n = \dfrac{(-1)^{n+1}}{2n-1}$
36. $a_n = \left(-\frac{1}{2}\right)^n$
37. $a_n = \sqrt{\dfrac{2n}{n+1}}$
38. $a_n = \dfrac{1}{(0.9)^n}$
39. $a_n = \sin\left(\dfrac{\pi}{2} + \dfrac{1}{n}\right)$
40. $a_n = n\pi\cos(n\pi)$
41. $a_n = \dfrac{\sin n}{n}$
42. $a_n = \dfrac{\sin^2 n}{2^n}$
43. $a_n = \dfrac{n}{2^n}$
44. $a_n = \dfrac{3^n}{n^3}$
45. $a_n = \dfrac{\ln(n+1)}{\sqrt{n}}$
46. $a_n = \dfrac{\ln n}{\ln 2n}$
47. $a_n = 8^{1/n}$
48. $a_n = (0.03)^{1/n}$
49. $a_n = \left(1 + \dfrac{7}{n}\right)^n$
50. $a_n = \left(1 - \dfrac{1}{n}\right)^n$
51. $a_n = \sqrt[n]{10n}$
52. $a_n = \sqrt[n]{n^2}$
53. $a_n = \left(\dfrac{3}{n}\right)^{1/n}$
54. $a_n = (n+4)^{1/(n+4)}$
55. $a_n = \dfrac{\ln n}{n^{1/n}}$
56. $a_n = \ln n - \ln(n+1)$
57. $a_n = \sqrt[n]{4^n n}$
58. $a_n = \sqrt[n]{3^{2n+1}}$
59. $a_n = \dfrac{n!}{n^n}$ (*Hint:* Compare with $1/n$.)

60. $a_n = \dfrac{(-4)^n}{n!}$

61. $a_n = \dfrac{n!}{10^{6n}}$

62. $a_n = \dfrac{n!}{2^n \cdot 3^n}$

63. $a_n = \left(\dfrac{1}{n}\right)^{1/(\ln n)}$

64. $a_n = \ln\left(1 + \dfrac{1}{n}\right)^n$

65. $a_n = \left(\dfrac{3n+1}{3n-1}\right)^n$

66. $a_n = \left(\dfrac{n}{n+1}\right)^n$

67. $a_n = \left(\dfrac{x^n}{2n+1}\right)^{1/n}, \quad x > 0$

68. $a_n = \left(1 - \dfrac{1}{n^2}\right)^n$

69. $a_n = \dfrac{3^n \cdot 6^n}{2^{-n} \cdot n!}$

70. $a_n = \dfrac{(10/11)^n}{(9/10)^n + (11/12)^n}$

71. $a_n = \tanh n$

72. $a_n = \sinh(\ln n)$

73. $a_n = \dfrac{n^2}{2n-1} \sin \dfrac{1}{n}$

74. $a_n = n\left(1 - \cos \dfrac{1}{n}\right)$

75. $a_n = \tan^{-1} n$

76. $a_n = \dfrac{1}{\sqrt{n}} \tan^{-1} n$

77. $a_n = \left(\dfrac{1}{3}\right)^n + \dfrac{1}{\sqrt{2^n}}$

78. $a_n = \sqrt[n]{n^2 + n}$

79. $a_n = \dfrac{(\ln n)^{200}}{n}$

80. $a_n = \dfrac{(\ln n)^5}{\sqrt{n}}$

81. $a_n = n - \sqrt{n^2 - n}$

82. $a_n = \dfrac{1}{\sqrt{n^2 - 1} - \sqrt{n^2 + n}}$

83. $a_n = \dfrac{1}{n}\displaystyle\int_1^n \dfrac{1}{x}\,dx$

84. $a_n = \displaystyle\int_1^n \dfrac{1}{x^p}\,dx, \quad p > 1$

Theory and Examples

85. The first term of a sequence is $x_1 = 1$. Each succeeding term is the sum of all those that come before it:

$$x_{n+1} = x_1 + x_2 + \cdots + x_n.$$

Write out enough early terms of the sequence to deduce a general formula for x_n that holds for $n \geq 2$.

86. A sequence of rational numbers is described as follows:

$$\frac{1}{1}, \frac{3}{2}, \frac{7}{5}, \frac{17}{12}, \ldots, \frac{a}{b}, \frac{a+2b}{a+b}, \ldots.$$

Here the numerators form one sequence, the denominators form a second sequence, and their ratios form a third sequence. Let x_n and y_n be, respectively, the numerator and the denominator of the nth fraction $r_n = x_n/y_n$.

a. Verify that $x_1^2 - 2y_1^2 = -1, x_2^2 - 2y_2^2 = +1$ and, more generally, that if $a^2 - 2b^2 = -1$ or $+1$, then

$$(a + 2b)^2 - 2(a + b)^2 = +1 \quad \text{or} \quad -1,$$

respectively.

b. The fractions $r_n = x_n/y_n$ approach a limit as n increases. What is that limit? (*Hint:* Use part (a) to show that $r_n^2 - 2 = \pm(1/y_n)^2$ and that y_n is not less than n.)

87. **Newton's method** The following sequences come from the recursion formula for Newton's method,

$$x_{n+1} = x_n - \frac{f(x_n)}{f'(x_n)}.$$

Do the sequences converge? If so, to what value? In each case, begin by identifying the function f that generates the sequence.

a. $x_0 = 1, \quad x_{n+1} = x_n - \dfrac{x_n^2 - 2}{2x_n} = \dfrac{x_n}{2} + \dfrac{1}{x_n}$

b. $x_0 = 1, \quad x_{n+1} = x_n - \dfrac{\tan x_n - 1}{\sec^2 x_n}$

c. $x_0 = 1, \quad x_{n+1} = x_n - 1$

88. **a.** Suppose that $f(x)$ is differentiable for all x in $[0, 1]$ and that $f(0) = 0$. Define the sequence $\{a_n\}$ by the rule $a_n = nf(1/n)$. Show that $\lim_{n\to\infty} a_n = f'(0)$.

Use the result in part (a) to find the limits of the following sequences $\{a_n\}$.

b. $a_n = n \tan^{-1} \dfrac{1}{n}$

c. $a_n = n(e^{1/n} - 1)$

d. $a_n = n \ln\left(1 + \dfrac{2}{n}\right)$

89. **Pythagorean triples** A triple of positive integers a, b, and c is called a **Pythagorean triple** if $a^2 + b^2 = c^2$. Let a be an odd positive integer and let

$$b = \left\lfloor \frac{a^2}{2} \right\rfloor \quad \text{and} \quad c = \left\lceil \frac{a^2}{2} \right\rceil$$

be, respectively, the integer floor and ceiling for $a^2/2$.

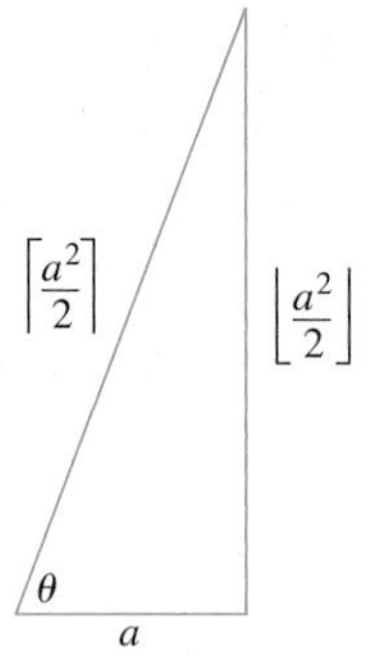

a. Show that $a^2 + b^2 = c^2$. (*Hint:* Let $a = 2n + 1$ and express b and c in terms of n.)

b. By direct calculation, or by appealing to the figure here, find

$$\lim_{a\to\infty} \frac{\left\lfloor \frac{a^2}{2} \right\rfloor}{\left\lceil \frac{a^2}{2} \right\rceil}.$$

90. The *n*th root of *n*!

a. Show that $\lim_{n\to\infty}(2n\pi)^{1/(2n)} = 1$ and hence, using Stirling's approximation (Chapter 8, Additional Exercise 50a), that

$$\sqrt[n]{n!} \approx \frac{n}{e} \quad \text{for large values of } n.$$

T **b.** Test the approximation in part (a) for $n = 40, 50, 60, \ldots$, as far as your calculator will allow.

91. a. Assuming that $\lim_{n\to\infty}(1/n^c) = 0$ if c is any positive constant, show that

$$\lim_{n\to\infty} \frac{\ln n}{n^c} = 0$$

if c is any positive constant.

b. Prove that $\lim_{n\to\infty}(1/n^c) = 0$ if c is any positive constant. (*Hint:* If $\epsilon = 0.001$ and $c = 0.04$, how large should N be to ensure that $|1/n^c - 0| < \epsilon$ if $n > N$?)

92. The zipper theorem Prove the "zipper theorem" for sequences: If $\{a_n\}$ and $\{b_n\}$ both converge to L, then the sequence

$$a_1, b_1, a_2, b_2, \ldots, a_n, b_n, \ldots$$

converges to L.

93. Prove that $\lim_{n\to\infty}\sqrt[n]{n} = 1$.

94. Prove that $\lim_{n\to\infty} x^{1/n} = 1, (x > 0)$.

95. Prove Theorem 2.

96. Prove Theorem 3.

In Exercises 97–100, determine if the sequence is nondecreasing and if it is bounded from above.

97. $a_n = \dfrac{3n+1}{n+1}$

98. $a_n = \dfrac{(2n+3)!}{(n+1)!}$

99. $a_n = \dfrac{2^n 3^n}{n!}$

100. $a_n = 2 - \dfrac{2}{n} - \dfrac{1}{2^n}$

Which of the sequences in Exercises 101–106 converge, and which diverge? Give reasons for your answers.

101. $a_n = 1 - \dfrac{1}{n}$

102. $a_n = n - \dfrac{1}{n}$

103. $a_n = \dfrac{2^n - 1}{2^n}$

104. $a_n = \dfrac{2^n - 1}{3^n}$

105. $a_n = ((-1)^n + 1)\left(\dfrac{n+1}{n}\right)$

106. The first term of a sequence is $x_1 = \cos(1)$. The next terms are $x_2 = x_1$ or $\cos(2)$, whichever is larger; and $x_3 = x_2$ or $\cos(3)$, whichever is larger (farther to the right). In general,

$$x_{n+1} = \max\{x_n, \cos(n+1)\}.$$

107. Nonincreasing sequences A sequence of numbers $\{a_n\}$ in which $a_n \ge a_{n+1}$ for every n is called a **nonincreasing sequence**. A sequence $\{a_n\}$ is **bounded from below** if there is a number M with $M \le a_n$ for every n. Such a number M is called a **lower bound** for the sequence. Deduce from Theorem 6 that a nonincreasing sequence that is bounded from below converges and that a nonincreasing sequence that is not bounded from below diverges.

(*Continuation of Exercise 107.*) Using the conclusion of Exercise 107, determine which of the sequences in Exercises 108–112 converge and which diverge.

108. $a_n = \dfrac{n+1}{n}$

109. $a_n = \dfrac{1 + \sqrt{2n}}{\sqrt{n}}$

110. $a_n = \dfrac{1 - 4^n}{2^n}$

111. $a_n = \dfrac{4^{n+1} + 3^n}{4^n}$

112. $a_1 = 1, \quad a_{n+1} = 2a_n - 3$

113. The sequence $\{n/(n+1)\}$ has a least upper bound of 1 Show that if M is a number less than 1, then the terms of $\{n/(n+1)\}$ eventually exceed M. That is, if $M < 1$ there is an integer N such that $n/(n+1) > M$ whenever $n > N$. Since $n/(n+1) < 1$ for every n, this proves that 1 is a least upper bound for $\{n/(n+1)\}$.

114. Uniqueness of least upper bounds Show that if M_1 and M_2 are least upper bounds for the sequence $\{a_n\}$, then $M_1 = M_2$. That is, a sequence cannot have two different least upper bounds.

115. Is it true that a sequence $\{a_n\}$ of positive numbers must converge if it is bounded from above? Give reasons for your answer.

116. Prove that if $\{a_n\}$ is a convergent sequence, then to every positive number ϵ there corresponds an integer N such that for all m and n,

$$m > N \quad \text{and} \quad n > N \quad \Rightarrow \quad |a_m - a_n| < \epsilon.$$

117. Uniqueness of limits Prove that limits of sequences are unique. That is, show that if L_1 and L_2 are numbers such that $a_n \to L_1$ and $a_n \to L_2$, then $L_1 = L_2$.

118. Limits and subsequences If the terms of one sequence appear in another sequence in their given order, we call the first sequence a **subsequence** of the second. Prove that if two subsequences of a sequence $\{a_n\}$ have different limits $L_1 \ne L_2$, then $\{a_n\}$ diverges.

119. For a sequence $\{a_n\}$ the terms of even index are denoted by a_{2k} and the terms of odd index by a_{2k+1}. Prove that if $a_{2k} \to L$ and $a_{2k+1} \to L$, then $a_n \to L$.

120. Prove that a sequence $\{a_n\}$ converges to 0 if and only if the sequence of absolute values $\{|a_n|\}$ converges to 0.

T Calculator Explorations of Limits

In Exercises 121–124, experiment with a calculator to find a value of N that will make the inequality hold for all $n > N$. Assuming that the inequality is the one from the formal definition of the limit of a sequence, what sequence is being considered in each case and what is its limit?

121. $|\sqrt[n]{0.5} - 1| < 10^{-3}$ **122.** $|\sqrt[n]{n} - 1| < 10^{-3}$

123. $(0.9)^n < 10^{-3}$ **124.** $2^n/n! < 10^{-7}$

125. Sequences generated by Newton's method Newton's method, applied to a differentiable function $f(x)$, begins with a starting value x_0 and constructs from it a sequence of numbers $\{x_n\}$ that under favorable circumstances converges to a zero of f. The recursion formula for the sequence is

$$x_{n+1} = x_n - \frac{f(x_n)}{f'(x_n)}.$$

a. Show that the recursion formula for $f(x) = x^2 - a, a > 0$, can be written as $x_{n+1} = (x_n + a/x_n)/2$.

b. Starting with $x_0 = 1$ and $a = 3$, calculate successive terms of the sequence until the display begins to repeat. What number is being approximated? Explain.

126. (*Continuation of Exercise 125.*) Repeat part (b) of Exercise 125 with $a = 2$ in place of $a = 3$.

127. A recursive definition of $\pi/2$ If you start with $x_1 = 1$ and define the subsequent terms of $\{x_n\}$ by the rule $x_n = x_{n-1} + \cos x_{n-1}$, you generate a sequence that converges rapidly to $\pi/2$. **a.** Try it. **b.** Use the accompanying figure to explain why the convergence is so rapid.

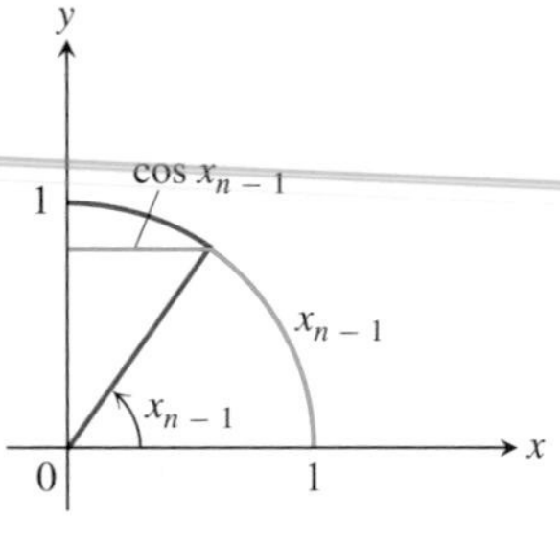

128. According to a front-page article in the December 15, 1992, issue of the *Wall Street Journal*, Ford Motor Company used about $7\frac{1}{4}$ hours of labor to produce stampings for the average vehicle, down from an estimated 15 hours in 1980. The Japanese needed only about $3\frac{1}{2}$ hours.

Ford's improvement since 1980 represents an average decrease of 6% per year. If that rate continues, then n years from 1992 Ford will use about

$$S_n = 7.25(0.94)^n$$

hours of labor to produce stampings for the average vehicle. Assuming that the Japanese continue to spend $3\frac{1}{2}$ hours per vehicle, how many more years will it take Ford to catch up? Find out two ways:

a. Find the first term of the sequence $\{S_n\}$ that is less than or equal to 3.5.

T **b.** Graph $f(x) = 7.25(0.94)^x$ and use Trace to find where the graph crosses the line $y = 3.5$.

COMPUTER EXPLORATIONS

Use a CAS to perform the following steps for the sequences in Exercises 129–140.

a. Calculate and then plot the first 25 terms of the sequence. Does the sequence appear to be bounded from above or below? Does it appear to converge or diverge? If it does converge, what is the limit L?

b. If the sequence converges, find an integer N such that $|a_n - L| \le 0.01$ for $n \ge N$. How far in the sequence do you have to get for the terms to lie within 0.0001 of L?

129. $a_n = \sqrt[n]{n}$ **130.** $a_n = \left(1 + \frac{0.5}{n}\right)^n$

131. $a_1 = 1, \quad a_{n+1} = a_n + \frac{1}{5^n}$

132. $a_1 = 1, \quad a_{n+1} = a_n + (-2)^n$

133. $a_n = \sin n$ **134.** $a_n = n \sin \frac{1}{n}$

135. $a_n = \frac{\sin n}{n}$ **136.** $a_n = \frac{\ln n}{n}$

137. $a_n = (0.9999)^n$ **138.** $a_n = 123456^{1/n}$

139. $a_n = \frac{8^n}{n!}$ **140.** $a_n = \frac{n^{41}}{19^n}$

141. Compound interest, deposits, and withdrawals If you invest an amount of money A_0 at a fixed annual interest rate r compounded m times per year, and if the constant amount b is added to the account at the end of each compounding period (or taken from the account if $b < 0$), then the amount you have after $n + 1$ compounding periods is

$$A_{n+1} = \left(1 + \frac{r}{m}\right)A_n + b. \qquad (1)$$

a. If $A_0 = 1000, r = 0.02015, m = 12$, and $b = 50$, calculate and plot the first 100 points (n, A_n). How much money is in your account at the end of 5 years? Does $\{A_n\}$ converge? Is $\{A_n\}$ bounded?

b. Repeat part (a) with $A_0 = 5000, r = 0.0589, m = 12$, and $b = -50$.

c. If you invest 5000 dollars in a certificate of deposit (CD) that pays 4.5% annually, compounded quarterly, and you make no further investments in the CD, approximately how many years will it take before you have 20,000 dollars? What if the CD earns 6.25%?

d. It can be shown that for any $k \geq 0$, the sequence defined recursively by Equation (1) satisfies the relation

$$A_k = \left(1 + \frac{r}{m}\right)^k \left(A_0 + \frac{mb}{r}\right) - \frac{mb}{r}. \qquad (2)$$

For the values of the constants A_0, r, m, and b given in part (a), validate this assertion by comparing the values of the first 50 terms of both sequences. Then show by direct substitution that the terms in Equation (2) satisfy the recursion formula in Equation (1).

142. Logistic difference equation The recursive relation

$$a_{n+1} = ra_n(1 - a_n)$$

is called the *logistic difference equation*, and when the initial value a_0 is given the equation defines the *logistic sequence* $\{a_n\}$. Throughout this exercise we choose a_0 in the interval $0 < a_0 < 1$, say $a_0 = 0.3$.

a. Choose $r = 3/4$. Calculate and plot the points (n, a_n) for the first 100 terms in the sequence. Does it appear to converge? What do you guess is the limit? Does the limit seem to depend on your choice of a_0?

b. Choose several values of r in the interval $1 < r < 3$ and repeat the procedures in part (a). Be sure to choose some points near the endpoints of the interval. Describe the behavior of the sequences you observe in your plots.

c. Now examine the behavior of the sequence for values of r near the endpoints of the interval $3 < r < 3.45$. The transition value $r = 3$ is called a **bifurcation value** and the new behavior of the sequence in the interval is called an **attracting 2-cycle**. Explain why this reasonably describes the behavior.

d. Next explore the behavior for r values near the endpoints of each of the intervals $3.45 < r < 3.54$ and $3.54 < r < 3.55$. Plot the first 200 terms of the sequences. Describe in your own words the behavior observed in your plots for each interval. Among how many values does the sequence appear to oscillate for each interval? The values $r = 3.45$ and $r = 3.54$ (rounded to two decimal places) are also called bifurcation values because the behavior of the sequence changes as r crosses over those values.

e. The situation gets even more interesting. There is actually an increasing sequence of bifurcation values $3 < 3.45 < 3.54 < \cdots < c_n < c_{n+1} \cdots$ such that for $c_n < r < c_{n+1}$ the logistic sequence $\{a_n\}$ eventually oscillates steadily among 2^n values, called an **attracting 2^n-cycle**. Moreover, the bifurcation sequence $\{c_n\}$ is bounded above by 3.57 (so it converges). If you choose a value of $r < 3.57$ you will observe a 2^n-cycle of some sort. Choose $r = 3.5695$ and plot 300 points.

f. Let us see what happens when $r > 3.57$. Choose $r = 3.65$ and calculate and plot the first 300 terms of $\{a_n\}$. Observe how the terms wander around in an unpredictable, chaotic fashion. You cannot predict the value of a_{n+1} from previous values of the sequence.

g. For $r = 3.65$ choose two starting values of a_0 that are close together, say, $a_0 = 0.3$ and $a_0 = 0.301$. Calculate and plot the first 300 values of the sequences determined by each starting value. Compare the behaviors observed in your plots. How far out do you go before the corresponding terms of your two sequences appear to depart from each other? Repeat the exploration for $r = 3.75$. Can you see how the plots look different depending on your choice of a_0? We say that the logistic sequence is *sensitive to the initial condition a_0*.

11.2 Infinite Series

An *infinite series* is the sum of an infinite sequence of numbers

$$a_1 + a_2 + a_3 + \cdots + a_n + \cdots$$

The goal of this section is to understand the meaning of such an infinite sum and to develop methods to calculate it. Since there are infinitely many terms to add in an infinite series, we cannot just keep adding to see what comes out. Instead we look at what we get by summing the first n terms of the sequence and stopping. The sum of the first n terms

$$s_n = a_1 + a_2 + a_3 + \cdots + a_n$$

is an ordinary finite sum and can be calculated by normal addition. It is called the *nth partial sum*. As n gets larger, we expect the partial sums to get closer and closer to a limiting value in the same sense that the terms of a sequence approach a limit, as discussed in Section 11.1.

For example, to assign meaning to an expression like

$$1 + \frac{1}{2} + \frac{1}{4} + \frac{1}{8} + \frac{1}{16} + \cdots$$

we add the terms one at a time from the beginning and look for a pattern in how these partial sums grow.

Partial sum		**Suggestive expression for partial sum**	**Value**
First:	$s_1 = 1$	$2 - 1$	1
Second:	$s_2 = 1 + \frac{1}{2}$	$2 - \frac{1}{2}$	$\frac{3}{2}$
Third:	$s_3 = 1 + \frac{1}{2} + \frac{1}{4}$	$2 - \frac{1}{4}$	$\frac{7}{4}$
$\vdots$	$\vdots$	$\vdots$	$\vdots$
nth:	$s_n = 1 + \frac{1}{2} + \frac{1}{4} + \cdots + \frac{1}{2^{n-1}}$	$2 - \frac{1}{2^{n-1}}$	$\frac{2^n - 1}{2^{n-1}}$

Indeed there is a pattern. The partial sums form a sequence whose nth term is

$$s_n = 2 - \frac{1}{2^{n-1}}.$$

This sequence of partial sums converges to 2 because $\lim_{n\to\infty}(1/2^n) = 0$. We say

$$\text{“the sum of the infinite series } 1 + \frac{1}{2} + \frac{1}{4} + \cdots + \frac{1}{2^{n-1}} + \cdots \text{ is 2.”}$$

Is the sum of any finite number of terms in this series equal to 2? No. Can we actually add an infinite number of terms one by one? No. But we can still define their sum by defining it to be the limit of the sequence of partial sums as $n \to \infty$, in this case 2 (Figure 11.5). Our knowledge of sequences and limits enables us to break away from the confines of finite sums.

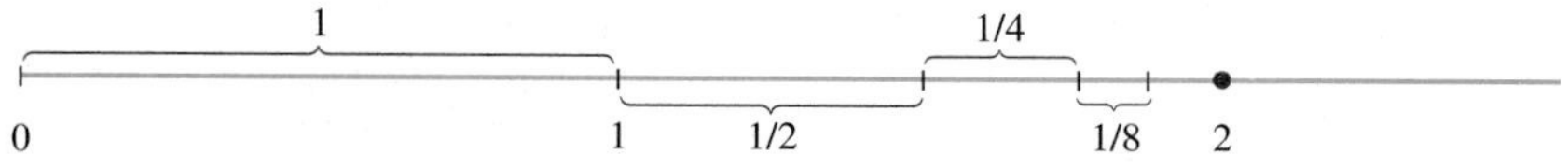

FIGURE 11.5 As the lengths 1, $\frac{1}{2}$, $\frac{1}{4}$, $\frac{1}{8}$, . . . are added one by one, the sum approaches 2.

Historical Biography

Blaise Pascal
(1623–1662)

DEFINITIONS Infinite Series, *n*th Term, Partial Sum, Converges, Sum

Given a sequence of numbers $\{a_n\}$, an expression of the form

$$a_1 + a_2 + a_3 + \cdots + a_n + \cdots$$

is an **infinite series**. The number a_n is the ***n*th term** of the series. The sequence $\{s_n\}$ defined by

$$\begin{aligned} s_1 &= a_1 \\ s_2 &= a_1 + a_2 \\ &\vdots \\ s_n &= a_1 + a_2 + \cdots + a_n = \sum_{k=1}^{n} a_k \\ &\vdots \end{aligned}$$

is the **sequence of partial sums** of the series, the number s_n being the ***n*th partial sum**. If the sequence of partial sums converges to a limit L, we say that the series **converges** and that its **sum** is L. In this case, we also write

$$a_1 + a_2 + \cdots + a_n + \cdots = \sum_{n=1}^{\infty} a_n = L.$$

If the sequence of partial sums of the series does not converge, we say that the series **diverges**.

When we begin to study a given series $a_1 + a_2 + \cdots + a_n + \cdots$, we might not know whether it converges or diverges. In either case, it is convenient to use sigma notation to write the series as

$$\sum_{n=1}^{\infty} a_n, \qquad \sum_{k=1}^{\infty} a_k, \qquad \text{or} \qquad \sum a_n$$

A useful shorthand when summation from 1 to ∞ is understood

Geometric Series

Geometric series are series of the form

$$a + ar + ar^2 + \cdots + ar^{n-1} + \cdots = \sum_{n=1}^{\infty} ar^{n-1}$$

in which a and r are fixed real numbers and $a \neq 0$. The series can also be written as $\sum_{n=0}^{\infty} ar^n$. The **ratio** r can be positive, as in

$$1 + \frac{1}{2} + \frac{1}{4} + \cdots + \left(\frac{1}{2}\right)^{n-1} + \cdots,$$

or negative, as in

$$1 - \frac{1}{3} + \frac{1}{9} - \cdots + \left(-\frac{1}{3}\right)^{n-1} + \cdots.$$

If $r = 1$, the nth partial sum of the geometric series is

$$s_n = a + a(1) + a(1)^2 + \cdots + a(1)^{n-1} = na,$$

and the series diverges because $\lim_{n\to\infty} s_n = \pm\infty$, depending on the sign of a. If $r = -1$, the series diverges because the nth partial sums alternate between a and 0. If $|r| \neq 1$, we can determine the convergence or divergence of the series in the following way:

$$s_n = a + ar + ar^2 + \cdots + ar^{n-1}$$

$$rs_n = ar + ar^2 + \cdots + ar^{n-1} + ar^n \qquad \text{Multiply } s_n \text{ by } r.$$

$$s_n - rs_n = a - ar^n \qquad \text{Subtract } rs_n \text{ from } s_n\text{. Most of the terms on the right cancel.}$$

$$s_n(1 - r) = a(1 - r^n) \qquad \text{Factor.}$$

$$s_n = \frac{a(1 - r^n)}{1 - r}, \quad (r \neq 1). \qquad \text{We can solve for } s_n \text{ if } r \neq 1.$$

If $|r| < 1$, then $r^n \to 0$ as $n \to \infty$ (as in Section 11.1) and $s_n \to a/(1 - r)$. If $|r| > 1$, then $|r^n| \to \infty$ and the series diverges.

> If $|r| < 1$, the geometric series $a + ar + ar^2 + \cdots + ar^{n-1} + \cdots$ converges to $a/(1 - r)$:
>
> $$\sum_{n=1}^{\infty} ar^{n-1} = \frac{a}{1 - r}, \qquad |r| < 1.$$
>
> If $|r| \geq 1$, the series diverges.

We have determined when a geometric series converges or diverges, and to what value. Often we can determine that a series converges without knowing the value to which it converges, as we will see in the next several sections. The formula $a/(1 - r)$ for the sum of a geometric series applies *only* when the summation index begins with $n = 1$ in the expression $\sum_{n=1}^{\infty} ar^{n-1}$ (or with the index $n = 0$ if we write the series as $\sum_{n=0}^{\infty} ar^n$).

EXAMPLE 1 Index Starts with $n = 1$

The geometric series with $a = 1/9$ and $r = 1/3$ is

$$\frac{1}{9} + \frac{1}{27} + \frac{1}{81} + \cdots = \sum_{n=1}^{\infty} \frac{1}{9}\left(\frac{1}{3}\right)^{n-1} = \frac{1/9}{1 - (1/3)} = \frac{1}{6}. \qquad \blacksquare$$

EXAMPLE 2 Index Starts with $n = 0$

The series

$$\sum_{n=0}^{\infty} \frac{(-1)^n 5}{4^n} = 5 - \frac{5}{4} + \frac{5}{16} - \frac{5}{64} + \cdots$$

is a geometric series with $a = 5$ and $r = -1/4$. It converges to

$$\frac{a}{1 - r} = \frac{5}{1 + (1/4)} = 4. \qquad \blacksquare$$

EXAMPLE 3 A Bouncing Ball

You drop a ball from a meters above a flat surface. Each time the ball hits the surface after falling a distance h, it rebounds a distance rh, where r is positive but less than 1. Find the total distance the ball travels up and down (Figure 11.6).

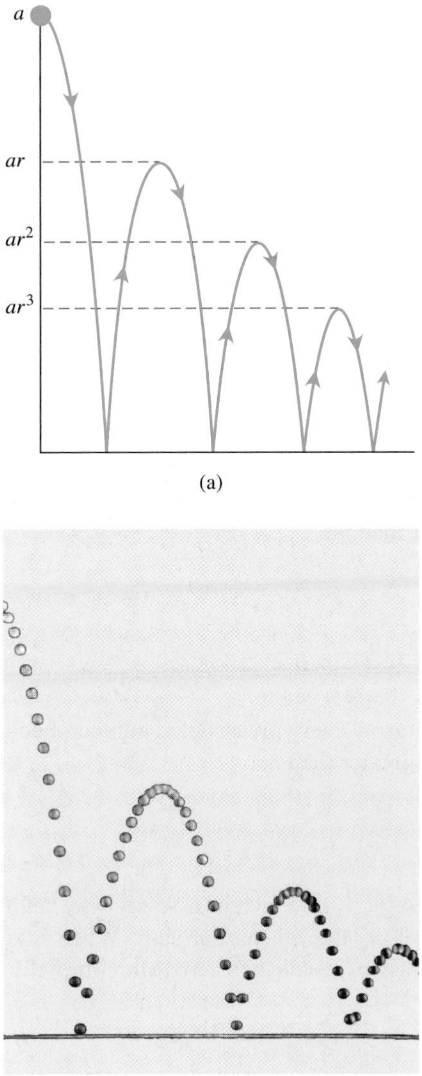

(a)

(b)

FIGURE 11.6 (a) Example 3 shows how to use a geometric series to calculate the total vertical distance traveled by a bouncing ball if the height of each rebound is reduced by the factor r. (b) A stroboscopic photo of a bouncing ball.

Solution The total distance is

$$s = a + \underbrace{2ar + 2ar^2 + 2ar^3 + \cdots}_{\text{This sum is } 2ar/(1-r).} = a + \frac{2ar}{1-r} = a\frac{1+r}{1-r}.$$

If $a = 6$ m and $r = 2/3$, for instance, the distance is

$$s = 6\frac{1 + (2/3)}{1 - (2/3)} = 6\left(\frac{5/3}{1/3}\right) = 30 \text{ m}.$$

■

EXAMPLE 4 Repeating Decimals

Express the repeating decimal 5.232323 . . . as the ratio of two integers.

Solution

$$\begin{aligned} 5.232323\ldots &= 5 + \frac{23}{100} + \frac{23}{(100)^2} + \frac{23}{(100)^3} + \cdots \\ &= 5 + \frac{23}{100}\underbrace{\left(1 + \frac{1}{100} + \left(\frac{1}{100}\right)^2 + \cdots\right)}_{1/(1-0.01)} \qquad a = 1,\ r = 1/100 \\ &= 5 + \frac{23}{100}\left(\frac{1}{0.99}\right) = 5 + \frac{23}{99} = \frac{518}{99} \end{aligned}$$

■

Unfortunately, formulas like the one for the sum of a convergent geometric series are rare and we usually have to settle for an estimate of a series' sum (more about this later). The next example, however, is another case in which we can find the sum exactly.

EXAMPLE 5 A Nongeometric but Telescoping Series

Find the sum of the series $\sum_{n=1}^{\infty} \frac{1}{n(n+1)}$.

Solution We look for a pattern in the sequence of partial sums that might lead to a formula for s_k. The key observation is the partial fraction decomposition

$$\frac{1}{n(n+1)} = \frac{1}{n} - \frac{1}{n+1},$$

so

$$\sum_{n=1}^{k} \frac{1}{n(n+1)} = \sum_{n=1}^{k} \left(\frac{1}{n} - \frac{1}{n+1}\right)$$

and

$$s_k = \left(\frac{1}{1} - \frac{1}{2}\right) + \left(\frac{1}{2} - \frac{1}{3}\right) + \left(\frac{1}{3} - \frac{1}{4}\right) + \cdots + \left(\frac{1}{k} - \frac{1}{k+1}\right).$$

Removing parentheses and canceling adjacent terms of opposite sign collapses the sum to

$$s_k = 1 - \frac{1}{k+1}.$$

We now see that $s_k \to 1$ as $k \to \infty$. The series converges, and its sum is 1:

$$\sum_{n=1}^{\infty} \frac{1}{n(n+1)} = 1.$$

■

Divergent Series

One reason that a series may fail to converge is that its terms don't become small.

EXAMPLE 6 Partial Sums Outgrow Any Number

(a) The series

$$\sum_{n=1}^{\infty} n^2 = 1 + 4 + 9 + \cdots + n^2 + \cdots$$

diverges because the partial sums grow beyond every number L. After $n = 1$, the partial sum $s_n = 1 + 4 + 9 + \cdots + n^2$ is greater than n^2.

(b) The series

$$\sum_{n=1}^{\infty} \frac{n+1}{n} = \frac{2}{1} + \frac{3}{2} + \frac{4}{3} + \cdots + \frac{n+1}{n} + \cdots$$

diverges because the partial sums eventually outgrow every preassigned number. Each term is greater than 1, so the sum of n terms is greater than n. ■

The nth-Term Test for Divergence

Observe that $\lim_{n\to\infty} a_n$ must equal zero if the series $\sum_{n=1}^{\infty} a_n$ converges. To see why, let S represent the series' sum and $s_n = a_1 + a_2 + \cdots + a_n$ the nth partial sum. When n is large, both s_n and s_{n-1} are close to S, so their difference, a_n, is close to zero. More formally,

$$a_n = s_n - s_{n-1} \quad \to \quad S - S = 0. \qquad \text{Difference Rule for sequences}$$

This establishes the following theorem.

Caution
Theorem 7 *does not say* that $\sum_{n=1}^{\infty} a_n$ converges if $a_n \to 0$. It is possible for a series to diverge when $a_n \to 0$.

THEOREM 7

If $\sum_{n=1}^{\infty} a_n$ converges, then $a_n \to 0$.

Theorem 7 leads to a test for detecting the kind of divergence that occurred in Example 6.

The nth-Term Test for Divergence

$\sum_{n=1}^{\infty} a_n$ diverges if $\lim_{n\to\infty} a_n$ fails to exist or is different from zero.

EXAMPLE 7 Applying the nth-Term Test

(a) $\sum_{n=1}^{\infty} n^2$ diverges because $n^2 \to \infty$

(b) $\sum_{n=1}^{\infty} \frac{n+1}{n}$ diverges because $\frac{n+1}{n} \to 1$

(c) $\sum_{n=1}^{\infty} (-1)^{n+1}$ diverges because $\lim_{n\to\infty}(-1)^{n+1}$ does not exist

(d) $\sum_{n=1}^{\infty} \frac{-n}{2n+5}$ diverges because $\lim_{n\to\infty} \frac{-n}{2n+5} = -\frac{1}{2} \neq 0.$ ■

EXAMPLE 8 $a_n \to 0$ but the Series Diverges

The series

$$1 + \underbrace{\frac{1}{2} + \frac{1}{2}}_{2 \text{ terms}} + \underbrace{\frac{1}{4} + \frac{1}{4} + \frac{1}{4} + \frac{1}{4}}_{4 \text{ terms}} + \cdots + \underbrace{\frac{1}{2^n} + \frac{1}{2^n} + \cdots + \frac{1}{2^n}}_{2^n \text{ terms}} + \cdots$$

diverges because the terms are grouped into clusters that add to 1, so the partial sums increase without bound. However, the terms of the series form a sequence that converges to 0. Example 1 of Section 11.3 shows that the harmonic series also behaves in this manner. ■

Combining Series

Whenever we have two convergent series, we can add them term by term, subtract them term by term, or multiply them by constants to make new convergent series.

THEOREM 8

If $\sum a_n = A$ and $\sum b_n = B$ are convergent series, then

1. *Sum Rule:* $\sum(a_n + b_n) = \sum a_n + \sum b_n = A + B$
2. *Difference Rule:* $\sum(a_n - b_n) = \sum a_n - \sum b_n = A - B$
3. *Constant Multiple Rule:* $\sum k a_n = k \sum a_n = kA$ (Any number k).

Proof The three rules for series follow from the analogous rules for sequences in Theorem 1, Section 11.1. To prove the Sum Rule for series, let

$$A_n = a_1 + a_2 + \cdots + a_n, \quad B_n = b_1 + b_2 + \cdots + b_n.$$

Then the partial sums of $\sum(a_n + b_n)$ are

$$\begin{aligned} s_n &= (a_1 + b_1) + (a_2 + b_2) + \cdots + (a_n + b_n) \\ &= (a_1 + \cdots + a_n) + (b_1 + \cdots + b_n) \\ &= A_n + B_n. \end{aligned}$$

Since $A_n \to A$ and $B_n \to B$, we have $s_n \to A + B$ by the Sum Rule for sequences. The proof of the Difference Rule is similar.

To prove the Constant Multiple Rule for series, observe that the partial sums of $\sum ka_n$ form the sequence

$$s_n = ka_1 + ka_2 + \cdots + ka_n = k(a_1 + a_2 + \cdots + a_n) = kA_n,$$

which converges to kA by the Constant Multiple Rule for sequences. ■

As corollaries of Theorem 8, we have

1. Every nonzero constant multiple of a divergent series diverges.
2. If $\sum a_n$ converges and $\sum b_n$ diverges, then $\sum(a_n + b_n)$ and $\sum(a_n - b_n)$ both diverge.

We omit the proofs.

CAUTION Remember that $\sum(a_n + b_n)$ can converge when $\sum a_n$ and $\sum b_n$ both diverge. For example, $\sum a_n = 1 + 1 + 1 + \cdots$ and $\sum b_n = (-1) + (-1) + (-1) + \cdots$ diverge, whereas $\sum(a_n + b_n) = 0 + 0 + 0 + \cdots$ converges to 0.

EXAMPLE 9 Find the sums of the following series.

(a)
$$\begin{aligned}\sum_{n=1}^{\infty} \frac{3^{n-1} - 1}{6^{n-1}} &= \sum_{n=1}^{\infty} \left(\frac{1}{2^{n-1}} - \frac{1}{6^{n-1}}\right) \\ &= \sum_{n=1}^{\infty} \frac{1}{2^{n-1}} - \sum_{n=1}^{\infty} \frac{1}{6^{n-1}} && \text{Difference Rule} \\ &= \frac{1}{1 - (1/2)} - \frac{1}{1 - (1/6)} && \text{Geometric series with } a = 1 \text{ and } r = 1/2, 1/6 \\ &= 2 - \frac{6}{5} \\ &= \frac{4}{5}\end{aligned}$$

(b)
$$\begin{aligned}\sum_{n=0}^{\infty} \frac{4}{2^n} &= 4\sum_{n=0}^{\infty} \frac{1}{2^n} && \text{Constant Multiple Rule} \\ &= 4\left(\frac{1}{1 - (1/2)}\right) && \text{Geometric series with } a = 1, r = 1/2 \\ &= 8\end{aligned}$$
■

Adding or Deleting Terms

We can add a finite number of terms to a series or delete a finite number of terms without altering the series' convergence or divergence, although in the case of convergence this will usually change the sum. If $\sum_{n=1}^{\infty} a_n$ converges, then $\sum_{n=k}^{\infty} a_n$ converges for any $k > 1$ and

$$\sum_{n=1}^{\infty} a_n = a_1 + a_2 + \cdots + a_{k-1} + \sum_{n=k}^{\infty} a_n.$$

Conversely, if $\sum_{n=k}^{\infty} a_n$ converges for any $k > 1$, then $\sum_{n=1}^{\infty} a_n$ converges. Thus,

$$\sum_{n=1}^{\infty} \frac{1}{5^n} = \frac{1}{5} + \frac{1}{25} + \frac{1}{125} + \sum_{n=4}^{\infty} \frac{1}{5^n}$$

and

$$\sum_{n=4}^{\infty} \frac{1}{5^n} = \left(\sum_{n=1}^{\infty} \frac{1}{5^n}\right) - \frac{1}{5} - \frac{1}{25} - \frac{1}{125}.$$

HISTORICAL BIOGRAPHY

Richard Dedekind
(1831–1916)

Reindexing

As long as we preserve the order of its terms, we can reindex any series without altering its convergence. To raise the starting value of the index h units, replace the n in the formula for a_n by $n - h$:

$$\sum_{n=1}^{\infty} a_n = \sum_{n=1+h}^{\infty} a_{n-h} = a_1 + a_2 + a_3 + \cdots.$$

To lower the starting value of the index h units, replace the n in the formula for a_n by $n + h$:

$$\sum_{n=1}^{\infty} a_n = \sum_{n=1-h}^{\infty} a_{n+h} = a_1 + a_2 + a_3 + \cdots.$$

It works like a horizontal shift. We saw this in starting a geometric series with the index $n = 0$ instead of the index $n = 1$, but we can use any other starting index value as well. We usually give preference to indexings that lead to simple expressions.

EXAMPLE 10 Reindexing a Geometric Series

We can write the geometric series

$$\sum_{n=1}^{\infty} \frac{1}{2^{n-1}} = 1 + \frac{1}{2} + \frac{1}{4} + \cdots$$

as

$$\sum_{n=0}^{\infty} \frac{1}{2^n}, \qquad \sum_{n=5}^{\infty} \frac{1}{2^{n-5}}, \qquad \text{or even} \qquad \sum_{n=-4}^{\infty} \frac{1}{2^{n+4}}.$$

The partial sums remain the same no matter what indexing we choose. ■

EXERCISES 11.2

Finding *n*th Partial Sums

In Exercises 1–6, find a formula for the nth partial sum of each series and use it to find the series' sum if the series converges.

1. $2 + \frac{2}{3} + \frac{2}{9} + \frac{2}{27} + \cdots + \frac{2}{3^{n-1}} + \cdots$

2. $\frac{9}{100} + \frac{9}{100^2} + \frac{9}{100^3} + \cdots + \frac{9}{100^n} + \cdots$

3. $1 - \frac{1}{2} + \frac{1}{4} - \frac{1}{8} + \cdots + (-1)^{n-1}\frac{1}{2^{n-1}} + \cdots$

4. $1 - 2 + 4 - 8 + \cdots + (-1)^{n-1}2^{n-1} + \cdots$

5. $\frac{1}{2\cdot 3} + \frac{1}{3\cdot 4} + \frac{1}{4\cdot 5} + \cdots + \frac{1}{(n+1)(n+2)} + \cdots$

6. $\frac{5}{1\cdot 2} + \frac{5}{2\cdot 3} + \frac{5}{3\cdot 4} + \cdots + \frac{5}{n(n+1)} + \cdots$

Series with Geometric Terms

In Exercises 7–14, write out the first few terms of each series to show how the series starts. Then find the sum of the series.

7. $\sum_{n=0}^{\infty} \frac{(-1)^n}{4^n}$

8. $\sum_{n=2}^{\infty} \frac{1}{4^n}$

9. $\sum_{n=1}^{\infty} \frac{7}{4^n}$

10. $\sum_{n=0}^{\infty} (-1)^n \frac{5}{4^n}$

11. $\sum_{n=0}^{\infty} \left(\frac{5}{2^n} + \frac{1}{3^n}\right)$

12. $\sum_{n=0}^{\infty} \left(\frac{5}{2^n} - \frac{1}{3^n}\right)$

13. $\sum_{n=0}^{\infty} \left(\frac{1}{2^n} + \frac{(-1)^n}{5^n}\right)$

14. $\sum_{n=0}^{\infty} \left(\frac{2^{n+1}}{5^n}\right)$

Telescoping Series

Find the sum of each series in Exercises 15–22.

15. $\sum_{n=1}^{\infty} \frac{4}{(4n-3)(4n+1)}$

16. $\sum_{n=1}^{\infty} \frac{6}{(2n-1)(2n+1)}$

17. $\sum_{n=1}^{\infty} \frac{40n}{(2n-1)^2(2n+1)^2}$

18. $\sum_{n=1}^{\infty} \frac{2n+1}{n^2(n+1)^2}$

19. $\sum_{n=1}^{\infty} \left(\frac{1}{\sqrt{n}} - \frac{1}{\sqrt{n+1}}\right)$

20. $\sum_{n=1}^{\infty} \left(\frac{1}{2^{1/n}} - \frac{1}{2^{1/(n+1)}}\right)$

21. $\sum_{n=1}^{\infty} \left(\frac{1}{\ln(n+2)} - \frac{1}{\ln(n+1)}\right)$

22. $\sum_{n=1}^{\infty} (\tan^{-1}(n) - \tan^{-1}(n+1))$

Convergence or Divergence

Which series in Exercises 23–40 converge, and which diverge? Give reasons for your answers. If a series converges, find its sum.

23. $\sum_{n=0}^{\infty} \left(\frac{1}{\sqrt{2}}\right)^n$

24. $\sum_{n=0}^{\infty} (\sqrt{2})^n$

25. $\sum_{n=1}^{\infty} (-1)^{n+1} \frac{3}{2^n}$

26. $\sum_{n=1}^{\infty} (-1)^{n+1} n$

27. $\sum_{n=0}^{\infty} \cos n\pi$

28. $\sum_{n=0}^{\infty} \frac{\cos n\pi}{5^n}$

29. $\sum_{n=0}^{\infty} e^{-2n}$

30. $\sum_{n=1}^{\infty} \ln \frac{1}{n}$

31. $\sum_{n=1}^{\infty} \frac{2}{10^n}$

32. $\sum_{n=0}^{\infty} \frac{1}{x^n}, \quad |x| > 1$

33. $\sum_{n=0}^{\infty} \frac{2^n - 1}{3^n}$

34. $\sum_{n=1}^{\infty} \left(1 - \frac{1}{n}\right)^n$

35. $\sum_{n=0}^{\infty} \frac{n!}{1000^n}$

36. $\sum_{n=1}^{\infty} \frac{n^n}{n!}$

37. $\sum_{n=1}^{\infty} \ln\left(\frac{n}{n+1}\right)$

38. $\sum_{n=1}^{\infty} \ln\left(\frac{n}{2n+1}\right)$

39. $\sum_{n=0}^{\infty} \left(\frac{e}{\pi}\right)^n$

40. $\sum_{n=0}^{\infty} \frac{e^{n\pi}}{\pi^{ne}}$

Geometric Series

In each of the geometric series in Exercises 41–44, write out the first few terms of the series to find a and r, and find the sum of the series. Then express the inequality $|r| < 1$ in terms of x and find the values of x for which the inequality holds and the series converges.

41. $\sum_{n=0}^{\infty} (-1)^n x^n$

42. $\sum_{n=0}^{\infty} (-1)^n x^{2n}$

43. $\sum_{n=0}^{\infty} 3\left(\frac{x-1}{2}\right)^n$

44. $\sum_{n=0}^{\infty} \frac{(-1)^n}{2}\left(\frac{1}{3+\sin x}\right)^n$

In Exercises 45–50, find the values of x for which the given geometric series converges. Also, find the sum of the series (as a function of x) for those values of x.

45. $\sum_{n=0}^{\infty} 2^n x^n$

46. $\sum_{n=0}^{\infty} (-1)^n x^{-2n}$

47. $\sum_{n=0}^{\infty} (-1)^n (x+1)^n$

48. $\sum_{n=0}^{\infty} \left(-\frac{1}{2}\right)^n (x-3)^n$

49. $\sum_{n=0}^{\infty} \sin^n x$

50. $\sum_{n=0}^{\infty} (\ln x)^n$

Repeating Decimals

Express each of the numbers in Exercises 51–58 as the ratio of two integers.

51. $0.\overline{23} = 0.23\ 23\ 23\ldots$

52. $0.\overline{234} = 0.234\ 234\ 234\ldots$

53. $0.\overline{7} = 0.7777\ldots$

54. $0.\overline{d} = 0.dddd\ldots,$ where d is a digit

55. $0.0\overline{6} = 0.06666\ldots$

56. $1.\overline{414} = 1.414\ 414\ 414\ldots$

57. $1.24\overline{123} = 1.24\ 123\ 123\ 123\ldots$

58. $3.\overline{142857} = 3.142857\ 142857\ldots$

Theory and Examples

59. The series in Exercise 5 can also be written as

$$\sum_{n=1}^{\infty} \frac{1}{(n+1)(n+2)} \quad \text{and} \quad \sum_{n=-1}^{\infty} \frac{1}{(n+3)(n+4)}.$$

Write it as a sum beginning with **(a)** $n = -2$, **(b)** $n = 0$, **(c)** $n = 5$.

60. The series in Exercise 6 can also be written as

$$\sum_{n=1}^{\infty} \frac{5}{n(n+1)} \quad \text{and} \quad \sum_{n=0}^{\infty} \frac{5}{(n+1)(n+2)}.$$

Write it as a sum beginning with **(a)** $n = -1$, **(b)** $n = 3$, **(c)** $n = 20$.

61. Make up an infinite series of nonzero terms whose sum is

a. 1 **b.** −3 **c.** 0.

62. (*Continuation of Exercise 61.*) Can you make an infinite series of nonzero terms that converges to any number you want? Explain.

63. Show by example that $\sum(a_n/b_n)$ may diverge even though $\sum a_n$ and $\sum b_n$ converge and no b_n equals 0.

64. Find convergent geometric series $A = \sum a_n$ and $B = \sum b_n$ that illustrate the fact that $\sum a_n b_n$ may converge without being equal to AB.

65. Show by example that $\sum (a_n/b_n)$ may converge to something other than A/B even when $A = \sum a_n$, $B = \sum b_n \neq 0$, and no b_n equals 0.

66. If $\sum a_n$ converges and $a_n > 0$ for all n, can anything be said about $\sum (1/a_n)$? Give reasons for your answer.

67. What happens if you add a finite number of terms to a divergent series or delete a finite number of terms from a divergent series? Give reasons for your answer.

68. If $\sum a_n$ converges and $\sum b_n$ diverges, can anything be said about their term-by-term sum $\sum (a_n + b_n)$? Give reasons for your answer.

69. Make up a geometric series $\sum ar^{n-1}$ that converges to the number 5 if

a. $a = 2$ **b.** $a = 13/2$.

70. Find the value of b for which

$$1 + e^b + e^{2b} + e^{3b} + \cdots = 9.$$

71. For what values of r does the infinite series

$$1 + 2r + r^2 + 2r^3 + r^4 + 2r^5 + r^6 + \cdots$$

converge? Find the sum of the series when it converges.

72. Show that the error $(L - s_n)$ obtained by replacing a convergent geometric series with one of its partial sums s_n is $ar^n/(1 - r)$.

73. A ball is dropped from a height of 4 m. Each time it strikes the pavement after falling from a height of h meters it rebounds to a height of $0.75h$ meters. Find the total distance the ball travels up and down.

74. (*Continuation of Exercise 73.*) Find the total number of seconds the ball in Exercise 73 is traveling. (*Hint:* The formula $s = 4.9t^2$ gives $t = \sqrt{s/4.9}$.)

75. The accompanying figure shows the first five of a sequence of squares. The outermost square has an area of 4 m^2. Each of the other squares is obtained by joining the midpoints of the sides of the squares before it. Find the sum of the areas of all the squares.

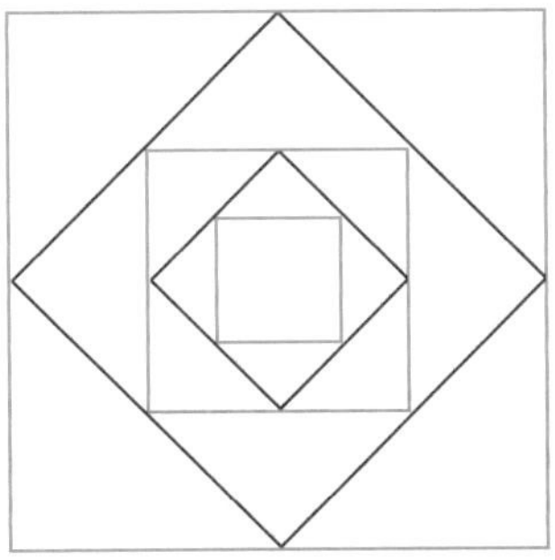

76. The accompanying figure shows the first three rows and part of the fourth row of a sequence of rows of semicircles. There are 2^n semicircles in the nth row, each of radius $1/2^n$. Find the sum of the areas of all the semicircles.

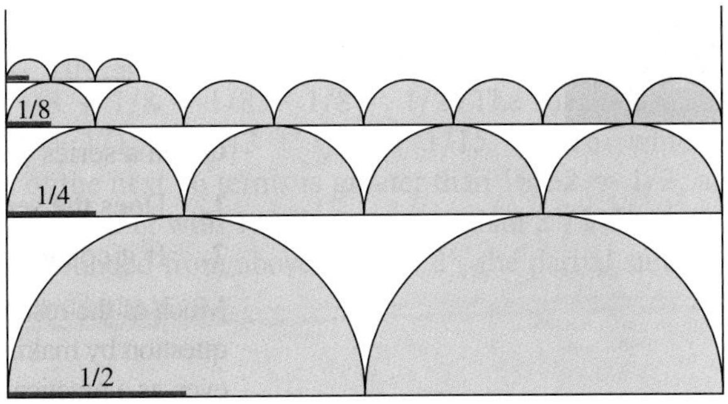

77. Helga von Koch's snowflake curve Helga von Koch's snowflake is a curve of infinite length that encloses a region of finite area. To see why this is so, suppose the curve is generated by starting with an equilateral triangle whose sides have length 1.

a. Find the length L_n of the nth curve C_n and show that $\lim_{n\to\infty} L_n = \infty$.

b. Find the area A_n of the region enclosed by C_n and calculate $\lim_{n\to\infty} A_n$.

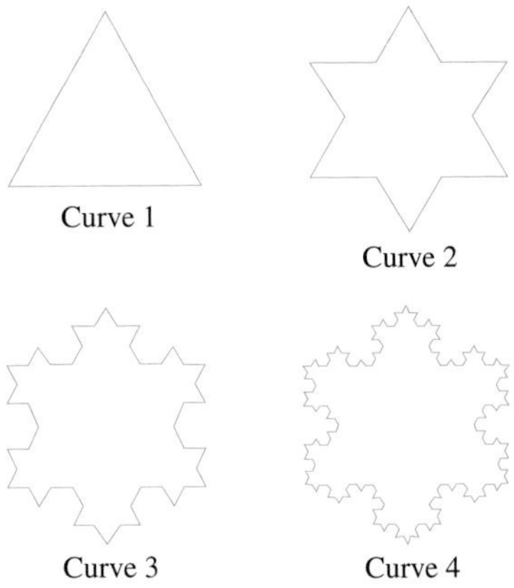

78. The accompanying figure provides an informal proof that $\sum_{n=1}^{\infty} (1/n^2)$ is less than 2. Explain what is going on. (*Source:* "Convergence with Pictures" by P. J. Rippon, *American Mathematical Monthly*, Vol. 93, No. 6, 1986, pp. 476–478.)

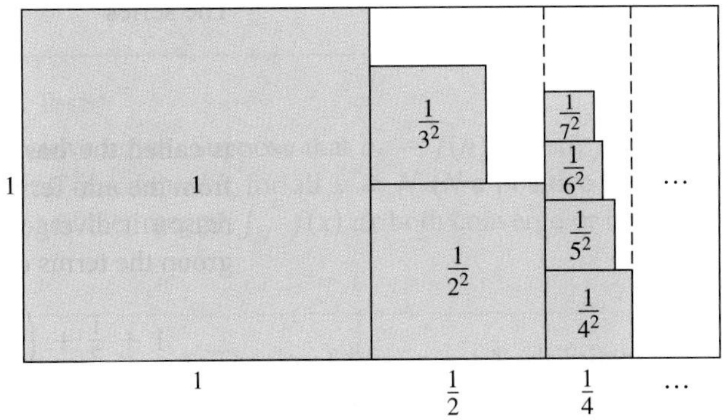

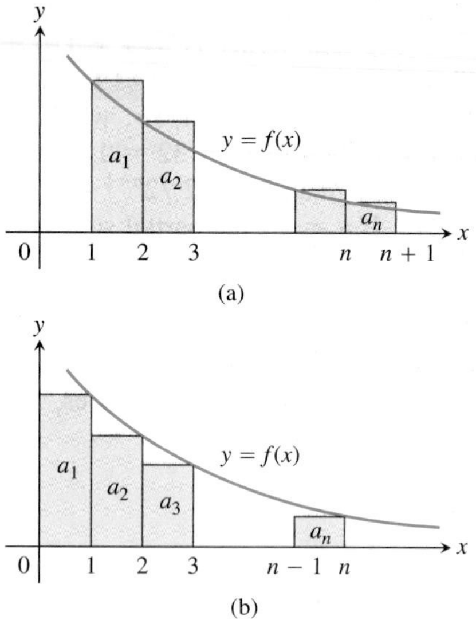

FIGURE 11.8 Subject to the conditions of the Integral Test, the series $\sum_{n=1}^{\infty} a_n$ and the integral $\int_1^{\infty} f(x)\, dx$ both converge or both diverge.

$a_1, a_2, \ldots, a_n$, collectively enclose more area than that under the curve $y = f(x)$ from $x = 1$ to $x = n + 1$. That is,

$$\int_1^{n+1} f(x)\, dx \leq a_1 + a_2 + \cdots + a_n.$$

In Figure 11.8b the rectangles have been faced to the left instead of to the right. If we momentarily disregard the first rectangle, of area a_1, we see that

$$a_2 + a_3 + \cdots + a_n \leq \int_1^{n} f(x)\, dx.$$

If we include a_1, we have

$$a_1 + a_2 + \cdots + a_n \leq a_1 + \int_1^{n} f(x)\, dx.$$

Combining these results gives

$$\int_1^{n+1} f(x)\, dx \leq a_1 + a_2 + \cdots + a_n \leq a_1 + \int_1^{n} f(x)\, dx.$$

These inequalities hold for each n, and continue to hold as $n \to \infty$.

If $\int_1^{\infty} f(x)\, dx$ is finite, the right-hand inequality shows that $\sum a_n$ is finite. If $\int_1^{\infty} f(x)\, dx$ is infinite, the left-hand inequality shows that $\sum a_n$ is infinite. Hence the series and the integral are both finite or both infinite. ■

EXAMPLE 3 The p-Series

Show that the ***p*-series**

$$\sum_{n=1}^{\infty} \frac{1}{n^p} = \frac{1}{1^p} + \frac{1}{2^p} + \frac{1}{3^p} + \cdots + \frac{1}{n^p} + \cdots$$

(p a real constant) converges if $p > 1$, and diverges if $p \leq 1$.

Solution If $p > 1$, then $f(x) = 1/x^p$ is a positive decreasing function of x. Since

$$\int_1^{\infty} \frac{1}{x^p}\, dx = \int_1^{\infty} x^{-p}\, dx = \lim_{b\to\infty} \left[\frac{x^{-p+1}}{-p+1}\right]_1^b$$

$$= \frac{1}{1-p} \lim_{b\to\infty} \left(\frac{1}{b^{p-1}} - 1\right)$$

$$= \frac{1}{1-p}(0 - 1) = \frac{1}{p-1},$$

$b^{p-1} \to \infty$ as $b \to \infty$ because $p - 1 > 0$.

the series converges by the Integral Test. We emphasize that the sum of the p-series is *not* $1/(p - 1)$. The series converges, but we don't know the value it converges to.

If $p < 1$, then $1 - p > 0$ and

$$\int_1^{\infty} \frac{1}{x^p}\, dx = \frac{1}{1-p} \lim_{b\to\infty} (b^{1-p} - 1) = \infty.$$

The series diverges by the Integral Test.

If $p = 1$, we have the (divergent) harmonic series

$$1 + \frac{1}{2} + \frac{1}{3} + \cdots + \frac{1}{n} + \cdots.$$

We have convergence for $p > 1$ but divergence for every other value of p. ■

The p-series with $p = 1$ is the **harmonic series** (Example 1). The p-Series Test shows that the harmonic series is just *barely* divergent; if we increase p to 1.000000001, for instance, the series converges!

The slowness with which the partial sums of the harmonic series approaches infinity is impressive. For instance, it takes about 178,482,301 terms of the harmonic series to move the partial sums beyond 20. It would take your calculator several weeks to compute a sum with this many terms. (See also Exercise 33b.)

EXAMPLE 4 A Convergent Series

The series

$$\sum_{n=1}^{\infty} \frac{1}{n^2 + 1}$$

converges by the Integral Test. The function $f(x) = 1/(x^2 + 1)$ is positive, continuous, and decreasing for $x \geq 1$, and

$$\begin{aligned} \int_1^{\infty} \frac{1}{x^2 + 1}\,dx &= \lim_{b\to\infty} \Big[\arctan x\Big]_1^b \\ &= \lim_{b\to\infty} [\arctan b - \arctan 1] \\ &= \frac{\pi}{2} - \frac{\pi}{4} = \frac{\pi}{4}. \end{aligned}$$

Again we emphasize that $\pi/4$ is *not* the sum of the series. The series converges, but we do not know the value of its sum. ■

Convergence of the series in Example 4 can also be verified by comparison with the series $\sum 1/n^2$. Comparison tests are studied in the next section.

EXERCISES 11.3

Determining Convergence or Divergence

Which of the series in Exercises 1–30 converge, and which diverge? Give reasons for your answers. (When you check an answer, remember that there may be more than one way to determine the series' convergence or divergence.)

1. $\sum_{n=1}^{\infty} \frac{1}{10^n}$ **2.** $\sum_{n=1}^{\infty} e^{-n}$ **3.** $\sum_{n=1}^{\infty} \frac{n}{n+1}$

4. $\sum_{n=1}^{\infty} \frac{5}{n+1}$ **5.** $\sum_{n=1}^{\infty} \frac{3}{\sqrt{n}}$ **6.** $\sum_{n=1}^{\infty} \frac{-2}{n\sqrt{n}}$

7. $\sum_{n=1}^{\infty} -\frac{1}{8^n}$ **8.** $\sum_{n=1}^{\infty} \frac{-8}{n}$ **9.** $\sum_{n=2}^{\infty} \frac{\ln n}{n}$

10. $\sum_{n=2}^{\infty} \frac{\ln n}{\sqrt{n}}$ **11.** $\sum_{n=1}^{\infty} \frac{2^n}{3^n}$ **12.** $\sum_{n=1}^{\infty} \frac{5^n}{4^n + 3}$

13. $\sum_{n=0}^{\infty} \frac{-2}{n+1}$ **14.** $\sum_{n=1}^{\infty} \frac{1}{2n-1}$ **15.** $\sum_{n=1}^{\infty} \frac{2^n}{n+1}$

16. $\sum_{n=1}^{\infty} \frac{1}{\sqrt{n}\left(\sqrt{n}+1\right)}$ **17.** $\sum_{n=2}^{\infty} \frac{\sqrt{n}}{\ln n}$ **18.** $\sum_{n=1}^{\infty} \left(1 + \frac{1}{n}\right)^n$

19. $\sum_{n=1}^{\infty} \frac{1}{(\ln 2)^n}$ **20.** $\sum_{n=1}^{\infty} \frac{1}{(\ln 3)^n}$

21. $\sum_{n=3}^{\infty} \frac{(1/n)}{(\ln n)\sqrt{\ln^2 n - 1}}$ **22.** $\sum_{n=1}^{\infty} \frac{1}{n(1 + \ln^2 n)}$

23. $\sum_{n=1}^{\infty} n \sin \frac{1}{n}$

24. $\sum_{n=1}^{\infty} n \tan \frac{1}{n}$

25. $\sum_{n=1}^{\infty} \frac{e^n}{1 + e^{2n}}$

26. $\sum_{n=1}^{\infty} \frac{2}{1 + e^n}$

27. $\sum_{n=1}^{\infty} \frac{8 \tan^{-1} n}{1 + n^2}$

28. $\sum_{n=1}^{\infty} \frac{n}{n^2 + 1}$

29. $\sum_{n=1}^{\infty} \operatorname{sech} n$

30. $\sum_{n=1}^{\infty} \operatorname{sech}^2 n$

Theory and Examples

For what values of a, if any, do the series in Exercises 31 and 32 converge?

31. $\sum_{n=1}^{\infty} \left(\frac{a}{n+2} - \frac{1}{n+4} \right)$

32. $\sum_{n=3}^{\infty} \left(\frac{1}{n-1} - \frac{2a}{n+1} \right)$

33. **a.** Draw illustrations like those in Figures 11.7 and 11.8 to show that the partial sums of the harmonic series satisfy the inequalities

$$\ln(n+1) = \int_1^{n+1} \frac{1}{x}\,dx \le 1 + \frac{1}{2} + \cdots + \frac{1}{n}$$
$$\le 1 + \int_1^n \frac{1}{x}\,dx = 1 + \ln n.$$

T **b.** There is absolutely no empirical evidence for the divergence of the harmonic series even though we know it diverges. The partial sums just grow too slowly. To see what we mean, suppose you had started with $s_1 = 1$ the day the universe was formed, 13 billion years ago, and added a new term every *second*. About how large would the partial sum s_n be today, assuming a 365-day year?

34. Are there any values of x for which $\sum_{n=1}^{\infty}(1/(nx))$ converges? Give reasons for your answer.

35. Is it true that if $\sum_{n=1}^{\infty} a_n$ is a divergent series of positive numbers then there is also a divergent series $\sum_{n=1}^{\infty} b_n$ of positive numbers with $b_n < a_n$ for every n? Is there a "smallest" divergent series of positive numbers? Give reasons for your answers.

36. (*Continuation of Exercise 35.*) Is there a "largest" convergent series of positive numbers? Explain.

37. **The Cauchy condensation test** The Cauchy condensation test says: Let $\{a_n\}$ be a nonincreasing sequence ($a_n \ge a_{n+1}$ for all n) of positive terms that converges to 0. Then $\sum a_n$ converges if and only if $\sum 2^n a_{2^n}$ converges. For example, $\sum(1/n)$ diverges because $\sum 2^n \cdot (1/2^n) = \sum 1$ diverges. Show why the test works.

38. Use the Cauchy condensation test from Exercise 37 to show that

a. $\sum_{n=2}^{\infty} \frac{1}{n \ln n}$ diverges;

b. $\sum_{n=1}^{\infty} \frac{1}{n^p}$ converges if $p > 1$ and diverges if $p \le 1$.

39. **Logarithmic *p*-series**

a. Show that

$$\int_2^{\infty} \frac{dx}{x(\ln x)^p} \quad (p \text{ a positive constant})$$

converges if and only if $p > 1$.

b. What implications does the fact in part (a) have for the convergence of the series

$$\sum_{n=2}^{\infty} \frac{1}{n(\ln n)^p}?$$

Give reasons for your answer.

40. (*Continuation of Exercise 39.*) Use the result in Exercise 39 to determine which of the following series converge and which diverge. Support your answer in each case.

a. $\sum_{n=2}^{\infty} \frac{1}{n(\ln n)}$

b. $\sum_{n=2}^{\infty} \frac{1}{n(\ln n)^{1.01}}$

c. $\sum_{n=2}^{\infty} \frac{1}{n \ln (n^3)}$

d. $\sum_{n=2}^{\infty} \frac{1}{n(\ln n)^3}$

41. **Euler's constant** Graphs like those in Figure 11.8 suggest that as n increases there is little change in the difference between the sum

$$1 + \frac{1}{2} + \cdots + \frac{1}{n}$$

and the integral

$$\ln n = \int_1^n \frac{1}{x}\,dx.$$

To explore this idea, carry out the following steps.

a. By taking $f(x) = 1/x$ in the proof of Theorem 9, show that

$$\ln(n+1) \le 1 + \frac{1}{2} + \cdots + \frac{1}{n} \le 1 + \ln n$$

or

$$0 < \ln(n+1) - \ln n \le 1 + \frac{1}{2} + \cdots + \frac{1}{n} - \ln n \le 1.$$

Thus, the sequence

$$a_n = 1 + \frac{1}{2} + \cdots + \frac{1}{n} - \ln n$$

is bounded from below and from above.

b. Show that

$$\frac{1}{n+1} < \int_n^{n+1} \frac{1}{x}\,dx = \ln(n+1) - \ln n,$$

and use this result to show that the sequence $\{a_n\}$ in part (a) is decreasing.

Since a decreasing sequence that is bounded from below converges (Exercise 107 in Section 11.1), the numbers a_n defined in part (a) converge:

$$1 + \frac{1}{2} + \cdots + \frac{1}{n} - \ln n \rightarrow \gamma.$$

The number γ, whose value is $0.5772\ldots$, is called *Euler's constant*. In contrast to other special numbers like π and e, no other expression with a simple law of formulation has ever been found for γ.

42. Use the integral test to show that

$$\sum_{n=0}^{\infty} e^{-n^2}$$

converges.

11.4 Comparison Tests

We have seen how to determine the convergence of geometric series, p-series, and a few others. We can test the convergence of many more series by comparing their terms to those of a series whose convergence is known.

THEOREM 10 The Comparison Test
Let $\sum a_n$ be a series with no negative terms.

(a) $\sum a_n$ converges if there is a convergent series $\sum c_n$ with $a_n \leq c_n$ for all $n > N$, for some integer N.

(b) $\sum a_n$ diverges if there is a divergent series of nonnegative terms $\sum d_n$ with $a_n \geq d_n$ for all $n > N$, for some integer N.

HISTORICAL BIOGRAPHY

Albert of Saxony
(ca. 1316–1390)

Proof In Part (a), the partial sums of $\sum a_n$ are bounded above by

$$M = a_1 + a_2 + \cdots + a_N + \sum_{n=N+1}^{\infty} c_n.$$

They therefore form a nondecreasing sequence with a limit $L \leq M$.

In Part (b), the partial sums of $\sum a_n$ are not bounded from above. If they were, the partial sums for $\sum d_n$ would be bounded by

$$M^* = d_1 + d_2 + \cdots + d_N + \sum_{n=N+1}^{\infty} a_n$$

and $\sum d_n$ would have to converge instead of diverge. ∎

EXAMPLE 1 Applying the Comparison Test

(a) The series

$$\sum_{n=1}^{\infty} \frac{5}{5n - 1}$$

diverges because its nth term

$$\frac{5}{5n - 1} = \frac{1}{n - \frac{1}{5}} > \frac{1}{n}$$

is greater than the nth term of the divergent harmonic series.

(b) The series

$$\sum_{n=0}^{\infty} \frac{1}{n!} = 1 + \frac{1}{1!} + \frac{1}{2!} + \frac{1}{3!} + \cdots$$

converges because its terms are all positive and less than or equal to the corresponding terms of

$$1 + \sum_{n=0}^{\infty} \frac{1}{2^n} = 1 + 1 + \frac{1}{2} + \frac{1}{2^2} + \cdots.$$

The geometric series on the left converges and we have

$$1 + \sum_{n=0}^{\infty} \frac{1}{2^n} = 1 + \frac{1}{1 - (1/2)} = 3.$$

The fact that 3 is an upper bound for the partial sums of $\sum_{n=0}^{\infty}(1/n!)$ does not mean that the series converges to 3. As we will see in Section 11.9, the series converges to e.

(c) The series

$$5 + \frac{2}{3} + \frac{1}{7} + 1 + \frac{1}{2 + \sqrt{1}} + \frac{1}{4 + \sqrt{2}} + \frac{1}{8 + \sqrt{3}} + \cdots + \frac{1}{2^n + \sqrt{n}} + \cdots$$

converges. To see this, we ignore the first three terms and compare the remaining terms with those of the convergent geometric series $\sum_{n=0}^{\infty}(1/2^n)$. The term $1/(2^n + \sqrt{n})$ of the truncated sequence is less than the corresponding term $1/2^n$ of the geometric series. We see that term by term we have the comparison,

$$1 + \frac{1}{2 + \sqrt{1}} + \frac{1}{4 + \sqrt{2}} + \frac{1}{8 + \sqrt{3}} + \cdots \leq 1 + \frac{1}{2} + \frac{1}{4} + \frac{1}{8} + \cdots$$

So the truncated series and the original series converge by an application of the Comparison Test. ■

The Limit Comparison Test

We now introduce a comparison test that is particularly useful for series in which a_n is a rational function of n.

THEOREM 11 Limit Comparison Test

Suppose that $a_n > 0$ and $b_n > 0$ for all $n \geq N$ (N an integer).

1. If $\lim_{n\to\infty} \frac{a_n}{b_n} = c > 0$, then $\sum a_n$ and $\sum b_n$ both converge or both diverge.
2. If $\lim_{n\to\infty} \frac{a_n}{b_n} = 0$ and $\sum b_n$ converges, then $\sum a_n$ converges.
3. If $\lim_{n\to\infty} \frac{a_n}{b_n} = \infty$ and $\sum b_n$ diverges, then $\sum a_n$ diverges.

Proof We will prove Part 1. Parts 2 and 3 are left as Exercises 37(a) and (b).

Since $c/2 > 0$, there exists an integer N such that for all n

$$n > N \Rightarrow \left|\frac{a_n}{b_n} - c\right| < \frac{c}{2}.$$

Limit definition with $\epsilon = c/2, L = c$, and a_n replaced by a_n/b_n

Thus, for $n > N$,

$$-\frac{c}{2} < \frac{a_n}{b_n} - c < \frac{c}{2},$$

$$\frac{c}{2} < \frac{a_n}{b_n} < \frac{3c}{2},$$

$$\left(\frac{c}{2}\right)b_n < a_n < \left(\frac{3c}{2}\right)b_n.$$

If $\sum b_n$ converges, then $\sum(3c/2)b_n$ converges and $\sum a_n$ converges by the Direct Comparison Test. If $\sum b_n$ diverges, then $\sum(c/2)b_n$ diverges and $\sum a_n$ diverges by the Direct Comparison Test. ■

EXAMPLE 2 Using the Limit Comparison Test

Which of the following series converge, and which diverge?

(a) $\frac{3}{4} + \frac{5}{9} + \frac{7}{16} + \frac{9}{25} + \cdots = \sum_{n=1}^{\infty} \frac{2n+1}{(n+1)^2} = \sum_{n=1}^{\infty} \frac{2n+1}{n^2+2n+1}$

(b) $\frac{1}{1} + \frac{1}{3} + \frac{1}{7} + \frac{1}{15} + \cdots = \sum_{n=1}^{\infty} \frac{1}{2^n - 1}$

(c) $\frac{1 + 2\ln 2}{9} + \frac{1 + 3\ln 3}{14} + \frac{1 + 4\ln 4}{21} + \cdots = \sum_{n=2}^{\infty} \frac{1 + n\ln n}{n^2 + 5}$

Solution

(a) Let $a_n = (2n+1)/(n^2+2n+1)$. For large n, we expect a_n to behave like $2n/n^2 = 2/n$ since the leading terms dominate for large n, so we let $b_n = 1/n$. Since

$$\sum_{n=1}^{\infty} b_n = \sum_{n=1}^{\infty} \frac{1}{n} \text{ diverges}$$

and

$$\lim_{n\to\infty} \frac{a_n}{b_n} = \lim_{n\to\infty} \frac{2n^2 + n}{n^2 + 2n + 1} = 2,$$

$\sum a_n$ diverges by Part 1 of the Limit Comparison Test. We could just as well have taken $b_n = 2/n$, but $1/n$ is simpler.

(b) Let $a_n = 1/(2^n - 1)$. For large n, we expect a_n to behave like $1/2^n$, so we let $b_n = 1/2^n$. Since

$$\sum_{n=1}^{\infty} b_n = \sum_{n=1}^{\infty} \frac{1}{2^n} \text{ converges}$$

and

$$\begin{aligned} \lim_{n\to\infty} \frac{a_n}{b_n} &= \lim_{n\to\infty} \frac{2^n}{2^n - 1} \\ &= \lim_{n\to\infty} \frac{1}{1 - (1/2^n)} \\ &= 1, \end{aligned}$$

Σa_n converges by Part 1 of the Limit Comparison Test.

(c) Let $a_n = (1 + n \ln n)/(n^2 + 5)$. For large n, we expect a_n to behave like $(n \ln n)/n^2 = (\ln n)/n$, which is greater than $1/n$ for $n \geq 3$, so we take $b_n = 1/n$. Since

$$\sum_{n=2}^{\infty} b_n = \sum_{n=2}^{\infty} \frac{1}{n} \text{ diverges}$$

and

$$\begin{aligned} \lim_{n\to\infty} \frac{a_n}{b_n} &= \lim_{n\to\infty} \frac{n + n^2 \ln n}{n^2 + 5} \\ &= \infty, \end{aligned}$$

Σa_n diverges by Part 3 of the Limit Comparison Test. ■

EXAMPLE 3 Does $\displaystyle\sum_{n=1}^{\infty} \frac{\ln n}{n^{3/2}}$ converge?

Solution Because $\ln n$ grows more slowly than n^c for any positive constant c (Section 11.1, Exercise 91), we would expect to have

$$\frac{\ln n}{n^{3/2}} < \frac{n^{1/4}}{n^{3/2}} = \frac{1}{n^{5/4}}$$

for n sufficiently large. Indeed, taking $a_n = (\ln n)/n^{3/2}$ and $b_n = 1/n^{5/4}$, we have

$$\begin{aligned} \lim_{n\to\infty} \frac{a_n}{b_n} &= \lim_{n\to\infty} \frac{\ln n}{n^{1/4}} \\ &= \lim_{n\to\infty} \frac{1/n}{(1/4)n^{-3/4}} && \text{l'Hôpital's Rule} \\ &= \lim_{n\to\infty} \frac{4}{n^{1/4}} = 0. \end{aligned}$$

Since $\Sigma b_n = \Sigma(1/n^{5/4})$ (a p-series with $p > 1$) converges, Σa_n converges by Part 2 of the Limit Comparison Test. ■

EXERCISES 11.4

Determining Convergence or Divergence

Which of the series in Exercises 1–36 converge, and which diverge? Give reasons for your answers.

1. $\sum_{n=1}^{\infty} \frac{1}{2\sqrt{n} + \sqrt[3]{n}}$ **2.** $\sum_{n=1}^{\infty} \frac{3}{n + \sqrt{n}}$ **3.** $\sum_{n=1}^{\infty} \frac{\sin^2 n}{2^n}$

4. $\sum_{n=1}^{\infty} \frac{1 + \cos n}{n^2}$ **5.** $\sum_{n=1}^{\infty} \frac{2n}{3n - 1}$ **6.** $\sum_{n=1}^{\infty} \frac{n + 1}{n^2\sqrt{n}}$

7. $\sum_{n=1}^{\infty} \left(\frac{n}{3n + 1}\right)^n$ **8.** $\sum_{n=1}^{\infty} \frac{1}{\sqrt{n^3 + 2}}$ **9.** $\sum_{n=3}^{\infty} \frac{1}{\ln(\ln n)}$

10. $\sum_{n=2}^{\infty} \frac{1}{(\ln n)^2}$ **11.** $\sum_{n=1}^{\infty} \frac{(\ln n)^2}{n^3}$ **12.** $\sum_{n=1}^{\infty} \frac{(\ln n)^3}{n^3}$

13. $\sum_{n=2}^{\infty} \frac{1}{\sqrt{n}\ln n}$ **14.** $\sum_{n=1}^{\infty} \frac{(\ln n)^2}{n^{3/2}}$ **15.** $\sum_{n=1}^{\infty} \frac{1}{1 + \ln n}$

16. $\sum_{n=1}^{\infty} \frac{1}{(1 + \ln n)^2}$ **17.** $\sum_{n=2}^{\infty} \frac{\ln(n + 1)}{n + 1}$ **18.** $\sum_{n=1}^{\infty} \frac{1}{(1 + \ln^2 n)}$

19. $\sum_{n=2}^{\infty} \frac{1}{n\sqrt{n^2 - 1}}$ **20.** $\sum_{n=1}^{\infty} \frac{\sqrt{n}}{n^2 + 1}$ **21.** $\sum_{n=1}^{\infty} \frac{1 - n}{n2^n}$

22. $\sum_{n=1}^{\infty} \frac{n + 2^n}{n^2 2^n}$ **23.** $\sum_{n=1}^{\infty} \frac{1}{3^{n-1} + 1}$ **24.** $\sum_{n=1}^{\infty} \frac{3^{n-1} + 1}{3^n}$

25. $\sum_{n=1}^{\infty} \sin\frac{1}{n}$ **26.** $\sum_{n=1}^{\infty} \tan\frac{1}{n}$

27. $\sum_{n=1}^{\infty} \frac{10n + 1}{n(n + 1)(n + 2)}$ **28.** $\sum_{n=3}^{\infty} \frac{5n^3 - 3n}{n^2(n - 2)(n^2 + 5)}$

29. $\sum_{n=1}^{\infty} \frac{\tan^{-1} n}{n^{1.1}}$ **30.** $\sum_{n=1}^{\infty} \frac{\sec^{-1} n}{n^{1.3}}$ **31.** $\sum_{n=1}^{\infty} \frac{\coth n}{n^2}$

32. $\sum_{n=1}^{\infty} \frac{\tanh n}{n^2}$ **33.** $\sum_{n=1}^{\infty} \frac{1}{n\sqrt[n]{n}}$ **34.** $\sum_{n=1}^{\infty} \frac{\sqrt[n]{n}}{n^2}$

35. $\sum_{n=1}^{\infty} \frac{1}{1 + 2 + 3 + \cdots + n}$ **36.** $\sum_{n=1}^{\infty} \frac{1}{1 + 2^2 + 3^2 + \cdots + n^2}$

Theory and Examples

37. Prove **(a)** Part 2 and **(b)** Part 3 of the Limit Comparison Test.

38. If $\sum_{n=1}^{\infty} a_n$ is a convergent series of nonnegative numbers, can anything be said about $\sum_{n=1}^{\infty}(a_n/n)$? Explain.

39. Suppose that $a_n > 0$ and $b_n > 0$ for $n \geq N$ (N an integer). If $\lim_{n\to\infty}(a_n/b_n) = \infty$ and $\sum a_n$ converges, can anything be said about $\sum b_n$? Give reasons for your answer.

40. Prove that if $\sum a_n$ is a convergent series of nonnegative terms, then $\sum a_n^2$ converges.

COMPUTER EXPLORATION

41. It is not yet known whether the series

$$\sum_{n=1}^{\infty} \frac{1}{n^3 \sin^2 n}$$

converges or diverges. Use a CAS to explore the behavior of the series by performing the following steps.

a. Define the sequence of partial sums

$$s_k = \sum_{n=1}^{k} \frac{1}{n^3 \sin^2 n}.$$

What happens when you try to find the limit of s_k as $k \to \infty$? Does your CAS find a closed form answer for this limit?

b. Plot the first 100 points (k, s_k) for the sequence of partial sums. Do they appear to converge? What would you estimate the limit to be?

c. Next plot the first 200 points (k, s_k). Discuss the behavior in your own words.

d. Plot the first 400 points (k, s_k). What happens when $k = 355$? Calculate the number $355/113$. Explain from your calculation what happened at $k = 355$. For what values of k would you guess this behavior might occur again?

You will find an interesting discussion of this series in Chapter 72 of *Mazes for the Mind* by Clifford A. Pickover, St. Martin's Press, Inc., New York, 1992.

11.5 The Ratio and Root Tests

The Ratio Test measures the rate of growth (or decline) of a series by examining the ratio a_{n+1}/a_n. For a geometric series $\sum ar^n$, this rate is a constant $((ar^{n+1})/(ar^n) = r)$, and the series converges if and only if its ratio is less than 1 in absolute value. The Ratio Test is a powerful rule extending that result. We prove it on the next page using the Comparison Test.

THEOREM 12 The Ratio Test
Let $\sum a_n$ be a series with positive terms and suppose that

$$\lim_{n\to\infty} \frac{a_{n+1}}{a_n} = \rho.$$

Then

(a) the series *converges* if $\rho < 1$,

(b) the series *diverges* if $\rho > 1$ or ρ is infinite,

(c) the test is *inconclusive* if $\rho = 1$.

Proof

(a) $\boldsymbol{\rho < 1}$**.** Let r be a number between ρ and 1. Then the number $\epsilon = r - \rho$ is positive. Since

$$\frac{a_{n+1}}{a_n} \rightarrow \rho,$$

a_{n+1}/a_n must lie within ϵ of ρ when n is large enough, say for all $n \geq N$. In particular

$$\frac{a_{n+1}}{a_n} < \rho + \epsilon = r, \qquad \text{when } n \geq N.$$

That is,

$$\begin{aligned} a_{N+1} &< ra_N, \\ a_{N+2} &< ra_{N+1} < r^2 a_N, \\ a_{N+3} &< ra_{N+2} < r^3 a_N, \\ &\vdots \\ a_{N+m} &< ra_{N+m-1} < r^m a_N. \end{aligned}$$

These inequalities show that the terms of our series, after the Nth term, approach zero more rapidly than the terms in a geometric series with ratio $r < 1$. More precisely, consider the series $\sum c_n$, where $c_n = a_n$ for $n = 1, 2, \ldots, N$ and $c_{N+1} = ra_N$, $c_{N+2} = r^2 a_N, \ldots, c_{N+m} = r^m a_N, \ldots$. Now $a_n \leq c_n$ for all n, and

$$\begin{aligned} \sum_{n=1}^{\infty} c_n &= a_1 + a_2 + \cdots + a_{N-1} + a_N + ra_N + r^2 a_N + \cdots \\ &= a_1 + a_2 + \cdots + a_{N-1} + a_N(1 + r + r^2 + \cdots). \end{aligned}$$

The geometric series $1 + r + r^2 + \cdots$ converges because $|r| < 1$, so $\sum c_n$ converges. Since $a_n \leq c_n$, $\sum a_n$ also converges.

(b) $\boldsymbol{1 < \rho \leq \infty}$**.** From some index M on,

$$\frac{a_{n+1}}{a_n} > 1 \qquad \text{and} \qquad a_M < a_{M+1} < a_{M+2} < \cdots.$$

The terms of the series do not approach zero as n becomes infinite, and the series diverges by the nth-Term Test.

(c) $\rho = 1$. The two series

$$\sum_{n=1}^{\infty} \frac{1}{n} \quad \text{and} \quad \sum_{n=1}^{\infty} \frac{1}{n^2}$$

show that some other test for convergence must be used when $\rho = 1$.

$$\text{For } \sum_{n=1}^{\infty} \frac{1}{n}: \qquad \frac{a_{n+1}}{a_n} = \frac{1/(n+1)}{1/n} = \frac{n}{n+1} \to 1.$$

$$\text{For } \sum_{n=1}^{\infty} \frac{1}{n^2}: \qquad \frac{a_{n+1}}{a_n} = \frac{1/(n+1)^2}{1/n^2} = \left(\frac{n}{n+1}\right)^2 \to 1^2 = 1.$$

In both cases, $\rho = 1$, yet the first series diverges, whereas the second converges. ■

The Ratio Test is often effective when the terms of a series contain factorials of expressions involving n or expressions raised to a power involving n.

EXAMPLE 1 Applying the Ratio Test

Investigate the convergence of the following series.

(a) $\displaystyle\sum_{n=0}^{\infty} \frac{2^n + 5}{3^n}$ **(b)** $\displaystyle\sum_{n=1}^{\infty} \frac{(2n)!}{n!n!}$ **(c)** $\displaystyle\sum_{n=1}^{\infty} \frac{4^n n! n!}{(2n)!}$

Solution

(a) For the series $\sum_{n=0}^{\infty} (2^n + 5)/3^n$,

$$\frac{a_{n+1}}{a_n} = \frac{(2^{n+1} + 5)/3^{n+1}}{(2^n + 5)/3^n} = \frac{1}{3} \cdot \frac{2^{n+1} + 5}{2^n + 5} = \frac{1}{3} \cdot \left(\frac{2 + 5 \cdot 2^{-n}}{1 + 5 \cdot 2^{-n}}\right) \to \frac{1}{3} \cdot \frac{2}{1} = \frac{2}{3}.$$

The series converges because $\rho = 2/3$ is less than 1. This does *not* mean that $2/3$ is the sum of the series. In fact,

$$\sum_{n=0}^{\infty} \frac{2^n + 5}{3^n} = \sum_{n=0}^{\infty} \left(\frac{2}{3}\right)^n + \sum_{n=0}^{\infty} \frac{5}{3^n} = \frac{1}{1 - (2/3)} + \frac{5}{1 - (1/3)} = \frac{21}{2}.$$

(b) If $a_n = \dfrac{(2n)!}{n!n!}$, then $a_{n+1} = \dfrac{(2n+2)!}{(n+1)!(n+1)!}$ and

$$\frac{a_{n+1}}{a_n} = \frac{n!n!(2n+2)(2n+1)(2n)!}{(n+1)!(n+1)!(2n)!}$$

$$= \frac{(2n+2)(2n+1)}{(n+1)(n+1)} = \frac{4n+2}{n+1} \to 4.$$

The series diverges because $\rho = 4$ is greater than 1.

(c) If $a_n = 4^n n! n!/(2n)!$, then

$$\frac{a_{n+1}}{a_n} = \frac{4^{n+1}(n+1)!(n+1)!}{(2n+2)(2n+1)(2n)!} \cdot \frac{(2n)!}{4^n n! n!}$$

$$= \frac{4(n+1)(n+1)}{(2n+2)(2n+1)} = \frac{2(n+1)}{2n+1} \to 1.$$

Because the limit is $\rho = 1$, we cannot decide from the Ratio Test whether the series converges. When we notice that $a_{n+1}/a_n = (2n + 2)/(2n + 1)$, we conclude that a_{n+1} is always greater than a_n because $(2n + 2)/(2n + 1)$ is always greater than 1. Therefore, all terms are greater than or equal to $a_1 = 2$, and the nth term does not approach zero as $n \to \infty$. The series diverges. ■

The Root Test

The convergence tests we have so far for Σa_n work best when the formula for a_n is relatively simple. But consider the following.

EXAMPLE 2 Let $a_n = \begin{cases} n/2^n, & n \text{ odd} \\ 1/2^n, & n \text{ even}. \end{cases}$ Does Σa_n converge?

Solution We write out several terms of the series:

$$\sum_{n=1}^{\infty} a_n = \frac{1}{2^1} + \frac{1}{2^2} + \frac{3}{2^3} + \frac{1}{2^4} + \frac{5}{2^5} + \frac{1}{2^6} + \frac{7}{2^7} + \cdots$$

$$= \frac{1}{2} + \frac{1}{4} + \frac{3}{8} + \frac{1}{16} + \frac{5}{32} + \frac{1}{64} + \frac{7}{128} + \cdots.$$

Clearly, this is not a geometric series. The nth term approaches zero as $n \to \infty$, so we do not know if the series diverges. The Integral Test does not look promising. The Ratio Test produces

$$\frac{a_{n+1}}{a_n} = \begin{cases} \dfrac{1}{2n}, & n \text{ odd} \\[2ex] \dfrac{n+1}{2}, & n \text{ even}. \end{cases}$$

As $n \to \infty$, the ratio is alternately small and large and has no limit.

A test that will answer the question (the series converges) is the Root Test. ■

THEOREM 13 The Root Test

Let Σa_n be a series with $a_n \geq 0$ for $n \geq N$, and suppose that

$$\lim_{n\to\infty} \sqrt[n]{a_n} = \rho.$$

Then

(a) the series *converges* if $\rho < 1$,

(b) the series *diverges* if $\rho > 1$ or ρ is infinite,

(c) the test is *inconclusive* if $\rho = 1$.

Proof

(a) $\boldsymbol{\rho < 1}$**.** Choose an $\epsilon > 0$ so small that $\rho + \epsilon < 1$. Since $\sqrt[n]{a_n} \to \rho$, the terms $\sqrt[n]{a_n}$ eventually get closer than ϵ to ρ. In other words, there exists an index $M \geq N$ such that

$$\sqrt[n]{a_n} < \rho + \epsilon \qquad \text{when } n \geq M.$$

Then it is also true that

$$a_n < (\rho + \epsilon)^n \qquad \text{for } n \geq M.$$

Now, $\sum_{n=M}^{\infty} (\rho + \epsilon)^n$, a geometric series with ratio $(\rho + \epsilon) < 1$, converges. By comparison, $\sum_{n=M}^{\infty} a_n$ converges, from which it follows that

$$\sum_{n=1}^{\infty} a_n = a_1 + \cdots + a_{M-1} + \sum_{n=M}^{\infty} a_n$$

converges.

(b) $\mathbf{1 < \rho \leq \infty}$**.** For all indices beyond some integer M, we have $\sqrt[n]{a_n} > 1$, so that $a_n > 1$ for $n > M$. The terms of the series do not converge to zero. The series diverges by the nth-Term Test.

(c) $\boldsymbol{\rho} = \mathbf{1}$**.** The series $\sum_{n=1}^{\infty} (1/n)$ and $\sum_{n=1}^{\infty} (1/n^2)$ show that the test is not conclusive when $\rho = 1$. The first series diverges and the second converges, but in both cases $\sqrt[n]{a_n} \rightarrow 1$. ■

EXAMPLE 3 Applying the Root Test

Which of the following series converges, and which diverges?

(a) $\displaystyle\sum_{n=1}^{\infty} \frac{n^2}{2^n}$ **(b)** $\displaystyle\sum_{n=1}^{\infty} \frac{2^n}{n^2}$ **(c)** $\displaystyle\sum_{n=1}^{\infty} \left(\frac{1}{1+n}\right)^n$

Solution

(a) $\displaystyle\sum_{n=1}^{\infty} \frac{n^2}{2^n}$ converges because $\displaystyle\sqrt[n]{\frac{n^2}{2^n}} = \frac{\sqrt[n]{n^2}}{\sqrt[n]{2^n}} = \frac{\left(\sqrt[n]{n}\right)^2}{2} \rightarrow \frac{1}{2} < 1.$

(b) $\displaystyle\sum_{n=1}^{\infty} \frac{2^n}{n^2}$ diverges because $\displaystyle\sqrt[n]{\frac{2^n}{n^2}} = \frac{2}{\left(\sqrt[n]{n}\right)^2} \rightarrow \frac{2}{1} > 1.$

(c) $\displaystyle\sum_{n=1}^{\infty} \left(\frac{1}{1+n}\right)^n$ converges because $\displaystyle\sqrt[n]{\left(\frac{1}{1+n}\right)^n} = \frac{1}{1+n} \rightarrow 0 < 1.$ ■

EXAMPLE 2 Revisited

Let $a_n = \begin{cases} n/2^n, & n \text{ odd} \\ 1/2^n, & n \text{ even.} \end{cases}$ Does $\sum a_n$ converge?

Solution We apply the Root Test, finding that

$$\sqrt[n]{a_n} = \begin{cases} \sqrt[n]{n}/2, & n \text{ odd} \\ 1/2, & n \text{ even.} \end{cases}$$

Therefore,

$$\frac{1}{2} \leq \sqrt[n]{a_n} \leq \frac{\sqrt[n]{n}}{2}.$$

Since $\sqrt[n]{n} \rightarrow 1$ (Section 11.1, Theorem 5), we have $\lim_{n\to\infty} \sqrt[n]{a_n} = 1/2$ by the Sandwich Theorem. The limit is less than 1, so the series converges by the Root Test. ■

EXERCISES 11.5

Determining Convergence or Divergence

Which of the series in Exercises 1–26 converge, and which diverge? Give reasons for your answers. (When checking your answers, remember there may be more than one way to determine a series' convergence or divergence.)

1. $\sum_{n=1}^{\infty} \frac{n^{\sqrt{2}}}{2^n}$

2. $\sum_{n=1}^{\infty} n^2 e^{-n}$

3. $\sum_{n=1}^{\infty} n! e^{-n}$

4. $\sum_{n=1}^{\infty} \frac{n!}{10^n}$

5. $\sum_{n=1}^{\infty} \frac{n^{10}}{10^n}$

6. $\sum_{n=1}^{\infty} \left(\frac{n-2}{n}\right)^n$

7. $\sum_{n=1}^{\infty} \frac{2+(-1)^n}{1.25^n}$

8. $\sum_{n=1}^{\infty} \frac{(-2)^n}{3^n}$

9. $\sum_{n=1}^{\infty} \left(1-\frac{3}{n}\right)^n$

10. $\sum_{n=1}^{\infty} \left(1-\frac{1}{3n}\right)^n$

11. $\sum_{n=1}^{\infty} \frac{\ln n}{n^3}$

12. $\sum_{n=1}^{\infty} \frac{(\ln n)^n}{n^n}$

13. $\sum_{n=1}^{\infty} \left(\frac{1}{n}-\frac{1}{n^2}\right)$

14. $\sum_{n=1}^{\infty} \left(\frac{1}{n}-\frac{1}{n^2}\right)^n$

15. $\sum_{n=1}^{\infty} \frac{\ln n}{n}$

16. $\sum_{n=1}^{\infty} \frac{n \ln n}{2^n}$

17. $\sum_{n=1}^{\infty} \frac{(n+1)(n+2)}{n!}$

18. $\sum_{n=1}^{\infty} e^{-n}(n^3)$

19. $\sum_{n=1}^{\infty} \frac{(n+3)!}{3!n!3^n}$

20. $\sum_{n=1}^{\infty} \frac{n2^n(n+1)!}{3^n n!}$

21. $\sum_{n=1}^{\infty} \frac{n!}{(2n+1)!}$

22. $\sum_{n=1}^{\infty} \frac{n!}{n^n}$

23. $\sum_{n=2}^{\infty} \frac{n}{(\ln n)^n}$

24. $\sum_{n=2}^{\infty} \frac{n}{(\ln n)^{(n/2)}}$

25. $\sum_{n=1}^{\infty} \frac{n! \ln n}{n(n+2)!}$

26. $\sum_{n=1}^{\infty} \frac{3^n}{n^3 2^n}$

Which of the series $\sum_{n=1}^{\infty} a_n$ defined by the formulas in Exercises 27–38 converge, and which diverge? Give reasons for your answers.

27. $a_1 = 2, \quad a_{n+1} = \frac{1+\sin n}{n} a_n$

28. $a_1 = 1, \quad a_{n+1} = \frac{1+\tan^{-1} n}{n} a_n$

29. $a_1 = \frac{1}{3}, \quad a_{n+1} = \frac{3n-1}{2n+5} a_n$

30. $a_1 = 3, \quad a_{n+1} = \frac{n}{n+1} a_n$

31. $a_1 = 2, \quad a_{n+1} = \frac{2}{n} a_n$

32. $a_1 = 5, \quad a_{n+1} = \frac{\sqrt[n]{n}}{2} a_n$

33. $a_1 = 1, \quad a_{n+1} = \frac{1+\ln n}{n} a_n$

34. $a_1 = \frac{1}{2}, \quad a_{n+1} = \frac{n+\ln n}{n+10} a_n$

35. $a_1 = \frac{1}{3}, \quad a_{n+1} = \sqrt[n]{a_n}$

36. $a_1 = \frac{1}{2}, \quad a_{n+1} = (a_n)^{n+1}$

37. $a_n = \frac{2^n n! n!}{(2n)!}$

38. $a_n = \frac{(3n)!}{n!(n+1)!(n+2)!}$

Which of the series in Exercises 39–44 converge, and which diverge? Give reasons for your answers.

39. $\sum_{n=1}^{\infty} \frac{(n!)^n}{(n^n)^2}$

40. $\sum_{n=1}^{\infty} \frac{(n!)^n}{n^{(n^2)}}$

41. $\sum_{n=1}^{\infty} \frac{n^n}{2^{(n^2)}}$

42. $\sum_{n=1}^{\infty} \frac{n^n}{(2^n)^2}$

43. $\sum_{n=1}^{\infty} \frac{1 \cdot 3 \cdot \cdots \cdot (2n-1)}{4^n 2^n n!}$

44. $\sum_{n=1}^{\infty} \frac{1 \cdot 3 \cdot \cdots \cdot (2n-1)}{[2 \cdot 4 \cdot \cdots \cdot (2n)](3^n+1)}$

Theory and Examples

45. Neither the Ratio nor the Root Test helps with p-series. Try them on

$$\sum_{n=1}^{\infty} \frac{1}{n^p}$$

and show that both tests fail to provide information about convergence.

46. Show that neither the Ratio Test nor the Root Test provides information about the convergence of

$$\sum_{n=2}^{\infty} \frac{1}{(\ln n)^p} \qquad (p \text{ constant}).$$

47. Let $a_n = \begin{cases} n/2^n, & \text{if } n \text{ is a prime number} \\ 1/2^n, & \text{otherwise.} \end{cases}$

Does $\sum a_n$ converge? Give reasons for your answer.

11.6 Alternating Series, Absolute and Conditional Convergence

A series in which the terms are alternately positive and negative is an **alternating series**. Here are three examples:

$$1 - \frac{1}{2} + \frac{1}{3} - \frac{1}{4} + \frac{1}{5} - \cdots + \frac{(-1)^{n+1}}{n} + \cdots \tag{1}$$

$$-2 + 1 - \frac{1}{2} + \frac{1}{4} - \frac{1}{8} + \cdots + \frac{(-1)^n 4}{2^n} + \cdots \tag{2}$$

$$1 - 2 + 3 - 4 + 5 - 6 + \cdots + (-1)^{n+1} n + \cdots \tag{3}$$

Series (1), called the **alternating harmonic series**, converges, as we will see in a moment. Series (2) a geometric series with ratio $r = -1/2$, converges to $-2/[1 + (1/2)] = -4/3$. Series (3) diverges because the nth term does not approach zero.

We prove the convergence of the alternating harmonic series by applying the Alternating Series Test.

THEOREM 14 The Alternating Series Test (Leibniz's Theorem)

The series

$$\sum_{n=1}^{\infty} (-1)^{n+1} u_n = u_1 - u_2 + u_3 - u_4 + \cdots$$

converges if all three of the following conditions are satisfied:

1. The u_n's are all positive.
2. $u_n \ge u_{n+1}$ for all $n \ge N$, for some integer N.
3. $u_n \to 0$.

Proof Assume $N = 1$. If n is an even integer, say $n = 2m$, then the sum of the first n terms is

$$\begin{aligned} s_{2m} &= (u_1 - u_2) + (u_3 - u_4) + \cdots + (u_{2m-1} - u_{2m}) \\ &= u_1 - (u_2 - u_3) - (u_4 - u_5) - \cdots - (u_{2m-2} - u_{2m-1}) - u_{2m}. \end{aligned}$$

The first equality shows that s_{2m} is the sum of m nonnegative terms, since each term in parentheses is positive or zero. Hence $s_{2m+2} \ge s_{2m}$, and the sequence $\{s_{2m}\}$ is nondecreasing. The second equality shows that $s_{2m} \le u_1$. Since $\{s_{2m}\}$ is nondecreasing and bounded from above, it has a limit, say

$$\lim_{m \to \infty} s_{2m} = L. \tag{4}$$

If n is an odd integer, say $n = 2m + 1$, then the sum of the first n terms is $s_{2m+1} = s_{2m} + u_{2m+1}$. Since $u_n \to 0$,

$$\lim_{m \to \infty} u_{2m+1} = 0$$

and, as $m \to \infty$,

$$s_{2m+1} = s_{2m} + u_{2m+1} \to L + 0 = L. \tag{5}$$

Combining the results of Equations (4) and (5) gives $\lim_{n \to \infty} s_n = L$ (Section 11.1, Exercise 119). ■

EXAMPLE 1 The alternating harmonic series

$$\sum_{n=1}^{\infty}(-1)^{n+1}\frac{1}{n} = 1 - \frac{1}{2} + \frac{1}{3} - \frac{1}{4} + \cdots$$

satisfies the three requirements of Theorem 14 with $N = 1$; it therefore converges. ■

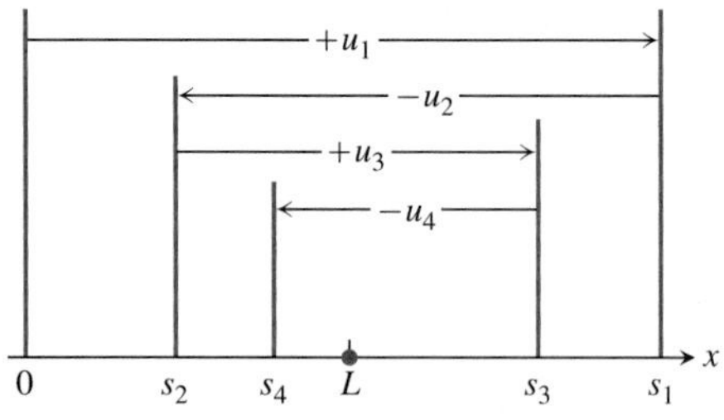

FIGURE 11.9 The partial sums of an alternating series that satisfies the hypotheses of Theorem 14 for $N = 1$ straddle the limit from the beginning.

A graphical interpretation of the partial sums (Figure 11.9) shows how an alternating series converges to its limit L when the three conditions of Theorem 14 are satisfied with $N = 1$. (Exercise 63 asks you to picture the case $N > 1$.) Starting from the origin of the x-axis, we lay off the positive distance $s_1 = u_1$. To find the point corresponding to $s_2 = u_1 - u_2$, we back up a distance equal to u_2. Since $u_2 \le u_1$, we do not back up any farther than the origin. We continue in this seesaw fashion, backing up or going forward as the signs in the series demand. But for $n \ge N$, each forward or backward step is shorter than (or at most the same size as) the preceding step, because $u_{n+1} \le u_n$. And since the nth term approaches zero as n increases, the size of step we take forward or backward gets smaller and smaller. We oscillate across the limit L, and the amplitude of oscillation approaches zero. The limit L lies between any two successive sums s_n and s_{n+1} and hence differs from s_n by an amount less than u_{n+1}.

Because

$$|L - s_n| < u_{n+1} \qquad \text{for } n \ge N,$$

we can make useful estimates of the sums of convergent alternating series.

THEOREM 15 The Alternating Series Estimation Theorem
If the alternating series $\sum_{n=1}^{\infty}(-1)^{n+1}u_n$ satisfies the three conditions of Theorem 14, then for $n \ge N$,

$$s_n = u_1 - u_2 + \cdots + (-1)^{n+1}u_n$$

approximates the sum L of the series with an error whose absolute value is less than u_{n+1}, the numerical value of the first unused term. Furthermore, the remainder, $L - s_n$, has the same sign as the first unused term.

We leave the verification of the sign of the remainder for Exercise 53.

EXAMPLE 2 We try Theorem 15 on a series whose sum we know:

$$\sum_{n=0}^{\infty}(-1)^n\frac{1}{2^n} = 1 - \frac{1}{2} + \frac{1}{4} - \frac{1}{8} + \frac{1}{16} - \frac{1}{32} + \frac{1}{64} - \frac{1}{128} \;\vdots + \frac{1}{256} - \cdots.$$

The theorem says that if we truncate the series after the eighth term, we throw away a total that is positive and less than $1/256$. The sum of the first eight terms is 0.6640625. The sum of the series is

$$\frac{1}{1-(-1/2)} = \frac{1}{3/2} = \frac{2}{3}.$$

The difference, $(2/3) - 0.6640625 = 0.0026041666\ldots$, is positive and less than $(1/256) = 0.00390625$. ■

Absolute and Conditional Convergence

DEFINITION Absolutely Convergent
A series Σa_n **converges absolutely** (is **absolutely convergent**) if the corresponding series of absolute values, $\Sigma |a_n|$, converges.

The geometric series

$$1 - \frac{1}{2} + \frac{1}{4} - \frac{1}{8} + \cdots$$

converges absolutely because the corresponding series of absolute values

$$1 + \frac{1}{2} + \frac{1}{4} + \frac{1}{8} + \cdots$$

converges. The alternating harmonic series does not converge absolutely. The corresponding series of absolute values is the (divergent) harmonic series.

DEFINITION Conditionally Convergent
A series that converges but does not converge absolutely **converges conditionally**.

The alternating harmonic series converges conditionally.

Absolute convergence is important for two reasons. First, we have good tests for convergence of series of positive terms. Second, if a series converges absolutely, then it converges. That is the thrust of the next theorem.

THEOREM 16 The Absolute Convergence Test
If $\sum_{n=1}^{\infty} |a_n|$ converges, then $\sum_{n=1}^{\infty} a_n$ converges.

Proof For each n,

$$-|a_n| \le a_n \le |a_n|, \qquad \text{so} \qquad 0 \le a_n + |a_n| \le 2|a_n|.$$

If $\sum_{n=1}^{\infty} |a_n|$ converges, then $\sum_{n=1}^{\infty} 2|a_n|$ converges and, by the Direct Comparison Test, the nonnegative series $\sum_{n=1}^{\infty} (a_n + |a_n|)$ converges. The equality $a_n = (a_n + |a_n|) - |a_n|$ now lets us express $\sum_{n=1}^{\infty} a_n$ as the difference of two convergent series:

$$\sum_{n=1}^{\infty} a_n = \sum_{n=1}^{\infty} (a_n + |a_n| - |a_n|) = \sum_{n=1}^{\infty} (a_n + |a_n|) - \sum_{n=1}^{\infty} |a_n|.$$

Therefore, $\sum_{n=1}^{\infty} a_n$ converges. ■

CAUTION We can rephrase Theorem 16 to say that every absolutely convergent series converges. However, the converse statement is false: Many convergent series do not converge absolutely (such as the alternating harmonic series in Example 1).

EXAMPLE 3 Applying the Absolute Convergence Test

(a) For $\sum_{n=1}^{\infty}(-1)^{n+1}\frac{1}{n^2} = 1 - \frac{1}{4} + \frac{1}{9} - \frac{1}{16} + \cdots$, the corresponding series of absolute values is the convergent series

$$\sum_{n=1}^{\infty}\frac{1}{n^2} = 1 + \frac{1}{4} + \frac{1}{9} + \frac{1}{16} + \cdots.$$

The original series converges because it converges absolutely.

(b) For $\sum_{n=1}^{\infty}\frac{\sin n}{n^2} = \frac{\sin 1}{1} + \frac{\sin 2}{4} + \frac{\sin 3}{9} + \cdots$, the corresponding series of absolute values is

$$\sum_{n=1}^{\infty}\left|\frac{\sin n}{n^2}\right| = \frac{|\sin 1|}{1} + \frac{|\sin 2|}{4} + \cdots,$$

which converges by comparison with $\sum_{n=1}^{\infty}(1/n^2)$ because $|\sin n| \le 1$ for every n. The original series converges absolutely; therefore it converges. ■

EXAMPLE 4 Alternating p-Series

If p is a positive constant, the sequence $\{1/n^p\}$ is a decreasing sequence with limit zero. Therefore the alternating p-series

$$\sum_{n=1}^{\infty}\frac{(-1)^{n-1}}{n^p} = 1 - \frac{1}{2^p} + \frac{1}{3^p} - \frac{1}{4^p} + \cdots, \qquad p > 0$$

converges.

If $p > 1$, the series converges absolutely. If $0 < p \le 1$, the series converges conditionally.

Conditional convergence: $\quad 1 - \frac{1}{\sqrt{2}} + \frac{1}{\sqrt{3}} - \frac{1}{\sqrt{4}} + \cdots$

Absolute convergence: $\quad 1 - \frac{1}{2^{3/2}} + \frac{1}{3^{3/2}} - \frac{1}{4^{3/2}} + \cdots$ ■

Rearranging Series

THEOREM 17 **The Rearrangement Theorem for Absolutely Convergent Series**

If $\sum_{n=1}^{\infty} a_n$ converges absolutely, and $b_1, b_2, \ldots, b_n, \ldots$ is any arrangement of the sequence $\{a_n\}$, then $\sum b_n$ converges absolutely and

$$\sum_{n=1}^{\infty} b_n = \sum_{n=1}^{\infty} a_n.$$

(For an outline of the proof, see Exercise 60.)

EXAMPLE 5 Applying the Rearrangement Theorem

As we saw in Example 3, the series

$$1 - \frac{1}{4} + \frac{1}{9} - \frac{1}{16} + \cdots + (-1)^{n-1}\frac{1}{n^2} + \cdots$$

converges absolutely. A possible rearrangement of the terms of the series might start with a positive term, then two negative terms, then three positive terms, then four negative terms, and so on: After k terms of one sign, take $k + 1$ terms of the opposite sign. The first ten terms of such a series look like this:

$$1 - \frac{1}{4} - \frac{1}{16} + \frac{1}{9} + \frac{1}{25} + \frac{1}{49} - \frac{1}{36} - \frac{1}{64} - \frac{1}{100} - \frac{1}{144} + \cdots.$$

The Rearrangement Theorem says that both series converge to the same value. In this example, if we had the second series to begin with, we would probably be glad to exchange it for the first, if we knew that we could. We can do even better: The sum of either series is also equal to

$$\sum_{n=1}^{\infty} \frac{1}{(2n - 1)^2} - \sum_{n=1}^{\infty} \frac{1}{(2n)^2}.$$

(See Exercise 61.) ■

If we rearrange infinitely many terms of a conditionally convergent series, we can get results that are far different from the sum of the original series. Here is an example.

EXAMPLE 6 Rearranging the Alternating Harmonic Series

The alternating harmonic series

$$\frac{1}{1} - \frac{1}{2} + \frac{1}{3} - \frac{1}{4} + \frac{1}{5} - \frac{1}{6} + \frac{1}{7} - \frac{1}{8} + \frac{1}{9} - \frac{1}{10} + \frac{1}{11} - \cdots$$

can be rearranged to diverge or to reach any preassigned sum.

(a) *Rearranging* $\sum_{n=1}^{\infty}(-1)^{n+1}/n$ *to diverge.* The series of terms $\sum[1/(2n - 1)]$ diverges to $+\infty$ and the series of terms $\sum(-1/2n)$ diverges to $-\infty$. No matter how far out in the sequence of odd-numbered terms we begin, we can always add enough positive terms to get an arbitrarily large sum. Similarly, with the negative terms, no matter how far out we start, we can add enough consecutive even-numbered terms to get a negative sum of arbitrarily large absolute value. If we wished to do so, we could start adding odd-numbered terms until we had a sum greater than $+3$, say, and then follow that with enough consecutive negative terms to make the new total less than -4. We could then add enough positive terms to make the total greater than $+5$ and follow with consecutive unused negative terms to make a new total less than -6, and so on. In this way, we could make the swings arbitrarily large in either direction.

(b) *Rearranging* $\sum_{n=1}^{\infty}(-1)^{n+1}/n$ *to converge to* 1. Another possibility is to focus on a particular limit. Suppose we try to get sums that converge to 1. We start with the first term, $1/1$, and then subtract $1/2$. Next we add $1/3$ and $1/5$, which brings the total back to 1 or above. Then we add consecutive negative terms until the total is less than 1. We continue in this manner: When the sum is less than 1, add positive terms until the total is 1 or more; then subtract (add negative) terms until the total is again less than 1. This process can be continued indefinitely. Because both the odd-numbered terms and the even-numbered terms of the original series approach zero as $n \to \infty$, the amount by which our partial sums exceed 1 or fall below it approaches zero. So the new series converges to 1. The rearranged series starts like this:

$$\frac{1}{1} - \frac{1}{2} + \frac{1}{3} + \frac{1}{5} - \frac{1}{4} + \frac{1}{7} + \frac{1}{9} - \frac{1}{6} + \frac{1}{11} + \frac{1}{13} - \frac{1}{8} + \frac{1}{15} + \frac{1}{17} - \frac{1}{10}$$
$$+ \frac{1}{19} + \frac{1}{21} - \frac{1}{12} + \frac{1}{23} + \frac{1}{25} - \frac{1}{14} + \frac{1}{27} - \frac{1}{16} + \cdots$$ ■

The kind of behavior illustrated by the series in Example 6 is typical of what can happen with any conditionally convergent series. Therefore we must always add the terms of a conditionally convergent series in the order given.

We have now developed several tests for convergence and divergence of series. In summary:

1. **The *n*th-Term Test:** Unless $a_n \to 0$, the series diverges.
2. **Geometric series:** $\sum ar^n$ converges if $|r| < 1$; otherwise it diverges.
3. ***p*-series:** $\sum 1/n^p$ converges if $p > 1$; otherwise it diverges.
4. **Series with nonnegative terms:** Try the Integral Test, Ratio Test, or Root Test. Try comparing to a known series with the Comparison Test.
5. **Series with some negative terms:** Does $\sum |a_n|$ converge? If yes, so does $\sum a_n$, since absolute convergence implies convergence.
6. **Alternating series:** $\sum a_n$ converges if the series satisfies the conditions of the Alternating Series Test.

EXERCISES 11.6

Determining Convergence or Divergence

Which of the alternating series in Exercises 1–10 converge, and which diverge? Give reasons for your answers.

1. $\sum_{n=1}^{\infty} (-1)^{n+1} \frac{1}{n^2}$
2. $\sum_{n=1}^{\infty} (-1)^{n+1} \frac{1}{n^{3/2}}$
3. $\sum_{n=1}^{\infty} (-1)^{n+1} \left(\frac{n}{10}\right)^n$
4. $\sum_{n=1}^{\infty} (-1)^{n+1} \frac{10^n}{n^{10}}$
5. $\sum_{n=2}^{\infty} (-1)^{n+1} \frac{1}{\ln n}$
6. $\sum_{n=1}^{\infty} (-1)^{n+1} \frac{\ln n}{n}$
7. $\sum_{n=2}^{\infty} (-1)^{n+1} \frac{\ln n}{\ln n^2}$
8. $\sum_{n=1}^{\infty} (-1)^{n} \ln\left(1 + \frac{1}{n}\right)$
9. $\sum_{n=1}^{\infty} (-1)^{n+1} \frac{\sqrt{n} + 1}{n + 1}$
10. $\sum_{n=1}^{\infty} (-1)^{n+1} \frac{3\sqrt{n + 1}}{\sqrt{n} + 1}$

Absolute Convergence

Which of the series in Exercises 11–44 converge absolutely, which converge, and which diverge? Give reasons for your answers.

11. $\sum_{n=1}^{\infty} (-1)^{n+1} (0.1)^n$
12. $\sum_{n=1}^{\infty} (-1)^{n+1} \frac{(0.1)^n}{n}$
13. $\sum_{n=1}^{\infty} (-1)^{n} \frac{1}{\sqrt{n}}$
14. $\sum_{n=1}^{\infty} \frac{(-1)^n}{1 + \sqrt{n}}$
15. $\sum_{n=1}^{\infty} (-1)^{n+1} \frac{n}{n^3 + 1}$
16. $\sum_{n=1}^{\infty} (-1)^{n+1} \frac{n!}{2^n}$
17. $\sum_{n=1}^{\infty} (-1)^{n} \frac{1}{n + 3}$
18. $\sum_{n=1}^{\infty} (-1)^{n} \frac{\sin n}{n^2}$
19. $\sum_{n=1}^{\infty} (-1)^{n+1} \frac{3 + n}{5 + n}$
20. $\sum_{n=2}^{\infty} (-1)^{n} \frac{1}{\ln (n^3)}$
21. $\sum_{n=1}^{\infty} (-1)^{n+1} \frac{1 + n}{n^2}$
22. $\sum_{n=1}^{\infty} \frac{(-2)^{n+1}}{n + 5^n}$
23. $\sum_{n=1}^{\infty} (-1)^{n} n^2 (2/3)^n$
24. $\sum_{n=1}^{\infty} (-1)^{n+1} \left(\sqrt[n]{10}\right)$
25. $\sum_{n=1}^{\infty} (-1)^{n} \frac{\tan^{-1} n}{n^2 + 1}$
26. $\sum_{n=2}^{\infty} (-1)^{n+1} \frac{1}{n \ln n}$

27. $\sum_{n=1}^{\infty}(-1)^n \frac{n}{n+1}$

28. $\sum_{n=1}^{\infty}(-1)^n \frac{\ln n}{n-\ln n}$

29. $\sum_{n=1}^{\infty} \frac{(-100)^n}{n!}$

30. $\sum_{n=1}^{\infty}(-5)^{-n}$

31. $\sum_{n=1}^{\infty} \frac{(-1)^{n-1}}{n^2+2n+1}$

32. $\sum_{n=2}^{\infty}(-1)^n\left(\frac{\ln n}{\ln n^2}\right)^n$

33. $\sum_{n=1}^{\infty} \frac{\cos n\pi}{n\sqrt{n}}$

34. $\sum_{n=1}^{\infty} \frac{\cos n\pi}{n}$

35. $\sum_{n=1}^{\infty} \frac{(-1)^n(n+1)^n}{(2n)^n}$

36. $\sum_{n=1}^{\infty} \frac{(-1)^{n+1}(n!)^2}{(2n)!}$

37. $\sum_{n=1}^{\infty}(-1)^n \frac{(2n)!}{2^n n! n}$

38. $\sum_{n=1}^{\infty}(-1)^n \frac{(n!)^2 3^n}{(2n+1)!}$

39. $\sum_{n=1}^{\infty}(-1)^n\left(\sqrt{n+1}-\sqrt{n}\right)$

40. $\sum_{n=1}^{\infty}(-1)^n\left(\sqrt{n^2+n}-n\right)$

41. $\sum_{n=1}^{\infty}(-1)^n\left(\sqrt{n+\sqrt{n}}-\sqrt{n}\right)$

42. $\sum_{n=1}^{\infty} \frac{(-1)^n}{\sqrt{n}+\sqrt{n+1}}$

43. $\sum_{n=1}^{\infty}(-1)^n \operatorname{sech} n$

44. $\sum_{n=1}^{\infty}(-1)^n \operatorname{csch} n$

Error Estimation

In Exercises 45–48, estimate the magnitude of the error involved in using the sum of the first four terms to approximate the sum of the entire series.

45. $\sum_{n=1}^{\infty}(-1)^{n+1} \frac{1}{n}$ It can be shown that the sum is ln 2.

46. $\sum_{n=1}^{\infty}(-1)^{n+1} \frac{1}{10^n}$

47. $\sum_{n=1}^{\infty}(-1)^{n+1} \frac{(0.01)^n}{n}$ As you will see in Section 11.7, the sum is ln (1.01).

48. $\frac{1}{1+t}=\sum_{n=0}^{\infty}(-1)^n t^n, \quad 0<t<1$

T Approximate the sums in Exercises 49 and 50 with an error of magnitude less than 5×10^{-6}.

49. $\sum_{n=0}^{\infty}(-1)^n \frac{1}{(2n)!}$ As you will see in Section 11.9, the sum is cos 1, the cosine of 1 radian.

50. $\sum_{n=0}^{\infty}(-1)^n \frac{1}{n!}$ As you will see in Section 11.9, the sum is e^{-1}.

Theory and Examples

51. **a.** The series

$$\frac{1}{3}-\frac{1}{2}+\frac{1}{9}-\frac{1}{4}+\frac{1}{27}-\frac{1}{8}+\cdots+\frac{1}{3^n}-\frac{1}{2^n}+\cdots$$

does not meet one of the conditions of Theorem 14. Which one?

b. Find the sum of the series in part (a).

T 52. The limit L of an alternating series that satisfies the conditions of Theorem 14 lies between the values of any two consecutive partial sums. This suggests using the average

$$\frac{s_n+s_{n+1}}{2}=s_n+\frac{1}{2}(-1)^{n+2}a_{n+1}$$

to estimate L. Compute

$$s_{20}+\frac{1}{2}\cdot\frac{1}{21}$$

as an approximation to the sum of the alternating harmonic series. The exact sum is $\ln 2 = 0.6931\ldots$.

53. **The sign of the remainder of an alternating series that satisfies the conditions of Theorem 14** Prove the assertion in Theorem 15 that whenever an alternating series satisfying the conditions of Theorem 14 is approximated with one of its partial sums, then the remainder (sum of the unused terms) has the same sign as the first unused term. (*Hint:* Group the remainder's terms in consecutive pairs.)

54. Show that the sum of the first $2n$ terms of the series

$$1-\frac{1}{2}+\frac{1}{2}-\frac{1}{3}+\frac{1}{3}-\frac{1}{4}+\frac{1}{4}-\frac{1}{5}+\frac{1}{5}-\frac{1}{6}+\cdots$$

is the same as the sum of the first n terms of the series

$$\frac{1}{1\cdot 2}+\frac{1}{2\cdot 3}+\frac{1}{3\cdot 4}+\frac{1}{4\cdot 5}+\frac{1}{5\cdot 6}+\cdots.$$

Do these series converge? What is the sum of the first $2n+1$ terms of the first series? If the series converge, what is their sum?

55. Show that if $\sum_{n=1}^{\infty} a_n$ diverges, then $\sum_{n=1}^{\infty}|a_n|$ diverges.

56. Show that if $\sum_{n=1}^{\infty} a_n$ converges absolutely, then

$$\left|\sum_{n=1}^{\infty} a_n\right| \leq \sum_{n=1}^{\infty}|a_n|.$$

57. Show that if $\sum_{n=1}^{\infty} a_n$ and $\sum_{n=1}^{\infty} b_n$ both converge absolutely, then so does

a. $\sum_{n=1}^{\infty}(a_n+b_n)$ **b.** $\sum_{n=1}^{\infty}(a_n-b_n)$

c. $\sum_{n=1}^{\infty} k a_n$ (k any number)

58. Show by example that $\sum_{n=1}^{\infty} a_n b_n$ may diverge even if $\sum_{n=1}^{\infty} a_n$ and $\sum_{n=1}^{\infty} b_n$ both converge.

T 59. In Example 6, suppose the goal is to arrange the terms to get a new series that converges to $-1/2$. Start the new arrangement with the first negative term, which is $-1/2$. Whenever you have a sum that is less than or equal to $-1/2$, start introducing positive terms, taken in order, until the new total is greater than $-1/2$. Then add negative terms until the total is less than or equal to $-1/2$ again. Continue this process until your partial sums have been above the target at least three times and finish at or below it. If s_n is the sum of the first n terms of your new series, plot the points (n, s_n) to illustrate how the sums are behaving.

60. Outline of the proof of the Rearrangement Theorem (Theorem 17)

a. Let ϵ be a positive real number, let $L = \sum_{n=1}^{\infty} a_n$, and let $s_k = \sum_{n=1}^{k} a_n$. Show that for some index N_1 and for some index $N_2 \geq N_1$,

$$\sum_{n=N_1}^{\infty} |a_n| < \frac{\epsilon}{2} \quad \text{and} \quad |s_{N_2} - L| < \frac{\epsilon}{2}.$$

Since all the terms $a_1, a_2, \ldots, a_{N_2}$ appear somewhere in the sequence $\{b_n\}$, there is an index $N_3 \geq N_2$ such that if $n \geq N_3$, then $\left(\sum_{k=1}^{n} b_k\right) - s_{N_2}$ is at most a sum of terms a_m with $m \geq N_1$. Therefore, if $n \geq N_3$,

$$\left|\sum_{k=1}^{n} b_k - L\right| \leq \left|\sum_{k=1}^{n} b_k - s_{N_2}\right| + |s_{N_2} - L|$$
$$\leq \sum_{k=N_1}^{\infty} |a_k| + |s_{N_2} - L| < \epsilon.$$

b. The argument in part (a) shows that if $\sum_{n=1}^{\infty} a_n$ converges absolutely then $\sum_{n=1}^{\infty} b_n$ converges and $\sum_{n=1}^{\infty} b_n = \sum_{n=1}^{\infty} a_n$. Now show that because $\sum_{n=1}^{\infty} |a_n|$ converges, $\sum_{n=1}^{\infty} |b_n|$ converges to $\sum_{n=1}^{\infty} |a_n|$.

61. Unzipping absolutely convergent series

a. Show that if $\sum_{n=1}^{\infty} |a_n|$ converges and

$$b_n = \begin{cases} a_n, & \text{if } a_n \geq 0 \\ 0, & \text{if } a_n < 0, \end{cases}$$

then $\sum_{n=1}^{\infty} b_n$ converges.

b. Use the results in part (a) to show likewise that if $\sum_{n=1}^{\infty} |a_n|$ converges and

$$c_n = \begin{cases} 0, & \text{if } a_n \geq 0 \\ a_n, & \text{if } a_n < 0, \end{cases}$$

then $\sum_{n=1}^{\infty} c_n$ converges.

In other words, if a series converges absolutely, its positive terms form a convergent series, and so do its negative terms. Furthermore,

$$\sum_{n=1}^{\infty} a_n = \sum_{n=1}^{\infty} b_n + \sum_{n=1}^{\infty} c_n$$

because $b_n = (a_n + |a_n|)/2$ and $c_n = (a_n - |a_n|)/2$.

62. What is wrong here?:

Multiply both sides of the alternating harmonic series

$$S = 1 - \frac{1}{2} + \frac{1}{3} - \frac{1}{4} + \frac{1}{5} - \frac{1}{6} + \frac{1}{7} - \frac{1}{8} + \frac{1}{9} - \frac{1}{10} + \frac{1}{11} - \frac{1}{12} + \cdots$$

by 2 to get

$$2S = 2 - 1 + \frac{2}{3} - \frac{1}{2} + \frac{2}{5} - \frac{1}{3} + \frac{2}{7} - \frac{1}{4} + \frac{2}{9} - \frac{1}{5} + \frac{2}{11} - \frac{1}{6} + \cdots.$$

Collect terms with the same denominator, as the arrows indicate, to arrive at

$$2S = 1 - \frac{1}{2} + \frac{1}{3} - \frac{1}{4} + \frac{1}{5} - \frac{1}{6} + \cdots.$$

The series on the right-hand side of this equation is the series we started with. Therefore, $2S = S$, and dividing by S gives $2 = 1$. (*Source:* "Riemann's Rearrangement Theorem" by Stewart Galanor, *Mathematics Teacher*, Vol. 80, No. 8, 1987, pp. 675–681.)

63. Draw a figure similar to Figure 11.9 to illustrate the convergence of the series in Theorem 14 when $N > 1$.

11.7 Power Series

Now that we can test infinite series for convergence we can study the infinite polynomials mentioned at the beginning of this chapter. We call these polynomials power series because they are defined as infinite series of powers of some variable, in our case x. Like polynomials, power series can be added, subtracted, multiplied, differentiated, and integrated to give new power series.

Power Series and Convergence

We begin with the formal definition.

DEFINITIONS Power Series, Center, Coefficients

A **power series about $x = 0$** is a series of the form

$$\sum_{n=0}^{\infty} c_n x^n = c_0 + c_1 x + c_2 x^2 + \cdots + c_n x^n + \cdots. \tag{1}$$

A **power series about $x = a$** is a series of the form

$$\sum_{n=0}^{\infty} c_n (x-a)^n = c_0 + c_1(x-a) + c_2(x-a)^2 + \cdots + c_n(x-a)^n + \cdots \tag{2}$$

in which the **center** a and the **coefficients** $c_0, c_1, c_2, \ldots, c_n, \ldots$ are constants.

Equation (1) is the special case obtained by taking $a = 0$ in Equation (2).

EXAMPLE 1 A Geometric Series

Taking all the coefficients to be 1 in Equation (1) gives the geometric power series

$$\sum_{n=0}^{\infty} x^n = 1 + x + x^2 + \cdots + x^n + \cdots.$$

This is the geometric series with first term 1 and ratio x. It converges to $1/(1-x)$ for $|x| < 1$. We express this fact by writing

$$\frac{1}{1-x} = 1 + x + x^2 + \cdots + x^n + \cdots, \qquad -1 < x < 1. \tag{3}$$

■

Up to now, we have used Equation (3) as a formula for the sum of the series on the right. We now change the focus: We think of the partial sums of the series on the right as polynomials $P_n(x)$ that approximate the function on the left. For values of x near zero, we need take only a few terms of the series to get a good approximation. As we move toward $x = 1$, or -1, we must take more terms. Figure 11.10 shows the graphs of $f(x) = 1/(1-x)$, and the approximating polynomials $y_n = P_n(x)$ for $n = 0, 1, 2$, and 8. The function $f(x) = 1/(1-x)$ is not continuous on intervals containing $x = 1$, where it has a vertical asymptote. The approximations do not apply when $x \geq 1$.

EXAMPLE 2 A Geometric Series

The power series

$$1 - \frac{1}{2}(x-2) + \frac{1}{4}(x-2)^2 + \cdots + \left(-\frac{1}{2}\right)^n (x-2)^n + \cdots \tag{4}$$

matches Equation (2) with $a = 2$, $c_0 = 1$, $c_1 = -1/2$, $c_2 = 1/4, \ldots, c_n = (-1/2)^n$. This is a geometric series with first term 1 and ratio $r = -\dfrac{x-2}{2}$. The series converges for

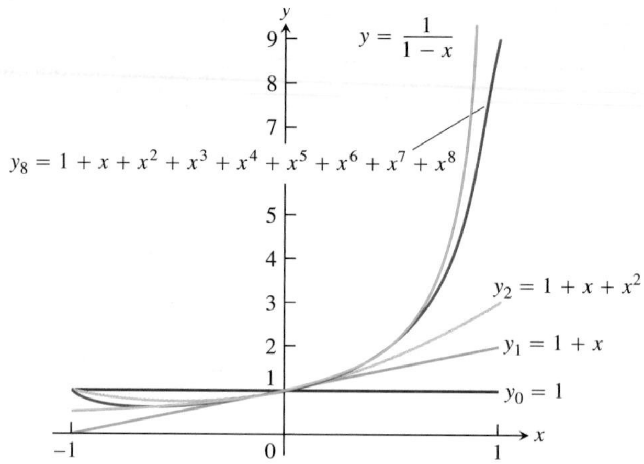

FIGURE 11.10 The graphs of $f(x) = 1/(1 - x)$ and four of its polynomial approximations (Example 1).

$\left|\frac{x-2}{2}\right| < 1$ or $0 < x < 4$. The sum is

$$\frac{1}{1-r} = \frac{1}{1 + \frac{x-2}{2}} = \frac{2}{x},$$

so

$$\frac{2}{x} = 1 - \frac{(x-2)}{2} + \frac{(x-2)^2}{4} - \cdots + \left(-\frac{1}{2}\right)^n (x-2)^n + \cdots, \qquad 0 < x < 4.$$

Series (4) generates useful polynomial approximations of $f(x) = 2/x$ for values of x near 2:

$$P_0(x) = 1$$

$$P_1(x) = 1 - \frac{1}{2}(x-2) = 2 - \frac{x}{2}$$

$$P_2(x) = 1 - \frac{1}{2}(x-2) + \frac{1}{4}(x-2)^2 = 3 - \frac{3x}{2} + \frac{x^2}{4},$$

and so on (Figure 11.11). ■

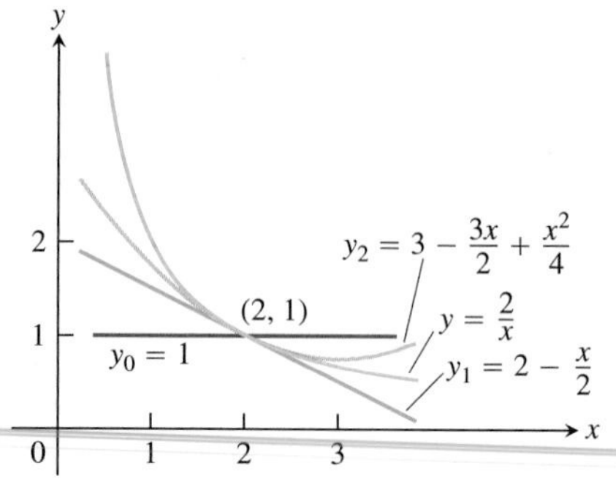

FIGURE 11.11 The graphs of $f(x) = 2/x$ and its first three polynomial approximations (Example 2).

EXAMPLE 3 Testing for Convergence Using the Ratio Test

For what values of x do the following power series converge?

(a) $\sum_{n=1}^{\infty} (-1)^{n-1} \frac{x^n}{n} = x - \frac{x^2}{2} + \frac{x^3}{3} - \cdots$

(b) $\sum_{n=1}^{\infty} (-1)^{n-1} \frac{x^{2n-1}}{2n-1} = x - \frac{x^3}{3} + \frac{x^5}{5} - \cdots$

(c) $\sum_{n=0}^{\infty} \frac{x^n}{n!} = 1 + x + \frac{x^2}{2!} + \frac{x^3}{3!} + \cdots$

(d) $\sum_{n=0}^{\infty} n!x^n = 1 + x + 2!x^2 + 3!x^3 + \cdots$

Solution Apply the Ratio Test to the series $\sum|u_n|$, where u_n is the nth term of the series in question.

(a) $$\left|\frac{u_{n+1}}{u_n}\right| = \frac{n}{n+1}|x| \rightarrow |x|.$$

The series converges absolutely for $|x| < 1$. It diverges if $|x| > 1$ because the nth term does not converge to zero. At $x = 1$, we get the alternating harmonic series $1 - 1/2 + 1/3 - 1/4 + \cdots$, which converges. At $x = -1$ we get $-1 - 1/2 - 1/3 - 1/4 - \cdots$, the negative of the harmonic series; it diverges. Series (a) converges for $-1 < x \leq 1$ and diverges elsewhere.

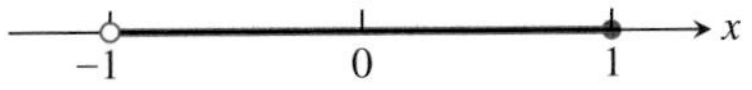

(b) $$\left|\frac{u_{n+1}}{u_n}\right| = \frac{2n-1}{2n+1}x^2 \rightarrow x^2.$$

The series converges absolutely for $x^2 < 1$. It diverges for $x^2 > 1$ because the nth term does not converge to zero. At $x = 1$ the series becomes $1 - 1/3 + 1/5 - 1/7 + \cdots$, which converges by the Alternating Series Theorem. It also converges at $x = -1$ because it is again an alternating series that satisfies the conditions for convergence. The value at $x = -1$ is the negative of the value at $x = 1$. Series (b) converges for $-1 \leq x \leq 1$ and diverges elsewhere.

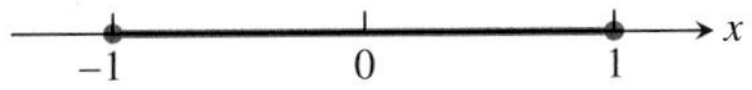

(c) $$\left|\frac{u_{n+1}}{u_n}\right| = \left|\frac{x^{n+1}}{(n+1)!} \cdot \frac{n!}{x^n}\right| = \frac{|x|}{n+1} \rightarrow 0 \text{ for every } x.$$

The series converges absolutely for all x.

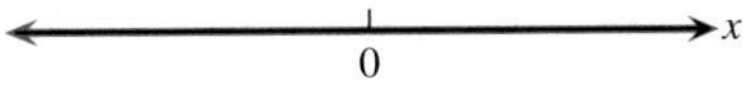

(d) $$\left|\frac{u_{n+1}}{u_n}\right| = \left|\frac{(n+1)!x^{n+1}}{n!x^n}\right| = (n+1)|x| \rightarrow \infty \text{ unless } x = 0.$$

The series diverges for all values of x except $x = 0$.

■

Example 3 illustrates how we usually test a power series for convergence, and the possible results.

THEOREM 18 The Convergence Theorem for Power Series

If the power series $\sum_{n=0}^{\infty} a_n x^n = a_0 + a_1 x + a_2 x^2 + \cdots$ converges for $x = c \neq 0$, then it converges absolutely for all x with $|x| < |c|$. If the series diverges for $x = d$, then it diverges for all x with $|x| > |d|$.

Proof Suppose the series $\sum_{n=0}^{\infty} a_n c^n$ converges. Then $\lim_{n\to\infty} a_n c^n = 0$. Hence, there is an integer N such that $|a_n c^n| < 1$ for all $n \geq N$. That is,

$$|a_n| < \frac{1}{|c|^n} \qquad \text{for } n \geq N. \tag{5}$$

Now take any x such that $|x| < |c|$ and consider

$$|a_0| + |a_1 x| + \cdots + |a_{N-1}x^{N-1}| + |a_N x^N| + |a_{N+1}x^{N+1}| + \cdots.$$

There are only a finite number of terms prior to $|a_N x^N|$, and their sum is finite. Starting with $|a_N x^N|$ and beyond, the terms are less than

$$\left|\frac{x}{c}\right|^N + \left|\frac{x}{c}\right|^{N+1} + \left|\frac{x}{c}\right|^{N+2} + \cdots \tag{6}$$

because of Inequality (5). But Series (6) is a geometric series with ratio $r = |x/c|$, which is less than 1, since $|x| < |c|$. Hence Series (6) converges, so the original series converges absolutely. This proves the first half of the theorem.

The second half of the theorem follows from the first. If the series diverges at $x = d$ and converges at a value x_0 with $|x_0| > |d|$, we may take $c = x_0$ in the first half of the theorem and conclude that the series converges absolutely at d. But the series cannot converge absolutely and diverge at one and the same time. Hence, if it diverges at d, it diverges for all x with $|x| > |d|$. ■

To simplify the notation, Theorem 18 deals with the convergence of series of the form $\sum a_n x^n$. For series of the form $\sum a_n(x - a)^n$ we can replace $x - a$ by x' and apply the results to the series $\sum a_n(x')^n$.

The Radius of Convergence of a Power Series

The theorem we have just proved and the examples we have studied lead to the conclusion that a power series $\sum c_n(x - a)^n$ behaves in one of three possible ways. It might converge only at $x = a$, or converge everywhere, or converge on some interval of radius R centered at $x = a$. We prove this as a Corollary to Theorem 18.

COROLLARY TO THEOREM 18

The convergence of the series $\sum c_n(x - a)^n$ is described by one of the following three possibilities:

1. There is a positive number R such that the series diverges for x with $|x - a| > R$ but converges absolutely for x with $|x - a| < R$. The series may or may not converge at either of the endpoints $x = a - R$ and $x = a + R$.
2. The series converges absolutely for every x $(R = \infty)$.
3. The series converges at $x = a$ and diverges elsewhere $(R = 0)$.

Proof We assume first that $a = 0$, so that the power series is centered at 0. If the series converges everywhere we are in Case 2. If it converges only at $x = 0$ we are in Case 3. Otherwise there is a nonzero number d such that $\sum c_n d^n$ diverges. The set S of values of x for which the series $\sum c_n x^n$ converges is nonempty because it contains 0 and a positive number p as well. By Theorem 18, the series diverges for all x with $|x| > |d|$, so $|x| \le |d|$ for all $x \in S$, and S is a bounded set. By the Completeness Property of the real numbers (see Appendix 4) a nonempty, bounded set has a least upper bound R. (The least upper bound is the smallest number with the property that the elements $x \in S$ satisfy $x \le R$.) If $|x| > R \ge p$, then $x \notin S$ so the series $\sum c_n x^n$ diverges. If $|x| < R$, then $|x|$ is not an upper bound for S (because it's smaller than the least upper bound) so there is a number $b \in S$ such that $b > |x|$. Since $b \in S$, the series $\sum c_n b^n$ converges and therefore the series $\sum c_n |x|^n$ converges by Theorem 18. This proves the Corollary for power series centered at $a = 0$.

For a power series centered at $a \ne 0$, we set $x' = (x - a)$ and repeat the argument with x'. Since $x' = 0$ when $x = a$, a radius R interval of convergence for $\sum c_n (x')^n$ centered at $x' = 0$ is the same as a radius R interval of convergence for $\sum c_n (x - a)^n$ centered at $x = a$. This establishes the Corollary for the general case. ■

R is called the **radius of convergence** of the power series and the interval of radius R centered at $x = a$ is called the **interval of convergence**. The interval of convergence may be open, closed, or half-open, depending on the particular series. At points x with $|x - a| < R$, the series converges absolutely. If the series converges for all values of x, we say its radius of convergence is infinite. If it converges only at $x = a$, we say its radius of convergence is zero.

How to Test a Power Series for Convergence

1. *Use the Ratio Test (or nth-Root Test) to find the interval where the series converges absolutely.* Ordinarily, this is an open interval

$$|x - a| < R \qquad \text{or} \qquad a - R < x < a + R.$$

2. *If the interval of absolute convergence is finite, test for convergence or divergence at each endpoint*, as in Examples 3a and b. Use a Comparison Test, the Integral Test, or the Alternating Series Test.
3. *If the interval of absolute convergence is $a - R < x < a + R$, the series diverges for $|x - a| > R$* (it does not even converge conditionally), because the nth term does not approach zero for those values of x.

Term-by-Term Differentiation

A theorem from advanced calculus says that a power series can be differentiated term by term at each interior point of its interval of convergence.

THEOREM 19 The Term-by-Term Differentiation Theorem
If $\sum c_n(x-a)^n$ converges for $a - R < x < a + R$ for some $R > 0$, it defines a function f:

$$f(x) = \sum_{n=0}^{\infty} c_n(x-a)^n, \qquad a - R < x < a + R.$$

Such a function f has derivatives of all orders inside the interval of convergence. We can obtain the derivatives by differentiating the original series term by term:

$$f'(x) = \sum_{n=1}^{\infty} nc_n(x-a)^{n-1}$$

$$f''(x) = \sum_{n=2}^{\infty} n(n-1)c_n(x-a)^{n-2},$$

and so on. Each of these derived series converges at every interior point of the interval of convergence of the original series.

EXAMPLE 4 Applying Term-by-Term Differentiation

Find series for $f'(x)$ and $f''(x)$ if

$$f(x) = \frac{1}{1-x} = 1 + x + x^2 + x^3 + x^4 + \cdots + x^n + \cdots$$
$$= \sum_{n=0}^{\infty} x^n, \qquad -1 < x < 1$$

Solution

$$f'(x) = \frac{1}{(1-x)^2} = 1 + 2x + 3x^2 + 4x^3 + \cdots + nx^{n-1} + \cdots$$
$$= \sum_{n=1}^{\infty} nx^{n-1}, \qquad -1 < x < 1$$
$$f''(x) = \frac{2}{(1-x)^3} = 2 + 6x + 12x^2 + \cdots + n(n-1)x^{n-2} + \cdots$$
$$= \sum_{n=2}^{\infty} n(n-1)x^{n-2}, \qquad -1 < x < 1$$

■

CAUTION Term-by-term differentiation might not work for other kinds of series. For example, the trigonometric series

$$\sum_{n=1}^{\infty} \frac{\sin(n!x)}{n^2}$$

converges for all x. But if we differentiate term by term we get the series

$$\sum_{n=1}^{\infty} \frac{n!\cos(n!x)}{n^2},$$

which diverges for all x. This is not a power series, since it is not a sum of positive integer powers of x.

Term-by-Term Integration

Another advanced calculus theorem states that a power series can be integrated term by term throughout its interval of convergence.

THEOREM 20 The Term-by-Term Integration Theorem
Suppose that

$$f(x) = \sum_{n=0}^{\infty} c_n(x-a)^n$$

converges for $a - R < x < a + R$ $(R > 0)$. Then

$$\sum_{n=0}^{\infty} c_n \frac{(x-a)^{n+1}}{n+1}$$

converges for $a - R < x < a + R$ and

$$\int f(x)\,dx = \sum_{n=0}^{\infty} c_n \frac{(x-a)^{n+1}}{n+1} + C$$

for $a - R < x < a + R$.

EXAMPLE 5 A Series for $\tan^{-1} x$, $-1 \le x \le 1$

Identify the function

$$f(x) = x - \frac{x^3}{3} + \frac{x^5}{5} - \cdots, \qquad -1 \le x \le 1.$$

Solution We differentiate the original series term by term and get

$$f'(x) = 1 - x^2 + x^4 - x^6 + \cdots, \qquad -1 < x < 1.$$

This is a geometric series with first term 1 and ratio $-x^2$, so

$$f'(x) = \frac{1}{1-(-x^2)} = \frac{1}{1+x^2}.$$

We can now integrate $f'(x) = 1/(1+x^2)$ to get

$$\int f'(x)\,dx = \int \frac{dx}{1+x^2} = \tan^{-1} x + C.$$

The series for $f(x)$ is zero when $x = 0$, so $C = 0$. Hence

$$f(x) = x - \frac{x^3}{3} + \frac{x^5}{5} - \frac{x^7}{7} + \cdots = \tan^{-1} x, \qquad -1 < x < 1. \tag{7}$$

In Section 11.10, we will see that the series also converges to $\tan^{-1} x$ at $x = \pm 1$. ■

Notice that the original series in Example 5 converges at both endpoints of the original interval of convergence, but Theorem 20 can guarantee the convergence of the differentiated series only inside the interval.

EXAMPLE 6 A Series for $\ln(1 + x)$, $-1 < x \le 1$

The series

$$\frac{1}{1+t} = 1 - t + t^2 - t^3 + \cdots$$

converges on the open interval $-1 < t < 1$. Therefore,

$$\begin{aligned}\ln(1+x) &= \int_0^x \frac{1}{1+t}\,dt = t - \frac{t^2}{2} + \frac{t^3}{3} - \frac{t^4}{4} + \cdots\Big]_0^x \qquad \text{Theorem 20}\\ &= x - \frac{x^2}{2} + \frac{x^3}{3} - \frac{x^4}{4} + \cdots, \qquad -1 < x < 1.\end{aligned}$$

It can also be shown that the series converges at $x = 1$ to the number ln 2, but that was not guaranteed by the theorem. ■

USING TECHNOLOGY Study of Series

Series are in many ways analogous to integrals. Just as the number of functions with explicit antiderivatives in terms of elementary functions is small compared to the number of integrable functions, the number of power series in x that agree with explicit elementary functions on x-intervals is small compared to the number of power series that converge on some x-interval. Graphing utilities can aid in the study of such series in much the same way that numerical integration aids in the study of definite integrals. The ability to study power series at particular values of x is built into most Computer Algebra Systems.

If a series converges rapidly enough, CAS exploration might give us an idea of the sum. For instance, in calculating the early partial sums of the series $\sum_{k=1}^{\infty}[1/(2^{k-1})]$ (Section 11.4, Example 2b), Maple returns $S_n = 1.6066\,95152$ for $31 \le n \le 200$. This suggests that the sum of the series is 1.6066 95152 to 10 digits. Indeed,

$$\sum_{k=201}^{\infty}\frac{1}{2^k - 1} = \sum_{k=201}^{\infty}\frac{1}{2^{k-1}(2 - (1/2^{k-1}))} < \sum_{k=201}^{\infty}\frac{1}{2^{k-1}} = \frac{1}{2^{199}} < 1.25 \times 10^{-60}.$$

The remainder after 200 terms is negligible.

However, CAS and calculator exploration cannot do much for us if the series converges or diverges very slowly, and indeed can be downright misleading. For example, try calculating the partial sums of the series $\sum_{k=1}^{\infty}[1/(10^{10}k)]$. The terms are tiny in comparison to the numbers we normally work with and the partial sums, even for hundreds of terms, are miniscule. We might well be fooled into thinking that the series converges. In fact, it diverges, as we can see by writing it as $(1/10^{10})\sum_{k=1}^{\infty}(1/k)$, a constant times the harmonic series.

We will know better how to interpret numerical results after studying error estimates in Section 11.9.

Multiplication of Power Series

Another theorem from advanced calculus states that absolutely converging power series can be multiplied the way we multiply polynomials. We omit the proof.

THEOREM 21 The Series Multiplication Theorem for Power Series
If $A(x) = \sum_{n=0}^{\infty} a_n x^n$ and $B(x) = \sum_{n=0}^{\infty} b_n x^n$ converge absolutely for $|x| < R$, and

$$c_n = a_0 b_n + a_1 b_{n-1} + a_2 b_{n-2} + \cdots + a_{n-1} b_1 + a_n b_0 = \sum_{k=0}^{n} a_k b_{n-k},$$

then $\sum_{n=0}^{\infty} c_n x^n$ converges absolutely to $A(x)B(x)$ for $|x| < R$:

$$\left(\sum_{n=0}^{\infty} a_n x^n\right) \cdot \left(\sum_{n=0}^{\infty} b_n x^n\right) = \sum_{n=0}^{\infty} c_n x^n.$$

EXAMPLE 7 Multiply the geometric series

$$\sum_{n=0}^{\infty} x^n = 1 + x + x^2 + \cdots + x^n + \cdots = \frac{1}{1-x}, \qquad \text{for } |x| < 1,$$

by itself to get a power series for $1/(1-x)^2$, for $|x| < 1$.

Solution Let

$$A(x) = \sum_{n=0}^{\infty} a_n x^n = 1 + x + x^2 + \cdots + x^n + \cdots = 1/(1-x)$$

$$B(x) = \sum_{n=0}^{\infty} b_n x^n = 1 + x + x^2 + \cdots + x^n + \cdots = 1/(1-x)$$

and

$$c_n = \underbrace{a_0 b_n + a_1 b_{n-1} + \cdots + a_k b_{n-k} + \cdots + a_n b_0}_{n+1 \text{ terms}}$$

$$= \underbrace{1 + 1 + \cdots + 1}_{n+1 \text{ ones}} = n + 1.$$

Then, by the Series Multiplication Theorem,

$$A(x) \cdot B(x) = \sum_{n=0}^{\infty} c_n x^n = \sum_{n=0}^{\infty} (n+1) x^n$$

$$= 1 + 2x + 3x^2 + 4x^3 + \cdots + (n+1)x^n + \cdots$$

is the series for $1/(1-x)^2$. The series all converge absolutely for $|x| < 1$.

Notice that Example 4 gives the same answer because

$$\frac{d}{dx}\left(\frac{1}{1-x}\right) = \frac{1}{(1-x)^2}.$$

■

EXERCISES 11.7

Intervals of Convergence

In Exercises 1–32, **(a)** find the series' radius and interval of convergence. For what values of x does the series converge **(b)** absolutely, **(c)** conditionally?

1. $\sum_{n=0}^{\infty} x^n$
2. $\sum_{n=0}^{\infty} (x+5)^n$
3. $\sum_{n=0}^{\infty} (-1)^n(4x+1)^n$
4. $\sum_{n=1}^{\infty} \frac{(3x-2)^n}{n}$
5. $\sum_{n=0}^{\infty} \frac{(x-2)^n}{10^n}$
6. $\sum_{n=0}^{\infty} (2x)^n$
7. $\sum_{n=0}^{\infty} \frac{nx^n}{n+2}$
8. $\sum_{n=1}^{\infty} \frac{(-1)^n(x+2)^n}{n}$
9. $\sum_{n=1}^{\infty} \frac{x^n}{n\sqrt{n}\,3^n}$
10. $\sum_{n=1}^{\infty} \frac{(x-1)^n}{\sqrt{n}}$
11. $\sum_{n=0}^{\infty} \frac{(-1)^n x^n}{n!}$
12. $\sum_{n=0}^{\infty} \frac{3^n x^n}{n!}$
13. $\sum_{n=0}^{\infty} \frac{x^{2n+1}}{n!}$
14. $\sum_{n=0}^{\infty} \frac{(2x+3)^{2n+1}}{n!}$
15. $\sum_{n=0}^{\infty} \frac{x^n}{\sqrt{n^2+3}}$
16. $\sum_{n=0}^{\infty} \frac{(-1)^n x^n}{\sqrt{n^2+3}}$
17. $\sum_{n=0}^{\infty} \frac{n(x+3)^n}{5^n}$
18. $\sum_{n=0}^{\infty} \frac{nx^n}{4^n(n^2+1)}$
19. $\sum_{n=0}^{\infty} \frac{\sqrt{n}x^n}{3^n}$
20. $\sum_{n=1}^{\infty} \sqrt[n]{n}(2x+5)^n$
21. $\sum_{n=1}^{\infty} \left(1+\frac{1}{n}\right)^n x^n$
22. $\sum_{n=1}^{\infty} (\ln n)x^n$
23. $\sum_{n=1}^{\infty} n^n x^n$
24. $\sum_{n=0}^{\infty} n!(x-4)^n$
25. $\sum_{n=1}^{\infty} \frac{(-1)^{n+1}(x+2)^n}{n2^n}$
26. $\sum_{n=0}^{\infty} (-2)^n(n+1)(x-1)^n$
27. $\sum_{n=2}^{\infty} \frac{x^n}{n(\ln n)^2}$ Get the information you need about $\sum 1/(n(\ln n)^2)$ from Section 11.3, Exercise 39.
28. $\sum_{n=2}^{\infty} \frac{x^n}{n \ln n}$ Get the information you need about $\sum 1/(n \ln n)$ from Section 11.3, Exercise 38.
29. $\sum_{n=1}^{\infty} \frac{(4x-5)^{2n+1}}{n^{3/2}}$
30. $\sum_{n=1}^{\infty} \frac{(3x+1)^{n+1}}{2n+2}$
31. $\sum_{n=1}^{\infty} \frac{(x+\pi)^n}{\sqrt{n}}$
32. $\sum_{n=0}^{\infty} \frac{(x-\sqrt{2})^{2n+1}}{2^n}$

In Exercises 33–38, find the series' interval of convergence and, within this interval, the sum of the series as a function of x.

33. $\sum_{n=0}^{\infty} \frac{(x-1)^{2n}}{4^n}$
34. $\sum_{n=0}^{\infty} \frac{(x+1)^{2n}}{9^n}$
35. $\sum_{n=0}^{\infty} \left(\frac{\sqrt{x}}{2}-1\right)^n$
36. $\sum_{n=0}^{\infty} (\ln x)^n$
37. $\sum_{n=0}^{\infty} \left(\frac{x^2+1}{3}\right)^n$
38. $\sum_{n=0}^{\infty} \left(\frac{x^2-1}{2}\right)^n$

Theory and Examples

39. For what values of x does the series

$$1 - \frac{1}{2}(x-3) + \frac{1}{4}(x-3)^2 + \cdots + \left(-\frac{1}{2}\right)^n(x-3)^n + \cdots$$

converge? What is its sum? What series do you get if you differentiate the given series term by term? For what values of x does the new series converge? What is its sum?

40. If you integrate the series in Exercise 39 term by term, what new series do you get? For what values of x does the new series converge, and what is another name for its sum?

41. The series

$$\sin x = x - \frac{x^3}{3!} + \frac{x^5}{5!} - \frac{x^7}{7!} + \frac{x^9}{9!} - \frac{x^{11}}{11!} + \cdots$$

converges to $\sin x$ for all x.

   a. Find the first six terms of a series for $\cos x$. For what values of x should the series converge?

   b. By replacing x by $2x$ in the series for $\sin x$, find a series that converges to $\sin 2x$ for all x.

   c. Using the result in part (a) and series multiplication, calculate the first six terms of a series for $2 \sin x \cos x$. Compare your answer with the answer in part (b).

42. The series

$$e^x = 1 + x + \frac{x^2}{2!} + \frac{x^3}{3!} + \frac{x^4}{4!} + \frac{x^5}{5!} + \cdots$$

converges to e^x for all x.

   a. Find a series for $(d/dx)e^x$. Do you get the series for e^x? Explain your answer.

   b. Find a series for $\int e^x\,dx$. Do you get the series for e^x? Explain your answer.

   c. Replace x by $-x$ in the series for e^x to find a series that converges to e^{-x} for all x. Then multiply the series for e^x and e^{-x} to find the first six terms of a series for $e^{-x} \cdot e^x$.

43. The series

$$\tan x = x + \frac{x^3}{3} + \frac{2x^5}{15} + \frac{17x^7}{315} + \frac{62x^9}{2835} + \cdots$$

converges to $\tan x$ for $-\pi/2 < x < \pi/2$.

a. Find the first five terms of the series for $\ln|\sec x|$. For what values of x should the series converge?

b. Find the first five terms of the series for $\sec^2 x$. For what values of x should this series converge?

c. Check your result in part (b) by squaring the series given for $\sec x$ in Exercise 44.

44. The series

$$\sec x = 1 + \frac{x^2}{2} + \frac{5}{24}x^4 + \frac{61}{720}x^6 + \frac{277}{8064}x^8 + \cdots$$

converges to $\sec x$ for $-\pi/2 < x < \pi/2$.

a. Find the first five terms of a power series for the function $\ln|\sec x + \tan x|$. For what values of x should the series converge?

b. Find the first four terms of a series for $\sec x \tan x$. For what values of x should the series converge?

c. Check your result in part (b) by multiplying the series for $\sec x$ by the series given for $\tan x$ in Exercise 43.

45. Uniqueness of convergent power series

a. Show that if two power series $\sum_{n=0}^{\infty} a_n x^n$ and $\sum_{n=0}^{\infty} b_n x^n$ are convergent and equal for all values of x in an open interval $(-c, c)$, then $a_n = b_n$ for every n. (*Hint:* Let $f(x) = \sum_{n=0}^{\infty} a_n x^n = \sum_{n=0}^{\infty} b_n x^n$. Differentiate term by term to show that a_n and b_n both equal $f^{(n)}(0)/(n!)$.)

b. Show that if $\sum_{n=0}^{\infty} a_n x^n = 0$ for all x in an open interval $(-c, c)$, then $a_n = 0$ for every n.

46. The sum of the series $\sum_{n=0}^{\infty}(n^2/2^n)$ To find the sum of this series, express $1/(1-x)$ as a geometric series, differentiate both sides of the resulting equation with respect to x, multiply both sides of the result by x, differentiate again, multiply by x again, and set x equal to $1/2$. What do you get? (*Source*: David E. Dobbs' letter to the editor, *Illinois Mathematics Teacher*, Vol. 33, Issue 4, 1982, p. 27.)

47. Convergence at endpoints Show by examples that the convergence of a power series at an endpoint of its interval of convergence may be either conditional or absolute.

48. Make up a power series whose interval of convergence is

a. $(-3, 3)$ **b.** $(-2, 0)$ **c.** $(1, 5)$.

11.8 Taylor and Maclaurin Series

This section shows how functions that are infinitely differentiable generate power series called Taylor series. In many cases, these series can provide useful polynomial approximations of the generating functions.

Series Representations

We know from Theorem 19 that within its interval of convergence the sum of a power series is a continuous function with derivatives of all orders. But what about the other way around? If a function $f(x)$ has derivatives of all orders on an interval I, can it be expressed as a power series on I? And if it can, what will its coefficients be?

We can answer the last question readily if we assume that $f(x)$ is the sum of a power series

$$\begin{aligned} f(x) &= \sum_{n=0}^{\infty} a_n(x-a)^n \\ &= a_0 + a_1(x-a) + a_2(x-a)^2 + \cdots + a_n(x-a)^n + \cdots \end{aligned}$$

with a positive radius of convergence. By repeated term-by-term differentiation within the interval of convergence I we obtain

$$\begin{aligned} f'(x) &= a_1 + 2a_2(x-a) + 3a_3(x-a)^2 + \cdots + na_n(x-a)^{n-1} + \cdots \\ f''(x) &= 1\cdot 2a_2 + 2\cdot 3a_3(x-a) + 3\cdot 4a_4(x-a)^2 + \cdots \\ f'''(x) &= 1\cdot 2\cdot 3a_3 + 2\cdot 3\cdot 4a_4(x-a) + 3\cdot 4\cdot 5a_5(x-a)^2 + \cdots, \end{aligned}$$

with the nth derivative, for all n, being

$$f^{(n)}(x) = n!a_n + \text{a sum of terms with } (x - a) \text{ as a factor.}$$

Since these equations all hold at $x = a$, we have

$$\begin{aligned} f'(a) &= a_1, \\ f''(a) &= 1 \cdot 2a_2, \\ f'''(a) &= 1 \cdot 2 \cdot 3a_3, \end{aligned}$$

and, in general,

$$f^{(n)}(a) = n!a_n.$$

These formulas reveal a pattern in the coefficients of any power series $\sum_{n=0}^{\infty} a_n(x - a)^n$ that converges to the values of f on I ("represents f on I"). If there *is* such a series (still an open question), then there is only one such series and its nth coefficient is

$$a_n = \frac{f^{(n)}(a)}{n!}.$$

If f has a series representation, then the series must be

$$f(x) = f(a) + f'(a)(x - a) + \frac{f''(a)}{2!}(x - a)^2 + \cdots + \frac{f^{(n)}(a)}{n!}(x - a)^n + \cdots. \tag{1}$$

But if we start with an arbitrary function f that is infinitely differentiable on an interval I centered at $x = a$ and use it to generate the series in Equation (1), will the series then converge to $f(x)$ at each x in the interior of I? The answer is maybe—for some functions it will but for other functions it will not, as we will see.

Taylor and Maclaurin Series

HISTORICAL BIOGRAPHIES

Brook Taylor
(1685–1731)

Colin Maclaurin
(1698–1746)

DEFINITIONS Taylor Series, Maclaurin Series

Let f be a function with derivatives of all orders throughout some interval containing a as an interior point. Then the **Taylor series generated by f at $x = a$ is**

$$\sum_{k=0}^{\infty} \frac{f^{(k)}(a)}{k!}(x - a)^k = f(a) + f'(a)(x - a) + \frac{f''(a)}{2!}(x - a)^2 + \cdots + \frac{f^{(n)}(a)}{n!}(x - a)^n + \cdots.$$

The **Maclaurin series generated by f is**

$$\sum_{k=0}^{\infty} \frac{f^{(k)}(0)}{k!}x^k = f(0) + f'(0)x + \frac{f''(0)}{2!}x^2 + \cdots + \frac{f^{(n)}(0)}{n!}x^n + \cdots,$$

the Taylor series generated by f at $x = 0$.

The Maclaurin series generated by f is often just called the Taylor series of f.

EXAMPLE 1 Finding a Taylor Series

Find the Taylor series generated by $f(x) = 1/x$ at $a = 2$. Where, if anywhere, does the series converge to $1/x$?

Solution We need to find $f(2), f'(2), f''(2), \ldots$. Taking derivatives we get

$$\begin{aligned}
f(x) &= x^{-1}, & f(2) &= 2^{-1} = \frac{1}{2},\\
f'(x) &= -x^{-2}, & f'(2) &= -\frac{1}{2^2},\\
f''(x) &= 2!x^{-3}, & \frac{f''(2)}{2!} &= 2^{-3} = \frac{1}{2^3},\\
f'''(x) &= -3!x^{-4}, & \frac{f'''(2)}{3!} &= -\frac{1}{2^4},\\
&\vdots & &\vdots\\
f^{(n)}(x) &= (-1)^n n! x^{-(n+1)}, & \frac{f^{(n)}(2)}{n!} &= \frac{(-1)^n}{2^{n+1}}.
\end{aligned}$$

The Taylor series is

$$\begin{aligned}
f(2) &+ f'(2)(x-2) + \frac{f''(2)}{2!}(x-2)^2 + \cdots + \frac{f^{(n)}(2)}{n!}(x-2)^n + \cdots\\
&= \frac{1}{2} - \frac{(x-2)}{2^2} + \frac{(x-2)^2}{2^3} - \cdots + (-1)^n \frac{(x-2)^n}{2^{n+1}} + \cdots.
\end{aligned}$$

This is a geometric series with first term $1/2$ and ratio $r = -(x-2)/2$. It converges absolutely for $|x-2| < 2$ and its sum is

$$\frac{1/2}{1 + (x-2)/2} = \frac{1}{2 + (x-2)} = \frac{1}{x}.$$

In this example the Taylor series generated by $f(x) = 1/x$ at $a = 2$ converges to $1/x$ for $|x-2| < 2$ or $0 < x < 4$. ■

Taylor Polynomials

The linearization of a differentiable function f at a point a is the polynomial of degree one given by

$$P_1(x) = f(a) + f'(a)(x-a).$$

In Section 3.8 we used this linearization to approximate $f(x)$ at values of x near a. If f has derivatives of higher order at a, then it has higher-order polynomial approximations as well, one for each available derivative. These polynomials are called the Taylor polynomials of f.

DEFINITION Taylor Polynomial of Order n

Let f be a function with derivatives of order k for $k = 1, 2, \dots, N$ in some interval containing a as an interior point. Then for any integer n from 0 through N, the **Taylor polynomial of order n** generated by f at $x = a$ is the polynomial

$$P_n(x) = f(a) + f'(a)(x - a) + \frac{f''(a)}{2!}(x-a)^2 + \cdots + \frac{f^{(k)}(a)}{k!}(x-a)^k + \cdots + \frac{f^{(n)}(a)}{n!}(x-a)^n.$$

We speak of a Taylor polynomial of *order n* rather than *degree n* because $f^{(n)}(a)$ may be zero. The first two Taylor polynomials of $f(x) = \cos x$ at $x = 0$, for example, are $P_0(x) = 1$ and $P_1(x) = 1$. The first-order Taylor polynomial has degree zero, not one.

Just as the linearization of f at $x = a$ provides the best linear approximation of f in the neighborhood of a, the higher-order Taylor polynomials provide the best polynomial approximations of their respective degrees. (See Exercise 32.)

EXAMPLE 2 Finding Taylor Polynomials for e^x

Find the Taylor series and the Taylor polynomials generated by $f(x) = e^x$ at $x = 0$.

Solution Since

$$f(x) = e^x, \quad f'(x) = e^x, \quad \dots, \quad f^{(n)}(x) = e^x, \quad \dots,$$

we have

$$f(0) = e^0 = 1, \quad f'(0) = 1, \quad \dots, \quad f^{(n)}(0) = 1, \quad \dots.$$

The Taylor series generated by f at $x = 0$ is

$$\begin{aligned} f(0) + f'(0)x + \frac{f''(0)}{2!}x^2 + \cdots + \frac{f^{(n)}(0)}{n!}x^n + \cdots &= 1 + x + \frac{x^2}{2} + \cdots + \frac{x^n}{n!} + \cdots \\ &= \sum_{k=0}^{\infty} \frac{x^k}{k!}. \end{aligned}$$

This is also the Maclaurin series for e^x. In Section 11.9 we will see that the series converges to e^x at every x.

The Taylor polynomial of order n at $x = 0$ is

$$P_n(x) = 1 + x + \frac{x^2}{2} + \cdots + \frac{x^n}{n!}.$$

See Figure 11.12. ■

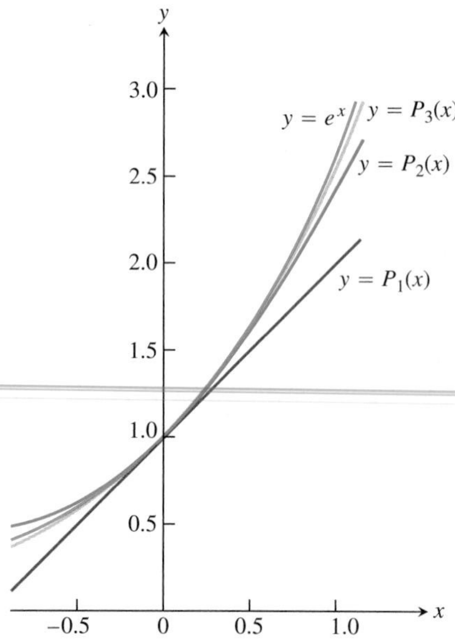

FIGURE 11.12 The graph of $f(x) = e^x$ and its Taylor polynomials
$P_1(x) = 1 + x$
$P_2(x) = 1 + x + (x^2/2!)$
$P_3(x) = 1 + x + (x^2/2!) + (x^3/3!)$.
Notice the very close agreement near the center $x = 0$ (Example 2).

EXAMPLE 3 Finding Taylor Polynomials for $\cos x$

Find the Taylor series and Taylor polynomials generated by $f(x) = \cos x$ at $x = 0$.

Solution The cosine and its derivatives are

$$\begin{aligned} f(x) &= \cos x, & f'(x) &= -\sin x, \\ f''(x) &= -\cos x, & f^{(3)}(x) &= \sin x, \\ &\vdots & &\vdots \\ f^{(2n)}(x) &= (-1)^n \cos x, & f^{(2n+1)}(x) &= (-1)^{n+1} \sin x. \end{aligned}$$

At $x = 0$, the cosines are 1 and the sines are 0, so

$$f^{(2n)}(0) = (-1)^n, \qquad f^{(2n+1)}(0) = 0.$$

The Taylor series generated by f at 0 is

$$\begin{aligned} f(0) + f'(0)x + \frac{f''(0)}{2!}x^2 + \frac{f'''(0)}{3!}x^3 + \cdots + \frac{f^{(n)}(0)}{n!}x^n + \cdots \\ = 1 + 0 \cdot x - \frac{x^2}{2!} + 0 \cdot x^3 + \frac{x^4}{4!} + \cdots + (-1)^n \frac{x^{2n}}{(2n)!} + \cdots \\ = \sum_{k=0}^{\infty} \frac{(-1)^k x^{2k}}{(2k)!}. \end{aligned}$$

This is also the Maclaurin series for $\cos x$. In Section 11.9, we will see that the series converges to $\cos x$ at every x.

Because $f^{(2n+1)}(0) = 0$, the Taylor polynomials of orders $2n$ and $2n + 1$ are identical:

$$P_{2n}(x) = P_{2n+1}(x) = 1 - \frac{x^2}{2!} + \frac{x^4}{4!} - \cdots + (-1)^n \frac{x^{2n}}{(2n)!}.$$

Figure 11.13 shows how well these polynomials approximate $f(x) = \cos x$ near $x = 0$. Only the right-hand portions of the graphs are given because the graphs are symmetric about the y-axis. ■

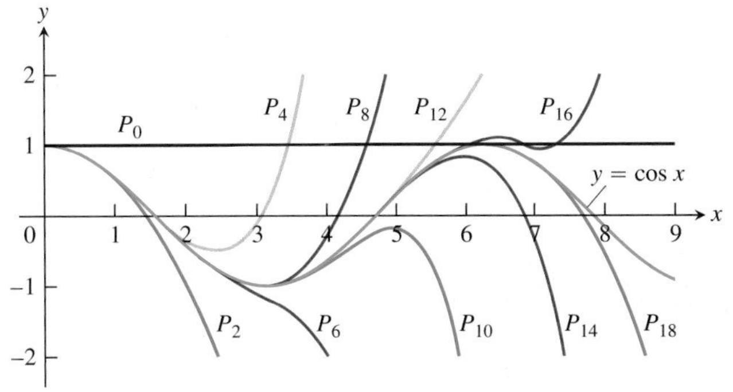

FIGURE 11.13 The polynomials

$$P_{2n}(x) = \sum_{k=0}^{n} \frac{(-1)^k x^{2k}}{(2k)!}$$

converge to $\cos x$ as $n \to \infty$. We can deduce the behavior of $\cos x$ arbitrarily far away solely from knowing the values of the cosine and its derivatives at $x = 0$ (Example 3).

EXAMPLE 4 A Function f Whose Taylor Series Converges at Every x but Converges to $f(x)$ Only at $x = 0$

It can be shown (though not easily) that

$$f(x) = \begin{cases} 0, & x = 0 \\ e^{-1/x^2}, & x \neq 0 \end{cases}$$

(Figure 11.14) has derivatives of all orders at $x = 0$ and that $f^{(n)}(0) = 0$ for all n. This means that the Taylor series generated by f at $x = 0$ is

$$\begin{aligned} f(0) + f'(0)x + \frac{f''(0)}{2!}x^2 + \cdots + \frac{f^{(n)}(0)}{n!}x^n + \cdots \\ = 0 + 0 \cdot x + 0 \cdot x^2 + \cdots + 0 \cdot x^n + \cdots \\ = 0 + 0 + \cdots + 0 + \cdots. \end{aligned}$$

The series converges for every x (its sum is 0) but converges to $f(x)$ only at $x = 0$. ■

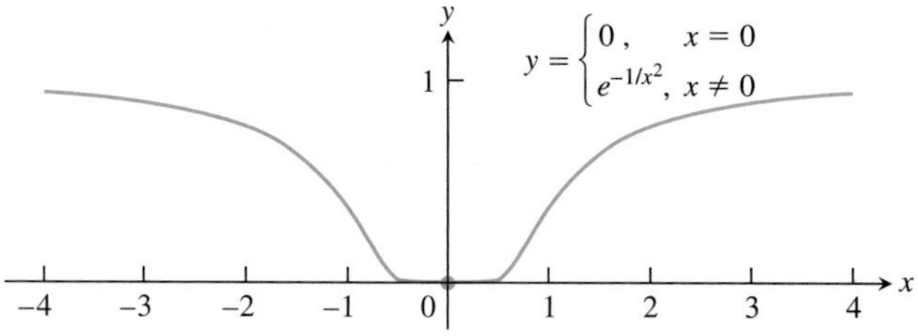

FIGURE 11.14 The graph of the continuous extension of $y = e^{-1/x^2}$ is so flat at the origin that all of its derivatives there are zero (Example 4).

Two questions still remain.

1. For what values of x can we normally expect a Taylor series to converge to its generating function?
2. How accurately do a function's Taylor polynomials approximate the function on a given interval?

The answers are provided by a theorem of Taylor in the next section.

EXERCISES 11.8

Finding Taylor Polynomials

In Exercises 1–8, find the Taylor polynomials of orders 0, 1, 2, and 3 generated by f at a.

1. $f(x) = \ln x, \quad a = 1$
2. $f(x) = \ln(1 + x), \quad a = 0$
3. $f(x) = 1/x, \quad a = 2$
4. $f(x) = 1/(x + 2), \quad a = 0$
5. $f(x) = \sin x, \quad a = \pi/4$
6. $f(x) = \cos x, \quad a = \pi/4$
7. $f(x) = \sqrt{x}, \quad a = 4$
8. $f(x) = \sqrt{x + 4}, \quad a = 0$

Finding Taylor Series at $x = 0$ (Maclaurin Series)

Find the Maclaurin series for the functions in Exercises 9–20.

9. e^{-x}
10. $e^{x/2}$
11. $\dfrac{1}{1 + x}$
12. $\dfrac{1}{1 - x}$
13. $\sin 3x$
14. $\sin \dfrac{x}{2}$

15. $7 \cos(-x)$

16. $5 \cos \pi x$

17. $\cosh x = \dfrac{e^x + e^{-x}}{2}$

18. $\sinh x = \dfrac{e^x - e^{-x}}{2}$

19. $x^4 - 2x^3 - 5x + 4$

20. $(x + 1)^2$

Finding Taylor Series

In Exercises 21–28, find the Taylor series generated by f at $x = a$.

21. $f(x) = x^3 - 2x + 4, \quad a = 2$

22. $f(x) = 2x^3 + x^2 + 3x - 8, \quad a = 1$

23. $f(x) = x^4 + x^2 + 1, \quad a = -2$

24. $f(x) = 3x^5 - x^4 + 2x^3 + x^2 - 2, \quad a = -1$

25. $f(x) = 1/x^2, \quad a = 1$

26. $f(x) = x/(1 - x), \quad a = 0$

27. $f(x) = e^x, \quad a = 2$

28. $f(x) = 2^x, \quad a = 1$

Theory and Examples

29. Use the Taylor series generated by e^x at $x = a$ to show that

$$e^x = e^a\left[1 + (x - a) + \frac{(x - a)^2}{2!} + \cdots\right].$$

30. (*Continuation of Exercise 29.*) Find the Taylor series generated by e^x at $x = 1$. Compare your answer with the formula in Exercise 29.

31. Let $f(x)$ have derivatives through order n at $x = a$. Show that the Taylor polynomial of order n and its first n derivatives have the same values that f and its first n derivatives have at $x = a$.

32. Of all polynomials of degree $\le n$, the Taylor polynomial of order n gives the best approximation Suppose that $f(x)$ is differentiable on an interval centered at $x = a$ and that $g(x) = b_0 + b_1(x - a) + \cdots + b_n(x - a)^n$ is a polynomial of degree n with constant coefficients $b_0, \ldots, b_n$. Let $E(x) = f(x) - g(x)$. Show that if we impose on g the conditions

a. $E(a) = 0$ — The approximation error is zero at $x = a$.

b. $\displaystyle\lim_{x \to a} \frac{E(x)}{(x - a)^n} = 0,$ — The error is negligible when compared to $(x - a)^n$.

then

$$g(x) = f(a) + f'(a)(x - a) + \frac{f''(a)}{2!}(x - a)^2 + \cdots + \frac{f^{(n)}(a)}{n!}(x - a)^n.$$

Thus, the Taylor polynomial $P_n(x)$ is the only polynomial of degree less than or equal to n whose error is both zero at $x = a$ and negligible when compared with $(x - a)^n$.

Quadratic Approximations

The Taylor polynomial of order 2 generated by a twice-differentiable function $f(x)$ at $x = a$ is called the **quadratic approximation** of f at $x = a$. In Exercises 33–38, find the **(a)** linearization (Taylor polynomial of order 1) and **(b)** quadratic approximation of f at $x = 0$.

33. $f(x) = \ln(\cos x)$

34. $f(x) = e^{\sin x}$

35. $f(x) = 1/\sqrt{1 - x^2}$

36. $f(x) = \cosh x$

37. $f(x) = \sin x$

38. $f(x) = \tan x$

11.9 Convergence of Taylor Series; Error Estimates

This section addresses the two questions left unanswered by Section 11.8:

1. When does a Taylor series converge to its generating function?
2. How accurately do a function's Taylor polynomials approximate the function on a given interval?

Taylor's Theorem

We answer these questions with the following theorem.

THEOREM 22 Taylor's Theorem

If f and its first n derivatives $f', f'', \ldots, f^{(n)}$ are continuous on the closed interval between a and b, and $f^{(n)}$ is differentiable on the open interval between a and b, then there exists a number c between a and b such that

$$f(b) = f(a) + f'(a)(b-a) + \frac{f''(a)}{2!}(b-a)^2 + \cdots + \frac{f^{(n)}(a)}{n!}(b-a)^n + \frac{f^{(n+1)}(c)}{(n+1)!}(b-a)^{n+1}.$$

Taylor's Theorem is a generalization of the Mean Value Theorem (Exercise 39). There is a proof of Taylor's Theorem at the end of this section.

When we apply Taylor's Theorem, we usually want to hold a fixed and treat b as an independent variable. Taylor's formula is easier to use in circumstances like these if we change b to x. Here is a version of the theorem with this change.

Taylor's Formula

If f has derivatives of all orders in an open interval I containing a, then for each positive integer n and for each x in I,

$$f(x) = f(a) + f'(a)(x-a) + \frac{f''(a)}{2!}(x-a)^2 + \cdots + \frac{f^{(n)}(a)}{n!}(x-a)^n + R_n(x), \quad (1)$$

where

$$R_n(x) = \frac{f^{(n+1)}(c)}{(n+1)!}(x-a)^{n+1} \quad \text{for some } c \text{ between } a \text{ and } x. \quad (2)$$

When we state Taylor's theorem this way, it says that for each $x \in I$,

$$f(x) = P_n(x) + R_n(x).$$

The function $R_n(x)$ is determined by the value of the $(n+1)$st derivative $f^{(n+1)}$ at a point c that depends on both a and x, and which lies somewhere between them. For any value of n we want, the equation gives both a polynomial approximation of f of that order and a formula for the error involved in using that approximation over the interval I.

Equation (1) is called **Taylor's formula**. The function $R_n(x)$ is called the **remainder of order n** or the **error term** for the approximation of f by $P_n(x)$ over I. If $R_n(x) \to 0$ as $n \to \infty$ for all $x \in I$, we say that the Taylor series generated by f at $x = a$ **converges** to f on I, and we write

$$f(x) = \sum_{k=0}^{\infty} \frac{f^{(k)}(a)}{k!}(x-a)^k.$$

Often we can estimate R_n without knowing the value of c, as the following example illustrates.

EXAMPLE 1 The Taylor Series for e^x Revisited

Show that the Taylor series generated by $f(x) = e^x$ at $x = 0$ converges to $f(x)$ for every real value of x.

Solution The function has derivatives of all orders throughout the interval $I = (-\infty, \infty)$. Equations (1) and (2) with $f(x) = e^x$ and $a = 0$ give

$$e^x = 1 + x + \frac{x^2}{2!} + \cdots + \frac{x^n}{n!} + R_n(x)$$

Polynomial from Section 11.8, Example 2

and

$$R_n(x) = \frac{e^c}{(n+1)!}x^{n+1} \qquad \text{for some } c \text{ between 0 and } x.$$

Since e^x is an increasing function of x, e^c lies between $e^0 = 1$ and e^x. When x is negative, so is c, and $e^c < 1$. When x is zero, $e^x = 1$ and $R_n(x) = 0$. When x is positive, so is c, and $e^c < e^x$. Thus,

$$|R_n(x)| \le \frac{|x|^{n+1}}{(n+1)!} \qquad \text{when } x \le 0,$$

and

$$|R_n(x)| < e^x \frac{x^{n+1}}{(n+1)!} \qquad \text{when } x > 0.$$

Finally, because

$$\lim_{n\to\infty} \frac{x^{n+1}}{(n+1)!} = 0 \qquad \text{for every } x,$$

Section 11.1

$\lim_{n\to\infty} R_n(x) = 0$, and the series converges to e^x for every x. Thus,

$$e^x = \sum_{k=0}^{\infty} \frac{x^k}{k!} = 1 + x + \frac{x^2}{2!} + \cdots + \frac{x^k}{k!} + \cdots. \tag{3}$$

■

Estimating the Remainder

It is often possible to estimate $R_n(x)$ as we did in Example 1. This method of estimation is so convenient that we state it as a theorem for future reference.

THEOREM 23 The Remainder Estimation Theorem
If there is a positive constant M such that $|f^{(n+1)}(t)| \le M$ for all t between x and a, inclusive, then the remainder term $R_n(x)$ in Taylor's Theorem satisfies the inequality

$$|R_n(x)| \le M\frac{|x-a|^{n+1}}{(n+1)!}.$$

If this condition holds for every n and the other conditions of Taylor's Theorem are satisfied by f, then the series converges to $f(x)$.

We are now ready to look at some examples of how the Remainder Estimation Theorem and Taylor's Theorem can be used together to settle questions of convergence. As you will see, they can also be used to determine the accuracy with which a function is approximated by one of its Taylor polynomials.

EXAMPLE 2 The Taylor Series for $\sin x$ at $x = 0$

Show that the Taylor series for $\sin x$ at $x = 0$ converges for all x.

Solution The function and its derivatives are

$$\begin{aligned} f(x) &= \sin x, & f'(x) &= \cos x, \\ f''(x) &= -\sin x, & f'''(x) &= -\cos x, \\ &\vdots & &\vdots \\ f^{(2k)}(x) &= (-1)^k \sin x, & f^{(2k+1)}(x) &= (-1)^k \cos x, \end{aligned}$$

so

$$f^{(2k)}(0) = 0 \quad \text{and} \quad f^{(2k+1)}(0) = (-1)^k.$$

The series has only odd-powered terms and, for $n = 2k + 1$, Taylor's Theorem gives

$$\sin x = x - \frac{x^3}{3!} + \frac{x^5}{5!} - \cdots + \frac{(-1)^k x^{2k+1}}{(2k+1)!} + R_{2k+1}(x).$$

All the derivatives of $\sin x$ have absolute values less than or equal to 1, so we can apply the Remainder Estimation Theorem with $M = 1$ to obtain

$$|R_{2k+1}(x)| \le 1 \cdot \frac{|x|^{2k+2}}{(2k+2)!}.$$

Since $(|x|^{2k+2}/(2k+2)!) \to 0$ as $k \to \infty$, whatever the value of x, $R_{2k+1}(x) \to 0$, and the Maclaurin series for $\sin x$ converges to $\sin x$ for every x. Thus,

$$\sin x = \sum_{k=0}^{\infty} \frac{(-1)^k x^{2k+1}}{(2k+1)!} = x - \frac{x^3}{3!} + \frac{x^5}{5!} - \frac{x^7}{7!} + \cdots. \tag{4}$$

■

EXAMPLE 3 The Taylor Series for $\cos x$ at $x = 0$ Revisited

Show that the Taylor series for $\cos x$ at $x = 0$ converges to $\cos x$ for every value of x.

Solution We add the remainder term to the Taylor polynomial for $\cos x$ (Section 11.8, Example 3) to obtain Taylor's formula for $\cos x$ with $n = 2k$:

$$\cos x = 1 - \frac{x^2}{2!} + \frac{x^4}{4!} - \cdots + (-1)^k \frac{x^{2k}}{(2k)!} + R_{2k}(x).$$

Because the derivatives of the cosine have absolute value less than or equal to 1, the Remainder Estimation Theorem with $M = 1$ gives

$$|R_{2k}(x)| \leq 1 \cdot \frac{|x|^{2k+1}}{(2k+1)!}.$$

For every value of x, $R_{2k} \to 0$ as $k \to \infty$. Therefore, the series converges to $\cos x$ for every value of x. Thus,

$$\cos x = \sum_{k=0}^{\infty} \frac{(-1)^k x^{2k}}{(2k)!} = 1 - \frac{x^2}{2!} + \frac{x^4}{4!} - \frac{x^6}{6!} + \cdots. \tag{5}$$

■

EXAMPLE 4 Finding a Taylor Series by Substitution

Find the Taylor series for $\cos 2x$ at $x = 0$.

Solution We can find the Taylor series for $\cos 2x$ by substituting $2x$ for x in the Taylor series for $\cos x$:

$$\begin{aligned}\cos 2x &= \sum_{k=0}^{\infty} \frac{(-1)^k (2x)^{2k}}{(2k)!} = 1 - \frac{(2x)^2}{2!} + \frac{(2x)^4}{4!} - \frac{(2x)^6}{6!} + \cdots \qquad \text{Equation (5) with } 2x \text{ for } x \\ &= 1 - \frac{2^2x^2}{2!} + \frac{2^4x^4}{4!} - \frac{2^6x^6}{6!} + \cdots \\ &= \sum_{k=0}^{\infty} (-1)^k \frac{2^{2k}x^{2k}}{(2k)!}.\end{aligned}$$

Equation (5) holds for $-\infty < x < \infty$, implying that it holds for $-\infty < 2x < \infty$, so the newly created series converges for all x. Exercise 45 explains why the series is in fact the Taylor series for $\cos 2x$. ■

EXAMPLE 5 Finding a Taylor Series by Multiplication

Find the Taylor series for $x \sin x$ at $x = 0$.

Solution We can find the Taylor series for $x \sin x$ by multiplying the Taylor series for $\sin x$ (Equation 4) by x:

$$\begin{aligned}x \sin x &= x\left(x - \frac{x^3}{3!} + \frac{x^5}{5!} - \frac{x^7}{7!} + \cdots\right) \\ &= x^2 - \frac{x^4}{3!} + \frac{x^6}{5!} - \frac{x^8}{7!} + \cdots.\end{aligned}$$

The new series converges for all x because the series for $\sin x$ converges for all x. Exercise 45 explains why the series is the Taylor series for $x \sin x$. ■

Truncation Error

The Taylor series for e^x at $x = 0$ converges to e^x for all x. But we still need to decide how many terms to use to approximate e^x to a given degree of accuracy. We get this information from the Remainder Estimation Theorem.

EXAMPLE 6 Calculate e with an error of less than 10^{-6}.

Solution We can use the result of Example 1 with $x = 1$ to write

$$e = 1 + 1 + \frac{1}{2!} + \cdots + \frac{1}{n!} + R_n(1),$$

with

$$R_n(1) = e^c \frac{1}{(n+1)!} \qquad \text{for some } c \text{ between 0 and 1.}$$

For the purposes of this example, we assume that we know that $e < 3$. Hence, we are certain that

$$\frac{1}{(n+1)!} < R_n(1) < \frac{3}{(n+1)!}$$

because $1 < e^c < 3$ for $0 < c < 1$.

By experiment we find that $1/9! > 10^{-6}$, while $3/10! < 10^{-6}$. Thus we should take $(n+1)$ to be at least 10, or n to be at least 9. With an error of less than 10^{-6},

$$e = 1 + 1 + \frac{1}{2} + \frac{1}{3!} + \cdots + \frac{1}{9!} \approx 2.718282.$$ ■

EXAMPLE 7 For what values of x can we replace $\sin x$ by $x - (x^3/3!)$ with an error of magnitude no greater than 3×10^{-4}?

Solution Here we can take advantage of the fact that the Taylor series for $\sin x$ is an alternating series for every nonzero value of x. According to the Alternating Series Estimation Theorem (Section 11.6), the error in truncating

$$\sin x = x - \frac{x^3}{3!} \,\Big|\, + \frac{x^5}{5!} - \cdots$$

after $(x^3/3!)$ is no greater than

$$\left|\frac{x^5}{5!}\right| = \frac{|x|^5}{120}.$$

Therefore the error will be less than or equal to 3×10^{-4} if

$$\frac{|x|^5}{120} < 3 \times 10^{-4} \quad \text{or} \quad |x| < \sqrt[5]{360 \times 10^{-4}} \approx 0.514. \qquad \text{Rounded down, to be safe}$$

The Alternating Series Estimation Theorem tells us something that the Remainder Estimation Theorem does not: namely, that the estimate $x - (x^3/3!)$ for $\sin x$ is an underestimate when x is positive because then $x^5/120$ is positive.

Figure 11.15 shows the graph of $\sin x$, along with the graphs of a number of its approximating Taylor polynomials. The graph of $P_3(x) = x - (x^3/3!)$ is almost indistinguishable from the sine curve when $-1 \le x \le 1$.

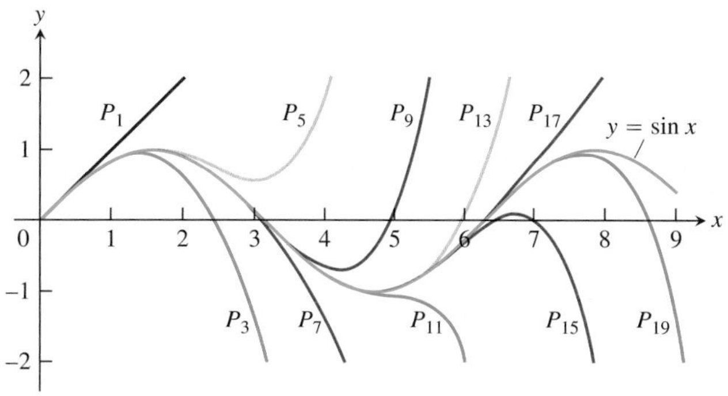

FIGURE 11.15 The polynomials

$$P_{2n+1}(x) = \sum_{k=0}^{n} \frac{(-1)^k x^{2k+1}}{(2k+1)!}$$

converge to $\sin x$ as $n \to \infty$. Notice how closely $P_3(x)$ approximates the sine curve for $x < 1$ (Example 7).

You might wonder how the estimate given by the Remainder Estimation Theorem compares with the one just obtained from the Alternating Series Estimation Theorem. If we write

$$\sin x = x - \frac{x^3}{3!} + R_3,$$

then the Remainder Estimation Theorem gives

$$|R_3| \le 1 \cdot \frac{|x|^4}{4!} = \frac{|x|^4}{24},$$

which is not as good. But if we recognize that $x - (x^3/3!) = 0 + x + 0x^2 - (x^3/3!) + 0x^4$ is the Taylor polynomial of order 4 as well as of order 3, then

$$\sin x = x - \frac{x^3}{3!} + 0 + R_4,$$

and the Remainder Estimation Theorem with $M = 1$ gives

$$|R_4| \le 1 \cdot \frac{|x|^5}{5!} = \frac{|x|^5}{120}.$$

This is what we had from the Alternating Series Estimation Theorem. ■

Combining Taylor Series

On the intersection of their intervals of convergence, Taylor series can be added, subtracted, and multiplied by constants, and the results are once again Taylor series. The Taylor series for $f(x) + g(x)$ is the sum of the Taylor series for $f(x)$ and $g(x)$ because the nth derivative of $f + g$ is $f^{(n)} + g^{(n)}$, and so on. Thus we obtain the Taylor series for $(1 + \cos 2x)/2$ by adding 1 to the Taylor series for $\cos 2x$ and dividing the combined results by 2, and the Taylor series for $\sin x + \cos x$ is the term-by-term sum of the Taylor series for $\sin x$ and $\cos x$.

Euler's Identity

As you may recall, a complex number is a number of the form $a + bi$, where a and b are real numbers and $i = \sqrt{-1}$. If we substitute $x = i\theta$ (θ real) in the Taylor series for e^x and use the relations

$$i^2 = -1, \qquad i^3 = i^2 i = -i, \qquad i^4 = i^2 i^2 = 1, \qquad i^5 = i^4 i = i,$$

and so on, to simplify the result, we obtain

$$\begin{aligned} e^{i\theta} &= 1 + \frac{i\theta}{1!} + \frac{i^2\theta^2}{2!} + \frac{i^3\theta^3}{3!} + \frac{i^4\theta^4}{4!} + \frac{i^5\theta^5}{5!} + \frac{i^6\theta^6}{6!} + \cdots \\ &= \left(1 - \frac{\theta^2}{2!} + \frac{\theta^4}{4!} - \frac{\theta^6}{6!} + \cdots\right) + i\left(\theta - \frac{\theta^3}{3!} + \frac{\theta^5}{5!} - \cdots\right) = \cos\theta + i\sin\theta. \end{aligned}$$

This does not *prove* that $e^{i\theta} = \cos\theta + i\sin\theta$ because we have not yet defined what it means to raise e to an imaginary power. Rather, it says how to define $e^{i\theta}$ to be consistent with other things we know.

DEFINITION

For any real number θ, $e^{i\theta} = \cos\theta + i\sin\theta$. (6)

Equation (6), called **Euler's identity**, enables us to define e^{a+bi} to be $e^a \cdot e^{bi}$ for any complex number $a + bi$. One consequence of the identity is the equation

$$e^{i\pi} = -1.$$

When written in the form $e^{i\pi} + 1 = 0$, this equation combines five of the most important constants in mathematics.

A Proof of Taylor's Theorem

We prove Taylor's theorem assuming $a < b$. The proof for $a > b$ is nearly the same.

The Taylor polynomial

$$P_n(x) = f(a) + f'(a)(x - a) + \frac{f''(a)}{2!}(x - a)^2 + \cdots + \frac{f^{(n)}(a)}{n!}(x - a)^n$$

and its first n derivatives match the function f and its first n derivatives at $x = a$. We do not disturb that matching if we add another term of the form $K(x - a)^{n+1}$, where K is any constant, because such a term and its first n derivatives are all equal to zero at $x = a$. The new function

$$\phi_n(x) = P_n(x) + K(x - a)^{n+1}$$

and its first n derivatives still agree with f and its first n derivatives at $x = a$.

We now choose the particular value of K that makes the curve $y = \phi_n(x)$ agree with the original curve $y = f(x)$ at $x = b$. In symbols,

$$f(b) = P_n(b) + K(b - a)^{n+1}, \qquad \text{or} \qquad K = \frac{f(b) - P_n(b)}{(b - a)^{n+1}}. \tag{7}$$

With K defined by Equation (7), the function

$$F(x) = f(x) - \phi_n(x)$$

measures the difference between the original function f and the approximating function ϕ_n for each x in $[a, b]$.

We now use Rolle's Theorem (Section 4.2). First, because $F(a) = F(b) = 0$ and both F and F' are continuous on $[a, b]$, we know that

$$F'(c_1) = 0 \qquad \text{for some } c_1 \text{ in } (a, b).$$

Next, because $F'(a) = F'(c_1) = 0$ and both F' and F'' are continuous on $[a, c_1]$, we know that

$$F''(c_2) = 0 \qquad \text{for some } c_2 \text{ in } (a, c_1).$$

Rolle's Theorem, applied successively to $F'', F''', \dots, F^{(n-1)}$ implies the existence of

$$\begin{aligned}
&c_3 \quad \text{in } (a, c_2) && \text{such that } F'''(c_3) = 0, \\
&c_4 \quad \text{in } (a, c_3) && \text{such that } F^{(4)}(c_4) = 0, \\
&\quad\vdots \\
&c_n \quad \text{in } (a, c_{n-1}) && \text{such that } F^{(n)}(c_n) = 0.
\end{aligned}$$

Finally, because $F^{(n)}$ is continuous on $[a, c_n]$ and differentiable on (a, c_n), and $F^{(n)}(a) = F^{(n)}(c_n) = 0$, Rolle's Theorem implies that there is a number c_{n+1} in (a, c_n) such that

$$F^{(n+1)}(c_{n+1}) = 0. \tag{8}$$

If we differentiate $F(x) = f(x) - P_n(x) - K(x - a)^{n+1}$ a total of $n + 1$ times, we get

$$F^{(n+1)}(x) = f^{(n+1)}(x) - 0 - (n + 1)!K. \tag{9}$$

Equations (8) and (9) together give

$$K = \frac{f^{(n+1)}(c)}{(n + 1)!} \qquad \text{for some number } c = c_{n+1} \text{ in } (a, b). \tag{10}$$

Equations (7) and (10) give

$$f(b) = P_n(b) + \frac{f^{(n+1)}(c)}{(n + 1)!}(b - a)^{n+1}.$$

This concludes the proof. ■

EXERCISES 11.9

Taylor Series by Substitution

Use substitution (as in Example 4) to find the Taylor series at $x = 0$ of the functions in Exercises 1–6.

1. e^{-5x} **2.** $e^{-x/2}$ **3.** $5 \sin(-x)$

4. $\sin\left(\frac{\pi x}{2}\right)$ **5.** $\cos\sqrt{x + 1}$ **6.** $\cos\left(x^{3/2}/\sqrt{2}\right)$

More Taylor Series

Find Taylor series at $x = 0$ for the functions in Exercises 7–18.

7. xe^x **8.** $x^2 \sin x$ **9.** $\frac{x^2}{2} - 1 + \cos x$

10. $\sin x - x + \frac{x^3}{3!}$ **11.** $x \cos \pi x$ **12.** $x^2 \cos(x^2)$

13. $\cos^2 x$ (*Hint:* $\cos^2 x = (1 + \cos 2x)/2$.)

14. $\sin^2 x$

15. $\dfrac{x^2}{1 - 2x}$

16. $x \ln(1 + 2x)$

17. $\dfrac{1}{(1 - x)^2}$

18. $\dfrac{2}{(1 - x)^3}$

Error Estimates

19. For approximately what values of x can you replace $\sin x$ by $x - (x^3/6)$ with an error of magnitude no greater than 5×10^{-4}? Give reasons for your answer.

20. If $\cos x$ is replaced by $1 - (x^2/2)$ and $|x| < 0.5$, what estimate can be made of the error? Does $1 - (x^2/2)$ tend to be too large, or too small? Give reasons for your answer.

21. How close is the approximation $\sin x = x$ when $|x| < 10^{-3}$? For which of these values of x is $x < \sin x$?

22. The estimate $\sqrt{1 + x} = 1 + (x/2)$ is used when x is small. Estimate the error when $|x| < 0.01$.

23. The approximation $e^x = 1 + x + (x^2/2)$ is used when x is small. Use the Remainder Estimation Theorem to estimate the error when $|x| < 0.1$.

24. (*Continuation of Exercise 23.*) When $x < 0$, the series for e^x is an alternating series. Use the Alternating Series Estimation Theorem to estimate the error that results from replacing e^x by $1 + x + (x^2/2)$ when $-0.1 < x < 0$. Compare your estimate with the one you obtained in Exercise 23.

25. Estimate the error in the approximation $\sinh x = x + (x^3/3!)$ when $|x| < 0.5$. (*Hint:* Use R_4, not R_3.)

26. When $0 \le h \le 0.01$, show that e^h may be replaced by $1 + h$ with an error of magnitude no greater than 0.6% of h. Use $e^{0.01} = 1.01$.

27. For what positive values of x can you replace $\ln(1 + x)$ by x with an error of magnitude no greater than 1% of the value of x?

28. You plan to estimate $\pi/4$ by evaluating the Maclaurin series for $\tan^{-1} x$ at $x = 1$. Use the Alternating Series Estimation Theorem to determine how many terms of the series you would have to add to be sure the estimate is good to two decimal places.

29. **a.** Use the Taylor series for $\sin x$ and the Alternating Series Estimation Theorem to show that

$$1 - \frac{x^2}{6} < \frac{\sin x}{x} < 1, \quad x \ne 0.$$

T **b.** Graph $f(x) = (\sin x)/x$ together with the functions $y = 1 - (x^2/6)$ and $y = 1$ for $-5 \le x \le 5$. Comment on the relationships among the graphs.

30. **a.** Use the Taylor series for $\cos x$ and the Alternating Series Estimation Theorem to show that

$$\frac{1}{2} - \frac{x^2}{24} < \frac{1 - \cos x}{x^2} < \frac{1}{2}, \quad x \ne 0.$$

(This is the inequality in Section 2.2, Exercise 52.)

T **b.** Graph $f(x) = (1 - \cos x)/x^2$ together with $y = (1/2) - (x^2/24)$ and $y = 1/2$ for $-9 \le x \le 9$. Comment on the relationships among the graphs.

Finding and Identifying Maclaurin Series

Recall that the Maclaurin series is just another name for the Taylor series at $x = 0$. Each of the series in Exercises 31–34 is the value of the Maclaurin series of a function $f(x)$ at some point. What function and what point? What is the sum of the series?

31. $(0.1) - \dfrac{(0.1)^3}{3!} + \dfrac{(0.1)^5}{5!} - \cdots + \dfrac{(-1)^k(0.1)^{2k+1}}{(2k + 1)!} + \cdots$

32. $1 - \dfrac{\pi^2}{4^2 \cdot 2!} + \dfrac{\pi^4}{4^4 \cdot 4!} - \cdots + \dfrac{(-1)^k(\pi)^{2k}}{4^{2k} \cdot (2k!)} + \cdots$

33. $\dfrac{\pi}{3} - \dfrac{\pi^3}{3^3 \cdot 3} + \dfrac{\pi^5}{3^5 \cdot 5} - \cdots + \dfrac{(-1)^k \pi^{2k+1}}{3^{2k+1}(2k + 1)} + \cdots$

34. $(\pi/4) - \dfrac{(\pi/4)^2}{2} + \dfrac{(\pi/4)^3}{3} - \cdots + (-1)^{k-1}\dfrac{(\pi/4)^k}{k} + \cdots$

35. Multiply the Maclaurin series for e^x and $\sin x$ together to find the first five nonzero terms of the Maclaurin series for $e^x \sin x$.

36. Multiply the Maclaurin series for e^x and $\cos x$ together to find the first five nonzero terms of the Maclaurin series for $e^x \cos x$.

37. Use the identity $\sin^2 x = (1 - \cos 2x)/2$ to obtain the Maclaurin series for $\sin^2 x$. Then differentiate this series to obtain the Maclaurin series for $2 \sin x \cos x$. Check that this is the series for $\sin 2x$.

38. (*Continuation of Exercise 37.*) Use the identity $\cos^2 x = \cos 2x + \sin^2 x$ to obtain a power series for $\cos^2 x$.

Theory and Examples

39. Taylor's Theorem and the Mean Value Theorem Explain how the Mean Value Theorem (Section 4.2, Theorem 4) is a special case of Taylor's Theorem.

40. Linearizations at inflection points Show that if the graph of a twice-differentiable function $f(x)$ has an inflection point at $x = a$, then the linearization of f at $x = a$ is also the quadratic approximation of f at $x = a$. This explains why tangent lines fit so well at inflection points.

41. The (second) second derivative test Use the equation

$$f(x) = f(a) + f'(a)(x - a) + \frac{f''(c_2)}{2}(x - a)^2$$

to establish the following test.

Let f have continuous first and second derivatives and suppose that $f'(a) = 0$. Then

a. f has a local maximum at a if $f'' \le 0$ throughout an interval whose interior contains a;

b. f has a local minimum at a if $f'' \ge 0$ throughout an interval whose interior contains a.

42. A cubic approximation Use Taylor's formula with $a = 0$ and $n = 3$ to find the standard cubic approximation of $f(x) = 1/(1 - x)$ at $x = 0$. Give an upper bound for the magnitude of the error in the approximation when $|x| \le 0.1$.

43. a. Use Taylor's formula with $n = 2$ to find the quadratic approximation of $f(x) = (1 + x)^k$ at $x = 0$ (k a constant).

b. If $k = 3$, for approximately what values of x in the interval $[0, 1]$ will the error in the quadratic approximation be less than $1/100$?

44. Improving approximations to π

a. Let P be an approximation of π accurate to n decimals. Show that $P + \sin P$ gives an approximation correct to $3n$ decimals. (*Hint:* Let $P = \pi + x$.)

T **b.** Try it with a calculator.

45. The Taylor series generated by $f(x) = \sum_{n=0}^{\infty} a_n x^n$ is $\sum_{n=0}^{\infty} a_n x^n$ A function defined by a power series $\sum_{n=0}^{\infty} a_n x^n$ with a radius of convergence $c > 0$ has a Taylor series that converges to the function at every point of $(-c, c)$. Show this by showing that the Taylor series generated by $f(x) = \sum_{n=0}^{\infty} a_n x^n$ is the series $\sum_{n=0}^{\infty} a_n x^n$ itself.

An immediate consequence of this is that series like

$$x \sin x = x^2 - \frac{x^4}{3!} + \frac{x^6}{5!} - \frac{x^8}{7!} + \cdots$$

and

$$x^2 e^x = x^2 + x^3 + \frac{x^4}{2!} + \frac{x^5}{3!} + \cdots,$$

obtained by multiplying Taylor series by powers of x, as well as series obtained by integration and differentiation of convergent power series, are themselves the Taylor series generated by the functions they represent.

46. Taylor series for even functions and odd functions (*Continuation of Section 11.7, Exercise 45.*) Suppose that $f(x) = \sum_{n=0}^{\infty} a_n x^n$ converges for all x in an open interval $(-c, c)$. Show that

a. If f is even, then $a_1 = a_3 = a_5 = \cdots = 0$, i.e., the Taylor series for f at $x = 0$ contains only even powers of x.

b. If f is odd, then $a_0 = a_2 = a_4 = \cdots = 0$, i.e., the Taylor series for f at $x = 0$ contains only odd powers of x.

47. Taylor polynomials of periodic functions

a. Show that every continuous periodic function $f(x)$, $-\infty < x < \infty$, is bounded in magnitude by showing that there exists a positive constant M such that $|f(x)| \le M$ for all x.

b. Show that the graph of every Taylor polynomial of positive degree generated by $f(x) = \cos x$ must eventually move away from the graph of $\cos x$ as $|x|$ increases. You can see this in Figure 11.13. The Taylor polynomials of $\sin x$ behave in a similar way (Figure 11.15).

T **48. a.** Graph the curves $y = (1/3) - (x^2)/5$ and $y = (x - \tan^{-1} x)/x^3$ together with the line $y = 1/3$.

b. Use a Taylor series to explain what you see. What is

$$\lim_{x \to 0} \frac{x - \tan^{-1} x}{x^3}?$$

Euler's Identity

49. Use Equation (6) to write the following powers of e in the form $a + bi$.

a. $e^{-i\pi}$ **b.** $e^{i\pi/4}$ **c.** $e^{-i\pi/2}$

50. Use Equation (6) to show that

$$\cos \theta = \frac{e^{i\theta} + e^{-i\theta}}{2} \quad \text{and} \quad \sin \theta = \frac{e^{i\theta} - e^{-i\theta}}{2i}.$$

51. Establish the equations in Exercise 50 by combining the formal Taylor series for $e^{i\theta}$ and $e^{-i\theta}$.

52. Show that

a. $\cosh i\theta = \cos \theta$, **b.** $\sinh i\theta = i \sin \theta$.

53. By multiplying the Taylor series for e^x and $\sin x$, find the terms through x^5 of the Taylor series for $e^x \sin x$. This series is the imaginary part of the series for

$$e^x \cdot e^{ix} = e^{(1+i)x}.$$

Use this fact to check your answer. For what values of x should the series for $e^x \sin x$ converge?

54. When a and b are real, we define $e^{(a+ib)x}$ with the equation

$$e^{(a+ib)x} = e^{ax} \cdot e^{ibx} = e^{ax}(\cos bx + i \sin bx).$$

Differentiate the right-hand side of this equation to show that

$$\frac{d}{dx} e^{(a+ib)x} = (a + ib)e^{(a+ib)x}.$$

Thus the familiar rule $(d/dx)e^{kx} = ke^{kx}$ holds for k complex as well as real.

55. Use the definition of $e^{i\theta}$ to show that for any real numbers θ, θ_1, and θ_2,

a. $e^{i\theta_1} e^{i\theta_2} = e^{i(\theta_1 + \theta_2)}$, **b.** $e^{-i\theta} = 1/e^{i\theta}$.

56. Two complex numbers $a + ib$ and $c + id$ are equal if and only if $a = c$ and $b = d$. Use this fact to evaluate

$$\int e^{ax} \cos bx \, dx \quad \text{and} \quad \int e^{ax} \sin bx \, dx$$

from

$$\int e^{(a+ib)x} \, dx = \frac{a - ib}{a^2 + b^2} e^{(a+ib)x} + C,$$

where $C = C_1 + iC_2$ is a complex constant of integration.

COMPUTER EXPLORATIONS

Linear, Quadratic, and Cubic Approximations

Taylor's formula with $n = 1$ and $a = 0$ gives the linearization of a function at $x = 0$. With $n = 2$ and $n = 3$ we obtain the standard quadratic and cubic approximations. In these exercises we explore the errors associated with these approximations. We seek answers to two questions:

a. For what values of x can the function be replaced by each approximation with an error less than 10^{-2}?

b. What is the maximum error we could expect if we replace the function by each approximation over the specified interval?

Using a CAS, perform the following steps to aid in answering questions (a) and (b) for the functions and intervals in Exercises 57–62.

Step 1: Plot the function over the specified interval.

Step 2: Find the Taylor polynomials $P_1(x)$, $P_2(x)$, and $P_3(x)$ at $x = 0$.

Step 3: Calculate the $(n + 1)$st derivative $f^{(n+1)}(c)$ associated with the remainder term for each Taylor polynomial. Plot the derivative as a function of c over the specified interval and estimate its maximum absolute value, M.

Step 4: Calculate the remainder $R_n(x)$ for each polynomial. Using the estimate M from Step 3 in place of $f^{(n+1)}(c)$, plot $R_n(x)$ over the specified interval. Then estimate the values of x that answer question (a).

Step 5: Compare your estimated error with the actual error $E_n(x) = |f(x) - P_n(x)|$ by plotting $E_n(x)$ over the specified interval. This will help answer question (b).

Step 6: Graph the function and its three Taylor approximations together. Discuss the graphs in relation to the information discovered in Steps 4 and 5.

57. $f(x) = \dfrac{1}{\sqrt{1 + x}}, \quad |x| \le \dfrac{3}{4}$

58. $f(x) = (1 + x)^{3/2}, \quad -\dfrac{1}{2} \le x \le 2$

59. $f(x) = \dfrac{x}{x^2 + 1}, \quad |x| \le 2$

60. $f(x) = (\cos x)(\sin 2x), \quad |x| \le 2$

61. $f(x) = e^{-x} \cos 2x, \quad |x| \le 1$

62. $f(x) = e^{x/3} \sin 2x, \quad |x| \le 2$

11.10 Applications of Power Series

This section introduces the binomial series for estimating powers and roots and shows how series are sometimes used to approximate the solution of an initial value problem, to evaluate nonelementary integrals, and to evaluate limits that lead to indeterminate forms. We provide a self-contained derivation of the Taylor series for $\tan^{-1} x$ and conclude with a reference table of frequently used series.

The Binomial Series for Powers and Roots

The Taylor series generated by $f(x) = (1 + x)^m$, when m is constant, is

$$1 + mx + \frac{m(m-1)}{2!}x^2 + \frac{m(m-1)(m-2)}{3!}x^3 + \cdots + \frac{m(m-1)(m-2)\cdots(m-k+1)}{k!}x^k + \cdots. \qquad (1)$$

This series, called the **binomial series**, converges absolutely for $|x| < 1$. To derive the series, we first list the function and its derivatives:

$$\begin{aligned}
f(x) &= (1 + x)^m \\
f'(x) &= m(1 + x)^{m-1} \\
f''(x) &= m(m - 1)(1 + x)^{m-2} \\
f'''(x) &= m(m - 1)(m - 2)(1 + x)^{m-3} \\
&\vdots \\
f^{(k)}(x) &= m(m - 1)(m - 2)\cdots(m - k + 1)(1 + x)^{m-k}.
\end{aligned}$$

We then evaluate these at $x = 0$ and substitute into the Taylor series formula to obtain Series (1).

If m is an integer greater than or equal to zero, the series stops after $(m + 1)$ terms because the coefficients from $k = m + 1$ on are zero.

If m is not a positive integer or zero, the series is infinite and converges for $|x| < 1$. To see why, let u_k be the term involving x^k. Then apply the Ratio Test for absolute convergence to see that

$$\left|\frac{u_{k+1}}{u_k}\right| = \left|\frac{m - k}{k + 1}x\right| \to |x| \qquad \text{as } k \to \infty.$$

Our derivation of the binomial series shows only that it is generated by $(1 + x)^m$ and converges for $|x| < 1$. The derivation does not show that the series converges to $(1 + x)^m$. It does, but we omit the proof.

The Binomial Series

For $-1 < x < 1$,

$$(1 + x)^m = 1 + \sum_{k=1}^{\infty} \binom{m}{k} x^k,$$

where we define

$$\binom{m}{1} = m, \qquad \binom{m}{2} = \frac{m(m - 1)}{2!},$$

and

$$\binom{m}{k} = \frac{m(m - 1)(m - 2)\cdots(m - k + 1)}{k!} \qquad \text{for } k \geq 3.$$

EXAMPLE 1 Using the Binomial Series

If $m = -1$,

$$\binom{-1}{1} = -1, \qquad \binom{-1}{2} = \frac{-1(-2)}{2!} = 1,$$

and

$$\binom{-1}{k} = \frac{-1(-2)(-3)\cdots(-1 - k + 1)}{k!} = (-1)^k \left(\frac{k!}{k!}\right) = (-1)^k.$$

With these coefficient values and with x replaced by $-x$, the binomial series formula gives the familiar geometric series

$$(1 + x)^{-1} = 1 + \sum_{k=1}^{\infty}(-1)^k x^k = 1 - x + x^2 - x^3 + \cdots + (-1)^k x^k + \cdots. \quad \blacksquare$$

EXAMPLE 2 Using the Binomial Series

We know from Section 3.8, Example 1, that $\sqrt{1 + x} \approx 1 + (x/2)$ for $|x|$ small. With $m = 1/2$, the binomial series gives quadratic and higher-order approximations as well, along with error estimates that come from the Alternating Series Estimation Theorem:

$$\begin{aligned}(1 + x)^{1/2} &= 1 + \frac{x}{2} + \frac{\left(\frac{1}{2}\right)\left(-\frac{1}{2}\right)}{2!}x^2 + \frac{\left(\frac{1}{2}\right)\left(-\frac{1}{2}\right)\left(-\frac{3}{2}\right)}{3!}x^3 \\ &\qquad + \frac{\left(\frac{1}{2}\right)\left(-\frac{1}{2}\right)\left(-\frac{3}{2}\right)\left(-\frac{5}{2}\right)}{4!}x^4 + \cdots \\ &= 1 + \frac{x}{2} - \frac{x^2}{8} + \frac{x^3}{16} - \frac{5x^4}{128} + \cdots.\end{aligned}$$

Substitution for x gives still other approximations. For example,

$$\sqrt{1 - x^2} \approx 1 - \frac{x^2}{2} - \frac{x^4}{8} \qquad \text{for } |x^2| \text{ small}$$

$$\sqrt{1 - \frac{1}{x}} \approx 1 - \frac{1}{2x} - \frac{1}{8x^2} \qquad \text{for } \left|\frac{1}{x}\right| \text{ small, that is, } |x| \text{ large}. \quad \blacksquare$$

Power Series Solutions of Differential Equations and Initial Value Problems

When we cannot find a relatively simple expression for the solution of an initial value problem or differential equation, we try to get information about the solution in other ways. One way is to try to find a power series representation for the solution. If we can do so, we immediately have a source of polynomial approximations of the solution, which may be all that we really need. The first example (Example 3) deals with a first-order linear differential equation that could be solved with the methods of Section 9.2. The example shows how, not knowing this, we can solve the equation with power series. The second example (Example 4) deals with an equation that cannot be solved analytically by previous methods.

EXAMPLE 3 Series Solution of an Initial Value Problem

Solve the initial value problem

$$y' - y = x, \qquad y(0) = 1.$$

Solution We assume that there is a solution of the form

$$y = a_0 + a_1 x + a_2 x^2 + \cdots + a_{n-1}x^{n-1} + a_n x^n + \cdots. \tag{2}$$

Our goal is to find values for the coefficients a_k that make the series and its first derivative

$$y' = a_1 + 2a_2 x + 3a_3 x^2 + \cdots + na_n x^{n-1} + \cdots \tag{3}$$

satisfy the given differential equation and initial condition. The series $y' - y$ is the difference of the series in Equations (2) and (3):

$$\begin{aligned} y' - y = (a_1 - a_0) + (2a_2 - a_1)x + (3a_3 - a_2)x^2 + \cdots \\ + (na_n - a_{n-1})x^{n-1} + \cdots. \end{aligned} \tag{4}$$

If y is to satisfy the equation $y' - y = x$, the series in Equation (4) must equal x. Since power series representations are unique (Exercise 45 in Section 11.7), the coefficients in Equation (4) must satisfy the equations

$$\begin{aligned} a_1 - a_0 &= 0 && \text{Constant terms} \\ 2a_2 - a_1 &= 1 && \text{Coefficients of } x \\ 3a_3 - a_2 &= 0 && \text{Coefficients of } x^2 \\ &\vdots && \vdots \\ na_n - a_{n-1} &= 0 && \text{Coefficients of } x^{n-1} \\ &\vdots && \vdots \end{aligned}$$

We can also see from Equation (2) that $y = a_0$ when $x = 0$, so that $a_0 = 1$ (this being the initial condition). Putting it all together, we have

$$a_0 = 1, \qquad a_1 = a_0 = 1, \qquad a_2 = \frac{1 + a_1}{2} = \frac{1+1}{2} = \frac{2}{2},$$

$$a_3 = \frac{a_2}{3} = \frac{2}{3 \cdot 2} = \frac{2}{3!}, \ldots, \qquad a_n = \frac{a_{n-1}}{n} = \frac{2}{n!}, \ldots$$

Substituting these coefficient values into the equation for y (Equation (2)) gives

$$\begin{aligned} y &= 1 + x + 2 \cdot \frac{x^2}{2!} + 2 \cdot \frac{x^3}{3!} + \cdots + 2 \cdot \frac{x^n}{n!} + \cdots \\ &= 1 + x + 2 \underbrace{\left(\frac{x^2}{2!} + \frac{x^3}{3!} + \cdots + \frac{x^n}{n!} + \cdots \right)}_{\text{the Taylor series for } e^x - 1 - x} \\ &= 1 + x + 2(e^x - 1 - x) = 2e^x - 1 - x. \end{aligned}$$

The solution of the initial value problem is $y = 2e^x - 1 - x$.

As a check, we see that

$$y(0) = 2e^0 - 1 - 0 = 2 - 1 = 1$$

and

$$y' - y = (2e^x - 1) - (2e^x - 1 - x) = x.$$

■

EXAMPLE 4 Solving a Differential Equation

Find a power series solution for

$$y'' + x^2 y = 0. \tag{5}$$

Solution We assume that there is a solution of the form

$$y = a_0 + a_1 x + a_2 x^2 + \cdots + a_n x^n + \cdots, \tag{6}$$

and find what the coefficients a_k have to be to make the series and its second derivative

$$y'' = 2a_2 + 3 \cdot 2a_3 x + \cdots + n(n-1)a_n x^{n-2} + \cdots \tag{7}$$

satisfy Equation (5). The series for $x^2 y$ is x^2 times the right-hand side of Equation (6):

$$x^2 y = a_0 x^2 + a_1 x^3 + a_2 x^4 + \cdots + a_n x^{n+2} + \cdots. \tag{8}$$

The series for $y'' + x^2 y$ is the sum of the series in Equations (7) and (8):

$$\begin{aligned} y'' + x^2 y &= 2a_2 + 6a_3 x + (12a_4 + a_0)x^2 + (20a_5 + a_1)x^3 \\ &\quad + \cdots + (n(n-1)a_n + a_{n-4})x^{n-2} + \cdots. \end{aligned} \tag{9}$$

Notice that the coefficient of x^{n-2} in Equation (8) is a_{n-4}. If y and its second derivative y'' are to satisfy Equation (5), the coefficients of the individual powers of x on the right-hand side of Equation (9) must all be zero:

$$2a_2 = 0, \qquad 6a_3 = 0, \qquad 12a_4 + a_0 = 0, \qquad 20a_5 + a_1 = 0, \tag{10}$$

and for all $n \ge 4$,

$$n(n-1)a_n + a_{n-4} = 0. \tag{11}$$

We can see from Equation (6) that

$$a_0 = y(0), \qquad a_1 = y'(0).$$

In other words, the first two coefficients of the series are the values of y and y' at $x = 0$. Equations in (10) and the recursion formula in Equation (11) enable us to evaluate all the other coefficients in terms of a_0 and a_1.

The first two of Equations (10) give

$$a_2 = 0, \qquad a_3 = 0.$$

Equation (11) shows that if $a_{n-4} = 0$, then $a_n = 0$; so we conclude that

$$a_6 = 0, \qquad a_7 = 0, \qquad a_{10} = 0, \qquad a_{11} = 0,$$

and whenever $n = 4k + 2$ or $4k + 3$, a_n is zero. For the other coefficients we have

$$a_n = \frac{-a_{n-4}}{n(n-1)}$$

so that

$$a_4 = \frac{-a_0}{4 \cdot 3}, \qquad a_8 = \frac{-a_4}{8 \cdot 7} = \frac{a_0}{3 \cdot 4 \cdot 7 \cdot 8}$$

$$a_{12} = \frac{-a_8}{11 \cdot 12} = \frac{-a_0}{3 \cdot 4 \cdot 7 \cdot 8 \cdot 11 \cdot 12}$$

and

$$a_5 = \frac{-a_1}{5 \cdot 4}, \qquad a_9 = \frac{-a_5}{9 \cdot 8} = \frac{a_1}{4 \cdot 5 \cdot 8 \cdot 9}$$

$$a_{13} = \frac{-a_9}{12 \cdot 13} = \frac{-a_1}{4 \cdot 5 \cdot 8 \cdot 9 \cdot 12 \cdot 13}.$$

The answer is best expressed as the sum of two separate series—one multiplied by a_0, the other by a_1:

$$y = a_0\left(1 - \frac{x^4}{3\cdot 4} + \frac{x^8}{3\cdot 4\cdot 7\cdot 8} - \frac{x^{12}}{3\cdot 4\cdot 7\cdot 8\cdot 11\cdot 12} + \cdots\right)$$
$$+ a_1\left(x - \frac{x^5}{4\cdot 5} + \frac{x^9}{4\cdot 5\cdot 8\cdot 9} - \frac{x^{13}}{4\cdot 5\cdot 8\cdot 9\cdot 12\cdot 13} + \cdots\right).$$

Both series converge absolutely for all x, as is readily seen by the Ratio Test. ■

Evaluating Nonelementary Integrals

Taylor series can be used to express nonelementary integrals in terms of series. Integrals like $\int \sin x^2\,dx$ arise in the study of the diffraction of light.

EXAMPLE 5 Express $\int \sin x^2\,dx$ as a power series.

Solution From the series for $\sin x$ we obtain

$$\sin x^2 = x^2 - \frac{x^6}{3!} + \frac{x^{10}}{5!} - \frac{x^{14}}{7!} + \frac{x^{18}}{9!} - \cdots.$$

Therefore,

$$\int \sin x^2\,dx = C + \frac{x^3}{3} - \frac{x^7}{7\cdot 3!} + \frac{x^{11}}{11\cdot 5!} - \frac{x^{15}}{15\cdot 7!} + \frac{x^{10}}{19\cdot 9!} - \cdots.$$ ■

EXAMPLE 6 Estimating a Definite Integral

Estimate $\int_0^1 \sin x^2\,dx$ with an error of less than 0.001.

Solution From the indefinite integral in Example 5,

$$\int_0^1 \sin x^2\,dx = \frac{1}{3} - \frac{1}{7\cdot 3!} + \frac{1}{11\cdot 5!} - \frac{1}{15\cdot 7!} + \frac{1}{19\cdot 9!} - \cdots.$$

The series alternates, and we find by experiment that

$$\frac{1}{11\cdot 5!} \approx 0.00076$$

is the first term to be numerically less than 0.001. The sum of the preceding two terms gives

$$\int_0^1 \sin x^2\,dx \approx \frac{1}{3} - \frac{1}{42} \approx 0.310.$$

With two more terms we could estimate

$$\int_0^1 \sin x^2\,dx \approx 0.310268$$

with an error of less than 10^{-6}. With only one term beyond that we have

$$\int_0^1 \sin x^2\,dx \approx \frac{1}{3} - \frac{1}{42} + \frac{1}{1320} - \frac{1}{75600} + \frac{1}{6894720} \approx 0.310268303,$$

with an error of about 1.08×10^{-9}. To guarantee this accuracy with the error formula for the Trapezoidal Rule would require using about 8000 subintervals. ■

Arctangents

In Section 11.7, Example 5, we found a series for $\tan^{-1} x$ by differentiating to get

$$\frac{d}{dx}\tan^{-1} x = \frac{1}{1 + x^2} = 1 - x^2 + x^4 - x^6 + \cdots$$

and integrating to get

$$\tan^{-1} x = x - \frac{x^3}{3} + \frac{x^5}{5} - \frac{x^7}{7} + \cdots.$$

However, we did not prove the term-by-term integration theorem on which this conclusion depended. We now derive the series again by integrating both sides of the finite formula

$$\frac{1}{1 + t^2} = 1 - t^2 + t^4 - t^6 + \cdots + (-1)^n t^{2n} + \frac{(-1)^{n+1} t^{2n+2}}{1 + t^2}, \tag{12}$$

in which the last term comes from adding the remaining terms as a geometric series with first term $a = (-1)^{n+1} t^{2n+2}$ and ratio $r = -t^2$. Integrating both sides of Equation (12) from $t = 0$ to $t = x$ gives

$$\tan^{-1} x = x - \frac{x^3}{3} + \frac{x^5}{5} - \frac{x^7}{7} + \cdots + (-1)^n \frac{x^{2n+1}}{2n + 1} + R_n(x),$$

where

$$R_n(x) = \int_0^x \frac{(-1)^{n+1} t^{2n+2}}{1 + t^2}\, dt.$$

The denominator of the integrand is greater than or equal to 1; hence

$$|R_n(x)| \le \int_0^{|x|} t^{2n+2}\, dt = \frac{|x|^{2n+3}}{2n + 3}.$$

If $|x| \le 1$, the right side of this inequality approaches zero as $n \to \infty$. Therefore $\lim_{n\to\infty} R_n(x) = 0$ if $|x| \le 1$ and

$$\begin{aligned} \tan^{-1} x &= \sum_{n=0}^{\infty} \frac{(-1)^n x^{2n+1}}{2n + 1}, \qquad |x| \le 1. \\ \tan^{-1} x &= x - \frac{x^3}{3} + \frac{x^5}{5} - \frac{x^7}{7} + \cdots, \qquad |x| \le 1 \end{aligned} \tag{13}$$

We take this route instead of finding the Taylor series directly because the formulas for the higher-order derivatives of $\tan^{-1} x$ are unmanageable. When we put $x = 1$ in Equation (13), we get **Leibniz's formula**:

$$\frac{\pi}{4} = 1 - \frac{1}{3} + \frac{1}{5} - \frac{1}{7} + \frac{1}{9} - \cdots + \frac{(-1)^n}{2n + 1} + \cdots.$$

Because this series converges very slowly, it is not used in approximating π to many decimal places. The series for $\tan^{-1} x$ converges most rapidly when x is near zero. For that reason, people who use the series for $\tan^{-1} x$ to compute π use various trigonometric identities.

For example, if

$$\alpha = \tan^{-1}\frac{1}{2} \qquad \text{and} \qquad \beta = \tan^{-1}\frac{1}{3},$$

then

$$\tan(\alpha + \beta) = \frac{\tan\alpha + \tan\beta}{1 - \tan\alpha\tan\beta} = \frac{\frac{1}{2} + \frac{1}{3}}{1 - \frac{1}{6}} = 1 = \tan\frac{\pi}{4}$$

and

$$\frac{\pi}{4} = \alpha + \beta = \tan^{-1}\frac{1}{2} + \tan^{-1}\frac{1}{3}.$$

Now Equation (13) may be used with $x = 1/2$ to evaluate $\tan^{-1}(1/2)$ and with $x = 1/3$ to give $\tan^{-1}(1/3)$. The sum of these results, multiplied by 4, gives π.

Evaluating Indeterminate Forms

We can sometimes evaluate indeterminate forms by expressing the functions involved as Taylor series.

EXAMPLE 7 Limits Using Power Series

Evaluate

$$\lim_{x\to 1}\frac{\ln x}{x - 1}.$$

Solution We represent $\ln x$ as a Taylor series in powers of $x - 1$. This can be accomplished by calculating the Taylor series generated by $\ln x$ at $x = 1$ directly or by replacing x by $x - 1$ in the series for $\ln(1 + x)$ in Section 11.7, Example 6. Either way, we obtain

$$\ln x = (x - 1) - \frac{1}{2}(x - 1)^2 + \cdots,$$

from which we find that

$$\lim_{x\to 1}\frac{\ln x}{x - 1} = \lim_{x\to 1}\left(1 - \frac{1}{2}(x - 1) + \cdots\right) = 1.$$ ■

EXAMPLE 8 Limits Using Power Series

Evaluate

$$\lim_{x\to 0}\frac{\sin x - \tan x}{x^3}.$$

Solution The Taylor series for $\sin x$ and $\tan x$, to terms in x^5, are

$$\sin x = x - \frac{x^3}{3!} + \frac{x^5}{5!} - \cdots, \qquad \tan x = x + \frac{x^3}{3} + \frac{2x^5}{15} + \cdots.$$

Hence,

$$\sin x - \tan x = -\frac{x^3}{2} - \frac{x^5}{8} - \cdots = x^3\left(-\frac{1}{2} - \frac{x^2}{8} - \cdots\right)$$

and

$$\lim_{x \to 0} \frac{\sin x - \tan x}{x^3} = \lim_{x \to 0}\left(-\frac{1}{2} - \frac{x^2}{8} - \cdots\right)$$

$$= -\frac{1}{2}.$$

If we apply series to calculate $\lim_{x \to 0}((1/\sin x) - (1/x))$, we not only find the limit successfully but also discover an approximation formula for $\csc x$.

EXAMPLE 9 Approximation Formula for csc x

Find $\lim_{x \to 0}\left(\frac{1}{\sin x} - \frac{1}{x}\right)$.

Solution

$$\frac{1}{\sin x} - \frac{1}{x} = \frac{x - \sin x}{x \sin x} = \frac{x - \left(x - \frac{x^3}{3!} + \frac{x^5}{5!} - \cdots\right)}{x \cdot \left(x - \frac{x^3}{3!} + \frac{x^5}{5!} - \cdots\right)}$$

$$= \frac{x^3\left(\frac{1}{3!} - \frac{x^2}{5!} + \cdots\right)}{x^2\left(1 - \frac{x^2}{3!} + \cdots\right)} = x\frac{\frac{1}{3!} - \frac{x^2}{5!} + \cdots}{1 - \frac{x^2}{3!} + \cdots}.$$

Therefore,

$$\lim_{x \to 0}\left(\frac{1}{\sin x} - \frac{1}{x}\right) = \lim_{x \to 0}\left(x\frac{\frac{1}{3!} - \frac{x^2}{5!} + \cdots}{1 - \frac{x^2}{3!} + \cdots}\right) = 0.$$

From the quotient on the right, we can see that if $|x|$ is small, then

$$\frac{1}{\sin x} - \frac{1}{x} \approx x \cdot \frac{1}{3!} = \frac{x}{6} \qquad \text{or} \qquad \csc x \approx \frac{1}{x} + \frac{x}{6}.$$

TABLE 11.1 Frequently used Taylor series

$$\frac{1}{1-x} = 1 + x + x^2 + \cdots + x^n + \cdots = \sum_{n=0}^{\infty} x^n, \qquad |x| < 1$$

$$\frac{1}{1+x} = 1 - x + x^2 - \cdots + (-x)^n + \cdots = \sum_{n=0}^{\infty} (-1)^n x^n, \qquad |x| < 1$$

$$e^x = 1 + x + \frac{x^2}{2!} + \cdots + \frac{x^n}{n!} + \cdots = \sum_{n=0}^{\infty} \frac{x^n}{n!}, \qquad |x| < \infty$$

$$\sin x = x - \frac{x^3}{3!} + \frac{x^5}{5!} - \cdots + (-1)^n \frac{x^{2n+1}}{(2n+1)!} + \cdots = \sum_{n=0}^{\infty} \frac{(-1)^n x^{2n+1}}{(2n+1)!}, \qquad |x| < \infty$$

$$\cos x = 1 - \frac{x^2}{2!} + \frac{x^4}{4!} - \cdots + (-1)^n \frac{x^{2n}}{(2n)!} + \cdots = \sum_{n=0}^{\infty} \frac{(-1)^n x^{2n}}{(2n)!}, \qquad |x| < \infty$$

$$\ln(1+x) = x - \frac{x^2}{2} + \frac{x^3}{3} - \cdots + (-1)^{n-1} \frac{x^n}{n} + \cdots = \sum_{n=1}^{\infty} \frac{(-1)^{n-1} x^n}{n}, \qquad -1 < x \le 1$$

$$\ln \frac{1+x}{1-x} = 2 \tanh^{-1} x = 2\left(x + \frac{x^3}{3} + \frac{x^5}{5} + \cdots + \frac{x^{2n+1}}{2n+1} + \cdots\right) = 2 \sum_{n=0}^{\infty} \frac{x^{2n+1}}{2n+1}, \qquad |x| < 1$$

$$\tan^{-1} x = x - \frac{x^3}{3} + \frac{x^5}{5} - \cdots + (-1)^n \frac{x^{2n+1}}{2n+1} + \cdots = \sum_{n=0}^{\infty} \frac{(-1)^n x^{2n+1}}{2n+1}, \qquad |x| \le 1$$

Binomial Series

$$(1+x)^m = 1 + mx + \frac{m(m-1)x^2}{2!} + \frac{m(m-1)(m-2)x^3}{3!} + \cdots + \frac{m(m-1)(m-2)\cdots(m-k+1)x^k}{k!} + \cdots$$

$$= 1 + \sum_{k=1}^{\infty} \binom{m}{k} x^k, \qquad |x| < 1,$$

where

$$\binom{m}{1} = m, \qquad \binom{m}{2} = \frac{m(m-1)}{2!}, \qquad \binom{m}{k} = \frac{m(m-1)\cdots(m-k+1)}{k!} \quad \text{for } k \ge 3.$$

Note: To write the binomial series compactly, it is customary to define $\binom{m}{0}$ to be 1 and to take $x^0 = 1$ (even in the usually excluded case where $x = 0$), yielding $(1+x)^m = \sum_{k=0}^{\infty} \binom{m}{k} x^k$. If m is a *positive integer*, the series terminates at x^m and the result converges for all x.

EXERCISES 11.10

Binomial Series

Find the first four terms of the binomial series for the functions in Exercises 1–10.

1. $(1+x)^{1/2}$ **2.** $(1+x)^{1/3}$ **3.** $(1-x)^{-1/2}$

4. $(1-2x)^{1/2}$ **5.** $\left(1+\frac{x}{2}\right)^{-2}$ **6.** $\left(1-\frac{x}{2}\right)^{-2}$

7. $(1+x^3)^{-1/2}$ **8.** $(1+x^2)^{-1/3}$

9. $\left(1+\frac{1}{x}\right)^{1/2}$ **10.** $\left(1-\frac{2}{x}\right)^{1/3}$

Find the binomial series for the functions in Exercises 11–14.

11. $(1+x)^4$ **12.** $(1+x^2)^3$

13. $(1 - 2x)^3$

14. $\left(1 - \frac{x}{2}\right)^4$

Initial Value Problems

Find series solutions for the initial value problems in Exercises 15–32.

15. $y' + y = 0, \quad y(0) = 1$

16. $y' - 2y = 0, \quad y(0) = 1$

17. $y' - y = 1, \quad y(0) = 0$

18. $y' + y = 1, \quad y(0) = 2$

19. $y' - y = x, \quad y(0) = 0$

20. $y' + y = 2x, \quad y(0) = -1$

21. $y' - xy = 0, \quad y(0) = 1$

22. $y' - x^2y = 0, \quad y(0) = 1$

23. $(1 - x)y' - y = 0, \quad y(0) = 2$

24. $(1 + x^2)y' + 2xy = 0, \quad y(0) = 3$

25. $y'' - y = 0, \quad y'(0) = 1$ and $y(0) = 0$

26. $y'' + y = 0, \quad y'(0) = 0$ and $y(0) = 1$

27. $y'' + y = x, \quad y'(0) = 1$ and $y(0) = 2$

28. $y'' - y = x, \quad y'(0) = 2$ and $y(0) = -1$

29. $y'' - y = -x, \quad y'(2) = -2$ and $y(2) = 0$

30. $y'' - x^2y = 0, \quad y'(0) = b$ and $y(0) = a$

31. $y'' + x^2y = x, \quad y'(0) = b$ and $y(0) = a$

32. $y'' - 2y' + y = 0, \quad y'(0) = 1$ and $y(0) = 0$

Approximations and Nonelementary Integrals

T In Exercises 33–36, use series to estimate the integrals' values with an error of magnitude less than 10^{-3}. (The answer section gives the integrals' values rounded to five decimal places.)

33. $\int_0^{0.2} \sin x^2 \, dx$

34. $\int_0^{0.2} \frac{e^{-x} - 1}{x} \, dx$

35. $\int_0^{0.1} \frac{1}{\sqrt{1 + x^4}} \, dx$

36. $\int_0^{0.25} \sqrt[3]{1 + x^2} \, dx$

T Use series to approximate the values of the integrals in Exercises 37–40 with an error of magnitude less than 10^{-8}.

37. $\int_0^{0.1} \frac{\sin x}{x} \, dx$

38. $\int_0^{0.1} e^{-x^2} \, dx$

39. $\int_0^{0.1} \sqrt{1 + x^4} \, dx$

40. $\int_0^{1} \frac{1 - \cos x}{x^2} \, dx$

41. Estimate the error if $\cos t^2$ is approximated by $1 - \frac{t^4}{2} + \frac{t^8}{4!}$ in the integral $\int_0^1 \cos t^2 \, dt$.

42. Estimate the error if $\cos \sqrt{t}$ is approximated by $1 - \frac{t}{2} + \frac{t^2}{4!} - \frac{t^3}{6!}$ in the integral $\int_0^1 \cos \sqrt{t} \, dt$.

In Exercises 43–46, find a polynomial that will approximate $F(x)$ throughout the given interval with an error of magnitude less than 10^{-3}.

43. $F(x) = \int_0^x \sin t^2 \, dt, \quad [0, 1]$

44. $F(x) = \int_0^x t^2 e^{-t^2} \, dt, \quad [0, 1]$

45. $F(x) = \int_0^x \tan^{-1} t \, dt, \quad$ **(a)** $[0, 0.5]$ **(b)** $[0, 1]$

46. $F(x) = \int_0^x \frac{\ln(1 + t)}{t} \, dt, \quad$ **(a)** $[0, 0.5]$ **(b)** $[0, 1]$

Indeterminate Forms

Use series to evaluate the limits in Exercises 47–56.

47. $\lim_{x \to 0} \frac{e^x - (1 + x)}{x^2}$

48. $\lim_{x \to 0} \frac{e^x - e^{-x}}{x}$

49. $\lim_{t \to 0} \frac{1 - \cos t - (t^2/2)}{t^4}$

50. $\lim_{\theta \to 0} \frac{\sin \theta - \theta + (\theta^3/6)}{\theta^5}$

51. $\lim_{y \to 0} \frac{y - \tan^{-1} y}{y^3}$

52. $\lim_{y \to 0} \frac{\tan^{-1} y - \sin y}{y^3 \cos y}$

53. $\lim_{x \to \infty} x^2(e^{-1/x^2} - 1)$

54. $\lim_{x \to \infty} (x + 1) \sin \frac{1}{x + 1}$

55. $\lim_{x \to 0} \frac{\ln(1 + x^2)}{1 - \cos x}$

56. $\lim_{x \to 2} \frac{x^2 - 4}{\ln(x - 1)}$

Theory and Examples

57. Replace x by $-x$ in the Taylor series for $\ln(1 + x)$ to obtain a series for $\ln(1 - x)$. Then subtract this from the Taylor series for $\ln(1 + x)$ to show that for $|x| < 1$,

$$\ln \frac{1 + x}{1 - x} = 2\left(x + \frac{x^3}{3} + \frac{x^5}{5} + \cdots\right).$$

58. How many terms of the Taylor series for $\ln(1 + x)$ should you add to be sure of calculating $\ln(1.1)$ with an error of magnitude less than 10^{-8}? Give reasons for your answer.

59. According to the Alternating Series Estimation Theorem, how many terms of the Taylor series for $\tan^{-1} 1$ would you have to add to be sure of finding $\pi/4$ with an error of magnitude less than 10^{-3}? Give reasons for your answer.

60. Show that the Taylor series for $f(x) = \tan^{-1} x$ diverges for $|x| > 1$.

T **61. Estimating Pi** About how many terms of the Taylor series for $\tan^{-1} x$ would you have to use to evaluate each term on the right-hand side of the equation

$$\pi = 48 \tan^{-1} \frac{1}{18} + 32 \tan^{-1} \frac{1}{57} - 20 \tan^{-1} \frac{1}{239}$$

with an error of magnitude less than 10^{-6}? In contrast, the convergence of $\sum_{n=1}^{\infty} (1/n^2)$ to $\pi^2/6$ is so slow that even 50 terms will not yield two-place accuracy.

62. Integrate the first three nonzero terms of the Taylor series for $\tan t$ from 0 to x to obtain the first three nonzero terms of the Taylor series for $\ln \sec x$.

63. a. Use the binomial series and the fact that

$$\frac{d}{dx}\sin^{-1} x = (1 - x^2)^{-1/2}$$

to generate the first four nonzero terms of the Taylor series for $\sin^{-1} x$. What is the radius of convergence?

b. Series for $\cos^{-1} x$ Use your result in part (a) to find the first five nonzero terms of the Taylor series for $\cos^{-1} x$.

64. a. Series for $\sinh^{-1} x$ Find the first four nonzero terms of the Taylor series for

$$\sinh^{-1} x = \int_0^x \frac{dt}{\sqrt{1 + t^2}}.$$

T **b.** Use the first *three* terms of the series in part (a) to estimate $\sinh^{-1} 0.25$. Give an upper bound for the magnitude of the estimation error.

65. Obtain the Taylor series for $1/(1 + x)^2$ from the series for $-1/(1 + x)$.

66. Use the Taylor series for $1/(1 - x^2)$ to obtain a series for $2x/(1 - x^2)^2$.

T **67. Estimating Pi** The English mathematician Wallis discovered the formula

$$\frac{\pi}{4} = \frac{2 \cdot 4 \cdot 4 \cdot 6 \cdot 6 \cdot 8 \cdot \cdots}{3 \cdot 3 \cdot 5 \cdot 5 \cdot 7 \cdot 7 \cdot \cdots}.$$

Find π to two decimal places with this formula.

T **68.** Construct a table of natural logarithms $\ln n$ for $n = 1, 2, 3, \ldots, 10$ by using the formula in Exercise 57, but taking advantage of the relationships $\ln 4 = 2 \ln 2$, $\ln 6 = \ln 2 + \ln 3$, $\ln 8 = 3 \ln 2$, $\ln 9 = 2 \ln 3$, and $\ln 10 = \ln 2 + \ln 5$ to reduce the job to the calculation of relatively few logarithms by series. Start by using the following values for x in Exercise 57:

$$\frac{1}{3}, \frac{1}{5}, \frac{1}{9}, \frac{1}{13}.$$

69. Series for $\sin^{-1} x$ Integrate the binomial series for $(1 - x^2)^{-1/2}$ to show that for $|x| < 1$,

$$\sin^{-1} x = x + \sum_{n=1}^{\infty} \frac{1 \cdot 3 \cdot 5 \cdot \cdots \cdot (2n - 1)}{2 \cdot 4 \cdot 6 \cdot \cdots \cdot (2n)} \frac{x^{2n+1}}{2n + 1}.$$

70. Series for $\tan^{-1} x$ for $|x| > 1$ Derive the series

$$\tan^{-1} x = \frac{\pi}{2} - \frac{1}{x} + \frac{1}{3x^3} - \frac{1}{5x^5} + \cdots, \quad x > 1$$

$$\tan^{-1} x = -\frac{\pi}{2} - \frac{1}{x} + \frac{1}{3x^3} - \frac{1}{5x^5} + \cdots, \quad x < -1,$$

by integrating the series

$$\frac{1}{1 + t^2} = \frac{1}{t^2} \cdot \frac{1}{1 + (1/t^2)} = \frac{1}{t^2} - \frac{1}{t^4} + \frac{1}{t^6} - \frac{1}{t^8} + \cdots$$

in the first case from x to ∞ and in the second case from $-\infty$ to x.

71. The value of $\sum_{n=1}^{\infty} \tan^{-1}(2/n^2)$

a. Use the formula for the tangent of the difference of two angles to show that

$$\tan(\tan^{-1}(n + 1) - \tan^{-1}(n - 1)) = \frac{2}{n^2}$$

b. Show that

$$\sum_{n=1}^{N} \tan^{-1}\frac{2}{n^2} = \tan^{-1}(N + 1) + \tan^{-1} N - \frac{\pi}{4}.$$

c. Find the value of $\sum_{n=1}^{\infty} \tan^{-1}\frac{2}{n^2}$.

11.11 Fourier Series

HISTORICAL BIOGRAPHY

Jean-Baptiste Joseph Fourier (1768–1830)

We have seen how Taylor series can be used to approximate a function f by polynomials. The Taylor polynomials give a close fit to f near a particular point $x = a$, but the error in the approximation can be large at points that are far away. There is another method that often gives good approximations on wide intervals, and often works with discontinuous functions for which Taylor polynomials fail. Introduced by Joseph Fourier, this method approximates functions with sums of sine and cosine functions. It is well suited for analyzing periodic functions, such as radio signals and alternating currents, for solving heat transfer problems, and for many other problems in science and engineering.

Suppose we wish to approximate a function f on the interval $[0, 2\pi]$ by a sum of sine and cosine functions,

$$\begin{aligned} f_n(x) &= a_0 + (a_1 \cos x + b_1 \sin x) + (a_2 \cos 2x + b_2 \sin 2x) + \cdots \\ &\quad + (a_n \cos nx + b_n \sin nx) \end{aligned}$$

or, in sigma notation,

$$f_n(x) = a_0 + \sum_{k=1}^{n} (a_k \cos kx + b_k \sin kx). \tag{1}$$

We would like to choose values for the constants $a_0, a_1, a_2, \ldots a_n$ and $b_1, b_2, \ldots, b_n$ that make $f_n(x)$ a "best possible" approximation to $f(x)$. The notion of "best possible" is defined as follows:

1. $f_n(x)$ and $f(x)$ give the same value when integrated from 0 to 2π.
2. $f_n(x) \cos kx$ and $f(x) \cos kx$ give the same value when integrated from 0 to 2π $(k = 1, \ldots, n)$.
3. $f_n(x) \sin kx$ and $f(x) \sin kx$ give the same value when integrated from 0 to 2π $(k = 1, \ldots, n)$.

Altogether we impose $2n + 1$ conditions on f_n:

$$\begin{aligned} \int_0^{2\pi} f_n(x)\, dx &= \int_0^{2\pi} f(x)\, dx, \\ \int_0^{2\pi} f_n(x) \cos kx\, dx &= \int_0^{2\pi} f(x) \cos kx\, dx, \qquad k = 1, \ldots, n, \\ \int_0^{2\pi} f_n(x) \sin kx\, dx &= \int_0^{2\pi} f(x) \sin kx\, dx, \qquad k = 1, \ldots, n. \end{aligned}$$

It is possible to choose $a_0, a_1, a_2, \ldots a_n$ and $b_1, b_2, \ldots, b_n$ so that all these conditions are satisfied, by proceeding as follows. Integrating both sides of Equation (1) from 0 to 2π gives

$$\int_0^{2\pi} f_n(x)\, dx = 2\pi a_0$$

since the integral over $[0, 2\pi]$ of $\cos kx$ equals zero when $k \geq 1$, as does the integral of $\sin kx$. Only the constant term a_0 contributes to the integral of f_n over $[0, 2\pi]$. A similar calculation applies with each of the other terms. If we multiply both sides of Equation (1) by $\cos x$ and integrate from 0 to 2π then we obtain

$$\int_0^{2\pi} f_n(x) \cos x\, dx = \pi a_1.$$

This follows from the fact that

$$\int_0^{2\pi} \cos px \cos px\, dx = \pi$$

and

$$\int_0^{2\pi} \cos px \cos qx\, dx = \int_0^{2\pi} \cos px \sin mx\, dx = \int_0^{2\pi} \sin px \sin qx\, dx = 0$$

whenever p, q and m are integers and p is not equal to q (Exercises 9–13). If we multiply Equation (1) by $\sin x$ and integrate from 0 to 2π we obtain

$$\int_0^{2\pi} f_n(x) \sin x \, dx = \pi b_1.$$

Proceeding in a similar fashion with

$$\cos 2x, \sin 2x, \ldots, \cos nx, \sin nx$$

we obtain only one nonzero term each time, the term with a sine-squared or cosine-squared term. To summarize,

$$\int_0^{2\pi} f_n(x) \, dx = 2\pi a_0$$

$$\int_0^{2\pi} f_n(x) \cos kx \, dx = \pi a_k, \qquad k = 1, \ldots, n$$

$$\int_0^{2\pi} f_n(x) \sin kx \, dx = \pi b_k, \qquad k = 1, \ldots, n$$

We chose f_n so that the integrals on the left remain the same when f_n is replaced by f, so we can use these equations to find $a_0, a_1, a_2, \ldots a_n$ and $b_1, b_2, \ldots, b_n$ from f:

$$a_0 = \frac{1}{2\pi} \int_0^{2\pi} f(x) \, dx \tag{2}$$

$$a_k = \frac{1}{\pi} \int_0^{2\pi} f(x) \cos kx \, dx, \qquad k = 1, \ldots, n \tag{3}$$

$$b_k = \frac{1}{\pi} \int_0^{2\pi} f(x) \sin kx \, dx, \qquad k = 1, \ldots, n \tag{4}$$

The only condition needed to find these coefficients is that the integrals above must exist. If we let $n \to \infty$ and use these rules to get the coefficients of an infinite series, then the resulting sum is called the **Fourier series for $f(x)$**,

$$a_0 + \sum_{k=1}^{\infty} (a_k \cos kx + b_k \sin kx). \tag{5}$$

EXAMPLE 1 Finding a Fourier Series Expansion

Fourier series can be used to represent some functions that cannot be represented by Taylor series; for example, the step function f shown in Figure 11.16a.

$$f(x) = \begin{cases} 1, & \text{if } 0 \le x \le \pi \\ 2, & \text{if } \pi < x \le 2\pi. \end{cases}$$

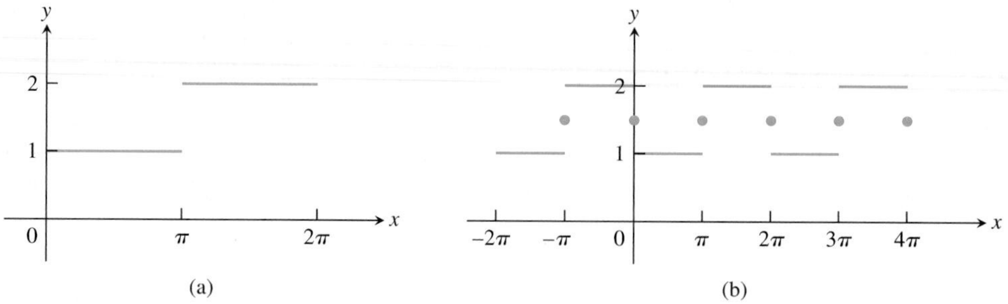

FIGURE 11.16 (a) The step function

$$f(x) = \begin{cases} 1, & 0 \le x \le \pi \\ 2, & \pi < x \le 2\pi \end{cases}$$

(b) The graph of the Fourier series for f is periodic and has the value $3/2$ at each point of discontinuity (Example 1).

The coefficients of the Fourier series of f are computed using Equations (2), (3), and (4).

$$\begin{aligned} a_0 &= \frac{1}{2\pi}\int_0^{2\pi} f(x)\,dx \\ &= \frac{1}{2\pi}\left(\int_0^{\pi} 1\,dx + \int_{\pi}^{2\pi} 2\,dx\right) = \frac{3}{2} \end{aligned}$$

$$\begin{aligned} a_k &= \frac{1}{\pi}\int_0^{2\pi} f(x)\cos kx\,dx \\ &= \frac{1}{\pi}\left(\int_0^{\pi} \cos kx\,dx + \int_{\pi}^{2\pi} 2\cos kx\,dx\right) \\ &= \frac{1}{\pi}\left(\left[\frac{\sin kx}{k}\right]_0^{\pi} + \left[\frac{2\sin kx}{k}\right]_{\pi}^{2\pi}\right) = 0, \qquad k \ge 1 \end{aligned}$$

$$\begin{aligned} b_k &= \frac{1}{\pi}\int_0^{2\pi} f(x)\sin kx\,dx \\ &= \frac{1}{\pi}\left(\int_0^{\pi} \sin kx\,dx + \int_{\pi}^{2\pi} 2\sin kx\,dx\right) \\ &= \frac{1}{\pi}\left(\left[-\frac{\cos kx}{k}\right]_0^{\pi} + \left[-\frac{2\cos kx}{k}\right]_{\pi}^{2\pi}\right) \\ &= \frac{\cos k\pi - 1}{k\pi} = \frac{(-1)^k - 1}{k\pi}. \end{aligned}$$

So

$$a_0 = \frac{3}{2}, \quad a_1 = a_2 = \cdots = 0,$$

and

$$b_1 = -\frac{2}{\pi}, \quad b_2 = 0, \quad b_3 = -\frac{2}{3\pi}, \quad b_4 = 0, \quad b_5 = -\frac{2}{5\pi}, \quad b_6 = 0, \ldots$$

The Fourier series is

$$\frac{3}{2} - \frac{2}{\pi}\left(\sin x + \frac{\sin 3x}{3} + \frac{\sin 5x}{5} + \cdots\right).$$

Notice that at $x = \pi$, where the function $f(x)$ jumps from 1 to 2, all the sine terms vanish, leaving 3/2 as the value of the series. This is not the value of f at π, since $f(\pi) = 1$. The Fourier series also sums to 3/2 at $x = 0$ and $x = 2\pi$. In fact, all terms in the Fourier series are periodic, of period 2π, and the value of the series at $x + 2\pi$ is the same as its value at x. The series we obtained represents the periodic function graphed in Figure 11.16b, with domain the entire real line and a pattern that repeats over every interval of width 2π. The function jumps discontinuously at $x = n\pi, n = 0, \pm 1, \pm 2, \ldots$ and at these points has value 3/2, the average value of the one-sided limits from each side. The convergence of the Fourier series of f is indicated in Figure 11.17. ■

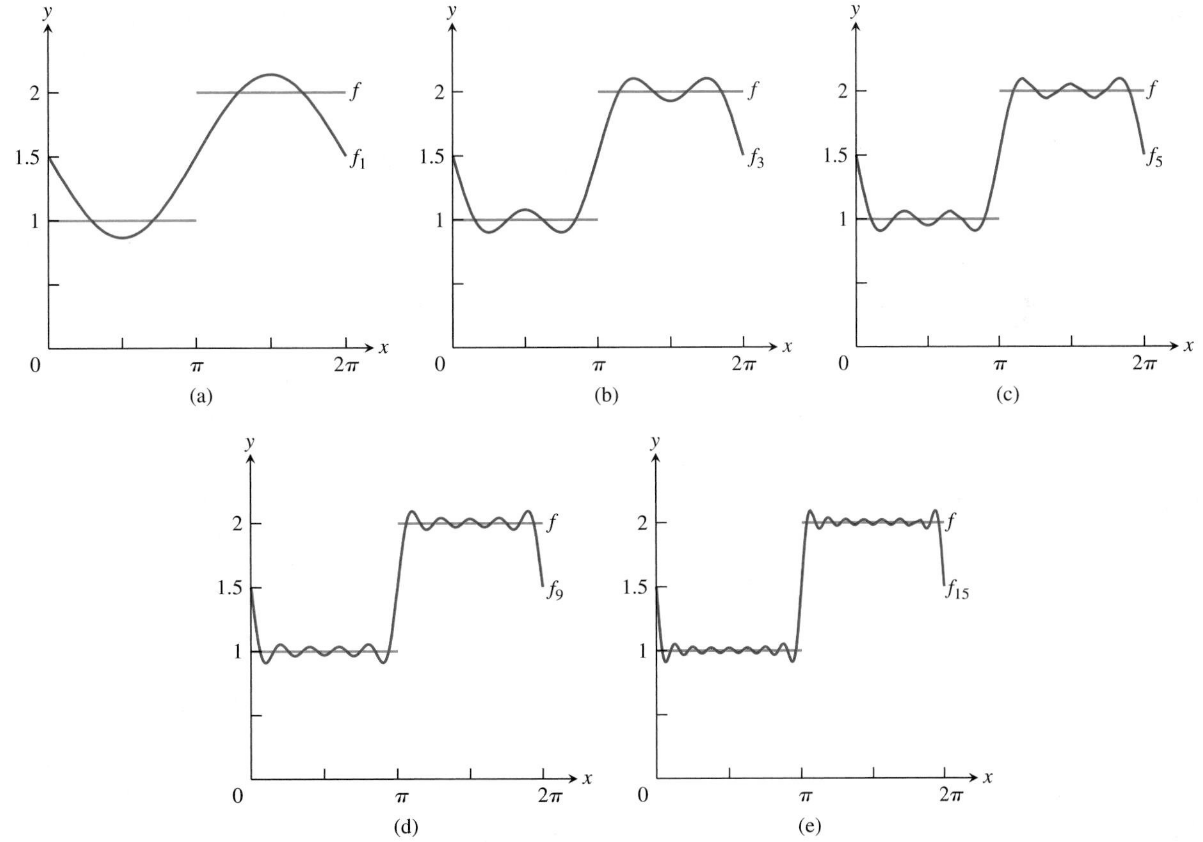

FIGURE 11.17 The Fourier approximation functions f_1, f_3, f_5, f_9, and f_{15} of the function $f(x) = \begin{cases} 1, & 0 \le x \le \pi \\ 2, & \pi < x \le 2\pi \end{cases}$ in Example 1.

Convergence of Fourier Series

Taylor series are computed from the value of a function and its derivatives at a single point $x = a$, and cannot reflect the behavior of a discontinuous function such as f in Example 1 past a discontinuity. The reason that a Fourier series can be used to represent such functions is that the Fourier series of a function depends on the existence of certain *integrals*, whereas the Taylor series depends on derivatives of a function near a single point. A function can be fairly "rough," even discontinuous, and still be integrable.

The coefficients used to construct Fourier series are precisely those one should choose to minimize the integral of the square of the error in approximating f by f_n. That is,

$$\int_0^{2\pi} [f(x) - f_n(x)]^2\, dx$$

is minimized by choosing $a_0, a_1, a_2, \ldots a_n$ and $b_1, b_2, \ldots, b_n$ as we did. While Taylor series are useful to approximate a function and its derivatives near a point, Fourier series minimize an error which is distributed over an interval.

We state without proof a result concerning the convergence of Fourier series. A function is **piecewise continuous** over an interval I if it has finitely many discontinuities on the interval, and at these discontinuities one-sided limits exist from each side. (See Chapter 5, Additional Exercises 11–18.)

THEOREM 24 Let $f(x)$ be a function such that f and f' are piecewise continuous on the interval $[0, 2\pi]$. Then f is equal to its Fourier series at all points where f is continuous. At a point c where f has a discontinuity, the Fourier series converges to

$$\frac{f(c^+) + f(c^-)}{2}$$

where $f(c^+)$ and $f(c^-)$ are the right- and left-hand limits of f at c.

EXERCISES 11.11

Finding Fourier Series

In Exercises 1–8, find the Fourier series associated with the given functions. Sketch each function.

1. $f(x) = 1 \quad 0 \le x \le 2\pi.$

2. $f(x) = \begin{cases} 1, & 0 \le x \le \pi \\ -1, & \pi < x \le 2\pi \end{cases}$

3. $f(x) = \begin{cases} x, & 0 \le x \le \pi \\ x - 2\pi, & \pi < x \le 2\pi \end{cases}$

4. $f(x) = \begin{cases} x^2, & 0 \le x \le \pi \\ 0, & \pi < x \le 2\pi \end{cases}$

5. $f(x) = e^x \quad 0 \le x \le 2\pi.$

6. $f(x) = \begin{cases} e^x, & 0 \le x \le \pi \\ 0, & \pi < x \le 2\pi \end{cases}$

7. $f(x) = \begin{cases} \cos x, & 0 \le x \le \pi \\ 0, & \pi < x \le 2\pi \end{cases}$

8. $f(x) = \begin{cases} 2, & 0 \le x \le \pi \\ -x, & \pi < x \le 2\pi \end{cases}$

Theory and Examples

Establish the results in Exercises 9–13, where p and q are positive integers.

9. $\displaystyle\int_0^{2\pi} \cos px\, dx = 0$ for all p.

10. $\int_0^{2\pi} \sin px\, dx = 0$ for all p.

11. $\int_0^{2\pi} \cos px \cos qx\, dx = \begin{cases} 0, & \text{if } p \neq q \\ \pi, & \text{if } p = q \end{cases}.$

(*Hint:* $\cos A \cos B = (1/2)[\cos(A + B) + \cos(A - B)]$.)

12. $\int_0^{2\pi} \sin px \sin qx\, dx = \begin{cases} 0, & \text{if } p \neq q \\ \pi, & \text{if } p = q \end{cases}.$

(*Hint:* $\sin A \sin B = (1/2)[\cos(A - B) - \cos(A + B)]$.)

13. $\int_0^{2\pi} \sin px \cos qx\, dx = 0$ for all p and q.

(*Hint*: $\sin A \cos B = (1/2)[\sin(A + B) + \sin(A - B)]$.)

14. **Fourier series of sums of functions** If f and g both satisfy the conditions of Theorem 24, is the Fourier series of $f + g$ on $[0, 2\pi]$ the sum of the Fourier series of f and the Fourier series of g? Give reasons for your answer.

15. **Term-by-term differentiation**

a. Use Theorem 24 to verify that the Fourier series for $f(x)$ in Exercise 3 converges to $f(x)$ for $0 < x < 2\pi$.

b. Although $f'(x) = 1$, show that the series obtained by term-by-term differentiation of the Fourier series in part (a) diverges.

16. Use Theorem 24 to find the Value of the Fourier series determined in Exercise 4 and show that $\frac{\pi^2}{6} = \sum_{n=1}^{\infty} \frac{1}{n^2}$.

Chapter 11 Questions to Guide Your Review

1. What is an infinite sequence? What does it mean for such a sequence to converge? To diverge? Give examples.
2. What is a nondecreasing sequence? Under what circumstances does such a sequence have a limit? Give examples.
3. What theorems are available for calculating limits of sequences? Give examples.
4. What theorem sometimes enables us to use l'Hôpital's Rule to calculate the limit of a sequence? Give an example.
5. What six sequence limits are likely to arise when you work with sequences and series?
6. What is an infinite series? What does it mean for such a series to converge? To diverge? Give examples.
7. What is a geometric series? When does such a series converge? Diverge? When it does converge, what is its sum? Give examples.
8. Besides geometric series, what other convergent and divergent series do you know?
9. What is the nth-Term Test for Divergence? What is the idea behind the test?
10. What can be said about term-by-term sums and differences of convergent series? About constant multiples of convergent and divergent series?
11. What happens if you add a finite number of terms to a convergent series? A divergent series? What happens if you delete a finite number of terms from a convergent series? A divergent series?
12. How do you reindex a series? Why might you want to do this?
13. Under what circumstances will an infinite series of nonnegative terms converge? Diverge? Why study series of nonnegative terms?
14. What is the Integral Test? What is the reasoning behind it? Give an example of its use.
15. When do p-series converge? Diverge? How do you know? Give examples of convergent and divergent p-series.
16. What are the Direct Comparison Test and the Limit Comparison Test? What is the reasoning behind these tests? Give examples of their use.
17. What are the Ratio and Root Tests? Do they always give you the information you need to determine convergence or divergence? Give examples.
18. What is an alternating series? What theorem is available for determining the convergence of such a series?
19. How can you estimate the error involved in approximating the sum of an alternating series with one of the series' partial sums? What is the reasoning behind the estimate?
20. What is absolute convergence? Conditional convergence? How are the two related?
21. What do you know about rearranging the terms of an absolutely convergent series? Of a conditionally convergent series? Give examples.
22. What is a power series? How do you test a power series for convergence? What are the possible outcomes?
23. What are the basic facts about

    a. term-by-term differentiation of power series?

    b. term-by-term integration of power series?

    c. multiplication of power series?

    Give examples.
24. What is the Taylor series generated by a function $f(x)$ at a point $x = a$? What information do you need about f to construct the series? Give an example.
25. What is a Maclaurin series?

26. Does a Taylor series always converge to its generating function? Explain.

27. What are Taylor polynomials? Of what use are they?

28. What is Taylor's formula? What does it say about the errors involved in using Taylor polynomials to approximate functions? In particular, what does Taylor's formula say about the error in a linearization? A quadratic approximation?

29. What is the binomial series? On what interval does it converge? How is it used?

30. How can you sometimes use power series to solve initial value problems?

31. How can you sometimes use power series to estimate the values of nonelementary definite integrals?

32. What are the Taylor series for $1/(1-x)$, $1/(1+x)$, e^x, $\sin x$, $\cos x$, $\ln(1+x)$, $\ln[(1+x)/(1-x)]$, and $\tan^{-1} x$? How do you estimate the errors involved in replacing these series with their partial sums?

33. What is a Fourier series? How do you calculate the Fourier coefficients $a_0, a_1, a_2, \ldots$ and $b_1, b_2, \ldots$ for a function $f(x)$ defined on the interval $[0, 2\pi]$?

34. State the theorem on convergence of the Fourier series for $f(x)$ when f and f' are piecewise continuous on $[0, 2\pi]$.

Chapter 11 Practice Exercises

Convergent or Divergent Sequences

Which of the sequences whose nth terms appear in Exercises 1–18 converge, and which diverge? Find the limit of each convergent sequence.

1. $a_n = 1 + \dfrac{(-1)^n}{n}$

2. $a_n = \dfrac{1-(-1)^n}{\sqrt{n}}$

3. $a_n = \dfrac{1-2^n}{2^n}$

4. $a_n = 1 + (0.9)^n$

5. $a_n = \sin\dfrac{n\pi}{2}$

6. $a_n = \sin n\pi$

7. $a_n = \dfrac{\ln(n^2)}{n}$

8. $a_n = \dfrac{\ln(2n+1)}{n}$

9. $a_n = \dfrac{n + \ln n}{n}$

10. $a_n = \dfrac{\ln(2n^3+1)}{n}$

11. $a_n = \left(\dfrac{n-5}{n}\right)^n$

12. $a_n = \left(1 + \dfrac{1}{n}\right)^{-n}$

13. $a_n = \sqrt[n]{\dfrac{3^n}{n}}$

14. $a_n = \left(\dfrac{3}{n}\right)^{1/n}$

15. $a_n = n(2^{1/n} - 1)$

16. $a_n = \sqrt[n]{2n+1}$

17. $a_n = \dfrac{(n+1)!}{n!}$

18. $a_n = \dfrac{(-4)^n}{n!}$

Convergent Series

Find the sums of the series in Exercises 19–24.

19. $\displaystyle\sum_{n=3}^{\infty} \frac{1}{(2n-3)(2n-1)}$

20. $\displaystyle\sum_{n=2}^{\infty} \frac{-2}{n(n+1)}$

21. $\displaystyle\sum_{n=1}^{\infty} \frac{9}{(3n-1)(3n+2)}$

22. $\displaystyle\sum_{n=3}^{\infty} \frac{-8}{(4n-3)(4n+1)}$

23. $\displaystyle\sum_{n=0}^{\infty} e^{-n}$

24. $\displaystyle\sum_{n=1}^{\infty} (-1)^n \frac{3}{4^n}$

Convergent or Divergent Series

Which of the series in Exercises 25–40 converge absolutely, which converge conditionally, and which diverge? Give reasons for your answers.

25. $\displaystyle\sum_{n=1}^{\infty} \frac{1}{\sqrt{n}}$

26. $\displaystyle\sum_{n=1}^{\infty} \frac{-5}{n}$

27. $\displaystyle\sum_{n=1}^{\infty} \frac{(-1)^n}{\sqrt{n}}$

28. $\displaystyle\sum_{n=1}^{\infty} \frac{1}{2n^3}$

29. $\displaystyle\sum_{n=1}^{\infty} \frac{(-1)^n}{\ln(n+1)}$

30. $\displaystyle\sum_{n=2}^{\infty} \frac{1}{n(\ln n)^2}$

31. $\displaystyle\sum_{n=1}^{\infty} \frac{\ln n}{n^3}$

32. $\displaystyle\sum_{n=3}^{\infty} \frac{\ln n}{\ln(\ln n)}$

33. $\displaystyle\sum_{n=1}^{\infty} \frac{(-1)^n}{n\sqrt{n^2+1}}$

34. $\displaystyle\sum_{n=1}^{\infty} \frac{(-1)^n 3n^2}{n^3+1}$

35. $\displaystyle\sum_{n=1}^{\infty} \frac{n+1}{n!}$

36. $\displaystyle\sum_{n=1}^{\infty} \frac{(-1)^n(n^2+1)}{2n^2+n-1}$

37. $\displaystyle\sum_{n=1}^{\infty} \frac{(-3)^n}{n!}$

38. $\displaystyle\sum_{n=1}^{\infty} \frac{2^n 3^n}{n^n}$

39. $\displaystyle\sum_{n=1}^{\infty} \frac{1}{\sqrt{n(n+1)(n+2)}}$

40. $\displaystyle\sum_{n=2}^{\infty} \frac{1}{n\sqrt{n^2-1}}$

Power Series

In Exercises 41–50, **(a)** find the series' radius and interval of convergence. Then identify the values of x for which the series converges **(b)** absolutely and **(c)** conditionally.

41. $\displaystyle\sum_{n=1}^{\infty} \frac{(x+4)^n}{n3^n}$

42. $\displaystyle\sum_{n=1}^{\infty} \frac{(x-1)^{2n-2}}{(2n-1)!}$

43. $\displaystyle\sum_{n=1}^{\infty} \frac{(-1)^{n-1}(3x-1)^n}{n^2}$

44. $\displaystyle\sum_{n=0}^{\infty} \frac{(n+1)(2x+1)^n}{(2n+1)2^n}$

45. $\displaystyle\sum_{n=1}^{\infty} \frac{x^n}{n^n}$

46. $\displaystyle\sum_{n=1}^{\infty} \frac{x^n}{\sqrt{n}}$

47. $\sum_{n=0}^{\infty} \frac{(n+1)x^{2n-1}}{3^n}$

48. $\sum_{n=0}^{\infty} \frac{(-1)^n(x-1)^{2n+1}}{2n+1}$

49. $\sum_{n=1}^{\infty} (\operatorname{csch} n)x^n$

50. $\sum_{n=1}^{\infty} (\coth n)x^n$

Maclaurin Series

Each of the series in Exercises 51–56 is the value of the Taylor series at $x = 0$ of a function $f(x)$ at a particular point. What function and what point? What is the sum of the series?

51. $1 - \frac{1}{4} + \frac{1}{16} - \cdots + (-1)^n \frac{1}{4^n} + \cdots$

52. $\frac{2}{3} - \frac{4}{18} + \frac{8}{81} - \cdots + (-1)^{n-1} \frac{2^n}{n3^n} + \cdots$

53. $\pi - \frac{\pi^3}{3!} + \frac{\pi^5}{5!} - \cdots + (-1)^n \frac{\pi^{2n+1}}{(2n+1)!} + \cdots$

54. $1 - \frac{\pi^2}{9 \cdot 2!} + \frac{\pi^4}{81 \cdot 4!} - \cdots + (-1)^n \frac{\pi^{2n}}{3^{2n}(2n)!} + \cdots$

55. $1 + \ln 2 + \frac{(\ln 2)^2}{2!} + \cdots + \frac{(\ln 2)^n}{n!} + \cdots$

56. $\frac{1}{\sqrt{3}} - \frac{1}{9\sqrt{3}} + \frac{1}{45\sqrt{3}} - \cdots$
$+ (-1)^{n-1} \frac{1}{(2n-1)(\sqrt{3})^{2n-1}} + \cdots$

Find Taylor series at $x = 0$ for the functions in Exercises 57–64.

57. $\frac{1}{1-2x}$

58. $\frac{1}{1+x^3}$

59. $\sin \pi x$

60. $\sin \frac{2x}{3}$

61. $\cos (x^{5/2})$

62. $\cos \sqrt{5x}$

63. $e^{(\pi x/2)}$

64. e^{-x^2}

Taylor Series

In Exercises 65–68, find the first four nonzero terms of the Taylor series generated by f at $x = a$.

65. $f(x) = \sqrt{3 + x^2}$ at $x = -1$

66. $f(x) = 1/(1 - x)$ at $x = 2$

67. $f(x) = 1/(x + 1)$ at $x = 3$

68. $f(x) = 1/x$ at $x = a > 0$

Initial Value Problems

Use power series to solve the initial value problems in Exercises 69–76.

69. $y' + y = 0, \quad y(0) = -1$

70. $y' - y = 0, \quad y(0) = -3$

71. $y' + 2y = 0, \quad y(0) = 3$

72. $y' + y = 1, \quad y(0) = 0$

73. $y' - y = 3x, \quad y(0) = -1$

74. $y' + y = x, \quad y(0) = 0$

75. $y' - y = x, \quad y(0) = 1$

76. $y' - y = -x, \quad y(0) = 2$

Nonelementary Integrals

Use series to approximate the values of the integrals in Exercises 77–80 with an error of magnitude less than 10^{-8}. (The answer section gives the integrals' values rounded to 10 decimal places.)

77. $\int_0^{1/2} e^{-x^3}\, dx$

78. $\int_0^1 x \sin (x^3)\, dx$

79. $\int_0^{1/2} \frac{\tan^{-1} x}{x}\, dx$

80. $\int_0^{1/64} \frac{\tan^{-1} x}{\sqrt{x}}\, dx$

Indeterminate Forms

In Exercises 81–86:

a. Use power series to evaluate the limit.

T **b.** Then use a grapher to support your calculation.

81. $\lim_{x \to 0} \frac{7 \sin x}{e^{2x} - 1}$

82. $\lim_{\theta \to 0} \frac{e^\theta - e^{-\theta} - 2\theta}{\theta - \sin \theta}$

83. $\lim_{t \to 0} \left(\frac{1}{2 - 2\cos t} - \frac{1}{t^2} \right)$

84. $\lim_{h \to 0} \frac{(\sin h)/h - \cos h}{h^2}$

85. $\lim_{z \to 0} \frac{1 - \cos^2 z}{\ln (1 - z) + \sin z}$

86. $\lim_{y \to 0} \frac{y^2}{\cos y - \cosh y}$

87. Use a series representation of $\sin 3x$ to find values of r and s for which

$$\lim_{x \to 0} \left(\frac{\sin 3x}{x^3} + \frac{r}{x^2} + s \right) = 0.$$

88. **a.** Show that the approximation $\csc x \approx 1/x + x/6$ in Section 11.10, Example 9, leads to the approximation $\sin x \approx 6x/(6 + x^2)$.

T **b.** Compare the accuracies of the approximations $\sin x \approx x$ and $\sin x \approx 6x/(6 + x^2)$ by comparing the graphs of $f(x) = \sin x - x$ and $g(x) = \sin x - (6x/(6 + x^2))$. Describe what you find.

Theory and Examples

89. **a.** Show that the series

$$\sum_{n=1}^{\infty} \left(\sin \frac{1}{2n} - \sin \frac{1}{2n+1} \right)$$

converges.

T **b.** Estimate the magnitude of the error involved in using the sum of the sines through $n = 20$ to approximate the sum of the series. Is the approximation too large, or too small? Give reasons for your answer.

90. **a.** Show that the series $\sum_{n=1}^{\infty} \left(\tan \frac{1}{2n} - \tan \frac{1}{2n+1} \right)$ converges.

T **b.** Estimate the magnitude of the error in using the sum of the tangents through $-\tan (1/41)$ to approximate the sum of the series. Is the approximation too large, or too small? Give reasons for your answer.

91. Find the radius of convergence of the series

$$\sum_{n=1}^{\infty} \frac{2 \cdot 5 \cdot 8 \cdot \cdots \cdot (3n-1)}{2 \cdot 4 \cdot 6 \cdot \cdots \cdot (2n)} x^n.$$

92. Find the radius of convergence of the series

$$\sum_{n=1}^{\infty} \frac{3 \cdot 5 \cdot 7 \cdot \cdots \cdot (2n+1)}{4 \cdot 9 \cdot 14 \cdot \cdots \cdot (5n-1)} (x-1)^n.$$

93. Find a closed-form formula for the nth partial sum of the series $\sum_{n=2}^{\infty} \ln(1 - (1/n^2))$ and use it to determine the convergence or divergence of the series.

94. Evaluate $\sum_{k=2}^{\infty} (1/(k^2 - 1))$ by finding the limits as $n \to \infty$ of the series' nth partial sum.

95. a. Find the interval of convergence of the series

$$y = 1 + \frac{1}{6}x^3 + \frac{1}{180}x^6 + \cdots + \frac{1 \cdot 4 \cdot 7 \cdot \cdots \cdot (3n-2)}{(3n)!} x^{3n} + \cdots.$$

b. Show that the function defined by the series satisfies a differential equation of the form

$$\frac{d^2y}{dx^2} = x^a y + b$$

and find the values of the constants a and b.

96. a. Find the Maclaurin series for the function $x^2/(1+x)$.

b. Does the series converge at $x = 1$? Explain.

97. If $\sum_{n=1}^{\infty} a_n$ and $\sum_{n=1}^{\infty} b_n$ are convergent series of nonnegative numbers, can anything be said about $\sum_{n=1}^{\infty} a_n b_n$? Give reasons for your answer.

98. If $\sum_{n=1}^{\infty} a_n$ and $\sum_{n=1}^{\infty} b_n$ are divergent series of nonnegative numbers, can anything be said about $\sum_{n=1}^{\infty} a_n b_n$? Give reasons for your answer.

99. Prove that the sequence $\{x_n\}$ and the series $\sum_{k=1}^{\infty} (x_{k+1} - x_k)$ both converge or both diverge.

100. Prove that $\sum_{n=1}^{\infty} (a_n/(1 + a_n))$ converges if $a_n > 0$ for all n and $\sum_{n=1}^{\infty} a_n$ converges.

101. (*Continuation of Section 4.7, Exercise 27.*) If you did Exercise 27 in Section 4.7, you saw that in practice Newton's method stopped too far from the root of $f(x) = (x-1)^{40}$ to give a useful estimate of its value, $x = 1$. Prove that nevertheless, for any starting value $x_0 \neq 1$, the sequence $x_0, x_1, x_2, \ldots, x_n, \ldots$ of approximations generated by Newton's method really does converge to 1.

102. Suppose that $a_1, a_2, a_3, \ldots, a_n$ are positive numbers satisfying the following conditions:

i. $a_1 \geq a_2 \geq a_3 \geq \cdots$;

ii. the series $a_2 + a_4 + a_8 + a_{16} + \cdots$ diverges.

Show that the series

$$\frac{a_1}{1} + \frac{a_2}{2} + \frac{a_3}{3} + \cdots$$

diverges.

103. Use the result in Exercise 102 to show that

$$1 + \sum_{n=2}^{\infty} \frac{1}{n \ln n}$$

diverges.

104. Suppose you wish to obtain a quick estimate for the value of $\int_0^1 x^2 e^x \, dx$. There are several ways to do this.

a. Use the Trapezoidal Rule with $n = 2$ to estimate $\int_0^1 x^2 e^x \, dx$.

b. Write out the first three nonzero terms of the Taylor series at $x = 0$ for $x^2 e^x$ to obtain the fourth Taylor polynomial $P(x)$ for $x^2 e^x$. Use $\int_0^1 P(x)\, dx$ to obtain another estimate for $\int_0^1 x^2 e^x \, dx$.

c. The second derivative of $f(x) = x^2 e^x$ is positive for all $x > 0$. Explain why this enables you to conclude that the Trapezoidal Rule estimate obtained in part (a) is too large. (*Hint:* What does the second derivative tell you about the graph of a function? How does this relate to the trapezoidal approximation of the area under this graph?)

d. All the derivatives of $f(x) = x^2 e^x$ are positive for $x > 0$. Explain why this enables you to conclude that all Maclaurin polynomial approximations to $f(x)$ for x in $[0, 1]$ will be too small. (*Hint:* $f(x) = P_n(x) + R_n(x)$.)

e. Use integration by parts to evaluate $\int_0^1 x^2 e^x \, dx$.

Fourier Series

Find the Fourier series for the functions in Exercises 105–108. Sketch each function.

105. $f(x) = \begin{cases} 0, & 0 \leq x \leq \pi \\ 1, & \pi < x \leq 2\pi \end{cases}$

106. $f(x) = \begin{cases} x, & 0 \leq x \leq \pi \\ 1, & \pi < x \leq 2\pi \end{cases}$

107. $f(x) = \begin{cases} \pi - x, & 0 \leq x \leq \pi \\ x - 2\pi, & \pi < x \leq 2\pi \end{cases}$

108. $f(x) = |\sin x|, \quad 0 \leq x \leq 2\pi$

Chapter 11 Additional and Advanced Exercises

Convergence or Divergence

Which of the series $\sum_{n=1}^{\infty} a_n$ defined by the formulas in Exercises 1–4 converge, and which diverge? Give reasons for your answers.

1. $\displaystyle\sum_{n=1}^{\infty} \frac{1}{(3n-2)^{n+(1/2)}}$

2. $\displaystyle\sum_{n=1}^{\infty} \frac{(\tan^{-1} n)^2}{n^2+1}$

3. $\displaystyle\sum_{n=1}^{\infty} (-1)^n \tanh n$

4. $\displaystyle\sum_{n=2}^{\infty} \frac{\log_n (n!)}{n^3}$

Which of the series $\sum_{n=1}^{\infty} a_n$ defined by the formulas in Exercises 5–8 converge, and which diverge? Give reasons for your answers.

5. $a_1 = 1, \quad a_{n+1} = \dfrac{n(n+1)}{(n+2)(n+3)} a_n$

(*Hint:* Write out several terms, see which factors cancel, and then generalize.)

6. $a_1 = a_2 = 7, \quad a_{n+1} = \dfrac{n}{(n-1)(n+1)} a_n \quad \text{if } n \ge 2$

7. $a_1 = a_2 = 1, \quad a_{n+1} = \dfrac{1}{1+a_n} \quad \text{if } n \ge 2$

8. $a_n = 1/3^n$ if n is odd, $a_n = n/3^n$ if n is even

Choosing Centers for Taylor Series

Taylor's formula

$$f(x) = f(a) + f'(a)(x-a) + \frac{f''(a)}{2!}(x-a)^2 + \cdots + \frac{f^{(n)}(a)}{n!}(x-a)^n + \frac{f^{(n+1)}(c)}{(n+1)!}(x-a)^{n+1}$$

expresses the value of f at x in terms of the values of f and its derivatives at $x = a$. In numerical computations, we therefore need a to be a point where we know the values of f and its derivatives. We also need a to be close enough to the values of f we are interested in to make $(x-a)^{n+1}$ so small we can neglect the remainder.

In Exercises 9–14, what Taylor series would you choose to represent the function near the given value of x? (There may be more than one good answer.) Write out the first four nonzero terms of the series you choose.

9. $\cos x$ near $x = 1$

10. $\sin x$ near $x = 6.3$

11. e^x near $x = 0.4$

12. $\ln x$ near $x = 1.3$

13. $\cos x$ near $x = 69$

14. $\tan^{-1} x$ near $x = 2$

Theory and Examples

15. Let a and b be constants with $0 < a < b$. Does the sequence $\{(a^n + b^n)^{1/n}\}$ converge? If it does converge, what is the limit?

16. Find the sum of the infinite series

$$1 + \frac{2}{10} + \frac{3}{10^2} + \frac{7}{10^3} + \frac{2}{10^4} + \frac{3}{10^5} + \frac{7}{10^6} + \frac{2}{10^7} + \frac{3}{10^8} + \frac{7}{10^9} + \cdots.$$

17. Evaluate

$$\sum_{n=0}^{\infty} \int_n^{n+1} \frac{1}{1+x^2}\, dx.$$

18. Find all values of x for which

$$\sum_{n=1}^{\infty} \frac{nx^n}{(n+1)(2x+1)^n}$$

converges absolutely.

19. Generalizing Euler's constant The accompanying figure shows the graph of a positive twice-differentiable decreasing function f whose second derivative is positive on $(0, \infty)$. For each n, the number A_n is the area of the lunar region between the curve and the line segment joining the points $(n, f(n))$ and $(n+1, f(n+1))$.

a. Use the figure to show that $\sum_{n=1}^{\infty} A_n < (1/2)(f(1) - f(2))$.

b. Then show the existence of

$$\lim_{n\to\infty} \left[\sum_{k=1}^{n} f(k) - \frac{1}{2}(f(1) + f(n)) - \int_1^n f(x)\, dx \right].$$

c. Then show the existence of

$$\lim_{n\to\infty} \left[\sum_{k=1}^{n} f(k) - \int_1^n f(x)\, dx \right].$$

If $f(x) = 1/x$, the limit in part (c) is Euler's constant (Section 11.3, Exercise 41). (*Source:* "Convergence with Pictures" by P. J. Rippon, *American Mathematical Monthly*, Vol. 93, No. 6, 1986, pp. 476–478.)

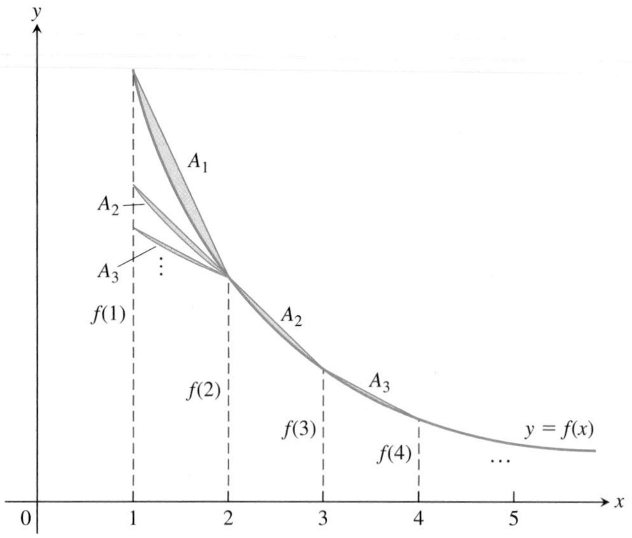

20. This exercise refers to the "right side up" equilateral triangle with sides of length $2b$ in the accompanying figure. "Upside down" equilateral triangles are removed from the original triangle as the sequence of pictures suggests. The sum of the areas removed from the original triangle forms an infinite series.

a. Find this infinite series.

b. Find the sum of this infinite series and hence find the total area removed from the original triangle.

c. Is every point on the original triangle removed? Explain why or why not.

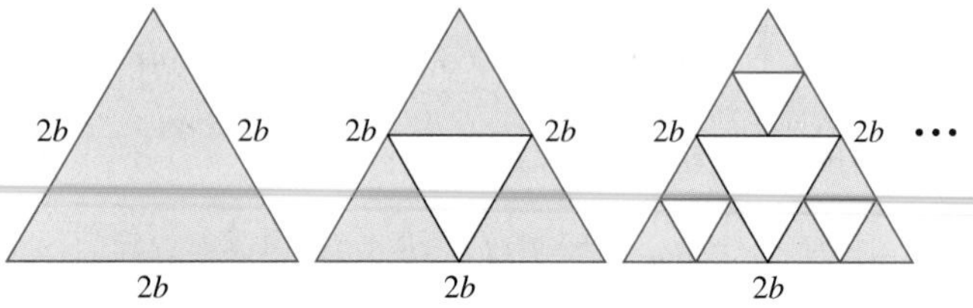

T **21. a.** Does the value of

$$\lim_{n\to\infty}\left(1-\frac{\cos(a/n)}{n}\right)^n, \quad a \text{ constant},$$

appear to depend on the value of a? If so, how?

b. Does the value of

$$\lim_{n\to\infty}\left(1-\frac{\cos(a/n)}{bn}\right)^n, \quad a \text{ and } b \text{ constant}, b \neq 0,$$

appear to depend on the value of b? If so, how?

c. Use calculus to confirm your findings in parts (a) and (b).

22. Show that if $\sum_{n=1}^{\infty} a_n$ converges, then

$$\sum_{n=1}^{\infty}\left(\frac{1+\sin(a_n)}{2}\right)^n$$

converges.

23. Find a value for the constant b that will make the radius of convergence of the power series

$$\sum_{n=2}^{\infty}\frac{b^n x^n}{\ln n}$$

equal to 5.

24. How do you know that the functions $\sin x$, $\ln x$, and e^x are not polynomials? Give reasons for your answer.

25. Find the value of a for which the limit

$$\lim_{x\to 0}\frac{\sin(ax)-\sin x - x}{x^3}$$

is finite and evaluate the limit.

26. Find values of a and b for which

$$\lim_{x\to 0}\frac{\cos(ax)-b}{2x^2} = -1.$$

27. Raabe's (or Gauss's) test The following test, which we state without proof, is an extension of the Ratio Test.

Raabe's test: If $\sum_{n=1}^{\infty} u_n$ is a series of positive constants and there exist constants C, K, and N such that

$$\frac{u_n}{u_{n+1}} = 1 + \frac{C}{n} + \frac{f(n)}{n^2}, \qquad (1)$$

where $|f(n)| < K$ for $n \geq N$, then $\sum_{n=1}^{\infty} u_n$ converges if $C > 1$ and diverges if $C \leq 1$.

Show that the results of Raabe's test agree with what you know about the series $\sum_{n=1}^{\infty}(1/n^2)$ and $\sum_{n=1}^{\infty}(1/n)$.

28. (*Continuation of Exercise 27.*) Suppose that the terms of $\sum_{n=1}^{\infty} u_n$ are defined recursively by the formulas

$$u_1 = 1, \quad u_{n+1} = \frac{(2n-1)^2}{(2n)(2n+1)}u_n.$$

Apply Raabe's test to determine whether the series converges.

29. If $\sum_{n=1}^{\infty} a_n$ converges, and if $a_n \neq 1$ and $a_n > 0$ for all n,

a. Show that $\sum_{n=1}^{\infty} a_n^{\,2}$ converges.

b. Does $\sum_{n=1}^{\infty} a_n/(1-a_n)$ converge? Explain.

30. (*Continuation of Exercise 29.*) If $\sum_{n=1}^{\infty} a_n$ converges, and if $1 > a_n > 0$ for all n, show that $\sum_{n=1}^{\infty}\ln(1-a_n)$ converges.

(*Hint:* First show that $|\ln(1-a_n)| \leq a_n/(1-a_n)$.)

31. Nicole Oresme's Theorem Prove Nicole Oresme's Theorem that

$$1 + \frac{1}{2}\cdot 2 + \frac{1}{4}\cdot 3 + \cdots + \frac{n}{2^{n-1}} + \cdots = 4.$$

(*Hint:* Differentiate both sides of the equation $1/(1-x) = 1 + \sum_{n=1}^{\infty} x^n$.)

32. a. Show that

$$\sum_{n=1}^{\infty} \frac{n(n+1)}{x^n} = \frac{2x^2}{(x-1)^3}$$

for $|x| > 1$ by differentiating the identity

$$\sum_{n=1}^{\infty} x^{n+1} = \frac{x^2}{1-x}$$

twice, multiplying the result by x, and then replacing x by $1/x$.

b. Use part (a) to find the real solution greater than 1 of the equation

$$x = \sum_{n=1}^{\infty} \frac{n(n+1)}{x^n}.$$

33. A fast estimate of $\pi/2$ As you saw if you did Exercise 127 in Section 11.1, the sequence generated by starting with $x_0 = 1$ and applying the recursion formula $x_{n+1} = x_n + \cos x_n$ converges rapidly to $\pi/2$. To explain the speed of the convergence, let $\epsilon_n = (\pi/2) - x_n$. (See the accompanying figure.) Then

$$\begin{aligned}\epsilon_{n+1} &= \frac{\pi}{2} - x_n - \cos x_n \\ &= \epsilon_n - \cos\left(\frac{\pi}{2} - \epsilon_n\right) \\ &= \epsilon_n - \sin \epsilon_n \\ &= \frac{1}{3!}(\epsilon_n)^3 - \frac{1}{5!}(\epsilon_n)^5 + \cdots.\end{aligned}$$

Use this equality to show that

$$0 < \epsilon_{n+1} < \frac{1}{6}(\epsilon_n)^3.$$

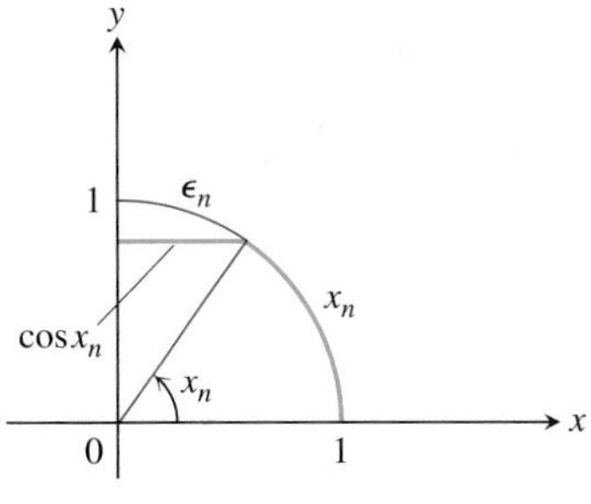

34. If $\sum_{n=1}^{\infty} a_n$ is a convergent series of positive numbers, can anything be said about the convergence of $\sum_{n=1}^{\infty} \ln(1 + a_n)$? Give reasons for your answer.

35. Quality control

a. Differentiate the series

$$\frac{1}{1-x} = 1 + x + x^2 + \cdots + x^n + \cdots$$

to obtain a series for $1/(1-x)^2$.

b. In one throw of two dice, the probability of getting a roll of 7 is $p = 1/6$. If you throw the dice repeatedly, the probability that a 7 will appear for the first time at the nth throw is $q^{n-1}p$, where $q = 1 - p = 5/6$. The expected number of throws until a 7 first appears is $\sum_{n=1}^{\infty} nq^{n-1}p$. Find the sum of this series.

c. As an engineer applying statistical control to an industrial operation, you inspect items taken at random from the assembly line. You classify each sampled item as either "good" or "bad." If the probability of an item's being good is p and of an item's being bad is $q = 1 - p$, the probability that the first bad item found is the nth one inspected is $p^{n-1}q$. The average number inspected up to and including the first bad item found is $\sum_{n=1}^{\infty} np^{n-1}q$. Evaluate this sum, assuming $0 < p < 1$.

36. Expected value Suppose that a random variable X may assume the values $1, 2, 3, \ldots$, with probabilities $p_1, p_2, p_3, \ldots$, where p_k is the probability that X equals k $(k = 1, 2, 3, \ldots)$. Suppose also that $p_k \geq 0$ and that $\sum_{k=1}^{\infty} p_k = 1$. The **expected value** of X, denoted by $E(X)$, is the number $\sum_{k=1}^{\infty} kp_k$, provided the series converges. In each of the following cases, show that $\sum_{k=1}^{\infty} p_k = 1$ and find $E(X)$ if it exists. (*Hint:* See Exercise 35.)

a. $p_k = 2^{-k}$ **b.** $p_k = \dfrac{5^{k-1}}{6^k}$

c. $p_k = \dfrac{1}{k(k+1)} = \dfrac{1}{k} - \dfrac{1}{k+1}$

T 37. Safe and effective dosage The concentration in the blood resulting from a single dose of a drug normally decreases with time as the drug is eliminated from the body. Doses may therefore need to be repeated periodically to keep the concentration from dropping below some particular level. One model for the effect of repeated doses gives the residual concentration just before the $(n+1)$st dose as

$$R_n = C_0 e^{-kt_0} + C_0 e^{-2kt_0} + \cdots + C_0 e^{-nkt_0},$$

where C_0 = the change in concentration achievable by a single dose (mg/mL), k = the *elimination constant* (h^{-1}), and t_0 = time between doses (h). See the accompanying figure.

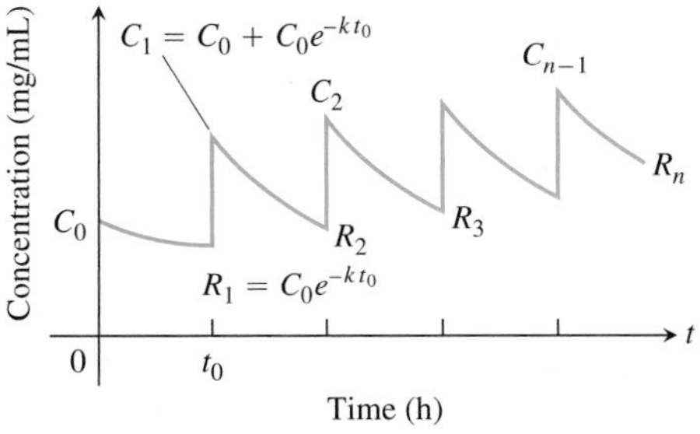

a. Write R_n in closed form as a single fraction, and find $R = \lim_{n\to\infty} R_n$.

b. Calculate R_1 and R_{10} for $C_0 = 1$ mg/mL, $k = 0.1\ \text{h}^{-1}$, and $t_0 = 10$ h. How good an estimate of R is R_{10}?

c. If $k = 0.01\ \text{h}^{-1}$ and $t_0 = 10$ h, find the smallest n such that $R_n > (1/2)R$.

(*Source: Prescribing Safe and Effective Dosage,* B. Horelick and S. Koont, COMAP, Inc., Lexington, MA.)

38. Time between drug doses (*Continuation of Exercise 37.*) If a drug is known to be ineffective below a concentration C_L and harmful above some higher concentration C_H, one needs to find values of C_0 and t_0 that will produce a concentration that is safe (not above C_H) but effective (not below C_L). See the accompanying figure. We therefore want to find values for C_0 and t_0 for which

$$R = C_L \quad \text{and} \quad C_0 + R = C_H.$$

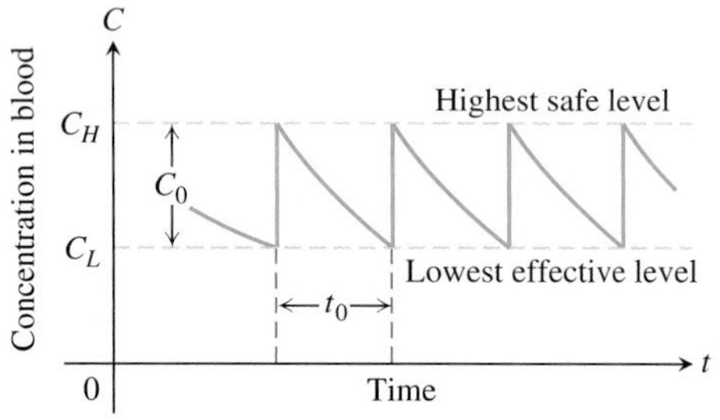

Thus $C_0 = C_H - C_L$. When these values are substituted in the equation for R obtained in part (a) of Exercise 37, the resulting equation simplifies to

$$t_0 = \frac{1}{k} \ln \frac{C_H}{C_L}.$$

To reach an effective level rapidly, one might administer a "loading" dose that would produce a concentration of C_H mg/mL. This could be followed every t_0 hours by a dose that raises the concentration by $C_0 = C_H - C_L$ mg/mL.

a. Verify the preceding equation for t_0.

b. If $k = 0.05\ \text{h}^{-1}$ and the highest safe concentration is e times the lowest effective concentration, find the length of time between doses that will assure safe and effective concentrations.

c. Given $C_H = 2$ mg/mL, $C_L = 0.5$ mg/mL, and $k = 0.02\ \text{h}^{-1}$, determine a scheme for administering the drug.

d. Suppose that $k = 0.2\ \text{h}^{-1}$ and that the smallest effective concentration is 0.03 mg/mL. A single dose that produces a concentration of 0.1 mg/mL is administered. About how long will the drug remain effective?

39. An infinite product The infinite product

$$\prod_{n=1}^{\infty}(1 + a_n) = (1 + a_1)(1 + a_2)(1 + a_3)\cdots$$

is said to converge if the series

$$\sum_{n=1}^{\infty} \ln(1 + a_n),$$

obtained by taking the natural logarithm of the product, converges. Prove that the product converges if $a_n > -1$ for every n and if $\sum_{n=1}^{\infty} |a_n|$ converges. (*Hint:* Show that

$$|\ln(1 + a_n)| \le \frac{|a_n|}{1 - |a_n|} \le 2|a_n|$$

when $|a_n| < 1/2$.)

40. If p is a constant, show that the series

$$1 + \sum_{n=3}^{\infty} \frac{1}{n \cdot \ln n \cdot [\ln(\ln n)]^p}$$

a. converges if $p > 1$, **b.** diverges if $p \le 1$. In general, if $f_1(x) = x$, $f_{n+1}(x) = \ln(f_n(x))$, and n takes on the values $1, 2, 3, \ldots$, we find that $f_2(x) = \ln x$, $f_3(x) = \ln(\ln x)$, and so on. If $f_n(a) > 1$, then

$$\int_a^{\infty} \frac{dx}{f_1(x) f_2(x) \cdots f_n(x)(f_{n+1}(x))^p}$$

converges if $p > 1$ and diverges if $p \le 1$.

41. a. Prove the following theorem: If $\{c_n\}$ is a sequence of numbers such that every sum $t_n = \sum_{k=1}^{n} c_k$ is bounded, then the series $\sum_{n=1}^{\infty} c_n/n$ converges and is equal to $\sum_{n=1}^{\infty} t_n/(n(n + 1))$.

Outline of proof: Replace c_1 by t_1 and c_n by $t_n - t_{n-1}$ for $n \ge 2$. If $s_{2n+1} = \sum_{k=1}^{2n+1} c_k/k$, show that

$$s_{2n+1} = t_1\left(1 - \frac{1}{2}\right) + t_2\left(\frac{1}{2} - \frac{1}{3}\right) + \cdots + t_{2n}\left(\frac{1}{2n} - \frac{1}{2n + 1}\right) + \frac{t_{2n+1}}{2n + 1}$$

$$= \sum_{k=1}^{2n} \frac{t_k}{k(k + 1)} + \frac{t_{2n+1}}{2n + 1}.$$

Because $|t_k| < M$ for some constant M, the series

$$\sum_{k=1}^{\infty} \frac{t_k}{k(k + 1)}$$

converges absolutely and s_{2n+1} has a limit as $n \to \infty$. Finally, if $s_{2n} = \sum_{k=1}^{2n} c_k/k$, then $s_{2n+1} - s_{2n} = c_{2n+1}/(2n + 1)$ approaches zero as $n \to \infty$ because $|c_{2n+1}| = |t_{2n+1} - t_{2n}| < 2M$. Hence the sequence of partial sums of the series $\sum c_k/k$ converges and the limit is $\sum_{k=1}^{\infty} t_k/(k(k + 1))$.

b. Show how the foregoing theorem applies to the alternating harmonic series

$$1-\frac{1}{2}+\frac{1}{3}-\frac{1}{4}+\frac{1}{5}-\frac{1}{6}+\cdots.$$

c. Show that the series

$$1-\frac{1}{2}-\frac{1}{3}+\frac{1}{4}+\frac{1}{5}-\frac{1}{6}-\frac{1}{7}+\cdots.$$

converges. (After the first term, the signs are two negative, two positive, two negative, two positive, and so on in that pattern.)

42. The convergence of $\sum_{n=1}^{\infty}[(-1)^{n-1}x^n]/n$ to $\ln(1+x)$ for $-1 < x \le 1$

a. Show by long division or otherwise that

$$\frac{1}{1+t}=1-t+t^2-t^3+\cdots+(-1)^n t^n+\frac{(-1)^{n+1}t^{n+1}}{1+t}.$$

b. By integrating the equation of part (a) with respect to t from 0 to x, show that

$$\ln(1+x)=x-\frac{x^2}{2}+\frac{x^3}{3}-\frac{x^4}{4}+\cdots$$
$$+(-1)^n\frac{x^{n+1}}{n+1}+R_{n+1}$$

where

$$R_{n+1}=(-1)^{n+1}\int_0^x \frac{t^{n+1}}{1+t}\,dt.$$

c. If $x \ge 0$, show that

$$|R_{n+1}| \le \int_0^x t^{n+1}\,dt=\frac{x^{n+2}}{n+2}.$$

(*Hint:* As t varies from 0 to x,

$$1+t \ge 1 \quad \text{and} \quad t^{n+1}/(1+t) \le t^{n+1},$$

and

$$\left|\int_0^x f(t)\,dt\right| \le \int_0^x |f(t)|\,dt.\Big)$$

d. If $-1 < x < 0$, show that

$$|R_{n+1}| \le \left|\int_0^x \frac{t^{n+1}}{1-|x|}\,dt\right|=\frac{|x|^{n+2}}{(n+2)(1-|x|)}.$$

(*Hint:* If $x < t \le 0$, then $|1+t| \ge 1-|x|$ and

$$\left|\frac{t^{n+1}}{1+t}\right| \le \frac{|t|^{n+1}}{1-|x|}.\Big)$$

e. Use the foregoing results to prove that the series

$$x-\frac{x^2}{2}+\frac{x^3}{3}-\frac{x^4}{4}+\cdots+\frac{(-1)^n x^{n+1}}{n+1}+\cdots$$

converges to $\ln(1+x)$ for $-1 < x \le 1$.

Chapter 11 Technology Application Projects

Mathematica/Maple Module
Bouncing Ball

The model predicts the height of a bouncing ball, and the time until it stops bouncing.

Mathematica/Maple Module
Taylor Polynomial Approximations of a Function

A graphical animation shows the convergence of the Taylor polynomials to functions having derivatives of all orders over an interval in their domains.

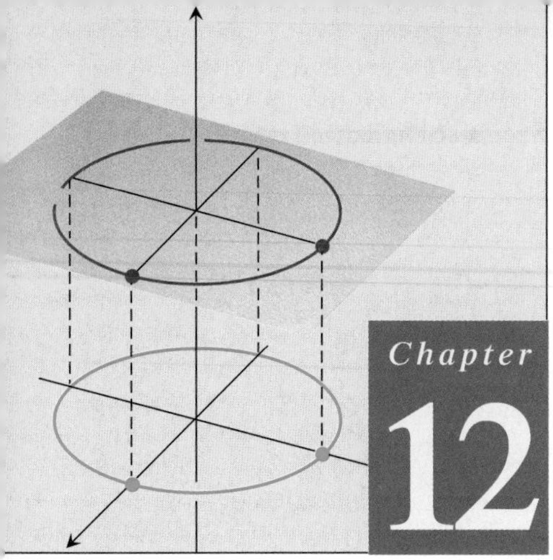

Chapter 12

VECTORS AND THE GEOMETRY OF SPACE

OVERVIEW To apply calculus in many real-world situations and in higher mathematics, we need a mathematical description of three-dimensional space. In this chapter we introduce three-dimensional coordinate systems and vectors. Building on what we already know about coordinates in the *xy*-plane, we establish coordinates in space by adding a third axis that measures distance above and below the *xy*-plane. Vectors are used to study the analytic geometry of space, where they give simple ways to describe lines, planes, surfaces, and curves in space. We use these geometric ideas in the rest of the book to study motion in space and the calculus of functions of several variables, with their many important applications in science, engineering, economics, and higher mathematics.

12.1 Three-Dimensional Coordinate Systems

To locate a point in space, we use three mutually perpendicular coordinate axes, arranged as in Figure 12.1. The axes shown there make a *right-handed* coordinate frame. When you hold your right hand so that the fingers curl from the positive x-axis toward the positive y-axis, your thumb points along the positive z-axis. So when you look down on the xy-plane from the positive direction of the z-axis, positive angles in the plane are measured counterclockwise from the positive x-axis and around the positive z-axis. (In a *left-handed* coordinate frame, the z-axis would point downward in Figure 12.1 and angles in the plane would be positive when measured clockwise from the positive x-axis. This is not the convention we have used for measuring angles in the xy-plane. Right-handed and left-handed coordinate frames are not equivalent.)

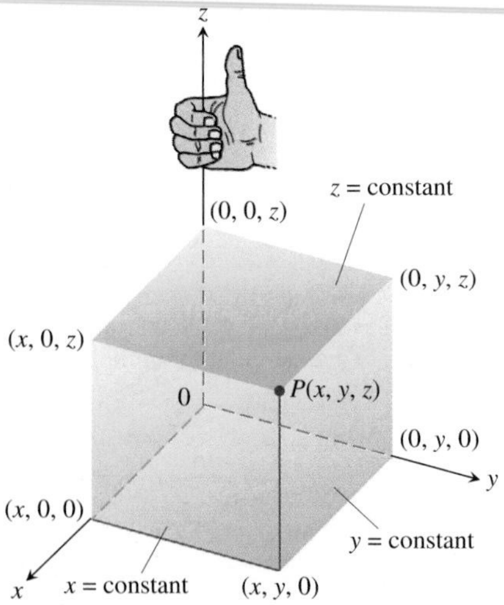

FIGURE 12.1 The Cartesian coordinate system is right-handed.

The Cartesian coordinates (x, y, z) of a point P in space are the numbers at which the planes through P perpendicular to the axes cut the axes. Cartesian coordinates for space are also called **rectangular coordinates** because the axes that define them meet at right angles. Points on the x-axis have y- and z-coordinates equal to zero. That is, they have coordinates of the form $(x, 0, 0)$. Similarly, points on the y-axis have coordinates of the form $(0, y, 0)$, and points on the z-axis have coordinates of the form $(0, 0, z)$.

The planes determined by the coordinates axes are the ***xy*-plane**, whose standard equation is $z = 0$; the ***yz*-plane**, whose standard equation is $x = 0$; and the ***xz*-plane**, whose standard equation is $y = 0$. They meet at the **origin** $(0, 0, 0)$ (Figure 12.2). The origin is also identified by simply 0 or sometimes the letter O.

The three **coordinate planes** $x = 0, y = 0$, and $z = 0$ divide space into eight cells called **octants**. The octant in which the point coordinates are all positive is called the **first octant**; there is no conventional numbering for the other seven octants.

The points in a plane perpendicular to the x-axis all have the same x-coordinate, this being the number at which that plane cuts the x-axis. The y- and z-coordinates can be any numbers. Similarly, the points in a plane perpendicular to the y-axis have a common y-coordinate and the points in a plane perpendicular to the z-axis have a common z-coordinate. To write equations for these planes, we name the common coordinate's value. The plane $x = 2$ is the plane perpendicular to the x-axis at $x = 2$. The plane $y = 3$ is the plane perpendicular to the y-axis at $y = 3$. The plane $z = 5$ is the plane perpendicular to the z-axis at $z = 5$. Figure 12.3 shows the planes $x = 2$, $y = 3$, and $z = 5$, together with their intersection point (2, 3, 5).

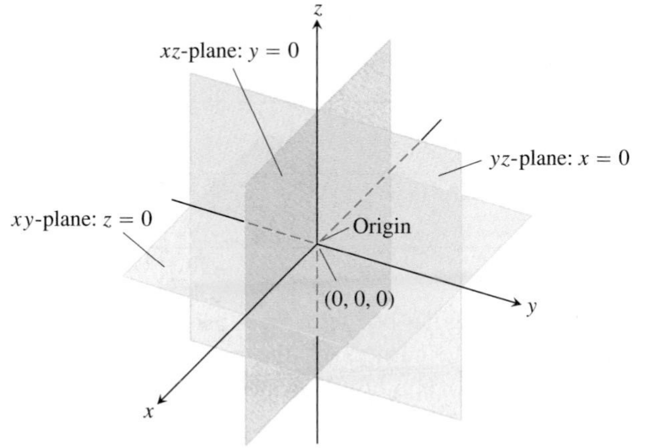

FIGURE 12.2 The planes $x = 0$, $y = 0$, and $z = 0$ divide space into eight octants.

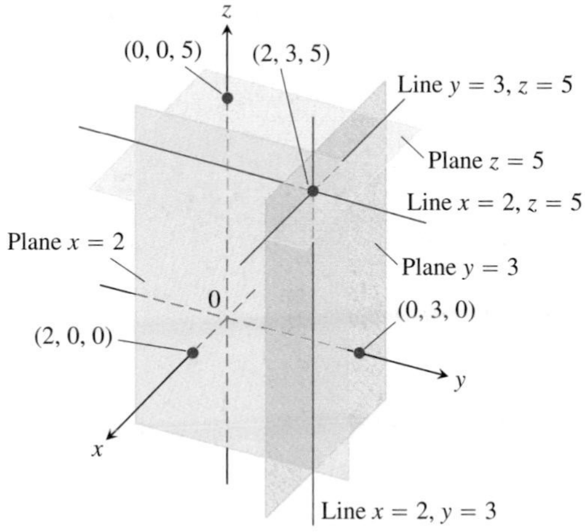

FIGURE 12.3 The planes $x = 2$, $y = 3$, and $z = 5$ determine three lines through the point (2, 3, 5).

The planes $x = 2$ and $y = 3$ in Figure 12.3 intersect in a line parallel to the z-axis. This line is described by the *pair* of equations $x = 2, y = 3$. A point (x, y, z) lies on the line if and only if $x = 2$ and $y = 3$. Similarly, the line of intersection of the planes $y = 3$ and $z = 5$ is described by the equation pair $y = 3, z = 5$. This line runs parallel to the x-axis. The line of intersection of the planes $x = 2$ and $z = 5$, parallel to the y-axis, is described by the equation pair $x = 2, z = 5$.

In the following examples, we match coordinate equations and inequalities with the sets of points they define in space.

EXAMPLE 1 Interpreting Equations and Inequalities Geometrically

(a) $z \geq 0$ — The half-space consisting of the points on and above the xy-plane.

(b) $x = -3$ — The plane perpendicular to the x-axis at $x = -3$. This plane lies parallel to the yz-plane and 3 units behind it.

(c) $z = 0, x \leq 0, y \geq 0$ — The second quadrant of the xy-plane.

(d) $x \geq 0, y \geq 0, z \geq 0$ — The first octant.

(e) $-1 \leq y \leq 1$ — The slab between the planes $y = -1$ and $y = 1$ (planes included).

(f) $y = -2, z = 2$ — The line in which the planes $y = -2$ and $z = 2$ intersect. Alternatively, the line through the point $(0, -2, 2)$ parallel to the x-axis. ■

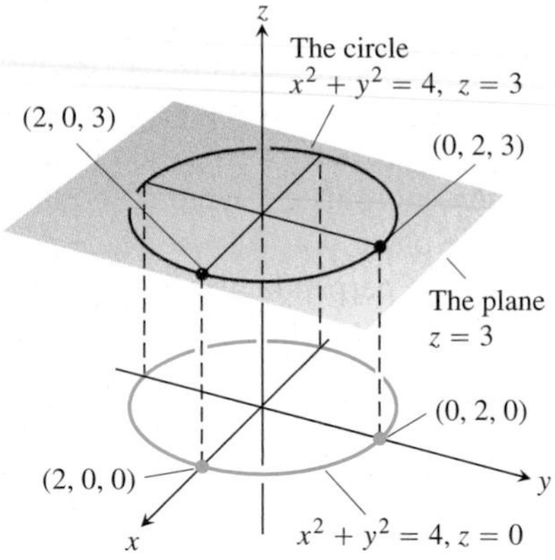

FIGURE 12.4 The circle $x^2 + y^2 = 4$ in the plane $z = 3$ (Example 2).

EXAMPLE 2 Graphing Equations

What points $P(x, y, z)$ satisfy the equations

$$x^2 + y^2 = 4 \quad \text{and} \quad z = 3?$$

Solution The points lie in the horizontal plane $z = 3$ and, in this plane, make up the circle $x^2 + y^2 = 4$. We call this set of points "the circle $x^2 + y^2 = 4$ in the plane $z = 3$" or, more simply, "the circle $x^2 + y^2 = 4, z = 3$" (Figure 12.4). ■

Distance and Spheres in Space

The formula for the distance between two points in the xy-plane extends to points in space.

The Distance Between $P_1(x_1, y_1, z_1)$ and $P_2(x_2, y_2, z_2)$ is

$$|P_1P_2| = \sqrt{(x_2 - x_1)^2 + (y_2 - y_1)^2 + (z_2 - z_1)^2}$$

Proof We construct a rectangular box with faces parallel to the coordinate planes and the points P_1 and P_2 at opposite corners of the box (Figure 12.5). If $A(x_2, y_1, z_1)$ and $B(x_2, y_2, z_1)$ are the vertices of the box indicated in the figure, then the three box edges P_1A, AB, and BP_2 have lengths

$$|P_1A| = |x_2 - x_1|, \qquad |AB| = |y_2 - y_1|, \qquad |BP_2| = |z_2 - z_1|.$$

Because triangles P_1BP_2 and P_1AB are both right-angled, two applications of the Pythagorean theorem give

$$|P_1P_2|^2 = |P_1B|^2 + |BP_2|^2 \quad \text{and} \quad |P_1B|^2 = |P_1A|^2 + |AB|^2$$

(see Figure 12.5).

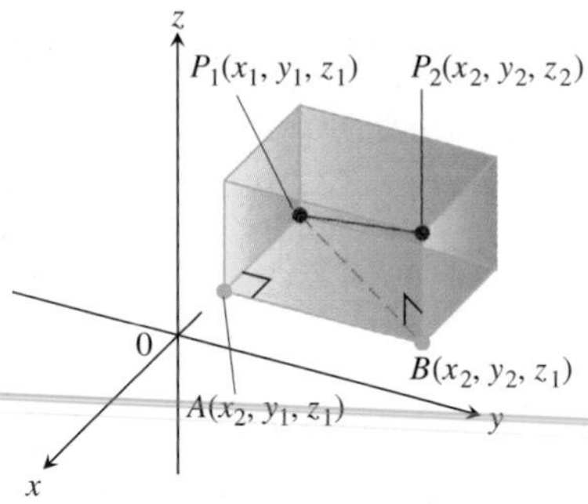

FIGURE 12.5 We find the distance between P_1 and P_2 by applying the Pythagorean theorem to the right triangles P_1AB and P_1BP_2.

So

$$\begin{aligned} |P_1P_2|^2 &= |P_1B|^2 + |BP_2|^2 \\ &= |P_1A|^2 + |AB|^2 + |BP_2|^2 && \text{Substitute } |P_1B|^2 = |P_1A|^2 + |AB|^2. \\ &= |x_2 - x_1|^2 + |y_2 - y_1|^2 + |z_2 - z_1|^2 \\ &= (x_2 - x_1)^2 + (y_2 - y_1)^2 + (z_2 - z_1)^2 \end{aligned}$$

Therefore

$$|P_1P_2| = \sqrt{(x_2 - x_1)^2 + (y_2 - y_1)^2 + (z_2 - z_1)^2}$$ ■

EXAMPLE 3 Finding the Distance Between Two Points

The distance between $P_1(2, 1, 5)$ and $P_2(-2, 3, 0)$ is

$$\begin{aligned} |P_1P_2| &= \sqrt{(-2 - 2)^2 + (3 - 1)^2 + (0 - 5)^2} \\ &= \sqrt{16 + 4 + 25} \\ &= \sqrt{45} \approx 6.708. \end{aligned}$$ ■

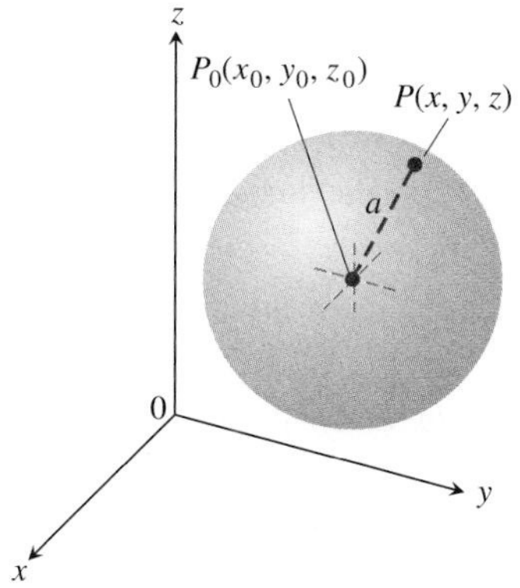

FIGURE 12.6 The standard equation of the sphere of radius a centered at the point (x_0, y_0, z_0) is

$$(x - x_0)^2 + (y - y_0)^2 + (z - z_0)^2 = a^2.$$

We can use the distance formula to write equations for spheres in space (Figure 12.6). A point $P(x, y, z)$ lies on the sphere of radius a centered at $P_0(x_0, y_0, z_0)$ precisely when $|P_0P| = a$ or

$$(x - x_0)^2 + (y - y_0)^2 + (z - z_0)^2 = a^2.$$

The Standard Equation for the Sphere of Radius a and Center (x_0, y_0, z_0)

$$(x - x_0)^2 + (y - y_0)^2 + (z - z_0)^2 = a^2$$

EXAMPLE 4 Finding the Center and Radius of a Sphere

Find the center and radius of the sphere

$$x^2 + y^2 + z^2 + 3x - 4z + 1 = 0.$$

Solution We find the center and radius of a sphere the way we find the center and radius of a circle: Complete the squares on the x-, y-, and z-terms as necessary and write each quadratic as a squared linear expression. Then, from the equation in standard form, read off the center and radius. For the sphere here, we have

$$x^2 + y^2 + z^2 + 3x - 4z + 1 = 0$$

$$(x^2 + 3x) + y^2 + (z^2 - 4z) = -1$$

$$\left(x^2 + 3x + \left(\frac{3}{2}\right)^2\right) + y^2 + \left(z^2 - 4z + \left(\frac{-4}{2}\right)^2\right) = -1 + \left(\frac{3}{2}\right)^2 + \left(\frac{-4}{2}\right)^2$$

$$\left(x + \frac{3}{2}\right)^2 + y^2 + (z - 2)^2 = -1 + \frac{9}{4} + 4 = \frac{21}{4}.$$

From this standard form, we read that $x_0 = -3/2$, $y_0 = 0$, $z_0 = 2$, and $a = \sqrt{21}/2$. The center is $(-3/2, 0, 2)$. The radius is $\sqrt{21}/2$. ■

EXAMPLE 5 Interpreting Equations and Inequalities

(a) $x^2 + y^2 + z^2 < 4$ — The interior of the sphere $x^2 + y^2 + z^2 = 4$.

(b) $x^2 + y^2 + z^2 \leq 4$ — The solid ball bounded by the sphere $x^2 + y^2 + z^2 = 4$. Alternatively, the sphere $x^2 + y^2 + z^2 = 4$ together with its interior.

(c) $x^2 + y^2 + z^2 > 4$ — The exterior of the sphere $x^2 + y^2 + z^2 = 4$.

(d) $x^2 + y^2 + z^2 = 4, z \leq 0$ — The lower hemisphere cut from the sphere $x^2 + y^2 + z^2 = 4$ by the xy-plane (the plane $z = 0$).

■

Just as polar coordinates give another way to locate points in the xy-plane (Section 10.5), alternative coordinate systems, different from the Cartesian coordinate system developed here, exist for three-dimensional space. We examine two of these coordinate systems in Section 15.6.

EXERCISES 12.1

Sets, Equations, and Inequalities

In Exercises 1–12, give a geometric description of the set of points in space whose coordinates satisfy the given pairs of equations.

1. $x = 2, \quad y = 3$

2. $x = -1, \quad z = 0$

3. $y = 0, \quad z = 0$

4. $x = 1, \quad y = 0$

5. $x^2 + y^2 = 4, \quad z = 0$

6. $x^2 + y^2 = 4, \quad z = -2$

7. $x^2 + z^2 = 4, \quad y = 0$

8. $y^2 + z^2 = 1, \quad x = 0$

9. $x^2 + y^2 + z^2 = 1, \quad x = 0$

10. $x^2 + y^2 + z^2 = 25, \quad y = -4$

11. $x^2 + y^2 + (z + 3)^2 = 25, \quad z = 0$

12. $x^2 + (y - 1)^2 + z^2 = 4, \quad y = 0$

In Exercises 13–18, describe the sets of points in space whose coordinates satisfy the given inequalities or combinations of equations and inequalities.

13. **a.** $x \geq 0, \quad y \geq 0, \quad z = 0$ **b.** $x \geq 0, \quad y \leq 0, \quad z = 0$

14. **a.** $0 \leq x \leq 1$ **b.** $0 \leq x \leq 1, \quad 0 \leq y \leq 1$

c. $0 \leq x \leq 1, \quad 0 \leq y \leq 1, \quad 0 \leq z \leq 1$

15. **a.** $x^2 + y^2 + z^2 \leq 1$ **b.** $x^2 + y^2 + z^2 > 1$

16. **a.** $x^2 + y^2 \leq 1, \quad z = 0$ **b.** $x^2 + y^2 \leq 1, \quad z = 3$

c. $x^2 + y^2 \leq 1$, no restriction on z

17. **a.** $x^2 + y^2 + z^2 = 1, \quad z \geq 0$

b. $x^2 + y^2 + z^2 \leq 1, \quad z \geq 0$

18. **a.** $x = y, \quad z = 0$ **b.** $x = y$, no restriction on z

In Exercises 19–28, describe the given set with a single equation or with a pair of equations.

19. The plane perpendicular to the

a. x-axis at $(3, 0, 0)$ **b.** y-axis at $(0, -1, 0)$

c. z-axis at $(0, 0, -2)$

20. The plane through the point $(3, -1, 2)$ perpendicular to the

a. x-axis **b.** y-axis **c.** z-axis

21. The plane through the point $(3, -1, 1)$ parallel to the

a. xy-plane **b.** yz-plane **c.** xz-plane

22. The circle of radius 2 centered at $(0, 0, 0)$ and lying in the

a. xy-plane **b.** yz-plane **c.** xz-plane

23. The circle of radius 2 centered at $(0, 2, 0)$ and lying in the

a. xy-plane **b.** yz-plane **c.** plane $y = 2$

24. The circle of radius 1 centered at $(-3, 4, 1)$ and lying in a plane parallel to the

a. xy-plane **b.** yz-plane **c.** xz-plane

25. The line through the point $(1, 3, -1)$ parallel to the

a. x-axis **b.** y-axis **c.** z-axis

26. The set of points in space equidistant from the origin and the point $(0, 2, 0)$

27. The circle in which the plane through the point $(1, 1, 3)$ perpendicular to the z-axis meets the sphere of radius 5 centered at the origin

28. The set of points in space that lie 2 units from the point $(0, 0, 1)$ and, at the same time, 2 units from the point $(0, 0, -1)$

Write inequalities to describe the sets in Exercises 29–34.

29. The slab bounded by the planes $z = 0$ and $z = 1$ (planes included)

30. The solid cube in the first octant bounded by the coordinate planes and the planes $x = 2$, $y = 2$, and $z = 2$

31. The half-space consisting of the points on and below the xy-plane

32. The upper hemisphere of the sphere of radius 1 centered at the origin

33. The **(a)** interior and **(b)** exterior of the sphere of radius 1 centered at the point $(1, 1, 1)$

34. The closed region bounded by the spheres of radius 1 and radius 2 centered at the origin. (*Closed* means the spheres are to be included. Had we wanted the spheres left out, we would have asked for the *open* region bounded by the spheres. This is analogous to the way we use *closed* and *open* to describe intervals: *closed* means endpoints included, *open* means endpoints left out. Closed sets include boundaries; open sets leave them out.)

Distance

In Exercises 35–40, find the distance between points P_1 and P_2.

35. $P_1(1, 1, 1), \quad P_2(3, 3, 0)$

36. $P_1(-1, 1, 5), \quad P_2(2, 5, 0)$

37. $P_1(1, 4, 5), \quad P_2(4, -2, 7)$

38. $P_1(3, 4, 5), \quad P_2(2, 3, 4)$

39. $P_1(0, 0, 0), \quad P_2(2, -2, -2)$

40. $P_1(5, 3, -2), \quad P_2(0, 0, 0)$

Spheres

Find the centers and radii of the spheres in Exercises 41–44.

41. $(x + 2)^2 + y^2 + (z - 2)^2 = 8$

42. $\left(x + \frac{1}{2}\right)^2 + \left(y + \frac{1}{2}\right)^2 + \left(z + \frac{1}{2}\right)^2 = \frac{21}{4}$

43. $(x - \sqrt{2})^2 + (y - \sqrt{2})^2 + (z + \sqrt{2})^2 = 2$

44. $x^2 + \left(y + \frac{1}{3}\right)^2 + \left(z - \frac{1}{3}\right)^2 = \frac{29}{9}$

Find equations for the spheres whose centers and radii are given in Exercises 45–48.

	Center	Radius
45.	$(1, 2, 3)$	$\sqrt{14}$
46.	$(0, -1, 5)$	2
47.	$(-2, 0, 0)$	$\sqrt{3}$
48.	$(0, -7, 0)$	7

Find the centers and radii of the spheres in Exercises 49–52.

49. $x^2 + y^2 + z^2 + 4x - 4z = 0$

50. $x^2 + y^2 + z^2 - 6y + 8z = 0$

51. $2x^2 + 2y^2 + 2z^2 + x + y + z = 9$

52. $3x^2 + 3y^2 + 3z^2 + 2y - 2z = 9$

Theory and Examples

53. Find a formula for the distance from the point $P(x, y, z)$ to the

a. x-axis **b.** y-axis **c.** z-axis

54. Find a formula for the distance from the point $P(x, y, z)$ to the

a. xy-plane **b.** yz-plane **c.** xz-plane

55. Find the perimeter of the triangle with vertices $A(-1, 2, 1)$, $B(1, -1, 3)$, and $C(3, 4, 5)$.

56. Show that the point $P(3, 1, 2)$ is equidistant from the points $A(2, -1, 3)$ and $B(4, 3, 1)$.

12.2 Vectors

Some of the things we measure are determined simply by their magnitudes. To record mass, length, or time, for example, we need only write down a number and name an appropriate unit of measure. We need more information to describe a force, displacement, or velocity. To describe a force, we need to record the direction in which it acts as well as how large it is. To describe a body's displacement, we have to say in what direction it moved as well as how far. To describe a body's velocity, we have to know where the body is headed as well as how fast it is going.

Component Form

A quantity such as force, displacement, or velocity is called a *vector* and is represented by a **directed line segment** (Figure 12.7). The arrow points in the direction of the action and its length gives the magnitude of the action in terms of a suitably chosen unit. For example, a force vector points in the direction in which the force acts; its length is a measure of the force's strength; a velocity vector points in the direction of motion and its length is the speed of the moving object. Figure 12.8 displays the velocity vector **v** at a specific location for a particle moving along a path in the plane or in space. (This application of vectors is studied in Chapter 13.)

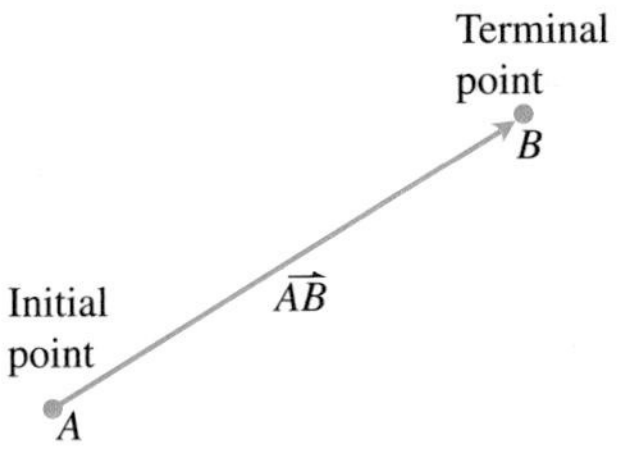

FIGURE 12.7 The directed line segment $\overrightarrow{AB}$.

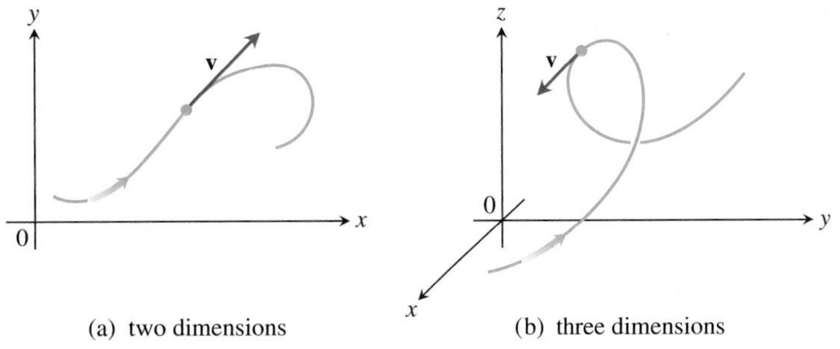

FIGURE 12.8 The velocity vector of a particle moving along a path (a) in the plane (b) in space. The arrowhead on the path indicates the direction of motion of the particle.

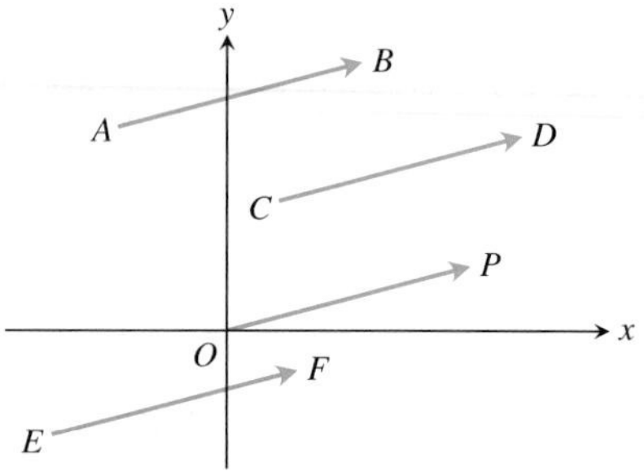

FIGURE 12.9 The four arrows in the plane (directed line segments) shown here have the same length and direction. They therefore represent the same vector, and we write $\overrightarrow{AB} = \overrightarrow{CD} = \overrightarrow{OP} = \overrightarrow{EF}$.

> **DEFINITIONS** Vector, Initial and Terminal Point, Length
>
> A **vector** is a directed line segment. The directed line segment $\overrightarrow{AB}$ has **initial point** A and **terminal point** B; its **length** is denoted by $|\overrightarrow{AB}|$. Two vectors are **equal** if they have the same length and direction.

The arrows we use when we draw vectors are understood to represent the same vector if they have the same length, are parallel, and point in the same direction (Figure 12.9) regardless of the initial point.

In textbooks, vectors are usually written in lowercase, boldface letters, for example **u**, **v**, and **w**. Sometimes we use uppercase boldface letters, such as **F**, to denote a force vector. In handwritten form, it is customary to draw small arrows above the letters, for example $\vec{u}$, $\vec{v}$, $\vec{w}$, and $\vec{F}$.

We need a way to represent vectors algebraically so that we can be more precise about the direction of a vector.

Let $\mathbf{v} = \overrightarrow{PQ}$. There is one directed line segment equal to $\overrightarrow{PQ}$ whose initial point is the origin (Figure 12.10). It is the representative of **v** in **standard position** and is the vector we normally use to represent **v**. We can specify **v** by writing the coordinates of its terminal point (v_1, v_2, v_3) when **v** is in standard position. If **v** is a vector in the plane its terminal point (v_1, v_2) has two coordinates.

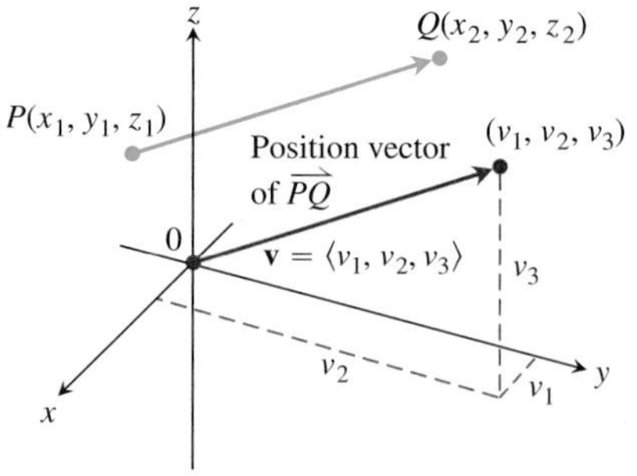

FIGURE 12.10 A vector $\overrightarrow{PQ}$ in standard position has its initial point at the origin. The directed line segments $\overrightarrow{PQ}$ and **v** are parallel and have the same length.

> **DEFINITION** Component Form
>
> If **v** is a **two-dimensional** vector in the plane equal to the vector with initial point at the origin and terminal point (v_1, v_2), then the **component form** of **v** is
>
> $$\mathbf{v} = \langle v_1, v_2 \rangle.$$
>
> If **v** is a **three-dimensional** vector equal to the vector with initial point at the origin and terminal point (v_1, v_2, v_3), then the **component form** of **v** is
>
> $$\mathbf{v} = \langle v_1, v_2, v_3 \rangle.$$

So a two-dimensional vector is an ordered pair $\mathbf{v} = \langle v_1, v_2 \rangle$ of real numbers, and a three-dimensional vector is an ordered triple $\mathbf{v} = \langle v_1, v_2, v_3 \rangle$ of real numbers. The numbers v_1, v_2, and v_3 are called the **components** of **v**.

Observe that if $\mathbf{v} = \langle v_1, v_2, v_3 \rangle$ is represented by the directed line segment $\overrightarrow{PQ}$, where the initial point is $P(x_1, y_1, z_1)$ and the terminal point is $Q(x_2, y_2, z_2)$, then $x_1 + v_1 = x_2$, $y_1 + v_2 = y_2$, and $z_1 + v_3 = z_2$ (see Figure 12.10). Thus, $v_1 = x_2 - x_1$, $v_2 = y_2 - y_1$, and $v_3 = z_2 - z_1$ are the components of $\overrightarrow{PQ}$.

In summary, given the points $P(x_1, y_1, z_1)$ and $Q(x_2, y_2, z_2)$, the standard position vector $\mathbf{v} = \langle v_1, v_2, v_3 \rangle$ equal to $\overrightarrow{PQ}$ is

$$\mathbf{v} = \langle x_2 - x_1, y_2 - y_1, z_2, -z_1 \rangle.$$

If **v** is two-dimensional with $P(x_1, y_1)$ and $Q(x_2, y_2)$ as points in the plane, then $\mathbf{v} = \langle x_2 - x_1, y_2 - y_1 \rangle$. There is no third component for planar vectors. With this understanding, we will develop the algebra of three-dimensional vectors and simply drop the third component when the vector is two-dimensional (a planar vector).

Two vectors are equal if and only if their standard position vectors are identical. Thus $\langle u_1, u_2, u_3 \rangle$ and $\langle v_1, v_2, v_3 \rangle$ are equal if and only if $u_1 = v_1$, $u_2 = v_2$, and $u_3 = v_3$.

The **magnitude** or **length** of the vector $\overrightarrow{PQ}$ is the length of any of its equivalent directed line segment representations. In particular, if $\mathbf{v} = \langle x_2 - x_1, y_2 - y_1, z_2 - z_1 \rangle$ is the standard position vector for $\overrightarrow{PQ}$, then the distance formula gives the magnitude or length of $\mathbf{v}$, denoted by the symbol $|\mathbf{v}|$ or $\|\mathbf{v}\|$.

The **magnitude** or **length** of the vector $\mathbf{v} = \overrightarrow{PQ}$ is the nonnegative number

$$|\mathbf{v}| = \sqrt{v_1^2 + v_2^2 + v_3^2} = \sqrt{(x_2 - x_1)^2 + (y_2 - y_1)^2 + (z_2 - z_1)^2}$$

(See Figure 12.10.)

The only vector with length 0 is the **zero vector** $\mathbf{0} = \langle 0, 0 \rangle$ or $\mathbf{0} = \langle 0, 0, 0 \rangle$. This vector is also the only vector with no specific direction.

EXAMPLE 1 Component Form and Length of a Vector

Find the **(a)** component form and **(b)** length of the vector with initial point $P(-3, 4, 1)$ and terminal point $Q(-5, 2, 2)$.

Solution

(a) The standard position vector $\mathbf{v}$ representing $\overrightarrow{PQ}$ has components

$$v_1 = x_2 - x_1 = -5 - (-3) = -2, \qquad v_2 = y_2 - y_1 = 2 - 4 = -2,$$

and

$$v_3 = z_2 - z_1 = 2 - 1 = 1.$$

The component form of $\overrightarrow{PQ}$ is

$$\mathbf{v} = \langle -2, -2, 1 \rangle.$$

(b) The length or magnitude of $\mathbf{v} = \overrightarrow{PQ}$ is

$$|\mathbf{v}| = \sqrt{(-2)^2 + (-2)^2 + (1)^2} = \sqrt{9} = 3.$$ ■

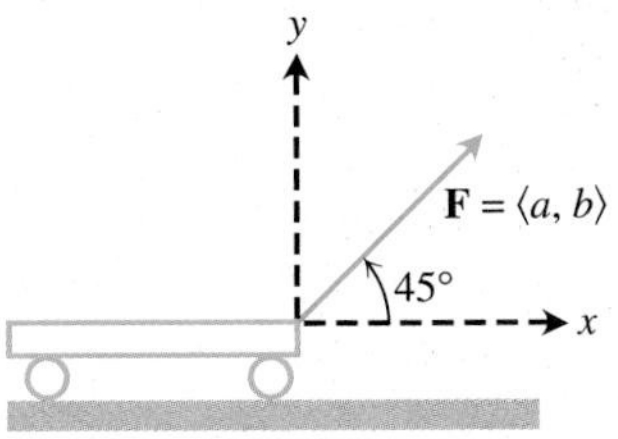

FIGURE 12.11 The force pulling the cart forward is represented by the vector **F** of magnitude 20 (pounds) making an angle of 45° with the horizontal ground (positive x-axis) (Example 2).

EXAMPLE 2 Force Moving a Cart

A small cart is being pulled along a smooth horizontal floor with a 20-lb force **F** making a 45° angle to the floor (Figure 12.11). What is the *effective* force moving the cart forward?

Solution The effective force is the horizontal component of $\mathbf{F} = \langle a, b \rangle$, given by

$$a = |\mathbf{F}| \cos 45° = (20)\left(\frac{\sqrt{2}}{2}\right) \approx 14.14 \text{ lb}.$$

Notice that **F** is a two-dimensional vector. ■

Vector Algebra Operations

Two principal operations involving vectors are *vector addition* and *scalar multiplication*. A **scalar** is simply a real number, and is called such when we want to draw attention to its differences from vectors. Scalars can be positive, negative, or zero.

DEFINITIONS Vector Addition and Multiplication of a Vector by a Scalar

Let $\mathbf{u} = \langle u_1, u_2, u_3 \rangle$ and $\mathbf{v} = \langle v_1, v_2, v_3 \rangle$ be vectors with k a scalar.

Addition: $\mathbf{u} + \mathbf{v} = \langle u_1 + v_1, u_2 + v_2, u_3 + v_3 \rangle$

Scalar multiplication: $k\mathbf{u} = \langle ku_1, ku_2, ku_3 \rangle$

We add vectors by adding the corresponding components of the vectors. We multiply a vector by a scalar by multiplying each component by the scalar. The definitions apply to planar vectors except there are only two components, $\langle u_1, u_2 \rangle$ and $\langle v_1, v_2 \rangle$.

The definition of vector addition is illustrated geometrically for planar vectors in Figure 12.12a, where the initial point of one vector is placed at the terminal point of the other. Another interpretation is shown in Figure 12.12b (called the **parallelogram law** of addition), where the sum, called the **resultant vector**, is the diagonal of the parallelogram. In physics, forces add vectorially as do velocities, accelerations, and so on. So the force acting on a particle subject to electric and gravitational forces is obtained by adding the two force vectors.

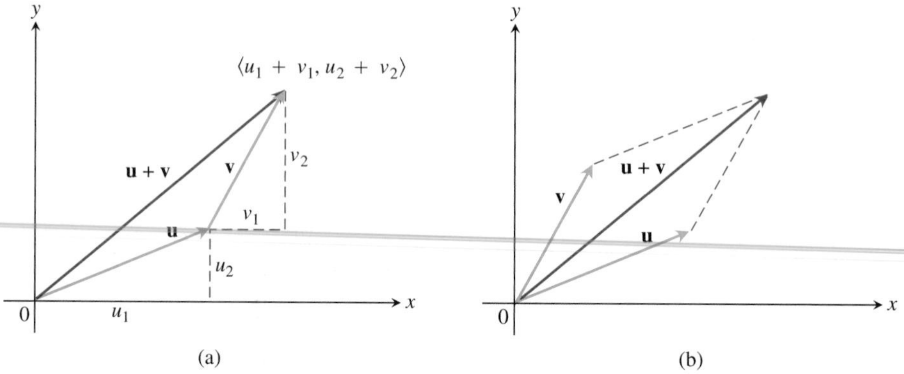

FIGURE 12.12 (a) Geometric interpretation of the vector sum. (b) The parallelogram law of vector addition.

Figure 12.13 displays a geometric interpretation of the product $k\mathbf{u}$ of the scalar k and vector $\mathbf{u}$. If $k > 0$, then $k\mathbf{u}$ has the same direction as $\mathbf{u}$; if $k < 0$, then the direction of $k\mathbf{u}$ is opposite to that of $\mathbf{u}$. Comparing the lengths of $\mathbf{u}$ and $k\mathbf{u}$, we see that

$$|k\mathbf{u}| = \sqrt{(ku_1)^2 + (ku_2)^2 + (ku_3)^2} = \sqrt{k^2(u_1^2 + u_2^2 + u_3^2)}$$

$$= \sqrt{k^2}\sqrt{u_1^2 + u_2^2 + u_3^2} = |k|\,|\mathbf{u}|.$$

The length of $k\mathbf{u}$ is the absolute value of the scalar k times the length of $\mathbf{u}$. The vector $(-1)\mathbf{u} = -\mathbf{u}$ has the same length as $\mathbf{u}$ but points in the opposite direction.

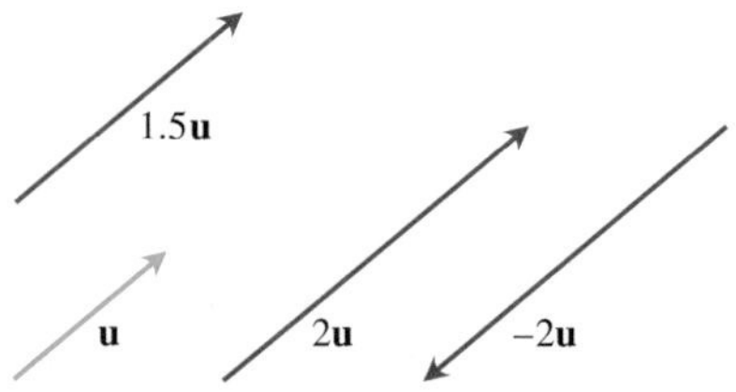

FIGURE 12.13 Scalar multiples of **u**.

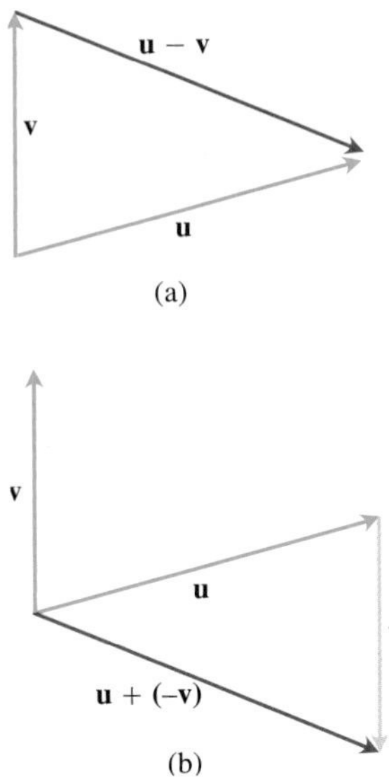

FIGURE 12.14 (a) The vector $\mathbf{u} - \mathbf{v}$, when added to $\mathbf{v}$, gives $\mathbf{u}$. (b) $\mathbf{u} - \mathbf{v} = \mathbf{u} + (-\mathbf{v})$.

By the **difference** $\mathbf{u} - \mathbf{v}$ of two vectors, we mean

$$\mathbf{u} - \mathbf{v} = \mathbf{u} + (-\mathbf{v}).$$

If $\mathbf{u} = \langle u_1, u_2, u_3 \rangle$ and $\mathbf{v} = \langle v_1, v_2, v_3 \rangle$, then

$$\mathbf{u} - \mathbf{v} = \langle u_1 - v_1, u_2 - v_2, u_3 - v_3 \rangle.$$

Note that $(\mathbf{u} - \mathbf{v}) + \mathbf{v} = \mathbf{u}$, so adding the vector $(\mathbf{u} - \mathbf{v})$ to $\mathbf{v}$ gives $\mathbf{u}$ (Figure 12.14a). Figure 12.14b shows the difference $\mathbf{u} - \mathbf{v}$ as the sum $\mathbf{u} + (-\mathbf{v})$.

EXAMPLE 3 Performing Operations on Vectors

Let $\mathbf{u} = \langle -1, 3, 1 \rangle$ and $\mathbf{v} = \langle 4, 7, 0 \rangle$. Find

(a) $2\mathbf{u} + 3\mathbf{v}$ **(b)** $\mathbf{u} - \mathbf{v}$ **(c)** $\left|\frac{1}{2}\mathbf{u}\right|$.

Solution

(a) $2\mathbf{u} + 3\mathbf{v} = 2\langle -1, 3, 1 \rangle + 3\langle 4, 7, 0 \rangle = \langle -2, 6, 2 \rangle + \langle 12, 21, 0 \rangle = \langle 10, 27, 2 \rangle$

(b) $\mathbf{u} - \mathbf{v} = \langle -1, 3, 1 \rangle - \langle 4, 7, 0 \rangle = \langle -1 - 4, 3 - 7, 1 - 0 \rangle = \langle -5, -4, 1 \rangle$

(c) $\left|\frac{1}{2}\mathbf{u}\right| = \left|\left\langle -\frac{1}{2}, \frac{3}{2}, \frac{1}{2} \right\rangle\right| = \sqrt{\left(-\frac{1}{2}\right)^2 + \left(\frac{3}{2}\right)^2 + \left(\frac{1}{2}\right)^2} = \frac{1}{2}\sqrt{11}.$ ■

Vector operations have many of the properties of ordinary arithmetic. These properties are readily verified using the definitions of vector addition and multiplication by a scalar.

> **Properties of Vector Operations**
>
> Let $\mathbf{u}$, $\mathbf{v}$, $\mathbf{w}$ be vectors and a, b be scalars.
>
> 1. $\mathbf{u} + \mathbf{v} = \mathbf{v} + \mathbf{u}$
> 2. $(\mathbf{u} + \mathbf{v}) + \mathbf{w} = \mathbf{u} + (\mathbf{v} + \mathbf{w})$
> 3. $\mathbf{u} + \mathbf{0} = \mathbf{u}$
> 4. $\mathbf{u} + (-\mathbf{u}) = \mathbf{0}$
> 5. $0\mathbf{u} = \mathbf{0}$
> 6. $1\mathbf{u} = \mathbf{u}$
> 7. $a(b\mathbf{u}) = (ab)\mathbf{u}$
> 8. $a(\mathbf{u} + \mathbf{v}) = a\mathbf{u} + a\mathbf{v}$
> 9. $(a + b)\mathbf{u} = a\mathbf{u} + b\mathbf{u}$

An important application of vectors occurs in navigation.

EXAMPLE 4 Finding Ground Speed and Direction

A Boeing® 767® airplane, flying due east at 500 mph in still air, encounters a 70-mph tailwind blowing in the direction 60° north of east. The airplane holds its compass heading due east but, because of the wind, acquires a new ground speed and direction. What are they?

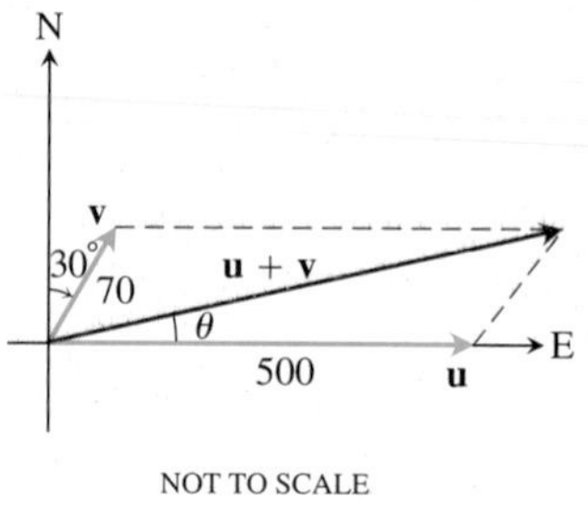

FIGURE 12.15 Vectors representing the velocities of the airplane **u** and tailwind **v** in Example 4.

Solution If **u** = the velocity of the airplane alone and **v** = the velocity of the tailwind, then $|\mathbf{u}| = 500$ and $|\mathbf{v}| = 70$ (Figure 12.15). The velocity of the airplane with respect to the ground is given by the magnitude and direction of the *resultant vector* $\mathbf{u} + \mathbf{v}$. If we let the positive x-axis represent east and the positive y-axis represent north, then the component forms of **u** and **v** are

$$\mathbf{u} = \langle 500, 0 \rangle \quad \text{and} \quad \mathbf{v} = \langle 70 \cos 60°, 70 \sin 60° \rangle = \langle 35, 35\sqrt{3} \rangle.$$

Therefore,

$$\mathbf{u} + \mathbf{v} = \langle 535, 35\sqrt{3} \rangle$$
$$|\mathbf{u} + \mathbf{v}| = \sqrt{535^2 + (35\sqrt{3})^2} \approx 538.4$$

and

$$\theta = \tan^{-1} \frac{35\sqrt{3}}{535} \approx 6.5°. \qquad \text{Figure 12.15}$$

The new ground speed of the airplane is about 538.4 mph, and its new direction is about 6.5° north of east. ■

Unit Vectors

A vector **v** of length 1 is called a **unit vector**. The **standard unit vectors** are

$$\mathbf{i} = \langle 1, 0, 0 \rangle, \quad \mathbf{j} = \langle 0, 1, 0 \rangle, \quad \text{and} \quad \mathbf{k} = \langle 0, 0, 1 \rangle.$$

Any vector $\mathbf{v} = \langle v_1, v_2, v_3 \rangle$ can be written as a *linear combination* of the standard unit vectors as follows:

$$\begin{aligned}\mathbf{v} = \langle v_1, v_2, v_3 \rangle &= \langle v_1, 0, 0 \rangle + \langle 0, v_2, 0 \rangle + \langle 0, 0, v_3 \rangle \\ &= v_1\langle 1, 0, 0 \rangle + v_2\langle 0, 1, 0 \rangle + v_3\langle 0, 0, 1 \rangle \\ &= v_1\mathbf{i} + v_2\mathbf{j} + v_3\mathbf{k}.\end{aligned}$$

We call the scalar (or number) v_1 the **i-component** of the vector **v**, v_2 the **j-component**, and v_3 the **k-component**. In component form, the vector from $P_1(x_1, y_1, z_1)$ to $P_2(x_2, y_2, z_2)$ is

$$\overrightarrow{P_1P_2} = (x_2 - x_1)\mathbf{i} + (y_2 - y_1)\mathbf{j} + (z_2 - z_1)\mathbf{k}$$

(Figure 12.16).

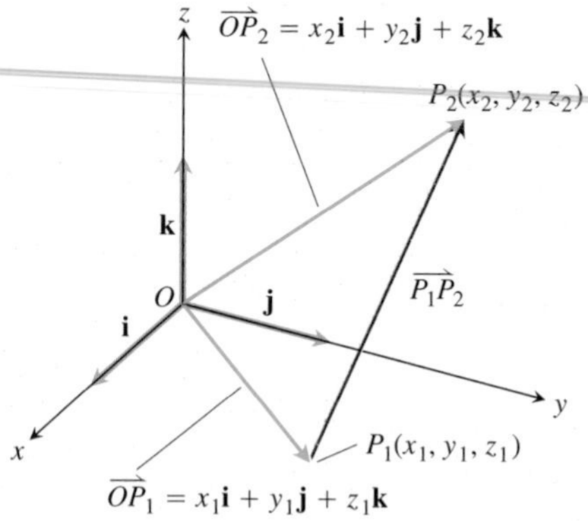

FIGURE 12.16 The vector from P_1 to P_2 is $\overrightarrow{P_1P_2} = (x_2 - x_1)\mathbf{i} + (y_2 - y_1)\mathbf{j} + (z_2 - z_1)\mathbf{k}$.

Whenever $\mathbf{v} \neq \mathbf{0}$, its length $|\mathbf{v}|$ is not zero and

$$\left| \frac{1}{|\mathbf{v}|}\mathbf{v} \right| = \frac{1}{|\mathbf{v}|}|\mathbf{v}| = 1.$$

That is, $\mathbf{v}/|\mathbf{v}|$ is a unit vector in the direction of **v**, called **the direction** of the nonzero vector **v**.

EXAMPLE 5 Finding a Vector's Direction

Find a unit vector **u** in the direction of the vector from $P_1(1, 0, 1)$ to $P_2(3, 2, 0)$.

Solution We divide $\overrightarrow{P_1P_2}$ by its length:

$$\overrightarrow{P_1P_2} = (3 - 1)\mathbf{i} + (2 - 0)\mathbf{j} + (0 - 1)\mathbf{k} = 2\mathbf{i} + 2\mathbf{j} - \mathbf{k}$$

$$|\overrightarrow{P_1P_2}| = \sqrt{(2)^2 + (2)^2 + (-1)^2} = \sqrt{4 + 4 + 1} = \sqrt{9} = 3$$

$$\mathbf{u} = \frac{\overrightarrow{P_1P_2}}{|\overrightarrow{P_1P_2}|} = \frac{2\mathbf{i} + 2\mathbf{j} - \mathbf{k}}{3} = \frac{2}{3}\mathbf{i} + \frac{2}{3}\mathbf{j} - \frac{1}{3}\mathbf{k}.$$

The unit vector $\mathbf{u}$ is the direction of $\overrightarrow{P_1P_2}$. ■

EXAMPLE 6 Expressing Velocity as Speed Times Direction

If $\mathbf{v} = 3\mathbf{i} - 4\mathbf{j}$ is a velocity vector, express $\mathbf{v}$ as a product of its speed times a unit vector in the direction of motion.

Solution Speed is the magnitude (length) of $\mathbf{v}$:

$$|\mathbf{v}| = \sqrt{(3)^2 + (-4)^2} = \sqrt{9 + 16} = 5.$$

HISTORICAL BIOGRAPHY

Hermann Grassmann
(1809–1877)

The unit vector $\mathbf{v}/|\mathbf{v}|$ has the same direction as $\mathbf{v}$:

$$\frac{\mathbf{v}}{|\mathbf{v}|} = \frac{3\mathbf{i} - 4\mathbf{j}}{5} = \frac{3}{5}\mathbf{i} - \frac{4}{5}\mathbf{j}.$$

So

$$\mathbf{v} = 3\mathbf{i} - 4\mathbf{j} = \underbrace{5}_{\text{Length (speed)}}\underbrace{\left(\frac{3}{5}\mathbf{i} - \frac{4}{5}\mathbf{j}\right)}_{\text{Direction of motion}}.$$ ■

In summary, we can express any nonzero vector $\mathbf{v}$ in terms of its two important features, length and direction, by writing $\mathbf{v} = |\mathbf{v}|\dfrac{\mathbf{v}}{|\mathbf{v}|}$.

> If $\mathbf{v} \neq \mathbf{0}$, then
>
> **1.** $\dfrac{\mathbf{v}}{|\mathbf{v}|}$ is a unit vector in the direction of $\mathbf{v}$;
>
> **2.** the equation $\mathbf{v} = |\mathbf{v}|\dfrac{\mathbf{v}}{|\mathbf{v}|}$ expresses $\mathbf{v}$ in terms of its length and direction.

EXAMPLE 7 A Force Vector

A force of 6 newtons is applied in the direction of the vector $\mathbf{v} = 2\mathbf{i} + 2\mathbf{j} - \mathbf{k}$. Express the force $\mathbf{F}$ as a product of its magnitude and direction.

Solution The force vector has magnitude 6 and direction $\dfrac{\mathbf{v}}{|\mathbf{v}|}$, so

$$\mathbf{F} = 6\frac{\mathbf{v}}{|\mathbf{v}|} = 6\frac{2\mathbf{i} + 2\mathbf{j} - \mathbf{k}}{\sqrt{2^2 + 2^2 + (-1)^2}} = 6\frac{2\mathbf{i} + 2\mathbf{j} - \mathbf{k}}{3}$$

$$= 6\left(\frac{2}{3}\mathbf{i} + \frac{2}{3}\mathbf{j} - \frac{1}{3}\mathbf{k}\right).$$ ■

Midpoint of a Line Segment

Vectors are often useful in geometry. For example, the coordinates of the midpoint of a line segment are found by averaging.

> The **midpoint** M of the line segment joining points $P_1(x_1, y_1, z_1)$ and $P_2(x_2, y_2, z_2)$ is the point
>
> $$\left(\frac{x_1 + x_2}{2}, \frac{y_1 + y_2}{2}, \frac{z_1 + z_2}{2}\right).$$

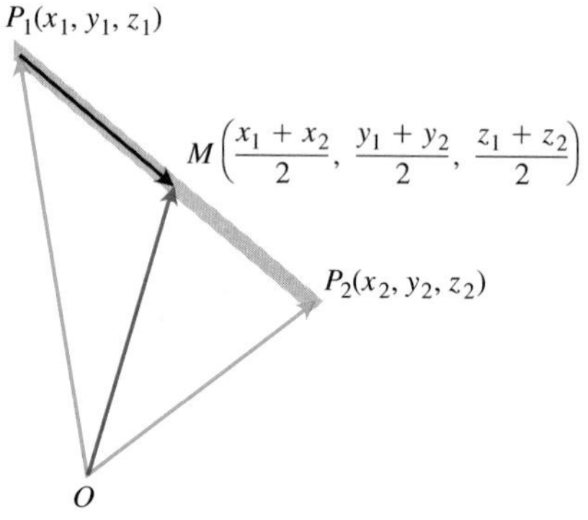

FIGURE 12.17 The coordinates of the midpoint are the averages of the coordinates of P_1 and P_2.

To see why, observe (Figure 12.17) that

$$\begin{aligned}\overrightarrow{OM} &= \overrightarrow{OP_1} + \frac{1}{2}(\overrightarrow{P_1P_2}) = \overrightarrow{OP_1} + \frac{1}{2}(\overrightarrow{OP_2} - \overrightarrow{OP_1}) \\ &= \frac{1}{2}(\overrightarrow{OP_1} + \overrightarrow{OP_2}) \\ &= \frac{x_1 + x_2}{2}\mathbf{i} + \frac{y_1 + y_2}{2}\mathbf{j} + \frac{z_1 + z_2}{2}\mathbf{k}.\end{aligned}$$

EXAMPLE 8 Finding Midpoints

The midpoint of the segment joining $P_1(3, -2, 0)$ and $P_2(7, 4, 4)$ is

$$\left(\frac{3 + 7}{2}, \frac{-2 + 4}{2}, \frac{0 + 4}{2}\right) = (5, 1, 2).$$ ■

EXERCISES 12.2

Vectors in the Plane

In Exercises 1–8, let $\mathbf{u} = \langle 3, -2 \rangle$ and $\mathbf{v} = \langle -2, 5 \rangle$. Find the **(a)** component form and **(b)** magnitude (length) of the vector.

1. $3\mathbf{u}$

2. $-2\mathbf{v}$

3. $\mathbf{u} + \mathbf{v}$

4. $\mathbf{u} - \mathbf{v}$

5. $2\mathbf{u} - 3\mathbf{v}$

6. $-2\mathbf{u} + 5\mathbf{v}$

7. $\frac{3}{5}\mathbf{u} + \frac{4}{5}\mathbf{v}$

8. $-\frac{5}{13}\mathbf{u} + \frac{12}{13}\mathbf{v}$

In Exercises 9–16, find the component form of the vector.

9. The vector $\overrightarrow{PQ}$, where $P = (1, 3)$ and $Q = (2, -1)$

10. The vector $\overrightarrow{OP}$ where O is the origin and P is the midpoint of segment RS, where $R = (2, -1)$ and $S = (-4, 3)$

11. The vector from the point $A = (2, 3)$ to the origin

12. The sum of $\overrightarrow{AB}$ and $\overrightarrow{CD}$, where $A = (1, -1)$, $B = (2, 0)$, $C = (-1, 3)$, and $D = (-2, 2)$

13. The unit vector that makes an angle $\theta = 2\pi/3$ with the positive x-axis

14. The unit vector that makes an angle $\theta = -3\pi/4$ with the positive x-axis

15. The unit vector obtained by rotating the vector $\langle 0, 1 \rangle$ 120° counterclockwise about the origin

16. The unit vector obtained by rotating the vector $\langle 1, 0 \rangle$ 135° counterclockwise about the origin

Vectors in Space

In Exercises 17–22, express each vector in the form $\mathbf{v} = v_1\mathbf{i} + v_2\mathbf{j} + v_3\mathbf{k}$.

17. $\overrightarrow{P_1P_2}$ if P_1 is the point $(5, 7, -1)$ and P_2 is the point $(2, 9, -2)$

18. $\overrightarrow{P_1P_2}$ if P_1 is the point $(1, 2, 0)$ and P_2 is the point $(-3, 0, 5)$

19. $\overrightarrow{AB}$ if A is the point $(-7, -8, 1)$ and B is the point $(-10, 8, 1)$

20. $\overrightarrow{AB}$ if A is the point $(1, 0, 3)$ and B is the point $(-1, 4, 5)$

21. $5\mathbf{u} - \mathbf{v}$ if $\mathbf{u} = \langle 1, 1, -1\rangle$ and $\mathbf{v} = \langle 2, 0, 3\rangle$

22. $-2\mathbf{u} + 3\mathbf{v}$ if $\mathbf{u} = \langle -1, 0, 2\rangle$ and $\mathbf{v} = \langle 1, 1, 1\rangle$

Geometry and Calculation

In Exercises 23 and 24, copy vectors **u**, **v**, and **w** head to tail as needed to sketch the indicated vector.

23.

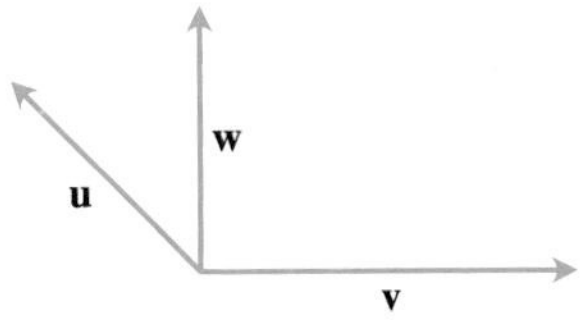

a. $\mathbf{u} + \mathbf{v}$ b. $\mathbf{u} + \mathbf{v} + \mathbf{w}$

c. $\mathbf{u} - \mathbf{v}$ d. $\mathbf{u} - \mathbf{w}$

24.

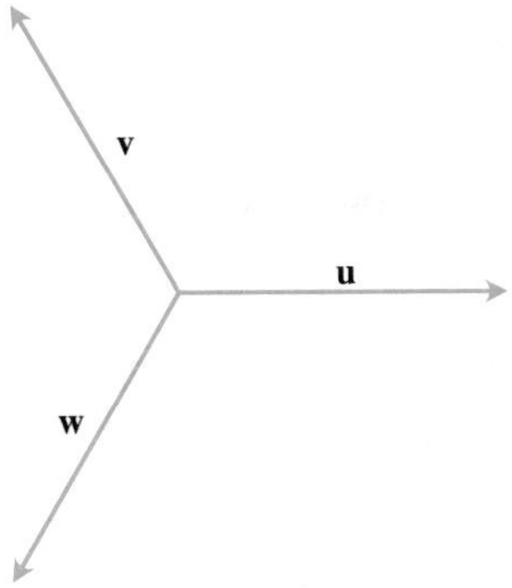

a. $\mathbf{u} - \mathbf{v}$ b. $\mathbf{u} - \mathbf{v} + \mathbf{w}$

c. $2\mathbf{u} - \mathbf{v}$ d. $\mathbf{u} + \mathbf{v} + \mathbf{w}$

Length and Direction

In Exercises 25–30, express each vector as a product of its length and direction.

25. $2\mathbf{i} + \mathbf{j} - 2\mathbf{k}$

26. $9\mathbf{i} - 2\mathbf{j} + 6\mathbf{k}$

27. $5\mathbf{k}$

28. $\frac{3}{5}\mathbf{i} + \frac{4}{5}\mathbf{k}$

29. $\frac{1}{\sqrt{6}}\mathbf{i} - \frac{1}{\sqrt{6}}\mathbf{j} - \frac{1}{\sqrt{6}}\mathbf{k}$

30. $\frac{\mathbf{i}}{\sqrt{3}} + \frac{\mathbf{j}}{\sqrt{3}} + \frac{\mathbf{k}}{\sqrt{3}}$

31. Find the vectors whose lengths and directions are given. Try to do the calculations without writing.

Length	Direction
a. 2	$\mathbf{i}$
b. $\sqrt{3}$	$-\mathbf{k}$
c. $\frac{1}{2}$	$\frac{3}{5}\mathbf{j} + \frac{4}{5}\mathbf{k}$
d. 7	$\frac{6}{7}\mathbf{i} - \frac{2}{7}\mathbf{j} + \frac{3}{7}\mathbf{k}$

32. Find the vectors whose lengths and directions are given. Try to do the calculations without writing.

Length	Direction
a. 7	$-\mathbf{j}$
b. $\sqrt{2}$	$-\frac{3}{5}\mathbf{i} - \frac{4}{5}\mathbf{k}$
c. $\frac{13}{12}$	$\frac{3}{13}\mathbf{i} - \frac{4}{13}\mathbf{j} - \frac{12}{13}\mathbf{k}$
d. $a > 0$	$\frac{1}{\sqrt{2}}\mathbf{i} + \frac{1}{\sqrt{3}}\mathbf{j} - \frac{1}{\sqrt{6}}\mathbf{k}$

33. Find a vector of magnitude 7 in the direction of $\mathbf{v} = 12\mathbf{i} - 5\mathbf{k}$.

34. Find a vector of magnitude 3 in the direction opposite to the direction of $\mathbf{v} = (1/2)\mathbf{i} - (1/2)\mathbf{j} - (1/2)\mathbf{k}$.

Vectors Determined by Points; Midpoints

In Exercises 35–38, find

a. the direction of $\overrightarrow{P_1P_2}$ and

b. the midpoint of line segment P_1P_2.

35. $P_1(-1, 1, 5)$ $P_2(2, 5, 0)$

36. $P_1(1, 4, 5)$ $P_2(4, -2, 7)$

37. $P_1(3, 4, 5)$ $P_2(2, 3, 4)$

38. $P_1(0, 0, 0)$ $P_2(2, -2, -2)$

39. If $\overrightarrow{AB} = \mathbf{i} + 4\mathbf{j} - 2\mathbf{k}$ and B is the point $(5, 1, 3)$, find A.

40. If $\overrightarrow{AB} = -7\mathbf{i} + 3\mathbf{j} + 8\mathbf{k}$ and A is the point $(-2, -3, 6)$, find B.

Theory and Applications

41. **Linear combination** Let $\mathbf{u} = 2\mathbf{i} + \mathbf{j}$, $\mathbf{v} = \mathbf{i} + \mathbf{j}$, and $\mathbf{w} = \mathbf{i} - \mathbf{j}$. Find scalars a and b such that $\mathbf{u} = a\mathbf{v} + b\mathbf{w}$.

42. **Linear combination** Let $\mathbf{u} = \mathbf{i} - 2\mathbf{j}$, $\mathbf{v} = 2\mathbf{i} + 3\mathbf{j}$, and $\mathbf{w} = \mathbf{i} + \mathbf{j}$. Write $\mathbf{u} = \mathbf{u}_1 + \mathbf{u}_2$, where $\mathbf{u}_1$ is parallel to $\mathbf{v}$ and $\mathbf{u}_2$ is parallel to $\mathbf{w}$. (See Exercise 41.)

43. **Force vector** You are pulling on a suitcase with a force **F** (pictured here) whose magnitude is $|\mathbf{F}| = 10$ lb. Find the **i**- and **j**-components of **F**.

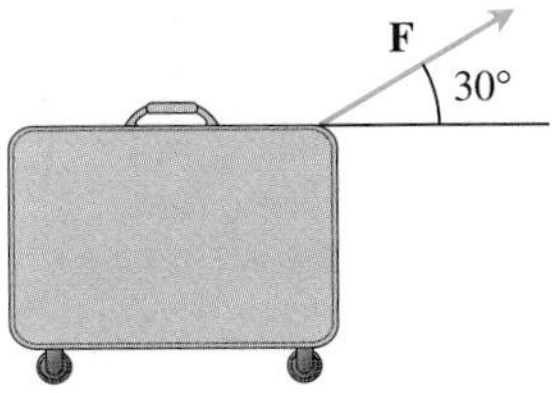

44. **Force vector** A kite string exerts a 12-lb pull ($|\mathbf{F}| = 12$) on a kite and makes a 45° angle with the horizontal. Find the horizontal and vertical components of **F**.

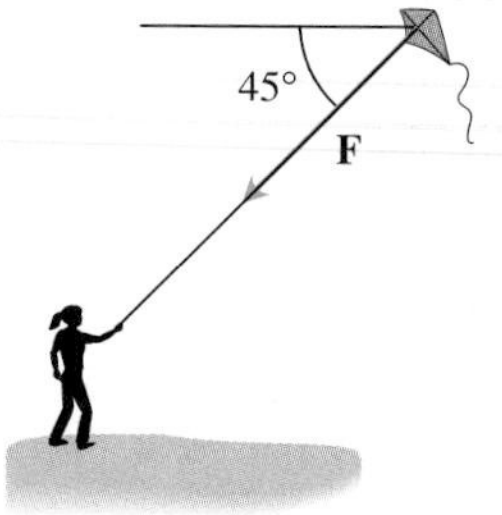

45. **Velocity** An airplane is flying in the direction 25° west of north at 800 km/h. Find the component form of the velocity of the airplane, assuming that the positive x-axis represents due east and the positive y-axis represents due north.

46. **Velocity** An airplane is flying in the direction 10° east of south at 600 km/h. Find the component form of the velocity of the airplane, assuming that the positive x-axis represents due east and the positive y-axis represents due north.

47. **Location** A bird flies from its nest 5 km in the direction 60° north of east, where it stops to rest on a tree. It then flies 10 km in the direction due southeast and lands atop a telephone pole. Place an xy-coordinate system so that the origin is the bird's nest, the x-axis points east, and the y-axis points north.
 a. At what point is the tree located?
 b. At what point is the telephone pole?

48. Use similar triangles to find the coordinates of the point Q that divides the segment from $P_1(x_1, y_1, z_1)$ to $P_2(x_2, y_2, z_2)$ into two lengths whose ratio is $p/q = r$.

49. **Medians of a triangle** Suppose that A, B, and C are the corner points of the thin triangular plate of constant density shown here.
 a. Find the vector from C to the midpoint M of side AB.
 b. Find the vector from C to the point that lies two-thirds of the way from C to M on the median CM.
 c. Find the coordinates of the point in which the medians of ΔABC intersect. According to Exercise 29, Section 6.4, this point is the plate's center of mass.

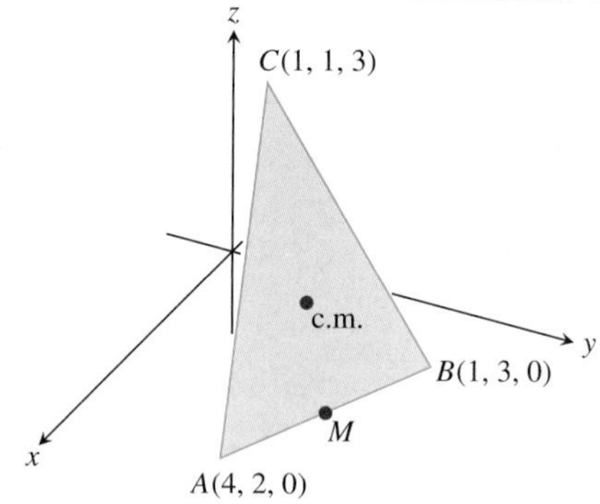

50. Find the vector from the origin to the point of intersection of the medians of the triangle whose vertices are

$$A(1, -1, 2), \quad B(2, 1, 3), \quad \text{and} \quad C(-1, 2, -1).$$

51. Let $ABCD$ be a general, not necessarily planar, quadrilateral in space. Show that the two segments joining the midpoints of opposite sides of $ABCD$ bisect each other. (*Hint:* Show that the segments have the same midpoint.)

52. Vectors are drawn from the center of a regular n-sided polygon in the plane to the vertices of the polygon. Show that the sum of the vectors is zero. (*Hint:* What happens to the sum if you rotate the polygon about its center?)

53. Suppose that A, B, and C are vertices of a triangle and that a, b, and c are, respectively, the midpoints of the opposite sides. Show that $\overrightarrow{Aa} + \overrightarrow{Bb} + \overrightarrow{Cc} = \mathbf{0}$.

54. **Unit vectors in the plane** Show that a unit vector in the plane can be expressed as $\mathbf{u} = (\cos\theta)\mathbf{i} + (\sin\theta)\mathbf{j}$, obtained by rotating $\mathbf{i}$ through an angle θ in the counterclockwise direction. Explain why this form gives *every* unit vector in the plane.

12.3 The Dot Product

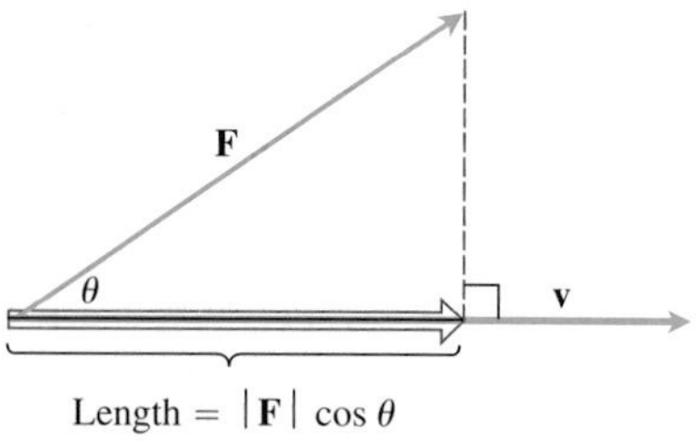

FIGURE 12.18 The magnitude of the force **F** in the direction of vector **v** is the length $|\mathbf{F}| \cos\theta$ of the projection of **F** onto **v**.

If a force **F** is applied to a particle moving along a path, we often need to know the magnitude of the force in the direction of motion. If **v** is parallel to the tangent line to the path at the point where **F** is applied, then we want the magnitude of **F** in the direction of **v**. Figure 12.18 shows that the scalar quantity we seek is the length $|\mathbf{F}| \cos\theta$, where θ is the angle between the two vectors **F** and **v**.

In this section, we show how to calculate easily the angle between two vectors directly from their components. A key part of the calculation is an expression called the *dot product*. Dot products are also called *inner* or *scalar* products because the product results in a scalar, not a vector. After investigating the dot product, we apply it to finding the projection of one vector onto another (as displayed in Figure 12.18) and to finding the work done by a constant force acting through a displacement.

Angle Between Vectors

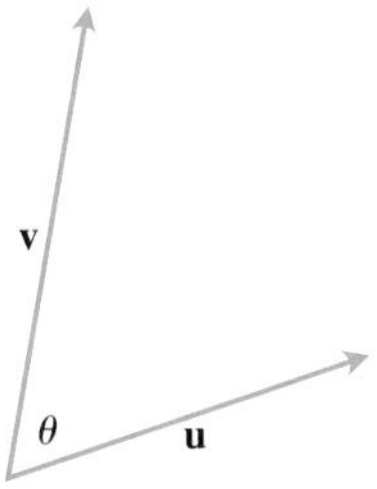

FIGURE 12.19 The angle between **u** and **v**.

When two nonzero vectors **u** and **v** are placed so their initial points coincide, they form an angle θ of measure $0 \leq \theta \leq \pi$ (Figure 12.19). If the vectors do not lie along the same line, the angle θ is measured in the plane containing both of them. If they do lie along the same line, the angle between them is 0 if they point in the same direction, and π if they point in opposite directions. The angle θ is the **angle between u** and **v**. Theorem 1 gives a formula to determine this angle.

THEOREM 1 Angle Between Two Vectors

The angle θ between two nonzero vectors $\mathbf{u} = \langle u_1, u_2, u_3 \rangle$ and $\mathbf{v} = \langle v_1, v_2, v_3 \rangle$ is given by

$$\theta = \cos^{-1}\left(\frac{u_1 v_1 + u_2 v_2 + u_3 v_3}{|\mathbf{u}||\mathbf{v}|}\right).$$

Before proving Theorem 1 (which is a consequence of the law of cosines), let's focus attention on the expression $u_1 v_1 + u_2 v_2 + u_3 v_3$ in the calculation for θ.

DEFINITION Dot Product

The **dot product u · v** ("**u** dot **v**") of vectors $\mathbf{u} = \langle u_1, u_2, u_3 \rangle$ and $\mathbf{v} = \langle v_1, v_2, v_3 \rangle$ is

$$\mathbf{u} \cdot \mathbf{v} = u_1 v_1 + u_2 v_2 + u_3 v_3.$$

EXAMPLE 1 Finding Dot Products

(a) $\langle 1, -2, -1 \rangle \cdot \langle -6, 2, -3 \rangle = (1)(-6) + (-2)(2) + (-1)(-3)$

$$= -6 - 4 + 3 = -7$$

(b) $\left(\frac{1}{2}\mathbf{i} + 3\mathbf{j} + \mathbf{k}\right) \cdot (4\mathbf{i} - \mathbf{j} + 2\mathbf{k}) = \left(\frac{1}{2}\right)(4) + (3)(-1) + (1)(2) = 1$ ■

The dot product of a pair of two-dimensional vectors is defined in a similar fashion:

$$\langle u_1, u_2 \rangle \cdot \langle v_1, v_2 \rangle = u_1 v_1 + u_2 v_2.$$

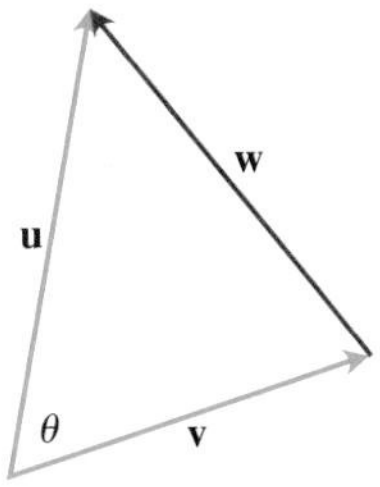

FIGURE 12.20 The parallelogram law of addition of vectors gives $\mathbf{w} = \mathbf{u} - \mathbf{v}$.

Proof of Theorem 1 Applying the law of cosines (Equation (6), Section 1.6) to the triangle in Figure 12.20, we find that

$$|\mathbf{w}|^2 = |\mathbf{u}|^2 + |\mathbf{v}|^2 - 2|\mathbf{u}||\mathbf{v}|\cos\theta \qquad \text{Law of cosines}$$

$$2|\mathbf{u}||\mathbf{v}|\cos\theta = |\mathbf{u}|^2 + |\mathbf{v}|^2 - |\mathbf{w}|^2.$$

EXAMPLE 6 Finding Vector Projections and Scalar Components

Find the vector projection of a force $\mathbf{F} = 5\mathbf{i} + 2\mathbf{j}$ onto $\mathbf{v} = \mathbf{i} - 3\mathbf{j}$ and the scalar component of $\mathbf{F}$ in the direction of $\mathbf{v}$.

Solution The vector projection is

$$\begin{aligned}\text{proj}_{\mathbf{v}}\,\mathbf{F} &= \left(\frac{\mathbf{F}\cdot\mathbf{v}}{|\mathbf{v}|^2}\right)\mathbf{v} \\ &= \frac{5-6}{1+9}(\mathbf{i}-3\mathbf{j}) = -\frac{1}{10}(\mathbf{i}-3\mathbf{j}) \\ &= -\frac{1}{10}\mathbf{i} + \frac{3}{10}\mathbf{j}.\end{aligned}$$

The scalar component of $\mathbf{F}$ in the direction of $\mathbf{v}$ is

$$|\mathbf{F}|\cos\theta = \frac{\mathbf{F}\cdot\mathbf{v}}{|\mathbf{v}|} = \frac{5-6}{\sqrt{1+9}} = -\frac{1}{\sqrt{10}}.$$ ■

Work

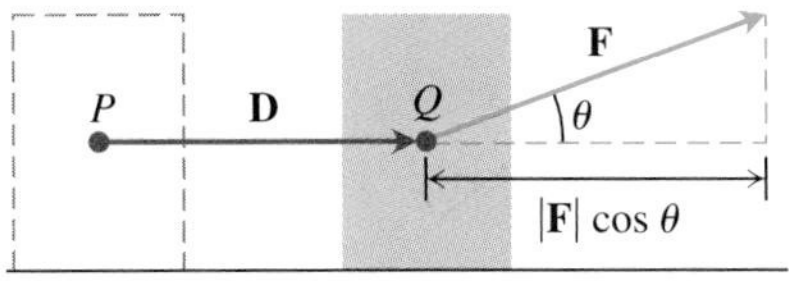

FIGURE 12.25 The work done by a constant force $\mathbf{F}$ during a displacement $\mathbf{D}$ is $(|\mathbf{F}|\cos\theta)|\mathbf{D}|$.

In Chapter 6, we calculated the work done by a constant force of magnitude F in moving an object through a distance d as $W = Fd$. That formula holds only if the force is directed along the line of motion. If a force $\mathbf{F}$ moving an object through a displacement $\mathbf{D} = \overrightarrow{PQ}$ has some other direction, the work is performed by the component of $\mathbf{F}$ in the direction of $\mathbf{D}$. If θ is the angle between $\mathbf{F}$ and $\mathbf{D}$ (Figure 12.25), then

$$\begin{aligned}\text{Work} &= \begin{pmatrix}\text{scalar component of } \mathbf{F} \\ \text{in the direction of } \mathbf{D}\end{pmatrix}(\text{length of } \mathbf{D}) \\ &= (|\mathbf{F}|\cos\theta)|\mathbf{D}| \\ &= \mathbf{F}\cdot\mathbf{D}.\end{aligned}$$

DEFINITION Work by Constant Force

The **work** done by a constant force $\mathbf{F}$ acting through a displacement $\mathbf{D} = \overrightarrow{PQ}$ is

$$W = \mathbf{F}\cdot\mathbf{D} = |\mathbf{F}||\mathbf{D}|\cos\theta,$$

where θ is the angle between $\mathbf{F}$ and $\mathbf{D}$.

EXAMPLE 7 Applying the Definition of Work

If $|\mathbf{F}| = 40$ N (newtons), $|\mathbf{D}| = 3$ m, and $\theta = 60°$, the work done by $\mathbf{F}$ in acting from P to Q is

$$\begin{aligned}\text{Work} &= |\mathbf{F}||\mathbf{D}|\cos\theta && \text{Definition} \\ &= (40)(3)\cos 60° && \text{Given values} \\ &= (120)(1/2) \\ &= 60\text{ J (joules)}.\end{aligned}$$ ■

We encounter more challenging work problems in Chapter 16 when we learn to find the work done by a variable force along a *path* in space.

Writing a Vector as a Sum of Orthogonal Vectors

We know one way to write a vector $\mathbf{u} = \langle u_1, u_2 \rangle$ or $\mathbf{u} = \langle u_1, u_2, u_3 \rangle$ as a sum of two orthogonal vectors:

$$\mathbf{u} = u_1\mathbf{i} + u_2\mathbf{j} \qquad \text{or} \qquad \mathbf{u} = u_1\mathbf{i} + (u_2\mathbf{j} + u_3\mathbf{k})$$

(since $\mathbf{i}\cdot\mathbf{j} = \mathbf{i}\cdot\mathbf{k} = \mathbf{j}\cdot\mathbf{k} = 0$).

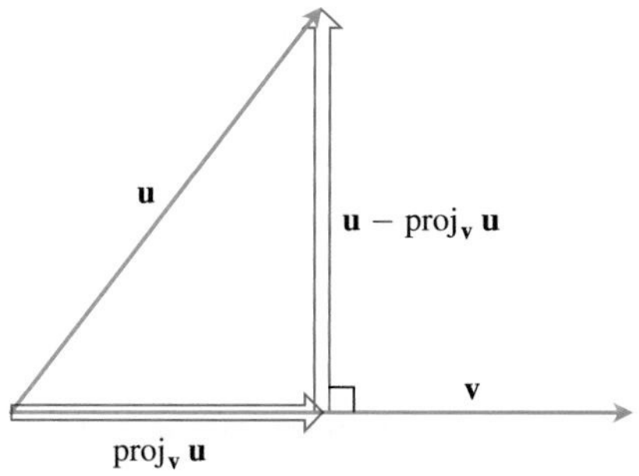

FIGURE 12.26 Writing **u** as the sum of vectors parallel and orthogonal to **v**.

Sometimes, however, it is more informative to express **u** as a different sum. In mechanics, for instance, we often need to write a vector **u** as a sum of a vector parallel to a given vector **v** and a vector orthogonal to **v**. As an example, in studying the motion of a particle moving along a path in the plane (or space), it is desirable to know the components of the acceleration vector in the direction of the tangent to the path (at a point) and of the normal to the path. (These *tangential* and *normal components* of acceleration are investigated in Section 13.4.) The acceleration vector can then be expressed as the sum of its (vector) tangential and normal components (which reflect important geometric properties about the nature of the path itself, such as *curvature*). Velocity and acceleration vectors are studied in the next chapter.

Generally, for vectors **u** and **v**, it is easy to see from Figure 12.26 that the vector

$$\mathbf{u} - \text{proj}_\mathbf{v}\,\mathbf{u}$$

is orthogonal to the projection vector $\text{proj}_\mathbf{v}\,\mathbf{u}$ (which has the same direction as **v**). The following calculation verifies this observation:

$$\begin{aligned}
(\mathbf{u} - \text{proj}_\mathbf{v}\,\mathbf{u})\cdot\text{proj}_\mathbf{v}\,\mathbf{u} &= \left(\mathbf{u} - \left(\frac{\mathbf{u}\cdot\mathbf{v}}{|\mathbf{v}|^2}\right)\mathbf{v}\right)\cdot\left(\frac{\mathbf{u}\cdot\mathbf{v}}{|\mathbf{v}|^2}\right)\mathbf{v} && \text{Equation (1)}\\
&= \left(\frac{\mathbf{u}\cdot\mathbf{v}}{|\mathbf{v}|^2}\right)(\mathbf{u}\cdot\mathbf{v}) - \left(\frac{\mathbf{u}\cdot\mathbf{v}}{|\mathbf{v}|^2}\right)^2(\mathbf{v}\cdot\mathbf{v}) && \text{Dot product properties 2 and 3}\\
&= \frac{(\mathbf{u}\cdot\mathbf{v})^2}{|\mathbf{v}|^2} - \frac{(\mathbf{u}\cdot\mathbf{v})^2}{|\mathbf{v}|^2} && \mathbf{v}\cdot\mathbf{v} = |\mathbf{v}|^2 \text{ cancels}\\
&= 0.
\end{aligned}$$

So the equation

$$\mathbf{u} = \text{proj}_\mathbf{v}\,\mathbf{u} + (\mathbf{u} - \text{proj}_\mathbf{v}\,\mathbf{u})$$

expresses **u** as a sum of orthogonal vectors.

How to Write u as a Vector Parallel to v Plus a Vector Orthogonal to v

$$\begin{aligned}
\mathbf{u} &= \text{proj}_\mathbf{v}\,\mathbf{u} + (\mathbf{u} - \text{proj}_\mathbf{v}\,\mathbf{u})\\
&= \underbrace{\left(\frac{\mathbf{u}\cdot\mathbf{v}}{|\mathbf{v}|^2}\right)\mathbf{v}}_{\text{Parallel to } \mathbf{v}} + \underbrace{\left(\mathbf{u} - \left(\frac{\mathbf{u}\cdot\mathbf{v}}{|\mathbf{v}|^2}\right)\mathbf{v}\right)}_{\text{Orthogonal to } \mathbf{v}}
\end{aligned}$$

EXAMPLE 8 Force on a Spacecraft

A force $\mathbf{F} = 2\mathbf{i} + \mathbf{j} - 3\mathbf{k}$ is applied to a spacecraft with velocity vector $\mathbf{v} = 3\mathbf{i} - \mathbf{j}$. Express $\mathbf{F}$ as a sum of a vector parallel to $\mathbf{v}$ and a vector orthogonal to $\mathbf{v}$.

Solution

$$\begin{aligned}
\mathbf{F} &= \text{proj}_{\mathbf{v}}\,\mathbf{F} + (\mathbf{F} - \text{proj}_{\mathbf{v}}\,\mathbf{F}) \\
&= \frac{\mathbf{F}\cdot\mathbf{v}}{\mathbf{v}\cdot\mathbf{v}}\mathbf{v} + \left(\mathbf{F} - \frac{\mathbf{F}\cdot\mathbf{v}}{\mathbf{v}\cdot\mathbf{v}}\mathbf{v}\right) \\
&= \left(\frac{6-1}{9+1}\right)\mathbf{v} + \left(\mathbf{F} - \left(\frac{6-1}{9+1}\right)\mathbf{v}\right) \\
&= \frac{5}{10}(3\mathbf{i} - \mathbf{j}) + \left(2\mathbf{i} + \mathbf{j} - 3\mathbf{k} - \frac{5}{10}(3\mathbf{i} - \mathbf{j})\right) \\
&= \left(\frac{3}{2}\mathbf{i} - \frac{1}{2}\mathbf{j}\right) + \left(\frac{1}{2}\mathbf{i} + \frac{3}{2}\mathbf{j} - 3\mathbf{k}\right).
\end{aligned}$$

The force $(3/2)\mathbf{i} - (1/2)\mathbf{j}$ is the effective force parallel to the velocity $\mathbf{v}$. The force $(1/2)\mathbf{i} + (3/2)\mathbf{j} - 3\mathbf{k}$ is orthogonal to $\mathbf{v}$. To check that this vector is orthogonal to $\mathbf{v}$, we find the dot product:

$$\left(\frac{1}{2}\mathbf{i} + \frac{3}{2}\mathbf{j} - 3\mathbf{k}\right)\cdot(3\mathbf{i} - \mathbf{j}) = \frac{3}{2} - \frac{3}{2} = 0.$$ ■

EXERCISES 12.3

Dot Product and Projections

In Exercises 1–8, find

a. $\mathbf{v}\cdot\mathbf{u}$, $|\mathbf{v}|$, $|\mathbf{u}|$

b. the cosine of the angle between $\mathbf{v}$ and $\mathbf{u}$

c. the scalar component of $\mathbf{u}$ in the direction of $\mathbf{v}$

d. the vector $\text{proj}_{\mathbf{v}}\,\mathbf{u}$.

1. $\mathbf{v} = 2\mathbf{i} - 4\mathbf{j} + \sqrt{5}\mathbf{k}$, $\mathbf{u} = -2\mathbf{i} + 4\mathbf{j} - \sqrt{5}\mathbf{k}$
2. $\mathbf{v} = (3/5)\mathbf{i} + (4/5)\mathbf{k}$, $\mathbf{u} = 5\mathbf{i} + 12\mathbf{j}$
3. $\mathbf{v} = 10\mathbf{i} + 11\mathbf{j} - 2\mathbf{k}$, $\mathbf{u} = 3\mathbf{j} + 4\mathbf{k}$
4. $\mathbf{v} = 2\mathbf{i} + 10\mathbf{j} - 11\mathbf{k}$, $\mathbf{u} = 2\mathbf{i} + 2\mathbf{j} + \mathbf{k}$
5. $\mathbf{v} = 5\mathbf{j} - 3\mathbf{k}$, $\mathbf{u} = \mathbf{i} + \mathbf{j} + \mathbf{k}$
6. $\mathbf{v} = -\mathbf{i} + \mathbf{j}$, $\mathbf{u} = \sqrt{2}\mathbf{i} + \sqrt{3}\mathbf{j} + 2\mathbf{k}$
7. $\mathbf{v} = 5\mathbf{i} + \mathbf{j}$, $\mathbf{u} = 2\mathbf{i} + \sqrt{17}\mathbf{j}$
8. $\mathbf{v} = \left\langle \frac{1}{\sqrt{2}}, \frac{1}{\sqrt{3}} \right\rangle$, $\mathbf{u} = \left\langle \frac{1}{\sqrt{2}}, -\frac{1}{\sqrt{3}} \right\rangle$

T Angles Between Vectors

Find the angles between the vectors in Exercises 9–12 to the nearest hundredth of a radian.

9. $\mathbf{u} = 2\mathbf{i} + \mathbf{j}$, $\mathbf{v} = \mathbf{i} + 2\mathbf{j} - \mathbf{k}$
10. $\mathbf{u} = 2\mathbf{i} - 2\mathbf{j} + \mathbf{k}$, $\mathbf{v} = 3\mathbf{i} + 4\mathbf{k}$
11. $\mathbf{u} = \sqrt{3}\mathbf{i} - 7\mathbf{j}$, $\mathbf{v} = \sqrt{3}\mathbf{i} + \mathbf{j} - 2\mathbf{k}$
12. $\mathbf{u} = \mathbf{i} + \sqrt{2}\mathbf{j} - \sqrt{2}\mathbf{k}$, $\mathbf{v} = -\mathbf{i} + \mathbf{j} + \mathbf{k}$
13. **Triangle** Find the measures of the angles of the triangle whose vertices are $A = (-1, 0)$, $B = (2, 1)$, and $C = (1, -2)$.
14. **Rectangle** Find the measures of the angles between the diagonals of the rectangle whose vertices are $A = (1, 0)$, $B = (0, 3)$, $C = (3, 4)$, and $D = (4, 1)$.
15. **Direction angles and direction cosines** The *direction angles* α, β, and γ of a vector $\mathbf{v} = a\mathbf{i} + b\mathbf{j} + c\mathbf{k}$ are defined as follows:

    α is the angle between $\mathbf{v}$ and the positive x-axis $(0 \le \alpha \le \pi)$

    β is the angle between $\mathbf{v}$ and the positive y-axis $(0 \le \beta \le \pi)$

    γ is the angle between $\mathbf{v}$ and the positive z-axis $(0 \le \gamma \le \pi)$.

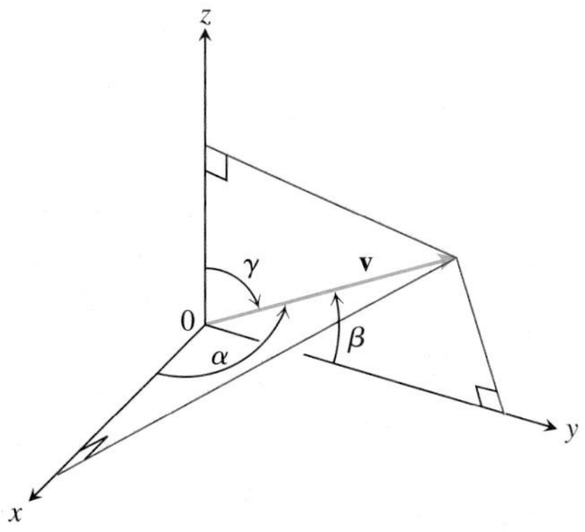

a. Show that

$$\cos\alpha = \frac{a}{|\mathbf{v}|}, \qquad \cos\beta = \frac{b}{|\mathbf{v}|}, \qquad \cos\gamma = \frac{c}{|\mathbf{v}|},$$

and $\cos^2\alpha + \cos^2\beta + \cos^2\gamma = 1$. These cosines are called the *direction cosines* of **v**.

b. **Unit vectors are built from direction cosines** Show that if $\mathbf{v} = a\mathbf{i} + b\mathbf{j} + c\mathbf{k}$ is a unit vector, then a, b, and c are the direction cosines of **v**.

16. **Water main construction** A water main is to be constructed with a 20% grade in the north direction and a 10% grade in the east direction. Determine the angle θ required in the water main for the turn from north to east.

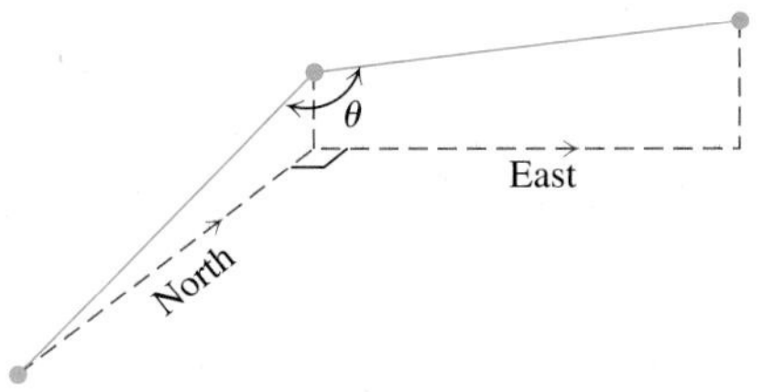

Decomposing Vectors

In Exercises 17–19, write **u** as the sum of a vector parallel to **v** and a vector orthogonal to **v**.

17. $\mathbf{u} = 3\mathbf{j} + 4\mathbf{k}, \quad \mathbf{v} = \mathbf{i} + \mathbf{j}$

18. $\mathbf{u} = \mathbf{j} + \mathbf{k}, \quad \mathbf{v} = \mathbf{i} + \mathbf{j}$

19. $\mathbf{u} = 8\mathbf{i} + 4\mathbf{j} - 12\mathbf{k}, \quad \mathbf{v} = \mathbf{i} + 2\mathbf{j} - \mathbf{k}$

20. **Sum of vectors** $\mathbf{u} = \mathbf{i} + (\mathbf{j} + \mathbf{k})$ is already the sum of a vector parallel to **i** and a vector orthogonal to **i**. If you use $\mathbf{v} = \mathbf{i}$, in the decomposition $\mathbf{u} = \text{proj}_{\mathbf{v}}\,\mathbf{u} + (\mathbf{u} - \text{proj}_{\mathbf{v}}\,\mathbf{u})$, do you get $\text{proj}_{\mathbf{v}}\,\mathbf{u} = \mathbf{i}$ and $(\mathbf{u} - \text{proj}_{\mathbf{v}}\,\mathbf{u}) = \mathbf{j} + \mathbf{k}$? Try it and find out.

Geometry and Examples

21. **Sums and differences** In the accompanying figure, it looks as if $\mathbf{v}_1 + \mathbf{v}_2$ and $\mathbf{v}_1 - \mathbf{v}_2$ are orthogonal. Is this mere coincidence, or are there circumstances under which we may expect the sum of two vectors to be orthogonal to their difference? Give reasons for your answer.

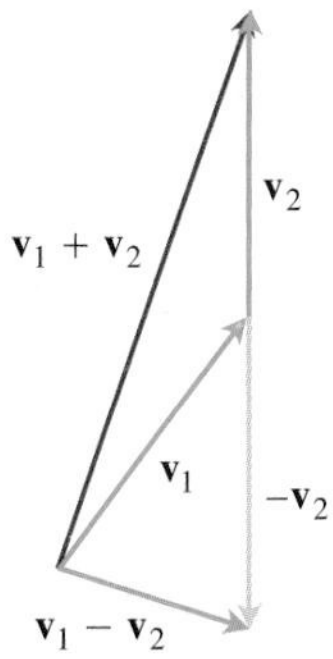

22. **Orthogonality on a circle** Suppose that AB is the diameter of a circle with center O and that C is a point on one of the two arcs joining A and B. Show that $\overrightarrow{CA}$ and $\overrightarrow{CB}$ are orthogonal.

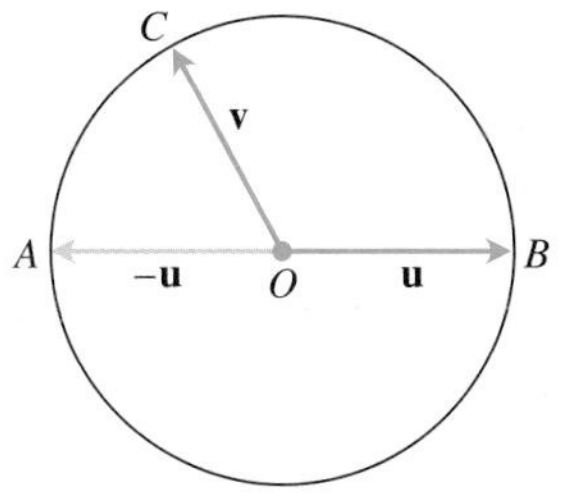

23. **Diagonals of a rhombus** Show that the diagonals of a rhombus (parallelogram with sides of equal length) are perpendicular.

24. **Perpendicular diagonals** Show that squares are the only rectangles with perpendicular diagonals.

25. **When parallelograms are rectangles** Prove that a parallelogram is a rectangle if and only if its diagonals are equal in length. (This fact is often exploited by carpenters.)

26. **Diagonal of parallelogram** Show that the indicated diagonal of the parallelogram determined by vectors **u** and **v** bisects the angle between **u** and **v** if $|\mathbf{u}| = |\mathbf{v}|$.

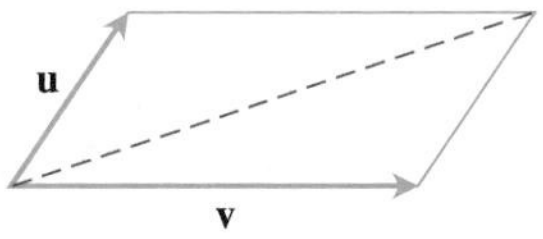

27. **Projectile motion** A gun with muzzle velocity of 1200 ft/sec is fired at an angle of 8° above the horizontal. Find the horizontal and vertical components of the velocity.

28. **Inclined plane** Suppose that a box is being towed up an inclined plane as shown in the figure. Find the force **w** needed to make the component of the force parallel to the inclined plane equal to 2.5 lb.

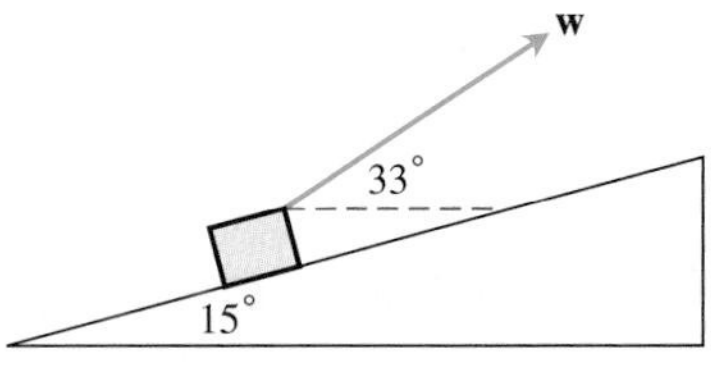

Theory and Examples

29. a. Cauchy-Schwartz inequality Use the fact that $\mathbf{u}\cdot\mathbf{v} = |\mathbf{u}||\mathbf{v}|\cos\theta$ to show that the inequality $|\mathbf{u}\cdot\mathbf{v}| \le |\mathbf{u}||\mathbf{v}|$ holds for any vectors **u** and **v**.

b. Under what circumstances, if any, does $|\mathbf{u}\cdot\mathbf{v}|$ equal $|\mathbf{u}||\mathbf{v}|$? Give reasons for your answer.

30. Copy the axes and vector shown here. Then shade in the points (x, y) for which $(x\mathbf{i} + y\mathbf{j})\cdot\mathbf{v} \le 0$. Justify your answer.

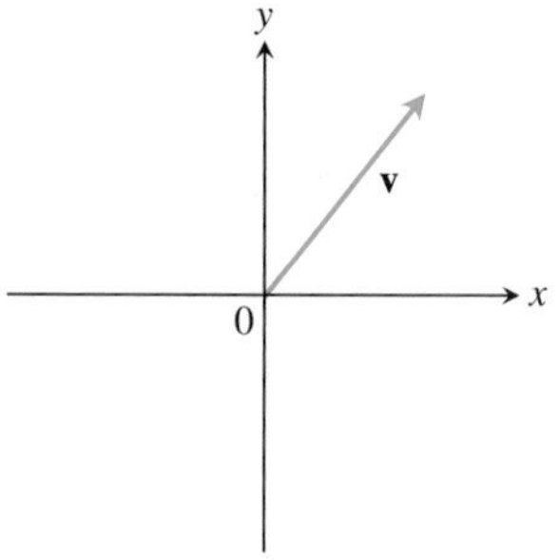

31. Orthogonal unit vectors If $\mathbf{u}_1$ and $\mathbf{u}_2$ are orthogonal unit vectors and $\mathbf{v} = a\mathbf{u}_1 + b\mathbf{u}_2$, find $\mathbf{v}\cdot\mathbf{u}_1$.

32. Cancellation in dot products In real-number multiplication, if $uv_1 = uv_2$ and $u \ne 0$, we can cancel the u and conclude that $v_1 = v_2$. Does the same rule hold for the dot product: If $\mathbf{u}\cdot\mathbf{v}_1 = \mathbf{u}\cdot\mathbf{v}_2$ and $\mathbf{u} \ne \mathbf{0}$, can you conclude that $\mathbf{v}_1 = \mathbf{v}_2$? Give reasons for your answer.

Equations for Lines in the Plane

33. Line perpendicular to a vector Show that the vector $\mathbf{v} = a\mathbf{i} + b\mathbf{j}$ is perpendicular to the line $ax + by = c$ by establishing that the slope of **v** is the negative reciprocal of the slope of the given line.

34. Line parallel to a vector Show that the vector $\mathbf{v} = a\mathbf{i} + b\mathbf{j}$ is parallel to the line $bx - ay = c$ by establishing that the slope of the line segment representing **v** is the same as the slope of the given line.

In Exercises 35–38, use the result of Exercise 33 to find an equation for the line through P perpendicular to **v**. Then sketch the line. Include **v** in your sketch *as a vector starting at the origin.*

35. $P(2, 1),\quad \mathbf{v} = \mathbf{i} + 2\mathbf{j}$

36. $P(-1, 2),\quad \mathbf{v} = -2\mathbf{i} - \mathbf{j}$

37. $P(-2, -7),\quad \mathbf{v} = -2\mathbf{i} + \mathbf{j}$

38. $P(11, 10),\quad \mathbf{v} = 2\mathbf{i} - 3\mathbf{j}$

In Exercises 39–42, use the result of Exercise 34 to find an equation for the line through P parallel to **v**. Then sketch the line. Include **v** in your sketch *as a vector starting at the origin.*

39. $P(-2, 1),\quad \mathbf{v} = \mathbf{i} - \mathbf{j}$

40. $P(0, -2),\quad \mathbf{v} = 2\mathbf{i} + 3\mathbf{j}$

41. $P(1, 2),\quad \mathbf{v} = -\mathbf{i} - 2\mathbf{j}$

42. $P(1, 3),\quad \mathbf{v} = 3\mathbf{i} - 2\mathbf{j}$

Work

43. Work along a line Find the work done by a force $\mathbf{F} = 5\mathbf{i}$ (magnitude 5 N) in moving an object along the line from the origin to the point $(1, 1)$ (distance in meters).

44. Locomotive The union Pacific's *Big Boy* locomotive could pull 6000-ton trains with a tractive effort (pull) of 602,148 N (135,375 lb). At this level of effort, about how much work did *Big Boy* do on the (approximately straight) 605-km journey from San Francisco to Los Angeles?

45. Inclined plane How much work does it take to slide a crate 20 m along a loading dock by pulling on it with a 200 N force at an angle of 30° from the horizontal?

46. Sailboat The wind passing over a boat's sail exerted a 1000-lb magnitude force **F** as shown here. How much work did the wind perform in moving the boat forward 1 mi? Answer in foot-pounds.

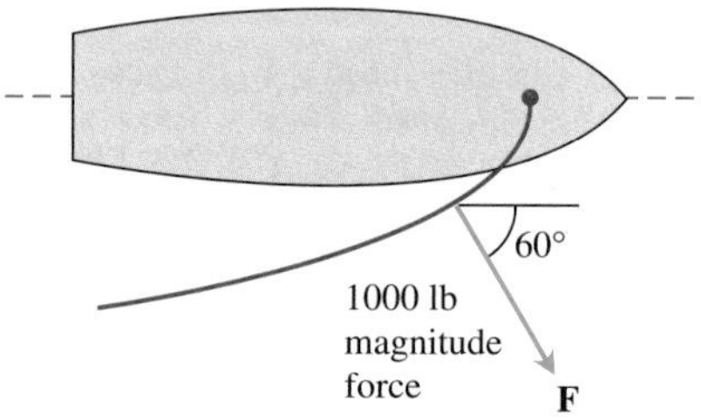

Angles Between Lines in the Plane

The acute angle between intersecting lines that do not cross at right angles is the same as the angle determined by vectors normal to the lines or by the vectors parallel to the lines.

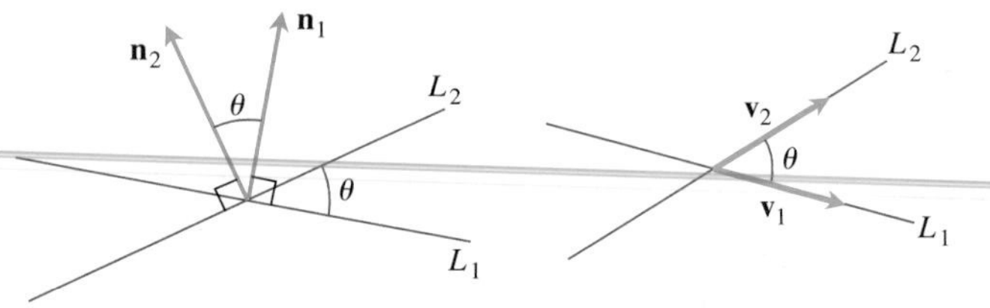

Use this fact and the results of Exercise 33 or 34 to find the acute angles between the lines in Exercises 47–52.

47. $3x + y = 5,\quad 2x - y = 4$

48. $y = \sqrt{3}x - 1,\quad y = -\sqrt{3}x + 2$

49. $\sqrt{3}x - y = -2,\quad x - \sqrt{3}y = 1$

50. $x + \sqrt{3}y = 1,\quad \left(1 - \sqrt{3}\right)x + \left(1 + \sqrt{3}\right)y = 8$

51. $3x - 4y = 3,\quad x - y = 7$

52. $12x + 5y = 1,\quad 2x - 2y = 3$

Angles Between Differentiable Curves

The angles between two differentiable curves at a point of intersection are the angles between the curves' tangent lines at these points. Find the angles between the curves in Exercises 53–56. Note that if $\mathbf{v} = a\mathbf{i} + b\mathbf{j}$ is a vector in the plane, then the vector has slope b/a provided $a \neq 0$.

53. $y = (3/2) - x^2, \quad y = x^2$ (two points of intersection)

54. $x = (3/4) - y^2, \quad x = y^2 - (3/4)$ (two points of intersection)

55. $y = x^3, \quad x = y^2$ (two points of intersection)

56. $y = -x^2, \quad y = \sqrt{x}$ (two points of intersection)

12.4 The Cross Product

In studying lines in the plane, when we needed to describe how a line was tilting, we used the notions of slope and angle of inclination. In space, we want a way to describe how a *plane* is tilting. We accomplish this by multiplying two vectors in the plane together to get a third vector perpendicular to the plane. The direction of this third vector tells us the "inclination" of the plane. The product we use to multiply the vectors together is the *vector* or *cross product*, the second of the two vector multiplication methods we study in calculus.

Cross products are widely used to describe the effects of forces in studies of electricity, magnetism, fluid flows, and orbital mechanics. This section presents the mathematical properties that account for the use of cross products in these fields.

The Cross Product of Two Vectors in Space

We start with two nonzero vectors **u** and **v** in space. If **u** and **v** are not parallel, they determine a plane. We select a unit vector **n** perpendicular to the plane by the **right-hand rule**. This means that we choose **n** to be the unit (normal) vector that points the way your right thumb points when your fingers curl through the angle θ from **u** to **v** (Figure 12.27). Then the **cross product u × v** ("**u** cross **v**") is the *vector* defined as follows.

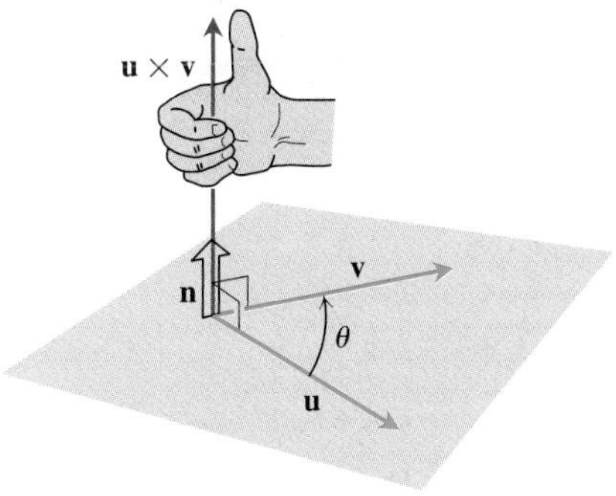

FIGURE 12.27 The construction of $\mathbf{u} \times \mathbf{v}$.

DEFINITION Cross Product

$$\mathbf{u} \times \mathbf{v} = (|\mathbf{u}||\mathbf{v}| \sin\theta)\,\mathbf{n}$$

Unlike the dot product, the cross product is a vector. For this reason it's also called the **vector product** of **u** and **v**, and applies *only* to vectors in space. The vector $\mathbf{u} \times \mathbf{v}$ is orthogonal to both **u** and **v** because it is a scalar multiple of **n**.

Since the sines of 0 and π are both zero, it makes sense to define the cross product of two parallel nonzero vectors to be **0**. If one or both of **u** and **v** are zero, we also define $\mathbf{u} \times \mathbf{v}$ to be zero. This way, the cross product of two vectors **u** and **v** is zero if and only if **u** and **v** are parallel or one or both of them are zero.

Parallel Vectors

Nonzero vectors **u** and **v** are parallel if and only if $\mathbf{u} \times \mathbf{v} = \mathbf{0}$.

The cross product obeys the following laws.

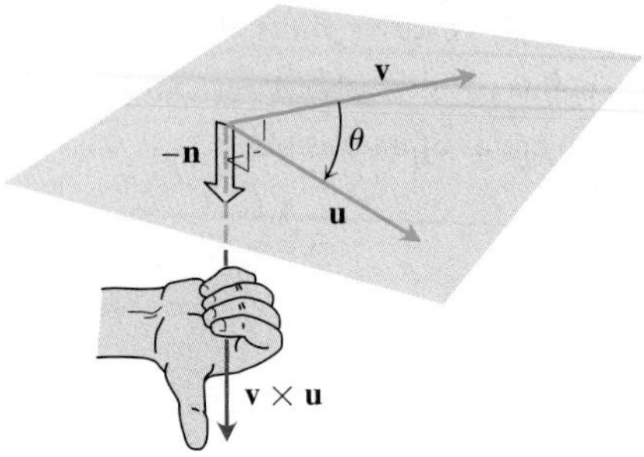

FIGURE 12.28 The construction of $\mathbf{v} \times \mathbf{u}$.

> **Properties of the Cross Product**
>
> If $\mathbf{u}$, $\mathbf{v}$, and $\mathbf{w}$ are any vectors and r, s are scalars, then
>
> 1. $(r\mathbf{u}) \times (s\mathbf{v}) = (rs)(\mathbf{u} \times \mathbf{v})$
> 2. $\mathbf{u} \times (\mathbf{v} + \mathbf{w}) = \mathbf{u} \times \mathbf{v} + \mathbf{u} \times \mathbf{w}$
> 3. $(\mathbf{v} + \mathbf{w}) \times \mathbf{u} = \mathbf{v} \times \mathbf{u} + \mathbf{w} \times \mathbf{u}$
> 4. $\mathbf{v} \times \mathbf{u} = -(\mathbf{u} \times \mathbf{v})$
> 5. $\mathbf{0} \times \mathbf{u} = \mathbf{0}$

To visualize Property 4, for example, notice that when the fingers of a right hand curl through the angle θ from $\mathbf{v}$ to $\mathbf{u}$, the thumb points the opposite way and the unit vector we choose in forming $\mathbf{v} \times \mathbf{u}$ is the negative of the one we choose in forming $\mathbf{u} \times \mathbf{v}$ (Figure 12.28).

Property 1 can be verified by applying the definition of cross product to both sides of the equation and comparing the results. Property 2 is proved in Appendix 6. Property 3 follows by multiplying both sides of the equation in Property 2 by -1 and reversing the order of the products using Property 4. Property 5 is a definition. As a rule, cross product multiplication is *not associative* so $(\mathbf{u} \times \mathbf{v}) \times \mathbf{w}$ does not generally equal $\mathbf{u} \times (\mathbf{v} \times \mathbf{w})$. (See Additional Exercise 15.)

When we apply the definition to calculate the pairwise cross products of $\mathbf{i}$, $\mathbf{j}$, and $\mathbf{k}$, we find (Figure 12.29)

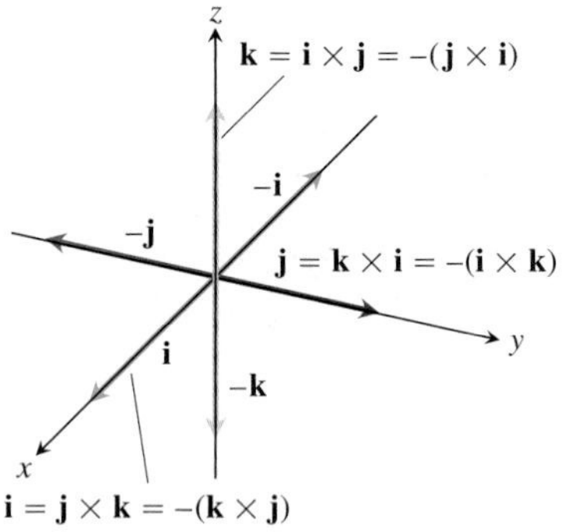

FIGURE 12.29 The pairwise cross products of $\mathbf{i}$, $\mathbf{j}$, and $\mathbf{k}$.

$$\mathbf{i} \times \mathbf{j} = -(\mathbf{j} \times \mathbf{i}) = \mathbf{k}$$

$$\mathbf{j} \times \mathbf{k} = -(\mathbf{k} \times \mathbf{j}) = \mathbf{i}$$

$$\mathbf{k} \times \mathbf{i} = -(\mathbf{i} \times \mathbf{k}) = \mathbf{j}$$

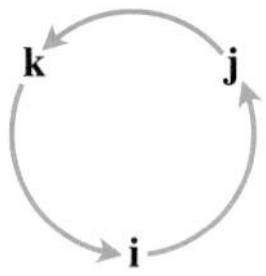

Diagram for recalling these products

and

$$\mathbf{i} \times \mathbf{i} = \mathbf{j} \times \mathbf{j} = \mathbf{k} \times \mathbf{k} = \mathbf{0}.$$

$|\mathbf{u} \times \mathbf{v}|$ Is the Area of a Parallelogram

Because $\mathbf{n}$ is a unit vector, the magnitude of $\mathbf{u} \times \mathbf{v}$ is

$$|\mathbf{u} \times \mathbf{v}| = |\mathbf{u}||\mathbf{v}|\,|\sin\theta|\,|\mathbf{n}| = |\mathbf{u}||\mathbf{v}|\sin\theta.$$

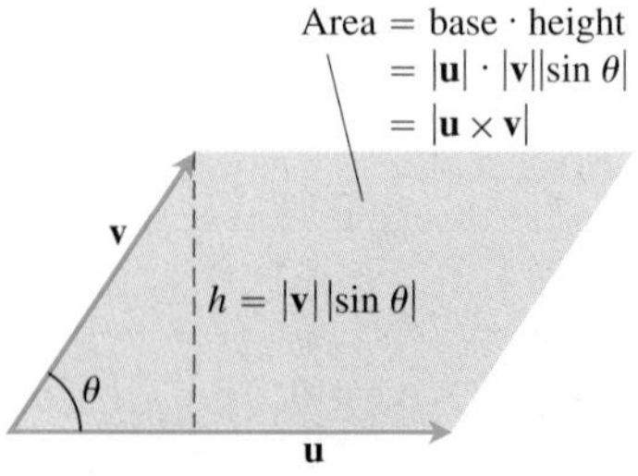

FIGURE 12.30 The parallelogram determined by $\mathbf{u}$ and $\mathbf{v}$.

This is the area of the parallelogram determined by $\mathbf{u}$ and $\mathbf{v}$ (Figure 12.30), $|\mathbf{u}|$ being the base of the parallelogram and $|\mathbf{v}||\sin\theta|$ the height.

Determinant Formula for $\mathbf{u} \times \mathbf{v}$

Our next objective is to calculate $\mathbf{u} \times \mathbf{v}$ from the components of $\mathbf{u}$ and $\mathbf{v}$ relative to a Cartesian coordinate system.

Determinants
2×2 and 3×3 determinants are evaluated as follows:

$$\begin{vmatrix} a & b \\ c & d \end{vmatrix} = ad - bc$$

EXAMPLE

$$\begin{vmatrix} 2 & 1 \\ -4 & 3 \end{vmatrix} = (2)(3) - (1)(-4) = 6 + 4 = 10$$

$$\begin{vmatrix} a_1 & a_2 & a_3 \\ b_1 & b_2 & b_3 \\ c_1 & c_2 & c_3 \end{vmatrix} = a_1\begin{vmatrix} b_2 & b_3 \\ c_2 & c_3 \end{vmatrix} - a_2\begin{vmatrix} b_1 & b_3 \\ c_1 & c_3 \end{vmatrix} + a_3\begin{vmatrix} b_1 & b_2 \\ c_1 & c_2 \end{vmatrix}$$

EXAMPLE

$$\begin{vmatrix} -5 & 3 & 1 \\ 2 & 1 & 1 \\ -4 & 3 & 1 \end{vmatrix} = (-5)\begin{vmatrix} 1 & 1 \\ 3 & 1 \end{vmatrix} - (3)\begin{vmatrix} 2 & 1 \\ -4 & 1 \end{vmatrix} + (1)\begin{vmatrix} 2 & 1 \\ -4 & 3 \end{vmatrix}$$
$$= -5(1 - 3) - 3(2 + 4) + 1(6 + 4)$$
$$= 10 - 18 + 10 = 2$$

(For more information, see the Web site at **www.aw-bc.com/thomas.**)

Suppose that

$$\mathbf{u} = u_1\mathbf{i} + u_2\mathbf{j} + u_3\mathbf{k}, \qquad \mathbf{v} = v_1\mathbf{i} + v_2\mathbf{j} + v_3\mathbf{k}.$$

Then the distributive laws and the rules for multiplying **i**, **j**, and **k** tell us that

$$\begin{aligned} \mathbf{u} \times \mathbf{v} &= (u_1\mathbf{i} + u_2\mathbf{j} + u_3\mathbf{k}) \times (v_1\mathbf{i} + v_2\mathbf{j} + v_3\mathbf{k}) \\ &= u_1v_1\mathbf{i} \times \mathbf{i} + u_1v_2\mathbf{i} \times \mathbf{j} + u_1v_3\mathbf{i} \times \mathbf{k} \\ &\quad + u_2v_1\mathbf{j} \times \mathbf{i} + u_2v_2\mathbf{j} \times \mathbf{j} + u_2v_3\mathbf{j} \times \mathbf{k} \\ &\quad + u_3v_1\mathbf{k} \times \mathbf{i} + u_3v_2\mathbf{k} \times \mathbf{j} + u_3v_3\mathbf{k} \times \mathbf{k} \\ &= (u_2v_3 - u_3v_2)\mathbf{i} - (u_1v_3 - u_3v_1)\mathbf{j} + (u_1v_2 - u_2v_1)\mathbf{k}. \end{aligned}$$

The terms in the last line are the same as the terms in the expansion of the symbolic determinant

$$\begin{vmatrix} \mathbf{i} & \mathbf{j} & \mathbf{k} \\ u_1 & u_2 & u_3 \\ v_1 & v_2 & v_3 \end{vmatrix}.$$

We therefore have the following rule.

> **Calculating Cross Products Using Determinants**
> If $\mathbf{u} = u_1\mathbf{i} + u_2\mathbf{j} + u_3\mathbf{k}$ and $\mathbf{v} = v_1\mathbf{i} + v_2\mathbf{j} + v_3\mathbf{k}$, then
> $$\mathbf{u} \times \mathbf{v} = \begin{vmatrix} \mathbf{i} & \mathbf{j} & \mathbf{k} \\ u_1 & u_2 & u_3 \\ v_1 & v_2 & v_3 \end{vmatrix}.$$

EXAMPLE 1 Calculating Cross Products with Determinants

Find $\mathbf{u} \times \mathbf{v}$ and $\mathbf{v} \times \mathbf{u}$ if $\mathbf{u} = 2\mathbf{i} + \mathbf{j} + \mathbf{k}$ and $\mathbf{v} = -4\mathbf{i} + 3\mathbf{j} + \mathbf{k}$.

Solution

$$\mathbf{u} \times \mathbf{v} = \begin{vmatrix} \mathbf{i} & \mathbf{j} & \mathbf{k} \\ 2 & 1 & 1 \\ -4 & 3 & 1 \end{vmatrix} = \begin{vmatrix} 1 & 1 \\ 3 & 1 \end{vmatrix}\mathbf{i} - \begin{vmatrix} 2 & 1 \\ -4 & 1 \end{vmatrix}\mathbf{j} + \begin{vmatrix} 2 & 1 \\ -4 & 3 \end{vmatrix}\mathbf{k}$$
$$= -2\mathbf{i} - 6\mathbf{j} + 10\mathbf{k}$$
$$\mathbf{v} \times \mathbf{u} = -(\mathbf{u} \times \mathbf{v}) = 2\mathbf{i} + 6\mathbf{j} - 10\mathbf{k}$$ ■

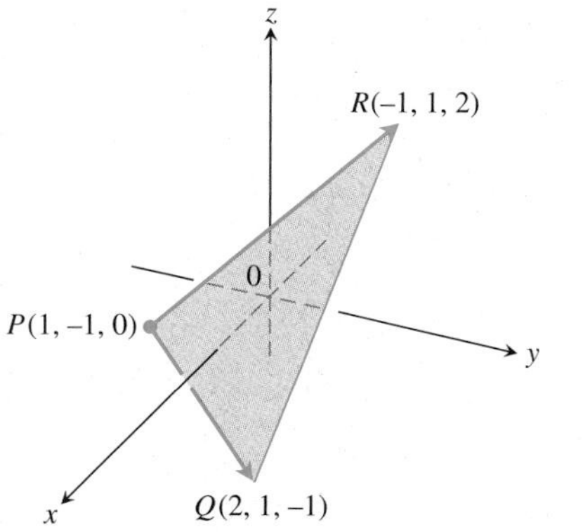

FIGURE 12.31 The area of triangle PQR is half of $|\overrightarrow{PQ} \times \overrightarrow{PR}|$ (Example 2).

EXAMPLE 2 Finding Vectors Perpendicular to a Plane

Find a vector perpendicular to the plane of $P(1, -1, 0)$, $Q(2, 1, -1)$, and $R(-1, 1, 2)$ (Figure 12.31).

Solution The vector $\overrightarrow{PQ} \times \overrightarrow{PR}$ is perpendicular to the plane because it is perpendicular to both vectors. In terms of components,

$$\overrightarrow{PQ} = (2 - 1)\mathbf{i} + (1 + 1)\mathbf{j} + (-1 - 0)\mathbf{k} = \mathbf{i} + 2\mathbf{j} - \mathbf{k}$$

$$\overrightarrow{PR} = (-1 - 1)\mathbf{i} + (1 + 1)\mathbf{j} + (2 - 0)\mathbf{k} = -2\mathbf{i} + 2\mathbf{j} + 2\mathbf{k}$$

$$\overrightarrow{PQ} \times \overrightarrow{PR} = \begin{vmatrix} \mathbf{i} & \mathbf{j} & \mathbf{k} \\ 1 & 2 & -1 \\ -2 & 2 & 2 \end{vmatrix} = \begin{vmatrix} 2 & -1 \\ 2 & 2 \end{vmatrix}\mathbf{i} - \begin{vmatrix} 1 & -1 \\ -2 & 2 \end{vmatrix}\mathbf{j} + \begin{vmatrix} 1 & 2 \\ -2 & 2 \end{vmatrix}\mathbf{k}$$

$$= 6\mathbf{i} + 6\mathbf{k}.$$

EXAMPLE 3 Finding the Area of a Triangle

Find the area of the triangle with vertices $P(1, -1, 0)$, $Q(2, 1, -1)$, and $R(-1, 1, 2)$ (Figure 12.31).

Solution The area of the parallelogram determined by P, Q, and R is

$$|\overrightarrow{PQ} \times \overrightarrow{PR}| = |6\mathbf{i} + 6\mathbf{k}| \qquad \text{Values from Example 2.}$$

$$= \sqrt{(6)^2 + (6)^2} = \sqrt{2 \cdot 36} = 6\sqrt{2}.$$

The triangle's area is half of this, or $3\sqrt{2}$.

EXAMPLE 4 Finding a Unit Normal to a Plane

Find a unit vector perpendicular to the plane of $P(1, -1, 0)$, $Q(2, 1, -1)$, and $R(-1, 1, 2)$.

Solution Since $\overrightarrow{PQ} \times \overrightarrow{PR}$ is perpendicular to the plane, its direction $\mathbf{n}$ is a unit vector perpendicular to the plane. Taking values from Examples 2 and 3, we have

$$\mathbf{n} = \frac{\overrightarrow{PQ} \times \overrightarrow{PR}}{|\overrightarrow{PQ} \times \overrightarrow{PR}|} = \frac{6\mathbf{i} + 6\mathbf{k}}{6\sqrt{2}} = \frac{1}{\sqrt{2}}\mathbf{i} + \frac{1}{\sqrt{2}}\mathbf{k}.$$

For ease in calculating the cross product using determinants, we usually write vectors in the form $\mathbf{v} = v_1\mathbf{i} + v_2\mathbf{j} + v_3\mathbf{k}$ rather than as ordered triples $\mathbf{v} = \langle v_1, v_2, v_3 \rangle$.

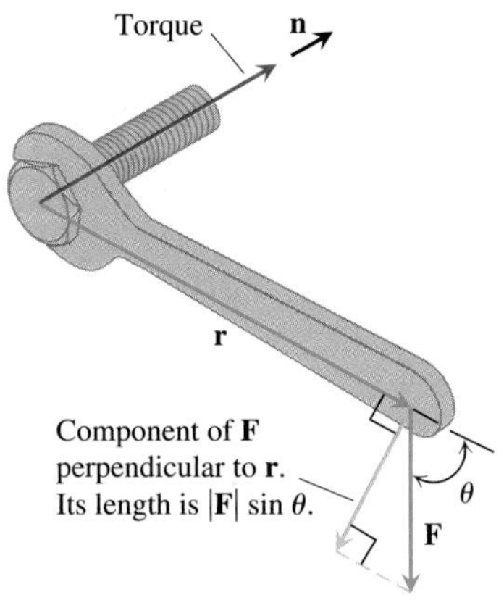

FIGURE 12.32 The torque vector describes the tendency of the force $\mathbf{F}$ to drive the bolt forward.

Torque

When we turn a bolt by applying a force $\mathbf{F}$ to a wrench (Figure 12.32), the torque we produce acts along the axis of the bolt to drive the bolt forward. The magnitude of the torque depends on how far out on the wrench the force is applied and on how much of the force is perpendicular to the wrench at the point of application. The number we use to measure the torque's magnitude is the product of the length of the lever arm $\mathbf{r}$ and the scalar component of $\mathbf{F}$ perpendicular to $\mathbf{r}$. In the notation of Figure 12.32,

$$\text{Magnitude of torque vector} = |\mathbf{r}||\mathbf{F}| \sin\theta,$$

or $|\mathbf{r} \times \mathbf{F}|$. If we let $\mathbf{n}$ be a unit vector along the axis of the bolt in the direction of the torque, then a complete description of the torque vector is $\mathbf{r} \times \mathbf{F}$, or

$$\text{Torque vector} = (|\mathbf{r}||\mathbf{F}| \sin\theta)\,\mathbf{n}.$$

Recall that we defined $\mathbf{u} \times \mathbf{v}$ to be $\mathbf{0}$ when $\mathbf{u}$ and $\mathbf{v}$ are parallel. This is consistent with the torque interpretation as well. If the force $\mathbf{F}$ in Figure 12.32 is parallel to the wrench, meaning that we are trying to turn the bolt by pushing or pulling along the line of the wrench's handle, the torque produced is zero.

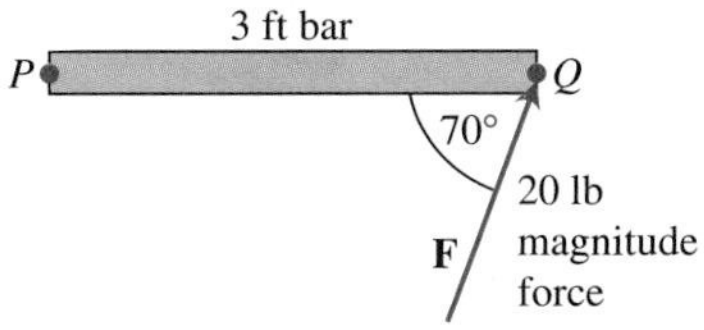

FIGURE 12.33 The magnitude of the torque exerted by **F** at *P* is about 56.4 ft-lb (Example 5).

EXAMPLE 5 Finding the Magnitude of a Torque

The magnitude of the torque generated by force $\mathbf{F}$ at the pivot point P in Figure 12.33 is

$$\begin{aligned} |\overrightarrow{PQ} \times \mathbf{F}| &= |\overrightarrow{PQ}||\mathbf{F}| \sin 70^\circ \\ &\approx (3)(20)(0.94) \\ &\approx 56.4 \text{ ft-lb}. \end{aligned}$$

■

Triple Scalar or Box Product

The product $(\mathbf{u} \times \mathbf{v}) \cdot \mathbf{w}$ is called the **triple scalar product** of $\mathbf{u}$, $\mathbf{v}$, and $\mathbf{w}$ (in that order). As you can see from the formula

$$|(\mathbf{u} \times \mathbf{v}) \cdot \mathbf{w}| = |\mathbf{u} \times \mathbf{v}||\mathbf{w}||\cos\theta|,$$

the absolute value of the product is the volume of the parallelepiped (parallelogram-sided box) determined by $\mathbf{u}$, $\mathbf{v}$, and $\mathbf{w}$ (Figure 12.34). The number $|\mathbf{u} \times \mathbf{v}|$ is the area of the base parallelogram. The number $|\mathbf{w}||\cos\theta|$ is the parallelepiped's height. Because of this geometry, $(\mathbf{u} \times \mathbf{v}) \cdot \mathbf{w}$ is also called the **box product** of $\mathbf{u}$, $\mathbf{v}$, and $\mathbf{w}$.

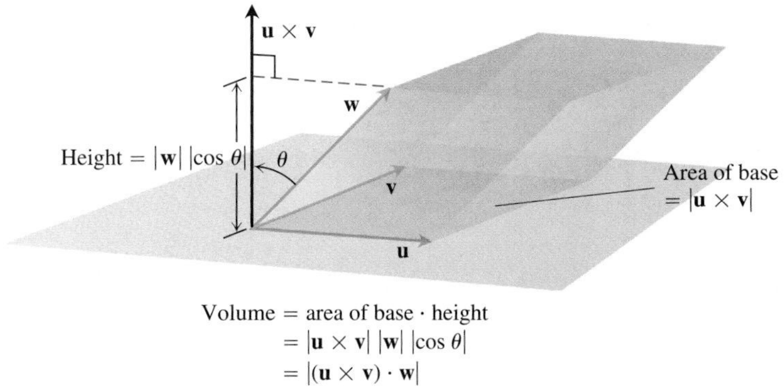

FIGURE 12.34 The number $|(\mathbf{u} \times \mathbf{v}) \cdot \mathbf{w}|$ is the volume of a parallelepiped.

The dot and cross may be interchanged in a triple scalar product without altering its value.

By treating the planes of $\mathbf{v}$ and $\mathbf{w}$ and of $\mathbf{w}$ and $\mathbf{u}$ as the base planes of the parallelepiped determined by $\mathbf{u}$, $\mathbf{v}$, and $\mathbf{w}$, we see that

$$(\mathbf{u} \times \mathbf{v}) \cdot \mathbf{w} = (\mathbf{v} \times \mathbf{w}) \cdot \mathbf{u} = (\mathbf{w} \times \mathbf{u}) \cdot \mathbf{v}.$$

Since the dot product is commutative, we also have

$$(\mathbf{u} \times \mathbf{v}) \cdot \mathbf{w} = \mathbf{u} \cdot (\mathbf{v} \times \mathbf{w}).$$

The triple scalar product can be evaluated as a determinant:

$$
\begin{aligned}
(\mathbf{u}\times\mathbf{v})\cdot\mathbf{w} &= \left[\begin{vmatrix} u_2 & u_3 \\ v_2 & v_3 \end{vmatrix}\mathbf{i} - \begin{vmatrix} u_1 & u_3 \\ v_1 & v_3 \end{vmatrix}\mathbf{j} + \begin{vmatrix} u_1 & u_2 \\ v_1 & v_2 \end{vmatrix}\mathbf{k}\right]\cdot\mathbf{w} \\
&= w_1\begin{vmatrix} u_2 & u_3 \\ v_2 & v_3 \end{vmatrix} - w_2\begin{vmatrix} u_1 & u_3 \\ v_1 & v_3 \end{vmatrix} + w_3\begin{vmatrix} u_1 & u_2 \\ v_1 & v_2 \end{vmatrix} \\
&= \begin{vmatrix} u_1 & u_2 & u_3 \\ v_1 & v_2 & v_3 \\ w_1 & w_2 & w_3 \end{vmatrix}.
\end{aligned}
$$

Calculating the Triple Scalar Product

$$
(\mathbf{u}\times\mathbf{v})\cdot\mathbf{w} = \begin{vmatrix} u_1 & u_2 & u_3 \\ v_1 & v_2 & v_3 \\ w_1 & w_2 & w_3 \end{vmatrix}
$$

EXAMPLE 6 Finding the Volume of a Parallelepiped

Find the volume of the box (parallelepiped) determined by $\mathbf{u} = \mathbf{i} + 2\mathbf{j} - \mathbf{k}$, $\mathbf{v} = -2\mathbf{i} + 3\mathbf{k}$, and $\mathbf{w} = 7\mathbf{j} - 4\mathbf{k}$.

Solution Using the rule for calculating determinants, we find

$$
(\mathbf{u}\times\mathbf{v})\cdot\mathbf{w} = \begin{vmatrix} 1 & 2 & -1 \\ -2 & 0 & 3 \\ 0 & 7 & -4 \end{vmatrix} = -23.
$$

The volume is $|(\mathbf{u}\times\mathbf{v})\cdot\mathbf{w}| = 23$ units cubed. ■

EXERCISES 12.4

Cross Product Calculations

In Exercises 1–8, find the length and direction (when defined) of $\mathbf{u}\times\mathbf{v}$ and $\mathbf{v}\times\mathbf{u}$.

1. $\mathbf{u} = 2\mathbf{i} - 2\mathbf{j} - \mathbf{k}, \quad \mathbf{v} = \mathbf{i} - \mathbf{k}$
2. $\mathbf{u} = 2\mathbf{i} + 3\mathbf{j}, \quad \mathbf{v} = -\mathbf{i} + \mathbf{j}$
3. $\mathbf{u} = 2\mathbf{i} - 2\mathbf{j} + 4\mathbf{k}, \quad \mathbf{v} = -\mathbf{i} + \mathbf{j} - 2\mathbf{k}$
4. $\mathbf{u} = \mathbf{i} + \mathbf{j} - \mathbf{k}, \quad \mathbf{v} = \mathbf{0}$
5. $\mathbf{u} = 2\mathbf{i}, \quad \mathbf{v} = -3\mathbf{j}$
6. $\mathbf{u} = \mathbf{i}\times\mathbf{j}, \quad \mathbf{v} = \mathbf{j}\times\mathbf{k}$
7. $\mathbf{u} = -8\mathbf{i} - 2\mathbf{j} - 4\mathbf{k}, \quad \mathbf{v} = 2\mathbf{i} + 2\mathbf{j} + \mathbf{k}$
8. $\mathbf{u} = \frac{3}{2}\mathbf{i} - \frac{1}{2}\mathbf{j} + \mathbf{k}, \quad \mathbf{v} = \mathbf{i} + \mathbf{j} + 2\mathbf{k}$

In Exercises 9–14, sketch the coordinate axes and then include the vectors $\mathbf{u}$, $\mathbf{v}$ and $\mathbf{u}\times\mathbf{v}$ as vectors starting at the origin.

9. $\mathbf{u} = \mathbf{i}, \quad \mathbf{v} = \mathbf{j}$
10. $\mathbf{u} = \mathbf{i} - \mathbf{k}, \quad \mathbf{v} = \mathbf{j}$
11. $\mathbf{u} = \mathbf{i} - \mathbf{k}, \quad \mathbf{v} = \mathbf{j} + \mathbf{k}$
12. $\mathbf{u} = 2\mathbf{i} - \mathbf{j}, \quad \mathbf{v} = \mathbf{i} + 2\mathbf{j}$
13. $\mathbf{u} = \mathbf{i} + \mathbf{j}, \quad \mathbf{v} = \mathbf{i} - \mathbf{j}$
14. $\mathbf{u} = \mathbf{j} + 2\mathbf{k}, \quad \mathbf{v} = \mathbf{i}$

Triangles in Space

In Exercises 15–18,

a. Find the area of the triangle determined by the points P, Q, and R.
b. Find a unit vector perpendicular to plane PQR.

15. $P(1, -1, 2)$, $Q(2, 0, -1)$, $R(0, 2, 1)$

16. $P(1, 1, 1)$, $Q(2, 1, 3)$, $R(3, -1, 1)$

17. $P(2, -2, 1)$, $Q(3, -1, 2)$, $R(3, -1, 1)$

18. $P(-2, 2, 0)$, $Q(0, 1, -1)$, $R(-1, 2, -2)$

Triple Scalar Products

In Exercises 19–22, verify that $(\mathbf{u} \times \mathbf{v}) \cdot \mathbf{w} = (\mathbf{v} \times \mathbf{w}) \cdot \mathbf{u} = (\mathbf{w} \times \mathbf{u}) \cdot \mathbf{v}$ and find the volume of the parallelepiped (box) determined by **u**, **v**, and **w**.

	u	**v**	**w**
19.	$2\mathbf{i}$	$2\mathbf{j}$	$2\mathbf{k}$
20.	$\mathbf{i} - \mathbf{j} + \mathbf{k}$	$2\mathbf{i} + \mathbf{j} - 2\mathbf{k}$	$-\mathbf{i} + 2\mathbf{j} - \mathbf{k}$
21.	$2\mathbf{i} + \mathbf{j}$	$2\mathbf{i} - \mathbf{j} + \mathbf{k}$	$\mathbf{i} + 2\mathbf{k}$
22.	$\mathbf{i} + \mathbf{j} - 2\mathbf{k}$	$-\mathbf{i} - \mathbf{k}$	$2\mathbf{i} + 4\mathbf{j} - 2\mathbf{k}$

Theory and Examples

23. Parallel and perpendicular vectors Let $\mathbf{u} = 5\mathbf{i} - \mathbf{j} + \mathbf{k}$, $\mathbf{v} = \mathbf{j} - 5\mathbf{k}$, $\mathbf{w} = -15\mathbf{i} + 3\mathbf{j} - 3\mathbf{k}$. Which vectors, if any, are **(a)** perpendicular? **(b)** Parallel? Give reasons for your answers.

24. Parallel and perpendicular vectors Let $\mathbf{u} = \mathbf{i} + 2\mathbf{j} - \mathbf{k}$, $\mathbf{v} = -\mathbf{i} + \mathbf{j} + \mathbf{k}$, $\mathbf{w} = \mathbf{i} + \mathbf{k}$, $\mathbf{r} = -(\pi/2)\mathbf{i} - \pi\mathbf{j} + (\pi/2)\mathbf{k}$. Which vectors, if any, are **(a)** perpendicular? **(b)** Parallel? Give reasons for your answers.

In Exercises 39 and 40, find the magnitude of the torque exerted by **F** on the bolt at P if $|\overrightarrow{PQ}| = 8$ in. and $|\mathbf{F}| = 30$ lb. Answer in foot-pounds.

25.

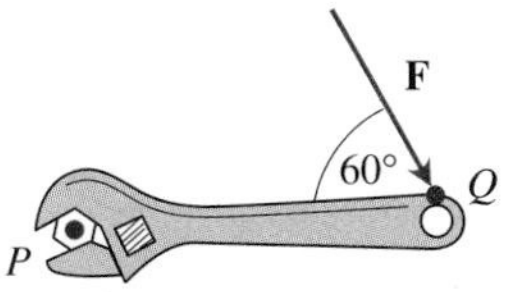

26.

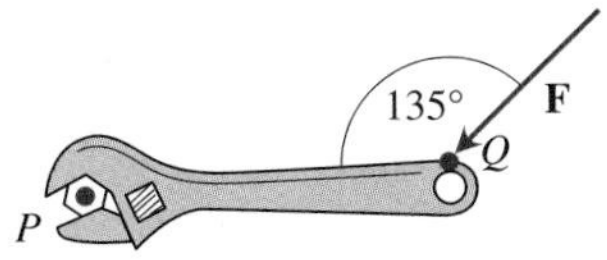

27. Which of the following are *always true*, and which are *not always true*? Give reasons for your answers.

a. $|\mathbf{u}| = \sqrt{\mathbf{u} \cdot \mathbf{u}}$ **b.** $\mathbf{u} \cdot \mathbf{u} = |\mathbf{u}|$

c. $\mathbf{u} \times \mathbf{0} = \mathbf{0} \times \mathbf{u} = \mathbf{0}$ **d.** $\mathbf{u} \times (-\mathbf{u}) = \mathbf{0}$

e. $\mathbf{u} \times \mathbf{v} = \mathbf{v} \times \mathbf{u}$

f. $\mathbf{u} \times (\mathbf{v} + \mathbf{w}) = \mathbf{u} \times \mathbf{v} + \mathbf{u} \times \mathbf{w}$

g. $(\mathbf{u} \times \mathbf{v}) \cdot \mathbf{v} = 0$

h. $(\mathbf{u} \times \mathbf{v}) \cdot \mathbf{w} = \mathbf{u} \cdot (\mathbf{v} \times \mathbf{w})$

28. Which of the following are *always true*, and which are *not always true*? Give reasons for your answers.

a. $\mathbf{u} \cdot \mathbf{v} = \mathbf{v} \cdot \mathbf{u}$ **b.** $\mathbf{u} \times \mathbf{v} = -(\mathbf{v} \times \mathbf{u})$

c. $(-\mathbf{u}) \times \mathbf{v} = -(\mathbf{u} \times \mathbf{v})$

d. $(c\mathbf{u}) \cdot \mathbf{v} = \mathbf{u} \cdot (c\mathbf{v}) = c(\mathbf{u} \cdot \mathbf{v})$ (any number c)

e. $c(\mathbf{u} \times \mathbf{v}) = (c\mathbf{u}) \times \mathbf{v} = \mathbf{u} \times (c\mathbf{v})$ (any number c)

f. $\mathbf{u} \cdot \mathbf{u} = |\mathbf{u}|^2$ **g.** $(\mathbf{u} \times \mathbf{u}) \cdot \mathbf{u} = 0$

h. $(\mathbf{u} \times \mathbf{v}) \cdot \mathbf{u} = \mathbf{v} \cdot (\mathbf{u} \times \mathbf{v})$

29. Given nonzero vectors **u**, **v**, and **w**, use dot product and cross product notation, as appropriate, to describe the following.

a. The vector projection of **u** onto **v**

b. A vector orthogonal to **u** and **v**

c. A vector orthogonal to $\mathbf{u} \times \mathbf{v}$ and **w**

d. The volume of the parallelepiped determined by **u**, **v**, and **w**

30. Given nonzero vectors **u**, **v**, and **w**, use dot product and cross product notation to describe the following.

a. A vector orthogonal to $\mathbf{u} \times \mathbf{v}$ and $\mathbf{u} \times \mathbf{w}$

b. A vector orthogonal to $\mathbf{u} + \mathbf{v}$ and $\mathbf{u} - \mathbf{v}$

c. A vector of length $|\mathbf{u}|$ in the direction of **v**

d. The area of the parallelogram determined by **u** and **w**

31. Let **u**, **v**, and **w** be vectors. Which of the following make sense, and which do not? Give reasons for your answers.

a. $(\mathbf{u} \times \mathbf{v}) \cdot \mathbf{w}$ **b.** $\mathbf{u} \times (\mathbf{v} \cdot \mathbf{w})$

c. $\mathbf{u} \times (\mathbf{v} \times \mathbf{w})$ **d.** $\mathbf{u} \cdot (\mathbf{v} \cdot \mathbf{w})$

32. Cross products of three vectors Show that except in degenerate cases, $(\mathbf{u} \times \mathbf{v}) \times \mathbf{w}$ lies in the plane of **u** and **v**, whereas $\mathbf{u} \times (\mathbf{v} \times \mathbf{w})$ lies in the plane of **v** and **w**. What *are* the degenerate cases?

33. Cancellation in cross products If $\mathbf{u} \times \mathbf{v} = \mathbf{u} \times \mathbf{w}$ and $\mathbf{u} \neq \mathbf{0}$, then does $\mathbf{v} = \mathbf{w}$? Give reasons for your answer.

34. Double cancellation If $\mathbf{u} \neq \mathbf{0}$ and if $\mathbf{u} \times \mathbf{v} = \mathbf{u} \times \mathbf{w}$ and $\mathbf{u} \cdot \mathbf{v} = \mathbf{u} \cdot \mathbf{w}$, then does $\mathbf{v} = \mathbf{w}$? Give reasons for your answer.

Area in the Plane

Find the areas of the parallelograms whose vertices are given in Exercises 35–38.

35. $A(1, 0)$, $B(0, 1)$, $C(-1, 0)$, $D(0, -1)$

36. $A(0, 0)$, $B(7, 3)$, $C(9, 8)$, $D(2, 5)$

37. $A(-1, 2)$, $B(2, 0)$, $C(7, 1)$, $D(4, 3)$

38. $A(-6, 0)$, $B(1, -4)$, $C(3, 1)$, $D(-4, 5)$

Find the areas of the triangles whose vertices are given in Exercises 39–42.

39. $A(0, 0)$, $B(-2, 3)$, $C(3, 1)$

40. $A(-1, -1)$, $B(3, 3)$, $C(2, 1)$

41. $A(-5, 3)$, $B(1, -2)$, $C(6, -2)$

42. $A(-6, 0)$, $B(10, -5)$, $C(-2, 4)$

43. Triangle area Find a formula for the area of the triangle in the xy-plane with vertices at $(0, 0)$, (a_1, a_2), and (b_1, b_2). Explain your work.

44. Triangle area Find a concise formula for the area of a triangle with vertices (a_1, a_2), (b_1, b_2), and (c_1, c_2).

12.5 Lines and Planes in Space

In the calculus of functions of a single variable, we used our knowledge of lines to study curves in the plane. We investigated tangents and found that, when highly magnified, differentiable curves were effectively linear.

To study the calculus of functions of more than one variable in the next chapter, we start with planes and use our knowledge of planes to study the surfaces that are the graphs of functions in space.

This section shows how to use scalar and vector products to write equations for lines, line segments, and planes in space.

Lines and Line Segments in Space

In the plane, a line is determined by a point and a number giving the slope of the line. In space a line is determined by a point and a *vector* giving the direction of the line.

Suppose that L is a line in space passing through a point $P_0(x_0, y_0, z_0)$ parallel to a vector $\mathbf{v} = v_1\mathbf{i} + v_2\mathbf{j} + v_3\mathbf{k}$. Then L is the set of all points $P(x, y, z)$ for which $\overrightarrow{P_0P}$ is parallel to $\mathbf{v}$ (Figure 12.35). Thus, $\overrightarrow{P_0P} = t\mathbf{v}$ for some scalar parameter t. The value of t depends on the location of the point P along the line, and the domain of t is $(-\infty, \infty)$. The expanded form of the equation $\overrightarrow{P_0P} = t\mathbf{v}$ is

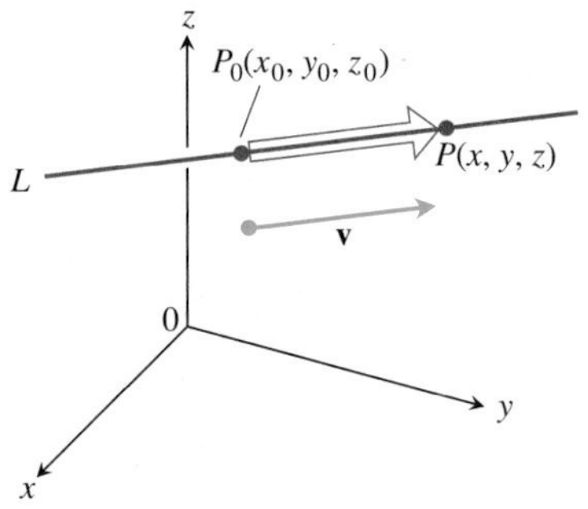

FIGURE 12.35 A point P lies on L through P_0 parallel to $\mathbf{v}$ if and only if $\overrightarrow{P_0P}$ is a scalar multiple of $\mathbf{v}$.

$$(x - x_0)\mathbf{i} + (y - y_0)\mathbf{j} + (z - z_0)\mathbf{k} = t(v_1\mathbf{i} + v_2\mathbf{j} + v_3\mathbf{k}),$$

which can be rewritten as

$$x\mathbf{i} + y\mathbf{j} + z\mathbf{k} = x_0\mathbf{i} + y_0\mathbf{j} + z_0\mathbf{k} + t(v_1\mathbf{i} + v_2\mathbf{j} + v_3\mathbf{k}). \tag{1}$$

If $\mathbf{r}(t)$ is the position vector of a point $P(x, y, z)$ on the line and $\mathbf{r}_0$ is the position vector of the point $P_0(x_0, y_0, z_0)$, then Equation (1) gives the following vector form for the equation of a line in space.

Vector Equation for a Line

A vector equation for the line L through $P_0(x_0, y_0, z_0)$ parallel to $\mathbf{v}$ is

$$\mathbf{r}(t) = \mathbf{r}_0 + t\mathbf{v}, \qquad -\infty < t < \infty, \tag{2}$$

where $\mathbf{r}$ is the position vector of a point $P(x, y, z)$ on L and $\mathbf{r}_0$ is the position vector of $P_0(x_0, y_0, z_0)$.

Equating the corresponding components of the two sides of Equation (1) gives three scalar equations involving the parameter t:

$$x = x_0 + tv_1, \qquad y = y_0 + tv_2, \qquad z = z_0 + tv_3.$$

These equations give us the standard parametrization of the line for the parameter interval $-\infty < t < \infty$.

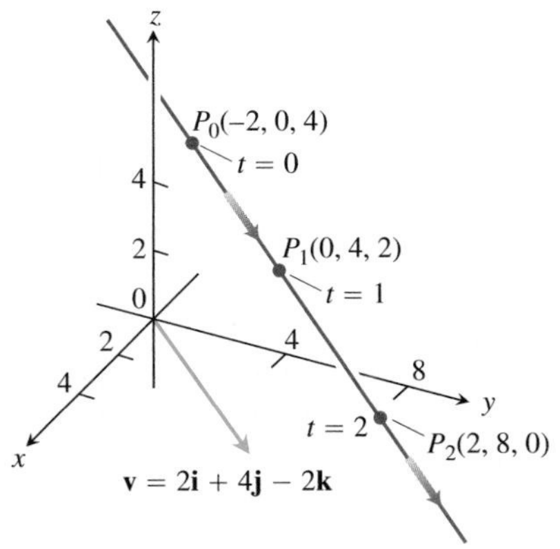

FIGURE 12.36 Selected points and parameter values on the line $x = -2 + 2t, y = 4t, z = 4 - 2t$. The arrows show the direction of increasing t (Example 1).

Parametric Equations for a Line

The standard parametrization of the line through $P_0(x_0, y_0, z_0)$ parallel to $\mathbf{v} = v_1\mathbf{i} + v_2\mathbf{j} + v_3\mathbf{k}$ is

$$x = x_0 + tv_1, \quad y = y_0 + tv_2, \quad z = z_0 + tv_3, \quad -\infty < t < \infty \tag{3}$$

EXAMPLE 1 Parametrizing a Line Through a Point Parallel to a Vector

Find parametric equations for the line through $(-2, 0, 4)$ parallel to $\mathbf{v} = 2\mathbf{i} + 4\mathbf{j} - 2\mathbf{k}$ (Figure 12.36).

Solution With $P_0(x_0, y_0, z_0)$ equal to $(-2, 0, 4)$ and $v_1\mathbf{i} + v_2\mathbf{j} + v_3\mathbf{k}$ equal to $2\mathbf{i} + 4\mathbf{j} - 2\mathbf{k}$, Equations (3) become

$$x = -2 + 2t, \qquad y = 4t, \qquad z = 4 - 2t.$$ ■

EXAMPLE 2 Parametrizing a Line Through Two Points

Find parametric equations for the line through $P(-3, 2, -3)$ and $Q(1, -1, 4)$.

Solution The vector

$$\begin{aligned}\overrightarrow{PQ} &= (1 - (-3))\mathbf{i} + (-1 - 2)\mathbf{j} + (4 - (-3))\mathbf{k} \\ &= 4\mathbf{i} - 3\mathbf{j} + 7\mathbf{k}\end{aligned}$$

is parallel to the line, and Equations (3) with $(x_0, y_0, z_0) = (-3, 2, -3)$ give

$$x = -3 + 4t, \qquad y = 2 - 3t, \qquad z = -3 + 7t.$$

We could have chosen $Q(1, -1, 4)$ as the "base point" and written

$$x = 1 + 4t, \quad y = -1 - 3t, \quad z = 4 + 7t.$$

These equations serve as well as the first; they simply place you at a different point on the line for a given value of t. ■

Notice that parametrizations are not unique. Not only can the "base point" change, but so can the parameter. The equations $x = -3 + 4t^3$, $y = 2 - 3t^3$, and $z = -3 + 7t^3$ also parametrize the line in Example 2.

To parametrize a line segment joining two points, we first parametrize the line through the points. We then find the t-values for the endpoints and restrict t to lie in the closed interval bounded by these values. The line equations together with this added restriction parametrize the segment.

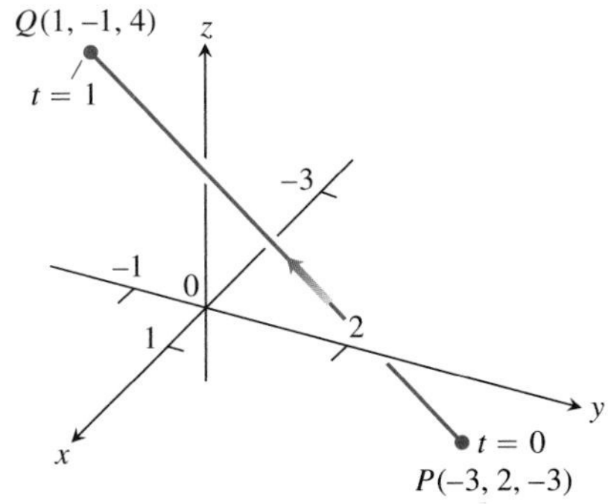

FIGURE 12.37 Example 3 derives a parametrization of line segment PQ. The arrow shows the direction of increasing t.

EXAMPLE 3 Parametrizing a Line Segment

Parametrize the line segment joining the points $P(-3, 2, -3)$ and $Q(1, -1, 4)$ (Figure 12.37).

Solution We begin with equations for the line through P and Q, taking them, in this case, from Example 2:

$$x = -3 + 4t, \qquad y = 2 - 3t, \qquad z = -3 + 7t.$$

We observe that the point

$$(x, y, z) = (-3 + 4t, 2 - 3t, -3 + 7t)$$

on the line passes through $P(-3, 2, -3)$ at $t = 0$ and $Q(1, -1, 4)$ at $t = 1$. We add the restriction $0 \le t \le 1$ to parametrize the segment:

$$x = -3 + 4t, \qquad y = 2 - 3t, \qquad z = -3 + 7t, \qquad 0 \le t \le 1. \qquad \blacksquare$$

The vector form (Equation (2)) for a line in space is more revealing if we think of a line as the path of a particle starting at position $P_0(x_0, y_0, z_0)$ and moving in the direction of vector $\mathbf{v}$. Rewriting Equation (2), we have

$$\begin{aligned} \mathbf{r}(t) &= \mathbf{r}_0 + t\mathbf{v} \\ &= \underbrace{\mathbf{r}_0}_{\text{Initial position}} + \underbrace{t}_{\text{Time}}\underbrace{|\mathbf{v}|}_{\text{Speed}}\underbrace{\frac{\mathbf{v}}{|\mathbf{v}|}}_{\text{Direction}}. \end{aligned} \tag{4}$$

In other words, the position of the particle at time t is its initial position plus its distance moved (speed $\times$ time) in the direction $\mathbf{v}/|\mathbf{v}|$ of its straight-line motion.

EXAMPLE 4 Flight of a Helicopter

A helicopter is to fly directly from a helipad at the origin in the direction of the point $(1, 1, 1)$ at a speed of 60 ft/sec. What is the position of the helicopter after 10 sec?

Solution We place the origin at the starting position (helipad) of the helicopter. Then the unit vector

$$\mathbf{u} = \frac{1}{\sqrt{3}}\mathbf{i} + \frac{1}{\sqrt{3}}\mathbf{j} + \frac{1}{\sqrt{3}}\mathbf{k}$$

gives the flight direction of the helicopter. From Equation (4), the position of the helicopter at any time t is

$$\begin{aligned} \mathbf{r}(t) &= \mathbf{r}_0 + t(\text{speed})\mathbf{u} \\ &= \mathbf{0} + t(60)\left(\frac{1}{\sqrt{3}}\mathbf{i} + \frac{1}{\sqrt{3}}\mathbf{j} + \frac{1}{\sqrt{3}}\mathbf{k}\right) \\ &= 20\sqrt{3}t(\mathbf{i} + \mathbf{j} + \mathbf{k}). \end{aligned}$$

When $t = 10$ sec,

$$\begin{aligned} \mathbf{r}(10) &= 200\sqrt{3}\,(\mathbf{i} + \mathbf{j} + \mathbf{k}) \\ &= \left\langle 200\sqrt{3}, 200\sqrt{3}, 200\sqrt{3} \right\rangle. \end{aligned}$$

After 10 sec of flight from the origin toward $(1, 1, 1)$, the helicopter is located at the point $(200\sqrt{3}, 200\sqrt{3}, 200\sqrt{3})$ in space. It has traveled a distance of $(60 \text{ ft/sec})(10 \text{ sec}) = 600$ ft, which is the length of the vector $\mathbf{r}(10)$. $\blacksquare$

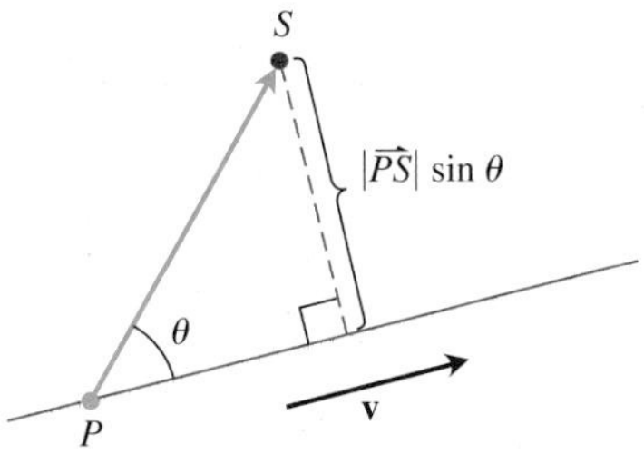

FIGURE 12.38 The distance from S to the line through P parallel to $\mathbf{v}$ is $|\overrightarrow{PS}| \sin \theta$, where θ is the angle between $\overrightarrow{PS}$ and $\mathbf{v}$.

The Distance from a Point to a Line in Space

To find the distance from a point S to a line that passes through a point P parallel to a vector $\mathbf{v}$, we find the absolute value of the scalar component of $\overrightarrow{PS}$ in the direction of a vector normal to the line (Figure 12.38). In the notation of the figure, the absolute value of the scalar component is, $|\overrightarrow{PS}| \sin \theta$, which is $\dfrac{|\overrightarrow{PS} \times \mathbf{v}|}{|\mathbf{v}|}$.

Distance from a Point S to a Line Through P Parallel to $\mathbf{v}$

$$d = \frac{|\overrightarrow{PS} \times \mathbf{v}|}{|\mathbf{v}|} \tag{5}$$

EXAMPLE 5 Finding Distance from a Point to a Line

Find the distance from the point $S(1, 1, 5)$ to the line

$$L: \quad x = 1 + t, \quad y = 3 - t, \quad z = 2t.$$

Solution We see from the equations for L that L passes through $P(1, 3, 0)$ parallel to $\mathbf{v} = \mathbf{i} - \mathbf{j} + 2\mathbf{k}$. With

$$\overrightarrow{PS} = (1 - 1)\mathbf{i} + (1 - 3)\mathbf{j} + (5 - 0)\mathbf{k} = -2\mathbf{j} + 5\mathbf{k}$$

and

$$\overrightarrow{PS} \times \mathbf{v} = \begin{vmatrix} \mathbf{i} & \mathbf{j} & \mathbf{k} \\ 0 & -2 & 5 \\ 1 & -1 & 2 \end{vmatrix} = \mathbf{i} + 5\mathbf{j} + 2\mathbf{k},$$

Equation (5) gives

$$d = \frac{|\overrightarrow{PS} \times \mathbf{v}|}{|\mathbf{v}|} = \frac{\sqrt{1 + 25 + 4}}{\sqrt{1 + 1 + 4}} = \frac{\sqrt{30}}{\sqrt{6}} = \sqrt{5}.$$ ∎

An Equation for a Plane in Space

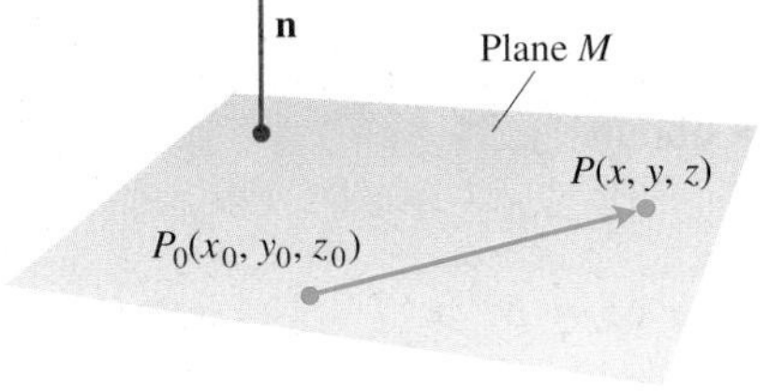

FIGURE 12.39 The standard equation for a plane in space is defined in terms of a vector normal to the plane: A point P lies in the plane through P_0 normal to $\mathbf{n}$ if and only if $\mathbf{n} \cdot \overrightarrow{P_0P} = 0$.

A plane in space is determined by knowing a point on the plane and its "tilt" or orientation. This "tilt" is defined by specifying a vector that is perpendicular or normal to the plane.

Suppose that plane M passes through a point $P_0(x_0, y_0, z_0)$ and is normal to the nonzero vector $\mathbf{n} = A\mathbf{i} + B\mathbf{j} + C\mathbf{k}$. Then M is the set of all points $P(x, y, z)$ for which $\overrightarrow{P_0P}$ is orthogonal to $\mathbf{n}$ (Figure 12.39). Thus, the dot product $\mathbf{n} \cdot \overrightarrow{P_0P} = 0$. This equation is equivalent to

$$(A\mathbf{i} + B\mathbf{j} + C\mathbf{k}) \cdot [(x - x_0)\mathbf{i} + (y - y_0)\mathbf{j} + (z - z_0)\mathbf{k}] = 0$$

or

$$A(x - x_0) + B(y - y_0) + C(z - z_0) = 0.$$

Equation for a Plane

The plane through $P_0(x_0, y_0, z_0)$ normal to $\mathbf{n} = A\mathbf{i} + B\mathbf{j} + C\mathbf{k}$ has

Vector equation:	$\mathbf{n} \cdot \overrightarrow{P_0P} = 0$
Component equation:	$A(x - x_0) + B(y - y_0) + C(z - z_0) = 0$
Component equation simplified:	$Ax + By + Cz = D$, where $D = Ax_0 + By_0 + Cz_0$

EXAMPLE 6 Finding an Equation for a Plane

Find an equation for the plane through $P_0(-3, 0, 7)$ perpendicular to $\mathbf{n} = 5\mathbf{i} + 2\mathbf{j} - \mathbf{k}$.

Solution The component equation is

$$5(x - (-3)) + 2(y - 0) + (-1)(z - 7) = 0.$$

Simplifying, we obtain

$$\begin{aligned} 5x + 15 + 2y - z + 7 &= 0 \\ 5x + 2y - z &= -22. \end{aligned}$$

■

Notice in Example 6 how the components of $\mathbf{n} = 5\mathbf{i} + 2\mathbf{j} - \mathbf{k}$ became the coefficients of x, y, and z in the equation $5x + 2y - z = -22$. The vector $\mathbf{n} = A\mathbf{i} + B\mathbf{j} + C\mathbf{k}$ is normal to the plane $Ax + By + Cz = D$.

EXAMPLE 7 Finding an Equation for a Plane Through Three Points

Find an equation for the plane through $A(0, 0, 1)$, $B(2, 0, 0)$, and $C(0, 3, 0)$.

Solution We find a vector normal to the plane and use it with one of the points (it does not matter which) to write an equation for the plane.

The cross product

$$\overrightarrow{AB} \times \overrightarrow{AC} = \begin{vmatrix} \mathbf{i} & \mathbf{j} & \mathbf{k} \\ 2 & 0 & -1 \\ 0 & 3 & -1 \end{vmatrix} = 3\mathbf{i} + 2\mathbf{j} + 6\mathbf{k}$$

is normal to the plane. We substitute the components of this vector and the coordinates of A(0, 0, 1) into the component form of the equation to obtain

$$\begin{aligned} 3(x - 0) + 2(y - 0) + 6(z - 1) &= 0 \\ 3x + 2y + 6z &= 6. \end{aligned}$$

■

Lines of Intersection

Just as lines are parallel if and only if they have the same direction, two planes are **parallel** if and only if their normals are parallel, or $\mathbf{n}_1 = k\mathbf{n}_2$ for some scalar k. Two planes that are not parallel intersect in a line.

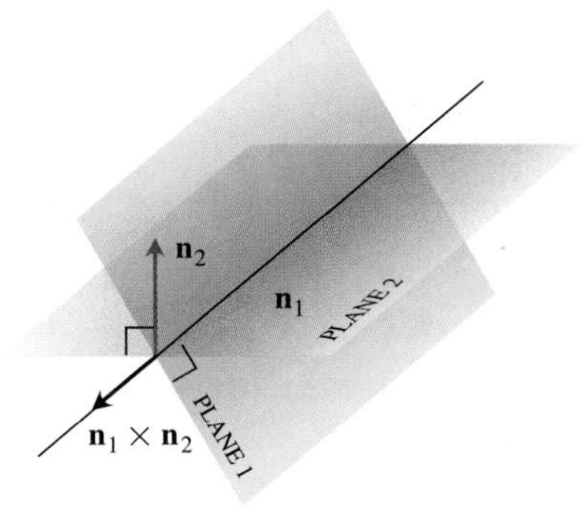

FIGURE 12.40 How the line of intersection of two planes is related to the planes' normal vectors (Example 8).

EXAMPLE 8 Finding a Vector Parallel to the Line of Intersection of Two Planes

Find a vector parallel to the line of intersection of the planes $3x - 6y - 2z = 15$ and $2x + y - 2z = 5$.

Solution The line of intersection of two planes is perpendicular to both planes' normal vectors $\mathbf{n}_1$ and $\mathbf{n}_2$ (Figure 12.40) and therefore parallel to $\mathbf{n}_1 \times \mathbf{n}_2$. Turning this around, $\mathbf{n}_1 \times \mathbf{n}_2$ is a vector parallel to the planes' line of intersection. In our case,

$$\mathbf{n}_1 \times \mathbf{n}_2 = \begin{vmatrix} \mathbf{i} & \mathbf{j} & \mathbf{k} \\ 3 & -6 & -2 \\ 2 & 1 & -2 \end{vmatrix} = 14\mathbf{i} + 2\mathbf{j} + 15\mathbf{k}.$$

Any nonzero scalar multiple of $\mathbf{n}_1 \times \mathbf{n}_2$ will do as well. ■

EXAMPLE 9 Parametrizing the Line of Intersection of Two Planes

Find parametric equations for the line in which the planes $3x - 6y - 2z = 15$ and $2x + y - 2z = 5$ intersect.

Solution We find a vector parallel to the line and a point on the line and use Equations (3).

Example 8 identifies $\mathbf{v} = 14\mathbf{i} + 2\mathbf{j} + 15\mathbf{k}$ as a vector parallel to the line. To find a point on the line, we can take any point common to the two planes. Substituting $z = 0$ in the plane equations and solving for x and y simultaneously identifies one of these points as $(3, -1, 0)$. The line is

$$x = 3 + 14t, \qquad y = -1 + 2t, \qquad z = 15t.$$

The choice $z = 0$ is arbitrary and we could have chosen $z = 1$ or $z = -1$ just as well. Or we could have let $x = 0$ and solved for y and z. The different choices would simply give different parametrizations of the same line. ■

Sometimes we want to know where a line and a plane intersect. For example, if we are looking at a flat plate and a line segment passes through it, we may be interested in knowing what portion of the line segment is hidden from our view by the plate. This application is used in computer graphics (Exercise 74).

EXAMPLE 10 Finding the Intersection of a Line and a Plane

Find the point where the line

$$x = \frac{8}{3} + 2t, \qquad y = -2t, \qquad z = 1 + t$$

intersects the plane $3x + 2y + 6z = 6$.

Solution The point

$$\left(\frac{8}{3} + 2t, -2t, 1 + t\right)$$

lies in the plane if its coordinates satisfy the equation of the plane, that is, if

$$\begin{aligned} 3\left(\frac{8}{3} + 2t\right) + 2(-2t) + 6(1 + t) &= 6 \\ 8 + 6t - 4t + 6 + 6t &= 6 \\ 8t &= -8 \\ t &= -1. \end{aligned}$$

The point of intersection is

$$(x, y, z)|_{t=-1} = \left(\frac{8}{3} - 2, 2, 1 - 1\right) = \left(\frac{2}{3}, 2, 0\right).$$

The Distance from a Point to a Plane

If P is a point on a plane with normal $\mathbf{n}$, then the distance from any point S to the plane is the length of the vector projection of $\overrightarrow{PS}$ onto $\mathbf{n}$. That is, the distance from S to the plane is

$$d = \left|\overrightarrow{PS} \cdot \frac{\mathbf{n}}{|\mathbf{n}|}\right| \tag{6}$$

where $\mathbf{n} = A\mathbf{i} + B\mathbf{j} + C\mathbf{k}$ is normal to the plane.

EXAMPLE 11 Finding the Distance from a Point to a Plane

Find the distance from $S(1, 1, 3)$ to the plane $3x + 2y + 6z = 6$.

Solution We find a point P in the plane and calculate the length of the vector projection of $\overrightarrow{PS}$ onto a vector $\mathbf{n}$ normal to the plane (Figure 12.41). The coefficients in the equation $3x + 2y + 6z = 6$ give

$$\mathbf{n} = 3\mathbf{i} + 2\mathbf{j} + 6\mathbf{k}.$$

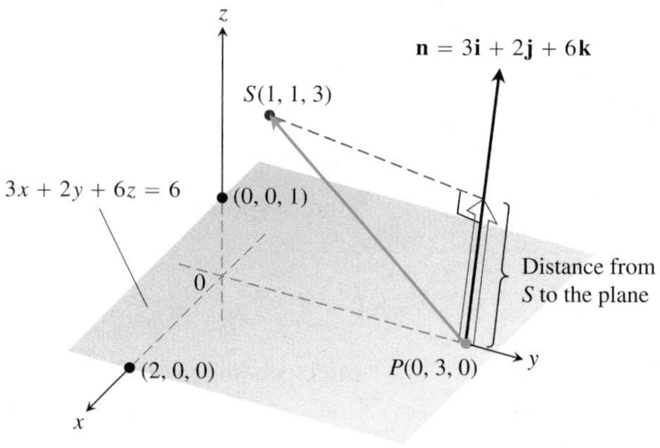

FIGURE 12.41 The distance from S to the plane is the length of the vector projection of $\overrightarrow{PS}$ onto $\mathbf{n}$ (Example 11).

The points on the plane easiest to find from the plane's equation are the intercepts. If we take P to be the y-intercept $(0, 3, 0)$, then

$$\begin{aligned}\overrightarrow{PS} &= (1-0)\mathbf{i} + (1-3)\mathbf{j} + (3-0)\mathbf{k} \\ &= \mathbf{i} - 2\mathbf{j} + 3\mathbf{k}, \\ |\mathbf{n}| &= \sqrt{(3)^2 + (2)^2 + (6)^2} = \sqrt{49} = 7.\end{aligned}$$

The distance from S to the plane is

$$\begin{aligned}d &= \left|\overrightarrow{PS}\cdot\frac{\mathbf{n}}{|\mathbf{n}|}\right| \qquad \text{length of } \mathrm{proj}_{\mathbf{n}}\,\overrightarrow{PS} \\ &= \left|(\mathbf{i} - 2\mathbf{j} + 3\mathbf{k})\cdot\left(\frac{3}{7}\mathbf{i} + \frac{2}{7}\mathbf{j} + \frac{6}{7}\mathbf{k}\right)\right| \\ &= \left|\frac{3}{7} - \frac{4}{7} + \frac{18}{7}\right| = \frac{17}{7}.\end{aligned}$$

■

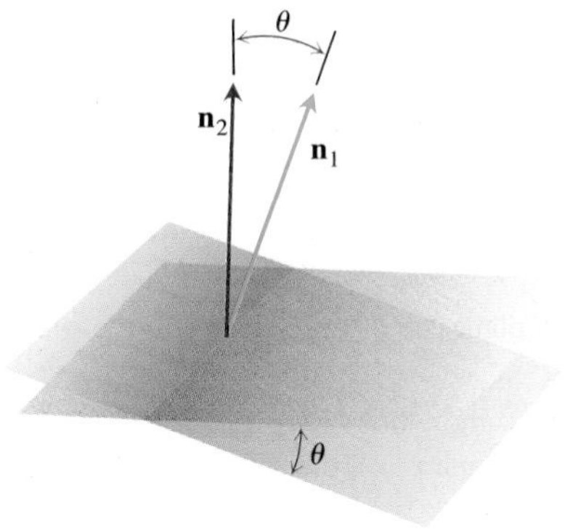

FIGURE 12.42 The angle between two planes is obtained from the angle between their normals.

Angles Between Planes

The angle between two intersecting planes is defined to be the (acute) angle determined by their normal vectors (Figure 12.42).

EXAMPLE 12 Find the angle between the planes $3x - 6y - 2z = 15$ and $2x + y - 2z = 5$.

Solution The vectors

$$\mathbf{n}_1 = 3\mathbf{i} - 6\mathbf{j} - 2\mathbf{k}, \qquad \mathbf{n}_2 = 2\mathbf{i} + \mathbf{j} - 2\mathbf{k}$$

are normals to the planes. The angle between them is

$$\begin{aligned}\theta &= \cos^{-1}\left(\frac{\mathbf{n}_1\cdot\mathbf{n}_2}{|\mathbf{n}_1||\mathbf{n}_2|}\right) \\ &= \cos^{-1}\left(\frac{4}{21}\right) \\ &\approx 1.38 \text{ radians.} \qquad \text{About 79 deg}\end{aligned}$$

■

EXERCISES 12.5

Lines and Line Segments

Find parametric equations for the lines in Exercises 1–12.

1. The line through the point $P(3, -4, -1)$ parallel to the vector $\mathbf{i} + \mathbf{j} + \mathbf{k}$
2. The line through $P(1, 2, -1)$ and $Q(-1, 0, 1)$
3. The line through $P(-2, 0, 3)$ and $Q(3, 5, -2)$
4. The line through $P(1, 2, 0)$ and $Q(1, 1, -1)$
5. The line through the origin parallel to the vector $2\mathbf{j} + \mathbf{k}$
6. The line through the point $(3, -2, 1)$ parallel to the line $x = 1 + 2t, y = 2 - t, z = 3t$
7. The line through $(1, 1, 1)$ parallel to the z-axis
8. The line through $(2, 4, 5)$ perpendicular to the plane $3x + 7y - 5z = 21$
9. The line through $(0, -7, 0)$ perpendicular to the plane $x + 2y + 2z = 13$

10. The line through (2, 3, 0) perpendicular to the vectors $\mathbf{u} = \mathbf{i} + 2\mathbf{j} + 3\mathbf{k}$ and $\mathbf{v} = 3\mathbf{i} + 4\mathbf{j} + 5\mathbf{k}$

11. The x-axis

12. The z-axis

Find parametrizations for the line segments joining the points in Exercises 13–20. Draw coordinate axes and sketch each segment, indicating the direction of increasing t for your parametrization.

13. $(0, 0, 0)$, $(1, 1, 3/2)$

14. $(0, 0, 0)$, $(1, 0, 0)$

15. $(1, 0, 0)$, $(1, 1, 0)$

16. $(1, 1, 0)$, $(1, 1, 1)$

17. $(0, 1, 1)$, $(0, -1, 1)$

18. $(0, 2, 0)$, $(3, 0, 0)$

19. $(2, 0, 2)$, $(0, 2, 0)$

20. $(1, 0, -1)$, $(0, 3, 0)$

Planes

Find equations for the planes in Exercises 21–26.

21. The plane through $P_0(0, 2, -1)$ normal to $\mathbf{n} = 3\mathbf{i} - 2\mathbf{j} - \mathbf{k}$

22. The plane through $(1, -1, 3)$ parallel to the plane

$$3x + y + z = 7$$

23. The plane through $(1, 1, -1)$, $(2, 0, 2)$, and $(0, -2, 1)$

24. The plane through $(2, 4, 5)$, $(1, 5, 7)$, and $(-1, 6, 8)$

25. The plane through $P_0(2, 4, 5)$ perpendicular to the line

$$x = 5 + t, \quad y = 1 + 3t, \quad z = 4t$$

26. The plane through $A(1, -2, 1)$ perpendicular to the vector from the origin to A

27. Find the point of intersection of the lines $x = 2t + 1$, $y = 3t + 2$, $z = 4t + 3$, and $x = s + 2$, $y = 2s + 4$, $z = -4s - 1$, and then find the plane determined by these lines.

28. Find the point of intersection of the lines $x = t$, $y = -t + 2$, $z = t + 1$, and $x = 2s + 2$, $y = s + 3$, $z = 5s + 6$, and then find the plane determined by these lines.

In Exercises 29 and 30, find the plane determined by the intersecting lines.

29. $L1$: $x = -1 + t$, $y = 2 + t$, $z = 1 - t$; $-\infty < t < \infty$

$L2$: $x = 1 - 4s$, $y = 1 + 2s$, $z = 2 - 2s$; $-\infty < s < \infty$

30. $L1$: $x = t$, $y = 3 - 3t$, $z = -2 - t$; $-\infty < t < \infty$

$L2$: $x = 1 + s$, $y = 4 + s$, $z = -1 + s$; $-\infty < s < \infty$

31. Find a plane through $P_0(2, 1, -1)$ and perpendicular to the line of intersection of the planes $2x + y - z = 3$, $x + 2y + z = 2$.

32. Find a plane through the points $P_1(1, 2, 3)$, $P_2(3, 2, 1)$ and perpendicular to the plane $4x - y + 2z = 7$.

Distances

In Exercises 33–38, find the distance from the point to the line.

33. $(0, 0, 12)$; $x = 4t$, $y = -2t$, $z = 2t$

34. $(0, 0, 0)$; $x = 5 + 3t$, $y = 5 + 4t$, $z = -3 - 5t$

35. $(2, 1, 3)$; $x = 2 + 2t$, $y = 1 + 6t$, $z = 3$

36. $(2, 1, -1)$; $x = 2t$, $y = 1 + 2t$, $z = 2t$

37. $(3, -1, 4)$; $x = 4 - t$, $y = 3 + 2t$, $z = -5 + 3t$

38. $(-1, 4, 3)$; $x = 10 + 4t$, $y = -3$, $z = 4t$

In Exercises 39–44, find the distance from the point to the plane.

39. $(2, -3, 4)$, $x + 2y + 2z = 13$

40. $(0, 0, 0)$, $3x + 2y + 6z = 6$

41. $(0, 1, 1)$, $4y + 3z = -12$

42. $(2, 2, 3)$, $2x + y + 2z = 4$

43. $(0, -1, 0)$, $2x + y + 2z = 4$

44. $(1, 0, -1)$, $-4x + y + z = 4$

45. Find the distance from the plane $x + 2y + 6z = 1$ to the plane $x + 2y + 6z = 10$.

46. Find the distance from the line $x = 2 + t$, $y = 1 + t$, $z = -(1/2) - (1/2)t$ to the plane $x + 2y + 6z = 10$.

Angles

Find the angles between the planes in Exercises 47 and 48.

47. $x + y = 1$, $2x + y - 2z = 2$

48. $5x + y - z = 10$, $x - 2y + 3z = -1$

T Use a calculator to find the acute angles between the planes in Exercises 49–52 to the nearest hundredth of a radian.

49. $2x + 2y + 2z = 3$, $2x - 2y - z = 5$

50. $x + y + z = 1$, $z = 0$ (the xy-plane)

51. $2x + 2y - z = 3$, $x + 2y + z = 2$

52. $4y + 3z = -12$, $3x + 2y + 6z = 6$

Intersecting Lines and Planes

In Exercises 53–56, find the point in which the line meets the plane.

53. $x = 1 - t$, $y = 3t$, $z = 1 + t$; $2x - y + 3z = 6$

54. $x = 2$, $y = 3 + 2t$, $z = -2 - 2t$; $6x + 3y - 4z = -12$

55. $x = 1 + 2t$, $y = 1 + 5t$, $z = 3t$; $x + y + z = 2$

56. $x = -1 + 3t$, $y = -2$, $z = 5t$; $2x - 3z = 7$

Find parametrizations for the lines in which the planes in Exercises 57–60 intersect.

57. $x + y + z = 1$, $x + y = 2$

58. $3x - 6y - 2z = 3$, $2x + y - 2z = 2$

59. $x - 2y + 4z = 2$, $x + y - 2z = 5$

60. $5x - 2y = 11$, $4y - 5z = -17$

Given two lines in space, either they are parallel, or they intersect, or they are skew (imagine, for example, the flight paths of two planes in the sky). Exercises 61 and 62 each give three lines. In each exercise, determine whether the lines, taken two at a time, are parallel, intersect, or are skew. If they intersect, find the point of intersection.

61. $L1$: $x = 3 + 2t$, $y = -1 + 4t$, $z = 2 - t$; $-\infty < t < \infty$

$L2$: $x = 1 + 4s$, $y = 1 + 2s$, $z = -3 + 4s$; $-\infty < s < \infty$

$L3$: $x = 3 + 2r$, $y = 2 + r$, $z = -2 + 2r$; $-\infty < r < \infty$

62. $L1$: $x = 1 + 2t$, $y = -1 - t$, $z = 3t$; $-\infty < t < \infty$

$L2$: $x = 2 - s$, $y = 3s$, $z = 1 + s$; $-\infty < s < \infty$

$L3$: $x = 5 + 2r$, $y = 1 - r$, $z = 8 + 3r$; $-\infty < r < \infty$

Theory and Examples

63. Use Equations (3) to generate a parametrization of the line through $P(2, -4, 7)$ parallel to $\mathbf{v}_1 = 2\mathbf{i} - \mathbf{j} + 3\mathbf{k}$. Then generate another parametrization of the line using the point $P_2(-2, -2, 1)$ and the vector $\mathbf{v}_2 = -\mathbf{i} + (1/2)\mathbf{j} - (3/2)\mathbf{k}$.

64. Use the component form to generate an equation for the plane through $P_1(4, 1, 5)$ normal to $\mathbf{n}_1 = \mathbf{i} - 2\mathbf{j} + \mathbf{k}$. Then generate another equation for the same plane using the point $P_2(3, -2, 0)$ and the normal vector $\mathbf{n}_2 = -\sqrt{2}\mathbf{i} + 2\sqrt{2}\mathbf{j} - \sqrt{2}\mathbf{k}$.

65. Find the points in which the line $x = 1 + 2t, y = -1 - t, z = 3t$ meets the coordinate planes. Describe the reasoning behind your answer.

66. Find equations for the line in the plane $z = 3$ that makes an angle of $\pi/6$ rad with $\mathbf{i}$ and an angle of $\pi/3$ rad with $\mathbf{j}$. Describe the reasoning behind your answer.

67. Is the line $x = 1 - 2t, y = 2 + 5t, z = -3t$ parallel to the plane $2x + y - z = 8$? Give reasons for your answer.

68. How can you tell when two planes $A_1x + B_1y + C_1z = D_1$ and $A_2x + B_2y + C_2z = D_2$ are parallel? Perpendicular? Give reasons for your answer.

69. Find two different planes whose intersection is the line $x = 1 + t, y = 2 - t, z = 3 + 2t$. Write equations for each plane in the form $Ax + By + Cz = D$.

70. Find a plane through the origin that meets the plane M: $2x + 3y + z = 12$ in a right angle. How do you know that your plane is perpendicular to M?

71. For any nonzero numbers a, b, and c, the graph of $(x/a) + (y/b) + (z/c) = 1$ is a plane. Which planes have an equation of this form?

72. Suppose L_1 and L_2 are disjoint (nonintersecting) nonparallel lines. Is it possible for a nonzero vector to be perpendicular to both L_1 and L_2? Give reasons for your answer.

Computer Graphics

73. Perspective in computer graphics In computer graphics and perspective drawing, we need to represent objects seen by the eye in space as images on a two-dimensional plane. Suppose that the eye is at $E(x_0, 0, 0)$ as shown here and that we want to represent a point $P_1(x_1, y_1, z_1)$ as a point on the yz-plane. We do this by projecting P_1 onto the plane with a ray from E. The point P_1 will be portrayed as the point $P(0, y, z)$. The problem for us as graphics designers is to find y and z given E and P_1.

a. Write a vector equation that holds between $\overrightarrow{EP}$ and $\overrightarrow{EP_1}$. Use the equation to express y and z in terms of x_0, x_1, y_1, and z_1.

b. Test the formulas obtained for y and z in part (a) by investigating their behavior at $x_1 = 0$ and $x_1 = x_0$ and by seeing what happens as $x_0 \to \infty$. What do you find?

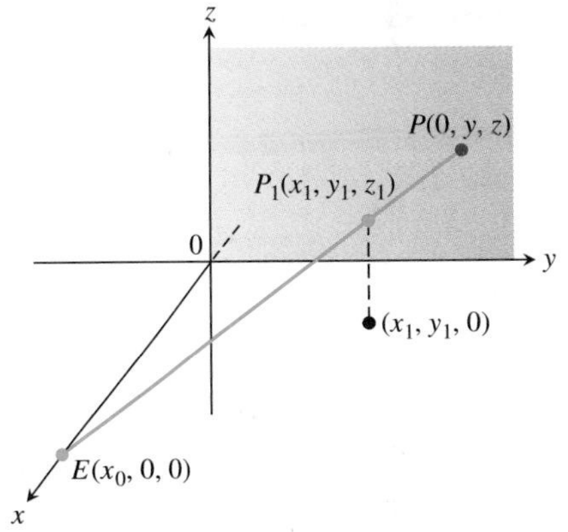

74. Hidden lines Here is another typical problem in computer graphics. Your eye is at $(4, 0, 0)$. You are looking at a triangular plate whose vertices are at $(1, 0, 1)$, $(1, 1, 0)$, and $(-2, 2, 2)$. The line segment from $(1, 0, 0)$ to $(0, 2, 2)$ passes through the plate. What portion of the line segment is hidden from your view by the plate? (This is an exercise in finding intersections of lines and planes.)

12.6 Cylinders and Quadric Surfaces

Up to now, we have studied two special types of surfaces: spheres and planes. In this section, we extend our inventory to include a variety of cylinders and quadric surfaces. Quadric surfaces are surfaces defined by second-degree equations in x, y, and z. Spheres are quadric surfaces, but there are others of equal interest.

Cylinders

A **cylinder** is a surface that is generated by moving a straight line along a given planar curve while holding the line parallel to a given fixed line. The curve is called a **generating curve** for the cylinder (Figure 12.43). In solid geometry, where *cylinder* means *circular* *cylinder*, the generating curves are circles, but now we allow generating curves of any kind. The cylinder in our first example is generated by a parabola.

When graphing a cylinder or other surface by hand or analyzing one generated by a computer, it helps to look at the curves formed by intersecting the surface with planes parallel to the coordinate planes. These curves are called **cross-sections** or **traces**.

EXAMPLE 1 The Parabolic Cylinder $y = x^2$

Find an equation for the cylinder made by the lines parallel to the z-axis that pass through the parabola $y = x^2, z = 0$ (Figure 12.44).

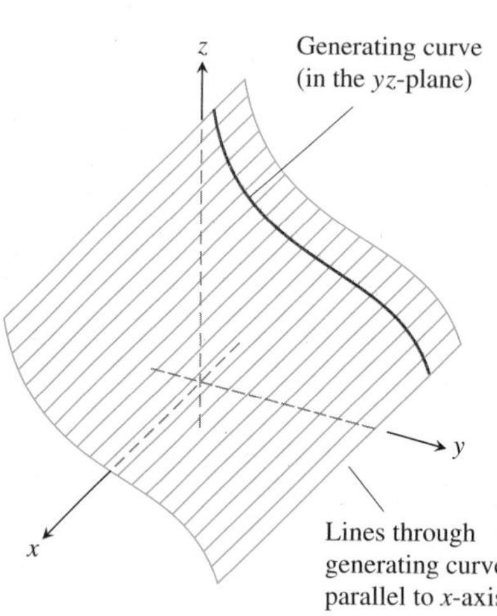

FIGURE 12.43 A cylinder and generating curve.

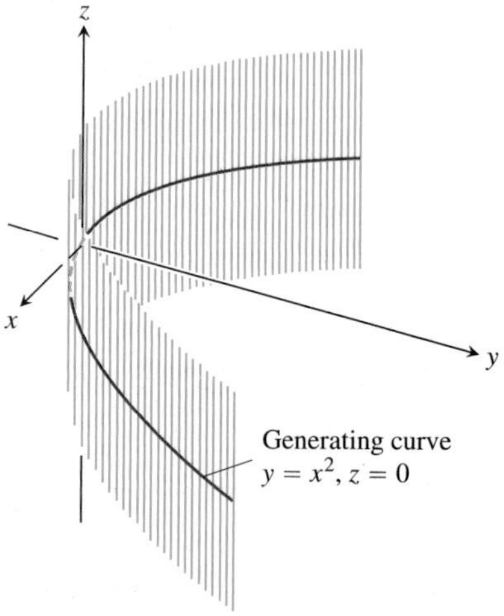

FIGURE 12.44 The cylinder of lines passing through the parabola $y = x^2$ in the xy-plane parallel to the z-axis (Example 1).

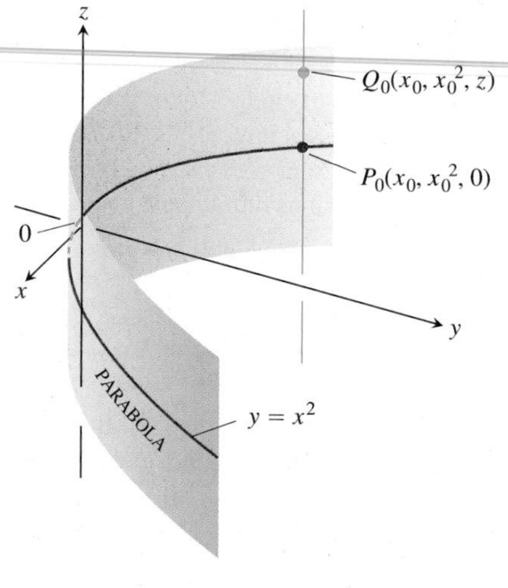

FIGURE 12.45 Every point of the cylinder in Figure 12.44 has coordinates of the form (x_0, x_0^2, z). We call it "the cylinder $y = x^2$."

Solution Suppose that the point $P_0(x_0, x_0^2, 0)$ lies on the parabola $y = x^2$ in the xy-plane. Then, for any value of z, the point $Q(x_0, x_0^2, z)$ will lie on the cylinder because it lies on the line $x = x_0, y = x_0^2$ through P_0 parallel to the z-axis. Conversely, any point $Q(x_0, x_0^2, z)$ whose y-coordinate is the square of its x-coordinate lies on the cylinder because it lies on the line $x = x_0, y = x_0^2$ through P_0 parallel to the z-axis (Figure 12.45).

Regardless of the value of z, therefore, the points on the surface are the points whose coordinates satisfy the equation $y = x^2$. This makes $y = x^2$ an equation for the cylinder. Because of this, we call the cylinder "the cylinder $y = x^2$." ■

As Example 1 suggests, any curve $f(x, y) = c$ in the xy-plane defines a cylinder parallel to the z-axis whose equation is also $f(x, y) = c$. The equation $x^2 + y^2 = 1$ defines the circular cylinder made by the lines parallel to the z-axis that pass through the circle $x^2 + y^2 = 1$ in the xy-plane. The equation $x^2 + 4y^2 = 9$ defines the elliptical cylinder made by the lines parallel to the z-axis that pass through the ellipse $x^2 + 4y^2 = 9$ in the xy-plane.

In a similar way, any curve $g(x, z) = c$ in the xz-plane defines a cylinder parallel to the y-axis whose space equation is also $g(x, z) = c$ (Figure 12.46). Any curve $h(y, z) = c$

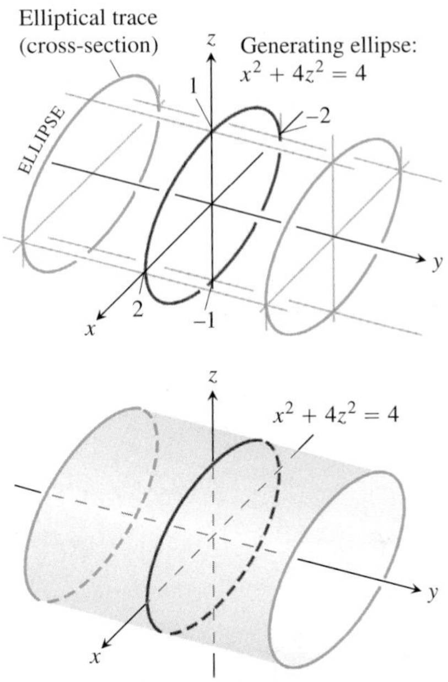

FIGURE 12.46 The elliptical cylinder $x^2 + 4z^2 = 4$ is made of lines parallel to the y-axis and passing through the ellipse $x^2 + 4z^2 = 4$ in the xz-plane. The cross-sections or "traces" of the cylinder in planes perpendicular to the y-axis are ellipses congruent to the generating ellipse. The cylinder extends along the entire y-axis.

defines a cylinder parallel to the x-axis whose space equation is also $h(y, z) = c$ (Figure 12.47). The axis of a cylinder need not be parallel to a coordinate axis, however.

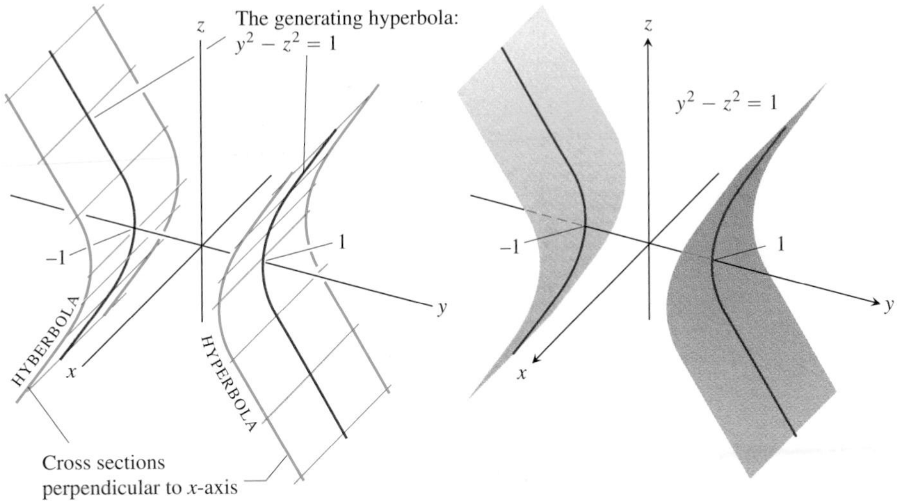

FIGURE 12.47 The hyperbolic cylinder $y^2 - z^2 = 1$ is made of lines parallel to the x-axis and passing through the hyperbola $y^2 - z^2 = 1$ in the yz-plane. The cross-sections of the cylinder in planes perpendicular to the x-axis are hyperbolas congruent to the generating hyperbola.

Quadric Surfaces

The next type of surface we examine is a *quadric* surface. These surfaces are the three-dimensional analogues of ellipses, parabolas, and hyperbolas.

A **quadric surface** is the graph in space of a second-degree equation in x, y, and z. The most general form is

$$Ax^2 + By^2 + Cz^2 + Dxy + Eyz + Fxz + Gx + Hy + Jz + K = 0,$$

where A, B, C, and so on are constants. However, this equation can be simplified by translation and rotation, as in the two-dimensional case. We will study only the simpler equations. Although defined differently, the cylinders in Figures 12.45 through 12.47 were also examples of quadric surfaces. The basic quadric surfaces are **ellipsoids**, **paraboloids**, **elliptical cones**, and **hyperboloids**. (We think of spheres as special ellipsoids.) We now present examples of each type.

EXAMPLE 2 Ellipsoids

The **ellipsoid**

$$\frac{x^2}{a^2} + \frac{y^2}{b^2} + \frac{z^2}{c^2} = 1 \qquad (1)$$

(Figure 12.48) cuts the coordinate axes at $(\pm a, 0, 0)$, $(0, \pm b, 0)$, and $(0, 0, \pm c)$. It lies within the rectangular box defined by the inequalities $|x| \le a$, $|y| \le b$, and $|z| \le c$. The surface is symmetric with respect to each of the coordinate planes because each variable in the defining equation is squared.

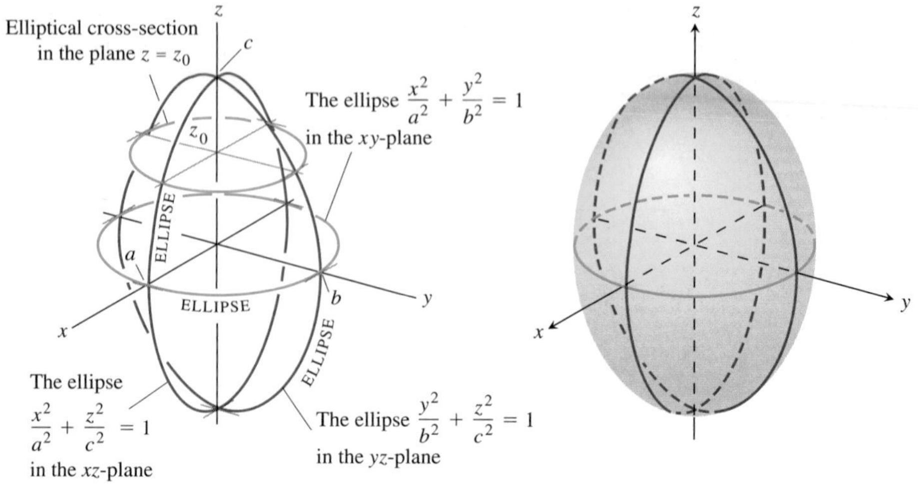

FIGURE 12.48 The ellipsoid

$$\frac{x^2}{a^2} + \frac{y^2}{b^2} + \frac{z^2}{c^2} = 1$$

in Example 2 has elliptical cross-sections in each of the three coordinate planes.

The curves in which the three coordinate planes cut the surface are ellipses. For example,

$$\frac{x^2}{a^2} + \frac{y^2}{b^2} = 1 \qquad \text{when} \qquad z = 0.$$

The section cut from the surface by the plane $z = z_0$, $|z_0| < c$, is the ellipse

$$\frac{x^2}{a^2(1 - (z_0/c)^2)} + \frac{y^2}{b^2(1 - (z_0/c)^2)} = 1.$$

If any two of the semiaxes a, b, and c are equal, the surface is an **ellipsoid of revolution**. If all three are equal, the surface is a sphere. ■

EXAMPLE 3 Paraboloids

The **elliptical paraboloid**

$$\frac{x^2}{a^2} + \frac{y^2}{b^2} = \frac{z}{c} \tag{2}$$

is symmetric with respect to the planes $x = 0$ and $y = 0$ (Figure 12.49). The only intercept on the axes is the origin. Except for this point, the surface lies above (if $c > 0$) or entirely below (if $c < 0$) the xy-plane, depending on the sign of c. The sections cut by the coordinate planes are

$x = 0$: the parabola $z = \frac{c}{b^2}y^2$

$y = 0$: the parabola $z = \frac{c}{a^2}x^2$

$z = 0$: the point $(0, 0, 0)$.

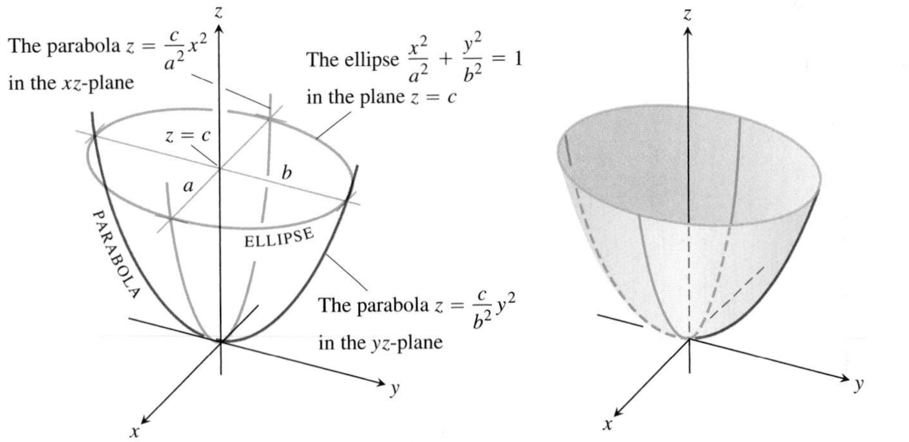

FIGURE 12.49 The elliptical paraboloid $(x^2/a^2) + (y^2/b^2) = z/c$ in Example 3, shown for $c > 0$. The cross-sections perpendicular to the z-axis above the xy-plane are ellipses. The cross-sections in the planes that contain the z-axis are parabolas.

Each plane $z = z_0$ above the xy-plane cuts the surface in the ellipse

$$\frac{x^2}{a^2} + \frac{y^2}{b^2} = \frac{z_0}{c}.$$

EXAMPLE 4 Cones

The **elliptical cone**

$$\frac{x^2}{a^2} + \frac{y^2}{b^2} = \frac{z^2}{c^2} \tag{3}$$

is symmetric with respect to the three coordinate planes (Figure 12.50). The sections cut

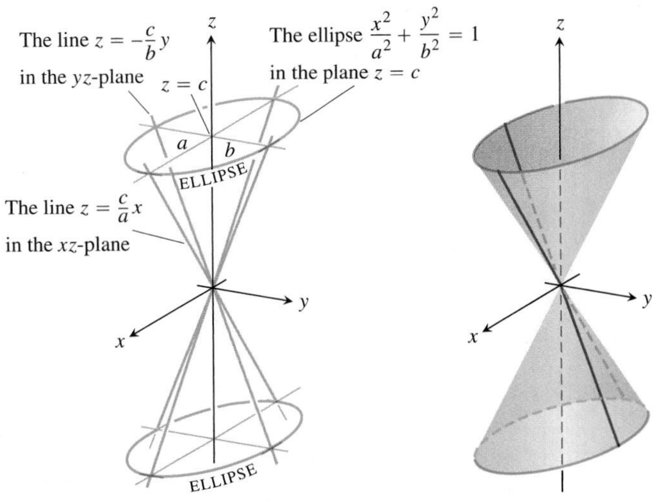

FIGURE 12.50 The elliptical cone $(x^2/a^2) + (y^2/b^2) = (z^2/c^2)$ in Example 4. Planes perpendicular to the z-axis cut the cone in ellipses above and below the xy-plane. Vertical planes that contain the z-axis cut it in pairs of intersecting lines.

by the coordinate planes are

$$x = 0: \quad \text{the lines } z = \pm \frac{c}{b} y$$

$$y = 0: \quad \text{the lines } z = \pm \frac{c}{a} x$$

$$z = 0: \quad \text{the point } (0, 0, 0).$$

The sections cut by planes $z = z_0$ above and below the xy-plane are ellipses whose centers lie on the z-axis and whose vertices lie on the lines given above.

If $a = b$, the cone is a right circular cone. ■

EXAMPLE 5 Hyperboloids

The **hyperboloid of one sheet**

$$\frac{x^2}{a^2} + \frac{y^2}{b^2} - \frac{z^2}{c^2} = 1 \tag{4}$$

is symmetric with respect to each of the three coordinate planes (Figure 12.51).

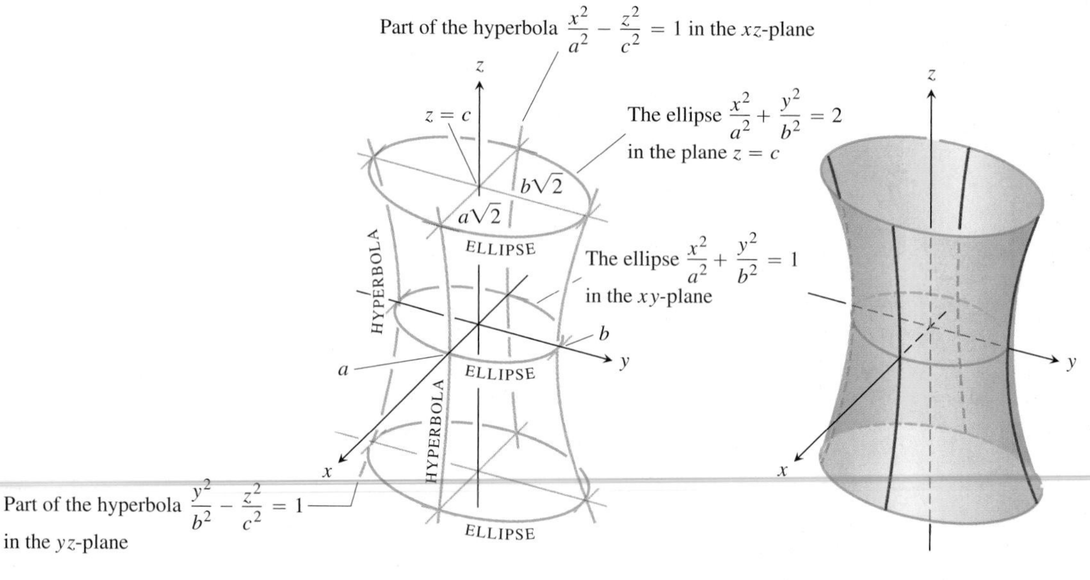

FIGURE 12.51 The hyperboloid $(x^2/a^2) + (y^2/b^2) - (z^2/c^2) = 1$ in Example 5. Planes perpendicular to the z-axis cut it in ellipses. Vertical planes containing the z-axis cut it in hyperbolas.

The sections cut out by the coordinate planes are

$$x = 0: \quad \text{the hyperbola } \frac{y^2}{b^2} - \frac{z^2}{c^2} = 1$$

$$y = 0: \quad \text{the hyperbola } \frac{x^2}{a^2} - \frac{z^2}{c^2} = 1$$

$$z = 0: \quad \text{the ellipse } \frac{x^2}{a^2} + \frac{y^2}{b^2} = 1.$$

The plane $z = z_0$ cuts the surface in an ellipse with center on the z-axis and vertices on one of the hyperbolic sections above.

The surface is connected, meaning that it is possible to travel from one point on it to any other without leaving the surface. For this reason, it is said to have *one* sheet, in contrast to the hyperboloid in the next example, which has two sheets.

If $a = b$, the hyperboloid is a surface of revolution. ■

EXAMPLE 6 Hyperboloids

The **hyperboloid of two sheets**

$$\frac{z^2}{c^2} - \frac{x^2}{a^2} - \frac{y^2}{b^2} = 1 \tag{5}$$

is symmetric with respect to the three coordinate planes (Figure 12.52). The plane $z = 0$ does not intersect the surface; in fact, for a horizontal plane to intersect the surface, we must have $|z| \geq c$. The hyperbolic sections

$$x = 0: \quad \frac{z^2}{c^2} - \frac{y^2}{b^2} = 1$$

$$y = 0: \quad \frac{z^2}{c^2} - \frac{x^2}{a^2} = 1$$

have their vertices and foci on the z-axis. The surface is separated into two portions, one above the plane $z = c$ and the other below the plane $z = -c$. This accounts for its name.

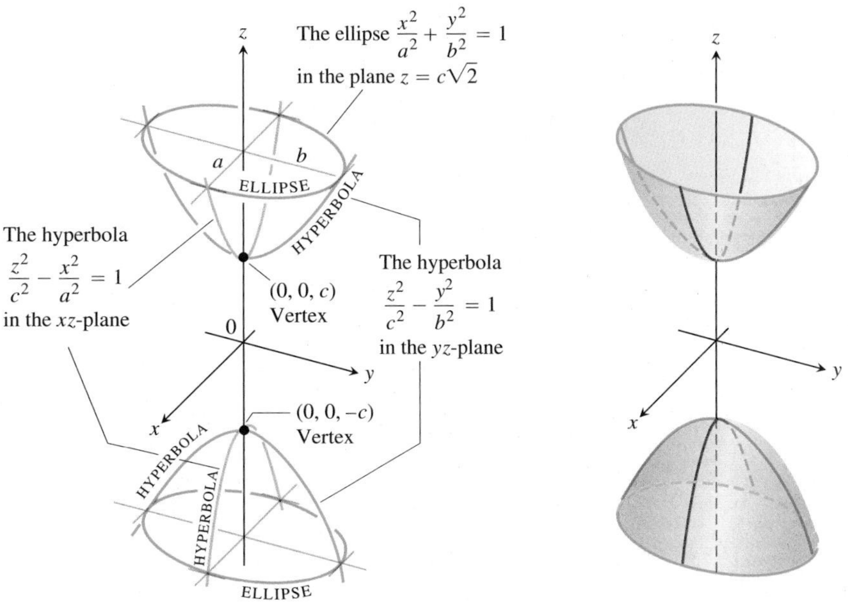

FIGURE 12.52 The hyperboloid $(z^2/c^2) - (x^2/a^2) - (y^2/b^2) = 1$ in Example 6. Planes perpendicular to the z-axis above and below the vertices cut it in ellipses. Vertical planes containing the z-axis cut it in hyperbolas.

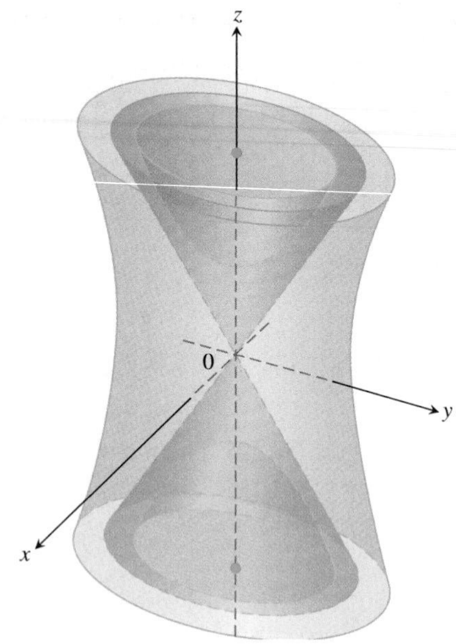

FIGURE 12.53 Both hyperboloids are asymptotic to the cone (Example 6).

Equations (4) and (5) have different numbers of negative terms. The number in each case is the same as the number of sheets of the hyperboloid. If we replace the 1 on the right side of either Equation (4) or Equation (5) by 0, we obtain the equation

$$\frac{x^2}{a^2} + \frac{y^2}{b^2} = \frac{z^2}{c^2}$$

for an elliptical cone (Equation 3). The hyperboloids are asymptotic to this cone (Figure 12.53) in the same way that the hyperbolas

$$\frac{x^2}{a^2} - \frac{y^2}{b^2} = \pm 1$$

are asymptotic to the lines

$$\frac{x^2}{a^2} - \frac{y^2}{b^2} = 0$$

in the xy-plane. ■

EXAMPLE 7 A Saddle Point

The **hyperbolic paraboloid**

$$\frac{y^2}{b^2} - \frac{x^2}{a^2} = \frac{z}{c}, \qquad c > 0 \tag{6}$$

has symmetry with respect to the planes $x = 0$ and $y = 0$ (Figure 12.54). The sections in these planes are

$$x = 0: \quad \text{the parabola } z = \frac{c}{b^2}y^2. \tag{7}$$

$$y = 0: \quad \text{the parabola } z = -\frac{c}{a^2}x^2. \tag{8}$$

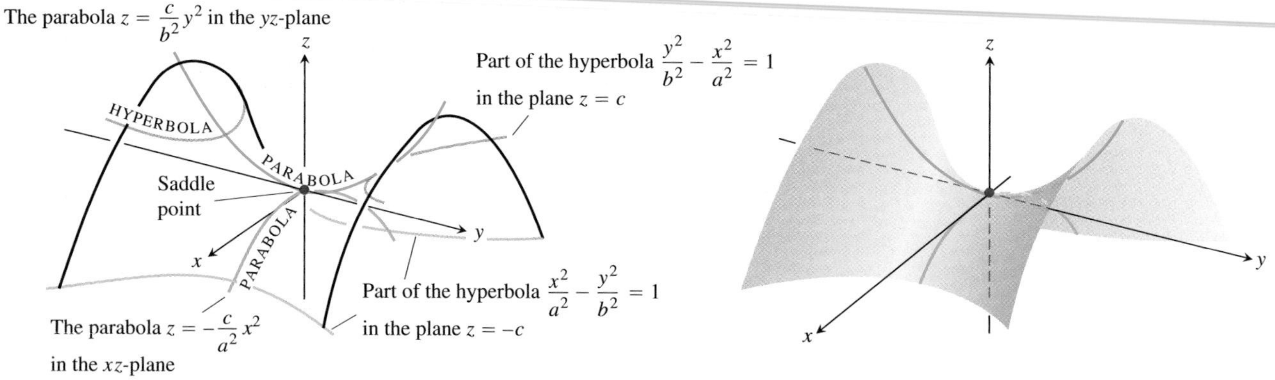

FIGURE 12.54 The hyperbolic paraboloid $(y^2/b^2) - (x^2/a^2) = z/c$, $c > 0$. The cross-sections in planes perpendicular to the z-axis above and below the xy-plane are hyperbolas. The cross-sections in planes perpendicular to the other axes are parabolas.

In the plane $x = 0$, the parabola opens upward from the origin. The parabola in the plane $y = 0$ opens downward.

If we cut the surface by a plane $z = z_0 > 0$, the section is a hyperbola,

$$\frac{y^2}{b^2} - \frac{x^2}{a^2} = \frac{z_0}{c},$$

with its focal axis parallel to the y-axis and its vertices on the parabola in Equation (7). If z_0 is negative, the focal axis is parallel to the x-axis and the vertices lie on the parabola in Equation (8).

Near the origin, the surface is shaped like a saddle or mountain pass. To a person traveling along the surface in the yz-plane the origin looks like a minimum. To a person traveling in the xz-plane the origin looks like a maximum. Such a point is called a **saddle point** of a surface. ■

USING TECHNOLOGY Visualizing in Space

A CAS or other graphing utility can help in visualizing surfaces in space. It can draw traces in different planes, and many computer graphing systems can rotate a figure so you can see it as if it were a physical model you could turn in your hand. Hidden-line algorithms (see Exercise 74, Section 12.5) are used to block out portions of the surface that you would not see from your current viewing angle. A system may require surfaces to be entered in parametric form, as discussed in Section 16.6 (see also CAS Exercises 57 through 60 in Section 14.1). Sometimes you may have to manipulate the grid mesh to see all portions of a surface.

EXERCISES 12.6

Matching Equations with Surfaces

In Exercises 1–12, match the equation with the surface it defines. Also, identify each surface by type (paraboloid, ellipsoid, etc.) The surfaces are labeled (a)–(1).

1. $x^2 + y^2 + 4z^2 = 10$
2. $z^2 + 4y^2 - 4x^2 = 4$
3. $9y^2 + z^2 = 16$
4. $y^2 + z^2 = x^2$
5. $x = y^2 - z^2$
6. $x = -y^2 - z^2$
7. $x^2 + 2z^2 = 8$
8. $z^2 + x^2 - y^2 = 1$
9. $x = z^2 - y^2$
10. $z = -4x^2 - y^2$
11. $x^2 + 4z^2 = y^2$
12. $9x^2 + 4y^2 + 2z^2 = 36$

a.

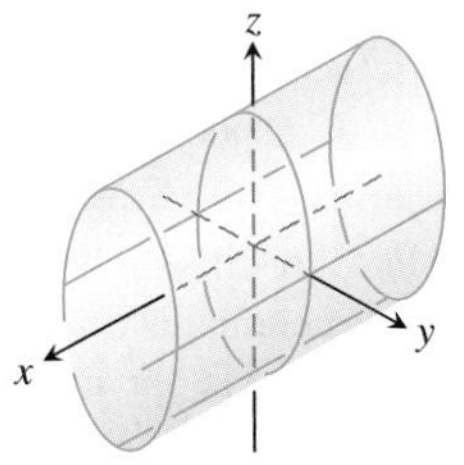

b.

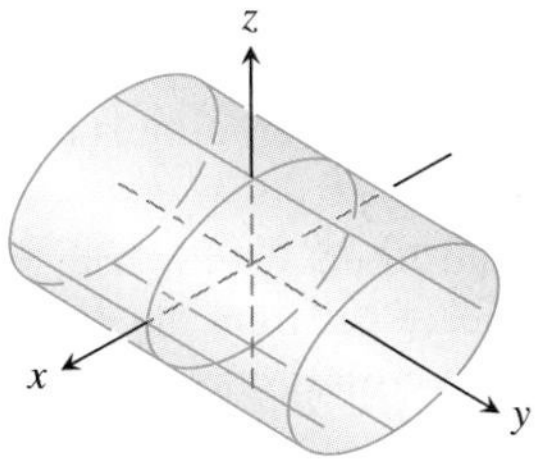

c.

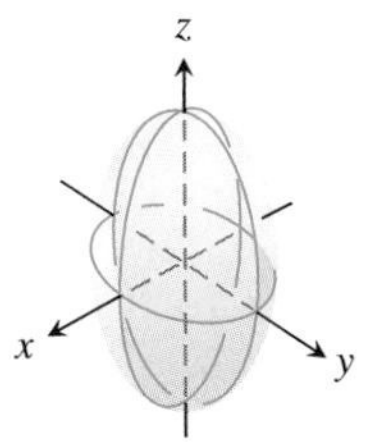

d.

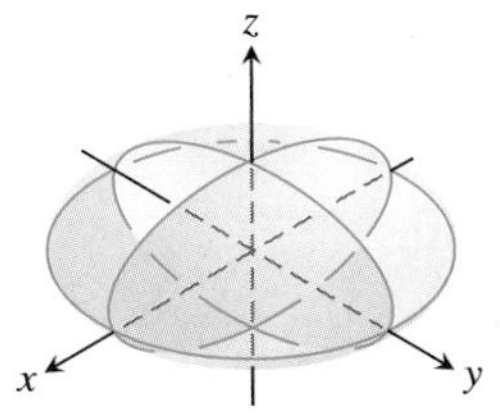

e.

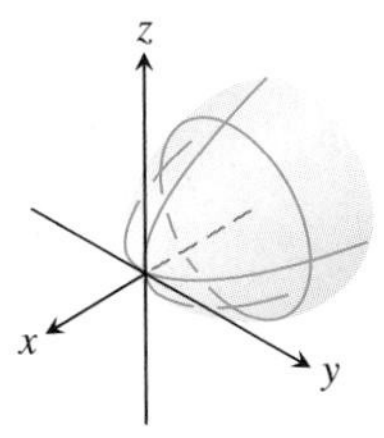

f.

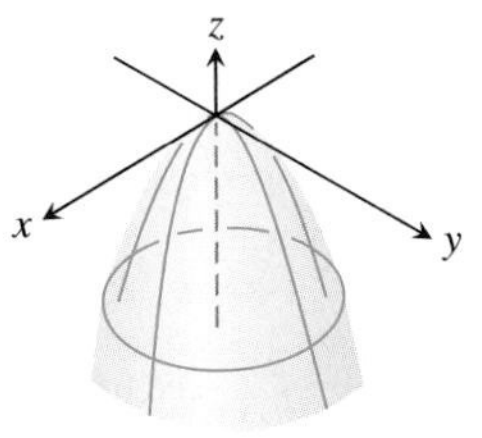

g.

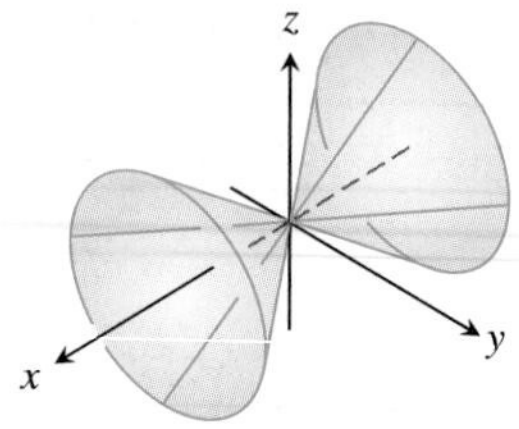

h.

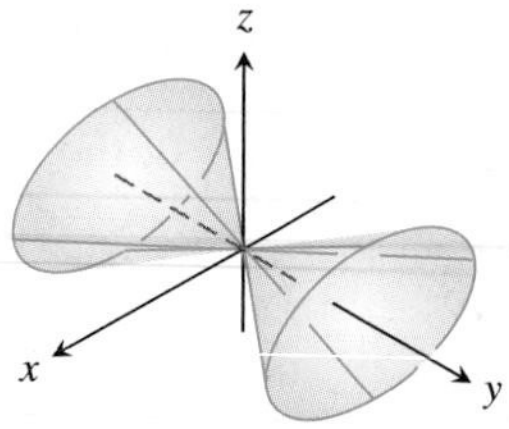

i.

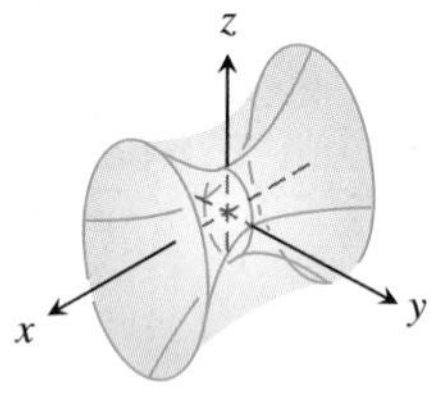

j.

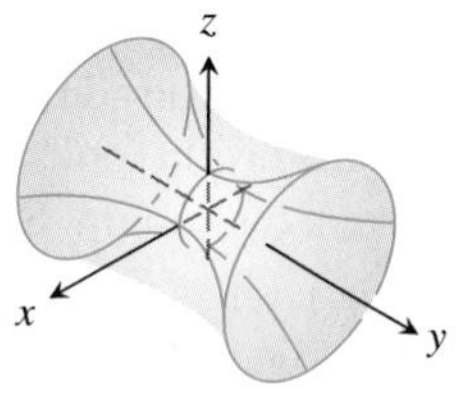

k.

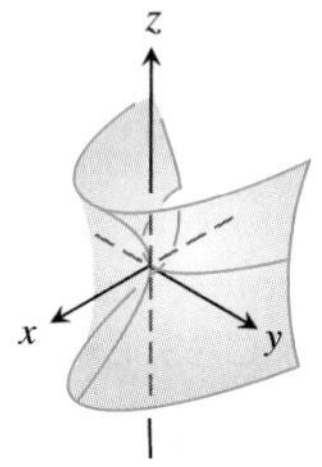

l.

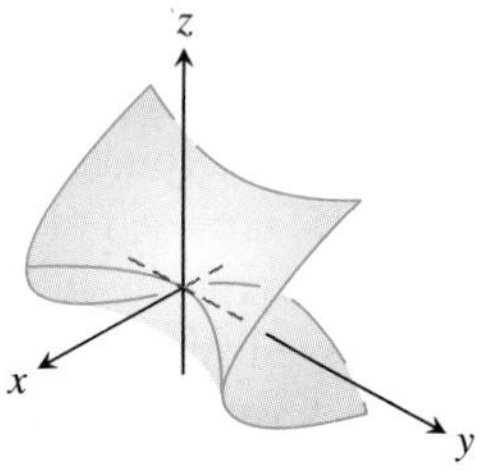

Drawing

Sketch the surfaces in Exercises 13–76.

CYLINDERS

13. $x^2 + y^2 = 4$
14. $x^2 + z^2 = 4$
15. $z = y^2 - 1$
16. $x = y^2$
17. $x^2 + 4z^2 = 16$
18. $4x^2 + y^2 = 36$
19. $z^2 - y^2 = 1$
20. $yz = 1$

ELLIPSOIDS

21. $9x^2 + y^2 + z^2 = 9$
22. $4x^2 + 4y^2 + z^2 = 16$
23. $4x^2 + 9y^2 + 4z^2 = 36$
24. $9x^2 + 4y^2 + 36z^2 = 36$

PARABOLOIDS

25. $z = x^2 + 4y^2$
26. $z = x^2 + 9y^2$
27. $z = 8 - x^2 - y^2$
28. $z = 18 - x^2 - 9y^2$
29. $x = 4 - 4y^2 - z^2$
30. $y = 1 - x^2 - z^2$

CONES

31. $x^2 + y^2 = z^2$
32. $y^2 + z^2 = x^2$
33. $4x^2 + 9z^2 = 9y^2$
34. $9x^2 + 4y^2 = 36z^2$

HYPERBOLOIDS

35. $x^2 + y^2 - z^2 = 1$
36. $y^2 + z^2 - x^2 = 1$
37. $(y^2/4) + (z^2/9) - (x^2/4) = 1$
38. $(x^2/4) + (y^2/4) - (z^2/9) = 1$
39. $z^2 - x^2 - y^2 = 1$
40. $(y^2/4) - (x^2/4) - z^2 = 1$
41. $x^2 - y^2 - (z^2/4) = 1$
42. $(x^2/4) - y^2 - (z^2/4) = 1$

HYPERBOLIC PARABOLOIDS

43. $y^2 - x^2 = z$
44. $x^2 - y^2 = z$

ASSORTED

45. $x^2 + y^2 + z^2 = 4$
46. $4x^2 + 4y^2 = z^2$
47. $z = 1 + y^2 - x^2$
48. $y^2 - z^2 = 4$
49. $y = -(x^2 + z^2)$
50. $z^2 - 4x^2 - 4y^2 = 4$
51. $16x^2 + 4y^2 = 1$
52. $z = x^2 + y^2 + 1$
53. $x^2 + y^2 - z^2 = 4$
54. $x = 4 - y^2$
55. $x^2 + z^2 = y$
56. $z^2 - (x^2/4) - y^2 = 1$
57. $x^2 + z^2 = 1$
58. $4x^2 + 4y^2 + z^2 = 4$
59. $16y^2 + 9z^2 = 4x^2$
60. $z = x^2 - y^2 - 1$
61. $9x^2 + 4y^2 + z^2 = 36$
62. $4x^2 + 9z^2 = y^2$
63. $x^2 + y^2 - 16z^2 = 16$
64. $z^2 + 4y^2 = 9$
65. $z = -(x^2 + y^2)$
66. $y^2 - x^2 - z^2 = 1$
67. $x^2 - 4y^2 = 1$
68. $z = 4x^2 + y^2 - 4$
69. $4y^2 + z^2 - 4x^2 = 4$
70. $z = 1 - x^2$
71. $x^2 + y^2 = z$
72. $(x^2/4) + y^2 - z^2 = 1$
73. $yz = 1$
74. $36x^2 + 9y^2 + 4z^2 = 36$
75. $9x^2 + 16y^2 = 4z^2$
76. $4z^2 - x^2 - y^2 = 4$

Theory and Examples

77. **a.** Express the area A of the cross-section cut from the ellipsoid

$$x^2 + \frac{y^2}{4} + \frac{z^2}{9} = 1$$

by the plane $z = c$ as a function of c. (The area of an ellipse with semiaxes a and b is πab.)

b. Use slices perpendicular to the z-axis to find the volume of the ellipsoid in part (a).

c. Now find the volume of the ellipsoid

$$\frac{x^2}{a^2} + \frac{y^2}{b^2} + \frac{z^2}{c^2} = 1.$$

Does your formula give the volume of a sphere of radius a if $a = b = c$?

78. The barrel shown here is shaped like an ellipsoid with equal pieces cut from the ends by planes perpendicular to the z-axis. The cross-sections perpendicular to the z-axis are circular. The barrel is $2h$ units high, its midsection radius is R, and its end radii are both r. Find a formula for the barrel's volume. Then check two things. First, suppose the sides of the barrel are straightened to turn the barrel into a cylinder of radius R and height $2h$. Does your formula give the cylinder's volume? Second, suppose $r = 0$ and $h = R$ so the barrel is a sphere. Does your formula give the sphere's volume?

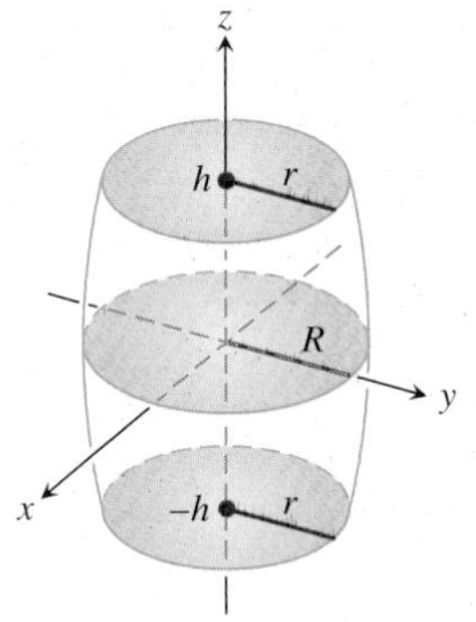

79. Show that the volume of the segment cut from the paraboloid

$$\frac{x^2}{a^2} + \frac{y^2}{b^2} = \frac{z}{c}$$

by the plane $z = h$ equals half the segment's base times its altitude. (Figure 12.49 shows the segment for the special case $h = c$.)

80. **a.** Find the volume of the solid bounded by the hyperboloid

$$\frac{x^2}{a^2} + \frac{y^2}{b^2} - \frac{z^2}{c^2} = 1$$

and the planes $z = 0$ and $z = h, h > 0$.

b. Express your answer in part (a) in terms of h and the areas A_0 and A_h of the regions cut by the hyperboloid from the planes $z = 0$ and $z = h$.

c. Show that the volume in part (a) is also given by the formula

$$V = \frac{h}{6}(A_0 + 4A_m + A_h),$$

where A_m is the area of the region cut by the hyperboloid from the plane $z = h/2$.

81. If the hyperbolic paraboloid $(y^2/b^2) - (x^2/a^2) = z/c$ is cut by the plane $y = y_1$, the resulting curve is a parabola. Find its vertex and focus.

82. Suppose you set $z = 0$ in the equation

$$Ax^2 + By^2 + Cz^2 + Dxy + Eyz + Fxz + Gx + Hy + Jz + K = 0$$

to obtain a curve in the xy-plane. What will the curve be like? Give reasons for your answer.

83. Every time we found the trace of a quadric surface in a plane parallel to one of the coordinate planes, it turned out to be a conic section. Was this mere coincidence? Did it have to happen? Give reasons for your answer.

84. Suppose you intersect a quadric surface with a plane that is *not* parallel to one of the coordinate planes. What will the trace in the plane be like? Give reasons for your answer.

T Computer Grapher Explorations

Plot the surfaces in Exercises 85–88 over the indicated domains. If you can, rotate the surface into different viewing positions.

85. $z = y^2, \quad -2 \le x \le 2, \quad -0.5 \le y \le 2$

86. $z = 1 - y^2, \quad -2 \le x \le 2, \quad -2 \le y \le 2$

87. $z = x^2 + y^2, \quad -3 \le x \le 3, \quad -3 \le y \le 3$

88. $z = x^2 + 2y^2$ over

a. $-3 \le x \le 3, \quad -3 \le y \le 3$

b. $-1 \le x \le 1, \quad -2 \le y \le 3$

c. $-2 \le x \le 2, \quad -2 \le y \le 2$

d. $-2 \le x \le 2, \quad -1 \le y \le 1$

COMPUTER EXPLORATIONS

Surface Plots

Use a CAS to plot the surfaces in Exercises 89–94. Identify the type of quadric surface from your graph.

89. $\frac{x^2}{9} + \frac{y^2}{36} = 1 - \frac{z^2}{25}$

90. $\frac{x^2}{9} - \frac{z^2}{9} = 1 - \frac{y^2}{16}$

91. $5x^2 = z^2 - 3y^2$

92. $\frac{y^2}{16} = 1 - \frac{x^2}{9} + z$

93. $\frac{x^2}{9} - 1 = \frac{y^2}{16} + \frac{z^2}{2}$

94. $y - \sqrt{4 - z^2} = 0$

Chapter 12 Questions to Guide Your Review

1. When do directed line segments in the plane represent the same vector?
2. How are vectors added and subtracted geometrically? Algebraically?
3. How do you find a vector's magnitude and direction?
4. If a vector is multiplied by a positive scalar, how is the result related to the original vector? What if the scalar is zero? Negative?

5. Define the *dot product* (*scalar product*) of two vectors. Which algebraic laws are satisfied by dot products? Give examples. When is the dot product of two vectors equal to zero?
6. What geometric interpretation does the dot product have? Give examples.
7. What is the vector projection of a vector **u** onto a vector **v**? How do you write **u** as the sum of a vector parallel to **v** and a vector orthogonal to **v**?
8. Define the *cross product* (*vector product*) of two vectors. Which algebraic laws are satisfied by cross products, and which are not? Give examples. When is the cross product of two vectors equal to zero?
9. What geometric or physical interpretations do cross products have? Give examples.
10. What is the determinant formula for calculating the cross product of two vectors relative to the Cartesian **i**, **j**, **k**-coordinate system? Use it in an example.
11. How do you find equations for lines, line segments, and planes in space? Give examples. Can you express a line in space by a single equation? A plane?
12. How do you find the distance from a point to a line in space? From a point to a plane? Give examples.
13. What are box products? What significance do they have? How are they evaluated? Give an example.
14. How do you find equations for spheres in space? Give examples.
15. How do you find the intersection of two lines in space? A line and a plane? Two planes? Give examples.
16. What is a cylinder? Give examples of equations that define cylinders in Cartesian coordinates.
17. What are quadric surfaces? Give examples of different kinds of ellipsoids, paraboloids, cones, and hyperboloids (equations and sketches).

Chapter 12 Practice Exercises

Vector Calculations in Two Dimensions

In Exercises 1–4, let $\mathbf{u} = \langle -3, 4\rangle$ and $\mathbf{v} = \langle 2, -5\rangle$. Find **(a)** the component form of the vector and **(b)** its magnitude.

1. $3\mathbf{u} - 4\mathbf{v}$
2. $\mathbf{u} + \mathbf{v}$
3. $-2\mathbf{u}$
4. $5\mathbf{v}$

In Exercises 5–8, find the component form of the vector.

5. The vector obtained by rotating $\langle 0, 1\rangle$ through an angle of $2\pi/3$ radians
6. The unit vector that makes an angle of $\pi/6$ radian with the positive x-axis
7. The vector 2 units long in the direction $4\mathbf{i} - \mathbf{j}$
8. The vector 5 units long in the direction opposite to the direction of $(3/5)\mathbf{i} + (4/5)\mathbf{j}$

Express the vectors in Exercises 9–12 in terms of their lengths and directions.

9. $\sqrt{2}\mathbf{i} + \sqrt{2}\mathbf{j}$
10. $-\mathbf{i} - \mathbf{j}$
11. Velocity vector $\mathbf{v} = (-2\sin t)\mathbf{i} + (2\cos t)\mathbf{j}$ when $t = \pi/2$.
12. Velocity vector $\mathbf{v} = (e^t\cos t - e^t\sin t)\mathbf{i} + (e^t\sin t + e^t\cos t)\mathbf{j}$ when $t = \ln 2$.

Vector Calculations in Three Dimensions

Express the vectors in Exercises 13 and 14 in terms of their lengths and directions.

13. $2\mathbf{i} - 3\mathbf{j} + 6\mathbf{k}$
14. $\mathbf{i} + 2\mathbf{j} - \mathbf{k}$
15. Find a vector 2 units long in the direction of $\mathbf{v} = 4\mathbf{i} - \mathbf{j} + 4\mathbf{k}$.
16. Find a vector 5 units long in the direction opposite to the direction of $\mathbf{v} = (3/5)\mathbf{i} + (4/5)\mathbf{k}$.

In Exercises 17 and 18, find $|\mathbf{v}|$, $|\mathbf{u}|$, $\mathbf{v}\cdot\mathbf{u}$, $\mathbf{u}\cdot\mathbf{v}$, $\mathbf{v}\times\mathbf{u}$, $\mathbf{u}\times\mathbf{v}$, $|\mathbf{v}\times\mathbf{u}|$, the angle between **v** and **u**, the scalar component of **u** in the direction of **v**, and the vector projection of **u** onto **v**.

17. $\mathbf{v} = \mathbf{i} + \mathbf{j}$
 $\mathbf{u} = 2\mathbf{i} + \mathbf{j} - 2\mathbf{k}$
18. $\mathbf{v} = \mathbf{i} + \mathbf{j} + 2\mathbf{k}$
 $\mathbf{u} = -\mathbf{i} - \mathbf{k}$

In Exercises 19 and 20, write **u** as the sum of a vector parallel to **v** and a vector orthogonal to **v**.

19. $\mathbf{v} = 2\mathbf{i} + \mathbf{j} - \mathbf{k}$
 $\mathbf{u} = \mathbf{i} + \mathbf{j} - 5\mathbf{k}$
20. $\mathbf{u} = \mathbf{i} - 2\mathbf{j}$
 $\mathbf{v} = \mathbf{i} + \mathbf{j} + \mathbf{k}$

In Exercises 21 and 22, draw coordinate axes and then sketch **u**, **v**, and $\mathbf{u}\times\mathbf{v}$ as vectors at the origin.

21. $\mathbf{u} = \mathbf{i}, \quad \mathbf{v} = \mathbf{i} + \mathbf{j}$
22. $\mathbf{u} = \mathbf{i} - \mathbf{j}, \quad \mathbf{v} = \mathbf{i} + \mathbf{j}$
23. If $|\mathbf{v}| = 2$, $|\mathbf{w}| = 3$, and the angle between **v** and **w** is $\pi/3$, find $|\mathbf{v} - 2\mathbf{w}|$.
24. For what value or values of a will the vectors $\mathbf{u} = 2\mathbf{i} + 4\mathbf{j} - 5\mathbf{k}$ and $\mathbf{v} = -4\mathbf{i} - 8\mathbf{j} + a\mathbf{k}$ be parallel?

In Exercises 25 and 26, find **(a)** the area of the parallelogram determined by vectors **u** and **v** and **(b)** the volume of the parallelepiped determined by the vectors **u**, **v**, and **w**.

25. $\mathbf{u} = \mathbf{i} + \mathbf{j} - \mathbf{k}, \quad \mathbf{v} = 2\mathbf{i} + \mathbf{j} + \mathbf{k}, \quad \mathbf{w} = -\mathbf{i} - 2\mathbf{j} + 3\mathbf{k}$
26. $\mathbf{u} = \mathbf{i} + \mathbf{j}, \quad \mathbf{v} = \mathbf{j}, \quad \mathbf{w} = \mathbf{i} + \mathbf{j} + \mathbf{k}$

Lines, Planes, and Distances

27. Suppose that **n** is normal to a plane and that **v** is parallel to the plane. Describe how you would find a vector **n** that is both perpendicular to **v** and parallel to the plane.

28. Find a vector in the plane parallel to the line $ax + by = c$.

In Exercises 29 and 30, find the distance from the point to the line.

29. $(2, 2, 0)$; $x = -t, \quad y = t, \quad z = -1 + t$

30. $(0, 4, 1)$; $x = 2 + t, \quad y = 2 + t, \quad z = t$

31. Parametrize the line that passes through the point $(1, 2, 3)$ parallel to the vector $\mathbf{v} = -3\mathbf{i} + 7\mathbf{k}$.

32. Parametrize the line segment joining the points $P(1, 2, 0)$ and $Q(1, 3, -1)$.

In Exercises 33 and 34, find the distance from the point to the plane.

33. $(6, 0, -6), \quad x - y = 4$

34. $(3, 0, 10), \quad 2x + 3y + z = 2$

35. Find an equation for the plane that passes through the point $(3, -2, 1)$ normal to the vector $\mathbf{n} = 2\mathbf{i} + \mathbf{j} + \mathbf{k}$.

36. Find an equation for the plane that passes through the point $(-1, 6, 0)$ perpendicular to the line $x = -1 + t, y = 6 - 2t, z = 3t$.

In Exercises 37 and 38, find an equation for the plane through points P, Q, and R.

37. $P(1, -1, 2), \quad Q(2, 1, 3), \quad R(-1, 2, -1)$

38. $P(1, 0, 0), \quad Q(0, 1, 0), \quad R(0, 0, 1)$

39. Find the points in which the line $x = 1 + 2t, y = -1 - t, z = 3t$ meets the three coordinate planes.

40. Find the point in which the line through the origin perpendicular to the plane $2x - y - z = 4$ meets the plane $3x - 5y + 2z = 6$.

41. Find the acute angle between the planes $x = 7$ and $x + y + \sqrt{2}z = -3$.

42. Find the acute angle between the planes $x + y = 1$ and $y + z = 1$.

43. Find parametric equations for the line in which the planes $x + 2y + z = 1$ and $x - y + 2z = -8$ intersect.

44. Show that the line in which the planes

$$x + 2y - 2z = 5 \quad \text{and} \quad 5x - 2y - z = 0$$

intersect is parallel to the line

$$x = -3 + 2t, \quad y = 3t, \quad z = 1 + 4t.$$

45. The planes $3x + 6z = 1$ and $2x + 2y - z = 3$ intersect in a line.

a. Show that the planes are orthogonal.

b. Find equations for the line of intersection.

46. Find an equation for the plane that passes through the point $(1, 2, 3)$ parallel to $\mathbf{u} = 2\mathbf{i} + 3\mathbf{j} + \mathbf{k}$ and $\mathbf{v} = \mathbf{i} - \mathbf{j} + 2\mathbf{k}$.

47. Is $\mathbf{v} = 2\mathbf{i} - 4\mathbf{j} + \mathbf{k}$ related in any special way to the plane $2x + y = 5$? Give reasons for your answer.

48. The equation $\mathbf{n} \cdot \overrightarrow{P_0P} = 0$ represents the plane through P_0 normal to $\mathbf{n}$. What set does the inequality $\mathbf{n} \cdot \overrightarrow{P_0P} > 0$ represent?

49. Find the distance from the point $P(1, 4, 0)$ to the plane through $A(0, 0, 0)$, $B(2, 0, -1)$ and $C(2, -1, 0)$.

50. Find the distance from the point $(2, 2, 3)$ to the plane $2x + 3y + 5z = 0$.

51. Find a vector parallel to the plane $2x - y - z = 4$ and orthogonal to $\mathbf{i} + \mathbf{j} + \mathbf{k}$.

52. Find a unit vector orthogonal to **A** in the plane of **B** and **C** if $\mathbf{A} = 2\mathbf{i} - \mathbf{j} + \mathbf{k}$, $\mathbf{B} = \mathbf{i} + 2\mathbf{j} + \mathbf{k}$, and $\mathbf{C} = \mathbf{i} + \mathbf{j} - 2\mathbf{k}$.

53. Find a vector of magnitude 2 parallel to the line of intersection of the planes $x + 2y + z - 1 = 0$ and $x - y + 2z + 7 = 0$.

54. Find the point in which the line through the origin perpendicular to the plane $2x - y - z = 4$ meets the plane $3x - 5y + 2z = 6$.

55. Find the point in which the line through $P(3, 2, 1)$ normal to the plane $2x - y + 2z = -2$ meets the plane.

56. What angle does the line of intersection of the planes $2x + y - z = 0$ and $x + y + 2z = 0$ make with the positive x-axis?

57. The line

$$L: \quad x = 3 + 2t, \quad y = 2t, \quad z = t$$

intersects the plane $x + 3y - z = -4$ in a point P. Find the coordinates of P and find equations for the line in the plane through P perpendicular to L.

58. Show that for every real number k the plane

$$x - 2y + z + 3 + k(2x - y - z + 1) = 0$$

contains the line of intersection of the planes

$$x - 2y + z + 3 = 0 \quad \text{and} \quad 2x - y - z + 1 = 0.$$

59. Find an equation for the plane through $A(-2, 0, -3)$ and $B(1, -2, 1)$ that lies parallel to the line through $C(-2, -13/5, 26/5)$ and $D(16/5, -13/5, 0)$.

60. Is the line $x = 1 + 2t, y = -2 + 3t, z = -5t$ related in any way to the plane $-4x - 6y + 10z = 9$? Give reasons for your answer.

61. Which of the following are equations for the plane through the points $P(1, 1, -1)$, $Q(3, 0, 2)$, and $R(-2, 1, 0)$?

a. $(2\mathbf{i} - 3\mathbf{j} + 3\mathbf{k}) \cdot ((x + 2)\mathbf{i} + (y - 1)\mathbf{j} + z\mathbf{k}) = 0$

b. $x = 3 - t, \quad y = -11t, \quad z = 2 - 3t$

c. $(x + 2) + 11(y - 1) = 3z$

d. $(2\mathbf{i} - 3\mathbf{j} + 3\mathbf{k}) \times ((x + 2)\mathbf{i} + (y - 1)\mathbf{j} + z\mathbf{k}) = \mathbf{0}$

e. $(2\mathbf{i} - \mathbf{j} + 3\mathbf{k}) \times (-3\mathbf{i} + \mathbf{k}) \cdot ((x + 2)\mathbf{i} + (y - 1)\mathbf{j} + z\mathbf{k}) = 0$

62. The parallelogram shown on page 902 has vertices at $A(2, -1, 4)$, $B(1, 0, -1)$, $C(1, 2, 3)$, and D. Find

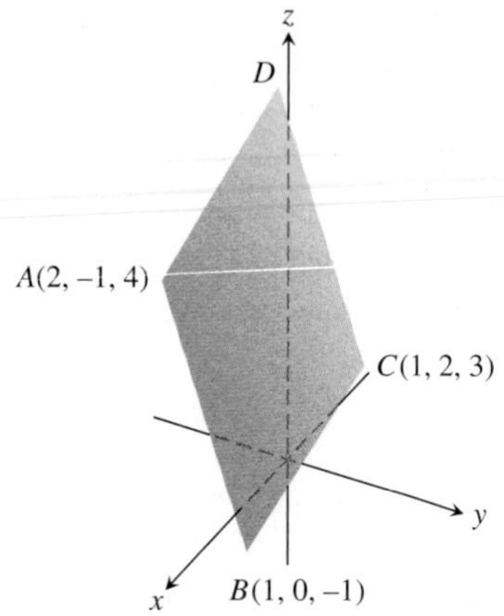

a. the coordinates of D,

b. the cosine of the interior angle at B,

c. the vector projection of $\overrightarrow{BA}$ onto $\overrightarrow{BC}$,

d. the area of the parallelogram,

e. an equation for the plane of the parallelogram,

f. the areas of the orthogonal projections of the parallelogram on the three coordinate planes.

63. Distance between lines Find the distance between the line L_1 through the points $A(1, 0, -1)$ and $B(-1, 1, 0)$ and the line L_2 through the points $C(3, 1, -1)$ and $D(4, 5, -2)$. The distance is to be measured along the line perpendicular to the two lines. First find a vector $\mathbf{n}$ perpendicular to both lines. Then project $\overrightarrow{AC}$ onto $\mathbf{n}$.

64. (*Continuation of Exercise 63.*) Find the distance between the line through $A(4, 0, 2)$ and $B(2, 4, 1)$ and the line through $C(1, 3, 2)$ and $D(2, 2, 4)$.

Quadric Surfaces

Identify and sketch the surfaces in Exercises 65–76.

65. $x^2 + y^2 + z^2 = 4$

66. $x^2 + (y - 1)^2 + z^2 = 1$

67. $4x^2 + 4y^2 + z^2 = 4$

68. $36x^2 + 9y^2 + 4z^2 = 36$

69. $z = -(x^2 + y^2)$

70. $y = -(x^2 + z^2)$

71. $x^2 + y^2 = z^2$

72. $x^2 + z^2 = y^2$

73. $x^2 + y^2 - z^2 = 4$

74. $4y^2 + z^2 - 4x^2 = 4$

75. $y^2 - x^2 - z^2 = 1$

76. $z^2 - x^2 - y^2 = 1$

Chapter 12 Additional and Advanced Exercises

1. Submarine hunting Two surface ships on maneuvers are trying to determine a submarine's course and speed to prepare for an aircraft intercept. As shown here, ship A is located at $(4, 0, 0)$, whereas ship B is located at $(0, 5, 0)$. All coordinates are given in thousands of feet. Ship A locates the submarine in the direction of the vector $2\mathbf{i} + 3\mathbf{j} - (1/3)\mathbf{k}$, and ship B locates it in the direction of the vector $18\mathbf{i} - 6\mathbf{j} - \mathbf{k}$. Four minutes ago, the submarine was located at $(2, -1, -1/3)$. The aircraft is due in 20 min. Assuming that the submarine moves in a straight line at a constant speed, to what position should the surface ships direct the aircraft?

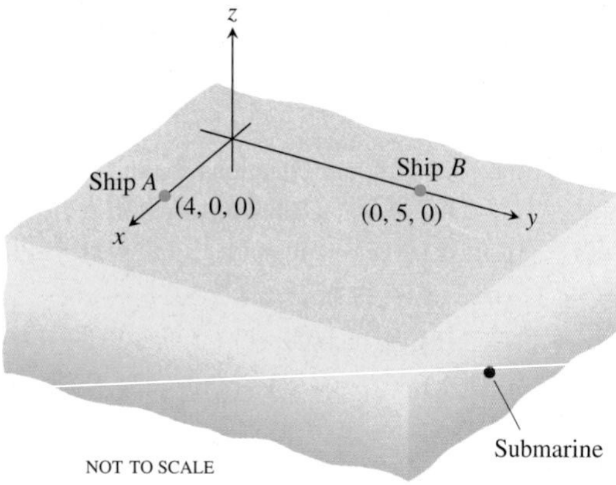

2. A helicopter rescue Two helicopters, H_1 and H_2, are traveling together. At time $t = 0$, they separate and follow different straight-line paths given by

$$H_1:\quad x = 6 + 40t, \quad y = -3 + 10t, \quad z = -3 + 2t$$

$$H_2:\quad x = 6 + 110t, \quad y = -3 + 4t, \quad z = -3 + t.$$

Time t is measured in hours and all coordinates are measured in miles. Due to system malfunctions, H_2 stops its flight at (446, 13, 1) and, in a negligible amount of time, lands at (446, 13, 0). Two hours later, H_1 is advised of this fact and heads toward H_2 at 150 mph. How long will it take H_1 to reach H_2?

3. Torque The operator's manual for the Toro® 21 in. lawnmower says "tighten the spark plug to 15 ft-lb (20.4 N · m)." If you are installing the plug with a 10.5-in. socket wrench that places the center of your hand 9 in. from the axis of the spark plug, about how hard should you pull? Answer in pounds.

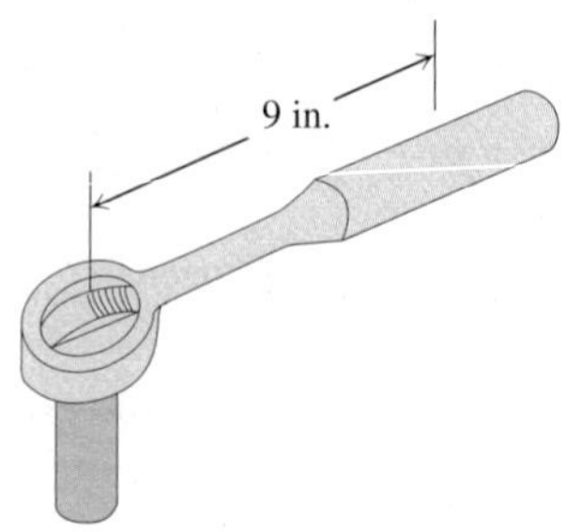

4. **Rotating body** The line through the origin and the point $A(1, 1, 1)$ is the axis of rotation of a right body rotating with a constant angular speed of $3/2$ rad/sec. The rotation appears to be clockwise when we look toward the origin from A. Find the velocity $\mathbf{v}$ of the point of the body that is at the position $B(1, 3, 2)$.

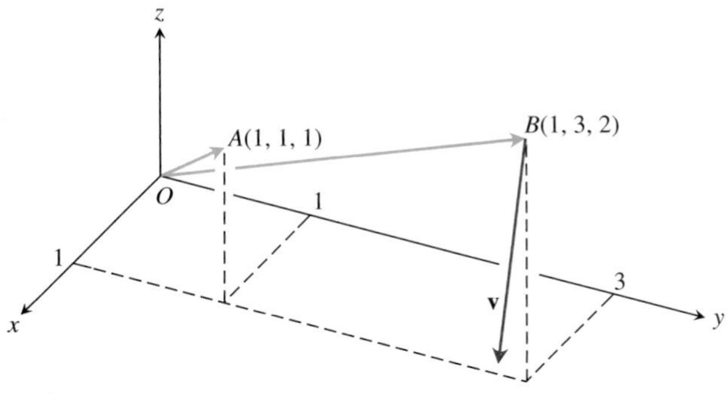

5. **Determinants and planes**
   a. Show that
$$\begin{vmatrix} x_1 - x & y_1 - y & z_1 - z \\ x_2 - x & y_2 - y & z_2 - z \\ x_3 - x & y_3 - y & z_3 - z \end{vmatrix} = 0$$
   is an equation for the plane through the three noncollinear points $P_1(x_1, y_1, z_1)$, $P_2(x_2, y_2, z_2)$, and $P_3(x_3, y_3, z_3)$.
   b. What set of points in space is described by the equation
$$\begin{vmatrix} x & y & z & 1 \\ x_1 & y_1 & z_1 & 1 \\ x_2 & y_2 & z_2 & 1 \\ x_3 & y_3 & z_3 & 1 \end{vmatrix} = 0?$$

6. **Determinants and lines** Show that the lines
$$x = a_1 s + b_1, y = a_2 s + b_2, z = a_3 s + b_3, -\infty < s < \infty,$$
and
$$x = c_1 t + d_1, y = c_2 t + d_2, z = c_3 t + d_3, -\infty < t < \infty,$$
intersect or are parallel if and only if
$$\begin{vmatrix} a_1 & c_1 & b_1 - d_1 \\ a_2 & c_2 & b_2 - d_2 \\ a_3 & c_3 & b_3 - d_3 \end{vmatrix} = 0.$$

7. **Parallelogram** The accompanying figure shows parallelogram $ABCD$ and the midpoint P of diagonal BD.
   a. Express $\overrightarrow{BD}$ in terms of $\overrightarrow{AB}$ and $\overrightarrow{AD}$.
   b. Express $\overrightarrow{AP}$ in terms of $\overrightarrow{AB}$ and $\overrightarrow{AD}$.
   c. Prove that P is also the midpoint of diagonal AC.

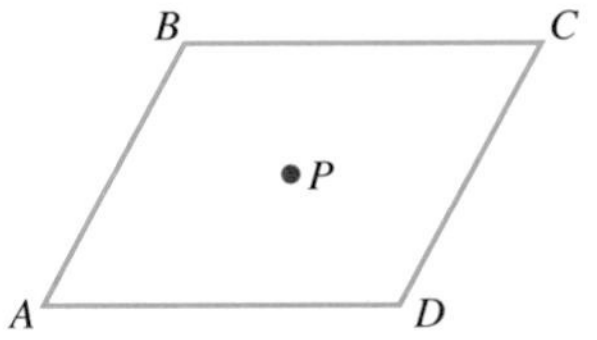

8. In the figure here, D is the midpoint of side AB of triangle ABC, and E is one-third of the way between C and B. Use vectors to prove that F is the midpoint of line segment CD.

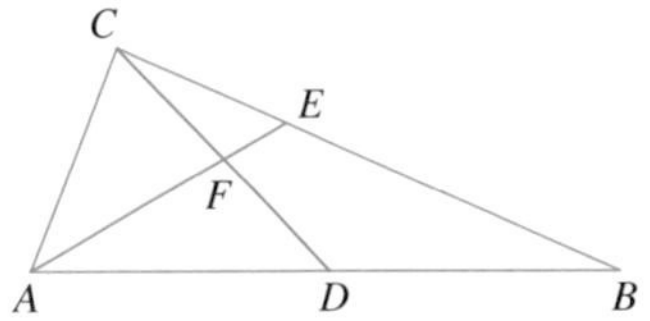

9. Use vectors to show that the distance from $P_1(x_1, y_1)$ to the line $ax + by = c$ is
$$d = \frac{|ax_1 + by_1 - c|}{\sqrt{a^2 + b^2}}.$$

10. a. Use vectors to show that the distance from $P_1(x_1, y_1, z_1)$ to the plane $Ax + By + Cz = D$ is
$$d = \frac{|Ax_1 + By_1 + Cz_1 - D|}{\sqrt{A^2 + B^2 + C^2}}.$$
   b. Find an equation for the sphere that is tangent to the planes $x + y + z = 3$ and $x + y + z = 9$ if the planes $2x - y = 0$ and $3x - z = 0$ pass through the center of the sphere.

11. a. Show that the distance between the parallel planes $Ax + By + Cz = D_1$ and $Ax + By + Cz = D_2$ is
$$d = \frac{|D_1 - D_2|}{|A\mathbf{i} + B\mathbf{j} + C\mathbf{k}|}.$$
   b. Find the distance between the planes $2x + 3y - z = 6$ and $2x + 3y - z = 12$.
   c. Find an equation for the plane parallel to the plane $2x - y + 2z = -4$ if the point $(3, 2, -1)$ is equidistant from the two planes.
   d. Write equations for the planes that lie parallel to and 5 units away from the plane $x - 2y + z = 3$.

12. Prove that four points A, B, C, and D are coplanar (lie in a common plane) if and only if $\overrightarrow{AD} \cdot (\overrightarrow{AB} \times \overrightarrow{BC}) = 0$.

13. **The projection of a vector on a plane** Let P be a plane in space and let $\mathbf{v}$ be a vector. The vector projection of $\mathbf{v}$ onto the plane P, $\text{proj}_P \mathbf{v}$, can be defined informally as follows. Suppose the sun is shining so that its rays are normal to the plane P. Then $\text{proj}_P \mathbf{v}$ is the "shadow" of $\mathbf{v}$ onto P. If P is the plane $x + 2y + 6z = 6$ and $\mathbf{v} = \mathbf{i} + \mathbf{j} + \mathbf{k}$, find $\text{proj}_P \mathbf{v}$.

14. The accompanying figure shows nonzero vectors $\mathbf{v}$, $\mathbf{w}$, and $\mathbf{z}$, with $\mathbf{z}$ orthogonal to the line L, and $\mathbf{v}$ and $\mathbf{w}$ making equal angles β with L. Assuming $|\mathbf{v}| = |\mathbf{w}|$, find $\mathbf{w}$ in terms of $\mathbf{v}$ and $\mathbf{z}$.

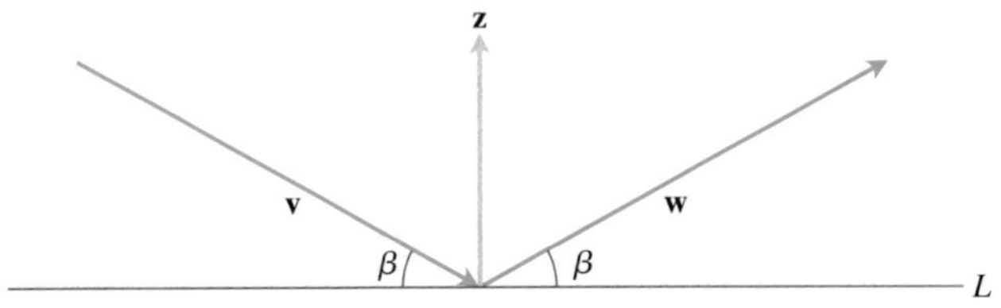

15. Triple vector products The *triple vector products* $(\mathbf{u} \times \mathbf{v}) \times \mathbf{w}$ and $\mathbf{u} \times (\mathbf{v} \times \mathbf{w})$ are usually not equal, although the formulas for evaluating them from components are similar:

$$(\mathbf{u} \times \mathbf{v}) \times \mathbf{w} = (\mathbf{u} \cdot \mathbf{w})\mathbf{v} - (\mathbf{v} \cdot \mathbf{w})\mathbf{u}.$$
$$\mathbf{u} \times (\mathbf{v} \times \mathbf{w}) = (\mathbf{u} \cdot \mathbf{w})\mathbf{v} - (\mathbf{u} \cdot \mathbf{v})\mathbf{w}.$$

Verify each formula for the following vectors by evaluating its two sides and comparing the results.

	u	**v**	**w**
a.	$2\mathbf{i}$	$2\mathbf{j}$	$2\mathbf{k}$
b.	$\mathbf{i} - \mathbf{j} + \mathbf{k}$	$2\mathbf{i} + \mathbf{j} - 2\mathbf{k}$	$-\mathbf{i} + 2\mathbf{j} - \mathbf{k}$
c.	$2\mathbf{i} + \mathbf{j}$	$2\mathbf{i} - \mathbf{j} + \mathbf{k}$	$\mathbf{i} + 2\mathbf{k}$
d.	$\mathbf{i} + \mathbf{j} - 2\mathbf{k}$	$-\mathbf{i} - \mathbf{k}$	$2\mathbf{i} + 4\mathbf{j} - 2\mathbf{k}$

16. Cross and dot products Show that if $\mathbf{u}$, $\mathbf{v}$, $\mathbf{w}$, and $\mathbf{r}$ are any vectors, then

a. $\mathbf{u} \times (\mathbf{v} \times \mathbf{w}) + \mathbf{v} \times (\mathbf{w} \times \mathbf{u}) + \mathbf{w} \times (\mathbf{u} \times \mathbf{v}) = \mathbf{0}$

b. $\mathbf{u} \times \mathbf{v} = (\mathbf{u} \cdot \mathbf{v} \times \mathbf{i})\mathbf{i} + (\mathbf{u} \cdot \mathbf{v} \times \mathbf{j})\mathbf{j} + (\mathbf{u} \cdot \mathbf{v} \times \mathbf{k})\mathbf{k}$

c. $(\mathbf{u} \times \mathbf{v}) \cdot (\mathbf{w} \times \mathbf{r}) = \begin{vmatrix} \mathbf{u} \cdot \mathbf{w} & \mathbf{v} \cdot \mathbf{w} \\ \mathbf{u} \cdot \mathbf{r} & \mathbf{v} \cdot \mathbf{r} \end{vmatrix}.$

17. Cross and dot products Prove or disprove the formula

$$\mathbf{u} \times (\mathbf{u} \times (\mathbf{u} \times \mathbf{v})) \cdot \mathbf{w} = -|\mathbf{u}|^2 \mathbf{u} \cdot \mathbf{v} \times \mathbf{w}.$$

18. By forming the cross product of two appropriate vectors, derive the trigonometric identity

$$\sin(A - B) = \sin A \cos B - \cos A \sin B.$$

19. Use vectors to prove that

$$(a^2 + b^2)(c^2 + d^2) \geq (ac + bd)^2$$

for any four numbers a, b, c, and d. (*Hint:* Let $\mathbf{u} = a\mathbf{i} + b\mathbf{j}$ and $\mathbf{v} = c\mathbf{i} + d\mathbf{j}$.)

20. Suppose that vectors $\mathbf{u}$ and $\mathbf{v}$ are not parallel and that $\mathbf{u} = \mathbf{w} + \mathbf{r}$, where $\mathbf{w}$ is parallel to $\mathbf{v}$ and $\mathbf{r}$ is orthogonal to $\mathbf{v}$. Express $\mathbf{w}$ and $\mathbf{r}$ in terms of $\mathbf{u}$ and $\mathbf{v}$.

21. Show that $|\mathbf{u} + \mathbf{v}| \leq |\mathbf{u}| + |\mathbf{v}|$ for any vectors $\mathbf{u}$ and $\mathbf{v}$.

22. Show that $\mathbf{w} = |\mathbf{v}|\mathbf{u} + |\mathbf{u}|\mathbf{v}$ bisects the angle between $\mathbf{u}$ and $\mathbf{v}$.

23. Show that $|\mathbf{v}|\mathbf{u} + |\mathbf{u}|\mathbf{v}$ and $|\mathbf{v}|\mathbf{u} - |\mathbf{u}|\mathbf{v}$ are orthogonal.

24. Dot multiplication is positive definite Show that dot multiplication of vectors is *positive definite*; that is, show that $\mathbf{u} \cdot \mathbf{u} \geq 0$ for every vector $\mathbf{u}$ and that $\mathbf{u} \cdot \mathbf{u} = 0$ if and only if $\mathbf{u} = \mathbf{0}$.

25. Point masses and gravitation In physics, the law of gravitation says that if P and Q are (point) masses with mass M and m, respectively, then P is attracted to Q by the force

$$\mathbf{F} = \frac{GMm\mathbf{r}}{|\mathbf{r}|^3},$$

where $\mathbf{r}$ is the vector from P to Q and G is the universal gravitational constant. Moreover, if $Q_1, \ldots, Q_k$ are (point) masses with mass $m_1, \ldots, m_k$, respectively, then the force on P due to all the Q_i's is

$$\mathbf{F} = \sum_{i=1}^{k} \frac{GMm_i}{|\mathbf{r}_i|^3}\mathbf{r}_i,$$

where $\mathbf{r}_i$ is the vector from P to Q_i.

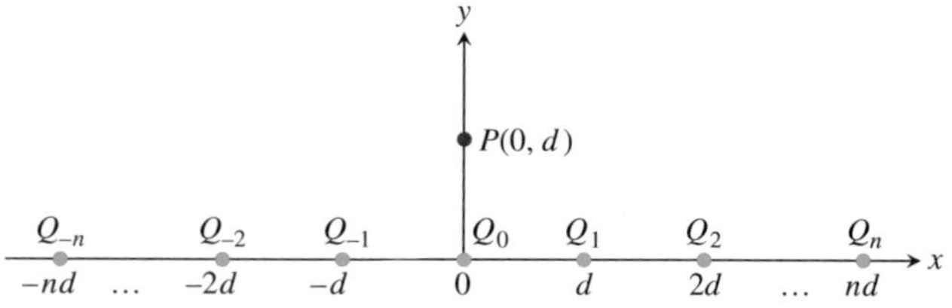

a. Let point P with mass M be located at the point $(0, d)$, $d > 0$, in the coordinate plane. For $i = -n, -n + 1, \ldots, -1, 0, 1, \ldots, n$, let Q_i be located at the point $(id, 0)$ and have mass mi. Find the magnitude of the gravitational force on P due to all the Q_i's.

b. Is the limit as $n \to \infty$ of the magnitude of the force on P finite? Why, or why not?

26. Relativistic sums Einstein's special theory of relativity roughly says that with respect to a reference frame (coordinate system) no material object can travel as fast as c, the speed of light. So, if $\vec{x}$ and $\vec{y}$ are two velocities such that $|\vec{x}| < c$ and $|\vec{y}| < c$, then the *relativistic sum* $\vec{x} \oplus \vec{y}$ of $\vec{x}$ and $\vec{y}$ must have length less than c. Einstein's special theory of relativity says that

$$\vec{x} \oplus \vec{y} = \frac{\vec{x} + \vec{y}}{1 + \dfrac{\vec{x} \cdot \vec{y}}{c^2}} + \frac{1}{c^2} \cdot \frac{\gamma_x}{\gamma_x + 1} \cdot \frac{\vec{x} \times (\vec{x} \times \vec{y})}{1 + \dfrac{\vec{x} \cdot \vec{y}}{c^2}},$$

where

$$\gamma_x = \frac{1}{\sqrt{1 - \dfrac{\vec{x} \cdot \vec{x}}{c^2}}}.$$

It can be shown that if $|\vec{x}| < c$ and $|\vec{y}| < c$, then $|\vec{x} \oplus \vec{y}| < c$. This exercise deals with two special cases.

a. Prove that if $\vec{x}$ and $\vec{y}$ are orthogonal, $|\vec{x}| < c$, $|\vec{y}| < c$, then $|\vec{x} \oplus \vec{y}| < c$.

b. Prove that if $\vec{x}$ and $\vec{y}$ are parallel, $|\vec{x}| < c$, $|\vec{y}| < c$, then $|\vec{x} \oplus \vec{y}| < c$.

c. Compute $\lim_{c \to \infty} \vec{x} \oplus \vec{y}$.

Chapter 12 Technology Application Projects

Mathematica/Maple Module
Using Vectors to Represent Lines and Find Distances

Parts I and II: Learn the advantages of interpreting lines as vectors.

Part III: Use vectors to find the distance from a point to a line.

Mathematica/Maple Module
Putting a Scene in Three Dimensions onto a Two-Dimensional Canvas

Use the concept of planes in space to obtain a two-dimensional image.

Mathematica/Maple Module
Getting Started in Plotting in 3D

Part I: Use the vector definition of lines and planes to generate graphs and equations, and to compare different forms for the equations of a single line.

Part II: Plot functions that are defined implicitly.

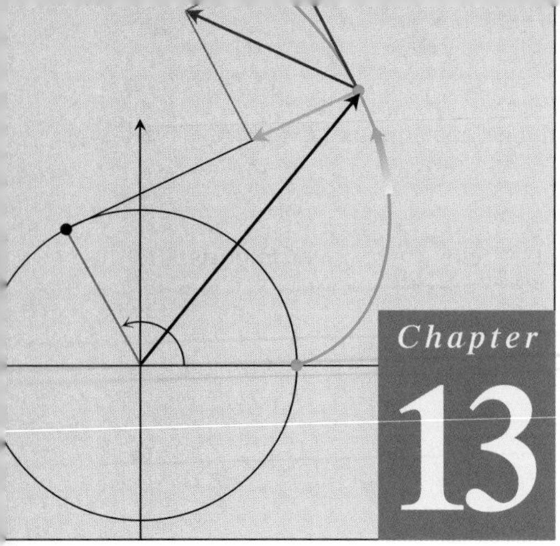

Chapter 13

VECTOR-VALUED FUNCTIONS AND MOTION IN SPACE

OVERVIEW When a body (or object) travels through space, the equations $x = f(t)$, $y = g(t)$, and $z = h(t)$ that give the body's coordinates as functions of time serve as parametric equations for the body's motion and path. With vector notation, we can condense these into a single equation $\mathbf{r}(t) = f(t)\mathbf{i} + g(t)\mathbf{j} + h(t)\mathbf{k}$ that gives the body's position as a vector function of time. For an object moving in the xy-plane, the component function $h(t)$ is zero for all time (that is, identically zero).

In this chapter, we use calculus to study the paths, velocities, and accelerations of moving bodies. As we go along, we will see how our work answers the standard questions about the paths and motions of projectiles, planets, and satellites. In the final section, we use our new vector calculus to derive Kepler's laws of planetary motion from Newton's laws of motion and gravitation.

13.1 Vector Functions

When a particle moves through space during a time interval I, we think of the particle's coordinates as functions defined on I:

$$x = f(t), \qquad y = g(t), \qquad z = h(t), \qquad t \in I. \tag{1}$$

The points $(x, y, z) = (f(t), g(t), h(t))$, $t \in I$, make up the **curve** in space that we call the particle's **path**. The equations and interval in Equation (1) **parametrize** the curve. A curve in space can also be represented in vector form. The vector

$$\mathbf{r}(t) = \overrightarrow{OP} = f(t)\mathbf{i} + g(t)\mathbf{j} + h(t)\mathbf{k} \tag{2}$$

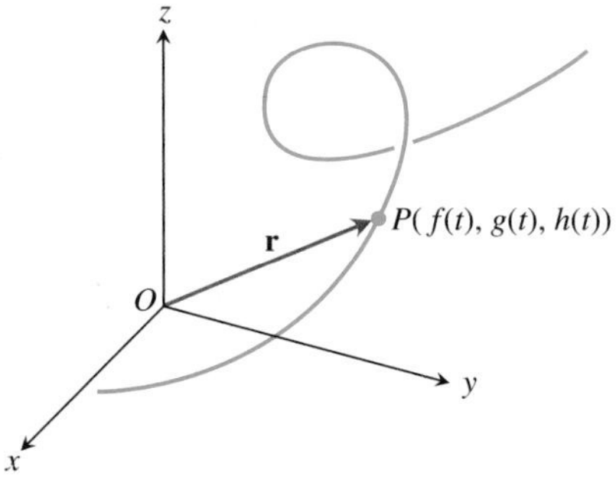

FIGURE 13.1 The position vector $\mathbf{r} = \overrightarrow{OP}$ of a particle moving through space is a function of time.

from the origin to the particle's **position** $P(f(t), g(t), h(t))$ at time t is the particle's **position vector** (Figure 13.1). The functions f, g, and h are the **component functions (components)** of the position vector. We think of the particle's path as the **curve traced by r** during the time interval I. Figure 13.2 displays several space curves generated by a computer graphing program. It would not be easy to plot these curves by hand.

Equation (2) defines $\mathbf{r}$ as a vector function of the real variable t on the interval I. More generally, a **vector function** or **vector-valued function** on a domain set D is a rule that assigns a vector in space to each element in D. For now, the domains will be intervals of real numbers resulting in a space curve. Later, in Chapter 16, the domains will be regions in the plane. Vector functions will then represent surfaces in space. Vector functions on a domain in the plane or space also give rise to "vector fields," which are important to the study of the flow of a fluid, gravitational fields, and electromagnetic phenomena. We investigate vector fields and their applications in Chapter 16.

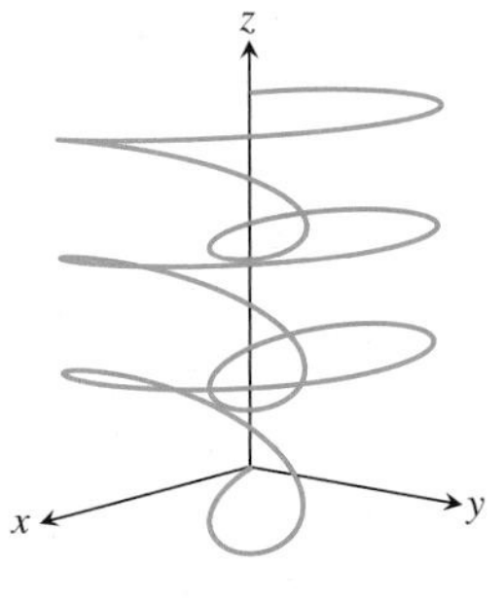

$\mathbf{r}(t) = (\sin 3t)(\cos t)\mathbf{i} + (\sin 3t)(\sin t)\mathbf{j} + t\mathbf{k}$

(a)

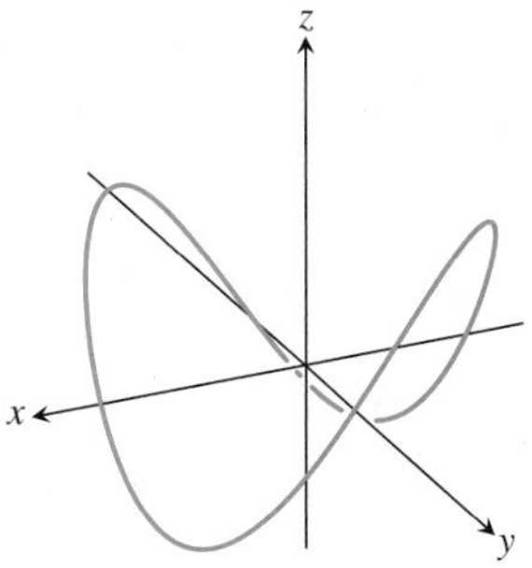

$\mathbf{r}(t) = (\cos t)\mathbf{i} + (\sin t)\mathbf{j} + (\sin 2t)\mathbf{k}$

(b)

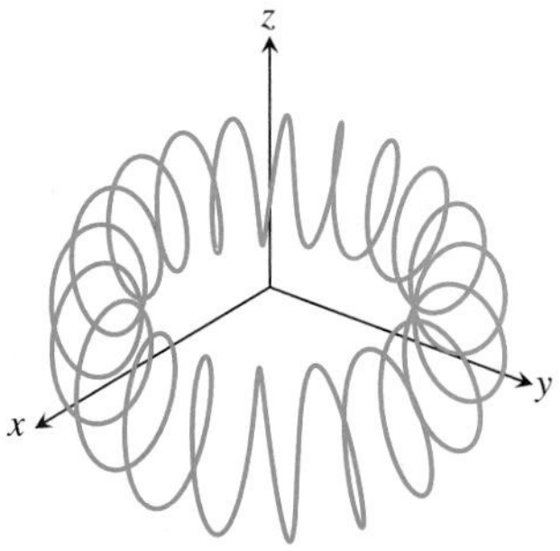

$\mathbf{r}(t) = (4 + \sin 20t)(\cos t)\mathbf{i} + (4 + \sin 20t)(\sin t)\mathbf{j} + (\cos 20t)\mathbf{k}$

(c)

FIGURE 13.2 Computer-generated space curves are defined by the position vectors $\mathbf{r}(t)$.

We refer to real-valued functions as **scalar functions** to distinguish them from vector functions. The components of $\mathbf{r}$ are scalar functions of t. When we define a vector-valued function by giving its component functions, we assume the vector function's domain to be the common domain of the components.

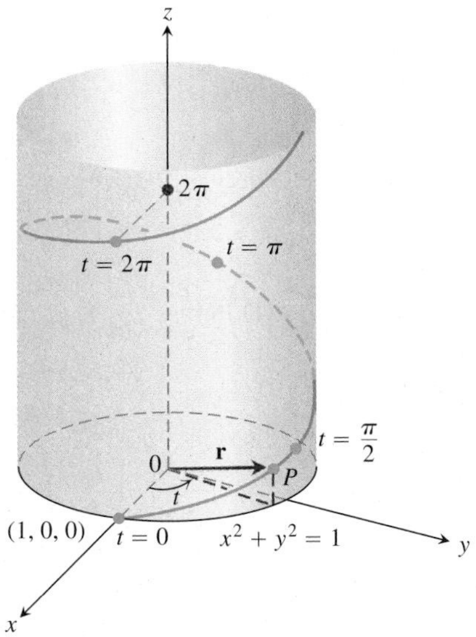

FIGURE 13.3 The upper half of the helix $\mathbf{r}(t) = (\cos t)\mathbf{i} + (\sin t)\mathbf{j} + t\mathbf{k}$ (Example 1).

EXAMPLE 1 Graphing a Helix

Graph the vector function

$$\mathbf{r}(t) = (\cos t)\mathbf{i} + (\sin t)\mathbf{j} + t\mathbf{k}.$$

Solution The vector function

$$\mathbf{r}(t) = (\cos t)\mathbf{i} + (\sin t)\mathbf{j} + t\mathbf{k}$$

is defined for all real values of t. The curve traced by $\mathbf{r}$ is a helix (from an old Greek word for "spiral") that winds around the circular cylinder $x^2 + y^2 = 1$ (Figure 13.3). The curve lies on the cylinder because the $\mathbf{i}$- and $\mathbf{j}$-components of $\mathbf{r}$, being the x- and y-coordinates of the tip of $\mathbf{r}$, satisfy the cylinder's equation:

$$x^2 + y^2 = (\cos t)^2 + (\sin t)^2 = 1.$$

The curve rises as the $\mathbf{k}$-component $z = t$ increases. Each time t increases by 2π, the curve completes one turn around the cylinder. The equations

$$x = \cos t, \qquad y = \sin t, \qquad z = t$$

parametrize the helix, the interval $-\infty < t < \infty$ being understood. You will find more helices in Figure 13.4. ■

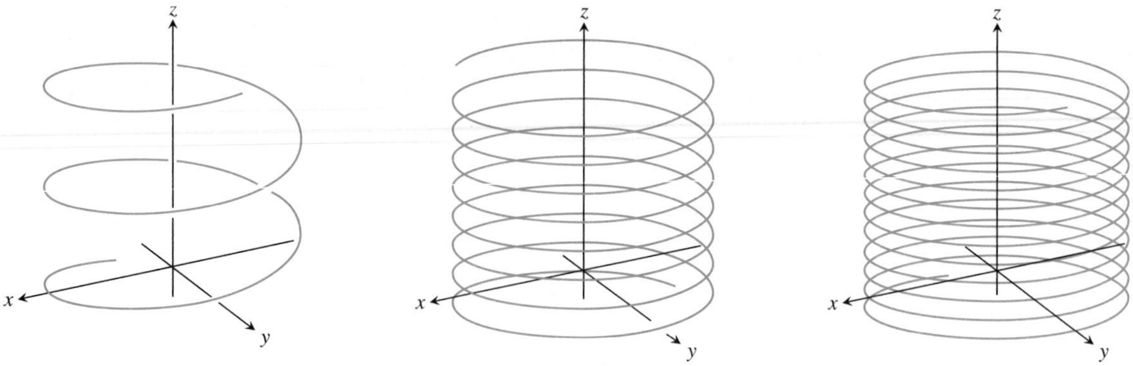

$\mathbf{r}(t) = (\cos t)\mathbf{i} + (\sin t)\mathbf{j} + t\mathbf{k}$ $\quad$ $\mathbf{r}(t) = (\cos t)\mathbf{i} + (\sin t)\mathbf{j} + 0.3t\mathbf{k}$ $\quad$ $\mathbf{r}(t) = (\cos 5t)\mathbf{i} + (\sin 5t)\mathbf{j} + t\mathbf{k}$

FIGURE 13.4 Helices drawn by computer.

Limits and Continuity

The way we define limits of vector-valued functions is similar to the way we define limits of real-valued functions.

> **DEFINITION Limit of Vector Functions**
>
> Let $\mathbf{r}(t) = f(t)\mathbf{i} + g(t)\mathbf{j} + h(t)\mathbf{k}$ be a vector function and $\mathbf{L}$ a vector. We say that $\mathbf{r}$ has **limit L** as t approaches t_0 and write
>
> $$\lim_{t\to t_0} \mathbf{r}(t) = \mathbf{L}$$
>
> if, for every number $\epsilon > 0$, there exists a corresponding number $\delta > 0$ such that for all t
>
> $$0 < |t - t_0| < \delta \quad \Rightarrow \quad |\mathbf{r}(t) - \mathbf{L}| < \epsilon.$$

If $\mathbf{L} = L_1\mathbf{i} + L_2\mathbf{j} + L_3\mathbf{k}$, then $\lim_{t\to t_0}\mathbf{r}(t) = \mathbf{L}$ precisely when

$$\lim_{t\to t_0} f(t) = L_1, \quad \lim_{t\to t_0} g(t) = L_2, \quad \text{and} \quad \lim_{t\to t_0} h(t) = L_3.$$

The equation

$$\lim_{t\to t_0} \mathbf{r}(t) = \left(\lim_{t\to t_0} f(t)\right)\mathbf{i} + \left(\lim_{t\to t_0} g(t)\right)\mathbf{j} + \left(\lim_{t\to t_0} h(t)\right)\mathbf{k} \tag{3}$$

provides a practical way to calculate limits of vector functions.

EXAMPLE 2 Finding Limits of Vector Functions

If $\mathbf{r}(t) = (\cos t)\mathbf{i} + (\sin t)\mathbf{j} + t\mathbf{k}$, then

$$\begin{aligned}\lim_{t\to\pi/4} \mathbf{r}(t) &= \left(\lim_{t\to\pi/4} \cos t\right)\mathbf{i} + \left(\lim_{t\to\pi/4} \sin t\right)\mathbf{j} + \left(\lim_{t\to\pi/4} t\right)\mathbf{k} \\ &= \frac{\sqrt{2}}{2}\mathbf{i} + \frac{\sqrt{2}}{2}\mathbf{j} + \frac{\pi}{4}\mathbf{k}.\end{aligned}$$ ■

We define continuity for vector functions the same way we define continuity for scalar functions.

DEFINITION Continuous at a Point

A vector function $\mathbf{r}(t)$ is **continuous at a point** $t = t_0$ in its domain if $\lim_{t \to t_0} \mathbf{r}(t) = \mathbf{r}(t_0)$. The function is **continuous** if it is continuous at every point in its domain.

From Equation (3), we see that $\mathbf{r}(t)$ is continuous at $t = t_0$ if and only if each component function is continuous there.

EXAMPLE 3 Continuity of Space Curves

(a) All the space curves shown in Figures 13.2 and 13.4 are continuous because their component functions are continuous at every value of t in $(-\infty, \infty)$.

(b) The function

$$\mathbf{g}(t) = (\cos t)\mathbf{i} + (\sin t)\mathbf{j} + \lfloor t \rfloor \mathbf{k}$$

is discontinuous at every integer, where the greatest integer function $\lfloor t \rfloor$ is discontinuous. ■

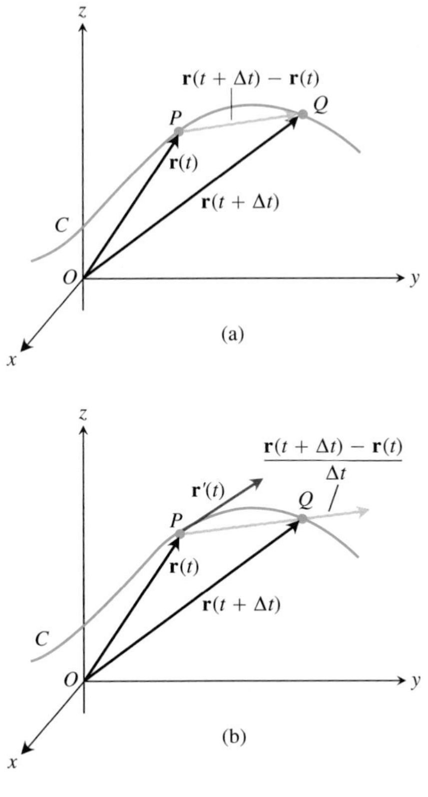

FIGURE 13.5 As $\Delta t \to 0$, the point Q approaches the point P along the curve C. In the limit, the vector $\overrightarrow{PQ}/\Delta t$ becomes the tangent vector $\mathbf{r}'(t)$.

Derivatives and Motion

Suppose that $\mathbf{r}(t) = f(t)\mathbf{i} + g(t)\mathbf{j} + h(t)\mathbf{k}$ is the position vector of a particle moving along a curve in space and that f, g, and h are differentiable functions of t. Then the difference between the particle's positions at time t and time $t + \Delta t$ is

$$\Delta \mathbf{r} = \mathbf{r}(t + \Delta t) - \mathbf{r}(t)$$

(Figure 13.5a). In terms of components,

$$\begin{aligned} \Delta \mathbf{r} &= \mathbf{r}(t + \Delta t) - \mathbf{r}(t) \\ &= [f(t + \Delta t)\mathbf{i} + g(t + \Delta t)\mathbf{j} + h(t + \Delta t)\mathbf{k}] \\ &\quad - [f(t)\mathbf{i} + g(t)\mathbf{j} + h(t)\mathbf{k}] \\ &= [f(t + \Delta t) - f(t)]\mathbf{i} + [g(t + \Delta t) - g(t)]\mathbf{j} + [h(t + \Delta t) - h(t)]\mathbf{k}. \end{aligned}$$

As Δt approaches zero, three things seem to happen simultaneously. First, Q approaches P along the curve. Second, the secant line PQ seems to approach a limiting position tangent to the curve at P. Third, the quotient $\Delta \mathbf{r}/\Delta t$ (Figure 13.5b) approaches the limit

$$\begin{aligned} \lim_{\Delta t \to 0} \frac{\Delta \mathbf{r}}{\Delta t} &= \left[\lim_{\Delta t \to 0} \frac{f(t + \Delta t) - f(t)}{\Delta t}\right]\mathbf{i} + \left[\lim_{\Delta t \to 0} \frac{g(t + \Delta t) - g(t)}{\Delta t}\right]\mathbf{j} \\ &\quad + \left[\lim_{\Delta t \to 0} \frac{h(t + \Delta t) - h(t)}{\Delta t}\right]\mathbf{k} \\ &= \left[\frac{df}{dt}\right]\mathbf{i} + \left[\frac{dg}{dt}\right]\mathbf{j} + \left[\frac{dh}{dt}\right]\mathbf{k}. \end{aligned}$$

We are therefore led by past experience to the following definition.

DEFINITION Derivative

The vector function $\mathbf{r}(t) = f(t)\mathbf{i} + g(t)\mathbf{j} + h(t)\mathbf{k}$ has a **derivative (is differentiable) at *t*** if f, g, and h have derivatives at t. The derivative is the vector function

$$\mathbf{r}'(t) = \frac{d\mathbf{r}}{dt} = \lim_{\Delta t \to 0} \frac{\mathbf{r}(t + \Delta t) - \mathbf{r}(t)}{\Delta t} = \frac{df}{dt}\mathbf{i} + \frac{dg}{dt}\mathbf{j} + \frac{dh}{dt}\mathbf{k}.$$

A vector function $\mathbf{r}$ is **differentiable** if it is differentiable at every point of its domain. The curve traced by $\mathbf{r}$ is **smooth** if $d\mathbf{r}/dt$ is continuous and never $\mathbf{0}$, that is, if f, g, and h have continuous first derivatives that are not simultaneously 0.

The geometric significance of the definition of derivative is shown in Figure 13.5. The points P and Q have position vectors $\mathbf{r}(t)$ and $\mathbf{r}(t + \Delta t)$, and the vector $\overrightarrow{PQ}$ is represented by $\mathbf{r}(t + \Delta t) - \mathbf{r}(t)$. For $\Delta t > 0$, the scalar multiple $(1/\Delta t)(\mathbf{r}(t + \Delta t) - \mathbf{r}(t))$ points in the same direction as the vector $\overrightarrow{PQ}$. As $\Delta t \to 0$, this vector approaches a vector that is tangent to the curve at P (Figure 13.5b). The vector $\mathbf{r}'(t)$, when different from $\mathbf{0}$, is defined to be the vector **tangent** to the curve at P. The **tangent line** to the curve at a point $(f(t_0), g(t_0), h(t_0))$ is defined to be the line through the point parallel to $\mathbf{r}'(t_0)$. We require $d\mathbf{r}/dt \neq \mathbf{0}$ for a smooth curve to make sure the curve has a continuously turning tangent at each point. On a smooth curve, there are no sharp corners or cusps.

A curve that is made up of a finite number of smooth curves pieced together in a continuous fashion is called **piecewise smooth** (Figure 13.6).

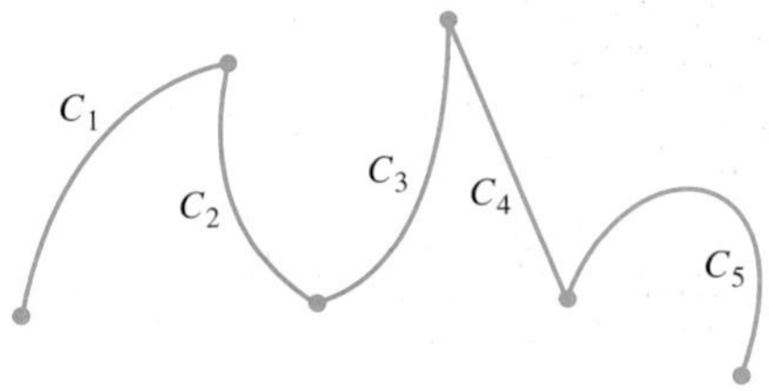

FIGURE 13.6 A piecewise smooth curve made up of five smooth curves connected end to end in continuous fashion.

Look once again at Figure 13.5. We drew the figure for Δt positive, so $\Delta\mathbf{r}$ points forward, in the direction of the motion. The vector $\Delta\mathbf{r}/\Delta t$, having the same direction as $\Delta\mathbf{r}$, points forward too. Had Δt been negative, $\Delta\mathbf{r}$ would have pointed backward, against the direction of motion. The quotient $\Delta\mathbf{r}/\Delta t$, however, being a negative scalar multiple of $\Delta\mathbf{r}$, would once again have pointed forward. No matter how $\Delta\mathbf{r}$ points, $\Delta\mathbf{r}/\Delta t$ points forward and we expect the vector $d\mathbf{r}/dt = \lim_{\Delta t \to 0} \Delta\mathbf{r}/\Delta t$, when different from $\mathbf{0}$, to do the same. This means that the derivative $d\mathbf{r}/dt$ is just what we want for modeling a particle's velocity. It points in the direction of motion and gives the rate of change of position with respect to time. For a smooth curve, the velocity is never zero; the particle does not stop or reverse direction.

DEFINITIONS Velocity, Direction, Speed, Acceleration

If $\mathbf{r}$ is the position vector of a particle moving along a smooth curve in space, then

$$\mathbf{v}(t) = \frac{d\mathbf{r}}{dt}$$

is the particle's **velocity vector**, tangent to the curve. At any time t, the direction of $\mathbf{v}$ is the **direction of motion**, the magnitude of $\mathbf{v}$ is the particle's **speed**, and the derivative $\mathbf{a} = d\mathbf{v}/dt$, when it exists, is the particle's **acceleration vector**. In summary,

1. Velocity is the derivative of position: $\mathbf{v} = \dfrac{d\mathbf{r}}{dt}$.
2. Speed is the magnitude of velocity: $\text{Speed} = |\mathbf{v}|$.
3. Acceleration is the derivative of velocity: $\mathbf{a} = \dfrac{d\mathbf{v}}{dt} = \dfrac{d^2\mathbf{r}}{dt^2}$.
4. The unit vector $\mathbf{v}/|\mathbf{v}|$ is the direction of motion at time t.

We can express the velocity of a moving particle as the product of its speed and direction:

$$\text{Velocity} = |\mathbf{v}|\left(\frac{\mathbf{v}}{|\mathbf{v}|}\right) = (\text{speed})(\text{direction}).$$

In Section 12.5, Example 4 we found this expression for velocity useful in locating, for example, the position of a helicopter moving along a straight line in space. Now let's look at an example of an object moving along a (nonlinear) space curve.

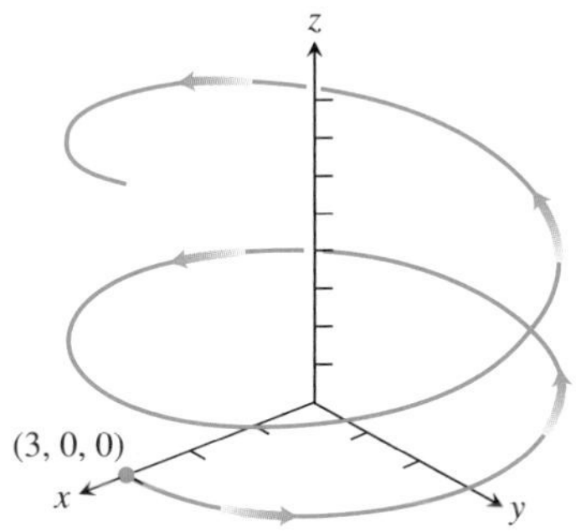

FIGURE 13.7 The path of a hang glider with position vector $\mathbf{r}(t) = (3\cos t)\mathbf{i} + (3\sin t)\mathbf{j} + t^2\mathbf{k}$ (Example 4).

EXAMPLE 4 Flight of a Hang Glider

A person on a hang glider is spiraling upward due to rapidly rising air on a path having position vector $\mathbf{r}(t) = (3\cos t)\mathbf{i} + (3\sin t)\mathbf{j} + t^2\mathbf{k}$. The path is similar to that of a helix (although it's *not* a helix, as you will see in Section 13.4) and is shown in Figure 13.7 for $0 \le t \le 4\pi$. Find

(a) the velocity and acceleration vectors,

(b) the glider's speed at any time t,

(c) the times, if any, when the glider's acceleration is orthogonal to its velocity.

Solution

(a) $\mathbf{r} = (3\cos t)\mathbf{i} + (3\sin t)\mathbf{j} + t^2\mathbf{k}$

$$\mathbf{v} = \frac{d\mathbf{r}}{dt} = -(3\sin t)\mathbf{i} + (3\cos t)\mathbf{j} + 2t\mathbf{k}$$

$$\mathbf{a} = \frac{d^2\mathbf{r}}{dt^2} = -(3\cos t)\mathbf{i} - (3\sin t)\mathbf{j} + 2\mathbf{k}$$

(b) Speed is the magnitude of $\mathbf{v}$:

$$\begin{aligned}|\mathbf{v}(t)| &= \sqrt{(-3\sin t)^2 + (3\cos t)^2 + (2t)^2}\\ &= \sqrt{9\sin^2 t + 9\cos^2 t + 4t^2}\\ &= \sqrt{9 + 4t^2}.\end{aligned}$$

The glider is moving faster and faster as it rises along its path.

(c) To find the times when $\mathbf{v}$ and $\mathbf{a}$ are orthogonal, we look for values of t for which

$$\mathbf{v}\cdot\mathbf{a} = 9\sin t\cos t - 9\cos t\sin t + 4t = 4t = 0.$$

Thus, the only time the acceleration vector is orthogonal to $\mathbf{v}$ is when $t = 0$. We study acceleration for motions along paths in more detail in Section 13.5. There we discover how the acceleration vector reveals the curving nature and tendency of the path to "twist" out of a certain plane containing the velocity vector. ■

Differentiation Rules

Because the derivatives of vector functions may be computed component by component, the rules for differentiating vector functions have the same form as the rules for differentiating scalar functions.

Differentiation Rules for Vector Functions

Let **u** and **v** be differentiable vector functions of t, **C** a constant vector, c any scalar, and f any differentiable scalar function.

1. *Constant Function Rule:* $\frac{d}{dt}\mathbf{C} = \mathbf{0}$
2. *Scalar Multiple Rules:* $\frac{d}{dt}[c\mathbf{u}(t)] = c\mathbf{u}'(t)$

   $\frac{d}{dt}[f(t)\mathbf{u}(t)] = f'(t)\mathbf{u}(t) + f(t)\mathbf{u}'(t)$
3. *Sum Rule:* $\frac{d}{dt}[\mathbf{u}(t) + \mathbf{v}(t)] = \mathbf{u}'(t) + \mathbf{v}'(t)$
4. *Difference Rule:* $\frac{d}{dt}[\mathbf{u}(t) - \mathbf{v}(t)] = \mathbf{u}'(t) - \mathbf{v}'(t)$
5. *Dot Product Rule:* $\frac{d}{dt}[\mathbf{u}(t)\cdot\mathbf{v}(t)] = \mathbf{u}'(t)\cdot\mathbf{v}(t) + \mathbf{u}(t)\cdot\mathbf{v}'(t)$
6. *Cross Product Rule:* $\frac{d}{dt}[\mathbf{u}(t)\times\mathbf{v}(t)] = \mathbf{u}'(t)\times\mathbf{v}(t) + \mathbf{u}(t)\times\mathbf{v}'(t)$
7. *Chain Rule:* $\frac{d}{dt}[\mathbf{u}(f(t))] = f'(t)\mathbf{u}'(f(t))$

When you use the Cross Product Rule, remember to preserve the order of the factors. If **u** comes first on the left side of the equation, it must also come first on the right or the signs will be wrong.

We will prove the product rules and Chain Rule but leave the rules for constants, scalar multiples, sums, and differences as exercises.

Proof of the Dot Product Rule Suppose that

$$\mathbf{u} = u_1(t)\mathbf{i} + u_2(t)\mathbf{j} + u_3(t)\mathbf{k}$$

and

$$\mathbf{v} = v_1(t)\mathbf{i} + v_2(t)\mathbf{j} + v_3(t)\mathbf{k}.$$

Then

$$\begin{aligned}\frac{d}{dt}(\mathbf{u}\cdot\mathbf{v}) &= \frac{d}{dt}(u_1v_1 + u_2v_2 + u_3v_3)\\ &= \underbrace{u_1'v_1 + u_2'v_2 + u_3'v_3}_{\mathbf{u}'\cdot\mathbf{v}} + \underbrace{u_1v_1' + u_2v_2' + u_3v_3'}_{\mathbf{u}\cdot\mathbf{v}'}.\end{aligned}$$

■

Proof of the Cross Product Rule We model the proof after the proof of the Product Rule for scalar functions. According to the definition of derivative,

$$\frac{d}{dt}(\mathbf{u}\times\mathbf{v}) = \lim_{h\to 0}\frac{\mathbf{u}(t+h)\times\mathbf{v}(t+h) - \mathbf{u}(t)\times\mathbf{v}(t)}{h}.$$

To change this fraction into an equivalent one that contains the difference quotients for the derivatives of **u** and **v**, we subtract and add $\mathbf{u}(t) \times \mathbf{v}(t+h)$ in the numerator. Then

$$\begin{aligned}
&\frac{d}{dt}(\mathbf{u}\times\mathbf{v}) \\
&\quad = \lim_{h\to 0} \frac{\mathbf{u}(t+h)\times\mathbf{v}(t+h) - \mathbf{u}(t)\times\mathbf{v}(t+h) + \mathbf{u}(t)\times\mathbf{v}(t+h) - \mathbf{u}(t)\times\mathbf{v}(t)}{h} \\
&\quad = \lim_{h\to 0} \left[\frac{\mathbf{u}(t+h)-\mathbf{u}(t)}{h}\times\mathbf{v}(t+h) + \mathbf{u}(t)\times\frac{\mathbf{v}(t+h)-\mathbf{v}(t)}{h}\right] \\
&\quad = \lim_{h\to 0} \frac{\mathbf{u}(t+h)-\mathbf{u}(t)}{h}\times\lim_{h\to 0}\mathbf{v}(t+h) + \lim_{h\to 0}\mathbf{u}(t)\times\lim_{h\to 0}\frac{\mathbf{v}(t+h)-\mathbf{v}(t)}{h}.
\end{aligned}$$

The last of these equalities holds because the limit of the cross product of two vector functions is the cross product of their limits if the latter exist (Exercise 52). As h approaches zero, $\mathbf{v}(t+h)$ approaches $\mathbf{v}(t)$ because **v**, being differentiable at t, is continuous at t (Exercise 53). The two fractions approach the values of $d\mathbf{u}/dt$ and $d\mathbf{v}/dt$ at t. In short,

$$\frac{d}{dt}(\mathbf{u}\times\mathbf{v}) = \frac{d\mathbf{u}}{dt}\times\mathbf{v} + \mathbf{u}\times\frac{d\mathbf{v}}{dt}. \quad ■$$

As an algebraic convenience, we sometimes write the product of a scalar c and a vector **v** as $\mathbf{v}c$ instead of $c\mathbf{v}$. This permits us, for instance, to write the Chain Rule in a familiar form:

$$\frac{d\mathbf{u}}{dt} = \frac{d\mathbf{u}}{ds}\frac{ds}{dt},$$

where $s = f(t)$.

Proof of the Chain Rule Suppose that $\mathbf{u}(s) = a(s)\mathbf{i} + b(s)\mathbf{j} + c(s)\mathbf{k}$ is a differentiable vector function of s and that $s = f(t)$ is a differentiable scalar function of t. Then a, b, and c are differentiable functions of t, and the Chain Rule for differentiable real-valued functions gives

$$\begin{aligned}
\frac{d}{dt}[\mathbf{u}(s)] &= \frac{da}{dt}\mathbf{i} + \frac{db}{dt}\mathbf{j} + \frac{dc}{dt}\mathbf{k} \\
&= \frac{da}{ds}\frac{ds}{dt}\mathbf{i} + \frac{db}{ds}\frac{ds}{dt}\mathbf{j} + \frac{dc}{ds}\frac{ds}{dt}\mathbf{k} \\
&= \frac{ds}{dt}\left(\frac{da}{ds}\mathbf{i} + \frac{db}{ds}\mathbf{j} + \frac{dc}{ds}\mathbf{k}\right) \\
&= \frac{ds}{dt}\frac{d\mathbf{u}}{ds} \\
&= f'(t)\mathbf{u}'(f(t)). && s = f(t) \quad ■
\end{aligned}$$

Vector Functions of Constant Length

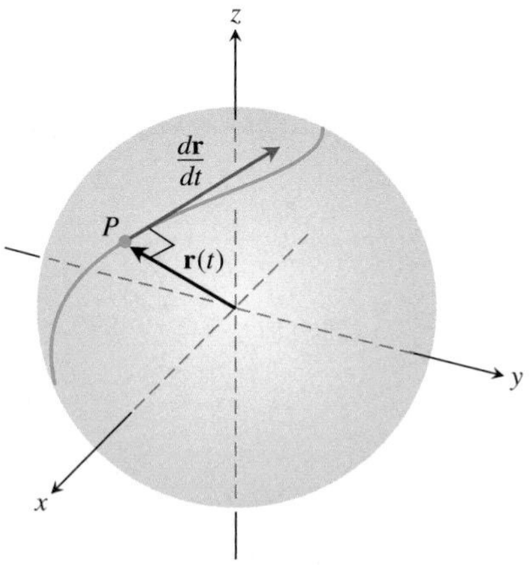

FIGURE 13.8 If a particle moves on a sphere in such a way that its position **r** is a differentiable function of time, then $\mathbf{r}\cdot(d\mathbf{r}/dt) = 0$.

When we track a particle moving on a sphere centered at the origin (Figure 13.8), the position vector has a constant length equal to the radius of the sphere. The velocity vector $d\mathbf{r}/dt$, tangent to the path of motion, is tangent to the sphere and hence perpendicular to **r**. This is always the case for a differentiable vector function of constant length: The vector and its first derivative are orthogonal. With the length constant, the change in the function is a change in direction only, and direction changes take place at right angles. We can also obtain this result by direct calculation:

$$\begin{aligned}
\mathbf{r}(t)\cdot\mathbf{r}(t) &= c^2 && |\mathbf{r}(t)| = c \text{ is constant.} \\
\frac{d}{dt}[\mathbf{r}(t)\cdot\mathbf{r}(t)] &= 0 && \text{Differentiate both sides.} \\
\mathbf{r}'(t)\cdot\mathbf{r}(t) + \mathbf{r}(t)\cdot\mathbf{r}'(t) &= 0 && \text{Rule 5 with } \mathbf{r}(t) = \mathbf{u}(t) = \mathbf{v}(t) \\
2\mathbf{r}'(t)\cdot\mathbf{r}(t) &= 0.
\end{aligned}$$

The vectors $\mathbf{r}'(t)$ and $\mathbf{r}(t)$ are orthogonal because their dot product is 0. In summary,

If $\mathbf{r}$ is a differentiable vector function of t of constant length, then

$$\mathbf{r} \cdot \frac{d\mathbf{r}}{dt} = 0. \tag{4}$$

We will use this observation repeatedly in Section 13.4.

EXAMPLE 5 Supporting Equation (4)

Show that $\mathbf{r}(t) = (\sin t)\mathbf{i} + (\cos t)\mathbf{j} + \sqrt{3}\mathbf{k}$ has constant length and is orthogonal to its derivative.

Solution

$$\mathbf{r}(t) = (\sin t)\mathbf{i} + (\cos t)\mathbf{j} + \sqrt{3}\mathbf{k}$$

$$|\mathbf{r}(t)| = \sqrt{(\sin t)^2 + (\cos t)^2 + \left(\sqrt{3}\right)^2} = \sqrt{1 + 3} = 2$$

$$\frac{d\mathbf{r}}{dt} = (\cos t)\mathbf{i} - (\sin t)\mathbf{j}$$

$$\mathbf{r} \cdot \frac{d\mathbf{r}}{dt} = \sin t \cos t - \sin t \cos t = 0$$

■

Integrals of Vector Functions

A differentiable vector function $\mathbf{R}(t)$ is an **antiderivative** of a vector function $\mathbf{r}(t)$ on an interval I if $d\mathbf{R}/dt = \mathbf{r}$ at each point of I. If $\mathbf{R}$ is an antiderivative of $\mathbf{r}$ on I, it can be shown, working one component at a time, that every antiderivative of $\mathbf{r}$ on I has the form $\mathbf{R} + \mathbf{C}$ for some constant vector $\mathbf{C}$ (Exercise 56). The set of all antiderivatives of $\mathbf{r}$ on I is the **indefinite integral** of $\mathbf{r}$ on I.

DEFINITION Indefinite Integral

The **indefinite integral** of $\mathbf{r}$ with respect to t is the set of all antiderivatives of $\mathbf{r}$, denoted by $\int \mathbf{r}(t)\,dt$. If $\mathbf{R}$ is any antiderivative of $\mathbf{r}$, then

$$\int \mathbf{r}(t)\,dt = \mathbf{R}(t) + \mathbf{C}.$$

The usual arithmetic rules for indefinite integrals apply.

EXAMPLE 6 Finding Indefinite Integrals

$$\int ((\cos t)\mathbf{i} + \mathbf{j} - 2t\mathbf{k})\,dt = \left(\int \cos t\,dt\right)\mathbf{i} + \left(\int dt\right)\mathbf{j} - \left(\int 2t\,dt\right)\mathbf{k} \tag{5}$$

$$= (\sin t + C_1)\mathbf{i} + (t + C_2)\mathbf{j} - (t^2 + C_3)\mathbf{k} \tag{6}$$

$$= (\sin t)\mathbf{i} + t\mathbf{j} - t^2\mathbf{k} + \mathbf{C} \qquad \mathbf{C} = C_1\mathbf{i} + C_2\mathbf{j} - C_3\mathbf{k}$$

As in the integration of scalar functions, we recommend that you skip the steps in Equations (5) and (6) and go directly to the final form. Find an antiderivative for each component and add a constant vector at the end. ■

Definite integrals of vector functions are best defined in terms of components.

DEFINITION Definite Integral

If the components of $\mathbf{r}(t) = f(t)\mathbf{i} + g(t)\mathbf{j} + h(t)\mathbf{k}$ are integrable over $[a, b]$, then so is $\mathbf{r}$, and the **definite integral** of $\mathbf{r}$ from a to b is

$$\int_a^b \mathbf{r}(t)\,dt = \left(\int_a^b f(t)\,dt\right)\mathbf{i} + \left(\int_a^b g(t)\,dt\right)\mathbf{j} + \left(\int_a^b h(t)\,dt\right)\mathbf{k}.$$

EXAMPLE 7 Evaluating Definite Integrals

$$\begin{aligned}\int_0^\pi ((\cos t)\mathbf{i} + \mathbf{j} - 2t\mathbf{k})\,dt &= \left(\int_0^\pi \cos t\,dt\right)\mathbf{i} + \left(\int_0^\pi dt\right)\mathbf{j} - \left(\int_0^\pi 2t\,dt\right)\mathbf{k} \\ &= \Big[\sin t\Big]_0^\pi \mathbf{i} + \Big[t\Big]_0^\pi \mathbf{j} - \Big[t^2\Big]_0^\pi \mathbf{k} \\ &= [0 - 0]\mathbf{i} + [\pi - 0]\mathbf{j} - [\pi^2 - 0^2]\mathbf{k} \\ &= \pi\mathbf{j} - \pi^2\mathbf{k}\end{aligned}$$

■

The Fundamental Theorem of Calculus for continuous vector functions says that

$$\int_a^b \mathbf{r}(t)\,dt = \mathbf{R}(t)\Big]_a^b = \mathbf{R}(b) - \mathbf{R}(a)$$

where $\mathbf{R}$ is any antiderivative of $\mathbf{r}$, so that $\mathbf{R}'(t) = \mathbf{r}(t)$ (Exercise 57).

EXAMPLE 8 Revisiting the Flight of a Glider

Suppose that we did not know the path of the glider in Example 4, but only its acceleration vector $\mathbf{a}(t) = -(3\cos t)\mathbf{i} - (3\sin t)\mathbf{j} + 2\mathbf{k}$. We also know that initially (at time $t = 0$), the glider departed from the point (3, 0, 0) with velocity $\mathbf{v}(0) = 3\mathbf{j}$. Find the glider's position as a function of t.

Solution Our goal is to find $\mathbf{r}(t)$ knowing

The differential equation: $\mathbf{a} = \dfrac{d^2\mathbf{r}}{dt^2} = -(3\cos t)\mathbf{i} - (3\sin t)\mathbf{j} + 2\mathbf{k}$

The initial conditions: $\mathbf{v}(0) = 3\mathbf{j}$ and $\mathbf{r}(0) = 3\mathbf{i} + 0\mathbf{j} + 0\mathbf{k}$.

Integrating both sides of the differential equation with respect to t gives

$$\mathbf{v}(t) = -(3\sin t)\mathbf{i} + (3\cos t)\mathbf{j} + 2t\mathbf{k} + \mathbf{C}_1.$$

We use $\mathbf{v}(0) = 3\mathbf{j}$ to find $\mathbf{C}_1$:

$$\begin{aligned}3\mathbf{j} &= -(3\sin 0)\mathbf{i} + (3\cos 0)\mathbf{j} + (0)\mathbf{k} + \mathbf{C}_1 \\ 3\mathbf{j} &= 3\mathbf{j} + \mathbf{C}_1 \\ \mathbf{C}_1 &= \mathbf{0}.\end{aligned}$$

The glider's velocity as a function of time is

$$\frac{d\mathbf{r}}{dt} = \mathbf{v}(t) = -(3\sin t)\mathbf{i} + (3\cos t)\mathbf{j} + 2t\mathbf{k}.$$

Integrating both sides of this last differential equation gives

$$\mathbf{r}(t) = (3\cos t)\mathbf{i} + (3\sin t)\mathbf{j} + t^2\mathbf{k} + \mathbf{C}_2.$$

We then use the initial condition $\mathbf{r}(0) = 3\mathbf{i}$ to find $\mathbf{C}_2$:

$$\begin{aligned} 3\mathbf{i} &= (3\cos 0)\mathbf{i} + (3\sin 0)\mathbf{j} + (0^2)\mathbf{k} + \mathbf{C}_2 \\ 3\mathbf{i} &= 3\mathbf{i} + (0)\mathbf{j} + (0)\mathbf{k} + \mathbf{C}_2 \\ \mathbf{C}_2 &= \mathbf{0}. \end{aligned}$$

The glider's position as a function of t is

$$\mathbf{r}(t) = (3\cos t)\mathbf{i} + (3\sin t)\mathbf{j} + t^2\mathbf{k}.$$

This is the path of the glider we know from Example 4 and is shown in Figure 13.7.

Note: It was peculiar to this example that both of the constant vectors of integration, $\mathbf{C}_1$ and $\mathbf{C}_2$, turned out to be $\mathbf{0}$. Exercises 31 and 32 give different results for these constants. ■

EXERCISES 13.1

Motion in the xy-plane

In Exercises 1–4, $\mathbf{r}(t)$ is the position of a particle in the xy-plane at time t. Find an equation in x and y whose graph is the path of the particle. Then find the particle's velocity and acceleration vectors at the given value of t.

1. $\mathbf{r}(t) = (t+1)\mathbf{i} + (t^2 - 1)\mathbf{j}, \quad t = 1$
2. $\mathbf{r}(t) = (t^2 + 1)\mathbf{i} + (2t - 1)\mathbf{j}, \quad t = 1/2$
3. $\mathbf{r}(t) = e^t\mathbf{i} + \frac{2}{9}e^{2t}\mathbf{j}, \quad t = \ln 3$
4. $\mathbf{r}(t) = (\cos 2t)\mathbf{i} + (3\sin 2t)\mathbf{j}, \quad t = 0$

Exercises 5–8 give the position vectors of particles moving along various curves in the xy-plane. In each case, find the particle's velocity and acceleration vectors at the stated times and sketch them as vectors on the curve.

5. **Motion on the circle $x^2 + y^2 = 1$**

$$\mathbf{r}(t) = (\sin t)\mathbf{i} + (\cos t)\mathbf{j}; \quad t = \pi/4 \text{ and } \pi/2$$

6. **Motion on the circle $x^2 + y^2 = 16$**

$$\mathbf{r}(t) = \left(4\cos\frac{t}{2}\right)\mathbf{i} + \left(4\sin\frac{t}{2}\right)\mathbf{j}; \quad t = \pi \text{ and } 3\pi/2$$

7. **Motion on the cycloid $x = t - \sin t$, $y = 1 - \cos t$**

$$\mathbf{r}(t) = (t - \sin t)\mathbf{i} + (1 - \cos t)\mathbf{j}; \quad t = \pi \text{ and } 3\pi/2$$

8. **Motion on the parabola $y = x^2 + 1$**

$$\mathbf{r}(t) = t\mathbf{i} + (t^2 + 1)\mathbf{j}; \quad t = -1, 0, \text{ and } 1$$

Velocity and Acceleration in Space

In Exercises 9–14, $\mathbf{r}(t)$ is the position of a particle in space at time t. Find the particle's velocity and acceleration vectors. Then find the particle's speed and direction of motion at the given value of t. Write the particle's velocity at that time as the product of its speed and direction.

9. $\mathbf{r}(t) = (t+1)\mathbf{i} + (t^2 - 1)\mathbf{j} + 2t\mathbf{k}, \quad t = 1$
10. $\mathbf{r}(t) = (1+t)\mathbf{i} + \frac{t^2}{\sqrt{2}}\mathbf{j} + \frac{t^3}{3}\mathbf{k}, \quad t = 1$
11. $\mathbf{r}(t) = (2\cos t)\mathbf{i} + (3\sin t)\mathbf{j} + 4t\mathbf{k}, \quad t = \pi/2$
12. $\mathbf{r}(t) = (\sec t)\mathbf{i} + (\tan t)\mathbf{j} + \frac{4}{3}t\mathbf{k}, \quad t = \pi/6$
13. $\mathbf{r}(t) = (2\ln(t+1))\mathbf{i} + t^2\mathbf{j} + \frac{t^2}{2}\mathbf{k}, \quad t = 1$
14. $\mathbf{r}(t) = (e^{-t})\mathbf{i} + (2\cos 3t)\mathbf{j} + (2\sin 3t)\mathbf{k}, \quad t = 0$

In Exercises 15–18, $\mathbf{r}(t)$ is the position of a particle in space at time t. Find the angle between the velocity and acceleration vectors at time $t = 0$.

15. $\mathbf{r}(t) = (3t + 1)\mathbf{i} + \sqrt{3}t\mathbf{j} + t^2\mathbf{k}$

16. $\mathbf{r}(t) = \left(\frac{\sqrt{2}}{2}t\right)\mathbf{i} + \left(\frac{\sqrt{2}}{2}t - 16t^2\right)\mathbf{j}$

17. $\mathbf{r}(t) = (\ln(t^2 + 1))\mathbf{i} + (\tan^{-1} t)\mathbf{j} + \sqrt{t^2 + 1}\,\mathbf{k}$

18. $\mathbf{r}(t) = \frac{4}{9}(1 + t)^{3/2}\mathbf{i} + \frac{4}{9}(1 - t)^{3/2}\mathbf{j} + \frac{1}{3}t\mathbf{k}$

In Exercises 19 and 20, $\mathbf{r}(t)$ is the position vector of a particle in space at time t. Find the time or times in the given time interval when the velocity and acceleration vectors are orthogonal.

19. $\mathbf{r}(t) = (t - \sin t)\mathbf{i} + (1 - \cos t)\mathbf{j}, \quad 0 \le t \le 2\pi$

20. $\mathbf{r}(t) = (\sin t)\mathbf{i} + t\mathbf{j} + (\cos t)\mathbf{k}, \quad t \ge 0$

Integrating Vector-Valued Functions

Evaluate the integrals in Exercises 21–26.

21. $\int_0^1 [t^3\mathbf{i} + 7\mathbf{j} + (t + 1)\mathbf{k}]\,dt$

22. $\int_1^2 \left[(6 - 6t)\mathbf{i} + 3\sqrt{t}\mathbf{j} + \left(\frac{4}{t^2}\right)\mathbf{k}\right] dt$

23. $\int_{-\pi/4}^{\pi/4} [(\sin t)\mathbf{i} + (1 + \cos t)\mathbf{j} + (\sec^2 t)\mathbf{k}]\,dt$

24. $\int_0^{\pi/3} [(\sec t \tan t)\mathbf{i} + (\tan t)\mathbf{j} + (2 \sin t \cos t)\mathbf{k}]\,dt$

25. $\int_1^4 \left[\frac{1}{t}\mathbf{i} + \frac{1}{5 - t}\mathbf{j} + \frac{1}{2t}\mathbf{k}\right] dt$

26. $\int_0^1 \left[\frac{2}{\sqrt{1 - t^2}}\mathbf{i} + \frac{\sqrt{3}}{1 + t^2}\mathbf{k}\right] dt$

Initial Value Problems for Vector-Valued Functions

Solve the initial value problems in Exercises 27–32 for $\mathbf{r}$ as a vector function of t.

27. Differential equation: $\frac{d\mathbf{r}}{dt} = -t\mathbf{i} - t\mathbf{j} - t\mathbf{k}$

Initial condition: $\mathbf{r}(0) = \mathbf{i} + 2\mathbf{j} + 3\mathbf{k}$

28. Differential equation: $\frac{d\mathbf{r}}{dt} = (180t)\mathbf{i} + (180t - 16t^2)\mathbf{j}$

Initial condition: $\mathbf{r}(0) = 100\mathbf{j}$

29. Differential equation: $\frac{d\mathbf{r}}{dt} = \frac{3}{2}(t + 1)^{1/2}\mathbf{i} + e^{-t}\mathbf{j} + \frac{1}{t + 1}\mathbf{k}$

Initial condition: $\mathbf{r}(0) = \mathbf{k}$

30. Differential equation: $\frac{d\mathbf{r}}{dt} = (t^3 + 4t)\mathbf{i} + t\mathbf{j} + 2t^2\mathbf{k}$

Initial condition: $\mathbf{r}(0) = \mathbf{i} + \mathbf{j}$

31. Differential equation: $\frac{d^2\mathbf{r}}{dt^2} = -32\mathbf{k}$

Initial conditions: $\mathbf{r}(0) = 100\mathbf{k}$ and

$\left.\frac{d\mathbf{r}}{dt}\right|_{t=0} = 8\mathbf{i} + 8\mathbf{j}$

32. Differential equation: $\frac{d^2\mathbf{r}}{dt^2} = -(\mathbf{i} + \mathbf{j} + \mathbf{k})$

Initial conditions: $\mathbf{r}(0) = 10\mathbf{i} + 10\mathbf{j} + 10\mathbf{k}$ and

$\left.\frac{d\mathbf{r}}{dt}\right|_{t=0} = \mathbf{0}$

Tangent Lines to Smooth Curves

As mentioned in the text, the tangent line to a smooth curve $\mathbf{r}(t) = f(t)\mathbf{i} + g(t)\mathbf{j} + h(t)\mathbf{k}$ at $t = t_0$ is the line that passes through the point $(f(t_0), g(t_0), h(t_0))$ parallel to $\mathbf{v}(t_0)$, the curve's velocity vector at t_0. In Exercises 33–36, find parametric equations for the line that is tangent to the given curve at the given parameter value $t = t_0$.

33. $\mathbf{r}(t) = (\sin t)\mathbf{i} + (t^2 - \cos t)\mathbf{j} + e^t\mathbf{k}, \quad t_0 = 0$

34. $\mathbf{r}(t) = (2 \sin t)\mathbf{i} + (2 \cos t)\mathbf{j} + 5t\mathbf{k}, \quad t_0 = 4\pi$

35. $\mathbf{r}(t) = (a \sin t)\mathbf{i} + (a \cos t)\mathbf{j} + bt\mathbf{k}, \quad t_0 = 2\pi$

36. $\mathbf{r}(t) = (\cos t)\mathbf{i} + (\sin t)\mathbf{j} + (\sin 2t)\mathbf{k}, \quad t_0 = \frac{\pi}{2}$

Motion on Circular Paths

37. Each of the following equations in parts (a)–(e) describes the motion of a particle having the same path, namely the unit circle $x^2 + y^2 = 1$. Although the path of each particle in parts (a)–(e) is the same, the behavior, or "dynamics," of each particle is different. For each particle, answer the following questions.

i. Does the particle have constant speed? If so, what is its constant speed?

ii. Is the particle's acceleration vector always orthogonal to its velocity vector?

iii. Does the particle move clockwise or counterclockwise around the circle?

iv. Does the particle begin at the point $(1, 0)$?

a. $\mathbf{r}(t) = (\cos t)\mathbf{i} + (\sin t)\mathbf{j}, \quad t \ge 0$

b. $\mathbf{r}(t) = \cos(2t)\mathbf{i} + \sin(2t)\mathbf{j}, \quad t \ge 0$

c. $\mathbf{r}(t) = \cos(t - \pi/2)\mathbf{i} + \sin(t - \pi/2)\mathbf{j}, \quad t \ge 0$

d. $\mathbf{r}(t) = (\cos t)\mathbf{i} - (\sin t)\mathbf{j}, \quad t \ge 0$

e. $\mathbf{r}(t) = \cos(t^2)\mathbf{i} + \sin(t^2)\mathbf{j}, \quad t \ge 0$

38. Show that the vector-valued function

$$\mathbf{r}(t) = (2\mathbf{i} + 2\mathbf{j} + \mathbf{k}) + \cos t\left(\frac{1}{\sqrt{2}}\mathbf{i} - \frac{1}{\sqrt{2}}\mathbf{j}\right) + \sin t\left(\frac{1}{\sqrt{3}}\mathbf{i} + \frac{1}{\sqrt{3}}\mathbf{j} + \frac{1}{\sqrt{3}}\mathbf{k}\right)$$

describes the motion of a particle moving in the circle of radius 1 centered at the point $(2, 2, 1)$ and lying in the plane $x + y - 2z = 2$.

Motion Along a Straight Line

39. At time $t = 0$, a particle is located at the point (1, 2, 3). It travels in a straight line to the point (4, 1, 4), has speed 2 at (1, 2, 3) and constant acceleration $3\mathbf{i} - \mathbf{j} + \mathbf{k}$. Find an equation for the position vector $\mathbf{r}(t)$ of the particle at time t.

40. A particle traveling in a straight line is located at the point $(1, -1, 2)$ and has speed 2 at time $t = 0$. The particle moves toward the point (3, 0, 3) with constant acceleration $2\mathbf{i} + \mathbf{j} + \mathbf{k}$. Find its position vector $\mathbf{r}(t)$ at time t.

Theory and Examples

41. Motion along a parabola A particle moves along the top of the parabola $y^2 = 2x$ from left to right at a constant speed of 5 units per second. Find the velocity of the particle as it moves through the point (2, 2).

42. Motion along a cycloid A particle moves in the xy-plane in such a way that its position at time t is

$$\mathbf{r}(t) = (t - \sin t)\mathbf{i} + (1 - \cos t)\mathbf{j}.$$

T **a.** Graph $\mathbf{r}(t)$. The resulting curve is a cycloid.

b. Find the maximum and minimum values of $|\mathbf{v}|$ and $|\mathbf{a}|$. (*Hint:* Find the extreme values of $|\mathbf{v}|^2$ and $|\mathbf{a}|^2$ first and take square roots later.)

43. Motion along an ellipse A particle moves around the ellipse $(y/3)^2 + (z/2)^2 = 1$ in the yz-plane in such a way that its position at time t is

$$\mathbf{r}(t) = (3\cos t)\mathbf{j} + (2\sin t)\mathbf{k}.$$

Find the maximum and minimum values of $|\mathbf{v}|$ and $|\mathbf{a}|$. (*Hint:* Find the extreme values of $|\mathbf{v}|^2$ and $|\mathbf{a}|^2$ first and take square roots later.)

44. A satellite in circular orbit A satellite of mass m is revolving at a constant speed v around a body of mass M (Earth, for example) in a circular orbit of radius r_0 (measured from the body's center of mass). Determine the satellite's orbital period T (the time to complete one full orbit), as follows:

a. Coordinatize the orbital plane by placing the origin at the body's center of mass, with the satellite on the x-axis at $t = 0$ and moving counterclockwise, as in the accompanying figure.

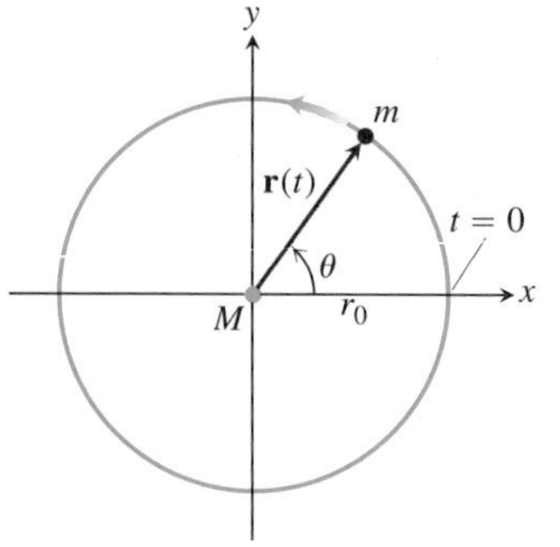

Let $\mathbf{r}(t)$ be the satellite's position vector at time t. Show that $\theta = vt/r_0$ and hence that

$$\mathbf{r}(t) = \left(r_0 \cos \frac{vt}{r_0}\right)\mathbf{i} + \left(r_0 \sin \frac{vt}{r_0}\right)\mathbf{j}.$$

b. Find the acceleration of the satellite.

c. According to Newton's law of gravitation, the gravitational force exerted on the satellite is directed toward M and is given by

$$\mathbf{F} = \left(-\frac{GmM}{r_0^2}\right)\frac{\mathbf{r}}{r_0},$$

where G is the universal constant of gravitation. Using Newton's second law, $\mathbf{F} = m\mathbf{a}$, show that $v^2 = GM/r_0$.

d. Show that the orbital period T satisfies $vT = 2\pi r_0$.

e. From parts (c) and (d), deduce that

$$T^2 = \frac{4\pi^2}{GM} r_0^3.$$

That is, the square of the period of a satellite in circular orbit is proportional to the cube of the radius from the orbital center.

45. Let $\mathbf{v}$ be a differentiable vector function of t. Show that if $\mathbf{v} \cdot (d\mathbf{v}/dt) = 0$ for all t, then $|\mathbf{v}|$ is constant.

46. Derivatives of triple scalar products

a. Show that if $\mathbf{u}$, $\mathbf{v}$, and $\mathbf{w}$ are differentiable vector functions of t, then

$$\frac{d}{dt}(\mathbf{u} \cdot \mathbf{v} \times \mathbf{w}) = \frac{d\mathbf{u}}{dt} \cdot \mathbf{v} \times \mathbf{w} + \mathbf{u} \cdot \frac{d\mathbf{v}}{dt} \times \mathbf{w} + \mathbf{u} \cdot \mathbf{v} \times \frac{d\mathbf{w}}{dt}. \tag{7}$$

b. Show that Equation (7) is equivalent to

$$\frac{d}{dt}\begin{vmatrix} u_1 & u_2 & u_3 \\ v_1 & v_2 & v_3 \\ w_1 & w_2 & w_3 \end{vmatrix} = \begin{vmatrix} \frac{du_1}{dt} & \frac{du_2}{dt} & \frac{du_3}{dt} \\ v_1 & v_2 & v_3 \\ w_1 & w_2 & w_3 \end{vmatrix} + \begin{vmatrix} u_1 & u_2 & u_3 \\ \frac{dv_1}{dt} & \frac{dv_2}{dt} & \frac{dv_3}{dt} \\ w_1 & w_2 & w_3 \end{vmatrix} + \begin{vmatrix} u_1 & u_2 & u_3 \\ v_1 & v_2 & v_3 \\ \frac{dw_1}{dt} & \frac{dw_2}{dt} & \frac{dw_3}{dt} \end{vmatrix}. \tag{8}$$

Equation (8) says that the derivative of a 3 by 3 determinant of differentiable functions is the sum of the three determinants obtained from the original by differentiating one row at a time. The result extends to determinants of any order.

47. (*Continuation of Exercise 46.*) Suppose that $\mathbf{r}(t) = f(t)\mathbf{i} + g(t)\mathbf{j} + h(t)\mathbf{k}$ and that f, g, and h have derivatives through order three. Use Equation (7) or (8) to show that

$$\frac{d}{dt}\left(\mathbf{r}\cdot\frac{d\mathbf{r}}{dt}\times\frac{d^2\mathbf{r}}{dt^2}\right) = \mathbf{r}\cdot\left(\frac{d\mathbf{r}}{dt}\times\frac{d^3\mathbf{r}}{dt^3}\right). \qquad (9)$$

(*Hint:* Differentiate on the left and look for vectors whose products are zero.)

48. Constant Function Rule Prove that if $\mathbf{u}$ is the vector function with the constant value $\mathbf{C}$, then $d\mathbf{u}/dt = \mathbf{0}$.

49. Scalar Multiple Rules

a. Prove that if $\mathbf{u}$ is a differentiable function of t and c is any real number, then

$$\frac{d(c\mathbf{u})}{dt} = c\frac{d\mathbf{u}}{dt}.$$

b. Prove that if $\mathbf{u}$ is a differentiable function of t and f is a differentiable scalar function of t, then

$$\frac{d}{dt}(f\mathbf{u}) = \frac{df}{dt}\mathbf{u} + f\frac{d\mathbf{u}}{dt}.$$

50. Sum and Difference Rules Prove that if $\mathbf{u}$ and $\mathbf{v}$ are differentiable functions of t, then

$$\frac{d}{dt}(\mathbf{u}+\mathbf{v}) = \frac{d\mathbf{u}}{dt} + \frac{d\mathbf{v}}{dt}$$

and

$$\frac{d}{dt}(\mathbf{u}-\mathbf{v}) = \frac{d\mathbf{u}}{dt} - \frac{d\mathbf{v}}{dt}.$$

51. Component Test for Continuity at a Point Show that the vector function $\mathbf{r}$ defined by $\mathbf{r}(t) = f(t)\mathbf{i} + g(t)\mathbf{j} + h(t)\mathbf{k}$ is continuous at $t = t_0$ if and only if f, g, and h are continuous at t_0.

52. Limits of cross products of vector functions Suppose that $\mathbf{r}_1(t) = f_1(t)\mathbf{i} + f_2(t)\mathbf{j} + f_3(t)\mathbf{k}$, $\mathbf{r}_2(t) = g_1(t)\mathbf{i} + g_2(t)\mathbf{j} + g_3(t)\mathbf{k}$, $\lim_{t\to t_0}\mathbf{r}_1(t) = \mathbf{A}$, and $\lim_{t\to t_0}\mathbf{r}_2(t) = \mathbf{B}$. Use the determinant formula for cross products and the Limit Product Rule for scalar functions to show that

$$\lim_{t\to t_0}(\mathbf{r}_1(t)\times\mathbf{r}_2(t)) = \mathbf{A}\times\mathbf{B}$$

53. Differentiable vector functions are continuous Show that if $\mathbf{r}(t) = f(t)\mathbf{i} + g(t)\mathbf{j} + h(t)\mathbf{k}$ is differentiable at $t = t_0$, then it is continuous at t_0 as well.

54. Establish the following properties of integrable vector functions.

a. The *Constant Scalar Multiple Rule:*

$$\int_a^b k\mathbf{r}(t)\,dt = k\int_a^b \mathbf{r}(t)\,dt \quad \text{(any scalar } k\text{)}$$

The *Rule for Negatives,*

$$\int_a^b (-\mathbf{r}(t))\,dt = -\int_a^b \mathbf{r}(t)\,dt,$$

is obtained by taking $k = -1$.

b. The *Sum and Difference Rules:*

$$\int_a^b (\mathbf{r}_1(t) \pm \mathbf{r}_2(t))\,dt = \int_a^b \mathbf{r}_1(t)\,dt \pm \int_a^b \mathbf{r}_2(t)\,dt$$

c. The *Constant Vector Multiple Rules:*

$$\int_a^b \mathbf{C}\cdot\mathbf{r}(t)\,dt = \mathbf{C}\cdot\int_a^b \mathbf{r}(t)\,dt \quad \text{(any constant vector } \mathbf{C}\text{)}$$

and

$$\int_a^b \mathbf{C}\times\mathbf{r}(t)\,dt = \mathbf{C}\times\int_a^b \mathbf{r}(t)\,dt \quad \text{(any constant vector } \mathbf{C}\text{)}$$

55. Products of scalar and vector functions Suppose that the scalar function $u(t)$ and the vector function $\mathbf{r}(t)$ are both defined for $a \le t \le b$.

a. Show that $u\mathbf{r}$ is continuous on $[a, b]$ if u and $\mathbf{r}$ are continuous on $[a, b]$.

b. If u and $\mathbf{r}$ are both differentiable on $[a, b]$, show that $u\mathbf{r}$ is differentiable on $[a, b]$ and that

$$\frac{d}{dt}(u\mathbf{r}) = u\frac{d\mathbf{r}}{dt} + \mathbf{r}\frac{du}{dt}.$$

56. Antiderivatives of vector functions

a. Use Corollary 2 of the Mean Value Theorem for scalar functions to show that if two vector functions $\mathbf{R}_1(t)$ and $\mathbf{R}_2(t)$ have identical derivatives on an interval I, then the functions differ by a constant vector value throughout I.

b. Use the result in part (a) to show that if $\mathbf{R}(t)$ is any antiderivative of $\mathbf{r}(t)$ on I, then any other antiderivative of $\mathbf{r}$ on I equals $\mathbf{R}(t) + \mathbf{C}$ for some constant vector $\mathbf{C}$.

57. The Fundamental Theorem of Calculus The Fundamental Theorem of Calculus for scalar functions of a real variable holds for vector functions of a real variable as well. Prove this by using the theorem for scalar functions to show first that if a vector function $\mathbf{r}(t)$ is continuous for $a \le t \le b$, then

$$\frac{d}{dt}\int_a^t \mathbf{r}(\tau)\,d\tau = \mathbf{r}(t)$$

at every point t of (a, b). Then use the conclusion in part (b) of Exercise 56 to show that if $\mathbf{R}$ is any antiderivative of $\mathbf{r}$ on $[a, b]$ then

$$\int_a^b \mathbf{r}(t)\,dt = \mathbf{R}(b) - \mathbf{R}(a).$$

COMPUTER EXPLORATIONS

Drawing Tangents to Space Curves

Use a CAS to perform the following steps in Exercises 58–61.

a. Plot the space curve traced out by the position vector $\mathbf{r}$.

b. Find the components of the velocity vector $d\mathbf{r}/dt$.

c. Evaluate $d\mathbf{r}/dt$ at the given point t_0 and determine the equation of the tangent line to the curve at $\mathbf{r}(t_0)$.

d. Plot the tangent line together with the curve over the given interval.

58. $\mathbf{r}(t) = (\sin t - t\cos t)\mathbf{i} + (\cos t + t\sin t)\mathbf{j} + t^2\mathbf{k}$, $0 \le t \le 6\pi$, $t_0 = 3\pi/2$

59. $\mathbf{r}(t) = \sqrt{2t}\,\mathbf{i} + e^t\mathbf{j} + e^{-t}\mathbf{k}$, $-2 \le t \le 3$, $t_0 = 1$

60. $\mathbf{r}(t) = (\sin 2t)\mathbf{i} + (\ln(1 + t))\mathbf{j} + t\mathbf{k}$, $0 \le t \le 4\pi$, $t_0 = \pi/4$

61. $\mathbf{r}(t) = (\ln(t^2 + 2))\mathbf{i} + (\tan^{-1} 3t)\mathbf{j} + \sqrt{t^2 + 1}\,\mathbf{k}$, $-3 \le t \le 5$, $t_0 = 3$

In Exercises 62 and 63, you will explore graphically the behavior of the helix

$$\mathbf{r}(t) = (\cos at)\mathbf{i} + (\sin at)\mathbf{j} + bt\mathbf{k}.$$

as you change the values of the constants a and b. Use a CAS to perform the steps in each exercise.

62. Set $b = 1$. Plot the helix $\mathbf{r}(t)$ together with the tangent line to the curve at $t = 3\pi/2$ for $a = 1, 2, 4$, and 6 over the interval $0 \le t \le 4\pi$. Describe in your own words what happens to the graph of the helix and the position of the tangent line as a increases through these positive values.

63. Set $a = 1$. Plot the helix $\mathbf{r}(t)$ together with the tangent line to the curve at $t = 3\pi/2$ for $b = 1/4, 1/2, 2$, and 4 over the interval $0 \le t \le 4\pi$. Describe in your own words what happens to the graph of the helix and the position of the tangent line as b increases through these positive values.

13.2 Modeling Projectile Motion

When we shoot a projectile into the air we usually want to know beforehand how far it will go (will it reach the target?), how high it will rise (will it clear the hill?), and when it will land (when do we get results?). We get this information from the direction and magnitude of the projectile's initial velocity vector, using Newton's second law of motion.

The Vector and Parametric Equations for Ideal Projectile Motion

To derive equations for projectile motion, we assume that the projectile behaves like a particle moving in a vertical coordinate plane and that the only force acting on the projectile during its flight is the constant force of gravity, which always points straight down. In practice, none of these assumptions really holds. The ground moves beneath the projectile as the earth turns, the air creates a frictional force that varies with the projectile's speed and altitude, and the force of gravity changes as the projectile moves along. All this must be taken into account by applying corrections to the predictions of the *ideal* equations we are about to derive. The corrections, however, are not the subject of this section.

We assume that the projectile is launched from the origin at time $t = 0$ into the first quadrant with an initial velocity $\mathbf{v}_0$ (Figure 13.9). If $\mathbf{v}_0$ makes an angle α with the horizontal, then

$$\mathbf{v}_0 = (|\mathbf{v}_0|\cos\alpha)\mathbf{i} + (|\mathbf{v}_0|\sin\alpha)\mathbf{j}.$$

If we use the simpler notation v_0 for the initial speed $|\mathbf{v}_0|$, then

$$\mathbf{v}_0 = (v_0\cos\alpha)\mathbf{i} + (v_0\sin\alpha)\mathbf{j}. \tag{1}$$

The projectile's initial position is

$$\mathbf{r}_0 = 0\mathbf{i} + 0\mathbf{j} = \mathbf{0}. \tag{2}$$

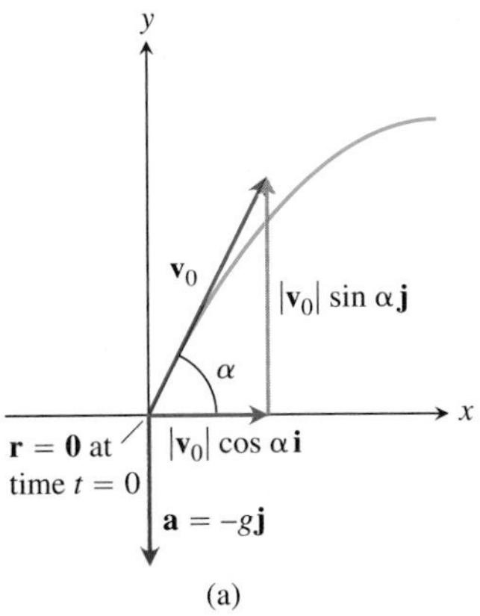

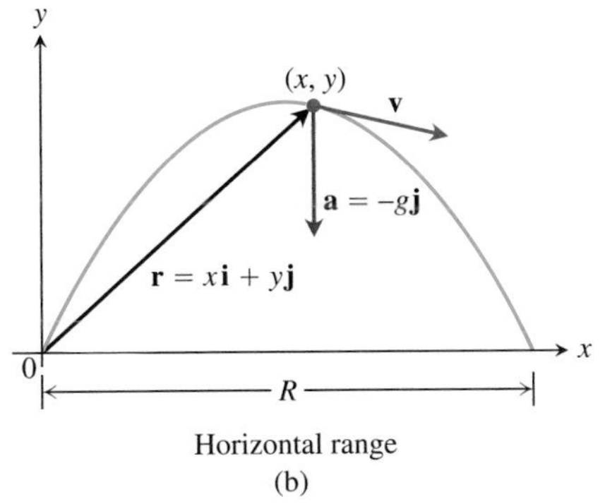

FIGURE 13.9 (a) Position, velocity, acceleration, and launch angle at $t = 0$. (b) Position, velocity, and acceleration at a later time t.

Newton's second law of motion says that the force acting on the projectile is equal to the projectile's mass m times its acceleration, or $m(d^2\mathbf{r}/dt^2)$ if $\mathbf{r}$ is the projectile's position vector and t is time. If the force is solely the gravitational force $-mg\mathbf{j}$, then

$$m\frac{d^2\mathbf{r}}{dt^2} = -mg\mathbf{j} \qquad \text{and} \qquad \frac{d^2\mathbf{r}}{dt^2} = -g\mathbf{j}.$$

We find $\mathbf{r}$ as a function of t by solving the following initial value problem.

Differential equation: $\dfrac{d^2\mathbf{r}}{dt^2} = -g\mathbf{j}$

Initial conditions: $\mathbf{r} = \mathbf{r}_0$ and $\dfrac{d\mathbf{r}}{dt} = \mathbf{v}_0$ when $t = 0$

The first integration gives

$$\frac{d\mathbf{r}}{dt} = -(gt)\mathbf{j} + \mathbf{v}_0.$$

A second integration gives

$$\mathbf{r} = -\frac{1}{2}gt^2\mathbf{j} + \mathbf{v}_0 t + \mathbf{r}_0.$$

Substituting the values of $\mathbf{v}_0$ and $\mathbf{r}_0$ from Equations (1) and (2) gives

$$\mathbf{r} = -\frac{1}{2}gt^2\mathbf{j} + \underbrace{(v_0\cos\alpha)t\mathbf{i} + (v_0\sin\alpha)t\mathbf{j}}_{\mathbf{v}_0 t} + \mathbf{0}$$

Collecting terms, we have

> **Ideal Projectile Motion Equation**
>
> $$\mathbf{r} = (v_0\cos\alpha)t\mathbf{i} + \left((v_0\sin\alpha)t - \frac{1}{2}gt^2\right)\mathbf{j}. \qquad (3)$$

Equation (3) is the *vector equation* for ideal projectile motion. The angle α is the projectile's **launch angle** (**firing angle**, **angle of elevation**), and v_0, as we said before, is the projectile's **initial speed**. The components of $\mathbf{r}$ give the parametric equations

$$x = (v_0\cos\alpha)t \qquad \text{and} \qquad y = (v_0\sin\alpha)t - \frac{1}{2}gt^2, \qquad (4)$$

where x is the distance downrange and y is the height of the projectile at time $t \geq 0$.

EXAMPLE 1 Firing an Ideal Projectile

A projectile is fired from the origin over horizontal ground at an initial speed of 500 m/sec and a launch angle of 60°. Where will the projectile be 10 sec later?

Solution We use Equation (3) with $v_0 = 500$, $\alpha = 60°$, $g = 9.8$, and $t = 10$ to find the projectile's components 10 sec after firing.

$$\begin{aligned}\mathbf{r} &= (v_0 \cos \alpha)t\mathbf{i} + \left((v_0 \sin \alpha)t - \frac{1}{2}gt^2\right)\mathbf{j} \\ &= (500)\left(\frac{1}{2}\right)(10)\mathbf{i} + \left((500)\left(\frac{\sqrt{3}}{2}\right)10 - \left(\frac{1}{2}\right)(9.8)(100)\right)\mathbf{j} \\ &\approx 2500\mathbf{i} + 3840\mathbf{j}.\end{aligned}$$

Ten seconds after firing, the projectile is about 3840 m in the air and 2500 m downrange. ■

Height, Flight Time, and Range

Equation (3) enables us to answer most questions about the ideal motion for a projectile fired from the origin.

The projectile reaches its highest point when its vertical velocity component is zero, that is, when

$$\frac{dy}{dt} = v_0 \sin \alpha - gt = 0, \qquad \text{or} \qquad t = \frac{v_0 \sin \alpha}{g}.$$

For this value of t, the value of y is

$$y_{\max} = (v_0 \sin \alpha)\left(\frac{v_0 \sin \alpha}{g}\right) - \frac{1}{2}g\left(\frac{v_0 \sin \alpha}{g}\right)^2 = \frac{(v_0 \sin \alpha)^2}{2g}.$$

To find when the projectile lands when fired over horizontal ground, we set the vertical component equal to zero in Equation (3) and solve for t.

$$\begin{aligned}(v_0 \sin \alpha)t - \frac{1}{2}gt^2 &= 0 \\ t\left(v_0 \sin \alpha - \frac{1}{2}gt\right) &= 0 \\ t = 0, \qquad t &= \frac{2v_0 \sin \alpha}{g}\end{aligned}$$

Since 0 is the time the projectile is fired, $(2v_0 \sin \alpha)/g$ must be the time when the projectile strikes the ground.

To find the projectile's **range** R, the distance from the origin to the point of impact on horizontal ground, we find the value of the horizontal component when $t = (2v_0 \sin \alpha)/g$.

$$\begin{aligned}x &= (v_0 \cos \alpha)t \\ R &= (v_0 \cos \alpha)\left(\frac{2v_0 \sin \alpha}{g}\right) = \frac{v_0^2}{g}(2 \sin \alpha \cos \alpha) = \frac{v_0^2}{g} \sin 2\alpha\end{aligned}$$

The range is largest when $\sin 2\alpha = 1$ or $\alpha = 45°$.

Height, Flight Time, and Range for Ideal Projectile Motion

For ideal projectile motion when an object is launched from the origin over a horizontal surface with initial speed v_0 and launch angle α:

Maximum height: $$y_{max} = \frac{(v_0 \sin \alpha)^2}{2g}$$

Flight time: $$t = \frac{2v_0 \sin \alpha}{g}$$

Range: $$R = \frac{v_0^2}{g} \sin 2\alpha .$$

EXAMPLE 2 Investigating Ideal Projectile Motion

Find the maximum height, flight time, and range of a projectile fired from the origin over horizontal ground at an initial speed of 500 m/sec and a launch angle of 60° (same projectile as Example 1).

Solution

$$\text{Maximum height:} \quad y_{max} = \frac{(v_0 \sin \alpha)^2}{2g} = \frac{(500 \sin 60°)^2}{2(9.8)} \approx 9566 \text{ m}$$

$$\text{Flight time:} \quad t = \frac{2v_0 \sin \alpha}{g} = \frac{2(500) \sin 60°}{9.8} \approx 88.4 \text{ sec}$$

$$\text{Range:} \quad R = \frac{v_0^2}{g} \sin 2\alpha = \frac{(500)^2 \sin 120°}{9.8} \approx 22{,}092 \text{ m}$$

From Equation (3), the position vector of the projectile is

$$\begin{aligned} \mathbf{r} &= (v_0 \cos \alpha)t\mathbf{i} + \left((v_0 \sin \alpha)t - \frac{1}{2}gt^2\right)\mathbf{j} \\ &= (500 \cos 60°)t\mathbf{i} + \left((500 \sin 60°)t - \frac{1}{2}(9.8)t^2\right)\mathbf{j} \\ &= 250t\mathbf{i} + \left(\left(250\sqrt{3}\right)t - 4.9t^2\right)\mathbf{j}. \end{aligned}$$

A graph of the projectile's path is shown in Figure 13.10. ■

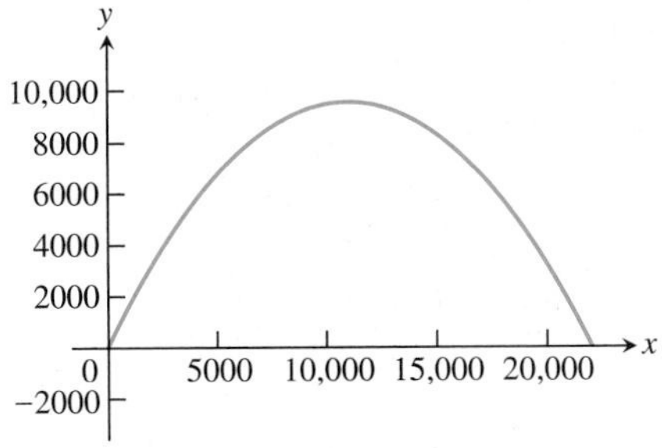

FIGURE 13.10 The graph of the projectile described in Example 2.

Ideal Trajectories Are Parabolic

It is often claimed that water from a hose traces a parabola in the air, but anyone who looks closely enough will see this is not so. The air slows the water down, and its forward progress is too slow at the end to keep pace with the rate at which it falls.

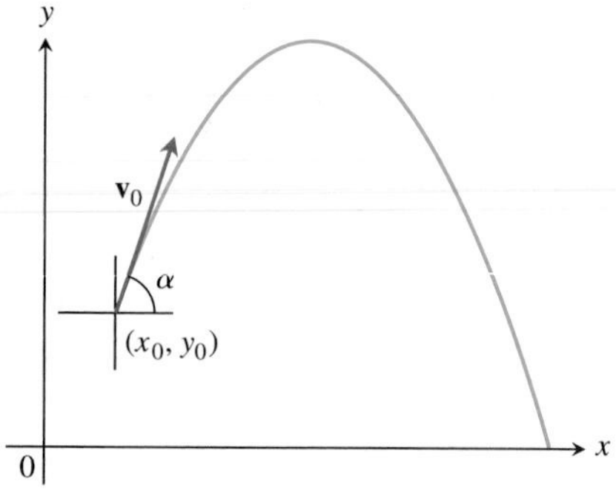

FIGURE 13.11 The path of a projectile fired from (x_0, y_0) with an initial velocity $\mathbf{v}_0$ at an angle of α degrees with the horizontal.

What is really being claimed is that ideal projectiles move along parabolas, and this we can see from Equations (4). If we substitute $t = x/(v_0 \cos \alpha)$ from the first equation into the second, we obtain the Cartesian-coordinate equation

$$y = -\left(\frac{g}{2v_0^2 \cos^2 \alpha}\right)x^2 + (\tan \alpha)x.$$

This equation has the form $y = ax^2 + bx$, so its graph is a parabola.

Firing from (x_0, y_0)

If we fire our ideal projectile from the point (x_0, y_0) instead of the origin (Figure 13.11), the position vector for the path of motion is

$$\mathbf{r} = (x_0 + (v_0 \cos \alpha)t)\mathbf{i} + \left(y_0 + (v_0 \sin \alpha)t - \frac{1}{2}gt^2\right)\mathbf{j}, \qquad (5)$$

as you are asked to show in Exercise 19.

EXAMPLE 3 Firing a Flaming Arrow

To open the 1992 Summer Olympics in Barcelona, bronze medalist archer Antonio Rebollo lit the Olympic torch with a flaming arrow (Figure 13.12). Suppose that Rebollo shot the arrow at a height of 6 ft above ground level 90 ft from the 70-ft-high cauldron, and he wanted the arrow to reach maximum height exactly 4 ft above the center of the cauldron (Figure 13.12).

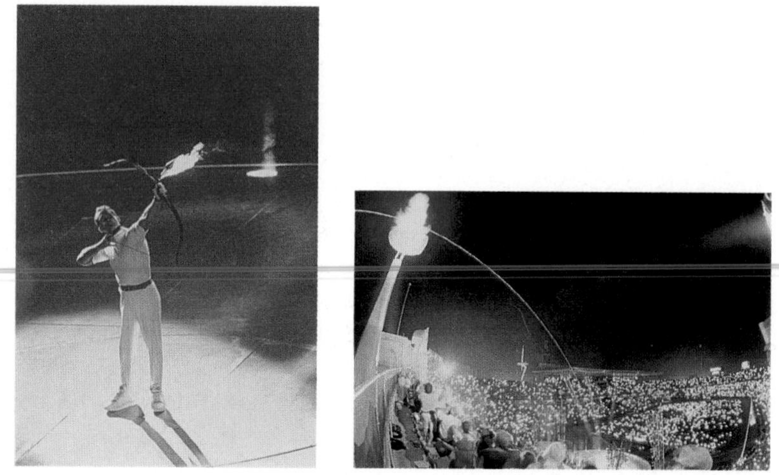

FIGURE 13.12 Spanish archer Antonio Rebollo lights the Olympic torch in Barcelona with a flaming arrow.

(a) Express $y_{\max}$ in terms of the initial speed v_0 and firing angle α.

(b) Use $y_{\max} = 74$ ft (Figure 13.13) and the result from part (a) to find the value of $v_0 \sin \alpha$.

(c) Find the value of $v_0 \cos \alpha$.

(d) Find the initial firing angle of the arrow.

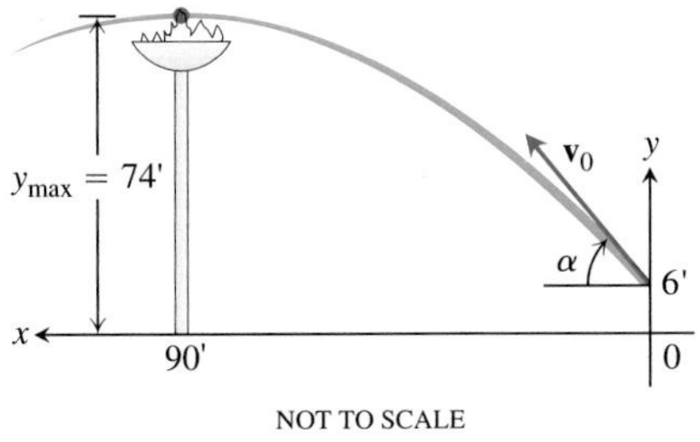

FIGURE 13.13 Ideal path of the arrow that lit the Olympic torch (Example 3).

Solution

(a) We use a coordinate system in which the positive x-axis lies along the ground toward the left (to match the second photograph in Figure 13.12) and the coordinates of the flaming arrow at $t = 0$ are $x_0 = 0$ and $y_0 = 6$ (Figure 13.13). We have

$$y = y_0 + (v_0 \sin \alpha)t - \frac{1}{2}gt^2 \qquad \text{Equation (5), } \mathbf{j}\text{-component}$$

$$= 6 + (v_0 \sin \alpha)t - \frac{1}{2}gt^2. \qquad y_0 = 6$$

We find the time when the arrow reaches its highest point by setting $dy/dt = 0$ and solving for t, obtaining

$$t = \frac{v_0 \sin \alpha}{g}.$$

For this value of t, the value of y is

$$y_{max} = 6 + (v_0 \sin \alpha)\left(\frac{v_0 \sin \alpha}{g}\right) - \frac{1}{2}g\left(\frac{v_0 \sin \alpha}{g}\right)^2$$

$$= 6 + \frac{(v_0 \sin \alpha)^2}{2g}.$$

(b) Using $y_{max} = 74$ and $g = 32$, we see from the preceeding equation in part (a) that

$$74 = 6 + \frac{(v_0 \sin \alpha)^2}{2(32)}$$

or

$$v_0 \sin \alpha = \sqrt{(68)(64)}.$$

(c) When the arrow reaches y_{max}, the horizontal distance traveled to the center of the cauldron is $x = 90$ ft. We substitute the time to reach y_{max} from part (a) and the horizontal distance $x = 90$ ft into the **i**-component of Equation (5) to obtain

$$x = x_0 + (v_0 \cos \alpha)t \qquad \text{Equation (5), } \mathbf{i}\text{-component}$$

$$90 = 0 + (v_0 \cos \alpha)t \qquad x = 90,\ x_0 = 0$$

$$= (v_0 \cos \alpha)\left(\frac{v_0 \sin \alpha}{g}\right). \qquad t = (v_0 \sin \alpha)/g$$

Solving this equation for $v_0 \cos \alpha$ and using $g = 32$ and the result from part (b), we have

$$v_0 \cos \alpha = \frac{90g}{v_0 \sin \alpha} = \frac{(90)(32)}{\sqrt{(68)(64)}}.$$

(d) Parts (b) and (c) together tell us that

$$\tan \alpha = \frac{v_0 \sin \alpha}{v_0 \cos \alpha} = \frac{\left(\sqrt{(68)(64)}\right)^2}{(90)(32)} = \frac{68}{45}$$

or

$$\alpha = \tan^{-1}\left(\frac{68}{45}\right) \approx 56.5°.$$

This is Rebollo's firing angle. ■

Projectile Motion with Wind Gusts

The next example shows how to account for another force acting on a projectile. We also assume that the path of the baseball in Example 4 lies in a vertical plane.

EXAMPLE 4 Hitting a Baseball

A baseball is hit when it is 3 ft above the ground. It leaves the bat with initial speed of 152 ft/sec, making an angle of 20° with the horizontal. At the instant the ball is hit, an instantaneous gust of wind blows in the horizontal direction directly opposite the direction the ball is taking toward the outfield, adding a component of $-8.8\mathbf{i}$ (ft/sec) to the ball's initial velocity (8.8 ft/sec = 6 mph).

(a) Find a vector equation (position vector) for the path of the baseball.

(b) How high does the baseball go, and when does it reach maximum height?

(c) Assuming that the ball is not caught, find its range and flight time.

Solution

(a) Using Equation (1) and accounting for the gust of wind, the initial velocity of the baseball is

$$\begin{aligned}\mathbf{v}_0 &= (v_0 \cos \alpha)\mathbf{i} + (v_0 \sin \alpha)\mathbf{j} - 8.8\mathbf{i} \\ &= (152 \cos 20°)\mathbf{i} + (152 \sin 20°)\mathbf{j} - (8.8)\mathbf{i} \\ &= (152 \cos 20° - 8.8)\mathbf{i} + (152 \sin 20°)\mathbf{j}.\end{aligned}$$

The initial position is $\mathbf{r}_0 = 0\mathbf{i} + 3\mathbf{j}$. Integration of $d^2\mathbf{r}/dt^2 = -g\mathbf{j}$ gives

$$\frac{d\mathbf{r}}{dt} = -(gt)\mathbf{j} + \mathbf{v}_0.$$

A second integration gives

$$\mathbf{r} = -\frac{1}{2}gt^2\mathbf{j} + \mathbf{v}_0 t + \mathbf{r}_0.$$

Substituting the values of $\mathbf{v}_0$ and $\mathbf{r}_0$ into the last equation gives the position vector of the baseball.

$$\begin{aligned}\mathbf{r} &= -\frac{1}{2}gt^2\mathbf{j} + \mathbf{v}_0 t + \mathbf{r}_0 \\ &= -16t^2\mathbf{j} + (152 \cos 20° - 8.8)t\mathbf{i} + (152 \sin 20°)t\mathbf{j} + 3\mathbf{j} \\ &= (152 \cos 20° - 8.8)t\mathbf{i} + \left(3 + (152 \sin 20°)t - 16t^2\right)\mathbf{j}.\end{aligned}$$

(b) The baseball reaches its highest point when the vertical component of velocity is zero, or

$$\frac{dy}{dt} = 152 \sin 20° - 32t = 0.$$

Solving for t we find

$$t = \frac{152 \sin 20°}{32} \approx 1.62 \text{ sec}.$$

Substituting this time into the vertical component for **r** gives the maximum height

$$\begin{aligned} y_{\max} &= 3 + (152 \sin 20°)(1.62) - 16(1.62)^2 \\ &\approx 45.2 \text{ ft}. \end{aligned}$$

That is, the maximum height of the baseball is about 45.2 ft, reached about 1.6 sec after leaving the bat.

(c) To find when the baseball lands, we set the vertical component for **r** equal to 0 and solve for t:

$$\begin{aligned} 3 + (152 \sin 20°)t - 16t^2 &= 0 \\ 3 + (51.99)t - 16t^2 &= 0. \end{aligned}$$

The solution values are about $t = 3.3$ sec and $t = -0.06$ sec. Substituting the positive time into the horizontal component for **r**, we find the range

$$\begin{aligned} R &= (152 \cos 20° - 8.8)(3.3) \\ &\approx 442 \text{ ft}. \end{aligned}$$

Thus, the horizontal range is about 442 ft, and the flight time is about 3.3 sec. ■

In Exercises 29 through 31, we consider projectile motion when there is air resistance slowing down the flight.

EXERCISES 13.2

Projectile flights in the following exercises are to be treated as ideal unless stated otherwise. All launch angles are assumed to be measured from the horizontal. All projectiles are assumed to be launched from the origin over a horizontal surface unless stated otherwise.

1. **Travel time** A projectile is fired at a speed of 840 m/sec at an angle of 60°. How long will it take to get 21 km downrange?

2. **Finding muzzle speed** Find the muzzle speed of a gun whose maximum range is 24.5 km.

3. **Flight time and height** A projectile is fired with an initial speed of 500 m/sec at an angle of elevation of 45°.

   a. When and how far away will the projectile strike?

   b. How high overhead will the projectile be when it is 5 km downrange?

   c. What is the greatest height reached by the projectile?

4. **Throwing a baseball** A baseball is thrown from the stands 32 ft above the field at an angle of 30° up from the horizontal. When and how far away will the ball strike the ground if its initial speed is 32 ft/sec?

5. **Shot put** An athlete puts a 16-lb shot at an angle of 45° to the horizontal from 6.5 ft above the ground at an initial speed of 44 ft/sec as suggested in the accompanying figure. How long after launch and how far from the inner edge of the stopboard does the shot land?

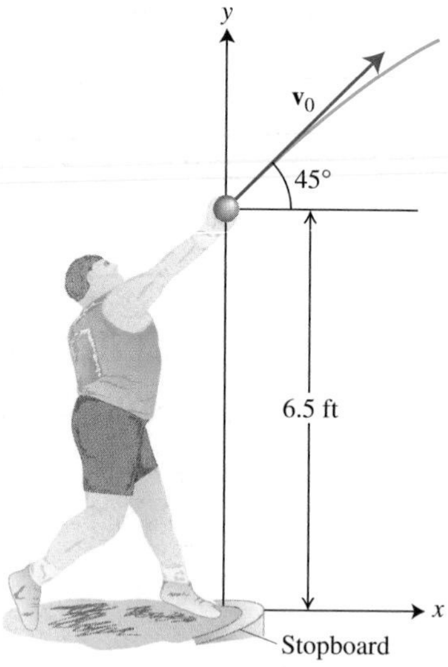

6. (*Continuation of Exercise* 5.) Because of its initial elevation, the shot in Exercise 5 would have gone slightly farther if it had been launched at a 40° angle. How much farther? Answer in inches.

7. **Firing golf balls** A spring gun at ground level fires a golf ball at an angle of 45°. The ball lands 10 m away.
   a. What was the ball's initial speed?
   b. For the same initial speed, find the two firing angles that make the range 6 m.

8. **Beaming electrons** An electron in a TV tube is beamed horizontally at a speed of 5×10^6 m/sec toward the face of the tube 40 cm away. About how far will the electron drop before it hits?

9. **Finding golf ball speed** Laboratory tests designed to find how far golf balls of different hardness go when hit with a driver showed that a 100-compression ball hit with a club-head speed of 100 mph at a launch angle of 9° carried 248.8 yd. What was the launch speed of the ball? (It was more than 100 mph. At the same time the club head was moving forward, the compressed ball was kicking away from the club face, adding to the ball's forward speed.)

10. A *human cannonball* is to be fired with an initial speed of $v_0 = 80\sqrt{10/3}$ ft/sec. The circus performer (of the right caliber, naturally) hopes to land on a special cushion located 200 ft downrange at the same height as the muzzle of the cannon. The circus is being held in a large room with a flat ceiling 75 ft higher than the muzzle. Can the performer be fired to the cushion without striking the ceiling? If so, what should the cannon's angle of elevation be?

11. A golf ball leaves the ground at a 30° angle at a speed of 90 ft/sec. Will it clear the top of a 30-ft tree that is in the way, 135 ft down the fairway? Explain.

12. **Elevated green** A golf ball is hit with an initial speed of 116 ft/sec at an angle of elevation of 45° from the tee to a green that is elevated 45 ft above the tee as shown in the diagram. Assuming that the pin, 369 ft downrange, does not get in the way, where will the ball land in relation to the pin?

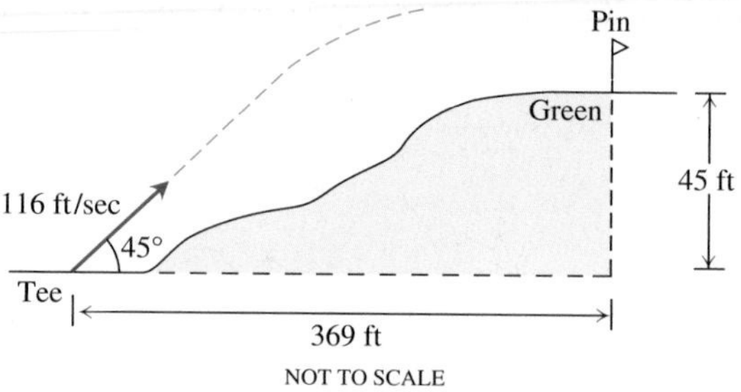

13. **The Green Monster** A baseball hit by a Boston Red Sox player at a 20° angle from 3 ft above the ground just cleared the left end of the "Green Monster," the left-field wall in Fenway Park. This wall is 37 ft high and 315 ft from home plate (see the accompanying figure).
    a. What was the initial speed of the ball?
    b. How long did it take the ball to reach the wall?

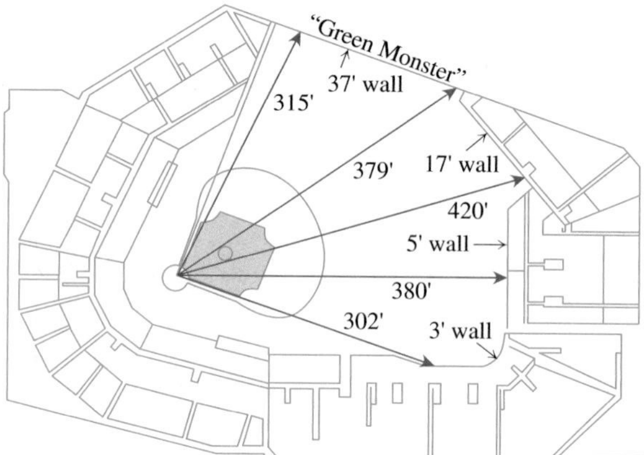

14. **Equal-range firing angles** Show that a projectile fired at an angle of α degrees, $0 < \alpha < 90$, has the same range as a projectile fired at the same speed at an angle of $(90 - \alpha)$ degrees. (In models that take air resistance into account, this symmetry is lost.)

15. **Equal-range firing angles** What two angles of elevation will enable a projectile to reach a target 16 km downrange on the same level as the gun if the projectile's initial speed is 400 m/sec?

16. **Range and height versus speed**
    a. Show that doubling a projectile's initial speed at a given launch angle multiplies its range by 4.
    b. By about what percentage should you increase the initial speed to double the height and range?

17. **Shot put** In Moscow in 1987, Natalya Lisouskaya set a women's world record by putting an 8 lb 13 oz shot 73 ft 10 in. Assuming that she launched the shot at a 40° angle to the horizontal from 6.5 ft above the ground, what was the shot's initial speed?

18. Height versus time Show that a projectile attains three-quarters of its maximum height in half the time it takes to reach the maximum height.

19. Firing from (x_0, y_0) Derive the equations

$$x = x_0 + (v_0 \cos \alpha)t,$$

$$y = y_0 + (v_0 \sin \alpha)t - \frac{1}{2}gt^2,$$

(see Equation (5) in the text) by solving the following initial value problem for a vector **r** in the plane.

Differential equation: $\frac{d^2\mathbf{r}}{dt^2} = -g\mathbf{j}$

Initial conditions: $\mathbf{r}(0) = x_0\mathbf{i} + y_0\mathbf{j}$

$$\frac{d\mathbf{r}}{dt}(0) = (v_0 \cos \alpha)\mathbf{i} + (v_0 \sin \alpha)\mathbf{j}$$

20. Flaming arrow Using the firing angle found in Example 3, find the speed at which the flaming arrow left Rebollo's bow. See Figure 13.13.

21. Flaming arrow The cauldron in Example 3 is 12 ft in diameter. Using Equation (5) and Example 3c, find how long it takes the flaming arrow to cover the horizontal distance to the rim. How high is the arrow at this time?

22. Describe the path of a projectile given by Equations (4) when $\alpha = 90°$.

23. Model train The accompanying multiflash photograph shows a model train engine moving at a constant speed on a straight horizontal track. As the engine moved along, a marble was fired into the air by a spring in the engine's smokestack. The marble, which continued to move with the same forward speed as the engine, rejoined the engine 1 sec after it was fired. Measure the angle the marble's path made with the horizontal and use the information to find how high the marble went and how fast the engine was moving.

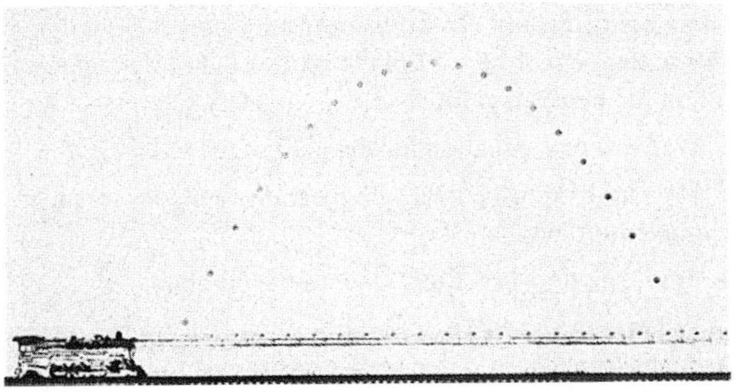

24. Colliding marbles The figure shows an experiment with two marbles. Marble A was launched toward marble B with launch angle α and initial speed v_0. At the same instant, marble B was released to fall from rest at $R \tan \alpha$ units directly above a spot R units downrange from A. The marbles were found to collide regardless of the value of v_0. Was this mere coincidence, or must this happen? Give reasons for your answer.

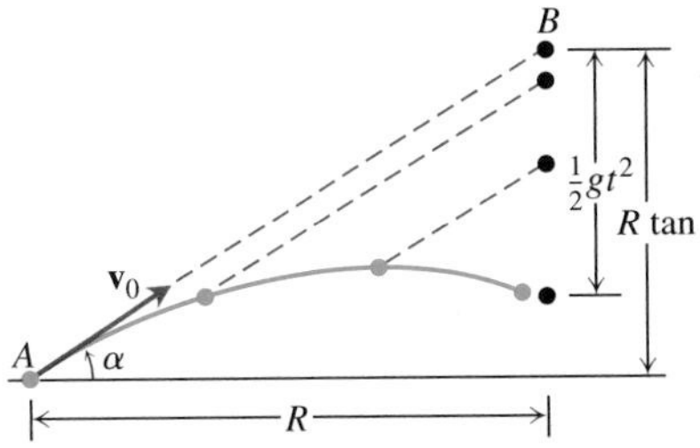

25. Launching downhill An ideal projectile is launched straight down an inclined plane as shown in the accompanying figure.

a. Show that the greatest downhill range is achieved when the initial velocity vector bisects angle AOR.

b. If the projectile were fired uphill instead of down, what launch angle would maximize its range? Give reasons for your answer.

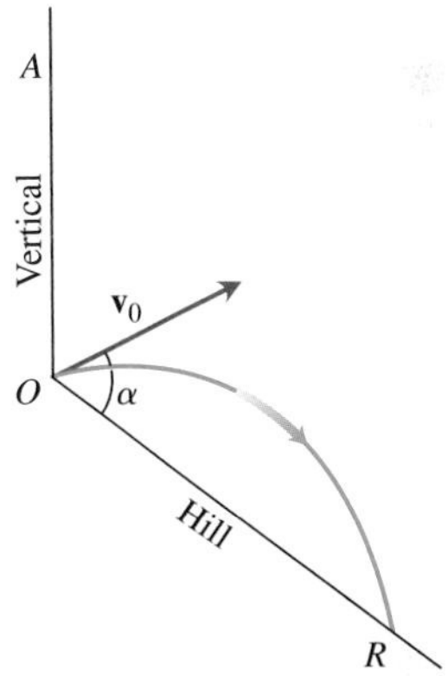

26. Hitting a baseball under a wind gust A baseball is hit when it is 2.5 ft above the ground. It leaves the bat with an initial velocity of 145 ft/sec at a launch angle of 23°. At the instant the ball is hit, an instantaneous gust of wind blows against the ball, adding a component of $-14\mathbf{i}$ (ft/sec) to the ball's initial velocity. A 15-ft-high fence lies 300 ft from home plate in the direction of the flight.

a. Find a vector equation for the path of the baseball.

b. How high does the baseball go, and when does it reach maximum height?

c. Find the range and flight time of the baseball, assuming that the ball is not caught.

d. When is the baseball 20 ft high? How far (ground distance) is the baseball from home plate at that height?

e. Has the batter hit a home run? Explain.

27. Volleyball A volleyball is hit when it is 4 ft above the ground and 12 ft from a 6-ft-high net. It leaves the point of impact with an initial velocity of 35 ft/sec at an angle of 27° and slips by the opposing team untouched.

a. Find a vector equation for the path of the volleyball.

b. How high does the volleyball go, and when does it reach maximum height?

c. Find its range and flight time.

d. When is the volleyball 7 ft above the ground? How far (ground distance) is the volleyball from where it will land?

e. Suppose that the net is raised to 8 ft. Does this change things? Explain.

28. Where trajectories crest For a projectile fired from the ground at launch angle α with initial speed v_0, consider α as a variable and v_0 as a fixed constant. For each α, $0 < \alpha < \pi/2$, we obtain a parabolic trajectory as shown in the accompanying figure. Show that the points in the plane that give the maximum heights of these parabolic trajectories all lie on the ellipse

$$x^2 + 4\left(y - \frac{v_0^2}{4g}\right)^2 = \frac{v_0^4}{4g^2},$$

where $x \geq 0$.

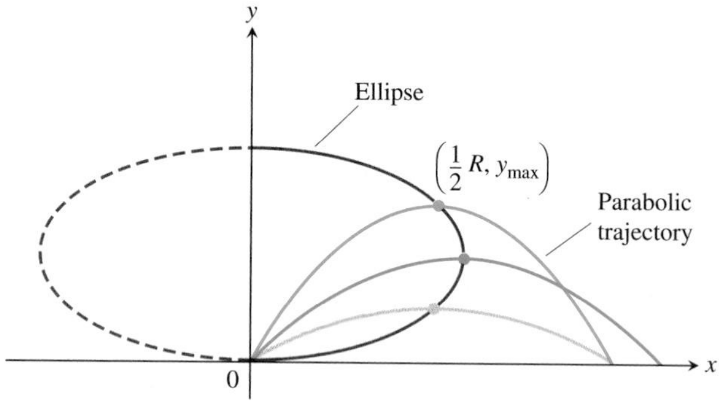

Projectile Motion with Linear Drag

The main force affecting the motion of a projectile, other than gravity, is air resistance. This slowing down force is **drag force**, and it acts in a direction *opposite* to the velocity of the projectile (see accompanying figure). For projectiles moving through the air at relatively low speeds, however, the drag force is (very nearly) proportional to the speed (to the first power) and so is called **linear**.

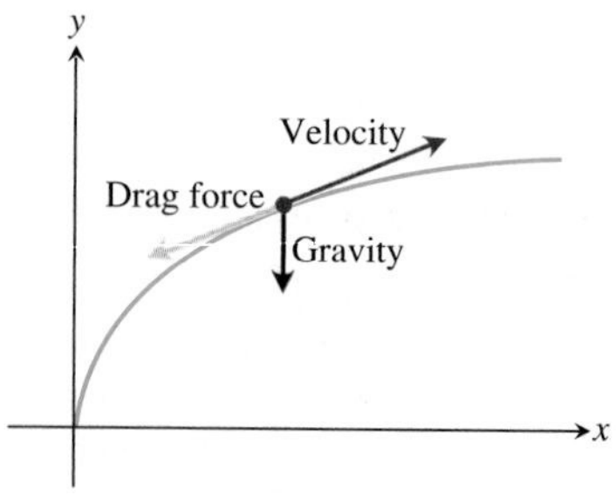

29. Linear drag Derive the equations

$$x = \frac{v_0}{k}(1 - e^{-kt})\cos\alpha$$

$$y = \frac{v_0}{k}(1 - e^{-kt})(\sin\alpha) + \frac{g}{k^2}(1 - kt - e^{-kt})$$

by solving the following initial value problem for a vector **r** in the plane.

Differential equation: $\dfrac{d^2\mathbf{r}}{dt^2} = -g\mathbf{j} - k\mathbf{v} = -g\mathbf{j} - k\dfrac{d\mathbf{r}}{dt}$

Initial conditions: $\mathbf{r}(0) = \mathbf{0}$

$$\left.\frac{d\mathbf{r}}{dt}\right|_{t=0} = \mathbf{v}_0 = (v_0\cos\alpha)\mathbf{i} + (v_0\sin\alpha)\mathbf{j}$$

The **drag coefficient** k is a positive constant representing resistance due to air density, v_0 and α are the projectile's initial speed and launch angle, and g is the acceleration of gravity.

30. Hitting a baseball with linear drag Consider the baseball problem in Example 4 when there is linear drag (see Exercise 29). Assume a drag coefficient $k = 0.12$, but no gust of wind.

a. From Exercise 29, find a vector form for the path of the baseball.

b. How high does the baseball go, and when does it reach maximum height?

c. Find the range and flight time of the baseball.

d. When is the baseball 30 ft high? How far (ground distance) is the baseball from home plate at that height?

e. A 10-ft-high outfield fence is 340 ft from home plate in the direction of the flight of the baseball. The outfielder can jump and catch any ball up to 11 ft off the ground to stop it from going over the fence. Has the batter hit a home run?

31. Hitting a baseball with linear drag under a wind gust Consider again the baseball problem in Example 4. This time assume a drag coefficient of 0.08 *and* an instantaneous gust of wind that adds a component of $-17.6\mathbf{i}$ (ft/sec) to the initial velocity at the instant the baseball is hit.

a. Find a vector equation for the path of the baseball.

b. How high does the baseball go, and when does it reach maximum height?

c. Find the range and flight time of the baseball.

d. When is the baseball 35 ft high? How far (ground distance) is the baseball from home plate at that height?

e. A 20-ft-high outfield fence is 380 ft from home plate in the direction of the flight of the baseball. Has the batter hit a home run? If "yes," what change in the horizontal component of the ball's initial velocity would have kept the ball in the park? If "no," what change would have allowed it to be a home run?

13.3 Arc Length and the Unit Tangent Vector T

Imagine the motions you might experience traveling at high speeds along a path through the air or space. Specifically, imagine the motions of turning to your left or right and the up-and-down motions tending to lift you from, or pin you down to, your seat. Pilots flying through the atmosphere, turning and twisting in flight acrobatics, certainly experience these motions. Turns that are too tight, descents or climbs that are too steep, or either one coupled with high and increasing speed can cause an aircraft to spin out of control, possibly even to break up in midair, and crash to Earth.

In this and the next two sections, we study the features of a curve's shape that describe mathematically the sharpness of its turning and its twisting perpendicular to the forward motion.

Arc Length Along a Space Curve

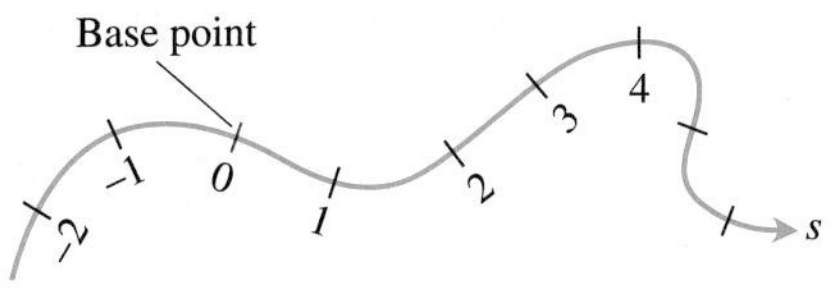

FIGURE 13.14 Smooth curves can be scaled like number lines, the coordinate of each point being its directed distance along the curve from a preselected base point.

One of the features of smooth space curves is that they have a measurable length. This enables us to locate points along these curves by giving their directed distance s along the curve from some **base point**, the way we locate points on coordinate axes by giving their directed distance from the origin (Figure 13.14). Time is the natural parameter for describing a moving body's velocity and acceleration, but s is the natural parameter for studying a curve's shape. Both parameters appear in analyses of space flight.

To measure distance along a smooth curve in space, we add a z-term to the formula we use for curves in the plane.

> **DEFINITION** Length of a Smooth Curve
>
> The **length** of a smooth curve $\mathbf{r}(t) = x(t)\mathbf{i} + y(t)\mathbf{j} + z(t)\mathbf{k}$, $a \le t \le b$, that is traced exactly once as t increases from $t = a$ to $t = b$, is
>
> $$L = \int_a^b \sqrt{\left(\frac{dx}{dt}\right)^2 + \left(\frac{dy}{dt}\right)^2 + \left(\frac{dz}{dt}\right)^2}\, dt. \tag{1}$$

Just as for plane curves, we can calculate the length of a curve in space from any convenient parametrization that meets the stated conditions. We omit the proof.

The square root in Equation (1) is $|\mathbf{v}|$, the length of a velocity vector $d\mathbf{r}/dt$. This enables us to write the formula for length a shorter way.

> **Arc Length Formula**
>
> $$L = \int_a^b |\mathbf{v}|\, dt \tag{2}$$

EXAMPLE 1 Distance Traveled by a Glider

A glider is soaring upward along the helix $\mathbf{r}(t) = (\cos t)\mathbf{i} + (\sin t)\mathbf{j} + t\mathbf{k}$. How far does the glider travel along its path from $t = 0$ to $t = 2\pi \approx 6.28$ sec?

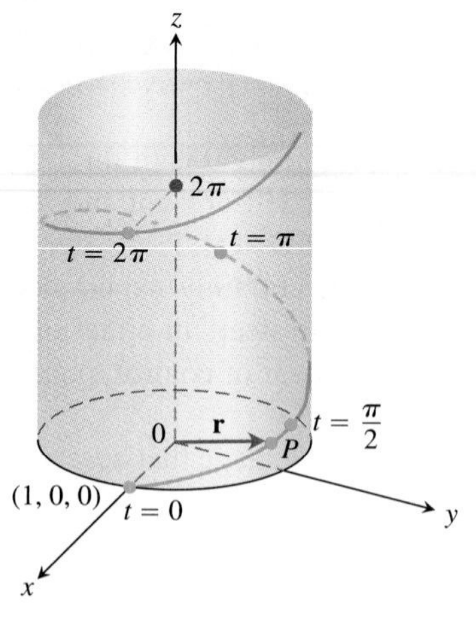

FIGURE 13.15 The helix $\mathbf{r}(t) = (\cos t)\mathbf{i} + (\sin t)\mathbf{j} + t\mathbf{k}$ in Example 1.

Solution The path segment during this time corresponds to one full turn of the helix (Figure 13.15). The length of this portion of the curve is

$$L = \int_a^b |\mathbf{v}|\,dt = \int_0^{2\pi} \sqrt{(-\sin t)^2 + (\cos t)^2 + (1)^2}\,dt$$

$$= \int_0^{2\pi} \sqrt{2}\,dt = 2\pi\sqrt{2} \text{ units of length.}$$

This is $\sqrt{2}$ times the length of the circle in the xy-plane over which the helix stands. ■

If we choose a base point $P(t_0)$ on a smooth curve C parametrized by t, each value of t determines a point $P(t) = (x(t), y(t), z(t))$ on C and a "directed distance"

$$s(t) = \int_{t_0}^{t} |\mathbf{v}(\tau)|\,d\tau,$$

measured along C from the base point (Figure 13.16). If $t > t_0$, $s(t)$ is the distance from $P(t_0)$ to $P(t)$. If $t < t_0$, $s(t)$ is the negative of the distance. Each value of s determines a point on C and this parametrizes C with respect to s. We call s an **arc length parameter** for the curve. The parameter's value increases in the direction of increasing t. The arc length parameter is particularly effective for investigating the turning and twisting nature of a space curve.

We use the Greek letter τ ("tau") as the variable of integration because the letter t is already in use as the upper limit.

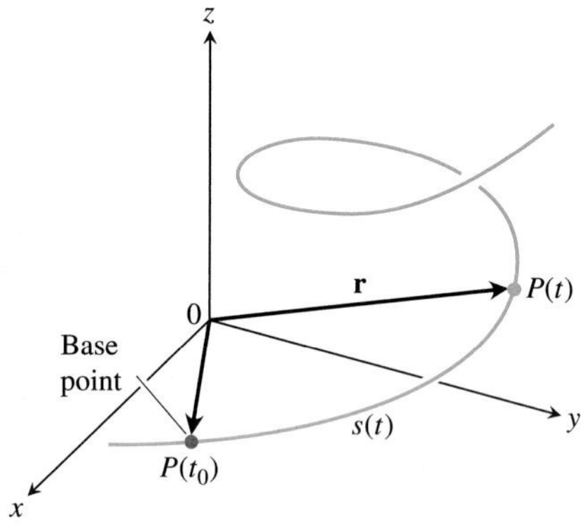

FIGURE 13.16 The directed distance along the curve from $P(t_0)$ to any point $P(t)$ is

$$s(t) = \int_{t_0}^{t} |\mathbf{v}(\tau)|\,d\tau.$$

Arc Length Parameter with Base Point $P(t_0)$

$$s(t) = \int_{t_0}^{t} \sqrt{[x'(\tau)]^2 + [y'(\tau)]^2 + [z'(\tau)]^2}\,d\tau = \int_{t_0}^{t} |\mathbf{v}(\tau)|\,d\tau \qquad (3)$$

If a curve $\mathbf{r}(t)$ is already given in terms of some parameter t and $s(t)$ is the arc length function given by Equation (3), then we may be able to solve for t as a function of s: $t = t(s)$. Then the curve can be reparametrized in terms of s by substituting for t: $\mathbf{r} = \mathbf{r}(t(s))$.

EXAMPLE 2 Finding an Arc Length Parametrization

If $t_0 = 0$, the arc length parameter along the helix

$$\mathbf{r}(t) = (\cos t)\mathbf{i} + (\sin t)\mathbf{j} + t\mathbf{k}$$

from t_0 to t is

$$s(t) = \int_{t_0}^{t} |\mathbf{v}(\tau)|\,d\tau \qquad \text{Equation (3)}$$

$$= \int_0^{t} \sqrt{2}\,d\tau \qquad \text{Value from Example 1}$$

$$= \sqrt{2}\,t.$$

Solving this equation for t gives $t = s/\sqrt{2}$. Substituting into the position vector $\mathbf{r}$ gives the following arc length parametrization for the helix:

$$\mathbf{r}(t(s)) = \left(\cos\frac{s}{\sqrt{2}}\right)\mathbf{i} + \left(\sin\frac{s}{\sqrt{2}}\right)\mathbf{j} + \frac{s}{\sqrt{2}}\mathbf{k}. \qquad \blacksquare$$

Unlike Example 2, the arc length parametrization is generally difficult to find analytically for a curve already given in terms of some other parameter t. Fortunately, however, we rarely need an exact formula for $s(t)$ or its inverse $t(s)$.

EXAMPLE 3 Distance Along a Line

Show that if $\mathbf{u} = u_1\mathbf{i} + u_2\mathbf{j} + u_3\mathbf{k}$ is a unit vector, then the arc length parameter along the line

$$\mathbf{r}(t) = (x_0 + tu_1)\mathbf{i} + (y_0 + tu_2)\mathbf{j} + (z_0 + tu_3)\mathbf{k}$$

from the point $P_0(x_0, y_0, z_0)$ where $t = 0$ is t itself.

Solution

$$\mathbf{v} = \frac{d}{dt}(x_0 + tu_1)\mathbf{i} + \frac{d}{dt}(y_0 + tu_2)\mathbf{j} + \frac{d}{dt}(z_0 + tu_3)\mathbf{k} = u_1\mathbf{i} + u_2\mathbf{j} + u_3\mathbf{k} = \mathbf{u},$$

so

$$s(t) = \int_0^t |\mathbf{v}|\,d\tau = \int_0^t |\mathbf{u}|\,d\tau = \int_0^t 1\,d\tau = t. \qquad \blacksquare$$

HISTORICAL BIOGRAPHY

Josiah Willard Gibbs
(1839–1903)

Speed on a Smooth Curve

Since the derivatives beneath the radical in Equation (3) are continuous (the curve is smooth), the Fundamental Theorem of Calculus tells us that s is a differentiable function of t with derivative

$$\frac{ds}{dt} = |\mathbf{v}(t)|. \tag{4}$$

As we already knew, the speed with which a particle moves along its path is the magnitude of $\mathbf{v}$.

Notice that although the base point $P(t_0)$ plays a role in defining s in Equation (3), it plays no role in Equation (4). The rate at which a moving particle covers distance along its path is independent of how far away it is from the base point.

Notice also that $ds/dt > 0$ since, by definition, $|\mathbf{v}|$ is never zero for a smooth curve. We see once again that s is an increasing function of t.

Unit Tangent Vector T

We already know the velocity vector $\mathbf{v} = d\mathbf{r}/dt$ is tangent to the curve and that the vector

$$\mathbf{T} = \frac{\mathbf{v}}{|\mathbf{v}|}$$

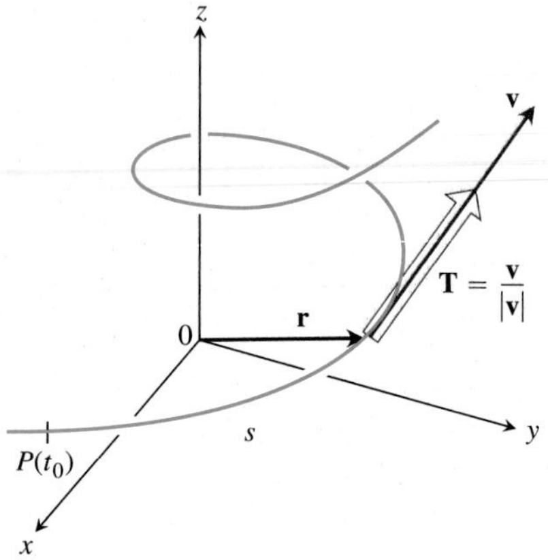

FIGURE 13.17 We find the unit tangent vector **T** by dividing **v** by $|\mathbf{v}|$.

is therefore a unit vector tangent to the (smooth) curve. Since $ds/dt > 0$ for the curves we are considering, s is one-to-one and has an inverse that gives t as a differentiable function of s (Section 7.1). The derivative of the inverse is

$$\frac{dt}{ds} = \frac{1}{ds/dt} = \frac{1}{|\mathbf{v}|}.$$

This makes **r** a differentiable function of s whose derivative can be calculated with the Chain Rule to be

$$\frac{d\mathbf{r}}{ds} = \frac{d\mathbf{r}}{dt}\frac{dt}{ds} = \mathbf{v}\frac{1}{|\mathbf{v}|} = \frac{\mathbf{v}}{|\mathbf{v}|} = \mathbf{T}.$$

This equation says that $d\mathbf{r}/ds$ is the unit tangent vector in the direction of the velocity vector **v** (Figure 13.17).

DEFINITION Unit Tangent Vector

The **unit tangent vector** of a smooth curve $\mathbf{r}(t)$ is

$$\mathbf{T} = \frac{d\mathbf{r}}{ds} = \frac{d\mathbf{r}/dt}{ds/dt} = \frac{\mathbf{v}}{|\mathbf{v}|}. \qquad (5)$$

The unit tangent vector **T** is a differentiable function of t whenever **v** is a differentiable function of t. As we see in Section 13.5, **T** is one of three unit vectors in a traveling reference frame that is used to describe the motion of space vehicles and other bodies traveling in three dimensions.

EXAMPLE 4 Finding the Unit Tangent Vector T

Find the unit tangent vector of the curve

$$\mathbf{r}(t) = (3\cos t)\mathbf{i} + (3\sin t)\mathbf{j} + t^2\mathbf{k}$$

representing the path of the glider in Example 4, Section 13.1.

Solution In that example, we found

$$\mathbf{v} = \frac{d\mathbf{r}}{dt} = -(3\sin t)\mathbf{i} + (3\cos t)\mathbf{j} + 2t\mathbf{k}$$

and

$$|\mathbf{v}| = \sqrt{9 + 4t^2}.$$

Thus,

$$\mathbf{T} = \frac{\mathbf{v}}{|\mathbf{v}|} = -\frac{3\sin t}{\sqrt{9+4t^2}}\mathbf{i} + \frac{3\cos t}{\sqrt{9+4t^2}}\mathbf{j} + \frac{2t}{\sqrt{9+4t^2}}\mathbf{k}.$$

■

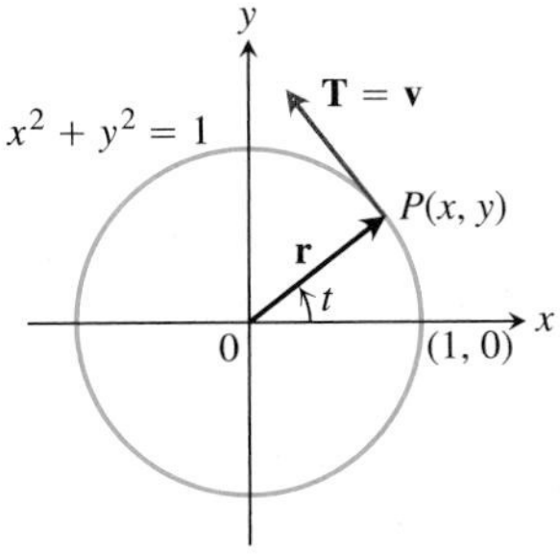

FIGURE 13.18 The motion $\mathbf{r}(t) = (\cos t)\mathbf{i} + (\sin t)\mathbf{j}$ (Example 5).

EXAMPLE 5 Motion on the Unit Circle

For the counterclockwise motion

$$\mathbf{r}(t) = (\cos t)\mathbf{i} + (\sin t)\mathbf{j}$$

around the unit circle,

$$\mathbf{v} = (-\sin t)\mathbf{i} + (\cos t)\mathbf{j}$$

is already a unit vector, so $\mathbf{T} = \mathbf{v}$ (Figure 13.18). ■

EXERCISES 13.3

Finding Unit Tangent Vectors and Lengths of Curves

In Exercises 1–8, find the curve's unit tangent vector. Also, find the length of the indicated portion of the curve.

1. $\mathbf{r}(t) = (2\cos t)\mathbf{i} + (2\sin t)\mathbf{j} + \sqrt{5}t\mathbf{k}, \quad 0 \le t \le \pi$
2. $\mathbf{r}(t) = (6\sin 2t)\mathbf{i} + (6\cos 2t)\mathbf{j} + 5t\mathbf{k}, \quad 0 \le t \le \pi$
3. $\mathbf{r}(t) = t\mathbf{i} + (2/3)t^{3/2}\mathbf{k}, \quad 0 \le t \le 8$
4. $\mathbf{r}(t) = (2 + t)\mathbf{i} - (t + 1)\mathbf{j} + t\mathbf{k}, \quad 0 \le t \le 3$
5. $\mathbf{r}(t) = (\cos^3 t)\mathbf{j} + (\sin^3 t)\mathbf{k}, \quad 0 \le t \le \pi/2$
6. $\mathbf{r}(t) = 6t^3\mathbf{i} - 2t^3\mathbf{j} - 3t^3\mathbf{k}, \quad 1 \le t \le 2$
7. $\mathbf{r}(t) = (t\cos t)\mathbf{i} + (t\sin t)\mathbf{j} + \left(2\sqrt{2}/3\right)t^{3/2}\mathbf{k}, \quad 0 \le t \le \pi$
8. $\mathbf{r}(t) = (t\sin t + \cos t)\mathbf{i} + (t\cos t - \sin t)\mathbf{j}, \quad \sqrt{2} \le t \le 2$
9. Find the point on the curve

$$\mathbf{r}(t) = (5\sin t)\mathbf{i} + (5\cos t)\mathbf{j} + 12t\mathbf{k}$$

at a distance 26π units along the curve from the point $(0, 5, 0)$ in the direction of increasing arc length.

10. Find the point on the curve

$$\mathbf{r}(t) = (12\sin t)\mathbf{i} - (12\cos t)\mathbf{j} + 5t\mathbf{k}$$

at a distance 13π units along the curve from the point $(0, -12, 0)$ in the direction opposite to the direction of increasing arc length.

Arc Length Parameter

In Exercises 11–14, find the arc length parameter along the curve from the point where $t = 0$ by evaluating the integral

$$s = \int_0^t |\mathbf{v}(\tau)|\, d\tau$$

from Equation (3). Then find the length of the indicated portion of the curve.

11. $\mathbf{r}(t) = (4\cos t)\mathbf{i} + (4\sin t)\mathbf{j} + 3t\mathbf{k}, \quad 0 \le t \le \pi/2$
12. $\mathbf{r}(t) = (\cos t + t\sin t)\mathbf{i} + (\sin t - t\cos t)\mathbf{j}, \quad \pi/2 \le t \le \pi$
13. $\mathbf{r}(t) = (e^t\cos t)\mathbf{i} + (e^t\sin t)\mathbf{j} + e^t\mathbf{k}, \quad -\ln 4 \le t \le 0$
14. $\mathbf{r}(t) = (1 + 2t)\mathbf{i} + (1 + 3t)\mathbf{j} + (6 - 6t)\mathbf{k}, \quad -1 \le t \le 0$

Theory and Examples

15. **Arc length** Find the length of the curve

$$\mathbf{r}(t) = \left(\sqrt{2}t\right)\mathbf{i} + \left(\sqrt{2}t\right)\mathbf{j} + (1 - t^2)\mathbf{k}$$

from $(0, 0, 1)$ to $\left(\sqrt{2}, \sqrt{2}, 0\right)$.

16. **Length of helix** The length $2\pi\sqrt{2}$ of the turn of the helix in Example 1 is also the length of the diagonal of a square 2π units on a side. Show how to obtain this square by cutting away and flattening a portion of the cylinder around which the helix winds.

17. **Ellipse**

  a. Show that the curve $\mathbf{r}(t) = (\cos t)\mathbf{i} + (\sin t)\mathbf{j} + (1 - \cos t)\mathbf{k}$, $0 \le t \le 2\pi$, is an ellipse by showing that it is the intersection of a right circular cylinder and a plane. Find equations for the cylinder and plane.

  b. Sketch the ellipse on the cylinder. Add to your sketch the unit tangent vectors at $t = 0, \pi/2, \pi$, and $3\pi/2$.

  c. Show that the acceleration vector always lies parallel to the plane (orthogonal to a vector normal to the plane). Thus, if you draw the acceleration as a vector attached to the ellipse, it will lie in the plane of the ellipse. Add the acceleration vectors for $t = 0, \pi/2, \pi$, and $3\pi/2$ to your sketch.

  d. Write an integral for the length of the ellipse. Do not try to evaluate the integral; it is nonelementary.

  T e. **Numerical integrator** Estimate the length of the ellipse to two decimal places.

18. **Length is independent of parametrization** To illustrate that the length of a smooth space curve does not depend on

the parametrization you use to compute it, calculate the length of one turn of the helix in Example 1 with the following parametrizations.

a. $\mathbf{r}(t) = (\cos 4t)\mathbf{i} + (\sin 4t)\mathbf{j} + 4t\mathbf{k}, \quad 0 \le t \le \pi/2$

b. $\mathbf{r}(t) = [\cos(t/2)]\mathbf{i} + [\sin(t/2)]\mathbf{j} + (t/2)\mathbf{k}, \quad 0 \le t \le 4\pi$

c. $\mathbf{r}(t) = (\cos t)\mathbf{i} - (\sin t)\mathbf{j} - t\mathbf{k}, \quad -2\pi \le t \le 0$

19. The involute of a circle If a string wound around a fixed circle is unwound while held taut in the plane of the circle, its end P traces an *involute* of the circle. In the accompanying figure, the circle in question is the circle $x^2 + y^2 = 1$ and the tracing point starts at $(1, 0)$. The unwound portion of the string is tangent to the circle at Q, and t is the radian measure of the angle from the positive x-axis to segment OQ. Derive the parametric equations

$$x = \cos t + t \sin t, \quad y = \sin t - t \cos t, \quad t > 0$$

of the point $P(x, y)$ for the involute.

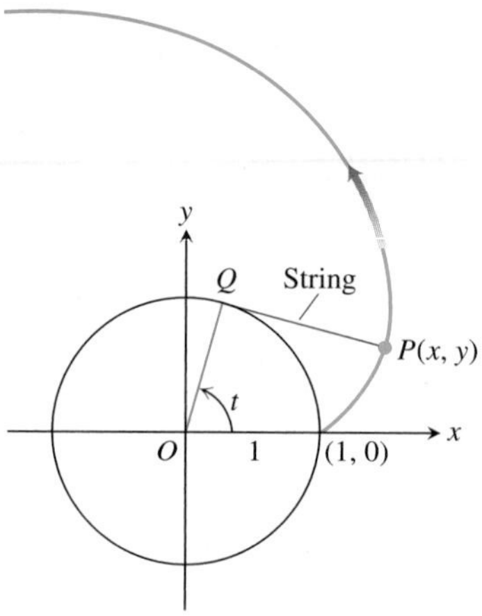

20. (*Continuation of Exercise 19.*) Find the unit tangent vector to the involute of the circle at the point $P(x, y)$.

13.4 Curvature and the Unit Normal Vector N

In this section we study how a curve turns or bends. We look first at curves in the coordinate plane, and then at curves in space.

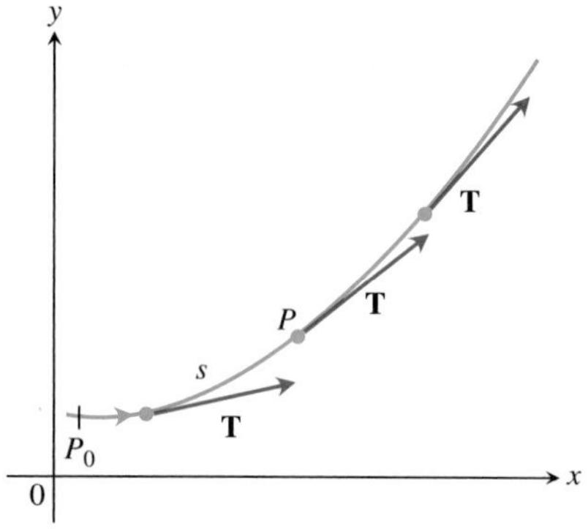

FIGURE 13.19 As P moves along the curve in the direction of increasing arc length, the unit tangent vector turns. The value of $|d\mathbf{T}/ds|$ at P is called the *curvature* of the curve at P.

Curvature of a Plane Curve

As a particle moves along a smooth curve in the plane, $\mathbf{T} = d\mathbf{r}/ds$ turns as the curve bends. Since $\mathbf{T}$ is a unit vector, its length remains constant and only its direction changes as the particle moves along the curve. The rate at which $\mathbf{T}$ turns per unit of length along the curve is called the *curvature* (Figure 13.19). The traditional symbol for the curvature function is the Greek letter κ ("kappa").

DEFINITION Curvature

If $\mathbf{T}$ is the unit vector of a smooth curve, the **curvature** function of the curve is

$$\kappa = \left|\frac{d\mathbf{T}}{ds}\right|.$$

If $|d\mathbf{T}/ds|$ is large, $\mathbf{T}$ turns sharply as the particle passes through P, and the curvature at P is large. If $|d\mathbf{T}/ds|$ is close to zero, $\mathbf{T}$ turns more slowly and the curvature at P is smaller.

If a smooth curve $\mathbf{r}(t)$ is already given in terms of some parameter t other than the arc length parameter s, we can calculate the curvature as

$$\kappa = \left|\frac{d\mathbf{T}}{ds}\right| = \left|\frac{d\mathbf{T}}{dt}\frac{dt}{ds}\right| \qquad \text{Chain Rule}$$

$$= \frac{1}{|ds/dt|}\left|\frac{d\mathbf{T}}{dt}\right|$$

$$= \frac{1}{|\mathbf{v}|}\left|\frac{d\mathbf{T}}{dt}\right|. \qquad \frac{ds}{dt} = |\mathbf{v}|$$

Formula for Calculating Curvature

If $\mathbf{r}(t)$ is a smooth curve, then the curvature is

$$\kappa = \frac{1}{|\mathbf{v}|}\left|\frac{d\mathbf{T}}{dt}\right|, \tag{1}$$

where $\mathbf{T} = \mathbf{v}/|\mathbf{v}|$ is the unit tangent vector.

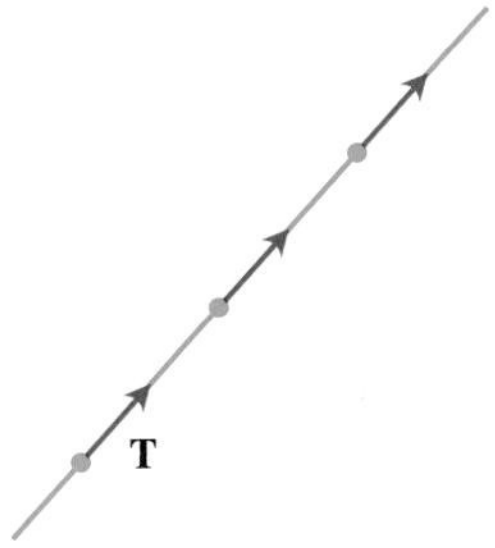

FIGURE 13.20 Along a straight line, **T** always points in the same direction. The curvature, $|d\mathbf{T}/ds|$, is zero (Example 1).

Testing the definition, we see in Examples 1 and 2 below that the curvature is constant for straight lines and circles.

EXAMPLE 1 The Curvature of a Straight Line Is Zero

On a straight line, the unit tangent vector **T** always points in the same direction, so its components are constants. Therefore, $|d\mathbf{T}/ds| = |\mathbf{0}| = 0$ (Figure 13.20). ■

EXAMPLE 2 The Curvature of a Circle of Radius a is $1/a$

To see why, we begin with the parametrization

$$\mathbf{r}(t) = (a\cos t)\mathbf{i} + (a\sin t)\mathbf{j}$$

of a circle of radius a. Then,

$$\mathbf{v} = \frac{d\mathbf{r}}{dt} = -(a\sin t)\mathbf{i} + (a\cos t)\mathbf{j}$$

$$|\mathbf{v}| = \sqrt{(-a\sin t)^2 + (a\cos t)^2} = \sqrt{a^2} = |a| = a.$$

Since $a > 0$, $|a| = a$.

From this we find

$$\mathbf{T} = \frac{\mathbf{v}}{|\mathbf{v}|} = -(\sin t)\mathbf{i} + (\cos t)\mathbf{j}$$

$$\frac{d\mathbf{T}}{dt} = -(\cos t)\mathbf{i} - (\sin t)\mathbf{j}$$

$$\left|\frac{d\mathbf{T}}{dt}\right| = \sqrt{\cos^2 t + \sin^2 t} = 1.$$

Hence, for any value of the parameter t,

$$\kappa = \frac{1}{|\mathbf{v}|}\left|\frac{d\mathbf{T}}{dt}\right| = \frac{1}{a}(1) = \frac{1}{a}.$$ ■

Although the formula for calculating κ in Equation (1) is also valid for space curves, in the next section we find a computational formula that is usually more convenient to apply.

Among the vectors orthogonal to the unit tangent vector **T** is one of particular significance because it points in the direction in which the curve is turning. Since **T** has constant length (namely, 1), the derivative $d\mathbf{T}/ds$ is orthogonal to **T** (Section 13.1). Therefore, if we divide $d\mathbf{T}/ds$ by its length κ, we obtain a *unit* vector **N** orthogonal to **T** (Figure 13.21).

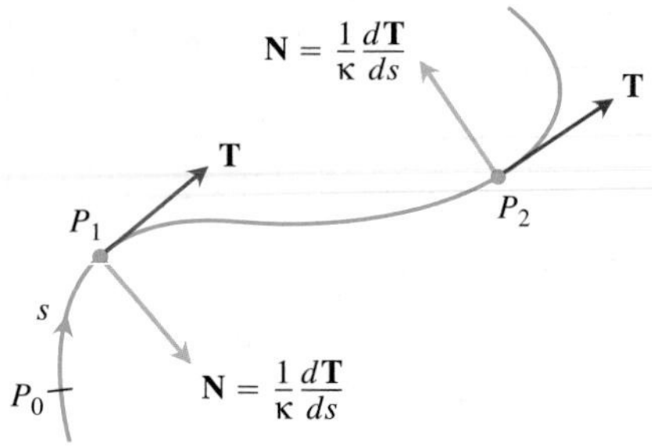

FIGURE 13.21 The vector $d\mathbf{T}/ds$, normal to the curve, always points in the direction in which $\mathbf{T}$ is turning. The unit normal vector $\mathbf{N}$ is the direction of $d\mathbf{T}/ds$.

DEFINITION Principal Unit Normal

At a point where $\kappa \neq 0$, the **principal unit normal** vector for a smooth curve in the plane is

$$\mathbf{N} = \frac{1}{\kappa}\frac{d\mathbf{T}}{ds}.$$

The vector $d\mathbf{T}/ds$ points in the direction in which $\mathbf{T}$ turns as the curve bends. Therefore, if we face in the direction of increasing arc length, the vector $d\mathbf{T}/ds$ points toward the right if $\mathbf{T}$ turns clockwise and toward the left if $\mathbf{T}$ turns counterclockwise. In other words, the principal normal vector $\mathbf{N}$ will point toward the concave side of the curve (Figure 13.21).

If a smooth curve $\mathbf{r}(t)$ is already given in terms of some parameter t other than the arc length parameter s, we can use the Chain Rule to calculate $\mathbf{N}$ directly:

$$\begin{aligned}\mathbf{N} &= \frac{d\mathbf{T}/ds}{|d\mathbf{T}/ds|} \\ &= \frac{(d\mathbf{T}/dt)(dt/ds)}{|d\mathbf{T}/dt||dt/ds|} \\ &= \frac{d\mathbf{T}/dt}{|d\mathbf{T}/dt|}. \qquad \frac{dt}{ds} = \frac{1}{ds/dt} > 0 \text{ cancels}\end{aligned}$$

This formula enables us to find $\mathbf{N}$ without having to find κ and s first.

Formula for Calculating N

If $\mathbf{r}(t)$ is a smooth curve, then the principal unit normal is

$$\mathbf{N} = \frac{d\mathbf{T}/dt}{|d\mathbf{T}/dt|}, \qquad (2)$$

where $\mathbf{T} = \mathbf{v}/|\mathbf{v}|$ is the unit tangent vector.

EXAMPLE 3 Finding T and N

Find $\mathbf{T}$ and $\mathbf{N}$ for the circular motion

$$\mathbf{r}(t) = (\cos 2t)\mathbf{i} + (\sin 2t)\mathbf{j}.$$

Solution We first find $\mathbf{T}$:

$$\begin{aligned}\mathbf{v} &= -(2\sin 2t)\mathbf{i} + (2\cos 2t)\mathbf{j} \\ |\mathbf{v}| &= \sqrt{4\sin^2 2t + 4\cos^2 2t} = 2 \\ \mathbf{T} &= \frac{\mathbf{v}}{|\mathbf{v}|} = -(\sin 2t)\mathbf{i} + (\cos 2t)\mathbf{j}.\end{aligned}$$

From this we find

$$\frac{d\mathbf{T}}{dt} = -(2\cos 2t)\mathbf{i} - (2\sin 2t)\mathbf{j}$$

$$\left|\frac{d\mathbf{T}}{dt}\right| = \sqrt{4\cos^2 2t + 4\sin^2 2t} = 2$$

and

$$\mathbf{N} = \frac{d\mathbf{T}/dt}{|d\mathbf{T}/dt|}$$

$$= -(\cos 2t)\mathbf{i} - (\sin 2t)\mathbf{j}. \qquad \text{Equation (2)}$$

Notice that $\mathbf{T} \cdot \mathbf{N} = 0$, verifying that **N** is orthogonal to **T**. Notice too, that for the circular motion here, **N** points from $\mathbf{r}(t)$ towards the circle's center at the origin. ∎

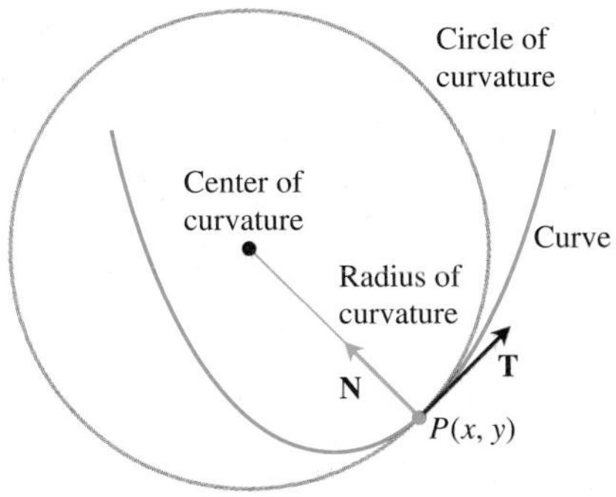

FIGURE 13.22 The osculating circle at $P(x, y)$ lies toward the inner side of the curve.

Circle of Curvature for Plane Curves

The **circle of curvature** or **osculating circle** at a point P on a plane curve where $\kappa \neq 0$ is the circle in the plane of the curve that

1. is tangent to the curve at P (has the same tangent line the curve has)
2. has the same curvature the curve has at P
3. lies toward the concave or inner side of the curve (as in Figure 13.22).

The **radius of curvature** of the curve at P is the radius of the circle of curvature, which, according to Example 2, is

$$\text{Radius of curvature} = \rho = \frac{1}{\kappa}.$$

To find ρ, we find κ and take the reciprocal. The **center of curvature** of the curve at P is the center of the circle of curvature.

EXAMPLE 4 Finding the Osculating Circle for a Parabola

Find and graph the osculating circle of the parabola $y = x^2$ at the origin.

Solution We parametrize the parabola using the parameter $t = x$ (Section 10.4, Example 1)

$$\mathbf{r}(t) = t\mathbf{i} + t^2\mathbf{j}.$$

First we find the curvature of the parabola at the origin, using Equation (1):

$$\mathbf{v} = \frac{d\mathbf{r}}{dt} = \mathbf{i} + 2t\mathbf{j}$$

$$|\mathbf{v}| = \sqrt{1 + 4t^2}$$

so that

$$\mathbf{T} = \frac{\mathbf{v}}{|\mathbf{v}|} = (1 + 4t^2)^{-1/2}\mathbf{i} + 2t(1 + 4t^2)^{-1/2}\mathbf{j}.$$

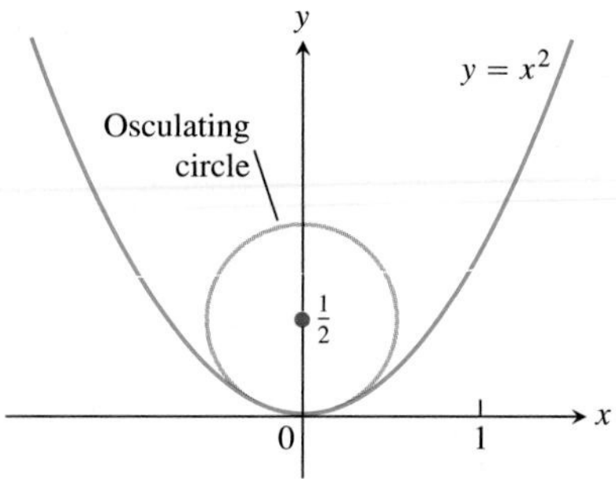

FIGURE 13.23 The osculating circle for the parabola $y = x^2$ at the origin (Example 4).

From this we find

$$\frac{d\mathbf{T}}{dt} = -4t(1 + 4t^2)^{-3/2}\mathbf{i} + [2(1 + 4t^2)^{-1/2} - 8t^2(1 + 4t^2)^{-3/2}]\,\mathbf{j}.$$

At the origin, $t = 0$, so the curvature is

$$\begin{aligned}\kappa(0) &= \frac{1}{|\mathbf{v}(0)|}\left|\frac{d\mathbf{T}}{dt}(0)\right| \qquad \text{Equation (1)}\\ &= \frac{1}{\sqrt{1}}|0\mathbf{i} + 2\mathbf{j}|\\ &= (1)\sqrt{0^2 + 2^2} = 2.\end{aligned}$$

Therefore, the radius of curvature is $1/\kappa = 1/2$ and the center of the circle is $(0, 1/2)$ (see Figure 13.23). The equation of the osculating circle is

$$(x - 0)^2 + \left(y - \frac{1}{2}\right)^2 = \left(\frac{1}{2}\right)^2$$

or

$$x^2 + \left(y - \frac{1}{2}\right)^2 = \frac{1}{4}.$$

You can see from Figure 13.23 that the osculating circle is a better approximation to the parabola at the origin than is the tangent line approximation $y = 0$. ■

Curvature and Normal Vectors for Space Curves

If a smooth curve in space is specified by the position vector $\mathbf{r}(t)$ as a function of some parameter t, and if s is the arc length parameter of the curve, then the unit tangent vector $\mathbf{T}$ is $d\mathbf{r}/ds = \mathbf{v}/|\mathbf{v}|$. The **curvature** in space is then defined to be

$$\kappa = \left|\frac{d\mathbf{T}}{ds}\right| = \frac{1}{|\mathbf{v}|}\left|\frac{d\mathbf{T}}{dt}\right| \tag{3}$$

just as for plane curves. The vector $d\mathbf{T}/ds$ is orthogonal to $\mathbf{T}$, and we define the **principal unit normal** to be

$$\mathbf{N} = \frac{1}{\kappa}\frac{d\mathbf{T}}{ds} = \frac{d\mathbf{T}/dt}{|d\mathbf{T}/dt|}. \tag{4}$$

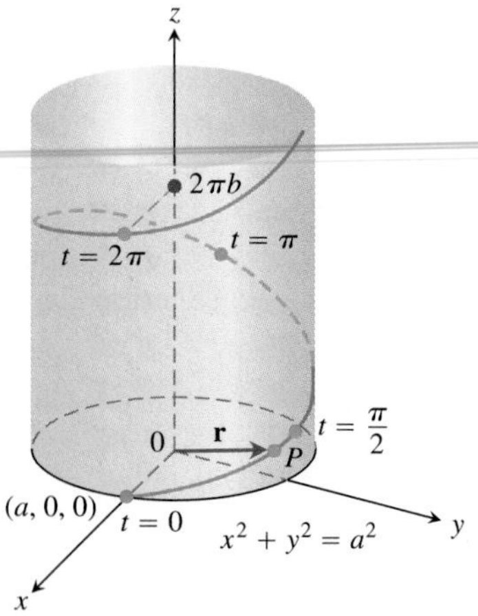

FIGURE 13.24 The helix

$$\mathbf{r}(t) = (a\cos t)\mathbf{i} + (a\sin t)\mathbf{j} + bt\mathbf{k},$$

drawn with a and b positive and $t \ge 0$ (Example 5).

EXAMPLE 5 Finding Curvature

Find the curvature for the helix (Figure 13.24)

$$\mathbf{r}(t) = (a\cos t)\mathbf{i} + (a\sin t)\mathbf{j} + bt\mathbf{k}, \qquad a, b \ge 0, \qquad a^2 + b^2 \ne 0.$$

Solution We calculate **T** from the velocity vector **v**:

$$\mathbf{v} = -(a \sin t)\mathbf{i} + (a \cos t)\mathbf{j} + b\mathbf{k}$$

$$|\mathbf{v}| = \sqrt{a^2 \sin^2 t + a^2 \cos^2 t + b^2} = \sqrt{a^2 + b^2}$$

$$\mathbf{T} = \frac{\mathbf{v}}{|\mathbf{v}|} = \frac{1}{\sqrt{a^2 + b^2}}[-(a \sin t)\mathbf{i} + (a \cos t)\mathbf{j} + b\mathbf{k}].$$

Then using Equation (3),

$$\begin{aligned}
\kappa &= \frac{1}{|\mathbf{v}|}\left|\frac{d\mathbf{T}}{dt}\right| \\
&= \frac{1}{\sqrt{a^2 + b^2}}\left|\frac{1}{\sqrt{a^2 + b^2}}[-(a \cos t)\mathbf{i} - (a \sin t)\mathbf{j}]\right| \\
&= \frac{a}{a^2 + b^2}|-(\cos t)\mathbf{i} - (\sin t)\mathbf{j}| \\
&= \frac{a}{a^2 + b^2}\sqrt{(\cos t)^2 + (\sin t)^2} = \frac{a}{a^2 + b^2}.
\end{aligned}$$

From this equation, we see that increasing b for a fixed a decreases the curvature. Decreasing a for a fixed b eventually decreases the curvature as well. Stretching a spring tends to straighten it.

If $b = 0$, the helix reduces to a circle of radius a and its curvature reduces to $1/a$, as it should. If $a = 0$, the helix becomes the z-axis, and its curvature reduces to 0, again as it should. ■

EXAMPLE 6 Finding the Principal Unit Normal Vector N

Find **N** for the helix in Example 5.

Solution We have

$$\frac{d\mathbf{T}}{dt} = -\frac{1}{\sqrt{a^2 + b^2}}[(a \cos t)\mathbf{i} + (a \sin t)\mathbf{j}] \qquad \text{Example 5}$$

$$\left|\frac{d\mathbf{T}}{dt}\right| = \frac{1}{\sqrt{a^2 + b^2}}\sqrt{a^2 \cos^2 t + a^2 \sin^2 t} = \frac{a}{\sqrt{a^2 + b^2}}$$

$$\begin{aligned}
\mathbf{N} &= \frac{d\mathbf{T}/dt}{|d\mathbf{T}/dt|} \qquad \text{Equatiion (4)} \\
&= -\frac{\sqrt{a^2 + b^2}}{a} \cdot \frac{1}{\sqrt{a^2 + b^2}}[(a \cos t)\mathbf{i} + (a \sin t)\mathbf{j}] \\
&= -(\cos t)\mathbf{i} - (\sin t)\mathbf{j}.
\end{aligned}$$

■

EXERCISES 13.4

Plane Curves

Find $\mathbf{T}$, $\mathbf{N}$, and κ for the plane curves in Exercises 1–4.

1. $\mathbf{r}(t) = t\mathbf{i} + (\ln \cos t)\mathbf{j}, \quad -\pi/2 < t < \pi/2$
2. $\mathbf{r}(t) = (\ln \sec t)\mathbf{i} + t\mathbf{j}, \quad -\pi/2 < t < \pi/2$
3. $\mathbf{r}(t) = (2t + 3)\mathbf{i} + (5 - t^2)\mathbf{j}$
4. $\mathbf{r}(t) = (\cos t + t \sin t)\mathbf{i} + (\sin t - t \cos t)\mathbf{j}, \quad t > 0$
5. **A formula for the curvature of the graph of a function in the *xy*-plane**
   a. The graph $y = f(x)$ in the xy-plane automatically has the parametrization $x = x, y = f(x)$, and the vector formula $\mathbf{r}(x) = x\mathbf{i} + f(x)\mathbf{j}$. Use this formula to show that if f is a twice-differentiable function of x, then
   $$\kappa(x) = \frac{|f''(x)|}{\left[1 + (f'(x))^2\right]^{3/2}}.$$
   b. Use the formula for κ in part (a) to find the curvature of $y = \ln(\cos x), -\pi/2 < x < \pi/2$. Compare your answer with the answer in Exercise 1.
   c. Show that the curvature is zero at a point of inflection.
6. **A formula for the curvature of a parametrized plane curve**
   a. Show that the curvature of a smooth curve $\mathbf{r}(t) = f(t)\mathbf{i} + g(t)\mathbf{j}$ defined by twice-differentiable functions $x = f(t)$ and $y = g(t)$ is given by the formula
   $$\kappa = \frac{|\dot{x}\ddot{y} - \dot{y}\ddot{x}|}{(\dot{x}^2 + \dot{y}^2)^{3/2}}.$$
   Apply the formula to find the curvatures of the following curves.
   b. $\mathbf{r}(t) = t\mathbf{i} + (\ln \sin t)\mathbf{j}, \quad 0 < t < \pi$
   c. $\mathbf{r}(t) = [\tan^{-1}(\sinh t)]\mathbf{i} + (\ln \cosh t)\mathbf{j}.$
7. **Normals to plane curves**
   a. Show that $\mathbf{n}(t) = -g'(t)\mathbf{i} + f'(t)\mathbf{j}$ and $-\mathbf{n}(t) = g'(t)\mathbf{i} - f'(t)\mathbf{j}$ are both normal to the curve $\mathbf{r}(t) = f(t)\mathbf{i} + g(t)\mathbf{j}$ at the point $(f(t), g(t))$.
   To obtain $\mathbf{N}$ for a particular plane curve, we can choose the one of $\mathbf{n}$ or $-\mathbf{n}$ from part (a) that points toward the concave side of the curve, and make it into a unit vector. (See Figure 13.21.) Apply this method to find $\mathbf{N}$ for the following curves.
   b. $\mathbf{r}(t) = t\mathbf{i} + e^{2t}\mathbf{j}$
   c. $\mathbf{r}(t) = \sqrt{4 - t^2}\,\mathbf{i} + t\mathbf{j}, \quad -2 \le t \le 2$
8. (*Continuation of Exercise 7.*)
   a. Use the method of Exercise 7 to find $\mathbf{N}$ for the curve $\mathbf{r}(t) = t\mathbf{i} + (1/3)t^3\mathbf{j}$ when $t < 0$; when $t > 0$.
   b. Calculate
   $$\mathbf{N} = \frac{d\mathbf{T}/dt}{|d\mathbf{T}/dt|}, \quad t \neq 0,$$
   for the curve in part (a). Does $\mathbf{N}$ exist at $t = 0$? Graph the curve and explain what is happening to $\mathbf{N}$ as t passes from negative to positive values.

Space Curves

Find $\mathbf{T}$, $\mathbf{N}$, and κ for the space curves in Exercises 9–16.

9. $\mathbf{r}(t) = (3 \sin t)\mathbf{i} + (3 \cos t)\mathbf{j} + 4t\mathbf{k}$
10. $\mathbf{r}(t) = (\cos t + t \sin t)\mathbf{i} + (\sin t - t \cos t)\mathbf{j} + 3\mathbf{k}$
11. $\mathbf{r}(t) = (e^t \cos t)\mathbf{i} + (e^t \sin t)\mathbf{j} + 2\mathbf{k}$
12. $\mathbf{r}(t) = (6 \sin 2t)\mathbf{i} + (6 \cos 2t)\mathbf{j} + 5t\mathbf{k}$
13. $\mathbf{r}(t) = (t^3/3)\mathbf{i} + (t^2/2)\mathbf{j}, \quad t > 0$
14. $\mathbf{r}(t) = (\cos^3 t)\mathbf{i} + (\sin^3 t)\mathbf{j}, \quad 0 < t < \pi/2$
15. $\mathbf{r}(t) = t\mathbf{i} + (a \cosh(t/a))\mathbf{j}, \quad a > 0$
16. $\mathbf{r}(t) = (\cosh t)\mathbf{i} - (\sinh t)\mathbf{j} + t\mathbf{k}$

More on Curvature

17. Show that the parabola $y = ax^2, a \neq 0$, has its largest curvature at its vertex and has no minimum curvature. (*Note:* Since the curvature of a curve remains the same if the curve is translated or rotated, this result is true for any parabola.)
18. Show that the ellipse $x = a \cos t, y = b \sin t, a > b > 0$, has its largest curvature on its major axis and its smallest curvature on its minor axis. (As in Exercise 17, the same is true for any ellipse.)
19. **Maximizing the curvature of a helix** In Example 5, we found the curvature of the helix $\mathbf{r}(t) = (a \cos t)\mathbf{i} + (a \sin t)\mathbf{j} + bt\mathbf{k}$ $(a, b \ge 0)$ to be $\kappa = a/(a^2 + b^2)$. What is the largest value κ can have for a given value of b? Give reasons for your answer.
20. **Total curvature** We find the **total curvature** of the portion of a smooth curve that runs from $s = s_0$ to $s = s_1 > s_0$ by integrating κ from s_0 to s_1. If the curve has some other parameter, say t, then the total curvature is
   $$K = \int_{s_0}^{s_1} \kappa\, ds = \int_{t_0}^{t_1} \kappa \frac{ds}{dt}\, dt = \int_{t_0}^{t_1} \kappa |\mathbf{v}|\, dt,$$
   where t_0 and t_1 correspond to s_0 and s_1. Find the total curvatures of
   a. The portion of the helix $\mathbf{r}(t) = (3 \cos t)\mathbf{i} + (3 \sin t)\mathbf{j} + t\mathbf{k}$, $0 \le t \le 4\pi$.
   b. The parabola $y = x^2, -\infty < x < \infty$.
21. Find an equation for the circle of curvature of the curve $\mathbf{r}(t) = t\mathbf{i} + (\sin t)\mathbf{j}$ at the point $(\pi/2, 1)$. (The curve parametrizes the graph of $y = \sin x$ in the xy-plane.)

22. Find an equation for the circle of curvature of the curve $\mathbf{r}(t) = (2 \ln t)\mathbf{i} - [t + (1/t)]\mathbf{j}, e^{-2} \le t \le e^2$, at the point $(0, -2)$, where $t = 1$.

T Grapher Explorations

The formula

$$\kappa(x) = \frac{|f''(x)|}{\left[1 + (f'(x))^2\right]^{3/2}},$$

derived in Exercise 5, expresses the curvature $\kappa(x)$ of a twice-differentiable plane curve $y = f(x)$ as a function of x. Find the curvature function of each of the curves in Exercises 23–26. Then graph $f(x)$ together with $\kappa(x)$ over the given interval. You will find some surprises.

23. $y = x^2, \quad -2 \le x \le 2$

24. $y = x^4/4, \quad -2 \le x \le 2$

25. $y = \sin x, \quad 0 \le x \le 2\pi$

26. $y = e^x, \quad -1 \le x \le 2$

COMPUTER EXPLORATIONS

Circles of Curvature

In Exercises 27–34 you will use a CAS to explore the osculating circle at a point P on a plane curve where $\kappa \ne 0$. Use a CAS to perform the following steps:

a. Plot the plane curve given in parametric or function form over the specified interval to see what it looks like.

b. Calculate the curvature κ of the curve at the given value t_0 using the appropriate formula from Exercise 5 or 6. Use the parametrization $x = t$ and $y = f(t)$ if the curve is given as a function $y = f(x)$.

c. Find the unit normal vector $\mathbf{N}$ at t_0. Notice that the signs of the components of $\mathbf{N}$ depend on whether the unit tangent vector $\mathbf{T}$ is turning clockwise or counterclockwise at $t = t_0$. (See Exercise 7.)

d. If $\mathbf{C} = a\mathbf{i} + b\mathbf{j}$ is the vector from the origin to the center (a, b) of the osculating circle, find the center $\mathbf{C}$ from the vector equation

$$\mathbf{C} = \mathbf{r}(t_0) + \frac{1}{\kappa(t_0)}\mathbf{N}(t_0).$$

The point $P(x_0, y_0)$ on the curve is given by the position vector $\mathbf{r}(t_0)$.

e. Plot implicitly the equation $(x - a)^2 + (y - b)^2 = 1/\kappa^2$ of the osculating circle. Then plot the curve and osculating circle together. You may need to experiment with the size of the viewing window, but be sure it is square.

27. $\mathbf{r}(t) = (3 \cos t)\mathbf{i} + (5 \sin t)\mathbf{j}, \quad 0 \le t \le 2\pi, \quad t_0 = \pi/4$

28. $\mathbf{r}(t) = (\cos^3 t)\mathbf{i} + (\sin^3 t)\mathbf{j}, \quad 0 \le t \le 2\pi, \quad t_0 = \pi/4$

29. $\mathbf{r}(t) = t^2\mathbf{i} + (t^3 - 3t)\mathbf{j}, \quad -4 \le t \le 4, \quad t_0 = 3/5$

30. $\mathbf{r}(t) = (t^3 - 2t^2 - t)\mathbf{i} + \dfrac{3t}{\sqrt{1 + t^2}}\mathbf{j}, \quad -2 \le t \le 5, \quad t_0 = 1$

31. $\mathbf{r}(t) = (2t - \sin t)\mathbf{i} + (2 - 2 \cos t)\mathbf{j}, \quad 0 \le t \le 3\pi,$ $t_0 = 3\pi/2$

32. $\mathbf{r}(t) = (e^{-t} \cos t)\mathbf{i} + (e^{-t} \sin t)\mathbf{j}, \quad 0 \le t \le 6\pi, \quad t_0 = \pi/4$

33. $y = x^2 - x, \quad -2 \le x \le 5, \quad x_0 = 1$

34. $y = x(1 - x)^{2/5}, \quad -1 \le x \le 2, \quad x_0 = 1/2$

13.5 Torsion and the Unit Binormal Vector B

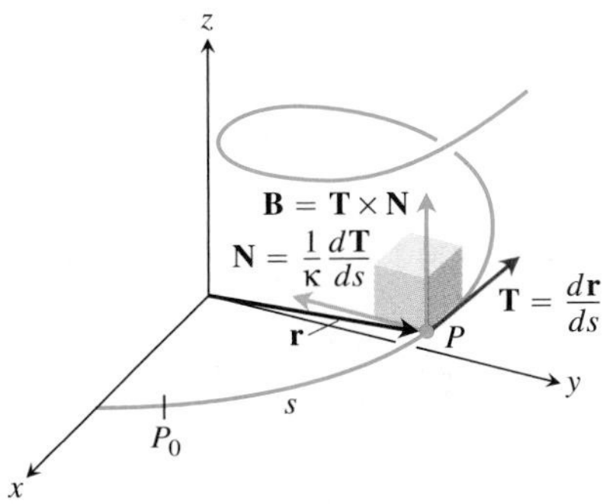

FIGURE 13.25 The **TNB** frame of mutually orthogonal unit vectors traveling along a curve in space.

If you are traveling along a space curve, the Cartesian **i**, **j**, and **k** coordinate system for representing the vectors describing your motion are not truly relevant to you. What is meaningful instead are the vectors representative of your forward direction (the unit tangent vector **T**), the direction in which your path is turning (the unit normal vector **N**), and the tendency of your motion to "twist" out of the plane created by these vectors in the direction perpendicular to this plane (defined by the *unit binormal vector* $\mathbf{B} = \mathbf{T} \times \mathbf{N}$). Expressing the acceleration vector along the curve as a linear combination of this **TNB** frame of mutually orthogonal unit vectors traveling with the motion (Figure 13.25) is particularly revealing of the nature of the path and motion along it.

Torsion

The **binormal vector** of a curve in space is $\mathbf{B} = \mathbf{T} \times \mathbf{N}$, a unit vector orthogonal to both **T** and **N** (Figure 13.26). Together **T**, **N**, and **B** define a moving right-handed vector frame that plays a significant role in calculating the paths of particles moving through space. It is

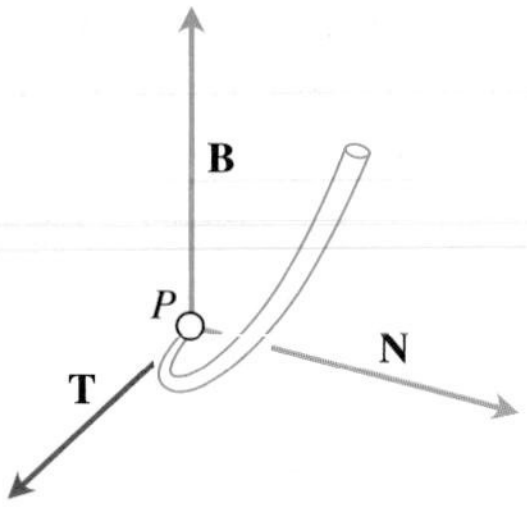

FIGURE 13.26 The vectors **T**, **N**, and **B** (in that order) make a right-handed frame of mutually orthogonal unit vectors in space.

called the **Frenet** ("fre-*nay*") **frame** (after Jean-Frédéric Frenet, 1816–1900), or the **TNB frame**.

How does $d\mathbf{B}/ds$ behave in relation to **T**, **N**, and **B**? From the rule for differentiating a cross product, we have

$$\frac{d\mathbf{B}}{ds} = \frac{d\mathbf{T}}{ds} \times \mathbf{N} + \mathbf{T} \times \frac{d\mathbf{N}}{ds}.$$

Since **N** is the direction of $d\mathbf{T}/ds$, $(d\mathbf{T}/ds) \times \mathbf{N} = \mathbf{0}$ and

$$\frac{d\mathbf{B}}{ds} = \mathbf{0} + \mathbf{T} \times \frac{d\mathbf{N}}{ds} = \mathbf{T} \times \frac{d\mathbf{N}}{ds}.$$

From this we see that $d\mathbf{B}/ds$ is orthogonal to **T** since a cross product is orthogonal to its factors.

Since $d\mathbf{B}/ds$ is also orthogonal to **B** (the latter has constant length), it follows that $d\mathbf{B}/ds$ is orthogonal to the plane of **B** and **T**. In other words, $d\mathbf{B}/ds$ is parallel to **N**, so $d\mathbf{B}/ds$ is a scalar multiple of **N**. In symbols,

$$\frac{d\mathbf{B}}{ds} = -\tau\mathbf{N}.$$

The negative sign in this equation is traditional. The scalar τ is called the *torsion* along the curve. Notice that

$$\frac{d\mathbf{B}}{ds}\cdot\mathbf{N} = -\tau\mathbf{N}\cdot\mathbf{N} = -\tau(1) = -\tau,$$

so that

$$\tau = -\frac{d\mathbf{B}}{ds}\cdot\mathbf{N}.$$

DEFINITION Torsion

Let $\mathbf{B} = \mathbf{T} \times \mathbf{N}$. The **torsion** function of a smooth curve is

$$\tau = -\frac{d\mathbf{B}}{ds}\cdot\mathbf{N}. \tag{1}$$

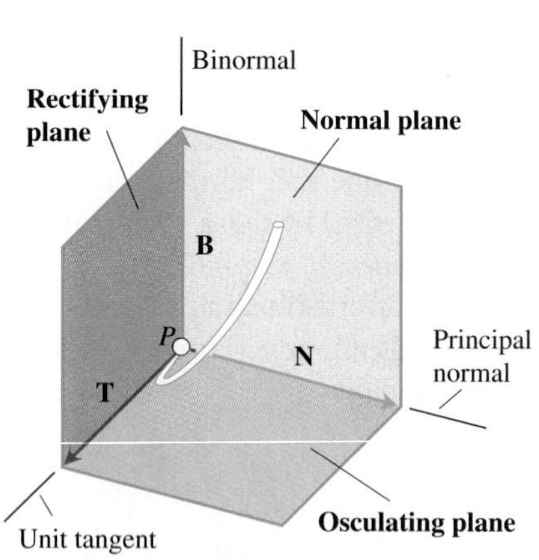

FIGURE 13.27 The names of the three planes determined by **T**, **N**, and **B**.

Unlike the curvature κ, which is never negative, the torsion τ may be positive, negative, or zero.

The three planes determined by **T**, **N**, and **B** are named and shown in Figure 13.27. The curvature $\kappa = |d\mathbf{T}/ds|$ can be thought of as the rate at which the normal plane turns as the point P moves along its path. Similarly, the torsion $\tau = -(d\mathbf{B}/ds)\cdot\mathbf{N}$ is the rate at which the osculating plane turns about **T** as P moves along the curve. Torsion measures how the curve twists.

If we think of the curve as the path of a moving body, then $|d\mathbf{T}/ds|$ tells how much the path turns to the left or right as the object moves along; it is called the *curvature* of the object's path. The number $-(d\mathbf{B}/ds)\cdot\mathbf{N}$ tells how much a body's path rotates or twists out of its plane of motion as the object moves along; it is called the *torsion* of the body's path. Look at Figure 13.28. If P is a train climbing up a curved track, the rate at which the headlight turns from side to side per unit distance is the curvature of the track. The rate at which the engine tends to twist out of the plane formed by **T** and **N** is the torsion.

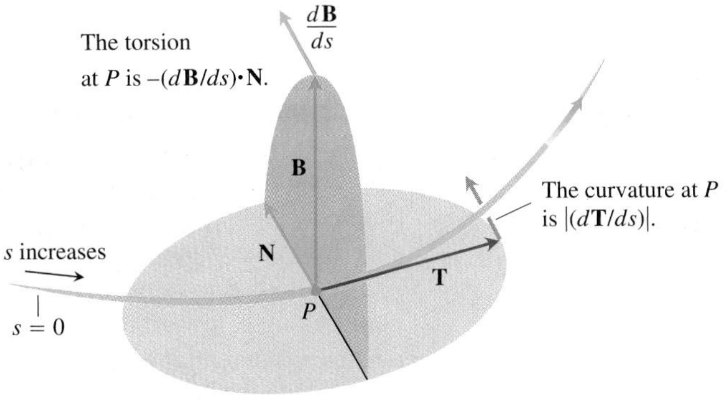

FIGURE 13.28 Every moving body travels with a **TNB** frame that characterizes the geometry of its path of motion.

Tangential and Normal Components of Acceleration

When a body is accelerated by gravity, brakes, a combination of rocket motors, or whatever, we usually want to know how much of the acceleration acts in the direction of motion, in the tangential direction **T**. We can calculate this using the Chain Rule to rewrite **v** as

$$\mathbf{v} = \frac{d\mathbf{r}}{dt} = \frac{d\mathbf{r}}{ds}\frac{ds}{dt} = \mathbf{T}\frac{ds}{dt}$$

and differentiating both ends of this string of equalities to get

$$\begin{aligned}\mathbf{a} &= \frac{d\mathbf{v}}{dt} = \frac{d}{dt}\left(\mathbf{T}\frac{ds}{dt}\right) = \frac{d^2s}{dt^2}\mathbf{T} + \frac{ds}{dt}\frac{d\mathbf{T}}{dt} \\ &= \frac{d^2s}{dt^2}\mathbf{T} + \frac{ds}{dt}\left(\frac{d\mathbf{T}}{ds}\frac{ds}{dt}\right) = \frac{d^2s}{dt^2}\mathbf{T} + \frac{ds}{dt}\left(\kappa\mathbf{N}\frac{ds}{dt}\right) \qquad \frac{d\mathbf{T}}{ds} = \kappa\mathbf{N} \\ &= \frac{d^2s}{dt^2}\mathbf{T} + \kappa\left(\frac{ds}{dt}\right)^2\mathbf{N}.\end{aligned}$$

DEFINITION **Tangential and Normal Components of Acceleration**

$$\mathbf{a} = a_{\mathrm{T}}\mathbf{T} + a_{\mathrm{N}}\mathbf{N}, \tag{2}$$

where

$$a_{\mathrm{T}} = \frac{d^2s}{dt^2} = \frac{d}{dt}|\mathbf{v}| \quad \text{and} \quad a_{\mathrm{N}} = \kappa\left(\frac{ds}{dt}\right)^2 = \kappa|\mathbf{v}|^2 \tag{3}$$

are the **tangential** and **normal** scalar components of acceleration.

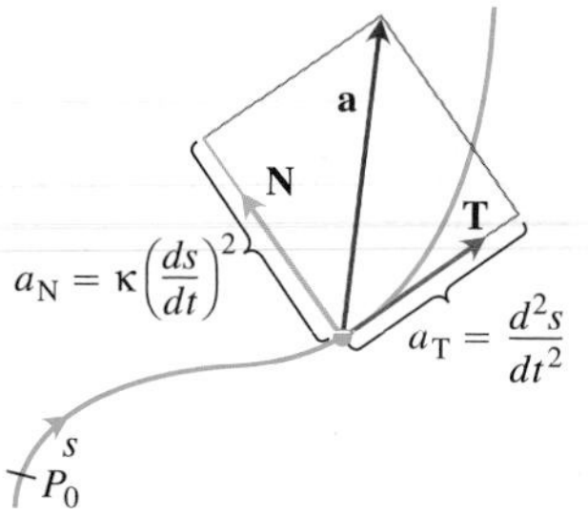

FIGURE 13.29 The tangential and normal components of acceleration. The acceleration **a** always lies in the plane of **T** and **N**, orthogonal to **B**.

Notice that the binormal vector **B** does not appear in Equation (2). No matter how the path of the moving body we are watching may appear to twist and turn in space, the acceleration **a** *always lies in the plane of* **T** and **N** orthogonal to **B**. The equation also tells us exactly how much of the acceleration takes place tangent to the motion (d^2s/dt^2) and how much takes place normal to the motion $[\kappa(ds/dt)^2]$ (Figure 13.29).

What information can we glean from Equations (3)? By definition, acceleration **a** is the rate of change of velocity **v**, and in general, both the length and direction of **v** change as a body moves along its path. The tangential component of acceleration a_T measures the rate of change of the *length* of **v** (that is, the change in the speed). The normal component of acceleration a_N measures the rate of change of the *direction* of **v**.

Notice that the normal scalar component of the acceleration is the curvature times the *square* of the speed. This explains why you have to hold on when your car makes a sharp (large κ), high-speed (large $|\mathbf{v}|$) turn. If you double the speed of your car, you will experience four times the normal component of acceleration for the same curvature.

If a body moves in a circle at a constant speed, d^2s/dt^2 is zero and all the acceleration points along **N** toward the circle's center. If the body is speeding up or slowing down, **a** has a nonzero tangential component (Figure 13.30).

To calculate a_N, we usually use the formula $a_N = \sqrt{|\mathbf{a}|^2 - a_T^2}$, which comes from solving the equation $|\mathbf{a}|^2 = \mathbf{a}\cdot\mathbf{a} = a_T^2 + a_N^2$ for a_N. With this formula, we can find a_N without having to calculate κ first.

Formula for Calculating the Normal Component of Acceleration

$$a_N = \sqrt{|\mathbf{a}|^2 - a_T^2} \qquad (4)$$

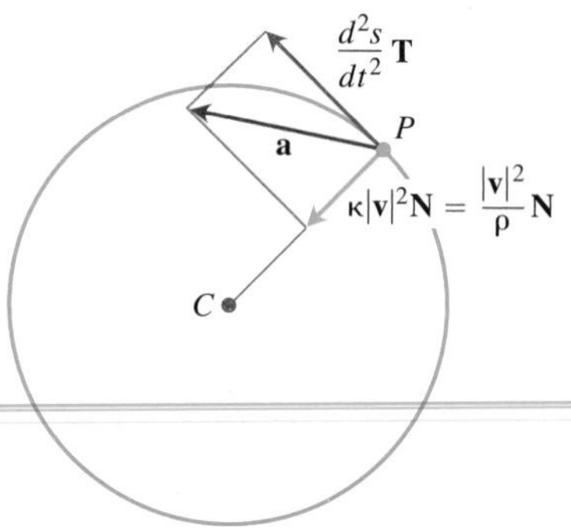

FIGURE 13.30 The tangential and normal components of the acceleration of a body that is speeding up as it moves counterclockwise around a circle of radius ρ.

EXAMPLE 1 Finding the Acceleration Scalar Components a_T, a_N

Without finding **T** and **N**, write the acceleration of the motion

$$\mathbf{r}(t) = (\cos t + t\sin t)\mathbf{i} + (\sin t - t\cos t)\mathbf{j}, \qquad t > 0$$

in the form $\mathbf{a} = a_T\mathbf{T} + a_N\mathbf{N}$. (The path of the motion is the involute of the circle in Figure 13.31.)

Solution We use the first of Equations (3) to find a_T:

$$\mathbf{v} = \frac{d\mathbf{r}}{dt} = (-\sin t + \sin t + t\cos t)\mathbf{i} + (\cos t - \cos t + t\sin t)\mathbf{j}$$
$$= (t\cos t)\mathbf{i} + (t\sin t)\mathbf{j}$$
$$|\mathbf{v}| = \sqrt{t^2\cos^2 t + t^2\sin^2 t} = \sqrt{t^2} = |t| = t \qquad t > 0$$
$$a_T = \frac{d}{dt}|\mathbf{v}| = \frac{d}{dt}(t) = 1. \qquad \text{Equation (3)}$$

Knowing a_T, we use Equation (4) to find a_N:

$$\mathbf{a} = (\cos t - t\sin t)\mathbf{i} + (\sin t + t\cos t)\mathbf{j}$$
$$|\mathbf{a}|^2 = t^2 + 1 \qquad \text{After some algebra}$$
$$a_N = \sqrt{|\mathbf{a}|^2 - a_T^2}$$
$$= \sqrt{(t^2 + 1) - (1)} = \sqrt{t^2} = t.$$

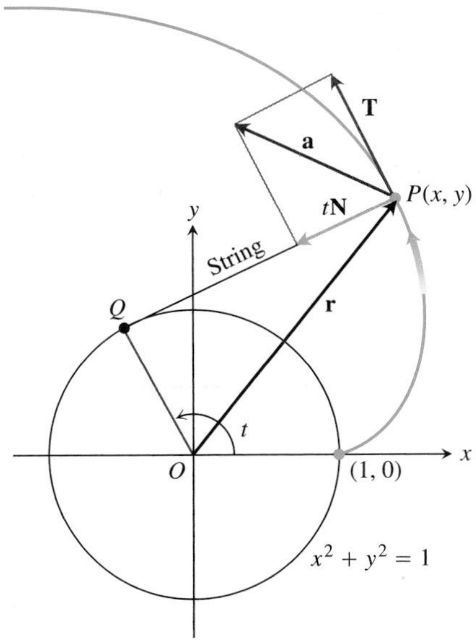

FIGURE 13.31 The tangential and normal components of the acceleration of the motion $\mathbf{r}(t) = (\cos t + t\sin t)\mathbf{i} + (\sin t - t\cos t)\mathbf{j}$, for $t > 0$. If a string wound around a fixed circle is unwound while held taut in the plane of the circle, its end P traces an involute of the circle (Example 1).

We then use Equation (2) to find **a**:

$$\mathbf{a} = a_{\mathrm{T}}\mathbf{T} + a_{\mathrm{N}}\mathbf{N} = (1)\mathbf{T} + (t)\mathbf{N} = \mathbf{T} + t\mathbf{N}. \qquad \blacksquare$$

Formulas for Computing Curvature and Torsion

We now give some easy-to-use formulas for computing the curvature and torsion of a smooth curve. From Equation (2), we have

$$\begin{aligned}\mathbf{v} \times \mathbf{a} &= \left(\frac{ds}{dt}\mathbf{T}\right) \times \left[\frac{d^2s}{dt^2}\mathbf{T} + \kappa\left(\frac{ds}{dt}\right)^2\mathbf{N}\right] && \mathbf{v} = d\mathbf{r}/dt = (ds/dt)\mathbf{T}\\ &= \left(\frac{ds}{dt}\frac{d^2s}{dt^2}\right)(\mathbf{T}\times\mathbf{T}) + \kappa\left(\frac{ds}{dt}\right)^3(\mathbf{T}\times\mathbf{N})\\ &= \kappa\left(\frac{ds}{dt}\right)^3\mathbf{B}. && \mathbf{T}\times\mathbf{T} = \mathbf{0} \text{ and } \mathbf{T}\times\mathbf{N} = \mathbf{B}\end{aligned}$$

It follows that

$$|\mathbf{v}\times\mathbf{a}| = \kappa\left|\frac{ds}{dt}\right|^3|\mathbf{B}| = \kappa|\mathbf{v}|^3. \qquad \frac{ds}{dt} = |\mathbf{v}| \text{ and } |\mathbf{B}| = 1$$

Solving for κ gives the following formula.

Vector Formula for Curvature

$$\kappa = \frac{|\mathbf{v}\times\mathbf{a}|}{|\mathbf{v}|^3} \tag{5}$$

Equation (5) calculates the curvature, a geometric property of the curve, from the velocity and acceleration of any vector representation of the curve in which $|\mathbf{v}|$ is different from zero. Take a moment to think about how remarkable this really is: From any formula for motion along a curve, no matter how variable the motion may be (as long as **v** is never zero), we can calculate a physical property of the curve that seems to have nothing to do with the way the curve is traversed.

The most widely used formula for torsion, derived in more advanced texts, is

$$\tau = \frac{\begin{vmatrix}\dot{x} & \dot{y} & \dot{z}\\ \ddot{x} & \ddot{y} & \ddot{z}\\ \dddot{x} & \dddot{y} & \dddot{z}\end{vmatrix}}{|\mathbf{v}\times\mathbf{a}|^2} \qquad (\text{if } \mathbf{v}\times\mathbf{a} \neq \mathbf{0}). \tag{6}$$

Newton's Dot Notation for Derivatives

The dots in Equation (6) denote differentiation with respect to t, one derivative for each dot. Thus, $\dot{x}$ ("x dot") means dx/dt, $\ddot{x}$ ("x double dot") means d^2x/dt^2, and $\dddot{x}$ ("x triple dot") means d^3x/dt^3. Similarly, $\dot{y} = dy/dt$, and so on.

This formula calculates the torsion directly from the derivatives of the component functions $x = f(t), y = g(t), z = h(t)$ that make up $\mathbf{r}$. The determinant's first row comes from $\mathbf{v}$, the second row comes from $\mathbf{a}$, and the third row comes from $\dot{\mathbf{a}} = d\mathbf{a}/dt$.

EXAMPLE 2 Finding Curvature and Torsion

Use Equations (5) and (6) to find κ and τ for the helix

$$\mathbf{r}(t) = (a\cos t)\mathbf{i} + (a\sin t)\mathbf{j} + bt\mathbf{k}, \qquad a, b \geq 0, \qquad a^2 + b^2 \neq 0.$$

Solution We calculate the curvature with Equation (5):

$$\mathbf{v} = -(a\sin t)\mathbf{i} + (a\cos t)\mathbf{j} + b\mathbf{k}$$

$$\mathbf{a} = -(a\cos t)\mathbf{i} - (a\sin t)\mathbf{j}$$

$$\mathbf{v}\times\mathbf{a} = \begin{vmatrix} \mathbf{i} & \mathbf{j} & \mathbf{k} \\ -a\sin t & a\cos t & b \\ -a\cos t & -a\sin t & 0 \end{vmatrix}$$

$$= (ab\sin t)\mathbf{i} - (ab\cos t)\mathbf{j} + a^2\mathbf{k}$$

$$\kappa = \frac{|\mathbf{v}\times\mathbf{a}|}{|\mathbf{v}|^3} = \frac{\sqrt{a^2b^2 + a^4}}{(a^2+b^2)^{3/2}} = \frac{a\sqrt{a^2+b^2}}{(a^2+b^2)^{3/2}} = \frac{a}{a^2+b^2}. \qquad (7)$$

Notice that Equation (7) agrees with the result in Example 5 in Section 13.4, where we calculated the curvature directly from its definition.

To evaluate Equation (6) for the torsion, we find the entries in the determinant by differentiating $\mathbf{r}$ with respect to t. We already have $\mathbf{v}$ and $\mathbf{a}$, and

$$\dot{\mathbf{a}} = \frac{d\mathbf{a}}{dt} = (a\sin t)\mathbf{i} - (a\cos t)\mathbf{j}.$$

Hence,

$$\tau = \frac{\begin{vmatrix} \dot{x} & \dot{y} & \dot{z} \\ \ddot{x} & \ddot{y} & \ddot{z} \\ \dddot{x} & \dddot{y} & \dddot{z} \end{vmatrix}}{|\mathbf{v}\times\mathbf{a}|^2} = \frac{\begin{vmatrix} -a\sin t & a\cos t & b \\ -a\cos t & -a\sin t & 0 \\ a\sin t & -a\cos t & 0 \end{vmatrix}}{\left(a\sqrt{a^2+b^2}\right)^2}$$

Value of $|\mathbf{v}\times\mathbf{a}|$ from Equation (7)

$$= \frac{b(a^2\cos^2 t + a^2\sin^2 t)}{a^2(a^2+b^2)}$$

$$= \frac{b}{a^2+b^2}.$$

From this last equation we see that the torsion of a helix about a circular cylinder is constant. In fact, constant curvature and constant torsion characterize the helix among all curves in space. ■

Formulas for Curves in Space

Unit tangent vector:	$\mathbf{T} = \dfrac{\mathbf{v}}{\lvert\mathbf{v}\rvert}$
Principal unit normal vector:	$\mathbf{N} = \dfrac{d\mathbf{T}/dt}{\lvert d\mathbf{T}/dt\rvert}$
Binormal vector:	$\mathbf{B} = \mathbf{T} \times \mathbf{N}$
Curvature:	$\kappa = \left\lvert \dfrac{d\mathbf{T}}{ds} \right\rvert = \dfrac{\lvert\mathbf{v} \times \mathbf{a}\rvert}{\lvert\mathbf{v}\rvert^3}$
Torsion:	$\tau = -\dfrac{d\mathbf{B}}{ds} \cdot \mathbf{N} = \dfrac{\begin{vmatrix} \dot{x} & \dot{y} & \dot{z} \\ \ddot{x} & \ddot{y} & \ddot{z} \\ \dddot{x} & \dddot{y} & \dddot{z} \end{vmatrix}}{\lvert\mathbf{v} \times \mathbf{a}\rvert^2}$
Tangential and normal scalar components of acceleration:	$\mathbf{a} = a_{\mathrm{T}}\mathbf{T} + a_{\mathrm{N}}\mathbf{N}$ $a_{\mathrm{T}} = \dfrac{d}{dt}\lvert\mathbf{v}\rvert$ $a_{\mathrm{N}} = \kappa\lvert\mathbf{v}\rvert^2 = \sqrt{\lvert\mathbf{a}\rvert^2 - a_{\mathrm{T}}^2}$

EXERCISES 13.5

Finding Torsion and the Binormal Vector

For Exercises 1–8 you found **T**, **N**, and κ in Section 13.4 (Exercises 9–16). Find now **B** and τ for these space curves.

1. $\mathbf{r}(t) = (3 \sin t)\mathbf{i} + (3 \cos t)\mathbf{j} + 4t\mathbf{k}$
2. $\mathbf{r}(t) = (\cos t + t \sin t)\mathbf{i} + (\sin t - t \cos t)\mathbf{j} + 3\mathbf{k}$
3. $\mathbf{r}(t) = (e^t \cos t)\mathbf{i} + (e^t \sin t)\mathbf{j} + 2\mathbf{k}$
4. $\mathbf{r}(t) = (6 \sin 2t)\mathbf{i} + (6 \cos 2t)\mathbf{j} + 5t\mathbf{k}$
5. $\mathbf{r}(t) = (t^3/3)\mathbf{i} + (t^2/2)\mathbf{j}, \quad t > 0$
6. $\mathbf{r}(t) = (\cos^3 t)\mathbf{i} + (\sin^3 t)\mathbf{j}, \quad 0 < t < \pi/2$
7. $\mathbf{r}(t) = t\mathbf{i} + (a \cosh (t/a))\mathbf{j}, \quad a > 0$
8. $\mathbf{r}(t) = (\cosh t)\mathbf{i} - (\sinh t)\mathbf{j} + t\mathbf{k}$

Tangential and Normal Components of Acceleration

In Exercises 9 and 10, write **a** in the form $a_{\mathrm{T}}\mathbf{T} + a_{\mathrm{N}}\mathbf{N}$ without finding **T** and **N**.

9. $\mathbf{r}(t) = (a \cos t)\mathbf{i} + (a \sin t)\mathbf{j} + bt\mathbf{k}$
10. $\mathbf{r}(t) = (1 + 3t)\mathbf{i} + (t - 2)\mathbf{j} - 3t\mathbf{k}$

In Exercises 11–14, write **a** in the form $\mathbf{a} = a_{\mathrm{T}}\mathbf{T} + a_{\mathrm{N}}\mathbf{N}$ at the given value of t without finding **T** and **N**.

11. $\mathbf{r}(t) = (t + 1)\mathbf{i} + 2t\mathbf{j} + t^2\mathbf{k}, \quad t = 1$
12. $\mathbf{r}(t) = (t \cos t)\mathbf{i} + (t \sin t)\mathbf{j} + t^2\mathbf{k}, \quad t = 0$
13. $\mathbf{r}(t) = t^2\mathbf{i} + (t + (1/3)t^3)\mathbf{j} + (t - (1/3)t^3)\mathbf{k}, \quad t = 0$
14. $\mathbf{r}(t) = (e^t \cos t)\mathbf{i} + (e^t \sin t)\mathbf{j} + \sqrt{2}e^t\mathbf{k}, \quad t = 0$

In Exercises 15 and 16, find **r**, **T**, **N**, and **B** at the given value of t. Then find equations for the osculating, normal, and rectifying planes at that value of t.

15. $\mathbf{r}(t) = (\cos t)\mathbf{i} + (\sin t)\mathbf{j} - \mathbf{k}, \quad t = \pi/4$
16. $\mathbf{r}(t) = (\cos t)\mathbf{i} + (\sin t)\mathbf{j} + t\mathbf{k}, \quad t = 0$

Physical Applications

17. The speedometer on your car reads a steady 35 mph. Could you be accelerating? Explain.
18. Can anything be said about the acceleration of a particle that is moving at a constant speed? Give reasons for your answer.
19. Can anything be said about the speed of a particle whose acceleration is always orthogonal to its velocity? Give reasons for your answer.

20. An object of mass m travels along the parabola $y = x^2$ with a constant speed of 10 units/sec. What is the force on the object due to its acceleration at $(0, 0)$? at $(2^{1/2}, 2)$? Write your answers in terms of $\mathbf{i}$ and $\mathbf{j}$. (Remember Newton's law, $\mathbf{F} = m\mathbf{a}$.)

21. The following is a quotation from an article in *The American Mathematical Monthly*, titled "Curvature in the Eighties" by Robert Osserman (October 1990, page 731):

> Curvature also plays a key role in physics. The magnitude of a force required to move an object at constant speed along a curved path is, according to Newton's laws, a constant multiple of the curvature of the trajectories.

Explain mathematically why the second sentence of the quotation is true.

22. Show that a moving particle will move in a straight line if the normal component of its acceleration is zero.

23. A sometime shortcut to curvature If you already know $|a_N|$ and $|\mathbf{v}|$, then the formula $a_N = \kappa|\mathbf{v}|^2$ gives a convenient way to find the curvature. Use it to find the curvature and radius of curvature of the curve

$$\mathbf{r}(t) = (\cos t + t\sin t)\mathbf{i} + (\sin t - t\cos t)\mathbf{j}, \quad t > 0.$$

(Take a_N and $|\mathbf{v}|$ from Example 1.)

24. Show that κ and τ are both zero for the line

$$\mathbf{r}(t) = (x_0 + At)\mathbf{i} + (y_0 + Bt)\mathbf{j} + (z_0 + Ct)\mathbf{k}.$$

Theory and Examples

25. What can be said about the torsion of a smooth plane curve $\mathbf{r}(t) = f(t)\mathbf{i} + g(t)\mathbf{j}$? Give reasons for your answer.

26. The torsion of a helix In Example 2, we found the torsion of the helix

$$\mathbf{r}(t) = (a\cos t)\mathbf{i} + (a\sin t)\mathbf{j} + bt\mathbf{k}, \quad a, b \geq 0$$

to be $\tau = b/(a^2 + b^2)$. What is the largest value τ can have for a given value of a? Give reasons for your answer.

27. Differentiable curves with zero torsion lie in planes That a sufficiently differentiable curve with zero torsion lies in a plane is a special case of the fact that a particle whose velocity remains perpendicular to a fixed vector $\mathbf{C}$ moves in a plane perpendicular to $\mathbf{C}$. This, in turn, can be viewed as the solution of the following problem in calculus.

Suppose $\mathbf{r}(t) = f(t)\mathbf{i} + g(t)\mathbf{j} + h(t)\mathbf{k}$ is twice differentiable for all t in an interval $[a, b]$, that $\mathbf{r} = 0$ when $t = a$, and that $\mathbf{v}\cdot\mathbf{k} = 0$ for all t in $[a, b]$. Then $h(t) = 0$ for all t in $[a, b]$.

Solve this problem. (*Hint:* Start with $\mathbf{a} = d^2\mathbf{r}/dt^2$ and apply the initial conditions in reverse order.)

28. A formula that calculates τ from B and v If we start with the definition $\tau = -(d\mathbf{B}/ds)\cdot\mathbf{N}$ and apply the Chain Rule to rewrite $d\mathbf{B}/ds$ as

$$\frac{d\mathbf{B}}{ds} = \frac{d\mathbf{B}}{dt}\frac{dt}{ds} = \frac{d\mathbf{B}}{dt}\frac{1}{|\mathbf{v}|},$$

we arrive at the formula

$$\tau = -\frac{1}{|\mathbf{v}|}\left(\frac{d\mathbf{B}}{dt}\cdot\mathbf{N}\right).$$

The advantage of this formula over Equation (6) is that it is easier to derive and state. The disadvantage is that it can take a lot of work to evaluate without a computer. Use the new formula to find the torsion of the helix in Example 2.

COMPUTER EXPLORATIONS

Curvature, Torsion, and the TNB Frame

Rounding the answers to four decimal places, use a CAS to find **v**, **a**, speed, **T**, **N**, **B**, κ, τ, and the tangential and normal components of acceleration for the curves in Exercises 29–32 at the given values of t.

29. $\mathbf{r}(t) = (t\cos t)\mathbf{i} + (t\sin t)\mathbf{j} + t\mathbf{k}, \quad t = \sqrt{3}$

30. $\mathbf{r}(t) = (e^t\cos t)\mathbf{i} + (e^t\sin t)\mathbf{j} + e^t\mathbf{k}, \quad t = \ln 2$

31. $\mathbf{r}(t) = (t - \sin t)\mathbf{i} + (1 - \cos t)\mathbf{j} + \sqrt{-t}\,\mathbf{k}, \quad t = -3\pi$

32. $\mathbf{r}(t) = (3t - t^2)\mathbf{i} + (3t^2)\mathbf{j} + (3t + t^3)\mathbf{k}, \quad t = 1$

13.6 Planetary Motion and Satellites

In this section, we derive Kepler's laws of planetary motion from Newton's laws of motion and gravitation and discuss the orbits of Earth satellites. The derivation of Kepler's laws from Newton's is one of the triumphs of calculus. It draws on almost everything we have studied so far, including the algebra and geometry of vectors in space, the calculus of vector functions, the solutions of differential equations and initial value problems, and the polar coordinate description of conic sections.

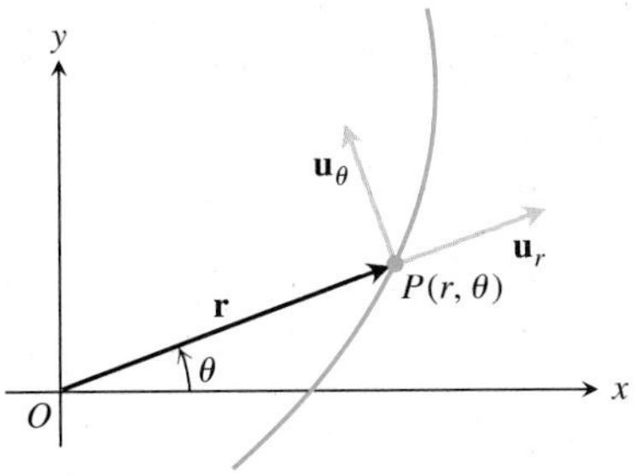

FIGURE 13.32 The length of $\mathbf{r}$ is the positive polar coordinate r of the point P. Thus, $\mathbf{u}_r$, which is $\mathbf{r}/|\mathbf{r}|$, is also $\mathbf{r}/r$. Equations (1) express $\mathbf{u}_r$ and $\mathbf{u}_\theta$ in terms of $\mathbf{i}$ and $\mathbf{j}$.

Motion in Polar and Cylindrical Coordinates

When a particle moves along a curve in the polar coordinate plane, we express its position, velocity, and acceleration in terms of the moving unit vectors

$$\mathbf{u}_r = (\cos\theta)\mathbf{i} + (\sin\theta)\mathbf{j}, \qquad \mathbf{u}_\theta = -(\sin\theta)\mathbf{i} + (\cos\theta)\mathbf{j}, \tag{1}$$

shown in Figure 13.32. The vector $\mathbf{u}_r$ points along the position vector $\overrightarrow{OP}$, so $\mathbf{r} = r\mathbf{u}_r$. The vector $\mathbf{u}_\theta$, orthogonal to $\mathbf{u}_r$, points in the direction of increasing θ.

We find from Equations (1) that

$$\begin{aligned} \frac{d\mathbf{u}_r}{d\theta} &= -(\sin\theta)\mathbf{i} + (\cos\theta)\mathbf{j} = \mathbf{u}_\theta \\ \frac{d\mathbf{u}_\theta}{d\theta} &= -(\cos\theta)\mathbf{i} - (\sin\theta)\mathbf{j} = -\mathbf{u}_r. \end{aligned} \tag{2}$$

When we differentiate $\mathbf{u}_r$ and $\mathbf{u}_\theta$ with respect to t to find how they change with time, the Chain Rule gives

$$\dot{\mathbf{u}}_r = \frac{d\mathbf{u}_r}{d\theta}\dot{\theta} = \dot{\theta}\mathbf{u}_\theta, \qquad \dot{\mathbf{u}}_\theta = \frac{d\mathbf{u}_\theta}{d\theta}\dot{\theta} = -\dot{\theta}\mathbf{u}_r. \tag{3}$$

Hence,

$$\mathbf{v} = \dot{\mathbf{r}} = \frac{d}{dt}\left(r\mathbf{u}_r\right) = \dot{r}\mathbf{u}_r + r\dot{\mathbf{u}}_r = \dot{r}\mathbf{u}_r + r\dot{\theta}\mathbf{u}_\theta. \tag{4}$$

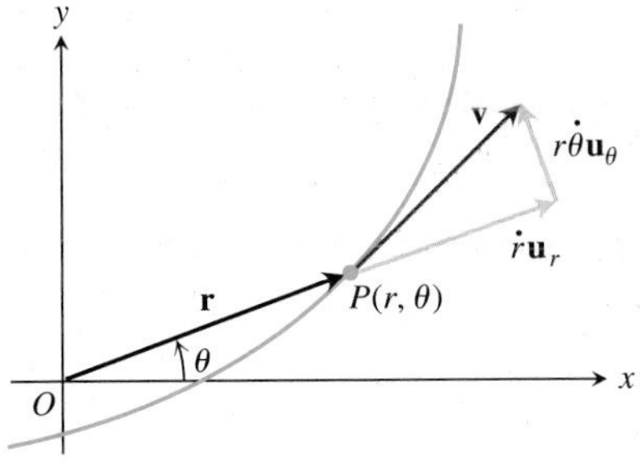

FIGURE 13.33 In polar coordinates, the velocity vector is

$$\mathbf{v} = \dot{r}\,\mathbf{u}_r + r\dot{\theta}\,\mathbf{u}_\theta$$

See Figure 13.33. As in the previous section, we use Newton's dot notation for time derivatives to keep the formulas as simple as we can: $\dot{\mathbf{u}}_r$ means $d\mathbf{u}_r/dt$, $\dot{\theta}$ means $d\theta/dt$, and so on.

The acceleration is

$$\mathbf{a} = \dot{\mathbf{v}} = (\ddot{r}\mathbf{u}_r + \dot{r}\dot{\mathbf{u}}_r) + (\dot{r}\dot{\theta}\mathbf{u}_\theta + r\ddot{\theta}\mathbf{u}_\theta + r\dot{\theta}\dot{\mathbf{u}}_\theta). \tag{5}$$

When Equations (3) are used to evaluate $\dot{\mathbf{u}}_r$ and $\dot{\mathbf{u}}_\theta$ and the components are separated, the equation for acceleration becomes

$$\mathbf{a} = (\ddot{r} - r\dot{\theta}^2)\mathbf{u}_r + (r\ddot{\theta} + 2\dot{r}\dot{\theta})\mathbf{u}_\theta. \tag{6}$$

To extend these equations of motion to space, we add $z\mathbf{k}$ to the right-hand side of the equation $\mathbf{r} = r\mathbf{u}_r$. Then, in these *cylindrical coordinates*,

Notice that $|\mathbf{r}| \neq r$ if $z \neq 0$.

$$\begin{aligned} \mathbf{r} &= r\mathbf{u}_r + z\mathbf{k} \\ \mathbf{v} &= \dot{r}\mathbf{u}_r + r\dot{\theta}\mathbf{u}_\theta + \dot{z}\mathbf{k} \\ \mathbf{a} &= (\ddot{r} - r\dot{\theta}^2)\mathbf{u}_r + (r\ddot{\theta} + 2\dot{r}\dot{\theta})\mathbf{u}_\theta + \ddot{z}\mathbf{k}. \end{aligned} \tag{7}$$

The vectors $\mathbf{u}_r$, $\mathbf{u}_\theta$, and $\mathbf{k}$ make a right-handed frame (Figure 13.34) in which

$$\mathbf{u}_r \times \mathbf{u}_\theta = \mathbf{k}, \qquad \mathbf{u}_\theta \times \mathbf{k} = \mathbf{u}_r, \qquad \mathbf{k} \times \mathbf{u}_r = \mathbf{u}_\theta. \tag{8}$$

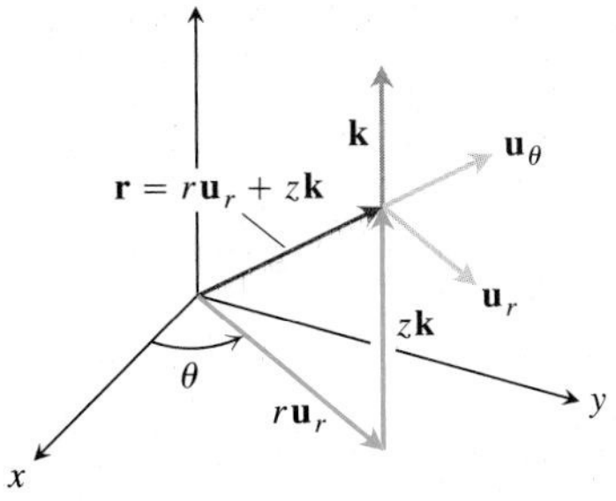

FIGURE 13.34 Position vector and basic unit vectors in cylindrical coordinates.

Planets Move in Planes

Newton's law of gravitation says that if $\mathbf{r}$ is the radius vector from the center of a sun of mass M to the center of a planet of mass m, then the force $\mathbf{F}$ of the gravitational attraction between the planet and sun is

$$\mathbf{F} = -\frac{GmM}{|\mathbf{r}|^2}\frac{\mathbf{r}}{|\mathbf{r}|} \tag{9}$$

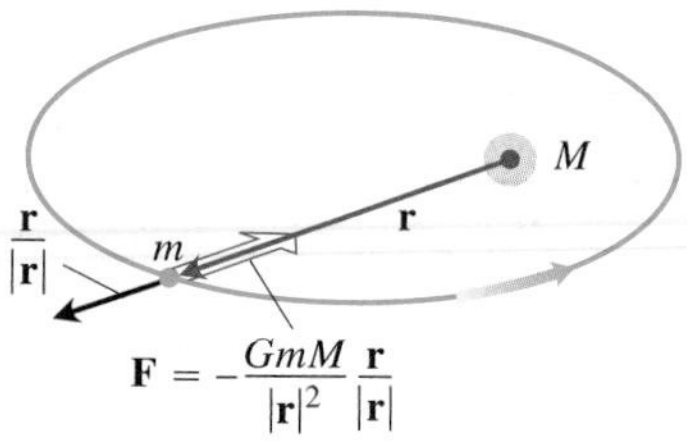

FIGURE 13.35 The force of gravity is directed along the line joining the centers of mass.

(Figure 13.35). The number G is the **universal gravitational constant**. If we measure mass in kilograms, force in newtons, and distance in meters, G is about $6.6726 \times 10^{-11}\ \text{Nm}^2\ \text{kg}^{-2}$.

Combining Equation (9) with Newton's second law, $\mathbf{F} = m\ddot{\mathbf{r}}$, for the force acting on the planet gives

$$m\ddot{\mathbf{r}} = -\frac{GmM}{|\mathbf{r}|^2}\frac{\mathbf{r}}{|\mathbf{r}|},$$

$$\ddot{\mathbf{r}} = -\frac{GM}{|\mathbf{r}|^2}\frac{\mathbf{r}}{|\mathbf{r}|}. \tag{10}$$

The planet is accelerated toward the sun's center at all times.

Equation (10) says that $\ddot{\mathbf{r}}$ is a scalar multiple of $\mathbf{r}$, so that

$$\mathbf{r} \times \ddot{\mathbf{r}} = \mathbf{0}. \tag{11}$$

A routine calculation shows $\mathbf{r} \times \ddot{\mathbf{r}}$ to be the derivative of $\mathbf{r} \times \dot{\mathbf{r}}$:

$$\frac{d}{dt}(\mathbf{r} \times \dot{\mathbf{r}}) = \underbrace{\dot{\mathbf{r}} \times \dot{\mathbf{r}}}_{0} + \mathbf{r} \times \ddot{\mathbf{r}} = \mathbf{r} \times \ddot{\mathbf{r}}. \tag{12}$$

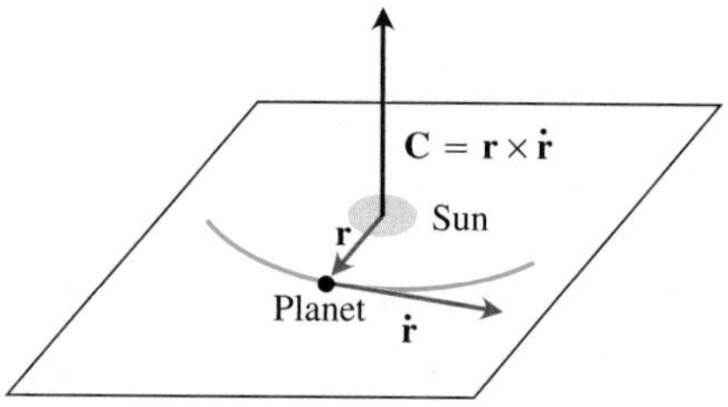

FIGURE 13.36 A planet that obeys Newton's laws of gravitation and motion travels in the plane through the sun's center of mass perpendicular to $\mathbf{C} = \mathbf{r} \times \dot{\mathbf{r}}$.

Hence Equation (11) is equivalent to

$$\frac{d}{dt}(\mathbf{r} \times \dot{\mathbf{r}}) = \mathbf{0}, \tag{13}$$

which integrates to

$$\mathbf{r} \times \dot{\mathbf{r}} = \mathbf{C} \tag{14}$$

for some constant vector $\mathbf{C}$.

Equation (14) tells us that $\mathbf{r}$ and $\dot{\mathbf{r}}$ always lie in a plane perpendicular to $\mathbf{C}$. Hence, the planet moves in a fixed plane through the center of its sun (Figure 13.36).

Coordinates and Initial Conditions

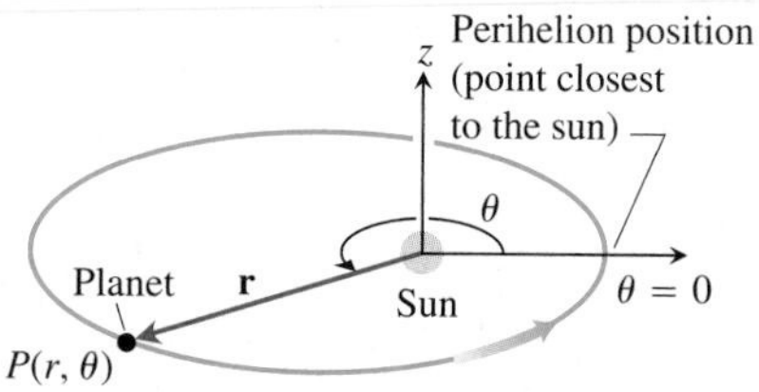

FIGURE 13.37 The coordinate system for planetary motion. The motion is counterclockwise when viewed from above, as it is here, and $\dot{\theta} > 0$.

We now introduce coordinates in a way that places the origin at the sun's center of mass and makes the plane of the planet's motion the polar coordinate plane. This makes $\mathbf{r}$ the planet's polar coordinate position vector and makes $|\mathbf{r}|$ equal to r and $\mathbf{r}/|\mathbf{r}|$ equal to $\mathbf{u}_r$. We also position the z-axis in a way that makes $\mathbf{k}$ the direction of $\mathbf{C}$. Thus, $\mathbf{k}$ has the same right-hand relation to $\mathbf{r} \times \dot{\mathbf{r}}$ that $\mathbf{C}$ does, and the planet's motion is counterclockwise when viewed from the positive z-axis. This makes θ increase with t, so that $\dot{\theta} > 0$ for all t. Finally, we rotate the polar coordinate plane about the z-axis, if necessary, to make the initial ray coincide with the direction $\mathbf{r}$ has when the planet is closest to the sun. This runs the ray through the planet's **perihelion** position (Figure 13.37).

If we measure time so that $t = 0$ at perihelion, we have the following initial conditions for the planet's motion.

1. $r = r_0$, the minimum radius, when $t = 0$
2. $\dot{r} = 0$ when $t = 0$ (because r has a minimum value then)
3. $\theta = 0$ when $t = 0$
4. $|\mathbf{v}| = v_0$ when $t = 0$

Since

$$\begin{aligned} v_0 &= |\mathbf{v}|_{t=0} \\ &= |\dot{r}\mathbf{u}_r + r\dot{\theta}\mathbf{u}_\theta|_{t=0} && \text{Equation (4)} \\ &= |r\dot{\theta}\mathbf{u}_\theta|_{t=0} && \dot{r} = 0 \text{ when } t = 0 \\ &= (|r\dot{\theta}||\mathbf{u}_\theta|)_{t=0} \\ &= |r\dot{\theta}|_{t=0} && |\mathbf{u}_\theta| = 1 \\ &= (r\dot{\theta})_{t=0}, && r \text{ and } \dot{\theta} \text{ both positive} \end{aligned}$$

we also know that

5. $r\dot{\theta} = v_0$ when $t = 0$.

HISTORICAL BIOGRAPHY

Johannes Kepler
(1571–1630)

Kepler's First Law (The Conic Section Law)

Kepler's first law says that a planet's path is a conic section with the sun at one focus. The eccentricity of the conic is

$$e = \frac{r_0 v_0^2}{GM} - 1 \tag{15}$$

and the polar equation is

$$r = \frac{(1 + e)r_0}{1 + e\cos\theta}. \tag{16}$$

The derivation uses Kepler's second law, so we will state and prove the second law before proving the first law.

Kepler's Second Law (The Equal Area Law)

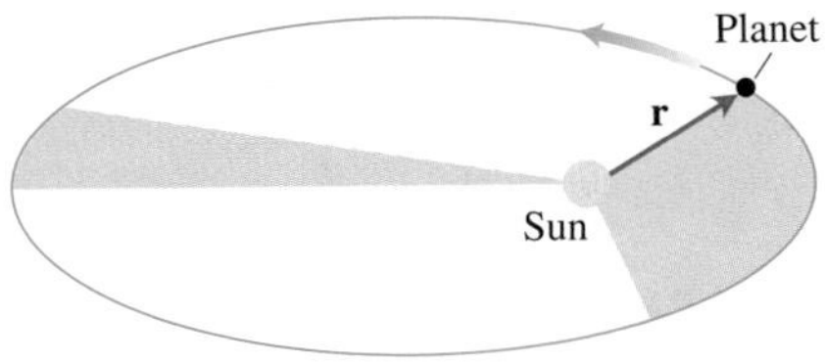

FIGURE 13.38 The line joining a planet to its sun sweeps over equal areas in equal times.

Kepler's second law says that the radius vector from the sun to a planet (the vector **r** in our model) sweeps out equal areas in equal times (Figure 13.38). To derive the law, we use Equation (4) to evaluate the cross product $\mathbf{C} = \mathbf{r} \times \dot{\mathbf{r}}$ from Equation (14):

$$\begin{aligned} \mathbf{C} &= \mathbf{r} \times \dot{\mathbf{r}} = \mathbf{r} \times \mathbf{v} \\ &= r\mathbf{u}_r \times (\dot{r}\mathbf{u}_r + r\dot{\theta}\mathbf{u}_\theta) && \text{Equation (4)} \\ &= r\dot{r}\underbrace{(\mathbf{u}_r \times \mathbf{u}_r)}_{\mathbf{0}} + r(r\dot{\theta})\underbrace{(\mathbf{u}_r \times \mathbf{u}_\theta)}_{\mathbf{k}} \\ &= r(r\dot{\theta})\mathbf{k}. \end{aligned} \tag{17}$$

Setting t equal to zero shows that

$$\mathbf{C} = [r(r\dot{\theta})]_{t=0}\,\mathbf{k} = r_0 v_0 \mathbf{k}. \tag{18}$$

Substituting this value for **C** in Equation (17) gives

$$r_0 v_0 \mathbf{k} = r^2\dot{\theta}\mathbf{k}, \qquad \text{or} \qquad r^2\dot{\theta} = r_0 v_0. \tag{19}$$

This is where the area comes in. The area differential in polar coordinates is

$$dA = \frac{1}{2}r^2\,d\theta$$

(Section 10.7). Accordingly, dA/dt has the constant value

$$\frac{dA}{dt} = \frac{1}{2}r^2\dot{\theta} = \frac{1}{2}r_0 v_0. \tag{20}$$

So dA/dt is constant, giving Kepler's second law.

For Earth, r_0 is about 150,000,000 km, v_0 is about 30 km/sec, and dA/dt is about 2,250,000,000 km^2/sec. Every time your heart beats, Earth advances 30 km along its orbit, and the radius joining Earth to the sun sweeps out 2,250,000,000 km^2 of area.

Proof of Kepler's First Law

To prove that a planet moves along a conic section with one focus at its sun, we need to express the planet's radius r as a function of θ. This requires a long sequence of calculations and some substitutions that are not altogether obvious.

We begin with the equation that comes from equating the coefficients of $\mathbf{u}_r = \mathbf{r}/|\mathbf{r}|$ in Equations (6) and (10):

$$\ddot{r} - r\dot{\theta}^2 = -\frac{GM}{r^2}. \tag{21}$$

We eliminate $\dot{\theta}$ temporarily by replacing it with $r_0 v_0/r^2$ from Equation (19) and rearrange the resulting equation to get

$$\ddot{r} = \frac{r_0^2 v_0^2}{r^3} - \frac{GM}{r^2}. \tag{22}$$

We change this into a first-order equation by a change of variable. With

$$p = \frac{dr}{dt}, \qquad \frac{d^2r}{dt^2} = \frac{dp}{dt} = \frac{dp}{dr}\frac{dr}{dt} = p\frac{dp}{dr}, \qquad \text{Chain Rule}$$

Equation (22) becomes

$$p\frac{dp}{dr} = \frac{r_0^2 v_0^2}{r^3} - \frac{GM}{r^2}. \tag{23}$$

Multiplying through by 2 and integrating with respect to r gives

$$p^2 = (\dot{r})^2 = -\frac{r_0^2 v_0^2}{r^2} + \frac{2GM}{r} + C_1. \tag{24}$$

The initial conditions that $r = r_0$ and $\dot{r} = 0$ when $t = 0$ determine the value of C_1 to be

$$C_1 = v_0^2 - \frac{2GM}{r_0}.$$

Accordingly, Equation (24), after a suitable rearrangement, becomes

$$\dot{r}^2 = v_0^2\left(1 - \frac{r_0^2}{r^2}\right) + 2GM\left(\frac{1}{r} - \frac{1}{r_0}\right). \tag{25}$$

The effect of going from Equation (21) to Equation (25) has been to replace a second-order differential equation in r by a first-order differential equation in r. Our goal is still to express r in terms of θ, so we now bring θ back into the picture. To accomplish this, we divide both sides of Equation (25) by the squares of the corresponding sides of the equation $r^2\dot{\theta} = r_0 v_0$ (Equation 19) and use the fact that $\dot{r}/\dot{\theta} = (dr/dt)/(d\theta/dt) = dr/d\theta$ to get

$$\begin{aligned}\frac{1}{r^4}\left(\frac{dr}{d\theta}\right)^2 &= \frac{1}{r_0{}^2} - \frac{1}{r^2} + \frac{2GM}{r_0{}^2 v_0{}^2}\left(\frac{1}{r} - \frac{1}{r_0}\right) \\ &= \frac{1}{r_0{}^2} - \frac{1}{r^2} + 2h\left(\frac{1}{r} - \frac{1}{r_0}\right). \qquad h = \frac{GM}{r_0{}^2 v_0{}^2}\end{aligned} \tag{26}$$

To simplify further, we substitute

$$u = \frac{1}{r}, \qquad u_0 = \frac{1}{r_0}, \qquad \frac{du}{d\theta} = -\frac{1}{r^2}\frac{dr}{d\theta}, \qquad \left(\frac{du}{d\theta}\right)^2 = \frac{1}{r^4}\left(\frac{dr}{d\theta}\right)^2,$$

obtaining

$$\left(\frac{du}{d\theta}\right)^2 = u_0{}^2 - u^2 + 2hu - 2hu_0 = (u_0 - h)^2 - (u - h)^2, \tag{27}$$

$$\frac{du}{d\theta} = \pm\sqrt{(u_0 - h)^2 - (u - h)^2}. \tag{28}$$

Which sign do we take? We know that $\dot{\theta} = r_0 v_0/r^2$ is positive. Also, r starts from a minimum value at $t = 0$, so it cannot immediately decrease, and $\dot{r} \geq 0$, at least for early positive values of t. Therefore,

$$\frac{dr}{d\theta} = \frac{\dot{r}}{\dot{\theta}} \geq 0 \qquad \text{and} \qquad \frac{du}{d\theta} = -\frac{1}{r^2}\frac{dr}{d\theta} \leq 0.$$

The correct sign for Equation (28) is the negative sign. With this determined, we rearrange Equation (28) and integrate both sides with respect to θ:

$$\begin{aligned}\frac{-1}{\sqrt{(u_0 - h)^2 - (u - h)^2}}\frac{du}{d\theta} &= 1 \\ \cos^{-1}\left(\frac{u - h}{u_0 - h}\right) &= \theta + C_2.\end{aligned} \tag{29}$$

The constant C_2 is zero because $u = u_0$ when $\theta = 0$ and $\cos^{-1}(1) = 0$. Therefore,

$$\frac{u - h}{u_0 - h} = \cos\theta$$

and

$$\frac{1}{r} = u = h + (u_0 - h)\cos\theta. \tag{30}$$

A few more algebraic maneuvers produce the final equation

$$r = \frac{(1 + e)r_0}{1 + e\cos\theta}, \tag{31}$$

where

$$e = \frac{1}{r_0 h} - 1 = \frac{r_0 v_0{}^2}{GM} - 1. \tag{32}$$

Together, Equations (31) and (32) say that the path of the planet is a conic section with one focus at the sun and with eccentricity $(r_0 v_0^2/GM) - 1$. This is the modern formulation of Kepler's first law.

Kepler's Third Law (The Time–Distance Law)

The time T it takes a planet to go around its sun once is the planet's **orbital period.** *Kepler's third law* says that T and the orbit's semimajor axis a are related by the equation

$$\frac{T^2}{a^3} = \frac{4\pi^2}{GM}. \tag{33}$$

Since the right-hand side of this equation is constant within a given solar system, the ratio of T^2 to a^3 *is the same for every planet in the system.*

Kepler's third law is the starting point for working out the size of our solar system. It allows the semimajor axis of each planetary orbit to be expressed in astronomical units, Earth's semimajor axis being one unit. The distance between any two planets at any time can then be predicted in astronomical units and all that remains is to find one of these distances in kilometers. This can be done by bouncing radar waves off Venus, for example. The astronomical unit is now known, after a series of such measurements, to be 149,597,870 km.

We derive Kepler's third law by combining two formulas for the area enclosed by the planet's elliptical orbit:

Formula 1: $\text{Area} = \pi ab$ — The geometry formula in which a is the semimajor axis and b is the semiminor axis

Formula 2:

$$\begin{aligned} \text{Area} &= \int_0^T dA \\ &= \int_0^T \frac{1}{2} r_0 v_0 \, dt \qquad \text{Equation (20)} \\ &= \frac{1}{2} T r_0 v_0. \end{aligned}$$

Equating these gives

$$T = \frac{2\pi ab}{r_0 v_0} = \frac{2\pi a^2}{r_0 v_0}\sqrt{1 - e^2}. \qquad \text{For any ellipse, } b = a\sqrt{1 - e^2} \tag{34}$$

It remains only to express a and e in terms of r_0, v_0, G, and M. Equation (32) does this for e. For a, we observe that setting θ equal to π in Equation (31) gives

$$r_{\max} = r_0 \frac{1 + e}{1 - e}.$$

Hence,

$$2a = r_0 + r_{\max} = \frac{2r_0}{1 - e} = \frac{2r_0 GM}{2GM - r_0 v_0^2}. \tag{35}$$

Squaring both sides of Equation (34) and substituting the results of Equations (32) and (35) now produces Kepler's third law (Exercise 15).

Orbit Data

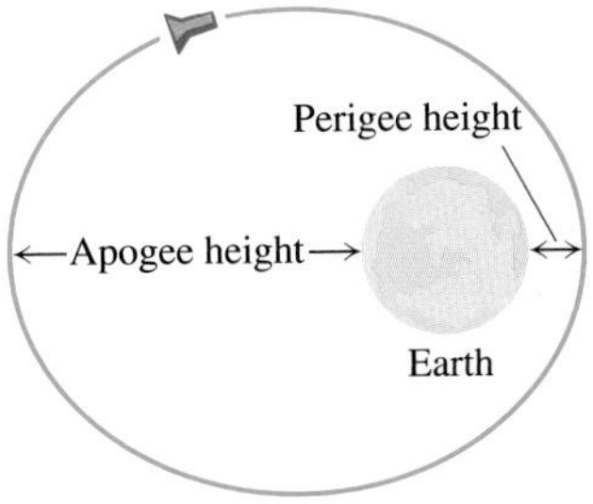

FIGURE 13.39 The orbit of an Earth satellite: $2a$ = diameter of Earth + perigee height + apogee height.

Although Kepler discovered his laws empirically and stated them only for the six planets known at the time, the modern derivations of Kepler's laws show that they apply to any body driven by a force that obeys an inverse square law like Equation (9). They apply to Halley's comet and the asteroid Icarus. They apply to the moon's orbit about Earth, and they applied to the orbit of the spacecraft *Apollo 8* about the moon.

Tables 13.1 through 13.3 give additional data for planetary orbits and for the orbits of seven of Earth's artificial satellites (Figure 13.39). *Vanguard 1* sent back data that revealed differences between the levels of Earth's oceans and provided the first determination of the precise locations of some of the more isolated Pacific islands. The data also verified that the gravitation of the sun and moon would affect the orbits of Earth's satellites and that solar radiation could exert enough pressure to deform an orbit.

TABLE 13.1 Values of a, e, and T for the major planets

Planet	Semimajor axis a^*	Eccentricity e	Period T
Mercury	57.95	0.2056	87.967 days
Venus	108.11	0.0068	224.701 days
Earth	149.57	0.0167	365.256 days
Mars	227.84	0.0934	1.8808 years
Jupiter	778.14	0.0484	11.8613 years
Saturn	1427.0	0.0543	29.4568 years
Uranus	2870.3	0.0460	84.0081 years
Neptune	4499.9	0.0082	164.784 years
Pluto	5909	0.2481	248.35 years

*Millions of kilometers.

TABLE 13.2 Data on Earth's satellites

Name	Launch date	Time or expected time aloft	Mass at launch (kg)	Period (min)	Perigee height (km)	Apogee height (km)	Semimajor axis a (km)	Eccentricity
Sputnik 1	Oct. 1957	57.6 days	83.6	96.2	215	939	6955	0.052
Vanguard 1	Mar. 1958	300 years	1.47	138.5	649	4340	8872	0.208
Syncom 3	Aug. 1964	$>10^6$ years	39	1436.2	35,718	35,903	42,189	0.002
Skylab 4	Nov. 1973	84.06 days	13,980	93.11	422	437	6808	0.001
Tiros II	Oct. 1978	500 years	734	102.12	850	866	7236	0.001
GOES 4	Sept. 1980	$>10^6$ years	627	1436.2	35,776	35,800	42,166	0.0003
Intelsat 5	Dec. 1980	$>10^6$ years	1928	1417.67	35,143	35,707	41,803	0.007

TABLE 13.3 Numerical data

Universal gravitational constant:	$G = 6.6726 \times 10^{-11}\ \text{Nm}^2\ \text{kg}^{-2}$
Sun's mass:	1.99×10^{30} kg
Earth's mass:	5.975×10^{24} kg
Equatorial radius of Earth:	6378.533 km
Polar radius of Earth:	6356.912 km
Earth's rotational period:	1436.1 min
Earth's orbital period:	1 year = 365.256 days

Syncom 3 is one of a series of U.S. Department of Defense telecommunications satellites. *Tiros II* (for "television infrared observation satellite") is one of a series of weather satellites. *GOES 4* (for "geostationary operational environmental satellite") is one of a series of satellites designed to gather information about Earth's atmosphere. Its orbital period, 1436.2 min, is nearly the same as Earth's rotational period of 1436.1 min, and its orbit is nearly circular ($e = 0.0003$). *Intelsat 5* is a heavy-capacity commercial telecommunications satellite.

EXERCISES 13.6

Reminder: When a calculation involves the gravitational constant G, express force in newtons, distance in meters, mass in kilograms, and time in seconds.

1. **Period of *Skylab 4*** Since the orbit of *Skylab 4* had a semimajor axis of $a = 6808$ km, Kepler's third law with M equal to Earth's mass should give the period. Calculate it. Compare your result with the value in Table 13.2.
2. **Earth's velocity at perihelion** Earth's distance from the sun at perihelion is approximately 149,577,000 km, and the eccentricity of Earth's orbit about the sun is 0.0167. Find the velocity v_0 of Earth in its orbit at perihelion. (Use Equation (15).)
3. **Semimajor axis of *Proton I*** In July 1965, the USSR launched *Proton I*, weighing 12,200 kg (at launch), with a perigee height of 183 km, an apogee height of 589 km, and a period of 92.25 min. Using the relevant data for the mass of Earth and the gravitational constant G, find the semimajor axis a of the orbit from Equation (3). Compare your answer with the number you get by adding the perigee and apogee heights to the diameter of the Earth.
4. **Semimajor axis of *Viking I*** The *Viking I* orbiter, which surveyed Mars from August 1975 to June 1976, had a period of 1639 min. Use this and the mass of Mars, 6.418×10^{23} kg, to find the semimajor axis of the *Viking I* orbit.
5. **Average diameter of Mars** (*Continuation of Exercise 4.*) The *Viking I* orbiter was 1499 km from the surface of Mars at its closest point and 35,800 km from the surface at its farthest point. Use this information together with the value you obtained in Exercise 4 to estimate the average diameter of Mars.
6. **Period of *Viking 2*** The *Viking 2* orbiter, which surveyed Mars from September 1975 to August 1976, moved in an ellipse whose semimajor axis was 22,030 km. What was the orbital period? (Express your answer in minutes.)
7. **Geosynchronous orbits** Several satellites in Earth's equatorial plane have nearly circular orbits whose periods are the same as Earth's rotational period. Such orbits are *geosynchronous* or *geostationary* because they hold the satellite over the same spot on the Earth's surface.
   a. Approximately what is the semimajor axis of a geosynchronous orbit? Give reasons for your answer.
   b. About how high is a geosynchronous orbit above Earth's surface?
   c. Which of the satellites in Table 13.2 have (nearly) geosynchronous orbits?
8. The mass of Mars is 6.418×10^{23} kg. If a satellite revolving about Mars is to hold a stationary orbit (have the same period as

the period of Mars's rotation, which is 1477.4 min), what must the semimajor axis of its orbit be? Give reasons for your answer.

9. **Distance from Earth to the moon** The period of the moon's rotation about Earth is 2.36055×10^6 sec. About how far away is the moon?

10. **Finding satellite speed** A satellite moves around Earth in a circular orbit. Express the satellite's speed as a function of the orbit's radius.

11. **Orbital period** If T is measured in seconds and a in meters, what is the value of T^2/a^3 for planets in our solar system? For satellites orbiting Earth? For satellites orbiting the moon? (The moon's mass is 7.354×10^{22} kg.)

12. **Type of orbit** For what values of v_0 in Equation (15) is the orbit in Equation (16) a circle? An ellipse? A parabola? A hyperbola?

13. **Circular orbits** Show that a planet in a circular orbit moves with a constant speed. (*Hint:* This is a consequence of one of Kepler's laws.)

14. Suppose that $\mathbf{r}$ is the position vector of a particle moving along a plane curve and dA/dt is the rate at which the vector sweeps out area. Without introducing coordinates, and assuming the necessary derivatives exist, give a geometric argument based on increments and limits for the validity of the equation

$$\frac{dA}{dt} = \frac{1}{2}|\mathbf{r} \times \dot{\mathbf{r}}|.$$

15. **Kepler's third law** Complete the derivation of Kepler's third law (the part following Equation (34)).

In Exercises 16 and 17, two planets, planet A and planet B, are orbiting their sun in circular orbits with A being the inner planet and B being farther away from the sun. Suppose the positions of A and B at time t are

$$\mathbf{r}_A(t) = 2\cos(2\pi t)\mathbf{i} + 2\sin(2\pi t)\mathbf{j}$$

and

$$\mathbf{r}_B(t) = 3\cos(\pi t)\mathbf{i} + 3\sin(\pi t)\mathbf{j},$$

respectively, where the sun is assumed to be located at the origin and distance is measured in astronomical units. (Notice that planet A moves faster than planet B.)

The people on planet A regard their planet, not the sun, as the center of their planetary system (their solar system).

16. Using planet A as the origin of a new coordinate system, give parametric equations for the location of planet B at time t. Write your answer in terms of $\cos(\pi t)$ and $\sin(\pi t)$.

T 17. Using planet A as the origin, graph the path of planet B.

This exercise illustrates the difficulty that people before Kepler's time, with an earth-centered (planet A) view of our solar system, had in understanding the motions of the planets (i.e., planet B = Mars). See D. G. Saari's article in the *American Mathematical Monthly*, Vol. 97 (Feb. 1990), pp. 105–119.

18. Kepler discovered that the path of Earth around the sun is an ellipse with the sun at one of the foci. Let $\mathbf{r}(t)$ be the position vector from the center of the sun to the center of Earth at time t. Let $\mathbf{w}$ be the vector from Earth's South Pole to North Pole. It is known that $\mathbf{w}$ is constant and not orthogonal to the plane of the ellipse (Earth's axis is tilted). In terms of $\mathbf{r}(t)$ and $\mathbf{w}$, give the mathematical meaning of (i) perihelion, (ii) aphelion, (iii) equinox, (iv) summer solstice, (v) winter solstice.

Chapter 13 Questions to Guide Your Review

1. State the rules for differentiating and integrating vector functions. Give examples.
2. How do you define and calculate the velocity, speed, direction of motion, and acceleration of a body moving along a sufficiently differentiable space curve? Give an example.
3. What is special about the derivatives of vector functions of constant length? Give an example.
4. What are the vector and parametric equations for ideal projectile motion? How do you find a projectile's maximum height, flight time, and range? Give examples.
5. How do you define and calculate the length of a segment of a smooth space curve? Give an example. What mathematical assumptions are involved in the definition?
6. How do you measure distance along a smooth curve in space from a preselected base point? Give an example.
7. What is a differentiable curve's unit tangent vector? Give an example.
8. Define curvature, circle of curvature (osculating circle), center of curvature, and radius of curvature for twice-differentiable curves in the plane. Give examples. What curves have zero curvature? Constant curvature?
9. What is a plane curve's principal normal vector? When is it defined? Which way does it point? Give an example.
10. How do you define $\mathbf{N}$ and κ for curves in space? How are these quantities related? Give examples.

11. What is a curve's binormal vector? Give an example. How is this vector related to the curve's torsion? Give an example.

12. What formulas are available for writing a moving body's acceleration as a sum of its tangential and normal components? Give an example. Why might one want to write the acceleration this way? What if the body moves at a constant speed? At a constant speed around a circle?

13. State Kepler's laws. To what phenomena do they apply?

Chapter 13 Practice Exercises

Motion in a Cartesian Plane

In Exercises 1 and 2, graph the curves and sketch their velocity and acceleration vectors at the given values of t. Then write $\mathbf{a}$ in the form $\mathbf{a} = a_T\mathbf{T} + a_N\mathbf{N}$ without finding $\mathbf{T}$ and $\mathbf{N}$, and find the value of κ at the given values of t.

1. $\mathbf{r}(t) = (4\cos t)\mathbf{i} + \left(\sqrt{2}\sin t\right)\mathbf{j}, \quad t = 0$ and $\pi/4$
2. $\mathbf{r}(t) = \left(\sqrt{3}\sec t\right)\mathbf{i} + \left(\sqrt{3}\tan t\right)\mathbf{j}, \quad t = 0$
3. The position of a particle in the plane at time t is

$$\mathbf{r} = \frac{1}{\sqrt{1+t^2}}\mathbf{i} + \frac{t}{\sqrt{1+t^2}}\mathbf{j}.$$

Find the particle's highest speed.

4. Suppose $\mathbf{r}(t) = (e^t\cos t)\mathbf{i} + (e^t\sin t)\mathbf{j}$. Show that the angle between $\mathbf{r}$ and $\mathbf{a}$ never changes. What *is* the angle?

5. **Finding curvature** At point P, the velocity and acceleration of a particle moving in the plane are $\mathbf{v} = 3\mathbf{i} + 4\mathbf{j}$ and $\mathbf{a} = 5\mathbf{i} + 15\mathbf{j}$. Find the curvature of the particle's path at P.

6. Find the point on the curve $y = e^x$ where the curvature is greatest.

7. A particle moves around the unit circle in the xy-plane. Its position at time t is $\mathbf{r} = x\mathbf{i} + y\mathbf{j}$, where x and y are differentiable functions of t. Find dy/dt if $\mathbf{v}\cdot\mathbf{i} = y$. Is the motion clockwise, or counterclockwise?

8. You send a message through a pneumatic tube that follows the curve $9y = x^3$ (distance in meters). At the point (3, 3), $\mathbf{v}\cdot\mathbf{i} = 4$ and $\mathbf{a}\cdot\mathbf{i} = -2$. Find the values of $\mathbf{v}\cdot\mathbf{j}$ and $\mathbf{a}\cdot\mathbf{j}$ at (3, 3).

9. **Characterizing circular motion** A particle moves in the plane so that its velocity and position vectors are always orthogonal. Show that the particle moves in a circle centered at the origin.

10. **Speed along a cycloid** A circular wheel with radius 1 ft and center C rolls to the right along the x-axis at a half-turn per second. (See the accompanying figure.) At time t seconds, the position vector of the point P on the wheel's circumference is

$$\mathbf{r} = (\pi t - \sin \pi t)\mathbf{i} + (1 - \cos \pi t)\mathbf{j}.$$

a. Sketch the curve traced by P during the interval $0 \le t \le 3$.

b. Find $\mathbf{v}$ and $\mathbf{a}$ at $t = 0, 1, 2,$ and 3 and add these vectors to your sketch.

c. At any given time, what is the forward speed of the topmost point of the wheel? Of C?

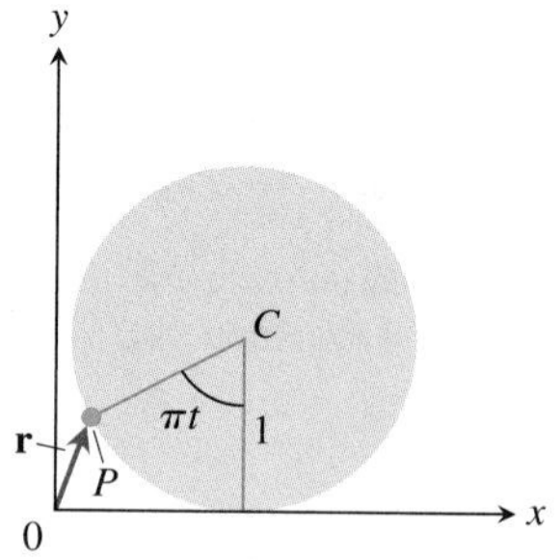

Projectile Motion and Motion in a Plane

11. **Shot put** A shot leaves the thrower's hand 6.5 ft above the ground at a 45° angle at 44 ft/sec. Where is it 3 sec later?

12. **Javelin** A javelin leaves the thrower's hand 7 ft above the ground at a 45° angle at 80 ft/sec. How high does it go?

13. A golf ball is hit with an initial speed v_0 at an angle α to the horizontal from a point that lies at the foot of a straight-sided hill that is inclined at an angle ϕ to the horizontal, where

$$0 < \phi < \alpha < \frac{\pi}{2}.$$

Show that the ball lands at a distance

$$\frac{2v_0^2\cos\alpha}{g\cos^2\phi}\sin(\alpha - \phi),$$

measured up the face of the hill. Hence, show that the greatest range that can be achieved for a given v_0 occurs when $\alpha = (\phi/2) + (\pi/4)$, i.e., when the initial velocity vector bisects the angle between the vertical and the hill.

14. The Dictator The Civil War mortar Dictator weighed so much (17,120 lb) that it had to be mounted on a railroad car. It had a 13-in. bore and used a 20-lb powder charge to fire a 200-lb shell. The mortar was made by Mr. Charles Knapp in his ironworks in Pittsburgh, Pennsylvania, and was used by the Union army in 1864 in the siege of Petersburg, Virginia. How far did it shoot? Here we have a difference of opinion. The ordnance manual claimed 4325 yd, while field officers claimed 4752 yd. Assuming a 45° firing angle, what muzzle speeds are involved here?

15. The World's record for popping a champagne cork

a. Until 1988, the world's record for popping a champagne cork was 109 ft. 6 in., once held by Captain Michael Hill of the British Royal Artillery (of course). Assuming Cpt. Hill held the bottle neck at ground level at a 45° angle, and the cork behaved like an ideal projectile, how fast was the cork going as it left the bottle?

b. A new world record of 177 ft. 9 in. was set on June 5, 1988, by Prof. Emeritus Heinrich of Rensselaer Polytechnic Institute, firing from 4 ft. above ground level at the Woodbury Vineyards Winery, New York. Assuming an ideal trajectory, what was the cork's initial speed?

16. Javelin In Potsdam in 1988, Petra Felke of (then) East Germany set a women's world record by throwing a javelin 262 ft 5 in.

a. Assuming that Felke launched the javelin at a 40° angle to the horizontal 6.5 ft above the ground, what was the javelin's initial speed?

b. How high did the javelin go?

17. Synchronous curves By eliminating α from the ideal projectile equations

$$x = (v_0 \cos\alpha)t, \quad y = (v_0 \sin\alpha)t - \frac{1}{2}gt^2,$$

show that $x^2 + (y + gt^2/2)^2 = v_0^2 t^2$. This shows that projectiles launched simultaneously from the origin at the same initial speed will, at any given instant, all lie on the circle of radius $v_0 t$ centered at $(0, -gt^2/2)$, regardless of their launch angle. These circles are the *synchronous curves* of the launching.

18. Radius of curvature Show that the radius of curvature of a twice-differentiable plane curve $\mathbf{r}(t) = f(t)\mathbf{i} + g(t)\mathbf{j}$ is given by the formula

$$\rho = \frac{\dot{x}^2 + \dot{y}^2}{\sqrt{\ddot{x}^2 + \ddot{y}^2 - \ddot{s}^2}}, \quad \text{where} \quad \ddot{s} = \frac{d}{dt}\sqrt{\dot{x}^2 + \dot{y}^2}.$$

19. Curvature Express the curvature of the curve

$$\mathbf{r}(t) = \left(\int_0^t \cos\left(\frac{1}{2}\pi\theta^2\right) d\theta\right)\mathbf{i} + \left(\int_0^t \sin\left(\frac{1}{2}\pi\theta^2\right) d\theta\right)\mathbf{j}$$

as a function of the directed distance s measured along the curve from the origin. (See the accompanying figure.)

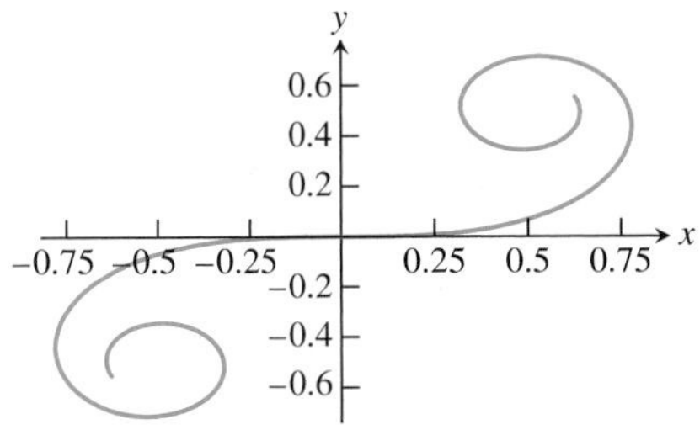

20. An alternative definition of curvature in the plane An alternative definition gives the curvature of a sufficiently differentiable plane curve to be $|d\phi/ds|$, where ϕ is the angle between $\mathbf{T}$ and $\mathbf{i}$ (Figure 13.40a). Figure 13.40b shows the distance s measured counterclockwise around the circle $x^2 + y^2 = a^2$ from the point $(a, 0)$ to a point P, along with the angle ϕ at P. Calculate the circle's curvature using the alternative definition. (*Hint:* $\phi = \theta + \pi/2$.)

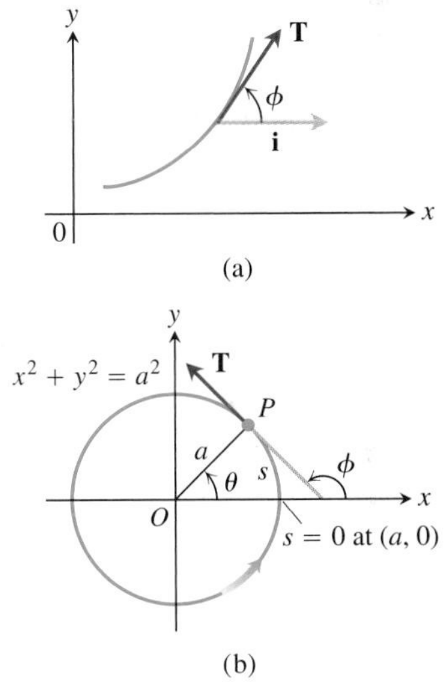

FIGURE 13.40 Figures for Exercise 20.

Motion in Space

Find the lengths of the curves in Exercises 21 and 22.

21. $\mathbf{r}(t) = (2\cos t)\mathbf{i} + (2\sin t)\mathbf{j} + t^2\mathbf{k}, \quad 0 \le t \le \pi/4$

22. $\mathbf{r}(t) = (3\cos t)\mathbf{i} + (3\sin t)\mathbf{j} + 2t^{3/2}\mathbf{k}, \quad 0 \le t \le 3$

In Exercises 23–26, find $\mathbf{T}$, $\mathbf{N}$, $\mathbf{B}$, κ, and τ at the given value of t.

23. $\mathbf{r}(t) = \frac{4}{9}(1 + t)^{3/2}\mathbf{i} + \frac{4}{9}(1 - t)^{3/2}\mathbf{j} + \frac{1}{3}t\mathbf{k}, \quad t = 0$

24. $\mathbf{r}(t) = (e^t \sin 2t)\mathbf{i} + (e^t \cos 2t)\mathbf{j} + 2e^t\mathbf{k}, \quad t = 0$

25. $\mathbf{r}(t) = t\mathbf{i} + \frac{1}{2}e^{2t}\mathbf{j}, \quad t = \ln 2$

26. $\mathbf{r}(t) = (3\cosh 2t)\mathbf{i} + (3\sinh 2t)\mathbf{j} + 6t\mathbf{k}, \quad t = \ln 2$

In Exercises 27 and 28, write $\mathbf{a}$ in the form $\mathbf{a} = a_T\mathbf{T} + a_N\mathbf{N}$ at $t = 0$ without finding $\mathbf{T}$ and $\mathbf{N}$.

27. $\mathbf{r}(t) = (2 + 3t + 3t^2)\mathbf{i} + (4t + 4t^2)\mathbf{j} - (6\cos t)\mathbf{k}$

28. $\mathbf{r}(t) = (2 + t)\mathbf{i} + (t + 2t^2)\mathbf{j} + (1 + t^2)\mathbf{k}$

29. Find $\mathbf{T}$, $\mathbf{N}$, $\mathbf{B}$, κ, and τ as functions of t if $\mathbf{r}(t) = (\sin t)\mathbf{i} + (\sqrt{2}\cos t)\mathbf{j} + (\sin t)\mathbf{k}$.

30. At what times in the interval $0 \le t \le \pi$ are the velocity and acceleration vectors of the motion $\mathbf{r}(t) = \mathbf{i} + (5\cos t)\mathbf{j} + (3\sin t)\mathbf{k}$ orthogonal?

31. The position of a particle moving in space at time $t \ge 0$ is

$$\mathbf{r}(t) = 2\mathbf{i} + \left(4\sin\frac{t}{2}\right)\mathbf{j} + \left(3 - \frac{t}{\pi}\right)\mathbf{k}.$$

Find the first time $\mathbf{r}$ is orthogonal to the vector $\mathbf{i} - \mathbf{j}$.

32. Find equations for the osculating, normal, and rectifying planes of the curve $\mathbf{r}(t) = t\mathbf{i} + t^2\mathbf{j} + t^3\mathbf{k}$ at the point $(1, 1, 1)$.

33. Find parametric equations for the line that is tangent to the curve $\mathbf{r}(t) = e^t\mathbf{i} + (\sin t)\mathbf{j} + \ln(1 - t)\mathbf{k}$ at $t = 0$.

34. Find parametric equations for the line tangent to the helix $\mathbf{r}(t) = (\sqrt{2}\cos t)\mathbf{i} + (\sqrt{2}\sin t)\mathbf{j} + t\mathbf{k}$ at the point where $t = \pi/4$.

35. The view from *Skylab 4* What percentage of Earth's surface area could the astronauts see when *Skylab 4* was at its apogee height, 437 km above the surface? To find out, model the visible surface as the surface generated by revolving the circular arc GT, shown here, about the y-axis. Then carry out these steps:

1. Use similar triangles in the figure to show that $y_0/6380 = 6380/(6380 + 437)$. Solve for y_0.
2. To four significant digits, calculate the visible area as
$$VA = \int_{y_0}^{6380} 2\pi x\sqrt{1 + \left(\frac{dx}{dy}\right)^2}\, dy.$$
3. Express the result as a percentage of Earth's surface area.

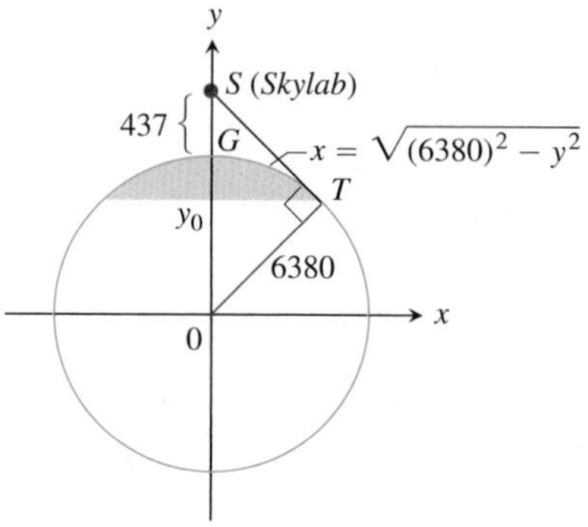

Chapter 13 Additional and Advanced Exercises

Applications

1. A straight river is 100 m wide. A rowboat leaves the far shore at time $t = 0$. The person in the boat rows at a rate of 20 m/min, always toward the near shore. The velocity of the river at (x, y) is

$$\mathbf{v} = \left(-\frac{1}{250}(y - 50)^2 + 10\right)\mathbf{i} \text{ m/min}, \quad 0 < y < 100.$$

a. Given that $\mathbf{r}(0) = 0\mathbf{i} + 100\mathbf{j}$, what is the position of the boat at time t?

b. How far downstream will the boat land on the near shore?

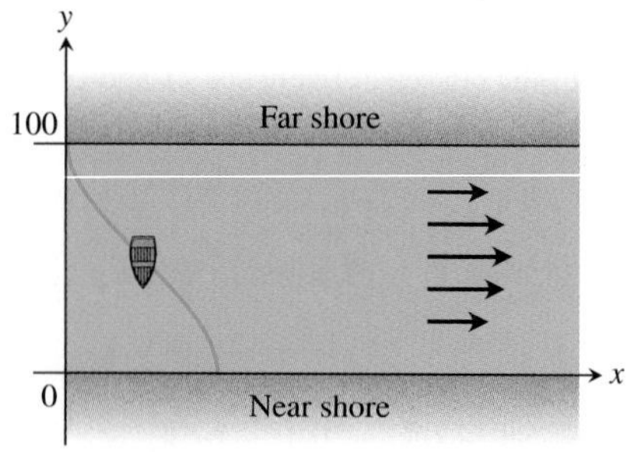

2. A straight river is 20 m wide. The velocity of the river at (x, y) is

$$\mathbf{v} = -\frac{3x(20 - x)}{100}\mathbf{j} \text{ m/min}, \quad 0 \le x \le 20.$$

A boat leaves the shore at $(0, 0)$ and travels through the water with a constant velocity. It arrives at the opposite shore at $(20, 0)$. The speed of the boat is always $\sqrt{20}$ m/min.

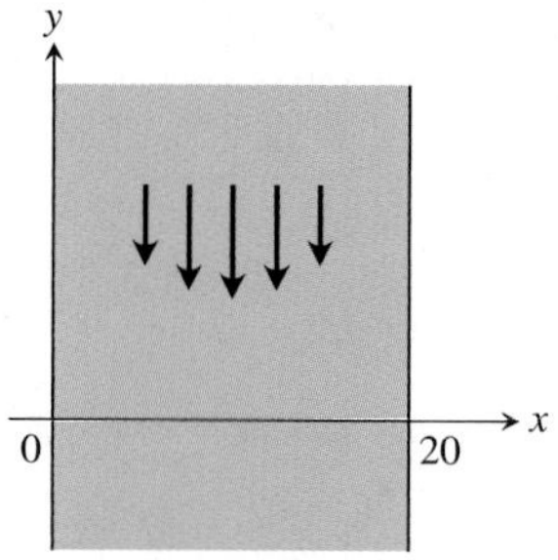

a. Find the velocity of the boat.

b. Find the location of the boat at time t.

c. Sketch the path of the boat.

3. A frictionless particle P, starting from rest at time $t = 0$ at the point $(a, 0, 0)$, slides down the helix

$$\mathbf{r}(\theta) = (a\cos\theta)\mathbf{i} + (a\sin\theta)\mathbf{j} + b\theta\mathbf{k} \quad (a, b > 0)$$

under the influence of gravity, as in the accompanying figure. The θ in this equation is the cylindrical coordinate θ and the helix is the curve $r = a, z = b\theta, \theta \geq 0$, in cylindrical coordinates. We assume θ to be a differentiable function of t for the motion. The law of conservation of energy tells us that the particle's speed after it has fallen straight down a distance z is $\sqrt{2gz}$, where g is the constant acceleration of gravity.

   a. Find the angular velocity $d\theta/dt$ when $\theta = 2\pi$.

   b. Express the particle's θ- and z-coordinates as functions of t.

   c. Express the tangential and normal components of the velocity $d\mathbf{r}/dt$ and acceleration $d^2\mathbf{r}/dt^2$ as functions of t. Does the acceleration have any nonzero component in the direction of the binormal vector $\mathbf{B}$?

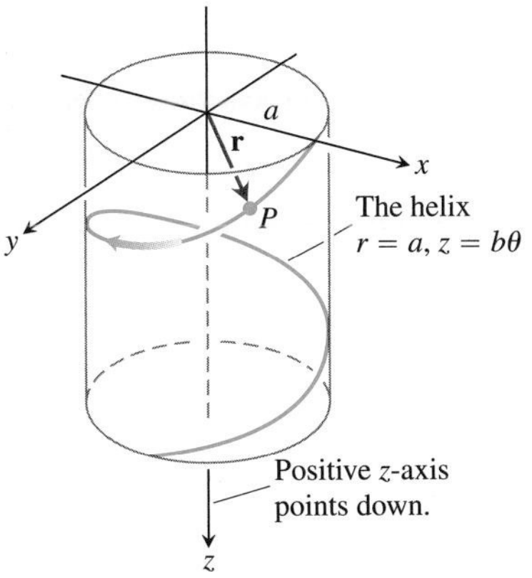

4. Suppose the curve in Exercise 3 is replaced by the conical helix $r = a\theta, z = b\theta$ shown in the accompanying figure.

   a. Express the angular velocity $d\theta/dt$ as a function of θ.

   b. Express the distance the particle travels along the helix as a function of θ.

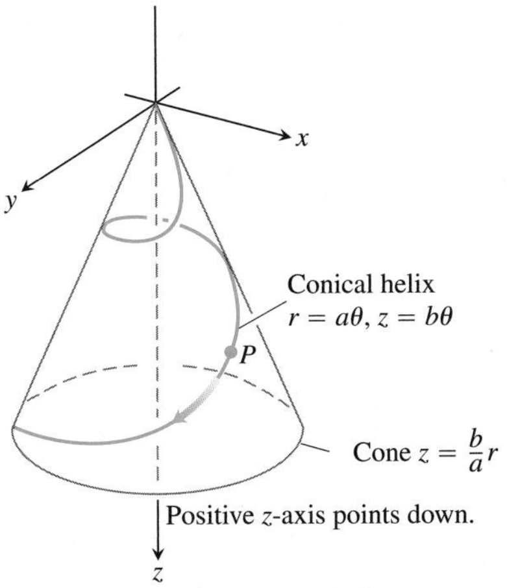

Polar Coordinate Systems and Motion in Space

5. Deduce from the orbit equation

$$r = \frac{(1 + e)r_0}{1 + e\cos\theta}$$

that a planet is closest to its sun when $\theta = 0$ and show that $r = r_0$ at that time.

T 6. **A Kepler equation** The problem of locating a planet in its orbit at a given time and date eventually leads to solving "Kepler" equations of the form

$$f(x) = x - 1 - \frac{1}{2}\sin x = 0.$$

   a. Show that this particular equation has a solution between $x = 0$ and $x = 2$.

   b. With your computer or calculator in radian mode, use Newton's method to find the solution to as many places as you can.

7. In Section 13.6, we found the velocity of a particle moving in the plane to be

$$\mathbf{v} = \dot{x}\,\mathbf{i} + \dot{y}\,\mathbf{j} = \dot{r}\,\mathbf{u}_r + r\dot{\theta}\,\mathbf{u}_\theta.$$

   a. Express $\dot{x}$ and $\dot{y}$ in terms of $\dot{r}$ and $r\dot{\theta}$ by evaluating the dot products $\mathbf{v}\cdot\mathbf{i}$ and $\mathbf{v}\cdot\mathbf{j}$.

   b. Express $\dot{r}$ and $r\dot{\theta}$ in terms of $\dot{x}$ and $\dot{y}$ by evaluating the dot products $\mathbf{v}\cdot\mathbf{u}_r$ and $\mathbf{v}\cdot\mathbf{u}_\theta$.

8. Express the curvature of a twice-differentiable curve $r = f(\theta)$ in the polar coordinate plane in terms of f and its derivatives.

9. A slender rod through the origin of the polar coordinate plane rotates (in the plane) about the origin at the rate of 3 rad/min. A beetle starting from the point (2, 0) crawls along the rod toward the origin at the rate of 1 in./min.

   a. Find the beetle's acceleration and velocity in polar form when it is halfway to (1 in. from) the origin.

   T **b.** To the nearest tenth of an inch, what will be the length of the path the beetle has traveled by the time it reaches the origin?

10. **Conservation of angular momentum** Let $\mathbf{r}(t)$ denote the position in space of a moving object at time t. Suppose the force acting on the object at time t is

$$\mathbf{F}(t) = -\frac{c}{|\mathbf{r}(t)|^3}\mathbf{r}(t),$$

where c is a constant. In physics the **angular momentum** of an object at time t is defined to be $\mathbf{L}(t) = \mathbf{r}(t) \times m\mathbf{v}(t)$, where m is the mass of the object and $\mathbf{v}(t)$ is the velocity. Prove that angular momentum is a conserved quantity; i.e., prove that $\mathbf{L}(t)$ is a constant vector, independent of time. Remember Newton's law $\mathbf{F} = m\mathbf{a}$. (This is a calculus problem, not a physics problem.)

Cylindrical Coordinate Systems

11. Unit vectors for position and motion in cylindrical coordinates When the position of a particle moving in space is given in cylindrical coordinates, the unit vectors we use to describe its position and motion are

$$\mathbf{u}_r = (\cos\theta)\mathbf{i} + (\sin\theta)\mathbf{j}, \qquad \mathbf{u}_\theta = -(\sin\theta)\mathbf{i} + (\cos\theta)\mathbf{j},$$

and $\mathbf{k}$ (see accompanying figure). The particle's position vector is then $\mathbf{r} = r\,\mathbf{u}_r + z\,\mathbf{k}$, where r is the positive polar distance coordinate of the particle's position.

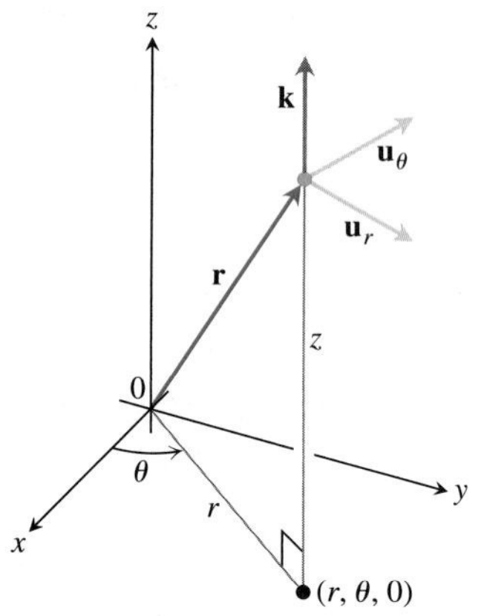

a. Show that $\mathbf{u}_r$, $\mathbf{u}_\theta$, and $\mathbf{k}$, in this order, form a right-handed frame of unit vectors.

b. Show that

$$\frac{d\mathbf{u}_r}{d\theta} = \mathbf{u}_\theta \quad \text{and} \quad \frac{d\mathbf{u}_\theta}{d\theta} = -\mathbf{u}_r.$$

c. Assuming that the necessary derivatives with respect to t exist, express $\mathbf{v} = \dot{\mathbf{r}}$ and $\mathbf{a} = \ddot{\mathbf{r}}$ in terms of $\mathbf{u}_r$, $\mathbf{u}_\theta$, $\mathbf{k}$, $\dot{r}$, and $\dot{\theta}$. (The dots indicate derivatives with respect to t: $\dot{\mathbf{r}}$ means $d\mathbf{r}/dt$, $\ddot{\mathbf{r}}$ means $d^2\mathbf{r}/dt^2$, and so on.) Section 13.6 derives these formulas and shows how the vectors mentioned here are used in describing planetary motion.

12. Arc length in cylindrical coordinates

a. Show that when you express $ds^2 = dx^2 + dy^2 + dz^2$ in terms of cylindrical coordinates, you get $ds^2 = dr^2 + r^2\,d\theta^2 + dz^2$.

b. Interpret this result geometrically in terms of the edges and a diagonal of a box. Sketch the box.

c. Use the result in part (a) to find the length of the curve $r = e^\theta, z = e^\theta, 0 \le \theta \le \theta \ln 8$.

Chapter 13 Technology Application Projects

Mathematica/Maple Module
Radar Tracking of a Moving Object

Visualize position, velocity, and acceleration vectors to analyze motion.

Mathematica/Maple Module
Parametric and Polar Equations with a Figure Skater

Visualize position, velocity, and acceleration vectors to analyze motion.

Mathematica/Maple Module
Moving in Three Dimensions

Compute distance traveled, speed, curvature, and torsion for motion along a space curve. Visualize and compute the tangential, normal, and binormal vectors associated with motion along a space curve.

Chapter 14

Partial Derivatives

OVERVIEW In studying a real-world phenomenon, a quantity being investigated usually depends on two or more independent variables. So we need to extend the basic ideas of the calculus of functions of a single variable to functions of several variables. Although the calculus rules remain essentially the same, the calculus is even richer. The derivatives of functions of several variables are more varied and more interesting because of the different ways in which the variables can interact. Their integrals lead to a greater variety of applications. The studies of probability, statistics, fluid dynamics, and electricity, to mention only a few, all lead in natural ways to functions of more than one variable.

14.1 Functions of Several Variables

Many functions depend on more than one independent variable. The function $V = \pi r^2 h$ calculates the volume of a right circular cylinder from its radius and height. The function $f(x, y) = x^2 + y^2$ calculates the height of the paraboloid $z = x^2 + y^2$ above the point $P(x, y)$ from the two coordinates of P. The temperature T of a point on Earth's surface depends on its latitude x and longitude y, expressed by writing $T = f(x, y)$. In this section, we define functions of more than one independent variable and discuss ways to graph them.

Real-valued functions of several independent real variables are defined much the way you would imagine from the single-variable case. The domains are sets of ordered pairs (triples, quadruples, n-tuples) of real numbers, and the ranges are sets of real numbers of the kind we have worked with all along.

DEFINITIONS Function of n Independent Variables

Suppose D is a set of n-tuples of real numbers $(x_1, x_2, \ldots, x_n)$. A **real-valued function** f on D is a rule that assigns a unique (single) real number

$$w = f(x_1, x_2, \ldots, x_n)$$

to each element in D. The set D is the function's **domain**. The set of w-values taken on by f is the function's **range**. The symbol w is the **dependent variable** of f, and f is said to be a function of the n **independent variables** x_1 to x_n. We also call the x_j's the function's **input variables** and call w the function's **output variable**.

If f is a function of two independent variables, we usually call the independent variables x and y and picture the domain of f as a region in the xy-plane. If f is a function of three independent variables, we call the variables x, y, and z and picture the domain as a region in space.

In applications, we tend to use letters that remind us of what the variables stand for. To say that the volume of a right circular cylinder is a function of its radius and height, we might write $V = f(r, h)$. To be more specific, we might replace the notation $f(r, h)$ by the formula that calculates the value of V from the values of r and h, and write $V = \pi r^2 h$. In either case, r and h would be the independent variables and V the dependent variable of the function.

As usual, we evaluate functions defined by formulas by substituting the values of the independent variables in the formula and calculating the corresponding value of the dependent variable.

EXAMPLE 1 Evaluating a Function

The value of $f(x, y, z) = \sqrt{x^2 + y^2 + z^2}$ at the point $(3, 0, 4)$ is

$$f(3, 0, 4) = \sqrt{(3)^2 + (0)^2 + (4)^2} = \sqrt{25} = 5.$$

From Section 12.1, we recognize f as the distance function from the origin to the point (x, y, z) in Cartesian space coordinates. ■

Domains and Ranges

In defining a function of more than one variable, we follow the usual practice of excluding inputs that lead to complex numbers or division by zero. If $f(x, y) = \sqrt{y - x^2}$, y cannot be less than x^2. If $f(x, y) = 1/(xy)$, xy cannot be zero. The domain of a function is assumed to be the largest set for which the defining rule generates real numbers, unless the domain is otherwise specified explicitly. The range consists of the set of output values for the dependent variable.

EXAMPLE 2(a) Functions of Two Variables

Function	Domain	Range
$w = \sqrt{y - x^2}$	$y \geq x^2$	$[0, \infty)$
$w = \dfrac{1}{xy}$	$xy \neq 0$	$(-\infty, 0) \cup (0, \infty)$
$w = \sin xy$	Entire plane	$[-1, 1]$

(b) Functions of Three Variables

Function	Domain	Range
$w = \sqrt{x^2 + y^2 + z^2}$	Entire space	$[0, \infty)$
$w = \dfrac{1}{x^2 + y^2 + z^2}$	$(x, y, z) \neq (0, 0, 0)$	$(0, \infty)$
$w = xy \ln z$	Half-space $z > 0$	$(-\infty, \infty)$

■

Functions of Two Variables

Regions in the plane can have interior points and boundary points just like intervals on the real line. Closed intervals $[a, b]$ include their boundary points, open intervals (a, b) don't include their boundary points, and intervals such as $[a, b)$ are neither open nor closed.

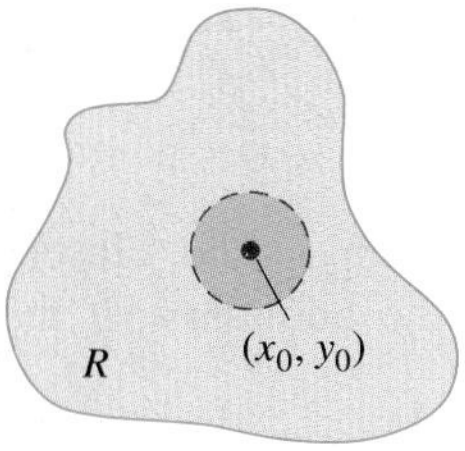

(a) Interior point

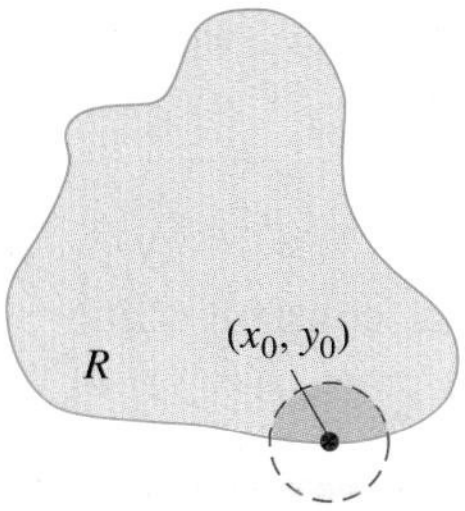

(b) Boundary point

FIGURE 14.1 Interior points and boundary points of a plane region R. An interior point is necessarily a point of R. A boundary point of R need not belong to R.

> **DEFINITIONS Interior and Boundary Points, Open, Closed**
> A point (x_0, y_0) in a region (set) R in the xy-plane is an **interior point** of R if it is the center of a disk of positive radius that lies entirely in R (Figure 14.1). A point (x_0, y_0) is a **boundary point** of R if every disk centered at (x_0, y_0) contains points that lie outside of R as well as points that lie in R. (The boundary point itself need not belong to R.)
>
> The interior points of a region, as a set, make up the **interior** of the region. The region's boundary points make up its **boundary**. A region is **open** if it consists entirely of interior points. A region is **closed** if it contains all its boundary points (Figure 14.2).

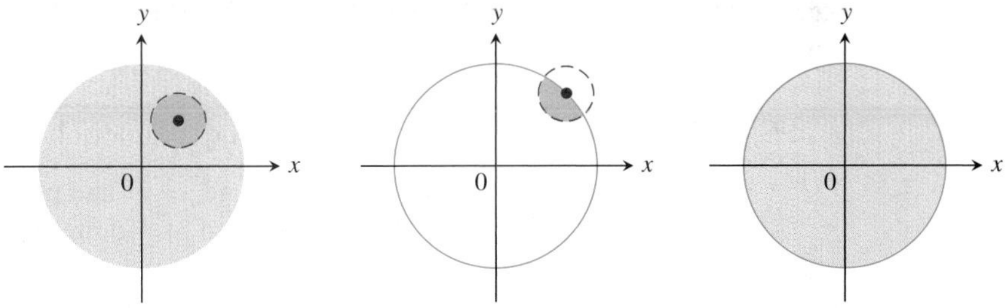

$\{(x, y) \mid x^2 + y^2 < 1\}$ Open unit disk. Every point an interior point.

$\{(x, y) \mid x^2 + y^2 = 1\}$ Boundary of unit disk. (The unit circle.)

$\{(x, y) \mid x^2 + y^2 \le 1\}$ Closed unit disk. Contains all boundary points.

FIGURE 14.2 Interior points and boundary points of the unit disk in the plane.

As with intervals of real numbers, some regions in the plane are neither open nor closed. If you start with the open disk in Figure 14.2 and add to it some of but not all its boundary points, the resulting set is neither open nor closed. The boundary points that *are* there keep the set from being open. The absence of the remaining boundary points keeps the set from being closed.

> **DEFINITIONS Bounded and Unbounded Regions in the Plane**
> A region in the plane is **bounded** if it lies inside a disk of fixed radius. A region is **unbounded** if it is not bounded.

Examples of *bounded* sets in the plane include line segments, triangles, interiors of triangles, rectangles, circles, and disks. Examples of *unbounded* sets in the plane include lines, coordinate axes, the graphs of functions defined on infinite intervals, quadrants, half-planes, and the plane itself.

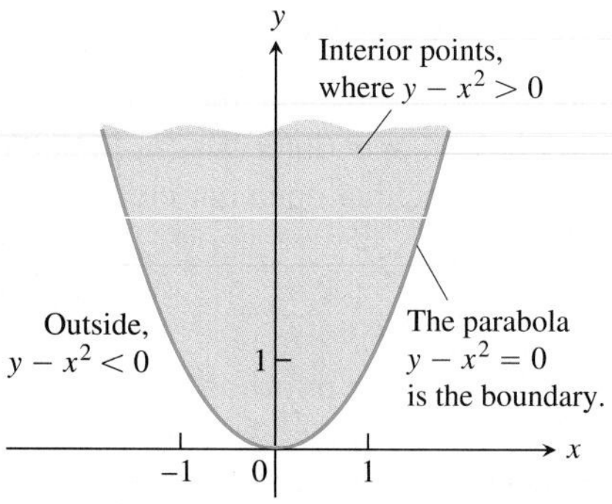

FIGURE 14.3 The domain of $f(x, y) = \sqrt{y - x^2}$ consists of the shaded region and its bounding parabola $y = x^2$ (Example 3).

EXAMPLE 3 Describing the Domain of a Function of Two Variables

Describe the domain of the function $f(x, y) = \sqrt{y - x^2}$.

Solution Since f is defined only where $y - x^2 \geq 0$, the domain is the closed, unbounded region shown in Figure 14.3. The parabola $y = x^2$ is the boundary of the domain. The points above the parabola make up the domain's interior. ■

Graphs, Level Curves, and Contours of Functions of Two Variables

There are two standard ways to picture the values of a function $f(x, y)$. One is to draw and label curves in the domain on which f has a constant value. The other is to sketch the surface $z = f(x, y)$ in space.

> **DEFINITIONS** Level Curve, Graph, Surface
>
> The set of points in the plane where a function $f(x, y)$ has a constant value $f(x, y) = c$ is called a **level curve** of f. The set of all points $(x, y, f(x, y))$ in space, for (x, y) in the domain of f, is called the **graph** of f. The graph of f is also called the **surface** $z = f(x, y)$.

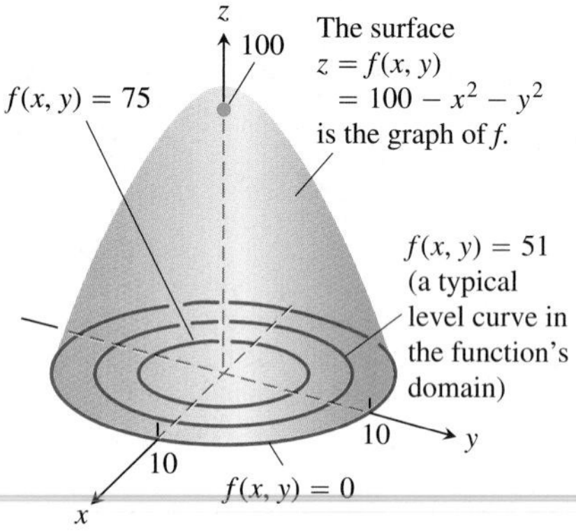

FIGURE 14.4 The graph and selected level curves of the function $f(x, y) = 100 - x^2 - y^2$ (Example 4).

EXAMPLE 4 Graphing a Function of Two Variables

Graph $f(x, y) = 100 - x^2 - y^2$ and plot the level curves $f(x, y) = 0$, $f(x, y) = 51$, and $f(x, y) = 75$ in the domain of f in the plane.

Solution The domain of f is the entire xy-plane, and the range of f is the set of real numbers less than or equal to 100. The graph is the paraboloid $z = 100 - x^2 - y^2$, a portion of which is shown in Figure 14.4.

The level curve $f(x, y) = 0$ is the set of points in the xy-plane at which

$$f(x, y) = 100 - x^2 - y^2 = 0, \qquad \text{or} \qquad x^2 + y^2 = 100,$$

which is the circle of radius 10 centered at the origin. Similarly, the level curves $f(x, y) = 51$ and $f(x, y) = 75$ (Figure 14.4) are the circles

$$f(x, y) = 100 - x^2 - y^2 = 51, \qquad \text{or} \qquad x^2 + y^2 = 49$$

$$f(x, y) = 100 - x^2 - y^2 = 75, \qquad \text{or} \qquad x^2 + y^2 = 25.$$

The level curve $f(x, y) = 100$ consists of the origin alone. (It is still a level curve.) ■

The curve in space in which the plane $z = c$ cuts a surface $z = f(x, y)$ is made up of the points that represent the function value $f(x, y) = c$. It is called the **contour curve** $f(x, y) = c$ to distinguish it from the level curve $f(x, y) = c$ in the domain of f. Figure 14.5 shows the contour curve $f(x, y) = 75$ on the surface $z = 100 - x^2 - y^2$ defined by the function $f(x, y) = 100 - x^2 - y^2$. The contour curve lies directly above the circle $x^2 + y^2 = 25$, which is the level curve $f(x, y) = 75$ in the function's domain.

Not everyone makes this distinction, however, and you may wish to call both kinds of curves by a single name and rely on context to convey which one you have in mind. On most maps, for example, the curves that represent constant elevation (height above sea level) are called contours, not level curves (Figure 14.6).

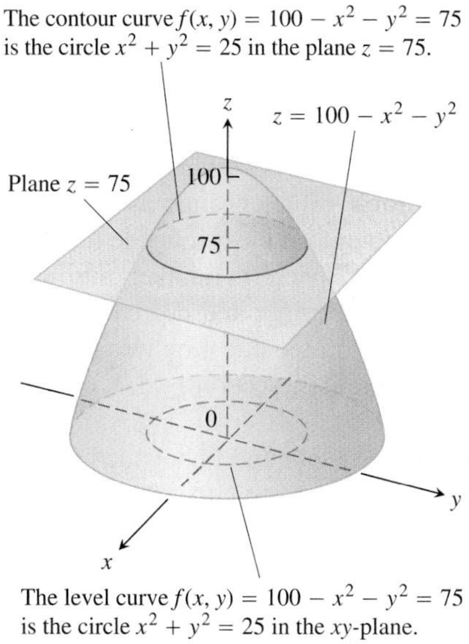

FIGURE 14.5 A plane $z = c$ parallel to the xy-plane intersecting a surface $z = f(x, y)$ produces a contour curve.

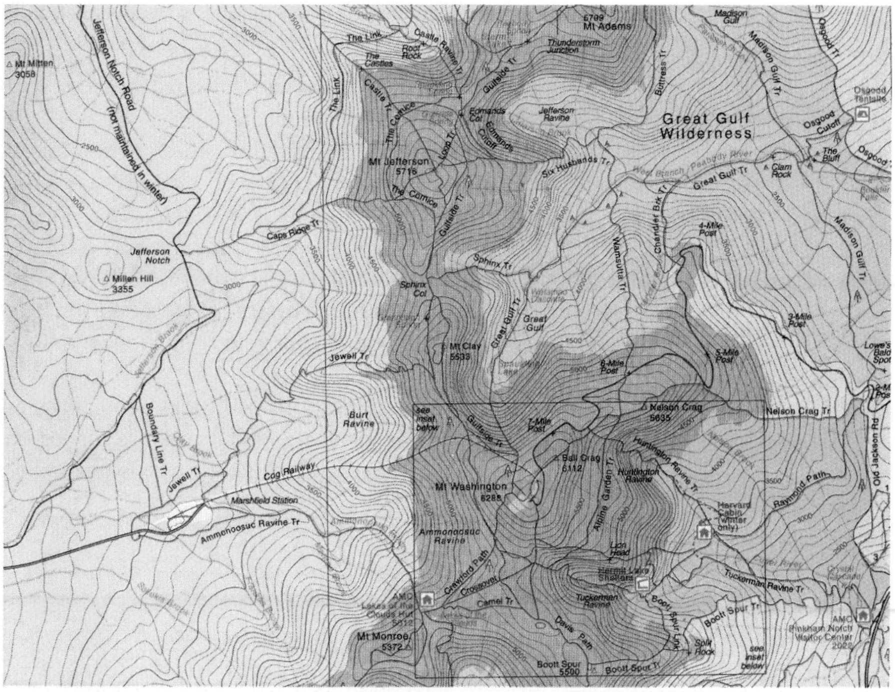

FIGURE 14.6 Contours on Mt. Washington in New Hampshire. (Reproduced by permission from the Appalachian Mountain Club.)

Functions of Three Variables

In the plane, the points where a function of two independent variables has a constant value $f(x, y) = c$ make a curve in the function's domain. In space, the points where a function of three independent variables has a constant value $f(x, y, z) = c$ make a surface in the function's domain.

> **DEFINITION** Level Surface
>
> The set of points (x, y, z) in space where a function of three independent variables has a constant value $f(x, y, z) = c$ is called a **level surface** of f.

Since the graphs of functions of three variables consist of points $(x, y, z, f(x, y, z))$ lying in a four-dimensional space, we cannot sketch them effectively in our three-dimensional frame of reference. We can see how the function behaves, however, by looking at its three-dimensional level surfaces.

EXAMPLE 5 Describing Level Surfaces of a Function of Three Variables

Describe the level surfaces of the function

$$f(x, y, z) = \sqrt{x^2 + y^2 + z^2}.$$

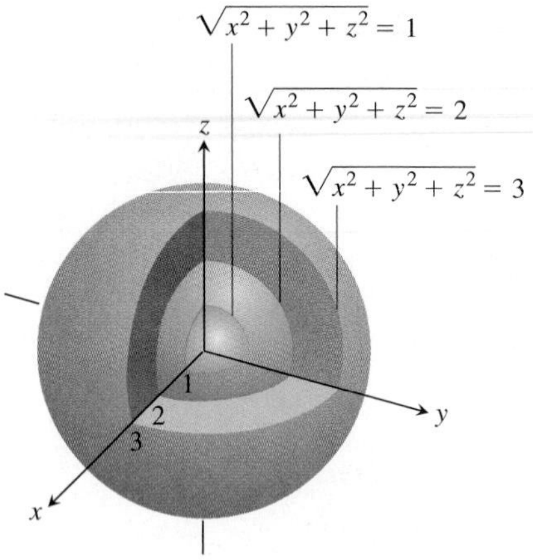

FIGURE 14.7 The level surfaces of $f(x, y, z) = \sqrt{x^2 + y^2 + z^2}$ are concentric spheres (Example 5).

Solution The value of f is the distance from the origin to the point (x, y, z). Each level surface $\sqrt{x^2 + y^2 + z^2} = c, c > 0$, is a sphere of radius c centered at the origin. Figure 14.7 shows a cutaway view of three of these spheres. The level surface $\sqrt{x^2 + y^2 + z^2} = 0$ consists of the origin alone.

We are not graphing the function here; we are looking at level surfaces in the function's domain. The level surfaces show how the function's values change as we move through its domain. If we remain on a sphere of radius c centered at the origin, the function maintains a constant value, namely c. If we move from one sphere to another, the function's value changes. It increases if we move away from the origin and decreases if we move toward the origin. The way the values change depends on the direction we take. The dependence of change on direction is important. We return to it in Section 14.5. ■

The definitions of interior, boundary, open, closed, bounded, and unbounded for regions in space are similar to those for regions in the plane. To accommodate the extra dimension, we use solid balls of positive radius instead of disks.

DEFINITIONS **Interior and Boundary Points for Space Regions**

A point (x_0, y_0, z_0) in a region R in space is an **interior point** of R if it is the center of a solid ball that lies entirely in R (Figure 14.8a). A point (x_0, y_0, z_0) is a **boundary point** of R if every sphere centered at (x_0, y_0, z_0) encloses points that lie outside of R as well as points that lie inside R (Figure 14.8b). The **interior** of R is the set of interior points of R. The **boundary** of R is the set of boundary points of R.

A region is **open** if it consists entirely of interior points. A region is **closed** if it contains its entire boundary.

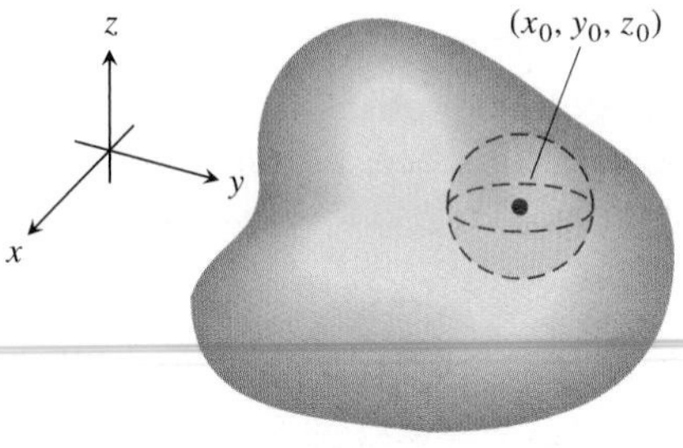

(a) Interior point

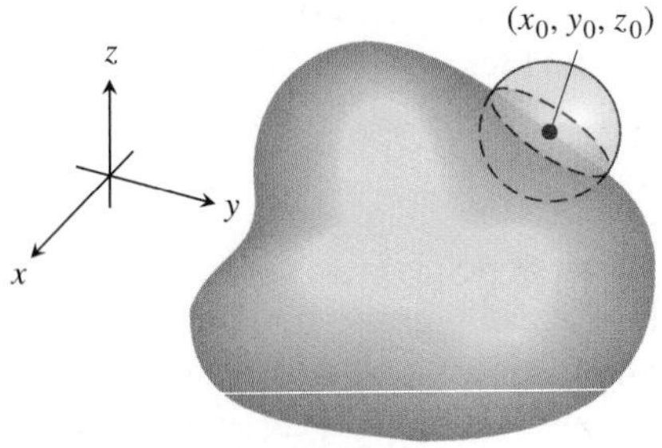

(b) Boundary point

FIGURE 14.8 Interior points and boundary points of a region in space.

Examples of *open* sets in space include the interior of a sphere, the open half-space $z > 0$, the first octant (where x, y, and z are all positive), and space itself.

Examples of *closed* sets in space include lines, planes, the closed half-space $z \geq 0$, the first octant together with its bounding planes, and space itself (since it has no boundary points).

A solid sphere with part of its boundary removed or a solid cube with a missing face, edge, or corner point would be *neither open nor closed.*

Functions of more than three independent variables are also important. For example, the temperature on a surface in space may depend not only on the location of the point $P(x, y, z)$ on the surface, but also on time t when it is visited, so we would write $T = f(x, y, z, t)$.

Computer Graphing

Three-dimensional graphing programs for computers and calculators make it possible to graph functions of two variables with only a few keystrokes. We can often get information more quickly from a graph than from a formula.

EXAMPLE 6 Modeling Temperature Beneath Earth's Surface

The temperature beneath the Earth's surface is a function of the depth x beneath the surface and the time t of the year. If we measure x in feet and t as the number of days elapsed from the expected date of the yearly highest surface temperature, we can model the variation in temperature with the function

$$w = \cos(1.7 \times 10^{-2}t - 0.2x)e^{-0.2x}.$$

(The temperature at 0 ft is scaled to vary from $+1$ to -1, so that the variation at x feet can be interpreted as a fraction of the variation at the surface.)

Figure 14.9 shows a computer-generated graph of the function. At a depth of 15 ft, the variation (change in vertical amplitude in the figure) is about 5% of the surface variation. At 30 ft, there is almost no variation during the year.

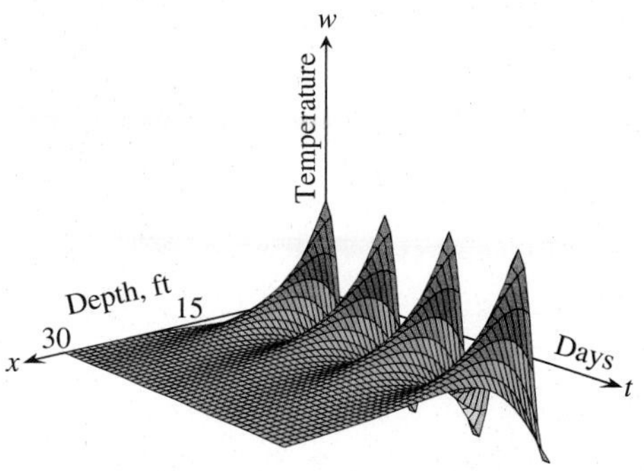

FIGURE 14.9 This computer-generated graph of

$$w = \cos(1.7 \times 10^{-2}t - 0.2x)e^{-0.2x}$$

shows the seasonal variation of the temperature belowground as a fraction of surface temperature. At $x = 15$ ft, the variation is only 5% of the variation at the surface. At $x = 30$ ft, the variation is less than 0.25% of the surface variation (Example 6). (Adapted from art provided by Norton Starr.)

The graph also shows that the temperature 15 ft below the surface is about half a year out of phase with the surface temperature. When the temperature is lowest on the surface (late January, say), it is at its highest 15 ft below. Fifteen feet below the ground, the seasons are reversed. ■

Figure 14.10 shows computer-generated graphs of a number of functions of two variables together with their level curves.

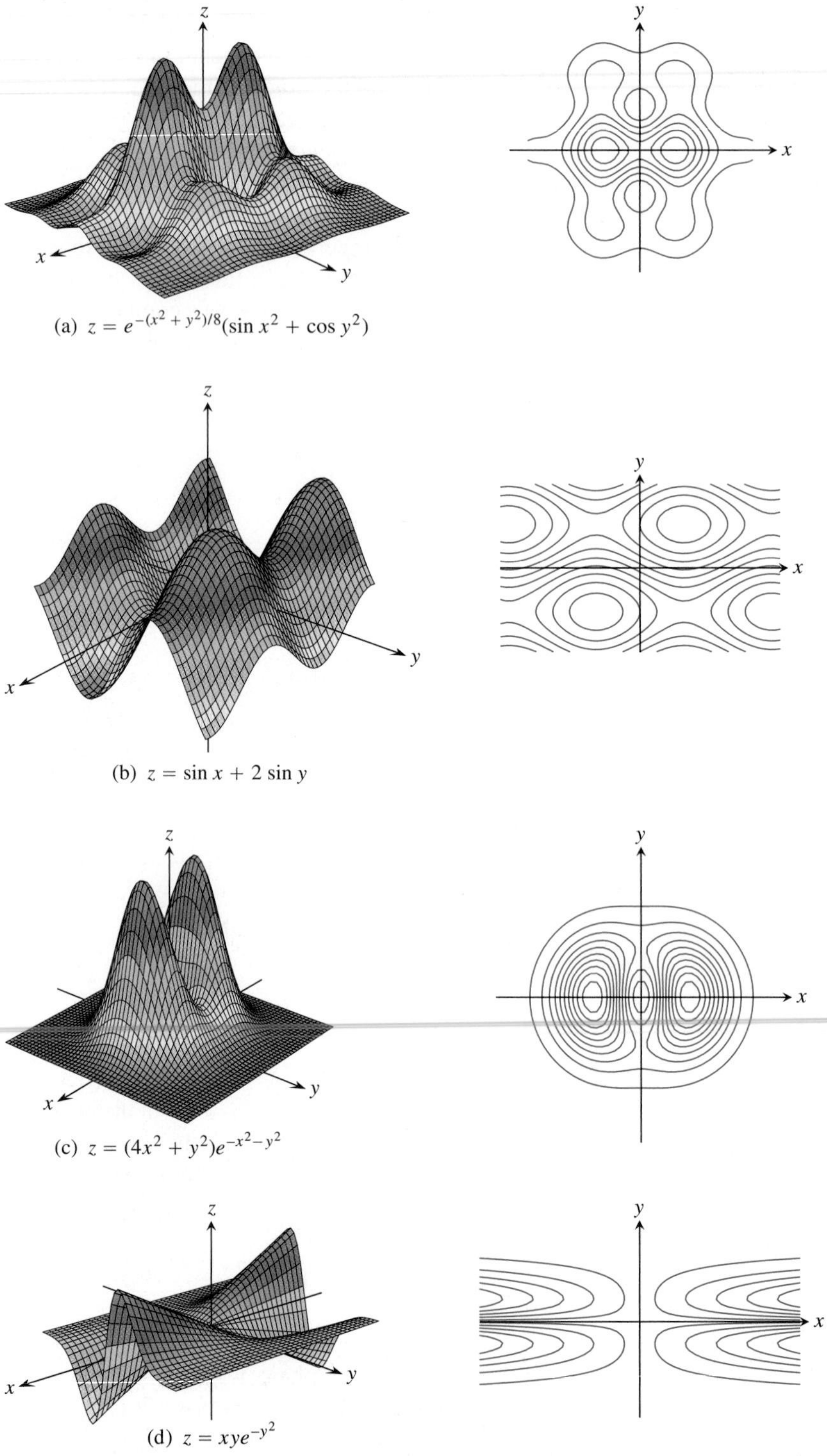

FIGURE 14.10 Computer-generated graphs and level curves of typical functions of two variables.

EXERCISES 14.1

Domain, Range, and Level Curves

In Exercises 1–12, **(a)** find the function's domain, **(b)** find the function's range, **(c)** describe the function's level curves, **(d)** find the boundary of the function's domain, **(e)** determine if the domain is an open region, a closed region, or neither, and **(f)** decide if the domain is bounded or unbounded.

1. $f(x, y) = y - x$

2. $f(x, y) = \sqrt{y - x}$

3. $f(x, y) = 4x^2 + 9y^2$

4. $f(x, y) = x^2 - y^2$

5. $f(x, y) = xy$

6. $f(x, y) = y/x^2$

7. $f(x, y) = \dfrac{1}{\sqrt{16 - x^2 - y^2}}$

8. $f(x, y) = \sqrt{9 - x^2 - y^2}$

9. $f(x, y) = \ln(x^2 + y^2)$

10. $f(x, y) = e^{-(x^2+y^2)}$

11. $f(x, y) = \sin^{-1}(y - x)$

12. $f(x, y) = \tan^{-1}\left(\dfrac{y}{x}\right)$

Identifying Surfaces and Level Curves

Exercises 13–18 show level curves for the functions graphed in (a)–(f). Match each set of curves with the appropriate function.

13.

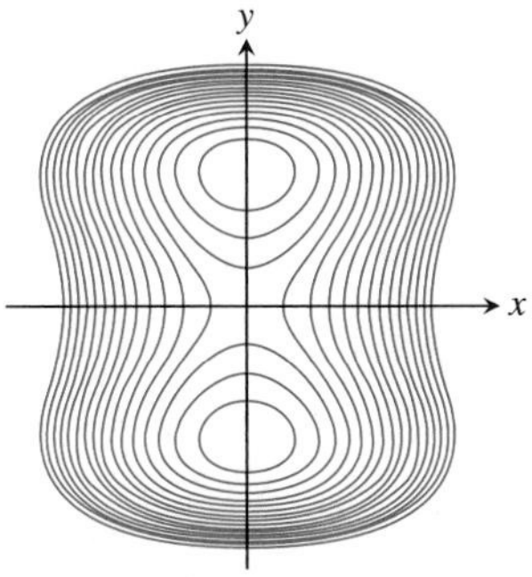

14.

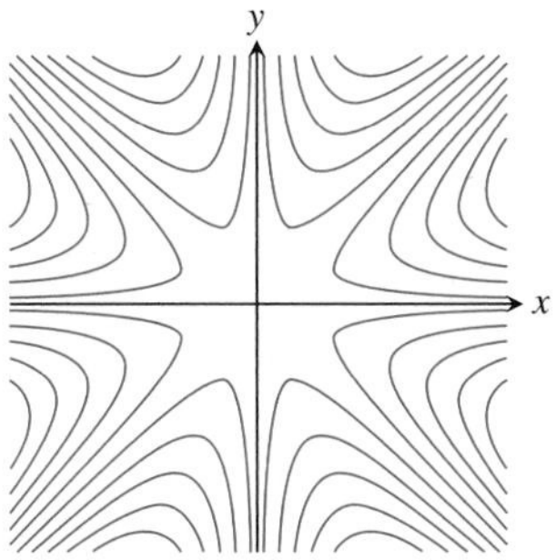

15.

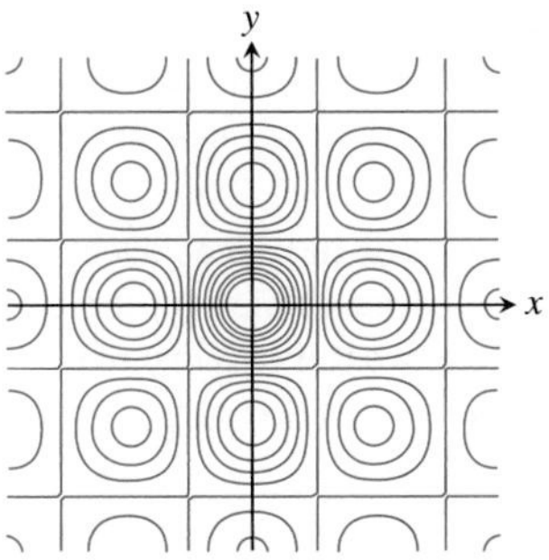

16.

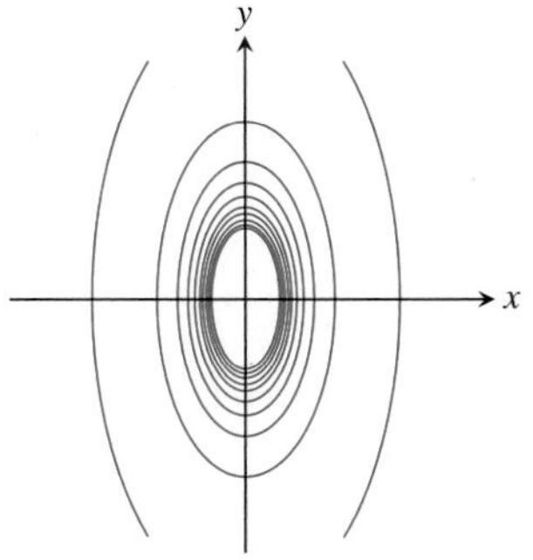

17.

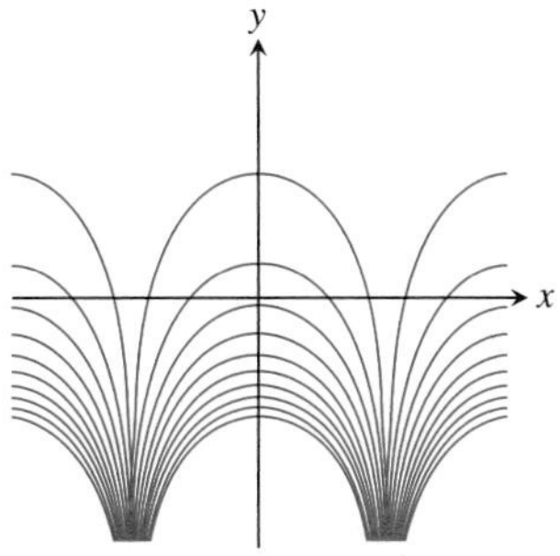

18.

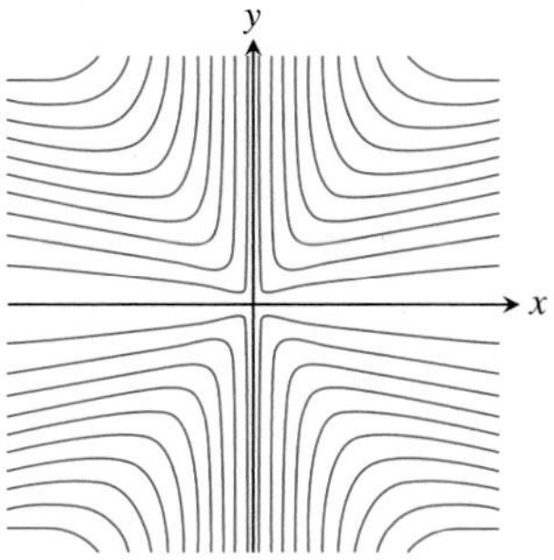

a.

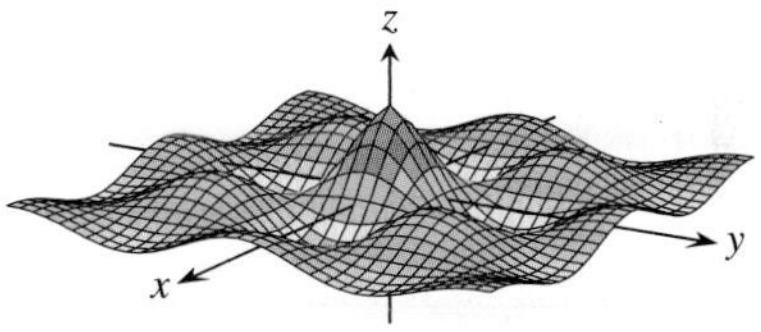

$z = (\cos x)(\cos y)\, e^{-\sqrt{x^2 + y^2}/4}$

b.

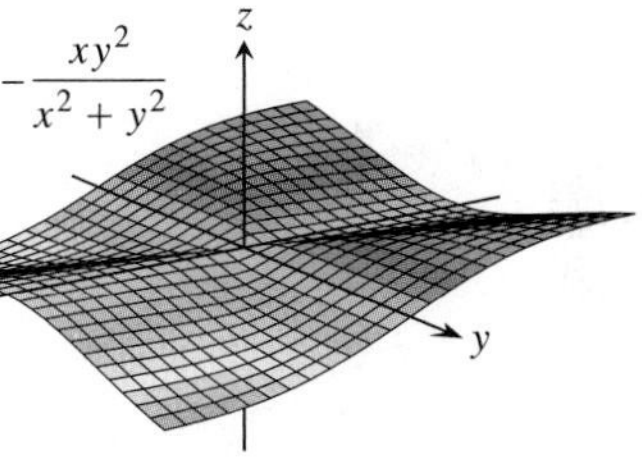

$z = -\dfrac{xy^2}{x^2 + y^2}$

c.

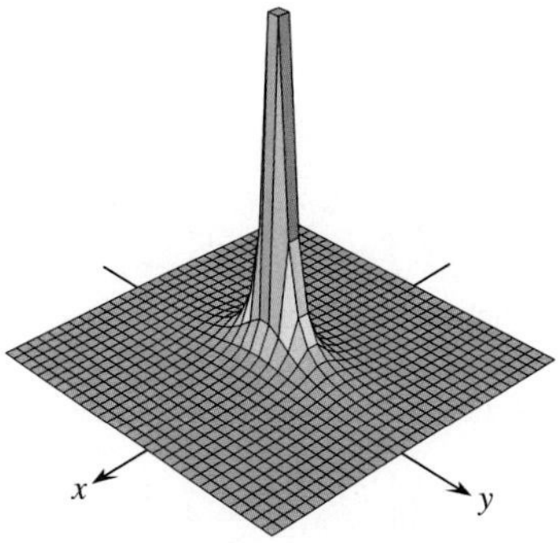

$z = \dfrac{1}{4x^2 + y^2}$

d.

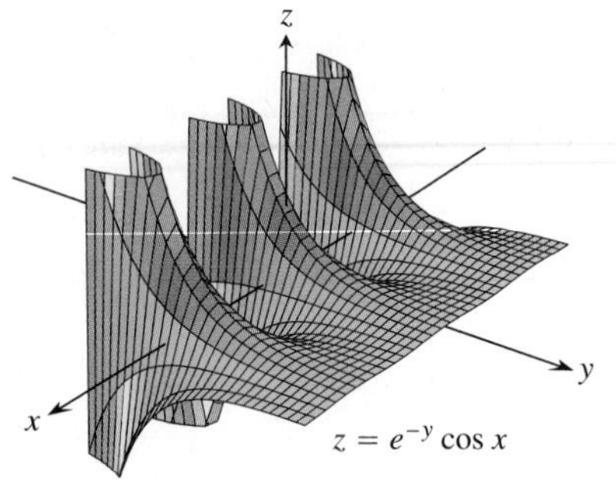

e.

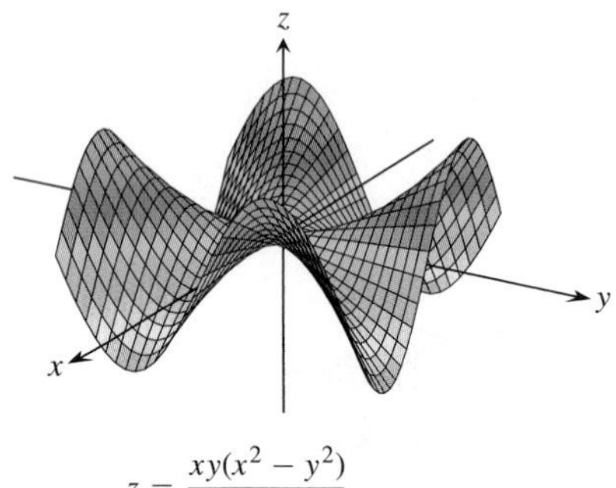

$$z = \frac{xy(x^2 - y^2)}{x^2 + y^2}$$

f.

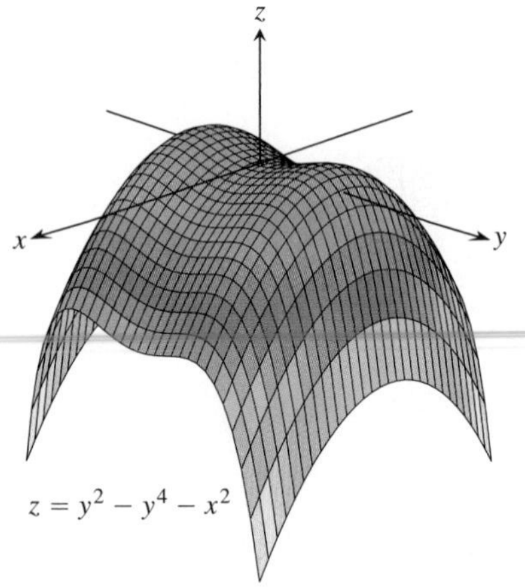

$z = y^2 - y^4 - x^2$

Identifying Functions of Two Variables

Display the values of the functions in Exercises 19–28 in two ways: **(a)** by sketching the surface $z = f(x, y)$ and **(b)** by drawing an assortment of level curves in the function's domain. Label each level curve with its function value.

19. $f(x, y) = y^2$ **20.** $f(x, y) = 4 - y^2$

21. $f(x, y) = x^2 + y^2$ **22.** $f(x, y) = \sqrt{x^2 + y^2}$

23. $f(x, y) = -(x^2 + y^2)$ **24.** $f(x, y) = 4 - x^2 - y^2$

25. $f(x, y) = 4x^2 + y^2$ **26.** $f(x, y) = 4x^2 + y^2 + 1$

27. $f(x, y) = 1 - |y|$ **28.** $f(x, y) = 1 - |x| - |y|$

Finding a Level Curve

In Exercises 29–32, find an equation for the level curve of the function $f(x, y)$ that passes through the given point.

29. $f(x, y) = 16 - x^2 - y^2, \quad \left(2\sqrt{2}, \sqrt{2}\right)$

30. $f(x, y) = \sqrt{x^2 - 1}, \quad (1, 0)$

31. $f(x, y) = \displaystyle\int_x^y \frac{dt}{1 + t^2}, \quad \left(-\sqrt{2}, \sqrt{2}\right)$

32. $f(x, y) = \displaystyle\sum_{n=0}^{\infty} \left(\frac{x}{y}\right)^n, \quad (1, 2)$

Sketching Level Surfaces

In Exercises 33–40, sketch a typical level surface for the function.

33. $f(x, y, z) = x^2 + y^2 + z^2$ **34.** $f(x, y, z) = \ln(x^2 + y^2 + z^2)$

35. $f(x, y, z) = x + z$ **36.** $f(x, y, z) = z$

37. $f(x, y, z) = x^2 + y^2$ **38.** $f(x, y, z) = y^2 + z^2$

39. $f(x, y, z) = z - x^2 - y^2$

40. $f(x, y, z) = (x^2/25) + (y^2/16) + (z^2/9)$

Finding a Level Surface

In Exercises 41–44, find an equation for the level surface of the function through the given point.

41. $f(x, y, z) = \sqrt{x - y} - \ln z, \quad (3, -1, 1)$

42. $f(x, y, z) = \ln(x^2 + y + z^2), \quad (-1, 2, 1)$

43. $g(x, y, z) = \displaystyle\sum_{n=0}^{\infty} \frac{(x + y)^n}{n!z^n}, \quad (\ln 2, \ln 4, 3)$

44. $g(x, y, z) = \displaystyle\int_x^y \frac{d\theta}{\sqrt{1 - \theta^2}} + \int_{\sqrt{2}}^z \frac{dt}{t\sqrt{t^2 - 1}}, \quad (0, 1/2, 2)$

Theory and Examples

45. The maximum value of a function on a line in space Does the function $f(x, y, z) = xyz$ have a maximum value on the line $x = 20 - t, y = t, z = 20$? If so, what is it? Give reasons for your answer. (*Hint:* Along the line, $w = f(x, y, z)$ is a differentiable function of t.)

46. The minimum value of a function on a line in space Does the function $f(x, y, z) = xy - z$ have a minimum value on the line $x = t - 1, y = t - 2, z = t + 7$? If so, what is it? Give reasons for your answer. (*Hint:* Along the line, $w = f(x, y, z)$ is a differentiable function of t.)

47. The Concorde's sonic booms Sound waves from the *Concorde* bend as the temperature changes above and below the altitude at which the plane flies. The sonic boom carpet is the region on the ground that receives shock waves directly from the plane, not reflected from the atmosphere or diffracted along the ground. The carpet is determined by the grazing rays striking the ground from the point directly under the plane. (See accompanying figure.)

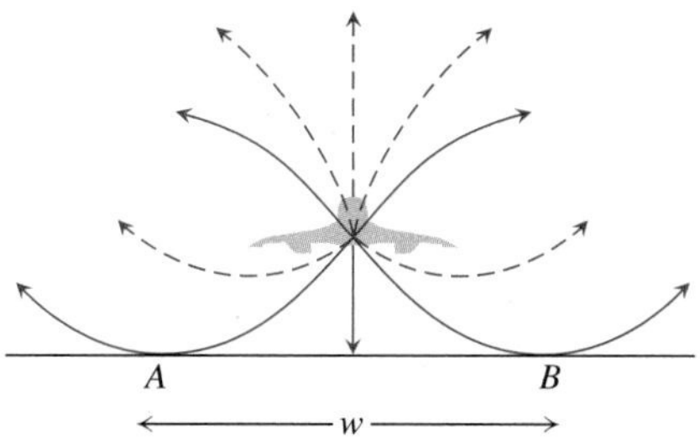

Sonic boom carpet

The width w of the region in which people on the ground hear the *Concorde*'s sonic boom directly, not reflected from a layer in the atmosphere, is a function of

T = air temperature at ground level (in degrees Kelvin)

h = the *Concorde*'s altitude (in kilometers)

d = the vertical temperature gradient (temperature drop in degrees Kelvin per kilometer).

The formula for w is

$$w = 4\left(\frac{Th}{d}\right)^{1/2}.$$

The Washington-bound *Concorde* approached the United States from Europe on a course that took it south of Nantucket Island at an altitude of 16.8 km. If the surface temperature is 290 K and the vertical temperature gradient is 5 K/km, how many kilometers south of Nantucket did the plane have to be flown to keep its sonic boom carpet away from the island? (From "Concorde Sonic Booms as an Atmospheric Probe" by N. K. Balachandra, W. L. Donn, and D. H. Rind, *Science*, Vol. 197 (July 1, 1977), pp. 47–49.)

48. As you know, the graph of a real-valued function of a single real variable is a set in a two-coordinate space. The graph of a real-valued function of two independent real variables is a set in a three-coordinate space. The graph of a real-valued function of three independent real variables is a set in a four-coordinate space. How would you define the graph of a real-valued function $f(x_1, x_2, x_3, x_4)$ of four independent real variables? How would you define the graph of a real-valued function $f(x_1, x_2, x_3, \dots, x_n)$ of n independent real variables?

COMPUTER EXPLORATIONS

Explicit Surfaces

Use a CAS to perform the following steps for each of the functions in Exercises 49–52.

a. Plot the surface over the given rectangle.

b. Plot several level curves in the rectangle.

c. Plot the level curve of f through the given point.

49. $f(x, y) = x \sin\frac{y}{2} + y \sin 2x, \quad 0 \le x \le 5\pi \quad 0 \le y \le 5\pi,$ $P(3\pi, 3\pi)$

50. $f(x, y) = (\sin x)(\cos y)e^{\sqrt{x^2+y^2}/8}, \quad 0 \le x \le 5\pi,$ $0 \le y \le 5\pi, \quad P(4\pi, 4\pi)$

51. $f(x, y) = \sin(x + 2\cos y), \quad -2\pi \le x \le 2\pi,$ $-2\pi \le y \le 2\pi, \quad P(\pi, \pi)$

52. $f(x, y) = e^{(x^{0.1} - y)} \sin(x^2 + y^2), \quad 0 \le x \le 2\pi,$ $-2\pi \le y \le \pi, \quad P(\pi, -\pi)$

Implicit Surfaces

Use a CAS to plot the level surfaces in Exercises 53–56.

53. $4 \ln(x^2 + y^2 + z^2) = 1$

54. $x^2 + z^2 = 1$

55. $x + y^2 - 3z^2 = 1$

56. $\sin\left(\frac{x}{2}\right) - (\cos y)\sqrt{x^2 + z^2} = 2$

Parametrized Surfaces

Just as you describe curves in the plane parametrically with a pair of equations $x = f(t), y = g(t)$ defined on some parameter interval I, you can sometimes describe surfaces in space with a triple of equations $x = f(u, v), y = g(u, v), z = h(u, v)$ defined on some parameter rectangle $a \le u \le b, c \le v \le d$. Many computer algebra systems permit you to plot such surfaces in *parametric mode*. (Parametrized surfaces are discussed in detail in Section 16.6.) Use a CAS to plot the surfaces in Exercises 57–60. Also plot several level curves in the xy-plane.

57. $x = u\cos v, \quad y = u \sin v, \quad z = u, \quad 0 \le u \le 2,$ $0 \le v \le 2\pi$

58. $x = u\cos v, \quad y = u \sin v, \quad z = v, \quad 0 \le u \le 2,$ $0 \le v \le 2\pi$

59. $x = (2 + \cos u)\cos v, \quad y = (2 + \cos u)\sin v, \quad z = \sin u,$ $0 \le u \le 2\pi, \quad 0 \le v \le 2\pi$

60. $x = 2\cos u \cos v, \quad y = 2\cos u \sin v, \quad z = 2 \sin u,$ $0 \le u \le 2\pi, \quad 0 \le v \le \pi$

14.2 Limits and Continuity in Higher Dimensions

This section treats limits and continuity for multivariable functions. The definition of the limit of a function of two or three variables is similar to the definition of the limit of a function of a single variable but with a crucial difference, as we now see.

Limits

If the values of $f(x, y)$ lie arbitrarily close to a fixed real number L for all points (x, y) sufficiently close to a point (x_0, y_0), we say that f approaches the limit L as (x, y) approaches (x_0, y_0). This is similar to the informal definition for the limit of a function of a single variable. Notice, however, that if (x_0, y_0) lies in the interior of f's domain, (x, y) can approach (x_0, y_0) from any direction. The direction of approach can be an issue, as in some of the examples that follow.

DEFINITION Limit of a Function of Two Variables

We say that a function $f(x, y)$ approaches the **limit** L as (x, y) approaches (x_0, y_0), and write

$$\lim_{(x, y)\to(x_0, y_0)} f(x, y) = L$$

if, for every number $\epsilon > 0$, there exists a corresponding number $\delta > 0$ such that for all (x, y) in the domain of f,

$$|f(x, y) - L| < \epsilon \qquad \text{whenever} \qquad 0 < \sqrt{(x - x_0)^2 + (y - y_0)^2} < \delta.$$

The definition of limit says that the distance between $f(x, y)$ and L becomes arbitrarily small whenever the distance from (x, y) to (x_0, y_0) is made sufficiently small (but not 0).

The definition of limit applies to boundary points (x_0, y_0) as well as interior points of the domain of f. The only requirement is that the point (x, y) remain in the domain at all times. It can be shown, as for functions of a single variable, that

$$\lim_{(x, y)\to(x_0, y_0)} x = x_0$$

$$\lim_{(x, y)\to(x_0, y_0)} y = y_0$$

$$\lim_{(x, y)\to(x_0, y_0)} k = k \qquad \text{(any number } k\text{)}.$$

For example, in the first limit statement above, $f(x, y) = x$ and $L = x_0$. Using the definition of limit, suppose that $\epsilon > 0$ is chosen. If we let δ equal this ϵ, we see that

$$0 < \sqrt{(x - x_0)^2 + (y - y_0)^2} < \delta = \epsilon$$

implies

$$\begin{aligned} 0 < \sqrt{(x - x_0)^2} &< \epsilon && (x - x_0)^2 \le (x - x_0)^2 + (y - y_0)^2 \\ |x - x_0| &< \epsilon && \sqrt{a^2} = |a| \\ |f(x, y) - x_0| &< \epsilon && x = f(x, y) \end{aligned}$$

That is,

$$|f(x, y) - x_0| < \epsilon \qquad \text{whenever} \qquad 0 < \sqrt{(x - x_0)^2 + (y - y_0)^2} < \delta.$$

So

$$\lim_{(x, y)\to(x_0, y_0)} f(x, y) = \lim_{(x, y)\to(x_0, y_0)} x = x_0.$$

It can also be shown that the limit of the sum of two functions is the sum of their limits (when they both exist), with similar results for the limits of the differences, products, constant multiples, quotients, and powers.

THEOREM 1 Properties of Limits of Functions of Two Variables

The following rules hold if L, M, and k are real numbers and

$$\lim_{(x, y)\to(x_0, y_0)} f(x, y) = L \qquad \text{and} \qquad \lim_{(x, y)\to(x_0, y_0)} g(x, y) = M.$$

1. *Sum Rule:* $\displaystyle\lim_{(x, y)\to(x_0, y_0)} (f(x, y) + g(x, y)) = L + M$
2. *Difference Rule:* $\displaystyle\lim_{(x, y)\to(x_0, y_0)} (f(x, y) - g(x, y)) = L - M$
3. *Product Rule:* $\displaystyle\lim_{(x, y)\to(x_0, y_0)} (f(x, y) \cdot g(x, y)) = L \cdot M$
4. *Constant Multiple Rule:* $\displaystyle\lim_{(x, y)\to(x_0, y_0)} (kf(x, y)) = kL$ (any number k)
5. *Quotient Rule:* $\displaystyle\lim_{(x, y)\to(x_0, y_0)} \frac{f(x, y)}{g(x, y)} = \frac{L}{M}$ $\quad M \neq 0$
6. *Power Rule:* If r and s are integers with no common factors, and $s \neq 0$, then

$$\lim_{(x, y)\to(x_0, y_0)} (f(x, y))^{r/s} = L^{r/s}$$

provided $L^{r/s}$ is a real number. (If s is even, we assume that $L > 0$.)

While we won't prove Theorem 1 here, we give an informal discussion of why it's true. If (x, y) is sufficiently close to (x_0, y_0), then $f(x, y)$ is close to L and $g(x, y)$ is close to M (from the informal interpretation of limits). It is then reasonable that $f(x, y) + g(x, y)$ is close to $L + M$; $f(x, y) - g(x, y)$ is close to $L - M$; $f(x, y)g(x, y)$ is close to LM; $kf(x, y)$ is close to kL; and that $f(x, y)/g(x, y)$ is close to L/M if $M \neq 0$.

When we apply Theorem 1 to polynomials and rational functions, we obtain the useful result that the limits of these functions as $(x, y) \to (x_0, y_0)$ can be calculated by evaluating the functions at (x_0, y_0). The only requirement is that the rational functions be defined at (x_0, y_0).

EXAMPLE 1 Calculating Limits

(a) $$\lim_{(x, y)\to(0,1)} \frac{x - xy + 3}{x^2y + 5xy - y^3} = \frac{0 - (0)(1) + 3}{(0)^2(1) + 5(0)(1) - (1)^3} = -3$$

(b) $$\lim_{(x, y)\to(3, -4)} \sqrt{x^2 + y^2} = \sqrt{(3)^2 + (-4)^2} = \sqrt{25} = 5$$ ■

EXAMPLE 2 Calculating Limits

Find

$$\lim_{(x,y)\to(0,0)} \frac{x^2 - xy}{\sqrt{x} - \sqrt{y}}.$$

Solution Since the denominator $\sqrt{x} - \sqrt{y}$ approaches 0 as $(x, y) \to (0, 0)$, we cannot use the Quotient Rule from Theorem 1. If we multiply numerator and denominator by $\sqrt{x} + \sqrt{y}$, however, we produce an equivalent fraction whose limit we *can* find:

$$\begin{aligned}
\lim_{(x,y)\to(0,0)} \frac{x^2 - xy}{\sqrt{x} - \sqrt{y}} &= \lim_{(x,y)\to(0,0)} \frac{(x^2 - xy)(\sqrt{x} + \sqrt{y})}{(\sqrt{x} - \sqrt{y})(\sqrt{x} + \sqrt{y})} \\
&= \lim_{(x,y)\to(0,0)} \frac{x(x - y)(\sqrt{x} + \sqrt{y})}{x - y} && \text{Algebra} \\
&= \lim_{(x,y)\to(0,0)} x(\sqrt{x} + \sqrt{y}) && \text{Cancel the nonzero factor } (x - y). \\
&= 0(\sqrt{0} + \sqrt{0}) = 0
\end{aligned}$$

We can cancel the factor $(x - y)$ because the path $y = x$ (along which $x - y = 0$) is *not* in the domain of the function

$$\frac{x^2 - xy}{\sqrt{x} - \sqrt{y}}.$$

■

EXAMPLE 3 Applying the Limit Definition

Find $\displaystyle\lim_{(x,y)\to(0,0)} \frac{4xy^2}{x^2 + y^2}$ if it exists.

Solution We first observe that along the line $x = 0$, the function always has value 0 when $y \neq 0$. Likewise, along the line $y = 0$, the function has value 0 provided $x \neq 0$. So if the limit does exist as (x, y) approaches $(0, 0)$, the value of the limit must be 0. To see if this is true, we apply the definition of limit.

Let $\epsilon > 0$ be given, but arbitrary. We want to find a $\delta > 0$ such that

$$\left|\frac{4xy^2}{x^2 + y^2} - 0\right| < \epsilon \quad \text{whenever} \quad 0 < \sqrt{x^2 + y^2} < \delta$$

or

$$\frac{4|x|y^2}{x^2 + y^2} < \epsilon \quad \text{whenever} \quad 0 < \sqrt{x^2 + y^2} < \delta.$$

Since $y^2 \leq x^2 + y^2$ we have that

$$\frac{4|x|y^2}{x^2 + y^2} \leq 4|x| = 4\sqrt{x^2} \leq 4\sqrt{x^2 + y^2}.$$

So if we choose $\delta = \epsilon/4$ and let $0 < \sqrt{x^2 + y^2} < \delta$, we get

$$\left| \frac{4xy^2}{x^2 + y^2} - 0 \right| \le 4\sqrt{x^2 + y^2} < 4\delta = 4\left(\frac{\epsilon}{4}\right) = \epsilon.$$

It follows from the definition that

$$\lim_{(x, y) \to (0,0)} \frac{4xy^2}{x^2 + y^2} = 0. \quad ■$$

Continuity

As with functions of a single variable, continuity is defined in terms of limits.

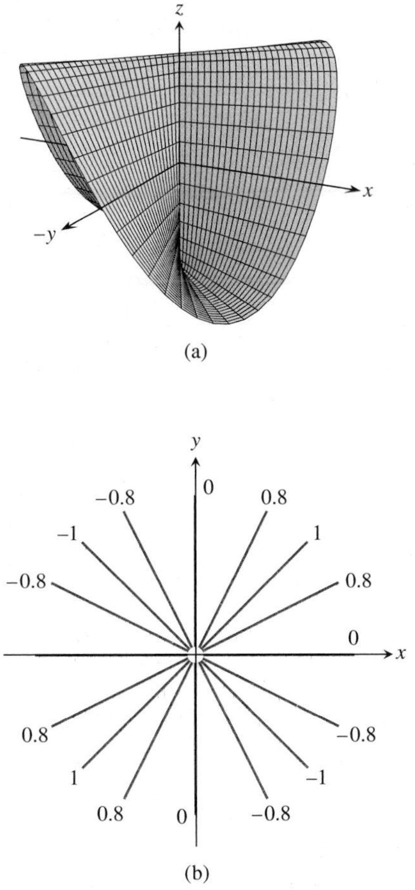

FIGURE 14.11 (a) The graph of

$$f(x, y) = \begin{cases} \dfrac{2xy}{x^2 + y^2}, & (x, y) \ne (0, 0) \\ 0, & (x, y) = (0, 0). \end{cases}$$

The function is continuous at every point except the origin. (b) The level curves of f (Example 4).

DEFINITION Continuous Function of Two Variables

A function $f(x, y)$ is **continuous at the point** (x_0, y_0) if

1. f is defined at (x_0, y_0),
2. $\lim_{(x, y) \to (x_0, y_0)} f(x, y)$ exists,
3. $\lim_{(x, y) \to (x_0, y_0)} f(x, y) = f(x_0, y_0)$.

A function is **continuous** if it is continuous at every point of its domain.

As with the definition of limit, the definition of continuity applies at boundary points as well as interior points of the domain of f. The only requirement is that the point (x, y) remain in the domain at all times.

As you may have guessed, one of the consequences of Theorem 1 is that algebraic combinations of continuous functions are continuous at every point at which all the functions involved are defined. This means that sums, differences, products, constant multiples, quotients, and powers of continuous functions are continuous where defined. In particular, polynomials and rational functions of two variables are continuous at every point at which they are defined.

EXAMPLE 4 A Function with a Single Point of Discontinuity

Show that

$$f(x, y) = \begin{cases} \dfrac{2xy}{x^2 + y^2}, & (x, y) \ne (0, 0) \\ 0, & (x, y) = (0, 0) \end{cases}$$

is continuous at every point except the origin (Figure 14.11).

Solution The function f is continuous at any point $(x, y) \ne (0, 0)$ because its values are then given by a rational function of x and y.

At $(0, 0)$, the value of f is defined, but f, we claim, has no limit as $(x, y) \to (0, 0)$. The reason is that different paths of approach to the origin can lead to different results, as we now see.

For every value of m, the function f has a constant value on the "punctured" line $y = mx, x \neq 0$, because

$$f(x, y)\Big|_{y=mx} = \frac{2xy}{x^2 + y^2}\Big|_{y=mx} = \frac{2x(mx)}{x^2 + (mx)^2} = \frac{2mx^2}{x^2 + m^2x^2} = \frac{2m}{1 + m^2}.$$

Therefore, f has this number as its limit as (x, y) approaches $(0, 0)$ along the line:

$$\lim_{\substack{(x, y)\to(0,0)\\ \text{along } y=mx}} f(x, y) = \lim_{(x, y)\to(0,0)} \left[f(x, y)\Big|_{y=mx}\right] = \frac{2m}{1 + m^2}.$$

This limit changes with m. There is therefore no single number we may call the limit of f as (x, y) approaches the origin. The limit fails to exist, and the function is not continuous. ■

Example 4 illustrates an important point about limits of functions of two variables (or even more variables, for that matter). For a limit to exist at a point, the limit must be the same along every approach path. This result is analogous to the single-variable case where both the left- and right-sided limits had to have the same value; therefore, for functions of two or more variables, if we ever find paths with different limits, we know the function has no limit at the point they approach.

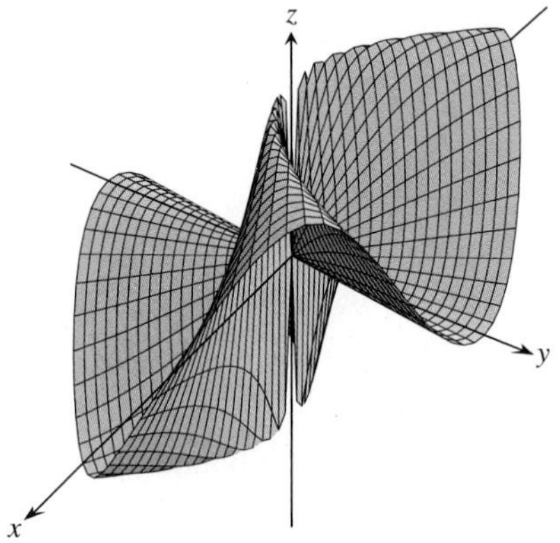

(a)

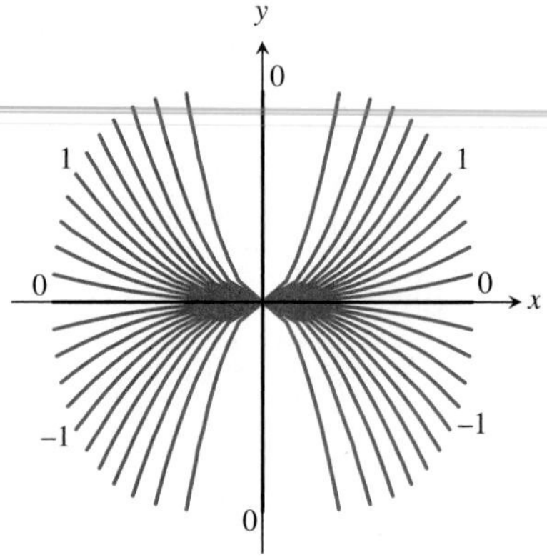

(b)

FIGURE 14.12 (a) The graph of $f(x, y) = 2x^2y/(x^4 + y^2)$. As the graph suggests and the level-curve values in part (b) confirm, $\lim_{(x, y)\to(0,0)} f(x, y)$ does not exist (Example 5).

Two-Path Test for Nonexistence of a Limit

If a function $f(x, y)$ has different limits along two different paths as (x, y) approaches (x_0, y_0), then $\lim_{(x, y)\to(x_0, y_0)} f(x, y)$ does not exist.

EXAMPLE 5 Applying the Two-Path Test

Show that the function

$$f(x, y) = \frac{2x^2y}{x^4 + y^2}$$

(Figure 14.12) has no limit as (x, y) approaches $(0, 0)$.

Solution The limit cannot be found by direct substitution, which gives the form $0/0$. We examine the values of f along curves that end at $(0, 0)$. Along the curve $y = kx^2, x \neq 0$, the function has the constant value

$$f(x, y)\Big|_{y=kx^2} = \frac{2x^2y}{x^4 + y^2}\Big|_{y=kx^2} = \frac{2x^2(kx^2)}{x^4 + (kx^2)^2} = \frac{2kx^4}{x^4 + k^2x^4} = \frac{2k}{1 + k^2}.$$

Therefore,

$$\lim_{\substack{(x, y)\to(0,0)\\ \text{along } y=kx^2}} f(x, y) = \lim_{(x, y)\to(0,0)} \left[f(x, y)\Big|_{y=kx^2}\right] = \frac{2k}{1 + k^2}.$$

This limit varies with the path of approach. If (x, y) approaches $(0, 0)$ along the parabola $y = x^2$, for instance, $k = 1$ and the limit is 1. If (x, y) approaches $(0, 0)$ along the x-axis, $k = 0$ and the limit is 0. By the two-path test, f has no limit as (x, y) approaches $(0, 0)$.

The language here may seem contradictory. You might well ask, "What do you mean f has no limit as (x, y) approaches the origin—it has lots of limits." But that is the point. There is no *single* path-independent limit, and therefore, by the definition, $\lim_{(x,y)\to(0,0)} f(x, y)$ does not exist. ■

Compositions of continuous functions are also continuous. The proof, omitted here, is similar to that for functions of a single variable (Theorem 10 in Section 2.6).

Continuity of Composites
If f is continuous at (x_0, y_0) and g is a single-variable function continuous at $f(x_0, y_0)$, then the composite function $h = g \circ f$ defined by $h(x, y) = g(f(x, y))$ is continuous at (x_0, y_0).

For example, the composite functions

$$e^{x-y}, \qquad \cos\frac{xy}{x^2+1}, \qquad \ln(1 + x^2y^2)$$

are continuous at every point (x, y).

As with functions of a single variable, the general rule is that composites of continuous functions are continuous. The only requirement is that each function be continuous where it is applied.

Functions of More Than Two Variables

The definitions of limit and continuity for functions of two variables and the conclusions about limits and continuity for sums, products, quotients, powers, and composites all extend to functions of three or more variables. Functions like

$$\ln(x + y + z) \qquad \text{and} \qquad \frac{y \sin z}{x - 1}$$

are continuous throughout their domains, and limits like

$$\lim_{P\to(1,0,-1)} \frac{e^{x+z}}{z^2 + \cos\sqrt{xy}} = \frac{e^{1-1}}{(-1)^2 + \cos 0} = \frac{1}{2},$$

where P denotes the point (x, y, z), may be found by direct substitution.

Extreme Values of Continuous Functions on Closed, Bounded Sets

We have seen that a function of a single variable that is continuous throughout a closed, bounded interval $[a, b]$ takes on an absolute maximum value and an absolute minimum value at least once in $[a, b]$. The same is true of a function $z = f(x, y)$ that is continuous on a closed, bounded set R in the plane (like a line segment, a disk, or a filled-in triangle). The function takes on an absolute maximum value at some point in R and an absolute minimum value at some point in R.

Theorems similar to these and other theorems of this section hold for functions of three or more variables. A continuous function $w = f(x, y, z)$, for example, must take on absolute maximum and minimum values on any closed, bounded set (solid ball or cube, spherical shell, rectangular solid) on which it is defined.

We learn how to find these extreme values in Section 14.7, but first we need to study derivatives in higher dimensions. That is the topic of the next section.

EXERCISES 14.2

Limits with Two Variables

Find the limits in Exercises 1–12.

1. $\lim_{(x,y)\to(0,0)} \dfrac{3x^2 - y^2 + 5}{x^2 + y^2 + 2}$

2. $\lim_{(x,y)\to(0,4)} \dfrac{x}{\sqrt{y}}$

3. $\lim_{(x,y)\to(3,4)} \sqrt{x^2 + y^2 - 1}$

4. $\lim_{(x,y)\to(2,-3)} \left(\dfrac{1}{x} + \dfrac{1}{y}\right)^2$

5. $\lim_{(x,y)\to(0,\pi/4)} \sec x \tan y$

6. $\lim_{(x,y)\to(0,0)} \cos \dfrac{x^2 + y^3}{x + y + 1}$

7. $\lim_{(x,y)\to(0,\ln 2)} e^{x-y}$

8. $\lim_{(x,y)\to(1,1)} \ln |1 + x^2 y^2|$

9. $\lim_{(x,y)\to(0,0)} \dfrac{e^y \sin x}{x}$

10. $\lim_{(x,y)\to(1,1)} \cos \sqrt[3]{|xy| - 1}$

11. $\lim_{(x,y)\to(1,0)} \dfrac{x \sin y}{x^2 + 1}$

12. $\lim_{(x,y)\to(\pi/2,0)} \dfrac{\cos y + 1}{y - \sin x}$

Limits of Quotients

Find the limits in Exercises 13–20 by rewriting the fractions first.

13. $\lim_{\substack{(x,y)\to(1,1)\\ x\neq y}} \dfrac{x^2 - 2xy + y^2}{x - y}$

14. $\lim_{\substack{(x,y)\to(1,1)\\ x\neq y}} \dfrac{x^2 - y^2}{x - y}$

15. $\lim_{\substack{(x,y)\to(1,1)\\ x\neq 1}} \dfrac{xy - y - 2x + 2}{x - 1}$

16. $\lim_{\substack{(x,y)\to(2,-4)\\ y\neq -4,\, x\neq x^2}} \dfrac{y + 4}{x^2 y - xy + 4x^2 - 4x}$

17. $\lim_{\substack{(x,y)\to(0,0)\\ x\neq y}} \dfrac{x - y + 2\sqrt{x} - 2\sqrt{y}}{\sqrt{x} - \sqrt{y}}$

18. $\lim_{\substack{(x,y)\to(2,2)\\ x+y\neq 4}} \dfrac{x + y - 4}{\sqrt{x + y} - 2}$

19. $\lim_{\substack{(x,y)\to(2,0)\\ 2x-y\neq 4}} \dfrac{\sqrt{2x - y} - 2}{2x - y - 4}$

20. $\lim_{\substack{(x,y)\to(4,3)\\ x\neq y+1}} \dfrac{\sqrt{x} - \sqrt{y + 1}}{x - y - 1}$

Limits with Three Variables

Find the limits in Exercises 21–26.

21. $\lim_{P\to(1,3,4)} \left(\dfrac{1}{x} + \dfrac{1}{y} + \dfrac{1}{z}\right)$

22. $\lim_{P\to(1,-1,-1)} \dfrac{2xy + yz}{x^2 + z^2}$

23. $\lim_{P\to(3,3,0)} (\sin^2 x + \cos^2 y + \sec^2 z)$

24. $\lim_{P\to(-1/4,\pi/2,2)} \tan^{-1} xyz$

25. $\lim_{P\to(\pi,0,3)} ze^{-2y} \cos 2x$

26. $\lim_{P\to(0,-2,0)} \ln \sqrt{x^2 + y^2 + z^2}$

Continuity in the Plane

At what points (x, y) in the plane are the functions in Exercises 27–30 continuous?

27. a. $f(x, y) = \sin (x + y)$ b. $f(x, y) = \ln (x^2 + y^2)$

28. a. $f(x, y) = \dfrac{x + y}{x - y}$ b. $f(x, y) = \dfrac{y}{x^2 + 1}$

29. a. $g(x, y) = \sin \dfrac{1}{xy}$ b. $g(x, y) = \dfrac{x + y}{2 + \cos x}$

30. a. $g(x, y) = \dfrac{x^2 + y^2}{x^2 - 3x + 2}$ b. $g(x, y) = \dfrac{1}{x^2 - y}$

Continuity in Space

At what points (x, y, z) in space are the functions in Exercises 31–34 continuous?

31. a. $f(x, y, z) = x^2 + y^2 - 2z^2$
    b. $f(x, y, z) = \sqrt{x^2 + y^2 - 1}$

32. a. $f(x, y, z) = \ln xyz$ b. $f(x, y, z) = e^{x+y} \cos z$

33. a. $h(x, y, z) = xy \sin \dfrac{1}{z}$ b. $h(x, y, z) = \dfrac{1}{x^2 + z^2 - 1}$

34. a. $h(x, y, z) = \dfrac{1}{|y| + |z|}$ b. $h(x, y, z) = \dfrac{1}{|xy| + |z|}$

No Limit at a Point

By considering different paths of approach, show that the functions in Exercises 35–42 have no limit as $(x, y) \to (0, 0)$.

35. $f(x, y) = -\dfrac{x}{\sqrt{x^2 + y^2}}$

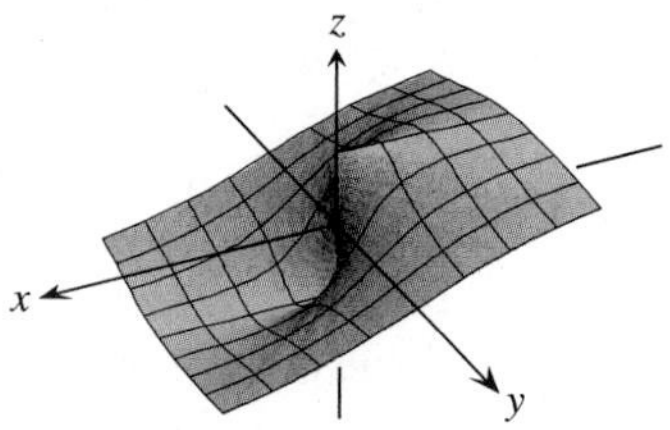

36. $f(x, y) = \dfrac{x^4}{x^4 + y^2}$

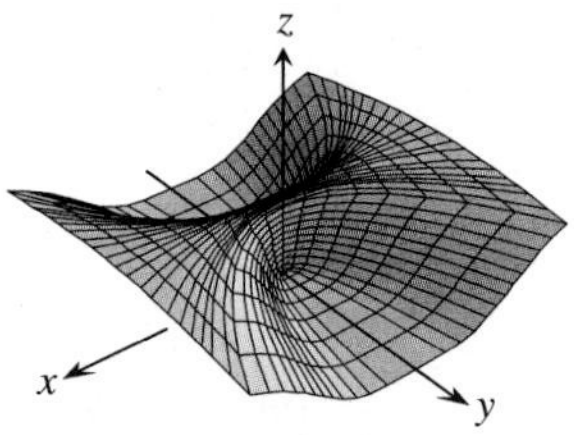

37. $f(x, y) = \dfrac{x^4 - y^2}{x^4 + y^2}$

38. $f(x, y) = \dfrac{xy}{|xy|}$

39. $g(x, y) = \dfrac{x - y}{x + y}$

40. $g(x, y) = \dfrac{x + y}{x - y}$

41. $h(x, y) = \dfrac{x^2 + y}{y}$

42. $h(x, y) = \dfrac{x^2}{x^2 - y}$

Theory and Examples

43. If $\lim_{(x,y)\to(x_0,y_0)} f(x,y) = L$, must f be defined at (x_0, y_0)? Give reasons for your answer.

44. If $f(x_0, y_0) = 3$, what can you say about

$$\lim_{(x,y)\to(x_0,y_0)} f(x,y)$$

if f is continuous at (x_0, y_0)? If f is not continuous at (x_0, y_0)? Give reasons for your answer.

The Sandwich Theorem for functions of two variables states that if $g(x,y) \le f(x,y) \le h(x,y)$ for all $(x,y) \ne (x_0, y_0)$ in a disk centered at (x_0, y_0) and if g and h have the same finite limit L as $(x,y) \to (x_0, y_0)$, then

$$\lim_{(x,y)\to(x_0,y_0)} f(x,y) = L.$$

Use this result to support your answers to the questions in Exercises 45–48.

45. Does knowing that

$$1 - \frac{x^2y^2}{3} < \frac{\tan^{-1} xy}{xy} < 1$$

tell you anything about

$$\lim_{(x,y)\to(0,0)} \frac{\tan^{-1} xy}{xy}?$$

Give reasons for your answer.

46. Does knowing that

$$2|xy| - \frac{x^2y^2}{6} < 4 - 4\cos\sqrt{|xy|} < 2|xy|$$

tell you anything about

$$\lim_{(x,y)\to(0,0)} \frac{4 - 4\cos\sqrt{|xy|}}{|xy|}?$$

Give reasons for your answer.

47. Does knowing that $|\sin(1/x)| \le 1$ tell you anything about

$$\lim_{(x,y)\to(0,0)} y \sin\frac{1}{x}?$$

Give reasons for your answer.

48. Does knowing that $|\cos(1/y)| \le 1$ tell you anything about

$$\lim_{(x,y)\to(0,0)} x \cos\frac{1}{y}?$$

Give reasons for your answer.

49. *(Continuation of Example 4.)*

a. Reread Example 4. Then substitute $m = \tan\theta$ into the formula

$$f(x,y)\Big|_{y=mx} = \frac{2m}{1+m^2}$$

and simplify the result to show how the value of f varies with the line's angle of inclination.

b. Use the formula you obtained in part (a) to show that the limit of f as $(x,y) \to (0,0)$ along the line $y = mx$ varies from -1 to 1 depending on the angle of approach.

50. Continuous extension Define $f(0,0)$ in a way that extends

$$f(x,y) = xy\frac{x^2 - y^2}{x^2 + y^2}$$

to be continuous at the origin.

Changing to Polar Coordinates

If you cannot make any headway with $\lim_{(x,y)\to(0,0)} f(x,y)$ in rectangular coordinates, try changing to polar coordinates. Substitute $x = r\cos\theta$, $y = r\sin\theta$, and investigate the limit of the resulting expression as $r \to 0$. In other words, try to decide whether there exists a number L satisfying the following criterion:

Given $\epsilon > 0$, there exists a $\delta > 0$ such that for all r and θ,

$$|r| < \delta \quad \Rightarrow \quad |f(r,\theta) - L| < \epsilon. \tag{1}$$

If such an L exists, then

$$\lim_{(x,y)\to(0,0)} f(x,y) = \lim_{r\to 0} f(r,\theta) = L.$$

For instance,

$$\lim_{(x,y)\to(0,0)} \frac{x^3}{x^2+y^2} = \lim_{r\to 0} \frac{r^3\cos^3\theta}{r^2} = \lim_{r\to 0} r\cos^3\theta = 0.$$

To verify the last of these equalities, we need to show that Equation (1) is satisfied with $f(r,\theta) = r\cos^3\theta$ and $L = 0$. That is, we need to show that given any $\epsilon > 0$ there exists a $\delta > 0$ such that for all r and θ,

$$|r| < \delta \quad \Rightarrow \quad |r\cos^3\theta - 0| < \epsilon.$$

Since

$$|r\cos^3\theta| = |r||\cos^3\theta| \le |r| \cdot 1 = |r|,$$

the implication holds for all r and θ if we take $\delta = \epsilon$.

In contrast,

$$\frac{x^2}{x^2+y^2} = \frac{r^2\cos^2\theta}{r^2} = \cos^2\theta$$

takes on all values from 0 to 1 regardless of how small $|r|$ is, so that $\lim_{(x,y)\to(0,0)} x^2/(x^2+y^2)$ does not exist.

In each of these instances, the existence or nonexistence of the limit as $r \to 0$ is fairly clear. Shifting to polar coordinates does not always help, however, and may even tempt us to false conclusions. For example, the limit may exist along every straight line (or ray) $\theta =$ constant and yet fail to exist in the broader sense. Example 4 illustrates this point. In polar coordinates, $f(x,y) = (2x^2y)/(x^4+y^2)$ becomes

$$f(r\cos\theta, r\sin\theta) = \frac{r\cos\theta\sin 2\theta}{r^2\cos^4\theta + \sin^2\theta}$$

for $r \neq 0$. If we hold θ constant and let $r \to 0$, the limit is 0. On the path $y = x^2$, however, we have $r \sin\theta = r^2 \cos^2\theta$ and

$$f(r\cos\theta, r\sin\theta) = \frac{r\cos\theta\sin 2\theta}{r^2\cos^4\theta + (r\cos^2\theta)^2}$$

$$= \frac{2r\cos^2\theta\sin\theta}{2r^2\cos^4\theta} = \frac{r\sin\theta}{r^2\cos^2\theta} = 1.$$

In Exercises 51–56, find the limit of f as $(x, y) \to (0, 0)$ or show that the limit does not exist.

51. $f(x, y) = \dfrac{x^3 - xy^2}{x^2 + y^2}$

52. $f(x, y) = \cos\left(\dfrac{x^3 - y^3}{x^2 + y^2}\right)$

53. $f(x, y) = \dfrac{y^2}{x^2 + y^2}$

54. $f(x, y) = \dfrac{2x}{x^2 + x + y^2}$

55. $f(x, y) = \tan^{-1}\left(\dfrac{|x| + |y|}{x^2 + y^2}\right)$

56. $f(x, y) = \dfrac{x^2 - y^2}{x^2 + y^2}$

In Exercises 57 and 58, define $f(0, 0)$ in a way that extends f to be continuous at the origin.

57. $f(x, y) = \ln\left(\dfrac{3x^2 - x^2y^2 + 3y^2}{x^2 + y^2}\right)$

58. $f(x, y) = \dfrac{3x^2y}{x^2 + y^2}$

Using the δ-ϵ Definition

Each of Exercises 59–62 gives a function $f(x, y)$ and a positive number ϵ. In each exercise, show that there exists a $\delta > 0$ such that for all (x, y),

$$\sqrt{x^2 + y^2} < \delta \quad\Rightarrow\quad |f(x, y) - f(0, 0)| < \epsilon.$$

59. $f(x, y) = x^2 + y^2, \quad \epsilon = 0.01$

60. $f(x, y) = y/(x^2 + 1), \quad \epsilon = 0.05$

61. $f(x, y) = (x + y)/(x^2 + 1), \quad \epsilon = 0.01$

62. $f(x, y) = (x + y)/(2 + \cos x), \quad \epsilon = 0.02$

Each of Exercises 63–66 gives a function $f(x, y, z)$ and a positive number ϵ. In each exercise, show that there exists a $\delta > 0$ such that for all (x, y, z),

$$\sqrt{x^2 + y^2 + z^2} < \delta \quad\Rightarrow\quad |f(x, y, z) - f(0, 0, 0)| < \epsilon.$$

63. $f(x, y, z) = x^2 + y^2 + z^2, \quad \epsilon = 0.015$

64. $f(x, y, z) = xyz, \quad \epsilon = 0.008$

65. $f(x, y, z) = \dfrac{x + y + z}{x^2 + y^2 + z^2 + 1}, \quad \epsilon = 0.015$

66. $f(x, y, z) = \tan^2 x + \tan^2 y + \tan^2 z, \quad \epsilon = 0.03$

67. Show that $f(x, y, z) = x + y - z$ is continuous at every point (x_0, y_0, z_0).

68. Show that $f(x, y, z) = x^2 + y^2 + z^2$ is continuous at the origin.

14.3 Partial Derivatives

The calculus of several variables is basically single-variable calculus applied to several variables one at a time. When we hold all but one of the independent variables of a function constant and differentiate with respect to that one variable, we get a "partial" derivative. This section shows how partial derivatives are defined and interpreted geometrically, and how to calculate them by applying the rules for differentiating functions of a single variable.

Partial Derivatives of a Function of Two Variables

If (x_0, y_0) is a point in the domain of a function $f(x, y)$, the vertical plane $y = y_0$ will cut the surface $z = f(x, y)$ in the curve $z = f(x, y_0)$ (Figure 14.13). This curve is the graph of the function $z = f(x, y_0)$ in the plane $y = y_0$. The horizontal coordinate in this plane is x; the vertical coordinate is z. The y-value is held constant at y_0, so y is not a variable.

We define the partial derivative of f with respect to x at the point (x_0, y_0) as the ordinary derivative of $f(x, y_0)$ with respect to x at the point $x = x_0$. To distinguish partial derivatives from ordinary derivatives we use the symbol ∂ rather than the d previously used.

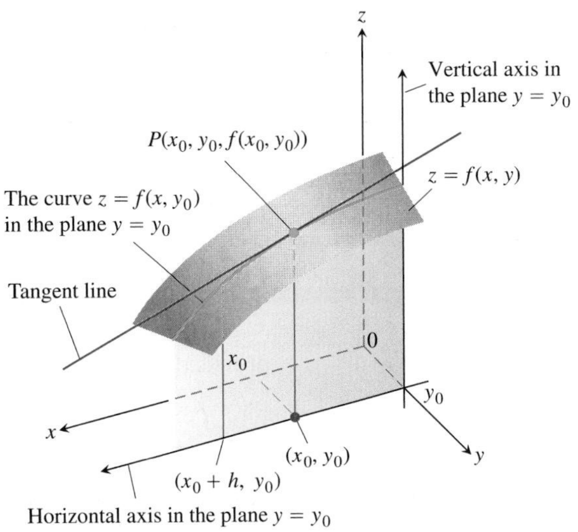

FIGURE 14.13 The intersection of the plane $y = y_0$ with the surface $z = f(x, y)$, viewed from above the first quadrant of the xy-plane.

DEFINITION Partial Derivative with Respect to x

The **partial derivative of $f(x, y)$ with respect to x** at the point (x_0, y_0) is

$$\left.\frac{\partial f}{\partial x}\right|_{(x_0, y_0)} = \lim_{h \to 0} \frac{f(x_0 + h, y_0) - f(x_0, y_0)}{h},$$

provided the limit exists.

An equivalent expression for the partial derivative is

$$\left.\frac{d}{dx} f(x, y_0)\right|_{x=x_0}.$$

The slope of the curve $z = f(x, y_0)$ at the point $P(x_0, y_0, f(x_0, y_0))$ in the plane $y = y_0$ is the value of the partial derivative of f with respect to x at (x_0, y_0). The tangent line to the curve at P is the line in the plane $y = y_0$ that passes through P with this slope. The partial derivative $\partial f/\partial x$ at (x_0, y_0) gives the rate of change of f with respect to x when y is held fixed at the value y_0. This is the rate of change of f in the direction of **i** at (x_0, y_0).

The notation for a partial derivative depends on what we want to emphasize:

$\frac{\partial f}{\partial x}(x_0, y_0)$ or $f_x(x_0, y_0)$	"Partial derivative of f with respect to x at (x_0, y_0)" or "f sub x at (x_0, y_0)." Convenient for stressing the point (x_0, y_0).
$\left.\frac{\partial z}{\partial x}\right\|_{(x_0, y_0)}$	"Partial derivative of z with respect to x at (x_0, y_0)." Common in science and engineering when you are dealing with variables and do not mention the function explicitly.
$f_x, \frac{\partial f}{\partial x}, z_x$, or $\frac{\partial z}{\partial x}$	"Partial derivative of f (or z) with respect to x." Convenient when you regard the partial derivative as a function in its own right.

The definition of the partial derivative of $f(x, y)$ with respect to y at a point (x_0, y_0) is similar to the definition of the partial derivative of f with respect to x. We hold x fixed at the value x_0 and take the ordinary derivative of $f(x_0, y)$ with respect to y at y_0.

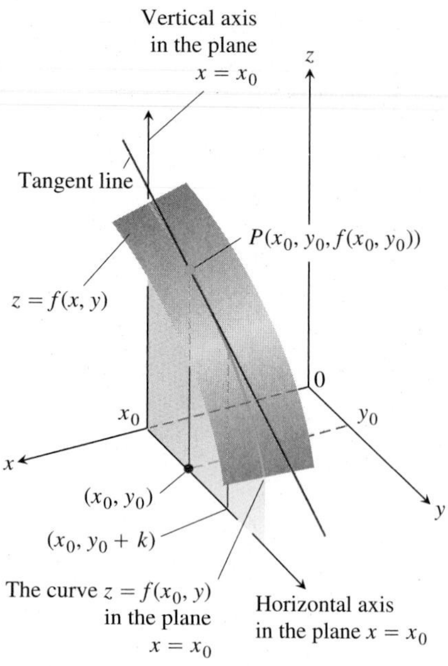

FIGURE 14.14 The intersection of the plane $x = x_0$ with the surface $z = f(x, y)$, viewed from above the first quadrant of the xy-plane.

DEFINITION Partial Derivative with Respect to y

The **partial derivative of $f(x, y)$ with respect to y** at the point (x_0, y_0) is

$$\left.\frac{\partial f}{\partial y}\right|_{(x_0, y_0)} = \left.\frac{d}{dy} f(x_0, y)\right|_{y=y_0} = \lim_{h \to 0} \frac{f(x_0, y_0 + h) - f(x_0, y_0)}{h},$$

provided the limit exists.

The slope of the curve $z = f(x_0, y)$ at the point $P(x_0, y_0, f(x_0, y_0))$ in the vertical plane $x = x_0$ (Figure 14.14) is the partial derivative of f with respect to y at (x_0, y_0). The tangent line to the curve at P is the line in the plane $x = x_0$ that passes through P with this slope. The partial derivative gives the rate of change of f with respect to y at (x_0, y_0) when x is held fixed at the value x_0. This is the rate of change of f in the direction of $\mathbf{j}$ at (x_0, y_0).

The partial derivative with respect to y is denoted the same way as the partial derivative with respect to x:

$$\frac{\partial f}{\partial y}(x_0, y_0), \qquad f_y(x_0, y_0), \qquad \frac{\partial f}{\partial y}, \qquad f_y.$$

Notice that we now have two tangent lines associated with the surface $z = f(x, y)$ at the point $P(x_0, y_0, f(x_0, y_0))$ (Figure 14.15). Is the plane they determine tangent to the surface at P? We will see that it is, but we have to learn more about partial derivatives before we can find out why.

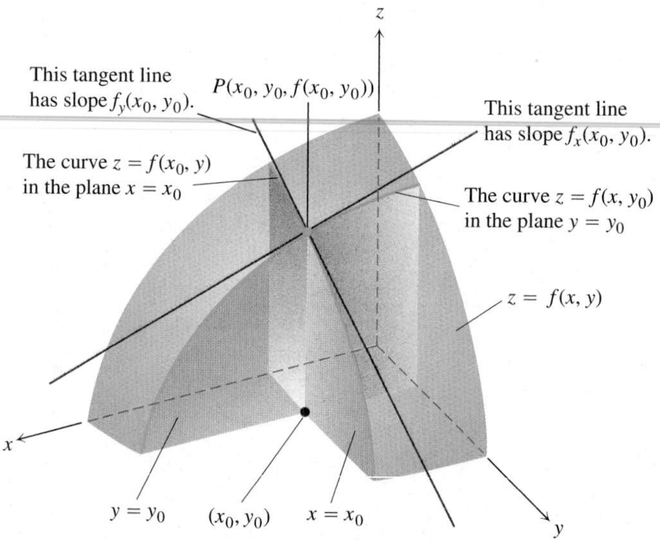

FIGURE 14.15 Figures 14.13 and 14.14 combined. The tangent lines at the point $(x_0, y_0, f(x_0, y_0))$ determine a plane that, in this picture at least, appears to be tangent to the surface.

Calculations

The definitions of $\partial f/\partial x$ and $\partial f/\partial y$ give us two different ways of differentiating f at a point: with respect to x in the usual way while treating y as a constant and with respect to y in the usual way while treating x as constant. As the following examples show, the values of these partial derivatives are usually different at a given point (x_0, y_0).

EXAMPLE 1 Finding Partial Derivatives at a Point

Find the values of $\partial f/\partial x$ and $\partial f/\partial y$ at the point $(4, -5)$ if

$$f(x, y) = x^2 + 3xy + y - 1.$$

Solution To find $\partial f/\partial x$, we treat y as a constant and differentiate with respect to x:

$$\frac{\partial f}{\partial x} = \frac{\partial}{\partial x}(x^2 + 3xy + y - 1) = 2x + 3 \cdot 1 \cdot y + 0 - 0 = 2x + 3y.$$

The value of $\partial f/\partial x$ at $(4, -5)$ is $2(4) + 3(-5) = -7$.

To find $\partial f/\partial y$, we treat x as a constant and differentiate with respect to y:

$$\frac{\partial f}{\partial y} = \frac{\partial}{\partial y}(x^2 + 3xy + y - 1) = 0 + 3 \cdot x \cdot 1 + 1 - 0 = 3x + 1.$$

The value of $\partial f/\partial y$ at $(4, -5)$ is $3(4) + 1 = 13$. ■

EXAMPLE 2 Finding a Partial Derivative as a Function

Find $\partial f/\partial y$ if $f(x, y) = y \sin xy$.

Solution We treat x as a constant and f as a product of y and $\sin xy$:

$$\frac{\partial f}{\partial y} = \frac{\partial}{\partial y}(y \sin xy) = y \frac{\partial}{\partial y} \sin xy + (\sin xy) \frac{\partial}{\partial y}(y)$$

$$= (y \cos xy) \frac{\partial}{\partial y}(xy) + \sin xy = xy \cos xy + \sin xy.$$ ■

USING TECHNOLOGY Partial Differentiation

A simple grapher can support your calculations even in multiple dimensions. If you specify the values of all but one independent variable, the grapher can calculate partial derivatives and can plot traces with respect to that remaining variable. Typically, a CAS can compute partial derivatives symbolically and numerically as easily as it can compute simple derivatives. Most systems use the same command to differentiate a function, regardless of the number of variables. (Simply specify the variable with which differentiation is to take place).

EXAMPLE 3 Partial Derivatives May Be Different Functions

Find f_x and f_y if

$$f(x, y) = \frac{2y}{y + \cos x}.$$

Solution We treat f as a quotient. With y held constant, we get

$$f_x = \frac{\partial}{\partial x}\left(\frac{2y}{y + \cos x}\right) = \frac{(y + \cos x)\frac{\partial}{\partial x}(2y) - 2y\frac{\partial}{\partial x}(y + \cos x)}{(y + \cos x)^2}$$

$$= \frac{(y + \cos x)(0) - 2y(-\sin x)}{(y + \cos x)^2} = \frac{2y \sin x}{(y + \cos x)^2}.$$

With x held constant, we get

$$f_y = \frac{\partial}{\partial y}\left(\frac{2y}{y + \cos x}\right) = \frac{(y + \cos x)\frac{\partial}{\partial y}(2y) - 2y\frac{\partial}{dy}(y + \cos x)}{(y + \cos x)^2}$$

$$= \frac{(y + \cos x)(2) - 2y(1)}{(y + \cos x)^2} = \frac{2 \cos x}{(y + \cos x)^2}.$$ ■

Implicit differentiation works for partial derivatives the way it works for ordinary derivatives, as the next example illustrates.

EXAMPLE 4 Implicit Partial Differentiation

Find $\partial z/\partial x$ if the equation

$$yz - \ln z = x + y$$

defines z as a function of the two independent variables x and y and the partial derivative exists.

Solution We differentiate both sides of the equation with respect to x, holding y constant and treating z as a differentiable function of x:

$$\frac{\partial}{\partial x}(yz) - \frac{\partial}{\partial x}\ln z = \frac{\partial x}{\partial x} + \frac{\partial y}{\partial x}$$

$$y\frac{\partial z}{\partial x} - \frac{1}{z}\frac{\partial z}{\partial x} = 1 + 0$$

With y constant, $\frac{\partial}{\partial x}(yz) = y\frac{\partial z}{\partial x}$.

$$\left(y - \frac{1}{z}\right)\frac{\partial z}{\partial x} = 1$$

$$\frac{\partial z}{\partial x} = \frac{z}{yz - 1}.$$ ■

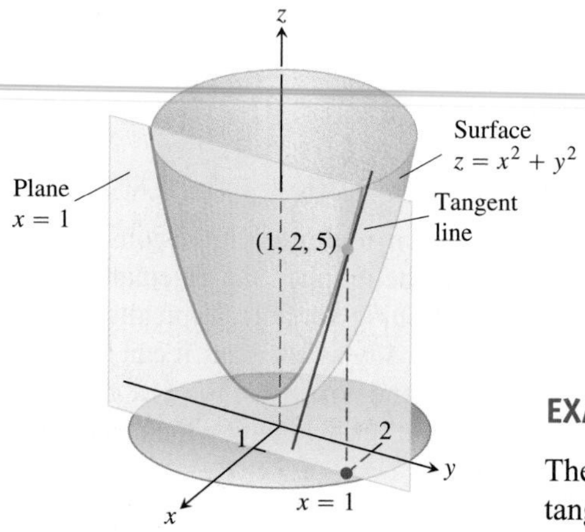

FIGURE 14.16 The tangent to the curve of intersection of the plane $x = 1$ and surface $z = x^2 + y^2$ at the point (1, 2, 5) (Example 5).

EXAMPLE 5 Finding the Slope of a Surface in the y-Direction

The plane $x = 1$ intersects the paraboloid $z = x^2 + y^2$ in a parabola. Find the slope of the tangent to the parabola at (1, 2, 5) (Figure 14.16).

Solution The slope is the value of the partial derivative $\partial z/\partial y$ at (1, 2):

$$\left.\frac{\partial z}{\partial y}\right|_{(1,2)} = \left.\frac{\partial}{\partial y}(x^2 + y^2)\right|_{(1,2)} = \left.2y\right|_{(1,2)} = 2(2) = 4.$$

As a check, we can treat the parabola as the graph of the single-variable function $z = (1)^2 + y^2 = 1 + y^2$ in the plane $x = 1$ and ask for the slope at $y = 2$. The slope, calculated now as an ordinary derivative, is

$$\left.\frac{dz}{dy}\right|_{y=2} = \left.\frac{d}{dy}(1 + y^2)\right|_{y=2} = \left.2y\right|_{y=2} = 4.$$

Functions of More Than Two Variables

The definitions of the partial derivatives of functions of more than two independent variables are like the definitions for functions of two variables. They are ordinary derivatives with respect to one variable, taken while the other independent variables are held constant.

EXAMPLE 6 A Function of Three Variables

If x, y, and z are independent variables and

$$f(x, y, z) = x \sin(y + 3z),$$

then

$$\begin{aligned}\frac{\partial f}{\partial z} &= \frac{\partial}{\partial z}[x \sin(y + 3z)] = x\frac{\partial}{\partial z}\sin(y + 3z)\\ &= x\cos(y + 3z)\frac{\partial}{\partial z}(y + 3z) = 3x\cos(y + 3z).\end{aligned}$$

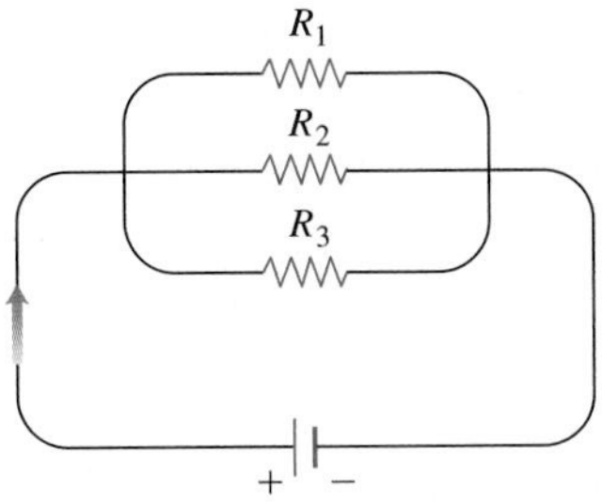

FIGURE 14.17 Resistors arranged this way are said to be connected in parallel (Example 7). Each resistor lets a portion of the current through. Their equivalent resistance R is calculated with the formula

$$\frac{1}{R} = \frac{1}{R_1} + \frac{1}{R_2} + \frac{1}{R_3}.$$

EXAMPLE 7 Electrical Resistors in Parallel

If resistors of R_1, R_2, and R_3 ohms are connected in parallel to make an R-ohm resistor, the value of R can be found from the equation

$$\frac{1}{R} = \frac{1}{R_1} + \frac{1}{R_2} + \frac{1}{R_3}$$

(Figure 14.17). Find the value of $\partial R/\partial R_2$ when $R_1 = 30$, $R_2 = 45$, and $R_3 = 90$ ohms.

Solution To find $\partial R/\partial R_2$, we treat R_1 and R_3 as constants and, using implicit differentiation, differentiate both sides of the equation with respect to R_2:

$$\begin{aligned}\frac{\partial}{\partial R_2}\left(\frac{1}{R}\right) &= \frac{\partial}{\partial R_2}\left(\frac{1}{R_1} + \frac{1}{R_2} + \frac{1}{R_3}\right)\\ -\frac{1}{R^2}\frac{\partial R}{\partial R_2} &= 0 - \frac{1}{R_2^{\,2}} + 0\\ \frac{\partial R}{\partial R_2} &= \frac{R^2}{R_2^{\,2}} = \left(\frac{R}{R_2}\right)^2.\end{aligned}$$

When $R_1 = 30$, $R_2 = 45$, and $R_3 = 90$,

$$\frac{1}{R} = \frac{1}{30} + \frac{1}{45} + \frac{1}{90} = \frac{3 + 2 + 1}{90} = \frac{6}{90} = \frac{1}{15},$$

so $R = 15$ and

$$\frac{\partial R}{\partial R_2} = \left(\frac{15}{45}\right)^2 = \left(\frac{1}{3}\right)^2 = \frac{1}{9}.$$ ■

Partial Derivatives and Continuity

A function $f(x, y)$ can have partial derivatives with respect to both x and y at a point without the function being continuous there. This is different from functions of a single variable, where the existence of a derivative implies continuity. If the partial derivatives of $f(x, y)$ exist and are continuous throughout a disk centered at (x_0, y_0), however, then f *is* continuous at (x_0, y_0), as we see at the end of this section.

EXAMPLE 8 Partials Exist, But f Discontinuous

Let

$$f(x, y) = \begin{cases} 0, & xy \neq 0 \\ 1, & xy = 0 \end{cases}$$

(Figure 14.18).

(a) Find the limit of f as (x, y) approaches $(0, 0)$ along the line $y = x$.

(b) Prove that f is not continuous at the origin.

(c) Show that both partial derivatives $\partial f/\partial x$ and $\partial f/\partial y$ exist at the origin.

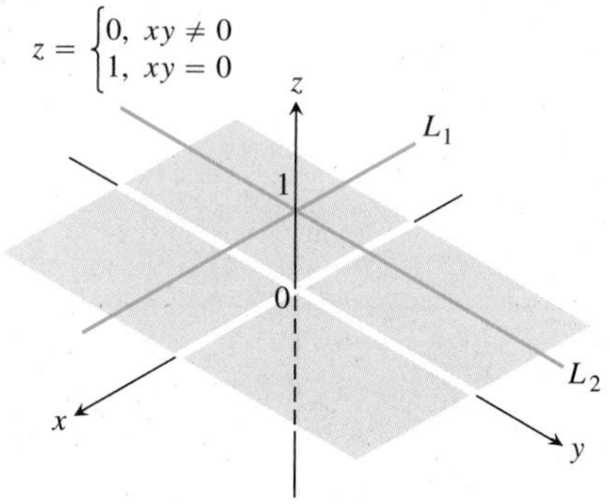

FIGURE 14.18 The graph of

$$f(x, y) = \begin{cases} 0, & xy \neq 0 \\ 1, & xy = 0 \end{cases}$$

consists of the lines L_1 and L_2 and the four open quadrants of the xy-plane. The function has partial derivatives at the origin but is not continuous there (Example 8).

Solution

(a) Since $f(x, y)$ is constantly zero along the line $y = x$ (except at the origin), we have

$$\lim_{(x, y)\to(0,0)} f(x, y)\Big|_{y=x} = \lim_{(x, y)\to(0,0)} 0 = 0.$$

(b) Since $f(0, 0) = 1$, the limit in part (a) proves that f is not continuous at $(0, 0)$.

(c) To find $\partial f/\partial x$ at $(0, 0)$, we hold y fixed at $y = 0$. Then $f(x, y) = 1$ for all x, and the graph of f is the line L_1 in Figure 14.18. The slope of this line at any x is $\partial f/\partial x = 0$. In particular, $\partial f/\partial x = 0$ at $(0, 0)$. Similarly, $\partial f/\partial y$ is the slope of line L_2 at any y, so $\partial f/\partial y = 0$ at $(0, 0)$. ■

Example 8 notwithstanding, it is still true in higher dimensions that *differentiability* at a point implies continuity. What Example 8 suggests is that we need a stronger requirement for differentiability in higher dimensions than the mere existence of the partial derivatives. We define differentiability for functions of two variables at the end of this section and revisit the connection to continuity.

Second-Order Partial Derivatives

When we differentiate a function $f(x, y)$ twice, we produce its second-order derivatives. These derivatives are usually denoted by

$$\frac{\partial^2 f}{\partial x^2} \quad \text{"}d\text{ squared }fdx\text{ squared"} \quad \text{or} \quad f_{xx} \quad \text{"}f\text{ sub }xx\text{"}$$

$$\frac{\partial^2 f}{\partial y^2} \quad \text{"}d\text{ squared }fdy\text{ squared"} \quad \text{or} \quad f_{yy} \quad \text{"}f\text{ sub }yy\text{"}$$

$$\frac{\partial^2 f}{\partial x \partial y} \quad \text{"}d\text{ squared } f\,dx\,dy\text{"} \quad \text{or} \quad f_{yx} \quad \text{"}f\text{ sub } yx\text{"}$$

$$\frac{\partial^2 f}{\partial y \partial x} \quad \text{"}d\text{ squared } f\,dy\,dx\text{"} \quad \text{or} \quad f_{xy} \quad \text{"}f\text{ sub } xy\text{"}$$

The defining equations are

$$\frac{\partial^2 f}{\partial x^2} = \frac{\partial}{\partial x}\left(\frac{\partial f}{\partial x}\right), \qquad \frac{\partial^2 f}{\partial x \partial y} = \frac{\partial}{\partial x}\left(\frac{\partial f}{\partial y}\right),$$

and so on. Notice the order in which the derivatives are taken:

$\frac{\partial^2 f}{\partial x \partial y}$ Differentiate first with respect to y, then with respect to x.

$f_{yx} = (f_y)_x$ Means the same thing.

HISTORICAL BIOGRAPHY

Pierre-Simon Laplace
(1749–1827)

EXAMPLE 9 Finding Second-Order Partial Derivatives

If $f(x, y) = x \cos y + ye^x$, find

$$\frac{\partial^2 f}{\partial x^2}, \quad \frac{\partial^2 f}{\partial y \partial x}, \quad \frac{\partial^2 f}{\partial y^2}, \quad \text{and} \quad \frac{\partial^2 f}{\partial x \partial y}.$$

Solution

$$\frac{\partial f}{\partial x} = \frac{\partial}{\partial x}(x \cos y + ye^x) = \cos y + ye^x$$

So

$$\frac{\partial^2 f}{\partial y \partial x} = \frac{\partial}{\partial y}\left(\frac{\partial f}{\partial x}\right) = -\sin y + e^x$$

$$\frac{\partial^2 f}{\partial x^2} = \frac{\partial}{\partial x}\left(\frac{\partial f}{\partial x}\right) = ye^x.$$

$$\frac{\partial f}{\partial y} = \frac{\partial}{\partial y}(x \cos y + ye^x) = -x \sin y + e^x$$

So

$$\frac{\partial^2 f}{\partial x \partial y} = \frac{\partial}{\partial x}\left(\frac{\partial f}{\partial y}\right) = -\sin y + e^x$$

$$\frac{\partial^2 f}{\partial y^2} = \frac{\partial}{\partial y}\left(\frac{\partial f}{\partial y}\right) = -x \cos y.$$

■

The Mixed Derivative Theorem

You may have noticed that the "mixed" second-order partial derivatives

$$\frac{\partial^2 f}{\partial y \partial x} \quad \text{and} \quad \frac{\partial^2 f}{\partial x \partial y}$$

in Example 9 were equal. This was not a coincidence. They must be equal whenever f, f_x, f_y, f_{xy}, and f_{yx} are continuous, as stated in the following theorem.

THEOREM 2 The Mixed Derivative Theorem

If $f(x, y)$ and its partial derivatives f_x, f_y, f_{xy}, and f_{yx} are defined throughout an open region containing a point (a, b) and are all continuous at (a, b), then

$$f_{xy}(a, b) = f_{yx}(a, b).$$

HISTORICAL BIOGRAPHY

Alexis Clairaut
(1713–1765)

Theorem 2 is also known as Clairaut's Theorem, named after the French mathematician Alexis Clairaut who discovered it. A proof is given in Appendix 7. Theorem 2 says that to calculate a mixed second-order derivative, we may differentiate in either order, provided the continuity conditions are satisfied. This can work to our advantage.

EXAMPLE 10 Choosing the Order of Differentiation

Find $\partial^2 w/\partial x \partial y$ if

$$w = xy + \frac{e^y}{y^2 + 1}.$$

Solution The symbol $\partial^2 w/\partial x \partial y$ tells us to differentiate first with respect to y and then with respect to x. If we postpone the differentiation with respect to y and differentiate first with respect to x, however, we get the answer more quickly. In two steps,

$$\frac{\partial w}{\partial x} = y \quad \text{and} \quad \frac{\partial^2 w}{\partial y \partial x} = 1.$$

If we differentiate first with respect to y, we obtain $\partial^2 w/\partial x \partial y = 1$ as well. ■

Partial Derivatives of Still Higher Order

Although we will deal mostly with first- and second-order partial derivatives, because these appear the most frequently in applications, there is no theoretical limit to how many times we can differentiate a function as long as the derivatives involved exist. Thus, we get third- and fourth-order derivatives denoted by symbols like

$$\frac{\partial^3 f}{\partial x \partial y^2} = f_{yyx}$$

$$\frac{\partial^4 f}{\partial x^2 \partial y^2} = f_{yyxx},$$

and so on. As with second-order derivatives, the order of differentiation is immaterial as long as all the derivatives through the order in question are continuous.

EXAMPLE 11 Calculating a Partial Derivative of Fourth-Order

Find f_{yxyz} if $f(x, y, z) = 1 - 2xy^2z + x^2y$.

Solution We first differentiate with respect to the variable y, then x, then y again, and finally with respect to z:

$$f_y = -4xyz + x^2$$

$$f_{yx} = -4yz + 2x$$

$$f_{yxy} = -4z$$

$$f_{yxyz} = -4$$

■

Differentiability

The starting point for differentiability is not Fermat's difference quotient but rather the idea of increment. You may recall from our work with functions of a single variable in Section 3.8 that if $y = f(x)$ is differentiable at $x = x_0$, then the change in the value of f that results from changing x from x_0 to $x_0 + \Delta x$ is given by an equation of the form

$$\Delta y = f'(x_0)\Delta x + \epsilon \Delta x$$

in which $\epsilon \to 0$ as $\Delta x \to 0$. For functions of two variables, the analogous property becomes the definition of differentiability. The Increment Theorem (from advanced calculus) tells us when to expect the property to hold.

THEOREM 3 The Increment Theorem for Functions of Two Variables

Suppose that the first partial derivatives of $f(x, y)$ are defined throughout an open region R containing the point (x_0, y_0) and that f_x and f_y are continuous at (x_0, y_0). Then the change

$$\Delta z = f(x_0 + \Delta x, y_0 + \Delta y) - f(x_0, y_0)$$

in the value of f that results from moving from (x_0, y_0) to another point $(x_0 + \Delta x, y_0 + \Delta y)$ in R satisfies an equation of the form

$$\Delta z = f_x(x_0, y_0)\Delta x + f_y(x_0, y_0)\Delta y + \epsilon_1 \Delta x + \epsilon_2 \Delta y,$$

in which each of $\epsilon_1, \epsilon_2 \to 0$ as both $\Delta x, \Delta y \to 0$.

You can see where the epsilons come from in the proof in Appendix 7. You will also see that similar results hold for functions of more than two independent variables.

DEFINITION Differentiable Function

A function $z = f(x, y)$ is **differentiable at** (x_0, y_0) if $f_x(x_0, y_0)$ and $f_y(x_0, y_0)$ exist and Δz satisfies an equation of the form

$$\Delta z = f_x(x_0, y_0)\Delta x + f_y(x_0, y_0)\Delta y + \epsilon_1 \Delta x + \epsilon_2 \Delta y,$$

in which each of $\epsilon_1, \epsilon_2 \to 0$ as both $\Delta x, \Delta y \to 0$. We call f **differentiable** if it is differentiable at every point in its domain.

In light of this definition, we have the immediate corollary of Theorem 3 that a function is differentiable if its first partial derivatives are *continuous*.

COROLLARY OF THEOREM 3 Continuity of Partial Derivatives Implies Differentiability

If the partial derivatives f_x and f_y of a function $f(x, y)$ are continuous throughout an open region R, then f is differentiable at every point of R.

If $z = f(x, y)$ is differentiable, then the definition of differentiability assures that $\Delta z = f(x_0 + \Delta x, y_0 + \Delta y) - f(x_0, y_0)$ approaches 0 as Δx and Δy approach 0. This tells us that a function of two variables is continuous at every point where it is differentiable.

THEOREM 4 Differentiability Implies Continuity
If a function $f(x, y)$ is differentiable at (x_0, y_0), then f is continuous at (x_0, y_0).

As we can see from Theorems 3 and 4, a function $f(x, y)$ must be continuous at a point (x_0, y_0) if f_x and f_y are continuous throughout an open region containing (x_0, y_0). Remember, however, that it is still possible for a function of two variables to be discontinuous at a point where its first partial derivatives exist, as we saw in Example 8. Existence alone of the partial derivative at a point is not enough.

EXERCISES 14.3

Calculating First-Order Partial Derivatives

In Exercises 1–22, find $\partial f/\partial x$ and $\partial f/\partial y$.

1. $f(x, y) = 2x^2 - 3y - 4$ **2.** $f(x, y) = x^2 - xy + y^2$

3. $f(x, y) = (x^2 - 1)(y + 2)$

4. $f(x, y) = 5xy - 7x^2 - y^2 + 3x - 6y + 2$

5. $f(x, y) = (xy - 1)^2$ **6.** $f(x, y) = (2x - 3y)^3$

7. $f(x, y) = \sqrt{x^2 + y^2}$ **8.** $f(x, y) = (x^3 + (y/2))^{2/3}$

9. $f(x, y) = 1/(x + y)$ **10.** $f(x, y) = x/(x^2 + y^2)$

11. $f(x, y) = (x + y)/(xy - 1)$ **12.** $f(x, y) = \tan^{-1}(y/x)$

13. $f(x, y) = e^{(x+y+1)}$ **14.** $f(x, y) = e^{-x}\sin(x + y)$

15. $f(x, y) = \ln(x + y)$ **16.** $f(x, y) = e^{xy}\ln y$

17. $f(x, y) = \sin^2(x - 3y)$ **18.** $f(x, y) = \cos^2(3x - y^2)$

19. $f(x, y) = x^y$ **20.** $f(x, y) = \log_y x$

21. $f(x, y) = \int_x^y g(t)\,dt$ (g continuous for all t)

22. $f(x, y) = \sum_{n=0}^{\infty} (xy)^n \quad (|xy| < 1)$

In Exercises 23–34, find f_x, f_y, and f_z.

23. $f(x, y, z) = 1 + xy^2 - 2z^2$ **24.** $f(x, y, z) = xy + yz + xz$

25. $f(x, y, z) = x - \sqrt{y^2 + z^2}$

26. $f(x, y, z) = (x^2 + y^2 + z^2)^{-1/2}$

27. $f(x, y, z) = \sin^{-1}(xyz)$ **28.** $f(x, y, z) = \sec^{-1}(x + yz)$

29. $f(x, y, z) = \ln(x + 2y + 3z)$

30. $f(x, y, z) = yz\ln(xy)$ **31.** $f(x, y, z) = e^{-(x^2+y^2+z^2)}$

32. $f(x, y, z) = e^{-xyz}$

33. $f(x, y, z) = \tanh(x + 2y + 3z)$

34. $f(x, y, z) = \sinh(xy - z^2)$

In Exercises 35–40, find the partial derivative of the function with respect to each variable.

35. $f(t, \alpha) = \cos(2\pi t - \alpha)$ **36.** $g(u, v) = v^2 e^{(2u/v)}$

37. $h(\rho, \phi, \theta) = \rho\sin\phi\cos\theta$ **38.** $g(r, \theta, z) = r(1 - \cos\theta) - z$

39. Work done by the heart (Section 3.8, Exercise 51)

$$W(P, V, \delta, v, g) = PV + \frac{V\delta v^2}{2g}$$

40. Wilson lot size formula (Section 4.5, Exercise 45)

$$A(c, h, k, m, q) = \frac{km}{q} + cm + \frac{hq}{2}$$

Calculating Second-Order Partial Derivatives

Find all the second-order partial derivatives of the functions in Exercises 41–46.

41. $f(x, y) = x + y + xy$ **42.** $f(x, y) = \sin xy$

43. $g(x, y) = x^2y + \cos y + y \sin x$

44. $h(x, y) = xe^y + y + 1$ **45.** $r(x, y) = \ln(x + y)$

46. $s(x, y) = \tan^{-1}(y/x)$

Mixed Partial Derivatives

In Exercises 47–50, verify that $w_{xy} = w_{yx}$.

47. $w = \ln(2x + 3y)$ **48.** $w = e^x + x \ln y + y \ln x$

49. $w = xy^2 + x^2y^3 + x^3y^4$ **50.** $w = x \sin y + y \sin x + xy$

51. Which order of differentiation will calculate f_{xy} faster: x first or y first? Try to answer without writing anything down.

a. $f(x, y) = x \sin y + e^y$

b. $f(x, y) = 1/x$

c. $f(x, y) = y + (x/y)$

d. $f(x, y) = y + x^2y + 4y^3 - \ln(y^2 + 1)$

e. $f(x, y) = x^2 + 5xy + \sin x + 7e^x$

f. $f(x, y) = x \ln xy$

52. The fifth-order partial derivative $\partial^5 f/\partial x^2 \partial y^3$ is zero for each of the following functions. To show this as quickly as possible, which variable would you differentiate with respect to first: x or y? Try to answer without writing anything down.

a. $f(x, y) = y^2x^4e^x + 2$

b. $f(x, y) = y^2 + y(\sin x - x^4)$

c. $f(x, y) = x^2 + 5xy + \sin x + 7e^x$

d. $f(x, y) = xe^{y^2/2}$

Using the Partial Derivative Definition

In Exercises 53 and 54, use the limit definition of partial derivative to compute the partial derivatives of the functions at the specified points.

53. $f(x, y) = 1 - x + y - 3x^2y,\quad \dfrac{\partial f}{\partial x}$ and $\dfrac{\partial f}{\partial y}$ at $(1, 2)$

54. $f(x, y) = 4 + 2x - 3y - xy^2,\quad \dfrac{\partial f}{\partial x}$ and $\dfrac{\partial f}{\partial y}$ at $(-2, 1)$

55. Three variables Let $w = f(x, y, z)$ be a function of three independent variables and write the formal definition of the partial derivative $\partial f/\partial z$ at (x_0, y_0, z_0). Use this definition to find $\partial f/\partial z$ at $(1, 2, 3)$ for $f(x, y, z) = x^2yz^2$.

56. Three variables Let $w = f(x, y, z)$ be a function of three independent variables and write the formal definition of the partial derivative $\partial f/\partial y$ at (x_0, y_0, z_0). Use this definition to find $\partial f/\partial y$ at $(-1, 0, 3)$ for $f(x, y, z) = -2xy^2 + yz^2$.

Differentiating Implicitly

57. Find the value of $\partial z/\partial x$ at the point $(1, 1, 1)$ if the equation

$$xy + z^3x - 2yz = 0$$

defines z as a function of the two independent variables x and y and the partial derivative exists.

58. Find the value of $\partial x/\partial z$ at the point $(1, -1, -3)$ if the equation

$$xz + y \ln x - x^2 + 4 = 0$$

defines x as a function of the two independent variables y and z and the partial derivative exists.

Exercises 59 and 60 are about the triangle shown here.

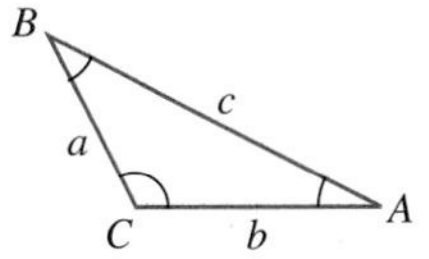

59. Express A implicitly as a function of a, b, and c and calculate $\partial A/\partial a$ and $\partial A/\partial b$.

60. Express a implicitly as a function of A, b, and B and calculate $\partial a/\partial A$ and $\partial a/\partial B$.

61. Two dependent variables Express v_x in terms of u and v if the equations $x = v \ln u$ and $y = u \ln v$ define u and v as functions of the independent variables x and y, and if v_x exists. (*Hint:* Differentiate both equations with respect to x and solve for v_x by eliminating u_x.)

62. Two dependent variables Find $\partial x/\partial u$ and $\partial y/\partial u$ if the equations $u = x^2 - y^2$ and $v = x^2 - y$ define x and y as functions of the independent variables u and v, and the partial derivatives exist. (See the hint in Exercise 61.) Then let $s = x^2 + y^2$ and find $\partial s/\partial u$.

Laplace Equations

The **three-dimensional Laplace equation**

$$\frac{\partial^2 f}{\partial x^2} + \frac{\partial^2 f}{\partial y^2} + \frac{\partial^2 f}{\partial z^2} = 0$$

is satisfied by steady-state temperature distributions $T = f(x, y, z)$ in space, by gravitational potentials, and by electrostatic potentials. The **two-dimensional Laplace equation**

$$\frac{\partial^2 f}{\partial x^2} + \frac{\partial^2 f}{\partial y^2} = 0,$$

obtained by dropping the $\partial^2 f/\partial z^2$ term from the previous equation, describes potentials and steady-state temperature distributions in a plane (see the accompanying figure). The plane (a) may be treated as a thin slice of the solid (b) perpendicular to the z-axis.

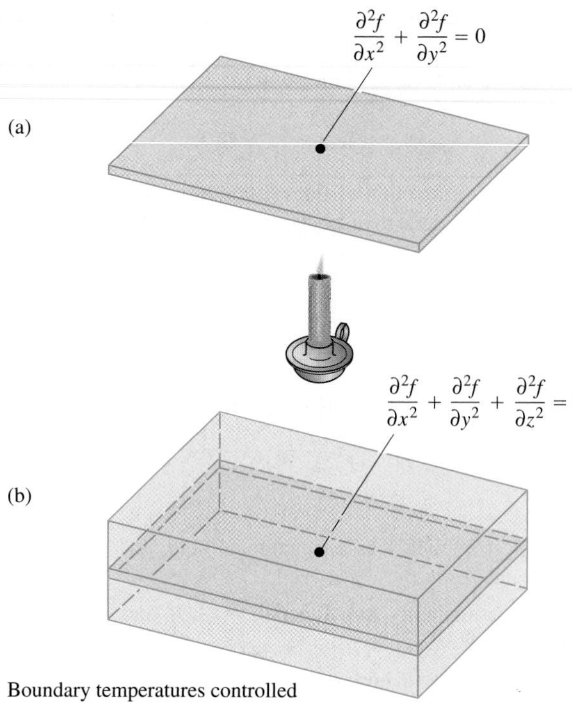

Show that each function in Exercises 63–68 satisfies a Laplace equation.

63. $f(x, y, z) = x^2 + y^2 - 2z^2$

64. $f(x, y, z) = 2z^3 - 3(x^2 + y^2)z$

65. $f(x, y) = e^{-2y} \cos 2x$

66. $f(x, y) = \ln \sqrt{x^2 + y^2}$

67. $f(x, y, z) = (x^2 + y^2 + z^2)^{-1/2}$

68. $f(x, y, z) = e^{3x+4y} \cos 5z$

The Wave Equation

If we stand on an ocean shore and take a snapshot of the waves, the picture shows a regular pattern of peaks and valleys in an instant of time. We see periodic vertical motion in space, with respect to distance. If we stand in the water, we can feel the rise and fall of the water as the waves go by. We see periodic vertical motion in time. In physics, this beautiful symmetry is expressed by the **one-dimensional wave equation**

$$\frac{\partial^2 w}{\partial t^2} = c^2 \frac{\partial^2 w}{\partial x^2},$$

where w is the wave height, x is the distance variable, t is the time variable, and c is the velocity with which the waves are propagated.

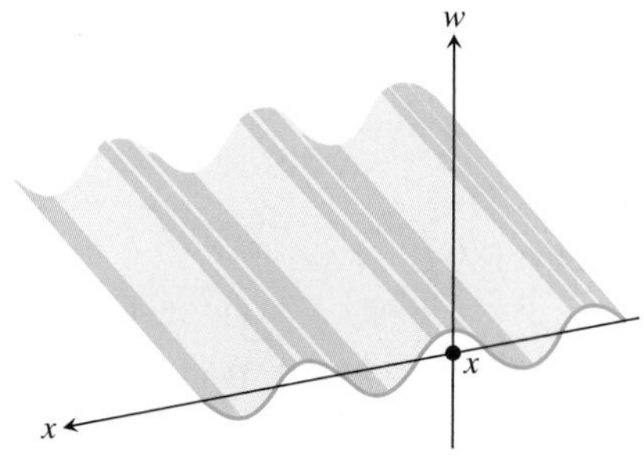

In our example, x is the distance across the ocean's surface, but in other applications, x might be the distance along a vibrating string, distance through air (sound waves), or distance through space (light waves). The number c varies with the medium and type of wave.

Show that the functions in Exercises 69–75 are all solutions of the wave equation.

69. $w = \sin(x + ct)$

70. $w = \cos(2x + 2ct)$

71. $w = \sin(x + ct) + \cos(2x + 2ct)$

72. $w = \ln(2x + 2ct)$

73. $w = \tan(2x - 2ct)$

74. $w = 5\cos(3x + 3ct) + e^{x+ct}$

75. $w = f(u)$, where f is a differentiable function of u, and $u = a(x + ct)$, where a is a constant

Continuous Partial Derivatives

76. Does a function $f(x, y)$ with continuous first partial derivatives throughout an open region R have to be continuous on R? Give reasons for your answer.

77. If a function $f(x, y)$ has continuous second partial derivatives throughout an open region R, must the first-order partial derivatives of f be continuous on R? Give reasons for your answer.

14.4 The Chain Rule

The Chain Rule for functions of a single variable studied in Section 3.5 said that when $w = f(x)$ was a differentiable function of x and $x = g(t)$ was a differentiable function of t, w became a differentiable function of t and dw/dt could be calculated with the formula

$$\frac{dw}{dt} = \frac{dw}{dx}\frac{dx}{dt}.$$

For functions of two or more variables the Chain Rule has several forms. The form depends on how many variables are involved but works like the Chain Rule in Section 3.5 once we account for the presence of additional variables.

Functions of Two Variables

The Chain Rule formula for a function $w = f(x, y)$ when $x = x(t)$ and $y = y(t)$ are both differentiable functions of t is given in the following theorem.

THEOREM 5 Chain Rule for Functions of Two Independent Variables
If $w = f(x, y)$ has continuous partial derivatives f_x and f_y and if $x = x(t), y = y(t)$ are differentiable functions of t, then the composite $w = f(x(t), y(t))$ is a differentiable function of t and

$$\frac{df}{dt} = f_x(x(t), y(t)) \cdot x'(t) + f_y(x(t), y(t)) \cdot y'(t),$$

or

$$\frac{dw}{dt} = \frac{\partial f}{\partial x}\frac{dx}{dt} + \frac{\partial f}{\partial y}\frac{dy}{dt}.$$

Proof The proof consists of showing that if x and y are differentiable at $t = t_0$, then w is differentiable at t_0 and

$$\left(\frac{dw}{dt}\right)_{t_0} = \left(\frac{\partial w}{\partial x}\right)_{P_0}\left(\frac{dx}{dt}\right)_{t_0} + \left(\frac{\partial w}{\partial y}\right)_{P_0}\left(\frac{dy}{dt}\right)_{t_0},$$

where $P_0 = (x(t_0), y(t_0))$. The subscripts indicate where each of the derivatives are to be evaluated.

To remember the Chain Rule picture the diagram below. To find dw/dt, start at w and read down each route to t, multiplying derivatives along the way. Then add the products.

Chain Rule

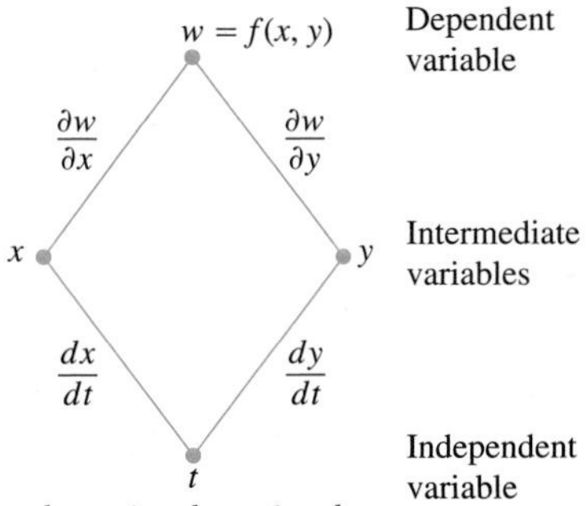

$$\frac{dw}{dt} = \frac{\partial w}{\partial x}\frac{dx}{dt} + \frac{\partial w}{\partial y}\frac{dy}{dt}$$

Let Δx, Δy, and Δw be the increments that result from changing t from t_0 to $t_0 + \Delta t$. Since f is differentiable (see the definition in Section 14.3),

$$\Delta w = \left(\frac{\partial w}{\partial x}\right)_{P_0}\Delta x + \left(\frac{\partial w}{\partial y}\right)_{P_0}\Delta y + \epsilon_1 \Delta x + \epsilon_2 \Delta y,$$

where $\epsilon_1, \epsilon_2 \to 0$ as $\Delta x, \Delta y \to 0$. To find dw/dt, we divide this equation through by Δt and let Δt approach zero. The division gives

$$\frac{\Delta w}{\Delta t} = \left(\frac{\partial w}{\partial x}\right)_{P_0}\frac{\Delta x}{\Delta t} + \left(\frac{\partial w}{\partial y}\right)_{P_0}\frac{\Delta y}{\Delta t} + \epsilon_1\frac{\Delta x}{\Delta t} + \epsilon_2\frac{\Delta y}{\Delta t}.$$

Letting Δt approach zero gives

$$\left(\frac{dw}{dt}\right)_{t_0} = \lim_{\Delta t \to 0}\frac{\Delta w}{\Delta t}$$

$$= \left(\frac{\partial w}{\partial x}\right)_{P_0}\left(\frac{dx}{dt}\right)_{t_0} + \left(\frac{\partial w}{\partial y}\right)_{P_0}\left(\frac{dy}{dt}\right)_{t_0} + 0 \cdot \left(\frac{dx}{dt}\right)_{t_0} + 0 \cdot \left(\frac{dy}{dt}\right)_{t_0}. \qquad \blacksquare$$

The **tree diagram** in the margin provides a convenient way to remember the Chain Rule. From the diagram, you see that when $t = t_0$, the derivatives dx/dt and dy/dt are evaluated at t_0. The value of t_0 then determines the value x_0 for the differentiable function x and the value y_0 for the differentiable function y. The partial derivatives $\partial w/\partial x$ and $\partial w/\partial y$ (which are themselves functions of x and y) are evaluated at the point $P_0(x_0, y_0)$ corresponding to t_0. The "true" independent variable is t, whereas x and y are *intermediate variables* (controlled by t) and w is the dependent variable.

A more precise notation for the Chain Rule shows how the various derivatives in Theorem 5 are evaluated:

$$\frac{dw}{dt}(t_0) = \frac{\partial f}{\partial x}(x_0, y_0) \cdot \frac{dx}{dt}(t_0) + \frac{\partial f}{\partial y}(x_0, y_0) \cdot \frac{dy}{dt}(t_0).$$

EXAMPLE 1 Applying the Chain Rule

Use the Chain Rule to find the derivative of

$$w = xy$$

with respect to t along the path $x = \cos t, y = \sin t$. What is the derivative's value at $t = \pi/2$?

Solution We apply the Chain Rule to find dw/dt as follows:

$$\begin{aligned}
\frac{dw}{dt} &= \frac{\partial w}{\partial x}\frac{dx}{dt} + \frac{\partial w}{\partial y}\frac{dy}{dt} \\
&= \frac{\partial(xy)}{\partial x} \cdot \frac{d}{dt}(\cos t) + \frac{\partial(xy)}{\partial y} \cdot \frac{d}{dt}(\sin t) \\
&= (y)(-\sin t) + (x)(\cos t) \\
&= (\sin t)(-\sin t) + (\cos t)(\cos t) \\
&= -\sin^2 t + \cos^2 t \\
&= \cos 2t.
\end{aligned}$$

In this example, we can check the result with a more direct calculation. As a function of t,

$$w = xy = \cos t \sin t = \frac{1}{2}\sin 2t,$$

so

$$\frac{dw}{dt} = \frac{d}{dt}\left(\frac{1}{2}\sin 2t\right) = \frac{1}{2} \cdot 2\cos 2t = \cos 2t.$$

In either case, at the given value of t,

$$\left(\frac{dw}{dt}\right)_{t=\pi/2} = \cos\left(2 \cdot \frac{\pi}{2}\right) = \cos \pi = -1. \qquad \blacksquare$$

Functions of Three Variables

You can probably predict the Chain Rule for functions of three variables, as it only involves adding the expected third term to the two-variable formula.

THEOREM 6 Chain Rule for Functions of Three Independent Variables
If $w = f(x, y, z)$ is differentiable and x, y, and z are differentiable functions of t, then w is a differentiable function of t and

$$\frac{dw}{dt} = \frac{\partial f}{\partial x}\frac{dx}{dt} + \frac{\partial f}{\partial y}\frac{dy}{dt} + \frac{\partial f}{\partial z}\frac{dz}{dt}.$$

Here we have three routes from w to t instead of two, but finding dw/dt is still the same. Read down each route, multiplying derivatives along the way; then add.

Chain Rule

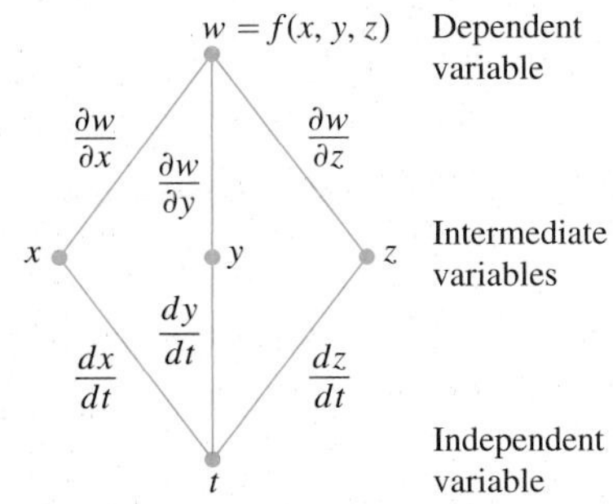

$$\frac{dw}{dt} = \frac{\partial w}{\partial x}\frac{dx}{dt} + \frac{\partial w}{\partial y}\frac{dy}{dt} + \frac{\partial w}{\partial z}\frac{dz}{dt}$$

The proof is identical with the proof of Theorem 5 except that there are now three intermediate variables instead of two. The diagram we use for remembering the new equation is similar as well, with three routes from w to t.

EXAMPLE 2 Changes in a Function's Values Along a Helix

Find dw/dt if

$$w = xy + z, \qquad x = \cos t, \qquad y = \sin t, \qquad z = t.$$

In this example the values of w are changing along the path of a helix (Section 13.1). What is the derivative's value at $t = 0$?

Solution

$$\begin{aligned}\frac{dw}{dt} &= \frac{\partial w}{\partial x}\frac{dx}{dt} + \frac{\partial w}{\partial y}\frac{dy}{dt} + \frac{\partial w}{\partial z}\frac{dz}{dt}\\ &= (y)(-\sin t) + (x)(\cos t) + (1)(1)\\ &= (\sin t)(-\sin t) + (\cos t)(\cos t) + 1 \qquad \text{Substitute for the intermediate variables.}\\ &= -\sin^2 t + \cos^2 t + 1 = 1 + \cos 2t.\end{aligned}$$

$$\left(\frac{dw}{dt}\right)_{t=0} = 1 + \cos(0) = 2.$$ ■

Here is a physical interpretation of change along a curve. If $w = T(x, y, z)$ is the temperature at each point (x, y, z) along a curve C with parametric equations $x = x(t)$, $y = y(t)$, and $z = z(t)$, then the composite function $w = T(x(t), y(t), z(t))$ represents the temperature relative to t along the curve. The derivative dw/dt is then the instantaneous rate of change of temperature along the curve, as calculated in Theorem 6.

Functions Defined on Surfaces

If we are interested in the temperature $w = f(x, y, z)$ at points (x, y, z) on a globe in space, we might prefer to think of x, y, and z as functions of the variables r and s that give the points' longitudes and latitudes. If $x = g(r, s)$, $y = h(r, s)$, and $z = k(r, s)$, we could then express the temperature as a function of r and s with the composite function

$$w = f(g(r, s), h(r, s), k(r, s)).$$

Under the right conditions, w would have partial derivatives with respect to both r and s that could be calculated in the following way.

THEOREM 7 Chain Rule for Two Independent Variables and Three Intermediate Variables

Suppose that $w = f(x, y, z)$, $x = g(r, s)$, $y = h(r, s)$, and $z = k(r, s)$. If all four functions are differentiable, then w has partial derivatives with respect to r and s, given by the formulas

$$\frac{\partial w}{\partial r} = \frac{\partial w}{\partial x}\frac{\partial x}{\partial r} + \frac{\partial w}{\partial y}\frac{\partial y}{\partial r} + \frac{\partial w}{\partial z}\frac{\partial z}{\partial r}$$

$$\frac{\partial w}{\partial s} = \frac{\partial w}{\partial x}\frac{\partial x}{\partial s} + \frac{\partial w}{\partial y}\frac{\partial y}{\partial s} + \frac{\partial w}{\partial z}\frac{\partial z}{\partial s}.$$

The first of these equations can be derived from the Chain Rule in Theorem 6 by holding s fixed and treating r as t. The second can be derived in the same way, holding r fixed and treating s as t. The tree diagrams for both equations are shown in Figure 14.19.

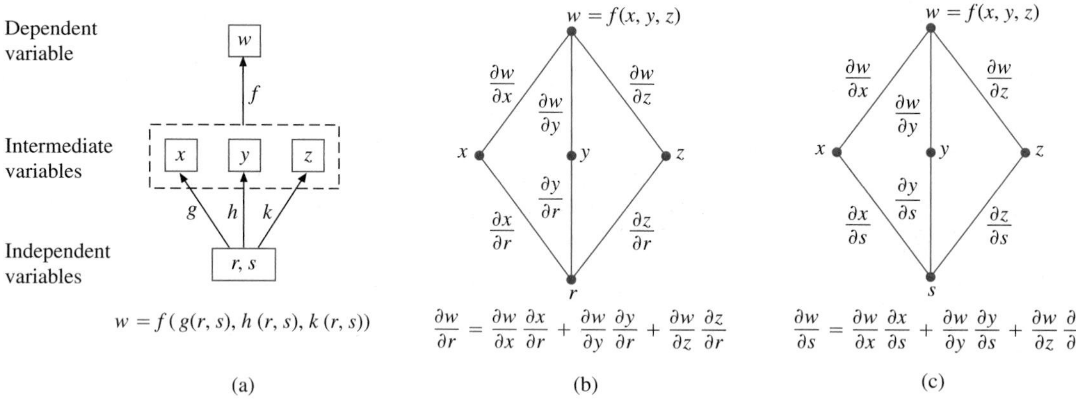

FIGURE 14.19 Composite function and tree diagrams for Theorem 7.

EXAMPLE 3 Partial Derivatives Using Theorem 7

Express $\partial w/\partial r$ and $\partial w/\partial s$ in terms of r and s if

$$w = x + 2y + z^2, \qquad x = \frac{r}{s}, \qquad y = r^2 + \ln s, \qquad z = 2r.$$

Solution

$$\begin{aligned}\frac{\partial w}{\partial r} &= \frac{\partial w}{\partial x}\frac{\partial x}{\partial r} + \frac{\partial w}{\partial y}\frac{\partial y}{\partial r} + \frac{\partial w}{\partial z}\frac{\partial z}{\partial r}\\ &= (1)\left(\frac{1}{s}\right) + (2)(2r) + (2z)(2)\\ &= \frac{1}{s} + 4r + (4r)(2) = \frac{1}{s} + 12r\end{aligned}$$

Substitute for intermediate variable z.

$$\begin{aligned}\frac{\partial w}{\partial s} &= \frac{\partial w}{\partial x}\frac{\partial x}{\partial s} + \frac{\partial w}{\partial y}\frac{\partial y}{\partial s} + \frac{\partial w}{\partial z}\frac{\partial z}{\partial s}\\ &= (1)\left(-\frac{r}{s^2}\right) + (2)\left(\frac{1}{s}\right) + (2z)(0) = \frac{2}{s} - \frac{r}{s^2}\end{aligned}$$ ■

If f is a function of two variables instead of three, each equation in Theorem 7 becomes correspondingly one term shorter.

Chain Rule

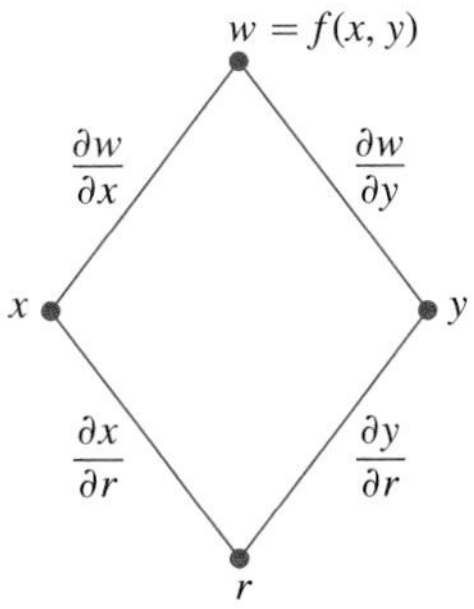

$$\frac{\partial w}{\partial r} = \frac{\partial w}{\partial x}\frac{\partial x}{\partial r} + \frac{\partial w}{\partial y}\frac{\partial y}{\partial r}$$

FIGURE 14.20 Tree diagram for the equation

$$\frac{\partial w}{\partial r} = \frac{\partial w}{\partial x}\frac{\partial x}{\partial r} + \frac{\partial w}{\partial y}\frac{\partial y}{\partial r}.$$

> If $w = f(x, y)$, $x = g(r, s)$, and $y = h(r, s)$, then
>
> $$\frac{\partial w}{\partial r} = \frac{\partial w}{\partial x}\frac{\partial x}{\partial r} + \frac{\partial w}{\partial y}\frac{\partial y}{\partial r} \quad \text{and} \quad \frac{\partial w}{\partial s} = \frac{\partial w}{\partial x}\frac{\partial x}{\partial s} + \frac{\partial w}{\partial y}\frac{\partial y}{\partial s}.$$

Figure 14.20 shows the tree diagram for the first of these equations. The diagram for the second equation is similar; just replace r with s.

EXAMPLE 4 More Partial Derivatives

Express $\partial w/\partial r$ and $\partial w/\partial s$ in terms of r and s if

$$w = x^2 + y^2, \qquad x = r - s, \qquad y = r + s.$$

Solution

$$\begin{aligned}\frac{\partial w}{\partial r} &= \frac{\partial w}{\partial x}\frac{\partial x}{\partial r} + \frac{\partial w}{\partial y}\frac{\partial y}{\partial r} \\ &= (2x)(1) + (2y)(1) \\ &= 2(r - s) + 2(r + s) \\ &= 4r\end{aligned} \qquad \begin{aligned}\frac{\partial w}{\partial s} &= \frac{\partial w}{\partial x}\frac{\partial x}{\partial s} + \frac{\partial w}{\partial y}\frac{\partial y}{\partial s} \\ &= (2x)(-1) + (2y)(1) \\ &= -2(r - s) + 2(r + s) \\ &= 4s\end{aligned}$$

Substitute for the intermediate variables.

■

If f is a function of x alone, our equations become even simpler.

Chain Rule

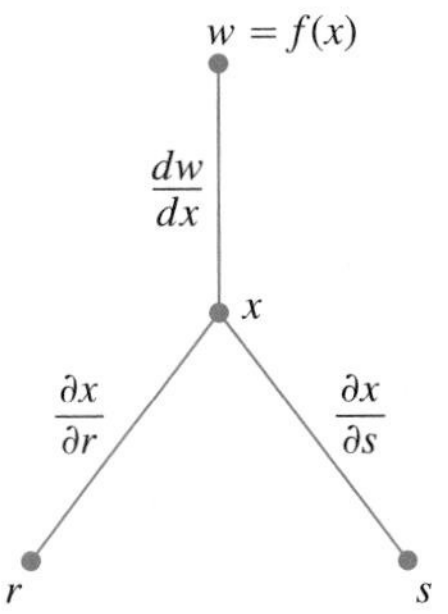

$$\frac{\partial w}{\partial r} = \frac{dw}{dx}\frac{\partial x}{\partial r}$$

$$\frac{\partial w}{\partial s} = \frac{dw}{dx}\frac{\partial x}{\partial s}$$

FIGURE 14.21 Tree diagram for differentiating f as a composite function of r and s with one intermediate variable.

> If $w = f(x)$ and $x = g(r, s)$, then
>
> $$\frac{\partial w}{\partial r} = \frac{dw}{dx}\frac{\partial x}{\partial r} \quad \text{and} \quad \frac{\partial w}{\partial s} = \frac{dw}{dx}\frac{\partial x}{\partial s}.$$

In this case, we can use the ordinary (single-variable) derivative, dw/dx. The tree diagram is shown in Figure 14.21.

Implicit Differentiation Revisited

The two-variable Chain Rule in Theorem 5 leads to a formula that takes most of the work out of implicit differentiation. Suppose that

1. The function $F(x, y)$ is differentiable and
2. The equation $F(x, y) = 0$ defines y implicitly as a differentiable function of x, say $y = h(x)$.

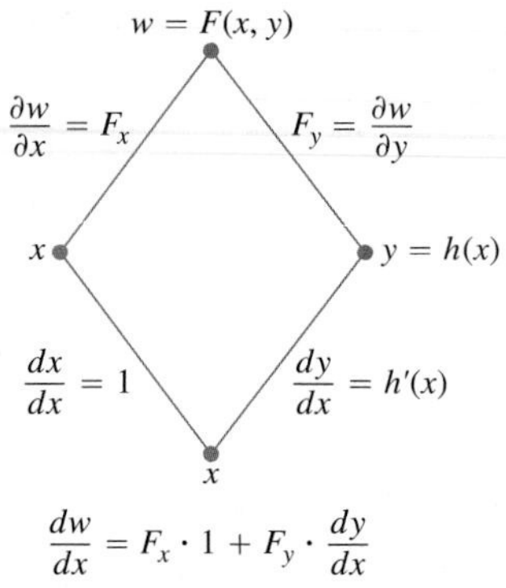

FIGURE 14.22 Tree diagram for differentiating $w = F(x, y)$ with respect to x. Setting $dw/dx = 0$ leads to a simple computational formula for implicit differentiation (Theorem 8).

Since $w = F(x, y) = 0$, the derivative dw/dx must be zero. Computing the derivative from the Chain Rule (tree diagram in Figure 14.22), we find

$$\begin{aligned} 0 = \frac{dw}{dx} &= F_x \frac{dx}{dx} + F_y \frac{dy}{dx} \qquad \text{Theorem 5 with } t = x \text{ and } f = F \\ &= F_x \cdot 1 + F_y \cdot \frac{dy}{dx}. \end{aligned}$$

If $F_y = \partial w/\partial y \neq 0$, we can solve this equation for dy/dx to get

$$\frac{dy}{dx} = -\frac{F_x}{F_y}.$$

This relationship gives a surprisingly simple shortcut to finding derivatives of implicitly defined functions, which we state here as a theorem.

THEOREM 8 A Formula for Implicit Differentiation
Suppose that $F(x, y)$ is differentiable and that the equation $F(x, y) = 0$ defines y as a differentiable function of x. Then at any point where $F_y \neq 0$,

$$\frac{dy}{dx} = -\frac{F_x}{F_y}.$$

EXAMPLE 5 Implicit Differentiation

Use Theorem 8 to find dy/dx if $y^2 - x^2 - \sin xy = 0$.

Solution Take $F(x, y) = y^2 - x^2 - \sin xy$. Then

$$\begin{aligned} \frac{dy}{dx} = -\frac{F_x}{F_y} &= -\frac{-2x - y\cos xy}{2y - x\cos xy} \\ &= \frac{2x + y\cos xy}{2y - x\cos xy}. \end{aligned}$$

This calculation is significantly shorter than the single-variable calculation with which we found dy/dx in Section 3.6, Example 3. ■

Functions of Many Variables

We have seen several different forms of the Chain Rule in this section, but you do not have to memorize them all if you can see them as special cases of the same general formula. When solving particular problems, it may help to draw the appropriate tree diagram by placing the dependent variable on top, the intermediate variables in the middle, and the selected independent variable at the bottom. To find the derivative of the dependent variable with respect to the selected independent variable, start at the dependent variable and read down each route of the tree to the independent variable, calculating and multiplying the derivatives along each route. Then add the products you found for the different routes.

In general, suppose that $w = f(x, y, \ldots, v)$ is a differentiable function of the variables $x, y, \ldots, v$ (a finite set) and the $x, y, \ldots, v$ are differentiable functions of $p, q, \ldots, t$ (another finite set). Then w is a differentiable function of the variables p through t and the partial derivatives of w with respect to these variables are given by equations of the form

$$\frac{\partial w}{\partial p} = \frac{\partial w}{\partial x}\frac{\partial x}{\partial p} + \frac{\partial w}{\partial y}\frac{\partial y}{\partial p} + \cdots + \frac{\partial w}{\partial v}\frac{\partial v}{\partial p}.$$

The other equations are obtained by replacing p by $q, \ldots, t$, one at a time.

One way to remember this equation is to think of the right-hand side as the dot product of two vectors with components

$$\underbrace{\left(\frac{\partial w}{\partial x}, \frac{\partial w}{\partial y}, \ldots, \frac{\partial w}{\partial v}\right)}_{\text{Derivatives of } w \text{ with respect to the intermediate variables}} \quad \text{and} \quad \underbrace{\left(\frac{\partial x}{\partial p}, \frac{\partial y}{\partial p}, \ldots, \frac{\partial v}{\partial p}\right)}_{\text{Derivatives of the intermediate variables with respect to the selected independent variable}}.$$

EXERCISES 14.4

Chain Rule: One Independent Variable

In Exercises 1–6, **(a)** express dw/dt as a function of t, both by using the Chain Rule and by expressing w in terms of t and differentiating directly with respect to t. Then **(b)** evaluate dw/dt at the given value of t.

1. $w = x^2 + y^2, \quad x = \cos t, \quad y = \sin t; \quad t = \pi$

2. $w = x^2 + y^2, \quad x = \cos t + \sin t, \quad y = \cos t - \sin t; \quad t = 0$

3. $w = \frac{x}{z} + \frac{y}{z}, \quad x = \cos^2 t, \quad y = \sin^2 t, \quad z = 1/t; \quad t = 3$

4. $w = \ln(x^2 + y^2 + z^2), \quad x = \cos t, \quad y = \sin t, \quad z = 4\sqrt{t};$ $t = 3$

5. $w = 2ye^x - \ln z, \quad x = \ln(t^2 + 1), \quad y = \tan^{-1} t, \quad z = e^t;$ $t = 1$

6. $w = z - \sin xy, \quad x = t, \quad y = \ln t, \quad z = e^{t-1}; \quad t = 1$

Chain Rule: Two and Three Independent Variables

In Exercises 7 and 8, **(a)** express $\partial z/\partial u$ and $\partial z/\partial v$ as functions of u and v both by using the Chain Rule and by expressing z directly in terms of u and v before differentiating. Then **(b)** evaluate $\partial z/\partial u$ and $\partial z/\partial v$ at the given point (u, v).

7. $z = 4e^x \ln y, \quad x = \ln(u \cos v), \quad y = u \sin v;$ $(u, v) = (2, \pi/4)$

8. $z = \tan^{-1}(x/y), \quad x = u \cos v, \quad y = u \sin v;$ $(u, v) = (1.3, \pi/6)$

In Exercises 9 and 10, **(a)** express $\partial w/\partial u$ and $\partial w/\partial v$ as functions of u and v both by using the Chain Rule and by expressing w directly in terms of u and v before differentiating. Then **(b)** evaluate $\partial w/\partial u$ and $\partial w/\partial v$ at the given point (u, v).

9. $w = xy + yz + xz, \quad x = u + v, \quad y = u - v, \quad z = uv;$ $(u, v) = (1/2, 1)$

10. $w = \ln(x^2 + y^2 + z^2), \quad x = ue^v \sin u, \quad y = ue^v \cos u,$ $z = ue^v; \quad (u, v) = (-2, 0)$

In Exercises 11 and 12, **(a)** express $\partial u/\partial x$, $\partial u/\partial y$, and $\partial u/\partial z$ as functions of x, y, and z both by using the Chain Rule and by expressing u directly in terms of x, y, and z before differentiating. Then **(b)** evaluate $\partial u/\partial x$, $\partial u/\partial y$, and $\partial u/\partial z$ at the given point (x, y, z).

11. $u = \frac{p - q}{q - r}, \quad p = x + y + z, \quad q = x - y + z,$ $r = x + y - z; \quad (x, y, z) = \left(\sqrt{3}, 2, 1\right)$

12. $u = e^{qr} \sin^{-1} p, \quad p = \sin x, \quad q = z^2 \ln y, \quad r = 1/z;$ $(x, y, z) = (\pi/4, 1/2, -1/2)$

Using a Tree Diagram

In Exercises 13–24, draw a tree diagram and write a Chain Rule formula for each derivative.

13. $\frac{dz}{dt}$ for $z = f(x, y), \quad x = g(t), \quad y = h(t)$

14. $\frac{dz}{dt}$ for $z = f(u, v, w), \quad u = g(t), \quad v = h(t), \quad w = k(t)$

15. $\frac{\partial w}{\partial u}$ and $\frac{\partial w}{\partial v}$ for $w = h(x, y, z), \quad x = f(u, v), \quad y = g(u, v),$ $z = k(u, v)$

16. $\frac{\partial w}{\partial x}$ and $\frac{\partial w}{\partial y}$ for $w = f(r, s, t), \quad r = g(x, y), \quad s = h(x, y), \quad t = k(x, y)$

17. $\frac{\partial w}{\partial u}$ and $\frac{\partial w}{\partial v}$ for $w = g(x, y), \quad x = h(u, v), \quad y = k(u, v)$

18. $\frac{\partial w}{\partial x}$ and $\frac{\partial w}{\partial y}$ for $w = g(u, v), \quad u = h(x, y), \quad v = k(x, y)$

19. $\frac{\partial z}{\partial t}$ and $\frac{\partial z}{\partial s}$ for $z = f(x, y), \quad x = g(t, s), \quad y = h(t, s)$

20. $\frac{\partial y}{\partial r}$ for $y = f(u), \quad u = g(r, s)$

21. $\frac{\partial w}{\partial s}$ and $\frac{\partial w}{\partial t}$ for $w = g(u), \quad u = h(s, t)$

22. $\frac{\partial w}{\partial p}$ for $w = f(x, y, z, v), \quad x = g(p, q), \quad y = h(p, q), \quad z = j(p, q), \quad v = k(p, q)$

23. $\frac{\partial w}{\partial r}$ and $\frac{\partial w}{\partial s}$ for $w = f(x, y), \quad x = g(r), \quad y = h(s)$

24. $\frac{\partial w}{\partial s}$ for $w = g(x, y), \quad x = h(r, s, t), \quad y = k(r, s, t)$

Implicit Differentiation

Assuming that the equations in Exercises 25–28 define y as a differentiable function of x, use Theorem 8 to find the value of dy/dx at the given point.

25. $x^3 - 2y^2 + xy = 0, \quad (1, 1)$

26. $xy + y^2 - 3x - 3 = 0, \quad (-1, 1)$

27. $x^2 + xy + y^2 - 7 = 0, \quad (1, 2)$

28. $xe^y + \sin xy + y - \ln 2 = 0, \quad (0, \ln 2)$

Three-Variable Implicit Differentiation

Theorem 8 can be generalized to functions of three variables and even more. The three-variable version goes like this: If the equation $F(x, y, z) = 0$ determines z as a differentiable function of x and y, then, at points where $F_z \neq 0$,

$$\frac{\partial z}{\partial x} = -\frac{F_x}{F_z} \quad \text{and} \quad \frac{\partial z}{\partial y} = -\frac{F_y}{F_z}.$$

Use these equations to find the values of $\partial z/\partial x$ and $\partial z/\partial y$ at the points in Exercises 29–32.

29. $z^3 - xy + yz + y^3 - 2 = 0, \quad (1, 1, 1)$

30. $\frac{1}{x} + \frac{1}{y} + \frac{1}{z} - 1 = 0, \quad (2, 3, 6)$

31. $\sin(x + y) + \sin(y + z) + \sin(x + z) = 0, \quad (\pi, \pi, \pi)$

32. $xe^y + ye^z + 2\ln x - 2 - 3\ln 2 = 0, \quad (1, \ln 2, \ln 3)$

Finding Specified Partial Derivatives

33. Find $\partial w/\partial r$ when $r = 1, s = -1$ if $w = (x + y + z)^2$, $x = r - s, y = \cos(r + s), z = \sin(r + s)$.

34. Find $\partial w/\partial v$ when $u = -1, v = 2$ if $w = xy + \ln z$, $x = v^2/u, y = u + v, z = \cos u$.

35. Find $\partial w/\partial v$ when $u = 0, v = 0$ if $w = x^2 + (y/x)$, $x = u - 2v + 1, y = 2u + v - 2$.

36. Find $\partial z/\partial u$ when $u = 0, v = 1$ if $z = \sin xy + x\sin y$, $x = u^2 + v^2, y = uv$.

37. Find $\partial z/\partial u$ and $\partial z/\partial v$ when $u = \ln 2, v = 1$ if $z = 5\tan^{-1} x$ and $x = e^u + \ln v$.

38. Find $\partial z/\partial u$ and $\partial z/\partial v$ when $u = 1$ and $v = -2$ if $z = \ln q$ and $q = \sqrt{v + 3}\tan^{-1} u$.

Theory and Examples

39. Changing voltage in a circuit The voltage V in a circuit that satisfies the law $V = IR$ is slowly dropping as the battery wears out. At the same time, the resistance R is increasing as the resistor heats up. Use the equation

$$\frac{dV}{dt} = \frac{\partial V}{\partial I}\frac{dI}{dt} + \frac{\partial V}{\partial R}\frac{dR}{dt}$$

to find how the current is changing at the instant when $R = 600$ ohms, $I = 0.04$ amp, $dR/dt = 0.5$ ohm/sec, and $dV/dt = -0.01$ volt/sec.

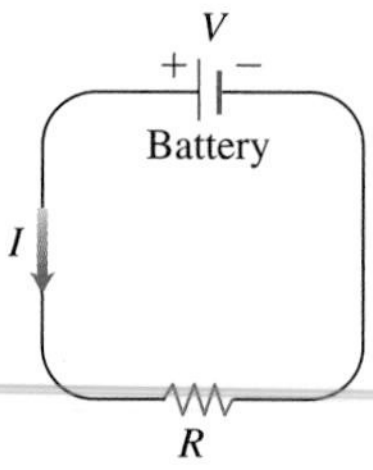

40. Changing dimensions in a box The lengths a, b, and c of the edges of a rectangular box are changing with time. At the instant in question, $a = 1$ m, $b = 2$ m, $c = 3$ m, $da/dt = db/dt = 1$ m/sec, and $dc/dt = -3$ m/sec. At what rates are the box's volume V and surface area S changing at that instant? Are the box's interior diagonals increasing in length or decreasing?

41. If $f(u, v, w)$ is differentiable and $u = x - y, v = y - z$, and $w = z - x$, show that

$$\frac{\partial f}{\partial x} + \frac{\partial f}{\partial y} + \frac{\partial f}{\partial z} = 0.$$

42. Polar coordinates Suppose that we substitute polar coordinates $x = r\cos\theta$ and $y = r\sin\theta$ in a differentiable function $w = f(x, y)$.

a. Show that

$$\frac{\partial w}{\partial r} = f_x \cos\theta + f_y \sin\theta$$

and

$$\frac{1}{r}\frac{\partial w}{\partial \theta} = -f_x \sin\theta + f_y \cos\theta .$$

b. Solve the equations in part (a) to express f_x and f_y in terms of $\partial w/\partial r$ and $\partial w/\partial \theta$.

c. Show that

$$(f_x)^2 + (f_y)^2 = \left(\frac{\partial w}{\partial r}\right)^2 + \frac{1}{r^2}\left(\frac{\partial w}{\partial \theta}\right)^2.$$

43. Laplace equations Show that if $w = f(u, v)$ satisfies the Laplace equation $f_{uu} + f_{vv} = 0$ and if $u = (x^2 - y^2)/2$ and $v = xy$, then w satisfies the Laplace equation $w_{xx} + w_{yy} = 0$.

44. Laplace equations Let $w = f(u) + g(v)$, where $u = x + iy$ and $v = x - iy$ and $i = \sqrt{-1}$. Show that w satisfies the Laplace equation $w_{xx} + w_{yy} = 0$ if all the necessary functions are differentiable.

Changes in Functions Along Curves

45. Extreme values on a helix Suppose that the partial derivatives of a function $f(x, y, z)$ at points on the helix $x = \cos t$, $y = \sin t$, $z = t$ are

$$f_x = \cos t, \qquad f_y = \sin t, \qquad f_z = t^2 + t - 2.$$

At what points on the curve, if any, can f take on extreme values?

46. A space curve Let $w = x^2 e^{2y} \cos 3z$. Find the value of dw/dt at the point $(1, \ln 2, 0)$ on the curve $x = \cos t$, $y = \ln(t + 2)$, $z = t$.

47. Temperature on a circle Let $T = f(x, y)$ be the temperature at the point (x, y) on the circle $x = \cos t$, $y = \sin t$, $0 \le t \le 2\pi$ and suppose that

$$\frac{\partial T}{\partial x} = 8x - 4y, \qquad \frac{\partial T}{\partial y} = 8y - 4x.$$

a. Find where the maximum and minimum temperatures on the circle occur by examining the derivatives dT/dt and d^2T/dt^2.

b. Suppose that $T = 4x^2 - 4xy + 4y^2$. Find the maximum and minimum values of T on the circle.

48. Temperature on an ellipse Let $T = g(x, y)$ be the temperature at the point (x, y) on the ellipse

$$x = 2\sqrt{2}\cos t, \qquad y = \sqrt{2}\sin t, \qquad 0 \le t \le 2\pi,$$

and suppose that

$$\frac{\partial T}{\partial x} = y, \qquad \frac{\partial T}{\partial y} = x.$$

a. Locate the maximum and minimum temperatures on the ellipse by examining dT/dt and d^2T/dt^2.

b. Suppose that $T = xy - 2$. Find the maximum and minimum values of T on the ellipse.

Differentiating Integrals

Under mild continuity restrictions, it is true that if

$$F(x) = \int_a^b g(t, x)\, dt,$$

then $F'(x) = \int_a^b g_x(t, x)\, dt$. Using this fact and the Chain Rule, we can find the derivative of

$$F(x) = \int_a^{f(x)} g(t, x)\, dt$$

by letting

$$G(u, x) = \int_a^u g(t, x)\, dt,$$

where $u = f(x)$. Find the derivatives of the functions in Exercises 49 and 50.

49. $F(x) = \int_0^{x^2} \sqrt{t^4 + x^3}\, dt$

50. $F(x) = \int_{x^2}^1 \sqrt{t^3 + x^2}\, dt$

14.5 Directional Derivatives and Gradient Vectors

If you look at the map (Figure 14.23) showing contours on the West Point Area along the Hudson River in New York, you will notice that the tributary streams flow perpendicular to the contours. The streams are following paths of steepest descent so the waters reach the Hudson as quickly as possible. Therefore, the instantaneous rate of change in a stream's altitude above sea level has a particular direction. In this section, you see why this direction, called the "downhill" direction, is perpendicular to the contours.

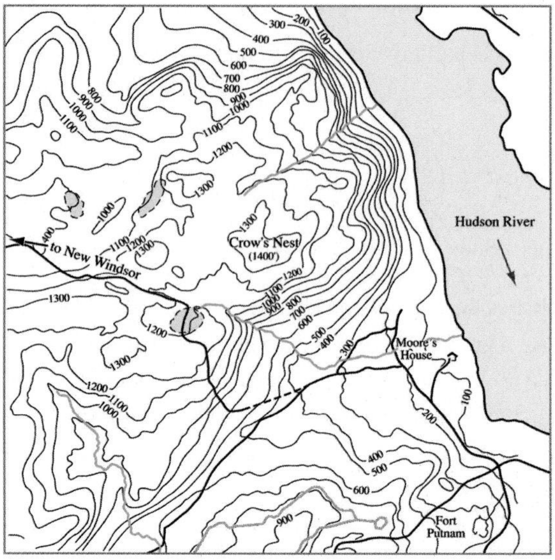

FIGURE 14.23 Contours of the West Point Area in New York show streams, which follow paths of steepest descent, running perpendicular to the contours.

Directional Derivatives in the Plane

We know from Section 14.4 that if $f(x, y)$ is differentiable, then the rate at which f changes with respect to t along a differentiable curve $x = g(t), y = h(t)$ is

$$\frac{df}{dt} = \frac{\partial f}{\partial x}\frac{dx}{dt} + \frac{\partial f}{\partial y}\frac{dy}{dt}.$$

At any point $P_0(x_0, y_0) = P_0(g(t_0), h(t_0))$, this equation gives the rate of change of f with respect to increasing t and therefore depends, among other things, on the direction of motion along the curve. If the curve is a straight line and t is the arc length parameter along the line measured from P_0 in the direction of a given unit vector $\mathbf{u}$, then df/dt is the rate of change of f with respect to distance in its domain in the direction of $\mathbf{u}$. By varying $\mathbf{u}$, we find the rates at which f changes with respect to distance as we move through P_0 in different directions. We now define this idea more precisely.

Suppose that the function $f(x, y)$ is defined throughout a region R in the xy-plane, that $P_0(x_0, y_0)$ is a point in R, and that $\mathbf{u} = u_1\mathbf{i} + u_2\mathbf{j}$ is a unit vector. Then the equations

$$x = x_0 + su_1, \qquad y = y_0 + su_2$$

parametrize the line through P_0 parallel to $\mathbf{u}$. If the parameter s measures arc length from P_0 in the direction of $\mathbf{u}$, we find the rate of change of f at P_0 in the direction of $\mathbf{u}$ by calculating df/ds at P_0 (Figure 14.24).

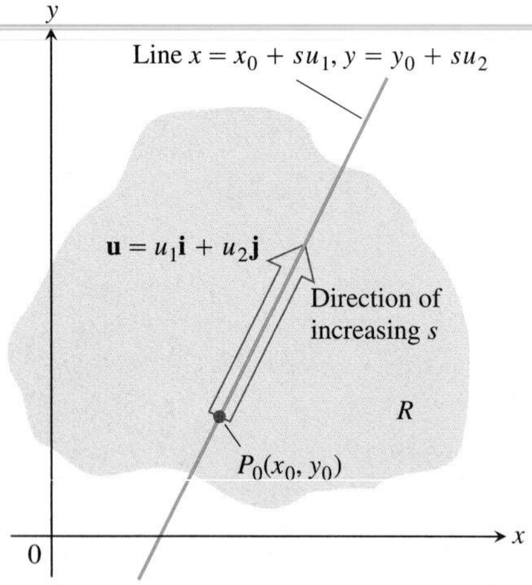

FIGURE 14.24 The rate of change of f in the direction of $\mathbf{u}$ at a point P_0 is the rate at which f changes along this line at P_0.

DEFINITION Directional Derivative

The **derivative of f at $P_0(x_0, y_0)$ in the direction of the unit vector $\mathbf{u} = u_1\mathbf{i} + u_2\mathbf{j}$** is the number

$$\left(\frac{df}{ds}\right)_{\mathbf{u},P_0} = \lim_{s\to 0} \frac{f(x_0 + su_1, y_0 + su_2) - f(x_0, y_0)}{s}, \tag{1}$$

provided the limit exists.

The directional derivative is also denoted by

$$(D_{\mathbf{u}} f)_{P_0}. \qquad \text{“The derivative of } f \text{ at } P_0 \text{ in the direction of } \mathbf{u}\text{”}$$

EXAMPLE 1 Finding a Directional Derivative Using the Definition

Find the derivative of

$$f(x, y) = x^2 + xy$$

at $P_0(1, 2)$ in the direction of the unit vector $\mathbf{u} = \left(1/\sqrt{2}\right)\mathbf{i} + \left(1/\sqrt{2}\right)\mathbf{j}$.

Solution

$$\left(\frac{df}{ds}\right)_{\mathbf{u},P_0} = \lim_{s\to 0} \frac{f(x_0 + su_1, y_0 + su_2) - f(x_0, y_0)}{s} \qquad \text{Equation (1)}$$

$$= \lim_{s\to 0} \frac{f\left(1 + s\cdot\frac{1}{\sqrt{2}}, 2 + s\cdot\frac{1}{\sqrt{2}}\right) - f(1, 2)}{s}$$

$$= \lim_{s\to 0} \frac{\left(1 + \frac{s}{\sqrt{2}}\right)^2 + \left(1 + \frac{s}{\sqrt{2}}\right)\left(2 + \frac{s}{\sqrt{2}}\right) - (1^2 + 1\cdot 2)}{s}$$

$$= \lim_{s\to 0} \frac{\left(1 + \frac{2s}{\sqrt{2}} + \frac{s^2}{2}\right) + \left(2 + \frac{3s}{\sqrt{2}} + \frac{s^2}{2}\right) - 3}{s}$$

$$= \lim_{s\to 0} \frac{\frac{5s}{\sqrt{2}} + s^2}{s} = \lim_{s\to 0}\left(\frac{5}{\sqrt{2}} + s\right) = \left(\frac{5}{\sqrt{2}} + 0\right) = \frac{5}{\sqrt{2}}.$$

The rate of change of $f(x, y) = x^2 + xy$ at $P_0(1, 2)$ in the direction $\mathbf{u} = \left(1/\sqrt{2}\right)\mathbf{i} + \left(1/\sqrt{2}\right)\mathbf{j}$ is $5/\sqrt{2}$. ■

Interpretation of the Directional Derivative

The equation $z = f(x, y)$ represents a surface S in space. If $z_0 = f(x_0, y_0)$, then the point $P(x_0, y_0, z_0)$ lies on S. The vertical plane that passes through P and $P_0(x_0, y_0)$ parallel to $\mathbf{u}$

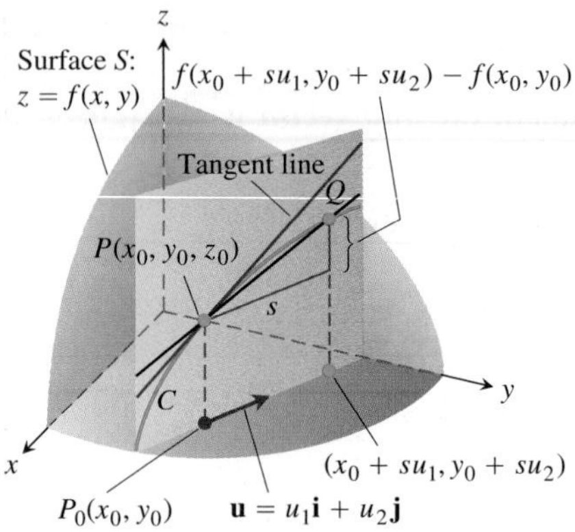

FIGURE 14.25 The slope of curve C at P_0 is $\lim_{Q \to P}$ slope (PQ); this is the directional derivative

$$\left(\frac{df}{ds}\right)_{\mathbf{u},P_0} = (D_{\mathbf{u}}f)_{P_0}.$$

intersects S in a curve C (Figure 14.25). The rate of change of f in the direction of $\mathbf{u}$ is the slope of the tangent to C at P.

When $\mathbf{u} = \mathbf{i}$, the directional derivative at P_0 is $\partial f/\partial x$ evaluated at (x_0, y_0). When $\mathbf{u} = \mathbf{j}$, the directional derivative at P_0 is $\partial f/\partial y$ evaluated at (x_0, y_0). The directional derivative generalizes the two partial derivatives. We can now ask for the rate of change of f in any direction $\mathbf{u}$, not just the directions $\mathbf{i}$ and $\mathbf{j}$.

Here's a physical interpretation of the directional derivative. Suppose that $T = f(x, y)$ is the temperature at each point (x, y) over a region in the plane. Then $f(x_0, y_0)$ is the temperature at the point $P_0(x_0, y_0)$ and $(D_{\mathbf{u}}f)_{P_0}$ is the instantaneous rate of change of the temperature at P_0 stepping off in the direction $\mathbf{u}$.

Calculation and Gradients

We now develop an efficient formula to calculate the directional derivative for a differentiable function f. We begin with the line

$$x = x_0 + su_1, \qquad y = y_0 + su_2, \tag{2}$$

through $P_0(x_0, y_0)$, parametrized with the arc length parameter s increasing in the direction of the unit vector $\mathbf{u} = u_1\mathbf{i} + u_2\mathbf{j}$. Then

$$\left(\frac{df}{ds}\right)_{\mathbf{u},P_0} = \left(\frac{\partial f}{\partial x}\right)_{P_0}\frac{dx}{ds} + \left(\frac{\partial f}{\partial y}\right)_{P_0}\frac{dy}{ds}$$

Chain Rule for differentiable f

$$= \left(\frac{\partial f}{\partial x}\right)_{P_0}\cdot u_1 + \left(\frac{\partial f}{\partial y}\right)_{P_0}\cdot u_2$$

From Equations (2), $dx/ds = u_1$ and $dy/ds = u_2$

$$= \underbrace{\left[\left(\frac{\partial f}{\partial x}\right)_{P_0}\mathbf{i} + \left(\frac{\partial f}{\partial y}\right)_{P_0}\mathbf{j}\right]}_{\text{Gradient of } f \text{ at } P_0}\cdot\underbrace{\left[u_1\mathbf{i} + u_2\mathbf{j}\right]}_{\text{Direction } \mathbf{u}}. \tag{3}$$

DEFINITION Gradient Vector

The **gradient vector (gradient)** of $f(x, y)$ at a point $P_0(x_0, y_0)$ is the vector

$$\nabla f = \frac{\partial f}{\partial x}\mathbf{i} + \frac{\partial f}{\partial y}\mathbf{j}$$

obtained by evaluating the partial derivatives of f at P_0.

The notation ∇f is read "grad f" as well as "gradient of f" and "del f." The symbol ∇ by itself is read "del." Another notation for the gradient is grad f, read the way it is written.

Equation (3) says that the derivative of a differentiable function f in the direction of $\mathbf{u}$ at P_0 is the dot product of $\mathbf{u}$ with the gradient of f at P_0.

THEOREM 9 The Directional Derivative Is a Dot Product
If $f(x, y)$ is differentiable in an open region containing $P_0(x_0, y_0)$, then

$$\left(\frac{df}{ds}\right)_{\mathbf{u},P_0} = (\nabla f)_{P_0} \cdot \mathbf{u}, \qquad (4)$$

the dot product of the gradient f at P_0 and $\mathbf{u}$.

EXAMPLE 2 Finding the Directional Derivative Using the Gradient

Find the derivative of $f(x, y) = xe^y + \cos(xy)$ at the point $(2, 0)$ in the direction of $\mathbf{v} = 3\mathbf{i} - 4\mathbf{j}$.

Solution The direction of $\mathbf{v}$ is the unit vector obtained by dividing $\mathbf{v}$ by its length:

$$\mathbf{u} = \frac{\mathbf{v}}{|\mathbf{v}|} = \frac{\mathbf{v}}{5} = \frac{3}{5}\mathbf{i} - \frac{4}{5}\mathbf{j}.$$

The partial derivatives of f are everywhere continuous and at $(2, 0)$ are given by

$$f_x(2, 0) = (e^y - y\sin(xy))_{(2,0)} = e^0 - 0 = 1$$

$$f_y(2, 0) = (xe^y - x\sin(xy))_{(2,0)} = 2e^0 - 2\cdot 0 = 2.$$

The gradient of f at $(2, 0)$ is

$$\nabla f|_{(2,0)} = f_x(2, 0)\mathbf{i} + f_y(2, 0)\mathbf{j} = \mathbf{i} + 2\mathbf{j}$$

(Figure 14.26). The derivative of f at $(2, 0)$ in the direction of $\mathbf{v}$ is therefore

$$(D_{\mathbf{u}}f)|_{(2,0)} = \nabla f|_{(2,0)} \cdot \mathbf{u} \qquad \text{Equation (4)}$$

$$= (\mathbf{i} + 2\mathbf{j}) \cdot \left(\frac{3}{5}\mathbf{i} - \frac{4}{5}\mathbf{j}\right) = \frac{3}{5} - \frac{8}{5} = -1.$$ ■

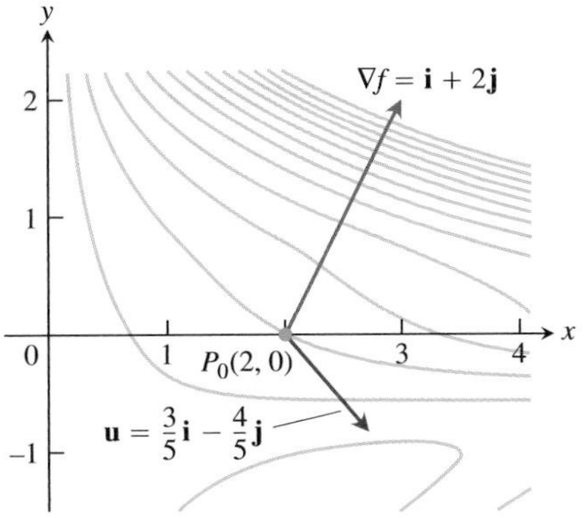

FIGURE 14.26 Picture ∇f as a vector in the domain of f. In the case of $f(x, y) = xe^y + \cos(xy)$, the domain is the entire plane. The rate at which f changes at $(2, 0)$ in the direction $\mathbf{u} = (3/5)\mathbf{i} - (4/5)\mathbf{j}$ is $\nabla f \cdot \mathbf{u} = -1$ (Example 2).

Evaluating the dot product in the formula

$$D_{\mathbf{u}}f = \nabla f \cdot \mathbf{u} = |\nabla f||\mathbf{u}|\cos\theta = |\nabla f|\cos\theta,$$

where θ is the angle between the vectors $\mathbf{u}$ and ∇f, reveals the following properties.

Properties of the Directional Derivative $D_{\mathbf{u}}f = \nabla f \cdot \mathbf{u} = |\nabla f|\cos\theta$

1. The function f increases most rapidly when $\cos\theta = 1$ or when $\mathbf{u}$ is the direction of ∇f. That is, at each point P in its domain, f increases most rapidly in the direction of the gradient vector ∇f at P. The derivative in this direction is
$$D_{\mathbf{u}}f = |\nabla f|\cos(0) = |\nabla f|.$$
2. Similarly, f decreases most rapidly in the direction of $-\nabla f$. The derivative in this direction is $D_{\mathbf{u}}f = |\nabla f|\cos(\pi) = -|\nabla f|$.
3. Any direction $\mathbf{u}$ orthogonal to a gradient $\nabla f \neq 0$ is a direction of zero change in f because θ then equals $\pi/2$ and
$$D_{\mathbf{u}}f = |\nabla f|\cos(\pi/2) = |\nabla f|\cdot 0 = 0.$$

As we discuss later, these properties hold in three dimensions as well as two.

EXAMPLE 3 Finding Directions of Maximal, Minimal, and Zero Change

Find the directions in which $f(x, y) = (x^2/2) + (y^2/2)$

(a) Increases most rapidly at the point (1, 1)

(b) Decreases most rapidly at (1, 1).

(c) What are the directions of zero change in f at (1, 1)?

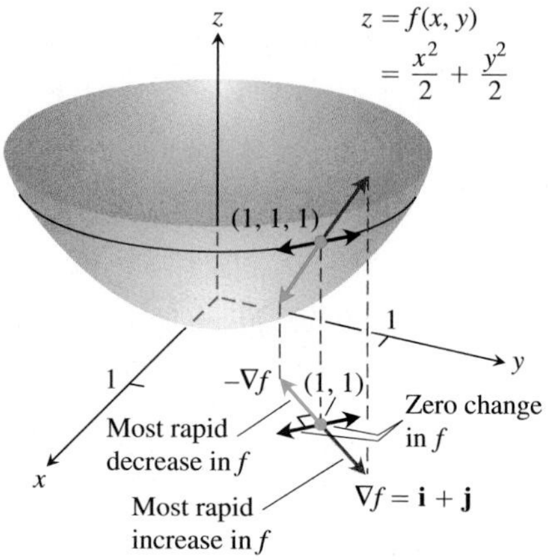

FIGURE 14.27 The direction in which $f(x, y) = (x^2/2) + (y^2/2)$ increases most rapidly at (1, 1) is the direction of $\nabla f|_{(1,1)} = \mathbf{i} + \mathbf{j}$. It corresponds to the direction of steepest ascent on the surface at (1, 1, 1) (Example 3).

Solution

(a) The function increases most rapidly in the direction of ∇f at (1, 1). The gradient there is

$$(\nabla f)_{(1,1)} = (x\mathbf{i} + y\mathbf{j})_{(1,1)} = \mathbf{i} + \mathbf{j}.$$

Its direction is

$$\mathbf{u} = \frac{\mathbf{i} + \mathbf{j}}{|\mathbf{i} + \mathbf{j}|} = \frac{\mathbf{i} + \mathbf{j}}{\sqrt{(1)^2 + (1)^2}} = \frac{1}{\sqrt{2}}\mathbf{i} + \frac{1}{\sqrt{2}}\mathbf{j}.$$

(b) The function decreases most rapidly in the direction of $-\nabla f$ at (1, 1), which is

$$-\mathbf{u} = -\frac{1}{\sqrt{2}}\mathbf{i} - \frac{1}{\sqrt{2}}\mathbf{j}.$$

(c) The directions of zero change at (1, 1) are the directions orthogonal to ∇f:

$$\mathbf{n} = -\frac{1}{\sqrt{2}}\mathbf{i} + \frac{1}{\sqrt{2}}\mathbf{j} \quad \text{and} \quad -\mathbf{n} = \frac{1}{\sqrt{2}}\mathbf{i} - \frac{1}{\sqrt{2}}\mathbf{j}.$$

See Figure 14.27. ■

Gradients and Tangents to Level Curves

If a differentiable function $f(x, y)$ has a constant value c along a smooth curve $\mathbf{r} = g(t)\mathbf{i} + h(t)\mathbf{j}$ (making the curve a level curve of f), then $f(g(t), h(t)) = c$. Differentiating both sides of this equation with respect to t leads to the equations

$$\frac{d}{dt} f(g(t), h(t)) = \frac{d}{dt}(c)$$

$$\frac{\partial f}{\partial x}\frac{dg}{dt} + \frac{\partial f}{\partial y}\frac{dh}{dt} = 0 \qquad \text{Chain Rule}$$

$$\underbrace{\left(\frac{\partial f}{\partial x}\mathbf{i} + \frac{\partial f}{\partial y}\mathbf{j}\right)}_{\nabla f} \cdot \underbrace{\left(\frac{dg}{dt}\mathbf{i} + \frac{dh}{dt}\mathbf{j}\right)}_{\frac{d\mathbf{r}}{dt}} = 0. \tag{5}$$

Equation (5) says that ∇f is normal to the tangent vector $d\mathbf{r}/dt$, so it is normal to the curve.

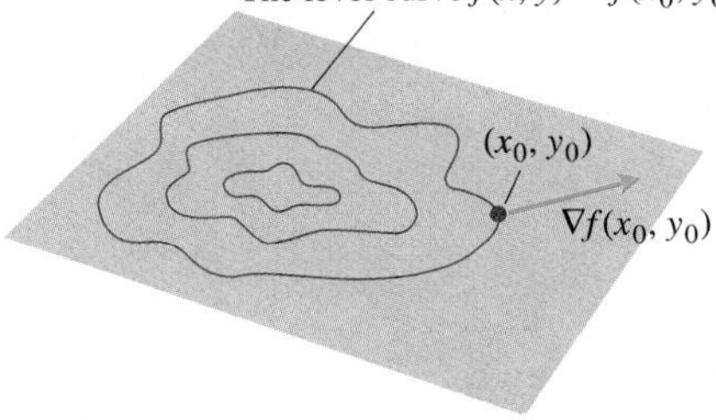

FIGURE 14.28 The gradient of a differentiable function of two variables at a point is always normal to the function's level curve through that point.

At every point (x_0, y_0) in the domain of a differentiable function $f(x, y)$, the gradient of f is normal to the level curve through (x_0, y_0) (Figure 14.28).

Equation (5) validates our observation that streams flow perpendicular to the contours in topographical maps (see Figure 14.23). Since the downflowing stream will reach its destination in the fastest way, it must flow in the direction of the negative gradient vectors from Property 2 for the directional derivative. Equation (5) tells us these directions are perpendicular to the level curves.

This observation also enables us to find equations for tangent lines to level curves. They are the lines normal to the gradients. The line through a point $P_0(x_0, y_0)$ normal to a vector $\mathbf{N} = A\mathbf{i} + B\mathbf{j}$ has the equation

$$A(x - x_0) + B(y - y_0) = 0$$

(Exercise 35). If $\mathbf{N}$ is the gradient $(\nabla f)_{(x_0, y_0)} = f_x(x_0, y_0)\mathbf{i} + f_y(x_0, y_0)\mathbf{j}$, the equation is the tangent line given by

$$f_x(x_0, y_0)(x - x_0) + f_y(x_0, y_0)(y - y_0) = 0. \tag{6}$$

EXAMPLE 4 Finding the Tangent Line to an Ellipse

Find an equation for the tangent to the ellipse

$$\frac{x^2}{4} + y^2 = 2$$

(Figure 14.29) at the point $(-2, 1)$.

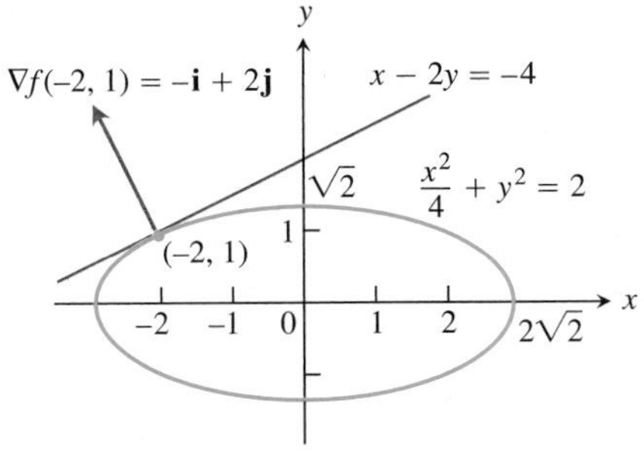

FIGURE 14.29 We can find the tangent to the ellipse $(x^2/4) + y^2 = 2$ by treating the ellipse as a level curve of the function $f(x, y) = (x^2/4) + y^2$ (Example 4).

Solution The ellipse is a level curve of the function

$$f(x, y) = \frac{x^2}{4} + y^2.$$

The gradient of f at $(-2, 1)$ is

$$\nabla f|_{(-2,1)} = \left(\frac{x}{2}\mathbf{i} + 2y\mathbf{j}\right)_{(-2,1)} = -\mathbf{i} + 2\mathbf{j}.$$

The tangent is the line

$$(-1)(x + 2) + (2)(y - 1) = 0 \qquad \text{Equation (6)}$$
$$x - 2y = -4.$$

■

If we know the gradients of two functions f and g, we automatically know the gradients of their constant multiples, sum, difference, product, and quotient. You are asked to establish the following rules in Exercise 36. Notice that these rules have the same form as the corresponding rules for derivatives of single-variable functions.

Algebra Rules for Gradients

1. *Constant Multiple Rule:* $\nabla(kf) = k\nabla f$ (any number k)
2. *Sum Rule:* $\nabla(f + g) = \nabla f + \nabla g$
3. *Difference Rule:* $\nabla(f - g) = \nabla f - \nabla g$
4. *Product Rule:* $\nabla(fg) = f\nabla g + g\nabla f$
5. *Quotient Rule:* $\nabla\left(\dfrac{f}{g}\right) = \dfrac{g\nabla f - f\nabla g}{g^2}$

EXAMPLE 5 Illustrating the Gradient Rules

We illustrate the rules with

$$f(x, y) = x - y \qquad g(x, y) = 3y$$
$$\nabla f = \mathbf{i} - \mathbf{j} \qquad \nabla g = 3\mathbf{j}.$$

We have

1. $\nabla(2f) = \nabla(2x - 2y) = 2\mathbf{i} - 2\mathbf{j} = 2\nabla f$
2. $\nabla(f + g) = \nabla(x + 2y) = \mathbf{i} + 2\mathbf{j} = \nabla f + \nabla g$
3. $\nabla(f - g) = \nabla(x - 4y) = \mathbf{i} - 4\mathbf{j} = \nabla f - \nabla g$
4. $$\begin{aligned}\nabla(fg) &= \nabla(3xy - 3y^2) = 3y\mathbf{i} + (3x - 6y)\mathbf{j} \\ &= 3y(\mathbf{i} - \mathbf{j}) + 3y\mathbf{j} + (3x - 6y)\mathbf{j} \\ &= 3y(\mathbf{i} - \mathbf{j}) + (3x - 3y)\mathbf{j} \\ &= 3y(\mathbf{i} - \mathbf{j}) + (x - y)3\mathbf{j} = g\nabla f + f\nabla g\end{aligned}$$
5. $$\begin{aligned}\nabla\left(\frac{f}{g}\right) &= \nabla\left(\frac{x - y}{3y}\right) = \nabla\left(\frac{x}{3y} - \frac{1}{3}\right) \\ &= \frac{1}{3y}\mathbf{i} - \frac{x}{3y^2}\mathbf{j} \\ &= \frac{3y\mathbf{i} - 3x\mathbf{j}}{9y^2} = \frac{3y(\mathbf{i} - \mathbf{j}) - (3x - 3y)\mathbf{j}}{9y^2} \\ &= \frac{3y(\mathbf{i} - \mathbf{j}) - (x - y)3\mathbf{j}}{9y^2} = \frac{g\nabla f - f\nabla g}{g^2}.\end{aligned}$$

■

Functions of Three Variables

For a differentiable function $f(x, y, z)$ and a unit vector $\mathbf{u} = u_1\mathbf{i} + u_2\mathbf{j} + u_3\mathbf{k}$ in space, we have

$$\nabla f = \frac{\partial f}{\partial x}\mathbf{i} + \frac{\partial f}{\partial y}\mathbf{j} + \frac{\partial f}{\partial z}\mathbf{k}$$

and

$$D_{\mathbf{u}}f = \nabla f \cdot \mathbf{u} = \frac{\partial f}{\partial x}u_1 + \frac{\partial f}{\partial y}u_2 + \frac{\partial f}{\partial z}u_3.$$

The directional derivative can once again be written in the form

$$D_{\mathbf{u}}f = \nabla f \cdot \mathbf{u} = |\nabla f||u|\cos\theta = |\nabla f|\cos\theta,$$

so the properties listed earlier for functions of two variables continue to hold. At any given point, f increases most rapidly in the direction of ∇f and decreases most rapidly in the direction of $-\nabla f$. In any direction orthogonal to ∇f, the derivative is zero.

EXAMPLE 6 Finding Directions of Maximal, Minimal, and Zero Change

(a) Find the derivative of $f(x, y, z) = x^3 - xy^2 - z$ at $P_0(1, 1, 0)$ in the direction of $\mathbf{v} = 2\mathbf{i} - 3\mathbf{j} + 6\mathbf{k}$.

(b) In what directions does f change most rapidly at P_0, and what are the rates of change in these directions?

Solution

(a) The direction of $\mathbf{v}$ is obtained by dividing $\mathbf{v}$ by its length:

$$|\mathbf{v}| = \sqrt{(2)^2 + (-3)^2 + (6)^2} = \sqrt{49} = 7$$

$$\mathbf{u} = \frac{\mathbf{v}}{|\mathbf{v}|} = \frac{2}{7}\mathbf{i} - \frac{3}{7}\mathbf{j} + \frac{6}{7}\mathbf{k}.$$

The partial derivatives of f at P_0 are

$$f_x = (3x^2 - y^2)_{(1,1,0)} = 2, \qquad f_y = -2xy|_{(1,1,0)} = -2, \qquad f_z = -1|_{(1,1,0)} = -1.$$

The gradient of f at P_0 is

$$\nabla f|_{(1,1,0)} = 2\mathbf{i} - 2\mathbf{j} - \mathbf{k}.$$

The derivative of f at P_0 in the direction of $\mathbf{v}$ is therefore

$$(D_{\mathbf{u}}f)_{(1,1,0)} = \nabla f|_{(1,1,0)} \cdot \mathbf{u} = (2\mathbf{i} - 2\mathbf{j} - \mathbf{k}) \cdot \left(\frac{2}{7}\mathbf{i} - \frac{3}{7}\mathbf{j} + \frac{6}{7}\mathbf{k}\right)$$

$$= \frac{4}{7} + \frac{6}{7} - \frac{6}{7} = \frac{4}{7}.$$

(b) The function increases most rapidly in the direction of $\nabla f = 2\mathbf{i} - 2\mathbf{j} - \mathbf{k}$ and decreases most rapidly in the direction of $-\nabla f$. The rates of change in the directions are, respectively,

$$|\nabla f| = \sqrt{(2)^2 + (-2)^2 + (-1)^2} = \sqrt{9} = 3 \quad \text{and} \quad -|\nabla f| = -3. \quad \blacksquare$$

EXERCISES 14.5

Calculating Gradients at Points

In Exercises 1–4, find the gradient of the function at the given point. Then sketch the gradient together with the level curve that passes through the point.

1. $f(x, y) = y - x, \quad (2, 1)$
2. $f(x, y) = \ln(x^2 + y^2), \quad (1, 1)$
3. $g(x, y) = y - x^2, \quad (-1, 0)$
4. $g(x, y) = \frac{x^2}{2} - \frac{y^2}{2}, \quad (\sqrt{2}, 1)$

In Exercises 5–8, find ∇f at the given point.

5. $f(x, y, z) = x^2 + y^2 - 2z^2 + z\ln x, \quad (1, 1, 1)$
6. $f(x, y, z) = 2z^3 - 3(x^2 + y^2)z + \tan^{-1} xz, \quad (1, 1, 1)$

7. $f(x, y, z) = (x^2 + y^2 + z^2)^{-1/2} + \ln(xyz)$, $(-1, 2, -2)$
8. $f(x, y, z) = e^{x+y}\cos z + (y + 1)\sin^{-1} x$, $(0, 0, \pi/6)$

Finding Directional Derivatives

In Exercises 9–16, find the derivative of the function at P_0 in the direction of $\mathbf{A}$.

9. $f(x, y) = 2xy - 3y^2$, $P_0(5, 5)$, $\mathbf{A} = 4\mathbf{i} + 3\mathbf{j}$
10. $f(x, y) = 2x^2 + y^2$, $P_0(-1, 1)$, $\mathbf{A} = 3\mathbf{i} - 4\mathbf{j}$
11. $g(x, y) = x - (y^2/x) + \sqrt{3}\sec^{-1}(2xy)$, $P_0(1, 1)$, $\mathbf{A} = 12\mathbf{i} + 5\mathbf{j}$
12. $h(x, y) = \tan^{-1}(y/x) + \sqrt{3}\sin^{-1}(xy/2)$, $P_0(1, 1)$, $\mathbf{A} = 3\mathbf{i} - 2\mathbf{j}$
13. $f(x, y, z) = xy + yz + zx$, $P_0(1, -1, 2)$, $\mathbf{A} = 3\mathbf{i} + 6\mathbf{j} - 2\mathbf{k}$
14. $f(x, y, z) = x^2 + 2y^2 - 3z^2$, $P_0(1, 1, 1)$, $\mathbf{A} = \mathbf{i} + \mathbf{j} + \mathbf{k}$
15. $g(x, y, z) = 3e^x \cos yz$, $P_0(0, 0, 0)$, $\mathbf{A} = 2\mathbf{i} + \mathbf{j} - 2\mathbf{k}$
16. $h(x, y, z) = \cos xy + e^{yz} + \ln zx$, $P_0(1, 0, 1/2)$, $\mathbf{A} = \mathbf{i} + 2\mathbf{j} + 2\mathbf{k}$

Directions of Most Rapid Increase and Decrease

In Exercises 17–22, find the directions in which the functions increase and decrease most rapidly at P_0. Then find the derivatives of the functions in these directions.

17. $f(x, y) = x^2 + xy + y^2$, $P_0(-1, 1)$
18. $f(x, y) = x^2y + e^{xy}\sin y$, $P_0(1, 0)$
19. $f(x, y, z) = (x/y) - yz$, $P_0(4, 1, 1)$
20. $g(x, y, z) = xe^y + z^2$, $P_0(1, \ln 2, 1/2)$
21. $f(x, y, z) = \ln xy + \ln yz + \ln xz$, $P_0(1, 1, 1)$
22. $h(x, y, z) = \ln(x^2 + y^2 - 1) + y + 6z$, $P_0(1, 1, 0)$

Tangent Lines to Curves

In Exercises 23–26, sketch the curve $f(x, y) = c$ together with ∇f and the tangent line at the given point. Then write an equation for the tangent line.

23. $x^2 + y^2 = 4$, $\left(\sqrt{2}, \sqrt{2}\right)$
24. $x^2 - y = 1$, $\left(\sqrt{2}, 1\right)$
25. $xy = -4$, $(2, -2)$
26. $x^2 - xy + y^2 = 7$, $(-1, 2)$

Theory and Examples

27. **Zero directional derivative** In what direction is the derivative of $f(x, y) = xy + y^2$ at $P(3, 2)$ equal to zero?
28. **Zero directional derivative** In what directions is the derivative of $f(x, y) = (x^2 - y^2)/(x^2 + y^2)$ at $P(1, 1)$ equal to zero?
29. Is there a direction $\mathbf{u}$ in which the rate of change of $f(x, y) = x^2 - 3xy + 4y^2$ at $P(1, 2)$ equals 14? Give reasons for your answer.
30. **Changing temperature along a circle** Is there a direction $\mathbf{u}$ in which the rate of change of the temperature function $T(x, y, z) = 2xy - yz$ (temperature in degrees Celsius, distance in feet) at $P(1, -1, 1)$ is $-3°\text{C/ft}$? Give reasons for your answer.
31. The derivative of $f(x, y)$ at $P_0(1, 2)$ in the direction of $\mathbf{i} + \mathbf{j}$ is $2\sqrt{2}$ and in the direction of $-2\mathbf{j}$ is -3. What is the derivative of f in the direction of $-\mathbf{i} - 2\mathbf{j}$? Give reasons for your answer.
32. The derivative of $f(x, y, z)$ at a point P is greatest in the direction of $\mathbf{v} = \mathbf{i} + \mathbf{j} - \mathbf{k}$. In this direction, the value of the derivative is $2\sqrt{3}$.
    a. What is ∇f at P? Give reasons for your answer.
    b. What is the derivative of f at P in the direction of $\mathbf{i} + \mathbf{j}$?
33. **Directional derivatives and scalar components** How is the derivative of a differentiable function $f(x, y, z)$ at a point P_0 in the direction of a unit vector $\mathbf{u}$ related to the scalar component of $(\nabla f)_{P_0}$ in the direction of $\mathbf{u}$? Give reasons for your answer.
34. **Directional derivatives and partial derivatives** Assuming that the necessary derivatives of $f(x, y, z)$ are defined, how are $D_{\mathbf{i}}f$, $D_{\mathbf{j}}f$, and $D_{\mathbf{k}}f$ related to f_x, f_y, and f_z? Give reasons for your answer.
35. **Lines in the *xy*-plane** Show that $A(x - x_0) + B(y - y_0) = 0$ is an equation for the line in the xy-plane through the point (x_0, y_0) normal to the vector $\mathbf{N} = A\mathbf{i} + B\mathbf{j}$.
36. **The algebra rules for gradients** Given a constant k and the gradients

$$\nabla f = \frac{\partial f}{\partial x}\mathbf{i} + \frac{\partial f}{\partial y}\mathbf{j} + \frac{\partial f}{\partial z}\mathbf{k}$$

and

$$\nabla g = \frac{\partial g}{\partial x}\mathbf{i} + \frac{\partial g}{\partial y}\mathbf{j} + \frac{\partial g}{\partial z}\mathbf{k},$$

use the scalar equations

$$\frac{\partial}{\partial x}(kf) = k\frac{\partial f}{\partial x}, \qquad \frac{\partial}{\partial x}(f \pm g) = \frac{\partial f}{\partial x} \pm \frac{\partial g}{\partial x},$$

$$\frac{\partial}{\partial x}(fg) = f\frac{\partial g}{\partial x} + g\frac{\partial f}{\partial x}, \qquad \frac{\partial}{\partial x}\left(\frac{f}{g}\right) = \frac{g\frac{\partial f}{\partial x} - f\frac{\partial g}{\partial x}}{g^2},$$

and so on, to establish the following rules.

a. $\nabla(kf) = k\nabla f$
b. $\nabla(f + g) = \nabla f + \nabla g$
c. $\nabla(f - g) = \nabla f - \nabla g$
d. $\nabla(fg) = f\nabla g + g\nabla f$
e. $\nabla\left(\frac{f}{g}\right) = \frac{g\nabla f - f\nabla g}{g^2}$

14.6 Tangent Planes and Differentials

In this section we define the tangent plane at a point on a smooth surface in space. We calculate an equation of the tangent plane from the partial derivatives of the function defining the surface. This idea is similar to the definition of the tangent line at a point on a curve in the coordinate plane for single-variable functions (Section 2.7). We then study the total differential and linearization of functions of several variables.

Tangent Planes and Normal Lines

If $\mathbf{r} = g(t)\mathbf{i} + h(t)\mathbf{j} + k(t)\mathbf{k}$ is a smooth curve on the level surface $f(x, y, z) = c$ of a differentiable function f, then $f(g(t), h(t), k(t)) = c$. Differentiating both sides of this equation with respect to t leads to

$$\frac{d}{dt} f(g(t), h(t), k(t)) = \frac{d}{dt}(c)$$

$$\frac{\partial f}{\partial x}\frac{dg}{dt} + \frac{\partial f}{\partial y}\frac{dh}{dt} + \frac{\partial f}{\partial z}\frac{dk}{dt} = 0 \qquad \text{Chain Rule}$$

$$\underbrace{\left(\frac{\partial f}{\partial x}\mathbf{i} + \frac{\partial f}{\partial y}\mathbf{j} + \frac{\partial f}{\partial z}\mathbf{k}\right)}_{\nabla f} \cdot \underbrace{\left(\frac{dg}{dt}\mathbf{i} + \frac{dh}{dt}\mathbf{j} + \frac{dk}{dt}\mathbf{k}\right)}_{d\mathbf{r}/dt} = 0. \tag{1}$$

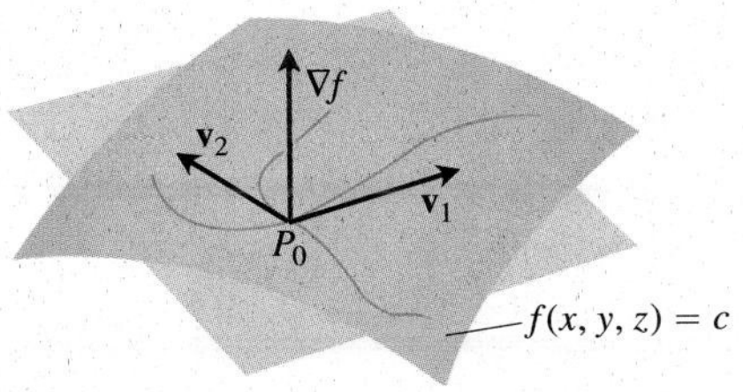

FIGURE 14.30 The gradient ∇f is orthogonal to the velocity vector of every smooth curve in the surface through P_0. The velocity vectors at P_0 therefore lie in a common plane, which we call the tangent plane at P_0.

At every point along the curve, ∇f is orthogonal to the curve's velocity vector.

Now let us restrict our attention to the curves that pass through P_0 (Figure 14.30). All the velocity vectors at P_0 are orthogonal to ∇f at P_0, so the curves' tangent lines all lie in the plane through P_0 normal to ∇f. We call this plane the tangent plane of the surface at P_0. The line through P_0 perpendicular to the plane is the surface's normal line at P_0.

DEFINITIONS Tangent Plane, Normal Line

The **tangent plane** at the point $P_0(x_0, y_0, z_0)$ on the level surface $f(x, y, z) = c$ of a differentiable function f is the plane through P_0 normal to $\nabla f|_{P_0}$.

The **normal line** of the surface at P_0 is the line through P_0 parallel to $\nabla f|_{P_0}$.

Thus, from Section 12.5, the tangent plane and normal line have the following equations:

Tangent Plane to $f(x, y, z) = c$ at $P_0(x_0, y_0, z_0)$

$$f_x(P_0)(x - x_0) + f_y(P_0)(y - y_0) + f_z(P_0)(z - z_0) = 0 \tag{2}$$

Normal Line to $f(x, y, z) = c$ at $P_0(x_0, y_0, z_0)$

$$x = x_0 + f_x(P_0)t, \qquad y = y_0 + f_y(P_0)t, \qquad z = z_0 + f_z(P_0)t \tag{3}$$

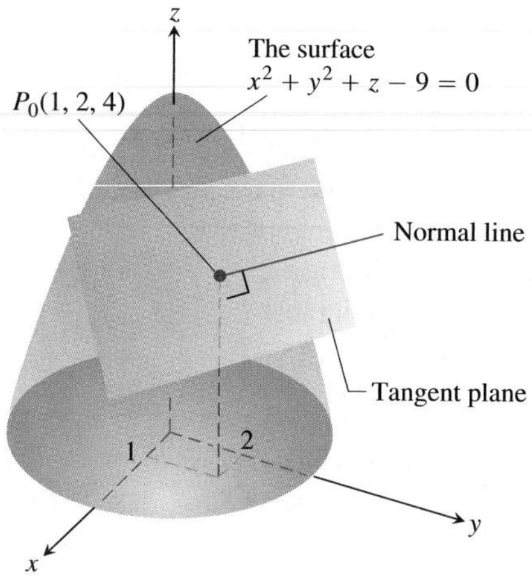

FIGURE 14.31 The tangent plane and normal line to the surface $x^2 + y^2 + z - 9 = 0$ at $P_0(1, 2, 4)$ (Example 1).

EXAMPLE 1 Finding the Tangent Plane and Normal Line

Find the tangent plane and normal line of the surface

$$f(x, y, z) = x^2 + y^2 + z - 9 = 0 \qquad \text{A circular paraboloid}$$

at the point $P_0(1, 2, 4)$.

Solution The surface is shown in Figure 14.31.

The tangent plane is the plane through P_0 perpendicular to the gradient of f at P_0. The gradient is

$$\nabla f|_{P_0} = (2x\mathbf{i} + 2y\mathbf{j} + \mathbf{k})_{(1,2,4)} = 2\mathbf{i} + 4\mathbf{j} + \mathbf{k}.$$

The tangent plane is therefore the plane

$$2(x - 1) + 4(y - 2) + (z - 4) = 0, \qquad \text{or} \qquad 2x + 4y + z = 14.$$

The line normal to the surface at P_0 is

$$x = 1 + 2t, \qquad y = 2 + 4t, \qquad z = 4 + t. \qquad \blacksquare$$

To find an equation for the plane tangent to a smooth surface $z = f(x, y)$ at a point $P_0(x_0, y_0, z_0)$ where $z_0 = f(x_0, y_0)$, we first observe that the equation $z = f(x, y)$ is equivalent to $f(x, y) - z = 0$. The surface $z = f(x, y)$ is therefore the zero level surface of the function $F(x, y, z) = f(x, y) - z$. The partial derivatives of F are

$$F_x = \frac{\partial}{\partial x}(f(x, y) - z) = f_x - 0 = f_x$$

$$F_y = \frac{\partial}{\partial y}(f(x, y) - z) = f_y - 0 = f_y$$

$$F_z = \frac{\partial}{\partial z}(f(x, y) - z) = 0 - 1 = -1.$$

The formula

$$F_x(P_0)(x - x_0) + F_y(P_0)(y - y_0) + F_z(P_0)(z - z_0) = 0$$

for the plane tangent to the level surface at P_0 therefore reduces to

$$f_x(x_0, y_0)(x - x_0) + f_y(x_0, y_0)(y - y_0) - (z - z_0) = 0.$$

Plane Tangent to a Surface $z = f(x, y)$ at $(x_0, y_0, f(x_0, y_0))$

The plane tangent to the surface $z = f(x, y)$ of a differentiable function f at the point $P_0(x_0, y_0, z_0) = (x_0, y_0, f(x_0, y_0))$ is

$$f_x(x_0, y_0)(x - x_0) + f_y(x_0, y_0)(y - y_0) - (z - z_0) = 0. \qquad (4)$$

EXAMPLE 2 Finding a Plane Tangent to a Surface $z = f(x, y)$

Find the plane tangent to the surface $z = x \cos y - ye^x$ at $(0, 0, 0)$.

Solution We calculate the partial derivatives of $f(x, y) = x\cos y - ye^x$ and use Equation (4):

$$f_x(0, 0) = (\cos y - ye^x)_{(0,0)} = 1 - 0 \cdot 1 = 1$$

$$f_y(0, 0) = (-x\sin y - e^x)_{(0,0)} = 0 - 1 = -1.$$

The tangent plane is therefore

$$1 \cdot (x - 0) - 1 \cdot (y - 0) - (z - 0) = 0, \qquad \text{Equation (4)}$$

or

$$x - y - z = 0.$$

■

EXAMPLE 3 Tangent Line to the Curve of Intersection of Two Surfaces

The surfaces

$$f(x, y, z) = x^2 + y^2 - 2 = 0 \qquad \text{A cylinder}$$

and

$$g(x, y, z) = x + z - 4 = 0 \qquad \text{A plane}$$

meet in an ellipse E (Figure 14.32). Find parametric equations for the line tangent to E at the point $P_0(1, 1, 3)$.

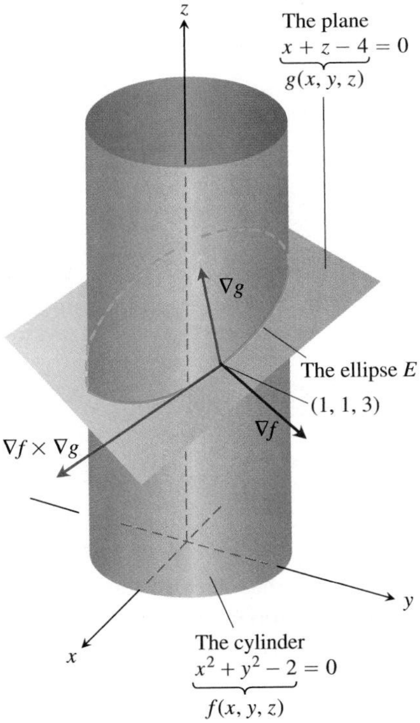

FIGURE 14.32 The cylinder $f(x, y, z) = x^2 + y^2 - 2 = 0$ and the plane $g(x, y, z) = x + z - 4 = 0$ intersect in an ellipse E (Example 3).

Solution The tangent line is orthogonal to both ∇f and ∇g at P_0, and therefore parallel to $\mathbf{v} = \nabla f \times \nabla g$. The components of $\mathbf{v}$ and the coordinates of P_0 give us equations for the line. We have

$$\nabla f\big|_{(1,1,3)} = (2x\mathbf{i} + 2y\mathbf{j})_{(1,1,3)} = 2\mathbf{i} + 2\mathbf{j}$$

$$\nabla g\big|_{(1,1,3)} = (\mathbf{i} + \mathbf{k})_{(1,1,3)} = \mathbf{i} + \mathbf{k}$$

$$\mathbf{v} = (2\mathbf{i} + 2\mathbf{j}) \times (\mathbf{i} + \mathbf{k}) = \begin{vmatrix} \mathbf{i} & \mathbf{j} & \mathbf{k} \\ 2 & 2 & 0 \\ 1 & 0 & 1 \end{vmatrix} = 2\mathbf{i} - 2\mathbf{j} - 2\mathbf{k}.$$

The tangent line is

$$x = 1 + 2t, \qquad y = 1 - 2t, \qquad z = 3 - 2t.$$

■

Estimating Change in a Specific Direction

The directional derivative plays the role of an ordinary derivative when we want to estimate how much the value of a function f changes if we move a small distance ds from a point P_0 to another point nearby. If f were a function of a single variable, we would have

$$df = f'(P_0)\,ds. \qquad \text{Ordinary derivative} \times \text{increment}$$

For a function of two or more variables, we use the formula

$$df = (\nabla f|_{P_0} \cdot \mathbf{u})\,ds, \qquad \text{Directional derivative} \times \text{increment}$$

where $\mathbf{u}$ is the direction of the motion away from P_0.

Estimating the Change in f in a Direction u

To estimate the change in the value of a differentiable function f when we move a small distance ds from a point P_0 in a particular direction $\mathbf{u}$, use the formula

$$df = \underbrace{(\nabla f|_{P_0} \cdot \mathbf{u})}_{\text{Directional derivative}} \cdot \underbrace{ds}_{\text{Distance increment}}$$

EXAMPLE 4 Estimating Change in the Value of $f(x, y, z)$

Estimate how much the value of

$$f(x, y, z) = y \sin x + 2yz$$

will change if the point $P(x, y, z)$ moves 0.1 unit from $P_0(0, 1, 0)$ straight toward $P_1(2, 2, -2)$.

Solution We first find the derivative of f at P_0 in the direction of the vector $\overrightarrow{P_0P_1} = 2\mathbf{i} + \mathbf{j} - 2\mathbf{k}$. The direction of this vector is

$$\mathbf{u} = \frac{\overrightarrow{P_0P_1}}{|\overrightarrow{P_0P_1}|} = \frac{\overrightarrow{P_0P_1}}{3} = \frac{2}{3}\mathbf{i} + \frac{1}{3}\mathbf{j} - \frac{2}{3}\mathbf{k}.$$

The gradient of f at P_0 is

$$\nabla f|_{(0,1,0)} = ((y \cos x)\mathbf{i} + (\sin x + 2z)\mathbf{j} + 2y\mathbf{k})_{(0,1,0)} = \mathbf{i} + 2\mathbf{k}.$$

Therefore,

$$\nabla f|_{P_0} \cdot \mathbf{u} = (\mathbf{i} + 2\mathbf{k}) \cdot \left(\frac{2}{3}\mathbf{i} + \frac{1}{3}\mathbf{j} - \frac{2}{3}\mathbf{k}\right) = \frac{2}{3} - \frac{4}{3} = -\frac{2}{3}.$$

The change df in f that results from moving $ds = 0.1$ unit away from P_0 in the direction of $\mathbf{u}$ is approximately

$$df = (\nabla f|_{P_0} \cdot \mathbf{u})(ds) = \left(-\frac{2}{3}\right)(0.1) \approx -0.067 \text{ unit.}$$ ■

How to Linearize a Function of Two Variables

Functions of two variables can be complicated, and we sometimes need to replace them with simpler ones that give the accuracy required for specific applications without being so difficult to work with. We do this in a way that is similar to the way we find linear replacements for functions of a single variable (Section 3.8).

Suppose the function we wish to replace is $z = f(x, y)$ and that we want the replacement to be effective near a point (x_0, y_0) at which we know the values of f, f_x, and f_y and at which f is differentiable. If we move from (x_0, y_0) to any point (x, y) by increments $\Delta x = x - x_0$ and $\Delta y = y - y_0$, then the definition of differentiability from Section 14.3 gives the change

$$f(x, y) - f(x_0, y_0) = f_x(x_0, y_0)\Delta x + f_y(x_0, y_0)\Delta y + \epsilon_1 \Delta x + \epsilon_2 \Delta y,$$

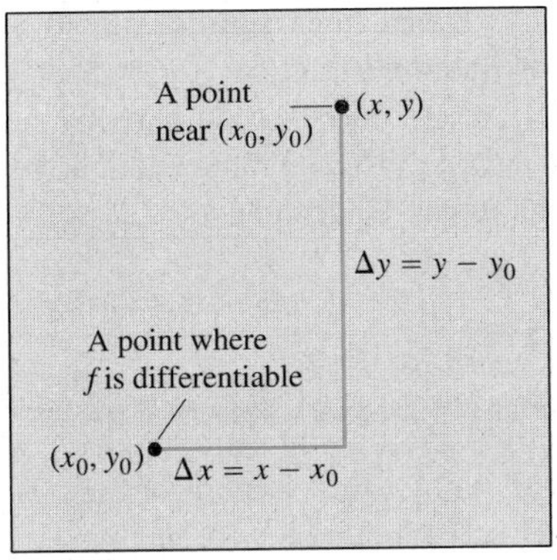

FIGURE 14.33 If f is differentiable at (x_0, y_0), then the value of f at any point (x, y) nearby is approximately $f(x_0, y_0) + f_x(x_0, y_0)\Delta x + f_y(x_0, y_0)\Delta y$.

where $\epsilon_1, \epsilon_2 \to 0$ as $\Delta x, \Delta y \to 0$. If the increments Δx and Δy are small, the products $\epsilon_1 \Delta x$ and $\epsilon_2 \Delta y$ will eventually be smaller still and we will have

$$f(x, y) \approx \underbrace{f(x_0, y_0) + f_x(x_0, y_0)(x - x_0) + f_y(x_0, y_0)(y - y_0)}_{L(x, y)}.$$

In other words, as long as Δx and Δy are small, f will have approximately the same value as the linear function L. If f is hard to use, and our work can tolerate the error involved, we may approximate f by L (Figure 14.33).

DEFINITIONS Linearization, Standard Linear Approximation

The **linearization** of a function $f(x, y)$ at a point (x_0, y_0) where f is differentiable is the function

$$L(x, y) = f(x_0, y_0) + f_x(x_0, y_0)(x - x_0) + f_y(x_0, y_0)(y - y_0). \qquad (5)$$

The approximation

$$f(x, y) \approx L(x, y)$$

is the **standard linear approximation** of f at (x_0, y_0).

From Equation (4), we see that the plane $z = L(x, y)$ is tangent to the surface $z = f(x, y)$ at the point (x_0, y_0). Thus, the linearization of a function of two variables is a tangent-*plane* approximation in the same way that the linearization of a function of a single variable is a tangent-*line* approximation.

EXAMPLE 5 Finding a Linearization

Find the linearization of

$$f(x, y) = x^2 - xy + \frac{1}{2}y^2 + 3$$

at the point (3, 2).

Solution We first evaluate f, f_x, and f_y at the point $(x_0, y_0) = (3, 2)$:

$$f(3, 2) = \left(x^2 - xy + \frac{1}{2}y^2 + 3\right)_{(3,2)} = 8$$

$$f_x(3, 2) = \frac{\partial}{\partial x}\left(x^2 - xy + \frac{1}{2}y^2 + 3\right)_{(3,2)} = (2x - y)_{(3,2)} = 4$$

$$f_y(3, 2) = \frac{\partial}{\partial y}\left(x^2 - xy + \frac{1}{2}y^2 + 3\right)_{(3,2)} = (-x + y)_{(3,2)} = -1,$$

giving

$$\begin{aligned} L(x, y) &= f(x_0, y_0) + f_x(x_0, y_0)(x - x_0) + f_y(x_0, y_0)(y - y_0) \\ &= 8 + (4)(x - 3) + (-1)(y - 2) = 4x - y - 2. \end{aligned}$$

The linearization of f at (3, 2) is $L(x, y) = 4x - y - 2$. ■

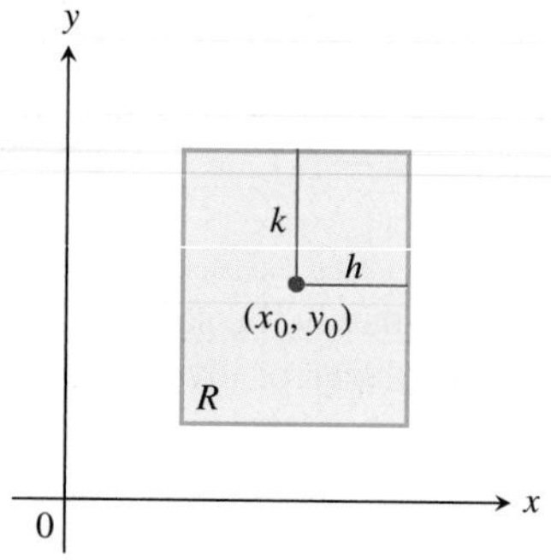

FIGURE 14.34 The rectangular region R: $|x - x_0| \le h, |y - y_0| \le k$ in the xy-plane.

When approximating a differentiable function $f(x, y)$ by its linearization $L(x, y)$ at (x_0, y_0), an important question is how accurate the approximation might be.

If we can find a common upper bound M for $|f_{xx}|, |f_{yy}|$, and $|f_{xy}|$ on a rectangle R centered at (x_0, y_0) (Figure 14.34), then we can bound the error E throughout R by using a simple formula (derived in Section 14.10). The **error** is defined by $E(x, y) = f(x, y) - L(x, y)$.

The Error in the Standard Linear Approximation

If f has continuous first and second partial derivatives throughout an open set containing a rectangle R centered at (x_0, y_0) and if M is any upper bound for the values of $|f_{xx}|, |f_{yy}|$, and $|f_{xy}|$ on R, then the error $E(x, y)$ incurred in replacing $f(x, y)$ on R by its linearization

$$L(x, y) = f(x_0, y_0) + f_x(x_0, y_0)(x - x_0) + f_y(x_0, y_0)(y - y_0)$$

satisfies the inequality

$$|E(x, y)| \le \frac{1}{2}M(|x - x_0| + |y - y_0|)^2.$$

To make $|E(x, y)|$ small for a given M, we just make $|x - x_0|$ and $|y - y_0|$ small.

EXAMPLE 6 Bounding the Error in Example 5

Find an upper bound for the error in the approximation $f(x, y) \approx L(x, y)$ in Example 5 over the rectangle

$$R: \quad |x - 3| \le 0.1, \qquad |y - 2| \le 0.1 .$$

Express the upper bound as a percentage of $f(3, 2)$, the value of f at the center of the rectangle.

Solution We use the inequality

$$|E(x, y)| \le \frac{1}{2}M(|x - x_0| + |y - y_0|)^2.$$

To find a suitable value for M, we calculate f_{xx}, f_{xy}, and f_{yy}, finding, after a routine differentiation, that all three derivatives are constant, with values

$$|f_{xx}| = |2| = 2, \qquad |f_{xy}| = |-1| = 1, \qquad |f_{yy}| = |1| = 1.$$

The largest of these is 2, so we may safely take M to be 2. With $(x_0, y_0) = (3, 2)$, we then know that, throughout R,

$$|E(x, y)| \le \frac{1}{2}(2)(|x - 3| + |y - 2|)^2 = (|x - 3| + |y - 2|)^2.$$

Finally, since $|x - 3| \le 0.1$ and $|y - 2| \le 0.1$ on R, we have

$$|E(x, y)| \le (0.1 + 0.1)^2 = 0.04.$$

As a percentage of $f(3, 2) = 8$, the error is no greater than

$$\frac{0.04}{8} \times 100 = 0.5\% .$$

■

Differentials

Recall from Section 3.8 that for a function of a single variable, $y = f(x)$, we defined the change in f as x changes from a to $a + \Delta x$ by

$$\Delta f = f(a + \Delta x) - f(a)$$

and the differential of f as

$$df = f'(a)\Delta x.$$

We now consider a function of two variables.

Suppose a differentiable function $f(x, y)$ and its partial derivatives exist at a point (x_0, y_0). If we move to a nearby point $(x_0 + \Delta x, y_0 + \Delta y)$, the change in f is

$$\Delta f = f(x_0 + \Delta x, y_0 + \Delta y) - f(x_0, y_0).$$

A straightforward calculation from the definition of $L(x, y)$, using the notation $x - x_0 = \Delta x$ and $y - y_0 = \Delta y$, shows that the corresponding change in L is

$$\begin{aligned}\Delta L &= L(x_0 + \Delta x, y_0 + \Delta y) - L(x_0, y_0)\\ &= f_x(x_0, y_0)\Delta x + f_y(x_0, y_0)\Delta y.\end{aligned}$$

The **differentials** dx and dy are independent variables, so they can be assigned any values. Often we take $dx = \Delta x = x - x_0$, and $dy = \Delta y = y - y_0$. We then have the following definition of the differential or *total* differential of f.

DEFINITION **Total Differential**

If we move from (x_0, y_0) to a point $(x_0 + dx, y_0 + dy)$ nearby, the resulting change

$$df = f_x(x_0, y_0)\, dx + f_y(x_0, y_0)\, dy$$

in the linearization of f is called the **total differential of f**.

EXAMPLE 7 Estimating Change in Volume

Suppose that a cylindrical can is designed to have a radius of 1 in. and a height of 5 in., but that the radius and height are off by the amounts $dr = +0.03$ and $dh = -0.1$. Estimate the resulting absolute change in the volume of the can.

Solution To estimate the absolute change in $V = \pi r^2 h$, we use

$$\Delta V \approx dV = V_r(r_0, h_0)\, dr + V_h(r_0, h_0)\, dh.$$

With $V_r = 2\pi rh$ and $V_h = \pi r^2$, we get

$$\begin{aligned}dV &= 2\pi r_0 h_0\, dr + \pi r_0^2\, dh = 2\pi(1)(5)(0.03) + \pi(1)^2(-0.1)\\ &= 0.3\pi - 0.1\pi = 0.2\pi \approx 0.63 \text{ in.}^3\end{aligned}$$

■

Instead of absolute change in the value of a function $f(x, y)$, we can estimate *relative change* or *percentage change* by

$$\frac{df}{f(x_0, y_0)} \quad \text{and} \quad \frac{df}{f(x_0, y_0)} \times 100,$$

respectively. In Example 7, the relative change is estimated by

$$\frac{dV}{V(r_0, h_0)} = \frac{0.2\pi}{\pi r_0^2 h_0} = \frac{0.2\pi}{\pi(1)^2(5)} = 0.04,$$

giving 4% as an estimate of the percentage change.

EXAMPLE 8 Sensitivity to Change

Your company manufactures right circular cylindrical molasses storage tanks that are 25 ft high with a radius of 5 ft. How sensitive are the tanks' volumes to small variations in height and radius?

Solution With $V = \pi r^2 h$, we have the approximation for the change in volume as

$$\begin{aligned} dV &= V_r(5, 25)\,dr + V_h(5, 25)\,dh \\ &= (2\pi rh)_{(5,25)}\,dr + (\pi r^2)_{(5,25)}\,dh \\ &= 250\pi\,dr + 25\pi\,dh. \end{aligned}$$

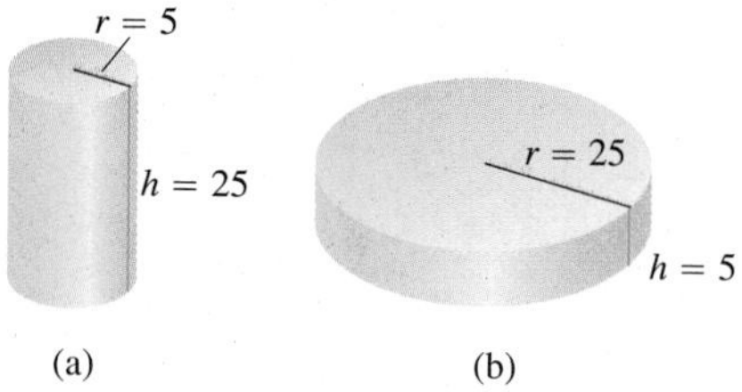

FIGURE 14.35 The volume of cylinder (a) is more sensitive to a small change in r than it is to an equally small change in h. The volume of cylinder (b) is more sensitive to small changes in h than it is to small changes in r (Example 8).

Thus, a 1-unit change in r will change V by about 250π units. A 1-unit change in h will change V by about 25π units. The tank's volume is 10 times more sensitive to a small change in r than it is to a small change of equal size in h. As a quality control engineer concerned with being sure the tanks have the correct volume, you would want to pay special attention to their radii.

In contrast, if the values of r and h are reversed to make $r = 25$ and $h = 5$, then the total differential in V becomes

$$dV = (2\pi rh)_{(25,5)}\,dr + (\pi r^2)_{(25,5)}\,dh = 250\pi\,dr + 625\pi\,dh.$$

Now the volume is more sensitive to changes in h than to changes in r (Figure 14.35).

The general rule is that functions are most sensitive to small changes in the variables that generate the largest partial derivatives. ■

EXAMPLE 9 Estimating Percentage Error

The volume $V = \pi r^2 h$ of a right circular cylinder is to be calculated from measured values of r and h. Suppose that r is measured with an error of no more than 2% and h with an error of no more than 0.5%. Estimate the resulting possible percentage error in the calculation of V.

Solution We are told that

$$\left|\frac{dr}{r} \times 100\right| \le 2 \quad \text{and} \quad \left|\frac{dh}{h} \times 100\right| \le 0.5.$$

Since

$$\frac{dV}{V} = \frac{2\pi rh\,dr + \pi r^2\,dh}{\pi r^2 h} = \frac{2\,dr}{r} + \frac{dh}{h},$$

we have

$$\begin{aligned}\left|\frac{dV}{V}\right| &= \left|2\frac{dr}{r} + \frac{dh}{h}\right| \\ &\le \left|2\frac{dr}{r}\right| + \left|\frac{dh}{h}\right| \\ &\le 2(0.02) + 0.005 = 0.045.\end{aligned}$$

We estimate the error in the volume calculation to be at most 4.5%. ■

Functions of More Than Two Variables

Analogous results hold for differentiable functions of more than two variables.

1. The **linearization** of $f(x, y, z)$ at a point $P_0(x_0, y_0, z_0)$ is

$$L(x, y, z) = f(P_0) + f_x(P_0)(x - x_0) + f_y(P_0)(y - y_0) + f_z(P_0)(z - z_0).$$

2. Suppose that R is a closed rectangular solid centered at P_0 and lying in an open region on which the second partial derivatives of f are continuous. Suppose also that $|f_{xx}|, |f_{yy}|, |f_{zz}|, |f_{xy}|, |f_{xz}|$, and $|f_{yz}|$ are all less than or equal to M throughout R. Then the **error** $E(x, y, z) = f(x, y, z) - L(x, y, z)$ in the approximation of f by L is bounded throughout R by the inequality

$$|E| \le \frac{1}{2}M(|x - x_0| + |y - y_0| + |z - z_0|)^2.$$

3. If the second partial derivatives of f are continuous and if x, y, and z change from $x_0, y_0,$ and z_0 by small amounts dx, dy, and dz, the **total differential**

$$df = f_x(P_0)\,dx + f_y(P_0)\,dy + f_z(P_0)\,dz$$

gives a good approximation of the resulting change in f.

EXAMPLE 10 Finding a Linear Approximation in 3-Space

Find the linearization $L(x, y, z)$ of

$$f(x, y, z) = x^2 - xy + 3\sin z$$

at the point $(x_0, y_0, z_0) = (2, 1, 0)$. Find an upper bound for the error incurred in replacing f by L on the rectangle

$$R:\quad |x - 2| \le 0.01, \qquad |y - 1| \le 0.02, \qquad |z| \le 0.01.$$

Solution A routine evaluation gives

$$f(2, 1, 0) = 2, \qquad f_x(2, 1, 0) = 3, \qquad f_y(2, 1, 0) = -2, \qquad f_z(2, 1, 0) = 3.$$

Thus,

$$L(x, y, z) = 2 + 3(x - 2) + (-2)(y - 1) + 3(z - 0) = 3x - 2y + 3z - 2.$$

Since

$$\begin{aligned} f_{xx} &= 2, & f_{yy} &= 0, & f_{zz} &= -3\sin z, \\ f_{xy} &= -1, & f_{xz} &= 0, & f_{yz} &= 0, \end{aligned}$$

we may safely take M to be $\max|-3 \sin z| = 3$. Hence, the error incurred by replacing f by L on R satisfies

$$|E| \leq \frac{1}{2}(3)(0.01 + 0.02 + 0.01)^2 = 0.0024.$$

The error will be no greater than 0.0024. ∎

EXERCISES 14.6

Tangent Planes and Normal Lines to Surfaces

In Exercises 1–8, find equations for the

(a) tangent plane and **(b)** normal line at the point P_0 on the given surface.

1. $x^2 + y^2 + z^2 = 3, \quad P_0(1, 1, 1)$

2. $x^2 + y^2 - z^2 = 18, \quad P_0(3, 5, -4)$

3. $2z - x^2 = 0, \quad P_0(2, 0, 2)$

4. $x^2 + 2xy - y^2 + z^2 = 7, \quad P_0(1, -1, 3)$

5. $\cos \pi x - x^2 y + e^{xz} + yz = 4, \quad P_0(0, 1, 2)$

6. $x^2 - xy - y^2 - z = 0, \quad P_0(1, 1, -1)$

7. $x + y + z = 1, \quad P_0(0, 1, 0)$

8. $x^2 + y^2 - 2xy - x + 3y - z = -4, \quad P_0(2, -3, 18)$

In Exercises 9–12, find an equation for the plane that is tangent to the given surface at the given point.

9. $z = \ln(x^2 + y^2), \quad (1, 0, 0)$ **10.** $z = e^{-(x^2+y^2)}, \quad (0, 0, 1)$

11. $z = \sqrt{y - x}, \quad (1, 2, 1)$ **12.** $z = 4x^2 + y^2, \quad (1, 1, 5)$

Tangent Lines to Curves

In Exercises 13–18, find parametric equations for the line tangent to the curve of intersection of the surfaces at the given point.

13. Surfaces: $x + y^2 + 2z = 4, \quad x = 1$

Point: $(1, 1, 1)$

14. Surfaces: $xyz = 1, \quad x^2 + 2y^2 + 3z^2 = 6$

Point: $(1, 1, 1)$

15. Surfaces: $x^2 + 2y + 2z = 4, \quad y = 1$

Point: $(1, 1, 1/2)$

16. Surfaces: $x + y^2 + z = 2, \quad y = 1$

Point: $(1/2, 1, 1/2)$

17. Surfaces: $x^3 + 3x^2y^2 + y^3 + 4xy - z^2 = 0, \quad x^2 + y^2 + z^2 = 11$

Point: $(1, 1, 3)$

18. Surfaces: $x^2 + y^2 = 4, \quad x^2 + y^2 - z = 0$

Point: $(\sqrt{2}, \sqrt{2}, 4)$

Estimating Change

19. By about how much will

$$f(x, y, z) = \ln\sqrt{x^2 + y^2 + z^2}$$

change if the point $P(x, y, z)$ moves from $P_0(3, 4, 12)$ a distance of $ds = 0.1$ unit in the direction of $3\mathbf{i} + 6\mathbf{j} - 2\mathbf{k}$?

20. By about how much will

$$f(x, y, z) = e^x \cos yz$$

change as the point $P(x, y, z)$ moves from the origin a distance of $ds = 0.1$ unit in the direction of $2\mathbf{i} + 2\mathbf{j} - 2\mathbf{k}$?

21. By about how much will

$$g(x, y, z) = x + x\cos z - y\sin z + y$$

change if the point $P(x, y, z)$ moves from $P_0(2, -1, 0)$ a distance of $ds = 0.2$ unit toward the point $P_1(0, 1, 2)$?

22. By about how much will

$$h(x, y, z) = \cos(\pi xy) + xz^2$$

change if the point $P(x, y, z)$ moves from $P_0(-1, -1, -1)$ a distance of $ds = 0.1$ unit toward the origin?

23. Temperature change along a circle Suppose that the Celsius temperature at the point (x, y) in the xy-plane is $T(x, y) = x \sin 2y$ and that distance in the xy-plane is measured in meters. A particle is moving *clockwise* around the circle of radius 1 m centered at the origin at the constant rate of 2 m/sec.

a. How fast is the temperature experienced by the particle changing in degrees Celsius per meter at the point $P(1/2, \sqrt{3}/2)$?

b. How fast is the temperature experienced by the particle changing in degrees Celsius per second at P?

24. Changing temperature along a space curve The Celsius temperature in a region in space is given by $T(x, y, z) = 2x^2 - xyz$. A particle is moving in this region and its position at time t is given by $x = 2t^2, y = 3t, z = -t^2$, where time is measured in seconds and distance in meters.

a. How fast is the temperature experienced by the particle changing in degrees Celsius per meter when the particle is at the point $P(8, 6, -4)$?

b. How fast is the temperature experienced by the particle changing in degrees Celsius per second at P?

Finding Linearizations

In Exercises 25–30, find the linearization $L(x, y)$ of the function at each point.

25. $f(x, y) = x^2 + y^2 + 1$ at **a.** $(0, 0)$, **b.** $(1, 1)$

26. $f(x, y) = (x + y + 2)^2$ at **a.** $(0, 0)$, **b.** $(1, 2)$

27. $f(x, y) = 3x - 4y + 5$ at **a.** $(0, 0)$, **b.** $(1, 1)$

28. $f(x, y) = x^3y^4$ at **a.** $(1, 1)$, **b.** $(0, 0)$

29. $f(x, y) = e^x \cos y$ at **a.** $(0, 0)$, **b.** $(0, \pi/2)$

30. $f(x, y) = e^{2y-x}$ at **a.** $(0, 0)$, **b.** $(1, 2)$

Upper Bounds for Errors in Linear Approximations

In Exercises 31–36, find the linearization $L(x, y)$ of the function $f(x, y)$ at P_0. Then find an upper bound for the magnitude $|E|$ of the error in the approximation $f(x, y) \approx L(x, y)$ over the rectangle R.

31. $f(x, y) = x^2 - 3xy + 5$ at $P_0(2, 1)$,
R: $|x - 2| \le 0.1$, $|y - 1| \le 0.1$

32. $f(x, y) = (1/2)x^2 + xy + (1/4)y^2 + 3x - 3y + 4$ at $P_0(2, 2)$,
R: $|x - 2| \le 0.1$, $|y - 2| \le 0.1$

33. $f(x, y) = 1 + y + x \cos y$ at $P_0(0, 0)$,
R: $|x| \le 0.2$, $|y| \le 0.2$
(Use $|\cos y| \le 1$ and $|\sin y| \le 1$ in estimating E.)

34. $f(x, y) = xy^2 + y \cos(x - 1)$ at $P_0(1, 2)$,
R: $|x - 1| \le 0.1$, $|y - 2| \le 0.1$

35. $f(x, y) = e^x \cos y$ at $P_0(0, 0)$,
R: $|x| \le 0.1$, $|y| \le 0.1$
(Use $e^x \le 1.11$ and $|\cos y| \le 1$ in estimating E.)

36. $f(x, y) = \ln x + \ln y$ at $P_0(1, 1)$,
R: $|x - 1| \le 0.2$, $|y - 1| \le 0.2$

Functions of Three Variables

Find the linearizations $L(x, y, z)$ of the functions in Exercises 37–42 at the given points.

37. $f(x, y, z) = xy + yz + xz$ at
a. $(1, 1, 1)$ **b.** $(1, 0, 0)$ **c.** $(0, 0, 0)$

38. $f(x, y, z) = x^2 + y^2 + z^2$ at
a. $(1, 1, 1)$ **b.** $(0, 1, 0)$ **c.** $(1, 0, 0)$

39. $f(x, y, z) = \sqrt{x^2 + y^2 + z^2}$ at
a. $(1, 0, 0)$ **b.** $(1, 1, 0)$ **c.** $(1, 2, 2)$

40. $f(x, y, z) = (\sin xy)/z$ at
a. $(\pi/2, 1, 1)$ **b.** $(2, 0, 1)$

41. $f(x, y, z) = e^x + \cos(y + z)$ at
a. $(0, 0, 0)$ **b.** $\left(0, \frac{\pi}{2}, 0\right)$ **c.** $\left(0, \frac{\pi}{4}, \frac{\pi}{4}\right)$

42. $f(x, y, z) = \tan^{-1}(xyz)$ at
a. $(1, 0, 0)$ **b.** $(1, 1, 0)$ **c.** $(1, 1, 1)$

In Exercises 43–46, find the linearization $L(x, y, z)$ of the function $f(x, y, z)$ at P_0. Then find an upper bound for the magnitude of the error E in the approximation $f(x, y, z) \approx L(x, y, z)$ over the region R.

43. $f(x, y, z) = xz - 3yz + 2$ at $P_0(1, 1, 2)$
R: $|x - 1| \le 0.01$, $|y - 1| \le 0.01$, $|z - 2| \le 0.02$

44. $f(x, y, z) = x^2 + xy + yz + (1/4)z^2$ at $P_0(1, 1, 2)$
R: $|x - 1| \le 0.01$, $|y - 1| \le 0.01$, $|z - 2| \le 0.08$

45. $f(x, y, z) = xy + 2yz - 3xz$ at $P_0(1, 1, 0)$
R: $|x - 1| \le 0.01$, $|y - 1| \le 0.01$, $|z| \le 0.01$

46. $f(x, y, z) = \sqrt{2} \cos x \sin(y + z)$ at $P_0(0, 0, \pi/4)$
R: $|x| \le 0.01$, $|y| \le 0.01$, $|z - \pi/4| \le 0.01$

Estimating Error; Sensitivity to Change

47. Estimating maximum error Suppose that T is to be found from the formula $T = x(e^y + e^{-y})$, where x and y are found to be 2 and ln 2 with maximum possible errors of $|dx| = 0.1$ and $|dy| = 0.02$. Estimate the maximum possible error in the computed value of T.

48. Estimating volume of a cylinder About how accurately may $V = \pi r^2 h$ be calculated from measurements of r and h that are in error by 1%?

49. Maximum percentage error If $r = 5.0$ cm and $h = 12.0$ cm to the nearest millimeter, what should we expect the maximum percentage error in calculating $V = \pi r^2 h$ to be?

50. Variation in electrical resistance The resistance R produced by wiring resistors of R_1 and R_2 ohms in parallel (see accompanying figure) can be calculated from the formula

$$\frac{1}{R} = \frac{1}{R_1} + \frac{1}{R_2}.$$

a. Show that

$$dR = \left(\frac{R}{R_1}\right)^2 dR_1 + \left(\frac{R}{R_2}\right)^2 dR_2.$$

b. You have designed a two-resistor circuit like the one shown on the next page to have resistances of $R_1 = 100$ ohms and $R_2 = 400$ ohms, but there is always some variation in manufacturing and the resistors received by your firm will probably not have these exact values. Will the value of R be more sensitive to variation in R_1 or to variation in R_2? Give reasons for your answer.

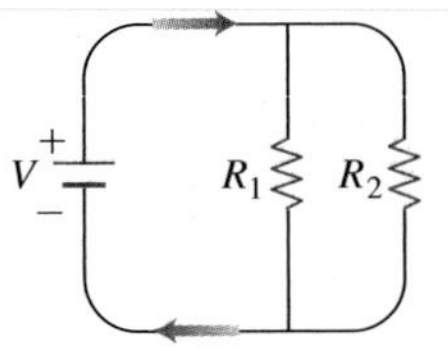

c. In another circuit like the one shown you plan to change R_1 from 20 to 20.1 ohms and R_2 from 25 to 24.9 ohms. By about what percentage will this change R?

51. You plan to calculate the area of a long, thin rectangle from measurements of its length and width. Which dimension should you measure more carefully? Give reasons for your answer.

52. a. Around the point $(1, 0)$, is $f(x, y) = x^2(y + 1)$ more sensitive to changes in x or to changes in y? Give reasons for your answer.

b. What ratio of dx to dy will make df equal zero at $(1, 0)$?

53. Error carryover in coordinate changes

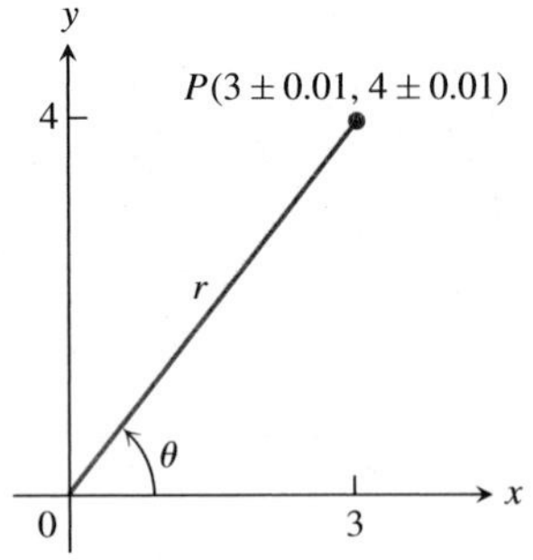

a. If $x = 3 \pm 0.01$ and $y = 4 \pm 0.01$, as shown here, with approximately what accuracy can you calculate the polar coordinates r and θ of the point $P(x, y)$ from the formulas $r^2 = x^2 + y^2$ and $\theta = \tan^{-1}(y/x)$? Express your estimates as percentage changes of the values that r and θ have at the point $(x_0, y_0) = (3, 4)$.

b. At the point $(x_0, y_0) = (3, 4)$, are the values of r and θ more sensitive to changes in x or to changes in y? Give reasons for your answer.

54. Designing a soda can A standard 12-fl oz can of soda is essentially a cylinder of radius $r = 1$ in. and height $h = 5$ in.

a. At these dimensions, how sensitive is the can's volume to a small change in radius versus a small change in height?

b. Could you design a soda can that *appears* to hold more soda but in fact holds the same 12-fl oz? What might its dimensions be? (There is more than one correct answer.)

55. Value of a 2 × 2 determinant If $|a|$ is much greater than $|b|$, $|c|$, and $|d|$, to which of a, b, c, and d is the value of the determinant

$$f(a, b, c, d) = \begin{vmatrix} a & b \\ c & d \end{vmatrix}$$

most sensitive? Give reasons for your answer.

56. Estimating maximum error Suppose that $u = xe^y + y \sin z$ and that x, y, and z can be measured with maximum possible errors of ± 0.2, ± 0.6, and $\pm \pi/180$, respectively. Estimate the maximum possible error in calculating u from the measured values $x = 2$, $y = \ln 3$, $z = \pi/2$.

57. The Wilson lot size formula The Wilson lot size formula in economics says that the most economical quantity Q of goods (radios, shoes, brooms, whatever) for a store to order is given by the formula $Q = \sqrt{2KM/h}$, where K is the cost of placing the order, M is the number of items sold per week, and h is the weekly holding cost for each item (cost of space, utilities, security, and so on). To which of the variables K, M, and h is Q most sensitive near the point $(K_0, M_0, h_0) = (2, 20, 0.05)$? Give reasons for your answer.

58. Surveying a triangular field The area of a triangle is $(1/2)ab \sin C$, where a and b are the lengths of two sides of the triangle and C is the measure of the included angle. In surveying a triangular plot, you have measured a, b, and C to be 150 ft, 200 ft, and 60°, respectively. By about how much could your area calculation be in error if your values of a and b are off by half a foot each and your measurement of C is off by 2°? See the accompanying figure. Remember to use radians.

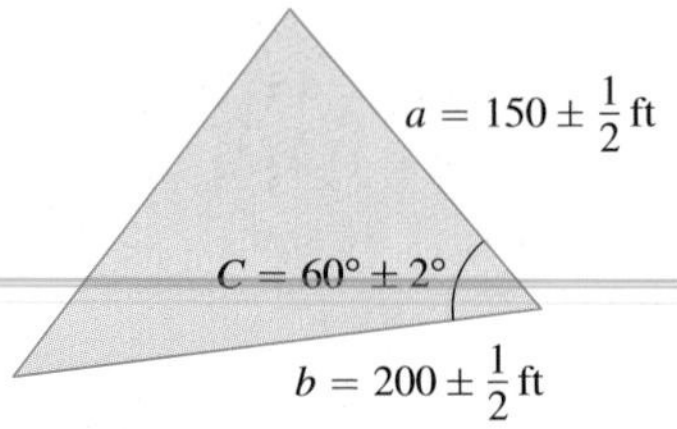

Theory and Examples

59. The linearization of $f(x, y)$ is a tangent-plane approximation Show that the tangent plane at the point $P_0(x_0, y_0, f(x_0, y_0))$ on the surface $z = f(x, y)$ defined by a differentiable function f is the plane

$$f_x(x_0, y_0)(x - x_0) + f_y(x_0, y_0)(y - y_0) - (z - f(x_0, y_0)) = 0$$

or

$$z = f(x_0, y_0) + f_x(x_0, y_0)(x - x_0) + f_y(x_0, y_0)(y - y_0).$$

Thus, the tangent plane at P_0 is the graph of the linearization of f at P_0 (see accompanying figure).

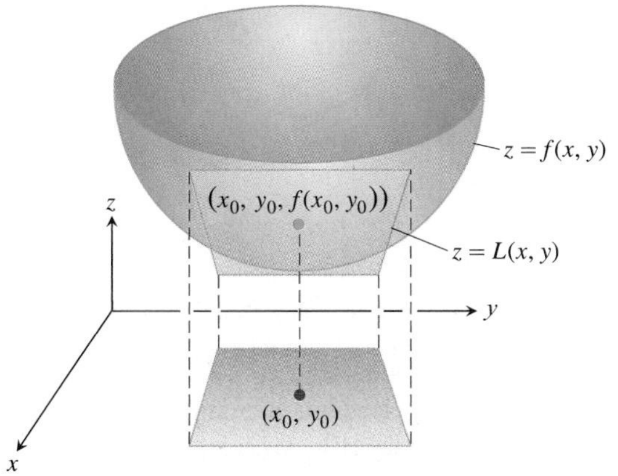

60. Change along the involute of a circle Find the derivative of $f(x, y) = x^2 + y^2$ in the direction of the unit tangent vector of the curve

$$\mathbf{r}(t) = (\cos t + t \sin t)\mathbf{i} + (\sin t - t \cos t)\mathbf{j}, \qquad t > 0.$$

61. Change along a helix Find the derivative of $f(x, y, z) = x^2 + y^2 + z^2$ in the direction of the unit tangent vector of the helix

$$\mathbf{r}(t) = (\cos t)\mathbf{i} + (\sin t)\mathbf{j} + t\mathbf{k}$$

at the points where $t = -\pi/4$, 0, and $\pi/4$. The function f gives the square of the distance from a point $P(x, y, z)$ on the helix to the origin. The derivatives calculated here give the rates at which the square of the distance is changing with respect to t as P moves through the points where $t = -\pi/4$, 0, and $\pi/4$.

62. Normal curves A smooth curve is *normal* to a surface $f(x, y, z) = c$ at a point of intersection if the curve's velocity vector is a nonzero scalar multiple of ∇f at the point.

Show that the curve

$$\mathbf{r}(t) = \sqrt{t}\mathbf{i} + \sqrt{t}\mathbf{j} - \frac{1}{4}(t + 3)\mathbf{k}$$

is normal to the surface $x^2 + y^2 - z = 3$ when $t = 1$.

63. Tangent curves A smooth curve is *tangent* to the surface at a point of intersection if its velocity vector is orthogonal to ∇f there.

Show that the curve

$$\mathbf{r}(t) = \sqrt{t}\mathbf{i} + \sqrt{t}\mathbf{j} + (2t - 1)\mathbf{k}$$

is tangent to the surface $x^2 + y^2 - z = 1$ when $t = 1$.

14.7 Extreme Values and Saddle Points

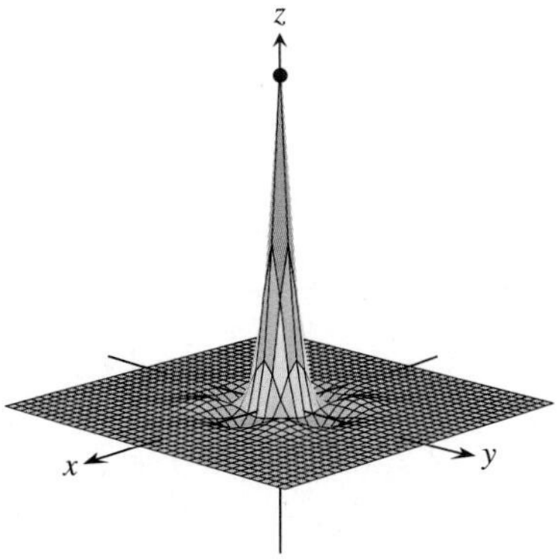

FIGURE 14.36 The function

$$z = (\cos x)(\cos y)e^{-\sqrt{x^2+y^2}}$$

has a maximum value of 1 and a minimum value of about -0.067 on the square region $|x| \le 3\pi/2, |y| \le 3\pi/2$.

Continuous functions of two variables assume extreme values on closed, bounded domains (see Figures 14.36 and 14.37). We see in this section that we can narrow the search for these extreme values by examining the functions' first partial derivatives. A function of two variables can assume extreme values only at domain boundary points or at interior domain points where both first partial derivatives are zero or where one or both of the first partial derivatives fails to exist. However, the vanishing of derivatives at an interior point (a, b) does not always signal the presence of an extreme value. The surface that is the graph of the function might be shaped like a saddle right above (a, b) and cross its tangent plane there.

Derivative Tests for Local Extreme Values

To find the local extreme values of a function of a single variable, we look for points where the graph has a horizontal tangent line. At such points, we then look for local maxima, local minima, and points of inflection. For a function $f(x, y)$ of two variables, we look for points where the surface $z = f(x, y)$ has a horizontal tangent *plane*. At such points, we then look for local maxima, local minima, and saddle points (more about saddle points in a moment).

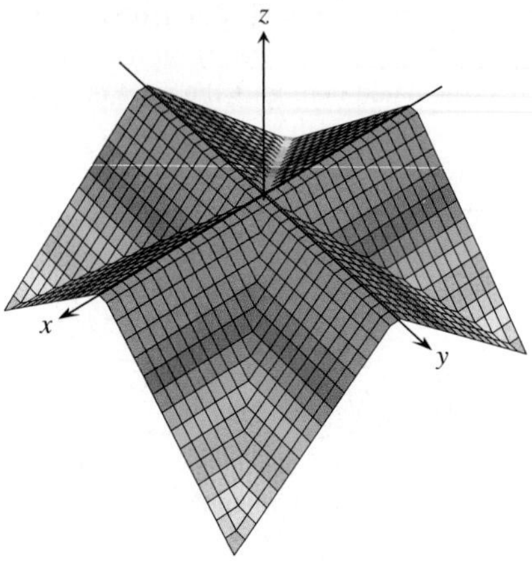

FIGURE 14.37 The "roof surface"

$$z = \frac{1}{2}\left(\left||x| - |y|\right| - |x| - |y|\right)$$

viewed from the point (10, 15, 20). The defining function has a maximum value of 0 and a minimum value of $-a$ on the square region $|x| \le a, |y| \le a$.

HISTORICAL BIOGRAPHY

Siméon-Denis Poisson
(1781–1840)

DEFINITIONS Local Maximum, Local Minimum

Let $f(x, y)$ be defined on a region R containing the point (a, b). Then

1. $f(a, b)$ is a **local maximum** value of f if $f(a, b) \ge f(x, y)$ for all domain points (x, y) in an open disk centered at (a, b).
2. $f(a, b)$ is a **local minimum** value of f if $f(a, b) \le f(x, y)$ for all domain points (x, y) in an open disk centered at (a, b).

Local maxima correspond to mountain peaks on the surface $z = f(x, y)$ and local minima correspond to valley bottoms (Figure 14.38). At such points the tangent planes, when they exist, are horizontal. Local extrema are also called **relative extrema**.

As with functions of a single variable, the key to identifying the local extrema is a first derivative test.

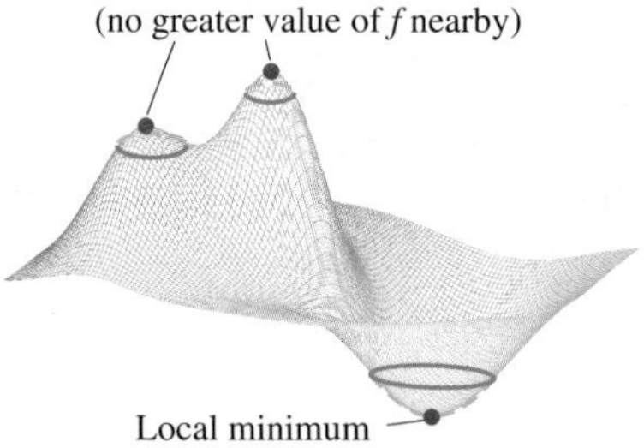

FIGURE 14.38 A local maximum is a mountain peak and a local minimum is a valley low.

THEOREM 10 First Derivative Test for Local Extreme Values

If $f(x, y)$ has a local maximum or minimum value at an interior point (a, b) of its domain and if the first partial derivatives exist there, then $f_x(a, b) = 0$ and $f_y(a, b) = 0$.

Proof If f has a local extremum at (a, b), then the function $g(x) = f(x, b)$ has a local extremum at $x = a$ (Figure 14.39). Therefore, $g'(a) = 0$ (Chapter 4, Theorem 2). Now $g'(a) = f_x(a, b)$, so $f_x(a, b) = 0$. A similar argument with the function $h(y) = f(a, y)$ shows that $f_y(a, b) = 0$. ■

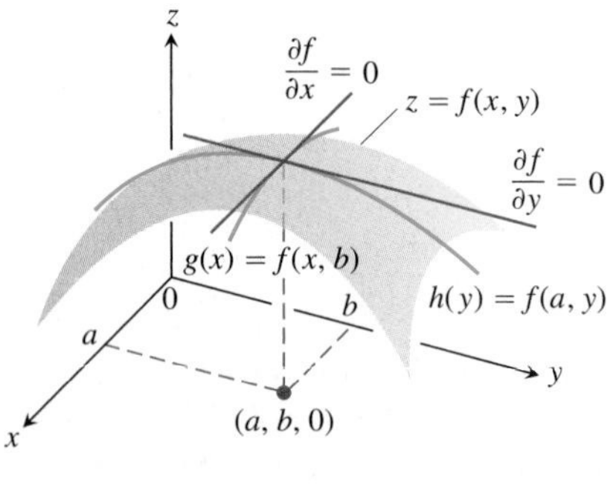

FIGURE 14.39 If a local maximum of f occurs at $x = a, y = b$, then the first partial derivatives $f_x(a, b)$ and $f_y(a, b)$ are both zero.

If we substitute the values $f_x(a, b) = 0$ and $f_y(a, b) = 0$ into the equation

$$f_x(a, b)(x - a) + f_y(a, b)(y - b) - (z - f(a, b)) = 0$$

for the tangent plane to the surface $z = f(x, y)$ at (a, b), the equation reduces to

$$0 \cdot (x - a) + 0 \cdot (y - b) - z + f(a, b) = 0$$

or

$$z = f(a, b).$$

Thus, Theorem 10 says that the surface does indeed have a horizontal tangent plane at a local extremum, provided there is a tangent plane there.

DEFINITION Critical Point

An interior point of the domain of a function $f(x, y)$ where both f_x and f_y are zero or where one or both of f_x and f_y do not exist is a **critical point** of f.

Theorem 10 says that the only points where a function $f(x, y)$ can assume extreme values are critical points and boundary points. As with differentiable functions of a single variable, not every critical point gives rise to a local extremum. A differentiable function of a single variable might have a point of inflection. A differentiable function of two variables might have a *saddle point*.

DEFINITION Saddle Point

A differentiable function $f(x, y)$ has a **saddle point** at a critical point (a, b) if in every open disk centered at (a, b) there are domain points (x, y) where $f(x, y) > f(a, b)$ and domain points (x, y) where $f(x, y) < f(a, b)$. The corresponding point $(a, b, f(a, b))$ on the surface $z = f(x, y)$ is called a saddle point of the surface (Figure 14.40).

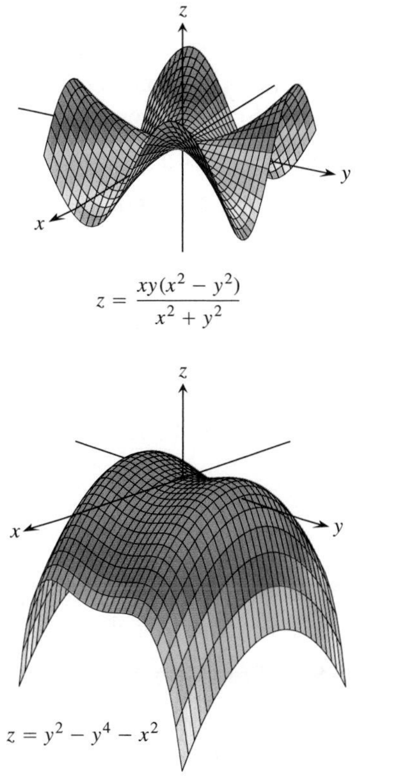

FIGURE 14.40 Saddle points at the origin.

EXAMPLE 1 Finding Local Extreme Values

Find the local extreme values of $f(x, y) = x^2 + y^2$.

Solution The domain of f is the entire plane (so there are no boundary points) and the partial derivatives $f_x = 2x$ and $f_y = 2y$ exist everywhere. Therefore, local extreme values can occur only where

$$f_x = 2x = 0 \qquad \text{and} \qquad f_y = 2y = 0.$$

The only possibility is the origin, where the value of f is zero. Since f is never negative, we see that the origin gives a local minimum (Figure 14.41). ■

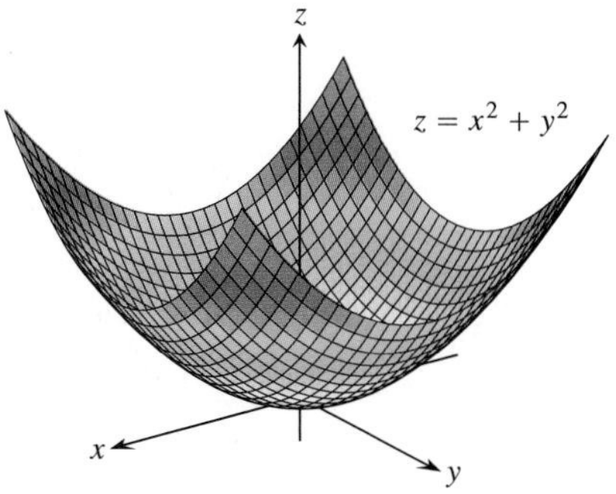

FIGURE 14.41 The graph of the function $f(x, y) = x^2 + y^2$ is the paraboloid $z = x^2 + y^2$. The function has a local minimum value of 0 at the origin (Example 1).

EXAMPLE 2 Identifying a Saddle Point

Find the local extreme values (if any) of $f(x, y) = y^2 - x^2$.

Solution The domain of f is the entire plane (so there are no boundary points) and the partial derivatives $f_x = -2x$ and $f_y = 2y$ exist everywhere. Therefore, local extrema can occur only at the origin $(0, 0)$. Along the positive x-axis, however, f has the value $f(x, 0) = -x^2 < 0$; along the positive y-axis, f has the value $f(0, y) = y^2 > 0$. Therefore, every open disk in the xy-plane centered at $(0, 0)$ contains points where the function is positive and points where it is negative. The function has a saddle point at the origin (Figure 14.42) instead of a local extreme value. We conclude that the function has no local extreme values. ■

That $f_x = f_y = 0$ at an interior point (a, b) of R does not guarantee f has a local extreme value there. If f and its first and second partial derivatives are continuous on R, however, we may be able to learn more from the following theorem, proved in Section 14.10.

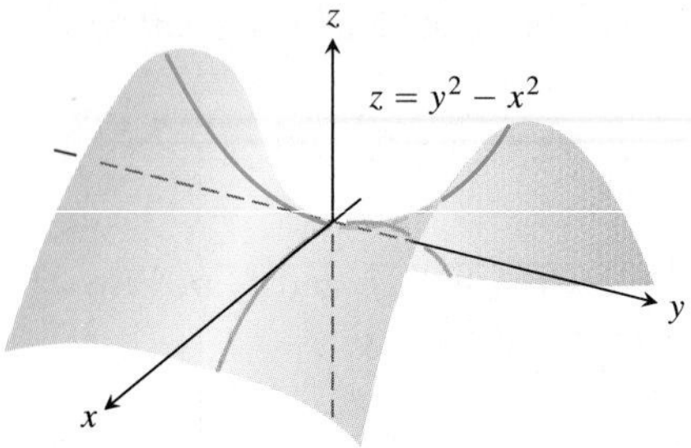

FIGURE 14.42 The origin is a saddle point of the function $f(x, y) = y^2 - x^2$. There are no local extreme values (Example 2).

THEOREM 11 Second Derivative Test for Local Extreme Values

Suppose that $f(x, y)$ and its first and second partial derivatives are continuous throughout a disk centered at (a, b) and that $f_x(a, b) = f_y(a, b) = 0$. Then

i. f has a **local maximum** at (a, b) if $f_{xx} < 0$ and $f_{xx}f_{yy} - f_{xy}^2 > 0$ at (a, b).
ii. f has a **local minimum** at (a, b) if $f_{xx} > 0$ and $f_{xx}f_{yy} - f_{xy}^2 > 0$ at (a, b).
iii. f has a **saddle point** at (a, b) if $f_{xx}f_{yy} - f_{xy}^2 < 0$ at (a, b).
iv. **The test is inconclusive** at (a, b) if $f_{xx}f_{yy} - f_{xy}^2 = 0$ at (a, b). In this case, we must find some other way to determine the behavior of f at (a, b).

The expression $f_{xx}f_{yy} - f_{xy}^2$ is called the **discriminant** or **Hessian** of f. It is sometimes easier to remember it in determinant form,

$$f_{xx}f_{yy} - f_{xy}^2 = \begin{vmatrix} f_{xx} & f_{xy} \\ f_{xy} & f_{yy} \end{vmatrix}.$$

Theorem 11 says that if the discriminant is positive at the point (a, b), then the surface curves the same way in all directions: downward if $f_{xx} < 0$, giving rise to a local maximum, and upward if $f_{xx} > 0$, giving a local minimum. On the other hand, if the discriminant is negative at (a, b), then the surface curves up in some directions and down in others, so we have a saddle point.

EXAMPLE 3 Finding Local Extreme Values

Find the local extreme values of the function

$$f(x, y) = xy - x^2 - y^2 - 2x - 2y + 4.$$

Solution The function is defined and differentiable for all x and y and its domain has no boundary points. The function therefore has extreme values only at the points where f_x and f_y are simultaneously zero. This leads to

$$f_x = y - 2x - 2 = 0, \qquad f_y = x - 2y - 2 = 0,$$

or

$$x = y = -2.$$

Therefore, the point $(-2, -2)$ is the only point where f may take on an extreme value. To see if it does so, we calculate

$$f_{xx} = -2, \qquad f_{yy} = -2, \qquad f_{xy} = 1.$$

The discriminant of f at $(a, b) = (-2, -2)$ is

$$f_{xx}f_{yy} - f_{xy}^2 = (-2)(-2) - (1)^2 = 4 - 1 = 3.$$

The combination

$$f_{xx} < 0 \qquad \text{and} \qquad f_{xx}f_{yy} - f_{xy}^2 > 0$$

tells us that f has a local maximum at $(-2, -2)$. The value of f at this point is $f(-2, -2) = 8$. ■

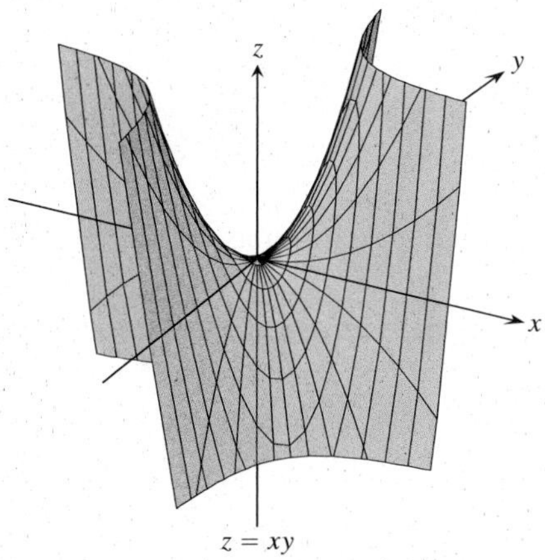

FIGURE 14.43 The surface $z = xy$ has a saddle point at the origin (Example 4).

EXAMPLE 4 Searching for Local Extreme Values

Find the local extreme values of $f(x, y) = xy$.

Solution Since f is differentiable everywhere (Figure 14.43), it can assume extreme values only where

$$f_x = y = 0 \quad \text{and} \quad f_y = x = 0.$$

Thus, the origin is the only point where f might have an extreme value. To see what happens there, we calculate

$$f_{xx} = 0, \quad f_{yy} = 0, \quad f_{xy} = 1.$$

The discriminant,

$$f_{xx}f_{yy} - f_{xy}{}^2 = -1,$$

is negative. Therefore, the function has a saddle point at (0, 0). We conclude that $f(x, y) = xy$ has no local extreme values. ■

Absolute Maxima and Minima on Closed Bounded Regions

We organize the search for the absolute extrema of a continuous function $f(x, y)$ on a closed and bounded region R into three steps.

1. *List the interior points of* R where f may have local maxima and minima and evaluate f at these points. These are the critical points of f.
2. *List the boundary points of* R where f has local maxima and minima and evaluate f at these points. We show how to do this shortly.
3. *Look through the lists* for the maximum and minimum values of f. These will be the absolute maximum and minimum values of f on R. Since absolute maxima and minima are also local maxima and minima, the absolute maximum and minimum values of f appear somewhere in the lists made in Steps 1 and 2.

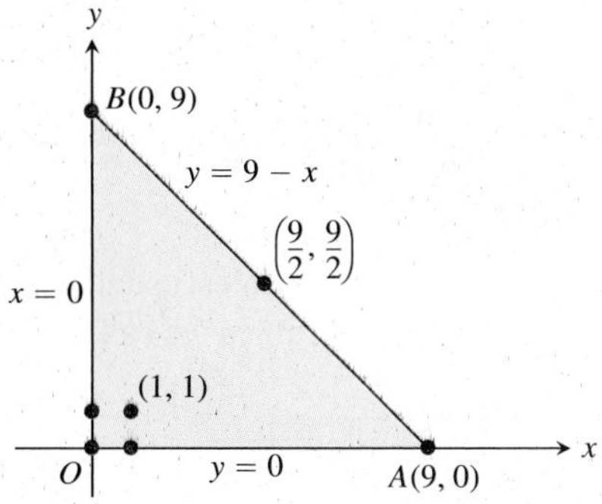

FIGURE 14.44 This triangular region is the domain of the function in Example 5.

EXAMPLE 5 Finding Absolute Extrema

Find the absolute maximum and minimum values of

$$f(x, y) = 2 + 2x + 2y - x^2 - y^2$$

on the triangular region in the first quadrant bounded by the lines $x = 0, y = 0, y = 9 - x$.

Solution Since f is differentiable, the only places where f can assume these values are points inside the triangle (Figure 14.44) where $f_x = f_y = 0$ and points on the boundary.

(a) Interior points. For these we have

$$f_x = 2 - 2x = 0, \quad f_y = 2 - 2y = 0,$$

yielding the single point $(x, y) = (1, 1)$. The value of f there is

$$f(1, 1) = 4.$$

(b) Boundary points. We take the triangle one side at a time:

(i) On the segment OA, $y = 0$. The function

$$f(x, y) = f(x, 0) = 2 + 2x - x^2$$

may now be regarded as a function of x defined on the closed interval $0 \le x \le 9$. Its extreme values (we know from Chapter 4) may occur at the endpoints

$$x = 0 \quad \text{where} \quad f(0, 0) = 2$$

$$x = 9 \quad \text{where} \quad f(9, 0) = 2 + 18 - 81 = -61$$

and at the interior points where $f'(x, 0) = 2 - 2x = 0$. The only interior point where $f'(x, 0) = 0$ is $x = 1$, where

$$f(x, 0) = f(1, 0) = 3.$$

(ii) On the segment OB, $x = 0$ and

$$f(x, y) = f(0, y) = 2 + 2y - y^2.$$

We know from the symmetry of f in x and y and from the analysis we just carried out that the candidates on this segment are

$$f(0, 0) = 2, \qquad f(0, 9) = -61, \qquad f(0, 1) = 3.$$

(iii) We have already accounted for the values of f at the endpoints of AB, so we need only look at the interior points of AB. With $y = 9 - x$, we have

$$f(x, y) = 2 + 2x + 2(9 - x) - x^2 - (9 - x)^2 = -61 + 18x - 2x^2.$$

Setting $f'(x, 9 - x) = 18 - 4x = 0$ gives

$$x = \frac{18}{4} = \frac{9}{2}.$$

At this value of x,

$$y = 9 - \frac{9}{2} = \frac{9}{2} \quad \text{and} \quad f(x, y) = f\left(\frac{9}{2}, \frac{9}{2}\right) = -\frac{41}{2}.$$

Summary We list all the candidates: 4, 2, −61, 3, −(41/2). The maximum is 4, which f assumes at (1, 1). The minimum is −61, which f assumes at (0, 9) and (9, 0). ■

Solving extreme value problems with algebraic constraints on the variables usually requires the method of Lagrange multipliers in the next section. But sometimes we can solve such problems directly, as in the next example.

EXAMPLE 6 Solving a Volume Problem with a Constraint

A delivery company accepts only rectangular boxes the sum of whose length and girth (perimeter of a cross-section) does not exceed 108 in. Find the dimensions of an acceptable box of largest volume.

Solution Let x, y, and z represent the length, width, and height of the rectangular box, respectively. Then the girth is $2y + 2z$. We want to maximize the volume $V = xyz$ of the

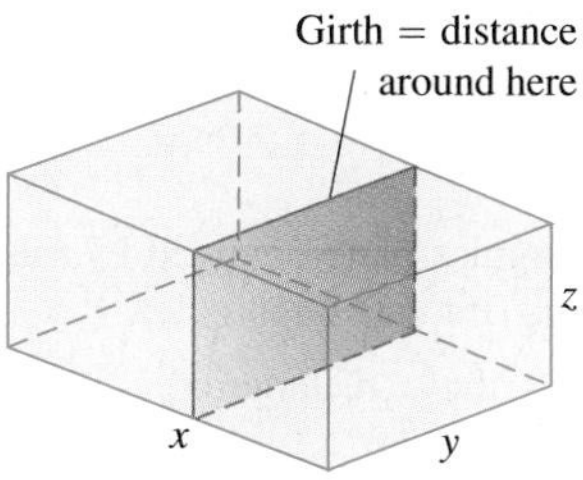

FIGURE 14.45 The box in Example 6.

box (Figure 14.45) satisfying $x + 2y + 2z = 108$ (the largest box accepted by the delivery company). Thus, we can write the volume of the box as a function of two variables.

$$\begin{aligned} V(y, z) &= (108 - 2y - 2z)yz \qquad V = xyz \text{ and } x = 108 - 2y - 2z \\ &= 108yz - 2y^2z - 2yz^2 \end{aligned}$$

Setting the first partial derivatives equal to zero,

$$\begin{aligned} V_y(y, z) &= 108z - 4yz - 2z^2 = (108 - 4y - 2z)z = 0 \\ V_z(y, z) &= 108y - 2y^2 - 4yz = (108 - 2y - 4z)y = 0, \end{aligned}$$

gives the critical points (0, 0), (0, 54), (54, 0), and (18, 18). The volume is zero at (0, 0), (0, 54), (54, 0), which are not maximum values. At the point (18, 18), we apply the Second Derivative Test (Theorem 11):

$$V_{yy} = -4z, \qquad V_{zz} = -4y, \qquad V_{yz} = 108 - 4y - 4z.$$

Then

$$V_{yy}V_{zz} - V_{yz}^2 = 16yz - 16(27 - y - z)^2.$$

Thus,

$$V_{yy}(18, 18) = -4(18) < 0$$

and

$$\left[V_{yy}V_{zz} - V_{yz}^2\right]_{(18,18)} = 16(18)(18) - 16(-9)^2 > 0$$

imply that (18, 18) gives a maximum volume. The dimensions of the package are $x = 108 - 2(18) - 2(18) = 36$ in., $y = 18$ in., and $z = 18$ in. The maximum volume is $V = (36)(18)(18) = 11{,}664 \text{ in.}^3$, or 6.75 ft^3. ■

Despite the power of Theorem 10, we urge you to remember its limitations. It does not apply to boundary points of a function's domain, where it is possible for a function to have extreme values along with nonzero derivatives. Also, it does not apply to points where either f_x or f_y fails to exist.

Summary of Max-Min Tests

The extreme values of $f(x, y)$ can occur only at

i. **boundary points** of the domain of f

ii. **critical points** (interior points where $f_x = f_y = 0$ or points where f_x or f_y fail to exist).

If the first- and second-order partial derivatives of f are continuous throughout a disk centered at a point (a, b) and $f_x(a, b) = f_y(a, b) = 0$, the nature of $f(a, b)$ can be tested with the **Second Derivative Test**:

i. $f_{xx} < 0$ and $f_{xx}f_{yy} - f_{xy}^2 > 0$ at (a, b) $\Rightarrow$ **local maximum**

ii. $f_{xx} > 0$ and $f_{xx}f_{yy} - f_{xy}^2 > 0$ at (a, b) $\Rightarrow$ **local minimum**

iii. $f_{xx}f_{yy} - f_{xy}^2 < 0$ at (a, b) $\Rightarrow$ **saddle point**

iv. $f_{xx}f_{yy} - f_{xy}^2 = 0$ at (a, b) $\Rightarrow$ **test is inconclusive**.

EXERCISES 14.7

Finding Local Extrema

Find all the local maxima, local minima, and saddle points of the functions in Exercises 1–30.

1. $f(x, y) = x^2 + xy + y^2 + 3x - 3y + 4$

2. $f(x, y) = x^2 + 3xy + 3y^2 - 6x + 3y - 6$

3. $f(x, y) = 2xy - 5x^2 - 2y^2 + 4x + 4y - 4$

4. $f(x, y) = 2xy - 5x^2 - 2y^2 + 4x - 4$

5. $f(x, y) = x^2 + xy + 3x + 2y + 5$

6. $f(x, y) = y^2 + xy - 2x - 2y + 2$

7. $f(x, y) = 5xy - 7x^2 + 3x - 6y + 2$

8. $f(x, y) = 2xy - x^2 - 2y^2 + 3x + 4$

9. $f(x, y) = x^2 - 4xy + y^2 + 6y + 2$

10. $f(x, y) = 3x^2 + 6xy + 7y^2 - 2x + 4y$

11. $f(x, y) = 2x^2 + 3xy + 4y^2 - 5x + 2y$

12. $f(x, y) = 4x^2 - 6xy + 5y^2 - 20x + 26y$

13. $f(x, y) = x^2 - y^2 - 2x + 4y + 6$

14. $f(x, y) = x^2 - 2xy + 2y^2 - 2x + 2y + 1$

15. $f(x, y) = x^2 + 2xy$

16. $f(x, y) = 3 + 2x + 2y - 2x^2 - 2xy - y^2$

17. $f(x, y) = x^3 - y^3 - 2xy + 6$

18. $f(x, y) = x^3 + 3xy + y^3$

19. $f(x, y) = 6x^2 - 2x^3 + 3y^2 + 6xy$

20. $f(x, y) = 3y^2 - 2y^3 - 3x^2 + 6xy$

21. $f(x, y) = 9x^3 + y^3/3 - 4xy$

22. $f(x, y) = 8x^3 + y^3 + 6xy$

23. $f(x, y) = x^3 + y^3 + 3x^2 - 3y^2 - 8$

24. $f(x, y) = 2x^3 + 2y^3 - 9x^2 + 3y^2 - 12y$

25. $f(x, y) = 4xy - x^4 - y^4$

26. $f(x, y) = x^4 + y^4 + 4xy$

27. $f(x, y) = \dfrac{1}{x^2 + y^2 - 1}$

28. $f(x, y) = \dfrac{1}{x} + xy + \dfrac{1}{y}$

29. $f(x, y) = y \sin x$

30. $f(x, y) = e^{2x} \cos y$

Finding Absolute Extrema

In Exercises 31–38, find the absolute maxima and minima of the functions on the given domains.

31. $f(x, y) = 2x^2 - 4x + y^2 - 4y + 1$ on the closed triangular plate bounded by the lines $x = 0, y = 2, y = 2x$ in the first quadrant

32. $D(x, y) = x^2 - xy + y^2 + 1$ on the closed triangular plate in the first quadrant bounded by the lines $x = 0, y = 4, y = x$

33. $f(x, y) = x^2 + y^2$ on the closed triangular plate bounded by the lines $x = 0, y = 0, y + 2x = 2$ in the first quadrant

34. $T(x, y) = x^2 + xy + y^2 - 6x$ on the rectangular plate $0 \le x \le 5, -3 \le y \le 3$

35. $T(x, y) = x^2 + xy + y^2 - 6x + 2$ on the rectangular plate $0 \le x \le 5, -3 \le y \le 0$

36. $f(x, y) = 48xy - 32x^3 - 24y^2$ on the rectangular plate $0 \le x \le 1, 0 \le y \le 1$

37. $f(x, y) = (4x - x^2) \cos y$ on the rectangular plate $1 \le x \le 3$, $-\pi/4 \le y \le \pi/4$ (see accompanying figure).

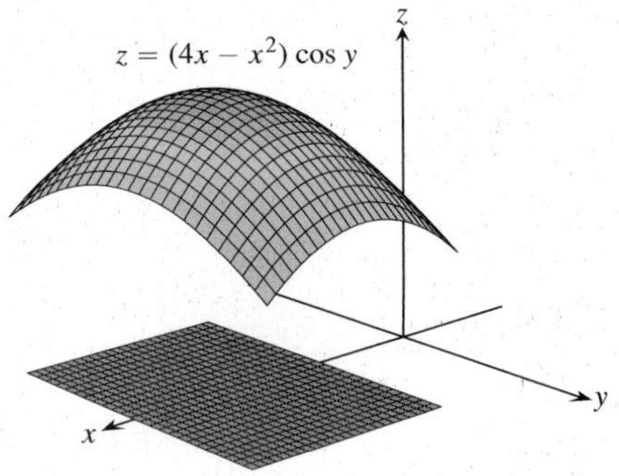

38. $f(x, y) = 4x - 8xy + 2y + 1$ on the triangular plate bounded by the lines $x = 0, y = 0, x + y = 1$ in the first quadrant

39. Find two numbers a and b with $a \le b$ such that

$$\int_a^b (6 - x - x^2)\,dx$$

has its largest value.

40. Find two numbers a and b with $a \le b$ such that

$$\int_a^b (24 - 2x - x^2)^{1/3}\,dx$$

has its largest value.

41. Temperatures The flat circular plate in Figure 14.46 has the shape of the region $x^2 + y^2 \le 1$. The plate, including the boundary where $x^2 + y^2 = 1$, is heated so that the temperature at the point (x, y) is

$$T(x, y) = x^2 + 2y^2 - x.$$

Find the temperatures at the hottest and coldest points on the plate.

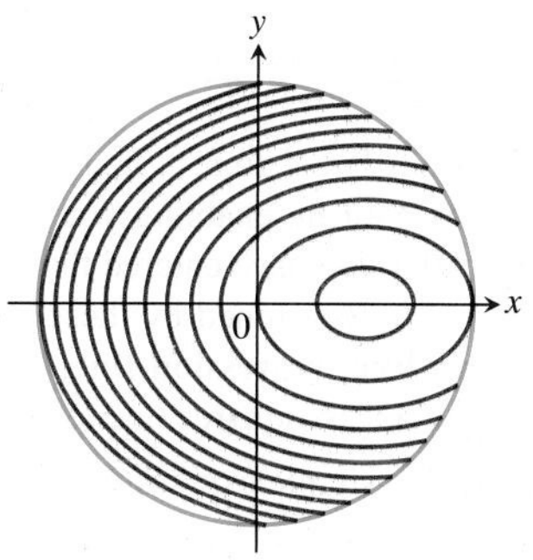

FIGURE 14.46 Curves of constant temperature are called isotherms. The figure shows isotherms of the temperature function $T(x, y) = x^2 + 2y^2 - x$ on the disk $x^2 + y^2 \leq 1$ in the xy-plane. Exercise 41 asks you to locate the extreme temperatures.

42. Find the critical point of

$$f(x, y) = xy + 2x - \ln x^2y$$

in the open first quadrant $(x > 0, y > 0)$ and show that f takes on a minimum there (Figure 14.47).

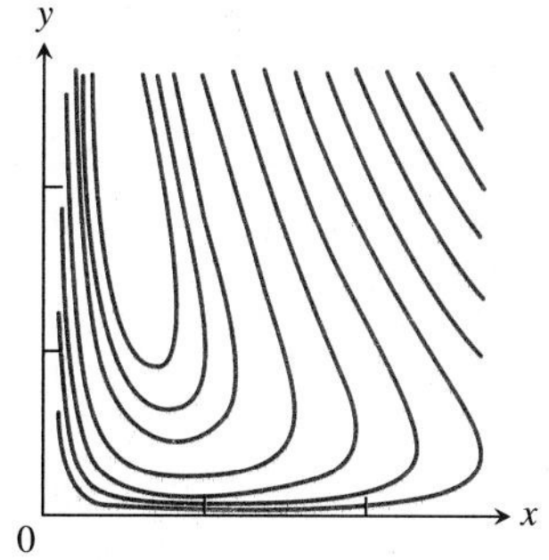

FIGURE 14.47 The function $f(x, y) = xy + 2x - \ln x^2y$ (selected level curves shown here) takes on a minimum value somewhere in the open first quadrant $x > 0, y > 0$ (Exercise 42).

Theory and Examples

43. Find the maxima, minima, and saddle points of $f(x, y)$, if any, given that

a. $f_x = 2x - 4y$ and $f_y = 2y - 4x$

b. $f_x = 2x - 2$ and $f_y = 2y - 4$

c. $f_x = 9x^2 - 9$ and $f_y = 2y + 4$

Describe your reasoning in each case.

44. The discriminant $f_{xx}f_{yy} - f_{xy}^2$ is zero at the origin for each of the following functions, so the Second Derivative Test fails there. Determine whether the function has a maximum, a minimum, or neither at the origin by imagining what the surface $z = f(x, y)$ looks like. Describe your reasoning in each case.

a. $f(x, y) = x^2y^2$ **b.** $f(x, y) = 1 - x^2y^2$

c. $f(x, y) = xy^2$ **d.** $f(x, y) = x^3y^2$

e. $f(x, y) = x^3y^3$ **f.** $f(x, y) = x^4y^4$

45. Show that $(0, 0)$ is a critical point of $f(x, y) = x^2 + kxy + y^2$ no matter what value the constant k has. (*Hint:* Consider two cases: $k = 0$ and $k \neq 0$.)

46. For what values of the constant k does the Second Derivative Test guarantee that $f(x, y) = x^2 + kxy + y^2$ will have a saddle point at $(0, 0)$? A local minimum at $(0, 0)$? For what values of k is the Second Derivative Test inconclusive? Give reasons for your answers.

47. If $f_x(a, b) = f_y(a, b) = 0$, must f have a local maximum or minimum value at (a, b)? Give reasons for your answer.

48. Can you conclude anything about $f(a, b)$ if f and its first and second partial derivatives are continuous throughout a disk centered at the critical point (a, b) and $f_{xx}(a, b)$ and $f_{yy}(a, b)$ differ in sign? Give reasons for your answer.

49. Among all the points on the graph of $z = 10 - x^2 - y^2$ that lie above the plane $x + 2y + 3z = 0$, find the point farthest from the plane.

50. Find the point on the graph of $z = x^2 + y^2 + 10$ nearest the plane $x + 2y - z = 0$.

51. The function $f(x, y) = x + y$ fails to have an absolute maximum value in the closed first quadrant $x \geq 0$ and $y \geq 0$. Does this contradict the discussion on finding absolute extrema given in the text? Give reasons for your answer.

52. Consider the function $f(x, y) = x^2 + y^2 + 2xy - x - y + 1$ over the square $0 \leq x \leq 1$ and $0 \leq y \leq 1$.

a. Show that f has an absolute minimum along the line segment $2x + 2y = 1$ in this square. What *is* the absolute minimum value?

b. Find the absolute maximum value of f over the square.

Extreme Values on Parametrized Curves

To find the extreme values of a function $f(x, y)$ on a curve $x = x(t), y = y(t)$, we treat f as a function of the single variable t and

TABLE 14.4 Sinkings of German submarines by U.S. during 16 consecutive months of WWII

Month	Guesses by U.S. (reported sinkings) x	Actual number y
1	3	3
2	2	2
3	4	6
4	2	3
5	5	4
6	5	3
7	9	11
8	12	9
9	8	10
10	13	16
11	14	13
12	3	5
13	4	6
14	13	19
15	10	15
16	16	15
	123	140

COMPUTER EXPLORATIONS

Exploring Local Extrema at Critical Points

In Exercises 65–70, you will explore functions to identify their local extrema. Use a CAS to perform the following steps:

a. Plot the function over the given rectangle.

b. Plot some level curves in the rectangle.

c. Calculate the function's first partial derivatives and use the CAS equation solver to find the critical points. How do the critical points relate to the level curves plotted in part (b)? Which critical points, if any, appear to give a saddle point? Give reasons for your answer.

d. Calculate the function's second partial derivatives and find the discriminant $f_{xx}f_{yy} - f_{xy}^2$.

e. Using the max-min tests, classify the critical points found in part (c). Are your findings consistent with your discussion in part (c)?

65. $f(x, y) = x^2 + y^3 - 3xy, \quad -5 \le x \le 5, \quad -5 \le y \le 5$

66. $f(x, y) = x^3 - 3xy^2 + y^2, \quad -2 \le x \le 2, \quad -2 \le y \le 2$

67. $f(x, y) = x^4 + y^2 - 8x^2 - 6y + 16, \quad -3 \le x \le 3, \quad -6 \le y \le 6$

68. $f(x, y) = 2x^4 + y^4 - 2x^2 - 2y^2 + 3, \quad -3/2 \le x \le 3/2, \quad -3/2 \le y \le 3/2$

69. $f(x, y) = 5x^6 + 18x^5 - 30x^4 + 30xy^2 - 120x^3, \quad -4 \le x \le 3, \quad -2 \le y \le 2$

70. $f(x, y) = \begin{cases} x^5 \ln(x^2 + y^2), & (x, y) \ne (0, 0) \\ 0, & (x, y) = (0, 0) \end{cases}, \quad -2 \le x \le 2, \quad -2 \le y \le 2$

14.8 Lagrange Multipliers

HISTORICAL BIOGRAPHY

Joseph Louis Lagrange (1736–1813)

Sometimes we need to find the extreme values of a function whose domain is constrained to lie within some particular subset of the plane—a disk, for example, a closed triangular region, or along a curve. In this section, we explore a powerful method for finding extreme values of constrained functions: the method of *Lagrange multipliers*.

Constrained Maxima and Minima

EXAMPLE 1 Finding a Minimum with Constraint

Find the point $P(x, y, z)$ closest to the origin on the plane $2x + y - z - 5 = 0$.

Solution The problem asks us to find the minimum value of the function

$$\begin{aligned} |\overrightarrow{OP}| &= \sqrt{(x - 0)^2 + (y - 0)^2 + (z - 0)^2} \\ &= \sqrt{x^2 + y^2 + z^2} \end{aligned}$$

subject to the constraint that

$$2x + y - z - 5 = 0.$$

Since $|\overrightarrow{OP}|$ has a minimum value wherever the function

$$f(x, y, z) = x^2 + y^2 + z^2$$

has a minimum value, we may solve the problem by finding the minimum value of $f(x, y, z)$ subject to the constraint $2x + y - z - 5 = 0$ (thus avoiding square roots). If we regard x and y as the independent variables in this equation and write z as

$$z = 2x + y - 5,$$

our problem reduces to one of finding the points (x, y) at which the function

$$h(x, y) = f(x, y, 2x + y - 5) = x^2 + y^2 + (2x + y - 5)^2$$

has its minimum value or values. Since the domain of h is the entire xy-plane, the First Derivative Test of Section 14.7 tells us that any minima that h might have must occur at points where

$$h_x = 2x + 2(2x + y - 5)(2) = 0, \qquad h_y = 2y + 2(2x + y - 5) = 0.$$

This leads to

$$10x + 4y = 20, \qquad 4x + 4y = 10,$$

and the solution

$$x = \frac{5}{3}, \qquad y = \frac{5}{6}.$$

We may apply a geometric argument together with the Second Derivative Test to show that these values minimize h. The z-coordinate of the corresponding point on the plane $z = 2x + y - 5$ is

$$z = 2\left(\frac{5}{3}\right) + \frac{5}{6} - 5 = -\frac{5}{6}.$$

Therefore, the point we seek is

$$\text{Closest point:} \qquad P\left(\frac{5}{3}, \frac{5}{6}, -\frac{5}{6}\right).$$

The distance from P to the origin is $5/\sqrt{6} \approx 2.04$. ■

Attempts to solve a constrained maximum or minimum problem by substitution, as we might call the method of Example 1, do not always go smoothly. This is one of the reasons for learning the new method of this section.

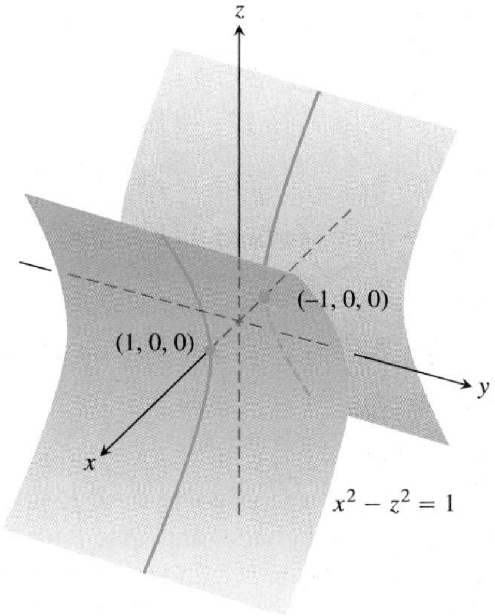

FIGURE 14.50 The hyperbolic cylinder $x^2 - z^2 - 1 = 0$ in Example 2.

EXAMPLE 2 Finding a Minimum with Constraint

Find the points closest to the origin on the hyperbolic cylinder $x^2 - z^2 - 1 = 0$.

Solution 1 The cylinder is shown in Figure 14.50. We seek the points on the cylinder closest to the origin. These are the points whose coordinates minimize the value of the function

$$f(x, y, z) = x^2 + y^2 + z^2 \qquad \text{Square of the distance}$$

subject to the constraint that $x^2 - z^2 - 1 = 0$. If we regard x and y as independent variables in the constraint equation, then

$$z^2 = x^2 - 1$$

and the values of $f(x, y, z) = x^2 + y^2 + z^2$ on the cylinder are given by the function

$$h(x, y) = x^2 + y^2 + (x^2 - 1) = 2x^2 + y^2 - 1.$$

To find the points on the cylinder whose coordinates minimize f, we look for the points in the xy-plane whose coordinates minimize h. The only extreme value of h occurs where

$$h_x = 4x = 0 \quad \text{and} \quad h_y = 2y = 0,$$

that is, at the point (0, 0). But there are no points on the cylinder where both x and y are zero. What went wrong?

What happened was that the First Derivative Test found (as it should have) the point *in the domain of h* where h has a minimum value. We, on the other hand, want the points *on the cylinder* where h has a minimum value. Although the domain of h is the entire xy-plane, the domain from which we can select the first two coordinates of the points (x, y, z) on the cylinder is restricted to the "shadow" of the cylinder on the xy-plane; it does not include the band between the lines $x = -1$ and $x = 1$ (Figure 14.51).

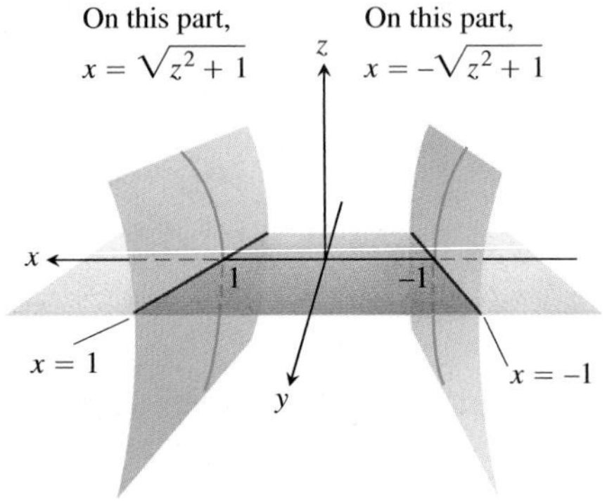

FIGURE 14.51 The region in the xy-plane from which the first two coordinates of the points (x, y, z) on the hyperbolic cylinder $x^2 - z^2 = 1$ are selected excludes the band $-1 < x < 1$ in the xy-plane (Example 2).

We can avoid this problem if we treat y and z as independent variables (instead of x and y) and express x in terms of y and z as

$$x^2 = z^2 + 1.$$

With this substitution, $f(x, y, z) = x^2 + y^2 + z^2$ becomes

$$k(y, z) = (z^2 + 1) + y^2 + z^2 = 1 + y^2 + 2z^2$$

and we look for the points where k takes on its smallest value. The domain of k in the yz-plane now matches the domain from which we select the y- and z-coordinates of the points (x, y, z) on the cylinder. Hence, the points that minimize k in the plane will have corresponding points on the cylinder. The smallest values of k occur where

$$k_y = 2y = 0 \quad \text{and} \quad k_z = 4z = 0,$$

or where $y = z = 0$. This leads to

$$x^2 = z^2 + 1 = 1, \qquad x = \pm 1.$$

The corresponding points on the cylinder are $(\pm 1, 0, 0)$. We can see from the inequality

$$k(y, z) = 1 + y^2 + 2z^2 \geq 1$$

that the points $(\pm 1, 0, 0)$ give a minimum value for k. We can also see that the minimum distance from the origin to a point on the cylinder is 1 unit.

Solution 2 Another way to find the points on the cylinder closest to the origin is to imagine a small sphere centered at the origin expanding like a soap bubble until it just touches the cylinder (Figure 14.52). At each point of contact, the cylinder and sphere have the same tangent plane and normal line. Therefore, if the sphere and cylinder are represented as the level surfaces obtained by setting

$$f(x, y, z) = x^2 + y^2 + z^2 - a^2 \quad \text{and} \quad g(x, y, z) = x^2 - z^2 - 1$$

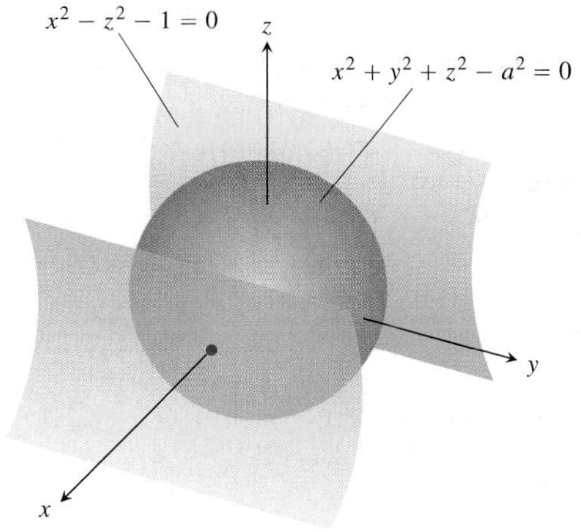

FIGURE 14.52 A sphere expanding like a soap bubble centered at the origin until it just touches the hyperbolic cylinder $x^2 - z^2 - 1 = 0$ (Example 2).

equal to 0, then the gradients ∇f and ∇g will be parallel where the surfaces touch. At any point of contact, we should therefore be able to find a scalar λ ("lambda") such that

$$\nabla f = \lambda \nabla g,$$

or

$$2x\mathbf{i} + 2y\mathbf{j} + 2z\mathbf{k} = \lambda(2x\mathbf{i} - 2z\mathbf{k}).$$

Thus, the coordinates x, y, and z of any point of tangency will have to satisfy the three scalar equations

$$2x = 2\lambda x, \qquad 2y = 0, \qquad 2z = -2\lambda z.$$

For what values of λ will a point (x, y, z) whose coordinates satisfy these scalar equations also lie on the surface $x^2 - z^2 - 1 = 0$? To answer this question, we use our knowledge that no point on the surface has a zero x-coordinate to conclude that $x \neq 0$. Hence, $2x = 2\lambda x$ only if

$$2 = 2\lambda, \qquad \text{or} \qquad \lambda = 1.$$

For $\lambda = 1$, the equation $2z = -2\lambda z$ becomes $2z = -2z$. If this equation is to be satisfied as well, z must be zero. Since $y = 0$ also (from the equation $2y = 0$), we conclude that the points we seek all have coordinates of the form

$$(x, 0, 0).$$

What points on the surface $x^2 - z^2 = 1$ have coordinates of this form? The answer is the points $(x, 0, 0)$ for which

$$x^2 - (0)^2 = 1, \qquad x^2 = 1, \qquad \text{or} \qquad x = \pm 1.$$

The points on the cylinder closest to the origin are the points $(\pm 1, 0, 0)$. ■

The Method of Lagrange Multipliers

In Solution 2 of Example 2, we used the **method of Lagrange multipliers**. The method says that the extreme values of a function $f(x, y, z)$ whose variables are subject to a constraint $g(x, y, z) = 0$ are to be found on the surface $g = 0$ at the points where

$$\nabla f = \lambda \nabla g$$

for some scalar λ (called a **Lagrange multiplier**).

To explore the method further and see why it works, we first make the following observation, which we state as a theorem.

THEOREM 12 The Orthogonal Gradient Theorem

Suppose that $f(x, y, z)$ is differentiable in a region whose interior contains a smooth curve

$$C: \quad \mathbf{r}(t) = g(t)\mathbf{i} + h(t)\mathbf{j} + k(t)\mathbf{k}.$$

If P_0 is a point on C where f has a local maximum or minimum relative to its values on C, then ∇f is orthogonal to C at P_0.

Proof We show that ∇f is orthogonal to the curve's velocity vector at P_0. The values of f on C are given by the composite $f(g(t), h(t), k(t))$, whose derivative with respect to t is

$$\frac{df}{dt} = \frac{\partial f}{\partial x}\frac{dg}{dt} + \frac{\partial f}{\partial y}\frac{dh}{dt} + \frac{\partial f}{\partial z}\frac{dk}{dt} = \nabla f \cdot \mathbf{v}.$$

At any point P_0 where f has a local maximum or minimum relative to its values on the curve, $df/dt = 0$, so

$$\nabla f \cdot \mathbf{v} = 0. \qquad \blacksquare$$

By dropping the z-terms in Theorem 12, we obtain a similar result for functions of two variables.

COROLLARY OF THEOREM 12

At the points on a smooth curve $\mathbf{r}(t) = g(t)\mathbf{i} + h(t)\mathbf{j}$ where a differentiable function $f(x, y)$ takes on its local maxima and minima relative to its values on the curve, $\nabla f \cdot \mathbf{v} = 0$, where $\mathbf{v} = d\mathbf{r}/dt$.

Theorem 12 is the key to the method of Lagrange multipliers. Suppose that $f(x, y, z)$ and $g(x, y, z)$ are differentiable and that P_0 is a point on the surface $g(x, y, z) = 0$ where f has a local maximum or minimum value relative to its other values on the surface. Then f takes on a local maximum or minimum at P_0 relative to its values on every differentiable curve through P_0 on the surface $g(x, y, z) = 0$. Therefore, ∇f is orthogonal to the velocity vector of every such differentiable curve through P_0. So is ∇g, moreover (because ∇g is orthogonal to the level surface $g = 0$, as we saw in Section 14.5). Therefore, at P_0, ∇f is some scalar multiple λ of ∇g.

> **The Method of Lagrange Multipliers**
> Suppose that $f(x, y, z)$ and $g(x, y, z)$ are differentiable and $\nabla g \neq \mathbf{0}$ when $g(x, y, z) = 0$. To find the local maximum and minimum values of f subject to the constraint $g(x, y, z) = 0$ (if these exist), find the values of x, y, z, and λ that simultaneously satisfy the equations
>
> $$\nabla f = \lambda \nabla g \quad \text{and} \quad g(x, y, z) = 0. \tag{1}$$
>
> For functions of two independent variables, the condition is similar, but without the variable z.

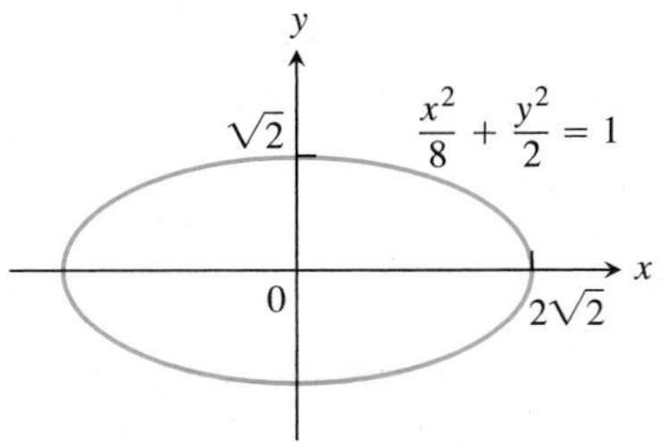

FIGURE 14.53 Example 3 shows how to find the largest and smallest values of the product xy on this ellipse.

EXAMPLE 3 Using the Method of Lagrange Multipliers

Find the greatest and smallest values that the function

$$f(x, y) = xy$$

takes on the ellipse (Figure 14.53)

$$\frac{x^2}{8} + \frac{y^2}{2} = 1.$$

Solution We want the extreme values of $f(x, y) = xy$ subject to the constraint

$$g(x, y) = \frac{x^2}{8} + \frac{y^2}{2} - 1 = 0.$$

To do so, we first find the values of x, y, and λ for which

$$\nabla f = \lambda \nabla g \quad \text{and} \quad g(x, y) = 0.$$

The gradient equation in Equations (1) gives

$$y\mathbf{i} + x\mathbf{j} = \frac{\lambda}{4}x\mathbf{i} + \lambda y\mathbf{j},$$

from which we find

$$y = \frac{\lambda}{4}x, \qquad x = \lambda y, \qquad \text{and} \qquad y = \frac{\lambda}{4}(\lambda y) = \frac{\lambda^2}{4}y,$$

so that $y = 0$ or $\lambda = \pm 2$. We now consider these two cases.

Case 1: If $y = 0$, then $x = y = 0$. But $(0, 0)$ is not on the ellipse. Hence, $y \neq 0$.
Case 2: If $y \neq 0$, then $\lambda = \pm 2$ and $x = \pm 2y$. Substituting this in the equation $g(x, y) = 0$ gives

$$\frac{(\pm 2y)^2}{8} + \frac{y^2}{2} = 1, \qquad 4y^2 + 4y^2 = 8 \qquad \text{and} \qquad y = \pm 1.$$

The function $f(x, y) = xy$ therefore takes on its extreme values on the ellipse at the four points $(\pm 2, 1)$, $(\pm 2, -1)$. The extreme values are $xy = 2$ and $xy = -2$.

The Geometry of the Solution

The level curves of the function $f(x, y) = xy$ are the hyperbolas $xy = c$ (Figure 14.54). The farther the hyperbolas lie from the origin, the larger the absolute value of f. We want

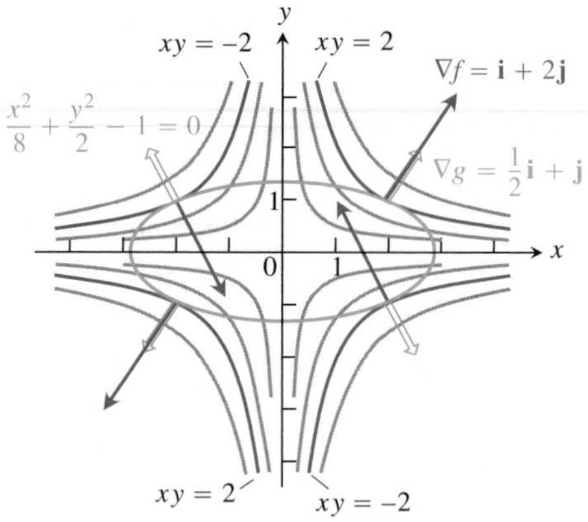

FIGURE 14.54 When subjected to the constraint $g(x, y) = x^2/8 + y^2/2 - 1 = 0$, the function $f(x, y) = xy$ takes on extreme values at the four points $(\pm 2, \pm 1)$. These are the points on the ellipse when ∇f (red) is a scalar multiple of ∇g (blue) (Example 3).

to find the extreme values of $f(x, y)$, given that the point (x, y) also lies on the ellipse $x^2 + 4y^2 = 8$. Which hyperbolas intersecting the ellipse lie farthest from the origin? The hyperbolas that just graze the ellipse, the ones that are tangent to it, are farthest. At these points, any vector normal to the hyperbola is normal to the ellipse, so $\nabla f = y\mathbf{i} + x\mathbf{j}$ is a multiple $(\lambda = \pm 2)$ of $\nabla g = (x/4)\mathbf{i} + y\mathbf{j}$. At the point (2, 1), for example,

$$\nabla f = \mathbf{i} + 2\mathbf{j}, \qquad \nabla g = \frac{1}{2}\mathbf{i} + \mathbf{j}, \qquad \text{and} \qquad \nabla f = 2\nabla g.$$

At the point (−2, 1),

$$\nabla f = \mathbf{i} - 2\mathbf{j}, \qquad \nabla g = -\frac{1}{2}\mathbf{i} + \mathbf{j}, \qquad \text{and} \qquad \nabla f = -2\nabla g.$$ ■

EXAMPLE 4 Finding Extreme Function Values on a Circle

Find the maximum and minimum values of the function $f(x, y) = 3x + 4y$ on the circle $x^2 + y^2 = 1$.

Solution We model this as a Lagrange multiplier problem with

$$f(x, y) = 3x + 4y, \qquad g(x, y) = x^2 + y^2 - 1$$

and look for the values of x, y, and λ that satisfy the equations

$$\nabla f = \lambda \nabla g: \quad 3\mathbf{i} + 4\mathbf{j} = 2x\lambda\mathbf{i} + 2y\lambda\mathbf{j}$$

$$g(x, y) = 0: \quad x^2 + y^2 - 1 = 0.$$

The gradient equation in Equations (1) implies that $\lambda \neq 0$ and gives

$$x = \frac{3}{2\lambda}, \qquad y = \frac{2}{\lambda}.$$

These equations tell us, among other things, that x and y have the same sign. With these values for x and y, the equation $g(x, y) = 0$ gives

$$\left(\frac{3}{2\lambda}\right)^2 + \left(\frac{2}{\lambda}\right)^2 - 1 = 0,$$

so

$$\frac{9}{4\lambda^2} + \frac{4}{\lambda^2} = 1, \qquad 9 + 16 = 4\lambda^2, \qquad 4\lambda^2 = 25, \qquad \text{and} \qquad \lambda = \pm\frac{5}{2}.$$

Thus,

$$x = \frac{3}{2\lambda} = \pm\frac{3}{5}, \qquad y = \frac{2}{\lambda} = \pm\frac{4}{5},$$

and $f(x, y) = 3x + 4y$ has extreme values at $(x, y) = \pm(3/5, 4/5)$.

By calculating the value of $3x + 4y$ at the points $\pm(3/5, 4/5)$, we see that its maximum and minimum values on the circle $x^2 + y^2 = 1$ are

$$3\left(\frac{3}{5}\right) + 4\left(\frac{4}{5}\right) = \frac{25}{5} = 5 \qquad \text{and} \qquad 3\left(-\frac{3}{5}\right) + 4\left(-\frac{4}{5}\right) = -\frac{25}{5} = -5.$$

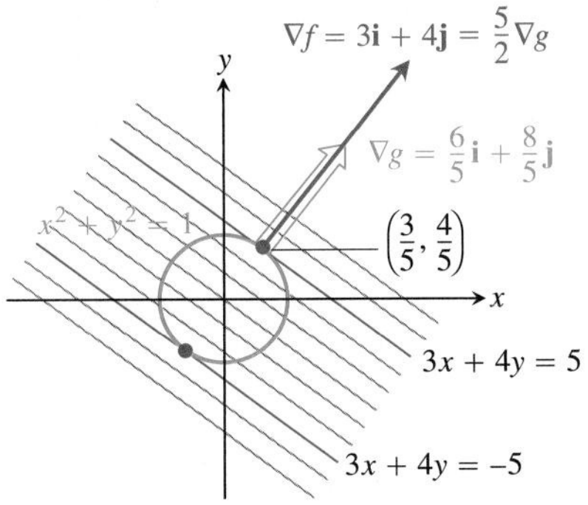

FIGURE 14.55 The function $f(x, y) = 3x + 4y$ takes on its largest value on the unit circle $g(x, y) = x^2 + y^2 - 1 = 0$ at the point $(3/5, 4/5)$ and its smallest value at the point $(-3/5, -4/5)$ (Example 4). At each of these points, ∇f is a scalar multiple of ∇g. The figure shows the gradients at the first point but not the second.

The Geometry of the Solution

The level curves of $f(x, y) = 3x + 4y$ are the lines $3x + 4y = c$ (Figure 14.55). The farther the lines lie from the origin, the larger the absolute value of f. We want to find the extreme values of $f(x, y)$ given that the point (x, y) also lies on the circle $x^2 + y^2 = 1$. Which lines intersecting the circle lie farthest from the origin? The lines tangent to the circle are farthest. At the points of tangency, any vector normal to the line is normal to the circle, so the gradient $\nabla f = 3\mathbf{i} + 4\mathbf{j}$ is a multiple ($\lambda = \pm 5/2$) of the gradient $\nabla g = 2x\mathbf{i} + 2y\mathbf{j}$. At the point $(3/5, 4/5)$, for example,

$$\nabla f = 3\mathbf{i} + 4\mathbf{j}, \qquad \nabla g = \frac{6}{5}\mathbf{i} + \frac{8}{5}\mathbf{j}, \qquad \text{and} \qquad \nabla f = \frac{5}{2}\nabla g. \qquad \blacksquare$$

Lagrange Multipliers with Two Constraints

Many problems require us to find the extreme values of a differentiable function $f(x, y, z)$ whose variables are subject to two constraints. If the constraints are

$$g_1(x, y, z) = 0 \qquad \text{and} \qquad g_2(x, y, z) = 0$$

and g_1 and g_2 are differentiable, with ∇g_1 not parallel to ∇g_2, we find the constrained local maxima and minima of f by introducing two Lagrange multipliers λ and μ (mu, pronounced "mew"). That is, we locate the points $P(x, y, z)$ where f takes on its constrained extreme values by finding the values of $x, y, z, \lambda,$ and μ that simultaneously satisfy the equations

$$\nabla f = \lambda\nabla g_1 + \mu\nabla g_2, \qquad g_1(x, y, z) = 0, \qquad g_2(x, y, z) = 0 \qquad (2)$$

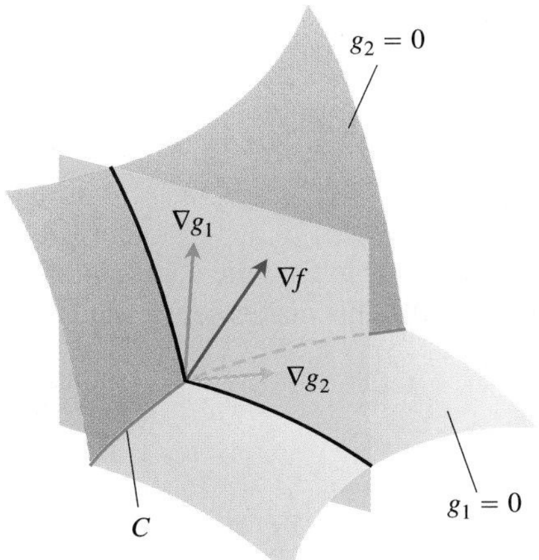

FIGURE 14.56 The vectors ∇g_1 and ∇g_2 lie in a plane perpendicular to the curve C because ∇g_1 is normal to the surface $g_1 = 0$ and ∇g_2 is normal to the surface $g_2 = 0$.

Equations (2) have a nice geometric interpretation. The surfaces $g_1 = 0$ and $g_2 = 0$ (usually) intersect in a smooth curve, say C (Figure 14.56). Along this curve we seek the points where f has local maximum and minimum values relative to its other values on the curve.

These are the points where ∇f is normal to C, as we saw in Theorem 12. But ∇g_1 and ∇g_2 are also normal to C at these points because C lies in the surfaces $g_1 = 0$ and $g_2 = 0$. Therefore, ∇f lies in the plane determined by ∇g_1 and ∇g_2, which means that $\nabla f = \lambda \nabla g_1 + \mu \nabla g_2$ for some λ and μ. Since the points we seek also lie in both surfaces, their coordinates must satisfy the equations $g_1(x, y, z) = 0$ and $g_2(x, y, z) = 0$, which are the remaining requirements in Equations (2).

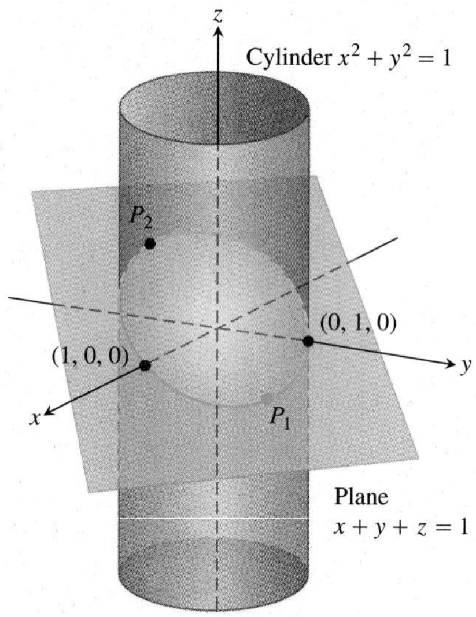

FIGURE 14.57 On the ellipse where the plane and cylinder meet, what are the points closest to and farthest from the origin? (Example 5)

EXAMPLE 5 Finding Extremes of Distance on an Ellipse

The plane $x + y + z = 1$ cuts the cylinder $x^2 + y^2 = 1$ in an ellipse (Figure 14.57). Find the points on the ellipse that lie closest to and farthest from the origin.

Solution We find the extreme values of

$$f(x, y, z) = x^2 + y^2 + z^2$$

(the square of the distance from (x, y, z) to the origin) subject to the constraints

$$g_1(x, y, z) = x^2 + y^2 - 1 = 0 \tag{3}$$

$$g_2(x, y, z) = x + y + z - 1 = 0. \tag{4}$$

The gradient equation in Equations (2) then gives

$$\begin{aligned}
\nabla f &= \lambda \nabla g_1 + \mu \nabla g_2 \\
2x\mathbf{i} + 2y\mathbf{j} + 2z\mathbf{k} &= \lambda(2x\mathbf{i} + 2y\mathbf{j}) + \mu(\mathbf{i} + \mathbf{j} + \mathbf{k}) \\
2x\mathbf{i} + 2y\mathbf{j} + 2z\mathbf{k} &= (2\lambda x + \mu)\mathbf{i} + (2\lambda y + \mu)\mathbf{j} + \mu\mathbf{k}
\end{aligned}$$

or

$$2x = 2\lambda x + \mu, \qquad 2y = 2\lambda y + \mu, \qquad 2z = \mu. \tag{5}$$

The scalar equations in Equations (5) yield

$$\begin{aligned}
2x &= 2\lambda x + 2z \Rightarrow (1 - \lambda)x = z, \\
2y &= 2\lambda y + 2z \Rightarrow (1 - \lambda)y = z.
\end{aligned} \tag{6}$$

Equations (6) are satisfied simultaneously if either $\lambda = 1$ and $z = 0$ or $\lambda \neq 1$ and $x = y = z/(1 - \lambda)$.

If $z = 0$, then solving Equations (3) and (4) simultaneously to find the corresponding points on the ellipse gives the two points $(1, 0, 0)$ and $(0, 1, 0)$. This makes sense when you look at Figure 14.57.

If $x = y$, then Equations (3) and (4) give

$$\begin{aligned}
x^2 + x^2 - 1 &= 0 & x + x + z - 1 &= 0 \\
2x^2 &= 1 & z &= 1 - 2x \\
x &= \pm\frac{\sqrt{2}}{2} & z &= 1 \mp \sqrt{2}.
\end{aligned}$$

The corresponding points on the ellipse are

$$P_1 = \left(\frac{\sqrt{2}}{2}, \frac{\sqrt{2}}{2}, 1 - \sqrt{2}\right) \quad \text{and} \quad P_2 = \left(-\frac{\sqrt{2}}{2}, -\frac{\sqrt{2}}{2}, 1 + \sqrt{2}\right).$$

Here we need to be careful, however. Although P_1 and P_2 both give local maxima of f on the ellipse, P_2 is farther from the origin than P_1.

The points on the ellipse closest to the origin are (1, 0, 0) and (0, 1, 0). The point on the ellipse farthest from the origin is P_2. ■

EXERCISES 14.8

Two Independent Variables with One Constraint

1. **Extrema on an ellipse** Find the points on the ellipse $x^2 + 2y^2 = 1$ where $f(x, y) = xy$ has its extreme values.

2. **Extrema on a circle** Find the extreme values of $f(x, y) = xy$ subject to the constraint $g(x, y) = x^2 + y^2 - 10 = 0$.

3. **Maximum on a line** Find the maximum value of $f(x, y) = 49 - x^2 - y^2$ on the line $x + 3y = 10$.

4. **Extrema on a line** Find the local extreme values of $f(x, y) = x^2 y$ on the line $x + y = 3$.

5. **Constrained minimum** Find the points on the curve $xy^2 = 54$ nearest the origin.

6. **Constrained minimum** Find the points on the curve $x^2 y = 2$ nearest the origin.

7. Use the method of Lagrange multipliers to find

   a. **Minimum on a hyperbola** The minimum value of $x + y$, subject to the constraints $xy = 16, x > 0, y > 0$

   b. **Maximum on a line** The maximum value of xy, subject to the constraint $x + y = 16$.

   Comment on the geometry of each solution.

8. **Extrema on a curve** Find the points on the curve $x^2 + xy + y^2 = 1$ in the xy-plane that are nearest to and farthest from the origin.

9. **Minimum surface area with fixed volume** Find the dimensions of the closed right circular cylindrical can of smallest surface area whose volume is 16π cm^3.

10. **Cylinder in a sphere** Find the radius and height of the open right circular cylinder of largest surface area that can be inscribed in a sphere of radius a. What *is* the largest surface area?

11. **Rectangle of greatest area in an ellipse** Use the method of Lagrange multipliers to find the dimensions of the rectangle of greatest area that can be inscribed in the ellipse $x^2/16 + y^2/9 = 1$ with sides parallel to the coordinate axes.

12. **Rectangle of longest perimeter in an ellipse** Find the dimensions of the rectangle of largest perimeter that can be inscribed in the ellipse $x^2/a^2 + y^2/b^2 = 1$ with sides parallel to the coordinate axes. What *is* the largest perimeter?

13. **Extrema on a circle** Find the maximum and minimum values of $x^2 + y^2$ subject to the constraint $x^2 - 2x + y^2 - 4y = 0$.

14. **Extrema on a circle** Find the maximum and minimum values of $3x - y + 6$ subject to the constraint $x^2 + y^2 = 4$.

15. **Ant on a metal plate** The temperature at a point (x, y) on a metal plate is $T(x, y) = 4x^2 - 4xy + y^2$. An ant on the plate walks around the circle of radius 5 centered at the origin. What are the highest and lowest temperatures encountered by the ant?

16. **Cheapest storage tank** Your firm has been asked to design a storage tank for liquid petroleum gas. The customer's specifications call for a cylindrical tank with hemispherical ends, and the tank is to hold 8000 m^3 of gas. The customer also wants to use the smallest amount of material possible in building the tank. What radius and height do you recommend for the cylindrical portion of the tank?

Three Independent Variables with One Constraint

17. **Minimum distance to a point** Find the point on the plane $x + 2y + 3z = 13$ closest to the point (1, 1, 1).

18. **Maximum distance to a point** Find the point on the sphere $x^2 + y^2 + z^2 = 4$ farthest from the point $(1, -1, 1)$.

19. **Minimum distance to the origin** Find the minimum distance from the surface $x^2 - y^2 - z^2 = 1$ to the origin.

20. **Minimum distance to the origin** Find the point on the surface $z = xy + 1$ nearest the origin.

21. **Minimum distance to the origin** Find the points on the surface $z^2 = xy + 4$ closest to the origin.

22. **Minimum distance to the origin** Find the point(s) on the surface $xyz = 1$ closest to the origin.

23. **Extrema on a sphere** Find the maximum and minimum values of

$$f(x, y, z) = x - 2y + 5z$$

on the sphere $x^2 + y^2 + z^2 = 30$.

24. **Extrema on a sphere** Find the points on the sphere $x^2 + y^2 + z^2 = 25$ where $f(x, y, z) = x + 2y + 3z$ has its maximum and minimum values.

25. **Minimizing a sum of squares** Find three real numbers whose sum is 9 and the sum of whose squares is as small as possible.

26. **Maximizing a product** Find the largest product the positive numbers x, y, and z can have if $x + y + z^2 = 16$.

27. **Rectangular box of longest volume in a sphere** Find the dimensions of the closed rectangular box with maximum volume that can be inscribed in the unit sphere.

28. **Box with vertex on a plane** Find the volume of the largest closed rectangular box in the first octant having three faces in the coordinate planes and a vertex on the plane $x/a + y/b + z/c = 1$, where $a > 0$, $b > 0$, and $c > 0$.

29. **Hottest point on a space probe** A space probe in the shape of the ellipsoid

$$4x^2 + y^2 + 4z^2 = 16$$

enters Earth's atmosphere and its surface begins to heat. After 1 hour, the temperature at the point (x, y, z) on the probe's surface is

$$T(x, y, z) = 8x^2 + 4yz - 16z + 600.$$

Find the hottest point on the probe's surface.

30. **Extreme temperatures on a sphere** Suppose that the Celsius temperature at the point (x, y, z) on the sphere $x^2 + y^2 + z^2 = 1$ is $T = 400xyz^2$. Locate the highest and lowest temperatures on the sphere.

31. **Maximizing a utility function: an example from economics** In economics, the usefulness or *utility* of amounts x and y of two capital goods G_1 and G_2 is sometimes measured by a function $U(x, y)$. For example, G_1 and G_2 might be two chemicals a pharmaceutical company needs to have on hand and $U(x, y)$ the gain from manufacturing a product whose synthesis requires different amounts of the chemicals depending on the process used. If G_1 costs a dollars per kilogram, G_2 costs b dollars per kilogram, and the total amount allocated for the purchase of G_1 and G_2 together is c dollars, then the company's managers want to maximize $U(x, y)$ given that $ax + by = c$. Thus, they need to solve a typical Lagrange multiplier problem.

Suppose that

$$U(x, y) = xy + 2x$$

and that the equation $ax + by = c$ simplifies to

$$2x + y = 30.$$

Find the maximum value of U and the corresponding values of x and y subject to this latter constraint.

32. **Locating a radio telescope** You are in charge of erecting a radio telescope on a newly discovered planet. To minimize interference, you want to place it where the magnetic field of the planet is weakest. The planet is spherical, with a radius of 6 units. Based on a coordinate system whose origin is at the center of the planet, the strength of the magnetic field is given by $M(x, y, z) = 6x - y^2 + xz + 60$. Where should you locate the radio telescope?

Extreme Values Subject to Two Constraints

33. Maximize the function $f(x, y, z) = x^2 + 2y - z^2$ subject to the constraints $2x - y = 0$ and $y + z = 0$.

34. Minimize the function $f(x, y, z) = x^2 + y^2 + z^2$ subject to the constraints $x + 2y + 3z = 6$ and $x + 3y + 9z = 9$.

35. **Minimum distance to the origin** Find the point closest to the origin on the line of intersection of the planes $y + 2z = 12$ and $x + y = 6$.

36. **Maximum value on line of intersection** Find the maximum value that $f(x, y, z) = x^2 + 2y - z^2$ can have on the line of intersection of the planes $2x - y = 0$ and $y + z = 0$.

37. **Extrema on a curve of intersection** Find the extreme values of $f(x, y, z) = x^2yz + 1$ on the intersection of the plane $z = 1$ with the sphere $x^2 + y^2 + z^2 = 10$.

38. a. **Maximum on line of intersection** Find the maximum value of $w = xyz$ on the line of intersection of the two planes $x + y + z = 40$ and $x + y - z = 0$.

    b. Give a geometric argument to support your claim that you have found a maximum, and not a minimum, value of w.

39. **Extrema on a circle of intersection** Find the extreme values of the function $f(x, y, z) = xy + z^2$ on the circle in which the plane $y - x = 0$ intersects the sphere $x^2 + y^2 + z^2 = 4$.

40. **Minimum distance to the origin** Find the point closest to the origin on the curve of intersection of the plane $2y + 4z = 5$ and the cone $z^2 = 4x^2 + 4y^2$.

Theory and Examples

41. **The condition $\nabla f = \lambda \nabla g$ is not sufficient** Although $\nabla f = \lambda \nabla g$ is a necessary condition for the occurrence of an extreme value of $f(x, y)$ subject to the condition $g(x, y) = 0$, it does not in itself guarantee that one exists. As a case in point, try using the method of Lagrange multipliers to find a maximum value of $f(x, y) = x + y$ subject to the constraint that $xy = 16$. The method will identify the two points $(4, 4)$ and $(-4, -4)$ as candidates for the location of extreme values. Yet the sum $(x + y)$ has no maximum value on the hyperbola $xy = 16$. The farther you go from the origin on this hyperbola in the first quadrant, the larger the sum $f(x, y) = x + y$ becomes.

42. **A least squares plane** The plane $z = Ax + By + C$ is to be "fitted" to the following points (x_k, y_k, z_k):

$$(0, 0, 0), \quad (0, 1, 1), \quad (1, 1, 1), \quad (1, 0, -1).$$

Find the values of A, B, and C that minimize

$$\sum_{k=1}^{4} (Ax_k + By_k + C - z_k)^2,$$

the sum of the squares of the deviations.

43. a. **Maximum on a sphere** Show that the maximum value of $a^2b^2c^2$ on a sphere of radius r centered at the origin of a Cartesian abc-coordinate system is $(r^2/3)^3$.

    b. **Geometric and arithmetic means** Using part (a), show that for nonnegative numbers a, b, and c,

$$(abc)^{1/3} \le \frac{a + b + c}{3};$$

that is, the *geometric mean* of three nonnegative numbers is less than or equal to their *arithmetic mean*.

44. **Sum of products** Let $a_1, a_2, \ldots, a_n$ be n positive numbers. Find the maximum of $\Sigma_{i=1}^{n} a_i x_i$ subject to the constraint $\Sigma_{i=1}^{n} x_i^2 = 1$.

COMPUTER EXPLORATIONS

Implementing the Method of Lagrange Multipliers

In Exercises 45–50, use a CAS to perform the following steps implementing the method of Lagrange multipliers for finding constrained extrema:

a. Form the function $h = f - \lambda_1 g_1 - \lambda_2 g_2$, where f is the function to optimize subject to the constraints $g_1 = 0$ and $g_2 = 0$.

b. Determine all the first partial derivatives of h, including the partials with respect to λ_1 and λ_2, and set them equal to 0.

c. Solve the system of equations found in part (b) for all the unknowns, including λ_1 and λ_2.

d. Evaluate f at each of the solution points found in part (c) and select the extreme value subject to the constraints asked for in the exercise.

45. Minimize $f(x, y, z) = xy + yz$ subject to the constraints $x^2 + y^2 - 2 = 0$ and $x^2 + z^2 - 2 = 0$.

46. Minimize $f(x, y, z) = xyz$ subject to the constraints $x^2 + y^2 - 1 = 0$ and $x - z = 0$.

47. Maximize $f(x, y, z) = x^2 + y^2 + z^2$ subject to the constraints $2y + 4z - 5 = 0$ and $4x^2 + 4y^2 - z^2 = 0$.

48. Minimize $f(x, y, z) = x^2 + y^2 + z^2$ subject to the constraints $x^2 - xy + y^2 - z^2 - 1 = 0$ and $x^2 + y^2 - 1 = 0$.

49. Minimize $f(x, y, z, w) = x^2 + y^2 + z^2 + w^2$ subject to the constraints $2x - y + z - w - 1 = 0$ and $x + y - z + w - 1 = 0$.

50. Determine the distance from the line $y = x + 1$ to the parabola $y^2 = x$. (*Hint:* Let (x, y) be a point on the line and (w, z) a point on the parabola. You want to minimize $(x - w)^2 + (y - z)^2$.)

14.9 Partial Derivatives with Constrained Variables

In finding partial derivatives of functions like $w = f(x, y)$, we have assumed x and y to be independent. In many applications, however, this is not the case. For example, the internal energy U of a gas may be expressed as a function $U = f(P, V, T)$ of pressure P, volume V, and temperature T. If the individual molecules of the gas do not interact, however, P, V, and T obey (and are constrained by) the ideal gas law

$$PV = nRT \qquad (n \text{ and } R \text{ constant}),$$

and fail to be independent. In this section we learn how to find partial derivatives in situations like this, which you may encounter in studying economics, engineering, or physics.†

Decide Which Variables Are Dependent and Which Are Independent

If the variables in a function $w = f(x, y, z)$ are constrained by a relation like the one imposed on x, y, and z by the equation $z = x^2 + y^2$, the geometric meanings and the numerical values of the partial derivatives of f will depend on which variables are chosen to be dependent and which are chosen to be independent. To see how this choice can affect the outcome, we consider the calculation of $\partial w/\partial x$ when $w = x^2 + y^2 + z^2$ and $z = x^2 + y^2$.

EXAMPLE 1 Finding a Partial Derivative with Constrained Independent Variables

Find $\partial w/\partial x$ if $w = x^2 + y^2 + z^2$ and $z = x^2 + y^2$.

†This section is based on notes written for MIT by Arthur P. Mattuck.

Solution We are given two equations in the four unknowns x, y, z, and w. Like many such systems, this one can be solved for two of the unknowns (the dependent variables) in terms of the others (the independent variables). In being asked for $\partial w/\partial x$, we are told that w is to be a dependent variable and x an independent variable. The possible choices for the other variables come down to

Dependent	*Independent*
w, z	x, y
w, y	x, z

In either case, we can express w explicitly in terms of the selected independent variables. We do this by using the second equation $z = x^2 + y^2$ to eliminate the remaining dependent variable in the first equation.

In the first case, the remaining dependent variable is z. We eliminate it from the first equation by replacing it by $x^2 + y^2$. The resulting expression for w is

$$\begin{aligned} w &= x^2 + y^2 + z^2 = x^2 + y^2 + (x^2 + y^2)^2 \\ &= x^2 + y^2 + x^4 + 2x^2y^2 + y^4 \end{aligned}$$

and

$$\frac{\partial w}{\partial x} = 2x + 4x^3 + 4xy^2. \tag{1}$$

This is the formula for $\partial w/\partial x$ when x and y are the independent variables.

In the second case, where the independent variables are x and z and the remaining dependent variable is y, we eliminate the dependent variable y in the expression for w by replacing y^2 in the second equation by $z - x^2$. This gives

$$w = x^2 + y^2 + z^2 = x^2 + (z - x^2) + z^2 = z + z^2$$

and

$$\frac{\partial w}{\partial x} = 0. \tag{2}$$

This is the formula for $\partial w/\partial x$ when x and z are the independent variables.

The formulas for $\partial w/\partial x$ in Equations (1) and (2) are genuinely different. We cannot change either formula into the other by using the relation $z = x^2 + y^2$. There is not just one $\partial w/\partial x$, there are two, and we see that the original instruction to find $\partial w/\partial x$ was incomplete. *Which* $\partial w/\partial x$? we ask.

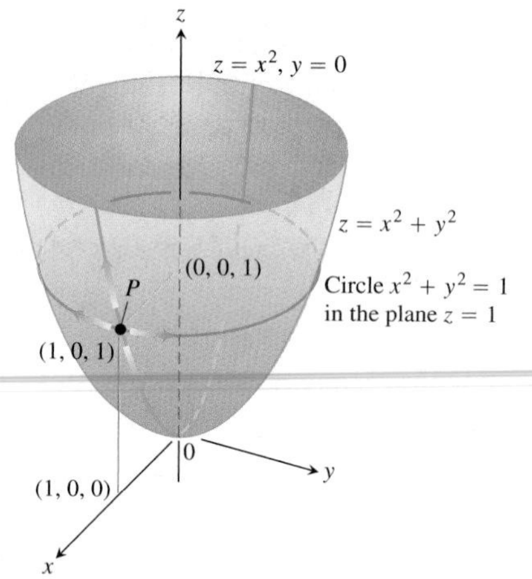

FIGURE 14.58 If P is constrained to lie on the paraboloid $z = x^2 + y^2$, the value of the partial derivative of $w = x^2 + y^2 + z^2$ with respect to x at P depends on the direction of motion (Example 1). (1) As x changes, with $y = 0$, P moves up or down the surface on the parabola $z = x^2$ in the xz-plane with $\partial w/\partial x = 2x + 4x^3$. (2) As x changes, with $z = 1$, P moves on the circle $x^2 + y^2 = 1$, $z = 1$, and $\partial w/\partial x = 0$.

The geometric interpretations of Equations (1) and (2) help to explain why the equations differ. The function $w = x^2 + y^2 + z^2$ measures the square of the distance from the point (x, y, z) to the origin. The condition $z = x^2 + y^2$ says that the point (x, y, z) lies on the paraboloid of revolution shown in Figure 14.58. What does it mean to calculate $\partial w/\partial x$ at a point $P(x, y, z)$ that can move only on this surface? What is the value of $\partial w/\partial x$ when the coordinates of P are, say, $(1, 0, 1)$?

If we take x and y to be independent, then we find $\partial w/\partial x$ by holding y fixed (at $y = 0$ in this case) and letting x vary. Hence, P moves along the parabola $z = x^2$ in the xz-plane. As P moves on this parabola, w, which is the square of the distance from P to the origin, changes. We calculate $\partial w/\partial x$ in this case (our first solution above) to be

$$\frac{\partial w}{\partial x} = 2x + 4x^3 + 4xy^2.$$

At the point $P(1, 0, 1)$, the value of this derivative is

$$\frac{\partial w}{\partial x} = 2 + 4 + 0 = 6.$$

If we take x and z to be independent, then we find $\partial w/\partial x$ by holding z fixed while x varies. Since the z-coordinate of P is 1, varying x moves P along a circle in the plane $z = 1$. As P moves along this circle, its distance from the origin remains constant, and w, being the square of this distance, does not change. That is,

$$\frac{\partial w}{\partial x} = 0,$$

as we found in our second solution. ■

How to Find $\partial w/\partial x$ When the Variables in $w = f(x, y, z)$ Are Constrained by Another Equation

As we saw in Example 1, a typical routine for finding $\partial w/\partial x$ when the variables in the function $w = f(x, y, z)$ are related by another equation has three steps. These steps apply to finding $\partial w/\partial y$ and $\partial w/\partial z$ as well.

1. *Decide* which variables are to be dependent and which are to be independent. (In practice, the decision is based on the physical or theoretical context of our work. In the exercises at the end of this section, we say which variables are which.)
2. *Eliminate* the other dependent variable(s) in the expression for w.
3. *Differentiate* as usual.

If we cannot carry out Step 2 after deciding which variables are dependent, we differentiate the equations as they are and try to solve for $\partial w/\partial x$ afterward. The next example shows how this is done.

EXAMPLE 2 Finding a Partial Derivative with Identified Constrained Independent Variables

Find $\partial w/\partial x$ at the point $(x, y, z) = (2, -1, 1)$ if

$$w = x^2 + y^2 + z^2, \qquad z^3 - xy + yz + y^3 = 1,$$

and x and y are the independent variables.

Solution It is not convenient to eliminate z in the expression for w. We therefore differentiate both equations implicitly with respect to x, treating x and y as independent variables and w and z as dependent variables. This gives

$$\frac{\partial w}{\partial x} = 2x + 2z\frac{\partial z}{\partial x} \tag{3}$$

and

$$3z^2 \frac{\partial z}{\partial x} - y + y \frac{\partial z}{\partial x} + 0 = 0. \qquad (4)$$

These equations may now be combined to express $\partial w/\partial x$ in terms of x, y, and z. We solve Equation (4) for $\partial z/\partial x$ to get

$$\frac{\partial z}{\partial x} = \frac{y}{y + 3z^2}$$

and substitute into Equation (3) to get

$$\frac{\partial w}{\partial x} = 2x + \frac{2yz}{y + 3z^2}.$$

The value of this derivative at $(x, y, z) = (2, -1, 1)$ is

$$\left(\frac{\partial w}{\partial x}\right)_{(2,-1,1)} = 2(2) + \frac{2(-1)(1)}{-1 + 3(1)^2} = 4 + \frac{-2}{2} = 3. \quad ■$$

HISTORICAL BIOGRAPHY

Sonya Kovalevsky
(1850–1891)

Notation

To show what variables are assumed to be independent in calculating a derivative, we can use the following notation:

$\left(\frac{\partial w}{\partial x}\right)_y$ $\quad$ $\partial w/\partial x$ with x and y independent

$\left(\frac{\partial f}{\partial y}\right)_{x,t}$ $\quad$ $\partial f/\partial y$ with y, x and t independent

EXAMPLE 3 Finding a Partial Derivative with Constrained Variables Notationally Identified

Find $(\partial w/\partial x)_{y,z}$ if $w = x^2 + y - z + \sin t$ and $x + y = t$.

Solution With x, y, z independent, we have

$$t = x + y, \qquad w = x^2 + y - z + \sin(x + y)$$

$$\left(\frac{\partial w}{\partial x}\right)_{y,z} = 2x + 0 - 0 + \cos(x + y)\frac{\partial}{\partial x}(x + y)$$

$$= 2x + \cos(x + y). \quad ■$$

Arrow Diagrams

In solving problems like the one in Example 3, it often helps to start with an arrow diagram that shows how the variables and functions are related. If

$$w = x^2 + y - z + \sin t \qquad \text{and} \qquad x + y = t$$

and we are asked to find $\partial w/\partial x$ when x, y, and z are independent, the appropriate diagram is one like this:

$$\begin{pmatrix} x \\ y \\ z \end{pmatrix} \to \begin{pmatrix} x \\ y \\ z \\ t \end{pmatrix} \to w \qquad (5)$$

Independent variables | Intermediate variables | Dependent variable

To avoid confusion between the independent and intermediate variables with the same symbolic names in the diagram, it is helpful to rename the intermediate variables (so they are seen as *functions* of the independent variables). Thus, let $u = x$, $v = y$, and $s = z$ denote the renamed intermediate variables. With this notation, the arrow diagram becomes

$$\begin{pmatrix} x \\ y \\ z \end{pmatrix} \to \begin{pmatrix} u \\ v \\ s \\ t \end{pmatrix} \to w \qquad (6)$$

Independent variables | Intermediate variables and relations $u = x$, $v = y$, $s = z$, $t = x + y$ | Dependent variable

The diagram shows the independent variables on the left, the intermediate variables and their relation to the independent variables in the middle, and the dependent variable on the right. The function w now becomes

$$w = u^2 + v - s + \sin t,$$

where

$$u = x, \qquad v = y, \qquad s = z, \qquad \text{and} \qquad t = x + y.$$

To find $\partial w/\partial x$, we apply the four-variable form of the Chain Rule to w, guided by the arrow diagram in Equation (6):

$$\begin{aligned}\frac{\partial w}{\partial x} &= \frac{\partial w}{\partial u}\frac{\partial u}{\partial x} + \frac{\partial w}{\partial v}\frac{\partial v}{\partial x} + \frac{\partial w}{\partial s}\frac{\partial s}{\partial x} + \frac{\partial w}{\partial t}\frac{\partial t}{\partial x} \\ &= (2u)(1) + (1)(0) + (-1)(0) + (\cos t)(1) \\ &= 2u + \cos t \\ &= 2x + \cos(x + y). \qquad \text{Substituting the original independent variables } u = x \text{ and } t = x + y.\end{aligned}$$

EXERCISES 14.9

Finding Partial Derivatives with Constrained Variables

In Exercises 1–3, begin by drawing a diagram that shows the relations among the variables.

1. If $w = x^2 + y^2 + z^2$ and $z = x^2 + y^2$, find

a. $\left(\dfrac{\partial w}{\partial y}\right)_z$ **b.** $\left(\dfrac{\partial w}{\partial z}\right)_x$ **c.** $\left(\dfrac{\partial w}{\partial z}\right)_y$.

2. If $w = x^2 + y - z + \sin t$ and $x + y = t$, find

a. $\left(\dfrac{\partial w}{\partial y}\right)_{x,z}$ **b.** $\left(\dfrac{\partial w}{\partial y}\right)_{z,t}$ **c.** $\left(\dfrac{\partial w}{\partial z}\right)_{x,y}$

d. $\left(\dfrac{\partial w}{\partial z}\right)_{y,t}$ **e.** $\left(\dfrac{\partial w}{\partial t}\right)_{x,z}$ **f.** $\left(\dfrac{\partial w}{\partial t}\right)_{y,z}$.

3. Let $U = f(P, V, T)$ be the internal energy of a gas that obeys the ideal gas law $PV = nRT$ (n and R constant). Find

 a. $\left(\frac{\partial U}{\partial P}\right)_V$ b. $\left(\frac{\partial U}{\partial T}\right)_V$.

4. Find

 a. $\left(\frac{\partial w}{\partial x}\right)_y$ b. $\left(\frac{\partial w}{\partial z}\right)_y$

 at the point $(x, y, z) = (0, 1, \pi)$ if

$$w = x^2 + y^2 + z^2 \quad \text{and} \quad y \sin z + z \sin x = 0.$$

5. Find

 a. $\left(\frac{\partial w}{\partial y}\right)_x$ b. $\left(\frac{\partial w}{\partial y}\right)_z$

 at the point $(w, x, y, z) = (4, 2, 1, -1)$ if

$$w = x^2y^2 + yz - z^3 \quad \text{and} \quad x^2 + y^2 + z^2 = 6.$$

6. Find $(\partial u/\partial y)_x$ at the point $(u, v) = \left(\sqrt{2}, 1\right)$, if $x = u^2 + v^2$ and $y = uv$.

7. Suppose that $x^2 + y^2 = r^2$ and $x = r\cos\theta$, as in polar coordinates. Find

$$\left(\frac{\partial x}{\partial r}\right)_\theta \quad \text{and} \quad \left(\frac{\partial r}{\partial x}\right)_y.$$

8. Suppose that

$$w = x^2 - y^2 + 4z + t \quad \text{and} \quad x + 2z + t = 25.$$

 Show that the equations

$$\frac{\partial w}{\partial x} = 2x - 1 \quad \text{and} \quad \frac{\partial w}{\partial x} = 2x - 2$$

 each give $\partial w/\partial x$, depending on which variables are chosen to be dependent and which variables are chosen to be independent. Identify the independent variables in each case.

Partial Derivatives Without Specific Formulas

9. Establish the fact, widely used in hydrodynamics, that if $f(x, y, z) = 0$, then

$$\left(\frac{\partial x}{\partial y}\right)_z \left(\frac{\partial y}{\partial z}\right)_x \left(\frac{\partial z}{\partial x}\right)_y = -1.$$

 (*Hint:* Express all the derivatives in terms of the formal partial derivatives $\partial f/\partial x$, $\partial f/\partial y$, and $\partial f/\partial z$.)

10. If $z = x + f(u)$, where $u = xy$, show that

$$x\frac{\partial z}{\partial x} - y\frac{\partial z}{\partial y} = x.$$

11. Suppose that the equation $g(x, y, z) = 0$ determines z as a differentiable function of the independent variables x and y and that $g_z \neq 0$. Show that

$$\left(\frac{\partial z}{\partial y}\right)_x = -\frac{\partial g/\partial y}{\partial g/\partial z}.$$

12. Suppose that $f(x, y, z, w) = 0$ and $g(x, y, z, w) = 0$ determine z and w as differentiable functions of the independent variables x and y, and suppose that

$$\frac{\partial f}{\partial z}\frac{\partial g}{\partial w} - \frac{\partial f}{\partial w}\frac{\partial g}{\partial z} \neq 0.$$

 Show that

$$\left(\frac{\partial z}{\partial x}\right)_y = -\frac{\dfrac{\partial f}{\partial x}\dfrac{\partial g}{\partial w} - \dfrac{\partial f}{\partial w}\dfrac{\partial g}{\partial x}}{\dfrac{\partial f}{\partial z}\dfrac{\partial g}{\partial w} - \dfrac{\partial f}{\partial w}\dfrac{\partial g}{\partial z}}$$

 and

$$\left(\frac{\partial w}{\partial y}\right)_x = -\frac{\dfrac{\partial f}{\partial z}\dfrac{\partial g}{\partial y} - \dfrac{\partial f}{\partial y}\dfrac{\partial g}{\partial z}}{\dfrac{\partial f}{\partial z}\dfrac{\partial g}{\partial w} - \dfrac{\partial f}{\partial w}\dfrac{\partial g}{\partial z}}.$$

14.10 Taylor's Formula for Two Variables

This section uses Taylor's formula to derive the Second Derivative Test for local extreme values (Section 14.7) and the error formula for linearizations of functions of two independent variables (Section 14.6). The use of Taylor's formula in these derivations leads to an extension of the formula that provides polynomial approximations of all orders for functions of two independent variables.

Derivation of the Second Derivative Test

Let $f(x, y)$ have continuous partial derivatives in an open region R containing a point $P(a, b)$ where $f_x = f_y = 0$ (Figure 14.59). Let h and k be increments small enough to put the

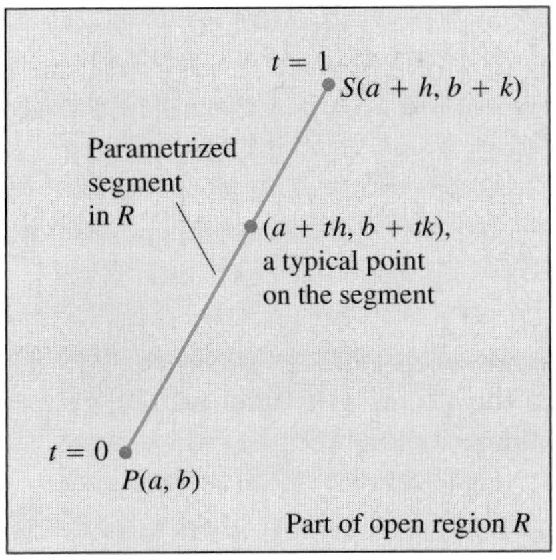

FIGURE 14.59 We begin the derivation of the second derivative test at $P(a, b)$ by parametrizing a typical line segment from P to a point S nearby.

point $S(a + h, b + k)$ and the line segment joining it to P inside R. We parametrize the segment PS as

$$x = a + th, \qquad y = b + tk, \qquad 0 \le t \le 1.$$

If $F(t) = f(a + th, b + tk)$, the Chain Rule gives

$$F'(t) = f_x \frac{dx}{dt} + f_y \frac{dy}{dt} = hf_x + kf_y.$$

Since f_x and f_y are differentiable (they have continuous partial derivatives), F' is a differentiable function of t and

$$F'' = \frac{\partial F'}{\partial x}\frac{dx}{dt} + \frac{\partial F'}{\partial y}\frac{dy}{dt} = \frac{\partial}{\partial x}(hf_x + kf_y) \cdot h + \frac{\partial}{\partial y}(hf_x + kf_y) \cdot k$$

$$= h^2 f_{xx} + 2hkf_{xy} + k^2 f_{yy}. \qquad f_{xy} = f_{yx}$$

Since F and F' are continuous on $[0, 1]$ and F' is differentiable on $(0, 1)$, we can apply Taylor's formula with $n = 2$ and $a = 0$ to obtain

$$F(1) = F(0) + F'(0)(1 - 0) + F''(c)\frac{(1 - 0)^2}{2}$$

$$F(1) = F(0) + F'(0) + \frac{1}{2}F''(c) \qquad (1)$$

for some c between 0 and 1. Writing Equation (1) in terms of f gives

$$f(a + h, b + k) = f(a, b) + hf_x(a, b) + kf_y(a, b)$$

$$+ \frac{1}{2}\left(h^2 f_{xx} + 2hkf_{xy} + k^2 f_{yy}\right)\Big|_{(a+ch,\, b+ck)}. \qquad (2)$$

Since $f_x(a, b) = f_y(a, b) = 0$, this reduces to

$$f(a + h, b + k) - f(a, b) = \frac{1}{2}\left(h^2 f_{xx} + 2hkf_{xy} + k^2 f_{yy}\right)\Big|_{(a+ch,\, b+ck)}. \qquad (3)$$

The presence of an extremum of f at (a, b) is determined by the sign of $f(a + h, b + k) - f(a, b)$. By Equation (3), this is the same as the sign of

$$Q(c) = (h^2 f_{xx} + 2hkf_{xy} + k^2 f_{yy})|_{(a+ch, b+ck)}.$$

Now, if $Q(0) \neq 0$, the sign of $Q(c)$ will be the same as the sign of $Q(0)$ for sufficiently small values of h and k. We can predict the sign of

$$Q(0) = h^2 f_{xx}(a, b) + 2hkf_{xy}(a, b) + k^2 f_{yy}(a, b) \qquad (4)$$

from the signs of f_{xx} and $f_{xx}f_{yy} - f_{xy}^2$ at (a, b). Multiply both sides of Equation (4) by f_{xx} and rearrange the right-hand side to get

$$f_{xx}Q(0) = (hf_{xx} + kf_{xy})^2 + (f_{xx}f_{yy} - f_{xy}^2)k^2. \qquad (5)$$

From Equation (5) we see that

1. If $f_{xx} < 0$ and $f_{xx}f_{yy} - f_{xy}^2 > 0$ at (a, b), then $Q(0) < 0$ for all sufficiently small nonzero values of h and k, and f has a *local maximum* value at (a, b).
2. If $f_{xx} > 0$ and $f_{xx}f_{yy} - f_{xy}^2 > 0$ at (a, b), then $Q(0) > 0$ for all sufficiently small nonzero values of h and k and f has a *local minimum* value at (a, b).

3. If $f_{xx}f_{yy} - f_{xy}^2 < 0$ at (a, b), there are combinations of arbitrarily small nonzero values of h and k for which $Q(0) > 0$, and other values for which $Q(0) < 0$. Arbitrarily close to the point $P_0(a, b, f(a, b))$ on the surface $z = f(x, y)$ there are points above P_0 and points below P_0, so f has a *saddle point* at (a, b).
4. If $f_{xx}f_{yy} - f_{xy}^2 = 0$, another test is needed. The possibility that $Q(0)$ equals zero prevents us from drawing conclusions about the sign of $Q(c)$.

The Error Formula for Linear Approximations

We want to show that the difference $E(x, y)$, between the values of a function $f(x, y)$, and its linearization $L(x, y)$ at (x_0, y_0) satisfies the inequality

$$|E(x, y)| \le \frac{1}{2}M(|x - x_0| + |y - y_0|)^2.$$

The function f is assumed to have continuous second partial derivatives throughout an open set containing a closed rectangular region R centered at (x_0, y_0). The number M is an upper bound for $|f_{xx}|, |f_{yy}|$, and $|f_{xy}|$ on R.

The inequality we want comes from Equation (2). We substitute x_0 and y_0 for a and b, and $x - x_0$ and $y - y_0$ for h and k, respectively, and rearrange the result as

$$f(x, y) = \underbrace{f(x_0, y_0) + f_x(x_0, y_0)(x - x_0) + f_y(x_0, y_0)(y - y_0)}_{\text{linearization } L(x, y)}$$

$$\underbrace{+ \frac{1}{2}\left((x - x_0)^2 f_{xx} + 2(x - x_0)(y - y_0)f_{xy} + (y - y_0)^2 f_{yy}\right)\Big|_{(x_0+c(x-x_0),\, y_0+c(y-y_0))}}_{\text{error } E(x, y)}.$$

This equation reveals that

$$|E| \le \frac{1}{2}\left(|x - x_0|^2|f_{xx}| + 2|x - x_0||y - y_0||f_{xy}| + |y - y_0|^2|f_{yy}|\right).$$

Hence, if M is an upper bound for the values of $|f_{xx}|, |f_{xy}|$, and $|f_{yy}|$ on R,

$$\begin{aligned} |E| &\le \frac{1}{2}\left(|x - x_0|^2 M + 2|x - x_0||y - y_0|M + |y - y_0|^2 M\right) \\ &= \frac{1}{2}M(|x - x_0| + |y - y_0|)^2. \end{aligned}$$

Taylor's Formula for Functions of Two Variables

The formulas derived earlier for F' and F'' can be obtained by applying to $f(x, y)$ the operators

$$\left(h\frac{\partial}{\partial x} + k\frac{\partial}{\partial y}\right) \quad \text{and} \quad \left(h\frac{\partial}{\partial x} + k\frac{\partial}{\partial y}\right)^2 = h^2\frac{\partial^2}{\partial x^2} + 2hk\frac{\partial^2}{\partial x\,\partial y} + k^2\frac{\partial^2}{\partial y^2}.$$

These are the first two instances of a more general formula,

$$F^{(n)}(t) = \frac{d^n}{dt^n}F(t) = \left(h\frac{\partial}{\partial x} + k\frac{\partial}{\partial y}\right)^n f(x, y), \tag{6}$$

which says that applying d^n/dt^n to $F(t)$ gives the same result as applying the operator

$$\left(h\frac{\partial}{\partial x} + k\frac{\partial}{\partial y}\right)^n$$

to $f(x, y)$ after expanding it by the Binomial Theorem.

If partial derivatives of f through order $n + 1$ are continuous throughout a rectangular region centered at (a, b), we may extend the Taylor formula for $F(t)$ to

$$F(t) = F(0) + F'(0)t + \frac{F''(0)}{2!}t^2 + \cdots + \frac{F^{(n)}(0)}{n!}t^{(n)} + \text{remainder},$$

and take $t = 1$ to obtain

$$F(1) = F(0) + F'(0) + \frac{F''(0)}{2!} + \cdots + \frac{F^{(n)}(0)}{n!} + \text{remainder}.$$

When we replace the first n derivatives on the right of this last series by their equivalent expressions from Equation (6) evaluated at $t = 0$ and add the appropriate remainder term, we arrive at the following formula.

Taylor's Formula for $f(x, y)$ at the Point (a, b)

Suppose $f(x, y)$ and its partial derivatives through order $n + 1$ are continuous throughout an open rectangular region R centered at a point (a, b). Then, throughout R,

$$\begin{aligned} f(a + h, b + k) = {} & f(a, b) + (hf_x + kf_y)\big|_{(a,b)} + \frac{1}{2!}(h^2 f_{xx} + 2hkf_{xy} + k^2 f_{yy})\big|_{(a,b)} \\ & + \frac{1}{3!}(h^3 f_{xxx} + 3h^2 k f_{xxy} + 3hk^2 f_{xyy} + k^3 f_{yyy})\big|_{(a,b)} + \cdots + \frac{1}{n!}\left(h\frac{\partial}{\partial x} + k\frac{\partial}{\partial y}\right)^n f\bigg|_{(a,b)} \\ & + \frac{1}{(n+1)!}\left(h\frac{\partial}{\partial x} + k\frac{\partial}{\partial y}\right)^{n+1} f\bigg|_{(a+ch,b+ck)}. \end{aligned} \tag{7}$$

The first n derivative terms are evaluated at (a, b). The last term is evaluated at some point $(a + ch, b + ck)$ on the line segment joining (a, b) and $(a + h, b + k)$.

If $(a, b) = (0, 0)$ and we treat h and k as independent variables (denoting them now by x and y), then Equation (7) assumes the following simpler form.

Taylor's Formula for $f(x, y)$ at the Origin

$$\begin{aligned} f(x, y) = {} & f(0, 0) + xf_x + yf_y + \frac{1}{2!}(x^2 f_{xx} + 2xyf_{xy} + y^2 f_{yy}) \\ & + \frac{1}{3!}(x^3 f_{xxx} + 3x^2 y f_{xxy} + 3xy^2 f_{xyy} + y^3 f_{yyy}) + \cdots + \frac{1}{n!}\left(x\frac{\partial}{\partial x} + y\frac{\partial}{\partial y}\right)^n f \\ & + \frac{1}{(n+1)!}\left(x\frac{\partial}{\partial x} + y\frac{\partial}{\partial y}\right)^{n+1} f\bigg|_{(cx,cy)} \end{aligned} \tag{8}$$

The first n derivative terms are evaluated at $(0, 0)$. The last term is evaluated at a point on the line segment joining the origin and (x, y).

Taylor's formula provides polynomial approximations of two-variable functions. The first n derivative terms give the polynomial; the last term gives the approximation error. The first three terms of Taylor's formula give the function's linearization. To improve on the linearization, we add higher power terms.

EXAMPLE 1 Finding a Quadratic Approximation

Find a quadratic approximation to $f(x, y) = \sin x \sin y$ near the origin. How accurate is the approximation if $|x| \leq 0.1$ and $|y| \leq 0.1$?

Solution We take $n = 2$ in Equation (8):

$$f(x, y) = f(0, 0) + (xf_x + yf_y) + \frac{1}{2}(x^2 f_{xx} + 2xyf_{xy} + y^2 f_{yy})$$
$$+ \frac{1}{6}(x^3 f_{xxx} + 3x^2 y f_{xxy} + 3xy^2 f_{xyy} + y^3 f_{yyy})_{(cx,cy)}$$

with

$$f(0, 0) = \sin x \sin y|_{(0,0)} = 0, \qquad f_{xx}(0, 0) = -\sin x \sin y|_{(0,0)} = 0,$$
$$f_x(0, 0) = \cos x \sin y|_{(0,0)} = 0, \qquad f_{xy}(0, 0) = \cos x \cos y|_{(0,0)} = 1,$$
$$f_y(0, 0) = \sin x \cos y|_{(0,0)} = 0, \qquad f_{yy}(0, 0) = -\sin x \sin y|_{(0,0)} = 0,$$

we have

$$\sin x \sin y \approx 0 + 0 + 0 + \frac{1}{2}(x^2(0) + 2xy(1) + y^2(0)),$$
$$\sin x \sin y \approx xy.$$

The error in the approximation is

$$E(x, y) = \frac{1}{6}(x^3 f_{xxx} + 3x^2 y f_{xxy} + 3xy^2 f_{xyy} + y^3 f_{yyy})|_{(cx,cy)}.$$

The third derivatives never exceed 1 in absolute value because they are products of sines and cosines. Also, $|x| \leq 0.1$ and $|y| \leq 0.1$. Hence

$$|E(x, y)| \leq \frac{1}{6}((0.1)^3 + 3(0.1)^3 + 3(0.1)^3 + (0.1)^3) = \frac{8}{6}(0.1)^3 \leq 0.00134$$

(rounded up). The error will not exceed 0.00134 if $|x| \leq 0.1$ and $|y| \leq 0.1$. ■

EXERCISES 14.10

Finding Quadratic and Cubic Approximations

In Exercises 1–10, use Taylor's formula for $f(x, y)$ at the origin to find quadratic and cubic approximations of f near the origin.

1. $f(x, y) = xe^y$
2. $f(x, y) = e^x \cos y$
3. $f(x, y) = y \sin x$
4. $f(x, y) = \sin x \cos y$
5. $f(x, y) = e^x \ln(1 + y)$
6. $f(x, y) = \ln(2x + y + 1)$
7. $f(x, y) = \sin(x^2 + y^2)$
8. $f(x, y) = \cos(x^2 + y^2)$

9. $f(x, y) = \dfrac{1}{1 - x - y}$

10. $f(x, y) = \dfrac{1}{1 - x - y + xy}$

11. Use Taylor's formula to find a quadratic approximation of $f(x, y) = \cos x \cos y$ at the origin. Estimate the error in the approximation if $|x| \le 0.1$ and $|y| \le 0.1$.

12. Use Taylor's formula to find a quadratic approximation of $e^x \sin y$ at the origin. Estimate the error in the approximation if $|x| \le 0.1$ and $|y| \le 0.1$.

Chapter 14 Questions to Guide Your Review

1. What is a real-valued function of two independent variables? Three independent variables? Give examples.
2. What does it mean for sets in the plane or in space to be open? Closed? Give examples. Give examples of sets that are neither open nor closed.
3. How can you display the values of a function $f(x, y)$ of two independent variables graphically? How do you do the same for a function $f(x, y, z)$ of three independent variables?
4. What does it mean for a function $f(x, y)$ to have limit L as $(x, y) \to (x_0, y_0)$? What are the basic properties of limits of functions of two independent variables?
5. When is a function of two (three) independent variables continuous at a point in its domain? Give examples of functions that are continuous at some points but not others.
6. What can be said about algebraic combinations and composites of continuous functions?
7. Explain the two-path test for nonexistence of limits.
8. How are the partial derivatives $\partial f/\partial x$ and $\partial f/\partial y$ of a function $f(x, y)$ defined? How are they interpreted and calculated?
9. How does the relation between first partial derivatives and continuity of functions of two independent variables differ from the relation between first derivatives and continuity for real-valued functions of a single independent variable? Give an example.
10. What is the Mixed Derivative Theorem for mixed second-order partial derivatives? How can it help in calculating partial derivatives of second and higher orders? Give examples.
11. What does it mean for a function $f(x, y)$ to be differentiable? What does the Increment Theorem say about differentiability?
12. How can you sometimes decide from examining f_x and f_y that a function $f(x, y)$ is differentiable? What is the relation between the differentiability of f and the continuity of f at a point?
13. What is the Chain Rule? What form does it take for functions of two independent variables? Three independent variables? Functions defined on surfaces? How do you diagram these different forms? Give examples. What pattern enables one to remember all the different forms?
14. What is the derivative of a function $f(x, y)$ at a point P_0 in the direction of a unit vector $\mathbf{u}$? What rate does it describe? What geometric interpretation does it have? Give examples.
15. What is the gradient vector of a differentiable function $f(x, y)$? How is it related to the function's directional derivatives? State the analogous results for functions of three independent variables.
16. How do you find the tangent line at a point on a level curve of a differentiable function $f(x, y)$? How do you find the tangent plane and normal line at a point on a level surface of a differentiable function $f(x, y, z)$? Give examples.
17. How can you use directional derivatives to estimate change?
18. How do you linearize a function $f(x, y)$ of two independent variables at a point (x_0, y_0)? Why might you want to do this? How do you linearize a function of three independent variables?
19. What can you say about the accuracy of linear approximations of functions of two (three) independent variables?
20. If (x, y) moves from (x_0, y_0) to a point $(x_0 + dx, y_0 + dy)$ nearby, how can you estimate the resulting change in the value of a differentiable function $f(x, y)$? Give an example.
21. How do you define local maxima, local minima, and saddle points for a differentiable function $f(x, y)$? Give examples.
22. What derivative tests are available for determining the local extreme values of a function $f(x, y)$? How do they enable you to narrow your search for these values? Give examples.
23. How do you find the extrema of a continuous function $f(x, y)$ on a closed bounded region of the xy-plane? Give an example.
24. Describe the method of Lagrange multipliers and give examples.
25. If $w = f(x, y, z)$, where the variables x, y, and z are constrained by an equation $g(x, y, z) = 0$, what is the meaning of the notation $(\partial w/\partial x)_y$? How can an arrow diagram help you calculate this partial derivative with constrained variables? Give examples.
26. How does Taylor's formula for a function $f(x, y)$ generate polynomial approximations and error estimates?

Chapter 14 Practice Exercises

Domain, Range, and Level Curves

In Exercises 1–4, find the domain and range of the given function and identify its level curves. Sketch a typical level curve.

1. $f(x, y) = 9x^2 + y^2$
2. $f(x, y) = e^{x+y}$
3. $g(x, y) = 1/xy$
4. $g(x, y) = \sqrt{x^2 - y}$

In Exercises 5–8, find the domain and range of the given function and identify its level surfaces. Sketch a typical level surface.

5. $f(x, y, z) = x^2 + y^2 - z$
6. $g(x, y, z) = x^2 + 4y^2 + 9z^2$
7. $h(x, y, z) = \dfrac{1}{x^2 + y^2 + z^2}$
8. $k(x, y, z) = \dfrac{1}{x^2 + y^2 + z^2 + 1}$

Evaluating Limits

Find the limits in Exercises 9–14.

9. $\lim_{(x,y)\to(\pi, \ln 2)} e^y \cos x$
10. $\lim_{(x,y)\to(0,0)} \dfrac{2 + y}{x + \cos y}$
11. $\lim_{(x,y)\to(1,1)} \dfrac{x - y}{x^2 - y^2}$
12. $\lim_{(x,y)\to(1,1)} \dfrac{x^3y^3 - 1}{xy - 1}$
13. $\lim_{P\to(1,-1,e)} \ln|x + y + z|$
14. $\lim_{P\to(1,-1,-1)} \tan^{-1}(x + y + z)$

By considering different paths of approach, show that the limits in Exercises 15 and 16 do not exist.

15. $\lim_{\substack{(x,y)\to(0,0)\\ y\neq x^2}} \dfrac{y}{x^2 - y}$
16. $\lim_{\substack{(x,y)\to(0,0)\\ xy\neq 0}} \dfrac{x^2 + y^2}{xy}$

17. **Continuous extension** Let $f(x, y) = (x^2 - y^2)/(x^2 + y^2)$ for $(x, y) \neq (0, 0)$. Is it possible to define $f(0, 0)$ in a way that makes f continuous at the origin? Why?

18. **Continuous extension** Let

$$f(x, y) = \begin{cases} \dfrac{\sin(x - y)}{|x| + |y|}, & |x| + |y| \neq 0 \\ 0, & (x, y) = (0, 0). \end{cases}$$

Is f continuous at the origin? Why?

Partial Derivatives

In Exercises 19–24, find the partial derivative of the function with respect to each variable.

19. $g(r, \theta) = r\cos\theta + r\sin\theta$
20. $f(x, y) = \dfrac{1}{2}\ln(x^2 + y^2) + \tan^{-1}\dfrac{y}{x}$
21. $f(R_1, R_2, R_3) = \dfrac{1}{R_1} + \dfrac{1}{R_2} + \dfrac{1}{R_3}$
22. $h(x, y, z) = \sin(2\pi x + y - 3z)$
23. $P(n, R, T, V) = \dfrac{nRT}{V}$ (the ideal gas law)
24. $f(r, l, T, w) = \dfrac{1}{2rl}\sqrt{\dfrac{T}{\pi w}}$

Second-Order Partials

Find the second-order partial derivatives of the functions in Exercises 25–28.

25. $g(x, y) = y + \dfrac{x}{y}$
26. $g(x, y) = e^x + y\sin x$
27. $f(x, y) = x + xy - 5x^3 + \ln(x^2 + 1)$
28. $f(x, y) = y^2 - 3xy + \cos y + 7e^y$

Chain Rule Calculations

29. Find dw/dt at $t = 0$ if $w = \sin(xy + \pi)$, $x = e^t$, and $y = \ln(t + 1)$.
30. Find dw/dt at $t = 1$ if $w = xe^y + y\sin z - \cos z$, $x = 2\sqrt{t}$, $y = t - 1 + \ln t$, and $z = \pi t$.
31. Find $\partial w/\partial r$ and $\partial w/\partial s$ when $r = \pi$ and $s = 0$ if $w = \sin(2x - y)$, $x = r + \sin s$, $y = rs$.
32. Find $\partial w/\partial u$ and $\partial w/\partial v$ when $u = v = 0$ if $w = \ln\sqrt{1 + x^2} - \tan^{-1}x$ and $x = 2e^u\cos v$.
33. Find the value of the derivative of $f(x, y, z) = xy + yz + xz$ with respect to t on the curve $x = \cos t$, $y = \sin t$, $z = \cos 2t$ at $t = 1$.
34. Show that if $w = f(s)$ is any differentiable function of s and if $s = y + 5x$, then

$$\frac{\partial w}{\partial x} - 5\frac{\partial w}{\partial y} = 0.$$

Implicit Differentiation

Assuming that the equations in Exercises 35 and 36 define y as a differentiable function of x, find the value of dy/dx at point P.

35. $1 - x - y^2 - \sin xy = 0$, $P(0, 1)$
36. $2xy + e^{x+y} - 2 = 0$, $P(0, \ln 2)$

Directional Derivatives

In Exercises 37–40, find the directions in which f increases and decreases most rapidly at P_0 and find the derivative of f in each direction. Also, find the derivative of f at P_0 in the direction of the vector $\mathbf{v}$.

37. $f(x, y) = \cos x\cos y$, $P_0(\pi/4, \pi/4)$, $\mathbf{v} = 3\mathbf{i} + 4\mathbf{j}$
38. $f(x, y) = x^2e^{-2y}$, $P_0(1, 0)$, $\mathbf{v} = \mathbf{i} + \mathbf{j}$
39. $f(x, y, z) = \ln(2x + 3y + 6z)$, $P_0(-1, -1, 1)$, $\mathbf{v} = 2\mathbf{i} + 3\mathbf{j} + 6\mathbf{k}$

40. $f(x, y, z) = x^2 + 3xy - z^2 + 2y + z + 4$, $P_0(0, 0, 0)$, $\mathbf{v} = \mathbf{i} + \mathbf{j} + \mathbf{k}$

41. Derivative in velocity direction Find the derivative of $f(x, y, z) = xyz$ in the direction of the velocity vector of the helix

$$\mathbf{r}(t) = (\cos 3t)\mathbf{i} + (\sin 3t)\mathbf{j} + 3t\mathbf{k}$$

at $t = \pi/3$.

42. Maximum directional derivative What is the largest value that the directional derivative of $f(x, y, z) = xyz$ can have at the point $(1, 1, 1)$?

43. Directional derivatives with given values At the point $(1, 2)$, the function $f(x, y)$ has a derivative of 2 in the direction toward $(2, 2)$ and a derivative of -2 in the direction toward $(1, 1)$.

a. Find $f_x(1, 2)$ and $f_y(1, 2)$.

b. Find the derivative of f at $(1, 2)$ in the direction toward the point $(4, 6)$.

44. Which of the following statements are true if $f(x, y)$ is differentiable at (x_0, y_0)? Give reasons for your answers.

a. If $\mathbf{u}$ is a unit vector, the derivative of f at (x_0, y_0) in the direction of $\mathbf{u}$ is $(f_x(x_0, y_0)\mathbf{i} + f_y(x_0, y_0)\mathbf{j}) \cdot \mathbf{u}$.

b. The derivative of f at (x_0, y_0) in the direction of $\mathbf{u}$ is a vector.

c. The directional derivative of f at (x_0, y_0) has its greatest value in the direction of ∇f.

d. At (x_0, y_0), vector ∇f is normal to the curve $f(x, y) = f(x_0, y_0)$.

Gradients, Tangent Planes, and Normal Lines

In Exercises 45 and 46, sketch the surface $f(x, y, z) = c$ together with ∇f at the given points.

45. $x^2 + y + z^2 = 0$; $(0, -1, \pm 1)$, $(0, 0, 0)$

46. $y^2 + z^2 = 4$; $(2, \pm 2, 0)$, $(2, 0, \pm 2)$

In Exercises 47 and 48, find an equation for the plane tangent to the level surface $f(x, y, z) = c$ at the point P_0. Also, find parametric equations for the line that is normal to the surface at P_0.

47. $x^2 - y - 5z = 0$, $P_0(2, -1, 1)$

48. $x^2 + y^2 + z = 4$, $P_0(1, 1, 2)$

In Exercises 49 and 50, find an equation for the plane tangent to the surface $z = f(x, y)$ at the given point.

49. $z = \ln(x^2 + y^2)$, $(0, 1, 0)$

50. $z = 1/(x^2 + y^2)$, $(1, 1, 1/2)$

In Exercises 51 and 52, find equations for the lines that are tangent and normal to the level curve $f(x, y) = c$ at the point P_0. Then sketch the lines and level curve together with ∇f at P_0.

51. $y - \sin x = 1$, $P_0(\pi, 1)$ **52.** $\dfrac{y^2}{2} - \dfrac{x^2}{2} = \dfrac{3}{2}$, $P_0(1, 2)$

Tangent Lines to Curves

In Exercises 53 and 54, find parametric equations for the line that is tangent to the curve of intersection of the surfaces at the given point.

53. Surfaces: $x^2 + 2y + 2z = 4$, $y = 1$
Point: $(1, 1, 1/2)$

54. Surfaces: $x + y^2 + z = 2$, $y = 1$
Point: $(1/2, 1, 1/2)$

Linearizations

In Exercises 55 and 56, find the linearization $L(x, y)$ of the function $f(x, y)$ at the point P_0. Then find an upper bound for the magnitude of the error E in the approximation $f(x, y) \approx L(x, y)$ over the rectangle R.

55. $f(x, y) = \sin x \cos y$, $P_0(\pi/4, \pi/4)$

R: $\left|x - \dfrac{\pi}{4}\right| \le 0.1$, $\left|y - \dfrac{\pi}{4}\right| \le 0.1$

56. $f(x, y) = xy - 3y^2 + 2$, $P_0(1, 1)$

R: $|x - 1| \le 0.1$, $|y - 1| \le 0.2$

Find the linearizations of the functions in Exercises 57 and 58 at the given points.

57. $f(x, y, z) = xy + 2yz - 3xz$ at $(1, 0, 0)$ and $(1, 1, 0)$

58. $f(x, y, z) = \sqrt{2} \cos x \sin(y + z)$ at $(0, 0, \pi/4)$ and $(\pi/4, \pi/4, 0)$

Estimates and Sensitivity to Change

59. Measuring the volume of a pipeline You plan to calculate the volume inside a stretch of pipeline that is about 36 in. in diameter and 1 mile long. With which measurement should you be more careful, the length or the diameter? Why?

60. Sensitivity to change Near the point $(1, 2)$, is $f(x, y) = x^2 - xy + y^2 - 3$ more sensitive to changes in x or to changes in y? How do you know?

61. Change in an electrical circuit Suppose that the current I (amperes) in an electrical circuit is related to the voltage V (volts) and the resistance R (ohms) by the equation $I = V/R$. If the voltage drops from 24 to 23 volts and the resistance drops from 100 to 80 ohms, will I increase or decrease? By about how much? Is the change in I more sensitive to change in the voltage or to change in the resistance? How do you know?

62. Maximum error in estimating the area of an ellipse If $a = 10$ cm and $b = 16$ cm to the nearest millimeter, what should you expect the maximum percentage error to be in the calculated area $A = \pi ab$ of the ellipse $x^2/a^2 + y^2/b^2 = 1$?

63. Error in estimating a product Let $y = uv$ and $z = u + v$, where u and v are positive independent variables.

a. If u is measured with an error of 2% and v with an error of 3%, about what is the percentage error in the calculated value of y?

b. Show that the percentage error in the calculated value of z is less than the percentage error in the value of y.

64. Cardiac index To make different people comparable in studies of cardiac output (Section 3.7, Exercise 25), researchers divide the measured cardiac output by the body surface area to find the *cardiac index C*:

$$C = \frac{\text{cardiac output}}{\text{body surface area}}.$$

The body surface area B of a person with weight w and height h is approximated by the formula

$$B = 71.84w^{0.425}h^{0.725},$$

which gives B in square centimeters when w is measured in kilograms and h in centimeters. You are about to calculate the cardiac index of a person with the following measurements:

Cardiac output:	7 L/min
Weight:	70 kg
Height:	180 cm

Which will have a greater effect on the calculation, a 1-kg error in measuring the weight or a 1-cm error in measuring the height?

Local Extrema

Test the functions in Exercises 65–70 for local maxima and minima and saddle points. Find each function's value at these points.

65. $f(x, y) = x^2 - xy + y^2 + 2x + 2y - 4$

66. $f(x, y) = 5x^2 + 4xy - 2y^2 + 4x - 4y$

67. $f(x, y) = 2x^3 + 3xy + 2y^3$

68. $f(x, y) = x^3 + y^3 - 3xy + 15$

69. $f(x, y) = x^3 + y^3 + 3x^2 - 3y^2$

70. $f(x, y) = x^4 - 8x^2 + 3y^2 - 6y$

Absolute Extrema

In Exercises 71–78, find the absolute maximum and minimum values of f on the region R.

71. $f(x, y) = x^2 + xy + y^2 - 3x + 3y$

R: The triangular region cut from the first quadrant by the line $x + y = 4$

72. $f(x, y) = x^2 - y^2 - 2x + 4y + 1$

R: The rectangular region in the first quadrant bounded by the coordinate axes and the lines $x = 4$ and $y = 2$

73. $f(x, y) = y^2 - xy - 3y + 2x$

R: The square region enclosed by the lines $x = \pm 2$ and $y = \pm 2$

74. $f(x, y) = 2x + 2y - x^2 - y^2$

R: The square region bounded by the coordinate axes and the lines $x = 2, y = 2$ in the first quadrant

75. $f(x, y) = x^2 - y^2 - 2x + 4y$

R: The triangular region bounded below by the x-axis, above by the line $y = x + 2$, and on the right by the line $x = 2$

76. $f(x, y) = 4xy - x^4 - y^4 + 16$

R: The triangular region bounded below by the line $y = -2$, above by the line $y = x$, and on the right by the line $x = 2$

77. $f(x, y) = x^3 + y^3 + 3x^2 - 3y^2$

R: The square region enclosed by the lines $x = \pm 1$ and $y = \pm 1$

78. $f(x, y) = x^3 + 3xy + y^3 + 1$

R: The square region enclosed by the lines $x = \pm 1$ and $y = \pm 1$

Lagrange Multipliers

79. Extrema on a circle Find the extreme values of $f(x, y) = x^3 + y^2$ on the circle $x^2 + y^2 = 1$.

80. Extrema on a circle Find the extreme values of $f(x, y) = xy$ on the circle $x^2 + y^2 = 1$.

81. Extrema in a disk Find the extreme values of $f(x, y) = x^2 + 3y^2 + 2y$ on the unit disk $x^2 + y^2 \le 1$.

82. Extrema in a disk Find the extreme values of $f(x, y) = x^2 + y^2 - 3x - xy$ on the disk $x^2 + y^2 \le 9$.

83. Extrema on a sphere Find the extreme values of $f(x, y, z) = x - y + z$ on the unit sphere $x^2 + y^2 + z^2 = 1$.

84. Minimum distance to origin Find the points on the surface $x^2 - zy = 4$ closest to the origin.

85. Minimizing cost of a box A closed rectangular box is to have volume V cm^3. The cost of the material used in the box is a cents/cm^2 for top and bottom, b cents/cm^2 for front and back, and c cents/cm^2 for the remaining sides. What dimensions minimize the total cost of materials?

86. Least volume Find the plane $x/a + y/b + z/c = 1$ that passes through the point $(2, 1, 2)$ and cuts off the least volume from the first octant.

87. Extrema on curve of intersecting surfaces Find the extreme values of $f(x, y, z) = x(y + z)$ on the curve of intersection of the right circular cylinder $x^2 + y^2 = 1$ and the hyperbolic cylinder $xz = 1$.

88. Minimum distance to origin on curve of intersecting plane and cone Find the point closest to the origin on the curve of intersection of the plane $x + y + z = 1$ and the cone $z^2 = 2x^2 + 2y^2$.

Partial Derivatives with Constrained Variables

In Exercises 89 and 90, begin by drawing a diagram that shows the relations among the variables.

89. If $w = x^2e^{yz}$ and $z = x^2 - y^2$ find

a. $\left(\frac{\partial w}{\partial y}\right)_z$ **b.** $\left(\frac{\partial w}{\partial z}\right)_x$ **c.** $\left(\frac{\partial w}{\partial z}\right)_y$.

90. Let $U = f(P, V, T)$ be the internal energy of a gas that obeys the ideal gas law $PV = nRT$ (n and R constant). Find

a. $\left(\frac{\partial U}{\partial T}\right)_P$ **b.** $\left(\frac{\partial U}{\partial V}\right)_T$.

Theory and Examples

91. Let $w = f(r, \theta)$, $r = \sqrt{x^2 + y^2}$, and $\theta = \tan^{-1}(y/x)$. Find $\partial w/\partial x$ and $\partial w/\partial y$ and express your answers in terms of r and θ.

92. Let $z = f(u, v)$, $u = ax + by$, and $v = ax - by$. Express z_x and z_y in terms of f_u, f_v, and the constants a and b.

93. If a and b are constants, $w = u^3 + \tanh u + \cos u$, and $u = ax + by$, show that

$$a\frac{\partial w}{\partial y} = b\frac{\partial w}{\partial x}.$$

94. Using the Chain Rule If $w = \ln(x^2 + y^2 + 2z)$, $x = r + s$, $y = r - s$, and $z = 2rs$, find w_r and w_s by the Chain Rule. Then check your answer another way.

95. Angle between vectors The equations $e^u \cos v - x = 0$ and $e^u \sin v - y = 0$ define u and v as differentiable functions of x and y. Show that the angle between the vectors

$$\frac{\partial u}{\partial x}\mathbf{i} + \frac{\partial u}{\partial y}\mathbf{j} \quad \text{and} \quad \frac{\partial v}{\partial x}\mathbf{i} + \frac{\partial v}{\partial y}\mathbf{j}$$

is constant.

96. Polar coordinates and second derivatives Introducing polar coordinates $x = r\cos\theta$ and $y = r\sin\theta$ changes $f(x, y)$ to $g(r, \theta)$. Find the value of $\partial^2 g/\partial\theta^2$ at the point $(r, \theta) = (2, \pi/2)$, given that

$$\frac{\partial f}{\partial x} = \frac{\partial f}{\partial y} = \frac{\partial^2 f}{\partial x^2} = \frac{\partial^2 f}{\partial y^2} = 1$$

at that point.

97. Normal line parallel to a plane Find the points on the surface

$$(y + z)^2 + (z - x)^2 = 16$$

where the normal line is parallel to the yz-plane.

98. Tangent plane parallel to xy-plane Find the points on the surface

$$xy + yz + zx - x - z^2 = 0$$

where the tangent plane is parallel to the xy-plane.

99. When gradient is parallel to position vector Suppose that $\nabla f(x, y, z)$ is always parallel to the position vector $x\mathbf{i} + y\mathbf{j} + z\mathbf{k}$. Show that $f(0, 0, a) = f(0, 0, -a)$ for any a.

100. One-sided directional derivative in all directions, but no gradient The one-sided directional derivative of f at $P(x_0, y_0, z_0)$ in the direction $\mathbf{u} = u_1\mathbf{i} + u_2\mathbf{j} + u_3\mathbf{k}$ is the number

$$\lim_{s\to 0^+} \frac{f(x_0 + su_1, y_0 + su_2, z_0 + su_3) - f(x_0, y_0, z_0)}{s}.$$

Show that the one-sided directional derivative of

$$f(x, y, z) = \sqrt{x^2 + y^2 + z^2}$$

at the origin equals 1 in any direction but that f has no gradient vector at the origin.

101. Normal line through origin Show that the line normal to the surface $xy + z = 2$ at the point $(1, 1, 1)$ passes through the origin.

102. Tangent plane and normal line

a. Sketch the surface $x^2 - y^2 + z^2 = 4$.

b. Find a vector normal to the surface at $(2, -3, 3)$. Add the vector to your sketch.

c. Find equations for the tangent plane and normal line at $(2, -3, 3)$.

Chapter 14 Additional and Advanced Exercises

Partial Derivatives

1. Function with saddle at the origin If you did Exercise 50 in Section 14.2, you know that the function

$$f(x, y) = \begin{cases} xy\dfrac{x^2 - y^2}{x^2 + y^2}, & (x, y) \neq (0, 0) \\ 0, & (x, y) = (0, 0) \end{cases}$$

(see the accompanying figure) is continuous at $(0, 0)$. Find $f_{xy}(0, 0)$ and $f_{yx}(0, 0)$.

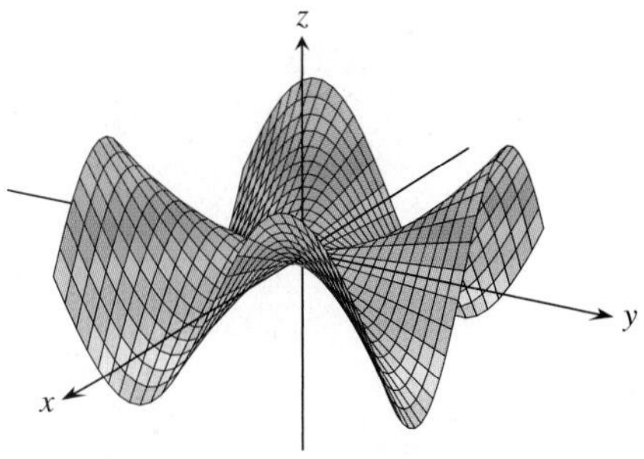

2. **Finding a function from second partials** Find a function $w = f(x, y)$ whose first partial derivatives are $\partial w/\partial x = 1 + e^x \cos y$ and $\partial w/\partial y = 2y - e^x \sin y$ and whose value at the point $(\ln 2, 0)$ is $\ln 2$.

3. **A proof of Leibniz's Rule** Leibniz's Rule says that if f is continuous on $[a, b]$ and if $u(x)$ and $v(x)$ are differentiable functions of x whose values lie in $[a, b]$, then

$$\frac{d}{dx}\int_{u(x)}^{v(x)} f(t)\,dt = f(v(x))\frac{dv}{dx} - f(u(x))\frac{du}{dx}.$$

Prove the rule by setting

$$g(u, v) = \int_u^v f(t)\,dt, \qquad u = u(x), \qquad v = v(x)$$

and calculating dg/dx with the Chain Rule.

4. **Finding a function with constrained second partials** Suppose that f is a twice-differentiable function of r, that $r = \sqrt{x^2 + y^2 + z^2}$, and that

$$f_{xx} + f_{yy} + f_{zz} = 0.$$

Show that for some constants a and b,

$$f(r) = \frac{a}{r} + b.$$

5. **Homogeneous functions** A function $f(x, y)$ is *homogeneous of degree n* (n a nonnegative integer) if $f(tx, ty) = t^n f(x, y)$ for all t, x, and y. For such a function (sufficiently differentiable), prove that

a. $x\dfrac{\partial f}{\partial x} + y\dfrac{\partial f}{\partial y} = nf(x, y)$

b. $x^2\left(\dfrac{\partial^2 f}{\partial x^2}\right) + 2xy\left(\dfrac{\partial^2 f}{\partial x \partial y}\right) + y^2\left(\dfrac{\partial^2 f}{\partial y^2}\right) = n(n-1)f.$

6. **Surface in polar coordinates** Let

$$f(r, \theta) = \begin{cases} \dfrac{\sin 6r}{6r}, & r \neq 0 \\ 1, & r = 0, \end{cases}$$

where r and θ are polar coordinates. Find

a. $\lim_{r \to 0} f(r, \theta)$ b. $f_r(0, 0)$ c. $f_\theta(r, \theta), \quad r \neq 0.$

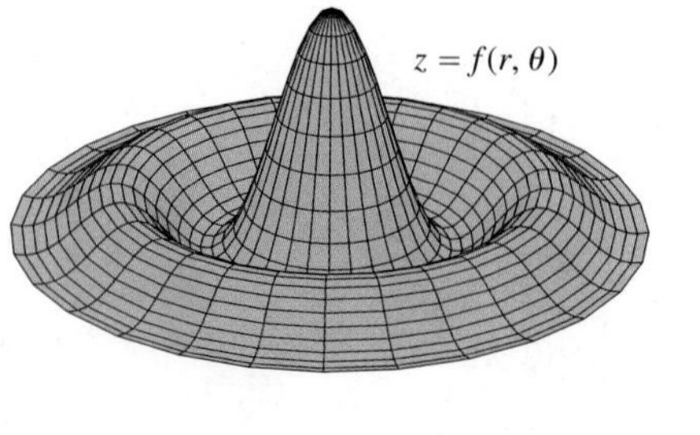

Gradients and Tangents

7. **Properties of position vectors** Let $\mathbf{r} = x\mathbf{i} + y\mathbf{j} + z\mathbf{k}$ and let $r = |\mathbf{r}|$.

a. Show that $\nabla r = \mathbf{r}/r$.

b. Show that $\nabla(r^n) = nr^{n-2}\mathbf{r}$.

c. Find a function whose gradient equals $\mathbf{r}$.

d. Show that $\mathbf{r} \cdot d\mathbf{r} = r\,dr$.

e. Show that $\nabla(\mathbf{A} \cdot \mathbf{r}) = \mathbf{A}$ for any constant vector $\mathbf{A}$.

8. **Gradient orthogonal to tangent** Suppose that a differentiable function $f(x, y)$ has the constant value c along the differentiable curve $x = g(t), y = h(t)$; that is

$$f(g(t), h(t)) = c$$

for all values of t. Differentiate both sides of this equation with respect to t to show that ∇f is orthogonal to the curve's tangent vector at every point on the curve.

9. **Curve tangent to a surface** Show that the curve

$$\mathbf{r}(t) = (\ln t)\mathbf{i} + (t \ln t)\mathbf{j} + t\mathbf{k}$$

is tangent to the surface

$$xz^2 - yz + \cos xy = 1$$

at $(0, 0, 1)$.

10. **Curve tangent to a surface** Show that the curve

$$\mathbf{r}(t) = \left(\frac{t^3}{4} - 2\right)\mathbf{i} + \left(\frac{4}{t} - 3\right)\mathbf{j} + \cos(t-2)\mathbf{k}$$

is tangent to the surface

$$x^3 + y^3 + z^3 - xyz = 0$$

at $(0, -1, 1)$.

Extreme Values

11. **Extrema on a surface** Show that the only possible maxima and minima of z on the surface $z = x^3 + y^3 - 9xy + 27$ occur at $(0, 0)$ and $(3, 3)$. Show that neither a maximum nor a minimum occurs at $(0, 0)$. Determine whether z has a maximum or a minimum at $(3, 3)$.

12. **Maximum in closed first quadrant** Find the maximum value of $f(x, y) = 6xye^{-(2x+3y)}$ in the closed first quadrant (includes the nonnegative axes).

13. **Minimum volume cut from first octant** Find the minimum volume for a region bounded by the planes $x = 0, y = 0, z = 0$ and a plane tangent to the ellipsoid

$$\frac{x^2}{a^2} + \frac{y^2}{b^2} + \frac{z^2}{c^2} = 1$$

at a point in the first octant.

14. Minimum distance from line to parabola in *xy*-plane By minimizing the function $f(x, y, u, v) = (x - u)^2 + (y - v)^2$ subject to the constraints $y = x + 1$ and $u = v^2$, find the minimum distance in the xy-plane from the line $y = x + 1$ to the parabola $y^2 = x$.

Theory and Examples

15. Boundedness of first partials implies continuity Prove the following theorem: If $f(x, y)$ is defined in an open region R of the xy-plane and if f_x and f_y are bounded on R, then $f(x, y)$ is continuous on R. (The assumption of boundedness is essential.)

16. Suppose that $\mathbf{r}(t) = g(t)\mathbf{i} + h(t)\mathbf{j} + k(t)\mathbf{k}$ is a smooth curve in the domain of a differentiable function $f(x, y, z)$. Describe the relation between df/dt, ∇f, and $\mathbf{v} = d\mathbf{r}/dt$. What can be said about ∇f and $\mathbf{v}$ at interior points of the curve where f has extreme values relative to its other values on the curve? Give reasons for your answer.

17. Finding functions from partial derivatives Suppose that f and g are functions of x and y such that

$$\frac{\partial f}{\partial y} = \frac{\partial g}{\partial x} \quad \text{and} \quad \frac{\partial f}{\partial x} = \frac{\partial g}{\partial y},$$

and suppose that

$$\frac{\partial f}{\partial x} = 0, \qquad f(1, 2) = g(1, 2) = 5 \quad \text{and} \quad f(0, 0) = 4.$$

Find $f(x, y)$ and $g(x, y)$.

18. Rate of change of the rate of change We know that if $f(x, y)$ is a function of two variables and if $\mathbf{u} = a\mathbf{i} + b\mathbf{j}$ is a unit vector, then $D_{\mathbf{u}}f(x, y) = f_x(x, y)a + f_y(x, y)b$ is the rate of change of $f(x, y)$ at (x, y) in the direction of $\mathbf{u}$. Give a similar formula for the rate of change *of the rate of change of* $f(x, y)$ at (x, y) in the direction $\mathbf{u}$.

19. Path of a heat-seeking particle A heat-seeking particle has the property that at any point (x, y) in the plane it moves in the direction of maximum temperature increase. If the temperature at (x, y) is $T(x, y) = -e^{-2y}\cos x$, find an equation $y = f(x)$ for the path of a heat-seeking particle at the point $(\pi/4, 0)$.

20. Velocity after a ricochet A particle traveling in a straight line with constant velocity $\mathbf{i} + \mathbf{j} - 5\mathbf{k}$ passes through the point $(0, 0, 30)$ and hits the surface $z = 2x^2 + 3y^2$. The particle ricochets off the surface, the angle of reflection being equal to the angle of incidence. Assuming no loss of speed, what is the velocity of the particle after the ricochet? Simplify your answer.

21. Directional derivatives tangent to a surface Let S be the surface that is the graph of $f(x, y) = 10 - x^2 - y^2$. Suppose that the temperature in space at each point (x, y, z) is $T(x, y, z) = x^2y + y^2z + 4x + 14y + z$.

a. Among all the possible directions tangential to the surface S at the point $(0, 0, 10)$, which direction will make the rate of change of temperature at $(0, 0, 10)$ a maximum?

b. Which direction tangential to S at the point $(1, 1, 8)$ will make the rate of change of temperature a maximum?

22. Drilling another borehole On a flat surface of land, geologists drilled a borehole straight down and hit a mineral deposit at 1000 ft. They drilled a second borehole 100 ft to the north of the first and hit the mineral deposit at 950 ft. A third borehole 100 ft east of the first borehole struck the mineral deposit at 1025 ft. The geologists have reasons to believe that the mineral deposit is in the shape of a dome, and for the sake of economy, they would like to find where the deposit is closest to the surface. Assuming the surface to be the xy-plane, in what direction from the first borehole would you suggest the geologists drill their fourth borehole?

The One-Dimensional Heat Equation

If $w(x, t)$ represents the temperature at position x at time t in a uniform conducting rod with perfectly insulated sides (see the accompanying figure), then the partial derivatives w_{xx} and w_t satisfy a differential equation of the form

$$w_{xx} = \frac{1}{c^2}w_t.$$

This equation is called the **one-dimensional heat equation**. The value of the positive constant c^2 is determined by the material from which the rod is made. It has been determined experimentally for a broad range of materials. For a given application, one finds the appropriate value in a table. For dry soil, for example, $c^2 = 0.19$ ft^2/day.

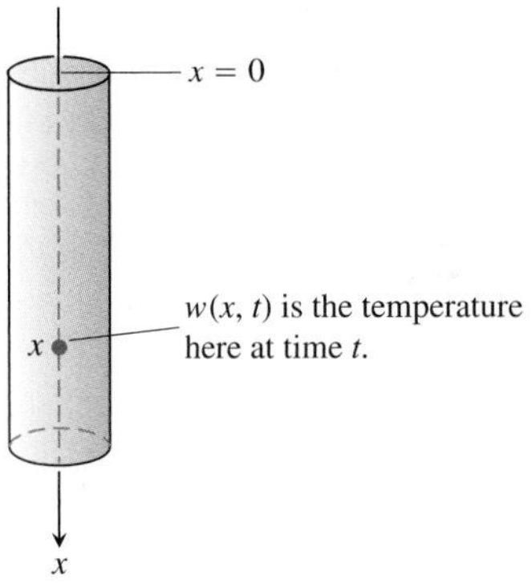

In chemistry and biochemistry, the heat equation is known as the **diffusion equation**. In this context, $w(x, t)$ represents the concentration of a dissolved substance, a salt for instance, diffusing along a tube filled with liquid. The value of $w(x, t)$ is the concentration at point x at time t. In other applications, $w(x, t)$ represents the diffusion of a gas down a long, thin pipe.

In electrical engineering, the heat equation appears in the forms

$$v_{xx} = RCv_t$$

and

$$i_{xx} = RCi_t.$$

These equations describe the voltage v and the flow of current i in a coaxial cable or in any other cable in which leakage and inductance are negligible. The functions and constants in these equations are

$v(x, t)$ = voltage at point x at time t

R = resistance per unit length

C = capacitance to ground per unit of cable length

$i(x, t)$ = current at point x at time t.

23. Find all solutions of the one-dimensional heat equation of the form $w = e^{rt} \sin \pi x$, where r is a constant.

24. Find all solutions of the one-dimensional heat equation that have the form $w = e^{rt} \sin kx$ and satisfy the conditions that $w(0, t) = 0$ and $w(L, t) = 0$. What happens to these solutions as $t \to \infty$?

Chapter 14 Technology Application Projects

Mathematica/Maple Module

Plotting Surfaces

Efficiently generate plots of surfaces, contours, and level curves.

Mathematica/Maple Module

Exploring the Mathematics Behind Skateboarding: Analysis of the Directional Derivative

The path of a skateboarder is introduced, first on a level plane, then on a ramp, and finally on a paraboloid. Compute, plot, and analyze the directional derivative in terms of the skateboarder.

Mathematica/Maple Module

Looking for Patterns and Applying the Method of Least Squares to Real Data

Fit a line to a set of numerical data points by choosing the line that minimizes the sum of the squares of the vertical distances from the points to the line.

Mathematica/Maple Module

Lagrange Goes Skateboarding: How High Does He Go?

Revisit and analyze the skateboarders' adventures for maximum and minimum heights from both a graphical and analytic perspective using Lagrange multipliers.

Chapter 15 Multiple Integrals

OVERVIEW In this chapter we consider the integral of a function of two variables $f(x, y)$ over a region in the plane and the integral of a function of three variables $f(x, y, z)$ over a region in space. These integrals are called *multiple integrals* and are defined as the limit of approximating Riemann sums, much like the single-variable integrals presented in Chapter 5. We can use multiple integrals to calculate quantities that vary over two or three dimensions, such as the total mass or the angular momentum of an object of varying density and the volumes of solids with general curved boundaries.

15.1 Double Integrals

In Chapter 5 we defined the definite integral of a continuous function $f(x)$ over an interval $[a, b]$ as a limit of Riemann sums. In this section we extend this idea to define the integral of a continuous function of two variables $f(x, y)$ over a bounded region R in the plane. In both cases the integrals are limits of approximating Riemann sums. The Riemann sums for the integral of a single-variable function $f(x)$ are obtained by partitioning a finite interval into thin subintervals, multiplying the width of each subinterval by the value of f at a point c_k inside that subinterval, and then adding together all the products. A similar method of partitioning, multiplying, and summing is used to construct double integrals. However, this time we pack a planar region R with small rectangles, rather than small subintervals. We then take the product of each small rectangle's area with the value of f at a point inside that rectangle, and finally sum together all these products. When f is continuous, these sums converge to a single number as each of the small rectangles shrinks in both width and height. The limit is the *double integral* of f over R. As with single integrals, we can evaluate multiple integrals via antiderivatives, which frees us from the formidable task of calculating a double integral directly from its definition as a limit of Riemann sums. The major practical problem that arises in evaluating multiple integrals lies in determining the limits of integration. While the integrals of Chapter 5 were evaluated over an interval, which is determined by its two endpoints, multiple integrals are evaluated over a region in the plane or in space. This gives rise to limits of integration which often involve variables, not just constants. Describing the regions of integration is the main new issue that arises in the calculation of multiple integrals.

Double Integrals over Rectangles

We begin our investigation of double integrals by considering the simplest type of planar region, a rectangle. We consider a function $f(x, y)$ defined on a rectangular region R,

$$R: \quad a \leq x \leq b, \quad c \leq y \leq d.$$

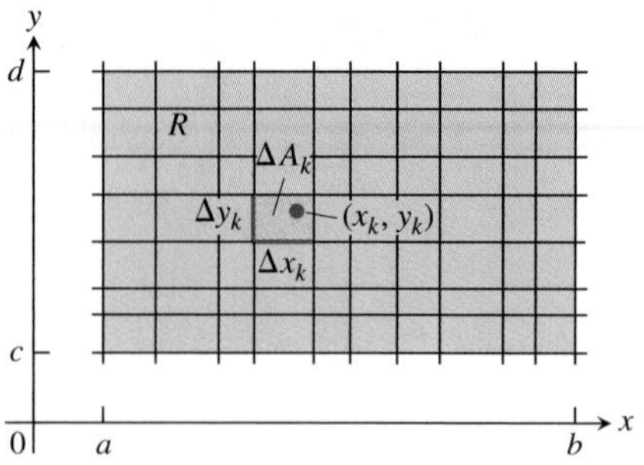

FIGURE 15.1 Rectangular grid partitioning the region R into small rectangles of area $\Delta A_k = \Delta x_k \, \Delta y_k$.

We subdivide R into small rectangles using a network of lines parallel to the x- and y-axes (Figure 15.1). The lines divide R into n rectangular pieces, where the number of such pieces n gets large as the width and height of each piece gets small. These rectangles form a **partition** of R. A small rectangular piece of width Δx and height Δy has area $\Delta A = \Delta x \Delta y$. If we number the small pieces partitioning R in some order, then their areas are given by numbers $\Delta A_1, \Delta A_2, \ldots, \Delta A_n$, where ΔA_k is the area of the kth small rectangle.

To form a Riemann sum over R, we choose a point (x_k, y_k) in the kth small rectangle, multiply the value of f at that point by the area ΔA_k, and add together the products:

$$S_n = \sum_{k=1}^{n} f(x_k, y_k)\, \Delta A_k.$$

Depending on how we pick (x_k, y_k) in the kth small rectangle, we may get different values for S_n.

We are interested in what happens to these Riemann sums as the widths and heights of all the small rectangles in the partition of R approach zero. The **norm** of a partition P, written $\|P\|$, is the largest width or height of any rectangle in the partition. If $\|P\| = 0.1$ then all the rectangles in the partition of R have width at most 0.1 and height at most 0.1. Sometimes the Riemann sums converge as the norm of P goes to zero, written $\|P\| \rightarrow 0$. The resulting limit is then written as

$$\lim_{\|P\|\rightarrow 0} \sum_{k=1}^{n} f(x_k, y_k)\, \Delta A_k.$$

As $\|P\| \rightarrow 0$ and the rectangles get narrow and short, their number n increases, so we can also write this limit as

$$\lim_{n\rightarrow\infty} \sum_{k=1}^{n} f(x_k, y_k)\, \Delta A_k.$$

with the understanding that $\Delta A_k \rightarrow 0$ as $n \rightarrow \infty$ and $\|P\| \rightarrow 0$.

There are many choices involved in a limit of this kind. The collection of small rectangles is determined by the grid of vertical and horizontal lines that determine a rectangular partition of R. In each of the resulting small rectangles there is a choice of an arbitrary point (x_k, y_k) at which f is evaluated. These choices together determine a single Riemann sum. To form a limit, we repeat the whole process again and again, choosing partitions whose rectangle widths and heights both go to zero and whose number goes to infinity.

When a limit of the sums S_n exists, giving the same limiting value no matter what choices are made, then the function f is said to be **integrable** and the limit is called the **double integral** of f over R, written as

$$\iint_R f(x, y)\, dA \qquad \text{or} \qquad \iint_R f(x, y)\, dx\, dy.$$

It can be shown that if $f(x, y)$ is a continuous function throughout R, then f is integrable, as in the single-variable case discussed in Chapter 5. Many discontinuous functions are also integrable, including functions which are discontinuous only on a finite number of points or smooth curves. We leave the proof of these facts to a more advanced text.

Double Integrals as Volumes

When $f(x, y)$ is a positive function over a rectangular region R in the xy-plane, we may interpret the double integral of f over R as the volume of the 3-dimensional solid region over the xy-plane bounded below by R and above by the surface $z = f(x, y)$ (Figure 15.2). Each term $f(x_k, y_k)\Delta A_k$ in the sum $S_n = \sum f(x_k, y_k)\Delta A_k$ is the volume of a vertical

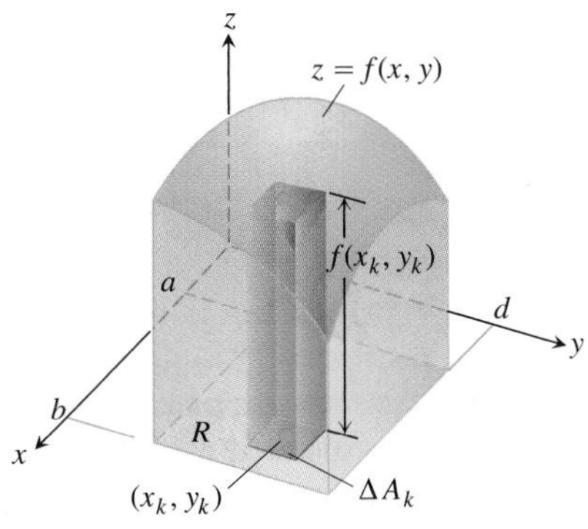

FIGURE 15.2 Approximating solids with rectangular boxes leads us to define the volumes of more general solids as double integrals. The volume of the solid shown here is the double integral of $f(x, y)$ over the base region R.

rectangular box that approximates the volume of the portion of the solid that stands directly above the base ΔA_k. The sum S_n thus approximates what we want to call the total volume of the solid. We *define* this volume to be

$$\text{Volume} = \lim_{n\to\infty} S_n = \iint_R f(x, y)\, dA,$$

where $\Delta A_k \to 0$ as $n \to \infty$.

As you might expect, this more general method of calculating volume agrees with the methods in Chapter 6, but we do not prove this here. Figure 15.3 shows Riemann sum approximations to the volume becoming more accurate as the number n of boxes increases.

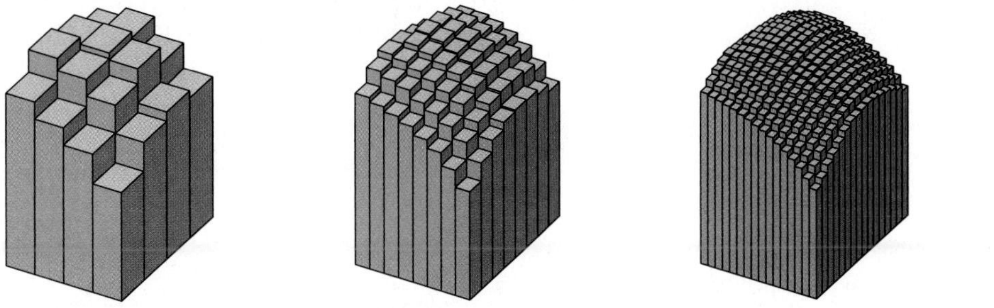

(a) $n = 16$ (b) $n = 64$ (c) $n = 256$

FIGURE 15.3 As n increases, the Riemann sum approximations approach the total volume of the solid shown in Figure 15.2.

Fubini's Theorem for Calculating Double Integrals

Suppose that we wish to calculate the volume under the plane $z = 4 - x - y$ over the rectangular region R: $0 \le x \le 2$, $0 \le y \le 1$ in the xy-plane. If we apply the method of slicing from Section 6.1, with slices perpendicular to the x-axis (Figure 15.4), then the volume is

$$\int_{x=0}^{x=2} A(x)\, dx, \tag{1}$$

where $A(x)$ is the cross-sectional area at x. For each value of x, we may calculate $A(x)$ as the integral

$$A(x) = \int_{y=0}^{y=1} (4 - x - y)\, dy, \tag{2}$$

which is the area under the curve $z = 4 - x - y$ in the plane of the cross-section at x. In calculating $A(x)$, x is held fixed and the integration takes place with respect to y. Combining Equations (1) and (2), we see that the volume of the entire solid is

$$\begin{aligned}
\text{Volume} &= \int_{x=0}^{x=2} A(x)\, dx = \int_{x=0}^{x=2} \left(\int_{y=0}^{y=1} (4 - x - y)\, dy \right) dx \\
&= \int_{x=0}^{x=2} \left[4y - xy - \frac{y^2}{2} \right]_{y=0}^{y=1} dx = \int_{x=0}^{x=2} \left(\frac{7}{2} - x \right) dx \\
&= \left[\frac{7}{2}x - \frac{x^2}{2} \right]_0^2 = 5.
\end{aligned} \tag{3}$$

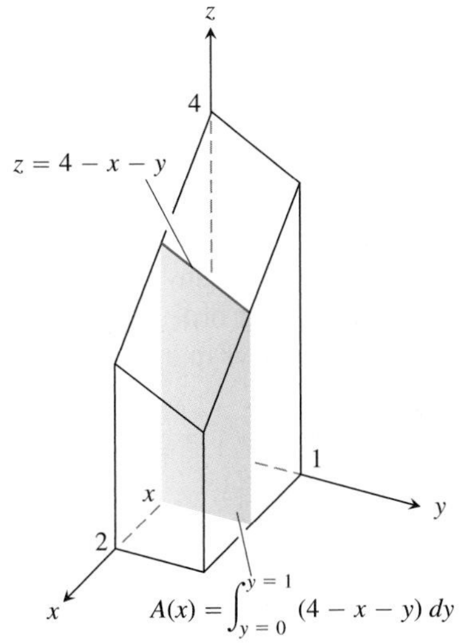

FIGURE 15.4 To obtain the cross-sectional area $A(x)$, we hold x fixed and integrate with respect to y.

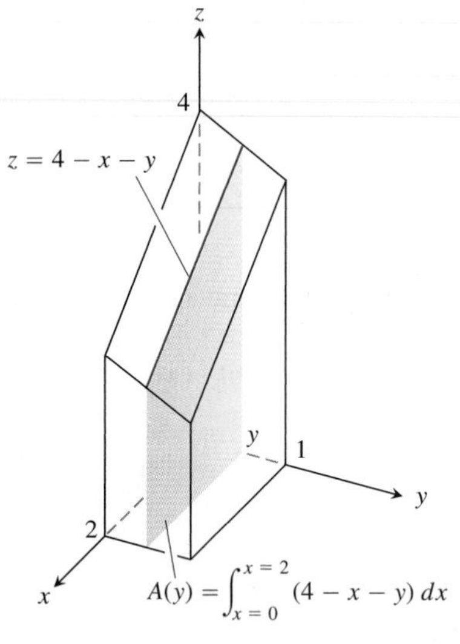

FIGURE 15.5 To obtain the cross-sectional area $A(y)$, we hold y fixed and integrate with respect to x.

If we just wanted to write a formula for the volume, without carrying out any of the integrations, we could write

$$\text{Volume} = \int_0^2 \int_0^1 (4 - x - y)\, dy\, dx.$$

The expression on the right, called an **iterated** or **repeated integral**, says that the volume is obtained by integrating $4 - x - y$ with respect to y from $y = 0$ to $y = 1$, holding x fixed, and then integrating the resulting expression in x with respect to x from $x = 0$ to $x = 2$. The limits of integration 0 and 1 are associated with y, so they are placed on the integral closest to dy. The other limits of integration, 0 and 2, are associated with the variable x, so they are placed on the outside integral symbol that is paired with dx.

What would have happened if we had calculated the volume by slicing with planes perpendicular to the y-axis (Figure 15.5)? As a function of y, the typical cross-sectional area is

$$A(y) = \int_{x=0}^{x=2} (4 - x - y)\, dx = \left[4x - \frac{x^2}{2} - xy\right]_{x=0}^{x=2} = 6 - 2y. \tag{4}$$

The volume of the entire solid is therefore

$$\text{Volume} = \int_{y=0}^{y=1} A(y)\, dy = \int_{y=0}^{y=1} (6 - 2y)\, dy = \left[6y - y^2\right]_0^1 = 5,$$

in agreement with our earlier calculation.

Again, we may give a formula for the volume as an iterated integral by writing

$$\text{Volume} = \int_0^1 \int_0^2 (4 - x - y)\, dx\, dy.$$

The expression on the right says we can find the volume by integrating $4 - x - y$ with respect to x from $x = 0$ to $x = 2$ as in Equation (4) and integrating the result with respect to y from $y = 0$ to $y = 1$. In this iterated integral, the order of integration is first x and then y, the reverse of the order in Equation (3).

What do these two volume calculations with iterated integrals have to do with the double integral

$$\iint_R (4 - x - y)\, dA$$

over the rectangle R: $0 \le x \le 2$, $0 \le y \le 1$? The answer is that both iterated integrals give the value of the double integral. This is what we would reasonably expect, since the double integral measures the volume of the same region as the two iterated integrals. A theorem published in 1907 by Guido Fubini says that the double integral of any continuous function over a rectangle can be calculated as an iterated integral in either order of integration. (Fubini proved his theorem in greater generality, but this is what it says in our setting.)

HISTORICAL BIOGRAPHY

Guido Fubini
(1879–1943)

THEOREM 1 Fubini's Theorem (First Form)
If $f(x, y)$ is continuous throughout the rectangular region R: $a \le x \le b$, $c \le y \le d$, then

$$\iint_R f(x, y)\, dA = \int_c^d \int_a^b f(x, y)\, dx\, dy = \int_a^b \int_c^d f(x, y)\, dy\, dx.$$

Fubini's Theorem says that double integrals over rectangles can be calculated as iterated integrals. Thus, we can evaluate a double integral by integrating with respect to one variable at a time.

Fubini's Theorem also says that we may calculate the double integral by integrating in *either* order, a genuine convenience, as we see in Example 3. When we calculate a volume by slicing, we may use either planes perpendicular to the x-axis or planes perpendicular to the y-axis.

EXAMPLE 1 Evaluating a Double Integral

Calculate $\iint_R f(x, y)\, dA$ for

$$f(x, y) = 1 - 6x^2y \qquad \text{and} \qquad R:\ 0 \le x \le 2, \quad -1 \le y \le 1.$$

Solution By Fubini's Theorem,

$$\iint_R f(x, y)\, dA = \int_{-1}^{1}\int_0^2 (1 - 6x^2y)\, dx\, dy = \int_{-1}^{1} \left[x - 2x^3y\right]_{x=0}^{x=2} dy$$

$$= \int_{-1}^{1} (2 - 16y)\, dy = \left[2y - 8y^2\right]_{-1}^{1} = 4.$$

Reversing the order of integration gives the same answer:

$$\int_0^2\int_{-1}^{1} (1 - 6x^2y)\, dy\, dx = \int_0^2 \left[y - 3x^2y^2\right]_{y=-1}^{y=1} dx$$

$$= \int_0^2 [(1 - 3x^2) - (-1 - 3x^2)]\, dx$$

$$= \int_0^2 2\, dx = 4.$$

■

USING TECHNOLOGY Multiple Integration

Most CAS can calculate both multiple and iterated integrals. The typical procedure is to apply the CAS integrate command in nested iterations according to the order of integration you specify.

Integral	Typical CAS Formulation
$\iint x^2y\, dx\, dy$	int (int (x ^ 2 * y, x), y) ;
$\int_{-\pi/3}^{\pi/4}\int_0^1 x\cos y\, dx\, dy$	int (int (x * cos (y), x = 0 . . 1), y = −Pi/3 . . Pi/4);

If a CAS cannot produce an exact value for a definite integral, it can usually find an approximate value numerically. Setting up a multiple integral for a CAS to solve can be a highly nontrivial task, and requires an understanding of how to describe the boundaries of the region and set up an appropriate integral.

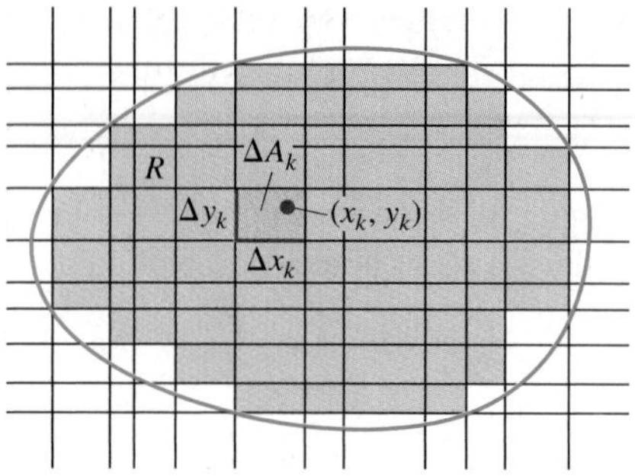

FIGURE 15.6 A rectangular grid partitioning a bounded nonrectangular region into rectangular cells.

Double Integrals over Bounded Nonrectangular Regions

To define the double integral of a function $f(x, y)$ over a bounded, nonrectangular region R, such as the one in Figure 15.6, we again begin by covering R with a grid of small rectangular cells whose union contains all points of R. This time, however, we cannot exactly fill R with a finite number of rectangles lying inside R, since its boundary is curved, and some of the small rectangles in the grid lie partly outside R. A partition of R is formed by taking the rectangles that lie completely inside it, not using any that are either partly or completely outside. For commonly arising regions, more and more of R is included as the norm of a partition (the largest width or height of any rectangle used) approaches zero.

Once we have a partition of R, we number the rectangles in some order from 1 to n and let ΔA_k be the area of the kth rectangle. We then choose a point (x_k, y_k) in the kth rectangle and form the Riemann sum

$$S_n = \sum_{k=1}^{n} f(x_k, y_k)\, \Delta A_k.$$

As the norm of the partition forming S_n goes to zero, $\|P\| \to 0$, the width and height of each enclosed rectangle goes to zero and their number goes to infinity. If $f(x, y)$ is a continuous function, then these Riemann sums converge to a limiting value, not dependent on any of the choices we made. This limit is called the **double integral** of $f(x, y)$ over R:

$$\lim_{\|P\|\to 0} \sum_{k=1}^{n} f(x_k, y_k)\, \Delta A_k = \iint_R f(x, y)\, dA.$$

The nature of the boundary of R introduces issues not found in integrals over an interval. When R has a curved boundary, the n rectangles of a partition lie inside R but do not cover all of R. In order for a partition to approximate R well, the parts of R covered by small rectangles lying partly outside R must become negligible as the norm of the partition approaches zero. This property of being nearly filled in by a partition of small norm is satisfied by all the regions that we will encounter. There is no problem with boundaries made from polygons, circles, ellipses, and from continuous graphs over an interval, joined end to end. A curve with a "fractal" type of shape would be problematic, but such curves are not relevant for most applications. A careful discussion of which type of regions R can be used for computing double integrals is left to a more advanced text.

Double integrals of continuous functions over nonrectangular regions have the same algebraic properties (summarized further on) as integrals over rectangular regions. The domain Additivity Property says that if R is decomposed into nonoverlapping regions R_1 and R_2 with boundaries that are again made of a finite number of line segments or smooth curves (see Figure 15.7 for an example), then

$$\iint_R f(x, y)\, dA = \iint_{R_1} f(x, y)\, dA + \iint_{R_2} f(x, y)\, dA.$$

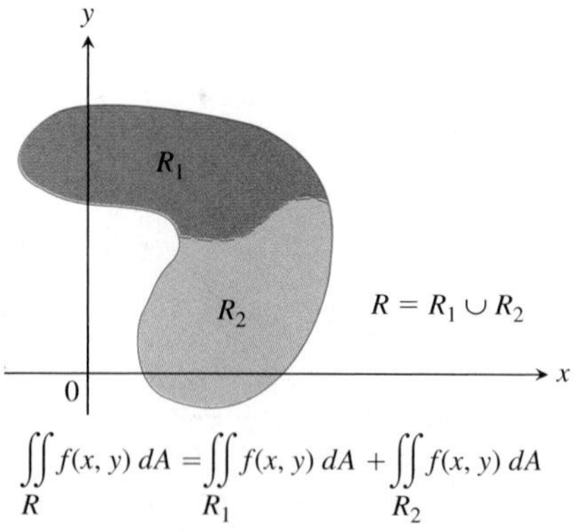

FIGURE 15.7 The Additivity Property for rectangular regions holds for regions bounded by continuous curves.

If $f(x, y)$ is positive and continuous over R we define the volume of the solid region between R and the surface $z = f(x, y)$ to be $\iint_R f(x, y)\, dA$, as before (Figure 15.8).

If R is a region like the one shown in the xy-plane in Figure 15.9, bounded "above" and "below" by the curves $y = g_2(x)$ and $y = g_1(x)$ and on the sides by the lines $x = a$, $x = b$, we may again calculate the volume by the method of slicing. We first calculate the cross-sectional area

$$A(x) = \int_{y=g_1(x)}^{y=g_2(x)} f(x, y)\, dy$$

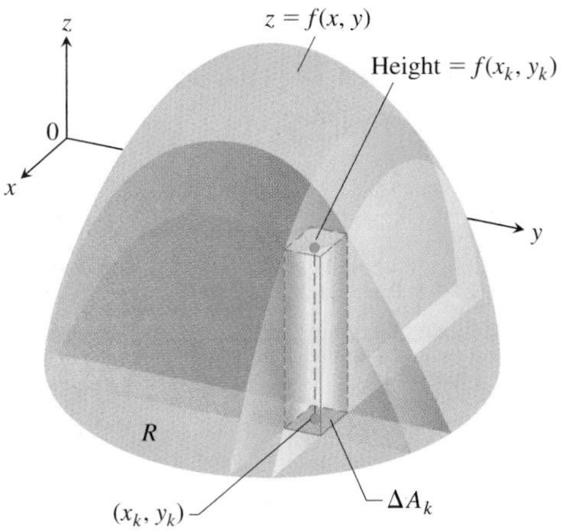

$$\text{Volume} = \lim \sum f(x_k, y_k)\,\Delta A_k = \iint_R f(x, y)\,dA$$

FIGURE 15.8 We define the volumes of solids with curved bases the same way we define the volumes of solids with rectangular bases.

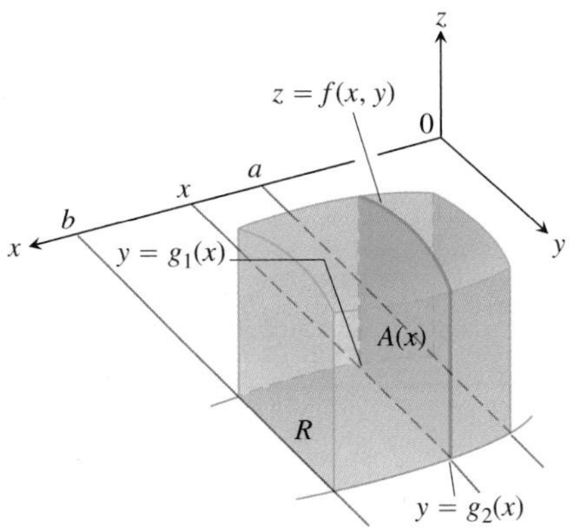

FIGURE 15.9 The area of the vertical slice shown here is

$$A(x) = \int_{g_1(x)}^{g_2(x)} f(x, y)\,dy.$$

To calculate the volume of the solid, we integrate this area from $x = a$ to $x = b$.

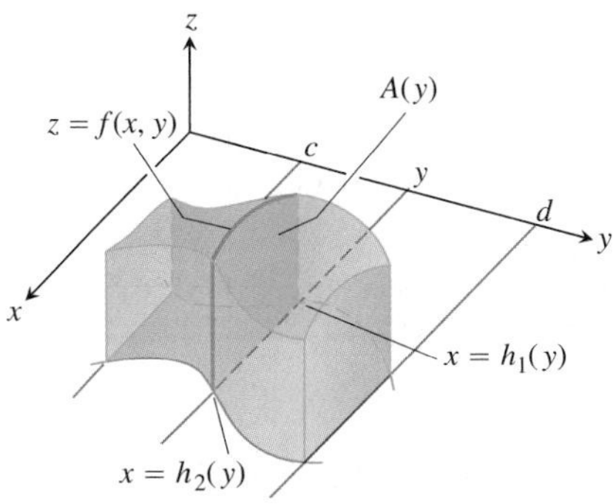

FIGURE 15.10 The volume of the solid shown here is

$$\int_c^d A(y)\,dy = \int_c^d \int_{h_1(y)}^{h_2(y)} f(x, y)\,dx\,dy.$$

and then integrate $A(x)$ from $x = a$ to $x = b$ to get the volume as an iterated integral:

$$V = \int_a^b A(x)\,dx = \int_a^b \int_{g_1(x)}^{g_2(x)} f(x, y)\,dy\,dx. \tag{5}$$

Similarly, if R is a region like the one shown in Figure 15.10, bounded by the curves $x = h_2(y)$ and $x = h_1(y)$ and the lines $y = c$ and $y = d$, then the volume calculated by slicing is given by the iterated integral

$$\text{Volume} = \int_c^d \int_{h_1(y)}^{h_2(y)} f(x, y)\,dx\,dy. \tag{6}$$

That the iterated integrals in Equations (5) and (6) both give the volume that we defined to be the double integral of f over R is a consequence of the following stronger form of Fubini's Theorem.

THEOREM 2 Fubini's Theorem (Stronger Form)

Let $f(x, y)$ be continuous on a region R.

1. If R is defined by $a \le x \le b$, $g_1(x) \le y \le g_2(x)$, with g_1 and g_2 continuous on $[a, b]$, then

$$\iint_R f(x, y)\,dA = \int_a^b \int_{g_1(x)}^{g_2(x)} f(x, y)\,dy\,dx.$$

2. If R is defined by $c \le y \le d$, $h_1(y) \le x \le h_2(y)$, with h_1 and h_2 continuous on $[c, d]$, then

$$\iint_R f(x, y)\,dA = \int_c^d \int_{h_1(y)}^{h_2(y)} f(x, y)\,dx\,dy.$$

EXAMPLE 2 Finding Volume

Find the volume of the prism whose base is the triangle in the xy-plane bounded by the x-axis and the lines $y = x$ and $x = 1$ and whose top lies in the plane

$$z = f(x, y) = 3 - x - y.$$

Solution See Figure 15.11 on page 1075. For any x between 0 and 1, y may vary from $y = 0$ to $y = x$ (Figure 15.11b). Hence,

$$V = \int_0^1 \int_0^x (3 - x - y)\, dy\, dx = \int_0^1 \left[3y - xy - \frac{y^2}{2}\right]_{y=0}^{y=x} dx$$

$$= \int_0^1 \left(3x - \frac{3x^2}{2}\right) dx = \left[\frac{3x^2}{2} - \frac{x^3}{2}\right]_{x=0}^{x=1} = 1.$$

When the order of integration is reversed (Figure 15.11c), the integral for the volume is

$$V = \int_0^1 \int_y^1 (3 - x - y)\, dx\, dy = \int_0^1 \left[3x - \frac{x^2}{2} - xy\right]_{x=y}^{x=1} dy$$

$$= \int_0^1 \left(3 - \frac{1}{2} - y - 3y + \frac{y^2}{2} + y^2\right) dy$$

$$= \int_0^1 \left(\frac{5}{2} - 4y + \frac{3}{2}y^2\right) dy = \left[\frac{5}{2}y - 2y^2 + \frac{y^3}{2}\right]_{y=0}^{y=1} = 1.$$

The two integrals are equal, as they should be. ■

Although Fubini's Theorem assures us that a double integral may be calculated as an iterated integral in either order of integration, the value of one integral may be easier to find than the value of the other. The next example shows how this can happen.

EXAMPLE 3 Evaluating a Double Integral

Calculate

$$\iint_R \frac{\sin x}{x}\, dA,$$

where R is the triangle in the xy-plane bounded by the x-axis, the line $y = x$, and the line $x = 1$.

Solution The region of integration is shown in Figure 15.12. If we integrate first with respect to y and then with respect to x, we find

$$\int_0^1 \left(\int_0^x \frac{\sin x}{x}\, dy\right) dx = \int_0^1 \left(y \frac{\sin x}{x}\right]_{y=0}^{y=x}\right) dx = \int_0^1 \sin x\, dx$$

$$= -\cos(1) + 1 \approx 0.46.$$

If we reverse the order of integration and attempt to calculate

$$\int_0^1 \int_y^1 \frac{\sin x}{x}\, dx\, dy,$$

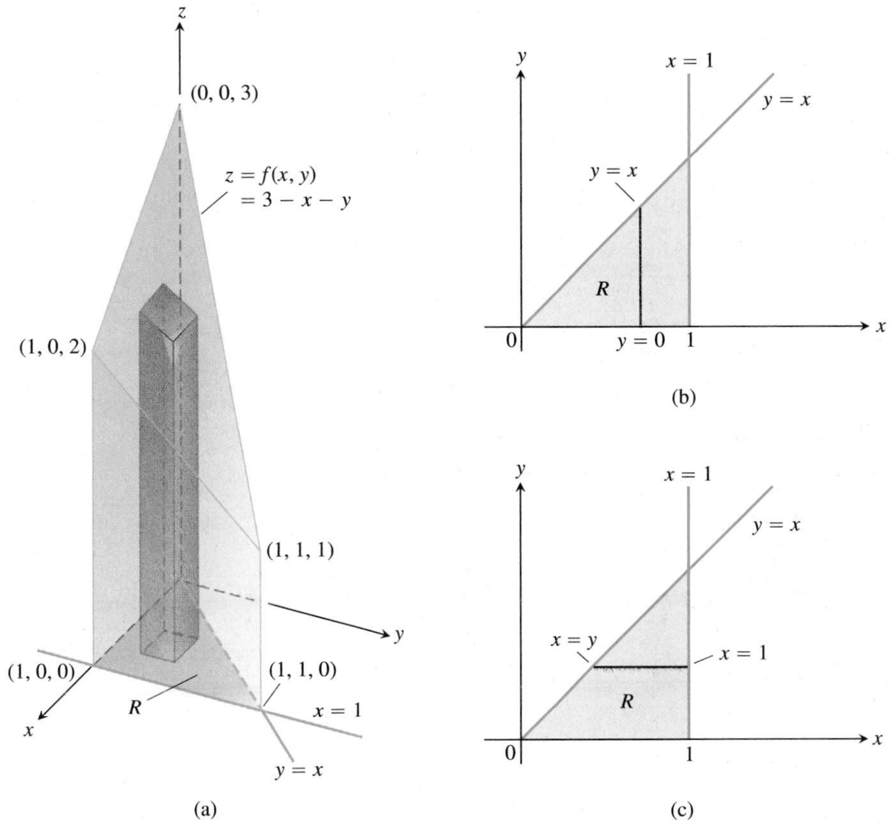

FIGURE 15.11 (a) Prism with a triangular base in the xy-plane. The volume of this prism is defined as a double integral over R. To evaluate it as an iterated integral, we may integrate first with respect to y and then with respect to x, or the other way around (Example 2).
(b) Integration limits of

$$\int_{x=0}^{x=1} \int_{y=0}^{y=x} f(x, y)\, dy\, dx.$$

If we integrate first with respect to y, we integrate along a vertical line through R and then integrate from left to right to include all the vertical lines in R.
(c) Integration limits of

$$\int_{y=0}^{y=1} \int_{x=y}^{x=1} f(x, y)\, dx\, dy.$$

If we integrate first with respect to x, we integrate along a horizontal line through R and then integrate from bottom to top to include all the horizontal lines in R.

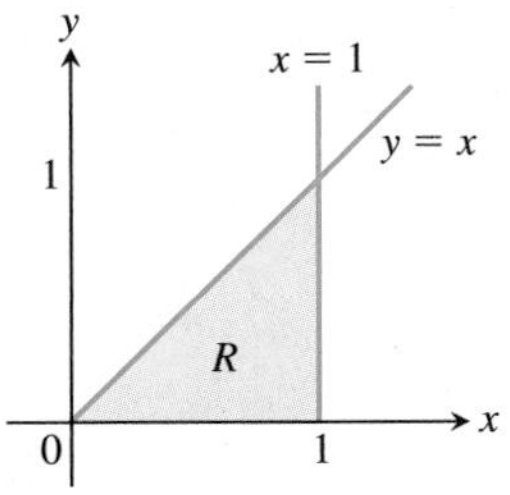

FIGURE 15.12 The region of integration in Example 3.

we run into a problem, because $\int ((\sin x)/x)\, dx$ cannot be expressed in terms of elementary functions (there is no simple antiderivative).

There is no general rule for predicting which order of integration will be the good one in circumstances like these. If the order you first choose doesn't work, try the other. Sometimes neither order will work, and then we need to use numerical approximations. ■

Finding Limits of Integration

We now give a procedure for finding limits of integration that applies for many regions in the plane. Regions that are more complicated, and for which this procedure fails, can often be split up into pieces on which the procedure works.

When faced with evaluating $\iint_R f(x, y)\,dA$, integrating first with respect to y and then with respect to x, do the following:

1. *Sketch.* Sketch the region of integration and label the bounding curves.

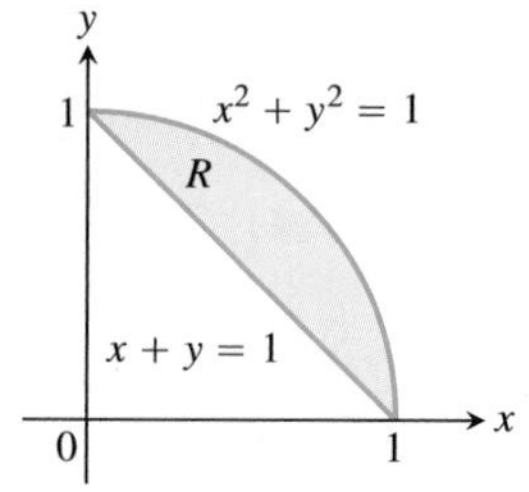

2. *Find the y-limits of integration.* Imagine a vertical line L cutting through R in the direction of increasing y. Mark the y-values where L enters and leaves. These are the y-limits of integration and are usually functions of x (instead of constants).

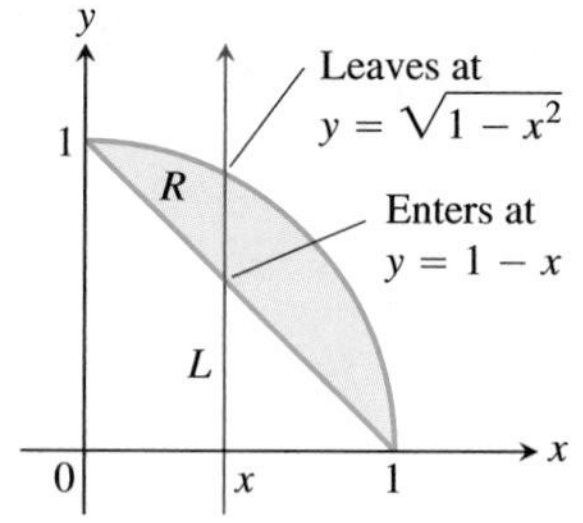

3. *Find the x-limits of integration.* Choose x-limits that include all the vertical lines through R. The integral shown here is

$$\iint_R f(x, y)\,dA =$$

$$\int_{x=0}^{x=1} \int_{y=1-x}^{y=\sqrt{1-x^2}} f(x, y)\,dy\,dx.$$

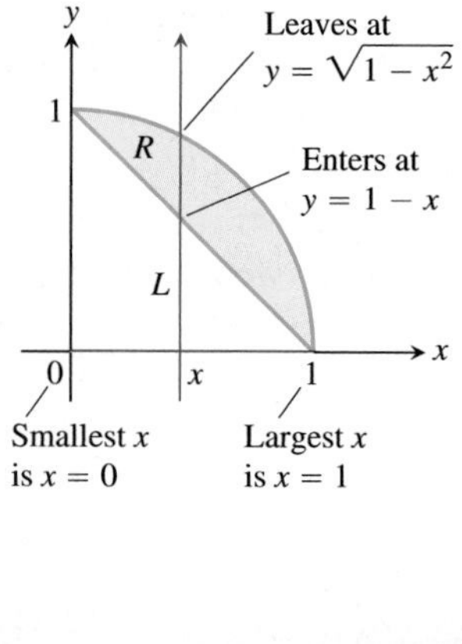

To evaluate the same double integral as an iterated integral with the order of integration reversed, use horizontal lines instead of vertical lines in Steps 2 and 3. The integral is

$$\iint_R f(x, y)\, dA = \int_0^1 \int_{1-y}^{\sqrt{1-y^2}} f(x, y)\, dx\, dy.$$

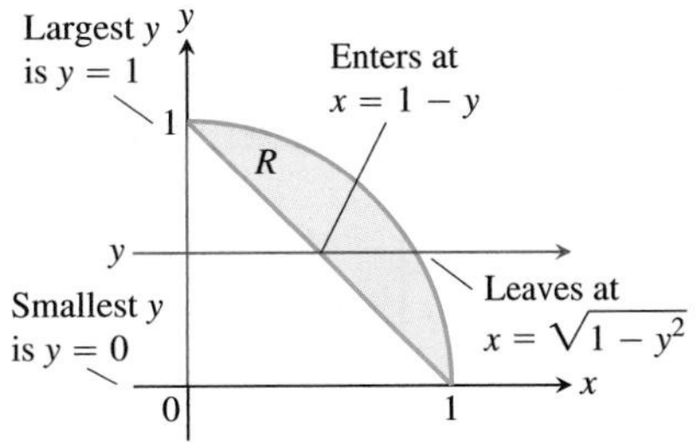

EXAMPLE 4 Reversing the Order of Integration

Sketch the region of integration for the integral

$$\int_0^2 \int_{x^2}^{2x} (4x + 2)\, dy\, dx$$

and write an equivalent integral with the order of integration reversed.

Solution The region of integration is given by the inequalities $x^2 \le y \le 2x$ and $0 \le x \le 2$. It is therefore the region bounded by the curves $y = x^2$ and $y = 2x$ between $x = 0$ and $x = 2$ (Figure 15.13a).

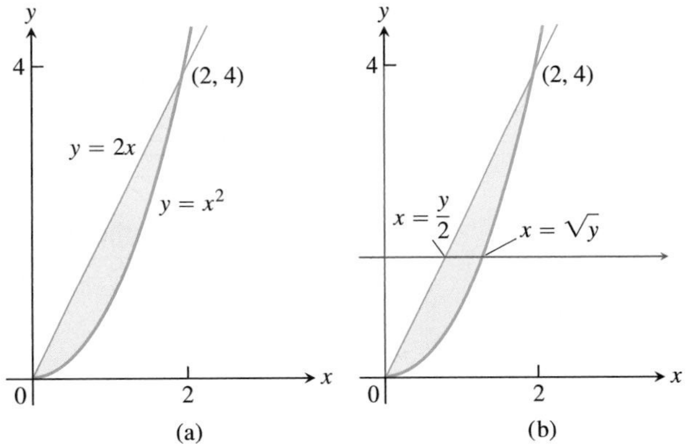

FIGURE 15.13 Region of integration for Example 4.

To find limits for integrating in the reverse order, we imagine a horizontal line passing from left to right through the region. It enters at $x = y/2$ and leaves at $x = \sqrt{y}$. To include all such lines, we let y run from $y = 0$ to $y = 4$ (Figure 15.13b). The integral is

$$\int_0^4 \int_{y/2}^{\sqrt{y}} (4x + 2)\, dx\, dy.$$

The common value of these integrals is 8. ■

Properties of Double Integrals

Like single integrals, double integrals of continuous functions have algebraic properties that are useful in computations and applications.

Properties of Double Integrals

If $f(x, y)$ and $g(x, y)$ are continuous, then

1. *Constant Multiple:* $\displaystyle\iint_R cf(x, y)\, dA = c\iint_R f(x, y)\, dA \quad \text{(any number } c\text{)}$

2. *Sum and Difference:*

$$\iint_R (f(x, y) \pm g(x, y))\, dA = \iint_R f(x, y)\, dA \pm \iint_R g(x, y)\, dA$$

3. *Domination:*

(a) $\displaystyle\iint_R f(x, y)\, dA \geq 0 \quad \text{if} \quad f(x, y) \geq 0 \text{ on } R$

(b) $\displaystyle\iint_R f(x, y)\, dA \geq \iint_R g(x, y)\, dA \quad \text{if} \quad f(x, y) \geq g(x, y) \text{ on } R$

4. *Additivity:* $\displaystyle\iint_R f(x, y)\, dA = \iint_{R_1} f(x, y)\, dA + \iint_{R_2} f(x, y)\, dA$

if R is the union of two nonoverlapping regions R_1 and R_2 (Figure 15.7).

The idea behind these properties is that integrals behave like sums. If the function $f(x, y)$ is replaced by its constant multiple $cf(x, y)$, then a Riemann sum for f

$$S_n = \sum_{k=1}^{n} f(x_k, y_k)\, \Delta A_k$$

is replaced by a Riemann sum for cf

$$\sum_{k=1}^{n} cf(x_k, y_k)\, \Delta A_k = c\sum_{k=1}^{n} f(x_k, y_k)\, \Delta A_k = cS_n .$$

Taking limits as $n \to \infty$ shows that $c \lim_{n\to\infty} S_n = c\iint_R f\, dA$ and $\lim_{n\to\infty} cS_n = \iint_R cf\, dA$ are equal. It follows that the constant multiple property carries over from sums to double integrals.

The other properties are also easy to verify for Riemann sums, and carry over to double integrals for the same reason. While this discussion gives the idea, an actual proof that these properties hold requires a more careful analysis of how Riemann sums converge.

EXERCISES 15.1

Finding Regions of Integration and Double Integrals

In Exercises 1–10, sketch the region of integration and evaluate the integral.

1. $\int_0^3 \int_0^2 (4 - y^2)\, dy\, dx$

2. $\int_0^3 \int_{-2}^0 (x^2 y - 2xy)\, dy\, dx$

3. $\int_{-1}^0 \int_{-1}^1 (x + y + 1)\, dx\, dy$

4. $\int_{\pi}^{2\pi} \int_0^{\pi} (\sin x + \cos y)\, dx\, dy$

5. $\int_0^{\pi} \int_0^{x} x \sin y\, dy\, dx$

6. $\int_0^{\pi} \int_0^{\sin x} y\, dy\, dx$

7. $\int_1^{\ln 8} \int_0^{\ln y} e^{x+y}\, dx\, dy$

8. $\int_1^2 \int_y^{y^2} dx\, dy$

9. $\int_0^1 \int_0^{y^2} 3y^3 e^{xy}\, dx\, dy$

10. $\int_1^4 \int_0^{\sqrt{x}} \frac{3}{2} e^{y/\sqrt{x}}\, dy\, dx$

In Exercises 11–16, integrate f over the given region.

11. **Quadrilateral** $f(x, y) = x/y$ over the region in the first quadrant bounded by the lines $y = x$, $y = 2x$, $x = 1$, $x = 2$

12. **Square** $f(x, y) = 1/(xy)$ over the square $1 \le x \le 2$, $1 \le y \le 2$

13. **Triangle** $f(x, y) = x^2 + y^2$ over the triangular region with vertices $(0, 0)$, $(1, 0)$, and $(0, 1)$

14. **Rectangle** $f(x, y) = y \cos xy$ over the rectangle $0 \le x \le \pi$, $0 \le y \le 1$

15. **Triangle** $f(u, v) = v - \sqrt{u}$ over the triangular region cut from the first quadrant of the uv-plane by the line $u + v = 1$

16. **Curved region** $f(s, t) = e^s \ln t$ over the region in the first quadrant of the st-plane that lies above the curve $s = \ln t$ from $t = 1$ to $t = 2$

Each of Exercises 17–20 gives an integral over a region in a Cartesian coordinate plane. Sketch the region and evaluate the integral.

17. $\int_{-2}^0 \int_v^{-v} 2\, dp\, dv$ (the pv-plane)

18. $\int_0^1 \int_0^{\sqrt{1-s^2}} 8t\, dt\, ds$ (the st-plane)

19. $\int_{-\pi/3}^{\pi/3} \int_0^{\sec t} 3 \cos t\, du\, dt$ (the tu-plane)

20. $\int_0^3 \int_1^{4-2u} \frac{4 - 2u}{v^2}\, dv\, du$ (the uv-plane)

Reversing the Order of Integration

In Exercises 21–30, sketch the region of integration and write an equivalent double integral with the order of integration reversed.

21. $\int_0^1 \int_2^{4-2x} dy\, dx$

22. $\int_0^2 \int_{y-2}^0 dx\, dy$

23. $\int_0^1 \int_y^{\sqrt{y}} dx\, dy$

24. $\int_0^1 \int_{1-x}^{1-x^2} dy\, dx$

25. $\int_0^1 \int_1^{e^x} dy\, dx$

26. $\int_0^{\ln 2} \int_{e^y}^2 dx\, dy$

27. $\int_0^{3/2} \int_0^{9-4x^2} 16x\, dy\, dx$

28. $\int_0^2 \int_0^{4-y^2} y\, dx\, dy$

29. $\int_0^1 \int_{-\sqrt{1-y^2}}^{\sqrt{1-y^2}} 3y\, dx\, dy$

30. $\int_0^2 \int_{-\sqrt{4-x^2}}^{\sqrt{4-x^2}} 6x\, dy\, dx$

Evaluating Double Integrals

In Exercises 31–40, sketch the region of integration, reverse the order of integration, and evaluate the integral.

31. $\int_0^{\pi} \int_x^{\pi} \frac{\sin y}{y}\, dy\, dx$

32. $\int_0^2 \int_x^2 2y^2 \sin xy\, dy\, dx$

33. $\int_0^1 \int_y^1 x^2 e^{xy}\, dx\, dy$

34. $\int_0^2 \int_0^{4-x^2} \frac{xe^{2y}}{4 - y}\, dy\, dx$

35. $\int_0^{2\sqrt{\ln 3}} \int_{y/2}^{\sqrt{\ln 3}} e^{x^2}\, dx\, dy$

36. $\int_0^3 \int_{\sqrt{x/3}}^1 e^{y^3}\, dy\, dx$

37. $\int_0^{1/16} \int_{y^{1/4}}^{1/2} \cos(16\pi x^5)\, dx\, dy$

38. $\int_0^8 \int_{\sqrt[3]{x}}^2 \frac{dy\, dx}{y^4 + 1}$

39. **Square region** $\iint_R (y - 2x^2)\, dA$ where R is the region bounded by the square $|x| + |y| = 1$

40. **Triangular region** $\iint_R xy\, dA$ where R is the region bounded by the lines $y = x$, $y = 2x$, and $x + y = 2$

Volume Beneath a Surface $z = f(x, y)$

41. Find the volume of the region bounded by the paraboloid $z = x^2 + y^2$ and below by the triangle enclosed by the lines $y = x$, $x = 0$, and $x + y = 2$ in the xy-plane.

42. Find the volume of the solid that is bounded above by the cylinder $z = x^2$ and below by the region enclosed by the parabola $y = 2 - x^2$ and the line $y = x$ in the xy-plane.

43. Find the volume of the solid whose base is the region in the xy-plane that is bounded by the parabola $y = 4 - x^2$ and the line $y = 3x$, while the top of the solid is bounded by the plane $z = x + 4$.

44. Find the volume of the solid in the first octant bounded by the coordinate planes, the cylinder $x^2 + y^2 = 4$, and the plane $z + y = 3$.

45. Find the volume of the solid in the first octant bounded by the coordinate planes, the plane $x = 3$, and the parabolic cylinder $z = 4 - y^2$.

46. Find the volume of the solid cut from the first octant by the surface $z = 4 - x^2 - y$.

47. Find the volume of the wedge cut from the first octant by the cylinder $z = 12 - 3y^2$ and the plane $x + y = 2$.

48. Find the volume of the solid cut from the square column $|x| + |y| \leq 1$ by the planes $z = 0$ and $3x + z = 3$.

49. Find the volume of the solid that is bounded on the front and back by the planes $x = 2$ and $x = 1$, on the sides by the cylinders $y = \pm 1/x$, and above and below by the planes $z = x + 1$ and $z = 0$.

50. Find the volume of the solid bounded on the front and back by the planes $x = \pm\pi/3$, on the sides by the cylinders $y = \pm\sec x$, above by the cylinder $z = 1 + y^2$, and below by the xy-plane.

Integrals over Unbounded Regions

Improper double integrals can often be computed similarly to improper integrals of one variable. The first iteration of the following improper integrals is conducted just as if they were proper integrals. One then evaluates an improper integral of a single variable by taking appropriate limits, as in Section 8.8. Evaluate the improper integrals in Exercises 51–54 as iterated integrals.

51. $\displaystyle\int_1^\infty \int_{e^{-x}}^1 \frac{1}{x^3 y}\, dy\, dx$

52. $\displaystyle\int_{-1}^1 \int_{-1/\sqrt{1-x^2}}^{1/\sqrt{1-x^2}} (2y + 1)\, dy\, dx$

53. $\displaystyle\int_{-\infty}^\infty \int_{-\infty}^\infty \frac{1}{(x^2 + 1)(y^2 + 1)}\, dx\, dy$

54. $\displaystyle\int_0^\infty \int_0^\infty xe^{-(x+2y)}\, dx\, dy$

Approximating Double Integrals

In Exercises 55 and 56, approximate the double integral of $f(x, y)$ over the region R partitioned by the given vertical lines $x = a$ and horizontal lines $y = c$. In each subrectangle, use (x_k, y_k) as indicated for your approximation.

$$\iint_R f(x, y)\, dA \approx \sum_{k=1}^{n} f(x_k, y_k)\, \Delta A_k$$

55. $f(x, y) = x + y$ over the region R bounded above by the semicircle $y = \sqrt{1 - x^2}$ and below by the x-axis, using the partition $x = -1, -1/2, 0, 1/4, 1/2, 1$ and $y = 0, 1/2, 1$ with (x_k, y_k) the lower left corner in the kth subrectangle (provided the subrectangle lies within R)

56. $f(x, y) = x + 2y$ over the region R inside the circle $(x - 2)^2 + (y - 3)^2 = 1$ using the partition $x = 1, 3/2, 2, 5/2, 3$ and $y = 2, 5/2, 3, 7/2, 4$ with (x_k, y_k) the center (centroid) in the kth subrectangle (provided the subrectangle lies within R)

Theory and Examples

57. Circular sector Integrate $f(x, y) = \sqrt{4 - x^2}$ over the smaller sector cut from the disk $x^2 + y^2 \leq 4$ by the rays $\theta = \pi/6$ and $\theta = \pi/2$.

58. Unbounded region Integrate $f(x, y) = 1/[(x^2 - x)(y - 1)^{2/3}]$ over the infinite rectangle $2 \leq x < \infty, 0 \leq y \leq 2$.

59. Noncircular cylinder A solid right (noncircular) cylinder has its base R in the xy-plane and is bounded above by the paraboloid $z = x^2 + y^2$. The cylinder's volume is

$$V = \int_0^1 \int_0^y (x^2 + y^2)\, dx\, dy + \int_1^2 \int_0^{2-y} (x^2 + y^2)\, dx\, dy.$$

Sketch the base region R and express the cylinder's volume as a single iterated integral with the order of integration reversed. Then evaluate the integral to find the volume.

60. Converting to a double integral Evaluate the integral

$$\int_0^2 (\tan^{-1}\pi x - \tan^{-1} x)\, dx.$$

(*Hint:* Write the integrand as an integral.)

61. Maximizing a double integral What region R in the xy-plane maximizes the value of

$$\iint_R (4 - x^2 - 2y^2)\, dA?$$

Give reasons for your answer.

62. Minimizing a double integral What region R in the xy-plane minimizes the value of

$$\iint_R (x^2 + y^2 - 9)\, dA?$$

Give reasons for your answer.

63. Is it possible to evaluate the integral of a continuous function $f(x, y)$ over a rectangular region in the xy-plane and get different answers depending on the order of integration? Give reasons for your answer.

64. How would you evaluate the double integral of a continuous function $f(x, y)$ over the region R in the xy-plane enclosed by the triangle with vertices (0, 1), (2, 0), and (1, 2)? Give reasons for your answer.

65. Unbounded region Prove that

$$\int_{-\infty}^\infty \int_{-\infty}^\infty e^{-x^2-y^2}\, dx\, dy = \lim_{b\to\infty} \int_{-b}^b \int_{-b}^b e^{-x^2-y^2}\, dx\, dy = 4\left(\int_0^\infty e^{-x^2}\, dx\right)^2.$$

66. Improper double integral Evaluate the improper integral

$$\int_0^1 \int_0^3 \frac{x^2}{(y - 1)^{2/3}}\, dy\, dx.$$

COMPUTER EXPLORATIONS

Evaluating Double Integrals Numerically

Use a CAS double-integral evaluator to estimate the values of the integrals in Exercises 67–70.

67. $\int_1^3 \int_1^x \frac{1}{xy}\, dy\, dx$

68. $\int_0^1 \int_0^1 e^{-(x^2+y^2)}\, dy\, dx$

69. $\int_0^1 \int_0^1 \tan^{-1} xy\, dy\, dx$

70. $\int_{-1}^1 \int_0^{\sqrt{1-x^2}} 3\sqrt{1 - x^2 - y^2}\, dy\, dx$

Use a CAS double-integral evaluator to find the integrals in Exercises 71–76. Then reverse the order of integration and evaluate, again with a CAS.

71. $\int_0^1 \int_{2y}^4 e^{x^2}\, dx\, dy$

72. $\int_0^3 \int_{x^2}^9 x \cos(y^2)\, dy\, dx$

73. $\int_0^2 \int_{y^3}^{4\sqrt{2y}} (x^2 y - xy^2)\, dx\, dy$

74. $\int_0^2 \int_0^{4-y^2} e^{xy}\, dx\, dy$

75. $\int_1^2 \int_0^{x^2} \frac{1}{x+y}\, dy\, dx$

76. $\int_1^2 \int_{y^3}^8 \frac{1}{\sqrt{x^2+y^2}}\, dx\, dy$

15.2 Area, Moments, and Centers of Mass

In this section, we show how to use double integrals to calculate the areas of bounded regions in the plane and to find the average value of a function of two variables. Then we study the physical problem of finding the center of mass of a thin plate covering a region in the plane.

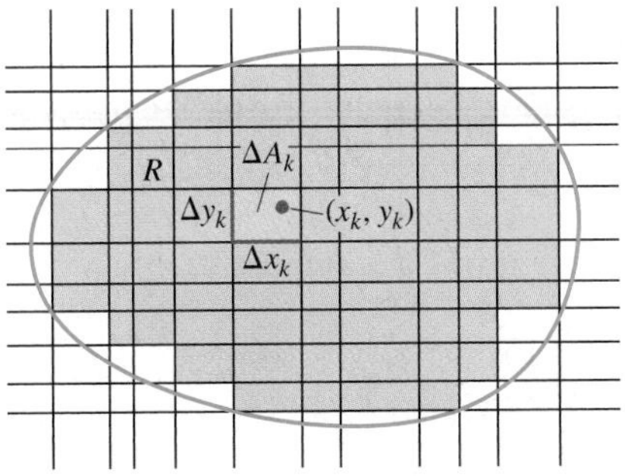

FIGURE 15.14 As the norm of a partition of the region R approaches zero, the sum of the areas ΔA_k gives the area of R defined by the double integral $\iint_R dA$.

Areas of Bounded Regions in the Plane

If we take $f(x, y) = 1$ in the definition of the double integral over a region R in the preceding section, the Riemann sums reduce to

$$S_n = \sum_{k=1}^{n} f(x_k, y_k)\, \Delta A_k = \sum_{k=1}^{n} \Delta A_k. \tag{1}$$

This is simply the sum of the areas of the small rectangles in the partition of R, and approximates what we would like to call the area of R. As the norm of a partition of R approaches zero, the height and width of all rectangles in the partition approach zero, and the coverage of R becomes increasingly complete (Figure 15.14). We define the area of R to be the limit

$$Area = \lim_{\|P\|\to 0} \sum_{k=1}^{n} \Delta A_k = \iint_R dA \tag{2}$$

DEFINITION Area

The **area** of a closed, bounded plane region R is

$$A = \iint_R dA.$$

As with the other definitions in this chapter, the definition here applies to a greater variety of regions than does the earlier single-variable definition of area, but it agrees with the earlier definition on regions to which they both apply. To evaluate the integral in the definition of area, we integrate the constant function $f(x, y) = 1$ over R.

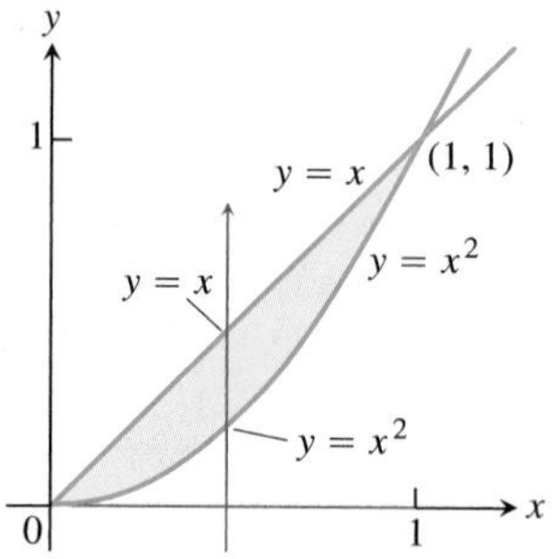

FIGURE 15.15 The region in Example 1.

EXAMPLE 1 Finding Area

Find the area of the region R bounded by $y = x$ and $y = x^2$ in the first quadrant.

Solution We sketch the region (Figure 15.15), noting where the two curves intersect, and calculate the area as

$$A = \int_0^1 \int_{x^2}^{x} dy\, dx = \int_0^1 \Big[y \Big]_{x^2}^{x} dx$$

$$= \int_0^1 (x - x^2)\, dx = \left[\frac{x^2}{2} - \frac{x^3}{3} \right]_0^1 = \frac{1}{6}.$$

Notice that the single integral $\int_0^1 (x - x^2)\, dx$, obtained from evaluating the inside iterated integral, is the integral for the area between these two curves using the method of Section 5.5. ■

EXAMPLE 2 Finding Area

Find the area of the region R enclosed by the parabola $y = x^2$ and the line $y = x + 2$.

Solution If we divide R into the regions R_1 and R_2 shown in Figure 15.16a, we may calculate the area as

$$A = \iint_{R_1} dA + \iint_{R_2} dA = \int_0^1 \int_{-\sqrt{y}}^{\sqrt{y}} dx\, dy + \int_1^4 \int_{y-2}^{\sqrt{y}} dx\, dy.$$

On the other hand, reversing the order of integration (Figure 15.16b) gives

$$A = \int_{-1}^{2} \int_{x^2}^{x+2} dy\, dx.$$

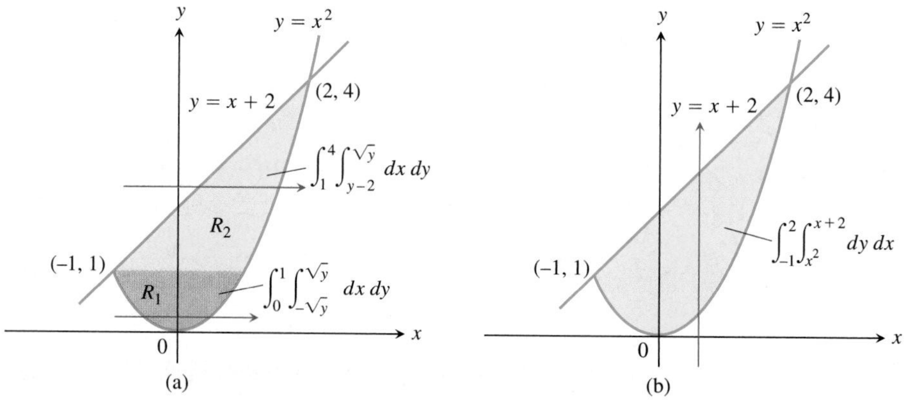

FIGURE 15.16 Calculating this area takes (a) two double integrals if the first integration is with respect to x, but (b) only one if the first integration is with respect to y (Example 2).

This second result, which requires only one integral, is simpler and is the only one we would bother to write down in practice. The area is

$$A = \int_{-1}^{2} \Big[y \Big]_{x^2}^{x+2} dx = \int_{-1}^{2} (x + 2 - x^2)\, dx = \left[\frac{x^2}{2} + 2x - \frac{x^3}{3} \right]_{-1}^{2} = \frac{9}{2}.$$

■

Average Value

The average value of an integrable function of one variable on a closed interval is the integral of the function over the interval divided by the length of the interval. For an integrable function of two variables defined on a bounded region in the plane, the average value is the integral over the region divided by the area of the region. This can be visualized by thinking of the function as giving the height at one instant of some water sloshing around in a tank whose vertical walls lie over the boundary of the region. The average height of the water in the tank can be found by letting the water settle down to a constant height. The height is then equal to the volume of water in the tank divided by the area of R. We are led to define the average value of an integrable function f over a region R to be

$$\textbf{Average value of } f \text{ over } R = \frac{1}{\text{area of } R} \iint_R f\, dA. \qquad (3)$$

If f is the temperature of a thin plate covering R, then the double integral of f over R divided by the area of R is the plate's average temperature. If $f(x, y)$ is the distance from the point (x, y) to a fixed point P, then the average value of f over R is the average distance of points in R from P.

EXAMPLE 3 Finding Average Value

Find the average value of $f(x, y) = x \cos xy$ over the rectangle R: $0 \le x \le \pi$, $0 \le y \le 1$.

Solution The value of the integral of f over R is

$$\int_0^{\pi} \int_0^1 x \cos xy\, dy\, dx = \int_0^{\pi} \Big[\sin xy \Big]_{y=0}^{y=1} dx \qquad \int x \cos xy\, dy = \sin xy + C$$

$$= \int_0^{\pi} (\sin x - 0)\, dx = -\cos x \Big]_0^{\pi} = 1 + 1 = 2.$$

The area of R is π. The average value of f over R is $2/\pi$. ■

Moments and Centers of Mass for Thin Flat Plates

In Section 6.4 we introduced the concepts of moments and centers of mass, and we saw how to compute these quantities for thin rods or strips and for plates of constant density. Using multiple integrals we can extend these calculations to a great variety of shapes with varying density. We first consider the problem of finding the center of mass of a thin flat plate: a disk of aluminum, say, or a triangular sheet of metal. We assume the distribution of mass in such a plate to be continuous. A material's *density* function, denoted by $\delta(x, y)$, is the mass per unit area. The *mass* of a plate is obtained by integrating the density function over the region R forming the plate. The first moment about an axis is calculated by integrating over R the distance from the axis times the density. The center of mass is found from the first moments. Table 15.1 gives the double integral formulas for mass, first moments, and center of mass.

TABLE 15.1 Mass and first moment formulas for thin plates covering a region R in the xy-plane

Mass:	$M = \iint_R \delta(x, y)\, dA$	$\delta(x, y)$ is the density at (x, y)
First moments:	$M_x = \iint_R y\delta(x, y)\, dA,$	$M_y = \iint_R x\delta(x, y)\, dA$
Center of mass:	$\bar{x} = \dfrac{M_y}{M},$	$\bar{y} = \dfrac{M_x}{M}$

EXAMPLE 4 Finding the Center of Mass of a Thin Plate of Variable Density

A thin plate covers the triangular region bounded by the x-axis and the lines $x = 1$ and $y = 2x$ in the first quadrant. The plate's density at the point (x, y) is $\delta(x, y) = 6x + 6y + 6$. Find the plate's mass, first moments, and center of mass about the coordinate axes.

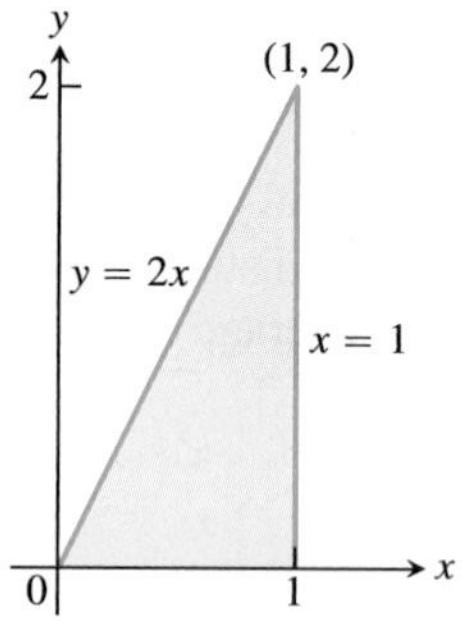

FIGURE 15.17 The triangular region covered by the plate in Example 4.

Solution We sketch the plate and put in enough detail to determine the limits of integration for the integrals we have to evaluate (Figure 15.17).

The plate's mass is

$$M = \int_0^1 \int_0^{2x} \delta(x, y)\, dy\, dx = \int_0^1 \int_0^{2x} (6x + 6y + 6)\, dy\, dx$$

$$= \int_0^1 \left[6xy + 3y^2 + 6y \right]_{y=0}^{y=2x} dx$$

$$= \int_0^1 (24x^2 + 12x)\, dx = \left[8x^3 + 6x^2 \right]_0^1 = 14.$$

The first moment about the x-axis is

$$M_x = \int_0^1 \int_0^{2x} y\delta(x, y)\, dy\, dx = \int_0^1 \int_0^{2x} (6xy + 6y^2 + 6y)\, dy\, dx$$

$$= \int_0^1 \left[3xy^2 + 2y^3 + 3y^2 \right]_{y=0}^{y=2x} dx = \int_0^1 (28x^3 + 12x^2)\, dx$$

$$= \left[7x^4 + 4x^3 \right]_0^1 = 11.$$

A similar calculation gives the moment about the y-axis:

$$M_y = \int_0^1 \int_0^{2x} x\delta(x, y)\, dy\, dx = 10.$$

The coordinates of the center of mass are therefore

$$\bar{x} = \frac{M_y}{M} = \frac{10}{14} = \frac{5}{7}, \qquad \bar{y} = \frac{M_x}{M} = \frac{11}{14}.$$

■

Moments of Inertia

A body's first moments (Table 15.1) tell us about balance and about the torque the body exerts about different axes in a gravitational field. If the body is a rotating shaft, however, we are more likely to be interested in how much energy is stored in the shaft or about how much energy it will take to accelerate the shaft to a particular angular velocity. This is where the second moment or moment of inertia comes in.

Think of partitioning the shaft into small blocks of mass Δm_k and let r_k denote the distance from the kth block's center of mass to the axis of rotation (Figure 15.18). If the shaft rotates at an angular velocity of $\omega = d\theta/dt$ radians per second, the block's center of mass will trace its orbit at a linear speed of

$$v_k = \frac{d}{dt}(r_k\theta) = r_k\frac{d\theta}{dt} = r_k\omega.$$

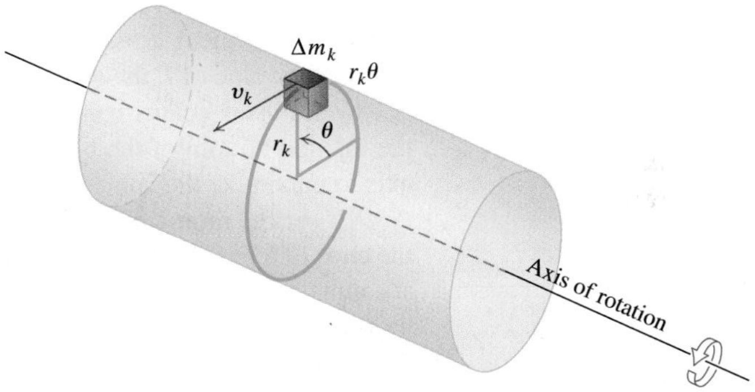

FIGURE 15.18 To find an integral for the amount of energy stored in a rotating shaft, we first imagine the shaft to be partitioned into small blocks. Each block has its own kinetic energy. We add the contributions of the individual blocks to find the kinetic energy of the shaft.

The block's kinetic energy will be approximately

$$\frac{1}{2}\Delta m_k v_k^2 = \frac{1}{2}\Delta m_k (r_k\omega)^2 = \frac{1}{2}\omega^2 r_k^2\, \Delta m_k.$$

The kinetic energy of the shaft will be approximately

$$\sum \frac{1}{2}\omega^2 r_k^2\, \Delta m_k.$$

The integral approached by these sums as the shaft is partitioned into smaller and smaller blocks gives the shaft's kinetic energy:

$$\mathrm{KE}_{\text{shaft}} = \int \frac{1}{2}\omega^2 r^2\, dm = \frac{1}{2}\omega^2 \int r^2\, dm. \tag{4}$$

The factor

$$I = \int r^2\, dm$$

is the *moment of inertia* of the shaft about its axis of rotation, and we see from Equation (4) that the shaft's kinetic energy is

$$\mathrm{KE}_{\text{shaft}} = \frac{1}{2} I\omega^2.$$

The moment of inertia of a shaft resembles in some ways the inertia of a locomotive. To start a locomotive with mass m moving at a linear velocity v, we need to provide a kinetic energy of $\mathrm{KE} = (1/2)mv^2$. To stop the locomotive we have to remove this amount of energy. To start a shaft with moment of inertia I rotating at an angular velocity ω, we need to provide a kinetic energy of $\mathrm{KE} = (1/2)I\omega^2$. To stop the shaft we have to take this amount of energy back out. The shaft's moment of inertia is analogous to the locomotive's mass. What makes the locomotive hard to start or stop is its mass. What makes the shaft hard to start or stop is its moment of inertia. The moment of inertia depends not only on the mass of the shaft, but also its distribution.

The moment of inertia also plays a role in determining how much a horizontal metal beam will bend under a load. The stiffness of the beam is a constant times I, the moment of inertia of a typical cross-section of the beam about the beam's longitudinal axis. The greater the value of I, the stiffer the beam and the less it will bend under a given load. That is why we use I-beams instead of beams whose cross-sections are square. The flanges at the top and bottom of the beam hold most of the beam's mass away from the longitudinal axis to maximize the value of I (Figure 15.19).

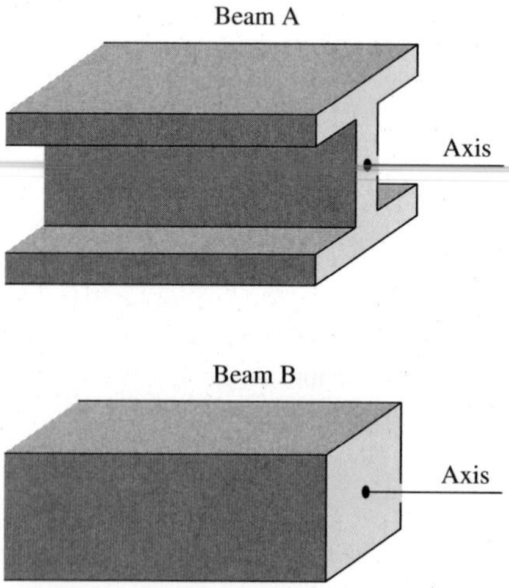

FIGURE 15.19 The greater the polar moment of inertia of the cross-section of a beam about the beam's longitudinal axis, the stiffer the beam. Beams A and B have the same cross-sectional area, but A is stiffer.

To see the moment of inertia at work, try the following experiment. Tape two coins to the ends of a pencil and twiddle the pencil about the center of mass. The moment of inertia accounts for the resistance you feel each time you change the direction of motion. Now move the coins an equal distance toward the center of mass and twiddle the pencil again. The system has the same mass and the same center of mass but now offers less resistance to the changes in motion. The moment of inertia has been reduced. The moment of inertia is what gives a baseball bat, golf club, or tennis racket its "feel." Tennis rackets that weigh the same, look the same, and have identical centers of mass will feel different and behave differently if their masses are not distributed the same way.

Computations of moments of inertia for thin plates in the plane lead to double integral formulas, which are summarized in Table 15.2. A small thin piece of mass Δm is equal to its small area ΔA multiplied by the density of a point in the piece. Computations of moments of inertia for objects occupying a region in space are discussed in Section 15.5.

The mathematical difference between the **first moments** M_x and M_y and the **moments of inertia**, or **second moments**, I_x and I_y is that the second moments use the *squares* of the "lever-arm" distances x and y.

The moment I_0 is also called the **polar moment** of inertia about the origin. It is calculated by integrating the density $\delta(x, y)$ (mass per unit area) times $r^2 = x^2 + y^2$, the square of the distance from a representative point (x, y) to the origin. Notice that $I_0 = I_x + I_y$; once we find two, we get the third automatically. (The moment I_0 is sometimes called I_z, for moment of inertia about the z-axis. The identity $I_z = I_x + I_y$ is then called the **Perpendicular Axis Theorem.**)

The **radius of gyration** R_x is defined by the equation

$$I_x = MR_x^2.$$

It tells how far from the x-axis the entire mass of the plate might be concentrated to give the same I_x. The radius of gyration gives a convenient way to express the moment of inertia in terms of a mass and a length. The radii R_y and R_0 are defined in a similar way, with

$$I_y = MR_y^2 \quad \text{and} \quad I_0 = MR_0^2.$$

We take square roots to get the formulas in Table 15.2, which gives the formulas for moments of inertia (second moments) as well as for radii of gyration.

TABLE 15.2 Second moment formulas for thin plates in the xy-plane

Moments of inertia (second moments):		
About the x-axis:	$I_x = \iint y^2\delta(x, y)\, dA$	
About the y-axis:	$I_y = \iint x^2\delta(x, y)\, dA$	
About a line L:	$I_L = \iint r^2(x, y)\delta(x, y)\, dA$, where $r(x, y) =$ distance from (x, y) to L	
About the origin (polar moment):	$I_0 = \iint (x^2 + y^2)\delta(x, y)\, dA = I_x + I_y$	
Radii of gyration:	About the x-axis:	$R_x = \sqrt{I_x/M}$
	About the y-axis:	$R_y = \sqrt{I_y/M}$
	About the origin:	$R_0 = \sqrt{I_0/M}$

EXAMPLE 5 Finding Moments of Inertia and Radii of Gyration

For the thin plate in Example 4 (Figure 15.17), find the moments of inertia and radii of gyration about the coordinate axes and the origin.

Solution Using the density function $\delta(x, y) = 6x + 6y + 6$ given in Example 4, the moment of inertia about the x-axis is

$$I_x = \int_0^1 \int_0^{2x} y^2\delta(x, y)\, dy\, dx = \int_0^1 \int_0^{2x} (6xy^2 + 6y^3 + 6y^2)\, dy\, dx$$

$$= \int_0^1 \left[2xy^3 + \frac{3}{2}y^4 + 2y^3\right]_{y=0}^{y=2x} dx = \int_0^1 (40x^4 + 16x^3)\, dx$$

$$= \left[8x^5 + 4x^4\right]_0^1 = 12.$$

Similarly, the moment of inertia about the y-axis is

$$I_y = \int_0^1 \int_0^{2x} x^2 \delta(x, y)\, dy\, dx = \frac{39}{5}.$$

Notice that we integrate y^2 times density in calculating I_x and x^2 times density to find I_y.

Since we know I_x and I_y, we do not need to evaluate an integral to find I_0; we can use the equation $I_0 = I_x + I_y$ instead:

$$I_0 = 12 + \frac{39}{5} = \frac{60 + 39}{5} = \frac{99}{5}.$$

The three radii of gyration are

$$R_x = \sqrt{I_x/M} = \sqrt{12/14} = \sqrt{6/7} \approx 0.93$$

$$R_y = \sqrt{I_y/M} = \sqrt{\left(\frac{39}{5}\right)/14} = \sqrt{39/70} \approx 0.75$$

$$R_0 = \sqrt{I_0/M} = \sqrt{\left(\frac{99}{5}\right)/14} = \sqrt{99/70} \approx 1.19. \qquad \blacksquare$$

Moments are also of importance in statistics. The first moment is used in computing the mean μ of a set of data, and the second moment is used in computing the variance $\left(\Sigma^2\right)$ and the standard deviation $\left(\Sigma\right)$. Third and fourth moments are used for computing statistical quantities known as skewness and kurtosis.

Centroids of Geometric Figures

When the density of an object is constant, it cancels out of the numerator and denominator of the formulas for $\bar{x}$ and $\bar{y}$ in Table 15.1. As far as $\bar{x}$ and $\bar{y}$ are concerned, δ might as well be 1. Thus, when δ is constant, the location of the center of mass becomes a feature of the object's shape and not of the material of which it is made. In such cases, engineers may call the center of mass the **centroid** of the shape. To find a centroid, we set δ equal to 1 and proceed to find $\bar{x}$ and $\bar{y}$ as before, by dividing first moments by masses.

EXAMPLE 6 Finding the Centroid of a Region

Find the centroid of the region in the first quadrant that is bounded above by the line $y = x$ and below by the parabola $y = x^2$.

Solution We sketch the region and include enough detail to determine the limits of integration (Figure 15.20). We then set δ equal to 1 and evaluate the appropriate formulas from Table 15.1:

$$M = \int_0^1 \int_{x^2}^{x} 1\, dy\, dx = \int_0^1 \left[y \right]_{y=x^2}^{y=x} dx = \int_0^1 (x - x^2)\, dx = \left[\frac{x^2}{2} - \frac{x^3}{3} \right]_0^1 = \frac{1}{6}$$

$$M_x = \int_0^1 \int_{x^2}^{x} y\, dy\, dx = \int_0^1 \left[\frac{y^2}{2} \right]_{y=x^2}^{y=x} dx$$

$$= \int_0^1 \left(\frac{x^2}{2} - \frac{x^4}{2} \right) dx = \left[\frac{x^3}{6} - \frac{x^5}{10} \right]_0^1 = \frac{1}{15}$$

$$M_y = \int_0^1 \int_{x^2}^{x} x\, dy\, dx = \int_0^1 \left[xy \right]_{y=x^2}^{y=x} dx = \int_0^1 (x^2 - x^3)\, dx = \left[\frac{x^3}{3} - \frac{x^4}{4} \right]_0^1 = \frac{1}{12}.$$

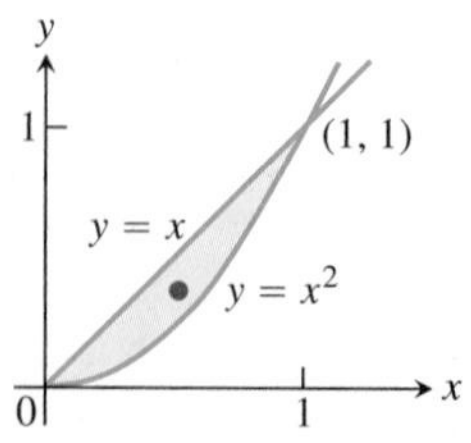

FIGURE 15.20 The centroid of this region is found in Example 6.

From these values of M, M_x, and M_y, we find

$$\bar{x} = \frac{M_y}{M} = \frac{1/12}{1/6} = \frac{1}{2} \quad \text{and} \quad \bar{y} = \frac{M_x}{M} = \frac{1/15}{1/6} = \frac{2}{5}.$$

The centroid is the point $(1/2, 2/5)$. ■

EXERCISES 15.2

Area by Double Integration

In Exercises 1–8, sketch the region bounded by the given lines and curves. Then express the region's area as an iterated double integral and evaluate the integral.

1. The coordinate axes and the line $x + y = 2$
2. The lines $x = 0$, $y = 2x$, and $y = 4$
3. The parabola $x = -y^2$ and the line $y = x + 2$
4. The parabola $x = y - y^2$ and the line $y = -x$
5. The curve $y = e^x$ and the lines $y = 0$, $x = 0$, and $x = \ln 2$
6. The curves $y = \ln x$ and $y = 2 \ln x$ and the line $x = e$, in the first quadrant
7. The parabolas $x = y^2$ and $x = 2y - y^2$
8. The parabolas $x = y^2 - 1$ and $x = 2y^2 - 2$

Identifying the Region of Integration

The integrals and sums of integrals in Exercises 9–14 give the areas of regions in the xy-plane. Sketch each region, label each bounding curve with its equation, and give the coordinates of the points where the curves intersect. Then find the area of the region.

9. $\displaystyle\int_0^6 \int_{y^2/3}^{2y} dx\, dy$

10. $\displaystyle\int_0^3 \int_{-x}^{x(2-x)} dy\, dx$

11. $\displaystyle\int_0^{\pi/4} \int_{\sin x}^{\cos x} dy\, dx$

12. $\displaystyle\int_{-1}^2 \int_{y^2}^{y+2} dx\, dy$

13. $\displaystyle\int_{-1}^0 \int_{-2x}^{1-x} dy\, dx + \int_0^2 \int_{-x/2}^{1-x} dy\, dx$

14. $\displaystyle\int_0^2 \int_{x^2-4}^{0} dy\, dx + \int_0^4 \int_0^{\sqrt{x}} dy\, dx$

Average Values

15. Find the average value of $f(x, y) = \sin(x + y)$ over
    a. the rectangle $0 \le x \le \pi$, $0 \le y \le \pi$
    b. the rectangle $0 \le x \le \pi$, $0 \le y \le \pi/2$
16. Which do you think will be larger, the average value of $f(x, y) = xy$ over the square $0 \le x \le 1$, $0 \le y \le 1$, or the average value of f over the quarter circle $x^2 + y^2 \le 1$ in the first quadrant? Calculate them to find out.
17. Find the average height of the paraboloid $z = x^2 + y^2$ over the square $0 \le x \le 2$, $0 \le y \le 2$.
18. Find the average value of $f(x, y) = 1/(xy)$ over the square $\ln 2 \le x \le 2 \ln 2$, $\ln 2 \le y \le 2 \ln 2$.

Constant Density

19. **Finding center of mass** Find a center of mass of a thin plate of density $\delta = 3$ bounded by the lines $x = 0$, $y = x$, and the parabola $y = 2 - x^2$ in the first quadrant.
20. **Finding moments of inertia and radii of gyration** Find the moments of inertia and radii of gyration about the coordinate axes of a thin rectangular plate of constant density δ bounded by the lines $x = 3$ and $y = 3$ in the first quadrant.
21. **Finding a centroid** Find the centroid of the region in the first quadrant bounded by the x-axis, the parabola $y^2 = 2x$, and the line $x + y = 4$.
22. **Finding a centroid** Find the centroid of the triangular region cut from the first quadrant by the line $x + y = 3$.
23. **Finding a centroid** Find the centroid of the semicircular region bounded by the x-axis and the curve $y = \sqrt{1 - x^2}$.
24. **Finding a centroid** The area of the region in the first quadrant bounded by the parabola $y = 6x - x^2$ and the line $y = x$ is $125/6$ square units. Find the centroid.
25. **Finding a centroid** Find the centroid of the region cut from the first quadrant by the circle $x^2 + y^2 = a^2$.
26. **Finding a centroid** Find the centroid of the region between the x-axis and the arch $y = \sin x$, $0 \le x \le \pi$.
27. **Finding moments of inertia** Find the moment of inertia about the x-axis of a thin plate of density $\delta = 1$ bounded by the circle $x^2 + y^2 = 4$. Then use your result to find I_y and I_0 for the plate.
28. **Finding a moment of inertia** Find the moment of inertia with respect to the y-axis of a thin sheet of constant density $\delta = 1$ bounded by the curve $y = (\sin^2 x)/x^2$ and the interval $\pi \le x \le 2\pi$ of the x-axis.
29. **The centroid of an infinite region** Find the centroid of the infinite region in the second quadrant enclosed by the coordinate axes and the curve $y = e^x$. (Use improper integrals in the mass-moment formulas.)

30. The first moment of an infinite plate Find the first moment about the y-axis of a thin plate of density $\delta(x, y) = 1$ covering the infinite region under the curve $y = e^{-x^2/2}$ in the first quadrant.

Variable Density

31. Finding a moment of inertia and radius of gyration Find the moment of inertia and radius of gyration about the x-axis of a thin plate bounded by the parabola $x = y - y^2$ and the line $x + y = 0$ if $\delta(x, y) = x + y$.

32. Finding mass Find the mass of a thin plate occupying the smaller region cut from the ellipse $x^2 + 4y^2 = 12$ by the parabola $x = 4y^2$ if $\delta(x, y) = 5x$.

33. Finding a center of mass Find the center of mass of a thin triangular plate bounded by the y-axis and the lines $y = x$ and $y = 2 - x$ if $\delta(x, y) = 6x + 3y + 3$.

34. Finding a center of mass and moment of inertia Find the center of mass and moment of inertia about the x-axis of a thin plate bounded by the curves $x = y^2$ and $x = 2y - y^2$ if the density at the point (x, y) is $\delta(x, y) = y + 1$.

35. Center of mass, moment of inertia, and radius of gyration Find the center of mass and the moment of inertia and radius of gyration about the y-axis of a thin rectangular plate cut from the first quadrant by the lines $x = 6$ and $y = 1$ if $\delta(x, y) = x + y + 1$.

36. Center of mass, moment of inertia, and radius of gyration Find the center of mass and the moment of inertia and radius of gyration about the y-axis of a thin plate bounded by the line $y = 1$ and the parabola $y = x^2$ if the density is $\delta(x, y) = y + 1$.

37. Center of mass, moment of inertia, and radius of gyration Find the center of mass and the moment of inertia and radius of gyration about the y-axis of a thin plate bounded by the x-axis, the lines $x = \pm 1$, and the parabola $y = x^2$ if $\delta(x, y) = 7y + 1$.

38. Center of mass, moment of inertia, and radius of gyration Find the center of mass and the moment of inertia and radius of gyration about the x-axis of a thin rectangular plate bounded by the lines $x = 0, x = 20, y = -1$, and $y = 1$ if $\delta(x, y) = 1 + (x/20)$.

39. Center of mass, moments of inertia, and radii of gyration Find the center of mass, the moment of inertia and radii of gyration about the coordinate axes, and the polar moment of inertia and radius of gyration of a thin triangular plate bounded by the lines $y = x, y = -x$, and $y = 1$ if $\delta(x, y) = y + 1$.

40. Center of mass, moments of inertia, and radii of gyration Repeat Exercise 39 for $\delta(x, y) = 3x^2 + 1$.

Theory and Examples

41. Bacterium population If $f(x, y) = (10{,}000e^y)/(1 + |x|/2)$ represents the "population density" of a certain bacterium on the xy-plane, where x and y are measured in centimeters, find the total population of bacteria within the rectangle $-5 \le x \le 5$ and $-2 \le y \le 0$.

42. Regional population If $f(x, y) = 100\,(y + 1)$ represents the population density of a planar region on Earth, where x and y are measured in miles, find the number of people in the region bounded by the curves $x = y^2$ and $x = 2y - y^2$.

43. Appliance design When we design an appliance, one of the concerns is how hard the appliance will be to tip over. When tipped, it will right itself as long as its center of mass lies on the correct side of the *fulcrum*, the point on which the appliance is riding as it tips. Suppose that the profile of an appliance of approximately constant density is parabolic, like an old-fashioned radio. It fills the region $0 \le y \le a(1 - x^2), -1 \le x \le 1$, in the xy-plane (see accompanying figure). What values of a will guarantee that the appliance will have to be tipped more than 45° to fall over?

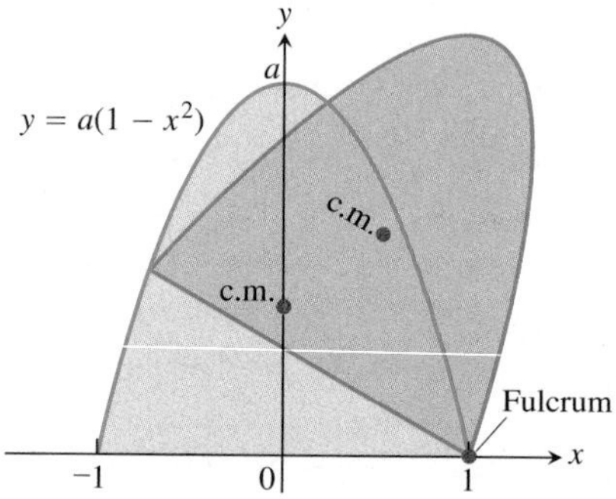

44. Minimizing a moment of inertia A rectangular plate of constant density $\delta(x, y) = 1$ occupies the region bounded by the lines $x = 4$ and $y = 2$ in the first quadrant. The moment of inertia I_a of the rectangle about the line $y = a$ is given by the integral

$$I_a = \int_0^4 \int_0^2 (y - a)^2 \, dy \, dx.$$

Find the value of a that minimizes I_a.

45. Centroid of unbounded region Find the centroid of the infinite region in the xy-plane bounded by the curves $y = 1/\sqrt{1 - x^2}$, $y = -1/\sqrt{1 - x^2}$, and the lines $x = 0, x = 1$.

46. Radius of gyration of slender rod Find the radius of gyration of a slender rod of constant linear density δ gm/cm and length L cm with respect to an axis

a. through the rod's center of mass perpendicular to the rod's axis.

b. perpendicular to the rod's axis at one end of the rod.

47. (*Continuation of Exercise 34.*) A thin plate of now constant density δ occupies the region R in the xy-plane bounded by the curves $x = y^2$ and $x = 2y - y^2$.

a. Constant density Find δ such that the plate has the same mass as the plate in Exercise 34.

b. Average value Compare the value of δ found in part (a) with the average value of $\delta(x, y) = y + 1$ over R.

48. Average temperature in Texas According to the *Texas Almanac*, Texas has 254 counties and a National Weather Service station in each county. Assume that at time t_0, each of the 254 weather stations recorded the local temperature. Find a formula that would give a reasonable approximation to the average temperature in Texas at time t_0. Your answer should involve information that you would expect to be readily available in the *Texas Almanac*.

The Parallel Axis Theorem

Let $L_{\text{c.m.}}$ be a line in the xy-plane that runs through the center of mass of a thin plate of mass m covering a region in the plane. Let L be a line in the plane parallel to and h units away from $L_{\text{c.m.}}$. The **Parallel Axis Theorem** says that under these conditions the moments of inertia I_L and $I_{\text{c.m.}}$ of the plate about L and $L_{\text{c.m.}}$ satisfy the equation

$$I_L = I_{\text{c.m.}} + mh^2.$$

This equation gives a quick way to calculate one moment when the other moment and the mass are known.

49. Proof of the Parallel Axis Theorem

a. Show that the first moment of a thin flat plate about any line in the plane of the plate through the plate's center of mass is zero. (*Hint:* Place the center of mass at the origin with the line along the y-axis. What does the formula $\bar{x} = M_y/M$ then tell you?)

b. Use the result in part (a) to derive the Parallel Axis Theorem. Assume that the plane is coordinatized in a way that makes $L_{\text{c.m.}}$ the y-axis and L the line $x = h$. Then expand the integrand of the integral for I_L to rewrite the integral as the sum of integrals whose values you recognize.

50. Finding moments of inertia

a. Use the Parallel Axis Theorem and the results of Example 4 to find the moments of inertia of the plate in Example 4 about the vertical and horizontal lines through the plate's center of mass.

b. Use the results in part (a) to find the plate's moments of inertia about the lines $x = 1$ and $y = 2$.

Pappus's Formula

Pappus knew that the centroid of the union of two nonoverlapping plane regions lies on the line segment joining their individual centroids. More specifically, suppose that m_1 and m_2 are the masses of thin plates P_1 and P_2 that cover nonoverlapping regions in the xy-plane. Let $\mathbf{c}_1$ and $\mathbf{c}_2$ be the vectors from the origin to the respective centers of mass of P_1 and P_2. Then the center of mass of the union $P_1 \cup P_2$ of the two plates is determined by the vector

$$\mathbf{c} = \frac{m_1\mathbf{c}_1 + m_2\mathbf{c}_2}{m_1 + m_2}. \tag{5}$$

Equation (5) is known as **Pappus's formula**. For more than two nonoverlapping plates, as long as their number is finite, the formula generalizes to

$$\mathbf{c} = \frac{m_1\mathbf{c}_1 + m_2\mathbf{c}_2 + \cdots + m_n\mathbf{c}_n}{m_1 + m_2 + \cdots + m_n}. \tag{6}$$

This formula is especially useful for finding the centroid of a plate of irregular shape that is made up of pieces of constant density whose centroids we know from geometry. We find the centroid of each piece and apply Equation (6) to find the centroid of the plate.

51. Derive Pappus's formula (Equation (5)). (*Hint:* Sketch the plates as regions in the first quadrant and label their centers of mass as $(\bar{x}_1, \bar{y}_1)$ and $(\bar{x}_2, \bar{y}_2)$. What are the moments of $P_1 \cup P_2$ about the coordinate axes?)

52. Use Equation (5) and mathematical induction to show that Equation (6) holds for any positive integer $n > 2$.

53. Let A, B, and C be the shapes indicated in the accompanying figure. Use Pappus's formula to find the centroid of

a. $A \cup B$ **b.** $A \cup C$

c. $B \cup C$ **d.** $A \cup B \cup C$.

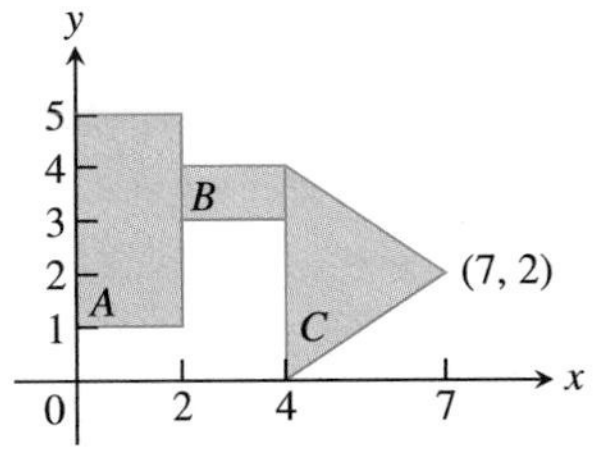

54. Locating center of mass Locate the center of mass of the carpenter's square, shown here.

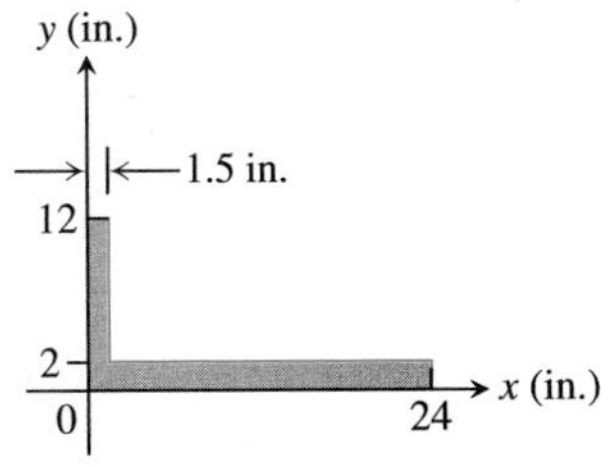

55. An isosceles triangle T has base $2a$ and altitude h. The base lies along the diameter of a semicircular disk D of radius a so that the two together make a shape resembling an ice cream cone. What relation must hold between a and h to place the centroid of $T \cup D$ on the common boundary of T and D? Inside T?

56. An isosceles triangle T of altitude h has as its base one side of a square Q whose edges have length s. (The square and triangle do not overlap.) What relation must hold between h and s to place the centroid of $T \cup Q$ on the base of the triangle? Compare your answer with the answer to Exercise 55.

15.3 Double Integrals in Polar Form

Integrals are sometimes easier to evaluate if we change to polar coordinates. This section shows how to accomplish the change and how to evaluate integrals over regions whose boundaries are given by polar equations.

Integrals in Polar Coordinates

When we defined the double integral of a function over a region R in the xy-plane, we began by cutting R into rectangles whose sides were parallel to the coordinate axes. These were the natural shapes to use because their sides have either constant x-values or constant y-values. In polar coordinates, the natural shape is a "polar rectangle" whose sides have constant r- and θ-values.

Suppose that a function $f(r, \theta)$ is defined over a region R that is bounded by the rays $\theta = \alpha$ and $\theta = \beta$ and by the continuous curves $r = g_1(\theta)$ and $r = g_2(\theta)$. Suppose also that $0 \le g_1(\theta) \le g_2(\theta) \le a$ for every value of θ between α and β. Then R lies in a fan-shaped region Q defined by the inequalities $0 \le r \le a$ and $\alpha \le \theta \le \beta$. See Figure 15.21.

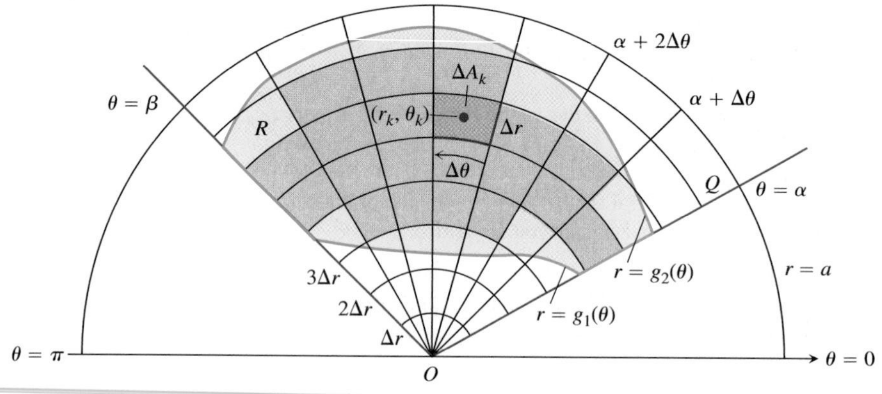

FIGURE 15.21 The region R: $g_1(\theta) \le r \le g_2(\theta)$, $\alpha \le \theta \le \beta$, is contained in the fan-shaped region Q: $0 \le r \le a$, $\alpha \le \theta \le \beta$. The partition of Q by circular arcs and rays induces a partition of R.

We cover Q by a grid of circular arcs and rays. The arcs are cut from circles centered at the origin, with radii $\Delta r, 2\Delta r, \ldots, m\Delta r$, where $\Delta r = a/m$. The rays are given by

$$\theta = \alpha, \qquad \theta = \alpha + \Delta\theta, \qquad \theta = \alpha + 2\Delta\theta, \qquad \ldots, \qquad \theta = \alpha + m'\Delta\theta = \beta,$$

where $\Delta\theta = (\beta - \alpha)/m'$. The arcs and rays partition Q into small patches called "polar rectangles."

We number the polar rectangles that lie inside R (the order does not matter), calling their areas $\Delta A_1, \Delta A_2, \ldots, \Delta A_n$. We let (r_k, θ_k) be any point in the polar rectangle whose area is ΔA_k. We then form the sum

$$S_n = \sum_{k=1}^{n} f(r_k, \theta_k)\, \Delta A_k.$$

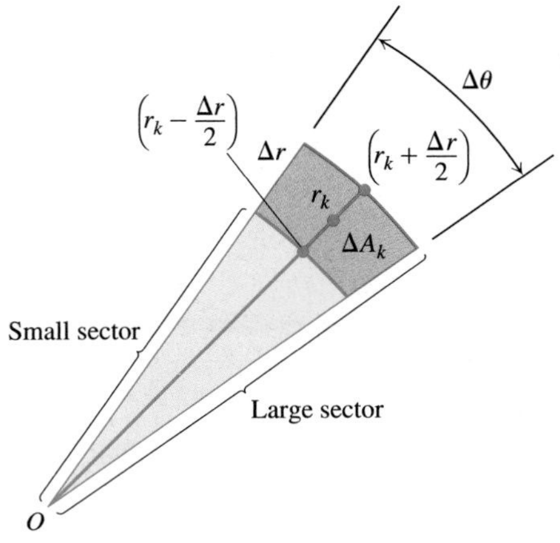

FIGURE 15.22 The observation that

$$\Delta A_k = \begin{pmatrix}\text{area of} \\ \text{large sector}\end{pmatrix} - \begin{pmatrix}\text{area of} \\ \text{small sector}\end{pmatrix}$$

leads to the formula $\Delta A_k = r_k\,\Delta r\,\Delta\theta$.

If f is continuous throughout R, this sum will approach a limit as we refine the grid to make Δr and $\Delta\theta$ go to zero. The limit is called the double integral of f over R. In symbols,

$$\lim_{n\to\infty} S_n = \iint_R f(r, \theta)\,dA.$$

To evaluate this limit, we first have to write the sum S_n in a way that expresses ΔA_k in terms of Δr and $\Delta\theta$. For convenience we choose r_k to be the average of the radii of the inner and outer arcs bounding the kth polar rectangle ΔA_k. The radius of the inner arc bounding ΔA_k is then $r_k - (\Delta r/2)$ (Figure 15.22). The radius of the outer arc is $r_k + (\Delta r/2)$.

The area of a wedge-shaped sector of a circle having radius r and angle θ is

$$A = \frac{1}{2}\theta \cdot r^2,$$

as can be seen by multiplying πr^2, the area of the circle, by $\theta/2\pi$, the fraction of the circle's area contained in the wedge. So the areas of the circular sectors subtended by these arcs at the origin are

$$\text{Inner radius:} \qquad \frac{1}{2}\left(r_k - \frac{\Delta r}{2}\right)^2 \Delta\theta$$

$$\text{Outer radius:} \qquad \frac{1}{2}\left(r_k + \frac{\Delta r}{2}\right)^2 \Delta\theta.$$

Therefore,

$$\begin{aligned}\Delta A_k &= \text{area of large sector} - \text{area of small sector} \\ &= \frac{\Delta\theta}{2}\left[\left(r_k + \frac{\Delta r}{2}\right)^2 - \left(r_k - \frac{\Delta r}{2}\right)^2\right] = \frac{\Delta\theta}{2}(2r_k\,\Delta r) = r_k\,\Delta r\,\Delta\theta.\end{aligned}$$

Combining this result with the sum defining S_n gives

$$S_n = \sum_{k=1}^{n} f(r_k, \theta_k) r_k\,\Delta r\,\Delta\theta.$$

As $n \to \infty$ and the values of Δr and $\Delta\theta$ approach zero, these sums converge to the double integral

$$\lim_{n\to\infty} S_n = \iint_R f(r, \theta)\,r\,dr\,d\theta.$$

A version of Fubini's Theorem says that the limit approached by these sums can be evaluated by repeated single integrations with respect to r and θ as

$$\iint_R f(r, \theta)\,dA = \int_{\theta=\alpha}^{\theta=\beta}\int_{r=g_1(\theta)}^{r=g_2(\theta)} f(r, \theta)\,r\,dr\,d\theta.$$

Finding Limits of Integration

The procedure for finding limits of integration in rectangular coordinates also works for polar coordinates. To evaluate $\iint_R f(r, \theta)\,dA$ over a region R in polar coordinates, integrating first with respect to r and then with respect to θ, take the following steps.

1. *Sketch*: Sketch the region and label the bounding curves.

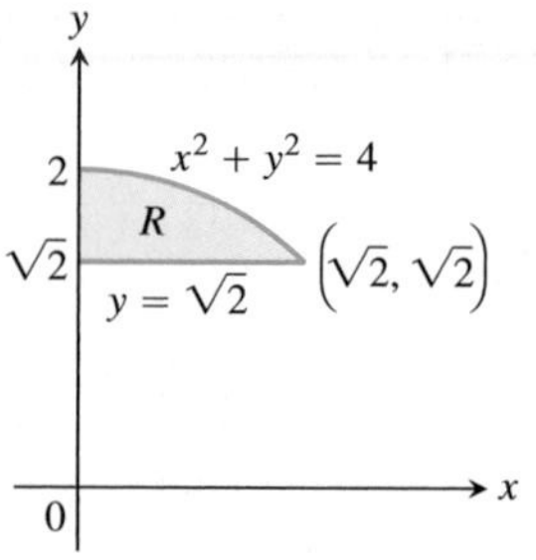

2. *Find the r-limits of integration*: Imagine a ray L from the origin cutting through R in the direction of increasing r. Mark the r-values where L enters and leaves R. These are the r-limits of integration. They usually depend on the angle θ that L makes with the positive x-axis.

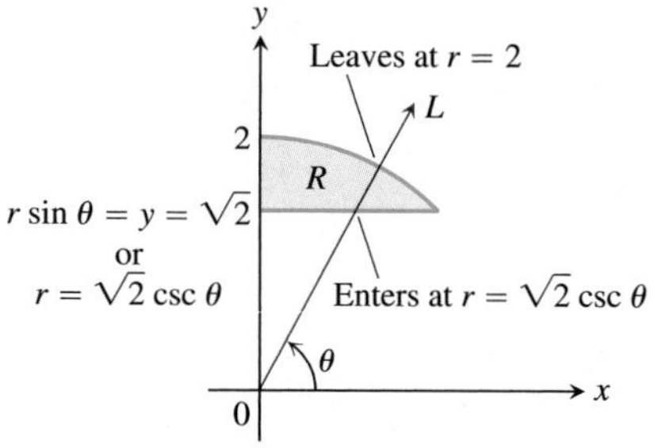

3. *Find the θ-limits of integration*: Find the smallest and largest θ-values that bound R. These are the θ-limits of integration.

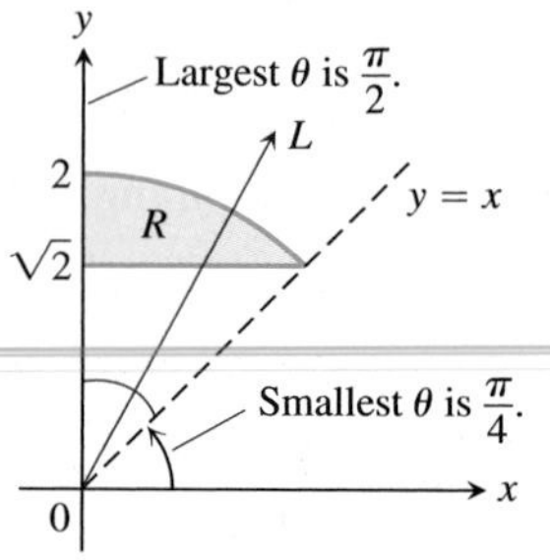

The integral is

$$\iint_R f(r, \theta)\, dA = \int_{\theta=\pi/4}^{\theta=\pi/2} \int_{r=\sqrt{2}\csc\theta}^{r=2} f(r, \theta)\, r\, dr\, d\theta.$$

EXAMPLE 1 Finding Limits of Integration

Find the limits of integration for integrating $f(r, \theta)$ over the region R that lies inside the cardioid $r = 1 + \cos\theta$ and outside the circle $r = 1$.

Solution

1. We first sketch the region and label the bounding curves (Figure 15.23).
2. Next we find the *r-limits of integration*. A typical ray from the origin enters R where $r = 1$ and leaves where $r = 1 + \cos\theta$.

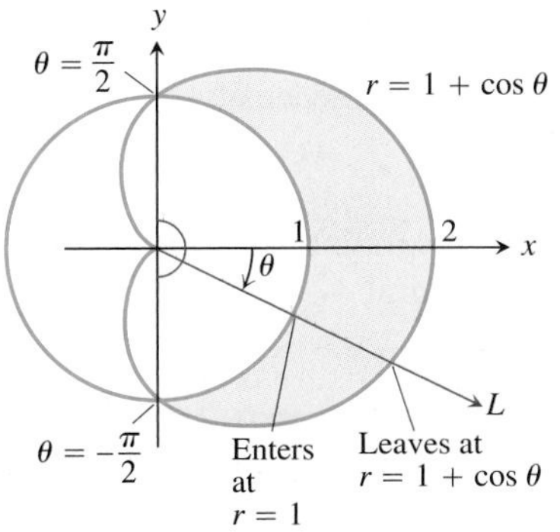

FIGURE 15.23 Finding the limits of integration in polar coordinates for the region in Example 1.

3. Finally we find the *θ-limits of integration*. The rays from the origin that intersect R run from $\theta = -\pi/2$ to $\theta = \pi/2$. The integral is

$$\int_{-\pi/2}^{\pi/2}\int_{1}^{1+\cos\theta} f(r,\theta)\, r\, dr\, d\theta.$$

If $f(r, \theta)$ is the constant function whose value is 1, then the integral of f over R is the area of R.

Area in Polar Coordinates

The area of a closed and bounded region R in the polar coordinate plane is

$$A = \iint_R r\, dr\, d\theta.$$

This formula for area is consistent with all earlier formulas, although we do not prove this fact.

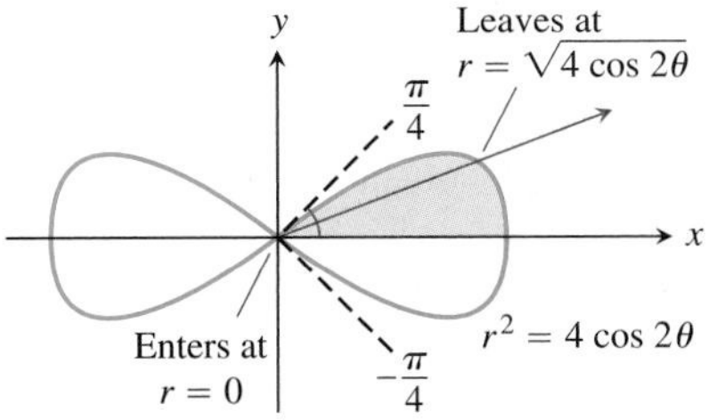

FIGURE 15.24 To integrate over the shaded region, we run r from 0 to $\sqrt{4\cos 2\theta}$ and θ from 0 to $\pi/4$ (Example 2).

EXAMPLE 2 Finding Area in Polar Coordinates

Find the area enclosed by the lemniscate $r^2 = 4\cos 2\theta$.

Solution We graph the lemniscate to determine the limits of integration (Figure 15.24) and see from the symmetry of the region that the total area is 4 times the first-quadrant portion.

$$A = 4\int_0^{\pi/4}\int_0^{\sqrt{4\cos 2\theta}} r\, dr\, d\theta = 4\int_0^{\pi/4}\left[\frac{r^2}{2}\right]_{r=0}^{r=\sqrt{4\cos 2\theta}} d\theta$$

$$= 4\int_0^{\pi/4} 2\cos 2\theta\, d\theta = 4\sin 2\theta\Big]_0^{\pi/4} = 4.$$

Changing Cartesian Integrals into Polar Integrals

The procedure for changing a Cartesian integral $\iint_R f(x, y)\, dx\, dy$ into a polar integral has two steps. First substitute $x = r\cos\theta$ and $y = r\sin\theta$, and replace $dx\, dy$ by $r\, dr\, d\theta$ in the Cartesian integral. Then supply polar limits of integration for the boundary of R.

The Cartesian integral then becomes

$$\iint_R f(x, y)\, dx\, dy = \iint_G f(r\cos\theta, r\sin\theta)\, r\, dr\, d\theta,$$

where G denotes the region of integration in polar coordinates. This is like the substitution method in Chapter 5 except that there are now two variables to substitute for instead of one. Notice that $dx\, dy$ is not replaced by $dr\, d\theta$ but by $r\, dr\, d\theta$. A more general discussion of changes of variables (substitutions) in multiple integrals is given in Section 15.7.

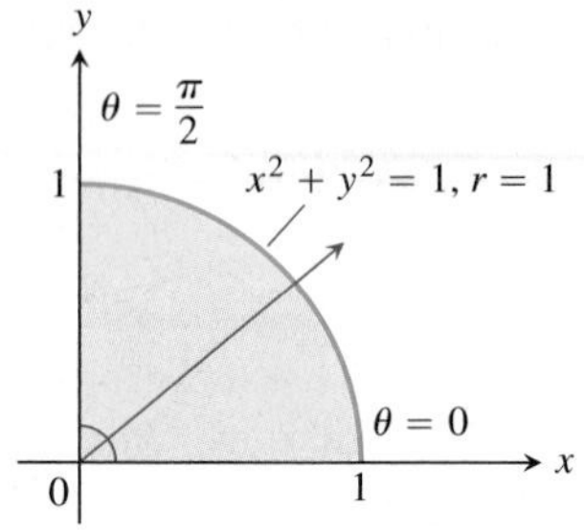

FIGURE 15.25 In polar coordinates, this region is described by simple inequalities:

$$0 \le r \le 1 \quad \text{and} \quad 0 \le \theta \le \pi/2$$

(Example 3).

EXAMPLE 3 Changing Cartesian Integrals to Polar Integrals

Find the polar moment of inertia about the origin of a thin plate of density $\delta(x, y) = 1$ bounded by the quarter circle $x^2 + y^2 = 1$ in the first quadrant.

Solution We sketch the plate to determine the limits of integration (Figure 15.25). In Cartesian coordinates, the polar moment is the value of the integral

$$\int_0^1 \int_0^{\sqrt{1-x^2}} (x^2 + y^2)\, dy\, dx.$$

Integration with respect to y gives

$$\int_0^1 \left(x^2\sqrt{1 - x^2} + \frac{(1 - x^2)^{3/2}}{3} \right) dx,$$

an integral difficult to evaluate without tables.

Things go better if we change the original integral to polar coordinates. Substituting $x = r\cos\theta$, $y = r\sin\theta$ and replacing $dx\, dy$ by $r\, dr\, d\theta$, we get

$$\int_0^1 \int_0^{\sqrt{1-x^2}} (x^2 + y^2)\, dy\, dx = \int_0^{\pi/2} \int_0^1 (r^2)\, r\, dr\, d\theta$$

$$= \int_0^{\pi/2} \left[\frac{r^4}{4} \right]_{r=0}^{r=1} d\theta = \int_0^{\pi/2} \frac{1}{4}\, d\theta = \frac{\pi}{8}.$$

Why is the polar coordinate transformation so effective here? One reason is that $x^2 + y^2$ simplifies to r^2. Another is that the limits of integration become constants. ■

EXAMPLE 4 Evaluating Integrals Using Polar Coordinates

Evaluate

$$\iint_R e^{x^2+y^2}\, dy\, dx,$$

where R is the semicircular region bounded by the x-axis and the curve $y = \sqrt{1 - x^2}$ (Figure 15.26).

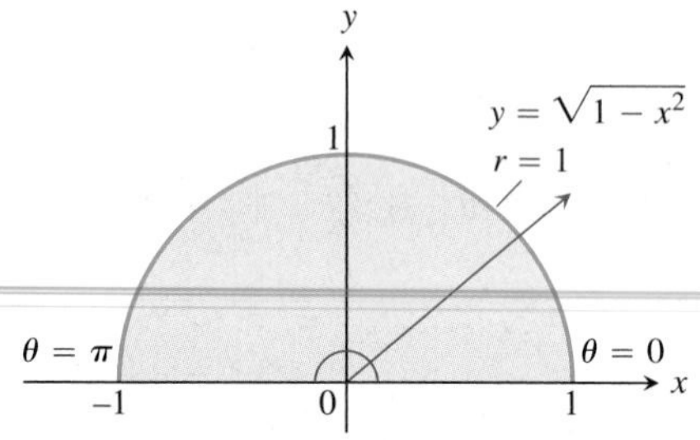

FIGURE 15.26 The semicircular region in Example 4 is the region

$$0 \le r \le 1, \qquad 0 \le \theta \le \pi.$$

Solution In Cartesian coordinates, the integral in question is a nonelementary integral and there is no direct way to integrate $e^{x^2+y^2}$ with respect to either x or y. Yet this integral and others like it are important in mathematics—in statistics, for example—and we need to find a way to evaluate it. Polar coordinates save the day. Substituting $x = r\cos\theta$, $y = r\sin\theta$ and replacing $dy\, dx$ by $r\, dr\, d\theta$ enables us to evaluate the integral as

$$\iint_R e^{x^2+y^2}\, dy\, dx = \int_0^{\pi} \int_0^1 e^{r^2}\, r\, dr\, d\theta = \int_0^{\pi} \left[\frac{1}{2} e^{r^2} \right]_0^1 d\theta$$

$$= \int_0^{\pi} \frac{1}{2}(e - 1)\, d\theta = \frac{\pi}{2}(e - 1).$$

The r in the $r\, dr\, d\theta$ was just what we needed to integrate e^{r^2}. Without it, we would have been unable to proceed. ■

EXERCISES 15.3

Evaluating Polar Integrals

In Exercises 1–16, change the Cartesian integral into an equivalent polar integral. Then evaluate the polar integral.

1. $\int_{-1}^{1}\int_{0}^{\sqrt{1-x^2}} dy\,dx$

2. $\int_{-1}^{1}\int_{-\sqrt{1-x^2}}^{\sqrt{1-x^2}} dy\,dx$

3. $\int_{0}^{1}\int_{0}^{\sqrt{1-y^2}} (x^2+y^2)\,dx\,dy$

4. $\int_{-1}^{1}\int_{-\sqrt{1-y^2}}^{\sqrt{1-y^2}} (x^2+y^2)\,dy\,dx$

5. $\int_{-a}^{a}\int_{-\sqrt{a^2-x^2}}^{\sqrt{a^2-x^2}} dy\,dx$

6. $\int_{0}^{2}\int_{0}^{\sqrt{4-y^2}} (x^2+y^2)\,dx\,dy$

7. $\int_{0}^{6}\int_{0}^{y} x\,dx\,dy$

8. $\int_{0}^{2}\int_{0}^{x} y\,dy\,dx$

9. $\int_{-1}^{0}\int_{-\sqrt{1-x^2}}^{0} \frac{2}{1+\sqrt{x^2+y^2}}\,dy\,dx$

10. $\int_{-1}^{1}\int_{-\sqrt{1-y^2}}^{0} \frac{4\sqrt{x^2+y^2}}{1+x^2+y^2}\,dx\,dy$

11. $\int_{0}^{\ln 2}\int_{0}^{\sqrt{(\ln 2)^2-y^2}} e^{\sqrt{x^2+y^2}}\,dx\,dy$

12. $\int_{0}^{1}\int_{0}^{\sqrt{1-x^2}} e^{-(x^2+y^2)}\,dy\,dx$

13. $\int_{0}^{2}\int_{0}^{\sqrt{1-(x-1)^2}} \frac{x+y}{x^2+y^2}\,dy\,dx$

14. $\int_{0}^{2}\int_{-\sqrt{1-(y-1)^2}}^{0} xy^2\,dx\,dy$

15. $\int_{-1}^{1}\int_{-\sqrt{1-y^2}}^{\sqrt{1-y^2}} \ln(x^2+y^2+1)\,dx\,dy$

16. $\int_{-1}^{1}\int_{-\sqrt{1-x^2}}^{\sqrt{1-x^2}} \frac{2}{(1+x^2+y^2)^2}\,dy\,dx$

Finding Area in Polar Coordinates

17. Find the area of the region cut from the first quadrant by the curve $r = 2(2 - \sin 2\theta)^{1/2}$.

18. **Cardioid overlapping a circle** Find the area of the region that lies inside the cardioid $r = 1 + \cos\theta$ and outside the circle $r = 1$.

19. **One leaf of a rose** Find the area enclosed by one leaf of the rose $r = 12\cos 3\theta$.

20. **Snail shell** Find the area of the region enclosed by the positive x-axis and spiral $r = 4\theta/3$, $0 \le \theta \le 2\pi$. The region looks like a snail shell.

21. **Cardioid in the first quadrant** Find the area of the region cut from the first quadrant by the cardioid $r = 1 + \sin\theta$.

22. **Overlapping cardioids** Find the area of the region common to the interiors of the cardioids $r = 1 + \cos\theta$ and $r = 1 - \cos\theta$.

Masses and Moments

23. **First moment of a plate** Find the first moment about the x-axis of a thin plate of constant density $\delta(x, y) = 3$, bounded below by the x-axis and above by the cardioid $r = 1 - \cos\theta$.

24. **Inertial and polar moments of a disk** Find the moment of inertia about the x-axis and the polar moment of inertia about the origin of a thin disk bounded by the circle $x^2 + y^2 = a^2$ if the disk's density at the point (x, y) is $\delta(x, y) = k(x^2 + y^2)$, k a constant.

25. **Mass of a plate** Find the mass of a thin plate covering the region outside the circle $r = 3$ and inside the circle $r = 6\sin\theta$ if the plate's density function is $\delta(x, y) = 1/r$.

26. **Polar moment of a cardioid overlapping circle** Find the polar moment of inertia about the origin of a thin plate covering the region that lies inside the cardioid $r = 1 - \cos\theta$ and outside the circle $r = 1$ if the plate's density function is $\delta(x, y) = 1/r^2$.

27. **Centroid of a cardioid region** Find the centroid of the region enclosed by the cardioid $r = 1 + \cos\theta$.

28. **Polar moment of a cardioid region** Find the polar moment of inertia about the origin of a thin plate enclosed by the cardioid $r = 1 + \cos\theta$ if the plate's density function is $\delta(x, y) = 1$.

Average Values

29. **Average height of a hemisphere** Find the average height of the hemisphere $z = \sqrt{a^2 - x^2 - y^2}$ above the disk $x^2 + y^2 \le a^2$ in the xy-plane.

30. **Average height of a cone** Find the average height of the (single) cone $z = \sqrt{x^2 + y^2}$ above the disk $x^2 + y^2 \le a^2$ in the xy-plane.

31. **Average distance from interior of disk to center** Find the average distance from a point $P(x, y)$ in the disk $x^2 + y^2 \le a^2$ to the origin.

32. **Average distance squared from a point in a disk to a point in its boundary** Find the average value of the *square* of the distance from the point $P(x, y)$ in the disk $x^2 + y^2 \le 1$ to the boundary point $A(1, 0)$.

Theory and Examples

33. **Converting to a polar integral** Integrate $f(x, y) = [\ln(x^2 + y^2)]/\sqrt{x^2 + y^2}$ over the region $1 \le x^2 + y^2 \le e$.

34. **Converting to a polar integral** Integrate $f(x, y) = [\ln(x^2 + y^2)]/(x^2 + y^2)$ over the region $1 \le x^2 + y^2 \le e^2$.

35. **Volume of noncircular right cylinder** The region that lies inside the cardioid $r = 1 + \cos\theta$ and outside the circle $r = 1$ is the base of a solid right cylinder. The top of the cylinder lies in the plane $z = x$. Find the cylinder's volume.

36. Volume of noncircular right cylinder The region enclosed by the lemniscate $r^2 = 2\cos 2\theta$ is the base of a solid right cylinder whose top is bounded by the sphere $z = \sqrt{2 - r^2}$. Find the cylinder's volume.

37. Converting to polar integrals

a. The usual way to evaluate the improper integral $I = \int_0^\infty e^{-x^2}\,dx$ is first to calculate its square:

$$I^2 = \left(\int_0^\infty e^{-x^2}\,dx\right)\left(\int_0^\infty e^{-y^2}\,dy\right) = \int_0^\infty\int_0^\infty e^{-(x^2+y^2)}\,dx\,dy.$$

Evaluate the last integral using polar coordinates and solve the resulting equation for I.

b. Evaluate

$$\lim_{x\to\infty} \operatorname{erf}(x) = \lim_{x\to\infty}\int_0^x \frac{2e^{-t^2}}{\sqrt{\pi}}\,dt.$$

38. Converting to a polar integral Evaluate the integral

$$\int_0^\infty\int_0^\infty \frac{1}{(1 + x^2 + y^2)^2}\,dx\,dy.$$

39. Existence Integrate the function $f(x, y) = 1/(1 - x^2 - y^2)$ over the disk $x^2 + y^2 \le 3/4$. Does the integral of $f(x, y)$ over the disk $x^2 + y^2 \le 1$ exist? Give reasons for your answer.

40. Area formula in polar coordinates Use the double integral in polar coordinates to derive the formula

$$A = \int_\alpha^\beta \frac{1}{2} r^2\,d\theta$$

for the area of the fan-shaped region between the origin and polar curve $r = f(\theta)$, $\alpha \le \theta \le \beta$.

41. Average distance to a given point inside a disk Let P_0 be a point inside a circle of radius a and let h denote the distance from P_0 to the center of the circle. Let d denote the distance from an arbitrary point P to P_0. Find the average value of d^2 over the region enclosed by the circle. (*Hint:* Simplify your work by placing the center of the circle at the origin and P_0 on the x-axis.)

42. Area Suppose that the area of a region in the polar coordinate plane is

$$A = \int_{\pi/4}^{3\pi/4}\int_{\csc\theta}^{2\sin\theta} r\,dr\,d\theta.$$

Sketch the region and find its area.

COMPUTER EXPLORATIONS

Coordinate Conversions

In Exercises 43–46, use a CAS to change the Cartesian integrals into an equivalent polar integral and evaluate the polar integral. Perform the following steps in each exercise.

a. Plot the Cartesian region of integration in the xy-plane.

b. Change each boundary curve of the Cartesian region in part (a) to its polar representation by solving its Cartesian equation for r and θ.

c. Using the results in part (b), plot the polar region of integration in the $r\theta$-plane.

d. Change the integrand from Cartesian to polar coordinates. Determine the limits of integration from your plot in part (c) and evaluate the polar integral using the CAS integration utility.

43. $\displaystyle\int_0^1\int_x^1 \frac{y}{x^2 + y^2}\,dy\,dx$

44. $\displaystyle\int_0^1\int_0^{x/2} \frac{x}{x^2 + y^2}\,dy\,dx$

45. $\displaystyle\int_0^1\int_{-y/3}^{y/3} \frac{y}{\sqrt{x^2 + y^2}}\,dx\,dy$

46. $\displaystyle\int_0^1\int_y^{2-y} \sqrt{x + y}\,dx\,dy$

15.4 Triple Integrals in Rectangular Coordinates

Just as double integrals allow us to deal with more general situations than could be handled by single integrals, triple integrals enable us to solve still more general problems. We use triple integrals to calculate the volumes of three-dimensional shapes, the masses and moments of solids of varying density, and the average value of a function over a three-dimensional region. Triple integrals also arise in the study of vector fields and fluid flow in three dimensions, as we will see in Chapter 16.

Triple Integrals

If $F(x, y, z)$ is a function defined on a closed bounded region D in space, such as the region occupied by a solid ball or a lump of clay, then the integral of F over D may be defined in

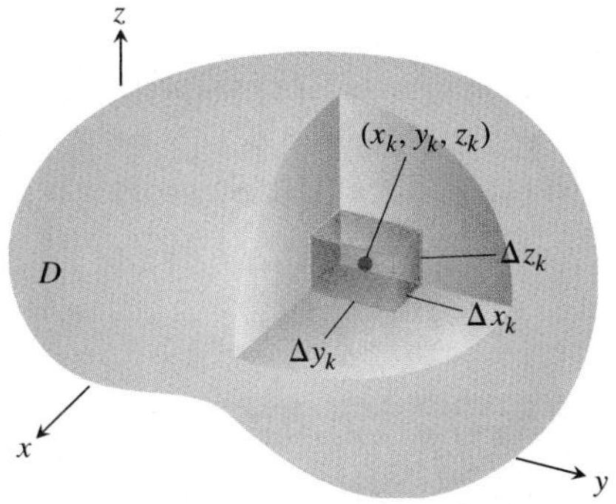

FIGURE 15.27 Partitioning a solid with rectangular cells of volume ΔV_k.

the following way. We partition a rectangular boxlike region containing D into rectangular cells by planes parallel to the coordinate axis (Figure 15.27). We number the cells that lie inside D from 1 to n in some order, the kth cell having dimensions Δx_k by Δy_k by Δz_k and volume $\Delta V_k = \Delta x_k \Delta y_k \Delta z_k$. We choose a point (x_k, y_k, z_k) in each cell and form the sum

$$S_n = \sum_{k=1}^{n} F(x_k, y_k, z_k)\,\Delta V_k. \tag{1}$$

We are interested in what happens as D is partitioned by smaller and smaller cells, so that Δx_k, Δy_k, Δz_k and the norm of the partition $\|P\|$, the largest value among Δx_k, Δy_k, Δz_k, all approach zero. When a single limiting value is attained, no matter how the partitions and points (x_k, y_k, z_k) are chosen, we say that F is **integrable** over D. As before, it can be shown that when F is continuous and the bounding surface of D is formed from finitely many smooth surfaces joined together along finitely many smooth curves, then F is integrable. As $\|P\| \to 0$ and the number of cells n goes to ∞, the sums S_n approach a limit. We call this limit the **triple integral of F over D** and write

$$\lim_{n\to\infty} S_n = \iiint_D F(x, y, z)\,dV \qquad \text{or} \qquad \lim_{\|P\|\to 0} S_n = \iiint_D F(x, y, z)\,dx\,dy\,dz.$$

The regions D over which continuous functions are integrable are those that can be closely approximated by small rectangular cells. Such regions include those encountered in applications.

Volume of a Region in Space

If F is the constant function whose value is 1, then the sums in Equation (1) reduce to

$$S_n = \sum F(x_k, y_k, z_k)\,\Delta V_k = \sum 1 \cdot \Delta V_k = \sum \Delta V_k.$$

As Δx_k, Δy_k, and Δz_k approach zero, the cells ΔV_k become smaller and more numerous and fill up more and more of D. We therefore define the volume of D to be the triple integral

$$\lim_{n\to\infty} \sum_{k=1}^{n} \Delta V_k = \iiint_D dV.$$

DEFINITION Volume

The **volume** of a closed, bounded region D in space is

$$V = \iiint_D dV.$$

This definition is in agreement with our previous definitions of volume, though we omit the verification of this fact. As we see in a moment, this integral enables us to calculate the volumes of solids enclosed by curved surfaces.

Finding Limits of Integration

We evaluate a triple integral by applying a three-dimensional version of Fubini's Theorem (Section 15.1) to evaluate it by three repeated single integrations. As with double integrals, there is a geometric procedure for finding the limits of integration for these single integrals.

To evaluate

$$\iiint_D F(x, y, z)\, dV$$

over a region D, integrate first with respect to z, then with respect to y, finally with x.

1. *Sketch:* Sketch the region D along with its "shadow" R (vertical projection) in the xy-plane. Label the upper and lower bounding surfaces of D and the upper and lower bounding curves of R.

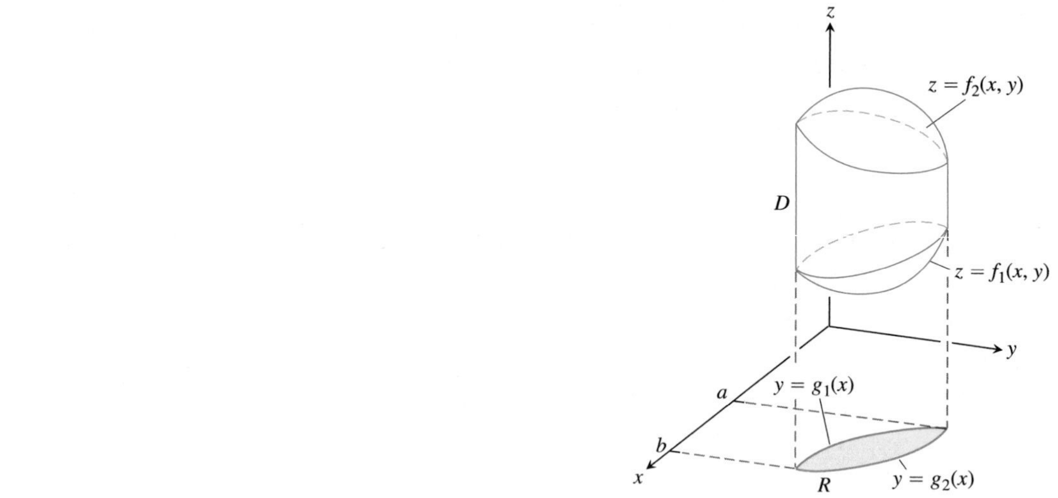

2. *Find the z-limits of integration:* Draw a line M passing through a typical point (x, y) in R parallel to the z-axis. As z increases, M enters D at $z = f_1(x, y)$ and leaves at $z = f_2(x, y)$. These are the z-limits of integration.

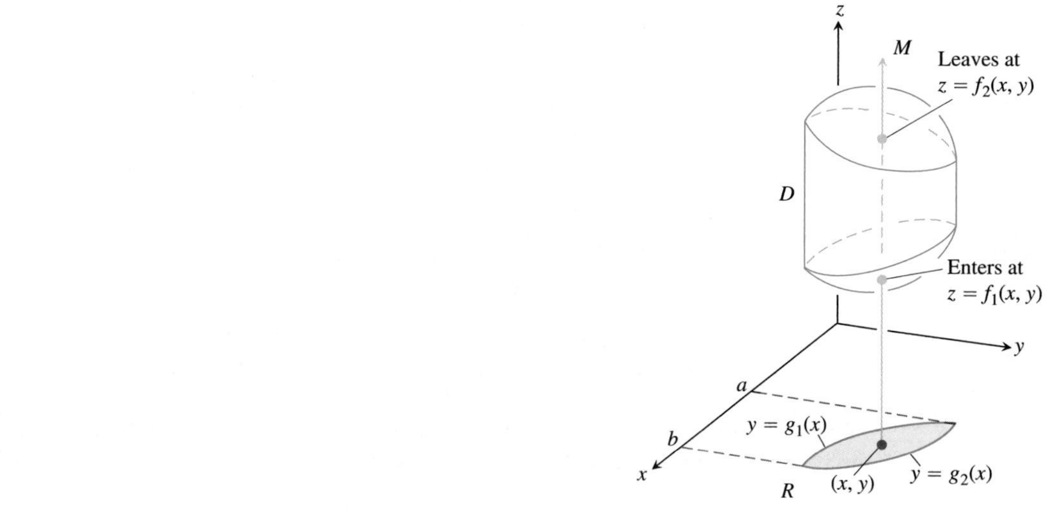

3. *Find the y-limits of integration:* Draw a line L through (x, y) parallel to the y-axis. As y increases, L enters R at $y = g_1(x)$ and leaves at $y = g_2(x)$. These are the y-limits of integration.

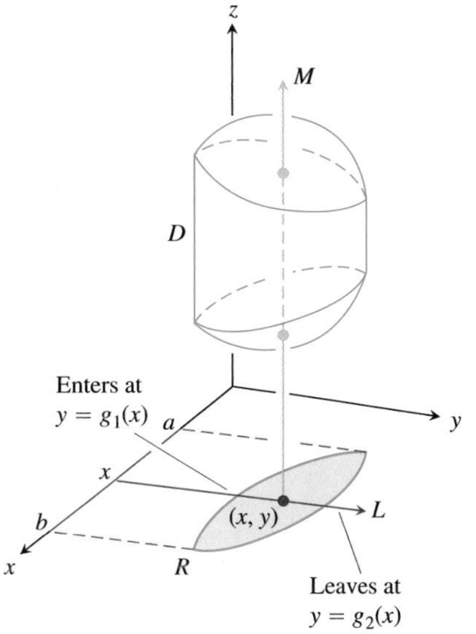

4. *Find the x-limits of integration:* Choose x-limits that include all lines through R parallel to the y-axis ($x = a$ and $x = b$ in the preceding figure). These are the x-limits of integration. The integral is

$$\int_{x=a}^{x=b} \int_{y=g_1(x)}^{y=g_2(x)} \int_{z=f_1(x,y)}^{z=f_2(x,y)} F(x, y, z)\, dz\, dy\, dx.$$

Follow similar procedures if you change the order of integration. The "shadow" of region D lies in the plane of the last two variables with respect to which the iterated integration takes place.

The above procedure applies whenever a solid region D is bounded above and below by a surface, and when the "shadow" region R is bounded by a lower and upper curve. It does not apply to regions with complicated holes through them, although sometimes such regions can be subdivided into simpler regions for which the procedure does apply.

EXAMPLE 1 Finding a Volume

Find the volume of the region D enclosed by the surfaces $z = x^2 + 3y^2$ and $z = 8 - x^2 - y^2$.

Solution The volume is

$$V = \iiint_D dz\, dy\, dx,$$

the integral of $F(x, y, z) = 1$ over D. To find the limits of integration for evaluating the integral, we first sketch the region. The surfaces (Figure 15.28) intersect on the elliptical cylinder $x^2 + 3y^2 = 8 - x^2 - y^2$ or $x^2 + 2y^2 = 4$, $z > 0$. The boundary of the region R, the projection of D onto the xy-plane, is an ellipse with the same equation: $x^2 + 2y^2 = 4$. The "upper" boundary of R is the curve $y = \sqrt{(4 - x^2)/2}$. The lower boundary is the curve $y = -\sqrt{(4 - x^2)/2}$.

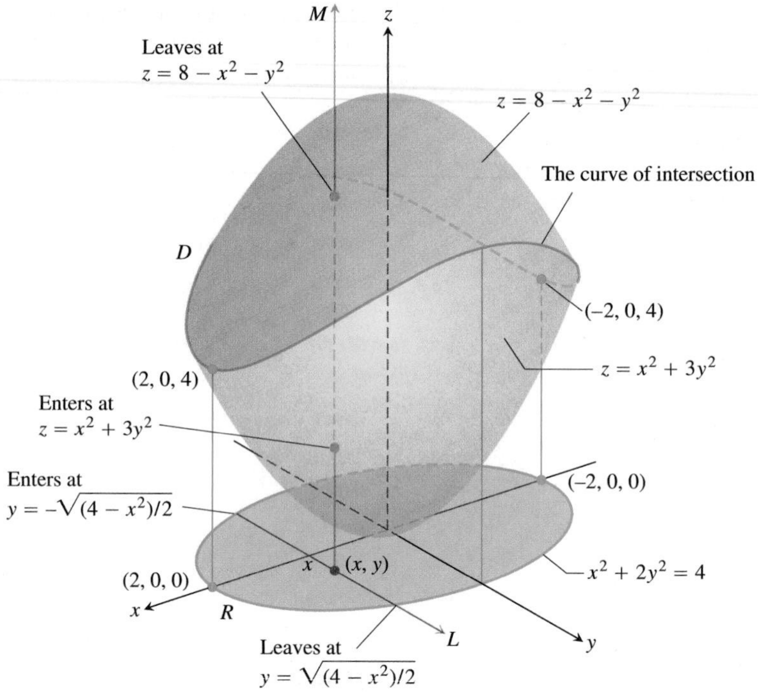

FIGURE 15.28 The volume of the region enclosed by two paraboloids, calculated in Example 1.

Now we find the z-limits of integration. The line M passing through a typical point (x, y) in R parallel to the z-axis enters D at $z = x^2 + 3y^2$ and leaves at $z = 8 - x^2 - y^2$.

Next we find the y-limits of integration. The line L through (x, y) parallel to the y-axis enters R at $y = -\sqrt{(4 - x^2)/2}$ and leaves at $y = \sqrt{(4 - x^2)/2}$.

Finally we find the x-limits of integration. As L sweeps across R, the value of x varies from $x = -2$ at $(-2, 0, 0)$ to $x = 2$ at $(2, 0, 0)$. The volume of D is

$$\begin{aligned}
V &= \iiint_D dz\,dy\,dx \\
&= \int_{-2}^{2}\int_{-\sqrt{(4-x^2)/2}}^{\sqrt{(4-x^2)/2}}\int_{x^2+3y^2}^{8-x^2-y^2} dz\,dy\,dx \\
&= \int_{-2}^{2}\int_{-\sqrt{(4-x^2)/2}}^{\sqrt{(4-x^2)/2}} (8 - 2x^2 - 4y^2)\,dy\,dx \\
&= \int_{-2}^{2}\left[(8 - 2x^2)y - \frac{4}{3}y^3\right]_{y=-\sqrt{(4-x^2)/2}}^{y=\sqrt{(4-x^2)/2}} dx \\
&= \int_{-2}^{2}\left(2(8 - 2x^2)\sqrt{\frac{4 - x^2}{2}} - \frac{8}{3}\left(\frac{4 - x^2}{2}\right)^{3/2}\right) dx \\
&= \int_{-2}^{2}\left[8\left(\frac{4 - x^2}{2}\right)^{3/2} - \frac{8}{3}\left(\frac{4 - x^2}{2}\right)^{3/2}\right] dx = \frac{4\sqrt{2}}{3}\int_{-2}^{2}(4 - x^2)^{3/2}\,dx \\
&= 8\pi\sqrt{2}. \qquad \text{After integration with the substitution } x = 2\sin u.
\end{aligned}$$

■

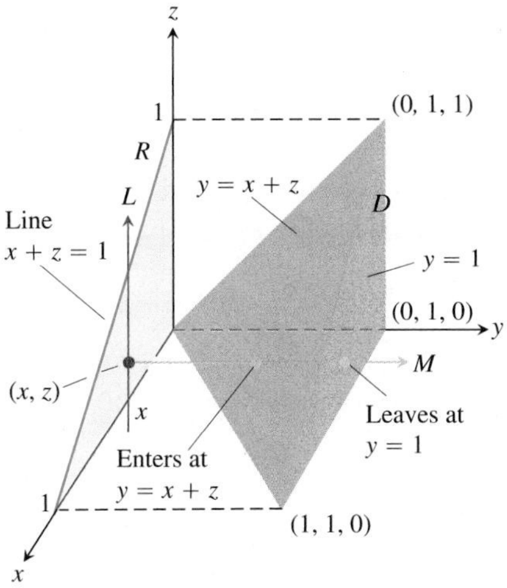

FIGURE 15.29 Finding the limits of integration for evaluating the triple integral of a function defined over the tetrahedron D (Example 2).

In the next example, we project D onto the xz-plane instead of the xy-plane, to show how to use a different order of integration.

EXAMPLE 2 Finding the Limits of Integration in the Order *dy dz dx*

Set up the limits of integration for evaluating the triple integral of a function $F(x, y, z)$ over the tetrahedron D with vertices (0, 0, 0), (1, 1, 0), (0, 1, 0), and (0, 1, 1).

Solution We sketch D along with its "shadow" R in the xz-plane (Figure 15.29). The upper (right-hand) bounding surface of D lies in the plane $y = 1$. The lower (left-hand) bounding surface lies in the plane $y = x + z$. The upper boundary of R is the line $z = 1 - x$. The lower boundary is the line $z = 0$.

First we find the y-limits of integration. The line through a typical point (x, z) in R parallel to the y-axis enters D at $y = x + z$ and leaves at $y = 1$.

Next we find the z-limits of integration. The line L through (x, z) parallel to the z-axis enters R at $z = 0$ and leaves at $z = 1 - x$.

Finally we find the x-limits of integration. As L sweeps across R, the value of x varies from $x = 0$ to $x = 1$. The integral is

$$\int_0^1 \int_0^{1-x} \int_{x+z}^1 F(x, y, z)\, dy\, dz\, dx.$$

■

EXAMPLE 3 Revisiting Example 2 Using the Order *dz dy dx*

To integrate $F(x, y, z)$ over the tetrahedron D in the order $dz\, dy\, dx$, we perform the steps in the following way.

First we find the z-limits of integration. A line parallel to the z-axis through a typical point (x, y) in the xy-plane "shadow" enters the tetrahedron at $z = 0$ and exits through the upper plane where $z = y - x$ (Figure 15.29).

Next we find the y-limits of integration. On the xy-plane, where $z = 0$, the sloped side of the tetrahedron crosses the plane along the line $y = x$. A line through (x, y) parallel to the y-axis enters the shadow in the xy-plane at $y = x$ and exits at $y = 1$.

Finally we find the x-limits of integration. As the line parallel to the y-axis in the previous step sweeps out the shadow, the value of x varies from $x = 0$ to $x = 1$ at the point (1, 1, 0). The integral is

$$\int_0^1 \int_x^1 \int_0^{y-x} F(x, y, z)\, dz\, dy\, dx.$$

For example, if $F(x, y, z) = 1$, we would find the volume of the tetrahedron to be

$$\begin{aligned} V &= \int_0^1 \int_x^1 \int_0^{y-x} dz\, dy\, dx \\ &= \int_0^1 \int_x^1 (y - x)\, dy\, dx \\ &= \int_0^1 \left[\frac{1}{2}y^2 - xy\right]_{y=x}^{y=1} dx \\ &= \int_0^1 \left(\frac{1}{2} - x + \frac{1}{2}x^2\right) dx \\ &= \left[\frac{1}{2}x - \frac{1}{2}x^2 + \frac{1}{6}x^3\right]_0^1 \\ &= \frac{1}{6}. \end{aligned}$$

We get the same result by integrating with the order $dy\,dz\,dx$,

$$V = \int_0^1 \int_0^{1-x} \int_{x+z}^{1} dy\,dz\,dx = \frac{1}{6}.$$ ■

As we have seen, there are sometimes (but not always) two different orders in which the iterated single integrations for evaluating a double integral may be worked. For triple integrals, there can be as many as six, since there are six ways of ordering dx, dy, and dz. Each ordering leads to a different description of the region of integration in space, and to different limits of integration.

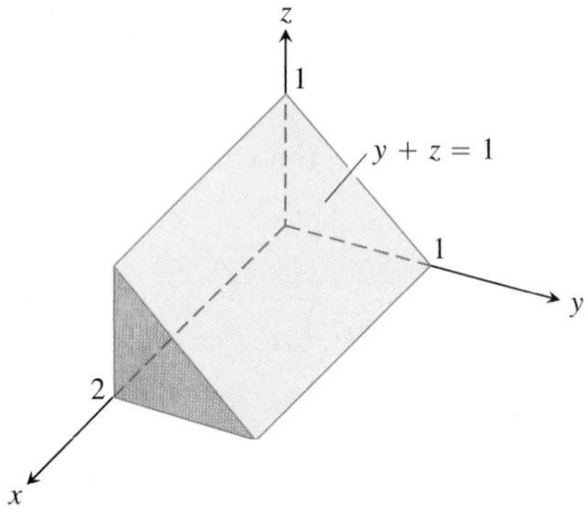

FIGURE 15.30 Example 4 gives six different iterated triple integrals for the volume of this prism.

EXAMPLE 4 Using Different Orders of Integration

Each of the following integrals gives the volume of the solid shown in Figure 15.30.

(a) $\displaystyle\int_0^1 \int_0^{1-z} \int_0^2 dx\,dy\,dz$ **(b)** $\displaystyle\int_0^1 \int_0^{1-y} \int_0^2 dx\,dz\,dy$

(c) $\displaystyle\int_0^1 \int_0^2 \int_0^{1-z} dy\,dx\,dz$ **(d)** $\displaystyle\int_0^2 \int_0^1 \int_0^{1-z} dy\,dz\,dx$

(e) $\displaystyle\int_0^1 \int_0^2 \int_0^{1-y} dz\,dx\,dy$ **(f)** $\displaystyle\int_0^2 \int_0^1 \int_0^{1-y} dz\,dy\,dx$

We work out the integrals in parts (b) and (c):

$$\begin{aligned}
V &= \int_0^1 \int_0^{1-y} \int_0^2 dx\,dz\,dy && \text{Integral in part (b)}\\
&= \int_0^1 \int_0^{1-y} 2\,dz\,dy\\
&= \int_0^1 \left[2z\right]_{z=0}^{z=1-y} dy\\
&= \int_0^1 2(1-y)\,dy\\
&= 1.
\end{aligned}$$

Also,

$$\begin{aligned}
V &= \int_0^1 \int_0^2 \int_0^{1-z} dy\,dx\,dz && \text{Integral in part (c)}\\
&= \int_0^1 \int_0^2 (1-z)\,dx\,dz\\
&= \int_0^1 \left[x - zx\right]_{x=0}^{x=2} dz
\end{aligned}$$

$$= \int_0^1 (2 - 2z)\, dz$$

$$= 1.$$

The integrals in parts (a), (d), (e), and (f) also give $V = 1$. ■

Average Value of a Function in Space

The average value of a function F over a region D in space is defined by the formula

$$\textbf{Average value} \text{ of } F \text{ over } D = \frac{1}{\text{volume of } D} \iiint_D F\, dV. \tag{2}$$

For example, if $F(x, y, z) = \sqrt{x^2 + y^2 + z^2}$, then the average value of F over D is the average distance of points in D from the origin. If $F(x, y, z)$ is the temperature at (x, y, z) on a solid that occupies a region D in space, then the average value of F over D is the average temperature of the solid.

EXAMPLE 5 Finding an Average Value

Find the average value of $F(x, y, z) = xyz$ over the cube bounded by the coordinate planes and the planes $x = 2$, $y = 2$, and $z = 2$ in the first octant.

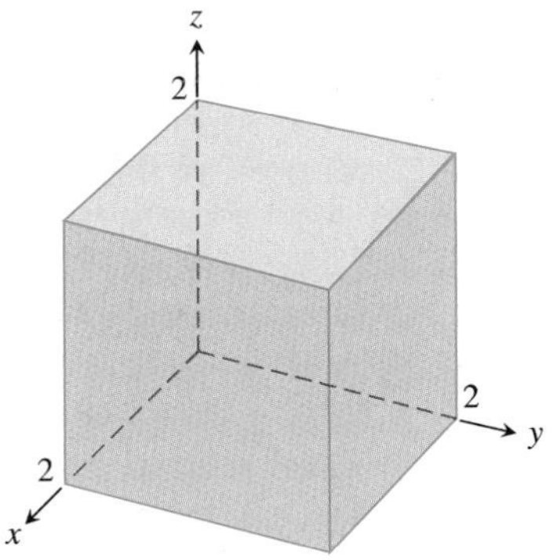

FIGURE 15.31 The region of integration in Example 5.

Solution We sketch the cube with enough detail to show the limits of integration (Figure 15.31). We then use Equation (2) to calculate the average value of F over the cube.

The volume of the cube is $(2)(2)(2) = 8$. The value of the integral of F over the cube is

$$\int_0^2\int_0^2\int_0^2 xyz\, dx\, dy\, dz = \int_0^2\int_0^2 \left[\frac{x^2}{2} yz\right]_{x=0}^{x=2} dy\, dz = \int_0^2\int_0^2 2yz\, dy\, dz$$

$$= \int_0^2 \left[y^2 z\right]_{y=0}^{y=2} dz = \int_0^2 4z\, dz = \left[2z^2\right]_0^2 = 8.$$

With these values, Equation (2) gives

$$\begin{array}{c}\text{Average value of} \\ xyz \text{ over the cube}\end{array} = \frac{1}{\text{volume}} \iiint_{\text{cube}} xyz\, dV = \left(\frac{1}{8}\right)(8) = 1.$$

In evaluating the integral, we chose the order $dx\, dy\, dz$, but any of the other five possible orders would have done as well. ■

Properties of Triple Integrals

Triple integrals have the same algebraic properties as double and single integrals.

Properties of Triple Integrals

If $F = F(x, y, z)$ and $G = G(x, y, z)$ are continuous, then

1. *Constant Multiple:* $\iiint_D kF\,dV = k\iiint_D F\,dV$ (any number k)

2. *Sum and Difference:* $\iiint_D (F \pm G)\,dV = \iiint_D F\,dV \pm \iiint_D G\,dV$

3. *Domination:*

   (a) $\iiint_D F\,dV \geq 0$ if $F \geq 0$ on D

   (b) $\iiint_D F\,dV \geq \iiint_D G\,dV$ if $F \geq G$ on D

4. *Additivity:* $\iiint_D F\,dV = \iiint_{D_1} E\,dV + \iiint_{D_2} F\,dV$

   if D is the union of two nonoverlapping regions D_1 and D_2.

EXERCISES 15.4

Evaluating Triple Integrals in Different Iterations

1. Evaluate the integral in Example 2 taking $F(x, y, z) = 1$ to find the volume of the tetrahedron.

2. **Volume of rectangular solid** Write six different iterated triple integrals for the volume of the rectangular solid in the first octant bounded by the coordinate planes and the planes $x = 1, y = 2$, and $z = 3$. Evaluate one of the integrals.

3. **Volume of tetrahedron** Write six different iterated triple integrals for the volume of the tetrahedron cut from the first octant by the plane $6x + 3y + 2z = 6$. Evaluate one of the integrals.

4. **Volume of solid** Write six different iterated triple integrals for the volume of the region in the first octant enclosed by the cylinder $x^2 + z^2 = 4$ and the plane $y = 3$. Evaluate one of the integrals.

5. **Volume enclosed by paraboloids** Let D be the region bounded by the paraboloids $z = 8 - x^2 - y^2$ and $z = x^2 + y^2$. Write six different triple iterated integrals for the volume of D. Evaluate one of the integrals.

6. **Volume inside paraboloid beneath a plane** Let D be the region bounded by the paraboloid $z = x^2 + y^2$ and the plane $z = 2y$. Write triple iterated integrals in the order $dz\,dx\,dy$ and $dz\,dy\,dx$ that give the volume of D. Do not evaluate either integral.

Evaluating Triple Iterated Integrals

Evaluate the integrals in Exercises 7–20.

7. $\int_0^1\int_0^1\int_0^1 (x^2 + y^2 + z^2)\,dz\,dy\,dx$

8. $\int_0^{\sqrt{2}}\int_0^{3y}\int_{x^2+3y^2}^{8-x^2-y^2} dz\,dx\,dy$

9. $\int_1^e\int_1^e\int_1^e \frac{1}{xyz}\,dx\,dy\,dz$

10. $\int_0^1\int_0^{3-3x}\int_0^{3-3x-y} dz\,dy\,dx$

11. $\int_0^1\int_0^{\pi}\int_0^{\pi} y\sin z\,dx\,dy\,dz$

12. $\int_{-1}^1\int_{-1}^1\int_{-1}^1 (x + y + z)\,dy\,dx\,dz$

13. $\int_0^3\int_0^{\sqrt{9-x^2}}\int_0^{\sqrt{9-x^2}} dz\,dy\,dx$

14. $\int_0^2\int_{-\sqrt{4-y^2}}^{\sqrt{4-y^2}}\int_0^{2x+y} dz\,dx\,dy$

15. $\int_0^1\int_0^{2-x}\int_0^{2-x-y} dz\,dy\,dx$

16. $\int_0^1\int_0^{1-x^2}\int_3^{4-x^2-y} x\,dz\,dy\,dx$

17. $\int_0^{\pi}\int_0^{\pi}\int_0^{\pi} \cos(u + v + w)\,du\,dv\,dw$ (uvw-space)

18. $\int_1^e\int_1^e\int_1^e \ln r \ln s \ln t\,dt\,dr\,ds$ (rst-space)

19. $\int_0^{\pi/4}\int_0^{\ln\sec v}\int_{-\infty}^{2t} e^x\,dx\,dt\,dv$ (tvx-space)

20. $\displaystyle\int_0^7\int_0^2\int_0^{\sqrt{4-q^2}} \frac{q}{r+1}\,dp\,dq\,dr$ (pqr-space)

Volumes Using Triple Integrals

21. Here is the region of integration of the integral

$$\int_{-1}^{1}\int_{x^2}^{1}\int_0^{1-y} dz\,dy\,dx.$$

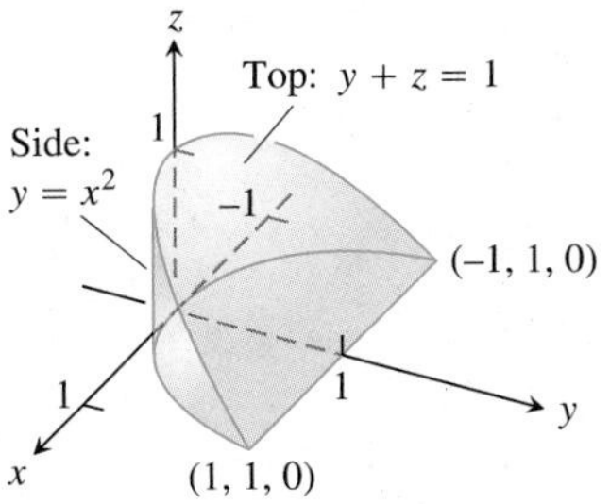

Rewrite the integral as an equivalent iterated integral in the order

a. $dy\,dz\,dx$ **b.** $dy\,dx\,dz$

c. $dx\,dy\,dz$ **d.** $dx\,dz\,dy$

e. $dz\,dx\,dy$.

22. Here is the region of integration of the integral

$$\int_0^1\int_{-1}^{0}\int_0^{y^2} dz\,dy\,dx.$$

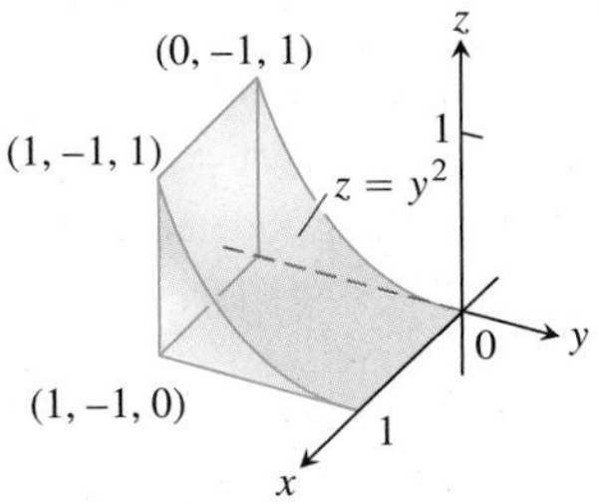

Rewrite the integral as an equivalent iterated integral in the order

a. $dy\,dz\,dx$ **b.** $dy\,dx\,dz$

c. $dx\,dy\,dz$ **d.** $dx\,dz\,dy$

e. $dz\,dx\,dy$.

Find the volumes of the regions in Exercises 23–36.

23. The region between the cylinder $z = y^2$ and the xy-plane that is bounded by the planes $x = 0, x = 1, y = -1, y = 1$

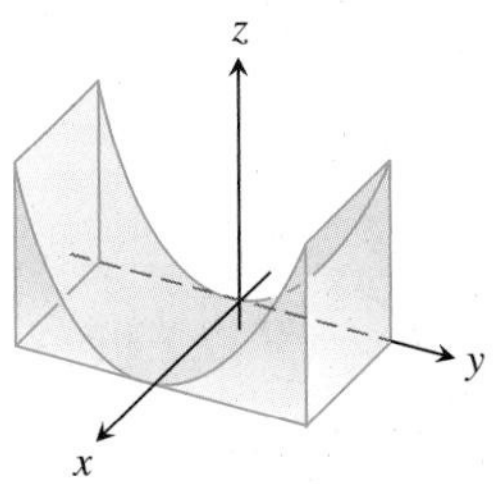

24. The region in the first octant bounded by the coordinate planes and the planes $x + z = 1, y + 2z = 2$

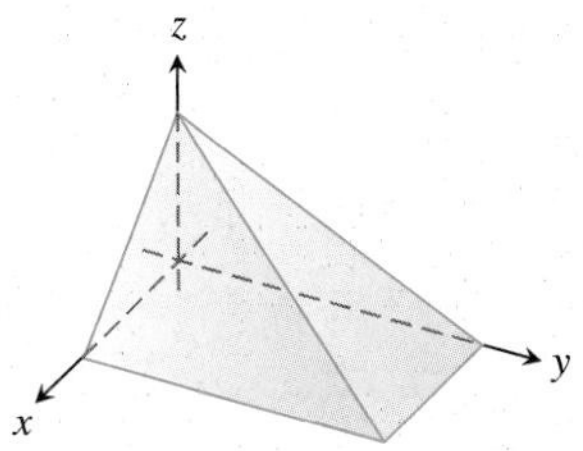

25. The region in the first octant bounded by the coordinate planes, the plane $y + z = 2$, and the cylinder $x = 4 - y^2$

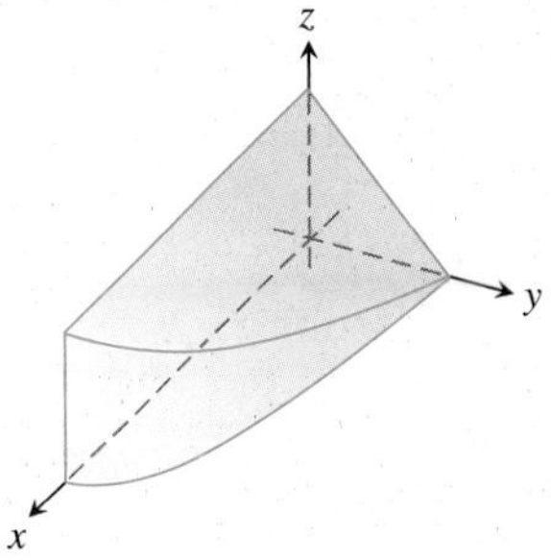

26. The wedge cut from the cylinder $x^2 + y^2 = 1$ by the planes $z = -y$ and $z = 0$

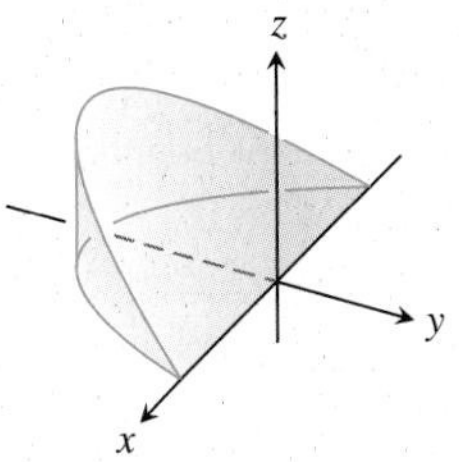

27. The tetrahedron in the first octant bounded by the coordinate planes and the plane passing through (1, 0, 0), (0, 2, 0), and (0, 0, 3).

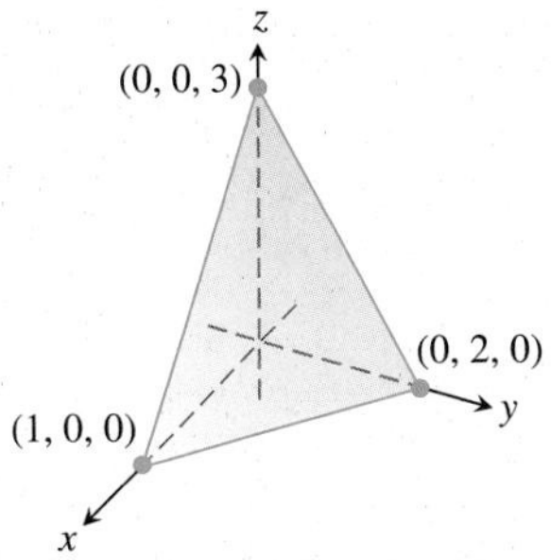

28. The region in the first octant bounded by the coordinate planes, the plane $y = 1 - x$, and the surface $z = \cos(\pi x/2)$, $0 \le x \le 1$

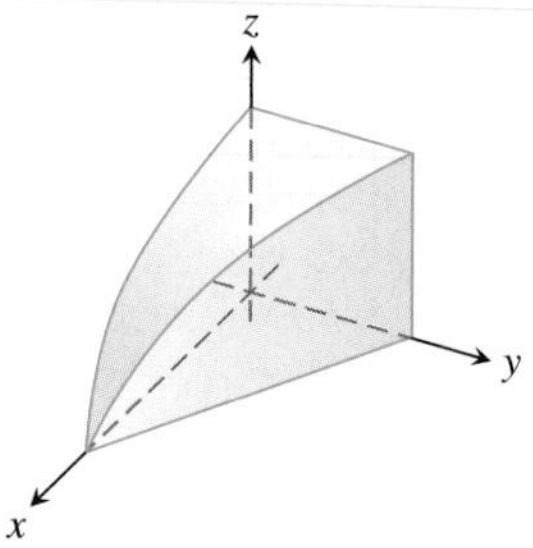

29. The region common to the interiors of the cylinders $x^2 + y^2 = 1$ and $x^2 + z^2 = 1$, one-eighth of which is shown in the accompanying figure.

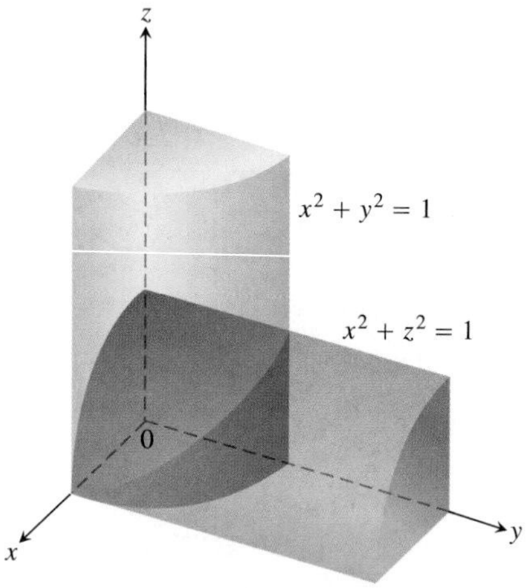

30. The region in the first octant bounded by the coordinate planes and the surface $z = 4 - x^2 - y$

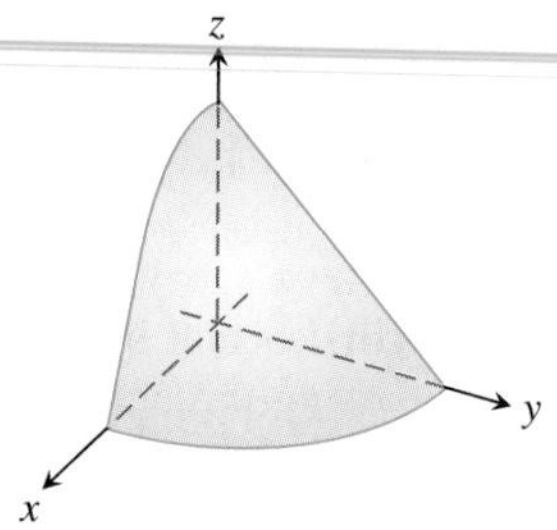

31. The region in the first octant bounded by the coordinate planes, the plane $x + y = 4$, and the cylinder $y^2 + 4z^2 = 16$

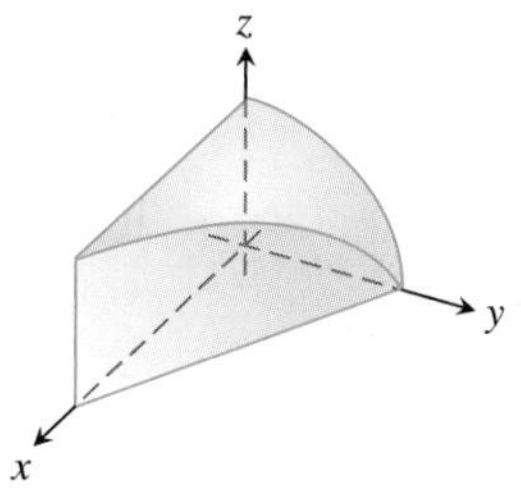

32. The region cut from the cylinder $x^2 + y^2 = 4$ by the plane $z = 0$ and the plane $x + z = 3$

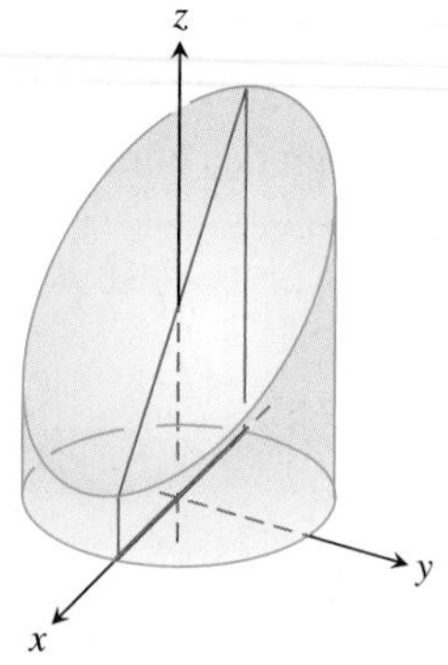

33. The region between the planes $x + y + 2z = 2$ and $2x + 2y + z = 4$ in the first octant

34. The finite region bounded by the planes $z = x$, $x + z = 8$, $z = y$, $y = 8$, and $z = 0$.

35. The region cut from the solid elliptical cylinder $x^2 + 4y^2 \le 4$ by the xy-plane and the plane $z = x + 2$

36. The region bounded in back by the plane $x = 0$, on the front and sides by the parabolic cylinder $x = 1 - y^2$, on the top by the paraboloid $z = x^2 + y^2$, and on the bottom by the xy-plane

Average Values

In Exercises 37–40, find the average value of $F(x, y, z)$ over the given region.

37. $F(x, y, z) = x^2 + 9$ over the cube in the first octant bounded by the coordinate planes and the planes $x = 2$, $y = 2$, and $z = 2$

38. $F(x, y, z) = x + y - z$ over the rectangular solid in the first octant bounded by the coordinate planes and the planes $x = 1$, $y = 1$, and $z = 2$

39. $F(x, y, z) = x^2 + y^2 + z^2$ over the cube in the first octant bounded by the coordinate planes and the planes $x = 1$, $y = 1$, and $z = 1$

40. $F(x, y, z) = xyz$ over the cube in the first octant bounded by the coordinate planes and the planes $x = 2$, $y = 2$, and $z = 2$

Changing the Order of Integration

Evaluate the integrals in Exercises 41–44 by changing the order of integration in an appropriate way.

41. $\displaystyle\int_0^4 \int_0^1 \int_{2y}^2 \frac{4\cos(x^2)}{2\sqrt{z}}\, dx\, dy\, dz$

42. $\displaystyle\int_0^1 \int_0^1 \int_{x^2}^1 12xze^{zy^2}\, dy\, dx\, dz$

43. $\displaystyle\int_0^1 \int_{\sqrt[3]{z}}^1 \int_0^{\ln 3} \frac{\pi e^{2x} \sin \pi y^2}{y^2}\, dx\, dy\, dz$

44. $\displaystyle\int_0^2 \int_0^{4-x^2} \int_0^x \frac{\sin 2z}{4 - z}\, dy\, dz\, dx$

Theory and Examples

45. **Finding upper limit of iterated integral** Solve for a:

$$\int_0^1 \int_0^{4-a-x^2} \int_a^{4-x^2-y} dz\, dy\, dx = \frac{4}{15}.$$

46. **Ellipsoid** For what value of c is the volume of the ellipsoid $x^2 + (y/2)^2 + (z/c)^2 = 1$ equal to 8π?

47. **Minimizing a triple integral** What domain D in space minimizes the value of the integral

$$\iiint_D (4x^2 + 4y^2 + z^2 - 4)\, dV?$$

Give reasons for your answer.

48. **Maximizing a triple integral** What domain D in space maximizes the value of the integral

$$\iiint_D (1 - x^2 - y^2 - z^2)\, dV?$$

Give reasons for your answer.

COMPUTER EXPLORATIONS

Numerical Evaluations

In Exercises 49–52, use a CAS integration utility to evaluate the triple integral of the given function over the specified solid region.

49. $F(x, y, z) = x^2y^2z$ over the solid cylinder bounded by $x^2 + y^2 = 1$ and the planes $z = 0$ and $z = 1$

50. $F(x, y, z) = |xyz|$ over the solid bounded below by the paraboloid $z = x^2 + y^2$ and above by the plane $z = 1$

51. $F(x, y, z) = \dfrac{z}{(x^2 + y^2 + z^2)^{3/2}}$ over the solid bounded below by the cone $z = \sqrt{x^2 + y^2}$ and above by the plane $z = 1$

52. $F(x, y, z) = x^4 + y^2 + z^2$ over the solid sphere $x^2 + y^2 + z^2 \le 1$

15.5 Masses and Moments in Three Dimensions

This section shows how to calculate the masses and moments of three-dimensional objects in Cartesian coordinates. The formulas are similar to those for two-dimensional objects. For calculations in spherical and cylindrical coordinates, see Section 15.6.

Masses and Moments

If $\delta(x, y, z)$ is the density of an object occupying a region D in space (mass per unit volume), the integral of δ over D gives the **mass** of the object. To see why, imagine partitioning the object into n mass elements like the one in Figure 15.32. The object's mass is the limit

$$M = \lim_{n\to\infty} \sum_{k=1}^{n} \Delta m_k = \lim_{n\to\infty} \sum_{k=1}^{n} \delta(x_k, y_k, z_k)\, \Delta V_k = \iiint_D \delta(x, y, z)\, dV.$$

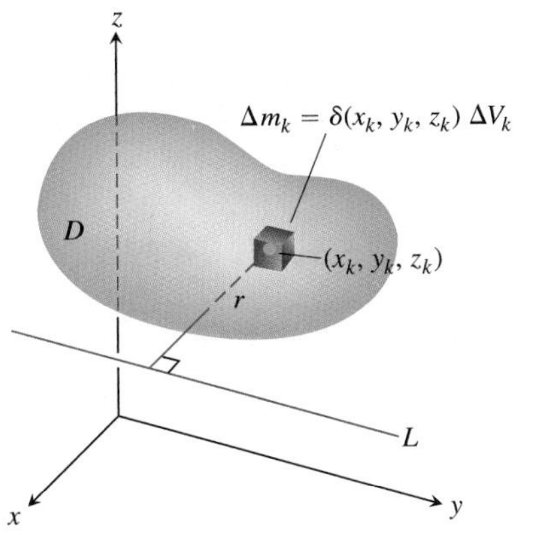

FIGURE 15.32 To define an object's mass and moment of inertia about a line, we first imagine it to be partitioned into a finite number of mass elements Δm_k.

We now derive a formula for the moment of inertia. If $r(x, y, z)$ is the distance from the point (x, y, z) in D to a line L, then the moment of inertia of the mass $\Delta m_k = \delta(x_k, y_k, z_k)\Delta V_k$ about the line L (shown in Figure 15.32) is approximately $\Delta I_k = r^2(x_k, y_k, z_k)\Delta m_k$. **The moment of inertia about L of the entire object is**

$$I_L = \lim_{n\to\infty} \sum_{k=1}^{n} \Delta I_k = \lim_{n\to\infty} \sum_{k=1}^{n} r^2(x_k, y_k, z_k)\, \delta(x_k, y_k, z_k)\, \Delta V_k = \iiint_D r^2\delta\, dV.$$

If L is the x-axis, then $r^2 = y^2 + z^2$ (Figure 15.33) and

$$I_x = \iiint_D (y^2 + z^2)\, \delta\, dV.$$

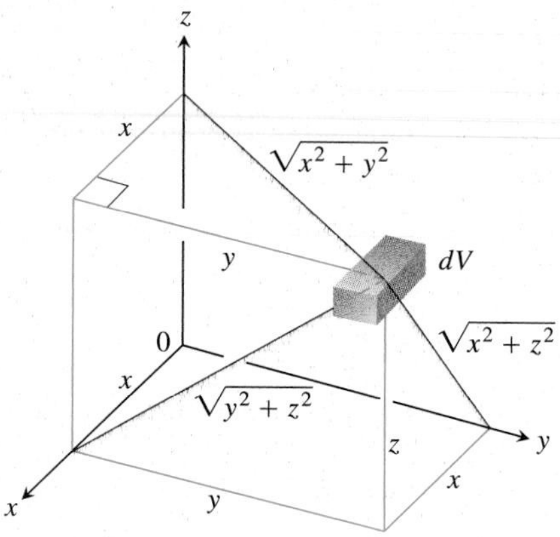

FIGURE 15.33 Distances from dV to the coordinate planes and axes.

Similarly, if L is the y-axis or z-axis we have

$$I_y = \iiint_D (x^2 + z^2)\,\delta\,dV \quad \text{and} \quad I_z = \iiint_D (x^2 + y^2)\,\delta\,dV.$$

Likewise, we can obtain the **first moments about the coordinate planes**. For example,

$$M_{yz} = \iiint_D x\delta(x, y, z)\,dV$$

gives the first moment about the yz-plane.

The mass and moment formulas in space analogous to those discussed for planar regions in Section 15.2 are summarized in Table 15.3.

TABLE 15.3 Mass and moment formulas for solid objects in space

Mass: $M = \iiint_D \delta\,dV \qquad (\delta = \delta(x, y, z) = \text{density})$

First moments about the coordinate planes:

$$M_{yz} = \iiint_D x\,\delta\,dV, \qquad M_{xz} = \iiint_D y\,\delta\,dV, \qquad M_{xy} = \iiint_D z\,\delta\,dV$$

Center of mass:

$$\bar{x} = \frac{M_{yz}}{M}, \qquad \bar{y} = \frac{M_{xz}}{M}, \qquad \bar{z} = \frac{M_{xy}}{M}$$

Moments of inertia (second moments) about the coordinate axes:

$$I_x = \iiint (y^2 + z^2)\,\delta\,dV$$

$$I_y = \iiint (x^2 + z^2)\,\delta\,dV$$

$$I_z = \iiint (x^2 + y^2)\,\delta\,dV$$

Moments of inertia about a line *L*:

$$I_L = \iiint r^2\,\delta\,dV \qquad (r(x, y, z) = \text{distance from the point } (x, y, z) \text{ to line } L)$$

Radius of gyration about a line *L*:

$$R_L = \sqrt{I_L/M}$$

EXAMPLE 1 Finding the Center of Mass of a Solid in Space

Find the center of mass of a solid of constant density δ bounded below by the disk $R: x^2 + y^2 \le 4$ in the plane $z = 0$ and above by the paraboloid $z = 4 - x^2 - y^2$ (Figure 15.34).

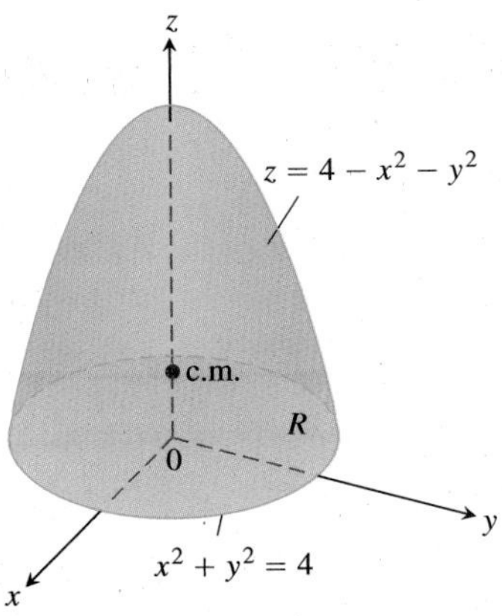

FIGURE 15.34 Finding the center of mass of a solid (Example 1).

Solution By symmetry $\bar{x} = \bar{y} = 0$. To find $\bar{z}$, we first calculate

$$M_{xy} = \iiint_R \int_{z=0}^{z=4-x^2-y^2} z\,\delta\,dz\,dy\,dx = \iint_R \left[\frac{z^2}{2}\right]_{z=0}^{z=4-x^2-y^2} \delta\,dy\,dx$$

$$= \frac{\delta}{2}\iint_R (4 - x^2 - y^2)^2\,dy\,dx$$

$$= \frac{\delta}{2}\int_0^{2\pi}\int_0^2 (4 - r^2)^2\,r\,dr\,d\theta \qquad \text{Polar coordinates}$$

$$= \frac{\delta}{2}\int_0^{2\pi}\left[-\frac{1}{6}(4 - r^2)^3\right]_{r=0}^{r=2} d\theta = \frac{16\delta}{3}\int_0^{2\pi} d\theta = \frac{32\pi\delta}{3}.$$

A similar calculation gives

$$M = \iint_R \int_0^{4-x^2-y^2} \delta\,dz\,dy\,dx = 8\pi\delta.$$

Therefore $\bar{z} = (M_{xy}/M) = 4/3$ and the center of mass is $(\bar{x}, \bar{y}, \bar{z}) = (0, 0, 4/3)$. ■

When the density of a solid object is constant (as in Example 1), the center of mass is called the **centroid** of the object (as was the case for two-dimensional shapes in Section 15.2).

EXAMPLE 2 Finding the Moments of Inertia About the Coordinate Axes

Find I_x, I_y, I_z for the rectangular solid of constant density δ shown in Figure 15.35.

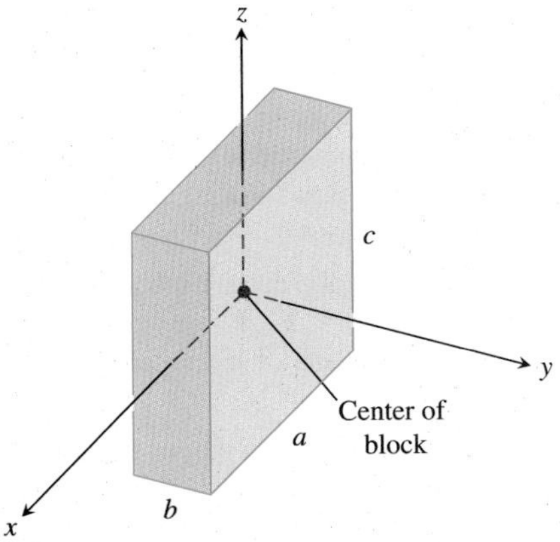

FIGURE 15.35 Finding I_x, I_y, and I_z for the block shown here. The origin lies at the center of the block (Example 2).

Solution The formula for I_x gives

$$I_x = \int_{-c/2}^{c/2}\int_{-b/2}^{b/2}\int_{-a/2}^{a/2} (y^2 + z^2)\,\delta\,dx\,dy\,dz.$$

We can avoid some of the work of integration by observing that $(y^2 + z^2)\delta$ is an even function of x, y, and z. The rectangular solid consists of eight symmetric pieces, one in each octant. We can evaluate the integral on one of these pieces and then multiply by 8 to get the total value.

$$I_x = 8\int_0^{c/2}\int_0^{b/2}\int_0^{a/2} (y^2 + z^2)\,\delta\,dx\,dy\,dz = 4a\delta\int_0^{c/2}\int_0^{b/2} (y^2 + z^2)\,dy\,dz$$

$$= 4a\delta\int_0^{c/2}\left[\frac{y^3}{3} + z^2 y\right]_{y=0}^{y=b/2} dz$$

$$= 4a\delta\int_0^{c/2}\left(\frac{b^3}{24} + \frac{z^2 b}{2}\right) dz$$

$$= 4a\delta\left(\frac{b^3 c}{48} + \frac{c^3 b}{48}\right) = \frac{abc\delta}{12}(b^2 + c^2) = \frac{M}{12}(b^2 + c^2).$$

Similarly,

$$I_y = \frac{M}{12}(a^2 + c^2) \qquad \text{and} \qquad I_z = \frac{M}{12}(a^2 + b^2).$$ ■

EXERCISES 15.5

Constant Density

The solids in Exercises 1–12 all have constant density $\delta = 1$.

1. (*Example 1 Revisited.*) Evaluate the integral for I_x in Table 15.3 directly to show that the shortcut in Example 2 gives the same answer. Use the results in Example 2 to find the radius of gyration of the rectangular solid about each coordinate axis.

2. **Moments of inertia** The coordinate axes in the figure run through the centroid of a solid wedge parallel to the labeled edges. Find I_x, I_y, and I_z if $a = b = 6$ and $c = 4$.

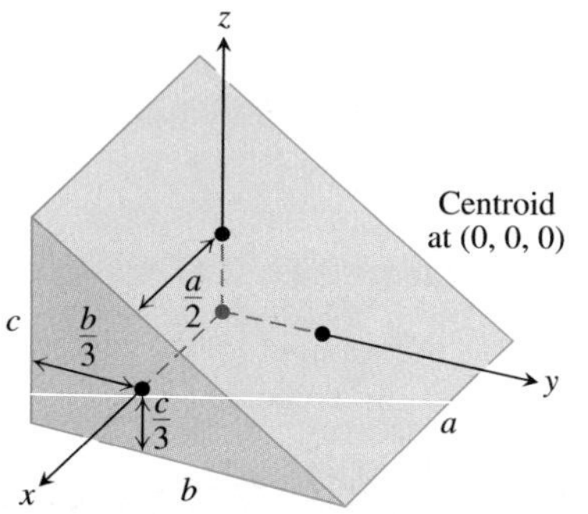

3. **Moments of inertia** Find the moments of inertia of the rectangular solid shown here with respect to its edges by calculating I_x, I_y, and I_z.

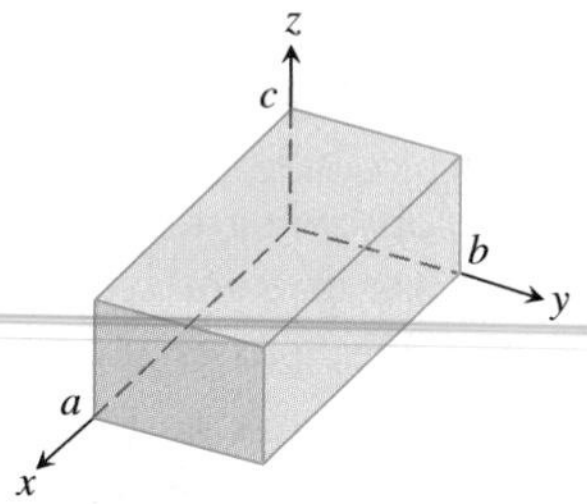

4. a. **Centroid and moments of inertia** Find the centroid and the moments of inertia I_x, I_y, and I_z of the tetrahedron whose vertices are the points (0, 0, 0), (1, 0, 0), (0, 1, 0), and (0, 0, 1).

   b. **Radius of gyration** Find the radius of gyration of the tetrahedron about the x-axis. Compare it with the distance from the centroid to the x-axis.

5. **Center of mass and moments of inertia** A solid "trough" of constant density is bounded below by the surface $z = 4y^2$, above by the plane $z = 4$, and on the ends by the planes $x = 1$ and $x = -1$. Find the center of mass and the moments of inertia with respect to the three axes.

6. **Center of mass** A solid of constant density is bounded below by the plane $z = 0$, on the sides by the elliptical cylinder $x^2 + 4y^2 = 4$, and above by the plane $z = 2 - x$ (see the accompanying figure).

   a. Find $\bar{x}$ and $\bar{y}$.

   b. Evaluate the integral

$$M_{xy} = \int_{-2}^{2}\int_{-(1/2)\sqrt{4-x^2}}^{(1/2)\sqrt{4-x^2}}\int_{0}^{2-x} z\,dz\,dy\,dx$$

   using integral tables to carry out the final integration with respect to x. Then divide M_{xy} by M to verify that $\bar{z} = 5/4$.

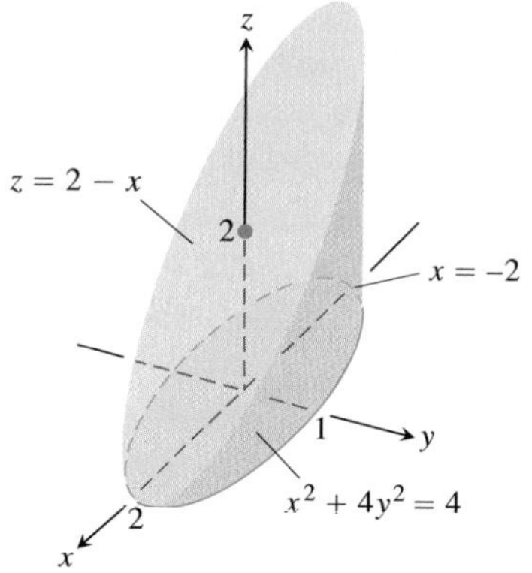

7. a. **Center of mass** Find the center of mass of a solid of constant density bounded below by the paraboloid $z = x^2 + y^2$ and above by the plane $z = 4$.

   b. Find the plane $z = c$ that divides the solid into two parts of equal volume. This plane does not pass through the center of mass.

8. **Moments and radii of gyration** A solid cube, 2 units on a side, is bounded by the planes $x = \pm 1$, $z = \pm 1$, $y = 3$, and $y = 5$. Find the center of mass and the moments of inertia and radii of gyration about the coordinate axes.

9. **Moment of inertia and radius of gyration about a line** A wedge like the one in Exercise 2 has $a = 4$, $b = 6$, and $c = 3$. Make a quick sketch to check for yourself that the square of the distance from a typical point (x, y, z) of the wedge to the line $L: z = 0, y = 6$ is $r^2 = (y - 6)^2 + z^2$. Then calculate the moment of inertia and radius of gyration of the wedge about L.

10. **Moment of inertia and radius of gyration about a line** A wedge like the one in Exercise 2 has $a = 4$, $b = 6$, and $c = 3$. Make a quick sketch to check for yourself that the square of the distance from a typical point (x, y, z) of the wedge to the line $L: x = 4, y = 0$ is $r^2 = (x - 4)^2 + y^2$. Then calculate the moment of inertia and radius of gyration of the wedge about L.

11. **Moment of inertia and radius of gyration about a line** A solid like the one in Exercise 3 has $a = 4$, $b = 2$, and $c = 1$. Make a quick sketch to check for yourself that the square of the distance between a typical point (x, y, z) of the solid and the line $L: y = 2, z = 0$ is $r^2 = (y - 2)^2 + z^2$. Then find the moment of inertia and radius of gyration of the solid about L.

12. Moment of inertia and radius of gyration about a line A solid like the one in Exercise 3 has $a = 4$, $b = 2$, and $c = 1$. Make a quick sketch to check for yourself that the square of the distance between a typical point (x, y, z) of the solid and the line $L: x = 4, y = 0$ is $r^2 = (x - 4)^2 + y^2$. Then find the moment of inertia and radius of gyration of the solid about L.

Variable Density

In Exercises 13 and 14, find

a. the mass of the solid.

b. the center of mass.

13. A solid region in the first octant is bounded by the coordinate planes and the plane $x + y + z = 2$. The density of the solid is $\delta(x, y, z) = 2x$.

14. A solid in the first octant is bounded by the planes $y = 0$ and $z = 0$ and by the surfaces $z = 4 - x^2$ and $x = y^2$ (see the accompanying figure). Its density function is $\delta(x, y, z) = kxy$, k a constant.

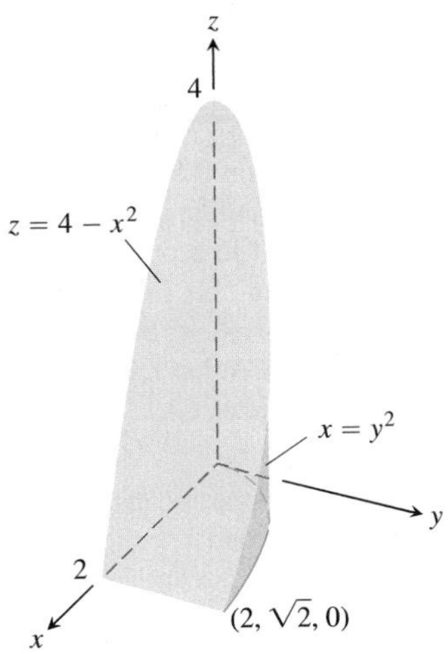

In Exercises 15 and 16, find

a. the mass of the solid.

b. the center of mass.

c. the moments of inertia about the coordinate axes.

d. the radii of gyration about the coordinate axes.

15. A solid cube in the first octant is bounded by the coordinate planes and by the planes $x = 1$, $y = 1$, and $z = 1$. The density of the cube is $\delta(x, y, z) = x + y + z + 1$.

16. A wedge like the one in Exercise 2 has dimensions $a = 2$, $b = 6$, and $c = 3$. The density is $\delta(x, y, z) = x + 1$. Notice that if the density is constant, the center of mass will be $(0, 0, 0)$.

17. Mass Find the mass of the solid bounded by the planes $x + z = 1$, $x - z = -1$, $y = 0$ and the surface $y = \sqrt{z}$. The density of the solid is $\delta(x, y, z) = 2y + 5$.

18. Mass Find the mass of the solid region bounded by the parabolic surfaces $z = 16 - 2x^2 - 2y^2$ and $z = 2x^2 + 2y^2$ if the density of the solid is $\delta(x, y, z) = \sqrt{x^2 + y^2}$.

Work

In Exercises 19 and 20, calculate the following.

a. The amount of work done by (constant) gravity g in moving the liquid filling in the container to the xy-plane. (*Hint:* Partition the liquid into small volume elements ΔV_i and find the work done (approximately) by gravity on each element. Summation and passage to the limit gives a triple integral to evaluate.)

b. The work done by gravity in moving the center of mass down to the xy-plane.

19. The container is a cubical box in the first octant bounded by the coordinate planes and the planes $x = 1$, $y = 1$, and $z = 1$. The density of the liquid filling the box is $\delta(x, y, z) = x + y + z + 1$ (see Exercise 15).

20. The container is in the shape of the region bounded by $y = 0$, $z = 0$, $z = 4 - x^2$, and $x = y^2$. The density of the liquid filling the region is $\delta(x, y, z) = kxy$, k a constant (see Exercise 14).

The Parallel Axis Theorem

The Parallel Axis Theorem (Exercises 15.2) holds in three dimensions as well as in two. Let $L_{\text{c.m.}}$ be a line through the center of mass of a body of mass m and let L be a parallel line h units away from $L_{\text{c.m.}}$. The **Parallel Axis Theorem** says that the moments of inertia $I_{\text{c.m.}}$ and I_L of the body about $L_{\text{c.m.}}$ and L satisfy the equation

$$I_L = I_{\text{c.m.}} + mh^2. \tag{1}$$

As in the two-dimensional case, the theorem gives a quick way to calculate one moment when the other moment and the mass are known.

21. Proof of the Parallel Axis Theorem

a. Show that the first moment of a body in space about any plane through the body's center of mass is zero. (*Hint:* Place the body's center of mass at the origin and let the plane be the yz-plane. What does the formula $\bar{x} = M_{yz}/M$ then tell you?)

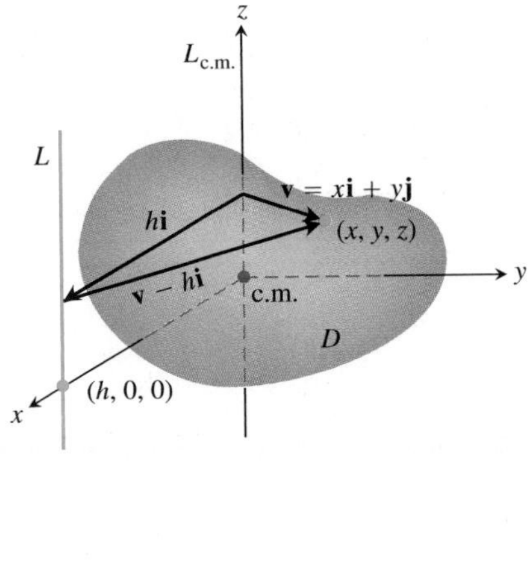

b. To prove the Parallel Axis Theorem, place the body with its center of mass at the origin, with the line $L_{\text{c.m.}}$ along the z-axis and the line L perpendicular to the xy-plane at the point $(h, 0, 0)$. Let D be the region of space occupied by the body. Then, in the notation of the figure,

$$I_L = \iiint_D |\mathbf{v} - h\mathbf{i}|^2\, dm.$$

Expand the integrand in this integral and complete the proof.

22. The moment of inertia about a diameter of a solid sphere of constant density and radius a is $(2/5)ma^2$, where m is the mass of the sphere. Find the moment of inertia about a line tangent to the sphere.

23. The moment of inertia of the solid in Exercise 3 about the z-axis is $I_z = abc(a^2 + b^2)/3$.

a. Use Equation (1) to find the moment of inertia and radius of gyration of the solid about the line parallel to the z-axis through the solid's center of mass.

b. Use Equation (1) and the result in part (a) to find the moment of inertia and radius of gyration of the solid about the line $x = 0, y = 2b$.

24. If $a = b = 6$ and $c = 4$, the moment of inertia of the solid wedge in Exercise 2 about the x-axis is $I_x = 208$. Find the moment of inertia of the wedge about the line $y = 4, z = -4/3$ (the edge of the wedge's narrow end).

Pappus's Formula

Pappus's formula (Exercises 15.2) holds in three dimensions as well as in two. Suppose that bodies B_1 and B_2 of mass m_1 and m_2, respectively, occupy nonoverlapping regions in space and that $\mathbf{c}_1$ and $\mathbf{c}_2$ are the vectors from the origin to the bodies' respective centers of mass. Then the center of mass of the union $B_1 \cup B_2$ of the two bodies is determined by the vector

$$\mathbf{c} = \frac{m_1\mathbf{c}_1 + m_2\mathbf{c}_2}{m_1 + m_2}.$$

As before, this formula is called **Pappus's formula**. As in the two-dimensional case, the formula generalizes to

$$\mathbf{c} = \frac{m_1\mathbf{c}_1 + m_2\mathbf{c}_2 + \cdots + m_n\mathbf{c}_n}{m_1 + m_2 + \cdots + m_n}$$

for n bodies.

25. Derive Pappus's formula. (*Hint:* Sketch B_1 and B_2 as nonoverlapping regions in the first octant and label their centers of mass $(\bar{x}_1, \bar{y}_1, \bar{z}_1)$ and $(\bar{x}_2, \bar{y}_2, \bar{z}_2)$. Express the moments of $B_1 \cup B_2$ about the coordinate planes in terms of the masses m_1 and m_2 and the coordinates of these centers.)

26. The accompanying figure shows a solid made from three rectangular solids of constant density $\delta = 1$. Use Pappus's formula to find the center of mass of

a. $A \cup B$ **b.** $A \cup C$
c. $B \cup C$ **d.** $A \cup B \cup C$.

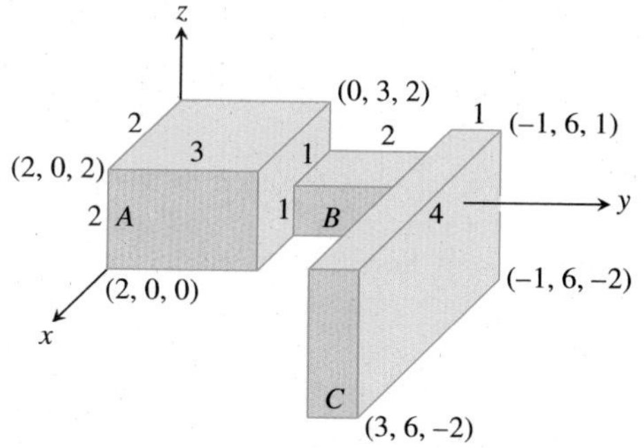

27. a. Suppose that a solid right circular cone C of base radius a and altitude h is constructed on the circular base of a solid hemisphere S of radius a so that the union of the two solids resembles an ice cream cone. The centroid of a solid cone lies one-fourth of the way from the base toward the vertex. The centroid of a solid hemisphere lies three-eighths of the way from the base to the top. What relation must hold between h and a to place the centroid of $C \cup S$ in the common base of the two solids?

b. If you have not already done so, answer the analogous question about a triangle and a semicircle (Section 15.2, Exercise 55). The answers are not the same.

28. A solid pyramid P with height h and four congruent sides is built with its base as one face of a solid cube C whose edges have length s. The centroid of a solid pyramid lies one-fourth of the way from the base toward the vertex. What relation must hold between h and s to place the centroid of $P \cup C$ in the base of the pyramid? Compare your answer with the answer to Exercise 27. Also compare it with the answer to Exercise 56 in Section 15.2.

15.6 Triple Integrals in Cylindrical and Spherical Coordinates

When a calculation in physics, engineering, or geometry involves a cylinder, cone, or sphere, we can often simplify our work by using cylindrical or spherical coordinates, which are introduced in this section. The procedure for transforming to these coordinates and evaluating the resulting triple integrals is similar to the transformation to polar coordinates in the plane studied in Section 15.3.

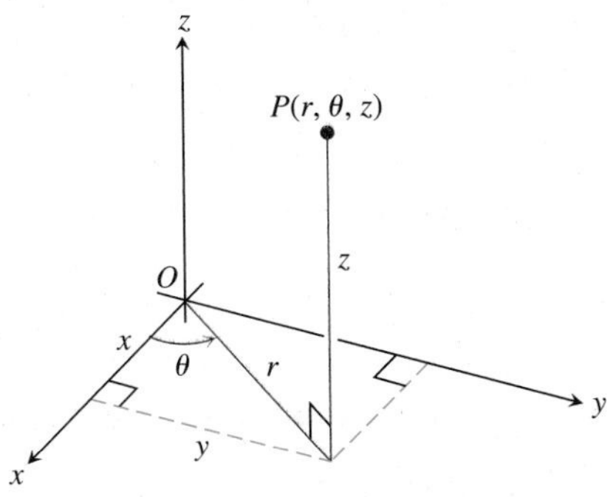

FIGURE 15.36 The cylindrical coordinates of a point in space are r, θ, and z.

Integration in Cylindrical Coordinates

We obtain cylindrical coordinates for space by combining polar coordinates in the xy-plane with the usual z-axis. This assigns to every point in space one or more coordinate triples of the form (r, θ, z), as shown in Figure 15.36.

> **DEFINITION Cylindrical Coordinates**
>
> **Cylindrical coordinates** represent a point P in space by ordered triples (r, θ, z) in which
>
> 1. r and θ are polar coordinates for the vertical projection of P on the xy-plane
> 2. z is the rectangular vertical coordinate.

The values of x, y, r, and θ in rectangular and cylindrical coordinates are related by the usual equations.

> **Equations Relating Rectangular (x, y, z) and Cylindrical (r, θ, z) Coordinates**
>
> $$x = r\cos\theta, \qquad y = r\sin\theta, \qquad z = z,$$
> $$r^2 = x^2 + y^2, \qquad \tan\theta = y/x$$

In cylindrical coordinates, the equation $r = a$ describes not just a circle in the xy-plane but an entire cylinder about the z-axis (Figure 15.37). The z-axis is given by $r = 0$. The equation $\theta = \theta_0$ describes the plane that contains the z-axis and makes an angle θ_0 with the positive x-axis. And, just as in rectangular coordinates, the equation $z = z_0$ describes a plane perpendicular to the z-axis.

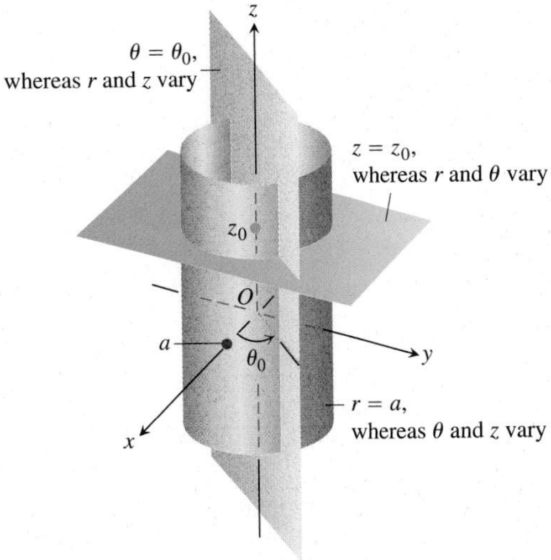

FIGURE 15.37 Constant-coordinate equations in cylindrical coordinates yield cylinders and planes.

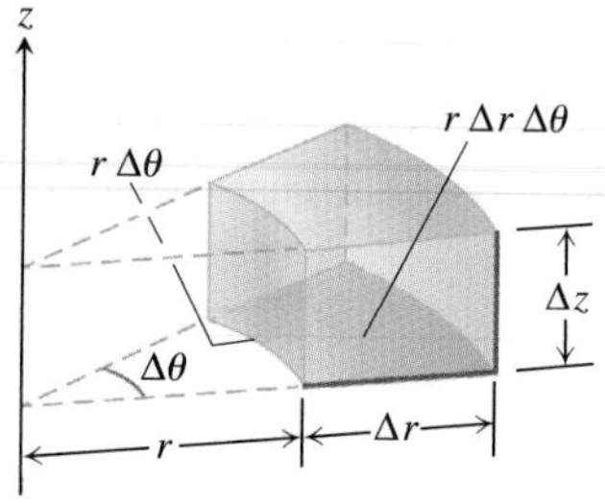

FIGURE 15.38 In cylindrical coordinates the volume of the wedge is approximated by the product $\Delta V = \Delta z\, r\, \Delta r\, \Delta\theta$.

Cylindrical coordinates are good for describing cylinders whose axes run along the z-axis and planes that either contain the z-axis or lie perpendicular to the z-axis. Surfaces like these have equations of constant coordinate value:

$$r = 4. \qquad \text{Cylinder, radius 4, axis the } z\text{-axis}$$

$$\theta = \frac{\pi}{3}. \qquad \text{Plane containing the } z\text{-axis}$$

$$z = 2. \qquad \text{Plane perpendicular to the } z\text{-axis}$$

When computing triple integrals over a region D in cylindrical coordinates, we partition the region into n small cylindrical wedges, rather than into rectangular boxes. In the kth cylindrical wedge, r, θ and z change by Δr_k, $\Delta\theta_k$, and Δz_k, and the largest of these numbers among all the cylindrical wedges is called the **norm** of the partition. We define the triple integral as a limit of Riemann sums using these wedges. The volume of such a cylindrical wedge ΔV_k is obtained by taking the area ΔA_k of its base in the $r\theta$-plane and multiplying by the height Δz (Figure 15.38).

For a point (r_k, θ_k, z_k) in the center of the kth wedge, we calculated in polar coordinates that $\Delta A_k = r_k\, \Delta r_k\, \Delta\theta_k$. So $\Delta V_k = \Delta z_k\, r_k\, \Delta r_k\, \Delta\theta_k$ and a Riemann sum for f over D has the form

$$S_n = \sum_{k=1}^{n} f(r_k, \theta_k, z_k)\, \Delta z_k\, r_k\, \Delta r_k\, \Delta\theta_k.$$

The triple integral of a function f over D is obtained by taking a limit of such Riemann sums with partitions whose norms approach zero

$$\lim_{n\to\infty} S_n = \iiint_D f\, dV = \iiint_D f\, dz\, r\, dr\, d\theta.$$

Triple integrals in cylindrical coordinates are then evaluated as iterated integrals, as in the following example.

EXAMPLE 1 Finding Limits of Integration in Cylindrical Coordinates

Find the limits of integration in cylindrical coordinates for integrating a function $f(r, \theta, z)$ over the region D bounded below by the plane $z = 0$, laterally by the circular cylinder $x^2 + (y-1)^2 = 1$, and above by the paraboloid $z = x^2 + y^2$.

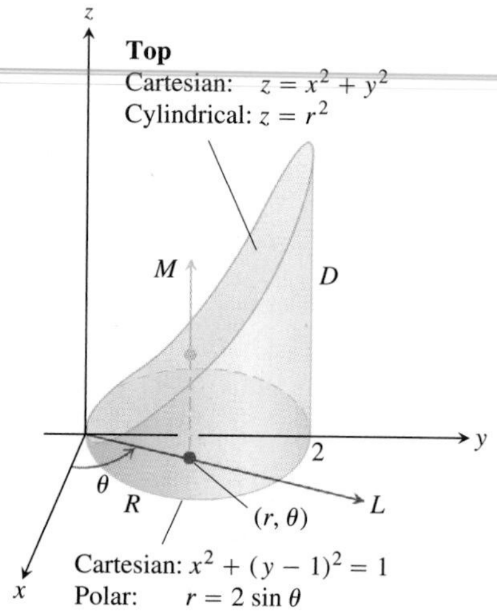

FIGURE 15.39 Finding the limits of integration for evaluating an integral in cylindrical coordinates (Example 1).

Solution The base of D is also the region's projection R on the xy-plane. The boundary of R is the circle $x^2 + (y-1)^2 = 1$. Its polar coordinate equation is

$$x^2 + (y-1)^2 = 1$$

$$x^2 + y^2 - 2y + 1 = 1$$

$$r^2 - 2r\sin\theta = 0$$

$$r = 2\sin\theta.$$

The region is sketched in Figure 15.39.

We find the limits of integration, starting with the z-limits. A line M through a typical point (r, θ) in R parallel to the z-axis enters D at $z = 0$ and leaves at $z = x^2 + y^2 = r^2$.

Next we find the r-limits of integration. A ray L through (r, θ) from the origin enters R at $r = 0$ and leaves at $r = 2\sin\theta$.

Finally we find the θ-limits of integration. As L sweeps across R, the angle θ it makes with the positive x-axis runs from $\theta = 0$ to $\theta = \pi$. The integral is

$$\iiint_D f(r, \theta, z)\, dV = \int_0^{\pi} \int_0^{2\sin\theta} \int_0^{r^2} f(r, \theta, z)\, dz\, r\, dr\, d\theta. \qquad \blacksquare$$

Example 1 illustrates a good procedure for finding limits of integration in cylindrical coordinates. The procedure is summarized as follows.

How to Integrate in Cylindrical Coordinates

To evaluate

$$\iiint_D f(r, \theta, z)\, dV$$

over a region D in space in cylindrical coordinates, integrating first with respect to z, then with respect to r, and finally with respect to θ, take the following steps.

1. *Sketch*. Sketch the region D along with its projection R on the xy-plane. Label the surfaces and curves that bound D and R.

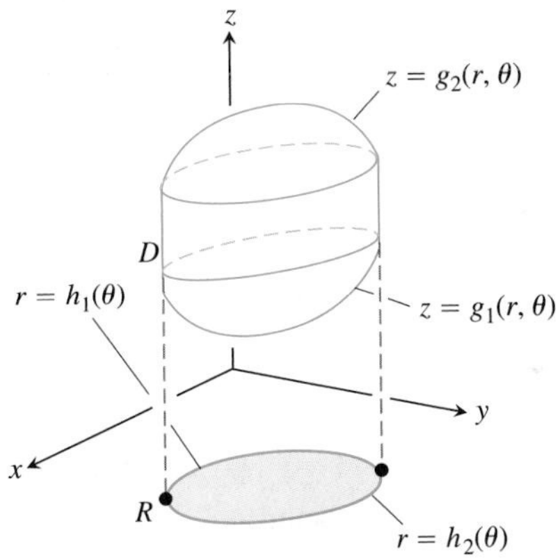

2. *Find the z-limits of integration*. Draw a line M through a typical point (r, θ) of R parallel to the z-axis. As z increases, M enters D at $z = g_1(r, \theta)$ and leaves at $z = g_2(r, \theta)$. These are the z-limits of integration.

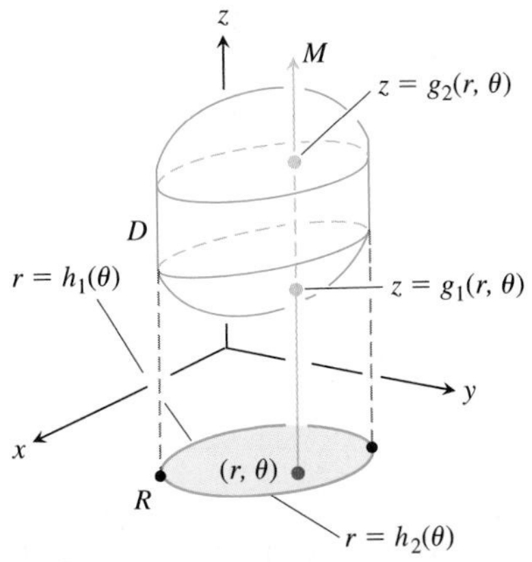

3. *Find the r-limits of integration.* Draw a ray L through (r, θ) from the origin. The ray enters R at $r = h_1(\theta)$ and leaves at $r = h_2(\theta)$. These are the r-limits of integration.

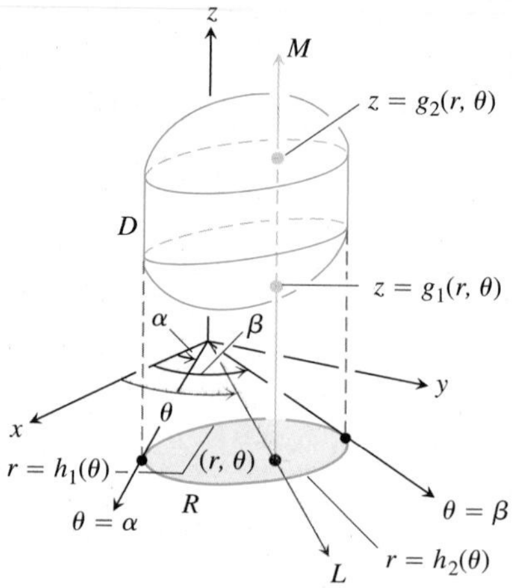

4. *Find the θ-limits of integration.* As L sweeps across R, the angle θ it makes with the positive x-axis runs from $\theta = \alpha$ to $\theta = \beta$. These are the θ-limits of integration. The integral is

$$\iiint_D f(r, \theta, z)\, dV = \int_{\theta=\alpha}^{\theta=\beta} \int_{r=h_1(\theta)}^{r=h_2(\theta)} \int_{z=g_1(r,\theta)}^{z=g_2(r,\theta)} f(r, \theta, z)\, dz\, r\, dr\, d\theta.$$

EXAMPLE 2 Finding a Centroid

Find the centroid ($\delta = 1$) of the solid enclosed by the cylinder $x^2 + y^2 = 4$, bounded above by the paraboloid $z = x^2 + y^2$, and bounded below by the xy-plane.

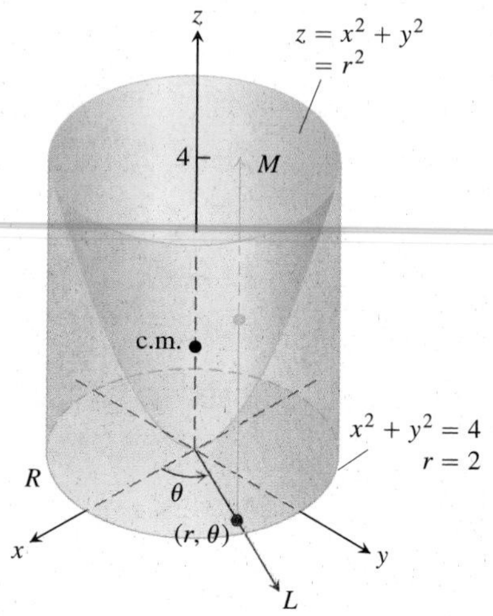

FIGURE 15.40 Example 2 shows how to find the centroid of this solid.

Solution We sketch the solid, bounded above by the paraboloid $z = r^2$ and below by the plane $z = 0$ (Figure 15.40). Its base R is the disk $0 \le r \le 2$ in the xy-plane.

The solid's centroid $(\bar{x}, \bar{y}, \bar{z})$ lies on its axis of symmetry, here the z-axis. This makes $\bar{x} = \bar{y} = 0$. To find $\bar{z}$, we divide the first moment M_{xy} by the mass M.

To find the limits of integration for the mass and moment integrals, we continue with the four basic steps. We completed our initial sketch. The remaining steps give the limits of integration.

The z-limits. A line M through a typical point (r, θ) in the base parallel to the z-axis enters the solid at $z = 0$ and leaves at $z = r^2$.

The r-limits. A ray L through (r, θ) from the origin enters R at $r = 0$ and leaves at $r = 2$.

The θ-limits. As L sweeps over the base like a clock hand, the angle θ it makes with the positive x-axis runs from $\theta = 0$ to $\theta = 2\pi$. The value of M_{xy} is

$$M_{xy} = \int_0^{2\pi} \int_0^2 \int_0^{r^2} z\, dz\, r\, dr\, d\theta = \int_0^{2\pi} \int_0^2 \left[\frac{z^2}{2}\right]_0^{r^2} r\, dr\, d\theta$$

$$= \int_0^{2\pi} \int_0^2 \frac{r^5}{2}\, dr\, d\theta = \int_0^{2\pi} \left[\frac{r^6}{12}\right]_0^2 d\theta = \int_0^{2\pi} \frac{16}{3}\, d\theta = \frac{32\pi}{3}.$$

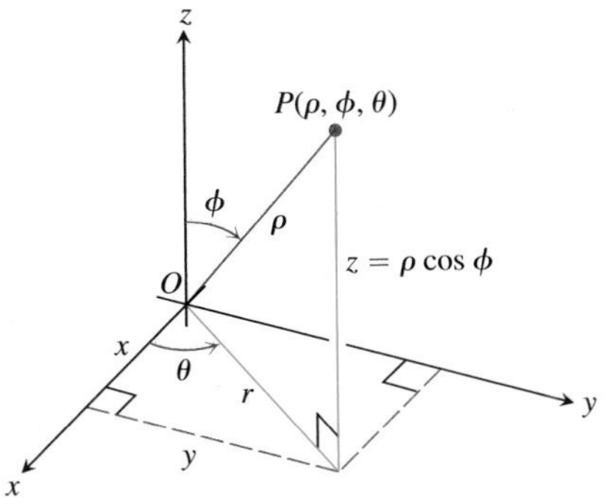

FIGURE 15.41 The spherical coordinates ρ, ϕ, and θ and their relation to x, y, z, and r.

The value of M is

$$M = \int_0^{2\pi}\int_0^2\int_0^{r^2} dz\, r\, dr\, d\theta = \int_0^{2\pi}\int_0^2 \Big[z\Big]_0^{r^2} r\, dr\, d\theta$$

$$= \int_0^{2\pi}\int_0^2 r^3\, dr\, d\theta = \int_0^{2\pi}\left[\frac{r^4}{4}\right]_0^2 d\theta = \int_0^{2\pi} 4\, d\theta = 8\pi.$$

Therefore,

$$\bar{z} = \frac{M_{xy}}{M} = \frac{32\pi}{3}\frac{1}{8\pi} = \frac{4}{3},$$

and the centroid is (0, 0, 4/3). Notice that the centroid lies outside the solid. ■

Spherical Coordinates and Integration

Spherical coordinates locate points in space with two angles and one distance, as shown in Figure 15.41. The first coordinate, $\rho = |\overrightarrow{OP}|$, is the point's distance from the origin. Unlike r, *the variable ρ is never negative*. The second coordinate, ϕ, is the angle $\overrightarrow{OP}$ makes with the positive z-axis. It is required to lie in the interval $[0, \pi]$. The third coordinate is the angle θ as measured in cylindrical coordinates.

DEFINITION Spherical Coordinates

Spherical coordinates represent a point P in space by ordered triples (ρ, ϕ, θ) in which

1. ρ is the distance from P to the origin.
2. ϕ is the angle $\overrightarrow{OP}$ makes with the positive z-axis $(0 \le \phi \le \pi)$.
3. θ is the angle from cylindrical coordinates.

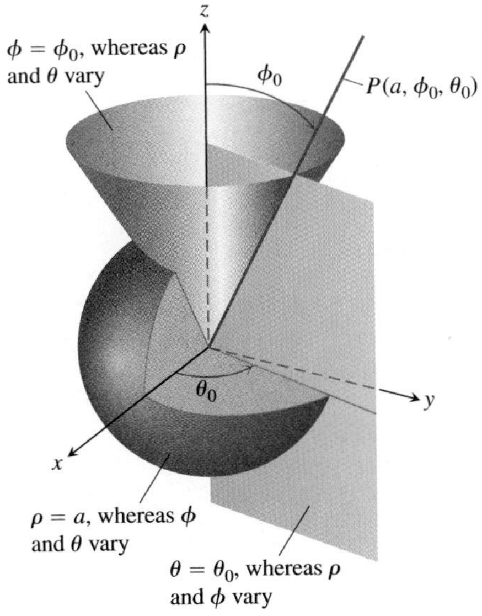

FIGURE 15.42 Constant-coordinate equations in spherical coordinates yield spheres, single cones, and half-planes.

On maps of the Earth, θ is related to the meridian of a point on the Earth and ϕ to its latitude, while ρ is related to elevation above the Earth's surface.

The equation $\rho = a$ describes the sphere of radius a centered at the origin (Figure 15.42). The equation $\phi = \phi_0$ describes a single cone whose vertex lies at the origin and whose axis lies along the z-axis. (We broaden our interpretation to include the xy-plane as the cone $\phi = \pi/2$.) If ϕ_0 is greater than $\pi/2$, the cone $\phi = \phi_0$ opens downward. The equation $\theta = \theta_0$ describes the half-plane that contains the z-axis and makes an angle θ_0 with the positive x-axis.

Equations Relating Spherical Coordinates to Cartesian and Cylindrical Coordinates

$$r = \rho\sin\phi, \qquad x = r\cos\theta = \rho\sin\phi\cos\theta,$$
$$z = \rho\cos\phi, \qquad y = r\sin\theta = \rho\sin\phi\sin\theta, \qquad (1)$$
$$\rho = \sqrt{x^2 + y^2 + z^2} = \sqrt{r^2 + z^2}.$$

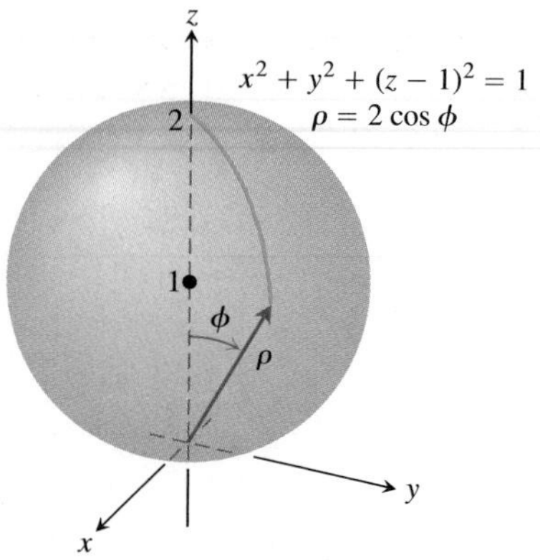

FIGURE 15.43 The sphere in Example 3.

EXAMPLE 3 Converting Cartesian to Spherical

Find a spherical coordinate equation for the sphere $x^2 + y^2 + (z-1)^2 = 1$.

Solution We use Equations (1) to substitute for x, y, and z:

$$x^2 + y^2 + (z-1)^2 = 1$$

$$\rho^2 \sin^2\phi \cos^2\theta + \rho^2 \sin^2\phi \sin^2\theta + (\rho\cos\phi - 1)^2 = 1 \qquad \text{Equations (1)}$$

$$\rho^2 \sin^2\phi \underbrace{(\cos^2\theta + \sin^2\theta)}_{1} + \rho^2\cos^2\phi - 2\rho\cos\phi + 1 = 1$$

$$\rho^2 \underbrace{(\sin^2\phi + \cos^2\phi)}_{1} = 2\rho\cos\phi$$

$$\rho^2 = 2\rho\cos\phi$$

$$\rho = 2\cos\phi.$$

See Figure 15.43. ■

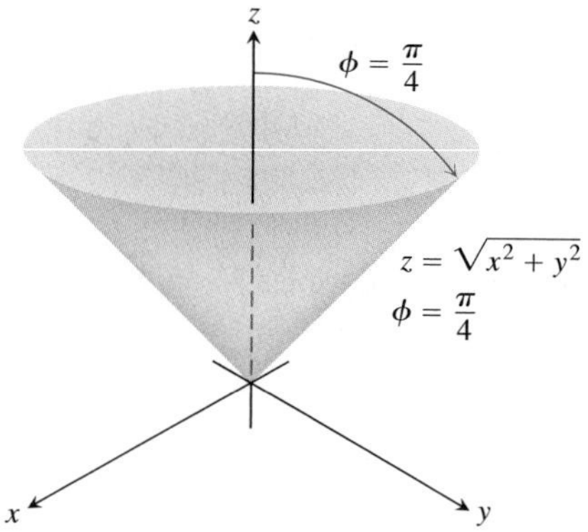

FIGURE 15.44 The cone in Example 4.

EXAMPLE 4 Converting Cartesian to Spherical

Find a spherical coordinate equation for the cone $z = \sqrt{x^2 + y^2}$ (Figure 15.44).

Solution 1 *Use geometry.* The cone is symmetric with respect to the z-axis and cuts the first quadrant of the yz-plane along the line $z = y$. The angle between the cone and the positive z-axis is therefore $\pi/4$ radians. The cone consists of the points whose spherical coordinates have ϕ equal to $\pi/4$, so its equation is $\phi = \pi/4$.

Solution 2 *Use algebra.* If we use Equations (1) to substitute for x, y, and z we obtain the same result:

$$z = \sqrt{x^2 + y^2}$$

$$\rho\cos\phi = \sqrt{\rho^2\sin^2\phi} \qquad \text{Example 3}$$

$$\rho\cos\phi = \rho\sin\phi \qquad \rho \geq 0,\ \sin\phi \geq 0$$

$$\cos\phi = \sin\phi$$

$$\phi = \frac{\pi}{4}. \qquad 0 \leq \phi \leq \pi$$

■

Spherical coordinates are good for describing spheres centered at the origin, half-planes hinged along the z-axis, and cones whose vertices lie at the origin and whose axes lie along the z-axis. Surfaces like these have equations of constant coordinate value:

$$\rho = 4 \qquad \text{Sphere, radius 4, center at origin}$$

$$\phi = \frac{\pi}{3} \qquad \text{Cone opening up from the origin, making an angle of } \pi/3 \text{ radians with the positive } z\text{-axis}$$

$$\theta = \frac{\pi}{3}. \qquad \text{Half-plane, hinged along the } z\text{-axis, making an angle of } \pi/3 \text{ radians with the positive } x\text{-axis}$$

When computing triple integrals over a region D in spherical coordinates, we partition the region into n spherical wedges. The size of the kth spherical wedge, which contains a point $(\rho_k, \phi_k, \theta_k)$, is given by changes by $\Delta\rho_k$, $\Delta\theta_k$, and $\Delta\phi_k$ in ρ, θ, and ϕ. Such a spherical wedge has one edge a circular arc of length $\rho_k\,\Delta\phi_k$, another edge a circular arc of

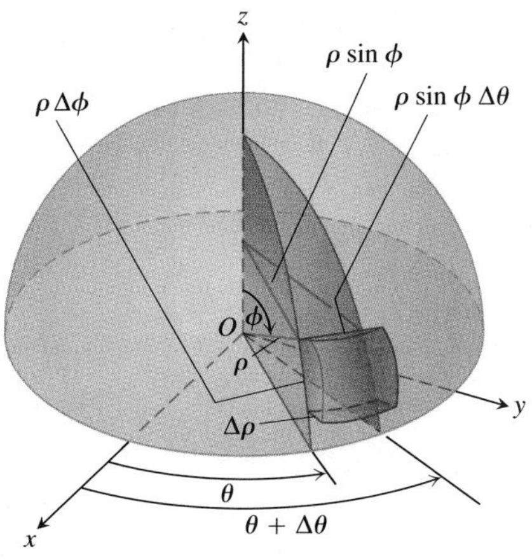

FIGURE 15.45 In spherical coordinates

$$dV = d\rho \cdot \rho\, d\phi \cdot \rho \sin\phi\, d\theta = \rho^2 \sin\phi\, d\rho\, d\phi\, d\theta.$$

length $\rho_k \sin\phi_k\, \Delta\theta_k$, and thickness $\Delta\rho_k$. The spherical wedge closely approximates a cube of these dimensions when $\Delta\rho_k$, $\Delta\theta_k$, and $\Delta\phi_k$ are all small (Figure 15.45). It can be shown that the volume of this spherical wedge ΔV_k is $\Delta V_k = \rho_k^2 \sin\phi_k\, \Delta\rho_k\, \Delta\phi_k\, \Delta\theta_k$ for $(\rho_k, \phi_k, \theta_k)$ a point chosen inside the wedge.

The corresponding Riemann sum for a function $F(\rho, \phi, \theta)$ is

$$S_n = \sum_{k=1}^{n} F(\rho_k, \phi_k, \theta_k)\, \rho_k^2 \sin\phi_k\, \Delta\rho_k\, \Delta\phi_k\, \Delta\theta_k.$$

As the norm of a partition approaches zero, and the spherical wedges get smaller, the Riemann sums have a limit when F is continuous:

$$\lim_{n\to\infty} S_n = \iiint_D F(\rho, \phi, \theta)\, dV = \iiint_D F(\rho, \phi, \theta)\, \rho^2 \sin\phi\, d\rho\, d\phi\, d\theta.$$

In spherical coordinates, we have

$$dV = \rho^2 \sin\phi\, d\rho\, d\phi\, d\theta.$$

To evaluate integrals in spherical coordinates, we usually integrate first with respect to ρ. The procedure for finding the limits of integration is shown below. We restrict our attention to integrating over domains that are solids of revolution about the z-axis (or portions thereof) and for which the limits for θ and ϕ are constant.

How to Integrate in Spherical Coordinates

To evaluate

$$\iiint_D f(\rho, \phi, \theta)\, dV$$

over a region D in space in spherical coordinates, integrating first with respect to ρ, then with respect to ϕ, and finally with respect to θ, take the following steps.

1. *Sketch.* Sketch the region D along with its projection R on the xy-plane. Label the surfaces that bound D.

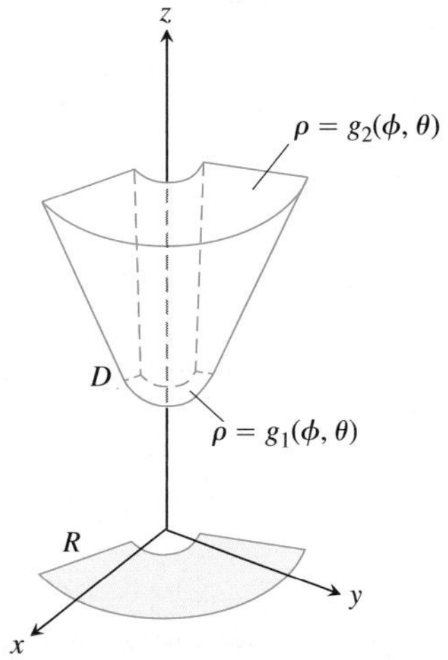

2. *Find the ρ-limits of integration.* Draw a ray M from the origin through D making an angle ϕ with the positive z-axis. Also draw the projection of M on the xy-plane (call the projection L). The ray L makes an angle θ with the positive x-axis. As ρ increases, M enters D at $\rho = g_1(\phi, \theta)$ and leaves at $\rho = g_2(\phi, \theta)$. These are the ρ-limits of integration.

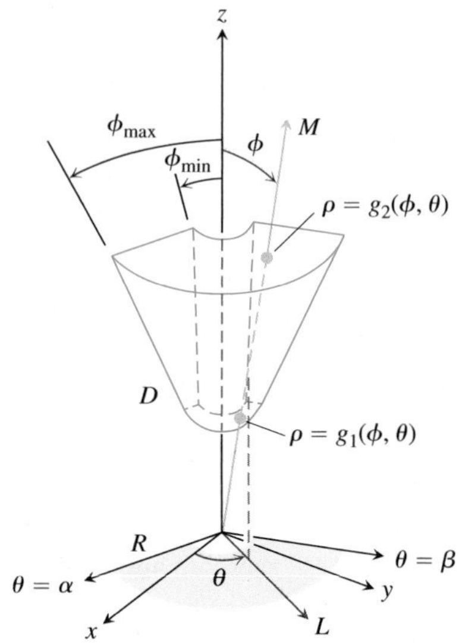

3. *Find the φ-limits of integration.* For any given θ, the angle ϕ that M makes with the z-axis runs from $\phi = \phi_{min}$ to $\phi = \phi_{max}$. These are the ϕ-limits of integration.
4. *Find the θ-limits of integration.* The ray L sweeps over R as θ runs from α to β. These are the θ-limits of integration. The integral is

$$\iiint_D f(\rho, \phi, \theta)\, dV = \int_{\theta=\alpha}^{\theta=\beta} \int_{\phi=\phi_{min}}^{\phi=\phi_{max}} \int_{\rho=g_1(\phi,\theta)}^{\rho=g_2(\phi,\theta)} f(\rho, \phi, \theta)\, \rho^2 \sin\phi \, d\rho \, d\phi \, d\theta.$$

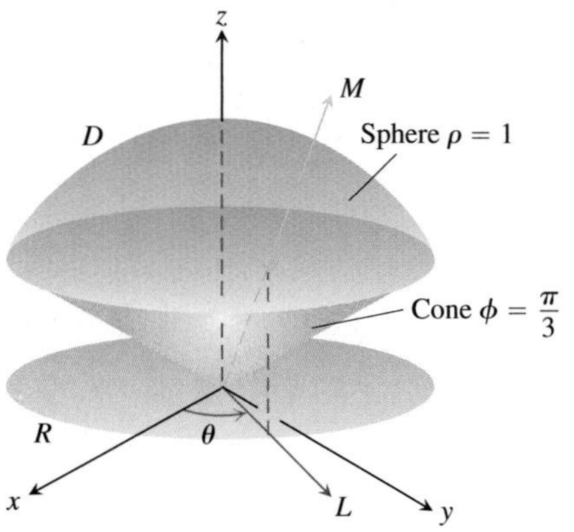

FIGURE 15.46 The ice cream cone in Example 5.

EXAMPLE 5 Finding a Volume in Spherical Coordinates

Find the volume of the "ice cream cone" D cut from the solid sphere $\rho \le 1$ by the cone $\phi = \pi/3$.

Solution The volume is $V = \iiint_D \rho^2 \sin\phi \, d\rho \, d\phi \, d\theta$, the integral of $f(\rho, \phi, \theta) = 1$ over D.

To find the limits of integration for evaluating the integral, we begin by sketching D and its projection R on the xy-plane (Figure 15.46).

The ρ-limits of integration. We draw a ray M from the origin through D making an angle ϕ with the positive z-axis. We also draw L, the projection of M on the xy-plane, along with the angle θ that L makes with the positive x-axis. Ray M enters D at $\rho = 0$ and leaves at $\rho = 1$.

The φ-limits of integration. The cone $\phi = \pi/3$ makes an angle of $\pi/3$ with the positive z-axis. For any given θ, the angle ϕ can run from $\phi = 0$ to $\phi = \pi/3$.

The θ-limits of integration. The ray L sweeps over R as θ runs from 0 to 2π. The volume is

$$\begin{aligned}
V &= \iiint_D \rho^2 \sin\phi \, d\rho \, d\phi \, d\theta = \int_0^{2\pi}\int_0^{\pi/3}\int_0^1 \rho^2 \sin\phi \, d\rho \, d\phi \, d\theta \\
&= \int_0^{2\pi}\int_0^{\pi/3}\left[\frac{\rho^3}{3}\right]_0^1 \sin\phi \, d\phi \, d\theta = \int_0^{2\pi}\int_0^{\pi/3} \frac{1}{3}\sin\phi \, d\phi \, d\theta \\
&= \int_0^{2\pi}\left[-\frac{1}{3}\cos\phi\right]_0^{\pi/3} d\theta = \int_0^{2\pi}\left(-\frac{1}{6}+\frac{1}{3}\right) d\theta = \frac{1}{6}(2\pi) = \frac{\pi}{3}.
\end{aligned}$$

■

EXAMPLE 6 Finding a Moment of Inertia

A solid of constant density $\delta = 1$ occupies the region D in Example 5. Find the solid's moment of inertia about the z-axis.

Solution In rectangular coordinates, the moment is

$$I_z = \iiint (x^2 + y^2)\, dV.$$

In spherical coordinates, $x^2 + y^2 = (\rho \sin\phi \cos\theta)^2 + (\rho \sin\phi \sin\theta)^2 = \rho^2 \sin^2\phi$. Hence,

$$I_z = \iiint (\rho^2 \sin^2\phi)\, \rho^2 \sin\phi \, d\rho \, d\phi \, d\theta = \iiint \rho^4 \sin^3\phi \, d\rho \, d\phi \, d\theta.$$

For the region in Example 5, this becomes

$$\begin{aligned}
I_z &= \int_0^{2\pi}\int_0^{\pi/3}\int_0^1 \rho^4 \sin^3\phi \, d\rho \, d\phi \, d\theta = \int_0^{2\pi}\int_0^{\pi/3}\left[\frac{\rho^5}{5}\right]_0^1 \sin^3\phi \, d\phi \, d\theta \\
&= \frac{1}{5}\int_0^{2\pi}\int_0^{\pi/3} (1 - \cos^2\phi)\sin\phi \, d\phi \, d\theta = \frac{1}{5}\int_0^{2\pi}\left[-\cos\phi + \frac{\cos^3\phi}{3}\right]_0^{\pi/3} d\theta \\
&= \frac{1}{5}\int_0^{2\pi}\left(-\frac{1}{2} + 1 + \frac{1}{24} - \frac{1}{3}\right) d\theta = \frac{1}{5}\int_0^{2\pi}\frac{5}{24}\, d\theta = \frac{1}{24}(2\pi) = \frac{\pi}{12}.
\end{aligned}$$

■

Coordinate Conversion Formulas

Cylindrical to Rectangular	Spherical to Rectangular	Spherical to Cylindrical
$x = r\cos\theta$	$x = \rho \sin\phi \cos\theta$	$r = \rho \sin\phi$
$y = r\sin\theta$	$y = \rho \sin\phi \sin\theta$	$z = \rho \cos\phi$
$z = z$	$z = \rho \cos\phi$	$\theta = \theta$

Corresponding formulas for dV in triple integrals:

$$\begin{aligned}
dV &= dx\, dy\, dz \\
&= dz\, r\, dr\, d\theta \\
&= \rho^2 \sin\phi \, d\rho \, d\phi \, d\theta
\end{aligned}$$

In the next section we offer a more general procedure for determining dV in cylindrical and spherical coordinates. The results, of course, will be the same.

EXERCISES 15.6

Evaluating Integrals in Cylindrical Coordinates

Evaluate the cylindrical coordinate integrals in Exercises 1–6.

1. $\int_0^{2\pi}\int_0^1\int_r^{\sqrt{2-r^2}} dz\, r\, dr\, d\theta$
2. $\int_0^{2\pi}\int_0^3\int_{r^2/3}^{\sqrt{18-r^2}} dz\, r\, dr\, d\theta$
3. $\int_0^{2\pi}\int_0^{\theta/2\pi}\int_0^{3+24r^2} dz\, r\, dr\, d\theta$
4. $\int_0^{\pi}\int_0^{\theta/\pi}\int_{-\sqrt{4-r^2}}^{3\sqrt{4-r^2}} z\, dz\, r\, dr\, d\theta$
5. $\int_0^{2\pi}\int_0^1\int_r^{1/\sqrt{2-r^2}} 3\, dz\, r\, dr\, d\theta$
6. $\int_0^{2\pi}\int_0^1\int_{-1/2}^{1/2} (r^2\sin^2\theta + z^2)\, dz\, r\, dr\, d\theta$

Changing Order of Integration in Cylindrical Coordinates

The integrals we have seen so far suggest that there are preferred orders of integration for cylindrical coordinates, but other orders usually work well and are occasionally easier to evaluate. Evaluate the integrals in Exercises 7–10.

7. $\int_0^{2\pi}\int_0^3\int_0^{z/3} r^3\, dr\, dz\, d\theta$
8. $\int_{-1}^1\int_0^{2\pi}\int_0^{1+\cos\theta} 4r\, dr\, d\theta\, dz$
9. $\int_0^1\int_0^{\sqrt{z}}\int_0^{2\pi} (r^2\cos^2\theta + z^2)\, r\, d\theta\, dr\, dz$
10. $\int_0^2\int_{r-2}^{\sqrt{4-r^2}}\int_0^{2\pi} (r\sin\theta + 1)\, r\, d\theta\, dz\, dr$

11. Let D be the region bounded below by the plane $z = 0$, above by the sphere $x^2 + y^2 + z^2 = 4$, and on the sides by the cylinder $x^2 + y^2 = 1$. Set up the triple integrals in cylindrical coordinates that give the volume of D using the following orders of integration.
    a. $dz\, dr\, d\theta$
    b. $dr\, dz\, d\theta$
    c. $d\theta\, dz\, dr$

12. Let D be the region bounded below by the cone $z = \sqrt{x^2 + y^2}$ and above by the paraboloid $z = 2 - x^2 - y^2$. Set up the triple integrals in cylindrical coordinates that give the volume of D using the following orders of integration.
    a. $dz\, dr\, d\theta$
    b. $dr\, dz\, d\theta$
    c. $d\theta\, dz\, dr$

13. Give the limits of integration for evaluating the integral
$$\iiint f(r, \theta, z)\, dz\, r\, dr\, d\theta$$
as an iterated integral over the region that is bounded below by the plane $z = 0$, on the side by the cylinder $r = \cos\theta$, and on top by the paraboloid $z = 3r^2$.

14. Convert the integral
$$\int_{-1}^1\int_0^{\sqrt{1-y^2}}\int_0^x (x^2 + y^2)\, dz\, dx\, dy$$
to an equivalent integral in cylindrical coordinates and evaluate the result.

Finding Iterated Integrals in Cylindrical Coordinates

In Exercises 15–20, set up the iterated integral for evaluating $\iiint_D f(r, \theta, z)\, dz\, r\, dr\, d\theta$ over the given region D.

15. D is the right circular cylinder whose base is the circle $r = 2\sin\theta$ in the xy-plane and whose top lies in the plane $z = 4 - y$.

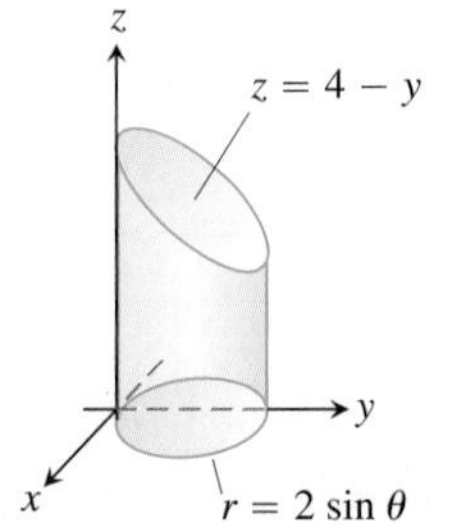

16. D is the right circular cylinder whose base is the circle $r = 3\cos\theta$ and whose top lies in the plane $z = 5 - x$.

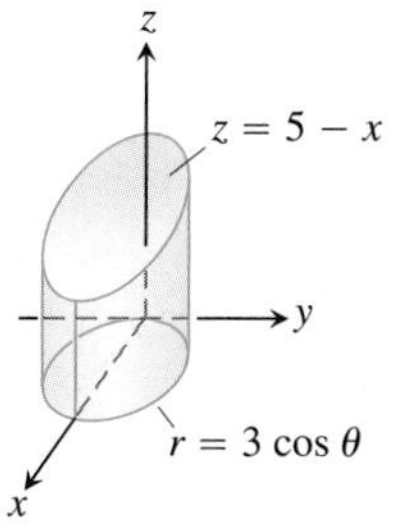

17. D is the solid right cylinder whose base is the region in the xy-plane that lies inside the cardioid $r = 1 + \cos\theta$ and outside the circle $r = 1$ and whose top lies in the plane $z = 4$.

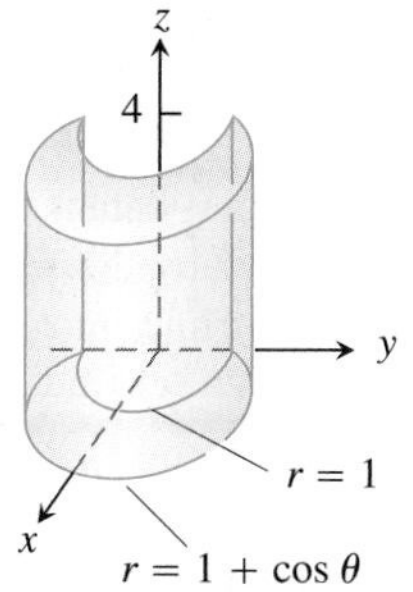

18. D is the solid right cylinder whose base is the region between the circles $r = \cos\theta$ and $r = 2\cos\theta$ and whose top lies in the plane $z = 3 - y$.

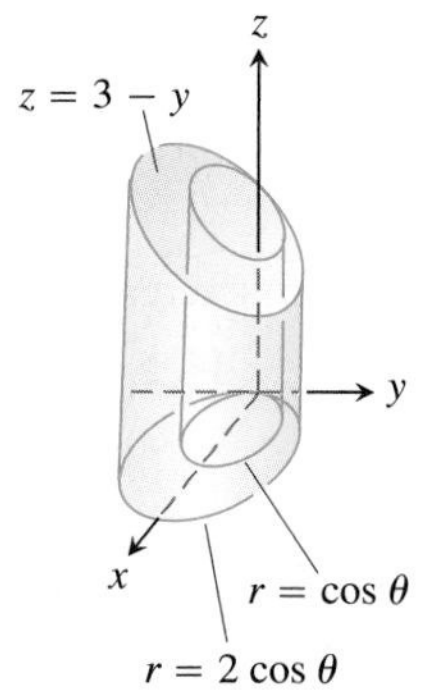

19. D is the prism whose base is the triangle in the xy-plane bounded by the x-axis and the lines $y = x$ and $x = 1$ and whose top lies in the plane $z = 2 - y$.

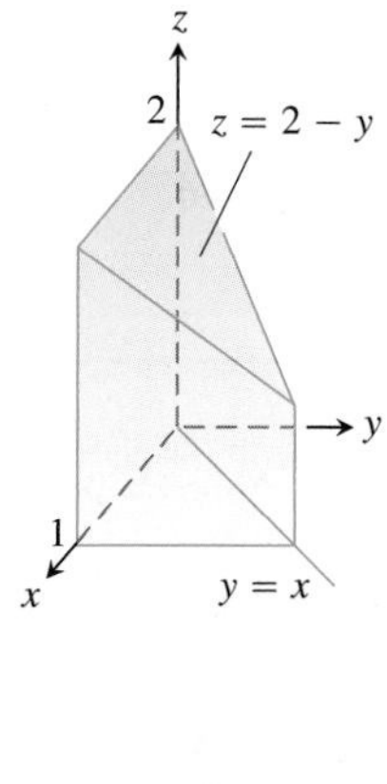

20. D is the prism whose base is the triangle in the xy-plane bounded by the y-axis and the lines $y = x$ and $y = 1$ and whose top lies in the plane $z = 2 - x$.

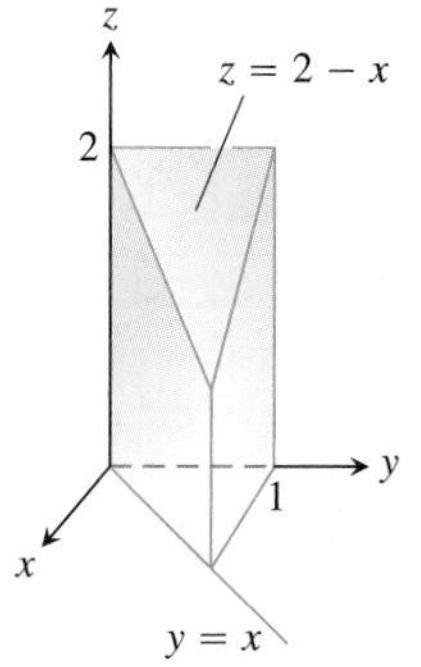

Evaluating Integrals in Spherical Coordinates

Evaluate the spherical coordinate integrals in Exercises 21–26.

21. $\displaystyle\int_0^{\pi}\int_0^{\pi}\int_0^{2\sin\phi} \rho^2 \sin\phi \, d\rho\, d\phi\, d\theta$

22. $\displaystyle\int_0^{2\pi}\int_0^{\pi/4}\int_0^{2} (\rho\cos\phi)\, \rho^2 \sin\phi \, d\rho\, d\phi\, d\theta$

23. $\displaystyle\int_0^{2\pi}\int_0^{\pi}\int_0^{(1-\cos\phi)/2} \rho^2 \sin\phi \, d\rho\, d\phi\, d\theta$

24. $\displaystyle\int_0^{3\pi/2}\int_0^{\pi}\int_0^{1} 5\rho^3 \sin^3\phi \, d\rho\, d\phi\, d\theta$

25. $\displaystyle\int_0^{2\pi}\int_0^{\pi/3}\int_{\sec\phi}^{2} 3\rho^2 \sin\phi \, d\rho\, d\phi\, d\theta$

26. $\displaystyle\int_0^{2\pi}\int_0^{\pi/4}\int_0^{\sec\phi} (\rho\cos\phi)\, \rho^2 \sin\phi \, d\rho\, d\phi\, d\theta$

Changing Order of Integration in Spherical Coordinates

The previous integrals suggest there are preferred orders of integration for spherical coordinates, but other orders are possible and occasionally easier to evaluate. Evaluate the integrals in Exercises 27–30.

27. $\displaystyle\int_0^{2}\int_{-\pi}^{0}\int_{\pi/4}^{\pi/2} \rho^3 \sin 2\phi \, d\phi\, d\theta\, d\rho$

28. $\displaystyle\int_{\pi/6}^{\pi/3}\int_{\csc\phi}^{2\csc\phi}\int_0^{2\pi} \rho^2 \sin\phi \, d\theta\, d\rho\, d\phi$

29. $\displaystyle\int_0^{1}\int_0^{\pi}\int_0^{\pi/4} 12\rho \sin^3\phi \, d\phi\, d\theta\, d\rho$

30. $\displaystyle\int_{\pi/6}^{\pi/2}\int_{-\pi/2}^{\pi/2}\int_{\csc\phi}^{2} 5\rho^4 \sin^3\phi \, d\rho\, d\theta\, d\phi$

31. Let D be the region in Exercise 11. Set up the triple integrals in spherical coordinates that give the volume of D using the following orders of integration.

a. $d\rho\, d\phi\, d\theta$ **b.** $d\phi\, d\rho\, d\theta$

32. Let D be the region bounded below by the cone $z = \sqrt{x^2 + y^2}$ and above by the plane $z = 1$. Set up the triple integrals in spherical coordinates that give the volume of D using the following orders of integration.

a. $d\rho\, d\phi\, d\theta$ **b.** $d\phi\, d\rho\, d\theta$

Finding Iterated Integrals in Spherical Coordinates

In Exercises 33–38, **(a)** find the spherical coordinate limits for the integral that calculates the volume of the given solid and **(b)** then evaluate the integral.

33. The solid between the sphere $\rho = \cos\phi$ and the hemisphere $\rho = 2, z \geq 0$

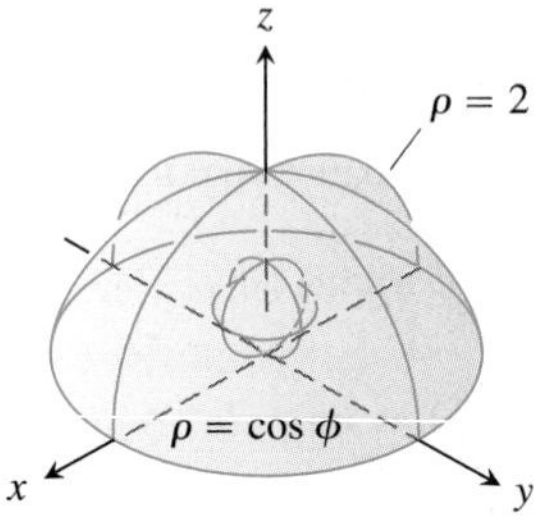

34. The solid bounded below by the hemisphere $\rho = 1, z \geq 0$, and above by the cardioid of revolution $\rho = 1 + \cos\phi$

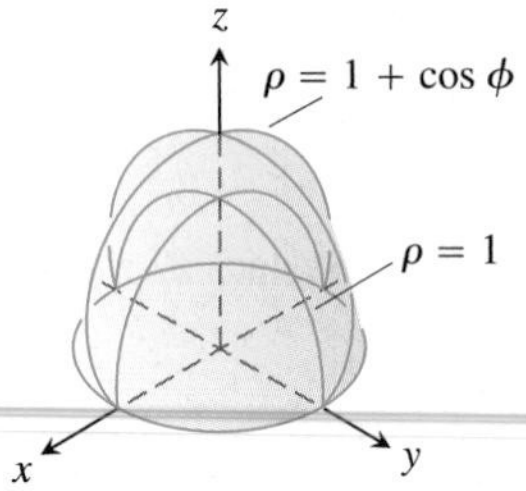

35. The solid enclosed by the cardioid of revolution $\rho = 1 - \cos\phi$

36. The upper portion cut from the solid in Exercise 35 by the xy-plane

37. The solid bounded below by the sphere $\rho = 2\cos\phi$ and above by the cone $z = \sqrt{x^2 + y^2}$

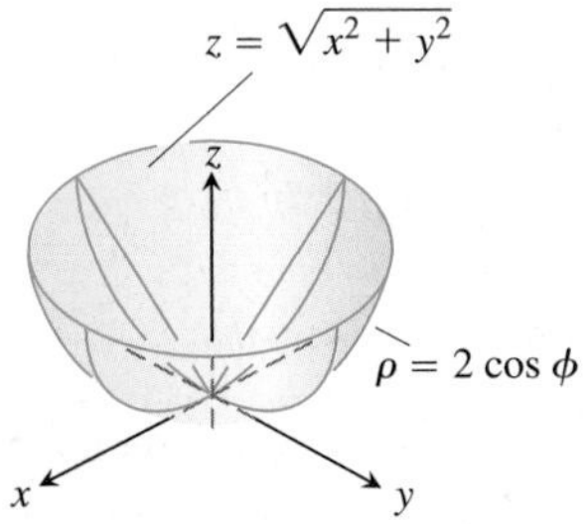

38. The solid bounded below by the xy-plane, on the sides by the sphere $\rho = 2$, and above by the cone $\phi = \pi/3$

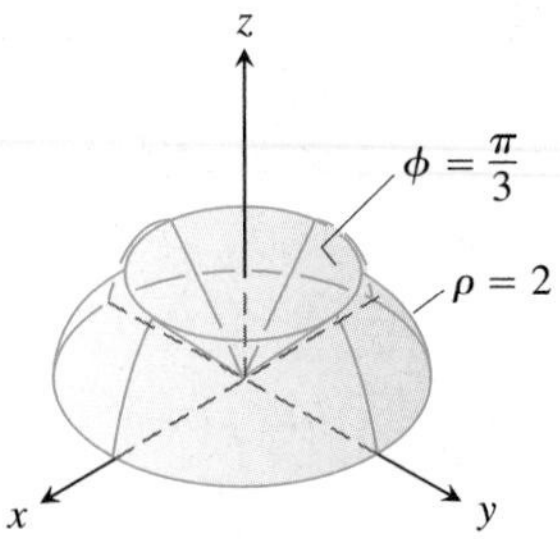

Rectangular, Cylindrical, and Spherical Coordinates

39. Set up triple integrals for the volume of the sphere $\rho = 2$ in **(a)** spherical, **(b)** cylindrical, and **(c)** rectangular coordinates.

40. Let D be the region in the first octant that is bounded below by the cone $\phi = \pi/4$ and above by the sphere $\rho = 3$. Express the volume of D as an iterated triple integral in **(a)** cylindrical and **(b)** spherical coordinates. Then **(c)** find V.

41. Let D be the smaller cap cut from a solid ball of radius 2 units by a plane 1 unit from the center of the sphere. Express the volume of D as an iterated triple integral in **(a)** spherical, **(b)** cylindrical, and **(c)** rectangular coordinates. Then **(d)** find the volume by evaluating one of the three triple integrals.

42. Express the moment of inertia I_z of the solid hemisphere $x^2 + y^2 + z^2 \leq 1, z \geq 0$, as an iterated integral in **(a)** cylindrical and **(b)** spherical coordinates. Then **(c)** find I_z.

Volumes

Find the volumes of the solids in Exercises 43–48.

43.

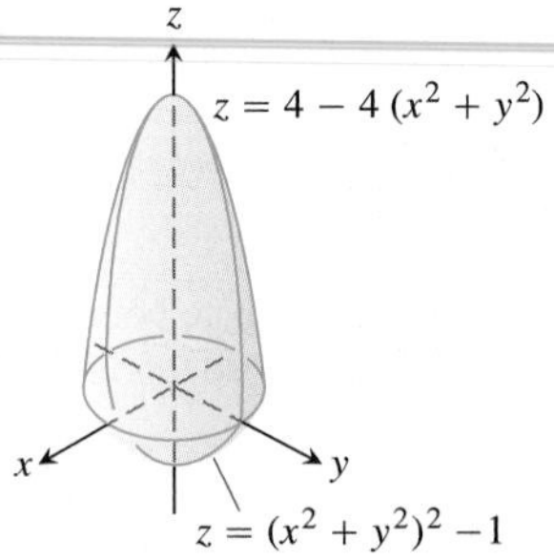

44.

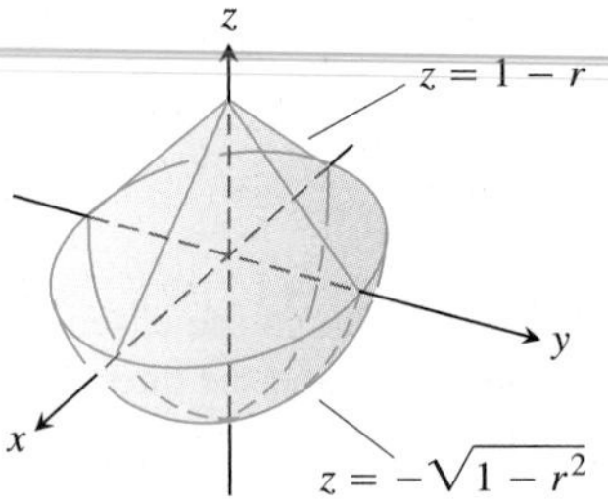

45.

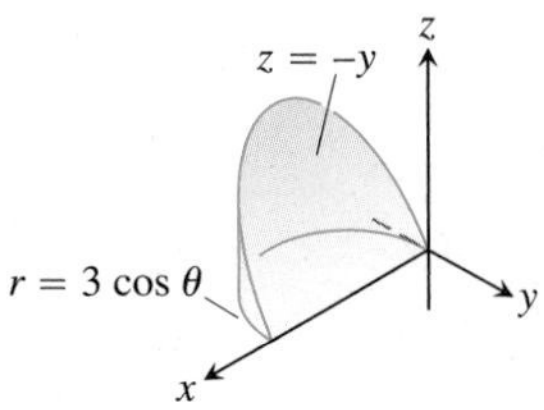

46.

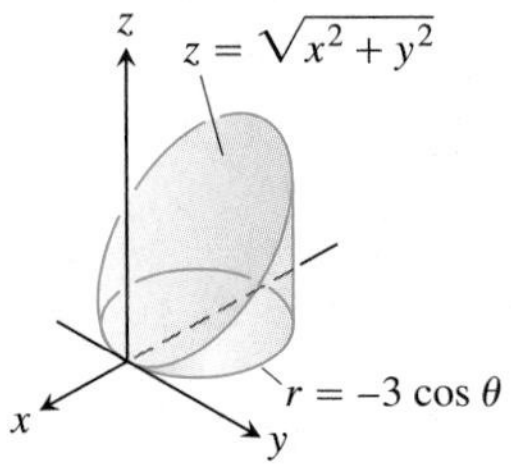

47.

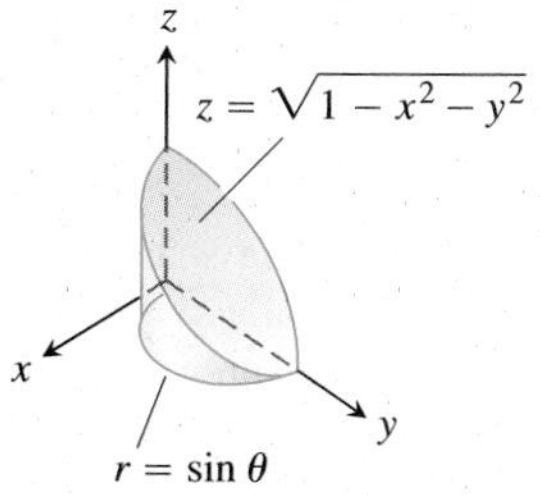

48.

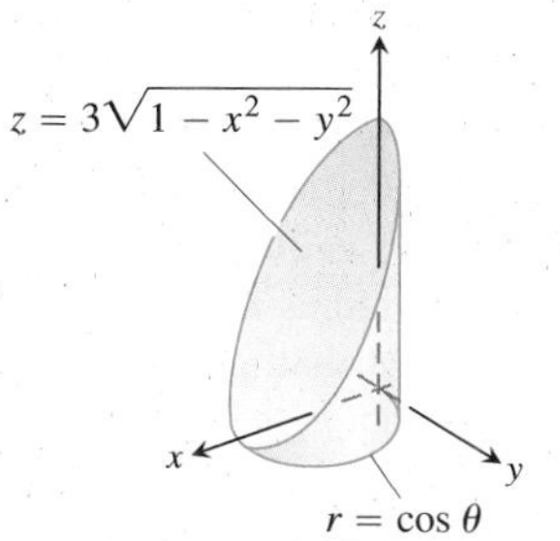

49. Sphere and cones Find the volume of the portion of the solid sphere $\rho \le a$ that lies between the cones $\phi = \pi/3$ and $\phi = 2\pi/3$.

50. Sphere and half-planes Find the volume of the region cut from the solid sphere $\rho \le a$ by the half-planes $\theta = 0$ and $\theta = \pi/6$ in the first octant.

51. Sphere and plane Find the volume of the smaller region cut from the solid sphere $\rho \le 2$ by the plane $z = 1$.

52. Cone and planes Find the volume of the solid enclosed by the cone $z = \sqrt{x^2 + y^2}$ between the planes $z = 1$ and $z = 2$.

53. Cylinder and paraboloid Find the volume of the region bounded below by the plane $z = 0$, laterally by the cylinder $x^2 + y^2 = 1$, and above by the paraboloid $z = x^2 + y^2$.

54. Cylinder and paraboloids Find the volume of the region bounded below by the paraboloid $z = x^2 + y^2$, laterally by the cylinder $x^2 + y^2 = 1$, and above by the paraboloid $z = x^2 + y^2 + 1$.

55. Cylinder and cones Find the volume of the solid cut from the thick-walled cylinder $1 \le x^2 + y^2 \le 2$ by the cones $z = \pm\sqrt{x^2 + y^2}$.

56. Sphere and cylinder Find the volume of the region that lies inside the sphere $x^2 + y^2 + z^2 = 2$ and outside the cylinder $x^2 + y^2 = 1$.

57. Cylinder and planes Find the volume of the region enclosed by the cylinder $x^2 + y^2 = 4$ and the planes $z = 0$ and $y + z = 4$.

58. Cylinder and planes Find the volume of the region enclosed by the cylinder $x^2 + y^2 = 4$ and the planes $z = 0$ and $x + y + z = 4$.

59. Region trapped by paraboloids Find the volume of the region bounded above by the paraboloid $z = 5 - x^2 - y^2$ and below by the paraboloid $z = 4x^2 + 4y^2$.

60. Paraboloid and cylinder Find the volume of the region bounded above by the paraboloid $z = 9 - x^2 - y^2$, below by the xy-plane, and lying *outside* the cylinder $x^2 + y^2 = 1$.

61. Cylinder and sphere Find the volume of the region cut from the solid cylinder $x^2 + y^2 \le 1$ by the sphere $x^2 + y^2 + z^2 = 4$.

62. Sphere and paraboloid Find the volume of the region bounded above by the sphere $x^2 + y^2 + z^2 = 2$ and below by the paraboloid $z = x^2 + y^2$.

Average Values

63. Find the average value of the function $f(r, \theta, z) = r$ over the region bounded by the cylinder $r = 1$ between the planes $z = -1$ and $z = 1$.

64. Find the average value of the function $f(r, \theta, z) = r$ over the solid ball bounded by the sphere $r^2 + z^2 = 1$. (This is the sphere $x^2 + y^2 + z^2 = 1$.)

65. Find the average value of the function $f(\rho, \phi, \theta) = \rho$ over the solid ball $\rho \le 1$.

66. Find the average value of the function $f(\rho, \phi, \theta) = \rho\cos\phi$ over the solid upper ball $\rho \le 1, 0 \le \phi \le \pi/2$.

Masses, Moments, and Centroids

67. Center of mass A solid of constant density is bounded below by the plane $z = 0$, above by the cone $z = r, r \ge 0$, and on the sides by the cylinder $r = 1$. Find the center of mass.

68. Centroid Find the centroid of the region in the first octant that is bounded above by the cone $z = \sqrt{x^2 + y^2}$, below by the plane $z = 0$, and on the sides by the cylinder $x^2 + y^2 = 4$ and the planes $x = 0$ and $y = 0$.

69. Centroid Find the centroid of the solid in Exercise 38.

70. Centroid Find the centroid of the solid bounded above by the sphere $\rho = a$ and below by the cone $\phi = \pi/4$.

71. Centroid Find the centroid of the region that is bounded above by the surface $z = \sqrt{r}$, on the sides by the cylinder $r = 4$, and below by the xy-plane.

72. Centroid Find the centroid of the region cut from the solid ball $r^2 + z^2 \le 1$ by the half-planes $\theta = -\pi/3, r \ge 0$, and $\theta = \pi/3, r \ge 0$.

73. Inertia and radius of gyration Find the moment of inertia and radius of gyration about the z-axis of a thick-walled right circular cylinder bounded on the inside by the cylinder $r = 1$, on the outside by the cylinder $r = 2$, and on the top and bottom by the planes $z = 4$ and $z = 0$. (Take $\delta = 1$.)

74. Moments of inertia of solid circular cylinder Find the moment of inertia of a solid circular cylinder of radius 1 and height 2 **(a)** about the axis of the cylinder and **(b)** about a line through the centroid perpendicular to the axis of the cylinder. (Take $\delta = 1$.)

75. Moment of inertia of solid cone Find the moment of inertia of a right circular cone of base radius 1 and height 1 about an axis through the vertex parallel to the base. (Take $\delta = 1$.)

76. Moment of inertia of solid sphere Find the moment of inertia of a solid sphere of radius a about a diameter. (Take $\delta = 1$.)

77. Moment of inertia of solid cone Find the moment of inertia of a right circular cone of base radius a and height h about its axis. (*Hint:* Place the cone with its vertex at the origin and its axis along the z-axis.)

78. Variable density A solid is bounded on the top by the paraboloid $z = r^2$, on the bottom by the plane $z = 0$, and on the sides by the cylinder $r = 1$. Find the center of mass and the moment of inertia and radius of gyration about the z-axis if the density is

a. $\delta(r, \theta, z) = z$

b. $\delta(r, \theta, z) = r$.

79. Variable density A solid is bounded below by the cone $z = \sqrt{x^2 + y^2}$ and above by the plane $z = 1$. Find the center of mass and the moment of inertia and radius of gyration about the z-axis if the density is

a. $\delta(r, \theta, z) = z$

b. $\delta(r, \theta, z) = z^2$.

80. Variable density A solid ball is bounded by the sphere $\rho = a$. Find the moment of inertia and radius of gyration about the z-axis if the density is

a. $\delta(\rho, \phi, \theta) = \rho^2$

b. $\delta(\rho, \phi, \theta) = r = \rho \sin \phi$.

81. Centroid of solid semiellipsoid Show that the centroid of the solid semiellipsoid of revolution $(r^2/a^2) + (z^2/h^2) \le 1, z \ge 0$, lies on the z-axis three-eighths of the way from the base to the top. The special case $h = a$ gives a solid hemisphere. Thus, the centroid of a solid hemisphere lies on the axis of symmetry three-eighths of the way from the base to the top.

82. Centroid of solid cone Show that the centroid of a solid right circular cone is one-fourth of the way from the base to the vertex. (In general, the centroid of a solid cone or pyramid is one-fourth of the way from the centroid of the base to the vertex.)

83. Variable density A solid right circular cylinder is bounded by the cylinder $r = a$ and the planes $z = 0$ and $z = h, h > 0$. Find the center of mass and the moment of inertia and radius of gyration about the z-axis if the density is $\delta(r, \theta, z) = z + 1$.

84. Mass of planet's atmosphere A spherical planet of radius R has an atmosphere whose density is $\mu = \mu_0 e^{-ch}$, where h is the altitude above the surface of the planet, μ_0 is the density at sea level, and c is a positive constant. Find the mass of the planet's atmosphere.

85. Density of center of a planet A planet is in the shape of a sphere of radius R and total mass M with spherically symmetric density distribution that increases linearly as one approaches its center. What is the density at the center of this planet if the density at its edge (surface) is taken to be zero?

Theory and Examples

86. Vertical circular cylinders in spherical coordinates Find an equation of the form $\rho = f(\phi)$ for the cylinder $x^2 + y^2 = a^2$.

87. Vertical planes in cylindrical coordinates

a. Show that planes perpendicular to the x-axis have equations of the form $r = a \sec \theta$ in cylindrical coordinates.

b. Show that planes perpendicular to the y-axis have equations of the form $r = b \csc \theta$.

88. (*Continuation of Exercise 87.*) Find an equation of the form $r = f(\theta)$ in cylindrical coordinates for the plane $ax + by = c$, $c \ne 0$.

89. Symmetry What symmetry will you find in a surface that has an equation of the form $r = f(z)$ in cylindrical coordinates? Give reasons for your answer.

90. Symmetry What symmetry will you find in a surface that has an equation of the form $\rho = f(\phi)$ in spherical coordinates? Give reasons for your answer.

15.7 Substitutions in Multiple Integrals

This section shows how to evaluate multiple integrals by substitution. As in single integration, the goal of substitution is to replace complicated integrals by ones that are easier to evaluate. Substitutions accomplish this by simplifying the integrand, the limits of integration, or both.

Substitutions in Double Integrals

The polar coordinate substitution of Section 15.3 is a special case of a more general substitution method for double integrals, a method that pictures changes in variables as transformations of regions.

Suppose that a region G in the uv-plane is transformed one-to-one into the region R in the xy-plane by equations of the form

$$x = g(u, v), \qquad y = h(u, v),$$

as suggested in Figure 15.47. We call R the **image** of G under the transformation, and G the **preimage** of R. Any function $f(x, y)$ defined on R can be thought of as a function

$f(g(u, v), h(u, v))$ defined on G as well. How is the integral of $f(x, y)$ over R related to the integral of $f(g(u, v), h(u, v))$ over G?

The answer is: If g, h, and f have continuous partial derivatives and $J(u, v)$ (to be discussed in a moment) is zero only at isolated points, if at all, then

$$\iint_R f(x, y)\, dx\, dy = \iint_G f(g(u, v), h(u, v))\,|J(u, v)|\, du\, dv. \tag{1}$$

HISTORICAL BIOGRAPHY

Carl Gustav Jacob Jacobi (1804–1851)

The factor $J(u, v)$, whose absolute value appears in Equation (1), is the *Jacobian* of the coordinate transformation, named after German mathematician Carl Jacobi. It measures how much the transformation is expanding or contracting the area around a point in G as G is transformed into R.

Definition Jacobian

The **Jacobian determinant** or **Jacobian** of the coordinate transformation $x = g(u, v)$, $y = h(u, v)$ is

$$J(u, v) = \begin{vmatrix} \dfrac{\partial x}{\partial u} & \dfrac{\partial x}{\partial v} \\[2mm] \dfrac{\partial y}{\partial u} & \dfrac{\partial y}{\partial v} \end{vmatrix} = \frac{\partial x}{\partial u}\frac{\partial y}{\partial v} - \frac{\partial y}{\partial u}\frac{\partial x}{\partial v}. \tag{2}$$

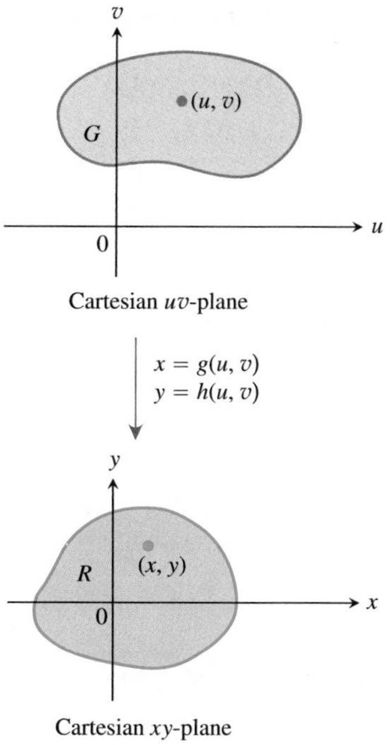

FIGURE 15.47 The equations $x = g(u, v)$ and $y = h(u, v)$ allow us to change an integral over a region R in the xy-plane into an integral over a region G in the uv-plane.

The Jacobian is also denoted by

$$J(u, v) = \frac{\partial(x, y)}{\partial(u, v)}$$

to help remember how the determinant in Equation (2) is constructed from the partial derivatives of x and y. The derivation of Equation (1) is intricate and properly belongs to a course in advanced calculus. We do not give the derivation here.

For polar coordinates, we have r and θ in place of u and v. With $x = r\cos\theta$ and $y = r\sin\theta$, the Jacobian is

$$J(r, \theta) = \begin{vmatrix} \dfrac{\partial x}{\partial r} & \dfrac{\partial x}{\partial \theta} \\[2mm] \dfrac{\partial y}{\partial r} & \dfrac{\partial y}{\partial \theta} \end{vmatrix} = \begin{vmatrix} \cos\theta & -r\sin\theta \\ \sin\theta & r\cos\theta \end{vmatrix} = r(\cos^2\theta + \sin^2\theta) = r.$$

Hence, Equation (1) becomes

$$\iint_R f(x, y)\, dx\, dy = \iint_G f(r\cos\theta, r\sin\theta)\,|r|\, dr\, d\theta$$

$$= \iint_G f(r\cos\theta, r\sin\theta)\, r\, dr\, d\theta, \qquad \text{If } r \ge 0 \tag{3}$$

which is the equation found in Section 15.3.

Figure 15.48 shows how the equations $x = r\cos\theta$, $y = r\sin\theta$ transform the rectangle G: $0 \le r \le 1$, $0 \le \theta \le \pi/2$ into the quarter circle R bounded by $x^2 + y^2 = 1$ in the first quadrant of the xy-plane.

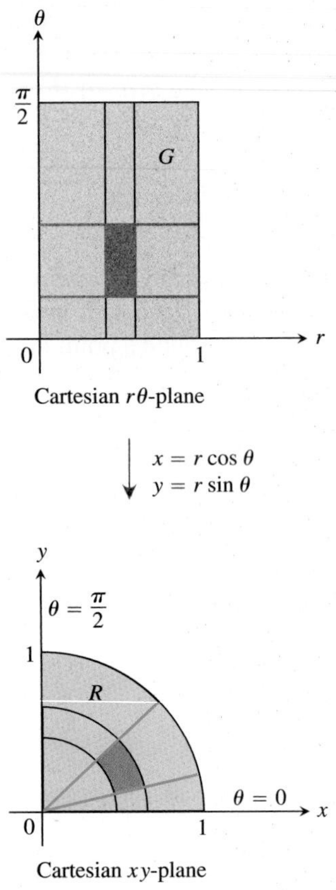

FIGURE 15.48 The equations $x = r\cos\theta$, $y = r\sin\theta$ transform G into R.

Notice that the integral on the right-hand side of Equation (3) is not the integral of $f(r\cos\theta, r\sin\theta)$ over a region in the polar coordinate plane. It is the integral of the product of $f(r\cos\theta, r\sin\theta)$ and r over a region G in the *Cartesian* $r\theta$-plane.

Here is an example of another substitution.

EXAMPLE 1 Applying a Transformation to Integrate

Evaluate

$$\int_0^4 \int_{x=y/2}^{x=(y/2)+1} \frac{2x - y}{2}\, dx\, dy$$

by applying the transformation

$$u = \frac{2x - y}{2}, \qquad v = \frac{y}{2} \tag{4}$$

and integrating over an appropriate region in the uv-plane.

Solution We sketch the region R of integration in the xy-plane and identify its boundaries (Figure 15.49).

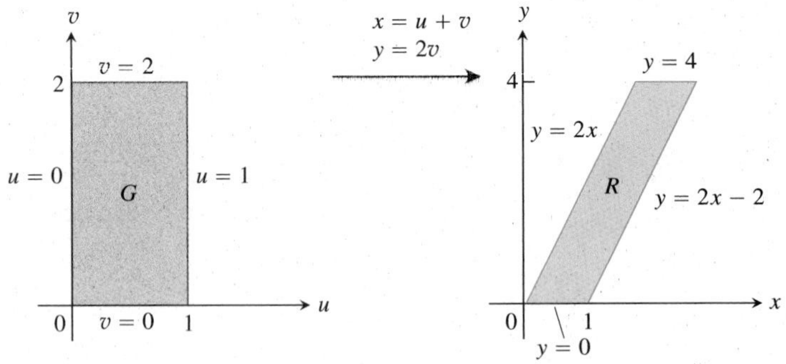

FIGURE 15.49 The equations $x = u + v$ and $y = 2v$ transform G into R. Reversing the transformation by the equations $u = (2x - y)/2$ and $v = y/2$ transforms R into G (Example 1).

To apply Equation (1), we need to find the corresponding uv-region G and the Jacobian of the transformation. To find them, we first solve Equations (4) for x and y in terms of u and v. Routine algebra gives

$$x = u + v \qquad y = 2v. \tag{5}$$

We then find the boundaries of G by substituting these expressions into the equations for the boundaries of R (Figure 15.49).

xy*-equations for the boundary of *R	**Corresponding *uv*-equations for the boundary of *G***	**Simplified *uv*-equations**
$x = y/2$	$u + v = 2v/2 = v$	$u = 0$
$x = (y/2) + 1$	$u + v = (2v/2) + 1 = v + 1$	$u = 1$
$y = 0$	$2v = 0$	$v = 0$
$y = 4$	$2v = 4$	$v = 2$

The Jacobian of the transformation (again from Equations (5)) is

$$J(u, v) = \begin{vmatrix} \dfrac{\partial x}{\partial u} & \dfrac{\partial x}{\partial v} \\ \dfrac{\partial y}{\partial u} & \dfrac{\partial y}{\partial v} \end{vmatrix} = \begin{vmatrix} \dfrac{\partial}{\partial u}(u + v) & \dfrac{\partial}{\partial v}(u + v) \\ \dfrac{\partial}{\partial u}(2v) & \dfrac{\partial}{\partial v}(2v) \end{vmatrix} = \begin{vmatrix} 1 & 1 \\ 0 & 2 \end{vmatrix} = 2.$$

We now have everything we need to apply Equation (1):

$$\int_0^4 \int_{x=y/2}^{x=(y/2)+1} \frac{2x - y}{2}\, dx\, dy = \int_{v=0}^{v=2} \int_{u=0}^{u=1} u\,|J(u, v)|\, du\, dv$$

$$= \int_0^2 \int_0^1 (u)(2)\, du\, dv = \int_0^2 \Big[u^2\Big]_0^1 dv = \int_0^2 dv = 2.$$

■

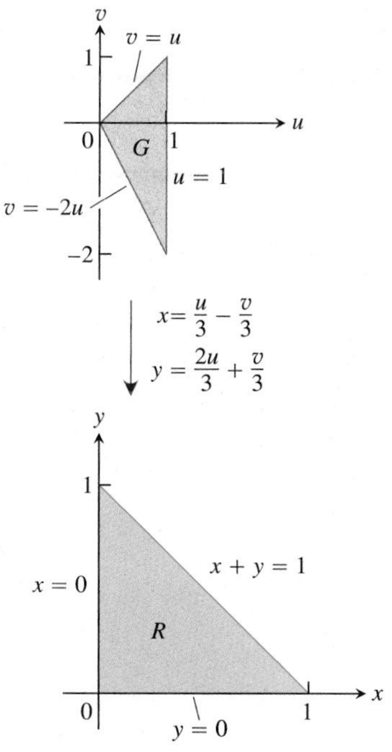

FIGURE 15.50 The equations $x = (u/3) - (v/3)$ and $y = (2u/3) + (v/3)$ transform G into R. Reversing the transformation by the equations $u = x + y$ and $v = y - 2x$ transforms R into G (Example 2).

EXAMPLE 2 Applying a Transformation to Integrate

Evaluate

$$\int_0^1 \int_0^{1-x} \sqrt{x + y}\,(y - 2x)^2\, dy\, dx.$$

Solution We sketch the region R of integration in the xy-plane and identify its boundaries (Figure 15.50). The integrand suggests the transformation $u = x + y$ and $v = y - 2x$. Routine algebra produces x and y as functions of u and v:

$$x = \frac{u}{3} - \frac{v}{3}, \qquad y = \frac{2u}{3} + \frac{v}{3}. \tag{6}$$

From Equations (6), we can find the boundaries of the uv-region G (Figure 15.50).

xy-equations for the boundary of R	Corresponding uv-equations for the boundary of G	Simplified uv-equations
$x + y = 1$	$\left(\frac{u}{3} - \frac{v}{3}\right) + \left(\frac{2u}{3} + \frac{v}{3}\right) = 1$	$u = 1$
$x = 0$	$\frac{u}{3} - \frac{v}{3} = 0$	$v = u$
$y = 0$	$\frac{2u}{3} + \frac{v}{3} = 0$	$v = -2u$

The Jacobian of the transformation in Equations (6) is

$$J(u, v) = \begin{vmatrix} \dfrac{\partial x}{\partial u} & \dfrac{\partial x}{\partial v} \\ \dfrac{\partial y}{\partial u} & \dfrac{\partial y}{\partial v} \end{vmatrix} = \begin{vmatrix} \dfrac{1}{3} & -\dfrac{1}{3} \\ \dfrac{2}{3} & \dfrac{1}{3} \end{vmatrix} = \frac{1}{3}.$$

Applying Equation (1), we evaluate the integral:

$$\int_0^1 \int_0^{1-x} \sqrt{x+y}\,(y-2x)^2\,dy\,dx = \int_{u=0}^{u=1} \int_{v=-2u}^{v=u} u^{1/2} v^2 |J(u, v)|\,dv\,du$$

$$= \int_0^1 \int_{-2u}^{u} u^{1/2} v^2 \left(\frac{1}{3}\right) dv\,du = \frac{1}{3}\int_0^1 u^{1/2} \left[\frac{1}{3} v^3\right]_{v=-2u}^{v=u} du$$

$$= \frac{1}{9}\int_0^1 u^{1/2}(u^3 + 8u^3)\,du = \int_0^1 u^{7/2}\,du = \frac{2}{9}u^{9/2}\Big]_0^1 = \frac{2}{9}.$$

■

Substitutions in Triple Integrals

The cylindrical and spherical coordinate substitutions in Section 15.6 are special cases of a substitution method that pictures changes of variables in triple integrals as transformations of three-dimensional regions. The method is like the method for double integrals except that now we work in three dimensions instead of two.

Suppose that a region G in uvw-space is transformed one-to-one into the region D in xyz-space by differentiable equations of the form

$$x = g(u, v, w), \qquad y = h(u, v, w), \qquad z = k(u, v, w),$$

as suggested in Figure 15.51. Then any function $F(x, y, z)$ defined on D can be thought of as a function

$$F(g(u, v, w), h(u, v, w), k(u, v, w)) = H(u, v, w)$$

defined on G. If g, h, and k have continuous first partial derivatives, then the integral of $F(x, y, z)$ over D is related to the integral of $H(u, v, w)$ over G by the equation

$$\iiint_D F(x, y, z)\,dx\,dy\,dz = \iiint_G H(u, v, w)\,|J(u, v, w)|\,du\,dv\,dw. \tag{7}$$

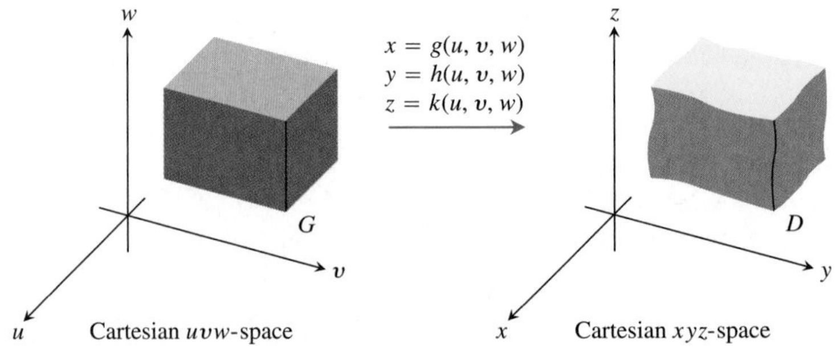

FIGURE 15.51 The equations $x = g(u, v, w)$, $y = h(u, v, w)$, and $z = k(u, v, w)$ allow us to change an integral over a region D in Cartesian xyz-space into an integral over a region G in Cartesian uvw-space.

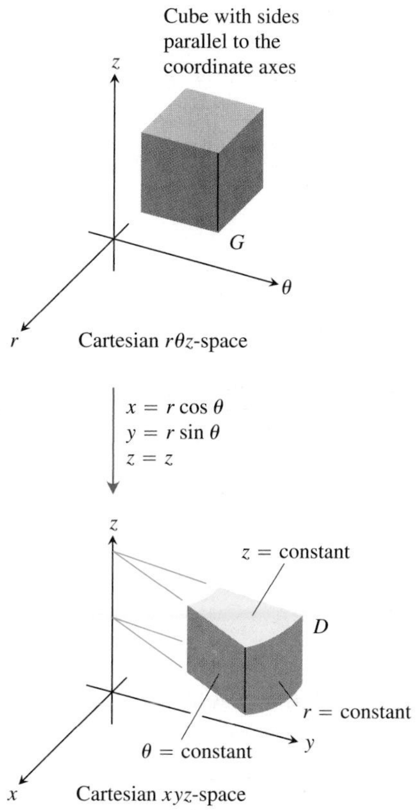

FIGURE 15.52 The equations $x = r\cos\theta, y = r\sin\theta$, and $z = z$ transform the cube G into a cylindrical wedge D.

The factor $J(u, v, w)$, whose absolute value appears in this equation, is the **Jacobian determinant**

$$J(u, v, w) = \begin{vmatrix} \dfrac{\partial x}{\partial u} & \dfrac{\partial x}{\partial v} & \dfrac{\partial x}{\partial w} \\[2ex] \dfrac{\partial y}{\partial u} & \dfrac{\partial y}{\partial v} & \dfrac{\partial y}{\partial w} \\[2ex] \dfrac{\partial z}{\partial u} & \dfrac{\partial z}{\partial v} & \dfrac{\partial z}{\partial w} \end{vmatrix} = \frac{\partial(x, y, z)}{\partial(u, v, w)}.$$

This determinant measures how much the volume near a point in G is being expanded or contracted by the transformation from (u, v, w) to (x, y, z) coordinates. As in the two-dimensional case, the derivation of the change-of-variable formula in Equation (7) is complicated and we do not go into it here.

For cylindrical coordinates, r, θ, and z take the place of u, v, and w. The transformation from *Cartesian* $r\theta z$*-space* to Cartesian xyz-space is given by the equations

$$x = r\cos\theta, \qquad y = r\sin\theta, \qquad z = z$$

(Figure 15.52). The Jacobian of the transformation is

$$J(r, \theta, z) = \begin{vmatrix} \dfrac{\partial x}{\partial r} & \dfrac{\partial x}{\partial \theta} & \dfrac{\partial x}{\partial z} \\[2ex] \dfrac{\partial y}{\partial r} & \dfrac{\partial y}{\partial \theta} & \dfrac{\partial y}{\partial z} \\[2ex] \dfrac{\partial z}{\partial r} & \dfrac{\partial z}{\partial \theta} & \dfrac{\partial z}{\partial z} \end{vmatrix} = \begin{vmatrix} \cos\theta & -r\sin\theta & 0 \\ \sin\theta & r\cos\theta & 0 \\ 0 & 0 & 1 \end{vmatrix}$$

$$= r\cos^2\theta + r\sin^2\theta = r.$$

The corresponding version of Equation (7) is

$$\iiint\limits_D F(x, y, z)\, dx\, dy\, dz = \iiint\limits_G H(r, \theta, z)\,|r|\, dr\, d\theta\, dz.$$

We can drop the absolute value signs whenever $r \geq 0$.

For spherical coordinates, ρ, ϕ, and θ take the place of u, v, and w. The transformation from Cartesian $\rho\phi\theta$-space to Cartesian xyz-space is given by

$$x = \rho\sin\phi\cos\theta, \qquad y = \rho\sin\phi\sin\theta, \qquad z = \rho\cos\phi$$

(Figure 15.53). The Jacobian of the transformation is

$$J(\rho, \phi, \theta) = \begin{vmatrix} \dfrac{\partial x}{\partial \rho} & \dfrac{\partial x}{\partial \phi} & \dfrac{\partial x}{\partial \theta} \\[2ex] \dfrac{\partial y}{\partial \rho} & \dfrac{\partial y}{\partial \phi} & \dfrac{\partial y}{\partial \theta} \\[2ex] \dfrac{\partial z}{\partial \rho} & \dfrac{\partial z}{\partial \phi} & \dfrac{\partial z}{\partial \theta} \end{vmatrix} = \rho^2\sin\phi$$

(Exercise 17). The corresponding version of Equation (7) is

$$\iiint_D F(x, y, z)\,dx\,dy\,dz = \iiint_G H(\rho, \phi, \theta)\,|\rho^2 \sin\phi|\,d\rho\,d\phi\,d\theta.$$

We can drop the absolute value signs because $\sin\phi$ is never negative for $0 \le \phi \le \pi$. Note that this is the same result we obtained in Section 15.6.

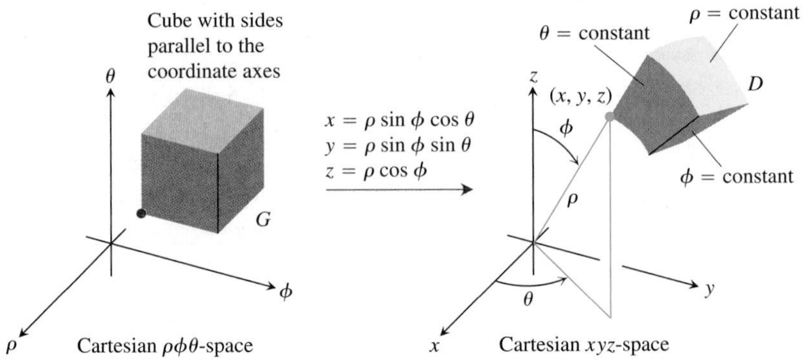

FIGURE 15.53 The equations $x = \rho \sin\phi \cos\theta$, $y = \rho \sin\phi \sin\theta$, and $z = \rho \cos\phi$ transform the cube G into the spherical wedge D.

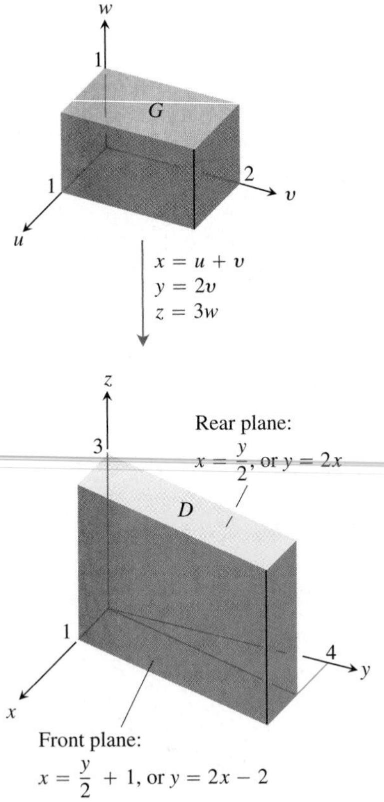

FIGURE 15.54 The equations $x = u + v$, $y = 2v$, and $z = 3w$ transform G into D. Reversing the transformation by the equations $u = (2x - y)/2$, $v = y/2$, and $w = z/3$ transforms D into G (Example 3).

Here is an example of another substitution. Although we could evaluate the integral in this example directly, we have chosen it to illustrate the substitution method in a simple (and fairly intuitive) setting.

EXAMPLE 3 Applying a Transformation to Integrate

Evaluate

$$\int_0^3 \int_0^4 \int_{x=y/2}^{x=(y/2)+1} \left(\frac{2x - y}{2} + \frac{z}{3}\right) dx\,dy\,dz$$

by applying the transformation

$$u = (2x - y)/2, \qquad v = y/2, \qquad w = z/3 \tag{8}$$

and integrating over an appropriate region in uvw-space.

Solution We sketch the region D of integration in xyz-space and identify its boundaries (Figure 15.54). In this case, the bounding surfaces are planes.

To apply Equation (7), we need to find the corresponding uvw-region G and the Jacobian of the transformation. To find them, we first solve Equations (8) for x, y, and z in terms of u, v, and w. Routine algebra gives

$$x = u + v, \qquad y = 2v, \qquad z = 3w. \tag{9}$$

We then find the boundaries of G by substituting these expressions into the equations for the boundaries of D:

xyz-equations for the boundary of *D*	Corresponding *uvw*-equations for the boundary of *G*	Simplified *uvw*-equations
$x = y/2$	$u + v = 2v/2 = v$	$u = 0$
$x = (y/2) + 1$	$u + v = (2v/2) + 1 = v + 1$	$u = 1$
$y = 0$	$2v = 0$	$v = 0$
$y = 4$	$2v = 4$	$v = 2$
$z = 0$	$3w = 0$	$w = 0$
$z = 3$	$3w = 3$	$w = 1$

The Jacobian of the transformation, again from Equations (9), is

$$J(u, v, w) = \begin{vmatrix} \frac{\partial x}{\partial u} & \frac{\partial x}{\partial v} & \frac{\partial x}{\partial w} \\ \frac{\partial y}{\partial u} & \frac{\partial y}{\partial v} & \frac{\partial y}{\partial w} \\ \frac{\partial z}{\partial u} & \frac{\partial z}{\partial v} & \frac{\partial z}{\partial w} \end{vmatrix} = \begin{vmatrix} 1 & 1 & 0 \\ 0 & 2 & 0 \\ 0 & 0 & 3 \end{vmatrix} = 6.$$

We now have everything we need to apply Equation (7):

$$\int_0^3 \int_0^4 \int_{x=y/2}^{x=(y/2)+1} \left(\frac{2x - y}{2} + \frac{z}{3}\right) dx\, dy\, dz$$

$$= \int_0^1 \int_0^2 \int_0^1 (u + w)|J(u, v, w)|\, du\, dv\, dw$$

$$= \int_0^1 \int_0^2 \int_0^1 (u + w)(6)\, du\, dv\, dw = 6\int_0^1 \int_0^2 \left[\frac{u^2}{2} + uw\right]_0^1 dv\, dw$$

$$= 6\int_0^1 \int_0^2 \left(\frac{1}{2} + w\right) dv\, dw = 6\int_0^1 \left[\frac{v}{2} + vw\right]_0^2 dw = 6\int_0^1 (1 + 2w)\, dw$$

$$= 6\left[w + w^2\right]_0^1 = 6(2) = 12.$$

The goal of this section was to introduce you to the ideas involved in coordinate transformations. A thorough discussion of transformations, the Jacobian, and multivariable substitution is best given in an advanced calculus course after a study of linear algebra.

EXERCISES 15.7

Finding Jacobians and Transformed Regions for Two Variables

1. **a.** Solve the system

$$u = x - y, \qquad v = 2x + y$$

for x and y in terms of u and v. Then find the value of the Jacobian $\partial(x, y)/\partial(u, v)$.

b. Find the image under the transformation $u = x - y$, $v = 2x + y$ of the triangular region with vertices $(0, 0)$, $(1, 1)$, and $(1, -2)$ in the xy-plane. Sketch the transformed region in the uv-plane.

2. **a.** Solve the system

$$u = x + 2y, \qquad v = x - y$$

for x and y in terms of u and v. Then find the value of the Jacobian $\partial(x, y)/\partial(u, v)$.

b. Find the image under the transformation $u = x + 2y$, $v = x - y$ of the triangular region in the xy-plane bounded by the lines $y = 0$, $y = x$, and $x + 2y = 2$. Sketch the transformed region in the uv-plane.

3. a. Solve the system

$$u = 3x + 2y, \qquad v = x + 4y$$

for x and y in terms of u and v. Then find the value of the Jacobian $\partial(x, y)/\partial(u, v)$.

b. Find the image under the transformation $u = 3x + 2y$, $v = x + 4y$ of the triangular region in the xy-plane bounded by the x-axis, the y-axis, and the line $x + y = 1$. Sketch the transformed region in the uv-plane.

4. a. Solve the system

$$u = 2x - 3y, \qquad v = -x + y$$

for x and y in terms of u and v. Then find the value of the Jacobian $\partial(x, y)/\partial(u, v)$.

b. Find the image under the transformation $u = 2x - 3y$, $v = -x + y$ of the parallelogram R in the xy-plane with boundaries $x = -3$, $x = 0$, $y = x$, and $y = x + 1$. Sketch the transformed region in the uv-plane.

Applying Transformations to Evaluate Double Integrals

5. Evaluate the integral

$$\int_0^4 \int_{x=y/2}^{x=(y/2)+1} \frac{2x - y}{2}\, dx\, dy$$

from Example 1 directly by integration with respect to x and y to confirm that its value is 2.

6. Use the transformation in Exercise 1 to evaluate the integral

$$\iint_R (2x^2 - xy - y^2)\, dx\, dy$$

for the region R in the first quadrant bounded by the lines $y = -2x + 4$, $y = -2x + 7$, $y = x - 2$, and $y = x + 1$.

7. Use the transformation in Exercise 3 to evaluate the integral

$$\iint_R (3x^2 + 14xy + 8y^2)\, dx\, dy$$

for the region R in the first quadrant bounded by the lines $y = -(3/2)x + 1$, $y = -(3/2)x + 3$, $y = -(1/4)x$, and $y = -(1/4)x + 1$.

8. Use the transformation and parallelogram R in Exercise 4 to evaluate the integral

$$\iint_R 2(x - y)\, dx\, dy.$$

9. Let R be the region in the first quadrant of the xy-plane bounded by the hyperbolas $xy = 1$, $xy = 9$ and the lines $y = x$, $y = 4x$. Use the transformation $x = u/v$, $y = uv$ with $u > 0$ and $v > 0$ to rewrite

$$\iint_R \left(\sqrt{\frac{y}{x}} + \sqrt{xy}\right) dx\, dy$$

as an integral over an appropriate region G in the uv-plane. Then evaluate the uv-integral over G.

10. a. Find the Jacobian of the transformation $x = u$, $y = uv$, and sketch the region G: $1 \le u \le 2$, $1 \le uv \le 2$ in the uv-plane.

b. Then use Equation (1) to transform the integral

$$\int_1^2 \int_1^2 \frac{y}{x}\, dy\, dx$$

into an integral over G, and evaluate both integrals.

11. Polar moment of inertia of an elliptical plate A thin plate of constant density covers the region bounded by the ellipse $x^2/a^2 + y^2/b^2 = 1$, $a > 0$, $b > 0$, in the xy-plane. Find the first moment of the plate about the origin. (*Hint:* Use the transformation $x = ar\cos\theta$, $y = br\sin\theta$.)

12. The area of an ellipse The area πab of the ellipse $x^2/a^2 + y^2/b^2 = 1$ can be found by integrating the function $f(x, y) = 1$ over the region bounded by the ellipse in the xy-plane. Evaluating the integral directly requires a trigonometric substitution. An easier way to evaluate the integral is to use the transformation $x = au$, $y = bv$ and evaluate the transformed integral over the disk G: $u^2 + v^2 \le 1$ in the uv-plane. Find the area this way.

13. Use the transformation in Exercise 2 to evaluate the integral

$$\int_0^{2/3} \int_y^{2-2y} (x + 2y)e^{(y-x)}\, dx\, dy$$

by first writing it as an integral over a region G in the uv-plane.

14. Use the transformation $x = u + (1/2)v$, $y = v$ to evaluate the integral

$$\int_0^2 \int_{y/2}^{(y+4)/2} y^3(2x - y)e^{(2x-y)^2}\, dx\, dy$$

by first writing it as an integral over a region G in the uv-plane.

Finding Jacobian Determinants

15. Find the Jacobian $\partial(x, y)/\partial(u, v)$ for the transformation

a. $x = u\cos v, \quad y = u\sin v$

b. $x = u\sin v, \quad y = u\cos v$.

16. Find the Jacobian $\partial(x, y, z)/\partial(u, v, w)$ of the transformation

a. $x = u\cos v, \quad y = u\sin v, \quad z = w$

b. $x = 2u - 1, \quad y = 3v - 4, \quad z = (1/2)(w - 4)$.

17. Evaluate the appropriate determinant to show that the Jacobian of the transformation from Cartesian $\rho\phi\theta$-space to Cartesian xyz-space is $\rho^2 \sin\phi$.

18. Substitutions in single integrals How can substitutions in single definite integrals be viewed as transformations of regions? What is the Jacobian in such a case? Illustrate with an example.

Applying Transformations to Evaluate Triple Integrals

19. Evaluate the integral in Example 3 by integrating with respect to x, y, and z.

20. Volume of an ellipsoid Find the volume of the ellipsoid

$$\frac{x^2}{a^2} + \frac{y^2}{b^2} + \frac{z^2}{c^2} = 1.$$

(*Hint:* Let $x = au$, $y = bv$, and $z = cw$. Then find the volume of an appropriate region in uvw-space.)

21. Evaluate

$$\iiint |xyz|\, dx\, dy\, dz$$

over the solid ellipsoid

$$\frac{x^2}{a^2} + \frac{y^2}{b^2} + \frac{z^2}{c^2} \le 1.$$

(*Hint:* Let $x = au$, $y = bv$, and $z = cw$. Then integrate over an appropriate region in uvw-space.)

22. Let D be the region in xyz-space defined by the inequalities

$$1 \le x \le 2, \quad 0 \le xy \le 2, \quad 0 \le z \le 1.$$

Evaluate

$$\iiint_D (x^2y + 3xyz)\, dx\, dy\, dz$$

by applying the transformation

$$u = x, \quad v = xy, \quad w = 3z$$

and integrating over an appropriate region G in uvw-space.

23. Centroid of a solid semiellipsoid Assuming the result that the centroid of a solid hemisphere lies on the axis of symmetry three-eighths of the way from the base toward the top, show, by transforming the appropriate integrals, that the center of mass of a solid semiellipsoid $(x^2/a^2) + (y^2/b^2) + (z^2/c^2) \le 1$, $z \ge 0$, lies on the z-axis three-eighths of the way from the base toward the top. (You can do this without evaluating any of the integrals.)

24. Cylindrical shells In Section 6.2, we learned how to find the volume of a solid of revolution using the shell method; namely, if the region between the curve $y = f(x)$ and the x-axis from a to b $(0 < a < b)$ is revolved about the y-axis, the volume of the resulting solid is $\int_a^b 2\pi x f(x)\, dx$. Prove that finding volumes by using triple integrals gives the same result. (*Hint:* Use cylindrical coordinates with the roles of y and z changed.)

Chapter 15 Questions to Guide Your Review

1. Define the double integral of a function of two variables over a bounded region in the coordinate plane.
2. How are double integrals evaluated as iterated integrals? Does the order of integration matter? How are the limits of integration determined? Give examples.
3. How are double integrals used to calculate areas, average values, masses, moments, centers of mass, and radii of gyration? Give examples.
4. How can you change a double integral in rectangular coordinates into a double integral in polar coordinates? Why might it be worthwhile to do so? Give an example.
5. Define the triple integral of a function $f(x, y, z)$ over a bounded region in space.
6. How are triple integrals in rectangular coordinates evaluated? How are the limits of integration determined? Give an example.
7. How are triple integrals in rectangular coordinates used to calculate volumes, average values, masses, moments, centers of mass, and radii of gyration? Give examples.
8. How are triple integrals defined in cylindrical and spherical coordinates? Why might one prefer working in one of these coordinate systems to working in rectangular coordinates?
9. How are triple integrals in cylindrical and spherical coordinates evaluated? How are the limits of integration found? Give examples.
10. How are substitutions in double integrals pictured as transformations of two-dimensional regions? Give a sample calculation.
11. How are substitutions in triple integrals pictured as transformations of three-dimensional regions? Give a sample calculation.

Chapter 15 Practice Exercises

Planar Regions of Integration

In Exercises 1–4, sketch the region of integration and evaluate the double integral.

1. $\int_1^{10}\int_0^{1/y} ye^{xy}\,dx\,dy$

2. $\int_0^1\int_0^{x^3} e^{y/x}\,dy\,dx$

3. $\int_0^{3/2}\int_{-\sqrt{9-4t^2}}^{\sqrt{9-4t^2}} t\,ds\,dt$

4. $\int_0^1\int_{\sqrt{y}}^{2-\sqrt{y}} xy\,dx\,dy$

Reversing the Order of Integration

In Exercises 5–8, sketch the region of integration and write an equivalent integral with the order of integration reversed. Then evaluate both integrals.

5. $\int_0^4\int_{-\sqrt{4-y}}^{(y-4)/2} dx\,dy$

6. $\int_0^1\int_{x^2}^{x} \sqrt{x}\,dy\,dx$

7. $\int_0^{3/2}\int_{-\sqrt{9-4y^2}}^{\sqrt{9-4y^2}} y\,dx\,dy$

8. $\int_0^2\int_0^{4-x^2} 2x\,dy\,dx$

Evaluating Double Integrals

Evaluate the integrals in Exercises 9–12.

9. $\int_0^1\int_{2y}^{2} 4\cos(x^2)\,dx\,dy$

10. $\int_0^2\int_{y/2}^{1} e^{x^2}\,dx\,dy$

11. $\int_0^8\int_{\sqrt[3]{x}}^{2} \frac{dy\,dx}{y^4+1}$

12. $\int_0^1\int_{\sqrt[3]{y}}^{1} \frac{2\pi\sin\pi x^2}{x^2}\,dx\,dy$

Areas and Volumes

13. **Area between line and parabola** Find the area of the region enclosed by the line $y = 2x + 4$ and the parabola $y = 4 - x^2$ in the xy-plane.

14. **Area bounded by lines and parabola** Find the area of the "triangular" region in the xy-plane that is bounded on the right by the parabola $y = x^2$, on the left by the line $x + y = 2$, and above by the line $y = 4$.

15. **Volume of the region under a paraboloid** Find the volume under the paraboloid $z = x^2 + y^2$ above the triangle enclosed by the lines $y = x$, $x = 0$, and $x + y = 2$ in the xy-plane.

16. **Volume of the region under parabolic cylinder** Find the volume under the parabolic cylinder $z = x^2$ above the region enclosed by the parabola $y = 6 - x^2$ and the line $y = x$ in the xy-plane.

Average Values

Find the average value of $f(x, y) = xy$ over the regions in Exercises 17 and 18.

17. The square bounded by the lines $x = 1$, $y = 1$ in the first quadrant

18. The quarter circle $x^2 + y^2 \le 1$ in the first quadrant

Masses and Moments

19. **Centroid** Find the centroid of the "triangular" region bounded by the lines $x = 2$, $y = 2$ and the hyperbola $xy = 2$ in the xy-plane.

20. **Centroid** Find the centroid of the region between the parabola $x + y^2 - 2y = 0$ and the line $x + 2y = 0$ in the xy-plane.

21. **Polar moment** Find the polar moment of inertia about the origin of a thin triangular plate of constant density $\delta = 3$ bounded by the y-axis and the lines $y = 2x$ and $y = 4$ in the xy-plane.

22. **Polar moment** Find the polar moment of inertia about the center of a thin rectangular sheet of constant density $\delta = 1$ bounded by the lines

    a. $x = \pm 2$, $y = \pm 1$ in the xy-plane

    b. $x = \pm a$, $y = \pm b$ in the xy-plane.

    (*Hint:* Find I_x. Then use the formula for I_x to find I_y and add the two to find I_0).

23. **Inertial moment and radius of gyration** Find the moment of inertia and radius of gyration about the x-axis of a thin plate of constant density δ covering the triangle with vertices $(0, 0)$, $(3, 0)$, and $(3, 2)$ in the xy-plane.

24. **Plate with variable density** Find the center of mass and the moments of inertia and radii of gyration about the coordinate axes of a thin plate bounded by the line $y = x$ and the parabola $y = x^2$ in the xy-plane if the density is $\delta(x, y) = x + 1$.

25. **Plate with variable density** Find the mass and first moments about the coordinate axes of a thin square plate bounded by the lines $x = \pm 1$, $y = \pm 1$ in the xy-plane if the density is $\delta(x, y) = x^2 + y^2 + 1/3$.

26. **Triangles with same inertial moment and radius of gyration** Find the moment of inertia and radius of gyration about the x-axis of a thin triangular plate of constant density δ whose base lies along the interval $[0, b]$ on the x-axis and whose vertex lies on the line $y = h$ above the x-axis. As you will see, it does not matter where on the line this vertex lies. All such triangles have the same moment of inertia and radius of gyration about the x-axis.

Polar Coordinates

Evaluate the integrals in Exercises 27 and 28 by changing to polar coordinates.

27. $\int_{-1}^1\int_{-\sqrt{1-x^2}}^{\sqrt{1-x^2}} \frac{2\,dy\,dx}{(1+x^2+y^2)^2}$

28. $\int_{-1}^1\int_{-\sqrt{1-y^2}}^{\sqrt{1-y^2}} \ln(x^2+y^2+1)\,dx\,dy$

29. **Centroid** Find the centroid of the region in the polar coordinate plane defined by the inequalities $0 \le r \le 3$, $-\pi/3 \le \theta \le \pi/3$.

30. Centroid Find the centroid of the region in the first quadrant bounded by the rays $\theta = 0$ and $\theta = \pi/2$ and the circles $r = 1$ and $r = 3$.

31. a. Centroid Find the centroid of the region in the polar coordinate plane that lies inside the cardioid $r = 1 + \cos\theta$ and outside the circle $r = 1$.

b. Sketch the region and show the centroid in your sketch.

32. a. Centroid Find the centroid of the plane region defined by the polar coordinate inequalities $0 \le r \le a$, $-\alpha \le \theta \le \alpha$ $(0 < \alpha \le \pi)$. How does the centroid move as $\alpha \to \pi^-$?

b. Sketch the region for $\alpha = 5\pi/6$ and show the centroid in your sketch.

33. Integrating over lemniscate Integrate the function $f(x, y) = 1/(1 + x^2 + y^2)^2$ over the region enclosed by one loop of the lemniscate $(x^2 + y^2)^2 - (x^2 - y^2) = 0$.

34. Integrate $f(x, y) = 1/(1 + x^2 + y^2)^2$ over

a. Triangular region The triangle with vertices (0, 0), (1, 0), $\left(1, \sqrt{3}\right)$.

b. First quadrant The first quadrant of the xy-plane.

Triple Integrals in Cartesian Coordinates

Evaluate the integrals in Exercises 35–38.

35. $\displaystyle\int_0^{\pi}\int_0^{\pi}\int_0^{\pi} \cos(x + y + z)\, dx\, dy\, dz$

36. $\displaystyle\int_{\ln 6}^{\ln 7}\int_0^{\ln 2}\int_{\ln 4}^{\ln 5} e^{(x+y+z)}\, dz\, dy\, dx$

37. $\displaystyle\int_0^1\int_0^{x^2}\int_0^{x+y} (2x - y - z)\, dz\, dy\, dx$

38. $\displaystyle\int_1^e\int_1^x\int_0^z \frac{2y}{z^3}\, dy\, dz\, dx$

39. Volume Find the volume of the wedge-shaped region enclosed on the side by the cylinder $x = -\cos y$, $-\pi/2 \le y \le \pi/2$, on the top by the plane $z = -2x$, and below by the xy-plane.

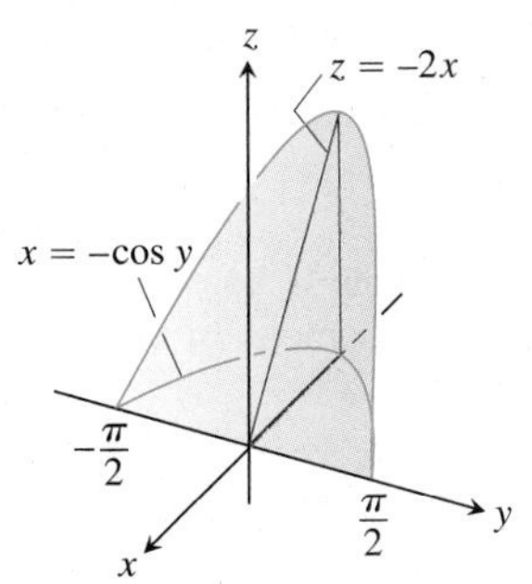

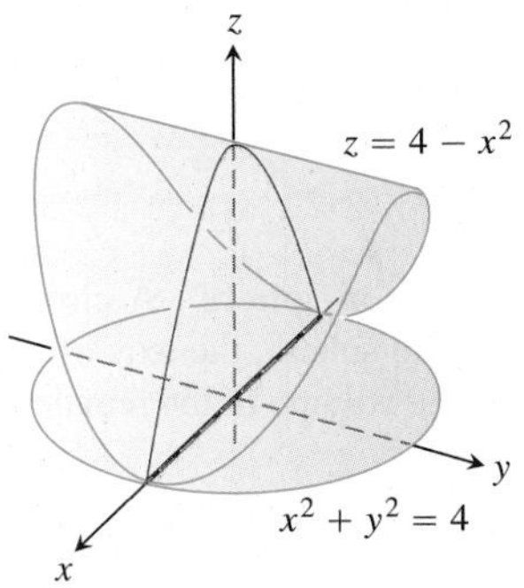

40. Volume Find the volume of the solid that is bounded above by the cylinder $z = 4 - x^2$, on the sides by the cylinder $x^2 + y^2 = 4$, and below by the xy-plane.

41. Average value Find the average value of $f(x, y, z) = 30xz\sqrt{x^2 + y}$ over the rectangular solid in the first octant bounded by the coordinate planes and the planes $x = 1, y = 3, z = 1$.

42. Average value Find the average value of ρ over the solid sphere $\rho \le a$ (spherical coordinates).

Cylindrical and Spherical Coordinates

43. Cylindrical to rectangular coordinates Convert

$$\int_0^{2\pi}\int_0^{\sqrt{2}}\int_r^{\sqrt{4-r^2}} 3\, dz\, r\, dr\, d\theta, \qquad r \ge 0$$

to **(a)** rectangular coordinates with the order of integration $dz\, dx\, dy$ and **(b)** spherical coordinates. Then **(c)** evaluate one of the integrals.

44. Rectangular to cylindrical coordinates **(a)** Convert to cylindrical coordinates. Then **(b)** evaluate the new integral.

$$\int_0^1\int_{-\sqrt{1-x^2}}^{\sqrt{1-x^2}}\int_{-(x^2+y^2)}^{(x^2+y^2)} 21xy^2\, dz\, dy\, dx$$

45. Rectangular to spherical coordinates **(a)** Convert to spherical coordinates. Then **(b)** evaluate the new integral.

$$\int_{-1}^1\int_{-\sqrt{1-x^2}}^{\sqrt{1-x^2}}\int_{\sqrt{x^2+y^2}}^{1} dz\, dy\, dx$$

46. Rectangular, cylindrical, and spherical coordinates Write an iterated triple integral for the integral of $f(x, y, z) = 6 + 4y$ over the region in the first octant bounded by the cone $z = \sqrt{x^2 + y^2}$, the cylinder $x^2 + y^2 = 1$, and the coordinate planes in **(a)** rectangular coordinates, **(b)** cylindrical coordinates, and **(c)** spherical coordinates. Then **(d)** find the integral of f by evaluating one of the triple integrals.

47. Cylindrical to rectangular coordinates Set up an integral in rectangular coordinates equivalent to the integral

$$\int_0^{\pi/2}\int_1^{\sqrt{3}}\int_1^{\sqrt{4-r^2}} r^3(\sin\theta\cos\theta)z^2\, dz\, dr\, d\theta.$$

Arrange the order of integration to be z first, then y, then x.

48. Rectangular to cylindrical coordinates The volume of a solid is

$$\int_0^2\int_0^{\sqrt{2x-x^2}}\int_{-\sqrt{4-x^2-y^2}}^{\sqrt{4-x^2-y^2}} dz\, dy\, dx.$$

a. Describe the solid by giving equations for the surfaces that form its boundary.

b. Convert the integral to cylindrical coordinates but do not evaluate the integral.

49. Spherical versus cylindrical coordinates Triple integrals involving spherical shapes do not always require spherical coordinates for convenient evaluation. Some calculations may be accomplished more easily with cylindrical coordinates. As a case in point, find the volume of the region bounded above by the sphere $x^2 + y^2 + z^2 = 8$ and below by the plane $z = 2$ by using **(a)** cylindrical coordinates and **(b)** spherical coordinates.

50. Finding I_z in spherical coordinates Find the moment of inertia about the z-axis of a solid of constant density $\delta = 1$ that is bounded above by the sphere $\rho = 2$ and below by the cone $\phi = \pi/3$ (spherical coordinates).

51. Moment of inertia of a "thick" sphere Find the moment of inertia of a solid of constant density δ bounded by two concentric spheres of radii a and b $(a < b)$ about a diameter.

52. Moment of inertia of an apple Find the moment of inertia about the z-axis of a solid of density $\delta = 1$ enclosed by the spherical coordinate surface $\rho = 1 - \cos\phi$. The solid is the red curve rotated about the z-axis in the accompanying figure.

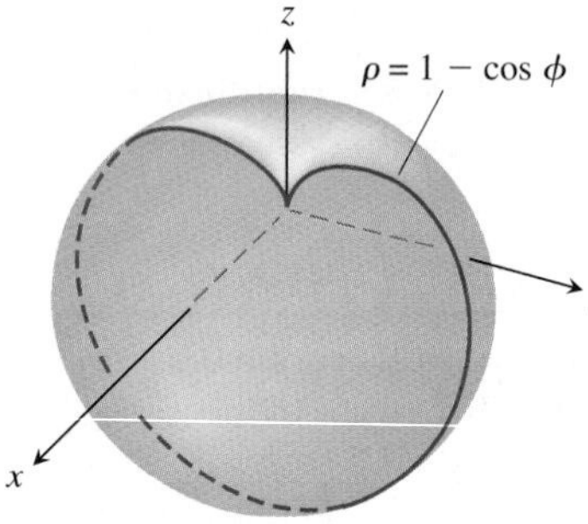

Substitutions

53. Show that if $u = x - y$ and $v = y$, then

$$\int_0^{\infty}\int_0^{x} e^{-sx} f(x-y, y)\, dy\, dx = \int_0^{\infty}\int_0^{\infty} e^{-s(u+v)} f(u, v)\, du\, dv.$$

54. What relationship must hold between the constants a, b, and c to make

$$\int_{-\infty}^{\infty}\int_{-\infty}^{\infty} e^{-(ax^2+2bxy+cy^2)}\, dx\, dy = 1?$$

(*Hint:* Let $s = \alpha x + \beta y$ and $t = \gamma x + \delta y$, where $(\alpha\delta - \beta\gamma)^2 = ac - b^2$. Then $ax^2 + 2bxy + cy^2 = s^2 + t^2$.)

Chapter 15 Additional and Advanced Exercises

Volumes

1. Sand pile: double and triple integrals The base of a sand pile covers the region in the xy-plane that is bounded by the parabola $x^2 + y = 6$ and the line $y = x$. The height of the sand above the point (x, y) is x^2. Express the volume of sand as **(a)** a double integral, **(b)** a triple integral. Then **(c)** find the volume.

2. Water in a hemispherical bowl A hemispherical bowl of radius 5 cm is filled with water to within 3 cm of the top. Find the volume of water in the bowl.

3. Solid cylindrical region between two planes Find the volume of the portion of the solid cylinder $x^2 + y^2 \le 1$ that lies between the planes $z = 0$ and $x + y + z = 2$.

4. Sphere and paraboloid Find the volume of the region bounded above by the sphere $x^2 + y^2 + z^2 = 2$ and below by the paraboloid $z = x^2 + y^2$.

5. Two paraboloids Find the volume of the region bounded above by the paraboloid $z = 3 - x^2 - y^2$ and below by the paraboloid $z = 2x^2 + 2y^2$.

6. Spherical coordinates Find the volume of the region enclosed by the spherical coordinate surface $\rho = 2\sin\phi$ (see accompanying figure).

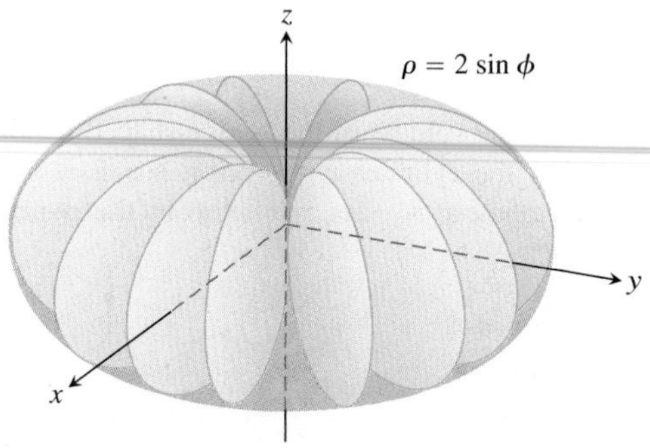

7. Hole in sphere A circular cylindrical hole is bored through a solid sphere, the axis of the hole being a diameter of the sphere. The volume of the remaining solid is

$$V = 2\int_0^{2\pi}\int_0^{\sqrt{3}}\int_1^{\sqrt{4-z^2}} r\, dr\, dz\, d\theta.$$

a. Find the radius of the hole and the radius of the sphere.

b. Evaluate the integral.

8. Sphere and cylinder Find the volume of material cut from the solid sphere $r^2 + z^2 \le 9$ by the cylinder $r = 3\sin\theta$.

9. Two paraboloids Find the volume of the region enclosed by the surfaces $z = x^2 + y^2$ and $z = (x^2 + y^2 + 1)/2$.

10. Cylinder and surface $z = xy$ Find the volume of the region in the first octant that lies between the cylinders $r = 1$ and $r = 2$ and that is bounded below by the xy-plane and above by the surface $z = xy$.

Changing the Order of Integration

11. Evaluate the integral

$$\int_0^\infty \frac{e^{-ax} - e^{-bx}}{x}\,dx.$$

(*Hint:* Use the relation

$$\frac{e^{-ax} - e^{-bx}}{x} = \int_a^b e^{-xy}\,dy$$

to form a double integral and evaluate the integral by changing the order of integration.)

12. a. Polar coordinates Show, by changing to polar coordinates, that

$$\int_0^{a\sin\beta}\int_{y\cot\beta}^{\sqrt{a^2-y^2}} \ln(x^2 + y^2)\,dx\,dy = a^2\beta\left(\ln a - \frac{1}{2}\right),$$

where $a > 0$ and $0 < \beta < \pi/2$.

b. Rewrite the Cartesian integral with the order of integration reversed.

13. Reducing a double to a single integral By changing the order of integration, show that the following double integral can be reduced to a single integral:

$$\int_0^x\int_0^u e^{m(x-t)} f(t)\,dt\,du = \int_0^x (x - t)e^{m(x-t)} f(t)\,dt.$$

Similarly, it can be shown that

$$\int_0^x\int_0^v\int_0^u e^{m(x-t)} f(t)\,dt\,du\,dv = \int_0^x \frac{(x-t)^2}{2} e^{m(x-t)} f(t)\,dt.$$

14. Transforming a double integral to obtain constant limits Sometimes a multiple integral with variable limits can be changed into one with constant limits. By changing the order of integration, show that

$$\int_0^1 f(x)\left(\int_0^x g(x - y)f(y)\,dy\right)dx$$
$$= \int_0^1 f(y)\left(\int_y^1 g(x - y)f(x)\,dx\right)dy$$
$$= \frac{1}{2}\int_0^1\int_0^1 g(|x - y|)f(x)f(y)\,dx\,dy.$$

Masses and Moments

15. Minimizing polar inertia A thin plate of constant density is to occupy the triangular region in the first quadrant of the xy-plane having vertices $(0, 0)$, $(a, 0)$, and $(a, 1/a)$. What value of a will minimize the plate's polar moment of inertia about the origin?

16. Polar inertia of triangular plate Find the polar moment of inertia about the origin of a thin triangular plate of constant density $\delta = 3$ bounded by the y-axis and the lines $y = 2x$ and $y = 4$ in the xy-plane.

17. Mass and polar inertia of a counterweight The counterweight of a flywheel of constant density 1 has the form of the smaller segment cut from a circle of radius a by a chord at a distance b from the center $(b < a)$. Find the mass of the counterweight and its polar moment of inertia about the center of the wheel.

18. Centroid of boomerang Find the centroid of the boomerang-shaped region between the parabolas $y^2 = -4(x - 1)$ and $y^2 = -2(x - 2)$ in the xy-plane.

Theory and Applications

19. Evaluate

$$\int_0^a\int_0^b e^{\max(b^2x^2,\, a^2y^2)}\,dy\,dx,$$

where a and b are positive numbers and

$$\max(b^2x^2, a^2y^2) = \begin{cases} b^2x^2 & \text{if } b^2x^2 \ge a^2y^2 \\ a^2y^2 & \text{if } b^2x^2 < a^2y^2. \end{cases}$$

20. Show that

$$\iint \frac{\partial^2 F(x, y)}{\partial x\,\partial y}\,dx\,dy$$

over the rectangle $x_0 \le x \le x_1, y_0 \le y \le y_1$, is

$$F(x_1, y_1) - F(x_0, y_1) - F(x_1, y_0) + F(x_0, y_0).$$

21. Suppose that $f(x, y)$ can be written as a product $f(x, y) = F(x)G(y)$ of a function of x and a function of y. Then the integral of f over the rectangle R: $a \le x \le b, c \le y \le d$ can be evaluated as a product as well, by the formula

$$\iint_R f(x, y)\,dA = \left(\int_a^b F(x)\,dx\right)\left(\int_c^d G(y)\,dy\right). \tag{1}$$

The argument is that

$$\iint_R f(x, y)\,dA = \int_c^d\left(\int_a^b F(x)G(y)\,dx\right)dy \tag{i}$$
$$= \int_c^d\left(G(y)\int_a^b F(x)\,dx\right)dy \tag{ii}$$
$$= \int_c^d\left(\int_a^b F(x)\,dx\right)G(y)\,dy \tag{iii}$$
$$= \left(\int_a^b F(x)\,dx\right)\int_c^d G(y)\,dy. \tag{iv}$$

a. Give reasons for steps (i) through (v).

When it applies, Equation (1) can be a time saver. Use it to evaluate the following integrals.

b. $\int_0^{\ln 2}\int_0^{\pi/2} e^x \cos y \, dy \, dx$ **c.** $\int_1^2\int_{-1}^1 \frac{x}{y^2}\, dx \, dy$

22. Let $D_{\mathbf{u}}f$ denote the derivative of $f(x, y) = (x^2 + y^2)/2$ in the direction of the unit vector $\mathbf{u} = u_1\mathbf{i} + u_2\mathbf{j}$.

a. Finding average value Find the average value of $D_{\mathbf{u}}f$ over the triangular region cut from the first quadrant by the line $x + y = 1$.

b. Average value and centroid Show in general that the average value of $D_{\mathbf{u}}f$ over a region in the xy-plane is the value of $D_{\mathbf{u}}f$ at the centroid of the region.

23. The value of $\Gamma(1/2)$ The gamma function,

$$\Gamma(x) = \int_0^\infty t^{x-1}e^{-t}\, dt,$$

extends the factorial function from the nonnegative integers to other real values. Of particular interest in the theory of differential equations is the number

$$\Gamma\left(\frac{1}{2}\right) = \int_0^\infty t^{(1/2)-1}e^{-t}\, dt = \int_0^\infty \frac{e^{-t}}{\sqrt{t}}\, dt. \qquad (2)$$

a. If you have not yet done Exercise 37 in Section 15.3, do it now to show that

$$I = \int_0^\infty e^{-y^2}\, dy = \frac{\sqrt{\pi}}{2}.$$

b. Substitute $y = \sqrt{t}$ in Equation (2) to show that $\Gamma(1/2) = 2I = \sqrt{\pi}$.

24. Total electrical charge over circular plate The electrical charge distribution on a circular plate of radius R meters is $\sigma(r, \theta) = kr(1 - \sin\theta)$ coulomb/m^2 (k a constant). Integrate σ over the plate to find the total charge Q.

25. A parabolic rain gauge A bowl is in the shape of the graph of $z = x^2 + y^2$ from $z = 0$ to $z = 10$ in. You plan to calibrate the bowl to make it into a rain gauge. What height in the bowl would correspond to 1 in. of rain? 3 in. of rain?

26. Water in a satellite dish A parabolic satellite dish is 2 m wide and 1/2 m deep. Its axis of symmetry is tilted 30 degrees from the vertical.

a. Set up, but do not evaluate, a triple integral in rectangular coordinates that gives the amount of water the satellite dish will hold. (*Hint:* Put your coordinate system so that the satellite dish is in "standard position" and the plane of the water level is slanted.) (*Caution:* The limits of integration are not "nice.")

b. What would be the smallest tilt of the satellite dish so that it holds no water?

27. An infinite half-cylinder Let D be the interior of the infinite right circular half-cylinder of radius 1 with its single-end face suspended 1 unit above the origin and its axis the ray from (0, 0, 1) to ∞. Use cylindrical coordinates to evaluate

$$\iiint_D z(r^2 + z^2)^{-5/2}\, dV.$$

28. Hypervolume We have learned that $\int_a^b 1\, dx$ is the length of the interval $[a, b]$ on the number line (one-dimensional space), $\iint_R 1\, dA$ is the area of region R in the xy-plane (two-dimensional space), and $\iiint_D 1\, dV$ is the volume of the region D in three-dimensional space (xyz-space). We could continue: If Q is a region in 4-space ($xyzw$-space), then $\iiiint_Q 1\, dV$ is the "hypervolume" of Q. Use your generalizing abilities and a Cartesian coordinate system of 4-space to find the hypervolume inside the unit 3-dimensional sphere $x^2 + y^2 + z^2 + w^2 = 1$.

Chapter 15 Technology Application Projects

Mathematica/Maple Module

Take Your Chances: Try the Monte Carlo Technique for Numerical Integration in Three Dimensions

Use the Monte Carlo technique to integrate numerically in three dimensions.

Mathematica/Maple Module

Means and Moments and Exploring New Plotting Techniques, Part II.

Use the method of moments in a form that makes use of geometric symmetry as well as multiple integration.

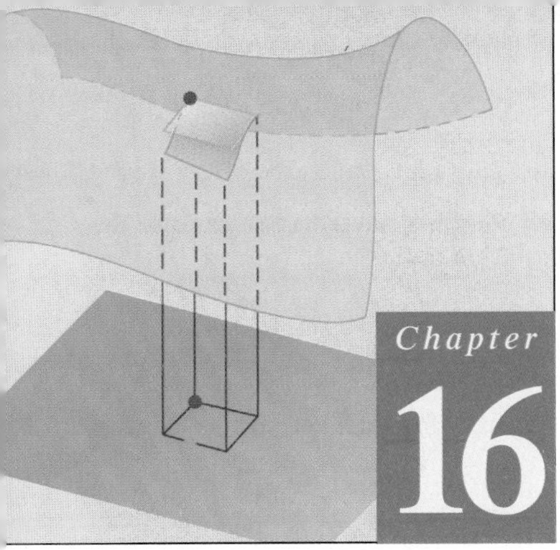

Chapter

16 INTEGRATION IN VECTOR FIELDS

OVERVIEW This chapter treats integration in vector fields. It is the mathematics that engineers and physicists use to describe fluid flow, design underwater transmission cables, explain the flow of heat in stars, and put satellites in orbit. In particular, we define line integrals, which are used to find the work done by a force field in moving an object along a path through the field. We also define surface integrals so we can find the rate that a fluid flows across a surface. Along the way we develop key concepts and results, such as *conservative* force fields and Green's Theorem, to simplify our calculations of these new integrals by connecting them to the single, double, and triple integrals we have already studied.

16.1 Line Integrals

In Chapter 5 we defined the definite integral of a function over a finite closed interval $[a, b]$ on the x-axis. We used definite integrals to find the mass of a thin straight rod, or the work done by a variable force directed along the x-axis. Now we would like to calculate the masses of thin rods or wires lying along a *curve* in the plane or space, or to find the work done by a variable force acting along such a curve. For these calculations we need a more general notion of a "line" integral than integrating over a line segment on the x-axis. Instead we need to integrate over a curve C in the plane or in space. These more general integrals are called *line integrals*, although "curve" integrals might be more descriptive. We make our definitions for space curves, remembering that curves in the xy-plane are just a special case with z-coordinate identically zero.

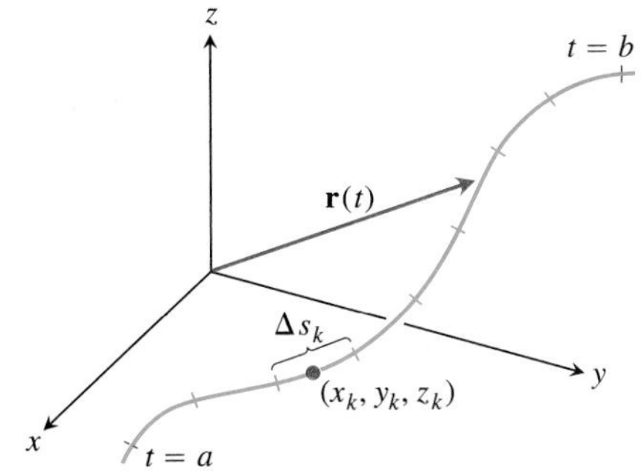

FIGURE 16.1 The curve $\mathbf{r}(t)$ partitioned into small arcs from $t = a$ to $t = b$. The length of a typical subarc is Δs_k.

Suppose that $f(x, y, z)$ is a real-valued function we wish to integrate over the curve $\mathbf{r}(t) = g(t)\mathbf{i} + h(t)\mathbf{j} + k(t)\mathbf{k}$, $a \le t \le b$, lying within the domain of f. The values of f along the curve are given by the composite function $f(g(t), h(t), k(t))$. We are going to integrate this composite with respect to arc length from $t = a$ to $t = b$. To begin, we first partition the curve into a finite number n of subarcs (Figure 16.1). The typical subarc has length Δs_k. In each subarc we choose a point (x_k, y_k, z_k) and form the sum

$$S_n = \sum_{k=1}^{n} f(x_k, y_k, z_k)\, \Delta s_k.$$

If f is continuous and the functions g, h, and k have continuous first derivatives, then these sums approach a limit as n increases and the lengths Δs_k approach zero. We call this limit the **line integral of f over the curve from a to b**. If the curve is denoted by a single letter, C for example, the notation for the integral is

$$\int_C f(x, y, z)\, ds \qquad \text{"The integral of } f \text{ over } C\text{"} \tag{1}$$

If $\mathbf{r}(t)$ is smooth for $a \le t \le b$ ($\mathbf{v} = d\mathbf{r}/dt$ is continuous and never $\mathbf{0}$), we can use the equation

$$s(t) = \int_a^t |\mathbf{v}(\tau)|\, d\tau \qquad \text{Equation (3) of Section 13.3 with } t_0 = a$$

to express ds in Equation (1) as $ds = |\mathbf{v}(t)|\, dt$. A theorem from advanced calculus says that we can then evaluate the integral of f over C as

$$\int_C f(x, y, z)\, ds = \int_a^b f(g(t), h(t), k(t))|\mathbf{v}(t)|\, dt.$$

Notice that the integral on the right side of this last equation is just an ordinary (single) definite integral, as defined in Chapter 5, where we are integrating with respect to the parameter t. The formula evaluates the line integral on the left side correctly no matter what parametrization is used, as long as the parametrization is smooth.

How to Evaluate a Line Integral

To integrate a continuous function $f(x, y, z)$ over a curve C:

1. Find a smooth parametrization of C,

$$\mathbf{r}(t) = g(t)\mathbf{i} + h(t)\mathbf{j} + k(t)\mathbf{k}, \qquad a \le t \le b$$

2. Evaluate the integral as

$$\int_C f(x, y, z)\, ds = \int_a^b f(g(t), h(t), k(t))|\mathbf{v}(t)|\, dt. \qquad (2)$$

If f has the constant value 1, then the integral of f over C gives the length of C.

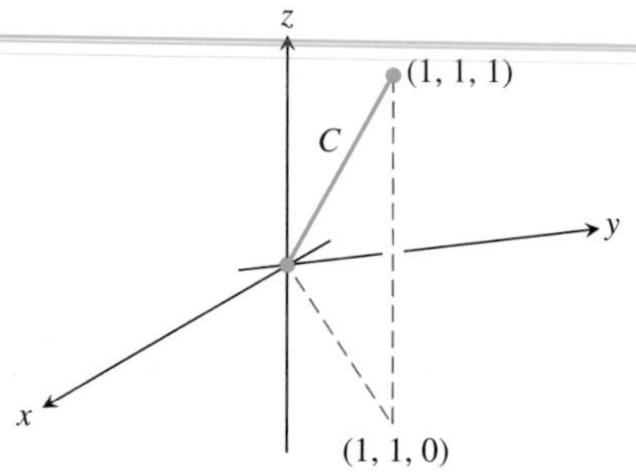

FIGURE 16.2 The integration path in Example 1.

EXAMPLE 1 Evaluating a Line Integral

Integrate $f(x, y, z) = x - 3y^2 + z$ over the line segment C joining the origin to the point (1, 1, 1) (Figure 16.2).

Solution We choose the simplest parametrization we can think of:

$$\mathbf{r}(t) = t\mathbf{i} + t\mathbf{j} + t\mathbf{k}, \qquad 0 \le t \le 1.$$

The components have continuous first derivatives and $|\mathbf{v}(t)| = |\mathbf{i} + \mathbf{j} + \mathbf{k}| = \sqrt{1^2 + 1^2 + 1^2} = \sqrt{3}$ is never 0, so the parametrization is smooth. The integral of f over C is

$$\begin{aligned}\int_C f(x, y, z)\, ds &= \int_0^1 f(t, t, t)\left(\sqrt{3}\right) dt \qquad \text{Equation (2)}\\ &= \int_0^1 (t - 3t^2 + t)\sqrt{3}\, dt\\ &= \sqrt{3}\int_0^1 (2t - 3t^2)\, dt = \sqrt{3}\left[t^2 - t^3\right]_0^1 = 0.\end{aligned}$$

■

Additivity

Line integrals have the useful property that if a curve C is made by joining a finite number of curves $C_1, C_2, \ldots, C_n$ end to end, then the integral of a function over C is the sum of the integrals over the curves that make it up:

$$\int_C f\,ds = \int_{C_1} f\,ds + \int_{C_2} f\,ds + \cdots + \int_{C_n} f\,ds. \tag{3}$$

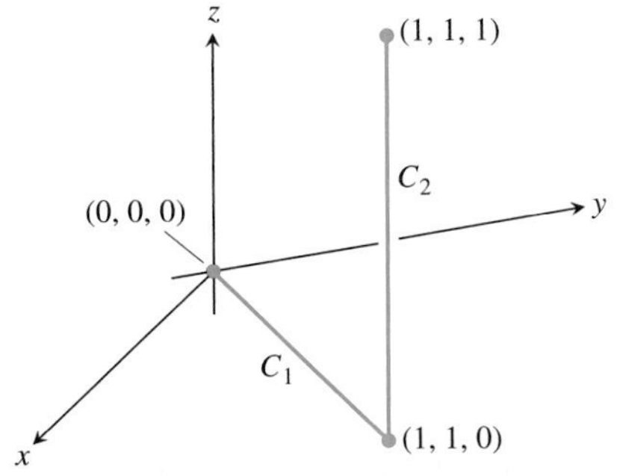

FIGURE 16.3 The path of integration in Example 2.

EXAMPLE 2 Line Integral for Two Joined Paths

Figure 16.3 shows another path from the origin to (1, 1, 1), the union of line segments C_1 and C_2. Integrate $f(x, y, z) = x - 3y^2 + z$ over $C_1 \cup C_2$.

Solution We choose the simplest parametrizations for C_1 and C_2 we can think of, checking the lengths of the velocity vectors as we go along:

$$C_1:\quad \mathbf{r}(t) = t\mathbf{i} + t\mathbf{j},\quad 0 \le t \le 1;\quad |\mathbf{v}| = \sqrt{1^2 + 1^2} = \sqrt{2}$$

$$C_2:\quad \mathbf{r}(t) = \mathbf{i} + \mathbf{j} + t\mathbf{k},\quad 0 \le t \le 1;\quad |\mathbf{v}| = \sqrt{0^2 + 0^2 + 1^2} = 1.$$

With these parametrizations we find that

$$\begin{aligned}
\int_{C_1 \cup C_2} f(x, y, z)\,ds &= \int_{C_1} f(x, y, z)\,ds + \int_{C_2} f(x, y, z)\,ds && \text{Equation (3)}\\
&= \int_0^1 f(t, t, 0)\sqrt{2}\,dt + \int_0^1 f(1, 1, t)(1)\,dt && \text{Equation (2)}\\
&= \int_0^1 (t - 3t^2 + 0)\sqrt{2}\,dt + \int_0^1 (1 - 3 + t)(1)\,dt\\
&= \sqrt{2}\left[\frac{t^2}{2} - t^3\right]_0^1 + \left[\frac{t^2}{2} - 2t\right]_0^1 = -\frac{\sqrt{2}}{2} - \frac{3}{2}.
\end{aligned}$$ ■

Notice three things about the integrations in Examples 1 and 2. First, as soon as the components of the appropriate curve were substituted into the formula for f, the integration became a standard integration with respect to t. Second, the integral of f over $C_1 \cup C_2$ was obtained by integrating f over each section of the path and adding the results. Third, the integrals of f over C and $C_1 \cup C_2$ had different values. For most functions, the value of the integral along a path joining two points changes if you change the path between them. For some functions, however, the value remains the same.

Mass and Moment Calculations

We treat coil springs and wires like masses distributed along smooth curves in space. The distribution is described by a continuous density function $\delta(x, y, z)$ (mass per unit length). The spring's or wire's mass, center of mass, and moments are then calculated with the formulas in Table 16.1. The formulas also apply to thin rods.

TABLE 16.1 Mass and moment formulas for coil springs, thin rods, and wires lying along a smooth curve C in space

Mass: $M = \int_C \delta(x, y, z)\, ds \qquad (\delta = \delta(x, y, z) = \text{density})$

First moments about the coordinate planes:

$$M_{yz} = \int_C x\,\delta\, ds, \qquad M_{xz} = \int_C y\,\delta\, ds, \qquad M_{xy} = \int_C z\,\delta\, ds$$

Coordinates of the center of mass:

$$\bar{x} = M_{yz}/M, \qquad \bar{y} = M_{xz}/M, \qquad \bar{z} = M_{xy}/M$$

Moments of inertia about axes and other lines:

$$I_x = \int_C (y^2 + z^2)\,\delta\, ds, \qquad I_y = \int_C (x^2 + z^2)\,\delta\, ds$$

$$I_z = \int_C (x^2 + y^2)\,\delta\, ds, \qquad I_L = \int_C r^2\,\delta\, ds$$

$$r(x, y, z) = \text{distance from the point } (x, y, z) \text{ to line } L$$

Radius of gyration about a line L: $R_L = \sqrt{I_L/M}$

EXAMPLE 3 Finding Mass, Center of Mass, Moment of Inertia, Radius of Gyration

A coil spring lies along the helix

$$\mathbf{r}(t) = (\cos 4t)\mathbf{i} + (\sin 4t)\mathbf{j} + t\mathbf{k}, \qquad 0 \le t \le 2\pi.$$

The spring's density is a constant, $\delta = 1$. Find the spring's mass and center of mass, and its moment of inertia and radius of gyration about the z-axis.

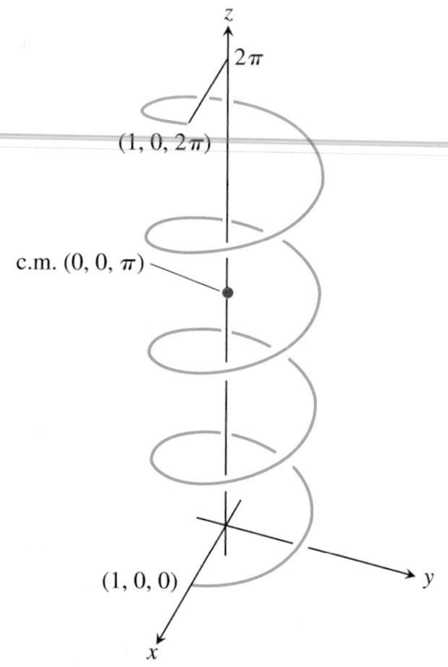

FIGURE 16.4 The helical spring in Example 3.

Solution We sketch the spring (Figure 16.4). Because of the symmetries involved, the center of mass lies at the point $(0, 0, \pi)$ on the z-axis.

For the remaining calculations, we first find $|\mathbf{v}(t)|$:

$$|\mathbf{v}(t)| = \sqrt{\left(\frac{dx}{dt}\right)^2 + \left(\frac{dy}{dt}\right)^2 + \left(\frac{dz}{dt}\right)^2}$$

$$= \sqrt{(-4\sin 4t)^2 + (4\cos 4t)^2 + 1} = \sqrt{17}.$$

We then evaluate the formulas from Table 16.1 using Equation (2):

$$M = \int_{\text{Helix}} \delta\, ds = \int_0^{2\pi} (1)\sqrt{17}\, dt = 2\pi\sqrt{17}$$

$$I_z = \int_{\text{Helix}} (x^2 + y^2)\delta\, ds = \int_0^{2\pi} (\cos^2 4t + \sin^2 4t)(1)\sqrt{17}\, dt$$

$$= \int_0^{2\pi} \sqrt{17}\, dt = 2\pi\sqrt{17}$$

$$R_z = \sqrt{I_z/M} = \sqrt{2\pi\sqrt{17}/(2\pi\sqrt{17})} = 1.$$

Notice that the radius of gyration about the z-axis is the radius of the cylinder around which the helix winds. ■

EXAMPLE 4 Finding an Arch's Center of Mass

A slender metal arch, denser at the bottom than top, lies along the semicircle $y^2 + z^2 = 1, z \geq 0$, in the yz-plane (Figure 16.5). Find the center of the arch's mass if the density at the point (x, y, z) on the arch is $\delta(x, y, z) = 2 - z$.

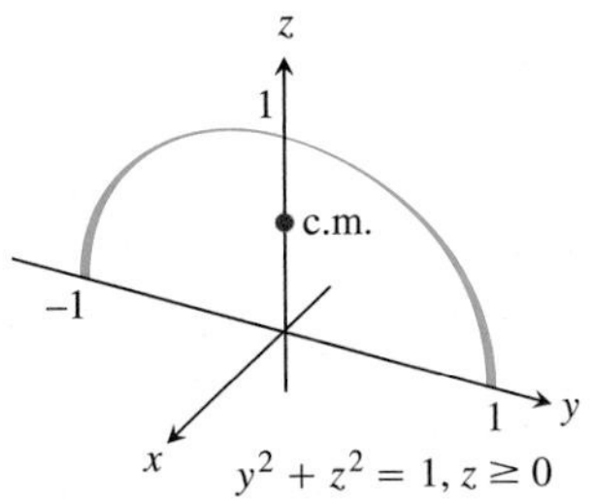

FIGURE 16.5 Example 4 shows how to find the center of mass of a circular arch of variable density.

Solution We know that $\bar{x} = 0$ and $\bar{y} = 0$ because the arch lies in the yz-plane with its mass distributed symmetrically about the z-axis. To find $\bar{z}$, we parametrize the circle as

$$\mathbf{r}(t) = (\cos t)\mathbf{j} + (\sin t)\mathbf{k}, \qquad 0 \leq t \leq \pi.$$

For this parametrization,

$$|\mathbf{v}(t)| = \sqrt{\left(\frac{dx}{dt}\right)^2 + \left(\frac{dy}{dt}\right)^2 + \left(\frac{dz}{dt}\right)^2} = \sqrt{(0)^2 + (-\sin t)^2 + (\cos t)^2} = 1.$$

The formulas in Table 16.1 then give

$$M = \int_C \delta\, ds = \int_C (2 - z)\, ds = \int_0^{\pi} (2 - \sin t)(1)\, dt = 2\pi - 2$$

$$M_{xy} = \int_C z\delta\, ds = \int_C z(2 - z)\, ds = \int_0^{\pi} (\sin t)(2 - \sin t)\, dt$$

$$= \int_0^{\pi} (2\sin t - \sin^2 t)\, dt = \frac{8 - \pi}{2}$$

$$\bar{z} = \frac{M_{xy}}{M} = \frac{8 - \pi}{2} \cdot \frac{1}{2\pi - 2} = \frac{8 - \pi}{4\pi - 4} \approx 0.57.$$

With $\bar{z}$ to the nearest hundredth, the center of mass is $(0, 0, 0.57)$. ■

EXERCISES 16.1

Graphs of Vector Equations

Match the vector equations in Exercises 1–8 with the graphs (a)–(h) given here.

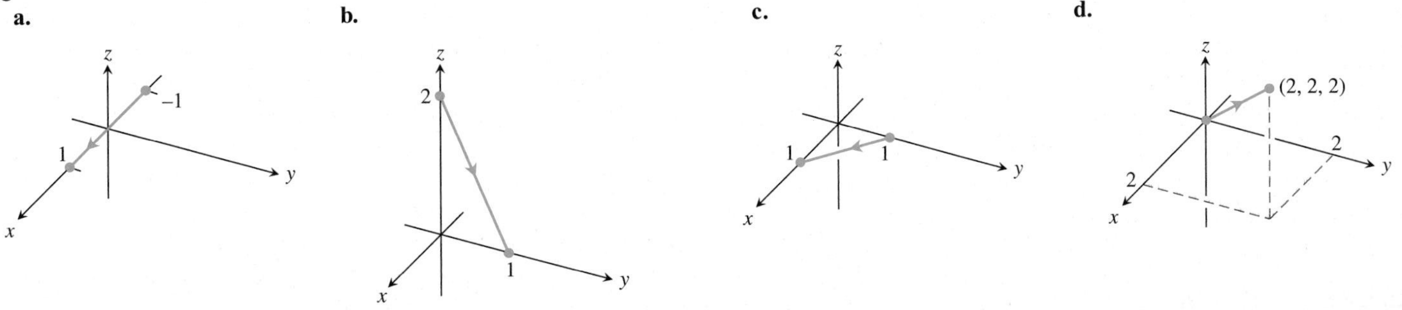

e.

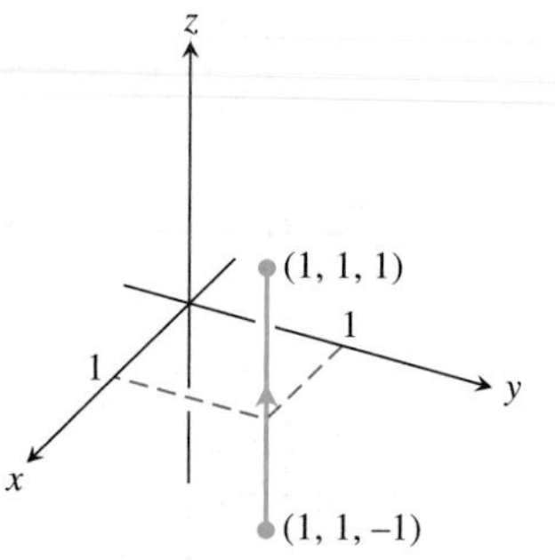

f.

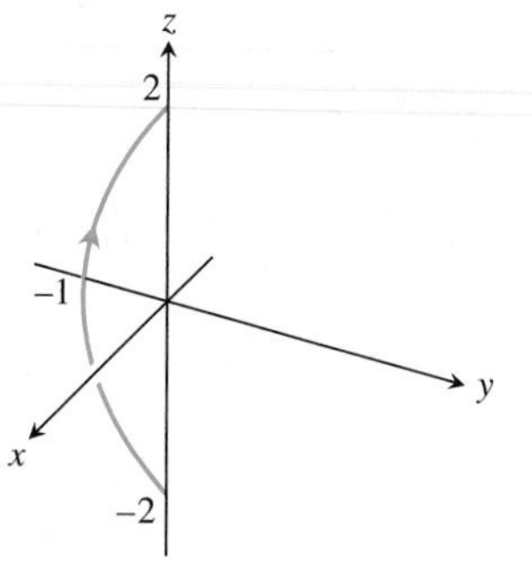

g.

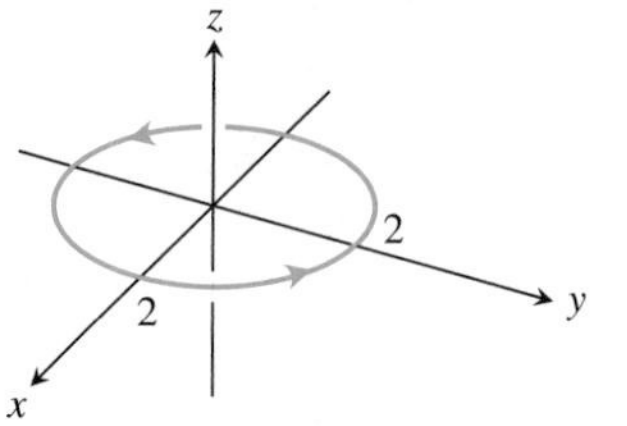

h.

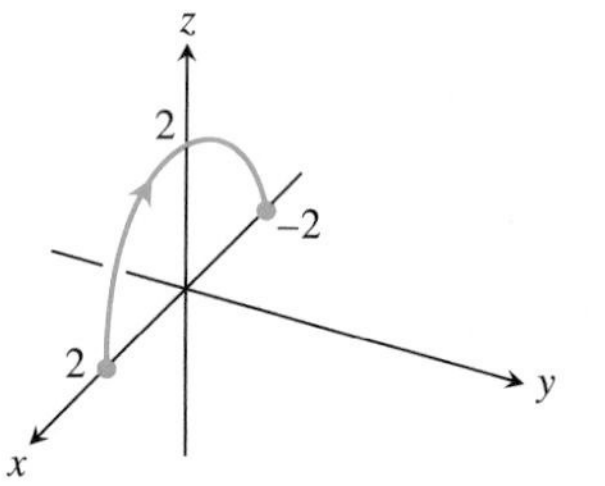

1. $\mathbf{r}(t) = t\mathbf{i} + (1 - t)\mathbf{j}, \quad 0 \le t \le 1$
2. $\mathbf{r}(t) = \mathbf{i} + \mathbf{j} + t\mathbf{k}, \quad -1 \le t \le 1$
3. $\mathbf{r}(t) = (2\cos t)\mathbf{i} + (2\sin t)\mathbf{j}, \quad 0 \le t \le 2\pi$
4. $\mathbf{r}(t) = t\mathbf{i}, \quad -1 \le t \le 1$
5. $\mathbf{r}(t) = t\mathbf{i} + t\mathbf{j} + t\mathbf{k}, \quad 0 \le t \le 2$
6. $\mathbf{r}(t) = t\mathbf{j} + (2 - 2t)\mathbf{k}, \quad 0 \le t \le 1$
7. $\mathbf{r}(t) = (t^2 - 1)\mathbf{j} + 2t\mathbf{k}, \quad -1 \le t \le 1$
8. $\mathbf{r}(t) = (2\cos t)\mathbf{i} + (2\sin t)\mathbf{k}, \quad 0 \le t \le \pi$

Evaluating Line Integrals over Space Curves

9. Evaluate $\int_C (x + y)\, ds$ where C is the straight-line segment $x = t, y = (1 - t), z = 0$, from $(0, 1, 0)$ to $(1, 0, 0)$.
10. Evaluate $\int_C (x - y + z - 2)\, ds$ where C is the straight-line segment $x = t, y = (1 - t), z = 1$, from $(0, 1, 1)$ to $(1, 0, 1)$.
11. Evaluate $\int_C (xy + y + z)\, ds$ along the curve $\mathbf{r}(t) = 2t\mathbf{i} + t\mathbf{j} + (2 - 2t)\mathbf{k}, 0 \le t \le 1$.
12. Evaluate $\int_C \sqrt{x^2 + y^2}\, ds$ along the curve $\mathbf{r}(t) = (4\cos t)\mathbf{i} + (4\sin t)\mathbf{j} + 3t\mathbf{k}, -2\pi \le t \le 2\pi$.
13. Find the line integral of $f(x, y, z) = x + y + z$ over the straight-line segment from $(1, 2, 3)$ to $(0, -1, 1)$.
14. Find the line integral of $f(x, y, z) = \sqrt{3}/(x^2 + y^2 + z^2)$ over the curve $\mathbf{r}(t) = t\mathbf{i} + t\mathbf{j} + t\mathbf{k}, 1 \le t \le \infty$.
15. Integrate $f(x, y, z) = x + \sqrt{y} - z^2$ over the path from $(0, 0, 0)$ to $(1, 1, 1)$ (Figure 16.6a) given by

$$C_1: \quad \mathbf{r}(t) = t\mathbf{i} + t^2\mathbf{j}, \quad 0 \le t \le 1$$

$$C_2: \quad \mathbf{r}(t) = \mathbf{i} + \mathbf{j} + t\mathbf{k}, \quad 0 \le t \le 1$$

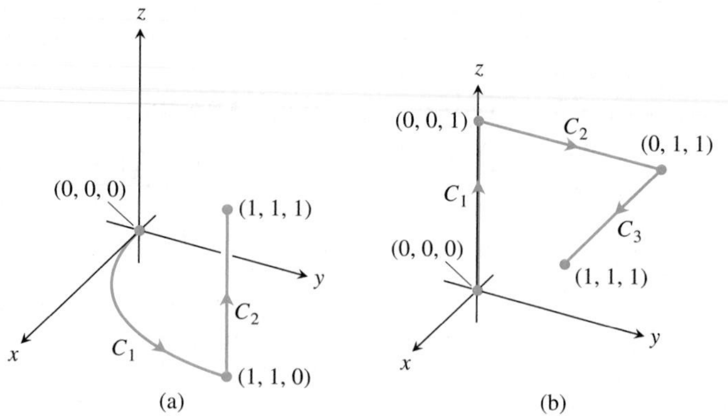

FIGURE 16.6 The paths of integration for Exercises 15 and 16.

16. Integrate $f(x, y, z) = x + \sqrt{y} - z^2$ over the path from $(0, 0, 0)$ to $(1, 1, 1)$ (Figure 16.6b) given by

$$C_1: \quad \mathbf{r}(t) = t\mathbf{k}, \quad 0 \le t \le 1$$

$$C_2: \quad \mathbf{r}(t) = t\mathbf{j} + \mathbf{k}, \quad 0 \le t \le 1$$

$$C_3: \quad \mathbf{r}(t) = t\mathbf{i} + \mathbf{j} + \mathbf{k}, \quad 0 \le t \le 1$$

17. Integrate $f(x, y, z) = (x + y + z)/(x^2 + y^2 + z^2)$ over the path $\mathbf{r}(t) = t\mathbf{i} + t\mathbf{j} + t\mathbf{k}, 0 < a \le t \le b$.
18. Integrate $f(x, y, z) = -\sqrt{x^2 + z^2}$ over the circle

$$\mathbf{r}(t) = (a\cos t)\mathbf{j} + (a\sin t)\mathbf{k}, \qquad 0 \le t \le 2\pi.$$

Line Integrals over Plane Curves

In Exercises 19–22, integrate f over the given curve.

19. $f(x, y) = x^3/y, \quad C: \; y = x^2/2, \quad 0 \le x \le 2$
20. $f(x, y) = (x + y^2)/\sqrt{1 + x^2}, \quad C: \; y = x^2/2$ from $(1, 1/2)$ to $(0, 0)$
21. $f(x, y) = x + y, \quad C: \; x^2 + y^2 = 4$ in the first quadrant from $(2, 0)$ to $(0, 2)$
22. $f(x, y) = x^2 - y, \quad C: \; x^2 + y^2 = 4$ in the first quadrant from $(0, 2)$ to $(\sqrt{2}, \sqrt{2})$

Mass and Moments

23. **Mass of a wire** Find the mass of a wire that lies along the curve $\mathbf{r}(t) = (t^2 - 1)\mathbf{j} + 2t\mathbf{k}, 0 \le t \le 1$, if the density is $\delta = (3/2)t$.
24. **Center of mass of a curved wire** A wire of density $\delta(x, y, z) = 15\sqrt{y + 2}$ lies along the curve $\mathbf{r}(t) = (t^2 - 1)\mathbf{j} + 2t\mathbf{k}, -1 \le t \le 1$. Find its center of mass. Then sketch the curve and center of mass together.
25. **Mass of wire with variable density** Find the mass of a thin wire lying along the curve $\mathbf{r}(t) = \sqrt{2}t\mathbf{i} + \sqrt{2}t\mathbf{j} + (4 - t^2)\mathbf{k}, 0 \le t \le 1$, if the density is **(a)** $\delta = 3t$ and **(b)** $\delta = 1$.

26. Center of mass of wire with variable density Find the center of mass of a thin wire lying along the curve $\mathbf{r}(t) = t\mathbf{i} + 2t\mathbf{j} + (2/3)t^{3/2}\mathbf{k}$, $0 \le t \le 2$, if the density is $\delta = 3\sqrt{5 + t}$.

27. Moment of inertia and radius of gyration of wire hoop A circular wire hoop of constant density δ lies along the circle $x^2 + y^2 = a^2$ in the xy-plane. Find the hoop's moment of inertia and radius of gyration about the z-axis.

28. Inertia and radii of gyration of slender rod A slender rod of constant density lies along the line segment $\mathbf{r}(t) = t\mathbf{j} + (2 - 2t)\mathbf{k}$, $0 \le t \le 1$, in the yz-plane. Find the moments of inertia and radii of gyration of the rod about the three coordinate axes.

29. Two springs of constant density A spring of constant density δ lies along the helix

$$\mathbf{r}(t) = (\cos t)\mathbf{i} + (\sin t)\mathbf{j} + t\mathbf{k}, \qquad 0 \le t \le 2\pi.$$

a. Find I_z and R_z.

b. Suppose that you have another spring of constant density δ that is twice as long as the spring in part (a) and lies along the helix for $0 \le t \le 4\pi$. Do you expect I_z and R_z for the longer spring to be the same as those for the shorter one, or should they be different? Check your predictions by calculating I_z and R_z for the longer spring.

30. Wire of constant density A wire of constant density $\delta = 1$ lies along the curve

$$\mathbf{r}(t) = (t\cos t)\mathbf{i} + (t\sin t)\mathbf{j} + \left(2\sqrt{2}/3\right)t^{3/2}\mathbf{k}, \qquad 0 \le t \le 1.$$

Find $\bar{z}$, I_z, and R_z.

31. The arch in Example 4 Find I_x and R_x for the arch in Example 4.

32. Center of mass, moments of inertia, and radii of gyration for wire with variable density Find the center of mass, and the moments of inertia and radii of gyration about the coordinate axes of a thin wire lying along the curve

$$\mathbf{r}(t) = t\mathbf{i} + \frac{2\sqrt{2}}{3}t^{3/2}\mathbf{j} + \frac{t^2}{2}\mathbf{k}, \qquad 0 \le t \le 2,$$

if the density is $\delta = 1/(t + 1)$.

COMPUTER EXPLORATIONS

Evaluating Line Integrals Numerically

In Exercises 33–36, use a CAS to perform the following steps to evaluate the line integrals.

a. Find $ds = |\mathbf{v}(t)|\,dt$ for the path $\mathbf{r}(t) = g(t)\mathbf{i} + h(t)\mathbf{j} + k(t)\mathbf{k}$.

b. Express the integrand $f(g(t), h(t), k(t))\,|\mathbf{v}(t)|$ as a function of the parameter t.

c. Evaluate $\int_C f\,ds$ using Equation (2) in the text.

33. $f(x, y, z) = \sqrt{1 + 30x^2 + 10y}$; $\mathbf{r}(t) = t\mathbf{i} + t^2\mathbf{j} + 3t^2\mathbf{k}$, $0 \le t \le 2$

34. $f(x, y, z) = \sqrt{1 + x^3 + 5y^3}$; $\mathbf{r}(t) = t\mathbf{i} + \frac{1}{3}t^2\mathbf{j} + \sqrt{t}\mathbf{k}$, $0 \le t \le 2$

35. $f(x, y, z) = x\sqrt{y} - 3z^2$; $\mathbf{r}(t) = (\cos 2t)\mathbf{i} + (\sin 2t)\mathbf{j} + 5t\mathbf{k}$, $0 \le t \le 2\pi$

36. $f(x, y, z) = \left(1 + \frac{9}{4}z^{1/3}\right)^{1/4}$; $\mathbf{r}(t) = (\cos 2t)\mathbf{i} + (\sin 2t)\mathbf{j} + t^{5/2}\mathbf{k}$, $0 \le t \le 2\pi$

16.2 Vector Fields, Work, Circulation, and Flux

When we study physical phenomena that are represented by vectors, we replace integrals over closed intervals by integrals over paths through vector fields. We use such integrals to find the work done in moving an object along a path against a variable force (such as a vehicle sent into space against Earth's gravitational field) or to find the work done by a vector field in moving an object along a path through the field (such as the work done by an accelerator in raising the energy of a particle). We also use line integrals to find the rates at which fluids flow along and across curves.

Vector Fields

Suppose a region in the plane or in space is occupied by a moving fluid such as air or water. Imagine that the fluid is made up of a very large number of particles, and that at any instant of time a particle has a velocity $\mathbf{v}$. If we take a picture of the velocities of some particles at

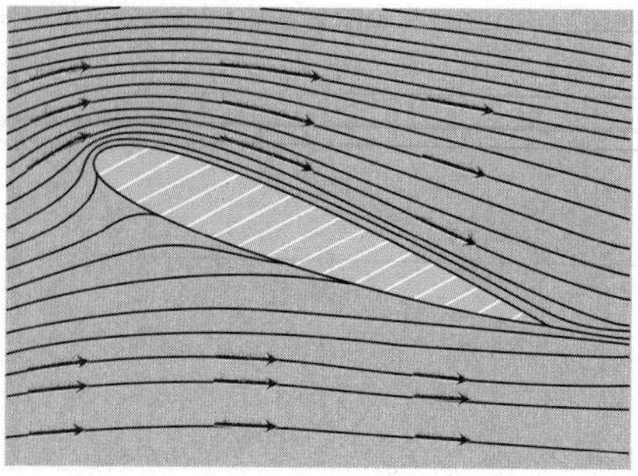

FIGURE 16.7 Velocity vectors of a flow around an airfoil in a wind tunnel. The streamlines were made visible by kerosene smoke.

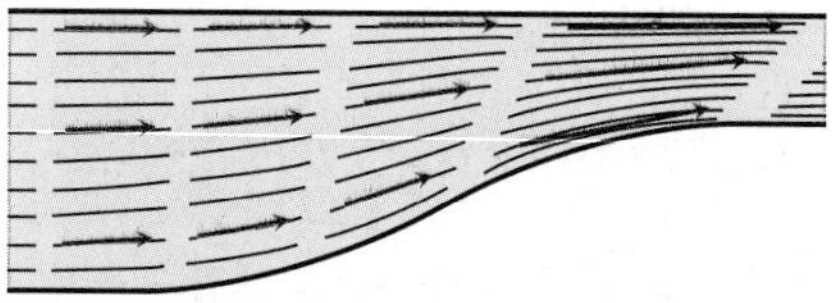

FIGURE 16.8 Streamlines in a contracting channel. The water speeds up as the channel narrows and the velocity vectors increase in length.

different position points at the same instant, we would expect to find that these velocities vary from position to position. We can think of a velocity vector as being attached to each point of the fluid. Such a fluid flow exemplifies a *vector field*. For example, Figure 16.7 shows a velocity vector field obtained by attaching a velocity vector to each point of air flowing around an airfoil in a wind tunnel. Figure 16.8 shows another vector field of velocity vectors along the streamlines of water moving through a contracting channel. In addition to vector fields associated with fluid flows, there are vector force fields that are associated with gravitational attraction (Figure 16.9), magnetic force fields, electric fields, and even purely mathematical fields.

Generally, a **vector field** on a domain in the plane or in space is a function that assigns a vector to each point in the domain. A field of three-dimensional vectors might have a formula like

$$\mathbf{F}(x, y, z) = M(x, y, z)\mathbf{i} + N(x, y, z)\mathbf{j} + P(x, y, z)\mathbf{k}.$$

The field is **continuous** if the **component functions** M, N, and P are continuous, **differentiable** if M, N, and P are differentiable, and so on. A field of two-dimensional vectors might have a formula like

$$\mathbf{F}(x, y) = M(x, y)\mathbf{i} + N(x, y)\mathbf{j}.$$

If we attach a projectile's velocity vector to each point of the projectile's trajectory in the plane of motion, we have a two-dimensional field defined along the trajectory. If we attach the gradient vector of a scalar function to each point of a level surface of the function, we have a three-dimensional field on the surface. If we attach the velocity vector to each point of a flowing fluid, we have a three-dimensional field defined on a region in space. These and other fields are illustrated in Figures 16.10–16.15. Some of the illustrations give formulas for the fields as well.

To sketch the fields that had formulas, we picked a representative selection of domain points and sketched the vectors attached to them. The arrows representing the vectors are drawn with their tails, not their heads, at the points where the vector functions are

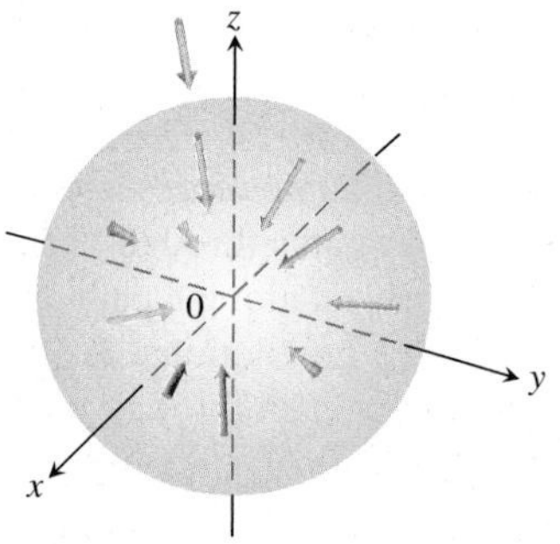

FIGURE 16.9 Vectors in a gravitational field point toward the center of mass that gives the source of the field.

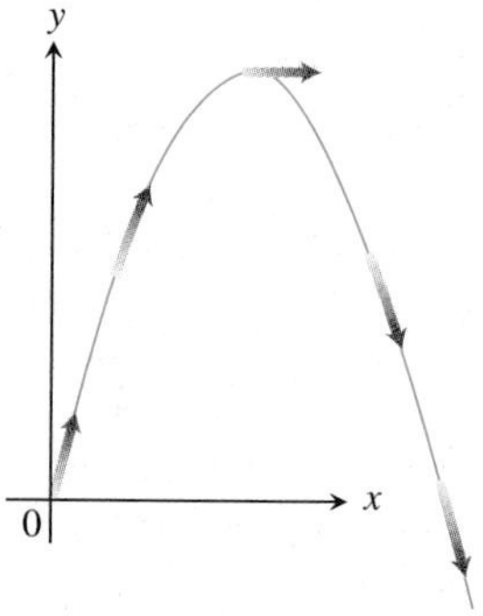

FIGURE 16.10 The velocity vectors $\mathbf{v}(t)$ of a projectile's motion make a vector field along the trajectory.

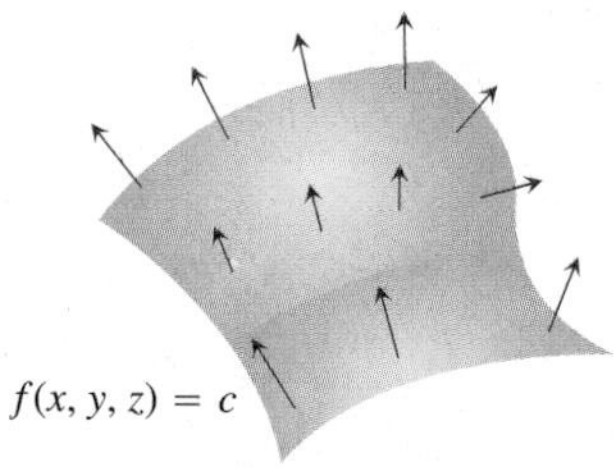

FIGURE 16.11 The field of gradient vectors ∇f on a surface $f(x, y, z) = c$.

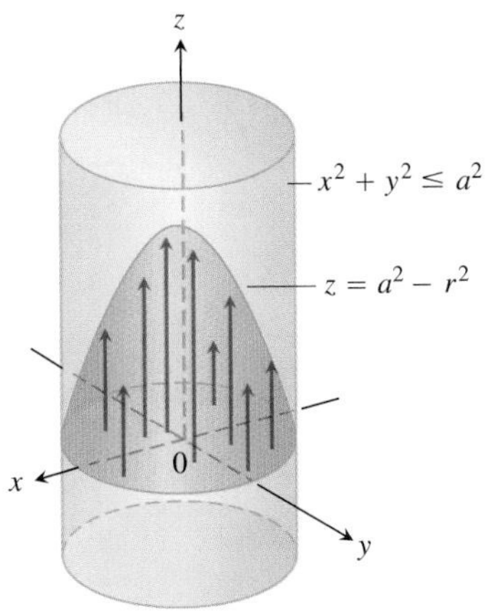

FIGURE 16.12 The flow of fluid in a long cylindrical pipe. The vectors $\mathbf{v} = (a^2 - r^2)\mathbf{k}$ inside the cylinder that have their bases in the xy-plane have their tips on the paraboloid $z = a^2 - r^2$.

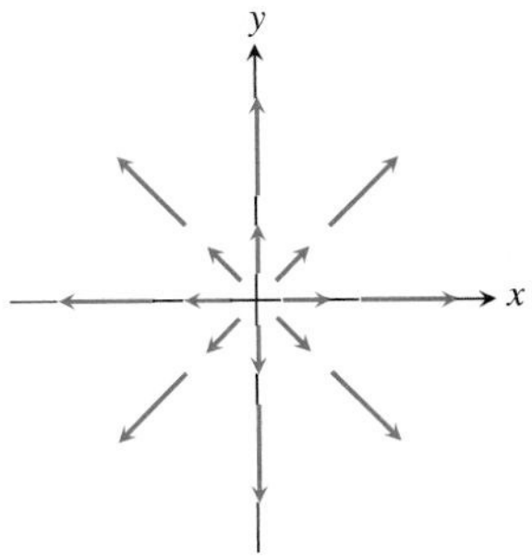

FIGURE 16.13 The radial field $\mathbf{F} = x\mathbf{i} + y\mathbf{j}$ of position vectors of points in the plane. Notice the convention that an arrow is drawn with its tail, not its head, at the point where $\mathbf{F}$ is evaluated.

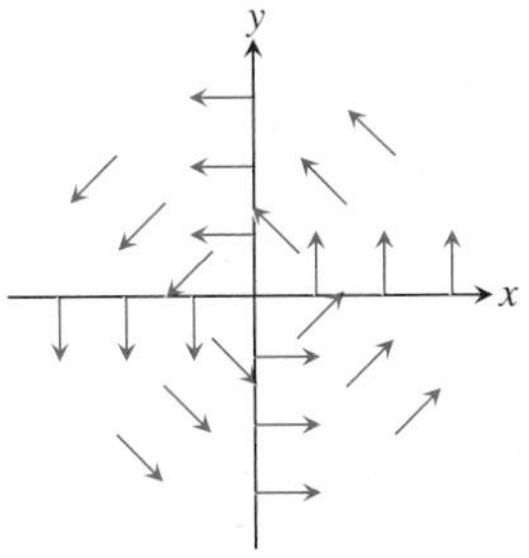

FIGURE 16.14 The circumferential or "spin" field of unit vectors

$$\mathbf{F} = (-y\mathbf{i} + x\mathbf{j})/(x^2 + y^2)^{1/2}$$

in the plane. The field is not defined at the origin.

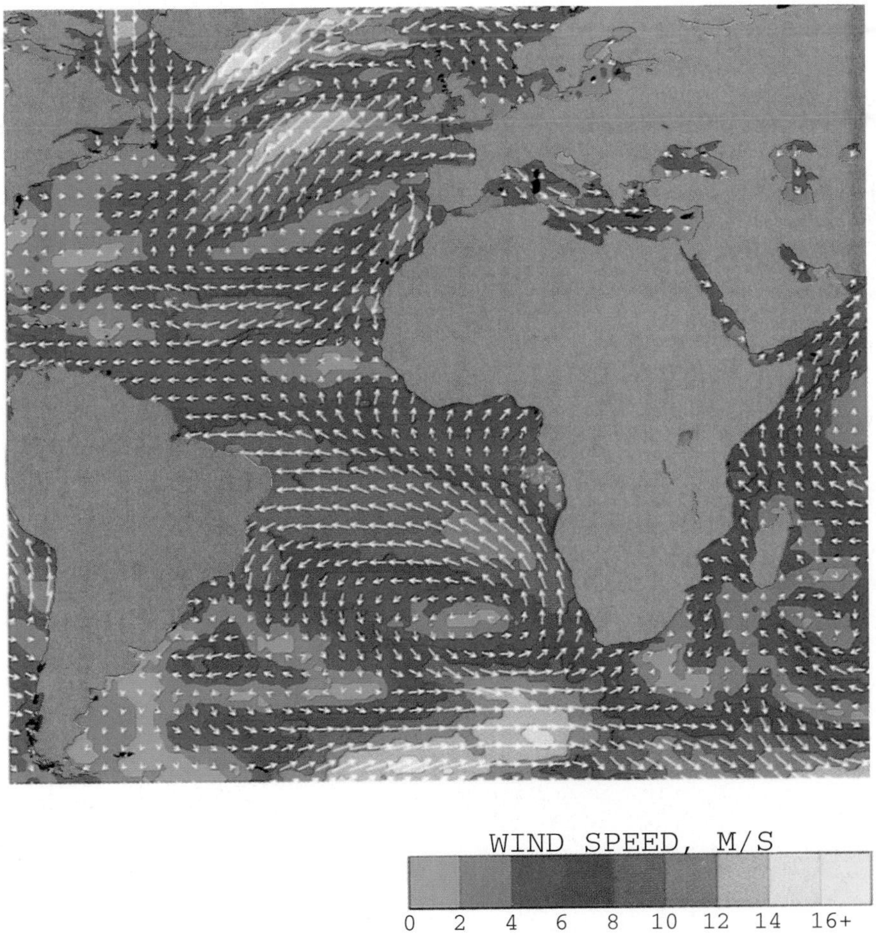

FIGURE 16.15 NASA's *Seasat* used radar to take 350,000 wind measurements over the world's oceans. The arrows show wind direction; their length and the color contouring indicate speed. Notice the heavy storm south of Greenland.

evaluated. This is different from the way we draw position vectors of planets and projectiles, with their tails at the origin and their heads at the planet's and projectile's locations.

Gradient Fields

> **DEFINITION Gradient Field**
> The **gradient field** of a differentiable function $f(x, y, z)$ is the field of gradient vectors
> $$\nabla f = \frac{\partial f}{\partial x}\mathbf{i} + \frac{\partial f}{\partial y}\mathbf{j} + \frac{\partial f}{\partial z}\mathbf{k}.$$

EXAMPLE 1 Finding a Gradient Field

Find the gradient field of $f(x, y, z) = xyz$.

Solution The gradient field of f is the field $\mathbf{F} = \nabla f = yz\mathbf{i} + xz\mathbf{j} + xy\mathbf{k}$. ■

As we will see in Section 16.3, gradient fields are of special importance in engineering, mathematics, and physics.

Work Done by a Force over a Curve in Space

Suppose that the vector field $\mathbf{F} = M(x, y, z)\mathbf{i} + N(x, y, z)\mathbf{j} + P(x, y, z)\mathbf{k}$ represents a force throughout a region in space (it might be the force of gravity or an electromagnetic force of some kind) and that

$$\mathbf{r}(t) = g(t)\mathbf{i} + h(t)\mathbf{j} + k(t)\mathbf{k}, \qquad a \le t \le b,$$

is a smooth curve in the region. Then the integral of $\mathbf{F} \cdot \mathbf{T}$, the scalar component of $\mathbf{F}$ in the direction of the curve's unit tangent vector, over the curve is called the work done by $\mathbf{F}$ over the curve from a to b (Figure 16.16).

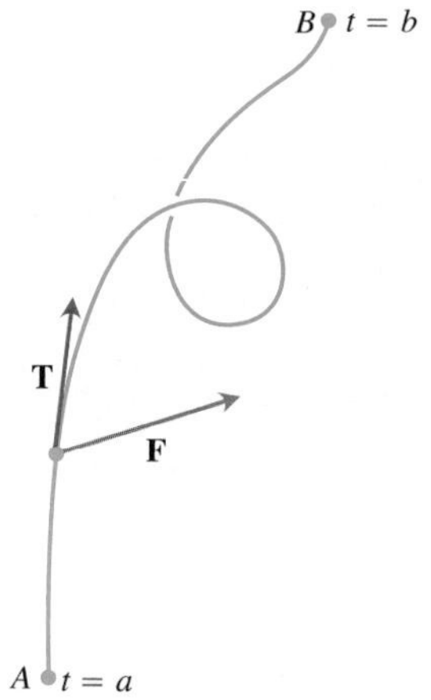

FIGURE 16.16 The work done by a force **F** is the line integral of the scalar component $\mathbf{F} \cdot \mathbf{T}$ over the smooth curve from A to B.

> **DEFINITION Work over a Smooth Curve**
> The **work** done by a force $\mathbf{F} = M\mathbf{i} + N\mathbf{j} + P\mathbf{k}$ over a smooth curve $\mathbf{r}(t)$ from $t = a$ to $t = b$ is
> $$W = \int_{t=a}^{t=b} \mathbf{F} \cdot \mathbf{T}\, ds. \qquad (1)$$

We motivate Equation (1) with the same kind of reasoning we used in Chapter 6 to derive the formula $W = \int_a^b F(x)\, dx$ for the work done by a continuous force of magnitude $F(x)$ directed along an interval of the x-axis. We divide the curve into short segments, apply the (constant-force) $\times$ (distance) formula for work to approximate the work over each curved segment, add the results to approximate the work over the entire curve, and calculate the work as the limit of the approximating sums as the segments become shorter and more numerous. To find exactly what the limiting integral should be, we partition the parameter interval $[a, b]$ in the usual way and choose a point c_k in each subinterval $[t_k, t_{k+1}]$. The partition of $[a, b]$ determines ("induces," we say) a partition of the curve, with the point P_k being the tip of the position vector $\mathbf{r}(t_k)$ and Δs_k being the length of the curve segment P_kP_{k+1} (Figure 16.17).

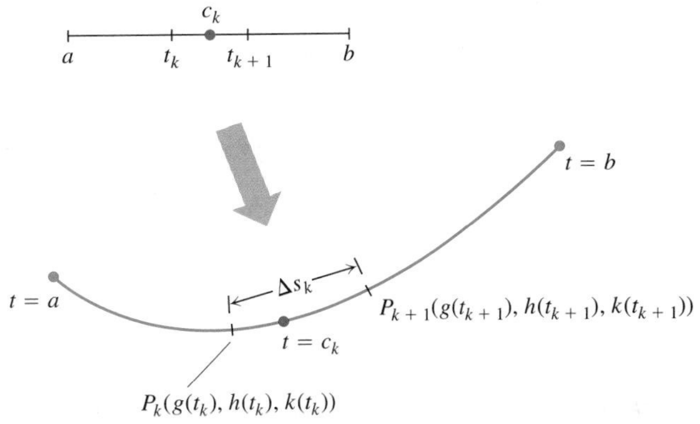

FIGURE 16.17 Each partition of $[a, b]$ induces a partition of the curve $\mathbf{r}(t) = g(t)\mathbf{i} + h(t)\mathbf{j} + k(t)\mathbf{k}$.

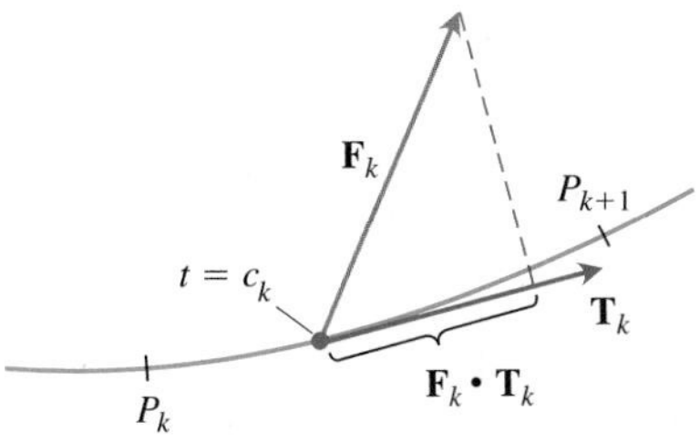

FIGURE 16.18 An enlarged view of the curve segment P_kP_{k+1} in Figure 16.17, showing the force and unit tangent vectors at the point on the curve where $t = c_k$.

If $\mathbf{F}_k$ denotes the value of $\mathbf{F}$ at the point on the curve corresponding to $t = c_k$ and $\mathbf{T}_k$ denotes the curve's unit tangent vector at this point, then $\mathbf{F}_k \cdot \mathbf{T}_k$ is the scalar component of $\mathbf{F}$ in the direction of $\mathbf{T}$ at $t = c_k$ (Figure 16.18). The work done by $\mathbf{F}$ along the curve segment P_kP_{k+1} is approximately

$$\begin{pmatrix}\text{Force component in}\\ \text{direction of motion}\end{pmatrix} \times \begin{pmatrix}\text{distance}\\ \text{applied}\end{pmatrix} = \mathbf{F}_k \cdot \mathbf{T}_k \, \Delta s_k.$$

The work done by $\mathbf{F}$ along the curve from $t = a$ to $t = b$ is approximately

$$\sum_{k=1}^{n} \mathbf{F}_k \cdot \mathbf{T}_k \, \Delta s_k.$$

As the norm of the partition of $[a, b]$ approaches zero, the norm of the induced partition of the curve approaches zero and these sums approach the line integral

$$\int_{t=a}^{t=b} \mathbf{F} \cdot \mathbf{T} \, ds.$$

The sign of the number we calculate with this integral depends on the direction in which the curve is traversed as t increases. If we reverse the direction of motion, we reverse the direction of $\mathbf{T}$ and change the sign of $\mathbf{F} \cdot \mathbf{T}$ and its integral.

Table 16.2 shows six ways to write the work integral in Equation (1). Despite their variety, the formulas in Table 16.2 are all evaluated the same way. In the table, $\mathbf{r}(t) = g(t)\mathbf{i} + h(t)\mathbf{j} + k(t)\mathbf{k} = x\mathbf{i} + y\mathbf{j} + z\mathbf{k}$ is a smooth curve, and

$$d\mathbf{r} = \frac{d\mathbf{r}}{dt}\,dt = dg\mathbf{i} + dh\mathbf{j} + dk\mathbf{k}$$

is its differential.

TABLE 16.2 Six different ways to write the work integral

$\mathbf{W} = \int_{t=a}^{t=b} \mathbf{F}\cdot\mathbf{T}\, ds$	The definition
$= \int_{t=a}^{t=b} \mathbf{F}\cdot d\mathbf{r}$	Compact differential form
$= \int_{a}^{b} \mathbf{F}\cdot\frac{d\mathbf{r}}{dt}\, dt$	Expanded to include dt; emphasizes the parameter t and velocity vector $d\mathbf{r}/dt$
$= \int_{a}^{b} \left(M\frac{dg}{dt} + N\frac{dh}{dt} + P\frac{dk}{dt}\right) dt$	Emphasizes the component functions
$= \int_{a}^{b} \left(M\frac{dx}{dt} + N\frac{dy}{dt} + P\frac{dz}{dt}\right) dt$	Abbreviates the components of $\mathbf{r}$
$= \int_{a}^{b} M\,dx + N\,dy + P\,dz$	dt's canceled; the most common form

Evaluating a Work Integral

To evaluate the work integral along a smooth curve $\mathbf{r}(t)$, take these steps:

1. Evaluate $\mathbf{F}$ on the curve as a function of the parameter t.
2. Find $d\mathbf{r}/dt$
3. Integrate $\mathbf{F}\cdot d\mathbf{r}/dt$ from $t = a$ to $t = b$.

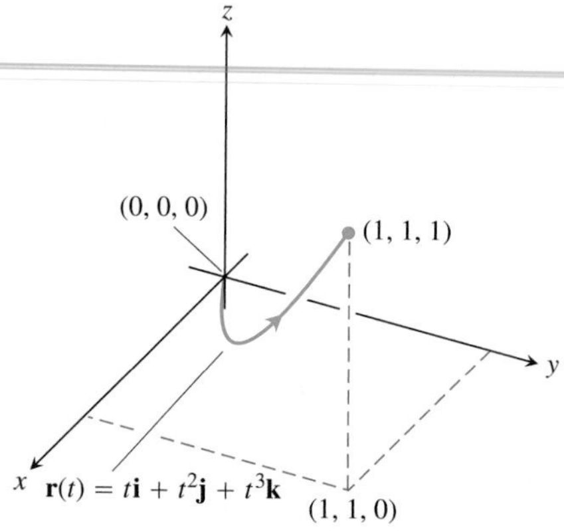

FIGURE 16.19 The curve in Example 2.

EXAMPLE 2 Finding Work Done by a Variable Force over a Space Curve

Find the work done by $\mathbf{F} = (y - x^2)\mathbf{i} + (z - y^2)\mathbf{j} + (x - z^2)\mathbf{k}$ over the curve $\mathbf{r}(t) = t\mathbf{i} + t^2\mathbf{j} + t^3\mathbf{k}$, $0 \le t \le 1$, from (0, 0, 0) to (1, 1, 1) (Figure 16.19).

Solution First we evaluate $\mathbf{F}$ on the curve:

$$\mathbf{F} = (y - x^2)\mathbf{i} + (z - y^2)\mathbf{j} + (x - z^2)\mathbf{k}$$
$$= \underbrace{(t^2 - t^2)}_{0}\mathbf{i} + (t^3 - t^4)\mathbf{j} + (t - t^6)\mathbf{k}$$

Then we find $d\mathbf{r}/dt$,

$$\frac{d\mathbf{r}}{dt} = \frac{d}{dt}(t\mathbf{i} + t^2\mathbf{j} + t^3\mathbf{k}) = \mathbf{i} + 2t\mathbf{j} + 3t^2\mathbf{k}$$

Finally, we find $\mathbf{F}\cdot d\mathbf{r}/dt$ and integrate from $t = 0$ to $t = 1$:

$$\mathbf{F}\cdot\frac{d\mathbf{r}}{dt} = [(t^3 - t^4)\mathbf{j} + (t - t^6)\mathbf{k}]\cdot(\mathbf{i} + 2t\mathbf{j} + 3t^2\mathbf{k})$$
$$= (t^3 - t^4)(2t) + (t - t^6)(3t^2) = 2t^4 - 2t^5 + 3t^3 - 3t^8,$$

so

$$\text{Work} = \int_0^1 (2t^4 - 2t^5 + 3t^3 - 3t^8)\,dt$$
$$= \left[\frac{2}{5}t^5 - \frac{2}{6}t^6 + \frac{3}{4}t^4 - \frac{3}{9}t^9\right]_0^1 = \frac{29}{60}.$$ ■

Flow Integrals and Circulation for Velocity Fields

Instead of being a force field, suppose that $\mathbf{F}$ represents the velocity field of a fluid flowing through a region in space (a tidal basin or the turbine chamber of a hydroelectric generator, for example). Under these circumstances, the integral of $\mathbf{F} \cdot \mathbf{T}$ along a curve in the region gives the fluid's flow along the curve.

DEFINITIONS **Flow Integral, Circulation**

If $\mathbf{r}(t)$ is a smooth curve in the domain of a continuous velocity field $\mathbf{F}$, the **flow** along the curve from $t = a$ to $t = b$ is

$$\text{Flow} = \int_a^b \mathbf{F} \cdot \mathbf{T}\,ds. \qquad (2)$$

The integral in this case is called a **flow integral**. If the curve is a closed loop, the flow is called the **circulation** around the curve.

We evaluate flow integrals the same way we evaluate work integrals.

EXAMPLE 3 Finding Flow Along a Helix

A fluid's velocity field is $\mathbf{F} = x\mathbf{i} + z\mathbf{j} + y\mathbf{k}$. Find the flow along the helix $\mathbf{r}(t) = (\cos t)\mathbf{i} + (\sin t)\mathbf{j} + t\mathbf{k}$, $0 \le t \le \pi/2$.

Solution We evaluate $\mathbf{F}$ on the curve,

$$\mathbf{F} = x\mathbf{i} + z\mathbf{j} + y\mathbf{k} = (\cos t)\mathbf{i} + t\mathbf{j} + (\sin t)\mathbf{k}$$

and then find $d\mathbf{r}/dt$:

$$\frac{d\mathbf{r}}{dt} = (-\sin t)\mathbf{i} + (\cos t)\mathbf{j} + \mathbf{k}.$$

Then we integrate $\mathbf{F} \cdot (d\mathbf{r}/dt)$ from $t = 0$ to $t = \dfrac{\pi}{2}$:

$$\mathbf{F} \cdot \frac{d\mathbf{r}}{dt} = (\cos t)(-\sin t) + (t)(\cos t) + (\sin t)(1)$$
$$= -\sin t \cos t + t \cos t + \sin t$$

so,

$$\text{Flow} = \int_{t=a}^{t=b} \mathbf{F} \cdot \frac{d\mathbf{r}}{dt}\, dt = \int_0^{\pi/2} (-\sin t \cos t + t \cos t + \sin t)\, dt$$

$$= \left[\frac{\cos^2 t}{2} + t \sin t\right]_0^{\pi/2} = \left(0 + \frac{\pi}{2}\right) - \left(\frac{1}{2} + 0\right) = \frac{\pi}{2} - \frac{1}{2}.$$

EXAMPLE 4 Finding Circulation Around a Circle

Find the circulation of the field $\mathbf{F} = (x - y)\mathbf{i} + x\mathbf{j}$ around the circle $\mathbf{r}(t) = (\cos t)\mathbf{i} + (\sin t)\mathbf{j}$, $0 \le t \le 2\pi$.

Solution On the circle, $\mathbf{F} = (x - y)\mathbf{i} + x\mathbf{j} = (\cos t - \sin t)\mathbf{i} + (\cos t)\mathbf{j}$, and

$$\frac{d\mathbf{r}}{dt} = (-\sin t)\mathbf{i} + (\cos t)\mathbf{j}.$$

Then

$$\mathbf{F} \cdot \frac{d\mathbf{r}}{dt} = -\sin t \cos t + \underbrace{\sin^2 t + \cos^2 t}_{1}$$

gives

$$\text{Circulation} = \int_0^{2\pi} \mathbf{F} \cdot \frac{d\mathbf{r}}{dt}\, dt = \int_0^{2\pi} (1 - \sin t \cos t)\, dt$$

$$= \left[t - \frac{\sin^2 t}{2}\right]_0^{2\pi} = 2\pi.$$

Flux Across a Plane Curve

To find the rate at which a fluid is entering or leaving a region enclosed by a smooth curve C in the xy-plane, we calculate the line integral over C of $\mathbf{F} \cdot \mathbf{n}$, the scalar component of the fluid's velocity field in the direction of the curve's outward-pointing normal vector. The value of this integral is the *flux* of $\mathbf{F}$ across C. *Flux* is Latin for *flow*, but many flux calculations involve no motion at all. If $\mathbf{F}$ were an electric field or a magnetic field, for instance, the integral of $\mathbf{F} \cdot \mathbf{n}$ would still be called the flux of the field across C.

DEFINITION Flux Across a Closed Curve in the Plane

If C is a smooth closed curve in the domain of a continuous vector field $\mathbf{F} = M(x, y)\mathbf{i} + N(x, y)\mathbf{j}$ in the plane and if $\mathbf{n}$ is the outward-pointing unit normal vector on C, the **flux** of $\mathbf{F}$ across C is

$$\text{Flux of } \mathbf{F} \text{ across } C = \int_C \mathbf{F} \cdot \mathbf{n}\, ds. \tag{3}$$

Notice the difference between flux and circulation. The flux of $\mathbf{F}$ across C is the line integral with respect to arc length of $\mathbf{F} \cdot \mathbf{n}$, the scalar component of $\mathbf{F}$ in the direction of the

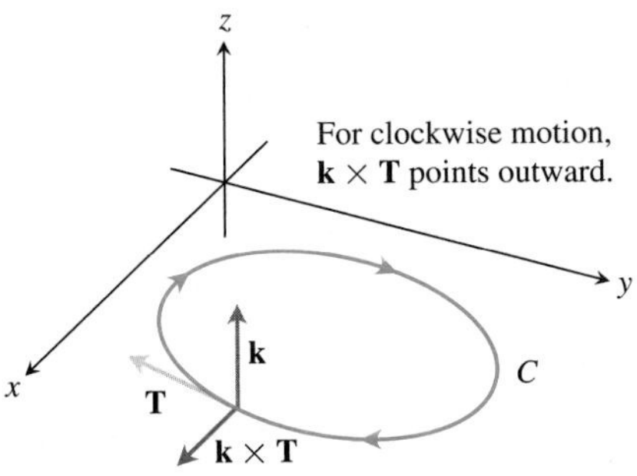

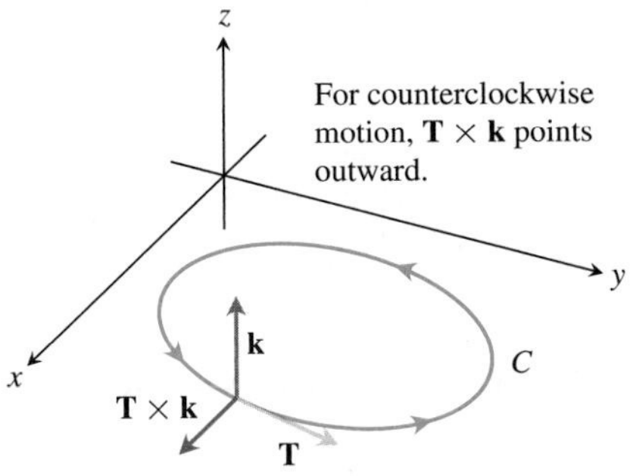

FIGURE 16.20 To find an outward unit normal vector for a smooth curve C in the xy-plane that is traversed counterclockwise as t increases, we take $\mathbf{n} = \mathbf{T} \times \mathbf{k}$. For clockwise motion, we take $\mathbf{n} = \mathbf{k} \times \mathbf{T}$.

outward normal. The circulation of $\mathbf{F}$ around C is the line integral with respect to arc length of $\mathbf{F} \cdot \mathbf{T}$, the scalar component of $\mathbf{F}$ in the direction of the unit tangent vector. Flux is the integral of the normal component of $\mathbf{F}$; circulation is the integral of the tangential component of $\mathbf{F}$.

To evaluate the integral in Equation (3), we begin with a smooth parametrization

$$x = g(t), \qquad y = h(t), \qquad a \le t \le b,$$

that traces the curve C exactly once as t increases from a to b. We can find the outward unit normal vector $\mathbf{n}$ by crossing the curve's unit tangent vector $\mathbf{T}$ with the vector $\mathbf{k}$. But which order do we choose, $\mathbf{T} \times \mathbf{k}$ or $\mathbf{k} \times \mathbf{T}$? Which one points outward? It depends on which way C is traversed as t increases. If the motion is clockwise, $\mathbf{k} \times \mathbf{T}$ points outward; if the motion is counterclockwise, $\mathbf{T} \times \mathbf{k}$ points outward (Figure 16.20). The usual choice is $\mathbf{n} = \mathbf{T} \times \mathbf{k}$, the choice that assumes counterclockwise motion. Thus, although the value of the arc length integral in the definition of flux in Equation (3) does not depend on which way C is traversed, the formulas we are about to derive for evaluating the integral in Equation (3) will assume counterclockwise motion.

In terms of components,

$$\mathbf{n} = \mathbf{T} \times \mathbf{k} = \left(\frac{dx}{ds}\mathbf{i} + \frac{dy}{ds}\mathbf{j}\right) \times \mathbf{k} = \frac{dy}{ds}\mathbf{i} - \frac{dx}{ds}\mathbf{j}.$$

If $\mathbf{F} = M(x, y)\mathbf{i} + N(x, y)\mathbf{j}$, then

$$\mathbf{F} \cdot \mathbf{n} = M(x, y)\frac{dy}{ds} - N(x, y)\frac{dx}{ds}.$$

Hence,

$$\int_C \mathbf{F} \cdot \mathbf{n}\, ds = \int_C \left(M\frac{dy}{ds} - N\frac{dx}{ds}\right) ds = \oint_C M\,dy - N\,dx.$$

We put a directed circle $\circlearrowleft$ on the last integral as a reminder that the integration around the closed curve C is to be in the counterclockwise direction. To evaluate this integral, we express M, dy, N, and dx in terms of t and integrate from $t = a$ to $t = b$. We do not need to know either $\mathbf{n}$ or ds to find the flux.

Calculating Flux Across a Smooth Closed Plane Curve

$$(\text{Flux of } \mathbf{F} = M\mathbf{i} + N\mathbf{j} \text{ across } C) = \oint_C M\,dy - N\,dx \qquad (4)$$

The integral can be evaluated from any smooth parametrization $x = g(t), y = h(t)$, $a \le t \le b$, that traces C counterclockwise exactly once.

EXAMPLE 5 Finding Flux Across a Circle

Find the flux of $\mathbf{F} = (x - y)\mathbf{i} + x\mathbf{j}$ across the circle $x^2 + y^2 = 1$ in the xy-plane.

Solution The parametrization $\mathbf{r}(t) = (\cos t)\mathbf{i} + (\sin t)\mathbf{j}$, $0 \le t \le 2\pi$, traces the circle counterclockwise exactly once. We can therefore use this parametrization in Equation (4). With

$$M = x - y = \cos t - \sin t, \qquad dy = d(\sin t) = \cos t\,dt$$
$$N = x = \cos t, \qquad dx = d(\cos t) = -\sin t\,dt,$$

We find

$$\text{Flux} = \int_C M\,dy - N\,dx = \int_0^{2\pi} (\cos^2 t - \sin t \cos t + \cos t \sin t)\,dt \qquad \text{Equation (4)}$$
$$= \int_0^{2\pi} \cos^2 t\,dt = \int_0^{2\pi} \frac{1 + \cos 2t}{2}\,dt = \left[\frac{t}{2} + \frac{\sin 2t}{4}\right]_0^{2\pi} = \pi.$$

The flux of $\mathbf{F}$ across the circle is π. Since the answer is positive, the net flow across the curve is outward. A net inward flow would have given a negative flux. ■

EXERCISES 16.2

Vector and Gradient Fields

Find the gradient fields of the functions in Exercises 1–4.

1. $f(x, y, z) = (x^2 + y^2 + z^2)^{-1/2}$
2. $f(x, y, z) = \ln\sqrt{x^2 + y^2 + z^2}$
3. $g(x, y, z) = e^z - \ln(x^2 + y^2)$
4. $g(x, y, z) = xy + yz + xz$
5. Give a formula $\mathbf{F} = M(x, y)\mathbf{i} + N(x, y)\mathbf{j}$ for the vector field in the plane that has the property that $\mathbf{F}$ points toward the origin with magnitude inversely proportional to the square of the distance from (x, y) to the origin. (The field is not defined at $(0, 0)$.)
6. Give a formula $\mathbf{F} = M(x, y)\mathbf{i} + N(x, y)\mathbf{j}$ for the vector field in the plane that has the properties that $\mathbf{F} = \mathbf{0}$ at $(0, 0)$ and that at any other point (a, b), $\mathbf{F}$ is tangent to the circle $x^2 + y^2 = a^2 + b^2$ and points in the clockwise direction with magnitude $|\mathbf{F}| = \sqrt{a^2 + b^2}$.

Work

In Exercises 7–12, find the work done by force $\mathbf{F}$ from $(0, 0, 0)$ to $(1, 1, 1)$ over each of the following paths (Figure 16.21):

a. The straight-line path C_1: $\mathbf{r}(t) = t\mathbf{i} + t\mathbf{j} + t\mathbf{k}$, $0 \le t \le 1$

b. The curved path C_2: $\mathbf{r}(t) = t\mathbf{i} + t^2\mathbf{j} + t^4\mathbf{k}$, $0 \le t \le 1$

c. The path $C_3 \cup C_4$ consisting of the line segment from $(0, 0, 0)$ to $(1, 1, 0)$ followed by the segment from $(1, 1, 0)$ to $(1, 1, 1)$

7. $\mathbf{F} = 3y\mathbf{i} + 2x\mathbf{j} + 4z\mathbf{k}$
8. $\mathbf{F} = [1/(x^2 + 1)]\mathbf{j}$
9. $\mathbf{F} = \sqrt{z}\mathbf{i} - 2x\mathbf{j} + \sqrt{y}\mathbf{k}$
10. $\mathbf{F} = xy\mathbf{i} + yz\mathbf{j} + xz\mathbf{k}$
11. $\mathbf{F} = (3x^2 - 3x)\mathbf{i} + 3z\mathbf{j} + \mathbf{k}$
12. $\mathbf{F} = (y + z)\mathbf{i} + (z + x)\mathbf{j} + (x + y)\mathbf{k}$

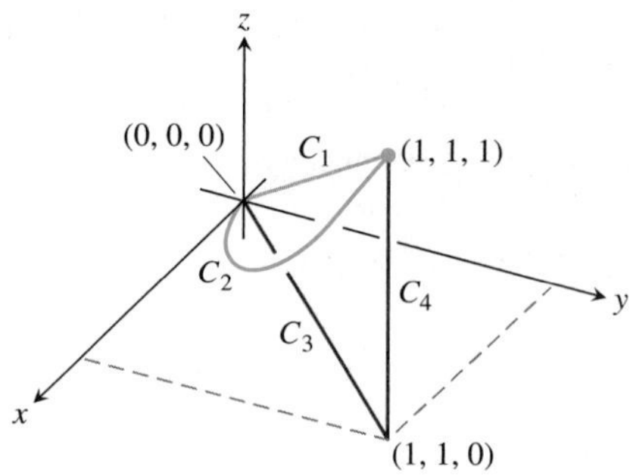

FIGURE 16.21 The paths from $(0, 0, 0)$ to $(1, 1, 1)$.

In Exercises 13–16, find the work done by $\mathbf{F}$ over the curve in the direction of increasing t.

13. $\mathbf{F} = xy\mathbf{i} + y\mathbf{j} - yz\mathbf{k}$
    $\mathbf{r}(t) = t\mathbf{i} + t^2\mathbf{j} + t\mathbf{k}$, $0 \le t \le 1$
14. $\mathbf{F} = 2y\mathbf{i} + 3x\mathbf{j} + (x + y)\mathbf{k}$
    $\mathbf{r}(t) = (\cos t)\mathbf{i} + (\sin t)\mathbf{j} + (t/6)\mathbf{k}$, $0 \le t \le 2\pi$
15. $\mathbf{F} = z\mathbf{i} + x\mathbf{j} + y\mathbf{k}$
    $\mathbf{r}(t) = (\sin t)\mathbf{i} + (\cos t)\mathbf{j} + t\mathbf{k}$, $0 \le t \le 2\pi$
16. $\mathbf{F} = 6z\mathbf{i} + y^2\mathbf{j} + 12x\mathbf{k}$
    $\mathbf{r}(t) = (\sin t)\mathbf{i} + (\cos t)\mathbf{j} + (t/6)\mathbf{k}$, $0 \le t \le 2\pi$

Line Integrals and Vector Fields in the Plane

17. Evaluate $\int_C xy\,dx + (x + y)\,dy$ along the curve $y = x^2$ from $(-1, 1)$ to $(2, 4)$.
18. Evaluate $\int_C (x - y)\,dx + (x + y)\,dy$ counterclockwise around the triangle with vertices $(0, 0)$, $(1, 0)$, and $(0, 1)$.

19. Evaluate $\int_C \mathbf{F}\cdot\mathbf{T}\,ds$ for the vector field $\mathbf{F} = x^2\mathbf{i} - y\mathbf{j}$ along the curve $x = y^2$ from $(4, 2)$ to $(1, -1)$.

20. Evaluate $\int_C \mathbf{F}\cdot d\mathbf{r}$ for the vector field $\mathbf{F} = y\mathbf{i} - x\mathbf{j}$ counterclockwise along the unit circle $x^2 + y^2 = 1$ from $(1, 0)$ to $(0, 1)$.

21. Work Find the work done by the force $\mathbf{F} = xy\mathbf{i} + (y - x)\mathbf{j}$ over the straight line from $(1, 1)$ to $(2, 3)$.

22. Work Find the work done by the gradient of $f(x, y) = (x + y)^2$ counterclockwise around the circle $x^2 + y^2 = 4$ from $(2, 0)$ to itself.

23. Circulation and flux Find the circulation and flux of the fields

$$\mathbf{F}_1 = x\mathbf{i} + y\mathbf{j} \quad \text{and} \quad \mathbf{F}_2 = -y\mathbf{i} + x\mathbf{j}$$

around and across each of the following curves.

a. The circle $\mathbf{r}(t) = (\cos t)\mathbf{i} + (\sin t)\mathbf{j}, \quad 0 \le t \le 2\pi$

b. The ellipse $\mathbf{r}(t) = (\cos t)\mathbf{i} + (4\sin t)\mathbf{j}, \quad 0 \le t \le 2\pi$

24. Flux across a circle Find the flux of the fields

$$\mathbf{F}_1 = 2x\mathbf{i} - 3y\mathbf{j} \quad \text{and} \quad \mathbf{F}_2 = 2x\mathbf{i} + (x - y)\mathbf{j}$$

across the circle

$$\mathbf{r}(t) = (a\cos t)\mathbf{i} + (a\sin t)\mathbf{j}, \quad 0 \le t \le 2\pi.$$

Circulation and Flux

In Exercises 25–28, find the circulation and flux of the field $\mathbf{F}$ around and across the closed semicircular path that consists of the semicircular arch $\mathbf{r}_1(t) = (a\cos t)\mathbf{i} + (a\sin t)\mathbf{j}$, $0 \le t \le \pi$, followed by the line segment $\mathbf{r}_2(t) = t\mathbf{i}$, $-a \le t \le a$.

25. $\mathbf{F} = x\mathbf{i} + y\mathbf{j}$

26. $\mathbf{F} = x^2\mathbf{i} + y^2\mathbf{j}$

27. $\mathbf{F} = -y\mathbf{i} + x\mathbf{j}$

28. $\mathbf{F} = -y^2\mathbf{i} + x^2\mathbf{j}$

29. Flow integrals Find the flow of the velocity field $\mathbf{F} = (x + y)\mathbf{i} - (x^2 + y^2)\mathbf{j}$ along each of the following paths from $(1, 0)$ to $(-1, 0)$ in the xy-plane.

a. The upper half of the circle $x^2 + y^2 = 1$

b. The line segment from $(1, 0)$ to $(-1, 0)$

c. The line segment from $(1, 0)$ to $(0, -1)$ followed by the line segment from $(0, -1)$ to $(-1, 0)$.

30. Flux across a triangle Find the flux of the field $\mathbf{F}$ in Exercise 29 outward across the triangle with vertices $(1, 0)$, $(0, 1)$, $(-1, 0)$.

Sketching and Finding Fields in the Plane

31. Spin field Draw the spin field

$$\mathbf{F} = -\frac{y}{\sqrt{x^2 + y^2}}\mathbf{i} + \frac{x}{\sqrt{x^2 + y^2}}\mathbf{j}$$

(see Figure 16.14) along with its horizontal and vertical components at a representative assortment of points on the circle $x^2 + y^2 = 4$.

32. Radial field Draw the radial field

$$\mathbf{F} = x\mathbf{i} + y\mathbf{j}$$

(see Figure 16.13) along with its horizontal and vertical components at a representative assortment of points on the circle $x^2 + y^2 = 1$.

33. A field of tangent vectors

a. Find a field $\mathbf{G} = P(x, y)\mathbf{i} + Q(x, y)\mathbf{j}$ in the xy-plane with the property that at any point $(a, b) \ne (0, 0)$, $\mathbf{G}$ is a vector of magnitude $\sqrt{a^2 + b^2}$ tangent to the circle $x^2 + y^2 = a^2 + b^2$ and pointing in the counterclockwise direction. (The field is undefined at $(0, 0)$.)

b. How is $\mathbf{G}$ related to the spin field $\mathbf{F}$ in Figure 16.14?

34. A field of tangent vectors

a. Find a field $\mathbf{G} = P(x, y)\mathbf{i} + Q(x, y)\mathbf{j}$ in the xy-plane with the property that at any point $(a, b) \ne (0, 0)$, $\mathbf{G}$ is a unit vector tangent to the circle $x^2 + y^2 = a^2 + b^2$ and pointing in the clockwise direction.

b. How is $\mathbf{G}$ related to the spin field $\mathbf{F}$ in Figure 16.14?

35. Unit vectors pointing toward the origin Find a field $\mathbf{F} = M(x, y)\mathbf{i} + N(x, y)\mathbf{j}$ in the xy-plane with the property that at each point $(x, y) \ne (0, 0)$, $\mathbf{F}$ is a unit vector pointing toward the origin. (The field is undefined at $(0, 0)$.)

36. Two "central" fields Find a field $\mathbf{F} = M(x, y)\mathbf{i} + N(x, y)\mathbf{j}$ in the xy-plane with the property that at each point $(x, y) \ne (0, 0)$, $\mathbf{F}$ points toward the origin and $|\mathbf{F}|$ is **(a)** the distance from (x, y) to the origin, **(b)** inversely proportional to the distance from (x, y) to the origin. (The field is undefined at $(0, 0)$.)

Flow Integrals in Space

In Exercises 37–40, $\mathbf{F}$ is the velocity field of a fluid flowing through a region in space. Find the flow along the given curve in the direction of increasing t.

37. $\mathbf{F} = -4xy\mathbf{i} + 8y\mathbf{j} + 2\mathbf{k}$
$\mathbf{r}(t) = t\mathbf{i} + t^2\mathbf{j} + \mathbf{k}, \quad 0 \le t \le 2$

38. $\mathbf{F} = x^2\mathbf{i} + yz\mathbf{j} + y^2\mathbf{k}$
$r(t) = 3t\mathbf{j} + 4t\mathbf{k}, \quad 0 \le t \le 1$

39. $\mathbf{F} = (x - z)\mathbf{i} + x\mathbf{k}$
$\mathbf{r}(t) = (\cos t)\mathbf{i} + (\sin t)\mathbf{k}, \quad 0 \le t \le \pi$

40. $\mathbf{F} = -y\mathbf{i} + x\mathbf{j} + 2\mathbf{k}$
$\mathbf{r}(t) = (-2\cos t)\mathbf{i} + (2\sin t)\mathbf{j} + 2t\mathbf{k}, \quad 0 \le t \le 2\pi$

41. Circulation Find the circulation of $\mathbf{F} = 2x\mathbf{i} + 2z\mathbf{j} + 2y\mathbf{k}$ around the closed path consisting of the following three curves traversed in the direction of increasing t:

C_1: $\mathbf{r}(t) = (\cos t)\mathbf{i} + (\sin t)\mathbf{j} + t\mathbf{k}, \quad 0 \le t \le \pi/2$

C_2: $\mathbf{r}(t) = \mathbf{j} + (\pi/2)(1 - t)\mathbf{k}, \quad 0 \le t \le 1$

C_3: $\mathbf{r}(t) = t\mathbf{i} + (1 - t)\mathbf{j}, \quad 0 \le t \le 1$

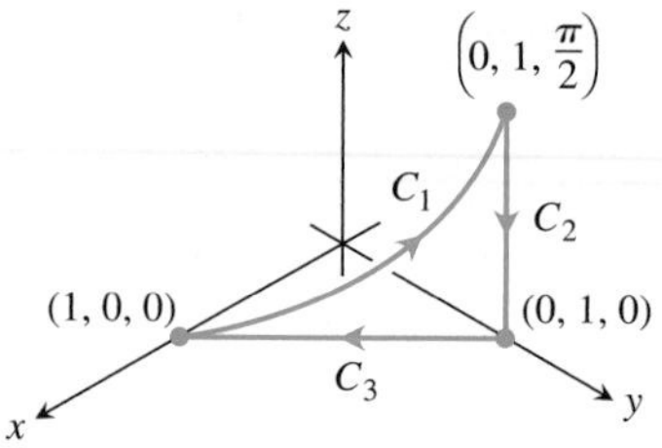

42. **Zero circulation** Let C be the ellipse in which the plane $2x + 3y - z = 0$ meets the cylinder $x^2 + y^2 = 12$. Show, without evaluating either line integral directly, that the circulation of the field $\mathbf{F} = x\mathbf{i} + y\mathbf{j} + z\mathbf{k}$ around C in either direction is zero.

43. **Flow along a curve** The field $\mathbf{F} = xy\mathbf{i} + y\mathbf{j} - yz\mathbf{k}$ is the velocity field of a flow in space. Find the flow from (0, 0, 0) to (1, 1, 1) along the curve of intersection of the cylinder $y = x^2$ and the plane $z = x$. (*Hint:* Use $t = x$ as the parameter.)

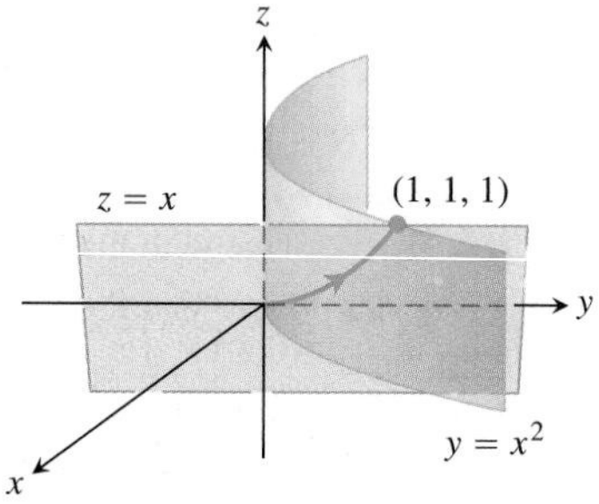

44. **Flow of a gradient field** Find the flow of the field $\mathbf{F} = \nabla(xy^2z^3)$:
   a. Once around the curve C in Exercise 42, clockwise as viewed from above
   b. Along the line segment from (1, 1, 1) to (2, 1, −1).

Theory and Examples

45. **Work and area** Suppose that $f(t)$ is differentiable and positive for $a \le t \le b$. Let C be the path $\mathbf{r}(t) = t\mathbf{i} + f(t)\mathbf{j}$, $a \le t \le b$, and $\mathbf{F} = y\mathbf{i}$. Is there any relation between the value of the work integral

$$\int_C \mathbf{F} \cdot d\mathbf{r}$$

and the area of the region bounded by the t-axis, the graph of f, and the lines $t = a$ and $t = b$? Give reasons for your answer.

46. **Work done by a radial force with constant magnitude** A particle moves along the smooth curve $y = f(x)$ from $(a, f(a))$ to $(b, f(b))$. The force moving the particle has constant magnitude k and always points away from the origin. Show that the work done by the force is

$$\int_C \mathbf{F} \cdot \mathbf{T}\, ds = k\left[(b^2 + (f(b))^2)^{1/2} - (a^2 + (f(a))^2)^{1/2}\right].$$

COMPUTER EXPLORATIONS

Finding Work Numerically

In Exercises 47–52, use a CAS to perform the following steps for finding the work done by force **F** over the given path:

a. Find $d\mathbf{r}$ for the path $\mathbf{r}(t) = g(t)\mathbf{i} + h(t)\mathbf{j} + k(t)\mathbf{k}$.
b. Evaluate the force **F** along the path.
c. Evaluate $\int_C \mathbf{F} \cdot d\mathbf{r}$.

47. $\mathbf{F} = xy^6\mathbf{i} + 3x(xy^5 + 2)\mathbf{j}$; $\mathbf{r}(t) = (2\cos t)\mathbf{i} + (\sin t)\mathbf{j}$, $0 \le t \le 2\pi$

48. $\mathbf{F} = \dfrac{3}{1 + x^2}\mathbf{i} + \dfrac{2}{1 + y^2}\mathbf{j}$; $\mathbf{r}(t) = (\cos t)\mathbf{i} + (\sin t)\mathbf{j}$, $0 \le t \le \pi$

49. $\mathbf{F} = (y + yz\cos xyz)\mathbf{i} + (x^2 + xz\cos xyz)\mathbf{j} + (z + xy\cos xyz)\mathbf{k}$; $\mathbf{r}(t) = (2\cos t)\mathbf{i} + (3\sin t)\mathbf{j} + \mathbf{k}$, $0 \le t \le 2\pi$

50. $\mathbf{F} = 2xy\mathbf{i} - y^2\mathbf{j} + ze^x\mathbf{k}$; $\mathbf{r}(t) = -t\mathbf{i} + \sqrt{t}\mathbf{j} + 3t\mathbf{k}$, $1 \le t \le 4$

51. $\mathbf{F} = (2y + \sin x)\mathbf{i} + (z^2 + (1/3)\cos y)\mathbf{j} + x^4\mathbf{k}$; $\mathbf{r}(t) = (\sin t)\mathbf{i} + (\cos t)\mathbf{j} + (\sin 2t)\mathbf{k}$, $-\pi/2 \le t \le \pi/2$

52. $\mathbf{F} = (x^2y)\mathbf{i} + \frac{1}{3}x^3\mathbf{j} + xy\mathbf{k}$; $\mathbf{r}(t) = (\cos t)\mathbf{i} + (\sin t)\mathbf{j} + (2\sin^2 t - 1)\mathbf{k}$, $0 \le t \le 2\pi$

16.3 Path Independence, Potential Functions, and Conservative Fields

In gravitational and electric fields, the amount of work it takes to move a mass or a charge from one point to another depends only on the object's initial and final positions and not on the path taken in between. This section discusses the notion of path independence of work integrals and describes the properties of fields in which work integrals are path independent. Work integrals are often easier to evaluate if they are path independent.

Path Independence

If A and B are two points in an open region D in space, the work $\int \mathbf{F} \cdot d\mathbf{r}$ done in moving a particle from A to B by a field $\mathbf{F}$ defined on D usually depends on the path taken. For some special fields, however, the integral's value is the same for all paths from A to B.

DEFINITIONS Path Independence, Conservative Field

Let $\mathbf{F}$ be a field defined on an open region D in space, and suppose that for any two points A and B in D the work $\int_A^B \mathbf{F} \cdot d\mathbf{r}$ done in moving from A to B is the same over all paths from A to B. Then the integral $\int \mathbf{F} \cdot d\mathbf{r}$ is **path independent in D** and the field $\mathbf{F}$ is **conservative on D.**

The word *conservative* comes from physics, where it refers to fields in which the principle of conservation of energy holds (it does, in conservative fields).

Under differentiability conditions normally met in practice, a field $\mathbf{F}$ is conservative if and only if it is the gradient field of a scalar function f; that is, if and only if $\mathbf{F} = \nabla f$ for some f. The function f then has a special name.

DEFINITION Potential Function

If $\mathbf{F}$ is a field defined on D and $\mathbf{F} = \nabla f$ for some scalar function f on D, then f is called a **potential function for F**.

An electric potential is a scalar function whose gradient field is an electric field. A gravitational potential is a scalar function whose gradient field is a gravitational field, and so on. As we will see, once we have found a potential function f for a field $\mathbf{F}$, we can evaluate all the work integrals in the domain of $\mathbf{F}$ over any path between A and B by

$$\int_A^B \mathbf{F} \cdot d\mathbf{r} = \int_A^B \nabla f \cdot d\mathbf{r} = f(B) - f(A). \tag{1}$$

If you think of ∇f for functions of several variables as being something like the derivative f' for functions of a single variable, then you see that Equation (1) is the vector calculus analogue of the Fundamental Theorem of Calculus formula

$$\int_a^b f'(x)\, dx = f(b) - f(a).$$

Conservative fields have other remarkable properties we will study as we go along. For example, saying that $\mathbf{F}$ is conservative on D is equivalent to saying that the integral of $\mathbf{F}$ around every closed path in D is zero. Naturally, certain conditions on the curves, fields, and domains must be satisfied for Equation (1) to be valid. We discuss these conditions below.

Assumptions in Effect from Now On: Connectivity and Simple Connectivity

We assume that all curves are **piecewise smooth**, that is, made up of finitely many smooth pieces connected end to end, as discussed in Section 13.1. We also assume that the components of **F** have continuous first partial derivatives. When $\mathbf{F} = \nabla f$, this continuity requirement guarantees that the mixed second derivatives of the potential function f are equal, a result we will find revealing in studying conservative fields **F**.

We assume D to be an *open* region in space. This means that every point in D is the center of an open ball that lies entirely in D. We assume D to be **connected**, which in an open region means that every point can be connected to every other point by a smooth curve that lies in the region. Finally, we assume D is **simply connected**, which means every loop in D can be contracted to a point in D without ever leaving D. (If D consisted of space with a line removed, for example, D would not be simply connected. There would be no way to contract a loop around the line to a point without leaving D.)

Connectivity and simple connectivity are not the same, and neither implies the other. Think of connected regions as being in "one piece" and simply connected regions as not having any "holes that catch loops." All of space itself is both connected and simply connected. Some of the results in this chapter can fail to hold if applied to domains where these conditions do not hold. For example, the component test for conservative fields, given later in this section, is not valid on domains that are not simply connected.

Line Integrals in Conservative Fields

The following result provides a convenient way to evaluate a line integral in a conservative field. The result establishes that the value of the integral depends only on the endpoints and not on the specific path joining them.

THEOREM 1 The Fundamental Theorem of Line Integrals

1. Let $\mathbf{F} = M\mathbf{i} + N\mathbf{j} + P\mathbf{k}$ be a vector field whose components are continuous throughout an open connected region D in space. Then there exists a differentiable function f such that

$$\mathbf{F} = \nabla f = \frac{\partial f}{\partial x}\mathbf{i} + \frac{\partial f}{\partial y}\mathbf{j} + \frac{\partial f}{\partial z}\mathbf{k}$$

if and only if for all points A and B in D the value of $\int_A^B \mathbf{F} \cdot d\mathbf{r}$ is independent of the path joining A to B in D.

2. If the integral is independent of the path from A to B, its value is

$$\int_A^B \mathbf{F} \cdot d\mathbf{r} = f(B) - f(A).$$

Proof that $\mathbf{F} = \nabla f$ Implies Path Independence of the Integral Suppose that A and B are two points in D and that C: $\mathbf{r}(t) = g(t)\mathbf{i} + h(t)\mathbf{j} + k(t)\mathbf{k}$, $a \le t \le b$, is a smooth curve in D joining A and B. Along the curve, f is a differentiable function of t and

$$\frac{df}{dt} = \frac{\partial f}{\partial x}\frac{dx}{dt} + \frac{\partial f}{\partial y}\frac{dy}{dt} + \frac{\partial f}{\partial z}\frac{dz}{dt} \qquad \text{Chain Rule with } x = g(t),\ y = h(t),\ z = k(t)$$

$$= \nabla f \cdot \left(\frac{dx}{dt}\mathbf{i} + \frac{dy}{dt}\mathbf{j} + \frac{dz}{dt}\mathbf{k}\right) = \nabla f \cdot \frac{d\mathbf{r}}{dt} = \mathbf{F} \cdot \frac{d\mathbf{r}}{dt}. \qquad \text{Because } \mathbf{F} = \nabla f$$

Therefore,

$$\int_C \mathbf{F} \cdot d\mathbf{r} = \int_{t=a}^{t=b} \mathbf{F} \cdot \frac{d\mathbf{r}}{dt}\, dt = \int_a^b \frac{df}{dt}\, dt$$

$$= f(g(t), h(t), k(t))\Big]_a^b = f(B) - f(A).$$

Thus, the value of the work integral depends only on the values of f at A and B and not on the path in between. This proves Part 2 as well as the forward implication in Part 1. We omit the more technical proof of the reverse implication. ∎

EXAMPLE 1 Finding Work Done by a Conservative Field

Find the work done by the conservative field

$$\mathbf{F} = yz\mathbf{i} + xz\mathbf{j} + xy\mathbf{k} = \nabla(xyz)$$

along any smooth curve C joining the point $A(-1, 3, 9)$ to $B(1, 6, -4)$.

Solution With $f(x, y, z) = xyz$, we have

$$\int_A^B \mathbf{F} \cdot d\mathbf{r} = \int_A^B \nabla f \cdot d\mathbf{r} \qquad \mathbf{F} = \nabla f$$

$$= f(B) - f(A) \qquad \text{Fundamental Theorem, Part 2}$$

$$= xyz|_{(1,6,-4)} - xyz|_{(-1,3,9)}$$

$$= (1)(6)(-4) - (-1)(3)(9)$$

$$= -24 + 27 = 3.$$ ∎

THEOREM 2 Closed-Loop Property of Conservative Fields

The following statements are equivalent.

1. $\oint \mathbf{F} \cdot d\mathbf{r} = 0$ around every closed loop in D.

2. The field $\mathbf{F}$ is conservative on D.

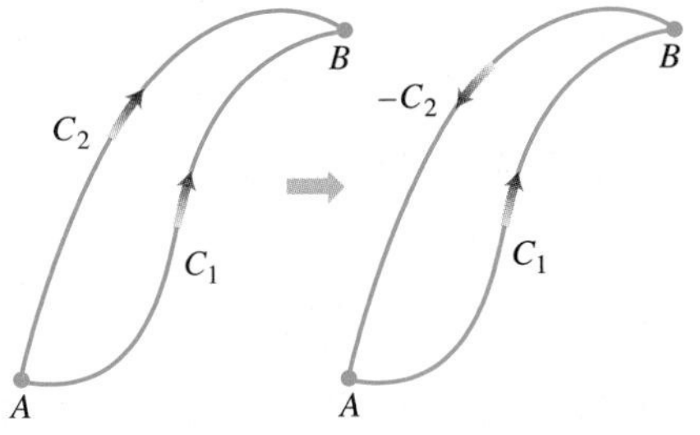

FIGURE 16.22 If we have two paths from A to B, one of them can be reversed to make a loop.

Proof that Part 1 ⇒ Part 2 We want to show that for any two points A and B in D, the integral of $\mathbf{F} \cdot d\mathbf{r}$ has the same value over any two paths C_1 and C_2 from A to B. We reverse the direction on C_2 to make a path $-C_2$ from B to A (Figure 16.22). Together, C_1 and $-C_2$ make a closed loop C, and

$$\int_{C_1} \mathbf{F} \cdot d\mathbf{r} - \int_{C_2} \mathbf{F} \cdot d\mathbf{r} = \int_{C_1} \mathbf{F} \cdot d\mathbf{r} + \int_{-C_2} \mathbf{F} \cdot d\mathbf{r} = \int_C \mathbf{F} \cdot d\mathbf{r} = 0.$$

Thus, the integrals over C_1 and C_2 give the same value. Note that the definition of line integral shows that changing the direction along a curve reverses the sign of the line integral.

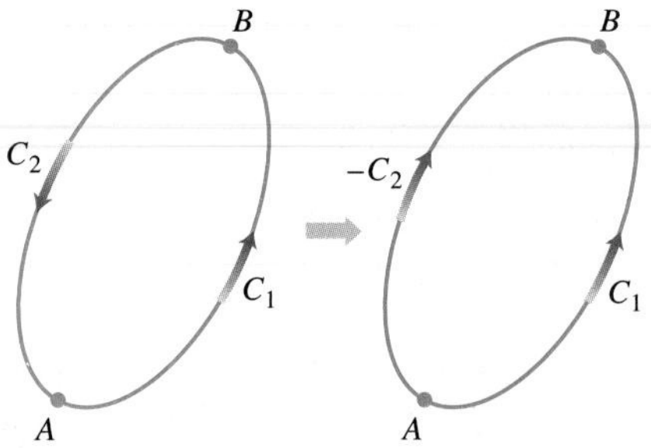

FIGURE 16.23 If A and B lie on a loop, we can reverse part of the loop to make two paths from A to B.

Proof that Part 2 $\Rightarrow$ Part 1 We want to show that the integral of $\mathbf{F}\cdot d\mathbf{r}$ is zero over any closed loop C. We pick two points A and B on C and use them to break C into two pieces: C_1 from A to B followed by C_2 from B back to A (Figure 16.23). Then

$$\oint_C \mathbf{F}\cdot d\mathbf{r} = \int_{C_1} \mathbf{F}\cdot d\mathbf{r} + \int_{C_2} \mathbf{F}\cdot d\mathbf{r} = \int_A^B \mathbf{F}\cdot d\mathbf{r} - \int_A^B \mathbf{F}\cdot d\mathbf{r} = 0. \qquad \blacksquare$$

The following diagram summarizes the results of Theorems 1 and 2.

	Theorem 1		Theorem 2	
$\mathbf{F} = \nabla f$ on D	$\Leftrightarrow$	$\mathbf{F}$ conservative on D	$\Leftrightarrow$	$\oint_C \mathbf{F}\cdot d\mathbf{r} = 0$ over any closed path in D

Now that we see how convenient it is to evaluate line integrals in conservative fields, two questions remain.

1. How do we know when a given field $\mathbf{F}$ is conservative?
2. If $\mathbf{F}$ is in fact conservative, how do we find a potential function f (so that $\mathbf{F} = \nabla f$)?

Finding Potentials for Conservative Fields

The test for being conservative is the following. Keep in mind our assumption that the domain of $\mathbf{F}$ is connected and simply connected.

> **Component Test for Conservative Fields**
> Let $\mathbf{F} = M(x, y, z)\mathbf{i} + N(x, y, z)\mathbf{j} + P(x, y, z)\mathbf{k}$ be a field whose component functions have continuous first partial derivatives. Then, $\mathbf{F}$ is conservative if and only if
>
> $$\frac{\partial P}{\partial y} = \frac{\partial N}{\partial z}, \quad \frac{\partial M}{\partial z} = \frac{\partial P}{\partial x}, \quad \text{and} \quad \frac{\partial N}{\partial x} = \frac{\partial M}{\partial y}. \qquad (2)$$

Proof that Equations (2) hold if F is conservative There is a potential function f such that

$$\mathbf{F} = M\mathbf{i} + N\mathbf{j} + P\mathbf{k} = \frac{\partial f}{\partial x}\mathbf{i} + \frac{\partial f}{\partial y}\mathbf{j} + \frac{\partial f}{\partial z}\mathbf{k}.$$

Hence,

$$\begin{aligned}\frac{\partial P}{\partial y} &= \frac{\partial}{\partial y}\left(\frac{\partial f}{\partial z}\right) = \frac{\partial^2 f}{\partial y\,\partial z} \\ &= \frac{\partial^2 f}{\partial z\,\partial y} \\ &= \frac{\partial}{\partial z}\left(\frac{\partial f}{\partial y}\right) = \frac{\partial N}{\partial z}.\end{aligned}$$

Continuity implies that the mixed partial derivatives are equal.

The others in Equations (2) are proved similarly. $\blacksquare$

The second half of the proof, that Equations (2) imply that $\mathbf{F}$ is conservative, is a consequence of Stokes' Theorem, taken up in Section 16.7, and requires our assumption that the domain of $\mathbf{F}$ be simply connected.

Once we know that **F** is conservative, we usually want to find a potential function for **F**. This requires solving the equation $\nabla f = \mathbf{F}$ or

$$\frac{\partial f}{\partial x}\mathbf{i} + \frac{\partial f}{\partial y}\mathbf{j} + \frac{\partial f}{\partial z}\mathbf{k} = M\mathbf{i} + N\mathbf{j} + P\mathbf{k}$$

for f. We accomplish this by integrating the three equations

$$\frac{\partial f}{\partial x} = M, \qquad \frac{\partial f}{\partial y} = N, \qquad \frac{\partial f}{\partial z} = P,$$

as illustrated in the next example.

EXAMPLE 2 Finding a Potential Function

Show that $\mathbf{F} = (e^x \cos y + yz)\mathbf{i} + (xz - e^x \sin y)\mathbf{j} + (xy + z)\mathbf{k}$ is conservative and find a potential function for it.

Solution We apply the test in Equations (2) to

$$M = e^x \cos y + yz, \qquad N = xz - e^x \sin y, \qquad P = xy + z$$

and calculate

$$\frac{\partial P}{\partial y} = x = \frac{\partial N}{\partial z}, \qquad \frac{\partial M}{\partial z} = y = \frac{\partial P}{\partial x}, \qquad \frac{\partial N}{\partial x} = -e^x \sin y + z = \frac{\partial M}{\partial y}.$$

Together, these equalities tell us that there is a function f with $\nabla f = \mathbf{F}$.

We find f by integrating the equations

$$\frac{\partial f}{\partial x} = e^x \cos y + yz, \qquad \frac{\partial f}{\partial y} = xz - e^x \sin y, \qquad \frac{\partial f}{\partial z} = xy + z. \tag{3}$$

We integrate the first equation with respect to x, holding y and z fixed, to get

$$f(x, y, z) = e^x \cos y + xyz + g(y, z).$$

We write the constant of integration as a function of y and z because its value may change if y and z change. We then calculate $\partial f/\partial y$ from this equation and match it with the expression for $\partial f/\partial y$ in Equations (3). This gives

$$-e^x \sin y + xz + \frac{\partial g}{\partial y} = xz - e^x \sin y,$$

so $\partial g/\partial y = 0$. Therefore, g is a function of z alone, and

$$f(x, y, z) = e^x \cos y + xyz + h(z).$$

We now calculate $\partial f/\partial z$ from this equation and match it to the formula for $\partial f/\partial z$ in Equations (3). This gives

$$xy + \frac{dh}{dz} = xy + z, \qquad \text{or} \qquad \frac{dh}{dz} = z,$$

so

$$h(z) = \frac{z^2}{2} + C.$$

Hence,

$$f(x, y, z) = e^x \cos y + xyz + \frac{z^2}{2} + C.$$

We have infinitely many potential functions of **F**, one for each value of C. ■

EXAMPLE 3 Showing That a Field Is Not Conservative

Show that $\mathbf{F} = (2x - 3)\mathbf{i} - z\mathbf{j} + (\cos z)\mathbf{k}$ is not conservative.

Solution We apply the component test in Equations (2) and find immediately that

$$\frac{\partial P}{\partial y} = \frac{\partial}{\partial y}(\cos z) = 0, \qquad \frac{\partial N}{\partial z} = \frac{\partial}{\partial z}(-z) = -1.$$

The two are unequal, so **F** is not conservative. No further testing is required. ■

Exact Differential Forms

As we see in the next section and again later on, it is often convenient to express work and circulation integrals in the "differential" form

$$\int_A^B M\,dx + N\,dy + P\,dz$$

mentioned in Section 16.2. Such integrals are relatively easy to evaluate if $M\,dx + N\,dy + P\,dz$ is the total differential of a function f. For then

$$\begin{aligned}\int_A^B M\,dx + N\,dy + P\,dz &= \int_A^B \frac{\partial f}{\partial x}dx + \frac{\partial f}{\partial y}dy + \frac{\partial f}{\partial z}dz \\ &= \int_A^B \nabla f \cdot d\mathbf{r} \\ &= f(B) - f(A). \qquad \text{Theorem 1}\end{aligned}$$

Thus,

$$\int_A^B df = f(B) - f(A),$$

just as with differentiable functions of a single variable.

DEFINITIONS **Exact Differential Form**

Any expression $M(x, y, z)\,dx + N(x, y, z)\,dy + P(x, y, z)\,dz$ is a **differential form**. A differential form is **exact** on a domain D in space if

$$M\,dx + N\,dy + P\,dz = \frac{\partial f}{\partial x}dx + \frac{\partial f}{\partial y}dy + \frac{\partial f}{\partial z}dz = df$$

for some scalar function f throughout D.

Notice that if $M\,dx + N\,dy + P\,dz = df$ on D, then $\mathbf{F} = M\mathbf{i} + N\mathbf{j} + P\mathbf{k}$ is the gradient field of f on D. Conversely, if $\mathbf{F} = \nabla f$, then the form $M\,dx + N\,dy + P\,dz$ is exact. The test for the form's being exact is therefore the same as the test for **F**'s being conservative.

Component Test for Exactness of $M\,dx + N\,dy + P\,dz$

The differential form $M\,dx + N\,dy + P\,dz$ is exact if and only if

$$\frac{\partial P}{\partial y} = \frac{\partial N}{\partial z}, \quad \frac{\partial M}{\partial z} = \frac{\partial P}{\partial x}, \quad \text{and} \quad \frac{\partial N}{\partial x} = \frac{\partial M}{\partial y}.$$

This is equivalent to saying that the field $\mathbf{F} = M\mathbf{i} + N\mathbf{j} + P\mathbf{k}$ is conservative.

EXAMPLE 4 Showing That a Differential Form Is Exact

Show that $y\,dx + x\,dy + 4\,dz$ is exact and evaluate the integral

$$\int_{(1,1,1)}^{(2,3,-1)} y\,dx + x\,dy + 4\,dz$$

over the line segment from $(1, 1, 1)$ to $(2, 3, -1)$.

Solution We let $M = y, N = x, P = 4$ and apply the Test for Exactness:

$$\frac{\partial P}{\partial y} = 0 = \frac{\partial N}{\partial z}, \quad \frac{\partial M}{\partial z} = 0 = \frac{\partial P}{\partial x}, \quad \frac{\partial N}{\partial x} = 1 = \frac{\partial M}{\partial y}.$$

These equalities tell us that $y\,dx + x\,dy + 4\,dz$ is exact, so

$$y\,dx + x\,dy + 4\,dz = df$$

for some function f, and the integral's value is $f(2, 3, -1) - f(1, 1, 1)$.

We find f up to a constant by integrating the equations

$$\frac{\partial f}{\partial x} = y, \quad \frac{\partial f}{\partial y} = x, \quad \frac{\partial f}{\partial z} = 4. \tag{4}$$

From the first equation we get

$$f(x, y, z) = xy + g(y, z).$$

The second equation tells us that

$$\frac{\partial f}{\partial y} = x + \frac{\partial g}{\partial y} = x, \quad \text{or} \quad \frac{\partial g}{\partial y} = 0.$$

Hence, g is a function of z alone, and

$$f(x, y, z) = xy + h(z).$$

The third of Equations (4) tells us that

$$\frac{\partial f}{\partial z} = 0 + \frac{dh}{dz} = 4, \quad \text{or} \quad h(z) = 4z + C.$$

Therefore,

$$f(x, y, z) = xy + 4z + C.$$

The value of the integral is

$$f(2, 3, -1) - f(1, 1, 1) = 2 + C - (5 + C) = -3.$$

■

EXERCISES 16.3

Testing for Conservative Fields

Which fields in Exercises 1–6 are conservative, and which are not?

1. $\mathbf{F} = yz\mathbf{i} + xz\mathbf{j} + xy\mathbf{k}$
2. $\mathbf{F} = (y \sin z)\mathbf{i} + (x \sin z)\mathbf{j} + (xy \cos z)\mathbf{k}$
3. $\mathbf{F} = y\mathbf{i} + (x + z)\mathbf{j} - y\mathbf{k}$
4. $\mathbf{F} = -y\mathbf{i} + x\mathbf{j}$
5. $\mathbf{F} = (z + y)\mathbf{i} + z\mathbf{j} + (y + x)\mathbf{k}$
6. $\mathbf{F} = (e^x \cos y)\mathbf{i} - (e^x \sin y)\mathbf{j} + z\mathbf{k}$

Finding Potential Functions

In Exercises 7–12, find a potential function f for the field $\mathbf{F}$.

7. $\mathbf{F} = 2x\mathbf{i} + 3y\mathbf{j} + 4z\mathbf{k}$
8. $\mathbf{F} = (y + z)\mathbf{i} + (x + z)\mathbf{j} + (x + y)\mathbf{k}$
9. $\mathbf{F} = e^{y+2z}(\mathbf{i} + x\mathbf{j} + 2x\mathbf{k})$
10. $\mathbf{F} = (y \sin z)\mathbf{i} + (x \sin z)\mathbf{j} + (xy \cos z)\mathbf{k}$
11. $\mathbf{F} = (\ln x + \sec^2(x + y))\mathbf{i} + \left(\sec^2(x + y) + \dfrac{y}{y^2 + z^2}\right)\mathbf{j} + \dfrac{z}{y^2 + z^2}\mathbf{k}$
12. $\mathbf{F} = \dfrac{y}{1 + x^2y^2}\mathbf{i} + \left(\dfrac{x}{1 + x^2y^2} + \dfrac{z}{\sqrt{1 - y^2z^2}}\right)\mathbf{j} + \left(\dfrac{y}{\sqrt{1 - y^2z^2}} + \dfrac{1}{z}\right)\mathbf{k}$

Evaluating Line Integrals

In Exercises 13–17, show that the differential forms in the integrals are exact. Then evaluate the integrals.

13. $\displaystyle\int_{(0,0,0)}^{(2,3,-6)} 2x\,dx + 2y\,dy + 2z\,dz$
14. $\displaystyle\int_{(1,1,2)}^{(3,5,0)} yz\,dx + xz\,dy + xy\,dz$
15. $\displaystyle\int_{(0,0,0)}^{(1,2,3)} 2xy\,dx + (x^2 - z^2)\,dy - 2yz\,dz$
16. $\displaystyle\int_{(0,0,0)}^{(3,3,1)} 2x\,dx - y^2\,dy - \frac{4}{1 + z^2}\,dz$
17. $\displaystyle\int_{(1,0,0)}^{(0,1,1)} \sin y \cos x\,dx + \cos y \sin x\,dy + dz$

Although they are not defined on all of space R^3, the fields associated with Exercises 18–22 have simply connected domains and the Component Test can be used to show they are conservative. Find a potential function for each field and evaluate the integrals as in Example 4.

18. $\displaystyle\int_{(0,2,1)}^{(1,\pi/2,2)} 2\cos y\,dx + \left(\frac{1}{y} - 2x \sin y\right)dy + \frac{1}{z}\,dz,\ y > 0$
19. $\displaystyle\int_{(1,1,1)}^{(1,2,3)} 3x^2\,dx + \frac{z^2}{y}\,dy + 2z \ln y\,dz,\ y > 0$
20. $\displaystyle\int_{(1,2,1)}^{(2,1,1)} (2x \ln y - yz)\,dx + \left(\frac{x^2}{y} - xz\right)dy - xy\,dz,\ y > 0$
21. $\displaystyle\int_{(1,1,1)}^{(2,2,2)} \frac{1}{y}\,dx + \left(\frac{1}{z} - \frac{x}{y^2}\right)dy - \frac{y}{z^2}\,dz,\ y > 0,\ z > 0$
22. $\displaystyle\int_{(-1,-1,-1)}^{(2,2,2)} \frac{2x\,dx + 2y\,dy + 2z\,dz}{x^2 + y^2 + z^2}$
23. **Revisiting Example 4** Evaluate the integral
$$\int_{(1,1,1)}^{(2,3,-1)} y\,dx + x\,dy + 4\,dz$$
from Example 4 by finding parametric equations for the line segment from $(1, 1, 1)$ to $(2, 3, -1)$ and evaluating the line integral of $\mathbf{F} = y\mathbf{i} + x\mathbf{j} + 4\mathbf{k}$ along the segment. Since $\mathbf{F}$ is conservative, the integral is independent of the path.
24. Evaluate
$$\int_C x^2\,dx + yz\,dy + (y^2/2)\,dz$$
along the line segment C joining $(0, 0, 0)$ to $(0, 3, 4)$.

Theory, Applications, and Examples

Independence of path Show that the values of the integrals in Exercises 25 and 26 do not depend on the path taken from A to B.

25. $\displaystyle\int_A^B z^2\,dx + 2y\,dy + 2xz\,dz$
26. $\displaystyle\int_A^B \frac{x\,dx + y\,dy + z\,dz}{\sqrt{x^2 + y^2 + z^2}}$

In Exercises 27 and 28, find a potential function for $\mathbf{F}$.

27. $\mathbf{F} = \dfrac{2x}{y}\mathbf{i} + \left(\dfrac{1 - x^2}{y^2}\right)\mathbf{j},\ y > 0$
28. $\mathbf{F} = (e^x \ln y)\mathbf{i} + \left(\dfrac{e^x}{y} + \sin z\right)\mathbf{j} + (y \cos z)\mathbf{k},\ y > 0$
29. **Work along different paths** Find the work done by $\mathbf{F} = (x^2 + y)\mathbf{i} + (y^2 + x)\mathbf{j} + ze^z\mathbf{k}$ over the following paths from $(1, 0, 0)$ to $(1, 0, 1)$.
    a. The line segment $x = 1, y = 0, 0 \le z \le 1$
    b. The helix $\mathbf{r}(t) = (\cos t)\mathbf{i} + (\sin t)\mathbf{j} + (t/2\pi)\mathbf{k},\ 0 \le t \le 2\pi$
    c. The x-axis from $(1, 0, 0)$ to $(0, 0, 0)$ followed by the parabola $z = x^2, y = 0$ from $(0, 0, 0)$ to $(1, 0, 1)$
30. **Work along different paths** Find the work done by $\mathbf{F} = e^{yz}\mathbf{i} + (xze^{yz} + z \cos y)\mathbf{j} + (xye^{yz} + \sin y)\mathbf{k}$ over the following paths from $(1, 0, 1)$ to $(1, \pi/2, 0)$.

a. The line segment $x = 1, y = \pi t/2, z = 1 - t, 0 \leq t \leq 1$

b. The line segment from $(1, 0, 1)$ to the origin followed by the line segment from the origin to $(1, \pi/2, 0)$

c. The line segment from $(1, 0, 1)$ to $(1, 0, 0)$, followed by the x-axis from $(1, 0, 0)$ to the origin, followed by the parabola $y = \pi x^2/2, z = 0$ from there to $(1, \pi/2, 0)$

31. Evaluating a work integral two ways Let $\mathbf{F} = \nabla(x^3y^2)$ and let C be the path in the xy-plane from $(-1, 1)$ to $(1, 1)$ that consists of the line segment from $(-1, 1)$ to $(0, 0)$ followed by the line segment from $(0, 0)$ to $(1, 1)$. Evaluate $\int_C \mathbf{F} \cdot d\mathbf{r}$ in two ways.

a. Find parametrizations for the segments that make up C and evaluate the integral.

b. Using $f(x, y) = x^3y^2$ as a potential function for $\mathbf{F}$.

32. Integral along different paths Evaluate $\int_C 2x \cos y\, dx - x^2 \sin y\, dy$ along the following paths C in the xy-plane.

a. The parabola $y = (x - 1)^2$ from $(1, 0)$ to $(0, 1)$

b. The line segment from $(-1, \pi)$ to $(1, 0)$

c. The x-axis from $(-1, 0)$ to $(1, 0)$

d. The astroid $\mathbf{r}(t) = (\cos^3 t)\mathbf{i} + (\sin^3 t)\mathbf{j}, 0 \leq t \leq 2\pi$, counterclockwise from $(1, 0)$ back to $(1, 0)$

33. a. Exact differential form How are the constants a, b, and c related if the following differential form is exact?

$$(ay^2 + 2czx)\, dx + y(bx + cz)\, dy + (ay^2 + cx^2)\, dz$$

b. Gradient field For what values of b and c will

$$\mathbf{F} = (y^2 + 2czx)\mathbf{i} + y(bx + cz)\mathbf{j} + (y^2 + cx^2)\mathbf{k}$$

be a gradient field?

34. Gradient of a line integral Suppose that $\mathbf{F} = \nabla f$ is a conservative vector field and

$$g(x, y, z) = \int_{(0,0,0)}^{(x,y,z)} \mathbf{F} \cdot d\mathbf{r}.$$

Show that $\nabla g = \mathbf{F}$.

35. Path of least work You have been asked to find the path along which a force field $\mathbf{F}$ will perform the least work in moving a particle between two locations. A quick calculation on your part shows $\mathbf{F}$ to be conservative. How should you respond? Give reasons for your answer.

36. A revealing experiment By experiment, you find that a force field $\mathbf{F}$ performs only half as much work in moving an object along path C_1 from A to B as it does in moving the object along path C_2 from A to B. What can you conclude about $\mathbf{F}$? Give reasons for your answer.

37. Work by a constant force Show that the work done by a constant force field $\mathbf{F} = a\mathbf{i} + b\mathbf{j} + c\mathbf{k}$ in moving a particle along any path from A to B is $W = \mathbf{F} \cdot \overrightarrow{AB}$.

38. Gravitational field

a. Find a potential function for the gravitational field

$$\mathbf{F} = -GmM \frac{x\mathbf{i} + y\mathbf{j} + z\mathbf{k}}{(x^2 + y^2 + z^2)^{3/2}} \qquad (G, m, \text{ and } M \text{ are constants}).$$

b. Let P_1 and P_2 be points at distance s_1 and s_2 from the origin. Show that the work done by the gravitational field in part (a) in moving a particle from P_1 to P_2 is

$$GmM\left(\frac{1}{s_2} - \frac{1}{s_1}\right).$$

16.4 Green's Theorem in the Plane

From Table 16.2 in Section 16.2, we know that every line integral $\int_C M\, dx + N\, dy$ can be written as a flow integral $\int_a^b \mathbf{F} \cdot \mathbf{T}\, ds$. If the integral is independent of path, so the field $\mathbf{F}$ is conservative (over a domain satisfying the basic assumptions), we can evaluate the integral easily from a potential function for the field. In this section we consider how to evaluate the integral if it is *not* associated with a conservative vector field, but is a flow or flux integral across a closed curve in the xy-plane. The means for doing so is a result known as Green's Theorem, which converts the line integral into a double integral over the region enclosed by the path.

We frame our discussion in terms of velocity fields of fluid flows because they are easy to picture. However, Green's Theorem applies to any vector field satisfying certain mathematical conditions. It does not depend for its validity on the field's having a particular physical interpretation.

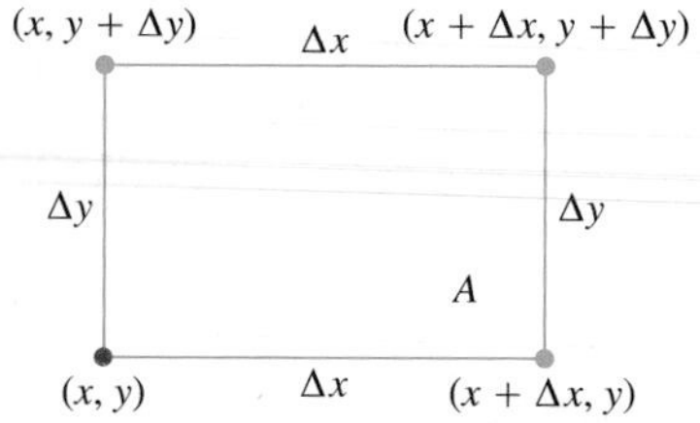

FIGURE 16.24 The rectangle for defining the divergence (flux density) of a vector field at a point (x, y).

Divergence

We need two new ideas for Green's Theorem. The first is the idea of the *divergence* of a vector field at a point, sometimes called the *flux density* of the vector field by physicists and engineers. We obtain it in the following way.

Suppose that $\mathbf{F}(x, y) = M(x, y)\mathbf{i} + N(x, y)\mathbf{j}$ is the velocity field of a fluid flow in the plane and that the first partial derivatives of M and N are continuous at each point of a region R. Let (x, y) be a point in R and let A be a small rectangle with one corner at (x, y) that, along with its interior, lies entirely in R (Figure 16.24). The sides of the rectangle, parallel to the coordinate axes, have lengths of Δx and Δy. The rate at which fluid leaves the rectangle across the bottom edge is approximately

$$\mathbf{F}(x, y) \cdot (-\mathbf{j})\, \Delta x = -N(x, y)\Delta x.$$

This is the scalar component of the velocity at (x, y) in the direction of the outward normal times the length of the segment. If the velocity is in meters per second, for example, the exit rate will be in meters per second times meters or square meters per second. The rates at which the fluid crosses the other three sides in the directions of their outward normals can be estimated in a similar way. All told, we have

Exit Rates:	Top:	$\mathbf{F}(x, y + \Delta y) \cdot \mathbf{j}\, \Delta x = N(x, y + \Delta y)\Delta x$
	Bottom:	$\mathbf{F}(x, y) \cdot (-\mathbf{j})\, \Delta x = -N(x, y)\Delta x$
	Right:	$\mathbf{F}(x + \Delta x, y) \cdot \mathbf{i}\, \Delta y = M(x + \Delta x, y)\Delta y$
	Left:	$\mathbf{F}(x, y) \cdot (-\mathbf{i})\, \Delta y = -M(x, y)\Delta y.$

Combining opposite pairs gives

Top and bottom: $$(N(x, y + \Delta y) - N(x, y))\Delta x \approx \left(\frac{\partial N}{\partial y}\Delta y\right)\Delta x$$

Right and left: $$(M(x + \Delta x, y) - M(x, y))\Delta y \approx \left(\frac{\partial M}{\partial x}\Delta x\right)\Delta y.$$

Adding these last two equations gives

$$\text{Flux across rectangle boundary} \approx \left(\frac{\partial M}{\partial x} + \frac{\partial N}{\partial y}\right)\Delta x \Delta y.$$

We now divide by $\Delta x \Delta y$ to estimate the total flux per unit area or flux density for the rectangle:

$$\frac{\text{Flux across rectangle boundary}}{\text{rectangle area}} \approx \left(\frac{\partial M}{\partial x} + \frac{\partial N}{\partial y}\right).$$

Finally, we let Δx and Δy approach zero to define what we call the *flux density* of $\mathbf{F}$ at the point (x, y). In mathematics, we call the flux density the *divergence* of $\mathbf{F}$. The symbol for it is div $\mathbf{F}$, pronounced "divergence of $\mathbf{F}$" or "div $\mathbf{F}$."

DEFINITION Divergence (Flux Density)

The **divergence (flux density)** of a vector field $\mathbf{F} = M\mathbf{i} + N\mathbf{j}$ at the point (x, y) is

$$\operatorname{div} \mathbf{F} = \frac{\partial M}{\partial x} + \frac{\partial N}{\partial y}. \tag{1}$$

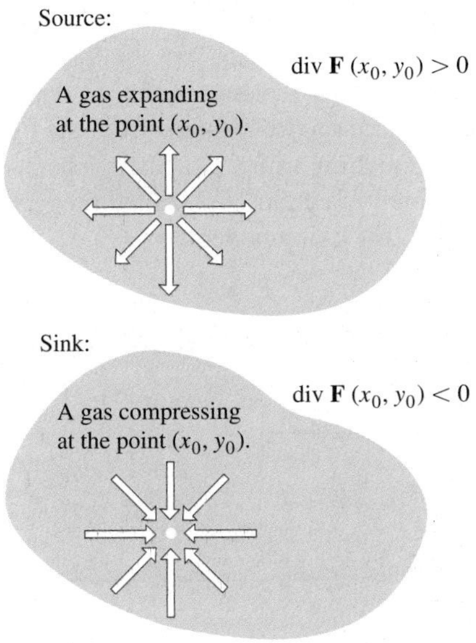

FIGURE 16.25 If a gas is expanding at a point (x_0, y_0), the lines of flow have positive divergence; if the gas is compressing, the divergence is negative.

Intuitively, if a gas is expanding at the point (x_0, y_0), the lines of flow would diverge there (hence the name) and, since the gas would be flowing out of a small rectangle about (x_0, y_0) the divergence of **F** at (x_0, y_0) would be positive. If the gas were compressing instead of expanding, the divergence would be negative (see Figure 16.25).

EXAMPLE 1 Finding Divergence

Find the divergence of $\mathbf{F}(x, y) = (x^2 - y)\mathbf{i} + (xy - y^2)\mathbf{j}$.

Solution We use the formula in Equation (1):

$$\begin{aligned}\text{div}\,\mathbf{F} &= \frac{\partial M}{\partial x} + \frac{\partial N}{\partial y} = \frac{\partial}{\partial x}(x^2 - y) + \frac{\partial}{\partial y}(xy - y^2)\\ &= 2x + x - 2y = 3x - 2y.\end{aligned}$$

■

Spin Around an Axis: The k-Component of Curl

The second idea we need for Green's Theorem has to do with measuring how a paddle wheel spins at a point in a fluid flowing in a plane region. This idea gives some sense of how the fluid is circulating around axes located at different points and perpendicular to the region. Physicists sometimes refer to this as the *circulation density* of a vector field **F** at a point. To obtain it, we return to the velocity field

$$\mathbf{F}(x, y) = M(x, y)\mathbf{i} + N(x, y)\mathbf{j}$$

and the rectangle A. The rectangle is redrawn here as Figure 16.26.

The counterclockwise circulation of **F** around the boundary of A is the sum of flow rates along the sides. For the bottom edge, the flow rate is approximately

$$\mathbf{F}(x, y) \cdot \mathbf{i}\,\Delta x = M(x, y)\Delta x.$$

This is the scalar component of the velocity $\mathbf{F}(x, y)$ in the direction of the tangent vector **i** times the length of the segment. The rates of flow along the other sides in the counterclockwise direction are expressed in a similar way. In all, we have

Top:	$\mathbf{F}(x, y + \Delta y) \cdot (-\mathbf{i})\,\Delta x = -M(x, y + \Delta y)\Delta x$
Bottom:	$\mathbf{F}(x, y) \cdot \mathbf{i}\,\Delta x = M(x, y)\Delta x$
Right:	$\mathbf{F}(x + \Delta x, y) \cdot \mathbf{j}\,\Delta y = N(x + \Delta x, y)\Delta y$
Left:	$\mathbf{F}(x, y) \cdot (-\mathbf{j})\,\Delta y = -N(x, y)\Delta y.$

We add opposite pairs to get

Top and bottom:

$$-(M(x, y + \Delta y) - M(x, y))\Delta x \approx -\left(\frac{\partial M}{\partial y}\Delta y\right)\Delta x$$

Right and left:

$$(N(x + \Delta x, y) - N(x, y))\Delta y \approx \left(\frac{\partial N}{\partial x}\Delta x\right)\Delta y.$$

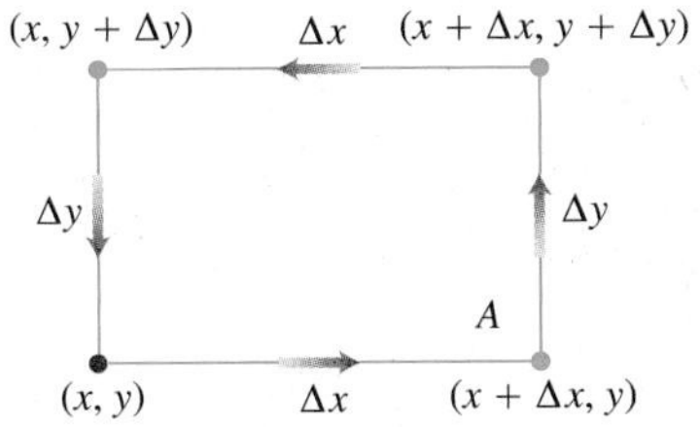

FIGURE 16.26 The rectangle for defining the curl (circulation density) of a vector field at a point (x, y).

Adding these last two equations and dividing by $\Delta x \Delta y$ gives an estimate of the circulation density for the rectangle:

$$\frac{\text{Circulation around rectangle}}{\text{rectangle area}} \approx \frac{\partial N}{\partial x} - \frac{\partial M}{\partial y}.$$

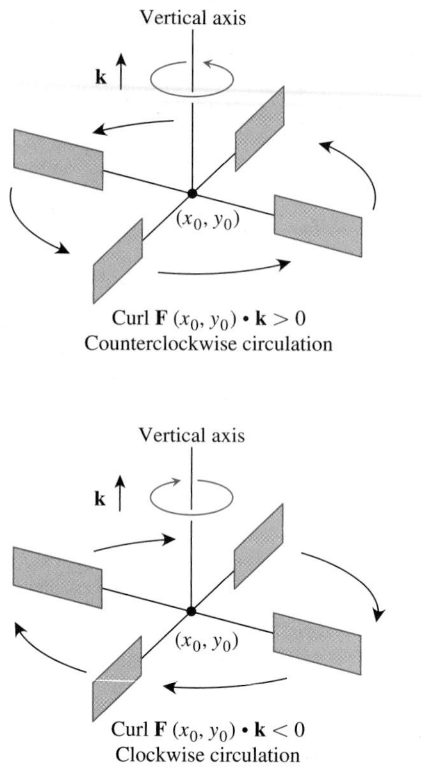

FIGURE 16.27 In the flow of an incompressible fluid over a plane region, the **k**-component of the curl measures the rate of the fluid's rotation at a point. The **k**-component of the curl is positive at points where the rotation is counterclockwise and negative where the rotation is clockwise.

We let Δx and Δy approach zero to define what we call the *circulation density* of **F** at the point (x, y).

The positive orientation of the circulation density for the plane is the *counterclockwise* rotation around the vertical axis, looking downward on the xy-plane from the tip of the (vertical) unit vector **k** (Figure 16.27). The circulation value is actually the **k**-component of a more general circulation vector we define in Section 16.7, called the *curl* of the vector field **F**. For Green's Theorem, we need only this **k**-component.

DEFINITION k-Component of Curl (Circulation Density)

The k-component of the curl (circulation density) of a vector field $\mathbf{F} = M\mathbf{i} + N\mathbf{j}$ at the point (x, y) is the scalar

$$(\text{curl } \mathbf{F}) \cdot \mathbf{k} = \frac{\partial N}{\partial x} - \frac{\partial M}{\partial y}. \tag{2}$$

If water is moving about a region in the xy-plane in a thin layer, then the **k**-component of the circulation, or curl, at a point (x_0, y_0) gives a way to measure how fast and in what direction a small paddle wheel will spin if it is put into the water at (x_0, y_0) with its axis perpendicular to the plane, parallel to **k** (Figure 16.27).

EXAMPLE 2 Finding the k-Component of the Curl

Find the **k**-component of the curl for the vector field

$$\mathbf{F}(x, y) = (x^2 - y)\mathbf{i} + (xy - y^2)\mathbf{j}.$$

Solution We use the formula in Equation (2):

$$(\text{curl } \mathbf{F}) \cdot \mathbf{k} = \frac{\partial N}{\partial x} - \frac{\partial M}{\partial y} = \frac{\partial}{\partial x}(xy - y^2) - \frac{\partial}{\partial y}(x^2 - y) = y + 1.$$ ■

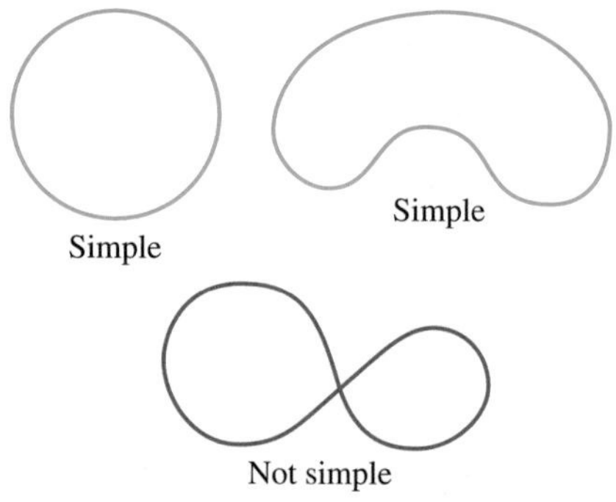

FIGURE 16.28 In proving Green's Theorem, we distinguish between two kinds of closed curves, simple and not simple. Simple curves do not cross themselves. A circle is simple but a figure 8 is not.

Two Forms for Green's Theorem

In one form, Green's Theorem says that under suitable conditions the outward flux of a vector field across a simple closed curve in the plane (Figure 16.28) equals the double integral of the divergence of the field over the region enclosed by the curve. Recall the formulas for flux in Equations (3) and (4) in Section 16.2.

THEOREM 3 Green's Theorem (Flux-Divergence or Normal Form)

The outward flux of a field $\mathbf{F} = M\mathbf{i} + N\mathbf{j}$ across a simple closed curve C equals the double integral of div **F** over the region R enclosed by C.

$$\underbrace{\oint_C \mathbf{F} \cdot \mathbf{n}\, ds}_{\text{Outward flux}} = \oint_C M\, dy - N\, dx = \underbrace{\iint_R \left(\frac{\partial M}{\partial x} + \frac{\partial N}{\partial y}\right) dx\, dy}_{\text{Divergence integral}} \tag{3}$$

In another form, Green's Theorem says that the counterclockwise circulation of a vector field around a simple closed curve is the double integral of the **k**-component of the curl of the field over the region enclosed by the curve. Recall the defining Equation (2) for circulation in Section 16.2.

THEOREM 4 Green's Theorem (Circulation-Curl or Tangential Form)
The counterclockwise circulation of a field $\mathbf{F} = M\mathbf{i} + N\mathbf{j}$ around a simple closed curve C in the plane equals the double integral of $(\operatorname{curl} \mathbf{F}) \cdot \mathbf{k}$ over the region R enclosed by C.

$$\underbrace{\oint_C \mathbf{F} \cdot \mathbf{T}\, ds = \oint_C M\,dx + N\,dy}_{\text{Counterclockwise circulation}} = \underbrace{\iint_R \left(\frac{\partial N}{\partial x} - \frac{\partial M}{\partial y}\right) dx\, dy}_{\text{Curl integral}} \tag{4}$$

The two forms of Green's Theorem are equivalent. Applying Equation (3) to the field $\mathbf{G}_1 = N\mathbf{i} - M\mathbf{j}$ gives Equation (4), and applying Equation (4) to $\mathbf{G}_2 = -N\mathbf{i} + M\mathbf{j}$ gives Equation (3).

Mathematical Assumptions

We need two kinds of assumptions for Green's Theorem to hold. First, we need conditions on M and N to ensure the existence of the integrals. The usual assumptions are that M, N, and their first partial derivatives are continuous at every point of some open region containing C and R. Second, we need geometric conditions on the curve C. It must be simple, closed, and made up of pieces along which we can integrate M and N. The usual assumptions are that C is piecewise smooth. The proof we give for Green's Theorem, however, assumes things about the shape of R as well. You can find proofs that are less restrictive in more advanced texts. First let's look at examples.

EXAMPLE 3 Supporting Green's Theorem

Verify both forms of Green's Theorem for the field

$$\mathbf{F}(x, y) = (x - y)\mathbf{i} + x\mathbf{j}$$

and the region R bounded by the unit circle

$$C: \quad \mathbf{r}(t) = (\cos t)\mathbf{i} + (\sin t)\mathbf{j}, \qquad 0 \le t \le 2\pi.$$

Solution We have

$$M = \cos t - \sin t, \qquad dx = d(\cos t) = -\sin t\, dt,$$

$$N = \cos t, \qquad dy = d(\sin t) = \cos t\, dt,$$

$$\frac{\partial M}{\partial x} = 1, \qquad \frac{\partial M}{\partial y} = -1, \qquad \frac{\partial N}{\partial x} = 1, \qquad \frac{\partial N}{\partial y} = 0.$$

The two sides of Equation (3) are

$$\oint_C M\,dy - N\,dx = \int_{t=0}^{t=2\pi} (\cos t - \sin t)(\cos t\,dt) - (\cos t)(-\sin t\,dt)$$

$$= \int_0^{2\pi} \cos^2 t\,dt = \pi$$

$$\iint_R \left(\frac{\partial M}{\partial x} + \frac{\partial N}{\partial y}\right) dx\,dy = \iint_R (1 + 0)\,dx\,dy$$

$$= \iint_R dx\,dy = \text{area inside the unit circle} = \pi.$$

The two sides of Equation (4) are

$$\oint_C M\,dx + N\,dy = \int_{t=0}^{t=2\pi} (\cos t - \sin t)(-\sin t\,dt) + (\cos t)(\cos t\,dt)$$

$$= \int_0^{2\pi} (-\sin t \cos t + 1)\,dt = 2\pi$$

$$\iint_R \left(\frac{\partial N}{\partial x} - \frac{\partial M}{dy}\right) dx\,dy = \iint_R (1 - (-1))\,dx\,dy = 2\iint_R dx\,dy = 2\pi.$$ ■

Using Green's Theorem to Evaluate Line Integrals

If we construct a closed curve C by piecing a number of different curves end to end, the process of evaluating a line integral over C can be lengthy because there are so many different integrals to evaluate. If C bounds a region R to which Green's Theorem applies, however, we can use Green's Theorem to change the line integral around C into one double integral over R.

EXAMPLE 4 Evaluating a Line Integral Using Green's Theorem

Evaluate the integral

$$\oint_C xy\,dy - y^2\,dx,$$

where C is the square cut from the first quadrant by the lines $x = 1$ and $y = 1$.

Solution We can use either form of Green's Theorem to change the line integral into a double integral over the square.

1. *With the Normal Form Equation* (3): Taking $M = xy$, $N = y^2$, and C and R as the square's boundary and interior gives

$$\oint_C xy\,dy - y^2\,dx = \iint_R (y + 2y)\,dx\,dy = \int_0^1\int_0^1 3y\,dx\,dy$$

$$= \int_0^1 \left[3xy\right]_{x=0}^{x=1} dy = \int_0^1 3y\,dy = \frac{3}{2}y^2\Big]_0^1 = \frac{3}{2}.$$

2. *With the Tangential Form Equation* (4): Taking $M = -y^2$ and $N = xy$ gives the same result:

$$\oint_C - y^2\,dx + xy\,dy = \iint_R (y - (-2y))\,dx\,dy = \frac{3}{2}.$$ ■

EXAMPLE 5 Finding Outward Flux

Calculate the outward flux of the field $\mathbf{F}(x, y) = x\mathbf{i} + y^2\mathbf{j}$ across the square bounded by the lines $x = \pm 1$ and $y = \pm 1$.

Solution Calculating the flux with a line integral would take four integrations, one for each side of the square. With Green's Theorem, we can change the line integral to one double integral. With $M = x$, $N = y^2$, C the square, and R the square's interior, we have

$$\begin{aligned}\text{Flux} &= \oint_C \mathbf{F}\cdot\mathbf{n}\,ds = \oint_C M\,dy - N\,dx \\ &= \iint_R \left(\frac{\partial M}{\partial x} + \frac{\partial N}{\partial y}\right) dx\,dy \qquad \text{Green's Theorem} \\ &= \int_{-1}^{1}\int_{-1}^{1} (1 + 2y)\,dx\,dy = \int_{-1}^{1} \Big[x + 2xy\Big]_{x=-1}^{x=1} dy \\ &= \int_{-1}^{1} (2 + 4y)\,dy = \Big[2y + 2y^2\Big]_{-1}^{1} = 4.\end{aligned}$$ ■

Proof of Green's Theorem for Special Regions

Let C be a smooth simple closed curve in the xy-plane with the property that lines parallel to the axes cut it in no more than two points. Let R be the region enclosed by C and suppose that M, N, and their first partial derivatives are continuous at every point of some open region containing C and R. We want to prove the circulation-curl form of Green's Theorem,

$$\oint_C M\,dx + N\,dy = \iint_R \left(\frac{\partial N}{\partial x} - \frac{\partial M}{\partial y}\right) dx\,dy. \tag{5}$$

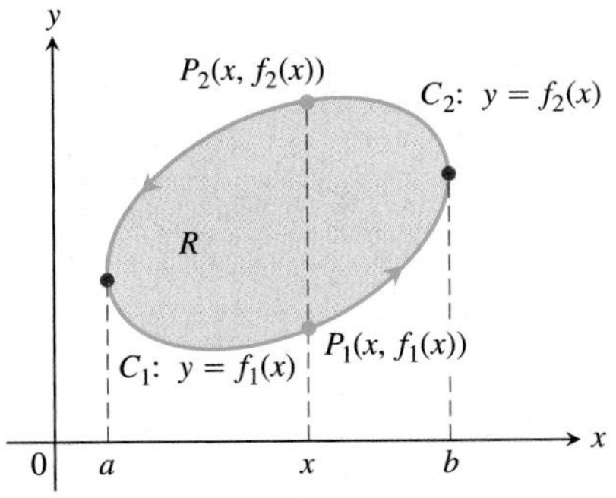

FIGURE 16.29 The boundary curve C is made up of C_1, the graph of $y = f_1(x)$, and C_2, the graph of $y = f_2(x)$.

Figure 16.29 shows C made up of two directed parts:

$$C_1:\quad y = f_1(x),\quad a \le x \le b, \qquad C_2:\quad y = f_2(x),\quad b \ge x \ge a.$$

For any x between a and b, we can integrate $\partial M/\partial y$ with respect to y from $y = f_1(x)$ to $y = f_2(x)$ and obtain

$$\int_{f_1(x)}^{f_2(x)} \frac{\partial M}{\partial y}\,dy = M(x, y)\Big]_{y=f_1(x)}^{y=f_2(x)} = M(x, f_2(x)) - M(x, f_1(x)).$$

We can then integrate this with respect to x from a to b:

$$\begin{aligned}\int_a^b \int_{f_1(x)}^{f_2(x)} \frac{\partial M}{\partial y}\, dy\, dx &= \int_a^b [M(x, f_2(x)) - M(x, f_1(x))]\, dx \\ &= -\int_b^a M(x, f_2(x))\, dx - \int_a^b M(x, f_1(x))\, dx \\ &= -\int_{C_2} M\, dx - \int_{C_1} M\, dx \\ &= -\oint_C M\, dx.\end{aligned}$$

Therefore

$$\oint_C M\, dx = \iint_R \left(-\frac{\partial M}{\partial y}\right) dx\, dy. \qquad (6)$$

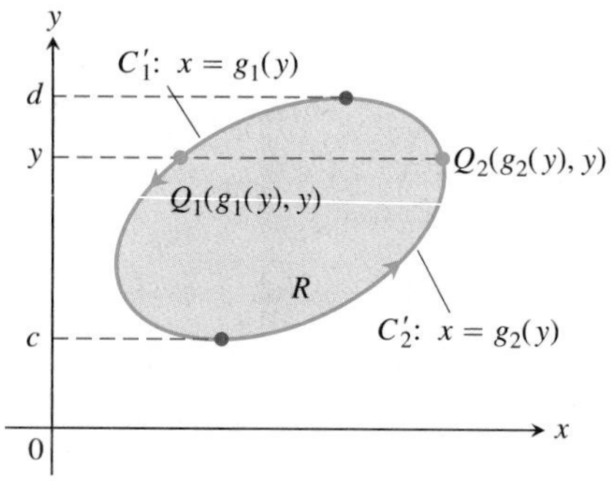

FIGURE 16.30 The boundary curve C is made up of C_1', the graph of $x = g_1(y)$, and C_2', the graph of $x = g_2(y)$.

Equation (6) is half the result we need for Equation (5). We derive the other half by integrating $\partial N/\partial x$ first with respect to x and then with respect to y, as suggested by Figure 16.30. This shows the curve C of Figure 16.29 decomposed into the two directed parts C_1': $x = g_1(y)$, $d \geq y \geq c$ and C_2': $x = g_2(y)$, $c \leq y \leq d$. The result of this double integration is

$$\oint_C N\, dy = \iint_R \frac{\partial N}{\partial x}\, dx\, dy. \qquad (7)$$

Summing Equations (6) and (7) gives Equation (5). This concludes the proof. ■

Extending the Proof to Other Regions

The argument we just gave does not apply directly to the rectangular region in Figure 16.31 because the lines $x = a, x = b, y = c$, and $y = d$ meet the region's boundary in more than two points. If we divide the boundary C into four directed line segments, however,

$$C_1:\quad y = c,\quad a \leq x \leq b, \qquad C_2:\quad x = b,\quad c \leq y \leq d$$

$$C_3:\quad y = d,\quad b \geq x \geq a, \qquad C_4:\quad x = a,\quad d \geq y \geq c,$$

we can modify the argument in the following way.

Proceeding as in the proof of Equation (7), we have

$$\begin{aligned}\int_c^d \int_a^b \frac{\partial N}{\partial x}\, dx\, dy &= \int_c^d (N(b, y) - N(a, y))\, dy \\ &= \int_c^d N(b, y)\, dy + \int_d^c N(a, y)\, dy \\ &= \int_{C_2} N\, dy + \int_{C_4} N\, dy. \qquad (8)\end{aligned}$$

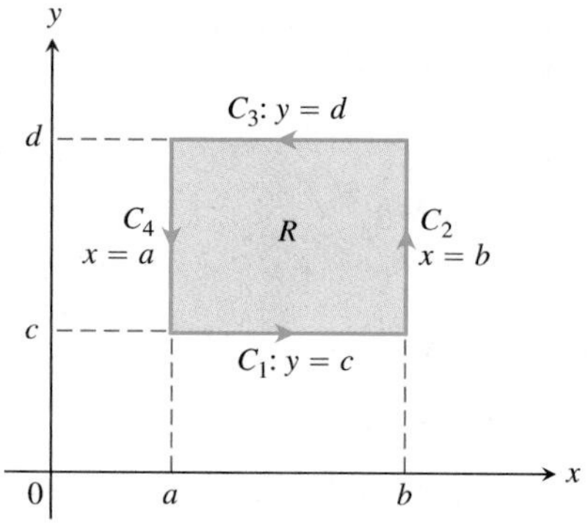

FIGURE 16.31 To prove Green's Theorem for a rectangle, we divide the boundary into four directed line segments.

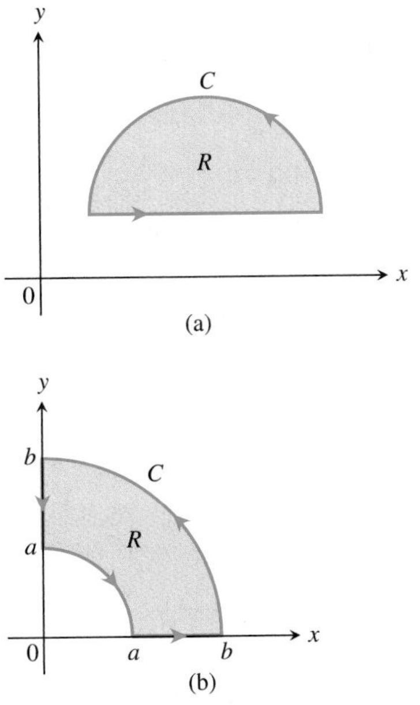

FIGURE 16.32 Other regions to which Green's Theorem applies.

Because y is constant along C_1 and C_3, $\int_{C_1} N\,dy = \int_{C_3} N\,dy = 0$, so we can add $\int_{C_1} N\,dy = \int_{C_3} N\,dy$ to the right-hand side of Equation (8) without changing the equality. Doing so, we have

$$\int_c^d \int_a^b \frac{\partial N}{\partial x}\,dx\,dy = \oint_C N\,dy. \tag{9}$$

Similarly, we can show that

$$\int_a^b \int_c^d \frac{\partial M}{\partial y}\,dy\,dx = -\oint_C M\,dx. \tag{10}$$

Subtracting Equation (10) from Equation (9), we again arrive at

$$\oint_C M\,dx + N\,dy = \iint_R \left(\frac{\partial N}{\partial x} - \frac{\partial M}{\partial y}\right) dx\,dy.$$

Regions like those in Figure 16.32 can be handled with no greater difficulty. Equation (5) still applies. It also applies to the horseshoe-shaped region R shown in Figure 16.33, as we see by putting together the regions R_1 and R_2 and their boundaries. Green's Theorem applies to C_1, R_1 and to C_2, R_2, yielding

$$\int_{C_1} M\,dx + N\,dy = \iint_{R_1} \left(\frac{\partial N}{\partial x} - \frac{\partial M}{\partial y}\right) dx\,dy$$

$$\int_{C_2} M\,dx + N\,dy = \iint_{R_2} \left(\frac{\partial N}{\partial x} - \frac{\partial M}{\partial y}\right) dx\,dy.$$

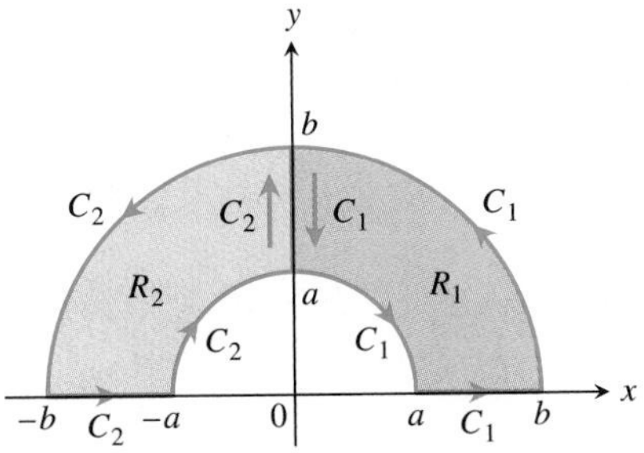

FIGURE 16.33 A region R that combines regions R_1 and R_2.

When we add these two equations, the line integral along the y-axis from b to a for C_1 cancels the integral over the same segment but in the opposite direction for C_2. Hence,

$$\oint_C M\,dx + N\,dy = \iint_R \left(\frac{\partial N}{\partial x} - \frac{\partial M}{\partial y}\right) dx\,dy,$$

where C consists of the two segments of the x-axis from $-b$ to $-a$ and from a to b and of the two semicircles, and where R is the region inside C.

The device of adding line integrals over separate boundaries to build up an integral over a single boundary can be extended to any finite number of subregions. In Figure 16.34a let C_1 be the boundary, oriented counterclockwise, of the region R_1 in the first quadrant. Similarly, for the other three quadrants, C_i is the boundary of the region R_i, $i = 2, 3, 4$. By Green's Theorem,

$$\oint_{C_i} M\,dx + N\,dy = \iint_{R_i} \left(\frac{\partial N}{\partial x} - \frac{\partial M}{\partial y}\right) dx\,dy. \tag{11}$$

We sum Equation (11) over $i = 1, 2, 3, 4$, and get (Figure 16.34b):

$$\oint_{r=b} (M\,dx + N\,dy) + \oint_{r=a} (M\,dx + N\,dy) = \iint_{\cup R_i} \left(\frac{\partial N}{\partial x} - \frac{\partial M}{\partial y}\right) dx\,dy. \tag{12}$$

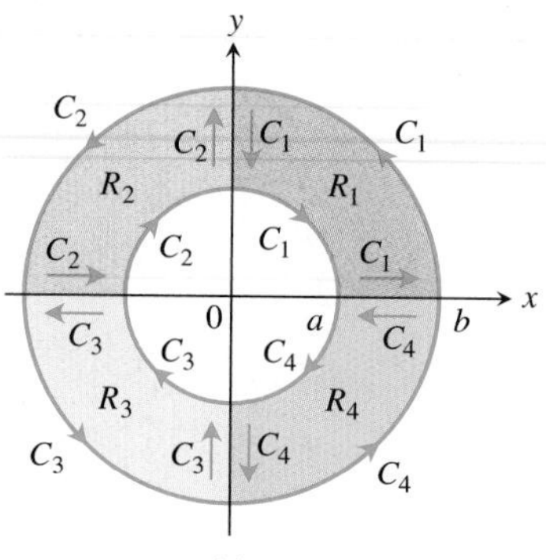

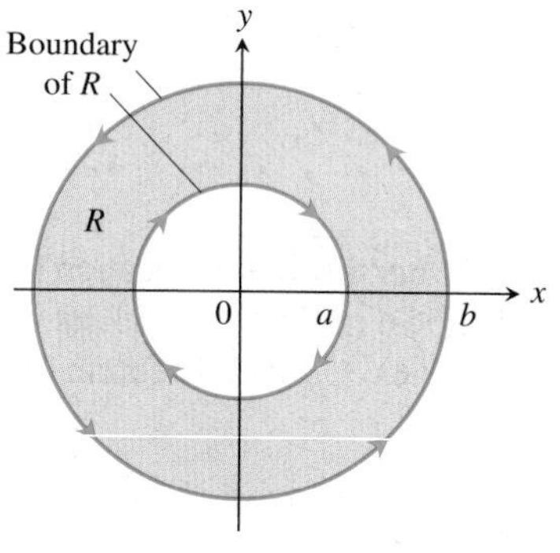

FIGURE 16.34 The annular region R combines four smaller regions. In polar coordinates, $r = a$ for the inner circle, $r = b$ for the outer circle, and $a \leq r \leq b$ for the region itself.

Equation (12) says that the double integral of $(\partial N/\partial x) - (\partial M/\partial y)$ over the annular ring R equals the line integral of $M\,dx + N\,dy$ over the complete boundary of R in the direction that keeps R on our left as we progress (Figure 16.34b).

EXAMPLE 6 Verifying Green's Theorem for an Annular Ring

Verify the circulation form of Green's Theorem (Equation 4) on the annular ring R: $h^2 \leq x^2 + y^2 \leq 1, 0 < h < 1$ (Figure 16.35), if

$$M = \frac{-y}{x^2 + y^2}, \qquad N = \frac{x}{x^2 + y^2}.$$

Here R is not simply connected.

Solution The boundary of R consists of the circle

$$C_1: \quad x = \cos t, \qquad y = \sin t, \qquad 0 \leq t \leq 2\pi,$$

traversed counterclockwise as t increases, and the circle

$$C_h: \quad x = h\cos\theta, \qquad y = -h\sin\theta, \qquad 0 \leq \theta \leq 2\pi,$$

traversed clockwise as θ increases. The functions M and N and their partial derivatives are continuous throughout R. Moreover,

$$\begin{aligned}\frac{\partial M}{\partial y} &= \frac{(x^2 + y^2)(-1) + y(2y)}{(x^2 + y^2)^2} \\ &= \frac{y^2 - x^2}{(x^2 + y^2)^2} = \frac{\partial N}{\partial x},\end{aligned}$$

so

$$\iint_R \left(\frac{\partial N}{\partial x} - \frac{\partial M}{\partial y}\right) dx\,dy = \iint_R 0\,dx\,dy = 0.$$

The integral of $M\,dx + N\,dy$ over the boundary of R is

$$\begin{aligned}\int_C M\,dx + N\,dy &= \oint_{C_1} \frac{x\,dy - y\,dx}{x^2 + y^2} + \oint_{C_h} \frac{x\,dy - y\,dx}{x^2 + y^2} \\ &= \int_0^{2\pi} (\cos^2 t + \sin^2 t)\,dt - \int_0^{2\pi} \frac{h^2(\cos^2\theta + \sin^2\theta)}{h^2}\,d\theta \\ &= 2\pi - 2\pi = 0.\end{aligned}$$

■

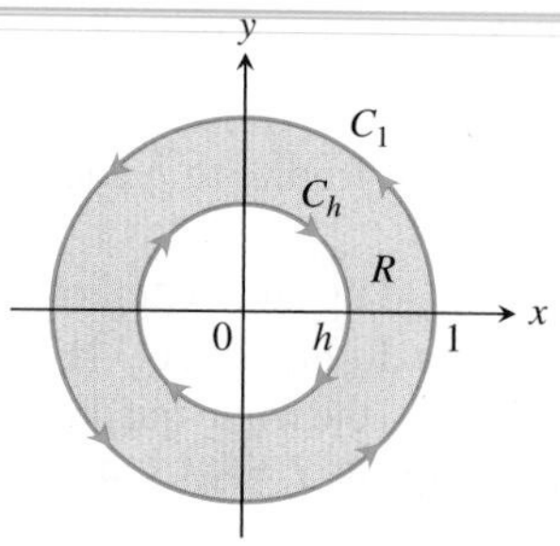

FIGURE 16.35 Green's Theorem may be applied to the annular region R by integrating along the boundaries as shown (Example 6).

The functions M and N in Example 6 are discontinuous at (0, 0), so we cannot apply Green's Theorem to the circle C_1 and the region inside it. We must exclude the origin. We do so by excluding the points interior to C_h.

We could replace the circle C_1 in Example 6 by an ellipse or any other simple closed curve K surrounding C_h (Figure 16.36). The result would still be

$$\oint_K (M\,dx + N\,dy) + \oint_{C_h} (M\,dx + N\,dy) = \iint_R \left(\frac{\partial N}{\partial x} - \frac{\partial M}{\partial y}\right) dy\,dx = 0,$$

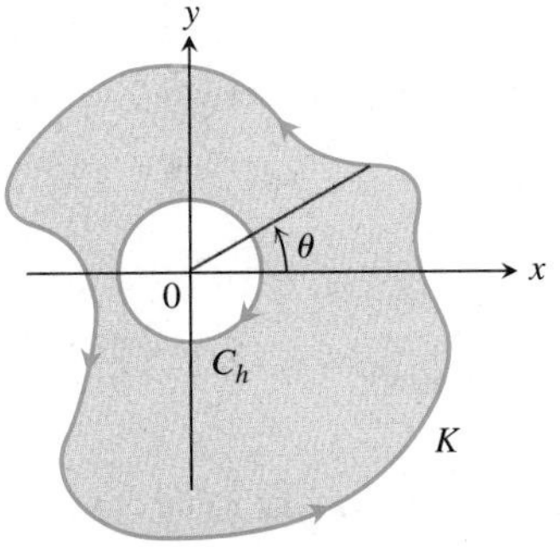

FIGURE 16.36 The region bounded by the circle C_h and the curve K.

which leads to the conclusion that

$$\oint_K (M\,dx + N\,dy) = 2\pi$$

for any such curve K. We can explain this result by changing to polar coordinates. With

$$x = r\cos\theta, \qquad y = r\sin\theta,$$
$$dx = -r\sin\theta\,d\theta + \cos\theta\,dr, \qquad dy = r\cos\theta\,d\theta + \sin\theta\,dr,$$

we have

$$\frac{x\,dy - y\,dx}{x^2 + y^2} = \frac{r^2(\cos^2\theta + \sin^2\theta)\,d\theta}{r^2} = d\theta,$$

and θ increases by 2π as we traverse K once counterclockwise.

EXERCISES 16.4

Verifying Green's Theorem

In Exercises 1–4, verify the conclusion of Green's Theorem by evaluating both sides of Equations (3) and (4) for the field $\mathbf{F} = M\mathbf{i} + N\mathbf{j}$. Take the domains of integration in each case to be the disk R: $x^2 + y^2 \le a^2$ and its bounding circle C: $\mathbf{r} = (a\cos t)\mathbf{i} + (a\sin t)\mathbf{j}$, $0 \le t \le 2\pi$.

1. $\mathbf{F} = -y\mathbf{i} + x\mathbf{j}$

2. $\mathbf{F} = y\mathbf{i}$

3. $\mathbf{F} = 2x\mathbf{i} - 3y\mathbf{j}$

4. $\mathbf{F} = -x^2y\mathbf{i} + xy^2\mathbf{j}$

Counterclockwise Circulation and Outward Flux

In Exercises 5–10, use Green's Theorem to find the counterclockwise circulation and outward flux for the field $\mathbf{F}$ and curve C.

5. $\mathbf{F} = (x - y)\mathbf{i} + (y - x)\mathbf{j}$

C: The square bounded by $x = 0, x = 1, y = 0, y = 1$

6. $\mathbf{F} = (x^2 + 4y)\mathbf{i} + (x + y^2)\mathbf{j}$

C: The square bounded by $x = 0, x = 1, y = 0, y = 1$

7. $\mathbf{F} = (y^2 - x^2)\mathbf{i} + (x^2 + y^2)\mathbf{j}$

C: The triangle bounded by $y = 0, x = 3$, and $y = x$

8. $\mathbf{F} = (x + y)\mathbf{i} - (x^2 + y^2)\mathbf{j}$

C: The triangle bounded by $y = 0, x = 1$, and $y = x$

9. $\mathbf{F} = (x + e^x\sin y)\mathbf{i} + (x + e^x\cos y)\mathbf{j}$

C: The right-hand loop of the lemniscate $r^2 = \cos 2\theta$

10. $\mathbf{F} = \left(\tan^{-1}\frac{y}{x}\right)\mathbf{i} + \ln(x^2 + y^2)\mathbf{j}$

C: The boundary of the region defined by the polar coordinate inequalities $1 \le r \le 2, 0 \le \theta \le \pi$

11. Find the counterclockwise circulation and outward flux of the field $\mathbf{F} = xy\mathbf{i} + y^2\mathbf{j}$ around and over the boundary of the region enclosed by the curves $y = x^2$ and $y = x$ in the first quadrant.

12. Find the counterclockwise circulation and the outward flux of the field $\mathbf{F} = (-\sin y)\mathbf{i} + (x\cos y)\mathbf{j}$ around and over the square cut from the first quadrant by the lines $x = \pi/2$ and $y = \pi/2$.

13. Find the outward flux of the field

$$\mathbf{F} = \left(3xy - \frac{x}{1 + y^2}\right)\mathbf{i} + (e^x + \tan^{-1}y)\mathbf{j}$$

across the cardioid $r = a(1 + \cos\theta)$, $a > 0$.

14. Find the counterclockwise circulation of $\mathbf{F} = (y + e^x\ln y)\mathbf{i} + (e^x/y)\mathbf{j}$ around the boundary of the region that is bounded above by the curve $y = 3 - x^2$ and below by the curve $y = x^4 + 1$.

Work

In Exercises 15 and 16, find the work done by $\mathbf{F}$ in moving a particle once counterclockwise around the given curve.

15. $\mathbf{F} = 2xy^3\mathbf{i} + 4x^2y^2\mathbf{j}$

C: The boundary of the "triangular" region in the first quadrant enclosed by the x-axis, the line $x = 1$, and the curve $y = x^3$

16. $\mathbf{F} = (4x - 2y)\mathbf{i} + (2x - 4y)\mathbf{j}$

C: The circle $(x - 2)^2 + (y - 2)^2 = 4$

Evaluating Line Integrals in the Plane

Apply Green's Theorem to evaluate the integrals in Exercises 17–20.

17. $\oint_C (y^2\,dx + x^2\,dy)$

C: The triangle bounded by $x = 0, x + y = 1, y = 0$

18. $\oint_C (3y\,dx + 2x\,dy)$

C: The boundary of $0 \le x \le \pi, 0 \le y \le \sin x$

19. $\oint_C (6y + x)\,dx + (y + 2x)\,dy$

C: The circle $(x - 2)^2 + (y - 3)^2 = 4$

20. $\oint_C (2x + y^2)\,dx + (2xy + 3y)\,dy$

C: Any simple closed curve in the plane for which Green's Theorem holds

Calculating Area with Green's Theorem

If a simple closed curve C in the plane and the region R it encloses satisfy the hypotheses of Green's Theorem, the area of R is given by

Green's Theorem Area Formula

$$\text{Area of } R = \frac{1}{2}\oint_C x\,dy - y\,dx \qquad (13)$$

The reason is that by Equation (3), run backward,

$$\text{Area of } R = \iint_R dy\,dx = \iint_R \left(\frac{1}{2} + \frac{1}{2}\right) dy\,dx$$

$$= \oint_C \frac{1}{2}x\,dy - \frac{1}{2}y\,dx.$$

Use the Green's Theorem area formula (Equation 13) to find the areas of the regions enclosed by the curves in Exercises 21–24.

21. The circle $\mathbf{r}(t) = (a\cos t)\mathbf{i} + (a\sin t)\mathbf{j}, \quad 0 \le t \le 2\pi$

22. The ellipse $\mathbf{r}(t) = (a\cos t)\mathbf{i} + (b\sin t)\mathbf{j}, \quad 0 \le t \le 2\pi$

23. The astroid $\mathbf{r}(t) = (\cos^3 t)\mathbf{i} + (\sin^3 t)\mathbf{j}, \quad 0 \le t \le 2\pi$

24. The curve $\mathbf{r}(t) = t^2\mathbf{i} + ((t^3/3) - t)\mathbf{j}, \quad -\sqrt{3} \le t \le \sqrt{3}$ (see accompanying figure).

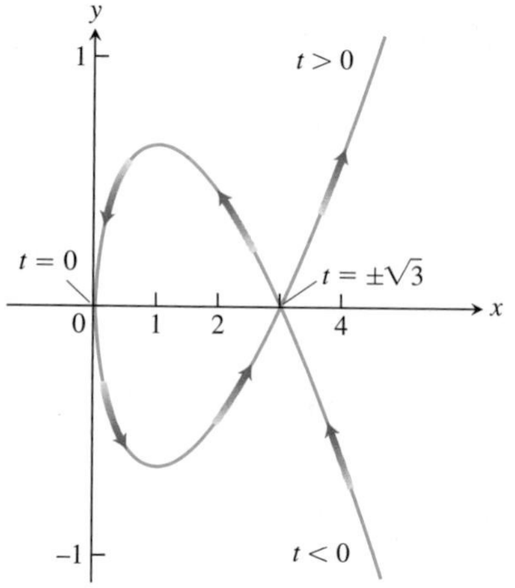

Theory and Examples

25. Let C be the boundary of a region on which Green's Theorem holds. Use Green's Theorem to calculate

a. $\oint_C f(x)\,dx + g(y)\,dy$

b. $\oint_C ky\,dx + hx\,dy$ (k and h constants).

26. Integral dependent only on area Show that the value of

$$\oint_C xy^2\,dx + (x^2y + 2x)\,dy$$

around any square depends only on the area of the square and not on its location in the plane.

27. What is special about the integral

$$\oint_C 4x^3y\,dx + x^4\,dy?$$

Give reasons for your answer.

28. What is special about the integral

$$\oint_C -y^3\,dy + x^3\,dx?$$

Give reasons for your answer.

29. Area as a line integral Show that if R is a region in the plane bounded by a piecewise-smooth simple closed curve C, then

$$\text{Area of } R = \oint_C x\,dy = -\oint_C y\,dx.$$

30. Definite integral as a line integral Suppose that a nonnegative function $y = f(x)$ has a continuous first derivative on $[a, b]$. Let C be the boundary of the region in the xy-plane that is bounded below by the x-axis, above by the graph of f, and on the sides by the lines $x = a$ and $x = b$. Show that

$$\int_a^b f(x)\,dx = -\oint_C y\,dx.$$

31. Area and the centroid Let A be the area and $\bar{x}$ the x-coordinate of the centroid of a region R that is bounded by a piecewise-smooth simple closed curve C in the xy-plane. Show that

$$\frac{1}{2}\oint_C x^2\,dy = -\oint_C xy\,dx = \frac{1}{3}\oint_C x^2\,dy - xy\,dx = A\bar{x}.$$

32. Moment of inertia Let I_y be the moment of inertia about the y-axis of the region in Exercise 31. Show that

$$\frac{1}{3}\oint_C x^3\,dy = -\oint_C x^2y\,dx = \frac{1}{4}\oint_C x^3\,dy - x^2y\,dx = I_y.$$

33. Green's Theorem and Laplace's equation Assuming that all the necessary derivatives exist and are continuous, show that if $f(x, y)$ satisfies the Laplace equation

$$\frac{\partial^2 f}{\partial x^2} + \frac{\partial^2 f}{\partial y^2} = 0,$$

then

$$\oint_C \frac{\partial f}{\partial y}\,dx - \frac{\partial f}{\partial x}\,dy = 0$$

for all closed curves C to which Green's Theorem applies. (The converse is also true: If the line integral is always zero, then f satisfies the Laplace equation.)

34. Maximizing work Among all smooth simple closed curves in the plane, oriented counterclockwise, find the one along which the work done by

$$\mathbf{F} = \left(\frac{1}{4}x^2y + \frac{1}{3}y^3\right)\mathbf{i} + x\mathbf{j}$$

is greatest. (*Hint:* Where is (curl $\mathbf{F}$) · $\mathbf{k}$ positive?)

35. Regions with many holes Green's Theorem holds for a region R with any finite number of holes as long as the bounding curves are smooth, simple, and closed and we integrate over each component of the boundary in the direction that keeps R on our immediate left as we go along (Figure 16.37).

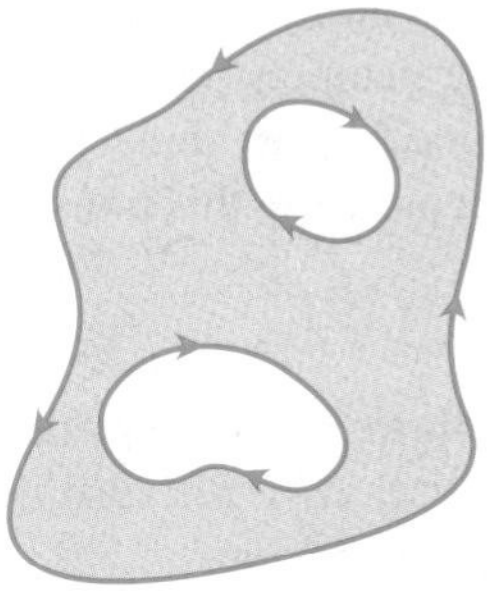

FIGURE 16.37 Green's Theorem holds for regions with more than one hole (Exercise 35).

a. Let $f(x, y) = \ln(x^2 + y^2)$ and let C be the circle $x^2 + y^2 = a^2$. Evaluate the flux integral

$$\oint_C \nabla f \cdot \mathbf{n}\,ds.$$

b. Let K be an arbitrary smooth simple closed curve in the plane that does not pass through (0, 0). Use Green's Theorem to show that

$$\oint_K \nabla f \cdot \mathbf{n}\,ds$$

has two possible values, depending on whether (0, 0) lies inside K or outside K.

36. Bendixson's criterion The *streamlines* of a planar fluid flow are the smooth curves traced by the fluid's individual particles. The vectors $\mathbf{F} = M(x, y)\mathbf{i} + N(x, y)\mathbf{j}$ of the flow's velocity field are the tangent vectors of the streamlines. Show that if the flow takes place over a simply connected region R (no holes or missing points) and that if $M_x + N_y \neq 0$ throughout R, then none of the streamlines in R is closed. In other words, no particle of fluid ever has a closed trajectory in R. The criterion $M_x + N_y \neq 0$ is called **Bendixson's criterion** for the nonexistence of closed trajectories.

37. Establish Equation (7) to finish the proof of the special case of Green's Theorem.

38. Establish Equation (10) to complete the argument for the extension of Green's Theorem.

39. Curl component of conservative fields Can anything be said about the curl component of a conservative two-dimensional vector field? Give reasons for your answer.

40. Circulation of conservative fields Does Green's Theorem give any information about the circulation of a conservative field? Does this agree with anything else you know? Give reasons for your answer.

COMPUTER EXPLORATIONS

Finding Circulation

In Exercises 41–44, use a CAS and Green's Theorem to find the counterclockwise circulation of the field $\mathbf{F}$ around the simple closed curve C. Perform the following CAS steps.

a. Plot C in the xy-plane.

b. Determine the integrand $(\partial N/\partial x) - (\partial M/\partial y)$ for the curl form of Green's Theorem.

c. Determine the (double integral) limits of integration from your plot in part (a) and evaluate the curl integral for the circulation.

41. $\mathbf{F} = (2x - y)\mathbf{i} + (x + 3y)\mathbf{j}$, C: The ellipse $x^2 + 4y^2 = 4$

42. $\mathbf{F} = (2x^3 - y^3)\mathbf{i} + (x^3 + y^3)\mathbf{j}$, C: The ellipse $\dfrac{x^2}{4} + \dfrac{y^2}{9} = 1$

43. $\mathbf{F} = x^{-1}e^y\mathbf{i} + (e^y \ln x + 2x)\mathbf{j}$,
C: The boundary of the region defined by $y = 1 + x^4$ (below) and $y = 2$ (above)

44. $\mathbf{F} = xe^y\mathbf{i} + 4x^2 \ln y\,\mathbf{j}$,
C: The triangle with vertices (0, 0), (2, 0), and (0, 4)

16.5 Surface Area and Surface Integrals

We know how to integrate a function over a flat region in a plane, but what if the function is defined over a curved surface? To evaluate one of these so-called surface integrals, we rewrite it as a double integral over a region in a coordinate plane beneath the surface (Figure 16.38). Surface integrals are used to compute quantities such as the flow of liquid across a membrane or the upward force on a falling parachute.

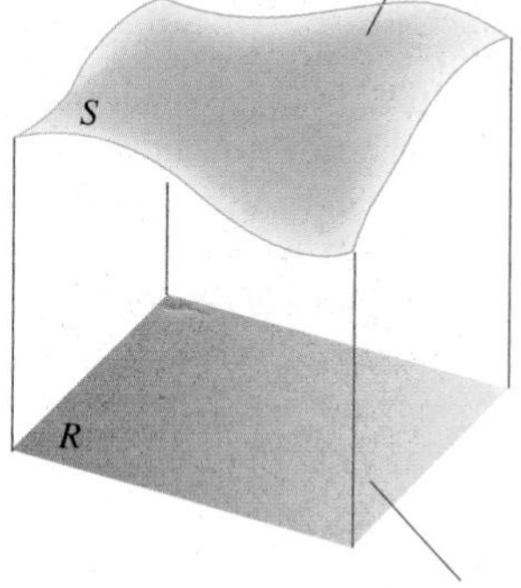

FIGURE 16.38 As we soon see, the integral of a function $g(x, y, z)$ over a surface S in space can be calculated by evaluating a related double integral over the vertical projection or "shadow" of S on a coordinate plane.

Surface Area

Figure 16.39 shows a surface S lying above its "shadow" region R in a plane beneath it. The surface is defined by the equation $f(x, y, z) = c$. If the surface is **smooth** (∇f is continuous and never vanishes on S), we can define and calculate its area as a double integral over R. We assume that this projection of the surface onto its shadow R is one-to-one. That is, each point in R corresponds to exactly one point (x, y, z) satisfying $f(x, y, z) = c$.

The first step in defining the area of S is to partition the region R into small rectangles ΔA_k of the kind we would use if we were defining an integral over R. Directly above each ΔA_k lies a patch of surface $\Delta\sigma_k$ that we may approximate by a parallelogram ΔP_k in the tangent plane to S at a point $T_k(x_k, y_k, z_k)$ in $\Delta\sigma_k$. This parallelogram in the tangent plane projects directly onto ΔA_k. To be specific, we choose the point $T_k(x_k, y_k, z_k)$ lying directly above the back corner C_k of ΔA_k, as shown in Figure 16.39. If the tangent plane is parallel to R, then ΔP_k will be congruent to ΔA_k. Otherwise, it will be a parallelogram whose area is somewhat larger than the area of ΔA_k.

Figure 16.40 gives a magnified view of $\Delta\sigma_k$ and ΔP_k, showing the gradient vector $\nabla f(x_k, y_k, z_k)$ at T_k and a unit vector $\mathbf{p}$ that is normal to R. The figure also shows the angle γ_k between ∇f and $\mathbf{p}$. The other vectors in the picture, $\mathbf{u}_k$ and $\mathbf{v}_k$, lie along the edges of the patch ΔP_k in the tangent plane. Thus, both $\mathbf{u}_k \times \mathbf{v}_k$ and ∇f are normal to the tangent plane.

We now need to know from advanced vector geometry that $|(\mathbf{u}_k \times \mathbf{v}_k) \cdot \mathbf{p}|$ is the area of the projection of the parallelogram determined by $\mathbf{u}_k$ and $\mathbf{v}_k$ onto any plane whose normal is $\mathbf{p}$. (A proof is given in Appendix 8.) In our case, this translates into the statement

$$|(\mathbf{u}_k \times \mathbf{v}_k) \cdot \mathbf{p}| = \Delta A_k.$$

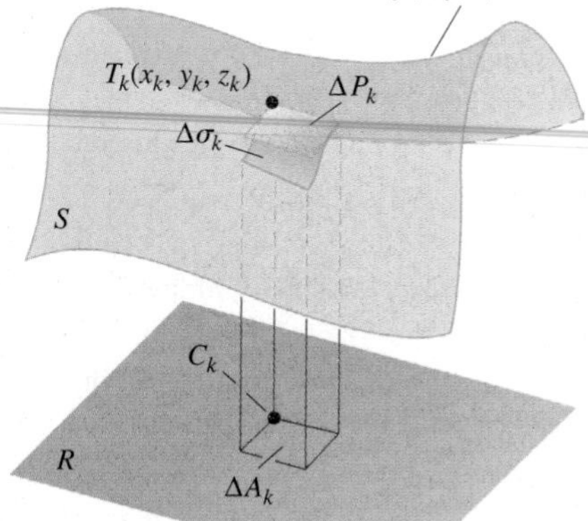

FIGURE 16.39 A surface S and its vertical projection onto a plane beneath it. You can think of R as the shadow of S on the plane. The tangent plane ΔP_k approximates the surface patch $\Delta\sigma_k$ above ΔA_k.

To simplify the notation in the derivation that follows, we are now denoting the *area* of the small rectangular region by ΔA_k as well. Likewise, ΔP_k will also denote the area of the portion of the tangent plane directly above this small region.

Now, $|\mathbf{u}_k \times \mathbf{v}_k|$ itself is the area ΔP_k (standard fact about cross products) so this last equation becomes

$$\underbrace{|\mathbf{u}_k \times \mathbf{v}_k|}_{\Delta P_k}\ \underbrace{|\mathbf{p}|}_{1}\ \underbrace{|\cos(\text{angle between } \mathbf{u}_k \times \mathbf{v}_k \text{ and } \mathbf{p})|}_{\substack{\text{Same as } |\cos\gamma_k| \text{ because } \nabla f \text{ and } \mathbf{u}_k \times \mathbf{v}_k \\ \text{are both normal to the tangent plane}}} = \Delta A_k$$

or

$$\Delta P_k |\cos\gamma_k| = \Delta A_k$$

or

$$\Delta P_k = \frac{\Delta A_k}{|\cos\gamma_k|},$$

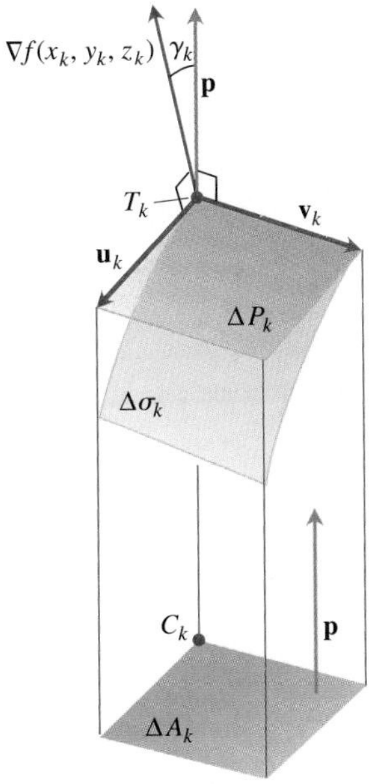

FIGURE 16.40 Magnified view from the preceding figure. The vector $\mathbf{u}_k \times \mathbf{v}_k$ (not shown) is parallel to the vector ∇f because both vectors are normal to the plane of ΔP_k.

provided $\cos \gamma_k \neq 0$. We will have $\cos \gamma_k \neq 0$ as long as ∇f is not parallel to the ground plane and $\nabla f \cdot \mathbf{p} \neq 0$.

Since the patches ΔP_k approximate the surface patches $\Delta \sigma_k$ that fit together to make S, the sum

$$\sum \Delta P_k = \sum \frac{\Delta A_k}{|\cos \gamma_k|} \tag{1}$$

looks like an approximation of what we might like to call the surface area of S. It also looks as if the approximation would improve if we refined the partition of R. In fact, the sums on the right-hand side of Equation (1) are approximating sums for the double integral

$$\iint_R \frac{1}{|\cos \gamma|}\, dA. \tag{2}$$

We therefore define the **area** of S to be the value of this integral whenever it exists. For any surface $f(x, y, z) = c$, we have $|\nabla f \cdot \mathbf{p}| = |\nabla f|\,|\mathbf{p}|\,|\cos \gamma|$, so

$$\frac{1}{|\cos \gamma|} = \frac{|\nabla f|}{|\nabla f \cdot \mathbf{p}|}.$$

This combines with Equation (2) to give a practical formula for surface area.

Formula for Surface Area

The area of the surface $f(x, y, z) = c$ over a closed and bounded plane region R is

$$\text{Surface area} = \iint_R \frac{|\nabla f|}{|\nabla f \cdot \mathbf{p}|}\, dA, \tag{3}$$

where $\mathbf{p}$ is a unit vector normal to R and $\nabla f \cdot \mathbf{p} \neq 0$.

Thus, the area is the double integral over R of the magnitude of ∇f divided by the magnitude of the scalar component of ∇f normal to R.

We reached Equation (3) under the assumption that $\nabla f \cdot \mathbf{p} \neq 0$ throughout R and that ∇f is continuous. Whenever the integral exists, however, we define its value to be the area of the portion of the surface $f(x, y, z) = c$ that lies over R. (Recall that the projection is assumed to be one-to-one.)

In the exercises (see Equation 11), we show how Equation (3) simplifies if the surface is defined by $z = f(x, y)$.

EXAMPLE 1 Finding Surface Area

Find the area of the surface cut from the bottom of the paraboloid $x^2 + y^2 - z = 0$ by the plane $z = 4$.

Solution We sketch the surface S and the region R below it in the xy-plane (Figure 16.41). The surface S is part of the level surface $f(x, y, z) = x^2 + y^2 - z = 0$, and R is the disk $x^2 + y^2 \leq 4$ in the xy-plane. To get a unit vector normal to the plane of R, we can take $\mathbf{p} = \mathbf{k}$.

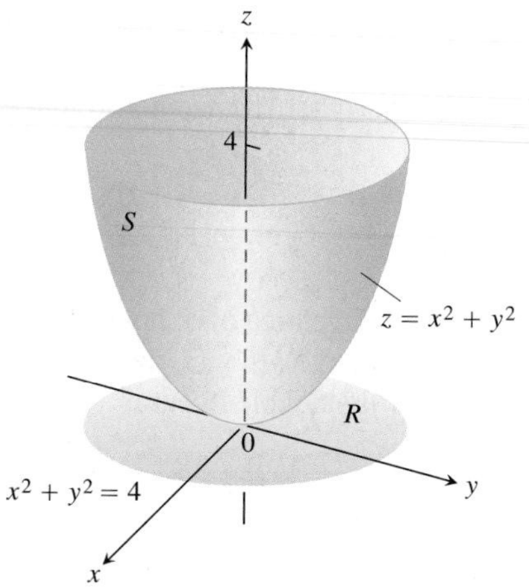

FIGURE 16.41 The area of this parabolic surface is calculated in Example 1.

At any point (x, y, z) on the surface, we have

$$\begin{aligned} f(x, y, z) &= x^2 + y^2 - z \\ \nabla f &= 2x\mathbf{i} + 2y\mathbf{j} - \mathbf{k} \\ |\nabla f| &= \sqrt{(2x)^2 + (2y)^2 + (-1)^2} \\ &= \sqrt{4x^2 + 4y^2 + 1} \\ |\nabla f \cdot \mathbf{p}| &= |\nabla f \cdot \mathbf{k}| = |-1| = 1. \end{aligned}$$

In the region R, $dA = dx\,dy$. Therefore,

$$\begin{aligned} \text{Surface area} &= \iint_R \frac{|\nabla f|}{|\nabla f \cdot \mathbf{p}|}\, dA && \text{Equation (3)} \\ &= \iint_{x^2+y^2\le 4} \sqrt{4x^2 + 4y^2 + 1}\, dx\, dy \\ &= \int_0^{2\pi}\int_0^2 \sqrt{4r^2 + 1}\, r\, dr\, d\theta && \text{Polar coordinates} \\ &= \int_0^{2\pi} \left[\frac{1}{12}(4r^2 + 1)^{3/2}\right]_0^2 d\theta \\ &= \int_0^{2\pi} \frac{1}{12}(17^{3/2} - 1)\, d\theta = \frac{\pi}{6}\left(17\sqrt{17} - 1\right). \end{aligned}$$

■

EXAMPLE 2 Finding Surface Area

Find the area of the cap cut from the hemisphere $x^2 + y^2 + z^2 = 2$, $z \ge 0$, by the cylinder $x^2 + y^2 = 1$ (Figure 16.42).

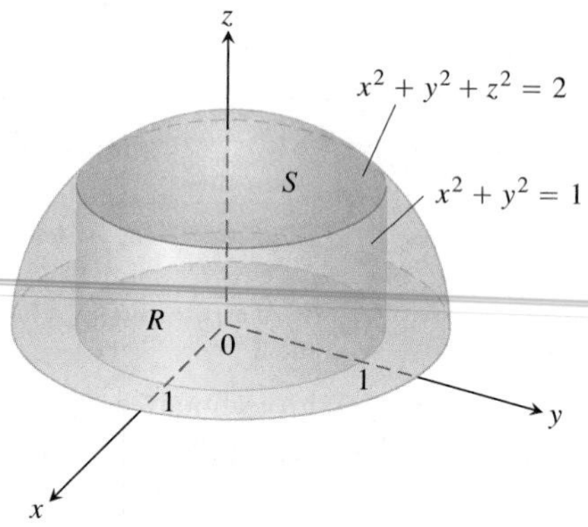

FIGURE 16.42 The cap cut from the hemisphere by the cylinder projects vertically onto the disk R: $x^2 + y^2 \le 1$ in the xy-plane (Example 2).

Solution The cap S is part of the level surface $f(x, y, z) = x^2 + y^2 + z^2 = 2$. It projects one-to-one onto the disk R: $x^2 + y^2 \le 1$ in the xy-plane. The unit vector $\mathbf{p} = \mathbf{k}$ is normal to the plane of R.

At any point on the surface,

$$\begin{aligned} f(x, y, z) &= x^2 + y^2 + z^2 \\ \nabla f &= 2x\mathbf{i} + 2y\mathbf{j} + 2z\mathbf{k} \\ |\nabla f| &= 2\sqrt{x^2 + y^2 + z^2} = 2\sqrt{2} && \text{Because } x^2 + y^2 + z^2 = 2 \text{ at points of } S \\ |\nabla f \cdot \mathbf{p}| &= |\nabla f \cdot \mathbf{k}| = |2z| = 2z. \end{aligned}$$

Therefore,

$$\text{Surface area} = \iint_R \frac{|\nabla f|}{|\nabla f \cdot \mathbf{p}|}\, dA = \iint_R \frac{2\sqrt{2}}{2z}\, dA = \sqrt{2}\iint_R \frac{dA}{z}. \qquad (4)$$

What do we do about the z?

Since z is the z-coordinate of a point on the sphere, we can express it in terms of x and y as

$$z = \sqrt{2 - x^2 - y^2}.$$

We continue the work of Equation (4) with this substitution:

$$\text{Surface area} = \sqrt{2}\iint_R \frac{dA}{z} = \sqrt{2}\iint_{x^2+y^2\le 1} \frac{dA}{\sqrt{2-x^2-y^2}}$$

$$= \sqrt{2}\int_0^{2\pi}\int_0^1 \frac{r\,dr\,d\theta}{\sqrt{2-r^2}} \qquad \text{Polar coordinates}$$

$$= \sqrt{2}\int_0^{2\pi}\left[-(2-r^2)^{1/2}\right]_{r=0}^{r=1} d\theta$$

$$= \sqrt{2}\int_0^{2\pi}\left(\sqrt{2}-1\right) d\theta = 2\pi\left(2-\sqrt{2}\right).$$

■

Surface Integrals

We now show how to integrate a function over a surface, using the ideas just developed for calculating surface area.

Suppose, for example, that we have an electrical charge distributed over a surface $f(x, y, z) = c$ like the one shown in Figure 16.43 and that the function $g(x, y, z)$ gives the charge per unit area (charge density) at each point on S. Then we may calculate the total charge on S as an integral in the following way.

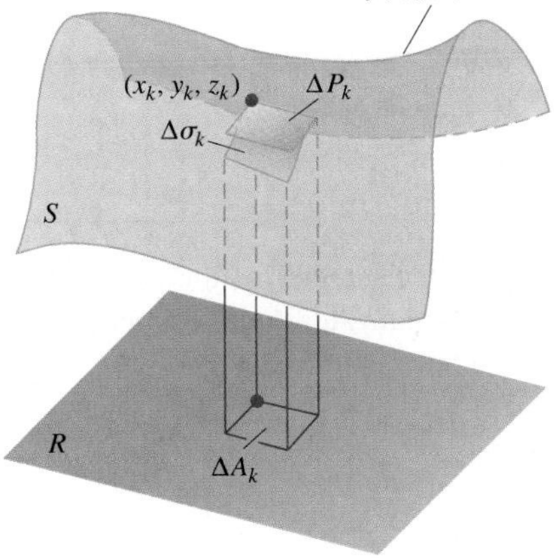

FIGURE 16.43 If we know how an electrical charge $g(x, y, z)$ is distributed over a surface, we can find the total charge with a suitably modified surface integral.

We partition the shadow region R on the ground plane beneath the surface into small rectangles of the kind we would use if we were defining the surface area of S. Then directly above each ΔA_k lies a patch of surface $\Delta\sigma_k$ that we approximate with a parallelogram-shaped portion of tangent plane, ΔP_k. (See Figure 16.43.)

Up to this point the construction proceeds as in the definition of surface area, but now we take an additional step: We evaluate g at (x_k, y_k, z_k) and approximate the total charge on the surface path $\Delta\sigma_k$ by the product $g(x_k, y_k, z_k)\,\Delta P_k$. The rationale is that when the partition of R is sufficiently fine, the value of g throughout $\Delta\sigma_k$ is nearly constant and ΔP_k is nearly the same as $\Delta\sigma_k$. The total charge over S is then approximated by the sum

$$\text{Total charge} \approx \sum g(x_k, y_k, z_k)\,\Delta P_k = \sum g(x_k, y_k, z_k)\frac{\Delta A_k}{|\cos\gamma_k|}.$$

If f, the function defining the surface S, and its first partial derivatives are continuous, and if g is continuous over S, then the sums on the right-hand side of the last equation approach the limit

$$\iint_R g(x, y, z)\frac{dA}{|\cos\gamma|} = \iint_R g(x, y, z)\frac{|\nabla f|}{|\nabla f\cdot\mathbf{p}|}\,dA \tag{5}$$

as the partition of R is refined in the usual way. This limit is called the integral of g over the surface S and is calculated as a double integral over R. The value of the integral is the total charge on the surface S.

As you might expect, the formula in Equation (5) defines the integral of *any* function g over the surface S as long as the integral exists.

DEFINITION Surface Integral

If R is the shadow region of a surface S defined by the equation $f(x, y, z) = c$, and g is a continuous function defined at the points of S, then the **integral of g over S** is the integral

$$\iint_R g(x, y, z) \frac{|\nabla f|}{|\nabla f \cdot \mathbf{p}|}\, dA, \tag{6}$$

where $\mathbf{p}$ is a unit vector normal to R and $\nabla f \cdot \mathbf{p} \neq 0$. The integral itself is called a **surface integral**.

The integral in Equation (6) takes on different meanings in different applications. If g has the constant value 1, the integral gives the area of S. If g gives the mass density of a thin shell of material modeled by S, the integral gives the mass of the shell.

We can abbreviate the integral in Equation (6) by writing $d\sigma$ for $(|\nabla f|/|\nabla f \cdot \mathbf{p}|)\, dA$.

The Surface Area Differential and the Differential Form for Surface Integrals

$$d\sigma = \frac{|\nabla f|}{|\nabla f \cdot \mathbf{p}|}\, dA \qquad \iint_S g\, d\sigma \tag{7}$$

Surface area differential — Differential formula for surface integrals

Surface integrals behave like other double integrals, the integral of the sum of two functions being the sum of their integrals and so on. The domain Additivity Property takes the form

$$\iint_S g\, d\sigma = \iint_{S_1} g\, d\sigma + \iint_{S_2} g\, d\sigma + \cdots + \iint_{S_n} g\, d\sigma.$$

The idea is that if S is partitioned by smooth curves into a finite number of nonoverlapping smooth patches (i.e., if S is **piecewise smooth**), then the integral over S is the sum of the integrals over the patches. Thus, the integral of a function over the surface of a cube is the sum of the integrals over the faces of the cube. We integrate over a turtle shell of welded plates by integrating one plate at a time and adding the results.

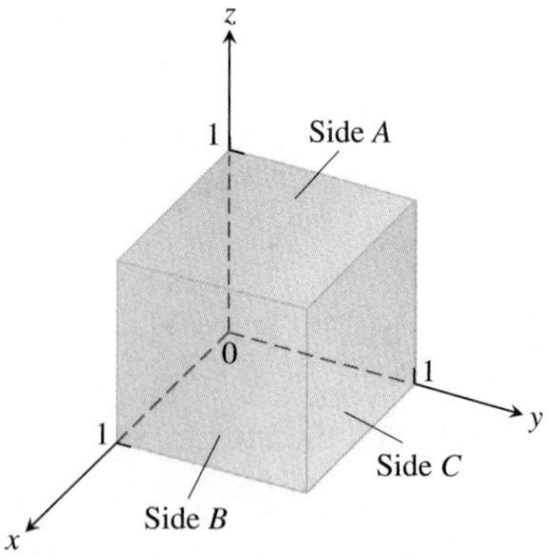

FIGURE 16.44 The cube in Example 3.

EXAMPLE 3 Integrating Over a Surface

Integrate $g(x, y, z) = xyz$ over the surface of the cube cut from the first octant by the planes $x = 1, y = 1$, and $z = 1$ (Figure 16.44).

Solution We integrate xyz over each of the six sides and add the results. Since $xyz = 0$ on the sides that lie in the coordinate planes, the integral over the surface of the cube reduces to

$$\iint_{\substack{\text{Cube}\\ \text{surface}}} xyz\, d\sigma = \iint_{\text{Side } A} xyz\, d\sigma + \iint_{\text{Side } B} xyz\, d\sigma + \iint_{\text{Side } C} xyz\, d\sigma.$$

Side A is the surface $f(x, y, z) = z = 1$ over the square region R_{xy}: $0 \le x \le 1$, $0 \le y \le 1$, in the xy-plane. For this surface and region,

$$\mathbf{p} = \mathbf{k}, \qquad \nabla f = \mathbf{k}, \qquad |\nabla f| = 1, \qquad |\nabla f \cdot \mathbf{p}| = |\mathbf{k} \cdot \mathbf{k}| = 1$$

$$d\sigma = \frac{|\nabla f|}{|\nabla f \cdot \mathbf{p}|} dA = \frac{1}{1} dx\, dy = dx\, dy$$

$$xyz = xy(1) = xy$$

and

$$\iint_{\text{Side } A} xyz\, d\sigma = \iint_{R_{xy}} xy\, dx\, dy = \int_0^1 \int_0^1 xy\, dx\, dy = \int_0^1 \frac{y}{2} dy = \frac{1}{4}.$$

Symmetry tells us that the integrals of xyz over sides B and C are also $1/4$. Hence,

$$\iint_{\substack{\text{Cube} \\ \text{surface}}} xyz\, d\sigma = \frac{1}{4} + \frac{1}{4} + \frac{1}{4} = \frac{3}{4}.$$

■

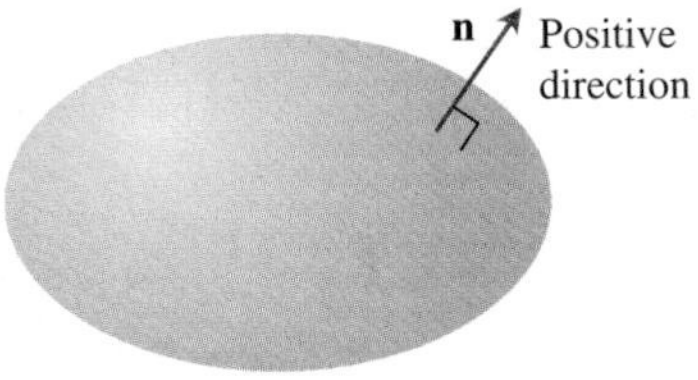

FIGURE 16.45 Smooth closed surfaces in space are orientable. The outward unit normal vector defines the positive direction at each point.

Orientation

We call a smooth surface S **orientable** or **two-sided** if it is possible to define a field $\mathbf{n}$ of unit normal vectors on S that varies continuously with position. Any patch or subportion of an orientable surface is orientable. Spheres and other smooth closed surfaces in space (smooth surfaces that enclose solids) are orientable. By convention, we choose $\mathbf{n}$ on a closed surface to point outward.

Once $\mathbf{n}$ has been chosen, we say that we have **oriented** the surface, and we call the surface together with its normal field an **oriented surface**. The vector $\mathbf{n}$ at any point is called the **positive direction** at that point (Figure 16.45).

The Möbius band in Figure 16.46 is not orientable. No matter where you start to construct a continuous-unit normal field (shown as the shaft of a thumbtack in the figure), moving the vector continuously around the surface in the manner shown will return it to the starting point with a direction opposite to the one it had when it started out. The vector at that point cannot point both ways and yet it must if the field is to be continuous. We conclude that no such field exists.

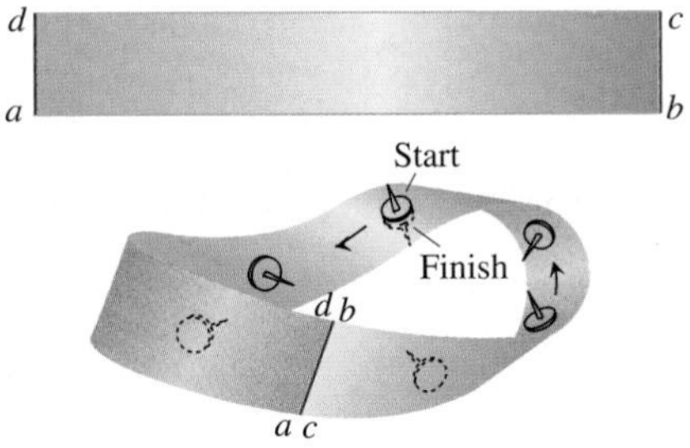

FIGURE 16.46 To make a Möbius band, take a rectangular strip of paper $abcd$, give the end bc a single twist, and paste the ends of the strip together to match a with c and b with d. The Möbius band is a nonorientable or one-sided surface.

Surface Integral for Flux

Suppose that $\mathbf{F}$ is a continuous vector field defined over an oriented surface S and that $\mathbf{n}$ is the chosen unit normal field on the surface. We call the integral of $\mathbf{F} \cdot \mathbf{n}$ over S the flux of $\mathbf{F}$ across S in the positive direction. Thus, the flux is the integral over S of the scalar component of $\mathbf{F}$ in the direction of $\mathbf{n}$.

DEFINITION Flux

The **flux** of a three-dimensional vector field $\mathbf{F}$ across an oriented surface S in the direction of $\mathbf{n}$ is

$$\text{Flux} = \iint_S \mathbf{F} \cdot \mathbf{n}\, d\sigma. \tag{8}$$

The definition is analogous to the flux of a two-dimensional field **F** across a plane curve C. In the plane (Section 16.2), the flux is

$$\int_C \mathbf{F}\cdot\mathbf{n}\,ds,$$

the integral of the scalar component of **F** normal to the curve.

If **F** is the velocity field of a three-dimensional fluid flow, the flux of **F** across S is the net rate at which fluid is crossing S in the chosen positive direction. We discuss such flows in more detail in Section 16.7.

If **S** is part of a level surface $g(x, y, z) = c$, then **n** may be taken to be one of the two fields

$$\mathbf{n} = \pm\frac{\nabla g}{|\nabla g|}, \tag{9}$$

depending on which one gives the preferred direction. The corresponding flux is

$$\begin{aligned}\text{Flux} &= \iint_S \mathbf{F}\cdot\mathbf{n}\,d\sigma \\ &= \iint_R \left(\mathbf{F}\cdot\frac{\pm\nabla g}{|\nabla g|}\right)\frac{|\nabla g|}{|\nabla g\cdot\mathbf{p}|}\,dA \qquad \text{Equations (9) and (7)} \qquad (8)\\ &= \iint_R \mathbf{F}\cdot\frac{\pm\nabla g}{|\nabla g\cdot\mathbf{p}|}\,dA. \qquad (10)\end{aligned}$$

EXAMPLE 4 Finding Flux

Find the flux of $\mathbf{F} = yz\mathbf{j} + z^2\mathbf{k}$ outward through the surface S cut from the cylinder $y^2 + z^2 = 1$, $z \geq 0$, by the planes $x = 0$ and $x = 1$.

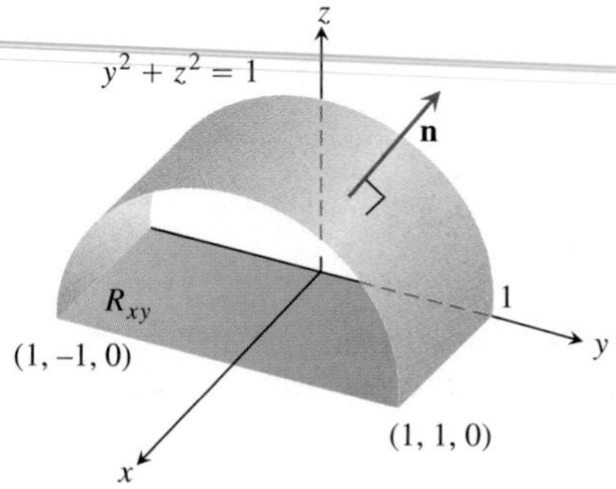

FIGURE 16.47 Calculating the flux of a vector field outward through this surface. The area of the shadow region R_{xy} is 2 (Example 4).

Solution The outward normal field on S (Figure 16.47) may be calculated from the gradient of $g(x, y, z) = y^2 + z^2$ to be

$$\mathbf{n} = +\frac{\nabla g}{|\nabla g|} = \frac{2y\mathbf{j} + 2z\mathbf{k}}{\sqrt{4y^2 + 4z^2}} = \frac{2y\mathbf{j} + 2z\mathbf{k}}{2\sqrt{1}} = y\mathbf{j} + z\mathbf{k}.$$

With $\mathbf{p} = \mathbf{k}$, we also have

$$d\sigma = \frac{|\nabla g|}{|\nabla g\cdot\mathbf{k}|}\,dA = \frac{2}{|2z|}\,dA = \frac{1}{z}\,dA.$$

We can drop the absolute value bars because $z \geq 0$ on S.

The value of $\mathbf{F}\cdot\mathbf{n}$ on the surface is

$$\begin{aligned}\mathbf{F}\cdot\mathbf{n} &= (yz\mathbf{j} + z^2\mathbf{k})\cdot(y\mathbf{j} + z\mathbf{k}) \\ &= y^2z + z^3 = z(y^2 + z^2) \\ &= z. \qquad y^2 + z^2 = 1 \text{ on } S\end{aligned}$$

Therefore, the flux of **F** outward through S is

$$\iint_S \mathbf{F}\cdot\mathbf{n}\,d\sigma = \iint_S (z)\left(\frac{1}{z}\,dA\right) = \iint_{R_{xy}} dA = \text{area}(R_{xy}) = 2.$$

■

Moments and Masses of Thin Shells

Thin shells of material like bowls, metal drums, and domes are modeled with surfaces. Their moments and masses are calculated with the formulas in Table 16.3.

TABLE 16.3 Mass and moment formulas for very thin shells

Mass: $M = \iint_S \delta(x, y, z)\, d\sigma \qquad (\delta(x, y, z) = \text{density at } (x, y, z), \text{ mass per unit area})$

First moments about the coordinate planes:

$$M_{yz} = \iint_S x\, \delta\, d\sigma, \qquad M_{xz} = \iint_S y\, \delta\, d\sigma, \qquad M_{xy} = \iint_S z\, \delta\, d\sigma$$

Coordinates of center of mass:

$$\bar{x} = M_{yz}/M, \qquad \bar{y} = M_{xz}/M, \qquad \bar{z} = M_{xy}/M$$

Moments of inertia about coordinate axes:

$$I_x = \iint_S (y^2 + z^2)\, \delta\, d\sigma, \qquad I_y = \iint_S (x^2 + z^2)\, \delta\, d\sigma,$$

$$I_z = \iint_S (x^2 + y^2)\, \delta\, d\sigma, \qquad I_L = \iint_S r^2 \delta\, d\sigma,$$

$$r(x, y, z) = \text{distance from point } (x, y, z) \text{ to line } L$$

Radius of gyration about a line L: $R_L = \sqrt{I_L/M}$

EXAMPLE 5 Finding Center of Mass

Find the center of mass of a thin hemispherical shell of radius a and constant density δ.

Solution We model the shell with the hemisphere

$$f(x, y, z) = x^2 + y^2 + z^2 = a^2, \qquad z \geq 0$$

(Figure 16.48). The symmetry of the surface about the z-axis tells us that $\bar{x} = \bar{y} = 0$. It remains only to find $\bar{z}$ from the formula $\bar{z} = M_{xy}/M$.

The mass of the shell is

$$M = \iint_S \delta\, d\sigma = \delta \iint_S d\sigma = (\delta)(\text{area of } S) = 2\pi a^2 \delta.$$

To evaluate the integral for M_{xy}, we take $\mathbf{p} = \mathbf{k}$ and calculate

$$|\nabla f| = |2x\mathbf{i} + 2y\mathbf{j} + 2z\mathbf{k}| = 2\sqrt{x^2 + y^2 + z^2} = 2a$$

$$|\nabla f \cdot \mathbf{p}| = |\nabla f \cdot \mathbf{k}| = |2z| = 2z$$

$$d\sigma = \frac{|\nabla f|}{|\nabla f \cdot \mathbf{p}|}\, dA = \frac{a}{z}\, dA.$$

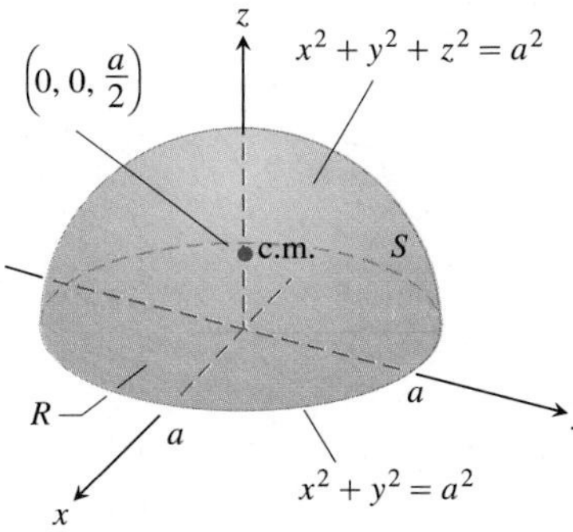

FIGURE 16.48 The center of mass of a thin hemispherical shell of constant density lies on the axis of symmetry halfway from the base to the top (Example 5).

Then

$$M_{xy} = \iint_S z\delta\, d\sigma = \delta \iint_R z\frac{a}{z}\, dA = \delta a \iint_R dA = \delta a(\pi a^2) = \delta\pi a^3$$

$$\bar{z} = \frac{M_{xy}}{M} = \frac{\pi a^3 \delta}{2\pi a^2 \delta} = \frac{a}{2}.$$

The shell's center of mass is the point $(0, 0, a/2)$. ∎

EXERCISES 16.5

Surface Area

1. Find the area of the surface cut from the paraboloid $x^2 + y^2 - z = 0$ by the plane $z = 2$.
2. Find the area of the band cut from the paraboloid $x^2 + y^2 - z = 0$ by the planes $z = 2$ and $z = 6$.
3. Find the area of the region cut from the plane $x + 2y + 2z = 5$ by the cylinder whose walls are $x = y^2$ and $x = 2 - y^2$.
4. Find the area of the portion of the surface $x^2 - 2z = 0$ that lies above the triangle bounded by the lines $x = \sqrt{3}$, $y = 0$, and $y = x$ in the xy-plane.
5. Find the area of the surface $x^2 - 2y - 2z = 0$ that lies above the triangle bounded by the lines $x = 2$, $y = 0$, and $y = 3x$ in the xy-plane.
6. Find the area of the cap cut from the sphere $x^2 + y^2 + z^2 = 2$ by the cone $z = \sqrt{x^2 + y^2}$.
7. Find the area of the ellipse cut from the plane $z = cx$ (c a constant) by the cylinder $x^2 + y^2 = 1$.
8. Find the area of the upper portion of the cylinder $x^2 + z^2 = 1$ that lies between the planes $x = \pm 1/2$ and $y = \pm 1/2$.
9. Find the area of the portion of the paraboloid $x = 4 - y^2 - z^2$ that lies above the ring $1 \le y^2 + z^2 \le 4$ in the yz-plane.
10. Find the area of the surface cut from the paraboloid $x^2 + y + z^2 = 2$ by the plane $y = 0$.
11. Find the area of the surface $x^2 - 2\ln x + \sqrt{15}y - z = 0$ above the square R: $1 \le x \le 2$, $0 \le y \le 1$, in the xy-plane.
12. Find the area of the surface $2x^{3/2} + 2y^{3/2} - 3z = 0$ above the square R: $0 \le x \le 1$, $0 \le y \le 1$, in the xy-plane.

Surface Integrals

13. Integrate $g(x, y, z) = x + y + z$ over the surface of the cube cut from the first octant by the planes $x = a$, $y = a$, $z = a$.
14. Integrate $g(x, y, z) = y + z$ over the surface of the wedge in the first octant bounded by the coordinate planes and the planes $x = 2$ and $y + z = 1$.
15. Integrate $g(x, y, z) = xyz$ over the surface of the rectangular solid cut from the first octant by the planes $x = a$, $y = b$, and $z = c$.
16. Integrate $g(x, y, z) = xyz$ over the surface of the rectangular solid bounded by the planes $x = \pm a$, $y = \pm b$, and $z = \pm c$.
17. Integrate $g(x, y, z) = x + y + z$ over the portion of the plane $2x + 2y + z = 2$ that lies in the first octant.
18. Integrate $g(x, y, z) = x\sqrt{y^2 + 4}$ over the surface cut from the parabolic cylinder $y^2 + 4z = 16$ by the planes $x = 0$, $x = 1$, and $z = 0$.

Flux Across a Surface

In Exercises 19 and 20, find the flux of the field **F** across the portion of the given surface in the specified direction.

19. $\mathbf{F}(x, y, z) = -\mathbf{i} + 2\mathbf{j} + 3\mathbf{k}$

    S: rectangular surface $z = 0$, $0 \le x \le 2$, $0 \le y \le 3$, direction $\mathbf{k}$

20. $\mathbf{F}(x, y, z) = yx^2\mathbf{i} - 2\mathbf{j} + xz\mathbf{k}$

    S: rectangular surface $y = 0$, $-1 \le x \le 2$, $2 \le z \le 7$, direction $-\mathbf{j}$

In Exercises 21–26, find the flux of the field **F** across the portion of the sphere $x^2 + y^2 + z^2 = a^2$ in the first octant in the direction away from the origin.

21. $\mathbf{F}(x, y, z) = z\mathbf{k}$
22. $\mathbf{F}(x, y, z) = -y\mathbf{i} + x\mathbf{j}$
23. $\mathbf{F}(x, y, z) = y\mathbf{i} - x\mathbf{j} + \mathbf{k}$
24. $\mathbf{F}(x, y, z) = zx\mathbf{i} + zy\mathbf{j} + z^2\mathbf{k}$
25. $\mathbf{F}(x, y, z) = x\mathbf{i} + y\mathbf{j} + z\mathbf{k}$
26. $\mathbf{F}(x, y, z) = \dfrac{x\mathbf{i} + y\mathbf{j} + z\mathbf{k}}{\sqrt{x^2 + y^2 + z^2}}$

27. Find the flux of the field $\mathbf{F}(x, y, z) = z^2\mathbf{i} + x\mathbf{j} - 3z\mathbf{k}$ outward through the surface cut from the parabolic cylinder $z = 4 - y^2$ by the planes $x = 0, x = 1$, and $z = 0$.

28. Find the flux of the field $\mathbf{F}(x, y, z) = 4x\mathbf{i} + 4y\mathbf{j} + 2\mathbf{k}$ outward (away from the z-axis) through the surface cut from the bottom of the paraboloid $z = x^2 + y^2$ by the plane $z = 1$.

29. Let S be the portion of the cylinder $y = e^x$ in the first octant that projects parallel to the x-axis onto the rectangle R_{yz}: $1 \le y \le 2$, $0 \le z \le 1$ in the yz-plane (see the accompanying figure). Let $\mathbf{n}$ be the unit vector normal to S that points away from the yz-plane. Find the flux of the field $\mathbf{F}(x, y, z) = -2\mathbf{i} + 2y\mathbf{j} + z\mathbf{k}$ across S in the direction of $\mathbf{n}$.

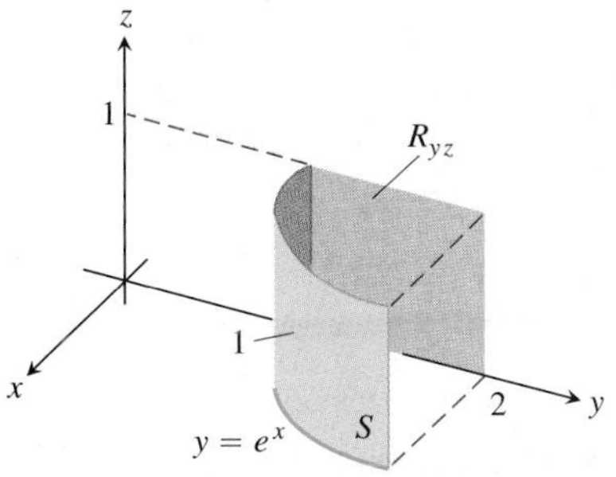

30. Let S be the portion of the cylinder $y = \ln x$ in the first octant whose projection parallel to the y-axis onto the xz-plane is the rectangle R_{xz}: $1 \le x \le e, 0 \le z \le 1$. Let $\mathbf{n}$ be the unit vector normal to S that points away from the xz-plane. Find the flux of $\mathbf{F} = 2y\mathbf{j} + z\mathbf{k}$ through S in the direction of $\mathbf{n}$.

31. Find the outward flux of the field $\mathbf{F} = 2xy\mathbf{i} + 2yz\mathbf{j} + 2xz\mathbf{k}$ across the surface of the cube cut from the first octant by the planes $x = a, y = a, z = a$.

32. Find the outward flux of the field $\mathbf{F} = xz\mathbf{i} + yz\mathbf{j} + \mathbf{k}$ across the surface of the upper cap cut from the solid sphere $x^2 + y^2 + z^2 \le 25$ by the plane $z = 3$.

Moments and Masses

33. Centroid Find the centroid of the portion of the sphere $x^2 + y^2 + z^2 = a^2$ that lies in the first octant.

34. Centroid Find the centroid of the surface cut from the cylinder $y^2 + z^2 = 9, z \ge 0$, by the planes $x = 0$ and $x = 3$ (resembles the surface in Example 4).

35. Thin shell of constant density Find the center of mass and the moment of inertia and radius of gyration about the z-axis of a thin shell of constant density δ cut from the cone $x^2 + y^2 - z^2 = 0$ by the planes $z = 1$ and $z = 2$,

36. Conical surface of constant density Find the moment of inertia about the z-axis of a thin shell of constant density δ cut from the cone $4x^2 + 4y^2 - z^2 = 0, z \ge 0$, by the circular cylinder $x^2 + y^2 = 2x$ (see the accompanying figure).

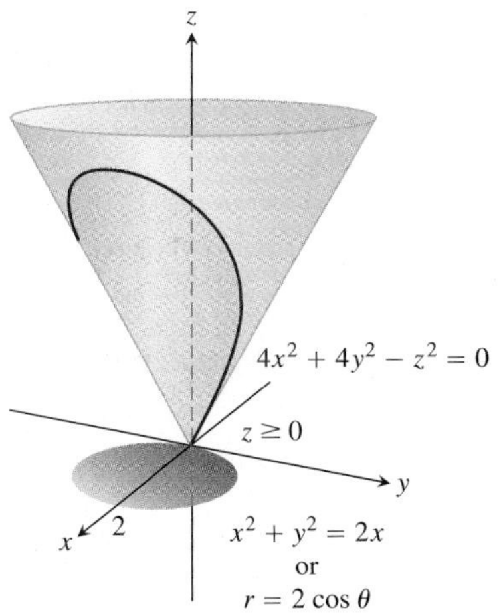

37. Spherical shells

a. Find the moment of inertia about a diameter of a thin spherical shell of radius a and constant density δ. (Work with a hemispherical shell and double the result.)

b. Use the Parallel Axis Theorem (Exercises 15.5) and the result in part (a) to find the moment of inertia about a line tangent to the shell.

38. a. Cones with and without ice cream Find the centroid of the lateral surface of a solid cone of base radius a and height h (cone surface minus the base).

b. Use Pappus's formula (Exercises 15.5) and the result in part (a) to find the centroid of the complete surface of a solid cone (side plus base).

c. A cone of radius a and height h is joined to a hemisphere of radius a to make a surface S that resembles an ice cream cone. Use Pappus's formula and the results in part (a) and Example 5 to find the centroid of S. How high does the cone have to be to place the centroid in the plane shared by the bases of the hemisphere and cone?

Special Formulas for Surface Area

If S is the surface defined by a function $z = f(x, y)$ that has continuous first partial derivatives throughout a region R_{xy} in the xy-plane (Figure 16.49), then S is also the level surface $F(x, y, z) = 0$ of the function $F(x, y, z) = f(x, y) - z$. Taking the unit normal to R_{xy} to be $\mathbf{p} = \mathbf{k}$ then gives

$$|\nabla F| = |f_x\mathbf{i} + f_y\mathbf{j} - \mathbf{k}| = \sqrt{f_x^{\,2} + f_y^{\,2} + 1}$$

$$|\nabla F \cdot \mathbf{p}| = |(f_x\mathbf{i} + f_y\mathbf{j} - \mathbf{k}) \cdot \mathbf{k}| = |-1| = 1$$

and

$$\iint_{R_{xy}} \frac{|\nabla F|}{|\nabla F \cdot \mathbf{p}|}\, dA = \iint_{R_{xy}} \sqrt{f_x^{\,2} + f_y^{\,2} + 1}\, dx\, dy, \tag{11}$$

Similarly, the area of a smooth surface $x = f(y, z)$ over a region R_{yz} in the yz-plane is

$$A = \iint_{R_{yz}} \sqrt{f_y^2 + f_z^2 + 1}\, dy\, dz, \qquad (12)$$

and the area of a smooth $y = f(x, z)$ over a region R_{xz} in the xz-plane is

$$A = \iint_{R_{xz}} \sqrt{f_x^2 + f_z^2 + 1}\, dx\, dz. \qquad (13)$$

Use Equations (11)–(13) to find the area of the surfaces in Exercises 39–44.

39. The surface cut from the bottom of the paraboloid $z = x^2 + y^2$ by the plane $z = 3$

40. The surface cut from the "nose" of the paraboloid $x = 1 - y^2 - z^2$ by the yz-plane

41. The portion of the cone $z = \sqrt{x^2 + y^2}$ that lies over the region between the circle $x^2 + y^2 = 1$ and the ellipse $9x^2 + 4y^2 = 36$ in the xy-plane. (*Hint:* Use formulas from geometry to find the area of the region.)

42. The triangle cut from the plane $2x + 6y + 3z = 6$ by the bounding planes of the first octant. Calculate the area three ways, once with each area formula

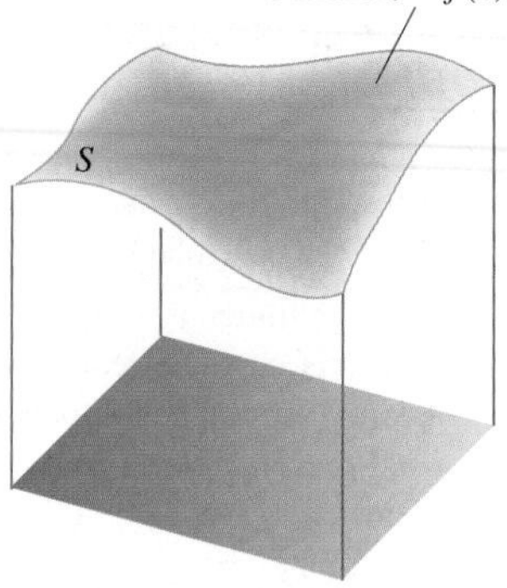

FIGURE 16.49 For a surface $z = f(x, y)$, the surface area formula in Equation (3) takes the form

$$A = \iint_{R_{xy}} \sqrt{f_x^2 + f_y^2 + 1}\, dx\, dy.$$

43. The surface in the first octant cut from the cylinder $y = (2/3)z^{3/2}$ by the planes $x = 1$ and $y = 16/3$

44. The portion of the plane $y + z = 4$ that lies above the region cut from the first quadrant of the xz-plane by the parabola $x = 4 - z^2$

16.6 Parametrized Surfaces

We have defined curves in the plane in three different ways:

Explicit form:	$y = f(x)$
Implicit form:	$F(x, y) = 0$
Parametric vector form:	$\mathbf{r}(t) = f(t)\mathbf{i} + g(t)\mathbf{j}, \quad a \le t \le b.$

We have analogous definitions of surfaces in space:

Explicit form:	$z = f(x, y)$
Implicit form:	$F(x, y, z) = 0.$

There is also a parametric form that gives the position of a point on the surface as a vector function of two variables. The present section extends the investigation of surface area and surface integrals to surfaces described parametrically.

Parametrizations of Surfaces

Let

$$\mathbf{r}(u, v) = f(u, v)\mathbf{i} + g(u, v)\mathbf{j} + h(u, v)\mathbf{k} \qquad (1)$$

be a continuous vector function that is defined on a region R in the uv-plane and one-to-one on the interior of R (Figure 16.50). We call the range of $\mathbf{r}$ the **surface** S defined or traced by $\mathbf{r}$. Equation (1) together with the domain R constitute a **parametrization** of the surface. The variables u and v are the **parameters**, and R is the **parameter domain**.

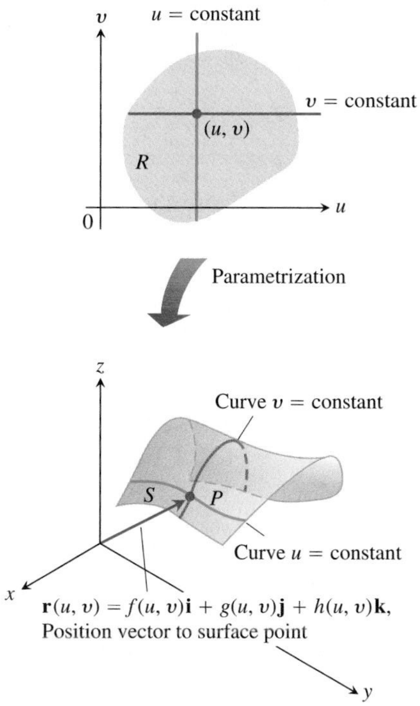

FIGURE 16.50 A parametrized surface S expressed as a vector function of two variables defined on a region R.

To simplify our discussion, we take R to be a rectangle defined by inequalities of the form $a \le u \le b, c \le v \le d$. The requirement that **r** be one-to-one on the interior of R ensures that S does not cross itself. Notice that Equation (1) is the vector equivalent of *three* parametric equations:

$$x = f(u, v), \qquad y = g(u, v), \qquad z = h(u, v).$$

EXAMPLE 1 Parametrizing a Cone

Find a parametrization of the cone

$$z = \sqrt{x^2 + y^2}, \qquad 0 \le z \le 1.$$

Solution Here, cylindrical coordinates provide everything we need. A typical point (x, y, z) on the cone (Figure 16.51) has $x = r\cos\theta, y = r\sin\theta$, and $z = \sqrt{x^2 + y^2} = r$, with $0 \le r \le 1$ and $0 \le \theta \le 2\pi$. Taking $u = r$ and $v = \theta$ in Equation (1) gives the parametrization

$$\mathbf{r}(r, \theta) = (r\cos\theta)\mathbf{i} + (r\sin\theta)\mathbf{j} + r\mathbf{k}, \qquad 0 \le r \le 1, \quad 0 \le \theta \le 2\pi.$$ ■

EXAMPLE 2 Parametrizing a Sphere

Find a parametrization of the sphere $x^2 + y^2 + z^2 = a^2$.

Solution Spherical coordinates provide what we need. A typical point (x, y, z) on the sphere (Figure 16.52) has $x = a\sin\phi\cos\theta, y = a\sin\phi\sin\theta$, and $z = a\cos\phi$, $0 \le \phi \le \pi$, $0 \le \theta \le 2\pi$. Taking $u = \phi$ and $v = \theta$ in Equation (1) gives the parametrization

$$\mathbf{r}(\phi, \theta) = (a\sin\phi\cos\theta)\mathbf{i} + (a\sin\phi\sin\theta)\mathbf{j} + (a\cos\phi)\mathbf{k},$$
$$0 \le \phi \le \pi, \quad 0 \le \theta \le 2\pi.$$ ■

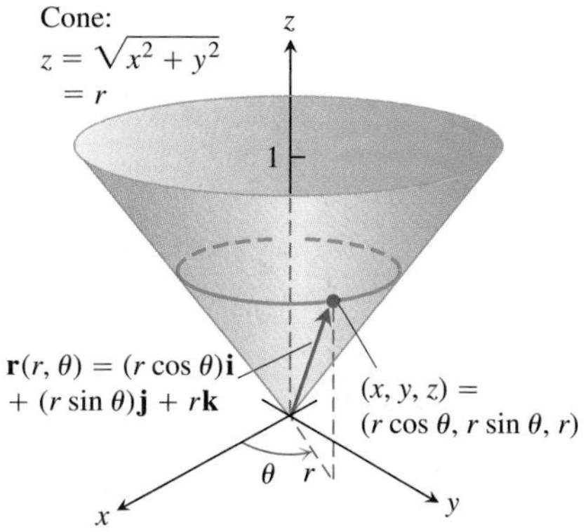

FIGURE 16.51 The cone in Example 1 can be parametrized using cylindrical coordinates.

EXAMPLE 3 Parametrizing a Cylinder

Find a parametrization of the cylinder

$$x^2 + (y - 3)^2 = 9, \qquad 0 \le z \le 5.$$

Solution In cylindrical coordinates, a point (x, y, z) has $x = r\cos\theta, y = r\sin\theta$, and $z = z$. For points on the cylinder $x^2 + (y - 3)^2 = 9$ (Figure 16.53), the equation is the same as the polar equation for the cylinder's base in the xy-plane:

$$x^2 + (y^2 - 6y + 9) = 9$$
$$r^2 - 6r\sin\theta = 0$$

or

$$r = 6\sin\theta, \qquad 0 \le \theta \le \pi.$$

A typical point on the cylinder therefore has

$$x = r\cos\theta = 6\sin\theta\cos\theta = 3\sin 2\theta$$
$$y = r\sin\theta = 6\sin^2\theta$$
$$z = z.$$

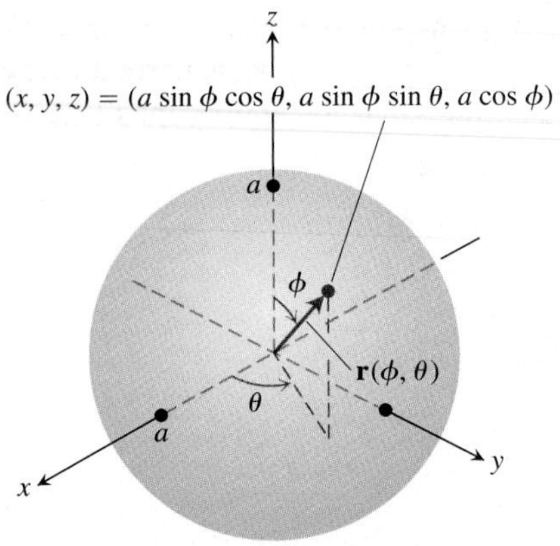

FIGURE 16.52 The sphere in Example 2 can be parametrized using spherical coordinates.

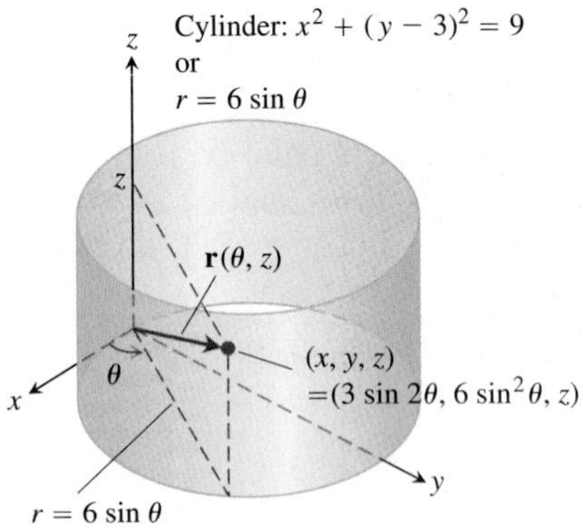

FIGURE 16.53 The cylinder in Example 3 can be parametrized using cylindrical coordinates.

Taking $u = \theta$ and $v = z$ in Equation (1) gives the parametrization

$$\mathbf{r}(\theta, z) = (3 \sin 2\theta)\mathbf{i} + (6 \sin^2 \theta)\mathbf{j} + z\mathbf{k}, \quad 0 \le \theta \le \pi, \quad 0 \le z \le 5.$$ ■

Surface Area

Our goal is to find a double integral for calculating the area of a curved surface S based on the parametrization

$$\mathbf{r}(u, v) = f(u, v)\mathbf{i} + g(u, v)\mathbf{j} + h(u, v)\mathbf{k}, \qquad a \le u \le b, \quad c \le v \le d.$$

We need S to be smooth for the construction we are about to carry out. The definition of smoothness involves the partial derivatives of $\mathbf{r}$ with respect to u and v:

$$\mathbf{r}_u = \frac{\partial \mathbf{r}}{\partial u} = \frac{\partial f}{\partial u}\mathbf{i} + \frac{\partial g}{\partial u}\mathbf{j} + \frac{\partial h}{\partial u}\mathbf{k}$$

$$\mathbf{r}_v = \frac{\partial \mathbf{r}}{\partial v} = \frac{\partial f}{\partial v}\mathbf{i} + \frac{\partial g}{\partial v}\mathbf{j} + \frac{\partial h}{\partial v}\mathbf{k}.$$

DEFINITION Smooth Parametrized Surface

A parametrized surface $\mathbf{r}(u, v) = f(u, v)\mathbf{i} + g(u, v)\mathbf{j} + h(u, v)\mathbf{k}$ is **smooth** if $\mathbf{r}_u$ and $\mathbf{r}_v$ are continuous and $\mathbf{r}_u \times \mathbf{r}_v$ is never zero on the parameter domain.

The condition that $\mathbf{r}_u \times \mathbf{r}_v$ is never the zero vector in the definition of smoothness means that the two vectors $\mathbf{r}_u$ and $\mathbf{r}_v$ are nonzero and never lie along the same line, so they always determine a plane tangent to the surface.

Now consider a small rectangle ΔA_{uv} in R with sides on the lines $u = u_0$, $u = u_0 + \Delta u$, $v = v_0$ and $v = v_0 + \Delta v$ (Figure 16.54). Each side of ΔA_{uv} maps to a curve on the surface S, and together these four curves bound a "curved area element" $\Delta\sigma_{uv}$. In the notation of the figure, the side $v = v_0$ maps to curve C_1, the side $u = u_0$ maps to C_2, and their common vertex (u_0, v_0) maps to P_0.

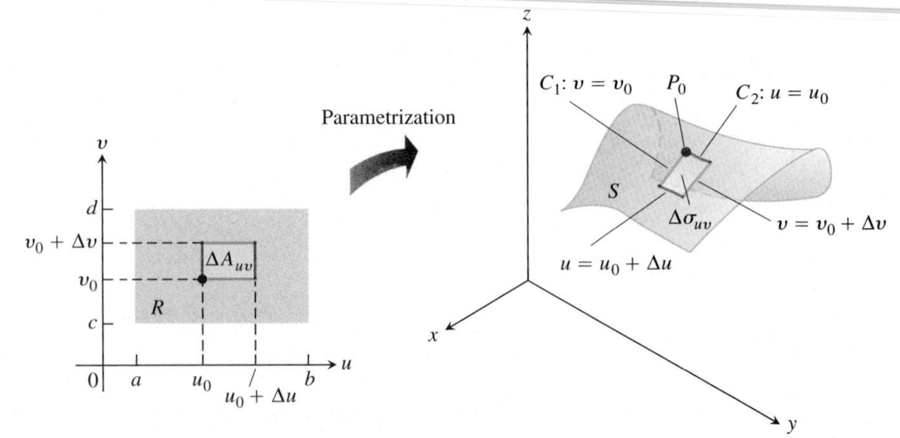

FIGURE 16.54 A rectangular area element ΔA_{uv} in the uv-plane maps onto a curved area element $\Delta\sigma_{uv}$ on S.

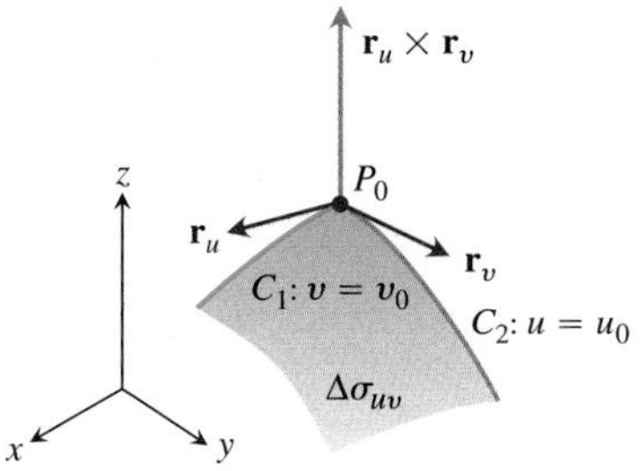

FIGURE 16.55 A magnified view of a surface area element $\Delta\sigma_{uv}$.

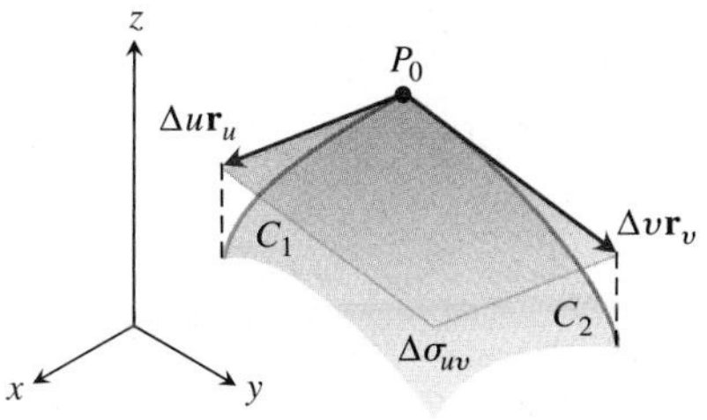

FIGURE 16.56 The parallelogram determined by the vectors $\Delta u\mathbf{r}_u$ and $\Delta v\mathbf{r}_v$ approximates the surface area element $\Delta\sigma_{uv}$.

Figure 16.55 shows an enlarged view of $\Delta\sigma_{uv}$. The vector $\mathbf{r}_u(u_0, v_0)$ is tangent to C_1 at P_0. Likewise, $\mathbf{r}_v(u_0, v_0)$ is tangent to C_2 at P_0. The cross product $\mathbf{r}_u \times \mathbf{r}_v$ is normal to the surface at P_0. (Here is where we begin to use the assumption that S is smooth. We want to be sure that $\mathbf{r}_u \times \mathbf{r}_v \neq \mathbf{0}$.)

We next approximate the surface element $\Delta\sigma_{uv}$ by the parallelogram on the tangent plane whose sides are determined by the vectors $\Delta u\mathbf{r}_u$ and $\Delta v\mathbf{r}_v$ (Figure 16.56). The area of this parallelogram is

$$|\Delta u\mathbf{r}_u \times \Delta v\mathbf{r}_v| = |\mathbf{r}_u \times \mathbf{r}_v|\ \Delta u\ \Delta v. \tag{2}$$

A partition of the region R in the uv-plane by rectangular regions ΔA_{uv} generates a partition of the surface S into surface area elements $\Delta\sigma_{uv}$. We approximate the area of each surface element $\Delta\sigma_{uv}$ by the parallelogram area in Equation (2) and sum these areas together to obtain an approximation of the area of S:

$$\sum_u \sum_v |\mathbf{r}_u \times \mathbf{r}_v|\ \Delta u\ \Delta v. \tag{3}$$

As Δu and Δv approach zero independently, the continuity of $\mathbf{r}_u$ and $\mathbf{r}_v$ guarantees that the sum in Equation (3) approaches the double integral $\int_c^d \int_a^b |\mathbf{r}_u \times \mathbf{r}_v|\ du\ dv$. This double integral defines the area of the surface S and agrees with previous definitions of area, though it is more general.

DEFINITION Area of a Smooth Surface

The **area** of the smooth surface

$$\mathbf{r}(u, v) = f(u, v)\mathbf{i} + g(u, v)\mathbf{j} + h(u, v)\mathbf{k}, \qquad a \le u \le b, \quad c \le v \le d$$

is

$$A = \int_c^d \int_a^b |\mathbf{r}_u \times \mathbf{r}_v|\ du\ dv. \tag{4}$$

As in Section 16.5, we can abbreviate the integral in Equation (4) by writing $d\sigma$ for $|\mathbf{r}_u \times \mathbf{r}_v|\ du\ dv$.

Surface Area Differential and Differential Formula for Surface Area

$$d\sigma = |\mathbf{r}_u \times \mathbf{r}_v|\ du\ dv \qquad \iint_S d\sigma \tag{5}$$

Surface area differential — Differential formula for surface area

EXAMPLE 4 Finding Surface Area (Cone)

Find the surface area of the cone in Example 1 (Figure 16.51).

Solution In Example 1, we found the parametrization

$$\mathbf{r}(r, \theta) = (r\cos\theta)\mathbf{i} + (r\sin\theta)\mathbf{j} + r\mathbf{k}, \qquad 0 \le r \le 1, \quad 0 \le \theta \le 2\pi.$$

To apply Equation (4), we first find $\mathbf{r}_r \times \mathbf{r}_\theta$:

$$\begin{aligned}\mathbf{r}_r \times \mathbf{r}_\theta &= \begin{vmatrix} \mathbf{i} & \mathbf{j} & \mathbf{k} \\ \cos\theta & \sin\theta & 1 \\ -r\sin\theta & r\cos\theta & 0 \end{vmatrix} \\ &= -(r\cos\theta)\mathbf{i} - (r\sin\theta)\mathbf{j} + \underbrace{(r\cos^2\theta + r\sin^2\theta)}_{r}\mathbf{k}.\end{aligned}$$

Thus, $|\mathbf{r}_r \times \mathbf{r}_\theta| = \sqrt{r^2\cos^2\theta + r^2\sin^2\theta + r^2} = \sqrt{2r^2} = \sqrt{2}r$. The area of the cone is

$$\begin{aligned}A &= \int_0^{2\pi}\int_0^1 |\mathbf{r}_r \times \mathbf{r}_\theta|\, dr\, d\theta \qquad \text{Equation (4) with } u = r,\ v = \theta \\ &= \int_0^{2\pi}\int_0^1 \sqrt{2}\, r\, dr\, d\theta = \int_0^{2\pi} \frac{\sqrt{2}}{2}\, d\theta = \frac{\sqrt{2}}{2}(2\pi) = \pi\sqrt{2} \text{ units squared.}\end{aligned}$$

■

EXAMPLE 5 Finding Surface Area (Sphere)

Find the surface area of a sphere of radius a.

Solution We use the parametrization from Example 2:

$$\mathbf{r}(\phi, \theta) = (a\sin\phi\cos\theta)\mathbf{i} + (a\sin\phi\sin\theta)\mathbf{j} + (a\cos\phi)\mathbf{k},$$
$$0 \le \phi \le \pi, \quad 0 \le \theta \le 2\pi.$$

For $\mathbf{r}_\phi \times \mathbf{r}_\theta$, we get

$$\begin{aligned}\mathbf{r}_\phi \times \mathbf{r}_\theta &= \begin{vmatrix} \mathbf{i} & \mathbf{j} & \mathbf{k} \\ a\cos\phi\cos\theta & a\cos\phi\sin\theta & -a\sin\phi \\ -a\sin\phi\sin\theta & a\sin\phi\cos\theta & 0 \end{vmatrix} \\ &= (a^2\sin^2\phi\cos\theta)\mathbf{i} + (a^2\sin^2\phi\sin\theta)\mathbf{j} + (a^2\sin\phi\cos\phi)\mathbf{k}.\end{aligned}$$

Thus,

$$\begin{aligned}|\mathbf{r}_\phi \times \mathbf{r}_\theta| &= \sqrt{a^4\sin^4\phi\cos^2\theta + a^4\sin^4\phi\sin^2\theta + a^4\sin^2\phi\cos^2\phi} \\ &= \sqrt{a^4\sin^4\phi + a^4\sin^2\phi\cos^2\phi} = \sqrt{a^4\sin^2\phi(\sin^2\phi + \cos^2\phi)} \\ &= a^2\sqrt{\sin^2\phi} = a^2\sin\phi,\end{aligned}$$

since $\sin\phi \ge 0$ for $0 \le \phi \le \pi$. Therefore, the area of the sphere is

$$\begin{aligned}A &= \int_0^{2\pi}\int_0^{\pi} a^2\sin\phi\, d\phi\, d\theta \\ &= \int_0^{2\pi}\Big[-a^2\cos\phi\Big]_0^{\pi} d\theta = \int_0^{2\pi} 2a^2\, d\theta = 4\pi a^2 \text{ units squared.}\end{aligned}$$

This agrees with the well-known formula for the surface area of a sphere. ■

Surface Integrals

Having found a formula for calculating the area of a parametrized surface, we can now integrate a function over the surface using the parametrized form.

DEFINITION Parametric Surface Integral

If S is a smooth surface defined parametrically as $\mathbf{r}(u, v) = f(u, v)\mathbf{i} + g(u, v)\mathbf{j} + h(u, v)\mathbf{k}$, $a \le u \le b$, $c \le v \le d$, and $G(x, y, z)$ is a continuous function defined on S, then the **integral of G over S** is

$$\iint_S G(x, y, z)\, d\sigma = \int_c^d \int_a^b G(f(u, v), g(u, v), h(u, v))\,|\mathbf{r}_u \times \mathbf{r}_v|\, du\, dv.$$

EXAMPLE 6 Integrating Over a Surface Defined Parametrically

Integrate $G(x, y, z) = x^2$ over the cone $z = \sqrt{x^2 + y^2}$, $0 \le z \le 1$.

Solution Continuing the work in Examples 1 and 4, we have $|\mathbf{r}_r \times \mathbf{r}_\theta| = \sqrt{2}r$ and

$$\begin{aligned}
\iint_S x^2\, d\sigma &= \int_0^{2\pi}\int_0^1 \left(r^2\cos^2\theta\right)\left(\sqrt{2}r\right) dr\, d\theta \qquad x = r\cos\theta \\
&= \sqrt{2}\int_0^{2\pi}\int_0^1 r^3\cos^2\theta\, dr\, d\theta \\
&= \frac{\sqrt{2}}{4}\int_0^{2\pi}\cos^2\theta\, d\theta = \frac{\sqrt{2}}{4}\left[\frac{\theta}{2} + \frac{1}{4}\sin 2\theta\right]_0^{2\pi} = \frac{\pi\sqrt{2}}{4}.
\end{aligned}$$

■

EXAMPLE 7 Finding Flux

Find the flux of $\mathbf{F} = yz\mathbf{i} + x\mathbf{j} - z^2\mathbf{k}$ outward through the parabolic cylinder $y = x^2$, $0 \le x \le 1$, $0 \le z \le 4$ (Figure 16.57).

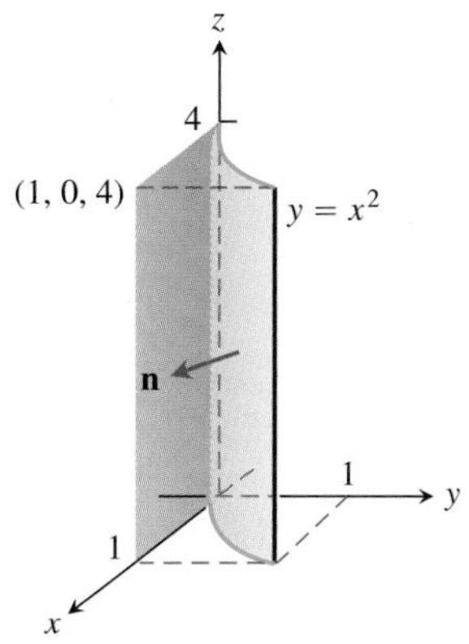

FIGURE 16.57 Finding the flux through the surface of a parabolic cylinder (Example 7).

Solution On the surface we have $x = x$, $y = x^2$, and $z = z$, so we automatically have the parametrization $\mathbf{r}(x, z) = x\mathbf{i} + x^2\mathbf{j} + z\mathbf{k}$, $0 \le x \le 1$, $0 \le z \le 4$. The cross product of tangent vectors is

$$\mathbf{r}_x \times \mathbf{r}_z = \begin{vmatrix} \mathbf{i} & \mathbf{j} & \mathbf{k} \\ 1 & 2x & 0 \\ 0 & 0 & 1 \end{vmatrix} = 2x\mathbf{i} - \mathbf{j}.$$

The unit normal pointing outward from the surface is

$$\mathbf{n} = \frac{\mathbf{r}_x \times \mathbf{r}_z}{|\mathbf{r}_x \times \mathbf{r}_z|} = \frac{2x\mathbf{i} - \mathbf{j}}{\sqrt{4x^2 + 1}}.$$

On the surface, $y = x^2$, so the vector field there is

$$\mathbf{F} = yz\mathbf{i} + x\mathbf{j} - z^2\mathbf{k} = x^2z\mathbf{i} + x\mathbf{j} - z^2\mathbf{k}.$$

Thus,

$$\begin{aligned}
\mathbf{F}\cdot\mathbf{n} &= \frac{1}{\sqrt{4x^2 + 1}}\left((x^2z)(2x) + (x)(-1) + (-z^2)(0)\right) \\
&= \frac{2x^3z - x}{\sqrt{4x^2 + 1}}.
\end{aligned}$$

The flux of **F** outward through the surface is

$$\begin{aligned}\iint_S \mathbf{F}\cdot\mathbf{n}\,d\sigma &= \int_0^4\int_0^1 \frac{2x^3z - x}{\sqrt{4x^2+1}}\,|\mathbf{r}_x \times \mathbf{r}_z|\,dx\,dz\\
&= \int_0^4\int_0^1 \frac{2x^3z - x}{\sqrt{4x^2+1}}\sqrt{4x^2+1}\,dx\,dz\\
&= \int_0^4\int_0^1 (2x^3z - x)\,dx\,dz = \int_0^4 \left[\frac{1}{2}x^4z - \frac{1}{2}x^2\right]_{x=0}^{x=1} dz\\
&= \int_0^4 \frac{1}{2}(z-1)\,dz = \frac{1}{4}(z-1)^2\bigg]_0^4\\
&= \frac{1}{4}(9) - \frac{1}{4}(1) = 2.\end{aligned}$$

■

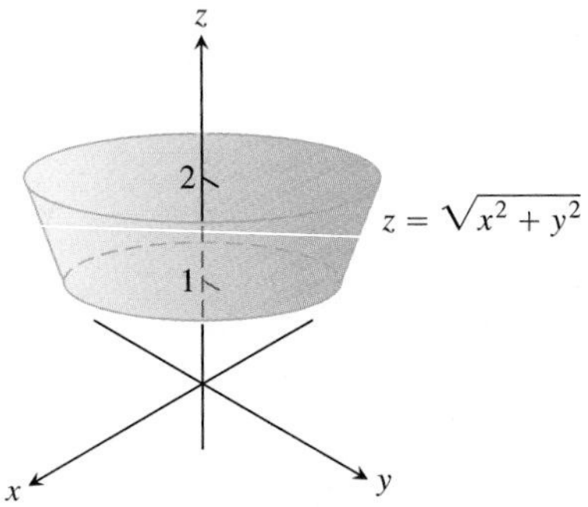

FIGURE 16.58 The cone frustum formed when the cone $z = \sqrt{x^2 + y^2}$ is cut by the planes $z = 1$ and $z = 2$ (Example 8).

EXAMPLE 8 Finding a Center of Mass

Find the center of mass of a thin shell of constant density δ cut from the cone $z = \sqrt{x^2 + y^2}$ by the planes $z = 1$ and $z = 2$ (Figure 16.58).

Solution The symmetry of the surface about the z-axis tells us that $\bar{x} = \bar{y} = 0$. We find $\bar{z} = M_{xy}/M$. Working as in Examples 1 and 4, we have

$$\mathbf{r}(r, \theta) = r\cos\theta\mathbf{i} + r\sin\theta\mathbf{j} + r\mathbf{k}, \qquad 1 \le r \le 2, \quad 0 \le \theta \le 2\pi,$$

and

$$|\mathbf{r}_r \times \mathbf{r}_\theta| = \sqrt{2}r.$$

Therefore,

$$\begin{aligned}M &= \iint_S \delta\,d\sigma = \int_0^{2\pi}\int_1^2 \delta\sqrt{2}r\,dr\,d\theta\\
&= \delta\sqrt{2}\int_0^{2\pi}\left[\frac{r^2}{2}\right]_1^2 d\theta = \delta\sqrt{2}\int_0^{2\pi}\left(2 - \frac{1}{2}\right)d\theta\\
&= \delta\sqrt{2}\left[\frac{3\theta}{2}\right]_0^{2\pi} = 3\pi\delta\sqrt{2}\\
M_{xy} &= \iint_S \delta z\,d\sigma = \int_0^{2\pi}\int_1^2 \delta r\sqrt{2}r\,dr\,d\theta\\
&= \delta\sqrt{2}\int_0^{2\pi}\int_1^2 r^2\,dr\,d\theta = \delta\sqrt{2}\int_0^{2\pi}\left[\frac{r^3}{3}\right]_1^2 d\theta\\
&= \delta\sqrt{2}\int_0^{2\pi}\frac{7}{3}\,d\theta = \frac{14}{3}\pi\delta\sqrt{2}\\
\bar{z} &= \frac{M_{xy}}{M} = \frac{14\pi\delta\sqrt{2}}{3\left(3\pi\delta\sqrt{2}\right)} = \frac{14}{9}.\end{aligned}$$

The shell's center of mass is the point (0, 0, 14/9). ■

EXERCISES 16.6

Finding Parametrizations for Surfaces

In Exercises 1–16, find a parametrization of the surface. (There are many correct ways to do these, so your answers may not be the same as those in the back of the book.)

1. The paraboloid $z = x^2 + y^2, z \leq 4$
2. The paraboloid $z = 9 - x^2 - y^2, z \geq 0$
3. **Cone frustum** The first-octant portion of the cone $z = \sqrt{x^2 + y^2}/2$ between the planes $z = 0$ and $z = 3$
4. **Cone frustum** The portion of the cone $z = 2\sqrt{x^2 + y^2}$ between the planes $z = 2$ and $z = 4$
5. **Spherical cap** The cap cut from the sphere $x^2 + y^2 + z^2 = 9$ by the cone $z = \sqrt{x^2 + y^2}$
6. **Spherical cap** The portion of the sphere $x^2 + y^2 + z^2 = 4$ in the first octant between the xy-plane and the cone $z = \sqrt{x^2 + y^2}$
7. **Spherical band** The portion of the sphere $x^2 + y^2 + z^2 = 3$ between the planes $z = \sqrt{3}/2$ and $z = -\sqrt{3}/2$
8. **Spherical cap** The upper portion cut from the sphere $x^2 + y^2 + z^2 = 8$ by the plane $z = -2$
9. **Parabolic cylinder between planes** The surface cut from the parabolic cylinder $z = 4 - y^2$ by the planes $x = 0, x = 2$, and $z = 0$
10. **Parabolic cylinder between planes** The surface cut from the parabolic cylinder $y = x^2$ by the planes $z = 0, z = 3$ and $y = 2$
11. **Circular cylinder band** The portion of the cylinder $y^2 + z^2 = 9$ between the planes $x = 0$ and $x = 3$
12. **Circular cylinder band** The portion of the cylinder $x^2 + z^2 = 4$ above the xy-plane between the planes $y = -2$ and $y = 2$
13. **Tilted plane inside cylinder** The portion of the plane $x + y + z = 1$
    a. Inside the cylinder $x^2 + y^2 = 9$
    b. Inside the cylinder $y^2 + z^2 = 9$
14. **Tilted plane inside cylinder** The portion of the plane $x - y + 2z = 2$
    a. Inside the cylinder $x^2 + z^2 = 3$
    b. Inside the cylinder $y^2 + z^2 = 2$
15. **Circular cylinder band** The portion of the cylinder $(x - 2)^2 + z^2 = 4$ between the planes $y = 0$ and $y = 3$
16. **Circular cylinder band** The portion of the cylinder $y^2 + (z - 5)^2 = 25$ between the planes $x = 0$ and $x = 10$

Areas of Parametrized Surfaces

In Exercises 17–26, use a parametrization to express the area of the surface as a double integral. Then evaluate the integral. (There are many correct ways to set up the integrals, so your integrals may not be the same as those in the back of the book. They should have the same values, however.)

17. **Titled plane inside cylinder** The portion of the plane $y + 2z = 2$ inside the cylinder $x^2 + y^2 = 1$
18. **Plane inside cylinder** The portion of the plane $z = -x$ inside the cylinder $x^2 + y^2 = 4$
19. **Cone frustum** The portion of the cone $z = 2\sqrt{x^2 + y^2}$ between the planes $z = 2$ and $z = 6$
20. **Cone frustum** The portion of the cone $z = \sqrt{x^2 + y^2}/3$ between the planes $z = 1$ and $z = 4/3$
21. **Circular cylinder band** The portion of the cylinder $x^2 + y^2 = 1$ between the planes $z = 1$ and $z = 4$
22. **Circular cylinder band** The portion of the cylinder $x^2 + z^2 = 10$ between the planes $y = -1$ and $y = 1$
23. **Parabolic cap** The cap cut from the paraboloid $z = 2 - x^2 - y^2$ by the cone $z = \sqrt{x^2 + y^2}$
24. **Parabolic band** The portion of the paraboloid $z = x^2 + y^2$ between the planes $z = 1$ and $z = 4$
25. **Sawed-off sphere** The lower portion cut from the sphere $x^2 + y^2 + z^2 = 2$ by the cone $z = \sqrt{x^2 + y^2}$
26. **Spherical band** The portion of the sphere $x^2 + y^2 + z^2 = 4$ between the planes $z = -1$ and $z = \sqrt{3}$

Integrals Over Parametrized Surfaces

In Exercises 27–34, integrate the given function over the given surface.

27. **Parabolic cylinder** $G(x, y, z) = x$, over the parabolic cylinder $y = x^2, 0 \leq x \leq 2, 0 \leq z \leq 3$
28. **Circular cylinder** $G(x, y, z) = z$, over the cylindrical surface $y^2 + z^2 = 4, z \geq 0, 1 \leq x \leq 4$
29. **Sphere** $G(x, y, z) = x^2$, over the unit sphere $x^2 + y^2 + z^2 = 1$
30. **Hemisphere** $G(x, y, z) = z^2$, over the hemisphere $x^2 + y^2 + z^2 = a^2, z \geq 0$
31. **Portion of plane** $F(x, y, z) = z$, over the portion of the plane $x + y + z = 4$ that lies above the square $0 \leq x \leq 1$, $0 \leq y \leq 1$, in the xy-plane
32. **Cone** $F(x, y, z) = z - x$, over the cone $z = \sqrt{x^2 + y^2}$, $0 \leq z \leq 1$
33. **Parabolic dome** $H(x, y, z) = x^2\sqrt{5 - 4z}$, over the parabolic dome $z = 1 - x^2 - y^2, z \geq 0$
34. **Spherical cap** $H(x, y, z) = yz$, over the part of the sphere $x^2 + y^2 + z^2 = 4$ that lies above the cone $z = \sqrt{x^2 + y^2}$

Flux Across Parametrized Surfaces

In Exercises 35–44, use a parametrization to find the flux $\iint_S \mathbf{F}\cdot\mathbf{n}\,d\sigma$ across the surface in the given direction.

35. **Parabolic cylinder** $\mathbf{F} = z^2\mathbf{i} + x\mathbf{j} - 3z\mathbf{k}$ outward (normal away from the x-axis) through the surface cut from the parabolic cylinder $z = 4 - y^2$ by the planes $x = 0, x = 1$, and $z = 0$

36. **Parabolic cylinder** $\mathbf{F} = x^2\mathbf{j} - xz\mathbf{k}$ outward (normal away from the yz-plane) through the surface cut from the parabolic cylinder $y = x^2, -1 \le x \le 1$, by the planes $z = 0$ and $z = 2$

37. **Sphere** $\mathbf{F} = z\mathbf{k}$ across the portion of the sphere $x^2 + y^2 + z^2 = a^2$ in the first octant in the direction away from the origin

38. **Sphere** $\mathbf{F} = x\mathbf{i} + y\mathbf{j} + z\mathbf{k}$ across the sphere $x^2 + y^2 + z^2 = a^2$ in the direction away from the origin

39. **Plane** $\mathbf{F} = 2xy\mathbf{i} + 2yz\mathbf{j} + 2xz\mathbf{k}$ upward across the portion of the plane $x + y + z = 2a$ that lies above the square $0 \le x \le a, 0 \le y \le a$, in the xy-plane

40. **Cylinder** $\mathbf{F} = x\mathbf{i} + y\mathbf{j} + z\mathbf{k}$ outward through the portion of the cylinder $x^2 + y^2 = 1$ cut by the planes $z = 0$ and $z = a$

41. **Cone** $\mathbf{F} = xy\mathbf{i} - z\mathbf{k}$ outward (normal away from the z-axis) through the cone $z = \sqrt{x^2 + y^2}, 0 \le z \le 1$

42. **Cone** $\mathbf{F} = y^2\mathbf{i} + xz\mathbf{j} - \mathbf{k}$ outward (normal away from the z-axis) through the cone $z = 2\sqrt{x^2 + y^2}, 0 \le z \le 2$

43. **Cone frustum** $\mathbf{F} = -x\mathbf{i} - y\mathbf{j} + z^2\mathbf{k}$ outward (normal away from the z-axis) through the portion of the cone $z = \sqrt{x^2 + y^2}$ between the planes $z = 1$ and $z = 2$

44. **Paraboloid** $\mathbf{F} = 4x\mathbf{i} + 4y\mathbf{j} + 2\mathbf{k}$ outward (normal way from the z-axis) through the surface cut from the bottom of the paraboloid $z = x^2 + y^2$ by the plane $z = 1$

Moments and Masses

45. Find the centroid of the portion of the sphere $x^2 + y^2 + z^2 = a^2$ that lies in the first octant.

46. Find the center of mass and the moment of inertia and radius of gyration about the z-axis of a thin shell of constant density δ cut from the cone $x^2 + y^2 - z^2 = 0$ by the planes $z = 1$ and $z = 2$.

47. Find the moment of inertia about the z-axis of a thin spherical shell $x^2 + y^2 + z^2 = a^2$ of constant density δ.

48. Find the moment of inertia about the z-axis of a thin conical shell $z = \sqrt{x^2 + y^2}, 0 \le z \le 1$, of constant density δ.

Planes Tangent to Parametrized Surfaces

The tangent plane at a point $P_0(f(u_0, v_0), g(u_0, v_0), h(u_0, v_0))$ on a parametrized surface $\mathbf{r}(u, v) = f(u, v)\mathbf{i} + g(u, v)\mathbf{j} + h(u, v)\mathbf{k}$ is the plane through P_0 normal to the vector $\mathbf{r}_u(u_0, v_0) \times \mathbf{r}_v(u_0, v_0)$, the cross product of the tangent vectors $\mathbf{r}_u(u_0, v_0)$ and $\mathbf{r}_v(u_0, v_0)$ at P_0. In Exercises 49–52, find an equation for the plane tangent to the surface at P_0. Then find a Cartesian equation for the surface and sketch the surface and tangent plane together.

49. **Cone** The cone $\mathbf{r}(r, \theta) = (r\cos\theta)\mathbf{i} + (r\sin\theta)\mathbf{j} + r\mathbf{k}, r \ge 0, 0 \le \theta \le 2\pi$ at the point $P_0(\sqrt{2}, \sqrt{2}, 2)$ corresponding to $(r, \theta) = (2, \pi/4)$

50. **Hemisphere** The hemisphere surface $\mathbf{r}(\phi, \theta) = (4\sin\phi\cos\theta)\mathbf{i} + (4\sin\phi\sin\theta)\mathbf{j} + (4\cos\phi)\mathbf{k}, 0 \le \phi \le \pi/2, 0 \le \theta \le 2\pi$, at the point $P_0(\sqrt{2}, \sqrt{2}, 2\sqrt{3})$ corresponding to $(\phi, \theta) = (\pi/6, \pi/4)$

51. **Circular cylinder** The circular cylinder $\mathbf{r}(\theta, z) = (3\sin 2\theta)\mathbf{i} + (6\sin^2\theta)\mathbf{j} + z\mathbf{k}, 0 \le \theta \le \pi$, at the point $P_0(3\sqrt{3}/2, 9/2, 0)$ corresponding to $(\theta, z) = (\pi/3, 0)$ (See Example 3.)

52. **Parabolic cylinder** The parabolic cylinder surface $\mathbf{r}(x, y) = x\mathbf{i} + y\mathbf{j} - x^2\mathbf{k}, -\infty < x < \infty, -\infty < y < \infty$, at the point $P_0(1, 2, -1)$ corresponding to $(x, y) = (1, 2)$

Further Examples of Parametrizations

53. **a.** A *torus of revolution* (doughnut) is obtained by rotating a circle C in the xz-plane about the z-axis in space. (See the accompanying figure.) If C has radius $r > 0$ and center $(R, 0, 0)$, show that a parametrization of the torus is

$$\mathbf{r}(u, v) = ((R + r\cos u)\cos v)\mathbf{i} + ((R + r\cos u)\sin v)\mathbf{j} + (r\sin u)\mathbf{k},$$

where $0 \le u \le 2\pi$ and $0 \le v \le 2\pi$ are the angles in the figure.

b. Show that the surface area of the torus is $A = 4\pi^2 Rr$.

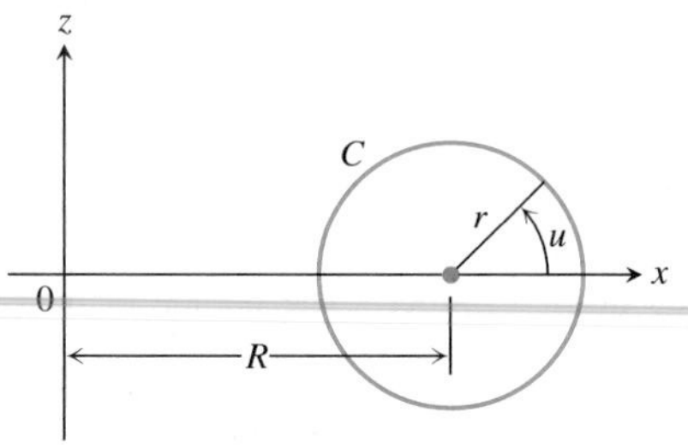

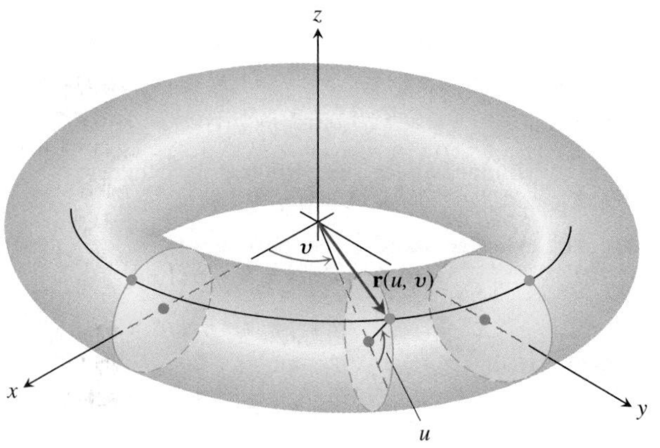

54. Parametrization of a surface of revolution Suppose that the parametrized curve C: $(f(u), g(u))$ is revolved about the x-axis, where $g(u) > 0$ for $a \le u \le b$.

a. Show that

$$\mathbf{r}(u, v) = f(u)\mathbf{i} + (g(u)\cos v)\mathbf{j} + (g(u)\sin v)\mathbf{k}$$

is a parametrization of the resulting surface of revolution, where $0 \le v \le 2\pi$ is the angle from the xy-plane to the point $\mathbf{r}(u, v)$ on the surface. (See the accompanying figure.) Notice that $f(u)$ measures distance *along* the axis of revolution and $g(u)$ measures distance *from* the axis of revolution.

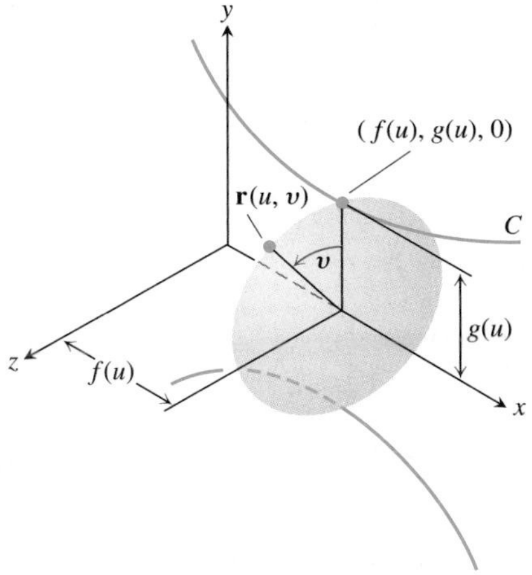

b. Find a parametrization for the surface obtained by revolving the curve $x = y^2, y \ge 0$, about the x-axis.

55. a. Parametrization of an ellipsoid Recall the parametrization $x = a\cos\theta, y = b\sin\theta, 0 \le \theta \le 2\pi$ for the ellipse $(x^2/a^2) + (y^2/b^2) = 1$ (Section 3.5, Example 13). Using the angles θ and ϕ in spherical coordinates, show that

$$\mathbf{r}(\theta, \phi) = (a\cos\theta\cos\phi)\mathbf{i} + (b\sin\theta\cos\phi)\mathbf{j} + (c\sin\phi)\mathbf{k}$$

is a parametrization of the ellipsoid $(x^2/a^2) + (y^2/b^2) + (z^2/c^2) = 1$.

b. Write an integral for the surface area of the ellipsoid, but do not evaluate the integral.

56. Hyperboloid of one sheet

a. Find a parametrization for the hyperboloid of one sheet $x^2 + y^2 - z^2 = 1$ in terms of the angle θ associated with the circle $x^2 + y^2 = r^2$ and the hyperbolic parameter u associated with the hyperbolic function $r^2 - z^2 = 1$. (See Section 7.8, Exercise 84.)

b. Generalize the result in part (a) to the hyperboloid $(x^2/a^2) + (y^2/b^2) - (z^2/c^2) = 1$.

57. (*Continuation of Exercise 56.*) Find a Cartesian equation for the plane tangent to the hyperboloid $x^2 + y^2 - z^2 = 25$ at the point $(x_0, y_0, 0)$, where $x_0^2 + y_0^2 = 25$.

58. Hyperboloid of two sheets Find a parametrization of the hyperboloid of two sheets $(z^2/c^2) - (x^2/a^2) - (y^2/b^2) = 1$.

16.7 Stokes' Theorem

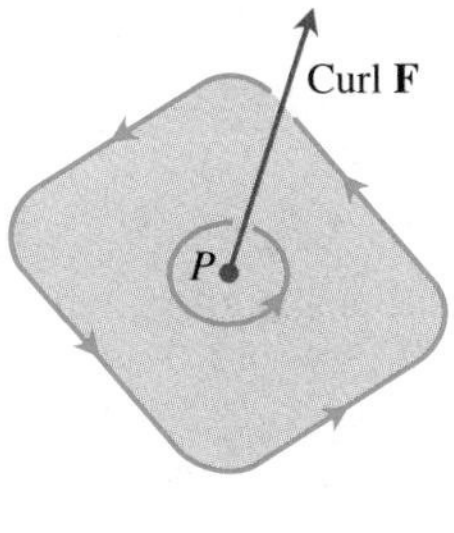

FIGURE 16.59 The circulation vector at a point P in a plane in a three-dimensional fluid flow. Notice its right-hand relation to the circulation line.

As we saw in Section 16.4, the circulation density or curl component of a two-dimensional field $\mathbf{F} = M\mathbf{i} + N\mathbf{j}$ at a point (x, y) is described by the scalar quantity $(\partial N/\partial x - \partial M/\partial y)$. In three dimensions, the circulation around a point P in a plane is described with a vector. This vector is normal to the plane of the circulation (Figure 16.59) and points in the direction that gives it a right-hand relation to the circulation line. The length of the vector gives the rate of the fluid's rotation, which usually varies as the circulation plane is tilted about P. It turns out that the vector of greatest circulation in a flow with velocity field $\mathbf{F} = M\mathbf{i} + N\mathbf{j} + P\mathbf{k}$ is the **curl vector**

$$\operatorname{curl}\mathbf{F} = \left(\frac{\partial P}{\partial y} - \frac{\partial N}{\partial z}\right)\mathbf{i} + \left(\frac{\partial M}{\partial z} - \frac{\partial P}{\partial x}\right)\mathbf{j} + \left(\frac{\partial N}{\partial x} - \frac{\partial M}{\partial y}\right)\mathbf{k}. \tag{1}$$

We get this information from Stokes' Theorem, the generalization of the circulation-curl form of Green's Theorem to space.

Notice that $(\operatorname{curl}\mathbf{F})\cdot\mathbf{k} = (\partial N/\partial x - \partial M/\partial y)$ is consistent with our definition in Section 16.4 when $\mathbf{F} = M(x, y)\mathbf{i} + N(x, y)\mathbf{j}$. The formula for curl $\mathbf{F}$ in Equation (1) is often written using the symbolic operator

$$\nabla = \mathbf{i}\frac{\partial}{\partial x} + \mathbf{j}\frac{\partial}{\partial y} + \mathbf{k}\frac{\partial}{\partial z}. \tag{2}$$

(The symbol ∇ is pronounced "del.") The curl of $\mathbf{F}$ is $\nabla \times \mathbf{F}$:

$$\begin{aligned}\nabla \times \mathbf{F} &= \begin{vmatrix} \mathbf{i} & \mathbf{j} & \mathbf{k} \\ \dfrac{\partial}{\partial x} & \dfrac{\partial}{\partial y} & \dfrac{\partial}{\partial z} \\ M & N & P \end{vmatrix} \\ &= \left(\frac{\partial P}{\partial y} - \frac{\partial N}{\partial z}\right)\mathbf{i} + \left(\frac{\partial M}{\partial z} - \frac{\partial P}{\partial x}\right)\mathbf{j} + \left(\frac{\partial N}{\partial x} - \frac{\partial M}{\partial y}\right)\mathbf{k} \\ &= \text{curl } \mathbf{F}.\end{aligned}$$

$$\text{curl } \mathbf{F} = \nabla \times \mathbf{F} \tag{3}$$

EXAMPLE 1 Finding Curl F

Find the curl of $\mathbf{F} = (x^2 - y)\mathbf{i} + 4z\mathbf{j} + x^2\mathbf{k}$.

Solution

$$\begin{aligned}\text{curl } \mathbf{F} &= \nabla \times \mathbf{F} \qquad \text{Equation (3)} \\ &= \begin{vmatrix} \mathbf{i} & \mathbf{j} & \mathbf{k} \\ \dfrac{\partial}{\partial x} & \dfrac{\partial}{\partial y} & \dfrac{\partial}{\partial z} \\ x^2 - y & 4z & x^2 \end{vmatrix} \\ &= \left(\frac{\partial}{\partial y}(x^2) - \frac{\partial}{\partial z}(4z)\right)\mathbf{i} - \left(\frac{\partial}{\partial x}(x^2) - \frac{\partial}{\partial z}(x^2 - y)\right)\mathbf{j} \\ &\quad + \left(\frac{\partial}{\partial x}(4z) - \frac{\partial}{\partial y}(x^2 - y)\right)\mathbf{k} \\ &= (0 - 4)\mathbf{i} - (2x - 0)\mathbf{j} + (0 + 1)\mathbf{k} \\ &= -4\mathbf{i} - 2x\mathbf{j} + \mathbf{k}\end{aligned}$$

■

As we will see, the operator ∇ has a number of other applications. For instance, when applied to a scalar function $f(x, y, z)$, it gives the gradient of f:

$$\nabla f = \frac{\partial f}{\partial x}\mathbf{i} + \frac{\partial f}{\partial y}\mathbf{j} + \frac{\partial f}{\partial z}\mathbf{k}.$$

This may now be read as "del f" as well as "grad f."

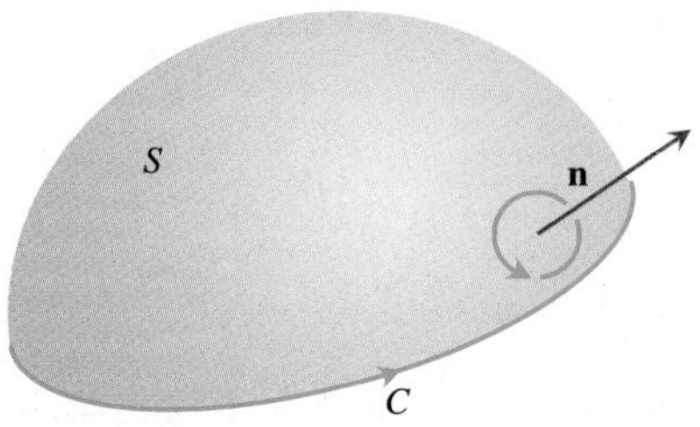

FIGURE 16.60 The orientation of the bounding curve C gives it a right-handed relation to the normal field $\mathbf{n}$.

Stokes' Theorem

Stokes' Theorem says that, under conditions normally met in practice, the circulation of a vector field around the boundary of an oriented surface in space in the direction counterclockwise with respect to the surface's unit normal vector field $\mathbf{n}$ (Figure 16.60) equals the integral of the normal component of the curl of the field over the surface.

THEOREM 5 Stokes' Theorem

The circulation of a vector field $\mathbf{F} = M\mathbf{i} + N\mathbf{j} + P\mathbf{k}$ around the boundary C of an oriented surface S in the direction counterclockwise with respect to the surface's unit normal vector $\mathbf{n}$ equals the integral of $\nabla \times \mathbf{F} \cdot \mathbf{n}$ over S.

$$\underbrace{\oint_C \mathbf{F} \cdot d\mathbf{r}}_{\text{Counterclockwise circulation}} = \underbrace{\iint_S \nabla \times \mathbf{F} \cdot \mathbf{n}\, d\sigma}_{\text{Curl integral}} \tag{4}$$

Notice from Equation (4) that if two different oriented surfaces S_1 and S_2 have the same boundary C, their curl integrals are equal:

$$\iint_{S_1} \nabla \times \mathbf{F} \cdot \mathbf{n}_1\, d\sigma = \iint_{S_2} \nabla \times \mathbf{F} \cdot \mathbf{n}_2\, d\sigma.$$

Both curl integrals equal the counterclockwise circulation integral on the left side of Equation (4) as long as the unit normal vectors $\mathbf{n}_1$ and $\mathbf{n}_2$ correctly orient the surfaces.

Naturally, we need some mathematical restrictions on $\mathbf{F}$, C, and S to ensure the existence of the integrals in Stokes' equation. The usual restrictions are that all functions, vector fields, and their derivatives be continuous.

If C is a curve in the xy-plane, oriented counterclockwise, and R is the region in the xy-plane bounded by C, then $d\sigma = dx\, dy$ and

$$(\nabla \times \mathbf{F}) \cdot \mathbf{n} = (\nabla \times \mathbf{F}) \cdot \mathbf{k} = \left(\frac{\partial N}{\partial x} - \frac{\partial M}{\partial y}\right).$$

Under these conditions, Stokes' equation becomes

$$\oint_C \mathbf{F} \cdot d\mathbf{r} = \iint_R \left(\frac{\partial N}{\partial x} - \frac{\partial M}{\partial y}\right) dx\, dy,$$

which is the circulation-curl form of the equation in Green's Theorem. Conversely, by reversing these steps we can rewrite the circulation-curl form of Green's Theorem for two-dimensional fields in del notation as

$$\oint_C \mathbf{F} \cdot d\mathbf{r} = \iint_R \nabla \times \mathbf{F} \cdot \mathbf{k}\, dA. \tag{5}$$

See Figure 16.61.

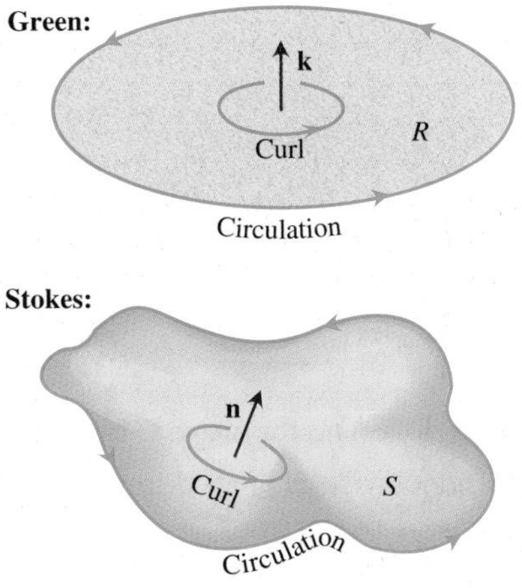

FIGURE 16.61 Comparison of Green's Theorem and Stokes' Theorem.

EXAMPLE 2 Verifying Stokes' Equation for a Hemisphere

Evaluate Equation (4) for the hemisphere $S: x^2 + y^2 + z^2 = 9, z \geq 0$, its bounding circle $C: x^2 + y^2 = 9, z = 0$, and the field $\mathbf{F} = y\mathbf{i} - x\mathbf{j}$.

Solution We calculate the counterclockwise circulation around C (as viewed from above) using the parametrization $\mathbf{r}(\theta) = (3\cos\theta)\mathbf{i} + (3\sin\theta)\mathbf{j},\ 0 \le \theta \le 2\pi$:

$$d\mathbf{r} = (-3\sin\theta\, d\theta)\mathbf{i} + (3\cos\theta\, d\theta)\mathbf{j}$$

$$\mathbf{F} = y\mathbf{i} - x\mathbf{j} = (3\sin\theta)\mathbf{i} - (3\cos\theta)\mathbf{j}$$

$$\mathbf{F}\cdot d\mathbf{r} = -9\sin^2\theta\, d\theta - 9\cos^2\theta\, d\theta = -9\, d\theta$$

$$\oint_C \mathbf{F}\cdot d\mathbf{r} = \int_0^{2\pi} -9\, d\theta = -18\pi.$$

For the curl integral of **F**, we have

$$\nabla\times\mathbf{F} = \left(\frac{\partial P}{\partial y} - \frac{\partial N}{\partial z}\right)\mathbf{i} + \left(\frac{\partial M}{\partial z} - \frac{\partial P}{\partial x}\right)\mathbf{j} + \left(\frac{\partial N}{\partial x} - \frac{\partial M}{\partial y}\right)\mathbf{k}$$

$$= (0-0)\mathbf{i} + (0-0)\mathbf{j} + (-1-1)\mathbf{k} = -2\mathbf{k}$$

$$\mathbf{n} = \frac{x\mathbf{i} + y\mathbf{j} + z\mathbf{k}}{\sqrt{x^2+y^2+z^2}} = \frac{x\mathbf{i} + y\mathbf{j} + z\mathbf{k}}{3} \qquad \text{Outer unit normal}$$

$$d\sigma = \frac{3}{z}\, dA \qquad \text{Section 16.5, Example 5, with } a = 3$$

$$\nabla\times\mathbf{F}\cdot\mathbf{n}\, d\sigma = -\frac{2z}{3}\frac{3}{z}\, dA = -2\, dA$$

and

$$\iint_S \nabla\times\mathbf{F}\cdot\mathbf{n}\, d\sigma = \iint_{x^2+y^2\le 9} -2\, dA = -18\pi.$$

The circulation around the circle equals the integral of the curl over the hemisphere, as it should. ■

EXAMPLE 3 Finding Circulation

Find the circulation of the field $\mathbf{F} = (x^2 - y)\mathbf{i} + 4z\mathbf{j} + x^2\mathbf{k}$ around the curve C in which the plane $z = 2$ meets the cone $z = \sqrt{x^2+y^2}$, counterclockwise as viewed from above (Figure 16.62).

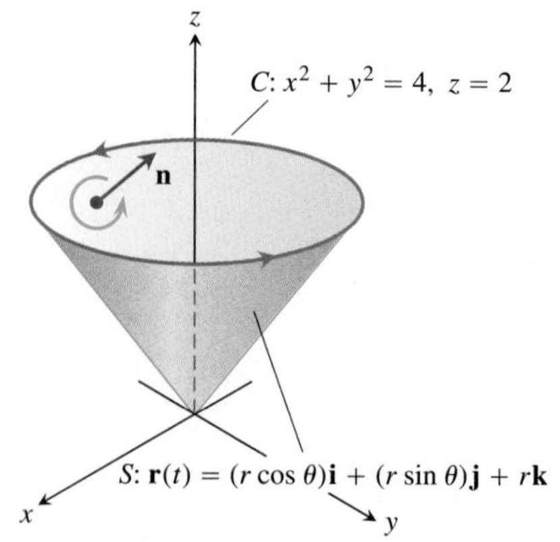

FIGURE 16.62 The curve C and cone S in Example 3.

Solution Stokes' Theorem enables us to find the circulation by integrating over the surface of the cone. Traversing C in the counterclockwise direction viewed from above corresponds to taking the *inner* normal **n** to the cone, the normal with a positive z-component.

We parametrize the cone as

$$\mathbf{r}(r,\theta) = (r\cos\theta)\mathbf{i} + (r\sin\theta)\mathbf{j} + r\mathbf{k}, \qquad 0 \le r \le 2, \quad 0 \le \theta \le 2\pi.$$

We then have

$$\mathbf{n} = \frac{\mathbf{r}_r\times\mathbf{r}_\theta}{|\mathbf{r}_r\times\mathbf{r}_\theta|} = \frac{-(r\cos\theta)\mathbf{i} - (r\sin\theta)\mathbf{j} + r\mathbf{k}}{r\sqrt{2}} \qquad \text{Section 16.6, Example 4}$$

$$= \frac{1}{\sqrt{2}}\left(-(\cos\theta)\mathbf{i} - (\sin\theta)\mathbf{j} + \mathbf{k}\right)$$

$$\begin{aligned} d\sigma &= r\sqrt{2}\, dr\, d\theta && \text{Section 16.6, Example 4} \\ \nabla \times \mathbf{F} &= -4\mathbf{i} - 2x\mathbf{j} + \mathbf{k} && \text{Example 1} \\ &= -4\mathbf{i} - 2r\cos\theta\mathbf{j} + \mathbf{k}. && x = r\cos\theta \end{aligned}$$

Accordingly,

$$\begin{aligned} \nabla \times \mathbf{F} \cdot \mathbf{n} &= \frac{1}{\sqrt{2}}\left(4\cos\theta + 2r\cos\theta\sin\theta + 1\right) \\ &= \frac{1}{\sqrt{2}}\left(4\cos\theta + r\sin 2\theta + 1\right) \end{aligned}$$

and the circulation is

$$\begin{aligned} \oint_C \mathbf{F} \cdot d\mathbf{r} &= \iint_S \nabla \times \mathbf{F} \cdot \mathbf{n}\, d\sigma && \text{Stokes' Theorem, Equation (4)} \\ &= \int_0^{2\pi}\int_0^2 \frac{1}{\sqrt{2}}\left(4\cos\theta + r\sin 2\theta + 1\right)\left(r\sqrt{2}\, dr\, d\theta\right) = 4\pi. \end{aligned}$$

■

Paddle Wheel Interpretation of $\nabla \times \mathbf{F}$

Suppose that $\mathbf{v}(x, y, z)$ is the velocity of a moving fluid whose density at (x, y, z) is $\delta(x, y, z)$ and let $\mathbf{F} = \delta\mathbf{v}$. Then

$$\oint_C \mathbf{F} \cdot d\mathbf{r}$$

is the circulation of the fluid around the closed curve C. By Stokes' Theorem, the circulation is equal to the flux of $\nabla \times \mathbf{F}$ through a surface S spanning C:

$$\oint_C \mathbf{F} \cdot d\mathbf{r} = \iint_S \nabla \times \mathbf{F} \cdot \mathbf{n}\, d\sigma.$$

Suppose we fix a point Q in the domain of $\mathbf{F}$ and a direction $\mathbf{u}$ at Q. Let C be a circle of radius ρ, with center at Q, whose plane is normal to $\mathbf{u}$. If $\nabla \times \mathbf{F}$ is continuous at Q, the average value of the $\mathbf{u}$-component of $\nabla \times \mathbf{F}$ over the circular disk S bounded by C approaches the $\mathbf{u}$-component of $\nabla \times \mathbf{F}$ at Q as $\rho \to 0$:

$$(\nabla \times \mathbf{F} \cdot \mathbf{u})_Q = \lim_{\rho \to 0} \frac{1}{\pi\rho^2} \iint_S \nabla \times \mathbf{F} \cdot \mathbf{u}\, d\sigma.$$

If we replace the surface integral in this last equation by the circulation, we get

$$(\nabla \times \mathbf{F} \cdot \mathbf{u})_Q = \lim_{\rho \to 0} \frac{1}{\pi\rho^2} \oint_C \mathbf{F} \cdot d\mathbf{r}. \tag{6}$$

The left-hand side of Equation (6) has its maximum value when $\mathbf{u}$ is the direction of $\nabla \times \mathbf{F}$. When ρ is small, the limit on the right-hand side of Equation (6) is approximately

$$\frac{1}{\pi\rho^2} \oint_C \mathbf{F} \cdot d\mathbf{r},$$

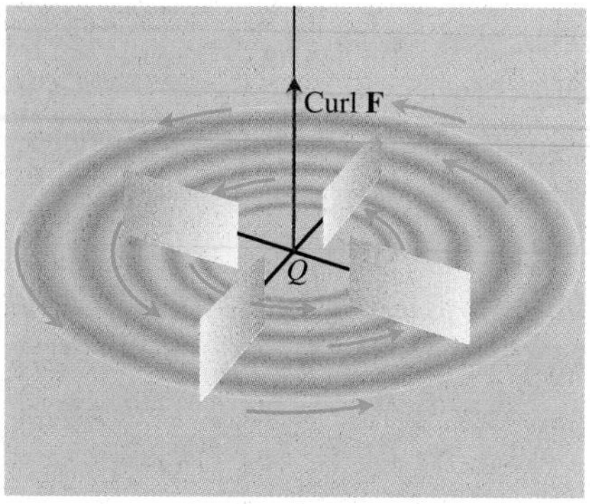

FIGURE 16.63 The paddle wheel interpretation of curl **F**.

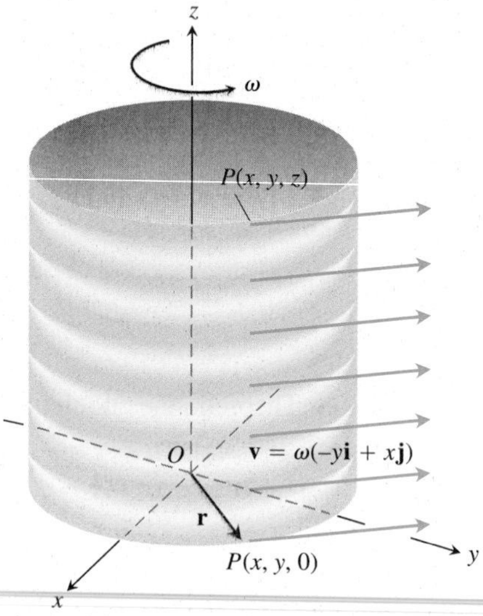

FIGURE 16.64 A steady rotational flow parallel to the xy-plane, with constant angular velocity ω in the positive (counterclockwise) direction (Example 4).

which is the circulation around C divided by the area of the disk (circulation density). Suppose that a small paddle wheel of radius ρ is introduced into the fluid at Q, with its axle directed along **u**. The circulation of the fluid around C will affect the rate of spin of the paddle wheel. The wheel will spin fastest when the circulation integral is maximized; therefore it will spin fastest when the axle of the paddle wheel points in the direction of $\nabla \times \mathbf{F}$ (Figure 16.63).

EXAMPLE 4 Relating $\nabla \times \mathbf{F}$ to Circulation Density

A fluid of constant density rotates around the z-axis with velocity $\mathbf{v} = \omega(-y\mathbf{i} + x\mathbf{j})$, where ω is a positive constant called the *angular velocity* of the rotation (Figure 16.64). If $\mathbf{F} = \mathbf{v}$, find $\nabla \times \mathbf{F}$ and relate it to the circulation density.

Solution With $\mathbf{F} = \mathbf{v} = -\omega y\mathbf{i} + \omega x\mathbf{j}$,

$$\nabla \times \mathbf{F} = \left(\frac{\partial P}{\partial y} - \frac{\partial N}{\partial z}\right)\mathbf{i} + \left(\frac{\partial M}{\partial z} - \frac{\partial P}{\partial x}\right)\mathbf{j} + \left(\frac{\partial N}{\partial x} - \frac{\partial M}{\partial y}\right)\mathbf{k}$$

$$= (0 - 0)\mathbf{i} + (0 - 0)\mathbf{j} + (\omega - (-\omega))\mathbf{k} = 2\omega\mathbf{k}.$$

By Stokes' Theorem, the circulation of **F** around a circle C of radius ρ bounding a disk S in a plane normal to $\nabla \times \mathbf{F}$, say the xy-plane, is

$$\oint_C \mathbf{F} \cdot d\mathbf{r} = \iint_S \nabla \times \mathbf{F} \cdot \mathbf{n}\, d\sigma = \iint_S 2\omega\mathbf{k} \cdot \mathbf{k}\, dx\, dy = (2\omega)(\pi\rho^2).$$

Thus,

$$(\nabla \times \mathbf{F}) \cdot \mathbf{k} = 2\omega = \frac{1}{\pi\rho^2}\oint_C \mathbf{F} \cdot d\mathbf{r},$$

consistent with Equation (6) when $\mathbf{u} = \mathbf{k}$. ■

EXAMPLE 5 Applying Stokes' Theorem

Use Stokes' Theorem to evaluate $\int_C \mathbf{F} \cdot d\mathbf{r}$, if $\mathbf{F} = xz\mathbf{i} + xy\mathbf{j} + 3xz\mathbf{k}$ and C is the boundary of the portion of the plane $2x + y + z = 2$ in the first octant, traversed counterclockwise as viewed from above (Figure 16.65).

Solution The plane is the level surface $f(x, y, z) = 2$ of the function $f(x, y, z) = 2x + y + z$. The unit normal vector

$$\mathbf{n} = \frac{\nabla f}{|\nabla f|} = \frac{(2\mathbf{i} + \mathbf{j} + \mathbf{k})}{|2\mathbf{i} + \mathbf{j} + \mathbf{k}|} = \frac{1}{\sqrt{6}}\left(2\mathbf{i} + \mathbf{j} + \mathbf{k}\right)$$

is consistent with the counterclockwise motion around C. To apply Stokes' Theorem, we find

$$\text{curl } \mathbf{F} = \nabla \times \mathbf{F} = \begin{vmatrix} \mathbf{i} & \mathbf{j} & \mathbf{k} \\ \dfrac{\partial}{\partial x} & \dfrac{\partial}{\partial y} & \dfrac{\partial}{\partial z} \\ xz & xy & 3xz \end{vmatrix} = (x - 3z)\mathbf{j} + y\mathbf{k}.$$

On the plane, z equals $2 - 2x - y$, so

$$\nabla \times \mathbf{F} = (x - 3(2 - 2x - y))\mathbf{j} + y\mathbf{k} = (7x + 3y - 6)\mathbf{j} + y\mathbf{k}$$

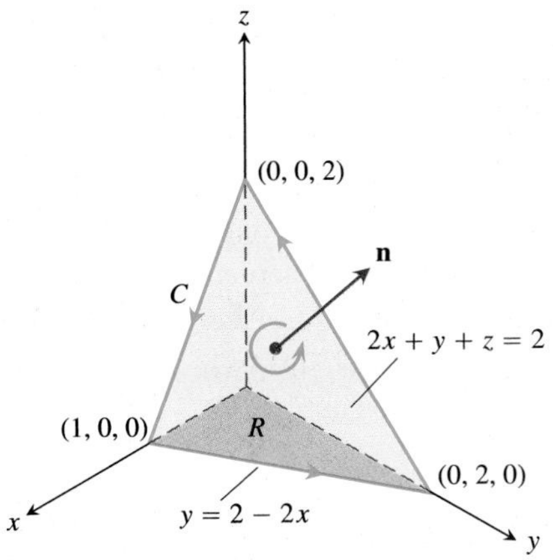

FIGURE 16.65 The planar surface in Example 5.

and

$$\nabla \times \mathbf{F} \cdot \mathbf{n} = \frac{1}{\sqrt{6}}\left(7x + 3y - 6 + y\right) = \frac{1}{\sqrt{6}}\left(7x + 4y - 6\right).$$

The surface area element is

$$d\sigma = \frac{|\nabla f|}{|\nabla f \cdot \mathbf{k}|}\, dA = \frac{\sqrt{6}}{1}\, dx\, dy.$$

The circulation is

$$\oint_C \mathbf{F} \cdot d\mathbf{r} = \iint_S \nabla \times \mathbf{F} \cdot \mathbf{n}\, d\sigma \qquad \text{Stokes' Theorem, Equation (4)}$$

$$= \int_0^1 \int_0^{2-2x} \frac{1}{\sqrt{6}}\left(7x + 4y - 6\right)\sqrt{6}\, dy\, dx$$

$$= \int_0^1 \int_0^{2-2x} (7x + 4y - 6)\, dy\, dx = -1.$$

Proof of Stokes' Theorem for Polyhedral Surfaces

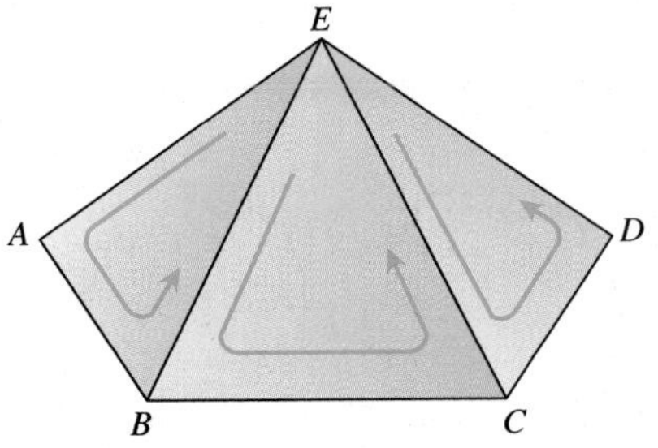

FIGURE 16.66 Part of a polyhedral surface.

Let S be a polyhedral surface consisting of a finite number of plane regions. (See Figure 16.66 for an example.) We apply Green's Theorem to each separate panel of S. There are two types of panels:

1. Those that are surrounded on all sides by other panels
2. Those that have one or more edges that are not adjacent to other panels.

The boundary Δ of S consists of those edges of the type 2 panels that are not adjacent to other panels. In Figure 16.66, the triangles EAB, BCE, and CDE represent a part of S, with $ABCD$ part of the boundary Δ. Applying Green's Theorem to the three triangles in turn and adding the results, we get

$$\left(\oint_{EAB} + \oint_{BCE} + \oint_{CDE}\right)\mathbf{F} \cdot d\mathbf{r} = \left(\iint_{EAB} + \iint_{BCE} + \iint_{CDE}\right)\nabla \times \mathbf{F} \cdot \mathbf{n}\, d\sigma. \qquad (7)$$

The three line integrals on the left-hand side of Equation (7) combine into a single line integral taken around the periphery $ABCDE$ because the integrals along interior segments cancel in pairs. For example, the integral along segment BE in triangle ABE is opposite in sign to the integral along the same segment in triangle EBC. The same holds for segment CE. Hence, Equation (7) reduces to

$$\oint_{ABCDE} \mathbf{F} \cdot d\mathbf{r} = \iint_{ABCDE} \nabla \times \mathbf{F} \cdot \mathbf{n}\, d\sigma.$$

When we apply Green's Theorem to all the panels and add the results, we get

$$\oint_{\Delta} \mathbf{F} \cdot d\mathbf{r} = \iint_S \nabla \times \mathbf{F} \cdot \mathbf{n}\, d\sigma.$$

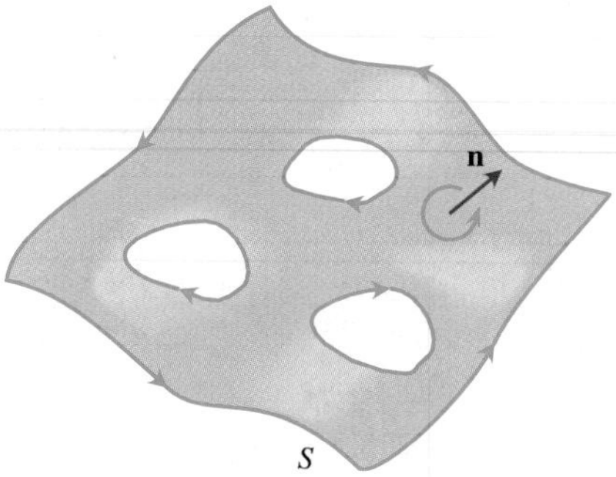

FIGURE 16.67 Stokes' Theorem also holds for oriented surfaces with holes.

This is Stokes' Theorem for a polyhedral surface S. You can find proofs for more general surfaces in advanced calculus texts.

Stokes' Theorem for Surfaces with Holes

Stokes' Theorem can be extended to an oriented surface S that has one or more holes (Figure 16.67), in a way analogous to the extension of Green's Theorem: The surface integral over S of the normal component of $\nabla \times \mathbf{F}$ equals the sum of the line integrals around all the boundary curves of the tangential component of $\mathbf{F}$, where the curves are to be traced in the direction induced by the orientation of S.

An Important Identity

The following identity arises frequently in mathematics and the physical sciences.

$$\text{curl grad } f = \mathbf{0} \qquad \text{or} \qquad \nabla \times \nabla f = \mathbf{0} \tag{8}$$

This identity holds for any function $f(x, y, z)$ whose second partial derivatives are continuous. The proof goes like this:

$$\nabla \times \nabla f = \begin{vmatrix} \mathbf{i} & \mathbf{j} & \mathbf{k} \\ \dfrac{\partial}{\partial x} & \dfrac{\partial}{\partial y} & \dfrac{\partial}{\partial z} \\ \dfrac{\partial f}{\partial x} & \dfrac{\partial f}{\partial y} & \dfrac{\partial f}{\partial z} \end{vmatrix} = (f_{zy} - f_{yz})\mathbf{i} - (f_{zx} - f_{xz})\mathbf{j} + (f_{yx} - f_{xy})\mathbf{k}.$$

If the second partial derivatives are continuous, the mixed second derivatives in parentheses are equal (Theorem 2, Section 14.3) and the vector is zero.

Conservative Fields and Stokes' Theorem

In Section 16.3, we found that a field $\mathbf{F}$ is conservative in an open region D in space is equivalent to the integral of $\mathbf{F}$ around every closed loop in D being zero. This, in turn, is equivalent in *simply connected* open regions to saying that $\nabla \times \mathbf{F} = \mathbf{0}$.

THEOREM 6 Curl F = 0 Related to the Closed-Loop Property
If $\nabla \times \mathbf{F} = \mathbf{0}$ at every point of a simply connected open region D in space, then on any piecewise-smooth closed path C in D,

$$\oint_C \mathbf{F} \cdot d\mathbf{r} = 0.$$

Sketch of a Proof Theorem 6 is usually proved in two steps. The first step is for simple closed curves. A theorem from topology, a branch of advanced mathematics, states that every differentiable simple closed curve C in a simply connected open region D is the boundary of a smooth two-sided surface S that also lies in D. Hence, by Stokes' Theorem,

$$\oint_C \mathbf{F} \cdot d\mathbf{r} = \iint_S \nabla \times \mathbf{F} \cdot \mathbf{n}\, d\sigma = 0.$$

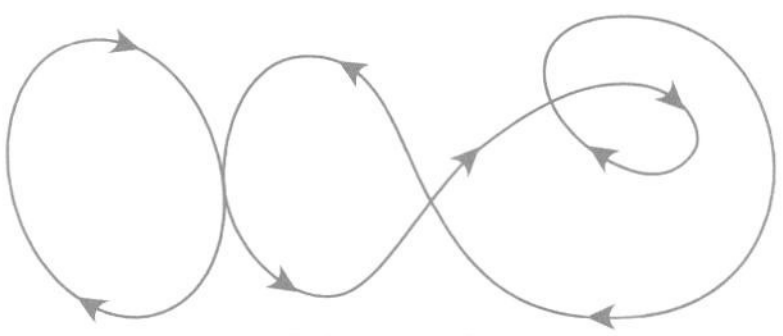

FIGURE 16.68 In a simply connected open region in space, differentiable curves that cross themselves can be divided into loops to which Stokes' Theorem applies.

The second step is for curves that cross themselves, like the one in Figure 16.68. The idea is to break these into simple loops spanned by orientable surfaces, apply Stokes' Theorem one loop at a time, and add the results. ■

The following diagram summarizes the results for conservative fields defined on connected, simply connected open regions.

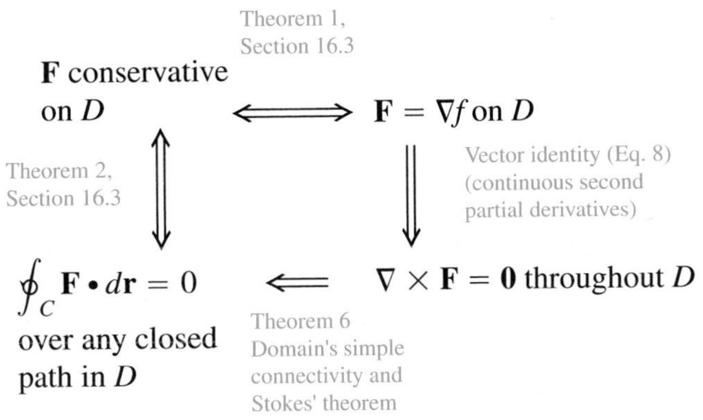

EXERCISES 16.7

Using Stokes' Theorem to Calculate Circulation

In Exercises 1–6, use the surface integral in Stokes' Theorem to calculate the circulation of the field $\mathbf{F}$ around the curve C in the indicated direction.

1. $\mathbf{F} = x^2\mathbf{i} + 2x\mathbf{j} + z^2\mathbf{k}$

   C: The ellipse $4x^2 + y^2 = 4$ in the xy-plane, counterclockwise when viewed from above

2. $\mathbf{F} = 2y\mathbf{i} + 3x\mathbf{j} - z^2\mathbf{k}$

   C: The circle $x^2 + y^2 = 9$ in the xy-plane, counterclockwise when viewed from above

3. $\mathbf{F} = y\mathbf{i} + xz\mathbf{j} + x^2\mathbf{k}$

   C: The boundary of the triangle cut from the plane $x + y + z = 1$ by the first octant, counterclockwise when viewed from above

4. $\mathbf{F} = (y^2 + z^2)\mathbf{i} + (x^2 + z^2)\mathbf{j} + (x^2 + y^2)\mathbf{k}$

   C: The boundary of the triangle cut from the plane $x + y + z = 1$ by the first octant, counterclockwise when viewed from above

5. $\mathbf{F} = (y^2 + z^2)\mathbf{i} + (x^2 + y^2)\mathbf{j} + (x^2 + y^2)\mathbf{k}$

   C: The square bounded by the lines $x = \pm 1$ and $y = \pm 1$ in the xy-plane, counterclockwise when viewed from above

6. $\mathbf{F} = x^2y^3\mathbf{i} + \mathbf{j} + z\mathbf{k}$

   C: The intersection of the cylinder $x^2 + y^2 = 4$ and the hemisphere $x^2 + y^2 + z^2 = 16, z \geq 0$, counterclockwise when viewed from above.

Flux of the Curl

7. Let $\mathbf{n}$ be the outer unit normal of the elliptical shell

   $$S:\quad 4x^2 + 9y^2 + 36z^2 = 36, \qquad z \geq 0,$$

   and let

   $$\mathbf{F} = y\mathbf{i} + x^2\mathbf{j} + (x^2 + y^4)^{3/2} \sin e^{\sqrt{xyz}}\,\mathbf{k}.$$

   Find the value of

   $$\iint_S \nabla \times \mathbf{F} \cdot \mathbf{n}\, d\sigma.$$

   (*Hint:* One parametrization of the ellipse at the base of the shell is $x = 3\cos t$, $y = 2\sin t$, $0 \leq t \leq 2\pi$.)

8. Let $\mathbf{n}$ be the outer unit normal (normal away from the origin) of the parabolic shell

   $$S:\quad 4x^2 + y + z^2 = 4, \qquad y \geq 0,$$

and let

$$\mathbf{F} = \left(-z + \frac{1}{2+x}\right)\mathbf{i} + (\tan^{-1} y)\mathbf{j} + \left(x + \frac{1}{4+z}\right)\mathbf{k}.$$

Find the value of

$$\iint_S \nabla \times \mathbf{F} \cdot \mathbf{n}\, d\sigma.$$

9. Let S be the cylinder $x^2 + y^2 = a^2, 0 \le z \le h$, together with its top, $x^2 + y^2 \le a^2, z = h$. Let $\mathbf{F} = -y\mathbf{i} + x\mathbf{j} + x^2\mathbf{k}$. Use Stokes' Theorem to find the flux of $\nabla \times \mathbf{F}$ outward through S.

10. Evaluate

$$\iint_S \nabla \times (y\mathbf{i}) \cdot \mathbf{n}\, d\sigma,$$

where S is the hemisphere $x^2 + y^2 + z^2 = 1, z \ge 0$.

11. Flux of curl F Show that

$$\iint_S \nabla \times \mathbf{F} \cdot \mathbf{n}\, d\sigma$$

has the same value for all oriented surfaces S that span C and that induce the same positive direction on C.

12. Let $\mathbf{F}$ be a differentiable vector field defined on a region containing a smooth closed oriented surface S and its interior. Let $\mathbf{n}$ be the unit normal vector field on S. Suppose that S is the union of two surfaces S_1 and S_2 joined along a smooth simple closed curve C. Can anything be said about

$$\iint_S \nabla \times \mathbf{F} \cdot \mathbf{n}\, d\sigma?$$

Give reasons for your answer.

Stokes' Theorem for Parametrized Surfaces

In Exercises 13–18, use the surface integral in Stokes' Theorem to calculate the flux of the curl of the field $\mathbf{F}$ across the surface S in the direction of the outward unit normal $\mathbf{n}$.

13. $\mathbf{F} = 2z\mathbf{i} + 3x\mathbf{j} + 5y\mathbf{k}$

S: $\mathbf{r}(r, \theta) = (r\cos\theta)\mathbf{i} + (r\sin\theta)\mathbf{j} + (4 - r^2)\mathbf{k}$, $0 \le r \le 2, \quad 0 \le \theta \le 2\pi$

14. $\mathbf{F} = (y - z)\mathbf{i} + (z - x)\mathbf{j} + (x + z)\mathbf{k}$

S: $\mathbf{r}(r, \theta) = (r\cos\theta)\mathbf{i} + (r\sin\theta)\mathbf{j} + (9 - r^2)\mathbf{k}$, $0 \le r \le 3, \quad 0 \le \theta \le 2\pi$

15. $\mathbf{F} = x^2y\mathbf{i} + 2y^3z\mathbf{j} + 3z\mathbf{k}$

S: $\mathbf{r}(r, \theta) = (r\cos\theta)\mathbf{i} + (r\sin\theta)\mathbf{j} + r\mathbf{k}$, $0 \le r \le 1, \quad 0 \le \theta \le 2\pi$

16. $\mathbf{F} = (x - y)\mathbf{i} + (y - z)\mathbf{j} + (z - x)\mathbf{k}$

S: $\mathbf{r}(r, \theta) = (r\cos\theta)\mathbf{i} + (r\sin\theta)\mathbf{j} + (5 - r)\mathbf{k}$, $0 \le r \le 5, \quad 0 \le \theta \le 2\pi$

17. $\mathbf{F} = 3y\mathbf{i} + (5 - 2x)\mathbf{j} + (z^2 - 2)\mathbf{k}$

S: $\mathbf{r}(\phi, \theta) = \left(\sqrt{3}\sin\phi\cos\theta\right)\mathbf{i} + \left(\sqrt{3}\sin\phi\sin\theta\right)\mathbf{j} + \left(\sqrt{3}\cos\phi\right)\mathbf{k}, \quad 0 \le \phi \le \pi/2, \quad 0 \le \theta \le 2\pi$

18. $\mathbf{F} = y^2\mathbf{i} + z^2\mathbf{j} + x\mathbf{k}$

S: $\mathbf{r}(\phi, \theta) = (2\sin\phi\cos\theta)\mathbf{i} + (2\sin\phi\sin\theta)\mathbf{j} + (2\cos\phi)\mathbf{k}$, $0 \le \phi \le \pi/2, \quad 0 \le \theta \le 2\pi$

Theory and Examples

19. Zero circulation Use the identity $\nabla \times \nabla f = \mathbf{0}$ (Equation (8) in the text) and Stokes' Theorem to show that the circulations of the following fields around the boundary of any smooth orientable surface in space are zero.

a. $\mathbf{F} = 2x\mathbf{i} + 2y\mathbf{j} + 2z\mathbf{k}$

b. $\mathbf{F} = \nabla(xy^2z^3)$

c. $\mathbf{F} = \nabla \times (x\mathbf{i} + y\mathbf{j} + z\mathbf{k})$

d. $\mathbf{F} = \nabla f$

20. Zero circulation Let $f(x, y, z) = (x^2 + y^2 + z^2)^{-1/2}$. Show that the clockwise circulation of the field $\mathbf{F} = \nabla f$ around the circle $x^2 + y^2 = a^2$ in the xy-plane is zero

a. by taking $\mathbf{r} = (a\cos t)\mathbf{i} + (a\sin t)\mathbf{j}, 0 \le t \le 2\pi$, and integrating $\mathbf{F} \cdot d\mathbf{r}$ over the circle.

b. by applying Stokes' Theorem.

21. Let C be a simple closed smooth curve in the plane $2x + 2y + z = 2$, oriented as shown here. Show that

$$\oint_C 2y\, dx + 3z\, dy - x\, dz$$

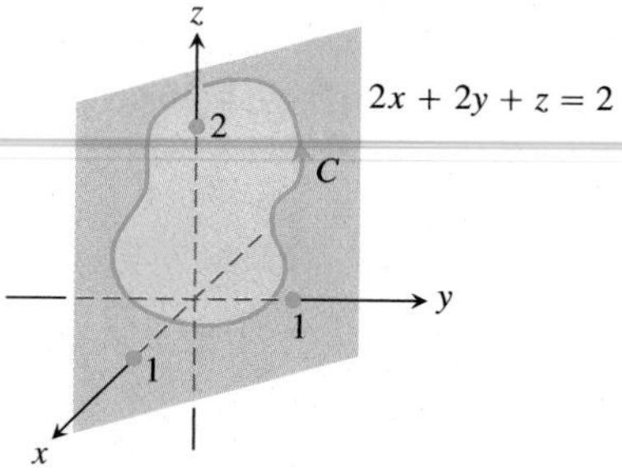

depends only on the area of the region enclosed by C and not on the position or shape of C.

22. Show that if $\mathbf{F} = x\mathbf{i} + y\mathbf{j} + z\mathbf{k}$, then $\nabla \times \mathbf{F} = \mathbf{0}$.

23. Find a vector field with twice-differentiable components whose curl is $x\mathbf{i} + y\mathbf{j} + z\mathbf{k}$ or prove that no such field exists.

24. Does Stokes' Theorem say anything special about circulation in a field whose curl is zero? Give reasons for your answer.

25. Let R be a region in the xy-plane that is bounded by a piecewise-smooth simple closed curve C and suppose that the moments of inertia of R about the x- and y-axes are known to be I_x and I_y. Evaluate the integral

$$\oint_C \nabla(r^4) \cdot \mathbf{n}\, ds,$$

where $r = \sqrt{x^2 + y^2}$, in terms of I_x and I_y.

26. Zero curl, yet field not conservative Show that the curl of

$$\mathbf{F} = \frac{-y}{x^2 + y^2}\mathbf{i} + \frac{x}{x^2 + y^2}\mathbf{j} + z\mathbf{k}$$

is zero but that

$$\oint_C \mathbf{F} \cdot d\mathbf{r}$$

is not zero if C is the circle $x^2 + y^2 = 1$ in the xy-plane. (Theorem 6 does not apply here because the domain of $\mathbf{F}$ is not simply connected. The field $\mathbf{F}$ is not defined along the z-axis so there is no way to contract C to a point without leaving the domain of $\mathbf{F}$.)

16.8 The Divergence Theorem and a Unified Theory

The divergence form of Green's Theorem in the plane states that the net outward flux of a vector field across a simple closed curve can be calculated by integrating the divergence of the field over the region enclosed by the curve. The corresponding theorem in three dimensions, called the Divergence Theorem, states that the net outward flux of a vector field across a closed surface in space can be calculated by integrating the divergence of the field over the region enclosed by the surface. In this section, we prove the Divergence Theorem and show how it simplifies the calculation of flux. We also derive Gauss's law for flux in an electric field and the continuity equation of hydrodynamics. Finally, we unify the chapter's vector integral theorems into a single fundamental theorem.

Divergence in Three Dimensions

The **divergence** of a vector field $\mathbf{F} = M(x, y, z)\mathbf{i} + N(x, y, z)\mathbf{j} + P(x, y, z)\mathbf{k}$ is the scalar function

$$\text{div}\,\mathbf{F} = \nabla \cdot \mathbf{F} = \frac{\partial M}{\partial x} + \frac{\partial N}{\partial y} + \frac{\partial P}{\partial z}. \tag{1}$$

The symbol "div **F**" is read as "divergence of **F**" or "div **F**." The notation $\nabla \cdot \mathbf{F}$ is read "del dot **F**."

Div **F** has the same physical interpretation in three dimensions that it does in two. If **F** is the velocity field of a fluid flow, the value of div **F** at a point (x, y, z) is the rate at which fluid is being piped in or drained away at (x, y, z). The divergence is the flux per unit volume or flux density at the point.

EXAMPLE 1 Finding Divergence

Find the divergence of $\mathbf{F} = 2xz\mathbf{i} - xy\mathbf{j} - z\mathbf{k}$.

Solution The divergence of **F** is

$$\nabla \cdot \mathbf{F} = \frac{\partial}{\partial x}(2xz) + \frac{\partial}{\partial y}(-xy) + \frac{\partial}{\partial z}(-z) = 2z - x - 1.$$

■

Divergence Theorem

The Divergence Theorem says that under suitable conditions, the outward flux of a vector field across a closed surface (oriented outward) equals the triple integral of the divergence of the field over the region enclosed by the surface.

THEOREM 7 Divergence Theorem

The flux of a vector field $\mathbf{F}$ across a closed oriented surface S in the direction of the surface's outward unit normal field $\mathbf{n}$ equals the integral of $\nabla \cdot \mathbf{F}$ over the region D enclosed by the surface:

$$\underbrace{\iint_S \mathbf{F} \cdot \mathbf{n}\, d\sigma}_{\text{Outward flux}} = \underbrace{\iiint_D \nabla \cdot \mathbf{F}\, dV.}_{\text{Divergence integral}} \tag{2}$$

EXAMPLE 2 Supporting the Divergence Theorem

Evaluate both sides of Equation (2) for the field $\mathbf{F} = x\mathbf{i} + y\mathbf{j} + z\mathbf{k}$ over the sphere $x^2 + y^2 + z^2 = a^2$.

Solution The outer unit normal to S, calculated from the gradient of $f(x, y, z) = x^2 + y^2 + z^2 - a^2$, is

$$\mathbf{n} = \frac{2(x\mathbf{i} + y\mathbf{j} + z\mathbf{k})}{\sqrt{4(x^2 + y^2 + z^2)}} = \frac{x\mathbf{i} + y\mathbf{j} + z\mathbf{k}}{a}.$$

Hence,

$$\mathbf{F} \cdot \mathbf{n}\, d\sigma = \frac{x^2 + y^2 + z^2}{a}\, d\sigma = \frac{a^2}{a}\, d\sigma = a\, d\sigma$$

because $x^2 + y^2 + z^2 = a^2$ on the surface. Therefore,

$$\iint_S \mathbf{F} \cdot \mathbf{n}\, d\sigma = \iint_S a\, d\sigma = a \iint_S d\sigma = a(4\pi a^2) = 4\pi a^3.$$

The divergence of $\mathbf{F}$ is

$$\nabla \cdot \mathbf{F} = \frac{\partial}{\partial x}(x) + \frac{\partial}{\partial y}(y) + \frac{\partial}{\partial z}(z) = 3,$$

so

$$\iiint_D \nabla \cdot \mathbf{F}\, dV = \iiint_D 3\, dV = 3\left(\frac{4}{3}\pi a^3\right) = 4\pi a^3. \quad \blacksquare$$

EXAMPLE 3 Finding Flux

Find the flux of $\mathbf{F} = xy\mathbf{i} + yz\mathbf{j} + xz\mathbf{k}$ outward through the surface of the cube cut from the first octant by the planes $x = 1$, $y = 1$, and $z = 1$.

Solution Instead of calculating the flux as a sum of six separate integrals, one for each face of the cube, we can calculate the flux by integrating the divergence

$$\nabla \cdot \mathbf{F} = \frac{\partial}{\partial x}(xy) + \frac{\partial}{\partial y}(yz) + \frac{\partial}{\partial z}(xz) = y + z + x$$

over the cube's interior:

$$\text{Flux} = \iint_{\substack{\text{Cube}\\ \text{surface}}} \mathbf{F} \cdot \mathbf{n}\, d\sigma = \iiint_{\substack{\text{Cube}\\ \text{interior}}} \nabla \cdot \mathbf{F}\, dV \qquad \text{The Divergence Theorem}$$

$$= \int_0^1 \int_0^1 \int_0^1 (x + y + z)\, dx\, dy\, dz = \frac{3}{2}. \qquad \text{Routine integration}$$

■

Proof of the Divergence Theorem for Special Regions

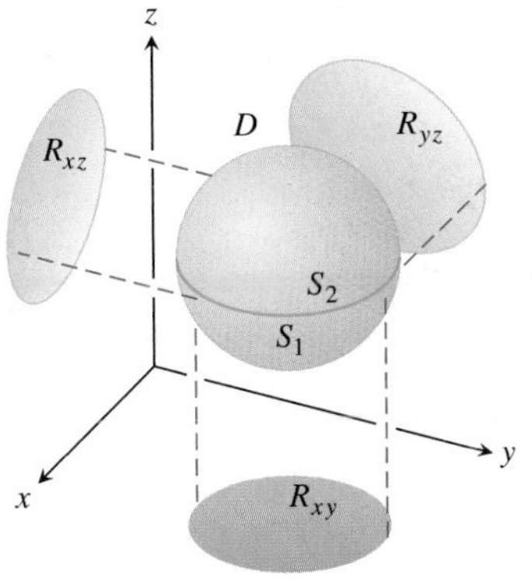

FIGURE 16.69 We first prove the Divergence Theorem for the kind of three-dimensional region shown here. We then extend the theorem to other regions.

To prove the Divergence Theorem, we assume that the components of **F** have continuous first partial derivatives. We also assume that D is a convex region with no holes or bubbles, such as a solid sphere, cube, or ellipsoid, and that S is a piecewise smooth surface. In addition, we assume that any line perpendicular to the xy-plane at an interior point of the region R_{xy} that is the projection of D on the xy-plane intersects the surface S in exactly two points, producing surfaces

$$S_1: \quad z = f_1(x, y), \qquad (x, y) \text{ in } R_{xy}$$

$$S_2: \quad z = f_2(x, y), \qquad (x, y) \text{ in } R_{xy},$$

with $f_1 \le f_2$. We make similar assumptions about the projection of D onto the other coordinate planes. See Figure 16.69.

The components of the unit normal vector $\mathbf{n} = n_1\mathbf{i} + n_2\mathbf{j} + n_3\mathbf{k}$ are the cosines of the angles α, β, and γ that $\mathbf{n}$ makes with **i**, **j**, and **k** (Figure 16.70). This is true because all the vectors involved are unit vectors. We have

$$n_1 = \mathbf{n} \cdot \mathbf{i} = |\mathbf{n}||\mathbf{i}| \cos\alpha = \cos\alpha$$

$$n_2 = \mathbf{n} \cdot \mathbf{j} = |\mathbf{n}||\mathbf{j}| \cos\beta = \cos\beta$$

$$n_3 = \mathbf{n} \cdot \mathbf{k} = |\mathbf{n}||\mathbf{k}| \cos\gamma = \cos\gamma$$

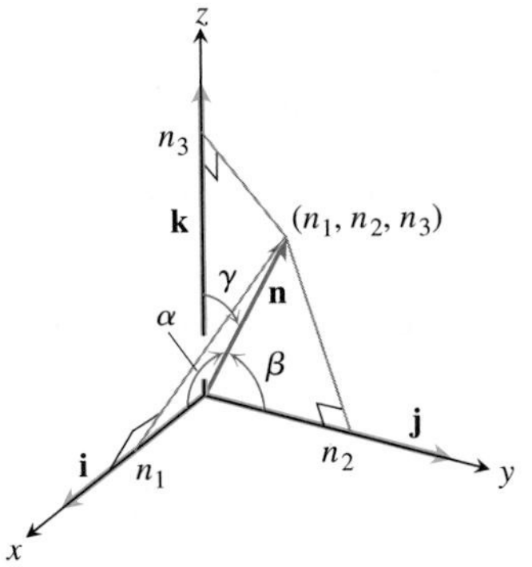

FIGURE 16.70 The scalar components of the unit normal vector **n** are the cosines of the angles α, β, and γ that it makes with **i**, **j**, and **k**.

Thus,

$$\mathbf{n} = (\cos\alpha)\mathbf{i} + (\cos\beta)\mathbf{j} + (\cos\gamma)\mathbf{k}$$

and

$$\mathbf{F} \cdot \mathbf{n} = M\cos\alpha + N\cos\beta + P\cos\gamma.$$

In component form, the Divergence Theorem states that

$$\iint_S (M\cos\alpha + N\cos\beta + P\cos\gamma)\, d\sigma = \iiint_D \left(\frac{\partial M}{\partial x} + \frac{\partial N}{\partial y} + \frac{\partial P}{\partial z}\right) dx\, dy\, dz.$$

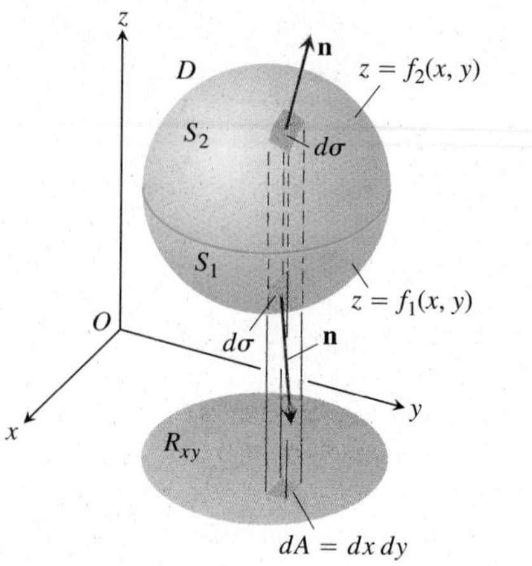

FIGURE 16.71 The three-dimensional region D enclosed by the surfaces S_1 and S_2 shown here projects vertically onto a two-dimensional region R_{xy} in the xy-plane.

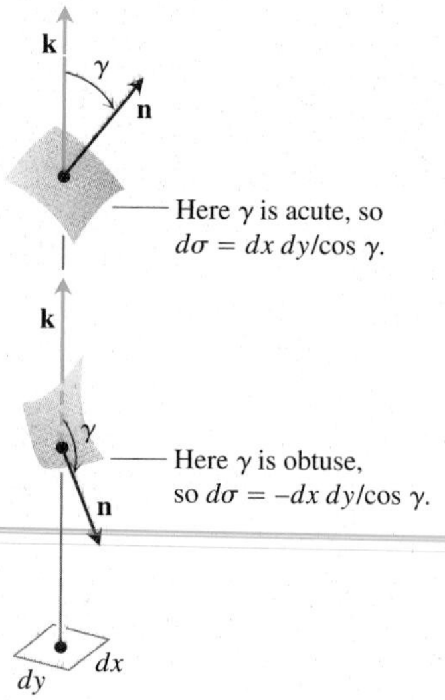

FIGURE 16.72 An enlarged view of the area patches in Figure 16.71. The relations $d\sigma = \pm dx\,dy/\cos\gamma$ are derived in Section 16.5.

We prove the theorem by proving the three following equalities:

$$\iint_S M\cos\alpha\,d\sigma = \iiint_D \frac{\partial M}{\partial x}\,dx\,dy\,dz \tag{3}$$

$$\iint_S N\cos\beta\,d\sigma = \iiint_D \frac{\partial N}{\partial y}\,dx\,dy\,dz \tag{4}$$

$$\iint_S P\cos\gamma\,d\sigma = \iiint_D \frac{\partial P}{\partial z}\,dx\,dy\,dz \tag{5}$$

Proof of Equation (5) We prove Equation (5) by converting the surface integral on the left to a double integral over the projection R_{xy} of D on the xy-plane (Figure 16.71). The surface S consists of an upper part S_2 whose equation is $z = f_2(x, y)$ and a lower part S_1 whose equation is $z = f_1(x, y)$. On S_2, the outer normal **n** has a positive **k**-component and

$$\cos\gamma\,d\sigma = dx\,dy \qquad \text{because} \qquad d\sigma = \frac{dA}{|\cos\gamma|} = \frac{dx\,dy}{\cos\gamma}.$$

See Figure 16.72. On S_1, the outer normal **n** has a negative **k**-component and

$$\cos\gamma\,d\sigma = -dx\,dy.$$

Therefore,

$$\begin{aligned}\iint_S P\cos\gamma\,d\sigma &= \iint_{S_2} P\cos\gamma\,d\sigma + \iint_{S_1} P\cos\gamma\,d\sigma\\ &= \iint_{R_{xy}} P(x, y, f_2(x, y))\,dx\,dy - \iint_{R_{xy}} P(x, y, f_1(x, y))\,dx\,dy\\ &= \iint_{R_{xy}} [P(x, y, f_2(x, y)) - P(x, y, f_1(x, y))]\,dx\,dy\\ &= \iint_{R_{xy}} \left[\int_{f_1(x,y)}^{f_2(x,y)} \frac{\partial P}{\partial z}\,dz\right] dx\,dy = \iiint_D \frac{\partial P}{\partial z}\,dz\,dx\,dy.\end{aligned}$$

This proves Equation (5). ■

The proofs for Equations (3) and (4) follow the same pattern; or just permute x, y, z; M, N, P; α, β, γ, in order, and get those results from Equation (5).

Divergence Theorem for Other Regions

The Divergence Theorem can be extended to regions that can be partitioned into a finite number of simple regions of the type just discussed and to regions that can be defined as limits of simpler regions in certain ways. For example, suppose that D is the region between two concentric spheres and that **F** has continuously differentiable components throughout D and on the bounding surfaces. Split D by an equatorial plane and apply the

Divergence Theorem to each half separately. The bottom half, D_1, is shown in Figure 16.73. The surface S_1 that bounds D_1 consists of an outer hemisphere, a plane washer-shaped base, and an inner hemisphere. The Divergence Theorem says that

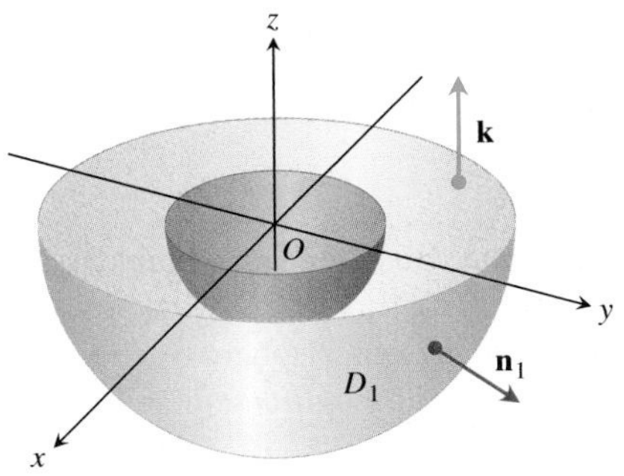

FIGURE 16.73 The lower half of the solid region between two concentric spheres.

$$\iint_{S_1} \mathbf{F} \cdot \mathbf{n}_1 \, d\sigma_1 = \iiint_{D_1} \nabla \cdot \mathbf{F} \, dV_1. \tag{6}$$

The unit normal $\mathbf{n}_1$ that points outward from D_1 points away from the origin along the outer surface, equals $\mathbf{k}$ along the flat base, and points toward the origin along the inner surface. Next apply the Divergence Theorem to D_2, and its surface S_2 (Figure 16.74):

$$\iint_{S_2} \mathbf{F} \cdot \mathbf{n}_2 \, d\sigma_2 = \iiint_{D_2} \nabla \cdot \mathbf{F} \, dV_2. \tag{7}$$

As we follow $\mathbf{n}_2$ over S_2, pointing outward from D_2, we see that $\mathbf{n}_2$ equals $-\mathbf{k}$ along the washer-shaped base in the xy-plane, points away from the origin on the outer sphere, and points toward the origin on the inner sphere. When we add Equations (6) and (7), the integrals over the flat base cancel because of the opposite signs of $\mathbf{n}_1$ and $\mathbf{n}_2$. We thus arrive at the result

$$\iint_{S} \mathbf{F} \cdot \mathbf{n} \, d\sigma = \iiint_{D} \nabla \cdot \mathbf{F} \, dV,$$

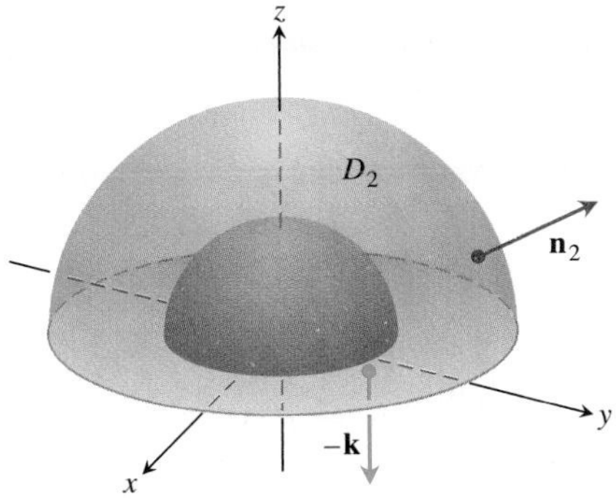

FIGURE 16.74 The upper half of the solid region between two concentric spheres.

with D the region between the spheres, S the boundary of D consisting of two spheres, and $\mathbf{n}$ the unit normal to S directed outward from D.

EXAMPLE 4 Finding Outward Flux

Find the net outward flux of the field

$$\mathbf{F} = \frac{x\mathbf{i} + y\mathbf{j} + z\mathbf{k}}{\rho^3}, \qquad \rho = \sqrt{x^2 + y^2 + z^2}$$

across the boundary of the region D: $0 < a^2 \le x^2 + y^2 + z^2 \le b^2$.

Solution The flux can be calculated by integrating $\nabla \cdot \mathbf{F}$ over D. We have

$$\frac{\partial \rho}{\partial x} = \frac{1}{2}(x^2 + y^2 + z^2)^{-1/2}(2x) = \frac{x}{\rho}$$

and

$$\frac{\partial M}{\partial x} = \frac{\partial}{\partial x}(x\rho^{-3}) = \rho^{-3} - 3x\rho^{-4}\frac{\partial \rho}{\partial x} = \frac{1}{\rho^3} - \frac{3x^2}{\rho^5}.$$

Similarly,

$$\frac{\partial N}{\partial y} = \frac{1}{\rho^3} - \frac{3y^2}{\rho^5} \quad \text{and} \quad \frac{\partial P}{\partial z} = \frac{1}{\rho^3} - \frac{3z^2}{\rho^5}.$$

Hence,

$$\text{div}\, \mathbf{F} = \frac{3}{\rho^3} - \frac{3}{\rho^5}(x^2 + y^2 + z^2) = \frac{3}{\rho^3} - \frac{3\rho^2}{\rho^5} = 0$$

and

$$\iiint\limits_D \nabla \cdot \mathbf{F}\, dV = 0.$$

So the integral of $\nabla \cdot \mathbf{F}$ over D is zero and the net outward flux across the boundary of D is zero. There is more to learn from this example, though. The flux leaving D across the inner sphere S_a is the negative of the flux leaving D across the outer sphere S_b (because the sum of these fluxes is zero). Hence, the flux of $\mathbf{F}$ across S_a in the direction away from the origin equals the flux of $\mathbf{F}$ across S_b in the direction away from the origin. Thus, the flux of $\mathbf{F}$ across a sphere centered at the origin is independent of the radius of the sphere. What is this flux?

To find it, we evaluate the flux integral directly. The outward unit normal on the sphere of radius a is

$$\mathbf{n} = \frac{x\mathbf{i} + y\mathbf{j} + z\mathbf{k}}{\sqrt{x^2 + y^2 + z^2}} = \frac{x\mathbf{i} + y\mathbf{j} + z\mathbf{k}}{a}.$$

Hence, on the sphere,

$$\mathbf{F} \cdot \mathbf{n} = \frac{x\mathbf{i} + y\mathbf{j} + z\mathbf{k}}{a^3} \cdot \frac{x\mathbf{i} + y\mathbf{j} + z\mathbf{k}}{a} = \frac{x^2 + y^2 + z^2}{a^4} = \frac{a^2}{a^4} = \frac{1}{a^2}$$

and

$$\iint\limits_{S_a} \mathbf{F} \cdot \mathbf{n}\, d\sigma = \frac{1}{a^2} \iint\limits_{S_a} d\sigma = \frac{1}{a^2}(4\pi a^2) = 4\pi.$$

The outward flux of $\mathbf{F}$ across any sphere centered at the origin is 4π. ■

Gauss's Law: One of the Four Great Laws of Electromagnetic Theory

There is still more to be learned from Example 4. In electromagnetic theory, the electric field created by a point charge q located at the origin is

$$\mathbf{E}(x, y, z) = \frac{1}{4\pi\epsilon_0} \frac{q}{|\mathbf{r}|^2} \left(\frac{\mathbf{r}}{|\mathbf{r}|} \right) = \frac{q}{4\pi\epsilon_0} \frac{\mathbf{r}}{|\mathbf{r}|^3} = \frac{q}{4\pi\epsilon_0} \frac{x\mathbf{i} + y\mathbf{j} + z\mathbf{k}}{\rho^3},$$

where ϵ_0 is a physical constant, $\mathbf{r}$ is the position vector of the point (x, y, z), and $\rho = |\mathbf{r}| = \sqrt{x^2 + y^2 + z^2}$. In the notation of Example 4,

$$\mathbf{E} = \frac{q}{4\pi\epsilon_0} \mathbf{F}.$$

The calculations in Example 4 show that the outward flux of $\mathbf{E}$ across any sphere centered at the origin is q/ϵ_0, but this result is not confined to spheres. The outward flux of $\mathbf{E}$ across any closed surface S that encloses the origin (and to which the Divergence Theorem applies) is also q/ϵ_0. To see why, we have only to imagine a large sphere S_a centered at the origin and enclosing the surface S. Since

$$\nabla \cdot \mathbf{E} = \nabla \cdot \frac{q}{4\pi\epsilon_0} \mathbf{F} = \frac{q}{4\pi\epsilon_0} \nabla \cdot \mathbf{F} = 0$$

when $\rho > 0$, the integral of $\nabla \cdot \mathbf{E}$ over the region D between S and S_a is zero. Hence, by the Divergence Theorem,

$$\iint_{\substack{\text{Boundary} \\ \text{of } D}} \mathbf{E} \cdot \mathbf{n} \, d\sigma = 0,$$

and the flux of $\mathbf{E}$ across S in the direction away from the origin must be the same as the flux of $\mathbf{E}$ across S_a in the direction away from the origin, which is q/ϵ_0. This statement, called *Gauss's Law*, also applies to charge distributions that are more general than the one assumed here, as you will see in nearly any physics text.

$$\text{Gauss's law:} \quad \iint_S \mathbf{E} \cdot \mathbf{n} \, d\sigma = \frac{q}{\epsilon_0}$$

Continuity Equation of Hydrodynamics

Let D be a region in space bounded by a closed oriented surface S. If $\mathbf{v}(x, y, z)$ is the velocity field of a fluid flowing smoothly through D, $\delta = \delta(t, x, y, z)$ is the fluid's density at (x, y, z) at time t, and $\mathbf{F} = \delta\mathbf{v}$, then the **continuity equation** of hydrodynamics states that

$$\nabla \cdot \mathbf{F} + \frac{\partial \delta}{\partial t} = 0.$$

If the functions involved have continuous first partial derivatives, the equation evolves naturally from the Divergence Theorem, as we now see.

First, the integral

$$\iint_S \mathbf{F} \cdot \mathbf{n} \, d\sigma$$

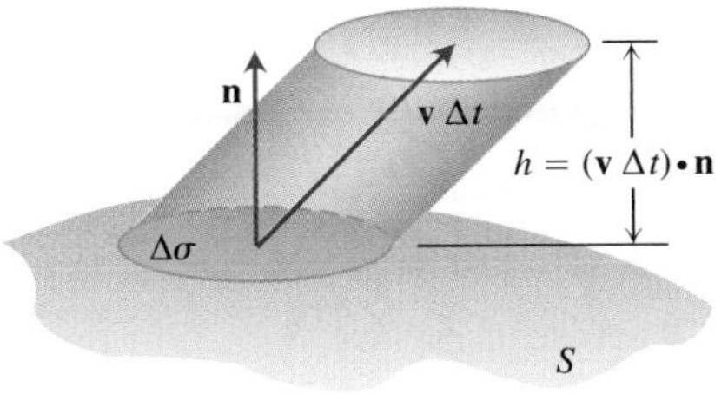

FIGURE 16.75 The fluid that flows upward through the patch $\Delta\sigma$ in a short time Δt fills a "cylinder" whose volume is approximately base $\times$ height $= \mathbf{v} \cdot \mathbf{n} \, \Delta\sigma \, \Delta t$.

is the rate at which mass leaves D across S (leaves because $\mathbf{n}$ is the outer normal). To see why, consider a patch of area $\Delta\sigma$ on the surface (Figure 16.75). In a short time interval Δt, the volume ΔV of fluid that flows across the patch is approximately equal to the volume of a cylinder with base area $\Delta\sigma$ and height $(\mathbf{v}\Delta t) \cdot \mathbf{n}$, where $\mathbf{v}$ is a velocity vector rooted at a point of the patch:

$$\Delta V \approx \mathbf{v} \cdot \mathbf{n} \, \Delta\sigma \, \Delta t.$$

The mass of this volume of fluid is about

$$\Delta m \approx \delta\mathbf{v} \cdot \mathbf{n} \, \Delta\sigma \, \Delta t,$$

so the rate at which mass is flowing out of D across the patch is about

$$\frac{\Delta m}{\Delta t} \approx \delta\mathbf{v} \cdot \mathbf{n} \, \Delta\sigma.$$

This leads to the approximation

$$\frac{\sum \Delta m}{\Delta t} \approx \sum \delta\mathbf{v} \cdot \mathbf{n} \, \Delta\sigma$$

as an estimate of the average rate at which mass flows across S. Finally, letting $\Delta\sigma \to 0$ and $\Delta t \to 0$ gives the instantaneous rate at which mass leaves D across S as

$$\frac{dm}{dt} = \iint_S \delta\mathbf{v}\cdot\mathbf{n}\,d\sigma,$$

which for our particular flow is

$$\frac{dm}{dt} = \iint_S \mathbf{F}\cdot\mathbf{n}\,d\sigma.$$

Now let B be a solid sphere centered at a point Q in the flow. The average value of $\nabla\cdot\mathbf{F}$ over B is

$$\frac{1}{\text{volume of } B}\iiint_B \nabla\cdot\mathbf{F}\,dV.$$

It is a consequence of the continuity of the divergence that $\nabla\cdot\mathbf{F}$ actually takes on this value at some point P in B. Thus,

$$\begin{aligned}(\nabla\cdot\mathbf{F})_P &= \frac{1}{\text{volume of } B}\iiint_B \nabla\cdot\mathbf{F}\,dV = \frac{\iint_S \mathbf{F}\cdot\mathbf{n}\,d\sigma}{\text{volume of } B}\\ &= \frac{\text{rate at which mass leaves } B \text{ across its surface } S}{\text{volume of } B}\end{aligned} \tag{8}$$

The fraction on the right describes decrease in mass per unit volume.

Now let the radius of B approach zero while the center Q stays fixed. The left side of Equation (8) converges to $(\nabla\cdot\mathbf{F})_Q$, the right side to $(-\partial\delta/\partial t)_Q$. The equality of these two limits is the continuity equation

$$\nabla\cdot\mathbf{F} = -\frac{\partial\delta}{\partial t}.$$

The continuity equation "explains" $\nabla\cdot\mathbf{F}$: The divergence of $\mathbf{F}$ at a point is the rate at which the density of the fluid is decreasing there.

The Divergence Theorem

$$\iint_S \mathbf{F}\cdot\mathbf{n}\,d\sigma = \iiint_D \nabla\cdot\mathbf{F}\,dV$$

now says that the net decrease in density of the fluid in region D is accounted for by the mass transported across the surface S. So, the theorem is a statement about conservation of mass (Exercise 31).

Unifying the Integral Theorems

If we think of a two-dimensional field $\mathbf{F} = M(x, y)\mathbf{i} + N(x, y)\mathbf{j}$ as a three-dimensional field whose $\mathbf{k}$-component is zero, then $\nabla\cdot\mathbf{F} = (\partial M/\partial x) + (\partial N/\partial y)$ and the normal form of Green's Theorem can be written as

$$\oint_C \mathbf{F}\cdot\mathbf{n}\,ds = \iint_R \left(\frac{\partial M}{\partial x} + \frac{\partial N}{\partial y}\right)dx\,dy = \iint_R \nabla\cdot\mathbf{F}\,dA.$$

Similarly, $\nabla \times \mathbf{F} \cdot \mathbf{k} = (\partial N/\partial x) - (\partial M/\partial y)$, so the tangential form of Green's Theorem can be written as

$$\oint_C \mathbf{F} \cdot d\mathbf{r} = \iint_R \left(\frac{\partial N}{\partial x} - \frac{\partial M}{\partial y}\right) dx\, dy = \iint_R \nabla \times \mathbf{F} \cdot \mathbf{k}\, dA.$$

With the equations of Green's Theorem now in del notation, we can see their relationships to the equations in Stokes' Theorem and the Divergence Theorem.

Green's Theorem and Its Generalization to Three Dimensions

Normal form of Green's Theorem: $\oint_C \mathbf{F} \cdot \mathbf{n}\, ds = \iint_R \nabla \cdot \mathbf{F}\, dA$

Divergence Theorem: $\iint_S \mathbf{F} \cdot \mathbf{n}\, d\sigma = \iiint_D \nabla \cdot \mathbf{F}\, dV$

Tangential form of Green's Theorem: $\oint_C \mathbf{F} \cdot d\mathbf{r} = \iint_R \nabla \times \mathbf{F} \cdot \mathbf{k}\, dA$

Stokes' Theorem: $\oint_C \mathbf{F} \cdot d\mathbf{r} = \iint_S \nabla \times \mathbf{F} \cdot \mathbf{n}\, d\sigma$

Notice how Stokes' Theorem generalizes the tangential (curl) form of Green's Theorem from a flat surface in the plane to a surface in three-dimensional space. In each case, the integral of the normal component of curl $\mathbf{F}$ over the interior of the surface equals the circulation of $\mathbf{F}$ around the boundary.

Likewise, the Divergence Theorem generalizes the normal (flux) form of Green's Theorem from a two-dimensional region in the plane to a three-dimensional region in space. In each case, the integral of $\nabla \cdot \mathbf{F}$ over the interior of the region equals the total flux of the field across the boundary.

There is still more to be learned here. All these results can be thought of as forms of a *single fundamental theorem*. Think back to the Fundamental Theorem of Calculus in Section 5.4. It says that if $f(x)$ is differentiable on (a, b) and continuous on $[a, b]$, then

FIGURE 16.76 The outward unit normals at the boundary of $[a, b]$ in one-dimensional space.

$$\int_a^b \frac{df}{dx}\, dx = f(b) - f(a).$$

If we let $\mathbf{F} = f(x)\mathbf{i}$ throughout $[a, b]$, then $(df/dx) = \nabla \cdot \mathbf{F}$. If we define the unit vector field $\mathbf{n}$ normal to the boundary of $[a, b]$ to be $\mathbf{i}$ at b and $-\mathbf{i}$ at a (Figure 16.76), then

$$\begin{aligned} f(b) - f(a) &= f(b)\mathbf{i} \cdot (\mathbf{i}) + f(a)\mathbf{i} \cdot (-\mathbf{i}) \\ &= \mathbf{F}(b) \cdot \mathbf{n} + \mathbf{F}(a) \cdot \mathbf{n} \\ &= \text{total outward flux of } \mathbf{F} \text{ across the boundary of } [a, b]. \end{aligned}$$

The Fundamental Theorem now says that

$$\mathbf{F}(b) \cdot \mathbf{n} + \mathbf{F}(a) \cdot \mathbf{n} = \int_{[a,b]} \nabla \cdot \mathbf{F}\, dx.$$

The Fundamental Theorem of Calculus, the normal form of Green's Theorem, and the Divergence Theorem all say that the integral of the differential operator $\nabla\cdot$ operating on a field $\mathbf{F}$ over a region equals the sum of the normal field components over the boundary of the region. (Here we are interpreting the line integral in Green's Theorem and the surface integral in the Divergence Theorem as "sums" over the boundary.)

Stokes' Theorem and the tangential form of Green's Theorem say that, when things are properly oriented, the integral of the normal component of the curl operating on a field equals the sum of the tangential field components on the boundary of the surface.

The beauty of these interpretations is the observance of a single unifying principle, which we might state as follows.

> The integral of a differential operator acting on a field over a region equals the sum of the field components appropriate to the operator over the boundary of the region.

EXERCISES 16.8

Calculating Divergence

In Exercises 1–4, find the divergence of the field.

1. The spin field in Figure 16.14.
2. The radial field in Figure 16.13.
3. The gravitational field in Figure 16.9.
4. The velocity field in Figure 16.12.

Using the Divergence Theorem to Calculate Outward Flux

In Exercises 5–16, use the Divergence Theorem to find the outward flux of $\mathbf{F}$ across the boundary of the region D.

5. **Cube** $\mathbf{F} = (y - x)\mathbf{i} + (z - y)\mathbf{j} + (y - x)\mathbf{k}$

   D: The cube bounded by the planes $x = \pm 1, y = \pm 1$, and $z = \pm 1$

6. $\mathbf{F} = x^2\mathbf{i} + y^2\mathbf{j} + z^2\mathbf{k}$

   a. **Cube** D: The cube cut from the first octant by the planes $x = 1, y = 1$, and $z = 1$

   b. **Cube** D: The cube bounded by the planes $x = \pm 1$, $y = \pm 1$, and $z = \pm 1$

   c. **Cylindrical can** D: The region cut from the solid cylinder $x^2 + y^2 \le 4$ by the planes $z = 0$ and $z = 1$

7. **Cylinder and paraboloid** $\mathbf{F} = y\mathbf{i} + xy\mathbf{j} - z\mathbf{k}$

   D: The region inside the solid cylinder $x^2 + y^2 \le 4$ between the plane $z = 0$ and the paraboloid $z = x^2 + y^2$

8. **Sphere** $\mathbf{F} = x^2\mathbf{i} + xz\mathbf{j} + 3z\mathbf{k}$

   D: The solid sphere $x^2 + y^2 + z^2 \le 4$

9. **Portion of sphere** $\mathbf{F} = x^2\mathbf{i} - 2xy\mathbf{j} + 3xz\mathbf{k}$

   D: The region cut from the first octant by the sphere $x^2 + y^2 + z^2 = 4$

10. **Cylindrical can** $\mathbf{F} = (6x^2 + 2xy)\mathbf{i} + (2y + x^2z)\mathbf{j} + 4x^2y^3\mathbf{k}$

    D: The region cut from the first octant by the cylinder $x^2 + y^2 = 4$ and the plane $z = 3$

11. **Wedge** $\mathbf{F} = 2xz\mathbf{i} - xy\mathbf{j} - z^2\mathbf{k}$

    D: The wedge cut from the first octant by the plane $y + z = 4$ and the elliptical cylinder $4x^2 + y^2 = 16$

12. **Sphere** $\mathbf{F} = x^3\mathbf{i} + y^3\mathbf{j} + z^3\mathbf{k}$

    D: The solid sphere $x^2 + y^2 + z^2 \le a^2$

13. **Thick sphere** $\mathbf{F} = \sqrt{x^2 + y^2 + z^2}\,(x\mathbf{i} + y\mathbf{j} + z\mathbf{k})$

    D: The region $1 \le x^2 + y^2 + z^2 \le 2$

14. **Thick sphere** $\mathbf{F} = (x\mathbf{i} + y\mathbf{j} + z\mathbf{k})/\sqrt{x^2 + y^2 + z^2}$

    D: The region $1 \le x^2 + y^2 + z^2 \le 4$

15. **Thick sphere** $\mathbf{F} = (5x^3 + 12xy^2)\mathbf{i} + (y^3 + e^y \sin z)\mathbf{j} + (5z^3 + e^y \cos z)\mathbf{k}$

    D: The solid region between the spheres $x^2 + y^2 + z^2 = 1$ and $x^2 + y^2 + z^2 = 2$

16. **Thick cylinder** $\mathbf{F} = \ln(x^2 + y^2)\mathbf{i} - \left(\frac{2z}{x}\tan^{-1}\frac{y}{x}\right)\mathbf{j} + z\sqrt{x^2 + y^2}\,\mathbf{k}$

    D: The thick-walled cylinder $1 \le x^2 + y^2 \le 2, \quad -1 \le z \le 2$

Properties of Curl and Divergence

17. div (curl G) is zero

a. Show that if the necessary partial derivatives of the components of the field $\mathbf{G} = M\mathbf{i} + N\mathbf{j} + P\mathbf{k}$ are continuous, then $\nabla \cdot \nabla \times \mathbf{G} = 0$.

b. What, if anything, can you conclude about the flux of the field $\nabla \times \mathbf{G}$ across a closed surface? Give reasons for your answer.

18. Let $\mathbf{F}_1$ and $\mathbf{F}_2$ be differentiable vector fields and let a and b be arbitrary real constants. Verify the following identities.

a. $\nabla \cdot (a\mathbf{F}_1 + b\mathbf{F}_2) = a\nabla \cdot \mathbf{F}_1 + b\nabla \cdot \mathbf{F}_2$

b. $\nabla \times (a\mathbf{F}_1 + b\mathbf{F}_2) = a\nabla \times \mathbf{F}_1 + b\nabla \times \mathbf{F}_2$

c. $\nabla \cdot (\mathbf{F}_1 \times \mathbf{F}_2) = \mathbf{F}_2 \cdot \nabla \times \mathbf{F}_1 - \mathbf{F}_1 \cdot \nabla \times \mathbf{F}_2$

19. Let $\mathbf{F}$ be a differentiable vector field and let $g(x, y, z)$ be a differentiable scalar function. Verify the following identities.

a. $\nabla \cdot (g\mathbf{F}) = g\nabla \cdot \mathbf{F} + \nabla g \cdot \mathbf{F}$

b. $\nabla \times (g\mathbf{F}) = g\nabla \times \mathbf{F} + \nabla g \times \mathbf{F}$

20. If $\mathbf{F} = M\mathbf{i} + N\mathbf{j} + P\mathbf{k}$ is a differentiable vector field, we define the notation $\mathbf{F} \cdot \nabla$ to mean

$$M\frac{\partial}{\partial x} + N\frac{\partial}{\partial y} + P\frac{\partial}{\partial z}.$$

For differentiable vector fields $\mathbf{F}_1$ and $\mathbf{F}_2$, verify the following identities.

a. $\nabla \times (\mathbf{F}_1 \times \mathbf{F}_2) = (\mathbf{F}_2 \cdot \nabla)\mathbf{F}_1 - (\mathbf{F}_1 \cdot \nabla)\mathbf{F}_2 + (\nabla \cdot \mathbf{F}_2)\mathbf{F}_1 - (\nabla \cdot \mathbf{F}_1)\mathbf{F}_2$

b. $\nabla(\mathbf{F}_1 \cdot \mathbf{F}_2) = (\mathbf{F}_1 \cdot \nabla)\mathbf{F}_2 + (\mathbf{F}_2 \cdot \nabla)\mathbf{F}_1 + \mathbf{F}_1 \times (\nabla \times \mathbf{F}_2) + \mathbf{F}_2 \times (\nabla \times \mathbf{F}_1)$

Theory and Examples

21. Let $\mathbf{F}$ be a field whose components have continuous first partial derivatives throughout a portion of space containing a region D bounded by a smooth closed surface S. If $|\mathbf{F}| \le 1$, can any bound be placed on the size of

$$\iiint_D \nabla \cdot \mathbf{F}\, dV?$$

Give reasons for your answer.

22. The base of the closed cubelike surface shown here is the unit square in the xy-plane. The four sides lie in the planes $x = 0, x = 1, y = 0$, and $y = 1$. The top is an arbitrary smooth surface whose identity is unknown. Let $\mathbf{F} = x\mathbf{i} - 2y\mathbf{j} + (z + 3)\mathbf{k}$ and suppose the outward flux of $\mathbf{F}$ through side A is 1 and through side B is -3. Can you conclude anything about the outward flux through the top? Give reasons for your answer.

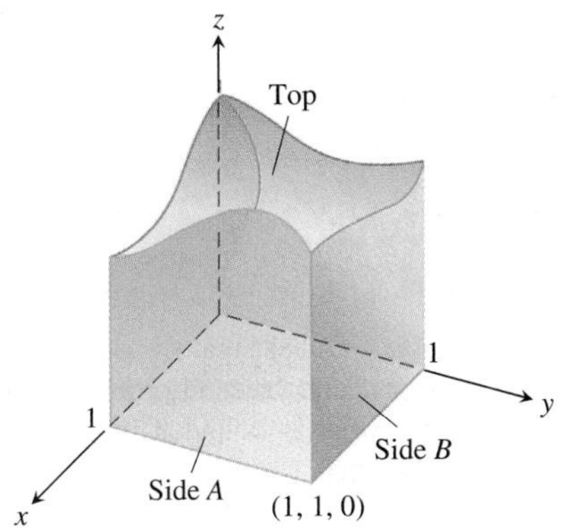

23. a. Show that the flux of the position vector field $\mathbf{F} = x\mathbf{i} + y\mathbf{j} + z\mathbf{k}$ outward through a smooth closed surface S is three times the volume of the region enclosed by the surface.

b. Let $\mathbf{n}$ be the outward unit normal vector field on S. Show that it is not possible for $\mathbf{F}$ to be orthogonal to $\mathbf{n}$ at every point of S.

24. Maximum flux Among all rectangular solids defined by the inequalities $0 \le x \le a, 0 \le y \le b, 0 \le z \le 1$, find the one for which the total flux of $\mathbf{F} = (-x^2 - 4xy)\mathbf{i} - 6yz\mathbf{j} + 12z\mathbf{k}$ outward through the six sides is greatest. What *is* the greatest flux?

25. Volume of a solid region Let $\mathbf{F} = x\mathbf{i} + y\mathbf{j} + z\mathbf{k}$ and suppose that the surface S and region D satisfy the hypotheses of the Divergence Theorem. Show that the volume of D is given by the formula

$$\text{Volume of } D = \frac{1}{3}\iint_S \mathbf{F} \cdot \mathbf{n}\, d\sigma.$$

26. Flux of a constant field Show that the outward flux of a constant vector field $\mathbf{F} = \mathbf{C}$ across any closed surface to which the Divergence Theorem applies is zero.

27. Harmonic functions A function $f(x, y, z)$ is said to be *harmonic* in a region D in space if it satisfies the Laplace equation

$$\nabla^2 f = \nabla \cdot \nabla f = \frac{\partial^2 f}{\partial x^2} + \frac{\partial^2 f}{\partial y^2} + \frac{\partial^2 f}{\partial z^2} = 0$$

throughout D.

a. Suppose that f is harmonic throughout a bounded region D enclosed by a smooth surface S and that $\mathbf{n}$ is the chosen unit normal vector on S. Show that the integral over S of $\nabla f \cdot \mathbf{n}$, the derivative of f in the direction of $\mathbf{n}$, is zero.

b. Show that if f is harmonic on D, then

$$\iint_S f\,\nabla f \cdot \mathbf{n}\, d\sigma = \iiint_D |\nabla f|^2\, dV.$$

28. **Flux of a gradient field** Let S be the surface of the portion of the solid sphere $x^2 + y^2 + z^2 \le a^2$ that lies in the first octant and let $f(x, y, z) = \ln\sqrt{x^2 + y^2 + z^2}$. Calculate

$$\iint_S \nabla f \cdot \mathbf{n}\, d\sigma.$$

($\nabla f \cdot \mathbf{n}$ is the derivative of f in the direction of $\mathbf{n}$.)

29. **Green's first formula** Suppose that f and g are scalar functions with continuous first- and second-order partial derivatives throughout a region D that is bounded by a closed piecewise-smooth surface S. Show that

$$\iint_S f\,\nabla g \cdot \mathbf{n}\, d\sigma = \iiint_D (f\,\nabla^2 g + \nabla f \cdot \nabla g)\, dV. \qquad (9)$$

Equation (9) is **Green's first formula**. (*Hint:* Apply the Divergence Theorem to the field $\mathbf{F} = f\,\nabla g$.)

30. **Green's second formula** (*Continuation of Exercise 29.*) Interchange f and g in Equation (9) to obtain a similar formula. Then subtract this formula from Equation (9) to show that

$$\iint_S (f\,\nabla g - g\nabla f) \cdot \mathbf{n}\, d\sigma = \iiint_D (f\,\nabla^2 g - g\nabla^2 f)\, dV. \qquad (10)$$

This equation is **Green's second formula**.

31. **Conservation of mass** Let $\mathbf{v}(t, x, y, z)$ be a continuously differentiable vector field over the region D in space and let $p(t, x, y, z)$ be a continuously differentiable scalar function. The variable t represents the time domain. The Law of Conservation of Mass asserts that

$$\frac{d}{dt}\iiint_D p(t, x, y, z)\, dV = -\iint_S p\mathbf{v} \cdot \mathbf{n}\, d\sigma,$$

where S is the surface enclosing D.

a. Give a physical interpretation of the conservation of mass law if $\mathbf{v}$ is a velocity flow field and p represents the density of the fluid at point (x, y, z) at time t.

b. Use the Divergence Theorem and Leibniz's Rule,

$$\frac{d}{dt}\iiint_D p(t, x, y, z)\, dV = \iiint_D \frac{\partial p}{\partial t}\, dV,$$

to show that the Law of Conservation of Mass is equivalent to the continuity equation,

$$\nabla \cdot p\mathbf{v} + \frac{\partial p}{\partial t} = 0.$$

(In the first term $\nabla \cdot p\mathbf{v}$, the variable t is held fixed, and in the second term $\partial p/\partial t$, it is assumed that the point (x, y, z) in D is held fixed.)

32. **The heat diffusion equation** Let $T(t, x, y, z)$ be a function with continuous second derivatives giving the temperature at time t at the point (x, y, z) of a solid occupying a region D in space. If the solid's heat capacity and mass density are denoted by the constants c and ρ, respectively, the quantity $c\rho T$ is called the solid's **heat energy per unit volume**.

a. Explain why $-\nabla T$ points in the direction of heat flow.

b. Let $-k\nabla T$ denote the **energy flux vector**. (Here the constant k is called the **conductivity**.) Assuming the Law of Conservation of Mass with $-k\nabla T = \mathbf{v}$ and $c\rho T = p$ in Exercise 31, derive the diffusion (heat) equation

$$\frac{\partial T}{\partial t} = K\nabla^2 T,$$

where $K = k/(c\rho) > 0$ is the *diffusivity* constant. (Notice that if $T(t, x)$ represents the temperature at time t at position x in a uniform conducting rod with perfectly insulated sides, then $\nabla^2 T = \partial^2 T/\partial x^2$ and the diffusion equation reduces to the one-dimensional heat equation in Chapter 14's Additional Exercises.)

Chapter 16 Questions to Guide Your Review

1. What are line integrals? How are they evaluated? Give examples.
2. How can you use line integrals to find the centers of mass of springs? Explain.
3. What is a vector field? A gradient field? Give examples.
4. How do you calculate the work done by a force in moving a particle along a curve? Give an example.
5. What are flow, circulation, and flux?
6. What is special about path independent fields?
7. How can you tell when a field is conservative?
8. What is a potential function? Show by example how to find a potential function for a conservative field.
9. What is a differential form? What does it mean for such a form to be exact? How do you test for exactness? Give examples.
10. What is the divergence of a vector field? How can you interpret it?
11. What is the curl of a vector field? How can you interpret it?
12. What is Green's theorem? How can you interpret it?
13. How do you calculate the area of a curved surface in space? Give an example.

14. What is an oriented surface? How do you calculate the flux of a three-dimensional vector field across an oriented surface? Give an example.
15. What are surface integrals? What can you calculate with them? Give an example.
16. What is a parametrized surface? How do you find the area of such a surface? Give examples.
17. How do you integrate a function over a parametrized surface? Give an example.
18. What is Stokes' Theorem? How can you interpret it?
19. Summarize the chapter's results on conservative fields.
20. What is the Divergence Theorem? How can you interpret it?
21. How does the Divergence Theorem generalize Green's Theorem?
22. How does Stokes' Theorem generalize Green's Theorem?
23. How can Green's Theorem, Stokes' Theorem, and the Divergence Theorem be thought of as forms of a single fundamental theorem?

Chapter 16 Practice Exercises

Evaluating Line Integrals

1. The accompanying figure shows two polygonal paths in space joining the origin to the point $(1, 1, 1)$. Integrate $f(x, y, z) = 2x - 3y^2 - 2z + 3$ over each path.

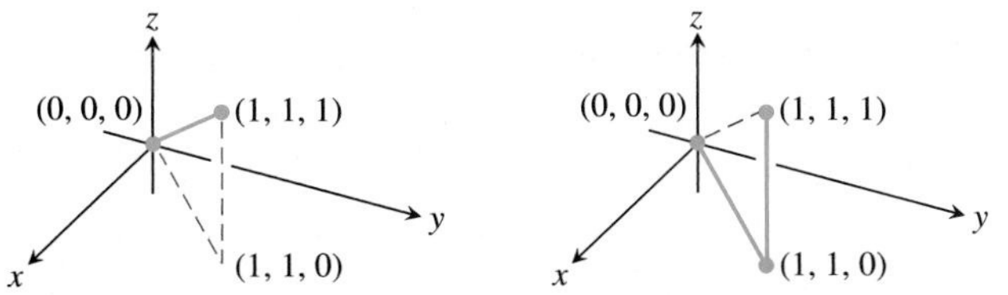

Path 1 Path 2

2. The accompanying figure shows three polygonal paths joining the origin to the point $(1, 1, 1)$. Integrate $f(x, y, z) = x^2 + y - z$ over each path.

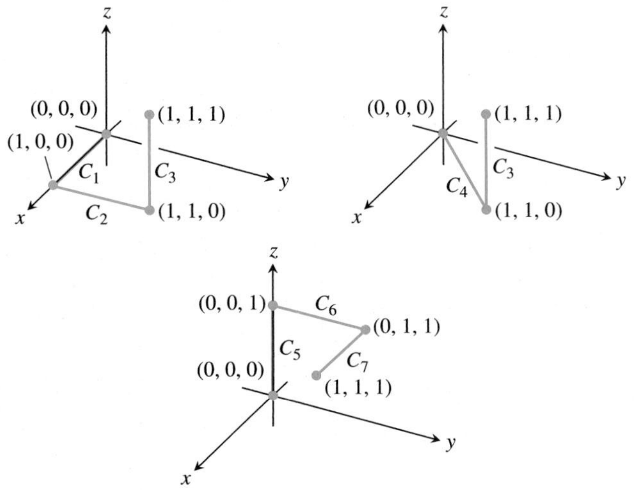

3. Integrate $f(x, y, z) = \sqrt{x^2 + z^2}$ over the circle

$$\mathbf{r}(t) = (a \cos t)\mathbf{j} + (a \sin t)\mathbf{k}, \qquad 0 \le t \le 2\pi.$$

4. Integrate $f(x, y, z) = \sqrt{x^2 + y^2}$ over the involute curve

$$\mathbf{r}(t) = (\cos t + t \sin t)\mathbf{i} + (\sin t - t \cos t)\mathbf{j}, \qquad 0 \le t \le \sqrt{3}.$$

Evaluate the integrals in Exercises 5 and 6.

5. $\displaystyle\int_{(-1,1,1)}^{(4,-3,0)} \frac{dx + dy + dz}{\sqrt{x + y + z}}$

6. $\displaystyle\int_{(1,1,1)}^{(10,3,3)} dx - \sqrt{\frac{z}{y}}\, dy - \sqrt{\frac{y}{z}}\, dz$

7. Integrate $\mathbf{F} = -(y \sin z)\mathbf{i} + (x \sin z)\mathbf{j} + (xy \cos z)\mathbf{k}$ around the circle cut from the sphere $x^2 + y^2 + z^2 = 5$ by the plane $z = -1$, clockwise as viewed from above.

8. Integrate $\mathbf{F} = 3x^2y\mathbf{i} + (x^3 + 1)\mathbf{j} + 9z^2\mathbf{k}$ around the circle cut from the sphere $x^2 + y^2 + z^2 = 9$ by the plane $x = 2$.

Evaluate the integrals in Exercises 9 and 10.

9. $\displaystyle\int_C 8x \sin y\, dx - 8y \cos x\, dy$

   C is the square cut from the first quadrant by the lines $x = \pi/2$ and $y = \pi/2$.

10. $\displaystyle\int_C y^2\, dx + x^2\, dy$

    C is the circle $x^2 + y^2 = 4$.

Evaluating Surface Integrals

11. **Area of an elliptical region** Find the area of the elliptical region cut from the plane $x + y + z = 1$ by the cylinder $x^2 + y^2 = 1$.

12. **Area of a parabolic cap** Find the area of the cap cut from the paraboloid $y^2 + z^2 = 3x$ by the plane $x = 1$.

13. **Area of a spherical cap** Find the area of the cap cut from the top of the sphere $x^2 + y^2 + z^2 = 1$ by the plane $z = \sqrt{2}/2$.

14. a. Hemisphere cut by cylinder Find the area of the surface cut from the hemisphere $x^2 + y^2 + z^2 = 4$, $z \geq 0$, by the cylinder $x^2 + y^2 = 2x$.

b. Find the area of the portion of the cylinder that lies inside the hemisphere. (*Hint:* Project onto the xz-plane. Or evaluate the integral $\int h\, ds$, where h is the altitude of the cylinder and ds is the element of arc length on the circle $x^2 + y^2 = 2x$ in the xy-plane.)

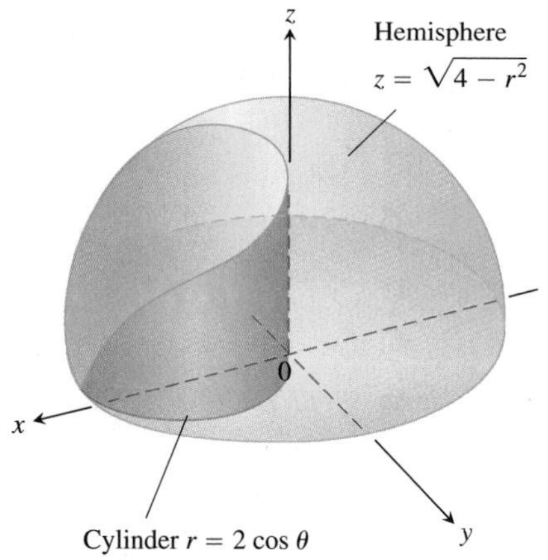

15. Area of a triangle Find the area of the triangle in which the plane $(x/a) + (y/b) + (z/c) = 1$ $(a, b, c > 0)$ intersects the first octant. Check your answer with an appropriate vector calculation.

16. Parabolic cylinder cut by planes Integrate

a. $g(x, y, z) = \dfrac{yz}{\sqrt{4y^2 + 1}}$ **b.** $g(x, y, z) = \dfrac{z}{\sqrt{4y^2 + 1}}$

over the surface cut from the parabolic cylinder $y^2 - z = 1$ by the planes $x = 0$, $x = 3$, and $z = 0$.

17. Circular cylinder cut by planes Integrate $g(x, y, z) = x^4y(y^2 + z^2)$ over the portion of the cylinder $y^2 + z^2 = 25$ that lies in the first octant between the planes $x = 0$ and $x = 1$ and above the plane $z = 3$.

18. Area of Wyoming The state of Wyoming is bounded by the meridians 111°3′ and 104°3′ west longitude and by the circles 41° and 45° north latitude. Assuming that Earth is a sphere of radius $R = 3959$ mi, find the area of Wyoming.

Parametrized Surfaces

Find the parametrizations for the surfaces in Exercises 19–24. (There are many ways to do these, so your answers may not be the same as those in the back of the book.)

19. Spherical band The portion of the sphere $x^2 + y^2 + z^2 = 36$ between the planes $z = -3$ and $z = 3\sqrt{3}$

20. Parabolic cap The portion of the paraboloid $z = -(x^2 + y^2)/2$ above the plane $z = -2$

21. Cone The cone $z = 1 + \sqrt{x^2 + y^2}$, $z \leq 3$

22. Plane above square The portion of the plane $4x + 2y + 4z = 12$ that lies above the square $0 \leq x \leq 2$, $0 \leq y \leq 2$ in the first quadrant

23. Portion of paraboloid The portion of the paraboloid $y = 2(x^2 + z^2)$, $y \leq 2$, that lies above the xy-plane

24. Portion of hemisphere The portion of the hemisphere $x^2 + y^2 + z^2 = 10$, $y \geq 0$, in the first octant

25. Surface area Find the area of the surface

$$\mathbf{r}(u, v) = (u + v)\mathbf{i} + (u - v)\mathbf{j} + v\mathbf{k},$$
$$0 \leq u \leq 1, \quad 0 \leq v \leq 1.$$

26. Surface integral Integrate $f(x, y, z) = xy - z^2$ over the surface in Exercise 25.

27. Area of a helicoid Find the surface area of the helicoid

$$\mathbf{r}(r, \theta) = (r\cos\theta)\mathbf{i} + (r\sin\theta)\mathbf{j} + \theta\mathbf{k}, \quad 0 \leq \theta \leq 2\pi, \quad 0 \leq r \leq 1,$$

in the accompanying figure.

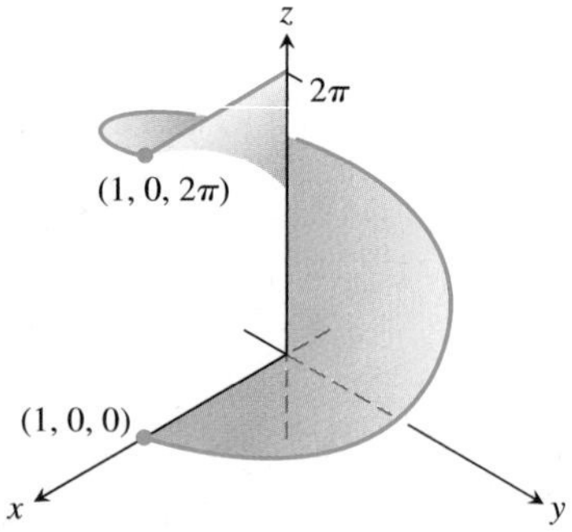

28. Surface integral Evaluate the integral $\iint_S \sqrt{x^2 + y^2 + 1}\, d\sigma$, where S is the helicoid in Exercise 27.

Conservative Fields

Which of the fields in Exercises 29–32 are conservative, and which are not?

29. $\mathbf{F} = x\mathbf{i} + y\mathbf{j} + z\mathbf{k}$

30. $\mathbf{F} = (x\mathbf{i} + y\mathbf{j} + z\mathbf{k})/(x^2 + y^2 + z^2)^{3/2}$

31. $\mathbf{F} = xe^y\mathbf{i} + ye^z\mathbf{j} + ze^x\mathbf{k}$

32. $\mathbf{F} = (\mathbf{i} + z\mathbf{j} + y\mathbf{k})/(x + yz)$

Find potential functions for the fields in Exercises 33 and 34.

33. $\mathbf{F} = 2\mathbf{i} + (2y + z)\mathbf{j} + (y + 1)\mathbf{k}$

34. $\mathbf{F} = (z\cos xz)\mathbf{i} + e^y\mathbf{j} + (x\cos xz)\mathbf{k}$

Work and Circulation

In Exercises 35 and 36, find the work done by each field along the paths from (0, 0, 0) to (1, 1, 1) in Exercise 1.

35. $\mathbf{F} = 2xy\mathbf{i} + \mathbf{j} + x^2\mathbf{k}$ **36.** $\mathbf{F} = 2xy\mathbf{i} + x^2\mathbf{j} + \mathbf{k}$

37. Finding work in two ways Find the work done by

$$\mathbf{F} = \frac{x\mathbf{i} + y\mathbf{j}}{(x^2 + y^2)^{3/2}}$$

over the plane curve $\mathbf{r}(t) = (e^t \cos t)\mathbf{i} + (e^t \sin t)\mathbf{j}$ from the point $(1, 0)$ to the point $(e^{2\pi}, 0)$ in two ways:

a. By using the parametrization of the curve to evaluate the work integral

b. By evaluating a potential function for **F**.

38. Flow along different paths Find the flow of the field $\mathbf{F} = \nabla(x^2 z e^y)$

a. Once around the ellipse C in which the plane $x + y + z = 1$ intersects the cylinder $x^2 + z^2 = 25$, clockwise as viewed from the positive y-axis

b. Along the curved boundary of the helicoid in Exercise 27 from $(1, 0, 0)$ to $(1, 0, 2\pi)$.

In Exercises 39 and 40, use the surface integral in Stokes' Theorem to find the circulation of the field **F** around the curve C in the indicated direction.

39. Circulation around an ellipse $\mathbf{F} = y^2\mathbf{i} - y\mathbf{j} + 3z^2\mathbf{k}$

C: The ellipse in which the plane $2x + 6y - 3z = 6$ meets the cylinder $x^2 + y^2 = 1$, counterclockwise as viewed from above

40. Circulation around a circle $\mathbf{F} = (x^2 + y)\mathbf{i} + (x + y)\mathbf{j} + (4y^2 - z)\mathbf{k}$

C: The circle in which the plane $z = -y$ meets the sphere $x^2 + y^2 + z^2 = 4$, counterclockwise as viewed from above

Mass and Moments

41. Wire with different densities Find the mass of a thin wire lying along the curve $\mathbf{r}(t) = \sqrt{2}t\mathbf{i} + \sqrt{2}t\mathbf{j} + (4 - t^2)\mathbf{k}$, $0 \le t \le 1$, if the density at t is **(a)** $\delta = 3t$ and **(b)** $\delta = 1$.

42. Wire with variable density Find the center of mass of a thin wire lying along the curve $\mathbf{r}(t) = t\mathbf{i} + 2t\mathbf{j} + (2/3)t^{3/2}\mathbf{k}$, $0 \le t \le 2$, if the density at t is $\delta = 3\sqrt{5 + t}$.

43. Wire with variable density Find the center of mass and the moments of inertia and radii of gyration about the coordinate axes of a thin wire lying along the curve

$$\mathbf{r}(t) = t\mathbf{i} + \frac{2\sqrt{2}}{3}t^{3/2}\mathbf{j} + \frac{t^2}{2}\mathbf{k}, \qquad 0 \le t \le 2,$$

if the density at t is $\delta = 1/(t + 1)$.

44. Center of mass of an arch A slender metal arch lies along the semicircle $y = \sqrt{a^2 - x^2}$ in the xy-plane. The density at the point (x, y) on the arch is $\delta(x, y) = 2a - y$. Find the center of mass.

45. Wire with constant density A wire of constant density $\delta = 1$ lies along the curve $\mathbf{r}(t) = (e^t \cos t)\mathbf{i} + (e^t \sin t)\mathbf{j} + e^t\mathbf{k}$, $0 \le t \le \ln 2$. Find $\bar{z}$, I_z, and R_z.

46. Helical wire with constant density Find the mass and center of mass of a wire of constant density δ that lies along the helix $\mathbf{r}(t) = (2 \sin t)\mathbf{i} + (2 \cos t)\mathbf{j} + 3t\mathbf{k}$, $0 \le t \le 2\pi$.

47. Inertia, radius of gyration, center of mass of a shell Find I_z, R_z, and the center of mass of a thin shell of density $\delta(x, y, z) = z$ cut from the upper portion of the sphere $x^2 + y^2 + z^2 = 25$ by the plane $z = 3$.

48. Moment of inertia of a cube Find the moment of inertia about the z-axis of the surface of the cube cut from the first octant by the planes $x = 1$, $y = 1$, and $z = 1$ if the density is $\delta = 1$.

Flux Across a Plane Curve or Surface

Use Green's Theorem to find the counterclockwise circulation and outward flux for the fields and curves in Exercises 49 and 50.

49. Square $\mathbf{F} = (2xy + x)\mathbf{i} + (xy - y)\mathbf{j}$

C: The square bounded by $x = 0, x = 1, y = 0, y = 1$

50. Triangle $\mathbf{F} = (y - 6x^2)\mathbf{i} + (x + y^2)\mathbf{j}$

C: The triangle made by the lines $y = 0, y = x$, and $x = 1$

51. Zero line integral Show that

$$\oint_C \ln x \sin y \, dy - \frac{\cos y}{x} \, dx = 0$$

for any closed curve C to which Green's Theorem applies.

52. a. Outward flux and area Show that the outward flux of the position vector field $\mathbf{F} = x\mathbf{i} + y\mathbf{j}$ across any closed curve to which Green's Theorem applies is twice the area of the region enclosed by the curve.

b. Let **n** be the outward unit normal vector to a closed curve to which Green's Theorem applies. Show that it is not possible for $\mathbf{F} = x\mathbf{i} + y\mathbf{j}$ to be orthogonal to **n** at every point of C.

In Exercises 53–56, find the outward flux of **F** across the boundary of D.

53. Cube $\mathbf{F} = 2xy\mathbf{i} + 2yz\mathbf{j} + 2xz\mathbf{k}$

D: The cube cut from the first octant by the planes $x = 1, y = 1, z = 1$

54. Spherical cap $\mathbf{F} = xz\mathbf{i} + yz\mathbf{j} + \mathbf{k}$

D: The entire surface of the upper cap cut from the solid sphere $x^2 + y^2 + z^2 \le 25$ by the plane $z = 3$

55. Spherical cap $\mathbf{F} = -2x\mathbf{i} - 3y\mathbf{j} + z\mathbf{k}$

D: The upper region cut from the solid sphere $x^2 + y^2 + z^2 \le 2$ by the paraboloid $z = x^2 + y^2$

56. Cone and cylinder $\mathbf{F} = (6x + y)\mathbf{i} - (x + z)\mathbf{j} + 4yz\mathbf{k}$

D: The region in the first octant bounded by the cone $z = \sqrt{x^2 + y^2}$, the cylinder $x^2 + y^2 = 1$, and the coordinate planes

57. Hemisphere, cylinder, and plane Let S be the surface that is bounded on the left by the hemisphere $x^2 + y^2 + z^2 = a^2, y \leq 0$, in the middle by the cylinder $x^2 + z^2 = a^2, 0 \leq y \leq a$, and on the right by the plane $y = a$. Find the flux of $\mathbf{F} = y\mathbf{i} + z\mathbf{j} + x\mathbf{k}$ outward across S.

58. Cylinder and planes Find the outward flux of the field $\mathbf{F} = 3xz^2\mathbf{i} + y\mathbf{j} - z^3\mathbf{k}$ across the surface of the solid in the first octant that is bounded by the cylinder $x^2 + 4y^2 = 16$ and the planes $y = 2z, x = 0$, and $z = 0$.

59. Cylindrical can Use the Divergence Theorem to find the flux of $\mathbf{F} = xy^2\mathbf{i} + x^2y\mathbf{j} + y\mathbf{k}$ outward through the surface of the region enclosed by the cylinder $x^2 + y^2 = 1$ and the planes $z = 1$ and $z = -1$.

60. Hemisphere Find the flux of $\mathbf{F} = (3z + 1)\mathbf{k}$ upward across the hemisphere $x^2 + y^2 + z^2 = a^2, z \geq 0$ **(a)** with the Divergence Theorem and **(b)** by evaluating the flux integral directly.

Chapter 16 Additional and Advanced Exercises

Finding Areas with Green's Theorem

Use the Green's Theorem area formula, Equation (13) in Exercises 16.4, to find the areas of the regions enclosed by the curves in Exercises 1–4.

1. The limaçon $x = 2\cos t - \cos 2t$, $y = 2\sin t - \sin 2t$, $0 \leq t \leq 2\pi$

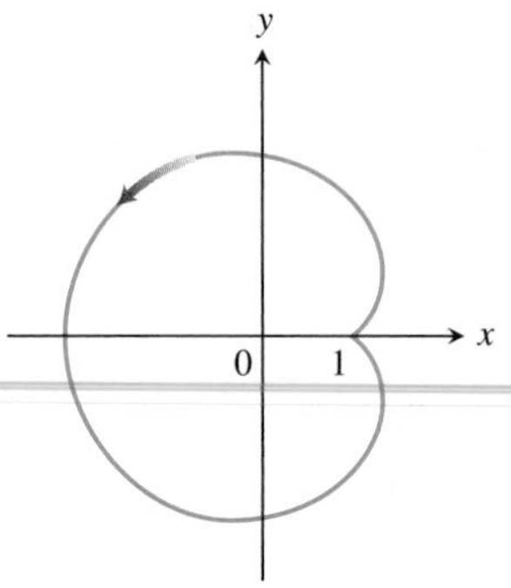

2. The deltoid $x = 2\cos t + \cos 2t$, $y = 2\sin t - \sin 2t$, $0 \leq t \leq 2\pi$

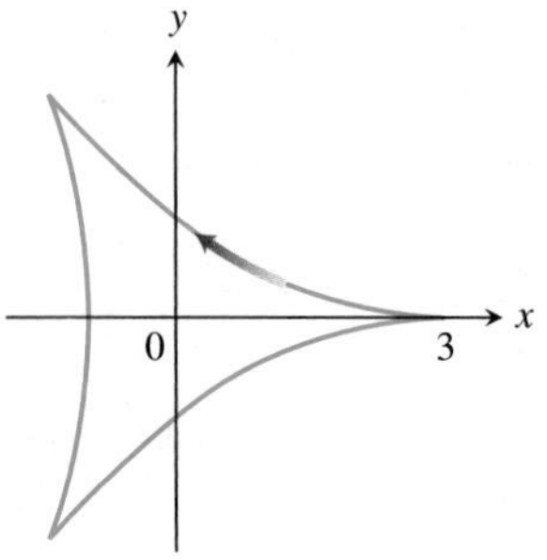

3. The eight curve $x = (1/2)\sin 2t, y = \sin t, 0 \leq t \leq \pi$ (one loop)

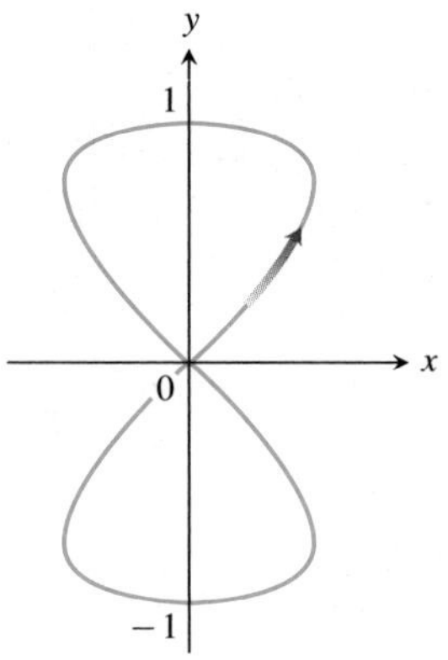

4. The teardrop $x = 2a\cos t - a\sin 2t, y = b\sin t, 0 \leq t \leq 2\pi$

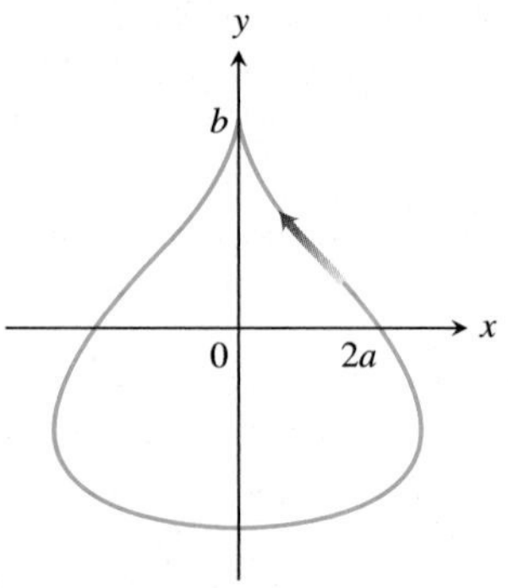

Theory and Applications

5. a. Give an example of a vector field $\mathbf{F}(x, y, z)$ that has value $\mathbf{0}$ at only one point and such that curl $\mathbf{F}$ is nonzero everywhere. Be sure to identify the point and compute the curl.

b. Give an example of a vector field $\mathbf{F}(x, y, z)$ that has value $\mathbf{0}$ on precisely one line and such that curl $\mathbf{F}$ is nonzero everywhere. Be sure to identify the line and compute the curl.

c. Give an example of a vector field $\mathbf{F}(x, y, z)$ that has value $\mathbf{0}$ on a surface and such that curl $\mathbf{F}$ is nonzero everywhere. Be sure to identify the surface and compute the curl.

6. Find all points (a, b, c) on the sphere $x^2 + y^2 + z^2 = R^2$ where the vector field $\mathbf{F} = yz^2\mathbf{i} + xz^2\mathbf{j} + 2xyz\mathbf{k}$ is normal to the surface and $\mathbf{F}(a, b, c) \neq \mathbf{0}$.

7. Find the mass of a spherical shell of radius R such that at each point (x, y, z) on the surface the mass density $\delta(x, y, z)$ is its distance to some fixed point (a, b, c) of the surface.

8. Find the mass of a helicoid

$$\mathbf{r}(r, \theta) = (r\cos\theta)\mathbf{i} + (r\sin\theta)\mathbf{j} + \theta\mathbf{k},$$

$0 \le r \le 1, 0 \le \theta \le 2\pi$, if the density function is $\delta(x, y, z) = 2\sqrt{x^2 + y^2}$. See Practice Exercise 27 for a figure.

9. Among all rectangular regions $0 \le x \le a, 0 \le y \le b$, find the one for which the total outward flux of $\mathbf{F} = (x^2 + 4xy)\mathbf{i} - 6y\mathbf{j}$ across the four sides is least. What *is* the least flux?

10. Find an equation for the plane through the origin such that the circulation of the flow field $\mathbf{F} = z\mathbf{i} + x\mathbf{j} + y\mathbf{k}$ around the circle of intersection of the plane with the sphere $x^2 + y^2 + z^2 = 4$ is a maximum.

11. A string lies along the circle $x^2 + y^2 = 4$ from (2, 0) to (0, 2) in the first quadrant. The density of the string is $\rho(x, y) = xy$

a. Partition the string into a finite number of subarcs to show that the work done by gravity to move the string straight down to the x-axis is given by

$$\text{Work} = \lim_{n\to\infty} \sum_{k=1}^{n} g\, x_k y_k^2 \Delta s_k = \int_C g\, xy^2\, ds,$$

where g is the gravitational constant.

b. Find the total work done by evaluating the line integral in part (a).

c. Show that the total work done equals the work required to move the string's center of mass $(\bar{x}, \bar{y})$ straight down to the x-axis.

12. A thin sheet lies along the portion of the plane $x + y + z = 1$ in the first octant. The density of the sheet is $\delta(x, y, z) = xy$.

a. Partition the sheet into a finite number of subpieces to show that the work done by gravity to move the sheet straight down to the xy-plane is given by

$$\text{Work} = \lim_{n\to\infty} \sum_{k=1}^{n} g\, x_k y_k z_k\, \Delta\sigma_k = \iint_S g\, xyz\, d\sigma,$$

where g is the gravitational constant.

b. Find the total work done by evaluating the surface integral in part (a).

c. Show that the total work done equals the work required to move the sheet's center of mass $(\bar{x}, \bar{y}, \bar{z})$ straight down to the xy-plane.

13. Archimedes' principle If an object such as a ball is placed in a liquid, it will either sink to the bottom, float, or sink a certain distance and remain suspended in the liquid. Suppose a fluid has constant weight density w and that the fluid's surface coincides with the plane $z = 4$. A spherical ball remains suspended in the fluid and occupies the region $x^2 + y^2 + (z - 2)^2 \le 1$.

a. Show that the surface integral giving the magnitude of the total force on the ball due to the fluid's pressure is

$$\text{Force} = \lim_{n\to\infty} \sum_{k=1}^{n} w(4 - z_k)\, \Delta\sigma_k = \iint_S w(4 - z)\, d\sigma.$$

b. Since the ball is not moving, it is being held up by the buoyant force of the liquid. Show that the magnitude of the buoyant force on the sphere is

$$\text{Buoyant force} = \iint_S w(z - 4)\mathbf{k}\cdot\mathbf{n}\, d\sigma,$$

where $\mathbf{n}$ is the outer unit normal at (x, y, z). This illustrates Archimedes' principle that the magnitude of the buoyant force on a submerged solid equals the weight of the displaced fluid.

c. Use the Divergence Theorem to find the magnitude of the buoyant force in part (b).

14. Fluid force on a curved surface A cone in the shape of the surface $z = \sqrt{x^2 + y^2}, 0 \le z \le 2$ is filled with a liquid of constant weight density w. Assuming the xy-plane is "ground level," show that the total force on the portion of the cone from $z = 1$ to $z = 2$ due to liquid pressure is the surface integral

$$F = \iint_S w(2 - z)\, d\sigma.$$

Evaluate the integral.

15. Faraday's Law If $\mathbf{E}(t, x, y, z)$ and $\mathbf{B}(t, x, y, z)$ represent the electric and magnetic fields at point (x, y, z) at time t, a basic principle of electromagnetic theory says that $\nabla \times \mathbf{E} = -\partial\mathbf{B}/\partial t$. In this expression $\nabla \times \mathbf{E}$ is computed with t held fixed and $\partial\mathbf{B}/\partial t$ is calculated with (x, y, z) fixed. Use Stokes' Theorem to derive Faraday's Law

$$\oint_C \mathbf{E}\cdot d\mathbf{r} = -\frac{\partial}{\partial t}\iint_S \mathbf{B}\cdot\mathbf{n}\, d\sigma,$$

where C represents a wire loop through which current flows counterclockwise with respect to the surface's unit normal $\mathbf{n}$, giving rise to the voltage

$$\oint_C \mathbf{E}\cdot d\mathbf{r}$$

around C. The surface integral on the right side of the equation is called the *magnetic flux*, and S is any oriented surface with boundary C.

16. Let

$$\mathbf{F} = -\frac{GmM}{|\mathbf{r}|^3}\mathbf{r}$$

be the gravitational force field defined for $\mathbf{r} \neq \mathbf{0}$. Use Gauss's Law in Section 16.8 to show that there is no continuously differentiable vector field $\mathbf{H}$ satisfying $\mathbf{F} = \nabla \times \mathbf{H}$.

17. If $f(x, y, z)$ and $g(x, y, z)$ are continuously differentiable scalar functions defined over the oriented surface S with boundary curve C, prove that

$$\iint_S (\nabla f \times \nabla g) \cdot \mathbf{n}\, d\sigma = \oint_C f\, \nabla g \cdot d\mathbf{r}.$$

18. Suppose that $\nabla \cdot \mathbf{F}_1 = \nabla \cdot \mathbf{F}_2$ and $\nabla \times \mathbf{F}_1 = \nabla \times \mathbf{F}_2$ over a region D enclosed by the oriented surface S with outward unit normal $\mathbf{n}$ and that $\mathbf{F}_1 \cdot \mathbf{n} = \mathbf{F}_2 \cdot \mathbf{n}$ on S. Prove that $\mathbf{F}_1 = \mathbf{F}_2$ throughout D.

19. Prove or disprove that if $\nabla \cdot \mathbf{F} = 0$ and $\nabla \times \mathbf{F} = \mathbf{0}$, then $\mathbf{F} = \mathbf{0}$.

20. Let S be an oriented surface parametrized by $\mathbf{r}(u, v)$. Define the notation $d\boldsymbol{\sigma} = \mathbf{r}_u\, du \times \mathbf{r}_v\, dv$ so that $d\boldsymbol{\sigma}$ is a vector normal to the surface. Also, the magnitude $d\sigma = |d\boldsymbol{\sigma}|$ is the element of surface area (by Equation 5 in Section 16.6). Derive the identity

$$d\sigma = (EG - F^2)^{1/2}\, du\, dv$$

where

$$E = |\mathbf{r}_u|^2, \quad F = \mathbf{r}_u \cdot \mathbf{r}_v, \quad \text{and} \quad G = |\mathbf{r}_v|^2.$$

21. Show that the volume V of a region D in space enclosed by the oriented surface S with outward normal $\mathbf{n}$ satisfies the identity

$$V = \frac{1}{3}\iint_S \mathbf{r} \cdot \mathbf{n}\, d\sigma,$$

where $\mathbf{r}$ is the position vector of the point (x, y, z) in D.

Chapter 16 Technology Application Projects

Mathematica/Maple Module

Work in Conservative and Nonconservative Force Fields

Explore integration over vector fields and experiment with conservative and nonconservative force functions along different paths in the field.

Mathematica/Maple Module

How Can You Visualize Green's Theorem?

Explore integration over vector fields and use parametrizations to compute line integrals. Both forms of Green's Theorem are explored.

Mathematica/Maple Module

Visualizing and Interpreting the Divergence Theorem

Verify the Divergence Theorem by formulating and evaluating certain divergence and surface integrals.

APPENDICES

A.1 Mathematical Induction

Many formulas, like

$$1 + 2 + \cdots + n = \frac{n(n+1)}{2},$$

can be shown to hold for every positive integer n by applying an axiom called the *mathematical induction principle*. A proof that uses this axiom is called a *proof by mathematical induction* or a *proof by induction*.

The steps in proving a formula by induction are the following:

1. Check that the formula holds for $n = 1$.
2. Prove that if the formula holds for any positive integer $n = k$, then it also holds for the next integer, $n = k + 1$.

The induction axiom says that once these steps are completed, the formula holds for all positive integers n. By Step 1 it holds for $n = 1$. By Step 2 it holds for $n = 2$, and therefore by Step 2 also for $n = 3$, and by Step 2 again for $n = 4$, and so on. If the first domino falls, and the kth domino always knocks over the $(k + 1)$st when it falls, all the dominoes fall.

From another point of view, suppose we have a sequence of statements S_1, $S_2, \ldots, S_n, \ldots$, one for each positive integer. Suppose we can show that assuming any one of the statements to be true implies that the next statement in line is true. Suppose that we can also show that S_1 is true. Then we may conclude that the statements are true from S_1 on.

EXAMPLE 1 Use mathematical induction to prove that for every positive integer n,

$$1 + 2 + \cdots + n = \frac{n(n+1)}{2}.$$

Solution We accomplish the proof by carrying out the two steps above.

1. The formula holds for $n = 1$ because

$$1 = \frac{1(1+1)}{2}.$$

2. If the formula holds for $n = k$, does it also hold for $n = k + 1$? The answer is yes, as we now show. If

$$1 + 2 + \cdots + k = \frac{k(k + 1)}{2},$$

then

$$\begin{aligned} 1 + 2 + \cdots + k + (k + 1) &= \frac{k(k + 1)}{2} + (k + 1) = \frac{k^2 + k + 2k + 2}{2} \\ &= \frac{(k + 1)(k + 2)}{2} = \frac{(k + 1)((k + 1) + 1)}{2}. \end{aligned}$$

The last expression in this string of equalities is the expression $n(n + 1)/2$ for $n = (k + 1)$.

The mathematical induction principle now guarantees the original formula for all positive integers n. ■

In Example 4 of Section 5.2 we gave another proof for the formula giving the sum of the first n integers. However, proof by mathematical induction is more general. It can be used to find the sums of the squares and cubes of the first n integers (Exercises 9 and 10). Here is another example.

EXAMPLE 2 Show by mathematical induction that for all positive integers n,

$$\frac{1}{2^1} + \frac{1}{2^2} + \cdots + \frac{1}{2^n} = 1 - \frac{1}{2^n}.$$

Solution We accomplish the proof by carrying out the two steps of mathematical induction.

1. The formula holds for $n = 1$ because

$$\frac{1}{2^1} = 1 - \frac{1}{2^1}.$$

2. If

$$\frac{1}{2^1} + \frac{1}{2^2} + \cdots + \frac{1}{2^k} = 1 - \frac{1}{2^k},$$

then

$$\begin{aligned} \frac{1}{2^1} + \frac{1}{2^2} + \cdots + \frac{1}{2^k} + \frac{1}{2^{k+1}} &= 1 - \frac{1}{2^k} + \frac{1}{2^{k+1}} = 1 - \frac{1 \cdot 2}{2^k \cdot 2} + \frac{1}{2^{k+1}} \\ &= 1 - \frac{2}{2^{k+1}} + \frac{1}{2^{k+1}} = 1 - \frac{1}{2^{k+1}}. \end{aligned}$$

Thus, the original formula holds for $n = (k + 1)$ whenever it holds for $n = k$.

With these steps verified, the mathematical induction principle now guarantees the formula for every positive integer n. ■

Other Starting Integers

Instead of starting at $n = 1$ some induction arguments start at another integer. The steps for such an argument are as follows.

1. Check that the formula holds for $n = n_1$ (the first appropriate integer).
2. Prove that if the formula holds for any integer $n = k \geq n_1$, then it also holds for $n = (k + 1)$.

Once these steps are completed, the mathematical induction principle guarantees the formula for all $n \geq n_1$.

EXAMPLE 3 Show that $n! > 3^n$ if n is large enough.

Solution How large is large enough? We experiment:

n	1	2	3	4	5	6	7
$n!$	1	2	6	24	120	720	5040
3^n	3	9	27	81	243	729	2187

It looks as if $n! > 3^n$ for $n \geq 7$. To be sure, we apply mathematical induction. We take $n_1 = 7$ in Step 1 and complete Step 2.

Suppose $k! > 3^k$ for some $k \geq 7$. Then

$$(k + 1)! = (k + 1)(k!) > (k + 1)3^k > 7 \cdot 3^k > 3^{k+1}.$$

Thus, for $k \geq 7$,

$$k! > 3^k \quad \text{implies} \quad (k + 1)! > 3^{k+1}.$$

The mathematical induction principle now guarantees $n! \geq 3^n$ for all $n \geq 7$. ■

EXERCISES A.1

1. Assuming that the triangle inequality $|a + b| \leq |a| + |b|$ holds for any two numbers a and b, show that
$$|x_1 + x_2 + \cdots + x_n| \leq |x_1| + |x_2| + \cdots + |x_n|$$
for any n numbers.

2. Show that if $r \neq 1$, then
$$1 + r + r^2 + \cdots + r^n = \frac{1 - r^{n+1}}{1 - r}$$
for every positive integer n.

3. Use the Product Rule, $\frac{d}{dx}(uv) = u\frac{dv}{dx} + v\frac{du}{dx}$, and the fact that $\frac{d}{dx}(x) = 1$ to show that $\frac{d}{dx}(x^n) = nx^{n-1}$ for every positive integer n.

4. Suppose that a function $f(x)$ has the property that $f(x_1x_2) = f(x_1) + f(x_2)$ for any two positive numbers x_1 and x_2. Show that
$$f(x_1x_2\cdots x_n) = f(x_1) + f(x_2) + \cdots + f(x_n)$$
for the product of any n positive numbers $x_1, x_2 \ldots, x_n$.

5. Show that
$$\frac{2}{3^1} + \frac{2}{3^2} + \cdots + \frac{2}{3^n} = 1 - \frac{1}{3^n}$$
for all positive integers n.

6. Show that $n! > n^3$ if n is large enough.

7. Show that $2^n > n^2$ if n is large enough.

8. Show that $2^n \geq 1/8$ for $n \geq -3$.

9. **Sums of squares** Show that the sum of the squares of the first n positive integers is
$$\frac{n\left(n + \frac{1}{2}\right)(n + 1)}{3}.$$

10. **Sums of cubes** Show that the sum of the cubes of the first n positive integers is $(n(n + 1)/2)^2$.

11. **Rules for finite sums** Show that the following finite sum rules hold for every positive integer n.

    a. $\sum_{k=1}^{n}(a_k + b_k) = \sum_{k=1}^{n} a_k + \sum_{k=1}^{n} b_k$

b. $\sum_{k=1}^{n}(a_k - b_k) = \sum_{k=1}^{n} a_k - \sum_{k=1}^{n} b_k$

c. $\sum_{k=1}^{n} ca_k = c \cdot \sum_{k=1}^{n} a_k$ (Any number c)

d. $\sum_{k=1}^{n} a_k = n \cdot c$ (if a_k has the constant value c)

12. Show that $|x^n| = |x|^n$ for every positive integer n and every real number x.

A.2 Proofs of Limit Theorems

This appendix proves Theorem 1, Parts 2–5, and Theorem 4 from Section 2.2.

THEOREM 1 Limit Laws

If L, M, c, and k are real numbers and

$$\lim_{x \to c} f(x) = L \quad \text{and} \quad \lim_{x \to c} g(x) = M, \quad \text{then}$$

1. *Sum Rule:* $\lim_{x \to c} \left(f(x) + g(x)\right) = L + M$
2. *Difference Rule:* $\lim_{x \to c} \left(f(x) - g(x)\right) = L - M$
3. *Product Rule:* $\lim_{x \to c} \left(f(x) \cdot g(x)\right) = L \cdot M$
4. *Constant Multiple Rule:* $\lim_{x \to c} \left(kf(x)\right) = kL$ (any number k)
5. *Quotient Rule:* $\lim_{x \to c} \dfrac{f(x)}{g(x)} = \dfrac{L}{M}$, if $M \neq 0$
6. *Power Rule:* If r and s are integers with no common factor and $s \neq 0$, then

$$\lim_{x \to c} \left(f(x)\right)^{r/s} = L^{r/s}$$

provided that $L^{r/s}$ is a real number. (If s is even, we assume that $L > 0$.)

We proved the Sum Rule in Section 2.3 and the Power Rule is proved in more advanced texts. We obtain the Difference Rule by replacing $g(x)$ by $-g(x)$ and M by $-M$ in the Sum Rule. The Constant Multiple Rule is the special case $g(x) = k$ of the Product Rule. This leaves only the Product and Quotient Rules.

Proof of the Limit Product Rule We show that for any $\epsilon > 0$ there exists a $\delta > 0$ such that for all x in the intersection D of the domains of f and g,

$$0 < |x - c| < \delta \quad \Rightarrow \quad |f(x)g(x) - LM| < \epsilon.$$

Suppose then that ϵ is a positive number, and write $f(x)$ and $g(x)$ as

$$f(x) = L + (f(x) - L), \qquad g(x) = M + (g(x) - M).$$

Multiply these expressions together and subtract LM:

$$\begin{aligned} f(x)\cdot g(x) - LM &= (L + (f(x) - L))(M + (g(x) - M)) - LM \\ &= LM + L(g(x) - M) + M(f(x) - L) \\ &\quad + (f(x) - L)(g(x) - M) - LM \\ &= L(g(x) - M) + M(f(x) - L) + (f(x) - L)(g(x) - M). \quad (1) \end{aligned}$$

Since f and g have limits L and M as $x \to c$, there exist positive numbers δ_1, δ_2, δ_3, and δ_4 such that for all x in D

$$\begin{aligned} 0 < |x - c| < \delta_1 &\quad \Rightarrow \quad |f(x) - L| < \sqrt{\epsilon/3} \\ 0 < |x - c| < \delta_2 &\quad \Rightarrow \quad |g(x) - M| < \sqrt{\epsilon/3} \\ 0 < |x - c| < \delta_3 &\quad \Rightarrow \quad |f(x) - L| < \epsilon/(3(1 + |M|)) \\ 0 < |x - c| < \delta_4 &\quad \Rightarrow \quad |g(x) - M| < \epsilon/(3(1 + |L|)) \end{aligned} \quad (2)$$

If we take δ to be the smallest numbers δ_1 through δ_4, the inequalities on the right-hand side of the Implications (2) will hold simultaneously for $0 < |x - c| < \delta$. Therefore, for all x in D, $0 < |x - c| < \delta$ implies

$$|f(x)\cdot g(x) - LM|$$

Triangle inequality applied to Equation (1)

$$\begin{aligned} &\leq |L||g(x) - M| + |M||f(x) - L| + |f(x) - L||g(x) - M| \\ &\leq (1 + |L|)|g(x) - M| + (1 + |M|)|f(x) - L| + |f(x) - L||g(x) - M| \\ &< \frac{\epsilon}{3} + \frac{\epsilon}{3} + \sqrt{\frac{\epsilon}{3}}\sqrt{\frac{\epsilon}{3}} = \epsilon. \end{aligned}$$

Values from (2)

This completes the proof of the Limit Product Rule. ■

Proof of the Limit Quotient Rule We show that $\lim_{x\to c}(1/g(x)) = 1/M$. We can then conclude that

$$\lim_{x\to c}\frac{f(x)}{g(x)} = \lim_{x\to c}\left(f(x)\cdot\frac{1}{g(x)}\right) = \lim_{x\to c} f(x)\cdot\lim_{x\to c}\frac{1}{g(x)} = L\cdot\frac{1}{M} = \frac{L}{M}$$

by the Limit Product Rule.

Let $\epsilon > 0$ be given. To show that $\lim_{x\to c}(1/g(x)) = 1/M$, we need to show that there exists a $\delta > 0$ such that for all x.

$$0 < |x - c| < \delta \quad \Rightarrow \quad \left|\frac{1}{g(x)} - \frac{1}{M}\right| < \epsilon.$$

Since $|M| > 0$, there exists a positive number δ_1 such that for all x

$$0 < |x - c| < \delta_1 \quad \Rightarrow \quad |g(x) - M| < \frac{M}{2}. \quad (3)$$

For any numbers A and B it can be shown that $|A| - |B| \leq |A - B|$ and $|B| - |A| \leq |A - B|$, from which it follows that $||A| - |B|| \leq |A - B|$. With $A = g(x)$ and $B = M$, this becomes

$$||g(x)| - |M|| \leq |g(x) - M|,$$

which can be combined with the inequality on the right in Implication (3) to get, in turn,

$$||g(x)| - |M|| < \frac{|M|}{2}$$

$$-\frac{|M|}{2} < |g(x)| - |M| < \frac{|M|}{2}$$

$$\frac{|M|}{2} < |g(x)| < \frac{3|M|}{2}$$

$$|M| < 2|g(x)| < 3|M|$$

$$\frac{1}{|g(x)|} < \frac{2}{|M|} < \frac{3}{|g(x)|} \tag{4}$$

Therefore, $0 < |x - c| < \delta_1$ implies that

$$\left|\frac{1}{g(x)} - \frac{1}{M}\right| = \left|\frac{M - g(x)}{Mg(x)}\right| \le \frac{1}{|M|} \cdot \frac{1}{|g(x)|} \cdot |M - g(x)|$$

$$< \frac{1}{|M|} \cdot \frac{2}{|M|} \cdot |M - g(x)|. \quad \text{Inequality (4)} \tag{5}$$

Since $(1/2)|M|^2\epsilon > 0$, there exists a number $\delta_2 > 0$ such that for all x

$$0 < |x - c| < \delta_2 \quad \Rightarrow \quad |M - g(x)| < \frac{\epsilon}{2}|M|^2. \tag{6}$$

If we take δ to be the smaller of δ_1 and δ_2, the conclusions in (5) and (6) both hold for all x such that $0 < |x - c| < \delta$. Combining these conclusions gives

$$0 < |x - c| < \delta \quad \Rightarrow \quad \left|\frac{1}{g(x)} - \frac{1}{M}\right| < \epsilon.$$

This concludes the proof of the Limit Quotient Rule. ■

THEOREM 4 The Sandwich Theorem

Suppose that $g(x) \le f(x) \le h(x)$ for all x in some open interval I containing c, except possibly at $x = c$ itself. Suppose also that $\lim_{x\to c} g(x) = \lim_{x\to c} h(x) = L$. Then $\lim_{x\to c} f(x) = L$.

Proof for Right-Hand Limits Suppose $\lim_{x\to c^+} g(x) = \lim_{x\to c^+} h(x) = L$. Then for any $\epsilon > 0$ there exists a $\delta > 0$ such that for all x the interval $c < x < c + \delta$ is contained in I and the inequality implies

$$L - \epsilon < g(x) < L + \epsilon \quad \text{and} \quad L - \epsilon < h(x) < L + \epsilon.$$

These inequalities combine with the inequality $g(x) \le f(x) \le h(x)$ to give

$$L - \epsilon < g(x) \le f(x) \le h(x) < L + \epsilon,$$

$$L - \epsilon < f(x) < L + \epsilon,$$

$$-\epsilon < f(x) - L < \epsilon.$$

Therefore, for all x, the inequality $c < x < c + \delta$ implies $|f(x) - L| < \epsilon$. ■

Proof for Left-Hand Limits Suppose $\lim_{x\to c^-} g(x) = \lim_{x\to c^-} h(x) = L$. Then for any $\epsilon > 0$ there exists a $\delta > 0$ such that for all x the interval $c - \delta < x < c$ is contained in I and the inequality implies

$$L - \epsilon < g(x) < L + \epsilon \quad \text{and} \quad L - \epsilon < h(x) < L + \epsilon.$$

We conclude as before that for all x, $c - \delta < x < c$ implies $|f(x) - L| < \epsilon$. ■

Proof for Two-Sided Limits If $\lim_{x\to c} g(x) = \lim_{x\to c} h(x) = L$, then $g(x)$ and $h(x)$ both approach L as $x \to c^+$ and as $x \to c^-$; so $\lim_{x\to c^+} f(x) = L$ and $\lim_{x\to c^-} f(x) = L$. Hence $\lim_{x\to c} f(x)$ exists and equals L. ■

EXERCISES A.2

1. Suppose that functions $f_1(x)$, $f_2(x)$, and $f_3(x)$ have limits L_1, L_2, and L_3, respectively, as $x \to c$. Show that their sum has limit $L_1 + L_2 + L_3$. Use mathematical induction (Appendix 1) to generalize this result to the sum of any finite number of functions.
2. Use mathematical induction and the Limit Product Rule in Theorem 1 to show that if functions $f_1(x), f_2(x), \ldots, f_n(x)$ have limits $L_1, L_2, \ldots, L_n$ as $x \to c$, then
$$\lim_{x\to c} f_1(x) f_2(x) \cdot \cdots \cdot f_n(x) = L_1 \cdot L_2 \cdot \cdots \cdot L_n.$$
3. Use the fact that $\lim_{x\to c} x = c$ and the result of Exercise 2 to show that $\lim_{x\to c} x^n = c^n$ for any integer $n > 1$.
4. **Limits of polynomials** Use the fact that $\lim_{x\to c}(k) = k$ for any number k together with the results of Exercises 1 and 3 to show that $\lim_{x\to c} f(x) = f(c)$ for any polynomial function
$$f(x) = a_n x^n + a_{n-1}x^{n-1} + \cdots + a_1 x + a_0.$$
5. **Limits of rational functions** Use Theorem 1 and the result of Exercise 4 to show that if $f(x)$ and $g(x)$ are polynomial functions and $g(c) \neq 0$, then
$$\lim_{x\to c} \frac{f(x)}{g(x)} = \frac{f(c)}{g(c)}.$$
6. **Composites of continuous functions** Figure A.1 gives the diagram for a proof that the composite of two continuous functions is continuous. Reconstruct the proof from the diagram. The statement to be proved is this: If f is continuous at $x = c$ and g is continuous at $f(c)$, then $g \circ f$ is continuous at c.

   Assume that c is an interior point of the domain of f and that $f(c)$ is an interior point of the domain of g. This will make the limits involved two-sided. (The arguments for the cases that involve one-sided limits are similar.)

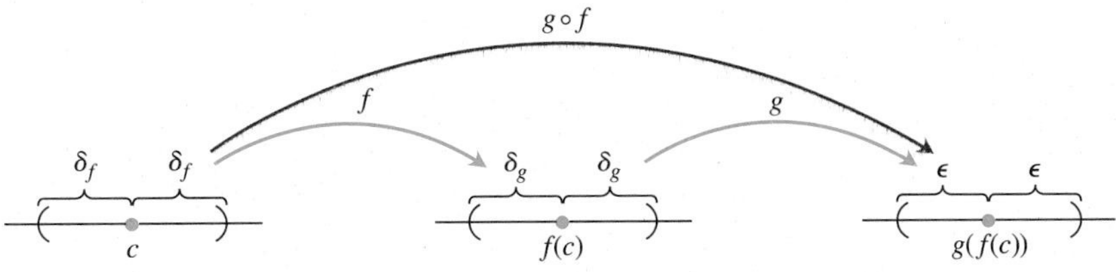

FIGURE A.1 The diagram for a proof that the composite of two continuous functions is continuous.

A.3 Commonly Occurring Limits

This appendix verifies limits (4)–(6) in Theorem 5 of Section 11.1.

Limit 4: If $|x| < 1$, $\lim_{n\to\infty} x^n = 0$ We need to show that to each $\epsilon > 0$ there corresponds an integer N so large that $|x^n| < \epsilon$ for all n greater than N. Since $\epsilon^{1/n} \to 1$, while

$|x| < 1$, there exists an integer N for which $\epsilon^{1/N} > |x|$. In other words,

$$|x^N| = |x|^N < \epsilon. \tag{1}$$

This is the integer we seek because, if $|x| < 1$, then

$$|x^n| < |x^N| \quad \text{for all } n > N. \tag{2}$$

Combining (1) and (2) produces $|x^n| < \epsilon$ for all $n > N$, concluding the proof. ∎

Limit 5: For any number x, $\lim_{n\to\infty}\left(1 + \frac{x}{n}\right)^n = e^x$ Let

$$a_n = \left(1 + \frac{x}{n}\right)^n.$$

Then

$$\ln a_n = \ln\left(1 + \frac{x}{n}\right)^n = n \ln\left(1 + \frac{x}{n}\right) \rightarrow x,$$

as we can see by the following application of l'Hôpital's Rule, in which we differentiate with respect to n:

$$\lim_{n\to\infty} n \ln\left(1 + \frac{x}{n}\right) = \lim_{n\to\infty} \frac{\ln(1 + x/n)}{1/n}$$

$$= \lim_{n\to\infty} \frac{\left(\dfrac{1}{1 + x/n}\right)\cdot\left(-\dfrac{x}{n^2}\right)}{-1/n^2} = \lim_{n\to\infty} \frac{x}{1 + x/n} = x.$$

Apply Theorem 4, Section 11.1, with $f(x) = e^x$ to conclude that

$$\left(1 + \frac{x}{n}\right)^n = a_n = e^{\ln a_n} \rightarrow e^x.$$

∎

Limit 6: For any number x, $\lim_{n\to\infty} \frac{x^n}{n!} = 0$ Since

$$-\frac{|x|^n}{n!} \le \frac{x^n}{n!} \le \frac{|x|^n}{n!},$$

all we need to show is that $|x|^n/n! \rightarrow 0$. We can then apply the Sandwich Theorem for Sequences (Section 11.1, Theorem 2) to conclude that $x^n/n! \rightarrow 0$.

The first step in showing that $|x|^n/n! \rightarrow 0$ is to choose an integer $M > |x|$, so that $(|x|/M) < 1$. By Limit 4, just proved, we then have $(|x|/M)^n \rightarrow 0$. We then restrict our attention to values of $n > M$. For these values of n, we can write

$$\frac{|x|^n}{n!} = \frac{|x|^n}{1\cdot 2\cdot \cdots \cdot M\cdot \underbrace{(M+1)(M+2)\cdot \cdots \cdot n}_{(n-M)\text{ factors}}}$$

$$\le \frac{|x|^n}{M!M^{n-M}} = \frac{|x|^n M^M}{M!M^n} = \frac{M^M}{M!}\left(\frac{|x|}{M}\right)^n.$$

Thus,

$$0 \le \frac{|x|^n}{n!} \le \frac{M^M}{M!}\left(\frac{|x|}{M}\right)^n.$$

Now, the constant $M^M/M!$ does not change as n increases. Thus the Sandwich Theorem tells us that $|x|^n/n! \to 0$ because $(|x|/M)^n \to 0$. ■

A.4 Theory of the Real Numbers

A rigorous development of calculus is based on properties of the real numbers. Many results about functions, derivatives, and integrals would be false if stated for functions defined only on the rational numbers. In this appendix we briefly examine some basic concepts of the theory of the reals that hint at what might be learned in a deeper, more theoretical study of calculus.

Three types of properties make the real numbers what they are. These are the **algebraic**, **order**, and **completeness** properties. The algebraic properties involve addition and multiplication, subtraction and division. They apply to rational or complex numbers as well as to the reals.

The structure of numbers is built around a set with addition and multiplication operations. The following properties are required of addition and multiplication.

A1 $a + (b + c) = (a + b) + c$ for all a, b, c.

A2 $a + b = b + a$ for all a, b, c.

A3 There is a number called "0" such that $a + 0 = a$ for all a.

A4 For each number a, there is a b such that $a + b = 0$.

M1 $a(bc) = (ab)c$ for all a, b, c.

M2 $ab = ba$ for all a, b.

M3 There is a number called "1" such that $a \cdot 1 = a$ for all a.

M4 For each nonzero a, there is a b such that $ab = 1$.

D $a(b + c) = ab + bc$ for all a, b, c.

A1 and M1 are *associative laws*, A2 and M2 are *commutativity laws*, A3 and M3 are *identity laws*, and D is the *distributive law*. Sets that have these algebraic properties are examples of **fields**, and are studied in depth in the area of theoretical mathematics called abstract algebra.

The **order** properties allow us to compare the size of any two numbers. The order properties are

O1 For any a and b, either $a \le b$ or $b \le a$ or both.

O2 If $a \le b$ and $b \le a$ then $a = b$.

O3 If $a \le b$ and $b \le c$ then $a \le c$.

O4 If $a \le b$ then $a + c \le b + c$.

O5 If $a \le b$ and $0 \le c$ then $ac \le bc$.

O3 is the *transitivity law*, and O4 and O5 relate ordering to addition and multiplication.

We can order the reals, the integers, and the rational numbers, but we cannot order the complex numbers (see Appendix A.5). There is no reasonable way to decide whether a number like $i = \sqrt{-1}$ is bigger or smaller than zero. A field in which the size of any two elements can be compared as above is called an **ordered field**. Both the rational numbers and the real numbers are ordered fields, and there are many others.

We can think of real numbers geometrically, lining them up as points on a line. The **completeness property** says that the real numbers correspond to all points on the line, with no "holes" or "gaps." The rationals, in contrast, omit points such as $\sqrt{2}$ and π, and the integers even leave out fractions like $1/2$. The reals, having the completeness property, omit no points.

What exactly do we mean by this vague idea of missing holes? To answer this we must give a more precise description of completeness. A number M is an **upper bound** for a set of numbers if all numbers in the set are smaller than or equal to M. M is a **least upper bound** if it is the smallest upper bound. For example, $M = 2$ is an upper bound for the negative numbers. So is $M = 1$, showing that 2 is not a least upper bound. The least upper bound for the set of negative numbers is $M = 0$. We define a **complete** ordered field to be one in which every nonempty set bounded above has a least upper bound.

If we work with just the rational numbers, the set of numbers less than $\sqrt{2}$ is bounded, but it does not have a rational least upper bound, since any rational upper bound M can be replaced by a slightly smaller rational number that is still larger than $\sqrt{2}$. So the rationals are not complete. In the real numbers, a set that is bounded above always has a least upper bound. The reals are a complete ordered field.

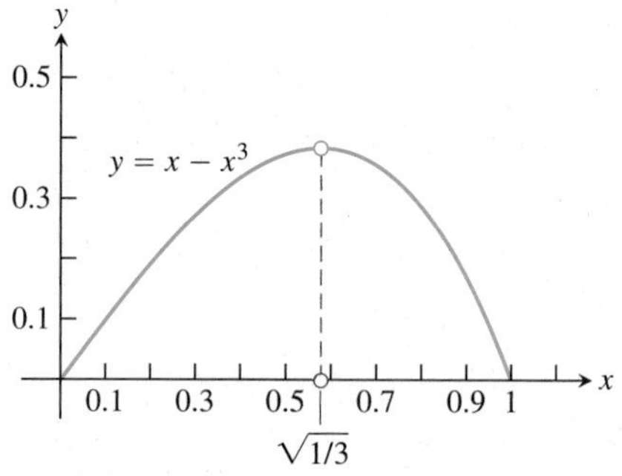

FIGURE A.2 The maximum value of $y = x - x^3$ on [0, 1] occurs at the irrational number $x = \sqrt{1/3}$.

The completeness property is at the heart of many results in calculus. One example occurs when searching for a maximum value for a function on a closed interval $[a, b]$, as in Section 4.1. The function $y = x - x^3$ has a maximum value on [0, 1] at the point x satisfying $1 - 3x^2 = 0$, or $x = \sqrt{1/3}$. If we limited our consideration to functions defined only on rational numbers, we would have to conclude that the function has no maximum, since $\sqrt{1/3}$ is irrational (Figure A.2). The Extreme Value Theorem (Section 4.1), which implies that continuous functions on closed intervals $[a, b]$ have a maximum value, is not true for functions defined only on the rationals.

The Intermediate Value Theorem implies that a continuous function f on an interval $[a, b]$ with $f(a) < 0$ and $f(b) > 0$ must be zero somewhere in $[a, b]$. The function values cannot jump from negative to positive without there being some point x in $[a, b]$ where $f(x) = 0$. The Intermediate Value Theorem also relies on the completeness of the real numbers and is false for continuous functions defined only on the rationals. The function $f(x) = 3x^2 - 1$ has $f(0) = -1$ and $f(1) = 2$, but if we consider f only on the rational numbers, it never equals zero. The only value of x for which $f(x) = 0$ is $x = \sqrt{1/3}$, an irrational number.

We have captured the desired properties of the reals by saying that the real numbers are a complete ordered field. But we're not quite finished. Greek mathematicians in the school of Pythagoras tried to impose another property on the numbers of the real line, the condition that all numbers are ratios of integers. They learned that their effort was doomed when they discovered irrational numbers such as $\sqrt{2}$. How do we know that our efforts to specify the real numbers are not also flawed, for some unseen reason? The artist Escher drew optical illusions of spiral staircases that went up and up until they rejoined themselves at the bottom. An engineer trying to build such a staircase would find that no structure realized the plans the architect had drawn. Could it be that our design for the reals contains some subtle contradiction, and that no construction of such a number system can be made?

We resolve this issue by giving a specific description of the real numbers and verifying that the algebraic, order, and completeness properties are satisfied in this model. This is called a **construction** of the reals, and just as stairs can be built with wood, stone, or steel, there are several approaches to constructing the reals. One construction treats the reals as all the infinite decimals,

$$a.d_1d_2d_3d_4\ldots$$

In this approach a real number is an integer a followed by a sequence of decimal digits $d_1, d_2, d_3, \ldots$, each between 0 and 9. This sequence may stop, or repeat in a periodic pattern, or keep going forever with no pattern. In this form, $2.00, 0.3333333\ldots$ and $3.1415926535898\ldots$ represent three familiar real numbers. The real meaning of the dots "$\ldots$" following these digits requires development of the theory of sequences and series, as in Chapter 11. Each real number is constructed as the limit of a sequence of rational numbers given by its finite decimal approximations. An infinite decimal is then the same as a series

$$a + \frac{d_1}{10} + \frac{d_2}{100} + \cdots.$$

This decimal construction of the real numbers is not entirely straightforward. It's easy enough to check that it gives numbers that satisfy the completeness and order properties, but verifying the algebraic properties is rather involved. Even adding or multiplying two numbers requires an infinite number of operations. Making sense of division requires a careful argument involving limits of rational approximations to infinite decimals.

A different approach was taken by Richard Dedekind (1831–1916), a German mathematician, who gave the first rigorous construction of the real numbers in 1872. Given any real number x, we can divide the rational numbers into two sets: those less than or equal to x and those greater. Dedekind cleverly reversed this reasoning and defined a real number to be a division of the rational numbers into two such sets. This seems like a strange approach, but such indirect methods of constructing new structures from old are common in theoretical mathematics.

These and other approaches (see Appendix A.5) can be used to construct a system of numbers having the desired algebraic, order, and completeness properties. A final issue that arises is whether all the constructions give the same thing. Is it possible that different constructions result in different number systems satisfying all the required properties? If yes, which of these is the real numbers? Fortunately, the answer turns out to be no. The reals are the only number system satisfying the algebraic, order, and completeness properties.

Confusion about the nature of real numbers and about limits caused considerable controversy in the early development of calculus. Calculus pioneers such as Newton, Leibniz, and their successors, when looking at what happens to the difference quotient

$$\frac{\Delta y}{\Delta x} = \frac{f(x + \Delta x) - f(x)}{\Delta x}$$

as each of Δy and Δx approach zero, talked about the resulting derivative being a quotient of two infinitely small quantities. These "infinitesimals," written dx and dy, were thought to be some new kind of number, smaller than any fixed number but not zero. Similarly, a definite integral was thought of as a sum of an infinite number of infinitesimals

$$f(x) \cdot dx$$

as x varied over a closed interval. While the approximating difference quotients $\Delta y/\Delta x$ were understood much as today, it was the quotient of infinitesimal quantities, rather than a limit, that was thought to encapsulate the meaning of the derivative. This way of thinking led to logical difficulties, as attempted definitions and manipulations of infinitesimals ran into contradictions and inconsistencies. The more concrete and computable difference quotients did not cause such trouble, but they were thought of merely as useful calculation tools. Difference quotients were used to work out the numerical value of the derivative and to derive general formulas for calculation, but were not considered to be at the heart of the question of what the derivative actually was. Today we realize that the logical problems associated with infinitesimals can be avoided by *defining* the derivative to be the limit of its approximating difference quotients. The ambiguities of the old approach are no longer present, and in the standard theory of calculus, infinitesimals are neither needed nor used.

A.5 Complex Numbers

Complex numbers are expressions of the form $a + ib$, where a and b are real numbers and i is a symbol for $\sqrt{-1}$. Unfortunately, the words "real" and "imaginary" have connotations that somehow place $\sqrt{-1}$ in a less favorable position in our minds than $\sqrt{2}$. As a matter of fact, a good deal of imagination, in the sense of *inventiveness*, has been required to construct the *real* number system, which forms the basis of the calculus (see Appendix A.4). In this appendix we review the various stages of this invention. The further invention of a complex number system is then presented.

The Development of the Real Numbers

The earliest stage of number development was the recognition of the **counting numbers** $1, 2, 3, \dots$, which we now call the **natural numbers** or the **positive integers**. Certain simple arithmetical operations can be performed with these numbers without getting outside the system. That is, the system of positive integers is **closed** under the operations of addition and multiplication. By this we mean that if m and n are any positive integers, then

$$m + n = p \qquad \text{and} \qquad mn = q \tag{1}$$

are also positive integers. Given the two positive integers on the left side of either equation in (1), we can find the corresponding positive integer on the right side. More than this, we can sometimes specify the positive integers m and p and find a positive integer n such that $m + n = p$. For instance, $3 + n = 7$ can be solved when the only numbers we know are the positive integers. But the equation $7 + n = 3$ cannot be solved unless the number system is enlarged.

The number zero and the negative integers were invented to solve equations like $7 + n = 3$. In a civilization that recognizes all the **integers**

$$\dots, -3, -2, -1, 0, 1, 2, 3, \dots, \tag{2}$$

an educated person can always find the missing integer that solves the equation $m + n = p$ when given the other two integers in the equation.

Suppose our educated people also know how to multiply any two of the integers in the list (2). If, in Equations (1), they are given m and q, they discover that sometimes they can find n and sometimes they cannot. Using their imagination, they may be

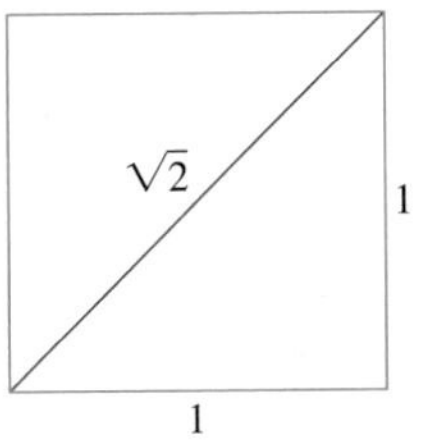

FIGURE A.3 With a straightedge and compass, it is possible to construct a segment of irrational length.

inspired to invent still more numbers and introduce fractions, which are just ordered pairs m/n of integers m and n. The number zero has special properties that may bother them for a while, but they ultimately discover that it is handy to have all ratios of integers m/n, excluding only those having zero in the denominator. This system, called the set of **rational numbers**, is now rich enough for them to perform the **rational operations** of arithmetic:

1. (a) addition **(b)** subtraction

2. (a) multiplication **(b)** division

on any two numbers in the system, *except that they cannot divide by zero* because it is meaningless.

The geometry of the unit square (Figure A.3) and the Pythagorean theorem showed that they could construct a geometric line segment that, in terms of some basic unit of length, has length equal to $\sqrt{2}$. Thus they could solve the equation

$$x^2 = 2$$

by a geometric construction. But then they discovered that the line segment representing $\sqrt{2}$ is an incommensurable quantity. This means that $\sqrt{2}$ cannot be expressed as the ratio of two *integer* multiples of some unit of length. That is, our educated people could not find a rational number solution of the equation $x^2 = 2$.

There *is* no rational number whose square is 2. To see why, suppose that there were such a rational number. Then we could find integers p and q with no common factor other than 1, and such that

$$p^2 = 2q^2. \tag{3}$$

Since p and q are integers, p must be even; otherwise its product with itself would be odd. In symbols, $p = 2p_1$, where p_1 is an integer. This leads to $2{p_1}^2 = q^2$ which says q must be even, say $q = 2q_1$, where q_1 is an integer. This makes 2 a factor of both p and q, contrary to our choice of p and q as integers with no common factor other than 1. Hence there is no rational number whose square is 2.

Although our educated people could not find a rational solution of the equation $x^2 = 2$, they could get a sequence of rational numbers

$$\frac{1}{1}, \quad \frac{7}{5}, \quad \frac{41}{29}, \quad \frac{239}{169}, \quad \ldots, \tag{4}$$

whose squares form a sequence

$$\frac{1}{1}, \quad \frac{49}{25}, \quad \frac{1681}{841}, \quad \frac{57{,}121}{28{,}561}, \quad \ldots, \tag{5}$$

that converges to 2 as its limit. This time their imagination suggested that they needed the concept of a limit of a sequence of rational numbers. If we accept the fact that an increasing sequence that is bounded from above always approaches a limit (Theorem 6, Section 11.1) and observe that the sequence in (4) has these properties, then we want it to have a limit L. This would also mean, from (5), that $L^2 = 2$, and hence L is *not* one of our rational numbers. If to the rational numbers we further add the limits of all bounded increasing sequences of rational numbers, we arrive at the system of all "real" numbers. The word *real* is placed in quotes because there is nothing that is either "more real" or "less real" about this system than there is about any other mathematical system.

The Complex Numbers

Imagination was called upon at many stages during the development of the real number system. In fact, the art of invention was needed at least three times in constructing the systems we have discussed so far:

1. The *first invented* system: the set of *all integers* as constructed from the counting numbers.
2. The *second invented* system: the set of *rational numbers* m/n as constructed from the integers.
3. The *third invented* system: the set of all *real numbers* x as constructed from the rational numbers.

These invented systems form a hierarchy in which each system contains the previous system. Each system is also richer than its predecessor in that it permits additional operations to be performed without going outside the system:

1. In the system of all integers, we can solve all equations of the form
$$x + a = 0, \tag{6}$$
where a can be any integer.
2. In the system of all rational numbers, we can solve all equations of the form
$$ax + b = 0, \tag{7}$$
provided a and b are rational numbers and $a \neq 0$.
3. In the system of all real numbers, we can solve all of Equations (6) and (7) and, in addition, all quadratic equations
$$ax^2 + bx + c = 0 \quad \text{having} \quad a \neq 0 \quad \text{and} \quad b^2 - 4ac \geq 0. \tag{8}$$

You are probably familiar with the formula that gives the solutions of Equation (8), namely,
$$x = \frac{-b \pm \sqrt{b^2 - 4ac}}{2a}, \tag{9}$$
and are familiar with the further fact that when the discriminant, $b^2 - 4ac$, is negative, the solutions in Equation (9) do *not* belong to any of the systems discussed above. In fact, the very simple quadratic equation
$$x^2 + 1 = 0$$
is impossible to solve if the only number systems that can be used are the three invented systems mentioned so far.

Thus we come to the *fourth invented* system, the set of *all complex numbers* $a + ib$. We could dispense entirely with the symbol i and use the ordered pair notation (a, b). Since, under algebraic operations, the numbers a and b are treated somewhat differently, it is essential to keep the *order* straight. We therefore might say that the **complex number system** consists of the set of all ordered pairs of real numbers (a, b), together with the rules by which they are to be equated, added, multiplied, and so on, listed below. We will use both the (a, b) notation and the notation $a + ib$ in the discussion that follows. We call a the **real part** and b the **imaginary part** of the complex number (a, b).

We make the following definitions.

Equality $a + ib = c + id$ if and only if $a = c$ and $b = d$.	Two complex numbers (a, b) and (c, d) are *equal* if and only if $a = c$ and $b = d$.
Addition $(a + ib) + (c + id)$ $= (a + c) + i(b + d)$	The *sum* of the two complex numbers (a, b) and (c, d) is the complex number $(a + c, b + d)$.
Multiplication $(a + ib)(c + id)$ $= (ac - bd) + i(ad + bc)$	The *product* of two complex numbers (a, b) and (c, d) is the complex number $(ac - bd, ad + bc)$.
$c(a + ib) = ac + i(bc)$	The product of a real number c and the complex number (a, b) is the complex number (ac, bc).

The set of all complex numbers (a, b) in which the second number b is zero has all the properties of the set of real numbers a. For example, addition and multiplication of $(a, 0)$ and $(c, 0)$ give

$$(a, 0) + (c, 0) = (a + c, 0),$$
$$(a, 0) \cdot (c, 0) = (ac, 0),$$

which are numbers of the same type with imaginary part equal to zero. Also, if we multiply a "real number" $(a, 0)$ and the complex number (c, d), we get

$$(a, 0) \cdot (c, d) = (ac, ad) = a(c, d).$$

In particular, the complex number (0, 0) plays the role of *zero* in the complex number system, and the complex number (1, 0) plays the role of *unity* or *one*.

The number pair (0, 1), which has real part equal to zero and imaginary part equal to one, has the property that its square,

$$(0, 1)(0, 1) = (-1, 0),$$

has real part equal to minus one and imaginary part equal to zero. Therefore, in the system of complex numbers (a, b) there is a number $x = (0, 1)$ whose square can be added to unity $= (1, 0)$ to produce zero $= (0, 0)$, that is,

$$(0, 1)^2 + (1, 0) = (0, 0).$$

The equation

$$x^2 + 1 = 0$$

therefore has a solution $x = (0, 1)$ in this new number system.

You are probably more familiar with the $a + ib$ notation than you are with the notation (a, b). And since the laws of algebra for the ordered pairs enable us to write

$$(a, b) = (a, 0) + (0, b) = a(1, 0) + b(0, 1),$$

while (1, 0) behaves like unity and (0, 1) behaves like a square root of minus one, we need not hesitate to write $a + ib$ in place of (a, b). The i associated with b is like a tracer element that tags the imaginary part of $a + ib$. We can pass at will from the realm of ordered pairs (a, b) to the realm of expressions $a + ib$, and conversely. But there is nothing less "real" about the symbol $(0, 1) = i$ than there is about the symbol $(1, 0) = 1$, once we have learned the laws of algebra in the complex number system of ordered pairs (a, b).

To reduce any rational combination of complex numbers to a single complex number, we apply the laws of elementary algebra, replacing i^2 wherever it appears by -1. Of course, we cannot divide by the complex number $(0, 0) = 0 + i0$. But if $a + ib \neq 0$, then we may carry out a division as follows:

$$\frac{c + id}{a + ib} = \frac{(c + id)(a - ib)}{(a + ib)(a - ib)} = \frac{(ac + bd) + i(ad - bc)}{a^2 + b^2}.$$

The result is a complex number $x + iy$ with

$$x = \frac{ac + bd}{a^2 + b^2}, \qquad y = \frac{ad - bc}{a^2 + b^2},$$

and $a^2 + b^2 \neq 0$, since $a + ib = (a, b) \neq (0, 0)$.

The number $a - ib$ that is used as multiplier to clear the i from the denominator is called the **complex conjugate** of $a + ib$. It is customary to use $\bar{z}$ (read "z bar") to denote the complex conjugate of z; thus

$$z = a + ib, \qquad \bar{z} = a - ib.$$

Multiplying the numerator and denominator of the fraction $(c + id)/(a + ib)$ by the complex conjugate of the denominator will always replace the denominator by a real number.

EXAMPLE 1 Arithmetic Operations with Complex Numbers

(a) $(2 + 3i) + (6 - 2i) = (2 + 6) + (3 - 2)i = 8 + i$

(b) $(2 + 3i) - (6 - 2i) = (2 - 6) + (3 - (-2))i = -4 + 5i$

(c) $(2 + 3i)(6 - 2i) = (2)(6) + (2)(-2i) + (3i)(6) + (3i)(-2i)$

$$= 12 - 4i + 18i - 6i^2 = 12 + 14i + 6 = 18 + 14i$$

(d) $$\frac{2 + 3i}{6 - 2i} = \frac{2 + 3i}{6 - 2i}\,\frac{6 + 2i}{6 + 2i} = \frac{12 + 4i + 18i + 6i^2}{36 + 12i - 12i - 4i^2} = \frac{6 + 22i}{40} = \frac{3}{20} + \frac{11}{20}i$$ ■

Argand Diagrams

There are two geometric representations of the complex number $z = x + iy$:

1. as the point $P(x, y)$ in the xy-plane
2. as the vector $\overrightarrow{OP}$ from the origin to P.

In each representation, the x-axis is called the **real axis** and the y-axis is the **imaginary axis**. Both representations are **Argand diagrams** for $x + iy$ (Figure A.4).

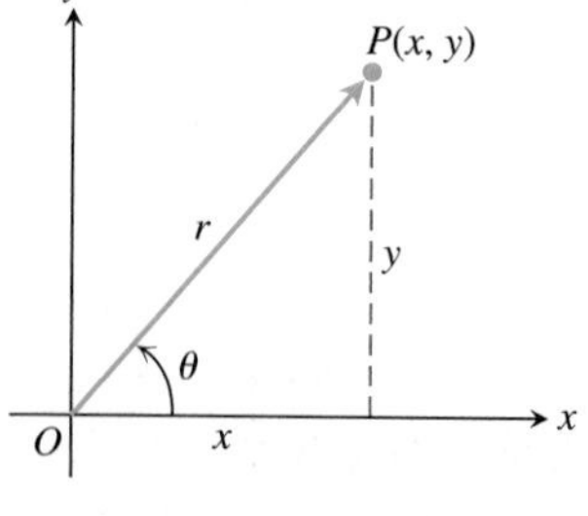

FIGURE A.4 This Argand diagram represents $z = x + iy$ both as a point $P(x, y)$ and as a vector $\overrightarrow{OP}$.

In terms of the polar coordinates of x and y, we have

$$x = r\cos\theta, \qquad y = r\sin\theta,$$

and

$$z = x + iy = r(\cos\theta + i\sin\theta). \tag{10}$$

We define the **absolute value** of a complex number $x + iy$ to be the length r of a vector $\overrightarrow{OP}$ from the origin to $P(x, y)$. We denote the absolute value by vertical bars; thus,

$$|x + iy| = \sqrt{x^2 + y^2}.$$

If we always choose the polar coordinates r and θ so that r is nonnegative, then

$$r = |x + iy|.$$

The polar angle θ is called the **argument** of z and is written $\theta = \arg z$. Of course, any integer multiple of 2π may be added to θ to produce another appropriate angle.

The following equation gives a useful formula connecting a complex number z, its conjugate $\bar{z}$, and its absolute value $|z|$, namely,

$$z \cdot \bar{z} = |z|^2.$$

Euler's Formula

The identity

$$e^{i\theta} = \cos\theta + i\sin\theta,$$

called **Euler's formula**, enables us to rewrite Equation (10) as

$$z = re^{i\theta}.$$

This formula, in turn, leads to the following rules for calculating products, quotients, powers, and roots of complex numbers. It also leads to Argand diagrams for $e^{i\theta}$. Since $\cos\theta + i\sin\theta$ is what we get from Equation (10) by taking $r = 1$, we can say that $e^{i\theta}$ is represented by a unit vector that makes an angle θ with the positive x-axis, as shown in Figure A.5.

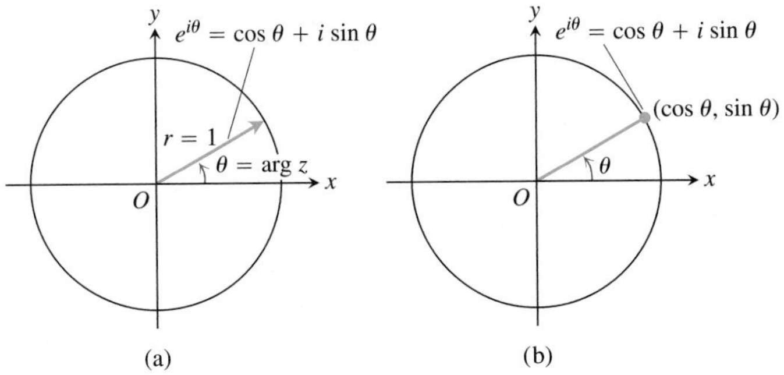

FIGURE A.5 Argand diagrams for $e^{i\theta} = \cos\theta + i\sin\theta$ (a) as a vector and (b) as a point.

Products

To multiply two complex numbers, we multiply their absolute values and add their angles. Let

$$z_1 = r_1 e^{i\theta_1}, \qquad z_2 = r_2 e^{i\theta_2}, \tag{11}$$

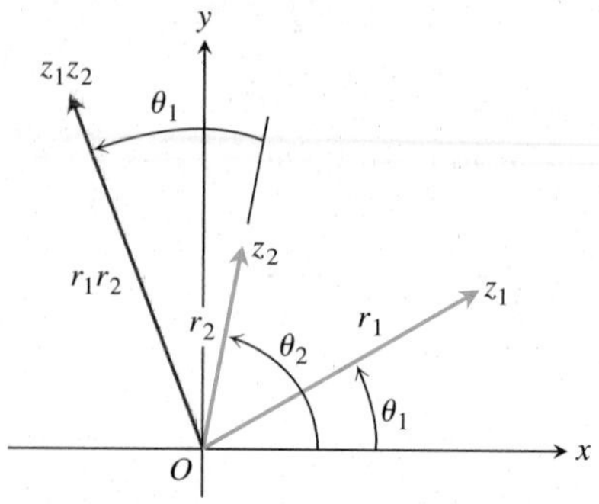

FIGURE A.6 When z_1 and z_2 are multiplied, $|z_1z_2| = r_1 \cdot r_2$ and $\arg(z_1z_2) = \theta_1 + \theta_2$.

so that

$$|z_1| = r_1, \qquad \arg z_1 = \theta_1; \qquad |z_2| = r_2, \qquad \arg z_2 = \theta_2.$$

Then

$$z_1z_2 = r_1e^{i\theta_1} \cdot r_2e^{i\theta_2} = r_1r_2e^{i(\theta_1+\theta_2)}$$

and hence

$$|z_1z_2| = r_1r_2 = |z_1|\cdot|z_2|$$
$$\arg(z_1z_2) = \theta_1 + \theta_2 = \arg z_1 + \arg z_2. \tag{12}$$

Thus, the product of two complex numbers is represented by a vector whose length is the product of the lengths of the two factors and whose argument is the sum of their arguments (Figure A.6). In particular, from Equation (12) a vector may be rotated counterclockwise through an angle θ by multiplying it by $e^{i\theta}$. Multiplication by i rotates 90°, by -1 rotates 180°, by $-i$ rotates 270°, and so on.

EXAMPLE 2 Finding a Product of Complex Numbers

Let $z_1 = 1 + i$, $z_2 = \sqrt{3} - i$. We plot these complex numbers in an Argand diagram (Figure A.7) from which we read off the polar representations

$$z_1 = \sqrt{2}e^{i\pi/4}, \qquad z_2 = 2e^{-i\pi/6}.$$

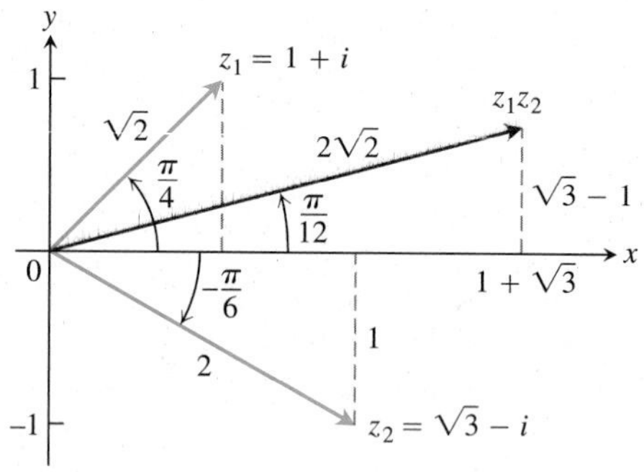

FIGURE A.7 To multiply two complex numbers, multiply their absolute values and add their arguments.

Then

$$z_1z_2 = 2\sqrt{2}\exp\left(\frac{i\pi}{4} - \frac{i\pi}{6}\right) = 2\sqrt{2}\exp\left(\frac{i\pi}{12}\right)$$
$$= 2\sqrt{2}\left(\cos\frac{\pi}{12} + i\sin\frac{\pi}{12}\right) \approx 2.73 + 0.73i.$$

The notation $\exp(A)$ stands for e^A. ■

Quotients

Suppose $r_2 \neq 0$ in Equation (11). Then

$$\frac{z_1}{z_2} = \frac{r_1e^{i\theta_1}}{r_2e^{i\theta_2}} = \frac{r_1}{r_2}e^{i(\theta_1-\theta_2)}.$$

Hence

$$\left|\frac{z_1}{z_2}\right| = \frac{r_1}{r_2} = \frac{|z_1|}{|z_2|} \quad \text{and} \quad \arg\left(\frac{z_1}{z_2}\right) = \theta_1 - \theta_2 = \arg z_1 - \arg z_2.$$

That is, we divide lengths and subtract angles for the quotient of complex numbers.

EXAMPLE 3 Let $z_1 = 1 + i$ and $z_2 = \sqrt{3} - i$, as in Example 2. Then

$$\frac{1+i}{\sqrt{3}-i} = \frac{\sqrt{2}e^{i\pi/4}}{2e^{-i\pi/6}} = \frac{\sqrt{2}}{2}e^{5\pi i/12} \approx 0.707\left(\cos\frac{5\pi}{12} + i\sin\frac{5\pi}{12}\right)$$
$$\approx 0.183 + 0.683i.$$

■

Powers

If n is a positive integer, we may apply the product formulas in Equation (12) to find

$$z^n = z \cdot z \cdot \cdots \cdot z. \qquad n \text{ factors}$$

With $z = re^{i\theta}$, we obtain

$$\begin{aligned} z^n = (re^{i\theta})^n &= r^n e^{i(\theta+\theta+\cdots+\theta)} \qquad n \text{ summands} \\ &= r^n e^{in\theta}. \end{aligned} \tag{13}$$

The length $r = |z|$ is raised to the nth power and the angle $\theta = \arg z$ is multiplied by n.

If we take $r = 1$ in Equation (13), we obtain De Moivre's Theorem.

De Moivre's Theorem

$$(\cos\theta + i\sin\theta)^n = \cos n\theta + i\sin n\theta. \tag{14}$$

If we expand the left side of De Moivre's equation above by the Binomial Theorem and reduce it to the form $a + ib$, we obtain formulas for $\cos n\theta$ and $\sin n\theta$ as polynomials of degree n in $\cos\theta$ and $\sin\theta$.

EXAMPLE 4 If $n = 3$ in Equation (14), we have

$$(\cos\theta + i\sin\theta)^3 = \cos 3\theta + i\sin 3\theta.$$

The left side of this equation expands to

$$\cos^3\theta + 3i\cos^2\theta\sin\theta - 3\cos\theta\sin^2\theta - i\sin^3\theta.$$

The real part of this must equal $\cos 3\theta$ and the imaginary part must equal $\sin 3\theta$. Therefore,

$$\cos 3\theta = \cos^3\theta - 3\cos\theta\sin^2\theta,$$

$$\sin 3\theta = 3\cos^2\theta\sin\theta - \sin^3\theta.$$

■

Roots

If $z = re^{i\theta}$ is a complex number different from zero and n is a positive integer, then there are precisely n different complex numbers $w_0, w_1, \ldots, w_{n-1}$, that are nth roots of z. To see why, let $w = \rho e^{i\alpha}$ be an nth root of $z = re^{i\theta}$, so that

$$w^n = z$$

or

$$\rho^n e^{in\alpha} = re^{i\theta}.$$

Then

$$\rho = \sqrt[n]{r}$$

is the real, positive nth root of r. For the argument, although we cannot say that $n\alpha$ and θ must be equal, we can say that they may differ only by an integer multiple of 2π. That is,

$$n\alpha = \theta + 2k\pi, \qquad k = 0, \pm 1, \pm 2, \ldots.$$

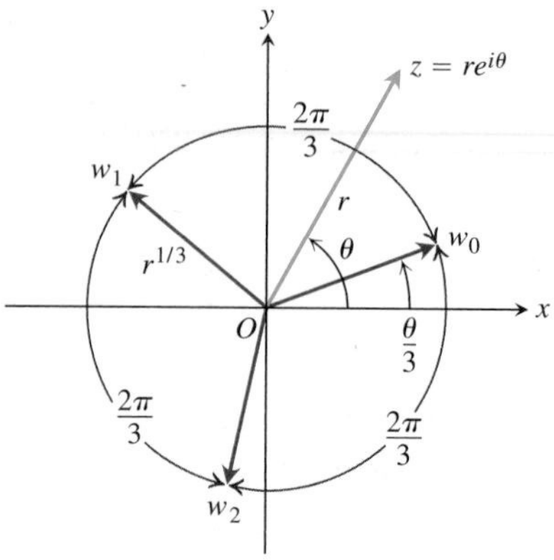

FIGURE A.8 The three cube roots of $z = re^{i\theta}$.

Therefore,

$$\alpha = \frac{\theta}{n} + k\frac{2\pi}{n}.$$

Hence, all the nth roots of $z = re^{i\theta}$ are given by

$$\sqrt[n]{re^{i\theta}} = \sqrt[n]{r}\exp i\left(\frac{\theta}{n} + k\frac{2\pi}{n}\right), \qquad k = 0, \pm 1, \pm 2, \ldots. \tag{15}$$

There might appear to be infinitely many different answers corresponding to the infinitely many possible values of k, but $k = n + m$ gives the same answer as $k = m$ in Equation (15). Thus, we need only take n consecutive values for k to obtain all the different nth roots of z. For convenience, we take

$$k = 0, 1, 2, \ldots, n - 1.$$

All the nth roots of $re^{i\theta}$ lie on a circle centered at the origin and having radius equal to the real, positive nth root of r. One of them has argument $\alpha = \theta/n$. The others are uniformly spaced around the circle, each being separated from its neighbors by an angle equal to $2\pi/n$. Figure A.8 illustrates the placement of the three cube roots, w_0, w_1, w_2, of the complex number $z = re^{i\theta}$.

EXAMPLE 5 Finding Fourth Roots

Find the four fourth roots of -16.

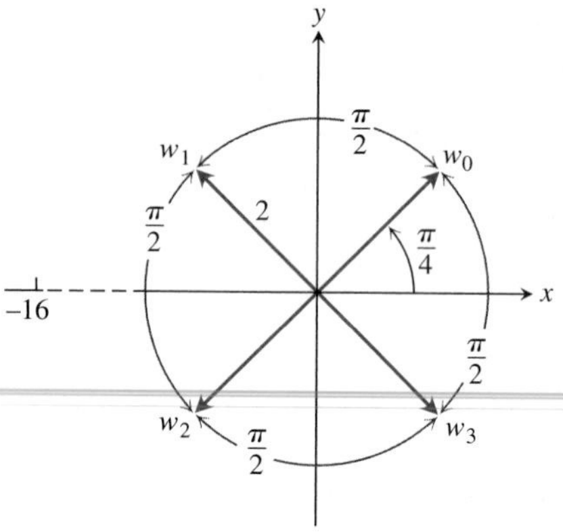

FIGURE A.9 The four fourth roots of -16.

Solution As our first step, we plot the number -16 in an Argand diagram (Figure A.9) and determine its polar representation $re^{i\theta}$. Here, $z = -16$, $r = +16$, and $\theta = \pi$. One of the fourth roots of $16e^{i\pi}$ is $2e^{i\pi/4}$. We obtain others by successive additions of $2\pi/4 = \pi/2$ to the argument of this first one. Hence,

$$\sqrt[4]{16\exp i\pi} = 2\exp i\left(\frac{\pi}{4}, \frac{3\pi}{4}, \frac{5\pi}{4}, \frac{7\pi}{4}\right),$$

and the four roots are

$$w_0 = 2\left[\cos\frac{\pi}{4} + i\sin\frac{\pi}{4}\right] = \sqrt{2}(1 + i)$$

$$w_1 = 2\left[\cos\frac{3\pi}{4} + i\sin\frac{3\pi}{4}\right] = \sqrt{2}(-1 + i)$$

$$w_2 = 2\left[\cos\frac{5\pi}{4} + i\sin\frac{5\pi}{4}\right] = \sqrt{2}(-1 - i)$$

$$w_3 = 2\left[\cos\frac{7\pi}{4} + i\sin\frac{7\pi}{4}\right] = \sqrt{2}(1 - i). \qquad \blacksquare$$

The Fundamental Theorem of Algebra

One might say that the invention of $\sqrt{-1}$ is all well and good and leads to a number system that is richer than the real number system alone; but where will this process end? Are we also going to invent still more systems so as to obtain $\sqrt[4]{-1}$, $\sqrt[6]{-1}$, and so on? But it turns out this is not necessary. These numbers are already expressible in terms of the complex number system $a + ib$. In fact, the Fundamental Theorem of Algebra says that with the introduction of the complex numbers we now have enough numbers to factor every polynomial into a product of linear factors and so enough numbers to solve every possible polynomial equation.

The Fundamental Theorem of Algebra

Every polynomial equation of the form

$$a_n z^n + a_{n-1} z^{n-1} + \cdots + a_1 z + a_0 = 0,$$

in which the coefficients $a_0, a_1, \ldots, a_n$ are any complex numbers, whose degree n is greater than or equal to one, and whose leading coefficient a_n is not zero, has exactly n roots in the complex number system, provided each multiple root of multiplicity m is counted as m roots.

A proof of this theorem can be found in almost any text on the theory of functions of a complex variable.

EXERCISES A.5

Operations with Complex Numbers

1. **How computers multiply complex numbers** Find $(a, b) \cdot (c, d) = (ac - bd, ad + bc)$.

   a. $(2, 3) \cdot (4, -2)$ b. $(2, -1) \cdot (-2, 3)$

   c. $(-1, -2) \cdot (2, 1)$

   (This is how complex numbers are multiplied by computers.)

2. Solve the following equations for the real numbers, x and y.

   a. $(3 + 4i)^2 - 2(x - iy) = x + iy$

   b. $\left(\dfrac{1 + i}{1 - i}\right)^2 + \dfrac{1}{x + iy} = 1 + i$

   c. $(3 - 2i)(x + iy) = 2(x - 2iy) + 2i - 1$

Graphing and Geometry

3. How may the following complex numbers be obtained from $z = x + iy$ geometrically? Sketch.

   a. $\bar{z}$ b. $\overline{(-z)}$

   c. $-z$ d. $1/z$

4. Show that the distance between the two points z_1 and z_2 in an Argand diagram is $|z_1 - z_2|$.

In Exercises 5–10, graph the points $z = x + iy$ that satisfy the given conditions.

5. a. $|z| = 2$ b. $|z| < 2$ c. $|z| > 2$

6. $|z - 1| = 2$

7. $|z + 1| = 1$

8. $|z + 1| = |z - 1|$

9. $|z + i| = |z - 1|$

10. $|z + 1| \geq |z|$

Express the complex numbers in Exercises 11–14 in the form $re^{i\theta}$, with $r \geq 0$ and $-\pi < \theta \leq \pi$. Draw an Argand diagram for each calculation.

11. $\left(1 + \sqrt{-3}\right)^2$

12. $\dfrac{1 + i}{1 - i}$

13. $\dfrac{1 + i\sqrt{3}}{1 - i\sqrt{3}}$

14. $(2 + 3i)(1 - 2i)$

Powers and Roots

Use De Moivre's Theorem to express the trigonometric functions in Exercises 15 and 16 in terms of $\cos\theta$ and $\sin\theta$.

15. $\cos 4\theta$

16. $\sin 4\theta$

17. Find the three cube roots of 1.

18. Find the two square roots of i.
19. Find the three cube roots of $-8i$.
20. Find the six sixth roots of 64.
21. Find the four solutions of the equation $z^4 - 2z^2 + 4 = 0$.
22. Find the six solutions of the equation $z^6 + 2z^3 + 2 = 0$.
23. Find all solutions of the equation $x^4 + 4x^2 + 16 = 0$.
24. Solve the equation $x^4 + 1 = 0$.

Theory and Examples

25. **Complex numbers and vectors in the plane** Show with an Argand diagram that the law for adding complex numbers is the same as the parallelogram law for adding vectors.
26. **Complex arithmetic with conjugates** Show that the conjugate of the sum (product, or quotient) of two complex numbers, z_1 and z_2, is the same as the sum (product, or quotient) of their conjugates.
27. **Complex roots of polynomials with real coefficients come in complex-conjugate pairs**
    a. Extend the results of Exercise 26 to show that $f(\bar{z}) = \overline{f(z)}$ if
    $$f(z) = a_n z^n + a_{n-1}z^{n-1} + \cdots + a_1 z + a_0$$
    is a polynomial with real coefficients $a_0, \ldots, a_n$.
    b. If z is a root of the equation $f(z) = 0$, where $f(z)$ is a polynomial with real coefficients as in part (a), show that the conjugate $\bar{z}$ is also a root of the equation. (*Hint:* Let $f(z) = u + iv = 0$; then both u and v are zero. Use the fact that $f(\bar{z}) = \overline{f(z)} = u - iv$.)
28. **Absolute value of a conjugate** Show that $|\bar{z}| = |z|$.
29. **When $\mathbf{z} = \bar{\mathbf{z}}$** If z and $\bar{z}$ are equal, what can you say about the location of the point z in the complex plane?
30. **Real and imaginary parts** Let Re(z) denote the real part of z and Im(z) the imaginary part. Show that the following relations hold for any complex numbers z, z_1, and z_2.
    a. $z + \bar{z} = 2\text{Re}(z)$
    b. $z - \bar{z} = 2i\text{Im}(z)$
    c. $|\text{Re}(z)| \le |z|$
    d. $|z_1 + z_2|^2 = |z_1|^2 + |z_2|^2 + 2\text{Re}(z_1\bar{z}_2)$
    e. $|z_1 + z_2| \le |z_1| + |z_2|$

A.6 The Distributive Law for Vector Cross Products

In this appendix, we prove the Distributive Law

$$\mathbf{u} \times (\mathbf{v} + \mathbf{w}) = \mathbf{u} \times \mathbf{v} + \mathbf{u} \times \mathbf{w}$$

which is Property 2 in Section 12.4.

Proof To derive the Distributive Law, we construct $\mathbf{u} \times \mathbf{v}$ a new way. We draw $\mathbf{u}$ and $\mathbf{v}$ from the common point O and construct a plane M perpendicular to $\mathbf{u}$ at O (Figure A.10). We then project $\mathbf{v}$ orthogonally onto M, yielding a vector $\mathbf{v}'$ with length $|\mathbf{v}| \sin\theta$. We rotate $\mathbf{v}'$ 90° about $\mathbf{u}$ in the positive sense to produce a vector $\mathbf{v}''$. Finally, we multiply $\mathbf{v}''$ by the

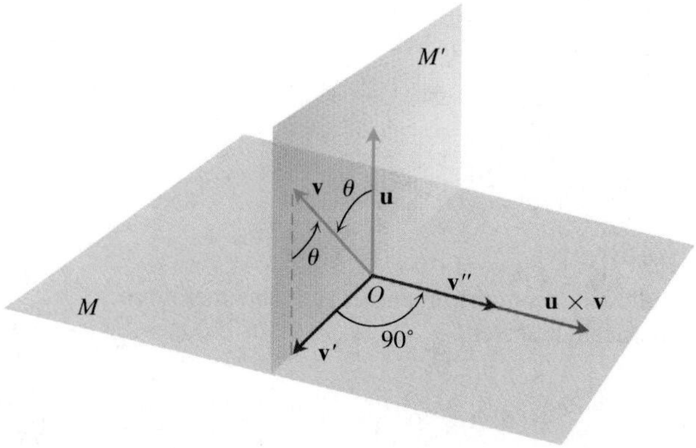

FIGURE A.10 As explained in the text, $\mathbf{u} \times \mathbf{v} = |\mathbf{u}|\mathbf{v}''$.

length of **u**. The resulting vector $|\mathbf{u}|\mathbf{v}''$ is equal to $\mathbf{u} \times \mathbf{v}$ since $\mathbf{v}''$ has the same direction as $\mathbf{u} \times \mathbf{v}$ by its construction (Figure A.10) and

$$|\mathbf{u}||\mathbf{v}''| = |\mathbf{u}||\mathbf{v}'| = |\mathbf{u}||\mathbf{v}|\sin\theta = |\mathbf{u} \times \mathbf{v}|.$$

Now each of these three operations, namely,

1. projection onto M
2. rotation about **u** through 90°
3. multiplication by the scalar $|\mathbf{u}|$

when applied to a triangle whose plane is not parallel to **u**, will produce another triangle. If we start with the triangle whose sides are **v**, **w**, and $\mathbf{v} + \mathbf{w}$ (Figure A.11) and apply these three steps, we successively obtain the following:

1. A triangle whose sides are $\mathbf{v}'$, $\mathbf{w}'$, and $(\mathbf{v} + \mathbf{w})'$ satisfying the vector equation
$$\mathbf{v}' + \mathbf{w}' = (\mathbf{v} + \mathbf{w})'$$
2. A triangle whose sides are $\mathbf{v}''$, $\mathbf{w}''$, and $(\mathbf{v} + \mathbf{w})''$ satisfying the vector equation
$$\mathbf{v}'' + \mathbf{w}'' = (\mathbf{v} + \mathbf{w})''$$
(the double prime on each vector has the same meaning as in Figure A.10)

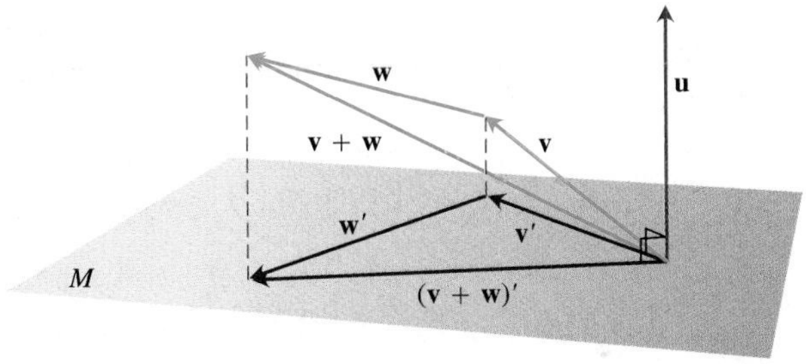

FIGURE A.11 The vectors, **v**, **w**, $\mathbf{v} + \mathbf{w}$, and their projections onto a plane perpendicular to **u**.

3. A triangle whose sides are $|\mathbf{u}|\mathbf{v}''$, $|\mathbf{u}|\mathbf{w}''$, and $|\mathbf{u}|(\mathbf{v} + \mathbf{w})''$ satisfying the vector equation
$$|\mathbf{u}|\mathbf{v}'' + |\mathbf{u}|\mathbf{w}'' = |\mathbf{u}|(\mathbf{v} + \mathbf{w})''.$$

Substituting $|\mathbf{u}|\mathbf{v}'' = \mathbf{u} \times \mathbf{v}$, $|\mathbf{u}|\mathbf{w}'' = \mathbf{u} \times \mathbf{w}$, and $|\mathbf{u}|(\mathbf{v} + \mathbf{w})'' = \mathbf{u} \times (\mathbf{v} + \mathbf{w})$ from our discussion above into this last equation gives

$$\mathbf{u} \times \mathbf{v} + \mathbf{u} \times \mathbf{w} = \mathbf{u} \times (\mathbf{v} + \mathbf{w}),$$

which is the law we wanted to establish. ■

A.7 The Mixed Derivative Theorem and the Increment Theorem

This appendix derives the Mixed Derivative Theorem (Theorem 2, Section 14.3) and the Increment Theorem for Functions of Two Variables (Theorem 3, Section 14.3). Euler first published the Mixed Derivative Theorem in 1734, in a series of papers he wrote on hydrodynamics.

THEOREM 2 The Mixed Derivative Theorem
If $f(x, y)$ and its partial derivatives f_x, f_y, f_{xy}, and f_{yx} are defined throughout an open region containing a point (a, b) and are all continuous at (a, b), then $f_{xy}(a, b) = f_{yx}(a, b)$.

Proof The equality of $f_{xy}(a, b)$ and $f_{yx}(a, b)$ can be established by four applications of the Mean Value Theorem (Theorem 4, Section 4.2). By hypothesis, the point (a, b) lies in the interior of a rectangle R in the xy-plane on which f, f_x, f_y, f_{xy}, and f_{yx} are all defined. We let h and k be the numbers such that the point $(a + h, b + k)$ also lies in R, and we consider the difference

$$\Delta = F(a + h) - F(a), \tag{1}$$

where

$$F(x) = f(x, b + k) - f(x, b). \tag{2}$$

We apply the Mean Value Theorem to F, which is continuous because it is differentiable. Then Equation (1) becomes

$$\Delta = hF'(c_1), \tag{3}$$

where c_1 lies between a and $a + h$. From Equation (2).

$$F'(x) = f_x(x, b + k) - f_x(x, b),$$

so Equation (3) becomes

$$\Delta = h[f_x(c_1, b + k) - f_x(c_1, b)]. \tag{4}$$

Now we apply the Mean Value Theorem to the function $g(y) = f_x(c_1, y)$ and have

$$g(b + k) - g(b) = kg'(d_1),$$

or

$$f_x(c_1, b + k) - f_x(c_1, b) = kf_{xy}(c_1, d_1)$$

for some d_1 between b and $b + k$. By substituting this into Equation (4), we get

$$\Delta = hkf_{xy}(c_1, d_1) \tag{5}$$

for some point (c_1, d_1) in the rectangle R' whose vertices are the four points (a, b), $(a + h, b)$, $(a + h, b + k)$, and $(a, b + k)$. (See Figure A.12.)

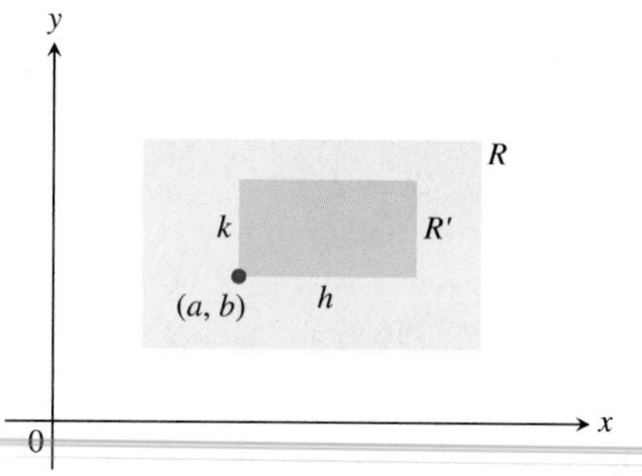

FIGURE A.12 The key to proving $f_{xy}(a, b) = f_{yx}(a, b)$ is that no matter how small R' is, f_{xy} and f_{yx} take on equal values somewhere inside R' (although not necessarily at the same point).

By substituting from Equation (2) into Equation (1), we may also write

$$\begin{aligned}\Delta &= f(a + h, b + k) - f(a + h, b) - f(a, b + k) + f(a, b)\\ &= [f(a + h, b + k) - f(a, b + k)] - [f(a + h, b) - f(a, b)]\\ &= \phi(b + k) - \phi(b),\end{aligned} \tag{6}$$

where

$$\phi(y) = f(a + h, y) - f(a, y). \tag{7}$$

The Mean Value Theorem applied to Equation (6) now gives

$$\Delta = k\phi'(d_2) \tag{8}$$

for some d_2 between b and $b + k$. By Equation (7),

$$\phi'(y) = f_y(a + h, y) - f_y(a, y). \tag{9}$$

Substituting from Equation (9) into Equation (8) gives

$$\Delta = k[f_y(a + h, d_2) - f_y(a, d_2)].$$

Finally, we apply the Mean Value Theorem to the expression in brackets and get

$$\Delta = khf_{yx}(c_2, d_2) \tag{10}$$

for some c_2 between a and $a + h$.

Together, Equations (5) and (10) show that

$$f_{xy}(c_1, d_1) = f_{yx}(c_2, d_2), \tag{11}$$

where (c_1, d_1) and (c_2, d_2) both lie in the rectangle R' (Figure A.12). Equation (11) is not quite the result we want, since it says only that f_{xy} has the same value at (c_1, d_1) that f_{yx} has at (c_2, d_2). The numbers h and k in our discussion, however, may be made as small as we wish. The hypothesis that f_{xy} and f_{yx} are both continuous at (a, b) means that $f_{xy}(c_1, d_1) = f_{xy}(a, b) + \epsilon_1$ and $f_{yx}(c_2, d_2) = f_{yx}(a, b) + \epsilon_2$, where each of $\epsilon_1, \epsilon_2 \to 0$ as both $h, k \to 0$. Hence, if we let h and $k \to 0$, we have $f_{xy}(a, b) = f_{yx}(a, b)$. ■

The equality of $f_{xy}(a, b)$ and $f_{yx}(a, b)$ can be proved with hypotheses weaker than the ones we assumed. For example, it is enough for f, f_x, and f_y to exist in R and for f_{xy} to be continuous at (a, b). Then f_{yx} will exist at (a, b) and equal f_{xy} at that point.

THEOREM 3 The Increment Theorem for Functions of Two Variables
Suppose that the first partial derivatives of $z = f(x, y)$ are defined throughout an open region R containing the point (x_0, y_0) and that f_x and f_y are continuous at (x_0, y_0). Then the change $\Delta z = f(x_0 + \Delta x, y_0 + \Delta y) - f(x_0, y_0)$ in the value of f that results from moving from (x_0, y_0) to another point $(x_0 + \Delta x, y_0 + \Delta y)$ in R satisfies an equation of the form

$$\Delta z = f_x(x_0, y_0)\Delta x + f_y(x_0, y_0)\Delta y + \epsilon_1 \Delta x + \epsilon_2 \Delta y,$$

in which each of $\epsilon_1, \epsilon_2 \to 0$ as both $\Delta x, \Delta y \to 0$.

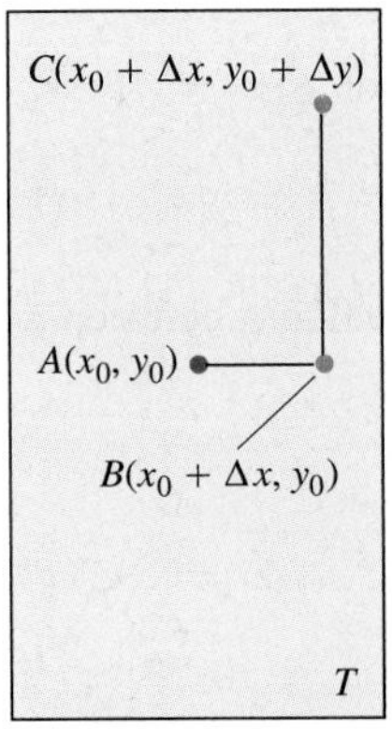

FIGURE A.13 The rectangular region T in the proof of the Increment Theorem. The figure is drawn for Δx and Δy positive, but either increment might be zero or negative.

Proof We work within a rectangle T centered at $A(x_0, y_0)$ and lying within R, and we assume that Δx and Δy are already so small that the line segment joining A to $B(x_0 + \Delta x, y_0)$ and the line segment joining B to $C(x_0 + \Delta x, y_0 + \Delta y)$ lie in the interior of T (Figure A.13).

We may think of Δz as the sum $\Delta z = \Delta z_1 + \Delta z_2$ of two increments, where

$$\Delta z_1 = f(x_0 + \Delta x, y_0) - f(x_0, y_0)$$

is the change in the value of f from A to B and

$$\Delta z_2 = f(x_0 + \Delta x, y_0 + \Delta y) - f(x_0 + \Delta x, y_0)$$

is the change in the value of f from B to C (Figure A.14).

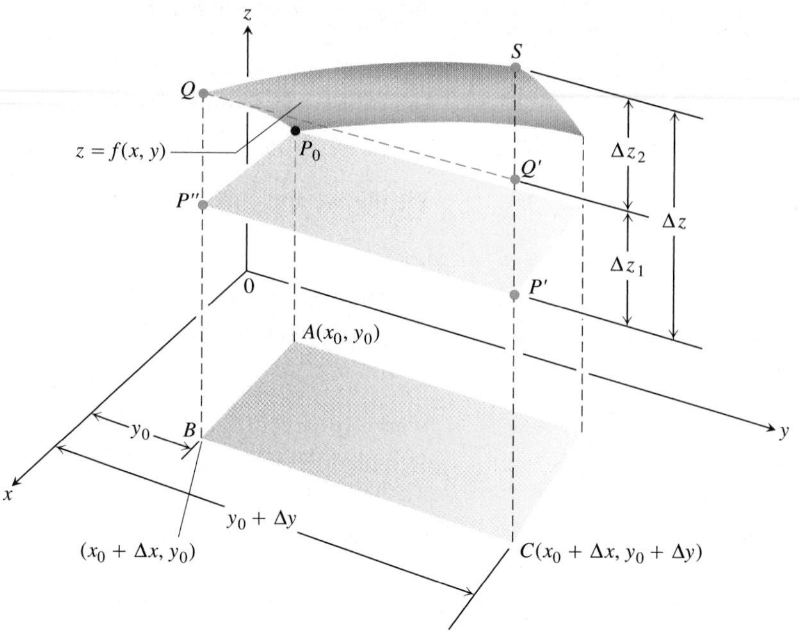

FIGURE A.14 Part of the surface $z = f(x, y)$ near $P_0(x_0, y_0, f(x_0, y_0))$. The points P_0, P', and P'' have the same height $z_0 = f(x_0, y_0)$ above the xy-plane. The change in z is $\Delta z = P'S$. The change

$$\Delta z_1 = f(x_0 + \Delta x, y_0) - f(x_0, y_0),$$

shown as $P''Q = P'Q'$, is caused by changing x from x_0 to $x_0 + \Delta x$ while holding y equal to y_0. Then, with x held equal to $x_0 + \Delta x$,

$$\Delta z_2 = f(x_0 + \Delta x, y_0 + \Delta y) - f(x_0 + \Delta x, y_0)$$

is the change in z caused by changing y_0 from $y_0 + \Delta y$, which is represented by $Q'S$? The total change in z is the sum of Δz_1 and Δz_2.

On the closed interval of x-values joining x_0 to $x_0 + \Delta x$, the function $F(x) = f(x, y_0)$ is a differentiable (and hence continuous) function of x, with derivative

$$F'(x) = f_x(x, y_0).$$

By the Mean Value Theorem (Theorem 4, Section 4.2), there is an x-value c between x_0 and $x_0 + \Delta x$ at which

$$F(x_0 + \Delta x) - F(x_0) = F'(c)\Delta x$$

or

$$f(x_0 + \Delta x, y_0) - f(x_0, y_0) = f_x(c, y_0)\Delta x$$

or

$$\Delta z_1 = f_x(c, y_0)\Delta x. \tag{12}$$

Similarly, $G(y) = f(x_0 + \Delta x, y)$ is a differentiable (and hence continuous) function of y on the closed y-interval joining y_0 and $y_0 + \Delta y$, with derivative

$$G'(y) = f_y(x_0 + \Delta x, y).$$

Hence, there is a y-value d between y_0 and $y_0 + \Delta y$ at which

$$G(y_0 + \Delta y) - G(y_0) = G'(d)\Delta y$$

or

$$f(x_0 + \Delta x, y_0 + \Delta y) - f(x_0 + \Delta x, y) = f_y(x_0 + \Delta x, d)\Delta y$$

or

$$\Delta z_2 = f_y(x_0 + \Delta x, d)\Delta y. \tag{13}$$

Now, as both Δx and $\Delta y \to 0$, we know that $c \to x_0$ and $d \to y_0$. Therefore, since f_x and f_y are continuous at (x_0, y_0), the quantities

$$\begin{aligned} \epsilon_1 &= f_x(c, y_0) - f_x(x_0, y_0), \\ \epsilon_2 &= f_y(x_0 + \Delta x, d) - f_y(x_0, y_0) \end{aligned} \tag{14}$$

both approach zero as both Δx and $\Delta y \to 0$.

Finally,

$$\begin{aligned} \Delta z &= \Delta z_1 + \Delta z_2 \\ &= f_x(c, y_0)\Delta x + f_y(x_0 + \Delta x, d)\Delta y && \text{From Equations (12) and (13)} \\ &= [f_x(x_0, y_0) + \epsilon_1]\Delta x + [f_y(x_0, y_0) + \epsilon_2]\Delta y && \text{From Equation (14)} \\ &= f_x(x_0, y_0)\Delta x + f_y(x_0, y_0)\Delta y + \epsilon_1\Delta x + \epsilon_2\Delta y, \end{aligned}$$

where both ϵ_1 and $\epsilon_2 \to 0$ as both Δx and $\Delta y \to 0$, which is what we set out to prove. ■

Analogous results hold for functions of any finite number of independent variables. Suppose that the first partial derivatives of $w = f(x, y, z)$ are defined throughout an open region containing the point (x_0, y_0, z_0) and that f_x, f_y, and f_z are continuous at (x_0, y_0, z_0). Then

$$\begin{aligned} \Delta w &= f(x_0 + \Delta x, y_0 + \Delta y, z_0 + \Delta z) - f(x_0, y_0, z_0) \\ &= f_x\Delta x + f_y\Delta y + f_z\Delta z + \epsilon_1\Delta x + \epsilon_2\Delta y + \epsilon_3\Delta z, \end{aligned} \tag{15}$$

where $\epsilon_1, \epsilon_2, \epsilon_3 \to 0$ as Δx, Δy, and $\Delta z \to 0$.

The partial derivatives f_x, f_y, f_z in Equation (15) are to be evaluated at the point (x_0, y_0, z_0).

Equation (15) can be proved by treating Δw as the sum of three increments,

$$\Delta w_1 = f(x_0 + \Delta x, y_0, z_0) - f(x_0, y_0, z_0) \tag{16}$$

$$\Delta w_2 = f(x_0 + \Delta x, y_0 + \Delta y, z_0) - f(x_0 + \Delta x, y_0, z_0) \tag{17}$$

$$\Delta w_3 = f(x_0 + \Delta x, y_0 + \Delta y, z_0 + \Delta z) - f(x_0 + \Delta x, y_0 + \Delta y, z_0), \tag{18}$$

and applying the Mean Value Theorem to each of these separately. Two coordinates remain constant and only one varies in each of these partial increments $\Delta w_1, \Delta w_2, \Delta w_3$. In Equation (17), for example, only y varies, since x is held equal to $x_0 + \Delta x$ and z is held equal to z_0. Since $f(x_0 + \Delta x, y, z_0)$ is a continuous function of y with a derivative f_y, it is subject to the Mean Value Theorem, and we have

$$\Delta w_2 = f_y(x_0 + \Delta x, y_1, z_0)\Delta y$$

for some y_1 between y_0 and $y_0 + \Delta y$.

A.8 The Area of a Parallelogram's Projection on a Plane

This appendix proves the result needed in Section 16.5 that $|(\mathbf{u} \times \mathbf{v}) \cdot \mathbf{p}|$ is the area of the projection of the parallelogram with sides determined by $\mathbf{u}$ and $\mathbf{v}$ onto any plane whose normal is $\mathbf{p}$. (See Figure A.15.)

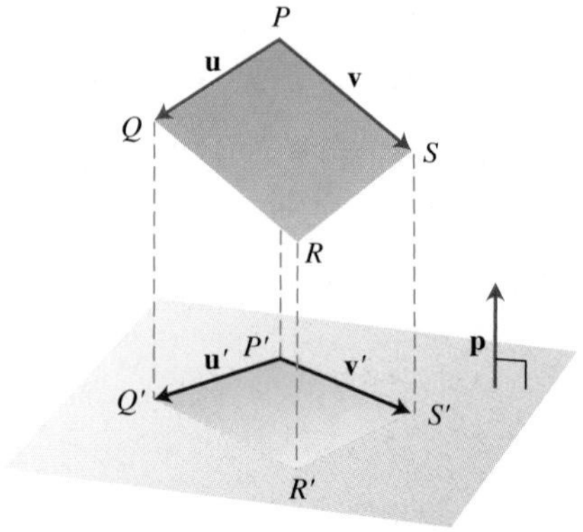

FIGURE A.15 The parallelogram determined by two vectors **u** and **v** in space and the orthogonal projection of the parallelogram onto a plane. The projection lines, orthogonal to the plane, lie parallel to the unit normal vector **p**.

THEOREM

The area of the orthogonal projection of the parallelogram determined by two vectors **u** and **v** in space onto a plane with unit normal vector **p** is

$$\text{Area} = |(\mathbf{u} \times \mathbf{v}) \cdot \mathbf{p}|.$$

Proof In the notation of Figure A.15, which shows a typical parallelogram determined by vectors **u** and **v** and its orthogonal projection onto a plane with unit normal vector **p**,

$$\begin{aligned}
\mathbf{u} &= \overrightarrow{PP'} + \mathbf{u}' + \overrightarrow{Q'Q} \\
&= \mathbf{u}' + \overrightarrow{PP'} - \overrightarrow{QQ'} && (\overrightarrow{Q'Q} = -\overrightarrow{QQ'}) \\
&= \mathbf{u}' + s\mathbf{p}. && \text{(For some scalar } s \text{ because } (\overrightarrow{PP'} - \overrightarrow{QQ'}) \text{ is parallel to } \mathbf{p})
\end{aligned}$$

Similarly,

$$\mathbf{v} = \mathbf{v}' + t\mathbf{p}$$

for some scalar t. Hence,

$$\begin{aligned}
\mathbf{u} \times \mathbf{v} &= (\mathbf{u}' + s\mathbf{p}) \times (\mathbf{v}' + t\mathbf{p}) \\
&= (\mathbf{u}' \times \mathbf{v}') + s(\mathbf{p} \times \mathbf{v}') + t(\mathbf{u}' \times \mathbf{p}) + st\underbrace{(\mathbf{p} \times \mathbf{p})}_{0}. \qquad (1)
\end{aligned}$$

The vectors $\mathbf{p} \times \mathbf{v}'$ and $\mathbf{u}' \times \mathbf{p}$ are both orthogonal to **p**. Hence, when we dot both sides of Equation (1) with **p**, the only nonzero term on the right is $(\mathbf{u}' \times \mathbf{v}') \cdot \mathbf{p}$. That is,

$$(\mathbf{u} \times \mathbf{v}) \cdot \mathbf{p} = (\mathbf{u}' \times \mathbf{v}') \cdot \mathbf{p}.$$

In particular,

$$|(\mathbf{u} \times \mathbf{v}) \cdot \mathbf{p}| = |(\mathbf{u}' \times \mathbf{v}') \cdot \mathbf{p}|. \qquad (2)$$

The absolute value on the right is the volume of the box determined by $\mathbf{u}'$, $\mathbf{v}'$, and **p**. The height of this particular box is $|\mathbf{p}| = 1$, so the box's volume is numerically the same as its base area, the area of parallelogram $P'Q'R'S'$. Combining this observation with Equation (2) gives

$$\text{Area of } P'Q'R'S' = |(\mathbf{u}' \times \mathbf{v}') \cdot \mathbf{p}| = |(\mathbf{u} \times \mathbf{v}) \cdot \mathbf{p}|,$$

which says that the area of the orthogonal projection of the parallelogram determined by **u** and **v** onto a plane with unit normal vector **p** is $|(\mathbf{u} \times \mathbf{v}) \cdot \mathbf{p}|$. This is what we set out to prove. ■

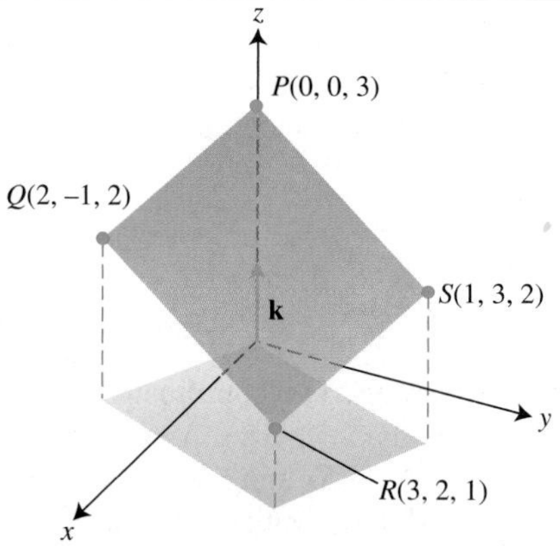

FIGURE A.16 Example 1 calculates the area of the orthogonal projection of parallelogram $PQRS$ on the xy-plane.

EXAMPLE 1 Finding the Area of a Projection

Find the area of the orthogonal projection onto the xy-plane of the parallelogram determined by the points $P(0, 0, 3)$, $Q(2, -1, 2)$, $R(3, 2, 1)$, and $S(1, 3, 2)$ (Figure A.16).

Solution With

$$\mathbf{u} = \overrightarrow{PQ} = 2\mathbf{i} - \mathbf{j} - \mathbf{k}, \qquad \mathbf{v} = \overrightarrow{PS} = \mathbf{i} + 3\mathbf{j} - \mathbf{k}, \qquad \text{and} \qquad \mathbf{p} = \mathbf{k},$$

we have

$$(\mathbf{u} \times \mathbf{v}) \cdot \mathbf{p} = \begin{vmatrix} 2 & -1 & -1 \\ 1 & 3 & -1 \\ 0 & 0 & 1 \end{vmatrix} = \begin{vmatrix} 2 & -1 \\ 1 & 3 \end{vmatrix} = 7,$$

so the area is $|(\mathbf{u} \times \mathbf{v}) \cdot \mathbf{p}| = |7| = 7$. ■

A.9 Basic Algebra, Geometry, and Trigonometry Formulas

Algebra

Arithmetic Operations

$$a(b + c) = ab + ac, \qquad \frac{a}{b} \cdot \frac{c}{d} = \frac{ac}{bd}$$

$$\frac{a}{b} + \frac{c}{d} = \frac{ad + bc}{bd}, \qquad \frac{a/b}{c/d} = \frac{a}{b} \cdot \frac{d}{c}$$

Laws of Signs

$$-(-a) = a, \qquad \frac{-a}{b} = -\frac{a}{b} = \frac{a}{-b}$$

Zero Division by zero is not defined.

$$\text{If } a \neq 0: \ \frac{0}{a} = 0, \quad a^0 = 1, \quad 0^a = 0$$

$$\text{For any number } a: \ a \cdot 0 = 0 \cdot a = 0$$

Laws of Exponents

$$a^m a^n = a^{m+n}, \qquad (ab)^m = a^m b^m, \qquad (a^m)^n = a^{mn}, \qquad a^{m/n} = \sqrt[n]{a^m} = \left(\sqrt[n]{a}\right)^m$$

If $a \neq 0$,

$$\frac{a^m}{a^n} = a^{m-n}, \quad a^0 = 1, \quad a^{-m} = \frac{1}{a^m}.$$

The Binomial Theorem For any positive integer n,

$$(a + b)^n = a^n + na^{n-1}b + \frac{n(n-1)}{1 \cdot 2} a^{n-2}b^2 + \frac{n(n-1)(n-2)}{1 \cdot 2 \cdot 3} a^{n-3}b^3 + \cdots + nab^{n-1} + b^n.$$

For instance,

$$(a+b)^2 = a^2 + 2ab + b^2, \qquad (a-b)^2 = a^2 - 2ab + b^2$$

$$(a+b)^3 = a^3 + 3a^2b + 3ab^2 + b^3, \qquad (a-b)^3 = a^2 - 3a^2b + 3ab^2 - b^3.$$

Factoring the Difference of Like Integer Powers, $n > 1$

$$a^n - b^n = (a-b)(a^{n-1} + a^{n-2}b + a^{n-3}b^2 + \cdots + ab^{n-2} + b^{n-1})$$

For instance,

$$a^2 - b^2 = (a-b)(a+b),$$
$$a^3 - b^3 = (a-b)(a^2 + ab + b^2),$$
$$a^4 - b^4 = (a-b)(a^3 + a^2b + ab^2 + b^3).$$

Completing the Square If $a \neq 0$,

$$\begin{aligned} ax^2 + bx + c &= a\left(x^2 + \frac{b}{a}x\right) + c \\ &= a\left(x^2 + \frac{b}{a}x + \frac{b^2}{4a^2} - \frac{b^2}{4a^2}\right) + c \\ &= a\left(x^2 + \frac{b}{a}x + \frac{b^2}{4a^2}\right) + a\left(-\frac{b^2}{4a^2}\right) + c \\ &= \underbrace{a\left(x^2 + \frac{b}{a}x + \frac{b^2}{4a^2}\right)}_{\text{This is } \left(x + \frac{b}{2a}\right)^2.} + \underbrace{c - \frac{b^2}{4a}}_{\text{Call this part } C.} \\ &= au^2 + C \qquad (u = x + (b/2a)) \end{aligned}$$

The Quadratic Formula If $a \neq 0$ and $ax^2 + bx + c = 0$, then

$$x = \frac{-b \pm \sqrt{b^2 - 4ac}}{2a}.$$

Geometry

Formulas for area, circumference, and volume: (A = area, B = area of base, C = circumference, S = lateral area or surface area, V = volume)

Triangle

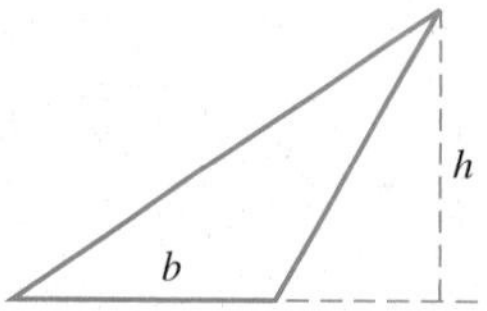

$A = \frac{1}{2}bh$

Similar Triangles

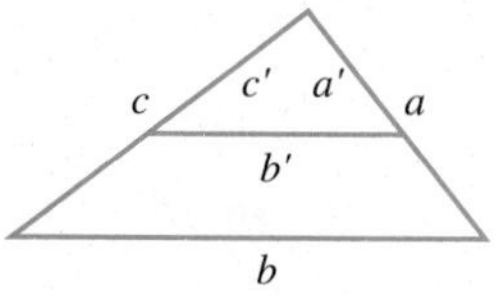

$\frac{a'}{a} = \frac{b'}{b} = \frac{c'}{c}$

Pythagorean Theorem

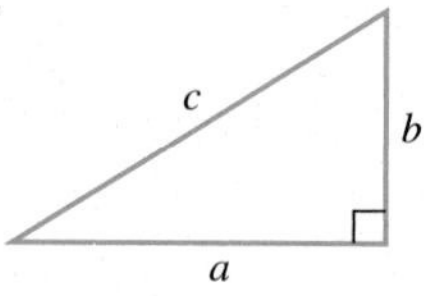

$a^2 + b^2 = c^2$

Parallelogram

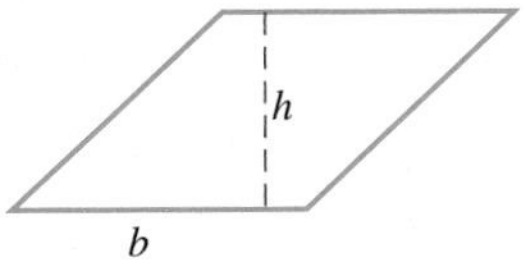

$A = bh$

Trapezoid

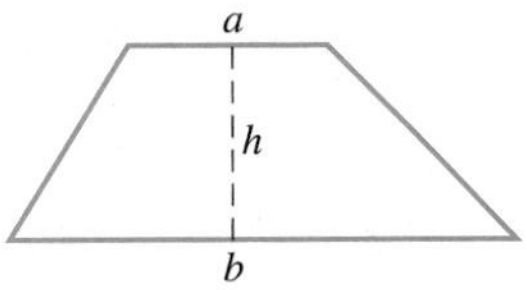

$A = \frac{1}{2}(a + b)h$

Circle

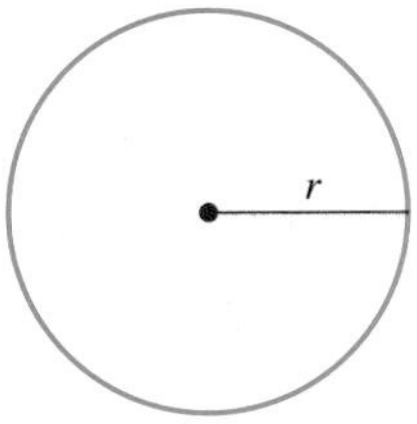

$A = \pi r^2,$
$C = 2\pi r$

Any Cylinder or Prism with Parallel Bases

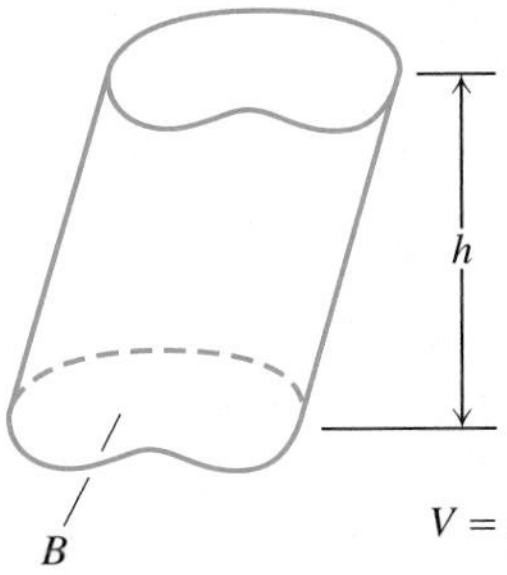

$V = Bh$

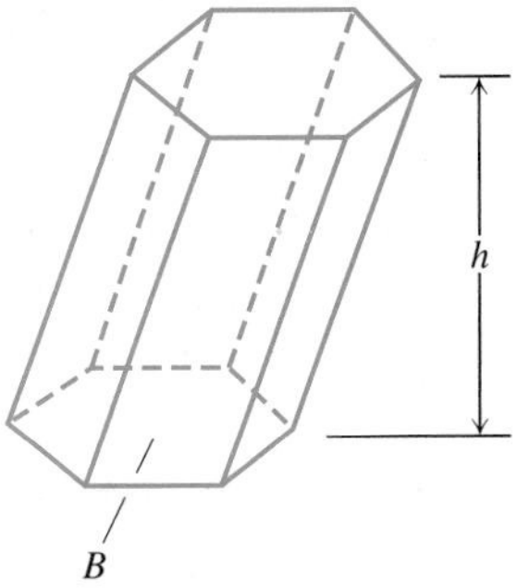

Right Circular Cylinder

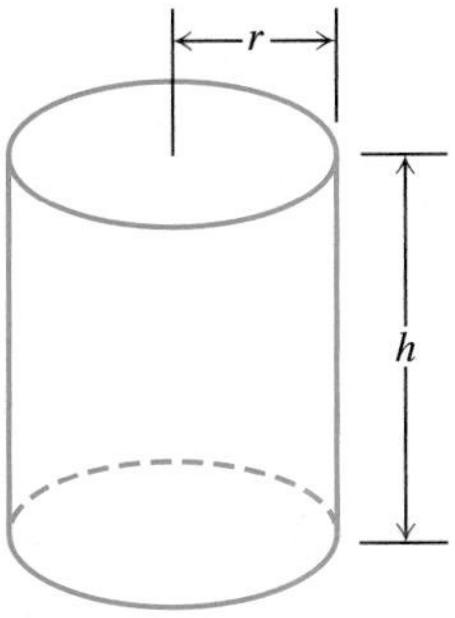

$V = \pi r^2 h$
$S = 2\pi rh =$ Area of side

Any Cone or Pyramid

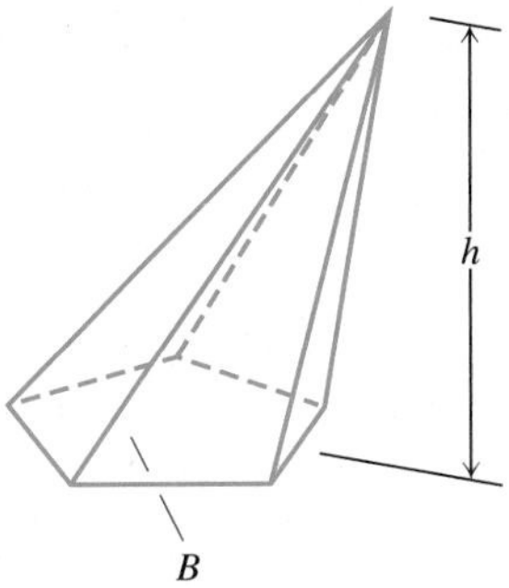

$V = \frac{1}{3}Bh$

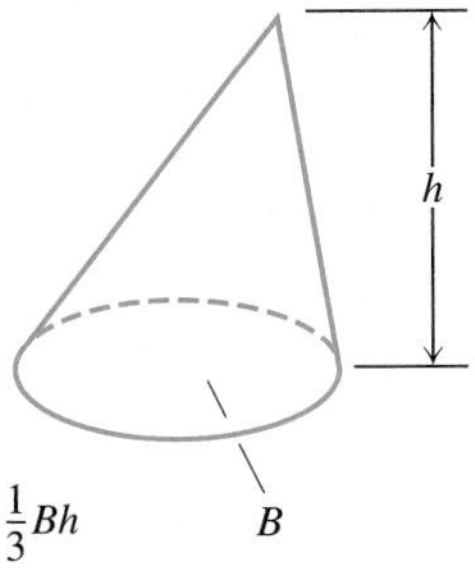

Right Circular Cone

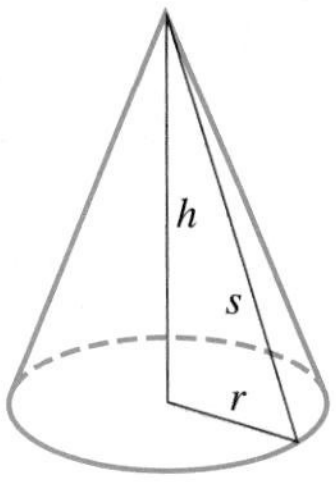

$V = \frac{1}{3}\pi r^2 h$
$S = \pi rs =$ Area of side

Sphere

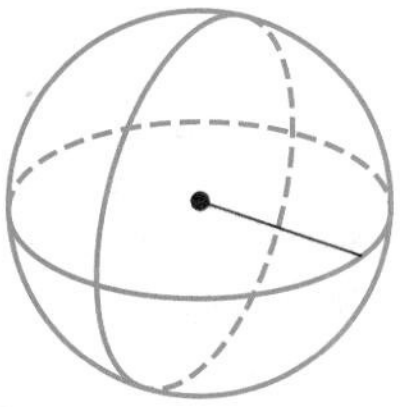

$V = \frac{4}{3}\pi r^3, S = 4\pi r^2$

Trigonometry Formulas

Definitions and Fundamental Identities

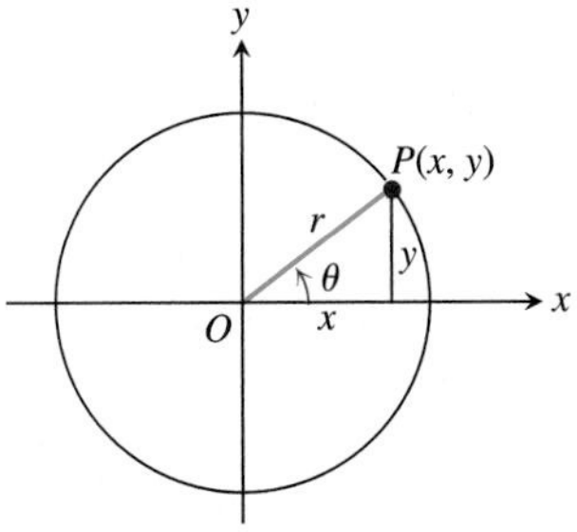

Sine: $\sin\theta = \frac{y}{r} = \frac{1}{\csc\theta}$

Cosine: $\cos\theta = \frac{x}{r} = \frac{1}{\sec\theta}$

Tangent: $\tan\theta = \frac{y}{x} = \frac{1}{\cot\theta}$

Identities

$$\sin(-\theta) = -\sin\theta, \quad \cos(-\theta) = \cos\theta$$

$$\sin^2\theta + \cos^2\theta = 1, \quad \sec^2\theta = 1 + \tan^2\theta, \quad \csc^2\theta = 1 + \cot^2\theta$$

$$\sin 2\theta = 2\sin\theta\cos\theta, \quad \cos 2\theta = \cos^2\theta - \sin^2\theta$$

$$\cos^2\theta = \frac{1 + \cos 2\theta}{2}, \quad \sin^2\theta = \frac{1 - \cos 2\theta}{2}$$

$$\sin(A + B) = \sin A\cos B + \cos A\sin B$$

$$\sin(A - B) = \sin A\cos B - \cos A\sin B$$

$$\cos(A + B) = \cos A\cos B - \sin A\sin B$$

$$\cos(A - B) = \cos A\cos B + \sin A\sin B$$

$$\tan(A + B) = \frac{\tan A + \tan B}{1 - \tan A\tan B}, \quad \tan(A - B) = \frac{\tan A - \tan B}{1 + \tan A\tan B}$$

$$\sin\left(A - \frac{\pi}{2}\right) = -\cos A, \quad \cos\left(A - \frac{\pi}{2}\right) = \sin A$$

$$\sin\left(A + \frac{\pi}{2}\right) = \cos A, \quad \cos\left(A + \frac{\pi}{2}\right) = -\sin A$$

$$\sin A\sin B = \frac{1}{2}\cos(A - B) - \frac{1}{2}\cos(A + B)$$

$$\cos A\cos B = \frac{1}{2}\cos(A - B) + \frac{1}{2}\cos(A + B)$$

$$\sin A\cos B = \frac{1}{2}\sin(A - B) + \frac{1}{2}\sin(A + B)$$

$$\sin A + \sin B = 2\sin\frac{1}{2}(A + B)\cos\frac{1}{2}(A - B)$$

$$\sin A - \sin B = 2\cos\frac{1}{2}(A + B)\sin\frac{1}{2}(A - B)$$

$$\cos A + \cos B = 2\cos\frac{1}{2}(A + B)\cos\frac{1}{2}(A - B)$$

$$\cos A - \cos B = -2\sin\frac{1}{2}(A + B)\sin\frac{1}{2}(A - B)$$

Trigonometric Functions

Radian Measure

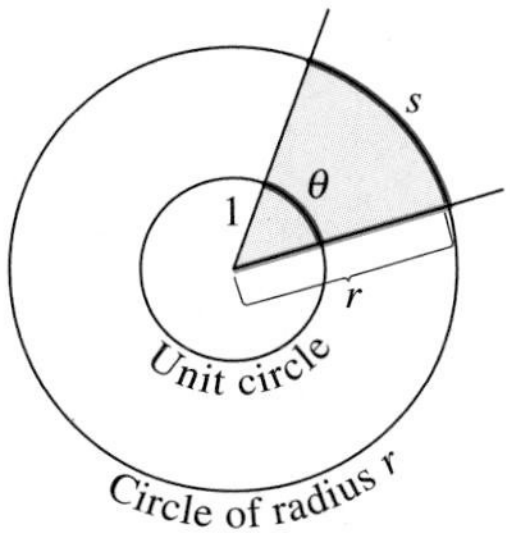

$$\frac{s}{r} = \frac{\theta}{1} = \theta \qquad \text{or} \qquad \theta = \frac{s}{r},$$

$180° = \pi$ radians.

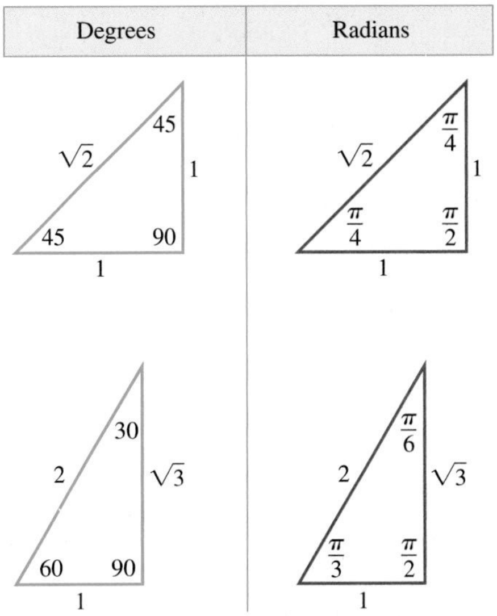

The angles of two common triangles, in degrees and radians.

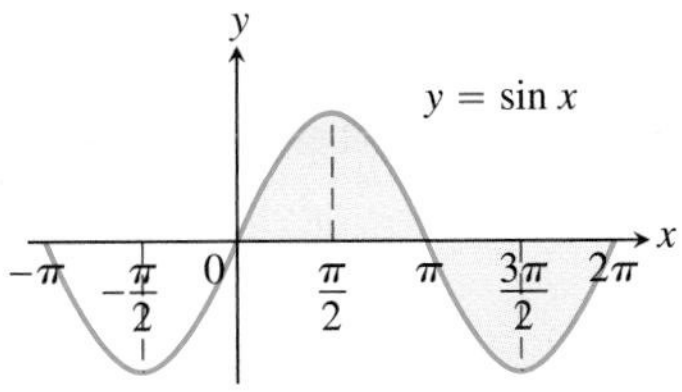

Domain: $(-\infty, \infty)$
Range: $[-1, 1]$

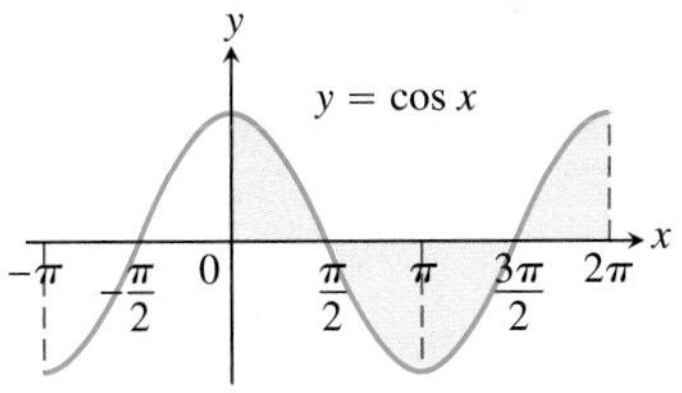

Domain: $(-\infty, \infty)$
Range: $[-1, 1]$

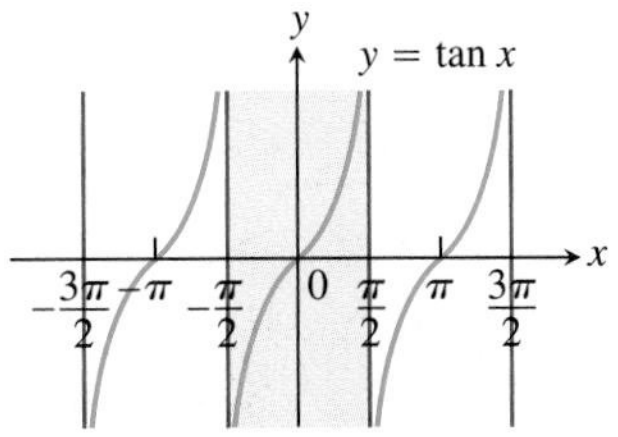

Domain: All real numbers except odd integer multiples of $\pi/2$
Range: $(-\infty, \infty)$

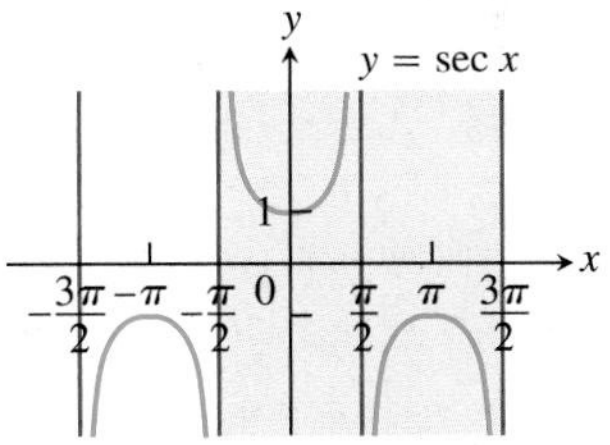

Domain: $x \neq \pm\frac{\pi}{2}, \pm\frac{3\pi}{2}, \ldots$
Range: $(-\infty, -1] \cup [1, \infty)$

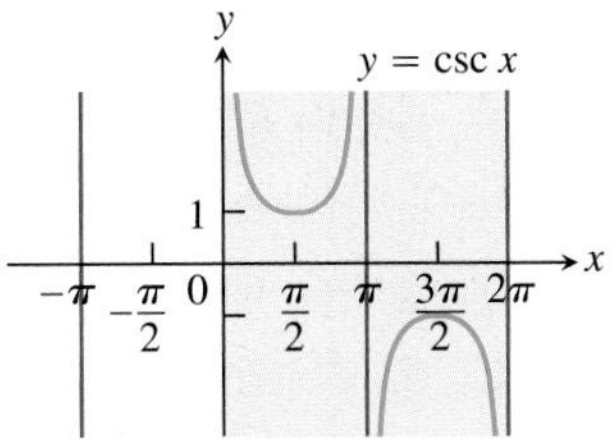

Domain: $x \neq 0, \pm\pi, \pm 2\pi, \ldots$
Range: $(-\infty, -1] \cup [1, \infty)$

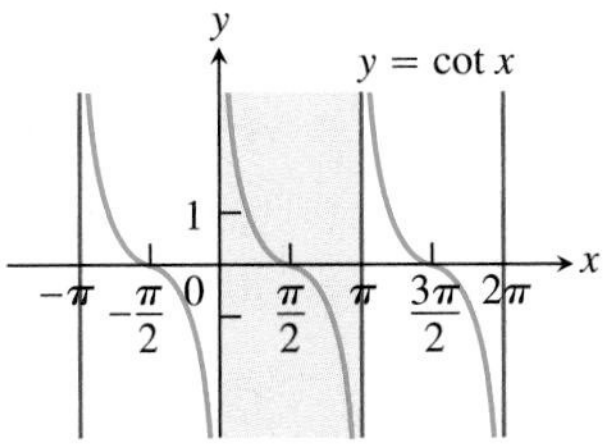

Domain: $x \neq 0, \pm\pi, \pm 2\pi, \ldots$
Range: $(-\infty, \infty)$

ANSWERS

CHAPTER 1

Section 1.1, pp. 7–8

1. $0.\bar{1}, 0.\bar{2}, 0.\bar{3}, 0.\bar{8}, 0.\bar{9}$ or 1
3. **(a)** Not necessarily true **(b)** True **(c)** True **(d)** True **(e)** True **(f)** True **(g)** True **(h)** True
5. $x < -2$ 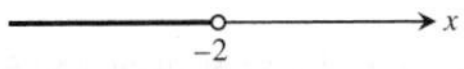

7. $x \le \frac{5}{4}$ 

9. $x \le -\frac{1}{3}$ 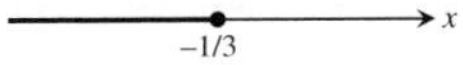

11. $x < -\frac{6}{7}$ 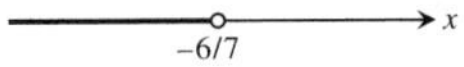

13. ± 3 **15.** $-\frac{1}{2}, -\frac{9}{2}$ **17.** $\frac{7}{6}, \frac{25}{6}$
19. $-2 < x < 2$ 

21. $-2 \le t \le 4$ 

23. $1 < y < \frac{11}{3}$ 

25. $0 \le z \le 10$ 

27. $\frac{2}{7} < x < \frac{2}{5}$ or $\frac{10}{35} < x < \frac{14}{35}$ 

29. $(-\infty, -2] \cup [2, \infty)$ 

31. $(-\infty, 0) \cup (2, \infty)$ 

33. $(-\infty, -3] \cup [1, \infty)$ 

35. $(-\sqrt{2}, \sqrt{2})$ **37.** $(-3, -2) \cup (2, 3)$ **39.** $(-1, 3)$
41. $(0, 1)$ **43.** $a \ge 0$; any negative real number
47. $-\frac{1}{2} < x \le 3$

Section 1.2, pp. 16–19

1. $2, -4; 2\sqrt{5}$ **3.** $-4.9, 0; 4.9$ **5.** Unit circle
7. The circle centered at the origin with points less than a radius of $\sqrt{3}$ and its interior
9. $m_\perp = -\frac{1}{3}$

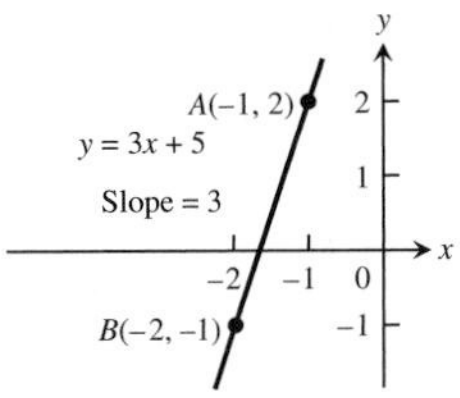

11. $m_\perp$ is undefined.

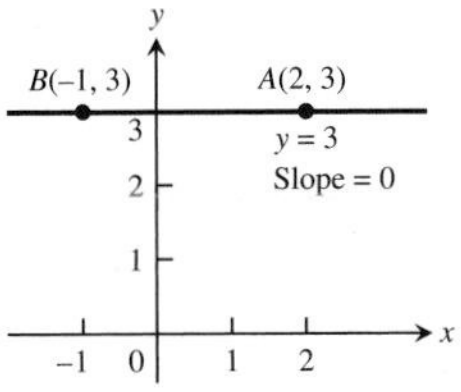

13. **(a)** $x = -1$ **(b)** $y = 4/3$
15. **(a)** $x = 0$ **(b)** $y = -\sqrt{2}$
17. $y = -x$ **19.** $y = -\frac{x}{5} + \frac{23}{5}$ **21.** $y = -\frac{5}{4}x + 6$
23. $y = -9$ **25.** $y = 4x + 4$ **27.** $y = -\frac{2}{5}x + 1$
29. $y = -\frac{x}{2} + 12$
31. x-intercept $= 4$, y-intercept $= 3$

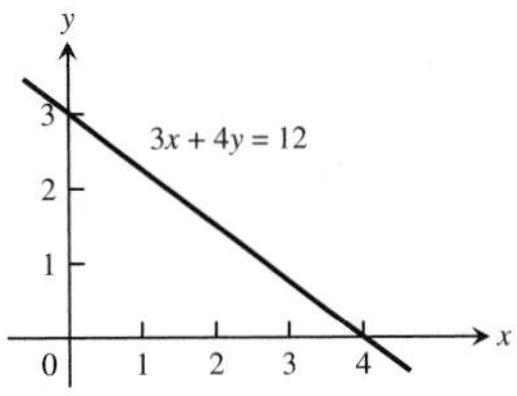

33. x-intercept $= \sqrt{3}$, y-intercept $= -\sqrt{2}$

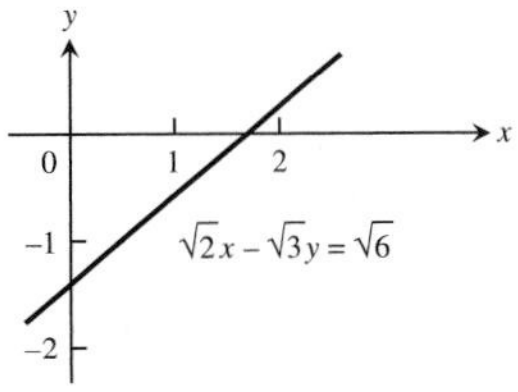

35. Yes. The lines are perpendicular because their slopes, $-A/B$ and B/A, are negative reciprocals of one another.

37. $(3, -3)$ **39.** $(-2, -9)$

41. $x^2 + (y - 2)^2 = 4$

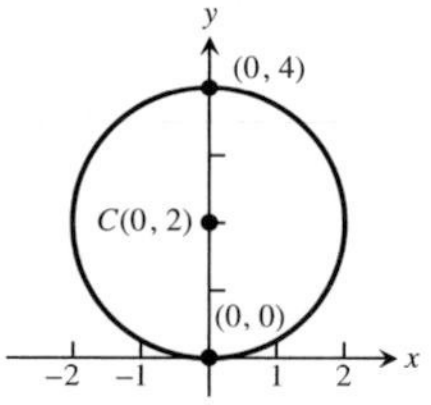

43. $(x + 1)^2 + (y - 5)^2 = 10$

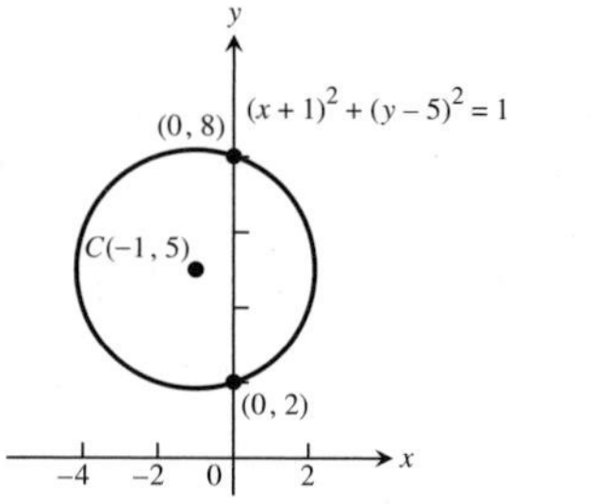

45. $(x + \sqrt{3})^2 + (y + 2)^2 = 4$

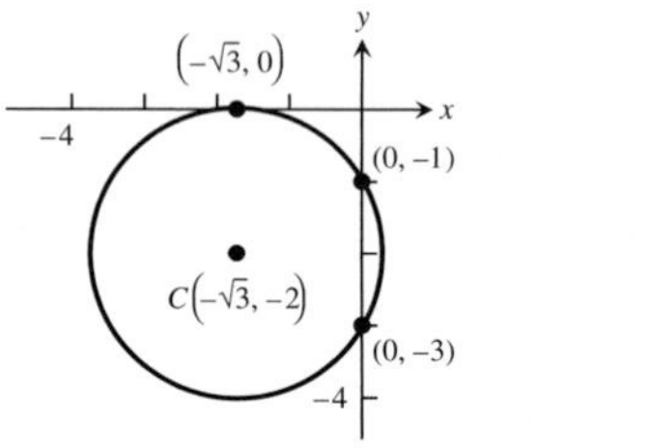

47. $(x + 2)^2 + (y - 2)^2 = 4$

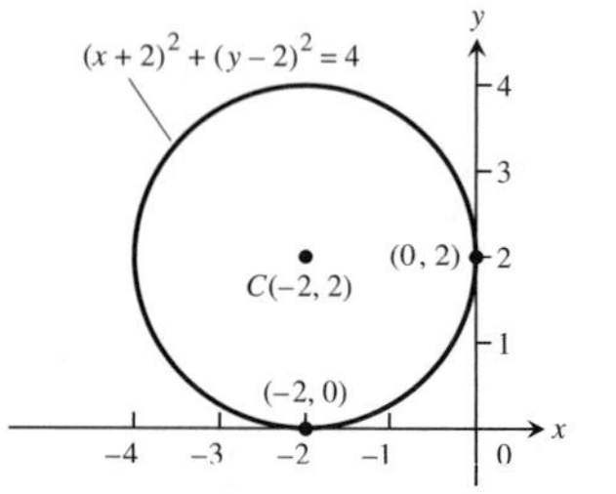

49. $x^2 + (y - 3/2)^2 = 25/4$

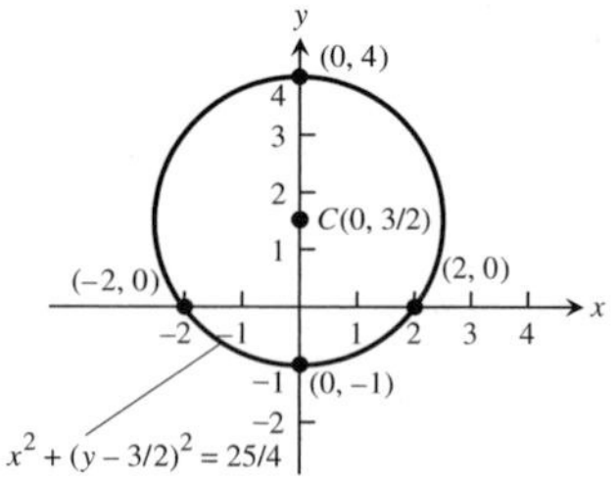

51. $(x - 2)^2 + (y + 2)^2 = 8$

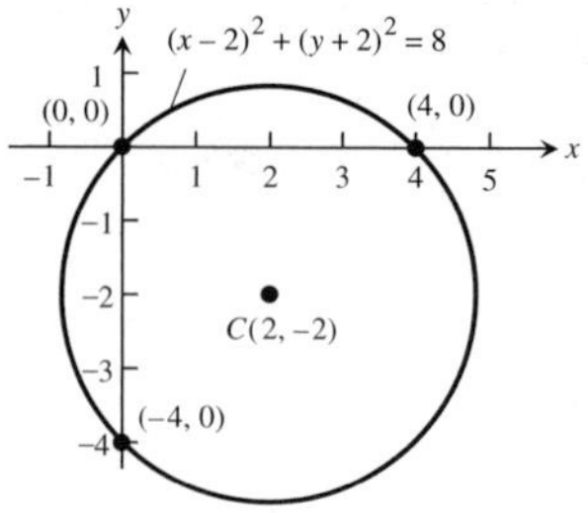

53.

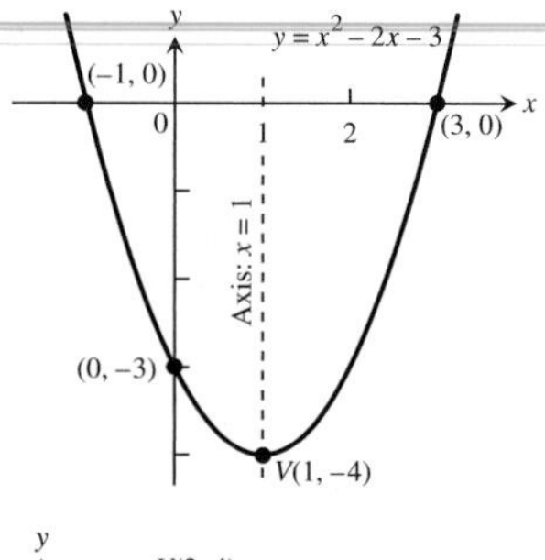

55.

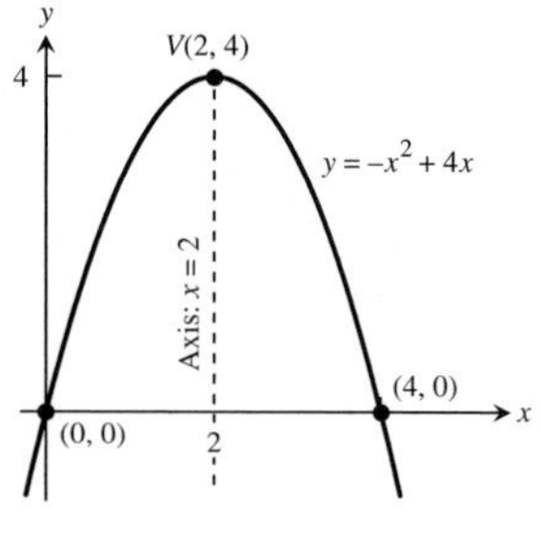

57.

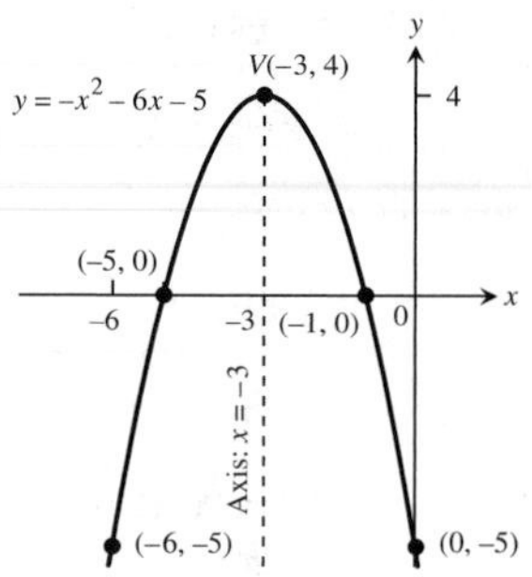

59.

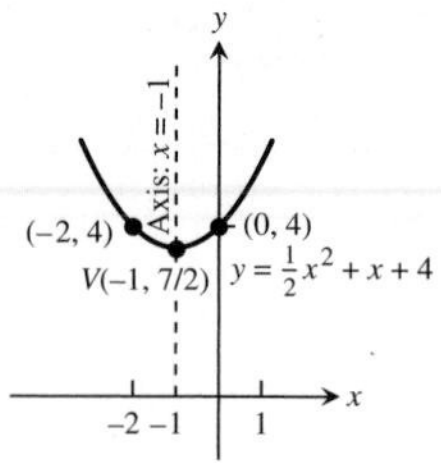

61. Exterior points of a circle of radius $\sqrt{7}$, centered at the origin

63. A circle of radius 2, centered at $(1, 0)$, together with its interior

65. The washer between the circles $x^2 + y^2 = 1$ and $x^2 + y^2 = 4$ (points with distance from the origin between 1 and 2)

67. The interior points of a circle centered at $(0, -3)$ with a radius of 3 that lie above the line $y = -3$

69. $(x + 2)^2 + (y - 1)^2 < 6$ **71.** $x^2 + y^2 \le 2, \quad x \ge 1$

73. $\left(\frac{1}{\sqrt{5}}, \frac{2}{\sqrt{5}}\right), \quad \left(-\frac{1}{\sqrt{5}}, -\frac{2}{\sqrt{5}}\right)$

75. $\left(\frac{1 - \sqrt{5}}{2}, \frac{3 - \sqrt{5}}{2}\right), \quad \left(\frac{1 + \sqrt{5}}{2}, \frac{3 + \sqrt{5}}{2}\right)$

77. $\left(-\frac{1}{\sqrt{3}}, -\frac{1}{3}\right), \quad \left(\frac{1}{\sqrt{3}}, -\frac{1}{3}\right)$

79. $\left(\frac{1}{2}, -\frac{\sqrt{3}}{2}\right), \quad \left(\frac{1}{2}, \frac{\sqrt{3}}{2}\right)$

81. (a) ≈ -2.5 degrees/inch **(b)** ≈ -16.1 degrees/inch **(c)** ≈ -8.3 degrees/inch **83.** 5.97 atm

85. Yes: $C = F = -40°$

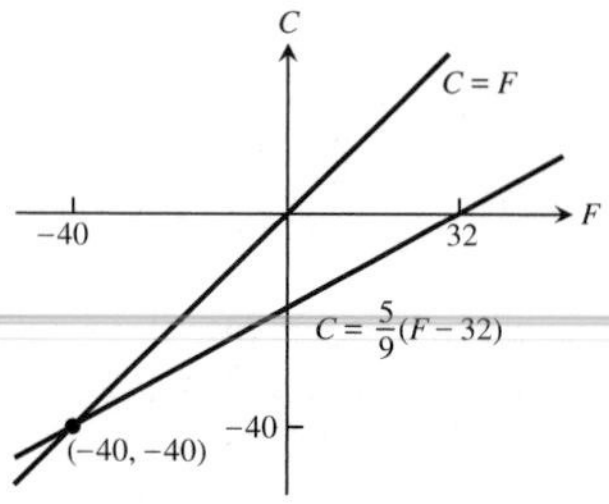

91.

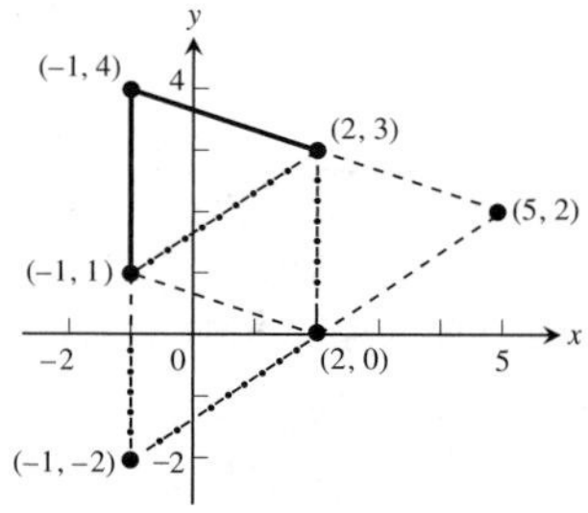

93. $k = -8, \quad k = 1/2$

Section 1.3, pp. 26–28

1. D: $(-\infty, \infty)$, R: $[1, \infty)$ **3.** D: $(0, \infty)$, R: $(0, \infty)$

5. D: $[-2, 2]$, R: $[0, 2]$

7. **(a)** Not a function of x because some values of x have two values of y **(b)** A function of x because for every x there is only one possible y

9. **(a)** No **(b)** No **(c)** No **(d)** $(0, 1]$

11. $A = \dfrac{\sqrt{3}}{4}x^2, \quad p = 3x$

13. $x = \dfrac{d}{\sqrt{3}}, \quad A = 2d^2, \quad V = \dfrac{d^3}{3\sqrt{3}}$

15. $(-\infty, \infty)$

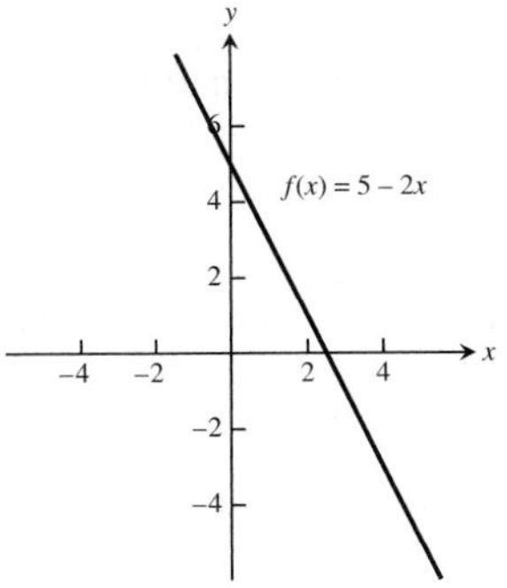

17. $(-\infty, \infty)$

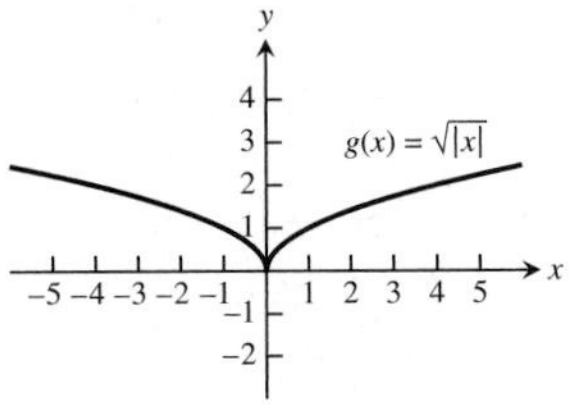

19. $(-\infty, 0) \cup (0, \infty)$

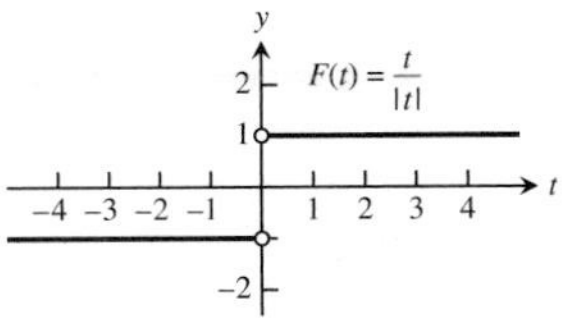

21. **(a)** For each positive value of x, there are two values of y.

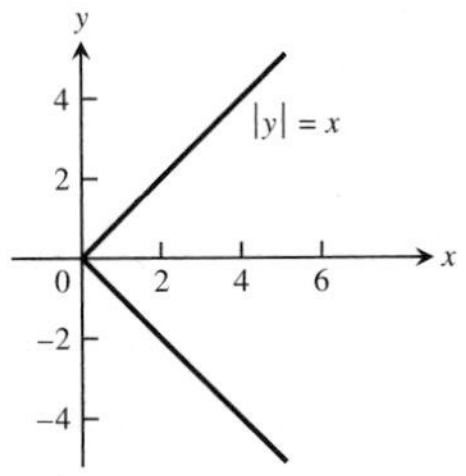

(b) For each value of $x \neq 0$, there are two values of y.

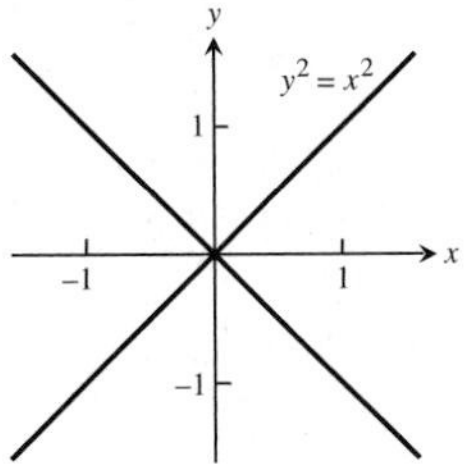

23.

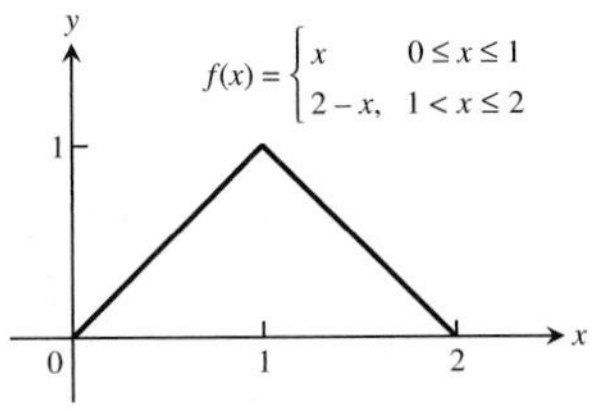

25.

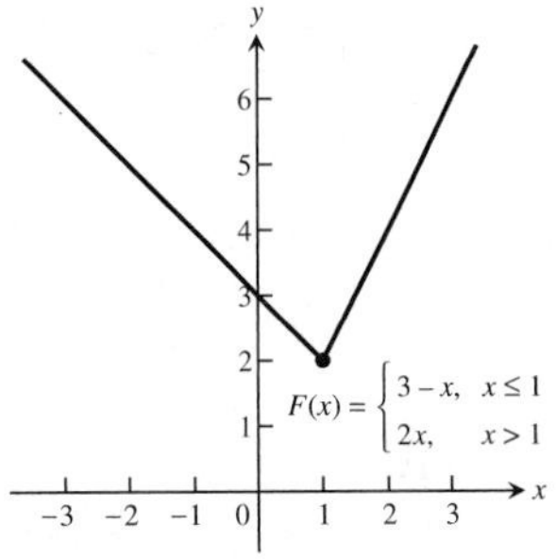

27. **(a)** $f(x) = \begin{cases} x, & 0 \le x \le 1 \\ -x + 2, & 1 < x \le 2 \end{cases}$

(b) $f(x) = \begin{cases} 2, & 0 \le x < 1 \\ 0, & 1 \le x < 2 \\ 2, & 2 \le x < 3 \\ 0, & 3 \le x \le 4 \end{cases}$

29. **(a)** $f(x) = \begin{cases} -x, & -1 \le x < 0 \\ 1, & 0 < x \le 1 \\ -\frac{1}{2}x + \frac{3}{2}, & 1 < x < 3 \end{cases}$

(b) $f(x) = \begin{cases} \frac{1}{2}x, & -2 \le x \le 0 \\ -2x + 2, & 0 < x \le 1 \\ -1, & 1 < x \le 3 \end{cases}$

31. **(a)** $(-2, 0) \cup (4, \infty)$

33. **(a)** $0 \le x < 1$ **(b)** $-1 < x \le 0$

35. Yes **37.** $V = x(14 - 2x)(22 - 2x)$

39. **(a)** Because the circumference of the original circle was 8π and a piece of length x was removed

(b) $r = \dfrac{8\pi - x}{2\pi} = 4 - \dfrac{x}{2\pi}$

(c) $h = \sqrt{16 - r^2} = \dfrac{\sqrt{16\pi x - x^2}}{2\pi}$

(d) $V = \dfrac{1}{3}\pi r^2 h = \dfrac{(8\pi - x)^2\sqrt{16\pi x - x^2}}{24\pi^2}$

Section 1.4, pp. 37–38

1. **(a)** linear, algebraic, polynomial of degree 1 **(b)** power, algebraic **(c)** rational, algebraic **(d)** exponential

3. **(a)** rational, algebraic **(b)** algebraic **(c)** trigonometric **(d)** logarithmic

5. **(a)** h **(b)** f **(c)** g

7. Symmetric about the origin

Dec. $-\infty < x < \infty$

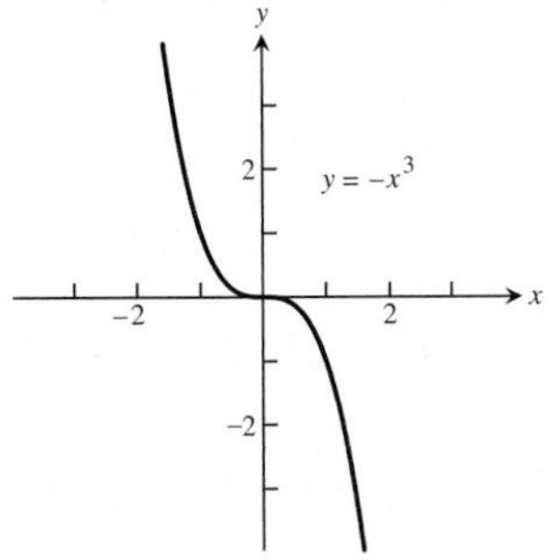

9. Symmetric about the origin

Inc. $-\infty < x < 0$ and $0 < x < \infty$

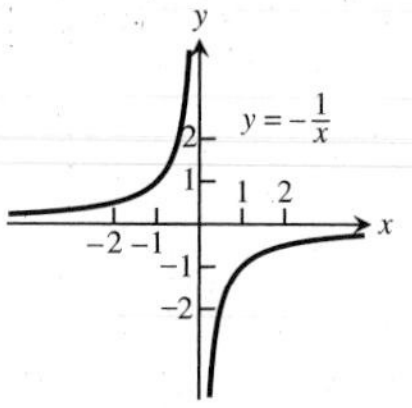

11. Symmetric about the y-axis

Dec. $-\infty < x \le 0$;
inc. $0 \le x < \infty$

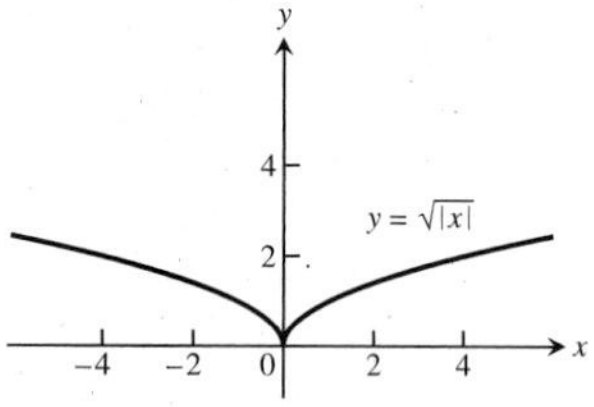

13. Symmetric about the origin

Inc. $-\infty < x < \infty$

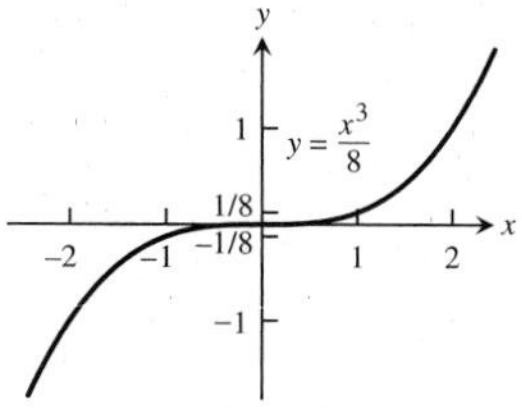

15. No symmetry

Dec. $0 \le x < \infty$

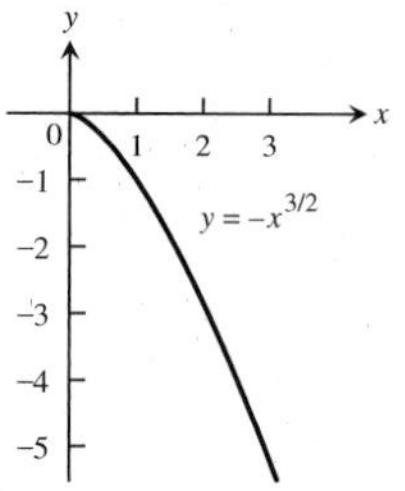

17. Symmetric about the y-axis

Dec. $-\infty < x \le 0$;
inc. $0 \le x < \infty$

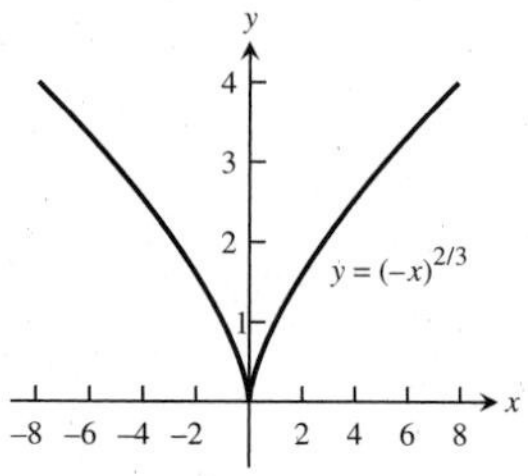

19. Even **21.** Even **23.** Odd **25.** Even **27.** Neither
29. Neither

31. (a) The graph supports the assumption that y is proportional to x. The constant of proportionality is estimated from the slope of the line, which is 0.166.

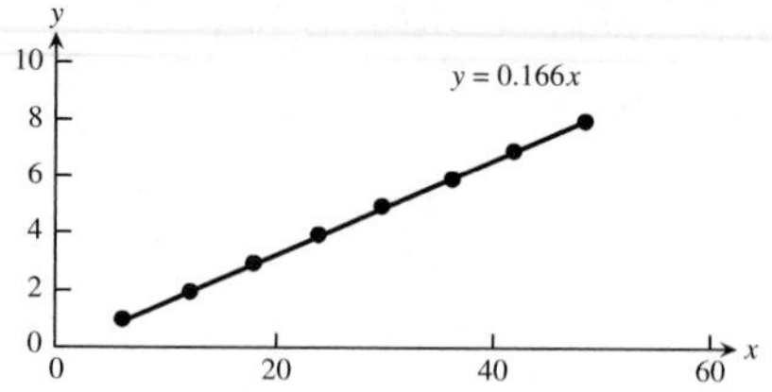

(b) The graph supports the assumption that y is proportional to $x^{1/2}$. The constant of proportionality is estimated from the slope of the line, which is 2.03.

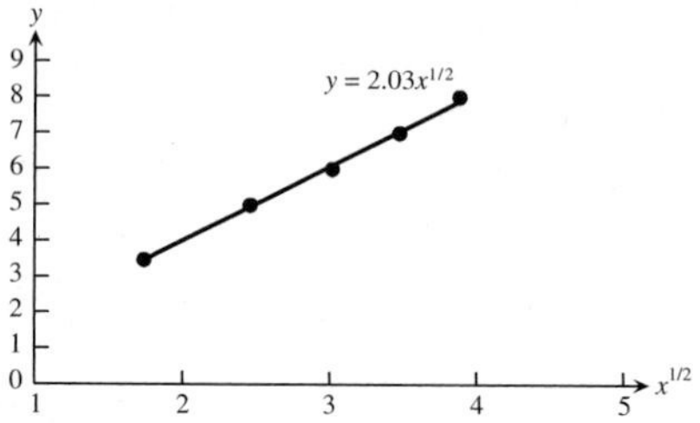

33. (a) $k \approx 1.1$ **(b)** $k \approx 0.059$

35. (a)

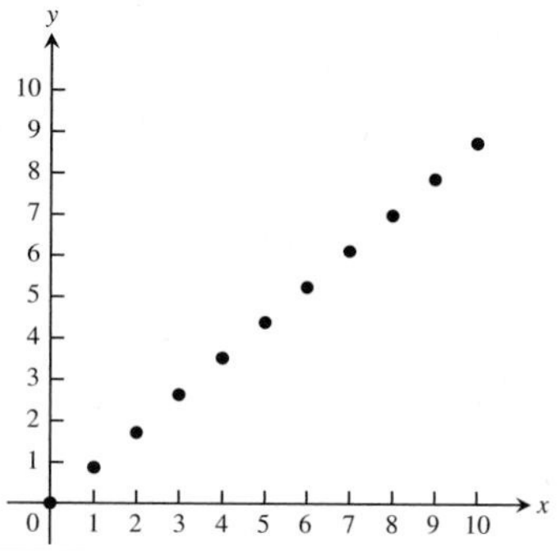

(b) $k \approx 0.87$
(c) Using $y = 0.87x$ with $x = 13$, we get $y = 11.31$.

Section 1.5, pp. 45–48

1. $D_f: -\infty < x < \infty$, $D_g: x \ge 1$, $R_f: -\infty < y < \infty$, $R_g: y \ge 0$, $D_{f+g} = D_{f\cdot g} = D_g$, $R_{f+g}: y \ge 1$, $R_{f\cdot g}: y \ge 0$

3. $D_f: -\infty < x < \infty$, $D_g: -\infty < x < \infty$, $R_f: y = 2$, $R_g: y \ge 1$, $D_{f/g}: -\infty < x < \infty$, $R_{f/g}: 0 < y \le 2$, $D_{g/f}: -\infty < x < \infty$, $R_{g/f}: y \ge 1/2$

5. (a) 2 **(b)** 22 **(c)** $x^2 + 2$ **(d)** $x^2 + 10x + 22$ **(e)** 5 **(f)** -2 **(g)** $x + 10$ **(h)** $x^4 - 6x^2 + 6$

7. (a) $\frac{4}{x^2} - 5$ **(b)** $\frac{4}{x^2} - 5$ **(c)** $\left(\frac{4}{x} - 5\right)^2$ **(d)** $\left(\frac{1}{4x - 5}\right)^2$ **(e)** $\frac{1}{4x^2 - 5}$ **(f)** $\frac{1}{(4x - 5)^2}$

9. **(a)** $f(g(x))$ **(b)** $j(g(x))$ **(c)** $g(g(x))$ **(d)** $j(j(x))$ **(e)** $g(h(f(x)))$ **(f)** $h(j(f(x)))$

11.

	$g(x)$	$f(x)$	$f \circ g(x)$
(a)	$x - 7$	$\sqrt{x}$	$\sqrt{x - 7}$
(b)	$x + 2$	$3x$	$3x + 6$
(c)	x^2	$\sqrt{x - 5}$	$\sqrt{x^2 - 5}$
(d)	$\dfrac{x}{x - 1}$	$\dfrac{x}{x - 1}$	x
(e)	$\dfrac{1}{x - 1}$	$1 + \dfrac{1}{x}$	x
(f)	$\dfrac{1}{x}$	$\dfrac{1}{x}$	x

13. **(a)** $f(g(x)) = \sqrt{\dfrac{1}{x} + 1}$, $g(f(x)) = \dfrac{1}{\sqrt{x + 1}}$

(b) $D_{f \circ g} = (0, \infty)$, $D_{g \circ f} = (-1, \infty)$

(c) $R_{f \circ g} = (1, \infty)$, $R_{g \circ f} = (0, \infty)$

15. **(a)** $y = -(x + 7)^2$ **(b)** $y = -(x - 4)^2$

17. **(a)** Position 4 **(b)** Position 1 **(c)** Position 2 **(d)** Position 3

19. $(x + 2)^2 + (y + 3)^2 = 49$

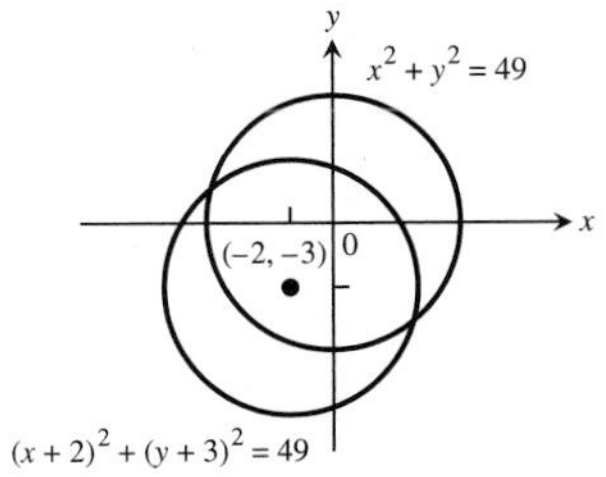

21. $y + 1 = (x + 1)^3$

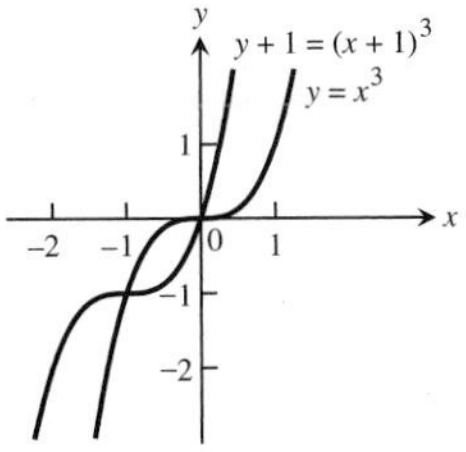

23. $y = \sqrt{x + 0.81}$

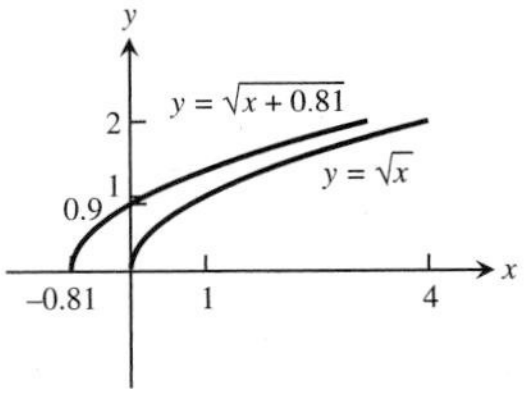

25. $y = 2x$

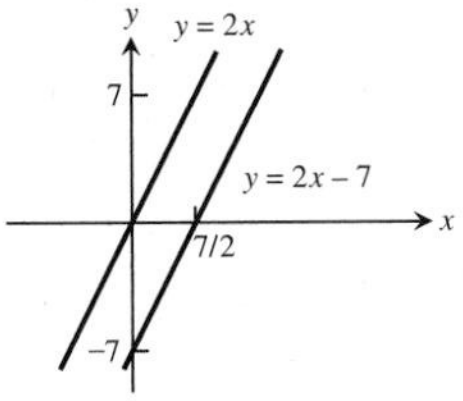

27. $y - 1 = \dfrac{1}{x - 1}$

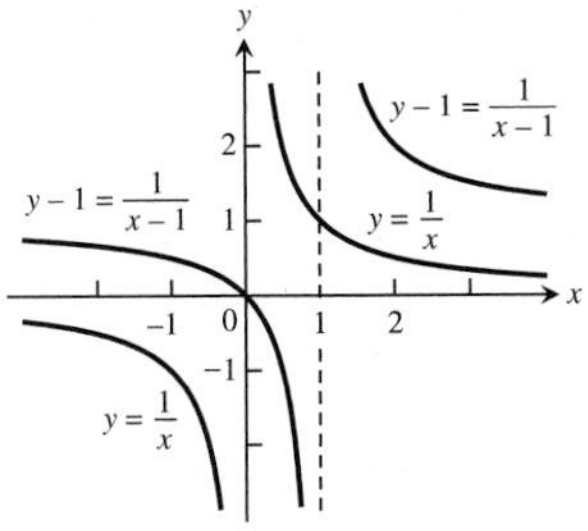

29.

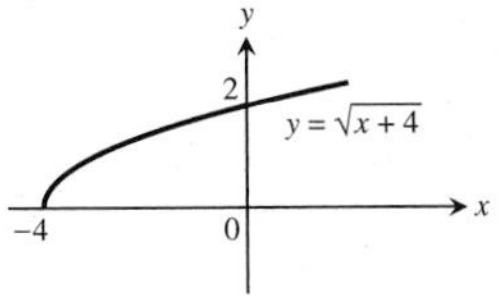

31.

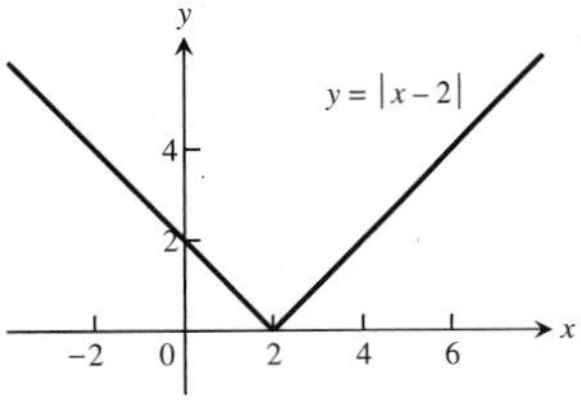

33.

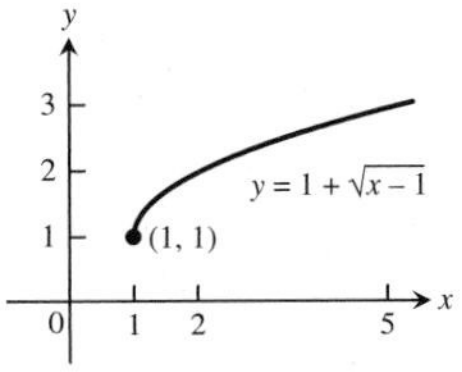

35.

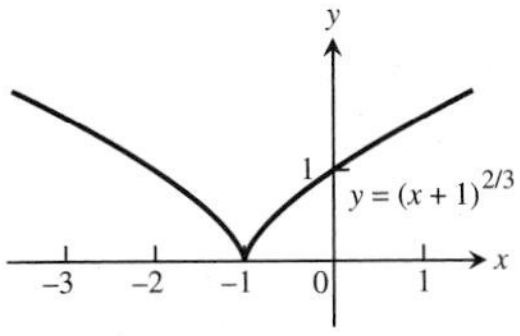

37.

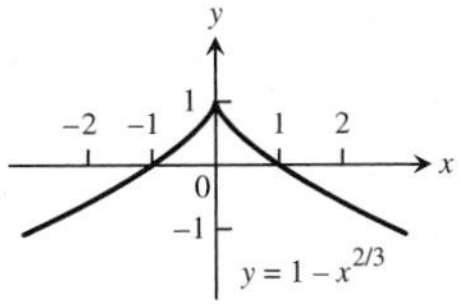

39.

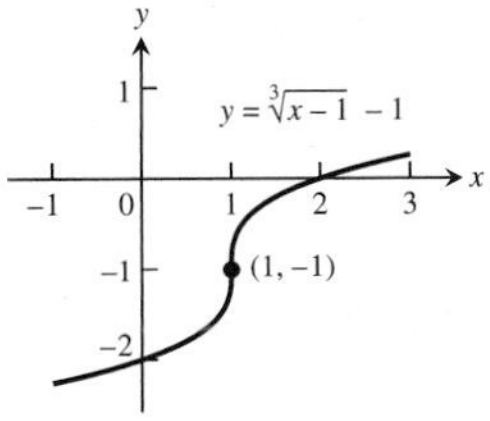

41.

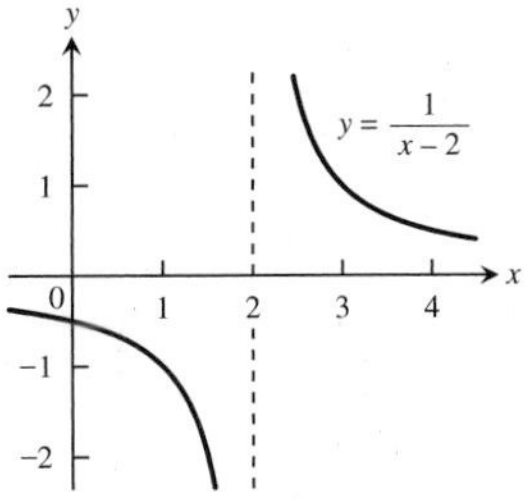

43.

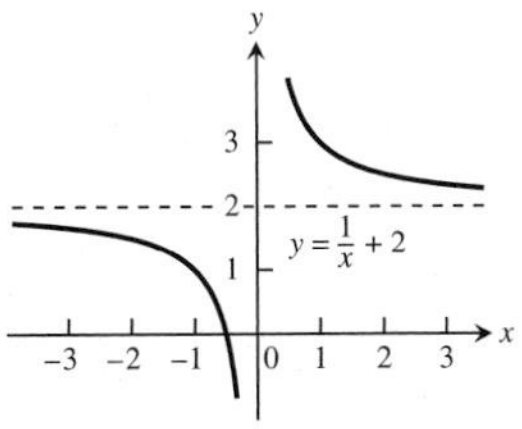

45.

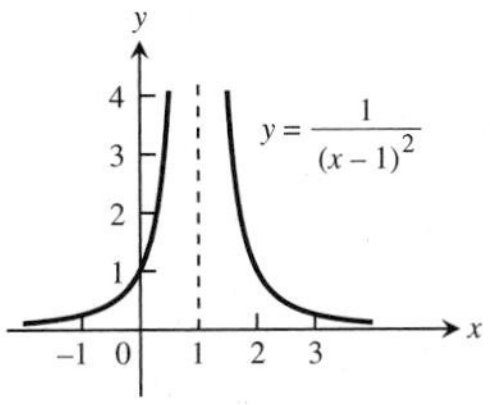

47.

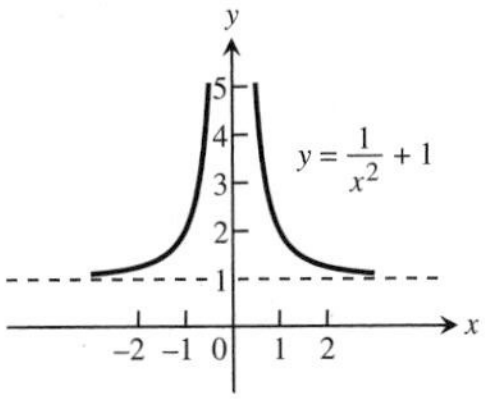

49. **(a)** $D: [0, 2]$, $R: [2, 3]$

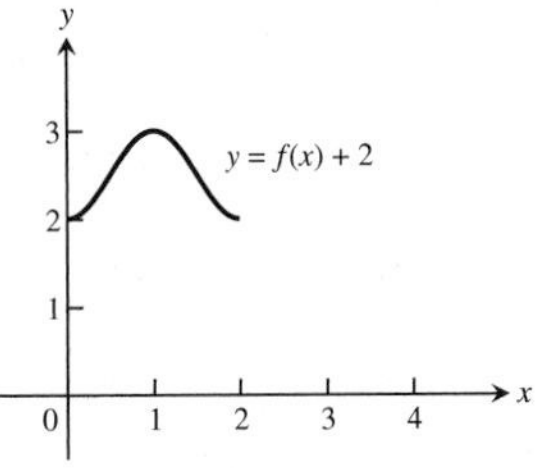

(b) $D: [0, 2]$, $R: [-1, 0]$

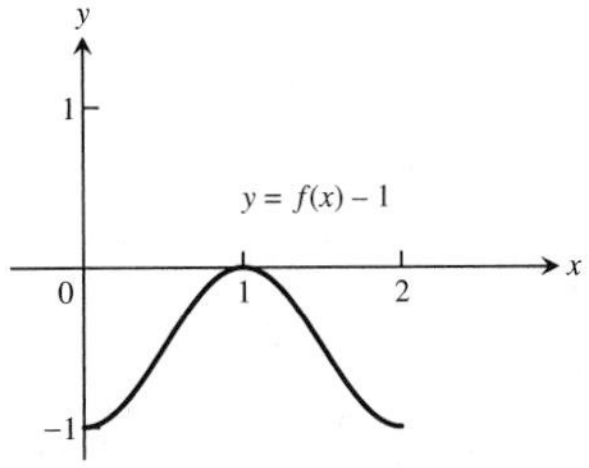

(c) $D: [0, 2], \quad R: [0, 2]$

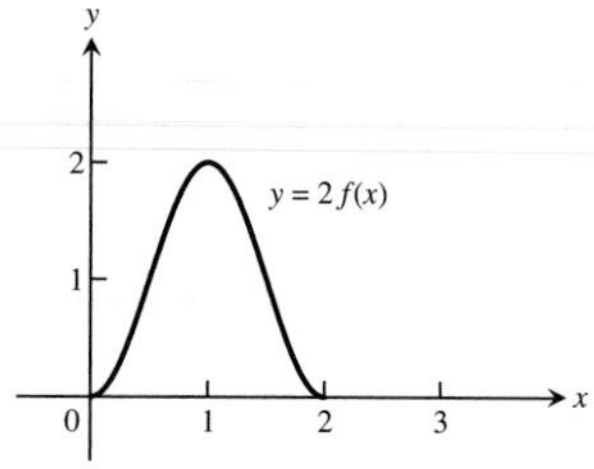

(d) $D: [0, 2], \quad R: [-1, 0]$

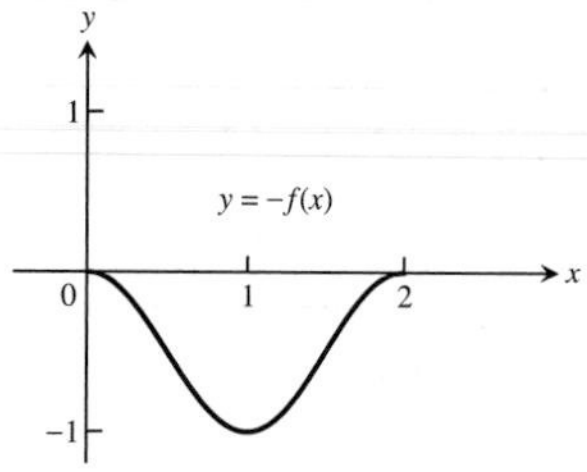

(e) $D: [-2, 0], \quad R: [0, 1]$

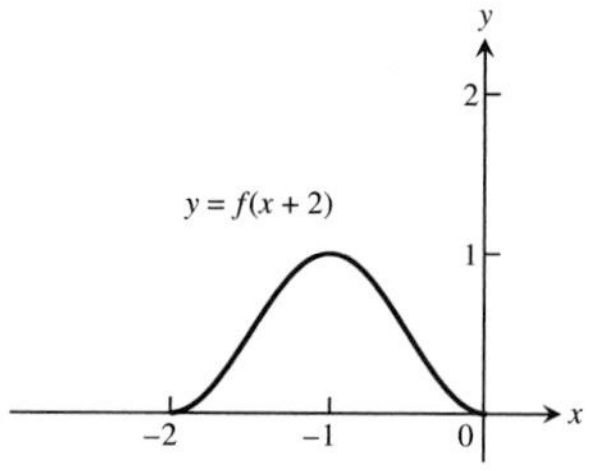

(f) $D: [1, 3], R: [0, 1]$

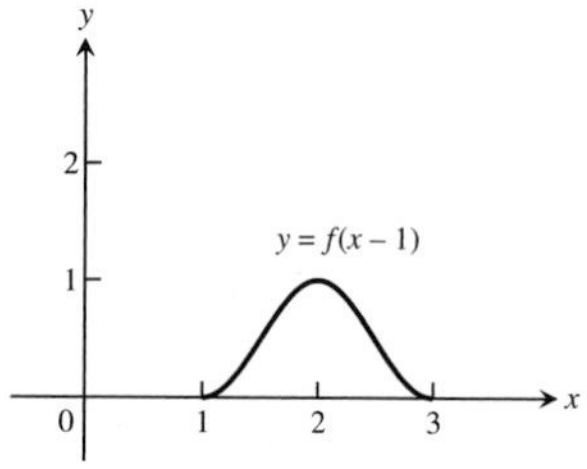

(g) $D: [-2, 0], \quad R: [0, 1]$

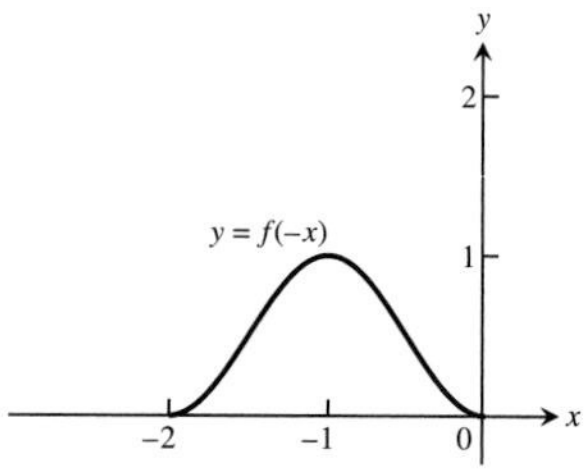

(h) $D: [-1, 1], \quad R: [0, 1]$

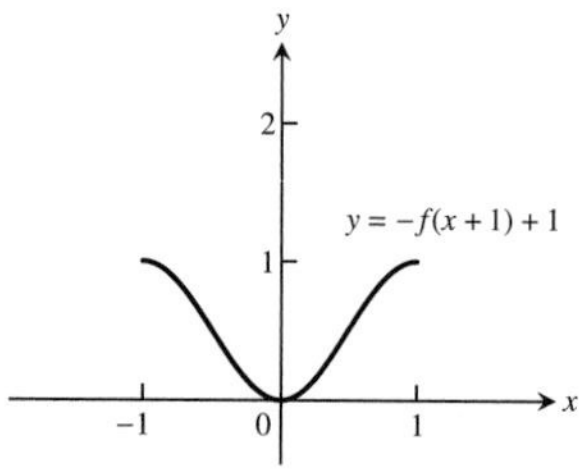

51. $y = 3x^2 - 3$ **53.** $y = \frac{1}{2} + \frac{1}{2x^2}$ **55.** $y = \sqrt{4x + 1}$

57. $y = \sqrt{4 - \frac{x^2}{4}}$ **59.** $y = 1 - 27x^3$

61.

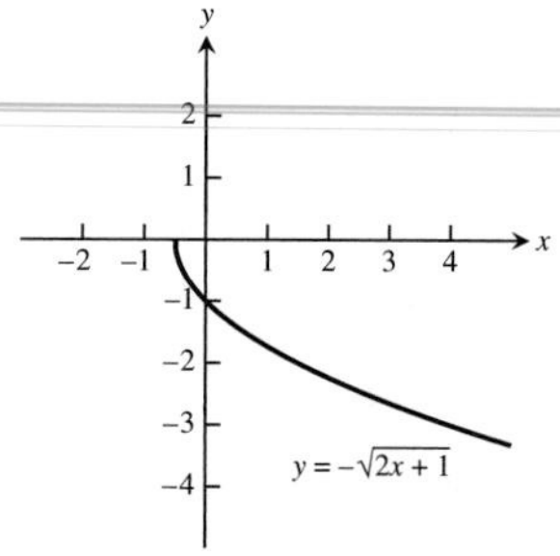

63.

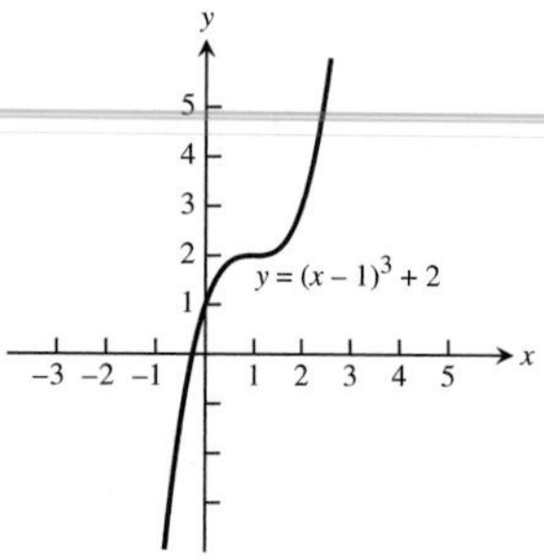

65.

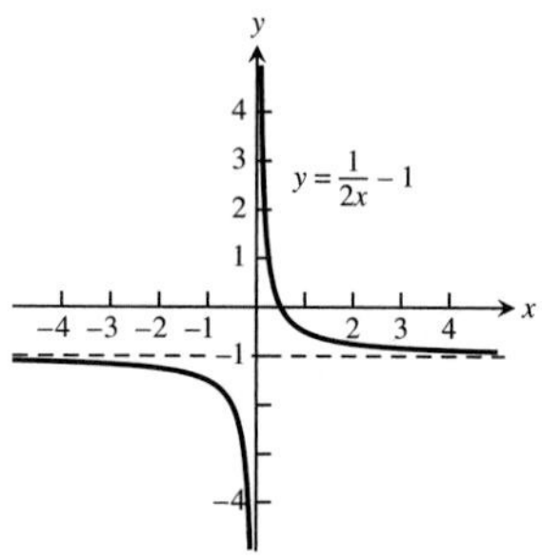

67.

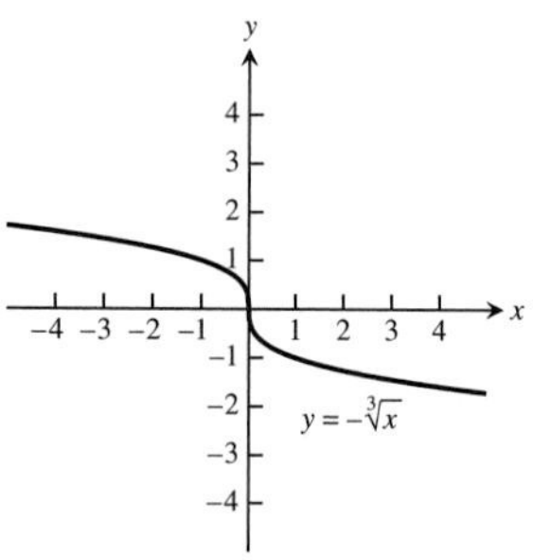

69.

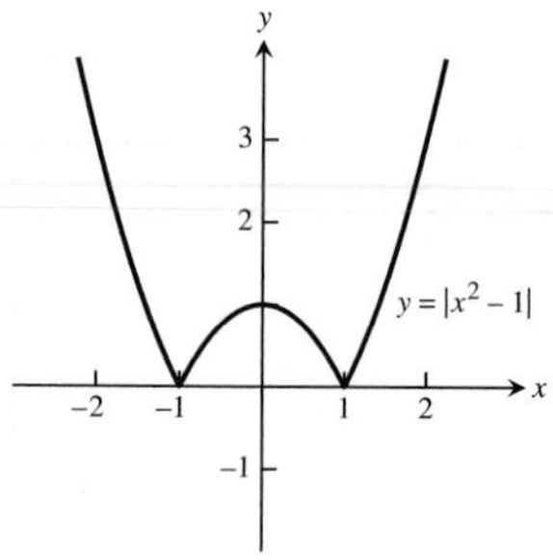

71.

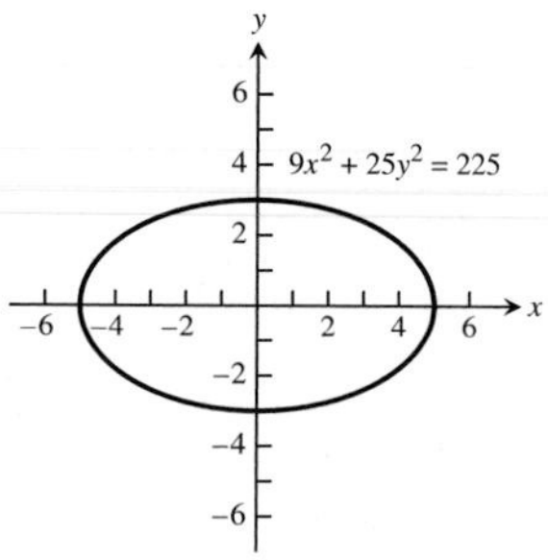

73.

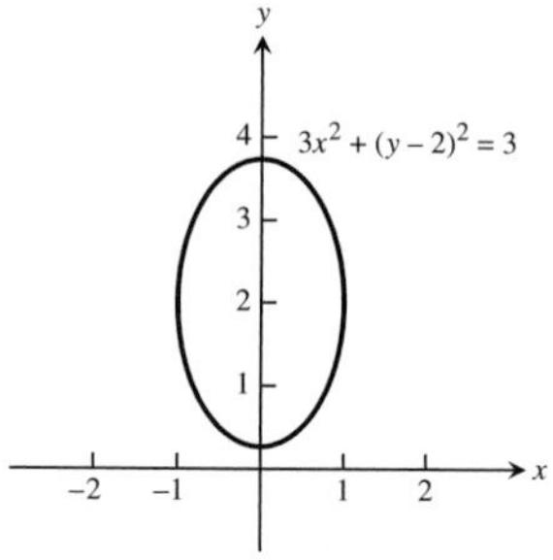

75.

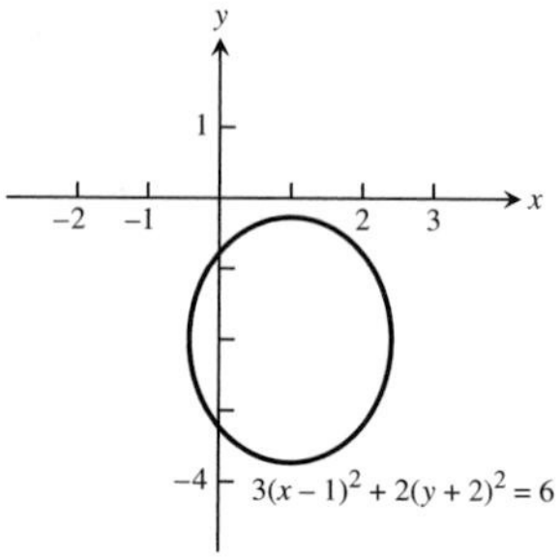

77. $\frac{(x + 4)^2}{16} + \frac{(y - 3)^2}{9} = 1$ Center: $(-4, 3)$

The major axis is the line segment between $(-8, 3)$ and $(0, 3)$.

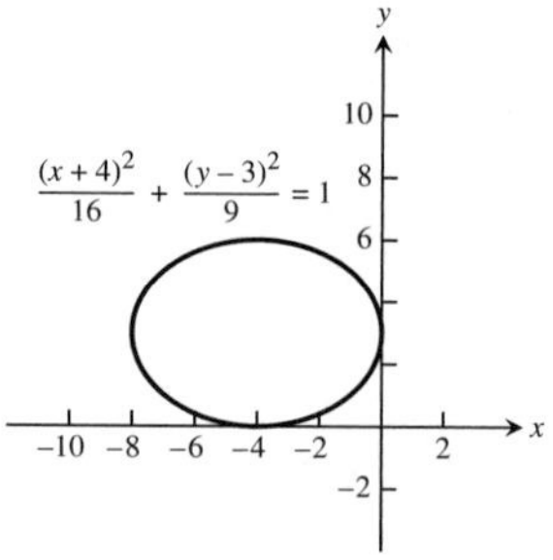

79. **(a)** Odd **(b)** Odd **(c)** Odd **(d)** Even **(e)** Even **(f)** Even **(g)** Even **(h)** Even **(i)** Odd

Section 1.6, pp. 56–58

1. **(a)** 8π m **(b)** $\frac{55\pi}{9}$ m **3.** 8.4 in.

5.

θ	$-\pi$	$-2\pi/3$	0	$\pi/2$	$3\pi/4$
$\sin\theta$	0	$-\frac{\sqrt{3}}{2}$	0	1	$\frac{1}{\sqrt{2}}$
$\cos\theta$	-1	$-\frac{1}{2}$	1	0	$-\frac{1}{\sqrt{2}}$
$\tan\theta$	0	$\sqrt{3}$	0	UND	-1
$\cot\theta$	UND	$\frac{1}{\sqrt{3}}$	UND	0	-1
$\sec\theta$	-1	-2	1	UND	$-\sqrt{2}$
$\csc\theta$	UND	$-\frac{2}{\sqrt{3}}$	UND	1	$\sqrt{2}$

7. $\cos x = -4/5, \tan x = -3/4$

9. $\sin x = -\dfrac{\sqrt{8}}{3}, \tan x = -\sqrt{8}$

11. $\sin x = -\dfrac{1}{\sqrt{5}}, \cos x = -\dfrac{2}{\sqrt{5}}$

13. Period π

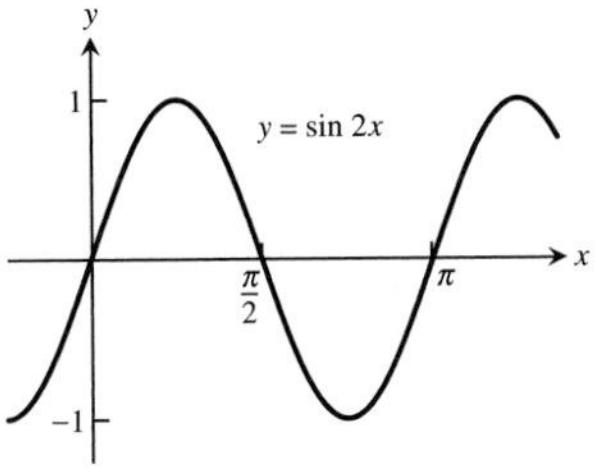

15. Period 2

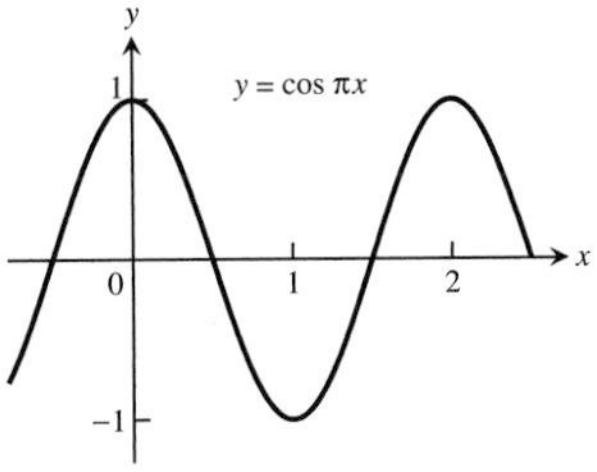

17. Period 6

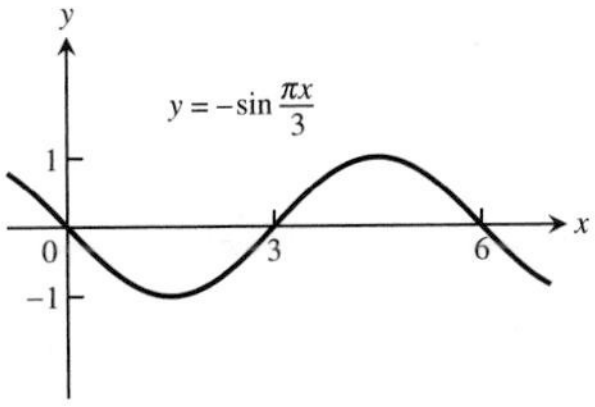

19. Period 2π

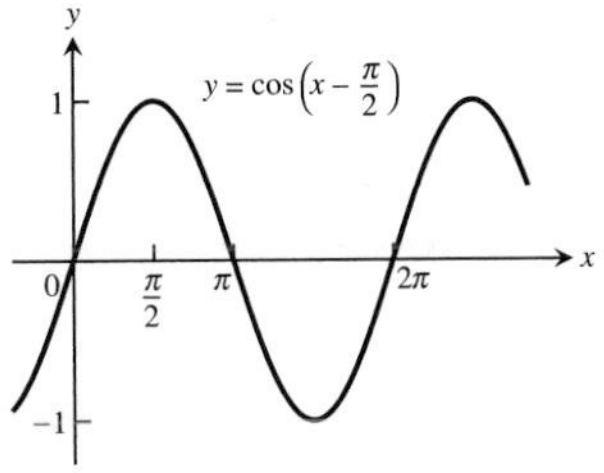

21. Period 2π

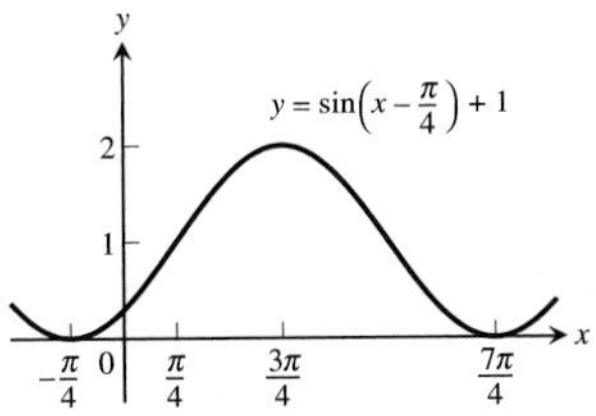

23. Period $\pi/2$, symmetric about the origin

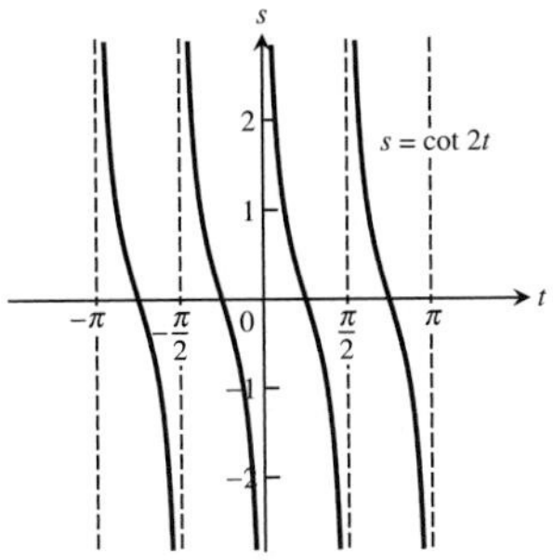

25. Period 4, symmetric about the y-axis

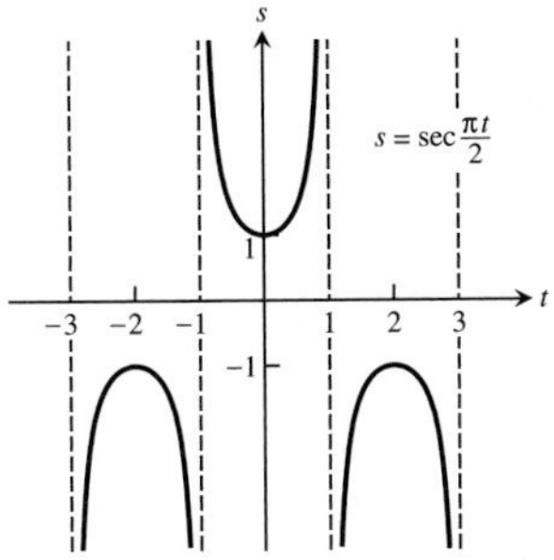

29. $D: (-\infty, \infty)$, $R: y = -1, 0, 1$

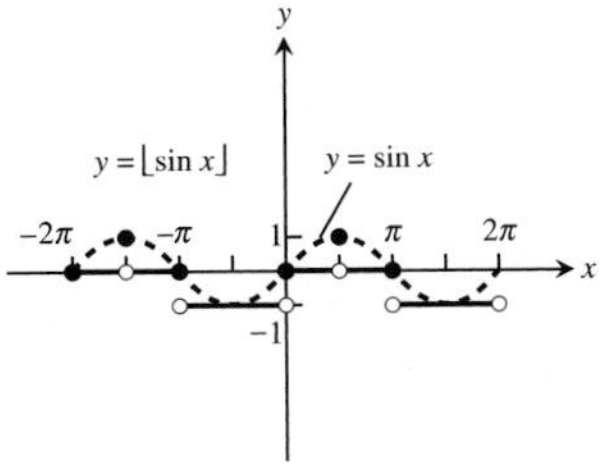

39. $-\cos x$ 41. $-\cos x$ 43. $\dfrac{\sqrt{6} + \sqrt{2}}{4}$ 45. $\dfrac{1 + \sqrt{3}}{2\sqrt{2}}$

47. $\dfrac{2 + \sqrt{2}}{4}$ 49. $\dfrac{2 - \sqrt{3}}{4}$ 55. $c = \sqrt{7} \approx 2.646$

59. $a = 1.464$

61. $A = 2, B = 2\pi, C = -\pi, D = -1$

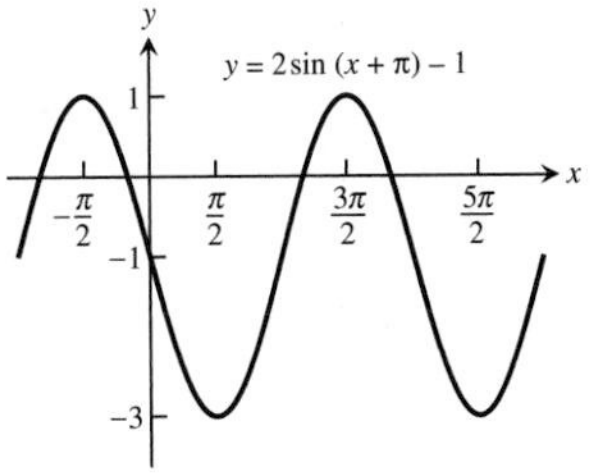

63. $A = -\dfrac{2}{\pi}, B = 4, C = 0, D = \dfrac{1}{\pi}$

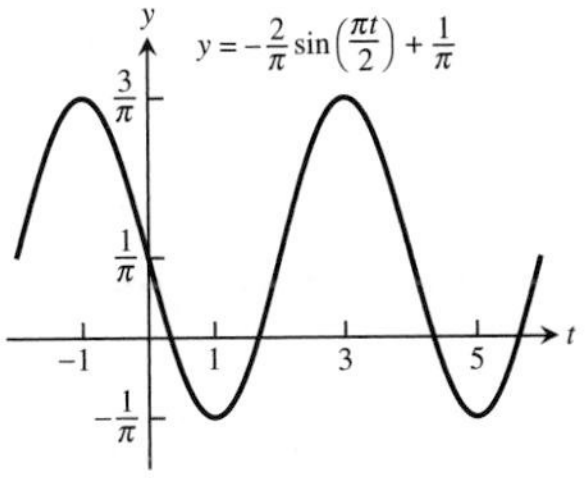

65. **(a)** 37 **(b)** 365 **(c)** Right 101 **(d)** Up 25

Section 1.7, pp. 66–68

1. **d)** 3. **d)**

5. $[-3, 5]$ by $[-15, 40]$

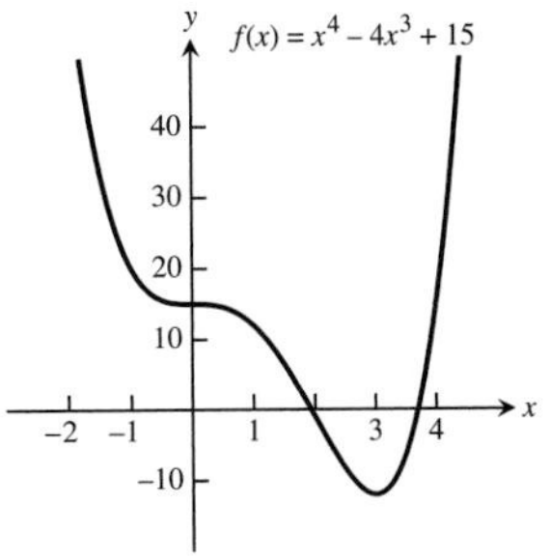

7. $[-3, 6]$ by $[-250, 50]$

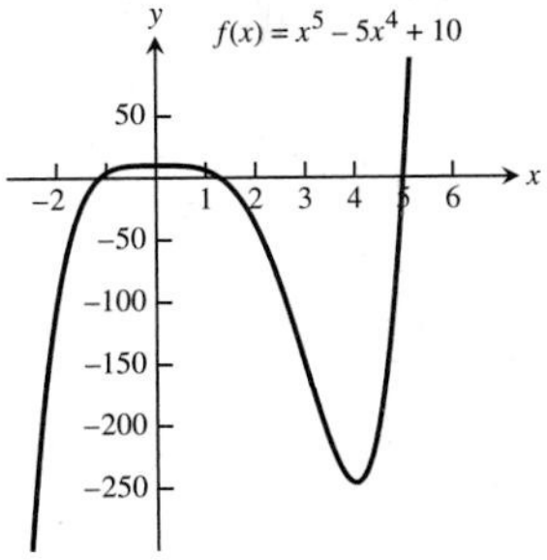

9. $[-3, 3]$ by $[-6, 6]$

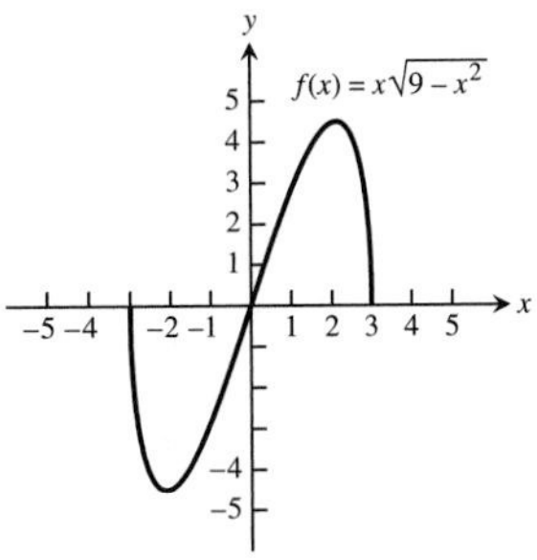

11. $[-2, 6]$ by $[-5, 4]$

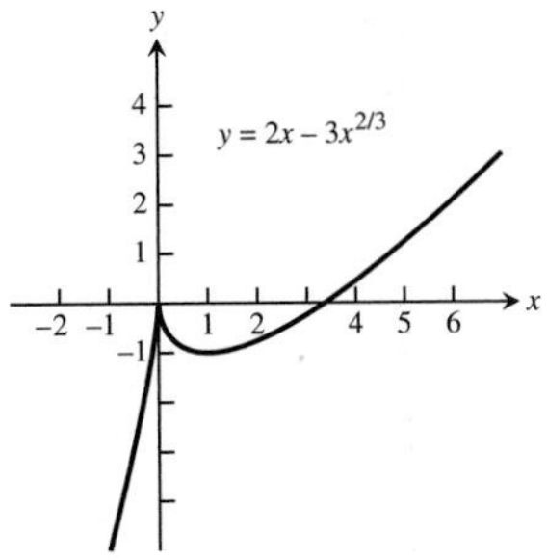

13. $[-2, 8]$ by $[-5, 10]$

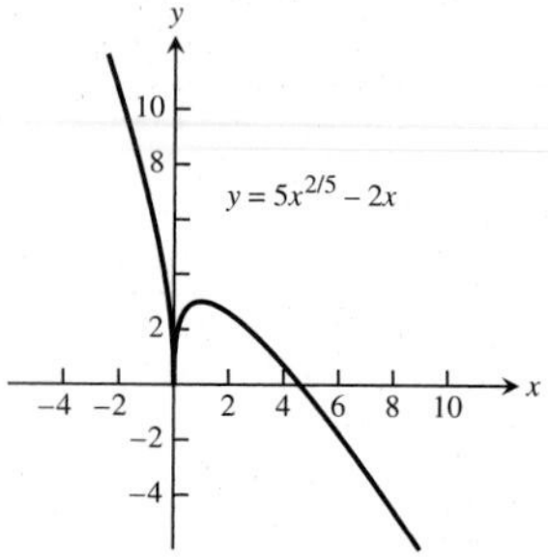

15. $[-3, 3]$ by $[0, 10]$

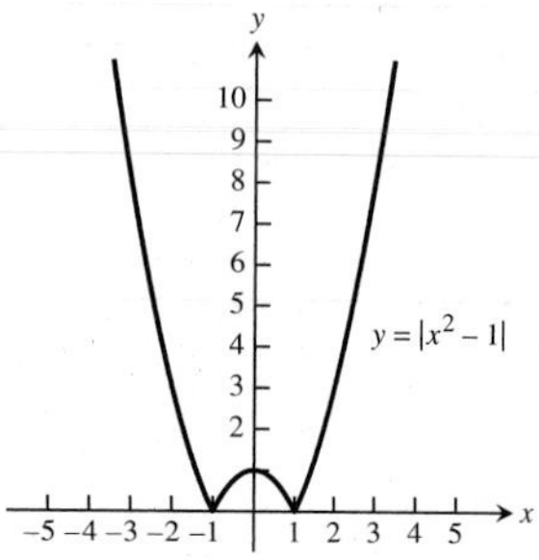

17. $[-15, 5]$ by $[-10, 10]$

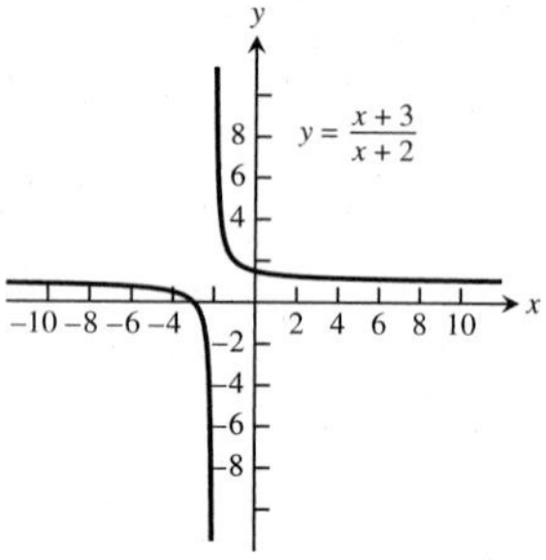

19. $[-4, 4]$ by $[0, 3]$

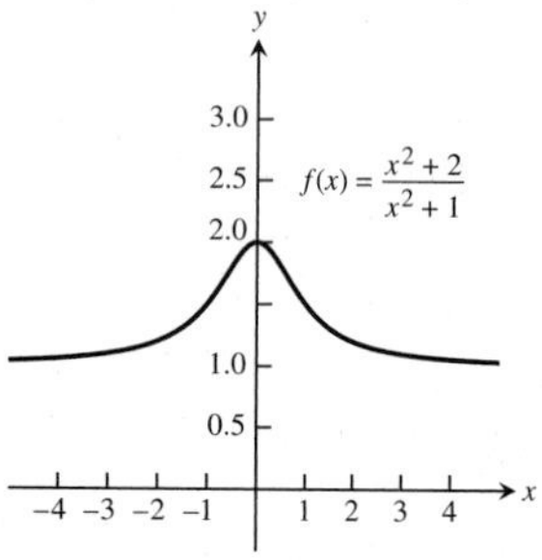

21. $[-10, 10]$ by $[-6, 6]$

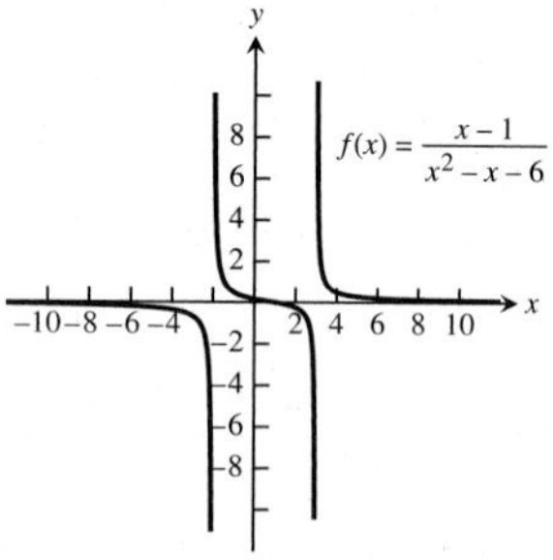

23. $[-6, 10]$ by $[-6, 6]$

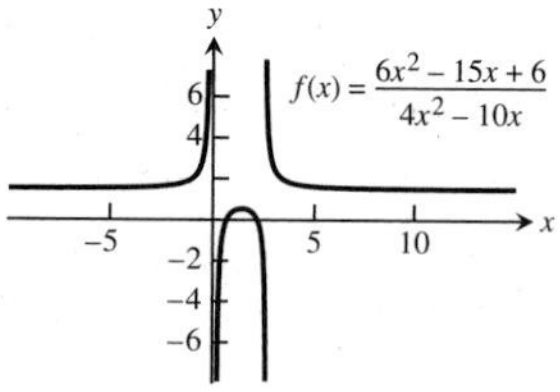

25. $\left[-\frac{\pi}{125}, \frac{\pi}{125}\right]$ by $[-1.25, 1.25]$

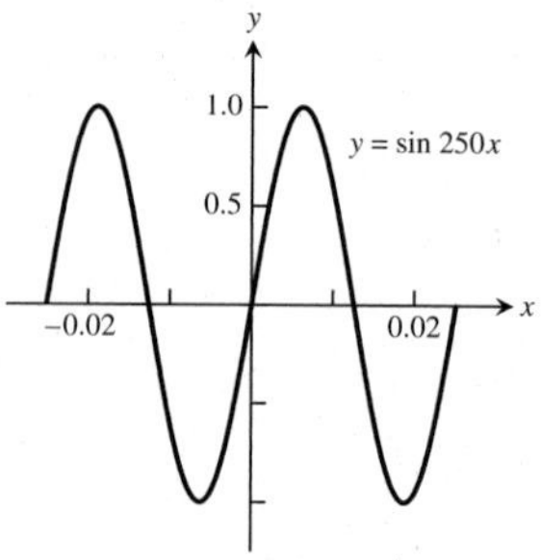

27. $[-100\pi, 100\pi]$ by $[-1.25, 1.25]$

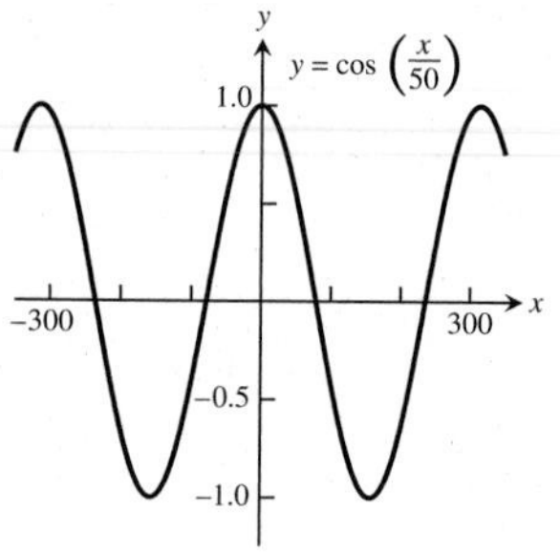

29. $\left[-\frac{\pi}{15}, \frac{\pi}{15}\right]$ by $[-0.25, 0.25]$

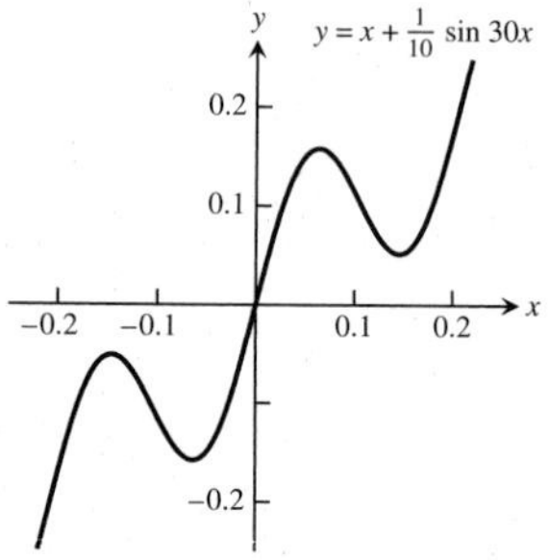

31.

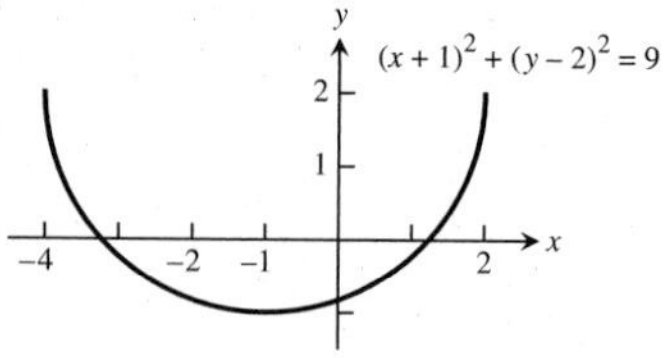

33.

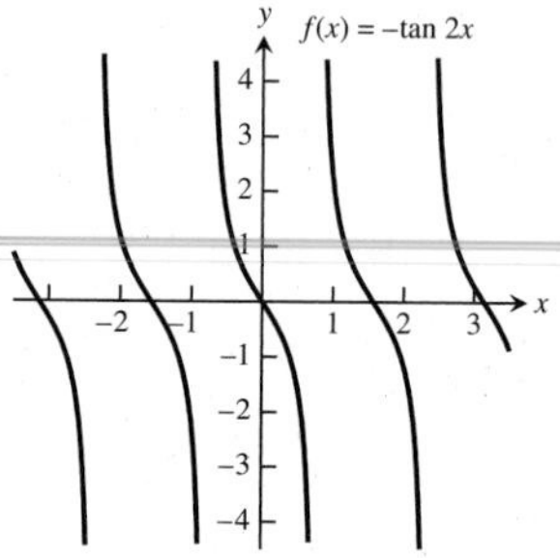

35.

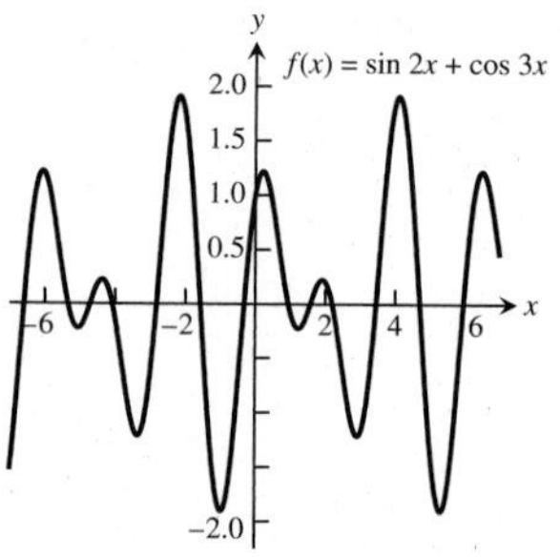

37.

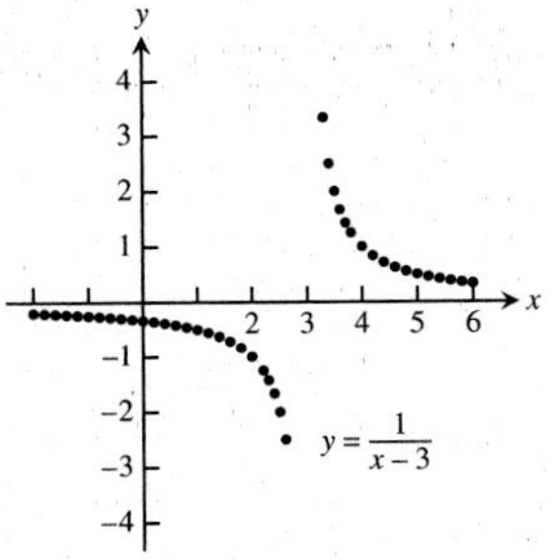

39.

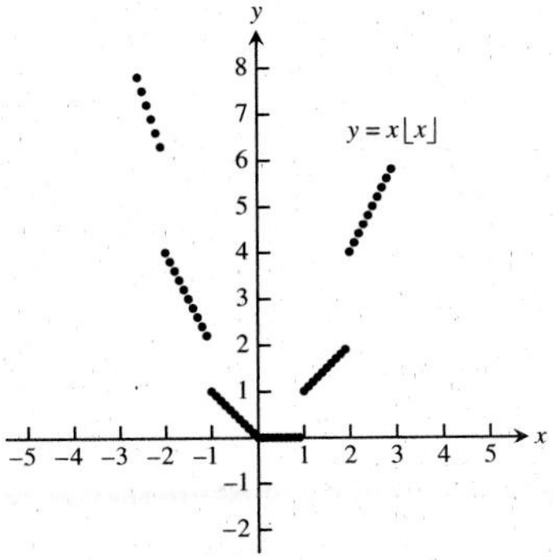

41. **(a)** $y = 1059.14x - 2074972.23$
(b) $m = 1059.14$; this is the amount the compensation will increase by each year.
(c)

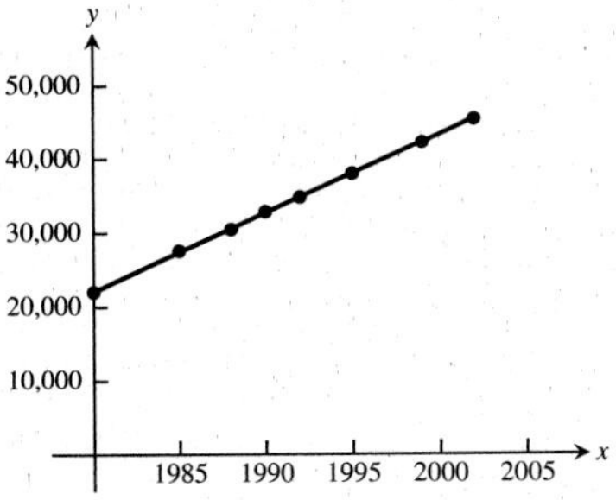

(d) \$53,899.17

43. **(a)** $y = 0.0866x^2 - 1.9701x + 50.0594$
(b)

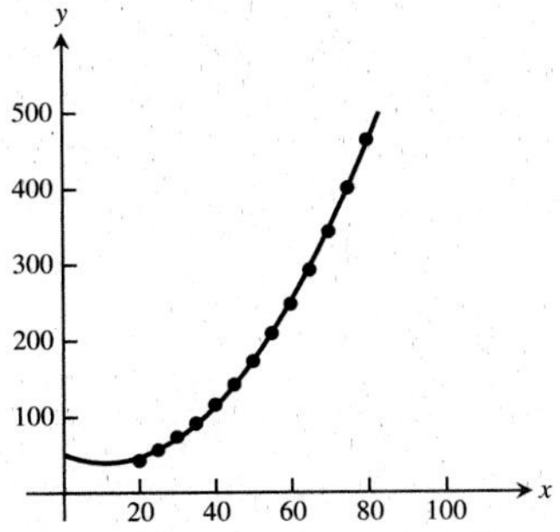

(c) If the speed is 72 mph, the approximate stopping distance is 367.50 ft. If the speed is 85 mph, the approximate stopping distance is 522.67 ft.

(d) $y = -140.4121 + 6.889x$

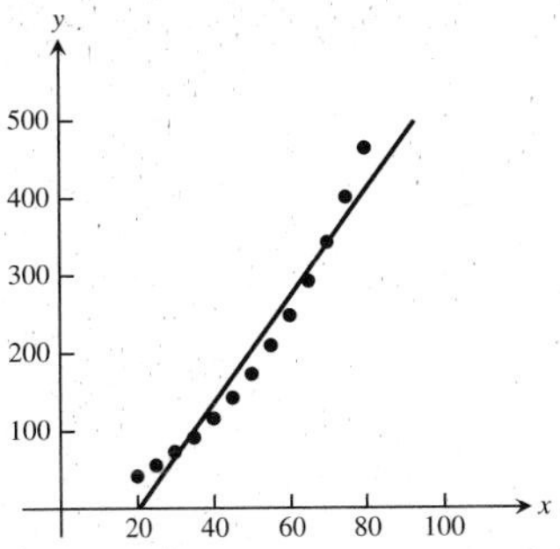

If the speed is 72 mph, the approximate stopping distance is 355.60 ft. If the speed is 85 mph, the approximate stopping distance is 445.15 ft
The quadratic regression equation is a better fit.

Practice Exercises, pp. 69–70

1. $x \geq -2$ 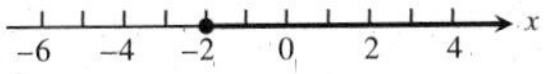

3. $x > 6$ 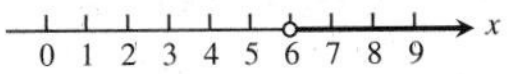

5. $-8, 6$ **7.** $x < -1$ or $x > 5$ **9.** $(0, 11)$
11. No, no two sides have the same length; no, no two sides are perpendicular
13. $y = 3x - 9$ **15.** $x = 0$ **17.** $y = 2$ **19.** $y = -3x + 3$
21. $y = -\frac{4}{3}x - \frac{20}{3}$ **23.** $y = \frac{2}{3}x + \frac{8}{3}$
25. $A = \pi r^2, C = 2\pi r, A = \frac{C^2}{4\pi}$ **27.** $x = \tan\theta, y = \tan^2\theta$
29. Origin **31.** Neither **33.** Even **35.** Even **37.** Odd
39. Neither **41.** **(a)** Domain: all reals **(b)** Range: $[-2, \infty)$
43. **(a)** Domain: $[-4, 4]$ **(b)** Range: $[0, 4]$
45. **(a)** Domain: all reals **(b)** Range: $(-3, \infty)$
47. **(a)** Domain: all reals **(b)** Range: $[-3, 1]$
49. **(a)** Domain: $(3, \infty)$ **(b)** Range: all reals
51. **(a)** Domain: $[-4, 4]$ **(b)** Range: $[0, 2]$
53. $f(x) = \begin{cases} 1 - x, & 0 \leq x < 1 \\ 2 - x, & 1 \leq x \leq 2 \end{cases}$
55. **(a)** 1 **(b)** $\frac{1}{\sqrt{2.5}} = \sqrt{\frac{2}{5}}$ **(c)** $x, x \neq 0$
(d) $\frac{1}{\sqrt{1/\sqrt{x+2}+2}}$
57. **(a)** $(f \circ g)(x) = -x, x \geq -2, (g \circ f)(x) = \sqrt{4 - x^2}$
(b) Domain $(f \circ g)$: $[-2, \infty)$, domain $(g \circ f)$: $[-2, 2]$
(c) Range $(f \circ g)$: $(-\infty, 2]$, range $(g \circ f)$: $[0, 2]$

59. Replace the portion for $x < 0$ with mirror image of the portion for $x > 0$ to make the new graph symmetric with respect to the y-axis.

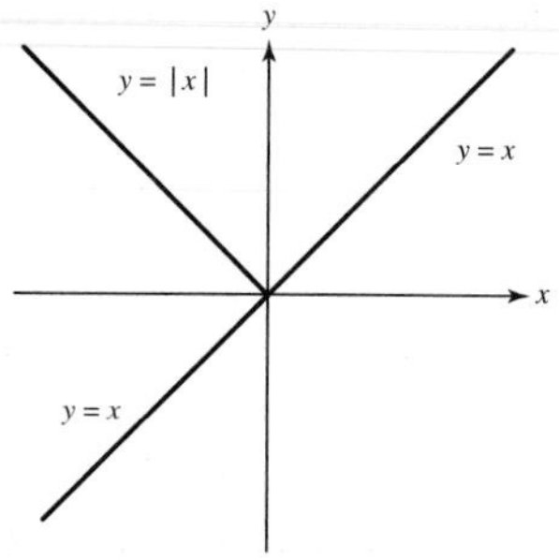

61. It does not change it.
63. Adds the mirror image of the portion for $x > 0$ to make the new graph symmetric with respect to the y-axis
65. Reflects the portion for $y < 0$ across the x-axis
67. Reflects the portion for $y < 0$ across the x-axis
69. Period π

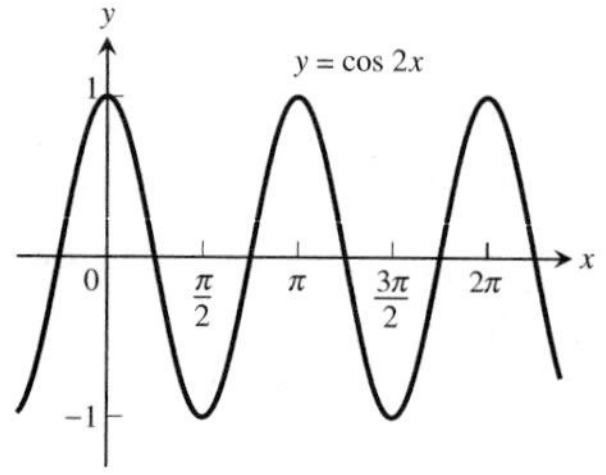

71. Period 2

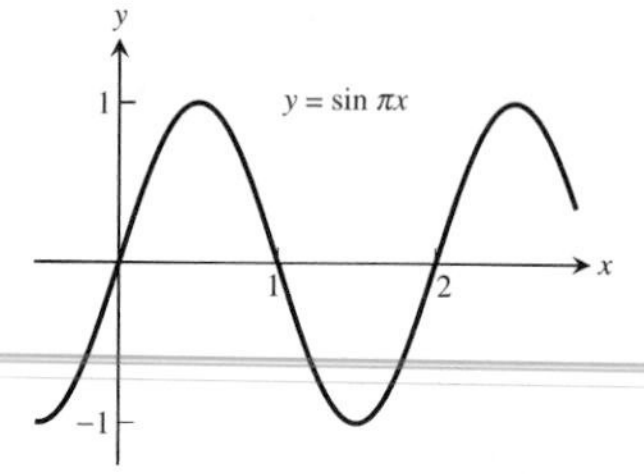

73.

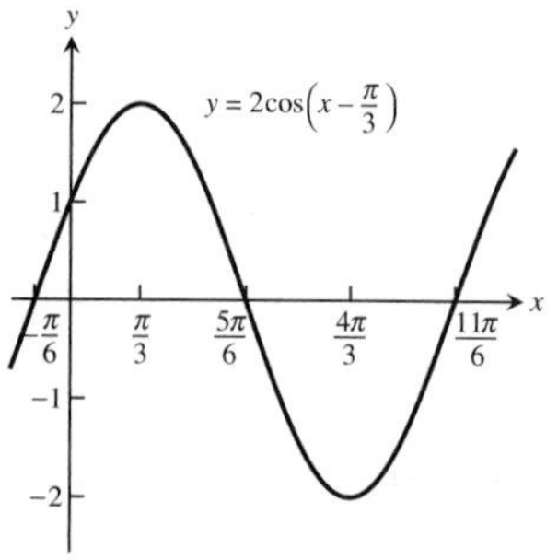

75. **(a)** $a = 1 \quad b = \sqrt{3}$ **(b)** $a = 2\sqrt{3}/3 \quad c = 4\sqrt{3}/3$

77. **(a)** $a = \dfrac{b}{\tan B}$ **(b)** $c = \dfrac{a}{\sin A}$

79. ≈ 16.98 m **81.** **(b)** 4π

Additional and Advanced Exercises, pp. 71–72

1. **(a)**

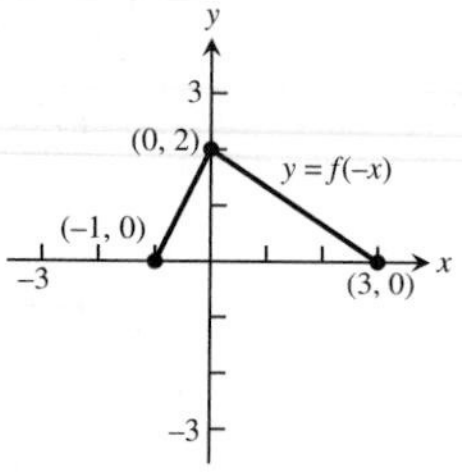

(b)

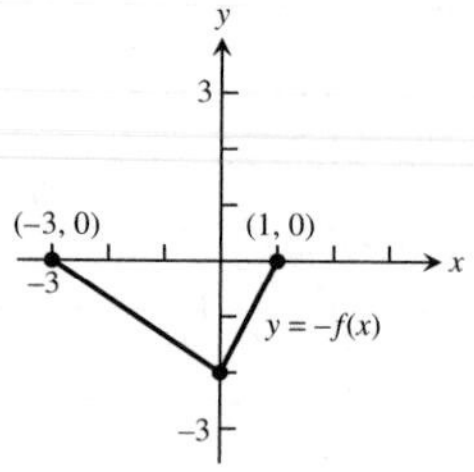

(c)

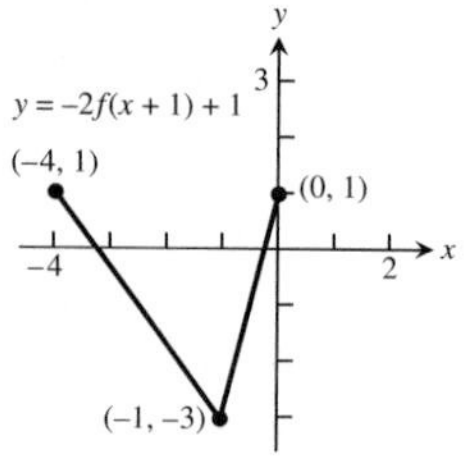

(d)

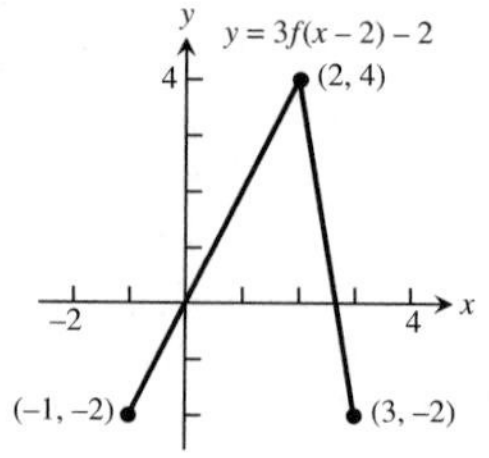

3. Yes. For instance: $f(x) = 1/x$ and $g(x) = 1/x$, or $f(x) = 2x$ and $g(x) = x/2$, or $f(x) = e^x$ and $g(x) = \ln x$.
5. If $f(x)$ is odd, then $g(x) = f(x) - 2$ is not odd. Nor is $g(x)$ even, unless $f(x) = 0$ for all x. If f is even, then $g(x) = f(x) - 2$ is also even.
7.

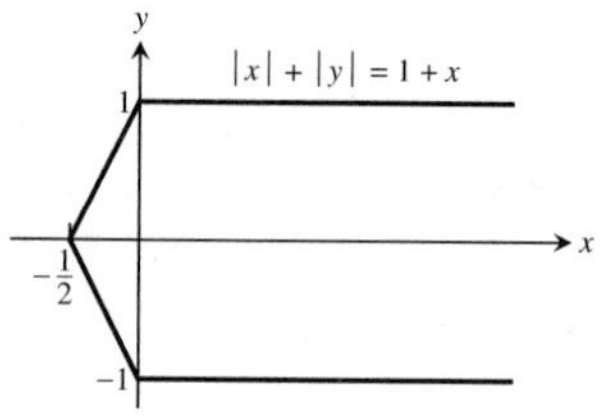

9. $\sqrt{2}$ **11.** $3/4$ **13.** $3\sqrt{15}/16$ **27.** $-4 < m < 0$

CHAPTER 2

Section 2.1, pp. 81–84

1. **(a)** Does not exist. As x approaches 1 from the right, $g(x)$ approaches 0. As x approaches 1 from the left, $g(x)$ approaches 1. There is no single number L that all the values $g(x)$ get arbitrarily close to as $x \to 1$. **(b)** 1 **(c)** 0
3. **(a)** True **(b)** True **(c)** False **(d)** False **(e)** False **(f)** True
5. As x approaches 0 from the left, $x/|x|$ approaches -1. As x approaches 0 from the right, $x/|x|$ approaches 1. There is no single number L that the function values all get arbitrarily close to as $x \to 0$.
7. Nothing can be said. **9.** No; no; no

11. **(a)** $f(x) = (x^2 - 9)/(x + 3)$

x	-3.1	-3.01	-3.001	-3.0001	-3.00001	-3.000001
$f(x)$	-6.1	-6.01	-6.001	-6.0001	-6.00001	-6.000001

x	-2.9	-2.99	-2.999	-2.9999	-2.99999	-2.999999
$f(x)$	-5.9	-5.99	-5.999	-5.9999	-5.99999	-5.999999

(c) $\lim_{x \to -3} f(x) = -6$

13. **(a)** $G(x) = (x + 6)/(x^2 + 4x - 12)$

x	-5.9	-5.99	-5.999	-5.9999	-5.99999	-5.999999
$G(x)$	$-.126582$	$-.1251564$	$-.1250156$	$-.1250015$	$-.1250001$	$-.1250000$

x	-6.1	-6.01	-6.001	-6.0001	-6.00001	-6.000001
$G(x)$	$-.123456$	$-.124843$	$-.124984$	$-.124998$	$-.124999$	$-.124999$

(c) $\lim_{x \to -6} G(x) = -1/8 = -0.125$

15. **(a)** $f(x) = (x^2 - 1)/(|x| - 1)$

x	-1.1	-1.01	-1.001	-1.0001	-1.00001	-1.000001
$f(x)$	2.1	2.01	2.001	2.0001	2.00001	2.000001

x	$-.9$	$-.99$	$-.999$	$-.9999$	$-.99999$	$-.999999$
$f(x)$	1.9	1.99	1.999	1.9999	1.99999	1.999999

(c) $\lim_{x \to -1} f(x) = 2$

17. **(a)** $g(\theta) = (\sin \theta)/\theta$

θ	$.1$	$.01$	$.001$	$.0001$	$.00001$	$.000001$
$g(\theta)$	$.998334$	$.999983$	$.999999$	$.999999$	$.999999$	$.999999$

θ	$-.1$	$-.01$	$-.001$	$-.0001$	$-.00001$	$-.000001$
$g(\theta)$	$.998334$	$.999983$	$.999999$	$.999999$	$.999999$	$.999999$

$\lim_{\theta \to 0} g(\theta) = 1$

19. **(a)** $f(x) = x^{1/(1-x)}$

x	$.9$	$.99$	$.999$	$.9999$	$.99999$	$.999999$
$f(x)$	$.348678$	$.366032$	$.367695$	$.367861$	$.367877$	$.367879$

x	1.1	1.01	1.001	1.0001	1.00001	1.000001
$f(x)$	$.385543$	$.369711$	$.368063$	$.367897$	$.367881$	$.367878$

$\lim_{x \to 1} f(x) \approx 0.36788$

21. 4 **23.** 0 **25.** 9 **27.** $\pi/2$ **29.** **(a)** 19 **(b)** 1

31. **(a)** $-\frac{4}{\pi}$ **(b)** $-\frac{3\sqrt{3}}{\pi}$ **33.** 1

35. Graphs can shift during a press run, so your estimates may not completely agree with these.

(a)

PQ_1	PQ_2	PQ_3	PQ_4
43	46	49	50

The appropriate units are m/sec.

(b) $\approx$ 50 m/sec or 180 km/h

37. **(a)**

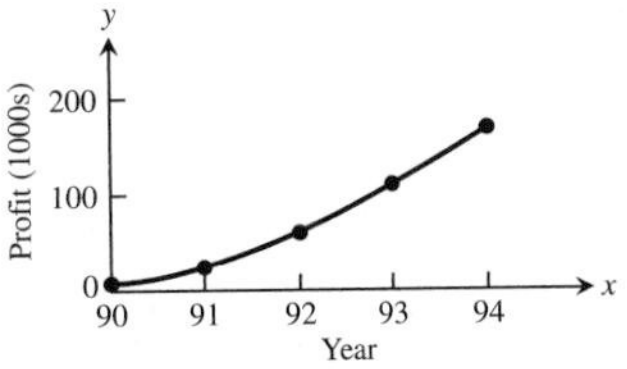

(b) $\approx$ \$56,000/year
(c) $\approx$ \$42,000/year

39. **(a)** 0.414213, 0.449489, $(\sqrt{1+h} - 1)/h$ **(b)** $g(x) = \sqrt{x}$

$1 + h$	1.1	1.01	1.001	1.0001	1.00001	1.000001
$\sqrt{1+h}$	1.04880	1.004987	1.0004998	1.0000499	1.000005	1.0000005
$(\sqrt{1+h} - 1)/h$	0.4880	0.4987	0.4998	0.499	0.5	0.5

(c) 0.5 **(d)** 0.5

Section 2.2, pp. 89–91

1. -9 **3.** 4 **5.** -8 **7.** 5/8 **9.** 5/2 **11.** 27 **13.** 16 **15.** 3/2 **17.** 3/2 **19.** 1/10 **21.** -7 **23.** 3/2 **25.** $-1/2$ **27.** 4/3 **29.** 1/6 **31.** 4 **33.** 1/2 **35.** 3/2

37. **(a)** Quotient Rule **(b)** Difference and Power Rules **(c)** Sum and Constant Multiple Rules

39. **(a)** -10 **(b)** -20 **(c)** -1 **(d)** 5/7

41. **(a)** 4 **(b)** -21 **(c)** -12 **(d)** $-7/3$

43. 2 **45.** 3 **47.** $1/(2\sqrt{7})$ **49.** $\sqrt{5}$

51. **(a)** The limit is 1. **53.** $c = 0, 1, -1$; The limit is 0 at $c = 0$, and 1 at $c = 1, -1$ **55.** 7 **57.** **(a)** 5 **(b)** 5

Section 2.3, pp. 98–101

1. $\delta = 2$

3. $\delta = 1/2$

5. $\delta = 1/18$

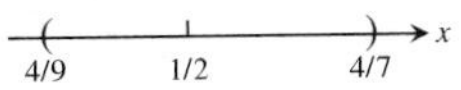

7. $\delta = 0.1$ **9.** $\delta = 7/16$ **11.** $\delta = \sqrt{5} - 2$ **13.** $\delta = 0.36$

15. $(3.99, 4.01)$, $\delta = 0.01$ **17.** $(-0.19, 0.21)$, $\delta = 0.19$

19. $(3, 15)$, $\delta = 5$ **21.** $(10/3, 5)$, $\delta = 2/3$

23. $(-\sqrt{4.5}, -\sqrt{3.5})$, $\delta = \sqrt{4.5} - 2 \approx 0.12$

25. $(\sqrt{15}, \sqrt{17})$, $\delta = \sqrt{17} - 4 \approx 0.12$

27. $\left(2 - \frac{0.03}{m}, 2 + \frac{0.03}{m}\right), \quad \delta = \frac{0.03}{m}$

29. $\left(\frac{1}{2} - \frac{c}{m}, \frac{c}{m} + \frac{1}{2}\right), \quad \delta = \frac{c}{m}$

31. $L = -3, \quad \delta = 0.01$ **33.** $L = 4, \quad \delta = 0.05$

35. $L = 4, \quad \delta = 0.75$

55. [3.384, 3.387]. To be safe, the left endpoint was rounded up and the right endpoint rounded down.

59. The limit does not exist as x approaches 3.

Section 2.4, pp. 111–114

1. **(a)** True **(b)** True **(c)** False **(d)** True **(e)** True **(f)** True **(g)** False **(h)** False **(i)** False **(j)** False **(k)** True **(l)** False

3. **(a)** 2, 1 **(b)** No, $\lim_{x\to 2^+} f(x) \neq \lim_{x\to 2^-} f(x)$ **(c)** 3, 3 **(d)** Yes, 3

5. **(a)** No **(b)** Yes, 0 **(c)** No

7. **(a)**

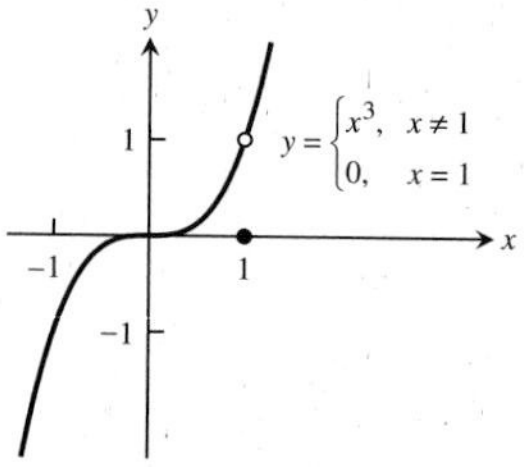

(b) 1, 1 **(c)** Yes, 1

9. **(a)** $D: 0 \le x \le 2, R: 0 < y \le 1$ and $y = 2$. **(b)** $(0, 1) \cup (1, 2)$ **(c)** $x = 2$ **(d)** $x = 0$

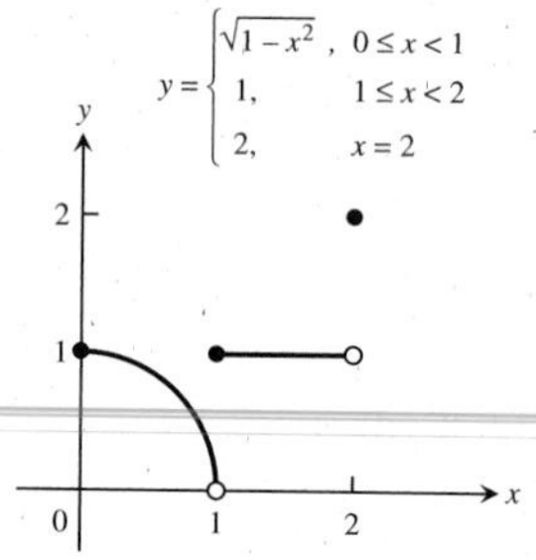

11. $\sqrt{3}$ **13.** 1 **15.** $2/\sqrt{5}$ **17.** **(a)** 1 **(b)** -1
19. **(a)** 1 **(b)** 2/3 **21.** 1 **23.** 3/4 **25.** 2 **27.** 1/2
29. 2 **31.** 1 **33.** 1/2 **35.** 3/8 **37.** **(a)** -3 **(b)** -3
39. **(a)** 1/2 **(b)** 1/2 **41.** **(a)** $-5/3$ **(b)** $-5/3$ **43.** 0
45. -1 **47.** **(a)** 2/5 **(b)** 2/5 **49.** **(a)** 0 **(b)** 0
51. **(a)** 7 **(b)** 7 **53.** **(a)** 0 **(b)** 0 **55.** **(a)** $-2/3$ **(b)** $-2/3$ **57.** 0 **59.** 1 **61.** ∞ **69.** 1
73. $\delta = \epsilon^2, \lim_{x\to 5^+} \sqrt{x - 5} = 0$
77. **(a)** 400 **(b)** 399 **(c)** The limit does not exist.
79. 1 **81.** 3/2 **83.** 3

Section 2.5, pp. 122–123

1. ∞ **3.** $-\infty$ **5.** $-\infty$ **7.** ∞ **9.** **(a)** ∞ **(b)** $-\infty$
11. ∞ **13.** ∞ **15.** $-\infty$ **17.** **(a)** ∞ **(b)** $-\infty$
(c) $-\infty$ **(d)** ∞ **19.** **(a)** $-\infty$ **(b)** ∞ **(c)** 0
(d) 3/2 **21.** **(a)** $-\infty$ **(b)** 1/4 **(c)** 1/4 **(d)** 1/4
(e) It will be $-\infty$. **23.** **(a)** $-\infty$ **(b)** ∞ **25.** **(a)** ∞
(b) ∞ **(c)** ∞ **(d)** ∞

27.

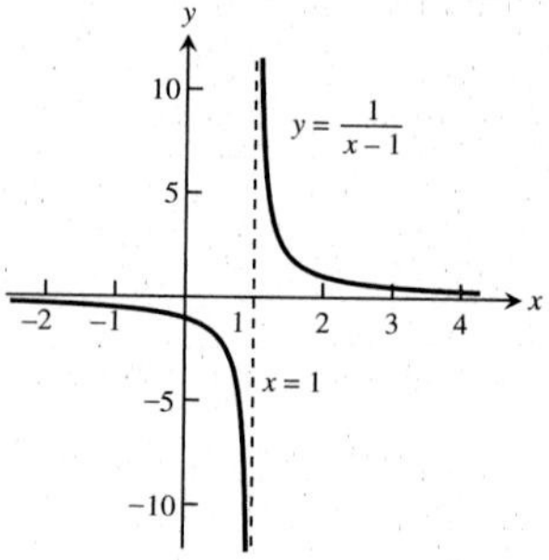

29.

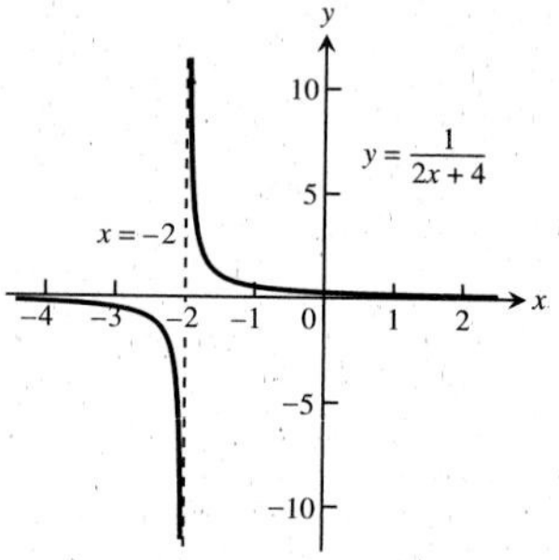

31.

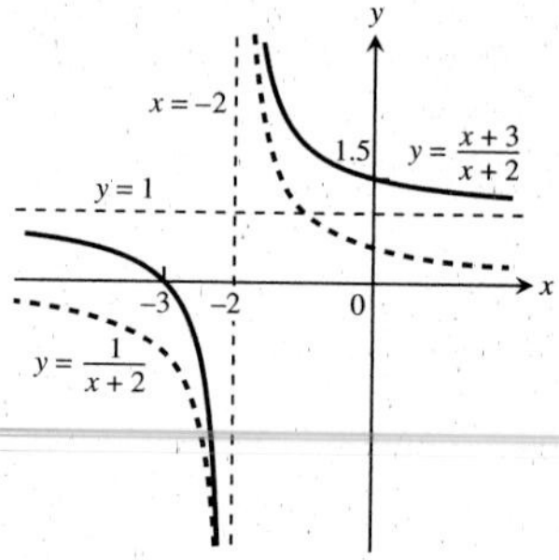

33.

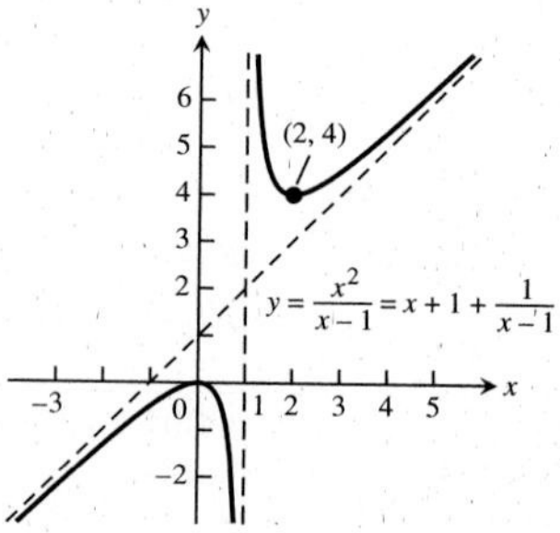

35.

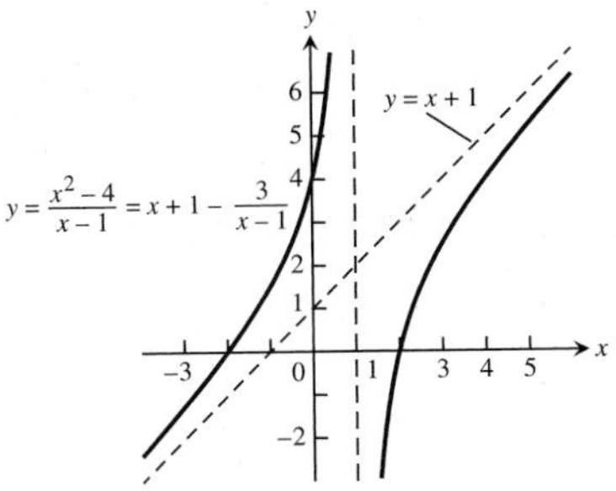

37.

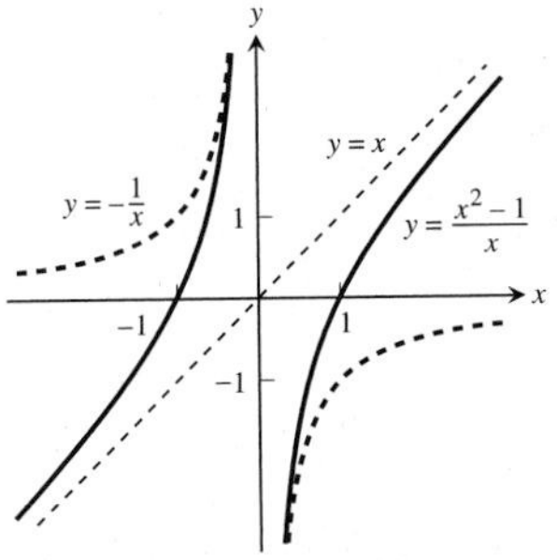

39. Here is one possibility.

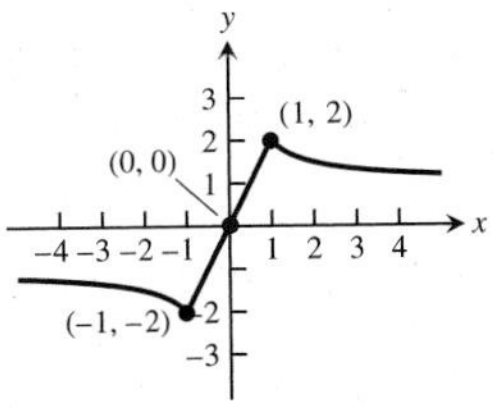

41. Here is one possibility.

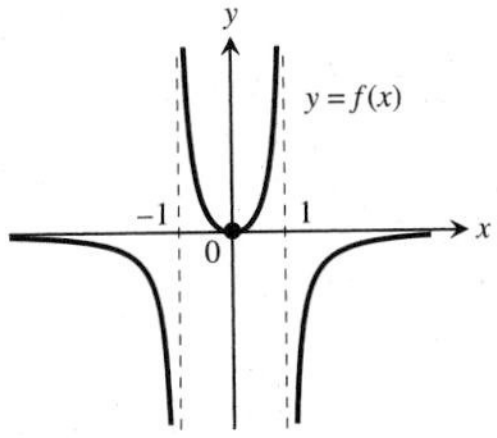

43. Here is one possibility.

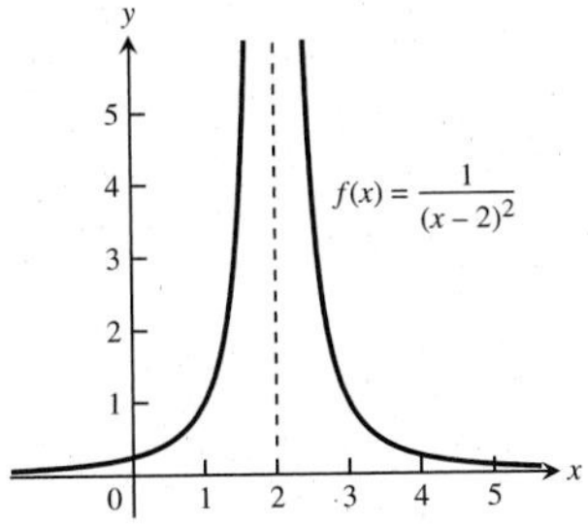

45. Here is one possibility.

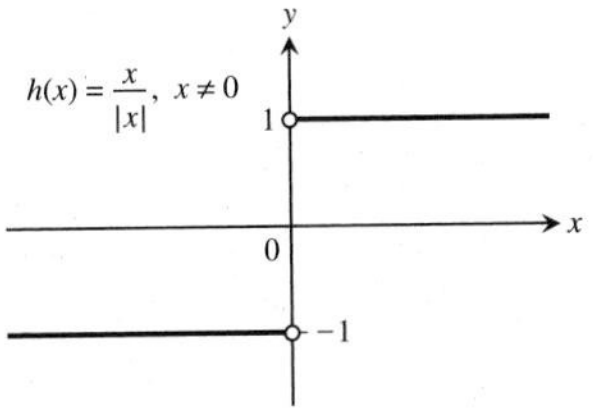

51. **(a)** For every positive real number B there exists a corresponding number $\delta > 0$ such that for all x

$$x_0 - \delta < x < x_0 \Rightarrow f(x) > B.$$

(b) For every negative real number $-B$ there exists a corresponding number $\delta > 0$ such that for all x

$$x_0 < x < x_0 + \delta \Rightarrow f(x) < -B.$$

(c) For every negative real number $-B$ there exists a corresponding number $\delta > 0$ such that for all x

$$x_0 - \delta < x < x_0 \Rightarrow f(x) < -B.$$

57.

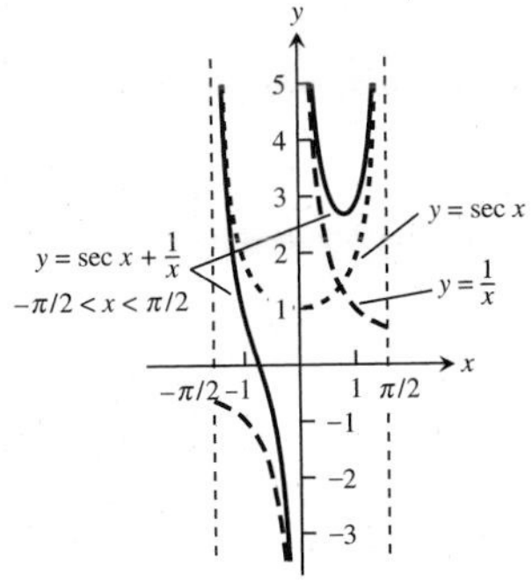

59.

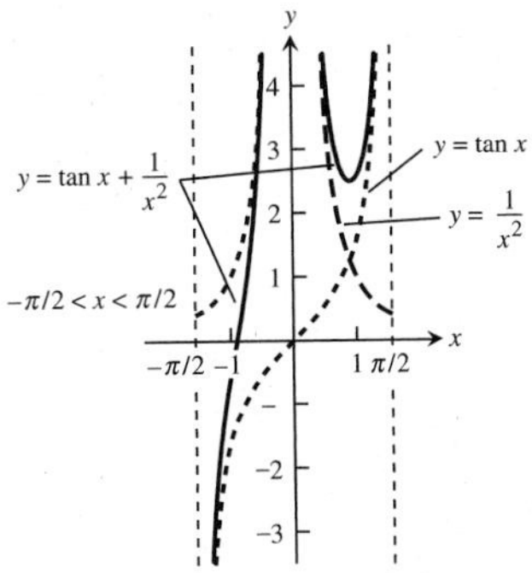

61.

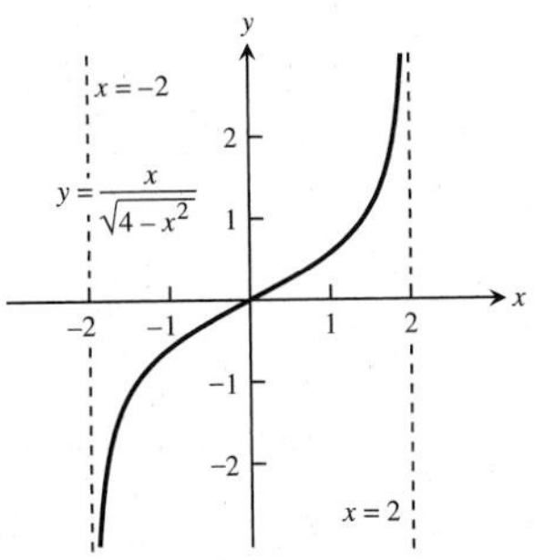

63.

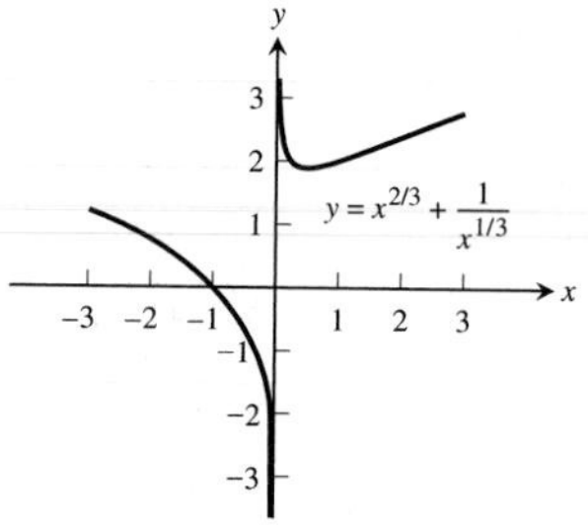

Section 2.6, pp. 132–134

1. No; discontinuous at $x = 2$; not defined at $x = 2$ **3.** Continuous **5. (a)** Yes **(b)** Yes **(c)** Yes **(d)** Yes **7. (a)** No **(b)** No **9.** 0 **11.** 1, nonremovable; 0, removable **13.** All x except $x = 2$ **15.** All x except $x = 3, x = 1$ **17.** All x **19.** All x except $x = 0$ **21.** All x except $x = n\pi/2$, n any integer **23.** All x except $n\pi/2$, n an odd integer **25.** All $x \geq -3/2$ **27.** All x **29.** 0; continuous at $x = \pi$ **31.** 1; continuous at $y = 1$ **33.** $\sqrt{2}/2$; continuous at $t = 0$ **35.** $g(3) = 6$ **37.** $f(1) = 3/2$ **39.** $a = 4/3$ **63.** $x \approx 1.8794, -1.5321, -0.3473$ **65.** $x \approx 1.7549$ **67.** $x \approx 3.5156$ **69.** $x \approx 0.7391$

Section 2.7, pp. 140–141

1. P_1: $m_1 = 1$, P_2: $m_2 = 5$ **3.** P_1: $m_1 = 5/2$, P_2: $m_2 = -1/2$ **5.** $y = 2x + 5$

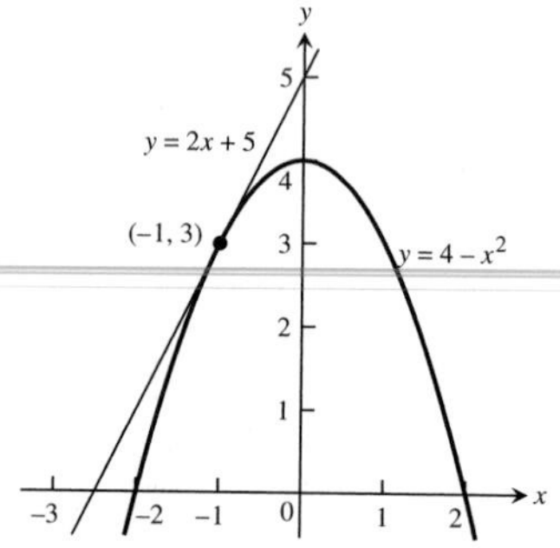

7. $y = x + 1$

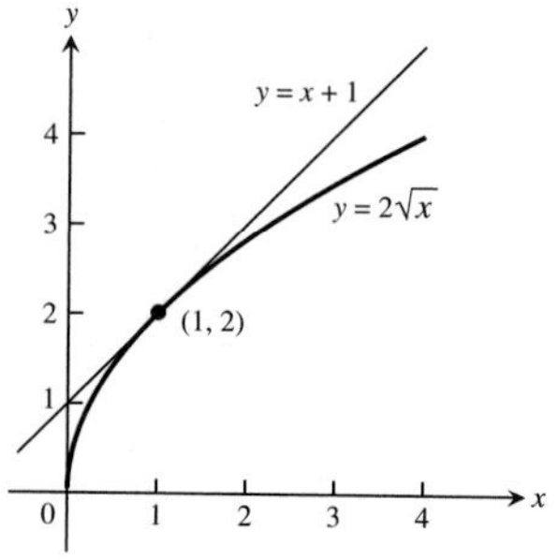

9. $y = 12x + 16$

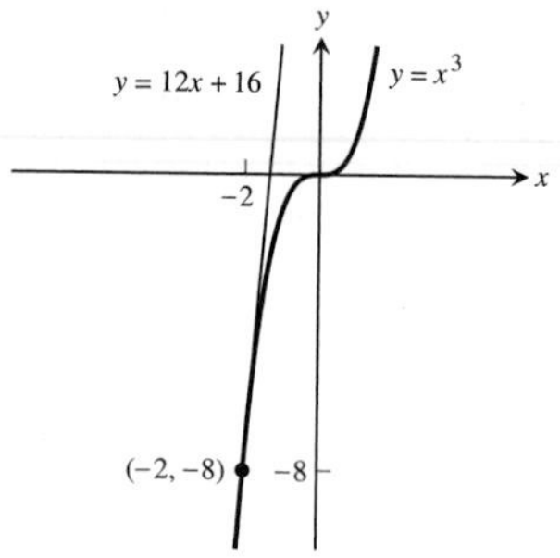

11. $m = 4, y - 5 = 4(x - 2)$ **13.** $m = -2, y - 3 = -2(x - 3)$ **15.** $m = 12, y - 8 = 12(t - 2)$ **17.** $m = \frac{1}{4}, y - 2 = \frac{1}{4}(x - 4)$ **19.** $m = -10$ **21.** $m = -1/4$ **23.** $(-2, -5)$ **25.** $y = -(x + 1), y = -(x - 3)$ **27.** 19.6 m/sec **29.** 6π **31.** Yes **33.** Yes **35. (a)** Nowhere **37. (a)** At $x = 0$ **39. (a)** Nowhere **41. (a)** At $x = 1$ **43. (a)** At $x = 0$

Practice Exercises, pp. 142–143

1. At $x = -1$: $\lim_{x\to -1^-} f(x) = \lim_{x\to -1^+} f(x) = 1$, so $\lim_{x\to -1} f(x) = 1 = f(-1)$; continuous at $x = -1$

At $x = 0$: $\lim_{x\to 0^-} f(x) = \lim_{x\to 0^+} f(x) = 0$, so $\lim_{x\to 0} f(x) = 0$. However, $f(0) \neq 0$, so f is discontinuous at $x = 0$. The discontinuity can be removed by redefining $f(0)$ to be 0.

At $x = 1$: $\lim_{x\to 1^-} f(x) = -1$ and $\lim_{x\to 1^+} f(x) = 1$, so $\lim_{x\to 1} f(x)$ does not exist. The function is discontinuous at $x = 1$, and the discontinuity is not removable.

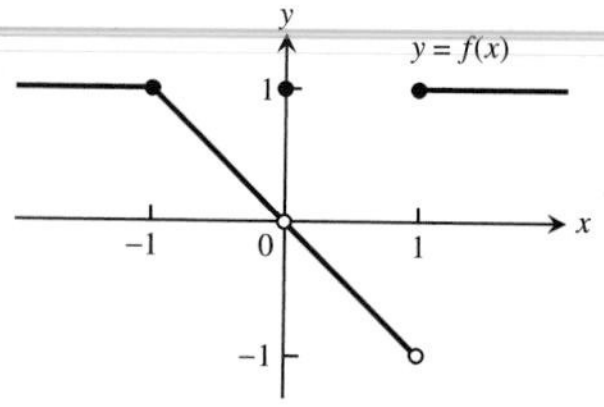

3. (a) -21 **(b)** 49 **(c)** 0 **(d)** 1 **(e)** 1 **(f)** 7 **(g)** -7 **(h)** $-\frac{1}{7}$ **5.** 4 **7. (a)** $(-\infty, +\infty)$ **(b)** $[0, \infty)$ **(c)** $(-\infty, 0)$ and $(0, \infty)$ **(d)** $(0, \infty)$ **9. (a)** Does not exist **(b)** 0 **11.** $\frac{1}{2}$ **13.** $2x$ **15.** $-\frac{1}{4}$ **17.** 2 **19.** 0 **21.** $\frac{2}{5}$ **23.** 0 **25.** $-\infty$ **27.** 0 **29.** 1 **31.** No in both cases, because $\lim_{x\to 1} f(x)$ does not exist, and $\lim_{x\to -1} f(x)$ does not exist. **37. (b)** 1.324717957

Additional and Advanced Exercises, pp. 144–146

3. 0; the left-hand limit was needed because the function is undefined for $v > c$. **5.** $65 < t < 75$; within 5°F

13. **(a)** B **(b)** A **(c)** A **(d)** A

21. **(a)** $\lim_{a\to 0} r_+(a) = 0.5$, $\lim_{a\to -1^+} r_+(a) = 1$

(b) $\lim_{a\to 0} r_-(a)$ does not exist, $\lim_{a\to -1^+} r_-(a) = 1$

25. 0 **27.** 1 **29.** 4

CHAPTER 3

Section 3.1, pp. 155–159

1. $-2x, 6, 0, -2$ **3.** $-\dfrac{2}{t^3}, 2, -\dfrac{1}{4}, -\dfrac{2}{3\sqrt{3}}$

5. $\dfrac{3}{2\sqrt{3\theta}}, \dfrac{3}{2\sqrt{3}}, \dfrac{1}{2}, \dfrac{3}{2\sqrt{2}}$ **7.** $6x^2$ **9.** $\dfrac{1}{(2t+1)^2}$

11. $\dfrac{-1}{2(q+1)\sqrt{q+1}}$ **13.** $1 - \dfrac{9}{x^2}, 0$ **15.** $3t^2 - 2t, 5$

17. $\dfrac{-4}{(x-2)\sqrt{x-2}}$, $y - 4 = -\dfrac{1}{2}(x - 6)$ **19.** 6

21. 1/8 **23.** $\dfrac{-1}{(x+2)^2}$ **25.** $\dfrac{-1}{(x-1)^2}$ **27.** b **29.** d

31. **(a)** $x = 0, 1, 4$ **(b)**

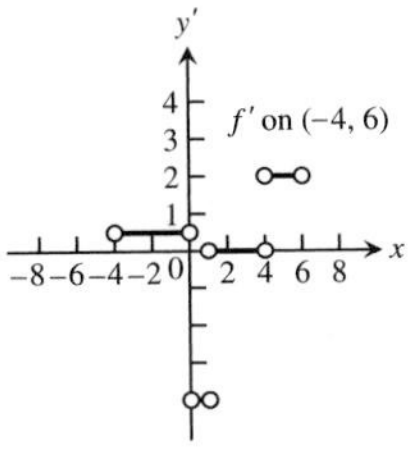

33.

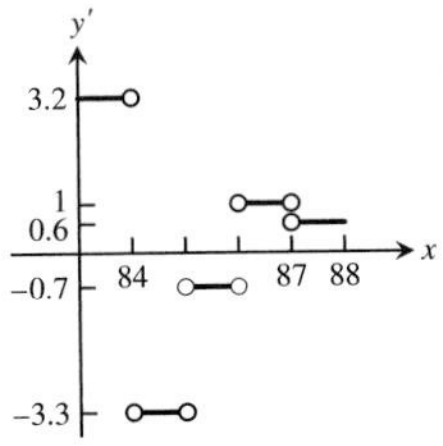

35. Since $\lim_{x\to 0^+} f'(x) = 1$ while $\lim_{x\to 0^-} f'(x) = 0$, $f(x)$ is not differentiable at $x = 0$.

37. Since $\lim_{x\to 1^+} f'(x) = 2$ while $\lim_{x\to 1^-} f'(x) = 1/2$, $f(x)$ is not differentiable at $x = 1$.

39. **(a)** $-3 \le x \le 2$ **(b)** None **(c)** None

41. **(a)** $-3 \le x < 0, 0 < x \le 3$ **(b)** None **(c)** $x = 0$

43. **(a)** $-1 \le x < 0, 0 < x \le 2$ **(b)** $x = 0$ **(c)** None

45. **(a)** $y' = -2x$ **(c)** $x < 0, x = 0, x > 0$

(d) $-\infty < x < 0, 0 < x < \infty$

47. **(a)** $y' = x^2$

(b)

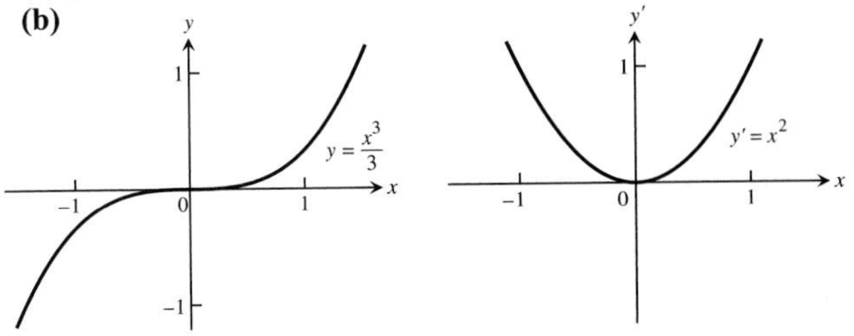

(c) $x \ne 0, x = 0$, none **(d)** $-\infty < x < \infty$, none

49. $y' = 3x^2$ is never negative

51. Yes, $y + 16 = -(x - 3)$ is tangent at $(3, -16)$

53. No, the function $y = \lfloor x \rfloor$ does not satisfy the intermediate value property of derivatives.

55. Yes, $(-f)'(x) = -(f'(x))$

57. For $g(t) = mt$ and $h(t) = t$, $\lim_{t\to 0} \dfrac{g(t)}{h(t)} = m$, which need not be zero.

Section 3.2, pp. 169–171

1. $\dfrac{dy}{dx} = -2x, \dfrac{d^2y}{dx^2} = -2$

3. $\dfrac{ds}{dt} = 15t^2 - 15t^4, \dfrac{d^2s}{dt^2} = 30t - 60t^3$

5. $\dfrac{dy}{dx} = 4x^2 - 1, \dfrac{d^2y}{dx^2} = 8x$

7. $\dfrac{dw}{dz} = -\dfrac{6}{z^3} + \dfrac{1}{z^2}, \dfrac{d^2w}{dz^2} = \dfrac{18}{z^4} - \dfrac{2}{z^3}$

9. $\dfrac{dy}{dx} = 12x - 10 + 10x^{-3}, \dfrac{d^2y}{dx^2} = 12 - 30x^{-4}$

11. $\dfrac{dr}{ds} = \dfrac{-2}{3s^3} + \dfrac{5}{2s^2}, \dfrac{d^2r}{ds^2} = \dfrac{2}{s^4} - \dfrac{5}{s^3}$

13. $y' = -5x^4 + 12x^2 - 2x - 3$

15. $y' = 3x^2 + 10x + 2 - \dfrac{1}{x^2}$ **17.** $y' = \dfrac{-19}{(3x-2)^2}$

19. $g'(x) = \dfrac{x^2 + x + 4}{(x + 0.5)^2}$ **21.** $\dfrac{dv}{dt} = \dfrac{t^2 - 2t - 1}{(1 + t^2)^2}$

23. $f'(s) = \dfrac{1}{\sqrt{s}(\sqrt{s} + 1)^2}$ **25.** $v' = -\dfrac{1}{x^2} + 2x^{-3/2}$

27. $y' = \dfrac{-4x^3 - 3x^2 + 1}{(x^2 - 1)^2(x^2 + x + 1)^2}$

29. $y' = 2x^3 - 3x - 1, y'' = 6x^2 - 3, y''' = 12x, y^{(4)} = 12, y^{(n)} = 0$ for $n \ge 5$

31. $y' = 2x - 7x^{-2}, y'' = 2 + 14x^{-3}$

33. $\dfrac{dr}{d\theta} = 3\theta^{-4}, \dfrac{d^2r}{d\theta^2} = -12\theta^{-5}$

35. $\dfrac{dw}{dz} = -z^{-2} - 1, \dfrac{d^2w}{dz^2} = 2z^{-3}$

37. $\dfrac{dp}{dq} = \dfrac{1}{6}q + \dfrac{1}{6}q^{-3} + q^{-5}, \dfrac{d^2p}{dq^2} = \dfrac{1}{6} - \dfrac{1}{2}q^{-4} - 5q^{-6}$

39. **(a)** 13 **(b)** −7 **(c)** 7/25 **(d)** 20

41. **(a)** $y = -\dfrac{x}{8} + \dfrac{5}{4}$ **(b)** $m = -4$ at (0, 1)

(c) $y = 8x - 15, y = 8x + 17$

43. $y = 4x, y = 2$ **45.** $a = 1, b = 1, c = 0$

47. **(a)** $y = 2x + 2$, **(c)** (2, 6)

49. $P'(x) = na_nx^{n-1} + (n-1)a_{n-1}x^{n-2} + \cdots + 2a_2x + a_1$

51. The Product Rule is then the Constant Multiple Rule, so the latter is a special case of the Product Rule.

53. **(a)** $\dfrac{d}{dx}(uvw) = uvw' + uv'w + u'vw$

(b) $\dfrac{d}{dx}(u_1u_2u_3u_4) = u_1u_2u_3u_4' + u_1u_2u_3'u_4 + u_1u_2'u_3u_4 + u_1'u_2u_3u_4$

(c) $\dfrac{d}{dx}(u_1\cdots u_n) = u_1u_2\cdots u_{n-1}u_n' + u_1u_2\cdots u_{n-2}u_{n-1}'u_n + \cdots + u_1'u_2\cdots u_n$

55. $\dfrac{dP}{dV} = -\dfrac{nRT}{(V-nb)^2} + \dfrac{2an^2}{V^3}$

Section 3.3, pp. 179–183

1. **(a)** −2 m, −1 m/sec **(b)** 3 m/sec, 1 m/sec; 2 m/sec^2, 2 m/sec^2 **(c)** changes direction at $t = 3/2$ sec

3. **(a)** −9 m, −3 m/sec **(b)** 3 m/sec, 12 m/sec; 6 m/sec^2, −12 m/sec^2 **(c)** no change in direction

5. **(a)** −20 m, −5 m/sec **(b)** 45 m/sec, (1/5) m/sec; 140 m/sec^2, (4/25) m/sec^2 **(c)** no change in direction

7. **(a)** $a(1) = -6$ m/sec^2, $a(3) = 6$ m/sec^2 **(b)** $v(2) = 3$ m/sec **(c)** 6 m

9. Mars: ≈ 7.5 sec, Jupiter: ≈ 1.2 sec **11.** $g_s = 0.75$ m/sec^2

13. **(a)** $v = -32t, |v| = 32t$ ft/sec, $a = -32$ ft/sec^2 **(b)** $t \approx 3.3$ sec **(c)** $v \approx -107.0$ ft/sec

15. **(a)** $t = 2, t = 7$ **(b)** $3 \le t \le 6$

(c)

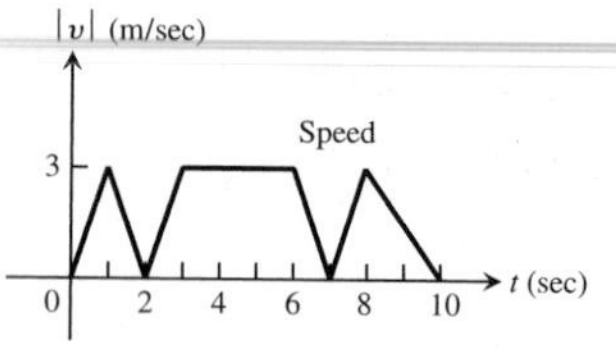

(d)

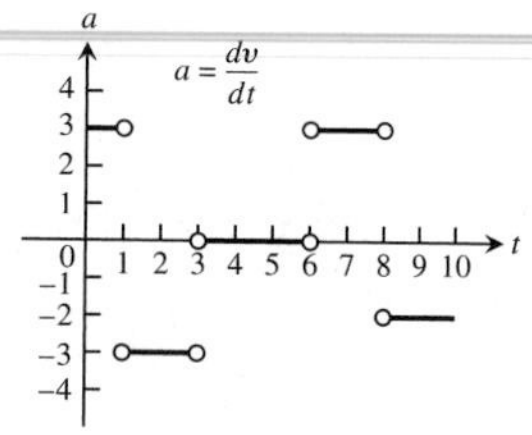

17. **(a)** 190 ft/sec **(b)** 2 sec **(c)** 8 sec, 0 ft/sec, **(d)** 10.8 sec, 90 ft/sec **(e)** 2.8 sec **(f)** greatest acceleration happens 2 sec after launch **(g)** constant acceleration between 2 and 10.8 sec, −32 ft/sec^2

19. **(a)** $\dfrac{4}{7}$ sec, 280 cm/sec **(b)** 560 cm/sec, 980 cm/sec^2 **(c)** 29.75 flashes/sec

21. C = position, A = velocity, B = acceleration

23. **(a)** \$110/machine **(b)** \$80 **(c)** \$79.90

25. **(a)** $b'(0) = 10^4$ bacteria/h **(b)** $b'(5) = 0$ bacteria/h **(c)** $b'(10) = -10^4$ bacteria/h

27. **(a)** $\dfrac{dy}{dt} = \dfrac{t}{12} - 1$ **(b)** The largest value of $\dfrac{dy}{dt}$ is 0 m/h when $t = 12$ and the smallest value of $\dfrac{dy}{dt}$ is −1 m/h, when $t = 0$.

(c)

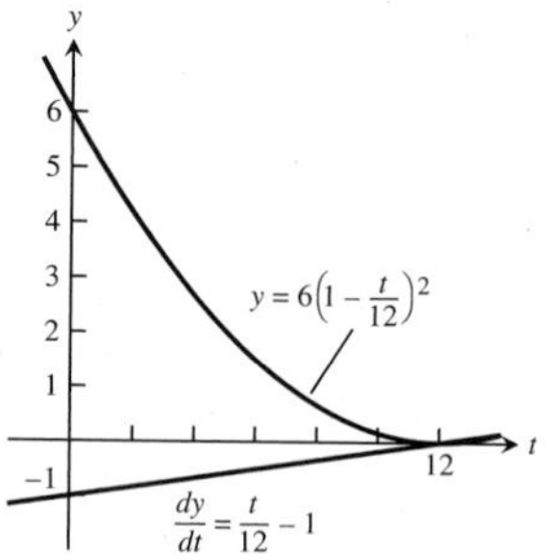

29. $t = 25$ sec $D = \dfrac{6250}{9}$ m

31.

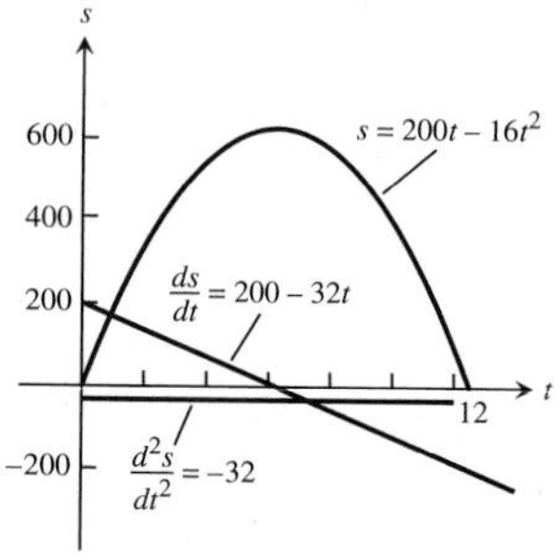

(a) $v = 0$ when $t = 6.25$ sec. **(b)** $v > 0$ when $0 \le t < 6.25 \Rightarrow$ body moves up; $v < 0$ when $6.25 < t \le 12.5 \Rightarrow$ body moves down. **(c)** Body changes direction at $t = 6.25$ sec. **(d)** Body speeds up on (6.25, 12.5] and slows down on [0, 6.25). **(e)** The body is moving fastest at the endpoints $t = 0$ and $t = 12.5$ when it is traveling 200 ft/sec. It's moving slowest at $t = 6.25$ when the speed is 0. **(f)** When $t = 6.25$ the body is $s = 625$ m from the origin and farthest away.

33.

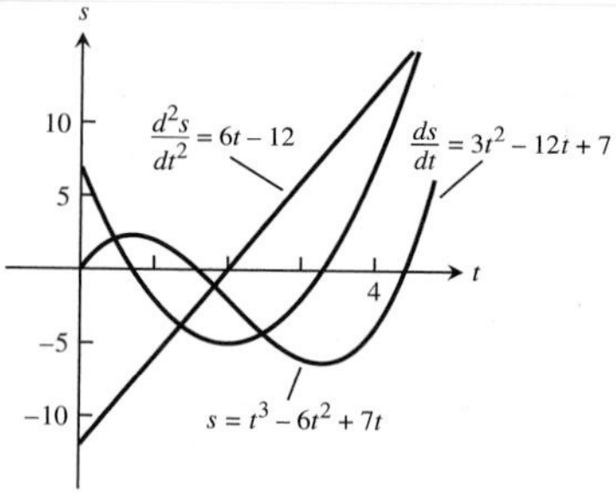

(a) $v = 0$ when $t = \dfrac{6 \pm \sqrt{15}}{3}$ sec.

(b) $v < 0$ when $\dfrac{6 - \sqrt{15}}{3} < t < \dfrac{6 + \sqrt{15}}{3} \Rightarrow$ body moves left; $v > 0$ when $0 \le t < \dfrac{6 - \sqrt{15}}{3}$ or $\dfrac{6 + \sqrt{15}}{3} < t \le 4 \Rightarrow$ body moves right.

(c) Body changes direction at $t = \frac{6 \pm \sqrt{15}}{3}$ sec.
(d) Body speeds up on $\left(\frac{6 - \sqrt{15}}{3}, 2\right) \cup \left(\frac{6 + \sqrt{15}}{3}, 4\right]$ and slows down on $\left[0, \frac{6 - \sqrt{15}}{3}\right) \cup \left(2, \frac{6 + \sqrt{15}}{3}\right)$.
(e) The body is moving fastest at $t = 0$ and $t = 4$ when it is moving 7 units/sec and slowest at $t = \frac{6 \pm \sqrt{15}}{3}$ sec.
(f) When $t = \frac{6 + \sqrt{15}}{3}$ the body is at position $s \approx -6.303$ units and farthest from the origin.

35. (a) It takes 135 sec. **(b)** Average speed $= \frac{\Delta F}{\Delta t} = \frac{5 - 0}{73 - 0} = \frac{5}{73} \approx 0.068$ furlongs/sec. **(c)** Using a centered difference quotient, the horse's speed is approximately (Section 3.4, Exercise 53) $\frac{\Delta F}{\Delta t} = \frac{4 - 2}{59 - 33} = \frac{2}{26} = \frac{1}{13} \approx 0.077$ furlongs/sec.
(d) The horse is running the fastest during the last furlong (between 9th and 10th furlong markers). This furlong takes only 11 sec to run, which is the least amount of time for a furlong.
(e) The horse accelerates the fastest during the first furlong (between markers 0 and 1).

Section 3.4, pp. 188–190

1. $-10 - 3 \sin x$ **3.** $-\csc x \cot x - \frac{2}{\sqrt{x}}$ **5.** 0
7. $\frac{-\csc^2 x}{(1 + \cot x)^2}$ **9.** $4 \tan x \sec x - \csc^2 x$ **11.** $x^2 \cos x$
13. $\sec^2 t - 1$ **15.** $\frac{-2 \csc t \cot t}{(1 - \csc t)^2}$ **17.** $-\theta (\theta \cos \theta + 2 \sin \theta)$
19. $\sec \theta \csc \theta (\tan \theta - \cot \theta) = \sec^2 \theta - \csc^2 \theta$
21. $\sec^2 q$ **23.** $\sec^2 q$ **25. (a)** $2 \csc^3 x - \csc x$ **(b)** $2 \sec^3 x - \sec x$
27.

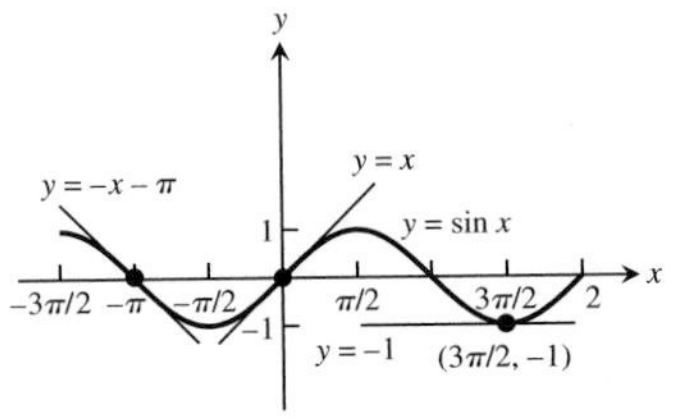

29.

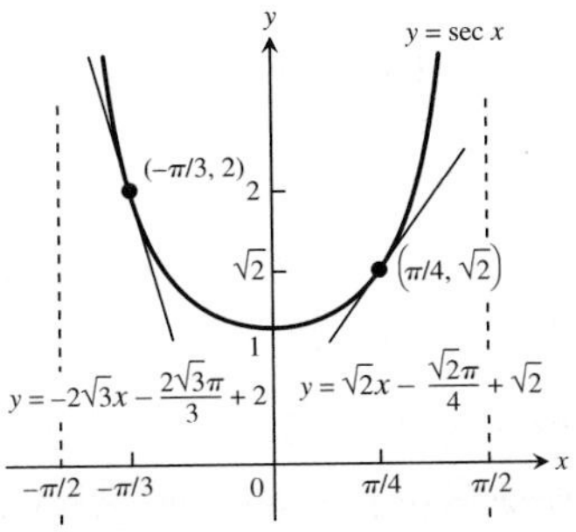

31. Yes, at $x = \pi$ **33.** No
35. $\left(-\frac{\pi}{4}, -1\right); \left(\frac{\pi}{4}, 1\right)$

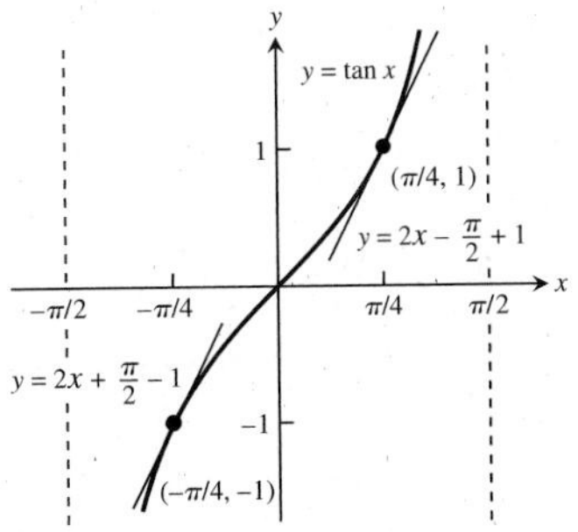

37. (a) $y = -x + \pi/2 + 2$ **(b)** $y = 4 - \sqrt{3}$
39. 0 **41.** -1 **43.** 0
45. $-\sqrt{2}$ m/sec, $\sqrt{2}$ m/sec, $\sqrt{2}$ m/sec^2, $\sqrt{2}$ m/sec^3
47. $c = 9$ **49.** $\sin x$

Section 3.5, pp. 201–205

1. $12x^3$ **3.** $3 \cos (3x + 1)$ **5.** $-\sin (\sin x) \cos x$
7. $10 \sec^2(10x - 5)$
9. With $u = (2x + 1), y = u^5: \frac{dy}{dx} = \frac{dy}{du}\frac{du}{dx} = 5u^4 \cdot 2 = 10(2x + 1)^4$
11. With $u = (1 - (x/7)), y = u^{-7}: \frac{dy}{dx} = \frac{dy}{du}\frac{du}{dx} = -7u^{-8} \cdot \left(-\frac{1}{7}\right) = \left(1 - \frac{x}{7}\right)^{-8}$
13. With $u = ((x^2/8) + x - (1/x)), y = u^4: \frac{dy}{dx} = \frac{dy}{du}\frac{du}{dx} = 4u^3 \cdot \left(\frac{x}{4} + 1 + \frac{1}{x^2}\right) = 4\left(\frac{x^2}{8} + x - \frac{1}{x}\right)^3 \left(\frac{x}{4} + 1 + \frac{1}{x^2}\right)$
15. With $u = \tan x, y = \sec u: \frac{dy}{dx} = \frac{dy}{du}\frac{du}{dx} = (\sec u \tan u)(\sec^2 x) = \sec (\tan x) \tan (\tan x) \sec^2 x$
17. With $u = \sin x, y = u^3: \frac{dy}{dx} = \frac{dy}{du}\frac{du}{dx} = 3u^2 \cos x = 3 \sin^2 x (\cos x)$
19. $-\frac{1}{2\sqrt{3 - t}}$ **21.** $\frac{4}{\pi}(\cos 3t - \sin 5t)$ **23.** $\frac{\csc \theta}{\cot \theta + \csc \theta}$
25. $2x \sin^4 x + 4x^2 \sin^3 x \cos x + \cos^{-2} x + 2x \cos^{-3} x \sin x$
27. $(3x - 2)^6 - \frac{1}{x^3\left(4 - \frac{1}{2x^2}\right)^2}$ **29.** $\frac{(4x + 3)^3(4x + 7)}{(x + 1)^4}$
31. $\sqrt{x} \sec^2 (2\sqrt{x}) + \tan (2\sqrt{x})$ **33.** $\frac{2 \sin \theta}{(1 + \cos \theta)^2}$
35. $\frac{dr}{d\theta} = -2 \sin (\theta^2) \sin 2\theta + 2\theta \cos (2\theta) \cos (\theta^2)$

37. $\dfrac{dq}{dt} = \left(\dfrac{t+2}{2(t+1)^{3/2}}\right)\cos\left(\dfrac{t}{\sqrt{t+1}}\right)$

39. $2\pi \sin(\pi t - 2)\cos(\pi t - 2)$ **41.** $\dfrac{8\sin(2t)}{(1+\cos 2t)^5}$

43. $-2\cos(\cos(2t-5))(\sin(2t-5))$

45. $\left(1+\tan^4\left(\dfrac{t}{12}\right)\right)^2\left(\tan^3\left(\dfrac{t}{12}\right)\sec^2\left(\dfrac{t}{12}\right)\right)$

47. $-\dfrac{t\sin(t^2)}{\sqrt{1+\cos(t^2)}}$ **49.** $\dfrac{6}{x^3}\left(1+\dfrac{1}{x}\right)\left(1+\dfrac{2}{x}\right)$

51. $2\csc^2(3x-1)\cot(3x-1)$ **53.** $5/2$ **55.** $-\pi/4$ **57.** 0

59. (a) $2/3$ **(b)** $2\pi+5$ **(c)** $15-8\pi$ **(d)** $37/6$ **(e)** -1 **(f)** $\sqrt{2}/24$ **(g)** $5/32$ **(h)** $-5/(3\sqrt{17})$ **61.** 5

63. (a) 1 **(b)** 1 **65. (a)** $y = \pi x + 2 - \pi$ **(b)** $\pi/2$

67.

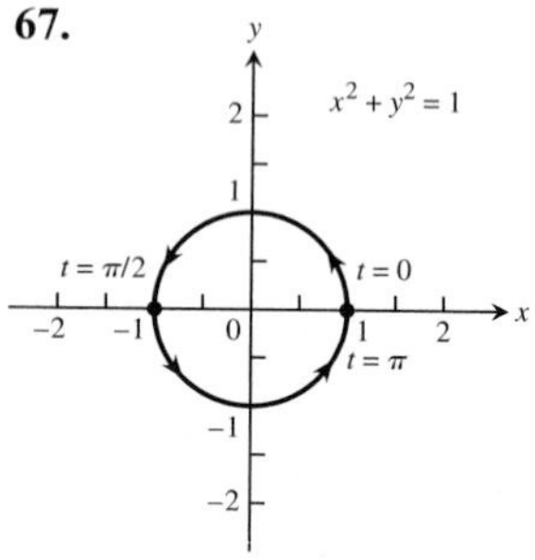

69.

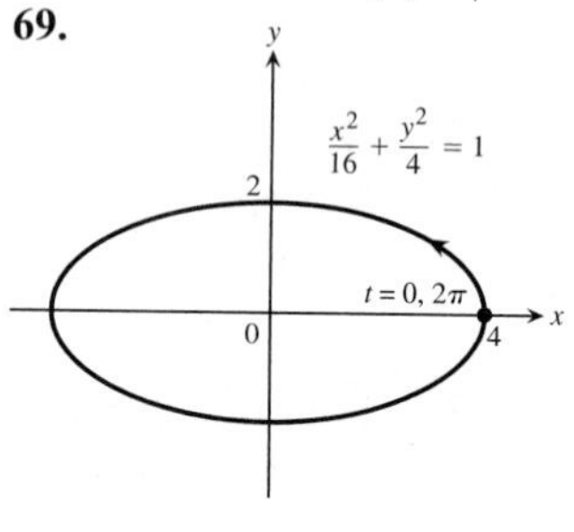

71.

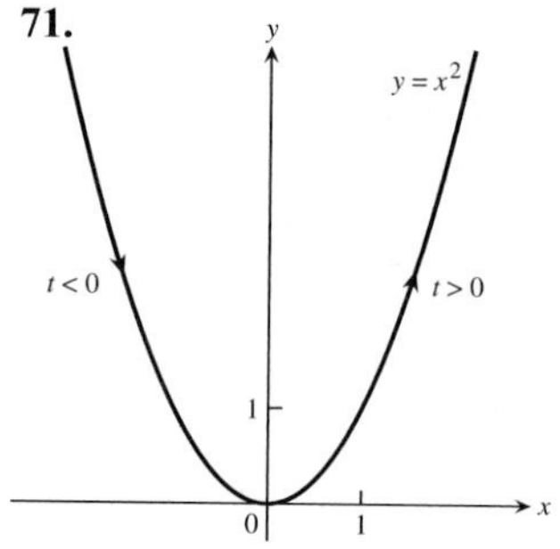

73.

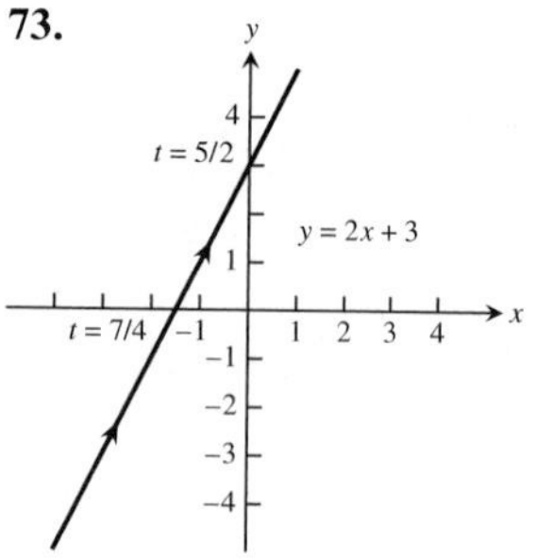

75.

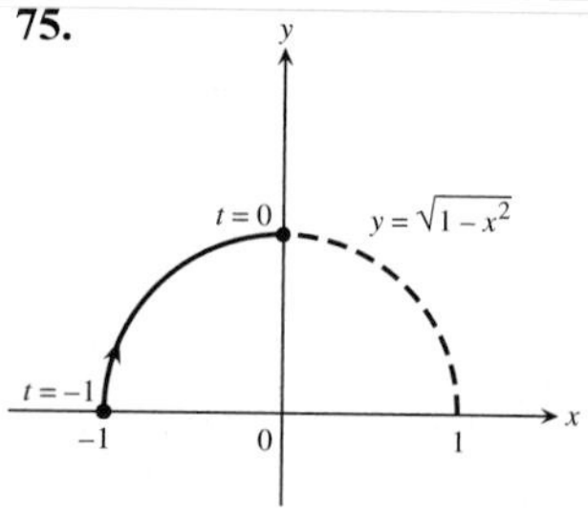

77.

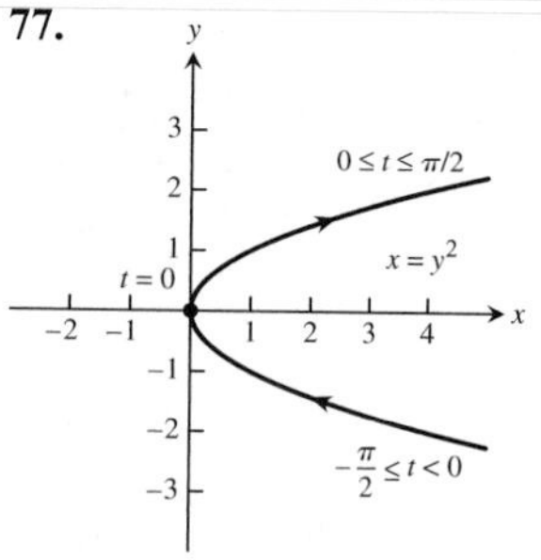

79. (a) $x = a\cos t,\ y = -a\sin t,\ 0 \le t \le 2\pi$
(b) $x = a\cos t,\ y = a\sin t,\ 0 \le t \le 2\pi$
(c) $x = a\cos t,\ y = -a\sin t,\ 0 \le t \le 4\pi$
(d) $x = a\cos t,\ y = a\sin t,\ 0 \le t \le 4\pi$

81. Possible answer: $x = -1 + 5t,\ y = -3 + 4t,\ 0 \le t \le 1$

83. Possible answer: $x = t^2 + 1,\ y = t,\ t \le 0$

85. Possible answer: $x = 2 - 3t,\ y = 3 - 4t,\ t \ge 0$

87. $y = -x + 2,\ \left.\dfrac{d^2y}{dx^2}\right|_{t=\pi/4} = -\sqrt{2}$

89. $y = x + \dfrac{1}{4},\ \left.\dfrac{d^2y}{dx^2}\right|_{t=1/4} = -2$

91. $y = x - 4,\ \left.\dfrac{d^2y}{dx^2}\right|_{t=-1} = \dfrac{1}{2}$

93. $y = 2,\ \left.\dfrac{d^2y}{dx^2}\right|_{t=\pi/2} = -1$

95. It multiplies the velocity, acceleration, and jerk by 2, 4, and 8, respectively.

97. $v(6) = \dfrac{2}{5}$ m/sec, $a(6) = -\dfrac{4}{125}$ m/sec^2

107. $\left(\dfrac{\sqrt{2}}{2}, 1\right)$, $y = 2x$ at $t = 0$, $y = -2x$ at $t = \pi$

Section 3.6, pp. 211–213

1. $\dfrac{9}{4}x^{5/4}$ **3.** $\dfrac{2^{1/3}}{3x^{2/3}}$ **5.** $\dfrac{7}{2(x+6)^{1/2}}$ **7.** $-(2x+5)^{-3/2}$

9. $\dfrac{2x^2+1}{(x^2+1)^{1/2}}$ **11.** $\dfrac{ds}{dt} = \dfrac{2}{7}t^{-5/7}$

13. $\dfrac{dy}{dt} = -\dfrac{4}{3}(2t+5)^{-5/3}\cos\left[(2t+5)^{-2/3}\right]$

15. $f'(x) = \dfrac{-1}{4\sqrt{x}(1-\sqrt{x})}$

17. $h'(\theta) = -\dfrac{2}{3}(\sin 2\theta)(1+\cos 2\theta)^{-2/3}$ **19.** $\dfrac{-2xy - y^2}{x^2 + 2xy}$

21. $\dfrac{1-2y}{2x+2y-1}$ **23.** $\dfrac{-2x^3 + 3x^2y - xy^2 + x}{x^2y - x^3 + y}$

25. $\dfrac{1}{y(x+1)^2}$ **27.** $\cos^2 y$ **29.** $\dfrac{-\cos^2(xy) - y}{x}$

31. $\dfrac{-y^2}{y\sin\left(\frac{1}{y}\right) - \cos\left(\frac{1}{y}\right) + xy}$ **33.** $-\dfrac{\sqrt{r}}{\sqrt{\theta}}$ **35.** $\dfrac{-r}{\theta}$

37. $y' = -\dfrac{x}{y},\ y'' = \dfrac{-y^2 - x^2}{y^3}$

39. $y' = \dfrac{x+1}{y},\ y'' = \dfrac{y^2 - (x+1)^2}{y^3}$

41. $y' = \dfrac{\sqrt{y}}{\sqrt{y}+1},\ y'' = \dfrac{1}{2(\sqrt{y}+1)^3}$

43. -2 **45.** $(-2, 1): m = -1,\ (-2, -1): m = 1$

47. (a) $y = \dfrac{7}{4}x - \dfrac{1}{2}$, **(b)** $y = -\dfrac{4}{7}x + \dfrac{29}{7}$

49. (a) $y = 3x + 6$, **(b)** $y = -\dfrac{1}{3}x + \dfrac{8}{3}$

51. (a) $y = \dfrac{6}{7}x + \dfrac{6}{7}$, **(b)** $y = -\dfrac{7}{6}x - \dfrac{7}{6}$

53. (a) $y = -\dfrac{\pi}{2}x + \pi$, **(b)** $y = \dfrac{2}{\pi}x - \dfrac{2}{\pi} + \dfrac{\pi}{2}$

55. **(a)** $y = 2\pi x - 2\pi$, **(b)** $y = -\dfrac{x}{2\pi} + \dfrac{1}{2\pi}$

57. Points: $(-\sqrt{7}, 0)$ and $(\sqrt{7}, 0)$, Slope: -2

59. $m = -1$ at $\left(\dfrac{\sqrt{3}}{4}, \dfrac{\sqrt{3}}{2}\right)$, $m = \sqrt{3}$ at $\left(\dfrac{\sqrt{3}}{4}, \dfrac{1}{2}\right)$

61. $(-3, 2): m = -\dfrac{27}{8}$; $(-3, -2): m = \dfrac{27}{8}$; $(3, 2): m = \dfrac{27}{8}$; $(3, -2): m = -\dfrac{27}{8}$

63. 0 **65.** -6 **67.** **(a)** False **(b)** True **(c)** True **(d)** True **69.** $(3, -1)$

73. $\dfrac{dy}{dx} = \dfrac{y^3 + 2xy}{x^2 + 3xy^2}$, $\dfrac{dx}{dy} = -\dfrac{x^2 + 3xy^2}{y^3 + 2xy}$, $\dfrac{dx}{dy} = \dfrac{1}{dy/dx}$

Section 3.7, pp. 218–221

1. $\dfrac{dA}{dt} = 2\pi r \dfrac{dr}{dt}$ **3.** **(a)** $\dfrac{dV}{dt} = \pi r^2 \dfrac{dh}{dt}$ **(b)** $\dfrac{dV}{dt} = 2\pi hr \dfrac{dr}{dt}$

(c) $\dfrac{dV}{dt} = \pi r^2 \dfrac{dh}{dt} + 2\pi hr \dfrac{dr}{dt}$ **5.** **(a)** 1 volt/sec

(b) $-\dfrac{1}{3}$ amp/sec **(c)** $\dfrac{dR}{dt} = \dfrac{1}{I}\left(\dfrac{dV}{dt} - \dfrac{V}{I}\dfrac{dI}{dt}\right)$

(d) 3/2 ohms/sec, R is increasing.

7. **(a)** $\dfrac{ds}{dt} = \dfrac{x}{\sqrt{x^2 + y^2}}\dfrac{dx}{dt}$

(b) $\dfrac{ds}{dt} = \dfrac{x}{\sqrt{x^2 + y^2}}\dfrac{dx}{dt} + \dfrac{y}{\sqrt{x^2 + y^2}}\dfrac{dy}{dt}$

(c) $\dfrac{dx}{dt} = -\dfrac{y}{x}\dfrac{dy}{dt}$

9. **(a)** $\dfrac{dA}{dt} = \dfrac{1}{2}ab\cos\theta \dfrac{d\theta}{dt}$

(b) $\dfrac{dA}{dt} = \dfrac{1}{2}ab\cos\theta \dfrac{d\theta}{dt} + \dfrac{1}{2}b\sin\theta \dfrac{da}{dt}$

(c) $\dfrac{dA}{dt} = \dfrac{1}{2}ab\cos\theta \dfrac{d\theta}{dt} + \dfrac{1}{2}b\sin\theta \dfrac{da}{dt} + \dfrac{1}{2}a\sin\theta \dfrac{db}{dt}$

11. **(a)** 14 cm^2/sec, increasing **(b)** 0 cm/sec, constant **(c)** $-14/13$ cm/sec, decreasing

13. **(a)** -12 ft/sec **(b)** -59.5 ft^2/sec **(c)** -1 rad/sec

15. 20 ft/sec

17. **(a)** $\dfrac{dh}{dt} = 11.19$ cm/min **(b)** $\dfrac{dr}{dt} = 14.92$ cm/min

19. **(a)** $\dfrac{-1}{24\pi}$ m/min **(b)** $r = \sqrt{26y - y^2}$ m,

(c) $\dfrac{dr}{dt} = -\dfrac{5}{288\pi}$ m/min

21. 1 ft/min, 40π ft^2/min **23.** 11 ft/sec **25.** Increasing at 466/1681 L/min **27.** 1 rad/sec **29.** -5 m/sec

31. -1500 ft/sec **33.** $\dfrac{5}{72\pi}$ in./min, $\dfrac{10}{3}$ in.^2/min

35. 7.1 in./min **37.** **(a)** $-32/\sqrt{13} \approx -8.875$ ft/sec, **(b)** $d\theta_1/dt = 8/65$ rad/sec, $d\theta_2/dt = -8/65$ rad/sec

(c) $d\theta_1/dt = 1/6$ rad/sec, $d\theta_2/dt = -1/6$ rad/sec

Section 3.8, pp. 231–234

1. $L(x) = 10x - 13$ **3.** $L(x) = 2$ **5.** $2x$ **7.** -5

9. $\dfrac{1}{12}x + \dfrac{4}{3}$ **11.** **(a)** $L(x) = x$ **(b)** $L(x) = \pi - x$

13. **(a)** $L(x) = 1$ **(b)** $L(x) = 2 - 2\sqrt{3}\left(x + \dfrac{\pi}{3}\right)$

15. $f(0) = 1$. Also, $f'(x) = k(1 + x)^{k-1}$, so $f'(0) = k$. This means the linearization at $x = 0$ is $L(x) = 1 + kx$.

17. **(a)** 1.01 **(b)** 1.003

19. $\left(3x^2 - \dfrac{3}{2\sqrt{x}}\right)dx$ **21.** $\dfrac{2 - 2x^2}{(1 + x^2)^2}dx$

23. $\dfrac{1 - y}{3\sqrt{y} + x}dx$ **25.** $\dfrac{5}{2\sqrt{x}}\cos(5\sqrt{x})\,dx$

27. $(4x^2)\sec^2\left(\dfrac{x^3}{3}\right)dx$

29. $\dfrac{3}{\sqrt{x}}(\csc(1 - 2\sqrt{x})\cot(1 - 2\sqrt{x}))\,dx$

31. **(a)** .41 **(b)** .4 **(c)** .01

33. **(a)** .231 **(b)** .2 **(c)** .031

35. **(a)** $-1/3$ **(b)** $-2/5$ **(c)** $1/15$

37. $dV = 4\pi r_0^2\,dr$ **39.** $dS = 12x_0\,dx$ **41.** $dV = 2\pi r_0 h\,dr$

43. **(a)** 0.08π m^2 **(b)** 2% **45.** $dV \approx 565.5$ in.^3

47. $\dfrac{1}{3}$% **49.** 0.05%

51. The ratio equals 37.87, so a change in the acceleration of gravity on the moon has about 38 times the effect that a change of the same magnitude has on Earth.

53. 3% **55.** 3%

59. $\lim\limits_{x\to 0}\dfrac{\sqrt{1 + x}}{1 + \left(\dfrac{x}{2}\right)} = \dfrac{\sqrt{1 + 0}}{1 + \left(\dfrac{0}{2}\right)} = \dfrac{1}{1} = 1$ **65.** $0.07c$

Practice Exercises, pp. 235–240

1. $5x^4 - 0.25x + 0.25$ **3.** $3x(x - 2)$

5. $2(x + 1)(2x^2 + 4x + 1)$

7. $3(\theta^2 + \sec\theta + 1)^2(2\theta + \sec\theta\tan\theta)$

9. $\dfrac{1}{2\sqrt{t}(1 + \sqrt{t})^2}$ **11.** $2\sec^2 x\tan x$

13. $8\cos^3(1 - 2t)\sin(1 - 2t)$ **15.** $5(\sec t)(\sec t + \tan t)^5$

17. $\dfrac{\theta\cos\theta + \sin\theta}{\sqrt{2\theta\sin\theta}}$ **19.** $\dfrac{\cos\sqrt{2\theta}}{\sqrt{2\theta}}$

21. $x\csc\left(\dfrac{2}{x}\right) + \csc\left(\dfrac{2}{x}\right)\cot\left(\dfrac{2}{x}\right)$

23. $\dfrac{1}{2}x^{1/2}\sec(2x)^2\left[16\tan(2x)^2 - x^{-2}\right]$

25. $-10x\csc^2(x^2)$ **27.** $8x^3\sin(2x^2)\cos(2x^2) + 2x\sin^2(2x^2)$

29. $\dfrac{-(t + 1)}{8t^3}$ **31.** $\dfrac{1 - x}{(x + 1)^3}$ **33.** $\dfrac{-1}{2x^2\left(1 + \dfrac{1}{x}\right)^{1/2}}$

35. $\dfrac{-2\sin\theta}{(\cos\theta - 1)^2}$ **37.** $3\sqrt{2x+1}$ **39.** $-9\left[\dfrac{5x+\cos 2x}{(5x^2+\sin 2x)^{5/2}}\right]$

41. $-\dfrac{y+2}{x+3}$ **43.** $\dfrac{-3x^2-4y+2}{4x-4y^{1/3}}$ **45.** $-\dfrac{y}{x}$

47. $\dfrac{1}{2y(x+1)^2}$ **49.** $\dfrac{dp}{dq} = \dfrac{6q-4p}{3p^2+4q}$

51. $\dfrac{dr}{ds} = (2r-1)(\tan 2s)$

53. **(a)** $\dfrac{d^2y}{dx^2} = \dfrac{-2xy^3-2x^4}{y^5}$ **(b)** $\dfrac{d^2y}{dx^2} = \dfrac{-2xy^2-1}{x^4y^3}$

55. **(a)** 7 **(b)** −2 **(c)** 5/12 **(d)** 1/4 **(e)** 12 **(f)** 9/2 **(g)** 3/4

57. 0 **59.** $\sqrt{3}$ **61.** $-\dfrac{1}{2}$ **63.** $\dfrac{-2}{(2t+1)^2}$

65. **(a)**

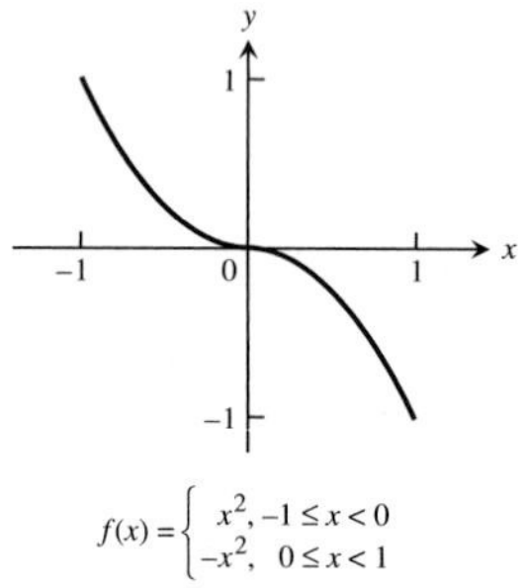

(b) Yes **(c)** Yes

67. **(a)**

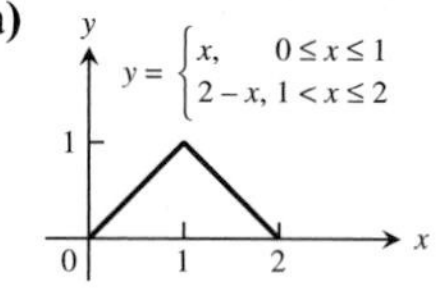

(b) Yes **(c)** No

69. $\left(\dfrac{5}{2}, \dfrac{9}{4}\right)$ and $\left(\dfrac{3}{2}, -\dfrac{1}{4}\right)$ **71.** $(-1, 27)$ and $(2, 0)$

73. **(a)** $(-2, 16)$, $(3, 11)$ **(b)** $(0, 20)$, $(1, 7)$

75.

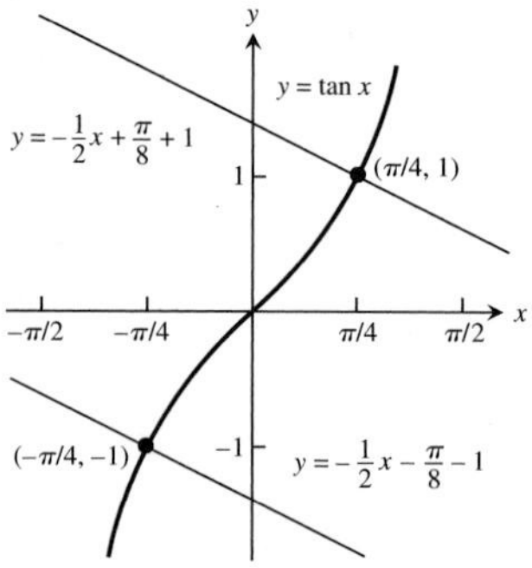

77. $\dfrac{1}{4}$ **79.** 4 **81.** Tangent: $y = -\dfrac{1}{4}x + \dfrac{9}{4}$, normal: $y = 4x - 2$

83. Tangent: $y = 2x - 4$, normal: $y = -\dfrac{1}{2}x + \dfrac{7}{2}$

85. Tangent: $y = -\dfrac{5}{4}x + 6$, normal: $y = \dfrac{4}{5}x - \dfrac{11}{5}$

87. $(1, 1)$: $m = -\dfrac{1}{2}$; $(1, -1)$: m not defined

89. $y = \left(\dfrac{\sqrt{3}}{2}\right)x + \dfrac{1}{4}, \dfrac{1}{4}$ **91.** B = graph of f, A = graph of f'

93.

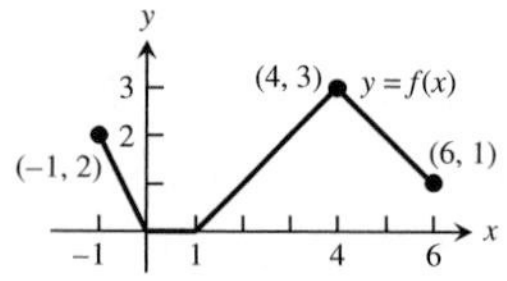

95. **(a)** 0, 0 **(b)** 1700 rabbits, ≈1400 rabbits

97. −1 **99.** 1/2 **101.** 4 **103.** 1

107. **(a)** $\dfrac{dS}{dt} = (4\pi r + 2\pi h)\dfrac{dr}{dt}$ **(b)** $\dfrac{dS}{dt} = 2\pi r\dfrac{dh}{dt}$

(c) $\dfrac{dS}{dt} = (4\pi r + 2\pi h)\dfrac{dr}{dt} + 2\pi r\dfrac{dh}{dt}$

(d) $\dfrac{dr}{dt} = -\dfrac{r}{2r+h}\dfrac{dh}{dt}$

109. $-40\ \text{m}^2/\text{sec}$ **111.** 0.02 ohm/sec **113.** 22 m/sec

115. **(a)** $r = \dfrac{2}{5}h$ **(b)** $-\dfrac{125}{144\pi}$ ft/min

117. **(a)** $\dfrac{3}{5}$ km/sec or 600 m/sec **(b)** $\dfrac{18}{\pi}$ rpm

119. **(a)** $L(x) = 2x + \dfrac{\pi - 2}{2}$

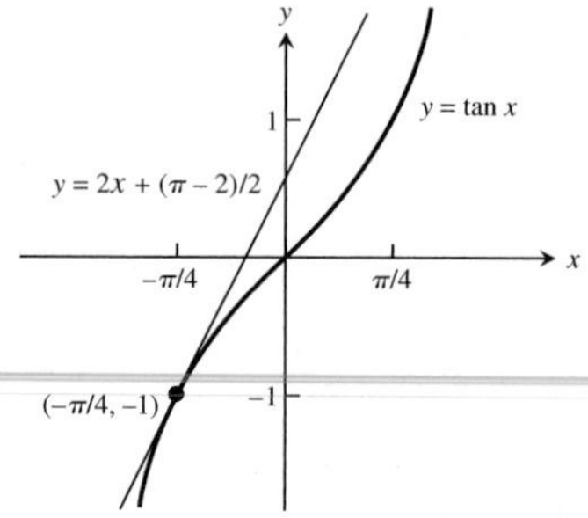

(b) $L(x) = -\sqrt{2}x + \dfrac{\sqrt{2}(4-\pi)}{4}$

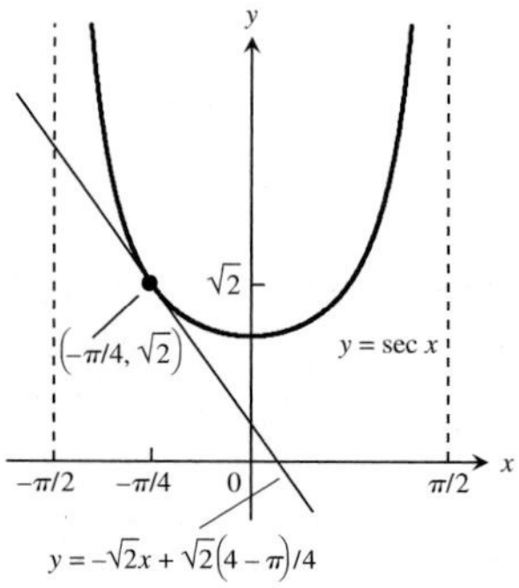

121. $L(x) = 1.5x + 0.5$ **123.** $dS = \dfrac{\pi r h_0}{\sqrt{r^2 + h_0^2}}\, dh$

125. (a) 4% **(b)** 8% **(c)** 12%

Additional and Advanced Exercises, pp. 240–243

1. (a) $\sin 2\theta = 2 \sin\theta \cos\theta$; $2 \cos 2\theta = 2 \sin\theta\,(-\sin\theta) + \cos\theta\,(2\cos\theta)$; $2\cos 2\theta = -2\sin^2\theta + 2\cos^2\theta$; $\cos 2\theta = \cos^2\theta - \sin^2\theta$
(b) $\cos 2\theta = \cos^2\theta - \sin^2\theta$; $-2\sin 2\theta = 2\cos\theta\,(-\sin\theta) - 2\sin\theta\,(\cos\theta)$; $\sin 2\theta = \cos\theta\sin\theta + \sin\theta\cos\theta$; $\sin 2\theta = 2\sin\theta\cos\theta$

3. (a) $a = 1, b = 0, c = -\dfrac{1}{2}$ **(b)** $b = \cos a, c = \sin a$

5. $h = -4, k = \dfrac{9}{2}, a = \dfrac{5\sqrt{5}}{2}$

7. (a) $0.09y$ **(b)** Increasing at 1% per year

9. Answers will vary. Here is one possibility.

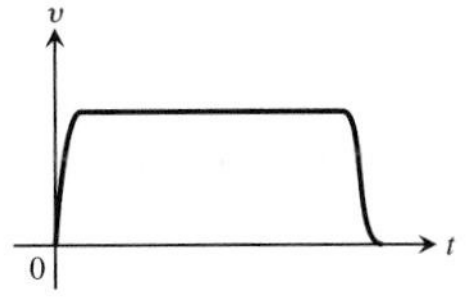

11. (a) 2 sec, 64 ft/sec **(b)** 12.31 sec, 393.85 ft

15. (a) $m = -\dfrac{b}{\pi}$ **(b)** $m = -1, b = \pi$

17. (a) $a = \dfrac{3}{4}, b = \dfrac{9}{4}$ **19.** f odd $\Rightarrow f'$ is even

23. h' is defined but not continuous at $x = 0$; k' is defined *and* continuous at $x = 0$.

27. (a) 0.8156 ft **(b)** 0.00613 sec **(c)** It will lose about 8.83 min/day.

CHAPTER 4

Section 4.1, pp. 252–255

1. Absolute minimum at $x = c_2$; absolute maximum at $x = b$

3. Absolute maximum at $x = c$; no absolute minimum

5. Absolute minimum at $x = a$; absolute maximum at $x = c$

7. Local minimum at $(-1, 0)$; local maximum at $(1, 0)$

9. Maximum at $(0, 5)$ **11.** (c) **13.** (d)

15. Absolute maximum: -3; absolute minimum: $-19/3$

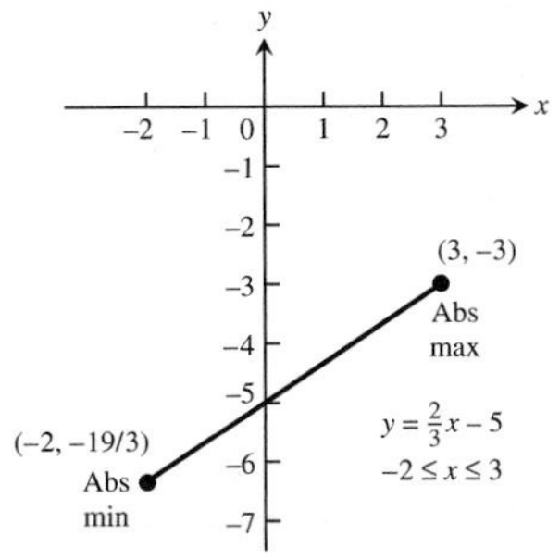

17. Absolute maximum: 3; absolute minimum: -1

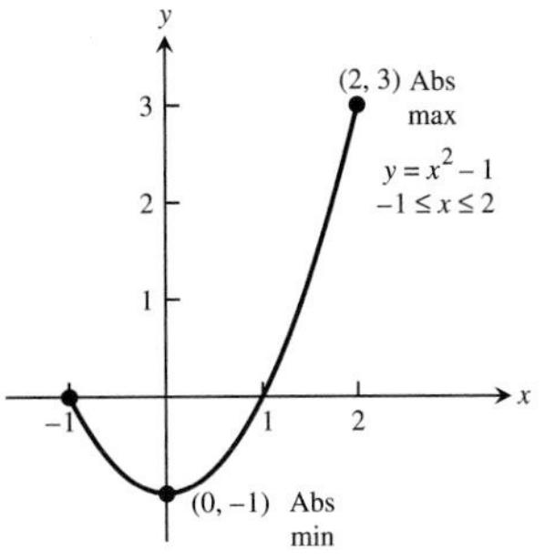

19. Absolute maximum: -0.25; absolute minimum: -4

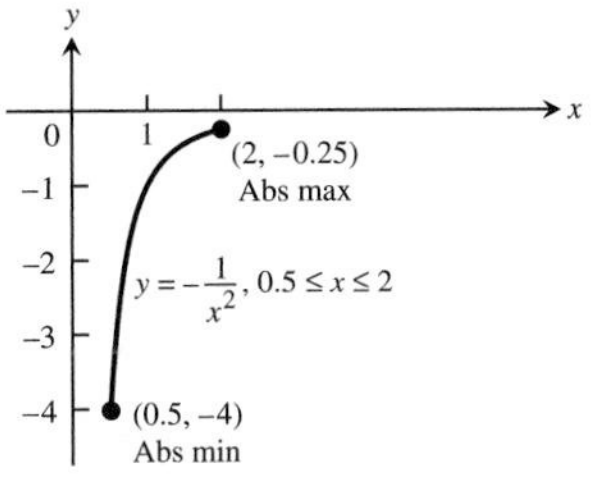

21. Absolute maximum: 2; absolute minimum: -1

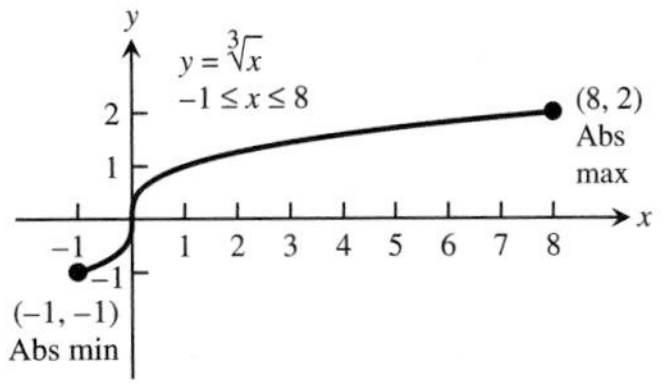

23. Absolute maximum: 2; absolute minimum: 0

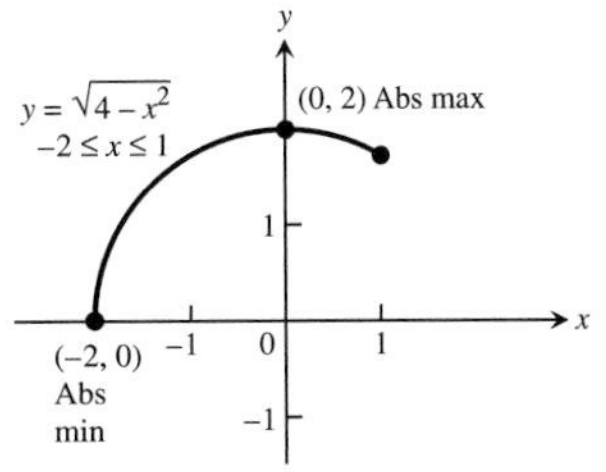

25. Absolute maximum: 1; absolute minimum: -1

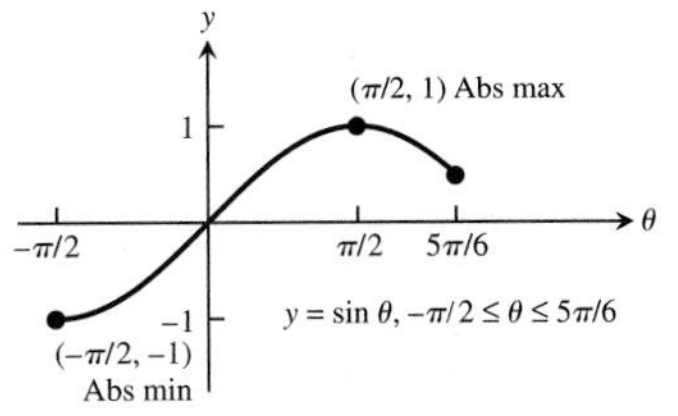

27. Absolute maximum: $2/\sqrt{3}$; absolute minimum: 1

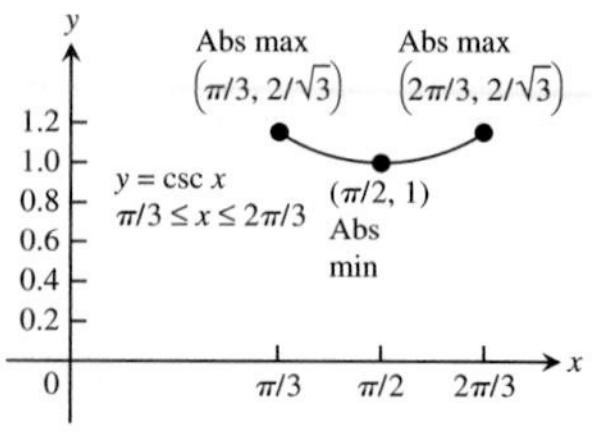

29. Absolute maximum: 2; absolute minimum: -1

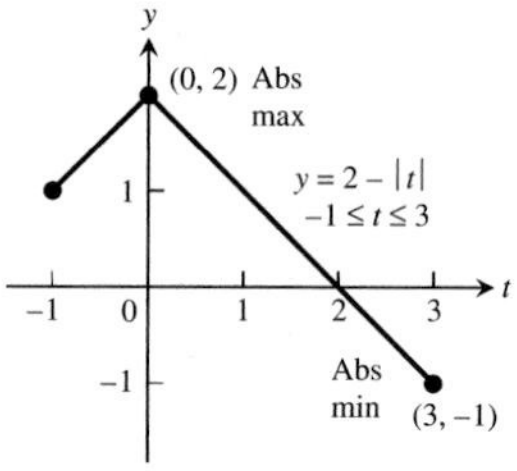

31. Increasing on (0, 8), decreasing on $(-1, 0)$; absolute maximum: 16 at $x = 8$; absolute minimum: 0 at $x = 0$

33. Increasing on $(-32, 1)$; absolute maximum: 1 at $\theta = 1$; absolute minimum: -8 at $\theta = -32$

35. Minimum value is 1 at $x = 2$.

37. Local maximum at $(-2, 17)$; local minimum at $\left(\frac{4}{3}, -\frac{41}{27}\right)$

39. Minimum value is 0 at $x = -1$ and $x = 1$.

41. There is a local minimum at (0, 1).

43. Maximum value is $\frac{1}{2}$ at $x = 1$; minimum value is $-\frac{1}{2}$ at $x = -1$.

45.

Critical point	Derivative	Extremum	Value
$x = -\frac{4}{5}$	0	Local max	$\frac{12}{25} 10^{1/3} = 1.034$
$x = 0$	Undefined	Local min	0

47.

Critical point	Derivative	Extremum	Value
$x = -2$	Undefined	Local max	0
$x = -\sqrt{2}$	0	Minimum	-2
$x = \sqrt{2}$	0	Maximum	2
$x = 2$	Undefined	Local min	0

49.

Critical point	Derivative	Extremum	Value
$x = 1$	Undefined	Minimum	2

51.

Critical point	Derivative	Extremum	Value
$x = -1$	0	Maximum	5
$x = 1$	Undefined	Local min	1
$x = 3$	0	Maximum	5

53. **(a)** No
(b) The derivative is defined and nonzero for $x \neq 2$. Also, $f(2) = 0$ and $f(x) > 0$ for all $x \neq 2$.
(c) No, because $(-\infty, \infty)$ is not a closed interval.
(d) The answers are the same as parts (a) and (b) with 2 replaced by a.

55. **(a)** $C(x) = 0.3\sqrt{16 + x^2} + 0.2(9 - x)$ million dollars, where $0 \le x \le 9$ mi. To minimize the cost of construction, the pipeline should be placed from the docking facility to point B, 3.58 mi along the shore from point A, and then along the shore from point B to the refinery.
(b) In theory, the underwater pipe cost per mile p would have to be infinite to justify running the pipe directly from the docking facility to point A (i.e., for x_c to be zero). For all values of $p > 0.218864$, there is always an x_c in (0, 9) that will give a minimum value for C. This is proved by looking at $C''(x_c) = \frac{16p}{(16 + x_c{}^2)^{3/2}}$, which is always positive for $p > 0$.

57. The length of pipeline is $L(x) = \sqrt{4 + x^2} + \sqrt{25 + (10 - x)^2}$ for $0 \le x \le 10$. $x = \frac{20}{7} \approx 2.857$ mi along the coast from Town A to Town B.

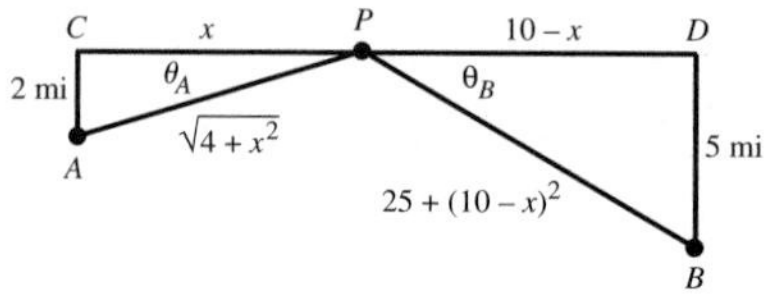

59. **(a)** Maximum value is 144 at $x = 2$.
(b) The largest volume of the box is 144 cubic units, and it occurs when $x = 2$.

61. The largest possible area is $A\left(\frac{5}{\sqrt{2}}\right) = \frac{25}{4}\text{ cm}^2$.

63. $\frac{v_0{}^2}{2g} + s_0$ **65.** Yes **67.** g assumes a local maximum at $-c$.

69. **(a)** $f'(x) = 3ax^2 + 2bx + c$ is a quadratic, so it can have 0, 1, or 2 zeros, which would be the critical points of f. Examples: The function $f(x) = x^3 - 3x$ has two critical points at $x = -1$ and $x = 1$.

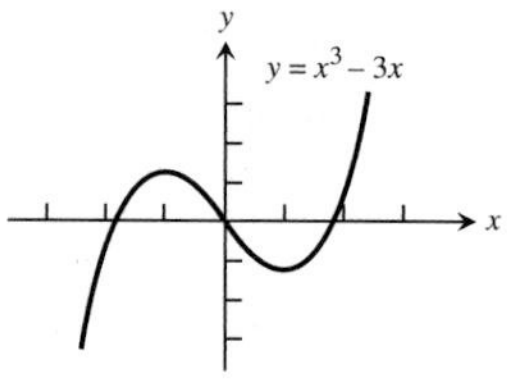

The function $f(x) = x^3 - 1$ has one critical point at $x = 0$.

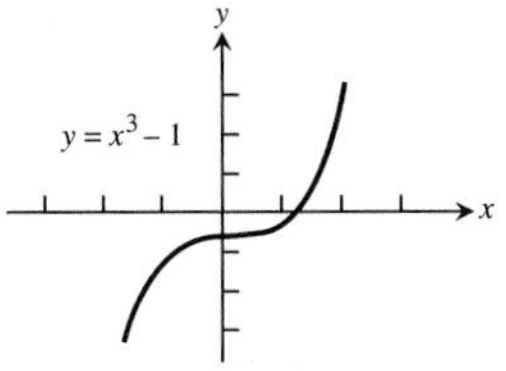

The function $f(x) = x^3 + x$ has no critical points.

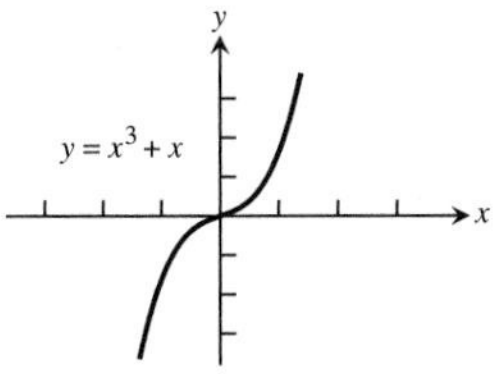

(b) Two or none

71. Maximum value is 11 at $x = 5$; minimum value is 5 on the interval $[-3, 2]$; local maximum at $(-5, 9)$.

73. Maximum value is 5 on the interval $[3, \infty)$; minimum value is -5 on the interval $(-\infty, -2]$.

Section 4.2, pp. 260–262

1. 1/2 **3.** 1 **5.** Does not; f is not differentiable at the interior domain point $x = 0$. **7.** Does

11. **(a)**

i)
ii)
iii)
iv)

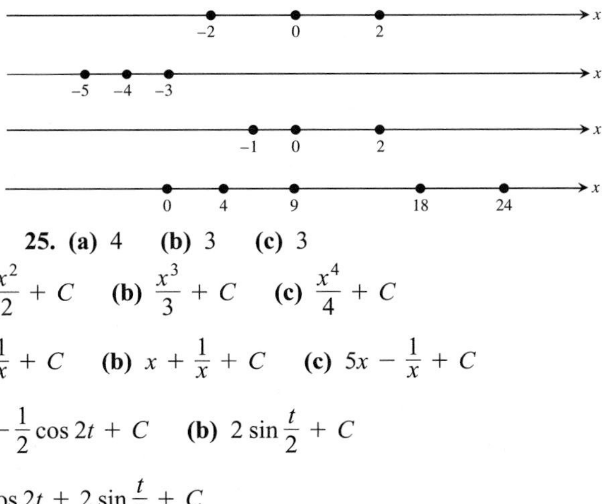

23. Yes **25.** **(a)** 4 **(b)** 3 **(c)** 3

27. **(a)** $\frac{x^2}{2} + C$ **(b)** $\frac{x^3}{3} + C$ **(c)** $\frac{x^4}{4} + C$

29. **(a)** $\frac{1}{x} + C$ **(b)** $x + \frac{1}{x} + C$ **(c)** $5x - \frac{1}{x} + C$

31. **(a)** $-\frac{1}{2}\cos 2t + C$ **(b)** $2\sin\frac{t}{2} + C$

(c) $-\frac{1}{2}\cos 2t + 2\sin\frac{t}{2} + C$

33. $f(x) = x^2 - x$ **35.** $r(\theta) = 8\theta + \cot\theta - 2\pi - 1$

37. $s = 4.9t^2 + 5t + 10$ **39.** $s = \dfrac{1 - \cos(\pi t)}{\pi}$

41. $s = 16t^2 + 20t + 5$ **43.** $s = \sin(2t) - 3$

45. If $T(t)$ is the temperature of the thermometer at time t, then $T(0) = -19\ °\text{C}$ and $T(14) = 100\ °\text{C}$. From the Mean Value Theorem, there exists a $0 < t_0 < 14$ such that $\dfrac{T(14) - T(0)}{14 - 0} = 8.5\ °\text{C/sec} = T'(t_0)$, the rate at which the temperature was changing at $t = t_0$ as measured by the rising mercury on the thermometer.

47. Because its average speed was approximately 7.667 knots, and by the Mean Value Theorem, it must have been going that speed at least once during the trip.

51. The conclusion of the Mean Value Theorem yields

$$\frac{\frac{1}{b} - \frac{1}{a}}{b - a} = -\frac{1}{c^2} \Rightarrow c^2\left(\frac{a - b}{ab}\right) = a - b \Rightarrow c = \sqrt{ab}.$$

55. $f(x)$ must be zero at least once between a and b by the Intermediate Value Theorem. Now suppose that $f(x)$ is zero twice between a and b. Then, by the Mean Value Theorem, $f'(x)$ would have to be zero at least once between the two zeros of $f(x)$, but this can't be true since we are given that $f'(x) \neq 0$ on this interval. Therefore, $f(x)$ is zero once and only once between a and b.

61. $1.09999 \leq f(0.1) \leq 1.1$

Section 4.3, pp. 266–267

1. **(a)** 0, 1 **(b)** increasing on $(-\infty, 0)$ and $(1, \infty)$, decreasing on $(0, 1)$ **(c)** local maximum at $x = 0$, local minimum at $x = 1$

3. **(a)** $-2, 1$ **(b)** increasing on $(-2, 1)$ and $(1, \infty)$, decreasing on $(-\infty, -2)$ **(c)** no local maximum, local minimum at $x = -2$

5. **(a)** $-2, 1, 3$ **(b)** increasing on $(-2, 1)$ and $(3, \infty)$, decreasing on $(-\infty, -2)$ and $(1, 3)$ **(c)** local maximum at $x = 1$, local minima at $x = -2, 3$

7. **(a)** $-2, 0$ **(b)** increasing on $(-\infty, -2)$ and $(0, \infty)$, decreasing on $(-2, 0)$ **(c)** local maximum at $x = -2$, local minimum at $x = 0$

9. **(a)** Increasing on $(-\infty, -1.5)$, decreasing on $(-1.5, \infty)$ **(b)** local maximum: 5.25 at $t = -1.5$ **(c)** absolute maximum: 5.25 at $t = -1.5$

11. **(a)** Decreasing on $(-\infty, 0)$, increasing on $(0, 4/3)$, decreasing on $(4/3, \infty)$ **(b)** local minimum at $x = 0$ $(0, 0)$, local maximum at $x = 4/3$ $(4/3, 32/27)$ **(c)** no absolute extrema

13. **(a)** Decreasing on $(-\infty, 0)$, increasing on $(0, 1/2)$, decreasing on $(1/2, \infty)$ **(b)** local minimum at $\theta = 0$ $(0, 0)$, local maximum at $\theta = 1/2$ $(1/2, 1/4)$ **(c)** no absolute extrema

15. **(a)** Increasing on $(-\infty, \infty)$, never decreasing **(b)** no local extrema **(c)** no absolute extrema

17. **(a)** Increasing on $(-2, 0)$ and $(2, \infty)$, decreasing on $(-\infty, -2)$ and $(0, 2)$ **(b)** local maximum: 16 at $x = 0$, local minimum: 0 at $x = \pm 2$ **(c)** no absolute maximum; absolute minimum: 0 at $x = \pm 2$

19. **(a)** Increasing on $(-\infty, -1)$, decreasing on $(-1, 0)$, increasing on $(0, 1)$, decreasing on $(1, \infty)$ **(b)** local maximum at $x = \pm 1$ $(1, 0.5), (-1, 0.5)$, local minimum at $x = 0$ $(0, 0)$ **(c)** absolute maximum: 1/2 at $x = \pm 1$; no absolute minimum

21. **(a)** Decreasing on $(-2\sqrt{2}, -2)$, increasing on $(-2, 2)$, decreasing on $(2, 2\sqrt{2})$ **(b)** local minima: $g(-2) = -4$, $g(2\sqrt{2}) = 0$; local maxima: $g(-2\sqrt{2}) = 0$, $g(2) = 4$ **(c)** absolute maximum: 4 at $x = 2$; absolute minimum: -4 at $x = -2$

23. **(a)** Increasing on $(-\infty, 1)$, decreasing when $1 < x < 2$, decreasing when $2 < x < 3$, discontinuous at $x = 2$, increasing on $(3, \infty)$ **(b)** local minimum at $x = 3$ $(3, 6)$, local maximum at $x = 1$ $(1, 2)$ **(c)** no absolute extrema

25. **(a)** Increasing on $(-2, 0)$ and $(0, \infty)$, decreasing on $(-\infty, -2)$ **(b)** local minimum: $-6\sqrt[3]{2}$ at $x = -2$ **(c)** no absolute maximum; absolute minimum: $-6\sqrt[3]{2}$ at $x = -2$

27. **(a)** Increasing on $\left(-\infty, -2/\sqrt{7}\right)$ and $\left(2/\sqrt{7}, \infty\right)$, decreasing on $\left(-2/\sqrt{7}, 0\right)$ and $\left(0, 2/\sqrt{7}\right)$ **(b)** local maximum: $24\sqrt[3]{2}/7^{7/6} \approx 3.12$ at $x = -2/\sqrt{7}$; local minimum: $-24\sqrt[3]{2}/7^{7/6} \approx -3.12$ at $x = 2/\sqrt{7}$ **(c)** no absolute extrema

29. **(a)** Local maximum: 1 at $x = 1$; local minimum: 0 at $x = 2$ **(b)** absolute maximum: 1 at $x = 1$; no absolute minimum

31. **(a)** Local maximum: 1 at $x = 1$; local minimum: 0 at $x = 2$ **(b)** no absolute maximum; absolute minimum: 0 at $x = 2$

33. **(a)** Local maxima: -9 at $t = -3$ and 16 at $t = 2$; local minimum: -16 at $t = -2$ **(b)** absolute maximum: 16 at $t = 2$; no absolute minimum

35. **(a)** Local minimum: 0 at $x = 0$ **(b)** no absolute maximum; absolute minimum: 0 at $x = 0$

37. **(a)** Local minimum: $(\pi/3) - \sqrt{3}$ at $x = 2\pi/3$; local maximum: 0 at $x = 0$; local maximum: π at $x = 2\pi$

39. **(a)** Local minimum: 0 at $x = \pi/4$

41. Local maximum: 3 at $\theta = 0$; local minimum: -3 at $\theta = 2\pi$

43.

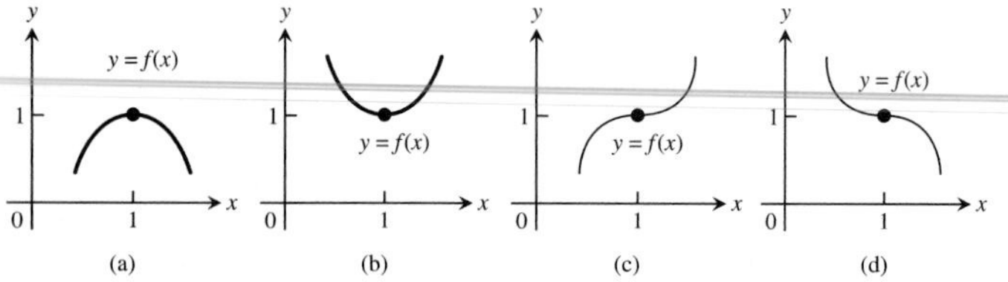

45. **(a)**

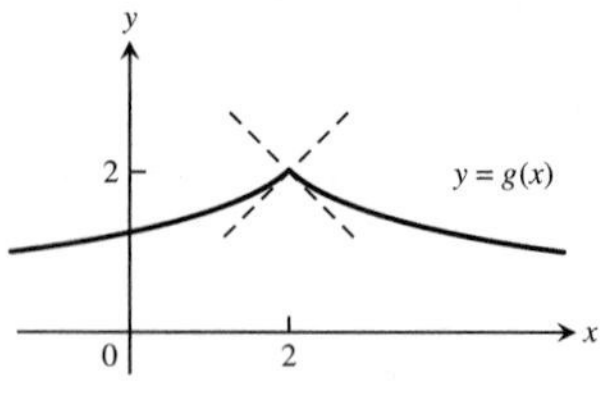

(b)

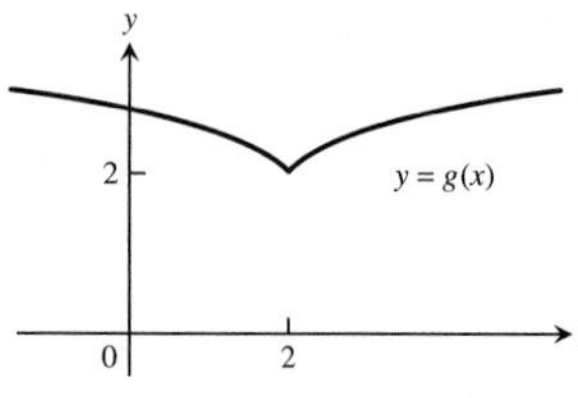

47. Rising

Section 4.4, pp. 274–277

1. Local maximum: 3/2 at $x = -1$, local minimum: -3 at $x = 2$, point of inflection at $(1/2, -3/4)$, rising on $(-\infty, -1)$ and $(2, \infty)$, falling on $(-1, 2)$, concave up on $(1/2, \infty)$, concave down on $(-\infty, 1/2)$

3. Local maximum: 3/4 at $x = 0$, local minimum: 0 at $x = \pm 1$, points of inflection at $\left(-\sqrt{3}, \frac{3\sqrt[3]{4}}{4}\right)$ and $\left(\sqrt{3}, \frac{3\sqrt[3]{4}}{4}\right)$, rising on $(-1, 0)$ and $(1, \infty)$, falling on $(-\infty, -1)$ and $(0, 1)$, concave up on $(-\infty, -\sqrt{3})$ and $(\sqrt{3}, \infty)$, concave down on $(-\sqrt{3}, \sqrt{3})$

5. Local maxima: $-2\pi/3 + \sqrt{3}/2$ at $x = -2\pi/3$; $\frac{\pi}{3} + \frac{\sqrt{3}}{2}$ at $x = \frac{\pi}{3}$, local minima: $-\frac{\pi}{3} - \frac{\sqrt{3}}{2}$ at $x = -\frac{\pi}{3}$; $2\pi/3 - \sqrt{3}/2$ at $x = \frac{2\pi}{3}$, points of inflection at $(-\pi/2, -\pi/2)$, $(0, 0)$, and $(\pi/2, \pi/2)$, rising on $(-\pi/3, \pi/3)$, falling on $(-2\pi/3, -\pi/3)$ and $(\pi/3, 2\pi/3)$, concave up on $(-\pi/2, 0)$ and $(\pi/2, 2\pi/3)$, concave down on $(-2\pi/3, -\pi/2)$ and $(0, \pi/2)$

7. Local maxima: 1 at $x = -\frac{\pi}{2}$ and $x = \frac{\pi}{2}$; 0 at $x = -2\pi$ and $x = 2\pi$; local minima: -1 at $x = -\frac{3\pi}{2}$ and $x = \frac{3\pi}{2}$, 0 at $x = 0$, points of inflection at $(-\pi, 0)$ and $(\pi, 0)$, rising on $(-3\pi/2, -\pi/2)$, $(0, \pi/2)$ and $(3\pi/2, 2\pi)$, falling on $(-2\pi, -3\pi/2)$, $(-\pi/2, 0)$ and $(\pi/2, 3\pi/2)$, concave up on $(-2\pi, -\pi)$ and $(\pi, 2\pi)$, concave down on $(-\pi, \pi)$

9.

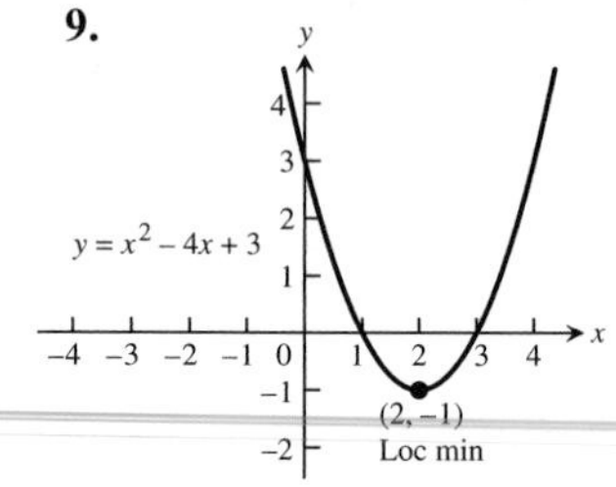

11.

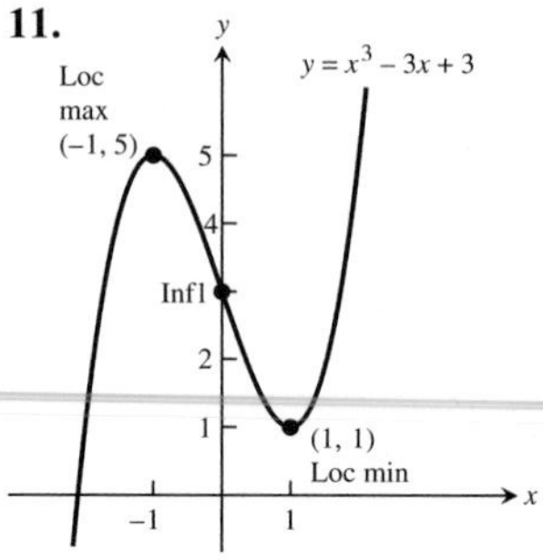

13.

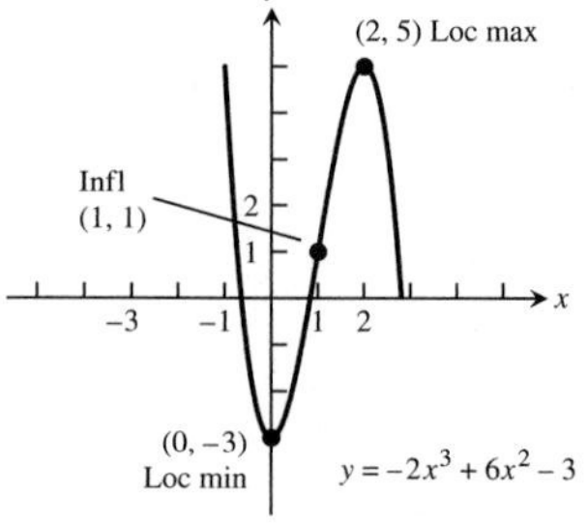

15.

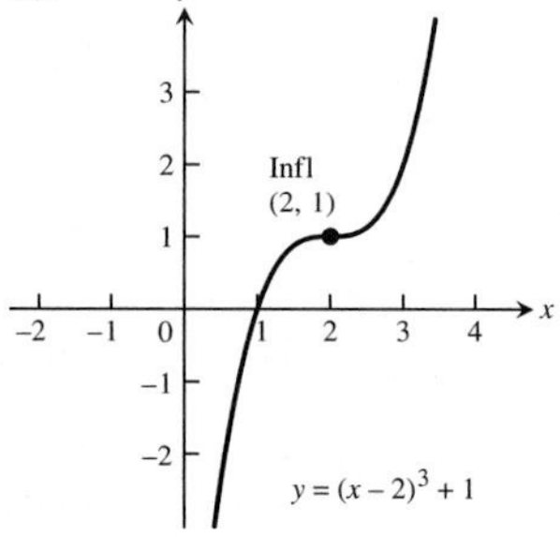

17.

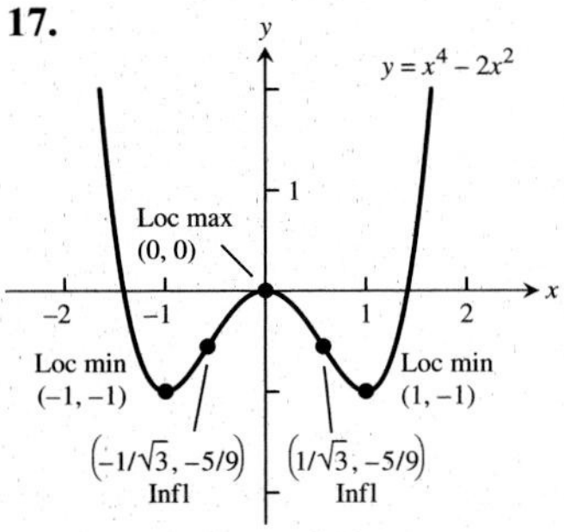

19.

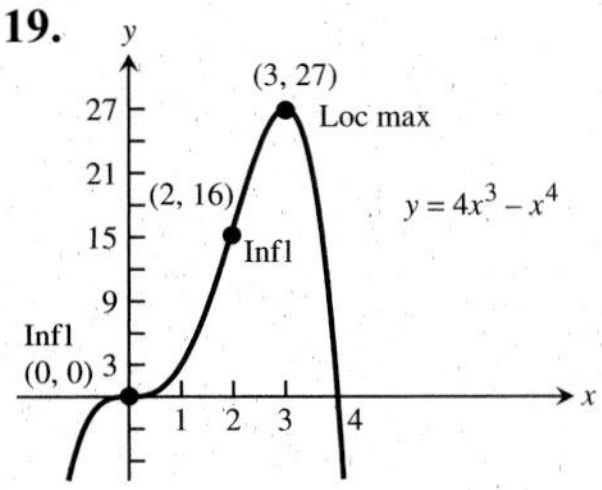

21.

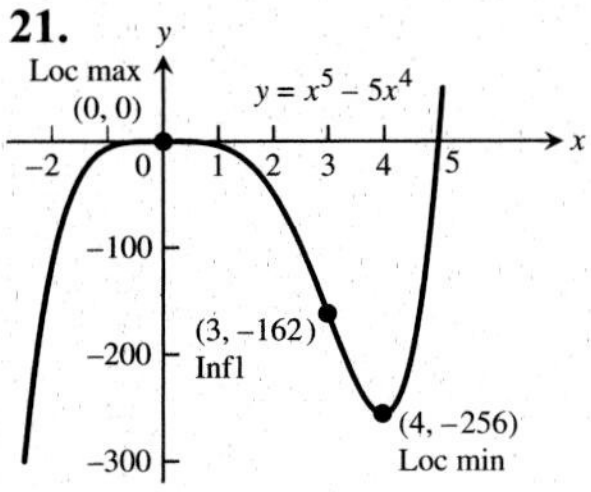

23.

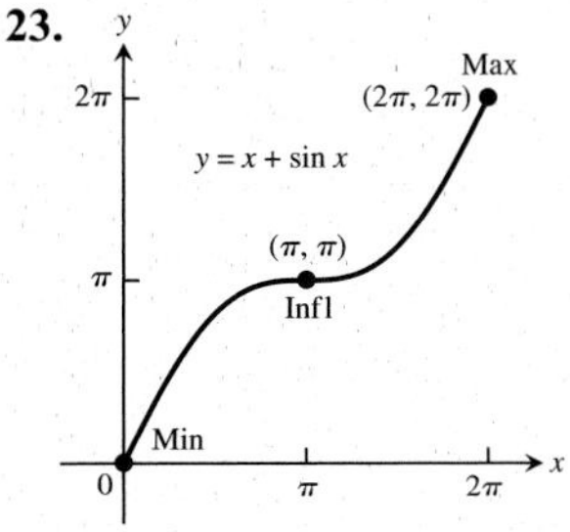

25.

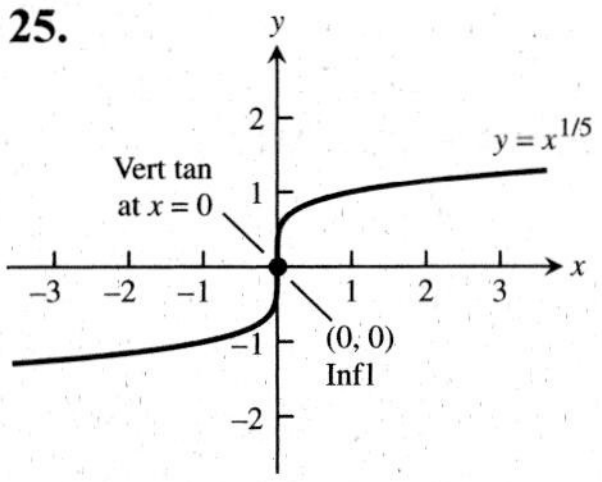

27.

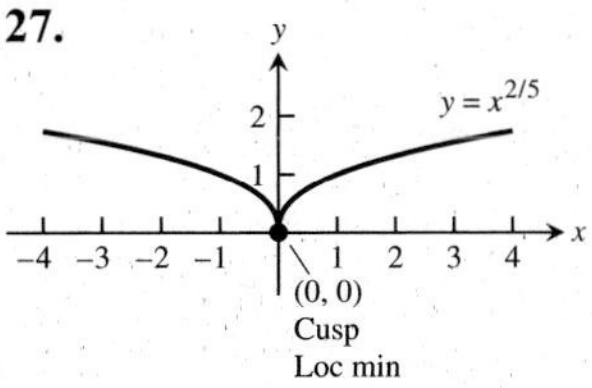

29.

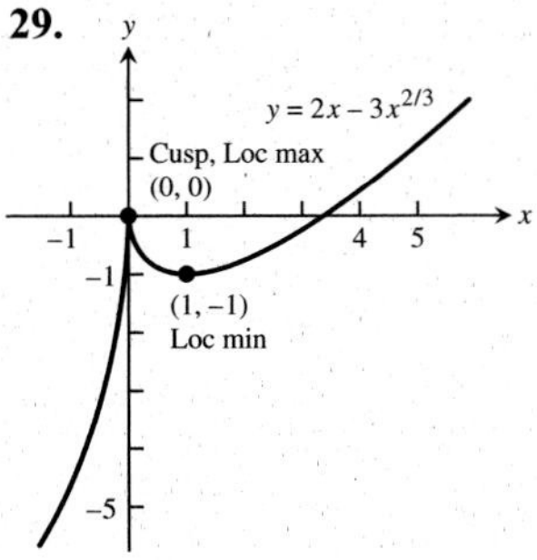

31.

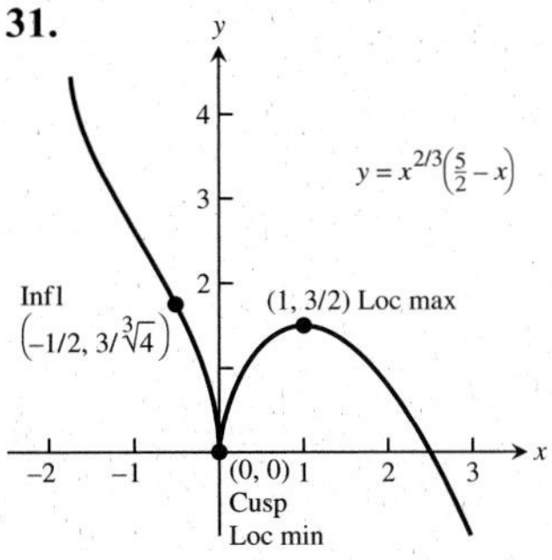

33.

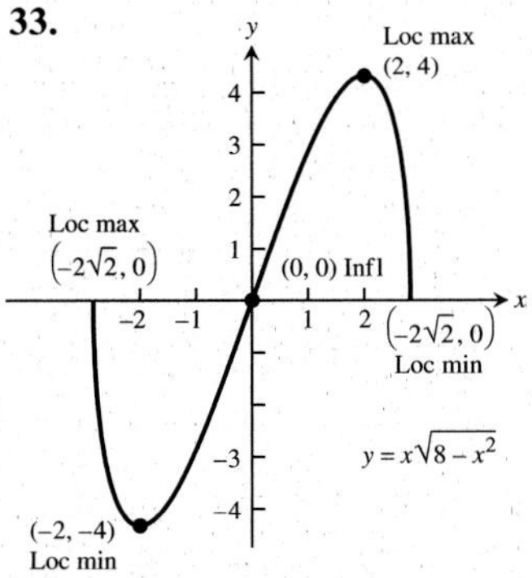

35.

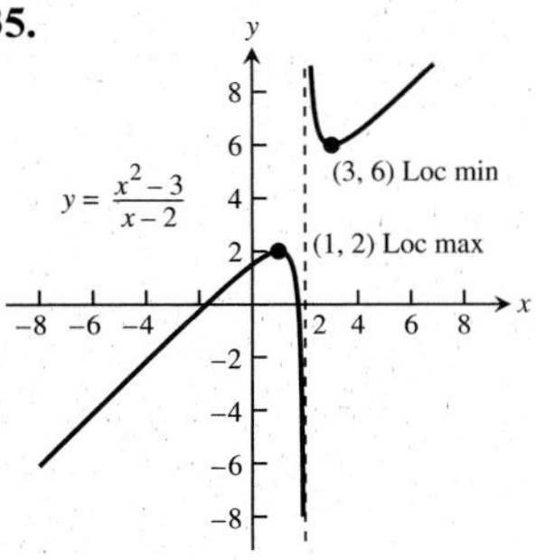

37.

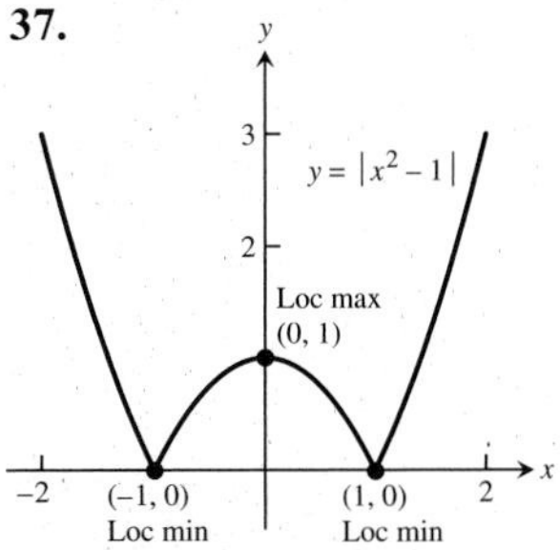

39.

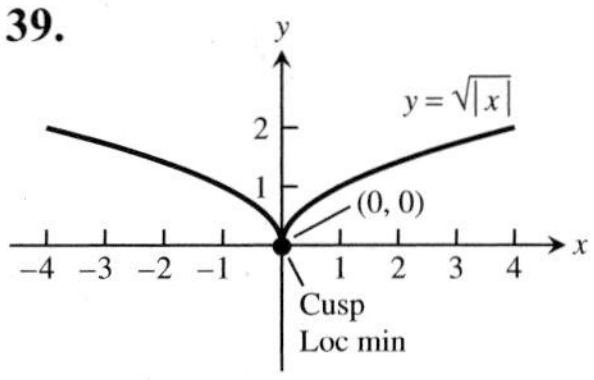

41. $y'' = 1 - 2x$

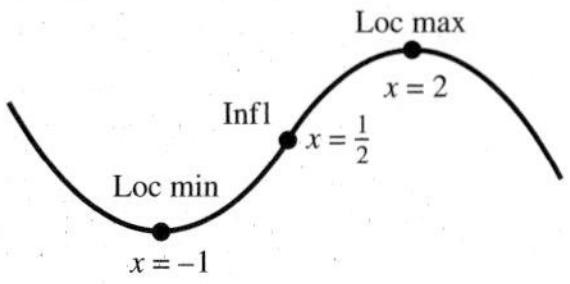

43. $y'' = 3(x - 3)(x - 1)$

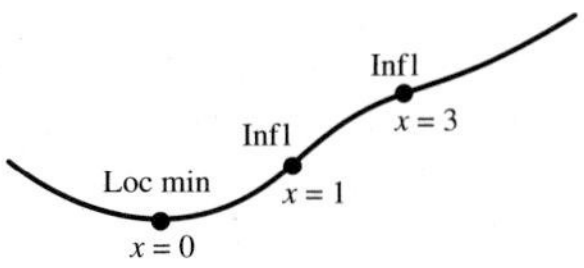

45. $y'' = 3(x - 2)(x + 2)$

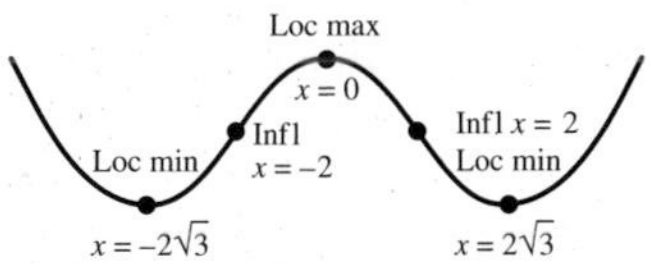

47. $y'' = 4(4 - x)(5x^2 - 16x + 8)$

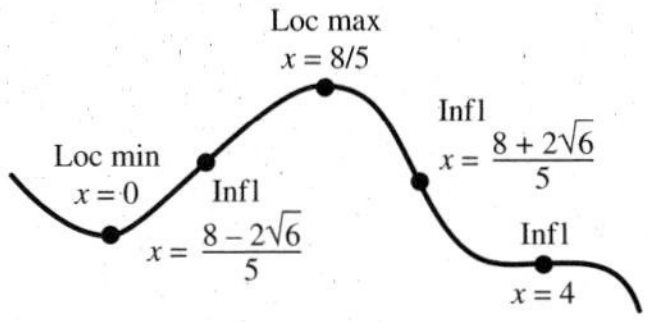

49. $y'' = 2\sec^2 x \tan x$

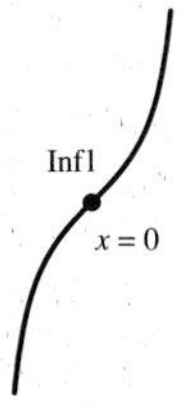

51. $y'' = -\dfrac{1}{2}\csc^2\dfrac{\theta}{2},\ 0 < \theta < 2\pi$

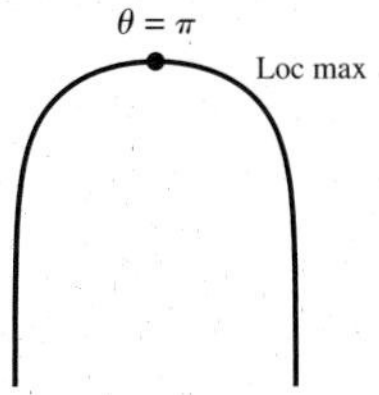

53. $y'' = 2\tan\theta\sec^2\theta,\ -\frac{\pi}{2} < \theta < \frac{\pi}{2}$

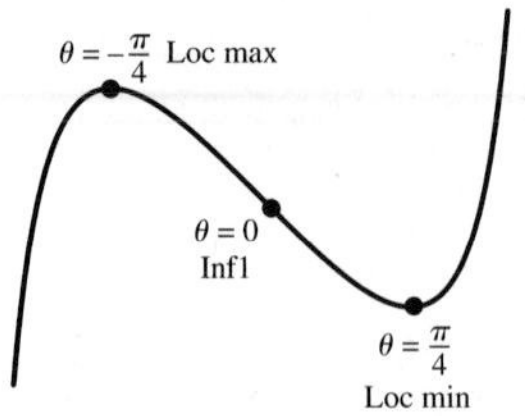

55. $y'' = -\sin t,\ 0 \le t \le 2\pi$

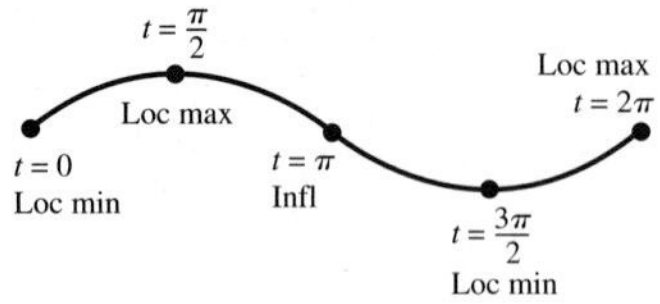

57. $y'' = -\frac{2}{3}(x+1)^{-5/3}$

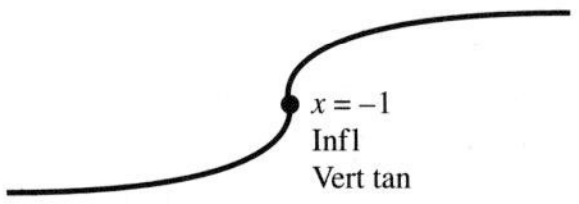

59. $y'' = \frac{1}{3}x^{-2/3} + \frac{2}{3}x^{-5/3}$

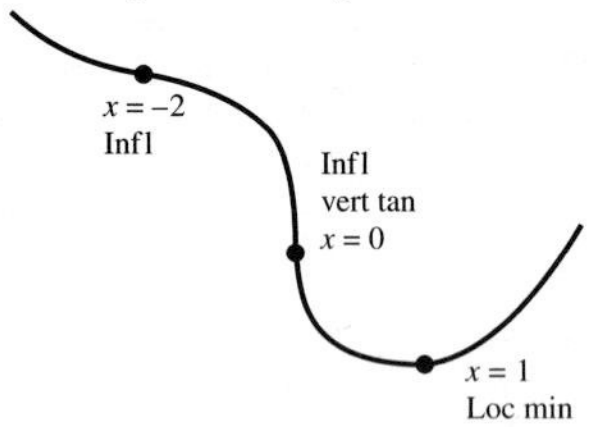

61. $y'' = \begin{cases} -2, & x < 0 \\ 2, & x > 0 \end{cases}$

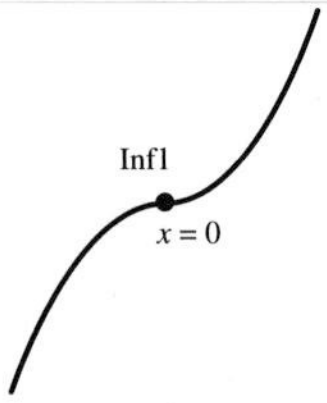

63.

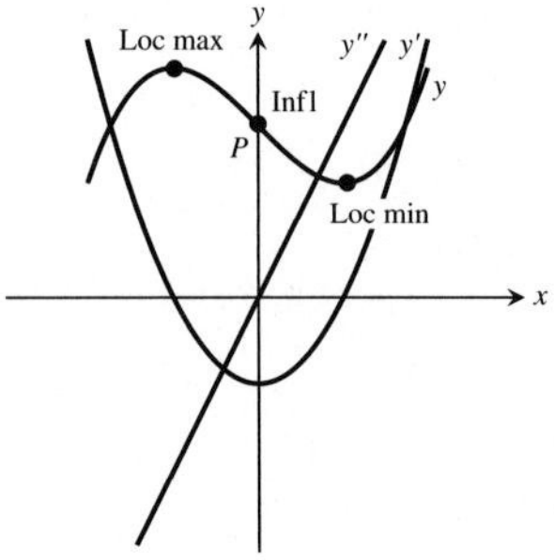

65.

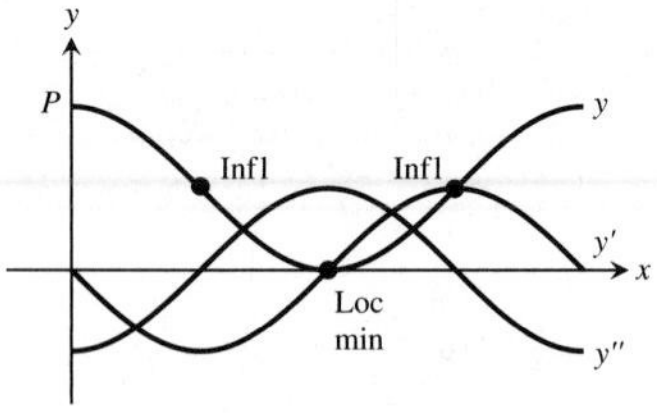

67.

Point	y'	y''
P	−	+
Q	+	0
R	+	−
S	0	−
T	−	−

69.

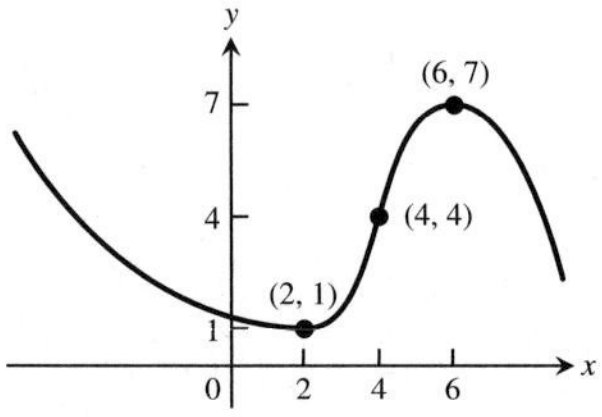

73. ≈ 60 thousand units **75.** Local minimum at $x = 2$, inflection points at $x = 1$ and $x = 5/3$ **79.** $b = -3$

81. (a) $\left(-\frac{b}{2a}, \frac{4ac - b^2}{4a}\right)$ **(b)** concave up if $a > 0$, concave down if $a < 0$

85. The zeros of $y' = 0$ and $y'' = 0$ are extrema and points of inflection, respectively. Inflection at $x = 3$, local maximum at $x = 0$, local minimum at $x = 4$.

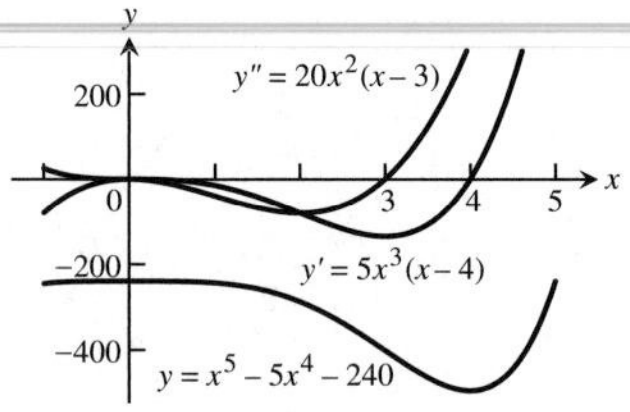

87. The zeros of $y' = 0$ and $y'' = 0$ are extrema and points of inflection, respectively. Inflection at $x = -\sqrt[3]{2}$, local maximum at $x = -2$, local minimum at $x = 0$.

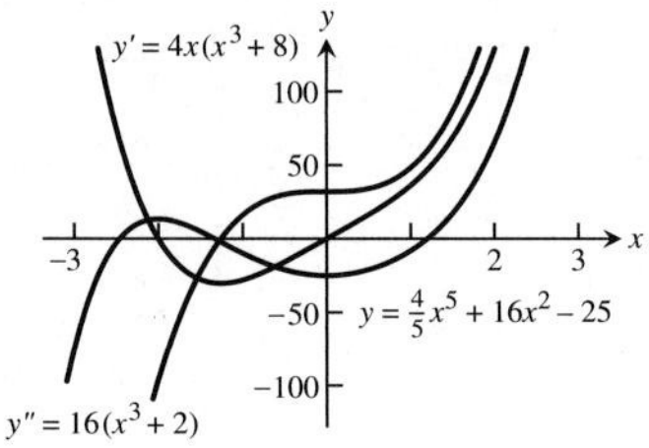

91. **(b)** $f'(x) = 3x^2 + k$; $-12k$; positive if $k < 0$, negative if $k > 0$, 0 if $k = 0$; f' has two zeros if $k < 0$, one zero if $k = 0$, no zeros if $k > 0$; that is, the sign of k controls the number of local extrema.

93. **(b)** A cusp since $\lim_{x\to 0^-} y' = \infty$ and $\lim_{x\to 0^+} y' = -\infty$

95. Yes, the graph of y' crosses through zero near -3, so y has a horizontal tangent near -3.

Section 4.5, pp. 285–292

1. 16 in., 4 in. by 4 in. **3.** **(a)** $(x, 1-x)$

(b) $A(x) = 2x(1-x)$ **(c)** $\frac{1}{2}$ square units, 1 by $\frac{1}{2}$

5. $\frac{14}{3} \times \frac{35}{3} \times \frac{5}{3}$ in., $\frac{2450}{27}$ in.^{3} **7.** 80,000 m^2; 400 m by 200 m

9. **(a)** The optimum dimensions of the tank are 10 ft on the base edges and 5 ft deep.

(b) Minimizing the surface area of the tank minimizes its weight for a given wall thickness. The thickness of the steel walls would likely be determined by other considerations such as structural requirements.

11. 9×18 in. **13.** $\frac{\pi}{2}$ **15.** $h : r = 8 : \pi$

17. **(a)** $V(x) = 2x(24 - 2x)(18 - 2x)$ **(b)** Domain: (0, 9)

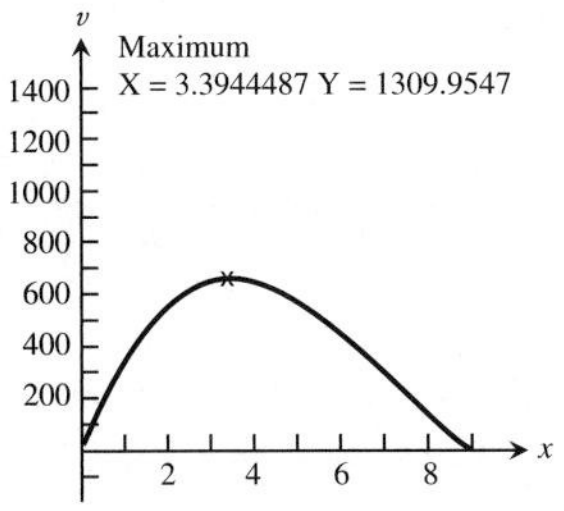

(c) Maximum volume ≈ 1309.95 in.^{3} when $x \approx 3.39$ in.

(d) $V'(x) = 24x^2 - 336x + 864$, so the critical point is at $x = 7 - \sqrt{13}$, which confirms the result in part (c).

(e) $x = 2$ in. or $x = 5$ in.

19. ≈ 2418.40 cm^3

21. **(a)** $h = 24$, $w = 18$

(b)

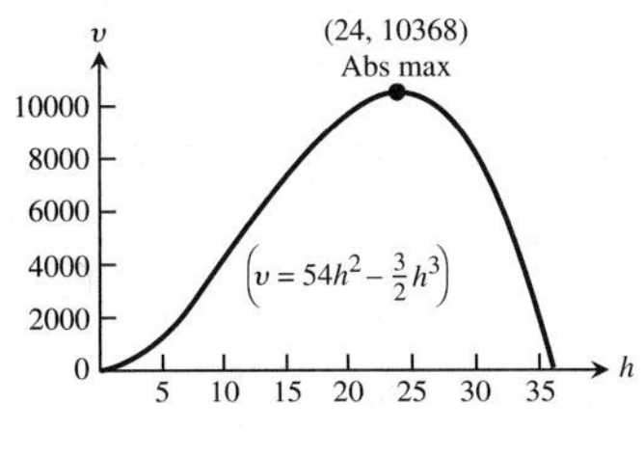

23. If r is the radius of the hemisphere, h the height of the cylinder, and V the volume, then $r = \left(\frac{3V}{8\pi}\right)^{1/3}$ and $h = \left(\frac{3V}{\pi}\right)^{1/3}$.

25. **(b)** $x = \frac{51}{8}$ **(c)** $L \approx 11$ in.

27. Radius $= \sqrt{2}$ m, height $= 1$ m, volume $\frac{2\pi}{3}$ m^3

31. **(a)** $v(0) = 96$ ft/sec **(b)** 256 ft at $t = 3$ sec

(c) Velocity when $s = 0$ is $v(7) = -128$ ft/sec

33. ≈ 46.87 ft **35.** **(a)** $6 \times 6\sqrt{3}$ in.

37. **(a)** $10\pi \approx 31.42$ cm/sec; when $t = 0.5$ sec, 1.5 sec, 2.5 sec, 3.5 sec; $s = 0$, acceleration is 0 **(b)** 10 cm from rest position; speed is 0

39. **(a)** $s = ((12 - 12t)^2 + 64t^2)^{1/2}$ **(b)** -12 knots, 8 knots

(c) No **(e)** $4\sqrt{13}$. This limit is the square root of the sums of the squares of the individual speeds.

41. $x = \frac{a}{2}$, $v = \frac{ka^2}{4}$ **43.** $\frac{c}{2} + 50$

45. **(a)** $\sqrt{\frac{2km}{h}}$ **(b)** $\sqrt{\frac{2km}{h}}$

49. **(a)** The cabinetmaker should order px units of material to have enough until the next delivery.

(c) Average cost per day $= \dfrac{\left(d + \frac{ps}{2}x^2\right)}{x} = \frac{d}{x} + \frac{ps}{2}x$;

$x^* = \sqrt{\frac{2d}{ps}}$; $px^* = \sqrt{\frac{2pd}{s}}$ gives a minimum.

(d) The line and hyperbola intersect when $\frac{d}{x} = \frac{ps}{2}x$. For $x > 0$, $x_{\text{intersection}} = \sqrt{\frac{2d}{ps}} = x^*$. The average cost per day is minimized when the average daily cost of delivery is equal to the average daily cost of storage.

51. $M = \frac{C}{2}$ **57.** **(a)** $y = -1$

59. **(a)** The minimum distance is $\frac{\sqrt{5}}{2}$.

(b) The minimum distance is from the point (3/2, 0) to the point (1, 1) on the graph of $y = \sqrt{x}$, and this occurs at the value $x = 1$, where $D(x)$, the distance squared, has its minimum value.

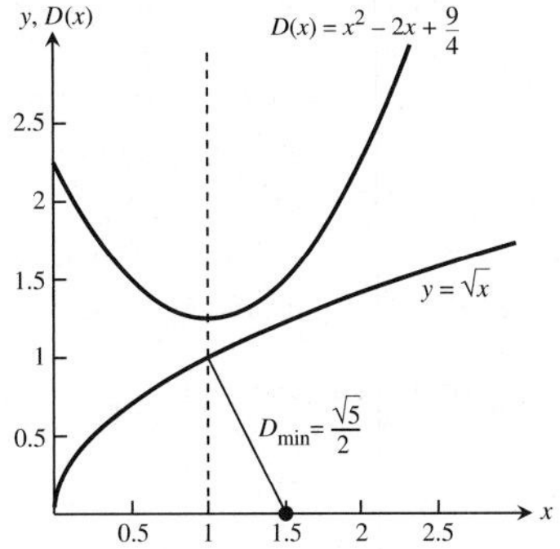

61. **(a)** $V(x) = \frac{\pi}{3}\left(\frac{2\pi a - x}{2\pi}\right)^2 \sqrt{a^2 - \left(\frac{2\pi a - x}{2\pi}\right)^2}$

(b) When $a = 4$: $r = \dfrac{4\sqrt{6}}{3}, h = \dfrac{4\sqrt{3}}{3}$; when $a = 5$: $r = \dfrac{5\sqrt{6}}{3}, h = \dfrac{5\sqrt{3}}{3}$; when $a = 6$: $r = 2\sqrt{6}, h = 2\sqrt{3}$; when $a = 8$: $r = \dfrac{8\sqrt{6}}{3}, h = \dfrac{8\sqrt{3}}{3}$

(c) Since $r = \dfrac{a\sqrt{6}}{3}$ and $h = \dfrac{a\sqrt{3}}{3}$, the relationship is $\dfrac{r}{h} = \sqrt{2}$.

Section 4.6, pp. 298–299

1. $\frac{1}{4}$ **3.** $\frac{5}{7}$ **5.** $\frac{1}{2}$ **7.** 0 **9.** 1 **11.** $\sqrt{2}$ **13.** $+1$

15. -1 **17.** $\frac{1}{2}$ **19.** 3 **21.** na **23.** $-\frac{1}{2}$ **25.** 0 **27.** 3

29. 1 **31.** (b) is correct, but (a) is not.

33. $c = \frac{27}{10}$ **35.** -1

Section 4.7, pp. 305–306

1. $x_2 = -\frac{5}{3}, \frac{13}{21}$ **3.** $x_2 = -\frac{51}{31}, \frac{5763}{4945}$ **5.** $x_2 = \frac{2387}{2000}$

7. x, and all later approximations will equal x_0.

9.

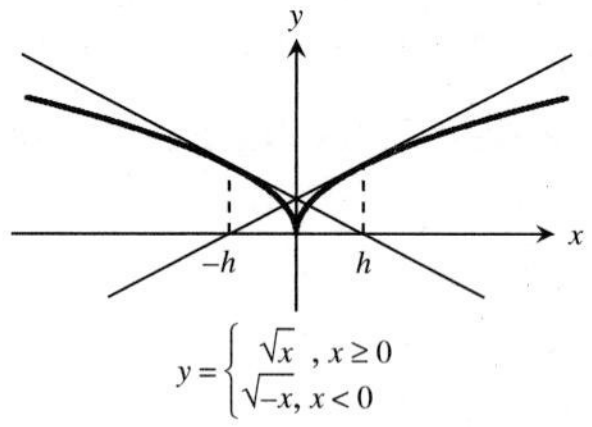

11. The points of intersection of $y = x^3$ and $y = 3x + 1$ or $y = x^3 - 3x$ and $y = 1$ have the same x-values as the roots of part (i) or the solutions of part (iv). **15.** 1.165561185

17. **(a)** Two **(b)** 0.35003501505249 and -1.0261731615301

19. ± 1.3065629648764, ± 0.5411961001462 **21.** $x \approx 0.45$

23. The root is 1.17951.

25. **(a)** For $x_0 = -2$ or $x_0 = -0.8$, $x_i \to -1$ as i gets large.
(b) For $x_0 = -0.5$ or $x_0 = 0.25$, $x_i \to 0$ as i gets large.
(c) For $x_0 = 0.8$ or $x_0 = 2$, $x_i \to 1$ as i gets large.
(d) For $x_0 = -\sqrt{21}/7$ or $x_0 = \sqrt{21}/7$, Newton's method does not converge. The values of x_i alternate between $-\sqrt{21}/7$ and $\sqrt{21}/7$ as i increases.

27. Answers will vary with machine speed. **29.** 2.45, 0.000245

Section 4.8, pp. 314–318

1. **(a)** x^2 **(b)** $\frac{x^3}{3}$ **(c)** $\frac{x^3}{3} - x^2 + x$

3. **(a)** x^{-3} **(b)** $-\frac{1}{3}x^{-3}$ **(c)** $-\frac{1}{3}x^{-3} + x^2 + 3x$

5. **(a)** $-\frac{1}{x}$ **(b)** $-\frac{5}{x}$ **(c)** $2x + \frac{5}{x}$

7. **(a)** $\sqrt{x^3}$ **(b)** $\sqrt{x}$ **(c)** $\frac{2\sqrt{x^3}}{3} + 2\sqrt{x}$

9. **(a)** $x^{2/3}$ **(b)** $x^{1/3}$ **(c)** $x^{-1/3}$

11. **(a)** $\cos(\pi x)$ **(b)** $-3\cos x$ **(c)** $-\frac{1}{\pi}\cos(\pi x) + \cos(3x)$

13. **(a)** $\tan x$ **(b)** $2\tan\left(\frac{x}{3}\right)$ **(c)** $-\frac{2}{3}\tan\left(\frac{3x}{2}\right)$

15. **(a)** $-\csc x$ **(b)** $\frac{1}{5}\csc(5x)$ **(c)** $2\csc\left(\frac{\pi x}{2}\right)$

17. $\frac{x^2}{2} + x + C$ **19.** $t^3 + \frac{t^2}{4} + C$ **21.** $\frac{x^4}{2} - \frac{5x^2}{2} + 7x + C$

23. $-\frac{1}{x} - \frac{x^3}{3} - \frac{x}{3} + C$ **25.** $\frac{3}{2}x^{2/3} + C$

27. $\frac{2}{3}x^{3/2} + \frac{3}{4}x^{4/3} + C$ **29.** $4y^2 - \frac{8}{3}y^{3/4} + C$

31. $x^2 + \frac{2}{x} + C$ **33.** $2\sqrt{t} - \frac{2}{\sqrt{t}} + C$ **35.** $-2\sin t + C$

37. $-21\cos\frac{\theta}{3} + C$ **39.** $3\cot x + C$ **41.** $-\frac{1}{2}\csc\theta + C$

43. $4\sec x - 2\tan x + C$ **45.** $-\frac{1}{2}\cos 2x + \cot x + C$

47. $\frac{t}{2} + \frac{\sin 4t}{8} + C$ **49.** $\tan\theta + C$

51. $-\cot x - x + C$ **53.** $-\cos\theta - \theta + C$

61. **(a)** Wrong: $\frac{d}{dx}\left(\frac{x^2}{2}\sin x + C\right) = \frac{2x}{2}\sin x + \frac{x^2}{2}\cos x = x\sin x + \frac{x^2}{2}\cos x$

(b) Wrong: $\frac{d}{dx}(-x\cos x + C) = -\cos x + x\sin x$

(c) Right: $\frac{d}{dx}(-x\cos x + \sin x + C) = -\cos x + x\sin x + \cos x = x\sin x$

63. **(a)** Wrong: $\frac{d}{dx}\left(\frac{(2x+1)^3}{3} + C\right) = \frac{3(2x+1)^2(2)}{3} = 2(2x+1)^2$

(b) Wrong: $\frac{d}{dx}((2x+1)^3 + C) = 3(2x+1)^2(2) = 6(2x+1)^2$

(c) Right: $\frac{d}{dx}((2x+1)^3 + C) = 6(2x+1)^2$

65. (b) **67.** $y = x^2 - 7x + 10$ **69.** $y = -\frac{1}{x} + \frac{x^2}{2} - \frac{1}{2}$

71. $y = 9x^{1/3} + 4$ **73.** $s = t + \sin t + 4$

75. $r = \cos(\pi\theta) - 1$ **77.** $v = \frac{1}{2}\sec t + \frac{1}{2}$

79. $y = x^2 - x^3 + 4x + 1$ **81.** $r = \frac{1}{t} + 2t - 2$

83. $y = x^3 - 4x^2 + 5$ **85.** $y = -\sin t + \cos t + t^3 - 1$

87. $y = 2x^{3/2} - 50$ **89.** $y = x - x^{4/3} + \frac{1}{2}$

91. $y = -\sin x - \cos x - 2$

93. **(a)** **(i)** 33.2 units, **(ii)** 33.2 units, **(iii)** 33.2 units **(b)** True

95. $t = 88/k,\ k = 16$

97. **(a)** $v = 10t^{3/2} - 6t^{1/2}$ **(b)** $s = 4t^{5/2} - 4t^{3/2}$

101. **(a)** $-\sqrt{x} + C$ **(b)** $x + C$ **(c)** $\sqrt{x} + C$
(d) $-x + C$ **(e)** $x - \sqrt{x} + C$ **(f)** $-x - \sqrt{x} + C$

Practice Exercises, pp. 318–321

1. No **3.** No minimum; absolute maximum: $f(1) = 16$; critical points: $x = 1$ and $11/3$ **5.** Yes, except at $x = 0$ **7.** No

11. **(b)** one

13. **(b)** 0.8555 99677 2 **19.** Global minimum value of $\frac{1}{2}$ at $x = 2$

21. **(a)** $t = 0, 6, 12$ **(b)** $t = 3, 9$ **(c)** $6 < t < 12$
(d) $0 < t < 6,\ 12 < t < 14$

23.

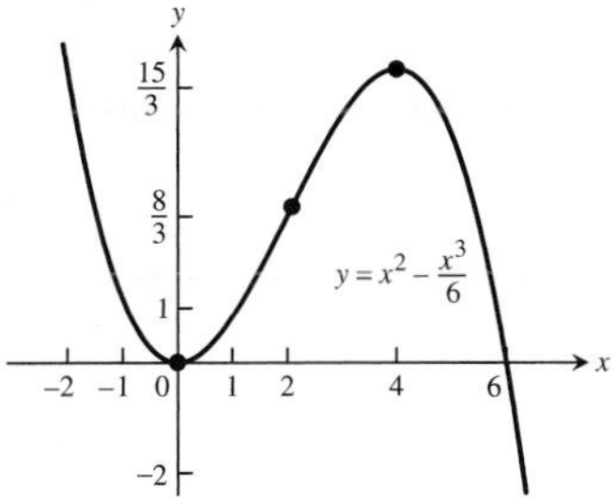

25.

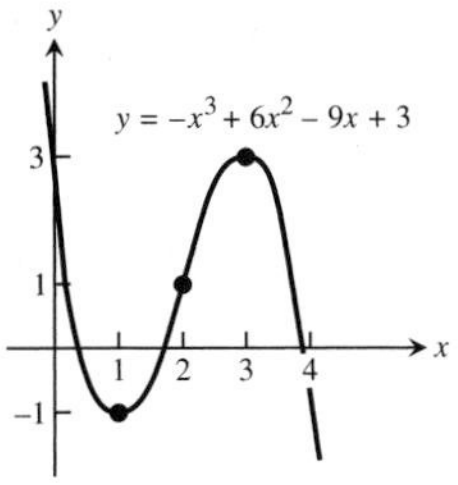

27.

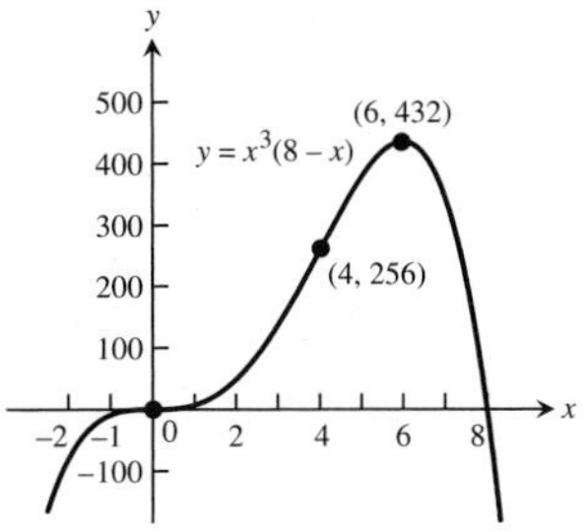

29.

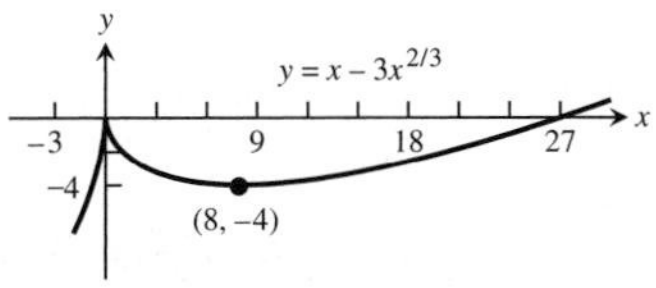

31.

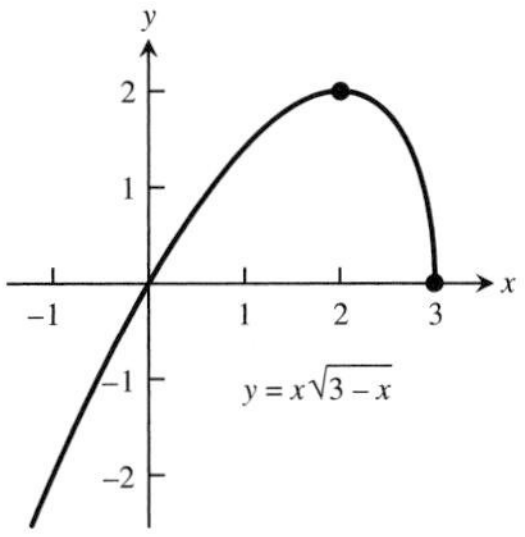

33. **(a)** Local maximum at $x = 4$, local minimum at $x = -4$, inflection point at $x = 0$

(b)

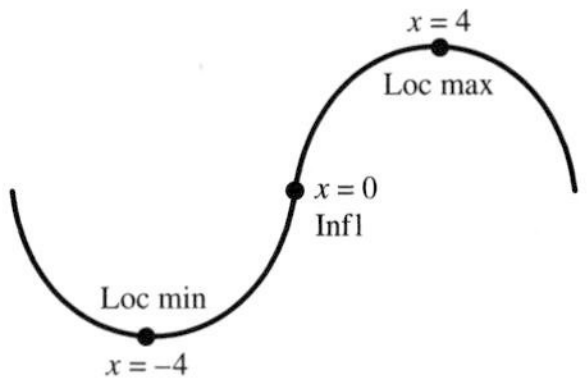

35. **(a)** Local maximum at $x = 0$, local minima at $x = -1$ and $x = 2$, inflection points at $x = \left(1 \pm \sqrt{7}\right)/3$

(b)

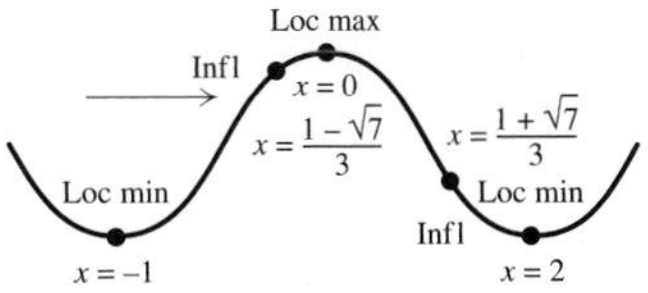

37. **(a)** Local maximum at $x = -\sqrt{2}$, local minimum at $x = \sqrt{2}$, inflection points at $x = \pm 1$ and 0

(b)

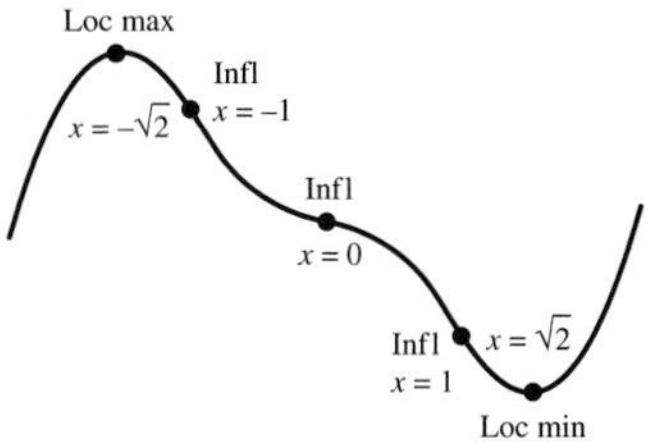

43.

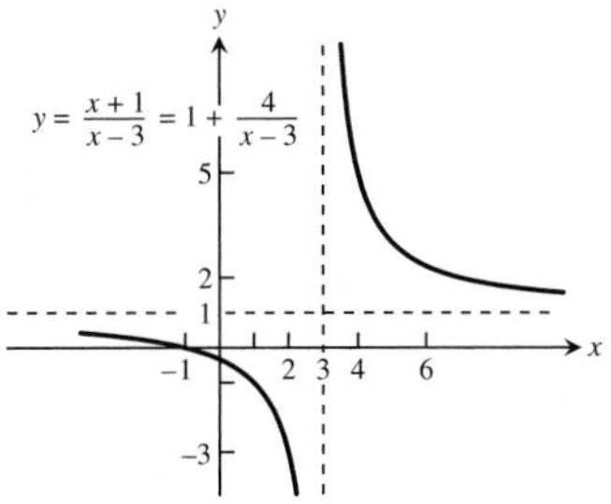

45.

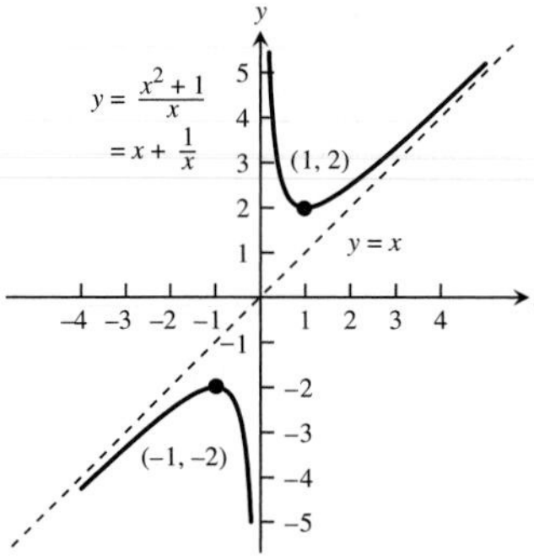

47.

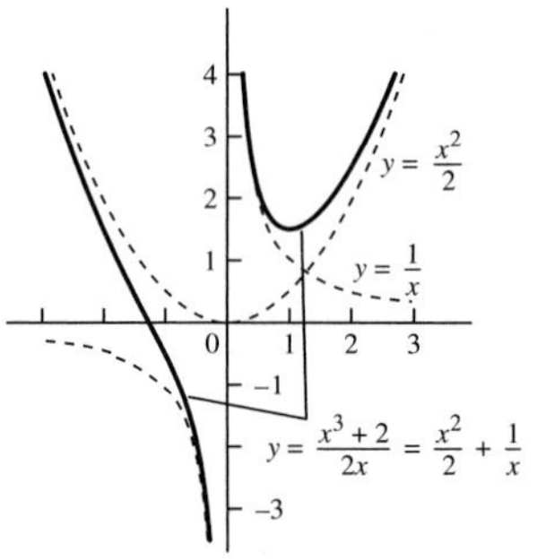

49.

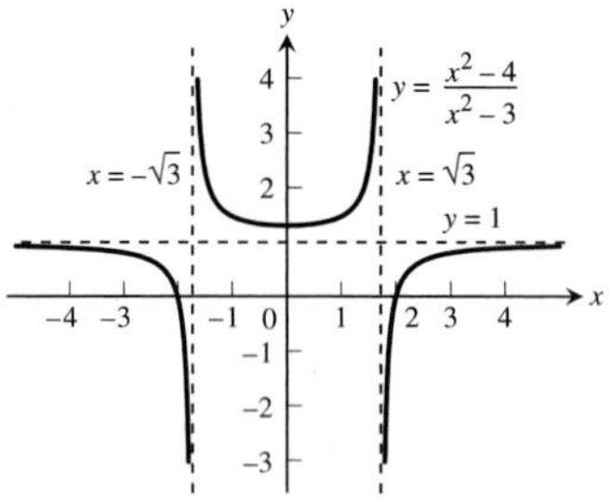

51. 5 **53.** 0 **55.** 1 **57.** 3/7 **59.** 0 **61.** 1
63. **(a)** 0, 36 **(b)** 18, 18 **65.** 54 square units
67. height = 2, radius = $\sqrt{2}$
69. $x = 5 - \sqrt{5}$ hundred ≈ 276 tires,
$y = 2(5 - \sqrt{5})$ hundred ≈ 553 tires
71. Dimensions: base is 6 in. by 12 in., height = 2 in.; maximum volume = 144 in.^{3}
73. $x_5 = 2.1958\,23345$ **75.** $\frac{x^4}{4} + \frac{5}{2}x^2 - 7x + C$
77. $2t^{3/2} - \frac{4}{t} + C$ **79.** $-\frac{1}{r+5} + C$
81. $(\theta^2 + 1)^{3/2} + C$ **83.** $\frac{1}{3}(1 + x^4)^{3/4} + C$
85. $10 \tan \frac{s}{10} + C$ **87.** $-\frac{1}{\sqrt{2}} \csc \sqrt{2}\,\theta + C$
89. $\frac{1}{2}x - \sin \frac{x}{2} + C$ **91.** $y = x - \frac{1}{x} - 1$
93. $r = 4t^{5/2} + 4t^{3/2} - 8t$

Additional and Advanced Exercises, pp. 322–324

1. The function is constant on the interval.
3. The extreme points will not be at the end of an open interval.
5. **(a)** A local minimum at $x = -1$, points of inflection at $x = 0$ and $x = 2$ **(b)** A local maximum at $x = 0$ and local minima at $x = -1$ and $x = 2$, points of inflection at $x = \frac{1 \pm \sqrt{7}}{3}$
9. No **11.** $a = 1, b = 0, c = 1$ **13.** Yes **15.** Drill the hole at $y = h/2$.
17. $r = \frac{RH}{2(H - R)}$ for $H > 2R$, $r = R$ if $H \le 2R$
19. **(a)** $\frac{10}{3}$ **(b)** $\frac{5}{3}$ **(c)** $\frac{1}{2}$ **(d)** 0 **(e)** $-\frac{1}{2}$ **(f)** 1 **(g)** $\frac{1}{2}$ **(h)** 3
21. **(a)** $\frac{c - b}{2e}$ **(b)** $\frac{c + b}{2}$ **(c)** $\frac{b^2 - 2bc + c^2 + 4ae}{4e}$
(d) $\frac{c + b + t}{2}$
23. $m_0 = 1 - \frac{1}{q}, m_1 = \frac{1}{q}$
25. **(a)** $k = -38.72$ **(b)** 25 ft
27. Yes, $y = x + C$ **29.** $v_0 = \frac{2\sqrt{2}}{3} b^{3/4}$

CHAPTER 5

Section 5.1, pp. 333–335

1. **(a)** 0.125 **(b)** 0.21875 **(c)** 0.625 **(d)** 0.46875
3. **(a)** 1.066667 **(b)** 1.283333 **(c)** 2.666667 **(d)** 2.083333
5. 0.3125, 0.328125 **7.** 1.5, 1.574603
9. **(a)** 87 in. **(b)** 87 in.
11. **(a)** 3490 ft **(b)** 3840 ft
13. **(a)** 74.65 ft/sec **(b)** 45.28 ft/sec **(c)** 146.59 ft
15. $\frac{31}{16}$ **17.** 1
19. **(a)** Upper = 758 gal, lower = 543 gal
(b) Upper = 2363 gal, lower = 1693 gal
(c) ≈ 31.4 h, ≈ 32.4 h
21. **(a)** 2 **(b)** $2\sqrt{2} \approx 2.828$
(c) $8 \sin\left(\frac{\pi}{8}\right) \approx 3.061$
(d) Each area is less than the area of the circle, π. As n increases, the polygon area approaches π.

Section 5.2, pp. 342–343

1. $\frac{6(1)}{1+1} + \frac{6(2)}{2+1} = 7$
3. $\cos(1)\pi + \cos(2)\pi + \cos(3)\pi + \cos(4)\pi = 0$
5. $\sin \pi - \sin \frac{\pi}{2} + \sin \frac{\pi}{3} = \frac{\sqrt{3} - 2}{2}$

7. All of them **9.** b **11.** $\sum_{k=1}^{6} k$ **13.** $\sum_{k=1}^{4} \frac{1}{2^k}$

15. $\sum_{k=1}^{5} (-1)^{k+1} \frac{1}{k}$

17. **(a)** -15 **(b)** 1 **(c)** 1 **(d)** -11 **(e)** 16

19. **(a)** 55 **(b)** 385 **(c)** 3025

21. -56 **23.** -73 **25.** 240 **27.** 3376

29. **(a)**

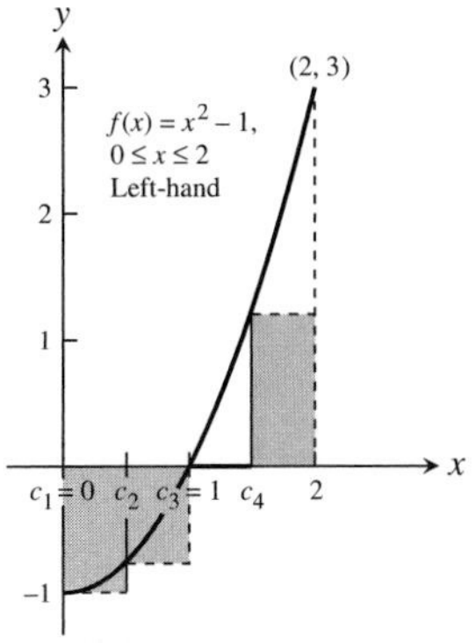

(b)

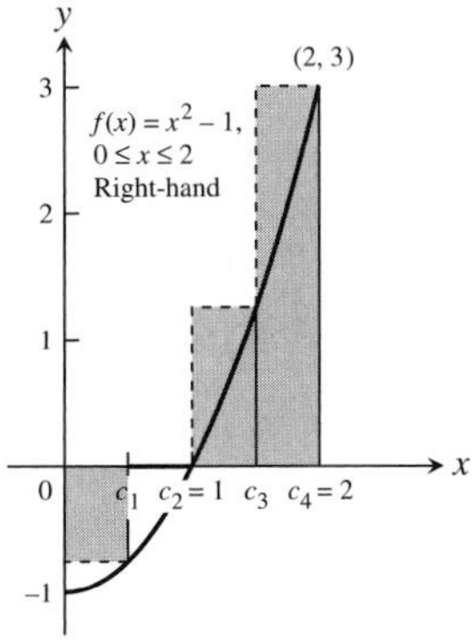

(c)

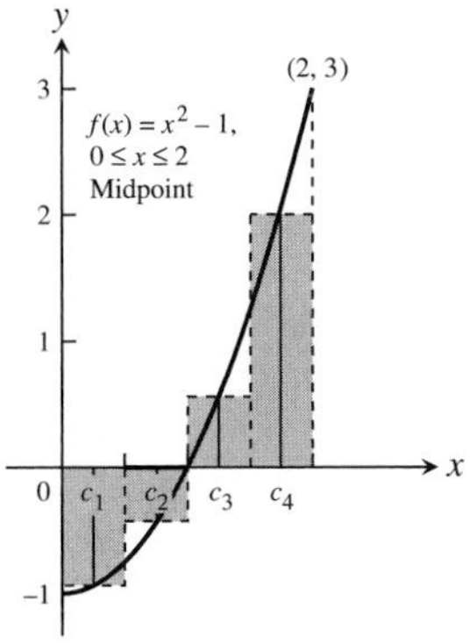

31. **(a)**

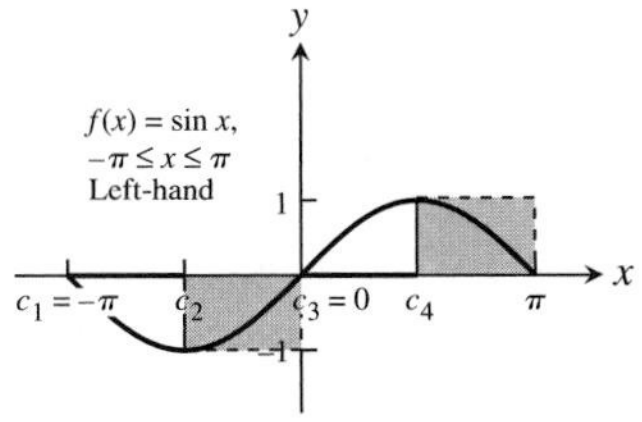

(b)

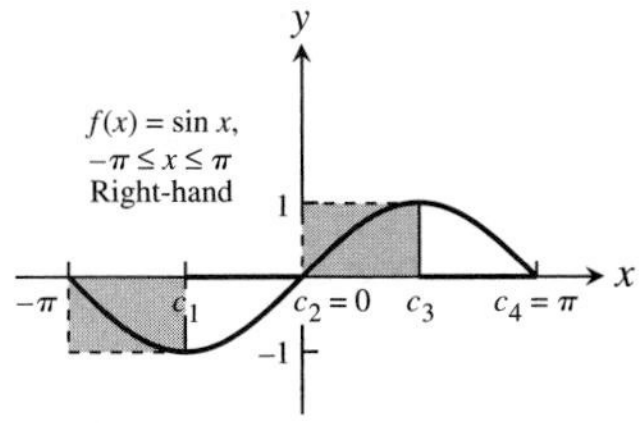

(c)

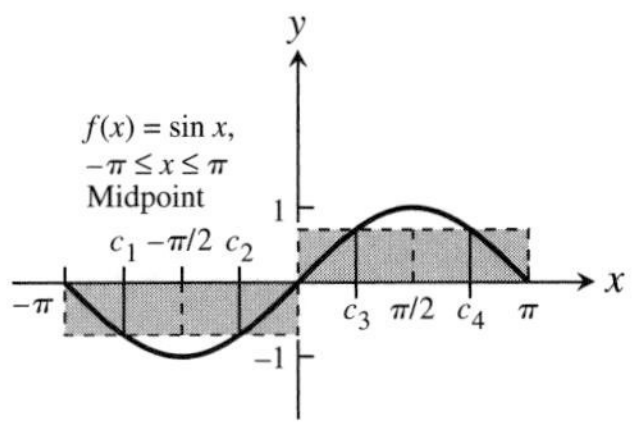

33. 1.2

35. $\frac{2}{3} + \frac{3n-1}{6n^2}, \frac{2}{3}$

37. $12 + \frac{27n+9}{2n^2}, 12$

39. $\frac{5}{6} + \frac{6n+1}{6n^2}, \frac{5}{6}$

Section 5.3, pp. 352–356

1. $\int_0^2 x^2\,dx$ **3.** $\int_{-7}^{5} (x^2 - 3x)\,dx$ **5.** $\int_2^3 \frac{1}{1-x}\,dx$

7. $\int_{-\pi/4}^{0} \sec x\,dx$

9. **(a)** 0 **(b)** -8 **(c)** -12 **(d)** 10 **(e)** -2 **(f)** 16

11. **(a)** 5 **(b)** $5\sqrt{3}$ **(c)** -5 **(d)** -5

13. **(a)** 4 **(b)** -4 **15.** Area $= 21$ square units

17. Area $= 9\pi/2$ square units **19.** Area $= 2.5$ square units

21. Area $= 3$ square units **23.** $b^2/4$ **25.** $b^2 - a^2$ **27.** $1/2$

29. $3\pi^2/2$ **31.** $7/3$ **33.** $1/24$ **35.** $3a^2/2$ **37.** $b/3$

39. -14 **41.** 10 **43.** -2 **45.** $-7/4$ **47.** 7 **49.** 0

51. Using n subintervals of length $\Delta x = b/n$ and right-endpoint values:

$$\text{Area} = \int_0^b 3x^2\,dx = b^3$$

53. Using n subintervals of length $\Delta x = b/n$ and right-endpoint values:

$$\text{Area} = \int_0^b 2x\,dx = b^2$$

55. $\text{av}(f) = 0$ **57.** $\text{av}(f) = -2$ **59.** $\text{av}(f) = 1$

61. **(a)** $\text{av}(g) = -1/2$ **(b)** $\text{av}(g) = 1$ **(c)** $\text{av}(g) = 1/4$

63. $a = 0$ and $b = 1$ maximize the integral.

65. Upper bound $= 1$, lower bound $= 1/2$

67. For example, $\int_0^1 \sin(x^2)\,dx \le \int_0^1 dx = 1$

69. $\int_a^b f(x)\,dx \ge \int_a^b 0\,dx = 0$ **71.** Upper bound $= 1/2$

Section 5.4, pp. 365–368

1. 6 **3.** 8 **5.** 1 **7.** $5/2$ **9.** 2 **11.** $2\sqrt{3}$ **13.** 0

15. $-\pi/4$ **17.** $\frac{2\pi^3}{3}$ **19.** $-8/3$ **21.** $-3/4$

23. $\sqrt{2} - \sqrt[4]{8} + 1$ **25.** 16 **27.** $(\cos\sqrt{x})\left(\frac{1}{2\sqrt{x}}\right)$

29. $4t^5$ **31.** $\sqrt{1+x^2}$ **33.** $-\frac{1}{2}x^{-1/2}\sin x$ **35.** 1

37. 28/3 **39.** 1/2 **41.** 51/4 **43.** π **45.** $\frac{\sqrt{2}\pi}{2}$

47. d, since $y' = \frac{1}{x}$ and $y(\pi) = \int_\pi^\pi \frac{1}{t}\,dt - 3 = -3$

49. b, since $y' = \sec x$ and $y(0) = \int_0^0 \sec t\,dt + 4 = 4$

51. $y = \int_2^x \sec t\,dt + 3$ **53.** $s = \int_{t_0}^t f(x)\,dx + s_0$

55. $\frac{2}{3}bh$ **57.** \$9.00

59. **(a)** $v = \frac{ds}{dt} = \frac{d}{dt}\int_0^t f(x)\,dx = f(t) \Rightarrow v(5) = f(5) = 2$ m/sec

(b) $a = df/dt$ is negative since the slope of the tangent line at $t = 5$ is negative.

(c) $s = \int_0^3 f(x)\,dx = \frac{1}{2}(3)(3) = \frac{9}{2}$ m since the integral is the area of the triangle formed by $y = f(x)$, the x-axis, and $x = 3$.

(d) $t = 6$ since after $t = 6$ to $t = 9$, the region lies below the x-axis.

(e) At $t = 4$ and $t = 7$, since there are horizontal tangents there.

(f) Toward the origin between $t = 6$ and $t = 9$ since the velocity is negative on this interval. Away from the origin between $t = 0$ and $t = 6$ since the velocity is positive there.

(g) Right or positive side, because the integral of f from 0 to 9 is positive, there being more area above the x-axis than below.

63. $2x - 2$ **65.** $-3x + 5$

67. **(a)** True. Since f is continuous, g is differentiable by Part 1 of the Fundamental Theorem of Calculus.

(b) True: g is continuous because it is differentiable.

(c) True, since $g'(1) = f(1) = 0$.

(d) False, since $g''(1) = f'(1) > 0$.

(e) True, since $g'(1) = 0$ and $g''(1) = f'(1) > 0$.

(f) False: $g''(x) = f'(x) > 0$, so g'' never changes sign.

(g) True, since $g'(1) = f(1) = 0$ and $g'(x) = f(x)$ is an increasing function of x (because $f'(x) > 0$).

Section 5.5, pp. 374–376

1. $-\frac{1}{3}\cos 3x + C$ **3.** $\frac{1}{2}\sec 2t + C$ **5.** $-(7x-2)^{-4} + C$

7. $-6(1 - r^3)^{1/2} + C$

9. $\frac{1}{3}(x^{3/2} - 1) - \frac{1}{6}\sin(2x^{3/2} - 2) + C$

11. **(a)** $-\frac{1}{4}(\cot^2 2\theta) + C$ **(b)** $-\frac{1}{4}(\csc^2 2\theta) + C$

13. $-\frac{1}{3}(3 - 2s)^{3/2} + C$ **15.** $\frac{2}{5}(5s + 4)^{1/2} + C$

17. $-\frac{2}{5}(1 - \theta^2)^{5/4} + C$ **19.** $-\frac{1}{3}(7 - 3y^2)^{3/2} + C$

21. $(-2/(1 + \sqrt{x})) + C$ **23.** $\frac{1}{3}\sin(3z + 4) + C$

25. $\frac{1}{3}\tan(3x + 2) + C$ **27.** $\frac{1}{2}\sin^6\left(\frac{x}{3}\right) + C$

29. $\left(\frac{r^3}{18} - 1\right)^6 + C$ **31.** $-\frac{2}{3}\cos(x^{3/2} + 1) + C$

33. $\sec\left(v + \frac{\pi}{2}\right) + C$ **35.** $\frac{1}{2\cos(2t + 1)} + C$

37. $-\frac{2}{3}(\cot^3 y)^{1/2} + C$ **39.** $-\sin\left(\frac{1}{t} - 1\right) + C$

41. $-\frac{\sin^2(1/\theta)}{2} + C$ **43.** $\frac{(s^3 + 2s^2 - 5s + 5)^2}{2} + C$

45. $\frac{1}{16}(1 + t^4)^4 + C$

47. $\frac{1}{5}(x^2 + 1)^{5/2} - \frac{1}{3}(x^2 + 1)^{3/2} + C$

49. **(a)** $-\frac{6}{2 + \tan^3 x} + C$ **(b)** $-\frac{6}{2 + \tan^3 x} + C$

(c) $-\frac{6}{2 + \tan^3 x} + C$

51. $\frac{1}{6}\sin\sqrt{3(2r - 1)^2 + 6} + C$

53. $s = \frac{1}{2}(3t^2 - 1)^4 - 5$

55. $s = 4t - 2\sin\left(2t + \frac{\pi}{6}\right) + 9$

57. $s = \sin\left(2t - \frac{\pi}{2}\right) + 100t + 1$

59. 6 m **63.** b) 399 Volts

Section 5.6, pp. 383–387

1. **(a)** 14/3 **(b)** 2/3
3. **(a)** 1/2 **(b)** $-1/2$
5. **(a)** 15/16 **(b)** 0
7. **(a)** 0 **(b)** 1/8
9. **(a)** 4 **(b)** 0
11. **(a)** 1/6 **(b)** 1/2
13. **(a)** 0 **(b)** 0
15. $2\sqrt{3}$ **17.** 3/4 **19.** $3^{5/2} - 1$ **21.** 3 **23.** $\pi/3$
25. 16/3 **27.** $2^{5/2}$ **29.** $\pi/2$ **31.** 128/15 **33.** 4/3
35. 5/6 **37.** 38/3 **39.** 49/6 **41.** 32/3 **43.** 48/5
45. 8/3 **47.** 8 **49.** 5/3 (There are three intersection points.)
51. 18 **53.** 243/8 **55.** 8/3 **57.** 2 **59.** 104/15
61. 56/15 **63.** 4 **65.** $\frac{4}{3} - \frac{4}{\pi}$ **67.** $\pi/2$ **69.** 2 **71.** 1/2
73. 1
75. **(a)** $(\pm\sqrt{c}, c)$ **(b)** $c = 4^{2/3}$ **(c)** $c = 4^{2/3}$
77. 11/3 **79.** 3/4 **81.** Neither **83.** $F(6) - F(2)$
85. **(a)** -3 **(b)** 3
87. $I = a/2$

Practice Exercises, pp. 388–391

1. **(a)** about 680 ft **(b)**

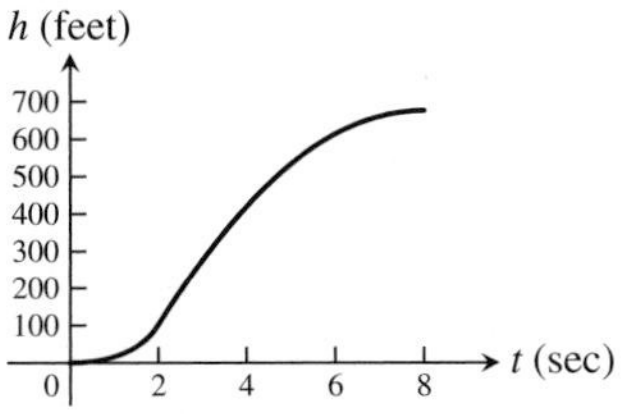

3. **(a)** $-1/2$ **(b)** 31 **(c)** 13 **(d)** 0

5. $\int_1^5 (2x-1)^{-1/2}\,dx = 2$ **7.** $\int_{-\pi}^{0} \cos\frac{x}{2}\,dx = 2$

9. **(a)** 4 **(b)** 2 **(c)** -2 **(d)** -2π **(e)** $8/5$

11. $8/3$ **13.** 62 **15.** 1 **17.** $1/6$ **19.** 18 **21.** $9/8$

23. $\frac{\pi^2}{32} + \frac{\sqrt{2}}{2} - 1$ **25.** 4 **27.** $\frac{8\sqrt{2}-7}{6}$

29. Min: -4, max: 0, area: $27/4$ **31.** $6/5$

35. $y = \int_5^x \left(\frac{\sin t}{t}\right) dt - 3$ **37.** $-4(\cos x)^{1/2} + C$

39. $\theta^2 + \theta + \sin(2\theta + 1) + C$ **41.** $\frac{t^3}{3} + \frac{4}{t} + C$

43. $-\frac{1}{3}\cos(2t^{3/2}) + C$ **45.** 16 **47.** 2 **49.** 1 **51.** 8

53. $27\sqrt{3}/160$ **55.** $\pi/2$ **57.** $\sqrt{3}$ **59.** $6\sqrt{3} - 2\pi$

61. -1 **63.** 2 **65.** -2 **67.** 1 **69.** $\sqrt{2} - 1$

71. **(a)** b **(b)** b

75. 25°F **77.** $\sqrt{2 + \cos^3 x}$ **79.** $\frac{-6}{3 + x^4}$ **81.** Yes

83. $-\sqrt{1 + x^2}$

85. cost $\approx$ \$10,899 using a lower sum estimate

87. 600, \$18.00 **89.** 300, \$6.00

Additional and Advanced Exercises, pp. 391–394

1. **(a)** Yes **(b)** No

5. **(a)** $1/4$ **(b)** $\sqrt[3]{12}$

7. $f(x) = \frac{x}{\sqrt{x^2 + 1}}$ **9.** $y = x^3 + 2x - 4$

11. $36/5$

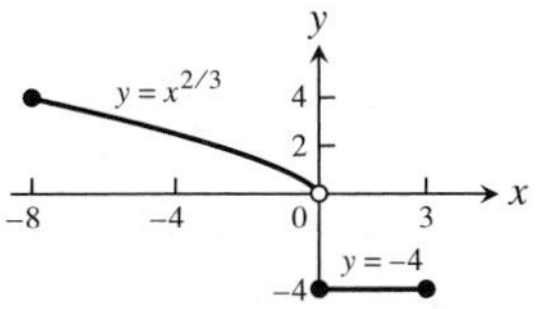

13. $\frac{1}{2} - \frac{2}{\pi}$

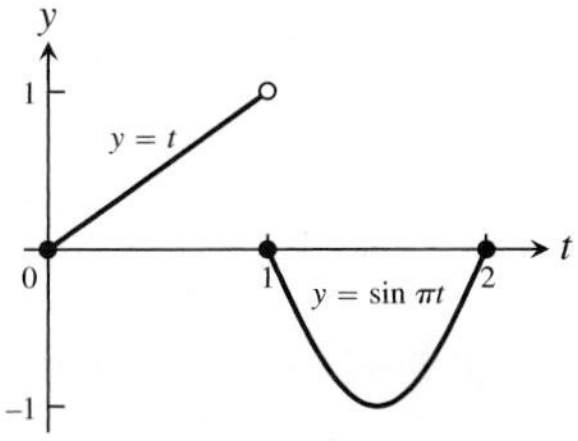

15. $13/3$

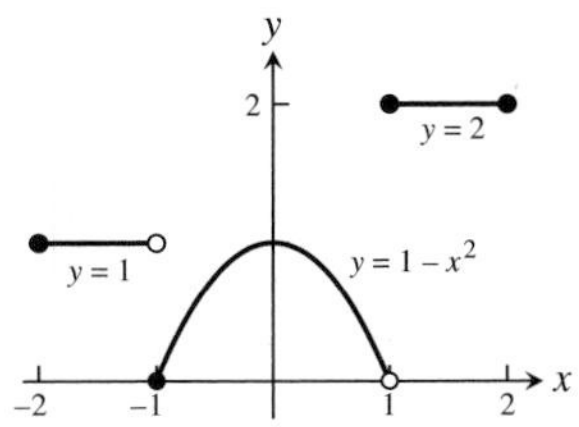

17. $1/2$ **19.** $2/x$

21. $\frac{\sin 4y}{\sqrt{y}} - \frac{\sin y}{2\sqrt{y}}$ **23.** $1/6$ **25.** $\int_0^1 f(x)\,dx$

29. **(a)** 0 **(b)** -1 **(c)** $-\pi$ **(d)** $x = 1$
(e) $y = 2x + 2 - \pi$ **(f)** $x = -1, x = 2$ **(g)** $[-2\pi, 0]$

CHAPTER 6

Section 6.1, pp. 405–409

1. **(a)** $A(x) = \pi(1 - x^2)$ **(b)** $A(x) = 4(1 - x^2)$
(c) $A(x) = 2(1 - x^2)$ **(d)** $A(x) = \sqrt{3}(1 - x^2)$

3. 16 **5.** $\frac{16}{3}$ **7.** **(a)** $2\sqrt{3}$ **(b)** 8 **9.** 8π

11. **(a)** s^2h **(b)** s^2h **13.** $\frac{2\pi}{3}$ **15.** $4 - \pi$ **17.** $\frac{32\pi}{5}$

19. 36π **21.** π **23.** $\pi\left(\frac{\pi}{2} + 2\sqrt{2} - \frac{11}{3}\right)$ **25.** 2π

27. 2π **29.** 3π **31.** $\pi^2 - 2\pi$ **33.** $\frac{2\pi}{3}$ **35.** $\frac{117\pi}{5}$

37. $\pi(\pi - 2)$ **39.** $\frac{4\pi}{3}$ **41.** 8π **43.** $\frac{7\pi}{6}$

45. **(a)** 8π **(b)** $\frac{32\pi}{5}$ **(c)** $\frac{8\pi}{3}$ **(d)** $\frac{224\pi}{15}$

47. **(a)** $\frac{16\pi}{15}$ **(b)** $\frac{56\pi}{15}$ **(c)** $\frac{64\pi}{15}$

49. $V = 2a^2b\pi^2$ **51.** **(a)** $V = \frac{\pi h^2(3a - h)}{3}$ **(b)** $\frac{1}{120\pi}$ m/sec

55. $V = 3308\ \text{cm}^3$

57. **(a)** $c = \frac{2}{\pi}$ **(b)** $c = 0$
(c)

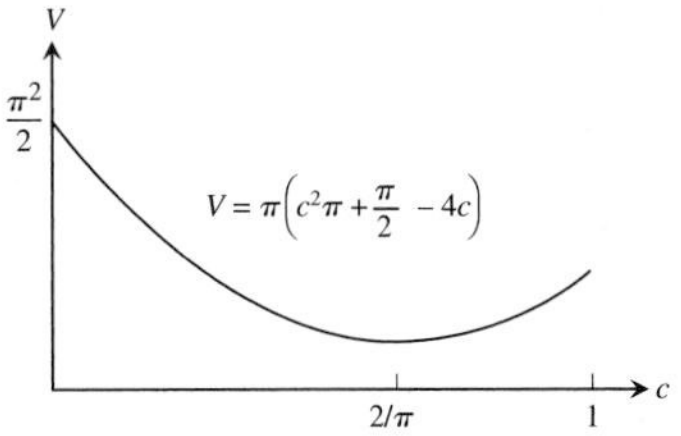

Section 6.2, pp. 414–416

1. 6π **3.** 2π **5.** $\frac{14\pi}{3}$ **7.** 8π **9.** $\frac{5\pi}{6}$ **11.** $\frac{7\pi}{15}$

13. **(b)** 4π **15.** $\frac{16\pi}{15}\left(3\sqrt{2} + 5\right)$ **17.** $\frac{8\pi}{3}$ **19.** $\frac{4\pi}{3}$

21. $\frac{16\pi}{3}$ **23.** **(a)** $\frac{6\pi}{5}$ **(b)** $\frac{4\pi}{5}$ **(c)** 2π **(d)** 2π

25. **(a)** About the x-axis: $V = \frac{2\pi}{15}$; about the y-axis: $V = \frac{\pi}{6}$
(b) About the x-axis: $V = \frac{2\pi}{15}$; about the y-axis: $V = \frac{\pi}{6}$
27. **(a)** $\frac{5\pi}{3}$ **(b)** $\frac{4\pi}{3}$ **(c)** 2π **(d)** $\frac{2\pi}{3}$
29. **(a)** $\frac{4\pi}{15}$ **(b)** $\frac{7\pi}{30}$ **31.** **(a)** $\frac{24\pi}{5}$ **(b)** $\frac{48\pi}{5}$
33. **(a)** $\frac{9\pi}{16}$ **(b)** $\frac{9\pi}{16}$
35. Disk: 2 integrals; washer: 2 integrals; shell: 1 integral

Section 6.3, pp. 423–424

1. $\frac{5\sqrt{10}}{3}$ **3.** 7 **5.** $\frac{21}{2}$ **7.** 12 **9.** $\frac{53}{6}$ **11.** $\frac{123}{32}$
13. $\frac{99}{8}$ **15.** 2 **17.** **(a)** $\int_{-1}^{2} \sqrt{1 + 4x^2}\, dx$ **(c)** ≈ 6.13
19. **(a)** $\int_0^{\pi} \sqrt{1 + \cos^2 y}\, dy$ **(c)** ≈ 3.82
21. **(a)** $\int_{-1}^{3} \sqrt{1 + (y + 1)^2}\, dy$ **(c)** ≈ 9.29
23. **(a)** $\int_0^{\pi/6} \sec x\, dx$ **(c)** ≈ 0.55
25. Yes, $f(x) = \pm x + C$ where C is any real number.
27. **(a)** $y = \sqrt{x}$ from (1, 1) to (4, 2)
(b) Only one. We know the derivative of the function and the value of the function at one value of x.
29. **(a)** π **(b)** π

Section 6.4, pp. 434–435

1. 4 ft **3.** $(L/4, L/4)$ **5.** $M_0 = 8, M = 8, \bar{x} = 1$
7. $M_0 = 15/2, M = 9/2, \bar{x} = 5/3$
9. $M_0 = 73/6, M = 5, \bar{x} = 73/30$ **11.** $M_0 = 3, M = 3, \bar{x} = 1$
13. $\bar{x} = 0, \bar{y} = 12/5$ **15.** $\bar{x} = 1, \bar{y} = -3/5$
17. $\bar{x} = 16/105, \bar{y} = 8/15$ **19.** $\bar{x} = 0, \bar{y} = \pi/8$
21. $\bar{x} = 1, \bar{y} = -2/5$ **23.** $\bar{x} = \bar{y} = \frac{2}{4 - \pi}$
25. $\bar{x} = 3/2, \bar{y} = 1/2$
27. **(a)** $\frac{224\pi}{3}$ **(b)** $\bar{x} = 2, \bar{y} = 0$
(c)

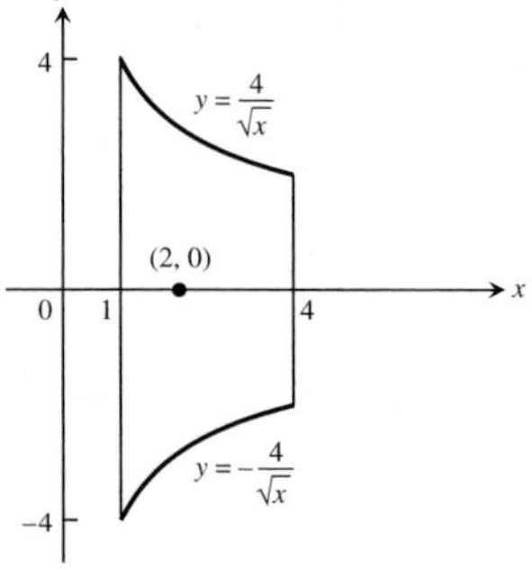

31. $\bar{x} = \bar{y} = 1/3$ **33.** $\bar{x} = a/3, \bar{y} = b/3$ **35.** $13\delta/6$
37. $\bar{x} = 0, \bar{y} = \frac{a\pi}{4}$

Section 6.5, pp. 444–447

1. **(a)** $2\pi \int_0^{\pi/4} \tan x \sqrt{1 + \sec^4 x}\, dx$ **(c)** ≈ 3.84
3. **(a)** $2\pi \int_1^{2} \frac{1}{y} \sqrt{1 + y^{-4}}\, dy$ **(c)** ≈ 5.02
5. **(a)** $2\pi \int_1^{4} (3 - \sqrt{x})^2 \sqrt{1 + (1 - 3x^{-1/2})^2}\, dx$ **(c)** ≈ 63.37
7. **(a)** $2\pi \int_0^{\pi/3} \left(\int_0^{y} \tan t\, dt \right) \sec y\, dy$ **(c)** ≈ 2.08
9. $4\pi\sqrt{5}$ **11.** $3\pi\sqrt{5}$ **13.** $98\pi/81$ **15.** 2π
17. $\pi(\sqrt{8} - 1)/9$ **19.** $35\pi\sqrt{5}/3$ **21.** $253\pi/20$
25. $2\pi \int_{-\pi/2}^{\pi/2} (\cos x) \sqrt{1 + \sin^2 x}\, dx$
27. Order 226.2 liters of each color. **31.** $5\sqrt{2}\pi$ **33.** $8\pi^2$
35. $52\pi/3$ **37.** $3\pi\sqrt{5}$ **41.** $V = 32\pi, S = 32\sqrt{2}\pi$
43. $4\pi^2$ **45.** $\bar{x} = 0, \bar{y} = \frac{2a}{\pi}$ **47.** $\bar{x} = 0, \bar{y} = \frac{4b}{3\pi}$
49. $\sqrt{2}\pi a^3(4 + 3\pi)/6$ **51.** $\frac{2a^3}{3}$

Section 6.6, pp. 452–455

1. 400 N·m **3.** 4 cm, 0.08 J
5. **(a)** 7238 lb/in. **(b)** 905 in.-lb, 2714 in.-lb **7.** 780 J
9. 72,900 ft-lb **13.** 160 ft-lb
15. **(a)** 1,497,600 ft-lb **(b)** 1 hr, 40 min **(d)** At 62.26 lb/ft^3: a) 1,494,240 ft-lb b) 1 hr, 40.1 min At 62.59 lb/ft^3: a) 1,502,160 ft-lb b) 1 hr, 40.1 min
17. 37,306 ft-lb **19.** 7,238,229.47 ft-lb
21. **(a)** 34,583 ft-lb **(b)** 53,483 ft-lb **23.** 15,073,100.75 J
27. 85.1 ft-lb **29.** 64.6 ft-lb **31.** 110.6 ft-lb
33. **(a)** $r(y) = 60 - \sqrt{50^2 - (y - 325)^2}$ for $325 \le y \le 375$ ft
(b) $\Delta V \approx \pi[60 - \sqrt{2500 - (y - 325)^2}]^2\, \Delta y$
(c) $W = 6.3358 \cdot 10^7$ ft-lb
35. 91.32 in.-oz **37.** 5.144×10^{10} J

Section 6.7, pp. 459–461

1. 1684.8 lb **3.** 2808 lb **5.** **(a)** 1164.8 lb **(b)** 1194.7 lb
7. 1309 lb **9.** 41.6 lb **11.** **(a)** 93.33 lb **(b)** 3 ft
13. 1035 ft^3 **15.** $wb/2$
17. No. The tank will overflow because the movable end will have moved only $3\frac{1}{3}$ ft by the time the tank is full.
19. 4.2 lb **21.** **(a)** 374.4 lb **(b)** 7.5 in. **(c)** No

Practice Exercises, pp. 461–464

1. $\frac{9\pi}{280}$ **3.** π^2 **5.** $\frac{72\pi}{35}$

7. **(a)** 2π **(b)** π **(c)** $12\pi/5$ **(d)** $26\pi/5$
9. **(a)** 8π **(b)** $1088\pi/15$ **(c)** $512\pi/15$
11. $\pi(3\sqrt{3} - \pi)/3$
13. **(a)** $16\pi/15$ **(b)** $8\pi/5$ **(c)** $8\pi/3$ **(d)** $32\pi/5$
15. $\frac{28\pi}{3}$ ft^3 **17.** $\frac{10}{3}$ **19.** $\frac{285}{8}$ **21.** 10 **23.** $\frac{9\pi}{2}$
25. $\bar{x} = 0, \bar{y} = 8/5$ **27.** $\bar{x} = 3/2, \bar{y} = 12/5$
29. $\bar{x} = 9/5, \bar{y} = 11/10$ **31.** $28\pi\sqrt{2}/3$ **33.** 4π
35. $76\pi/3$ **37.** 4640 J **39.** 10 ft-lb, 30 ft-lb
41. 418,208.81 ft-lb **43.** $22{,}500\pi$ ft-lb, 257 sec
45. 332.8 lb **47.** 2196.48 lb **49.** $216w_1 + 360w_2$

Additional and Advanced Exercises, pp. 464–465

1. $f(x) = \sqrt{\frac{2x - a}{\pi}}$ **3.** $f(x) = \sqrt{C^2 - 1}\,x + a$, where $C \geq 1$
5. $\bar{x} = 0, \bar{y} = \frac{n}{2n + 1}$, $(0, 1/2)$
9. **(a)** $\bar{x} = \bar{y} = 4(a^2 + ab + b^2)/(3\pi(a + b))$
(b) $(2a/\pi, 2a/\pi)$
11. 28/3 **13.** $\frac{4h\sqrt{3mh}}{3}$ **15.** ≈ 2329.6 lb
17. **(a)** $2h/3$ **(b)** $(6a^2 + 8ah + 3h^2)/(6a + 4h)$

CHAPTER 7

Section 7.1, 473–475

1. One-to-one **3.** Not one-to-one **5.** One-to-one
7. D: (0, 1] R: $[0, \infty)$

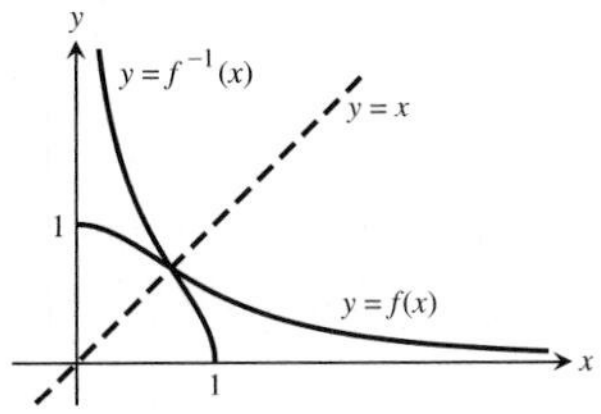

9. D: $[-1, 1]$ R: $[-\pi/2, \pi/2]$

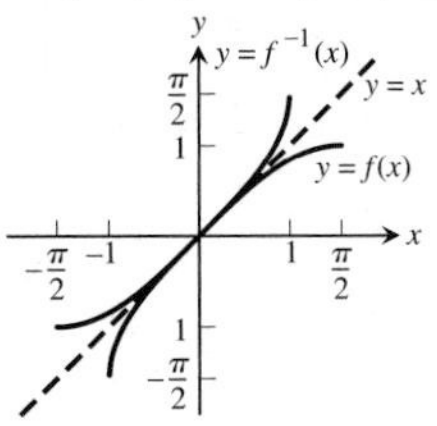

11. **(a)** Symmetric about the line $y = x$

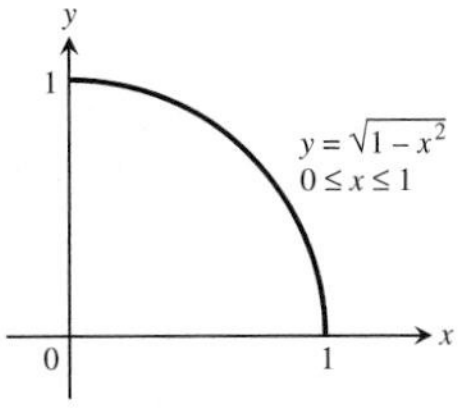

13. $f^{-1}(x) = \sqrt{x - 1}$ **15.** $f^{-1}(x) = \sqrt[3]{x + 1}$
17. $f^{-1}(x) = \sqrt{x} - 1$
19. $f^{-1}(x) = \sqrt[5]{x}$;domain: $-\infty < x < \infty$; range: $-\infty < y < \infty$
21. $f^{-1}(x) = 5\sqrt{x - 1}$; domain: $-\infty < x < \infty$; range: $-\infty < y < \infty$
23. $f^{-1}(x) = \frac{1}{\sqrt{x}}$; domain: $x > 0$; range: $y > 0$
25. **(a)** $f^{-1}(x) = \frac{x}{2} - \frac{3}{2}$
(b)

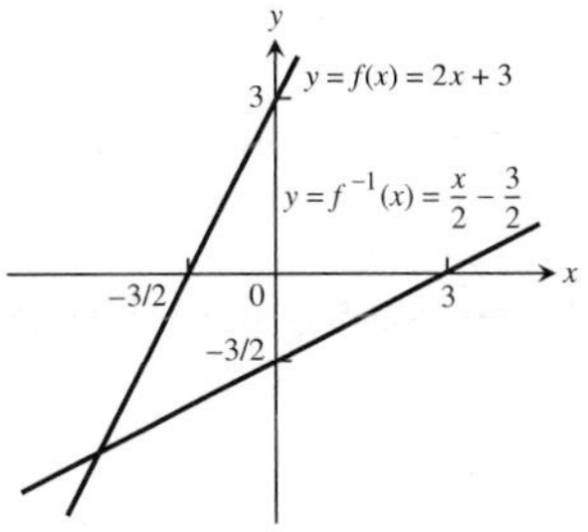

(c) 2, 1/2
27. **(a)** $f^{-1}(x) = -\frac{x}{4} + \frac{5}{4}$
(b)

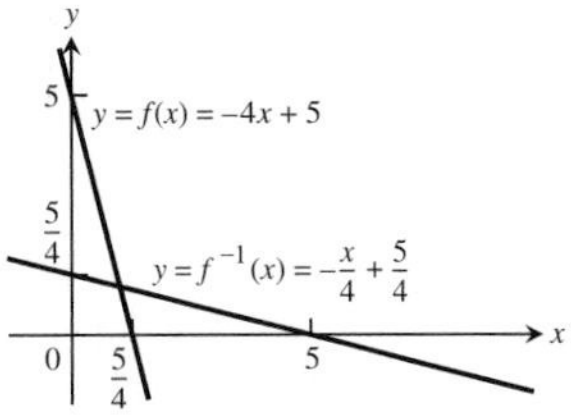

(c) $-4, -1/4$
29. **(b)**

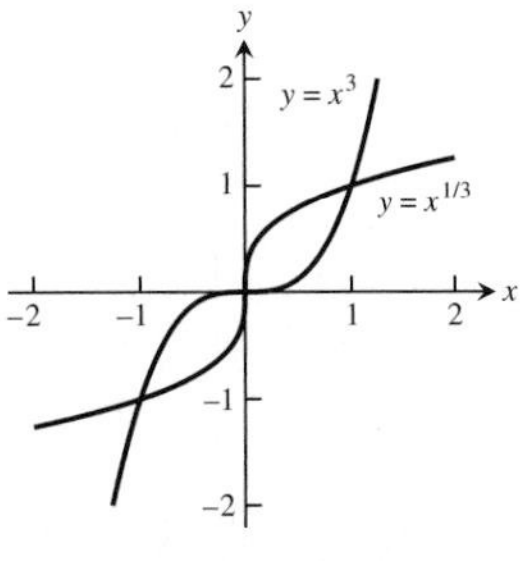

(c) Slope of f at (1, 1) : 3; slope of g at (1, 1): 1/3; slope of f at $(-1, -1)$: 3; slope of g at $(-1, -1)$: 1/3
(d) $y = 0$ is tangent to $y = x^3$ at $x = 0$; $x = 0$ is tangent to $y = \sqrt[3]{x}$ at $x = 0$

31. 1/9 **33.** 3

35. **(a)** $f^{-1}(x) = \frac{1}{m}x$

(b) The graph of f^{-1} is the line through the origin with slope $1/m$.

37. **(a)** $f^{-1}(x) = x - 1$

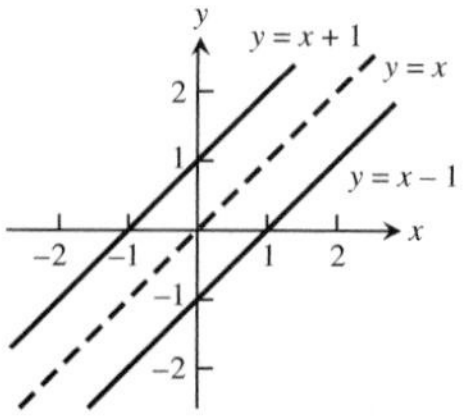

(b) $f^{-1}(x) = x - b$. The graph of f^{-1} is a line parallel to the graph of f. The graphs of f and f^{-1} lie on opposite sides of the line $y = x$ and are equidistant from that line.

(c) Their graphs will be parallel to one another and lie on opposite sides of the line $y = x$ equidistant from that line.

41. Increasing, therefore one-to-one; $df^{-1}/dx = \frac{1}{9}x^{-2/3}$

43. Decreasing, therefore one-to-one; $df^{-1}/dx = -\frac{1}{3}x^{-2/3}$

Section 7.2, pp. 484–485

1. **(a)** $\ln 3 - 2\ln 2$ **(b)** $2(\ln 2 - \ln 3)$ **(c)** $-\ln 2$, **(d)** $\frac{2}{3}\ln 3$ **(e)** $\ln 3 + \frac{1}{2}\ln 2$ **(f)** $\frac{1}{2}(3\ln 3 - \ln 2)$

3. **(a)** $\ln 5$ **(b)** $\ln(x - 3)$ **(c)** $\ln(t^2)$

5. $1/x$ **7.** $2/t$ **9.** $-1/x$ **11.** $\frac{1}{\theta + 1}$ **13.** $3/x$

15. $2(\ln t) + (\ln t)^2$ **17.** $x^3 \ln x$ **19.** $\frac{1 - \ln t}{t^2}$

21. $\frac{1}{x(1 + \ln x)^2}$ **23.** $\frac{1}{x\ln x}$ **25.** $2\cos(\ln\theta)$

27. $-\frac{3x + 2}{2x(x + 1)}$ **29.** $\frac{2}{t(1 - \ln t)^2}$ **31.** $\frac{\tan(\ln\theta)}{\theta}$

33. $\frac{10x}{x^2 + 1} + \frac{1}{2(1 - x)}$ **35.** $2x\ln|x| - x\ln\frac{|x|}{\sqrt{2}}$ **37.** $\ln\left(\frac{2}{3}\right)$

39. $\ln|y^2 - 25| + C$ **41.** $\ln 3$ **43.** $(\ln 2)^2$ **45.** $\frac{1}{\ln 4}$

47. $\ln|6 + 3\tan t| + C$ **49.** $\ln 2$ **51.** $\ln 27$

53. $\ln(1 + \sqrt{x}) + C$

55. $\left(\frac{1}{2}\right)\sqrt{x(x + 1)}\left(\frac{1}{x} + \frac{1}{x + 1}\right) = \frac{2x + 1}{2\sqrt{x(x + 1)}}$

57. $\left(\frac{1}{2}\right)\sqrt{\frac{t}{t + 1}}\left(\frac{1}{t} - \frac{1}{t + 1}\right) = \frac{1}{2\sqrt{t}(t + 1)^{3/2}}$

59. $\sqrt{\theta + 3}(\sin\theta)\left(\frac{1}{2(\theta + 3)} + \cot\theta\right)$

61. $t(t + 1)(t + 2)\left[\frac{1}{t} + \frac{1}{t + 1} + \frac{1}{t + 2}\right] = 3t^2 + 6t + 2$

63. $\frac{\theta + 5}{\theta\cos\theta}\left[\frac{1}{\theta + 5} - \frac{1}{\theta} + \tan\theta\right]$

65. $\frac{x\sqrt{x^2 + 1}}{(x + 1)^{2/3}}\left[\frac{1}{x} + \frac{x}{x^2 + 1} - \frac{2}{3(x + 1)}\right]$

67. $\frac{1}{3}\sqrt[3]{\frac{x(x - 2)}{x^2 + 1}}\left(\frac{1}{x} + \frac{1}{x - 2} - \frac{2x}{x^2 + 1}\right)$

69. **(a)** Max $= 0$ at $x = 0$, min $= -\ln 2$ at $x = \pi/3$
(b) Max $= 1$ at $x = 1$, min $= \cos(\ln 2)$ at $x = 1/2$ and $x = 2$

71. $\ln 16$ **73.** $4\pi\ln 4$ **75.** $\pi\ln 16$

77. **(a)** $6 + \ln 2$ **(b)** $8 + \ln 9$

79. **(a)** $\bar{x} \approx 1.44, \bar{y} \approx 0.36$
(b)

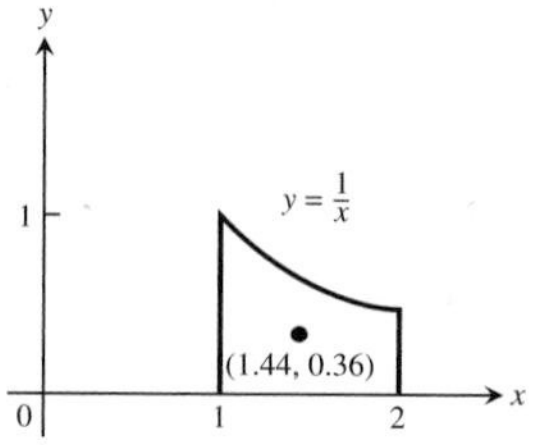

81. $y = x + \ln|x| + 2$ **83.** **(b)** 0.00469

Section 7.3, pp. 493–495

1. **(a)** 7.2 **(b)** $\frac{1}{x^2}$ **(c)** $\frac{x}{y}$

3. **(a)** 1 **(b)** 1 **(c)** $-x^2 - y^2$

5. e^{2t+4} **7.** $e^{5t} + 40$ **9.** $y = 2xe^x + 1$

11. **(a)** $k = \ln 2$ **(b)** $k = (1/10)\ln 2$ **(c)** $k = 1000\ln a$

13. **(a)** $t = -10\ln 3$ **(b)** $t = -\frac{\ln 2}{k}$ **(c)** $t = \frac{\ln .4}{\ln .2}$

15. $4(\ln x)^2$ **17.** $-5e^{-5x}$ **19.** $-7e^{(5-7x)}$ **21.** xe^x

23. x^2e^x **25.** $2e^\theta\cos\theta$ **27.** $2\theta e^{-\theta^2}\sin(e^{-\theta^2})$

29. $\frac{1 - t}{t}$ **31.** $1/(1 + e^\theta)$ **33.** $e^{\cos t}(1 - t\sin t)$

35. $(\sin x)/x$ **37.** $\frac{ye^y\cos x}{1 - ye^y\sin x}$ **39.** $\frac{2e^{2x} - \cos(x + 3y)}{3\cos(x + 3y)}$

41. $\frac{1}{3}e^{3x} - 5e^{-x} + C$ **43.** 1 **45.** $8e^{(x+1)} + C$ **47.** 2

49. $2e^{\sqrt{r}} + C$ **51.** $-e^{-t^2} + C$ **53.** $-e^{1/x} + C$ **55.** e

57. $\frac{1}{\pi}e^{\sec\pi t} + C$ **59.** 1 **61.** $\ln(1 + e^r) + C$

63. $y = 1 - \cos(e^t - 2)$ **65.** $y = 2(e^{-x} + x) - 1$

67. Maximum: 1 at $x = 0$, minimum: $2 - 2\ln 2$ at $x = \ln 2$

69. Abs max of $1/(2e)$ assumed at $x = 1/\sqrt{e}$

71. 2 **73.** $y = e^{x/2} - 1$

75. **(a)** $\frac{d}{dx}(x\ln x - x + C) = x\cdot\frac{1}{x} + \ln x - 1 + 0 = \ln x$
(b) $\frac{1}{e - 1}$

77. **(b)** $|\text{error}| \approx 0.02140$ **79.** 2.71828183

Section 7.4, pp. 500–502

1. **(a)** 7 **(b)** $\sqrt{2}$ **(c)** 75 **(d)** 2 **(e)** 0.5 **(f)** -1

3. **(a)** $\sqrt{x}$ **(b)** x^2 **(c)** $\sin x$ **5.** **(a)** $\frac{\ln 3}{\ln 2}$ **(b)** 3 **(c)** 2

7. $x = 12$ **9.** $x = 3$ or $x = 2$ **11.** $2^x\ln x$

13. $\left(\frac{\ln 5}{2\sqrt{s}}\right)5^{\sqrt{s}}$ **15.** $\pi x^{(\pi-1)}$ **17.** $-\sqrt{2}\cos\theta^{(\sqrt{2}-1)}\sin\theta$

19. $7^{\sec\theta}(\ln 7)^2(\sec\theta\tan\theta)$ **21.** $(3\cos 3t)(2^{\sin 3t})\ln 2$

23. $\frac{1}{\theta\ln 2}$ **25.** $\frac{3}{x\ln 4}$ **27.** $\frac{2(\ln r)}{r(\ln 2)(\ln 4)}$

29. $\frac{-2}{(x+1)(x-1)}$ **31.** $\sin(\log_7\theta) + \frac{1}{\ln 7}\cos(\log_7\theta)$

33. $\frac{1}{\ln 5}$ **35.** $\frac{1}{t}(\log_2 3)3^{\log_2 t}$ **37.** $\frac{1}{t}$

39. $(x+1)^x\left(\frac{x}{x+1} + \ln(x+1)\right)$ **41.** $(\sqrt{t})^t\left(\frac{\ln t}{2} + \frac{1}{2}\right)$

43. $(\sin x)^x(\ln\sin x + x\cot x)$ **45.** $(x^{\ln x})\left(\frac{\ln x^2}{x}\right)$

47. $\frac{5^x}{\ln 5} + C$ **49.** $\frac{1}{2\ln 2}$ **51.** $\frac{1}{\ln 2}$ **53.** $\frac{6}{\ln 7}$ **55.** 32760

57. $\frac{3x^{(\sqrt{3}+1)}}{\sqrt{3}+1} + C$ **59.** $3^{\sqrt{2}+1}$ **61.** $\frac{1}{\ln 10}\left(\frac{(\ln x)^2}{2}\right) + C$

63. $2(\ln 2)^2$ **65.** $\frac{3\ln 2}{2}$ **67.** $\ln 10$ **69.** $(\ln 10)\ln|\ln x| + C$

71. $\ln(\ln x), x > 1$ **73.** $-\ln x$ **75.** $2\ln 5$

77. $[10^{-7.44}, 10^{-7.37}]$ **79.** $k = 10$

81. **(a)** 10^{-7} **(b)** 7 **(c)** 1 : 1 **83.** $x \approx -0.76666$

85. **(a)** $L(x) = 1 + (\ln 2)x \approx 0.69x + 1$

87. **(a)** 1.89279 **(b)** −0.35621 **(c)** 0.94575 **(d)** −2.80735 **(e)** 5.29595 **(f)** 0.97041 **(g)** −1.03972 **(h)** −1.61181

91. $1/e$ **93.** 1 **95.** $1/e$ **97.** $e^{1/2}$ **99.** 1

Section 7.5, pp. 508–511

1. **(a)** −0.00001 **(b)** 10,536 years **(c)** 82%
3. 54.88 g **5.** 59.8 ft **7.** 2.8147498×10^{14}
9. **(a)** 8 years **(b)** 32.02 years **11.** 15.28 years
13. **(a)** $A_0e^{0.2}$ **(b)** 17.33 years; 27.47 years
15. 4.50% **17.** 0.585 days
21. **(a)** 17.5 min. **(b)** 13.26 min.
23. −3°C **25.** About 6659 years **27.** 41 years old

Section 7.6, pp. 515–517

1. **(a)** slower **(b)** slower **(c)** slower **(d)** faster **(e)** slower **(f)** slower **(g)** same **(h)** slower
3. **(a)** same **(b)** faster **(c)** same **(d)** same **(e)** slower **(f)** faster **(g)** slower **(h)** same
5. **(a)** same **(b)** same **(c)** same **(d)** faster **(e)** faster **(f)** same **(g)** slower **(h)** faster
7. d, a, c, b
9. **(a)** false **(b)** false **(c)** true **(d)** true **(e)** true **(f)** true **(g)** false **(h)** true
13. When the degree of f is less than or equal to the degree of g.
15. 1, 1
21. **(b)** $\ln(e^{17000000}) = 17{,}000{,}000 < (e^{17\times 10^6})^{1/10^6} = e^{17} \approx 24{,}154{,}952.75$
(c) $x \approx 3.4306311 \times 10^{15}$
(d) They cross at $x \approx 3.4306311 \times 10^{15}$

23. **(a)** The algorithm that takes $O(n\log_2 n)$ steps
25. It could take one million for a sequential search; at most 20 steps for a binary search.

Section 7.7, pp. 530–534

1. **(a)** $\pi/4$ **(b)** $-\pi/3$ **(c)** $\pi/6$
3. **(a)** $-\pi/6$ **(b)** $\pi/4$ **(c)** $-\pi/3$
5. **(a)** $\pi/3$ **(b)** $3\pi/4$ **(c)** $\pi/6$
7. **(a)** $3\pi/4$ **(b)** $\pi/6$ **(c)** $2\pi/3$
9. **(a)** $\pi/4$ **(b)** $-\pi/3$ **(c)** $\pi/6$
11. **(a)** $3\pi/4$ **(b)** $\pi/6$ **(c)** $2\pi/3$

13. $\cos\alpha = \frac{12}{13}, \tan\alpha = \frac{5}{12}, \sec\alpha = \frac{13}{12}, \csc\alpha = \frac{13}{5}, \cot\alpha = \frac{12}{5}$

15. $\sin\alpha = \frac{2}{\sqrt{5}}, \cos\alpha = -\frac{1}{\sqrt{5}}, \tan\alpha = -2, \csc\alpha = \frac{\sqrt{5}}{2}, \cot\alpha = -\frac{1}{2}$

17. $1/\sqrt{2}$ **19.** $-1/\sqrt{3}$ **21.** $\frac{4+\sqrt{3}}{2\sqrt{3}}$ **23.** 1

25. $-\sqrt{2}$ **27.** $\pi/6$ **29.** $\frac{\sqrt{x^2+4}}{2}$ **31.** $\sqrt{9y^2-1}$

33. $\sqrt{1-x^2}$ **35.** $\frac{\sqrt{x^2-2x}}{x-1}$ **37.** $\frac{\sqrt{9-4y^2}}{3}$

39. $\frac{\sqrt{x^2-16}}{x}$ **41.** $\pi/2$ **43.** $\pi/2$ **45.** $\pi/2$

47. 0 **49.** $\frac{-2x}{\sqrt{1-x^4}}$ **51.** $\frac{\sqrt{2}}{\sqrt{1-2t^2}}$

53. $\frac{1}{|2s+1|\sqrt{s^2+s}}$ **55.** $\frac{-2x}{(x^2+1)\sqrt{x^4+2x^2}}$

57. $\frac{-1}{\sqrt{1-t^2}}$ **59.** $\frac{-1}{2\sqrt{t}(1+t)}$ **61.** $\frac{1}{(\tan^{-1}x)(1+x^2)}$

63. $\frac{-e^t}{|e^t|\sqrt{(e^t)^2-1}} = \frac{-1}{\sqrt{e^{2t}-1}}$ **65.** $\frac{-2s^n}{\sqrt{1-s^2}}$ **67.** 0

69. $\sin^{-1}x$ **71.** $\sin^{-1}\frac{x}{7} + C$ **73.** $\frac{1}{\sqrt{17}}\tan^{-1}\frac{x}{\sqrt{17}} + C$

75. $\frac{1}{\sqrt{2}}\sec^{-1}\left|\frac{5x}{\sqrt{2}}\right| + C$ **77.** $2\pi/3$ **79.** $\pi/16$

81. $-\pi/12$ **83.** $\frac{3}{2}\sin^{-1}2(r-1) + C$

85. $\frac{\sqrt{2}}{2}\tan^{-1}\left(\frac{x-1}{\sqrt{2}}\right) + C$ **87.** $\frac{1}{4}\sec^{-1}\left|\frac{2x-1}{2}\right| + C$

89. π **91.** $\pi/12$ **93.** $\frac{1}{2}\sin^{-1}y^2 + C$ **95.** $\sin^{-1}(x-2) + C$

97. π **99.** $\frac{1}{2}\tan^{-1}\left(\frac{y-1}{2}\right) + C$ **101.** 2π

103. $\sec^{-1}|x+1| + C$ **105.** $e^{\sin^{-1}x} + C$ **107.** $\frac{1}{3}(\sin^{-1}(x))^3 + C$

109. $\ln|\tan^{-1}y| + C$ **111.** $\sqrt{3} - 1$ **113.** 5 **115.** 2

121. $y = \sin^{-1}x$ **123.** $y = \sec^{-1}x + \frac{2\pi}{3}, x > 1$

127. $\theta = \cos^{-1}\left(\frac{1}{\sqrt{3}}\right) \approx 54.7°$

133. **(a)** Defined; there is an angle whose tangent is 2.
(b) Not defined; there is no angle whose cosine is 2.

135. **(a)** Not defined; no angle has secant 0.
(b) Not defined; no angle has sine $\sqrt{2}$.

137. $3\sqrt{5}$ ft.

139. Yes, $\sin^{-1}(x)$ and $-\cos^{-1}(x)$ differ by the constant $\pi/2$.

147. $\pi^2/2$ **149.** **(a)** $\pi^2/2$ **(b)** 2π

151. **(a)** 0.84107 **(b)** −0.72973 **(c)** 0.46365

153. **(a)** Domain: all real numbers except those having the form $\frac{\pi}{2} + k\pi$ where k is an integer; range: $-\pi/2 < y < \pi/2$.
(b) Domain: $-\infty < x < \infty$; range: $-\infty < y < \infty$

155. **(a)** Domain: $-\infty < x < \infty$; range: $0 \le y \le \pi$
(b) Domain: $-1 \le x \le 1$; range: $-1 \le y \le 1$

157. The graphs are identical.

Section 7.8, pp. 542–546

1. $\cosh x = 5/4$, $\tanh x = -3/5$, $\coth x = -5/3$, $\operatorname{sech} x = 4/5$, $\operatorname{csch} x = -4/3$

3. $\sinh x = 8/15$, $\tanh x = 8/17$, $\coth x = 17/8$, $\operatorname{sech} x = 15/17$, $\operatorname{csch} x = 15/8$

5. $x + \frac{1}{x}$ **7.** e^{5x} **9.** e^{4x} **13.** $2\cosh\frac{x}{3}$

15. $\operatorname{sech}^2\sqrt{t} + \frac{\tanh\sqrt{t}}{\sqrt{t}}$ **17.** $\coth z$

19. $(\ln \operatorname{sech}\theta)(\operatorname{sech}\theta\tanh\theta)$ **21.** $\tanh^3 v$ **23.** 2

25. $\frac{1}{2\sqrt{x}(1+x)}$ **27.** $\frac{1}{1+\theta} - \tanh^{-1}\theta$

29. $\frac{1}{2\sqrt{t}} - \coth^{-1}\sqrt{t}$ **31.** $-\operatorname{sech}^{-1} x$ **33.** $\frac{\ln 2}{\sqrt{1 + \left(\frac{1}{2}\right)^{2\theta}}}$

35. $|\sec x|$ **41.** $\frac{\cosh 2x}{2} + C$

43. $12\sinh\left(\frac{x}{2} - \ln 3\right) + C$ **45.** $7\ln|e^{x/7} + e^{-x/7}| + C$

47. $\tanh\left(x - \frac{1}{2}\right) + C$ **49.** $-2\operatorname{sech}\sqrt{t} + C$ **51.** $\ln\frac{5}{2}$

53. $\frac{3}{32} + \ln 2$ **55.** $e - e^{-1}$ **57.** 3/4 **59.** $\frac{3}{8} + \ln\sqrt{2}$

61. $\ln(2/3)$ **63.** $\frac{-\ln 3}{2}$ **65.** $\ln 3$

67. **(a)** $\sinh^{-1}(\sqrt{3})$ **(b)** $\ln(\sqrt{3} + 2)$

69. **(a)** $\coth^{-1}(2) - \coth^{-1}(5/4)$ **(b)** $\left(\frac{1}{2}\right)\ln\left(\frac{1}{3}\right)$

71. **(a)** $-\operatorname{sech}^{-1}\left(\frac{12}{13}\right) + \operatorname{sech}^{-1}\left(\frac{4}{5}\right)$

(b) $-\ln\left(\frac{1 + \sqrt{1 - (12/13)^2}}{(12/13)}\right) + \ln\left(\frac{1 + \sqrt{1 - (4/5)^2}}{(4/5)}\right) = -\ln\left(\frac{3}{2}\right) + \ln(2) = \ln(4/3)$

73. **(a)** 0 **(b)** 0

75. **(b)** i) $f(x) = \frac{2f(x)}{2} + 0 = f(x)$, ii) $f(x) = 0 + \frac{2f(x)}{2} = f(x)$

77. **(b)** $\sqrt{\frac{mg}{k}}$ **(c)** $80\sqrt{5} \approx 178.89$ ft/sec

79. $y = \operatorname{sech}^{-1}(x) - \sqrt{1 - x^2}$ **81.** 2π **83.** $\frac{6}{5}$

85. $16\pi\ln 6 + \frac{455\pi}{9}$

89. **(c)** $a \approx 0.0417525$ **(d)** ≈ 47.90 lb

Practice Exercises, pp. 547–550

1. $-2e^{-x/5}$ **3.** xe^{4x} **5.** $\frac{2\sin\theta\cos\theta}{\sin^2\theta} = 2\cot\theta$ **7.** $\frac{2}{(\ln 2)x}$

9. $-8^{-t}(\ln 8)$ **11.** $18x^{2.6}$

13. $(x+2)^{x+2}(\ln(x+2) + 1)$ **15.** $-\frac{1}{\sqrt{1 - u^2}}$

17. $\frac{-1}{\sqrt{1 - x^2}\cos^{-1} x}$ **19.** $\tan^{-1}(t) + \frac{t}{1+t^2} - \frac{1}{2t}$

21. $\frac{1 - z}{\sqrt{z^2 - 1}} + \sec^{-1} z$ **23.** −1

25. $\frac{2(x^2+1)}{\sqrt{\cos 2x}}\left[\frac{2x}{x^2+1} + \tan 2x\right]$

27. $5\left[\frac{(t+1)(t-1)}{(t-2)(t+3)}\right]^5\left[\frac{1}{t+1} + \frac{1}{t-1} - \frac{1}{t-2} - \frac{1}{t+3}\right]$

29. $\frac{1}{\sqrt{\theta}}(\sin\theta)^{\sqrt{\theta}}\left(\frac{\ln\sqrt{\sin\theta}}{2} + \theta\cot\theta\right)$ **31.** $-\cos e^x + C$

33. $\tan(e^x - 7) + C$ **35.** $e^{\tan x} + C$ **37.** $\frac{-\ln 7}{3}$

39. $\ln 8$ **41.** $\ln(9/25)$ **43.** $-[\ln|\cos(\ln v)|] + C$

45. $-\frac{1}{2}(\ln x)^{-2} + C$ **47.** $-\cot(1 + \ln r) + C$

49. $\frac{1}{2\ln 3}(3^{x^2}) + C$ **51.** $3\ln 7$ **53.** $15/16 + \ln 2$

55. $e - 1$ **57.** 1/6 **59.** 9/14

61. $\frac{1}{3}\left[(\ln 4)^3 - (\ln 2)^3\right]$ or $\frac{7}{3}(\ln 2)^3$ **63.** $\frac{9\ln 2}{4}$ **65.** π

67. $\pi/\sqrt{3}$ **69.** $\sec^{-1}|2y| + C$ **71.** $\pi/12$

73. $\sin^{-1}(x+1) + C$ **75.** $\pi/2$ **77.** $\frac{1}{3}\sec^{-1}\left(\frac{t+1}{3}\right) + C$

79. $y = \frac{\ln 2}{\ln(3/2)}$ **81.** $y = \ln x - \ln 3$ **83.** $y = \frac{1}{1 - e^x}$

85. $\ln 10$ **87.** $\ln 2$ **89.** 5 **91.** $-\infty$ **93.** 1 **95.** e^3

97. **(a)** same rate **(b)** same rate **(c)** faster **(d)** faster **(e)** same rate **(f)** same rate

99. **(a)** true **(b)** false **(c)** false **(d)** true **(e)** true **(f)** true

101. 1/3

103. Absolute maximum = 0 at $x = e/2$, absolute minimum = −0.5 at $x = 0.5$

105. 1 **107.** $1/e$ m/sec

109. $1/\sqrt{2}$ units long by $1/\sqrt{e}$ units high, $A = 1/\sqrt{2e} \approx 0.43$ units^2

111. $\ln 5x - \ln 3x = \ln(5/3)$ **113.** 1/2

115. (a) Absolute maximum of $2/e$ at $x = e^2$, inflection point $(e^{8/3}, (8/3)e^{-4/3})$, concave up on $(e^{8/3}, \infty)$, concave down on $(0, e^{8/3})$
(b) Absolute maximum of 1 at $x = 0$, inflection points $(\pm 1/\sqrt{2}, 1/\sqrt{e})$, concave up on $(-\infty, -1/\sqrt{2}) \cup (1/\sqrt{2}, \infty)$, concave down on $(-1/\sqrt{2}, 1/\sqrt{2})$
(c) Absolute maximum of 1 at $x = 0$, inflection point $(1, 2/e)$, concave up on $(1, \infty)$, concave down on $(-\infty, 1)$

117. 18,935 years **119.** $20(5 - \sqrt{17})$ m

Additional and Advanced Exercises, pp. 550–552

1. $\pi/2$ **3.** $1/\sqrt{e}$ **5.** $\ln 2$ **7. (a)** 1 **(b)** $\pi/2$ **(c)** π
9. $\frac{1}{\ln 2}, \frac{1}{2\ln 2}, 2:1$ **11.** $x = 2$ **13.** $2/17$
21. $\bar{x} = \frac{\ln 4}{\pi}, \bar{y} = 0$ **25. (b)** 61°

CHAPTER 8

Section 8.1, pp. 558–560

1. $2\sqrt{8x^2 + 1} + C$ **3.** $2(\sin v)^{3/2} + C$ **5.** $\ln 5$
7. $2\ln\left(\sqrt{x} + 1\right) + C$ **9.** $-\frac{1}{7}\ln|\sin(3 - 7x)| + C$
11. $-\ln|\csc(e^\theta + 1) + \cot(e^\theta + 1)| + C$
13. $3\ln\left|\sec\frac{t}{3} + \tan\frac{t}{3}\right| + C$
15. $-\ln|\csc(s - \pi) + \cot(s - \pi)| + C$ **17.** 1
19. $e^{\tan v} + C$ **21.** $\frac{3^{(x+1)}}{\ln 3} + C$ **23.** $\frac{2^{\sqrt{w}}}{\ln 2} + C$
25. $3\tan^{-1} 3u + C$ **27.** $\pi/18$ **29.** $\sin^{-1} s^2 + C$
31. $6\sec^{-1}|5x| + C$ **33.** $\tan^{-1} e^x + C$ **35.** $\ln\left(2 + \sqrt{3}\right)$
37. 2π **39.** $\sin^{-1}(t - 2) + C$
41. $\sec^{-1}|x + 1| + C$, when $|x + 1| > 1$
43. $\tan x - 2\ln|\csc x + \cot x| - \cot x - x + C$
45. $x + \sin 2x + C$ **47.** $x - \ln|x + 1| + C$ **49.** $7 + \ln 8$
51. $2t^2 - t + 2\tan^{-1}\left(\frac{t}{2}\right) + C$ **53.** $\sin^{-1} x + \sqrt{1 - x^2} + C$
55. $\sqrt{2}$ **57.** $\tan x - \sec x + C$ **59.** $\ln|1 + \sin\theta| + C$
61. $\cot x + x + \csc x + C$ **63.** 4 **65.** $\sqrt{2}$ **67.** 2
69. $\ln\left|\sqrt{2} + 1\right| - \ln\left|\sqrt{2} - 1\right|$ **71.** $4 - \frac{\pi}{2}$
73. $-\ln|\csc(\sin\theta) + \cot(\sin\theta)| + C$
75. $\ln|\sin x| + \ln|\cos x| + C$ **77.** $12\tan^{-1}\left(\sqrt{y}\right) + C$
79. $\sec^{-1}\left|\frac{x - 1}{7}\right| + C$ **81.** $\ln|\sec(\tan t)| + C$

83. (a) $\sin\theta - \frac{1}{3}\sin^3\theta + C$
(b) $\sin\theta - \frac{2}{3}\sin^3\theta + \frac{1}{5}\sin^5\theta + C$
(c) $\int \cos^9\theta\, d\theta = \int \cos^8\theta\,(\cos\theta)\, d\theta = \int (1 - \sin^2\theta)^4(\cos\theta)\, d\theta$

85. (a) $\int \tan^3\theta\, d\theta = \frac{1}{2}\tan^2\theta - \int \tan\theta\, d\theta = \frac{1}{2}\tan^2\theta + \ln|\cos\theta| + C$
(b) $\int \tan^5\theta\, d\theta = \frac{1}{4}\tan^4\theta - \int \tan^3\theta\, d\theta$
(c) $\int \tan^7\theta\, d\theta = \frac{1}{6}\tan^6\theta - \int \tan^5\theta\, d\theta$
(d) $\int \tan^{2k+1}\theta\, d\theta = \frac{1}{2k}\tan^{2k}\theta - \int \tan^{2k-1}\theta\, d\theta$

87. $2\sqrt{2} - \ln\left(3 + 2\sqrt{2}\right)$ **89.** π^2
91. $\ln\left(2 + \sqrt{3}\right)$ **93.** $\bar{x} = 0, \bar{y} = \frac{1}{\ln\left(2\sqrt{2} + 3\right)}$

Section 8.2, pp. 568–570

1. $-2x\cos(x/2) + 4\sin(x/2) + C$
3. $t^2\sin t + 2t\cos t - 2\sin t + C$ **5.** $\ln 4 - \frac{3}{4}$
7. $y\tan^{-1}(y) - \ln\sqrt{1 + y^2} + C$
9. $x\tan x + \ln|\cos x| + C$
11. $(x^3 - 3x^2 + 6x - 6)e^x + C$
13. $(x^2 - 7x + 7)e^x + C$
15. $(x^5 - 5x^4 + 20x^3 - 60x^2 + 120x - 120)e^x + C$
17. $\frac{\pi^2 - 4}{8}$ **19.** $\frac{5\pi - 3\sqrt{3}}{9}$
21. $\frac{1}{2}(-e^\theta\cos\theta + e^\theta\sin\theta) + C$
23. $\frac{e^{2x}}{13}(3\sin 3x + 2\cos 3x) + C$
25. $\frac{2}{3}\left(\sqrt{3s + 9}\,e^{\sqrt{3s+9}} - e^{\sqrt{3s+9}}\right) + C$
27. $\frac{\pi\sqrt{3}}{3} - \ln(2) - \frac{\pi^2}{18}$
29. $\frac{1}{2}[-x\cos(\ln x) + x\sin(\ln x)] + C$
31. (a) π **(b)** 3π **(c)** 5π **(d)** $(2n + 1)\pi$
33. $2\pi(1 - \ln 2)$ **35. (a)** $\pi(\pi - 2)$ **(b)** 2π
37. $\frac{1}{2\pi}(1 - e^{-2\pi})$ **39.** $u = x^n, dv = \cos x\, dx$
41. $u = x^n, dv = e^{ax}\, dx$ **43.** $x\sin^{-1} x + \cos(\sin^{-1} x) + C$

45. $x\sec^{-1}x - \ln\left|x + \sqrt{x^2-1}\right| + C$ **47.** Yes

49. **(a)** $x\sinh^{-1}x - \cosh(\sinh^{-1}x) + C$
(b) $x\sinh^{-1}x - (1+x^2)^{1/2} + C$

Section 8.3, pp. 579–581

1. $\dfrac{2}{x-3} + \dfrac{3}{x-2}$ **3.** $\dfrac{1}{x+1} + \dfrac{3}{(x+1)^2}$

5. $\dfrac{-2}{z} + \dfrac{-1}{z^2} + \dfrac{2}{z-1}$ **7.** $1 + \dfrac{17}{t-3} + \dfrac{-12}{t-2}$

9. $\dfrac{1}{2}[\ln|1+x| - \ln|1-x|] + C$

11. $\dfrac{1}{7}\ln|(x+6)^2(x-1)^5| + C$ **13.** $(\ln 15)/2$

15. $-\dfrac{1}{2}\ln|t| + \dfrac{1}{6}\ln|t+2| + \dfrac{1}{3}\ln|t-1| + C$ **17.** $3\ln 2 - 2$

19. $\dfrac{1}{4}\ln\left|\dfrac{x+1}{x-1}\right| - \dfrac{x}{2(x^2-1)} + C$ **21.** $(\pi + 2\ln 2)/8$

23. $\tan^{-1}y - \dfrac{1}{y^2+1} + C$

25. $-(s-1)^{-2} + (s-1)^{-1} + \tan^{-1}s + C$

27. $\dfrac{-1}{\theta^2+2\theta+2} + \ln(\theta^2+2\theta+2) - \tan^{-1}(\theta+1) + C$

29. $x^2 + \ln\left|\dfrac{x-1}{x}\right| + C$

31. $9x + 2\ln|x| + \dfrac{1}{x} + 7\ln|x-1| + C$

33. $\dfrac{y^2}{2} - \ln|y| + \dfrac{1}{2}\ln(1+y^2) + C$ **35.** $\ln\left(\dfrac{e^t+1}{e^t+2}\right) + C$

37. $\dfrac{1}{5}\ln\left|\dfrac{\sin y - 2}{\sin y + 3}\right| + C$

39. $\dfrac{(\tan^{-1}2x)^2}{4} - 3\ln|x-2| + \dfrac{6}{x-2} + C$

41. $x = \ln|t-2| - \ln|t-1| + \ln 2$ **43.** $x = \dfrac{6t}{t+2} - 1$

45. $3\pi\ln 25$ **47.** 1.10

49. **(a)** $x = \dfrac{1000e^{4t}}{499+e^{4t}}$ **(b)** 1.55 days

51. **(a)** $\dfrac{22}{7} - \pi$ **(b)** 0.04% **(c)** The area is less than 0.003.

Section 8.4, pp. 585–586

1. 8/15 **3.** 4/3 **5.** 16/35 **7.** 3π **9.** π **11.** 2
13. 1 **15.** 4 **17.** 2 **19.** $2\ln\left(1+\sqrt{2}\right)$ **21.** $\sqrt{2}$

23. $2\sqrt{3} + \ln\left(2+\sqrt{3}\right)$ **25.** 4/3 **27.** 4/3

29. $2(1-\ln 2)$ **31.** $\dfrac{4}{3} - \ln\sqrt{3}$ **33.** $-6/5$ **35.** π

37. 0 **39.** $\dfrac{2\pi\left(\left(9\sqrt[3]{4}+1\right)^{3/2} - 1\right)}{27}$ **41.** $\ln\left(1+\sqrt{2}\right)$

43. $\pi^2/2$

Section 8.5, pp. 591–592

1. $\ln\left|\sqrt{9+y^2} + y\right| + C$ **3.** $\pi/4$ **5.** $\pi/6$

7. $\dfrac{25}{2}\sin^{-1}\left(\dfrac{t}{5}\right) + \dfrac{t\sqrt{25-t^2}}{2} + C$

9. $\dfrac{1}{2}\ln\left|\dfrac{2x}{7} + \dfrac{\sqrt{4x^2-49}}{7}\right| + C$

11. $7\left[\dfrac{\sqrt{y^2-49}}{7} - \sec^{-1}\left(\dfrac{y}{7}\right)\right] + C$ **13.** $\dfrac{\sqrt{x^2-1}}{x} + C$

15. $\dfrac{1}{3}(x^2+4)^{3/2} - 4\sqrt{x^2+4} + C$ **17.** $\dfrac{-2\sqrt{4-w^2}}{w} + C$

19. $4\sqrt{3} - 4\pi/3$ **21.** $-\dfrac{x}{\sqrt{x^2-1}} + C$

23. $-\dfrac{1}{5}\left(\dfrac{\sqrt{1-x^2}}{x}\right)^5 + C$ **25.** $2\tan^{-1}2x + \dfrac{4x}{(4x^2+1)} + C$

27. $\dfrac{1}{3}\left(\dfrac{v}{\sqrt{1-v^2}}\right)^3 + C$ **29.** $\ln 9 - \ln\left(1+\sqrt{10}\right)$

31. $\pi/6$ **33.** $\sec^{-1}|x| + C$ **35.** $\sqrt{x^2-1} + C$

37. $y = 2\left[\dfrac{\sqrt{x^2-4}}{2} - \sec^{-1}\left(\dfrac{x}{2}\right)\right]$

39. $y = \dfrac{3}{2}\tan^{-1}\left(\dfrac{x}{2}\right) - \dfrac{3\pi}{8}$ **41.** $3\pi/4$

43. $\dfrac{2}{1-\tan(x/2)} + C$ **45.** 1 **47.** $\dfrac{\sqrt{3}\pi}{9}$

49. $\dfrac{1}{\sqrt{2}}\ln\left|\dfrac{\tan(t/2) + 1 - \sqrt{2}}{\tan(t/2) + 1 + \sqrt{2}}\right| + C$

51. $\ln\left|\dfrac{1+\tan(\theta/2)}{1-\tan(\theta/2)}\right| + C$

Section 8.6, pp. 600–603

1. $\dfrac{2}{\sqrt{3}}\left(\tan^{-1}\sqrt{\dfrac{x-3}{3}}\right) + C$

3. $\sqrt{x-2}\left(\dfrac{2(x-2)}{3} + 4\right) + C$

5. $\dfrac{(2x-3)^{3/2}(x+1)}{5} + C$

7. $\dfrac{-\sqrt{9-4x}}{x} - \dfrac{2}{3}\ln\left|\dfrac{\sqrt{9-4x}-3}{\sqrt{9-4x}+3}\right| + C$

9. $\dfrac{(x+2)(2x-6)\sqrt{4x-x^2}}{6} + 4\sin^{-1}\left(\dfrac{x-2}{2}\right) + C$

11. $-\dfrac{1}{\sqrt{7}}\ln\left|\dfrac{\sqrt{7}+\sqrt{7+x^2}}{x}\right| + C$

13. $\sqrt{4-x^2} - 2\ln\left|\dfrac{2+\sqrt{4-x^2}}{x}\right| + C$

15. $\dfrac{p}{2}\sqrt{25-p^2} + \dfrac{25}{2}\sin^{-1}\dfrac{p}{5} + C$

17. $2\sin^{-1}\frac{r}{2} - \frac{1}{2}r\sqrt{4-r^2} + C$

19. $-\frac{1}{3}\tan^{-1}\left[\frac{1}{3}\tan\left(\frac{\pi}{4}-\theta\right)\right] + C$

21. $\frac{e^{2t}}{13}(2\cos 3t + 3\sin 3t) + C$

23. $\frac{x^2}{2}\cos^{-1}(x) + \frac{1}{4}\sin^{-1}(x) - \frac{1}{4}x\sqrt{1-x^2} + C$

25. $\frac{s}{18(9-s^2)} + \frac{1}{108}\ln\left|\frac{s+3}{s-3}\right| + C$

27. $-\frac{\sqrt{4x+9}}{x} + \frac{2}{3}\ln\left|\frac{\sqrt{4x+9}-3}{\sqrt{4x+9}+3}\right| + C$

29. $2\sqrt{3t-4} - 4\tan^{-1}\sqrt{\frac{3t-4}{4}} + C$

31. $\frac{x^3}{3}\tan^{-1}x - \frac{x^2}{6} + \frac{1}{6}\ln(1+x^2) + C$

33. $-\frac{\cos 5x}{10} - \frac{\cos x}{2} + C$ **35.** $8\left[\frac{\sin(7t/2)}{7} - \frac{\sin(9t/2)}{9}\right] + C$

37. $6\sin(\theta/12) + \frac{6}{7}\sin(7\theta/12) + C$

39. $\frac{1}{2}\ln(x^2+1) + \frac{x}{2(1+x^2)} + \frac{1}{2}\tan^{-1}x + C$

41. $\left(x-\frac{1}{2}\right)\sin^{-1}\sqrt{x} + \frac{1}{2}\sqrt{x-x^2} + C$

43. $\sin^{-1}\sqrt{x} - \sqrt{x-x^2} + C$

45. $\sqrt{1-\sin^2 t} - \ln\left|\frac{1+\sqrt{1-\sin^2 t}}{\sin t}\right| + C$

47. $\ln\left|\ln y + \sqrt{3+(\ln y)^2}\right| + C$

49. $\ln\left|3r + \sqrt{9r^2-1}\right| + C$

51. $x\cos^{-1}\sqrt{x} + \frac{1}{2}\sin^{-1}\sqrt{x} - \frac{1}{2}\sqrt{x-x^2} + C$

53. $-\frac{\sin^4 2x\cos 2x}{10} - \frac{2\sin^2 2x\cos 2x}{15} - \frac{4\cos 2x}{15} + C$

55. $\frac{\cos^3 2\pi t\sin 2\pi t}{\pi} + \frac{3}{2}\frac{\cos 2\pi t\sin 2\pi t}{\pi} + 3t + C$

57. $\frac{\sin^3 2\theta\cos^2 2\theta}{10} + \frac{\sin^3 2\theta}{15} + C$ **59.** $\frac{2}{3}\tan^3 t + C$

61. $\tan^2 2x - 2\ln|\sec 2x| + C$

63. $8\left[-\frac{1}{3}\cot^3 t + \cot t + t\right] + C$

65. $\frac{(\sec\pi x)(\tan\pi x)}{\pi} + \frac{1}{\pi}\ln|\sec\pi x + \tan\pi x| + C$

67. $\frac{\sec^2 3x\tan 3x}{3} + \frac{2}{3}\tan 3x + C$

69. $\frac{-\csc^3 x\cot x}{4} - \frac{3\csc x\cot x}{8} - \frac{3}{8}\ln|\csc x + \cot x| + C$

71. $4x^4(\ln x)^2 - 2x^4(\ln x) + \frac{x^2}{2} + C$ **73.** $\frac{e^{3x}}{9}(3x-1) + C$

75. $2x^3e^{x/2} - 12x^2e^{x/2} + 96e^{x/2}\left(\frac{x}{2}-1\right) + C$

77. $\frac{x^2 2^x}{\ln 2} - \frac{2}{\ln 2}\left[\frac{x2^x}{\ln 2} - \frac{2^x}{(\ln 2)^2}\right] + C$ **79.** $\frac{x\pi^x}{\ln\pi} - \frac{\pi^x}{(\ln\pi)^2} + C$

81. $\frac{1}{2}[\sec(e^t-1)\tan(e^t-1) + \ln|\sec(e^t-1) + \tan(e^t-1)|] + C$

83. $\sqrt{2} + \ln\left(\sqrt{2}+1\right)$ **85.** $\pi/3$

87. $\frac{1}{120}\sinh^4 3x\cosh 3x - \frac{1}{90}\sinh^2 3x\cosh 3x + \frac{1}{45}\cosh 3x + C$

89. $\frac{x^2}{3}\sinh 3x - \frac{2x}{9}\cosh 3x + \frac{2}{27}\sinh 3x + C$

91. $-\frac{\operatorname{sech}^7 x}{7} + C$ **101.** $2\pi\sqrt{3} + \pi\sqrt{2}\ln(\sqrt{2}+\sqrt{3})$

103. $\bar{x} = 4/3, \bar{y} = \ln\sqrt{2}$ **105.** 7.62 **107.** $\pi/8$ **111.** $\pi/4$

Section 8.7, pp. 613–619

1. I: (a) 1.5, 0 **(b)** 1.5, 0 **(c)** 0%
II: (a) 1.5, 0 **(b)** 1.5, 0 **(c)** 0%

3. I: (a) 2.75, 0.08 **(b)** 2.67, 0.08 **(c)** $0.0312 \approx 3\%$
II: (a) 2.67, 0 **(b)** 2.67, 0 **(c)** 0%

5. I: (a) 6.25, 0.5 **(b)** 6, 0.25 **(c)** $0.0417 \approx 4\%$
II: (a) 6, 0 **(b)** 6, 0 **(c)** 0%

7. I: (a) 0.509, 0.03125 **(b)** 0.5, 0.009 **(c)** $0.018 \approx 2\%$
II: (a) 0.5, 0.002604 **(b)** 0.5, 0.0004 **(c)** 0%

9. I: (a) 1.8961, 0.161 **(b)** 2, 0.1039 **(c)** $0.052 \approx 5\%$
II: (a) 2.0045, 0.0066 **(b)** 2, 0.00454 **(c)** 0%

11. (a) 0.31929 **(b)** 0.32812 **(c)** 1/3, 0.01404, 0.00521

13. (a) 1.95643 **(b)** 2.00421 **(c)** 2, 0.04357, −0.00421

15. (a) 1 **(b)** 2

17. (a) 116 **(b)** 2

19. (a) 283 **(b)** 2

21. (a) 71 **(b)** 10

23. (a) 76 **(b)** 12

25. (a) 82 **(b)** 8

27. 15,990 ft^3 **29.** 5166.346 ft ≈ 0.9785 mi **31.** ≈10.63 ft

33. (a) ≈0.00021 **(b)** ≈1.37079 **(c)** ≈0.015%

35. (a) 3.11571 **(b)** 0.02588
(c) With $M = 3.11$, we get $|E_T| \le (\pi^3/1200)(3.11) < 0.081$

39. 1.08943 **41.** 0.82812

43. (a) $T_{10} \approx 1.983523538$, $T_{100} \approx 1.999835504$, $T_{1000} \approx 1.999998355$

(b)

n	$\lvert E_T\rvert = 2 - T_n$
10	$0.016476462 = 1.6476462 \times 10^{-2}$
100	1.64496×10^{-4}
1000	1.645×10^{-6}

(c) $|E_{10n}| \approx 10^{-2}|E_n|$

(d) $b - a = \pi, h^2 = \frac{\pi^2}{n^2}, M = 1$

$$|E_n| \le \frac{\pi}{12}\left(\frac{\pi^2}{n^2}\right) = \frac{\pi^3}{12n^2}$$

$$|E_{10n}| \le \frac{\pi^3}{12(10n)^2} = 10^{-2}|E_n|$$

45. **(a)** $f''(x) = 2\cos(x^2) - 4x^2 \sin(x^2)$

(b)

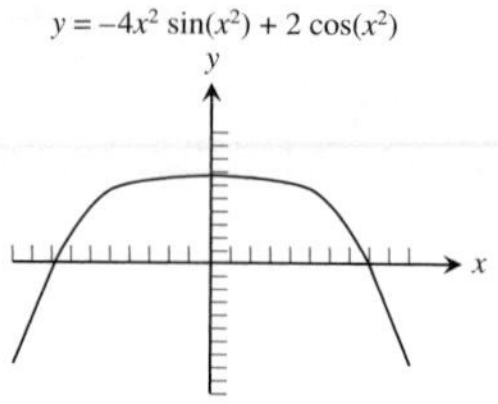

(c) The graph shows that $-3 \le f''(x) \le 2$ for $-1 \le x \le 1$.

(d) $|E_T| \le \frac{1-(-1)}{12}(\Delta x^2)(3) = \frac{\Delta x^2}{2}$

(e) $|E_T| \le \frac{\Delta x^2}{2} \le \frac{0.1^2}{2} < 0.01$ **(f)** $n \ge 20$

47. **(a)** 2.3, 1.6, 1.5, 2.1, 3.2, 4.8, 7.0, 9.3, 10.7, 10.7, 9.3, 6.4, 3.2

(b) $\frac{1}{4\pi}\int_0^6 (C(y))^2\, dy$ **(c)** ≈ 34.7 in.^{3}

(d) $V \approx 34.79$ in.^{3} by Simpson's Rule. Simpson's Rule estimate should be more accurate than the trapezoid estimate. The error in Simpson's Rule estimate is proportional to $\Delta x^4 = 0.0625$, whereas the error in the trapezoid estimate is proportional to $\Delta x^2 = 0.25$, a larger number when $\Delta x = 0.5$ in.

49. **(a)** ≈ 5.870 **(b)** $|E_T| \le 0.0032$

51. 21.07 in. **53.** 14.4 **55.** 54.9

Section 8.8, pp. 631–633

1. $\pi/2$ **3.** 2 **5.** 6 **7.** $\pi/2$ **9.** ln 3 **11.** ln 4 **13.** 0

15. $\sqrt{3}$ **17.** π **19.** $\ln\left(1 + \frac{\pi}{2}\right)$ **21.** -1 **23.** 1

25. $-1/4$ **27.** $\pi/2$ **29.** $\pi/3$ **31.** 6 **33.** ln 2

35. Diverges **37.** Converges **39.** Converges **41.** Converges

43. Diverges **45.** Converges **47.** Converges **49.** Diverges

51. Converges **53.** Converges **55.** Diverges **57.** Converges

59. Diverges **61.** Converges **63.** Converges

65. **(a)** Converges when $p < 1$ **(b)** Converges when $p > 1$

67. 1 **69.** 2π **71.** ln 2 **73.** **(b)** ≈ 0.88621

75. **(a)**

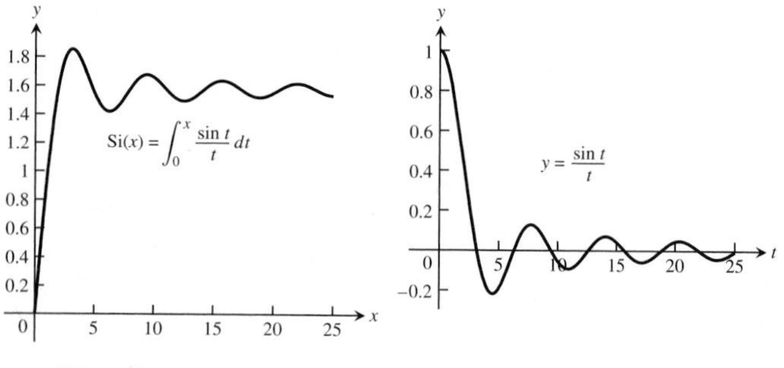

(b) $\pi/2$

77. **(a)**

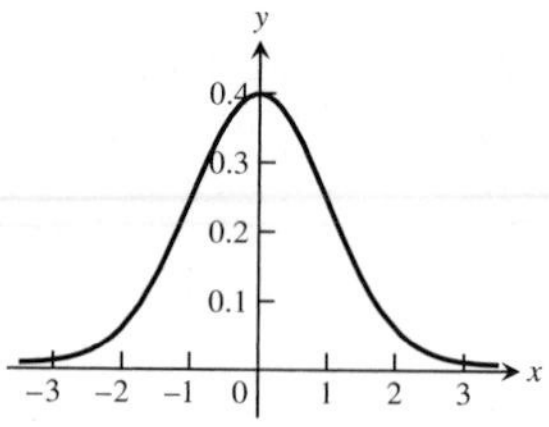

(b) ≈ 0.683, ≈ 0.954, ≈ 0.997

81. Diverges **83.** Converges **85.** Converges **87.** Diverges

Practice Exercises, pp. 634–638

1. $\frac{1}{12}(4x^2 - 9)^{3/2} + C$ **3.** $\frac{(2x+1)^{5/2}}{10} - \frac{(2x+1)^{3/2}}{6} + C$

5. $\frac{\sqrt{8x^2+1}}{8} + C$ **7.** $\frac{1}{2}\ln(25 + y^2) + C$

9. $\frac{-\sqrt{9 - 4t^4}}{8} + C$ **11.** $\frac{9}{25}(z^{5/3} + 1)^{5/3} + C$

13. $-\frac{1}{2(1 - \cos 2\theta)} + C$ **15.** $-\frac{1}{4}\ln|3 + 4\cos t| + C$

17. $-\frac{1}{2}e^{\cos 2x} + C$ **19.** $-\frac{1}{3}\cos^3(e^\theta) + C$ **21.** $\frac{2^{x-1}}{\ln 2} + C$

23. $\ln|\ln v| + C$ **25.** $\ln|2 + \tan^{-1} x| + C$

27. $\sin^{-1}(2x) + C$ **29.** $\frac{1}{3}\sin^{-1}\left(\frac{3t}{4}\right) + C$

31. $\frac{1}{3}\tan^{-1}\left(\frac{t}{3}\right) + C$ **33.** $\frac{1}{5}\sec^{-1}\left|\frac{5x}{4}\right| + C$

35. $\sin^{-1}\left(\frac{x-2}{2}\right) + C$ **37.** $\frac{1}{2}\tan^{-1}\left(\frac{y-2}{2}\right) + C$

39. $\sec^{-1}|x - 1| + C$ **41.** $\frac{x}{2} - \frac{\sin 2x}{4} + C$

43. $\frac{2}{3}\cos^3\left(\frac{\theta}{2}\right) - 2\cos\left(\frac{\theta}{2}\right) + C$

45. $\frac{\tan^2(2t)}{4} - \frac{1}{2}\ln|\sec 2t| + C$

47. $-\frac{1}{2}\ln|\csc(2x) + \cot(2x)| + C$ **49.** $\ln\sqrt{2}$ **51.** 2

53. $2\sqrt{2}$ **55.** $x - 2\tan^{-1}\left(\frac{x}{2}\right) + C$

57. $x + x^2 + 2\ln|2x - 1| + C$

59. $\ln(y^2 + 4) - \frac{1}{2}\tan^{-1}\left(\frac{y}{2}\right) + C$

61. $-\sqrt{4 - t^2} + 2\sin^{-1}\left(\frac{t}{2}\right) + C$ **63.** $x - \tan x + \sec x + C$

65. $-\frac{1}{3}\ln|\sec(5 - 3x) + \tan(5 - 3x)| + C$

67. $4\ln\left|\sin\left(\frac{x}{4}\right)\right| + C$

69. $-2\left(\frac{(\sqrt{1-x})^3}{3} - \frac{(\sqrt{1-x})^5}{5}\right) + C$

71. $\frac{1}{2}\left(z\sqrt{z^2+1}+\ln\left|z+\sqrt{z^2+1}\right|\right)+C$

73. $\ln\left|y+\sqrt{25+y^2}\right|+C$ **75.** $\frac{-\sqrt{1-x^2}}{x}+C$

77. $\frac{\sin^{-1}x}{2}-\frac{x\sqrt{1-x^2}}{2}+C$ **79.** $\ln\left|\frac{x}{3}+\frac{\sqrt{x^2-9}}{3}\right|+C$

81. $\sqrt{w^2-1}-\sec^{-1}(w)+C$

83. $(x+1)(\ln(x+1))-(x+1)+C$

85. $x\tan^{-1}(3x)-\frac{1}{6}\ln(1+9x^2)+C$

87. $(x+1)^2e^x-2(x+1)e^x+2e^x+C$

89. $\frac{2e^x\sin 2x}{5}+\frac{e^x\cos 2x}{5}+C$

91. $2\ln|x-2|-\ln|x-1|+C$

93. $\ln|x|-\ln|x+1|+\frac{1}{x+1}+C$

95. $-\frac{1}{3}\ln\left|\frac{\cos\theta-1}{\cos\theta+2}\right|+C$

97. $4\ln|x|-\frac{1}{2}\ln(x^2+1)+4\tan^{-1}x+C$

99. $\frac{1}{16}\ln\left|\frac{(v-2)^5(v+2)}{v^6}\right|+C$

101. $\frac{1}{2}\tan^{-1}t-\frac{\sqrt{3}}{6}\tan^{-1}\frac{t}{\sqrt{3}}+C$

103. $\frac{x^2}{2}+\frac{4}{3}\ln|x+2|+\frac{2}{3}\ln|x-1|+C$

105. $\frac{x^2}{2}-\frac{9}{2}\ln|x+3|+\frac{3}{2}\ln|x+1|+C$

107. $\frac{1}{3}\ln\left|\frac{\sqrt{x+1}-1}{\sqrt{x+1}+1}\right|+C$ **109.** $\ln|1-e^{-s}|+C$

111. $-\sqrt{16-y^2}+C$ **113.** $-\frac{1}{2}\ln|4-x^2|+C$

115. $\ln\frac{1}{\sqrt{9-x^2}}+C$ **117.** $\frac{1}{6}\ln\left|\frac{x+3}{x-3}\right|+C$

119. $-\frac{\cos^5 x}{5}+\frac{\cos^7 x}{7}+C$ **121.** $\frac{\tan^5 x}{5}+C$

123. $\frac{\cos\theta}{2}-\frac{\cos 11\theta}{22}+C$ **125.** $4\sqrt{1-\cos(t/2)}+C$

127. At least 16 **129.** $T=\pi, S=\pi$ **131.** 25°F

133. (a) $\approx$2.42 gal **(b)** $\approx$24.83 mi/gal

135. $\pi/2$ **137.** 6 **139.** ln 3 **141.** 2 **143.** $\pi/6$

145. Diverges **147.** Diverges **149.** Converges

151. $\frac{2x^{3/2}}{3}-x+2\sqrt{x}-2\ln(\sqrt{x}+1)+C$

153. $\ln\left|\frac{\sqrt{x}}{\sqrt{x^2+1}}\right|-\frac{1}{2}\left(\frac{x}{\sqrt{x^2+1}}\right)^2+C$

155. $\sin^{-1}(x+1)+C$ **157.** $\ln\left|u+\sqrt{1+u^2}\right|+C$

159. $-2\cot x-\ln|\csc x+\cot x|+\csc x+C$

161. $\frac{1}{12}\ln\left|\frac{3+v}{3-v}\right|+\frac{1}{6}\tan^{-1}\frac{v}{3}+C$

163. $\frac{\theta\sin(2\theta+1)}{2}+\frac{\cos(2\theta+1)}{4}+C$

165. $\frac{x^2}{2}+2x+3\ln|x-1|-\frac{1}{x-1}+C$

167. $-\cos(2\sqrt{x})+C$ **169.** $-\ln|\csc(2y)+\cot(2y)|+C$

171. $\frac{1}{2}\tan^2 x+C$ **173.** $-\sqrt{4-(r+2)^2}+C$

175. $\frac{1}{4}\sec^2\theta+C$ **177.** $\frac{\sqrt{2}}{2}$

179. $2\left(\frac{(\sqrt{2-x})^3}{3}-2\sqrt{2-x}\right)+C$ **181.** $\tan^{-1}(y-1)+C$

183. $\frac{1}{3}\ln|\sec\theta^3|+C$

185. $\frac{1}{4}\ln|z|-\frac{1}{4z}-\frac{1}{4}\left[\frac{1}{2}\ln(z^2+4)+\frac{1}{2}\tan^{-1}\left(\frac{z}{2}\right)\right]+C$

187. $-\frac{1}{4}\sqrt{9-4t^2}+C$ **189.** $\ln|\sin\theta|-\frac{1}{2}\ln(1+\sin^2\theta)+C$

191. $\ln|\sec\sqrt{y}|+C$ **193.** $-\theta\ln\left|\frac{\theta+2}{\theta-2}\right|+C$ **195.** $x+C$

197. $-\frac{\cos x}{2}+C$ **199.** $\ln(1+e^t)+C$ **201.** 1/4

203. $\ln|\ln\sin v|+C$ **205.** $\frac{2}{3}x^{3/2}+C$

207. $-\frac{1}{5}\tan^{-1}\cos(5t)+C$ **209.** $\frac{1}{3}\left(\frac{27^{3\theta+1}}{\ln 27}\right)+C$

211. $2\sqrt{r}-2\ln(1+\sqrt{r})+C$

213. $\ln\left|\frac{y}{y+2}\right|+\frac{2}{y}-\frac{2}{y^2}+C$ **215.** $4\sec^{-1}\left(\frac{7m}{2}\right)+C$

217. $\frac{\sqrt{8}-1}{6}$ **219.** $\frac{\pi}{2}(3b-a)+2$

Additional and Advanced Exercises, pp. 638–641

1. $x(\sin^{-1}x)^2+2(\sin^{-1}x)\sqrt{1-x^2}-2x+C$

3. $\frac{x^2\sin^{-1}x}{2}+\frac{x\sqrt{1-x^2}-\sin^{-1}x}{4}+C$

5. $\frac{\ln|\sec 2\theta+\tan 2\theta|+2\theta}{4}+C$

7. $\frac{1}{2}\left(\ln\left(t-\sqrt{1-t^2}\right)-\sin^{-1}t\right)+C$

9. $\frac{1}{16}\ln\left|\frac{x^2+2x+2}{x^2-2x+2}\right|+\frac{1}{8}(\tan^{-1}(x+1)+\tan^{-1}(x-1))+C$

11. 0 **13.** $\ln(4)-1$ **15.** 1 **17.** $32\pi/35$ **19.** 2π

21. (a) π **(b)** $\pi(2e-5)$

23. (b) $\pi\left(\frac{8(\ln 2)^2}{3}-\frac{16(\ln 2)}{9}+\frac{16}{27}\right)$ **25.** $\left(\frac{e^2+1}{4},\frac{e-2}{2}\right)$

27. $\sqrt{1+e^2}-\ln\left(\frac{\sqrt{1+e^2}}{e}+\frac{1}{e}\right)-\sqrt{2}+\ln\left(1+\sqrt{2}\right)$

29. 6 **31.** $y=\sqrt{x},\ 0\le x\le 4$ **33. (b)** 1

37. $a=\frac{1}{2},-\frac{\ln 2}{4}$ **39.** $\frac{1}{2}<p\le 1$

41. $\frac{e^{2x}}{13}(3\sin 3x + 2\cos 3x) + C$

43. $\frac{\cos x \sin 3x - 3\sin x \cos 3x}{8} + C$

45. $\frac{e^{ax}}{a^2 + b^2}(a\sin bx - b\cos bx) + C$ **47.** $x\ln(ax) - x + C$

CHAPTER 9

Section 9.1, pp. 648–650

9. $\frac{2}{3}y^{3/2} - x^{1/2} = C$ **11.** $e^y - e^x = C$

13. $-x + 2\tan\sqrt{y} = C$ **15.** $e^{-y} + 2e^{\sqrt{x}} = C$

17. $y = \sin(x^2 + C)$ **19.** (d) **21.** (a)

23.

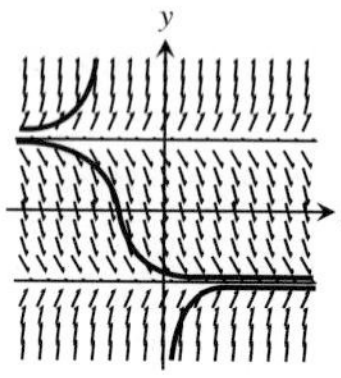

25.

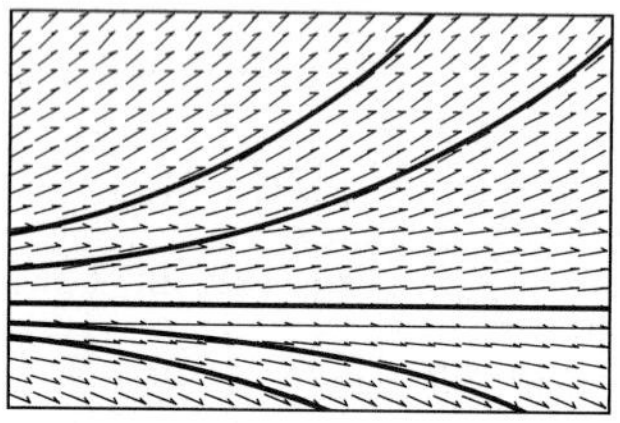

27.

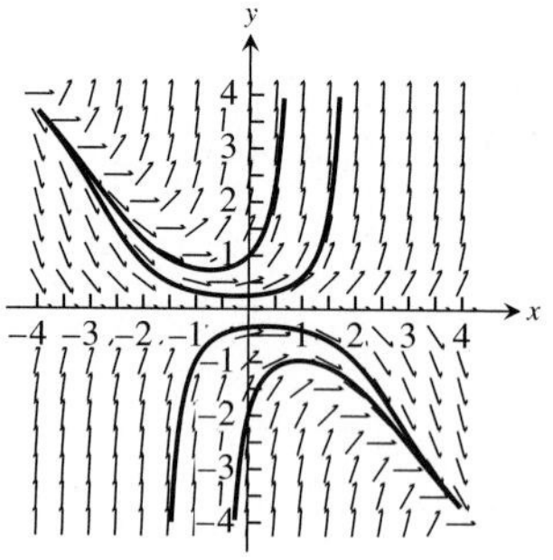

29.

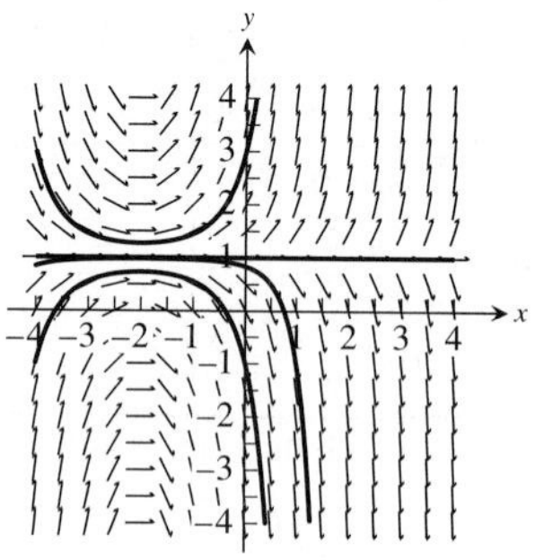

Section 9.2, pp. 657–659

1. $y = \frac{e^x + C}{x}, \quad x > 0$ **3.** $y = \frac{C - \cos x}{x^3}, \quad x > 0$

5. $y = \frac{1}{2} - \frac{1}{x} + \frac{C}{x^2}, \quad x > 0$ **7.** $y = \frac{1}{2}xe^{x/2} + Ce^{x/2}$

9. $y = x(\ln x)^2 + Cx$

11. $s = \frac{t^3}{3(t-1)^4} - \frac{t}{(t-1)^4} + \frac{C}{(t-1)^4}$

13. $r = (\csc\theta)(\ln|\sec\theta| + C), \quad 0 < \theta < \pi/2$

15. $y = \frac{3}{2} - \frac{1}{2}e^{-2t}$ **17.** $y = -\frac{1}{\theta}\cos\theta + \frac{\pi}{2\theta}$

19. $y = 6e^{x^2} - \frac{e^{x^2}}{x+1}$ **21.** $y = y_0 e^{kt}$

23. **(b)** is correct, but **(a)** is not.

25. **(a)** 10 lb/min **(b)** $(100 + t)$ gal

(c) $4\left(\frac{y}{100 + t}\right)$ lb/min

(d) $\frac{dy}{dt} = 10 - \frac{4y}{100 + t}, \quad y(0) = 50,$

$y = 2(100 + t) - \frac{150}{\left(1 + \frac{t}{100}\right)^4}$

(e) Concentration $= \frac{y(25)}{\text{amt. brine in tank}} = \frac{188.6}{125} \approx 1.5$ lb/gal

27. $y(27.8) \approx 14.8$ lb, $t \approx 27.8$ min **29.** $t = \frac{L}{R}\ln 2$ sec

31. **(a)** $i = \frac{V}{R} - \frac{V}{R}e^{-3} = \frac{V}{R}(1 - e^{-3}) \approx 0.95\frac{V}{R}$ amp **(b)** 86%

33. $y = \frac{1}{1 + Ce^{-x}}$ **35.** $y^3 = 1 + Cx^{-3}$

Section 9.3, pp. 664–665

1. $y\text{ (exact)} = \frac{x}{2} - \frac{4}{x}, \ y_1 = -0.25, \ y_2 = 0.3, \ y_3 = 0.75$

3. $y\text{ (exact)} = 3e^{x(x+2)}, \ y_1 = 4.2, \ y_2 = 6.216, \ y_3 = 9.697$

5. $y\text{ (exact)} = e^{x^2} + 1, \ y_1 = 2.0, \ y_2 = 2.0202, \ y_3 = 2.0618$

7. $y \approx 2.48832$, exact value is e

9. $y \approx -0.2272$, exact value is $1/(1 - 2\sqrt{5}) \approx -0.2880$

11.

x	z	y-approx.	y-exact	Error
0	1	3	3	0
0.2	4.2	4.608	4.658122	0.050122
0.4	6.81984	7.623475	7.835089	0.211614
0.6	11.89262	13.56369	14.27646	0.712777

13. Euler's method gives $y \approx 3.45835$; the exact solution is $y = 1 + e \approx 3.71828$

15. $y \approx 1.5000$; exact value is 1.5275.

17. **(a)** $y = \frac{1}{x^2 - 2x + 2}, \quad y(3) = -0.2$

(b) -0.1851, error ≈ 0.0149 **(c)** -0.1929, error ≈ 0.0071

(d) -0.1965, error ≈ 0.0035

19. The exact solution in $y = \frac{1}{x^2 - 2x + 2}$, so $y(3) = -0.2$. To find the approximation, let $z_n = y_{n-1} + 2y_{n-1}(x_{n-1} - 1)\,dx$ and $y_n = y_{n-1} + (y_{n-1}^2 \ (x_{n-1} - 1) + z_n^2 \ (x_n^2 - 1))\,dx$ with initial values $x_0 = 2$ and $y_0 = -\frac{1}{2}$. Use a spreadsheet, calculator, or CAS as indicated in parts (a) through (d).

(a) -0.2024, error ≈ 0.0024

(b) -0.2005, error ≈ 0.0005

(c) -0.2001, error ≈ 0.0001

(d) Each time the step size is cut in half, the error is reduced to approximately one-fourth of what it was for the larger step size.

Section 9.4, pp. 671–672

1. $y' = (y + 2)(y - 3)$

(a) $y = -2$ is a stable equilibrium value and $y = 3$ is an unstable equilibrium.

(b) $y'' = 2(y+2)\left(y - \frac{1}{2}\right)(y-3)$

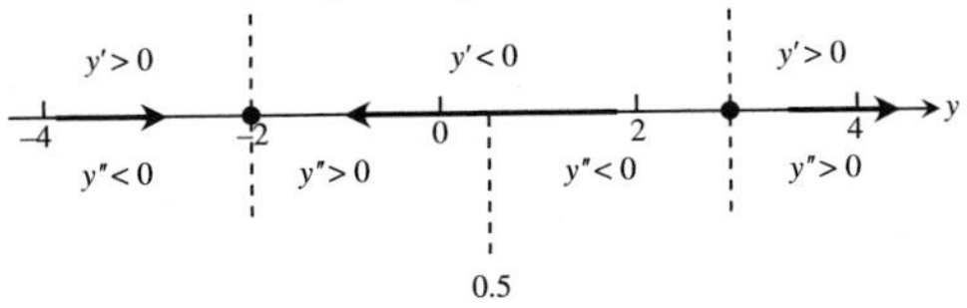

(c)

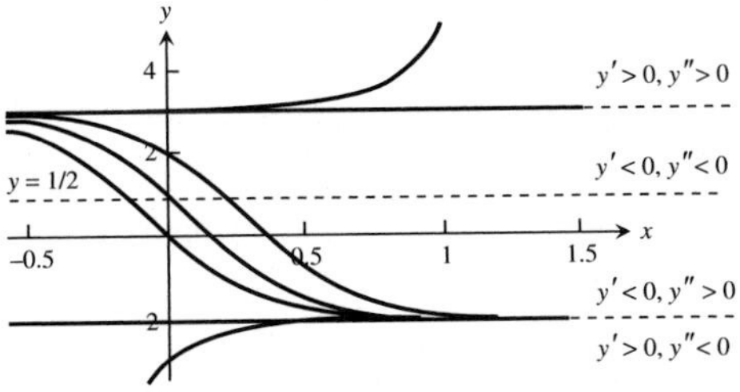

3. $y' = y^3 - y = (y+1)y(y-1)$

(a) $y = -1$ and $y = 1$ are unstable equilibria and $y = 0$ is a stable equilibrium.

(b) $y'' = (3y^2 - 1)y' = 3(y+1)\left(y + 1/\sqrt{3}\right)y\left(y - 1/\sqrt{3}\right)(y-1)$

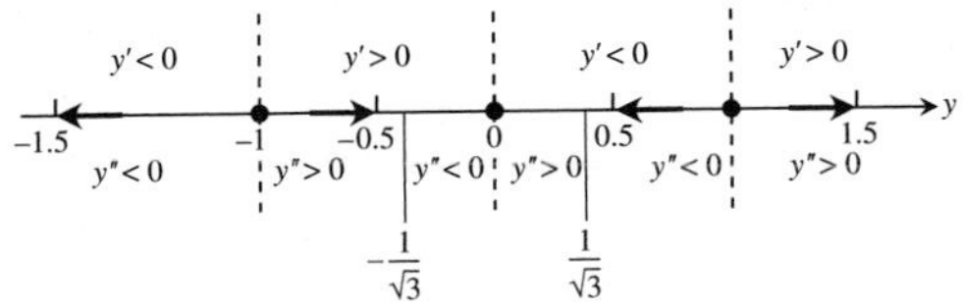

(c)

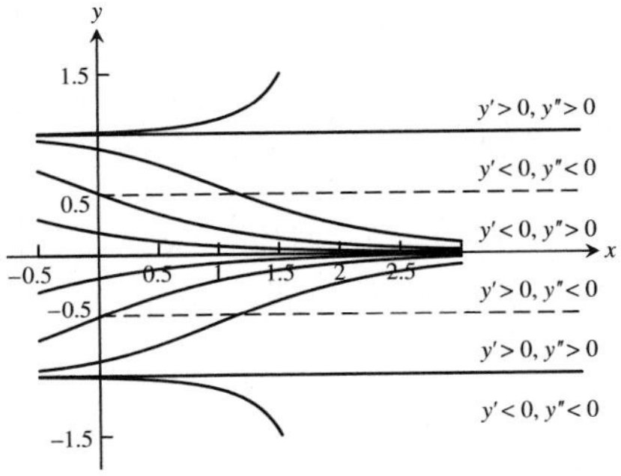

5. $y' = \sqrt{y}, y > 0$

(a) There are no equilibrium values.

(b) $y'' = \frac{1}{2}$

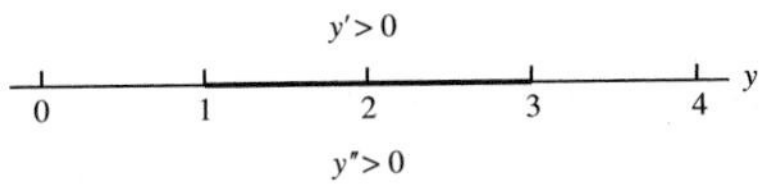

(c)

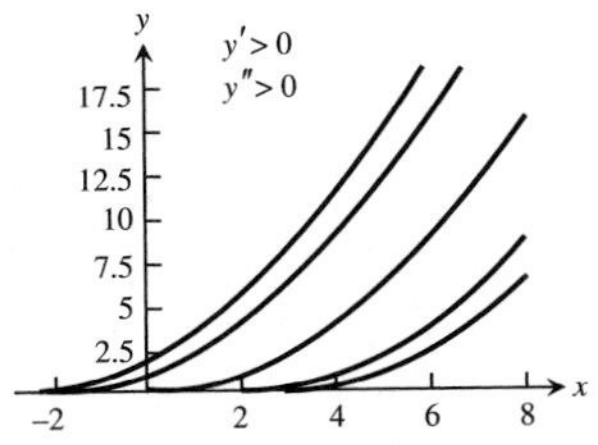

7. $y' = (y-1)(y-2)(y-3)$

(a) $y = 1$ and $y = 3$ are unstable equilibria and $y = 2$ is a stable equilibrium.

(b) $y'' = (3y^2 - 12y + 11)(y-1)(y-2)(y-3) = (y-1)\left(y - \frac{6-\sqrt{3}}{3}\right)(y-2)\left(y - \frac{6+\sqrt{3}}{3}\right)(y-3)$

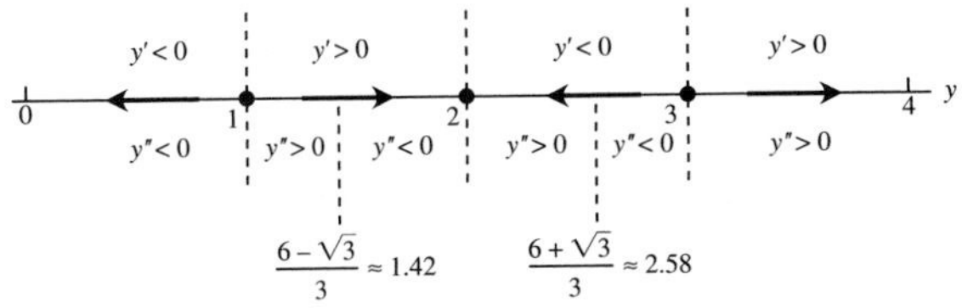

(c)

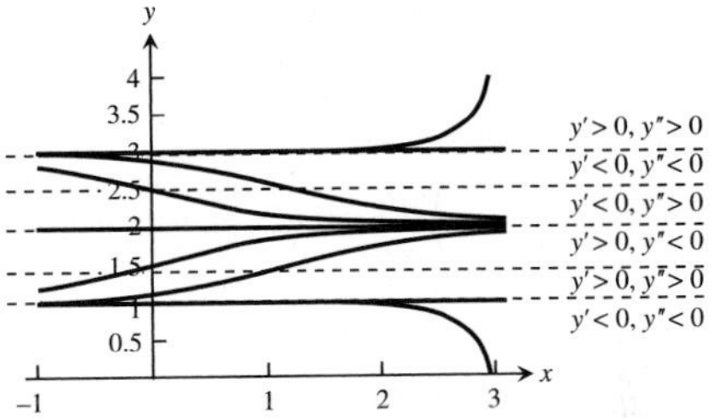

9. $\frac{dP}{dt} = 1 - 2P$ has a stable equilibrium at $P = \frac{1}{2}$; $\frac{d^2P}{dt^2} = -2\frac{dP}{dt} = -2(1-2P)$.

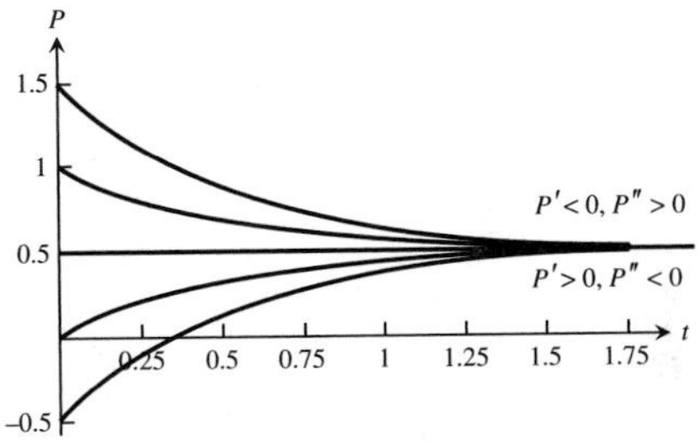

11. $\frac{dP}{dt} = 2P(P-3)$ has a stable equilibrium at $P = 0$ and an unstable equilibrium at $P = 3$; $\frac{d^2P}{dt^2} = 2(2P-3)\frac{dP}{dt} = 4P(2P-3)(P-3)$

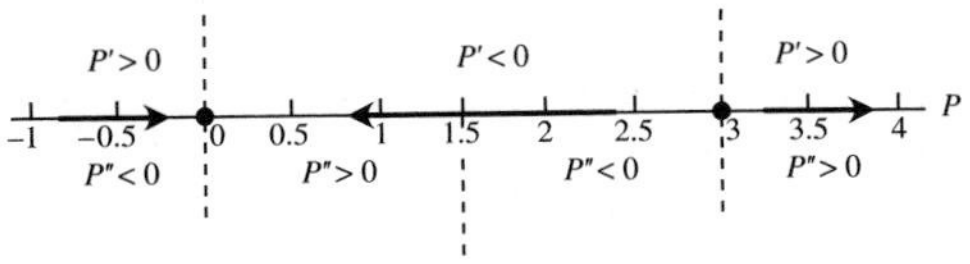

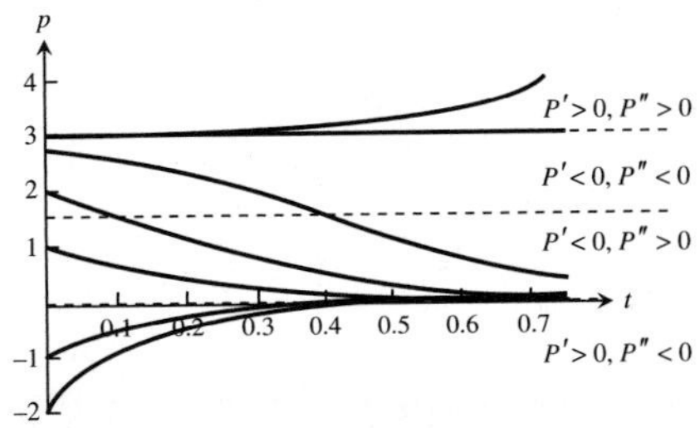

13. Before the catastrophe, the population exhibits logistic growth and $P(t)$ increases toward M_0, the stable equilibrium. After the catastrophe, the population declines logistically and $P(t)$ decreases toward M_1, the new stable equilibrium.

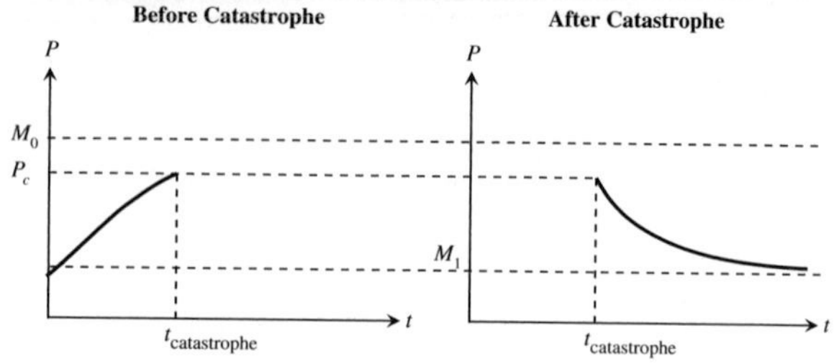

15. $\frac{dv}{dt} = g - \frac{k}{m}v^2, \quad g, k, m > 0$ and $v(t) \geq 0$

Equilibrium: $\frac{dv}{dt} = g - \frac{k}{m}v^2 = 0 \Rightarrow v = \sqrt{\frac{mg}{k}}$

Concavity: $\frac{d^2v}{dt^2} = -2\left(\frac{k}{m}v\right)\frac{dv}{dt} = -2\left(\frac{k}{m}v\right)\left(g - \frac{k}{m}v^2\right)$

(a)

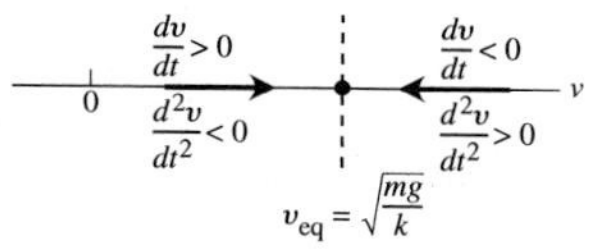

(b)

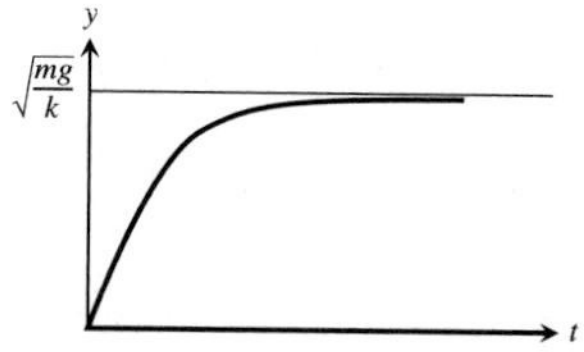

(c) $v_{\text{terminal}} = \sqrt{\frac{160}{0.005}} = 178.9$ ft/sec $= 122$ mph

17. $F = F_p - F_r$; $ma = 50 - 5|v|$; $\frac{dv}{dt} = \frac{1}{m}(50 - 5|v|)$. The maximum velocity occurs when $\frac{dv}{dt} = 0$ or $v = 10$ ft/sec.

19. Phase line:

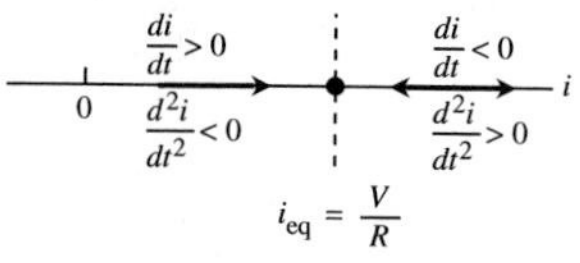

If the switch is closed at $t = 0$, then $i(0) = 0$, and the graph of the solution looks like this:

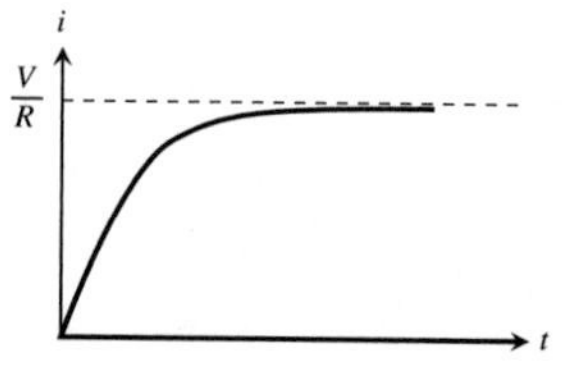

As $t \to \infty$, $i(t) \to i_{\text{steady state}} = \frac{V}{R}$.

Section 9.5, pp. 680–681

1. **(a)** 168.5 m **(b)** 41.13 sec

3. $s(t) = 4.91(1 - e^{-(22.36/39.92)t})$

5. **(a)** $P(t) = \frac{150}{1 + 24e^{-0.225t}}$

(b) About 17.21 weeks; 21.28 weeks

7. **(a)** $y(t) = \frac{8 \times 10^7}{1 + 4e^{-0.71t}} \Rightarrow y(1) \approx 2.69671 \times 10^7$ kg

(b) $t \approx 1.95253$ years

9. **(a)** $y = 2e^t - 1$ **(b)** $y(t) = \frac{400}{1 + 199e^{-200t}}$

11. **(a)** $P(t) = \frac{P_0}{1 - kP_0 t}$ **(b)** Vertical asymptote at $t = \frac{1}{kP_0}$

13. $x^2 + y^2 = C$

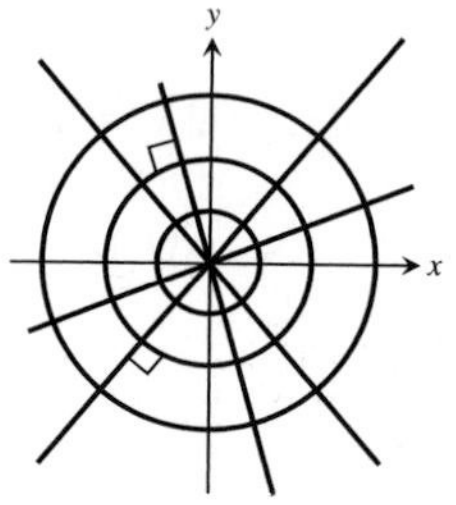

15. $\ln|y| - \frac{1}{2}y^2 = \frac{1}{2}x^2 + C$

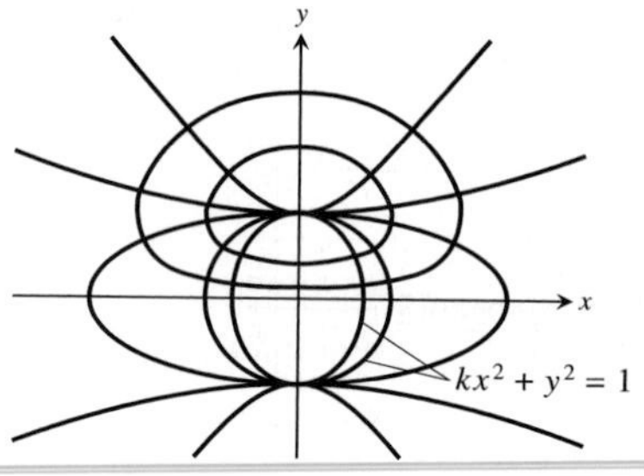

17. $y = \pm\sqrt{2x + C}$

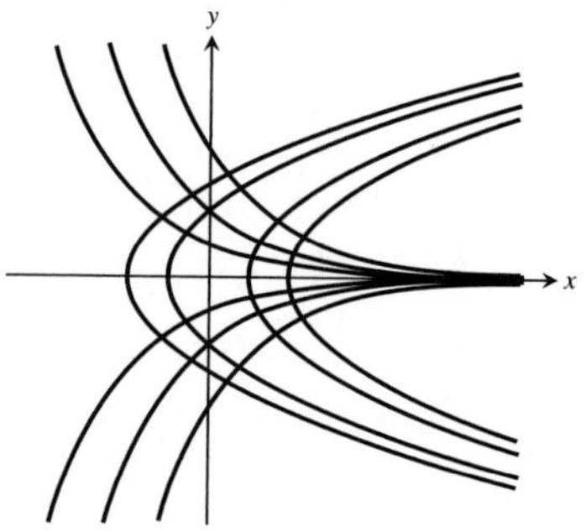

Practice Exercises, pp. 682–683

1. $y = \left(\tan^{-1}\left(\frac{x + C}{2}\right)\right)^2$ **3.** $y^2 = \sin^{-1}(2\tan x + C)$

5. $y = -\ln\left(C - \frac{2}{5}(x - 2)^{5/2} - \frac{4}{3}(x - 2)^{3/2}\right)$

7. $\tan y = -x\sin x - \cos x + C$ **9.** $(y + 1)e^{-y} = -\ln|x| + C$

11. $y = C\frac{x - 1}{x}$ **13.** $y = \frac{x^4}{4}e^{x/2} + Ce^{x/2}$

15. $y = \frac{x^2 - 2x + C}{2x^2}$ **17.** $y = \frac{e^{-x} + C}{1 + e^x}$ **19.** $xy + y^3 = C$

21. $y = -2 + \ln(2 - e^{-x})$ **23.** $y = \frac{2x^3 + 3x^2 + 6}{6(x + 1)^2}$

25. $y = \frac{1}{3}(1 - 4e^{-x^3})$ **27.** $y = 4x - 4\sqrt{x} + 1$

29. $y = e^{-x}(3x^3 - 3x^2)$

31.

x	y
0	0
0.1	0.1000
0.2	0.2095
0.3	0.3285
0.4	0.4568
0.5	0.5946
0.6	0.7418
0.7	0.8986
0.8	1.0649
0.9	1.2411
1.0	1.4273

x	y
1.1	1.6241
1.2	1.8319
1.3	2.0513
1.4	2.2832
1.5	2.5285
1.6	2.7884
1.7	3.0643
1.8	3.3579
1.9	3.6709
2.0	4.0057

33. $y(3) \approx 0.9063$

35.

(a)

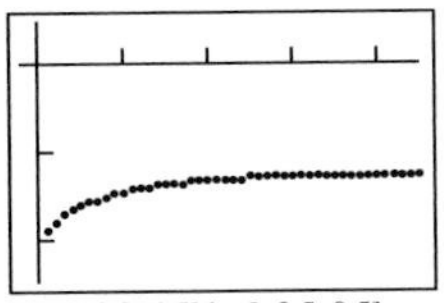

[−0.2, 4.5] by [−2.5, 0.5]

(b) Note that we choose a small interval of x-values because the y-values decrease very rapidly and our calculator cannot handle the calculations for $x \le -1$. (This occurs because the analytic solution is $y = -2 + \ln(2 - e^{-x})$, which has an asymptote at $x = -\ln 2 \approx -0.69$. Obviously, the Euler approximations are misleading for $x \le -0.7$.)

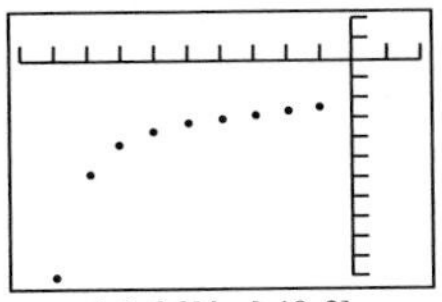

[−1, 0.2] by [−10, 2]

37. $y(\text{exact}) = \frac{1}{2}x^2 - \frac{3}{2}$; $y(2) \approx 0.4$; exact value is $\frac{1}{2}$

39. $y(\text{exact}) = -e^{(x^2-1)/2}$; $y(2) \approx -3.4192$; exact value is $-e^{3/2} \approx -4.4817$.

41. **(a)** $y = -1$ is stable and $y = 1$ is unstable.

(b) $\frac{d^2y}{dx^2} = 2y\frac{dy}{dx} = 2y(y^2 - 1)$

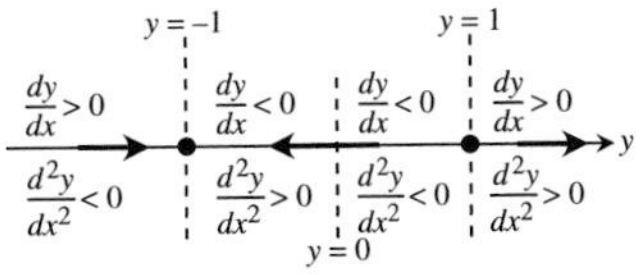

(c)

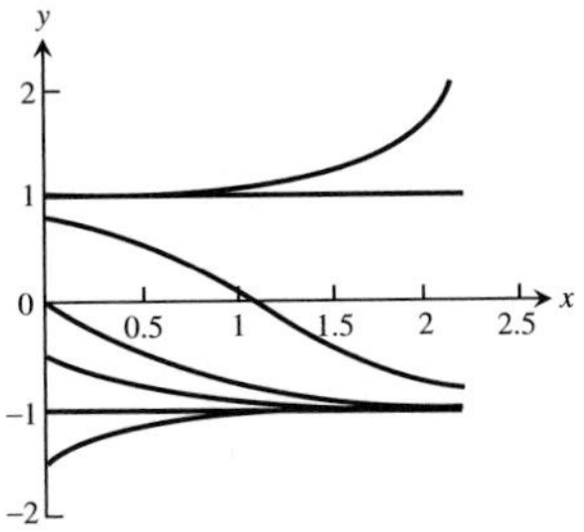

Additional and Advanced Exercises, pp. 683–684

1. **(a)** $y = c + (y_0 - c)e^{-k(A/V)t}$

(b) Steady-state solution: $y_\infty = c$

3. 0.179%

CHAPTER 10

Section 10.1, pp. 693–697

1. $y^2 = 8x$, $F(2, 0)$, directrix: $x = -2$

3. $x^2 = -6y$, $F(0, -3/2)$, directrix: $y = 3/2$

5. $\frac{x^2}{4} - \frac{y^2}{9} = 1$, $F(\pm\sqrt{13}, 0)$, $V(\pm 2, 0)$, asymptotes: $y = \pm\frac{3}{2}x$

7. $\frac{x^2}{2} + y^2 = 1$, $F(\pm 1, 0)$, $V(\pm\sqrt{2}, 0)$

9.

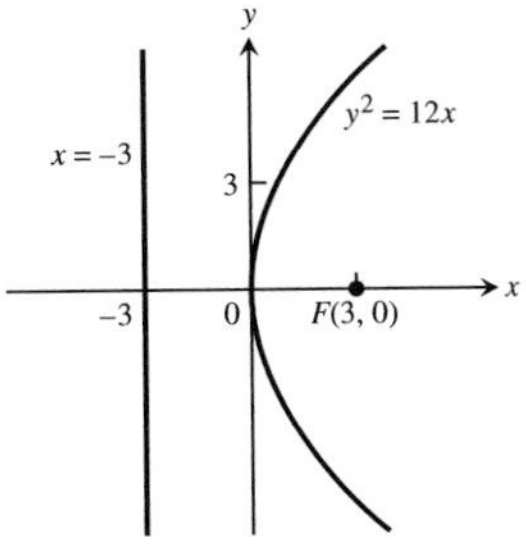

11.

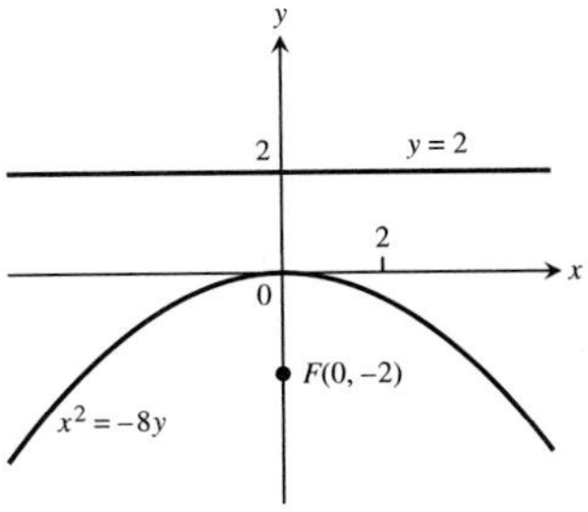

13.

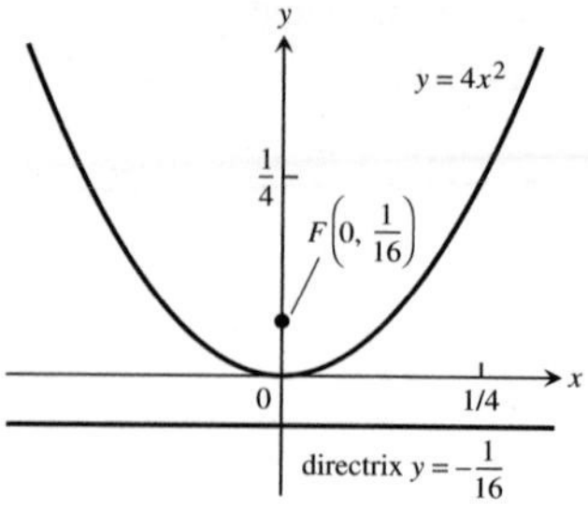

15.

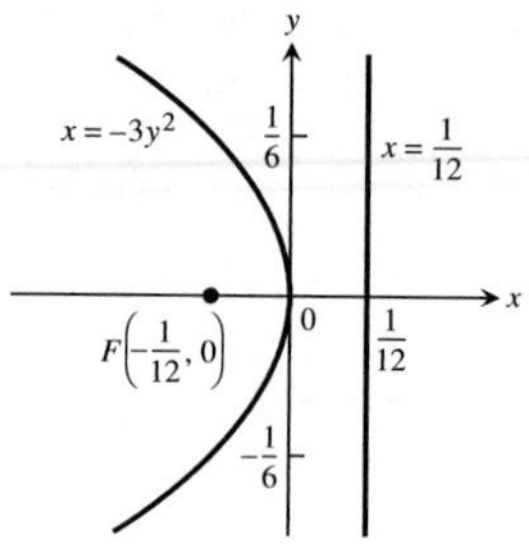

17.

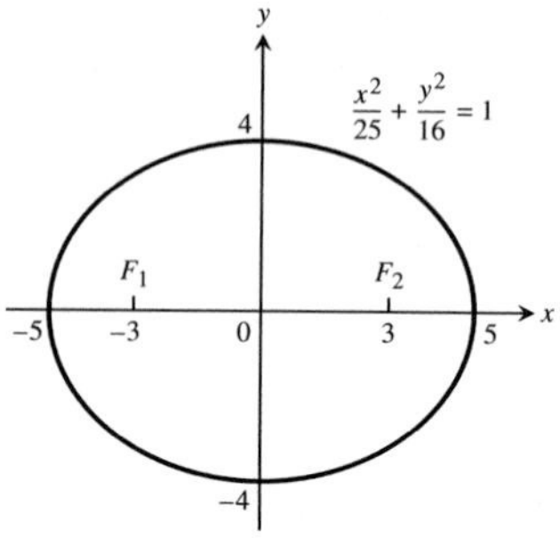

19.

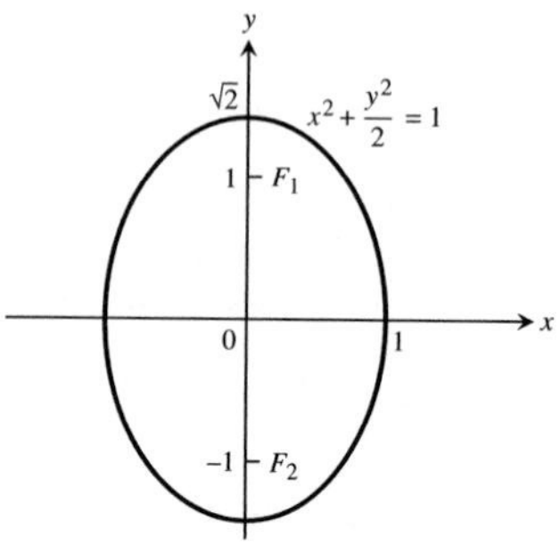

21.

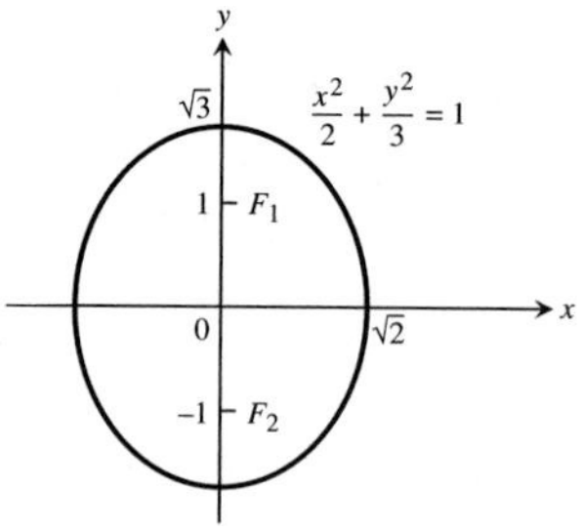

23.

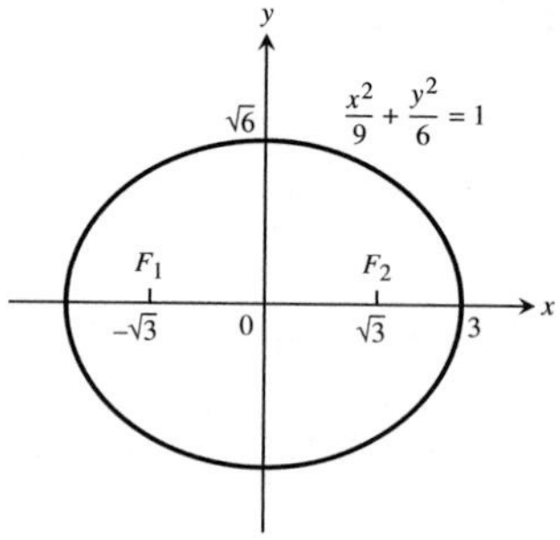

25. $\dfrac{x^2}{4} + \dfrac{y^2}{2} = 1$

27. Asymptotes: $y = \pm x$

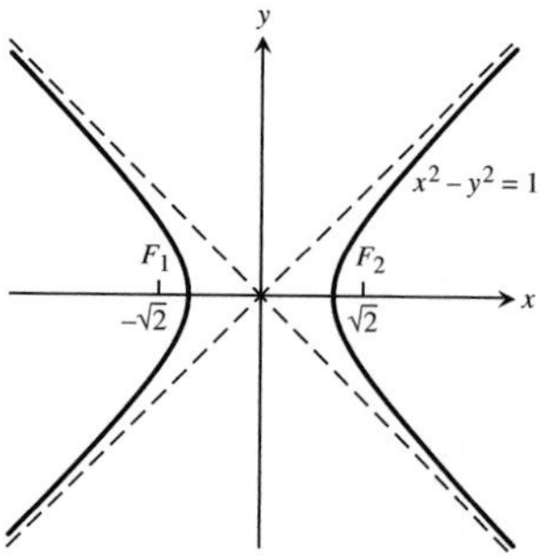

29. Asymptotes: $y = \pm x$

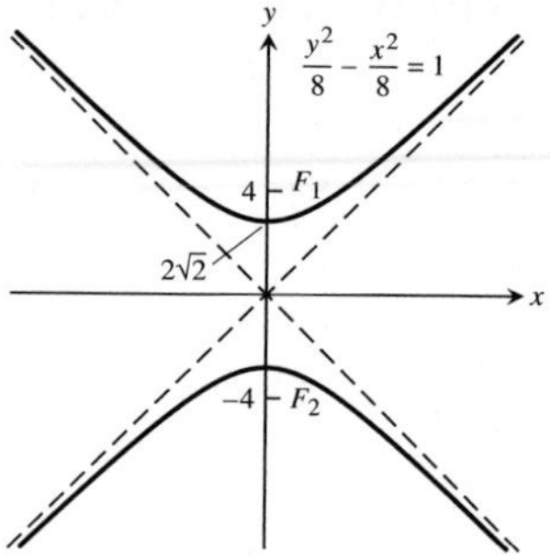

31. Asymptotes: $y = \pm 2x$

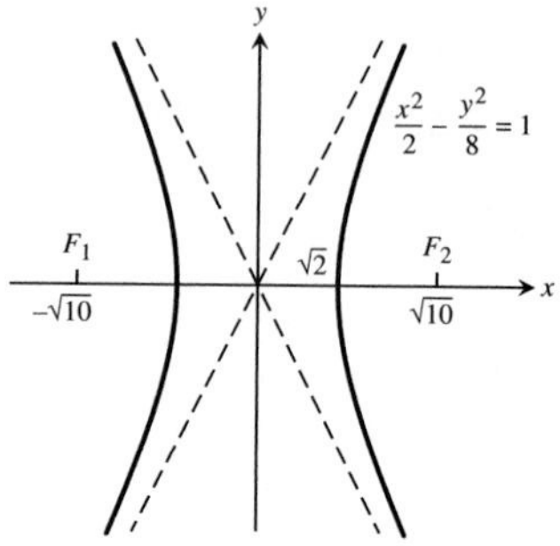

33. Asymptotes: $y = \pm\dfrac{x}{2}$

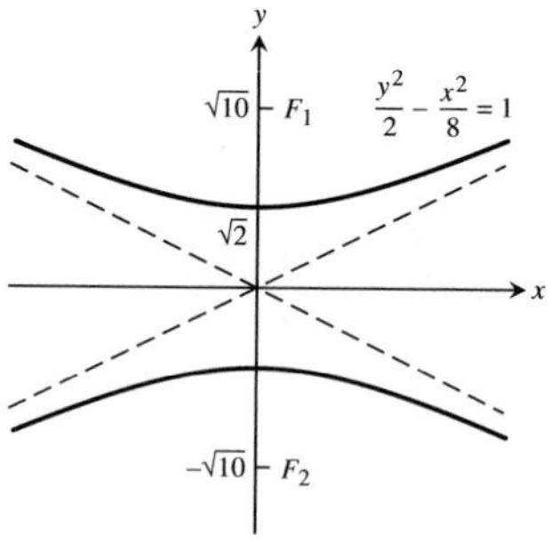

35. $y^2 - x^2 = 1$ **37.** $\dfrac{x^2}{9} - \dfrac{y^2}{16} = 1$

39. (a) Vertex: $(1, -2)$; focus: $(3, -2)$; directrix: $x = -1$

(b)

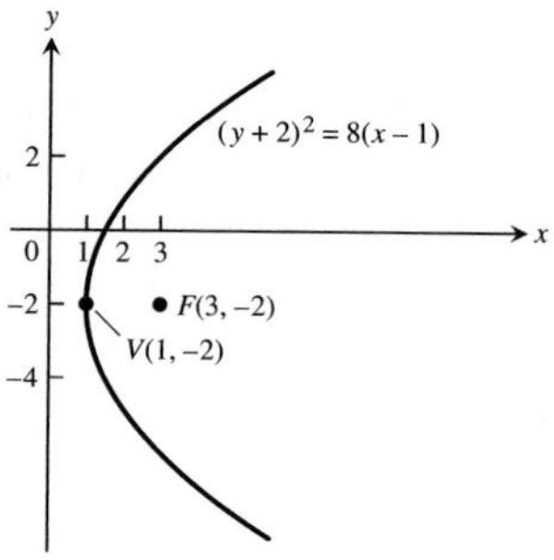

41. (a) Foci: $\left(4 \pm \sqrt{7}, 3\right)$; vertices: (8, 3) and (0, 3); center: (4, 3)
(b)

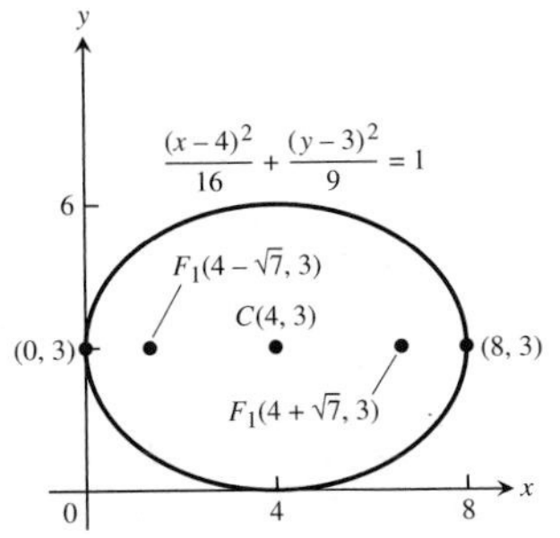

43. (a) Center: (2, 0); foci: (7, 0) and (−3, 0); vertices: (6, 0) and (−2, 0); asymptotes: $y = \pm\frac{3}{4}(x - 2)$
(b)

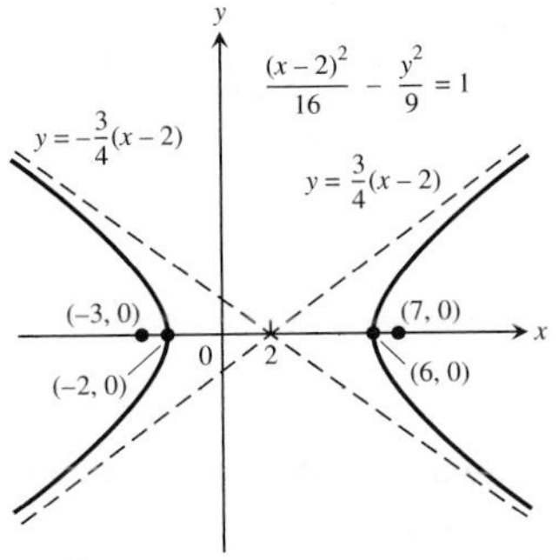

45. $(y + 3)^2 = 4(x + 2)$, $V(-2, -3)$, $F(-1, -3)$, directrix: $x = -3$
47. $(x - 1)^2 = 8(y + 7)$, $V(1, -7)$, $F(1, -5)$, directrix: $y = -9$
49. $\frac{(x + 2)^2}{6} + \frac{(y + 1)^2}{9} = 1$, $F\left(-2, \pm\sqrt{3} - 1\right)$, $V(-2, \pm 3 - 1)$, $C(-2, -1)$
51. $\frac{(x - 2)^2}{3} + \frac{(y - 3)^2}{2} = 1$, $F(3, 3)$ and $F(1, 3)$, $V\left(\pm\sqrt{3} + 2, 3\right)$, $C(2, 3)$
53. $\frac{(x - 2)^2}{4} - \frac{(y - 2)^2}{5} = 1$, $C(2, 2)$, $F(5, 2)$ and $F(-1, 2)$, $V(4, 2)$ and $V(0, 2)$; asymptotes: $(y - 2) = \pm\frac{\sqrt{5}}{2}(x - 2)$
55. $(y + 1)^2 - (x + 1)^2 = 1$, $C(-1, -1)$, $F\left(-1, \sqrt{2} - 1\right)$ and $F\left(-1, -\sqrt{2} - 1\right)$, $V(-1, 0)$ and $V(-1, -2)$; asymptotes $(y + 1) = \pm(x + 1)$
57. $C(-2, 0)$, $a = 4$ **59.** $V(-1, 1)$, $F(-1, 0)$
61. Ellipse: $\frac{(x + 2)^2}{5} + y^2 = 1$, $C(-2, 0)$, $F(0, 0)$ and $F(-4, 0)$, $V\left(\sqrt{5} - 2, 0\right)$ and $V\left(-\sqrt{5} - 2, 0\right)$
63. Ellipse: $\frac{(x - 1)^2}{2} + (y - 1)^2 = 1$, $C(1, 1)$, $F(2, 1)$ and $F(0, 1)$, $V\left(\sqrt{2} + 1, 1\right)$ and $V\left(-\sqrt{2} + 1, 1\right)$
65. Hyperbola: $(x - 1)^2 - (y - 2)^2 = 1$, $C(1, 2)$, $F\left(1 + \sqrt{2}, 2\right)$ and $F\left(1 - \sqrt{2}, 2\right)$, $V(2, 2)$ and $V(0, 2)$; asymptotes: $(y - 2) = \pm(x - 1)$
67. Hyperbola: $\frac{(y - 3)^2}{6} - \frac{x^2}{3} = 1$, $C(0, 3)$, $F(0, 6)$ and $F(0, 0)$, $V\left(0, \sqrt{6} + 3\right)$ and $V\left(0, -\sqrt{6} + 3\right)$; asymptotes: $y = \sqrt{2}x + 3$ or $y = -\sqrt{2}x + 3$

69.

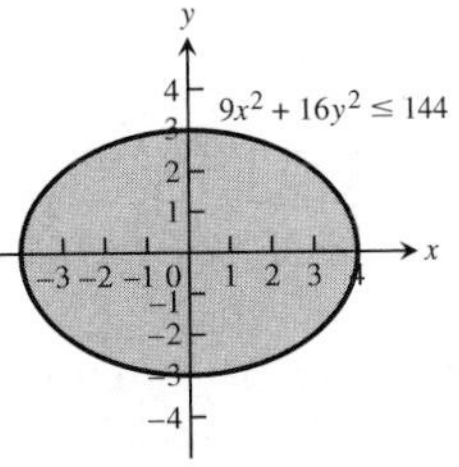

71.

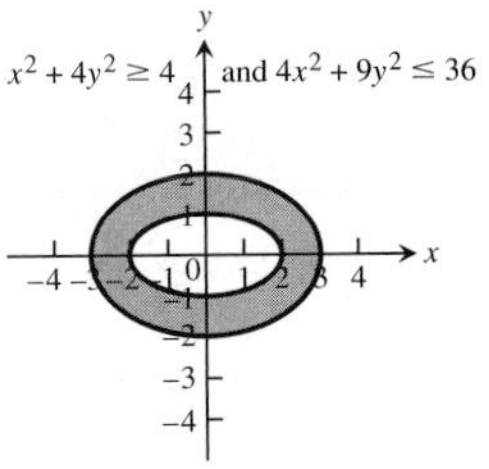

73.

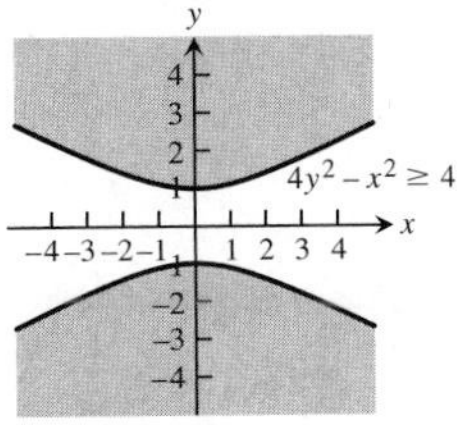

77. $3x^2 + 3y^2 - 7x - 7y + 4 = 0$
79. $(x + 2)^2 + (y - 1)^2 = 13$. The point is inside the circle.
81. (b) 1 : 1 **83.** Length $= 2\sqrt{2}$, width $= \sqrt{2}$, area $= 4$
85. 24π **87.** $(0, 16/(3\pi))$

Section 10.2, pp. 701–702

1. $e = 3/5$, $F(\pm 3, 0)$, $x = \pm 25/3$
3. $e = 1/\sqrt{2}$, $F(0, \pm 1)$, $y = \pm 2$
5. $e = 1/\sqrt{3}$, $F(0, \pm 1)$, $y = \pm 3$
7. $e = \sqrt{3}/3$, $F\left(\pm\sqrt{3}, 0\right)$, $x = \pm 3\sqrt{3}$ **9.** $\frac{x^2}{27} + \frac{y^2}{36} = 1$
11. $\frac{x^2}{4851} + \frac{y^2}{4900} = 1$ **13.** $e = \frac{\sqrt{5}}{3}$, $\frac{x^2}{9} + \frac{y^2}{4} = 1$
15. $e = 1/2$, $\frac{x^2}{64} + \frac{y^2}{48} = 1$
19. $\frac{(x - 1)^2}{4} + \frac{(y - 4)^2}{9} = 1$, $F\left(1, 4 \pm \sqrt{5}\right)$, $e = \sqrt{5}/3$, $y = 4 \pm \left(9\sqrt{5}/5\right)$
21. $a = 0$, $b = -4$, $c = 0$, $e = \sqrt{3}/2$
23. $e = \sqrt{2}$, $F\left(\pm\sqrt{2}, 0\right)$, $x = \pm 1/\sqrt{2}$
25. $e = \sqrt{2}$, $F(0, \pm 4)$, $y = \pm 2$
27. $e = \sqrt{5}$, $F\left(\pm\sqrt{10}, 0\right)$, $x = \pm 2/\sqrt{10}$
29. $e = \sqrt{5}$, $F\left(0, \pm\sqrt{10}\right)$, $y = \pm 2/\sqrt{10}$
31. $y^2 - \frac{x^2}{8} = 1$ **33.** $x^2 - \frac{y^2}{8} = 1$
35. $e = \sqrt{2}$, $\frac{x^2}{8} - \frac{y^2}{8} = 1$ **37.** $e = 2$, $x^2 - \frac{y^2}{3} = 1$

39. $\dfrac{(y-6)^2}{36} - \dfrac{(x-1)^2}{45} = 1$

Section 10.3, pp. 707–709

1. Hyperbola **3.** Ellipse **5.** Parabola **7.** Parabola
9. Hyperbola **11.** Hyperbola **13.** Ellipse **15.** Ellipse
17. $x'^2 - y'^2 = 4$, hyperbola **19.** $4x'^2 + 16y' = 0$, parabola
21. $y'^2 = 1$, parallel lines **23.** $2\sqrt{2}x'^2 + 8\sqrt{2}y' = 0$, parabola
25. $4x'^2 + 2y'^2 = 19$, ellipse
27. $\sin\alpha = 1/\sqrt{5}$, $\cos\alpha = 2/\sqrt{5}$
29. $A' = 0.88, B' = 0.00, C' = 3.10, D' = 0.74, E' = -1.20, F' = -3, 0.88x'^2 + 3.10y'^2 + 0.74x' - 1.20y' - 3 = 0$, ellipse
31. $A' = 0.00, B' = 0.00, C' = 5.00, D' = 0, E' = 0, F' = -5, 5.00y'^2 - 5 = 0$ or $y' = \pm 1.00$, parallel lines
33. $A' = 5.05, B' = 0.00, C' = -0.05, D' = -5.07, E' = -6.18, F' = -1, 5.05x'^2 - 0.05y'^2 - 5.07x' - 6.18y' - 1 = 0$, hyperbola
35. **(a)** $\dfrac{x'^2}{b^2} + \dfrac{y'^2}{a^2} = 1$ **(b)** $\dfrac{y'^2}{a^2} - \dfrac{x'^2}{b^2} = 1$
(c) $x'^2 + y'^2 = a^2$ **(d)** $y' = -\dfrac{1}{m}x'$
(e) $y' = -\dfrac{1}{m}x' + \dfrac{b}{m}$
37. **(a)** $x'^2 - y'^2 = 2$ **(b)** $x'^2 - y'^2 = 2a$ **43.** **(a)** Parabola
45. **(a)** Hyperbola
(b)

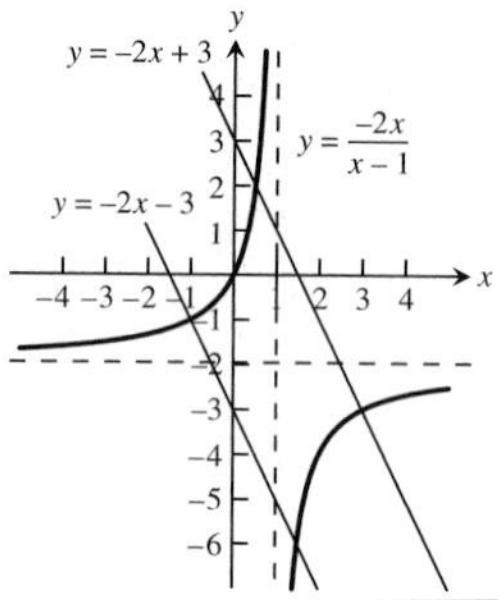

(c) $y = -2x - 3$, $y = -2x + 3$

Section 10.4, pp. 712–713

1.

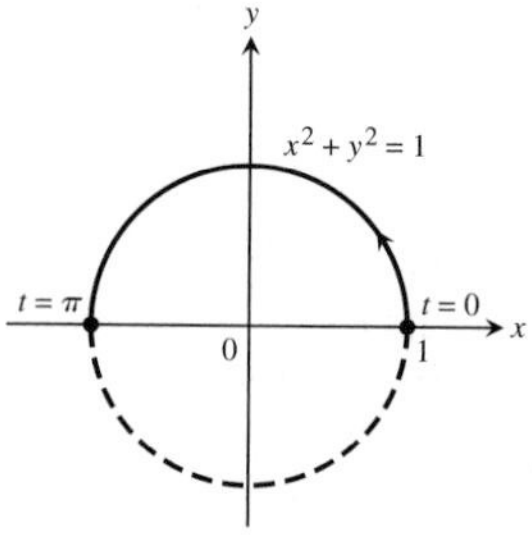

3.

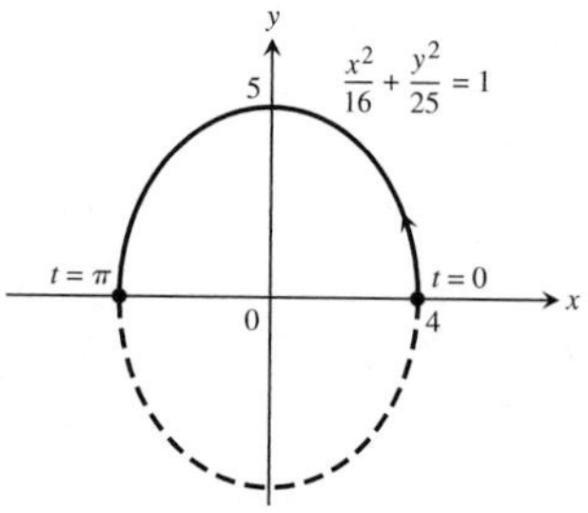

5.

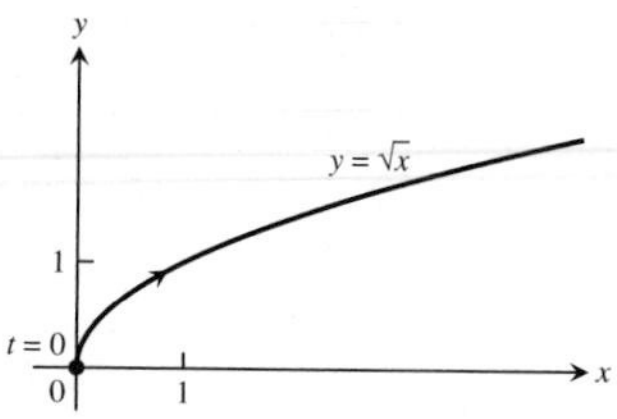

7.

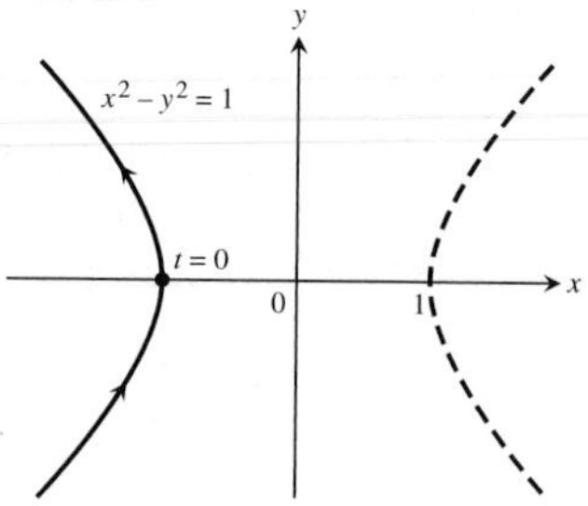

9.

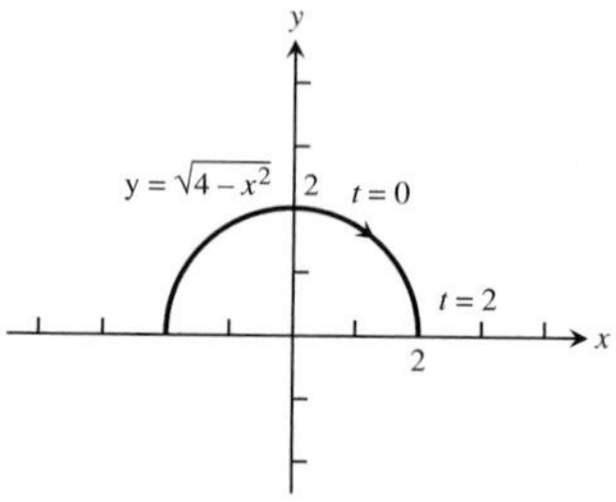

11.

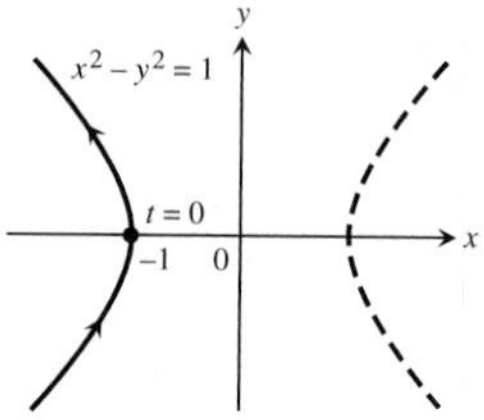

13. $x = (a-b)\cos\theta + b\cos\left(\dfrac{a-b}{b}\theta\right)$,
$y = (a-b)\sin\theta - b\sin\left(\dfrac{a-b}{b}\theta\right)$

15. $x = a\sin^2 t\tan t$, $y = a\sin^2 t$ **17.** $(1, 1)$

Section 10.5, pp. 718–719

1. a, e; b, g; c, h; d, f

3.

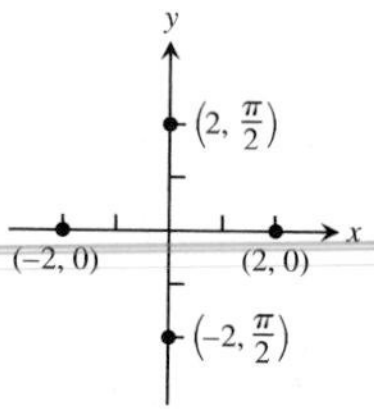

(a) $\left(2, \dfrac{\pi}{2} + 2n\pi\right)$ and $\left(-2, \dfrac{\pi}{2} + (2n+1)\pi\right)$, n an integer
(b) $(2, 2n\pi)$ and $(-2, (2n+1)\pi)$, n an integer
(c) $\left(2, \dfrac{3\pi}{2} + 2n\pi\right)$ and $\left(-2, \dfrac{3\pi}{2} + (2n+1)\pi\right)$, n an integer
(d) $(2, (2n+1)\pi)$ and $(-2, 2n\pi)$, n an integer

5. **(a)** $(3, 0)$ **(b)** $(-3, 0)$ **(c)** $(-1, \sqrt{3})$ **(d)** $(1, \sqrt{3})$
(e) $(3, 0)$ **(f)** $(1, \sqrt{3})$ **(g)** $(-3, 0)$ **(h)** $(-1, \sqrt{3})$

7.
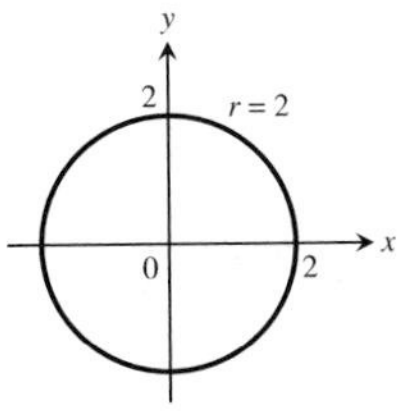

9.
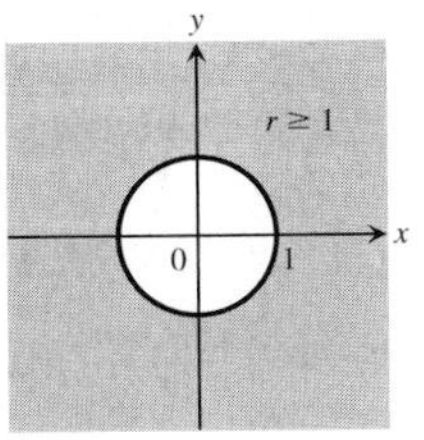

11.
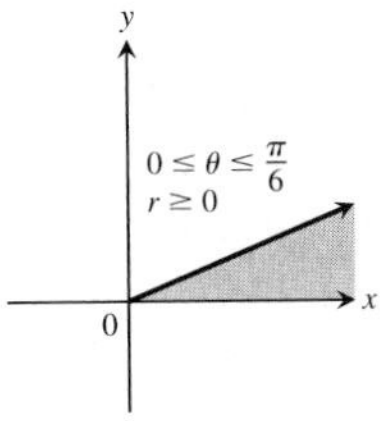

13.
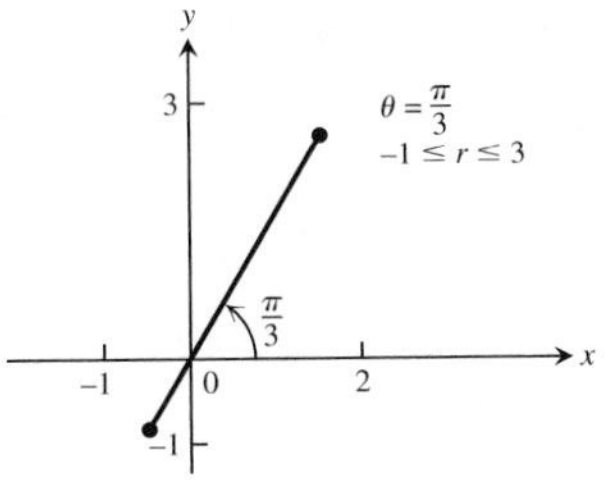

15.
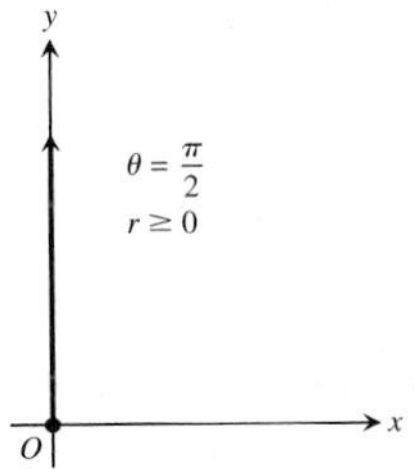

17.
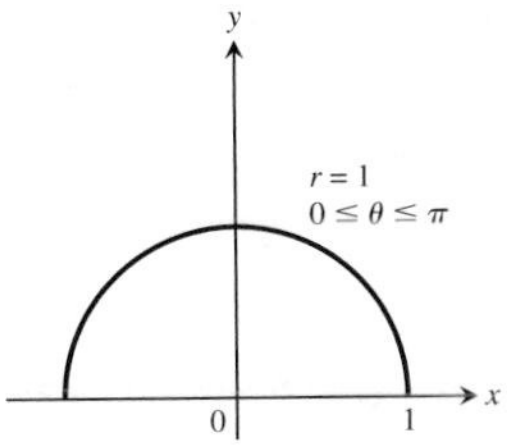

19.
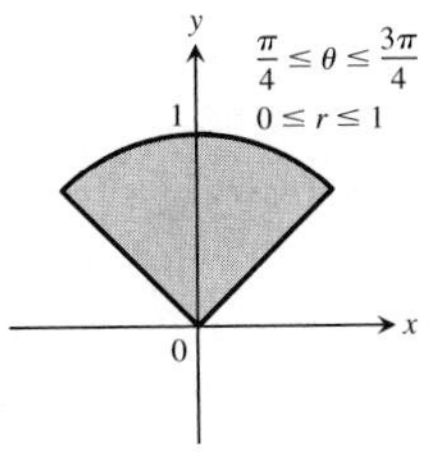

21.
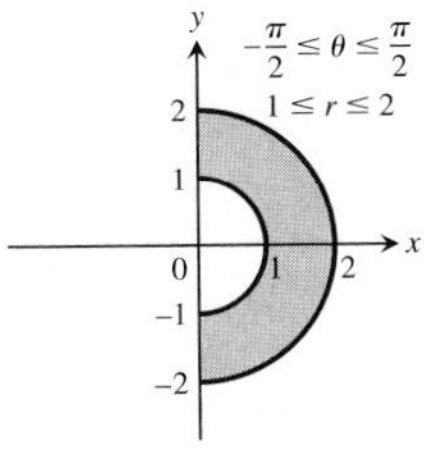

23. $x = 2$, vertical line through $(2, 0)$
25. $y = 0$, the x-axis
27. $y = 4$, horizontal line through $(0, 4)$
29. $x + y = 1$, line, $m = -1, b = 1$
31. $x^2 + y^2 = 1$, circle, $C(0, 0)$, radius 1
33. $y - 2x = 5$, line; $m = 2, b = 5$
35. $y^2 = x$, parabola, vertex $(0, 0)$, opens right
37. $y = e^x$, graph of natural exponential function
39. $x + y = \pm 1$, two straight lines of slope -1, y-intercepts $b = \pm 1$
41. $(x + 2)^2 + y^2 = 4$, circle, $C(-2, 0)$, radius 2
43. $x^2 + (y - 4)^2 = 16$, circle, $C(0, 4)$, radius 4
45. $(x - 1)^2 + (y - 1)^2 = 2$, circle, $C(1, 1)$, radius $\sqrt{2}$
47. $\sqrt{3}y + x = 4$ **49.** $r\cos\theta = 7$ **51.** $\theta = \pi/4$
53. $r = 2$ or $r = -2$ **55.** $4r^2\cos^2\theta + 9r^2\sin^2\theta = 36$
57. $r\sin^2\theta = 4\cos\theta$ **59.** $r = 4\sin\theta$
61. $r^2 = 6r\cos\theta - 2r\sin\theta - 6$
63. $(0, \theta)$, where θ is any angle

Section 10.6, pp. 725–726

1. x-axis
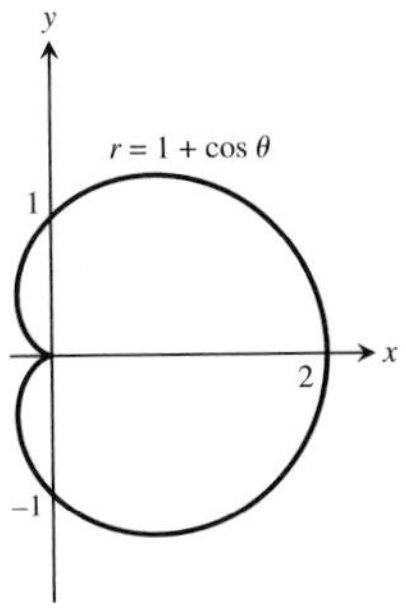

3. y-axis
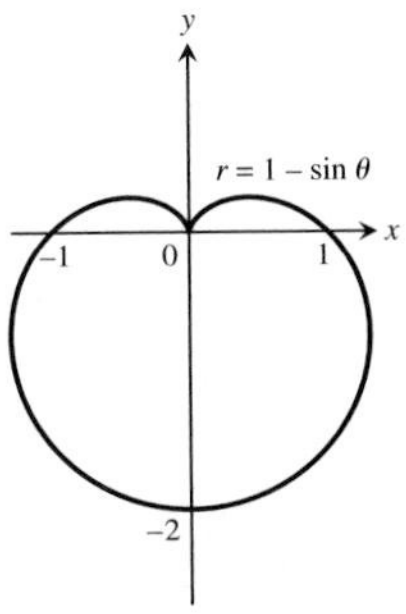

5. y-axis
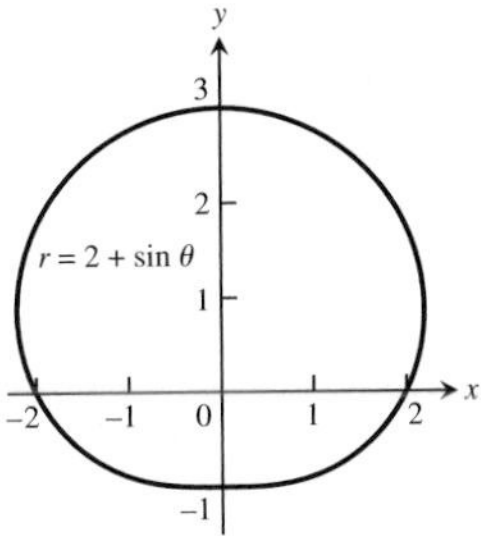

7. x-axis
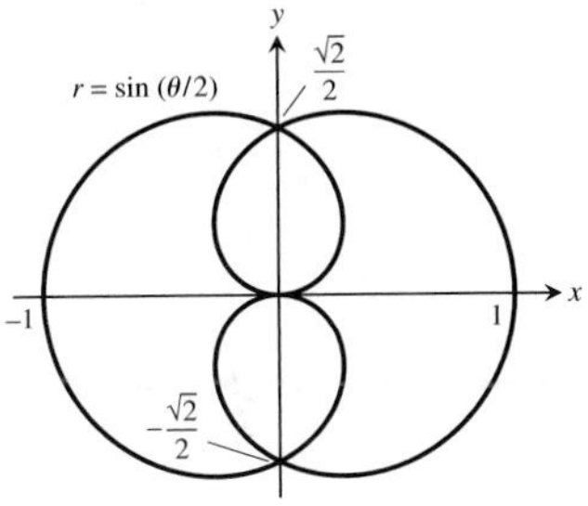

9. x-axis, y-axis, origin
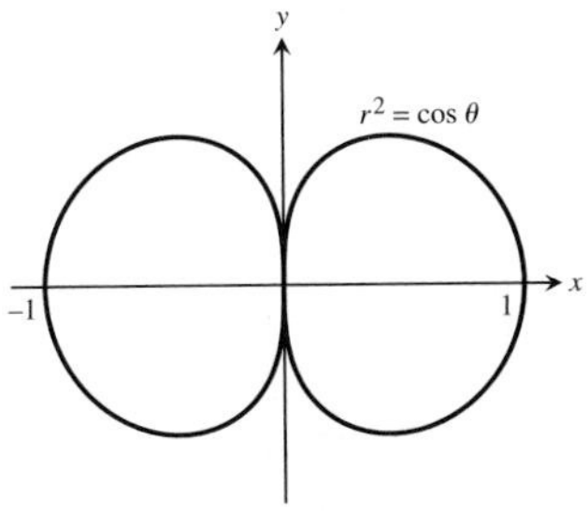

11. y-axis, x-axis, origin
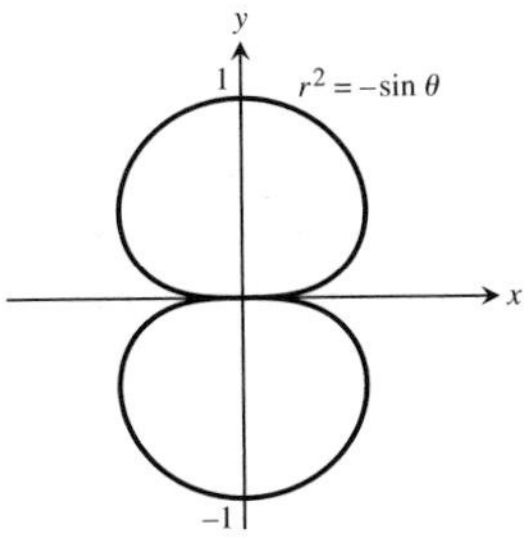

13. x-axis, y-axis, origin **15.** Origin

17. The slope at $(-1, \pi/2)$ is -1, at $(-1, -\pi/2)$ is 1.

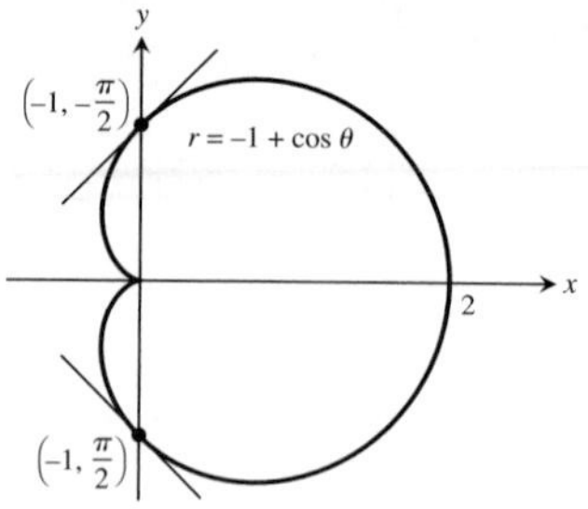

19. The slope at $(1, \pi/4)$ is -1, at $(-1, -\pi/4)$ is 1, at $(-1, 3\pi/4)$ is 1, at $(1, -3\pi/4)$ is -1.

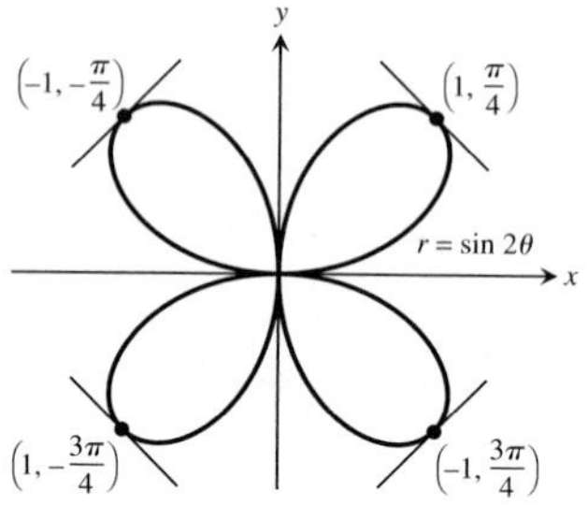

21. (a)

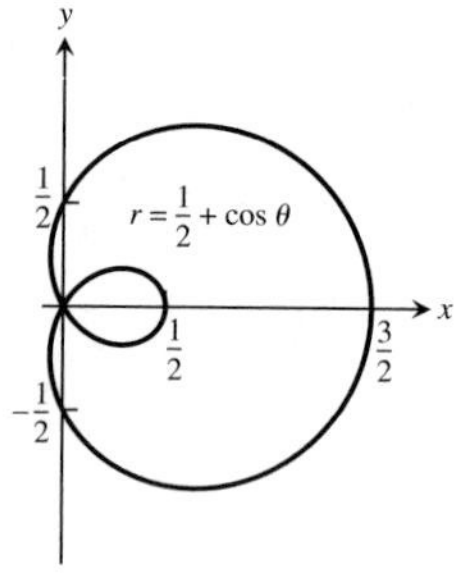

(b)

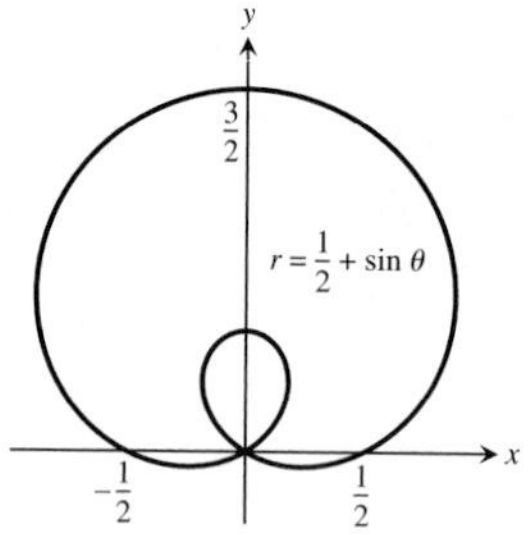

23. (a)

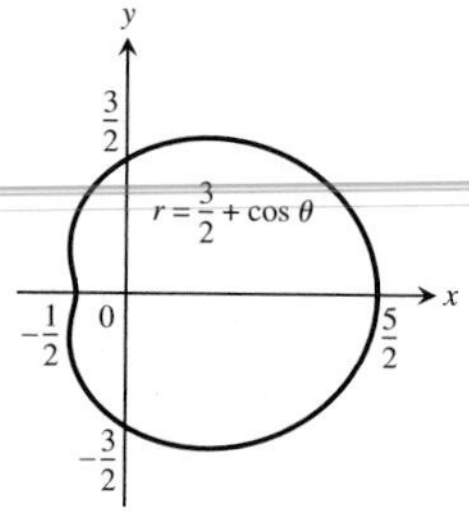

(b)

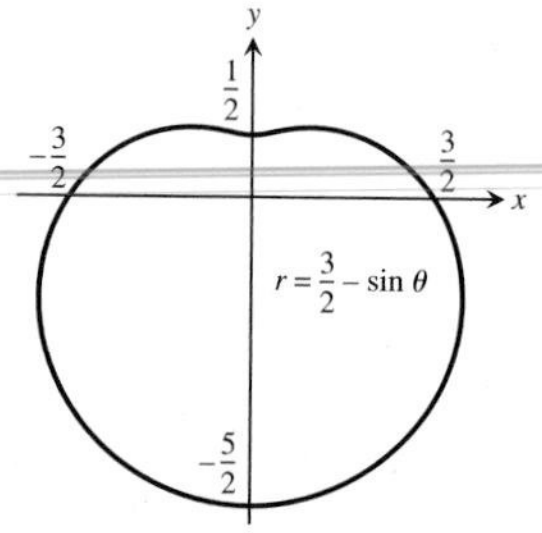

25.

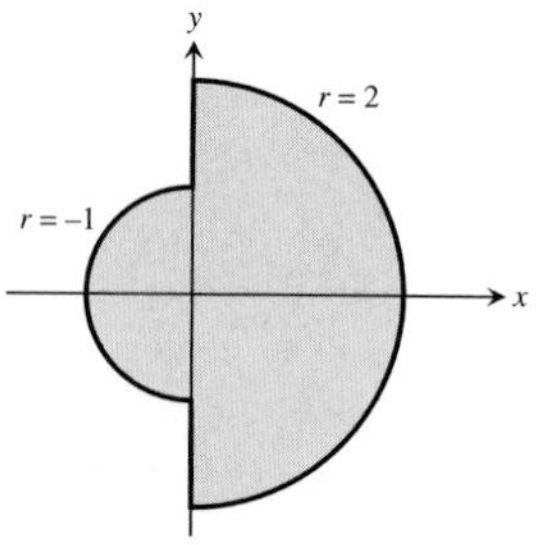

27.

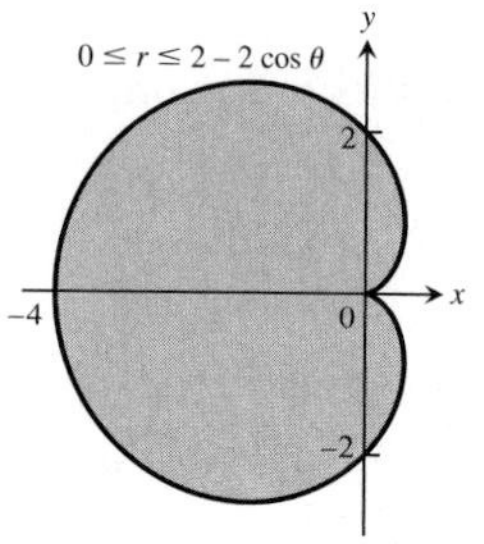

31. $(0, 0), (1, \pi/2), (1, 3\pi/2)$

33. $(0, 0), (\sqrt{3}, \pi/3), (-\sqrt{3}, -\pi/3)$

35. $(\sqrt{2}, \pm\pi/6), (\sqrt{2}, \pm 5\pi/6)$

37. $(1, \pi/12), (1, 5\pi/12), (1, 13\pi/12), (1, 17\pi/12)$

43. (a) **51.** $2y = \dfrac{2\sqrt{6}}{9}$

Section 10.7, pp. 730–731

1. 18π **3.** $\pi/8$ **5.** 2 **7.** $\dfrac{\pi}{2} - 1$ **9.** $5\pi - 8$

11. $3\sqrt{3} - \pi$ **13.** $\dfrac{\pi}{3} + \dfrac{\sqrt{3}}{2}$ **15.** $12\pi - 9\sqrt{3}$

17. (a) $\dfrac{3}{2} - \dfrac{\pi}{4}$ **19.** $19/3$ **21.** 8

23. $3(\sqrt{2} + \ln(1 + \sqrt{2}))$ **25.** $\dfrac{\pi}{8} + \dfrac{3}{8}$ **27.** 2π

29. $\pi\sqrt{2}$ **31.** $2\pi(2 - \sqrt{2})$ **37.** $\left(\dfrac{5}{6}a, 0\right)$

Section 10.8, pp. 737–739

1. $r\cos(\theta - \pi/6) = 5, y = -\sqrt{3}x + 10$

3. $r\cos(\theta - 4\pi/3) = 3, y = -(\sqrt{3}/3)x - 2\sqrt{3}$

5. $y = 2 - x$ **7.** $y = (\sqrt{3}/3)x + 2\sqrt{3}$

9. $r\cos\left(\theta - \dfrac{\pi}{4}\right) = 3$ **11.** $r\cos\left(\theta + \dfrac{\pi}{2}\right) = 5$

13. $r = 8\cos\theta$ **15.** $r = 2\sqrt{2}\sin\theta$ **17.** $C(2, 0)$, radius $= 2$

19. $C(1, \pi)$, radius $= 1$ **21.** $(x - 6)^2 + y^2 = 36, r = 12\cos\theta$

23. $x^2 + (y - 5)^2 = 25, r = 10\sin\theta$

25. $(x + 1)^2 + y^2 = 1, r = -2\cos\theta$

27. $x^2 + (y + 1/2)^2 = 1/4, r = -\sin\theta$ **29.** $r = 2/(1 + \cos\theta)$

31. $r = 30/(1 - 5\sin\theta)$ **33.** $r = 1/(2 + \cos\theta)$

35. $r = 10/(5 - \sin\theta)$

37.

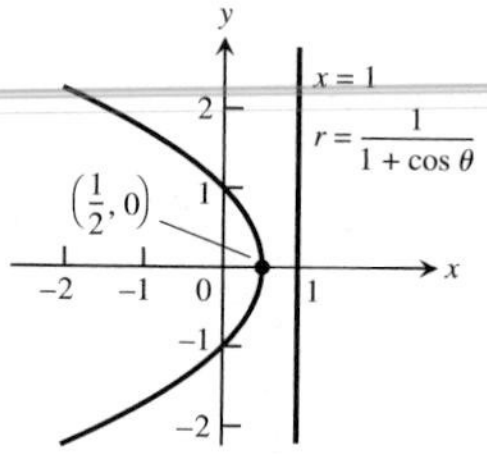

39.

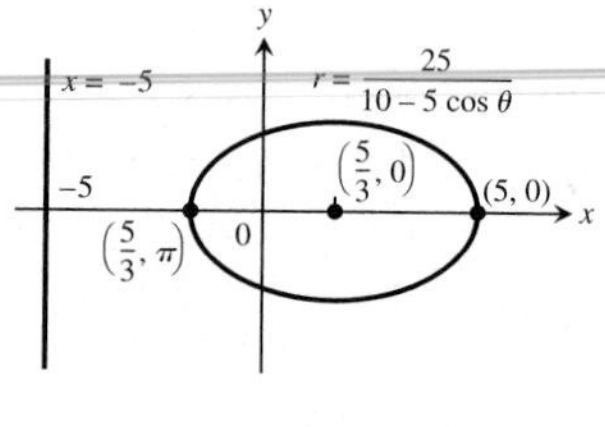

41.

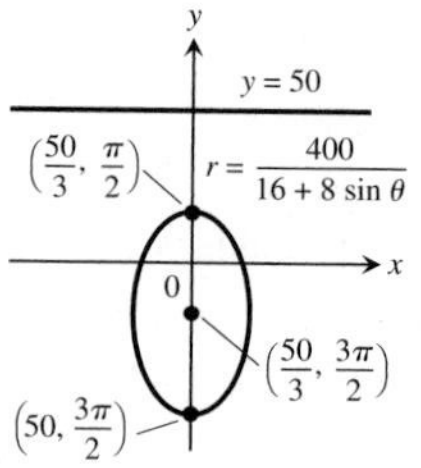

43.

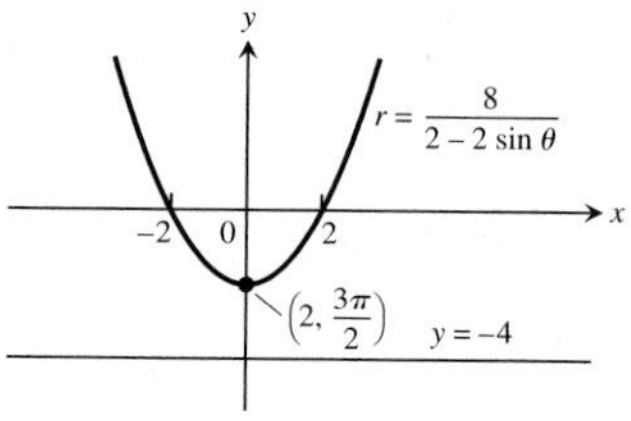

45.

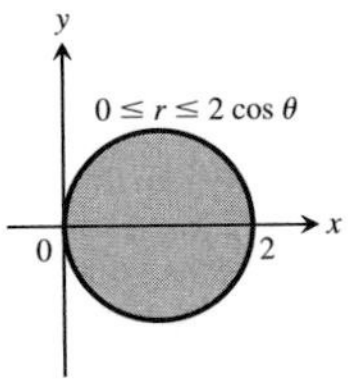

57. (b)

Planet	Perihelion	Aphelion
Mercury	0.3075 AU	0.4667 AU
Venus	0.7184 AU	0.7282 AU
Earth	0.9833 AU	1.0167 AU
Mars	1.3817 AU	1.6663 AU
Jupiter	4.9512 AU	5.4548 AU
Saturn	9.0210 AU	10.0570 AU
Uranus	18.2977 AU	20.0623 AU
Neptune	29.8135 AU	30.3065 AU
Pluto	29.6549 AU	49.2251 AU

59. (a) $x^2 + (y - 2)^2 = 4, x = \sqrt{3}$

(b)

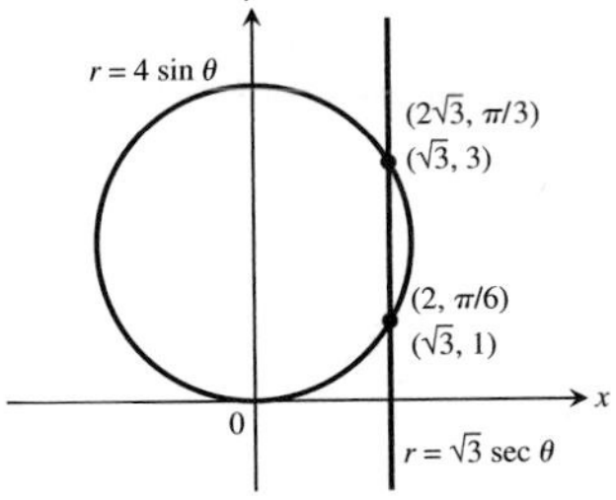

61. $r = 4/(1 + \cos\theta)$ **63. (b)** The pins should be 2 in. apart.

65. $r = 2a\sin\theta$ (a circle) **67.** $r\cos(\theta - \alpha) = p$ (a line)

Practice Exercises, pp. 740–743

1.

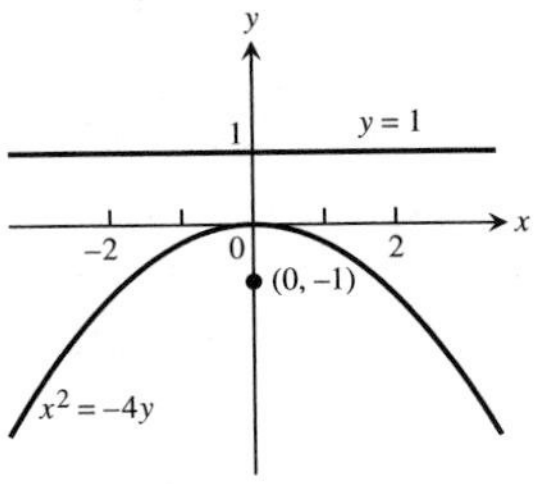

3.

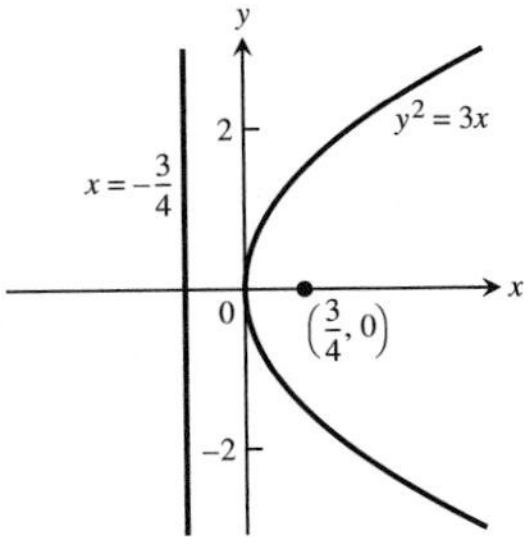

5. $e = 3/4$

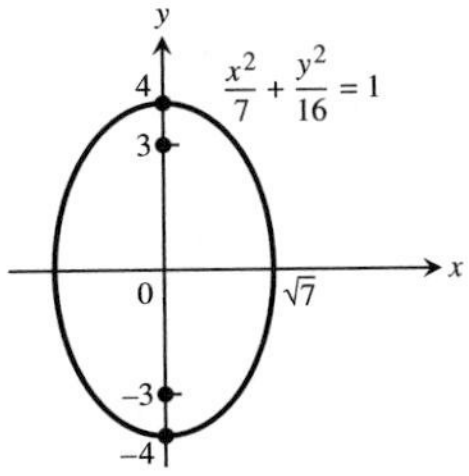

7. $e = 2$

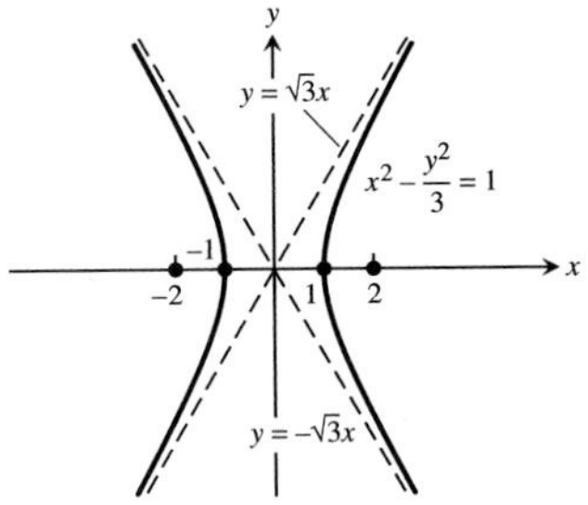

9. $(x - 2)^2 = -12(y - 3)$, $V(2, 3)$, $F(2, 0)$; directrix: $y = 6$

11. $\dfrac{(x + 3)^2}{9} + \dfrac{(y + 5)^2}{25} = 1$, $C(-3, -5)$, $V(-3, 0)$ and $V(-3, -10)$, $F(-3, -1)$ and $F(-3, -9)$

13. $\dfrac{(y - 2\sqrt{2})^2}{8} - \dfrac{(x - 2)^2}{2} = 1$, $C(2, 2\sqrt{2})$, $V(2, 4\sqrt{2})$ and $V(2, 0)$, $F(2, \sqrt{10} + 2\sqrt{2})$ and $F(2, -\sqrt{10} + 2\sqrt{2})$; asymptotes: $y = 2x - 4 + 2\sqrt{2}$ and $y = -2x + 4 + 2\sqrt{2}$

15. Hyperbola: $\dfrac{(x - 2)^2}{4} - y^2 = 1$, $F(2 \pm \sqrt{5}, 0)$, $V(2 \pm 2, 0)$, $C(2, 0)$; asymptotes: $y = \pm\frac{1}{2}(x - 2)$

17. Parabola: $(y - 1)^2 = -16(x + 3)$, $V(-3, 1)$, $F(-7, 1)$; directrix: $x = 1$

19. Ellipse: $\dfrac{(x + 3)^2}{16} + \dfrac{(y - 2)^2}{9} = 1$, $F(\pm\sqrt{7} - 3, 2)$, $V(\pm 4 - 3, 2)$, $C(-3, 2)$

21. Circle: $(x - 1)^2 + (y - 1)^2 = 2$, $C(1, 1)$, radius $= \sqrt{2}$

23. Ellipse **25.** Hyperbola **27.** Line

29. Ellipse, $5x'^2 + 3y'^2 = 30$ **31.** Hyperbola, $y'^2 - x'^2 = 2$

33.

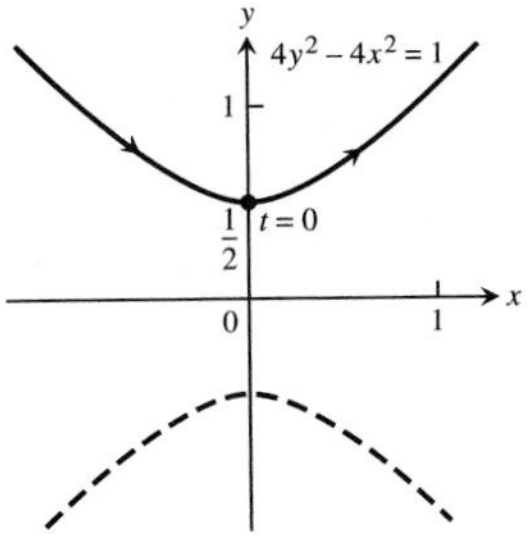

35.

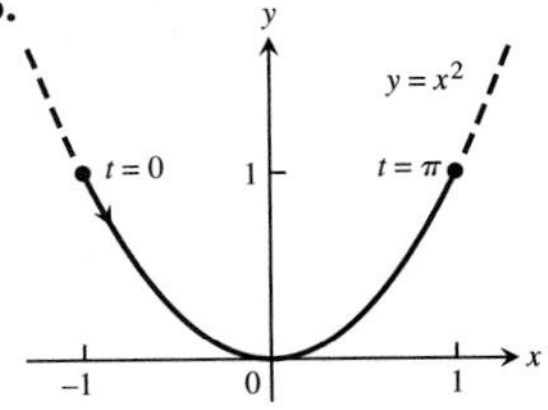

37.

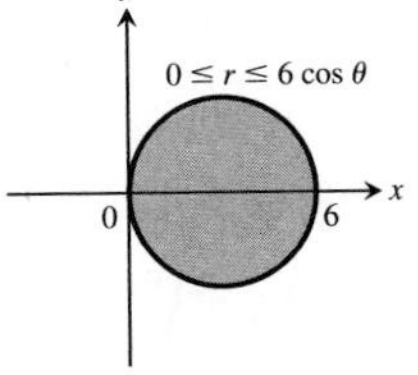

39. (d) **41.** (l) **43.** (k)

45. (i) **47.** (0, 0) **49.** $(0, 0), (1, \pm\pi/2)$

51. The graphs coincide. **53.** $(\sqrt{2}, \pi/4)$

55. $y = (\sqrt{3}/3)x - 4$ **57.** $x = 2$ **59.** $y = -3/2$
61. $x^2 + (y + 2)^2 = 4$ **63.** $(x - \sqrt{2})^2 + y^2 = 2$
65. $r = -5 \sin\theta$ **67.** $r = 3\cos\theta$
69.

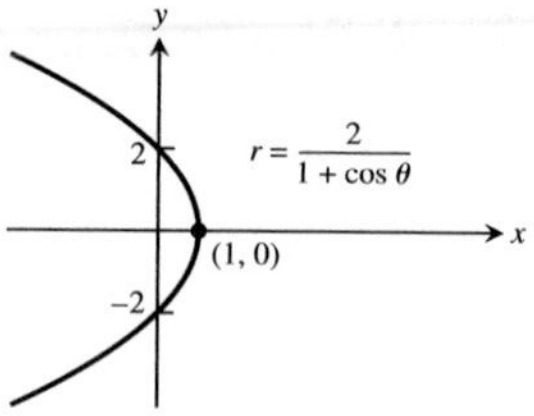

71.

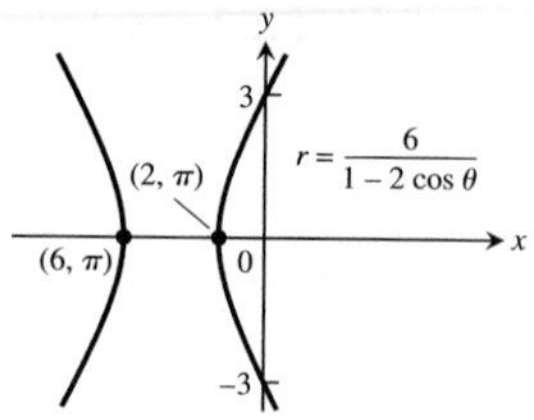

73. $r = \dfrac{4}{1 + 2\cos\theta}$ **75.** $r = \dfrac{2}{2 + \sin\theta}$ **77.** $9\pi/2$
79. $2 + \pi/4$ **81.** 8 **83.** $\pi - 3$ **85.** $(2 - \sqrt{2})\pi$
87. **(a)** 24π **(b)** 16π

Additional and Advanced Exercises, pp. 743–745

1.

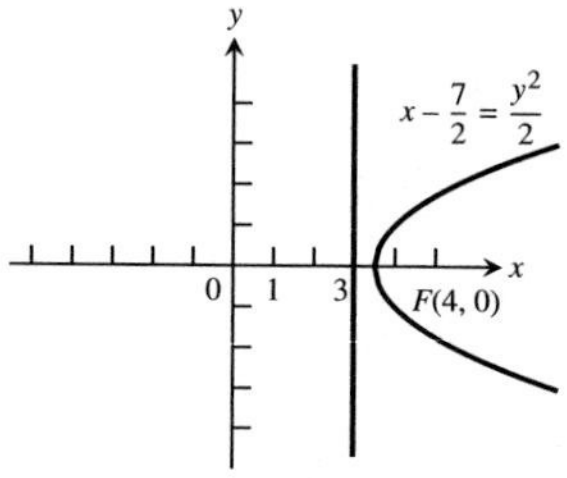

3. $3x^2 + 3y^2 - 8y + 4 = 0$

5. $(0, \pm 1)$

7. **(a)** $\dfrac{(y - 1)^2}{16} - \dfrac{x^2}{48} = 1$ **(b)** $\dfrac{16\left(y + \frac{3}{4}\right)^2}{25} - \dfrac{2x^2}{75} = 1$

11.

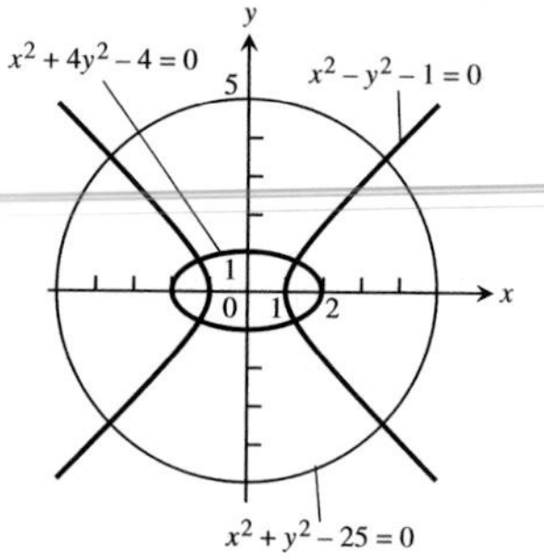

13.

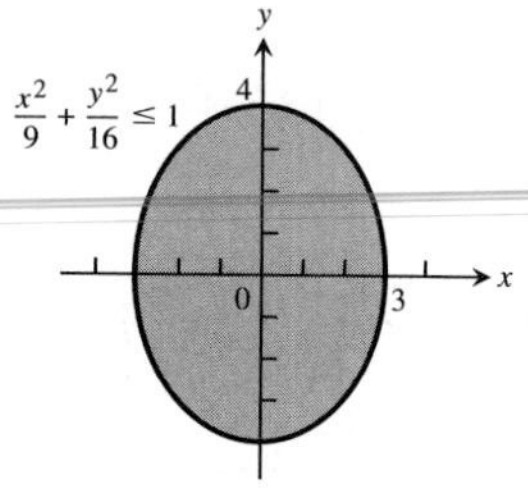

15.

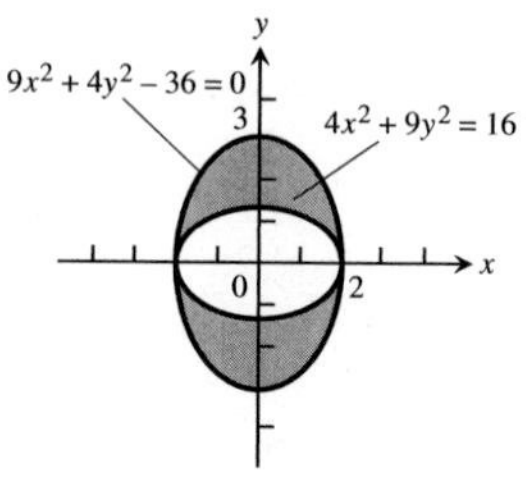

17.

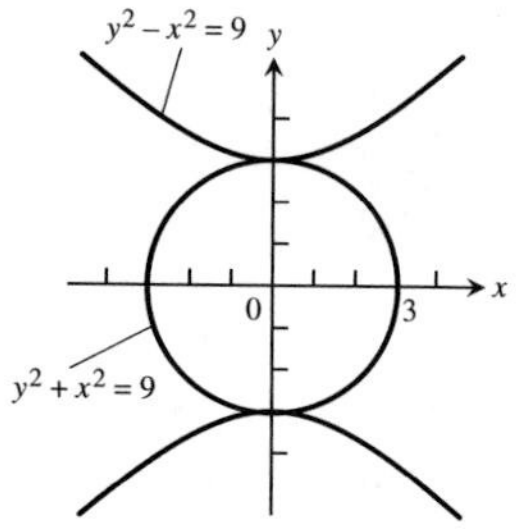

19. $x = (a + b)\cos\theta - b\cos\left(\dfrac{a + b}{b}\theta\right)$,

$y = (a + b)\sin\theta - b\sin\left(\dfrac{a + b}{b}\theta\right)$

21. **(a)** $r = e^{2\theta}$ **(b)** $\dfrac{\sqrt{5}}{2}(e^{4\pi} - 1)$ **23.** $\dfrac{32\pi - 4\pi\sqrt{2}}{5}$

25. $r = \dfrac{4}{1 + 2\cos\theta}$ **27.** $r = \dfrac{2}{2 + \sin\theta}$ **29.** **(a)** 120°

31. 1×10^7 miles **33.** $e = \sqrt{2/3}$ **35.** Yes, a parabola

37. **(a)** $r = \dfrac{2a}{1 + \cos\left(\theta - \dfrac{\pi}{4}\right)}$ **(b)** $r = \dfrac{8}{3 - \cos\theta}$

(c) $r = \dfrac{3}{1 + 2\sin\theta}$

43. $\pi/2$ **47.** $\left(2, \pm\dfrac{\pi}{3}\right), \dfrac{\pi}{2}$ **51.** $\pi/2$ **53.** $\pi/4$

CHAPTER 11

Section 11.1, pp. 757–761

1. $a_1 = 0, a_2 = -1/4, a_3 = -2/9, a_4 = -3/16$
3. $a_1 = 1, a_2 = -1/3, a_3 = 1/5, a_4 = -1/7$
5. $a_1 = 1/2, a_2 = 1/2, a_3 = 1/2, a_4 = 1/2$
7. $1, \dfrac{3}{2}, \dfrac{7}{4}, \dfrac{15}{8}, \dfrac{31}{16}, \dfrac{63}{32}, \dfrac{127}{64}, \dfrac{255}{128}, \dfrac{511}{256}, \dfrac{1023}{512}$
9. $2, 1, -\dfrac{1}{2}, -\dfrac{1}{4}, \dfrac{1}{8}, \dfrac{1}{16}, -\dfrac{1}{32}, -\dfrac{1}{64}, \dfrac{1}{128}, \dfrac{1}{256}$
11. 1, 1, 2, 3, 5, 8, 13, 21, 34, 55 **13.** $a_n = (-1)^{n+1}, n \ge 1$
15. $a_n = (-1)^{n+1}(n)^2, n \ge 1$ **17.** $a_n = n^2 - 1, n \ge 1$
19. $a_n = 4n - 3, n \ge 1$ **21.** $a_n = \dfrac{1 + (-1)^{n+1}}{2}, n \ge 1$
23. Converges, 2 **25.** Converges, −1 **27.** Converges, −5
29. Diverges **31.** Diverges **33.** Converges, 1/2
35. Converges, 0 **37.** Converges, $\sqrt{2}$ **39.** Converges, 1
41. Converges, 0 **43.** Converges, 0 **45.** Converges, 0
47. Converges, 1 **49.** Converges, e^7 **51.** Converges, 1
53. Converges, 1 **55.** Diverges **57.** Converges, 4
59. Converges, 0 **61.** Diverges **63.** Converges, e^{-1}
65. Converges, $e^{2/3}$ **67.** Converges, x $(x > 0)$
69. Converges, 0 **71.** Converges, 1 **73.** Converges, 1/2
75. Converges, $\pi/2$ **77.** Converges, 0 **79.** Converges, 0
81. Converges, 1/2 **83.** Converges, 0 **85.** $x_n = 2^{n-2}$
87. **(a)** $f(x) = x^2 - 2$, $1.414213562 \approx \sqrt{2}$
(b) $f(x) = \tan(x) - 1$, $0.7853981635 \approx \pi/4$
(c) $f(x) = e^x$, diverges
89. **(b)** 1 **97.** Nondecreasing, bounded
99. Not nondecreasing, bounded
101. Converges, nondecreasing sequence theorem
103. Converges, nondecreasing sequence theorem
105. Diverges, definition of divergence **109.** Converges
111. Converges **121.** $N = 692$, $a_n = \sqrt[n]{0.5}$, $L = 1$
123. $N = 65$, $a_n = (0.9)^n$, $L = 0$ **125.** **(b)** $\sqrt{3}$

Section 11.2, pp. 769–771

1. $s_n = \dfrac{2(1-(1/3)^n)}{1-(1/3)}, 3$ **3.** $s_n = \dfrac{1-(-1/2)^n}{1-(-1/2)}, 2/3$

5. $s_n = \dfrac{1}{2} - \dfrac{1}{n+2}, \dfrac{1}{2}$ **7.** $1 - \dfrac{1}{4} + \dfrac{1}{16} - \dfrac{1}{64} + \cdots, \dfrac{4}{5}$

9. $\dfrac{7}{4} + \dfrac{7}{16} + \dfrac{7}{64} + \cdots, \dfrac{7}{3}$

11. $(5+1) + \left(\dfrac{5}{2} + \dfrac{1}{3}\right) + \left(\dfrac{5}{4} + \dfrac{1}{9}\right) + \left(\dfrac{5}{8} + \dfrac{1}{27}\right) + \cdots, \dfrac{23}{2}$

13. $(1+1) + \left(\dfrac{1}{2} - \dfrac{1}{5}\right) + \left(\dfrac{1}{4} + \dfrac{1}{25}\right) + \left(\dfrac{1}{8} - \dfrac{1}{125}\right) + \cdots, \dfrac{17}{6}$

15. 1 **17.** 5 **19.** 1 **21.** $-\dfrac{1}{\ln 2}$ **23.** Converges, $2 + \sqrt{2}$

25. Converges, 1 **27.** Diverges **29.** Converges, $\dfrac{e^2}{e^2 - 1}$

31. Converges, 2/9 **33.** Converges, 3/2 **35.** Diverges

37. Diverges **39.** Converges, $\dfrac{\pi}{\pi - e}$

41. $a = 1, r = -x$; converges to $1/(1+x)$ for $|x| < 1$

43. $a = 3, r = (x-1)/2$; converges to $6/(3-x)$ for x in $(-1, 3)$

45. $|x| < \dfrac{1}{2}, \dfrac{1}{1-2x}$ **47.** $-2 < x < 0, \dfrac{1}{2+x}$

49. $x \neq (2k+1)\dfrac{\pi}{2}$, k an integer; $\dfrac{1}{1 - \sin x}$

51. 23/99 **53.** 7/9 **55.** 1/15 **57.** 41333/33300

59. (a) $\displaystyle\sum_{n=-2}^{\infty} \frac{1}{(n+4)(n+5)}$ **(b)** $\displaystyle\sum_{n=0}^{\infty} \frac{1}{(n+2)(n+3)}$

(c) $\displaystyle\sum_{n=5}^{\infty} \frac{1}{(n-3)(n-2)}$

69. (a) $r = 3/5$ **(b)** $r = -3/10$

71. $|r| < 1, \dfrac{1+2r}{1-r^2}$ **73.** 28 m **75.** 8 m^2

77. (a) $3\left(\dfrac{4}{3}\right)^{n-1}$

(b) $A_n = A + \dfrac{1}{3}A + \dfrac{1}{3}\left(\dfrac{4}{9}\right)A + \cdots + \dfrac{1}{3}\left(\dfrac{4}{9}\right)^{n-2}A,$

$A = \dfrac{\sqrt{3}}{4}, \lim_{n\to\infty} A_n = 2\sqrt{3}/5$

Section 11.3, pp. 775–777

1. Converges; geometric series, $r = \dfrac{1}{10} < 1$

3. Diverges; $\lim_{n\to\infty} \dfrac{n}{n+1} = 1 \neq 0$ **5.** Diverges; p-series, $p < 1$

7. Converges; geometric series, $r = \dfrac{1}{8} < 1$

9. Diverges; Integral Test

11. Converges; geometric series, $r = 2/3 < 1$

13. Diverges; Integral Test **15.** Diverges; $\lim_{n\to\infty} \dfrac{2^n}{n+1} \neq 0$

17. Diverges; $\lim_{n\to\infty}\left(\sqrt{n}/\ln n\right) \neq 0$

19. Diverges; geometric series, $r = \dfrac{1}{\ln 2} > 1$

21. Converges; Integral Test **23.** Diverges; nth-Term Test

25. Converges; Integral Test **27.** Converges; Integral Test

29. Converges; Integral Test **31.** $a = 1$ **33. (b)** About 41.55

35. True

Section 11.4, p. 781

1. Diverges; limit comparison with $\sum\left(1/\sqrt{n}\right)$

3. Converges; compare with $\sum\left(1/2^n\right)$

5. Diverges; nth-Term Test

7. Converges; $\left(\dfrac{n}{3n+1}\right)^n < \left(\dfrac{n}{3n}\right)^n = \left(\dfrac{1}{3}\right)^n$

9. Diverges; direct comparison with $\sum(1/n)$

11. Converges; limit comparison with $\sum(1/n^2)$

13. Diverges; limit comparison with $\sum(1/n)$

15. Diverges; limit comparison with $\sum(1/n)$

17. Diverges; Integral Test

19. Converges; compare with $\sum(1/n^{3/2})$

21. Converges; $\dfrac{1}{n2^n} \leq \dfrac{1}{2^n}$ **23.** Converges; $\dfrac{1}{3^{n-1}+1} < \dfrac{1}{3^{n-1}}$

25. Diverges; limit comparison with $\sum(1/n)$

27. Converges; compare with $\sum(1/n^2)$

29. Converges; $\dfrac{\tan^{-1} n}{n^{1.1}} < \dfrac{\pi/2}{n^{1.1}}$

31. Converges; compare with $\sum(1/n^2)$

33. Diverges; limit comparison with $\sum(1/n)$

35. Converges; limit comparison with $\sum(1/n^2)$

Section 11.5, p. 786

1. Converges; Ratio Test **3.** Diverges; Ratio Test

5. Converges; Ratio Test

7. Converges; compare with $\sum(3/(1.25)^n)$

9. Diverges; $\lim_{n\to\infty}\left(1 - \dfrac{3}{n}\right)^n = e^{-3} \neq 0$

11. Converges; compare with $\sum(1/n^2)$

13. Diverges; compare with $\sum(1/(2n))$

15. Diverges; compare with $\sum(1/n)$ **17.** Converges; Ratio Test

19. Converges; Ratio Test **21.** Converges; Ratio Test

23. Converges; Root Test

25. Converges; compare with $\sum(1/n^2)$

27. Converges; Ratio Test **29.** Diverges; Ratio Test

31. Converges; Ratio Test **33.** Converges; Ratio Test

35. Diverges; $a_n = \left(\dfrac{1}{3}\right)^{(1/n!)} \to 1$ **37.** Converges; Ratio Test

39. Diverges; Root Test **41.** Converges; Root Test

43. Converges; Ratio Test **47.** Yes

Section 11.6, pp. 792–794

1. Converges by Theorem 16
3. Diverges; $a_n \nrightarrow 0$
5. Converges by Theorem 16
7. Diverges; $a_n \to 1/2$
9. Converges by Theorem 16
11. Converges absolutely. Series of absolute values is a convergent geometric series.
13. Converges conditionally. $1/\sqrt{n} \to 0$ but $\sum_{n=1}^{\infty} \frac{1}{\sqrt{n}}$ diverges.
15. Converges absolutely. Compare with $\sum_{n=1}^{\infty}(1/n^2)$.
17. Converges conditionally. $1/(n+3) \to 0$ but $\sum_{n=1}^{\infty} \frac{1}{n+3}$ diverges (compare with $\sum_{n=1}^{\infty}(1/n)$).
19. Diverges; $\frac{3+n}{5+n} \to 1$
21. Converges conditionally; $\left(\frac{1}{n^2} + \frac{1}{n}\right) \to 0$ but $(1+n)/n^2 > 1/n$
23. Converges absolutely; Ratio Test
25. Converges absolutely by Integral Test
27. Diverges; $a_n \nrightarrow 0$
29. Converges absolutely by the Ratio Test
31. Converges absolutely; $\frac{1}{n^2 + 2n + 1} < \frac{1}{n^2}$
33. Converges absolutely since $\left|\frac{\cos n\pi}{n\sqrt{n}}\right| = \left|\frac{(-1)^{n+1}}{n^{3/2}}\right| = \frac{1}{n^{3/2}}$ (convergent p-series)
35. Converges absolutely by Root Test
37. Diverges; $a_n \to \infty$
39. Converges conditionally; $\sqrt{n+1} - \sqrt{n} = 1/\left(\sqrt{n} + \sqrt{n+1}\right) \to 0$, but series of absolute values diverges (compare with $\sum\left(1/\sqrt{n}\right)$)
41. Diverges, $a_n \to 1/2 \neq 0$
43. Converges absolutely; $\operatorname{sech} n = \frac{2}{e^n + e^{-n}} = \frac{2e^n}{e^{2n}+1} < \frac{2e^n}{e^{2n}} = \frac{2}{e^n}$, a term from a convergent geometric series.
45. $|\text{Error}| < 0.2$
47. $|\text{Error}| < 2 \times 10^{-11}$
49. 0.54030
51. (a) $a_n \geq a_{n+1}$ (b) $-1/2$

Section 11.7, pp. 804–805

1. (a) $1, -1 < x < 1$ (b) $-1 < x < 1$ (c) none
3. (a) $1/4, -1/2 < x < 0$ (b) $-1/2 < x < 0$ (c) none
5. (a) $10, -8 < x < 12$ (b) $-8 < x < 12$ (c) none
7. (a) $1, -1 < x < 1$ (b) $-1 < x < 1$ (c) none
9. (a) $3, -3 \leq x \leq 3$ (b) $-3 \leq x \leq 3$ (c) none
11. (a) ∞, for all x (b) for all x (c) none
13. (a) ∞, for all x (b) for all x (c) none
15. (a) $1, -1 \leq x < 1$ (b) $-1 < x < 1$ (c) $x = -1$
17. (a) $5, -8 < x < 2$ (b) $-8 < x < 2$ (c) none
19. (a) $3, -3 < x < 3$ (b) $-3 < x < 3$ (c) none
21. (a) $1, -1 < x < 1$ (b) $-1 < x < 1$ (c) none
23. (a) $0, x = 0$ (b) $x = 0$ (c) none
25. (a) $2, -4 < x \leq 0$ (b) $-4 < x < 0$ (c) $x = 0$
27. (a) $1, -1 \leq x \leq 1$ (b) $-1 \leq x \leq 1$ (c) none
29. (a) $1/4, 1 \leq x \leq 3/2$ (b) $1 \leq x \leq 3/2$ (c) none
31. (a) $1, (-1-\pi) \leq x < (1-\pi)$ (b) $(-1-\pi) < x < (1-\pi)$ (c) $x = -1-\pi$
33. $-1 < x < 3$, $4/(3 + 2x - x^2)$
35. $0 < x < 16$, $2/\left(4 - \sqrt{x}\right)$
37. $-\sqrt{2} < x < \sqrt{2}$, $3/(2 - x^2)$
39. $1 < x < 5$, $2/(x-1)$, $1 < x < 5$, $-2/(x-1)^2$
41. (a) $\cos x = 1 - \frac{x^2}{2!} + \frac{x^4}{4!} - \frac{x^6}{6!} + \frac{x^8}{8!} - \frac{x^{10}}{10!} + \cdots$; converges for all x
(b) and
(c) $2x - \frac{2^3x^3}{3!} + \frac{2^5x^5}{5!} - \frac{2^7x^7}{7!} + \frac{2^9x^9}{9!} - \frac{2^{11}x^{11}}{11!} + \cdots$
43. (a) $\frac{x^2}{2} + \frac{x^4}{12} + \frac{x^6}{45} + \frac{17x^8}{2520} + \frac{31x^{10}}{14175}$, $-\frac{\pi}{2} < x < \frac{\pi}{2}$
(b) $1 + x^2 + \frac{2x^4}{3} + \frac{17x^6}{45} + \frac{62x^8}{315} + \cdots$, $-\frac{\pi}{2} < x < \frac{\pi}{2}$

Section 11.8, pp. 810–811

1. $P_0(x) = 0$, $P_1(x) = x - 1$, $P_2(x) = (x-1) - \frac{1}{2}(x-1)^2$, $P_3(x) = (x-1) - \frac{1}{2}(x-1)^2 + \frac{1}{3}(x-1)^3$
3. $P_0(x) = \frac{1}{2}$, $P_1(x) = \frac{1}{2} - \frac{1}{4}(x-2)$, $P_2(x) = \frac{1}{2} - \frac{1}{4}(x-2) + \frac{1}{8}(x-2)^2$, $P_3(x) = \frac{1}{2} - \frac{1}{4}(x-2) + \frac{1}{8}(x-2)^2 - \frac{1}{16}(x-2)^3$
5. $P_0(x) = \frac{\sqrt{2}}{2}$, $P_1(x) = \frac{\sqrt{2}}{2} + \frac{\sqrt{2}}{2}\left(x - \frac{\pi}{4}\right)$, $P_2(x) = \frac{\sqrt{2}}{2} + \frac{\sqrt{2}}{2}\left(x - \frac{\pi}{4}\right) - \frac{\sqrt{2}}{4}\left(x - \frac{\pi}{4}\right)^2$, $P_3(x) = \frac{\sqrt{2}}{2} + \frac{\sqrt{2}}{2}\left(x - \frac{\pi}{4}\right) - \frac{\sqrt{2}}{4}\left(x - \frac{\pi}{4}\right)^2 - \frac{\sqrt{2}}{12}\left(x - \frac{\pi}{4}\right)^3$
7. $P_0(x) = 2$, $P_1(x) = 2 + \frac{1}{4}(x-4)$, $P_2(x) = 2 + \frac{1}{4}(x-4) - \frac{1}{64}(x-4)^2$, $P_3(x) = 2 + \frac{1}{4}(x-4) - \frac{1}{64}(x-4)^2 + \frac{1}{512}(x-4)^3$
9. $\sum_{n=0}^{\infty} \frac{(-x)^n}{n!} = 1 - x + \frac{x^2}{2!} - \frac{x^3}{3!} + \frac{x^4}{4!} - \cdots$
11. $\sum_{n=0}^{\infty}(-1)^n x^n = 1 - x + x^2 - x^3 + \cdots$
13. $\sum_{n=0}^{\infty} \frac{(-1)^n 3^{2n+1} x^{2n+1}}{(2n+1)!}$
15. $7\sum_{n=0}^{\infty} \frac{(-1)^n x^{2n}}{(2n)!}$
17. $\sum_{n=0}^{\infty} \frac{x^{2n}}{(2n)!}$
19. $x^4 - 2x^3 - 5x + 4$
21. $8 + 10(x-2) + 6(x-2)^2 + (x-2)^3$
23. $21 - 36(x+2) + 25(x+2)^2 - 8(x+2)^3 + (x+2)^4$

25. $\sum_{n=0}^{\infty}(-1)^n(n+1)(x-1)^n$ **27.** $\sum_{n=0}^{\infty}\frac{e^2}{n!}(x-2)^n$

33. $L(x)=0, Q(x)=-x^2/2$ **35.** $L(x)=1, Q(x)=1+x^2/2$

37. $L(x)=x, Q(x)=x$

Section 11.9, pp. 819–822

1. $\sum_{n=0}^{\infty}\frac{(-5x)^n}{n!}=1-5x+\frac{5^2x^2}{2!}-\frac{5^3x^3}{3!}+\cdots$

3. $\sum_{n=0}^{\infty}\frac{5(-1)^n(-x)^{2n+1}}{(2n+1)!}=\sum_{n=0}^{\infty}\frac{5(-1)^{n+1}x^{2n+1}}{(2n+1)!}$
$=-5x+\frac{5x^3}{3!}-\frac{5x^5}{5!}+\frac{5x^7}{7!}+\cdots$

5. $\sum_{n=0}^{\infty}\frac{(-1)^n(x+1)^n}{(2n)!}$

7. $\sum_{n=0}^{\infty}\frac{x^{n+1}}{n!}=x+x^2+\frac{x^3}{2!}+\frac{x^4}{3!}+\frac{x^5}{4!}+\cdots$

9. $\sum_{n=2}^{\infty}\frac{(-1)^nx^{2n}}{(2n)!}=\frac{x^4}{4!}-\frac{x^6}{6!}+\frac{x^8}{8!}-\frac{x^{10}}{10!}+\cdots$

11. $x-\frac{\pi^2x^3}{2!}+\frac{\pi^4x^5}{4!}-\frac{\pi^6x^7}{6!}+\cdots=\sum_{n=0}^{\infty}\frac{(-1)^n\pi^{2n}x^{2n+1}}{(2n)!}$

13. $1+\sum_{n=1}^{\infty}\frac{(-1)^n(2x)^{2n}}{2\cdot(2n)!}=$
$1-\frac{(2x)^2}{2\cdot 2!}+\frac{(2x)^4}{2\cdot 4!}-\frac{(2x)^6}{2\cdot 6!}+\frac{(2x)^8}{2\cdot 8!}-\cdots$

15. $x^2\sum_{n=0}^{\infty}(2x)^n=x^2+2x^3+4x^4+\cdots$

17. $\sum_{n=1}^{\infty}nx^{n-1}=1+2x+3x^2+4x^3+\cdots$

19. $|x|<(0.06)^{1/5}<0.56968$

21. $|\text{Error}|<(10^{-3})^3/6<1.67\times10^{-10}$, $-10^{-3}<x<0$

23. $|\text{Error}|<(3^{0.1})(0.1)^3/6<1.87\times10^{-4}$ **25.** 0.000293653

27. $|x|<0.02$ **31.** $\sin x$, $x=0.1$; $\sin(0.1)$

33. $\tan^{-1}x, x=\pi/3$

35. $e^x\sin x=x+x^2+\frac{x^3}{3}-\frac{x^5}{30}-\frac{x^6}{90}\cdots$

43. **(a)** $Q(x)=1+kx+\frac{k(k-1)}{2}x^2$ **(b)** for $0\le x<100^{-1/3}$

49. **(a)** -1 **(b)** $(1/\sqrt{2})(1+i)$ **(c)** $-i$

53. $x+x^2+\frac{1}{3}x^3-\frac{1}{30}x^5\cdots$; will converge for all x

Section 11.10, pp. 831–833

1. $1+\frac{x}{2}-\frac{x^2}{8}+\frac{x^3}{16}$ **3.** $1+\frac{1}{2}x+\frac{3}{8}x^2+\frac{5}{16}x^3+\cdots$

5. $1-x+\frac{3x^2}{4}-\frac{x^3}{2}$ **7.** $1-\frac{x^3}{2}+\frac{3x^6}{8}-\frac{5x^9}{16}$

9. $1+\frac{1}{2x}-\frac{1}{8x^2}+\frac{1}{16x^3}$

11. $(1+x)^4=1+4x+6x^2+4x^3+x^4$

13. $(1-2x)^3=1-6x+12x^2-8x^3$

15. $y=\sum_{n=0}^{\infty}\frac{(-1)^n}{n!}x^n=e^{-x}$ **17.** $y=\sum_{n=1}^{\infty}(x^n/n!)=e^x-1$

19. $y=\sum_{n=2}^{\infty}(x^n/n!)=e^x-x-1$ **21.** $y=\sum_{n=0}^{\infty}\frac{x^{2n}}{2^nn!}=e^{x^2/2}$

23. $y=\sum_{n=0}^{\infty}2x^n=\frac{2}{1-x}$ **25.** $y=\sum_{n=0}^{\infty}\frac{x^{2n+1}}{(2n+1)!}=\sinh x$

27. $y=2+x-2\sum_{n=1}^{\infty}\frac{(-1)^{n+1}x^{2n}}{(2n)!}$

29. $y=x-2\sum_{n=0}^{\infty}\frac{x^{2n}}{(2n)!}-3\sum_{n=0}^{\infty}\frac{x^{2n+1}}{(2n+1)!}$

31. $y=a+bx+\frac{1}{6}x^3-\frac{ax^4}{3\cdot4}-\frac{bx^5}{4\cdot5}-\frac{x^7}{6\cdot6\cdot7}+$
$\frac{ax^8}{3\cdot4\cdot7\cdot8}+\frac{bx^9}{4\cdot5\cdot8\cdot9}\cdots$.

33. 0.00267 **35.** 0.1 **37.** 0.0999444611 **39.** 0.100001

41. $1/(13\cdot 6!)\approx 0.00011$ **43.** $\frac{x^3}{3}-\frac{x^7}{7\cdot3!}+\frac{x^{11}}{11\cdot5!}$

45. **(a)** $\frac{x^2}{2}-\frac{x^4}{12}$

(b) $\frac{x^2}{2}-\frac{x^4}{3\cdot4}+\frac{x^6}{5\cdot6}-\frac{x^8}{7\cdot8}+\cdots+(-1)^{15}\frac{x^{32}}{31\cdot32}$

47. 1/2 **49.** $-1/24$ **51.** 1/3 **53.** -1 **55.** 2

59. 500 terms **61.** 4 terms

63. **(a)** $x+\frac{x^3}{6}+\frac{3x^5}{40}+\frac{5x^7}{112}$, radius of convergence = 1

(b) $\frac{\pi}{2}-x-\frac{x^3}{6}-\frac{3x^5}{40}-\frac{5x^7}{112}$

65. $1-2x+3x^2-4x^3+\cdots$ **71.** **(c)** $3\pi/4$

Section 11.11, pp. 838–839

1. $f(x)=1$

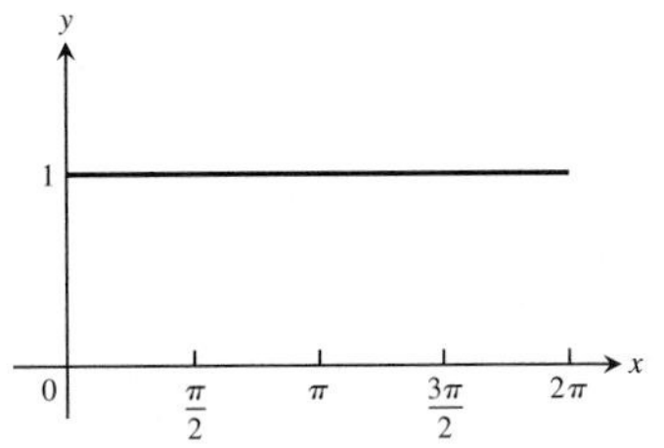

3. $f(x) = \sum_{n=1}^{\infty} \frac{2(-1)^{n+1} \sin(nx)}{n}$

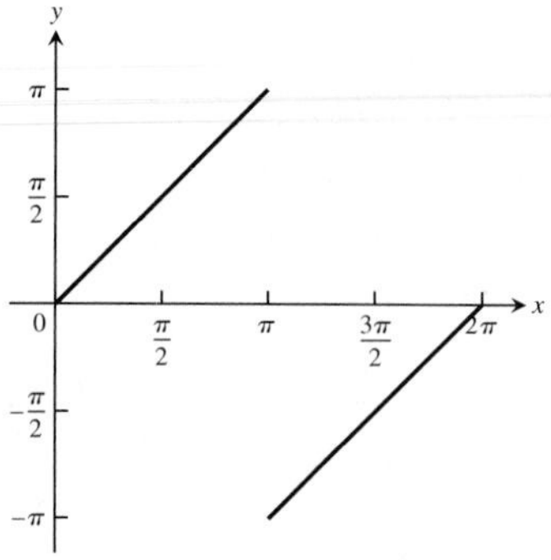

5. $\frac{e^{2\pi} - 1}{\pi}\left(\frac{1}{2} + \sum_{n=1}^{\infty} \frac{\cos(nx)}{n^2 + 1} - \sum_{n=1}^{\infty} \frac{n \sin(nx)}{n^2 + 1}\right)$

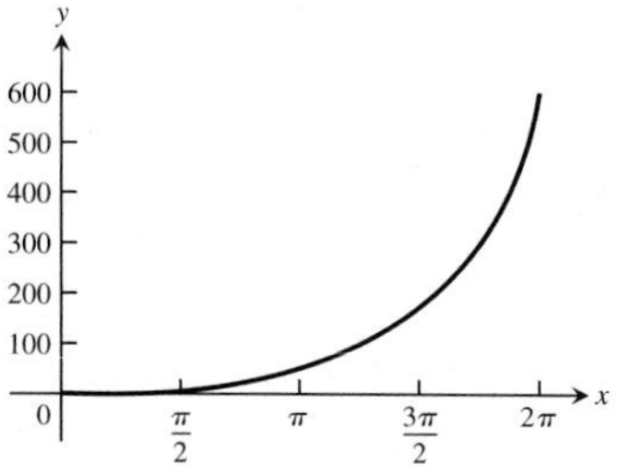

7. $f(x) = \frac{1}{2}\cos x + \frac{1}{\pi}\sum_{n=2}^{\infty} \frac{n(1 + (-1)^n)}{n^2 - 1} \sin nx$

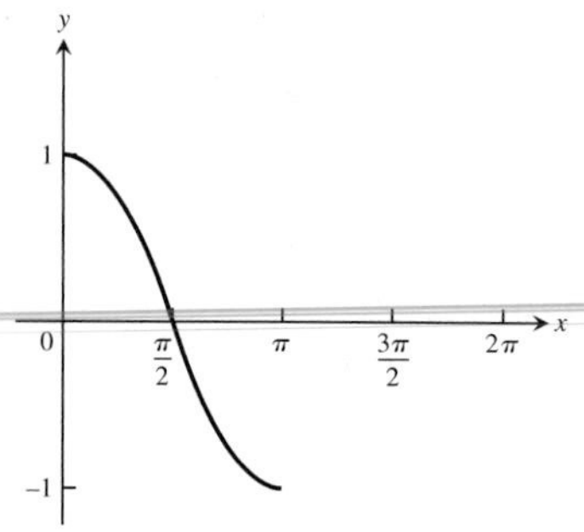

Practice Exercises, pp. 840–842

1. Converges to 1 **3.** Converges to -1 **5.** Diverges
7. Converges to 0 **9.** Converges to 1 **11.** Converges to e^{-5}
13. Converges to 3 **15.** Converges to ln 2 **17.** Diverges
19. $1/6$ **21.** $3/2$ **23.** $e/(e - 1)$ **25.** Diverges
27. Converges conditionally **29.** Converges conditionally
31. Converges absolutely **33.** Converges absolutely
35. Converges absolutely **37.** Converges absolutely
39. Converges absolutely
41. **(a)** $3, -7 \le x < -1$ **(b)** $-7 < x < -1$ **(c)** $x = -7$
43. **(a)** $1/3, 0 \le x \le 2/3$ **(b)** $0 \le x \le 2/3$ **(c)** none
45. **(a)** ∞, for all x **(b)** for all x **(c)** none
47. **(a)** $\sqrt{3}, -\sqrt{3} < x < \sqrt{3}$ **(b)** $-\sqrt{3} < x < \sqrt{3}$ **(c)** none
49. **(a)** $e, -e < x < e$ **(b)** $-e < x < e$ **(c)** empty set
51. $\frac{1}{1 + x}, \frac{1}{4}, \frac{4}{5}$ **53.** $\sin x, \pi, 0$ **55.** e^x, ln 2, 2 **57.** $\sum_{n=0}^{\infty} 2^n x^n$
59. $\sum_{n=0}^{\infty} \frac{(-1)^n \pi^{2n+1} x^{2n+1}}{(2n + 1)!}$ **61.** $\sum_{n=0}^{\infty} \frac{(-1)^n x^{5n}}{(2n)!}$
63. $\sum_{n=0}^{\infty} \frac{((\pi x)/2)^n}{n!}$
65. $2 - \frac{(x + 1)}{2 \cdot 1!} + \frac{3(x + 1)^2}{2^3 \cdot 2!} + \frac{9(x + 1)^3}{2^5 \cdot 3!} + \cdots$
67. $\frac{1}{4} - \frac{1}{4^2}(x - 3) + \frac{1}{4^3}(x - 3)^2 - \frac{1}{4^4}(x - 3)^3$
69. $y = \sum_{n=0}^{\infty} \frac{(-1)^{n+1}}{n!} x^n = -e^{-x}$
71. $y = 3\sum_{n=0}^{\infty} \frac{(-1)^n 2^n}{n!} x^n = 3e^{-2x}$
73. $y = -1 - x + 2\sum_{n=2}^{\infty} (x^n/n!) = 2e^x - 3x - 3$
75. $y = 1 + x + 2\sum_{n=0}^{\infty} (x^n/n!) = 2e^x - 1 - x$
77. 0.4849171431 **79.** ≈ 0.4872223583 **81.** $7/2$
83. $1/12$ **85.** -2 **87.** $r = -3, s = 9/2$
89. **(b)** $|\text{error}| < |\sin(1/42)| < 0.02381$; an underestimate because the remainder is positive
91. $2/3$ **93.** $\ln\left(\frac{n + 1}{2n}\right)$; the series converges to $\ln\left(\frac{1}{2}\right)$.
95. **(a)** ∞ **(b)** $a = 1, \quad b = 0$
97. It converges.
105. $\frac{1}{2} - \sum_{n=1}^{\infty} \frac{2 \sin((2n - 1)x)}{(2n - 1)\pi}$

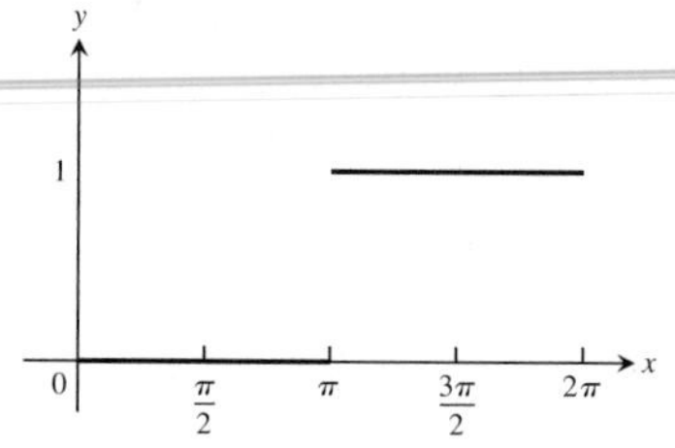

107. $\sum_{n=1}^{\infty} \frac{4\cos((2n-1)x)}{\pi(2n-1)^2} + \sum_{n=1}^{\infty} \frac{2\sin((2n-1)x)}{2n-1}$

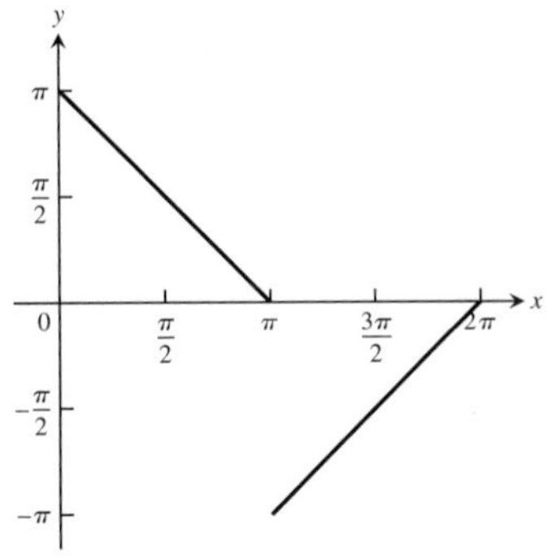

Additional and Advanced Exercises, pp. 843–847

1. Converges; Comparison Test **3.** Diverges; nth-Term Test
5. Converges; Comparison Test **7.** Diverges; nth-Term Test
9. With $a = \pi/3$, $\cos x = \frac{1}{2} - \frac{\sqrt{3}}{2}(x - \pi/3) - \frac{1}{4}(x - \pi/3)^2 + \frac{\sqrt{3}}{12}(x - \pi/3)^3 + \cdots$
11. With $a = 0$, $e^x = 1 + x + \frac{x^2}{2!} + \frac{x^3}{3!} + \cdots$
13. With $a = 22\pi$, $\cos x = 1 - \frac{1}{2}(x - 22\pi)^2 + \frac{1}{4!}(x - 22\pi)^4 - \frac{1}{6!}(x - 22\pi)^6 + \cdots$
15. Converges, limit $= b$ **17.** $\pi/2$ **23.** $b = \pm\frac{1}{5}$
25. $a = 2, L = -7/6$ **29. (b)** Yes
35. (a) $\sum_{n=1}^{\infty} nx^{n-1}$ **(b)** 6 **(c)** $1/q$
37. (a) $R_n = C_0 e^{-kt_0}(1 - e^{-nkt_0})/(1 - e^{-kt_0})$, $R = C_0(e^{-kt_0})/(1 - e^{-kt_0}) = C_0/(e^{kt_0} - 1)$
(b) $R_1 = 1/e \approx 0.368$, $R_{10} = R(1 - e^{-10}) \approx R(0.9999546) \approx 0.58195$; $R \approx 0.58198$; $0 < (R - R_{10})/R < 0.0001$
(c) 7

CHAPTER 12

Section 12.1, pp. 852–853

1. The line through the point (2, 3, 0) parallel to the z-axis
3. The x-axis **5.** The circle $x^2 + y^2 = 4$ in the xy-plane
7. The circle $x^2 + z^2 = 4$ in the xz-plane
9. The circle $y^2 + z^2 = 1$ in the yz-plane
11. The circle $x^2 + y^2 = 16$ in the xy-plane
13. (a) The first quadrant of the xy-plane
(b) The fourth quadrant of the xy-plane
15. (a) The ball of radius 1 centered at the origin
(b) All points greater than 1 unit from the origin
17. (a) The upper hemisphere of radius 1 centered at the origin
(b) The solid upper hemisphere of radius 1 centered at the origin
19. (a) $x = 3$ **(b)** $y = -1$ **(c)** $z = -2$
21. (a) $z = 1$ **(b)** $x = 3$ **(c)** $y = -1$
23. (a) $x^2 + (y - 2)^2 = 4, z = 0$
(b) $(y - 2)^2 + z^2 = 4, x = 0$ **(c)** $x^2 + z^2 = 4, y = 2$
25. (a) $y = 3, z = -1$ **(b)** $x = 1, z = -1$ **(c)** $x = 1, y = 3$
27. $x^2 + y^2 + z^2 = 25, z = 3$ **29.** $0 \le z \le 1$ **31.** $z \le 0$
33. (a) $(x - 1)^2 + (y - 1)^2 + (z - 1)^2 < 1$
(b) $(x - 1)^2 + (y - 1)^2 + (z - 1)^2 > 1$
35. 3 **37.** 7 **39.** $2\sqrt{3}$ **41.** $C(-2, 0, 2), a = 2\sqrt{2}$
43. $C(\sqrt{2}, \sqrt{2}, -\sqrt{2}), a = \sqrt{2}$
45. $(x - 1)^2 + (y - 2)^2 + (z - 3)^2 = 14$
47. $(x + 2)^2 + y^2 + z^2 = 3$ **49.** $C(-2, 0, 2), a = \sqrt{8}$
51. $C\left(-\frac{1}{4}, -\frac{1}{4}, -\frac{1}{4}\right), a = \frac{5\sqrt{3}}{4}$
53. (a) $\sqrt{y^2 + z^2}$ **(b)** $\sqrt{x^2 + z^2}$ **(c)** $\sqrt{x^2 + y^2}$
55. $\sqrt{17} + \sqrt{33} + 6$

Section 12.2, pp. 860–862

1. (a) $\langle 9, -6 \rangle$ **(b)** $3\sqrt{13}$ **3. (a)** $\langle 1, 3 \rangle$ **(b)** $\sqrt{10}$
5. (a) $\langle 12, -19 \rangle$ **(b)** $\sqrt{505}$ **7. (a)** $\left\langle \frac{1}{5}, \frac{14}{5} \right\rangle$ **(b)** $\frac{\sqrt{197}}{5}$
9. $\langle 1, -4 \rangle$ **11.** $\langle -2, -3 \rangle$ **13.** $\left\langle -\frac{1}{2}, \frac{\sqrt{3}}{2} \right\rangle$
15. $\left\langle -\frac{\sqrt{3}}{2}, -\frac{1}{2} \right\rangle$ **17.** $\langle -3\mathbf{i} + 2\mathbf{j} - \mathbf{k} \rangle$ **19.** $\langle -3\mathbf{i} + 16\mathbf{j} \rangle$
21. $\langle 3\mathbf{i} + 5\mathbf{j} - 8\mathbf{k} \rangle$
23. The vector **v** is horizontal and 1 in. long. The vectors **u** and **w** are $\frac{11}{16}$ in. long. **w** is vertical and **u** makes a 45° angle with the horizontal. All vectors must be drawn to scale.

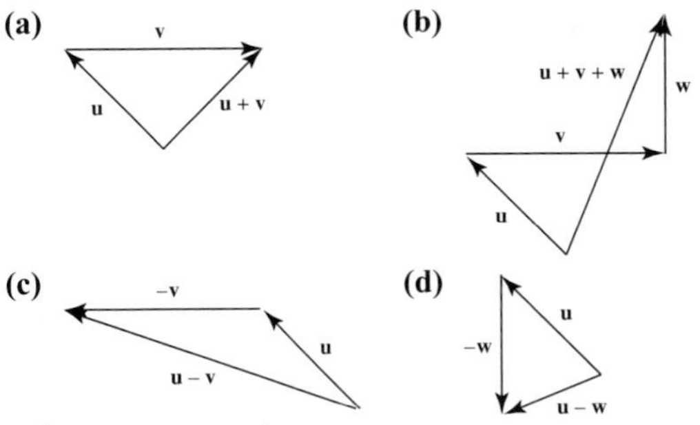

25. $3\left(\frac{2}{3}\mathbf{i} + \frac{1}{3}\mathbf{j} - \frac{2}{3}\mathbf{k}\right)$ **27.** $5(\mathbf{k})$
29. $\sqrt{\frac{1}{2}}\left(\frac{1}{\sqrt{3}}\mathbf{i} - \frac{1}{\sqrt{3}}\mathbf{j} - \frac{1}{\sqrt{3}}\mathbf{k}\right)$
31. (a) $2\mathbf{i}$ **(b)** $-\sqrt{3}\mathbf{k}$ **(c)** $\frac{3}{10}\mathbf{j} + \frac{2}{5}\mathbf{k}$ **(d)** $6\mathbf{i} - 2\mathbf{j} + 3\mathbf{k}$
33. $\frac{7}{13}(12\mathbf{i} - 5\mathbf{k})$
35. (a) $\frac{3}{5\sqrt{2}}\mathbf{i} + \frac{4}{5\sqrt{2}}\mathbf{j} - \frac{1}{\sqrt{2}}\mathbf{k}$ **(b)** $(1/2, 3, 5/2)$

37. **(a)** $-\frac{1}{\sqrt{3}}\mathbf{i} - \frac{1}{\sqrt{3}}\mathbf{j} - \frac{1}{\sqrt{3}}\mathbf{k}$ **(b)** $\left(\frac{5}{2}, \frac{7}{2}, \frac{9}{2}\right)$

39. $A(4, -3, 5)$ **41.** $a = \frac{3}{2}, b = \frac{1}{2}$ **43.** $5\sqrt{3}\mathbf{i}, 5\mathbf{j}$

45. $\approx\langle -338.095, 725.046\rangle$

47. **(a)** $\langle 5\cos 60°, 5\sin 60°\rangle = \left\langle\frac{5}{2}, \frac{5\sqrt{3}}{2}\right\rangle$

(b) $\langle 5\cos 60° + 10\cos 315°, 5\sin 60° + 10\sin 315°\rangle = \left\langle\frac{5 + 10\sqrt{2}}{2}, \frac{5\sqrt{3} - 10\sqrt{2}}{2}\right\rangle$

49. **(a)** $\frac{3}{2}\mathbf{i} + \frac{3}{2}\mathbf{j} - 3\mathbf{k}$ **(b)** $\mathbf{i} + \mathbf{j} - 2\mathbf{k}$ **(c)** $(2, 2, 1)$

Section 12.3, pp. 870–873

1. **(a)** $-25, 5, 5$ **(b)** -1 **(c)** -5 **(d)** $-2\mathbf{i} + 4\mathbf{j} - \sqrt{5}\mathbf{k}$

3. **(a)** $25, 15, 5$ **(b)** $\frac{1}{3}$ **(c)** $\frac{5}{3}$ **(d)** $\frac{1}{9}(10\mathbf{i} + 11\mathbf{j} - 2\mathbf{k})$

5. **(a)** $2, \sqrt{34}, \sqrt{3}$ **(b)** $\frac{2}{\sqrt{3}\sqrt{34}}$ **(c)** $\frac{2}{\sqrt{34}}$

(d) $\frac{1}{17}(5\mathbf{j} - 3\mathbf{k})$

7. **(a)** $10 + \sqrt{17}, \sqrt{26}, \sqrt{21}$ **(b)** $\frac{10 + \sqrt{17}}{\sqrt{546}}$

(c) $\frac{10 + \sqrt{17}}{\sqrt{26}}$ **(d)** $\frac{10 + \sqrt{17}}{26}(5\mathbf{i} + \mathbf{j})$

9. 0.75 rad **11.** 1.77 rad

13. Angle at $A = \cos^{-1}\left(\frac{1}{\sqrt{5}}\right) \approx 63.435$ degrees, angle at $B = \cos^{-1}\left(\frac{3}{5}\right) \approx 53.130$ degrees, angle at $C = \cos^{-1}\left(\frac{1}{\sqrt{5}}\right) \approx 63.435$ degrees.

17. $\left(\frac{3}{2}\mathbf{i} + \frac{3}{2}\mathbf{j}\right) + \left(-\frac{3}{2}\mathbf{i} + \frac{3}{2}\mathbf{j} + 4\mathbf{k}\right)$

19. $\left(\frac{14}{3}\mathbf{i} + \frac{28}{3}\mathbf{j} - \frac{14}{3}\mathbf{k}\right) + \left(\frac{10}{3}\mathbf{i} - \frac{16}{3}\mathbf{j} - \frac{22}{3}\mathbf{k}\right)$

21. The sum of two vectors of equal length is *always* orthogonal to their difference, as we can see from the equation $(\mathbf{v}_1 + \mathbf{v}_2)\cdot(\mathbf{v}_1 - \mathbf{v}_2) = \mathbf{v}_1\cdot\mathbf{v}_1 + \mathbf{v}_2\cdot\mathbf{v}_1 - \mathbf{v}_1\cdot\mathbf{v}_2 - \mathbf{v}_2\cdot\mathbf{v}_2 = |\mathbf{v}_1|^2 - |\mathbf{v}_2|^2 = 0$.

27. Horizontal component: ≈ 1188 ft/sec, vertical component: ≈ 167 ft/sec

29. **(a)** Since $|\cos\theta| \le 1$, we have $|\mathbf{u}\cdot\mathbf{v}| = |\mathbf{u}||\mathbf{v}||\cos\theta| \le |\mathbf{u}||\mathbf{v}|(1) = |\mathbf{u}||\mathbf{v}|$.

(b) We have equality precisely when $|\cos\theta| = 1$ or when one or both of $\mathbf{u}$ and $\mathbf{v}$ are $\mathbf{0}$. In the case of nonzero vectors, we have equality when $\theta = 0$ or π, that is, when the vectors are parallel.

31. a

35. $x + 2y = 4$

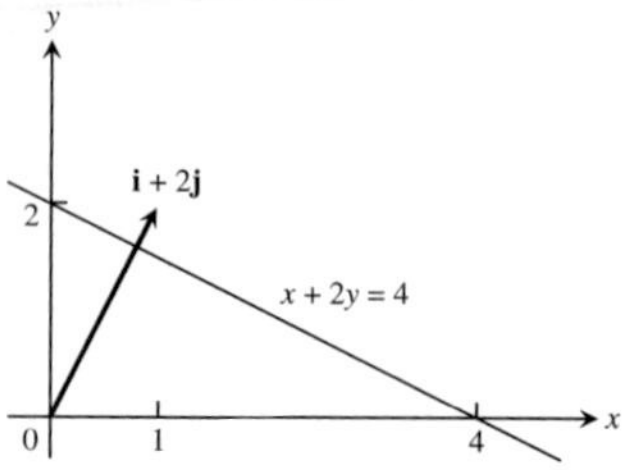

37. $-2x + y = -3$

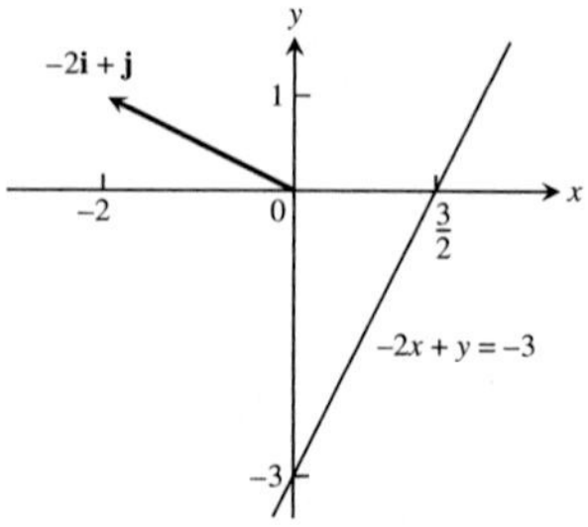

39. $x + y = -1$

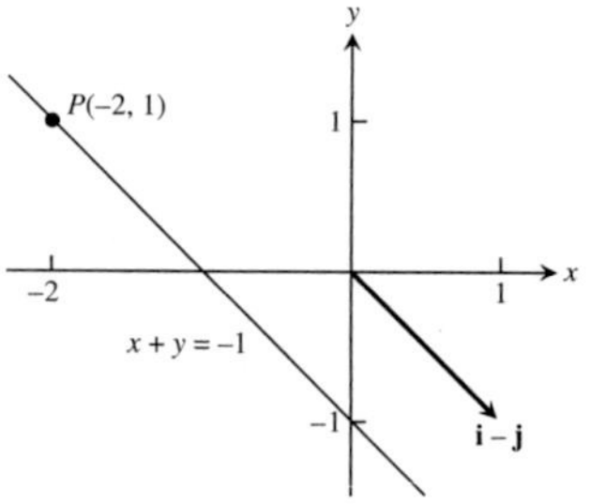

41. $2x - y = 0$

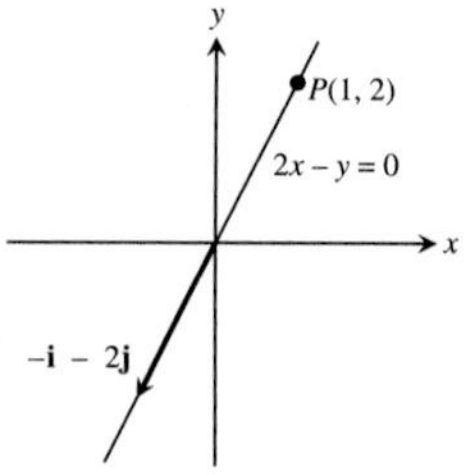

43. 5 J **45.** 3464 J **47.** $\frac{\pi}{4}$ **49.** $\frac{\pi}{6}$ **51.** 0.14

53. $\frac{\pi}{3}$ and $\frac{2\pi}{3}$ at each point **55.** At (0, 0): $\frac{\pi}{2}$; at (1, 1): $\frac{\pi}{4}$ and $\frac{3\pi}{4}$

Section 12.4, pp. 878–879

1. $|\mathbf{u}\times\mathbf{v}| = 3$, direction is $\frac{2}{3}\mathbf{i} + \frac{1}{3}\mathbf{j} + \frac{2}{3}\mathbf{k}$; $|\mathbf{v}\times\mathbf{u}| = 3$, direction is $-\frac{2}{3}\mathbf{i} - \frac{1}{3}\mathbf{j} - \frac{2}{3}\mathbf{k}$

3. $|\mathbf{u}\times\mathbf{v}| = 0$, no direction; $|\mathbf{v}\times\mathbf{u}| = 0$, no direction

5. $|\mathbf{u}\times\mathbf{v}| = 6$, direction is $-\mathbf{k}$; $|\mathbf{v}\times\mathbf{u}| = 6$, direction is $\mathbf{k}$

7. $|\mathbf{u}\times\mathbf{v}| = 6\sqrt{5}$, direction is $\frac{1}{\sqrt{5}}\mathbf{i} - \frac{2}{\sqrt{5}}\mathbf{k}$; $|\mathbf{v}\times\mathbf{u}| = 6\sqrt{5}$, direction is $-\frac{1}{\sqrt{5}}\mathbf{i} + \frac{2}{\sqrt{5}}\mathbf{k}$

9.

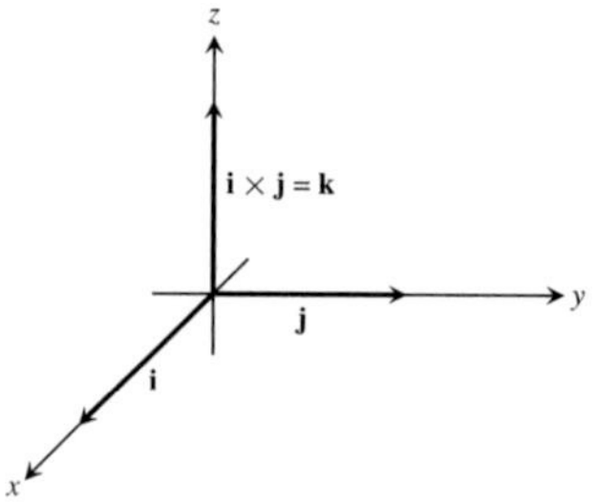

11.

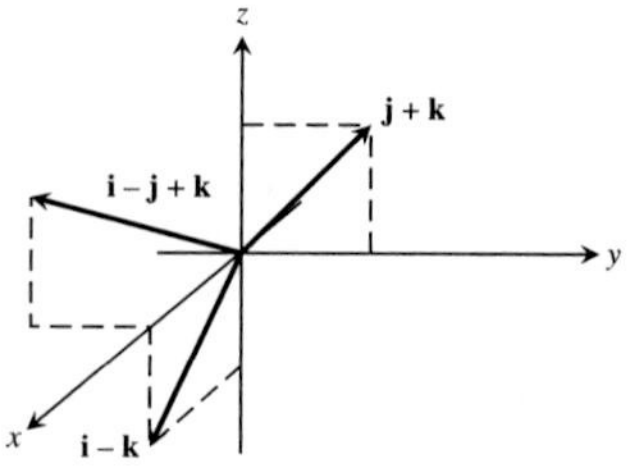

13.

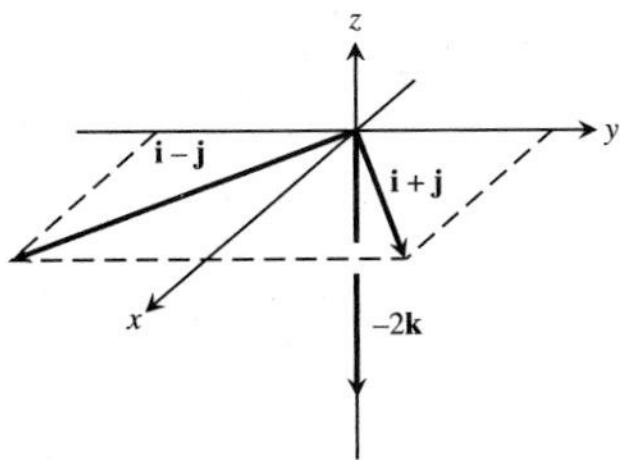

15. **(a)** $2\sqrt{6}$ **(b)** $\pm\frac{1}{\sqrt{6}}(2\mathbf{i}+\mathbf{j}+\mathbf{k})$

17. **(a)** $\frac{\sqrt{2}}{2}$ **(b)** $\pm\frac{1}{\sqrt{2}}(\mathbf{i}-\mathbf{j})$

19. 8 **21.** 7 **23.** **(a)** None **(b)** $\mathbf{u}$ and $\mathbf{w}$ **25.** $10\sqrt{3}$ ft-lb

27. **(a)** True **(b)** Not always true **(c)** True **(d)** True
(e) Not always true **(f)** True **(g)** True **(h)** True

29. **(a)** $\text{proj}_{\mathbf{v}}\,\mathbf{u} = \frac{\mathbf{u}\cdot\mathbf{v}}{\mathbf{v}\cdot\mathbf{v}}\mathbf{v}$ **(b)** $\pm\,\mathbf{u}\times\mathbf{v}$ **(c)** $\pm\,(\mathbf{u}\times\mathbf{v})\times\mathbf{w}$
(d) $|(\mathbf{u}\times\mathbf{v})\cdot\mathbf{w}|$

31. **(a)** Yes **(b)** No **(c)** Yes **(d)** No

33. No, $\mathbf{v}$ need not equal $\mathbf{w}$. For example, $\mathbf{i}+\mathbf{j}\neq -\mathbf{i}+\mathbf{j}$, but
$\mathbf{i}\times(\mathbf{i}+\mathbf{j}) = \mathbf{i}\times\mathbf{i}+\mathbf{i}\times\mathbf{j} = \mathbf{0}+\mathbf{k} = \mathbf{k}$ and
$\mathbf{i}\times(-\mathbf{i}+\mathbf{j}) = -\mathbf{i}\times\mathbf{i}+\mathbf{i}\times\mathbf{j} = \mathbf{0}+\mathbf{k} = \mathbf{k}$.

35. 2 **37.** 13 **39.** 11/2 **41.** 25/2

43. If $\mathbf{A} = a_1\mathbf{i}+a_2\mathbf{j}$ and $\mathbf{B} = b_1\mathbf{i}+b_2\mathbf{j}$, then

$$\mathbf{A}\times\mathbf{B} = \begin{vmatrix} \mathbf{i} & \mathbf{j} & \mathbf{k} \\ a_1 & a_2 & 0 \\ b_1 & b_2 & 0 \end{vmatrix} = \begin{vmatrix} a_1 & a_2 \\ b_1 & b_2 \end{vmatrix}\mathbf{k}$$

and the triangle's area is

$$\frac{1}{2}\left|\mathbf{A}\times\mathbf{B}\right| = \pm\frac{1}{2}\begin{vmatrix} a_1 & a_2 \\ b_1 & b_2 \end{vmatrix}.$$

The applicable sign is (+) if the acute angle from $\mathbf{A}$ to $\mathbf{B}$ runs counterclockwise in the xy-plane, and (–) if it runs clockwise.

Section 12.5, pp. 887–889

1. $x = 3+t, y = -4+t, z = -1+t$

3. $x = -2+5t, y = 5t, z = 3-5t$ **5.** $x = 0, y = 2t, z = t$

7. $x = 1, y = 1, z = 1+t$ **9.** $x = t, y = -7+2t, z = 2t$

11. $x = t, y = 0, z = 0$

13. $x = t, y = t, z = \frac{3}{2}t, 0 \leq t \leq 1$

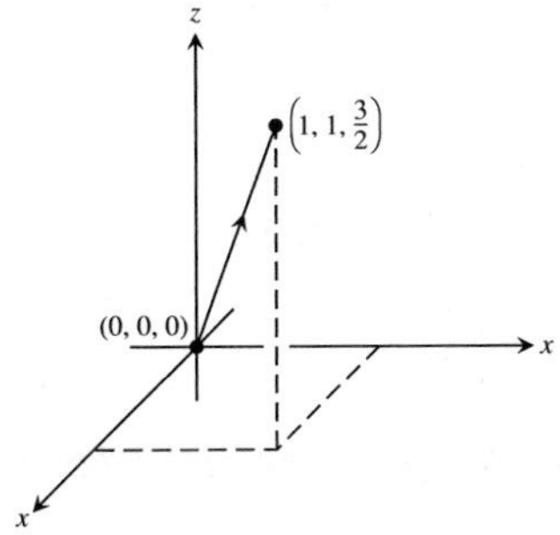

15. $x = 1, y = 1+t, z = 0, -1 \leq t \leq 0$

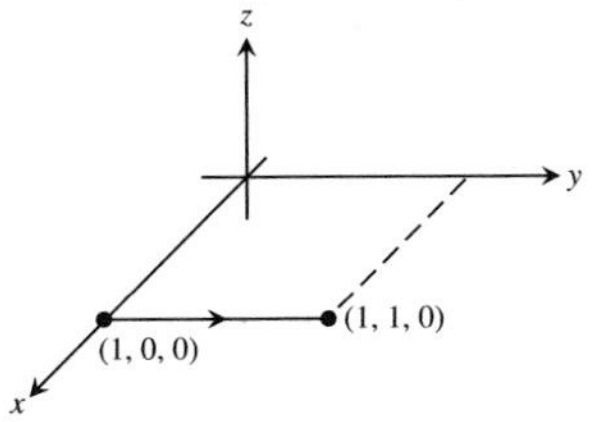

17. $x = 0, y = 1-2t, z = 1, 0 \leq t \leq 1$

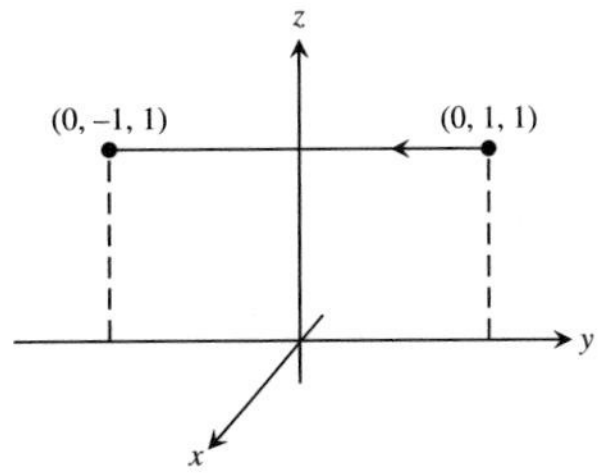

19. $x = 2-2t, y = 2t, z = 2-2t, 0 \leq t \leq 1$

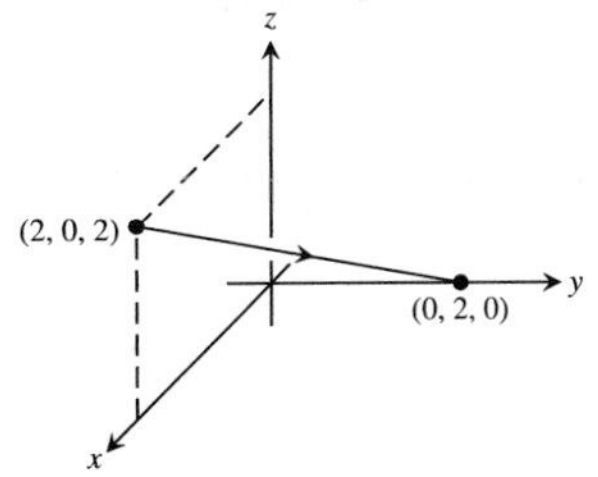

21. $3x-2y-z = -3$ **23.** $7x-5y-4z = 6$

25. $x+3y+4z = 34$ **27.** $(1, 2, 3), -20x+12y+z = 7$

29. $y+z = 3$ **31.** $x-y+z = 0$ **33.** $2\sqrt{30}$ **35.** 0

37. $\frac{9\sqrt{42}}{7}$ **39.** 3 **41.** 19/5 **43.** 5/3 **45.** $9/\sqrt{41}$

47. $\pi/4$ **49.** 1.76 rad **51.** 0.82 rad **53.** $\left(\frac{3}{2}, -\frac{3}{2}, \frac{1}{2}\right)$

55. $(1, 1, 0)$ **57.** $x = 1-t, y = 1+t, z = -1$

59. $x = 4, y = 3+6t, z = 1+3t$

61. $L1$ intersects $L2$; $L2$ is parallel to $L3$; $L1$ and $L3$ are skew.

63. $x = 2+2t, y = -4-t, z = 7+3t; x = -2-t,$
$y = -2+(1/2)t, z = 1-(3/2)t$

65. $\left(0, -\frac{1}{2}, -\frac{3}{2}\right), (-1, 0, -3), (1, -1, 0)$

69. Many possible answers. One possibility: $x+y = 3$ and $2y+z = 7$

71. $(x/a)+(y/b)+(z/c) = 1$ describes all planes *except* those through the origin or parallel to a coordinate axis.

Section 12.6, pp. 897–899

1. **(d),** ellipsoid **3.** **(a),** cylinder **5.** **(l),** hyperbolic paraboloid
7. **(b),** cylinder **9.** **(k),** hyperbolic paraboloid **11.** **(h),** cone

13.

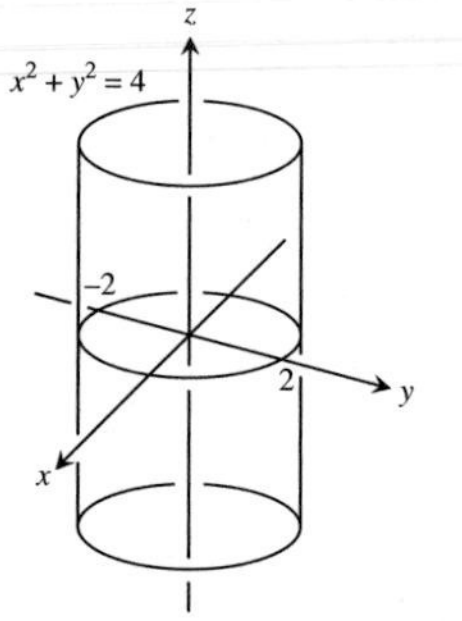

15.

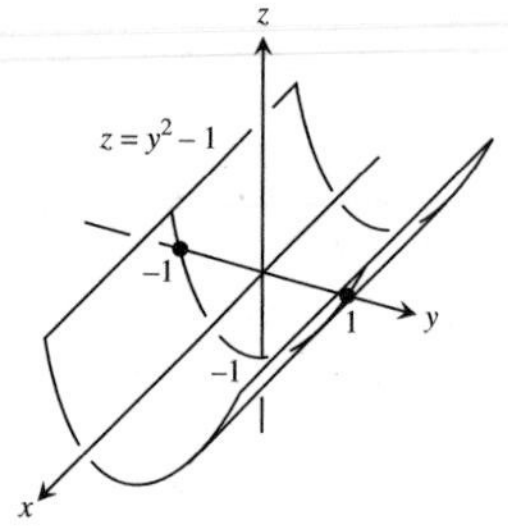

17.

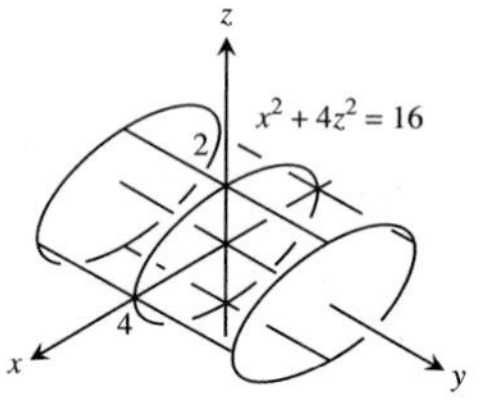

19.

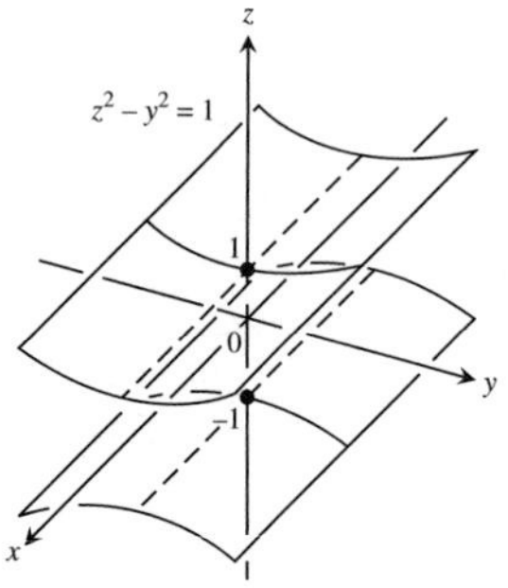

21.

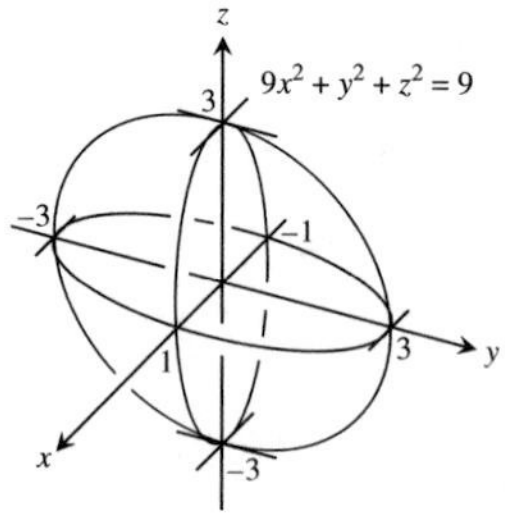

23.

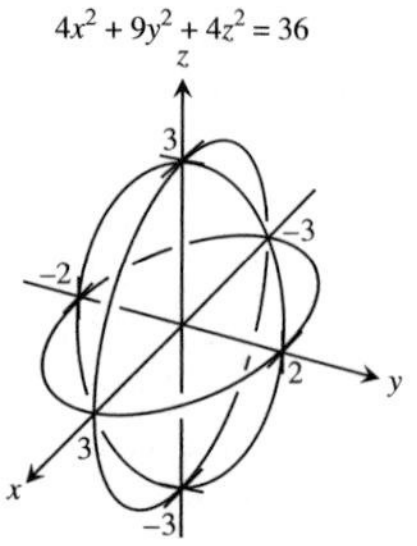

25.

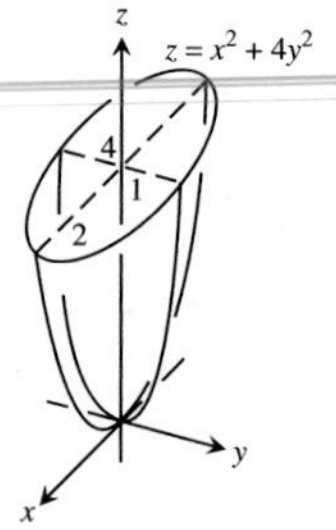

27.

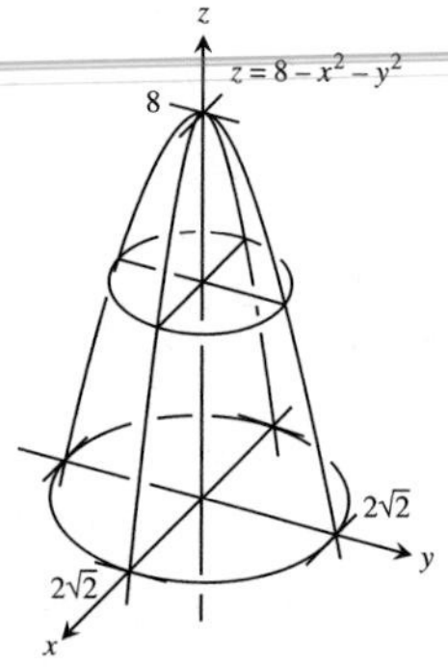

29.

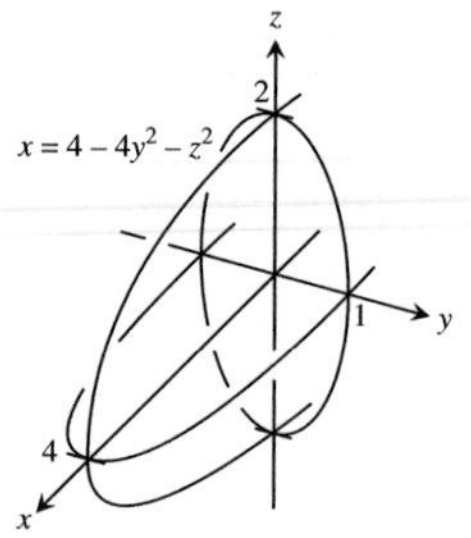

31.

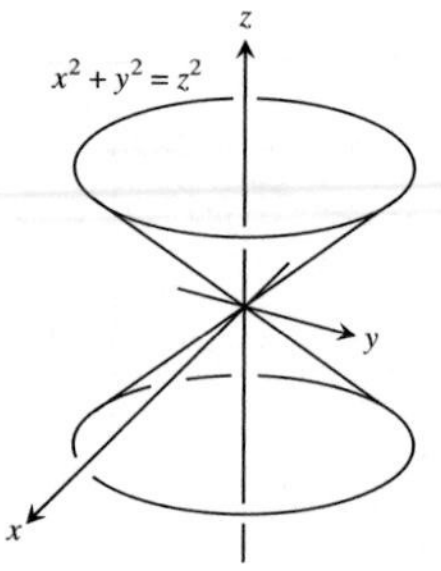

33.

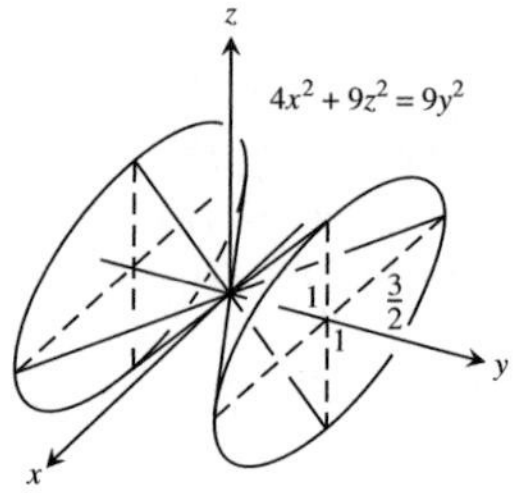

35.

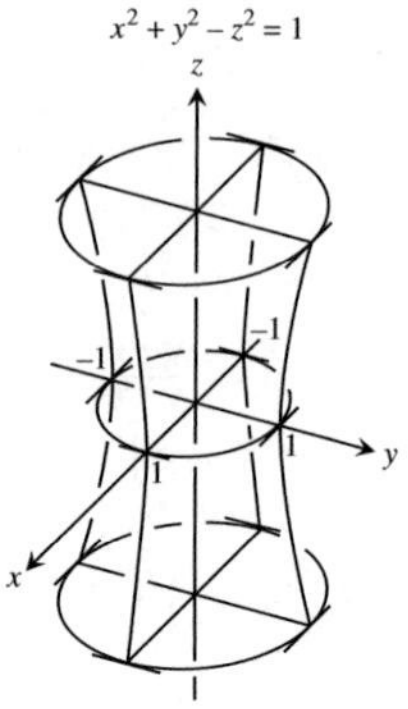

37.

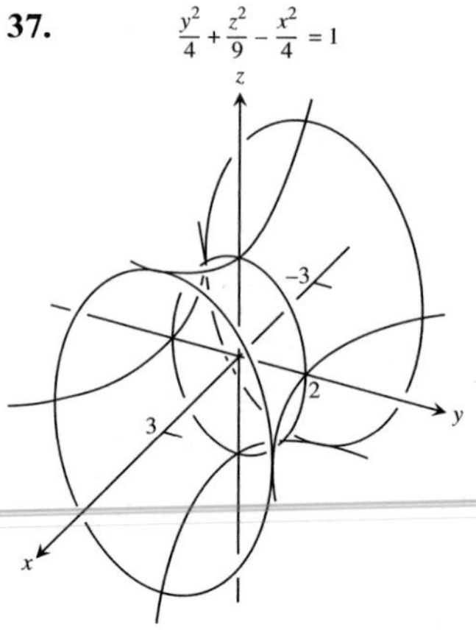

39.

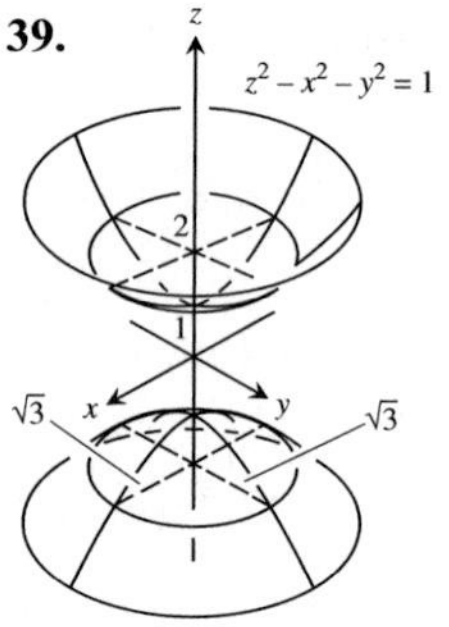

41.

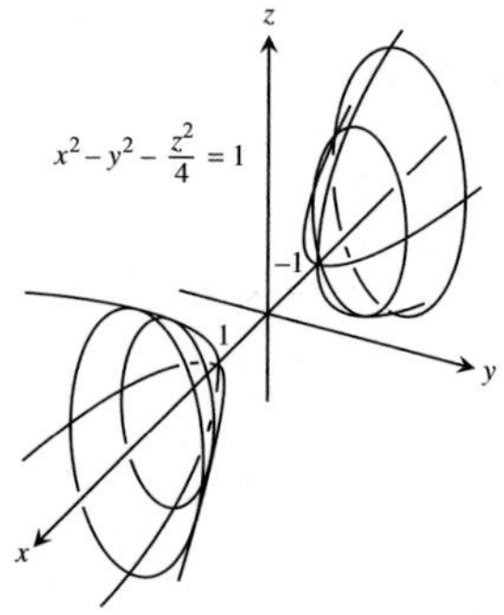

43.

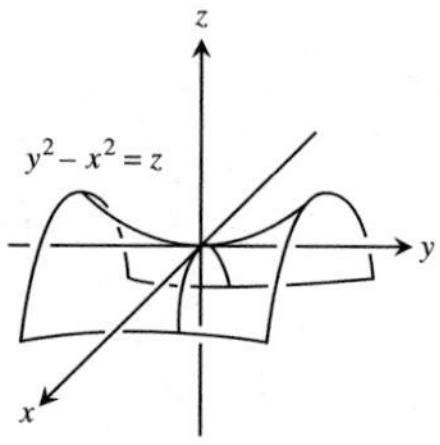

45.

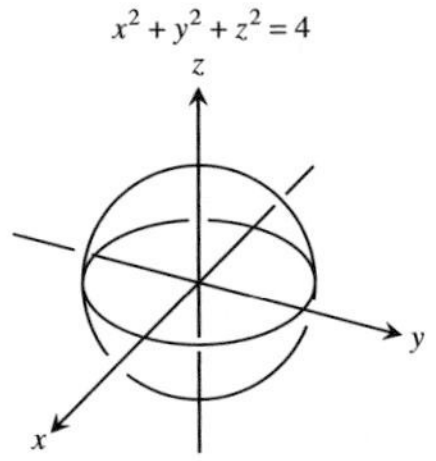

47.

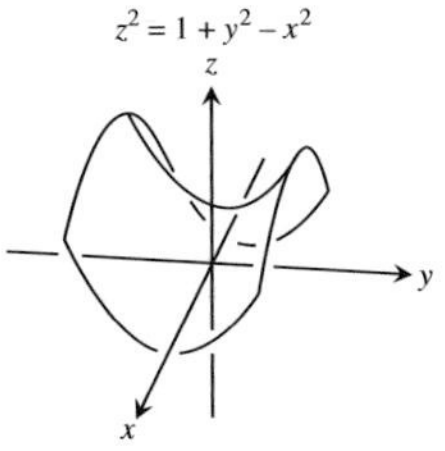

49.

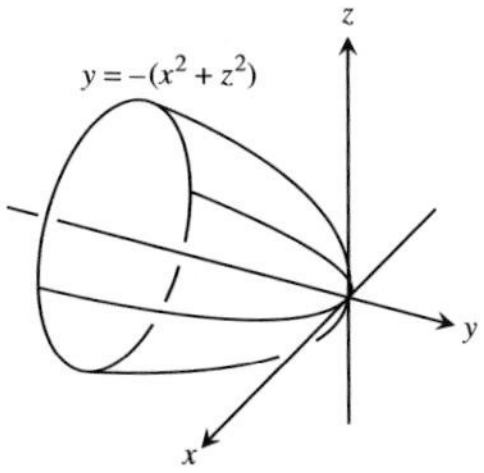

51.

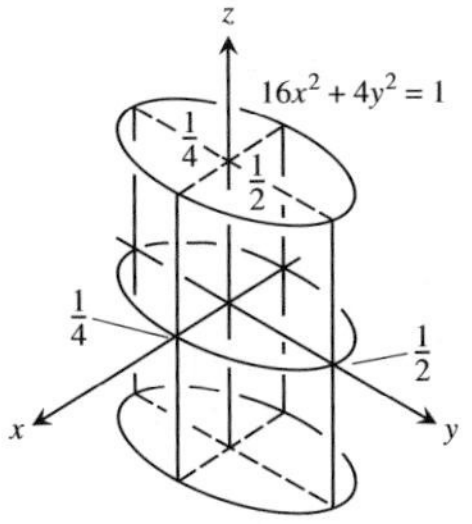

53.

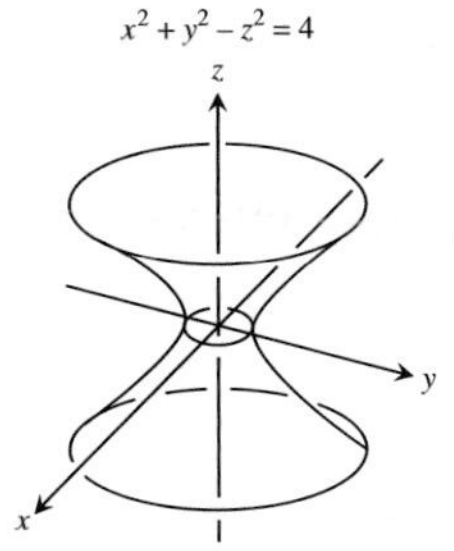

55.

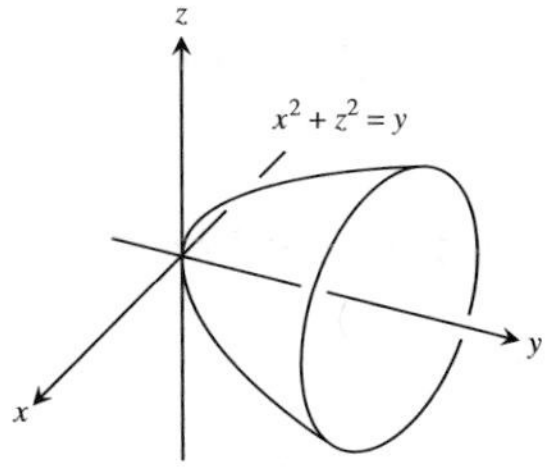

57.

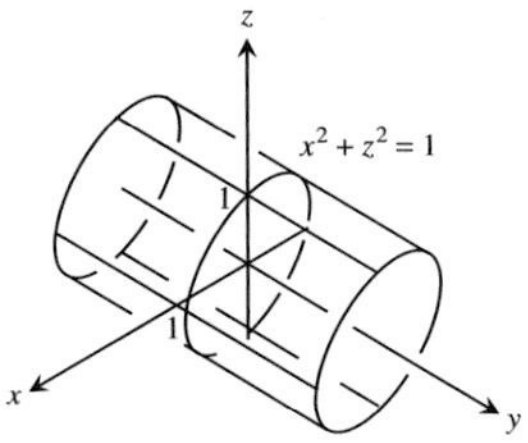

59.

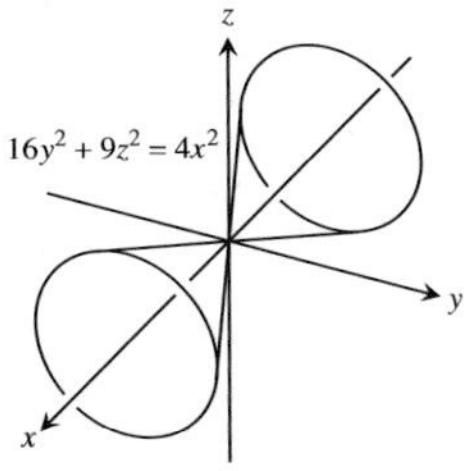

61.

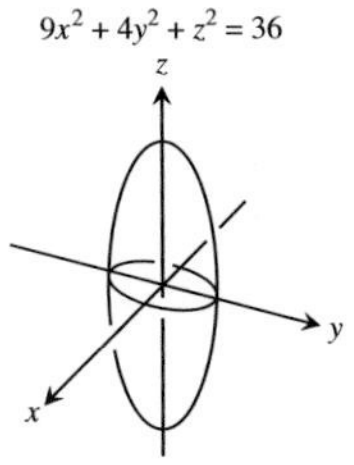

63.

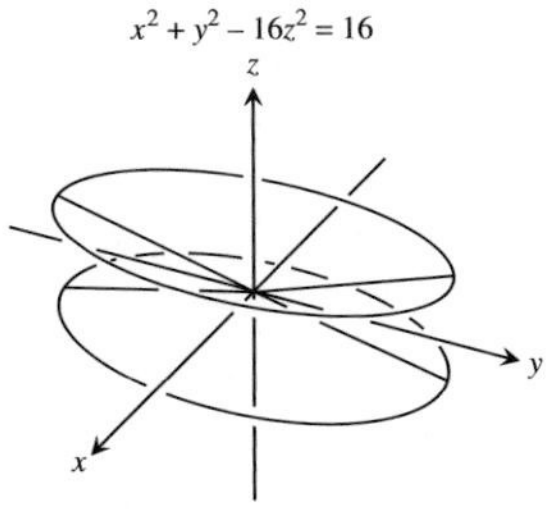

65.

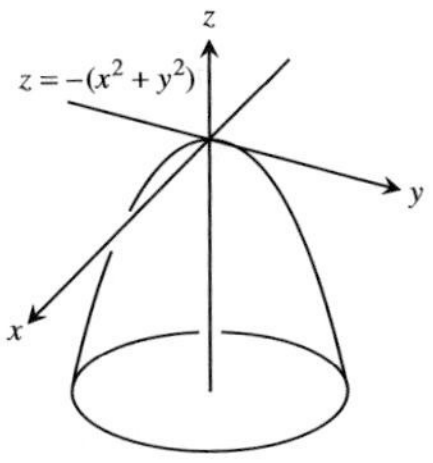

67.

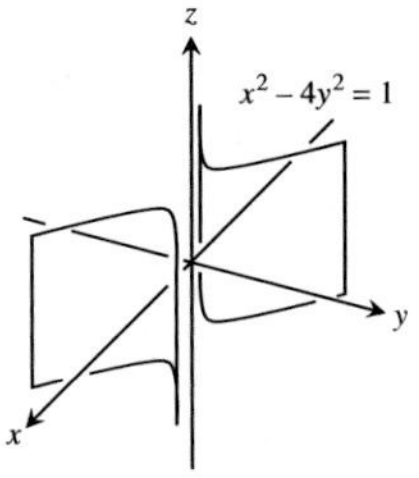

69.

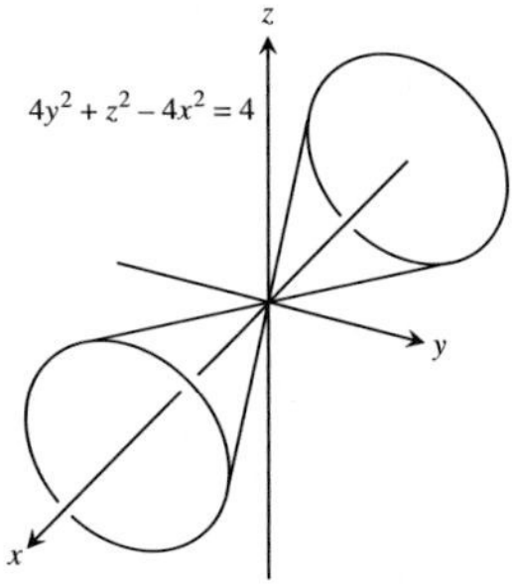

71.

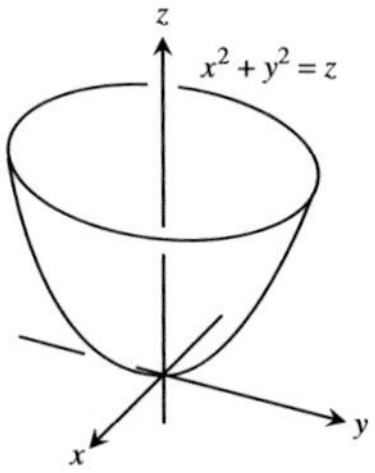

73.

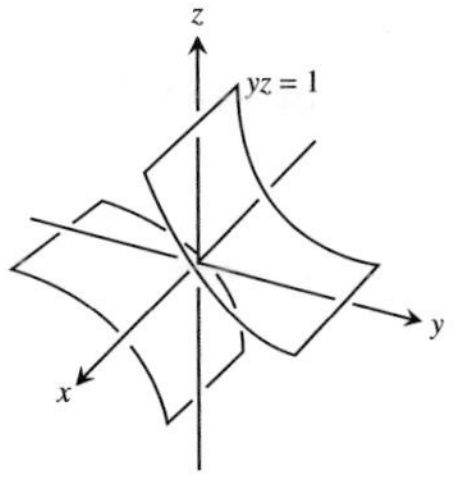

75.

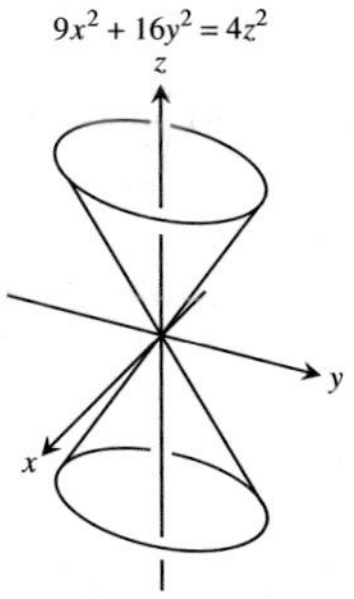

77. **(a)** $\dfrac{2\pi(9 - c^2)}{9}$ **(b)** 8π **(c)** $\dfrac{4\pi abc}{3}$

81. Vertex $(0, y_1, cy_1{}^2/b^2)$, focus $(0, y_1, c(y_1{}^2/b^2) - a^2/(4c))$

Practice Exercises, pp. 900–902

1. **(a)** $\langle -17, 32 \rangle$ **(b)** $\sqrt{1313}$ **3.** **(a)** $\langle 6, -8 \rangle$ **(b)** 10

5. $\left\langle -\dfrac{\sqrt{3}}{2}, -\dfrac{1}{2} \right\rangle$ [assuming counterclockwise]

7. $\left\langle \dfrac{8}{\sqrt{17}}, -\dfrac{2}{\sqrt{17}} \right\rangle$

9. Length $= 2$, direction is $\dfrac{1}{\sqrt{2}}\mathbf{i} + \dfrac{1}{\sqrt{2}}\mathbf{j}$.

11. $\mathbf{v}(\pi/2) = 2(-\mathbf{i})$

13. Length $= 7$, direction is $\dfrac{2}{7}\mathbf{i} - \dfrac{3}{7}\mathbf{j} + \dfrac{6}{7}\mathbf{k}$.

15. $\dfrac{8}{\sqrt{33}}\mathbf{i} - \dfrac{2}{\sqrt{33}}\mathbf{j} + \dfrac{8}{\sqrt{33}}\mathbf{k}$

17. $|\mathbf{v}| = \sqrt{2}, |\mathbf{u}| = 3, \mathbf{v}\cdot\mathbf{u} = \mathbf{u}\cdot\mathbf{v} = 3, \mathbf{v}\times\mathbf{u} = -2\mathbf{i} + 2\mathbf{j} - \mathbf{k}$,
$\mathbf{u}\times\mathbf{v} = 2\mathbf{i} - 2\mathbf{j} + \mathbf{k}, |\mathbf{v}\times\mathbf{u}| = 3, \theta = \cos^{-1}\left(\frac{1}{\sqrt{2}}\right) = \frac{\pi}{4}$,
$|\mathbf{u}|\cos\theta = \frac{3}{\sqrt{2}}, \text{proj}_{\mathbf{v}}\,\mathbf{u} = \frac{3}{2}(\mathbf{i} + \mathbf{j})$

19. $\frac{4}{3}(2\mathbf{i} + \mathbf{j} - \mathbf{k}) - \frac{1}{3}(5\mathbf{i} + \mathbf{j} + 11\mathbf{k})$

21. $\mathbf{u}\times\mathbf{v} = \mathbf{k}$

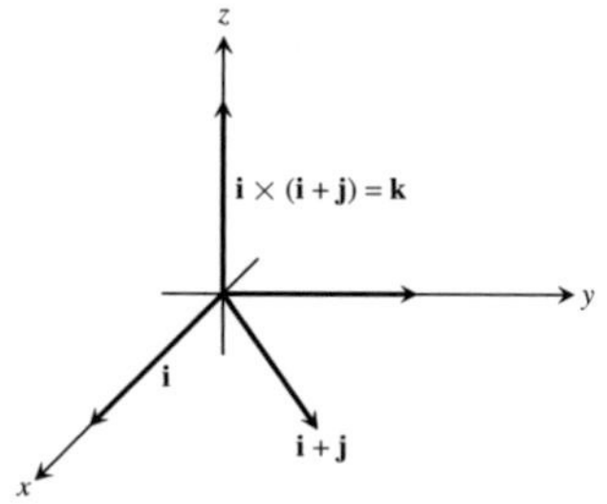

23. $2\sqrt{7}$ **25. (a)** $\sqrt{14}$ **(b)** 1 **29.** $\sqrt{78}/3$
31. $x = 1 - 3t, y = 2, z = 3 + 7t$ **33.** $\sqrt{2}$
35. $2x + y + z = 5$ **37.** $-9x + y + 7z = 4$
39. $\left(0, -\frac{1}{2}, -\frac{3}{2}\right), (-1, 0, -3), (1, -1, 0)$ **41.** $\pi/3$
43. $x = -5 + 5t, y = 3 - t, z = -3t$
45. (b) $x = -12t, y = 19/12 + 15t, z = 1/6 + 6t$
47. Yes; $\mathbf{v}$ is parallel to the plane. **49.** 3 **51.** $-3\mathbf{j} + 3\mathbf{k}$
53. $\frac{2}{\sqrt{35}}\left(5\mathbf{i} - \mathbf{j} - 3\mathbf{k}\right)$ **55.** $\left(\frac{11}{9}, \frac{26}{9}, \frac{7}{9}\right)$
57. $(1, -2, -1); x = 1 - 5t, y = -2 + 3t, z = -1 + 4t$
59. $2x + 7y + 2z + 10 = 0$
61. (a) no **(b)** no **(c)** no **(d)** no **(e)** yes
63. $11/\sqrt{107}$

65.

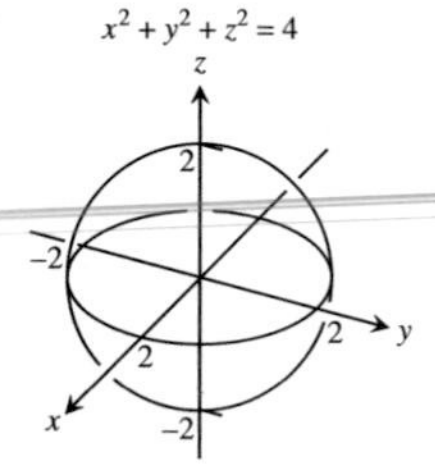

67.

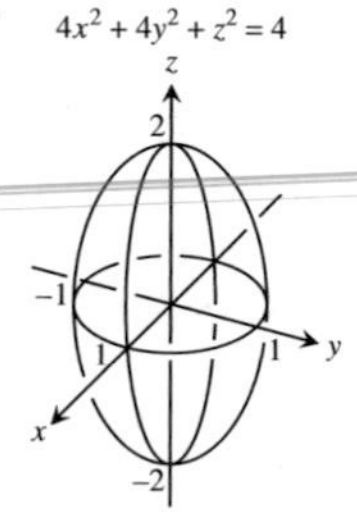

69.

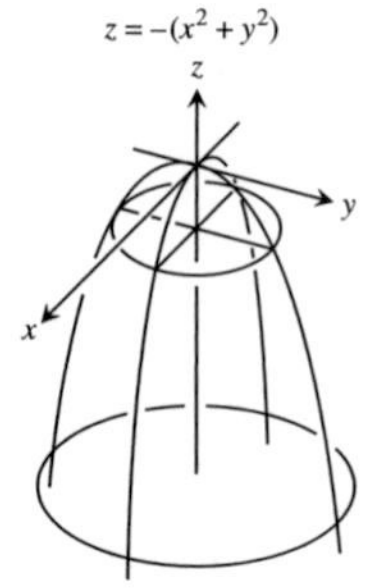

71.

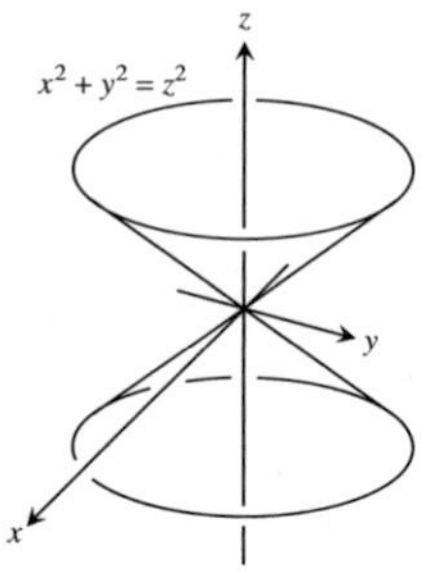

73.

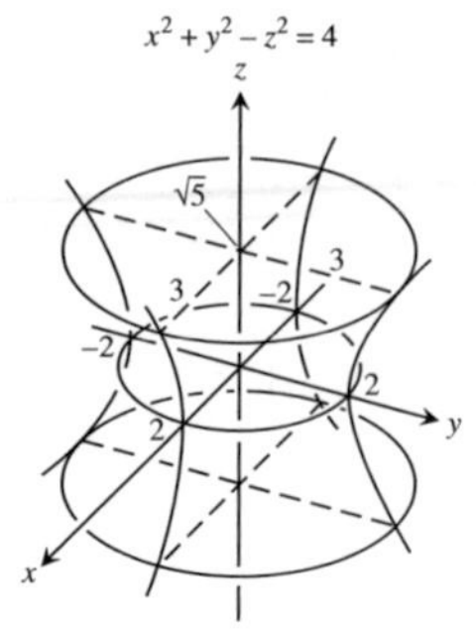

75.

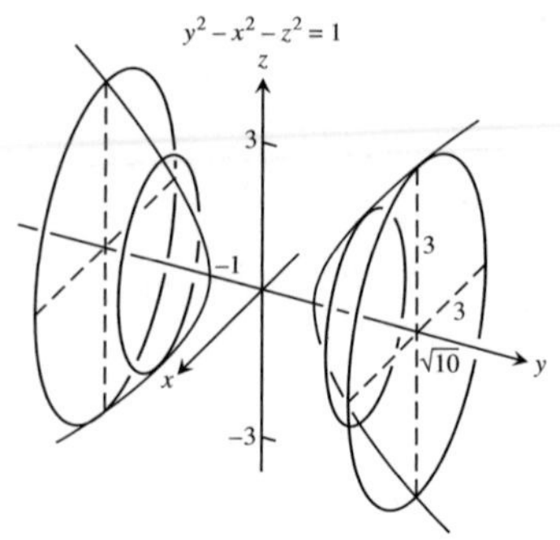

Additional and Advanced Exercises, pp. 902–904

1. $(26, 23, -1/3)$ **3.** $|\mathbf{F}| = 20$ lb
7. (a) $\overrightarrow{BD} = \overrightarrow{AD} - \overrightarrow{AB}$ **(b)** $\overrightarrow{AP} = \frac{1}{2}\overrightarrow{AB} + \frac{1}{2}\overrightarrow{AD}$
13. $\frac{32}{41}\mathbf{i} + \frac{23}{41}\mathbf{j} - \frac{13}{41}\mathbf{k}$
15. (a) 0, 0 **(b)** $-10\mathbf{i} - 2\mathbf{j} + 6\mathbf{k}, -9\mathbf{i} - 2\mathbf{j} + 7\mathbf{k}$
(c) $-4\mathbf{i} - 6\mathbf{j} + 2\mathbf{k}, \mathbf{i} - 2\mathbf{j} - 4\mathbf{k}$
(d) $-10\mathbf{i} - 10\mathbf{k}, -12\mathbf{i} - 4\mathbf{j} - 8\mathbf{k}$
25. (a) $|\mathbf{F}| = \frac{GMm}{d^2}\left(1 + \sum_{i=1}^{n}\frac{2}{(i^2 + 1)^{3/2}}\right)$ **(b)** Yes

CHAPTER 13

Section 13.1, pp. 916–920

1. $y = x^2 - 2x, \mathbf{v} = \mathbf{i} + 2\mathbf{j}, \mathbf{a} = 2\mathbf{j}$
3. $y = \frac{2}{9}x^2, \mathbf{v} = 3\mathbf{i} + 4\mathbf{j}, \mathbf{a} = 3\mathbf{i} + 8\mathbf{j}$
5. $t = \frac{\pi}{4}$: $\mathbf{v} = \frac{\sqrt{2}}{2}\mathbf{i} - \frac{\sqrt{2}}{2}\mathbf{j}, \mathbf{a} = \frac{-\sqrt{2}}{2}\mathbf{i} - \frac{\sqrt{2}}{2}\mathbf{j}$;
$t = \pi/2$: $\mathbf{v} = -\mathbf{j}, \mathbf{a} = -\mathbf{i}$

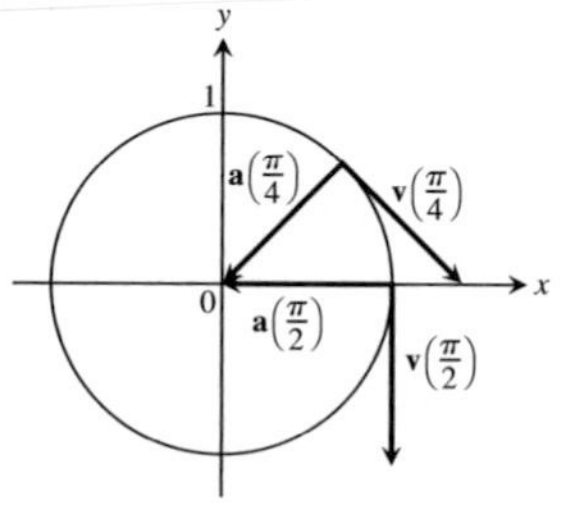

7. $t = \pi$: $\mathbf{v} = 2\mathbf{i}, \mathbf{a} = -\mathbf{j}$; $t = \frac{3\pi}{2}$: $\mathbf{v} = \mathbf{i} - \mathbf{j}, \mathbf{a} = -\mathbf{i}$

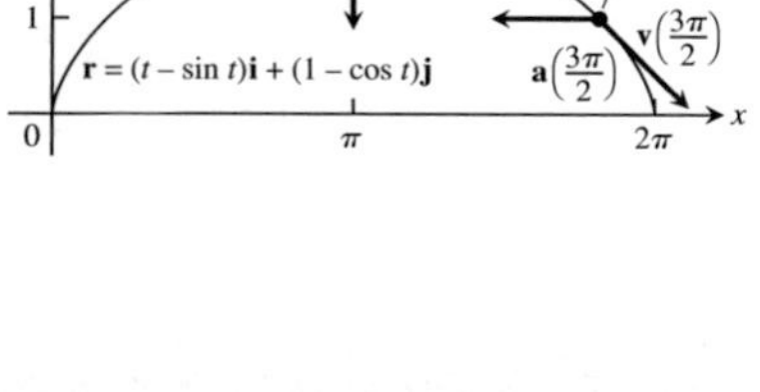

9. $\mathbf{v} = \mathbf{i} + 2t\mathbf{j} + 2\mathbf{k}$; $\mathbf{a} = 2\mathbf{j}$; speed: 3; direction: $\frac{1}{3}\mathbf{i} + \frac{2}{3}\mathbf{j} + \frac{2}{3}\mathbf{k}$; $\mathbf{v}(1) = 3\left(\frac{1}{3}\mathbf{i} + \frac{2}{3}\mathbf{j} + \frac{2}{3}\mathbf{k}\right)$

11. $\mathbf{v} = (-2\sin t)\mathbf{i} + (3\cos t)\mathbf{j} + 4\mathbf{k}$;
$\mathbf{a} = (-2\cos t)\mathbf{i} - (3\sin t)\mathbf{j}$; speed: $2\sqrt{5}$;
direction: $\left(-1/\sqrt{5}\right)\mathbf{i} + \left(2/\sqrt{5}\right)\mathbf{k}$;
$\mathbf{v}(\pi/2) = 2\sqrt{5}\left[\left(-1/\sqrt{5}\right)\mathbf{i} + \left(2/\sqrt{5}\right)\mathbf{k}\right]$

13. $\mathbf{v} = \left(\frac{2}{t+1}\right)\mathbf{i} + 2t\mathbf{j} + t\mathbf{k}$; $\mathbf{a} = \left(\frac{-2}{(t+1)^2}\right)\mathbf{i} + 2\mathbf{j} + \mathbf{k}$;
speed: $\sqrt{6}$; direction: $\frac{1}{\sqrt{6}}\mathbf{i} + \frac{2}{\sqrt{6}}\mathbf{j} + \frac{1}{\sqrt{6}}\mathbf{k}$;
$\mathbf{v}(1) = \sqrt{6}\left(\frac{1}{\sqrt{6}}\mathbf{i} + \frac{2}{\sqrt{6}}\mathbf{j} + \frac{1}{\sqrt{6}}\mathbf{k}\right)$

15. $\pi/2$ **17.** $\pi/2$ **19.** $t = 0, \pi, 2\pi$

21. $(1/4)\mathbf{i} + 7\mathbf{j} + (3/2)\mathbf{k}$ **23.** $\left(\frac{\pi + 2\sqrt{2}}{2}\right)\mathbf{j} + 2\mathbf{k}$

25. $(\ln 4)\mathbf{i} + (\ln 4)\mathbf{j} + (\ln 2)\mathbf{k}$

27. $\mathbf{r}(t) = \left(\frac{-t^2}{2} + 1\right)\mathbf{i} + \left(\frac{-t^2}{2} + 2\right)\mathbf{j} + \left(\frac{-t^2}{2} + 3\right)\mathbf{k}$

29. $\mathbf{r}(t) = ((t+1)^{3/2} - 1)\mathbf{i} + (-e^{-t} + 1)\mathbf{j} + (\ln(t+1) + 1)\mathbf{k}$

31. $\mathbf{r}(t) = 8t\mathbf{i} + 8t\mathbf{j} + (-16t^2 + 100)\mathbf{k}$

33. $x = t, y = -1, z = 1 + t$ **35.** $x = at, y = a, z = 2\pi b + bt$

37. **(a)** (i): It has constant speed 1 (ii): Yes
(iii): Counterclockwise (iv): Yes
(b) (i): It has constant speed 2 (ii): Yes
(iii): Counterclockwise (iv): Yes
(c) (i): It has constant speed 1 (ii): Yes
(iii): Counterclockwise
(iv): It starts at $(0, -1)$ instead of $(1, 0)$
(d) (i): It has constant speed 1 (ii): Yes
(iii): Clockwise (iv): Yes
(e) (i): It has variable speed (ii): No
(iii): Counterclockwise (iv): Yes

39. $\mathbf{r}(t) = \left(\frac{3}{2}t^2 + \frac{6}{\sqrt{11}}t + 1\right)\mathbf{i} - \left(\frac{1}{2}t^2 + \frac{2}{\sqrt{11}}t - 2\right)\mathbf{j} + \left(\frac{1}{2}t^2 + \frac{2}{\sqrt{11}}t + 3\right)\mathbf{k} = \left(\frac{1}{2}t^2 + \frac{2t}{\sqrt{11}}\right)(3\mathbf{i} - \mathbf{j} + \mathbf{k}) + (\mathbf{i} + 2\mathbf{j} + 3\mathbf{k})$

41. $\mathbf{v} = 2\sqrt{5}\mathbf{i} + \sqrt{5}\mathbf{j}$

43. $\max|\mathbf{v}| = 3$, $\min|\mathbf{v}| = 2$, $\max|\mathbf{a}| = 3$, $\min|\mathbf{a}| = 2$

Section 13.2, pp. 927–930

1. 50 sec

3. **(a)** 72.2 sec; 25,510 m **(b)** 4020 m **(c)** 6378 m

5. $t \approx 2.135$ sec, $x \approx 66.43$ ft

7. **(a)** $v_0 \approx 9.9$ m/sec **(b)** $\alpha \approx 18.4°$ or $71.6°$

9. 190 mph **11.** The golf ball will clip the leaves at the top.

13. **(a)** 149 ft/sec **(b)** 2.25 sec **15.** 39.3° or 50.7°

17. 46.6 ft/sec **21.** 1.92 sec, 73.7 ft (approx.)

23. 4.00 ft, 7.80 ft/sec **25.** **(b)** $\mathbf{v}_0$ would bisect $\angle AOR$

27. **(a)** (Assuming that "x" is zero at the point of impact.)
$\mathbf{r}(t) = (x(t))\mathbf{i} + (y(t))\mathbf{j}$, where $x(t) = (35\cos 27°)t$ and $y(t) = 4 + (35\sin 27°)t - 16t^2$.
(b) At $t \approx 0.497$ sec, it reaches its maximum height of about 7.945 ft.
(c) Range ≈ 37.45 ft; flight time ≈ 1.201 sec
(d) At $t \approx 0.254$ and $t \approx 0.740$ sec, when it is ≈ 29.532 and ≈ 14.376 ft from where it will land.
(e) Yes. It changes things because the ball won't clear the net.

31. **(a)** $\mathbf{r}(t) = (x(t))\mathbf{i} + (y(t))\mathbf{j}$, where $x(t) = \left(\frac{1}{0.08}\right)(1 - e^{-0.08t})\cdot(152\cos 20° - 17.6)$ and $y(t) = 3 + \left(\frac{152}{0.08}\right)(1 - e^{-0.08t})\cdot(\sin 20°) + \left(\frac{32}{0.08^2}\right)(1 - 0.08t - e^{-0.08t})$
(b) At $t \approx 1.527$ sec, it reaches its maximum height of about 41.893 ft.
(c) Range ≈ 351.734 ft, flight time ≈ 3.181 sec.
(d) At $t \approx 0.877$ and $t \approx 2.190$ sec, when it is about 106.028 and 251.530 ft from home plate.
(e) No. The wind gust would need to be greater than 12.846 ft/sec in the direction of the hit for the ball to clear the fence for a home run.

Section 13.3, pp. 935–936

1. $\mathbf{T} = \left(-\frac{2}{3}\sin t\right)\mathbf{i} + \left(\frac{2}{3}\cos t\right)\mathbf{j} + \frac{\sqrt{5}}{3}\mathbf{k}$, 3π

3. $\mathbf{T} = \frac{1}{\sqrt{1+t}}\mathbf{i} + \frac{\sqrt{t}}{\sqrt{1+t}}\mathbf{k}$, $\frac{52}{3}$

5. $\mathbf{T} = -\cos t\mathbf{j} + \sin t\mathbf{k}$, $\frac{3}{2}$

7. $\mathbf{T} = \left(\frac{\cos t - t\sin t}{t+1}\right)\mathbf{i} + \left(\frac{\sin t + t\cos t}{t+1}\right)\mathbf{j} + \left(\frac{\sqrt{2}t^{1/2}}{t+1}\right)\mathbf{k}$, $\frac{\pi^2}{2} + \pi$

9. $(0, 5, 24\pi)$ **11.** $s(t) = 5t$, $L = \frac{5\pi}{2}$

13. $s(t) = \sqrt{3}e^t - \sqrt{3}$, $L = \frac{3\sqrt{3}}{4}$ **15.** $\sqrt{2} + \ln\left(1 + \sqrt{2}\right)$

17. **(a)** Cylinder is $x^2 + y^2 = 1$, plane is $x + z = 1$.
(b) and **(c)**

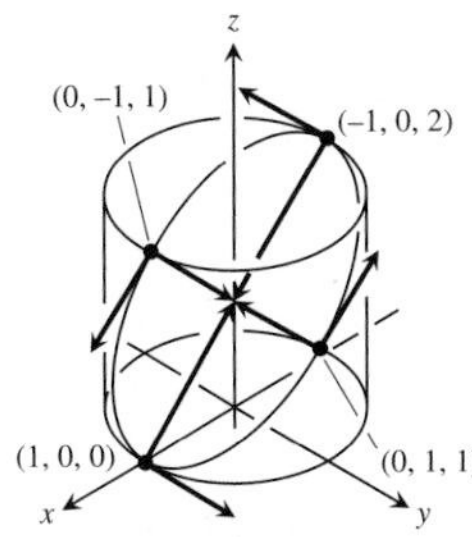

(d) $L = \int_0^{2\pi} \sqrt{1 + \sin^2 t}\, dt$ **(e)** $L \approx 7.64$

Section 13.4, pp. 942–943

1. $\mathbf{T} = (\cos t)\mathbf{i} - (\sin t)\mathbf{j}$, $\mathbf{N} = (-\sin t)\mathbf{i} - (\cos t)\mathbf{j}$, $\kappa = \cos t$

3. $\mathbf{T} = \frac{1}{\sqrt{1+t^2}}\mathbf{i} - \frac{t}{\sqrt{1+t^2}}\mathbf{j}$, $\mathbf{N} = \frac{-t}{\sqrt{1+t^2}}\mathbf{i} - \frac{1}{\sqrt{1+t^2}}\mathbf{j}$,

$\kappa = \frac{1}{2\left(\sqrt{1+t^2}\right)^3}$

5. (b) $\cos x$

7. (b) $\mathbf{N} = \frac{-2e^{2t}}{\sqrt{1+4e^{4t}}}\mathbf{i} + \frac{1}{\sqrt{1+4e^{4t}}}\mathbf{j}$

(c) $\mathbf{N} = -\frac{1}{2}\left(\sqrt{4-t^2}\,\mathbf{i} + t\mathbf{j}\right)$

9. $\mathbf{T} = \frac{3\cos t}{5}\mathbf{i} - \frac{3\sin t}{5}\mathbf{j} + \frac{4}{5}\mathbf{k}$, $\mathbf{N} = (-\sin t)\mathbf{i} - (\cos t)\mathbf{j}$,

$\kappa = \frac{3}{25}$

11. $\mathbf{T} = \left(\frac{\cos t - \sin t}{\sqrt{2}}\right)\mathbf{i} + \left(\frac{\cos t + \sin t}{\sqrt{2}}\right)\mathbf{j}$,

$\mathbf{N} = \left(\frac{-\cos t - \sin t}{\sqrt{2}}\right)\mathbf{i} + \left(\frac{-\sin t + \cos t}{\sqrt{2}}\right)\mathbf{j}$,

$\kappa = \frac{1}{e^t\sqrt{2}}$

13. $\mathbf{T} = \frac{t}{\sqrt{t^2+1}}\mathbf{i} + \frac{1}{\sqrt{t^2+1}}\mathbf{j}$, $\mathbf{N} = \frac{\mathbf{i}}{\sqrt{t^2+1}} - \frac{t\mathbf{j}}{\sqrt{t^2+1}}$,

$\kappa = \frac{1}{t(t^2+1)^{3/2}}$

15. $\mathbf{T} = \left(\operatorname{sech}\frac{t}{a}\right)\mathbf{i} + \left(\tanh\frac{t}{a}\right)\mathbf{j}$,

$\mathbf{N} = \left(-\tanh\frac{t}{a}\right)\mathbf{i} + \left(\operatorname{sech}\frac{t}{a}\right)\mathbf{j}$,

$\kappa = \frac{1}{a}\operatorname{sech}^2\frac{t}{a}$

19. $1/(2b)$ **21.** $\left(x - \frac{\pi}{2}\right)^2 + y^2 = 1$

23. $\kappa(x) = 2/(1+4x^2)^{3/2}$ **25.** $\kappa(x) = |\sin x|/(1+\cos^2 x)^{3/2}$

Section 13.5, pp. 949–950

1. $\mathbf{B} = \left(\frac{4}{5}\cos t\right)\mathbf{i} - \left(\frac{4}{5}\sin t\right)\mathbf{j} - \frac{3}{5}\mathbf{k}$, $\tau = -\frac{4}{25}$

3. $\mathbf{B} = \mathbf{k}$, $\tau = 0$ **5.** $\mathbf{B} = -\mathbf{k}$, $\tau = 0$ **7.** $\mathbf{B} = \mathbf{k}$, $\tau = 0$

9. $\mathbf{a} = |a|\mathbf{N}$ **11.** $\mathbf{a}(1) = \frac{4}{3}\mathbf{T} + \frac{2\sqrt{5}}{3}\mathbf{N}$ **13.** $\mathbf{a}(0) = 2\mathbf{N}$

15. $\mathbf{r}\left(\frac{\pi}{4}\right) = \frac{\sqrt{2}}{2}\mathbf{i} + \frac{\sqrt{2}}{2}\mathbf{j} - \mathbf{k}$, $\mathbf{T}\left(\frac{\pi}{4}\right) = -\frac{\sqrt{2}}{2}\mathbf{i} + \frac{\sqrt{2}}{2}\mathbf{j}$,

$\mathbf{N}\left(\frac{\pi}{4}\right) = -\frac{\sqrt{2}}{2}\mathbf{i} - \frac{\sqrt{2}}{2}\mathbf{j}$, $\mathbf{B}\left(\frac{\pi}{4}\right) = \mathbf{k}$; osculating plane: $z = -1$; normal plane: $-x + y = 0$; rectifying plane: $x + y = \sqrt{2}$

17. Yes. If the car is moving on a curved path ($\kappa \neq 0$), then $a_N = \kappa|\mathbf{v}|^2 \neq 0$ and $\mathbf{a} \neq \mathbf{0}$.

21. $|\mathbf{F}| = \kappa\left(m\left(\frac{ds}{dt}\right)^2\right)$ **23.** $\kappa = \frac{1}{t}$, $\rho = t$

29. Components of $\mathbf{v}$: -1.8701, 0.7089, 1.0000
Components of $\mathbf{a}$: -1.6960, -2.0307, 0
Speed: 2.2361; Components of $\mathbf{T}$: -0.8364, 0.3170, 0.4472
Components of $\mathbf{N}$: -0.4143, -0.8998, -0.1369
Components of $\mathbf{B}$: 0.3590, -0.2998, 0.8839; Curvature: 0.5060
Torsion: 0.2813; Tangential component of acceleration: 0.7746
Normal component of acceleration: 2.5298

31. Components of $\mathbf{v}$: 2.0000, 0, 0.1629
Components of $\mathbf{a}$: 0, -1.0000, 0.0086; Speed: 2.0066
Components of $\mathbf{T}$: 0.9967, 0, 0.0812
Components of $\mathbf{N}$: -0.0007, -1.0000, 0.0086
Components of $\mathbf{B}$: 0.0812, -0.0086, -0.9967;
Curvature: 0.2484
Torsion: -0.0411; Tangential component of acceleration: 0.0007
Normal component of acceleration: 1.0000

Section 13.6, pp. 958–959

1. $T = 93.2$ min **3.** $a = 6764$ km **5.** $D = 6501$ km

7. (a) 42,168 km **(b)** 35,789 km

(c) *Syncom 3, GOES 4,* and *Intelsat 5*

9. $a = 383{,}200$ km from the center of Earth, or about 376,821 km from the surface

11. $2.97 \times 10^{-19}\ \text{sec}^2/\text{m}^3$, $9.902 \times 10^{-14}\ \text{sec}^2/\text{m}^3$, $8.045 \times 10^{-12}\ \text{sec}^2/\text{m}^3$

Practice Exercises, pp. 960–962

1. $\frac{x^2}{16} + \frac{y^2}{2} = 1$

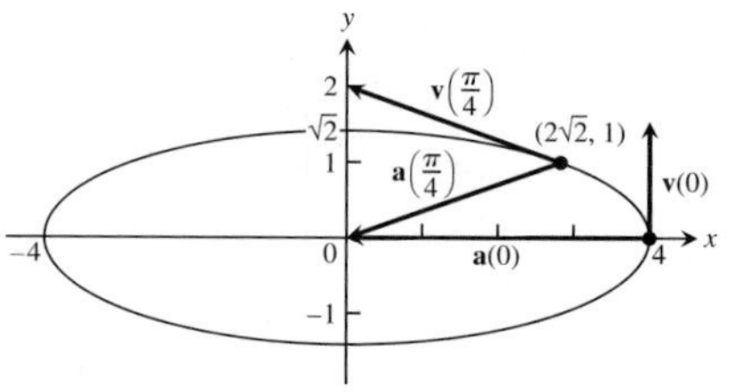

At $t = 0$: $a_{\mathbf{T}} = 0$, $a_{\mathbf{N}} = 4$, $\kappa = 2$;

At $t = \frac{\pi}{4}$: $a_{\mathbf{T}} = \frac{7}{3}$, $a_{\mathbf{N}} = \frac{4\sqrt{2}}{3}$, $\kappa = \frac{4\sqrt{2}}{27}$

3. $|\mathbf{v}|_{\max} = 1$ **5.** $\kappa = 1/5$ **7.** $dy/dt = -x$; clockwise
11. Shot put is on the ground, about 66 ft, 3 in. from the stopboard.
15. (a) 59.19 ft/sec **(b)** 74.58 ft/sec **19.** $\kappa = \pi s$

21. $\text{Length} = \frac{\pi}{4}\sqrt{1 + \frac{\pi^2}{16}} + \ln\left(\frac{\pi}{4} + \sqrt{1 + \frac{\pi^2}{16}}\right)$

23. $\mathbf{T}(0) = \frac{2}{3}\mathbf{i} - \frac{2}{3}\mathbf{j} + \frac{1}{3}\mathbf{k}$; $\mathbf{N}(0) = \frac{1}{\sqrt{2}}\mathbf{i} + \frac{1}{\sqrt{2}}\mathbf{j}$;

$\mathbf{B}(0) = -\frac{1}{3\sqrt{2}}\mathbf{i} + \frac{1}{3\sqrt{2}}\mathbf{j} + \frac{4}{3\sqrt{2}}\mathbf{k}$; $\kappa = \frac{\sqrt{2}}{3}$; $\tau = \frac{1}{6}$

25. $\mathbf{T}(\ln 2) = \frac{1}{\sqrt{17}}\mathbf{i} + \frac{4}{\sqrt{17}}\mathbf{j}$; $\mathbf{N}(\ln 2) = -\frac{4}{\sqrt{17}}\mathbf{i} + \frac{1}{\sqrt{17}}\mathbf{j}$;

$\mathbf{B}(\ln 2) = \mathbf{k}$; $\kappa = \frac{8}{17\sqrt{17}}$; $\tau = 0$

27. $\mathbf{a}(0) = 10\mathbf{T} + 6\mathbf{N}$

29. $\mathbf{T} = \left(\frac{1}{\sqrt{2}}\cos t\right)\mathbf{i} - (\sin t)\mathbf{j} + \left(\frac{1}{\sqrt{2}}\cos t\right)\mathbf{k}$;

$\mathbf{N} = \left(-\frac{1}{\sqrt{2}}\sin t\right)\mathbf{i} - (\cos t)\mathbf{j} - \left(\frac{1}{\sqrt{2}}\sin t\right)\mathbf{k}$;

$\mathbf{B} = \frac{1}{\sqrt{2}}\mathbf{i} - \frac{1}{\sqrt{2}}\mathbf{k}$; $\kappa = \frac{1}{\sqrt{2}}$; $\tau = 0$

31. $\pi/3$ **33.** $x = 1 + t, y = t, z = -t$

35. 5971 km, 1.639×10^7 km^2, 3.21% visible

Additional and Advanced Exercises, pp. 962–964

1. (a) $\mathbf{r}(t) = \left(-\frac{8}{15}t^3 + 4t^2\right)\mathbf{i} + (-20t + 100)\mathbf{j}$; **(b)** $\frac{100}{3}$ m

3. (a) $\left.\frac{d\theta}{dt}\right|_{\theta=2\pi} = 2\sqrt{\frac{\pi g b}{a^2 + b^2}}$

(b) $\theta = \frac{gbt^2}{2(a^2 + b^2)}, z = \frac{gb^2t^2}{2(a^2 + b^2)}$

(c) $\mathbf{v}(t) = \frac{gbt}{\sqrt{a^2 + b^2}}\mathbf{T}$; $\frac{d^2\mathbf{r}}{dt^2} = \frac{bg}{\sqrt{a^2 + b^2}}\mathbf{T} + a\left(\frac{bgt}{a^2 + b^2}\right)^2\mathbf{N}$

There is no component in the direction of **B**.

7. (a) $\frac{dx}{dt} = \dot{r}\cos\theta - r\dot{\theta}\sin\theta, \frac{dy}{dt} = \dot{r}\sin\theta + r\dot{\theta}\cos\theta$

(b) $\frac{dr}{dt} = \dot{x}\cos\theta + \dot{y}\sin\theta, r\frac{d\theta}{dt} = -\dot{x}\sin\theta + \dot{y}\cos\theta$

9. (a) $\mathbf{a}(1) = -9\mathbf{u}_r - 6\mathbf{u}_\theta, \mathbf{v}(1) = -\mathbf{u}_r + 3\mathbf{u}_\theta$ **(b)** 6.5 in.

11. (c) $\mathbf{v} = \dot{r}\mathbf{u}_r + r\dot{\theta}\mathbf{u}_\theta + \dot{z}\mathbf{k}, \mathbf{a} = (\ddot{r} - r\dot{\theta}^2)\mathbf{u}_r + (r\ddot{\theta} + 2\dot{r}\dot{\theta})\mathbf{u}_\theta + \ddot{z}\mathbf{k}$

CHAPTER 14

Section 14.1, pp. 973–975

1. (a) All points in the xy-plane **(b)** All reals
(c) The lines $y - x = c$ **(d)** No boundary points
(e) Both open and closed **(f)** Unbounded

3. (a) All points in the xy-plane **(b)** $z \geq 0$
(c) For $f(x, y) = 0$, the origin; for $f(x, y) \neq 0$, ellipses with the center (0, 0), and major and minor axes, along the x- and y-axes, respectively
(d) No boundary points **(e)** Both open and closed
(f) Unbounded

5. (a) All points in the xy-plane **(b)** All reals
(c) For $f(x, y) = 0$, the x- and y-axes; for $f(x, y) \neq 0$, hyperbolas with the x- and y-axes as asymptotes
(d) No boundary points **(e)** Both open and closed
(f) Unbounded

7. (a) All (x, y) satisfying $x^2 + y^2 < 16$ **(b)** $z \geq 1/4$
(c) Circles centered at the origin with radii $r < 4$
(d) Boundary is the circle $x^2 + y^2 = 16$
(e) Open **(f)** Bounded

9. (a) $(x, y) \neq (0, 0)$ **(b)** All reals
(c) The circles with center (0, 0) and radii $r > 0$
(d) Boundary is the single point (0, 0)
(e) Open **(f)** Unbounded

11. (a) All (x, y) satisfying $-1 \leq y - x \leq 1$
(b) $-\pi/2 \leq z \leq \pi/2$
(c) Straight lines of the form $y - x = c$ where $-1 \leq c \leq 1$
(d) Boundary is two straight lines $y = 1 + x$ and $y = -1 + x$
(e) Closed **(f)** Unbounded

13. (f) **15.** (a) **17.** (d)

19. (a)

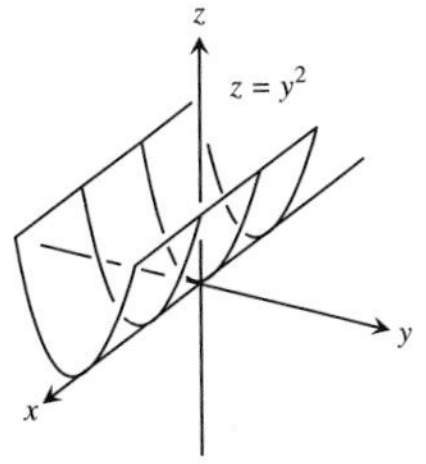

(b)

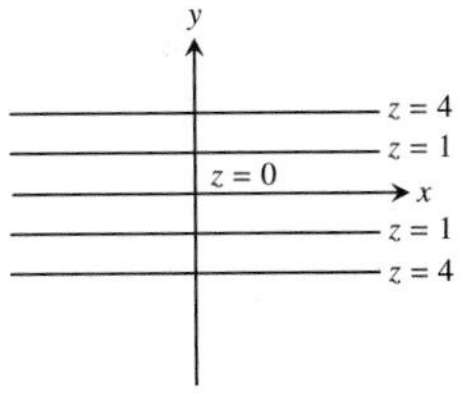

21. (a)

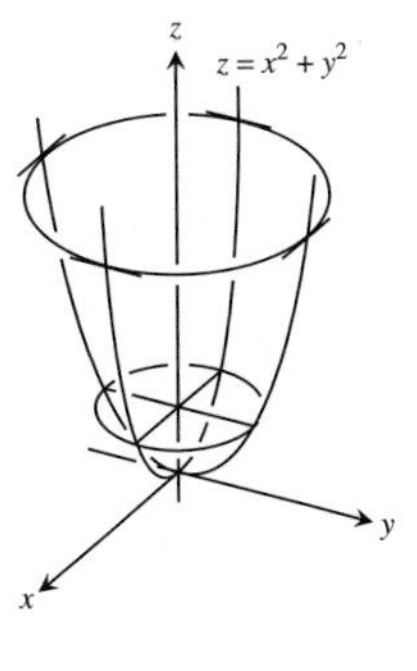

(b)

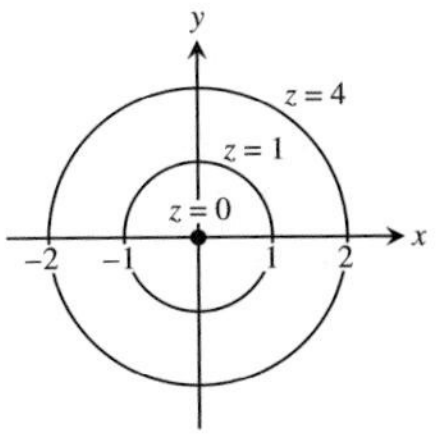

23. (a)

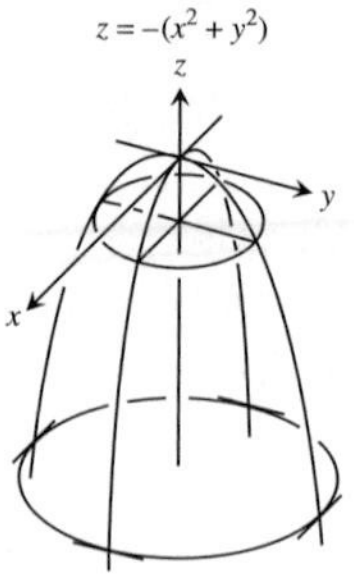

(b)

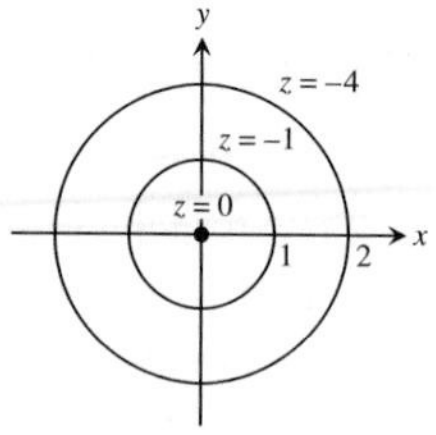

25. (a)

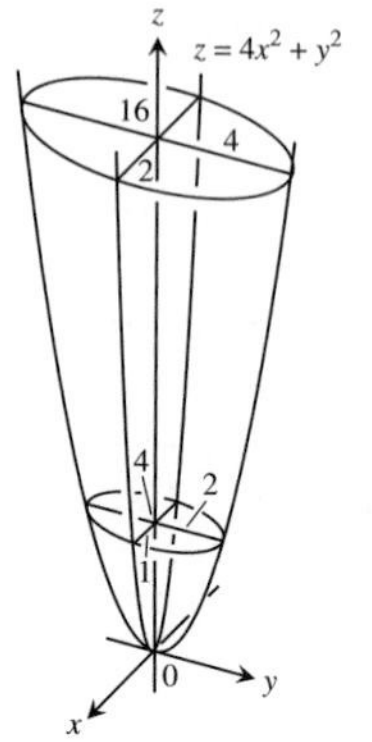

(b)

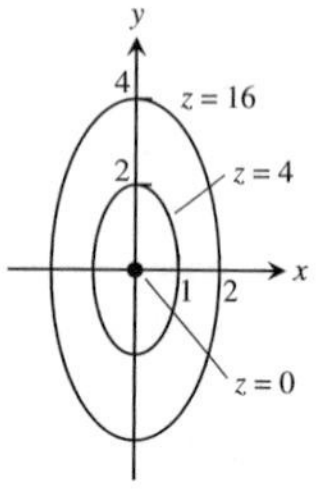

27. (a)

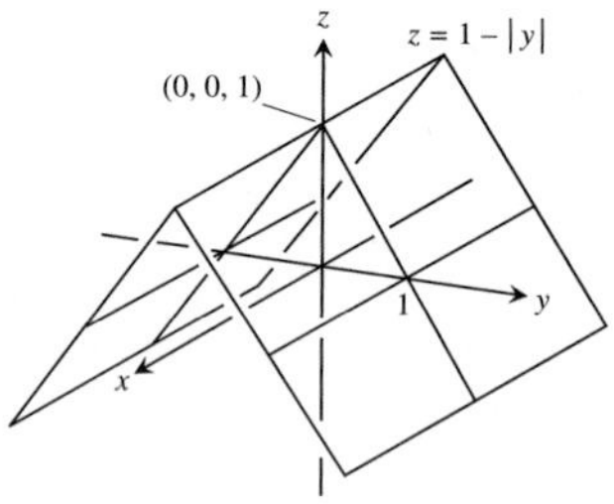

(b)

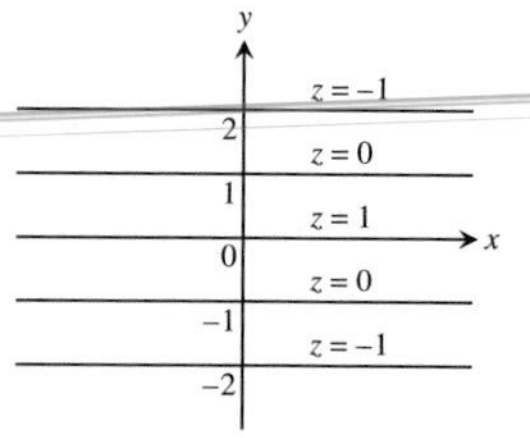

29. $x^2 + y^2 = 10$ **31.** $\tan^{-1} y - \tan^{-1} x = 2 \tan^{-1} \sqrt{2}$

33.

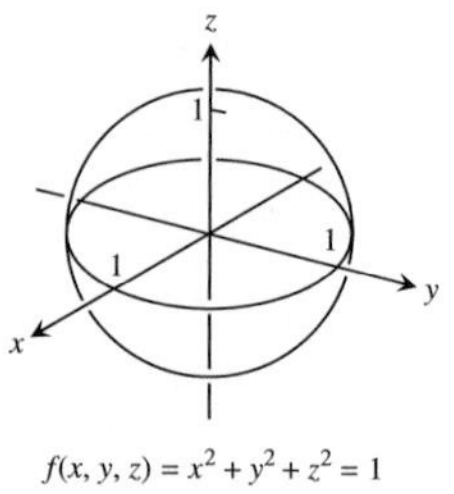

35.

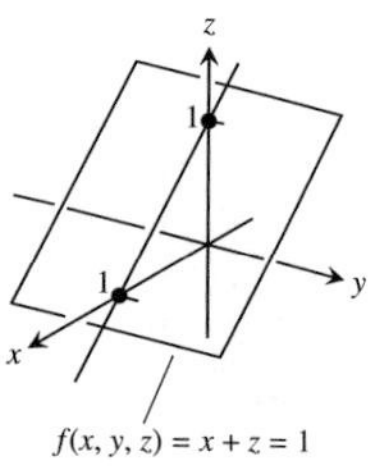

37.

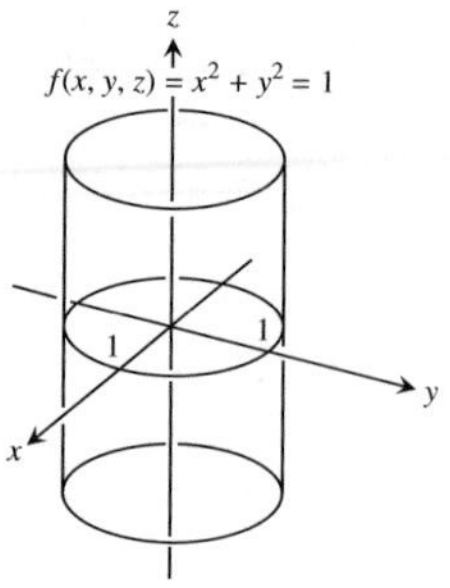

39.

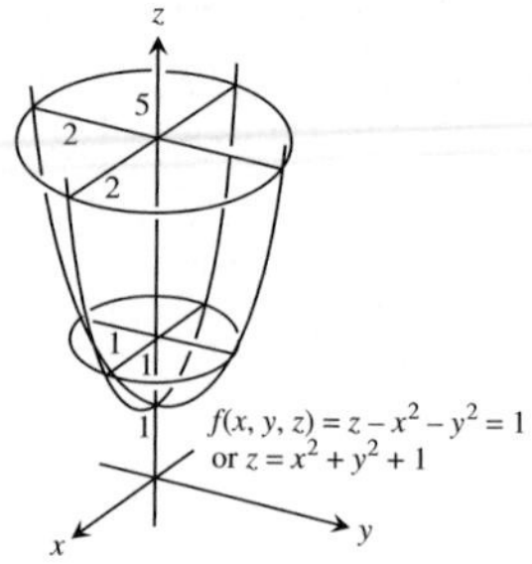

41. $\sqrt{x - y} - \ln z = 2$ **43.** $\dfrac{x + y}{z} = \ln 2$ **45.** Yes, 2000
47. 63 km

Section 14.2, pp. 982–984

1. $5/2$ **3.** $2\sqrt{6}$ **5.** 1 **7.** $1/2$ **9.** 1 **11.** 0 **13.** 0
15. -1 **17.** 2 **19.** $1/4$ **21.** $19/12$ **23.** 2 **25.** 3
27. (a) All (x, y) **(b)** All (x, y) except $(0, 0)$
29. (a) All (x, y) except where $x = 0$ or $y = 0$ **(b)** All (x, y)
31. (a) All (x, y, z) **(b)** All (x, y, z) except the interior of the cylinder $x^2 + y^2 = 1$
33. (a) All (x, y, z) with $z \neq 0$ **(b)** All (x, y, z) with $x^2 + z^2 \neq 1$
35. Consider paths along $y = x, x > 0$, and along $y = x, x < 0$
37. Consider the paths $y = kx^2$, k a constant
39. Consider the paths $y = mx$, m a constant, $m \neq -1$
41. Consider the paths $y = kx^2$, k a constant, $k \neq 0$
43. No **45.** The limit is 1 **47.** The limit is 0
49. (a) $f(x, y)|_{y=mx} = \sin 2\theta$ where $\tan \theta = m$ **51.** 0
53. Does not exist **55.** $\pi/2$ **57.** $f(0, 0) = \ln 3$
59. $\delta = 0.1$ **61.** $\delta = 0.005$ **63.** $\delta = \sqrt{0.015}$
65. $\delta = 0.005$

Section 14.3, pp. 994–996

1. $\dfrac{\partial f}{\partial x} = 4x, \dfrac{\partial f}{\partial y} = -3$ **3.** $\dfrac{\partial f}{\partial x} = 2x(y + 2), \dfrac{\partial f}{\partial y} = x^2 - 1$

5. $\dfrac{\partial f}{\partial x} = 2y(xy - 1), \dfrac{\partial f}{\partial y} = 2x(xy - 1)$

7. $\dfrac{\partial f}{\partial x} = \dfrac{x}{\sqrt{x^2 + y^2}}, \dfrac{\partial f}{\partial y} = \dfrac{y}{\sqrt{x^2 + y^2}}$

9. $\dfrac{\partial f}{\partial x} = \dfrac{-1}{(x + y)^2}, \dfrac{\partial f}{\partial y} = \dfrac{-1}{(x + y)^2}$

11. $\dfrac{\partial f}{\partial x} = \dfrac{-y^2 - 1}{(xy - 1)^2}, \dfrac{\partial f}{\partial y} = \dfrac{-x^2 - 1}{(xy - 1)^2}$

13. $\dfrac{\partial f}{\partial x} = e^{x+y+1}, \dfrac{\partial f}{\partial y} = e^{x+y+1}$ **15.** $\dfrac{\partial f}{\partial x} = \dfrac{1}{x + y}, \dfrac{\partial f}{\partial y} = \dfrac{1}{x + y}$

17. $\dfrac{\partial f}{\partial x} = 2 \sin (x - 3y) \cos (x - 3y),$

$\dfrac{\partial f}{\partial y} = -6 \sin (x - 3y) \cos (x - 3y)$

19. $\frac{\partial f}{\partial x} = yx^{y-1}, \frac{\partial f}{\partial y} = x^y \ln x$ 21. $\frac{\partial f}{\partial x} = -g(x), \frac{\partial f}{\partial y} = g(y)$

23. $f_x = y^2, f_y = 2xy, f_z = -4z$

25. $f_x = 1, f_y = -y(y^2 + z^2)^{-1/2}, f_z = -z(y^2 + z^2)^{-1/2}$

27. $f_x = \frac{yz}{\sqrt{1 - x^2y^2z^2}}, f_y = \frac{xz}{\sqrt{1 - x^2y^2z^2}}, f_z = \frac{xy}{\sqrt{1 - x^2y^2z^2}}$

29. $f_x = \frac{1}{x + 2y + 3z}, f_y = \frac{2}{x + 2y + 3z}, f_z = \frac{3}{x + 2y + 3z}$

31. $f_x = -2xe^{-(x^2+y^2+z^2)}, f_y = -2ye^{-(x^2+y^2+z^2)}, f_z = -2ze^{-(x^2+y^2+z^2)}$

33. $f_x = \operatorname{sech}^2(x + 2y + 3z), f_y = 2 \operatorname{sech}^2(x + 2y + 3z),$
$f_z = 3 \operatorname{sech}^2(x + 2y + 3z)$

35. $\frac{\partial f}{\partial t} = -2\pi \sin(2\pi t - \alpha), \frac{\partial f}{\partial \alpha} = \sin(2\pi t - \alpha)$

37. $\frac{\partial h}{\partial \rho} = \sin\phi \cos\theta, \frac{\partial h}{\partial \phi} = \rho \cos\phi \cos\theta, \frac{\partial h}{\partial \theta} = -\rho \sin\phi \sin\theta$

39. $W_P(P, V, \delta, v, g) = V, W_V(P, V, \delta, v, g) = P + \frac{\delta v^2}{2g},$
$W_\delta(P, V, \delta, v, g) = \frac{Vv^2}{2g}, W_v(P, V, \delta, v, g) = \frac{V\delta v}{g},$
$W_g(P, V, \delta, v, g) = -\frac{V\delta v^2}{2g^2}$

41. $\frac{\partial f}{\partial x} = 1 + y, \frac{\partial f}{\partial y} = 1 + x, \frac{\partial^2 f}{\partial x^2} = 0, \frac{\partial^2 f}{\partial y^2} = 0,$
$\frac{\partial^2 f}{\partial y\,\partial x} = \frac{\partial^2 f}{\partial x\,\partial y} = 1$

43. $\frac{\partial g}{\partial x} = 2xy + y\cos x, \frac{\partial g}{\partial y} = x^2 - \sin y + \sin x,$
$\frac{\partial^2 g}{\partial x^2} = 2y - y\sin x, \frac{\partial^2 g}{\partial y^2} = -\cos y,$
$\frac{\partial^2 g}{\partial y\,\partial x} = \frac{\partial^2 g}{\partial x\,\partial y} = 2x + \cos x$

45. $\frac{\partial r}{\partial x} = \frac{1}{x + y}, \frac{\partial r}{\partial y} = \frac{1}{x + y}, \frac{\partial^2 r}{\partial x^2} = \frac{-1}{(x + y)^2}, \frac{\partial^2 r}{\partial y^2} = \frac{-1}{(x + y)^2},$
$\frac{\partial^2 r}{\partial y\,\partial x} = \frac{\partial^2 r}{\partial x\,\partial y} = \frac{-1}{(x + y)^2}$

47. $\frac{\partial w}{\partial x} = \frac{2}{2x + 3y}, \frac{\partial w}{\partial y} = \frac{3}{2x + 3y}, \frac{\partial^2 w}{\partial y\,\partial x} = \frac{\partial^2 w}{\partial x\,\partial y} = \frac{-6}{(2x + 3y)^2}$

49. $\frac{\partial w}{\partial x} = y^2 + 2xy^3 + 3x^2y^4, \frac{\partial w}{\partial y} = 2xy + 3x^2y^2 + 4x^3y^3,$
$\frac{\partial^2 w}{\partial y\,\partial x} = \frac{\partial^2 w}{\partial x\,\partial y} = 2y + 6xy^2 + 12x^2y^3$

51. **(a)** x first **(b)** y first **(c)** x first
(d) x first **(e)** y first **(f)** y first

53. $f_x(1, 2) = -13, f_y(1, 2) = -2$ 55. 12 57. -2

59. $\frac{\partial A}{\partial a} = \frac{a}{bc \sin A}, \frac{\partial A}{\partial b} = \frac{c\cos A - b}{bc \sin A}$ 61. $v_x = \frac{\ln v}{(\ln u)(\ln v) - 1}$

77. Yes

Section 14.4, pp. 1003–1005

1. **(a)** $\frac{dw}{dt} = 0$, **(b)** $\frac{dw}{dt}(\pi) = 0$

3. **(a)** $\frac{dw}{dt} = 1$, **(b)** $\frac{dw}{dt}(3) = 1$

5. **(a)** $\frac{dw}{dt} = 4t\tan^{-1} t + 1$, **(b)** $\frac{dw}{dt}(1) = \pi + 1$

7. **(a)** $\frac{\partial z}{\partial u} = 4\cos v \ln(u \sin v) + 4\cos v,$
$\frac{\partial z}{\partial v} = -4u \sin v \ln(u \sin v) + \frac{4u\cos^2 v}{\sin v}$
(b) $\frac{\partial z}{\partial u} = \sqrt{2}(\ln 2 + 2), \frac{\partial z}{\partial v} = -2\sqrt{2}(\ln 2 - 2)$

9. **(a)** $\frac{\partial w}{\partial u} = 2u + 4uv, \frac{\partial w}{\partial v} = -2v + 2u^2$
(b) $\frac{\partial w}{\partial u} = 3, \frac{\partial w}{\partial v} = -\frac{3}{2}$

11. **(a)** $\frac{\partial u}{\partial x} = 0, \frac{\partial u}{\partial y} = \frac{z}{(z - y)^2}, \frac{\partial u}{\partial z} = \frac{-y}{(z - y)^2}$
(b) $\frac{\partial u}{\partial x} = 0, \frac{\partial u}{\partial y} = 1, \frac{\partial u}{\partial z} = -2$

13. $\frac{dz}{dt} = \frac{\partial z}{\partial x}\frac{dx}{dt} + \frac{\partial z}{\partial y}\frac{dy}{dt}$

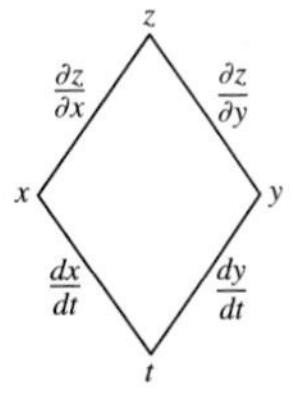

15. $\frac{\partial w}{\partial u} = \frac{\partial w}{\partial x}\frac{\partial x}{\partial u} + \frac{\partial w}{\partial y}\frac{\partial y}{\partial u} + \frac{\partial w}{\partial z}\frac{\partial z}{\partial u},$
$\frac{\partial w}{\partial v} = \frac{\partial w}{\partial x}\frac{\partial x}{\partial v} + \frac{\partial w}{\partial y}\frac{\partial y}{\partial v} + \frac{\partial w}{\partial z}\frac{\partial z}{\partial v}$

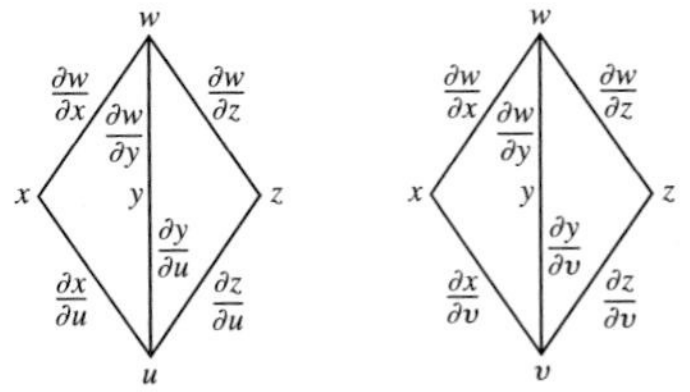

17. $\frac{\partial w}{\partial u} = \frac{\partial w}{\partial x}\frac{\partial x}{\partial u} + \frac{\partial w}{\partial y}\frac{\partial y}{\partial u}, \frac{\partial w}{\partial v} = \frac{\partial w}{\partial x}\frac{\partial x}{\partial v} + \frac{\partial w}{\partial y}\frac{\partial y}{\partial v}$

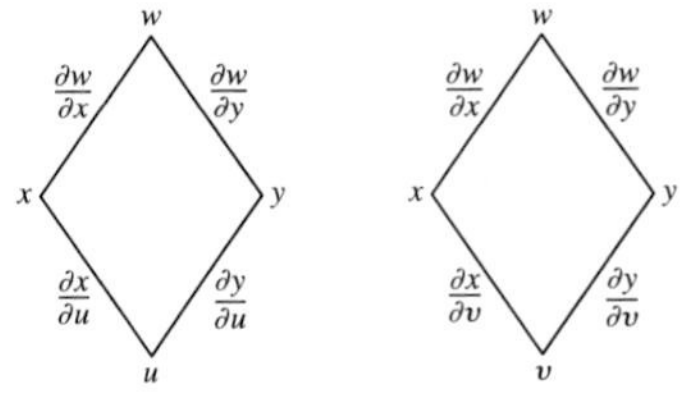

19. $\dfrac{\partial z}{\partial t} = \dfrac{\partial z}{\partial x}\dfrac{\partial x}{\partial t} + \dfrac{\partial z}{\partial y}\dfrac{\partial y}{\partial t}, \dfrac{\partial z}{\partial s} = \dfrac{\partial z}{\partial x}\dfrac{\partial x}{\partial s} + \dfrac{\partial z}{\partial y}\dfrac{\partial y}{\partial s}$

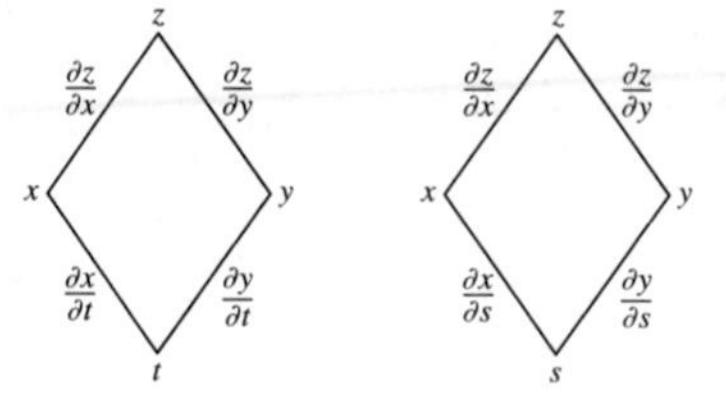

21. $\dfrac{\partial w}{\partial s} = \dfrac{dw}{du}\dfrac{\partial u}{\partial s}, \dfrac{\partial w}{\partial t} = \dfrac{dw}{du}\dfrac{\partial u}{\partial t}$

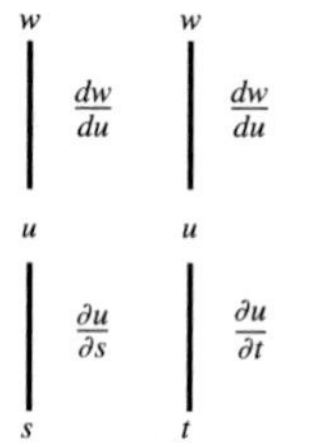

23. $\dfrac{\partial w}{\partial r} = \dfrac{\partial w}{\partial x}\dfrac{dx}{dr} + \dfrac{\partial w}{\partial y}\dfrac{dy}{dr} = \dfrac{\partial w}{\partial x}\dfrac{dx}{dr}$ since $\dfrac{dy}{dr} = 0$,

$\dfrac{\partial w}{\partial s} = \dfrac{\partial w}{\partial x}\dfrac{dx}{ds} + \dfrac{\partial w}{\partial y}\dfrac{dy}{ds} = \dfrac{\partial w}{\partial y}\dfrac{dy}{ds}$ since $\dfrac{dx}{ds} = 0$

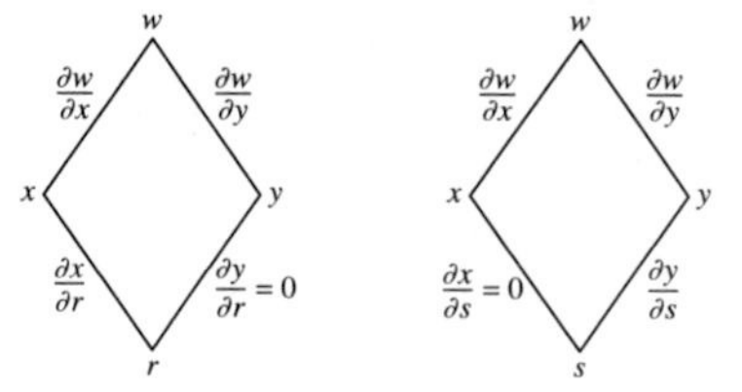

25. 4/3 **27.** −4/5 **29.** $\dfrac{\partial z}{\partial x} = \dfrac{1}{4}, \dfrac{\partial z}{\partial y} = -\dfrac{3}{4}$

31. $\dfrac{\partial z}{\partial x} = -1, \dfrac{\partial z}{\partial y} = -1$ **33.** 12 **35.** −7

37. $\dfrac{\partial z}{\partial u} = 2, \dfrac{\partial z}{\partial v} = 1$ **39.** −0.00005 amps/sec

45. $(\cos 1, \sin 1, 1)$ and $(\cos(-2), \sin(-2), -2)$

47. **(a)** Maximum at $\left(-\dfrac{\sqrt{2}}{2}, \dfrac{\sqrt{2}}{2}\right)$ and $\left(\dfrac{\sqrt{2}}{2}, -\dfrac{\sqrt{2}}{2}\right)$; minimum at $\left(\dfrac{\sqrt{2}}{2}, \dfrac{\sqrt{2}}{2}\right)$ and $\left(-\dfrac{\sqrt{2}}{2}, -\dfrac{\sqrt{2}}{2}\right)$

(b) Max = 6, min = 2

49. $2x\sqrt{x^8 + x^3} + \displaystyle\int_0^{x^2} \frac{3x^2}{2\sqrt{t^4 + x^3}}\,dt$

Section 14.5, pp. 1013–1014

1.

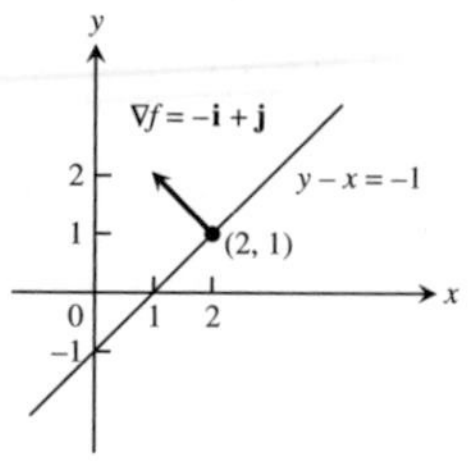

3.

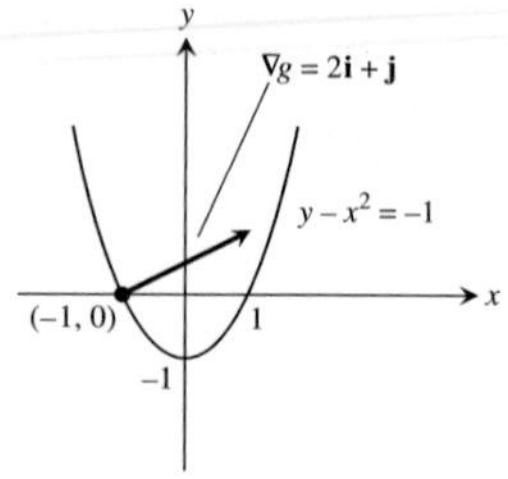

5. $\nabla f = 3\mathbf{i} + 2\mathbf{j} - 4\mathbf{k}$ **7.** $\nabla f = -\dfrac{26}{27}\mathbf{i} + \dfrac{23}{54}\mathbf{j} - \dfrac{23}{54}\mathbf{k}$

9. −4 **11.** 31/13 **13.** 3 **15.** 2

17. $\mathbf{u} = -\dfrac{1}{\sqrt{2}}\mathbf{i} + \dfrac{1}{\sqrt{2}}\mathbf{j}, (D_{\mathbf{u}}f)_{P_0} = \sqrt{2}; -\mathbf{u} = \dfrac{1}{\sqrt{2}}\mathbf{i} - \dfrac{1}{\sqrt{2}}\mathbf{j},$

$(D_{-\mathbf{u}}f)_{P_0} = -\sqrt{2}$

19. $\mathbf{u} = \dfrac{1}{3\sqrt{3}}\mathbf{i} - \dfrac{5}{3\sqrt{3}}\mathbf{j} - \dfrac{1}{3\sqrt{3}}\mathbf{k}, (D_{\mathbf{u}}f)_{P_0} = 3\sqrt{3};$

$-\mathbf{u} = -\dfrac{1}{3\sqrt{3}}\mathbf{i} + \dfrac{5}{3\sqrt{3}}\mathbf{j} + \dfrac{1}{3\sqrt{3}}\mathbf{k}, (D_{-\mathbf{u}}f)_{P_0} = -3\sqrt{3}$

21. $\mathbf{u} = \dfrac{1}{\sqrt{3}}(\mathbf{i} + \mathbf{j} + \mathbf{k}), (D_{\mathbf{u}}f)_{P_0} = 2\sqrt{3};$

$-\mathbf{u} = -\dfrac{1}{\sqrt{3}}(\mathbf{i} + \mathbf{j} + \mathbf{k}), (D_{-\mathbf{u}}f)_{P_0} = -2\sqrt{3}$

23.

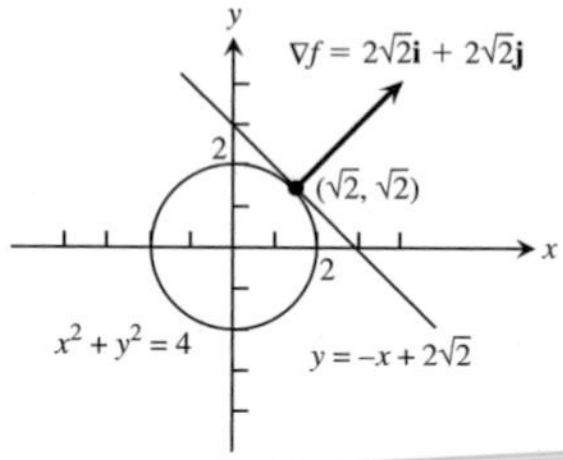

25.

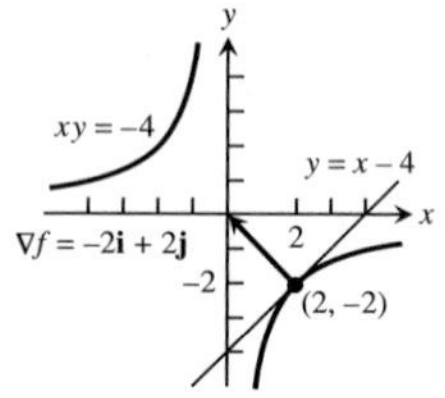

27. $\mathbf{u} = \dfrac{7}{\sqrt{53}}\mathbf{i} - \dfrac{2}{\sqrt{53}}\mathbf{j}, -\mathbf{u} = -\dfrac{7}{\sqrt{53}}\mathbf{i} + \dfrac{2}{\sqrt{53}}\mathbf{j}$

29. No, the maximum rate of change is $\sqrt{185} < 14$.

31. $-\dfrac{7}{\sqrt{5}}$

Section 14.6, pp. 1024–1027

1. **(a)** $x + y + z = 3$
(b) $x = 1 + 2t, y = 1 + 2t, z = 1 + 2t$

3. **(a)** $2x - z - 2 = 0$ **(b)** $x = 2 - 4t, y = 0, z = 2 + 2t$

5. **(a)** $2x + 2y + z - 4 = 0$
(b) $x = 2t, y = 1 + 2t, z = 2 + t$

7. **(a)** $x + y + z - 1 = 0$ **(b)** $x = t, y = 1 + t, z = t$

9. $2x - z - 2 = 0$ **11.** $x - y + 2z - 1 = 0$

13. $x = 1, y = 1 + 2t, z = 1 - 2t$

15. $x = 1 - 2t, y = 1, z = \frac{1}{2} + 2t$

17. $x = 1 + 90t, y = 1 - 90t, z = 3$

19. $df = \frac{9}{11{,}830} \approx 0.0008$ **21.** $dg = 0$

23. **(a)** $\frac{\sqrt{3}}{2}\sin\sqrt{3} - \frac{1}{2}\cos\sqrt{3} \approx 0.935°\text{C/ft}$
(b) $\sqrt{3}\sin\sqrt{3} - \cos\sqrt{3} \approx 1.87°\text{C/sec}$

25. **(a)** $L(x, y) = 1$ **(b)** $L(x, y) = 2x + 2y - 1$

27. **(a)** $L(x, y) = 3x - 4y + 5$ **(b)** $L(x, y) = 3x - 4y + 5$

29. **(a)** $L(x, y) = 1 + x$ **(b)** $L(x, y) = -y + \frac{\pi}{2}$

31. $L(x, y) = 7 + x - 6y; 0.06$ **33.** $L(x, y) = x + y + 1; 0.08$

35. $L(x, y) = 1 + x; 0.0222$

37. **(a)** $L(x, y, z) = 2x + 2y + 2z - 3$ **(b)** $L(x, y, z) = y + z$
(c) $L(x, y, z) = 0$

39. **(a)** $L(x, y, z) = x$ **(b)** $L(x, y, z) = \frac{1}{\sqrt{2}}x + \frac{1}{\sqrt{2}}y$
(c) $L(x, y, z) = \frac{1}{3}x + \frac{2}{3}y + \frac{2}{3}z$

41. **(a)** $L(x, y, z) = 2 + x$
(b) $L(x, y, z) = x - y - z + \frac{\pi}{2} + 1$
(c) $L(x, y, z) = x - y - z + \frac{\pi}{2} + 1$

43. $L(x, y, z) = 2x - 6y - 2z + 6, 0.0024$

45. $L(x, y, z) = x + y - z - 1, 0.00135$

47. Maximum error (estimate) ≤ 0.31 in magnitude

49. Maximum percentage error $= \pm 4.83\%$

51. Pay more attention to the smaller of the two dimensions. It will generate the larger partial derivative.

53. **(a)** 0.30% **55.** f is most sensitive to a change in d.

57. Q is most sensitive to changes in h.

61. At $-\frac{\pi}{4}, -\frac{\pi}{2\sqrt{2}}$; at 0, 0; at $\frac{\pi}{4}, \frac{\pi}{2\sqrt{2}}$

Section 14.7, pp. 1034–1038

1. $f(-3, 3) = -5$, local minimum

3. $f\left(\frac{2}{3}, \frac{4}{3}\right) = 0$, local maximum **5.** $f(-2, 1)$, saddle point

7. $f\left(\frac{6}{5}, \frac{69}{25}\right)$, saddle point **9.** $f(2, 1)$, saddle point

11. $f(2, -1) = -6$, local minimum **13.** $f(1, 2)$, saddle point

15. $f(0, 0)$, saddle point

17. $f(0, 0)$, saddle point; $f\left(-\frac{2}{3}, \frac{2}{3}\right) = \frac{170}{27}$, local maximum

19. $f(0, 0) = 0$, local minimum; $f(1, -1)$, saddle point

21. $f(0, 0)$, saddle point; $f\left(\frac{4}{9}, \frac{4}{3}\right) = -\frac{64}{81}$, local minimum

23. $f(0, 0)$, saddle point; $f(0, 2) = -12$, local minimum; $f(-2, 0) = -4$, local maximum; $f(-2, 2)$, saddle point

25. $f(0, 0)$, saddle point; $f(1, 1) = 2$, $f(-1, -1) = 2$, local maxima

27. $f(0, 0) = -1$, local maximum

29. $f(n\pi, 0)$, saddle point; $f(n\pi, 0) = 0$ for every n

31. Absolute maximum: 1 at (0, 0); absolute minimum: -5 at (1, 2)

33. Absolute maximum: 4 at (0, 2); absolute minimum: 0 at (0, 0)

35. Absolute maximum: 11 at (0, −3); absolute minimum: −10 at (4, −2)

37. Absolute maximum: 4 at (2, 0); absolute minimum: $\frac{3\sqrt{2}}{2}$ at $\left(3, -\frac{\pi}{4}\right), \left(3, \frac{\pi}{4}\right), \left(1, -\frac{\pi}{4}\right)$, and $\left(1, \frac{\pi}{4}\right)$

39. $a = -3, b = 2$

41. Hottest: $2\frac{1}{4}°$ at $\left(-\frac{1}{2}, \frac{\sqrt{3}}{2}\right)$ and $\left(-\frac{1}{2}, -\frac{\sqrt{3}}{2}\right)$; coldest: $-\frac{1}{4}°$ at $\left(\frac{1}{2}, 0\right)$

43. **(a)** $f(0, 0)$, saddle point **(b)** $f(1, 2)$, local minimum
(c) $f(1, -2)$, local minimum; $f(-1, -2)$, saddle point

49. $\left(\frac{1}{6}, \frac{1}{3}, \frac{355}{36}\right)$

53. **(a)** On the semicircle, max $f = 2\sqrt{2}$ at $t = \pi/4$, min $f = -2$ at $t = \pi$. On the quarter circle, max $f = 2\sqrt{2}$ at $t = \pi/4$, min $f = 2$ at $t = 0, \pi/2$.
(b) On the semicircle, max $g = 2$ at $t = \pi/4$, min $g = -2$ at $t = 3\pi/4$. On the quarter circle, max $g = 2$ at $t = \pi/4$, min $g = 0$ at $t = 0, \pi/2$.
(c) On the semicircle, max $h = 8$ at $t = 0, \pi$; min $h = 4$ at $t = \pi/2$. On the quarter circle, max $h = 8$ at $t = 0$, min $h = 4$ at $t = \pi/2$.

55. i) min $f = -1/2$ at $t = -1/2$; no max ii) max $f = 0$ at $t = -1, 0$; min $f = -1/2$ at $t = -1/2$ iii) max $f = 4$ at $t = 1$; min $f = 0$ at $t = 0$

57. $y = -\frac{20}{13}x + \frac{9}{13}, y|_{x=4} = -\frac{71}{13}$

59. $y = \frac{3}{2}x + \frac{1}{6}, y|_{x=4} = \frac{37}{6}$

61. $y = 0.122x + 3.59$

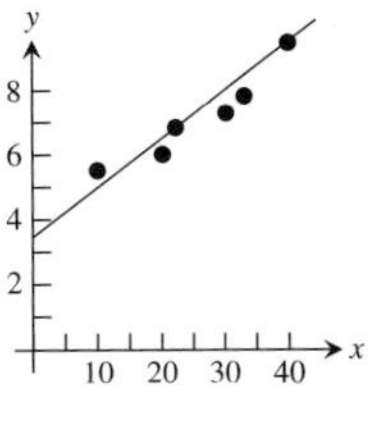

63. **(a)**

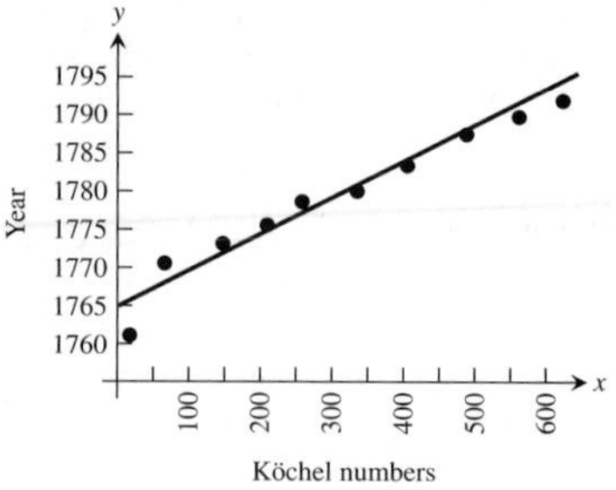

(b) $y = 0.0427K + 1764.8$ **(c)** 1780

Section 14.8, pp. 1047–1049

1. $\left(\pm\frac{1}{\sqrt{2}}, \frac{1}{2}\right), \left(\pm\frac{1}{\sqrt{2}}, -\frac{1}{2}\right)$ **3.** 39 **5.** $\left(3, \pm 3\sqrt{2}\right)$

7. **(a)** 8 **(b)** 64

9. $r = 2$ cm, $h = 4$ cm **11.** Length $= 4\sqrt{2}$, width $= 3\sqrt{2}$

13. $f(0, 0) = 0$ is minimum, $f(2, 4) = 20$ is maximum.

15. Lowest $= 0°$, highest $= 125°$

17. $\left(\frac{3}{2}, 2, \frac{5}{2}\right)$ **19.** 1 **21.** $(0, 0, 2), (0, 0, -2)$

23. $f(1, -2, 5) = 30$ is maximum, $f(-1, 2, -5) = -30$ is minimum.

25. 3, 3, 3 **27.** $\frac{2}{\sqrt{3}}$ by $\frac{2}{\sqrt{3}}$ by $\frac{2}{\sqrt{3}}$ units

29. $(\pm 4/3, -4/3, -4/3)$ **31.** $U(8, 14) = \$128$

33. $f(2/3, 4/3, -4/3) = \frac{4}{3}$ **35.** $(2, 4, 4)$

37. Maximum is $1 + 6\sqrt{3}$ at $\left(\pm\sqrt{6}, \sqrt{3}, 1\right)$, minimum is $1 - 6\sqrt{3}$ at $\left(\pm\sqrt{6}, -\sqrt{3}, 1\right)$.

39. Maximum is 4 at $(0, 0, \pm 2)$, minimum is 2 at $\left(\pm\sqrt{2}, \pm\sqrt{2}, 0\right)$.

Section 14.9, pp. 1053–1054

1. **(a)** 0 **(b)** $1 + 2z$ **(c)** $1 + 2z$

3. **(a)** $\frac{\partial U}{\partial P} + \frac{\partial U}{\partial T}\left(\frac{V}{nR}\right)$ **(b)** $\frac{\partial U}{\partial P}\left(\frac{nR}{V}\right) + \frac{\partial U}{\partial T}$

5. **(a)** 5 **(b)** 5

7. $\left(\frac{\partial x}{\partial r}\right)_\theta = \cos\theta$

$\left(\frac{\partial r}{\partial x}\right)_y = \frac{x}{\sqrt{x^2 + y^2}}$

Section 14.10, pp. 1058–1059

1. Quadratic: $x + xy$; cubic: $x + xy + \frac{1}{2}xy^2$

3. Quadratic: xy; cubic: xy

5. Quadratic: $y + \frac{1}{2}(2xy - y^2)$;

cubic: $y + \frac{1}{2}(2xy - y^2) + \frac{1}{6}(3x^2y - 3xy^2 + 2y^3)$

7. Quadratic: $\frac{1}{2}(2x^2 + 2y^2) = x^2 + y^2$; cubic: $x^2 + y^2$

9. Quadratic: $1 + (x + y) + (x + y)^2$;
cubic: $1 + (x + y) + (x + y)^2 + (x + y)^3$

11. Quadratic: $1 - \frac{1}{2}x^2 - \frac{1}{2}y^2$; $E(x, y) \le 0.00134$

Practice Exercises, pp. 1060–1063

1. Domain: all points in the xy-plane; range: $z \ge 0$. Level curves are ellipses with major axis along the y-axis and minor axis along the x-axis.

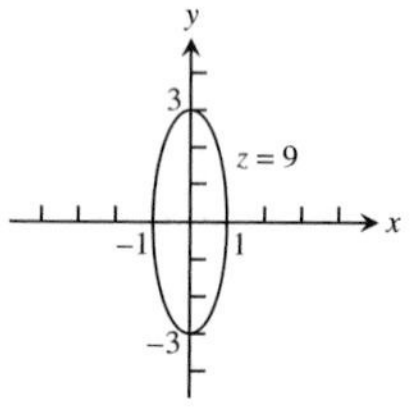

3. Domain: all (x, y) such that $x \ne 0$ and $y \ne 0$; range: $z \ne 0$. Level curves are hyperbolas with the x- and y-axes as asymptotes.

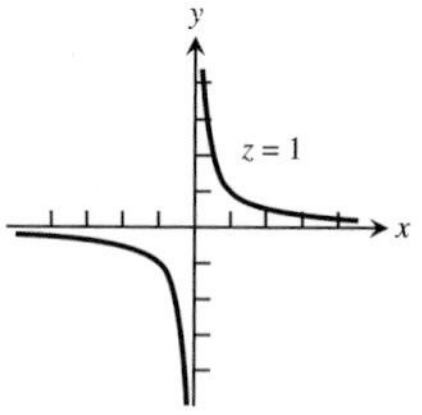

5. Domain: all points in xyz-space; range: all real numbers. Level surfaces are paraboloids of revolution with the z-axis as axis.

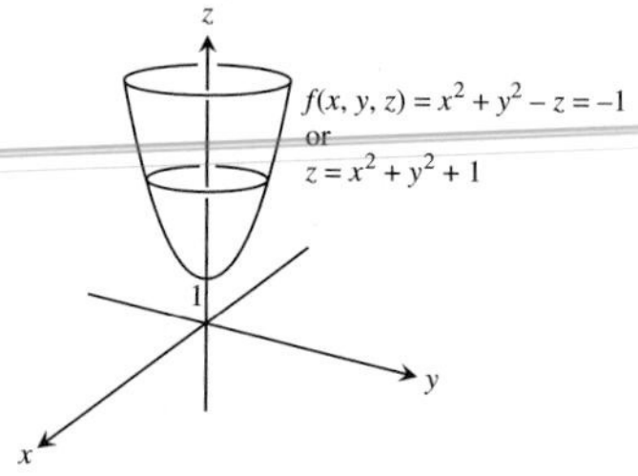

7. Domain: all (x, y, z) such that $(x, y, z) \ne (0, 0, 0)$; range: positive real numbers. Level surfaces are spheres with center $(0, 0, 0)$ and radius $r > 0$.

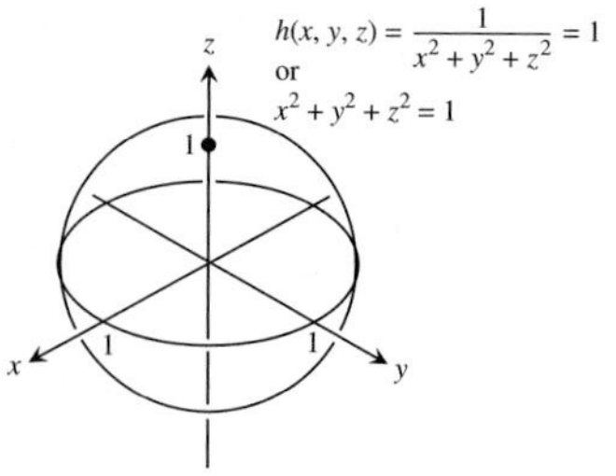

9. -2 **11.** $1/2$ **13.** 1 **15.** Let $y = kx^2, k \neq 1$

17. No; $\lim_{(x,y)\to(0,0)} f(x, y)$ does not exist.

19. $\dfrac{\partial g}{\partial r} = \cos\theta + \sin\theta, \dfrac{\partial g}{\partial \theta} = -r\sin\theta + r\cos\theta$

21. $\dfrac{\partial f}{\partial R_1} = -\dfrac{1}{R_1^2}, \dfrac{\partial f}{\partial R_2} = -\dfrac{1}{R_2^2}, \dfrac{\partial f}{\partial R_3} = -\dfrac{1}{R_3^2}$

23. $\dfrac{\partial P}{\partial n} = \dfrac{RT}{V}, \dfrac{\partial P}{\partial R} = \dfrac{nT}{V}, \dfrac{\partial P}{\partial T} = \dfrac{nR}{V}, \dfrac{\partial P}{\partial V} = -\dfrac{nRT}{V^2}$

25. $\dfrac{\partial^2 g}{\partial x^2} = 0, \dfrac{\partial^2 g}{\partial y^2} = \dfrac{2x}{y^3}, \dfrac{\partial^2 g}{\partial y\,\partial x} = \dfrac{\partial^2 g}{\partial x\,\partial y} = -\dfrac{1}{y^2}$

27. $\dfrac{\partial^2 f}{\partial x^2} = -30x + \dfrac{2 - 2x^2}{(x^2 + 1)^2}, \dfrac{\partial^2 f}{\partial y^2} = 0, \dfrac{\partial^2 f}{\partial y\,\partial x} = \dfrac{\partial^2 f}{\partial x\,\partial y} = 1$

29. $\left.\dfrac{dw}{dt}\right|_{t=0} = -1$

31. $\left.\dfrac{\partial w}{\partial r}\right|_{(r,s)=(\pi,0)} = 2, \left.\dfrac{\partial w}{\partial s}\right|_{(r,s)=(\pi,0)} = 2 - \pi$

33. $\left.\dfrac{df}{dt}\right|_{t=1} = -(\sin 1 + \cos 2)(\sin 1) + (\cos 1 + \cos 2)(\cos 1)$
$- 2(\sin 1 + \cos 1)(\sin 2)$

35. $\left.\dfrac{dy}{dx}\right|_{(x,y)=(0,1)} = -1$

37. Increases most rapidly in the direction $\mathbf{u} = -\dfrac{\sqrt{2}}{2}\mathbf{i} - \dfrac{\sqrt{2}}{2}\mathbf{j}$; decreases most rapidly in the direction $-\mathbf{u} = \dfrac{\sqrt{2}}{2}\mathbf{i} + \dfrac{\sqrt{2}}{2}\mathbf{j}$; $D_{\mathbf{u}}f = \dfrac{\sqrt{2}}{2}$; $D_{-\mathbf{u}}f = -\dfrac{\sqrt{2}}{2}$; $D_{\mathbf{u}_1}f = -\dfrac{7}{10}$ where $\mathbf{u}_1 = \dfrac{\mathbf{v}}{|\mathbf{v}|}$

39. Increases most rapidly in the direction $\mathbf{u} = \dfrac{2}{7}\mathbf{i} + \dfrac{3}{7}\mathbf{j} + \dfrac{6}{7}\mathbf{k}$; decreases most rapidly in the direction $-\mathbf{u} = -\dfrac{2}{7}\mathbf{i} - \dfrac{3}{7}\mathbf{j} - \dfrac{6}{7}\mathbf{k}$; $D_{\mathbf{u}}f = 7$; $D_{-\mathbf{u}}f = -7$; $D_{\mathbf{u}_1}f = 7$ where $\mathbf{u}_1 = \dfrac{\mathbf{v}}{|\mathbf{v}|}$

41. $\pi/\sqrt{2}$

43. **(a)** $f_x(1, 2) = f_y(1, 2) = 2$ **(b)** $14/5$

45.

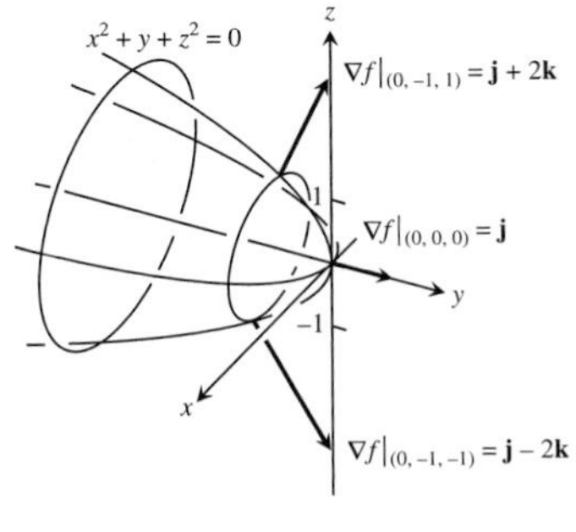

47. Tangent: $4x - y - 5z = 4$; normal line: $x = 2 + 4t, y = -1 - t, z = 1 - 5t$

49. $2y - z - 2 = 0$

51. Tangent: $x + y = \pi + 1$; normal line: $y = x - \pi + 1$

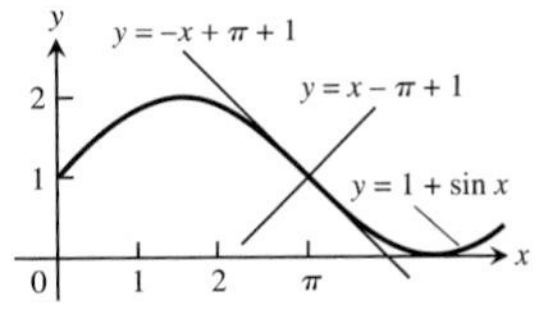

53. $x = 1 - 2t, y = 1, z = 1/2 + 2t$

55. Answers will depend on the upper bound used for $|f_{xx}|, |f_{xy}|, |f_{yy}|$. With $M = \sqrt{2}/2$, $|E| \leq 0.0142$. With $M = 1$, $|E| \leq 0.02$.

57. $L(x, y, z) = y - 3z, L(x, y, z) = x + y - z - 1$

59. Be more careful with the diameter.

61. $dI = 0.038$, % change in $I = 15.83\%$, more sensitive to voltage change

63. **(a)** 5%

65. Local minimum of -8 at $(-2, -2)$

67. Saddle point at $(0, 0)$, $f(0, 0) = 0$; local maximum of $1/4$ at $(-1/2, -1/2)$

69. Saddle point at $(0, 0)$, $f(0, 0) = 0$; local minimum of -4 at $(0, 2)$; local maximum of 4 at $(-2, 0)$; saddle point at $(-2, 2)$, $f(-2, 2) = 0$

71. Absolute maximum: 28 at $(0, 4)$; absolute minimum: $-9/4$ at $(3/2, 0)$

73. Absolute maximum: 18 at $(2, -2)$; absolute minimum: $-17/4$ at $(-2, 1/2)$

75. Absolute maximum: 8 at $(-2, 0)$; absolute minimum: -1 at $(1, 0)$

77. Absolute maximum: 4 at $(1, 0)$; absolute minimum: -4 at $(0, -1)$

79. Absolute maximum: 1 at $(0, \pm 1)$ and $(1, 0)$; absolute minimum: -1 at $(-1, 0)$

81. Maximum: 5 at $(0, 1)$; minimum: $-1/3$ at $(0, -1/3)$

83. Maximum: $\sqrt{3}$ at $\left(\dfrac{1}{\sqrt{3}}, -\dfrac{1}{\sqrt{3}}, \dfrac{1}{\sqrt{3}}\right)$; minimum: $-\sqrt{3}$ at $\left(-\dfrac{1}{\sqrt{3}}, \dfrac{1}{\sqrt{3}}, -\dfrac{1}{\sqrt{3}}\right)$

85. Width $= \left(\dfrac{c^2V}{ab}\right)^{1/3}$, depth $= \left(\dfrac{b^2V}{ac}\right)^{1/3}$, height $= \left(\dfrac{a^2V}{bc}\right)^{1/3}$

87. Maximum: $\dfrac{3}{2}$ at $\left(\dfrac{1}{\sqrt{2}}, \dfrac{1}{\sqrt{2}}, \sqrt{2}\right)$ and $\left(-\dfrac{1}{\sqrt{2}}, -\dfrac{1}{\sqrt{2}}, -\sqrt{2}\right)$; minimum: $\dfrac{1}{2}$ at $\left(-\dfrac{1}{\sqrt{2}}, \dfrac{1}{\sqrt{2}}, -\sqrt{2}\right)$ and $\left(\dfrac{1}{\sqrt{2}}, -\dfrac{1}{\sqrt{2}}, \sqrt{2}\right)$

89. **(a)** $(2y + x^2z)e^{yz}$ **(b)** $x^2e^{yz}\left(y - \dfrac{z}{2y}\right)$ **(c)** $(1 + x^2y)e^{yz}$

91. $\dfrac{\partial w}{\partial x} = \cos\theta\dfrac{\partial w}{\partial r} - \dfrac{\sin\theta}{r}\dfrac{\partial w}{\partial \theta}, \dfrac{\partial w}{\partial y} = \sin\theta\dfrac{\partial w}{\partial r} + \dfrac{\cos\theta}{r}\dfrac{\partial w}{\partial \theta}$

97. $(t, -t \pm 4, t)$, t a real number

Additional and Advanced Exercises, pp. 1063–1066

1. $f_{xy}(0, 0) = -1$, $f_{yx}(0, 0) = 1$

7. **(c)** $\dfrac{r^2}{2} = \dfrac{1}{2}(x^2 + y^2 + z^2)$ **13.** $V = \dfrac{\sqrt{3}abc}{2}$

17. $f(x, y) = \dfrac{y}{2} + 4$, $g(x, y) = \dfrac{x}{2} + \dfrac{9}{2}$

19. $y = 2\ln|\sin x| + \ln 2$

21. **(a)** $\dfrac{1}{\sqrt{53}}(2\mathbf{i} + 7\mathbf{j})$ **(b)** $\dfrac{-1}{\sqrt{29{,}097}}(98\mathbf{i} - 127\mathbf{j} + 58\mathbf{k})$

23. $w = e^{-c^2\pi^2 t}\sin \pi x$

CHAPTER 15

Section 15.1, pp. 1079–1081

1. 16

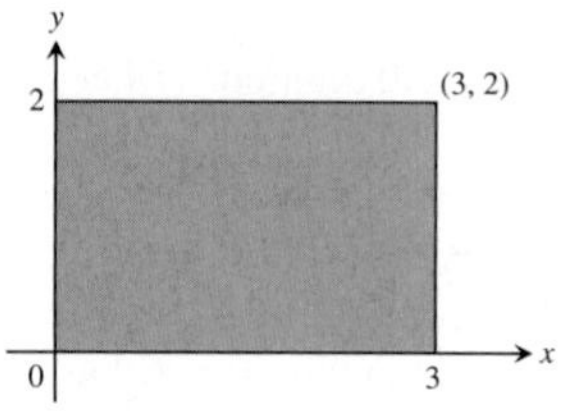

3. 1

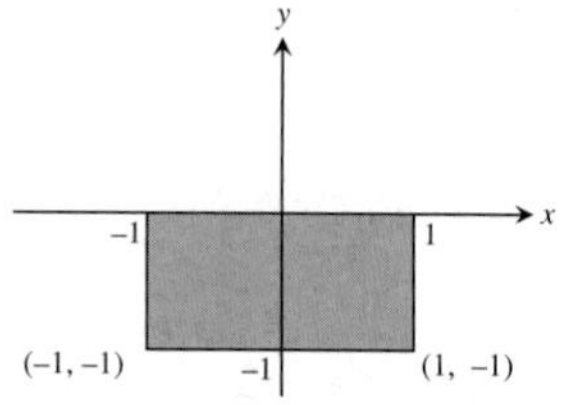

5. $\dfrac{\pi^2}{2} + 2$

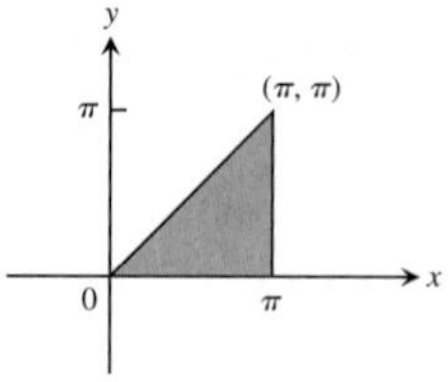

7. $8\ln 8 - 16 + e$

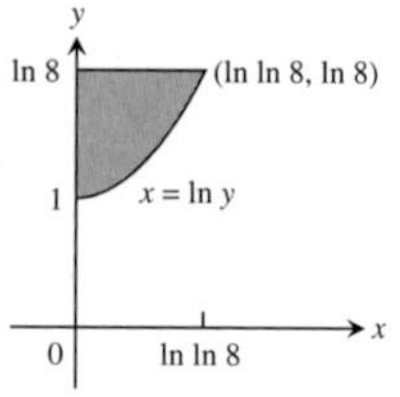

9. $e - 2$

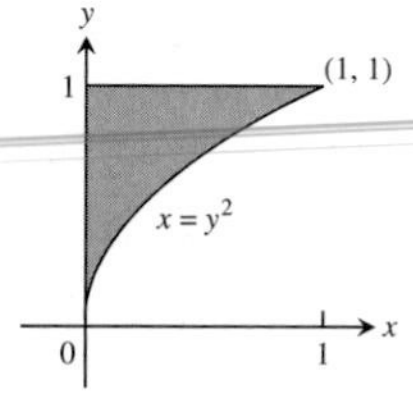

11. $\dfrac{3}{2}\ln 2$ **13.** $1/6$ **15.** $-1/10$

17. 8

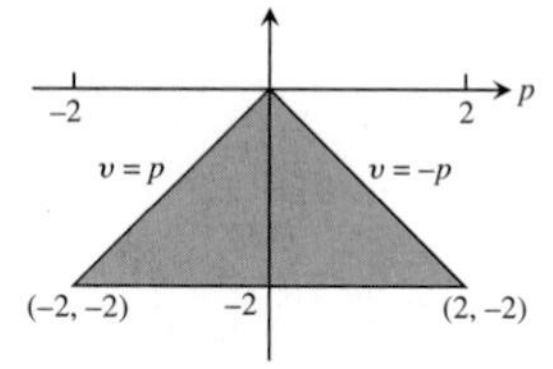

19. 2π

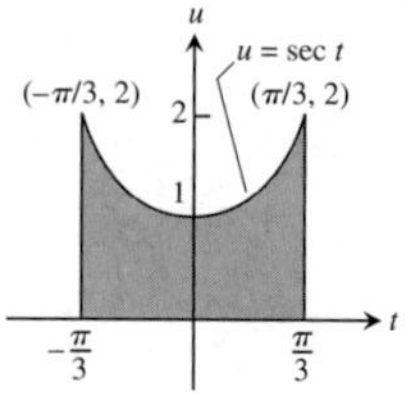

21. $\displaystyle\int_2^4 \int_0^{(4-y)/2} dx\, dy$

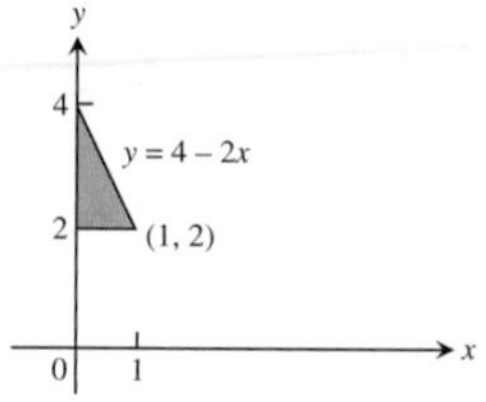

23. $\displaystyle\int_0^1 \int_{x^2}^{x} dy\, dx$

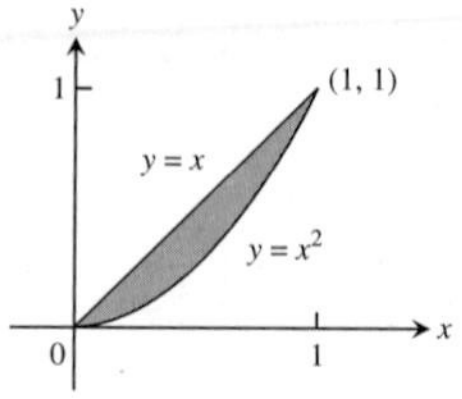

25. $\displaystyle\int_1^e \int_{\ln y}^{1} dx\, dy$

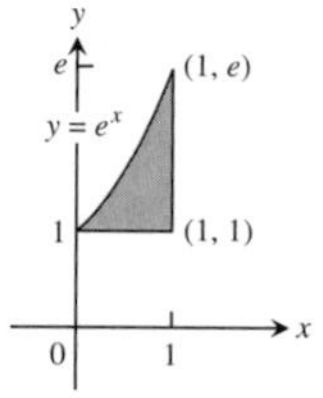

27. $\displaystyle\int_0^9 \int_0^{(\sqrt{9-y})/2} 16x\, dx\, dy$

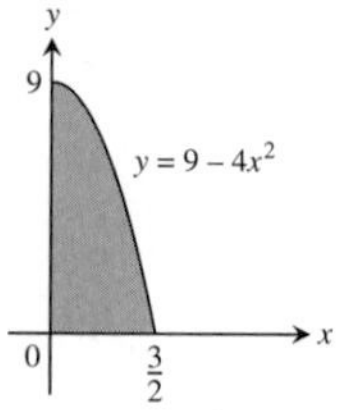

29. $\displaystyle\int_{-1}^1 \int_0^{\sqrt{1-x^2}} 3y\, dy\, dx$

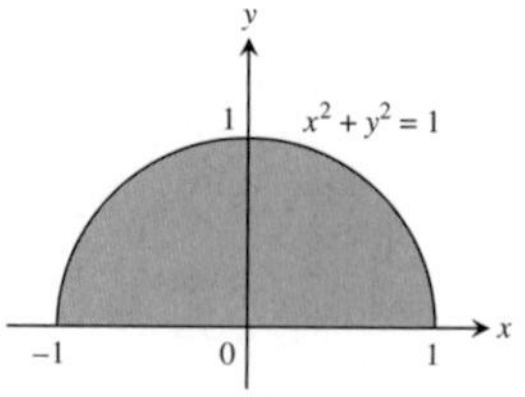

31. 2

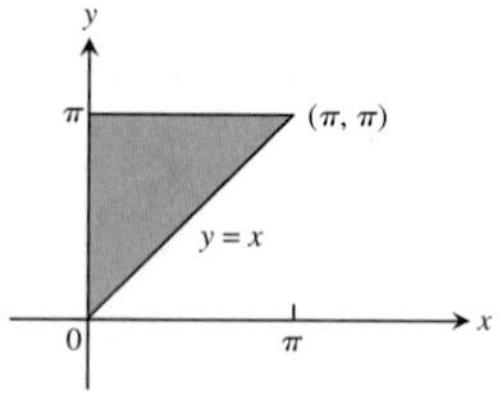

33. $\dfrac{e - 2}{2}$

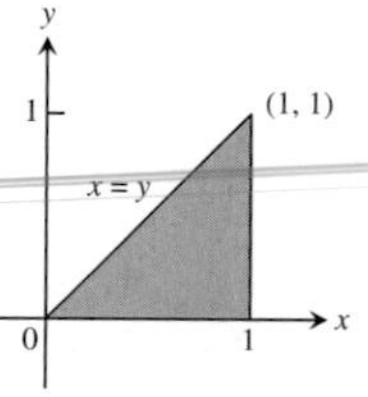

35. 2

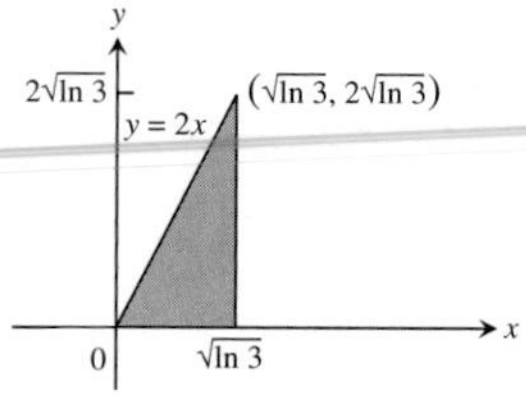

37. $1/(80\pi)$

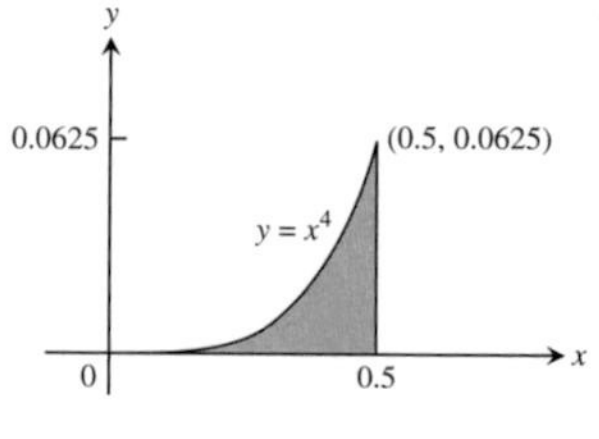

39. $-2/3$

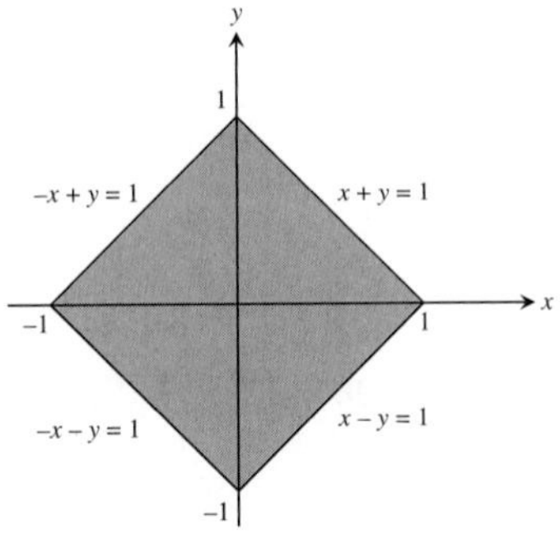

41. $4/3$ **43.** $625/12$ **45.** 16 **47.** 20 **49.** $2(1 + \ln 2)$

51. 1 **53.** π^2 **55.** $-\dfrac{3}{32}$ **57.** $\dfrac{20\sqrt{3}}{9}$

59. $\displaystyle\int_0^1 \int_x^{2-x} (x^2 + y^2)\,dy\,dx = \frac{4}{3}$

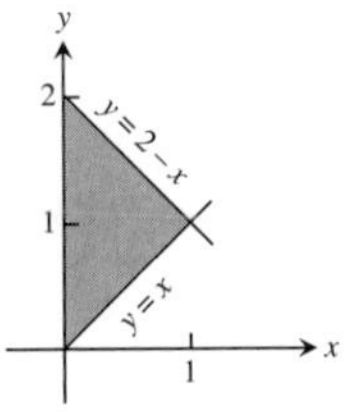

61. R is the set of points (x, y) such that $x^2 + 2y^2 < 4$

63. No, by Fubini's Theorem, the two orders of integration must give the same result.

67. 0.603 **69.** 0.233

Section 15.2, pp. 1089–1091

1. $\displaystyle\int_0^2 \int_0^{2-x} dy\,dx = 2$ or $\displaystyle\int_0^2 \int_0^{2-y} dx\,dy = 2$

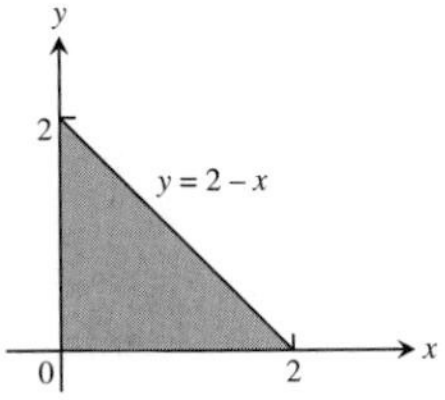

3. $\displaystyle\int_{-2}^{1} \int_{y-2}^{-y^2} dx\,dy = \frac{9}{2}$ **5.** $\displaystyle\int_0^{\ln 2} \int_0^{e^x} dy\,dx = 1$

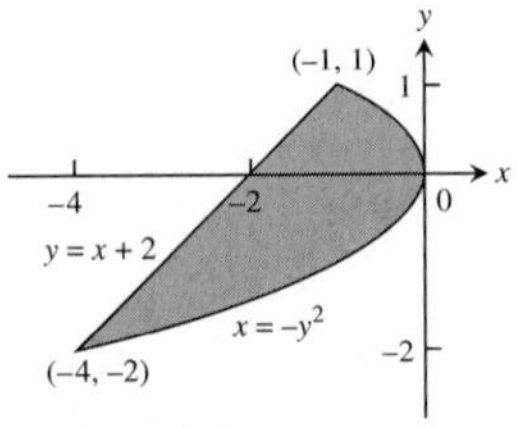

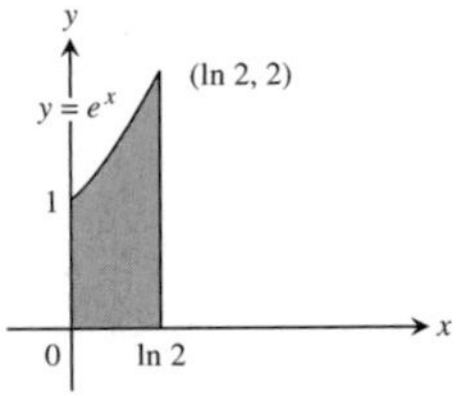

7. $\displaystyle\int_0^1 \int_{y^2}^{2y-y^2} dx\,dy = \frac{1}{3}$ **9.** 12

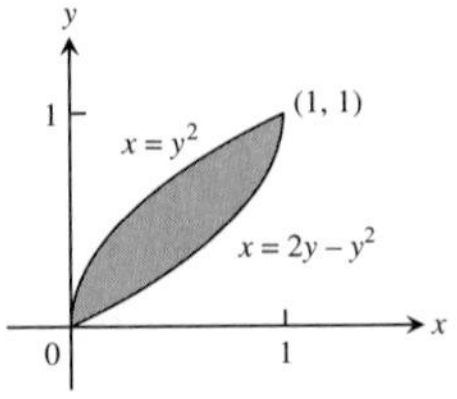

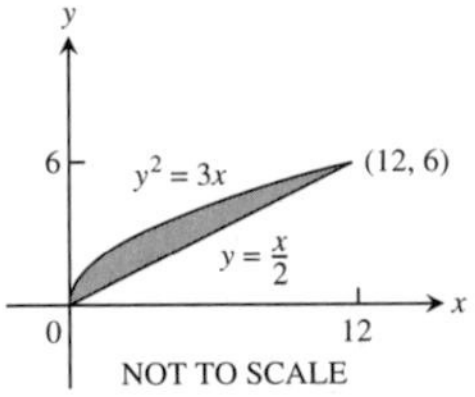

11. $\sqrt{2} - 1$ **13.** $\dfrac{3}{2}$

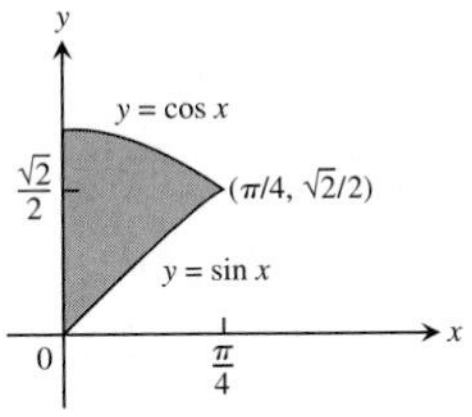

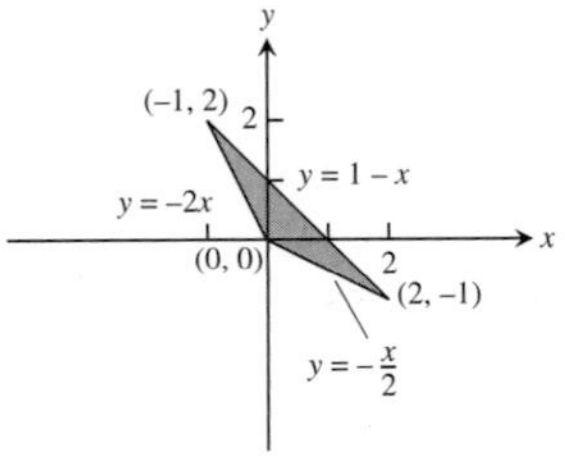

15. **(a)** 0 **(b)** $4/\pi^2$ **17.** $8/3$ **19.** $\bar{x} = 5/14, \bar{y} = 38/35$

21. $\bar{x} = 64/35, \bar{y} = 5/7$ **23.** $\bar{x} = 0, \bar{y} = 4/(3\pi)$

25. $\bar{x} = \bar{y} = 4a/(3\pi)$ **27.** $I_x = I_y = 4\pi, I_0 = 8\pi$

29. $\bar{x} = -1, \bar{y} = 1/4$ **31.** $I_x = 64/105, R_x = 2\sqrt{2/7}$

33. $\bar{x} = 3/8, \bar{y} = 17/16$

35. $\bar{x} = 11/3, \bar{y} = 14/27, I_y = 432, R_y = 4$

37. $\bar{x} = 0, \bar{y} = 13/31, I_y = 7/5, R_y = \sqrt{21/31}$

39. $\bar{x} = 0, \bar{y} = 7/10; I_x = 9/10, I_y = 3/10, I_0 = 6/5;$
$R_x = 3\sqrt{6}/10, R_y = 3\sqrt{2}/10, R_0 = 3\sqrt{2}/5$

41. $40{,}000(1 - e^{-2})\ln(7/2) \approx 43{,}329$

43. If $0 < a \le 5/2$, then the appliance will have to be tipped more than 45° to fall over.

45. $(\bar{x}, \bar{y}) = (2/\pi, 0)$ **47.** **(a)** $3/2$ **(b)** They are the same.

53. **(a)** $(7/5, 31/10)$ **(b)** $(19/7, 18/7)$ **(c)** $(9/2, 19/8)$
(d) $(11/4, 43/16)$

55. For the center of mass to be on the common boundary, $h = a\sqrt{2}$. For the center of mass to be inside T, $h > a\sqrt{2}$.

Section 15.3, pp. 1097–1098

1. $\pi/2$ **3.** $\pi/8$ **5.** πa^2 **7.** 36 **9.** $(1 - \ln 2)\pi$

11. $(2\ln 2 - 1)(\pi/2)$ **13.** $(\pi/2) + 1$ **15.** $\pi(\ln 4 - 1)$

17. $2(\pi - 1)$ **19.** 12π **21.** $(3\pi/8) + 1$ **23.** 4

25. $6\sqrt{3} - 2\pi$ **27.** $\bar{x} = 5/6, \bar{y} = 0$ **29.** $\dfrac{2a}{3}$ **31.** $\dfrac{2a}{3}$

33. $2\pi\left(2 - \sqrt{e}\right)$ **35.** $\dfrac{4}{3} + \dfrac{5\pi}{8}$ **37.** **(a)** $\dfrac{\sqrt{\pi}}{2}$ **(b)** 1

39. $\pi \ln 4$, no **41.** $\dfrac{1}{2}(a^2 + 2h^2)$

Section 15.4, pp. 1106–1109

1. $1/6$

3. $\int_0^1 \int_0^{2-2x} \int_0^{3-3x-3y/2} dz\,dy\,dx$, $\int_0^2 \int_0^{1-y/2} \int_0^{3-3x-3y/2} dz\,dx\,dy$,
$\int_0^1 \int_0^{3-3x} \int_0^{2-2x-2z/3} dy\,dz\,dx$, $\int_0^3 \int_0^{1-z/3} \int_0^{2-2x-2z/3} dy\,dx\,dz$,
$\int_0^2 \int_0^{3-3y/2} \int_0^{1-y/2-z/3} dx\,dz\,dy$,
$\int_0^3 \int_0^{2-2z/3} \int_0^{1-y/2-z/3} dx\,dy\,dz$.
The value of all six integrals is 1.

5. $\int_{-2}^2 \int_{-\sqrt{4-x^2}}^{\sqrt{4-x^2}} \int_{x^2+y^2}^{8-x^2-y^2} 1\,dz\,dx\,dy$, $\int_{-2}^2 \int_{-\sqrt{4-y^2}}^{\sqrt{4-y^2}} \int_{x^2+y^2}^{8-x^2-y^2} 1\,dz\,dx\,dy$,
$\int_{-2}^2 \int_4^{8-y^2} \int_{-\sqrt{8-z-y^2}}^{\sqrt{8-z-y^2}} 1\,dx\,dz\,dy + \int_{-2}^2 \int_{y^2}^4 \int_{-\sqrt{z-y^2}}^{\sqrt{z-y^2}} 1\,dx\,dz\,dy$,
$\int_4^8 \int_{-\sqrt{8-z}}^{\sqrt{8-z}} \int_{-\sqrt{8-z-y^2}}^{\sqrt{8-z-y^2}} 1\,dx\,dy\,dz + \int_0^4 \int_{-\sqrt{z}}^{\sqrt{z}} \int_{-\sqrt{z-y^2}}^{\sqrt{z-y^2}} 1\,dx\,dy\,dz$,
$\int_{-2}^2 \int_4^{8-x^2} \int_{-\sqrt{8-z-x^2}}^{\sqrt{8-z-x^2}} 1\,dy\,dz\,dx + \int_{-2}^2 \int_{x^2}^4 \int_{-\sqrt{z-x^2}}^{\sqrt{z-x^2}} 1\,dy\,dz\,dx$,
$\int_4^8 \int_{-\sqrt{8-z}}^{\sqrt{8-z}} \int_{-\sqrt{8-z-x^2}}^{\sqrt{8-z-x^2}} 1\,dy\,dx\,dz + \int_0^4 \int_{-\sqrt{z}}^{\sqrt{z}} \int_{-\sqrt{z-x^2}}^{\sqrt{z-x^2}} 1\,dy\,dx\,dz$.
The value of all six integrals is 16π.

7. 1 **9.** 1 **11.** $\frac{\pi^3}{2}(1 - \cos 1)$ **13.** 18 **15.** 7/6

17. 0 **19.** $\frac{1}{2} - \frac{\pi}{8}$

21. (a) $\int_{-1}^1 \int_0^{1-x^2} \int_{x^2}^{1-z} dy\,dz\,dx$ **(b)** $\int_0^1 \int_{-\sqrt{1-z}}^{\sqrt{1-z}} \int_{x^2}^{1-z} dy\,dx\,dz$
(c) $\int_0^1 \int_0^{1-z} \int_{-\sqrt{y}}^{\sqrt{y}} dx\,dy\,dz$ **(d)** $\int_0^1 \int_0^{1-y} \int_{-\sqrt{y}}^{\sqrt{y}} dx\,dz\,dy$
(e) $\int_0^1 \int_{-\sqrt{y}}^{\sqrt{y}} \int_0^{1-y} dz\,dx\,dy$

23. 2/3 **25.** 20/3 **27.** 1 **29.** 16/3 **31.** $8\pi - \frac{32}{3}$
33. 2 **35.** 4π **37.** 31/3 **39.** 1 **41.** 2 sin 4 **43.** 4
45. $a = 3$ or $a = 13/3$
47. The domain is the set of all point (x, y, z) such that $4x^2 + 4y^2 + z^2 \le 4$.

Section 15.5, pp. 1112–1114

1. $R_x = \sqrt{\frac{b^2 + c^2}{12}}$, $R_y = \sqrt{\frac{a^2 + c^2}{12}}$, $R_z = \sqrt{\frac{a^2 + b^2}{12}}$

3. $I_x = \frac{M}{3}(b^2 + c^2)$, $I_y = \frac{M}{3}(a^2 + c^2)$, $I_z = \frac{M}{3}(a^2 + b^2)$

5. $\bar{x} = \bar{y} = 0, \bar{z} = 12/5$, $I_x = 7904/105 \approx 75.28$, $I_y = 4832/63 \approx 76.70$, $I_z = 256/45 \approx 5.69$

7. (a) $\bar{x} = \bar{y} = 0, \bar{z} = 8/3$ **(b)** $c = 2\sqrt{2}$

9. $I_L = 1386$, $R_L = \sqrt{\frac{77}{2}}$ **11.** $I_L = \frac{40}{3}$, $R_L = \sqrt{\frac{5}{3}}$

13. (a) 4/3 **(b)** $\bar{x} = 4/5, \bar{y} = \bar{z} = 2/5$
15. (a) 5/2 **(b)** $\bar{x} = \bar{y} = \bar{z} = 8/15$ **(c)** $I_x = I_y = I_z = 11/6$
(d) $R_x = R_y = R_z = \sqrt{\frac{11}{15}}$ **17.** 3 **19. (a)** $\frac{4}{3}g$ **(b)** $\frac{4}{3}g$

23. (a) $I_{\text{c.m.}} = \frac{abc(a^2 + b^2)}{12}$, $R_{\text{c.m.}} = \sqrt{\frac{a^2 + b^2}{12}}$
(b) $I_L = \frac{abc(a^2 + 7b^2)}{3}$, $R_L = \sqrt{\frac{a^2 + 7b^2}{3}}$

27. (a) $h = a\sqrt{3}$ **(b)** $h = a\sqrt{2}$

Section 15.6, pp. 1124–1128

1. $\frac{4\pi(\sqrt{2} - 1)}{3}$ **3.** $\frac{17\pi}{5}$ **5.** $\pi(6\sqrt{2} - 8)$ **7.** $\frac{3\pi}{10}$
9. $\pi/3$

11. (a) $\int_0^{2\pi} \int_0^1 \int_0^{\sqrt{4-r^2}} r\,dz\,dr\,d\theta$
(b) $\int_0^{2\pi} \int_0^{\sqrt{3}} \int_0^1 r\,dr\,dz\,d\theta + \int_0^{2\pi} \int_{\sqrt{3}}^2 \int_0^{\sqrt{4-z^2}} r\,dr\,dz\,d\theta$
(c) $\int_0^1 \int_0^{\sqrt{4-r^2}} \int_0^{2\pi} r\,d\theta\,dz\,dr$

13. $\int_{-\pi/2}^{\pi/2} \int_0^{\cos\theta} \int_0^{3r^2} f(r, \theta, z)\,dz\,r\,dr\,d\theta$

15. $\int_0^{\pi} \int_0^{2\sin\theta} \int_0^{4-r\sin\theta} f(r, \theta, z)\,dz\,r\,dr\,d\theta$

17. $\int_{-\pi/2}^{\pi/2} \int_1^{1+\cos\theta} \int_0^4 f(r, \theta, z)\,dz\,r\,dr\,d\theta$

19. $\int_0^{\pi/4} \int_0^{\sec\theta} \int_0^{2-r\sin\theta} f(r, \theta, z)\,dz\,r\,dr\,d\theta$ **21.** π^2 **23.** $\pi/3$

25. 5π **27.** 2π **29.** $\left(\frac{8 - 5\sqrt{2}}{2}\right)\pi$

31. (a) $\int_0^{2\pi} \int_0^{\pi/6} \int_0^2 \rho^2 \sin\phi\,d\rho\,d\phi\,d\theta +$
$\int_0^{2\pi} \int_{\pi/6}^{\pi/2} \int_0^{\csc\phi} \rho^2 \sin\phi\,d\rho\,d\phi\,d\theta$

(b) $\int_0^{2\pi}\int_1^2\int_{\pi/6}^{\sin^{-1}(1/\rho)} \rho^2 \sin\phi\, d\phi\, d\rho\, d\theta + \int_0^{2\pi}\int_0^2\int_0^{\pi/6} \rho^2 \sin\phi\, d\phi\, d\rho\, d\theta$

33. $\int_0^{2\pi}\int_0^{\pi/2}\int_{\cos\phi}^{2} \rho^2 \sin\phi\, d\rho\, d\phi\, d\theta = \frac{31\pi}{6}$

35. $\int_0^{2\pi}\int_0^{\pi}\int_0^{1-\cos\phi} \rho^2 \sin\phi\, d\rho\, d\phi\, d\theta = \frac{8\pi}{3}$

37. $\int_0^{2\pi}\int_{\pi/4}^{\pi/2}\int_0^{2\cos\phi} \rho^2 \sin\phi\, d\rho\, d\phi\, d\theta = \frac{\pi}{3}$

39. **(a)** $8\int_0^{\pi/2}\int_0^{\pi/2}\int_0^{2} \rho^2 \sin\phi\, d\rho\, d\phi\, d\theta$

(b) $8\int_0^{\pi/2}\int_0^{2}\int_0^{\sqrt{4-r^2}} r\, dz\, dr\, d\theta$

(c) $8\int_0^{2}\int_0^{\sqrt{4-x^2}}\int_0^{\sqrt{4-x^2-y^2}} dz\, dy\, dx$

41. **(a)** $\int_0^{2\pi}\int_0^{\pi/3}\int_{\sec\phi}^{2} \rho^2 \sin\phi\, d\rho\, d\phi\, d\theta$

(b) $\int_0^{2\pi}\int_0^{\sqrt{3}}\int_1^{\sqrt{4-r^2}} r\, dz\, dr\, d\theta$

(c) $\int_{-\sqrt{3}}^{\sqrt{3}}\int_{-\sqrt{3-x^2}}^{\sqrt{3-x^2}}\int_1^{\sqrt{4-x^2-y^2}} dz\, dy\, dx$ **(d)** $5\pi/3$

43. $8\pi/3$ **45.** $9/4$ **47.** $\frac{3\pi-4}{18}$ **49.** $\frac{2\pi a^3}{3}$ **51.** $5\pi/3$

53. $\pi/2$ **55.** $\frac{4(2\sqrt{2}-1)\pi}{3}$ **57.** 16π **59.** $5\pi/2$

61. $\frac{4\pi(8-3\sqrt{3})}{3}$ **63.** $2/3$ **65.** $3/4$

67. $\bar{x} = \bar{y} = 0, \bar{z} = 3/8$ **69.** $(\bar{x}, \bar{y}, \bar{z}) = (0, 0, 3/8)$

71. $\bar{x} = \bar{y} = 0, \bar{z} = 5/6$ **73.** $I_z = 30\pi, R_z = \sqrt{\frac{5}{2}}$

75. $I_x = \pi/4$ **77.** $\frac{a^4 h\pi}{10}$

79. **(a)** $(\bar{x}, \bar{y}, \bar{z}) = \left(0, 0, \frac{4}{5}\right), I_z = \frac{\pi}{12}, R_z = \sqrt{\frac{1}{3}}$

(b) $(\bar{x}, \bar{y}, \bar{z}) = \left(0, 0, \frac{5}{6}\right), I_z = \frac{\pi}{14}, R_z = \sqrt{\frac{5}{14}}$

83. $(\bar{x}, \bar{y}, \bar{z}) = \left(0, 0, \frac{2h^2+3h}{3h+6}\right), I_z = \frac{\pi a^4(h^2+2h)}{4}, R_z = \frac{a}{\sqrt{2}}$

85. $\frac{3M}{\pi R^3}$

89. The surface's equation $r = f(z)$ tells us that the point $(r, \theta, z) = (f(z), \theta, z)$ will lie on the surface for all θ. In particular, $(f(z), \theta + \pi, z)$ lies on the surface whenever $(f(z), \theta, z)$ lies on the surface, so the surface is symmetric with respect to the z-axis.

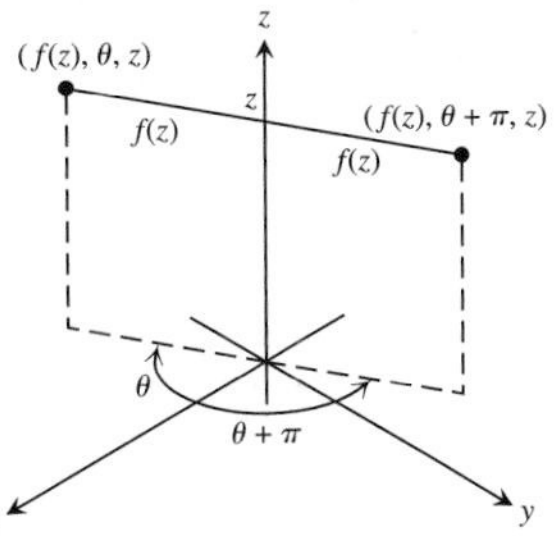

Section 15.7, pp. 1135–1137

1. **(a)** $x = \frac{u+v}{3}, y = \frac{v-2u}{3}; \frac{1}{3}$
(b) Triangular region with boundaries $u = 0, v = 0$, and $u + v = 3$

3. **(a)** $x = \frac{1}{5}(2u - v), y = \frac{1}{10}(3v - u); \frac{1}{10}$
(b) Triangular region with boundaries $3v = u, v = 2u$, and $3u + v = 10$

7. $64/5$ **9.** $\int_1^2\int_1^3 (u+v)\frac{2u}{v}\, du\, dv = 8 + \frac{52}{3}\ln 2$

11. $\frac{\pi ab(a^2+b^2)}{4}$ **13.** $\frac{1}{3}\left(1 + \frac{3}{e^2}\right) \approx 0.4687$

15. **(a)** $\begin{vmatrix} \cos v & -u\sin v \\ \sin v & u\cos v \end{vmatrix} = u\cos^2 v + u\sin^2 v = u$

(b) $\begin{vmatrix} \sin v & u\cos v \\ \cos v & -u\sin v \end{vmatrix} = -u\sin^2 v - u\cos^2 v = -u$

19. 12 **21.** $\frac{a^2b^2c^2}{6}$

Practice Exercises, pp. 1138–1140

1. $9e - 9$

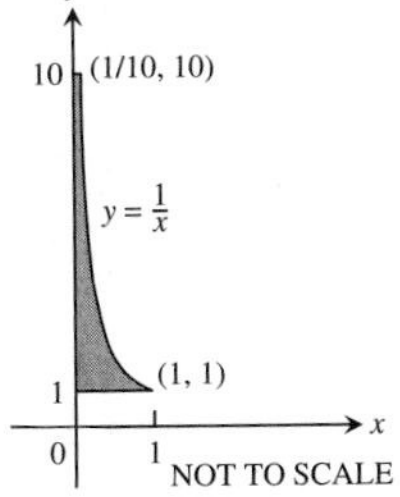

3. $9/2$

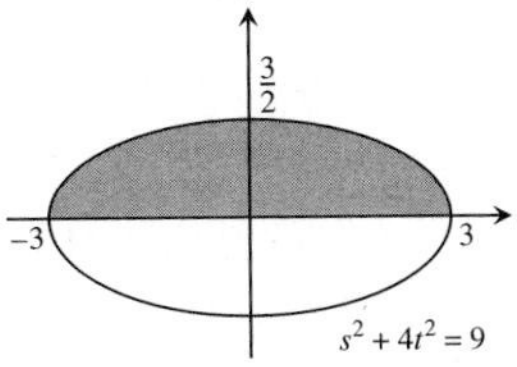

5. $\int_{-2}^{0}\int_{2x+4}^{4-x^2} dy\,dx = \frac{4}{3}$

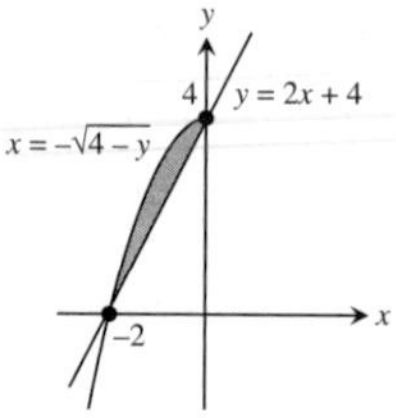

7. $\int_{-3}^{3}\int_{0}^{(1/2)\sqrt{9-x^2}} y\,dy\,dx = \frac{9}{2}$

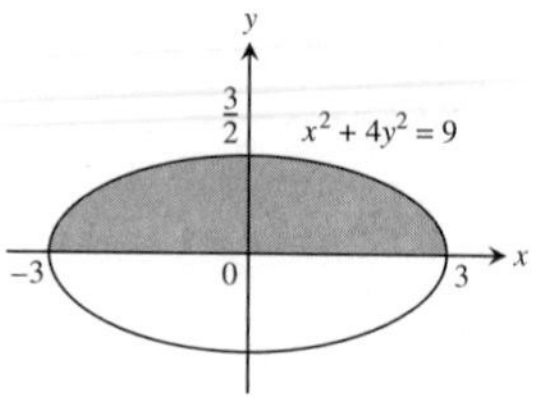

9. $\sin 4$ **11.** $\frac{\ln 17}{4}$ **13.** $4/3$ **15.** $4/3$ **17.** $1/4$

19. $\bar{x} = \bar{y} = \frac{1}{2 - \ln 4}$ **21.** $I_0 = 104$ **23.** $I_x = 2\delta, R_x = \sqrt{\frac{2}{3}}$

25. $M = 4, M_x = 0, M_y = 0$ **27.** π **29.** $\bar{x} = \frac{3\sqrt{3}}{\pi}, \bar{y} = 0$

31. **(a)** $\bar{x} = \frac{15\pi + 32}{6\pi + 48}, \bar{y} = 0$

(b)

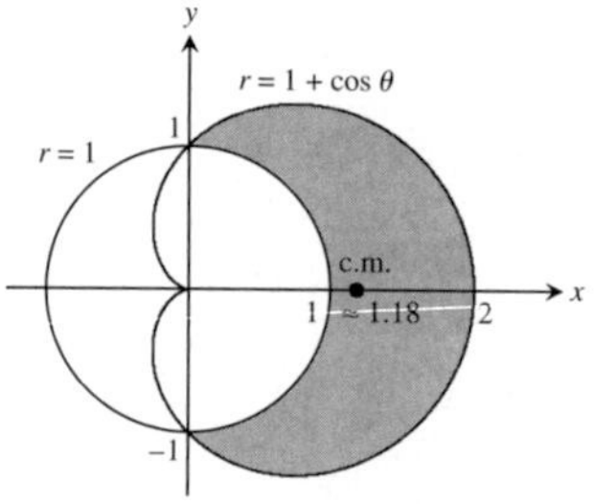

33. $\frac{\pi - 2}{4}$ **35.** 0 **37.** $8/35$ **39.** $\pi/2$ **41.** $\frac{2(31 - 3^{5/2})}{3}$

43. **(a)** $\int_{-\sqrt{2}}^{\sqrt{2}}\int_{-\sqrt{2-y^2}}^{\sqrt{2-y^2}}\int_{\sqrt{x^2+y^2}}^{\sqrt{4-x^2-y^2}} 3\,dz\,dx\,dy$

(b) $\int_0^{2\pi}\int_0^{\pi/4}\int_0^{2} 3\,\rho^2 \sin\phi\,d\rho\,d\phi\,d\theta$ **(c)** $2\pi\left(8 - 4\sqrt{2}\right)$

45. $\int_0^{2\pi}\int_0^{\pi/4}\int_0^{\sec\phi} \rho^2 \sin\phi\,d\rho\,d\phi\,d\theta = \frac{\pi}{3}$

47. $\int_0^1\int_{\sqrt{1-x^2}}^{\sqrt{3-x^2}}\int_1^{\sqrt{4-x^2-y^2}} z^2 xy\,dz\,dy\,dx$

$+ \int_1^{\sqrt{3}}\int_0^{\sqrt{3-x^2}}\int_1^{\sqrt{4-x^2-y^2}} z^2 xy\,dz\,dy\,dx$

49. **(a)** $\frac{8\pi\left(4\sqrt{2} - 5\right)}{3}$ **(b)** $\frac{8\pi\left(4\sqrt{2} - 5\right)}{3}$

51. $I_z = \frac{8\pi\delta(b^5 - a^5)}{15}$

Additional and Advanced Exercises, pp. 1140–1142

1. **(a)** $\int_{-3}^{2}\int_{x}^{6-x^2} x^2\,dy\,dx$ **(b)** $\int_{-3}^{2}\int_{x}^{6-x^2}\int_0^{x^2} dz\,dy\,dx$

(c) $125/4$

3. 2π **5.** $3\pi/2$ **7.** **(a)** Hole radius $= 1$, sphere radius $= 2$

(b) $4\sqrt{3}\pi$ **9.** $\pi/4$ **11.** $\ln\left(\frac{b}{a}\right)$ **15.** $1/\sqrt[4]{3}$

17. Mass $= a^2 \cos^{-1}\left(\frac{b}{a}\right) - b\sqrt{a^2 - b^2}$,

$I_0 = \frac{a^4}{2}\cos^{-1}\left(\frac{b}{a}\right) - \frac{b^3}{2}\sqrt{a^2 - b^2} - \frac{b^3}{6}(a^2 - b^2)^{3/2}$

19. $\frac{1}{ab}(e^{a^2b^2} - 1)$ **21.** **(b)** 1 **(c)** 0

25. $h = \sqrt{20}$ in., $h = \sqrt{60}$ in. **27.** $2\pi\left[\frac{1}{3} - \left(\frac{1}{3}\right)\frac{\sqrt{2}}{2}\right]$

CHAPTER 16

Section 16.1, pp. 1147–1149

1. Graph (c) **3.** Graph (g) **5.** Graph (d) **7.** Graph (f)

9. $\sqrt{2}$ **11.** $\frac{13}{2}$ **13.** $3\sqrt{14}$ **15.** $\frac{1}{6}\left(5\sqrt{5} + 9\right)$

17. $\sqrt{3}\ln\left(\frac{b}{a}\right)$ **19.** $\frac{10\sqrt{5} - 2}{3}$ **21.** 8 **23.** $2\sqrt{2} - 1$

25. **(a)** $4\sqrt{2} - 2$ **(b)** $\sqrt{2} + \ln\left(1 + \sqrt{2}\right)$

27. $I_z = 2\pi\delta a^3, R_z = a$

29. **(a)** $I_z = 2\pi\sqrt{2}\delta, R_z = 1$ **(b)** $I_z = 4\pi\sqrt{2}\delta, R_z = 1$

31. $I_x = 2\pi - 2, R_x = 1$

Section 16.2, pp. 1158–1160

1. $\nabla f = -(x\mathbf{i} + y\mathbf{j} + z\mathbf{k})(x^2 + y^2 + z^2)^{-3/2}$

3. $\nabla g = -\left(\frac{2x}{x^2 + y^2}\right)\mathbf{i} - \left(\frac{2y}{x^2 + y^2}\right)\mathbf{j} + e^z\mathbf{k}$

5. $\mathbf{F} = -\frac{kx}{(x^2 + y^2)^{3/2}}\mathbf{i} - \frac{ky}{(x^2 + y^2)^{3/2}}\mathbf{j}$, any $k > 0$

7. **(a)** $9/2$ **(b)** $13/3$ **(c)** $9/2$

9. **(a)** $1/3$ **(b)** $-1/5$ **(c)** 0

11. **(a)** 2 **(b)** $3/2$ **(c)** $1/2$

13. $1/2$ **15.** $-\pi$ **17.** $69/4$ **19.** $-39/2$ **21.** $25/6$

23. **(a)** $\text{Circ}_1 = 0, \text{circ}_2 = 2\pi, \text{flux}_1 = 2\pi, \text{flux}_2 = 0$

(b) $\text{Circ}_1 = 0, \text{circ}_2 = 8\pi, \text{flux}_1 = 8\pi, \text{flux}_2 = 0$

25. Circ $= 0$, flux $= a^2\pi$ **27.** Circ $= a^2\pi$, flux $= 0$

29. **(a)** $-\frac{\pi}{2}$ **(b)** 0 **(c)** 1

31.

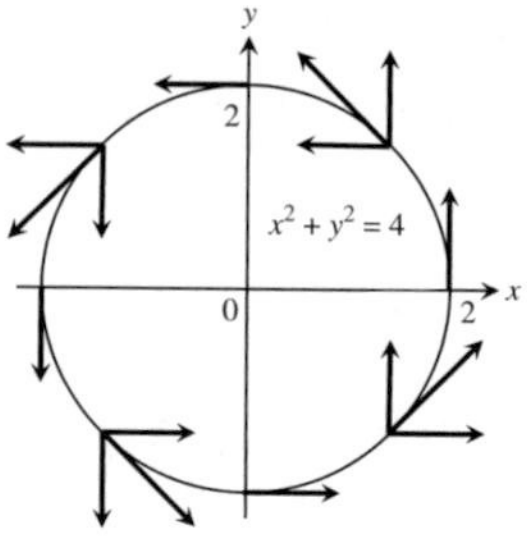

33. **(a)** $\mathbf{G} = -y\mathbf{i} + x\mathbf{j}$ **(b)** $\mathbf{G} = \sqrt{x^2 + y^2}\,\mathbf{F}$

35. $\mathbf{F} = -\dfrac{x\mathbf{i} + y\mathbf{j}}{\sqrt{x^2 + y^2}}$ **37.** 48 **39.** π **41.** 0 **43.** $\dfrac{1}{2}$

Section 16.3, pp. 1168–1169

1. Conservative **3.** Not conservative **5.** Not conversative

7. $f(x, y, z) = x^2 + \dfrac{3y^2}{2} + 2z^2 + C$

9. $f(x, y, z) = xe^{y+2z} + C$

11. $f(x, y, z) = x \ln x - x + \tan(x + y) + \dfrac{1}{2}\ln(y^2 + z^2) + C$

13. 49 **15.** -16 **17.** 1 **19.** $9 \ln 2$ **21.** 0 **23.** -3

27. $\mathbf{F} = \nabla\left(\dfrac{x^2 - 1}{y}\right)$ **29.** **(a)** 1 **(b)** 1 **(c)** 1

31. **(a)** 2 **(b)** 2 **33.** **(a)** $c = b = 2a$ **(b)** $c = b = 2$

35. It does not matter what path you use. The work will be the same on any path because the field is conservative.

37. The force $\mathbf{F}$ is conservative because all partial derivatives of M, N, and P are zero. $f(x, y, z) = ax + by + cz + C$; $A = (xa, ya, za)$ and $B = (xb, yb, zb)$. Therefore, $\int \mathbf{F} \cdot d\mathbf{r} = f(B) - f(A) = a(xb - xa) + b(yb - ya) + c(zb - za) = \mathbf{F} \cdot \overrightarrow{AB}$.

Section 16.4, pp. 1179–1181

1. Flux $= 0$, circ $= 2\pi a^2$ **3.** Flux $= -\pi a^2$, circ $= 0$

5. Flux $= 2$, circ $= 0$ **7.** Flux $= -9$, circ $= 9$

9. Flux $= 1/2$, circ $= 1/2$ **11.** Flux $= 1/5$, circ $= -1/12$

13. 0 **15.** $2/33$ **17.** 0 **19.** -16π **21.** πa^2 **23.** $\dfrac{3}{8}\pi$

25. **(a)** 4π if C is traversed counterclockwise
(b) $(h - k)$(area of the region) **35.** **(a)** 0

Section 16.5, pp. 1190–1192

1. $\dfrac{13}{3}\pi$ **3.** 4 **5.** $6\sqrt{6} - 2\sqrt{2}$ **7.** $\pi\sqrt{c^2 + 1}$

9. $\dfrac{\pi}{6}\left(17\sqrt{17} - 5\sqrt{5}\right)$ **11.** $3 + 2 \ln 2$ **13.** $9a^3$

15. $\dfrac{abc}{4}(ab + ac + bc)$ **17.** 2 **19.** 18 **21.** $\dfrac{\pi a^3}{6}$

23. $\dfrac{\pi a^2}{4}$ **25.** $\dfrac{\pi a^3}{2}$ **27.** -32 **29.** -4 **31.** $3a^4$

33. $\left(\dfrac{a}{2}, \dfrac{a}{2}, \dfrac{a}{2}\right)$

35. $(\bar{x}, \bar{y}, \bar{z}) = \left(0, 0, \dfrac{14}{9}\right)$, $I_z = \dfrac{15\pi\sqrt{2}}{2}\delta$, $R_z = \dfrac{\sqrt{10}}{2}$

37. **(a)** $\dfrac{8\pi}{3}a^4\delta$ **(b)** $\dfrac{20\pi}{3}a^4\delta$ **39.** $\dfrac{\pi}{6}\left(13\sqrt{13} - 1\right)$

41. $5\pi\sqrt{2}$ **43.** $\dfrac{2}{3}\left(5\sqrt{5} - 1\right)$

Section 16.6, pp. 1199–1201

1. $\mathbf{r}(r, \theta) = (r\cos\theta)\mathbf{i} + (r\sin\theta)\mathbf{j} + r^2\mathbf{k}$, $0 \le r \le 2$, $0 \le \theta \le 2\pi$

3. $\mathbf{r}(r, \theta) = (r\cos\theta)\mathbf{i} + (r\sin\theta)\mathbf{j} + (r/2)\mathbf{k}$, $0 \le r \le 6$, $0 \le \theta \le \pi/2$

5. $\mathbf{r}(r, \theta) = (r\cos\theta)\mathbf{i} + (r\sin\theta)\mathbf{j} + \sqrt{9 - r^2}\,\mathbf{k}$, $0 \le r \le 3\sqrt{2}/2$, $0 \le \theta \le 2\pi$; Also: $\mathbf{r}(\phi, \theta) = (3\sin\phi\cos\theta)\mathbf{i} + (3\sin\phi\sin\theta)\mathbf{j} + (3\cos\phi)\mathbf{k}$, $0 \le \phi \le \pi/4$, $0 \le \theta \le 2\pi$

7. $\mathbf{r}(\phi, \theta) = \left(\sqrt{3}\sin\phi\cos\theta\right)\mathbf{i} + \left(\sqrt{3}\sin\phi\sin\theta\right)\mathbf{j} + \left(\sqrt{3}\cos\phi\right)\mathbf{k}$, $\pi/3 \le \phi \le 2\pi/3$, $0 \le \theta \le 2\pi$

9. $\mathbf{r}(x, y) = x\mathbf{i} + y\mathbf{j} + (4 - y^2)\mathbf{k}$, $0 \le x \le 2$, $-2 \le y \le 2$

11. $\mathbf{r}(u, v) = u\mathbf{i} + (3\cos v)\mathbf{j} + (3\sin v)\mathbf{k}$, $0 \le u \le 3$, $0 \le v \le 2\pi$

13. **(a)** $\mathbf{r}(r, \theta) = (r\cos\theta)\mathbf{i} + (r\sin\theta)\mathbf{j} + (1 - r\cos\theta - r\sin\theta)\mathbf{k}$, $0 \le r \le 3$, $0 \le \theta \le 2\pi$
(b) $\mathbf{r}(u, v) = (1 - u\cos v - u\sin v)\mathbf{i} + (u\cos v)\mathbf{j} + (u\sin v)\mathbf{k}$, $0 \le u \le 3$, $0 \le v \le 2\pi$

15. $\mathbf{r}(u, v) = (4\cos^2 v)\mathbf{i} + u\mathbf{j} + (4\cos v\sin v)\mathbf{k}$, $0 \le u \le 3$, $-(\pi/2) \le v \le (\pi/2)$; Another way: $\mathbf{r}(u, v) = (2 + 2\cos v)\mathbf{i} + u\mathbf{j} + (2\sin v)\mathbf{k}$, $0 \le u \le 3$, $0 \le v \le 2\pi$

17. $\displaystyle\int_0^{2\pi}\int_0^1 \frac{\sqrt{5}}{2} r\,dr\,d\theta = \frac{\pi\sqrt{5}}{2}$

19. $\displaystyle\int_0^{2\pi}\int_1^3 r\sqrt{5}\,dr\,d\theta = 8\pi\sqrt{5}$ **21.** $\displaystyle\int_0^{2\pi}\int_1^4 1\,du\,dv = 6\pi$

23. $\displaystyle\int_0^{2\pi}\int_0^1 u\sqrt{4u^2 + 1}\,du\,dv = \frac{\left(5\sqrt{5} - 1\right)}{6}\pi$

25. $\displaystyle\int_0^{2\pi}\int_{\pi/4}^{\pi} 2\sin\phi\,d\phi\,d\theta = \left(4 + 2\sqrt{2}\right)\pi$

27. $\displaystyle\iint_S x\,d\sigma = \int_0^3\int_0^2 u\sqrt{4u^2 + 1}\,du\,dv = \frac{17\sqrt{17} - 1}{4}$

29. $\displaystyle\iint_S x^2\,d\sigma = \int_0^{2\pi}\int_0^{\pi} \sin^3\phi\cos^2\theta\,d\phi\,d\theta = \frac{4\pi}{3}$

31. $\displaystyle\iint_S z\,d\sigma = \int_0^1\int_0^1 (4 - u - v)\sqrt{3}\,dv\,du = 3\sqrt{3}$
(for $x = u, y = v$)

33. $\displaystyle\iint_S x^2\sqrt{5 - 4z}\,d\sigma = \int_0^1\int_0^{2\pi} u^2\cos^2 v \cdot \sqrt{4u^2 + 1} \cdot u\sqrt{4u^2 + 1}\,dv\,du = \int_0^1\int_0^{2\pi} u^3(4u^2 + 1)\cos^2 v\,dv\,du = \frac{11\pi}{12}$

35. -32 **37.** $\pi a^3/6$ **39.** $13a^4/6$ **41.** $2\pi/3$ **43.** $-73\pi/6$

45. $(a/2, a/2, a/2)$ **47.** $8\delta\pi a^4/3$

49.

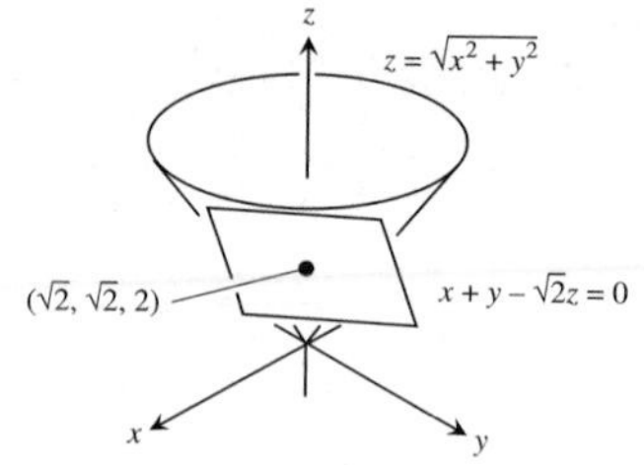

51.

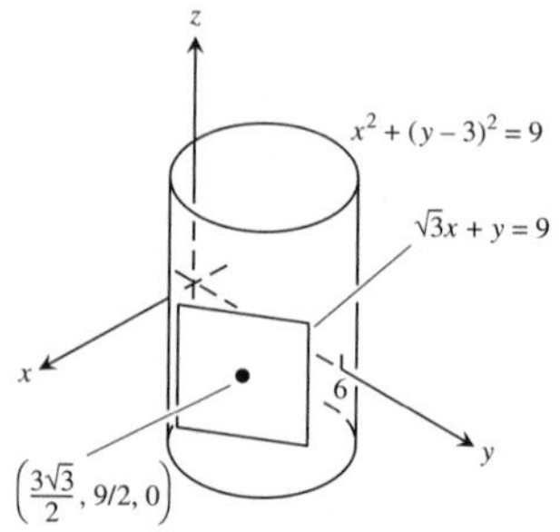

55. (b) $A = \int_0^{2\pi}\int_0^{\pi} [a^2b^2 \sin^2\phi \cos^2\phi + b^2c^2 \cos^4\phi \cos^2\theta + a^2c^2 \cos^4\phi \sin^2\theta]^{1/2}\, d\phi\, d\theta$

57. $x_0 x + y_0 y = 25$

Section 16.7, pp. 1209–1211

1. 4π **3.** $-5/6$ **5.** 0 **7.** -6π **9.** $2\pi a^2$ **13.** 12π
15. $-\pi/4$ **17.** -15π **25.** $16I_y + 16I_x$

Section 16.8, pp. 1220–1222

1. 0 **3.** 0 **5.** -16 **7.** -8π **9.** 3π **11.** $-40/3$
13. 12π **15.** $12\pi(4\sqrt{2} - 1)$
21. The integral's value never exceeds the surface area of S.

Practice Exercises, pp. 1223–1226

1. Path 1: $2\sqrt{3}$; path 2: $1 + 3\sqrt{2}$ **3.** $4a^2$ **5.** 0
7. $8\pi \sin(1)$ **9.** 0 **11.** $\pi\sqrt{3}$ **13.** $2\pi\left(1 - \frac{1}{\sqrt{2}}\right)$
15. $\frac{abc}{2}\sqrt{\frac{1}{a^2} + \frac{1}{b^2} + \frac{1}{c^2}}$ **17.** 50
19. $\mathbf{r}(\phi, \theta) = (6 \sin\phi \cos\theta)\mathbf{i} + (6 \sin\phi \sin\theta)\mathbf{j} + (6 \cos\phi)\mathbf{k}$, $\frac{\pi}{6} \le \phi \le \frac{2\pi}{3}, 0 \le \theta \le 2\pi$
21. $\mathbf{r}(r, \theta) = (r \cos\theta)\mathbf{i} + (r \sin\theta)\mathbf{j} + (1 + r)\mathbf{k}, 0 \le r \le 2$, $0 \le \theta \le 2\pi$
23. $\mathbf{r}(u, v) = (u \cos v)\mathbf{i} + 2u^2\mathbf{j} + (u \sin v)\mathbf{k}, 0 \le u \le 1$, $0 \le v \le \pi$
25. $\sqrt{6}$ **27.** $\pi\left[\sqrt{2} + \ln\left(1 + \sqrt{2}\right)\right]$ **29.** Conservative
31. Not conservative **33.** $f(x, y, z) = y^2 + yz + 2x + z$
35. Path 1: 2; path 2: 8/3 **37. (a)** $1 - e^{-2\pi}$ **(b)** $1 - e^{-2\pi}$
39. 0 **41. (a)** $4\sqrt{2} - 2$ **(b)** $\sqrt{2} + \ln\left(1 + \sqrt{2}\right)$
43. $(\bar{x}, \bar{y}, \bar{z}) = \left(1, \frac{16}{15}, \frac{2}{3}\right)$; $I_x = \frac{232}{45}, I_y = \frac{64}{15}, I_z = \frac{56}{9}$; $R_x = \frac{2\sqrt{29}}{3\sqrt{5}}, R_y = \frac{4\sqrt{2}}{\sqrt{15}}, R_z = \frac{2\sqrt{7}}{3}$
45. $\bar{z} = \frac{3}{2}, I_z = \frac{7\sqrt{3}}{3}, R_z = \sqrt{\frac{7}{3}}$
47. $(\bar{x}, \bar{y}, \bar{z}) = (0, 0, 49/12), I_z = 640\pi, R_z = 2\sqrt{2}$
49. Flux: 3/2; circ: $-1/2$ **53.** 3 **55.** $\frac{2\pi}{3}\left(7 - 8\sqrt{2}\right)$
57. 0 **59.** π

Additional and Advanced Exercises, pp. 1226–1228

1. 6π **3.** 2/3
5. (a) $\mathbf{F}(x, y, z) = z\mathbf{i} + x\mathbf{j} + y\mathbf{k}$ **(b)** $\mathbf{F}(x, y, z) = z\mathbf{i} + y\mathbf{k}$
(c) $\mathbf{F}(x, y, z) = z\mathbf{i}$
7. $\frac{16\pi R^3}{3}$ **9.** $a = 2, b = 1$. The minimum flux is -4.
11. (b) $\frac{16}{3}g$
(c) Work $= \left(\int_C gxy\, ds\right)\bar{y} = g\int_C xy^2\, ds = \frac{16}{3}g$
13. (c) $\frac{4}{3}\pi w$ **19.** False if $\mathbf{F} = y\mathbf{i} + x\mathbf{j}$

APPENDICES

Appendix A.5

1. (a) (14, 8) **(b)** $(-1, 8)$ **(c)** $(0, -5)$
3. (a) By reflecting z across the real axis
(b) By reflecting z across the imaginary axis
(c) By reflecting z in the real axis and then multiplying the length of the vector by $1/|z|^2$
5. (a) Points on the circle $x^2 + y^2 = 4$
(b) Points inside the circle $x^2 + y^2 = 4$
(c) Points outside the circle $x^2 + y^2 = 4$
7. Points on a circle of radius 1, center $(-1, 0)$
9. Points on the line $y = -x$ **11.** $4e^{2\pi i/3}$ **13.** $1e^{2\pi i/3}$
15. $\cos^4\theta - 6\cos^2\theta \sin^2\theta + \sin^4\theta$ **17.** $1, -\frac{1}{2} \pm \frac{\sqrt{3}}{2}i$
19. $2i, -\sqrt{3} - i, \sqrt{3} - i$ **21.** $\frac{\sqrt{6}}{2} \pm \frac{\sqrt{2}}{2}i, -\frac{\sqrt{6}}{2} \pm \frac{\sqrt{2}}{2}i$
23. $1 \pm \sqrt{3}i, -1 \pm \sqrt{3}i$

INDEX

Note: Page numbers in *italics* indicate figures; "t" indicates a table; "e" indicates an exercise.

Trigonometry Formulas

1. Definitions and Fundamental Identities

Sine: $\sin\theta = \dfrac{y}{r} = \dfrac{1}{\csc\theta}$

Cosine: $\cos\theta = \dfrac{x}{r} = \dfrac{1}{\sec\theta}$

Tangent: $\tan\theta = \dfrac{y}{x} = \dfrac{1}{\cot\theta}$

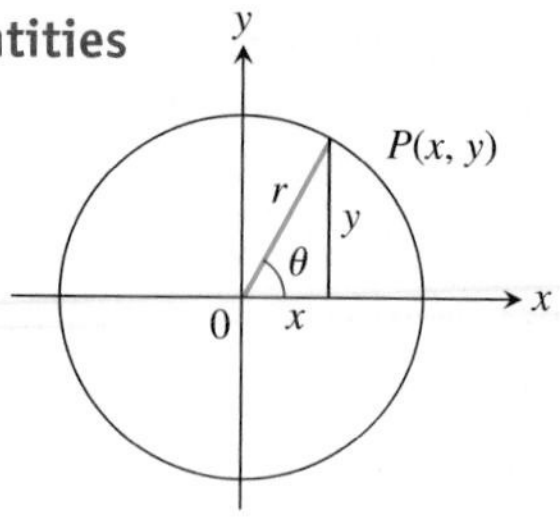

2. Identities

$\sin(-\theta) = -\sin\theta, \quad \cos(-\theta) = \cos\theta$

$\sin^2\theta + \cos^2\theta = 1, \quad \sec^2\theta = 1 + \tan^2\theta, \quad \csc^2\theta = 1 + \cot^2\theta$

$\sin 2\theta = 2\sin\theta\cos\theta, \quad \cos 2\theta = \cos^2\theta - \sin^2\theta$

$\cos^2\theta = \dfrac{1 + \cos 2\theta}{2}, \quad \sin^2\theta = \dfrac{1 - \cos 2\theta}{2}$

$\sin(A + B) = \sin A\cos B + \cos A\sin B$

$\sin(A - B) = \sin A\cos B - \cos A\sin B$

$\cos(A + B) = \cos A\cos B - \sin A\sin B$

$\cos(A - B) = \cos A\cos B + \sin A\sin B$

$\tan(A + B) = \dfrac{\tan A + \tan B}{1 - \tan A\tan B}$

$\tan(A - B) = \dfrac{\tan A - \tan B}{1 + \tan A\tan B}$

$\sin\left(A - \dfrac{\pi}{2}\right) = -\cos A, \quad \cos\left(A - \dfrac{\pi}{2}\right) = \sin A$

$\sin\left(A + \dfrac{\pi}{2}\right) = \cos A, \quad \cos\left(A + \dfrac{\pi}{2}\right) = -\sin A$

$\sin A\sin B = \dfrac{1}{2}\cos(A - B) - \dfrac{1}{2}\cos(A + B)$

$\cos A\cos B = \dfrac{1}{2}\cos(A - B) + \dfrac{1}{2}\cos(A + B)$

$\sin A\cos B = \dfrac{1}{2}\sin(A - B) + \dfrac{1}{2}\sin(A + B)$

$\sin A + \sin B = 2\sin\dfrac{1}{2}(A + B)\cos\dfrac{1}{2}(A - B)$

$\sin A - \sin B = 2\cos\dfrac{1}{2}(A + B)\sin\dfrac{1}{2}(A - B)$

$\cos A + \cos B = 2\cos\dfrac{1}{2}(A + B)\cos\dfrac{1}{2}(A - B)$

$\cos A - \cos B = -2\sin\dfrac{1}{2}(A + B)\sin\dfrac{1}{2}(A - B)$

Trigonometric Functions

Radian Measure

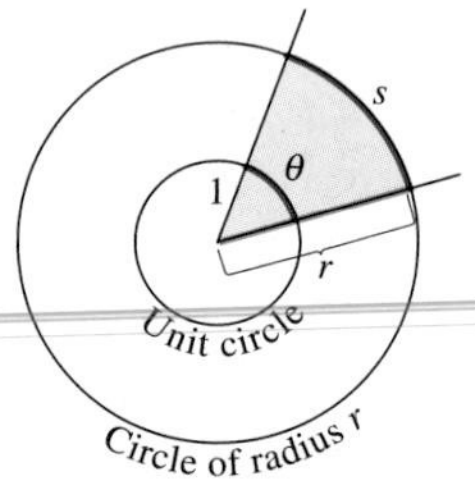

$\dfrac{s}{r} = \dfrac{\theta}{1} = \theta \quad \text{or} \quad \theta = \dfrac{s}{r},$

$180° = \pi$ radians.

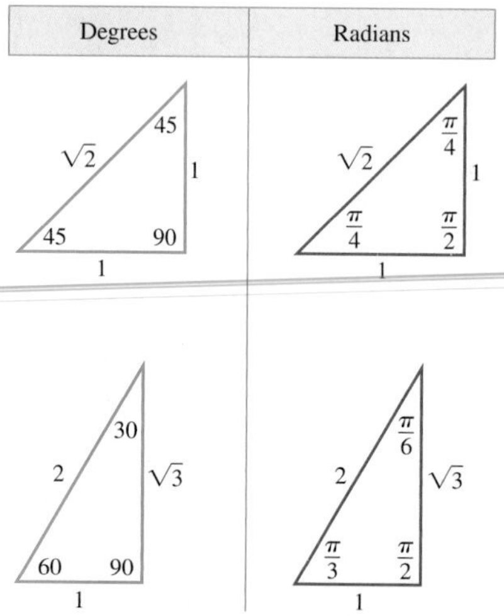

The angles of two common triangles, in degrees and radians.

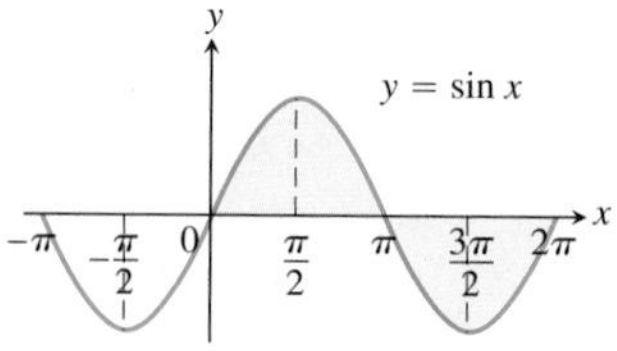

Domain: $(-\infty, \infty)$
Range: $[-1, 1]$

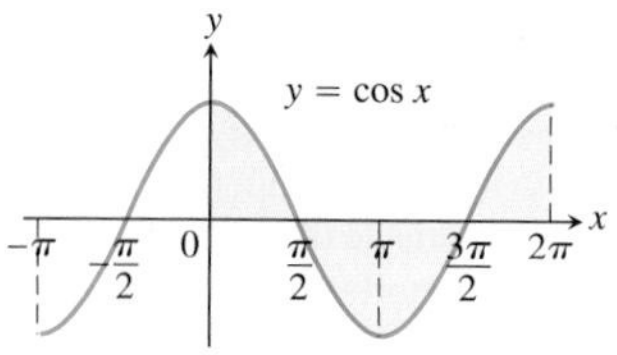

Domain: $(-\infty, \infty)$
Range: $[-1, 1]$

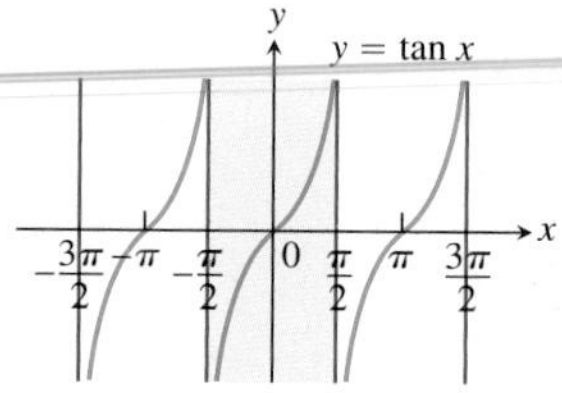

Domain: All real numbers except odd integer multiples of $\pi/2$
Range: $(-\infty, \infty)$

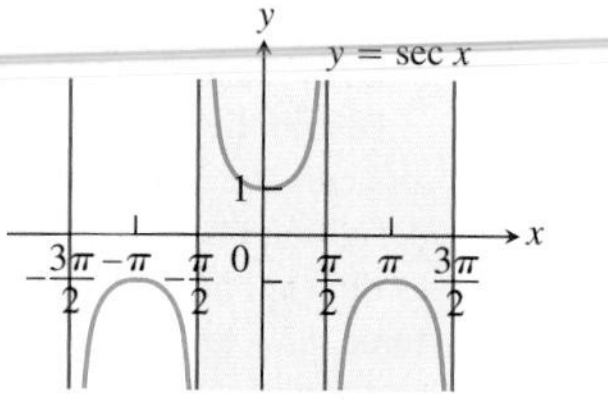

Domain: All real numbers except odd integer multiples of $\pi/2$
Range: $(-\infty, -1] \cup [1, \infty)$

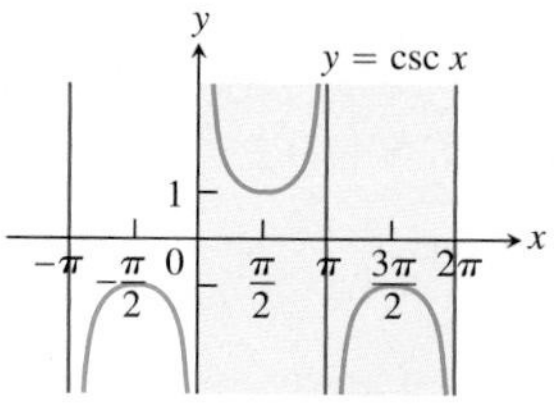

Domain: $x \neq 0, \pm\pi, \pm 2\pi, \ldots$
Range: $(-\infty, -1] \cup [1, \infty)$

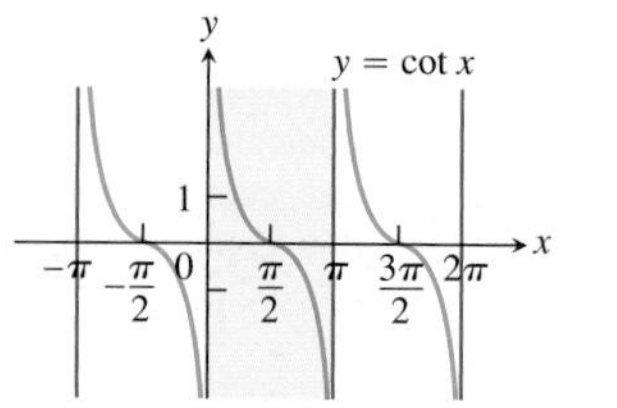

Domain: $x \neq 0, \pm\pi, \pm 2\pi, \ldots$
Range: $(-\infty, \infty)$

A Brief Table of Integrals

1. $\int u\,dv = uv - \int v\,du$

2. $\int a^u\,du = \frac{a^u}{\ln a} + C, \quad a \neq 1, \quad a > 0$

3. $\int \cos u\,du = \sin u + C$

4. $\int \sin u\,du = -\cos u + C$

5. $\int (ax+b)^n\,dx = \frac{(ax+b)^{n+1}}{a(n+1)} + C, \quad n \neq -1$

6. $\int (ax+b)^{-1}\,dx = \frac{1}{a}\ln|ax+b| + C$

7. $\int x(ax+b)^n\,dx = \frac{(ax+b)^{n+1}}{a^2}\left[\frac{ax+b}{n+2} - \frac{b}{n+1}\right] + C, \quad n \neq -1, -2$

8. $\int x(ax+b)^{-1}\,dx = \frac{x}{a} - \frac{b}{a^2}\ln|ax+b| + C$

9. $\int x(ax+b)^{-2}\,dx = \frac{1}{a^2}\left[\ln|ax+b| + \frac{b}{ax+b}\right] + C$

10. $\int \frac{dx}{x(ax+b)} = \frac{1}{b}\ln\left|\frac{x}{ax+b}\right| + C$

11. $\int \left(\sqrt{ax+b}\right)^n dx = \frac{2}{a}\frac{\left(\sqrt{ax+b}\right)^{n+2}}{n+2} + C, \quad n \neq -2$

12. $\int \frac{\sqrt{ax+b}}{x}\,dx = 2\sqrt{ax+b} + b\int \frac{dx}{x\sqrt{ax+b}}$

13. (a) $\int \frac{dx}{x\sqrt{ax-b}} = \frac{2}{\sqrt{b}}\tan^{-1}\sqrt{\frac{ax-b}{b}} + C$ (b) $\int \frac{dx}{x\sqrt{ax+b}} = \frac{1}{\sqrt{b}}\ln\left|\frac{\sqrt{ax+b}-\sqrt{b}}{\sqrt{ax+b}+\sqrt{b}}\right| + C$

14. $\int \frac{\sqrt{ax+b}}{x^2}\,dx = -\frac{\sqrt{ax+b}}{x} + \frac{a}{2}\int \frac{dx}{x\sqrt{ax+b}} + C$

15. $\int \frac{dx}{x^2\sqrt{ax+b}} = -\frac{\sqrt{ax+b}}{bx} - \frac{a}{2b}\int \frac{dx}{x\sqrt{ax+b}} + C$

16. $\int \frac{dx}{a^2+x^2} = \frac{1}{a}\tan^{-1}\frac{x}{a} + C$

17. $\int \frac{dx}{(a^2+x^2)^2} = \frac{x}{2a^2(a^2+x^2)} + \frac{1}{2a^3}\tan^{-1}\frac{x}{a} + C$

18. $\int \frac{dx}{a^2-x^2} = \frac{1}{2a}\ln\left|\frac{x+a}{x-a}\right| + C$

19. $\int \frac{dx}{(a^2-x^2)^2} = \frac{x}{2a^2(a^2-x^2)} + \frac{1}{4a^3}\ln\left|\frac{x+a}{x-a}\right| + C$

20. $\int \frac{dx}{\sqrt{a^2+x^2}} = \sinh^{-1}\frac{x}{a} + C = \ln\left(x + \sqrt{a^2+x^2}\right) + C$

21. $\int \sqrt{a^2+x^2}\,dx = \frac{x}{2}\sqrt{a^2+x^2} + \frac{a^2}{2}\ln\left(x + \sqrt{a^2+x^2}\right) + C$

22. $\int x^2\sqrt{a^2+x^2}\,dx = \frac{x}{8}(a^2+2x^2)\sqrt{a^2+x^2} - \frac{a^4}{8}\ln\left(x + \sqrt{a^2+x^2}\right) + C$

23. $\displaystyle\int \frac{\sqrt{a^2+x^2}}{x}\,dx = \sqrt{a^2+x^2} - a\ln\left|\frac{a+\sqrt{a^2+x^2}}{x}\right| + C$

24. $\displaystyle\int \frac{\sqrt{a^2+x^2}}{x^2}\,dx = \ln\left(x+\sqrt{a^2+x^2}\right) - \frac{\sqrt{a^2+x^2}}{x} + C$

25. $\displaystyle\int \frac{x^2}{\sqrt{a^2+x^2}}\,dx = -\frac{a^2}{2}\ln\left(x+\sqrt{a^2+x^2}\right) + \frac{x\sqrt{a^2+x^2}}{2} + C$

26. $\displaystyle\int \frac{dx}{x\sqrt{a^2+x^2}} = -\frac{1}{a}\ln\left|\frac{a+\sqrt{a^2+x^2}}{x}\right| + C$

27. $\displaystyle\int \frac{dx}{x^2\sqrt{a^2+x^2}} = -\frac{\sqrt{a^2+x^2}}{a^2x} + C$

28. $\displaystyle\int \frac{dx}{\sqrt{a^2-x^2}} = \sin^{-1}\frac{x}{a} + C$

29. $\displaystyle\int \sqrt{a^2-x^2}\,dx = \frac{x}{2}\sqrt{a^2-x^2} + \frac{a^2}{2}\sin^{-1}\frac{x}{a} + C$

30. $\displaystyle\int x^2\sqrt{a^2-x^2}\,dx = \frac{a^4}{8}\sin^{-1}\frac{x}{a} - \frac{1}{8}x\sqrt{a^2-x^2}\,(a^2-2x^2) + C$

31. $\displaystyle\int \frac{\sqrt{a^2-x^2}}{x}\,dx = \sqrt{a^2-x^2} - a\ln\left|\frac{a+\sqrt{a^2-x^2}}{x}\right| + C$

32. $\displaystyle\int \frac{\sqrt{a^2-x^2}}{x^2}\,dx = -\sin^{-1}\frac{x}{a} - \frac{\sqrt{a^2-x^2}}{x} + C$

33. $\displaystyle\int \frac{x^2}{\sqrt{a^2-x^2}}\,dx = \frac{a^2}{2}\sin^{-1}\frac{x}{a} - \frac{1}{2}x\sqrt{a^2-x^2} + C$

34. $\displaystyle\int \frac{dx}{x\sqrt{a^2-x^2}} = -\frac{1}{a}\ln\left|\frac{a+\sqrt{a^2-x^2}}{x}\right| + C$

35. $\displaystyle\int \frac{dx}{x^2\sqrt{a^2-x^2}} = -\frac{\sqrt{a^2-x^2}}{a^2x} + C$

36. $\displaystyle\int \frac{dx}{\sqrt{x^2-a^2}} = \cosh^{-1}\frac{x}{a} + C = \ln\left|x+\sqrt{x^2-a^2}\right| + C$

37. $\displaystyle\int \sqrt{x^2-a^2}\,dx = \frac{x}{2}\sqrt{x^2-a^2} - \frac{a^2}{2}\ln\left|x+\sqrt{x^2-a^2}\right| + C$

38. $\displaystyle\int \left(\sqrt{x^2-a^2}\right)^n dx = \frac{x\left(\sqrt{x^2-a^2}\right)^n}{n+1} - \frac{na^2}{n+1}\int\left(\sqrt{x^2-a^2}\right)^{n-2}dx, \quad n \neq -1$

39. $\displaystyle\int \frac{dx}{\left(\sqrt{x^2-a^2}\right)^n} = \frac{x\left(\sqrt{x^2-a^2}\right)^{2-n}}{(2-n)a^2} - \frac{n-3}{(n-2)a^2}\int\frac{dx}{\left(\sqrt{x^2-a^2}\right)^{n-2}}, \quad n \neq 2$

40. $\displaystyle\int x\left(\sqrt{x^2-a^2}\right)^n dx = \frac{\left(\sqrt{x^2-a^2}\right)^{n+2}}{n+2} + C, \quad n \neq -2$

41. $\displaystyle\int x^2\sqrt{x^2-a^2}\,dx = \frac{x}{8}(2x^2-a^2)\sqrt{x^2-a^2} - \frac{a^4}{8}\ln\left|x+\sqrt{x^2-a^2}\right| + C$

42. $\displaystyle\int \frac{\sqrt{x^2-a^2}}{x}\,dx = \sqrt{x^2-a^2} - a\sec^{-1}\left|\frac{x}{a}\right| + C$

43. $\displaystyle\int \frac{\sqrt{x^2-a^2}}{x^2}\,dx = \ln\left|x+\sqrt{x^2-a^2}\right| - \frac{\sqrt{x^2-a^2}}{x} + C$

44. $\displaystyle\int \frac{x^2}{\sqrt{x^2-a^2}}\,dx = \frac{a^2}{2}\ln\left|x+\sqrt{x^2-a^2}\right| + \frac{x}{2}\sqrt{x^2-a^2} + C$

45. $\displaystyle\int \frac{dx}{x\sqrt{x^2-a^2}} = \frac{1}{a}\sec^{-1}\left|\frac{x}{a}\right| + C = \frac{1}{a}\cos^{-1}\left|\frac{a}{x}\right| + C$

46. $\displaystyle\int \frac{dx}{x^2\sqrt{x^2-a^2}} = \frac{\sqrt{x^2-a^2}}{a^2x} + C$

47. $\displaystyle\int \frac{dx}{\sqrt{2ax - x^2}} = \sin^{-1}\left(\frac{x-a}{a}\right) + C$

48. $\displaystyle\int \sqrt{2ax - x^2}\,dx = \frac{x-a}{2}\sqrt{2ax - x^2} + \frac{a^2}{2}\sin^{-1}\left(\frac{x-a}{a}\right) + C$

49. $\displaystyle\int \left(\sqrt{2ax - x^2}\right)^n dx = \frac{(x-a)\left(\sqrt{2ax - x^2}\right)^n}{n+1} + \frac{na^2}{n+1}\int \left(\sqrt{2ax - x^2}\right)^{n-2} dx$

50. $\displaystyle\int \frac{dx}{\left(\sqrt{2ax - x^2}\right)^n} = \frac{(x-a)\left(\sqrt{2ax - x^2}\right)^{2-n}}{(n-2)a^2} + \frac{n-3}{(n-2)a^2}\int \frac{dx}{\left(\sqrt{2ax - x^2}\right)^{n-2}}$

51. $\displaystyle\int x\sqrt{2ax - x^2}\,dx = \frac{(x+a)(2x-3a)\sqrt{2ax - x^2}}{6} + \frac{a^3}{2}\sin^{-1}\left(\frac{x-a}{a}\right) + C$

52. $\displaystyle\int \frac{\sqrt{2ax - x^2}}{x}\,dx = \sqrt{2ax - x^2} + a\sin^{-1}\left(\frac{x-a}{a}\right) + C$

53. $\displaystyle\int \frac{\sqrt{2ax - x^2}}{x^2}\,dx = -2\sqrt{\frac{2a-x}{x}} - \sin^{-1}\left(\frac{x-a}{a}\right) + C$

54. $\displaystyle\int \frac{x\,dx}{\sqrt{2ax - x^2}} = a\sin^{-1}\left(\frac{x-a}{a}\right) - \sqrt{2ax - x^2} + C$

55. $\displaystyle\int \frac{dx}{x\sqrt{2ax - x^2}} = -\frac{1}{a}\sqrt{\frac{2a-x}{x}} + C$

56. $\displaystyle\int \sin ax\,dx = -\frac{1}{a}\cos ax + C$

57. $\displaystyle\int \cos ax\,dx = \frac{1}{a}\sin ax + C$

58. $\displaystyle\int \sin^2 ax\,dx = \frac{x}{2} - \frac{\sin 2ax}{4a} + C$

59. $\displaystyle\int \cos^2 ax\,dx = \frac{x}{2} + \frac{\sin 2ax}{4a} + C$

60. $\displaystyle\int \sin^n ax\,dx = -\frac{\sin^{n-1} ax \cos ax}{na} + \frac{n-1}{n}\int \sin^{n-2} ax\,dx$

61. $\displaystyle\int \cos^n ax\,dx = \frac{\cos^{n-1} ax \sin ax}{na} + \frac{n-1}{n}\int \cos^{n-2} ax\,dx$

62. (a) $\displaystyle\int \sin ax \cos bx\,dx = -\frac{\cos(a+b)x}{2(a+b)} - \frac{\cos(a-b)x}{2(a-b)} + C, \quad a^2 \neq b^2$

(b) $\displaystyle\int \sin ax \sin bx\,dx = \frac{\sin(a-b)x}{2(a-b)} - \frac{\sin(a+b)x}{2(a+b)} + C, \quad a^2 \neq b^2$

(c) $\displaystyle\int \cos ax \cos bx\,dx = \frac{\sin(a-b)x}{2(a-b)} + \frac{\sin(a+b)x}{2(a+b)} + C, \quad a^2 \neq b^2$

63. $\displaystyle\int \sin ax \cos ax\,dx = -\frac{\cos 2ax}{4a} + C$

64. $\displaystyle\int \sin^n ax \cos ax\,dx = \frac{\sin^{n+1} ax}{(n+1)a} + C, \quad n \neq -1$

65. $\displaystyle\int \frac{\cos ax}{\sin ax}\,dx = \frac{1}{a}\ln|\sin ax| + C$

66. $\displaystyle\int \cos^n ax \sin ax\,dx = -\frac{\cos^{n+1} ax}{(n+1)a} + C, \quad n \neq -1$

67. $\displaystyle\int \frac{\sin ax}{\cos ax}\,dx = -\frac{1}{a}\ln|\cos ax| + C$

68. $\int \sin^n ax \cos^m ax\,dx = -\frac{\sin^{n-1} ax \cos^{m+1} ax}{a(m+n)} + \frac{n-1}{m+n}\int \sin^{n-2} ax \cos^m ax\,dx, \quad n \neq -m \quad (\text{reduces } \sin^n ax)$

69. $\int \sin^n ax \cos^m ax\,dx = \frac{\sin^{n+1} ax \cos^{m-1} ax}{a(m+n)} + \frac{m-1}{m+n}\int \sin^n ax \cos^{m-2} ax\,dx, \quad m \neq -n \quad (\text{reduces } \cos^m ax)$

70. $\int \frac{dx}{b + c\sin ax} = \frac{-2}{a\sqrt{b^2 - c^2}}\tan^{-1}\left[\sqrt{\frac{b-c}{b+c}}\tan\left(\frac{\pi}{4} - \frac{ax}{2}\right)\right] + C, \quad b^2 > c^2$

71. $\int \frac{dx}{b + c\sin ax} = \frac{-1}{a\sqrt{c^2 - b^2}}\ln\left|\frac{c + b\sin ax + \sqrt{c^2 - b^2}\cos ax}{b + c\sin ax}\right| + C, \quad b^2 < c^2$

72. $\int \frac{dx}{1 + \sin ax} = -\frac{1}{a}\tan\left(\frac{\pi}{4} - \frac{ax}{2}\right) + C$

73. $\int \frac{dx}{1 - \sin ax} = \frac{1}{a}\tan\left(\frac{\pi}{4} + \frac{ax}{2}\right) + C$

74. $\int \frac{dx}{b + c\cos ax} = \frac{2}{a\sqrt{b^2 - c^2}}\tan^{-1}\left[\sqrt{\frac{b-c}{b+c}}\tan\frac{ax}{2}\right] + C, \quad b^2 > c^2$

75. $\int \frac{dx}{b + c\cos ax} = \frac{1}{a\sqrt{c^2 - b^2}}\ln\left|\frac{c + b\cos ax + \sqrt{c^2 - b^2}\sin ax}{b + c\cos ax}\right| + C, \quad b^2 < c^2$

76. $\int \frac{dx}{1 + \cos ax} = \frac{1}{a}\tan\frac{ax}{2} + C$

77. $\int \frac{dx}{1 - \cos ax} = -\frac{1}{a}\cot\frac{ax}{2} + C$

78. $\int x\sin ax\,dx = \frac{1}{a^2}\sin ax - \frac{x}{a}\cos ax + C$

79. $\int x\cos ax\,dx = \frac{1}{a^2}\cos ax + \frac{x}{a}\sin ax + C$

80. $\int x^n \sin ax\,dx = -\frac{x^n}{a}\cos ax + \frac{n}{a}\int x^{n-1}\cos ax\,dx$

81. $\int x^n \cos ax\,dx = \frac{x^n}{a}\sin ax - \frac{n}{a}\int x^{n-1}\sin ax\,dx$

82. $\int \tan ax\,dx = \frac{1}{a}\ln|\sec ax| + C$

83. $\int \cot ax\,dx = \frac{1}{a}\ln|\sin ax| + C$

84. $\int \tan^2 ax\,dx = \frac{1}{a}\tan ax - x + C$

85. $\int \cot^2 ax\,dx = -\frac{1}{a}\cot ax - x + C$

86. $\int \tan^n ax\,dx = \frac{\tan^{n-1} ax}{a(n-1)} - \int \tan^{n-2} ax\,dx, \quad n \neq 1$

87. $\int \cot^n ax\,dx = -\frac{\cot^{n-1} ax}{a(n-1)} - \int \cot^{n-2} ax\,dx, \quad n \neq 1$

88. $\int \sec ax\,dx = \frac{1}{a}\ln|\sec ax + \tan ax| + C$

89. $\int \csc ax\,dx = -\frac{1}{a}\ln|\csc ax + \cot ax| + C$

90. $\int \sec^2 ax\,dx = \frac{1}{a}\tan ax + C$

91. $\int \csc^2 ax\,dx = -\frac{1}{a}\cot ax + C$

92. $\int \sec^n ax\,dx = \frac{\sec^{n-2} ax \tan ax}{a(n-1)} + \frac{n-2}{n-1}\int \sec^{n-2} ax\,dx, \quad n \neq 1$

93. $\int \csc^n ax\,dx = -\frac{\csc^{n-2} ax \cot ax}{a(n-1)} + \frac{n-2}{n-1}\int \csc^{n-2} ax\,dx, \quad n \neq 1$

94. $\int \sec^n ax \tan ax\,dx = \frac{\sec^n ax}{na} + C, \quad n \neq 0$

95. $\int \csc^n ax \cot ax\,dx = -\frac{\csc^n ax}{na} + C, \quad n \neq 0$

96. $\int \sin^{-1} ax\, dx = x \sin^{-1} ax + \frac{1}{a}\sqrt{1 - a^2x^2} + C$

97. $\int \cos^{-1} ax\, dx = x \cos^{-1} ax - \frac{1}{a}\sqrt{1 - a^2x^2} + C$

98. $\int \tan^{-1} ax\, dx = x \tan^{-1} ax - \frac{1}{2a} \ln(1 + a^2x^2) + C$

99. $\int x^n \sin^{-1} ax\, dx = \frac{x^{n+1}}{n+1} \sin^{-1} ax - \frac{a}{n+1} \int \frac{x^{n+1}\, dx}{\sqrt{1 - a^2x^2}}, \quad n \neq -1$

100. $\int x^n \cos^{-1} ax\, dx = \frac{x^{n+1}}{n+1} \cos^{-1} ax + \frac{a}{n+1} \int \frac{x^{n+1}\, dx}{\sqrt{1 - a^2x^2}}, \quad n \neq -1$

101. $\int x^n \tan^{-1} ax\, dx = \frac{x^{n+1}}{n+1} \tan^{-1} ax - \frac{a}{n+1} \int \frac{x^{n+1}\, dx}{1 + a^2x^2}, \quad n \neq -1$

102. $\int e^{ax}\, dx = \frac{1}{a} e^{ax} + C$

103. $\int b^{ax}\, dx = \frac{1}{a} \frac{b^{ax}}{\ln b} + C, \quad b > 0, b \neq 1$

104. $\int x e^{ax}\, dx = \frac{e^{ax}}{a^2}(ax - 1) + C$

105. $\int x^n e^{ax}\, dx = \frac{1}{a} x^n e^{ax} - \frac{n}{a} \int x^{n-1} e^{ax}\, dx$

106. $\int x^n b^{ax}\, dx = \frac{x^n b^{ax}}{a \ln b} - \frac{n}{a \ln b} \int x^{n-1} b^{ax}\, dx, \quad b > 0, b \neq 1$

107. $\int e^{ax} \sin bx\, dx = \frac{e^{ax}}{a^2 + b^2}(a \sin bx - b \cos bx) + C$

108. $\int e^{ax} \cos bx\, dx = \frac{e^{ax}}{a^2 + b^2}(a \cos bx + b \sin bx) + C$

109. $\int \ln ax\, dx = x \ln ax - x + C$

110. $\int x^n (\ln ax)^m\, dx = \frac{x^{n+1}(\ln ax)^m}{n+1} - \frac{m}{n+1} \int x^n (\ln ax)^{m-1}\, dx, \quad n \neq -1$

111. $\int x^{-1} (\ln ax)^m\, dx = \frac{(\ln ax)^{m+1}}{m+1} + C, \quad m \neq -1$

112. $\int \frac{dx}{x \ln ax} = \ln |\ln ax| + C$

113. $\int \sinh ax\, dx = \frac{1}{a} \cosh ax + C$

114. $\int \cosh ax\, dx = \frac{1}{a} \sinh ax + C$

115. $\int \sinh^2 ax\, dx = \frac{\sinh 2ax}{4a} - \frac{x}{2} + C$

116. $\int \cosh^2 ax\, dx = \frac{\sinh 2ax}{4a} + \frac{x}{2} + C$

117. $\int \sinh^n ax\, dx = \frac{\sinh^{n-1} ax \cosh ax}{na} - \frac{n-1}{n} \int \sinh^{n-2} ax\, dx, \quad n \neq 0$

118. $\int \cosh^n ax\, dx = \frac{\cosh^{n-1} ax \sinh ax}{na} + \frac{n-1}{n} \int \cosh^{n-2} ax\, dx, \quad n \neq 0$

119. $\int x \sinh ax\, dx = \frac{x}{a} \cosh ax - \frac{1}{a^2} \sinh ax + C$

120. $\int x \cosh ax\, dx = \frac{x}{a} \sinh ax - \frac{1}{a^2} \cosh ax + C$

121. $\int x^n \sinh ax\, dx = \frac{x^n}{a} \cosh ax - \frac{n}{a} \int x^{n-1} \cosh ax\, dx$

122. $\int x^n \cosh ax\, dx = \frac{x^n}{a} \sinh ax - \frac{n}{a} \int x^{n-1} \sinh ax\, dx$

123. $\int \tanh ax\, dx = \frac{1}{a} \ln(\cosh ax) + C$

124. $\int \coth ax\, dx = \frac{1}{a} \ln |\sinh ax| + C$

125. $\int \tanh^2 ax\,dx = x - \frac{1}{a}\tanh ax + C$

126. $\int \coth^2 ax\,dx = x - \frac{1}{a}\coth ax + C$

127. $\int \tanh^n ax\,dx = -\frac{\tanh^{n-1} ax}{(n-1)a} + \int \tanh^{n-2} ax\,dx, \quad n \neq 1$

128. $\int \coth^n ax\,dx = -\frac{\coth^{n-1} ax}{(n-1)a} + \int \coth^{n-2} ax\,dx, \quad n \neq 1$

129. $\int \operatorname{sech} ax\,dx = \frac{1}{a}\sin^{-1}(\tanh ax) + C$

130. $\int \operatorname{csch} ax\,dx = \frac{1}{a}\ln\left|\tanh\frac{ax}{2}\right| + C$

131. $\int \operatorname{sech}^2 ax\,dx = \frac{1}{a}\tanh ax + C$

132. $\int \operatorname{csch}^2 ax\,dx = -\frac{1}{a}\coth ax + C$

133. $\int \operatorname{sech}^n ax\,dx = \frac{\operatorname{sech}^{n-2} ax \tanh ax}{(n-1)a} + \frac{n-2}{n-1}\int \operatorname{sech}^{n-2} ax\,dx, \quad n \neq 1$

134. $\int \operatorname{csch}^n ax\,dx = -\frac{\operatorname{csch}^{n-2} ax \coth ax}{(n-1)a} - \frac{n-2}{n-1}\int \operatorname{csch}^{n-2} ax\,dx, \quad n \neq 1$

135. $\int \operatorname{sech}^n ax \tanh ax\,dx = -\frac{\operatorname{sech}^n ax}{na} + C, \quad n \neq 0$

136. $\int \operatorname{csch}^n ax \coth ax\,dx = -\frac{\operatorname{csch}^n ax}{na} + C, \quad n \neq 0$

137. $\int e^{ax}\sinh bx\,dx = \frac{e^{ax}}{2}\left[\frac{e^{bx}}{a+b} - \frac{e^{-bx}}{a-b}\right] + C, \quad a^2 \neq b^2$

138. $\int e^{ax}\cosh bx\,dx = \frac{e^{ax}}{2}\left[\frac{e^{bx}}{a+b} + \frac{e^{-bx}}{a-b}\right] + C, \quad a^2 \neq b^2$

139. $\int_0^\infty x^{n-1}e^{-x}\,dx = \Gamma(n) = (n-1)!, \quad n > 0$

140. $\int_0^\infty e^{-ax^2}\,dx = \frac{1}{2}\sqrt{\frac{\pi}{a}}, \quad a > 0$

141. $\int_0^{\pi/2} \sin^n x\,dx = \int_0^{\pi/2} \cos^n x\,dx = \begin{cases} \dfrac{1\cdot3\cdot5\cdots(n-1)}{2\cdot4\cdot6\cdots n}\cdot\dfrac{\pi}{2}, & \text{if } n \text{ is an even integer} \geq 2 \\ \dfrac{2\cdot4\cdot6\cdots(n-1)}{3\cdot5\cdot7\cdots n}, & \text{if } n \text{ is an odd integer} \geq 3 \end{cases}$

Credits

Page 181, Section 3.3, photo for Exercise 19, *PSSC Physics*, 2nd ed., DC Heath & Co. with Education Development Center, Inc.; **Page 238, Figure 3.53,** NCPMF "Differentiation" by W. U. Walton et al., Project CALC, Education Development Center, Inc.; **Page 241, Chapter 3 Additional and Advanced Exercises, photo for Exercise 9,** AP/Wide World Photos; **Page 304, Figure 4.53,** Scott A. Burns; **Page 425, Figure 6.29a,** *PSSC Physics*, 2nd ed., DC Heath & Co. with Education Development Center, Inc.; **Page 615, Section 8.7, photo for Exercise 30,** Marshall Henrichs; **Page 696, Section 10.1, photo for Exercise 89,** *PSSC Physics*, 2nd ed., DC Heath & Co. with Education Development Center, Inc.; **Page 765, Figure 11.6,** *PSSC Physics*, 2nd ed., DC Heath & Co. with Education Development Center, Inc.; **Page 924, Figure 13.12,** Corbis; **Page 929, Section 13.2, photo for Exercise 23,** *PSSC Physics*, 2nd ed., DC Heath & Co. with Education Development Center, Inc.; **Page 969, Figure 14.6,** Reproduced by permission from Appalachian Mountain Club; **Page 1006, Figure 14.23,** Department of History, U.S. Military Academy, West Point, New York; **Page 1150, Figures 16.7 and 16.8,** *NCFMF Book of Film Notes*, 1974, MIT Press with Education Development Center, Inc.; **Page 1151, Figure 16.15,** Inter-Network Media, Inc., and NASA/JPL.

INTEGRATION RULES

General Formulas

Zero: $\displaystyle\int_a^a f(x)\,dx = 0$

Order of Integration: $\displaystyle\int_b^a f(x)\,dx = -\int_a^b f(x)\,dx$

Constant Multiples: $\displaystyle\int_a^b kf(x)\,dx = k\int_a^b f(x)\,dx$ (Any number k)

$$\int_a^b -f(x)\,dx = -\int_a^b f(x)\,dx \qquad (k = -1)$$

Sums and Differences: $\displaystyle\int_a^b (f(x) \pm g(x))\,dx = \int_a^b f(x)\,dx \pm \int_a^b g(x)\,dx$

Additivity: $\displaystyle\int_a^b f(x)\,dx + \int_b^c f(x)\,dx = \int_a^c f(x)\,dx$

Max-Min Inequality: If $\max f$ and $\min f$ are the maximum and minimum values of f on $[a, b]$, then

$$\min f \cdot (b - a) \le \int_a^b f(x)\,dx \le \max f \cdot (b - a).$$

Domination: $f(x) \ge g(x)$ on $[a, b]$ implies $\displaystyle\int_a^b f(x)\,dx \ge \int_a^b g(x)\,dx$

$f(x) \ge 0$ on $[a, b]$ implies $\displaystyle\int_a^b f(x)\,dx \ge 0$

The Fundamental Theorem of Calculus

Part 1 If f is continuous on $[a, b]$, then $F(x) = \int_a^x f(t)\,dt$ is continuous on $[a, b]$ and differentiable on (a, b) and its derivative is $f(x)$;

$$F'(x) = \frac{d}{dx}\int_a^x f(t)\,dt = f(x).$$

Part 2 If f is continuous at every point of $[a, b]$ and F is any antiderivative of f on $[a, b]$, then

$$\int_a^b f(x)\,dx = F(b) - F(a).$$

Substitution in Definite Integrals

$$\int_a^b f(g(x)) \cdot g'(x)\,dx = \int_{g(a)}^{g(b)} f(u)\,du$$

Integration by Parts

$$\int_a^b f(x)g'(x)\,dx = f(x)g(x)\Big]_a^b - \int_a^b f'(x)g(x)\,dx$$